COASTAL WETLAN

COASTAL WETLANDS

AN INTEGRATED ECOSYSTEM APPROACH

SECOND EDITION

Edited by

Gerardo M.E. Perillo
Eric Wolanski
Donald R. Cahoon
Charles S. Hopkinson

Elsevier
Radarweg 29, PO Box 211, 1000 AE Amsterdam, Netherlands
The Boulevard, Langford Lane, Kidlington, Oxford OX5 1GB, United Kingdom
50 Hampshire Street, 5th Floor, Cambridge, MA 02139, United States

Library of Congress Cataloging-in-Publication Data
A catalog record for this book is available from the Library of Congress

British Library Cataloguing-in-Publication Data
A catalogue record for this book is available from the British Library

ISBN: 978-0-444-63893-9

Publisher: Candice Janco
Acquisition Editor: Louisa Hutchins
Editorial Project Manager: Emily Thomson
Production Project Manager: Nilesh Kumar Shah
Cover Designer: Mark Rogers

Typeset by TNQ Technologies

Dedication

The editors wish to dedicate the present book to the memory of Dr. Mark M. Brinson, our friend, colleague, and member of the editorial team of the first edition of Coastal Wetlands: An integrated Ecosystem Approach.

Gerardo M.E. Perillo, Eric Wolanski, Donald R. Cahoon, and Charles S. Hopkinson

Dr. Mark M. Brinson retired in September 2010 with the title of Distinguished Research Professor, after more than 35 years at East Carolina University. We were looking forward to several more active years of research and professional service when he unexpectedly passed away a few months later, on January 3, 2011. In words of his dear friend and colleague, Dr. Robert Christian, we lost a thoughtful, hardworking, and creative Ecosystem Naturalist that day.

Mark devoted his research to understanding how wetlands function, unraveling the intricate relationships among the physical, chemical, and biological components of wetland ecosystems. His unique perspective made substantial contributions not only to wetland science but also to the environmental management of wetlands. Mark's work had a significant impact on the hydrogeomorphic classification of wetlands, central to the functional assessment of wetlands and mitigation procedures based on functional loss. This functional approach along with the use of reference states that one of his most significant contributions greatly changed mitigation policies and targets for restoration strategies.

Throughout his career, Mark taught numerous courses and workshops on wetland ecology within the United States and abroad. He has edited, authored, and coauthored many publications on wetland ecology and served as a technical consultant to the US Environmental Protection Agency, US Fish and Wildlife Service, and the Smithsonian Institution. He was also President of the Society of Wetland Scientists and served on its Board of Directors for several years. In addition, Mark has received many honors and awards, including the Thomas Harriot College of Arts and Sciences Distinguished Professor Award, Lifetime Achievement Award, a National Wetlands Award for Science Research cosponsored by the Environmental Law Institute and the Environmental Protection Agency, and Fellowship of the Society of Wetland Scientists.

On a more personal note, I met Mark in 2003, when I was a graduate student at Universidad de Buenos Aires. By that time he was using a Fulbright Foreign Scholarship Award to aid in the development of a national wetland inventory in Argentina. I was a clumsy and unfocused student who knew little about Mark then, but working with him gave me the opportunity of a lifetime to become a wetland ecologist. Mark was one of the most respectful people I ever knew, but I could not describe Mark better than himself. As he said in his retirement essay, he did not like "being shrouded by the transparent tapestry of ego." And that was Dr. Mark Brinson, the Senior Scientist and Distinguished Professor who did not miss a single field trip to my marsh sites. He wrote my field notes, helped me out with the peat sampler, and cooked dinner.

In his retirement speech he also said "Having fun is being hardwired to your profession – half of the time you don't realize how much fun it is." From all the pictures that came to my mind since I started to write this note, I choose to share this one: Mark and his big smile standing at the marsh, playing at being my field assistant, and having fun.

Dr. Paula G Pratolongo

Contents

PART II

PHYSICAL PROCESSES

PART III

TIDAL FLATS

18. Tidal Freshwater Wetlands

DENNIS F. WHIGHAM, ANDREW H. BALDWIN, AND AAT BARENDREGT

19. Biogeochemistry of Tidal Freshwater Wetlands

J. PATRICK MEGONIGAL AND SCOTT C. NEUBAUER

Part V

MANGROVES

20. Biogeomorphology of Mangroves

JOANNA C. ELLISON

21. Mangrove Biogeochemistry at Local to Global Scales Using Ecogeomorphic Approaches

ROBERT R. TWILLEY, VICTOR H. RIVERA-MONROY, ANDRE S. ROVAI, EDWARD CASTAÑEDA-MOYA, AND STEPHEN DAVIS

Part VI

COASTAL WETLAND RESTORATION AND MANAGEMENT

22. Tidal Marsh Creation

STEPHEN W. BROOME, CHRISTOPHER B. CRAFT, AND MICHAEL R. BURCHELL

23. Salt Marsh Restoration
PAUL ADAM

24. Methods and Criteria for Successful Mangrove Forest Rehabilitation
ROY R. LEWIS III, BENJAMIN M. BROWN, AND LAURA L. FLYNN

25. Evaluating Restored Tidal Freshwater Wetlands
ANDREW H. BALDWIN, RICHARD S. HAMMERSCHLAG, AND DONALD R. CAHOON

Part VII
COASTAL WETLAND SUSTAINABILITY

26. The Shifting Saltmarsh-Mangrove Ecotone in Australasia and the Americas
NEIL SAINTILAN, KERRYLEE ROGERS, AND KAREN L. MCKEE

27. The Value of Coastal Wetland Ecosystem Services
EDWARD B. BARBIER

28. Conservation of Blue Carbon Ecosystems for Climate Change Mitigation and Adaptation
OSCAR SERRANO, JEFFREY J. KELLEWAY, CATHERINE LOVELOCK, AND PAUL S. LAVERY

List of Contributors

Kenneth F. Abraham Trent University, Peterborough, ON, Canada

Paul Adam School of Biological Earth and Environmental Science, UNSW, Sydney, NSW, Australia

S. Ahmerkamp Max Planck Institute for Marine Microbiology, Bremen, Germany

Rebecca J. Aspden Scottish Oceans Institute, School of Biology, University of St Andrews, Fife, Scotland

Andrew H. Baldwin Department of Environmental Science and Technology, University of Maryland, College Park, MD, United States

Donald M. Baltz Department of Oceanography and Coastal Sciences, Louisiana State University, Baton Rouge, LA, United States

Edward B. Barbier Department of Economics, Colorado State University, Fort Collins, Colorado, United States

Aat Barendregt Utrecht University, Utrecht, The Netherlands

Kevin S. Black Partrac, Glasgow, Scotland

Laurence A. Boorman L A B Coastal, Cambridgeshire, United Kingdom

Mark M. Brinson†

Stephen W. Broome Department of Crop and Soil Sciences, North Carolina State University, Raleigh, NC, United States

Benjamin M. Brown Charles Darwin University, Research Institute for Environment and Livelihoods (RIEL), Darwin, NT, Australia

Michael R. Burchell Department of Biological and Agricultural Engineering, North Carolina State University, Raleigh, NC, United States

†Deceased.

Donald R. Cahoon United States Geological Survey, Patuxent Wildlife Research Center, Laurel, MD, United States

L. Carniello Department of Civil, Environmental and Architectural Engineering, University of Padua, Padova, Italy

Edward Castañeda-Moya Southeast Environmental Research Center, Florida International University, Miami, FL, United States

Elizabeth Christie Cambridge Coastal Research Unit, Department of Geography, University of Cambridge, Cambridge, United Kingdom

P.L.M. Cook Water Studies Centre, Monash University, Clayton, VIC, Australia

Christopher B. Craft School of Public and Environmental Affairs, Indiana University, Bloomington, IN, United States

Carolyn A. Currin NOAA, National Centers for Coastal Ocean Science, Beaufort Lab, Beaufort, NC, United States

Andrea D'Alpaos Department of Geosciences, University of Padova, PD, Italy

L. D'Alpaos Department of Civil, Environmental and Architectural Engineering, University of Padua, Padova, Italy

Stephen Davis Everglades Foundation, Palmetto Bay, FL, United States

Dirk de Beer Max Planck Institute for Marine Microbiology, Bremen, Germany

A. Defina Department of Civil, Environmental and Architectural Engineering, University of Padua, Padova, Italy

Joanna C. Ellison Discipline of Geography and Spatial Sciences, School of Technology, Environments and Design, University of Tasmania, Launceston, TAS, Australia

Laura L. Flynn Coastal Resources Group, Inc., Venice, FL, United States

Irene Fortune Scottish Oceans Institute, School of Biology, University of St Andrews, Fife, Scotland

Jon French Coastal and Estuarine Research Unit, UCL Department of Geography, University College London, Gower Street, London, United Kingdom

Shu Gao State Key Laboratory for Estuarine and Coastal Research, East China Normal University, Shanghai, China

Christopher Haight New York City Department of Parks & Recreation, New York, NY, United States

Richard S. Hammerschlag United States Geological Survey, Patuxent Wildlife Research Center, Laurel, MD, United States

Ellen Kracauer Hartig New York City Department of Parks & Recreation, New York, NY, United States

Marianne Holmer Department of Biology, University of Southern Denmark, Odense, Denmark

Charles S. Hopkinson Department of Marine Sciences, University of Georgia, Athens, GA, United States

Robert L. Jefferies[†]

S.B. Joye Department of Marine Sciences, University of Georgia, Athens, GA, United States

Jeffrey J. Kelleway Department of Environmental Sciences, Macquarie University, Sydney, NSW, Australia

Jason R. Kirby Liverpool John Moores University, School of Natural Sciences & Psychology, Liverpool, United Kingdom

Stefano Lanzoni Department of Civil, Environmental, and Architectural Engineering, University of Padova, PD, Italy

Marit Larson New York City Department of Parks & Recreation, New York, NY, United States

Paul S. Lavery School of Science & Centre for Marine Ecosystems Research, Edith Cowan University, Joondalup, WA, Australia; Centro de Estudios Avanzados de Blanes, Consejo Superior de Investigaciones Científicas, Blanes, Spain

Nicoletta Leonardi University of Liverpool, School of Environmental Sciences, Department of Geography and Planning, Liverpool, United Kingdom

Roy R. Lewis III Coastal Resources Group, Inc., Salt Springs, FL, United States

Catherine Lovelock The School of Biological Sciences, The University of Queensland, St Lucia, QLD, Australia; Global Change Institute, The University of Queensland, St Lucia, QLD, Australia

Marco Marani Department of Civil, Environmental, and Architectural Engineering, University of Padova, PD, Italy

I. Peter Martini School of Environmental Sciences, University of Guelph, Guelph, ON, Canada

Karen L. McKee U.S. Geological Survey, Wetland and Aquatic Research Center, Lafayette, LA, United States

J. Patrick Megonigal Smithsonian Environmental Research Center, Edgewater, MD, United States

Stephen Midway Department of Oceanography and Coastal Sciences, Louisiana State University, Baton Rouge, LA, United States

Iris Möller Cambridge Coastal Research Unit, Department of Geography, University of Cambridge, Cambridge, United Kingdom

R.I. Guy Morrison National Wildlife Research Centre, Environment and Climate Change Canada, Ottawa, ON, Canada

Scott C. Neubauer Department of Biology, Virginia Commonwealth University, Richmond, VA, United States

David M. Paterson Scottish Oceans Institute, School of Biology, University of St Andrews, Fife, Scotland

[†]Deceased.

Gerardo M.E. Perillo Instituto Argentino de Oceanografía (CONICET – UNS), Bahía Blanca, Argentina; Departamento de Geología, Universidad Nacional del Sur, Bahía Blanca, Argentina

Maria Cintia Piccolo Instituto Argentino de Oceanografia - Universidad Nacional del Sur, Bahia Blanca, Buenos Aires, Argentina

Andrew Plater University of Liverpool, School of Environmental Sciences, Department of Geography and Planning, Liverpool, United Kingdom

Paula Pratolongo Universidad Nacional del Sur, Dto. de Biología Bioquímica y Farmacia CONICET, Instituto Argentino de Oceanografía, Bahía Blanca, Argentina

Andrea Rinaldo Department of Civil, Environmental, and Architectural Engineering, University of Padova, PD, Italy; Laboratory of Ecohydrology, Ecole Polytechnique Fèdèrale Lausanne, Lausanne, Switzerland

Victor H. Rivera-Monroy Department of Oceanography and Coastal Sciences, College of the Coast and the Environment, Louisiana State University, Baton Rouge, LA, United States

Kerrylee Rogers Geoquest, University of Wollongong, Wollongong, NSW, Australia

Andre S. Rovai Department of Oceanography and Coastal Sciences, College of the Coast and the Environment, Louisiana State University, Baton Rouge, LA, United States; Programa de Pós-Graduação em Oceanografia, Universidade Federal de Santa Catarina, Florianópolis, Brazil

Neil Saintilan Department of Environmental Sciences, Macquarie University, Sydney, NSW, Australia

Charles E. Sasser Department of Oceanography and Coastal Sciences, Louisiana State University, Baton Rouge, LA, United States

C.A. Schutte Louisiana Universities Marine Consortium (LUMCON), Chauvin, LA, United States

M. Seidel Institute for Chemistry and Biology of the Marine Environment (ICBM), University of Oldenburg, Oldenburg, Germany

Liudmila A. Sergienko Department of Botany and Plant Physiology, Petrozavodsk State University, Petrozavodsk, Russia

Oscar Serrano School of Science & Centre for Marine Ecosystems Research, Edith Cowan University, Joondalup, WA, Australia

Daniel O. Suman Rosenstiel School of Marine and Atmospheric Science, University of Miami, Miami, FL, United States

Rebecca K. Swadek New York City Department of Parks & Recreation, New York, NY, United States

Craig Tobias University of Connecticut, Groton, CT, United States

Robert R. Twilley Department of Oceanography and Coastal Sciences, College of the Coast and the Environment, Louisiana State University, Baton Rouge, LA, United States

Jenneke M. Visser Institute for Coastal and Water Research, and School of Geosciences, University of Louisiana at Lafayette, Lafayette, LA, United States

Dennis F. Whigham Smithsonian Environmental Research Center, Edgewater, MD, United States

Eric Wolanski TropWATER and College of Science and Engineering, James Cook University, Townsville, QLD, Australia; Australian Institute of Marine Science, Townsville, QLD, Australia

Colin D. Woodroffe School of Earth and Environmental Sciences, University of Wollongong, Wollongong, NSW, Australia

C.S. Wu Max Planck Institute for Marine Microbiology, Bremen, Germany

Preface to the First Edition

Why coastal wetlands? What is so important about them that a whole book is required to try to review and explain their large variety of properties? Of all the coastal habitats, wetlands are the least depicted in the tourist brochures because they lack those paradisiacal long, white sandy beaches backed by palm trees or expensive resort hotels close to transparent blue waters. In fact, most coastal wetlands are quite muddy and are more likely to be inhabited by crabs and worms than by charismatic fish, birds and mammals. Hence, most inhabitants of our world either have never thought about coastal wetlands or may consider them a nuisance, not realizing that their seafood dinner likely had its origin as a detrital food web in a salt marsh or mangrove swamp. Bahía Blanca (Argentina) inhabitants are a classical example: a city of over 300,000 people living within 10 km of a 2300-km^2 wetland, the largest of Argentina, but fewer than 40% have any idea that they are so close to the sea and a short distance of places that are globally unique (Perillo and Iribarne, 2003, in Chapter 6).

Similarly, there are many other coastlines dominated by wetlands, yet they are only seen as areas to exploit in an unsustainable fashion. For example, mangroves have served local communities for generations in many Asian tropical countries for harvesting wood and fish in contrast to their wholesale replacement for rice cultivation and shrimp farming.

Even though management guidelines have been available for decades, the negative consequences of uninformed exploitation have resulted in poor or even total lack of management criteria by most governments at all levels. Even local stakeholders fail to act in their own best interest without consideration of the ecosystem goods and services that the nearby wetlands provide.

Coastal wetlands best develop along passive-margin coasts with low-gradient coastal plains and wide continental shelves. The combination of low hydraulic energy and gentle slope provides an ideal setting for the wetland development. Also passive margins are less prone to receive large episodic events like tsunamis. Tsunamis and storm surges, in particular, are major coast modifiers, but when they act on low coasts their effects are more far reaching than they are on higher relief coasts. For a wetland to form, there is a need for a particular geomorphological setting such as an embayment or estuary providing a relatively low-energy environment favoring sediment settling, deposition and preservation. However, that is only the beginning of a large and complex "life" where many geological (i.e., sediment supply, geological setting and isostasy), physical (i.e., oceanographic, atmospheric, fluvial, groundwater processes and sea level changes), chemical (i.e., nutrients, pollutants), biological (i.e., intervening flora and fauna), and anthropic factors play a wide spectra of roles. Coastal wetlands are areas that have combined physical sources and biological processes to develop structure that continues to take advantage of natural energy inputs.

This book has been planned to address in an integrated way all these processes and their consequences on the characterization and evolution of coastal wetlands. It aims to provide an integrated perspective on coastal wetlands as ecosystems for the public, engineers, scientists and resources managers. It is only after acquiring this perspective that scientists can confidently propose ecohydrologic solutions for managing these environments in an ecologically sustainable way. This is but one small step toward encouraging humanity to look beyond purely technological, and often failed, solutions to complex environmental problems.

This is done by focusing on the principal components considering the full range of environments from freshwater to subtidal and from polar to tropical systems. The book has been divided into seven parts starting from a synthesis chapter that integrates the whole book. Part I covers, in three chapters, the general description of the wetlands structured according to broad climatic regions and introduces the most important physical processes that are common to all coastal wetlands including some geomorphologic and modeling principles. Part II are specific to each particular type of wetland (tidal flats, marshes and seagrasses, and mangroves). Within each part (Parts III to V), there are chapters dealing with their particular geomorphology, sedimentology, biology and biogeochemistry. Finally Parts VI and VII provide insight into the restoration and management and sustainability and landscape dynamics.

As editors, our work was greatly facilitated by the tremendous cooperation and enthusiasm from each of the authors to complete this process that began mid 2006. Each author, an authority in his or her specialty, was specifically invited to write a review chapter. Therefore, the challenges were much larger than in the case of typical contributed articles. But the reward, we think, is much more beneficial for the student, professor, or researcher employing this book for his or her particular interest. Readers will not only be able to find a specific topic but will find related information to complement and enhance the understanding of the topic.

We are in debt to the more than 50 reviewers (most of them are not authors in the book) who have agreed and provided graciously and unselfishly their valuable time. Some took on the responsibility of two chapters, and their efforts are rewarded with improvements of each contribution. In many cases, reviewers gave us interesting ideas that helped in the general structure of the book. A list of the reviewers is provided.

We also thank Elsevier Science and the various Publishing Editors who were in charge of our book along the period since we first proposed our idea to the final result that you are reading now. First of all to Kristien van Lunen who first believed that our proposal was realizable and then to the important contributions and patience of Jennifer Hele, and also to Pauline Riebeek, Linda Versteeg and lastly Sara Pratt. Stalin Viswanathan did an excellent job copyediting the whole book.

Gerardo M.E. Perillo
Eric Wolanski
Donald R. Cahoon
Mark M. Brinson
July, 2008

This document is based on work partially supported by the U.S. National Science Foundation under Grants No. BSR-8702333-06, DEB-9211772, DEB-9411974, DEB-0080381 and DEB-0621014 and to SCOR under Grant No. OCE-0608600. Any opinions, findings, and conclusions or recommendations expressed in this material are those of the authors and do not necessarily reflect the views of the U.S. National Science Foundation (NSF).

Preface to the Second Edition

As we wrote this preface, we realized that exactly 10 years have passed since the time we wrote the original preface for the first edition. Three and half years ago, Elsevier—with a surprising mail by Candice Janco—invited us to look into the possibility of a second edition. Although initially the thought of getting back into all the effort that editing a book of these characteristics demands seemed an unhealthy adventure, all of us actually jumped right in with the idea. Unfortunately, as we present in the dedicatory, our good friend Mark Brinson passed away in the period between both editions. However, Chuck Hopkinson has been brave enough to come on board playing a major role in the outcome of the book.

All the comments given in the original preface are as true today as they were then. Unfortunately, many management issues and complex situations that coastal wetlands were suffering are still present today and, graver, they have become worse. Therefore, they will not be repeated here.

In this sense, to the original structure of the book, we added new chapters pointing out specifically new views of coastal wetlands management and valuation. Nevertheless, each chapter was significantly improved and updated to reflect on the new advances achieved by the wetland community in the last 10 years.

Many of the original lead authors from the first edition gladly accepted the challenge to get back to their old notes and redo them for the new edition. A few could not commit themselves to the task for various reasons. In the cases where it was possible, we found new adventurers who were willing to take the "relay baton" and write a completely new chapter on the subject.

In all cases, we must remember, authors were asked to make a thorough review on a specific subject rather than describe their own personal work. However, we relied on the outstanding capabilities and worldwide recognition each of the lead authors brought to the book, accompanied by no lesser coauthors.

Again, we could not have presented a book of this level of quality without the unselfish input of over 50 reviewers. They were essential in providing us with their expertise for those chapters in which we may not be as confident as them. A book is as good as its authors, and also it is improved by the quality of its reviewers. We cannot thank enough each and every one of them for their time and support.

Finally, again we cannot thank enough the original invitation of Candice Janco—who left us not only for another position within Elsevier almost at the beginning of the project but also in the very capable hands, the amenability, and the good humor of Emily Joy Grace Thomson. She helped us through all the steps—even those initial failures—with quick responses and many times going out of her way to look for solutions, some almost impossible. We also want to thank the invaluable contributions by Nilesh Shah and Rajesh Manohar for his capable and professional work in doing the copyediting of the book.

Gerardo M.E. Perillo
Eric Wolanski
Donald R. Cahoon
Charles S. Hopkinson
July 2018

List of Reviewers

9 anonymous
Iris Anderson
Donald R. Cahoon
John Callaway
Giovanni Coco
Steve Davis
John Day
Luiz Drude de Lacerda
Steven Dundas
Carl Fitz
Duncan Fitzgerald
Shu Gao
John Grant
Charles S. Hopkinson
Chris Kennedy
Mary Kentula
Ken W. Krauss
Megan LaPeyre
Christopher Madden
Marco Marani
Beth Middleton
Kathleen Onorevole
Michael Osland
Morten Pejrup
Steven Pennings
Gerardo M. E. Perillo
Stephen Polasky
Charles Roman
Rochelle Seitz
Tom Spencer
Craig Tobias
Raymond Torres
Reginald Uncles
Bethney Ward
Eric Wolanski

CHAPTER

1

Coastal Wetlands: A Synthesis

Charles S. Hopkinson[1], *Eric Wolanski*[2,3], *Mark M. Brinson*[†], *Donald R. Cahoon*[4], *Gerardo M.E. Perillo*[5,6]

[1]Department of Marine Sciences, University of Georgia, Athens, GA, United States; [2]TropWATER and College of Science and Engineering, James Cook University, Townsville, QLD, Australia; [3]Australian Institute of Marine Science, Townsville, QLD, Australia; [4]United States Geological Survey, Patuxent Wildlife Research Center, Laurel, MD, United States; [5]Instituto Argentino de Oceanografía (CONICET – UNS), Bahía Blanca, Argentina; [6]Departamento de Geología, Universidad Nacional del Sur, Bahía Blanca, Argentina

1. INTRODUCTION

What are coastal wetland ecosystems, what are their limits of distribution, and where do they exist in the overall coastal landscape? There are several general definitions for wetlands, but the Ramsar definition is likely the most broadly encompassing (http://www.ramsar.org/), whereas others are more focused definitions tailored to country-specific protection and management policies (Mitsch and Gosselink, 2006). We offer a very general approach rather than a precise definition: coastal wetlands are ecosystems that are found within an elevation gradient that ranges between subtidal depths where light penetrates to support photosynthesis of benthic plants to the landward edge where the sea passes its hydrologic influence to groundwater and atmospheric processes. At the seaward margin, biofilms, benthic algae, and seagrasses are representative biotic components. At the landward margin, vegetation boundaries range from those located on groundwater seeps or fens in humid climates to relatively barren salt flats in arid climates.

Tidal wetlands are a critical component of the coastal ocean landscape, which consists of a continuum of landscape elements or ecosystems stretching from where rivers enter the coastal zone, through the estuary, and onto the continental shelf (Fig. 1.1). In addition to tidal wetlands, the coastal ecosystems include seagrass meadows, rivers, tidal creeks, estuarine waters and unvegetated subtidal bottoms, tidal flats, coral reefs, and continental shelf

[†]Deceased.

Coastal Wetlands
https://doi.org/10.1016/B978-0-444-63893-9.00001-0

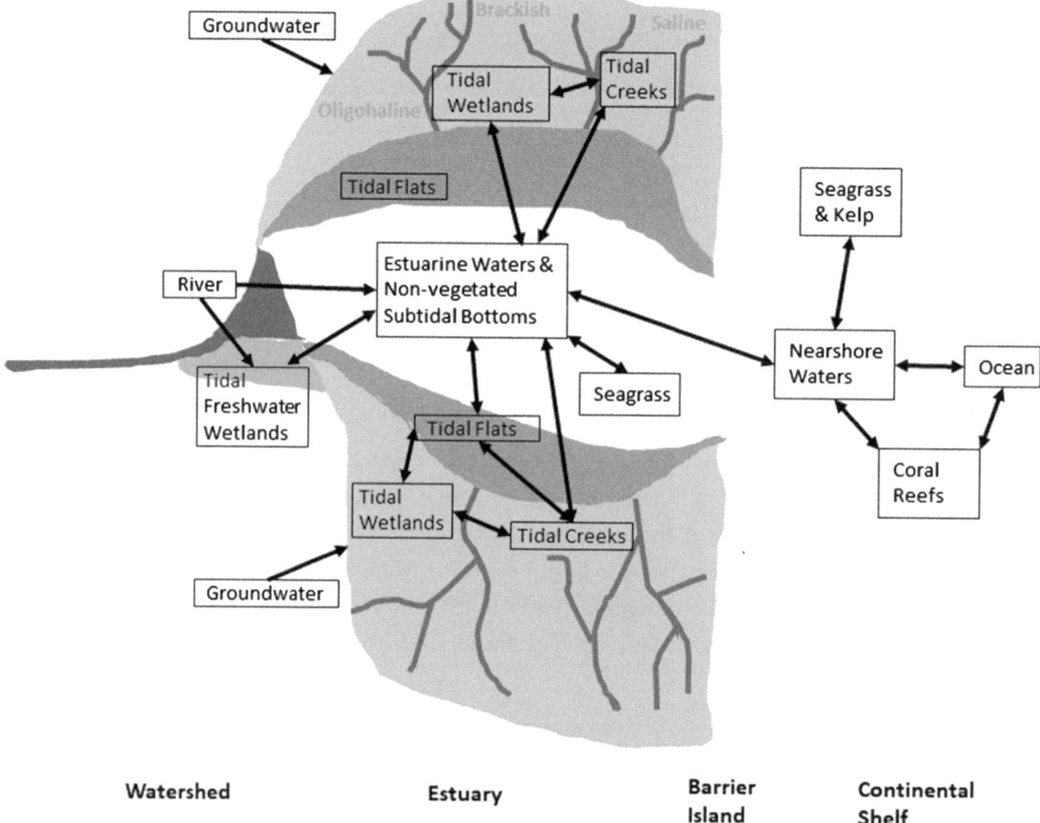

FIGURE 1.1 The coastal ocean landscape highlighting a tidal wetland-dominated estuary and its linkages to the adjacent watershed/river, the continental shelf, and the open ocean. *Arrows* illustrate key hydrologic and material exchanges between the various landscape elements.

waters and bottoms. Quite often there are extremely sharp transitions between landscape elements in the coastal zone, for example, between open water and marsh, marsh and uplands, and seagrass and mangroves (Fig. 1.2A and B). Groundwater (freshwater draining from uplands) is another source of upland-derived waters and materials. The exchange and mixing of water and materials entering from rivers and the ocean defines the overall structure and distribution of landscape elements. The distribution and deposition of sediments from land and the ocean establishes the overall bathymetry. Bathymetry in combination with tidal range and the spatial gradient in salinity are the primary determinants of ecosystem distribution within the coastal ocean landscape. Emergent and submerged vegetation, once established, exert an ecogeomorphic feedback on the fate of river and ocean sources of various materials. For example, vegetation slows the movement of water, promotes the settling of sediment particles, and accelerates estuarine infilling and tidal wetland expansion and accretion.

FIGURE 1.2 Examples of transitions and interactions between adjacent landscape components. Sharp transitions are typical across the coastal landscape (A) between the tidal salt marsh and the upland forest (Virginia Coast Reserve, USA) and (B) between seagrass and mangroves (Palau). Impacts in one landscape component can adversely affect adjacent components; for example, excess N-enrichment has led to (C) the invasion of ferns in the Mai Pai mangroves (Hong Kong) and (D) the accumulation of plant litter on the mangrove sediment surface. Compare the accumulation of litter here with the absence of litter in healthy mangroves shown in Fig. 1.8 H–J.

The movement of water links all the coastal ocean ecosystems and promotes the exchange of materials between them, thereby increasing the productivity of individual ecosystems and the entire system. The tidal wetlands serve as a valuable filter of watershed-derived materials, such as nitrogen, which lessens the potential for estuarine N-enrichment and eutrophication (Hopkinson and Giblin, 2008). Tidal wetlands also are an important source of organic carbon for estuarine and nearshore ecosystems, enhancing their secondary production including important coastal fisheries. The outwelling of organic matter from coastal wetlands and its importance in subsidizing secondary production was one aspect of coastal ecosystem linkages first studied (Odum and Heald, 1972, for mangroves; Odum, 1980, for saltmarshes, but also see Odum, 1985). Other connections that have been documented include biomass transfer by water currents of plant litter between different components of the coastal ecosystem (De Boer, 2000), the acceleration of organic matter and nutrient transfer by migrating fish and shrimp that use different ecosystems in different stages of their life and during different times of the day or season (Deegan, 1993; Mumby et al., 2004), and the protection that tidal wetlands offer to seagrass and coral reefs by trapping riverine nutrients and sediments, thereby enhancing water clarity (Wolanski and Elliott, 2015).

Inputs of freshwater from groundwater and upstream tidal freshwater wetlands can mitigate the effects of watershed drought by slowly releasing freshwater that ward off estuarine hypersalinity. This is the case, for instance, of freshwater tidal wetlands adjoining mangroves in Micronesia. There the mangroves were spared the stress of hypersalinity during a severe El Nino drought because the supply of freshwater from groundwater and the freshwater tidal wetlands continued 6 months into the drought even after the watercourses had dried (Drexler and Ewel, 2001; Drexler and de Carlo, 2002). On the negative side, this connectivity between coastal ocean ecosystems can lead to the degradation of the whole ecosystem if one component has been severely impacted by human activities. For example, the destruction of coastal wetlands can lead to a degradation of adjacent seagrass and coral reefs in coastal waters (Duke and Wolanski, 2001). As another example, a degraded estuary can in turn degrade adjoining coastal wetlands. For instance, the discharge of treated sewage from more than one million people in the small bay draining the Mai Po mangrove reserve in Hong Kong has resulted in an excess of nutrients in the bay waters and sediment. Ferns and weeds have invaded the substrate of these mangroves (Fig. 1.2C). There is so much plant litter that the detritivores cannot consume it all (Fig. 1.2D; Lee, 1990); the thick fern vegetation along the banks prevents the flushing of this plant litter; the plant litter accumulates and decays, releasing hydrogen sulphide (H_2S), which in turn further reduces the crab population, which stresses the mangrove trees by inhibiting the aeration of the soil and the flushing of excess salt from the soil. The stressed trees generate less tannin (Tong et al., 2006), and borers use this weakness to attack the trees, resulting in stunted tree growth (Wolanski et al., 2009; Wolanski and Elliott, 2015). The coastal landscape continues to evolve, especially in response to human activities that alter the magnitude and timing of water and material inputs from watersheds and contribute to climate change, including sea level rise (SLR). Saltwater intrusion is shifting the distribution of estuarine and tidal wetland ecosystems upstream, and SLR is causing tidal wetlands to transgress upland regions.

This book focuses on commonly recognized ecosystems along this hydrologic gradient: seagrass meadows, intertidal flats, tidal saltmarshes, mangrove forests, and tidal freshwater wetlands. Coral reefs are not covered at all because they are so physically and biologically distinct from the foregoing list, as well as in part because they have received research attention equivalent to the totality of all of the wetlands covered in this book (Duarte et al., 2008). Little direct reference is made in this book to lagoons that are intermittently connected to the sea; regardless, all five of the ecosystem types can and do occur in these and other more specialized geomorphic settings. They would all comprise the array that we recognize as coastal wetlands.

This book addresses the pressing need to quantify the ecological services provided by coastal wetlands as a tool to guide better management and conservation worldwide because coastal wetlands are disappearing worldwide at an alarming rate; in some countries the loss is 70%–80% in the last 50 years (Frayer et al., 1983; Duarte, 2002; Hily et al., 2003; Bernier et al., 2006; Duke et al., 2007; Wolanski, 2007).

2. A SYNTHESIS OF COASTAL WETLANDS SCIENCE

In this section we provide an overview of the structure and functioning of coastal wetlands with emphasis on key forces and processes that interact with their coastal geographic location. It is difficult to discuss these forces in isolation because, for example, climate change influences sea level, sea level forces changes in vegetation structure, disturbance, and herbivory affect vegetation, and so forth. As such, comparison across major ecosystem types can provide insight into the differences and the relative importance of both extrinsic forces and intrinsic structure. We also discuss the role of modeling in elucidating the relative importance of key processes and in predicting the effects of alteration by humans. This last effect presents us with challenges of how to best protect and manage coastal wetlands for their attributes of life support and, importantly, for their contribution to esthetic and cultural values. To this end, we outline key research needs that recognize the usefulness of working beyond "single factor cause and effect."

2.1 Geography

Coastal wetlands include seagrass meadows, intertidal flats, tidal saltmarshes, mangrove forests, and tidal freshwater wetlands. They are found in six continents and all but extreme polar latitudes. Cliffs and rocky shores are probably the only coasts with minimal wetlands. Worldwide, wetlands are heavily impacted by increasing population and coastal development. The area occupied by wetlands has been greatly reduced over the past 100 years and it will likely decrease substantially throughout the 21st century as population pressure and the rate of SLR accelerate. There is considerable uncertainty in estimates of the global area covered by all tidal wetlands. Much of the uncertainty is the result of insufficient detail in satellite imagery (e.g., 1 km) and the integration of disparate and incompatible geospatial and statistical data sources (FAO, 2007). Consistent utilization of the US Geological Survey (USGS) compiled Landsat archive, which is now freely available, could likely decrease the uncertainty in estimates that are currently in the literature.

Estimates of mangrove cover differ widely, from about 80,000 km^2 to about 230,000 km^2 worldwide (Diop, 2003; Duke et al., 2007). One of the most recent estimates, based on 30 m resolution Landsat imagery, is 137,760 km^2 (Giri et al., 2011). Using a slightly different approach, however, Hamilton and Casey (2016) estimate global mangrove area of only 83,495 km^2. The largest extent of mangroves is in Indonesia, and about 75% are located in just 15 countries (Giri et al., 2011). Although the greatest distribution is between 5°N and 5°S, mangroves extend from 31°20′ N to 38°59′ S (Fig. 1.3). The distribution is controlled by the combination of continental (no frosts or only very rarely, typically less than 1 frost every 10 years; Lugo and Patterson Zucca, 1977) and oceanic climates (warm waters; Duke et al., 2007). Recent compilations estimate global saltmarsh area at about 54,950 km^2 (Mcowen et al., 2017) The area of saltmarshes in Canada and the United States alone is about 19,600 km^2 according to Chmura et al. (2003) and 20,000 km^2 according to Mcowen et al. (2017). Large areas of saltmarshes (~7000–13,000 km^2) might also exist in northern Russia and Australia (Mcowen et al. (2017). Saltmarshes are also found scattered in the mangrove belt, usually in the upper intertidal areas landward of mangroves. Seagrasses cover was most recently estimated to be 177,000 km^2 worldwide (Green and Short, 2003; Waycott et al., 2016), ranging from at least 165,000 km^2

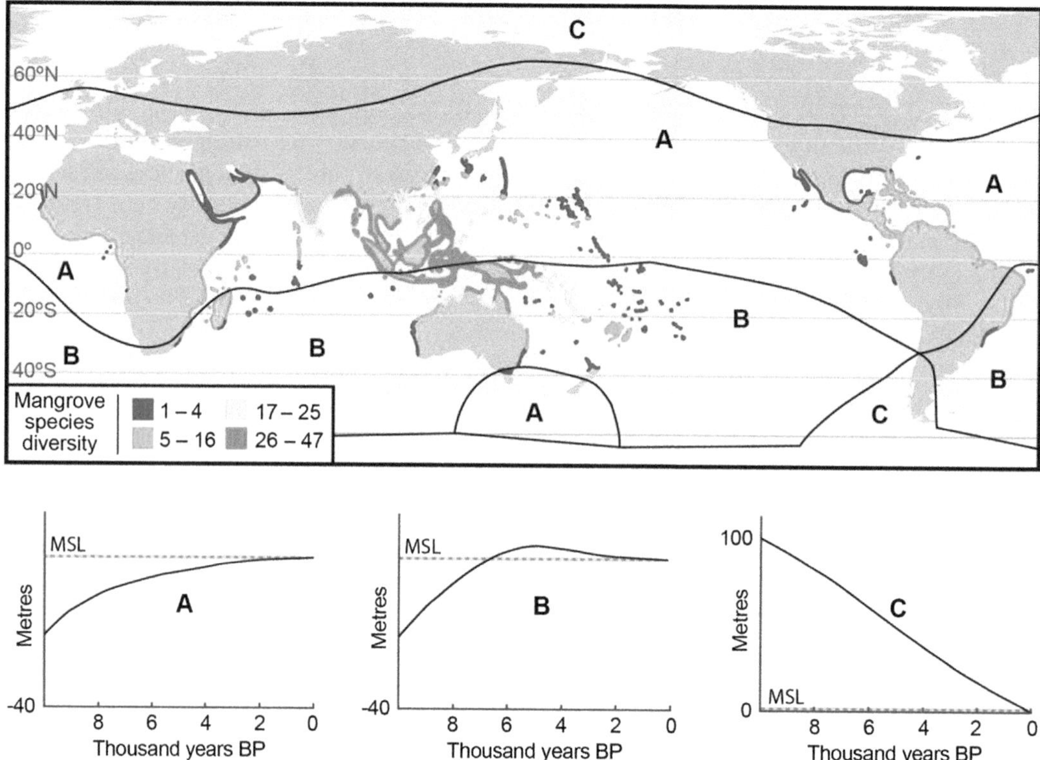

FIGURE 1.3 World distribution of mangroves (*shaded areas* along tropical and subtropical shorelines) and the approximate distribution of coasts that follow one of three relative sea level trajectories for the past 10,000 years. Zone A has seen a continual rise, zone B experienced a more rapid rise initially followed by a slow decrease in SL, while zone C experienced a continued decrease in sea level primarily as a result of tectonic activity and rebounding from glacial coverage. *Redrawn from Ellison, J., 2009. Geomorphology and sedimentology of mangroves. In: Perillo, G.M.E., Wolanski, E., Cahoon, D.R., Brinson, M.M. (Eds.), Coastal Wetlands: An Integrated Ecosystem Approach. Elsevier, Amsterdam. 565–591 and adapted from; Woodroffe, C.D., 1992. Mangrove sediments and geomorphology. In: Robertson, A.I., Alongi, D.M. (Eds.), Tropical Mangrove Ecosystems. American Geophysical Union, Washington D.C. 7–42; Pirazzoli, P.A., 1996. Sea Level Changes: The Last 20 000 Years. J Wiley & Sons, Chichester; Lambeck, K., Woodroffe, C.D., Antonioli, F., Anzidei, M., Gehrels, W.R., Laborel, J., Wright, A.J., 2010. Paleoenvironmental records, geophysical modelling, and reconstruction of sea level trends and variability on centennial and longer timescales. In: Church, J.A., Woodworth, P.L. Aarup, T., Wilson, W.S. (Eds.), Understanding Sea Level Rise and Variability, Wiley-Blackwell, Chichester, 61–121. See Chapter 20.*

up to 600,000 km^2 (Nellemann et al., 2009a; Hopkinson et al., 2012). The worldwide area of tidal flats and freshwater coastal wetlands seems unknown, though back-of-the-envelope estimates suggest that they may reach 300,000 km^2.

2.2 Geomorphic Evolution Under Past Climate Change: How Present Coastal Wetlands Came to Be

Wetlands continuously evolve in time and space. The story of present-day coastal wetlands starts 120,000 years ago, which is an interglacial period that lasted about 15,000 years and when the mean sea level (MSL) was at a level comparable with the present one (Fig. 1.4A). Where natural conditions allowed, wetlands existed in refugia with a less biodiverse flora than at present (DiMichele, 2014). The human population was tiny at that time and had a negligible impact on the evolution of the coastal wetlands.

Then came a 100,000-year-long ice age that buried much of the present temperate coasts and continental shelves (although mostly concentrated in the Northern Hemisphere) under hundreds of meters, sometimes several kilometers, of ice. Water was taken away from the ocean and stored on the continents. As a result, the MSL decreased and was at its lowest, about 120 m below the present MSL, about 20,000 years ago (Fig. 1.4A). At that time, coastal wetlands would only have existed along those coasts that were not buried by permanent ice, mainly along the then "tropical" belt. Even there the coastal wetlands could only have existed on the upper continental slope, which is where at that time the sea met the land (Fig. 1.5). Because the inclination of the continental slope is much steeper than the continental shelf where the coastal wetlands are now located, the space for accommodating coastal wetlands would have been very limited. At that time (20,000 years ago), the area occupied by coastal wetlands would have been much smaller and in many cases restricted to estuaries that developed along present-day submarine canyons (Perillo, 1995). Taking all of this into account, coastal wetlands would have occupied perhaps as little as 5% of the area covered just prior to human alterations.

Then, 20,000 years ago, came a period of rapid change for coastal wetlands. In response to melting of the glaciers, sea surface height rose rapidly, much faster than it fell during the previous 100,000 years (Fig. 1.4A) (Kemp et al., 2011). Absolute sea level rose rapidly until about 6000 years ago and then continued to rise slowly, for most of the world's oceans, until the mid-1800s (Fig. 1.4B). With the exception of equatorial regions, the absolute rate of SLR decreased to less than 0.4 mm year^{-1} (Fig. 1.4C). From the perspective of coastal systems and tidal wetlands, it is not only the sea surface height that affects water level but also changes in land elevation. Compression of the land by the weight of glaciers, subsidence of the land by the weight of overlying sediments or sinking because of the subsurface extraction of oil and gas, and tectonic activity can also affect the relative sea level. Equatorial ocean siphoning also affected the apparent rate of SLR in equatorial regions. This was the result of the process of glacial isostatic adjustment, in particular the collapse of forebulges at the periphery of previously glaciated regions (Mitrovica and Milne, 2002). In the equatorial region, sea level peaked a few meters above current levels about 5000 years ago.

About 6000 years ago, the bulk of the continental ice outside of polar and near-polar regions had melted, which released a huge weight from those continental shelves that were burdened by ice. In those areas the continental shelf rose by isostatic rebound of the land, with the largest rebound occurring where the ice burden was greatest. In zone C of Fig. 1.3, the land rose faster than the sea, and the relative MSL decreased over the last 10,000 years, the coast prograded, and new land emerged. In many places where this occurred, there is evidence of marine clays stranded tens of meters higher than where the sea is now (e.g., coastal Massachusetts, USA (McIntire and Morgan, 1962)). In other areas, the relative sea level rose about 20 m during the last 10,000 years. In some areas (zone A in Fig. 1.3) this increase was asymptotic, being largest 10,000 years ago and minimal at present; in such areas the coast retreated from an advancing sea. In the other areas (zone B in Fig. 1.3) the relative sea level reached a maximum about 5000 years ago typically 2–3 m

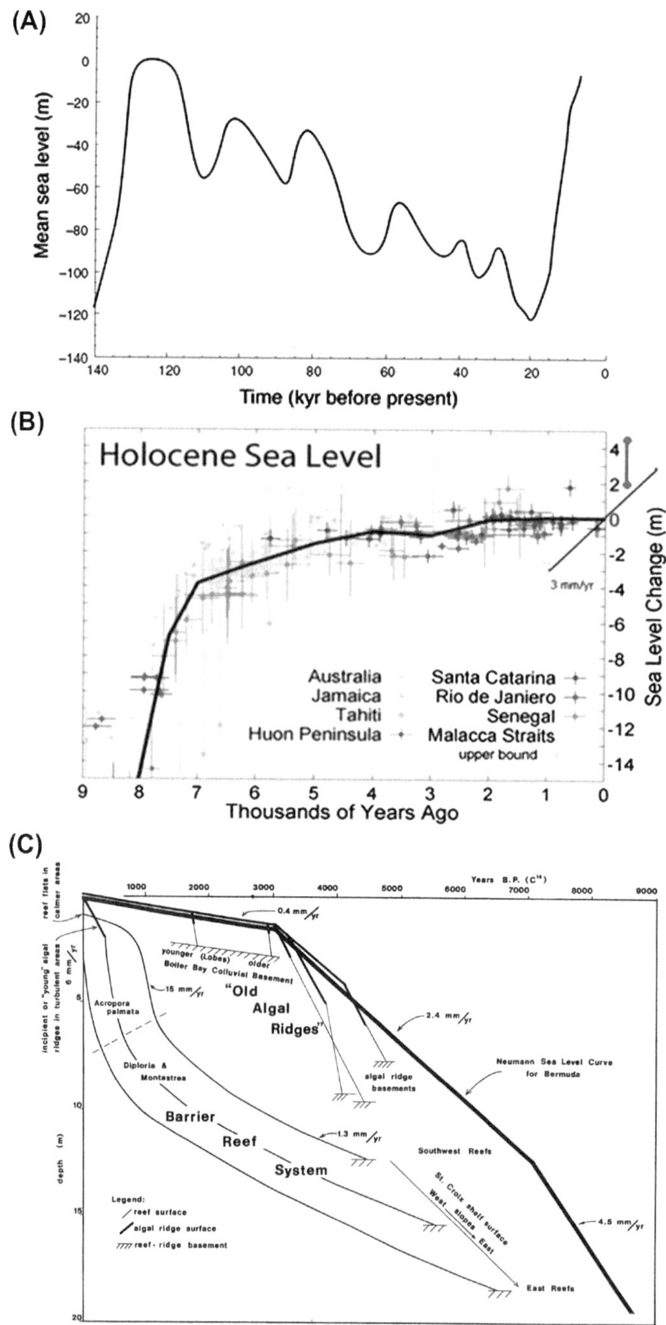

FIGURE 1.4 Time series of sea level and the rate of sea level rise over three time frames: (A) absolute mean sea level during the last 140,000 years, (B) the rate of sea level rise decreased rapidly over the past 9000 years. Rates decreased to about 0.4 mm year^{-1} until the mid-1800s. The current 3–3.5 mm year^{-1} rate is illustrated with the line at point 0,0. (C) Tidal wetlands as we know them today developed in the past 4–6k years after rates of sea level rise decreased to less than 1–2 mm year^{-1}. *(A) Reproduced from Wolanski, E., 2007. Protective functions of coastal forests and trees against natural hazards. In: Braatz, S., Fortuna, J., Broadhead, R., Leslie, R. (Eds.), Coastal Protection in the Aftermath of the Indian Ocean Tsunami. What Role for Forests and Trees? FAO, Bangkok. 157–179. (B) From Kemp, A., Horton, B., Donnelly, J., Mann, M., Vermeer, M., Rahmstorf, W., 2011. Climate related sea-level variations over the past two millennia. Proceedings of the National Academy of Sciences 108, 11017–11022 and Robert A. Rohde, Global warming art project. (C) From Adey, W., Burke, R., 1976. Holocene bioherms (algal ridges and bank-barrier reefs) of the eastern Caribbean. The Geological Society of America Bulletin 87, 95–109. Doc. No. 60112 – Fig. 3, page 100.*

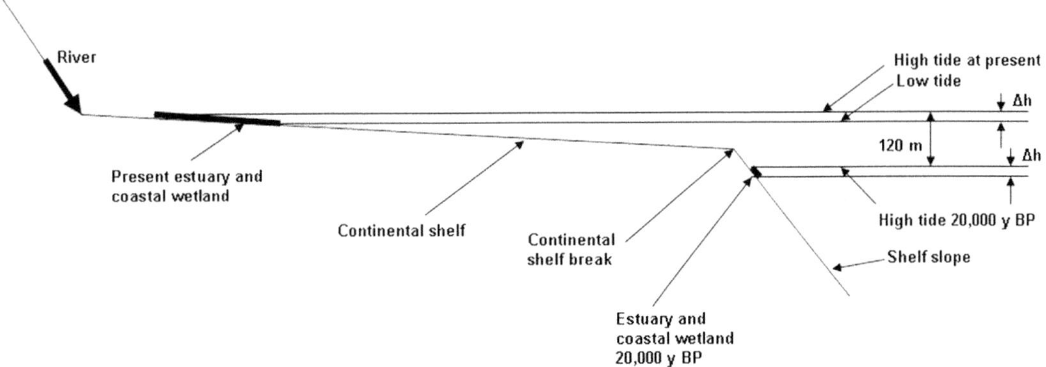

FIGURE 1.5 Sketch of the continental shelf and continental slope, and the location of the estuary and the coastal wetlands at present and 20,000 years BP (years before present).

higher (Isla, 1989) and up to 7 m higher than at present (Gómez and Perillo, 1995), and it decreased smoothly to its present elevation until about 2000 years ago. In those areas where sea level has dropped, the coast initially receded from an advancing sea and then prograded slightly during the last 5000 years. Within each of these zones, there were local exceptions (Sweet et al., 2017), especially in deltaic regions where sediment supply was sufficient to compensate for coastal retreat.

Worldwide the estuaries responded to these changing conditions of sea level, river and sediment discharge, changing sea ice, water currents, storms, and waves brought about by a new climate. The elevation of the MSL relative to land level determined at any time the location of the estuary. The net sediment budget, the balance between the sediment inflow (from the river and import from the sea), and the sediment outflow from the estuary to the sea determined the evolution of the estuary. Where the relative sea level fell rapidly (zone C in Fig. 1.3), new land emerged constantly and the estuary migrated seaward, continually reinventing itself; the "old" estuary rose above sea level and became part of the landscape; this evolution is still proceeding in zone C (Fig. 1.6A). Where sea level rose, "old" estuaries were drowned and new estuaries formed landward where the land met the sea (Fig. 1.6B).

In zone C (Fig. 1.3), no steady state has yet been reached. The estuary is still moving seaward as new land emerges. In zone A, a quasi-steady state was nearly reached by the end of the 1800s as the relative rate of SLR was only about 0.4 mm year^{-1} for the previous 1000 years or so. In zone B, the estuary is still evolving, albeit much slower than in the past because the changes in relative MSL slowed down considerably during the last 6000 years. The downward trajectory in the absolute rate of SLR began to reverse worldwide during the mid-1800s (Kemp et al., 2011). Regardless of whether estuaries were in steady state or not at that time, changes that began then have only become greater since. This will likely move all tidal wetlands out of steady state, so coastal wetlands will continue to migrate.

2.3 The Influence of Vegetation on the Geomorphic Evolution With Climate Change

Estuaries trap some portion of the riverine sediment input (Perillo, 1995; Wolanski and Elliott, 2015). Estuaries in zone C are thus silting as soon as they form in a new position; small

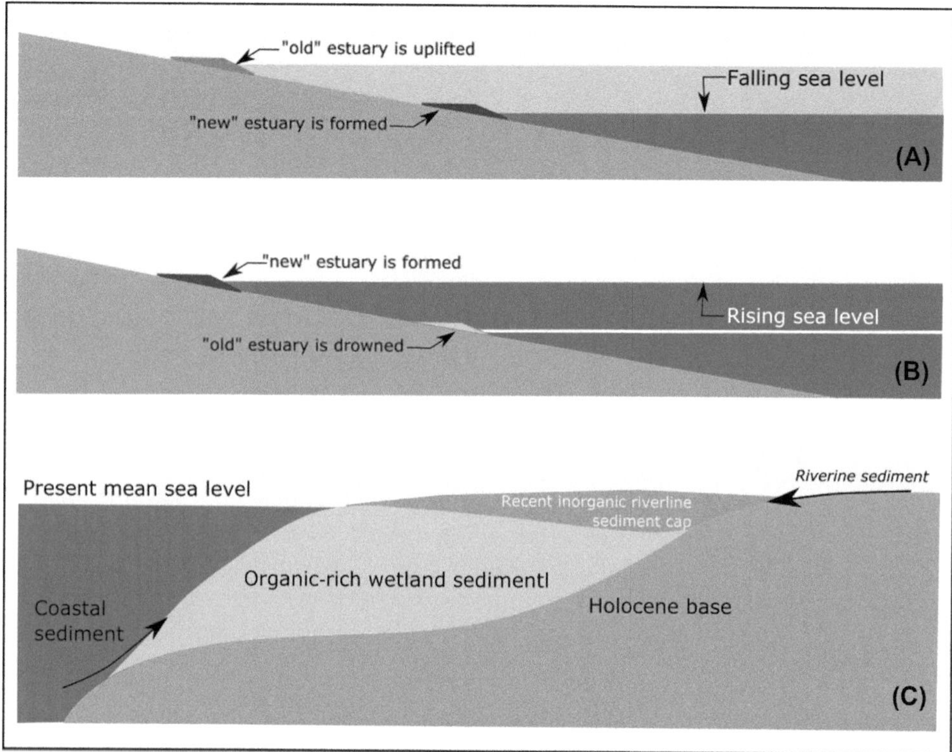

FIGURE 1.6 Sketch of the migration of an estuary with (A) falling and (B) rising sea level; (C) in macrotidal areas with large riverine sediment inflow, a vast coastal wetland often formed in the estuary that kept up with sea level rise until the last few thousand years when the wetland died as it was capped by inorganic riverine sediment; (D) in other open estuaries with a smaller riverine sediment inflow, rising seas drowned the wetlands, leaving behind organic-rich wetland mud. This mud was capped by inorganic sediment of oceanic and riverine origin, until, a few thousand years ago, it was close enough to the water surface that the coastal wetland could expand on it, creating a near-surface organic-rich mud layer; (E) in basins that were submerged by rising seas to depths too deep to support a wetland, riverine mud accumulated in the basin until depths became small enough that a costal wetland was established and accumulated organic-rich mud; (F) in lagoons with little riverine sediment inflow, a coastal wetland formed originally as the area was flooded by rising seas; the wetland could not keep up with rising seas; the wetland died and was capped by calcareous sediment of oceanic origin that can support seagrass; (G) in other lagoons, a coastal wetland was formed, as the area was flooded by rising seas; it died when it could not keep up with sea level rise and it was capped by inorganic sediment of terrestrial and oceanic origin. When this sediment surface neared the water surface, a coastal wetland developed over this sediment, creating a near-surface layer of organic-rich mud. Note that figure letters are not to be confused with zonation described in Fig. 3. *Examples taken from Ellison, J., 2009. Geomorphology and sedimentology of mangroves. In: Perillo, G.M.E., Wolanski, E., Cahoon, D.R., Brinson, M.M. (Eds.), Coastal Wetlands: An Integrated Ecosystem Approach. Elsevier, Amsterdam. 565–591; Lara, R.J., Szlafsztein, C.F., Cohen, M.C.L., Oxmann, J., Schmitt, B.B., Filho, P.W.M.S., 2009. Geomorphology and sedimentology of mangroves and salt marshes: the formation of geobotanical units. In: Perillo, G.M.R., Wolanski, E., Cahoon, D.R., Brinson, M.M. (Eds.), Coastal Wetlands: An Integrated Ecosystem Approach. Elsevier Science, Amsterdam, 593–614; Woodroffe, C.D., Davies, G., 2009. The morphology and development of tropical coastal wetlands. In: Perillo, G.M.R., Wolanski, E., Cahoon, D.R., Brinson, M.M. (Eds.), Coastal Wetlands: An Integrated Ecosystem Approach. Elsevier Science, Amsterdam, 65–88.*

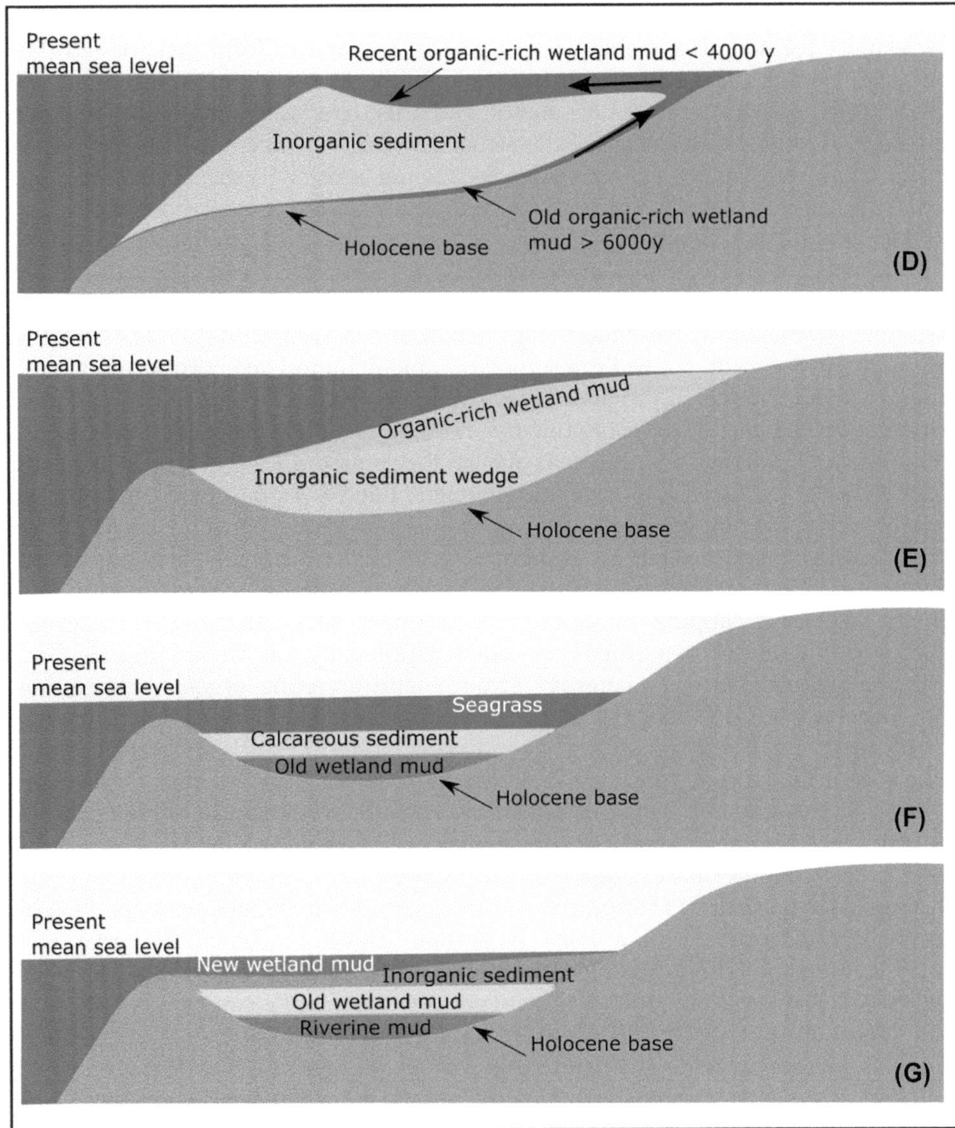

FIGURE 1.6 cont'd

estuaries not flushed by seasonal freshets of large rivers are turning into mud flats that are stabilized by biofilms and can rapidly be colonized by saltmarsh vegetation, which creates new high-latitude saltmarshes (Martini et al., Chapter 4). The evolution of estuaries in zones A and B was controlled by the geomorphology when the rising seas flooded new land and by the tides, river runoff, and oceanographic conditions at the coast. Because these conditions varied from site to site around the world, a large number of estuarine evolution pathways occurred (Woodroffe Chapter 2 and Ellison Chapter 20). At some locations (Fig. 1.6C) a wide, shallow estuary formed; this enabled a vast coastal wetland to form that fringed the main channel as the estuary filled with sediment from both the river and the sea (through processes such as tidal pumping). Because of vegetation and adequate sediment supplies, this wetland type accelerated siltation and enabled the estuarine system to keep up with SLR. Later this estuarine, saline sediment was capped by freshwater sediment; the estuary had by then reached "old age" because it had used all the accommodating space on land and could then only evolve further by moving offshore to form a delta, or it can occasionally be rejuvenated by opening a new channel or by isostatic or structural movements. In other cases (Fig. 1.6D), the estuary moved upland with a rising sea level until about 6000 years ago when sea level stabilized; as the estuary migrated, so did the coastal wetlands, leaving behind their signature in the form of a well-preserved organic-rich mud layer. This mud was capped by inorganic sediment, which provided a substrate on which the wetland prograded, and in the process formed a near-surface, organic-rich mud layer. Still in other cases (Fig. 1.6E), principally in microtidal areas and in areas where the geomorphology formed semienclosed coastal waters as sea level rose, riverine sediment filled a lagoon-type environment and this area was colonized by wetlands, leaving behind an organic-rich mud layer. The relative sea level evolution over the last few thousand years has created, in sediment-starved estuaries of zone B, a varied succession of wetland habitats, including (Fig. 1.6F) the die-off of mangroves and their capping by calcareous sediment later colonized by seagrass, and (Fig. 1.6G) the die-off of mangroves and their capping by calcareous sediment with an even later capping by a new mangrove swamp with accreting organic-rich mud. In the case of temperate, sediment-starved estuaries in zone B, the lack of sediment input led to erosion of the original wetland sediments and an evolution in wetland types. For example, in Bahía Blanca Estuary, Argentina, some older *Sarcocornia* high marshes eroded (Minkoff et al., 2005, 2006; Escapa et al., 2007), which lowered their position in relation to MSL and led to their evolution to tidal flats, which later were colonized by *Spartina* (Pratolongo et al., 2013; Pratolongo et al., Chapter 3).

The marshes in Barnstable Harbor in the northeastern United States are a good example of tidal wetland development in a zone A system. Development of the tidal wetlands commenced when the rate of SLR slowed to about 0.4 mm $year^{-1}$ about 2000 to 3000 years ago (Fig. 1.4C). Redfield (1965) took sediment cores to basement coarse sands and, on the basis of ^{14}C dating the peat at various depths in the cores, reconstructed the processes of harbor infilling, wetland transgression, wetland progradation, and vertical elevation gain (Fig. 1.7A). The initial marsh was first established along the upland-tidal flat shoreline between MSL and mean high water (MHW). Once marsh vegetation became established, the marsh gained elevation through the combined processes of enhanced sediment trapping during tidal inundation and the accumulation of undecomposed marsh plant roots and rhizomes (Fig. 1.7B). Concomitantly with marsh elevation gain, tidal flats gained elevation because of deposition of ocean-derived sediment through the process of tidal pumping (Woodroffe, Chapter 2 and Ellison Chapter 20). Marsh plants migrate onto these newly created tidal flats when elevation reaches MSL, which enables the marsh to prograde toward open water. As sea level rose (Fig. 1.7A and B), uplands flooded, enabling marsh plants to transgress up across the now flooded uplands. This process continues to this day as sea level continues to rise and sediments enter the harbor by tidal pumping from the ocean or runoff from uplands.

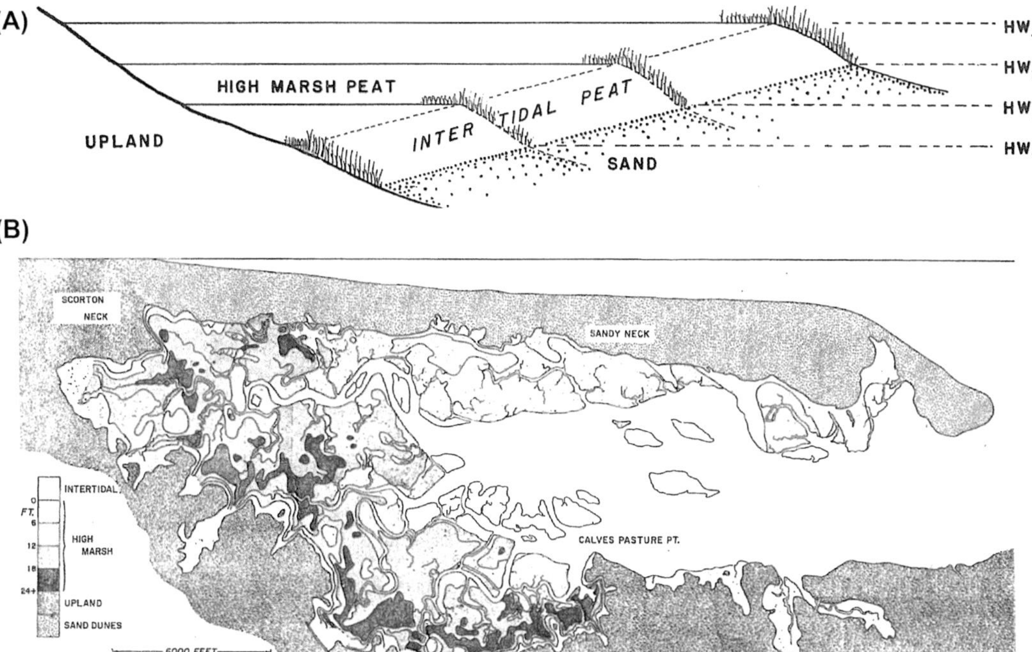

FIGURE 1.7 Salt marsh development over the past several thousand years. (A) The distribution of peat depths in the marshes of the Barnstable Harbor Estuary. Contours are 6 ft (1.6 m). (B) Salt marsh development in Barnstable Harbor Estuary demonstrating the processes of elevation gain, progradation toward open water and transgression onto uplands in response to sediment inputs and sea level rise. *From Redfield, A., 1965. Ontogeny of a salt marsh. Science 145, 50–55 – Figs. 1 and 2 page 52. Reprinted and used with permission of AAAS and ESA.*

Thus, coastal wetlands have been subjected to large changes in drivers over the past 20,000 years, yet they have adapted, greatly expanded in areal distribution, especially during the last 4000 years, and survived. Sea level evolution, changes in sediment and water input from the continent and from the ocean, and tidal range modifications, in addition to the evolution of wetland plants and fauna, are the controlling factors that acted on the various types of wetlands. Therefore, coastal wetlands provide a significant example of one of the most resilient ecosystems—an ecosystem that can withstand major modifications but still evolve in a sustainable manner. Coastal wetlands are constantly evolving by feedback mechanisms, especially with respect to rising sea level and to disturbances. The state of a particular wetland at any one time affects its future and is determined by the accumulated history of previous states. What we see in a wetland is a snapshot in time.

2.4 The Stabilizing Role of Vegetation

There is a strong interplay between tidal hydrodynamics, sediment infilling and redistribution in intertidal regions, wetland plants, and the geomorphic development of drainage networks and tidal wetlands in estuaries (Woodroffe, Chapter 2; Gao, Chapters 10 and Ellison, 20; D'Alpaos et al., Chapter 5). As long as mud flat elevation remains below MSL, generally the mud flat remains fairly flat and uneventful and tidal creeks may not form (Fig. 1.8A).

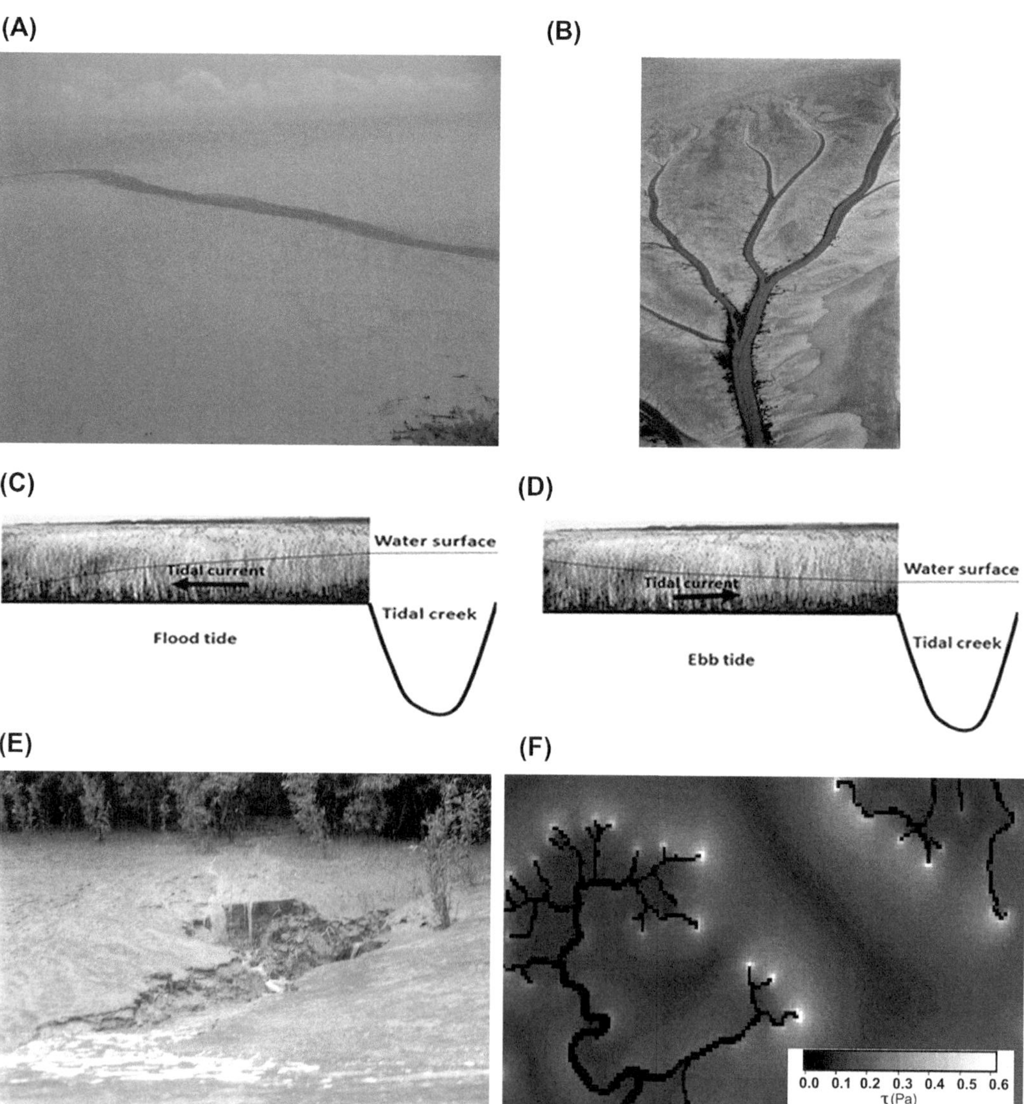

FIGURE 1.8 Aerial photographs of (A) the 6-km wide mud flat at the mouth of the Mary River, Australia's Northern Territory, showing the river channel and no tidal creeks, and (B) a dendritic tidal creek in an unvegetated tidal flat (King Sound, Western Australia; mangrove trees are only present in the lower reaches of the creek where they form a one-tree wide vegetation strip). (C) Sketch of the asymmetry of the water surface and the tidal currents in a coastal wetland at flood tide and (D) ebb tide. (E) The tip of a tidal creek in an unvegetated mud bank stops at the mangrove vegetation (Darwin Harbour, Australia). (F) Numerical prediction of the distribution of frictional stresses in a dendritic tidal creek in an unvegetated mud flat (D'Alpaos et al., 2009). (G) A dense mangrove root network locks the soil together to a depth of several meters and inhibits wave erosion of the bank. (H–J) High near-bottom vegetation density due to tree trunks, prop roots, buttresses, and pneumatophores in mangroves.

FIGURE 1.8 cont'd

However, once the mud flat rises above MSL, the tidal hydrodynamics change profoundly. A large volume of water floods and drains the mud flat in a short time compared with the entire tidal period. This flow becomes funneled in a natural depression generating large currents, which in turn quickly erode the bottom and the banks of the depression forming a tidal course (i.e., a tidal channel, creek, gully, groove, or rill; Fig. 1.8B; Perillo, 2009; Perillo, Chapter 6). The tidal creek grows by tidal current asymmetry, with larger ebb than flood tidal currents, due mainly to the higher mean depth of flood tides—hence lesser friction—at flood tide than at ebb tide (Fig. 1.8C and D). This process of tidal asymmetry operates both in mangroves and saltmarshes (Wolanski et al., 1980; Kjerfve et al., 1991; Mazda et al., 1995; D'Alpaos et al., Chapter 5). In addition, there can be intense erosion at the head of the tidal creek driven by small waterfalls that form there at ebb tide. Other forms of erosion also help in the growth of the tidal courses (*sensu* Perillo, 2009) over unvegetated mud flats (Perillo, Chapter 6). At ebb tide, frictional forces retard the flow over the mud flat more than the ebb flow is retarded in the tidal creek (Fig. 1.8E). A tidal creek forming in an unvegetated mud flat develops a dendritic pattern with several erosional "hot spots" at the tips and at the external flanks of tight meanders (Fig. 1.8F).

Before a mud flat becomes vegetated, it can be stabilized by biofilms or destabilized by bioturbation. Once the mud flat is vegetated by pioneer species of mangroves or saltmarsh vegetation, a quasi-steady state develops characterized by a sharp discontinuity separating the vegetated wetland from the unvegetated tidal course (Fig. 1.9A–C; Perillo, Chapter 6); the exception is Arctic coastal wetlands because scouring by ice reshapes the bathymetry yearly (Fig. 1.9D). The patterns of erosion and siltation change profoundly once vegetation is established for three main reasons. First, the plant roots hold the soil together and this slows down or even stops bank erosion by tidal currents and waves to the depth of the rooting depth. Rooting depth varies but generally ranges from about 15 cm in seagrass to over a meter in creek bank mangroves and saltmarshes (Fig. 1.8G–J; Holmer, Chapter 13). Second, the wave erosion of the vegetated substrate is decreased because hydraulic energy of waves is dissipated by vegetation (Fig. 1.10A–D). Third, the tidal currents are slowed 95% by the

FIGURE 1.9 Tidal channels through coastal wetlands are generally unvegetated and can persist for long periods (years to hundreds of years) such as in these photographs at low tide of (A) a salt marsh channel (Mont-Saint-Michel, France) and (B) a mangrove channel (Darwin, Australia). (C) Tidal channels start to form as soon as pioneer vegetation is established, even sparsely (Mont-Saint-Michel, France). (D) In the Arctic coastal wetlands, general mesotidal conditions and yearly scouring by ice tend to impede the formation of long-term tidal channels, especially on relatively flat coasts, though ice does lead to the formation of sculptured marshes ("jigsaw" marshes) (James Bay, Canada). *(A) Courtesy of E. Langlois-Saliou. (C) Courtesy of E. Langlois-Saliou. (D) Courtesy of I.P. Martini.*

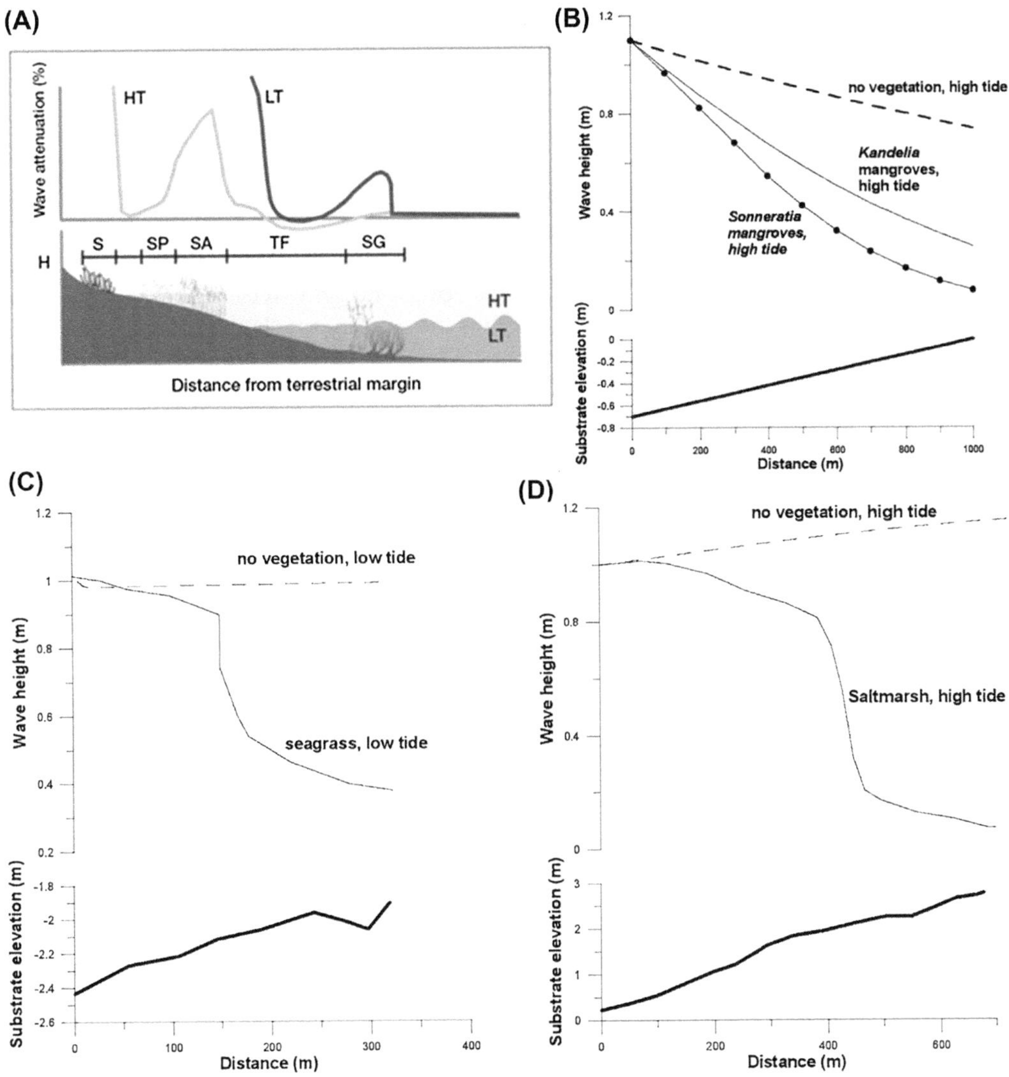

FIGURE 1.10 Effects of vegetation on the dissipation of wave height and energy. (A) A generic representation of wave attenuation (%) at high (HT) and low (LT) tide over an idealized temperate coastal wetland during nonextreme events. Biotic structure contributes to wave attenuation only when submersed. When marshes, mangroves, and seagrasses are exposed at low tide, they do not attenuate waves. Even so, during low tide, they may still contribute to coastal protection via sediment stabilization. In this figure, we have incorporated the geomorphologic effect, resulting in wave shoaling, wave regeneration, and wave breaking, which leads to 100% wave attenuation. In the present case, we are not considering the effects of wave surf, run-up/backwash, or the dampening of potential high turbidity situations, i.e., fluid mud. (*S*, *Salicornia* spp.; *SA*, *Spartina alterniflora*; *SG*, seagrass; *SP*, *Spartina patens*; *TF*, tidal flat) (B) Swell waves intruding in mangroves along the coast of the Gulf of Tonkin, Vietnam, decrease in amplitude with distance into the mangroves and this decrease is a function of the tidal height and the mangrove species. Hydrodynamics model predictions of the attenuation of swell waves (C) propagating over a seagrass meadow at low tide and (D) intruding in a salt marsh at high tide, together with the predicted wave height if there was no vegetation. In all cases, the vegetation significantly reduces the wave height. *(A) From Koch, E.W., Barbier, E.D., Silliman, B.R., Reed, D.J., Perillo, G.M.E., Hacker, S.D., Granek, E.F., Primavera, J.H., Muthiga, N., Polasky, S., Halpern, B.S., Kennedy, C.J., Wolanski, E., Kappel, C.V., Aswani, S., Cramer, L.A., Bael, D., Stoms, D.M., 2009. Non-linearity in ecosystem services: temporal and spatial variability in coastal protection. Frontiers in Ecology and the Environment 7, 29–37. (B) Drawn from data in Mazda, Y., Wolanski, E., 2009. Hydrodynamics and modeling of water flow in mangrove areas. In: Perillo, G.M.E., Wolanski, E., Cahoon, D.R., Brinson, M.M. (Eds.), Coastal Wetlands: An Integrated Ecosystem Approach. first ed., Elsevier, Amsterdam, 231–261.*

vegetation compared with currents in the tidal creek; also, the currents around the vegetation generate eddies and stagnation zones where the suspended sediment imported with the rising tide settles; the tidal currents are too small at falling tide to resuspend all that sediment; thus the vegetated tidal flat silts (Furukawa et al., 1997; Wolanski and Elliott, 2015; D'Alpaos et al., Chapter 5). From the wetland edge to the interior, waves are attenuated, thus reducing their ability to erode and transport sediments.

Wave attenuation and wave action across a subtidal to supratidal transect vary depending on tidal stage (Fig. 1.10A) (Koch et al., 2009). Except for ocean waves acting along wetlands open to the sea, most waves are generated within the estuary by (1) direct wave generation, (2) wind–current interaction, or (3) boat wakes (Perillo and Sequeira, 1989). The degree of wave activity and its extension within the wetland is fully dependent on the wave characteristics and tidal stage. At low tide, waves are only generated within the channels by processes (1) and (2), (process (3) is only important in highly navigated channels) and require intense winds. These waves only act against the channel flanks generating erosion because they tend to be very steep waves (Perillo, Chapter 6). As the tide advances, the fetch for wave generation increases, as does wave number and frequency, and their area of influence increases. Except for hypertidal estuaries, when tides cover the wetland, waves can be active in all dimensions. At that point, different plant types and the shoaling effect of wetland topography become important in defining how much sediment trapping or resuspension/erosion can occur on the wetland.

Tidal creek drainage patterns in fully developed marsh systems are very similar to their terrestrial counterparts (Novakowski et al., 2004). Both have mostly dendritic creek patterns (that is, for the mature marsh), and in the downstream (ebb) direction channel reaches converge toward a main channel. Youthful marshes on the other hand have a reticulated pattern. While only a subset of marsh systems has been examined from this perspective, the highest stream order observed is five, in the Strahler system (Horton, 1945). Both terrestrial and marsh drainage networks follow Horton's law of stream order, which says that stream reach length increases exponentially with order (Rinaldo et al., 1999). In coastal wetlands, the relation between drainage basin area (A_W) and total creek length is similar to the same relation in terrestrial basins. For the North Inlet (South Carolina, USA) marsh system, there is a power relation of $L = 0.08A_W^{0.73}$ (Fig. 1.11). This relation is in agreement to the L-A_W relation proposed by Hack (1957) for terrestrial drainage networks.

To quantify the evolution of an estuary, models have been developed to compute the sediment dynamics over vegetated and unvegetated mud flats. The aim of these models is to simulate the evolving interaction between currents and sediment transport because this interaction determines the evolution of the bathymetry (i.e., D'Alpaos et al., Chapter 5).

Kirwan and Murray (2007) successfully captured the interactive sediment transport processes with the dynamics of vegetation biomass and productivity that lead to tidal wetland and drainage network development in an estuary using a 3D simulation model. The model explores the interactive effects of tidal amplitude, sediment availability, and the rate of SLR. At a constant 1 mm year^{-1} rate of SLR and a suspended sediment concentration of 20 mg L^{-1} (held constant throughout the domain), open water areas shallow, a marsh platform and drainage network develop, and accretion rates on the marsh platform come to equilibrium with the rate of SLR (Fig. 1.12 left to right). Decreases in suspended solids lead to expansion of the tidal creek network, as does an increase in the rate of SLR at constant

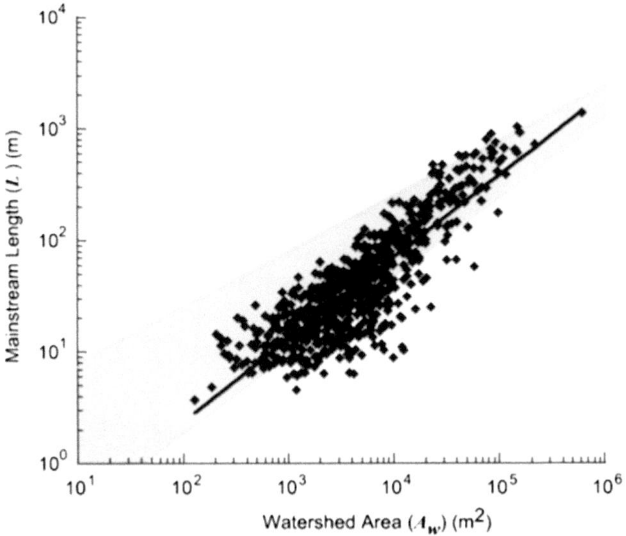

FIGURE 1.11 Relation between length of tidal creeks and the watershed area they drain for a salt marsh (*dots*) and terrestrial systems (*shaded area*). *From Novakowski, K., Torres, R., Gardner, R., Voulgaris, G., 2004. Geomorphic Analysis of Tidal Creek Networks. WRR 40, W05401. https://doi.org/10.1029/2003WR002722 – Fig. 6, page 8.*

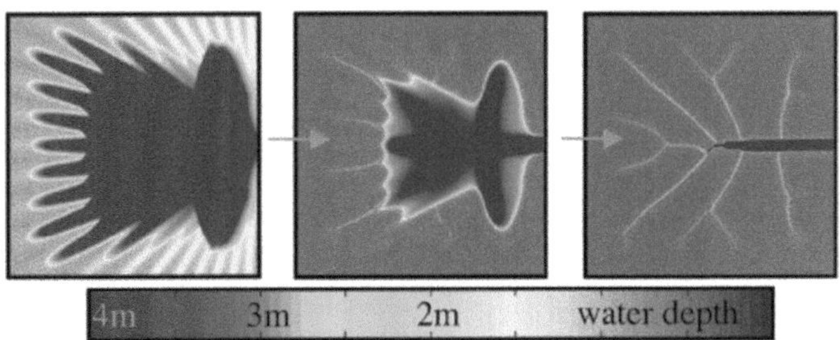

FIGURE 1.12 Simulated evolution of a marsh platform and drainage network under constant 1 mm year^{-1} sea level rise, 20 mg L^{-1} suspended solids concentration, and 4-m tidal range. This demonstrates the dendritic nature of the drainage network that develops over successional time. Time proceeds from left to right. *From Kirwan, M., and Murray, B., 2007. A coupled geomorphic and ecological model of tidal marsh evolution. Proceedings of the National Academy of Sciences 104, 6118–6122, https://doi.org/10.1073/pnas.0700958104 –Fig. 1, page 6120.*

marsh productivity. The development of a dendritic tidal creek drainage network as the marsh matures is in agreement with field observations (Wadsworth, 1980; Novakowski et al., 2004). The ability of models to predict the future evolution of the linked sedimentology and ecology for vegetated wetlands (saltmarshes and mangroves) remains an active and challenging area of research.

2.5 Coastal Evolution and State Change

Estuaries we know today evolved slowly over the past 2–4000 years (Frey and Basan, 1978). For saltmarshes of the southeastern United States, three stages of evolution or

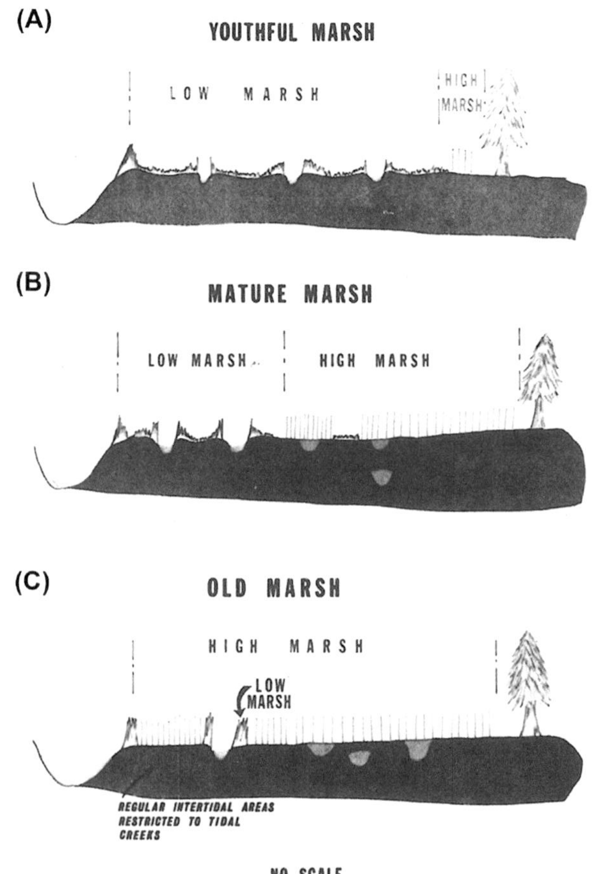

FIGURE 1.13 Three stages of marsh development (A—youthful, B—mature, C—old) showing the changes over time in the ratio of low to high marsh and the filling of tidal creeks. *From Frey, R.W., Basan, P., 1978. Coastal salt marshes. In: Davis R.A. (Ed.), Coastal Sedimentary Environments. Springer, New York, NY. doi-org.proxy-remote.galib.uga.edu/10.1007/978-1-4684-0056-4_4 – Fig. 6, page 114. Reprinted with permission from Springer Nature.*

succession have been identified: youthful, mature, and old age (Fig. 1.13A–C). A youthful marsh-dominated estuary has a low area of marsh relative to water. The marsh itself has a high density of tidal creeks that are often reticulated in nature (Fig. 1.14 top). The marsh has a low area of high marsh and a high amount of low marsh. Marsh grass biomass and productivity are high in the low marsh because of the frequency of tidal flooding (and draining). Sediment deposition and marsh surface accretion/elevation gain are rapid because of both the high frequency of flooding and the high plant density and biomass, which trap sediments. On the high marsh, productivity and biomass are low because of infrequent tidal flooding and as a result sediment deposition is relatively slow. In the mature stage, the ratio of marsh area to open water area is increased, while the ratio of low marsh to high marsh is decreased. Many of the former tidal creeks have infilled (Fig. 1.13B). The overall productivity of the mature marsh ecosystem is lower than in the youthful stage because tidal flooding is reduced

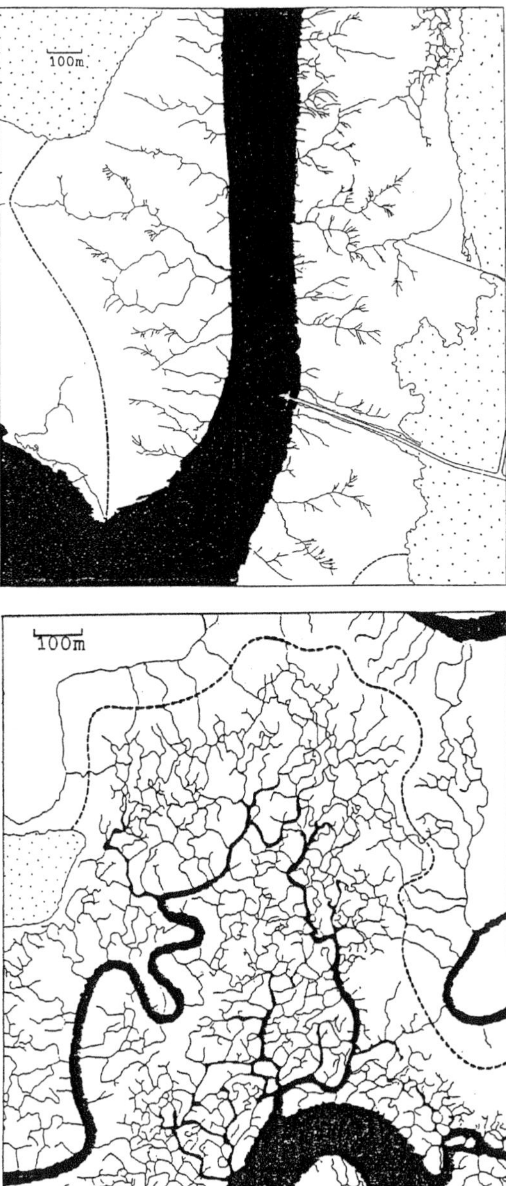

FIGURE 1.14 Development of tidal marsh drainage networks. Examples of dendritic (top) and reticulated (bottom) drainage networks in the Duplin River marsh system along Sapelo Island GA, USA. The dendritic pattern is common in old age marshes with a low drainage density and large expanse of high marsh platform. The reticulated network reflects a youthful marsh system with high drainage density and a high ratio of low marsh to high marsh. *From Wadsworth, J., 1980. Geomorphic Characteristics of Tidal Drainage Networks in the Duplin River System, Sapelo Island, GA. (PhD thesis), University of Georgia, Athens – Figs. 15 and18, pages 47 and 50.*

and the amount of creek bank marsh has decreased. Marsh sediment porewater exchange is reduced relative to the youthful marsh, thus salts and toxic metabolic products accumulate. Finally, in the old stage, the overall area of marsh to water is at its highest and changes only slowly from this point on—in relation to the rate of SLR and/or the availability of external sources of sediment. The ratio of high marsh to low marsh has reached a maximum and most tidal creeks have infilled. Tidal creeks demonstrate a dendritic pattern and the drainage density is greatly decreased (Fig. 1.13C and 1.14 bottom). The high marsh platform may only flood during spring tides. With a low drainage density of tidal creeks and infrequent flooding of the marsh platform, salts and toxic compounds accumulate in marsh porewaters and plant primary production is relatively low. The overall pattern of succession is not uniform across the marsh; rather it proceeds at different rates even within large tidal creek systems (e.g., the Duplin River tidal creek system in coastal Georgia USA, which is about 12 × 3 km). In the Duplin River, approximately one-third of the system is characterized as old and one-third as youthful (Wadsworth, 1980). While sediment availability remains high in the Duplin system, a rapidly increasing rate of SLR in the future may greatly slow or reverse this sequence of development.

A rising relative MSL and the geomorphologic evolution that it induces generate changes in the distribution of coastal wetland ecosystems (see Fig. 1.1)—also known as ecological state changes. These changes modulate the upslope transgression from upland forest to high marsh, high marsh to low marsh, and low marsh to subtidal ecosystems (Fig. 1.15A), which is similar to the scenario shown in Fig. 1.6C. During the 21st century, the historical evolution of coastal wetlands described above may be altered in response to the projected increased rate of SLR because of global warming. Transition from one type of wetland to another or from one state to another is the result of interactions among sediment availability, SLR, and the ecogeomorphic feedback with wetland vegetation. With adequate sediment supply, existing wetlands will be able to maintain elevation relative to SLR. When there is an imbalance, however, erosion along wetland edges will lead to conversion of wetland to unvegetated tidal flat below MSL (Mariotti et al., 2010; Fagherazzi et al., 2013). Lower salinity tidal wetlands may transition to saline wetlands as a result of saltwater intrusion (Williams, 2014). Saltwater intrusion can result from changes in hydrology (e.g., precipitation, river runoff, evapotranspiration) and increases in sea level. Finally, low-lying upland ecosystems (e.g., forests, grasslands, human land use) will experience salinity and water stress as SLRs, flooding frequency increases, and salinity increases.

Ecological succession of vegetation occurs within each of the states as a result of such disturbances as long as hydrology and sediments do not change substantially. For example, freshwater wetland forest landward of high marsh may dieback after a fire, but unless salinity has increased, ecological succession may return the vegetation to forest. Alternatively, an increase in salinity will prevent forest re-establishment and subsequently change the state to emergent saltmarsh. Often, the vegetation is out of phase with physical conditions so that disturbance such as fire or wrack deposition becomes the trigger to initiate change (Brinson et al., 1995).

Over larger time scales and more divergent conditions, ecosystem types may change from one to another (Fig. 1.15A). These coastal ecosystems have different flooding and aeration regimes that control plant dynamics (Fig. 1.15B). Submerged aquatic vegetation (SAV, such as seagrass) has persistent flooding, and mud/sand flats are intermittently exposed, as are wetlands and both will experience some degree of water table drop during low tides.

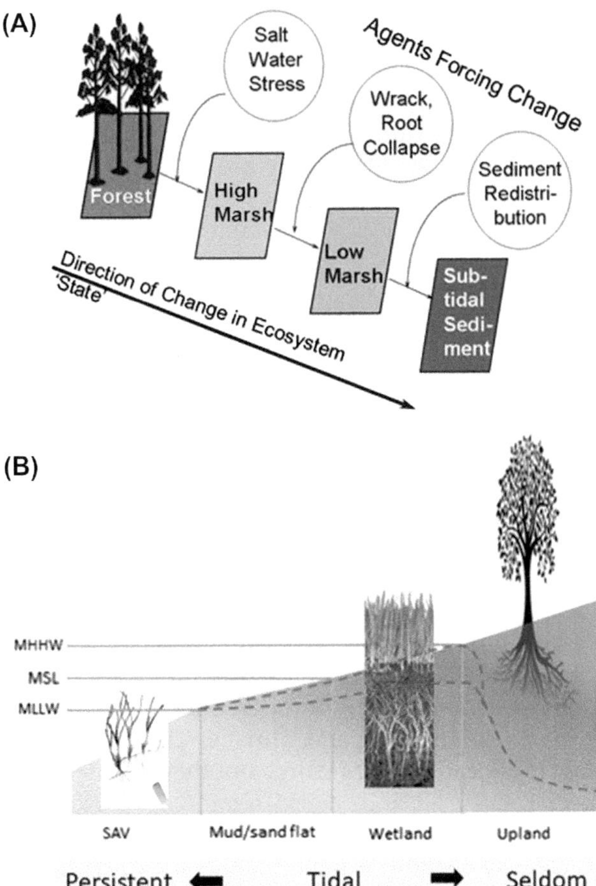

FIGURE 1.15 (A) Predicted state change model showing response of subtidal, tidal, and upland ecosystems to a change in the rate of sea level rise in combination with other agents of change, such as saltwater stress, wrack deposition and smothering of wetland vegetation, and sediment redistribution, such as edge erosion. Ecogeomorphic feedback and the availability of sediments will dictate whether tidal wetlands expand in the future (through upland transgression) or contract (through edge erosion and tidal wetland shoreline retreat). (B) Differences in flooding duration and distribution of saline groundwater (light blue) and fresh groundwater (light brown) across a subtidal to intertidal to upland gradient or transect. The dashed line indicates the saline water table at high and low tides. Sediments below the groundwater lines are 100% saturated, whereas those above the line range from fully saturated to field capacity or lower, depending on the balance between precipitation/flooding inputs and water losses associated with evapotranspiration and drainage. Rising sea level will result in a shift of zones up into uplands, with saltwater intrusion and flooding resulting in the death and loss of upland vegetation. *MHHW*, mean higher high water; *MLLW*, mean lower low water tide levels; *MSL*, mean sea level; *SAV*, submerged aquatic vegetation or seagrass.

Nevertheless, soils will always remain saturated with minimal aeration of the root zone. In contrast, uplands are not under the influence of sea level and rarely experience soil saturation or lack of aeration. SAVs differ from flats by the presence of roots in the sediment and a requirement for transparency in the water column.

2.6 The Role of Physical Disturbances

Because of their location at the land–sea interface, tidal wetlands and estuaries are some of the most physically disturbed ecosystems on Earth. They are subject to several orders of magnitude variation in river runoff or precipitation over annual cycles, they have the once to twice daily inundation from tides, and many are hit by severe ocean-generated storms. The role of such physical disturbances in coastal wetland dynamics and evolution is recognized but seldom quantified. At high latitudes, saltmarshes are scoured by sea ice, creating a network of patches of rich vegetation, degraded vegetation, and unvegetated mud flats (Fig. 1.9D; Martini et al., Chapter 4). Furthermore, floating ice imbedded with sediments is often transported elsewhere for deposition (e.g., marsh surfaces). At temperate latitudes, crabs in saltmarshes can excavate local depressions that can develop into tidal courses or become unvegetated muddy patches (Perillo and Iribarne, 2003a, b; Wilson et al., 2012; Perillo, Chapter 6). In temperate and tropical climates, however, the most common natural disturbances are storms. In mangrove areas, hurricanes can destroy the vegetation by the force of wind and waves (Fig. 1.16A).

People living along the edge of coastal wetlands are particularly threatened by storm surges (Fig. 1.17A–C) because storm surge flood waves occur at long time scales (hours to days) and are therefore little attenuated by vegetation (Highfield et al., 2018). However, coastal wetlands do serve a coastal protection role. Mangroves can attenuate a storm surge by 8–20 cm km^{-1} (Zhang et al., 2012; Spalding et al., 2012). Although storm surge attenuation can be negligible, saltmarshes can significantly attenuate the wind waves; indeed, a 0.9 m wind wave is halved in about 200 m after entering a saltmarsh (depth < 2 m) during a storm surge (Moller et al., 2014; Möller and Christie, Chapter 8). Seagrass meadows probably also have a negligible effect on a storm surge but they do attenuate wind waves, though to a smaller extent.

Tsunamis, although rare, generate a major disturbance over shorter periods of time. The presence of extensive areas of shallow waters, such as intertidal flats, is important because

FIGURE 1.16 (A) Mangroves defoliated by a hurricane in Florida. (B) A 5–10-m wide strip of mangroves along the estuary banks was destroyed by the 2004 Indian Ocean tsunami, and the remaining mangrove forest remained intact (Khao Lak, Thailand). *(A) Courtesy of USGS.*

FIGURE 1.17 People living along the edges of coastal wetlands such as in (A) the Fly River Estuary, Papua New Guinea, (B) Chumphon mangroves, Thailand, and (C) the mangrove inlets at Cedar Key, Florida, USA, are particularly at threat from natural hazards including tsunamis and storm surges.

it presents a wide area for dissipation of energy before the wave hits the vegetation. In mangroves for small (<6 m) tsunamis waves, the damaged area may be limited (Fig. 1.16B) to a 5–10 m wide strip along the tidal creek and another, possibly much wider, strip along the coast (Wolanski, 2007; Wolanski and Elliott, 2015). This occurs because the vegetation is an obstacle to swift flows and thus steers the tsunami wave along the tidal creek. People (fishermen) living along the tidal creek—with no vegetation between them and the creek—are not protected against a small tsunami; however, people living behind the mangrove belt receive some protection (Alongi, 2008; Wolanski and Elliott, 2015). Provided that the tsunami wave is less than 6 m and the trees are fully developed, the vegetation can survive and a wave 500 m inland where it is transformed from rising in 10 s to rising in 3 min; this gives a chance for people to take shelter or flee the area, and the attenuation considerably reduces damage to property and infrastructure (Wolanski and Elliott, 2015). A large tsunami (wave > 6 m) flattens and uproots the vegetation. In some cases, the energy extracted from the wave and used for this destruction reduces the wave energy sufficiently to provide some protection to people living in the coastal zone inland from the mangroves (Fig. 1.18A–B). However, for a large tsunami wave, mangroves and coastal forests are known to significantly contribute to human mortality from the physical damage of water-borne debris (Cochard et al., 2008). In climates with distinct wet and dry seasons, the evolution of the bathymetry

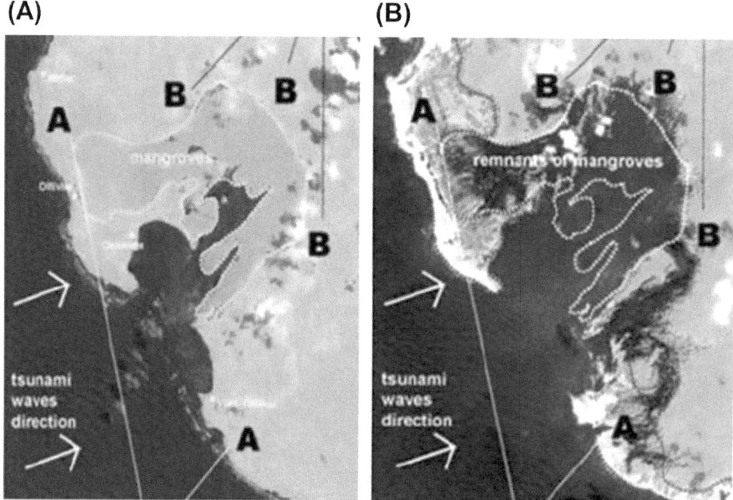

FIGURE 1.18 Satellite images of Kitchall Island (Nicobars, Indian Ocean) (A) before and (B) after the December 26, 2004 tsunami. The likely tsunami wave direction is shown by arrows. The white *dotted line* shows the bay. Before the tsunami, the bay was covered by mangroves. After the tsunami, no mangroves survived as the tsunami uprooted or snapped the trees. The vegetation and the villages in the area marked "B" behind the mangroves were not badly impacted by the tsunami disaster because their high elevation together with the tsunami wave attenuation by the vegetation prevented flooding; villages and the agricultural areas marked "A" were not in the lee of mangroves and were destroyed or severely damaged by the tsunami. Thus, the land behind the mangroves was partially protected by the sacrificial belt of mangroves. At longer time scales, the surviving mangroves died because tectonic movements changed the substrate elevation and the tidal hydrology. *Courtesy of Y. Mazda.*

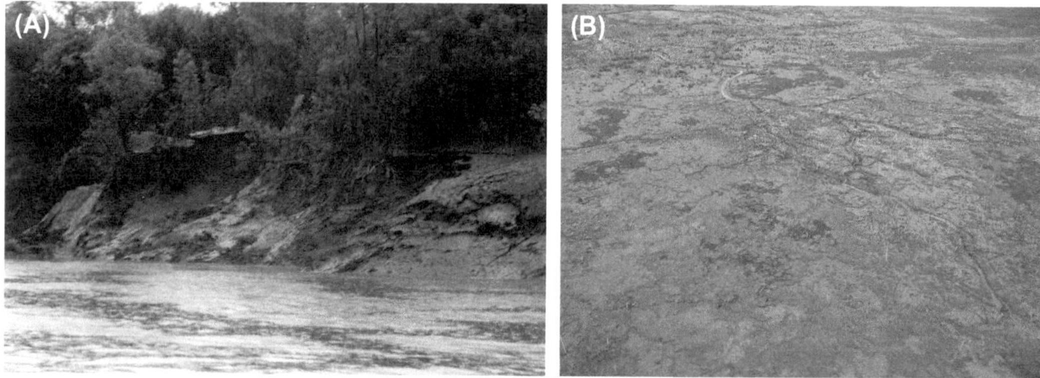

FIGURE 1.19 (A) Seasonal estuarine bank slumping at the end of the wet season (Daly Estuary, NT, Australia). (B) The growth of a tidal creek in freshwater flood plains is facilitated by the destruction of the protective vegetation by saltwater intruding in the dry season and creating bare mud flats that may be colonized by mangroves years later (Mary Estuary, NT, Australia).

of a tidal creek may be determined by seasonal bank slumping principally at the end of the wet season, which is a large-scale process that vegetation usually cannot stop (Fig. 1.19A) and which many estuarine geomorphology evolution models ignore so far. In arid and semi-arid areas, bank slumping depends on underwater erosion and failures in the security factor due to steep flanks (Ginsberg and Perillo, 1990). In other cases, the tidal creek develops from headward erosion at its tip during floods; during the subsequent dry season, saline water floods the eroded area, kills the freshwater vegetation, and creates bare soils that are unprotected by vegetation and thus readily erode; this allows the creek to grow in the tidal freshwater flood plains (Williams, 2014, Fig. 1.19B). This process is still largely ignored by estuarine geomorphology evolution models, although recently a model has been developed (Gong et al., 2018).

There is a continuum of water level variations within estuarine systems ranging from tides to severe ocean-generated storms and tsunami surges that differ tremendously on their ecological and geomorphic structure and function. These water level disturbances also occur with varied frequency and predictability. At the low end of the disturbance continuum, Odum et al. (1979) hypothesized that the twice daily tidal flooding was one of the factors contributing to their extremely high levels of production. Tides bring in inorganic nutrients and sediments from the ocean/estuary and remove toxic metabolic products for sediment porewaters. In the tidal marshes of Louisiana, rapid land loss has led to many human actions to retard loss, including the installation of weirs, which perch water levels, and placing spoil banks along canals and natural tidal creeks, which also limit tidal exchange. These actions tend to limit transport of nutrients, sediment, and toxic compounds, however, (e.g., Cahoon and Turner, 1989) and as a result marsh grass production generally decreases (Spalding and Hester, 2007). Although not a rigorous testing of Odum's hypothesis, these changes are consistent with predictions.

The effects of hurricanes and typhoons on tidal wetland structure and function remain an active area of research. H.T. Odum talked about mangroves as being pulse-stabilized systems meaning that the regular exposure to severe hurricanes helped maintain high levels of primary and secondary production (Odum et al., 1995). But there are records in the literature where rapid sediment deposits of more than 5 cm (during severe storms) can kill mangroves (Ellison, 1998). Research on the effects of hurricanes on mangroves in the Everglades has been a major component of the Florida Coastal Everglades Long-term Ecological Research Project (FCE-LTER). Smith et al. (2009) examined the effects of five major storms to pass through the Everglades mangroves and concluded that hurricane damage did occur and that it was related to the hydrogeomorphic type of mangrove. Damage was much greater in basin mangroves than in riverine or island mangroves. Some forests recovered while others did not. Those that did not transitioned into other ecosystem types, such as tidal flats. On the other hand, another study in the same region came to different conclusions, at least with respect to the wetlands that were not so heavily damaged. Danielson et al. (2017) observed a pattern where major defoliation led to significant declines in net primary production, but that often in less than 5 years, primary production returned to prestorm levels or higher (Fig. 1.20). The inputs of sediments were seen not as damaging to mangrove function but rather to be an important source of inorganic P, in these P-limited carbonate systems.

Saltmarsh plants also seem able to cope well and plant growth is even accelerated following major storm events (Garbutt and Boorman, 2009; Boorman, Chapter 17). The subsidy–stress hypothesis of Odum explains well the impacts of severe ocean-generated storms on tidal wetlands. Up to a point (threshold), storms subsidize tidal wetlands via their removal of dead vegetation and the input of valuable inorganic nutrients, often associated

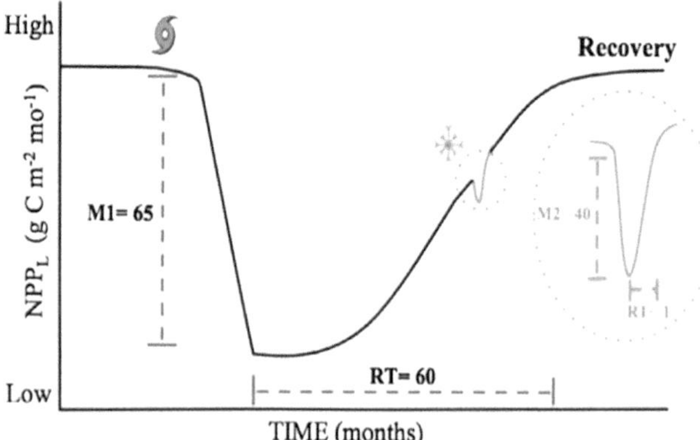

FIGURE 1.20 Results of a disturbance simulation model showing the recovery time (RT) of mangrove net primary production (NPP) following disturbances of varying intensity (hurricane Wilma and a cold snap). RT varies from 60 months following Hurricane Wilma to less than 1 month for the cold snap. *From Danielson, T., Rivera-Monroy, V., Casteneda-Moya, E., Briceno, H., Travieso, R., Marx, B., Gaiser, E., Farfan, L., 2017. Assessment of everglades mangrove forest resilience: implications for above-ground net primary productivity and carbon dynamics. Wetland Ecology and Management 303, 115–125 – Fig. 2, page 118 for Shark River Slough as part of the Florida Coastal Everglades Project: FCWE-LTER.*

with sediments. But there is a threshold, which varies on the initial hydrogeomorphic site condition, where the impact of storms is so severe that recovery is not possible and the wetland transitions to another ecosystem type, at least until conditions become favorable sometime in the future, such as a response to SLR, for recolonization of the original ecosystem vegetation.

Both saltmarshes and mangroves suffer mortality through hydrologic regime change, particularly from sustained flooding or loss of structural support through sediment erosion. The belowground biomass of mangroves, seagrass meadows, and saltmarshes offer protection against erosion, thus providing a more stable medium for plant growth. The soft sediments of unvegetated intertidal flats are highly mobile and generally inhospitable to emergent plant growth. The establishment of emergent vegetation is rapid once the right hydraulic conditions occur and from that time onward the vegetation modifies the water and sediment circulation and promotes further biomass accumulation and the creation of a new ecosystem. The establishment of pioneer vegetation over a soft sediment bottom appears to be controlled for saltmarsh primarily by vegetative fragments establishing themselves and forming patches that over time become connected, a process promoted by the growth of rhizomes (Sánchez et al., 2001). For mangroves pioneer seedlings that are often *Avicenna marina* stabilize the soils through their roots and pneumatophores (Saintilan et al., 2009).

Once initial colonization has been achieved, vegetation may develop, being possibly a seagrass meadow, a mangrove swamp, or a saltmarsh. This is commonly achieved by the arrival of species of fauna and flora with strong dispersal characteristics, followed by corresponding changes in food web dynamics, sedimentology, and hydrology. Humanity has accelerated this process by inadvertently or purposefully encouraging the spread of exotic invasive species and by the nutritive effects of eutrophication (e.g., Ellison and Sheehan, 2014; Sheehan and Ellison, 2014).

2.7 The Role of Animals (Herbivores and Carnivores)

Crabs are generally abundant in mangroves; there are over 40 species in Indo-Pacific mangroves alone, dominated by two families, the Grapsidae and Ocypodidae, and each family by one genus, *Sesarma* and *Uca*, respectively (Jones, 1984). In mangroves, crabs dig burrows often at high densities (Fig. 1.21A). These burrows and their occupants facilitate the

FIGURE 1.21 (A) Crab burrows are numerous in healthy mangrove soils (Congo River Delta). (B) Mangroves permanently destroyed by groundwater hypersalinity and acid sulfate leaching as a result of poor land use practices in reclaimed mangrove soils (Konkoure River delta, Guinea). (C) Nearly all mangrove seedlings planted in this abandoned shrimp pond died due to drowning as the tidal hydrology was not properly restored (Surat Thani, Thailand). (D) This mangrove seedling was destroyed by being physically brought down by the weight of algae due to local eutrophication from cattle dung (Iriomote Island, Japan). A bamboo curtain (E) and a seawall (F) were constructed as a wave barrier to protect mangrove seedlings in a mangrove restoration site in the upper Gulf of Thailand.

circulation of porewaters with tidal floodwaters, which permit salt to be flushed from sediments, thus reducing hypersalinity (Alongi, 2009; Mazda and Wolanski, 2009), which would otherwise occur from mangroves that selectively avoid salt uptake and increase porewater salinity of the sediments. When crab populations are absent, the groundwater becomes hypersaline and the vegetation either dies or becomes stunted. Human activities can overwhelm and destroy this relationship. For instance, aquaculture on sulfur-rich soils drains and acidifies the soils (Dent and Pons, 1995); the vegetation is destroyed and cannot be recovered (Fig. 1.21B–F) without massive additions of lime. Even with lime additions, it is not guaranteed that farming can be resumed because of the resilience of disease vectors buried in the mud from intensive aquaculture (Stevenson et al., 1999).

The fate of plant litter in coastal wetlands is largely controlled by the activities of benthic fauna and by the circulation of water. Crabs, bivalves, and snails can remove and process up to 50%–80% of mangrove leaf litter (Smith et al., 1991). This enables the nutrients to be recycled within the ecosystem. In turn the distribution of these animals is linked to the sedimentology, topography, and hydrology of the wetland. Birds are particularly important in also consuming and exporting wetland plant and animal biomass; they also connect various components of the coastal wetlands with each other and the coastal wetlands with the estuary and with surrounding ecosystems (Daborn et al., 1993; Mander et al., 2013; Bocher et al., 2015; Buelow and Sheaves, 2015; Wolanski and Elliott, 2015). In temperate wetlands, crabs and other burrowing animals are major bioengineers through their reworking of the surface and upper meter of tidal flat and marsh sediments. For instance, the grazing amphipod, *Corophium volutator*, not only bioturbates sediments but also indirectly influences the erosion threshold of sediments via their grazing of benthic diatoms, which excrete exopolysaccharides, a "sticky" organic compound that binds sediment particles. Changes in amphipod densities can lead to differences in sediment conditions and sediment transport in less than a couple of days (Daborn et al., 1993). As another example, high density of crab burrows can develop a groundwater circulation forced by the tides that in some cases creates surface depressions that induce the formation of ponds (Perillo, Chapter 6).

The fate of dead organic matter is important in determining the evolution of a coastal wetland. There are large differences between the fate of aboveground plant biomass and belowground plant biomass. Live aboveground biomass can be consumed by crabs, snails, shrimps, and cone shells (Fig. 1.22A and B). For instance, shrimps can remove 50%–80% of seagrass leaf production (Kneer et al., 2008) and crabs and snails climb trees to harvest mangrove leaves (Vannini and Ruwa, 1994; Vannini et al., 2008). Another fraction can be exported as particulate organic matter outwelling that in turn enhances detrital food webs in the adjoining estuarine and coastal waters, and the remaining fraction may be accumulated on the substrate for eventual burial by sediment deposition. Belowground organic biomass is generally not exported in particulate form but rather is lost through decomposition by microbes as carbon dioxide and methane and dissolved nutrients. Decomposition takes several months to years for labile organic matter, whereas refractory organic matter may take hundreds of years to decompose or become permanently buried through peat accumulation.

There are feedback between the vegetation and the fauna. For instance, coastal wetlands with an aerial canopy of vegetation generate a microclimate at the sediment surface, which

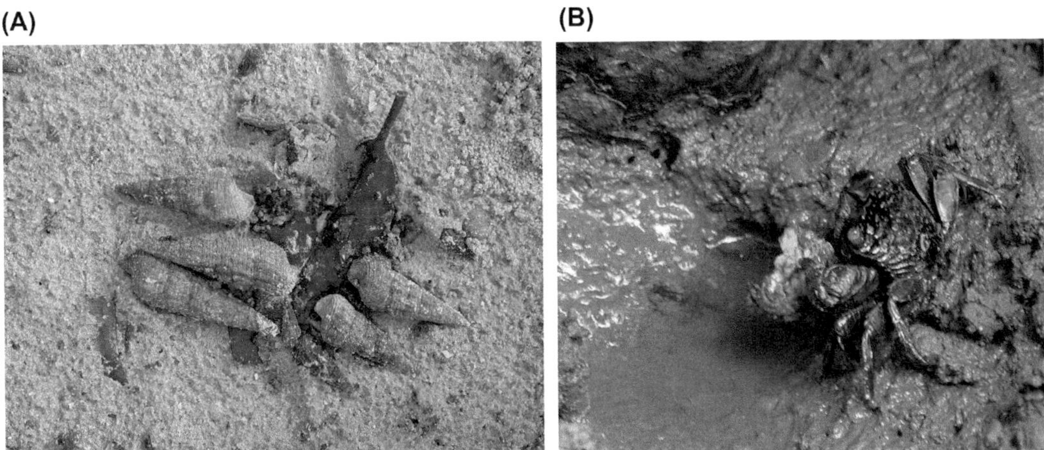

FIGURE 1.22 (A) Cone shells eating leaves in mangroves (Iriomote Island, Japan). (B) Crabs consume mangrove litter that falls on the ground; some crabs also climb on trees, principally at night, and eat the young leaves that have little tannin in them (Australia mangroves). *(A) Courtesy of K. Furukawa. (B) Courtesy of N. Duke.*

in turn affects the fauna and flora. As an example, at mid-latitudes the daily temperature variation in stagnant pools of water in wetlands exceeds 5°C if the pools are shaded by mangroves and 15°C if the pools are not shaded; this in turn affects the population dynamics of fish and mosquitoes in these pools (Knight, 2008).

The fate of dead, aboveground organic matter differs widely among the types of coastal wetlands. Much of the litter fall is consumed in mangroves (Smith et al., 1991; Kiwango et al., 2015). In some saltmarshes, a large portion of live plant biomass may be consumed and subsequently exported as birds migrate away from sites of grazing (Canada geese in North America; Rivers and Short, 2007). Usually a large fraction of dead-standing saltmarsh vegetation is decomposed in place by fungi, which are farmed and consumed by snails, such as *Littoraria irrorata* (Silliman and Newell, 2003). In other saltmarshes, consumption of plant litter by fungi and herbivores is not a major process; instead large wracks of dead grass form in winter as the aboveground components die (Fig. 1.23). The amount of aboveground organic matter can appear to be enormous, when small amounts exported from many km^2 accumulate on in small areas. The same occurs in seagrass meadows where storms can leave massive accumulations on beaches (Holmer, Chapter 13). Some of this material remains trapped in the wetland, redistributed during extreme tides, or exported as an outwelling event during a single storm (see Fig. 4.11 in Wolanski and Elliott, 2015). Large accumulations of wrack can produce large dieback of *Spartina,* which then may induce the formation of tidal depressions within the marsh (Perillo, Chapter 6). Fish and crabs can also consume biomass on wetland surfaces at high tide and then facilitate export as they move toward the estuary with ebbing tides (Deegan, 1993). Thus, mangroves, saltmarshes, and seagrass meadows can be net exporters of organic matter depending on local physical conditions. This link between flora and fauna in the energy flows and nutrient fluxes between ecosystem types is a rich, largely unexplored, field of research.

FIGURE 1.23 A wrack of dead salt marsh plants and some plastic litter after a storm (Bahía Blanca Estuary, Argentina). The plant litter is composed of dried stems of *Spartina alterniflora*. *Sarcocornia perennis* plants die when covered by the dead plants.

2.8 Observations Across Ecosystem Types

Surface and underground water circulation are vital to the sedimentation processes, chemistry, and biology of coastal wetlands. There is a strong feedback between water circulation and sediment dynamics and the constantly changing geomorphology of coastal wetlands. The circulation of surface and sediment porewaters in coastal wetlands also plays a key role in the adjoining estuarine and coastal ecosystems, as well as their biology and biodiversity. The processes vary spatially from small to large scales and temporally at time scales from that of individual mixing events to tidal to geomorphological scales.

Almost every day of field work in coastal wetlands reveals new aspects of the complex interplay between hydrodynamics, sedimentology, and flora and fauna. To cite a few examples, we compare attributes of the five major ecosystem types (Table 1.1) in terms of their geographic distribution, physical limits, food web dynamics, and responses to human activities. In making such comparisons, there is a tendency to overgeneralize and to gloss over the diversity of patterns found within each of the types. We make these comparisons in the spirit that they may lead to insight not obvious by working within or examining a single ecosystem type (the remainder of this section refers to the contents of Table 1.1).

Some patterns are obvious, such as climatic differences between the latitudinal ranges of saltmarshes and mangroves. Although the geographic distribution is reasonably well-established, inventories of the area covered by each ecosystem type are less reliable. For example, tidal freshwater wetlands have not been much studied outside of the North American and European continents, so their distribution has not been mapped to our knowledge (Conner et al., 2007; Whigham et al., Chapter 18). In fact, most of these ecosystems are associated with the large deltas of the world where dominance by freshwater discharge results in salinities lower than is often associated with the term "coastal."

TABLE 1.1 Comparison of Coastal Wetland Ecosystem Characteristics for Five Major Coastal Ecosystem Types

Characteristics	Seagrass Meadows	Intertidal Flats	Salt Marshes	Mangrove Forests	Freshwater Tidal Wetlands
GEOGRAPHIC DISTRIBUTION					
Geographic range[a]	Absent in polar regions	Unlimited	Mid to high latitudes; replaced by mangroves in subtropics	Mean monthly temps >20 C	Low to high latitudes
Global abundance[b]	180,000 km^2	Widespread; undocumented	60,000 km^2	150,000 km^2	Grossly underestimated
PHYSICAL LIMITS					
Maximum water depth limitation/ minimum elevation[c]	Approximately 20% incident light	Light limitation of net primary production where $P/R < 1$	Varies with amplitude and temperature	Unresolved	Unresolved
Maximum elevation/ minimum hydroperiod[d]	Intertidal conditions (exposure to some drying)	Supratidal during storms	Groundwater discharge zone; salt accumulation in arid climates	Groundwater discharge zone; salt accumulation in arid climates	No information
Salinity range[e]	Euryhaline to polyhaline	Variable	Hyperhaline to oligohaline	Hyperhaline to oligohaline	Fresh to oligohaline
Source of material for vertical accretion[f]	Also organic accretion	Allochthonous inorganic only (biofilm stabilizing effect)	Also organic accretion	Also organic accretion	Also organic accretion
Interannual variation in sea level; ENSO cycles[g]	Not applicable	Not applicable	Demonstrated effect on primary production	Fish nursery variation	Not known
Hurricanes and cyclones[h]	Temporary unless burial by sediments or severe erosion	Likely short-term effects	Local shore erosion and sediment redeposition	Decades-long effects after blowdown of trees	Surge may affect soil salinity
FOOD WEB DYNAMICS					
Typical dominant primary producers[i]	Submersed vascular plants	Epipelic microalgae (esp. diatoms) and macroalgae	Emergent grasses and forbs	Trees, shrubs	Ranges from herbaceous to forest dominance
Grazing food webs[j]	Large mammals and turtles	Infauna, epifauna	Snails and insects	Tree crabs	Rodents
Detrital food webs[k]	Invertebrate dominated	Indistinguishable from grazing	Invertebrate dominated	Burrowing crabs	Invertebrate dominated
Openness of organic matter fluxes[l]	Exports and imports well-documented	No information; presumed open	Evidence of net exports	Evidence of net exports	Few studies

(*Continued*)

TABLE 1.1 Comparison of Coastal Wetland Ecosystem Characteristics for Five Major Coastal Ecosystem Types—cont'd

Characteristics	Seagrass Meadows	Intertidal Flats	Salt Marshes	Mangrove Forests	Freshwater Tidal Wetlands
RESPONSE TO HUMAN ACTIVITIES—GLOBAL CHANGE					
Response to climate warming: ambient temperatures[m]	Thresholds not established	No information	Possible distribution to higher latitudes	Possible distribution to higher latitudes lacking frost	No information
Response to climate warming: acceleration of rising sea level[n]	Thresholds not established	No information	Strong sediment sources needed for survival (glacial rebound areas excluded)	Strong sediment sources needed for survival	Strong sediment sources needed for survival (little information)
Altered salinities in response to climate drying or wetting[o]	Hypersalinity in seasonally isolated lagoons; freshening in others	No information	Expansion or reduction of salt flats	Expansion or contraction of salt flats	Transformation to greater or lesser salinity tolerant species
RESPONSE TO HUMAN ACTIVITIES—LOCAL AND UPSTREAM					
Increased salinity from reduced freshwater flows (upstream withdrawals, irrigation, etc.)[p]	Hypersalinity in seasonally isolated lagoons	Increasingly stressful	Expansion of salt flats	Expansion of salt flats	Transformation to salinity tolerant species
Reduced sediment supply due to reduced freshwater flows (dams, etc.)[q]	No information (reduced suspended sediments beneficial to water clarity)	Reduced development of tidal flats	Erosion	Erosion; hypersaline conditions	Erosion
Enhanced or excessive sediment supply[r]	Reduced water clarity limits primary production	Accretion with shift to marsh and mangrove colonization	Accretion self-limiting	Sensitive to burial of pneumatophores	Massive development of marshes historically (Chesapeake Bay, USA)
Tidal barriers—dyking and bulkheads[s]	Not applicable	Not applicable	Eliminates physical and biotic exchanges	Eliminates physical and biotic exchanges	Eliminates physical and biotic exchanges
Eutrophication[t]	High sensitivity to nutrients; macroalgal smothering	Proliferation of macroalgae	Increased herbivory	No information	No information

TABLE 1.1 Comparison of Coastal Wetland Ecosystem Characteristics for Five Major Coastal Ecosystem Types—cont'd

Characteristics	Seagrass Meadows	Intertidal Flats	Salt Marshes	Mangrove Forests	Freshwater Tidal Wetlands
Harvesting of plants and animals[u]	Trawling for benthic organisms is extremely destructive	Shellfish harvesting (e.g., hand tonging for oysters and clams), disruption of substrate	Harvesting of marsh hay (can be done sustainably)	Logging for firewood (can be done sustainably); aquaculture impoundments (see#22, very destructive)	Cypress logging
Bottom disturbance: channel dredging, fish/shrimp ponds, etc.[v]	Extremely destructive, direct loss, and declining water clarity	Deepening beyond euphotic zone	Extremely destructive	Extremely destructive	Extremely destructive
Fragmentation within habitat[w]	No information	No information	No information	No information	No information
Loss of connectivity with other habitats[x]	No information	No information	No information	No information	No information
RESPONSE TO HUMAN ACTIVITIES—BIOTIC					
Invasive species[y]	*Caulerpa taxifolia* aggressive clones	No information	*Phragmites* in spring tidal and fresh zones	No information	*Phragmites* throughout
Diseases[z]	Wasting disease historically	No information	No information	No information	No information

Superscript alphabets corresponds to footnotes that refer to chapters in this volume and additional sources listed at the bottom of the table.

[a]*Chapters 2–4 and 28; Conner et al., 2007.*

[b]*Duarte et al. (2008) for all but intertidal flats and tidal freshwater wetlands, the latter grossly underestimated (Chapter 18). Mcowen et al. (2017) for saltmarshes and Giri et al. (2011) and Hamilton and Casey (2016) for mangroves. Green and Short (2003), Nellemann et al. (2009a) and Hopkinson et al. (2012) for seagrasses.*

[c]*Chapter 13; McKee and Patrick (1988) for saltmarshes, Gallegos and Kenworthy (1996) for seagrass.*

[d]*Chapters 3 and 16.*

[e]*Chapters 2, 3 and 4.*

[f]*Chapters 2, 3, and 4.*

[g]*For saltmarshes, Morris et al., 1990; for mangroves, Rehage and Loftus, 2007.*

[h]*For mangroves (Smith et al., 1994); for marshes and mangroves (Cahoon, 2006); for marshes (van de Plassche et al., 2004), for seagrasses Holmer Chapter 13.*

[i]*Chapters 2, 3, and 4.*

[j]*Chapters 2, 3, and 4; Chapter 11 for tidal flats; Chapter 15 for saltmarshes.*

[k]*Chapters 2, 3, and 4; Chapter 11 for tidal flats; Chapter 15 for saltmarshes.*

[l]*Chapters 11, 13, 15, 19.*

[m]*Chapters 2, 20, 22, 28.*

[n]*Chapters 2, 3, 4, and 28.*

[o]*Predictions depend upon local effects of precipitation/evapotranspiration change as they may affect continental sediment supplies.*

[p]*No documented examples in chapters.*

[q]*Pasternack et al., 2001.*

[r]*Pasternack et al., 2001; van Katwijk et al., 2016.*

[s]*Chapter 19.*

[t]*Seagrasses particularly vulnerable: Chapter 13, van Katwijk et al., 2016 and Waycott et al., 2016.*

[u]*Chapter 24 (Holmer) and van Katwijk et al., 2016 for seagrasses; Buchsbaum et al., 2009 and Holden et al., 2013 for saltmarsh haying; Perkins, 2017 and Mancil, 1972 for cypress tree logging; Goessens et al., 2014 for mangrove logging.*

[v]*Chapter 24.*

[w]*No information available.*

[x]*No information available.*

[y]*Chapter 13.*

[z]*Chapter 13 and Waycott et al., 2016.*

At a particular site, physical limits and factors (water depth, wave energy, salinity range, rate of rising sea level, etc.) determine the type of coastal wetland. Because of the high diversity in life forms across ecosystem types, ranging from diatoms to trees for just the primary producers, there is great variability in the capacity of biotic structure to influence physical processes. For tidal flats and seagrasses, both of which are dominated by obligate aquatic taxa, maximum depth distribution (or lowest elevation) is limited by light availability when flooded. The upper elevation of disturbance is limited by desiccation when the ecosystem is exposed. For emergent life forms (marsh grasses, shrubs, and trees), the vertical range is highly influenced by tidal amplitude, i.e., greater tidal amplitudes allow a greater elevational range of distribution (McKee and Patrick, 1988).

Variations in intertidal soil salinity are largely a consequence of climate (precipitation—evapotranspiration intensity and periodicity) and freshwater discharge, when significant. In some humid climates, groundwater discharge establishes the upper boundary of mangroves and marshes (Plater and Kirby, 2006), whereas in some arid climates, high evapotranspiration leads to high soil salinities that restrict the landward extent of coastal wetlands (Pratolongo et al., Chapter 3). Flooding frequency and bioturbation contribute to porewater exchange and soil salinity. With regular tidal flooding, porewater exchange maintains salinity (in psu) so that it is rarely above 50 in a mangrove forest or a saltmarsh, compared to 32—37 in an adjacent tidal creek and 100 in salt pan porewater (Sam and Ridd, 1998; Gardner, 2005; Boorman, Chapter 17). The intensity of the groundwater salinity intrusion in a wetland depends on the tidal range, occurrence of an impermeable clay/silt layer underneath the wetland soils penetrated by roots, and the fresh groundwater discharge from the upland (Barlow, 2003; Wilson and Morris, 2012). With changes in the relative position of sea level, the boundary at which coastal effects, such as salinity, are no longer apparent—and this boundary controls the vegetation—changes over time as wetlands migrate landward in response to rising sea level (most regions; see Fig. 1.3) and regress as sea level drops (mainly at high latitudes) (Figs. 1.3 and 1.6).

Virtually all coastal wetlands respond to and rely on sediment sources and exchanges. The effects of suspended sediment on light transmission in the water column are critical for survival of seagrasses and likely have short-term effects on the primary productivity of tidal flats. For marsh and forest ecosystems, however, sediment accumulation through deposition is a critical process because it maintains the relationship between a wetland surface and sea level change. This dynamic interaction is to some degree self-maintaining because too much accretion places the sediment surface too high for flooding and sediment deposition, whereas too little accretion has the opposite effect (Rybczyk and Callaway, 2009; Morris, 2016).

Interannual variation in sea level stand adds another dimension of complexity, but it has been demonstrated only for primary production rates in saltmarshes (Morris et al., 1990). Of course, hurricanes, cyclones, and tsunamis, where they occur, can produce short-term disruptions of tidal flats but highly variable effects locally for other coastal wetland types (Cahoon, 2006). For saltmarshes, sediment erosion is a shore phenomenon that in extreme cases can remove large areas in a single storm. In an example studied by van de Plassche et al. (2004) in Connecticut (USA), massive removal of saltmarsh sediments could be returned to intertidal status only with substantial infilling and regrowth of vegetation. In contrast, mangroves can experience massive blowdown with return to full growth forests only on decadal time scales (Cahoon et al., 2003; but see also Danielson et al. 2017) where mangrove recovery is seen in relation to the degree of disturbance. The effects of soil salinity on tidal freshwater swamps in temperate zones have been shown to cause tree mortality and replacement with marsh vegetation (Conner et al., 1997). Fire is likely a disturbance only in the upper portions of tidal wetlands where mortality of trees can accelerate the landward movement of marsh vegetation (Poulter, 2005).

Food web dynamics vary greatly among coastal ecosystem types. Tidal flats represent the largest departure from other ecosystem types in that the grazing food web is dominated by deposit-feeders rather than the more typical herbivory of higher plants. This is somewhat deceiving because marshes and swamps also have epiphytic communities that support substantial food webs. Algal production in both seagrasses and saltmarshes can be substantial, especially with regard to macroelements, such as N. This is not to diminish the role of living plant tissue in supporting grazing food webs because there are examples of substantial herbivory, ranging from sea turtles and dugongs for seagrasses and invertebrates for mangroves and saltmarshes. Regardless, plant tissue is often unpalatable to many potential consumers resulting in the majority of primary production being entrained in the detrital food webs (Visser et al., Chapter 15). In spite of the openness of most tidal ecosystems to organic matter exchanges, both particulate and dissolved, there is general agreement that greater tidal amplitudes not only facilitate imports and exports but also that associated currents serve to amplify nutrient cycling and associated primary productivity. Even seagrasses that are completely open to exchange have the capacity to trap particulate organic matter through their baffling effect in comparison with bare sediments (Holmer, Chapter 13). Nevertheless, it should be pointed out that shallow bare sediments used for comparison are similar to the intertidal flats that fully contribute to the habitat complexity of coastal wetlands. Migrating birds also provide a connection, through biomass transfer, between coastal wetlands that can be adjoining or widely separated, and even in different hemispheres.

2.9 The Human Impact

Starting about 7760 years ago in China (Zong et al., 2007) and typically a few hundred years ago in most other coastal areas, humanity has profoundly impacted, degraded, or destroyed many coastal wetlands by direct physical degradation and pollution. Ironically, reduced coastal wetland area increases the threat to human safety at the same time that shoreline development exposes populations to coastal hazards such as tsunamis, erosion, flooding, storm waves, and surges.

As a result, nearly all saltmarshes have been destroyed by land reclamation in several highly populated temperate countries, including Japan, China, and the Netherlands (Wolanski and Elliott, 2015). Recent wetland destruction in the United States has led to over 9 million hectares of wetlands lost between 1950 and 1970, the greatest loss occurring in coastal wetlands of the Gulf of Mexico (Frayer et al., 1983). Despite the "swampbuster" 1985 Food Securities Act and the "no net loss" of wetlands policy that emerged in 1989, wetland loss along that USA coast continues (specifically Louisiana), though at a slower rate (Streever, 2001; Entwistle et al., 2017). Seagrass area is also declining as a result of human activities. Since 1879, 29% of the known areal extent of seagrasses was lost and rates of loss since 1990 have accelerated to 7% $year^{-1}$ (Waycott et al., 2016).

This loss in turn is economically significant because coastal wetlands are critical habitats for many fishery species in the Gulf of Mexico (Heck et al., 2003; O'Connor and Matlock, 2005). The wetlands remaining are sinking and shrinking as they are not replenished by sediment because rivers are diverted elsewhere, exacerbating flooding by river floods and storm surges (Streever, 2001; Day et al., 2000). Human disturbance now threatens tropical coastal wetlands; indeed, we face the prospect of a world without mangroves this century (Diop, 2003; Duke et al., 2007). The main threats are clear-cutting mangroves for charcoal and conversion of the land for urbanization, salt ponds, rice farms, and shrimp ponds (Fig. 1.24A–D). Interestingly, the Sundarbans of India and Bangladesh lost only 1.2% of their mangroves in the past 25 years. These mangrove areas have the highest human population density in the world living on their periphery. Government actions are largely responsible, which is an excellent example of coexistence of coastal wetlands and humans (Giri et al., 2007) Arctic coastal wetlands remain the least impacted, simply because there are fewer people living in these areas.

Geomorphology teaches us that all coastal wetlands suffer mortality if mineral and organic sediment accumulations are unable to keep pace with SLR. This is important to consider

FIGURE 1.24 The ongoing worldwide destruction or degradation of mangroves is mainly due to (A) urbanization (Mai Po, Hong Kong), (B) shrimp ponds (Surat Thani, Thailand), (C) salt ponds (Wami Estuary, Tanzania), (D) rice farms (Konkoure Estuary, Guinea), as well as (not shown) clear-cutting for charcoal production.

when planning the future of coastal wetlands in view of a predicted SLR during this century. Most modern-day tidal wetlands formed over the past several thousand years when rates of SLR were much lower than the current 3.2 mm $year^{-1}$ rate (Rahmstorf et al., 2012). Due to ocean warming and glacial melting, sea level may increase up to 2 m by 2100 (Sweet et al., 2017). Decreasing rates of sediment supply to the coastal zone further compromise the ability of tidal wetlands to maintain elevation relative to SLR (Weston, 2013). Better land management, reforestation, and damming of rivers have substantially reduced sediment delivery worldwide (Milliman and Syvitski, 1992). Overall, rising rates of SLR and decreased sediment delivery to the coast have been linked to tidal wetland loss globally (Reed, 1995), witness wetland loss in the Mississippi River delta (Blum and Roberts, 2009), the Blackwater Creek marshes in Chesapeake Bay (Ganju et al., 2015), and Venice Lagoon in northern Italy (Day et al., 1998).

The impacts of human activities are also seen in loss of tidal flat ecosystems with great implications for migrating waterbird populations. SLR, declining sediment inputs, and upland tidal flat barriers, such as the building of rock walls and urban development, are contributing to tidal flat erosion, permanent flooding, and general wetland loss in area. Tidal flats along the 13,800 km Yellow Sea coastline of China, the Democratic People's Republic of Korea, and the Republic of Korea have seen some of the largest declines in tidal flat area. Murray et al. (2012, 2014, and 2015) report a decline of between 50% and 80% of tidal flat extent over the past 50 years, with a current loss rate of about 1.2% annually. This has led to declines in migratory bird populations that use the East Asian–Australasian Flyway, with 5%–9% per year losses in waterbird numbers (MacKinnon et al., 2012). Shorebird populations in Japan that use this flyway are rapidly declining, strongly suggesting the loss of Yellow Sea tidal flats as the likely driver of declines (Murray et al., 2015). The same trend of decreasing shorebird populations that use this flyway has been observed in Australia (Clemens et al., 2016).

Impacts from human activities are invariably destructive to coastal ecosystems as a general principle, whether inadvertent or in the course of ecosystem management. However, human activities often provide insight into the workings of relatively unaltered or pristine ecosystems. (The term "pristine" in this context is a relative term, and scientists studying global change would argue that there are no remaining examples that warrant this status.) We summarize human activities under three categories: those resulting from global change, effects from more localized physical and chemical changes, and effects from biotic sources (Table 1.1).

Three aspects of global change that are having a major impact on tidal wetland dynamics include a warming that may expand the range of frost-intolerant species (i.e., mangroves), accelerating rise in sea level that may exceed the capacity of some wetlands to keep pace, and the drying or wetting of climates that may alter coastal salinity and fluvial sediment delivery patterns (Kundzewicz et al., 2008). In comparison with more local types of human impacts, global changes tend to be longer term and subtler, but this feature is offset by the large geographic reach of such effects. Furthermore, there is great uncertainty in actual rates of change. Some reports have suggested that increased interannual variation in precipitation and temperatures, as well as increased "storminess," should be considered to have ecological consequences beyond more linear time-averaged projections (Michener et al., 1997). For terrestrial ecosystems, reconstruction of past climate changes indicates that forests did not shift latitudinally as intact communities, but rather individual species responded independently resulting in community mixtures not recognizable in today's assemblages (Clark et al., 1998). It is unlikely that the same occurred for coastal wetlands because of their relatively low plant diversity. Regardless, examples of asynchrony in migratory bird activity relative to the supply of seasonal foods are examples of global change that warrant scrutiny in coastal wetlands (Michener et al., 1997). We are in our infancy in the understanding of possible interacting effects of changing salinity, temperature, and diseases for wetlands.

Reduced freshwater flows resulting from domestic and agricultural extractions have the potential to increase coastal salinities and impact the existence of tidal wetlands. These extractions could and have increased the salinity in already hypersaline lagoons that have seagrasses and marshes (Buskey et al., 1997). The decline in freshwater supply was the fourth most common cause of mangrove decline in Southeast Asia between 2000 and 2012 (Giri et al., 2014). Increased estuarine salinization and turbidity from decreased freshwater flow and increased land clearing in river basins are at present the greatest threat to tropical estuarine ecosystems and their coastal wetlands in Africa (Kiwango et al., 2015; Jennerjahn et al., 2017). Furthermore, associated irrigation return flows and domestic wastewater discharges have additional influences in changing nutrient loading and introducing pesticides and other toxicants. Seagrasses are the most likely to be affected by nutrient enrichment, as discussed below.

Land clearing activities can increase sediment delivery to coastal wetlands, whereas impoundments have the opposite effect; both have the potential of changing sediment balances or imbalances locally. Seagrasses are particularly susceptible to effects of suspended sediments on light attenuation in the water column (Waycott et al., 2016). Mangrove mortality has resulted from excessive sedimentation (Ellison, 1998). Saltmarshes and tidal freshwater marshes along the east coast of United States, however, owe their current existence to past increases in sediment supply (Pasternack et al., 2001). The effects of sediment starvation have been well-documented as the cause of tidal marsh losses (Streever, 2001; Bernier et al., 2006), and a similar process is also apparent for mangroves, e.g., in the upper Gulf of Thailand (Thampanya et al., 2006; Wolanski and Elliott, 2015).

Barriers to tidal exchange interfere with the most fundamental processes in coastal ecosystems dependent on sediment dynamics. Barriers include dykes and various forms of seawalls (Adam, Chapter 23). For example, tidal marsh areas isolated from sediment supplies and astronomic tides often undergo major subsidence. Seagrasses generally may be immune to these subsidence effects.

Nutrient enrichment affects seagrass meadows in multiple ways, including accumulations of excessive macroalgal growth that depletes night time oxygen to the detriment of the plants, stimulation of sulfide production through organic enrichment resulting in toxicity to roots, and excessive periphyton growth having shading effects on plant photosynthesis (Holmer, Chapter 13). Approximately 54% of seagrass loss since 1879 has been attributed to water quality deterioration, chiefly eutrophication (van Katwijk et al., 2016).

Experimental fertilization of saltmarshes at high levels has been shown to increase insect herbivory (Vince et al., 1981). Experimental fertilization can differentially affect biogenic soil formation and, consequently, soil elevation trajectories across geomorphic settings of mangrove forests in mineral- and sediment-poor settings (McKee et al., 2007). Deegan et al. (2012) found that high levels of N-enrichment led to creek bank slumping of macrotidal saltmarshes in Northeastern USA, presumably through weakening soil structure via enhanced organic matter decomposition and possibly reduced belowground root and rhizome production.

Bottom disturbance from trawling activities is particularly destructive to seagrass beds, resulting in scars that can take decades to heal; these are the object of some restoration projects (Paling et al., 2009; van Katwijk et al., 2016). While trawling disturbance does not occur in saltmarshes and mangroves, more intense bottom disturbances (e.g., canal dredging and creation of aquaculture ponds) can completely alter the structure and function of these coastal wetlands. Effects often extend beyond the boundaries of these activities as a result of altering tidal reaches and organic matter fluxes, not to mention possible nutrient enrichment and pesticide loading often associated with discharges from aquaculture practices.

Fragmentation within a particular habitat has been little studied in coastal ecosystems. Fragmentation is a component of the placement of tidal barriers already discussed. The loss of "between-habitat" connections may have far-reaching effects for the exchange of larvae and the other nursery functions that many coastal wetland habitats provide (Davis et al., 2014).

Biotic effects include invasive species and diseases. Although not all examples are documented to have been initiated by human activities, human agents may be responsible for accelerating their introduction, and stressful conditions created by eutrophication may magnify their effects. The overharvesting of oysters in Chesapeake Bay, for example, is believed to have completely changed the trophic dynamics of the estuary. The extent to which this has interfered with oyster recovery, now hampered by the MSX and Dermo oyster parasites (Harvell et al., 1999; http://hatchery.hpl.umces.edu/oysters/history/), remains an important research question. While oyster habitat would be classified more as reef structures than a coastal wetland type, influence of the species on sediment dynamics, biogeochemistry, and species composition make it a keystone species.

Both native species and exotics can be aggressive invaders that rapidly change the structure and dynamics of coastal wetlands. *Phragmites australis*, for example, is capable of overtaking saltmarshes by outcompeting the shorter *Spartina patens*, the original dominant (Windham and Lathrop, 1999). The aggressiveness of *P. australis* is attributed to the development of an ecotype genetically dissimilar to the tamer native type (Vasquez et al., 2005). In the case of *Caulerpa taxifolia*, a submerged macrophyte, clones in the Mediterranean and California (USA) change the structure of benthic communities for both seagrasses and tidal flats. This not only interferes with fishing practices and recreation but also fundamentally alters food webs of lobsters (in California) and displaces eelgrass beds (Williams and Grosholz, 2002).

As with oyster diseases, the role of human influence on the incidence of other diseases is uncertain. Nevertheless, it is worthwhile to reveal examples that may be facilitated by human activity. A major loss of eelgrass beds that occurred in the 1930s has been attributed to a fungal pathogen that caused "wasting disease" (Orth et al., 2006). This had effects over large areas along the coasts of Europe and North America (Holmer, Chapter 13).

2.10 Tidal Wetland Metabolism and the Global Coastal Ocean Carbon Cycle

Ecosystems that comprise the coastal ocean (Fig. 1.1) play a disproportionately large role in the overall global C cycle relative to the global area the coastal ocean occupies (Bauer et al., 2013; Regnier et al., 2013). The combined net CO_2 uptake from the atmosphere by the coastal ocean ecosystems is 0.45 Pg year^{-1}. The coastal ocean receives 0.45 Pg of organic carbon year^{-1} from rivers, buries 0.3 Pg organic carbon (OC) year^{-1}, and exports about 0.25 Pg OC year^{-1} to the open ocean. Tidal wetlands dominate the coastal ocean carbon budget with a net CO_2 uptake (net ecosystem production [NEP]) of 0.35 Pg year^{-1}. Of this approximately 15%–30% is buried in wetland sediments and the remainder is exported to adjacent tidal creeks and estuarine waters and bottoms. The allochthonous OC inputs from rivers and tidal wetlands drive estuarine waters and bottoms toward net heterotrophy, with NEP of about −0.2 Pg year^{-1} (Bauer et al., 2013).

Estuarine net metabolism and the overall balance between gross primary production and ecosystem respiration are primarily a function of the relative balance between organic matter and inorganic nutrient loading (Hopkinson and Vallino, 1995; Smith and Hollibaugh, 1997; Testa and Kemp, 2008; Testa et al., 2012). Watersheds typically are the primary source of inorganic nutrients (with the exception of karst estuaries like the Everglades coastal wetlands in the United States where the ocean is the primary source), while watersheds and tidal wetlands are the primary sources of organic matter in estuaries. Results from coupled hydrologic/biogeochemical simulation models suggest that land use, land cover, and land management practices are primary factors controlling long-term trends in dissolved organic carbon (DOC) export from major river basins, while decadal and short-term variability are more controlled by year-to-year climate variability and extreme flooding events (Ren et al., 2016). Long-term trends in inorganic nutrient runoff are also largely controlled by land use and land use management (Yang et al., 2015). For instance, long-term trends in NO_3^- loading to the Louisiana coastal zone are related most directly to fertilizer use in the Mississippi River basin (Goolsby et al., 1997; Turner and Rabalais, 2003). Agricultural land abandonment and reforestation generally result in reduced export of all forms of N (Yang et al., 2015).

The estuarine metabolic response to watershed-derived organic matter and nutrients is also related to the estuarine water residence time, which in some estuaries is closely coupled to freshwater discharge (Hopkinson and Vallino, 1995). Residence time coupled to the C:N stoichiometry and lability of allochthonous organic matter can further control estuarine net metabolism. Watershed- and tidal wetland–derived organic matter is much less labile than phytoplankton or algal-derived organic matter and the C:N ratio is considerably higher as well (e.g., 6:1 for phytoplankton and 50:1 for watershed or wetland-derived organic matter). The turnover time of inorganic nutrients is extremely short as estuarine primary producers are often N-limited. Turnover time of wetland or watershed-derived organic matter is substantially longer—ranging from days to years (Uhlenhopp et al., 1995). The degree to which primary production or microbial decomposition of allochthonous organic matter is stimulated in estuarine systems is therefore a function of the relative balance between water residence time in the estuary and inorganic/organic turnover times. Holding the inorganic:organic loading constant, decreasing residence time pushes a system toward increased autotrophy, whereas decreasing residence time pushes a system toward heterotrophy (Hopkinson and Vallino, 1995). Stoichiometry also influences the metabolic balance of an estuary. High C:N ratio organic matter, such as that derived from saltmarshes or mangroves, regenerates relatively little inorganic nitrogen when decomposed, thus the primary production that can be supported by the regenerated N (such as by phytoplankton) is trivial compared to the amount of organic matter decomposed (such as of saltmarsh origin). As an example, consider that 50 units of saltmarsh organic matter decomposition releases

50 units of CO_2 and 1 unit of inorganic N. The 1 unit of inorganic N only supports 6 units of phytoplankton C uptake. Hence the ratio of production to respiration is 6:50—very heterotrophic. It is interesting to speculate whether there are general relationships between the relative area of tidal wetlands (and/or seagrasses) to open waters (bays, sounds, creeks) and estuarine metabolism. Does the presence of wetlands decrease estuarine water residence time? Does increasing wetland area linearly increase organic matter loading? Presumably, the greater the relative area of wetlands, the greater the degree of estuarine heterotrophy.

Our understanding of tidal wetland carbon cycling has changed substantially in the past several years with the advent of new technologies for measuring metabolism. In the past, most budgets were developed by measuring the primary production and respiration of all the major functional groups of either the donor system (e.g., the tidal wetland) or the receiving system (e.g., the estuary or nearshore ocean) (Hopkinson, 1985, 1988). The balance between production and respiration is typically small relative to either P or R (close to zero) and statistically it may not be different from zero. Many early budgets for saltmarshes, however, showed very high rates of primary production and considerably smaller rates for respiration. The conclusion was that tidal wetlands exported large quantities of organic matter (e.g., Hopkinson, 1988). Recently, scientists have been employing the eddy covariance technique to measure primary production and respiration of large expanses of tidal wetland (e.g., up to 1 km^2). Eddy covariance directly quantifies the vertical exchange of selected gases, such as CO_2 and CH_4, between the ground surface (e.g., saltmarsh) and the atmosphere by rapid, simultaneous measures of vertical air movement and gas concentration. It underestimates net ecosystem metabolism to the extent that dissolved gases and inorganic carbon can also be exchanged laterally by tides, but it can be close. While eddy covariance measured rates of respiration per unit area of tidal wetland have not changed substantially from the older component methods, rates of primary production are lower, and the resultant net metabolic balance is much smaller. For example, in the Plum Island Sound saltmarsh, eddy covariance results (Forbrich et al., 2018) show that primary production averages about 818 $gC\ m^{-2}\ year^{-1}$ and respiration averages about 650 $gC\ m^{-2}\ year^{-1}$. The net ecosystem exchange (NEE—exchange with the atmosphere only—not including dissolved inorganic carbon or DOC export/import via tidal waters) is only 168 $gC\ m^{-2}\ year^{-1}$. After accounting for an OC burial at about 114 $gC\ m^{-2}\ year^{-1}$, only 91 $gC\ m^{-2}\ year^{-1}$ are potentially available for export. This contrasts with budgets created in the 70s and 80s for another New England tidal wetland, Sippewissett marsh, where net primary production was estimated at about 1400 $gC\ m^{-2}\ year^{-1}$, heterotrophic respiration at 560 $gC\ m^{-2}\ year^{-1}$, and NEP in excess of 840 $gC\ m^{-2}\ year^{-1}$ (Hopkinson, 1988). This suggests that the organic carbon subsidy that tidal wetlands provide for estuarine waters and bottoms is much less than previously estimated. New global ocean carbon budgets have yet to be published that incorporate these new findings based on eddy covariance measures of gross primary production (GPP), R, and NEP.

Long-term deployments of the eddy covariance approach promise to greatly increase our understanding of the overall effect of tidal flooding, SLR, precipitation, and temperature on the fate of coastal wetland primary production. Conducting these measurements over multidecadal time frames that encompass long- and short-term changes in drivers should enable us to statistically unravel the relative contributions of each of these factors in controlling primary production, ecosystem respiration, and the net ecosystem carbon balance of tidal wetlands. Examining a 4-year record of saltmarsh metabolism, Forbrich et al. (2018) showed precipitation early in the growing season was the primary determinant of year-to-year variability in NEE and this was primarily due to the effect of salinity (as influenced by precipitation) on primary production, as opposed to salinity or temperature effects on respiration.

2.11 Modeling and Predictions

In view of the perennial changes in wetlands and the acceleration of these changes because of human impacts and climate change, there is a need to obtain models that can predict the future evolution of coastal wetlands. The models must be able to distinguish between two evolutionary pathways, namely a "state change" in contrast to "community succession." Community succession repeats vegetation patterns after disturbance (the same community), while state change leads to new patterns or new communities normally as a result of some underlying physical change in the ecosystem (Hayden et al., 1991). Brinson et al. (1995), for instance, showed that disturbances including sea level changes have generated state changes in coastal communities, resulting in a zonation of adjoining upland forest, high marsh, low marsh, and intertidal flat communities. Such models need elaboration to transition from being qualitative to being quantitative. The result should allow prediction in changes of coastal wetlands to climate change so that adaptation strategies can be put in place.

Predictive models of the evolution of coastal wetlands are needed and remain in early stages of meeting goals for management; this is particularly true for landscape-scale models with multiple communities or ecosystem types. Such models need the input of oceanographers, sedimentologists, chemists, biologists, ecologists, hydrologists, pedologists, dendrologists, entomologists, and geneticists. Clearly models to be successful and practical must avoid getting bogged down in details while recognizing that the system is complex and intrinsically biogeomorphic with equally important and connected biological, chemical, and physical components. Understanding the functioning of coastal wetland ecosystems depends both on the collection of good long-term sets of real-time data on the levels and fluxes of each of the significant plant nutrients and on the development of functional models of the magnitude and direction of all major organic and inorganic material fluxes. Cumulative effects on coastal wetlands are very difficult to predict. Much has already been revealed by various models; these have, among many findings, revealed the seasonal and interannual variations in these fluxes. In view of global climatic change, the baseline of environmental parameters (i.e., extreme climatic conditions) is changing. As a result, the adequacy of the existing data sets on coastal wetlands to address climate change needs to be re-examined. Thus, the problem of a shifting database of climate, a key problem for humanity in planning a reliable water supply (Kundzewicz et al., 2008; Milly et al., 2008), applies equally forcefully to the survival of coastal ecosystems. Models should also consider the interaction between adjoining coastal wetlands, such as between mangroves and seagrass, or saltmarshes and seagrass, or mangroves and saltmarshes, as these frequently coexist side by side, though very rarely are intermixed.

Scientists may choose to study pristine coastal wetlands to understand and quantify the dominant processes shaping the ecosystem. They then hope to apply this knowledge to quantify the human impact in other wetlands. However, we are rapidly losing our scientific control sites because pristine coastal wetlands are increasingly a rarity worldwide (Mcowen et al., 2017)

Scott Hagen and colleagues (Alizad et al., 2016a;b) studied tidal wetland response to SLR and sediment supply using coupled hydrodynamic/ecological models with sediment transport. They modeled drowning of saltmarshes, saltwater intrusion and conversion of tidal freshwater marsh to brackish or saline marsh, and transgression of marshes into uplands. They found substantial differences in marsh response to increased sea level that reflected differences in sediment availability. They found tipping points in marsh response, where marsh macrophyte biomass and stem density actually increased up to a certain rate of SLR and beyond that biomass declined, accretion decreased, and ultimately the marsh drowned. The models were useful in helping to assess the complex spatial dynamics of saltmarshes over time as the rate of SLR increased.

In the Mississippi River delta of coastal Louisiana, coastal models based on detailed sediment diversion modeling and marsh development have been used to predict the future extent of tidal wetlands. This modeling effort resulted in the development of the $50 billion Louisiana Master Plan 2017, which is a state plan to save coastal Louisiana, its people, and its economy (see http://coastal.la.gov/our-plan/2017-coastal-master-plan/).

3. HUMAN HEALTH AND COASTAL WETLAND SOCIOECONOMICS

Coastal wetlands provide numerous ecosystem services to humanity. They protect the coast against erosion and guard against loss of capital infrastructure and human lives. They are habitats that support seasonal or perennial fisheries and are vital for migratory and resident birds. In addition, they provide ecological services that have socioeconomic benefits to the human population, including, according to location, fuel, forage, building material, timber, fisheries, and protection of commercial, recreational, and naval vessels (Williams et al., 2007). Mangroves provide another important ecosystem service to the population living in the hinterland, by sheltering it from storm winds, and capturing salt spray (Fig. 1.25), thus improving crop production in arid coastal areas (Wolanski and Elliott, 2015).

Throughout history, coastal wetland management, in most countries in the world, has been a failure. The failure has led to the destruction and degradation of these ecosystems, resulting in the loss for humanity of the ecosystem services they provide and to a lowering of the human quality of life (Millennium Ecosystem Assessment reports; http://www.millenniumassessment.org/en/index.aspx; Frayer et al., 1983; Duarte, 2002; Hily et al., 2003; Bernier et al., 2006; Duke et al., 2007; Wolanski and Elliott, 2015). Destruction of coastal wetlands is often the result of management for other goals (e.g., flood control in the Mississippi River, which has prevented river sediments from being deposited within the tidal wetlands), which do not consider the effects on tidal wetlands. However, there are other examples from within the Mississippi River deltaic plain where management was directed specifically at protecting tidal wetlands, such as putting in weirs and installing small levees along marsh shorelines. Again, the result led to the destruction of the tidal wetlands, which require regular flooding and drying to trap sediment and build soil strength. The levees prevent regular tidal flooding thereby limiting the input of sediments from tidal floodwaters. The undesired outcome was increased wetland subsidence, decreased wetland biomass, and general marsh erosion and drowning.

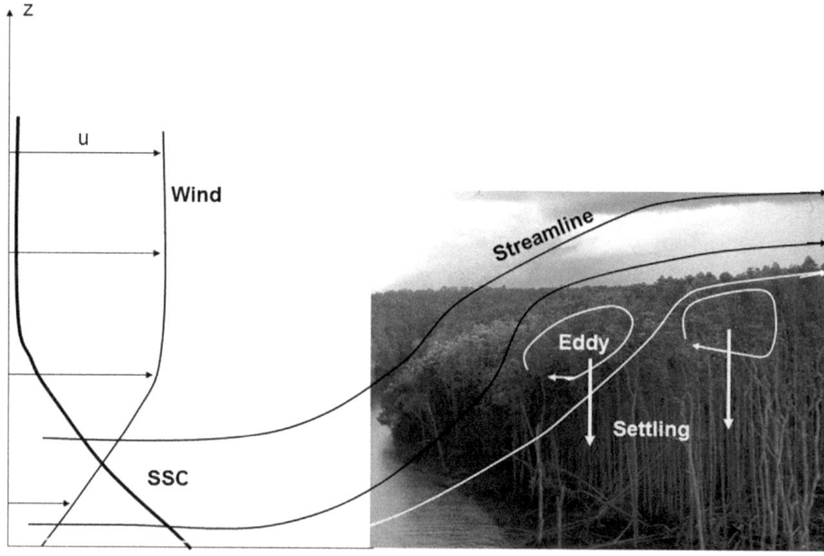

FIGURE 1.25 Sketch of the trajectory of the landward wind facing a mangrove belt and the fate of salt particles in suspension. The wind slows down as the air flows through the mangroves, the bulk of the flow is deflected upward over the vegetation, and a turbulent wake forms behind the trees. *Down arrows* indicate deposition of salt particles. *SSC*, suspended salt concentration. *Redrawn from Wolanski, E., 2007. Protective functions of coastal forests and trees against natural hazards. In: S. Braatz, S. Fortuna, J. Broadhead, R. Leslie, R. (Ed.), Coastal Protection in the Aftermath of the Indian Ocean Tsunami. What Role for Forests and Trees? FAO, Bangkok. 157–179.*

Scientists could help in the preservation of coastal wetlands by finding ways to more effectively communicate with stakeholders, managers, and policy-makers (Pielke, 2007). Communication is difficult because of the multitude of stakeholders in coastal wetlands and fragmented governance. Communication of coastal wetlands science results and their implications for humanity could be improved by quantifying the economic value of ecosystem services. This is very much an emerging science and there have been several attempts to do that (see, for example, Costanza et al., 1997 for wetland ecosystems in general; Balmford et al., 2002, for the case of tropical rainforests, mangroves, wetlands, and coral reefs; Barbier et al., 2008, for the case of mangroves; van den Belt and Costanza, 2011, for the case of estuaries without, however, separating the various components of the estuary such as its wetlands). As Barbier shows in Chapter 27 shows, the core of the wetland loss problem is the failure to consider the various values or benefits provided by coastal wetlands when deciding on their current and future development and use. Wetland ecosystems are even more undervalued if they are already degraded and thus have lost their capacity to generate goods and services (Valiela and Fox, 2008). Failing to account for the full range of ecosystem services can often lead to overexploitation of coastal wetlands and their deterioration (Ghermandi et al., 2011). This problem is exacerbated when using linear economics (i.e., as a linear interaction between production—e.g., labor, land, built, and financial capital—and consumption—e.g., individual utility and social welfare) that fails to distinguish between desirable and undesirable economic activities (de Groot, 2011), as well as when managing single aspects of coastal ecosystems independently of others (Ghermandi et al., 2011). Even if the socioeconomic services of coastal wetlands were fully quantified and communicated, it does not mean that these wetlands will be preserved when job-producing development opportunities arise. This, and economic forces operating at global scales, may be the principal causes for the ongoing wetland loss worldwide.

There is an urgent need to quantify the socioeconomic value of coastal wetlands because they may be perceived as a threat to people as they are a breeding ground for mosquitoes that transmit diseases to humans (Dale and Knight, 2008). Some such diseases can be life-threatening such as malaria, yellow fever, dengue, and forms of encephalitis. Other diseases are debilitating such as Ross River virus and West Nile virus, though these diseases occasionally also can be lethal. The link between coastal wetlands (both fresh and saline) and mosquitoes is undeniable. The saltwater mosquito *Aedes vigilax* (Fig. 1.26A) breeds in mangroves (Knight, 2008). Eggs are laid on most substrates, eggs mature under dry conditions (no tidal flooding) in as few as 3 days, eggs hatch when they are flooded by the tides, and the larvae develop in small water holes that occur in localized areas of poorly flushed mangroves. When the mosquitoes emerge, they may fly many kilometers. For combating this mosquito, it may be sufficient to increase the drainage pattern to avoid slow-flowing or stagnant waters (Dale and Knight, 2006); this requires careful engineering

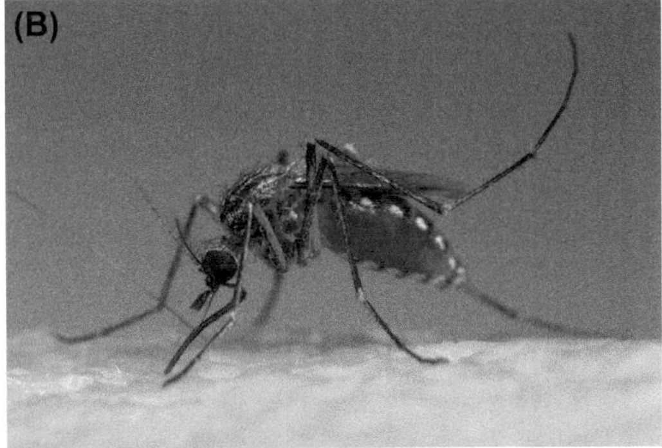

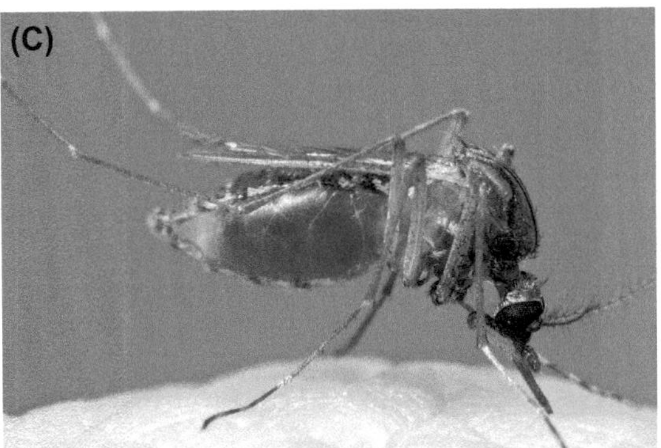

FIGURE 1.26 Disease-carrying mosquito species include (A) *Aedes vigilax*, (B) *Aedes Coquillettidia linealis*, and (C) *Aedes Culex annulirostris*. *Courtesy of Stephen Doggett, Westmead Hospital, Australia.*

to avoid creating acid sulfate soils. There are a number of disease-carrying mosquito species (Fig. 1.26A–C) and the management strategy to combat them in coastal wetlands varies from species to species because of their different breeding strategies. For instance, to control the saltmarsh mosquitoes *Ochlerotatus taeniorhynchus* and *Ochlerotatus sollicitans*, which do not lay their eggs on standing water, impoundments can be built with earthen dikes in saltmarsh or mangroves and kept flooded during the breeding season to prevent oviposition. The rest of the time the impoundment should be open through culverts in the dikes to allow access by estuarine plankton communities and fish (Rey et al., 1991; Brockmeyer et al., 1997). Mosquitoes can thus be managed without having to use pesticides.

There is clearly a need for mosquito-control managers and wetland managers to communicate and integrate their efforts to sustain both wetland functions and human health. A variety of chemical, biological, and physical methods is possible (Dale, 1994; White and Beumer, 1997; Dale and Knight, 2008; Wolanski and Elliott, 2015). Physical control methods that minimize the conditions favorable for breeding include modifying the hydrology by constructing runnels to connect depressions with tidal exchange, and this has been successful in saltmarshes. Historically, physical methods to control mosquitoes were very destructive of the coastal wetlands (i.e., filling or impounding) but they are increasingly discarded because of their destructive impact on flora and fauna. Biological control methods include introducing fish and larval predators (White and Beumer, 1997). Historically DDT was the preferred chemical to control mosquitoes; because of its impact on the environment, it has been replaced in many (but not all) countries by mosquito larvicides, including methoprene and temephos and *Bacillus thuringiensis* (EPA, 2002). Complete larval control of *Anopheline* and *Culicine* mosquitoes can be achieved using the mosquito larvicide Vectobac GR (Djenontin et al., 2014). Another less costly but proven method is to use mosquito-eating fish such as the guppy, as was demonstrated by Kusumawathie et al. (2008) in Sri Lanka. The cost of various mosquito control methods can be readily evaluated using the method of Dale et al. (2018).

Wetlands have often been blamed by the public for the proliferation of *Vibrio cholerae*, the etiological agent of cholera that, despite progress in medicine, still exists in at least 38 countries in 2016, and improvement over 90 countries a decade earlier (WHO, 1998; http://www.who.int/gho/epidemic_diseases/cholera/cases_text/en/). The combination of a brackish wetland environment, rich in organic matter and with a high density of human population, represents ideal conditions for *Vibrio cholerae*. Although the high human density near some wetlands amplifies the transmission, it may not be its primary cause because *V. cholerae* is part of the autochthonous flora of brackish and estuarine environments (Constantin de Magny et al., 2011). Cholera epidemics can also be linked to plankton blooms, rise in temperature, and El Niño Southern Oscillation. Outbreaks can occur after natural disasters (Colwell, 1996; Neogi et al., 2014). These additional links may explain why the disease is established along the Bay of Bengal and not along the Amazonia coast, even though human and wetland settings in these two regions are similar (OCHA, 2007; R. Lara, pers. com. IADO, Argentina).

Climate change may exacerbate the problem of mosquito-borne diseases (WHO, 2000). Increased temperature may speed up the development time for mosquitoes and pathogen cycles. This may extend their distribution to higher latitudes so that presently cooler wetlands may become mosquito-vectored disease habitats. Some infectious diseases, such as malaria, yellow fever, and dengue, are believed to be associated mainly with wetland conditions.

Dengue and yellow fever were common from the 17th century onward in the United States, with yellow fever killing tens of thousands of people as far north as New York City (Reiter, 1996). Temperature and favorable habitat for vectors may be a necessary but not sufficient condition for disease prevalence in human populations. Public health practices and lifestyles have a greater influence on the spread of diseases than the temperature tolerance of their vectors (Marshall, 1997). Increased incidence of West Nile virus in recent years (Bourgeade and Marchou, 2003) is apparently unrelated to changes in temperature. In a worst-case scenario, the melting of permafrost in northern latitudes would be likely to create larger areas suitable for mosquitoes.

Coastal wetlands ecosystems are important not only for their ecosystem services and values but also as a key component of the total ecosystem composed of the river basin, the river, the estuary, and coastal waters. Thus, for human needs, an ecohydrology approach could integrate management across ecosystem types (Wolanski and Elliott, 2015). We have not found examples in the world of such an approach being implemented, or if they do occur, we have not found outcomes of their success or failure widely communicated. Instead, we have found management regulated by social and political boundaries that often ignore the realities of the watershed geography and hydrology or by humanity's historical habits of managing individual users (water resources, water supply, irrigation, hydroelectricity, forestry, farming, and fisheries) without integration. We have found that this type of management invariably leads to environmental degradation and the loss of ecosystem services. A new approach could use an ecohydrology (i.e., holistic) management strategy.

The restoration of functional coastal wetlands is a combination of evolving multidisciplinary science and engineering practices based on analysis of past failures and success. The aim is to restore degraded coastal wetlands to have many of the same functions and ecosystem services, if not all, of natural wetlands. Some success has been achieved, but not more than 50% or so of documented case studies, for saltmarsh and tidal freshwater wetlands, in part because hydrology (i.e., sea level and tides) is predictable and practices have been ongoing for decades Baldwin et al., Chapter 22 and Broom et al., Chapter 25. However, there are still a number of failures for a number of reasons, such as coastal sediment being imported by the currents into the created wetlands and rapidly filling up the system to an elevation rarely flooded by the tides (Elliott et al., 2016). Success has been even less for mangroves but is improving, although it is costly and the prevalence of monoculture in practice does not necessarily benefit the ecology (Asaeda et al., 2016; Lewis et al., Chapter 24). The reasons for failure of coastal wetland restoration vary from site to site. Failures can be as simple as restoring incorrectly the hydrology (Fig. 1.21C), or coastal sediment being advected into the wetlands and rapidly filling up the system to an elevation rarely flooded by the tides, or the weight of mats of algae physically bringing down emerging plants and seedlings (Fig. 1.21D), or simply planting the wrong species at the wrong places (Elliott et al., 2016; Primavera and Estaban, 2008). Other reasons for failures can be more complex and involve the unintended interaction between a number of ecological processes and/or invasive species. Restoring coastal wetlands in sites exposed to high wave energy remains a challenge. Often, a temporary form of shoreline armoring is engineered (e.g., using bamboo curtains as in Fig. 1.21E or seawalls as in Fig. 1.21F) and installed to enable the establishment of wetland vegetation against wave and erosion until it is developed enough to be able to survive and attenuate waves alone. For seagrass, success is even less but is improving, though the size and location of projects are greatly constrained by high costs and appropriate environmental conditions (van Katwijk et al., 2016).

4. COASTAL WETLANDS ARE ESSENTIAL FOR OUR QUALITY OF LIFE

It may be unrealistic to imagine that humanity would act to preserve and cherish coastal wetlands simply because they are beautiful, but recent research suggests that nature's intrinsic beauty may be an important factor (Bekessy et al., 2018). As editors, we recognize that science alone, regardless of how compelling the evidence, is unlikely to motivate societies to protect coastal wetlands at the expense of numerous competing short-term gains. The environmental movement is not only facilitated by necessity based on the science but also by a sense of beauty that is vital for maintaining the quality of life of the human population (Bekessy et al., 2018). The human need for beauty should not be underestimated as a prime factor in the preservation of coastal wetlands (Dorst, 1965; translated from French):

> Nature will only be preserved if man loves it a little, simply because it is beautiful, and because we need beauty whatever is the form we are sensitive to as a result of our culture and intellectual formation. Indeed this is an integral part of the human soul

Coastal wetlands offer spectacular landscapes for humanity to enjoy (Fig. 1.27). Intangibles such as existence value relate to some of the more popular images of coastal wetlands including birds and other charismatic wildlife. Coastal wetlands are becoming increasingly important as a refuge to charismatic wildlife (Figs. 1.28 and 1.29) faced with the human coastal squeeze. Thus, coastal wetlands are increasingly becoming the last remaining coastal biodiversity hotspots worldwide. At less disturbed sites, large, charismatic animals still freely migrate between the land and the coastal wetlands, such as along the Arctic coast (Fig. 1.30).

5. BLUE CARBON AND HUMAN WELL-BEING

In the past decade, interest in tidal wetlands has increased greatly with the realization that these systems play a significant role in sequestering atmospheric CO_2 and that rising levels of CO_2 in the atmosphere (because of anthropogenic activities such as fossil fuel combustion and deforestation) are contributing to climate change and SLR (Nellemann et al., 2009b, Serrano et al., Chapter 28). The sequestration of atmospheric CO_2 by tidal wetlands is because of the accumulation and burial of organic carbon in their sediments (Hopkinson, 2018). Organic carbon accumulates and is buried in tidal wetlands as sediments settle on the marsh surface at a rate proportional to SLR. Sediment deposition on the marsh surface is often the primary mechanism whereby marshes keep up with SLR. In sediment-poor regions, the accumulation of dead belowground parts of plants is the primary mechanism of marsh elevation gain (Cahoon et al., 2009). Regardless, the primary source of organic matter that accumulates over time and becomes buried is root and rhizome tissues that escape decomposition in the anaerobic, poorly drained tidal wetland soils. Collectively the tidal wetlands that accumulate

FIGURE 1.27 Coastal wetlands offer spectacular landscapes such as (A) the mangrove-fringed Hinchinbrook channel, Australia, and (B) the salt marshes of the Virginia Coast Reserve, USA. *(A) Courtesy of H. Yorkston.*

and bury organic carbon are called blue carbon ecosystems and the organic carbon that accumulates in their soils over time is called blue carbon (Windham-Myers et al., 2018).

Blue carbon sequestration worldwide is calculated as the product of the areal extent of these systems, the organic carbon density of blue carbon sediments, and the accretion of the wetland sediment surface. Blue carbon sequestration is estimated betwee 215 Tg OC year^{-1} (Hopkinson et al., 2012). Lack of precision in estimates areal extent of these systems and variability in organic carbon density ments are the factors contributing to the wide range in sequestratio estimates of the areal extent of seagrasses range from 177,000 to 600,000 km^{2} (van Katwijk et al., 2016). For saltmarshes, by a factor of two.

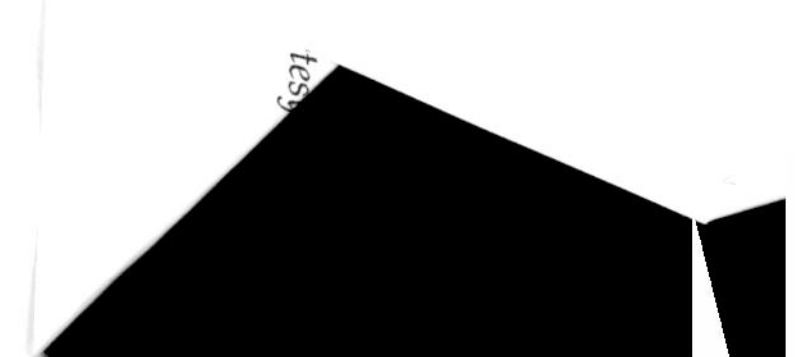

28 Charismatic wildlife found in coastal wetlands include (A) chimpanzee (Conkouati mangrove
Brazzaville), (B) crab-eating Macaque in *Rhizophora apiculata* mangroves (Ranong, Thailand),
rass along the Queensland coast, (D) mudskipper and saltwater crocodile (Queensland man-
s (Chwaka Bay, Zanzibar), (F) spotted deer (Sundarban, Bangladesh), (G) Sundarban tiger
popotamus (Wami Estuary, Tanzania). *(A) Courtesy of S. Thomas. (B) Courtesy of N. Duke. (C)*
eund. (D) Courtesy of M. Read. (E) Courtesy of Farhat Mbarouk. (F) Courtesy of P. Dyas. (G)
of H. Kiwango.

FIGURE 1.27 Coastal wetlands offer spectacular landscapes such as (A) the mangrove-fringed Hinchinbrook channel, Australia, and (B) the salt marshes of the Virginia Coast Reserve, USA. *(A) Courtesy of H. Yorkston.*

and bury organic carbon are called blue carbon ecosystems and the organic carbon that accumulates in their soils over time is called blue carbon (Windham-Myers et al., 2018).

Blue carbon sequestration worldwide is calculated as the product of the areal extent of these systems, the organic carbon density of blue carbon sediments, and the accretion rate of the wetland sediment surface. Blue carbon sequestration is estimated between 65 and 215 Tg OC $year^{-1}$ (Hopkinson et al., 2012). Lack of precision in estimates of the global areal extent of these systems and variability in organic carbon density of wetland sediments are the factors contributing to the wide range in sequestration rates. For instance, estimates of the areal extent of seagrasses range from 177,000 (Green and Short, 2003) to 600,000 km^2 (van Katwijk et al., 2016). For saltmarshes, estimates of areal extent vary by a factor of two.

FIGURE 1.28 Charismatic wildlife found in coastal wetlands include (A) chimpanzee (Conkouati mangrove reserve, Congo-Brazzaville), (B) crab-eating Macaque in *Rhizophora apiculata* mangroves (Ranong, Thailand), (C) dugong in seagrass along the Queensland coast, (D) mudskipper and saltwater crocodile (Queensland mangroves), (E) red colobus (Chwaka Bay, Zanzibar), (F) spotted deer (Sundarban, Bangladesh), (G) Sundarban tiger (Bangladesh), and (H) hippopotamus (Wami Estuary, Tanzania). *(A) Courtesy of S. Thomas. (B) Courtesy of N. Duke. (C) Courtesy of J. Freund and S. Freund. (D) Courtesy of M. Read. (E) Courtesy of Farhat Mbarouk. (F) Courtesy of P. Dyas. (G) Courtesy of P. Dyas. (H) Courtesy of H. Kiwango.*

FIGURE 1.28 cont'd

Tidal wetlands and seagrasses are some of the most threatened ecosystems in the world and their loss or declines in performance could impact not only the annual rate of blue carbon sequestration but also the fate of organic carbon stores that have accumulated over the past several thousand years (Chmura et al., 2003). Canals, wetland reclamation, sediment starvation from dams and diversion, and conversion to other land uses are the primary drivers of change in the United States. Worldwide, mangrove loss due to reclamation is

FIGURE 1.29 Coastal wetlands support an enormous diversity of resident and migratory birds including (A) snow geese, often found in large flocks on coastal wetlands where they raise their broods (Alaska, USA), (B) heron (Cedar Key, Florida, USA), (C) yellow-billed storks (Wami Estuary, Tanzania), (D) fish eagle (Congo Estuary, Angola), (E) lesser flamingoes (Wami Estuary, Tanzania), (F) diverse shorebirds (Mai Po, Hong Kong), (G) 20,000 waders (mainly knot, dunlin, short-billed dowitcher, and semipalmated sandpiper in the foreground with a few laughing gulls in the background) feeding on horseshoe crab eggs (Delaware Bay, USA), and (H) a flock of knots and oystercatchers over the Wash (UK). *(A) Courtesy of W. Streever. (C) Courtesy of H. Kiwango. (G) Courtesy of N. Clark. (H) Courtesy of N. Davidson.*

expanding at greater rates than that for saltmarshes. Giri et al. (2008) report that major causes of deforestation are agriculture (81%), aquaculture (12%), and urban development (2%). Mangroves are cut heavily in certain parts of the world for charcoal production and diked for fish farming in many other regions. Seagrasses are also being lost at high rates but

FIGURE 1.29 cont'd

FIGURE 1.29 cont'd

primarily because of N-enrichment leading to eutrophication and the resultant loss in water clarity because of increased turbidity (phytoplankton and suspended sediments) (Holmer, Chapter 13).

Decreased river export of sediments, climate change, and SLR are three additional factors that will affect the ability of tidal wetlands to continue to sequester CO_2 (Weston, 2013; Hopkinson et al., 2012). These factors can affect sequestration in multiple ways through affecting their (1) areal extent, (2) organic carbon density, and (3) vertical rates of accumulation. Worldwide, there has been a decrease in river sediment load that is the result of agricultural abandonment and reforestation, better land management (e.g., contour plowing), and damming of rivers (Milliman and Syvitski, 1992; Syvitski et al., 2005a,b). Sediment supply to estuaries controls the rate of bay bottom filling and marsh progradation. Sediment deposition onto the tidal wetland surface contributes to elevation gain and the ability of wetlands to maintain elevation relative to SLR (Cahoon et al., 2009).

FIGURE 1.30 American and Canadian Arctic coastal wetlands are probably among the least human-impacted coastal wetland areas of the world. Large charismatic wildlife such as (A) caribou (Alaska) can move back and forth between inland and coastal wetlands, and (B) polar bears (James Bay, Canada) use coastal wetlands during the summer when the sea ice has melted. *(A) Courtesy of W. Streever. (B) Courtesy of R.I.G. Morrison.*

Climate change influences CO_2 sequestration through its control over biological processes, including primary production and organic matter decomposition. The organic carbon buried in wetland sediments is primarily because of the accumulation of root and rhizome material that does not decompose (i.e., tidal wetland NEP). Changing temperature and salinity (controlled by the balance of freshwater supply from rivers, precipitation, and evaporation) directly influence rates of primary production and organic matter decomposition. The relative susceptibility of production and decomposition to changes in temperature and salinity is an active area of current research. Theoretically, respiration should be enhanced to a greater degree by a unit increase in temperature than primary production, as the activation energy is greater for respiration (Yvon-Durocher et al., 2010). Thus, global warming should lead to less organic carbon remaining for burial. However, field results from different studies often contrast with theoretical predictions. A decline in organic matter accumulation and its role in marsh elevation gain in a New England saltmarsh was attributed to an increase in water temperature over the past 30 years (Carey et al., 2017). In contrast, mangrove soil carbon density and OC accumulation were unrelated to temperature rise. At the same time, however, mangrove productivity and biomass increased in response to rising temperature, at least near the cold-temperature limits of their growth, where they typically transition into saltmarshes (Feher et al., 2017).

SLR will influence the ability of tidal wetlands to continue to sequester CO_2. The great expanse of wetlands we see today is largely the result of geomorphic processes that played out over the past 4000 to 6000 years, with the onset of the Holocene deceleration of SLR (Fig. 1.4B and C) (Redfield, 1965; Donnelly, 2006). Ocean volume remained nearly constant since about 2500 years ago at rates less than 0.4–0.6 mm year^{-1} (Kemp et al., 2011; Lambeck et al., 2014). Slow rates of SLR enabled sediment inputs to infill shallow bay bottoms and for wetlands to prograde outward into areas reaching MSL in elevation and it enabled existing wetlands to gain elevation through sediment deposition and accretion (as well as organic carbon accumulation of undecomposed roots and rhizomes). However, rates of SLR increased suddenly between 1865 and 1892 to about 2.1 mm year^{-1} (Fig. 1.31) (Kemp et al., 2011; Lambeck et al., 2014). Rates are now about 3.2 mm year^{-1} (Rahmstorf et al., 2012). Sea level is projected to increase by up to 2 m by 2100 based on high CO_2 emission scenarios (Fig. 1.31) (Walsh et al., 2014; Sweet et al., 2017). SLR can affect tidal wetland CO_2 sequestration through its combined effects on areal extent, organic carbon density, and vertical accretion rate.

Coupled hydrodynamic, geomorphic, and ecological models are becoming increasingly sophisticated and have become critical tools for predicting the fate of tidal wetlands, their

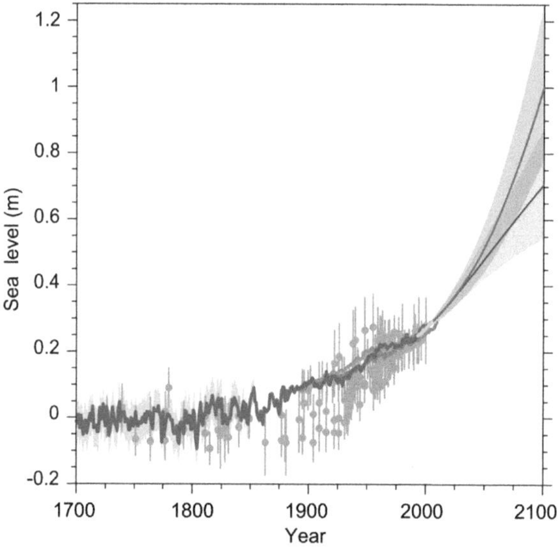

FIGURE 1.31 Sea level since 1700 and projected through the 21st century. For the past proxy data are shown in blue and tide gage data in purple. Future sea levels are based on IPCC projections for high emissions (red) and very low emissions (blue). *From Church, A., et al. 2013. Sea level change. In: Stocker, T., Qin, D., Plattner, G., Tignor, M., Allen, S., Boschung, J., Nauels, A., Xia, Y., Bex, V., Midgley, P. (Eds.), Climate Change 2013: The Physical Science Basis. Contribution of Working Group I to the Fifth Assessment Report of the Intergovernmental Panel on Climate Change (IPCC). Cambridge University Press, Cambridge, UK, IPCC AR5 Fig. 13.27, page 1204.*

rate of CO_2 sequestration, and for guiding coastal management primarily for protecting human populations and their economies (see, for example, the $50 billion 2017 Louisiana Coastal Master Plan). Models are addressing the issue of blue carbon sequestration in multiple ways: areal extent both in terms of horizontal expansion across uplands (transgression), through horizontal contraction through tidal wetland shoreline erosion, and through vertical elevation gain and organic carbon burial.

As SLRs and uplands flood, terrestrial vegetation will become stressed and die, and wetland vegetation will invade (Fig. 1.32). In terms of blue carbon, wetland transgression will need to be examined and modeled in terms of how upland productivity and existing above- and belowground OC stocks change during the onset and completion of wetland invasion. Blue carbon associated with the transgressing wetlands will also have to be examined as the production and organic carbon accumulation of the invading vegetation will be considerably lower in the stressful conditions at the upland-wetland ecotone. The potential increase in areal extent will reflect the slope of the upland-estuarine shoreline and the rate of SLR (Kirwan et al., 2016). Estimates are that tidal wetlands could increase in area by 38.7×10^3 km^2 with 2 m of SLR in the United States, which is considerably more than the current 21.7×10^3 km^2. In Chesapeake Bay, on the east coast of the United States, about 400 km^2 of uplands were transgressed by tidal marshes since the rate of SLR accelerated in the late 1800s. This represents about a third of all current tidal wetlands in this system (Schieder et al., 2017). A critical social question will also have to be addressed to accurately predict transgression—will current land owners armor their coastline to prevent flooding? Shoreline armoring is quite extensive in the United States (14% of total coastline), especially adjacent to developed urban lands (Gittman et al., 2015).

The fate of existing tidal wetlands is largely controlled by the relative availability of sediment and the rate of SLR (Fig. 1.33). As rates of SLR accelerate and sediment becomes

FIGURE 1.32 An example of marsh transgression into upland forest. This photo demonstrates the saltwater and tidal flooding stress on upland vegetation resulting in its death and the migration of salt marsh vegetation, including *Spartina alterniflora, Juncus roemerianus, and Sarcocornia virginica* into the former upland forest. *Photo courtesy Karen Sundberg, Univ. South Carolina, Georgetown, SC USA.*

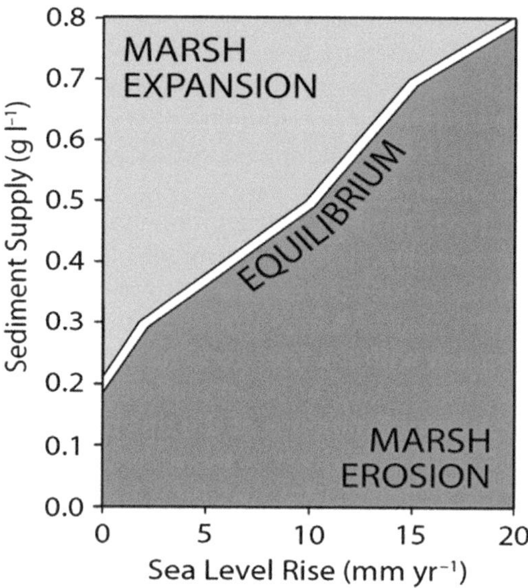

FIGURE 1.33 The relation between the rate of sea level rise (SLR), sediment supply, and marsh areal extent. Marsh expansion or progradation is only possible within a given range of SLR and sediment supply conditions, otherwise tidal wetland. *From Fagherazzi, S., Mariotti, G., Wiberg, P., McGlathery, K., 2013. Marsh collapse does not require sea level rise. Oceanography 26, 70–77 – Fig. 4, page 74.*

FIGURE 1.34 Shoreline erosion of the tidal marsh adjacent to Plum Island Sound, Massachusetts, USA. *Photo courtesy Sergio Fagherazzi, Boston University, Boston, MA USA.*

increasingly scarce, wetlands will transition from prograding to eroding (Fagherazzi et al., 2013) (Fig. 1.34), especially in microtidal systems (Kirwan et al., 2010). Tidal flats play a critical role in controlling wetland shoreline erosion (Mariotti and Fagherazzi, 2012, 2013). The effect of SLR is to increase tidal flat erosion, which will deepen them. Deepening increases water depth, however, so a positive feedback is established. Increased tidal flat

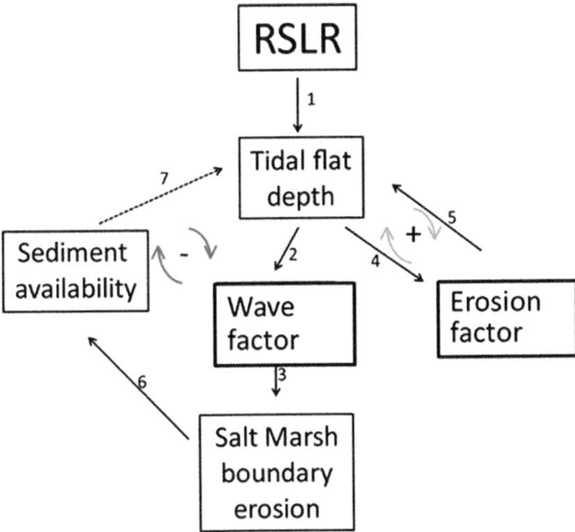

FIGURE 1.35 The relation between the relative rate of sea level rise and tidal flat depth on the erosion of tidal wetland shorelines. Both the rate of sea level rise and sediment availability influence tidal flat depth. *RSLR*, relative sea level rise. *From Mariotti, G.Fagherazzi, S., Wiberg, P., McGlathery, K., Carmiello, L.,Defina, A., 2010. Influence of storm surges and sea level on shallow tidal basin erosive processes. Journal of Geophysical Research 115, C11012 – Fig. 14, page 15.*

depth will lead to increased wave energy hitting the tidal flat-wetland boundary; hence wetland shoreline erosion will also increase. Here a negative feedback enters the equation, as erosion contributes sediment to the water column, which could decrease tidal flat depth if eroded sediment deposits there (Fig. 1.35). The fate of the marshes depends on the relative strength of the two feedback. If tidal flat erosion is greater than the increased sediment deposition, then there is an overall positive feedback in effect and marsh shoreline erosion will increase until all marshes disappear (Fig. 1.35). In the Plum Island Sound estuary in the northeastern United States, the rate of SLR and sediment availability has reached the critical point, and marsh shoreline erosion is accelerating (Leonardi and Fagherazzi, 2014, 2015). Interestingly, the sediment made available from shoreline erosion now contributes 30% of the annual mineral sediment required for the marsh to keep up with the recent rate of SLR (Hopkinson et al., 2018). Thus, the erosion of the marsh edge is helping to save the interior marsh platform. However, this process cannot last forever. It could be that the short-term resilience of the marsh platform at Plum Island is at the long-term expense of tidal flats and the marsh (Mariotti and Carr, 2014).

A more complex set of factors controls the vertical elevation gain of tidal wetlands and the magnitude of blue carbon burial per unit area (Fig. 1.36). All models show a positive relationship between sediment availability and the likelihood of tidal wetlands maintaining elevation relative to SLR (Kirwan et al., 2010; Fagherazzi et al., 2012). Net organic matter production (NEP) also plays into the equation of elevation gain, and the interaction between sediment availability and organic carbon preservation is an active research area. Wetland plants play a critical role in not only contributing organic matter belowground, which can increase overall elevation, but also in enhancing sediment accretion by slowing down water flow over the wetland surface. The extent to which they trap sediments is related to their aboveground

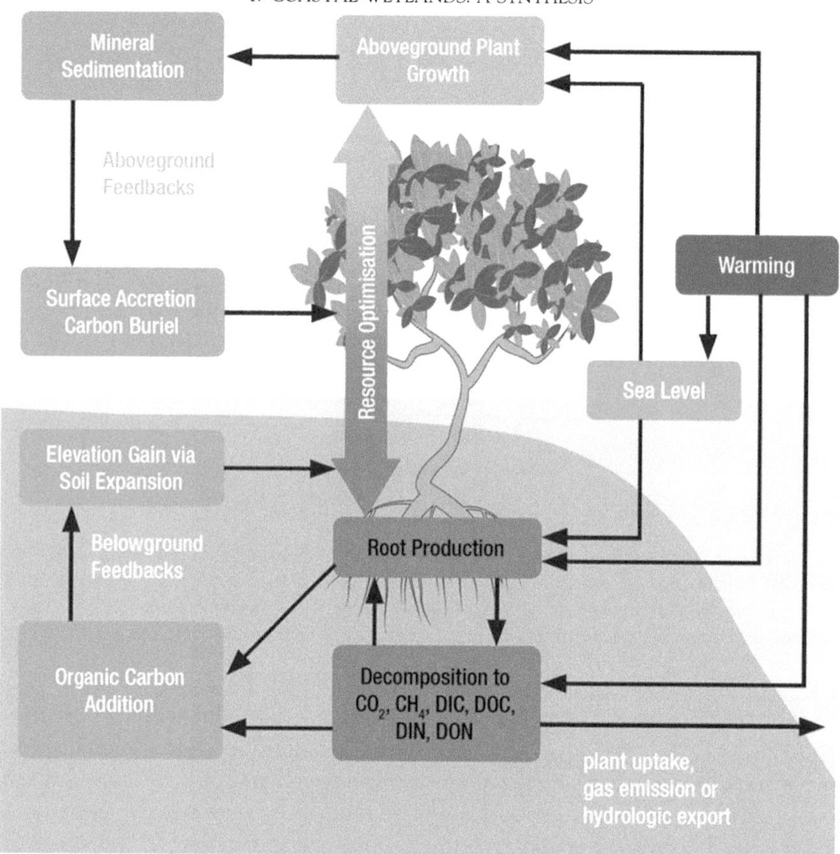

FIGURE 1.36 Conceptual model showing the major processes contributing to tidal wetland elevation gain and the effects of sea level and warming. *From Megonigal, P., Chapman, S., Crooks, S., Dijkstra, P., Kirwan, M., Langley, A., 2016. Impacts and effects of ocean warming on tidal marsh and tidal freshwater forest ecosystems. Pages 107–120. In: Laffoley, D., Baxter, M. (Eds.), Explaining Ocean Warming: Causes, Scale, Effects and Consequences. Full Report. IUCN, Gland, Switzerland. 456 p. – Fig. 3.4.2, page 109.*

biomass and density, which is hyperbolically related to tidal flooding depth and duration. The marsh equilibrium model (MEM—Morris et al., 2002; Morris, 2016) describes a hyperbolic relation between flooding depth and wetland plant biomass and above- and belowground productivity (Fig. 1.37). Productivity and biomass are optimal at an intermediate level of flooding and reduced at higher and lower levels. As SLRs, plants that are above the optimal flooding depth will see an increase in biomass production. Increased biomass promotes sediment deposition, which enhances elevation gain (Fig. 1.37). At the optimal flooding depth, root and rhizome production is maximized. Although the fate of the belowground production is an area of active research, models often allocate a certain percentage of belowground production to a very recalcitrant lignocellulosic material that decomposes very slowly (Morris, 2016; Kirwan et al., 2016). Enhanced surface deposition leads to enhanced rates of burial, which reduces decay of the recalcitrant material, further contributing to elevation gain.

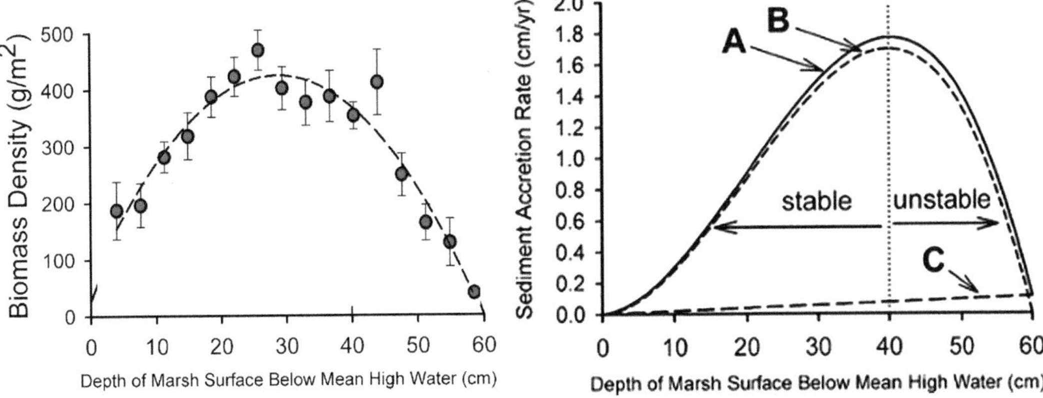

FIGURE 1.37 Left: Hyperbolic response of marsh biomass production to flooding depth as determined by plants grown in pipes held at various elevations in the macrotidal Plum Island Sound estuary, Massachusetts, MA USA. Right: Effect of marsh plant biomass and flooding depth on marsh surface sediment accretion. Line C shows accretion in the absence of vegetation trapping, line B shows accretion attributable to vegetation only and line A shows the combined total accretion. *From Morris, J.T., Sundareshwar, P.V., Nietch, C., Kjerfve, B., Cahoon, D., 2002. Responses of coastal wetlands to rising sea level. Ecology 83, 2869–2877. https://doi.org//10.1890/0012-9658(2002)083[2869:ROCWTRJ2.0.CO:2; Morris, J.T., 2007. Ecological engineering in intertidal saltmarshes. Hydrobiologia 577, 161–168. https://doi.org/10.1007/s10750-006-0425-4.; Morris, JT., 2016. Marsh equilibrium theory. In: 4th International Conference on Invasive Spartina, ICI-Spartina 2014. : University of Rennes Press, Rennes, France. https://doi.org//10.1007/s12237-014-9937-8. 67–71 – Used permission of Springer Nature. modified from Figs. 1 and 2, page 164.*

There is, however, a threshold where the rates of SLR and sediment nonavailability exceed the rate at which wetlands can gain elevation. If the rate of SLR exceeds this maximum, then the rate of elevation gain will decrease, and wetland survival will be compromised. Long-term simulations will be required to predict the trajectory of tidal wetland blue carbon sequestration and whether tidal wetlands will remain a net sink or a net source of atmospheric CO_2.

There have been several recent attempts to assemble mechanistic models that integrate vertical and horizontal dimensions of tidal wetland survival and to move away from "bathtub" approaches (Passerl et al., 2015). Kirwan et al. (2016) have coupled models of marsh transgression, with models of wetland edge erosion and with models of vertical soil accretion to examine how ecosystem connectivity influences tidal wetland size, the impact of anthropogenic drivers, and anthropogenic barriers that limit transgression, e.g., armoring. Hagen (e.g., Alizad et al., 2016a,b) has developed a spatially explicit hydro-MEM model that his team has used to explore the effects of various rates of future SLR on marsh grass productivity in several coastal regions of the United States. In northeastern Florida marshes, given current suspended solids availability, it is predicted that tidal wetlands can increase productivity and blue carbon storage with a 11 cm rise in sea level by 2050, but not a 48 cm rise by then (Alizad et al., 2016a,b). The SLR threshold for sustaining wetlands is somewhere between 11 and 48 cm by 2050. In Apalachicola wetlands, similar results were found, highlighting the importance of upland transgression under the highest rate of SLR (>50 cm by 2050), which partially reversed edge erosion (Alizad et al., 2016a,b).

With this background behind us, we now move into the crux of this second edition of Coastal Wetlands: an integrated ecosystem approach. Chapters in the rest of this book provide in-depth reviews of the state-of-the-science of coastal wetlands by leading scientists from around the world.

Acknowledgments

The authors want to acknowledge the continued intellectual contributions of our co-author, Mark Brinson, who unfortunately passed away prior to the rewriting of this chapter and the book. We will carry his memories with us the rest of our lives. The authors also wish to acknowledge several funding sources that supported us in undertaking the rewriting of this chapter. For GMEP-CONICET, ANPCYT, and Universidad Nacional del Sur. For CSH-NSF (PIE-LTER OCE-1237733, OCE-1237140, OCE-1637630, DEB-0614282, BCS-0709685) and NASA NNX10AU06G. We thank Dr. Walter D. Melo for upgrading some of the figures in the chapter. Any use of trade, firm, or product names is for descriptive purposes only and does not imply endorsement by the US Government or any other government.

References

Adey, W., Burke, R., 1976. Holocene bioherms (algal ridges and bank-barrier reefs) of the eastern Caribbean. The Geological Society of America Bulletin 87, 95–109. Doc. No. 60112.

Alizad, K., Hagen, S., Morris, J., Cacopoulos, P., Bilskie, M., Weishampel, J., 2016a. Coastal wetland response to sea-level rise in a fluvial estuarine system. Earth's Future 4, 483–497.

Alizad, K., Hagen, S., Morris, J., Cacopoulos, P., Bilskie, M., Weishampel, J., Medeiros, S., 2016b. Coupled, two-dimensional hydrodynamic-marsh model with biological feedback. Ecological Modelling 327, 29–43.

Alongi, D.M., 2008. Mangrove forests: resilience, protection from tsunamis, and responses to global climate change. Estuarine, Coastal and Shelf Science 76, 1–13.

Alongi, D.M., 2009. Paradigm shifts in mangrove biology. In: Perillo, G.M.E., Wolanski, E., Cahoon, D.R., Brinson, M.M. (Eds.), Coastal Wetlands: An Integrated Ecosystem Approach, first ed. Elsevier, Amsterdam, pp. 615–640.

Asaeda, T., Barnuevo, A., Sanjaya, K., Fortes, M.D., Kanesaka, Y., Wolanski, E., 2016. Mangrove plantation over a limestone reef – good for the ecology? Estuarine, Coastal and Shelf Science 173, 57–64.

Balmford, A., Bruner, A., Cooper, P., Costanza, R., Farber, S., Green, R.E., Jenkins, M., Jefferiss, P., Jessamy, V., Madden, J., Munro, K., Myers, N., Naeem, S., Paavola, J., Rayment, M., Rosendo, S., Roughgarden, J., Trumper, K., Turner, R.K., 2002. Economic reasons for conserving wild nature. Science 297, 950–953.

Barbier, E.B., Koch, E.W., Silliman, B.R., Hacker, S.D., Wolanski, E., Primavera, J., Granek, E.F., Polasky, S., Aswani, S., Cramer, L.A., Stoms, D.M., Kennedy, C.J., Bael, D., Kappel, C.V., Perillo, G.M.E., Reed, D.J., 2008. Coastal ecosystem-based management with non-linear ecological functions and values. Science 319, 321–323.

Barlow, P., 2003. Ground Water in Freshwater-saltwater Environments of the Atlantic Coast. http://pubs.usgs.gov/circ/2003/circ1262/.

Bauer, J., Raymond, P., Cai, W.-J., Bianchi, T., Hopkinson, C.S., Regnier, P., 2013. The changing C cycle of the coastal ocean. Nature 504, 61–70. https://doi.org/10.1038/nature12857.

Bekessy, S.A., Runge, M.C., Kusmanoff, A.M., Keith, D.A., Wintle, B.A., 2018. Ask not what nature can do for you: a critique of ecosystem services as a communication strategy. Biological Conservation 224, 71–74.

Bernier, J.C., Morton, R.A., Barras, J.A., 2006. Constraining rates and trends of historical wetland loss, Mississippi delta plain, South-central Louisiana. In: Xu, Y.J., Singh, V.P. (Eds.), Coastal Environment and Water Quality, Water Resources Publications, Highlands Ranch, Colorado, U.S.A., pp. 371–382.

Blum, M., Roberts, H., 2009. Drowning of the Mississippi Delta due to insufficient sediment supply and global sea-level rise. Nature Geoscience 2, 488–491.

Bocher, P., Robin, F., Kojadinovic, J., Delaporte, P., Rousseau, P., Dupuy, C., Bustamante, P., 2014. Tropic resource partitioning within a shorebird community feeding on intertidal mudflat habitats. Journal of Sea Research 92, 115–124.

Bourgeade, A., Marchou, B., 2003. Yellow fever, Dengue, Japanese encephalitis and West Nile virus infection: four major arbovirus diseases. Médecine et Maladies Infectieuses 33, 385–395.

Brinson, M.M., Christian, R.R., Blum, L.K., 1995. Multiple stages in the sea-level induced transition from terrestrial forest to estuary. Estuaries 18, 648–659.

Brockmeyer, R.E., Rey, J.R., Virnstein, R.W., Gilmore, R.G., Earnest, L., 1997. Rehabilitation of impounded estuarine wetlands by hydrologic reconnection to the Indian River Lagoon, Florida. Wetlands Ecology and Management 4, 93–109.

Buchsbaum, R., Deegan, L., Horowitz, J., Garritt, R., Giblin, A., Ludlam, J., Shull, D., 2009. Effects of regular salt marsh haying on marsh plants, algae, invertebrates and birds at Plum Island Sound, Massachusetts. Wetlands Ecology and Management 17, 469–487.

Buelow, C., Sheaves, M., 2015. A birds-eye view of biological connectivity in mangrove systems. Estuarine, Coastal and Shelf Science 152, 33–43.

Buskey, E.J., Montagna, P.A., Amos, A.F., Whitledge, T.E., 1997. Disruption of grazer populations as a contributing factor to the initiation of the Texas brown tide algal bloom. Limnology and Oceanography 42, 1215–1222.

Cahoon, D.R., 2006. A review of major storm impacts on coastal wetland elevations. Estuaries and Coasts 29, 889–898.

Cahoon, D.R., Hensel, P.F., Rybczyk, J., McKee, K.L., Proffitt, C.E., Perez, B.C., 2003. Mass tree mortality leads to mangrove peat collapse at Bay Islands, Honduras after Hurricane Mitch. Journal of Ecology 91, 1093–1105.

Cahoon, D.R., Turner, R., 1989. Accretion and canal impacts in a rapidly subsiding wetland. II. Feldspar marker horizon technique. Estuaries 12, 260–268.

Cahoon, D.R., Reed, D.J., Kolker, A., Brinson, M.M., Stevenson, J.C., Riggs, S., Christian, R., Reyes, E., Voss, C., Kunz, D., 2009. Coastal wetland sustainability. In: Coastal Sensitivity to Sea-level Rise: A Focus on the Mid-Atlantic Region. A Report by the U.S. Climate Change Science Program and the Subcommittee on Global Change Research. U.S. Environmental Protection Agency, Washington D.C., USA, pp. 57–72 [James G. Titus (Coordinating Lead Author), K. Eric Anderson, Donald R. Cahoon, Stephen Gill, E. Robert Thieler, S. Jeffress Williams (Lead Authors)] 784 p.

Carey, J.C., Moran, S., Kelly, R., Kolker, A., Fulweler, R.W., 2017. The declining role of organic matter in New England salt marshes. Estuaries and Coasts 40, 626–639.

Chmura, G., Anisfeld, S., Cahoon, D., Lynch, J., 2003. Global carbon sequestration in tidal, saline wetland soils. Global Biogeochemical Cycles 17, 1111–1123.

Church, A., et al., 2013. Sea level change. In: Stocker, T., Qin, D., Plattner, G., Tignor, M., Allen, S., Boschung, J., Nauels, A., Xia, Y., Bex, V., Midgley, P. (Eds.), Climate Change 2013: The Physical Science Basis. Contribution of Working Group I to the Fifth Assessment Report of the Intergovernmental Panel on Climate Change (IPCC). Cambridge University Press, Cambridge, UK.

Clark, J.S., Fastie, F., Hurtt, G., Jackson, G., Johnson, S.T., King, C., Lewis, G.A., Lynch, J., Pacala, S., Prentice, C., Schupp, E.W., Webb III, T., Wyckoff, P., 1998. Reid's paradox of rapid plant migration. BioScience 48, 13–24.

Clemens, R.S., Rogers, D.I., Hansen, B.D., et al., 2016. Continental-scale decreases in shorebird populations in Australia. Emu 116, 119–135.

Cochard, R., Ranamukhaarachchi, S.L., Shivakoti, G.P., Shipin, O.V., Edwards, P.J., Seeland, K.T., 2008. The 2004 tsunami in Aceh and Southern Thailand: a review on coastal ecosystems, wave hazards and vulnerability. Perspectives in Plant Ecology, Evolution and Systematics 10, 3–40.

Colwell, R.R., 1996. Global climate and infectious disease: the cholera paradigm. Science 274, 2025–2031.

Conner, W.H., Doyle, T.W., Krauss, K.W., 2007. Ecology of Tidal Freshwater Forested Wetland of the Southeastern United States. Springer, Dordrecht, The Netherlands.

Conner, W.H., McLeod, K.W., McCarron, J.K., 1997. Flooding and salinity effects on growth and survival of four common forested wetland species. Wetlands Ecology and Management 5, 99–109.

Constantin de Magny, G., Mozumder, P.K., Grim, C.J., Hasan, N.A., Naser, N.M., Alam, M., et al., 2011. Role of zooplankton diversity in *Vibrio cholera*population dynamics and in the incidence of cholera in the Bangladesh Sundarbans. Applied Environmental Micriobiology 77, 6125–6132.

Costanza, R., d'Arge, R., de Groot, R., Farber, S., Grasso, M., Hannon, B., Limburg, K., Naeem, S., O'Neil, R., Paruelo, J., Raskin, R., Sutton, P., van den Belt, M., 1997. The value of the world's ecosystem services and natural capital. Nature 387, 253–260.

D'Alpaos, A., Lanzoni, S., Rinaldo, A., Marani, M., 2009. Intertidal eco-geomorphological dynamics and hydrodynamic circulation. In: Perillo, G.M.E., Wolanski, E., Cahoon, D.R., Brinson, M.M. (Eds.), Coastal Wetlands: An Integrated Ecosystem Approach. Elsevier, Amsterdam, pp. 159–184.

Daborn, G.R., Amos, C.L., Brylinsky, M., Christian, H., Drapeau, G., Faas, R.W., Grant, J., Long, B., Paterson, D., Perillo, G.M.E., Piccolo, M.C., 1993. An ecological "cascade" effect: migratory birds affect stability of intertidal sediments. Limnology and Oceanography 38, 225–231.

Dale, P.E.R., 1994. An Australian perspective on coastal wetland management and vector borne diseases. In: Mitsch, W.J. (Ed.), Wetlands of the World. Elsevier, Utrecht, pp. 771–781.

Dale, P.E.R., Knight, J.M., 2006. Managing salt marshes for mosquito control: impacts of runnelling, open marsh water management and grid ditching in sub-tropical Australia. Wetlands Ecology and Management 14, 211–220.

Dale, P.E.R., Knight, J.M., 2008. Wetlands and mosquitoes: a review. Wetlands Ecology and Management 16, 255–276.

Dale, P.E.R., Knight, J.M., Daniels, P.L., 2018. Using present value as a simple approach to compare mosquito larval control methods. Journal of the American Mosquito Control Association 34, 25–33.

Danielson, T., Rivera-Monroy, V., Casteneda-Moya, E., Briceno, H., Travieso, R., Marx, B., Gaiser, E., Farfan, L., 2017. Assessment of Everglades mangrove forest resilience: implications for above-ground net primary productivity and carbon dynamics. Wetland Ecology and Management 303, 115–125.

Davis, B., Baker, R., Sheaves, M., 2014. Seascape and metacommunity processes regulate fish assemblage structure in coastal wetlands. Marine Ecology Progress Series 500, 187–202.

Day, J.W., Rismondo, A., Scarton, F., Are, D., Cecconi, G., 1998. Relative sea-level rise and Venice lagoon wetlands. Journal of Coastal Conservation 4, 27–34.

Day, J.W., Britsch, L.D., Hawes, S.R., et al., 2000. Pattern and process of land loss in the Mississippi Delta: a Spatial and temporal analysis of wetland habitat change. Estuaries 23, 425–438.

De Boer, E.F., 2000. Biomass dynamics of seagrasses and the role of mangrove and seagrass vegetation as different nutrient sources for an intertidal ecosystem. Aquatic Botany 66, 225–239.

de Groot, R., 2011. What are ecosystem services? In: van den Belt, M., Costanza, R. (Eds.), Ecological Economics of Estuaries and Coasts. Treatise on Coastal and Coastal Science, vol. 12. Elsevier, Amsterdam, pp. 16–34.

Deegan, L., 1993. Nutrient and energy transport between estuaries and coastal marine ecosystems by fish migration. Canadian Journal of Fisheries and Aquatic Sciences 50, 74–79. https://doi.org/10.1139/f93-009.

Deegan, L., Johnson, D., Warren, R.S., Peterson, B., Fleeger, J., Fagherazzi, S., Wollheim, W., 2012. Coastal eutrophication as a driver of salt marsh loss. Nature 490. https://doi.org/10.1038/nature11533.

Dent, D.L., Pons, L.J., 1995. A world perspective on acid sulphate soils. Geoderma 67, 263–276.

DiMichele, W.A., 2014. Wetland-dryland vegetational dynamics in the Pennsylvanian ice age tropics. International Journal of Plant Sciences 175, 123–164.

Diop, S., 2003. Vulnerability assessments of mangroves to environmental change. Estuarine, Coastal and Shelf Science 58, 1–3.

Djenontin, A., Pennetier, C., Zogo, B., Soukou, K.B., Ole-Sangba, M., Akogbeto, M., Chandre, F., Yadav, R., Corbel, V., 2014. Field efficacy of Vectobac GR as a mosquito larvicide for the control of Anopheline and Culicine mosquitoes in natural habitats in Benin, West Africa. PLoS One 9, e87934. https://doi.org/10.1371/journal.pone.0087934.

Donnelly, J., 2006. A revised late Holocene sea-level record for northern Massachusetts, USA. Journal of Coastal Research 22, 1051–1061. https://doi.org/10.2012/04-0207.1.

Dorst, J., 1965. Avant Que Nature Meure. Delachaux and Niestle, Neuchatel, Switzerland, p. 424.

Drexler, J.Z., de Carlo, E.W., 2002. Source water partitioning as a means of characterizing hydrologic function in mangroves. Wetlands Ecology and Management 10, 103–113.

Drexler, J.Z., Ewel, K.C., 2001. Effect of the 1997-1998 ENSO-related drought on hydrology and salinity in a Micronesian wetland complex. Estuaries 24, 347–356.

Duarte, C.M., 2002. The future of seagrass meadows. Environmental Conservation 29, 192–206.

Duarte, C.M., Dennison, W.C., Orth, R.J.W., Carruthers, T.J.B., 2008. The charisma of coastal ecosystems: addressing the imbalance. Estuaries and Coasts 31, 233–238.

Duke, N.C., Meynecke, J.-O., Dittmann, S., Ellison, A.M., Anger, K., Berger, U., Cannicci, S., Diele, K., Ewel, K.C., Field, C.D., Koedam, N., Lee, S.Y., Marchand, C., Nordhaus, I., Dahdouh-Guebas, F., 2007. A world without mangroves? Science 317, 41–42.

Duke, N.C., Wolanski, E., 2001. Muddy coastal waters and depleted mangrove coastlines - depleted seagrass and coral reefs. In: Wolanski, E. (Ed.), Oceanographic Processes of Coral Reefs: Physical and Biological Links in the Great Barrier Reef. CRC Press, Boca Raton, Florida, pp. 77–91.

Elliott, M., Mander, L., Mazik, K., Simenstad, C., Valesini, F., Whitfield, A., Wolanski, E., 2016. Ecoengineering with Ecohydrology: successes and failures in estuarine restoration. Estuarine, Coastal and Shelf Science 176, 12–35.

Ellison, J., 2009. Geomorphology and sedimentology of mangroves. In: Perillo, G.M.E., Wolanski, E., Cahoon, D.R., Brinson, M.M. (Eds.), Coastal Wetlands: An Integrated Ecosystem Approach. Elsevier, Amsterdam, pp. 565–591.

Ellison, J., 1998. Impacts of sediment burial on mangroves. Marine Pollution Bulletin 37, 420–426.

Ellison, J.C., Sheehan, M.R., 2014. Past, present and futures of the Tamar estuary, Tasmania. In: Wolanski, E. (Ed.), Estuaries of Australia in 2050 and beyond. Springer, The Netherlands, pp. 69–89.

Entwistle, C., Mora, M.A., Knight, R., 2017. Estimating Coastal Wetland Gain and Losses in Galveston County and Cameron County, Texas, USA, vol. 14. Integrated Environmental Assessment and Management, pp. 120–129.

EPA (Environment Protection Agency), 2002. Larvicides for Mosquito Control. US Environment Protection Agency, Washington, DC, USA.

Escapa, C.M., Minkoff, D.R., Perillo, G.M.E., Iribarne, O.O., 2007. Direct and indirect effects of burrowing crab activities on erosion of Southwest Atlantic Sarcocornia-dominated marshes. Limnology and Oceanography 52, 2340–2349.

Fagherazzi, S., Kirwan, M., Mudd, S., Guntenspergen, G., Temmerman, S., D'Alpaos, A., van de Koppel, J., Rybczyk, J., Reyes, E., Craft, C., Clough, J., 2012. Numerical models of salt marsh evolution: ecological, geomorphic, and climatic factors. In: Environmental Science, Paper 10.

Fagherazzi, S., Mariotti, G., Wiberg, P., McGlathery, K., 2013. Marsh collapse does not require sea level rise. Oceanography 26, 70–77.

FAO, 2007. The World's Mangroves 1980-2005. FAO Forestry Paper 153. FAO, Rome.

Feher, L., Osland, M., Griffith, K., et al., 2017. Linear and nonlinear effects of temperature and precipitation on ecosystem properties in tidal saline wetlands. Ecosphere 8, e01956. https://doi.org/10.1002/ecs2.1956.

Forbrich, I., Giblin, A., Hopkinson, C.S., 2018. Using atmospheric CO_2 exchange with carbon burial rates to constrain salt marsh carbon budgets. Journal of Geophysical Research-Biogeoscience 123. https://doi.org/10.1002/2017JG004336.

Frayer, W.E., Monahan, T.J., Bowden, D.C., Graybill, F.A., 1983. Status and Trends of Wetlands and Deepwater Habitats in the Conterminous United States, 1950's to 1970's. Colorado State University, Ft. Collins, Colorado, USA.

Frey, R.W., Basan, P., 1978. Coastal salt marshes. In: Davis, R.A. (Ed.), Coastal Sedimentary Environments. Springer, New York, NY. doi-org.proxy-remote.galib.uga.edu/10.1007/978-1-4684-0056-4_4.

Furukawa, K., Wolanski, E., Muller, H., 1997. Currents and sediment transport in mangrove forests. Estuarine, Coastal and Shelf Science 44, 301–310.

Gallegos, C.L., Kenworthy, W.J., 1996. Seagrass depth limits in the Indian River Lagoon (Florida, U.S.A.): Application of an optical water quality model. Estuarine, Coastal and Shelf Science 42, 267–288.

Ganju, N., Kirwan, M., Dickhudt, P., Guntenspergen, G., Cahoon, D., Kroeger, K., 2015. Sediment transport-based metrics of wetland stability. Geophysical Research Letters 42, 7992–8000. https://doi.org/10.1002/2015GL065980.

Garbutt, A., Boorman, L.A., 2009. Managed realignment: re-creating intertidal habitats on formerly reclaimed land. In: Perillo, G.M.E., Wolanski, E., Cahoon, D.R., Brinson, M.M. (Eds.), Coastal Wetlands: An Integrated Ecosystem Approach, first ed. Elsevier, Amsterdam, pp. 763–781.

Gardner, L.R., 2005. A modeling study of the dynamics of pore water seepage from intertidal marsh sediments. Estuarine, Coastal and Shelf Science 62, 691–698.

Ghermandi, A., Nunes, P.A.L.D., Portela, R., Rao, N., Teelucksingh, S.S., 2011. Recreational, cultural, and aesthetic services from estuarine and coastal ecosystems. In: van den Belt, M., Costanza, R. (Eds.), Ecological Economics of Estuaries and Coasts. Treatise on Coastal and Coastal Science, vol. 12. Elsevier, Amsterdam, pp. 217–237.

Ginsberg, S.S., Perillo, G.M.E., 1990. Channel bank recession in the Bahía Blanca Estuary, Argentina. Journal of Coastal Research 6, 999–1010.

Giri, C., Ochieng, E., Tieszen, L., Zhu, A., Singh, A., Loveland, T., Masek, J., Duke, D., 2011. Status and distribution of mangrove forests of the world using earth observation satellite data (version 1.3, updated by UNEP-WCMC). Global Ecology and Biogeography 20, 154–159. https://doi.org/10.1111/j.1466-8238.2010.00584.x. http://data.unep-wcmc.org/datasets/4. Data URL:

Giri, C., Long, J., Abbas, S., Murali, R., Qamer, F., Pengra, B., Thau, D., 2014. Distribution and dynamics of mangrove forests of South Asia. Journal of Environmental Economics and Management 148, 101–111.

Giri, C., Pengra, B., Zhu, Z., Singh, A., Tieszen, L., 2007. Monitoring mangrove forest dynamics of the Sundarbans in Bangladesh and India using multi-temporal satellite data from 1973 to 2000. Estuarine Coastal and Shelf Sciences 73, 91–100.

Giri, C., Zhu, Z., Tieszen, L., Singh, A., Gilletee, S., Kelmelis, J., 2008. Mangrove forest distributions and dynamics (1975-2005) of the tsunami-affected region of Asia. Journal of Biogeography 35, 519–528.

Gittman, R., Fodrie, J., Popowich, A., Keller, D., Bruno, J., Currin, C., Peterson, C., Piehler, M., 2015. Engineering away our natural defenses: an analysis of shoreline hardening in the US. Frontiers in Ecology and the Environment 13, 301–307.

Goessens, A., Satyanarayana, B., Van der Stocken, T., Quispe Zuniga, M., Mohd-Lokman, H., Sulong, I., 2014. Is Matang Mangrove Forest in Malaysia Sustainably Rejuvenating after More than a Century of Conservation and Harvesting Management? PLoS One 9 (8), e105069. https://doi.org/10.1371/journal.pone.0105069.

Gómez, E.A., Perillo, G.M.E., 1995. Sediment outcrops underneath shoreface-connected sand ridges, outer Bahía Blanca estuary, Argentina. Quaternary of South America and Antarctica Peninsula 9, 27–42.

Gong, Z., Zhao, K., Zhang, C., Dai, D., Coco, C., Zhou, Z., 2018. The role of bank collapse on tidal creek ontogeny: A novel process-based model for bank retreat. Geomorphology 311, 13–26.

Goolsby, D., Battaglin, W., Hooper, R., 1997. In: Sources and Transport of Nitrogen in the Mississippi River Basin. American Farm Bureau Federation Workshop. July 14, 1997. St. Louis, MO. http://wwwrcolka.cr.usgs.gov/midconherb/st.louis.hypoxia.html.

Green, E.P., Short, F.T., 2003. World Atlas of Seagrasses. University of California Press, Berkeley, USA, p. 298.

Hack, J.T., 1957. Studies of Longitudinal Profiles in Virginia and Maryland, U.S. Geol. Surv. Prof. Pap., 294-B.

Hamilton, S., Casey, D., 2016. Creation of a high spatio-temporal resolution global database of continuous mangrove forest cover for the 21st century (CGMFC-21). Global Ecology and Biogeography 25, 729–738. https://doi.org/10.1111/geb.12449.

Harvell, C.D., Kim, K., Burkholder, J.M., Colwell, R.R., Epstein, P.R., Grimes, D.J., Hofmann, E.E., Lipp, E.K., Osterhaus, A.D.M.E., Overstreet, R.M., Porter, J.W., Smith, G.W., Vasta, G.R., 1999. Emerging marine diseases-climate links and anthropogenic factors. Science 285, 1505–1510.

Hayden, B.P., Dueser, R.D., Callahan, J.T., Shugart, H.H., 1991. Long-term research at the Virginia Coast Reserve: Modeling a highly dynamic environment. BioScience 41, 310–318.

Heck Jr., K.L., Hays, G., Orth, R.J., 2003. Critical evaluation of the nursery role hypothesis for seagrass meadows. Marine Progress Series 253, 123–136.

Highfield, W.E., Brody, S.D., Shepard, C., 2018. The effects of estuarine wetlands on flood losses associated with storm surge. Ocean and Coastal Management 157, 50–55.

Hily, C., van Katwijk, M.M., den Hartog, C., 2003. The seagrasses of Western Europe. In: Green, E.P., Short, F. (Eds.), World Atlas of Seagrasses. UNEP, University of California Press, Berkeley, USA, pp. 48–58.

Holden, E., Newman, J., Thistle, I., Whiteman, E., 2013. Traditional Uses of the Great Marsh, a Review of Less-known Resources in the Massachusetts' Great Salt Marsh. Report to Tufts University, Urban and Environmental Policy and Planning, 66 pages.

Hopkinson, C.S., 1985. Shallow-water benthic and pelagic metabolism - Evidence of net heterotrophy in the nearshore Georgia Bight. Marine Biology 87, 19–32.

Hopkinson, C.S., 1988. Patterns of organic carbon exchange between coastal ecosystems: The mass balance approach in salt marsh ecosystems. In: Jansson, B.-O. (Ed.), Coastal-offshore Ecosystem Interactions. Lecture Notes on Coastal and Estuarine Studies, vol. 22. Springer-Verlag, pp. 122–154.

Hopkinson, C.S., 2018. Net ecosystem carbon balance of coastal wetland-dominated estuaries: where's the blue carbon. In: Windham-Myers, L., Crooks, S., Troxler, T. (Eds.), A Blue Carbon Primer: The State of Coastal Wetland Carbon Science, Policy, and Practice. CRC Press, Boca Raton, FL.

Hopkinson, C.S., Vallino, J., 1995. The nature of watershed perturbations and their influence on estuarine metabolism. Estuaries 18, 598–621.

Hopkinson, C.S., Cai, W-J., Hu, X., 2012. Carbon Sequestration in Wetland Dominated Coastal Systems - A Global Sink of Rapidly Diminishing Magnitude. Current Opinion on Environmental Sustainability 4, 1–9. https://doi.org/10.1016/j.cosust.2012.03.005.

Hopkinson, C., Giblin, A., 2008. Salt marsh N Cycling. In: R. Capone, D. Bronk, M. Mulholland, E. Carpenter (Eds.), Nitrogen in the Marine Environment, second ed. Elsevier Publ., pp. 991–1036

Hopkinson, C., Morris, J., Fagherazzi, S., Wollheim, W., Raymond, P., 2018. Lateral marsh edge erosion as a source of sediments for vertical marsh accretion. JGR-Biogeosciences. https://doi.org/10.1029/2017JG004358.

Horton, R.E., 1945. Erosional development of streams and their drainage basins; hydrophysical approach to quantitative morphology. Bulletin of the Geological Society of America 56, 275–370.

Isla, F.I., 1989. The Southern Hemisphere sea-level fluctuation. Quaternary Science Reviews 8, 359–368.

Jennerjahn, T.C., Gilman, E., Krauss, K.W., Lacerda, L.D., Nordhaus, I., Wolanski, E., 2017. Mangrove ecosystems under climate change. In: Rivera-Monroy, V.H., Lee, S.Y., Kristensen, E., Twilley, R.W. (Eds.), Mangrove Ecosystems: A Global Biogeographic Perspective. Springer International Publishing, pp. 211–244. https://doi.org/10.1007/978-3-319-62206-4_7.

Jones, D., 1984. Crabs of the mangal ecosystems. In: Por, F., Dor, I. (Eds.), Hydrobiology of the Mangal. Junk Publisher, The Hague, Netherlands.

Kemp, A., Horton, B., Donnelly, J., Mann, M., Vermeer, M., Rahmstorf, W., 2011. Climate related sea-level variations over the past two millennia. Proceedings of the National Academy of Sciences 108, 11017–11022.

Kirwan, M., Murray, B., 2007. A coupled geomorphic and ecological model of tidal marsh evolution. Proceedings of the National Academy of Sciences 104, 6118–6122. https://doi.org/10.1073/pnas.0700958104.

Kirwan, M., Guntenspergen, G., D' Alpaos, A., Morris, J., Mudd, S., Temmerman, S., 2010. Limits on the adaptability of coastal marshes to rising sea level. Geophysical Research Letters 37, L23401. https://doi.org/10.1020/2010G:L045489.

Kirwan, M., Walters, D., Reay, W., Carr, J., 2016. Sea level driven marsh expansion in a coupled model of marsh erosion and migration. Geophysical Research Letters 4366–4373.

Kiwango, H., Njau, K., Wolanski, E., 2015. The need to enforce minimum environmental flow requirements in Tanzania to preserve estuaries: case study of the mangrove-fringed Wami River estuary. Ecohydrology and Hydrobiology 15, 171–181.

Kjerfve, B., Miranda, L.B., Wolanski, E., 1991. Modelling water circulation in an estuary and intertidal salt marsh system. Netherlands Journal of Sea Research 28, 141–147.

Kneer, D., Asmus, H., Vonk, J.A., 2008. Seagrass as the main food source of *Neaxius acanthus* (Thalassinidea: Strahlaxiidae), its burrow associates, and of *Corallianassa coutierei* (Thalassinidea: Callianassidae). Estuarine, Coastal and Shelf Science 79, 620–630.

Knight, J.M., 2008. Characterizing the Biophysical Properties of a Mangrove Forest to Inform Mosquito Control (PhD thesis). University of Queensland, St. Lucia, 228 p.

Koch, E.W., Barbier, E.D., Silliman, B.R., Reed, D.J., Perillo, G.M.E., Hacker, S.D., Granek, E.F., Primavera, J.H., Muthiga, N., Polasky, S., Halpern, B.S., Kennedy, C.J., Wolanski, E., Kappel, C.V., Aswani, S., Cramer, L.A., Bael, D., Stoms, D.M., 2009. Non-linearity in ecosystem services: temporal and spatial variability in coastal protection. Frontiers in Ecology and the Environment 7, 29–37.

Kundzewicz, Z.P., Mata, L.J., Arnell, N.W., Doll, P., Jimenez, B., Miller, K., Oki, T., Sen, Z., Shiklomanov, V., 2008. The implications of projected climate change for freshwater resources and their management. Hydrological Sciences 53, 3–10.

Kusumawathie, P.H.D., Wickremasinghe, A.R., Karunaweera, N.D., Wijeyaratne, M.J.S., 2008. Costs and effectiveness of application of *Poecilia reticulata* (guppy) and temephos in anopheline mosquito control in river basins below the major dams of Sri Lanka. Transactions of the Royal Society of Tropical Medicine and Hygiene 102, 705–711.

Lambeck, K., Rouby, H., Purcell, A., Sun, Y., Sambridge, M., 2014. Sea level and global ice volumes from the Last Glacial Maximum to the Holocene. Proceedings of the National Academy of Sciences 111, 15296–15303 doi:10/1073/pnas.1411762111.

Lambeck, K., Woodroffe, C.D., Antonioli, F., Anzidei, M., Gehrels, W.R., Laborel, J., Wright, A.J., 2010. Paleoenvironmental records, geophysical modelling, and reconstruction of sea level trends and variability on centennial and longer timescales. In: Church, J.A., Woodworth, P.L., Aarup, T., Wilson, W.S. (Eds.), Understanding Sea Level Rise and Variability. Wiley-Blackwell, Chichester, pp. 61–121.

Lara, R.J., Szlafsztein, C.F., Cohen, M.C.L., Oxmann, J., Schmitt, B.B., Filho, P.W.M.S., 2009. Geomorphology and sedimentology of mangroves and salt marshes: the formation of geobotanical units. In: Perillo, G.M.R., Wolanski, E., Cahoon, D.R., Brinson, M.M. (Eds.), Coastal Wetlands: An Integrated Ecosystem Approach. Elsevier Science, Amsterdam, pp. 593–614.

Lee, S.Y., 1990. Net aerial primary productivity, litter production and decomposition of the reed Phragmites *communis* in a nature reserve in Hong Kong: management implications. Marine Ecology Progress Series 66, 161–173.

Leonardi, N., Fagherazzi, S., 2014. How waves shape salt marshes. Geology 42, 887–890.

Leonardi, N., Fagherazzi, S., 2015. Effect of local variability in erosional resistance on large-scale morphodynamic response of salt marshes to wind waves and extreme events. Geophysical Research Letters 42, 5872–5879.

Lugo, A.E., Patterson Zucca, C., 1977. The impact of low temperature stress on mangrove structure and growth. Tropical Ecology 18, 149–161.

MacKinnon, J., Verkuil, Y., Murray, N., 2012. IUCN situation analysis on East and Southeast Asian intertidal habitats, with particular reference to the Yellow Sea (including the Bohai Sea). In: Occasional Paper of the IUCN Species Survival Commission No. 47. IUCN, Gland, Switzerland, 70 p.

Mancil, E., 1972. LSU Historical Dissertations and Theses. An Historical Geography of Industrial Cypress Lumbering in Louisiana, vols. 1&2, p. 2296.

Mander, L., Marie-Orleach, L., Elliott, M., 2013. The value of wader foraging behaviour study to assess the success of restored intertidal areas. Estuarine, Coastal and Shelf Science 131, 1–5.

Mariotti, G., Fagherazzi, S., 2013. Critical width of tidal flats triggers marsh collapse in the absence of sea-level rise. Proceedings of the National Academy of Sciences 110, 5353–5356.

Mariotti, G., Carr, J., 2014. Dual role of salt marsh retreat: Long-term loss and short-term resilience. Water Resources Research 50, 2963–2974.

Mariotti, G., Fagherazzi, S., 2012. Modeling the effect of tides and waves on benthic biofilms. Journal of Geophysical Research - Biogeosciences 117, G04010. https://doi.org/10.1029/2012JG002064.

Mariotti, G., Fagherazzi, S., Wiberg, P., McGlathery, K., Carmiello, L., Defina, A., 2010. Influence of storm surges and sea level on shallow tidal basin erosive processes. Journal of Geophysical Research 115, C11012.

Marshall, E., 1997. Apocalypse not (News and comment section). Science 278, 1004–1006.

Mazda, Y., Kanazawa, N., Wolanski, E., 1995. Tidal asymmetry in mangrove swamps. Hydrobiologia 295, 51–58.

Mazda, Y., Wolanski, E., 2009. Hydrodynamics and modeling of water flow in mangrove areas. In: Perillo, G.M.E., Wolanski, E., Cahoon, D.R., Brinson, M.M. (Eds.), Coastal Wetlands: An Integrated Ecosystem Approach, first ed. Elsevier, Amsterdam, pp. 231–261.

McIntire, W., Morgan, J., 1962. Recent Geomorphic History of Plum Island, Massachusetts and Adjacent Coastal. Atlantic Coastal Studies, Technical Report No. 19, Part A, Report under Project No. Nonr 1575(03). LSU Coastal Studies Institute. Contribution No. 62-7. 57 p.

Mcowen, C., Weatherdon, L., Bochove, J., Sullivan, E., Blyth, S., Zockler, C., Stanwell-Smith, D., Kingston, N., Martin, C., Spalding, M., Fletcher, S., 2017. A global map of saltmarshes. Biodiversity Data Journal 5, e11764. https://doi.org/10.3897/BDJ.5.e11764.

McKee, K.L., Patrick Jr., W.H., 1988. The relationship of smooth cordgrass (*Spartina alterniflora*) to tidal datums: a review. Estuaries 11, 143–151.

McKee, K.L., Cahoon, D.R., Feller, I.C., 2007. Caribbean mangroves adjust to rising sea level through biotic controls on change in soil elevation. Global Ecology and Biogeography. https://doi.org/10.1111/j.1466-8238.2007.00317.x.

Megonigal, P., Chapman, S., Crooks, S., Dijkstra, P., Kirwan, M., Langley, A., 2016. Impacts and effects of ocean warming on tidal marsh and tidal freshwater forest ecosystems. In: Laffoley, D., Baxter, M. (Eds.), Explaining Ocean Warming: Causes, Scale, Effects and Consequences. Full Report. IUCN, Gland, Switzerland, pp. 107–120, 456 p.

Michener, W.K., Blood, E.R., Bildstein, K.L., Brinson, M.M., Gardner, L.R., 1997. Climate change, hurricanes and tropical storms, and rising sea level in coastal wetlands. Ecological Applications 7, 770–801.

Milliman, J.D., Syvitski, J.P.M., 1992. Geomorphic/tectonic control of sediment discharge to the ocean: the importance of small mountainous rivers. The Journal of Geology 100, 525–544.

Milly, P.C.D., Betancourt, J., Falkenmark, M., Hirsch, R.M., Kundzewicz, Z.W., Lettenmaier, D.P., Stouffer, R.J., 2008. Stationarity is dead: whither water management? Science 319, 573–574.

Minkoff, D.R., Escapa, C.M., Ferramola, F.E., Maraschin, S., Pierini, J.O., Perillo, G.M.E., Delrieux, C., 2006. Effects of crab-halophytic plant interactions on creek growth in a S.W. Atlantic salt marsh: a cellular automata model. Estuarine, Coastal and Shelf Science 69, 403–413.

Minkoff, D.R., Escapa, C.M., Ferramola, F.E., Perillo, G.M.E., 2005. Erosive processes due to physical - biological interactions based in a cellular automata model. Latin American Journal of Sedimentology and Basin Analysis 12, 25–34.

Mitrovica, J., Milne, G., 2002. On the origin of late Holocene sea-level highstands within equatorial ocean basins. Quaternary Science Reviews 21, 20179–22190.

Mitsch, W.J., Gosselink, J.G., 2006. Wetlands, 4rth edition. John Wiley and Sons, New York, NY, USA. 582 p.

Moller, I., Kudella, M., Rupprecht, F., Spencer, T., Paul, M., van Wesenbeeck, B.K., Wolters, G., Jensen, K., Bouma, T.J., Miaranda-Lange, M., Schimmels, S., 2014. Wave attenuation over coastal salt marshes under storm surge conditions. Nature Geoscience 7, 727–731.

Morris, J.T., 2016. Marsh equilibrium theory. In: 4th International Conference on Invasive Spartina, ICI-Spartina 2014. University of Rennes Press, Rennes, France, pp. 67–71. https://doi.org/10.1007/s12237-014-9937-8.

Morris, J.T., Kjerfve, B., Dean, J.M., 1990. Dependence of estuarine productivity on anomalies in mean sea level. Limnology and Oceanography 35, 926–930.

Morris, J.T., Sundareshwar, P.V., Nietch, C., Kjerfve, B., Cahoon, D., 2002. Responses of coastal wetlands to rising sea level. Ecology 83, 2869–2877. https://doi.org/10.1890/0012-9658(2002)083[2869:ROCWTR]2.0.CO:2.

Morris, J.T., 2007. Ecological engineering in intertidal saltmarshes. Hydrobiologia 577, 161–168. https://doi.org/10.1007/s10750-006-0425-4.

Mumby, P.J., Edwards, A.J., Arias-Gonzalez, J.E., Lindeman, K.C., Blackwell, P.G., Gall, A., Gorczynska, M.I., Harborne, A.R., Pescod, C.L., Renken, H., Wabnitz, C.C.C., Llewellyn, G., 2004. Mangroves enhance the biomass of coral reef fish communities in the Caribbean. Nature 427, 523–536.

Murray, N., Clemens, R., Phinn, S., Possingham, H., Fuller, R., 2014. Frontiers in Ecology and the Environment 12, 267–272. https://doi.org/10.1890/130260.

Murray, N., Ma, Z., Fuller, R., 2015. Tidal flats of the Yellow Sea: a review of ecosystem status and anthropogenic threats. Austral Ecology. https://doi.org/10.1111/aec.12211.

Murray, N., Phinn, S., Clemens, R., Roelfsema, C., Fuller, R., 2012. Continual scale mapping of tidal flats across East Asia using the Landsat archive. Remote Sensors 4, 2426–3417. https://doi.org/10.3390/rs4113417.

Nellemann, C., Corcoran, E., Duarte, C., Valdés, L., Young, C., Fonseca, L., Grimsditch, G., 2009a. Blue Carbon — the Role of Healthy Oceans in Binding Carbon. GRID-Arendal: United Nations Environment Programme.

Nellemann, C., Corcoran, E., Duarte, C.M., Valdes, L., DeYoung, C., Fonseca, L., Grimsditch, G. (Eds.), 2009b. Blue Carbon. A Rapid Response Assessment. United Nations Environment Programme, GRID-Arendal.

Neogi, S.B., Yamasaki, S., Alam, M., Lara, R., 2014. The role of wetlands microinvertebartes in spreading human diseases. Wetlands Ecology and Management 22, 469–491.

Novakowski, K., Torres, R., Gardner, R., Voulgaris, G., 2004. Geomorphic Analysis of Tidal Creek Networks. WRR 40, W05401. https://doi.org/10.1029/2003WR002722.

O'Connor, T.P., Matlock, G.C., 2005. Shrimp landing trends as indicators of estuarine habitat quality. Gulf of Mexico Science 2, 192–196.

OCHA, 2007. Bangladesh Cyclone Sidr. United Nations Office for the Coordination of Humanitarian Affairs. http://ochaonline.un.org/News/Emergencies/Bangladesh/tabid/2707/Default.aspx.

Odum, E.P., 1980. The Status of Three Ecosystem-level Hypotheses Regarding Salt Marsh Estuaries: Tidal Subsidy, Outwelling, and Detritus-based Food Chains. Estuarine Perspectives 485-495. Academic press.

Odum, E.P., 1985. Trends expected in stressed ecosystems. BioScience 35, 419–422.

Odum, E.P., Finn, J., Franz, E., 1979. Perturbation theory and the subsidy-stress gradient. BioScince 29, 349–352.

Odum, W.E., Odum, E.P., Odum, H.T., 1995. Nature's pulsing paradigm. Estuaries 18, 547–555.

Odum, W.E., Heald, E.J., 1972. Trophic analyses of an estuarine mangrove community. Bulletin of Marine Science 22, 671–738.

Orth, R.J., Carruthers, T.J.B., Dennison, W.C., Duarte, C.M., Fourqurean, J.W., Heck Jr., K.L., Hughes, A.R., Kendrick, G.A., Kenworthy, W.J., Olyarnik, S., Short, F.T., Waycott, M., Williams, S.L., 2006. A global crisis for seagrass ecosystems. BioScience 56, 987–996.

Paling, E.I., Fonseca, M., van Katwijk, M.M., van Keulen, M., 2009. Seagrass Restoration. In: Perillo, G.M.E., Wolanski, E., Cahoon, D.R., Brinson, M.M. (Eds.), Coastal Wetlands: An Integrated Ecosystem Approach, first ed. Elsevier, Amsterdam, pp. 687–713.

Passerl, D., Hagen, S., Medeiros, S., Bilskie, M., Alizad, K., Wang, D., 2015. The dynamic effects of sea-level rise on low-gradient coastal landscapes: a review. Earth's Future 3, 159–181.

Pasternack, G.B., Brush, G.S., Hilgartner, W.B., 2001. Impact of historic land-use change on sediment delivery to a Chesapeake Bay subestuarine delta. Earth Surface Processes and Landforms 26, 409–427.

Perillo, G.M.E. (Ed.), 1995. Geomorphology and Sedimentology of Estuaries. Elsevier, Amsterdam, The Netherlands, 471 p.

Perillo, G.M.E., 2009. Tidal courses: classification, origin and functionality. In: Perillo, G.M.E., Wolanski, E., Cahoon, D.R., Brinson, M.M. (Eds.), Coastal Wetlands: An Integrated Ecosystem Approach. Elsevier, Amsterdam, pp. 185–210.

Perillo, G.M.E., Iribarne, O.O., 2003a. New mechanisms studied for creek formation in tidal flats: from crabs to tidal channels. EOS American Geophysical Union Transactions 84 (1), 1–5.

Perillo, G.M.E., Iribarne, O.O., 2003b. Processes of tidal channels develop in salt and freshwater marshes. Earth Surface Processes and Landforms 28, 1473–1482.

Perillo, G.M.E., Sequeira, M.E., 1989. Geomorphologic and sediment transport characteristics of the middle reach of the Bahía Blanca Estuary, Argentina. Journal of Geophysical Research-Oceans 94, 14351–14362.

Perkins, S., 2017. Inner workings: how saving some of the Southeast's oldest trees might help scientists monitor climate change. Proceedings of the National Academy of Sciences 114, 6875–6876.

Pielke Jr., R.S., 2007. The Honest Broker: Making Sense of Science in Policy and Politics Cambridge. Cambridge University Press, 190 p.

Pirazzoli, P.A., 1996. Sea Level Changes: The Last 20 000 Years. J Wiley & Sons, Chichester.

Plater, A.J., Kirby, J.R., 2006. The potential for perimarine wetlands as an ecohydrological and phytotechnological management tool in the Guadiana estuary, Portugal. Estuarine, Coastal and Shelf Science 70, 98–108.

Poulter, B., 2005. Interactions between Landscape Disturbance and Gradual Environmental Change: Plant Community Migration in Response to Fire and Sea Level Rise. Duke University, Durham, NC, USA., 237 p.

Pratolongo, P., Mazzon, C., Zapperi, G., Piovan, M.J., Brinson, M.M., 2013. Land cover 1607 changes in tidal salt marshes of the Bahia Blanca Estuary (Argentina) during the past 1608 40 years. Estuarine Coastal and Shelf Sciences 133, 23–31.

Primavera, J.H., Estaban, J.M.A., 2008. A review of mangrove rehabilitation in the Philippines – successes, failures and future prospects. Wetlands Ecology and Management 16, 345–358.

Rahmstorf, S., Foster, G., Cazenave, A., 2012. Comparing climate projections to observations up to 2011. Environmental Research Letters 7, 1–5. https://doi.org/10.1088/1748-9326/7/4/044035.

Redfield, A., 1965. Ontogeny of a salt marsh. Science 145, 50–55.

Reed, D.J., 1995. The response of coastal marshes to sea-level rise: survival or submergence. Earth Surface Processes and Landforms 20, 39–48.

Regnier, P., Friedlingstein, P., Ciais, P., et al., 2013. Anthropogenic perturbation of the carbon fluxes from land to ocean. Nature Geosciences 6, 597–607.

Rehage, J.S., Loftus, W.F., 2007. Seasonal fish community variation in headwater mangrove creeks in the southwestern Everglades: An examination of their role as dry-down refuges. Bulletin of Marine Science 80, 625–645.

Reiter, P., 1996. Global warming and mosquito-borne disease in the USA. The Lancet 348, 622.

Ren, W., Tian, H., Cai, W.-J., Lohrenz, S., Hopkinson, C.S., Huang, W.-J., Yang, J., Tao, B., Pan, S., He, R., 2016. Century-long increasing trend and variability of dissolved organic carbon export from the Mississippi River basin driven by natural and anthropogenic forcing. Global Biogeochemical Cycles 30. https://doi.org/10.1002/2016GB005395.

Rey, J.R., Kain, T., Crossman, R., Peterson, M., Shaffer, J., Vose, F., 1991. Zooplankton of impounded marshes and shallow areas of a subtropical lagoon. Florida Scientist 54, 191–203.

Rinaldo, A., Fagherazzi, S., Lanzoni, S., Marani, M., Dietrich, W.E., 1999. Tidal networks 2. Watershed delineation and comparative network morphology. Water Resources Research 35, 3905–3917.

Rivers, D.S., Short, F.T., 2007. Effect of grazing by Canada geese *Branta canadensis* on an intertidal eelgrass *Zostera marina* meadow. Marine Ecology Progress Series 333, 271–279.

Rybczyk, J.M., Callaway, J.C., 2009. Surface elevation models. In: Perillo, G.M.E., Wolanski, E., Cahoon, D.R., Brinson, M.M. (Eds.), Coastal Wetlands: An Integrated Ecosystem Approach, first ed. Elsevier, Amsterdam, pp. 835–853.

Saintilan, N., Rogers, K., McKee, K., 2009. Saltmarsh-mangrove interactions in Australasia and the Americas. In: Perillo, et al. (Eds.), Coastal Wetlands. An Ecosystem Approach. Elsevier, Amsterdam, pp. 855–883.

Sam, R., Ridd, P.V., 1998. Spatial variation of groundwater salinity in a mangrove-salt flat system, Cocoa Creek, Australia. Mangrove Salt MarshSalt Marshes 2, 121–132.

Sánchez, J.M., SanLeon, D.G., Izco, J., 2001. Primary colonisation of mudflat estuaries by Spartina maritima (Curtis) Fernald in Northwest Spain: vegetation structure and sediment accretion. Aquatic Botany 69, 15–25.

Schieder, N., Walters, D., Kirwan, M., 2017. Massive upland to wetland conversion compensated for historical marsh loss in Chesapeake Bay, USA. Estuaries and Coasts. https://doi.org/10.1008/s12237 017 036 9.

Sheehan, M.R., Ellison, J.C., 2014. Intertidal morphology change following *Spartina anglica* introduction, Tamar Estuary, Tasmania. Estuarine, Coastal and Shelf Science 149, 24–37.

Silliman, B., Newell, S.Y., 2003. Fungal farming in a snail. Proceedings of the National Academy of Sciences 100, 15643–15648. https://doi.org/10.1073/pnas.2535227100.

Smith III, T.J., Robblee, M.B., Wanless, H.R., Doyle, T.W., 1994. Mangroves, hurricanes, and lightning strikes. BioScience 44, 256–262.

Smith, S.V., Hollibaugh, J.T., 1997. Annual cycle and interannual variability of ecosystem metabolism in a temperate climate embayment. Ecological Monographs 67, 509–533.

Smith, T., Anderson, G., Balentine, K., Tiling, G., Ward, G., Whelan, K., 2009. Cumulative impacts of hurricanses on Florida mangrove ecosystems: sediment deposition, storm surges and vegetation. Wetlands 29, 24–34.

Smith, T.J., Boto, K.G., Frusher, S.D., Giddins, R.L., 1991. Keystone species and mangrove forest dynamics: the influence of burrowing by crabs on soil nutrient status and forest productivity. Estuarine, Coastal and Shelf Science 33, 419–432.

Spalding, E., Hester, M., 2007. Interactive effects of hydrology and salinity on oligohaline plant species productivity: implications of relative sea-level rise. Estuaries and Coasts 30, 214–225.

Spalding, M., McIvor, C., Beck, M., Koch, E., Moller, I., Reed, D., Rubinoff, P., Spencer, T., Tolhurst, T., Wamlsey, T., van Wesenbeeck, B., Wolanski, E., Woodroffe, C., 2012. Coastal ecosystems: a critical element of risk reduction. Conservation Letters 11, 1–9.

Stevenson, N.J., Lewis, R.R., Burbridge, P.R., 1999. Disused shrimp ponds and mangrove rehabilitation. In: Streever, W. (Ed.), An International Perspective on Wetland Rehabilitation. Springer, Dordrecht, pp. 277–297.

Streever, W., 2001. Saving Louisiana? The Battle for Coastal Wetlands. University Press of Mississippi, Jackson, USA., 189 p.

Sweet, W., Kopp, R., Weaver, C., Obeysekera, J., Horton, R., Thieler, E.R., Zervas, C., 2017. Global and Regional Sea Level Rise Scenarios for the United States. NOAA Technical Report. NOS O-OPS 083. Silver Spring, Maryland, USA.

Syvitski, J.M.P., Harvey, N., Wolanski, E., Burnett, W.C., Perillo, G.M.E., Gornitz, V., Bokuniewicz, H., Huettel, M., Moore, W.S., Saito, Y., Taniguchi, M., Hesp, P., Yim, W.W.-S., Salisbury, J., Campbell, J., Snoussi, M., Haida, S., Arthurton, R., Gao, S., 2005a. Dynamics of the coastal zone. In: Crossland, C.J., Kremer, H.H., Lindeboom, H.J., Crossland, J.I.M., Le Tissier, M.D.A. (Eds.), Coastal Fluxes in the Anthropocene. Springer-Verlag, Berlín, pp. 39–94.

Syvitski, J., Vorosmartyy, C.J., Kettner, A., Green, P., 2005b. Impact of humans on the flux of terrestrial sediment to the global coastal ocean. The Journal of Geology 115, 1–19.

Testa, J., Kemp, W.M., Hopkinson, C.S., Smith, S.V., 2012. Ecosystem metabolism. In: Day, J., Kemp, W.M., Yanez, A., Crump, B. (Eds.), Estuarine Ecology. Wiley, pp. 381–416.

Testa, J., Kemp, W., 2008. Variability of biogeochemical processes and physical transport in a partially stratified estuary: a box-modeling approach. Marine Ecology Progress Series 356, 63–79.

Thampanya, U., Vermaat, J.E., Sinsakul, S., Panapitukku, N., 2006. Coastal erosion and mangrove progradation of Southern Thailand. Estuarine, Coastal and Shelf Science 68, 75–85.

Tong, Y.F., Lee, S.Y., Morton, B., 2006. The herbivore assemblage, herbivory and leaf chemistry of the mangrove *Kandelia obovata* in two contrasting forests in Hong Kong. Wetlands Ecology and Management 14, 39–52.

Turner, R., Rabalais, N., 2003. Linking landscape and water quality in the Mississippi river basin for 200 years. BioScience 53, 563–572.

Uhlenhopp, A., Hobbie, J., Vallino, J., 1995. Effects of land use on the degradability of dissolved organic matter in three watersheds of the Plum Island Sound Estuary. Biological Bulletin 189, 256–257.

Valiela, I., Fox, S.E., 2008. Managing coastal wetlands. Science 319, 290–291.

van de Plassche, O., Wright, A.J., van der Borg, K., de Jong, A.F.M., 2004. On the erosive trail of a 14th and 15th century hurricane in Connecticut (USA) salt marshes. Radiocarbon 46, 775–784.

van den Belt, M., Costanza, R., 2011. Ecological economics of estuaries and coasts. In: Wolanski, E., McLusky, D. (Eds.), Treatise on Estuarine and Coastal Science, vol. 12. Elsevier, Amsterdam, 483 p.

van Katwijk, M.M., Thorhaug, A., Marbà, N., et al., 2016. Global analysis of seagrass restoration: the importance of large-scale planting. Journal of Applied Ecology 53, 567–578.

Vannini, M., Coffa, C., Lori, E., Fratini, S., 2008. Vertical migrations of the mangrove snail *Cerithidea decollata* (L.) (Potamididae) through a synodic month. Estuarine, Coastal and Shelf Science 78, 644–648.

Vannini, M., Ruwa, R.K., 1994. Vertical migrations in the tree crab *Sesarma leptosoma* (Decapoda, Grapsidae). Marine Biology 118, 271–278.

Vasquez, E.A., Glenn, E.P., Brown, J.J., Guntenspergen, G.R., Nelson, S.G., 2005. Salt tolerance underlies the cryptic invasion of North American salt marshes by an introduced haplotype of the common reed *Phragmites australis* (Poaceae). Marine Ecology Progress Series 298, 1–8.

Vince, S.W., Valiela, I., Teal, J.M., 1981. An experimental study of the structure of herbivorous insect communities in a salt marsh. Ecology 62, 1662–1678.

Wadsworth, J., 1980. Geomorphic Characteristics of Tidal Drainage Networks in the Duplin River System, Sapelo Island, GA (PhD thesis). University of Georgia, Athens.

Walsh, J., Wuebbles, D., Hayhoe, K., Kossin, et al., 2014. Chapter 2: our changing climate. In: Melillo, J., Richmond, T., Yohe, G. (Eds.), Climate Change Impacts in the United states: The Third National Climate Assessment. U.S. Global Change Research Program, pp. 19–67. https://doi.org/10.7930/JOKW5CXT.

Waycott, M., Duarte, C., Carruthers, T., et al., 2016. Accelerating loss of seagrasses across the globe threatens coastal ecosystems. Proceedings of the National Academy of Sciences 106, 12377–12381.

Weston, N., 2013. Declining sediments and rising seas: an unfortunate convergence for tidal wetlands. Estuaries and Coasts 37, 1–23. https://doi.org/10.1007/212237-013-9654-8.

White, M., Beumer, J., 1997. Estuarine insect pest control from a fisheries group perspective. Arbovirus Research in Australia 7, 346–356.

WHO (World Health Organization), 1998. Weekly Epidemiological Record, vol. 73, pp. 201–208.

WHO (World Health Organization), 2000. Climate Change and Health: Impact and Adaptation. Available at: http://www.who.int/environmental_information/Climate/climchange.pdf.

Williams, D., 2014. Recent, rapid evolution of the lower Mary River estuary and flood plains. In: Wolanski, E. (Ed.), Estuaries of Australia in 2050 and beyond. Springer, Dordrecht, pp. 277–287, 292 p.

Williams, M.J., Coles, R., Primavera, J.H., 2007. A lesson from cyclone Larry: an untold story of the success of good coastal planning. Estuarine, Coastal and Shelf Science 71, 364–367.

Williams, S.L., Grosholz, E.D., 2002. Preliminary reports from the *Caulerpa taxifolia* invasion in southern California. Marine Ecology Progress Series 233, 307–310.

Wilson, A.M., Morris, J.T., 2012. The influence of tidal forcing on groundwater flow and nutrient exchange in a salt marsh-dominated estuary. Biogeochemistry 108, 27–38.

Wilson, C.A., Hughes, Z.J., FitzGerald, D.M., 2012. The effects of crab bioturbation on Mid-Atlantic salt marsh creek extension: geotechnical and geochemical changes. Estuarine, Coastal and Shelf Science 106, 33–44.

Windham, L., Lathrop Jr., R.G., 1999. Effects of *Phragmites australis* (Common Reed) invasion on aboveground biomass and soil properties in brackish tidal marsh of the Mullica River, New Jersey. Estuaries 22, 927–935.

Windham-Myers, L., Crooks, S., Troxler, T., 2018. A Blue Carbon Primer: The State of Coastal Wetland Carbon Science, Policy, and Practice. CRC Press, Boca Raton, FL.

Wolanski, E., 2007. Protective functions of coastal forests and trees against natural hazards. In: Braatz, S., Fortuna, S., Broadhead, J., Leslie, R. (Eds.), Coastal Protection in the Aftermath of the Indian Ocean Tsunami. What Role for Forests and Trees? FAO, Bangkok, pp. 157–179.

Wolanski, E., Brinson, M., Cahoon, D.M.E., Perillo, G.M.E., 2009. Coastal wetlands. A synthesis. In: Perillo, G.M.E., Wolanski, E., Cahoon, D.M.E., Brinson, M.M. (Eds.), Coastal Wetlands. An Integrated Ecosystem Approach, first ed. Elsevier, Amsterdam, pp. 1–62.

Wolanski, E., Elliott, M., 2015. Estuarine Ecohydrology. An Introduction. Elsevier, Amsterdam, 322 p.

Wolanski, E., Jones, M., Bunt, J.S., 1980. Hydrodynamics of a tidal creek-mangrove swamp system. Australian Journal of Marine and Freshwater Research 31, 431–450.

Woodroffe, C.D., 1992. Mangrove sediments and geomorphology. In: Robertson, A.I., Alongi, D.M. (Eds.), Tropical Mangrove Ecosystems. American Geophysical Union, Washington D.C., pp. 7–42

Woodroffe, C.D., Davies, G., 2009. The morphology and development of tropical coastal wetlands. In: Perillo, G.M.R., Wolanski, E., Cahoon, D.R., Brinson, M.M. (Eds.), Coastal Wetlands: An Integrated Ecosystem Approach. Elsevier Science, Amsterdam, pp. 65–88.

Yang, Q., Tian, H., Friedrichs, M.A.M., Hopkinson, C.S., Lu, C., Najjar, R., 2015. Increased nitrogen export from eastern North America to the Atlantic Ocean due to climatic and anthropogenic changes during 1901–2008. Journal of Geophysical Research: Biogeosciences 120, 1046–1068. https://doi.org/10.1002/2014JG002763.

Yvon-Durocher, G., Jones, J., Trimmer, M., Woodward, G., Montoya, J., 2010. Warming alters the metabolic balance of ecosystems. Philosophical Transactions of the Royal Society of London 365, 2117–2126.

Zhang, K., Liu, H., Li, Y., Xu, H., Shen, J., Rhome, J., Smith, T.J., 2012. The role of mangroves in attenuating storm surges. Estuarine, Coastal and Shelf Science 102–103, 11–23.

Zong, Y., Chen, Z., Innes, J.B., Chen, C., Wang, Z., Wang, H., 2007. Fire and flood management of coastal swamp enabled first rice paddy cultivation in east China. Nature 449, 459–462.

PART I

COASTAL WETLANDS AS ECOSYSTEMS

CHAPTER

2

The Morphology and Development of Coastal Wetlands in the Tropics

Colin D. Woodroffe

School of Earth and Environmental Sciences, University of Wollongong, Wollongong, NSW, Australia

1. INTRODUCTION

Tropical coastal wetlands represent particularly productive ecosystems that contribute to both terrestrial and marine biodiversity. In this review, the principal wetlands that will be considered are associated with mangrove shorelines. Mangroves characterize the upper intertidal zone on many low energy tropical coasts, often with salt marsh and associated wetlands that can form landward of such halophytic vegetation (Robertson and Alongi, 1992; Alongi, 1998). Seagrass, which can be extensive seaward of mangroves and in other nearshore settings, is beyond the scope of this overview.

There are several factors that have favored development of extensive coastal wetlands over the past 7000 years (the mid-late Holocene). In particular, the formation of low-lying coastal plains has occurred in response to the pattern of relative sea level change. For such wetlands to develop, it is generally necessary for extensive near-horizontal topography to be available at sea level. This occurs most frequently as a consequence of a particular set of geomorphological conditions that have accompanied a relatively stable period of sea level. It is sea level, or the level of the water table that is generally closely related to (though slightly above) sea level, which constrains wetland development. In some cases, modern sea level fortuitously coincides with a horizontal substrate at or close to sea level. In other instances, the near-horizontal topography is the result of landforms built during former Pleistocene sea level highstands or a complex of habitats resulting from deltaic-estuarine sediment deposition and Holocene coastal plain development. Coincidence of sea level with suitable topography is crucial for extensive mangrove forests to develop.

Coastal Wetlands
https://doi.org/10.1016/B978-0-444-63893-9.00002-2

2. MANGROVE AND ASSOCIATED WETLANDS

Mangroves can be defined as trees, shrubs, or palms, exceeding half a meter in height, which occur in the upper intertidal zone. They comprise a diverse range of plants and are not a single taxonomic group. Each has adaptations enabling survival in this otherwise inhospitable saline and anaerobic environment. Adaptations include viviparous propagules; for example, in many genera seeds remain attached and germinate on the tree and then are buoyant during a short aquatic dispersal phase (Woodroffe et al., 2014). Many mangroves have developed mechanisms to tolerate salt, and the majority have root systems that enable the plants to respire, despite being anchored in saturated, nonporous soils depleted in oxygen. Above-ground root systems include pneumatophores, prop roots, and buttresses, some of which provide structural support and most of which are covered with lenticels that promote gas exchange.

There is a significant climatic control on the distribution of tropical coastal wetlands (Hamilton and Casey, 2016). Mangroves are predominantly tropical, although extending into subtropical regions along the eastern coasts of major continents where ocean currents ameliorate temperatures. Although salt marsh does occur in the tropics, it usually represents a minor component in comparison with mangroves but becomes increasingly important toward the poleward limits of mangroves (Alongi, 1998). Temperature controls distribution in a broad sense, mangroves occur where mean annual sea surface temperatures exceed 20°C, and minimum air temperatures rarely fall below 0°C (Woodroffe and Grindrod, 1991; Osland et al., 2016). Exposure to winter frosts constrains the latitudinal limit (Cavanaugh et al., 2014; Saintilan et al., 2014). The genetically distinct population of the southeast Australian and New Zealand mangrove *Avicennia marina* var. *australasica* reaches its southernmost limit in Corner Inlet (38° 45′ S) in Victoria (Duke, 1992).

Coastal wetlands are more extensive, with greater areal extent and species diversity, in the wetter parts of the tropics for several reasons. First, deltas are more extensive at the mouths of rivers in high rainfall regions because greater sediment loads are eroded from the catchment and carried to the coast. Second, inundation is more common and widespread, both because of the river-fed flood waters and also as a result of direct precipitation. Mangroves do occur on arid coasts, however, and there are wetlands in arid and semiarid settings, but these appear to be limited by strongly hypersaline groundwaters (Cresswell and Semeniuk, 2011). For example, the wetter east coast of Australia has 20 species of mangrove compared with only 4 at comparable latitude on the drier west coast (Duke et al., 1998).

The global distribution of mangroves comprises two provinces, one centered in the West Indies, but including the Pacific coast of Latin America (termed Atlantic-East Pacific, AEP, by Duke et al., 1998), and the other in the Indo-West Pacific (IWP). The IWP covers a larger area, extending from the east coast of Africa to the central Pacific Ocean. It is more diverse in terms of number of mangrove taxa, as well as most other groups of organisms, such as corals. By contrast, only four mangrove genera dominate in the West Indies. *Rhizophora mangle* is the most distinctive species, although there are other *Rhizophora* species that occur in West Africa and South America. *Avicennia germinans*, termed the black mangrove, is the second most significant mangrove, generally occurring landward of *Rhizophora*. A second species, *Avicennia schaueriana*, extends to the southern limit of mangroves in Brazil. *Laguncularia* and *Conocarpus* are found at the landward margin and are typically less extensive. Mangrove associates in the AEP include the succulent *Batis maritima* and the fern *Acrostichum aurem* (also found in parts of the IWP). In South America, the genus *Pelliciera* also occurs in wetter areas and upstream in estuaries. Grass and sedgelands occur landward of mangroves in the Everglades of Florida, as well as along much of the northern coast of South America.

The structure of mangrove forests is relatively simple in the West Indies (Fig. 2.1). Based on ecological studies in Florida and islands in the Caribbean, Lugo and Snedaker (1974) and Lugo et al. (1976) defined five principal types of mangrove: overwash, fringe, basin, scrub, and riverine. Their research indicated functional differences between these mangrove types, and the classification has been widely adopted for the AEP region (Bacon, 1994). Overwash mangrove occurs on islands that are overtopped by the tide, as in the Bahamas and on the Belize barrier reef. Fringe mangrove comprises the seaward *Rhizophora* zone, flooded during each tidal cycle. Basin mangroves consist of more inland stands that are less regularly inundated or drained; in some situations, the flooding may be seasonal rather than tidal. Scrub

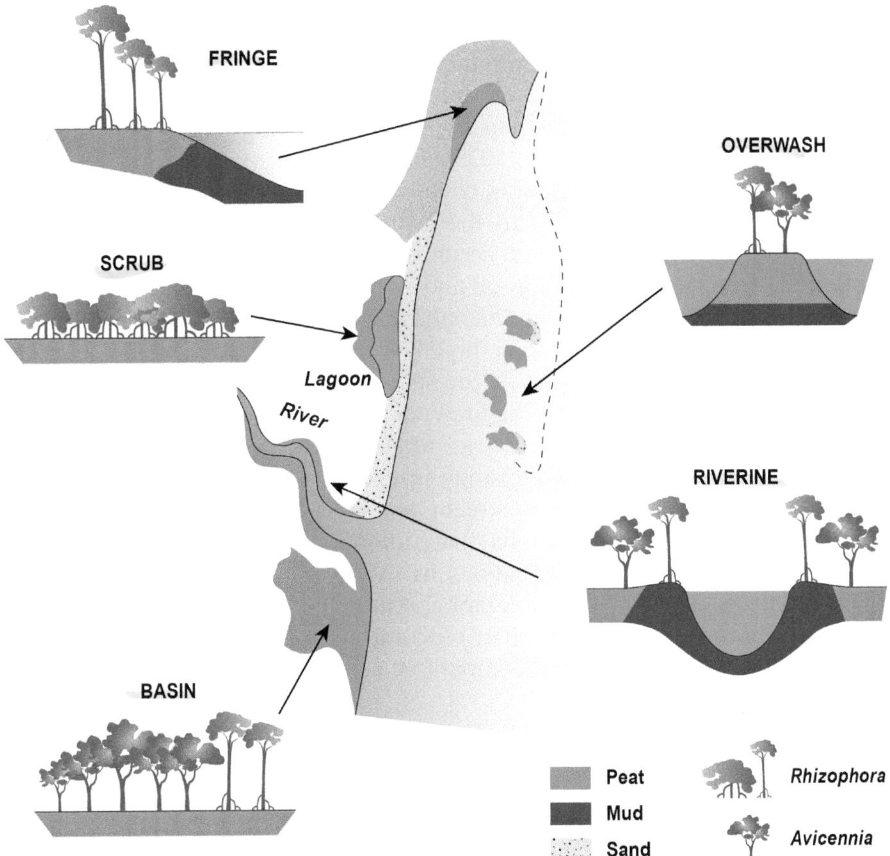

FIGURE 2.1 Mangrove forest types distinguished in the West Indies, illustrated schematically, and their typical occurrence in the landscape (following Lugo et al., 1976 and Bacon, 1994).

mangrove comprises dwarfed stands, often of *Rhizophora*, that appear to be nutrient limited. Riverine mangrove occurs along the banks of rivers and has the largest trees and the highest productivity.

Mangroves are much more diverse in the IWP; the center of diversity lies in Indonesia and northern Australia where there are more than 30 species. There are not only several species of *Rhizophora* in the IWP but also several other genera in the Rhizophoraceae, such as *Ceriops* and *Bruguiera*. Although there is considerably greater diversity of habitats in the IWP than in the AEP and many of these show complex mosaics of mangrove species, a general zonation pattern can be recognized on open shorelines (Macnae, 1969). Usually, this comprises a seaward zone of *Sonneratia* and then a prominent zone of *Rhizophora*, often with *Ceriops* and *Bruguiera*, and a landward zone of other mangroves such as *Lumnitzera* and *Excoecaria*. There are also several species of *Avicennia* in the IWP, and this genus demonstrates the broadest tolerance to environmental factors, especially salinity, being able to grow throughout the upper intertidal zone.

There is greatest diversity of mangrove habitats in southeast Asia and northern Australia as a consequence of geomorphological development of deltaic-estuarine and coastal plains during the mid- and late Holocene under relatively stable sea level. The vegetation landward of the mangroves varies, determined primarily by regional climatic trends. In arid and semi-arid areas, as flanking the macrotidal estuarine systems of Western Australia, there are extensive saline flats, and the only vegetation that survives is low shrubs and herbs, such as samphire (Jennings, 1975; Thom et al., 1975). In the wet-dry tropics of northern Australia by contrast, there are broad alluvial plains, covered by seasonally inundated grass and sedgelands (Woodroffe et al., 1989, 1993). In the perhumid tropics of much of Malaysia and Indonesia, similar plains support domed accumulations of woody peat beneath peat swamp forest (Anderson, 1964; Morley, 1981; Staub and Esterle, 1994). Brackish communities of *Nipa fruticans* and *Heritiera littoralis* often dominate the transition to inland peat swamp forest, as in the Mahakam delta. Prior to its clearing in the 1970s for intensive rice cultivation, the plains of the Mekong delta were largely covered by forests of *Melaleuca cajuputi*, with *Casuarina* on the sandy beach ridges; these species provide firewood and other services and reestablishment is being encouraged, especially in *Melaleuca* rice forest farming (Douglas, 2005).

Proceeding up estuaries, there is a change in species occurrence. In northern Australia, *Sonneratia lanceolata* is restricted to upstream locations where it grows in the intertidal zone, but at higher absolute elevations than intertidal mangroves further downstream (Finlayson and Woodroffe, 1996). Variations in salinity along the estuaries are important physiological controls. The most salt-tolerant species tend to be the slowest growing under low salinity conditions and thus are often outcompeted by less-tolerant species; evidently salt tolerance occurs at the expense of competitive ability (Ball, 1998).

3. ENVIRONMENTAL SETTINGS

It is important to discriminate particular environmental settings within which mangroves occur. Mangrove forests form complex mosaics in some geomorphological settings, such as on sediment-rich deltas, in contrast to simple stands of mangrove that may be sediment-starved in the lee of coral reefs. Within each setting, there may be a range of hydrodynamic conditions, flux of sediments, and organic and geochemical characteristics of the substrate (Thom, 1982; Woodroffe, 1992). For example, there are functional differences between stands of *R. mangle*, which are regularly and effectively flushed by tides, and more inland mangroves that are flooded seasonally or by only the highest tides, favoring accumulation of mangrove-derived organic carbon (Twilley, 1985). An ecomorphodynamic classification, combining aspects of mangrove growth form and hydrodynamics, was developed by Twilley and Rivera-Monroy (2005, 2009).

Fig. 2.2 outlines a framework, across several scales, within which to consider mangrove ecosystems (Woodroffe et al., 2016). The broadest scale is the macroscale appropriate for entire deltas or estuaries. At this scale, climate and relative sea level changes are important factors in explaining the complexity of environments such as deltas, estuaries, lagoons, and carbonate banks or reefs. Climate not only determines the overall floristic composition of the wetland and adjacent terrestrial ecosystems (e.g., mangrove or salt marsh) but also affects delivery of sediment to the coast. Relative sea level change captures the combined effect of local changes in sea level, together with net movement of the land, such as subsidence or other physical and biological processes, which in combination constrain the accommodation space within which wetlands form.

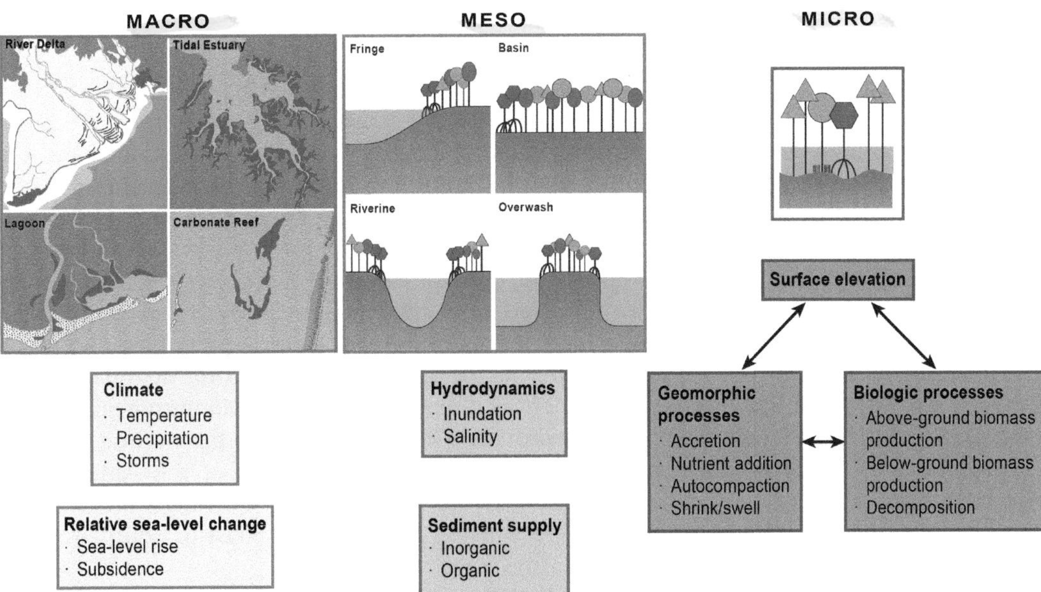

FIGURE 2.2 A scale-based framework for describing mangrove systems, comprising macroscale at which regional boundary conditions (climate and relative sea level change) are important; mesoscale at which hydrodynamics and sediment supply influence mangroves; and microscale at-a-site interactions within a mangrove stand including surface and subsurface processes. *After Woodroffe, C.D., Rogers, K., McKee, K.L., Lovelock, C.E., Mendelssohn, I.A., Saintilan, N., 2016. Mangrove sedimentation and response to relative sea level rise. Annual Review of Marine Science 8, 243–266.*

At the mesoscale, it is usually possible to distinguish distinctive mangrove habitats, often associated with different landforms or microtopography. Hydroperiod (frequency/duration of inundation), which is related to substrate elevation in the tidal frame, influences how a particular stand of mangroves functions. The availability of sediment is also important, whether this is terrigenous sediment or mangrove-derived peat. The functional forest types described by Lugo and Snedaker (1974) and as shown in Fig. 2.1 represent a classification at this mesoscale, which has been widely applied in the Americas but is less appropriate for the more structurally complex mangrove forests of the IWP where a wider range of geomorphologically defined habitats have been described (e.g., Semeniuk, 1985; Cresswell and Semeniuk, 2011).

At a microscale, involving an individual site, a range of geomorphological and biological surface and subsurface processes operate (Fig. 2.2). Hydroperiod influences the rate of sediment accretion and root production with negative feedbacks, as accommodation space is gradually reduced. Rates of surface elevation change measured over decades are extremely variable and often independent of the rate of sea level rise (Rogers et al., 2006). However, the upper limit to mangrove growth is usually limited by highest tides, and accumulation rates measured over centuries to millennia are constrained by relative sea level changes and often appear to track sea level. Mangrove forests are replaced by freshwater wetland vegetation once their substrate has built beyond this upper threshold but may be replaced by salt flats on arid coasts.

4. SEDIMENTATION AND THE DEVELOPMENT OF WETLANDS

Tropical coastal wetlands form near the water table under low energy conditions and can grow on substrates ranging from mud (silt and clay) to fibrous peat (derived primarily from mangrove roots). The source and supply of sediment determine the nature of the substrate, and an organic-rich mud is typical where there is both a large supply of sediment and prolific mangrove growth. Those extensive coastal wetlands associated with large river deltas are built on terrigenous sediment that is delivered from the catchments (Lovelock et al., 2010; Swales et al., 2015). The largest area of mangroves in the world, the Sundarbans, occurs in the abandoned delta of the former mouths of the Ganges-Brahmaputra-Meghna system, down which over a billion tonnes of sediment is delivered each year. Mangrove and associated wetlands have also been extensive, until large-scale clearance, at the mouths of other megadeltas in Asia (Woodroffe et al., 2006). By contrast, the mouth of the Amazon has not developed a progradational delta, but riverine muds carried by longshore currents nourish a suite of rapidly migrating landforms along the northeast coast of South America (Nascimento et al., 2013; Anthony et al., 2014; Gensac et al., 2015).

Considerable volumes of mud can also be advected into coastal wetlands from offshore by tidal processes. This is particularly true of macrotidal systems, such as those along the coast of northern and northwestern Australia, where tidal flows are large enough to entrain considerable volumes of mud and carry them in suspension into mangrove forests where they are deposited when velocities are reduced (Woodroffe et al., 1989; Wolanski, 2006a). The effectiveness of tidal pumping is demonstrated on the Ord River where reduction of river discharge, as a result of dam construction to form Lake Argyle, has been accompanied by rapid accumulation of tidal sediments in the upper part of the estuary (Wolanski et al., 2001).

Mangroves also occur in carbonate settings. The calcareous substrates on which mangroves establish may be either biologically produced, as in the case of coral reefs, or precipitated, as in the case of muds on the Great Bahama Bank. Mangroves can colonize a near-horizontal reef platform, and successive stages of mangrove colonization and consolidation have been described in response to the evolutionary stage of reef flat development on islands of the Great Barrier Reef (Stoddart, 1980). Sediment dynamics is a function of reef growth by corals and associated organisms, but sediment accumulation beneath mangroves also incorporates organic matter produced by the mangroves themselves.

Some mangrove forests grow on sediments that are primarily organic, indicating the considerable productivity of these ecosystems. In these systems where there is only limited inorganic sediment supplied from outside the wetland, the substrate is generally peaty, comprising the fibrous root material of the mangroves (McKee et al., 2007). These highly organic peats characterize sediment-starved settings, such as limestone islands in the West Indies, the oceanward margin of the Everglades, and parts of the coast of Latin America, including Belize (Woodroffe, 1992; Ramcharan, 2004). These sites around the Caribbean have experienced a gradual, but slowing, rise of sea level over the past 6000 years, in contrast to most of the rest of the tropics, because they lie in the forebulge region, relatively close to the former North American ice sheets, and have been experiencing ongoing glacial isostatic adjustment (Milne and Peros, 2013). It appears that this relative sea level rise has made accommodation space available, which has enabled several meters of mangrove-derived peat to accumulate over several millennia. The pattern of sea level change, observed in the Caribbean (Khan et al., 2017), contrasts with that which has been experienced by most mangrove shorelines. Elsewhere, including the east coast of South America, sea level was close to, or above, present 5000 years ago, providing little opportunity for comparable thicknesses of organic-rich substrate over this period. This seems to likely explain why highly biogenic peats are rare under mangroves in other parts of the world; by contrast, many northern hemisphere salt marshes do occur in similar forebulge settings and have been able to accumulate thick biogenic soils (Allen, 2000).

It is therefore essential to consider the nature of the substrate beneath mangrove forests in terms of sediment supply, distinguishing mud from outside, termed allochthonous, from the in situ production of sediment, termed autochthonous (mangrove-derived peat). While deep thixotropic muds are typical of the many turbid mangrove shorelines of the Indo-Pacific, these adaptable halophytes can also grow above several meters of fibrous peat produced by earlier generations of mangroves. Fig. 2.3 shows the interrelationships schematically. The elevation of the mangrove substrate is a function of the rate of accretion that is, in turn, partly a response to hydroperiod and the frequency and duration of inundation. Increase in surface elevation comes about through import of largely inorganic sediment from land or marine sources or the production of plant matter, primarily root material. This dichotomy is another useful distinction to consider in relation to the associated wetlands that form in the hinterland of mangroves. For example, peat swamp forests of Southeast Asia grow on autochthonous peats formed beneath forests of trees in ombrogenous wetlands. Other rain-fed wetlands occur on alluvial floodplains behind subtle levees built from the sediment that seasonal tropical rivers bring from their catchments.

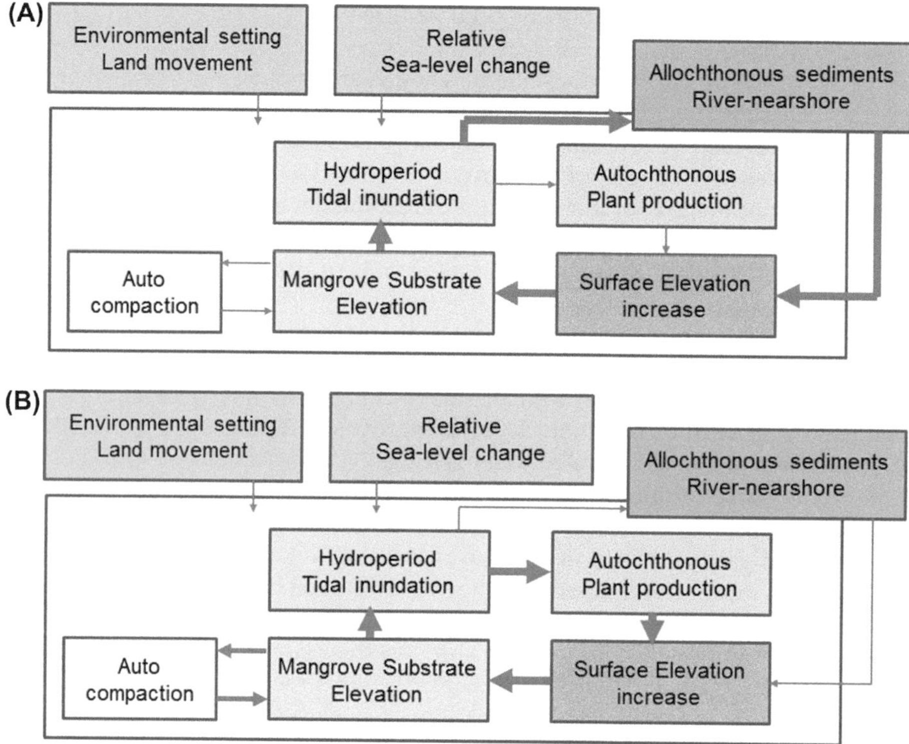

FIGURE 2.3 Schematic illustration of the relative importance of processes influencing sedimentation in mangrove systems: (A) A muddy coast on which allochthonous sediment derived from outside the forest is important. (B) A sediment-starved system in which autochthonous plant production contributes to the substrate. Environmental setting and the relative pattern of sea level change are important boundary conditions.

5. SEA LEVEL CONTROLS ON WETLAND DEVELOPMENT

Tropical coastal wetlands develop where near-horizontal topography coincides with sea level, and it is clear that past sea level changes have influenced their distribution and extent. Paleoenvironmental reconstruction provides insights into the nature of this response and may enable a clearer understanding of how such wetlands might respond to future environmental change, including sea level rise.

At a global scale, plate tectonic setting exerts a first-order control on the distribution of coastal wetlands. Shorelines along a plate margin, such as the west coast of North and South America, are generally experiencing gradual uplift and are backed by steep mountain ranges from which short, steep rivers drain, carrying sediment into deeper ocean waters. Such coasts provide relatively few areas suitable for development of wetlands (for example, isolated stands of tall mangroves in Ecuador, Hamilton and Lovette, 2015). By contrast, the passive, or trailing-edge, coasts that occur on the east of the American continents receive large sediment loads from the long river systems that traverse the continents, such as those of the Mississippi, Orinoco, and Amazon rivers. There are more extensive low-lying areas associated with the sedimentary basins on these coasts, which are more conducive to the development of suites of landforms within which wetlands can form.

Large sediment loads are delivered by the Ganges-Brahmaputra-Meghna, Irrawaddy, Mekong, Red, and Pearl rivers into the tropical waters of the IWP region from the tectonically active Himalayan massif (Ericson et al., 2006; Woodroffe et al., 2006). Sediment deposition has formed broad, flat delta plains in the lower reaches. Human impact on these catchments has been considerable, with land use change often resulting in increased loss of soil from the landscape. Elsewhere, the construction of dams has been reducing sediment delivery to the coast (Syvitski et al., 2005). Steep catchments along the tectonically active island arcs, such as Indonesia and Papua New Guinea, contribute a disproportionately large sediment load to the oceans, accompanied by delta development along many of the coasts of these islands (Milliman and Syvitski, 1992; Walsh and Nittrouer, 2004). Coastal progradation has formed near-horizontal topography, providing suitable habitat for extensive wetland development.

If sea level changes, then there are ramifications throughout the wetland system. Wetlands can best form where there has been sufficient time for near-horizontal substrates to have been deposited, such as has occurred in many parts of the tropics over the past few millennia as a result of a relatively stable period of sea level and its coincidence with near-horizontal underlying substrates (Woodroffe et al., 2016). The earth's climate is presently in a warm period, an interglacial, and sea level has consequently regained a level occupied during previous interglacials, implying that some landform complexes are partially inherited from previous highstands of the sea.

The peak of the last ice age occurred about 20,000 years ago, and sea level at that time was around 120 m below its present level (Lambeck and Chappell, 2001). Ice melted relatively quickly, and the postglacial sea appears to have risen to its present level by around 7000 years ago, with only a minor amount of ice melt since that time (Murray-Wallace and Woodroffe, 2014). Those areas that were covered by ice have experienced rapid uplift since the ice load has been removed and are continuing to uplift. Areas that were marginal to the large ice sheets have also experienced isostatic land movements; in many cases, they are continuing to undergo subsidence as the mantle adjusts to replenish the ice-loaded regions. Most tropical coastal wetlands are sufficiently remote from former ice sheets that the pattern of sea level change experienced is similar to the global eustatic pattern, namely a postglacial sea level rise up to ~7000 years ago but little change since. This is in contrast to temperate (or polar) latitudes, where there are ongoing isostatic movements of the land as a result of the redistribution of mass as ice has melted and ocean volume increased.

Knowledge of Holocene sea level trajectories, combined with paleoenvironmental information on mangrove forests, can be used to infer how such forests adjust to different rates of sea level change. This approach has been most effectively developed in the case of reefs because fossil reefs are better preserved than mangrove sediments. Several different responses, including drowning, backstepping, catch up, keep up, prograding, and emergence, have been identified for reefs (Neumann and Macintyre, 1985; Hubbard, 1997). Similar types of response are shown schematically for mangrove forests in Fig. 2.4. It can be inferred that although mangroves lined shorelines when the sea was at its lowest, mangrove substrates have not been able to keep pace with the most rapid rates of postglacial sea level rise that have been experienced. If they had, then mangrove shorelines would not have shifted since the last glacial maximum (when the coastline was up to several hundred kilometers seaward

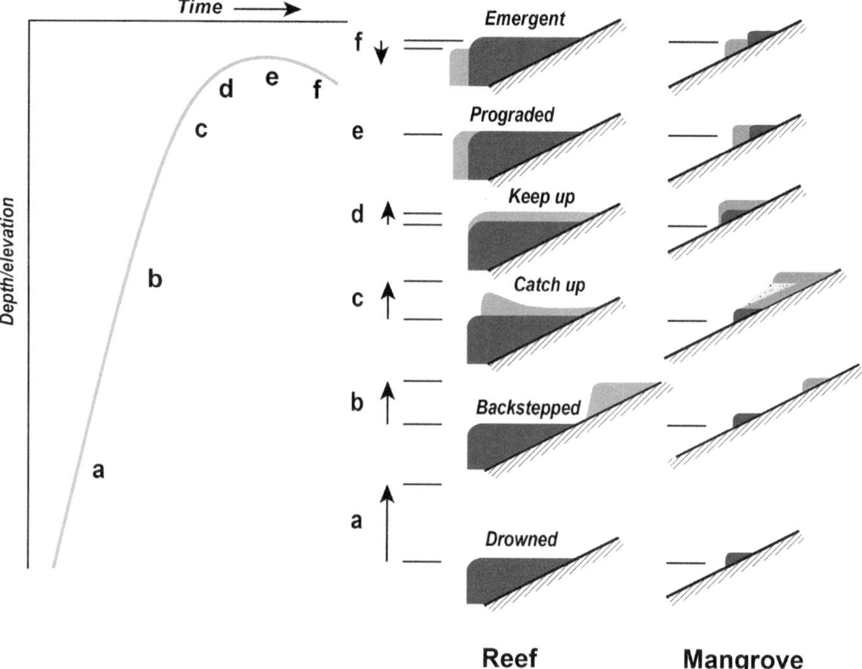

FIGURE 2.4 The response of mangrove shorelines to sea level change. A schematic sea level *curve* is shown with rapid sea level rise, decelerating, and then stabilizing before a gradual fall to present. The typical response, as indicated by the stratigraphy of mangrove sediments, is indicated, following a broader scheme that has been developed for coral reefs. *From Woodroffe C.D., 2018. Mangrove response to sea-level rise: palaeo-ecological insights from macrotidal systems in northern Australia. Marine and Freshwater Research 69, 917–932.*

of the present one), and mangrove sediments 120 m or more thick would have been deposited across continental shelves such as the Timor Shelf (Yokoyama et al., 2001).

Postglacial mangrove sediments have been identified on the broad Sahul and Sunda shelves, but these have been drowned by sea level rise (Hanebuth et al., 2000, 2011). In shallower waters, and when rates of sea level rise were slightly less, it seems that mangrove forests may have backstepped. In this case, mangrove sediments evidently did not keep pace with sea level rise but mangrove reestablished at a more landward location at higher elevation. This response can be inferred for some of the deeper parts of both the Belize reefs and the Great Barrier Reef where mangrove peat is encountered on the lagoon floor. Basal transgressive mangrove sediments, generally organic-rich peaty muds at the base of cores, are found beneath coastal plains in northern Australia (Woodroffe et al., 1993). It appears that mangrove sedimentation was unable to keep pace with sea level rise in these open-coastal plain settings, with the mangrove forests retreating landward and forming a fringe along the margin of the hinterland about 7000 years ago when sea level stabilized close to its present level. Progradation has occurred subsequently with the build-out of northern Australian mangrove shorelines often for several kilometers and mangrove mud overlying nearshore sediments, indicating a period of "catch up" before coastal wetlands reestablished (Woodroffe, 2018). Within the more shelter estuaries, however, contemporaneous mangrove sediments were able to keep pace with sea level rise as it slowed prior to reaching present level, and the mangrove forests appear to have persisted into the "big swamp" phase that characterized these estuarine plains (discussed below).

In the case of sea level fall, mangrove substrate is left emergent, as shown, for example, by the highly oxidized muds found in Malaysia (Geyh et al., 1979). Detailed evidence of relative sea level change from many tropical locations reveals a slight fall of sea level during the past 7000 years (Lewis et al., 2013). Subtle adjustments can be inferred related either to local isostatic flexure (primarily hydroisostatic flexure, the response of ocean basins to loading of water) or sediment-isostasy (flexural response to the loading of sediment, such as seems likely to have occurred where large loads of sediment have been deposited, as in the Asian megadeltas), or more generally responding to changes in the total volume of the ocean basins. There are, of course, exceptions to the general pattern of sea level change, especially for areas that are undergoing vertical movement. For example, fossil reefs on uplifting shorelines, such as along the Indonesian arc, have been repeatedly raised by coseismic activity. In the West Indies and Florida, evidence indicates that sea level has continued to rise at a decelerating rate in those regions, reflecting the ongoing adjustment of the glacial forebulge region in eastern North America. It has become apparent that the West Indies has thus experienced a different sea level history from that of other tropical areas (Murray-Wallace and Woodroffe, 2014; Khan et al., 2017), a fact that underpins the differences between the coastal wetlands of the New and Old World tropics.

6. SEA LEVEL CHANGE AND THE DIVERSIFICATION OF WEST INDIAN MANGROVES

The mangrove forests of Florida have been subject to a number of detailed studies (Bowman, 1917; Lugo and Snedaker, 1974). The mangroves of Florida Bay are adjacent to shallow marine carbonate banks and show the typical zonation with a seaward fringe of *R. mangle* and a more landward zone of *Avicennia*, flanking the extensive sedge- and grass-dominated freshwater wetlands of the Everglades. In an important paper, Davis (1940) considered that mangroves promoted sedimentation and had substantial land-building capability and were building out into Florida Bay. By contrast, Egler, an ecologist, regarded the deep peats beneath the mangrove forests as having accreted during a period of sea level rise and concluded that mangroves must have invaded previously freshwater environments (Egler, 1952). More detailed stratigraphic and palynological studies confirmed the latter interpretation and mangrove peat was used to reconstruct early sea level curves for the region (Scholl and Stuiver, 1967).

The stratigraphy typical of carbonate banks in the region is shown in Fig. 2.5. Such stratigraphy indicates that in the late Holocene, the sea was rising, as recorded by a basal mangrove peat that underlies most of the marine carbonate sediments. Such a sequence of sediments is found in Florida Bay, comprising an intertidal peat overlain by shallow water carbonate, and is transgressive, recording a relative rise of sea level (Enos and Perkins, 1979). On the other hand, mangroves are observed to be now spreading over marine carbonate banks, as observed by Davis (1940). This is a regressive stratigraphy, recording the progradation of mangroves (Parkinson, 1989). A similar stratigraphy is found elsewhere in the region, for

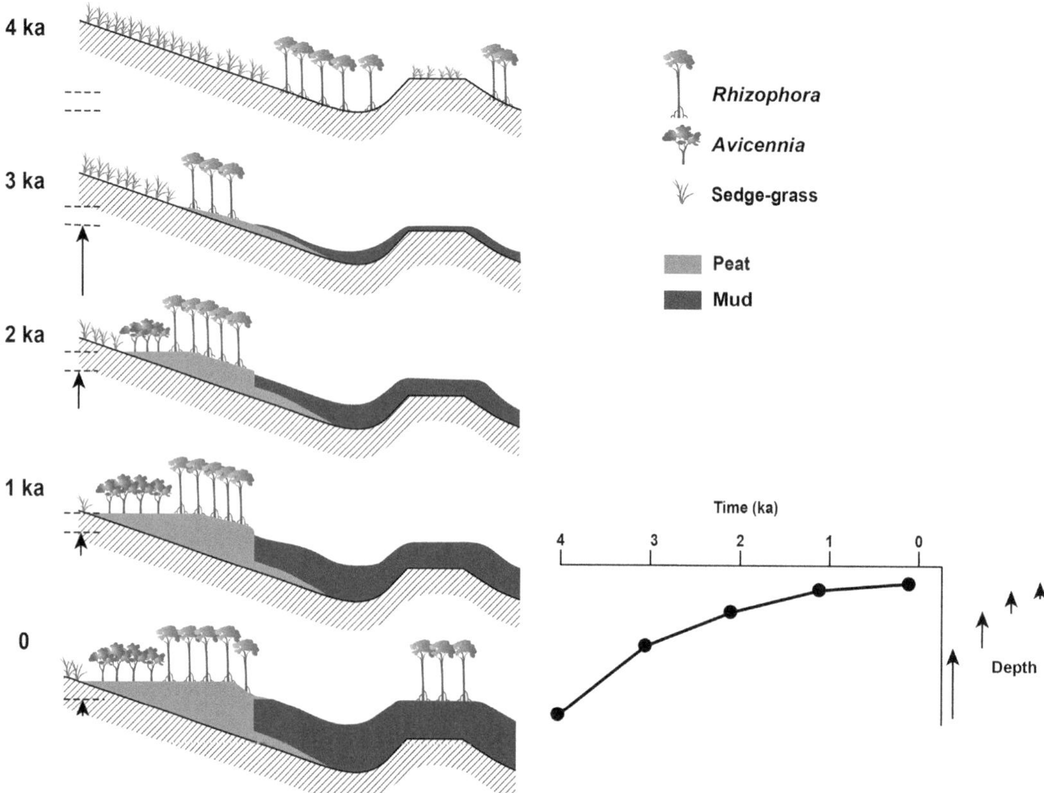

FIGURE 2.5 Typical sea level *curve* for the Florida and West Indian region and the schematic response of mangrove peat and carbonate environments as recorded in the stratigraphy of Florida Bay and several islands of the West Indies.

example, beneath the extensive mangrove forests of the Cayman Islands (Woodroffe, 1981) and Jamaica (Hendry and Digerfeldt, 1989). Such a pattern of decelerating sea level rise is observed widely throughout the Caribbean (Toscano and Macintyre, 2003; Cohen et al., 2016) and represents the gradual response of this region to the rebound from ice melt of the North American ice sheets. However, a slightly different sea level history is inferred for Belize (Gischler, 2006), and in South America it appears from the stratigraphy of mangrove forests that sea level reached its present level considerably earlier (Cohen et al., 2005), as it has in the IWP.

7. SEA LEVEL CHANGE AND MANGROVE HABITAT EVOLUTION IN THE PACIFIC

It appears that much of the Sunda Shelf experienced greater seasonality during glacial periods; wide, multichanneled, or braided sand and gravel-bedded rivers carried large quantities of coarse sediment to shorelines on the exposed shelf. More subtle changes of climate are recorded by interbedded sands and freshwater peat horizons, such as those at Pantai Remis, south of Ipoh (Kamaludin et al., 1993). As sea level rose, marine sediments were deposited (Verstappen, 1975), this being recorded in numerous pollen sequences (Woodroffe, 1993; Proske et al., 2011). The stratigraphy and pattern of habitat change in response to the general sea level history of the southeast Asian and northern Australian region is shown in Fig. 2.6. Transgressive sediments record the rapid rise of sea level across much of the region. Mangrove deposits on the Sunda Shelf indicate that a fringe of mangrove forests must have traversed much of the shelf as it was flooded following ice melt (Hanebuth et al., 2000).

The Holocene evolution of a number of systems in northern Australia has been interpreted, primarily on the basis of stratigraphic and radiocarbon dating analyses, with particular emphasis not only on depositional environments but also including the evolution of

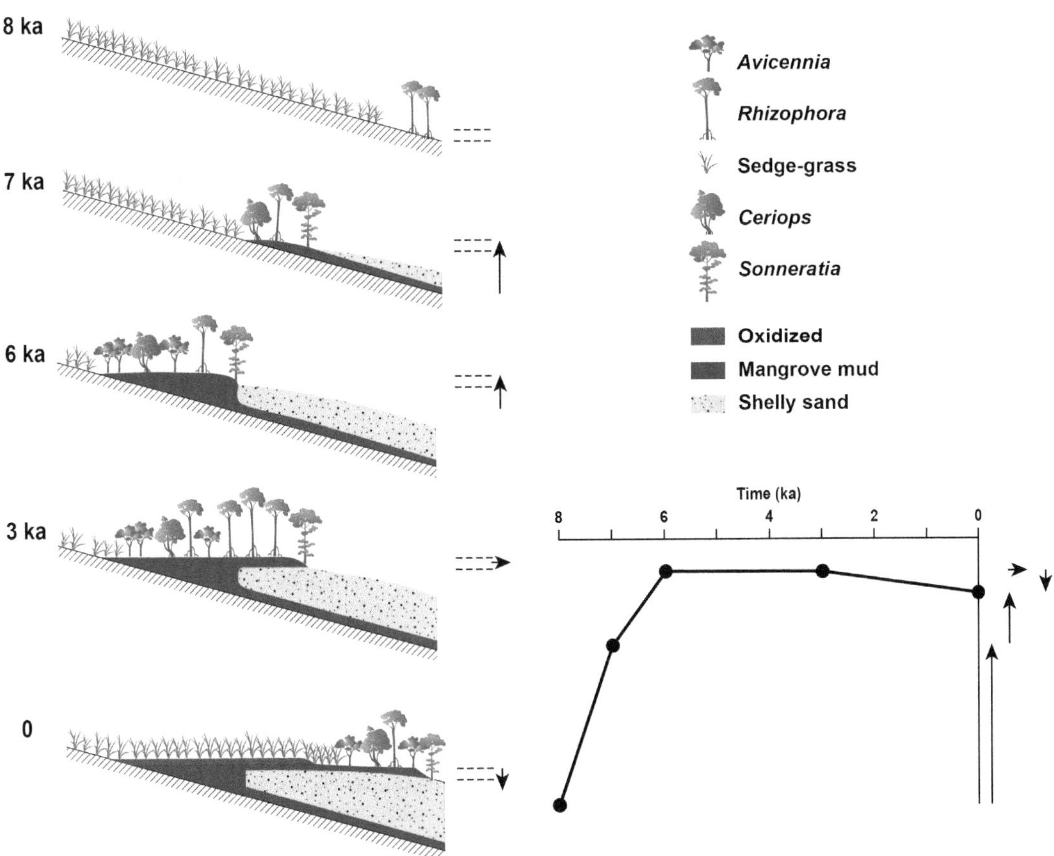

FIGURE 2.6 Typical sea level *curve* for the southeast Asian and northern Australian region, and schematic response of mangrove mud and adjacent environments. The slight emergence is a result of late Holocene relative sea level fall; such a pattern is seen in Malaysia.

channel form. Sea level rise drowned the valleys, resulting in the landward transgression of mangroves over terrestrial environments and the formation of broad, open embayments, as sediment supply did not keep pace with the rapid creation of accommodation space (Fig. 2.7). Sea level stabilized in northern Australia around 7000 years BP (Lewis et al., 2013),

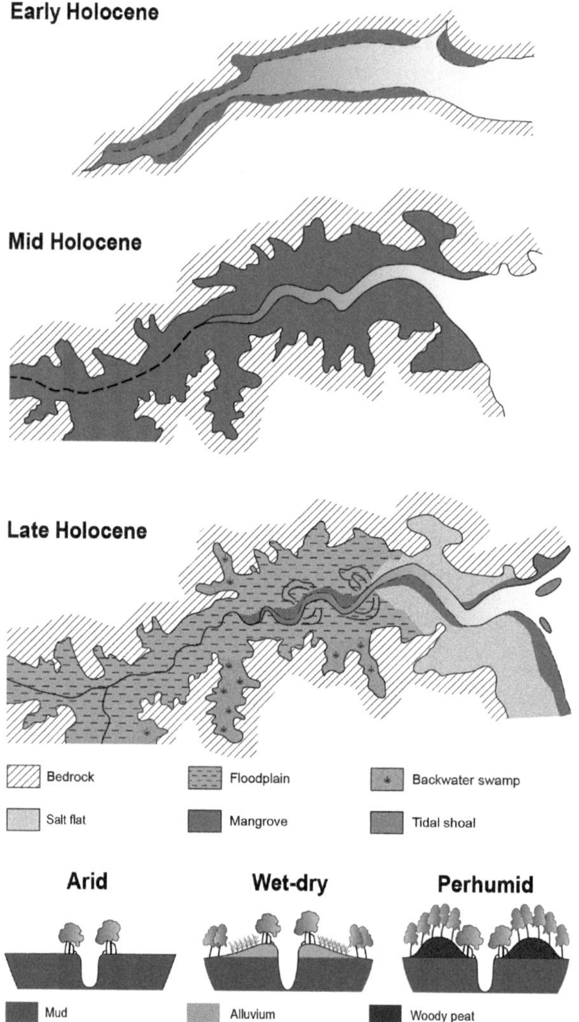

FIGURE 2.7 Three Holocene stages of estuarine evolution that occurred throughout much of northern Australia, but that can be inferred to have typified much of southeast Asia (Woodroffe, 1993). The morphology of the plains that have developed during the late Holocene depends on climate as shown by the schematic cross sections. In arid areas, a saline flat remains over mangrove sediments; in the wet-dry tropics (as in the Top End of Australia), seasonally inundated sedge-grassland dominates convex floodplains; whereas in the perhumid tropics, such as in Malaysia and Indonesia, domed peat occurs covered by peat swamp forest. *Based on Woodroffe, C.D., Chappell, J., Thom, B.G., Wallensky, E., 1989. Depositional model of a macrotidal estuary and floodplain, South Alligator River, Northern Australia. Sedimentology 36, 737–756; Woodroffe, C.D., Mulrennan, M.E., Chappell, J., 1993. Estuarine infill and coastal progradation, southern van Diemen Gulf, northern Australia. Sedimentary Geology 83, 257–275.*

and mangroves were widespread throughout many macrotidal estuaries for 1000 years or more in what has been called the "big swamp" (Woodroffe et al., 1985).

Mangrove deposits have been encountered in cores, and mangrove stumps occur along eroding banks in the estuary of the South Alligator system in particular (Woodroffe et al., 1985, 1989). Similar stratigraphy occurs in neighboring systems (Chappell, 1993; Woodroffe et al., 1993) and can be inferred for the Fitzroy and Ord rivers in Western Australia, as well as more generally along the north Australian coast (Jennings, 1975; Thom et al., 1975; Chappell and Thom, 1986; Crowley, 1996; Eliot and Eliot, 2013). Within the estuary where there is 10–15 m of "big swamp" mangrove mud, it appears that mangrove forests were able to keep up with the final decelerating rise in sea level. By contrast, along the adjacent coastal plain, the stratigraphy comprises a basal transgressive organic-rich mangrove mud overlying pre-Holocene terrestrial sediments. This is overlain by a shelly and muddy nearshore facies that in turn transitions upward into a regressive mangrove mud. In this case mangroves were initially drowned and subsequently reestablished indicating that sedimentation caught up with sea level (Woodroffe, 2018).

The nature of the sedimentary infill that has been deposited following sea level stabilization varies depending on the availability of material and the ability of the local environment to transport it, but typically it consists of either sands transported by wave or fluviotidal processes or muds associated with lower-energy mangrove environments (Semeniuk, 1985). Within the estuaries themselves, vertical accretion has continued, but pollen analysis indicates a transition from mangrove to freshwater vegetation dominated by grasses and sedges as alluvial sediments accreted to form the floodplain. In many cases mangrove forests had been replaced by ~4000 years BP, although in other places they persisted until more recently (Clark and Guppy, 1988). Simultaneously with the transition to freshwater plains, progradation at the coast has built out broad coastal plains.

Despite coeval transgressive-regressive stratigraphy, modern environments along the coast of northern Australia show considerable variability. Mangroves are generally of limited extent in the more arid environments of Western Australia (Semeniuk, 1996). High salinity restricts mangrove colonization, and the tidal mudflats that form accrete into salt flats. In those areas that receive sufficient rainfall, floodplains may continue to accrete, with levees developing that restrict tidal waters to the major channels during the dry season but impound freshwater on the floodplains during the wet season (Woodroffe et al., 1993). Subtle variations in topography across the plains (associated with low-elevation cheniers, scroll bars, and levees) can be very important in determining this balance; for example, the Mary River comprises a series of plains from which the wet season flood waters would have evaporated, without a fluvial connection to the sea until recent expansion of tidal systems reconnected the river to the sea (Mulrennan and Woodroffe, 1998a; Williams, 2014). Fluvial and tidal channel deposits associated with channel migration can also become abundant at these elevations if rates of channel migration are high enough and the supply of sediment is adequate, as shown by a series of deltas flanking the Gulf of Carpentaria, formed over this late Holocene period (Jones et al., 2003; Nanson et al., 2013). Tropical forests have replaced mangroves in the wettest regions of Queensland (Crowley and Gagan, 1995).

Distinct mangrove assemblages occur in particular geomorphological habitat settings along the coast of northern and northwestern Australia, and those mangroves along tidal creeks have been distinguished from those on the adjacent tidal flat and hinterland margin by Semeniuk (1985). Diversity varies, reducing from the seasonally wet in the north to the semiarid in the west. The influence of habitat type results from the different soil and groundwater regimes associated with each habitat, as these factors exert a strong effect on mangrove competition, and environmental gradients are reflected in corresponding species gradients (Semeniuk, 1985). The evolution of these landforms entails a change in the distribution of each habitat that will depend on the relative influence of different geomorphological processes. For example, alluvial fans are more extensive in those tidal embayments with greater fluvial inputs. Similarly, spits and cheniers are built more frequently in systems exposed to higher wave energies. Such patterns are likely to be associated with a corresponding shift in the distribution of mangrove assemblages, and the mangrove types that occur in these complex mangrove forests are more diverse than those summarized in the West Indian situation in Fig. 2.1.

The tidal river systems of northern Australia began as open embayments drowned by the rapidly rising postglacial sea level. Following sea level stabilization, the nature of sedimentation appears to have varied with distance from the coast. Seaward regions of the embayments filled with a mixture of shelly fine sand and mud, presumably delivered from offshore as a result of pumping by tidal currents (Chappell and Woodroffe, 1994; Wolanski, 2006a,b). Around 6000 years ago, many of the embayments contained substantial intertidal mangrove forests, interspersed with tidal channels (Fig. 2.7). These environments infilled rapidly with fine-grained sediment that appears to have been derived from offshore as the volume deposited far exceeds the estimated rate of fluvial sediment delivery (Chappell and Woodroffe, 1994). As sediments accreted and as indicated by the pollen record, lower intertidal species (*Sonneratia*) were replaced by mid-intertidal species (*Rhizophora/Ceriops*), in turn replaced by higher intertidal species (*Avicennia*) that were subsequently replaced by freshwater floodplain species (Woodroffe et al., 1986). In the case of the South Alligator and Daly rivers, the infilling of the mangrove swamps was also associated with the onset of channel migration throughout the estuarine and coastal plains, resulting in reworking of big swamp sediments and lateral accretion of laminated channel margin and shoal sediments (Chappell, 1993; Woodroffe et al., 1993). In smaller systems, the banks near the mouth may be reworked by waves, but fluvial discharges appear less efficient at eroding the banks.

In southeast Asia, a similar pattern of stratigraphy and chronology has been recorded (Woodroffe, 1993; Staub and Gastaldo, 2003; Hope, 2005; Proske et al., 2011; Li et al., 2012). Early studies of extensive peat swamp forests in the region had shown that these were formed on domed peat deposits in the wettest settings (Fig. 2.7) and that they had formed over mangrove muds (Anderson, 1964). Pollen analysis has also revealed the transition from mangrove to peat swamp forest and transition through successional stages of that forest as the peat has built up beneath the trees (Anderson and Muller, 1975). The Anderson model of peat swamp development is widely applicable in southeast Asia (Hope et al., 2004).

8. IMPACT OF FUTURE CLIMATE AND SEA LEVEL CHANGE

It is clear that mangrove forests have undergone, and survived, major geomorphological transformations during the Holocene, primarily as a consequence of global sea level rise. The paleoenvironmental record of past response to sea level change holds important lessons as to how these systems are likely to respond to future climate and sea level change. While temperature and carbon dioxide concentrations may have some effects on mangrove growth, this is already highly spatially variable and climate drivers seem likely to have minor impacts on ecosystem functioning (Saenger, 2002) and range extensions (Osland et al., 2016). Changes in storm intensity may affect mangrove distribution, but it seems clear that one of the most major impacts is likely to be from any acceleration of sea level rise (Semeniuk, 1994).

The stratigraphic record indicates that mangrove systems have survived rapid rates of sea level rise in the past, and it provides a rich archive of information on the nature of the response (Woodroffe, 1990; Ellison, 2015). At the ecosystem level, it is clear that mangroves do not face total extinction as a result of sea level rise, as even where such wetlands have been drowned by rapid phases of sea level rise during postglacial times, they can reestablish when conditions are again favorable. Stratigraphy may offer insights into the rates at which mangroves can keep up, as opposed to catch up, with sea level rise. Ellison and Stoddart (1991) considered that collapse of mangrove ecosystems might be imminent on small islands where there is not a large supply of inorganic sediment, but their view is not shared by other researchers familiar with these environments (Bacon, 1994; Snedaker et al., 1994).

The paleoenvironmental record could potentially yield insight into the critical rate of sea level rise beyond which mangrove systems change from "keep up" to "catch up," indicating a threshold beyond which they might lag future sea level rise. It is important, however, to note that the distinction is primarily about whether the seaward fringe has kept pace. Stratigraphy confirms that the most fundamental response of mangroves to sea level rise is incursion into more landward habitats. It will not always be the case that the space within which this can happen will be available in future. Loss of intertidal habitat where landward areas have been embanked or reclaimed for alternative uses is called coastal squeeze (Woodroffe, 1990; Delafontaine et al., 2000; Gilman et al., 2006; Phan et al., 2015). Perhaps less well investigated is the question of whether there is hinterland to invade. The presence of broad prograded plains, such as those developed on many coasts in the IWP region, implies that rapid sea level rise in these areas would lead to inundation of these plains. There has not been the detail of stratigraphic reconstruction to demonstrate whether such episodes have recurred in the past. It is clear that increased tidal processes will widen tidal channels (Wolanski and Chappell, 1996), but the intertidal systems can be anticipated to have only a limited ability to continue to build up and to accentuate the levees before sea level rise exceeds this capacity and widespread inundation of the plains occurs (Woodroffe, 2007). This has been likened to reestablishment of the big swamp.

In the case of salt marshes, stratigraphy of coastal plains around much of Europe and eastern North America contains interbedded marine muds and freshwater peats that are interpreted to record alternation of freshwater wetlands and more marine environments in the past (Allen, 2000). Managed realignment is already practiced in the United Kingdom, such reversion being promoted as a management strategy in the face of sea level rise (Wolters et al., 2005), and similar reintroduction of saline wetlands may become widespread elsewhere (Rogers et al., 2013). There will need to be much more research focused on the ways to manage mangrove wetlands. Preliminary efforts have focused on mangrove planting and design of dykes and protective structures to promote mangrove rehabilitation, for example, in the southern Mekong delta in Vietnam (e.g., Albers and Schmitt, 2015); but alternative designs yield different results on the predominantly erosional western shore of the delta (Thornton and Johnstone, 2015).

It is important to recognize that mangrove shorelines are highly sensitive to changes in sea level, and modifications in the distribution of mangroves may be one sign of sea level rise (Blasco et al., 1996). On the other hand, it is evident that mangroves are naturally dynamic and that substantial changes in mangrove distribution are already occurring as a part of the geomorphological behavior of these systems (Lucas et al., 2001; Nascimento et al., 2013). Mangrove incursion into more landward salt marsh habitats, and loss of salt marsh as a result, has been observed along a series of estuarine systems in southern New South Wales (Saintilan and Williams, 1999; Oliver et al., 2012), but this cannot be unequivocally attributed to sea level change alone and may be a more complex response to other anthropogenic pressures such as altered land use (Harty, 2004).

Widespread inundation of plains might occur under future rapid sea level rise (Lovelock et al., 2015), but more insidious might be saline incursion into freshwater wetlands through an expanding network of tidal creeks. Tidal creek systems are a component of the natural dynamics of these complex coastal systems, but increased salinization through creek extension is also occurring on several systems already. Freshwater wetlands in northern Australia are separated from tidal waters by grasslands, low chenier ridges, and channel levees that reach elevations only a few centimeters above high tide level. They act as barriers to saltwater intrusion, experiencing occasional inundation by storm surges and by seasonal freshwater floods (Woodroffe et al., 1993; Winn et al., 2006).

In recent decades, erosion of mangrove habitats and the growth of tidal channels have been recorded in several parts of northern Australia, including the Pilbara (Semeniuk, 1980, 1993) and Alligator River regions (Winn et al., 2006). Saline intrusion has occurred most dramatically on the Mary River plains (Knighton et al., 1991, 1992; Mulrennan and Woodroffe, 1998b). As the tidal channels extend and deepen, they create local increases in the tidal range and transfer saltwater to regions previously protected from tidal influence. This has resulted in the salinization of large areas of freshwater wetland, death of *Melaleuca*, and upstream extension of mangrove species. Channel networks have undergone exponential growth in terms of number and length, although the causes remain uncertain (Mulrennan and Woodroffe, 1998b; Williams, 2014). While it has not been demonstrated unambiguously that tidal creek extension is related to sea level change, these examples do demonstrate the speed with which saline intrusion can proceed by this mechanism, which is likely to become much more widespread as sea level rise accelerates.

Low-lying plains, such as those developed in much of the IWP region during the past few millennia, are likely to be subject to subsidence and compaction (Swales et al., 2018). This is accelerated in some regions through extraction of groundwater, which, for example, has been shown to have led to rapid subsidence beneath Bangkok (Phienwej and Nutalaya, 2005; Syvitski et al., 2009). Groundwater can influence other aspects of the behavior of wetlands (Perillo et al., 2005; Gallardo and Marui, 2006). Whether similar processes are involved in vegetation change on the relatively unpopulated plains of northern Australia is unclear, although direct evidence for this is weak. Nevertheless, detailed measurements of ground surface elevation in mangroves not only demonstrate considerable spatial variation in rates of sedimentation, with the most rapid potential rates accentuated by the presence of prop roots (Krauss et al., 2003, 2014), but also seem to imply that surface elevation changes are only partially a record of sediment deposition. Using controlled sedimentation tables, substrate elevation has been shown to also incorporate much more subtle changes of ground surface in response to groundwater changes within the substrate (Rogers et al., 2006, 2014).

These complexities imply that the response of mangrove shorelines to sea level change will differ between, or even within, systems (Semeniuk, 1994; Woodroffe, 1995; Woodroffe et al., 2016). There may be other subtle changes as well; for example, gap regeneration appears to be an important process in the cycle of mangrove tree replacement (Duke, 2001), although to differing extents in terms of which trees remain standing (Pinzón et al., 2003). These and other processes may adjust in response to other aspects of climate change, such as changes in storm frequency and intensity.

9. SUMMARY AND CONCLUDING REMARKS

Mangroves have often been interpreted as "trees that reclaim land from the sea," in view of their vivipary and unique aerial root systems (Carlton, 1974). This builds on ideas of Davis (1940) that mangrove peats in Florida overlie marine carbonate sediments. In fact, stratigraphic studies have indicated that in many cases these calcareous marls were laid down in more freshwater environments, as now found in the Everglades, and within the stratigraphy there is a complex pattern of initial transgression followed by subsequent regression (Parkinson, 1989).

Mangrove forests are some of the most conspicuously zoned of plant communities (Watson, 1928; Chapman, 1976), and this has often been interpreted in terms of a succession of species that might subsequently become replaced by a more landward, climax plant community. While pollen studies do provide evidence of such gradual replacement, for example, throughout much of the southern Asia and northern Australia region, reevaluation of areas in which these ideas were initially applied shows little apparent change over the past half a century (Alleng, 1998). In geomorphologically dynamic areas such as the Niger delta or the delta of the Grijalva in Mexico, mangroves appear to opportunistically colonize habitats (Allen, 1965; Thom, 1967). However, even where sediment supply might appear ample, mangrove shorelines can demonstrate stability and a morphological resilience, as in Sherbro Bay in West Africa, implying that they are in equilibrium with the hydrodynamics and sediment transport through the system (Anthony, 2004).

Geomorphologically, mangrove shorelines have adapted to past patterns of sea level change, some of which have occurred more rapidly than at rates occurring at present or anticipated in the immediate future. Mangrove systems are adjusted to substantial sediment loads, and sedimentation rates can be high and often spatially variable. Climate change threatens many ecosystems with conditions to which they are not adapted. Mangrove forests, by comparison, are relatively well adjusted to extreme environmental conditions, whether in terms of salinity tolerance or inundation by saltwater. Sea level rise, at the rates presently experienced or at accelerated rates being predicted, should not represent undue stress for mangrove forests. However, geomorphological adjustments are inevitable. Mangroves are generally well able to adapt to changing conditions, and the interactions with adjacent systems and the landward migration of mangroves may seem unexpected or undesirable. Under the most extreme conditions presently foreseeable, such as melting of the West Antarctic ice shelf, more rapid sea level rise may challenge mangrove systems and trigger more rapid changes than currently seen, but these seem unlikely to pose greater threats than are already caused by man. These anthropogenic pressures, combined with additional climate change, seem certain to lead to continued decline of mangrove ecosystems, unless the economic and ecological values of these systems are more broadly appreciated, and opportunities to preserve and extend their distribution by conservation and planting are undertaken with greater urgency.

References

Albers, T., Schmitt, K., 2015. Dyke design, floodplain restoration and mangrove co-management as parts of an area coastal protection strategy for the mud coasts of the Mekong Delta, Vietnam. Wetlands Ecology and Management 23, 991–1004.

Allen, J.R.L., 1965. Coastal geomorphology of eastern Nigeria: beach-ridge barrier islands and vegetated tidal flats. Geologie en Mijnbouw 44, 1–21.

Allen, J.R.L., 2000. Morphodynamics of Holocene salt marshes: a review sketch from the Atlantic and Southern North Sea coasts of Europe. Quaternary Science Reviews 19, 1155–1231.

Alleng, G.P., 1998. Historical development of the Port Royal Mangrove Wetland, Jamaica. Journal of Coastal Research 14, 951–959.

Alongi, D.M., 1998. Coastal Ecosystem Processes. CRC Press, Boca Raton, Florida, 419p.

Anderson, J.A.R., 1964. The structure and development of the peat swamps of Sarawak and Brunei. Journal of Tropical Geography 18, 7–16.

Anderson, J.A.R., Muller, J., 1975. Palynological study of a Holocene peat and a Miocene coal deposit from NW Borneo. Palaeobotany and Palynology 19, 291–351.

Anthony, E.J., 2004. Sediment dynamics and morphological stability of estuarine mangrove swamps in Shebro Bay, West Africa. Marine Geology 208, 207–224.

Anthony, E.J., Gardel, A., Gratiot, N., 2014. Fluvial sediment supply, mud banks, cheniers and the morphodynamics of the coast of South America between the Amazon and Orinoco river mouths. In: Martini, I.P., Wanless, H.R. (Eds.), Sedimentary Coastal Zones from High to Low Latitudes: Similarities and Differences. Geological Society, London, Special Publications, London, pp. 533–560.

Bacon, P.R., 1994. Template for evaluation of impacts of sea level rise on Caribbean coastal wetlands. Ecological Engineering 3, 171–186.

Ball, M.C., 1998. Mangrove species richness in relation to salinity and waterlogging: a case study along the Adelaide River floodplain, northern Australia. Global Ecology and Biogeography Letters 7, 73–82.

Blasco, F., Saenger, P., Janodet, E., 1996. Mangroves as indicators of coastal change. Catena 27, 167–178.

Bowman, H.H.M., 1917. Ecology and physiology of the red mangrove. Proceedings of the American Philosophical Society 56, 589–672.

Carlton, J.M., 1974. Land-building and stabilization by mangroves. Environmental Conservation 1, 285–294.

Cavanaugh, K.C., Kellner, J.R., Forde, A.J., Gruner, D.S., Parker, J.D., Rodriguez, W., Feller, I.C., 2014. Poleward expansion of mangroves is a threshold response to decreased frequency of extreme cold events. Proceedings of the National Academy of Sciences 111, 723–727.

Chapman, V.J., 1976. Mangrove Vegetation. J. Cramer, Germany, 447 p.

Chappell, J., 1993. Contrasting Holocene sedimentary geologies of lower Daly River, northern Australia, and lower Sepik-Ramu, Papua New Guinea. Sedimentary Geology 83, 339–358.

Chappell, J., Thom, B.G., 1986. Coastal morphodynamics in north Australia: review and prospect. Australian Geographical Studies 24, 110–127.

Chappell, J., Woodroffe, C.D., 1994. Macrotidal estuaries. In: Carter, R.G., Woodroffe, C.D. (Eds.), Coastal Evolution: Late Quaternary Shoreline Morphodynamics. Cambridge University Press, pp. 187–218.

Clark, R.L., Guppy, J.C., 1988. A transition from mangrove forest to freshwater wetland in the monsoon tropics of Australia. Journal of Biogeography 15, 665–684.

Cohen, M.C.L., Souza Filho, P.W.M., Lara, R.J., Behling, H., Angulo, R.J., 2005. A model of Holocene mangrove development and relative sea-level changes on Braganca Peninsula (northern Brazil). Wetlands Ecology and Management 13, 433–443.

Cohen, M.C.L., Lara, R.J., Cuevas, E., Oliveras, E.M., Stemberg, L.S., 2016. Effects of sea-level rise and climatic changes on mangroves from southwestern littoral of Puerto Rico during the middle and late Holocene. Catena 143, 187–200.

Cresswell, I.D., Semeniuk, V., 2011. Mangroves of the Kimberley Coast: ecological patterns in a tropical ria coast setting. Journal of the Royal Society of Western Australia 94, 213–237.

Crowley, G.M., 1996. Late quaternary mangrove distribution in Northern Australia. Australian Systematic Botany 9, 219–225.

Crowley, G.M., Gagan, M.K., 1995. Holocene evolution of coastal wetlands in wet-tropical northeastern Australia. The Holocene 5, 385–399.

Davis, J.H., 1940. The ecology and geologic role of mangroves in Florida. Papers from Tortugas Laboratory 32, 307–412.

Delafontaine, M.T., Flemming, B.W., Mai, S., 2000. The Wadden Sea squeeze as a cause of decreasing sedimentary organic loading. In: Flemming, B.W., Delafontaine, M.T., Liebezeit, G. (Eds.), Muddy Coast Dynamics and Resource Management. Elsevier, Amsterdam, pp. 273–286.

Douglas, I., 2005. The Mekong river basin. In: Gupta, A. (Ed.), The Physical Geography of Southeast Asia. Oxford University Press, pp. 193–218.

Duke, N.C., 1992. Mangrove floristics and biogeography. In: Alongi, D., Robertson, A. (Eds.), Tropical Mangrove Ecosystems. American Geophysical Union, Coastal and Estuarine Studies, pp. 63–100.

Duke, N.C., 2001. Gap creation and regeneration processes driving diversity and structure of mangrove ecosystems. Wetlands Ecology and Management 9, 257–269.

Duke, N.C., Ball, M.C., Ellison, J.C., 1998. Factors influencing biodiversity and distributional gradients in mangroves. Global Ecology and Biogeography Letters 7, 27–47.

Egler, F.E., 1952. Southeast saline everglades vegetation, Florida: and its management. Vegetatio 3, 213–265.

Eliot, M., Eliot, I., 2013. Interpreting estuarine change in northern Australia: physical responses to changing conditions. Hydrobiologia 708, 3–21.

Ellison, J.C., 2015. Vulnerability assessment of mangroves to climate change and sea-level rise impacts. Wetlands Ecology and Management 23, 115–137.

Ellison, J.C., Stoddart, D.R., 1991. Mangrove ecosystem collapse during predicted sea-level rise: Holocene analogues and implications. Journal of Coastal Research 7, 151–165.

Enos, P., Perkins, R.D., 1979. Evolution of Florida Bay from island stratigraphy. The Geological Society of America Bulletin 90, 59–83.

Ericson, J.P., Vorosmarty, C.J., Dingman, S.L., Ward, L.G., Meybeck, M., 2006. Effective sea- level rise and deltas: causes of change and human dimension implications. Global and Planetary Change 50, 63–82.

Finlayson, C.M., Woodroffe, C.D., 1996. Wetland vegetation of the Alligator Rivers region. In: Finlayson, C.M., von Oertzen, I. (Eds.), Landscape and Vegetation Ecology of the Kakadu Region, Northern Australia. Geobotany. Kluwer, Dordrecht, pp. 81–112.

Gallardo, A.H., Marui, A., 2006. Submarine groundwater discharge: an outlook of recent advances and current knowledge. Geo-Marine Letters 26, 102–113.

Gensac, E., Gardel, A., Lesourd, S., Brutier, L., 2015. Morphodynamic evolution of an intertidal mudflat under the influence of Amazon sediment supply – Kourou mud bank, French Guiana, South America. Estuarine, Coastal and Shelf Science 158, 53–62.

Geyh, M.A., Kudrass, H.-R., Streif, H., 1979. sea-level changes during the late Pleistocene and Holocene in the strait of Malacca. Nature 278, 441–443.

Gilman, E., van Lavieren, H., Ellison, J., Jungblut, V., Wilson, L., Areki, F., Brighouse, G., Bungitak, J., Dus, E., Henry, M., Kilman, M., Matthews, E., Sauni, I., Teariki-Ruata, N., Tukia, S., Yuknavage, K., 2006. Pacific Island mangroves in a changing climate and rising sea. UNEP Regional Seas Reports and Studies 179, 1–58.

Gischler, E., 2006. Comment on "Corrected western Atlantic sea-level curve for the last 11,000 years based on calibrated ^{14}C dates from Acropora palmata framework and intertidal mangrove peat" by Toscano and Macintyre. Coral Reefs 22:257–270 (2003), and their response in Coral Reefs 24:187–190(2005). Coral Reefs 25, 273–279.

Hamilton, S.E., Casey, D., 2016. Creation of a high spatio-temporal resolution global database of continuous mangrove forest cover for the 21st century (CGMFC-21). Global Ecology and Biogeography 25, 729–738.

Hamilton, S.E., Lovette, J., 2015. Ecuador's mangrove forest carbon stocks: a spatiotemporal analysis of living carbon holdings and their depletion since the advent of commercial aquaculture. PLoS One 10, e0118880. https://doi.org/10.1371/journal.pone.0118880.

Hanebuth, T., Stattegger, K., Grootes, P.M., 2000. Rapid flooding of the Sunda Shelf: a late-glacial sea-level record. Science 288, 1033–1035.

Hanebuth, T., Voris, H.K., Yokoyama, Y., Saito, Y., Okuno, J., 2011. Formation and fate of sedimentary depocentres on Southeast Asia's Sunda Shelf over the past sea-level cycle and biogeographic implications. Earth-Science Reviews 104, 92–110.

Harty, C., 2004. Planning strategies for mangrove and salt marsh changes in southeastern Australia. Coastal Management 32, 405–415.

Hendry, M., Digerfeldt, G., 1989. Paleogeography and palaeoenvironments of a tropical coastal wetland and offshore shelf during Holocen submergence, Jamaica. Palaeogeography, Palaeoclimatology, Palaeoecology 73, 1–10.

Hope, G., 2005. The quaternary in Southeast Asia. In: Gupta, A. (Ed.), The Physical Geography of Southeast Asia. Oxford University Press, pp. 24–37.

Hope, G., Kershaw, A.P., van der Kaars, S., Xiangjun, S., Liew, P.-M., Heusser, L.E., Takahara, H., McGlone, M., Miyoshi, N., Moss, P.T., 2004. History of vegetation and habitat change in the Austral-Asian region. Quaternary International 118–119, 103–126.

Hubbard, D.K., 1997. Reefs as dynamic systems. In: Birkeland, C. (Ed.), Life and Death of Coral Reefs. Chapman and Hall, New York, pp. 43–67.

Jennings, J.N., 1975. Desert dunes and estuarine fill in the Fitzroy estuary, north-western Australia. Catena 2, 215–262.

Jones, B.G., Woodroffe, C.D., Martin, G.R., 2003. Deltas in the Gulf of Carpentaria, Australia: forms, processes and products. In: Sidi, F.H., Nummedal, D., Imbert, P., Darman, H., Posamentier, H.W. (Eds.), Tropical Deltas of Southeast Asia: Sedimentology, Stratigraphy, and Petroleum Geology. SEPM Special Publication, Tulsa, OK, pp. 21–43.

Kamaludin, B.H., Nakamura, T., Price, D.M., Woodroffe, C.D., Fujii, S., 1993. Radiocarbon and thermoluminescence dating of the Old Alluvium from a coastal site in Perak, Malaysia. Sedimentary Geology 83, 199–210.

Khan, N., Ashe, E., Horton, B.P., Dutton, A., Kopp, R.E., Brocard, G., Engelhart, S.E., Hill, D.F., Peltier, W.R., Vane, C.H., Scatena, F.N., 2017. Drivers of Holocene sea-level change in the Caribbean. Quaternary Science Reviews 155, 13–36.

Knighton, A.D., Mills, K., Woodroffe, C.D., 1991. Tidal creek extension and saltwater intrusion in northern Australia. Geology 19, 831–834.

Knighton, A.D., Woodroffe, C.D., Mills, K., 1992. The evolution of tidal creek networks, Mary River, Northern Australia. Earth Surface Processes and Landforms 17, 167–190.

Krauss, K.W., Allen, J.A., Cahoon, D.R., 2003. Differential rates of vertical accretion and elevation change among aerial root types in Micronesian mangrove forests. Estuarine, Coastal and Shelf Science 56, 251–259.

Krauss, K.W., McKee, K.L., Lovelock, C.E., Cahoon, D.R., Saintilan, N., Reef, R., Chen, L., 2014. How mangrove forests adjust to rising sea level. New Phytologist 202, 19–34.

Lambeck, K., Chappell, J., 2001. Sea level change through the last glacial cycle. Science 292, 679–686.

Lewis, S.E., Sloss, C.R., Murray- Wallace, C.V., Woodroffe, C.D., Smithers, S.G., 2013. Post-glacial sea-level changes around the Australian margin: a review. Quaternary Science Reviews 75, 1–24.

Li, Z., Saito, Y., Mao, L., Tamura, T., Li, Z., Song, B., Zhang, Y., Lu, A., Sieng, S., Li, J., 2012. Mid-Holocene mangrove succession and its response to sea-level change in the upper Mekong River delta, Cambodia. Quaternary Research 78, 386–399.

Lovelock, C.E., Cahoon, D.R., Friess, D.A., Guntenspergen, G.R., Kruass, K.W., Reef, R., Rogers, K., Saunders, M.L., Sidik, F., Swales, A., Saintilan, N., Thuyen, L.X., Triet, T., 2015. The vulnerability of Indo-Pacific mangrove forests to sea-level rise. Nature 526, 559–563.

Lovelock, C.E., Sorrell, B., Hancock, N., Hua, Q., Swales, A., 2010. Mangrove forest and soil development on a rapidly accreting shore in New Zealand. Ecosystems 13, 437–451.

Lucas, R., Ellison, J.C., Mitchell, A., Donelly, B., Finlayson, M., Milne, A.K., 2001. Use of stereo aerial photography for quantifying changes in the extent and height of mangroves in tropical Australia. Wetlands Ecology and Management 10, 161–175.

Lugo, A.E., Snedaker, S.C., 1974. The ecology of mangroves. Annual Review of Ecology and Systematics 5, 39–64.

Lugo, A.E., Sell, M., Snedaker, S.C., 1976. Mangrove ecosystem analysis. In: Patten, B.C. (Ed.), Systems Analysis and Simulation in Ecology. Academic Press, New York, pp. 113–145.

Macnae, W., 1969. A general account of the fauna and flora of mangrove swamps and forests in the Indo-West-Pacific region. Advances in Marine Biology 6, 73–270.

McKee, K.L., Cahoon, D.R., Feller, I.C., 2007. Caribbean mangroves adjust to rising sea level through biotic controls on change in soil elevation. Global Ecology and Biogeography 16, 545–556.

Milliman, J.D., Syvitski, J.P.M., 1992. Geomorphic/tectonic control of sediment discharge to the ocean: the importance of small mountainous rivers. The Journal of Geology 100, 525–544.

Milne, G.A., Peros, M., 2013. Data–model comparison of Holocene sea-level change in the circum-Caribbean region. Global and Planetary Change 107, 119–131.

Morley, R.J., 1981. Development and vegetation dynamics of a lowland ombrogenous peat swamp in Kalimantan Tengah, Indonesia. Journal of Biogeography 8, 383–404.

Mulrennan, M.E., Woodroffe, C.D., 1998a. Holocene development of the lower Mary River plains, Northern Territory, Australia. The Holocene 8, 565–579.

Mulrennan, M.E., Woodroffe, C.D., 1998b. Saltwater intrusion into coastal plains of the Lower Mary River, Northern Territory, Australia. Journal of Environmental Management 54, 169–188.

Murray-Wallace, C.V., Woodroffe, C.D., 2014. Quaternary Sea-Level Changes: A Global Perspective. Cambridge Universitry Press, Cambridge.

Nanson, R.A., Valcarelov, B.K., Ainsworth, R.B., Williams, F., Price, D., 2013. Evolution of a Holocene, mixed-process, forced regressive shoreline: the Mitchell River delta, Queensland, Australia. Marine Geology 339, 22–43.

Nascimento, W.R., Souza Filho, P.W.M., Proisy, C., Lucas, R.M., Rosenqvist, A., 2013. Mapping changes in the largest continuous Amazonian mangrove belt using object-based classification of multisensor satellite imagery. Estuarine, Coastal and Shelf Science 117, 83–93.

Neumann, A.C., Macintyre, I., 1985. Reef response to sea level rise: keep-up, catch-up or give-up. Proceedings of the 5th International Coral Reef Congress 3, 105–110.

Oliver, T., Rogers, K., Woodroffe, C.D., 2012. Measuring, mapping and modelling: an integrated approach to the management of mangrove and saltmarsh in the Minnamurra River estuary, southeast Australia. Wetlands Ecology and Management 20, 353–371.

Osland, M.J., Enwright, N.M., Day, R.H., Gabler, C.A., Stagg, C.L., Grace, J.B., 2016. Beyond just sea-level rise: considering macroclimatic drivers within coastal wetland vulnerability assessments to climate change. Global Change Biology 22, 1–11.

Parkinson, R.W., 1989. Decelerating Holocene sea-level rise and its influence on southwest Florida coastal evolution: a transgressive/regressive stratigraphy. Journal of Sedimentary Petrology 59, 960–972.

Perillo, G.M.E., Minkoff, D.R., Piccolo, M.C., 2005. Novel mechanism of stream formation in coastal wetlands by crab-fish-groundwater interaction. Geo-Marine Letters 25, 214–220.

Phan, L.K., van Thiel de Vries, J.S.M., Stive, M.J.F., 2015. Coastal mangrove squeeze in the Mekong delta. Journal of Coastal Research 31, 233–243.

Phienwej, N., Nutalaya, P., 2005. Subsidence and flooding in Bangkok. In: Gupta, A. (Ed.), The Physical Geography of Southeast Asia. Oxford University Press, pp. 358–378.

Pinzón, Z.S., Ewel, K.C., Putz, F.E., 2003. Gap formation and forest regeneration in a Micronesian mangrove forest. Journal of Tropical Ecology 19, 143–153.

Proske, U., Hanebuth, T.J.J., Gröger, J., Diem, B.P., 2011. Late Holocene sedimentary and environmental development of the northern Mekong River Delta, Vietnam. Quaternary International 230, 57–66.

Ramcharan, E.K., 2004. Mid-to-late Holocene sea level influence on coastal wetland development in Trinidad. Quaternary International 120, 145–151.

Robertson, A.I., Alongi, D.M., 1992. Tropical mangrove ecosystems. In: Coastal and Estuarine Studies. American Geophysical Union, Washington, DC, 330 p.

Rogers, K., Saintilan, N., Copeland, C., 2013. Managed retreat of saline coastal wetlands: challenges and opportunities identified from the Hunter River Estuary, Australia. Estuaries and Coasts 37, 67–78.

Rogers, K., Saintilan, N., Woodoffe, C.D., 2014. Surface elevation change and vegetation distribution dynamics in a subtropical coastal wetland: implications for coastal wetland response to climate change. Estuarine, Coastal and Shelf Science 149, 46–56.

Rogers, K., Wilton, K.M., Saintilan, N., 2006. Vegetation change and surface elevation dynamics in estuarine wetlands of southeast Australia. Estuarine, Coastal and Shelf Science 66, 559–569.

Saenger, P., 2002. Mangrove Ecology, Silviculture and Conservation. Kluwer, Dordrecht, 360 p.

Saintilan, N., Williams, R.J., 1999. Mangrove transgression into saltmarsh environments in south-east Australia. Global Ecology and Biogeography 8, 117–124.

Saintilan, N., Wilson, N.C., Rogers, K., Rajkaran, A., Krauss, K.W., 2014. Mangrove expansion and salt marsh decline at mangrove poleward limits. Global Change Biology 20, 147–157.

Scholl, D.W., Stuiver, M., 1967. Recent submergence of southern Florida: a comparison with adjacent coasts and other eustatic data. The Geological Society of America Bulletin 78, 437–454.

Semeniuk, V., 1980. Mangrove zonation along an eroding coastline in King Sound, northwestern Australia. Journal of Ecology 68, 789–812.

Semeniuk, V., 1985. Development of mangrove habitats along ria shorelines in north and northwestern tropical Australia. Vegetatio 60, 3–23.

Semeniuk, V., 1993. The Pilbara Coast: a riverine coastal plain in a tropical arid setting, northwestern Australia. Sedimentary Geology 83, 235–256.

Semeniuk, V., 1994. Predicting the effect of sea-level rise on mangroves in northwestern Australia. Journal of Coastal Research 10, 1050–1076.

Semeniuk, V., 1996. Coastal forms and Quaternary processes along the arid Pilbara coast of northwestern Australia. Palaeogeography, Palaeoclimatology, Palaeoecology 123, 49–84.

Snedaker, S.C., Meeder, J.F., Ross, M.S., Ford, R.G., 1994. Discussion of Ellison, J.C. and Stoddart, D.R., 1991. Mangrove ecosystem collapse during predicted sea-level rise: Holocene analogues and implications. Journal of Coastal Research 7 (1), 151–165. Journal of Coastal Research 10, 497–498.

Staub, J.R., Esterle, J.S., 1994. Peat-accumulating depositional systems of Sarawak, east Malaysia. Sedimentary Geology 89, 91–106.

Staub, J.R., Gastaldo, R.A., 2003. Late quaternary sedimentation and peat development in the Rajang River Delta, Sarawak, East Malaysia. In: Sidi, F.H., Nummedal, D., Imbert, P., Darman, H., Posamentier, H.W. (Eds.), Tropical Deltas of Southeast Asia: Sedimentology, Stratigraphy, and Petroleum Geology. SEPM Special Publication, Tulsa, OK, pp. 71–87.

Stoddart, D.R., 1980. Mangroves as successional stages, inner reefs of the northern Great Barrier Reef. Journal of Biogeography 7, 269–284.

Swales, A., Bentley, S.J., Lovelock, C.E., 2015. Mangrove-forest evolution in a sediment-rich estuarine system: opportunists or agents of geomorphic change? Earth Surface Processes and Landforms 40, 1672–1687.

Swales, A., Denys, P., Pickett, V.I., Lovelock, C.E., 2018. Evaluating deep subsidence in a rapidly-accreting mangrove forest using GPS monitoring of surface-elevation benchmarks and sedimentary records. Marine Geology (in press).

Syvitski, J.P.M., Vörösmarty, C.J., Kettner, A.J., Green, P., 2005. Impact of humans on the flux of terrestrial sediment to the global coastal ocean. Science 308, 376–380.

Syvitski, J.P.M., Kettner, A.J., Overeem, I., Hutton, E.W.H., Hannon, M.T., Brakenridge, G.R., Day, J., Vorosmarty, C.J., Saito, Y., Giosan, L., Nicholls, R.J., 2009. Sinking deltas due to human activities. Nature Geoscience 2, 681–686.

Thom, B.G., 1967. Mangrove ecology and deltaic geomorphology: Tabasco, Mexico. Journal of Ecology 55, 301–343.

Thom, B.G., 1982. Mangrove ecology: a geomorphological perspective. In: Clough, B.F. (Ed.), Mangrove Ecosystems in Australia, Structure, Function and Management. ANU Press, Canberra, pp. 3–17.

Thom, B.G., Wright, L.D., Coleman, J.M., 1975. Mangrove ecology and deltaic-estuarine geomorphology, Cambridge Gulf-Ord River, Western Australia. Journal of Ecology 63, 203–222.

Thornton, S.R., Johnstone, R.W., 2015. Mangrove rehabilitation in high erosion areas: assessment using bioindicators. Estuarine, Coastal and Shelf Science 165, 176–184.

Toscano, M.A., Macintyre, I.G., 2003. Corrected western Atlantic sea-level curve for the last 11,000 years based on calibrated ^{14}C dates from *Acropora palmata* framework and intertidal mangrove peat. Coral Reefs 22, 257–270.

Twilley, R.R., 1985. The exchange of organic carbon in basin mangrove forests in a southwest Florida estuary. Estuarine, Coastal and Shelf Science 20, 543–557.

Twilley, R.R., Rivera-Monroy, V.H., 2005. Developing performance measures of mangrove wetlands using simulation models of hydrology, nutrient biogeochemistry, and community dynamics. Journal of Coastal Research 40, 79–93. Special Issue.

Twilley, R.R., Rivera-Monroy, V.H., 2009. Ecogeomorphic models of nutrient biogeochemistry for mangrove wetlands. In: Perillo, G.M.E., Wolanski, E., Cahoon, D.R., Brinson, M.M. (Eds.), Coastal Wetlands: An Integrated Ecosystem Approach. Elsevier, Amsterdam, pp. 641–683.

Verstappen, H.T., 1975. On paleoclimates and landform development in Malesia. Modern Quaternary Research in Southeast Asia 1, 3–35.

Walsh, J.P., Nittrouer, C.A., 2004. Mangrove-bank sedimentation in a mesotidal environment with large sediment supply, Gulf of Papua. Marine Geology 208, 225–248.

Watson, J.G., 1928. Mangrove forests of the Malay Peninsula. Malayan Forestry Records 6, 1–275.

Williams, D., 2014. Recent, rapid evolution of the lower Mary River estuary and flood plains. In: Wolanski, E. (Ed.), Estuaries of Australia in 2050 and beyond. Springer Science, pp. 277–287.

Winn, K.O., Saynor, M.J., Eliot, M.J., Eliot, I., 2006. Saltwater intrusion and morphological change at the mouth of the East Alligator River, Northern Territory. Journal of Coastal Research 22, 137–149.

Wolanski, E., 2006a. The evolution time scale of macro-tidal estuaries: examples from the Pacific Rim. Estuarine, Coastal and Shelf Science 66, 544–549.

Wolanski, E., 2006b. The sediment trapping efficiency of the macro-tidal Daly Estuary, tropical Australia. Estuarine, Coastal and Shelf Science 69, 291–298.

Wolanski, E., Chappell, J., 1996. The response of tropical Australian estuaries to a sea level rise. Journal of Marine Systems 7, 267–279.

Wolanski, E., Moore, K., Spagnol, S., D'Adamo, N., Pattiaratchi, C., 2001. Rapid, human-induced siltation of the macro-tidal Ord River Estuary, Western Australia. Estuarine, Coastal and Shelf Science 53, 717–732.

Wolters, M., Bakker, J.P., Bertness, M.D., Jefferies, R.L., Möller, I., 2005. Saltmarsh erosion and restoration in south-east England: squeezing the evidence requires realignment. Journal of Applied Ecology 42, 844–851.

Woodroffe, C.D., 1981. Mangrove swamp stratigraphy and Holocene transgression, Grand Cayman Island, West Indies. Marine Geology 41, 271–294.

Woodroffe, C.D., 1990. The impact of sea-level rise on mangrove shorelines. Progress in Physical Geography 14, 483–520.

Woodroffe, C.D., 1992. Mangrove sediments and geomorphology. In: Alongi, D., Robertson, A. (Eds.), Tropical Mangrove Ecosystems. American Geophysical Union, Coastal and Estuarine Studies, pp. 7–41.

Woodroffe, C.D., 1993. Late Quaternary evolution of coastal and lowland riverine plains of Southeast Asia and northern Australia: an overview. Sedimentary Geology 83, 163–173.

Woodroffe, C.D., 1995. Response of tide-dominated mangrove shorelines in northern Australia to anticipated sea-level rise. Earth Surface Processes and Landforms 20, 65–85.

Woodroffe, C.D., 2007. Critical thresholds and the vulnerability of Australian tropical coastal ecosystems to the impacts of climate change. Journal of Coastal Research 50, 469–473. Special Issue.

Woodroffe, C.D., 2018. Mangrove response to sea-level rise: palaeo-ecological insights from macrotidal systems in northern Australia. Marine and Freshwater Research 69, 917–932.

Woodroffe, C.D., Thom, B.G., Chappell, J., 1985. Development of widespread mangrove swamps in mid-Holocene times in northern Australia. Nature 317, 711–713.

Woodroffe, C.D., Grindrod, J., 1991. Mangrove biogeography: the role of Quaternary environmental and sea-level fluctuations. Journal of Biogeography 18, 479–492.

Woodroffe, C.D., Chappell, J., Thom, B.G., Wallensky, E., 1986. Geomorphological Dynamics and Evolution of the South Alligator River and Plains. N.T. North Australia Research Unit Monograph. ANU Press.

Woodroffe, C.D., Chappell, J., Thom, B.G., Wallensky, E., 1989. Depositional model of a macrotidal estuary and floodplain, South Alligator River, Northern Australia. Sedimentology 36, 737–756.

Woodroffe, C.D., Mulrennan, M.E., Chappell, J., 1993. Estuarine infill and coastal progradation, southern van Diemen Gulf, northern Australia. Sedimentary Geology 83, 257–275.

Woodroffe, C.D., Nicholls, R.J., Saito, Y., Chen, Z., Goodbred, S.L., 2006. Landscape variability and the response of Asian megadeltas to environmental change. In: Harvey, N. (Ed.), Global Change and Integrated Coastal Management: The Asia-Pacific Region. Springer, London, New York, pp. 277–314.

Woodroffe, C.D., Lovelock, C.E., Rogers, K., 2014. Mangrove shorelines. In: Masselink, G., Gehrels, R. (Eds.), Coastal Environments and Global Change. John Wiley and Sons, Ltd., pp. 251–267

Woodroffe, C.D., Rogers, K., McKee, K.L., Lovelock, C.E., Mendelssohn, I.A., Saintilan, N., 2016. Mangrove sedimentation and response to relative sea-level rise. Annual Review of Marine Science 8, 243–266.

Yokoyama, Y., De Deckker, P., Lambeck, K., Johnston, P., Fifield, L.K., 2001. Sea-level at the Last Glacial Maximum: evidence from northwestern Australia to constrain ice volumes for oxygen isotope stage 2. Palaeogeography, Palaeoclimatology, Palaeoecology 165, 281–297.

CHAPTER

3

Temperate Coastal Wetlands: Morphology, Sediment Processes, and Plant Communities

Paula Pratolongo[1], *Nicoletta Leonardi*[2], *Jason R. Kirby*[3], *Andrew Plater*[2]

[1]Universidad Nacional del Sur, Dto. de Biología Bioquímica y Farmacia CONICET, Instituto Argentino de Oceanografía, Bahía Blanca, Argentina; [2]University of Liverpool, School of Environmental Sciences, Department of Geography and Planning, Liverpool, United Kingdom; [3]Liverpool John Moores University, School of Natural Sciences & Psychology, Liverpool, United Kingdom

1. INTRODUCTION

Temperate climates of the Earth are characterized by relatively moderate mean annual temperatures, with average monthly temperatures above 10°C in their warmest months and above −3°C in their colder months (Trewartha and Horn, 1980). Most regions with a temperate climate present four seasons, and temperatures can change greatly between summer and winter (McColl, 2005). Most people live in temperate zones, and human population densities in coastal regions are about three times higher than the global average (Small and Nicholls, 2003). Globally, nearly all temperate coastal regions experienced net immigration during the last century (Neumann et al., 2015), and the increasing population associated with rapid economic growth (Hugo, 2011; Smith, 2011) has led to extensive conversion of natural coastal wetlands to agriculture, aquaculture, and silviculture, as well as industrial and residential uses (Valiela, 2006).

Humans both influence and depend on the extensive ecosystem services that temperate coastal wetlands provide, sustained by biological populations and their dynamic interaction with physical and chemical properties of the environment (Kirwan and Megonigal, 2013). The interaction of biota, hydrology, and sediments is clearly evident in the ecological and geomorphologic characteristics of temperate coastal wetlands. Living organisms respond to abiotic

Coastal Wetlands
https://doi.org/10.1016/B978-0-444-63893-9.00003-4

factors such as tidal inundation, climate, groundwater, accommodation space (the space between sediment surface and mean sea level), and sediment dynamics, as well as human interventions. The paleoenvironmental records preserved in coastal wetlands provide both ecological and chronological information on their evolution in response to many of these factors and reveal valuable information concerning past climate change, vegetation history, paleohydrology, sea level trends, and alteration by human activities (Tooley, 1986).

Coastal salt marshes are the dominant land cover type in the intertidal zone of many low energy temperate coasts (see Adam, 1990; Allen and Pye, 1992), instead of the extensive mangroves that characterize tropical latitude coasts (Odum et al., 1982). Salt marsh plants, mostly herbaceous halophytes, are adapted to regular inundation by saltwater, especially those species occupying the lowest portion of a marsh. In this zone, the vegetation type and distribution is controlled primarily by hydroperiod, defined as the product of the frequency and duration of tidal flooding (French, 1993). At higher elevations where tidal inundation becomes less frequent, evaporation is more intense and a different type of vegetation, composed of succulents, salt excreters, and salt excluders, may appear in response to salt stress (Cronk and Fenessy, 2001). In arid and semiarid coastal locations, high evaporation rates may produce muddy salt flats instead of a typical high marsh and, at the other end of the scale, large volumes of freshwater discharging into the upper salt marsh may support less salt-tolerant wetland plants, such as *Phragmites* reeds.

A broader definition of coastal wetlands includes a wide variety of environments, covering a spatial extent that may spread over several kilometers, from tidal flats, salt marshes, and mangroves to freshwater tidal and nontidal wetlands forming a transition to terrestrial systems, whose hydrology is still influenced by mean sea level. Hageman (1969) first used the term *perimarine* zone to refer to the area where freshwater wetlands persist under the control of relative mean sea level. In humid climates, freshwater seepage and high groundwater levels provide the waterlogged conditions necessary for the development and persistence of perimarine freshwater swamps, nontidal marshes, and fens (Gardner et al., 2000; Waller et al., 1999). In the perimarine zone of arid climates, under a combination of high evaporation and low freshwater inputs, soils develop extremely high salinities that eliminate all but the most tolerant plants (Fig. 3.1).

Tidal flats are intertidal, nonvegetated, soft sediment habitats (Dyer et al., 2000), globally distributed along protected, gentle sloping coastlines. In the temperate zone, tidal flats are often backed by salt marshes that grow at higher elevations within the intertidal fringe. Extensive tidal flats are commonly associated with deltas and estuaries. At the mouth of the Amazon River, riverine muds are carried northward by longshore currents and form nearshore mud banks that migrate along the coast of French Guiana (Eisma et al., 1991). In the Chinese coast, the massive suspended sediment load carried by the Yellow River supports extensive mudflats (Wang, 1983), especially in the Jiangsu province, known for having the widest mudflats on Earth (Chen et al., 2016). Tidal flats are also conspicuous coastal features in sheltered embayments of many estuaries in northwestern Europe, including the British Isles, France, and the Netherlands (Flemming, 2002; Carling et al., 2009). In the Wadden Sea, wide mudflats develop in sheltered environments behind the Frisian Islands. In this case, sediments are mainly provided by the North Sea rather than river input (Pejrup et al., 1997). Tidal flats comprise roughly 7% of total coastal shelf areas (Stutz and Pilkey, 2002), but they have a primary role in the recycling of organic matter and nutrients from both terrestrial and marine sources (Mann, 2009).

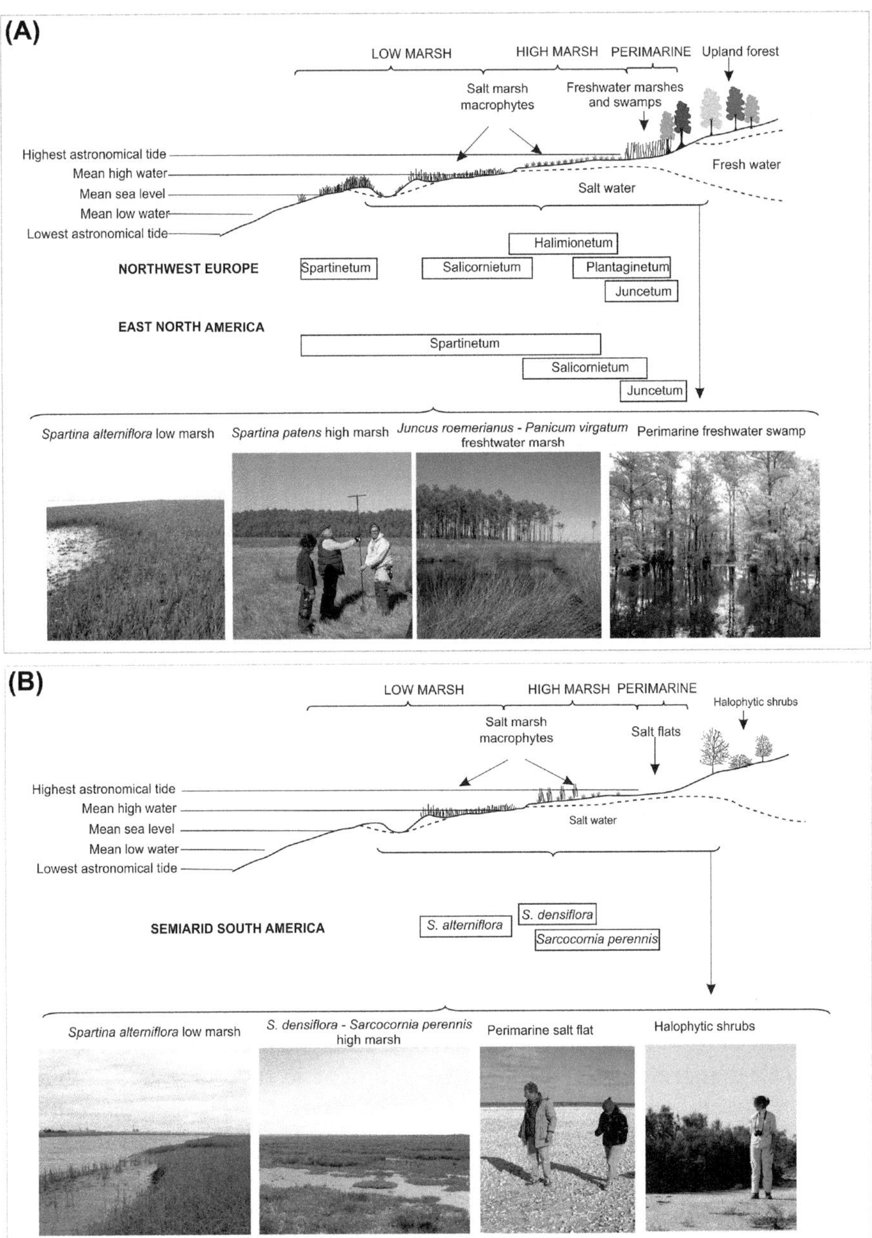

FIGURE 3.1 (A) Species distribution in temperate coastal wetlands of northwestern Europe and eastern North America, under humid climates. (B) Species distribution in temperate coastal wetlands of the Bahia Blanca Estuary, southeastern South America, under semiarid climate. *(A) After Carter, 1988. (B) After Pratolongo, P., Piovan, M.J., Cuadrado, D.G., 2016. Coastal environments in the Bahía Blanca estuary, Argentina. In: Khan, A.M., Boër, B., Özturk, M., Clüsener-Godt, M., Gul, B., Breckle, S.-W. (Eds.), Sabkha Ecosystems Volume V: The Americas. Springer International Publishing, Geneva, Switzerland, pp. 205–224. Photographs by Paula Pratolongo.*

Tidal freshwater marshes are a distinctive type of coastal wetland, restricted to particular locations within the tidal portion of some rivers, where there is bidirectional flow of freshwater. These conditions only occur when there is sufficient freshwater flow from a river, combined with significant tidal amplitude, commonly magnified by geomorphological constrains (Odum et al., 1984). Tidal freshwater marshes have been poorly studied, and it is difficult to assess their worldwide extent (Whigham et al., 2009). In North America, these marshes commonly appear along the temperate humid Atlantic coast, from southern New England to Florida (Odum et al., 1984). They were also common in northwestern Europe, but most of the original tidal freshwater habitats have disappeared after centuries of human activities (Barendregt et al., 2006). Extensive tidal freshwater marshes have been described on the Atlantic coast of South America, associated to the Paraná River Delta (Kandus and Malvárez, 2004; Pratolongo et al., 2007, 2008), and large marsh areas dominated by *Phragmites australis* occupy the tidal portion of the Yangtze River, China (Zhang et al., 2014). The presence of tidal freshwater marshes in other regions is less documented.

In the following sections, we aim to illustrate the main geomorphic and sedimentary processes shaping temperate coastal wetlands, explore the factors that control plant zonation and their change over time, and describe their geographic variation globally. The impacts of climate change are discussed along with new developments in numerical modeling of salt marsh development and response to sea level rise. Lastly, modifications by human activities are presented.

2. FACTORS CONTROLLING SALT MARSH DEVELOPMENT

If we consider salt marshes and contiguous tidal flats as a slope or ramp that increases in elevation landward, there is an evident decrease in both the frequency and duration of tidal inundation (hydroperiod) as elevation of the sediment surface approaches that of the highest tides (French, 1993). Inundation duration is the period of immersion either side of the high tide, whereas the frequency of inundation is a response to temporal variations in tidal range. Essentially, the higher the sediment surface elevation, the fewer the number of tides that are high enough to accomplish immersion. In this "ramp model," the deposition of tidal sediments in relation to hydroperiod is perhaps best considered with reference to the phase lag between idealized sinusoids of flow velocity and tidal height (Rahman and Plater, 2014). An increase in salt marsh elevation through net sediment accretion reduces both hydroperiod and incremental sediment accretion (e.g., Pethick, 1981). This simplistic ramp model is influenced by vegetation in a number of ways:

Increased bed roughness results from the presence of leaves and stems, thus increasing frictional drag and near-bed turbulence and enhancing sediment deposition potential (see Leonard and Luther, 1995; Leonard and Croft, 2006). Increasing biomass also increases the surface area available for sediment capture, e.g., leaf surfaces (e.g., Stumpf, 1983), and the presence of roots operates to bind and further protect sediment once it has been deposited (Garofalo, 1980). A protective binding function is also provided by filamentous algae that cover the mudflat surface, thus providing the initial stability for halophytes to become established in the first instance (Underwood and Paterson, 1993).

The result of the environmental gradients imposed by elevation is to produce a shore-parallel zonation of plants which, in reality, is made more complex and spatially variable by the micromorphology of the salt marsh surface. For example, enhanced sediment accretion has been observed on hummocks covered by *Puccinellia maritima* (Langlois et al., 2003). These local increases in sedimentation may be a function of vegetation height and stem density (Boorman et al., 1998). Another aspect of micromorphology is the drainage capability of the salt marsh surface, where subhorizontal sediment surfaces encourage the storage of salt-water in the form of salt pans. These may be observed as patches of bare mud, particularly in the mid- to upper salt marsh, where the enhanced salinity induced by evaporation from the pan leads to die back.

Micromorphology is further reflected in the "creek" model of salt marsh sedimentation, whereby the network of drainage channels that cross the salt marsh acts to capture the incoming tide. The rising tide then only spills onto much of the salt marsh surface when it exceeds the capacity of the creeks. The overspilling waters tend to deposit the majority of their sediment load in close proximity to channel margins. Furthermore, an observed increase in sediment grain size with proximity to the creek margin is a function of the competence of the "over-creek" flood waters decreasing with distance from the creek margin (Christiansen et al., 2000; Leonard et al., 2002; Temmerman et al., 2004). Consequently, levees develop on creek banks where more of the sediment load is deposited. Process observations and sedimentary records support the viability of both models for salt marsh accretion, but their operational scales in time and space are very different: the ramp model accounting for widespread and gradual vertical accretion, the creek model being more focused on the local scale; and potentially rapid development of three-dimensional sedimentary features (French and Spencer, 1993; French et al., 1995).

While the ramp and creek models provide a highly simplified view of the landscape, salt marsh evolution and accretion depend on several factors, such as sediment availability, vegetation cover, salt marsh platform elevation, and hydrodynamic forces (e.g., Fagherazzi et al., 2012; Kirwan et al., 2016; Leonardi et al., 2016). Fig. 3.2 represents a sketch of the physical and ecological processes that many models aim to reproduce by using different levels of simplification or by only focusing on specific dynamics (e.g., accretion, lateral migration, or a combination of both).

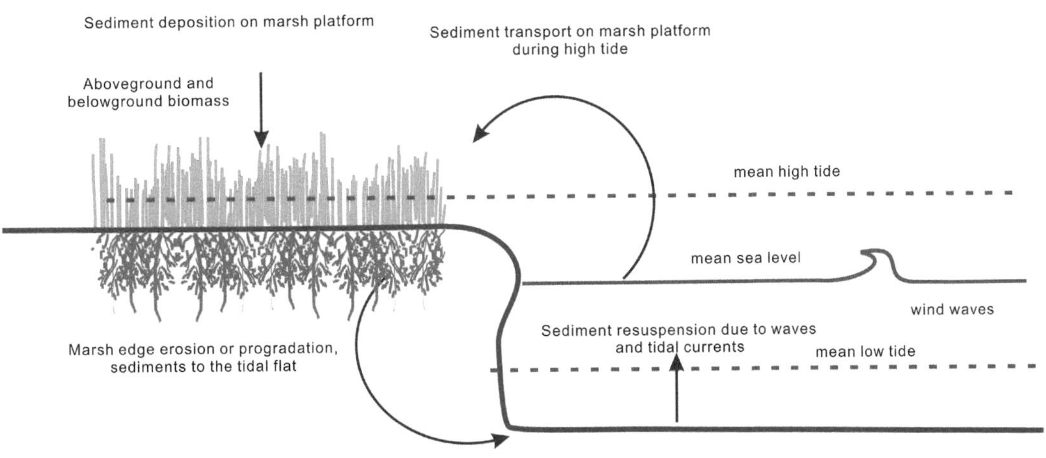

FIGURE 3.2 Mechanisms and sediment fluxes responsible for salt marsh vertical and horizontal dynamics.

2.1 Models for Salt Marsh Formation (Transition From Tidal Flat to Salt Marsh)

The evolution of a salt marsh from mudflats and sandflats can be described using a simplified model which explains the typical bimodal distribution of elevations found in shallow tidal basins (Fagherazzi et al., 2006). In fact, tidal flats and salt marshes generally lie within two different ranges of elevation, and intermediate conditions are rare (French, 1993; Fagherazzi et al., 2006; Defina et al., 2007). According to the model, the vertical evolution of tidal flats is governed by the following equation:

$$(1-n)\frac{dy}{dt} = R_D - R = R_D - A(\tau_b - \tau_{cr})^{\alpha} \tag{3.1}$$

where y is the elevation of the tidal flat, n is the porosity, R is the rate of sediment erosion, R_D is the rate of sediment deposition, A is a constant of proportionality, τ_b is the wave-induced bottom shear stress, and τ_{cr} is the critical shear stress for erosion. According to Eq. (3.1), tidal flats reach an equilibrium configuration $\left(\frac{dy}{dt} = 0\right)$ when the rate of deposition is equal to the rate of erosion, which occurs for

$$\tau_b = \left(\frac{R_D}{A}\right) + \tau_{cr} \tag{3.2}$$

In Fig. 3.3, the blue line corresponds to τ_b and represents erosional forces; the red line corresponds to the right-hand side of Eq. (3.2) and represents depositional forces. Note that the blue curve of erosion peaks for intermediate water depth values. In fact, wave action is limited at the two end-member configurations of very shallow and very deep water. For shallow water, the limitation comes from dissipative processes; for deep water, the limitation comes from the fact that waves cannot influence the bottom (Fredsøe and Deigaard, 1992). The intersection point on the right represents a stable equilibrium because the system can recover from small perturbations in elevation, whereas the equilibrium point on the left is unstable. For instance, starting from an unstable configuration, small depth increments will cause the transition to areas where erosion is higher than deposition, which will in turn cause greater depths leading to even higher erosion rates such that the system will move toward the more stable tidal flat configuration corresponding to the equilibrium point to the right. Conversely, if small perturbations cause a decrease in depth, deposition rates become higher and the system transitions toward a salt marsh configuration.

2.2 Models for Vertical Dynamics

Several models have been developed to represent vertical salt marsh dynamics and surface processes. Many of these models use different formulations in terms of both sediment and vegetation dynamics, but nonetheless they mostly predict that salt marsh accretion rates and sedimentation rates are higher under the following situations: (1) near channel and in areas that are frequently flooded because these are locations that can be easily

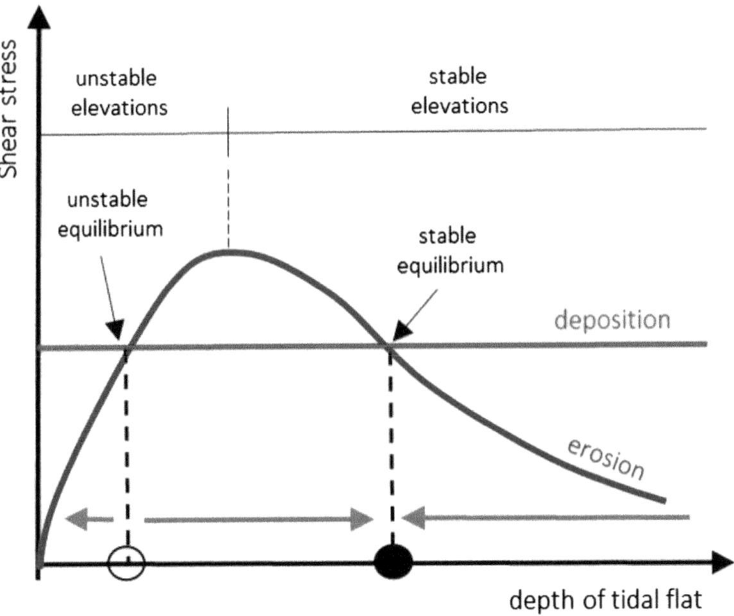

FIGURE 3.3 Equilibrium condition for tidal flats. The *blue line* is the bottom shear stress induced by waves and represents erosion; the *red line* represents deposition. Horizontal axis corresponds to tidal flat depth with values increasing from left (salt marsh condition) to right (deep tidal flat). The intersection points between *red and blue line* correspond to equilibrium configurations. The black filled point on the right is a stable equilibrium because small perturbations are recovered by the system. The point to the left is unstable. *After Fagherazzi, S., Carniello, L., D'Alpaos, L., Defina, A., 2006. Critical bifurcation of shallow microtidal landforms in tidal flats and salt marshes. Proceedings of the National Academy of Sciences 103, 8337–8341.*

reached by sediment-laden waters and (2) in the presence of dense vegetation because dense vegetation generally leads to increased biomass production and higher sedimentation rates created by slower flow and sediment trapping by leaves (e.g., Temmerman et al., 2003; Mudd et al., 2009; Mariotti et al., 2010; Fagherazzi et al., 2012; Kirwan et al., 2016).

For instance, salt marsh accretion through sedimentation and biological feedback has been explained using the following zero-dimensional model (Morris et al., 2002):

$$\frac{dy}{dt} = (q + kB)D \tag{3.3}$$

where $\frac{dy}{dt}$ is the change in elevation of the salt marsh surface, q represents sediment loading, D is the water depth above the salt marsh surface and below high tide, B and k represent aboveground production and the efficiency of vegetation as a sediment trap.

The model takes into account the dependence of aboveground biomass from water depth. This is because vegetation primary production has been found to increase with vegetation submergence, up to the point when hypoxia limits plant survival (Morris et al., 2002).

Temmerman et al. (2003) also developed a similar zero-dimensional model initially aimed to reproduce salt marsh growth in the Scheldt Estuary, which is more sophisticated in terms of suspended sediment transport dynamics. According to this model, early changes in surface elevation (y) can be expressed as

$$\frac{dy}{dt} = \frac{dS_{min}}{dt} + \frac{dS_{org}}{dt} - \frac{dP}{dt} \tag{3.4}$$

where $\frac{dS_{min}}{dt}$ is the rate of mineral sediment deposition, $\frac{dS_{org}}{dt}$ refers to organic sediment deposition, and $\frac{dP}{dt}$ is the compaction rate. The rate of mineral sediment deposition depends on the depth-averaged concentrations of the suspended sediments above the salt marsh platform ($C(t)$) whose variations during a tidal cycle are modeled as follows:

$$\frac{d[h(t) - y]}{dt} C(t) = -w_s C(t) + C(0)\frac{dh}{dt} \tag{3.5}$$

where $h(t)$ is the water level, w_s is the settling velocity, and $C(0)$ is the sediment concentration in the flooding water. Eq. (3.5) describes variations in sediment concentrations above the salt marsh platform caused by sediment settling (first right-hand side term) and sediment fluxes coming with the flooding water (second right-hand side term). While accounting for organic sedimentation, the model does not account for the dependence of biomass production from water depth, which leads to reduced ecogeomorphic feedbacks when dealing with biomass response to increasing rates of sea level rise (e.g., Kirwan and Temmerman, 2009).

The OIMAS-N zero-dimensional model for salt marsh accretion has also been used to explore the influence of salt marsh accretion rates on organic carbon accumulation under different sea level and sediment supply scenarios (e.g., Mudd et al., 2009). The model accounts for both aboveground and belowground biomass production, plant growth and mortality, and sediment compaction. Following results from Morris et al. (2002), a dependence of biomass production from inundation depth is considered, and the ratio between belowground and aboveground biomass is approximated as linearly decreasing with the depth below mean higher high water. Decomposition rates of shoots and roots are modeled, and rates of change of organic carbon linearly depend on decay coefficients, which in turn decrease exponentially with depth. For the inorganic sedimentation component, a linear relationship between submergence time and sedimentation rate is assumed, and for the compaction rate, a classic formulation is used, which depends on compaction coefficients and local stresses.

Full three-dimensional hydrodynamic and sediment transport models have been used to model salt marsh platform dynamics as well. As an example, three-dimensional hydrodynamic models such as Delft3D, Regional Ocean Modeling System, or sECOM (variation of the Princeton Ocean Model) can be coupled with sediment transport and vegetation modules and can therefore be used to investigate sediment patterns and flow dynamics over salt marsh surfaces (Blumberg and Mellor, 1987; Lesser et al., 2004; Warner et al., 2010). The effect of vegetation on the velocity field and on the vertical velocity structure is generally taken into account by considering an additional source term for friction, $F(z)$, and two additional terms for turbulent kinetic energy generation and dissipation by plants. In this regard, plants are generally schematized as vertical cylinders, whose most important characteristics include average stem diameter, stem density, and height above the bottom (Rodi, 1993; Baptist et al., 2007).

Applications of these methodologies include salt marshes in the Netherlands (Temmerman et al., 2005), Kingsport salt marshes in the Minas Basin in the Bay of Fundy (Ashall et al., 2016), and salt marshes in Jamaica Bay, NY (Marsooli et al., 2016). Results suggest that due to the frictional differences between vegetated and unvegetated surfaces, preferential flow routing is established in unvegetated areas, which results in higher velocities in tidal creeks and in between vegetation patches.

2.3 Models for Lateral Dynamics

The lateral migration of salt marshes is a very common process as salt marshes have frequently been found to be inherently unstable in the horizontal direction (Francalanci et al., 2013; Leonardi et al., 2016; Marani et al., 2011; Schwimmer, 2001). For example, salt marsh lateral erosion by wind waves is one of the principal causes for salt marsh losses worldwide, and by using numerical models and field data, it has been shown that this lateral migration is mainly caused by average weather conditions rather than by extreme storms (Leonardi et al., 2016).

Physically based and stochastic models have been developed to describe the evolution of marsh boundaries. Mariotti and Fagherazzi (2010) propose a one-dimensional transect model for the coupled evolution of salt marshes and tidal flats. For the tidal flat and salt marsh surface, the model considers the combined action of wind waves and tides. Wave energy is described following the one-dimensional equation for wave energy conservation at steady state, and tidal currents are calculated through the continuity equation. Vegetation dynamics are also considered following results from Morris et al. (2002) and Mudd et al. (2009). Bottom shear stresses are considered for wind waves and currents, as well as for wave breaking. In the model, wave power is the only governing parameter for the lateral migration of salt marsh boundaries, which are eroded proportionally to the excess power with respect to a critical value. Different formulations are used for the erosion of salt marsh boundaries with respect to the ones of the tidal flat. This is because the latter only depends on tangential shear stresses, whereas for a vertical scarp, normal stresses are present as well. Moreover, in contrast to tidal flats, sediment removal from salt marsh boundaries is a discontinuous process, which is also generally accompanied by superimposed mass failures (e.g., Mariotti and Fagherazzi, 2010; Francalanci et al., 2013; Leonardi and Fagherazzi, 2014, 2015). The model reproduces realistic salt marsh profiles and behaviors under a variety of external drivers.

By using similar boundary dynamics, it has been also shown that vertical and horizontal dynamics respond differently to wave erosive processes (Mariotti and Carr, 2014; Kirwan et al., 2016). In fact, horizontal erosion has been found to be higher with large fetch values and higher soil erodibility. However, the same conditions also promote sediment reworking in the mudflat, transport of sediments on the salt marsh surface during high tide and, as a consequence, higher accretion rates. On the contrary, in case of small fetch values and low mudflat erodibility, salt marsh drowning due to sea level rise is most likely to occur unless a high supply of allochthonous sediments prevails (e.g., Mariotti and Carr, 2014; Kirwan et al., 2016).

A two-dimensional cellular automata (CA) model has also been used to investigate the response of salt marshes to different wave climates and extreme events and to test whether these environments could reach self-organized criticality (SOC) (Leonardi and Fagherazzi, 2014, 2015). The model is the first one directly accounting for the fact that salt marshes are not uniform, and variability in erosional resistance is present across salt marshes. This is caused by natural heterogeneities including biological and ecological factors and different soil properties. The use of CA and simplified stochastic models is relevant in geomorphology as they can help to reproduce emergent properties, nonlinearities, and complex dynamics, which are common in natural systems (Bak, 1996; Rodríguez-Iturbe and Rinaldo, 2001; Dearing et al., 2006). The model has been found to be in agreement with field data from five different salt marsh locations in the Unites States and was calibrated by using real soil shear strength measurements (Leonardi and Fagherazzi, 2014, 2015). The model consists of a two-dimensional square grid representing a salt marsh. Each grid element i has a randomly distributed resistance value r_i, and each cell has an erosion rate:

$$E_i = \alpha P^{\beta} \exp\left(\frac{r_i}{P}\right) \tag{3.6}$$

where P is the wave power, and α and β are nondimensional constants equal to 0.35 and 1.1, respectively (Schwimmer, 2001). The first part of Eq. (3.6) follows classical theoretical and empirical investigations for salt marsh erosion. According to these, the retreat rate is proportional to wave power (Schwimmer, 2001; Marani et al., 2011; Priestas et al., 2015; Leonardi et al., 2016). The second part of the equation is meant to take into account the variability in erosional resistance induced by the variety of biological and ecological processes affecting each portion of the salt marsh. The erosion rate has two extreme limits such that for low-wave power values, the system is highly disordered (each element i has a different erosion rate), whereas for very high-wave power, only a weak disorder is present (all elements have similar erosion rates). The equation is such that when the wave power is very low, variability in erosional resistance is more important. For example, during low-wave energy conditions, areas with denser vegetation could erode significantly less.

On the contrary, when the wave energy is very high, for example, under the continuous action of very energetic storms, different salt marsh portions are eroded at the same rate because the variability in resistance is small when compared with the main external agent (the term in parenthesis in Eq. (3.6) goes to zero, and the exponential goes to one). At every simulation time step, cells at the boundary are eroded at random with a probability

$$p_i = \frac{E_i}{\Sigma E_i}, \tag{3.7}$$

where the sum refers to all the cells that can be potentially removed. The time interval before a cell is removed with a probability of 1 is thus given by the following equation (Leonardi and Fagherazzi, 2014, 2015):

$$t = \frac{1}{\Sigma E_i} \tag{3.8}$$

Results suggest that for high-wave energy conditions, erosion proceeds uniformly with the generation of a smooth salt marsh boundary profile because each cell has similar resistance when compared with the main external driver. On the contrary, when exposed to low-wave energy, salt marsh boundaries are rough and jagged because differential erosion rates apply and affect the global system behavior (Fig. 3.4).

In the model, the magnitude of erosion events is defined as the number of eroded cells for a given time interval; when collecting field measurements, erosion events correspond to the erosion rate of relatively short stretches of shoreline. Statistically, high-wave energy conditions correspond to a Gaussian distribution of erosion events, whereas for low-wave energy conditions, the frequency distribution of erosion events follows a long-tailed power law distribution with many low magnitude events, accompanied by occasional failures of large salt marsh blocks (Fig. 3.4G). In fact, for low-wave energy, a long time period might be required to erode resistant cells, but once the very resistant cells are eroded, new weak sites are uncovered and rapidly disintegrate with the consequent generation of large failures. Model results agree well with field measurements and suggest that under low-wave energy conditions the system might reach SOC (Leonardi and Fagherazzi, 2014). SOC refers to the state at which a system develops patterns and behaviors independently from any fine external tuning such that all members of the system influence each other so that local instabilities can affect the entire system behavior (Bak et al., 1987, Bak, 1996; Dearing et al., 2006). For salt marsh boundaries, the critical state could be the one promoting salt marsh "armoring" by removal of the weakest sites and exposure of the more uniform and resistant marsh portions. Internal physical mechanisms possibly promoting the establishment of a critical state are geotechnical and connected to the development of tensional cracks in the presence of which the system approaches the minimum energy state independently from any external tuning (Leonardi and Fagherazzi, 2014).

3. SPATIAL AND TEMPORAL PATTERNS OF VARIATION

3.1 Spatial Zonation of Vegetation

Salt marsh macrophytes commonly exhibit clear patterns of zonation, driven by the individual species tolerance to physical stress and biological interactions acting across abiotic gradients (Pennings and Bertness, 2001). Abiotic factors linked to salt marsh ecological zonation include soil moisture content, redox state, nutrient limitation, pH, and salinity (Thibodeau et al., 1998; Silvestri et al., 2005; Moffett et al., 2012), all of which are commonly correlated with elevation and hydroperiod across the salt marsh. Key biotic factors, such as interspecies competition, also operate in close interaction with abiotic stress (Bertness, 1991; Pennings et al., 2005), resulting in the typical shore-parallel zonation of plants.

The lower limit of a salt marsh is unambiguously defined as the seaward margin of emergent vascular plants (Adam, 1990). These areas are regularly inundated by saltwater and vegetation is usually restricted to a few hardy pioneer genera such as *Salicornia*, *Suaeda*, *Aster*, and *Spartina* (Doody, 1992). In low marsh areas, salinity is comparatively low because of regular tidal flushing that prevents the accumulation of salt. Instead, substrate stability, oxygenation, and sulfide toxicity are key factors controlling plant establishment at the seaward margin (Adam, 1990; Mendelssohn and Morris, 1999). Salinity levels vary across marsh elevations, but the shape of the salt gradient depends on climate. In humid warm regions, freshwater input from rain and upland sources moderates salinity at the terrestrial border,

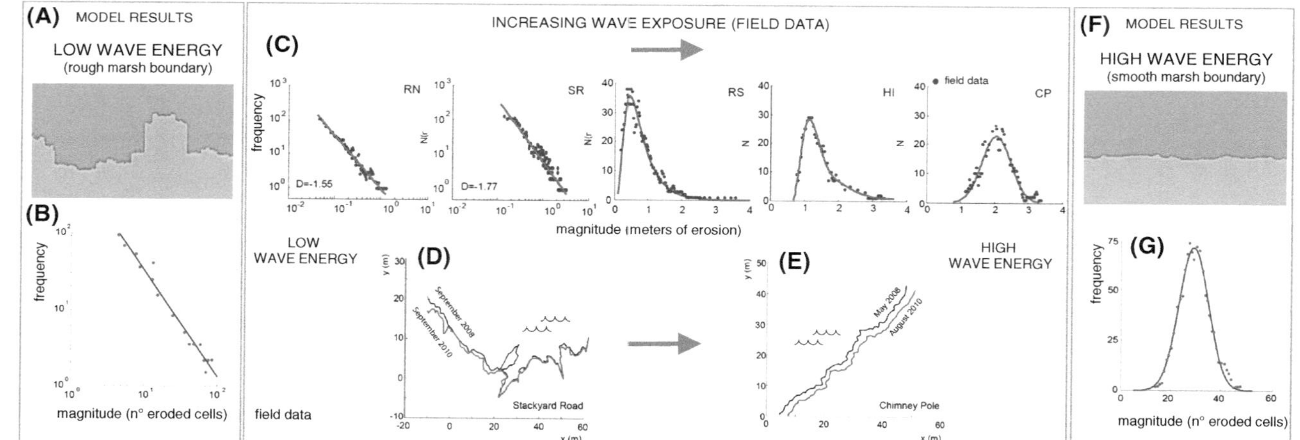

FIGURE 3.4 Results from two-dimensional (2D) cellular automata model and field measurements. (A) Salt marsh boundary profile resulting from the 2D cellular automata model under low-wave energy conditions. Note the rough boundary profile. (B) Model results; logarithmic frequency–magnitude distribution of erosion events under low-wave energy conditions. (C) Field data; frequency magnitude distribution of erosion events for five field sites in the United States. Note the passage from a logarithmic to a Gaussian distribution for increasing wave energy. The first three sites are in Plum Island Sound, MA (Refuge North, Stackyard Road, Refuge South), and the last two are in the Virginia Coast Reserve (Chimney Pole, Hog Island). Blue points are field data on salt marsh boundary erosion. In Plum Island, sound data were collected every year form 2008–13. In Virginia Coast Reserve, data were collected form 2008–10. (D) Salt marsh boundary profile at one of the sites subject to low-wave energy. (E) Salt marsh boundary profile for one site subject to high-wave energy; note that under high-wave energy conditions, the salt marsh boundary is smoother. (F) Model results; salt marsh boundary subject to high-wave energy. (G) Model results. Gaussian magnitude/frequency distribution of erosion events under high-wave energy conditions. *After Leonardi et al. (2014, 2015).*

but salts can concentrate by evaporation at intermediate marsh elevations (Davy and Costa, 1992). In colder climates, salinity declines monotonically with increasing elevation (Pennings and Bertness, 1999). Mid-marshes characterized by higher salinities support a more salt-tolerant flora, and salt accumulations may also lead to the development of bare areas known as salt pans. Despite these general patterns, salinity can be highly variable, depending on rainfall seasonality, interannual variations, or freshwater discharges (Callaway and Sabraw, 1994; Noe and Zedler, 2001).

In coastal settings with sufficient freshwater inputs, there would be a gradational transition from the upper salt marsh limit into freshwater wetland communities, where waterlogging is sustained by freshwater seepage, rainfall, and high groundwater levels due to proximity to the coast. Here, depending on the amount and nutrient status of the receiving water, an array of floristically different wetland communities may develop, including freshwater marshes and swamps further described in other chapters of this book. This transition from high salt marshes to freshwater wetland communities, which creates a clear zonation of plant species at different elevations, is often restricted because of anthropogenic activities such as drainage, reclamation, and emplacement of coastal defenses.

In arid climates, under a combination of high evaporation and low freshwater inputs, salt accumulations increase with elevation, and extremely high salinities eliminate all but the most tolerant plants from the upper intertidal and supratidal zones. *Sabkha* is an Arabic term for flat salt-crusted desert. The local terminology of the Arabian Gulf region describes extensive, barren, salt-encrusted flats that characterize arid coastal lowlands in those regions where evaporation by far exceeds precipitation (Barth and Böer, 2002). Barren sabkhas covered by evaporite accumulations are most common within the tropical and subtropical zones, but they also appear in arid coastal settings within temperate climates (e.g., Bahía Blanca Estuary in eastern South America; Shark Bay, in Western Australia). From a hydrogeomorphic point of view, coastal sabkhas, frequently inundated by seawater, occupy the landscape position of a high marsh. The equivalent to perimarine freshwater fens and swamps in these arid environments are inland sabkhas in the coastal zone, which are never flooded by tides but where sea level keeps a shallow water table near the surface (Pratolongo et al., 2016).

Besides zonation of vascular plants, other biota also occupy distinct zones across the elevation gradient. For sessile or low-mobility species (e.g., diatoms, foraminifera, macroinvertebrates), their position reflects the relative ecological tolerance to a combination of abiotic and biotic stressors (Scott and Medioli, 1978; Zong and Horton, 1998; Zapperi et al., 2016). The dominant control on the distribution of microorganisms on salt marshes is flooding duration as a function of elevation (Gehrels, 2000), allowing for paleoecological studies and reconstructions of past relative sea levels (Edwards and Horton, 2000, 2006). The elevation gradient across the salt marsh also constrains the distributions of mobile animals by creating different stress gradients to aquatic and terrestrial species, which must either tolerate or avoid submergence and emergence (Pennings and Bertness, 2001).

3.2 Changes Over Time

The ecological development of coastal wetlands is a product of both autogenic and allogenic factors (Waller et al., 1999). These factors determine spatial patterns in wetland plant communities and also their change over time. Autogenic processes drive plant succession until vegetation reaches the "equilibrium" with the environment (Schofield and Bunting, 2005). Furthermore, autogenic factors above and beyond the influence of morphodynamic controls have been shown to enable lateral expansion and elevation change of pioneer salt marshes (Silinski et al., 2016). However, patterns and pathways of successional change within coastal wetlands are largely controlled by allogenic factors such as sea level change and storms. As such, ecological succession, controlled by autogenic processes, gives way to state change principally controlled by changes in hydrology and salinity, which in turn lead to changes in vegetation (Brinson et al., 1995).

A dominant process in coastal wetland evolution is ecosystem response to relative sea level changes. A change in relative sea level produces an alteration in the ecological state of wetlands, and the different plant associations within the coastal wetlands continuum are expected to migrate in response to different hydrologic conditions. While the globally averaged sea level has been rising from the Last Glacial Maximum (LGM) to the present, the relative sea level (relative height of the sea with respect to land) can vary from place to place due to the effects of eustatic, isostatic, tectonic, ocean dynamic, and local factors (Khan et al., 2015).

The process of wetlands migration upslope under a rising relative sea level during the Holocene has been largely studied along the eastern coast of North America and northern Europe. Hageman (1969) described the evolution of freshwater swamps in the western Rhine/Meuse Delta, and there are examples of sedimentary records showing that peat-forming perimarine wetlands accumulated deep layers of organic matter between around 6000 and 2000 years BP (Waller, 1994; Kirby, 2001). Palynological analysis of these peat deposits showed sequences of salt marshes, reed swamps, fens, and woodland carr communities that developed under a rising sea level that maintained a near-surface groundwater table (Waller et al., 1999), enabling fen carr vegetation communities to regenerate for several thousand years. A similar model of transgression was described for a typical coastal barrier ecosystems extending along the seaward margin of the Delmarva Peninsula on the Atlantic coast of North America (Oertel et al., 1989). In this system, a sustained sea level rise during the Holocene set up a similar sequence of state changes in wetlands along the mainland edge (Brinson et al., 1995). As transgression occurred, upland forest was replaced with high marsh, high marsh with low marsh, low marsh with mudflats, and mudflats with open water (Christian et al., 2000).

In contrast to these well-studied examples of continuously rising relative sea level, little is known about wetland response in coastal environments, which developed under different conditions after the LGM. Some examples from the Gulf of Bothnia describe a downward migration of plant zones in response to the continuous sea level drop (Vartiainen, 1988; Ecke and Rydin, 2000) and the seaward expansion of pioneer plant communities (Zobel and Kont, 1992). A more complex dynamic characterizes wetland environments where the relative sea level reached a transgressive maximum during the Holocene. In these systems, the late Holocene marine regression resulted in wide low-lying coastal landforms inherited from the former estuarine setting. These coastal environments are commonly occupied by perimarine wetlands, which undergo increasing inundation under the current rising trends in relative sea level. In the Bahía Blanca Estuary, a rising sea level is a major cause of wetlands loss in elevated Holocene surfaces, but the accelerated erosion of soft sediments is also the main source of suspended solids to the tidal sediment budget, allowing deposition and seaward expansion of low salt marshes (Pratolongo et al., 2010, 2013).

4. GEOGRAPHIC VARIATION

Given that all tidal marshes share a similar range of geomorphologic settings and hydrological driving forces, they would be comparable in terms of ecosystem structure and function (Brinson, 1993). However, even within the temperate zone, substantial climatic differences arise among major geographic regions due to precipitation regimes. Typical patterns of salt marsh plant zonation largely reflect rainfall amount and seasonality, along with the biogeographic distribution of species. In addition, human activities have also exerted profound changes in coastal wetlands. Worldwide, it has been estimated that about 50% of salt marshes have been either lost or degraded with wetland losses exceeding 90% in some coastal areas from developed countries (Barbier et al., 2011). However, these processes have been quite variable across regions, leading to major differences in the extent and ecological integrity of remaining wetland areas. Introduction of nonnative species that rapidly expand into native communities has contributed to this trend and, in some cases, may have far-reaching consequences on both biotic and physical processes.

The geographical grouping used here is based on the work of Chapman (1960). In his early attempt to classify coastal systems, salt marshes of the world fall into distinct groups characterized by different types of vegetation. In this updated review, we also incorporate climate zones according to the Köppen-Geiger classification after the update by Peel et al. (2007), regional geomorphic settings that affect the distribution of vegetation, and major human modifications.

4.1 Northern Europe

Temperate tidal marshes in Northern Europe zone occur along the Atlantic and Channel coasts from the Iberian Peninsula northward to Denmark, through the barrier complex of the Friesian Islands and the Wadden Sea, and including the southern coasts of Great Britain (Fig. 3.5A). Marshes in this group develop under an oceanic (Cfb) climate (also known as marine or maritime), typical of west coasts in higher middle latitudes of continents. It is characterized by cool summers and winters, with a relatively narrow range of annual temperatures. Precipitation is evenly distributed throughout the year, without a dry season.

Some of the better-known salt marshes of Europe are found at the Mont Saint-Michel Bay, in the south of the Gulf of Normandy. The system has a very large tidal range (spring tidal range up to 16 m) creating an intertidal zone that expands over 220 km^2. Since the 11th century, 133 km^2 of salt marshes have been converted to agriculture. Nowadays, salt marshes extend over 40 km^2 with most of the area under sheep grazing (Lefeuvre et al., 2000). Typical vegetation pattern includes a pioneer zone with a sparse cover of *Spartina anglica* and *Salicornia dolichostachya/fragilis*. Landward, *Puccinellia maritima* appears as small clones, and a middle marsh establishes with higher vegetation cover and diversity. *Aster tripolium, Salicornia ramosissima, Suaeda maritima, Halimione portulacoides*, and *Atriplex portulacoides* may also appear in this zone, whereas *Festuca rubra, Juncus gerardii*, and *Elymus athericus* are common dominants in the

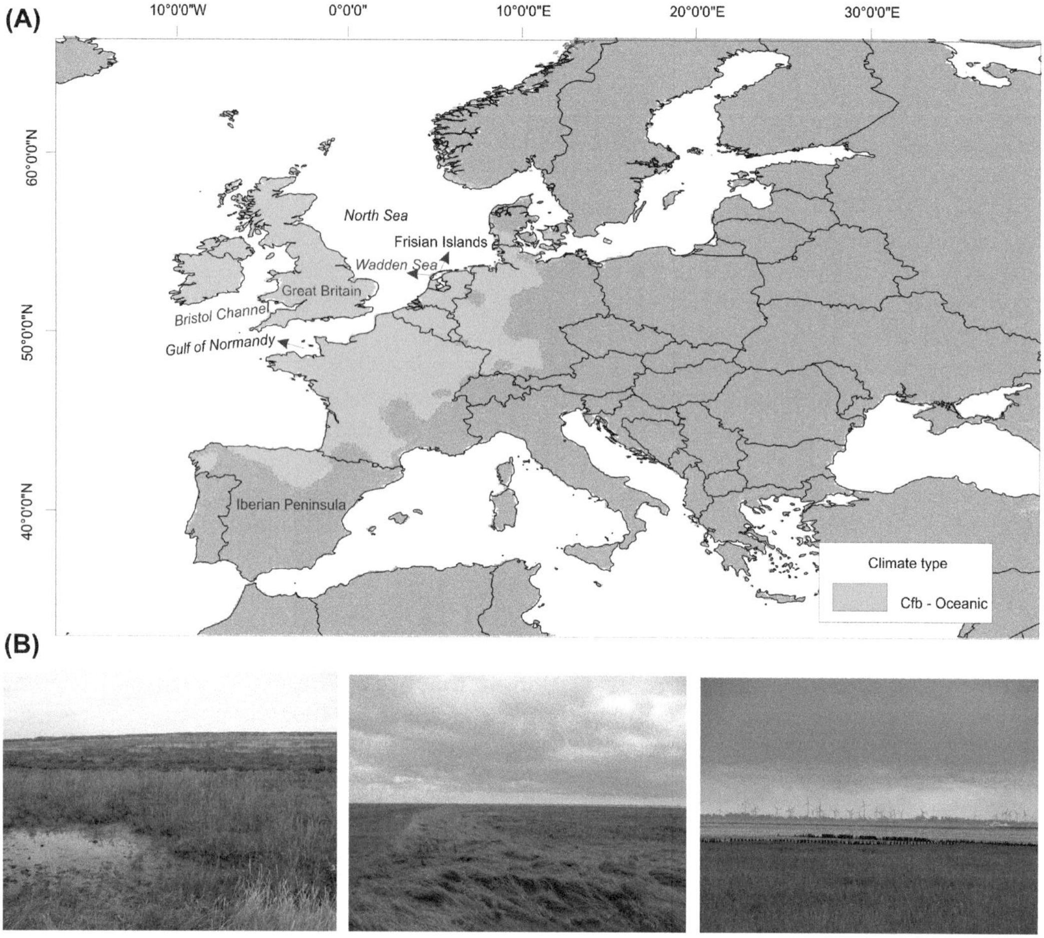

FIGURE 3.5 (A) Climate classification within the temperate zone of northern Europe. (B) Salt marshes on Langeoog Island, Germany after de-embankment of summer polders. *(A) Modified from Peel, M.C., Finlayson, B.L., McMahon, T.A., 2007. Updated world map of the Köppen-Geiger climate classification. Hydrology and Earth System Sciences 11, 1633–1644. (B) Photographs by Alejandro Loydi.*

upper marsh (Langlois et al., 2003; Valéry et al., 2004). Since the mid-1980, the native clonal grass *Elymus athericus* has been aggressively invading the middle and low marshes, where it often forms dense monospecific stands replacing the natural *Atriplex portulacoides* marshes. This pattern of expansion has been reported elsewhere in northern Europe, and the enrichment of water by nitrogenous compounds facilitates *Elymus* invasions (Valéry et al., 2016).

Well-studied tidal marshes also occur in the Severn Estuary, in the head of the Bristol Channel, the back-barrier environments along the Frisian Islands, and the open coast of the Wadden Sea. Mainland marshes in the Wadden have been embanked for centuries, but the aim of the Wadden Sea Plan is to maintain and, where possible, increase the total area of marshes through de-embankments of summer polders. After de-embankment on

Langeoog Island, Germany (Fig. 3.5B), nitrophytic species such as *Suaeda maritima*, *Atriplex prostrata*, and *Artemisia maritima* replaced the former glycophytic vegetation. As soil nitrogen decreased, *Atriplex portulacoides* and *Elymus athericus* became dominants in the lower and higher marsh (Barkowski et al., 2009).

In the remaining salt marshes on these complexes, typical vegetation zones and plant associations are similar to those described for northern France, with the recurring presence of *Spartina anglica* in the pioneer zone. This fertile allopolyploid arose by the end of the 1880s in Southampton Water, UK, by chromosomal doubling from *Spartina* × *townsendii*, a hybrid between the introduced *Spartina alterniflora* and the native *Spartina maritima* (Ayres and Strong, 2001). Although *Spartina anglica* invaded large areas during the first 30 years after being reported (Raybould, 1997), during the late 1920s and early 1930s, the species began to decline at some locations. In southern and southeastern England in particular, widespread diebacks of *Spartina anglica* have occurred for unknown reasons, resulting in the erosion of substantial areas of salt marsh (Baily and Pearson, 2007). However, this is not the case in the northeast or northwest coasts of England, where the species still seems to be expanding, despite control methods employed (Lacambra et al., 2004). An increase in abundance and accelerated spread of *Spartina anglica* were also observed in the Wadden Sea, possibly promoted by warmer spring temperatures (Nehring and Hesse, 2008).

4.2 Eastern North America

In the temperate zone of eastern North America, salt marshes dominate the intertidal zone from southern Canada to the northern Gulf Coast, excluding subtropical coastlines of Florida, south of approximately 30°N, where salt marshes are replaced by a mixture of mangrove forests and tidal marshes (Savage, 1972). The presence of barrier islands is a conspicuous feature all along the Atlantic coast of North America, occurring from Maine to Texas. Barrier islands are formed and maintained by changing sea level that makes them extremely dynamic and ephemeral landforms on a geological timescale. In eastern North America, different barrier complexes share a common origin, closely related to the sustained sea level rise during the last 4500–5000 years (Oertel et al., 1992; Stutz and Pilkey, 2011). The Holocene landward migration of shorelines led to the transgression of older barrier islands and dune ridges, and the inundation of lowlands behind them created extensive protected back-barrier environments suitable for wetland development (Oertel et al., 1989).

Despite similarities in the Holocene sea level history, there are several structural differences between North American and European coastal marshes. In eastern North America, *Spartina alterniflora* grows in the lower intertidal zone, with the lower elevational limit of occurrence near the tidal datum of mean low tide (Mckee and Patrick, 1988). In contrast, vegetation of European salt marshes is typically confined to the upper intertidal zone, at elevations between mean high tide and spring tide (Lefeuvre, 1996). Moreover, pioneer species in European salt marshes form a short prairie instead of those dense tall stands of *Spartina alterniflora* described on the Atlantic coast of North America, leading to a more organogenic substrate of peat.

Following Chapman's geographical groups, tidal marshes of eastern North America can be divided into three subgroups: the Bay of Fundy, New England, and Coastal Plain, which closely reflect climatic differences across the latitudinal gradient (Fig. 3.6).

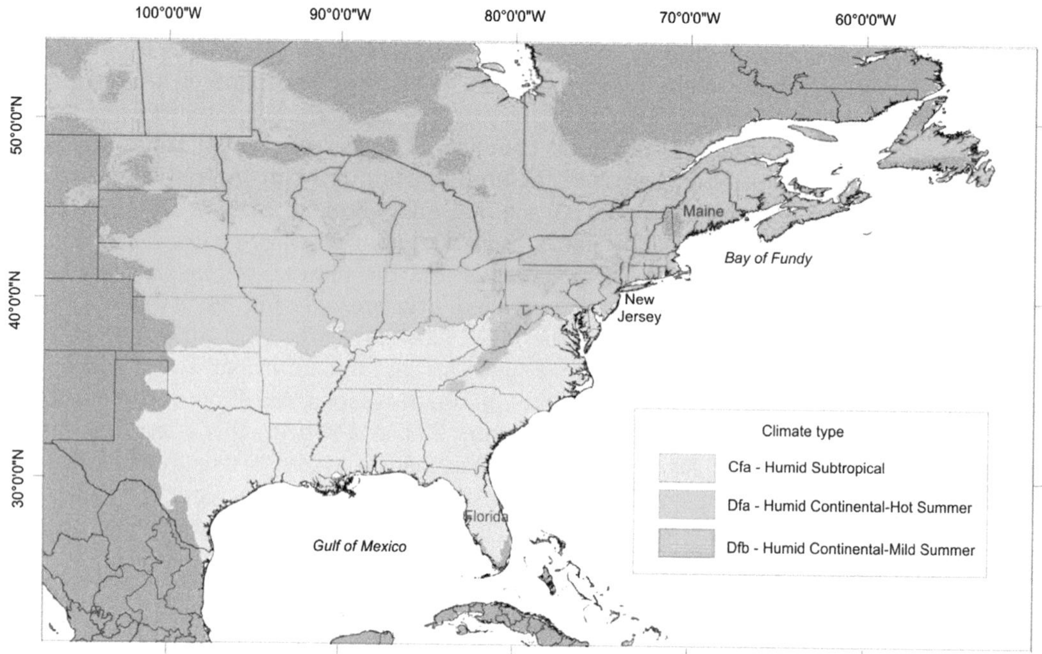

FIGURE 3.6 Climate classification within the temperate zone of eastern North America. *Modified from Peel, M.C., Finlayson, B.L., McMahon, T.A., 2007. Updated world map of the Köppen-Geiger climate classification. Hydrology and Earth System Sciences 11, 1633–1644.*

4.2.1 Bay of Fundy

The Bay of Fundy is located on the east coast of southern Canada, where it forms an extension of the Gulf of Maine. The region has a humid continental mild summer climate (Dfb). Humid continental climates are characterized by large seasonal temperature differences, with warm and often humid summers and cold (sometimes severely cold) winters. The average temperatures in the coldest month are generally below −3°C, and frost-free periods typically last 3–5 months. Precipitation is usually well distributed through the year.

Semidiurnal tides of about 4 m near the mouth amplify to an average range of about 11 m in the upper reaches of the Bay. The most remarkable feature in this area is that marshes here grow under the strong tidal currents generated by some of the largest tides in the world. Characteristic marsh topography results in a narrow low marsh fringing a wider high marsh area (Miller and Egler, 1950). At some locations, undercutting of creek banks generates a sharp topographic division separating the two zones. These escarpments also appear in southern New England marshes (Redfield, 1972) but, being a transition between temperate and subarctic regions, marsh geomorphology in the upper Bay of Fundy is more strongly affected by ice formation. The movement of blocks of frozen sediment causes both erosion of and sedimentation in the intertidal zone (Dionne, 1989).

Typical low marsh vegetation is flooded frequently by water as much as 4 m deep and consists almost exclusively of *Spartina alterniflora*. The high marsh zone is flooded infrequently for short periods by only extreme high tides, and vegetation is more diverse, usually dominated by *Spartina patens*, although different species such as *Salicornia europea*, *Triglochin maritima*, and *Juncus gerardii* may be also dominant (Chmura et al., 1997). Common species also found in these high marshes include *Atriplex patula, Glaux maritima, Suaeda maritima, Spergularia canadensis, Limonium nashii, Hierochloe odorata, Elymus arenarius, Distichlis spicata*, and *Puccinellia maritima* (van Proosdij et al., 1999). At some locations, Chmura et al. (1997) described a middle zone between the high and low marsh where *Spartina alterniflora* and *Plantago maritima* are codominants. The presence of *Puccinellia maritima* and extensive stands of *Juncus balticus*, common species at higher latitudes, indicates the transition from the sub-arctic to the more temperate marshes of New England. *Phragmites australis* and *Iva frutescens* often mark the upper limit of the high marsh (Jacobson Jr. and Jacobson, 1989).

4.2.2 New England

Tidal marshes in New England occur along the Atlantic coast from Maine to New Jersey. This zone is a climatic transition from humid continental mild summer climate (Dfb) to humid continental hot summer (Dfa) climate through the south. This southern area is characterized by an average temperature of at least 22°C in the warmest month and frost-free periods normally lasting more than 5 months.

In the words of Chapman (1960), the distinctive feature of these marshes is that they are built of a marine peat. The particular processes driving marsh peat development in New England raised an early scientific theme. After Redfield (1972), a new comprehensive model was accepted, highlighting the relevance of Holocene sea level history on the evolution of New England marshes. In Redfield's model with a sufficient sediment supply, the marsh would have responded to slowly rising sea level (c. 1 mm/year) by extending landward and covering uplands with marsh peat. Marshes prograded seaward through fine sediment accumulation and transgressed landward through high marsh development over the older intertidal peat.

Similarly to marshes in the Bay of Fundy, vertical zonation includes a low marsh flooded daily by tides and covered by dense monospecific stands of *Spartina alterniflora*. At higher elevations, marshes are dominated by *Spartina patens*, with the upland border dominated by monocultures of *J. gerardii*. The occurrence of a narrow strip of *Iva frutescens* usually delineates the terrestrial border (Bertness et al., 2002), and where a brackish transition exists, given sufficient freshwater input, *Zizania aquatica, Scirpus americanus, Typha* spp., and *Phragmites australis* are often dominants (Nixon, 1982). Although New England marshes are typically described as a regional unit, the more severe winter temperatures in northern marshes and the high frequency of icing events may cause severe damage to low marsh vegetation, preventing the establishment of dense and pronounced monocultures of *Spartina alterniflora* common in southern marshes (Ewanchuk and Bertness, 2003).

Besides climate, northern and southern marshes differ in the recent history of human alteration. During 1960–90, the human population along the Long Island Sound increased at twice the rate experienced in the Gulf of Maine (Shriver et al., 2004). As a result, remaining marshes in southern New England are smaller and well drained, whereas the larger marshes in the Gulf of Maine are more likely to have waterlogged pans. Vegetation of these pans comprises forbs such as *Agalinis maritima, Atriplex patula, Glaux maritima, Limonium nashii, Plantago maritima, Salicornia europaea, Suaeda linearis*, and *Triglochin maritima*, many of which occur at low abundances further south (Shumway and Bertness, 1992) but form a persistent community in the north.

An expansion of *Phragmites australis* has been observed on the landward edges of New England salt marshes, due to the introduction of an aggressive European haplotype that has largely replaced native Phragmites (Saltonstall, 2002). Although phenotypically similar, the native haplotype would be a superior competitor under nutrient-limited conditions, whereas the exotic Phragmites exhibits a vigorous aboveground response under high nitrogen availability. The global increase in ecosystem nitrogen input from human activity has given the exotic haplotype a competitive advantage in freshwater and brackish wetlands. Additionally, the exotic haplotype has a higher salt tolerance, outcompeting a low competitor salt marsh species such as *Spartina alterniflora* (Minchinton and Bertness, 2003). Disturbance and periodic long-term changes in tidal regime, in addition to nutrient enrichment, appear to play an important role in *Phragmites* expansion (Ravit et al., 2007).

4.2.3 Coastal Plain

Marshes of the Coastal Plain of eastern North America occur in the broad coastal zone from New Jersey to the Gulf Coast, under a humid subtropical (Cfa) climate. Still humid, such as in northern coastal locations, this climate has long hot summers and short mild winters, with annual rainfall concentrated in summer, and relatively dry winters. Along the southeastern Atlantic Coast, salt marshes usually develop in extensive shallow areas lying behind Pleistocene barrier islands (Wiegert and Freeman, 1990, Fig. 3.7A). In the Gulf Coast, extensive salt and freshwater tidal marshes are associated with the Mississippi River Delta (Gosselink and Pendleton, 1984).

South of the Chesapeake Bay to northern Florida, *Spartina alterniflora* dominates extensive intertidal low marshes that cover most of the coastal marsh area (Wiegert and Freeman, 1990, Fig. 3.7B). On levees and creek banks, the tallest and most productive form of the species appears as a narrow band. In the back levee low marsh, where tidal inundation is still several hours a day, a wide zone of vigorous plants extends up to higher elevations, where tidal inundation is just a couple of hours a day, and the short form of *Spartina alterniflora* becomes dominant (DeLaune et al., 1983). At even higher elevations, where flooding by tides is irregular, high marshes dominated by *Juncus roemerianus* and *Spartina patens* appear, along with inclusions of *Salicornia* spp., *Distichlis spicata*, and *Limonium* spp.; bare salt pans may also occur. In the brackish transitions to freshwater communities, *Spartina cynosuroides* commonly forms extensive stands that may mix with smaller stands of *Scirpus americanus* and *Pontederia cordata* (Wiegert and Freeman, 1990).

Tidal freshwater wetlands are particularly well developed in this region, south from New Jersey to Georgia. Extensive tidal freshwater marshes occur in the Delaware River, the Chesapeake Bay, South Carolina, and Georgia. Vegetation is generally more heterogeneous than in salt marshes, with species such as *Nuphar luteum*, *Zizania aquatica*, and *Zizaniopsis miliacea* in the low marsh and *Typha* spp., *Polygonum* spp., *Hibiscus moscheutos*, *Acorus calamus*, and many others at higher elevations (Odum et al., 1984). In the Albemarle-Pamlico Sound, North Carolina, coastal rivers flow into large sounds with negligible tides. Although most rivers present both tidal and freshwater conditions, the irregularly pattern of flooding determines that typical tidal freshwater marshes are often replaced by tidal swamps (Moorhead and Brinson, 1995).

Pocosins comprise the largest extent of true bogs in the Coastal Plain of North America, which once covered nearly 1 million ha only in North Carolina (Richardson, 1983). *Pocosin* is a local term used to describe raised bogs dominated by woody vegetation (*Pinus serotine*,

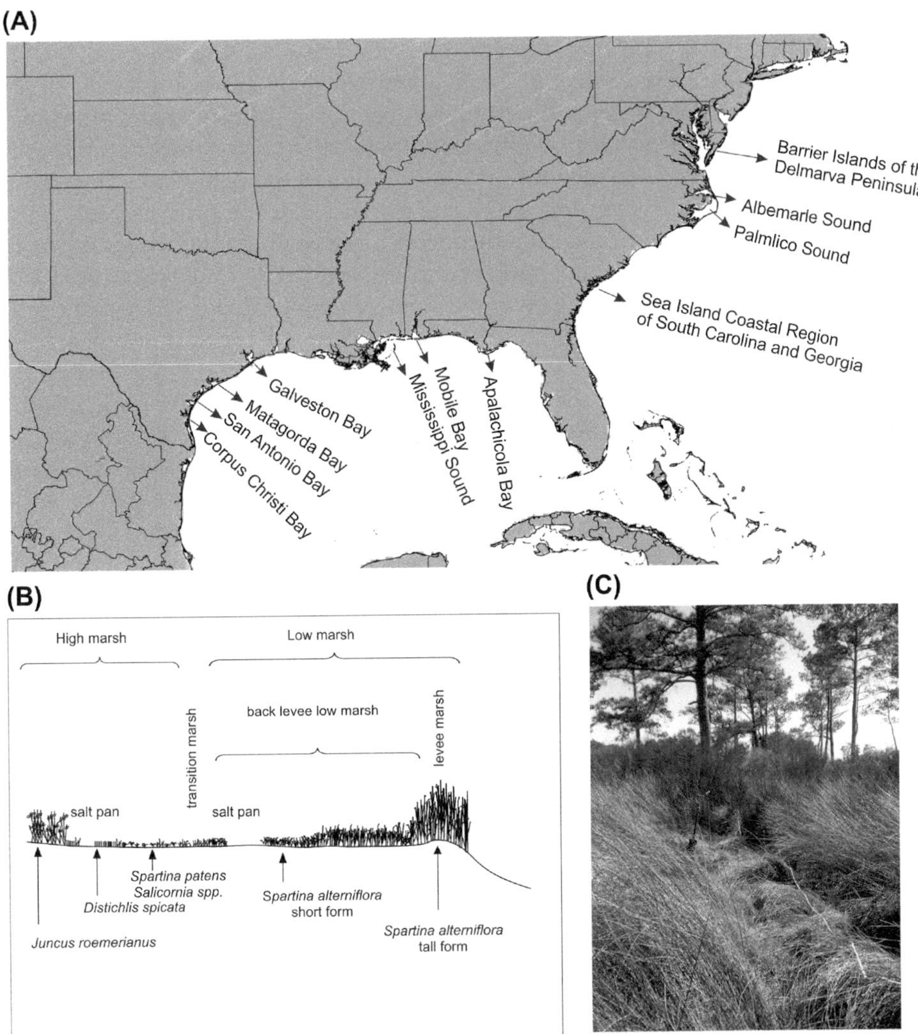

FIGURE 3.7 (A) Location of main wetland complexes associated to barrier island systems along the temperate Coastal Plain of the United States. (B) Cross section of a typical Southeast Atlantic coast salt marsh. (C) Transition zone between a brackish marsh and pocosin wetland in the Albemarle Sound, North Carolina. *(B) After Wiegert, R.G., Freeman, B.J., 1990. Tidal Salt Marshes of the Southeast Atlantic Coast: A Community Profile. (Washington, D.C., USA). (C) Photograph by Paula Pratolongo.*

Pinus taeda, Pinus palustris), with deep, acidic, and saturated peat soils. Groundwater seeps and often perched water tables cause the inundation, and for pocosins in the perimarine zone, sea level influences their hydrology and impedes drainage (Fig. 3.7C). In some areas of the Pamlico Sound, North Carolina, brackish marshes invade pocosins in response to a rising sea level. A transition zone commonly develops with a *Juncus roemerianus* marsh grading into fresher marshes (*Cladium jamaicense* and *Panicum virgatum*). Inland from the marsh is a shrub zone (*Juniperus virginiana*, *Myrica cerifera*), and the peat surface of pocosins becomes the platform for inland migration of the wetland continuum (Brinson, 1991).

In the Gulf Coast, extensive salt and freshwater tidal marshes are associated with the Mississippi River Delta, with a general pattern of decreasing salinity inland from the coast. In a typical delta cycle (since Scruton, 1960), once a new delta begins its progradation, broad coastal freshwater marshes develop, until the river switches the course to a different site of deposition. When deprived of the sediment load, marshes become progressively inundated by marine waters, mainly due to a rapid subsidence. Marsh deterioration increases rapidly with saltwater intrusion, ponds open, and the bay enlarges. The final stage on this cycle is an open coastal system with a large bay and a relative small area of salt marsh remaining (Day et al., 2000). Visser et al. (1998) recognized nine different vegetation types: (1) two in the polyhaline zone, one occupied by black mangrove (*Avicennia germinans* with *Spartina alterniflora* and *Batis maritima)* and the other by oyster grass (*Spartina alterniflora* sometimes in association with *Juncus roemerianus*); (2) in the mesohaline zone, a codominant mixture of *Spartina alterniflora, Spartina patens,* and *Distichlis spicata*; (3) mesohaline wiregrass dominated by *Spartina patens;* (4) oligohaline wiregrass dominated by *Spartina patens,* but with a higher species richness; (5) an oligohaline mix dominated by *Sagittaria lancifolia*; and (6) a freshwater zone, comprised of three different types—the fresh bulltongue dominated by *Sagittaria lancifolia* with the presence of ferns, the fresh maidencane dominated by *Panicum hemitomon*, and the fresh cutgrass dominated by *Zizaniopsis miliacea*.

Spartina alterniflora marshes along the Coastal Plain have undergone extensive diebacks, characterized by the premature browning of this species. In 2000, a large dieback event (termed "brown marsh") affected over 100,000 ha of *Spartina alterniflora* marshes along the Mississippi River deltaic plain (Lindstedt and Swenson, 2006). In 2001, a dieback that affected both *Spartina alterniflora* and *Juncus roemerianus* salt marshes was reported in Georgia (Ogburn and Alber, 2006). By 2002, dieback sites extended throughout Georgia into South Carolina, Long Island, and New England, where *Spartina alterniflora*, *Spartina patens*, *Distichlis spicata*, and *Juncus gerardii* marshes were all affected to varying degrees (Smith and Carullo, 2007). New diebacks of *Spartina alterniflora* continued to be reported in Delaware Inland Bays, the Virginia Coast Reserve, the Chesapeake Marshlands Blackwater National Wildlife Refuge, Louisiana, and Georgia (Crawford and Stone, 2015). There are multiple hypothesized drivers for diebacks, including elevated salinity, changes in herbivory, soil drying, and altered nutrient uptake, suggesting that there is no single answer able to explain all dieback occurrences (Alber et al., 2008).

4.3 Western North America

Coastal wetlands on the Pacific coast of North America are less extensive because of the morphology of the shoreline. Tidal regime is also different from the Atlantic coast, with a more pronounced mixed tide, which likely influences low and high marsh distribution. Main wetland areas occur in San Francisco Bay Estuary, a drowned river valley, formed by crustal movements during the late Pliocene and later inundated by rising Holocene sea level (Atwater et al., 1977). The area is under Mediterranean warm summer (Csb) climate, also called dry summer temperate under the Köppen climate classification (Fig. 3.8A). This climate has warm summers and short mild winters and, compared with the subtropical

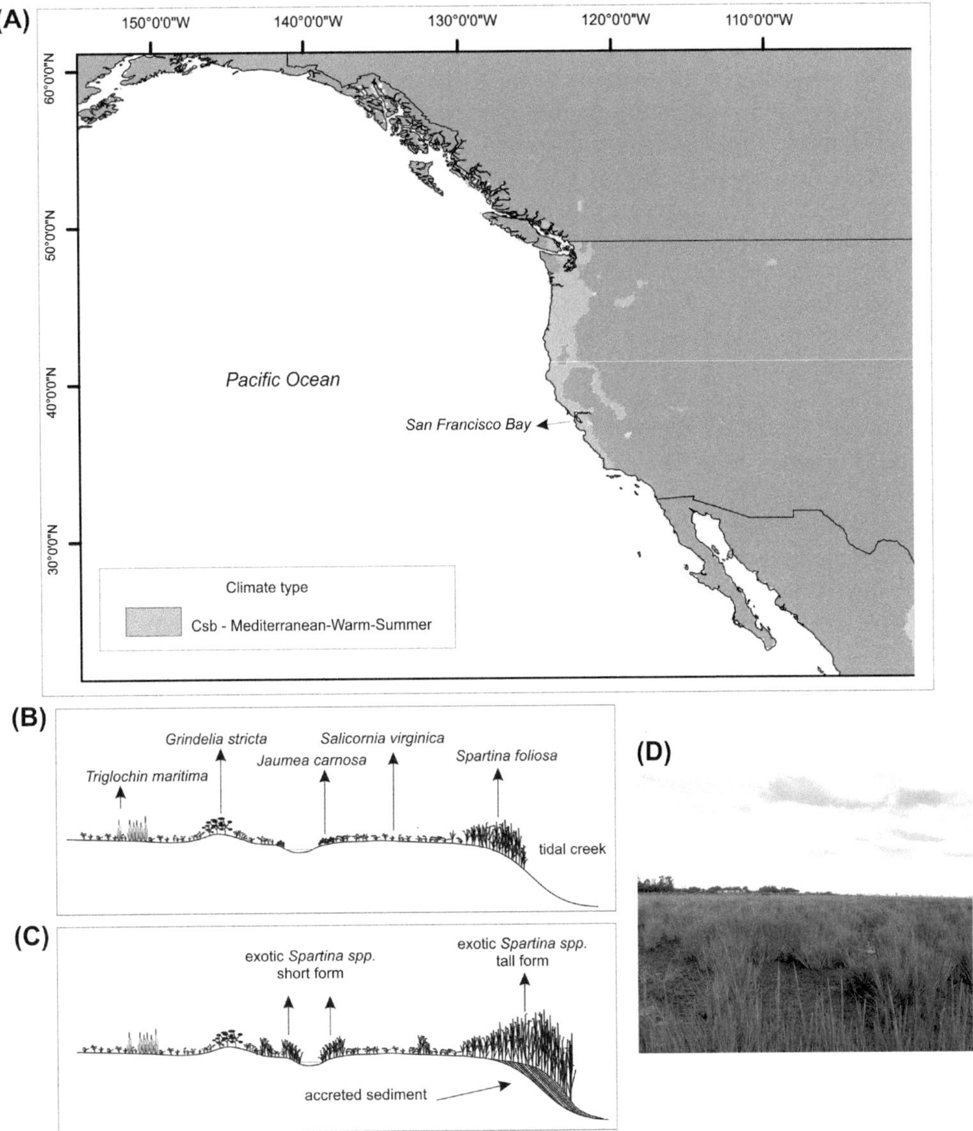

FIGURE 3.8 (A) Climate classification within the temperate zone of western North America. (B) Cross section of natural salt marshes in the lower portion of the San Francisco Bay Estuary and (C) modifications introduced by the invasive exotic species of *Spartina*. (D) Invasive *Spartina densiflora*. *(A) Modified from Peel, M.C., Finlayson, B.L., McMahon, T.A., 2007. Updated world map of the Köppen-Geiger climate classification. Hydrology and Earth System Sciences 11, 1633–1644. (B) After Josselyn, M., 1983. The ecology of San Francisco Bay tidal marshes: a community profile. In: Fish and Wildlife Service. Division of Biological Services, Washington, DC, USA. (C) After California State Coastal Conservancy/U.S. Fish and Wildlife Service, 2003. (D) Photograph by Paula Pratolongo.*

humid type, the seasonal rainfall imposed a marked restriction in water availability, with a rainfall peak in winter and dry summers.

Natural salt marshes in the estuary were early described by Atwater and Hedel (1976). Typically, the low marsh was dominated by *Spartina foliosa*, where ambient water salinity was greater than 15, and by *Scirpus californicus*, *Phragmites australis*, or *Typha* sp. under lower salinities. West of Carquinez Strait, *Salicornia virginica* is a common dominant salt marsh species. Brackish marshes appear in the upper reaches of the estuary, at Suisun Bay. Typical zonation includes a low marsh, dominated by *Schoenoplectus californicus*, a middle marsh with *Typha* spp., *Scirpus olneyi*, and *Scirpus robustus*, and a high marsh dominated by *Juncus balticus*. Many other species also occur in this zone like the endangered species *Lilaeopsis masonii*, and where salinities are high, *Salicornia virginica*, *Distichlis spicata*, and *Grindelia humilus* may also occur (Josselyn, 1983). Additional species such as *Atriplex patula*, *Cotula coronopifolia*, *Cuscuta salina*, *Frankenia grandifolia*, *Oenanthe sarmentosa*, and *Pontentilla egedii* occurred along the inland fringe of the marsh, rarely inundated by tides (Barbour et al., 1973). Tidal influence from San Francisco estuary extends upstream into the Sacramento–San Joaquin Delta. Islands in this delta formed by the deposition of natural levees along channels, and the further filling of their central areas with sediments. The regularly flooded lowlands originally supported dense tidal freshwater marshes, dominated by *Scirpus* spp., *Phragmites australis*, and *Typha latifolia* (Ingebritsen et al., 2000). These natural marshes presently occupy a relatively small area because wetlands in San Francisco Bay have been converted to salt ponds, urban and industrial land uses, and agriculture on a large scale. *Spartina foliosa* is the only cordgrass species native to the San Francisco Estuary. It has a narrow range of distribution, limited to the Pacific coast of North America, from Humboldt Bay to Baja California. Similarly to *Spartina alterniflora*, *Spartina foliosa* takes on a tall or "robust" form, which grows at lower elevations, and a "dwarf" form at elevations closer to mean high water. By 2000, four nonnative species of *Spartina* had been reported in San Francisco Bay and nearby estuaries: *Spartina alterniflora*, *Spartina densiflora*, *Spartina patens*, and *Spartina anglica*, as well as the exotic hybrid *Spartina alterniflora* × *Spartina foliosa* (Fig. 3.8B). Within the bay, this hybrid has the largest distribution, but control practices have reduced their cover during the past years (Batzer and Baldwin, 2012).

In Willapa Bay, nearly one-third of the area originally covered by mudflats replaced by *Spartina alterniflora* marshes. This species was introduced into Willapa Bay during the late 1800s but was not identified until the 1940s. During the first 50 years, the population expanded slowly, but from 1945 to 1988 the plant spread rapidly throughout the bay, resulting in severe habitat alteration as unvegetated mudflats were converted to salt marshes (Simenstad and Thom, 1995). Herbicides have been effectively applied in an aggressive eradication campaign that has virtually eliminated *Spartina alterniflora* from Willapa Bay (Strong and Ayres, 2016).

4.4 Mediterranean

Similar to western North America, the area is under Mediterranean climate, but mostly the Mediterranean hot summer (Csa) subtype, also called dry summer subtropical. Compared with dry summer temperate (Csb) climates, this zone is characterized by warmer summer months and decreased water availability (Fig. 3.9A). The Tagus Estuary is the largest estuary

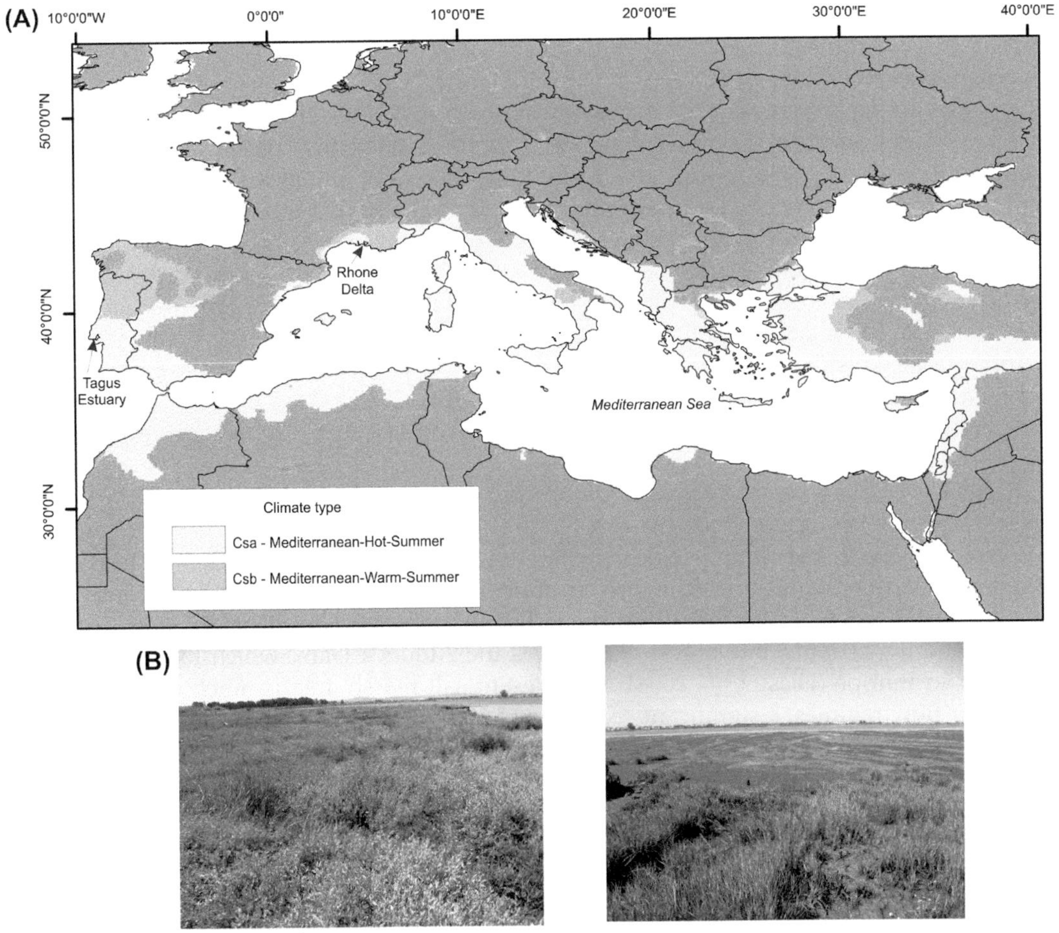

(A) 10°0'0"W 0°0'0" 10°0'0"E 20°0'0"E 30°0'0"E 40°0'0"E
50°0'0"N
40°0'0"N
30°0'0"N
Rhone Delta
Tagus Estuary
Mediterranean Sea
Climate type
Csa - Mediterranean-Hot-Summer
Csb - Mediterranean-Warm-Summer
(B)

FIGURE 3.9 (A) Climate classification within the Mediterranean zone of Europe and North Africa. (B) Rosário salt marsh, in the southern shore of the Tagus Estuary, Portugal. *(A) Modified from Peel, M.C., Finlayson, B.L., McMahon, T.A., 2007. Updated world map of the Köppen-Geiger climate classification. Hydrology and Earth System Sciences 11, 1633–1644. (B) Photograph by Vanesa Negrin.*

on the west coast of Europe. Almost 40% of the estuarine area is covered by mudflats, and salt marshes occur in the southern and eastern shores. Typical zonation in salt marshes of the Tagus Estuary includes homogeneous stands of *Spartina maritima* in the lower marsh area, with pure stands of *Sarcocornia perennis* at higher elevations. *Sarcocornia fruticosa* and *Halimione portulacoides* are commonly found in the upper salt marsh (Duarte et al., 2010). The Rosário salt marsh (Fig. 3.9B) covers an area of 200 ha in the southern shoreline of the estuary. The area is located near a heavy industrialized zone, and over the past decades chemical plants discharged effluents enriched in various contaminants into the estuary that were transported toward the Rosário marsh by tides. Several studies showed that Tagus salt marshes, including Rosário, incorporate large quantities of anthropogenic metals into the marsh sediments. Retention of contaminated particles and root–sediment interactions appear to contribute to the metal enrichment of sediments and belowground biomass (Santos-Echeandía et al., 2010).

Extensive temperate coastal wetlands are also found in Southwest Spain. The Doñana National Park comprises 54 ha of wetlands extending along the coastal delta plain of the Guadalquivir River. This mosaic of swamps, ponds, and marshes formed on estuarine sediment, but this maritime influence has declined so that only high tides inundate a small area, giving the marshes a continental character. Espinar et al. (2002) identify two main vegetation groups: a low marsh with *Bolboschoenus (Scirpus) maritimus* and *Schoenoplectus (Scirpus) litoralis* and a higher salt marsh with *Arthrocnemum macrostachyum* and *Juncus subulatus*. The Mediterranean climate, characterized by major seasonal rainfall and temperature contrasts, results in significant hydrological variability with maximum inundation in winter and completes drying in July (Green et al., 2016). The Doñana National Park protects this unique ecosystem in recognition of its biodiversity, ecology, and importance for migrating birds.

Further west along the coastal plain of the Gulf of Cadiz, the Guadiana Estuary forms the southwestern boundary of Portugal and Spain. The Portuguese (western) shore comprises and ebb tidal delta and dunes, which shelter salt marshes. On the Spanish (eastern) shore, a series of barrier islands and spits are separated by wide salt marshes. As a consequence of interannual rainfall variability, the river discharge of the Guadiana River is regulated by more than 100 dams, the largest of which is the Alqueva Dam, which forms the largest reservoir in Europe (Dias et al., 2004). The present salt marsh area is much reduced due to the lower flow regimes, a significant reduction in sediment supply and other anthropogenic pressures such as urbanization and aquaculture (Boski et al., 2008; Plater and Kirby, 2006; Sampath and Boski, 2016).

Coastal wetlands along the Mediterranean Sea are limited in distribution, with only very fragmented and patchy occurrences of marshes along protected shorelines (e.g., Eastern Adriatic coast of Croatia (Pandža et al., 2007)). Instead, salt marshes in the Mediterranean are mainly associated with deltas, such as those of the Rhône, Ebro, Nile, and Po, where the deceleration in the rise of sea level from 7000–6000 years BP favored a rapid deltaic progradation, dependent on alluvial sediment supplies (Vella et al., 2005). Similar to the Mississippi Delta, Mediterranean systems are subsiding, and many coastal wetlands in this region might presently face elevated rates of relative sea level rise (Day et al., 1995). The Rhône Delta (La Camargue) is one of the most important natural areas remaining in this region. Natural halophytic vegetation is composed of low shrubby species, instead of the herbaceous ones of the genus *Spartina*. Typical salt marsh communities include species such as *Sarcocornia fruticosa* and *Arthrocnemum macrostachyum* in the low marsh, an intermediate zone with *Limonium virgatum*, *Limonium girardianum*, *Frankenia pulverulenta*, and *Artemisia galla*, and a high marsh with *Juncus* spp. (Chapman, 1960).

However, intensive agriculture and human settlements have eliminated most of the original Mediterranean coastal wetland habitats (Benito et al., 2014). Changes in land and water uses, leading to artificial freshwater inputs and eutrophication, have largely altered habitat structure and functions of coastal wetlands (Marco-Barba et al., 2013; Prado et al., 2014). Rice fields, although their hydrology is no longer controlled by sea level, have become the main remaining wetland areas, especially in the northern Mediterranean, where they may offer a valuable habitat for waterbirds (Tourenq et al., 2001). Sustainable management of Mediterranean coastal wetlands through traditional aquaculture and capture fisheries is presently proposed as a regional strategy for habitat restoration and biodiversity conservation (Cataudella et al., 2015).

4.5 Eastern Asia

Major areas of marshes in eastern Asia occur along the temperate coasts of the East China Sea and the Bohai Sea where large rivers such as the Changjian (Yangtze River) and the Huanghe (Yellow River) develop extensive deltas. Coastal areas in this region have a humid subtropical climate, and the monsoon influence results in a modified subtype (Cwa), defined by a dry winter and larger annual temperature range than Cfa types (Fig. 3.10A). The high precipitation due to a monsoonal climate, the presence of large rivers with high sediment loads, and the Holocene relative sea level history are major reasons for the presence of extensive deltas in Asia (Saito, 2001). These monsoon-controlled coasts have been formed under a stable or slightly falling sea level over the past 6000 years. In most coastal regions of eastern Asia, the middle Holocene highstand was 2–4 m above the present relative sea level, and deltas began to build by sediment deposition and shoreline progradation as sea level fell to the present position (Song et al., 2013). The Yangtze River, the third largest in the world, transports 4.1×10^8 tons of sediments every year into the East China Sea. Most of these sediments are deposited in the estuary developing an extensive delta with a high freshwater influence (Chu et al., 2013).

The intertidal zone in the Yangtze Delta extends over a total area of 10,160 km^2 outside the seawall of 1980 (Li et al., 2014), and the most representative salt marshes in the estuary cover 242 km^2, in the Chongming Dongtan Nature Reserve, and 127 km^2 in the Jiuduansha Nature Reserve (Ge et al., 2016). The lower intertidal zone is typically characterized by mudflats lacking any vegetation cover, and *Scirpus mariqueter* is the typical low marsh species that appears at elevations higher than the neap high tide level. *Scirpus triqueter* and *Zizania aquatica* can be also found in locations with lower salinity, near the main channel of the Yangtze River (Gao and Zhang, 2006). In the high marsh, *Phragmites australis* is the common dominant and a variety of species such as *Imperata cylindrical, Suaeda glauca, Juncus setchuensis,* and *Carex scabriflora* may also appear in small patches (Gao and Zhang, 2006).

The Chongming Island is the largest island at the mouth of the Yangtze River Estuary. At the west end of this island, the Chongxi Wetland comprises more than 300 km^2 of freshwater tidal marshes. At this site, monospecific stands of *Phragmites australis* occupy the upper portions of the intertidal zone and make a significant contribution to the estuarine carbon budget through their extremely high net primary production (Zhang et al., 2014).

In 1979, *Spartina alterniflora* was transplanted into tidal marshes of coastal China to stabilize tidal flats. Since the mid-1990s, this species began its expansion in the Yangtze Estuary (Ouyang et al., 2013), and it had gradually invaded the former *Phragmites australis* communities and the upper *Scirpus mariqueter* zone, aggressively replacing these native species within the next 15 years (Cheng et al., 2006). *Spartina alterniflora* marshes presently occupy most of the upper intertidal zone on the coasts of Jiangsu, Shanghai, Zhejiang, and Fujian Provinces (Gao et al., 2014). Before the introduction of *Spartina alterniflora,*

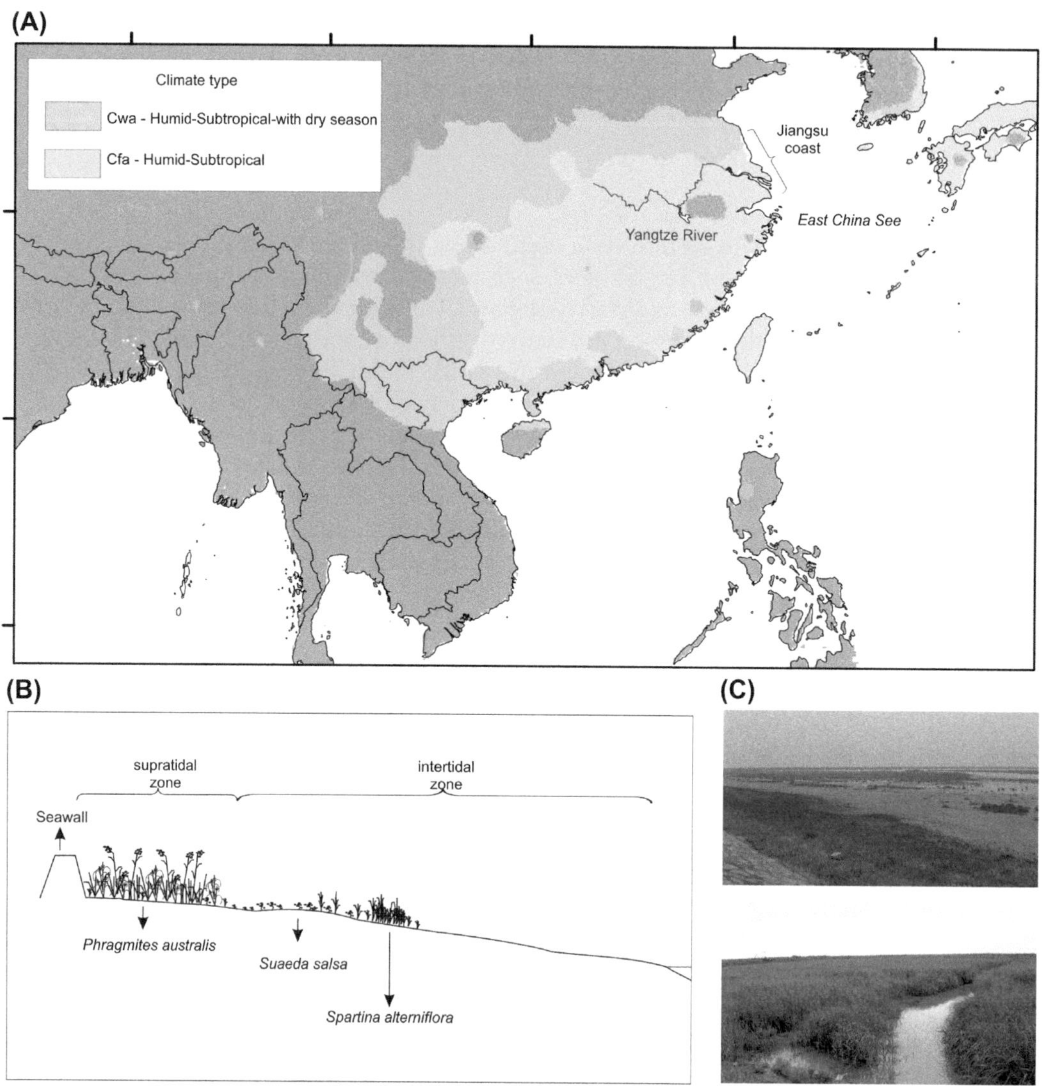

FIGURE 3.10 (A) Climate classification within the temperate zone of eastern Asia. (B) Typical sequence of plant species in coastal salt marshes of the Jiangsu Province, China. (C) Salt marshes in the Jiangsu coast, seaward from coastal defenses. *(A) Modified from Peel, M.C., Finlayson, B.L., McMahon, T.A., 2007. Updated world map of the Köppen-Geiger climate classification. Hydrology and Earth System Sciences 11, 1633–1644. (B) After Zhang et al. (2004). (C) Photograph by Benwei Shi.*

coastal wetlands in the Jiangsu Province typically included *Phragmites australis* in the supratidal zone and *Suaeda salsa* in the upper marsh, with bare mudflats at lower elevations. At present (Fig. 3.10B), the plant zonation of coastal wetlands includes *Spartina alterniflora* as the pioneer species dominating the upper intertidal zone, but marshes of *Suaeda salsa and*

Phragmites australis persist landward. In estuarine areas such as the Yangtze Delta, *Spartina alterniflora* occupies an elevation range that overlaps with *Scirpus mariqueter* and *Phragmites australis* marshes, and the expansion of *Spartina alterniflora* often leads to disappearance of the native marshes (Gao et al., 2014).

4.6 Australasia

Similar to eastern Asia, the coasts of Australia and New Zealand have also experienced a stable or slightly falling relative sea level since the mid-Holocene highstand. However, the internally arid continent in Australia constrains the sediment supply to coastal areas, precluding the formation of large deltas and allowing the persistence of unfilled estuaries behind barriers. Regarding climate, there is a sharp gradient in Australia between wet southeastern coasts (Fig. 3.11A), with humid subtropical to oceanic climates (Cfa to Cfb), and the dry southwestern counterparts, with Mediterranean to semiarid climates (Csb, Csa, and Bsh). The arid climate on the southwestern coast determines a more open vegetation cover in the high marshes and a more shrubby physiognomy, whereas on the wetter eastern coasts denser vegetation occurs, with a higher presence of grasses and sedges (Adam, 1990).

Although mangroves are associated with tropical climates worldwide, in Australia they also occur in temperate coastlines, where the southernmost mangroves in the world (*Avicennia marina*) occur in Corner Inlet, Victoria (38°54′25″S) (Saintilan et al., 2013). Mangrove expansion is a consistent pattern within the humid coasts of Australia and New Zealand (Stokes et al., 2010). An estimated 30% of salt marshes have been lost to mangroves across Australia (Saintilan et al., 2014), but rates of loss are lower (up to 10%) toward the southern limit in Victoria.

In Victoria, two different climatic patterns occur, with a drier central-west coastal section (Fig. 3.11B). In the humid southeastern Australia, salt marshes commonly develop to the landward side of mangroves, with *Sarcocornia quinqueflora* occupying the low marsh, *Sporobolus virginicus* and *Juncus kraussii* in the middle marsh, and *Phragmites australis* as the dominant reed in less saline areas (Boon et al., 2015). The only exceptions to this pattern occur in Tasmania, where mangroves are absent, and in the southern coasts of Victoria, where the introduced *Spartina anglica* grows at lower elevations than mangroves. In central-west coastal Victoria, under Mediterranean climate, the low summer rainfall and high temperatures can lead to intensely hypersaline conditions in the high marsh, where *Tecticornia pergranulata* and *Tecticornia halocnemoides* become dominant species (Boon et al., 2015). In the southwestern coastlines of Australia, under Mediterranean to semiarid climates, extremely saline systems, such as saline lakes and salt flats, commonly appear landward of mangroves. In Shark Bay, extensive supratidal flats with gypsum develop by evaporation of a shallow saline groundwater table (Burek and Prosser, 2008).

Most estuarine areas in New Zealand are shallow lagoons protected by sand bars. Typical marsh zonation, under a humid oceanic climate, includes a pioneer zone dominated by *Sarcocornia quinqueflora* and *Juncus maritimus* and a middle marsh with *Juncus articulatus*. At higher elevations, beyond mean high tides, a salt meadow develops, where species such as *Selliera radicans*, *Cotula coronopifolia*, and *Plagianthus divaricatus* commonly appear (Bishop, 1992).

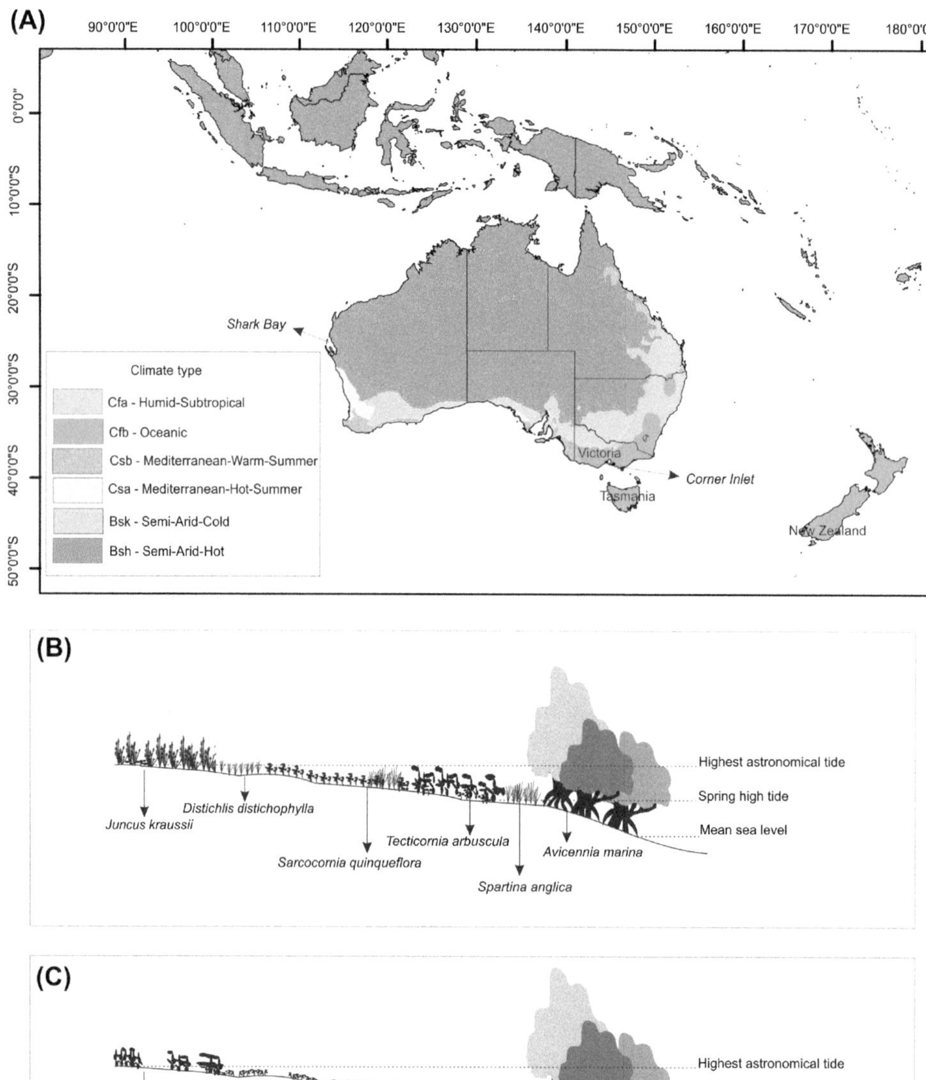

FIGURE 3.11 (A) Climate classification within the temperate zone of Australia and New Zealand. (B) Typical salt marshes in humid southeastern coastal Victoria. (C) Typical salt marshes in central-west coastal Victoria under Mediterranean climate. *(A) Modified from Peel, M.C., Finlayson, B.L., McMahon, T.A., 2007. Updated world map of the Köppen-Geiger climate classification. Hydrology and Earth System Sciences 11, 1633–1644. (C) After Boon, P.I., Allen, T., Carr, G., Frood, D., Harty, C., Mcmahon, A., Mathews, S., Rosengren, N., Sinclair, S., White, M., 2015. Coastal wetlands of Victoria, south-eastern Australia: providing the inventory and condition information needed for their effective management and conservation. Aquatic Conservation: Marine and Freshwater Ecosystems 25, 454–479.*

4.7 South America

Similar to North America, wetland development along the rugged western coasts of South America is restricted mostly to small barrier lagoons and fjord-head deltas in the south of Chile. In contrast, the Atlantic coast is occupied by a large number of coastal salt and freshwater marshes, from the temperate southern coasts of Brazil to Tierra del Fuego Island in Argentina. Although similar in the geographical configuration of the coast, lying along wide and shallow continental shelves, a falling or fluctuating relative sea level characterized South American late Holocene (Isla, 1989), instead of the rising sea level described in North American Atlantic coasts. In the northern coasts of Argentina, several authors locate a highstand about 6000 years BP, when the relative sea level reached around 6 m above present (Gómez and Perillo, 1995; Violante and Parker, 2000; Cavallotto et al., 2004). Southward, along the Patagonian coasts, several Holocene marine shorelines can be found at different elevations, with a southward increase in terrace elevation for terraces of the same age, suggesting a tectonic uplift effect (Rostami et al., 2000). The late Holocene falling trend in relative sea level has resulted in wide low-lying areas of former estuarine environments and typical regressive forms, such as extensive plains composed of beach ridge and lagoonal deposits. This regressive landscape, which is at present unconnected to tides, would become increasingly inundated under the recent rising trends in relative sea level. A zone called the arid diagonal crosses southern South America from the northwest to southeast, creating a sharp latitudinal gradient along the temperate Atlantic coasts and distinguishing the northern oceanic to humid subtropical climates (Cfb to Cfa) from the semiarid and arid counterparts (Bsk and BWk) through the south (Fig. 3.12A).

Two different patterns of salt marsh zonation have been described for Atlantic South America, and their boundary occurs approximately between 38°S and 39°S (Isacch et al., 2006). In temperate salt marshes from Southern Brazil to the coastal plain extending southward from the Río de la Plata Estuary, on the northern coasts of Argentina, *Spartina alterniflora* is the dominant species in the low marsh, typically forming monospecific stands. At higher elevations within this humid region, the middle salt marshes are dominated by mixed and monospecific stands of the southern cordgrass *Spartina densiflora* (Costa, 1997; Cagnoni, 1999; Costa et al., 2003) (Fig. 3.12B). Although extensive salt marshes of *Spartina densiflora* dominate coastal wetlands under the humid conditions of northern marshes, its presence south from 40°S would be strongly dependent on freshwater inputs, which are limited to the few rivers crossing the Patagonian plateau (Isacch et al., 2006).

The Río de la Plata Estuary is a major feature dominating the landscape in northern coasts of Argentina. The area comprises the Paraná River Delta, located at the head of the estuary, and extensive coastal plains that form a complex mosaic of freshwater wetlands, covering an area of about 17,500 km^2 (Malvarez, 1997). The Lower Delta, in the lower reaches of the Paraná River, is a 2700 km^2 freshwater tidal system, which is aggrading into Rio de la Plata Estuary at an estimated rate of 70 m/year (Parker and Marcolini, 1992). *Schoenoplectus californicus* is the pioneer species in the new bars, stabilizing sediments and establishing monospecific marshes. As these incipient islands mature, a mixture of forbs such as *Panicum grumosum, Ludwigia* spp., and *Senecio bonariensis* replace *Schoenoplectus californicus* and a surrounding levee begins to develop. Mature islands in the frontal zone finally assume the form of a central depression occupied by marshes, where *Scirpus giganteus* grows in nearly

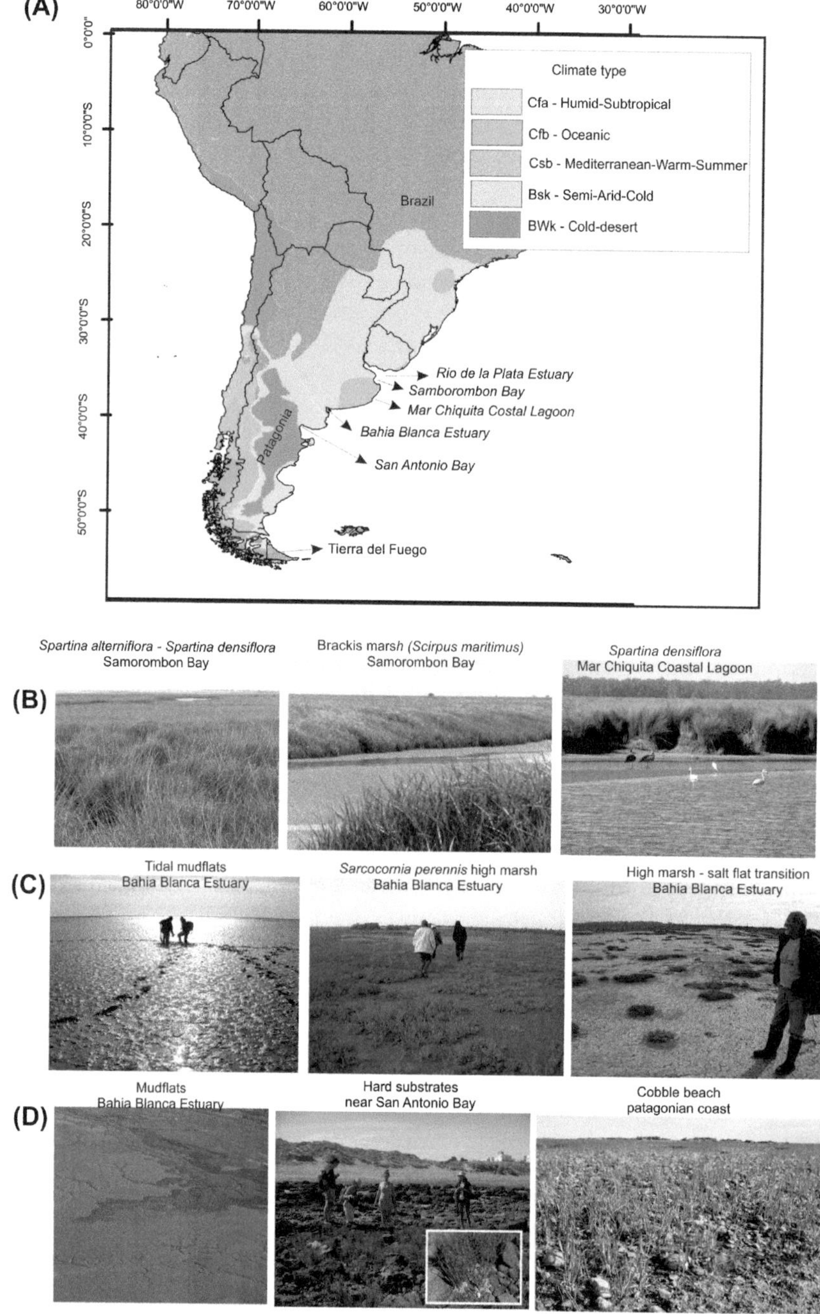

FIGURE 3.12 (A) Climate classification within the temperate zone of South America. (B) Typical salt marshes in coastal settings under humid climates in northern Argentina. (C) Coastal marshes in the Bahia Blanca Estuary in a transition to semiarid climates. (D) Expansions of *Spartina alterniflora* salt marshes over different substrates. *(A) Modified from Peel, M.C., Finlayson, B.L., McMahon, T.A., 2007. Updated world map of the Köppen-Geiger climate classification. Hydrology and Earth System Sciences 11, 1633–1644. (B) Photographs by Gabriela Gonzalez Trilla. (C, D) Photographs by Paula Pratolongo.*

monospecific stands and a surrounding levee covered with *Erythrina crista-galli* forest (Kandus and Malvárez, 2004; Pratolongo et al., 2005). As tidal influences attenuate upstream, the pulsing hydrology of the Paraná River becomes dominant (Neiff, 1999), and the tidal freshwater marshes in the Lower Delta progressively give place to a large corridor of inland freshwater wetlands, which extends along 3400 km of dam free rivers, from the Paraná Delta to the Gran Pantanal, through the Paraná River alluvial valley and the Paraguay River lowlands (Peteán and Cappato, 2005).

In a gradient of increasing aridity southward, vegetation becomes sparse in the intertidal zone. In the Bahía Blanca Estuary, at the northern limit of the Patagonian desert, most of the intertidal fringe is covered by extensive barren mudflats. Pure stands of *Spartina alterniflora* are commonly restricted to lower marshes in the middle reach of the estuary but do not appear in the inner zone (Pratolongo et al., 2010). Through the shallow inner section of the estuary, seasonally hypersaline conditions commonly develop because of the higher evaporation rates, and vegetated marshes in this zone are restricted to elevations close to the mean high tide level, with *Sarcocornia perennis* as the dominant species (Pratolongo et al., 2016). Above the elevation of the mean high tide in a zone irregularly inundated by seawater, vegetation is composed of a mosaic of salt flats, halophytic steppes, and shrubs (Fig. 3.12C).

Beyond tidal influence, an area of almost 195 km^2 is occupied by salt flats. In addition, more than 273 km^2 are covered by halophyte steppes, mainly composed of barren soil between vegetated patches. Vegetation in these supratidal patches is composed of typical southern species such as *Heterostachys ritteriana, Suaeda patagonica, Cressa truxillensis*, and *Allenrolfea patagonica*. Inland and coastal sabkhas of considerable extension have also been described further south, in Bahia San Blas, which develop under temperate semiarid and arid climates (Pratolongo et al., 2016). However, despite the wide area that they cover, salt flats in the region are virtually absent from the literature.

Noteworthy, *Spartina alterniflora* increased in both regional extent and abundance in Atlantic South America over the 20th century. Early records and indirect observations indicate that *Spartina alterniflora* was originally less abundant and more restricted in distribution in Suriname, Guyana, and Brazil. In the Bahia Blanca Estuary, Argentina, its area increased from 215 to 774 ha between 1967 and 2005 (Pratolongo et al., 2013), and similar expansions had been reported in Peninsula Valdes (42°24′ S) at the southernmost limit of its distribution (Bortolus et al., 2015) (Fig. 3.12D). Based on indirect evidence, like the species appearance in new regions, its association with anthropic mechanisms of dispersal, and the relatively isolated populations with restricted distribution, it has been recently proposed that *Spartina alterniflora* would be a nonnative species to Atlantic South America, which was accidentally introduced from either North America or Europe, sometime prior to 1817. If this is indeed the case, it would be one of the largest biological invasions involving this species, with these alien marshes entirely reshaping coastal systems (Bortolus et al., 2015).

5. HUMAN IMPACTS AND CLIMATE CHANGE

In coastal environments continuous change is the reference state or "normal" pattern. However, human activities can modify rates of natural changes, and there is evidence of anthropogenic actions that have significantly enhanced changes driven by natural agents. This section considers not just ecological change driven by global atmospheric and climate alterations but also coastal change created by human use of water on land and increased erosion of terrestrial sediments, as well as direct human destruction of coastal habitats.

5.1 Climate Change and Sea Level Rise

During the latter part of the 20th century, Earth's climate unequivocally warmed considerably faster than during the previous millennium. Long records of global mean temperature show that there was an increasing warming during the 20th century, particularly after the 1980s (Jones et al., 2001), and there is scientific consensus that temperatures will continue to rise. Projected warming for the end of the 21st century depends on greenhouse gas emissions, and for most scenarios, the global surface temperature increase is likely to exceed 1.5°C relative to the 1850–1900 period (IPCC, 2014).

Climate warming has the potential to completely alter the structure and function of coastal wetlands. The net primary production of wetland plants is higher at lower and warmer latitudes (Turner and Gosselink, 1975), and under warmer temperatures, the limits of tropical and subtropical mangrove communities are expected to migrate to higher latitudes (Snedaker, 1995). However, there are large uncertainties regarding the impacts of a warmer climate on coastal wetland function and the reciprocal effects of wetlands response on climate. The increase in atmospheric CO_2 is a major driver of climate change, and coastal wetlands have shown a great potential as sinks of blue carbon (Chmura et al., 2003; Beaumont et al., 2014). Increased atmospheric CO_2 could increase net primary production and carbon sequestration if nutrients, precipitation, and other factors are not limiting to plant growth (Rozema et al., 1991). Although growth enhancement would be common effect of elevated atmospheric CO_2 changes to the entire carbon cycle may alter the sequestration in ways that are still unclear.

Climate-enhanced sea level rise is a major consequence of global warming, and there has been considerable discussion as to how temperate coastal wetlands will respond in the future (Reed, 1990; Simas et al., 2001). A key determinant of coastal wetland vulnerability is whether its surface elevation can keep pace with rising sea level (Webb et al., 2013). Early studies (Titus, 1987; Boorman et al., 1989) predicted the large-scale loss of coastal wetlands as a consequence of sea level rise exceeding sediment supply. However, there are many feedbacks between plant growth and geomorphology, which allow salt marshes to cope with sea level rise (Kirwan and Megonigal, 2013). Measurements of salt marsh elevation worldwide indicate that marshes are generally building elevation at rates similar to or exceeding present sea level rise, and process-based models predict survival under a wide range of future sea level scenarios (Kirwan et al., 2016).

In addition to vertical accretion and progradation resulting from sedimentation, the different plant associations within the coastal wetlands continuum are expected to migrate landward in the future in response to rising sea levels (Christian et al., 2000). There are good examples of coastal marshes and mangroves through the world that possess the ability to move inland in response to sea level rise, but there are also many exceptions. Wetlands growing on islands within estuaries have no land to migrate (Kearney and Stevenson, 1991; Wray et al., 1995). Similarly, wetlands cannot migrate inland in places where the landward slope is too steep or where hard barriers were constructed landward. In these cases where transgression stalls, low sediment supply results in erosion at the seaward margin and a progressive loss of intertidal area. Under these situations, coastal wetlands may disappear by erosion over time through a process known as coastal squeeze (Moorhead and Brinson, 1995).

According to recent field observations and numerical modeling, salt marsh vulnerability would be largely dependent on biophysical feedback processes that accelerate soil building under a rising sea level. However, human activities interfere with these feedbacks through alteration of climate, nutrient inputs, sediment dynamics, and subsidence rates. Thus, the future development of coastal wetlands under a rapid sea level rise is strongly conditioned by human impacts and socioeconomic factors that influence transgression into adjacent uplands (Kirwan and Megonigal, 2013).

5.2 Alterations in Sediment Transport

Humans have altered fluvial transport of sediments to the coast, mainly through soil erosion enhanced by deforestation and agriculture, which in turn increase sediment loads in rivers (Douglas, 1990). On the other hand, the increasing freshwater requirements for human consumption and irrigation resulted in widespread construction of dams, dikes, and canals. Suspended solids accumulate behind dams, and there is a lower amount of sediments being exported to the coastal zone. As a common pattern, intensification of agriculture increases sediment loads in rivers. Dam construction is necessary to further augment agricultural production, which results in decreasing sediment loads entering coastal waters. In many deltas, there is a positive relationship between suspended sediments and land area, and variations in suspended sediments, higher or lower, would likely lead to proportional wetland changes in the coastal zone. Indeed, the operation of the Three Gorges Dam on the Yangtze has been linked to the decline in deltaic progradation rate—and even coastal erosion in some locations—due to reduced fluvial outflow and associated sediment load (Dai et al., 2014).

A well-known example of human-mediated changes in sediment transport is the catchment basin of the Mississippi River. Agriculture expanded in the area from the 1800s through the 1900s and enhanced soil erosion (Turner and Rabalais, 2003). In the 20th century, widespread reservoir construction and dikes built to control floods greatly reduced sediment transport downriver (Meade and Moody, 2010). Changes in area of the Mississippi delta clearly reflect these changes in sediment loads (Wells, 1996). In the mid-1800s, a period of rapid progradation resulted in 560 km^2 of new land (Wells and Coleman, 1987), but large wetland areas appearing in maps of the 1930s have been more recently eroded.

Anthropogenic changes in sediment transport have become the dominant factor determining whether relative sea level rise is positive or negative in many deltas (Syvitski and Kettner, 2011). Deltaic regions are particularly vulnerable to a relative sea level rise because of a rapid subsidence (Day et al., 1995). Under this scenario, river sediment supply is one of the most important agents in shaping deltaic evolution, and human-induced changes in sedimentary fluxes are thought to be a major cause of deltaic degradation when coupled with subsidence (Stanley and Warne, 1993). Holocene sedimentary evidence indicates that perimarine wetlands are particularly well suited to such a combination of reduced terrestrial sediment supply and increased sea level rise through autogenic sediment accumulation processes (Plater and Kirby, 2006).

5.3 Direct Human Transformations

Human impacts on wetlands have been registered since the end of the last glacial age, when the combination of changing climates and increasing pressures of expanding and migrating human populations extinguished a substantial number of wetland species (Martin and Wright, 1967). In the North and Wadden Seas, salt marshes have been used for livestock grazing since the Neolithic (Meier, 2004; Knottnerus, 2005), but large-scale wetland conversion to upland has been mostly undertaken for agriculture (Glover and Higham, 1996). Salt marsh reclamation for agriculture began in the Netherlands and France by the 11th century and probably earlier in China (Yoshinobu, 1998). Diking for flood protection is the most striking transformation of wetland areas and results in wetland conversion to uplands (Mendelssohn and Morris, 1999). Many areas that were originally reclaimed for agriculture were transformed into urban, residential, and industrial land. Large coastal cities such as Boston, San Francisco, Amsterdam, Rotterdam, Venice, and Tokyo expanded on former coastal wetlands (Pinder and Witherick, 1990).

The fact that many former and extensive perimarine wetlands are largely absent from European coastal lowlands is primarily due to land claim (e.g., the Somerset Levels, Romney Marsh, and Fenland systems in the United Kingdom (Rippon, 2000)). Former marshes in northern Europe have been extensively diked over the past 2000 years, and the few surviving areas are reduced in size and strongly modified. On the German coasts of the Wadden Sea, approximately 1000 km^2 of former coastal marshes have been diked over the last millennium. Continuous embankments of newly accreted land have shaped the mainland coasts of the northern Netherlands, where anthropogenic salt marshes extend over 190 km^2, and virtually no natural tidal marshes remain (Lotze, 2004). In the Severn estuary, about 840 km^2 of marshes have been impounded since the end of the Roman occupation, whereas only 14 km^2 of active marsh remains (Allen and Duffy, 1998), and similar relations between active and embanked marshes would also apply for France and Denmark (Allen, 2000). As a result of past and ongoing loss of intertidal habitat, managed realignment is increasingly favored as a soft coastal management tool as a means of salt marsh habitat creation in efforts to mitigate loss of intertidal area due to sea level rise and meteorological effects of climate change.

Prior to European settlement in the Bay of Fundy, marshes covered wide areas of the Minas Basin and the upper reaches of Chignecto Bay, where large amounts of fine sediments accumulate. However, during the past 400 years, lowlands have been intensely diked and reclaimed, with an estimated reduction of about 70% of the former marsh area (Gordon and Cranford, 1994). Agricultural expansion is the major cause of wetland losses. Since the early 1800s, wetland conversion to agriculture is estimated at over 20 million ha, including 65% of the coastal marshes of Atlantic Canada.

In Eastern Asia, several human activities, such as dredging and deepening of navigation channels, water diversions to northern China, and large-scale land reclamation are rapidly changing the Yangtze Delta environment (Xiqing, 1998). In the past 3 decades, the total area of salt marshes and mudflats has decreased from 1647 km^2 in 1985 to 1047 km^2 in 2014. The total area of land claim through sea dike construction has reached 1077 km^2, a value that exceeds the area of the remaining tidal flats (Chen et al., 2016). Since cultivation in this region dates back to more than 7000 years, human activities have played a major role in shaping present landscapes. With more than 50% of the world's population living in Asia, and most of the world's rice fields occurring in Asian deltas (Galloway and Melillo, 1998), human pressures on the contributing watersheds are expected to increase. In China, the 1999 Land Administration Law stressed a dynamic balance farmland policy (no net loss in farmland) and encouraged local governments to reclaim tidal flats for arable land compensation. The Twelfth Five-Year National Plan in 2010 proposed the claiming of 587 km^2 of tidal flats in Shanghai till 2020. In Jiangsu, the Million Land Project aims to enclose a further 1800 km^2 of tidal flats according to the Jiangsu Marine Function Zoning (2006–20). Given that human pressures have greatly influenced the past and present coastal processes, the future development of big Asian deltas may be driven, to a great extent, by large-scale engineering projects.

In the temperate coasts of Australia, reclamation for industrial and human development has been concentrated on the southeastern coasts, but losses have been small, compared with the original wetland area (Adam, 1990). Besides reclaimed areas, the main threats to salt marshes are ecosystem degradation due to alterations in hydrologic regimes, pollution, and weed invasion. The presence of large wetland complexes, comparatively little affected by human alterations, is an attribute that distinguishes South American marshes. Salt marshes have commonly been used as pasture for livestock, and the seasonal burning of vegetation to improve fodder quality is the most common management practice. However, the rapidly growing coastal population, the massive clearance of upland forests for agriculture, the poorly controlled industrial effluents, and the untreated sewage discharge from coastal cities are major threatens to the coastal environment (Pratolongo et al., 2013).

6. SUMMARY

Temperate coastal wetlands include a large variety of environments, from tidal flats and salt marshes to nontidal wetlands at the landward edge, whose hydrology is still influenced by sea level. Salt marsh evolution and accretion depends on several factors such as sediment availability, vegetation cover, marsh platform elevation, and hydrodynamic forces. Different models aim to reproduce the physical and ecological processes driving marsh evolution by using different levels of simplification. The evolution of a salt marsh from mudflats can be described using a simplified model that considers the differences in elevation between mudflats and salt marshes. Changes in elevation result from the difference between erosion and deposition rates. Models developed to represent vertical salt marsh dynamics use different formulations in terms of both sediment and vegetation dynamics. These models mostly predict that salt marsh accretion rates would increase in more frequently flooded areas and also in the presence of dense vegetation. Finally, the lateral migration of salt marshes due to wind wave erosion is one of the principal causes for salt marsh losses worldwide. By using numerical models and field data, it has been shown that this lateral migration is mainly caused by average weather conditions rather than by extreme storms.

Salt marsh plants commonly exhibit clear patterns of zonation, driven by the individual species tolerance to physical stress and biological interactions acting across the elevation gradient. The result is a shore-parallel zonation of plants which is made more complex and spatially variable by the micromorphology of the marsh surface. In low marsh areas, salinity is comparatively low because of regular tidal flushing, but soil salinity levels vary across marsh elevations. In humid warm regions, freshwater input from rain and upland sources moderates salinity at the terrestrial border, but salts can concentrate by evaporation at intermediate marsh elevations. Mid-marshes characterized by higher salinities support a more salt-tolerant flora, and salt accumulations may also lead to the development of bare areas known as salt pans. In humid climates, the upper salt marsh commonly grades into freshwater communities. In arid climates, however, salinities can exceed the limits of even the most tolerant halophytes, and salt flats devoid of vascular vegetation develop near the upland boundary.

Tidal marshes throughout the temperate zone would be comparable in terms of ecosystem structure and function. However, substantial differences arise among major geographic regions due to precipitation regimes. Typical patterns of salt marsh plant zonation largely reflect rainfall amount and seasonality, along with the biogeographic distribution of species. In addition, human activities have also exerted profound changes in coastal wetlands, and the processes of wetland loss and degradation have been quite variable in space and time, leading to major regional differences in the extent and ecological integrity of remaining wetland areas.

The evolution of salt marshes over time is strongly influenced by climate change and human-induced alterations. Major human impacts are associated with land claims and alterations of fluvial sediment transport. Considering a future climate-enhanced sea level rise, early studies predicted the large-scale loss of coastal wetlands as a consequence of sea level rise exceeding sediment supply. However, salt marsh vulnerability would be largely dependent on biophysical feedback processes that accelerate soil building under a rising sea level. Thus, the future development of coastal wetlands under a rapid sea level rise would be strongly conditioned by human activities that interfere with these feedbacks.

References

Adam, P., 1990. Saltmarsh Ecology. Cambridge University Press, Cambridge, England.

Alber, M., Swenson, E., Adamowicz, S., Mendelssohn, I., 2008. Salt Marsh Dieback: an overview of recent events in the US. Estuarine, Coastal and Shelf Science 80, 1–11.

Allen, J., 2000. Morphodynamics of Holocene salt marshes: a review sketch from the Atlantic and Southern North Sea coasts of Europe. Quaternary Science Reviews 19, 1155–1231.

Allen, J.R.L., Pye, K., 1992. Saltmarshes: Morphodynamics, Conservation and Engineering Significance. Cambridge University Press, Cambridge, UK.

Allen, J.R.L., Duffy, M.J., 1998. Temporal and spatial depositional patterns in the Severn Estuary, southwest Britain: intertidal studies at spring-neap and seasonal scales 1991-1993. Marine Geology 146, 147–171.

Ashall, L.M., Mulligan, R.P., Law, B.A., 2016. Variability in suspended sediment concentration in the Minas Basin, Bay of Fundy, and implications for changes due to tidal power extraction. Coastal Engineering 107, 102–115.

Atwater, B.F., Hedel, C.W., 1976. Distribution of Seed Plants with Respect to Top Water Levels and Water Salinity in the Natural Tidal Marshes of the Northern San Francisco Bay Estuary, California. U.S. Geological Survey Open File Report 76-0389, Reston, USA.

Atwater, B.F., Hedel, C.W., Helley, E.J., 1977. Late quaternary depositional history, Holocene sea-level changes, and vertical crustal movement, southern San Francisco Bay, California. In: U.S. Geological Survey Professional Paper 1014, Washington, D.C., USA.

Ayres, D.R., Strong, D.R., 2001. Origin and genetic diversity of *Spartina anglica* (Poaceae) using nuclear DNA markers. American Journal of Botany 88, 1863–1867.

Baily, B., Pearson, A.W., 2007. Change detection mapping and analysis of salt marsh areas of central southern England from Hurst Castle Spit to Pagham Harbour. Journal of Coastal Research 1549–1564.

Bak, P., 1996. How nature works: the science of self-organized criticality. Nature 383, 772–773.

Bak, P., Tang, C., Wiesenfeld, K., 1987. Self-organized criticality: An explanation of the 1/f noise. Physical Review Letters 59 (4), 381.

Baptist, M.J., Babovic, V., Rodríguez Uthurburu, J., Keijzer, M., Uittenbogaard, R.E., Mynett, A., Verwey, A., 2007. On inducing equations for vegetation resistance. Journal of Hydraulic Research 45, 435–450.

Barbier, E.B., Hacker, S.D., Kennedy, C., Koch, E.W., Stier, A.C., Silliman, B.R., 2011. The value of estuarine and coastal ecosystem services. Ecological Monographs 81, 169–193.

Barbour, M.G., Craig, R.B., Drysdale, F.R., Ghiselin, M.T., 1973. Coastal Ecology, Bodega Head. University of California Press, Berkeley, USA.

Barendregt, A., Whigham, D.F., Meire, P., Baldwin, A.H., Van Damme, S., 2006. Wetlands in the tidal freshwater zone. In: Bobbink, R., Beltman, B., Verhoeven, J.T.A., Whigham, D.F. (Eds.), Wetlands: Functioning, Biodiversity Conservation, and Restoration. Springer Berlin Heidelberg, Berlin, Germany, pp. 117–148.

Barkowski, J.W., Kolditz, K., Brumsack, H., Freund, H., 2009. The impact of tidal inundation on salt marsh vegetation after de-embankment on Langeoog Island, Germany—six years time series of permanent plots. Journal of Coastal Conservation 13, 185–206.

Barth, H.J., Böer, B., 2002. Introduction. In: Barth, H.J., Böer, B. (Eds.), Sabkha Ecosystems: Volume I: The Arabian Peninsula and Adjacent Countries. Kluwer Academic Publishers, Dordrecht, Netherlands, pp. 1–5.

Batzer, D.P., Baldwin, A.H., 2012. Wetland Habitats of North America: Ecology and Conservation Concerns. Univ of California Press, San Diego, USA.

Beaumont, N.J., Jones, L., Garbutt, A., Hanson, J.D., Toberman, M., 2014. The value of carbon sequestration and storage in coastal habitats. Estuarine, Coastal and Shelf Science 137, 32–40.

Benito, X., Trobajo, R., Ibáñez, C., 2014. Modelling habitat distribution of Mediterranean coastal wetlands: the Ebro delta as case study. Wetlands 34, 775–785.

Bertness, M.D., 1991. Zonation of Spartina patens and *Spartina alterniflora* in a New England salt marsh. Ecology 72, 138–148.

Bertness, M.D., Ewanchuk, P.J., Silliman, B.R., 2002. Anthropogenic modification of New England salt marsh landscapes. Proceedings of the National Academy of Sciences 99, 1395–1398.

Bishop, N., 1992. Natural History of New Zealand. Hodder and Stoughton, Auckland, New Zaeland.

Blumberg, A.F., Mellor, G.L., 1987. A description of a three-dimensional coastal ocean circulation model. In: Three-Dimensional Coastal Ocean Models. American Geophysical Union, pp. 1–16.

Boon, P.I., Allen, T., Carr, G., Frood, D., Harty, C., Mcmahon, A., Mathews, S., Rosengren, N., Sinclair, S., White, M., 2015. Coastal wetlands of Victoria, south-eastern Australia: providing the inventory and condition information needed for their effective management and conservation. Aquatic Conservation: Marine and Freshwater Ecosystems 25, 454–479.

Boorman, L.A., Goss-Custard, J.D., McGrorty, S., 1989. Climate Change, Rising Sea-Level and the British Coast. ITE Research Publication No. 1, HMSO, London.

Boorman, L.A., Garbutt, A., Barratt, D., 1998. The role of vegetation in determining patterns of the accretion of salt marsh sediments. In: Black, K.S., Patterson, D.M., Cramp, A. (Eds.), Sedimentary Processes in the Intertidal Zone. Geological Society, London, pp. 389–399.

Bortolus, A., Carlton, J.T., Schwindt, E., 2015. Reimagining South American coasts: unveiling the hidden invasion history of an iconic ecological engineer. Diversity and Distributions 21, 1267–1283.

Boski, T., Camacho, S., Moura, D., Flewtcher, W., Wilamowski, A., Veiga-Pires, C., Correira, V., Loureira, C., Santana, P., 2008. Chronology of the sedimentary processes during the postglacial sea level rise in two estuaries of the Algarve coast, Southern Portugal. Estuarine, Coastal and Shelf Science 77, 230–244.

Brinson, M.M., 1991. Landscape properties of pocosins and associated wetlands. Wetlands 11, 441–465.

Brinson, M.M., 1993. A hydrogeomorphic classification for wetlands. In: Wetlands Research Program Technical Report WRP-DE-4. US Army Corps of Engineers, Washington D.C., USA.

Brinson, M.M., Christian, R.R., Blum, L.K., 1995. Multiple states in the sea-level induced transition from terrestrial forest to estuary. Estuaries 18, 648–659.

Burek, C.D., Prosser, C.V. (Eds.), 2008. The History of Geoconservation. Geological Society of London, London, UK.

Cagnoni, M., 1999. Espartillares de la costa bonaerense de la República Argentina. Un caso de humedales costeros. In: Malvárez, A.I. (Ed.), Tópicos Sobre Humedales Subtropicales Y Templados de Sudamérica. MAB-UNESCO, Montevideo, pp. 55–69.

Callaway, R.M., Sabraw, C.S., 1994. Effects of Variable precipitation on the structure and diversity of a California salt marsh community. Journal of Vegetation Science 5, 433–438.

Carling, P.A., Williams, J.J., Croudace, I.W., Amos, C.L., 2009. Formation of mud ridge and runnels in the intertidal zone of the Severn estuary, UK. Continental Shelf Research 29, 1913–1926.

Carter, R.W.G., 1988. Coastal Environments. Academic Press, London.

Cataudella, S., Crosetti, D., Massa, F., 2015. Mediterranean Coastal Lagoons: Sustainable Management and Interactions Among Aquaculture, Capture Fisheries and the Environment. FAO Studies and Reviews No. 95. General Fisheries Commission for the Mediterranean, Rome, Italy.

Cavallotto, J.L., Violante, R.A., Parker, G., 2004. Sea-level fluctuations during the last 8600 years in the de la Plata river (Argentina). Quaternary International 114, 155–165.

Chapman, V.J., 1960. Salt Marshes and Deserts of the World. Intersciences, New York, USA.

Chen, Y., Dong, J., Xiao, X., Zhang, M., Tian, B., Zhou, Y., Li, B., Ma, Z., 2016. Land claim and loss of tidal flats in the Yangtze Estuary. Scientific Reports 6, 24018.

Cheng, X., Luo, Y., Chen, J., Lin, G., Li, B., 2006. Short-term C4 plant *Spartina alterniflora* invasions change the soil carbon in C3 plant-dominated tidal wetlands on a growing estuarine Island. Soil Biology and Biogeochemistry 38, 3380–3386.

Chmura, G., Chase, P., Bercovitch, J., 1997. Climatic controls of the middle marsh zone in the Bay of Fundy. Estuaries 20, 689–699.

Chmura, G.L., Anisfeld, S.C., Cahoon, D.R., Lynch, J.C., 2003. Global carbon sequestration in tidal, saline wetland soils. Global Biogeochemical Cycles 17, 1111–1133.

Christian, R.R., Stasavich, L., Thomas, C.R., Brinson, M.M., 2000. Reference is a moving target in sea-level controlled wetlands. In: Weinstein, M.P., Kreeger, D.A. (Eds.), Concepts and Controversies in Tidal Marsh Ecology. Kluwer Press, The Netherlands, pp. 805–825.

Chu, Z., Yang, X., Feng, X., Fan, D., Li, Y., Shen, X., Miao, A., 2013. Temporal and spatial changes in coastline movement of the Yangtze delta during 1974–2010. Journal of Asian Earth Sciences 66, 166–174.

Costa, C.S.B., 1997. Subtropical convergence environments: the coast and sea in the warm-temperate southwestern Atlantic. In: Seeliger, U., Odebrecht, C., Castello, J.P. (Eds.), Tidal Marshes and Wetlands. Springer-Verlag, Berlin, pp. 24–26.

Costa, C.S.B., Marangoni, J.C., Azevedo, A.M.G., 2003. Plant zonation in irregularly flooded salt marshes: relative importance of stress tolerance and biological interactions. Journal of Ecology 91, 951–965.

Crawford, J.T., Stone, A.G., 2015. Relationships between soil composition and *Spartina alterniflora* dieback in an Atlantic salt marsh. Wetlands 35, 13–20.

Christiansen, T., Wiberg, P.L., Milligan, T.G., 2000. Flow and sediment transport on a tidal salt marsh. Estuarine, Coastal and Shelf Science 50, 315–331.

Cronk, J.K., Fenessy, M.S., 2001. Wetland Plants Biology and Ecology. Lewis Publishers, Florida.

Dai, Z., Liu, J.T., Wei, E., Chen, J., 2014. Detection of the three Gorges dam influence on the Changjiang (Yangtze River) submerged delta. Scientific Reports 4, 6600. https://doi.org/10.1038/srep06600.

Davy, A.J., Costa, C.S.B., 1992. Development and organization of saltmarsh communities. In: Seeliger, U. (Ed.), Coastal Plant Communities of Latin America. Academic Press, San Diego, USA, pp. 157–178.

Day, J., Didier, P., Hensel, P., Ibañez, C., 1995. Impacts of sea-level rise on deltas in the Gulf of Mexico and the Mediterranean: the importance of pulsing events to sustainability. Estuaries and Coasts 18, 636–647.

Day, J., Shaffer, G., Britsch, L., Reed, D., Hawes, S., Cahoon, D., 2000. Pattern and process of land loss in the Mississippi delta: a spatial and temporal analysis of wetland habitat change. Estuaries 23, 425–438.

DeLaune, R.D., Smith, C.J., Patrick Jr., W.H., 1983. Relationship of marsh elevation, redox potential, and sulfide to Spartina alterniflora productivity. Soil Science Society of America Journal 47, 930–935.

Dearing, J.A., Richmond, N., Plater, A.J., Wolf, J., Prandle, D., Coulthard, T.J., 2006. Modelling approaches for coastal simulation based on cellular automata: the need and potential. Philosophical Transactions of the Royal Society A: Mathematical, Physical and Engineering Sciences 364, 1051–1071.

Defina, A., Carniello, L., Fagherazzi, S., D'Alpaos, L., 2007. Self-organization of shallow basins in tidal flats and salt marshes. Journal of Geophysical Research: Earth Surface 112, F03001.

Dias, J.M.A., Gonzalez, R., Ferreira, O., 2004. Natural versus anthropic causes in variations of sand export from river basins: an example from the Guadiana river mouth (Southwestern Iberia). In: Rapid Transgression into Semi-enclosed Basins, p. 95e102. Polish Geological Institute Special Papers, Gdansk.

Dionne, J.C., 1989. An estimate of shore ice action in a *Spartina* tidal marsh, St. Lawrence Estuary, Quebec, Canada. Journal of Coastal Research 5, 281–293.

Doody, J.P., 1992. The conservation of British saltmarshes. In: Allen, J.R.L., Pye, K. (Eds.), Saltmarshes: Morphodynamics, Conservation and Engineering Significance. Cambridge University Press, Cambridge, pp. 80–114.

Douglas, I., 1990. Sediment transfer and siltation. In: Turner, B.L., Clark, W.C., Kates, R.W., Richards, J.F., Matthews, J.T., Meyer, W.B. (Eds.), The Earth as Transformed by Human Action. Cambridge University Press, Cambridge, UK, pp. 215–234.

Duarte, B., Caetano, M., Almeida, P.R., Vale, C., Caçador, I., 2010. Accumulation and biological cycling of heavy metal in four salt marsh species, from Tagus estuary (Portugal). Environmental Pollution 158, 1661–1668.

Dyer, K.R., Christie, M.C., Wright, E.W., 2000. The classification of intertidal mudflats. Continental Shelf Research 20, 1039–1060.

Ecke, F., Rydin, H., 2000. Succession on a land uplift coast in relation to plant strategy theory. Annales Botanici Fennici 37, 163–171.

Edwards, R.J., Horton, B.P., 2000. Reconstructing relative sea-level change using UK salt-marsh foraminifera. Marine Geology 169, 41–56.

Edwards, R.J., Horton, B.P., 2006. Developing detailed records of relative sea-level change using a foraminiferal transfer function: an example from North Norfolk, UK. Philosophical Transactions of the Royal Society A: Mathematical, Physical and Engineering Sciences 364, 973–991.

Eisma, D., Augustinus, P.G., Alexanderm, C., 1991. Recent and subrecent changes in the dispersal of Amazon mud. Netherlands Journal of Sea Research 28, 181–192.

Espinar, J.L., García, L.V., García-Murillo, P., Toja, J., 2002. Submerged macrophyte zonation in a Mediterranean salt marsh: a facilitation effect from established helophytes? Journal of Vegetation Science 13, 831–840.

Ewanchuk, P.J., Bertness, M.D., 2003. Recovery of a northern New England salt marsh plant community from winter icing. Oecologia 136, 616–626.

Fagherazzi, S., Carniello, L., D'Alpaos, L., Defina, A., 2006. Critical bifurcation of shallow microtidal landforms in tidal flats and salt marshes. Proceedings of the National Academy of Sciences 103, 8337–8341.

Fagherazzi, S., Kirwan, M.L., Mudd, S.M., Guntenspergen, G.R., Temmerman, S., D'Alpaos, A., van de Koppel, J., Rybczyk, J.M., Reyes, E., Craft, C., Clough, J., 2012. Numerical models of salt marsh evolution: ecological, geomorphic, and climatic factors. Reviews of Geophysics 50, RG1002.

Flemming, B.W., 2002. Geographic distribution of muddy coasts. In: Healy, T., Wang, Y., Healy, A. (Eds.), Muddy Coasts of the World: Processes, Deposits and Function. Elsevier, Amsterdam, The Netherlands, pp. 99–201.

Francalanci, S., Bendoni, M., Rinaldi, M., Solari, L., 2013. Ecomorphodynamic evolution of salt marshes: experimental observations of bank retreat processes. Geomorphology 195, 53–65.

Fredsøe, J., Deigaard, R., 1992. Mechanics of Coastal Sediment Transport. World scientific, Singapore.

French, J.R., 1993. Numerical simulation of vertical marsh growth and adjustment to accelerated sea level rise, North Norfolk, UK. Earth Surface Processes and Landforms 81, 63–81.

French, J.R., Spencer, T., 1993. Dynamics of sedimentation in a tide-dominated backbarrier salt marsh, Norfolk, UK. Marine Geology 110, 315–331.

French, J.R., Spencer, T., Murray, A.L., Arnold, N.S., 1995. Geostatistical analysis of sediment deposition in two small tidal wetlands, Norfolk, UK. Journal of Coastal Research 11, 308–321.

Galloway, J.N., Melillo, J.M., 1998. Asian Change in the Context of Global Climate Change. Cambridge University Press, Cambridge, UK.

Gao, S., Du, Y., Xie, W., Gao, W., Wang, D., Wu, X., 2014. Environment-ecosystem dynamic processes of Spartina alterniflora salt-marshes along the eastern China coastlines. Science China Earth Sciences 57, 2567–2586.

Gao, Z.G., Zhang, L.Q., 2006. Multi-seasonal spectral characteristics analysis of coastal salt marsh vegetation in Shanghai, China. Estuarine, Coastal and Shelf Science 69, 217–224.

Gardner, L.R., Reeves, H.W., Thibodeau, P.M., 2000. Groundwater dynamics along forest-marsh transects in a southeastern salt marsh, USA: description, interpretation, and challenges for numerical modeling. Wetlands Ecology and Management 10, 145–159.

Garofalo, D., 1980. The influence of wetland vegetation in tidal stream migration and morphology. Estuaries 3, 258–270.

Ge, Z.-M., Wang, H., Cao, H.-B., Zhao, B., Zhou, X., Peltola, H., Cui, L.-F., Li, X.-Z., Zhang, L.-Q., 2016. Responses of eastern Chinese coastal salt marshes to sea-level rise combined with vegetative and sedimentary processes. Scientific Reports 6, 28466.

Gehrels, W.R., 2000. Using foraminiferal transfer functions to produce high-resolution sea-level records from salt-marsh deposits, Maine, USA. The Holocene 10, 367–376.

Glover, I.A., Higham, C.F.W., 1996. New evidence for early rice cultivation in south, southeast and East Asia. In: Harris, D.R. (Ed.), The Origins and Spread of Agriculture and Pastoralism in Eurasia. UCL Press, London, UK, pp. 413–441.

Gómez, E.A., Perillo, G.M.E., 1995. Submarine Outcrops underneath Shoreface- Connected Sand Ridges, outer Bahía Blanca Estuary, Argentina. Quaternary of South America and Antarctic Peninsula 9, 23–37.

Gordon, D.C., Cranford, P.J., 1994. Export of organic matter from macrotidal salt marshes in the upper Bay of Fundy, Canada. In: Mitsch, W.J. (Ed.), Global Wetlands: Old World and New. Elsevier, Amsterdam, The Netherlands, pp. 257–264.

Gosselink, J., Pendleton, E.C., 1984. The Ecology of Delta Marshes of Coastal Louisiana: A Community Profile. Louisiana State University Baton Rouge Center for Wetland Resources.

Green, A.J., Bustamante, J., Janss, G.F.E., Fernández-Zamudio, R., Díaz-Paniagua, C., 2016. In: Finlayson, C.M., et al. (Eds.), The Wetland Book. Springer, Dordrecht.

Hageman, B.P., 1969. Development of the western part of The Netherlands during the Holocene. Geologie en Mijnbouw 48, 373–386.

Hugo, G., 2011. Future demographic change and its interactions with migration and climate change. Global Environmental Change 21, S21–S33.

Ingebritsen, S.E., Ikehara, M.E., Galloway, D.L., Jones, D.R., 2000. Delta Subsidence in California: The Sinking Heart of the State. U.S. Geological Survey FS-005-00.

IPCC, 2014. Climate Change 2014: Synthesis Report. Contribution of Working Groups I, II and III to the Fifth Assessment Report of the Intergovernmental Panel on Climate Change. Intergovernmental Panel on Climate Change, Geneva, Switzerland.

Isacch, J., Costa, C., Rodriguez-Gallego, L., Conde, D., Escapa, M., Gagliardini, D., Iribarne, O., 2006. Distribution of salt marsh plant communities associated with environmental factors along a latitudinal gradient on the southwest Atlantic coast. Journal of Biogeography 33, 888–900.

Isla, F.I., 1989. Holocene sea-level fluctuation in the southern hemisphere. Quaternary Science Reviews 8, 359–368.

Jacobson Jr., G.L., Jacobson, H.A., 1989. An inventory of distribution and variation in salt marshes from different settings along the Maine coast. In: Anderson, W.A., Borns Jr., H.W. (Eds.), Neotectonics of Maine, pp. 69–83.

Jones, P.D., Osborn, T.J., Briffa, K.R., 2001. The evolution of climate over the last millennium. Science 292, 662–667.

Josselyn, M., 1983. The ecology of San Francisco Bay tidal marshes: a community profile. In: Fish and Wildlife Service. Division of Biological Services, Washington, DC, USA.

Kandus, P., Malvárez, A.I., 2004. Vegetation patterns and change analysis in the lower delta islands of the Parana River (Argentina). Wetlands 24, 620–632.

Kearney, M.S., Stevenson, J.C., 1991. Island land loss and marsh vertical accretion rate evidence for historical sea-level changes in Chesapeake Bay. Journal of Coastal Research 7, 403–415.

Khan, N.S., Ashe, E., Shaw, T.A., Vacchi, M., Walker, J., Peltier, W.R., Kopp, R.E., Horton, B.P., 2015. Holocene relative sea-level changes from near, intermediate, and far-field locations. Current Climate Change Reports 1, 247–262.

Kirby, J.R., 2001. Regional late quaternary marine and perimarine record. In: Bateman, M.D., Buckland, P.C., Frederick, C.D., Whitehouse, N.J. (Eds.), The Quaternary of East Yorkshire and North Lincolnshire. Field Guide. Quaternary Research Association, London, pp. 25–34.

Kirwan, M., Temmerman, S., 2009. Coastal marsh response to historical and future sea-level acceleration. Quaternary Science Reviews 28, 1801–1808.

Kirwan, M.L., Megonigal, J.P., 2013. Tidal wetland stability in the face of human impacts and sea-level rise. Nature 504, 53–60.

Kirwan, M.L., Temmerman, S., Skeehan, E.E., Guntenspergen, G.R., Fagherazzi, S., 2016. Overestimation of marsh vulnerability to sea level rise. Nature Climate Change 6, 253–260.

Knottnerus, O.S., 2005. History of human settlement, cultural change and interference with the marine environment. Helgoland Marine Research 59, 2–8.

Lacambra, C., Cutts, N., Allen, J., Burd, F., Elliott, M., 2004. Spartina Anglica: A Review of Its Status, Dynamics and Management. English Nature Research Reports No. 527. English Nature, Peterborough, UK.

Langlois, E., Bonis, A., Bouzillé, J.B., 2003. Sediment and plant dynamics in saltmarshes pioneer zone: *Puccinellia maritima* as a key species? Estuarine, Coastal and Shelf Science 56, 239–249.

Lefeuvre, J.C. (Ed.), 1996. Effect of Environmental Changes on Salt Marsh Processes. Commission of the European Community. Laboratoire d'Evolution des Systèmes Naturels et Modifiés, University of Rennes, France. EEC Contrat No. E5V-CT92-0098. Final report.

Lefeuvre, J.C., Bouchard, V., Feunteun, E., Grare, S., Laffaille, P., Radureau, A., 2000. European salt marshes diversity and functioning: the case study of the Mont Saint-Michel bay, France. Wetlands Ecology and Management 8, 147–161.

Leonard, L.A., Croft, A.L., 2006. The effect of standing biomass on flow velocity and turbulence in *Spartina alterniflora* canopies. Estuarine, Coastal and Shelf Science 69, 325–336.

Leonard, L.A., Luther, M.E., 1995. Flow hydrodynamics in tidal marsh canopies. Limnology and Oceanography 40, 1474–1484.

Leonard, L.A., Wren, P.A., Beavers, R.L., 2002. Flow dynamics and sedimentation in *Spartina alterniflora* and *Phragmites australis* marshes of Chesapeake Bay. Wetlands 22 (2), 415–424.

Leonardi, N., Fagherazzi, S., 2014. How waves shape salt marshes. Geology 42, 887–890.

Leonardi, N., Fagherazzi, S., 2015. Effect of local variability in erosional resistance on large-scale morphodynamic response of salt marshes to wind waves and extreme events. Geophysical Research Letters 42, 5872–5879.

Leonardi, N., Ganju, N.K., Fagherazzi, S., 2016. A linear relationship between wave power and erosion determines salt-marsh resilience to violent storms and hurricanes. Proceedings of the National Academy of Sciences 113, 64–68.

Lesser, G.R., Roelvink, J.A., Van Kester, J., Stelling, G.S., 2004. Development and validation of a three-dimensional morphological model. Coastal Engineering 51, 883–915.

Li, X., Ren, L., Liu, Y., Craft, C., Mander, Ü., Yang, S., 2014. The impact of the change in vegetation structure on the ecological functions of salt marshes: the example of the Yangtze estuary. Regional Environmental Change 14, 623–632.

Lindstedt, D.M., Swenson, E.M., 2006. The case of the dying marsh grass. Scientific-Technical Committee of the Barataria-Terrabonne National Estuary Program and Louisiana Department of Natural Resources, Baton Rouge, LA.

Lotze, H.K., 2004. Ecological history of the Wadden Sea: 2,000 years of human-induced change in a unique coastal ecosystem. Wadden Sea Newsletter 1, 22–23.

Malvarez, A.I., 1997. Las comunidades vegetales del Delta del Río Paraná. Su relación con factores ambientales y patrones de paisaje (Thesis). Universidad de Buenos Aires.

Mann, K.H., 2009. Ecology of Coastal Waters: With Implications for Management. John Wiley & Sons.

Marani, M., D'Alpaos, A., Lanzoni, S., Santalucia, M., 2011. Understanding and predicting wave erosion of marsh edges. Geophysical Research Letters 38, L21401.

Marco-Barba, J., Mesquita-Joanes, F., Miracle, M.R., 2013. Ostracod palaeolimnological analysis reveals drastic historical changes in salinity, eutrophication and biodiversity loss in a coastal Mediterranean lake. The Holocene 23, 556–567.

Mariotti, G., Carr, J., 2014. Dual role of salt marsh retreat: long-term loss and short-term resilience. Water Resources Research 50, 2963–2974.

Mariotti, G., Fagherazzi, S., 2010. A numerical model for the coupled long-term evolution of salt marshes and tidal flats. Journal of Geophysical Research: Earth Surface 115 (F1).

Mariotti, G., Fagherazzi, S., Wiberg, P.L., McGlathery, K.J., Carniello, L., Defina, A., 2010. Influence of storm surges and sea level on shallow tidal basin erosive processes. Journal of Geophysical Research: Oceans 115, C11012.

Marsooli, R., Orton, P.M., Georgas, N., Blumberg, A.F., 2016. Three-dimensional hydrodynamic modeling of coastal flood mitigation by wetlands. Coastal Engineering 111, 83–94.

Martin, P.S., Wright Jr., H.E., 1967. Pleistocene Extinctions. The Search for a Cause. Yale University Press, New Heaven, USA.

McColl, R.W., 2005. Encyclopedia of World Geography. Facts On File, Inc., New York, USA.

Mckee, K.L., Patrick, W.H., 1988. The relationship of smooth cordgrass (*Spartina alterniflora*) to tidal datums: a review. Estuaries 11, 143–151.

Meade, R.H., Moody, J.A., 2010. Causes for the decline of suspended-sediment discharge in the Mississippi River system 1940-2007. Hydrological Processes 24, 35–49.

Meier, D., 2004. Man and environment in the marsh area of Schleswig-Holstein from Roman until late Medieval times. Quaternary International 112, 55–69.

Mendelssohn, I.A., Morris, J.T., 1999. Ecophysiological controls on the productivity of *Spartina alterniflora*. In: Weinstein, M., Kreeger, D.A. (Eds.), Concepts and Controversies in Tidal Marsh Ecology. Kluwer Academic Publishers, Boston, USA, pp. 59–80.

Miller, W.R., Egler, F.E., 1950. Vegetation of the Wequetequock-Pawcatuck tidal-marshes, Connecticut. Ecological Monographs 20, 143–172.

Minchinton, T.E., Bertness, M.D., 2003. Disturbance-mediated competition and the spread of *Phragmites australis* in a coastal marsh. Ecological Applications 13, 1400–1416.

Moffett, K.B., Gorelick, S.M., McLaren, R.G., Sudicky, E.A., 2012. Salt marsh ecohydrological zonation due to heterogeneous vegetation–groundwater–surface water interactions. Water Resources Research 48, W02516.

Moorhead, K.K., Brinson, M.M., 1995. Response of wetlands to rising sea level in the lower coastal plain of North Carolina. Ecological Applications 5, 261–271.

Morris, J.T., Sundareshwar, P.V., Nietch, C.T., Kjerfve, B., Cahoon, D.R., 2002. Responses of coastal wetlands to rising sea level. Ecology 83, 2869–2877.

Mudd, S.M., Howell, S.M., Morris, J.T., 2009. Impact of dynamic feedbacks between sedimentation, sea-level rise, and biomass production on near-surface marsh stratigraphy and carbon accumulation. Estuarine, Coastal and Shelf Science 82, 377–389.

Nehring, S., Hesse, K.-J., 2008. Invasive alien plants in marine protected areas: the *Spartina anglica* affair in the European Wadden Sea. Biological Invasions 10, 937–950.

Neiff, J.J., 1999. El régimen de pulsos en ríos y grandes humedales de Sudamérica. In: Malvarez, A.I., Kandus, P. (Eds.), Tópicos Sobre Grandes Humedales Sudamericanos. ORCYT-MAB (UNESCO), Buenos Aires, Argentina, pp. 97–145.

Neumann, B., Vafeidis, A.T., Zimmermann, J., Nicholls, R.J., 2015. Future coastal population growth and exposure to sea-level rise and coastal flooding-a global assessment. PLoS One 10, e0118571.

Nixon, S.W., 1982. The Ecology of New England High Salt Marshes: A Community Profile. U.S. Fish and Wildlife Service, FWS/OBS-81/55, Washington, D.C., USA.

Noe, G.B., Zedler, J.B., 2001. Variable rainfall limits germination of upper intertidal marsh plants in Southern California. Estuaries 24, 30–40.

Odum, W.E., McIvor, C.C., Smith III, T.J., 1982. The Ecology of the Mangroves of South Florida: A Community Profile. U.S. Fish and Wildlife Service, Office of Biological Services, FWS/OBS-81/24, Washington, D.C.

Odum, W.E., Smith III, T.J., Hoover, J.K., McIvor, C.C., 1984. The ecology of tidal freshwater marshes of the United States east coast: a community profile. U.S. Fish and Wildlife Service, Office of Biological Services, Washington, D.C. FWS/OBS-83/17.

Oertel, G.F., Keaney, M.S., Leatherman, S.P., 1989. Anatomy of a barrier platform: outer barrier lagoon, Southern Delmarva peninsula, Virginia. Marine Geology 88, 303–318.

Oertel, G.F., Kraft, J.C., Kearney, M.S., Woo, H.J., 1992. A rational theory for barrier lagoon development. In: Soc Econ Paleontol Miner, 48, pp. 77–87. Spec Pub.

Ogburn, M.B., Alber, M., 2006. An investigation of salt marsh dieback in Georgia using field transplants. Estuaries and Coasts 29, 54–62.

Ouyang, Z.-T., Gao, Y., Xie, X., Guo, H.-Q., Zhang, T.-T., Zhao, B., 2013. Spectral discrimination of the invasive plant *Spartina alterniflora* at multiple phenological stages in a saltmarsh wetland. PLoS One 8, e67315.

Parker, G., Marcolini, S., 1992. Geomorfología del Delta del Paraná y su extensión hacia el Río de la Plata. Revista de la Asociación Geológica Argentina 47, 243–249.

Peel, M.C., Finlayson, B.L., McMahon, T.A., 2007. Updated world map of the Köppen-Geiger climate classification. Hydrology and Earth System Sciences 11, 1633–1644.

Pejrup, M., Larsen, M., Edelvang, K., 1997. A fine grained sediment budget for the Sylt-Rømø tidal basin. Helgoländer Meeresuntersunchungen 51, 253–268.

Pennings, S.C., Bertness, M.D., 1999. Using latitudinal variation to examine effects of climate on coastal salt marsh pattern and process. Current Topics in Wetland Biogeochemistry 3, 100–111.

Pennings, S.C., Bertness, M.D., 2001. Salt marsh communities. Marsh Community Ecology 289–316.

Pennings, S.C., Grant, M.B., Bertness, M.D., 2005. Plant zonation in low-latitude salt marshes: disentangling the roles of flooding, salinity and competition. Journal of Ecology 93, 159–167.

Peteán, J., Cappato, J., 2005. Humedales fluviales de América del Sur. Hacia un manejo Sustentable. Imprenta Lux S.A., Santa Fé, Argentina.

Pethick, J.S., 1981. Long-term accretion rates on tidal marshes. Journal of Sedimentary Petrology 51, 571–577.

Pinder, D.A., Witherick, M.E., 1990. Port industrialization, urbanization and wetland loss. In: Williams, M. (Ed.), Wetlands: A Threatened Landscape. Blackwell, Oxford, UK, pp. 235–266.

Plater, A., Kirby, J., 2006. The potential for perimarine wetlands as an ecohydrological and phytotechnological management tool in the Guadiana estuary, Portugal. Estuarine, Coastal and Shelf Science 70, 98–108.

Prado, P., Caiola, N., Ibáñez, C., 2014. Freshwater inflows and seasonal forcing strongly influence macrofaunal assemblages in Mediterranean coastal lagoons. Estuarine, Coastal and Shelf Science 147, 68–77.

Pandža, M., Franjić, J., Škvorc, Ž., 2007. The salt marsh vegetation on the East Adriatic coast. Biologia 62, 24–31.

Pratolongo, P., Kandus, P., Brinson, M.M., 2007. Net Aboveground Primary Production and Soil Properties of Floating and Attached Freshwater Tidal Marshes in the Río de la Plata Estuary, Argentina. Estuaries and Coasts 30, 618–626.

Pratolongo, P., Kandus, P., Brinson, M.M., 2008. Net aboveground primary production and biomass dynamics of *Schoenoplectus californicus* (Cyperaceae) marshes growing under different hydrological conditions. Darwiniana 46, 258–269.

Pratolongo, P., Mazzon, C., Zapperi, G., Piovan, M.J., Brinson, M.M., 2013. Land cover changes in tidal salt marshes of the Bahia Blanca Estuary (Argentina) during the past 40 years. Estuarine, Coastal and Shelf Science 133, 23–31.

Pratolongo, P., Perillo, G.M.E., Píccolo, M.C., 2010. Combined effects of waves and plants on a mud deposition event at a mudflat-saltmarsh edge in the Bahía Blanca estuary. Estuarine, Coastal and Shelf Science 87, 207–212.

Pratolongo, P., Piovan, M.J., Cuadrado, D.G., 2016. Coastal environments in the Bahía Blanca Estuary, Argentina. In: Khan, A.M., Boër, B., Özturk, M., Clüsener-Godt, M., Gul, B., Breckle, S.-W. (Eds.), Sabkha Ecosystems Volume V: The Americas. Springer International Publishing, Geneva, Switzerland, pp. 205–224.

Pratolongo, P., Vicari, R., Kandus, P., Malvárez, I., 2005. A new method for evaluating net aboveground primary production (NAPP) of Scirpus giganteus (Kunth). Wetlands 25, 228–232.

Priestas, M.A., Mariotti, G., Leonardi, N., Fagherazzi, S., 2015. Coupled wave energy and erosion dynamics along a salt marsh boundary, Hog Island Bay, Virginia, USA. Journal of Marine Science and Engineering 3 (3), 1041–1065.

Rahman, R., Plater, A.J., 2014. Particle-size evidence of estuary evolution: a rapid and diagnostic tool for determining the nature of recent saltmarsh accretion. Geomorphology 213, 139–152.

Ravit, B., Ehrenfeld, J.G., Häggblom, M.M., Bartels, M., 2007. The effects of drainage and nitrogen enrichment on *Phragmites australis, Spartina alterniflora*, and their root-associated microbial communities. Wetlands 27, 915–927.

Raybould, A.F., 1997. The history and ecology of *Spartina anglica* in Poole Harbour. Proceedings - Dorset Natural History and Archaeological Society 119, 147–158.

Richardson, C.J., 1983. Pocosins: vanishing wastelands or valuable wetlands? BioScience 33, 626–633.

Redfield, A.C., 1972. Development of a New England salt marhs. Ecological Monographs 42, 201–237.

Reed, D.J., 1990. The impact of sea-level rise on coastal saltmarshes. Progress in Physical Geography 14, 465–481.

Rippon, S., 2000. The Transformation of Coastal Wetlands: Exploitation and Management of Marshland Landscapes of North West Europe during the Roman and Medieval Periods. Oxford University Press, Oxford.

Rodi, W., 1993. Turbulence Models and Their Application in Hydraulics. CRC Press, Rotterdam, Netherlands.

Rodríguez-Iturbe, I., Rinaldo, A., 2001. Fractal River Basins: Chance and Self-organization. Cambridge University Press, Cambridge, UK.

Rostami, K., Peltier, W.R., Mangini, A., 2000. Quaternary marine, sea level changes and uplift history of Patagonia, Argentina: comparisons with predictions of the ICE-4G (VM2) model of the global process of glacialisostatic adjustment. Quaternary Science Reviews 19, 1495–1525.

Rozema, J., Dorel, F., Janissen, R., Lenssen, G., Broekman, R., Arp, W., Drake, B., 1991. Effect of elevated atmospheric CO_2 on growth, photosynthesis and water relations of salt marsh grass species. Aquatic Botany 39, 45–55.

Saintilan, N., Rogers, K., Mazumder, D., Woodroffe, C., 2013. Allochthonous and autochthonous contributions to carbon accumulation and carbon store in southeastern Australian coastal wetlands. Estuarine, Coastal and Shelf Science 128, 84–92.

Saintilan, N., Wilson, N.C., Rogers, K., Rajkaran, A., Krauss, K.W., 2014. Mangrove expansion and salt marsh decline at mangrove poleward limits. Global Change Biology 20, 147–157.

Saito, Y., 2001. Deltas in southeast and east Asia: their evolution and current problems. In: Mimura, N., Yokoki, H. (Eds.), Global Change and Asia Pacific Coasts. Proceedings of APN/SURVAS/LOICZ Joint Conference on Coastal Impacts of Climate Change and Adaptation in the Asia-Pacific Region. Kobe, Japan, pp. 185–191.

Saltonstall, K., 2002. Cryptic invasion by a non-native genotype of the common reed, Phragmites australis, into North America. Proceedings of the National Academy of Sciences 99, 2445–2449.

Sampath, D.M.R., Boski, T., 2016. Morphological response of the saltmarsh habitats of the Guadiana estuary due to flow regulation and sea-level rise. Estuarine, Coastal and Shelf Science 183, 314–326.

Santos-Echeandía, J., Vale, C., Caetano, M., Pereira, P., Prego, R., 2010. Effect of tidal flooding on metal distribution in pore waters of marsh sediments and its transport to water column (Tagus estuary, Portugal). Marine Environmental Research 70, 358–367.

Savage, T., 1972. Florida Mangroves as Shoreline Stabilizers. Florida Department of Natural Resources. Professional Papers Series 19. St. Petersburg.

Schofield, J.E., Bunting, M.J., 2005. Mid-Holocene presence of water chestnut (*Trapa natans L.*) in the meres of Holderness, East Yorkshire, UK. The Holocene 15, 687–697.

Schwimmer, R.A., 2001. Rates and processes of marsh shoreline erosion in Rehoboth Bay, Delaware, U.S.A. Journal of Coastal Research 17, 672–683.

Scott, D.S., Medioli, F.S., 1978. Vertical zonations of marsh foraminifera as accurate indicators of former sea-levels. Nature 272, 528–531.

Scruton, P.C., 1960. Delta building and the deltaic sequence. In: Shepard, F.P., Phleger, F.B., Van Andel, T.H. (Eds.), Recent Sediments, Northwest Gulf of Mexico. American Association of Petroleum Geologists, Tulsa, OK, pp. 82–102.

Shriver, W.G., Hodgman, T.P., Gibbs, J.P., Vickery, P.D., 2004. Landscape context influences salt marsh bird diversity and area requirements in New England. Biological Conservation 119, 545–553.

Shumway, S.W., Bertness, M.D., 1992. Salt stress limitation of seedling recruitment in a salt marsh plant community. Oecologia 92, 490–497.

Silinski, A., Fransen, E., Bouma, T.J., Meire, P., Temmerman, S., 2016. Unravelling the controls of lateral expansion and elevation change of pioneer tidal marshes. Geomorphology 274, 106–115.

Silvestri, S., Defina, A., Marani, M., 2005. Tidal regime, salinity and salt marsh plant zonation. Estuarine, Coastal and Shelf Science 62, 119–130.

Simas, T., Nunes, J.P., Ferreira, J.G., 2001. Effects of global climate change on coastal saltmarshes. Ecological Modelling 139, 1–15.

Simenstad, C.A., Thom, R.M., 1995. *Spartina alterniflora* (smooth cordgrass) as an invasive halophyte in Pacific Northwest estuaries. Hortus Northwest 6, 9–12.

Small, C., Nicholls, R.J., 2003. A global analysis of human settlement in coastal zones. Journal of Coastal Research 19, 584–599.

Smith, K., 2011. We are seven billion. Nature Climate Change 1, 331–335.

Smith, J.P., Carullo, M., 2007. Survey of potential marsh dieback sites in coastal Massachusetts. In: Massachusttes Bays National Estuary Program and Massachusttes A. Office of Coastal Zone Management, Boston, USA.

Snedaker, S.C., 1995. Mangroves and climate change in the Florida and Caribbean region: Scenarios and hypotheses. Hydrobiologia 295, 43–49.

Song, B., Li, Z., Saito, Y., Okuno, J., Li, Z., Lu, A., Hua, D., Li, J., Li, Y., Nakashima, R., 2013. Initiation of the Changjiang (Yangtze) delta and its response to the mid-Holocene sea level change. Palaeogeography, Palaeoclimatology, Palaeoecology 388, 81–97.

Stanley, D.J., Warne, A., 1993. Nile delta: recent geological evolution and human impacts. Science 260, 628–634.

Stokes, D.J., Healy, T.R., Cooke, P.J., 2010. Expansion dynamics of monospecific, temperate mangroves and sedimentation in two embayments of a barrier-enclosed lagoon, Tauranga Harbour, New Zealand. Journal of Coastal Research 26, 113–122.

Strong, D.R., Ayres, D.A., 2016. Control and consequences of *Spartina spp.* invasions with focus upon San Francisco Bay. Biological Invasions 18, 2237–2246.

Stumpf, R.P., 1983. The process of sedimentation on the surface of a saltmarsh. Estuarine, Coastal and Shelf Science 17, 495–508.

Stutz, M.L., Pilkey, O.H., 2002. Global distribution and morphology of deltaic barrier island systems. Journal of Coastal Research 36, 694–707.

Stutz, M.L., Pilkey, O.H., 2011. Open-Ocean Barrier Islands: Global Influence of Climatic, Oceanographic, and Depositional Settings. Journal of Coastal Research 27, 207–222.

Syvitski, J.P.M., Kettner, A.J., 2011. Sediment flux and the Anthropocene. Philosophical Transactions of the Royal Society A 369, 957–975.

Temmerman, S., Bouma, T.J., Govers, G., Wang, Z.B., De Vries, M.B., Herman, P.M.J., 2005. Impact of vegetation on flow routing and sedimentation patterns: three-dimensional modeling for a tidal marsh. Journal of Geophysical Research: Earth Surface 110, 1–18.

Temmerman, S., Govers, G., Meire, P., Wartel, S., 2003. Modelling long-term tidal marsh growth under changing tidal conditions and suspended sediment concentrations, Scheldt estuary, Belgium. Marine Geology 193, 151–169.

Temmerman, S., Govers, G., Meire, P., Wartel, S., 2004. Simulating the long-term development of levee-basin topography of tidal marshes. Geomorphology 63, 39–55.

Thibodeau, P.M., Gardner, L.R., Reeves, H.W., 1998. The role of groundwater flow in controlling the spatial distribution of soil salinity and rooted macrophytes in a southeastern salt marsh, USA. Mangroves and Salt Marshes 2, 1–13.

Titus, J.G., 1987. The greenhouse effect, rising sea-level, and society's response. In: Devoy, R.J.N. (Ed.), Sea Surface Studies. Croom Helm, London, pp. 499–528.

Tooley, M.J., 1986. Sea levels. Progress in Physical Geography 10, 120–129.

Tourenq, C., Bennetts, R.E., Kowalski, H., Vialet, E., Lucchesi, J.-L., Kayser, Y., Isenmann, P., 2001. Are ricefields a good alternative to natural marshes for waterbird communities in the Camargue, southern France? Biological Conservation 100, 335–343.

Trewartha, G.T., Horn, L.H., 1980. Introduction to Climate, fifth ed. McGraw Hill, New York, NY.

Turner, R.E., Gosselink, J.G., 1975. A note on standing crops of *Spartina alterniflora* in Texas and Florida. Contributions in Marine Science - The University of Texas at Austin 19, 113–118.

Turner, R.E., Rabalais, N.N., 2003. Linking landscape and water quality in the Mississippi River basin for 200 years. BioScience 53, 563–572.

Underwood, G.J.C., Paterson, D.M., 1993. Seasonal changes in diatom biomass, sediment stability and biogenic stabilization in the Severn Estuary. Journal of the Marine Biological Association of the United Kingdom 73 (4), 871–887.

Valéry, L., Bouchard, V., Lefeuvre, J.-C., 2004. Impact of the invasive native species *Elymus athericus* on carbon pools in a salt marsh. Wetlands 24, 268–276.

Valéry, L., Radureau, A., Lefeuvre, J.-C., 2016. Spread of the native *grass Elymus athericus* in salt marshes of Mont-Saint-Michel bay as an unusual case of coastal eutrophication. Journal of Coastal Conservation 1–13.

Valiela, I., 2006. Global Coastal Change. Blackwell Publishing, Malden, USA.

van Proosdij, D., Ollerhead, J., Davidson-Arnott, R.G.D., Schostak, L.E., 1999. Allen creek marsh, Bay of Fundy: a macro-tidal coastal saltmarsh. Canadian Geographer 43, 316–322.

Vartiainen, T., 1988. Vegetation development on the outer islands of the Bothnian Bay. Vegetatio 77, 149–158.

Vella, C., Fleury, T.-J., Raccasi, G., Provansal, M., Sabatier, F., Bourcier, M., 2005. Evolution of the Rhône delta plain in the Holocene. Marine Geology 222, 235–265.

Violante, R.A., Parker, G., 2000. El Holoceno en las regiones marinas y costeras del nordeste de Buenos Aires. The Revista de la Asociación Geológica Argentina 55, 337–351.

Visser, J.M., Sasser, C.E., Chabreck, R.H., Linscombe, R.G., 1998. Marsh vegetation types of the Mississippi River deltaic plain. Estuaries 21, 818–828.

Waller, M.P., 1994. The Fenland Project, Number 9: Flandrian Environmental Change in Fenland, East Anglian Archaeology Report No. 70. Cambridgeshire Archaeological Committee, Cambridge.

Waller, M.P., Long, a. J., Long, D., Innes, J.B., 1999. Patterns and processes in the development of coastal mire vegetation: Multi-site investigations from Walland Marsh, Southeast England. Quaternary Science Reviews 18, 1419–1444.

Wang, Y., 1983. The mudflat system of China. Canadian Journal of Fisheries and Aquatic Sciences 40, 160–171.

Warner, J.C., Armstrong, B., He, R., Zambon, J.B., 2010. Development of a coupled ocean–atmosphere–wave–sediment transport (COAWST) modeling system. Ocean Modelling 35, 230–244.

Webb, E.L., Friess, D.A., Krauss, K.W., Cahoon, D.R., Guntenspergen, G.R., Phelps, J., 2013. A global standard for monitoring coastal wetland vulnerability to accelerated sea-level rise. Nature Climate Change 3, 458–465.

Wells, J.T., 1996. Subsidence, sea-level rise, and wetland loss in the Lower Mississippi River delta. In: Milliman, J.D., Haq, B.U. (Eds.), Sea-Level Rise and Coastal Subsidence. Kluwer Academic Publishers, Dordrecht, the Netherlands, pp. 281–312.

Wells, J.T., Coleman, J.M., 1987. Wetland loss and the subdelta life cycle. Estuarine, Coastal and Shelf Science 25, 111–125.

Whigham, D.F., Baldwin, A.H., Barendregt, A., 2009. Tidal freshwater wetlands. In: Perillo, G.M.E., Wolanksi, E., Cahoon, D.R., Brinson, M.M. (Eds.), Coastal Wetlands: An Integrated Ecosystem Approach. Elsevier, Oxford, UK, pp. 515–533.

Wiegert, R.G., Freeman, B.J., 1990. Tidal Salt Marshes of the Southeast Atlantic Coast: A Community Profile (Washington, D.C., USA).

Wray, R.D., Leatherman, S.P., Nicholls, R.J., 1995. Historic and future land loss for upland and marsh islands in the Chesapeake Bay, Maryland, U.S.A. Journal of Coastal Research 11, 1195–1203.

Xiqing, C., 1998. Changjian (Yangtze) River Delta, China. Journal of Coastal Research 14, 838–858.

Yoshinobu, S., 1998. Environment versus water control: The case of the southern Hangzhou Bay area from the mid-Tang through the Qing. In: Elvin, M., Ts'ui-jung, L. (Eds.), Sediments of Time: Environment and Society in Chinese History. Cambridge University Press, New York, USA, pp. 135–164.

Zapperi, G., Pratolongo, P., Piovan, M.J., Marcovecchio, J.E., 2016. Benthic-Pelagic coupling in an Intertidal Mudflat in the Bahía Blanca Estuary (SW Atlantic). Journal of Coastal Research 32, 629–637.

Zhang, J., Jørgensen, S.E., Lu, J., Nielsen, S.N., Wang, Q., 2014. A model for the contribution of macrophyte-derived organic carbon in harvested tidal freshwater marshes to surrounding estuarine and oceanic ecosystems and its response to global warming. Ecology Modelling 294, 105–116.

Zhang, R.S., Shen, Y.M., Lu, L.Y., Yan, S.G., Wang, Y.H., Li, J.L., Zhang, Z.L., 2004. Formation of Spartina alterniflora salt marshes on the coast of Jiangsu Province, China. Ecology Engineering 23, 95–105.

Zobel, M., Kont, A., 1992. Formation and succession of alvar communities in the Baltic land uplift area. Nordic Journal of Botany 12, 249–256.

Zong, Y., Horton, B.P., 1998. Diatom zones across intertidal flats and coastal saltmarshes in Britain. Diatom Research 13, 375–394.

CHAPTER

4

Northern Polar Coastal Wetlands: Development, Structure, and Land Use

I. Peter Martini[1], *Robert L. Jefferies*[†], *R.I. Guy Morrison*[2], *Kenneth F. Abraham*[3], *Liudmila A. Sergienko*[4]

[1]School of Environmental Sciences, University of Guelph, Guelph, ON, Canada; [2]National Wildlife Research Centre, Environment and Climate Change Canada, Ottawa, ON, Canada; [3]Trent University, Peterborough, ON, Canada; [4]Department of Botany and Plant Physiology, Petrozavodsk State University, Petrozavodsk, Russia

1. INTRODUCTION

Polar coasts differ from others in terms of landscape, climate, oceanographic conditions (ice cover), soils (permafrost with seasonally variable surface active layer), type, rate and degree of weathering, variable growing season (flora and fauna), migratory animals (primarily birds), and geographic isolation (limiting environmental damage) (Forbes, 2011; Scheffers et al., 2012; Martini, 2014). Arctic coastal wetlands consist mostly of salt-, brackish-, and freshwater-marshes and tundra inundated by seawater during high tides and storms. Lowland wetlands may be backed by tundra plains with numerous ponds and shallow lakes in Arctic zones with continuous permafrost and by fens, bogs, and taiga in sub-Arctic to boreal zones with discontinuous permafrost (Fig. 4.1). Cold, vegetated wetlands are found along most northern low-lying marine coastlines, although they are often poorly or not developed in the High Arctic (Figs. 4.1 and 4.2) and are absent or extremely rare on Antarctic coasts. Northern polar deserts have a very cold climate (very low temperatures, <10°C average during the warmest month), very low precipitation (from 45 to 250 mm/year), and extreme poverty of life (Callaghan et al., 2005). Extensive, multiyear, multidisciplinary studies of cold coastal wetlands and their uses have been carried out along the northern marine shores

[†]Deceased.

Coastal Wetlands
https://doi.org/10.1016/B978-0-444-63893-9.00004-6

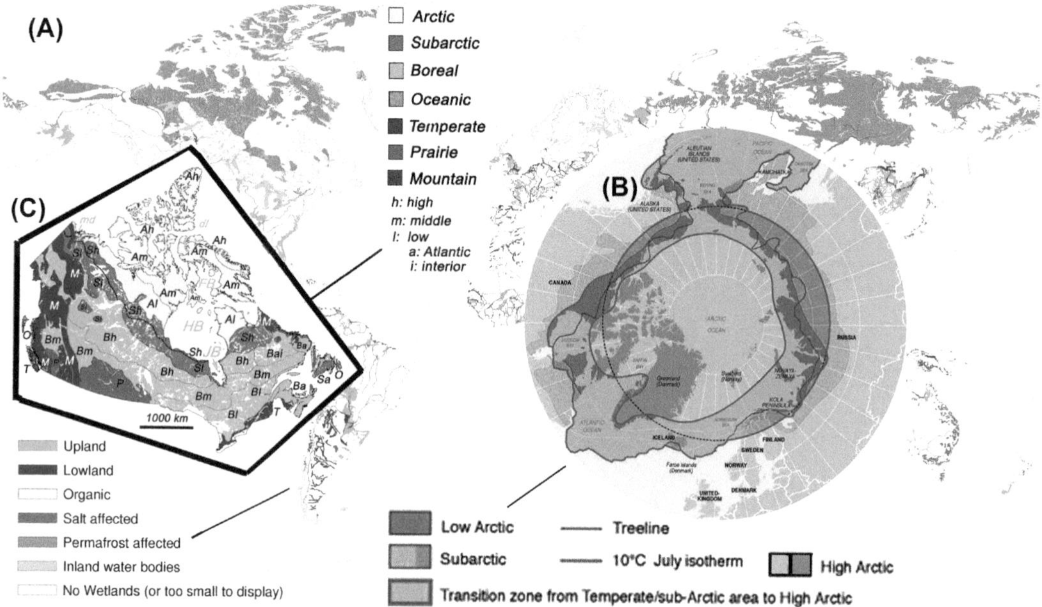

FIGURE 4.1 Wetlands. (A) Distribution of wetlands on Earth; polar wetlands are characterized by the presence of permafrost. (B) Distribution of Arctic and sub-Arctic zones. (C) Generalized distribution of the major types of wetlands regions of Canada (some zones are further subdivided into sectors, such as High, Middle, and Low). Note narrow sub-Arctic conditions extending along the western coast of James Bay. (*dl*, Devon Island; *FB*, Foxe Basin; *HB*, Hudson Bay; *JB*, James Bay; *md*, Mackenzie River Delta). *(B) Image credit P. Rekacewicz, UNEP/GRID-Arendal, 2005a. Distribution of Arctic and Sub-Arctic Zones. http://www.grida.no/resources/7010; Compilation of data from U.S. Dept of Agriculture, 1996. Global Distribution of Wetlands – Map. Natural Resources Conservation Service, Soil Survey Division, World Soil Resources, Washington, DC. https://www.nrcs.usda.gov/wps/portal/nrcs/detail/soils/use/worldsoils/?cid=nrcs142p2_054021; and modified from Zoltai, S.C., 1980. An outline of the wetland regions of Canada. In: Rubec, C.D.A., Pollett, F.C. (Eds.), Proceedings of a Workshop on Canadian Wetlands. Environment Canada, Lands Directorate, Ecological Land Classification, Series, No. 12, pp. 1–8. [Trace: Minister of Supply and Services Canada, 1980. Cat. No. En 73-3/12; ISBN 0-662-50919-6]. PDF available in http://www.cfs.nrcan.gc.ca/bookstore_pdfs/19315.pdf; and Tarnocai, C., 2009. The impact of climate change on Canadian Peatlands. Canadian Water Resources Journal 34, 453–466.*

of North America and Russia (this chapter), while there are fewer along the sub-Antarctic coasts of South America (Scott et al., 2014).

2. GEOLOGY/GEOMORPHOLOGY

The Arctic regions have been affected by several progenies (mountain-building episodes) and are mountainous over large tracts. These include the Mesozoic mountain ranges of northeastern Russia, Alaska (such as the Brooks Range [BR]), Canadian Arctic (such as the Richardson [RM] and Innuitian Mountains), and northern Greenland (Fig. 4.3;

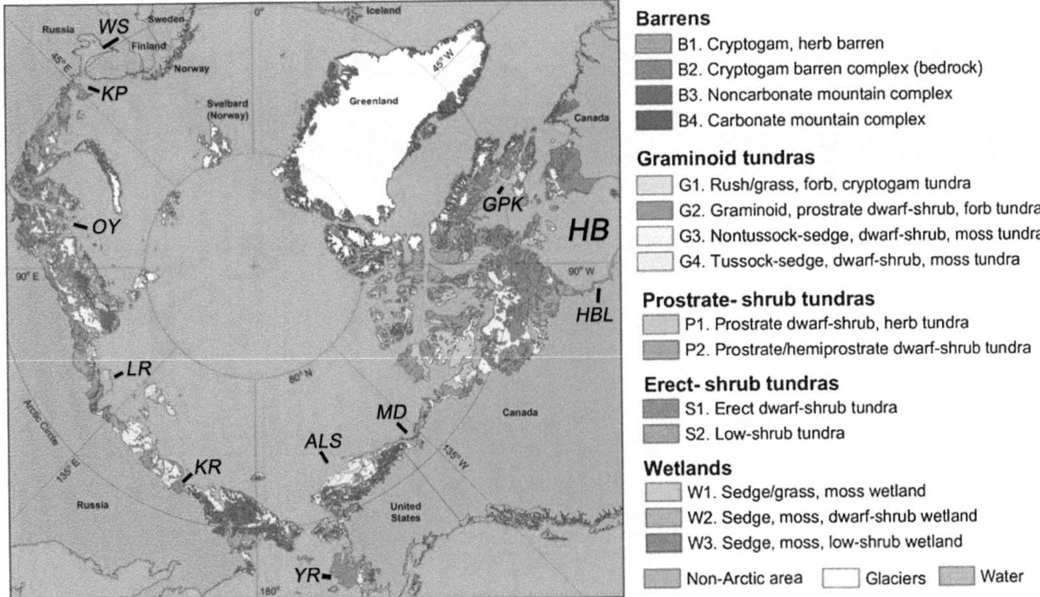

FIGURE 4.2 Generalized map of circumpolar Arctic vegetation. (A) Map and Legend of some of the largest coastal wetlands/coastal plains. Existing coastal wetlands of the White Sea are not shown. (*ALS*, Alaska North Slope; *GPK*, Great Plain of the Koukdjuak; *HB*, Hudson Bay; *HBL*, Hudson Bay Lowland; *KP*, Kanin Peninsula; *KR*, Kolyma River; *LR*, Lena River; *MD*, Mackenzie River; *OY*, Ob and Yenissei rivers; *WS*, White Sea; *YR*, Yukon River). *Modified from CAVM Team, 2003. Circumpolar Arctic Vegetation Map (1:7,500,000 Scale), Conservation of Arctic Flora and Fauna (CAFF) Map No. 1. U.S. Fish and Wildlife Service, Anchorage, Alaska. ISBN: 0-9767525-0-6, ISBN-13: 978-0-9767525-0-9 and Walker, D.A. et al., (Members of the CAVM Team), 2005. The circumpolar Arctic vegetation map. Journal of Vegetation Science 16, 267–282.*

Fulton, 1989). Older Caledonian mountain belts (Paleozoic) occur in eastern Greenland, northern Europe, and the central parts of Russia (Ural Mountains). The Kamchatka Peninsula in northeast Russia and the Aleutian Islands west of Alaska are active tectonic areas where continuing subduction of the Pacific tectonic plate has and is still generating active volcanoes and rugged terrain. Nevertheless, extensive circumpolar Arctic lowlands occur between or in front of these mountain chains along the Arctic Coastal Plain of northern Alaska (ALS), in the Mackenzie River Delta (MD) in northwest Canada, and in parts of the Russian Arctic west and east of the Ural Mountains (Figs. 4.2 and 4.3). In northeast Canada, extensive Arctic and sub-Arctic coastal plains extend southward along the continental coasts down to approximately location 51° 09′N, 79°48′W along the western coasts of Hudson and James bays, which are underlain by low-lying Paleozoic rocks of old inland basins bounded by the North American Precambrian Shield.

Most polar areas were glaciated during Late Pleistocene except in a few cold dry areas in northernmost North America and Russia (Fig. 4.4; Andrews, 1973). The ice sheets were thickest at their epicenters (domes) and thinner toward saddle and peripheral areas.

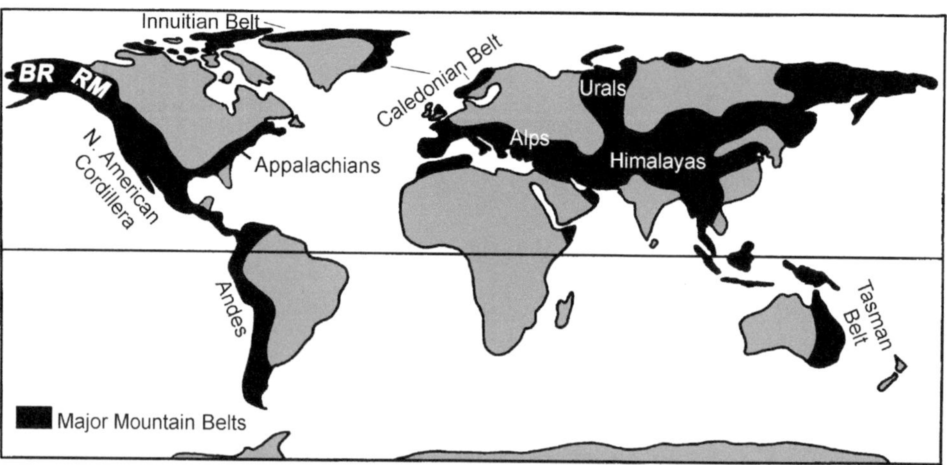

FIGURE 4.3 Distribution of major mountain chains on Earth (*BR*, Brook Range; *RM*, Richardson Mountains).

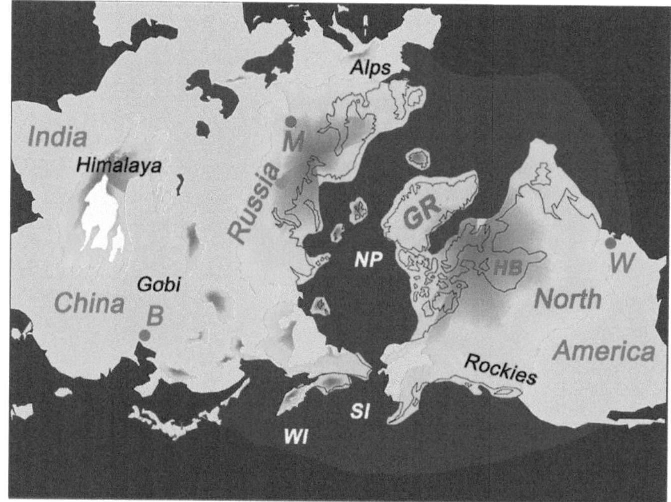

FIGURE 4.4 Generalized map of maximum Upper Pleistocene glaciations (light blue). (*B*, Beijing; *GR*, Greenland; *HB*, Hudson Bay; *M*, Moscow; *NP*, North Pole; *SI*, Summer sea ice; *W*, Washington; *WI*, Winter sea ice). *Modified from Grobe, H., 2000. Northern Hemisphere Glaciation During the Last Ice Ages. Creative Commons CC-BY-2.5, Wikipedia: https://en.wikipedia.org/wiki/Quaternary_glaciation#/media/File:Northern_icesheet_hg.png. Modified from Schlee, H.S. 2000. Our Changing Continent. USGS General Interest Publication https://pubs.usgs.gov/gip/continents/.*

The weight of the glaciers depressed the Earth's crust, and as the glaciers melted, differential isostatic rebound of up of 200 m has occurred. Continuing rebound rates still vary from minimal to about 1 m per century where the ice was thicker. Isostatic uplift has led to land emersion from large lakes and seas that had formed in front of the glaciers, resulting in extensive raised coastal plains. Land emersion continues in areas where uplift is more rapid than current sea level rise, for instance, along the western coasts of Hudson and James bays, the Gulf of Bothnia between Finland and Sweden, and the White Sea in northwest Russia (Jeglum et al., 2003). Where land uplift was small and no longer occurs and where there is no active neotectonism, marine transgression takes place, such as along the Arctic Coastal Plain of Alaska (ALS), northwestern Canada, and parts of the Siberian coastal plains. The extent of glaciation in central-north Russia remains unresolved; two major hypotheses (Velichko et al., 1997; Grosswald, 1998) have been proposed involving (1) a wide, thick glaciation during the Late Pleistocene, which could have generated postglacial isostatic uplift similar to that of North America and Fennoscandia and (2) a restricted glaciation that generated smaller areas of post-glacial isostatic uplift.

3. OCEANOGRAPHY

The inshore Arctic seas are generally shallow, particularly along the Russian coasts, Foxe Basin, and James Bay. They and their respective coasts are subjected to a harsh climate and a seasonally variable ice cover. At the northernmost latitudes, ice is present throughout the year. Sea ice influences marine currents, tides, and waves that, in turn, affect the stability of coastlines. The marine currents of the Arctic Ocean and adjoining inland seas are complex (Fig. 4.5). In North America, Arctic waters enter the Canadian inland

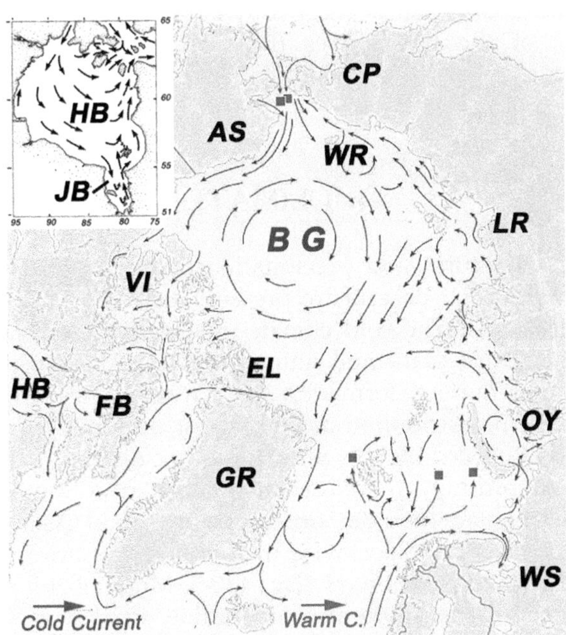

FIGURE 4.5 Marine currents (*arrows*) in the Arctic Ocean and adjacent seas. Insert: surface water currents in Hudson and James bays. (*AS*, Alaska; *BG*, Beaufort Gyre; *CP*, Chukotka Peninsula; *EL*, Ellesmere Island; *FB*, Foxe Basin; *GR*, Greenland; *HB*, Hudson Bay; *JB*, James Bay; *LR*, Lena River; *OY*, Ob-Yenisei rivers; *VR*, Victoria Island; *WI*, Wrangel Island; *WS*, White Sea). *Modified from AMAP, 1998. Arctic Marine Currents. http://neba.arcticresponsetechnology.org/media/1006/image013.png.*

seas from the Fury–Hecla Strait in the northwest corner of Foxe Basin and flow south into Hudson and James bays to latitudes of about 51°N, cooling the surrounding land and allowing the development of sub-Arctic coastal zones to such low latitudes (Fig. 4.5 insert). Conversely, the northern lands of Fennoscandia are warmed by a branch of the North Atlantic Drift Current north to approximate latitude 70°N, well above the Arctic Circle.

The northern seas are generally microtidal (<2 m in amplitude), but several shores experience mesotidal conditions, such as those of Hudson Bay, James Bay, and the Barents Sea. Macrotidal conditions (tides >5 m and locally >10 m) occur in some embayment, such as Bristol Bay in west Alaska, Bowman Bay in southeast Foxe Basin, Frobisher Bay in southeast Baffin Island, parts of Ungava Bay along the Hudson Strait, Mezen Gulf in east White Sea, and the Gulf of Shelikov in the northern Okhotsk Sea.

The salinity of the Arctic Seas may in some places vary drastically seasonally owing to the formation and melting of the ice cover and to the variable input of fluvial freshwater during spring–summer freshets (floods). While tropical oceans are temperature-stratified (thermocline), the Arctic Ocean and adjacent seas are mostly salinity-stratified (halocline), although this is less stable at high latitudes. The temperature profile of Arctic waters is nearly uniform at 0–1°C (Linacre and Geerts, 1998).

Brackish water conditions develop in and near the estuaries of major rivers during the spring–summer floods, and more marine saline conditions are reestablished later when the river discharge decreases. One dramatic case occurs along the shallow coast of southern James Bay, where warm freshwater is injected by northward flowing rivers. The anticlockwise marine current of the bay moves these waters northeastward, warming and freshening the eastern coast of the bay significantly more than the western coast (Fig. 4.5 insert; Prinsenberg, 1986).

4. CLIMATE

Average annual temperatures and precipitation decrease substantially from south to north, particularly where polar deserts are present, such as in the northernmost Russian and Canadian Arctic Islands. The cold climate and highly variable seasonal day length greatly affect assemblages of plant and animal species. A "cryosphere" has developed in northern areas, involving ice formation over waterbodies and within the ground (permafrost). The northern distribution of permafrost does not follow regular latitudinal gradients; rather, it is influenced by heat redistribution caused by atmospheric movements and marine currents. Accordingly, there is a southern dip of continuous and discontinuous permafrost in continental and mountainous areas of central Asia and a latitudinal dip in central-east Canada where anticlockwise marine currents bring cold Arctic waters to low latitudes in Hudson and James bays (Fig. 4.6A). Much of the permafrost is a legacy of colder climates, but new formation also occurs now in actively emerging High Arctic lands. Permafrost has undergone adjustment to the variable warming Holocene climates, such as during the postglacial temperature maximum (approximately 8000–7000 years BP) and at present. This has led to the development of various "periglacial" geomorphologic structures, including numerous modern thermokarst features (Washburn, 1973; Ruz et al., 1992). Ongoing climate amelioration is predicted to lead to considerable

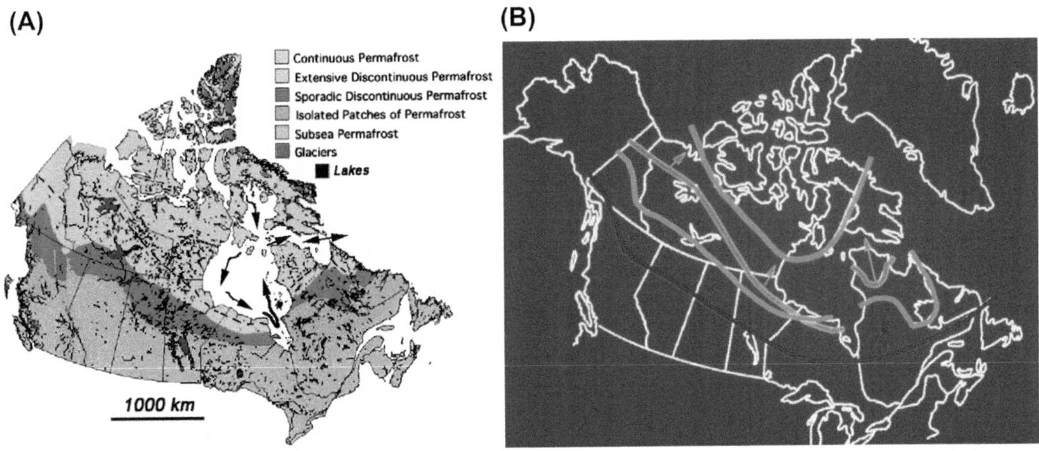

FIGURE 4.6 Permafrost in Canada. (A) Present distribution of permafrost in Northern Canada and marine currents in the inland seas (*arrows*: surficial water current). (b) Possible northward shift of permafrost boundaries due to climatic warming (green and blue lines indicate, respectively, present and predicted future southern limit of continuous permafrost; red and orange lines indicate modern and predicted future southern margin of permafrost, respectively). *(A) Modified from Heginbottom, J.A., Dubreuil, M.A., Harker, P.A., 1995. Canada – Permafrost in: National Atlas of Canada, fifth ed., National Atlas Information Service, Natural Resources Canada, MCR 4177; Koerner, R.M., 2002. Glaciers of Canada- glaciers of the Arctic Islands. In: Willliams, R.S., Ferrigno J.G. (Eds.), Satellite Image Atlas of the Glaciers of the World. U.S. Geological Survey Professional Paper 1386-J-1. J111–J146. https://pubs.usgs.gov/pp/p1386j/hiarctic/hiarctic-lores.pdf; Kettles, I.M., Tarnocai, C., 1999. Development of a model for estimating the sensitivity of Canadian peatlands to climate warming. Géographie Physique et Quaternaire 53, 323–338; and Tarnocai, C., 2009. The impact of climate change on Canadian Peatlands. Canadian Water Resources Journal 34, 453–466.*

northward shift of the southern limit of the permafrost and of the continuous permafrost boundary (Fig. 4.6B; Kettles and Tarnocai, 1999; Tarnocai, 2009).

5. STRUCTURE OF ARCTIC COASTAL WETLANDS

The characteristics of polar coastal wetlands are determined by abiotic and biotic conditions. They range from extensive seashore meadows (marshes), showing a transition from saltwater to freshwater on wide, open, coastal plains, to narrow wetland strips mostly developed in swales between beach ridges and brackish water systems in deltaic areas. Some of the best-developed coastal wetlands occur along the Low Arctic to sub-Arctic coasts of southwestern Hudson and James bays (Jefferies et al., 1979; Martini, 2006a,b). These have developed in the last 300–400 years, as the Hudson Bay Lowland coastal plain emerged as a result of isostatic uplift (Andrews, 1973). The widest salt marshes occur inland from open, extensive sand- and mudflats (Fig. 4.7A and B), and brackish marshes are formed in estuarine areas and on mainland coasts freshened by fluvial plumes (Fig. 4.7C and D). Less well-developed marshes form on steeper shores where higher waves develop beach ridges (Fig. 4.7E and F) and in areas of low sedimentation where limited fine-grained deposits occur.

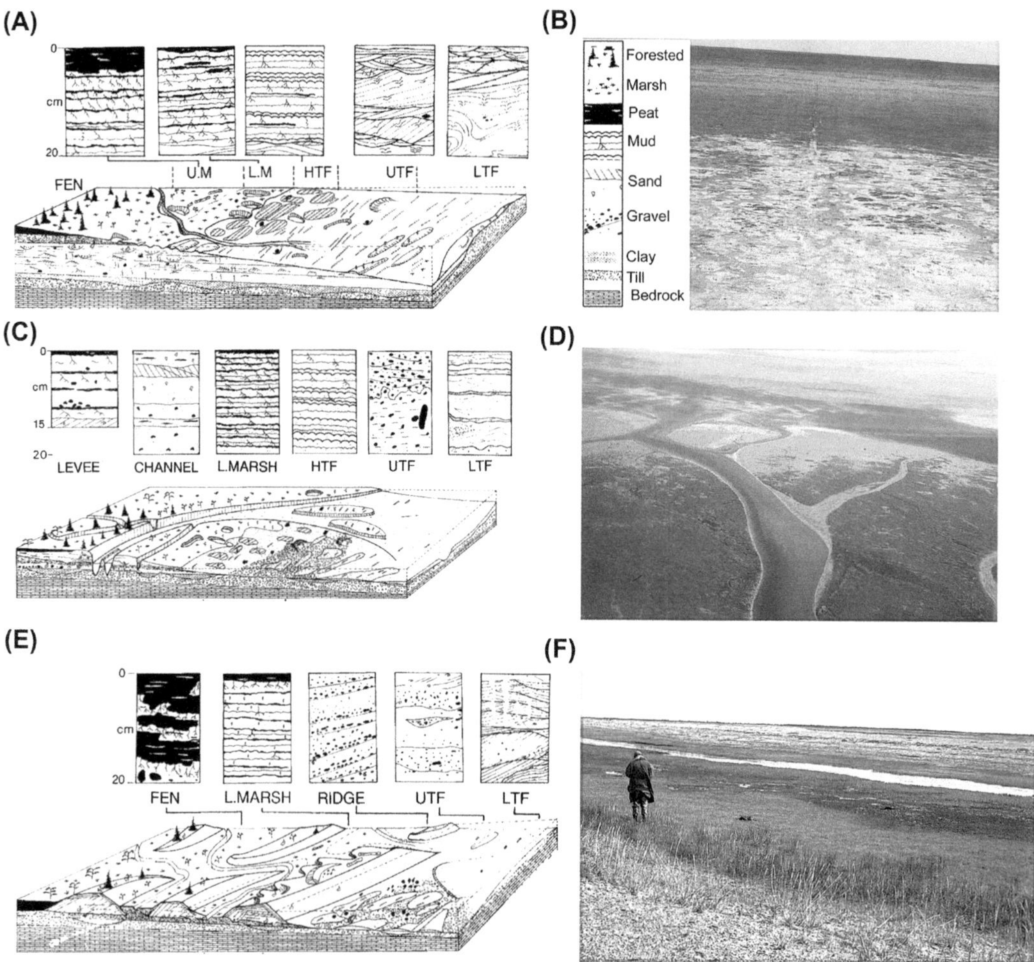

FIGURE 4.7 Structure of sub-Arctic coastal marshes, western James Bay. (A and B) Marsh of open coasts with well-developed algal high tidal flats at the shore grading into salt and freshwater marshes and inland peat-bearing fens. (C and D) Extensive brackish- to freshwater marshes of river-influenced coasts. (E and F) Coast with beach ridges and bilaterally structured coastal wetlands in the interridge swales. (*HTF*, high tidal flat; *LM*, lower marsh; *LTF*, lower tidal flat; *UM*, upper marsh; *UTF*, upper tidal flat). The diagrams show cross sections of the substrate stratigraphy, surficial depositional features (barbed short lines indicate sandy ripple marks), erosional features by ice floes, and internal laminations in the 20 cm long cores (inclined laminae: cross-laminations, and peat deposits; additional symbol explanations in vertical profile in "b"). The diagrams and the photographs also show the progressive colonization of the raised coastal plain by grassy vegetation, shrubs, and trees. *Modified from Martini, I.P., Broockfield, M.E., Sadura, S., 2001. Principles of Glacial Geomorphology and Geology. Prentice Hall, Upper Saddle River, NJ, 381 pp. Photo credit: I.P. Martini.*

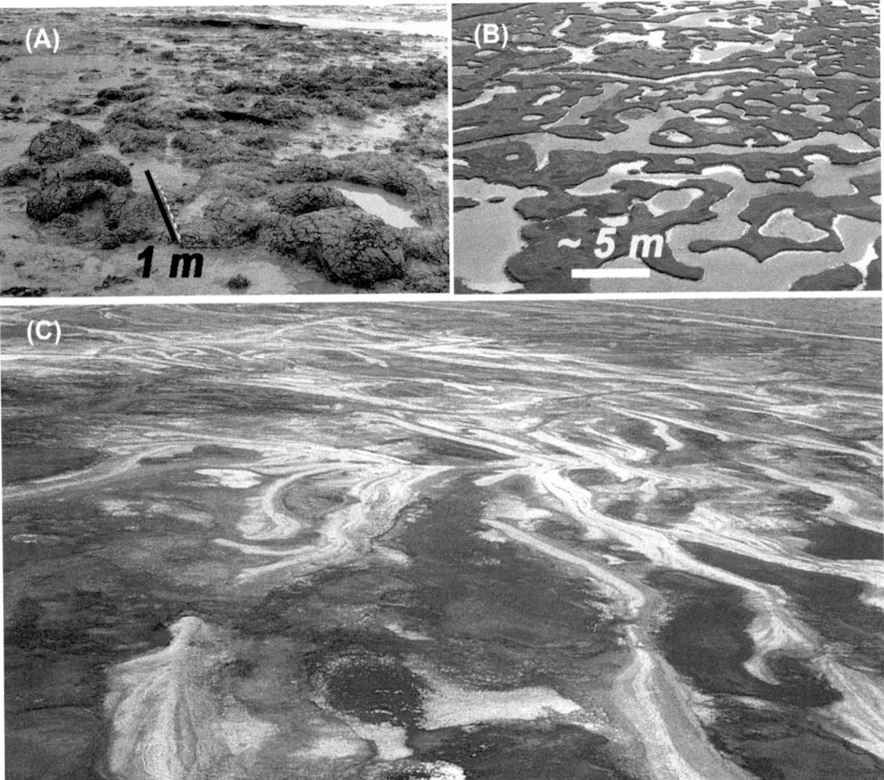

FIGURE 4.8 Typical Arctic coastal landscapes. (A and B) Effect of ice push and erosion on salt marshes: (A) Freshly developed ice push structures in high tidal flats, precursors to a (B) inland, raised, more mature, pond-riddle, jigsaw-pattern, salt marsh (scale applies to forefront area only). (C) Series of beach ridges in the Middle Arctic zone of Canada alternating with wetlands (Foxe Basin, Canada) (width of the immediate foreground is about 300 m). *Photo credit: I.P. Martini.*

The tidal salt marshes open to the sea are much impacted by seasonal ice action. Ice floes are repeatedly grounded, lifted, and removed by tides, and ice-pressured ridges may form (Fig. 4.8A). Marsh material frozen to the underside of the floes may be removed or scoured by ice push. As the land emerges further and becomes more vegetated, typical "jigsaw" patterns of pools develop through such ice action (Fig. 4.8B) and, locally, by geese trampling and grubbing below ground plant parts. Marshes formed in interridge swales are inundated by high tides and acquire a bilateral vegetation distribution pattern with salt species closer to the deeper, muddy, midswale tidal creek, as well as brackish and freshwater species farther away on the sandy beachridge flanks. Locally, the central muddy marsh deposits of the swales may also have some well-sorted sand grains (dispersed or in thin lenses) washed or blown-in from the bounding beach ridges (Fig. 4.7E and F).

In Low Arctic to sub-Arctic zones, the generally muddy, wet sediments of marshes are modified (ripened) into incipient soils. In the southernmost sub-Arctic western shores of Hudson and James bays, for example, incipient horizons with grayish brown colors

(2.5Y5/2) occur in the upper marshes, and ferrans (iron precipitates) may form around plant roots in better-drained parts of the system (Protz, 1982). Along transects from the shoreline inland, salt marsh soils show a gradual increase in thickness of the surficial organic layer (never reaching the 30–40 cm thickness to qualify as peatlands), a decrease in sodium and chloride concentrations with an associated drop in electrical conductivity (from a seasonal average of 6.1 mS cm^{-1} in the lower marsh to 2.2 mS cm^{-1} in the upper marsh), a marked decrease in pH in the upper brackish/freshwater marshes, and a decrease in calcium carbonate equivalents. In some cases, a landward salinity inversion occurs with brackish marshes formed near the coast and saltwater marshes developing farther inland (Martini, 2006a). Where there is inflow of freshwater from rivers, the more seaward sections of marshes are less saline than the landward sections where disturbance, lack of tidal cover in summer, overgrazing by geese, and drying out of the terrain can produce extreme hypersaline soils (about 120 g of solutes per liter), which are devoid of vascular plants (Iacobelli and Jefferies, 1991). Saline marshes located beyond the reach of storm surges may locally, as in southern James Bay, derive salt from groundwater desalinating marine argillaceous silts of the substrate (Price and Woo, 1988).

In the Middle to High Arctic zones, salt marshes open to the sea are generally poorly developed. Coasts that are affected by storm waves during the period of open seawater commonly develop beach ridges, spits, and barrier beaches. On isostatically rising lands, the beach ridges show various heights and spacing. They alternate with interridge lows occupied by shallow lakes and ponds (Fig. 4.8C). The vegetation that grows in these interridge coastal wetlands may constitute true oases in the northernmost colder areas (Martini and Morrison, 2014). The muddy sediments of the marshes can develop incipient soils characterized by a thin organic surface layer underlain by variously gleyed sediment and permanent frost table at less than 2 m depth. Typical features of Arctic soils (cryosols) may consist of deformations caused by freeze and thaw processes and development of periglacial features such as ice wedges and mud boils, leading to the development of patterned grounds (Walker and Peters, 1977; Kimble, 2004; French, 2011; Tarnocai and Bockheim, 2010).

Wetlands with numerous interlaced channels and lakes separated by patterned ground develop on Arctic deltas, particularly along the Beaufort Sea (such as those of the Mackenzie River Delta; Fig. 4.9A) and along the Arctic Russian coast (such as those of the Lena River; Fig. 4.9C and D). Pingos (small conical hills with a core of solid ice) may develop in shallow thermokarst lakes where the coastal areas are low-lying, such as in the Mackenzie River Delta (Fig. 4.9B). Thermokarst (local subsidence of the ground caused by partial melting of the underlying permafrost) also affects Arctic coastal zones, but the change in the landscape differs depending on rates of uplift relative to sea level rise. For example, the Arctic Coastal Plain of Alaska (ALS), the Mackenzie River (MD) and Lena (LR) deltas, and the Great Plain of the Koukdjuak (GPK) of southeast Foxe Basin are characterized by numerous thermokarst lakes (Fig. 4.2). Where sea level rise outpaces residual postglacial isostatic rebound, the ensuing marine transgression leads to considerable coastal erosion and the coastal thermokarst lakes are breached and invaded by the sea (Fig. 4.10A and B; Ruz et al., 1992; Wolfe et al., 1998); and local salt marshes can develop inside the newly formed embayments. Where the land is still undergoing isostatic uplift, such as in the GPK, new thaw lakes develop as the land emerges and remain isolated; farther inland in the raised plain; some are breached and

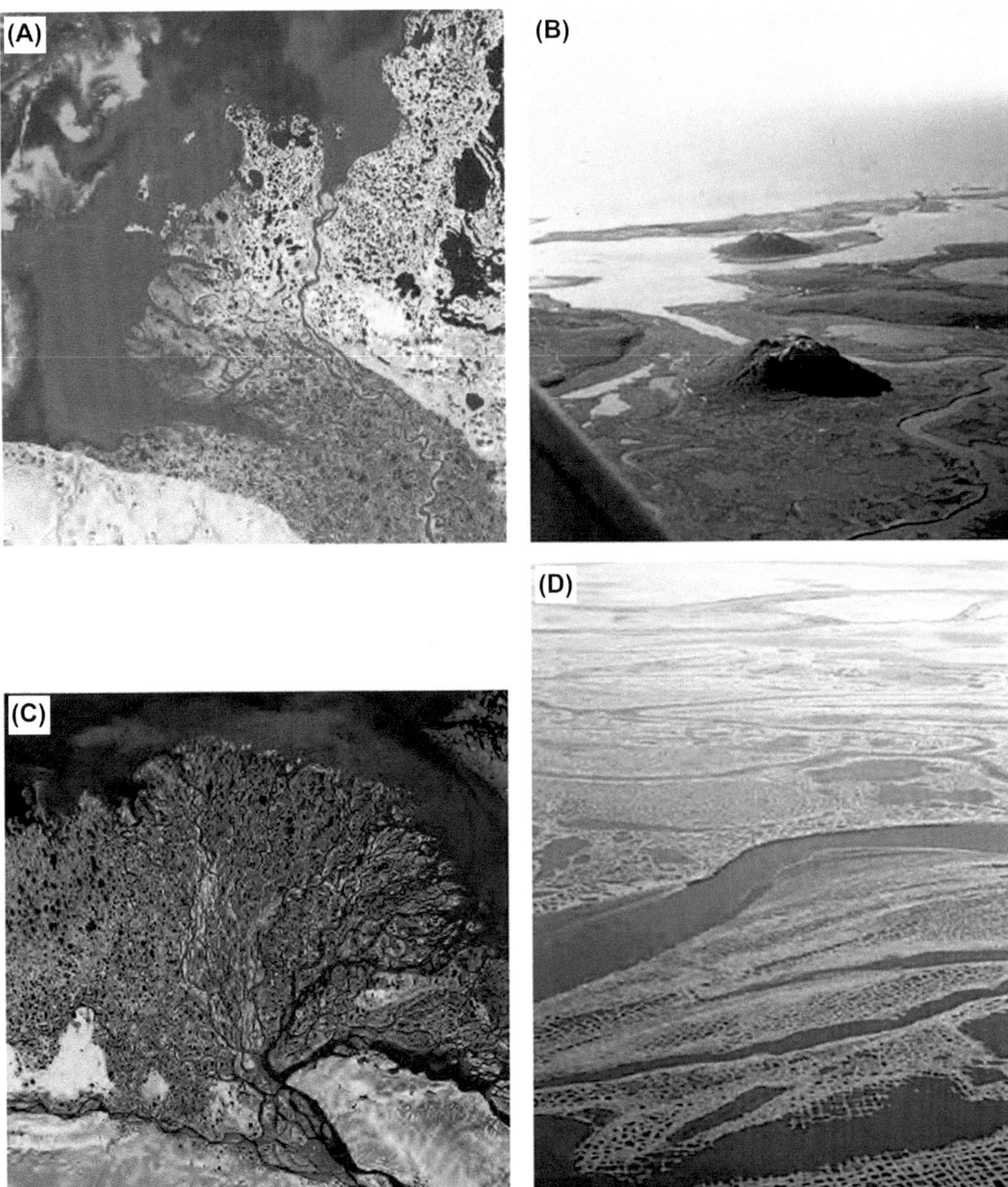

FIGURE 4.9 Coastal wetlands associated with Arctic river deltas. (A and B) Mackenzie River Delta (~12,170 km^2): (A) Interlaced channels and lakes (several are thermokarst lakes). (B) Aerial view of part of the delta with pingos (up to 70 m high hills) developed in a coastal lake (Wikipedia, 2016a). (C and D) Lena River Delta (~30,000 km^2): (C) interlaced channels, thaw lakes; (D) aerial view of well-developed patterned ground in interchannel areas. *(A and C) From NASA (D) Modified from Williams, L., 1994. Ust-Lensky. Russia. Wild Russia.*

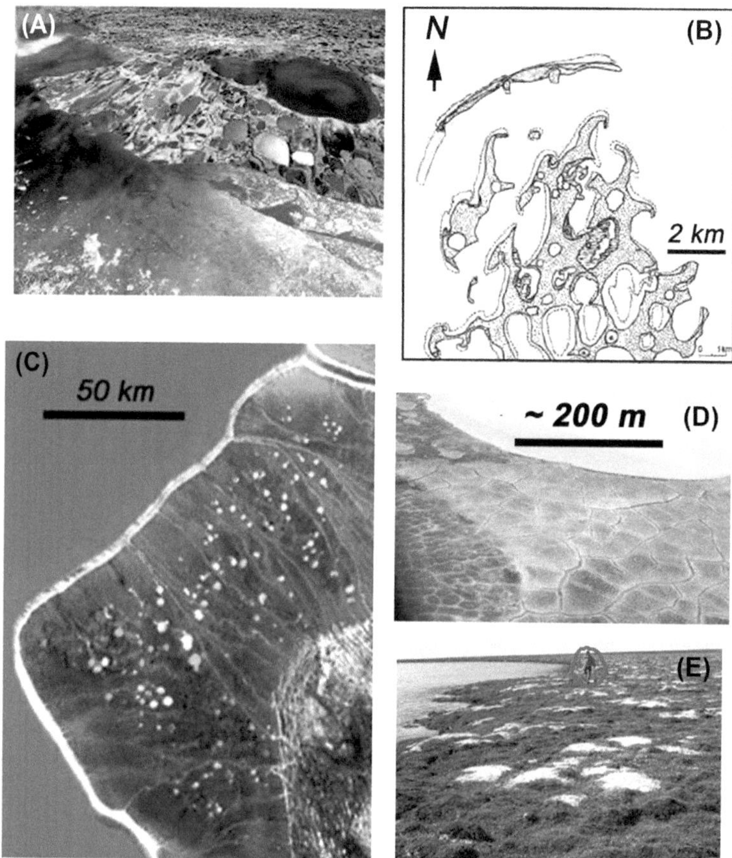

FIGURE 4.10 Periglacial structures in coastal zones. (A) Thaw lakes in the Arctic Coastal Plain, Cape Harlett, Alaska. (B) Breached thaw lakes in the Mackenzie River Delta. (C–E) Isolated lakes in the Great Plain of the Koukdjuak, southeastern Foxe Basin, Canada: (C) satellite image of the entire plain (Google image), (D) ice-wedges patterned ground associated with thaw lakes (scale applies to lakeshore only), (E) mud boils along the margin of a thaw lake (researcher in red circle). *(A) Modified from Bowen, W., 2005. Alaska Atlas of Panoramic Aerial Images. http://californiageographicalsurvey.com/alaska_panorama_atlas/index.html. (B) From Ruz, M.-H., Héquette, A., Hill, P.R., 1992. A model of coastal evolution in a transgressed thermokarst topography, Canadian Beaufort Sea. Marine Geology 106, 2512–3278. (D and E) Photo credit: I.P. Martini.*

joined by creeks (Fig. 4.10C; Martini and Morrison, 2014). These lakes are associated with patterned ground mostly caused by ice wedges and mud boils (Fig. 4.10D and E).

6. VEGETATION OF NORTHERN POLAR COASTAL WETLANDS

Severe environmental conditions (climate, ice) within the immediate coastal zones often restrict plant species richness. Unbroken sea ice persists until late spring (mid-June) on many shores or may occur as blocks driven by winds and currents. This keeps temperatures cool and restricts plant growth in the early growing season. Sub-Arctic salt marsh soils, such as those in southern Hudson Bay, are also severely nitrogen-limited for plant growth (Cargill and Jefferies, 1984; Ngai and Jefferies, 2004). For the immediate coastal freshwater mires, mostly poor fens, plant growth is limited by both nitrogen and phosphorus (Ngai and Jefferies, 2004). Consequently, the plant species richness of Arctic and sub-Arctic coastal salt marshes is low compared with temperate marshes, and prostrate graminoids dominate the vegetation. The most common vascular species that colonizes low-lying coastal sites throughout circumpolar regions is the grass *Puccinellia phryganodes* (Fig. 4.11A and B; Hultén, 1968). In Arctic North America, plants are triploid and although they flower, seeds are not produced. In northern Fennoscandia and in the White Sea region of the Russian Federation, there are reports of tetraploid races of this grass, but it is not known if seed set occurs (R.M.M. Crawford, personal communication). Hence, at least in North America and northern Russia, plants are dispersed by clonal propagation. Individual leaves, shoots, and tillers are able to establish in soft sediment and develop into plants (Chou et al., 1992). Another widespread circumpolar species is *Carex subspathacea* (Fig. 4.11C) and the closely related species (or variants) *Carex salina* and *Carex ramenskii*. These species appear to be less salt tolerant than *P. phryganodes* and tend to occur in areas that receive fresh or brackish drainage water from adjacent lowlands. *Carex subspathacea* can grow in anoxic soils where drainage is impeded. Seed set is episodic and most growth occurs via clonal reproduction.

Other common species include *Triglochin palustris, Triglochin maritima, Cochlearia officinalis, Plantago eriopoda, Potentilla egedii, Ranunculus cymbalaria, Stellaria humifusa, Carex ursina, Carex maritima*, and *Festuca rubra* (Kershaw, 1976; Jefferies, 1977; Jefferies et al., 1979). All flower infrequently, but seed set is uncommon and depends on prevailing local and seasonal weather conditions. Weather conditions are also important in the previous year when flower buds are laid down in most species. The few available studies of the seed bank in Arctic and sub-Arctic salt marshes (Staniforth et al., 1998; Chang et al., 2001) show that the composition of the vegetation and of the soil seed banks are only loosely correlated (approximately 50%), reflecting the poor seed contribution of the dominant graminoid species. Some Low Arctic weedy species typical of degraded or disturbed soils are overrepresented in the soil seed bank compared with their abundance in the vegetation. Within salt marshes, three species are represented: *Salicornia borealis, Koenigia islandica*, and *Atriplex patula*, which mostly grow on an organic substrate in the upper levels of salt marshes or in supratidal marshes (flooded by seawater only during storm surges) but only set seed in favorable years. *S. borealis* does not have a well-developed long-term seed bank, and annuals are not the primary colonizers of exposed mudflats, as in temperate salt marshes. Salt marshes may grade into beach ridges or dunes, and Leymus mollis var. *arenarius* is widespread in well-drained, disturbed, sandy habitats in the Arctic.

Brackish conditions often prevail in the tidal reaches of estuaries and on open marine coasts near river mouths, where extensive freshwater outflow produces salinities ranging from about 3 g of solutes per liter up to 12 g L^{-1}. High rates of sedimentation and brackish conditions result in soft sediments that are readily colonized by plants intolerant of full salinity, such as *Hippuris tetraphylla, Hippuris vulgaris* (less salt tolerant), *T. palustris, Potamogeton filiformis, P. pectinatus, C. officinalis*, and *R. cymbalaria*. This flora can also occur above the intertidal zone in coastal areas where isostatic uplift has left relict salt or brackish ponds. *Hippuris* species develop an extensive rhizome system resulting in stands that are very

FIGURE 4.11 Typical plants of Arctic and sub-Arctic coastal marshes. (A) *Puccinellia phryganodes* sub-Arctic salt marsh in James Bay. (B) *P. phryganodes* a detailed view. (C) *Carex subspathacea (Insert:* close-up of inflorescence). (D) *Hippuris tetraphilla* brackish marsh. (E) Thick silt deposits trapped by *Hippuris tetraphilla. (A) Photo credit: K.F Abraham. (C and D) From Canadian Arctic Archipelago: Adapted from Aiken, S.G., Dallwitz, M.J., Consaul, L.L., McJannet, C.L., Boles, R.L., Argus, G.W., Gillett, J.M., Scott, P.J., Elven, R., LeBlanc, M.C., Gillespie, L.J., Brysting, A.K., Solstad, H., Harris, J.G., 2007. Flora of the Canadian Arctic Archipelago: Descriptions, Illustrations, Identification, and Information Retrieval. NRC Research Press, National Research Council of Canada, Ottawa. https://nature.ca/aaflora/images/b1325090.jpg. (D and E) From sub-Arctic James Bay coasts; Photo credit: I.P. Martini.*

effective at trapping soft sediment, which may lead to a rapid change in coastal topography (Fig. 4.11D and E).

Arctic and sub-Arctic coastal wetlands have many similar plants. Variations in abundance and species occurrence may be related to geological history, local substrate, and

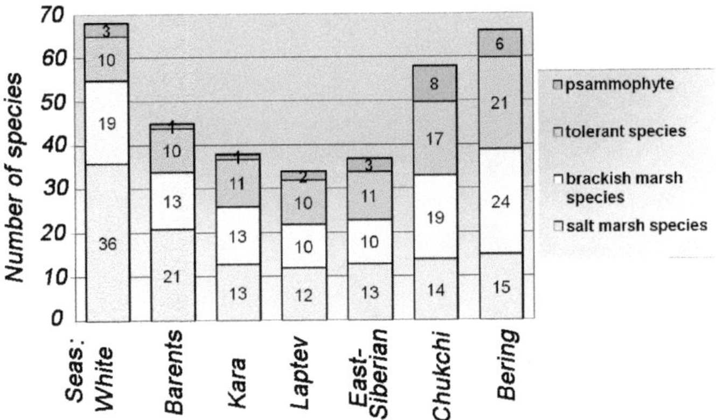

FIGURE 4.12 Distribution of ecological-coenotic groups of species in the coastal flora of the Russian Arctic. *Courtesy of L.A. Sergienko.*

environmental conditions, as illustrated along the salt marshes of the low-lying Arctic coasts of Russia (Sergienko, 2008, 2013a; Spiridonov, 2011). In predominantly muddy and sandy coasts, the number of families, genera, and species in the marsh flora consistently decrease regionally from the western (White and Barents seas) and eastern (Bering and Chukchi seas) areas toward the central zone (Kara, Laptev, and East-Siberian seas) (Figs. 4.12 and 4.13; Sergienko, 2013b). Taxonomic composition of the marsh flora differs between sectors of the Russian Arctic and along the coasts of the Arctic tidal seas (Alexandrova, 1977; Koroleva et al., 2011, 2012). Four typical marsh transects illustrate plant successions: two from western localities (coasts of the White and Barents seas), one from the central area (East-Siberian Sea), and one from an easternmost locality (Chukchi Sea) (Fig. 4.14). Beside differences that may be related to local elevation and the presence of beach ridges, a major regional variation occurs between the predominant species occurring along the White Sea transect and the others (Sergienko et al., 2012). In the White Sea case, Arctic species form the lowest percentage (Fig. 4.14A left circular diagram), whereas boreal species with an Atlantic and European source (Fig. 4.14A right circular diagram), such as sea aster (*Tripolium vulgare*), sea plantain (*Plantago maritima*), and seaside arrowgrass (*Triglochin maritima*), form the highest percentage. European species such as *Atriplex lapponica*, subpolar plantain (*Plantago subpolaris*), alkali grass (*Puccinellia capillaris*), and European glasswort (*Salicornia europaea*) are common. The Barents Sea transect has a typical Arctic profile, including the early salt marsh colonizers *Puccinellia phryganodes* and *Carex subspathacea*, but still contains boreal species of European origin (Fig. 4.14B). The influence of the European source is not felt to the east. The central transects of the Kara, Laptev, and Eastern-Siberian (Fig. 4.14C) seas have typical Arctic communities of circumpolar species, including *Puccinellia phryganodes*, *Carex subspathacea*, *Ranunculus tricrenatus*, *Dupontia psilosantha*, and *Arctophila fulva*, as well as vegetation of Siberian sources. The easternmost transects of the Chukchi have a well-defined Arctic composition as well as vegetation of American sources (Fig. 4.14D). The American influence is absent west of the Chukchi Sea.

The central Taymyr Peninsula acts as a major barrier for the distribution of plant species originating from western and eastern areas. The Peninsula also was the center of Upper

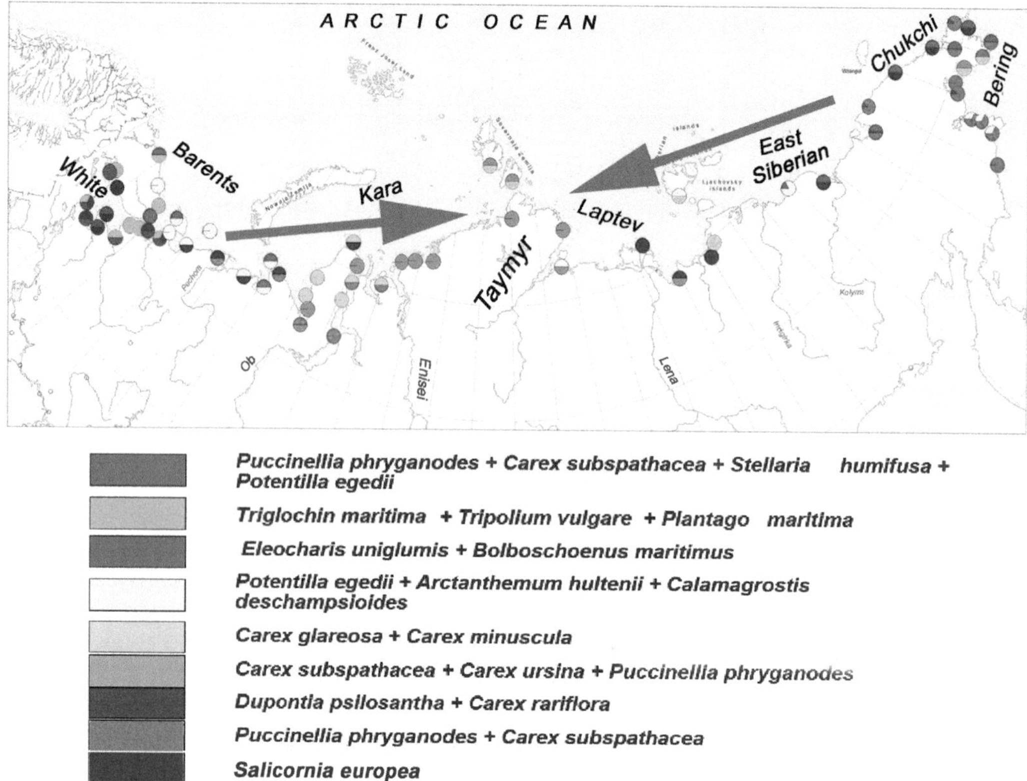

FIGURE 4.13 Typical marsh plant communities along the Russian Arctic coasts. Note that almost all species mentioned have a circumpolar Arctic distribution, but some exception occurs, notably, *Salicornia europea* that along the Russian coast is common only in the White Sea area. *Modified from Sergienko, L.A., 2013b. Analysis of the salt marsh flora of the Russian Arctic coasts. Annals of Valahia University of Targoviste. Geographical Series Tome 13/Issues 2, 69–75. http://fsu.valahia.ro/images/avutgs/1/2013/2013130201.html.*

Pleistocene glaciations and of glacier persistence (Arkhipov, 1997). Its inland polar desert still indicates a colder region in respect to the warmer westernmost and easternmost areas that have their climates ameliorated by Atlantic and Pacific marine warm currents, respectively (Fig. 4.5). This has led to an overall centripetal distribution and evolution of plant communities along the Russian Arctic coasts (Fig. 4.13).

7. FAUNA OF NORTHERN POLAR COASTAL WETLANDS

Although species diversity in the Arctic is considered low compared with other regions, the Arctic terrestrial fauna nevertheless contains some 6000 species (about 2% of the global total; Chernov, 1995; Callaghan et al., 2005). The fauna of polar coastal wetlands can be divided into two basic ecological categories: (1) an infauna consisting of organisms that spend their life cycles within or in close association with the wetland and (2) a migratory or nomadic

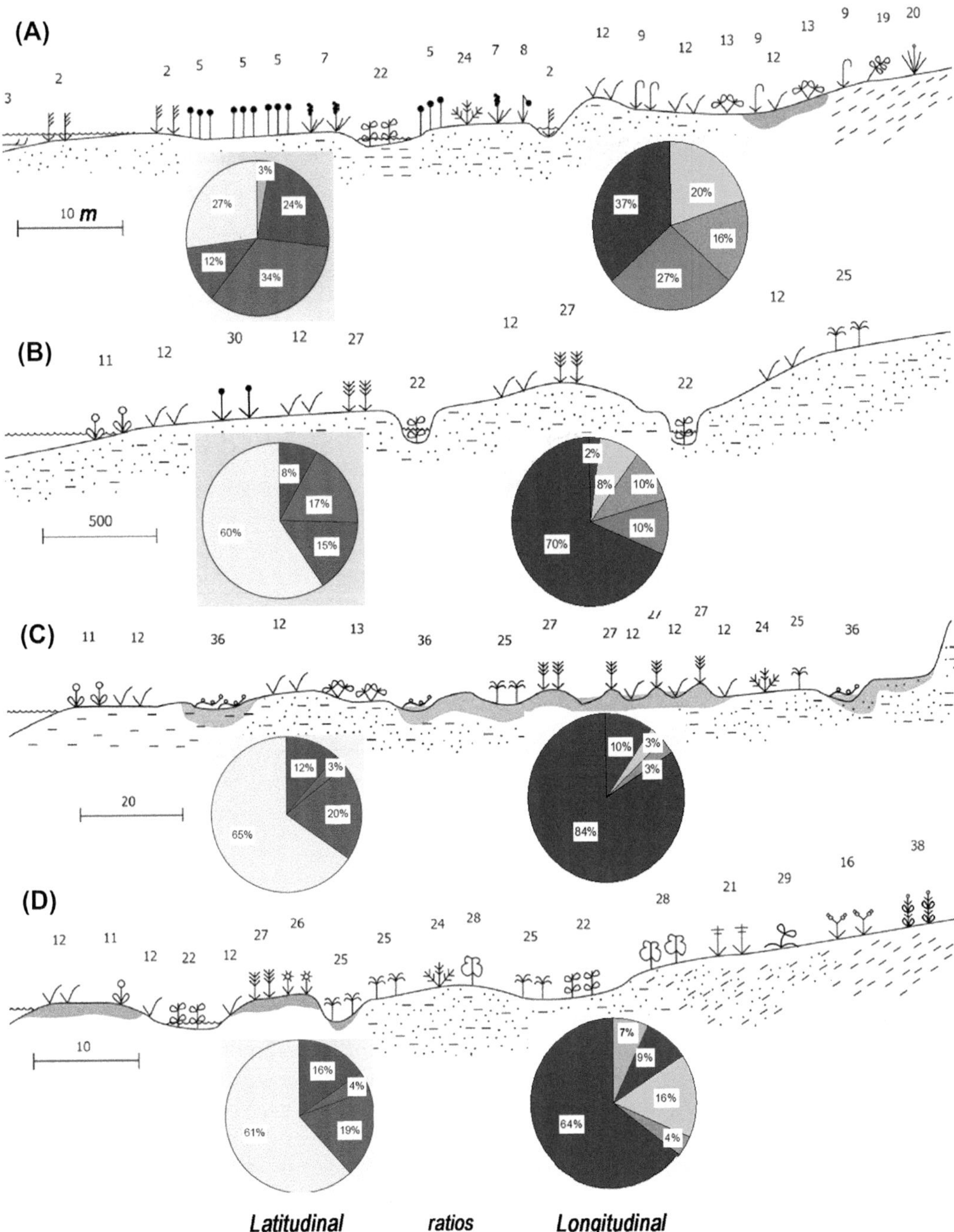

FIGURE 4.14 Landscape structure of coastal wetlands and ratios of latitudinal and longitudinal geographic elements of partial floras of (A) White Sea (B) Barents Sea, (C) Eastern-Siberian Sea, (D) Chukchi Sea (northern coast), and associated Legend. The circular left diagrams show the latitudinal ratio of the plant components of the area, whereas the right circular diagrams show the longitudinal ratio of the plant provenances. *Modified from Sergienko, L., 2013a. Salt marsh flora and vegetation of the Russian Arctic coasts. Czech Polar Reports 3, 30–37. http://www.sci.muni.cz/CPR/5cislo/Sergienko-web.pdf.*

No	Species	No	Species
2	*Bolboschoenus maritimus*	21	*Festuca rubra*
3	*Zostera marina,*	22	*Hippuris tetraphylla*
5	*Eleocharis uniglumis,*	24	*Stellaria humifusa*
7	*Triglochin maritima*	25	*Dupontia psilosantha*
8	*Tripolium vulgare,*	26	*Arctanthemum hultenii*
9	*Juncus gerardii*	27	*Calamagrostis deschampsioides*
11	*Puccinellia phryganodes*	28	*Salix ovalifolia*
12	*Carex subspathacea*	29	*Carex glareosa*
13	*Potentilla egedei*	30	*Rhodiola integrifolia*
16	*Leymus villosissimus*	36	*Plantago schrenkii*
19	*Lathyrus japonicus* ssp. *Pubescens*	38	*Ranunculus tricrenatus*
20	*Phragmites australis*		

LEGEND
LATITUDINAL RATIO
Plurizonal
Boreal
Arctic
Hypoarctic
Arcto-boreal
LONGITUDINAL RATIO
Preferably American
Preferably Siberian
Amphioceanic
Eurasian
European
Circumpolar
organic rich
clay
mud
sand

FIGURE 4.14 cont'd

exfauna that uses the wetland on a seasonal basis. Polar coastal wetlands play a critical role in the life cycles of many migratory animals, particularly birds. Many species of waterfowl and shorebirds use Arctic and sub-Arctic coastal wetlands both for breeding and/or as stopovers on migration to and from their breeding areas. The flora and infauna of the wetlands provide the food resources on which the birds depend to complete their annual cycles.

7.1 Invertebrate Fauna

Among invertebrates, primitive groups such as springtails (*Collembola*) (400 species, 6% of the global total) are more heavily represented than advanced groups such as spiders (300 species, 0.1% of the global total). Numbers of invertebrate species and families typically decrease with increasing latitude, and the distribution of some groups, such as spiders, is patchy (Chernov, 1995; Pickavance, 2006). In the far north, common invertebrate species tend to be widely distributed, with a few dominant species (such as 12 species of springtail in the northern Taymyr Peninsula, Russia; Chernov and Matveyeva, 1997). Coastal wetlands and their fauna fall into two broad categories: (1) those at or near the shore involving marine or saltwater-influenced habitats and (2) those occurring slightly inland involving mostly brackish and freshwater habitats.

7.1.1 Invertebrate Fauna of Coastal Saline Areas

Although poorly studied, marine intertidal invertebrates and other organisms are highly important food sources for a variety of birds, fish, and even mammals. Ecological conditions are harsh, including severe climate and physical disruption of habitats caused by ice action in near-shore areas. In Hudson and James bays, for instance, much of the intertidal stock of the common bivalve *Macoma balthica* is removed annually by ice and wave action but is replenished by spatfall (larval production) from subtidal populations during the summer. *M. balthica* is widely distributed and forms a prominent food resource for birds and fish in intertidal areas in Iceland, Hudson and James bays, and Alaska. Densities in James Bay average 2000–3700 individuals m^{-2}, with highest recorded densities of up to 12,800 individuals m^{-2} in zones of eelgrass *Zostera marina* (Martini and Morrison, 1987). Similar densities are found elsewhere in Alaska (maximum 4000 individuals m^{-2}; Powers et al., 2002), St. Lawrence Estuary, Canada (maximum 2700 individuals m^{-2}; Azouzi et al., 2002), and Dutch Wadden Sea (3250 individuals m^{-2}; Piersma and Koolhaas, 1997).

Sub-Arctic intertidal mudflats typically have fairly high densities but low species diversity of infauna, such as *M. balthica* in James Bay (Martini et al., 1980), and *M. balthica*, the amphipod *Corophium salmonis*, and the polychete *Eteone longa* on the Copper River Delta (CR), Alaska (Fig. 4.15; Powers et al., 2002). Other organisms that occur include gastropods, mussels, limpets, nematodes, oligochaetes, and polychaetes, as well as foraminifera, copepods, ostracods, amphipods, cladocerans, ectoprocta, and barnacles (Martini et al., 1980). Oligochete worms and Dipteran larvae are numerous along the edge of the short-grass salt marsh and are important food items for shorebirds. Oligochaetes (family Naididae, genus *Paranais*) numerically account for about 63% of the macrobenthos in the salt marshes, and their distribution is strongly correlated with electrical conductivity and the organic carbon content of the sediments. In the coastal ponds in western James Bay, Dipteran larvae of the families Chironomidae, Heleidae, and Tipulidae occur in densities of up to 5500 m^{-2} (Clarke, 1980).

Intertidal mudflats in High Arctic locations have low numbers and variety of organisms, reflecting increasingly severe environmental conditions. At Zackenberg (ZA) in central northeast Greenland (Fig. 4.15), mudflats contain low to moderate densities of nematodes, tardigrades, and crustaceans, which are preyed on by shorebirds during July and August (Caning and Rausch, 2001; Meltofte and Lahrmann, 2006). Red knots (*Calidris canutus*) and ruddy turnstones (*Arenaria interpres*) have been observed feeding on crustaceans on intertidal mudflats on the central east and northern coasts of nethermost Ellesmere Island (EL), Canada, during the postbreeding period (R.I.G. Morrison, personal observation).

Few studies are available of intertidal areas in sub-Antarctic wetlands and mudflats in the southern hemisphere. The mainland Atlantic coast of Tierra del Fuego contains important intertidal habitats, including the huge mudflats in Bahía Lomas, Chile, and Bahía San Sebastian, Argentina, and the rocky intertidal *restinga* habitats, which are used heavily by shorebirds, gulls, and terns. Although surrounded by the sub-Antarctic oceanographic zone, the area is cool-temperate in terms of climate and vegetation. At Bahía Lomas, polychaetes, bivalve mollusks, isopods, and amphipods predominate in sandy habitats, their abundance dependent on sediment size and type (Ponce et al., 2003). An ecological monitoring program was initiated at this site in 2003 (led by Prof. C. Espoz, Universidad Santo Tomás, Santiago, Chile). Espoz et al. (2008) published information on intertidal invertebrates and their role in the trophic ecology of the endangered red knot *Calidris canutus rufa* in Bahia Lomas, which supports over 95% of the southern wintering population of knots (R.I.G.Morrison, unpublished results).

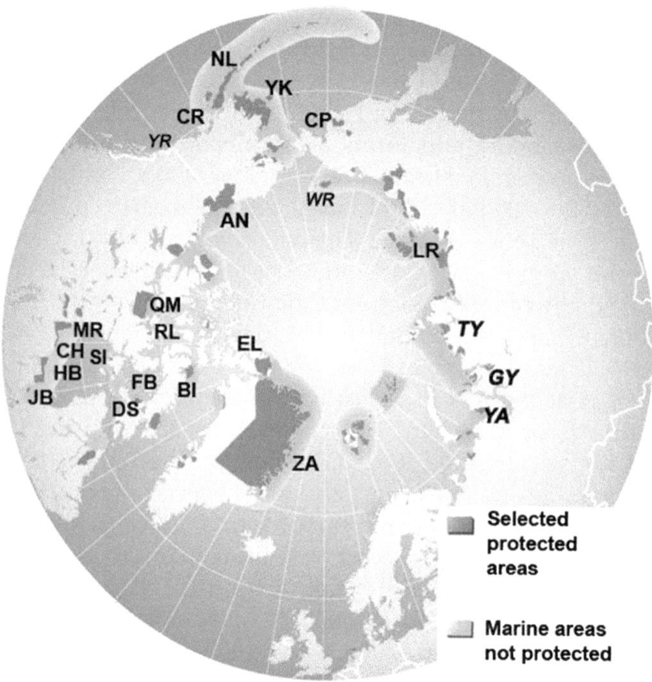

FIGURE 4.15 Major protected Arctic areas important for wildlife (*AN*, Arctic National Wildlife Reserve, Alaska; *BI*, Bylot Island; *CH*, Churchill, Manitoba Canada; *CP*, Chukotka Peninsula; *CR*, Copper River Delta, Alaska; *DS*, Dewey Soper Migratory Bird Sanctuary, Baffin Island; *EI*, Ellesmere Island; *FB*, Foxe Basin; *GY*, Gydan Peninsula; *HB*, Hudson Bay; *JB*, James Bay; *LR*, Lena River Delta, Russia; *MR*, McConnell River Migratory Bird Sanctuary; *NL*, Nelson Lagoon, Alaska; *QM*, Queen Maud Gulf Migratory Bird Sanctuary; *RL*, Rasmussen Lowlands; *SI*, Southampton Island; *TY*, Taymyr Peninula; *WR*, Wrangel Island; *YA*, Yamal Peninsula; *YF*, Yakutat Forelands, Alaska; *YK*,Yukon and Kuskokwim river deltas Alaska; *ZA*, Zackenberg, Greenland). *Modified from H. Ahlenius; UNEP/GRID-Arendal, 2005b. Protected Areas, Arctic. http://www.grida.no/resources/7002.*

7.1.2 Invertebrate Fauna of Near-Coast, Freshwater Areas

Areas farther inland or less regularly inundated by the tide can develop a rich insect fauna. In southern James Bay, Kakonge et al. (1979) identified 318 species of invertebrates (105 families, 14 orders) among which mosquitoes and biting flies were a prominent component. The insects play an important role in ecological processes in the marsh, including contributing to soil fertility, litter breakdown (by springtails, mites, nematodes, rotifers, and some exotic earthworms), as major secondary producers and as a food resource for migrating birds. Mosquitoes were reported to occur in densities of 5 million per acre (13.35 m ha^{-1}) on the coast of Hudson Bay (West, 1951). Chironomids often reach densities of many thousand per square meter in freshwater and brackish water (Pinder, 1983), and they are a major component of the macrobenthos in ponds at northern latitudes (Andersen, 1946; Butler et al., 2001) as well as being an important component of the diet of shorebirds (Summerhayes and Elton, 1923) and waterfowl (Bergman and Derksen, 1977; Danell and Sjöberg, 1977).

7.2 Vertebrate Fauna Using Coastal Wetlands

7.2.1 *Avifauna*

Birds, particularly waterbirds (ducks, geese, swans, loons, shorebirds, gulls, and terns), form a prominent component of the fauna using cold coastal wetlands. Within the Arctic, 450 species of birds have been recorded breeding (Callaghan et al., 2005); most migrate to southern latitudes where many inhabit coastal wetlands (Schmiegelow and Mönkkönen, 2002).

Polar wetland areas play a key role as migration sites enabling the birds to complete their long journeys (Pollock et al., 2012). These migration systems are generally known as flyways. Worldwide, there are some 8 recognizable flyway systems for waterfowl (Fig. 4.16) and 10 for shorebirds (Fig. 4.17). Some birds travel between Arctic and sub-Arctic wetlands to sub-Antarctic wetlands during the course of their annual travels. One population of the red knot (*C. canutus rufa*) (Fig. 4.18A), for instance, undertakes an annual round-trip journey of some 30,000 km between the central Canadian Arctic and Tierra del Fuego in southern South America (Morrison, 1984; Morrison and Harrington, 1992). Areas supporting important concentrations of shorebirds worldwide are located near coastal regions of high productivity (Butler et al., 2001).

Bird breeding areas. Recent population estimates for waterbirds breeding in polar coastal wetlands are presented in Table 4.1 (Zöckler et al., 2012). Some 175 species, involving 368 populations, totaled an estimated 123.2 million birds and included divers, grebes, swans, geese, ducks, cranes, shorebird, gulls, terns, and skuas. Geese form a prominent component (Fig. 4.18B), with important nesting areas in the North American and Eurasian Arctic and sub-Arctic, as well as in sub-Arctic wetlands in Iceland (Rowell and Hearn, 2005). Recent estimates differ somewhat. Fox and Leafloor (2018) report over 22 million geese of six species breeding in North American Arctic and nine species totaling an estimated 12 million birds

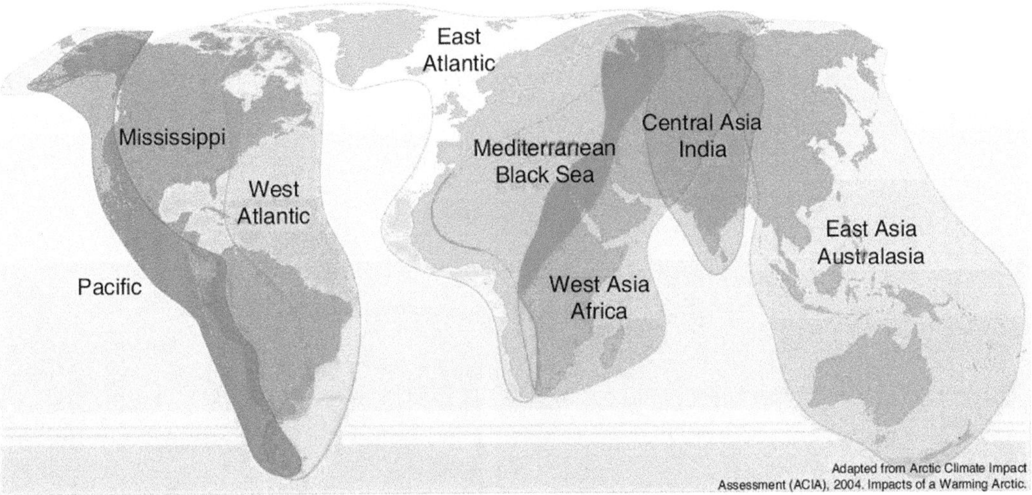

FIGURE 4.16 Major global flyway systems used by waterfowl. *Image credit: H. Ahlenius; UNEP/GRID-Arendal, 2005c. Major Global Waterfowl Flyways. http://www.grida.no/resources/7022.*

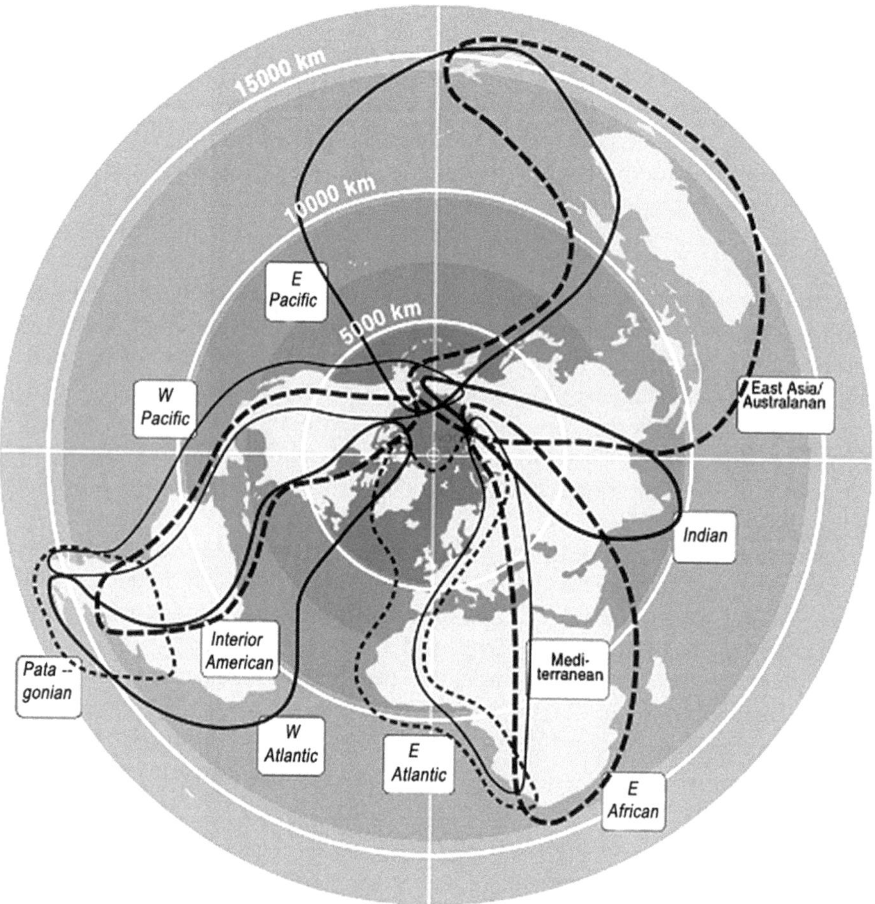

FIGURE 4.17 Major global shorebird flyway systems linking Arctic breeding wetlands with "wintering areas" some of which are sub-Antarctic wetlands. *From Piersna, T., Lindström, À., 2004. Migrating shorebirds as integrative sentinels of global environmental change. Ibis 146 (s1), 61–69.*

FIGURE 4.18 Birds utilizing the polar coastal wetlands. (A) Red knots in flight. (B) Lesser snow geese: light-colored, flightless adults in the background during breeding period and darker colored goslings in the foreground. *Photo credit: R.I.G. Morrison.*

TABLE 4.1 Populations of Waterbirds in Circumpolar Coastal Wetlands (Population Estimates in 1000 Individuals)

Group	No. of Species	No. of Populations	Popul.est. × 1000 ind.
Divers, grebes	7	18	2395
Swans	3	13	513
Geese	12	49	17,139
Ducks	30	87	30,558
Cranes	2	4	453
Shorebirds	71	170	48,760
Gulls, terns, skuas	28	46	23,350
Total	153	387	123,168

Modified From Zöckler, C., Lanctot, R., Brown, S., Syroechkovskiy. E., 2012. Waders (shorebirds). In: Arctic Report Card 2012, NOAA's Arctic Program, pp. 92–102. ftp://ftp.oar.noaa.gov/arctic/documents/ArcticReportCard_full_report2012.pdf.

breeding in the Eurasian Arctic; Zöckler et al. (2012) indicated that in the circumpolar Arctic, some 12 species involving 49 populations of geese totaled 17.1 million. Arctic goose populations are considered generally to be increasing in North America, in contrast to shorebirds (North American Bird Conservation Initiative Canada, 2016). Some geese occupy a relatively restricted range, such as the red-breasted goose (*Branta ruficollis*) in Siberia (Taymyr Peninsula [TY] and the Gydan [GY] and Yamal [YA] peninsulas (Dereliev, 2006)), whereas others have a much broader range, such as snow geese (*Chen caerulescens*), which ranges from northwest Greenland to Wrangel Island (WR) and the Chukotka Peninsula (CP) in Russia (Figs. 4.15 and 4.18B; Mowbray et al., 2000).

Shorebirds form a prominent component of the breeding avifauna of sub-Arctic and Arctic wetlands. Previous estimates by Zöckler (1998) (8.1 million individuals of 13 species [calidris sandpipers] in North America and 6.3 million individuals of 13 species in the Eurasian Arctic) were updated for the circumpolar Arctic, resulting in estimates for 71 species of shorebirds, involving 170 populations, totaling 48.8 million (Table 4.1, Zöckler et al., 2012). Increased population estimates generally represent better survey coverage. Some 27 species of shorebirds use central Arctic and sub-Arctic areas of Canada (Morrison and Gaston, 1986). Some species nest directly in wetland habitats, such as red phalarope (*Phalaropus fulicarius*) and dunlin (*Calidris alpina*), whereas others, such as ruddy turnstone and red knot, nest in nearby upland habitats, which are often found in close association with wetter habitats where the birds feed and raise their young. Shorebirds breed to the northern limit of land in North America and Eurasia, although in reduced numbers and density compared with areas farther south. Factors affecting distribution may include predation risk: predation risk decreased more than twofold along a 3350 km latitudinal transect from Akimiski Island (53°N) in James Bay to Alert (82°N), northern Ellesmere Island (McKinnon et al., 2010).

Bird staging areas. During migration, sub-Arctic and Arctic coastal habitats support large populations of waterfowl and shorebirds. For waterfowl, examples include geese migrating northward through Hudson and James bays, Iceland, and northern Norway *en route* to

Arctic breeding grounds (Jefferies et al., 2003; Ward et al., 2005; Glahder et al., 2006). During migration, geese and other waterfowl acquire nutrients that they bring to the breeding grounds in the form of body stores, which are used to form eggs or enhance survival (Alisauskas and Ankney, 1992).

In recent decades, rapidly increasing populations of lesser snow geese and Ross's geese in the Canadian Arctic have caused heavy damage to salt marshes and freshwater fens in migration and breeding areas (Abraham et al., 2012). Geometric population growth of these geese in North America (Abraham et al., 2005) and of the barnacle goose (*Branta leucopsis*) in Europe (Van Eerden et al., 2005; Jefferies et al., 2006a) is thought to result primarily from agricultural changes in southern areas, the availability of refugia from hunting, lower harvest rates, and possible climate change on the breeding grounds (Jefferies et al., 2003; Kerbes et al., 2006). Damage to northern coastal marshes depends on the densities of the birds and their foraging behavior (grazing, grubbing, and shoot pulling of sedges). Grubbing, in which the geese pull up the roots and rhizomes of the plants, and shoot pulling may lead to total loss of the selected graminoid plants of coastal marshes. The resulting physical and chemical changes in the exposed sediments and continued exposure to foraging geese alter habitat succession and recovery (Abraham et al., 2005). Vegetation loss in the Canadian Arctic has been so extensive that it can be readily detected by remote sensing (Jano et al., 1998; Didiuk and Ferguson, 2005; Jefferies et al., 2006b; Alisauskas et al., 2006). Reestablishment of vegetation is long-term and requires erosion of hypersaline, consolidated sediment, and the buildup of unconsolidated soft sediment, in which plants can reestablish themselves (Jefferies et al., 2003). A similar loss of vegetation is occurring in coastal freshwater marshes, although the abiotic and biotic processes are different (Jefferies et al., 2003; Alisauskas et al., 2006). Erosion of peat, following loss of vegetation, can lead to exposure of underlying glacial gravels and marine clays in coastal locations and can alter the trajectory of succession (Handa et al., 2002).

Migration areas are also critically important in shorebird migration. In Iceland, for instance, red knots *en route* to the eastern Canadian High Arctic from European wintering quarters depend on coastal habitats to enable a physiological transformation into virtual "flying machines," involving accumulation of fat, an increase in the size of organs and muscles used for flying, and a decrease in body components less important for flight (Piersma et al., 1999). Body stores remaining on arrival in the Arctic enable the birds to retransform into a state suitable for breeding or may be used for early season survival (Morrison and Hobson, 2004; Morrison et al., 2005). These aspects of bird migration emphasize the interconnected nature of the Arctic, sub-Arctic, and other wetlands farther south. The ability of birds to acquire the needed stores during migration has important survival implications. Shorebirds departing Iceland in better than average condition were shown to have a higher survival when faced with difficult weather conditions in the Arctic (Morrison, 2006; Morrison et al., 2007), and conversely, shorebirds prevented from reaching adequate departure weights at the final spring stopover area in North America suffered significantly decreased survival (Baker et al., 2004).

During autumn migration, many shorebirds and waterfowl use the coasts of Hudson and James bays to buildup body reserves for the flight south (Morrison and Harrington, 1979; Morrison and Gaston, 1986; Pollock et al., 2012). Shorebird distribution in this region was directly related to food abundance for several species, including semipalmated sandpiper (*Calidris pusilla*), red knot, and Hudsonian godwit (*Limosa haemastica*), at both local and regional scales (Morrison, 1983, 1984; Morrison and Gaston, 1986). In Alaska, the bar-tailed godwit (*Limosa lapponica*) accumulates up to 55% of its body weight in fat and undergoes critical physiological changes before a spectacular migration across the Pacific Ocean, which can involve nonstop flights of 11,000 km lasting six or more days to wintering areas in New Zealand and eastern Australia (Piersma and Gill, 1998; Gill et al., 2005).

7.2.2 *Mammalian Fauna*

A variety of mammals occurs in polar coastal wetlands, including bears (polar bears, grizzly bears, and brown bears), Arctic and red foxes, wolves, wolverines, caribou, Arctic hare, mink, weasels, lemmings, and voles. The cyclical patterns in abundance of small mammals such as lemmings and the associated responses of predators such as Arctic foxes (*Alopex lagopus*) influence the breeding success of other birds and animals in Arctic habitats. Breeding success of shorebirds is decreased in low lemming years owing to increased predation by foxes to the extent that lemming cycles in the Arctic can be detected by observing the number of shorebird young reaching migration and wintering areas (Underhill, 1987; Underhill et al., 1989; Blomqvist et al., 2002). Jaegers (*Stercorarius* spp.), which depend on lemmings for food, may not breed at all in some areas in low lemming years (Maher, 1970). Over the past 20–30 years, it has appeared that changing climate has led to suppression of some lemming cycles, which has the potential to greatly affect nesting success of Arctic birds (Hörnfeldt et al., 2005) including shorebirds (Fraser et al. 2013).

Polar bears (*Ursus maritimus*) are generally highly dependent on sea ice, which provides habitat for their principal prey, ringed seals (*Phoca hispida*) (Stirling et al., 1999; Derocher et al., 2004), but during the summer when the sea ice melts, polar bears come ashore in coastal areas. Polar bears can become significant predators of colonial nesting seabirds and waterfowl (Iverson et al., 2014). Polar bears are increasingly likely to seek alternative food sources as the extent of sea ice declines and onshore time increases under climate warming (Stirling and Parkinson, 2006; Gormezano and Rockwell, 2013).

In North America, wild caribou (*Rangifer tarandus*) herds use coastal plains primarily for calving in spring. In the summer, coastal areas are important for avoiding predators and biting and parasitic insects, and caribou diets shift in summer from the lichen-dominated winter diets to vascular plants, including wetland sedges, grasses, and other species.

8. ENVIRONMENTAL HAZARDS

Although polar coastal wetlands can be readily impacted by natural disturbances and human activities, they have been subject to relatively few anthropogenic influences compared with their temperate counterparts.

Climate change is expected to affect polar wetlands directly and more significantly than those in more temperate areas (Collins et al., 2013). Rises in temperature and precipitation will affect growth and reproductive success of plant populations. Other indirect effects are associated with increased melting of sea ice and permafrost and changes in salinity and hydrology. Although melting of Arctic sea ice is well documented (IPCC 2013), the rate of disappearance of the ice is proceeding far more quickly than most models predict

(Wadhams 2017). Low-lying coastal wetlands are vulnerable to increasing damage from more frequent storm surges and the destructive effects of wave action in the absence of sea ice, as well as from rise in sea level (Callaghan et al., 2005; Loeng et al., 2005; Cahoon et al., 2006; Anisimov et al., 2007; Ross and Adam, 2013). A warming climate may lead to mis-match of timing between production of young and their insect food resources, leading increased mortality, a phenomenon already noted for shorebirds (Senner 2012; van Gils et al. 2016).

Human activities may drastically impact polar coastal wetlands. Rapid adverse effects are associated with hydrological megaprojects, resource extraction (hydrocarbon exploration and exploitation, mining for lead, zinc, gold, and diamonds), ecotourism, fishing, and increased hunting and gathering by indigenous people (Anisimov et al., 2001). Large-scale hydrocarbon exploration and production is ongoing in on- and offshore areas in the Arctic Coastal Plain of Alaska, the Mackenzie River Delta, the Pechora Basin, the Lower Ob Basin, and the Western Siberian Plain (Ulmishek, 2003). Major mining projects, such as the Baffinland iron ore mine, are coming onstream in the eastern Canadian Arctic (Wikipedia, 2016b). Adverse effects on wetland ecosystems and wildlife result from infrastructure and disturbance. A more ominous impact is related to the extensive pollution that might derive from oil spills during exploration and transport of petroleum through a possible ice-free Northwest or Northeast Passage (Anisimov et al., 2001). Heavy metals and organic pollutants are transported by water and air, and both bioaccumulate in trophic food webs and in wetland soils, thereby posing environmental risks to wildlife and human populations (AMAP, 2002).

9. CONCLUSIONS AND RESEARCH PRIORITIES

Polar coastal wetlands contain unique habitats of great importance to wildlife and human populations. While Earth has changed repeatedly over geological time, the current rapid rate of global change is considered exceptional and likely to be felt most keenly in the Arctic and Polar Regions. Climatic and geochemical (pollution) changes are already impacting Polar Regions and will continue to do so increasingly. Rising temperatures can be expected to reduce the cover of both sea ice and permafrost inland. The resulting changes in albedo and heat storage could lead to a global redistribution of heat and modify atmospheric properties and oceanic currents; predicted effects include increased precipitation, increased frequency of storms, and a global sea level rise. Coastal Arctic and sub-Arctic environments and their associated biota are particularly vulnerable to these climatic changes. This, for example, may greatly impact birds that migrate over entire continents. Human settlements located on Arctic low-lying coasts will also be adversely affected by storms and tidal surges, and climate change has already affected traditional ways of life. The opening of Arctic sea routes presents further possible hazards for low-lying coastal regions and their biota. Pollution, physical disturbance, and increased access to these remote localities are likely to result in indirect changes, many of which are unforeseen at this stage.

Research priorities for polar coastal wetlands include the following:

1. Improve determination and monitoring of rates of isostatic land uplift and sea level rise to predict relative sea level changes along coasts, especially where severe coastal erosion is taking place.

2. Monitor rates of permafrost loss and document the northward shift of the boundary between continuous and discontinuous permafrost zones. These changes are likely to lead to extensive drying of peatlands and the release of greenhouse gasses (carbon dioxide and methane) and thus to an increase in the rate of global warming.
3. Monitor the effects of global change on oceanic and coastal environments. Assess and model effects of sea ice melting on weather patterns, coastal erosion, and associated phenomena (Werner et al., 2016). Determine the ability of vegetation and resident and migratory wildlife to adjust to these changes. Record spread of invasive species associated with climate change. Integrate traditional knowledge into assessments of effects of climate change on the environment and wildlife through more direct involvement of indigenous people.
4. Expand monitoring to understand the impact of local and long-range human activities.

References

Abraham, K.F., Jefferies, R.L., Alisauskas, R.T., 2005. The dynamics of landscape change and snow geese in mid-continent North America. Global Change Biology 11, 841–855.

Abraham, K.F., Jefferies, R.L., Alisauskas, R.T., Rockwell, R.F., 2012. Northern wetland ecosystems and their response to high densities of lesser snow geese and Ross's geese. In: Leafloor, J.O., Moser, T.J., Batt, B.D.J. (Eds.), Evaluation of Special Management Measures for Midcontinent Lesser Snow Geese and Ross's Geese, pp. 9–45. Arctic Goose Joint Venture Special Publication. U.S. Fish and Wildlife Service, Washington, DC and Canadian Wildlife Service, Ottawa, Ontario. http://www.agjv.ca/images/stories/pdf/AGJV_SNOW_GOOSE_RPT_2012_FINAL.pdf.

Aiken, S.G., Dallwitz, M.J., Consaul, L.L., McJannet, C.L., Boles, R.L., Argus, G.W., Gillett, J.M., Scott, P.J., Elven, R., LeBlanc, M.C., Gillespie, L.J., Brysting, A.K., Solstad, H., Harris, J.G., 2007. Flora of the Canadian Arctic Archipelago: Descriptions, Illustrations, Identification, and Information Retrieval. NRC Research Press, National Research Council of Canada, Ottawa. http://nature.ca/aaflora/data.

Alexandrova, V.C., 1977. Globotanicbeskoye rayonirovaniye Arktiki i Antarktiki (Geobotanical Zonation of the Arctic and Antarctic). Komarovskaye Chteniya 19, Leningrad (in Russian).

Alisauskas, R.T., Ankney, C.D., 1992. Egg laying and nutrient reserves. In: Batt, B.D.J., Afton, A.D., Anderson, M.G., Ankney, C.D., Johnson, D.H., Kadlec, J.A., Krapu, G.L. (Eds.), Ecology and Management of Breeding Waterfowl. University of Minnesota Press, Minneapolis, pp. 30–61.

Alisauskas, R.T., Charlwood, J.W., Kellett, D.K., 2006. Vegetation correlates of the history and density of nesting by Ross's geese and lesser snow geese at Karrak Lake. Nunavut. Arctic 59, 201–210.

AMAP, 1998. Arctic Marine Currents. http://neba.arcticresponsetechnology.org/media/1006/image013.png.

AMAP, 2002. Arctic pollution 2002: Persistent organic pollutants, heavy metals, radioactivity, human health, changing pathways. In: Arctic Monitoring and Assessment Programme. Oslo, Norway, pp. 1–137.

Andersen, F.S., 1946. East Greenland lakes as habitats for Chironomid larvae. Studies on the systematics and biology of Chironomidae, II. Meddeleseröm Grønland 100, 1–65.

Andrews, J.T., 1973. The Wisconsin Laurentide Ice-Sheet: dispersal centres, problems of rates of retreat and climatic interpretations. Arctic and Alpine Research 2, 115–134. http://www.grida.no/climate/ipcc_tar/wg2/592.htm.

Anisimov, O., Fitzharris, B., Hagen, J.O., Jefferies, R.L., Marchant, H., Nelson, F., Prowse, T., Vaughan, D.G., 2001. Polar Regions (Arctic and Antarctic). In: Climate Change 2001: Impacts Polar Coastal Wetlands 149 Adaptation, and Vulnerability IPCC Third Assessment. Cambridge University Press, Cambridge, pp. 801–841. http://www.grida.no/climate/ipcc_tar/wg2/592.htm.

Anisimov, O.A., Vaughan, D.G., Callaghan, T.V., Furgal, C., Marchant, H., Prowse, T.D., Vilhjálmsson, H., Walsh, J.E., 2007. Chapter 15: polar regions (Arctic and Antarctic). Climate change 2007: impacts, adaptation and vulnerability. In: Parry, M.L., Canziani, O.F., Palutikof, J.P., van der Linden, P.J., Hanson, C.E. (Eds.), Contribution of Working Group II to the Fourth Assessment Report of the Intergovernmental Panel on Climate Change. Cambridge University Press, Cambridge, pp. 653–685. https://www.ipcc.ch/pdf/assessment-report/ar4/wg2/ar4_wg2_full_report.pdf.

Arkhipov, A.A., 1997. Environment and climate of Sartan maximum and late glacial in Siberia. In: Martini, I.P. (Ed.), Late Glacial and Postglacial Environmental Changes–Quaternary, Carboniferous-Permian, and Proterozoic. Oxford University Press, New York, pp. 53–60.

Azouzi, L., Bourget, E., Borcard, D., 2002. Spatial variation in the intertidal bivalve *Macoma balthica*: biotic variables in relation to density and abiotic factors. Marine Ecology Progress Series 234, 159–170.

Baker, A.J., Gonzalez, P.M., Piersma, T., Niles, L.J., Nascimento, I.L.S., Atkinson, P.W., Clark, N.A., Minton, C.D.T., Peck, M., Aarts, G., 2004. Rapid population decline in red knots: fitness consequences of decreased refuelling rates and late arrival in Delaware Bay. Proceeding of the Royal Society B-Biological 271, 875–882.

Bergman, R.D., Derksen, D.V., 1977. Observations on arctic and red-throated loons at Storkensen Point, Alaska. Arctic 3, 41–51.

Blomqvist, S., Holmgren, N., Ákesson, S., Hedenström, A., Pettersson, J., 2002. Indirect effects of lemming cycles on sandpiper dynamics: 50 years of counts from southern Sweden. Oecologia 133, 146–158.

Bowen, W., 2005. Alaska Atlas of Panoramic Aerial Images. http://californiageographicalsurvey.com/alaska_panorama_atlas/index.html.

Butler, R.W., Davidson, N.C., Morrison, R.I.G., 2001. Global-scale shorebird distribution in relation to productivity of near-shore ocean waters. Waterbirds 24, 224–232. https://www.researchgate.net/publication/270439698_Global-Scale_Shorebird_Distribution_in_Relation_to_Productivity_of_Near-Shore_Ocean_Waters.

Cahoon, D.R., Hensel, P.F., Spencer, T., Reed, D.J., McKee, K.L., Saintilan, N., 2006. Coastal wetland vulnerability to relative sea-level rise: wetland elevation trends and process controls. In: Verhoeven, J.T.A., Beltman, B., Bobbink, R., Whigham, D.F. (Eds.), Wetlands and Natural Resource Management. Ecological Studies 190. Springer, Berlin, pp. 271–292.

Callaghan, T.V., Björn, L.O., Chapin III, F.S., Chernov, Y.I., Christensen, T.R., Huntley, B., Ims, R., Johansson, M., Riedlinger, D.J., Jonasson, S., Matveyeva, N.V., Ochel, W., Panikov, N., Shaver, G., 2005. Arctic Tundra and Polar Desert Ecosystems. Arctic Climate Impact Assessment. Cambridge University Press, Cambridge, pp. 243–352.

Caning, M., Rausch, M., 2001. Zackenberg Ecological Research Operations, 6th Annual Report, 2000. Danish Polar Centre, Ministry of Research and Information Technology, Copenhagen.

Cargill, S.M., Jefferies, R.L., 1984. Nutrient limitation of primary productivity in a sub-arctic salt marsh. Journal of Applied Ecology 21, 657–668.

CAVM Team, 2003. Circumpolar Arctic Vegetation Map (1:7,500,000 Scale), Conservation of Arctic Flora and Fauna (CAFF) Map No. 1. U.S. Fish and Wildlife Service, Anchorage, Alaska. ISBN: 0-9767525-0-6, ISBN-13: 978-0-9767525-0-9.

Chang, E.R., Jefferies, R.L., Carleton, T.J., 2001. Relationship betweengetation and soil seed banks in an arctic coastal marsh. Journal of Ecology 89, 367–384.

Chernov, Y., 1995. Diversity of the Arctic terrestrial fauna. In: Chapin III, F.S., Körner, C. (Eds.), Arctic and Alpine Biodiversity: Patterns, Causes, and Ecosystem Consequences. Springer Verlag, Berlin, pp. 81–95. http://www.academia.edu/11835206/Arctic_and_alpine_biodiversity_Patterns_causes_and_ecosystem_consequences.

Chernov, Y.I., Matveyeva, N.V., 1997. Arctic ecosystems in Russia. In: Wielgolaski, F.E. (Ed.), Polar and Alpine Tundra. Ecosystems of the World 3. Elsevier, Amsterdam, pp. 361–507.

Chou, R., Vardy, C., Jefferies, R.L., 1992. Establishment of leaves and other plant fragments produced by the foraging activities of geese. Functional Ecology 6, 297–301.

Clarke, K.E., 1980. Ecology of a Subarctic Coastal System, North Point, James Bay, Ontario (M.Sc. thesis). Department Land Resource Science, University of Guelph, Guelph, ON, Canada, 232 pp.

Collins, M., et al., 2013. Long-term climate change: projections, commitments and irreversibility. In: Stocker, T.F., et al. (Eds.), Climate Change 2013: The Physical Science Basis. Contribution of Working Group I to the Fifth Assessment Report of the Intergovernmental Panel on Climate Change. Cambridge University Press, Cambridge, United Kingdom and New York, NY, USA. http://www.climatechange2013.org/images/report/WG1AR5_Chapter12_FINAL.pdf.

Danell, K., Sjöberg, K., 1977. Seasonal emergence of chironomids in relation to egg laying and hatching of ducks in a restored lake (northern Sweden). Wildfowl 28, 129–135.

Dereliev, S.G., 2006. The Red-breasted Goose *Branta ruficollis* in the new millennium: a thriving species or a species on the brink of extinction? In: Boere, G.C., Galbraith, C.A., Stroud, D.A. (Eds.), Waterbirds Around the World. The Stationery Office, Edinburgh, pp. 619–623.

Derocher, A.E., Lunn, N.J., Stirling, I., 2004. Polar bears in a warming climate. Integrative and Comparative Biology 44, 163–176.

Didiuk, A.B., Ferguson, R.S., 2005. Land Cover Mapping of Queen Maud Gulf Migratory Bird Sanctuary, Nunavut. Occasional Paper, Number 111. Canadian Wildlife Service, Ottawa, 32 pp.

Espoz, A., Ponce, C., Matus, A., Blank, R., Rozbaczylo, O., Sitters, N., Rodriguez, H.P., Dey, S., Niles, A.D., Niles, L.J., 2008. Trophic ecology of the Red Knot *Calidris canutus rufa* at Bahía Lomas, Tierra del Fuego, Chile. Wader Study Group Bulletin 115, 69–76.

Fraser, J.D., Karpanty, S.M., Cohen, J.B., Truitt, B.R., 2013. The Red Knot (Calidris canutus rufa) decline in the western hemisphere: is there a lemming connection? Canadian. Journal of Zoology 91, 13–16.

French, H., 2011. Frozen sediments and previously-frozen sediments. In: Martini, I.P., French, H.M., Perez Alberti, A. (Eds.), Ice-marginal and Periglacial Processes and Sediments: Ancient and Modern. Geological Society, London, pp. 153–166. Special Publications 354.

Fulton, R.J. (Ed.), 1989. Geology of Canada and Greenland. Geology of Canada. No. 1. Geological Survey of Canada, Ottawa.

Forbes, D.L. (Ed.), 2011. State of the Arctic Coast 2010 – Scientific Review and Outlook. International Arctic Science Committee, Land-Ocean Interactions in the Coastal Zone, Arctic Monitoring and Assessment Programme. International Permafrost Association. Helmholtz-Zentrum, Geesthacht, Germany, 178 pp. http://arcticcoasts.org.

Fox, A.D., Leafloor, J.O., 2018. A Global Audit of the Status and Trends of Arctic and Northern Hemisphere Goose Populations. (Component 1). Conservation of Arctic Flora and Fauna report, 30 pp.

Gill, R.E., Piersma, T., Hufford, G., Servranckx, R., Riegen, A., 2005. Crossing the ultimate ecological barrier: evidence for an 11,000-km-long nonstop flight from Alaska to New Zealand and eastern Australia by Bar-tailed Godwits. Condor 107, 1–20.

Glahder, C.M., Fox, A.D., Hübner, C.E., Madsen, J., Tombre, I.M., 2006. Pre-nesting site use of nesting satellite transmitter tagged Svalbard Pink-footed Geese. Ardea 94, 679–690.

Gormezano, L.J., Rockwell, R.F., 2013. What to eat now? Shifts in polar bear diet in the ice-free season in western Hudson Bay. Ecology and Evolution 3509–3523.

Grobe H., 2000. Northern Hemisphere Glaciation During the Last Ice Ages. Creative Commons CC-BY-2.5. Wikipedia: https://en.wikipedia.org/wiki/Quaternary_glaciation#/media/File:Northern_icesheet_hg.png (Modified from Schlee, H.S., 2000. Our Changing Continent. USGS General Interest Publication. https://pubs.usgs.gov/gip/continents/).

Grosswald, M.G., 1998. Late-weichselian ice sheets in arctic and Pacific Siberia. Quaternary International 45, 3–18.

Handa, I.T., Harmsen, R., Jefferies, R.L., 2002. Patterns of vegetation change and the recovery potential of degraded areas in a coastal marsh system of the Hudson Bay Lowland. Journal of Ecology 90, 86–99.

Heginbottom, J.A., Dubreuil, M.A., Harker, P.A., 1995. Canada – Permafrost in: National Atlas of Canada, fifth ed. National Atlas Information Service, Natural Resources Canada, MCR 4177.

Hörnfeldt, B., Hipkiss, T., Eklund, U., 2005. Fading out of vole and predator cycles? Proceedings of the Royal Society B-Biological Sciences 272, 2045–2049. https://www.researchgate.net/publication/7572399_Fading_out_of_vole_and_predator_cycles.

Hultén, E., 1968. Flora of Alaska and Neighboring Territories. Stanford University Press, Stanford, California.

Iacobelli, A., Jefferies, R.L., 1991. Inverse salinity gradients in coastal marshes and the death of Salix; the effects of grubbing by geese. Journal of Ecology 79, 61–73.

IPCC (Intergovernmental Panel on Climate Change), 2013. Climate Change 2013. The Physical Science Basis. Working Group 1 Contribution to the Fifth Assessment Report of the Intergovernmental Panel on Climate Change. Summary for Policy-makers. Cambridge University Press, Cambridge.

Iverson, S.A., Gilchrist, H.G., Smith, P.A., Gaston, A.J., Forbes, M.R., 2014. Longer ice-free seasons increase the risk of nest depredation by polar bears for colonial breeding birds in the Canadian Arctic. Proceeding Royal Society of B, Biological Sciences 281, 2013–3128.

Jano, A.P., Jefferies, R.L., Rockwell, R.F., 1998. The detection of change by multitemporal analysis of LANDSAT data: the effects of goose foraging. Journal of Ecology 86, 93–99.

Jeglum, J.K., Salomonsson, A., Svensson, J. (Eds.), 2003. Primary Succession and Ecological Monitoring on Rising Coastlines, Proceedings a Workshop and Seminars, 17 to 20 April, 2002. Swedish University of Agricultural Sciences, Faculty of Forestry, Department of Forest Ecology, Umeå, Sweden. Stencilserie. No. 94. 36 pp.

Jefferies, R.L., 1977. The vegetation of salt marshes at some coastal sites in Arctic North America. Journal of Ecology 65, 661–672.

Jefferies, R.L., Drent, R.H., Bakker, J.P., 2006a. Connecting Arctic and temperate wetlands and agricultural landscapes: the dynamics of goose populations in response to global change. In: Verhoeven, J.T.A., Beltman, B., Bobbink, R., Whigham, D.F. (Eds.), Wetlands and Natural Resource Management. Ecological Studies 190. Springer, Berlin, pp. 293–314.

Jefferies, R.L., Jano, A.P., Abraham, K.F., 2006b. A biotic agent promotes large-scale catastrophic change in the coastal marshes of Hudson Bay. Journal of Ecology 94, 234–242.

Jefferies, R.L., Jensen, A., Abraham, K.F., 1979. Vegetational development and the effect of geese on the vegetation at La Pérouse Bay, Manitoba. Canadian Journal of Botany 57, 1439–1450.

Jefferies, R.L., Rockwell, R.F., Abraham, K.F., 2003. The embarrassment of riches: agricultural food subsidies, high goose numbers, and loss of Arctic wetlands – a continuing saga. Environmental Reviews 11, 193–232.

Kakonge, S.A.K., Morrison, R.I.G., Campbell, B.A., 1979. An Annotated List of Salt Marsh Invertebrates from North Point, Ontario. Unpublished Report. Canadian Wildlife Service, Ottawa, 200 pp.

Kerbes, R.H., Meeres, K.M., Alisauskas, R.T., Caswell, F.D., Abraham, K.F., Ross, R.K., 2006. Inventory of Nesting Mid-Continent Lesser Snow and Ross's Geese in Eastern and Central Arctic Canada, 1997–1998. Canadian Wildlife Service Technical Report Series, Winnipeg, MB, 54 pp.

Kershaw, K.A., 1976. The vegetational zonation of the east Pen Island salt marshes, Hudson Bay. Canadian Journal of Botany 54, 5–13.

Kettles, I.M., Tarnocai, C., 1999. Development of a model for estimating the sensitivity of Canadian peatlands to climate warming. Géographie physique et Quaternaire 53, 323–338.

Kimble, J.M., 2004. Cryosols: Permafrost-Affected Soils. Springer-Verlag, Berlin, 727 pp. https://books.google.ca/books/about/Cryosols.html?id=m33J44F-MMsC&hl=en.

Koerner, R.M., 2002. Glaciers of Canada- glaciers of the Arctic Islands. In: Willliams, R.S., Ferrigno, J.G. (Eds.), Satellite Image Atlas of the Glaciers of the World, pp. J111–J146. U.S. Geological Survey Professional Paper 1386-J-1. https://pubs.usgs.gov/pp/p1386j/hiarctic/hiarctic-lores.pdf.

Koroleva, T.M., Zverev, A.A., Katenin, A.E., Petrovskii, V.V., Pospelova, E.B., Rebristaya, O.V., Khitun, O.V., Khodachek, E.A., Chinenko, S.V., 2008 and 2011. Longitudinal geographic structure of local and regional floras of Asian Arctic. Paper 1 and Paper 2. Botan. Zhurn. 93, 193–220 and 96 145–169 (in Russian).

Koroleva, T.M., Zverev, A.A., Katenin, A.E., Petrovskii, V.V., Pospelova, E.B., Pospelov, I.N., Rebristaya, O.V., Khitun, O.V., Chinenko, S.V., 2012. Latitudinal geographical structure of local floras of the Asian Arctic: analysis of distribution of groups and fractions. Botan. Zhurn 97, 69–89 (in Russian).

Linacre, E., Geerts, B., 1998. The Arctic: The Ocean, Sea Ice, Icebergs, and Climate. http://www-das.uwyo.edu/~geerts/cwx/notes/chap17/arctic.html.

Loeng, H., Brander, K., Carmack, E., Denisenko, S., Drinkwater, K., Hansen, B., Kovacs, K., Livingston, P., McLaughlin, F., Sakshaug, E., 2005. Marine Systems. Arctic Climate Impact Assessment. Cambridge University Press, Cambridge, pp. 453–538.

Maher, W.J., 1970. Ecology of the long-tailed Jaeger at lake hazen, Ellesmere Island. Arctic 23, 112–129.

Martini, I.P., 2006a. Hudson–James Bays. (A Website of the Western Coasts). http://www.uoguelph.ca/geology/hudsonbay/.

Martini, I.P., 2006b. The cold-climate peatlands of the Hudson Bay Lowland, Canada: brief review of recent work. In: Martini, I.P., Cortizas, A.M., Chesworth, W. (Eds.), Peatlands: Evolution and Records of Environmental and Climatic Change. Elsevier B.V., pp. 53–84

Martini, I.P., 2014. General considerations and highlights of low-lying coastal zones: passive continental margins from the poles to the tropics. In: Martini, I.P., Wanless, H.R. (Eds.), Sedimentary Coastal Zones from High to Low Latitudes: Similarities and Differences. Geological Society, London, pp. 1–32. Special Publications 388.

Martini, I.P., Morrison, R.I.G., 1987. Regional distribution of *Macoma balthica* and *Hydrobia minuta* on the subarctic coasts of Hudson bay and James bay, ontario, Canada. Estuarine, Coastal and Shelf Science 24, 47–68.

Martini, I.P., Morrison, R.I.G., 2014. Coasts of Foxe Basin, Arctic Canada. In: Martini, I.P., Wanless, H.R. (Eds.), Sedimentary Coastal Zones from High to Low Latitudes: Similarities and Differences. Geological Society, London, pp. 165–198. Special Publications 388.

Martini, I.P., Morrison, R.I.G., Glooschenko, W.A., Protz, R., 1980. Coastal studies in James bay, ontario. Geoscience Canada 7, 11–21.

Martini, I.P., Broockfield, M.E., Sadura, S., 2001. Principles of Glacial Geomorphology and Geology. Prentice Hall, Upper Saddle River, NJ., 381 pp.

Meltofte, H., Lahrmann, D.P., 2006. Time allocation in Greenland high-arctic waders during summer. Dansk Ornitologisk Forenings Tidsskrift 100, 75–87.

McKinnon, L., Smith, P.A., Nol, E., Martin, J.-L., Abraham, K.F., Gilchrist, H.G., Morrison, R.I.G., Bety, J., 2010. Lower predation risk for migratory birds at high latitudes. Science 327, 326–327.

Morrison, R.I.G., 1983. A hemispheric perspective on the distribution and migration of some shorebirds in North and South America. In: Boyd, H. (Ed.), Proceedings of the First Western Hemisphere Waterfowl and Waterbird Symposium. Canadian Wildlife Service, Ottawa, pp. 84–94.

Morrison, R.I.G., 1984. Migration systems of some New World shorebirds. In: Burger, J., Olla, B.L. (Eds.), Shorebirds: Migration and Foraging Behavior. Plenum Press, New York, pp. 125–202.

Morrison, R.I.G., 2006. Body transformations, condition, and survival in red knots *Calidris canutus* travelling to breed at Alert, Ellesmere Island, Canada. Ardea 94, 607–618.

Morrison, R.I.G., Gaston, A.J., 1986. Marine and coastal birds of James Bay, Hudson Bay and Foxe Basin. In: Martini, I.P. (Ed.), Canadian Inland Seas. Elsevier, Amsterdam, pp. 355–386.

Morrison, R.I.G., Harrington, B.A., 1979. Critical shorebird resources in James Bay and eastern North America. Transactions of the North American Wildlife Natural Resources Conference 44, 498–507.

Morrison, R.I.G., Harrington, B.A., 1992. The migration system of the red knot *Calidris canutus rufa* in the new world. Wader Study Group Bulletin 64 (Suppl.), 71–84.

Morrison, R.I.G., Hobson, K.A., 2004. Use of body stores in shorebirds after arrival on High-Arctic breeding grounds. Auk 121, 333–344.

Morrison, R.I.G., Davidson, N.C., Piersma, T., 2005. Transformations at high latitudes: why do Red Knots bring body stores to the breeding grounds? Condor 107, 449–457.

Morrison, R.I.G., Davidson, N.C., Wilson, J.R., 2007. Survival of the fattest: body stores on migration and survival in red knots *Calidris canutus islandica*. Journal of Avian Biology 38, 479–487.

Mowbray, T.B., Cooke, F., Ganter, B., 2000. Snow goose (*Chen caerulescens*). In: Poole, A., Gill, F. (Eds.), The Birds of North America, No. 514. The Birds of North America, Inc., Philadelphia, PA.

Ngai, J., Jefferies, R.L., 2004. Nutrient limitation of plant growth and forage quality in arctic coastal marshes. Journal of Ecology 92, 1001–1010.

North American Bird Conservation Initiative Canada, 2016. The State of North America's Birds 2016. Environment and Climate Change Canada, Ottawa, Ontario, 8 pp. www.stateofthebirds.org.

Pickavance, J.R., 2006. The spiders of East Bay, Southampton Island, Nunavut, Canada. Arctic 59, 276–282.

Piersma, T., Gill, R.E., 1998. Guts don't fly: small digestive organs in obese bar-tailed godwits. Auk 115, 196–203.

Piersma, T., Koolhaas, A., 1997. Shorebirds, Shellfish(eries) and Sediments Around Griend, Western Wadden Sea, 1988–1996. NIOZ-RAPPORT 1997 – 7, Netherlands Instituut voor Onderzoek der Zee, Texel, 118 pp.

Piersna, T., Lindström, À., 2004. Migrating shorebirds as integrative sentinels of global environmental change. Ibis 146 (s1), 61–69.

Piersma, T., Gudmundsson, G.A., Lilliendahl, K., 1999. Rapid changes in the size of different functional organ and muscle groups during refueling in a long-distance migrating shorebird. Physiological and Biochemical Zoology 72, 405–415.

Pinder, L.C.V., 1983. The larvae of Chironomidae (Diptera) of the Holarctic region: introduction. Entomologica Scandinavica Supplement 19, 7–10.

Pollock, L.A., Abraham, K.F., Nol, E., 2012. Migrant shorebird use of Akimiski Island, Nunavut as a sub-arctic stopover site. Polar Biology 35, 1691–1701.

Ponce, A., Espoz, C., Rodriguez, S.R., Rozbaczylo, N., 2003. Macroinfauna de la playa de arena de Bahia Lomas, Tierra del Fuego, Chile. XLVI Reunion anual de la Sociedad de Biologia de Chile, XII reunion anual de la Sociedad de Ecologia de Chile. Biological Research 36, 3–4.

Powers, S.P., Bishop, M.A., Grabowski, J.H., Peterson, C.H., 2002. Intertidal benthic resources of the Copper River Delta, Alaska, USA. Journal of Sea Research 47, 13–23.

Price, J.S., Woo, M.-K., 1988. Studies of a subarctic coastal marsh, II. Salinity. Journal of Hydrology 103, 293–307.

Prinsenberg, S.J., 1986. Circulation patterns and current structure of Hudson Bay. In: Martini, I.P. (Ed.), Canadian Inland Seas. Elsevier, Amsterdam, pp. 187–204.

Protz, R., 1982. Development of gleysolic soils in the Hudson and James bay coastal zone, ontario. Naturaliste Canadien 109, 491–500.

Ross, P.M., Adam, P., 2013. Climate change and intertidal wetland. Biology (Basel) 2, 455–480. https://www.ncbi.nlm.nih.gov/pmc/articles/PMC4009871/.

Rowell, H.E., Hearn, R.D., 2005. The 2003 Icelandic-breeding Goose Census. The Wildfowl and Wetlands Trust/Joint Nature, Conservation Committee, Slimbridge.

Ruz, M.-H., Héquette, A., Hill, P.R., 1992. A model of coastal evolution in a transgressed thermokarst topography, Canadian Beaufort Sea. Marine Geology 106, 251–278.

Scheffers, A.M., Scheffers, S.R., Kelletat, D.H., 2012. The Coastlines of the World with Google Earth, vol. 2. Coastal Research Library, Springer, Dordrecht. http://www.springer.com/earth+sciences+and+geography/earth+system+sciences/book/978-94-007-0737-5.

Schmiegelow, F.K.A., Mönkkönen, M., 2002. Habitat loss and fragmentation in dynamic landscapes: avian perspectives from the boreal forest. Ecological Applications 12, 375–389.

Scott, D.B., Ferai-Gauthier, J., Mudie, P.J., 2014. Coastal Wetlands of the World: Geology, Ecology, Distribution and Applications. Cambridge University Press, Cambridge. https://books.google.ca/books?id=b1AHAwAAQBAJ&pg=PR7&lpg=PR7&dq=south+america+coastal+wetlands&source=bl&ots=CPOAsvZ95P&sig=susR4xmmqhJUooDuGw_Bn4oGqh4&hl=en&sa=X&ved=0ahUKEwirsoyQtZ3MAhVEwj4KHeqXBukQ6AEIODAI#v=onepage&q=south%20america%20coastal%20wetlands&f=false. http://ebooks.cambridge.org/ebook.jsf?bid=CBO9781107296916.

Senner, N.R., 2012. One species but two patterns: populations of the Hudsonian Godwit (Limosa haemastica) differ in spring migration timing. Auk 129, 670–682.

Sergienko, L.A., 2008. Flora and Vegetation of the Arctic Coasts and Adjacent Territories. Petrozavodsk State University, Petrozavodsk, 225 pp.

Sergienko, L., Markovskaya, E., Starodubtceva, A., 2012. Distribution of vascular plants in the coastal ecosystems of the White Sea. In: Gâstescu, P., Lewis Jr., W., Bretcan, P. (Eds.), Water Resources and Wetlands. Conference Proceedings, 14–16 September 2012, Tulcea – Romania, ISBN 978-606-605-038-8. In: http://www.limnology.ro/water2012/Proceedings/060.pdf.

Sergienko, L., 2013a. Salt marsh flora and vegetation of the Russian Arctic coasts. Czech Polar Reports 3, 30–37. http://www.sci.muni.cz/CPR/5cislo/Sergienko-web.pdf.

Sergienko, L.A., 2013b. Analysis of the salt marsh flora of the Russian arctic coasts. Annals of Valahia University of Targoviste. Geographical Series Tome 13 (2), 69–75. http://fsu.valahia.ro/images/avutgs/1/2013/2013130201.html.

Spiridonov, V.A., Gavrillo, M.V., Krasnova, B.D., Nikolaeva, N.G. (Eds.), 2011. Atlas of Marine and Coastal Biological Diversity of the Russian Arctic. WWF Russia, Moscow, 64 pp.

Staniforth, R.J., Griller, N., Lajzerowicz, C., 1998. Soil seed banks from coastal subarctic ecosystems of Bird Cove, Hudson Bay. Ecoscience 5, 241–249.

Stirling, I., Parkinson, C.L., 2006. Possible effects of climate warming on selected populations of polar bears (*Ursus maritimus*) in the Canadian Arctic. Arctic 59, 261–275.

Stirling, I., Lunn, N.J., Iacozza, J., 1999. Long-term trends in the population ecology of polar bears in western Hudson Bay in relation to climate change. Arctic 52, 294–306.

Summerhayes, V.S., Elton, C.S., 1923. Contributions to the ecology of Spitsbergen and Bear Island. Journal of Ecology 11, 214–286.

Tarnocai, C., 2009. The impact of climate change on Canadian Peatlands. Canadian Water Resources Journal 34, 453–466.

Tarnocai, C., Bockheim, J.G., 2010. Cryosolic soils of Canada: genesis, distribution, and classification. Canadian Journal of Soil Science 91, 749–762.

Ulmishek, G.F., 2003. Petroleum Geology and Resources of the West Siberian Basin, Russia. U.S. Geological Survey Bulletin 2201-G. USGS, 48 pp. http://pubs.usgs.gov/bul/2201/G/B2201-G.pdf.

Underhill, L.G., 1987. Changes in the age structure of curlew sandpiper populations at Langebaan Lagoon, South Africa, in relation to lemming cycles in Siberia. Transactions of the Royal Society of South Africa 46, 209–214.

Underhill, L.G., Waltner, M., Summers, R.W., 1989. Three-year cycles in breeding productivity of knots *Calidris canutus* wintering in southern Africa suggest Taimyr Peninsula provenance. Bird Study 36, 83–87.

UNEP/GRID-Arendal, 2005a. Distribution of Arctic and Sub-Arctic Zones. http://www.grida.no/resources/7010.

UNEP/GRID-Arendal, 2005b. Protected Areas, Arctic. http://www.grida.no/resources/7002.

UNEP/GRID-Arendal, 2005c. Major Global Waterfowl Flyways. http://www.grida.no/resources/7022.

U.S. Dept of Agriculture, 1996. Global Distribution of Wetlands – Map. Natural Resources Conservation service, Soil Survey Division, World Soil Resources, Washington DC. https://www.nrcs.usda.gov/wps/portal/nrcs/detail/soils/use/worldsoils/?cid=nrcs142p2_054021.

Van Eerden, M.R., Drent, R.H., Stahl, J., Bakker, J.P., 2005. Connecting seas: Western Palearctic continental flyway for water birds in the perspective of changing land use and climate. Global Change Biology 11, 894–908.

van Gils, J.A., Lisovski, S., Lok, T., Meissner, W., Ozarowska, A., de Fouvw, J., Rakhimberdiev, E., Soloviev, M.Y., Piersma, T., Klaassen, M., 2016. Body shrinkage due to Arctic warming reduces red knot fitness in tropical wintering range. Science 352, 819–821.

Velichko, A.A., Kononov, Y.M., Faustova, M.A., 1997. The last glaciation of Earth: size and volume of ice sheets. Quaternary International 41–42, 43–52.

Wadhams, P., 2017. A Farewell to Ice. A Report from the Arctic. Penguin Random House, UK.

Walker, B.D., Peters, T.W., 1977. Soils of truelove lowland and plateau. In: Bliss, L.C. (Ed.), Truelove Lowland, Devon Island, Canada: A High Arctic Ecosystem. University of Alberta Press, Edmonton, pp. 31–62. https://books.google.ca/books?id=GnLHYSUzxOIC&pg=PA31&lpg=PA31&dq=truelove+lowland+devon+island+soils&source=bl&ots=bWJM5_fyFm&sig=QN8V9A_seTog7Wzb9DxcedXhPIo&hl=en&sa=X&ved=0CDMQ6AEwA2oVChMIoMHvss38yAIVhaweCh34aAJo#v=onepage&q=truelove%20lowland%20devon%20island%20soils&f=true.

Walker, D.A., et al., 2005. The circumpolar Arctic vegetation map. Members of the CAVM Team Journal of Vegetation Science 16, 267–282.

Ward, D.H., Reed, A., Sedinger, J.S., Blacks, J.M., Derksen, D.V., Castelli, P.M., 2005. North American Brant: effects of changes in habitat and climate on population dynamics. Global Change Biology 11, 869–880.

Washburn, A.L., 1973. Periglacial Processes and Environments. Edward Arnold, London, 320 pp.

Werner, K., et al., 2016. Arctic in Rapid Transition: Priorities for the Future of Marine and Coastal Research in the Arctic, Polar Science. http://www.sciencedirect.com/science/article/pii/S1873965216300196.

West, A.S., 1951. The Canadian arctic and sub-arctic mosquito problems. In: Proceeding of 1951 NJMCA Annual Meeting. New Jersey Mosquito Extermination Control Association, pp. 105–110.

Wikipedia, 2016a. Pingo. https://en.wikipedia.org/wiki/Pingo.

Wikipedia, 2016b. Baffinland Iron Mine. https://en.wikipedia.org/wiki/Baffinland_Iron_Mine.

Williams, L., 1994. Ust-Lensky. Russia. Wild Russia. http://www.wild-russia.org/bioregion1/Ust-Lensky/1_ustlen.htm.

Wolfe, S.A., Dallimore, S.R., Solomon, S.M., 1998. Coastal permafrost investigations along a rapidly eroding shoreline, Tuktoyaktuk, N.W.T., Canada. In: Lewkowicz, A.G., Allard, M. (Eds.), Proceedings of the Seventh International Conference on Permafrost, Yellowknife, N.W.T. 23–27 June 1998. Collection Nordicanna No. 57. Centre d'Etudes Nordiques, Université Laval, Québec, Canada, pp. 1125–1131.

Zöckler, C., 1998. Patterns of diversity in Arctic birds. WCMC Biodiversity Bulletin 3, 1–16.

Zöckler, C., Lanctot, R., Brown, S., Syroechkovskiy, E., 2012. Waders (shorebirds). In: Arctic Report Card 2012, NOAA's Arctic Program, pp. 92–102. ftp://ftp.oar.noaa.gov/arctic/documents/ArcticReportCard_full_report2012.pdf.

Zoltai, S.C., 1980. An outline of the wetland regions of Canada. In: Rubec, C.D.A., Pollett, F.C. (Eds.), Proceedings of a Workshop on Canadian Wetlands. Environment Canada, Lands Directorate, Ecological Land Classification, Series, No. 12, pp. 1–8 [Trace: Minister of Supply and Services Canada, 1980. Cat. No. En 73-3/12; ISBN 0-662-50919-6]. PDF available in. http://www.cfs.nrcan.gc.ca/bookstore_pdfs/19315.pdf.

Further Reading

Abraham, K.F., Jefferies, R.L., 1997. High goose populations: causes, impacts, and implications. In: Batt, B.D.J. (Ed.), Arctic Ecosystems in Peril: A Report of the Arctic Goose Habitat. Working Group. U.S. Fish and Wildlife Service, Washington. DC/Canadian Wildlife Service, Ottawa, pp. 7–72. https://www.fws.gov/migratorybirds/pdf/management/arctic-goose/part2.pdf.

ACIA (Arctic Climate Impact Assessment), 2004. Impacts of a Warming Arctic. Cambridge University Press, Cambridge, UK, 139 pp.

Fox, A.D., Abraham, K.F., 2017. Why geese benefit from the transition from natural vegetation to agriculture. Ambio 46 (Suppl. 2), S188–S197. https://www.ncbi.nlm.nih.gov/pmc/articles/PMC5316322/.

Morrison, R.I.G., Ross, R.K., 1989. Atlas of Nearctic Shorebirds on the Coast of South America. Canadian Wildlife Service Special Publication, Ottawa, 325 pp. https://books.google.ca/books/about/Atlas_of_Nearctic_Shorebirds_on_the_Coas.html?id=HKblSQAACAAJ&redir_esc=y.

Tape, K.D., Flint P.L., Meixell, B.W., Gaglioti, B.V., 2013. Inundation, sedimentation, and subsidence creates goose habitat along the Arctic coast of Alaska. Environmental Research Letter, 8, N. 4, 9 pp. http://iopscience.iop.org/article/10.1088/1748-9326/8/4/045031/pdf.

Zöckler, C., Douglas, T., Contributing authors: Collen, B., Barry, T., Forbes, D., Loh, J., Gill, M., McRae, L., Sergienko, L., In: Forbes, D.L. (Ed.), 2011. State of the Arctic Coast 2010 – scientific review and outlook. International arctic science committee, Land-Ocean Interactions in the Coastal Zone, Arctic Monitoring and Assessment Programme, International Permafrost Association. Helmholtz- Zentrum, Geesthacht, Germany, 178 pp. http://library.arcticportal.org/1277/1/state_of_the_arctic_coast_2010.pdf.

PART II

PHYSICAL PROCESSES

CHAPTER

5

Salt-Marsh Ecogeomorphological Dynamics and Hydrodynamic Circulation

Andrea D'Alpaos[1], *Stefano Lanzoni*[2], *Andrea Rinaldo*[2,3], *Marco Marani*[2]

[1]Department of Geosciences, University of Padova, PD, Italy; [2]Department of Civil, Environmental, and Architectural Engineering, University of Padova, PD, Italy; [3]Laboratory of Ecohydrology, Ecole Polytechnique Fèdèrale Lausanne, Lausanne, Switzerland

1. INTRODUCTION

The strong dynamic coupling of intertidal platforms and tidal channel networks cutting through them, mediated by vegetation growth, gives rise to a complex system, whose nonlinear dynamics is arguably one of the most fascinating examples of ecomorphodynamics: The collective temporal evolution emerging from the mutual interactions and adjustments among hydrodynamic, morphological, and biological processes. Ecomorphodynamics is a fascinating and interdisciplinary research area that has recently emerged at the interface between ecological, hydrological, and geomorphological studies. Ecomorphodynamics, by accounting for the mutual role and interactions between water fluxes, sediment transport and morphology, on one side, and biological dynamics, on the other side, highlights the crucial role of ecogeomorphic feedbacks on the dynamics of Earth's landscapes (e.g., Murray et al., 2008; Reinhardt et al., 2010; D'Alpaos et al., 2016; Zhou et al., 2017).

Improving our understanding of the chief land-forming processes, of physical and biological nature, which drive intertidal system morphogenesis and long-term evolution, is an intriguing problem and a critical step to preserve such delicate systems, exposed to the effects of climate changes and human interference (Day et al., 2000; Marani et al., 2007; Temmerman and Kirwan, 2015). Wetland ecosystems host an extremely high biodiversity, exhibit one of the highest rates of primary production in the world, and play a fundamental role in determining the evolution of coastal lagoons and estuaries (Mitsch and Gosselink, 2000; Zedler

Coastal Wetlands
https://doi.org/10.1016/B978-0-444-63893-9.00005-8

and Kercher, 2005; Marani et al., 2006b). The decline of wetland areas worldwide and their potential sensitivity to abrupt sea-level fluctuations highlight their global importance and call for a deeper understanding of their dynamics.

This has motivated several researchers who have produced a large literature, especially in the last two decades (e.g., see Allen, 2000; Friedrichs and Perry, 2001; Marani et al., 2006a; Kirwan and Megongial, 2013; Saco and Rodríguez, 2013; Wolanski and Elliott, 2015; Larsen et al., 2016 for thorough reviews) to describe the evolution of estuarine systems. Most existing works, however, concentrate on specific aspects of intertidal dynamics, such as tidal propagation in estuarine channels (e.g., Friedrichs and Aubrey, 1994; Lanzoni and Seminara, 1998; Savenije, 2001; Savenije and Veling, 2005); tidal asymmetries and sediment dynamics in tidal channels (e.g., French and Stoddart, 1992; Friedrichs, 1995; Schuttelaars and de Swart, 2000; Lanzoni and Seminara, 2002); morphometric analyses of tidal networks (e.g., Steel and Pye, 1997; Fagherazzi et al., 1999; Rinaldo et al., 1999a,b; Marani et al., 2002, 2003; Rinaldo et al., 2004; Feola et al., 2005; Marani et al., 2006b); sedimentation and accretion patterns over vegetated marsh platforms (e.g., French and Spencer, 1993; Christiansen et al., 2000; Leonard and Reed, 2002; Neubauer, 2008); salt marsh ecological dynamics and patterns (e.g., Adam, 1990; Yallop et al., 1994; Marani et al., 2004; Silvestri et al., 2005; Belluco et al., 2006); saturated and unsaturated subsurface flows in salt marshes and their relationships with vegetation patterns (Ursino et al., 2004; Marani et al., 2005, 2006a; Cao et al., 2012; Xin et al., 2013; Boaga et al., 2014); and the influence of wind waves on the hydrodynamics of shallow tidal areas (Carniello et al., 2005; Fagherazzi et al., 2006). Even though significant advances have been achieved in all these fields, the understanding of the collective ecomorphological behavior of intertidal systems still lacks a comprehensive and predictive theory, due to the strongly intertwined interactions of their physical and ecological components. A deeper understanding may thus be achieved only by elucidating the detailed feedbacks between ecological and geomorphological processes, which, in turn, require a holistic approach encompassing the governing biomorphological processes over the wide range of spatial scales involved (Rinaldo et al., 1999a,b; Marani et al., 2003, 2006b).

To arrive at a mathematical description explicitly including intertidal biotic and abiotic processes, it is useful to provide a brief review of some modeling results that addressed, separately or jointly, the different components of the system (see Fagherazzi et al., 2012 for a thorough review).

A number of zero-dimensional models have been proposed to investigate the long-term vertical growth of salt marshes by assuming their accretion rate as a function of sediment supply and either marsh elevation or biomass (e.g., Randerson, 1979; French, 1993; Allen, 1997; Morris et al., 2002; Temmerman et al., 2003; Mudd et al., 2009; Kirwan et al., 2010; D'Alpaos et al., 2011). These models consider the evolution of a salt marsh point as a representative of the whole platform and, although providing helpful insights into the response of the marsh surface to tidal forcing and sea-level variations, they are unable to represent important space-dependent features.

The modeling of the differential accretion of the marsh surface induced by the spatial variability of sediment deposition rates has been relatively attempted only recently in a one-dimensional setting (e.g., Woolnough et al., 1995), whereas vegetation dynamics in a spatially explicit framework was first incorporated by Mudd et al. (2004). Three-dimensional analyses of sedimentation patterns in tidal marsh landscapes have been carried out both in the very short period (single inundation event) through complete hydrodynamic models (Temmerman et al., 2005) and in view of a long-term evolution through simplified process-oriented models (D'Alpaos et al., 2007a; Kirwan and Murray, 2007; Temmerman et al., 2007; D'Alpaos, 2011), thus emphasizing the strong control exerted by ecological processes on marsh morphodynamics. The purely geomorphological equilibria of unchanneled subtidal areas have been recently studied through conceptual (Fagherazzi et al., 2006) and numerical modeling (Defina et al., 2007). Marani et al. (2007, 2010) analyzed the fully coupled dynamics of landforms and biota in the intertidal zone, through a model of the coupled tidal physical and biological processes. They proved the existence of multiple equilibria, and transitions among them, governed by vegetation type, disturbances of the benthic biofilm, sediment availability and marine transgressions, or regressions, thus emphasizing the importance of the coupling between biological and sediment transport processes in determining the evolution of a tidal system as a whole. In the context of fully coupled modeling efforts, Marani et al. (2013) and Da Lio et al. (2013) emphasized that zonation patterns are the result of two-way feedbacks between biomass production and soil accretion, and that vegetation species are indeed capable of actively tuning marsh elevations within ranges of optimal adaptation. Zonation patterns are shown to be biogeomorphic features of salt marsh systems, i.e., they are the manifestation of multiple stable states, generated by competing vegetation species adapted to different elevation ranges.

In spite of the fundamental control exerted by tidal channels on the hydrodynamics and sediment dynamics within intertidal systems, and their importance for nutrient circulations within intertidal habitats, the literature on the morphogenesis and long-term morphological evolution of tidal channel networks is not as well developed. Field and laboratory observations (see, e.g., Pestrong, 1965; Redfield, 1965; Tambroni et al., 2005; Stefanon et al., 2010, 2012; Vlaswinkel and Cantelli, 2011) and conceptual models (e.g., Yapp et al., 1916; Beeftink, 1966; French and Stoddart, 1992; Allen, 2000) have, however, been developed. Laboratory observations have indeed highlighted the possibility of providing new insights on tidal network dynamics that can be used to benchmark numerical models that conceptualize and simplify the actual governing processes (e.g., Zhou et al., 2014a,b).

In the last 15 years, mathematical and numerical models of the morphogenesis and long-term morphological evolution of tidal channels have also been proposed. Schuttelaars and de Swart (2000) and Lanzoni and Seminara (2002) developed, within different theoretical frameworks, one-dimensional models that allow the investigation of the equilibrium configurations of estuaries and tidal channels. In particular, Lanzoni and Seminara (2002) observed that equilibrium configurations, allowing a vanishing net along-channel sediment flux, tend to be reached asymptotically. Fagherazzi and Furbish (2001) analyzed the long-term morphodynamic evolution of a reference cross section composed by an incipient channel zone and a marsh surface zone, through a model-simulating aspect of initial channel formation over an existing tidal flat. D'Alpaos et al. (2006) extended the analysis of Fagherazzi and Furbish (2001) tracking the channel cross-sectional morphodynamic evolution coupled with the vertical growth of the adjacent emerging marsh platform, with particular emphasis on the role played by the hydroperiod and halophytic vegetation. They found that channel cross sections tend to adapt quite rapidly to changes in the flow. Townend (2010) proposed a theoretical framework to provide a three-dimensional description of the equilibrium morphology of a tidal channel and of the adjacent platform, on the basis of a behavior-oriented model consisting of a planform described by an exponentially converging width and cross-sectional area imposed a priori, a low-water channel cross section parabolic in shape, and an intertidal flat profile. Indeed, mathematical models capable to describe the three-dimensional equilibrium morphology of a tidal channel and of the adjacent platform without imposing a priori channel properties such as longitudinal variations of channel width and/or depth, are quite rare (Canestrelli et al., 2007; van der Wegen et al., 2008; Lanzoni and D'Alpaos, 2015). In particular, Lanzoni and D'Alpaos, (2015) set up a simplified theoretical framework to analyze the three-dimensional equilibrium configuration of a tidal channel dissecting a short, unvegetated tidal flat in microtidal systems, allowing both channel bed and width to reach an equilibrium altimetric and planimetric configuration.

The morphogenesis and long-term evolution of channel networks have been recently studied through the use of simplified and more sophisticated behavior- and process-oriented models (Fagherazzi and Sun, 2004; D'Alpaos et al., 2005; Marciano et al., 2005; Kirwan and Murray, 2007; Temmerman et al., 2007; Coco et al., 2013; Zhou et al., 2014a; Belliard et al., 2015, 2016). A number of these models are based on the Poisson hydrodynamic model proposed by Rinaldo et al. (1999a,b). Fagherazzi and Sun (2004) developed a stochastic model for channel network formation in which water surface gradients drive the process of network incision. D'Alpaos et al. (2005) set up a mathematical model of tidal network ontogeny describing channel initiation and progressive headward extension through the carving of incised cross sections where the local shear stress—controlled by water surface gradients—exceeds a predefined, possibly site-dependent, threshold value. In agreement with observational evidence and conceptual models of marsh evolution, these approaches decouple the initial channel formation from the evolution of the adjacent marsh platform (Steers, 1960; Pestrong, 1965; French and Stoddart, 1992). However, contrary to the model proposed by Fagherazzi and Sun (2004), D'Alpaos et al. (2005) account for feedbacks existing between channel geometry and local hydrodynamic conditions, instantaneously adapting network configuration to the local discharge (or to the local tidal prism), in accordance with observational evidence and modeling (Friedrichs, 1995; Rinaldo et al., 1999b; Lanzoni and Seminara, 2002; D'Alpaos et al., 2006; D'Alpaos et al., 2010) and with laboratory experiments (Stefanon et al., 2010, 2012; Vlaswinkel and Cantelli, 2011). Moreover, D'Alpaos et al. (2007b) have recently tested the channel network model by simulating the rapid development of small creek networks within a newly constructed artificial salt marsh in the Venice Lagoon. They showed that the synthetic creeks tend to originate at locations that match those of the actual ones, thus supporting the assumption of the strong control exerted by the water surface elevation gradients in the process of channel incision. On the other hand, Kirwan and Murray (2007) proposed a model of the long-term evolution of channel networks through a simplified treatment of flow, sediment dynamics, and vegetation productivity. Water routing across the marsh platform is again based on the local gradients of a Poisson-parameterized surface (Rinaldo et al., 1999a), but part of the procedure used to represent channel erosion seems somewhat artificial.

Marciano et al. (2005) used the Delft3D hydrodynamic and sediment transport model to produce channel patterns in a short tidal basin. The results seem to be strongly influenced by the initial conditions specified, and only when an initial bottom configuration close to the expected equilibrium basin hypsometry is assigned, the model produces well-developed branching structures. Moreover, model validation is not conclusive in comparing generated and observed structures as it is carried out on the basis of Horton's hierarchical analysis, a formalism shown to be unable to discriminate different network statistics (Kirchner, 1993; Rinaldo et al., 1998). D'Alpaos et al. (2007a) discussed the interplay of erosion, sedimentation, and vegetation dynamics and their effects on the inter-twined ecomorphodynamic processes governing the evolution of the marsh platform and of the tidal channels cutting through it. Temmerman et al. (2007) developed a coupled morphodynamic and plant growth model, simulating plant colonization and tidal channel formation on an initially bare flat marsh surface. The interaction of different biotic and abiotic processes in particular environments was also addressed (Perillo et al., 2005; Minkoff et al., 2006; Hughes et al., 2009). A simplified cellular automaton model for the development of tidal creeks, accounting for observed bioturbation effects linked to crab—halophytic plant interactions, shows that, in the particular setting of the Bahía Blanca Estuary, this interaction exerts a relevant role in driving the development of tidal creeks, overcoming the role of water surface gradients (Minkoff et al., 2006). Likewise, Hughes et al. (2009) studied the dynamic behavior of tidal channel networks cutting through the salt marshes of the Santee Delta (SC, USA), and suggested that burrowing and herbivory by crabs weakens the soils at channel tips thus promoting faster channel headward growth than in other vegetated marsh platforms where crab—vegetation interactions are not observed. Hood (2006) suggested that in the particular environment represented by a rapidly prograding delta dominated by river discharge, tidal channels might be the result of depositional rather than erosional processes. Interestingly, Belliard et al. (2015, 2016) set up a modeling framework that describes the coevolution of the marsh platform and the embedded tidal networks in response to changes in the environmental forcing and suggested that erosion- and deposition-driven tidal channel development indeed coexist. Although erosional processes favor channel initiation over short temporal scales, depositional processes are mostly responsible for the slower elaboration of the channel network form and structure, thus playing a major role over long temporal scales. van Maanen et al. (2013) used a three-dimensional hydrodynamic model based on the unsteady Reynolds-averaged Navier—Stokes equations (ELCOM, Hodges et al., 2000) to study the role of environmental conditions, such as the range, on the long-term morphological evolution of tidal embayments, and showed, e.g., that increasing tidal ranges promoted faster channel network formation furthermore affecting final basin hypsometry and channel network characteristics. Interestingly, Coco et al. (2013) highlighted in their thorough review that different models using the same configurations and parameterizations can indeed produce tidal networks with different geomorphological features and structures. Finally, Zhou et al. (2014a,b) compared the results of laboratory experiments and numerical models exploring the possibility of reaching long-term morphodynamic equilibrium configurations, an issue that would deserve careful screening (Zhou et al., 2017).

It is worth at this point to remark that most of the contributions to the modeling of the morphogenesis and evolution of channel marsh systems discussed above do not rigorously address the problem of model validation. Very seldom a quantitative validation of models against observed morphologies is attempted, and the evaluation of model results is rather performed by qualitative visual appraisal or on the basis of lenient geomorphic measures (e.g., the traditional Hortonian measures considered by Marciano et al., 2005). Different from this common approach, D'Alpaos et al. (2005, 2007a,b) and Zhou et al. (2014a,b) used distinctive network statistics for a quantitative validation of model results, showing that the synthetic network structures generated by their models indeed reproduced several observed characteristics of geomorphic relevance such as, among others, unchanneled length distributions (Marani et al., 2003). Other studies have tested the possibility of using numerical models to study real world morphodynamics (*sensu* Zhou et al., 2017) based on the capability of these models to reproduce relevant geomorphic features such as the tidal—prism channel area relationship (e.g., Lanzoni and D'Alpaos, 2015; van der Wegen et al., 2010; van Maanen et al., 2013; Zhou et al., 2014a,b).

In the following, we describe a comprehensive theoretical framework aimed at extending our current understanding of the coupled ecogeomorphic evolution of intertidal environments, and our abilities to model it quantitatively, as defined by the literature discussed.

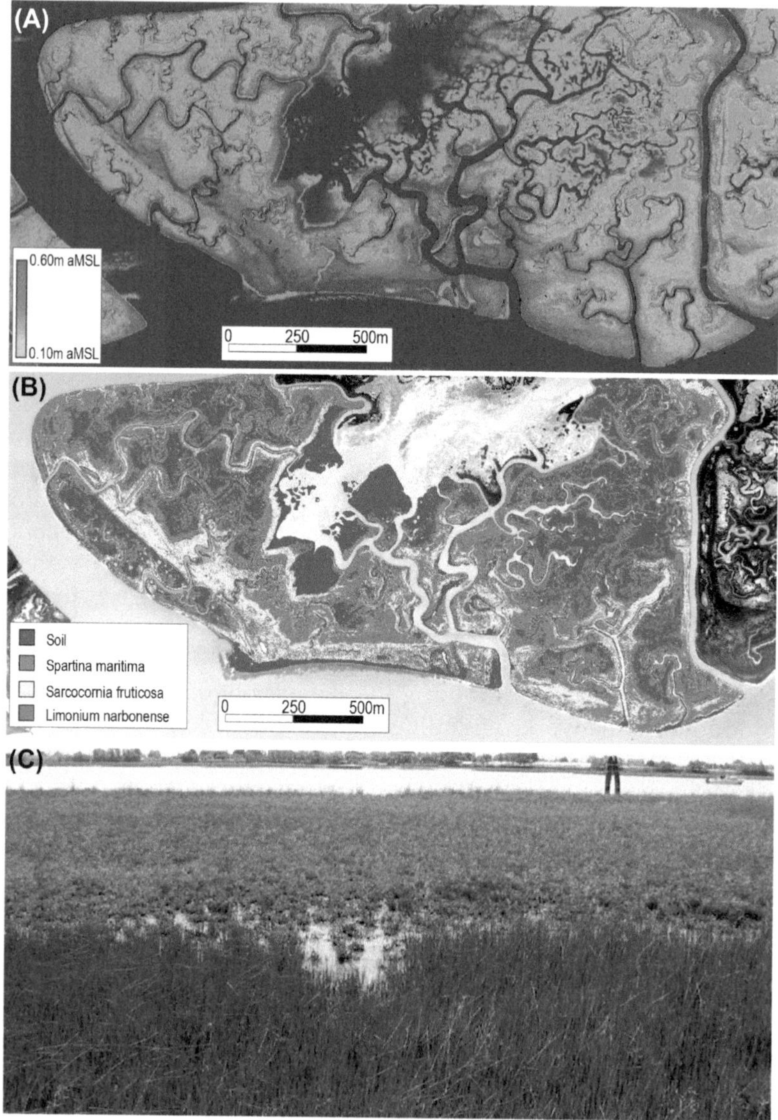

FIGURE 5.1 (A) Topography of the San Felice salt marsh in the Venice Lagoon obtained from a LiDAR survey. Lower elevations are coded in shades of blue and higher elevations in red. (B) A vegetation map of the same marsh is overlapped to the true-color representation of the multispectral remote sensing image from which it was created (Marani et al., 2006c). The Figure shows the typical patchy distribution emerging from the zonation phenomenon of different vegetation types (*Limonium narbonense*, *Sarcocornia fruticosa*, and *Spartina maritima*). (C) Example of a zonation pattern in the San Felice Salt marsh formed by *S. maritima* on the lower portion of the picture and *L. narbonense* on the upper portion).

2. INTERTIDAL ECOGEOMORPHOLOGICAL EVOLUTION

The chief morphological processes involved in the evolution of an intertidal area include the incision and subsequent elaboration of a channel network within a platform that may be evolving from a tidal flat to a salt marsh state. As mentioned before, the interaction between these processes is also coupled through the influence of biotic processes, such as vegetation or microphytobenthos colonization, affecting the sediment transport and stability.

Tidal channel initiation can be ascribed to the concentration of tidal fluxes over a surface, for example a mudflat, possibly induced by the presence of small perturbations in the topography (e.g., D'Alpaos et al., 2006; Belliard et al., 2015). Patches of pioneer vegetation species (e.g., Temmerman et al., 2007) or vegetation disturbance by crabs (e.g., Perillo et al., 2005; Hughes et al., 2009) can also favor the concentration of tidal fluxes over some portions of the marsh surface. In any case, flux concentration resulting from the space-dependent resistance encountered by tidal flows, produces local scour as a consequence of the excess shear stress exerted at the bottom. Channel incision favors a further flux concentration, generating a positive feedback mechanism that leads to the development of the observed tidal patterns (Yapp et al., 1917; Beeftink, 1966; French and Stoddart, 1992; Allen, 1997; Fagherazzi and Furbish, 2001; D'Alpaos et al., 2006). It is generally agreed that the process of network incision is a rather rapid one (Steers, 1960; Pestrong, 1965; Pethick, 1969; French and Stoddart, 1992; D'Alpaos et al., 2007b; Hughes et al., 2009): a permanent imprinting is likely to be given to the tidal environment, possibly later followed by a slower elaboration of the network structure, for example, by meandering and by the adjustment of channel geometry and stratal architecture to variations in the local tidal prism due to the vertical accretion of the flanking intertidal surface (Gabet, 1998; Marani et al., 2002; Stefanon et al., 2012; Brivio et al., 2016).

The transformation of a tidal flat into a salt marsh requires sediment deposition over the tidal flat to be larger than erosion and sea level rise effects. As soon as the local platform elevation exceeds a threshold for halophytic plant development, the surface is colonized by vegetation, which promotes sediment settling by reducing turbulent kinetic energy (Leonard and Croft, 2006; Mudd et al., 2010), direct sediment capture by vegetation during submersion periods (Leonard and Luther, 1995; Christiansen et al., 2000; Li and Yang, 2009; Mudd et al., 2010) and contributes organic material (e.g., Randerson, 1979; Morris et al., 2002; Nyman et al., 2006; Neubauer, 2008; Mudd et al., 2009). When vegetation extensively encroaches the marsh surface, the increased drag caused by plants influences tidal velocity profiles and the rate at which water floods into and drains from the platform adjacent to a channel (an increasing function of plant density, e.g., see Leonard and Luther, 1995; Nepf, 1999). The presence of vegetation also influences the planimetric evolution of tidal channels due to its stabilizing effects on surface sediments and channel banks (Garofalo, 1980; Marani et al., 2002). The influence of benthic fauna on erosion/deposition processes and on sediment characteristics through bioturbation and biodeposition has been observed as well (Yallop et al., 1994; Wood and Widdows, 2002). It is worthwhile emphasizing that observational evidence and modeling support the concept of inheritance of the major features of channelized patterns from sand flat or mudflat to a salt marsh (e.g., Allen, 2000; Friedrichs and Perry, 2001; Marani et al., 2003; Stefanon et al., 2012).

A numerically feasible description of such complex interactions, particularly in the context of a long-term model, requires the formulation of simplified model components retaining the most relevant features of the governing processes. Such a description of the key hydrodynamic properties of the flow over an intertidal platform can be obtained on the basis of the hydrodynamic model proposed by Rinaldo et al. (1999a,b), which we recall in the following.

2.1 Poisson Hydrodynamic Model

Under the assumption that a balance holds in the momentum equations between water surface slope and the linearized friction term, Rinaldo et al. (1999a) suitably simplified the two-dimensional shallow water equations to a Poisson equation:

$$\nabla^2 \eta_1 = \frac{\lambda}{(\eta_0 - z_0)^2} \frac{\partial \eta_0}{\partial t} \tag{5.1}$$

where $\eta_1(x; t)$ is the local deviation of the water surface from its instantaneous average value, $\eta_0(t)$, referenced to the mean sea level (hereinafter MSL); z_0 is the average marsh bottom elevation, referenced to the MSL; and λ is a bottom friction coefficient (Rinaldo et al., 1999a; Marani et al., 2003 for a detailed description). Further assuming tidal propagation to be much faster within the channel network than over the shallower flanking marsh areas, that is, considering a flat water level, $\eta_1 = 0$, within the network, allows one to determine the field of free surface elevations over the unchanneled marsh platform, at any instant t of the tidal cycle, by solving the Poisson boundary value problem Eq. (5.1).

On the basis of the resulting water surface topography, flow directions can be obtained at any location on the intertidal areas by determining the steepest descent direction, and watersheds related to any channel cross section may be thus identified. The above-simplified Poisson model applies, in principle, to relatively short tidal basins, that is, when the length of the basin is much smaller than the frictionless tidal wavelength (Lanzoni and Seminara, 1998). Nevertheless, as thoroughly discussed by Marani et al. (2003) by comparison with observations and complete hydrodynamic simulations, the Poisson model leads to quite robust estimates of drainage directions and watersheds, and, through the use of the continuity equation (Rinaldo et al., 1999b), of the landscape-forming discharges, even when the hypothesis of a short tidal basin is not strictly met.

On the basis of the water surface elevation field, the distribution of bottom shear stresses due to tidal currents at every point $\mathbf{x}$ on unchanneled areas can be determined as follows:

$$\tau = -\gamma \cdot D \cdot \nabla \eta_1 \tag{5.2}$$

where $\tau(\mathbf{x}; t)$ is the local value of the bottom shear stress, γ is the specific weight of water, and D is the local water depth.

The analysis of the spatial distribution of $\tau(\mathbf{x})$ for our case study sites in the Venice lagoon (Fig. 5.2A) is valuable in suggesting possible general features. It emerges that the higher values of the shear stress usually occur at the tips of the channel network and near pronounced channel bends. This observation is confirmed by Fig. 5.2B that shows the probability

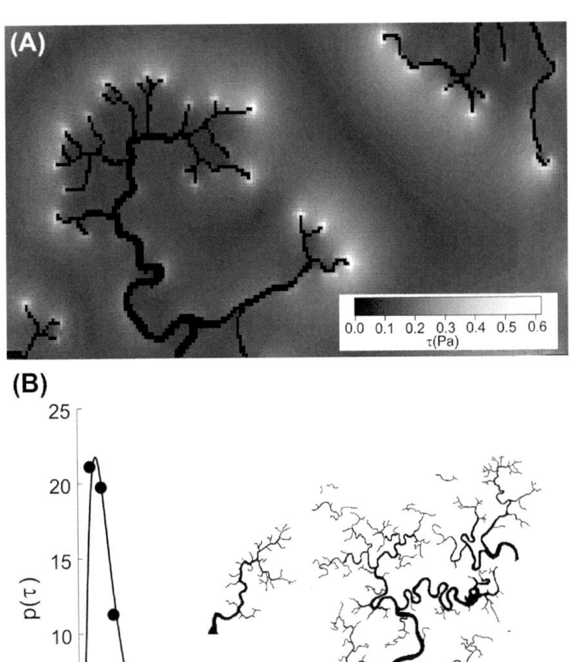

FIGURE 5.2 (A) Example of the spatial distribution of the bottom shear stress $\tau(x)$ attained on the intertidal areas adjacent to the network dissecting the southern part of the San Felice salt marsh in the Venice lagoon. The higher values of the shear stress usually occur at the tips of the channel network and in correspondence of quite pronounced channel bends; (B) probability density function of the bottom shear stress, $p(\tau)$, both at the tips (τ_{tips}) and in the remaining part of the sites adjacent to the tidal network (τ_{others}) characterizing San Felice channel network. The mean stress value acting at the tips of the network for the investigated zone is $\tau_{tips,mean} = 0.12$ Pa.

density function of the shear stresses at the channel network heads, τ_{tips}, and at all other adjacent sites with the exception of the tips, τ_{others}. Such observations corroborate the speculation that headward erosion and tributary addition (possibly originating at sites where the stress increases along bends) are the main processes responsible for channel elaboration during its early development (Pethick, 1969; Steel and Pye, 1997; Allen, 2000; Hughes et al., 2009). We thus suggest that channel headward growth, driven by the spatial distribution of local shear stress, is the chief land-forming agent for network formation on real marsh platforms. Under the assumption of approximately stable network configurations, the observed probability distributions provide useful information on critical shear stress values, which can be used in numerical simulations. The notion that erosional activities can be primarily expected in those parts of the basin where the local value $\tau(x)$ exceeds a threshold value for erosion, τ_c (Rinaldo et al., 1993, 1995; Rigon et al., 1994), is found to produce reasonable structures of tidal drainage densities and associated features within tidal landscapes (D'Alpaos et al., 2005, 2007b).

2.2 Model of Channel Network Early Development

We briefly review here the model of channel network development, presenting its more relevant features, and refer the reader to the paper by D'Alpaos et al. (2005) and D'Alpaos et al. (2007b) for a more detailed description.

Field evidence supports the main assumptions, in particular the hypothesis—adopted also in a number of conceptual models of salt marsh growth and supported by field observations (Pethick, 1969; French and Stoddart, 1992; French, 1993; Allen, 1997; Steel and Pye, 1997; D'Alpaos et al., 2007b; Hughes et al., 2009)—that during its initial development stage a tidal network quickly cuts down through the intertidal areas, acquiring a permanent basic structure (in analogy with the case of fluvial settings, e.g., Rodrìguez-Iturbe and Rinaldo, 1997). Such a quick initial network incision is later followed by elaboration through meandering and further branching (Garofalo, 1980; Marani et al., 2002; Fagherazzi et al., 2004). This later elaboration is deemed to produce minor changes compared to the initial network growth and is closely coupled (Brivio et al., 2016) to the vertical accretion of the adjacent marsh platform driven by the deposition of inorganic sediment and the accumulation of organic soil (Belliard et al., 2015). Nevertheless, the lateral migration of meandering tidal channels can induce neck-cutoff events (D'Alpaos et al., 2018) and channel piracy that are likely to produce important changes in network structure. These considerations indicate the existence of different timescales characteristic of the various processes and justify the choice of decoupling the initial rapid network incision from its subsequent slower elaboration and from the ecomorphological evolution of the adjacent marsh platforms (D'Alpaos et al., 2007a; Kirwan and Murray, 2007; D'Alpaos and Marani, 2016).

Furthermore, based on the computed spatial distribution of bottom shear stresses (Fig. 5.2), which displays higher values at channel tips, we assume that the mechanism dominating channel network development is headward growth (Hughes et al., 2009) driven by the exceedances of a critical shear stress, τ_c, which we take to coincide with a stability shear stress required to maintain an incised cross section through repeated tidal cycles (Friedrichs, 1995). Depending on the spatial heterogeneity of sediment, vegetation, and microphytobenthos, which influences channel network dynamics, τ_c may be assumed as constant or space dependent. Whenever the local bottom shear stress, $\tau(\mathbf{x})$, exceeds τ_c anywhere on the border of the channels, erosional activity and network development may be expected: The model of channel network incision is thus based on the evaluation of the bottom shear stress distribution.

According to the model proposed, the evolution of the network proceeds as follows. (1) For a given configuration of the channel network (initially consisting of a single-channeled site), Eq. (5.1) is solved using representative values of η_0 and $\partial\eta_0/\partial t$ and the $\tau(\mathbf{x})$ distribution is computed from Eq. (5.2). (2) One of the sites where $\tau(\mathbf{x})$ exceeds the fixed threshold for erosion, τ_c, is selected on the basis of a suitable procedure governed by a parameter, T (which may be considered as "temperature," in analogy with the simulated annealing procedure proposed by Kirkpatrick et al., 1983), expressing the possibly spatially heterogeneous distribution of the critical stress and becomes part of the network. (3) The new channel pixel is considered to be part of the channel axis and channel cross sections are instantaneously adapted to the tidal prism, P, flowing through them, defined as the total volume of water exchanged through any cross section between low-water slack and the following high water slack, that is, during flood or ebb phases. In fact, it has long been recognized that a power–law relation holds between the tidal prism, P, and the minimum cross-sectional area, Ω, for a large number of tidal systems believed to have achieved dynamic equilibrium (O'Brien, 1969; Jarrett, 1976; Marchi, 1990). More recently, Friedrichs (1995), Rinaldo et al. (1999b), Lanzoni and Seminara (2002), van der Wegen et al. (2008), and D'Alpaos et al. (2010) explored, in several tidal systems, the relationship between Ω and spring (i.e., maximum astronomical) peak discharge, Q, which is directly related to the tidal prism, finding that a near proportionality between Ω and Q also exists for sheltered sections. Friedrichs (1995) explains the existence of such relationship by relating the equilibrium cross-sectional geometry to the total bottom shear stress necessary to maintain a null along-channel gradient in net sediment transport, the so-called stability shear stress. In addition, D'Alpaos et al. (2010) verified, both through field evidence and numerical modeling applied to the Venice Lagoon, the broad applicability of tidal prism cross-sectional area relations to arbitrary sheltered cross sections within complex lagoonal configurations and embedded tidal networks. They found that values of the exponent α of the relation $\Omega = k\ P^{\alpha}$ nicely meet the value $\alpha = 6/7$, empirically observed by O'Brien (1969) and theoretically derived by Marchi (1990). Therefore, on the basis of the O'Brien-Jarrett-Marchi (OBJM) "law" (D'Alpaos et al., 2009) we consider the cross-sectional area, Ω, to be related to the landscape-forming tidal fluxes responsible for shaping network geometry (expressed through the tidal prism, P) on the basis of the relationship $\Omega = k\ P^{\alpha}$, with $k = 1.4\ 10^{-3}\ \text{m}^{2-3\alpha}$ and $\alpha = 6/7$. Such an assumption allows one to describe the evolution of the channel network in response to changes in the tidal prism, P, possibly due to variations in the elevation of the marsh platform or in relative mean sea level, as also supported by the experimental results provided by Stefanon et al. (2010, 2012). (4) Once the cross-sectional area has been determined, channel width is assigned based on a fixed value of the width-to-depth ratio, β (Marani et al., 2002; Lawrence et al., 2004; Lanzoni and D'Alpaos, 2015), in which we summarize the complex morphodynamic processes responsible for channel cross-sectional shape. Because the flow field has now varied due to the inclusion in the network structure of newly channelized pixels, Eq. (5.1) is solved again using the new boundary conditions reflecting the updated channel configuration and steps (1)–(4), which represent a model time step, are repeated iteratively. As the channel network extends into the intertidal area, the reference water surface and its gradients are progressively lowered and the procedure is repeated until the critical shear stress is nowhere exceeded.

2.3 Model of Marsh Platform Evolution

We briefly review here the ecomorphodynamic model proposed by D'Alpaos et al. (2007a) and modified by D'Alpaos and Marani (2016) to describe marsh platform evolution due to inorganic sediment transport, erosion, and deposition and accounting for vegetation competition and organic soil production as described by Marani et al. (2013) and Da Lio et al. (2013). A basic description of the models is provided below and the reader is referred to the original papers for full derivations.

The model assumes that the cohesive and nearly uniform bottom sediment particles are transported mainly in suspension. Evolution of bed topography is governed by the sediment continuity equation, which reads

$$(1-p)\frac{\partial z_b}{\partial t} = Q_d - Q_e - R \tag{5.3}$$

here z_b is the local bottom elevation with reference to the MSL, $P = .4$ is void fraction in the bed, and Q_d and Q_e are the local deposition and erosion fluxes, respectively, representing sediment volume exchange rates, per unit area, between the water column and the bed, and R is the rate of relative sea level rise (RSLR) (i.e., the algebraic sum of SLR and local subsidence).

We evaluate the erosion flux, Q_e, by a relationship that can be applied when bed properties are relatively uniform over the depth and the bed is consolidated (Mehta, 1984)

$$Q_e = \frac{Q_{e0}}{\rho_b}\left(\frac{\tau}{\tau_e} - 1\right)H(\tau - \tau_e) \tag{5.4}$$

here $Q_{e0} = 5.0 \times 10^{-4}$ kg m^{-1} s^{-1} is an empirical erosion rate; $\rho_b = (1-p)\rho_s$ is sediment bulk density after compaction has taken place, and $\rho_s = 2650$ kg m^{-3} is sediment bed porosity and density, respectively; τ is the absolute value of the local bottom shear stress evaluated through Eq. (5.2); $\tau_e = 0.4$ N m^{-3} is the cohesive shear stress strength with respect to erosion; and H is the Heaviside step function. We assume that Q_e vanishes as vegetation encroaches the marsh surface, in accordance with field observations emphasizing that tidal currents are unable to produce excess shear stress over vegetated marshes (Christiansen et al., 2000).

The total deposition flux, Q_d, is the sum of the local inorganic deposition fluxes due to sediment settling, Q_s, direct particle capture by plants, Q_c, and of the local organic soil production, Q_o, mainly associated with belowground biomass production:

$$Q_d = Q_s + Q_c + Q_o \tag{5.5}$$

If the marsh is not vegetated, both Q_c and Q_o are equal to zero, and the total deposition flux is equal to Q_s. According to the feedback mechanism existing between morphology, hydrodynamics, and sediment dynamics, the settling and trapping rates can be determined only when the equation for suspended sediment concentration (hereinafter SSC) has been solved. However, because bottom topography evolves on a much longer timescale with respect to the hydrodynamic circulation, one can decouple the solution of the hydrodynamic field from the morphological evolution. Under the assumption that the flow is fully turbulent, the equation for the conservation of sediment transported as a dilute suspension takes the form of a two-dimensional advection–diffusion equation for depth-averaged volumetric sediment concentration, $C(\mathbf{x}; t)$, which reads

$$\frac{\partial(CD)}{\partial t} + \nabla\cdot(\mathbf{U}CD - k_d D\nabla C) = Q_e - Q_d \tag{5.6}$$

where D is the local water depth, $\mathbf{U}$ is the local depth-averaged velocity field, and $k_d = 0.3\ \mathrm{m}^2\ \mathrm{s}^{-1}$ is a constant dispersion coefficient that accounts for dispersive effects associated with vertical variations in both flow velocity and sediment concentration. We estimate the deposition due to settling, Q_s, through the empirical relationships proposed by Einstein and Krone (1962), usually employed to describe cohesive sediment deposition in coastal environments

$$Q_s = w_s \frac{C}{\rho_b}\left(1 - \frac{\tau}{\tau_d}\right) H(\tau_d - \tau) \tag{5.7}$$

where $w_s = 1\ 10^{-4}\ \mathrm{m\ s}^{-1}$ is sediment settling velocity for silt and $\tau_d = 0.1\ \mathrm{N\ m}^{-3}$ is a critical shear stress below which all initially suspended sediment eventually deposits.

Vegetation encroachment at the surface, for emergent marsh platforms, increases sediment deposition rates as a consequence of particle capture by plants, Q_c, and of organic accretion rate, Q_o, which both depend on local plant biomass, B. Although several biotic and abiotic factors may be relevant in determining plant productivity (Silvestri et al., 2005, and references therein), locally, biomass production can, however, be related mainly to the elevation of the marsh platform encroached by plants. Such a relationship is the result of differences in soil aeration resulting from marsh flooding by the tide, and its form fundamentally depends on the biodiversity typical of the tidal environment considered.

The model assumes local biomass to be at all times in equilibrium with the local current soil elevation, i.e., $B = B[z(\mathbf{x},t)]$, on the basis of an "equilibrium vegetation model" (Marani et al., 2010) suggesting that annual vegetation productivity adjusts over a much faster timescale than the evolution of marsh surface elevation.

At this point is worth recalling that D'Alpaos et al. (2007a), in their original model formulation, addressed two scenarios of vegetation growth. The first scenario considered marshes characterized by a prevailing presence of *Spartina* spp., as typically occurs in many North European and North American marshes (e.g., Morris and Haskin, 1990; Morris et al., 2002). This scenario considered one vegetation species with biomass expressed as a linearly decreasing function of soil elevation between MSL and mean high water level (MHWL). The increased pore water salinity caused by evapotranspiration (enhanced by the progressive reduction of the duration and frequency of inundation, as the platform elevation increases) can in fact limit the growth of, or be fatal to, salt marsh macrophytes (Phleger, 1971). The second scenario addresses the situation in the Venice Lagoon where a mosaic of vegetation patches is observed (Marani et al., 2004; Silvestri et al., 2005; Marani et al., 2006a,b). This scenario considers multiple vegetation species for which, as a result of competition and individual species adaptations, biomass production linearly increases with soil elevation between MSL and MHWL. As soil elevation increases, in fact, because *Spartina* is not well adapted to more aerated soil conditions, it is outcompeted by other species (e.g., *Sarcocornia* or *Limonium* in the Venice Lagoon) which thus take over. These vegetation models were used in later modeling studies (e.g., D'Alpaos, 2011; Marani et al., 2007, 2010; Belliard et al., 2015). In such models, however, the competition among different species and its implications for organic soil production were not explicitly addressed. We therefore employ here the model proposed by D'Alpaos and Marani (2016), who modified D'Alpaos et al.'s (2007a) model following Marani et al. (2013). The latter authors considered a set of competing vegetation species, each adapted to different elevation ranges, to effectively represent the combined effects of environmental stressors, such as soil hypoxia and salinity concentration.

Previous analyses and studies suggest the amount of sediment directly captured by plants to be proportional to the local SSC and to the number of plant stems that can both reduce the turbulent energy and capture sediment particles (Leonard and Luther, 1995; Nepf, 1999; Leonard and Reed, 2002). In analogy with D'Alpaos et al. (2006, 2007a), sediment deposition due to particle capture (Palmer et al., 2004), reads

$$Q_c = \frac{C}{\rho_b}\alpha_c B_i^{\beta_c} U^{\gamma_c} \tag{5.8}$$

where $\alpha_c = 1.02 \times 10^6\, d_{50}{}^2\, (\mathrm{m\ s^{-1}})^{1-\gamma}(\mathrm{m^2\ g^{-1}})^{\beta c}$, $\beta_c = 0.382$, and $\gamma_c = 1.7$ are empirical coefficients (see D'Alpaos et al. 2006, 2007b for details); $d_{50} = 50\ \mu\mathrm{m}$ is the median sediment grain size; U is the magnitude of the local depth-averaged velocity; $B_i(x,t)$ is the annually averaged biomass production of vegetation species "i," which happens to colonize site $\mathbf{x}$ at time t (aboveground biomass, responsible for suspended sediment trapping, and belowground biomass, responsible for organic soil production, are assumed here to be both equal to $B_i(\mathbf{x},t)$). Based on the data collected by Morris and Haskin (1990) at North Inlet Estuary (South Carolina, USA), we have assumed that vegetation characteristics can be expressed as a function of plant biomass (see D'Alpaos et al., 2006, 2007a for a detailed description).

Finally, the organic accretion rate, Q_o, is linked to the annually averaged biomass production of vegetation species "i" (Randerson, 1979; Mudd et al., 2004; D'Alpaos et al., 2007a) as follows:

$$Q_o = Q_{o0} B_i \tag{5.9}$$

where $Q_{o0} = 2.5 \times 10^{-6}\ \mathrm{m^3\ year^{-1}\ g^{-1}}$ is a constant that incorporates typical vegetation characteristics and the density (after compaction and partial decomposition) of the organic soil produced.

Aboveground biomass is one of the main factors through which the control of vegetation on hydrodynamics and sediment deposition is exerted. The above recalled relationships were thus required to couple geomorphic and ecological models. Generally, the aboveground storage of organic material in salt marshes is an extremely complex process, which depends on vegetation characteristics and involves root production, microbial decomposition, and edaphic factors such as nutrient availability and salinity (e.g., Silvestri et al., 2005).

Following Marani et al. (2013), we express the local annually averaged biomass production, $B_i(x,t) = B_{\max} \cdot f_i[z(x, t)]$, as a fraction of the maximum annual biomass, $B_{\max} = 10^3\ \mathrm{g\ m^{-2}}$, on the basis of a species-specific fitness function, $0 < f_i(z) < 1$. The fitness function defines how biomass production and competitive abilities of each species vary with soil elevation, and hence describes the degree of adaptation of a species i to the local elevation z and to the related edaphic conditions (see Marani et al., 2013 and Da Lio et al., 2013 for further details). It is, however, worthwhile emphasizing that the fitness function not only regulates biomass production but also species competitive abilities, thus incorporating a competitive displacement mechanism. Following Marani et al. (2013) we adopt the following analytical relationship for the fitness function $f_i(z)$:

$$f_i = \frac{2}{e^{\lambda(\xi-\overline{\xi}_i)} + e^{-\lambda(\xi-\overline{\xi}_i)}} \tag{5.10}$$

where H is the tidal amplitude, $\xi = z/H$, $\overline{\xi}_i = z_i/H$ represents the dimensionless elevation at which the fitness function for species i reaches its maximum value ($f(\xi_i) = 1$), and λ is a scale parameter that expresses the rate at which the fitness function tends to zero (for $z \rightarrow \pm\infty$) as elevation deviates from the species-specific optimal value (Fig. 5.2). Eq. (5.10) accounts for the observation that halophytic vegetation species are maximally productive within specific ranges of optimal adaptation (Pennings and Callaway, 1992; Morris et al., 2002; Morris, 2006), whereas species competitive abilities decrease as elevation deviates from the species-specific elevation value providing optimal environmental conditions. Large values of λ allow one to mimic the behavior of specialized vegetation species, well adapted to a narrow range of elevations. Conversely, small values of λ are characteristic of species that are relatively well adapted to a broader range of marsh elevations.

As to the modeling of the changes in species distribution due to interspecific competition, which strongly affect the distribution of topographic elevations over the marsh platform, Marani et al. (2013) analyzed two competition mechanisms. These mechanisms are based on either (1) selecting, at each site $\boldsymbol{x}_k$ the species i for which $f_i(z_k)$ is maximum ("fittest takes all"), or (2) randomly selecting species i with a probability $p(i, x_k) = f_i(z_k)/\sum_j f_j(z_k)$ ("stochastic competition" mechanism), to account for the fact that biomorphodynamics in the real world are affected by stochastic forcings, stochasticity in competition mechanisms, and heterogenous edaphic conditions. Although this second criterion appears to more realistically account for real-life stochastic conditions, the first has the advantage of more clearly illustrating vegetation controls on marsh morphology, and therefore we adopt here, for the sake of simplicity, the "fittest takes all" mechanism (Marani et al., 2013). We therefore assume that a vegetation species j, which at time t colonizes a site x_k with elevation $z_k(\mathrm{x}; t)$, is replaced by another species i if $f_i(z_k) > f_j(z_k)$ for every $j \neq i$ (i.e., species i is best adapted to the current value of the elevation). The "fittest takes all" mechanism selects at each time step and at each site the species i whose fitness f_i is largest, allowing one to analyze system equilibria and patterns in the ideal case in which the outcome of vegetation dynamics can be isolated from the effects of stochasticity (in the environment and in the organisms).

According to our formulation, vegetation distribution influences sediment dynamics and local organic and inorganic accretion rates, thus affecting the patterns of net deposition, which, in turn, determine the change in soil topography. The latter determines changes in the spatial distribution of biomass, thus closing the feedback that is fully described in the model.

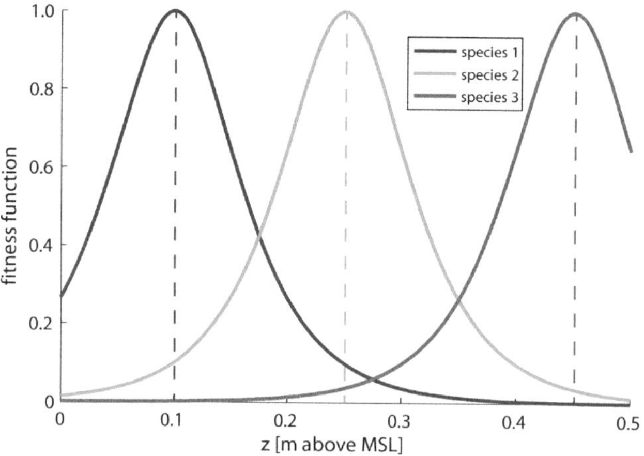

FIGURE 5.3 Fitness functions for three vegetation species with the same degree of specialization ($\lambda = 10$, see Eq. 5.10), characterized by different optimal elevations $\hat{z}_1 = 0.10$ m, $\hat{z}_2 = 0.25$ m, and $\hat{z}_3 = 0.45$ m above mean sea level (MSL) for the "blue," "green," and "red" species, respectively.

3. RESULTS

The model of channel network development (D'Alpaos et al., 2005) makes it possible to analyze both the initiation of a channel network over an undissected tidal embayment and the further elaboration of an already incised channel structure. A variety of experiments were performed starting from different initial conditions to analyze the effects related to the position of single or multiple inlets, the shape of the tidal basin, different values of the width-to-depth ratio, and different values of the critical shear stress for erosion, τ_c, and of the temperature, T. The model was also applied to simulate the evolution of a channel network within an actual catchment, emphasizing its noteworthy capabilities to reproduce real-life features (D'Alpaos et al., 2005, 2007b).

Here we present the results of numerical simulations aimed at studying the competition among tidal creeks to drain the marsh platform adjacent to a larger tidal channel. Fig. 5.4 shows some snapshots portraying the progressive development of creek networks within an idealized rectangular domain, limited by impermeable boundaries except for the bottom side, flanking a larger tidal channel. The marsh platform is characterized by an average elevation $z_0 = -0.20$ m above MSL. Channel network formation is a result of the dynamics of the system. Creeks are initiated at sites along the bottom channel where the first incision, initially due to chance, further grows because of the progressive flux concentration caused by creek development. At the beginning of the simulation, the domain is entirely drained by the boundary channel on the lower side and all of the boundary channel points drain the same amount of the watershed area. As soon as the networks start to develop, the drainage area associated with each of the growing networks, as well as their width, are relatively small. When the networks further develop and dissect the unchanneled domain, their watersheds and tidal prisms increase, thus causing their cross-sectional areas to increase as well, to accommodate the swelling tidal prism (van der Wegen et al., 2008; D'Alpaos et al., 2009, 2010).

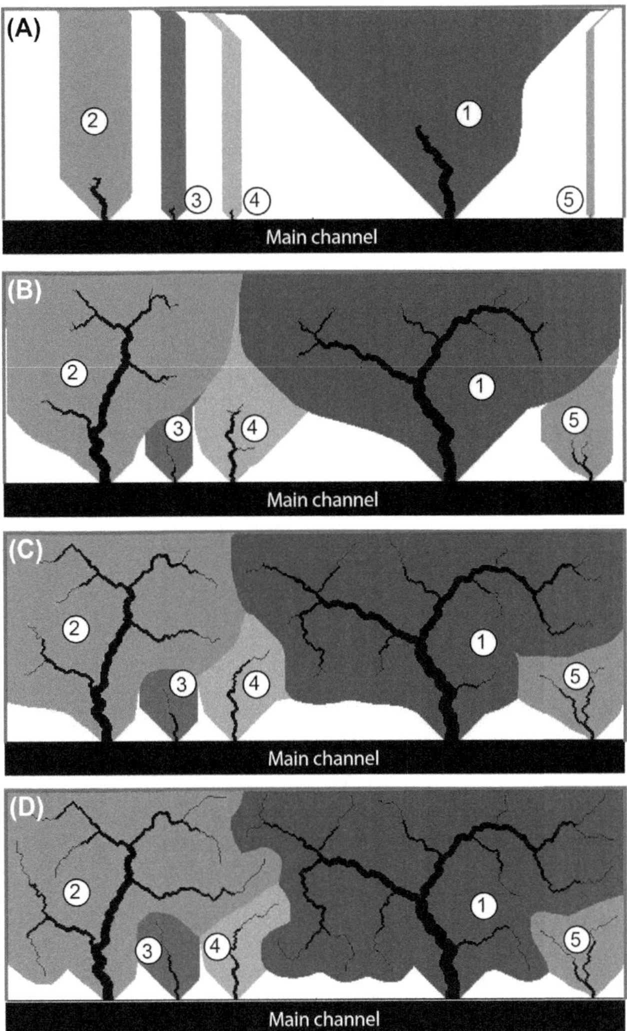

FIGURE 5.4 Evolution in time of planar network configurations and of the related watersheds within a rectangular domain (on which a 200 × 600 lattice is superimposed, the size of the pixel being equal to 1 m) limited by a tidal channel at the bottom and by otherwise impermeable boundaries (red lines in figure). Snapshots from (A–D) represent the evolution in time of the creek networks starting from the very beginning of the process to its end. Configurations (A–D) are obtained after 20, 200, 600, and 1000 model iterations, respectively. Shaded areas represent the watersheds associated with the five emerging creek networks, the portion of the domain in white is drained by the boundary channel at the bottom side.

The dynamics of the system is characterized by a "competition" among developing networks to capture the available watershed area. Stages of incision and retreat are observed, as well as situations in which divides migrate as a consequence of channel competition (Fig. 5.4).

To verify the validity of the proposed modeling approach, we compare relevant geomorphic features of the synthetic networks to those observed in actual tidal networks. The geomorphic characterizations necessary to compare synthetic morphologies with observed ones are provided by previous theoretical and observational analyses of the drainage density in tidal networks, relying on the statistics of unchanneled flow lengths, ℓ, that is, unchanneled flow paths from any unchanneled site to the nearest channel (Marani et al., 2003). Such statistics make it possible to capture site-specific features of network development and important morphological differences, providing a dynamically based geomorphic description, which proves distinctive of network aggregation features (contrary to traditional Hortonian measures). Indeed, the analysis of a great number of actual marsh systems in the Venice lagoon showed a clear tendency to develop watersheds characterized by exponential decays of the probability distributions of ℓ, and thereby a pointed absence of scale-free features (Marani et al., 2003). Interestingly, the probability distributions of ℓ, for the synthetic networks generated by the model, display a linear semilog trend of the type observed in the case of actual tidal patterns. Fig. 5.5 portrays the evolution in time of the probability

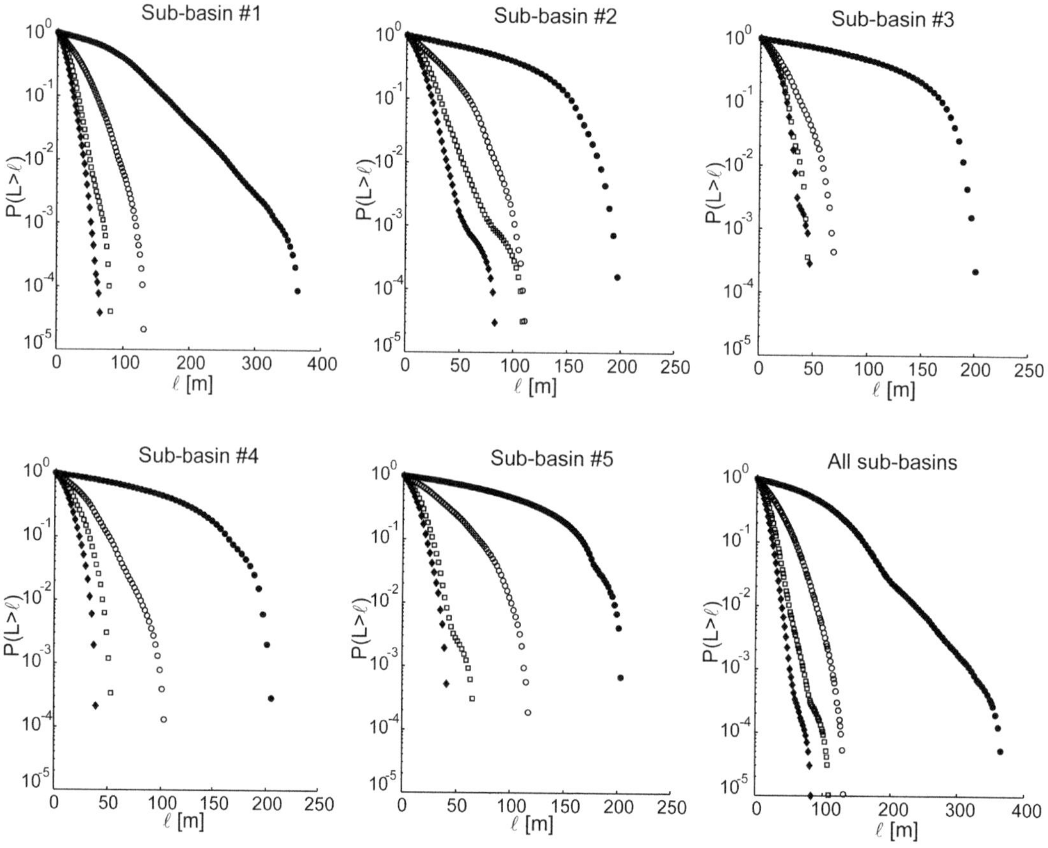

FIGURE 5.5 Semilog plots of the exceedance probability of unchanneled lengths $P(L > \ell)$ versus the current value of length ℓ, for the different subbasins represented in Fig. 5.4. *Black circles* refer to configuration (A) in Fig. 5.4; *empty circles* to configuration (B); *squares* to configuration (C); and *diamonds* to configuration (D).

distribution of unchanneled lengths, $P(L > \ell)$, as the synthetic networks cut through the undissected domain, moving from configuration (A) to configuration (D) in Fig. 5.4. The distribution of unchanneled lengths changes considerably as the network develops. The initial stages of network development are associated with larger values of the mean unchanneled length and with probability distributions that are quite far from an exponential form. At later stages, the mean unchanneled length decreases and the probability distributions tend to become exponential. This emphasizes that the model of network development is capable of providing complex structures and reproducing distinctive geometrical properties of geomorphic relevance (D'Alpaos et al., 2005, 2007b).

We then analyze the long-term morphological evolution of the marsh platform dissected by the (abovegenerated) synthetic creek networks by applying the model proposed by D'Alpaos and Marani (2016), which describes the mutual interaction and adjustment between tidal flows, sediment transport, morphology, and vegetation distribution, thus allowing one to study the biomorphodynamic evolution of salt marsh platforms. The model allows also one to investigate the response of tidal morphologies to different scenarios of sediment supply, colonization by halophytes, and changing sea level. To this end, we consider an idealized initial topographic configuration represented by a network structure in equilibrium with a flat salt marsh surface with elevation equal to 0.30 m above MSL in a microtidal system. The initial network structure is in equilibrium with the tidal prism computed on the basis of the initial assigned topography (*sensu* D'Alpaos et al., 2010). Furthermore, we assume the volumetric SSC within the channel network, C_0, to be constant in space and time and the system forced by a sinusoidal tide with tidal amplitude of 0.5 m. We considered fine cohesive and uniform sediments characterized by density $\rho_s = 2600\ \mathrm{kg\ m^{-3}}$; particle diameter $d_{50} = 50\ \mu\mathrm{m}$; settling velocity $w_s = 2.0 \times 10^{-4}\ \mathrm{m\ s^{-1}}$; porosity $P = 0.4$; and erosion rate parameter $Q_{e0} = 1/\rho_s = 3.0\ 10^{-4}\ \mathrm{m\ s^{-1}}$. The critical bottom shear stress for erosion, $\tau_e = 0.4\ \mathrm{N\ m^{-2}}$, and deposition, $\tau_d = 0.1\ \mathrm{N\ m^{-2}}$, are characteristic of fully consolidated mud (D'Alpaos et al., 2007a). We have further assumed that the marsh can be populated by three vegetation species with the same degree of specialization ($\lambda = 10$), characterized by different optimal elevations $\hat{z}_1 = 0.10$ m, $\hat{z}_2 = 0.25$ m, and $\hat{z}_3 = 0.45$ m above MSL for the "blue," "green," and "red" species, respectively, as depicted in Fig. 5.3. All species are characterized by equal maximum fitness and therefore by an equal maximum annually averaged biomass production $B_{max} = 10^3\ \mathrm{g\ m^{-2}}$.

Fig. 5.6 shows marsh topographies in equilibrium with different prescribed rates of RSLR and suspended sediment (SS) ($R = 5\ \mathrm{mm\ year^{-1}}$ and $C_0 = 10\ \mathrm{mg\ L^{-1}}$ for Fig. 5.5A; $R = 5\ \mathrm{mm\ year^{-1}}$ and $C_0 = 20\ \mathrm{mg\ L^{-1}}$ for Fig. 5.5B; $R = 2.5\ \mathrm{mm\ year^{-1}}$ and $C_0 = 10\ \mathrm{mg\ L^{-1}}$ for Fig. 5.5C), and allows one to distinguish the sedimentation patterns that characterize the nonuniform marsh topographies when the "fittest takes all" mechanism is considered. Fig. 5.6 shows the spatial distributions of the different vegetation species corresponding to the marsh topographic patterns of Fig. 5.5.

The three modeled scenarios present common evolutionary features, but important and interesting differences emerge.

The magnitude of deposition processes and, therefore, the local vertical growth of the marsh platform decrease with distance from the creeks. Marsh topographies are indeed characterized by the formation of higher levees paralleling channel banks and by bottom

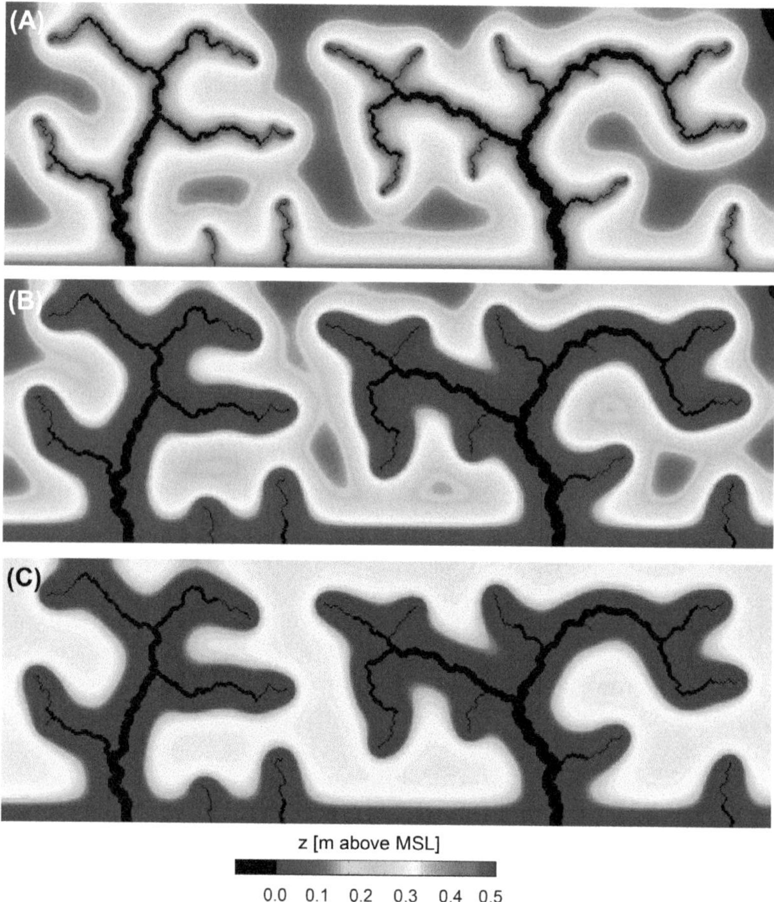

FIGURE 5.6 Comparison of marsh surface topographies. Color-coded representation of marsh surface elevations (referenced to MSL) in equilibrium with different rates of relative SLR and different values of the SSC within the channel network C_0, when the marsh is forced by a semidiurnal tide of amplitude 0.5 m and populated by three different vegetation species with fitness functions shown in Fig. 5.4. (A) $R = 5$ mm year^{-1} and $C_0 = 10$ mg L^{-1}; (B) $R = 5$ mm year^{-1} and $C_0 = 20$ mg L^{-1}; (C) $R = 2.5$ mm year^{-1} and $C_0 = 10$ mg L^{-1}.

elevations that progressively decrease toward the inner portion of the marsh. Such a behavior can be explained by considering the reduction in the SSC with distance from the creeks due to settling and direct particle capture by plant stems and to the progressive decrease of advective transport as prescribed by Eq. (5.6). These two processes promote the development of typical concave-up profiles at the marsh scale (see Fig. 5.1 for a qualitative comparison with actual marsh topography), in agreement with observational evidence (e.g., Temmerman et al., 2003; Roner et al., 2016) and with a number of numerical models describing salt marsh vertical accretion within the tidal frame (e.g., Allen, 2000).

It is, however, interesting to note that at a smaller scale a marked transition between neighboring gently sloping terrace-like structures emerges, which indeed results from the coupled

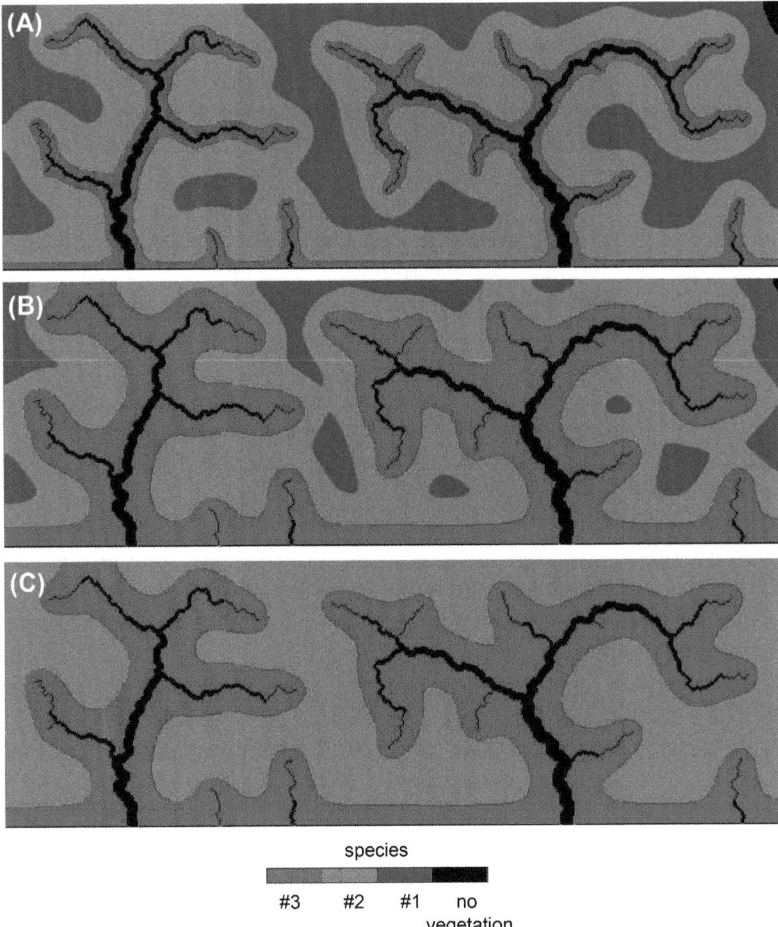

FIGURE 5.7 Comparison of vegetation distributions. Color-coded representation of vegetation patterns in equilibrium with topographic elevations of Fig. 5.6, when the marsh is forced by different rates of relative sea level rise (in the range 2.5–5.0 mm $year^{-1}$) and different values of the suspended sediment concentration within the channel network (in the range 10–20 mg L^{-1}).

evolution of salt marsh elevations and vegetation cover: This is the result of the interaction and adjustment between geomorphic and biological processes. Fig. 5.7 indeed shows the spatial distributions of the different vegetation species corresponding to the marsh topographic patterns of Fig. 5.6. Single vegetation species colonize gently sloping areas (Fig. 5.7), which display sharp transitions among them and are quite reminiscent of the zonation patterns observed across marshes worldwide (e.g., Adam, 1990; Pennings and Callaway, 1992; Silvestri et al., 2005; Marani et al., 2013).

Figs. 5.6 and 5.7 also allow one to analyze the spatially extended impacts of changes in the rates of RSLR and SS on both topography and vegetation dynamics.

An increase in SS from 10 to 20 mg L^{-1} (compare Fig. 5.6A and B) mostly increases marsh elevation in areas closer to the creeks and produces an increase in mean marsh elevation, as suggested by a number of morphodynamic models (see e.g., Fagherazzi et al., 2012 for a thorough review). The related change in vegetation cover (compare Fig. 5.7A and B) concerns an expansion of the "red" species that live at higher elevations, and a decrease in the marsh area occupied by the "green" and "blue" species that live at lower elevations.

A decrease in the rate of RSLR from 5 to 2.5 mm $year^{-1}$ (compare Fig. 5.6A and C) leads to an increase in marsh elevations in areas closer to the creeks and also produces an increase in mean marsh elevation, in analogy with the previous case. However, model results show that halving the rate of RSLR (from 5 to 2.5 mm $year^{-1}$, see Fig. 5.6A and C) produces a stronger general increase in marsh elevations than doubling the available SS (from 10 to 20 mg L^{-1}, see Fig. 5.6A and B) and moreover leads to the disappearance of the "blue" species that seems to be the most sensitive to changes in the environmental forcing (compare Fig. 5.6A and C).

These model results confirm those obtained by D'Alpaos and Marani (2016) who modeled the biomorphodynamic evolution of a tidal watershed whose geometry and shape was reminiscent of those displayed by an actual tidal watershed in the Venice Lagoon. Also in this case, where a synthetic domain is considered, we observe that biodiversity is strongly influenced by the environmental forcings (rate of RSLR and SS, in our case). Changes in the rate of RSLR and/or in SS may, in fact, result in the selective disappearance of some stable biogeomorphic equilibria associated with marsh biogeomorphic patterns, with consequent reductions in the biodiversity (Figs. 5.6 and Fig. 5.7).

Patterns emerging from the dynamics of marsh surface elevations and vegetation cover (Figs. 5.6 and 5.7) indicate that the coupled evolution of vegetation and morphology gives rise to different system properties. The critical role of biogeomorphic feedbacks in determining the coupled topographic and vegetation patterns can also be highlighted by analyzing the frequency distributions of topographic marsh elevations associated with the different vegetation species (Fig. 5.8). These frequency distributions display a multimodal behavior

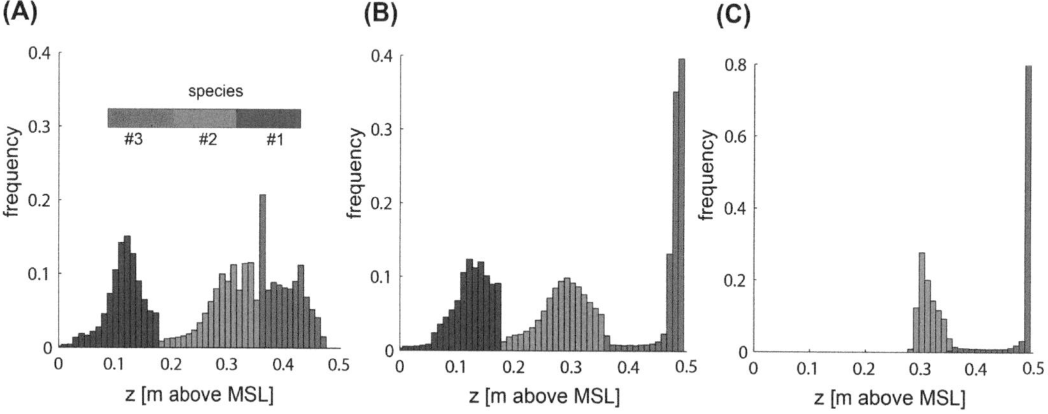

FIGURE 5.8 Comparison of frequency distribution of topographic elevations in which color codes represent the different species populating different elevation intervals. Panels (A–C) represent the frequency distributions for the topographic configurations and vegetation distributions of Figs. 5.6A–C and 5.7A–C, respectively.

(as shown by color-coded bars according to the most abundant species colonizing each elevation interval) where each frequency maximum is robustly associated with a single vegetation species, thus highlighting the fundamental role of biogeomorphic feedbacks in determining the observed coupled topographic and vegetation patterns. Indeed, the close relationship between maxima in the frequency distribution and vegetation species, observed for actual marshes in the Venice Lagoon (Marani et al., 2013; Da Lio et al., 2013), is found to be a characteristic signature of the underlying and intertwined physical and biological processes in marsh landscapes. The correspondence between species presence and the peaks in the frequency density of marsh elevations represents a detectable fingerprint of the landscape-constructing role of marsh plants (Marani et al., 2013; D'Alpaos and Marani, 2016).

4. DISCUSSION

The theoretical framework and modeling described here appears to reproduce geomorphologically relevant features of the ecomorphodynamic evolution of tidal networks and of the marsh platform they dissect. The choice to decouple the process of network initiation and early development (which appears to be quite a rapid one) from the subsequent slower network elaboration and evolution of the marsh platform is supported by field evidence (e.g., Steers, 1960; Collins et al., 1987; Wallace et al., 2005; D'Alpaos et al., 2007b), by conceptual and numerical models describing the evolution of channel marsh systems (Allen, 2000; D'Alpaos et al., 2005; Minkoff et al., 2006; Hughes et al., 2009; Vandenbruwaene et al., 2012; Zhou et al., 2014a,b), and by laboratory experiments (Stefanon et al., 2010, 2012; Vlaswinkel and Cantelli, 2011). Steers (1960) reported a channel headcut migration of up to 5–7 m year^{-1}, Collins et al. (1987) observed a headward erosion of more than 200 m in 130 years (i.e., more than 1.5 m year^{-1}), and Wallace et al. (2005) related a mean extension rate of 6.2 m year^{-1}. D'Alpaos et al. (2007b) described the rapid development of a network of volunteer creeks, branching from an artificial channel within a newly restored microtidal salt marsh, characterized by mean and maximum annual headward growth rates of 11 and 18 m year^{-1}, respectively. Vandenbruwaene et al. (2012) measured the formation and evolution of a tidal channel network in a newly constructed macrotidal marsh and observed a headward erosion rate of about 40 m year^{-1}. Interactions between physical and biological processes driving the formation and evolution of tidal channel networks have also been observed. Perillo et al. (2005) observed the interaction between the *Chasmagnathus granulata* crab and groundwater seepage responsible for channel initiation; Minkoff et al. (2006) modeled crab–vegetation dynamics and their effect on creek growth observed to occur in the field at maximum rates of 50 cm month^{-1}; Hughes et al. (2009) related headward erosion rates of 1.9 m year^{-1} promoted by vegetation dieback coupled with intense burrowing by crabs.

These observations and the comparison between actual and modeled geomorphic network features (D'Alpaos et al., 2005, 2007b) also substantiate the assumption concerning the strong control exerted by the water surface elevation gradients, and by the related bottom shear stresses, in driving the process of channel incision (see also Fagherazzi and Sun, 2004). It is worthwhile noting that our modeling approach does not contradict conceptual models of "depositional network development" (Redfield, 1965; Hood, 2006), which describe network development as the consequence of the vertical accretion and horizontal progradation of the vegetated marsh platform. Feedbacks that shape the network are fundamentally similar in both cases. According to the picture provided by our modeling approach, network incision and development take place at locations where the threshold shear stress is exceeded. In the vegetated platform progradation case, the extension of the channels is determined by the location where the bottom shear stress at channel heads is greater than the critical one (Redfield, 1965; Hood, 2006). Feedbacks between the evolution of the marsh platform and the channel network may lead to variations in network geometry due to the differential accretion between the platform and the channels: stronger tidal fluxes within the channel may promote erosion or maintenance, whereas weak fluxes on the marsh may allow deposition fluxes to overcome erosion lading to marsh vertical/horizontal accretion. Such an observation supports the known concept of inheritance of the major features of channelized patterns from sand flat or mudflat to a salt marsh (Allen, 2000). Moreover, the positive feedback between channel incision and flux concentration has recently been described by a number of numerical models of tidal landform evolution (D'Alpaos et al., 2006, 2007b; Kirwan and Murray, 2007; Temmerman et al., 2007; Zhou et al., 2014a,b; Belliard et al., 2015). The recent modeling results by Belliard et al. (2015, 2016) suggest that erosion- and deposition-driven tidal channel development coexist, although acting at different timescales. Erosional processes promote channel initiation, whereas depositional processes are mostly responsible for the elaboration of the channel network form and structure.

Numerical modeling of the cross-sectional evolution of tidal channels (D'Alpaos et al., 2006; van der Wegen et al., 2010; Zhou et al., 2014a,b; Lanzoni and D'Alpaos, 2015) together with field observations (D'Alpaos et al., 2010) and laboratory experiments (Stefanon et al., 2010, 2012) support the assumption of rapidly adapting cross-sectional areas to the flowing tidal prisms, on the basis of a deterministic power—law relation (O'Brien, 1969; Jarrett, 1976; Marchi, 1990; Friedrichs, 1995; Rinaldo et al., 1999b; Lanzoni and D'Alpaos, 2015) that D'Alpaos et al. (2009) proposed to term it the O'Brien-Jarrett-Marchi "law." The use of a constant width-to-depth ratio, which summarizes the complex morphodynamic processes responsible for channel cross-sectional shape, is supported by observational evidence (Marani et al., 2002; Lawrence et al., 2004) and modeling (D'Alpaos et al., 2006; Lanzoni and D'Alpaos, 2015). As a note, we observe that tidal flat and salt marsh channels display different values of the width-to-depth ratio, β, because they are seen to respond to different erosional processes resulting in different types of incision (Marani et al., 2002). Indeed, the presence of halophytic vegetation on the marsh platform is likely to strongly affect bank failure mechanisms, and therefore salt marsh creeks tend to be more deeply incised ($5 < \beta < 7$) than tidal flat channels ($8 < \beta < 20$). Moreover, D'Alpaos et al. (2006) showed that the width-to-depth ratio, β, decreases as the adjacent platform evolves from a tidal flat to a salt marsh, a result that was recently confirmed by Lanzoni and D'Alpaos (2015) on the basis of a different modeling approach.

Results from the ecogeomorphic model of marsh platform evolution concerning the equilibrium elevation reached by the platform within the tidal frame as a function of changes in the forcings agree with observational evidence (Allen, 200) and with a number of numerical models describing salt marsh vertical accretion within the tidal frame (Kirwan et al., 2010; Fagherazzi et al., 2012). The formation of marsh levees paralleling channel banks, which later broaden toward the inner part of the platform that exhibits a typical concave-up profile, is in accordance with field observations and modeling (e.g., Mudd et al., 2004; Silvestri et al., 2005; Temmerman et al., 2005; D'Alpaos et al., 2007a). The model accounts for the emergence of biogeomorphic patterns (e.g., Pennings and Callaway, 1992; Silvestri et al., 2005; Marani et al., 2013) due to the two-way interactions between physical and biological forcing, producing a chief effect on the long-term ecomorphodynamic evolution of salt marsh platforms. The response and the resilience of salt marsh landscapes to changes in the environmental forcings are critically affected by these biogeomorphic feedbacks.

5. CONCLUSIONS

We have shown here that long-term modeling of intertidal biogeomorphic systems is feasible by suitably simplifying the description of the governing processes and yet retaining their physically relevant features. We have also shown that biological–physical interactions are key in determining the observed spatial patterns both in the biological and geomorphic domains.

The models presented generate network structures that are quantitatively close to observed ones, whereas the topographic and vegetation spatial patterns produced are realistic and qualitatively similar to observed ones. The quantitative validation of the spatial patterns obtained from a biogeomorphic model is, however, always difficult to achieve. This is due to a difficulty in obtaining observations of the time evolution of the system, the long characteristic timescales, and the problematic definition of objective landscape metrics.

This problem has been overcome with reference to the spatial organization of a channel network and to the development of biogeomorphic patterns. The synthetic network structures have been compared to actual ones on the basis of statistics of unchanneled path lengths, which provide a quantitative characterization of the relationship between channels and the intertidal landscape they dissect. Spatial patterns of soil elevation and vegetation distribution have been qualitatively compared with zonation biogeomorphic patterns observed in marshes worldwide. In addition, a quantitative comparison between synthetic and actual pattern properties has been carried out on the basis of the frequency density of marsh elevations for single vegetation species. The frequency distributions of modeled marsh elevations display a multimodal behavior, with frequency maxima robustly associated with a single species, and nicely agree with observed ones. In addition, it is worthwhile emphasizing that the close relationship between maxima in the frequency distribution and vegetation species provides a characteristic signature of the intertwined biophysical processes that sculpt marsh landscapes. The correspondence between peaks in the frequency distribution of elevations and species distribution in the vertical frame represents a detectable fingerprint of the landscape-constructing role of marsh vegetation.

Although a number ecogeomorphological studies accounting for the mutual influence between hydrodynamics, sediment transport, and morphological and biological dynamics have been recently developed, the field of biogeomorphological modeling of intertidal systems is still in its infancy. Several issues remain to be tackled, including the clarification of the response of vegetation to changes in soil aeration, a more complete quantitative description of biological effects on sediment mechanical properties (e.g., physical factors regulating the onset and the time-space variability of microphytobenthos influence on sediment erosion), and the incorporation of the action of wind waves on the margins of intertidal platforms strengthened by plant roots. Many other important issues have not been listed or may have been overlooked here. However, the results presented certainly point to the fact that a predictive model of the evolution of intertidal landscapes must be based on the recognition that the system cannot be decomposed into its biological and physical components and that its dynamics is intrinsically a biogeomorphic one.

Acknowledgments

Funding from the CARIPARO Project titled "Reading signatures of the past to predict the future: 1000 years of stratigraphic record as a key for the future of the Venice Lagoon" is gratefully acknowledged.

References

Adam, P., 1990. Salt Marsh Ecology. Cambridge University Press, Cambridge.

Allen, J.R.L., 1997. Simulation models of salt-marsh morphodynamics: some implications for high- intertidal sediment couplets related to sea-level change. Sedimentary Geology 113, 211–223.

Allen, J.R.L., 2000. Morphodynamics of holocene salt marshes: a review sketch from the Atlantic and Southern North sea coasts of Europe. Quaternary Science Reviews 19 (17–18), 1155–1231.

Belliard, J.P., Toffolon, M., Carniello, L., D'Alpaos, A., 2015. An ecogeomorphic model of tidal channel initiation and elaboration in progressive marsh accretional contexts. Journal of Geophysical Research: Earth Surface 120, 1040–1064. https://doi.org/10.1002/2015JF003445.

Belliard, J.P., Di Marco, N., Carniello, L., Toffolon, M., 2016. Sediment and vegetation spatial dynamics facing sea-level rise in microtidal saltmarshes: insights from an ecogeomorphic model. Advances in Water Resources 93 (2016), 249–264. https://doi.org/10.1016/j.advwatres.2015.11.020.

Belluco, E., Camuffo, M., Ferrari, S., Modenese, L., Silvestri, S., Marani, A., Marani, M., 2006. Mapping salt-marsh vegetation by multispectral and hyperspectral remote sensing. Remote Sensing of Environment 105, 54–67.

Beeftink, W.G., 1966. Vegetation and habitat of the salt marshes and beach plains in the south-western part of the Nederlands. Wentia 15, 83–108.

Boaga, J., D'Alpaos, A., Cassiani, G., Marani, M., Putti, M., 2014. Plant-soil interactions in salt marsh environments: experimental evidence from electrical resistivity tomography in the Venice Lagoon. Geophysical Research Letters 41, 6160–6166. https://doi.org/10. 1002/2014GL060983.

Brivio, L., Ghinassi, M., D'Alpaos, A., Finotello, A., Fontana, A., Roner, M., Howes, N., 2016. Aggradation and lateral migration shaping geometry of a tidal point bar: an example from salt marshes of the Northern Venice Lagoon (Italy). Sedimentary Geology 343, 141–155.

Canestrelli, A., Defina, A., Lanzoni, S., D'Alpaos, L., 2007. Long-term evolution of tidal channels flanked by tidal flats. In: Dohmen, C., Hulscher, S. (Eds.), Proceedings of the 5th IAHR Symposium on River, Coastal and Estuarine Morphodynamics, Enschede, Netherlands, 17–21 Sept. Taylor and Francis, London, pp. 145–153.

Carniello, L., Defina, A., Fagherazzi, S., D'Alpaos, L., 2005. A combined wind wave-tidal model for the Venice lagoon, Italy. Journal of Geophysical Research: Earth Surface 110, F04007. https://doi.org/10.1029/2004JF000232.

Cao, M., Xin, P., Jin, G., Li, L., 2012. A field study on groundwater dynamics in a salt marsh – Chongming Dongtan wetland. Ecological Engineering 40, 61–69. https://doi.org/10.1016/j.ecoleng.2011.12.018.

Christiansen, T., Wiberg, P.L., Milligan, T.G., 2000. Flow and sediment transport on a tidal saltmarsh surface. Estuarine, Coastal and Shelf Science 50, 315–331.

Coco, G., Zhou, Z., van Maanen, B., Olabarrieta, M., Tinoco, R., Townend, I., 2013. Morphodynamics of tidal networks: advances and challenges. Marine Geology 346, 1–16. https://doi.org/10.1016/j.margeo.2013.08.005.

Collins, L.M., Collins, J.N., Leopold, L.B., 1987. Geomorphic processes of an estuarine marsh: preliminary results and hypotheses. In: Gardner, V. (Ed.), International Geomorphology 1986, Part I. John Wiley, Hoboken, NJ, pp. 1049–1072.

D'Alpaos, A., Lanzoni, S., Marani, M., Fagherazzi, S., Rinaldo, A., 2005. Tidal network ontogeny: channel initiation and early development. Journal of Geophysical Research 110, F02001. https://doi.org/10.1029/2004JF000182.

D'Alpaos, A., Lanzoni, S., Mudd, S.M., Fagherazzi, S., 2006. Modelling the influence of hydroperiodand vegetation on the cross-sectional formation of tidal channels. Estuarine, Coastal and Shelf Science 69, 311–324. https://doi.org/10.1016/j.ecss.2006.05.02.

D'Alpaos, A., Lanzoni, S., Marani, M., Rinaldo, A., 2007a. Landscape evolution in tidal embayments: modelling the interplay of erosion, sedimentation, and vegetation dynamics. Journal of Geophysical Research 112, F01008. https://doi.org/10.1029/2006JF000537.

D'Alpaos, A., Lanzoni, S., Marani, M., Bonometto, A., Cecconi, G., Rinaldo, A., 2007b. Spontaneous tidal network formation within a constructed salt marsh: observations and morphodynamic modelling. Geomorphology 91, 186–197. https://doi.org/10.1016/j.geomorph.2007.04.013.

D'Alpaos, A., Lanzoni, S., Marani, M., Rinaldo, A., 2009. On the O'Brien-Jarrett-Marchi law. Rend. Fis. Accad. Lincei. 20, 225–236. https://doi.org/10.1007/s12210-009-0052-x.

D'Alpaos, A., Lanzoni, S., Marani, M., Rinaldo, A., 2010. On the tidal prism-channel area relations. Journal of Geophysical Research: Earth Surface 115 (F1). https://doi.org/10.1029/2008JF001243.

D'Alpaos, A., 2011. The mutual influence of biotic and abiotic components on the long-term ecomorphodynamic evolution of salt-marsh ecosystems. Geomorphology 126, 269–278.

D'Alpaos, A., Mudd, S., Carniello, L., 2011. Dynamic response of marshes to perturbations in suspended sediment concentrations and rates of relative sea level rise. Journal of Geophysical Research 116, F04020.

D'Alpaos, A., Marani, M., 2016. Reading the signatures of biologic-geomorphic feedbacks in salt-marsh landscapes. Advances in Water Resources 93, 265–275. https://doi.org/10.1016/j.advwatres.2015.09.004.

D'Alpaos, A., Toffolon, M., Camporeale, C., 2016. Ecogeomorphological feedbacks of water fluxes, sediment transport and vegetation dynamics in rivers and estuaries. Advances in Water Resources 93, 151–155. https://doi.org/10.1016/j.advwatres.2016.05.019.

D'Alpaos, A., Ghinassi, M., Finotello, A., Brivio, L., Bellucci, L.G., Marani, M., 2017. Tidal meander migration and dynamics: a case study from the Venice Lagoon. Marine and Petroleum Geology 87, 80–90. https://doi.org/10.1016/j.marpetgeo.2017.04.012.

Da Lio, C., D'Alpaos, A., Marani, M., 2013. The secret gardener: vegetation and the emergence of biogeomorphic patterns in tidal environments. Philosophical Transactions of the Royal Society A 371, 20120367. https://doi.org/10.1098/rsta.2012.0367.

Day, J.W., Shaffer, G.P., Britsch, L.D., Reed, D.J., Hawes, S.R., Cahoon, D., 2000. Pattern and process of land loss in the Mississippi Delta: a spatial and temporal analysis of wetland habitat change. Estuaries 23, 425–438.

Defina, A., Carniello, L., D'Alpaos, L., Fagherazzi, S., 2007. Self organization of shallow basins intidal flats and salt marshes. Journal of Geophysical Research 112, F03001. https://doi.org/10.1029/2006JF000550.

Einstein, H.A., Krone, R.B., 1962. Experiments to determine modes of cohesive sediment transport in salt water. Journal of Geophysical Research 67 (4), 1451–1461.

Fagherazzi, S., Bortoluzzi, A., Dietrich, W.E., Adami, A., Lanzoni, S., Marani, M., Rinaldo, A., 1999. Tidal networks 1. Automatic network extraction and preliminary scaling features from Digital Elevation Maps. Water Resources Research 35 (12), 3891–3904.

Fagherazzi, S., Furbish, D.J., 2001. On the shape and widening of saltmarsh creeks. Journal of Geophysical Research 106 (C1), 991–1003.

Fagherazzi, S., Sun, T., 2004. A stochastic model for the formation of channel networks in tidal marshes. Geophysical Research Letters 31, L21503. https://doi.org/10.1029/2004GL020965.

Fagherazzi, S., Gabet, E.J., Furbish, D.J., 2004. The effect of bidirectional flow on tidal channel planforms. Earth Surf. Proc. Land 29 (3), 295–309.

Fagherazzi, S., Carniello, L., D'Alpaos, L., Defina, A., 2006. Critical bifurcation of shallow microtidallandforms in tidal flats and salt marshes. Proceedings of the National Academy of Sciences of the United States of America 103, 8337–8341. https://doi.org/10.1073/pnas.0508379103.

Fagherazzi, S., Kirwan, L., Mudd, S., Guntenspergen, G., Temmerman, S., D'Alpaos, A., van de Koppel, J., Rybczyk, J.M., Reyes, E., Craft, C., Clough, J., 2012. Numerical models of salt marsh evolution: ecological, geomorphic, and climatic factors. Reviews of Geophysics 50, RG1002.

Feola, A., Belluco, E., D'Alpaos, A., Lanzoni, S., Marani, M., Rinaldo, A., 2005. A geomorphic study of lagoonal landforms. Water Resources Research 41, W06019. https://doi.org/10.1029/2004WR003811.

French, J.R., Spencer, T., 1993. Dynamics of sedimentation in a tide-dominated back barrier salt marsh Norfolk, UK. Marine Geology 110, 315–331.

French, J.R., Stoddart, D.R., 1992. Hydrodynamics of saltmarsh creek systems: implications for marsh morphological development and material exchange. Earth Surface Processes and Landforms 17, 235–252.

French, J.R., 1993. Numerical simulation of vertical marsh growth and adjustment to accelerated sea-level rise, North Norfolk, UK. Earth Surface Processes and Landforms 18, 63–81.

Friedrichs, C.T., Aubrey, D.G., 1994. Tidal propagation in strongly convergent channels. Journal of Geophysical Research 99, 3321–3336.

Friedrichs, C.T., 1995. Stability shear stress and equilibrium cross-sectional geometry of sheltered tidal channels. Journal of Coastal Research 11 (4), 1062–1074.

Friedrichs, C.T., Perry, J.E., 2001. Tidal salt marsh morphodynamics. Journal of Coastal Research 27, 6–36.

Gabet, E.J., 1998. Lateral migration and bank erosion in a salt marsh tidal channel in San Francisco Bay, California. Estuaries 21 (4B), 745–753.

Garofalo, D., 1980. The influence of wetland vegetation on tidal stream migration and morphology. Estuaries 3, 258–270.

Hodges, B.R., Imberger, J., Saggio, A., Winters, K.B., 2000. Modeling basin-scale internal waves in a stratified lake. Limnology and Oceanography 45, 1603–1620.

Hood, W.G., 2006. A conceptual model of depositional, rather than erosional, tidal channel development in the rapidly prograding Skagit River Delta (Washington, USA). Earth Surface Processes and Landforms 31, 1824–1838.

Hughes, Z.J., FitzGerald, D.M., Wilson, C.A., Pennings, S.C., Wieski, K., Mahadevan, A., 2009. Rapid headward erosion of marsh creeks in response to relative sea level rise. Geophysical Research Letters 36, L03602. https://doi.org/10.1029/2008GL036000.

Jarrett, J.T., 1976. Tidal Prism-Inlet Area Relationships. General Investigation of Tidal Inlets Report 3. U.S. Army Coastal Engineering Research Center, Fort Belvoir, VA and U.S. Army Engineer Water. Exper. Stat., Vicksburg, MS, 32 pp.

Kirchner, J.W., 1993. Statistical inevitability of Horton's laws and the apparent randomness of stream channel networks. Geology 21, 591–594.

Kirkpatrick, S., Gelatt, G.D., Vecchi, M.P., 1983. Optimization by simulated annealing. Science 220, 671–680.

Kirwan, M.L., Murray, A.B., 2007. A coupled geomorphic and ecological model of tidal marsh evolution. Proceedings of the National Academy of Sciences of the United States of America 104, 6118–6122. https://doi.org/10.1073/pnas.0700958104.

Kirwan, M., Guntenspergen, G., D'Alpaos, A., Morris, J., Mudd, S., Temmerman, S., 2010. Limits on the adaptability of coastal marshes to rising sea level. Geophysical Research Letters 37, L23401.

Kirwan, M.L., Megongial, J.P., 2013. Tidal wetland stability in the face of human impacts and sea-level rise. Nature 504, 53–60. https://doi.org/10.1038/nature12856.

Lanzoni, S., Seminara, G., 1998. On tide propagation in convergent estuaries. Journal of Geophysical Research 103, 30793–30812.

Lanzoni, S., Seminara, G., 2002. Long term evolution and morphodynamic equilibrium of tidal channels. Journal of Geophysical Research 107 (C1), 3001. https://doi.org/10.1029/2000JC000468.

Lanzoni, S., D'Alpaos, A., 2015. On funneling of tidal channels. Journal of Geophysical Research: Earth Surface 120 (3), 433–452. https://doi.org/10.1002/2014JF003203.

Larsen, L.G., Eppinga, M.B., Passalacqua, P., Getz, W.M., Rose, K.A., Liang, M., 2016. Appropriate complexity landscape modeling. Earth-Science Reviews 160, 111–130.

Lawrence, D.S.L., Allen, J.R.L., Havelock, G.M., 2004. Salt marsh morphodynamics: an investigation on tidal flows and marsh channel equilibrium. Journal of Coastal Research 20 (1), 301–316.

Leonard, L., Luther, M., 1995. Flow hydrodynamics in tidal marsh canopies. Limnology and Oceanography 40, 1474–1484.

Leonard, L.A., Reed, D.J., 2002. Hydrodynamics and sediment transport through tidal marsh canopies. Journal of Coastal Research 36, 459–469.

Leonard, L.A., Croft, A.L., 2006. The effect of standing biomass on flow velocity and turbulence in *Spartina alterniflora* canopies. Estuarine, Coastal and Shelf Science 69, 325–336.

Li, H., Yang, S.L., 2009. Trapping effect of tidal marsh vegetation on suspended sediment, Yangtze Delta. Journal of Coastal Research 25, 915–936.

Marani, M., Lanzoni, S., Zandolin, D., Seminara, G., Rinaldo, A., 2002. Tidal meanders. Water Resources Research 38 (11), 1225–1233.

Marani, M., Belluco, E., D'Alpaos, A., Defina, A., Lanzoni, S., Rinaldo, A., 2003. On the drainage density of tidal networks. Water Resources Research 39, 1040. https://doi.org/10.1029/2001WR001051.

Marani, M., Lanzoni, S., Silvestri, S., Rinaldo, A., 2004. Tidal landforms, patterns of halophytic vegetation and the fate of the lagoon of Venice. Journal of Marine Systems 51, 191–210. https://doi.org/10.1016/j.jmarsys.2004.05.012.

Marani, M., Ursino, N., Silvestri, S., 2005. Reply to comment by Alicia M. Wilson and Leonard R. Gardner on "Subsurface flow and vegetation patterns in tidal environments". Water Resources Research 41 (W07022), 920. https://doi.org/10.1029/2004WR003722.

Marani, M., Silvestri, S., Belluco, E., Ursino, N., Comerlati, A., Tosatto, O., Putti, M., 2006a. Spatial organization and ecohydrological interactions in oxygen-limited vegetation ecosystems. Water Resources Research 42, W07S06. https://doi.org/10.1029/2005WR004582.

Marani, M., Belluco, E., Ferrari, S., Silvestri, S., D'Alpaos, A., Lanzoni, S., Feola, A., Rinaldo, A., 2006b. Analysis, synthesis, and modelling of high-resolution observations of salt-marsh ecogeomorphological patterns in the Venice lagoon. Estuarine, Coastal and Shelf Science 69, 414–426. https://doi.org/10.1016/j.ecss.2006.05.021.

Marani, M., Zillio, T., Belluco, E., Silvestri, S., Maritan, A., 2006c. Non-neutral vegetation dynamics. PLoS One 1 (1), 5.

Marani, M., D'Alpaos, A., Lanzoni, S., Carniello, L., Rinaldo, A., 2007. Biologically-controlled multiple equilibria of tidal landforms and the fate of the Venice lagoon. Geophysical Research Letters 34, L11402. https://doi.org/10.1029/2007GL030178.

Marani, M., D'Alpaos, A., Lanzoni, S., Carniello, L., Rinaldo, A., 2010. The importance of being coupled: stable states and catastrophic shifts in tidal biomorphodynamics. Journal of Geophysical Research 115, F04004.

Marani, M., Da Lio, C., D'Alpaos, A., 2013. Vegetation engineers marsh morphology through competing multiple stable states. Proceedings of the National Academy of Sciences of the United States of America 110, 3259–3263.

Marciano, R., Wang, Z.B., Hibma, A., de Vriend, H.J., Defina, A., 2005. Modeling of channel patterns in short tidal basins. Journal of Geophysical Research 110, F01001. https://doi.org/10.1029/2003JF000092.

Marchi, E., 1990. Sulla stabilità delle bocche lagunari a marea. Rend. Fis. Mat. Accad. Lincei. 1 (2), 137–150. https://doi.org/10.1007/BF03001888.

Mehta, A.J., 1984. Characterization of cohesive sediment properties and transport processes in estuaries. In: Mehta, A.J. (Ed.), Estuarine Cohesive Sediment Dynamics, Lecture Notes on Coastal and Estuarine Studies, vol. 14. Springer-Verlag, Berlin, pp. 290–315.

Mitsch, W.J., Gosselink, J.G., 2000. Wetlands. Wiley, New York.

Minkoff, D.R., Escapa, M., Ferramola, F.E., Maraschn, S.D., Pierini, J.O., Perillo, G.M.E., Delrieux, C., 2006. Effects of crab halophytic plant interactions on creek growth in a S.W. Atlantic salt marsh: a Cellular Automata model. Estuarine, Coastal and Shelf Science 69, 403–413. https://doi.org/10.1016/j.ecss.2006.05.008.

Morris, J.T., Haskin, B., 1990. A 5-yr record of aerial primary production and stand characteristics of *Spartina alterniflora*. Ecology 7 (16), 2209–2217.

Morris, J.T., Sundareshwar, P.V., Nietch, C.T., Kjerfve, B., Cahoon, D.R., 2002. Responses of coastal wetlands to rising sea level. Ecology 83, 2869–2877.

Morris, J., 2006. Competition among marsh macrophytes by means of geomorphological displacement in the intertidal zone. Estuarine, Coastal and Shelf Science 69, 395–402. https://doi.org/10.1016/j.ecss.2006.05.025.

Mudd, S.M., Fagherazzi, S., Morris, J.T., Furbish, D.J., 2004. Flow, sedimentation, and biomass production on a vegetated salt marsh in South Carolina: toward a predictive model of marsh morphologic and ecologic evolution. In: Fagherazzi, S., Marani, M., Blum, L.K. (Eds.), The Ecogeomorphology of Tidal Marshes, vol. 59. American Geophysical Union, Washington DC, pp. 165–188. Coastal and Estuarine Studies.

Mudd, S., Howell, S., Morris, J., 2009. Impact of dynamic feedbacks between sedimentation, sea-level rise, and biomass production on near surface marsh stratigraphy and carbon accumulation. Estuarine. Coastal and Shelf Science 82 (3), 377–389.

Mudd, S., D'Alpaos, A., Morris, J., 2010. How does vegetation affect sedimentation on tidal marshes? Investigating particle capture and hydrodynamic controls on biologically mediated sedimentation. Journal of Geophysical Research 115 (115), F03029. https://doi.org/10.1029/2009JF001566.

Murray, A.B., Knaapen, M.A.F., Tal, M., Kirwan, M.L., 2008. Biomorphodynamics: physical-biological feedbacks that shape landscapes. Water Resources Research 44, W11301. https://doi.org/10.1029/2007WR006410.

Nepf, H.M., 1999. Drag, turbulence, and diffusion in flow through emergent vegetation. Water Resources Research 35 (2), 479–489.

Neubauer, S.C., 2008. Contributions of mineral and organic components to tidal freshwater marsh accretion. Estuarine, Coastal and Shelf Science 78 (1), 78–88. https://doi.org/10.1016/j.ecss.2007.11.011.

Nyman, J.A., Walters, R.J., Delaune, R.D., Patrick, W.H., 2006. Marsh vertical accretion via vegetative growth. Estuarine, Coastal and Shelf Science 69.

O'Brien, M.P., 1969. Equilibrium flow areas of inlets in sandy coasts. Journal of the Waterways, Harbors, and Coast Engineering Division ASCE 95, 43–52.

Palmer, M.R., Nepf, H.M., Pettersson, T.J.R., Ackerman, J.D., 2004. Observations of particle capture on a cylindrical collector: implications for particle accumulation and removal in aquatic systems. Limnology and Oceanography 49 (1), 76–85.

Pennings, S., Callaway, R., 1992. Salt marsh plant zonation: the relative importance of competition and physical factors. Ecology 73, 681–690.

Perillo, G.M.E., Minkoff, D.R., Piccolo, M.C., 2005. Novel mechanism of stream formation in coastal wetlands by crab-fish-groundwater interaction. Geo-marine Letters 25, 214220. https://doi.org/10.1007/s00367-005-0209-2.

Pestrong, R., 1965. The Development of Drainage Patterns on Tidal Marshes. Technical Report 10. Stanford University Publications in the Geological Sciences, 87 pp.

Pethick, J.S., 1969. Drainage in tidal marshes. In: Steers, J.R. (Ed.), The Coastline of England and Wales, third ed. Cambridge University Press, Cambridge, pp. 725–730.

Phleger, C.F., 1971. Effect of salinity on growth of a salt marsh grass. Ecology 52 (5), 908–911.

Randerson, P.F., 1979. A simulation model of salt-marsh development and plant ecology. In: Knights, B., Phillips, A.J. (Eds.), Estuarine and Coastal Land Reclamation and Water Storage. Saxon House, Farnborough, pp. 48–67.

Redfield, A.C., 1965. Ontogeny of a salt marsh estuary. Science 147, 50–55.

Reinhardt, L., Jerolmack, D., Cardinale, B., Vanacker, V., Wright, J., 2010. Dynamic interactions of life and its landscape: feedbacks at the interface of geomorphology and ecology. Earth Surface Processes and Landforms 35, 78–101. https://doi.org/10.1002/esp.1912.

Rigon, R., Rinaldo, A., Rodrìguez-Iturbe, I., 1994. On landscape self organization. Journal of Geophysical Research 99, 11971–11993.

Rinaldo, A., Rodrìguez-Iturbe, I., Rigon, R., Bras, R.L., Ijjasz-Vasquez, E., 1993. Self-organized fractal river networks. Physical Review Letters 70, 1222–1226.

Rinaldo, A., Dietrich, W.E., Vogel, G., Rigon, R., Rodrìguez-Iturbe, I., 1995. Geomorphological signatures of varying climate. Nature 374, 632–636.

Rinaldo, A., Rigon, R., Rodrìguez-Iturbe, I., 1998. Channel networks. Annual Review of Earth and Planetary Sciences 26, 289–327.

Rinaldo, A., Fagherazzi, S., Lanzoni, S., Marani, M., Dietrich, W.E., 1999a. Tidal networks 2. Watershed delineation and comparative network morphology. Water Resources Research 35, 3905–3917.

Rinaldo, A., Fagherazzi, S., Lanzoni, S., Marani, M., Dietrich, W.E., 1999b. Tidal networks 3. Landscape-forming discharges and studies in empirical geomorphic relationships. Water Resources Research 35, 3919–3929.

Rinaldo, A., Belluco, E., D'ALpaos, A., Feola, A., Lanzoni, S., Marani, M., 2004. Tidal networks: form and function. In: Fagherazzi, S., Marani, M., Blum, L. (Eds.), Ecogeomorphology of Tidal Marshes, vol. 59. American Geophysical Union, Washington. Coastal and Estuarine Monograph Series, 266 pp.

Rodrìguez-Iturbe, I., Rinaldo, A., 1997. Fractal River Basins: Chance and Self-Organization. Cambridge University Press, New York.

Roner, M., D'Alpaos, A., Ghinassi, M., Marani, M., Silvestri, S., Franceschinis, E., Realdon, N., 2016. Spatial variation of salt-marsh organic and inorganic deposition and organic carbon accumulation: inferences from the Venice lagoon, Italy. Advances in Water Resources. https://doi.org/10.1016/j.advwatres.2015.11.011.

Saco, P.M., Rodríguez, J.F., 2013. Modeling ecogeomorphic systems. In: Shroder, J., Baas, A.C.W. (Eds.), Treatise on Geomorphology, vol. 2. Academic Press, San Diego, CA, pp. 201–220. Quantitative Modeling of Geomorphology.

Savenije, H.H.G., 2001. A simple analytical expression to describe tidal damping or amplification. Journal of Hydrology 243 (3–4), 205–215.

Savenije, H.H.G., Veling, E.J.M., 2005. Relation between tidal damping and wave celerity in estuaries. Journal of Geophysical Research 110, C04007. https://doi.org/10.1029/2004JC002278.

Schuttelaars, H.M., de Swart, H.E., 2000. Multiple morphodynamic equilibria in tidal embayments. Journal of Geophysical Research 105 (24), 105-124,118.

Silvestri, S., Defina, A., Marani, M., 2005. Tidal regime, salinity and salt-marsh plant zonation. Estuarine, Coastal and Shelf Science 62, 119–130.

Steel, T.J., Pye, K., 1997. The development of saltmarsh tidal creek networks: evidence from the UK. In: Proc. Canadian Coastal Conference, Can. Coastal Sci. and Eng. Assoc., Guelph, Ontario, vol. 1, pp. 267–280.

Steers, J.A., 1960. Physiography and evolution. In: Steers, J.A. (Ed.), Scolt Head Island, second ed. Heffer, Cambridge, pp. 12–26.

Stefanon, L., Carniello, L., D'Alpaos, A., Lanzoni, S., 2010. Experimental analysis of tidal network growth and development. Continental Shelf Research 30 (8), 950–962. https://doi.org/10.1016/j.csr.2009.08.018.

Stefanon, L., Carniello, L., D'Alpaos, A., Rinaldo, A., 2012. Signatures of sea level changes on tidal geomorphology: experiments on network incision and retreat. Geophysical Research Letters 39, L12402. https://doi.org/10.1029/2012GL051953.

Tambroni, N., Bolla Pittaluga, M., Seminara, G., 2005. Laboratory observations of the morphody-namic evolution of tidal channels and tidal inlets. Journal of Geophysical Research 110, F04009. https://doi.org/10.1029/2004JF000243.

Temmerman, S., Govers, G., Meire, P., Wartel, S., 2003. Modelling long-term tidal marsh growth under changing tidal conditions and suspended sediment concentrations, Scheldt estuary, Belgium. Marine Geology 193 (1–2), 151–169.

Temmerman, S., Bouma, T.J., Govers, G., Wang, Z.B., De Vries, M.B., Herman, P.M.J., 2005. Impact of vegetation on flow routing and sedimentation patterns: three-dimensional modeling for a tidal marsh. Journal of Geophysical Research 110, F04019. https://doi.org/10.1029/2005JF000301.

Temmerman, S., Bouma, T.J., van de Koppel, J., van der Wal, D., De Vries, M.B., Herman, P.M.J., 2007. Vegetation causes channel erosion in a tidal landscape. Geology 35 (7), 631–634.

Temmerman, S., Kirwan, M.L., 2015. Building land with a rising sea. Science 349, 588–589. https://doi.org/10.1126/science.aac8312.

Townend, I., 2010. An exploration of equilibrium in Venice Lagoon using an idealized form model. Continental Shelf Research 30 (8), 984–999. https://doi.org/10.1016/j.csr.2009.10.012.

Ursino, N., Silvestri, S., Marani, M., 2004. Subsurface flow and vegetation patterns in tidal environments. Water Resources Research 40, W05115. https://doi.org/10.1029/2003WR002702.

Vandenbruwaene, W., Meire, P., Temmerman, S., 2012. Formation and evolution of a tidal channel network within a constructed tidal marsh. Geomorphology 151–152, 114–125.

van der Wegen, M., Wang, Z.B., Savenije, H., Roelvink, J., 2008. Long-term morphodynamic evolution and energy dissipation in a coastal plain, tidal embayment. Journal of Geophysical Research 113 (F3). https://doi.org/10.1029/2007JF000898.

van der Wegen, M., Dastgheib, A., Roelvink, J., 2010. Morphodynamic modeling of tidal channel evolution in comparison to empirical PA relationship. Coastal Engineering 57 (9), 827–837. https://doi.org/10.1016/j.coastaleng.2010.04.003.

van Maanen, B., Coco, G., Bryan, K., 2013. Modelling the effects of tidal range and initial bathymetry on the morphological evolution of tidal embayments. Geomorphology 191, 23–34. https://doi.org/10.1016/j.geomorph.2013.02.023.

Vlaswinkel, B., Cantelli, A., 2011. Geometric characteristics and evolution of a tidal channel network in experimental setting. Earth Surface Processes and Landforms 36 (6), 739–752.

Wallace, K.J., Callaway, J.C., Zedler, J.B., 2005. Evolution of tidal creek networks in a high sedimentation environment: a 5-year experiment at Tijuana Estuary, California. Estuaries 28 (6), 795–811.

Wolanski, E., Elliott, M., 2015. Estuarine Ecohydrology, An Introduction, second ed. Elsevier Science. 322 pp.

Wood, R., Widdows, J., 2002. A model of sediment transport over an intertidal transect, comparing the influences of biological and physical factors. Limnology and Oceanography 47 (3), 848–855.

Woolnough, S.J., Allen, J.R.L., Wood, W.L., 1995. An exploratory numerical model of sediment deposition over tidal salt marshes. Estuarine, Coastal and Shelf Science 41, 515–543.

Xin, P., Kong, J., Li, L., Barry, D.A., 2013. Modelling of groundwater–vegetation interactions in a tidal marsh. Advances in Water Resources 57, 52–68.

Yallop, M.C., de Winder, B., Paterson, D.M., Stal, L.J., 1994. Comparative structure, primary production and biogenic stabilisation of cohesive and non-cohesive marine sediments inhabited by microphytobenthos. Estuarine, Coastal and Shelf Science 39, 565–582.

Yapp, R.H., Johns, D., Jones, O.T., 1916. The salt marshes of the Dovey Estuary. Part I. Introductory. Journal of Ecology 4, 27–42.

Yapp, R.H., Johns, D., Jones, O.T., 1917. The salt marshes of the Dovey Estuary. Part II. The salt marshes. Journal of Ecology 5, 65–103.

Zedler, J., Kercher, S., 2005. Wetland resources: status, trends, ecosystem services, and restorability. Annual Review of Environment and Resources Magazine 30, 39–74. https://doi.org/10.1146/annurev.energy.30.050504.144248.

Zhou, Z., Olabarrieta, M., Stefanon, L., D'Alpaos, A., Carniello, L., Coco, G., 2014a. A comparative study of physical and numerical modeling of tidal network ontogeny. Journal of Geophysical Research 119 (4), 892–912. https://doi.org/10.1002/2014JF003092.

Zhou, Z., Stefanon, L., Olabarrieta, M., D'Alpaos, A., Carniello, L., Coco, G., 2014b. Analysis of the drainage density of experimental and modelled tidal networks. Earth Surface Dynamics 2, 105–116.

Zhou, Z., Coco, G., Townend, I., et al., 2017. Is "morphodynamic equilibrium" an oxymoron? Earth-Science Reviews 165, 257–267. https://doi.org/10.1016/j.earscirev.2016.12.002.

CHAPTER

6

Geomorphology of Tidal Courses and Depressions

Gerardo M.E. Perillo[1,2]

[1]Instituto Argentino de Oceanografía (CONICET – UNS), Bahía Blanca, Argentina;
[2]Departamento de Geología, Universidad Nacional del Sur, Bahía Blanca, Argentina

1. INTRODUCTION

Wetlands are unique environments along the coasts of the world since their origin, and their preservation is required for the conjunction of a series of factors. First of all, the geomorphology of the initial coast must have enough accommodation space to allow the retention, minimizing the bypass, of the sediment being provided by the various possible sources. Sediment supply must also be important; therefore, the sources must provide a continuous input to build the wetland both vertically and horizontally, and keep it in equilibrium with the energy conditions proper of the coast, and potential modifications due to mean sea level variations. The location of the coastal setting should be such as to minimize the energy of the sea; for instance, protected coasts have a better chance to retain sediment than those found in the open shores. Furthermore, progradation and/or aggradation of the wetland, as it happens in deltaic coasts, requires a positive balance between the sediment input from the different sources and the material removed by waves, tides, and their corresponding currents (Perillo and Piccolo, 2011). This balance must also be positive for longer periods because the wetland needs to aggrade at least to the same rate to counterbalance the potential compaction/subsidence of the sediments and the increase in mean sea level.

Tidal courses and depressions are incisions or negative elevations in an otherwise level or slightly seaward-inclined surface represented by a typical coastal wetland. From the initial stage of wetland formation, courses and depressions are common features of the environment morphology. Even the smoothest topography has depressions where tidal water is retained during low tide. In fact, the irregularities of the surface when connected develop courses that, for a part of the tidal cycle, conduct the water and the substances and organism that it mobilizes. Therefore, courses are necessary features for the circulation of water and sediment during the preliminary stages of the formation of the wetland.

Coastal Wetlands
https://doi.org/10.1016/B978-0-444-63893-9.00006-X

Tidal courses (otherwise known as channels, creeks, or gullies) are the most distinctive and important features of coastal environments even with a minimum tidal influence. They represent the basic circulatory system through which water, sediments, organic matter, nutrients, and pollutants are transported in and out of these wetlands. The tide pumps water through the courses allowing the exchange with the open ocean and the exportation of materials (including pollutants) from it. As tide enters any of these environments such as tidal flats, salt marshes, or mangroves, it first becomes channelized until water level overflows channel banks and levees and develops a sheet flow. Ebbing is the reverse process, first water recedes as sheet flow, but final drainage is through the courses.

Without tidal courses, life in coastal wetlands could not be possible as they provide nourishment, protection for local fauna, a place for reproduction, growth of juveniles and, finally, a way out to the open ocean, when mature, for numerous species. In fact, tidal courses, together with tidal depressions, are one of the first features that appear on the formation of a coastal wetland either by modification of earlier fluvial networks or by the direct action of the tides, groundwater, and precipitation. As the wetland evolves in time and space, courses follow up and, many times, set the pass of this evolution because they, being the most energetic environment, are those that are most sensitive to possible changes in the external variables that influence the system.

Despite their importance, tidal courses have been taken for granted as a feature always present but little studied in comparison with the associated flat, marsh, and mangrove areas, except for the pioneer works of Yapp et al. (1917) and Pestrong (1965) among others. Later on, when studied, measurements focus on particular features such as biological or chemical composition and circulation (especially in meanders). Only in the last 30 years, a relatively small group of researchers have started to address the geomorphologic issues regarding tidal courses intensively. In most cases, their work concentrates in specific wetlands in western Europe (i.e., United Kingdom, Italy, the Netherlands), the Americas (i.e., Argentina, Canada, and United States), Australia, and China, but their ecological relevance is undervalued as reflected by the lack of research on these systems relative to other parts of the wetlands.

Tidal courses are widespread and abundant in estuarine ecosystems (Mallin, 2004). There have been some initial attempts to describe the geomorphologic characteristics of tidal courses and even some basic assumption about the mechanism by which they originate. Some relatively recent reviews (i.e., Eisma, 1997; Allen, 2000; Perillo, 2009; Hughes, 2012) provide an integrated approach from the classical geomorphologic point of view of our present knowledge about tidal courses. Work by the Italian group on the salt marshes of Venetia Lagoon (i.e., Fagherazzi et al., 1999, 2004; Rinaldo et al., 1999, 2004; Marani et al., 2002; D'Alpaos, 2005; D'Alpaos et al., 2010; Coco et al., 2013) have introduced new concepts about their function and evolution.

Furthermore, knowledge of their geomorphologic and sedimentological characteristics is essential to the interpretation of the stratigraphic record. Until the seminal work of Dalrymple et al. (1992), estuaries were seldom found in geological papers, as there were no adequate correlations between present day conditions and what was observed in geological outcrops (Perillo, 1995). Estuaries in general and specifically tidal courses have relatively small area distributions that are difficult to find in stratigraphic outcrops. In many cases, their imprint may be even confused with unidirectional flows if other criteria (i.e., estuarine fossils) are not present.

The actual natural mechanisms that originate the courses are largely unknown (Perillo, 2009) in spite of the numerous modeling efforts (both physical and numerical) that have been attempted in the last 10 years (i.e., D'Alpaos et al., 2010; Vlaswinkel and Cantelli, 2011; Fagherazzi et al., 2012; Vandenbruwaene et al., 2012; Kleinhans et al., 2014; Zhou et al., 2014). Although this seems to be a very simple concept, there is no real agreement as to how and when they initiate, not even if there are only one or multiple processes that develop them. In most cases, course initiation is an underwater process occurring over the intertidal area that is commonly occluded from direct observation by suspended sediments. Even if water transparency were not a problem, there is no way to predict when and where a rill will form, or if the courses will persist in time when they are flooded during the next tide. Nevertheless, there are some cases, where anthropic activities may produce courses either by dredging, ditching, or other mechanical action.

Probably, many of the uncertainties arise from the simplistic although inaccurate idea that tidal courses are somewhat the bidirectional flow counterpart of fluvial streams. The original studies made by pioneers such as Leopold et al. (1964) have specifically compared fluvial and tidal channels trying to apply various classification and statistical criteria employed in the terrestrial counterpart to the marine course networks. Preliminary studies, especially those related to drainage networks (Pethick, 1992; Pye and French, 1993), have considered fluvial terminology to describe tidal course networks although there are clear differences related to hydraulic and geomorphic considerations.

Despite of over 150 years of research in fluvial systems, there are still no adequate explanations for many of their basic problems, including the origin of gullies and rills (Poesen et al., 2003; Tucker et al., 2006; Nichols et al., 2016), which are comparable features to tidal gullies and rills. Still, less is expected from tidal courses when the length of time that they have been studied is only a minimum portion of the previous ones and, for the most part, the field study conditions are much more difficult. In this regard, the difficulties in studying both origin and evolution of tidal courses in physical models are much larger than fluvial ones because of the lack of widespread laboratories that have tidal-emulating facilities, together with the additional problem of resolving the sediment scaling (which is also common for the fluvial studies). This is not a minor issue because the large majority of coastal wetlands are dominated by fine (silt and clays) sediments whose behavior varies depending on the local environmental factors and interaction with the surrounding material. As we will show later on, biological factors play a significant role both affecting sediment transport (Cuadrado et al., 2014) but also in the mechanism for the formation of courses that cannot be reproduced in physical models so far.

On the other hand, tidal depressions are features present in all wetlands, but they have received significant attention only in the early stages of wetland research (i.e., Harshberger, 1916; Yapp et al., 1917; Chapman, 1960; Pestrong, 1965; Redfield, 1972, among others). However, in the last few years, interest for tidal depressions have started again (i.e., Stribling et al., 2007; Wilson et al., 2009, 2010, 2014; Goudie, 2013; Escapa et al., 2015; Linhoss and Underwood, 2016; Revollo Sarmiento et al., 2016) as there is a realization that these features can provide significant insight into the origin and evolution of coastal wetlands. Nevertheless, these authors provided a wide variety of names (i.e., pans, ponds, pools, panne, etc.) that are both confusing and related more with their origin than their geomorphology.

In all cases but one, their description concentrates in vegetated marshes (salt marshes) and, basically, there is no research that considers their presence in unvegetated wetlands (tidal flats).

Therefore, the objective of the chapter is to provide a review of the present knowledge about tidal courses and depression characteristics. We will consider networking and drainage systems to present some ideas and examples of mechanism for origin and evolution of courses and depressions. We also provide a possible classification of both features as a way to simplify the existing confusion based on the indiscriminate use of various common names. For the sake of a better description, we are dividing the chapter into three parts, one dedicated to their geomorphologic classification, second one corresponding to tidal courses, and a third one for tidal depressions. Even though they are analyzed separately, both features are quite related to each other and their mutual influence will be presented throughout the text.

2. CLASSIFICATIONS

On scanning the literature on wetland morphology, there is a remarkable confusion about the variety of names given to the different valleys and depressions observed on wetlands. In the case of courses, names as tidal channels and tidal creeks are very common and used interchangeably even in the same publication. Other widespread terms are gullies, rills, canals, and so on. A similar confusion occurs in the fluvial literature, especially regarding the differentiation between rivers and creeks. In the latter case, the problem appears to be related, usually, to local denominations without consideration of geomorphologic features notably distorted by the variety of definitions that exist in dictionaries and encyclopedias. On the other hand, classifications about depressions are based mostly on genetic issues. As research on these geoforms increases, also the number of processes that originate depressions will increase accordingly; therefore, there is a need to establish a classification based only on geomorphology and/or hydrologic conditions that will avoid future errors in their adequate identification.

2.1 Proposed Tidal Course Classification

In the present context, we propose a basic definition and classification of tidal courses to provide a common descriptive ground based on size and persistence of water on the course during low tide conditions. As a first step, we need to define a tidal course as follows:

Tidal course is any elongated indentation or valley in a wetland either originated by tidal processes or another origin, through which water flows primarily driven by tidal influence.

Tidal course is a general denomination that includes a series of indentations within a wide spectrum of sizes (width and depth) and with at least two levels of inundation (Table 6.1 and Fig. 6.1). The proposed classification, based on size ranges, also provides a descriptive terminology for all tidal courses. Only depth and width of a cross section estimated to the bankfull level are considered because length could have large variations, especially if there are artificial or major geomorphologic constrains (i.e., dikes, cliffs, etc.). Both the values for depth and width are average estimates of the widest and deepest portion of the course.

TABLE 6.1 General Classification and Size Range for Tidal Courses

Name	Water in Low Tide	Depth (cm)	Width (cm)	Cross-Section Area (cm^2)
Tidal rills	NO	<1	<2	<2
Tidal grooves	NO	1–5	2–10	<50
Tidal gullies	NO	5–100	10–100	50–1000
Tidal creeks	YES	10–200	10–200	100–4000
Tidal channels	YES	>100	>200	>2000

Depth is the mean vertical distance from the thalweg to the bankfull border of the indentation. Width is the mean horizontal distance measured across the indentation between the bankfull borders.

FIGURE 6.1 Examples of end members in the tidal course classification. (A) Tidal rills (tr) and tidal grooves (tg), (B) tidal gullies, (C) tidal creeks, and (D) tidal channels. *All photos from the Bahia Blanca Estuary, Argentina by the author.*

Tidal Rills (Fig. 6.1A—tr) are very small superficial indentations developing in the latter ebb stages along the unvegetated, sloping margins of larger courses or marsh fronts. To form, they require a small veneer of fine sediments that is dissected by slow flowing waters. Depending on the surface slope and sediment characteristics (i.e., size, cohesiveness, depth), rills vary in shape from linear to sinuous and may even develop braided conditions (both distributary and convergent). Another common mechanism of rill formation, as it is also frequent on sandy beaches, is due to groundwater discharge resulting in a large number of rills having a variety of shapes. Rills may be considered as the first step in the initiation of a tidal course, and their preservation and further evolution depend on the depth of the indentation, the soil characteristics and, most important, the processes occurring when tides inundate the course margin again. Being surficial and in normally soft, recently deposited sediments, reworking by the tidal currents may obliterate them. If the indentation progresses in depth to more than 1 cm, the rill becomes a tidal groove.

Tidal Grooves (Fig. 6.1A—tg) tend to be 1—5 m long or exceptionally longer courses, 1—5 cm deep and up to 10 cm wide. Normally they are linear to sinuous in sectors. They develop along channel margins and marsh fronts with relatively high (3—7 degrees) slopes due to strong ebb flows or groundwater outflows. Tidal grooves are more prone to resist the following tidal inundation as they produce a much deeper indentation in the soil. It is common to observe parallel groves along the margin of typical tidal channels. The concentration of the flow in some of those grooves may produce the formation of larger courses such as gullies or creeks.

Tidal gullies (Fig. 6.1B) are similar to those observed in continental drylands or those formed at the top of cliffs. They are deep indentations that may reach up to 1 m and much wider than rills and grooves. Gullies are preserved and enhanced by tidal inundation. They could easily superimpose in size with tidal creeks, the difference lies in the lack of tidal water during low tides although some flow may be observed, normally due to rainwater retained in the flats or marshes, or groundwater outflows. In all last three cases, the larger relative relief is found near the head of the course becoming shallower as it approaches the mouth. Their course is also linear to sinuous and seldom they develop meanders, but, if they do, those meanders appear at their mouth.

The major dynamic structures in wetlands are tidal creeks and channels. Probably the main difference with the other courses is the permanent inundation by tidally driven water in at least part of the course even during the lowermost tides. *Tidal Creeks* (Fig. 6.1C) range in size from few tens of cm in depth to up to 2 m (in macrotidal wetlands depths may be larger), whereas width has the same values. They normally have water during low tide; however, water depth during this time varies from practically none at the head to about 10—30% of bankfull depth at the mouth. Creeks, as well as the previous courses, are the tributaries of the tidal channels and form at their banks but they are distributed over tidal flats and marshes/mangroves as well. Rills and grooves never reach the level of the tidal flats and marshes, and gullies seldom do.

On the other hand, *Tidal Channels* (Fig. 6.1D) are the largest features in most wetlands as they clearly stand out in maps and satellite images. Channels always have water along their whole course even during the lowest spring tides. Their depth is greater than 2 m and maximum values are highly dependent on the characteristics of the wetland, but on average they have as much as 10 m, reaching in many cases up to 30—40 m deep as in the case of coastal lagoons inlets. Regarding channel width, the degree of variability is much larger than in the case of depth. Tidal channel widths start at about 2 m and may reach several kilometers (i.e., Bahía Blanca and Ord estuaries).

2.2 Proposed Tidal Depression Classification

As is also observed for the tidal courses, there is a wide variety of names and descriptions regarding tidal depressions on coastal wetlands. They are called indistinctly as ponds, pans, pannes, pools, hollow, to name only a few detected among the literature reviewed. Our aim is to provide a simple definition and simple classification of these features based on their geomorphologic and hydrologic characteristics rather than their origin, as it was common in the literature. We then define them as:

> Tidal depression is any surface concavity in a wetland either originated by tidal processes or another origin, which evolution depends on the geomorphologic and biologic characteristics and tidal-influenced dynamic processes acting on the wetland.

Tidal depression is a general denomination that includes a series of concavities that, for the most part, have been described for salt marshes (i.e., Harshberger, 1916; Yapp et al., 1917; Chapman, 1960; Pestrong, 1965, among others) and more recently by Collins et al. (1987); Adamowicz and Roman (2005); Perillo and Iribarne (2003a); Wilson et al. (2014); Escapa et al. (2015); Revollo Sarmiento et al. (2016), among others. Although they are common in tidal flats (Revollo Sarmiento et al., 2016) and carbonate platforms (i.e., Shinn et al., 1969; Rankey and Morgan, 2002; Rankey and Berkeley, 2012), depressions in them are seldom discussed in detail. Adam (1997), on the other hand, indicates that they are absent in many temperate salt marshes in Australia. Furthermore, there are other depressions that basically have not been studied so far. They are for the most cases subtidal and are described here as Tidal Scour Holes.

One of the problems in reaching a convenient classification of tidal depressions based on their geomorphologic features derives from the fact that they have different shapes ranging from circular to elongate including sinusoidal ones. Some depressions are isolated, but they can also have creeks attached. All these variables imply a significant complexity at the time to identify and separate them from pure creeks and other features on the surface employing automatic methods of classification from remote sensing sources (Cipolletti et al., 2012, 2014; Revollo Sarmiento et al., 2016).

Within the wide spectrum of sizes (area and depth) and with at least two levels of inundation (Fig. 6.2), the proposed classification, based on size ranges, also provides a descriptive terminology for all tidal depressions. Only depth and area are considered because shape could suffer large variations.

Tidal Pans (Fig. 6.2A) are shallow intertidal depressions ($\leq$5 cm), normally with small areas ($\leq$2 m^2), which tend to have water for shorter periods after tidal inundation. In cases of pans in high marshes, they have water due to precipitation or derived from groundwater only.

Tidal Ponds (Fig. 6.2B) are intertidal depressions with areas between 2 and 20 m^2 and maximum depth up to 50 cm. They preserve water during the whole tidal cycle except in regions with high evaporation.

Tidal Pools (Fig. 6.2C) are the largest intertidal depressions deeper than 50 cm and with areas larger than 20 m^2, being always inundated.

Tidal Scour Holes (Fig. 6.2D) are depressions commonly formed at the intersection of tidal courses with a relative relief nearly about twice as much as the depth of the tributary courses. They can be either intertidal or subtidal.

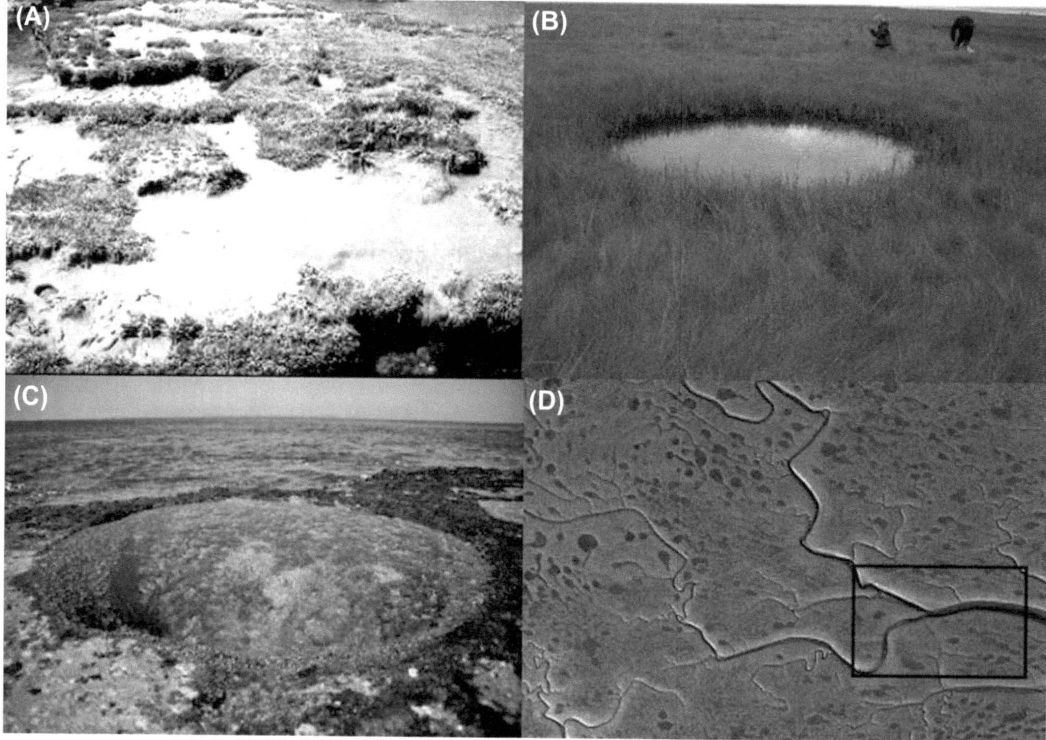

FIGURE 6.2 Examples of end members in the tidal depression classification. (A) Tidal pan (Loyola Bay, Argentina) (B) tidal pond (Minas Basin, Canada), (C) tidal pool at Kaena Point State Park, Oahu, Hawaii, (B) Tidal scour hole at the intersection of two tidal creeks (within the *rectangle*). *(A, B and D) Photos by the author. (C) Photo kindly provided by Mr Joel Metlen.*

All these types of depressions occur both in tidal flats and marshes/mangroves; therefore, there is no direct correlation between the type of wetland and depression. Some differences may occur about the type of sediment involved. Pools, for instance, are common in carbonate platform, where they tend to be deeper, but rare in sandy flats; whereas pans, ponds, and holes are observed in any ecosystem.

3. TIDAL COURSES

3.1 Geomorphology of Tidal Courses

At first sight, tidal courses resemble concentrated fluvial networks having also similar course shapes both in plan view and cross section. However, when analyzed in further detail, numerous differences appear that set them apart. The most obvious ones are the flow characteristics (roughly unidirectional vs. bidirectional), relative relief, the degree of course inundation, evolution pattern, among-course interactions, and so on. Although there are wetlands developed in sandy substrates, most of them are dominated by fine sediments; therefore,

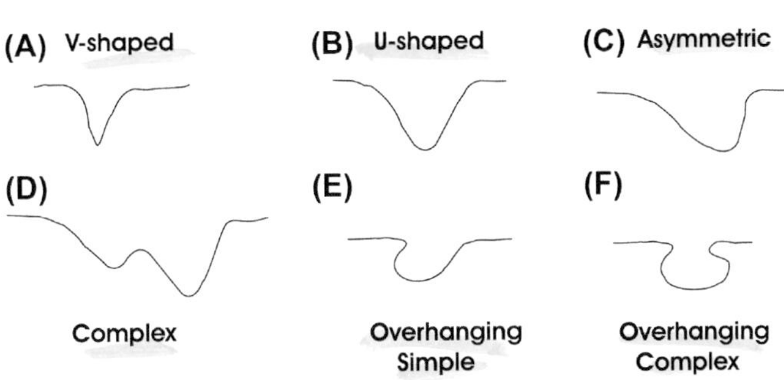

FIGURE 6.3 Description of the various possible cross sections of tidal courses. V- (A) and U-shaped (B) normally correspond to longitudinal portions of courses. Asymmetric (C) and Complex (D) are found commonly along meanders, whereas overhanging (E and F) represents differential erosion processes controlled by variations in sediment characteristics, water level in the course, and/or differential material strength as well as to marsh plants.

braided patterns are rare, and most valleys are single courses with roughly linear, sinuous or meandering patterns.

Various authors have partially analyzed the cross-section shape of courses in wetlands and tidal inlets (e.g., Pestrong, 1965; Vincent, 1976; Collins et al., 1987; French and Stoddart, 1992; Allen, 2000; Fagherazzi et al., 2004). The cross-section form depends on wetland stratigraphy, tidal range, their associated plant cover, and sinuosity (Allen, 2000). They can be classified as V-shaped (Fig. 6.3A), U-shaped (Fig. 6.3B), asymmetric (Fig. 6.3C), complex (Fig. 6.3D), or with one or both banks overhanging significantly (Fig. 6.3E and F). Beyond a general description of the shapes, there are no actual studies dedicated to relating cross-section forms and the course conditions, evolution, and size. There is, nevertheless, one example of a mathematical description of tidal inlet cross sections using empirical orthogonal functions (Vincent, 1976).

Both V- and U-shaped (Fig. 6.3A and B) are common in small courses and even in the linear portion of channels and creeks, whereas meanders have asymmetric cross sections. Many linear channels also show asymmetries (Fig. 6.3C), especially if they are subject to lateral migration (Ginsberg and Perillo, 2004). These types of cross sections appear during the initiation of the course when the linear pattern is still present and most of the flow strength is used to deepen and lengthen the course.

Complex cross sections (Fig. 6.3D) appear when bars are present along the course. Kjerfve (1978) described them as bimodal with two channels separated by a shallow area or bar having contrasting residual circulation associated with differential depths (flood and ebb dominance). However, there is still no clear indication whether the differential circulation is due to the bathymetry or the geomorphology is a consequence of the change in direction of the residual circulation often found on vertically homogeneous estuaries (Dyer, 1998) as there are no reports describing their formation process.

Overhanging banks (Fig. 6.3E and F) are the result of differential sedimentary characteristics as they are the remnant of flow undercutting within the course by tidal currents. The overhanging portion, due to higher sediment compaction, resistance, or sustained by plant roots, has been less affected by the currents and preserved while a lower portion of the bank is undercut. Preservation of overhanging banks is poor as they are intrinsically unstable features that will eventually collapse, and the falling material is then reworked and transported by the tidal currents.

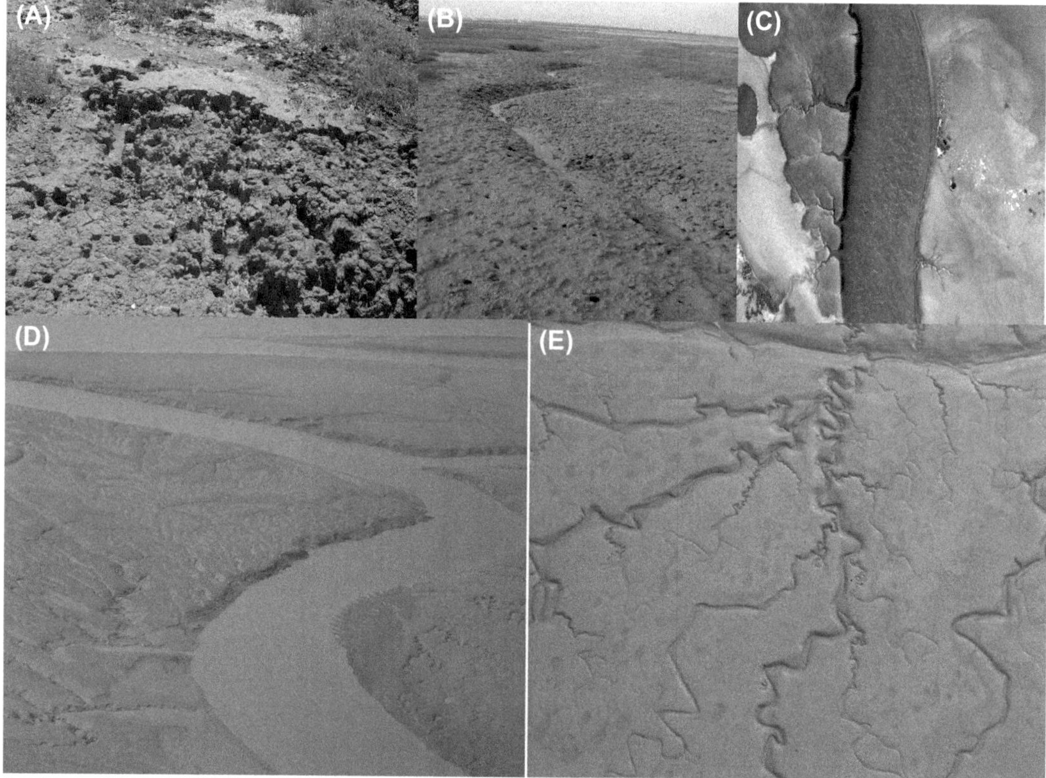

FIGURE 6.4 Various examples of tidal course characteristics. (A) complex structure of a gully head, (B) view of a tidal creek morphology, (C) marked differences in tributaries along a tidal channel, (D) typical meander in a tidal channel at low tide showing the formation of a point bar in the inner flank of the curve and the cliff on the outer flank, (E) variability of the meander pattern over a tidal flat showing that longer curvature radius occur on the flatter portion of the flat. *All photos from the Bahia Blanca Estuary by the author. (C) Photo is a part of an Ikonos high resolution image.*

In most cases, longitudinal profiles of rills, grooves, and gullies tend to be concave at the close-ended head becoming shallower and convex for a larger part of their course (Fig. 6.1A). At the head of these courses, there is fast erosion by headward retreat, which induces a relatively high relief represented by a microcliff (Minkoff, 2007). Ebbing water cascades at the microcliff, resulting in a deeper thalweg at the head. In areas where sediment is compact, the head, flanks, and bottom of the course at and near the head are marked by dislodged sediment clasts, flakes, crumbs and plant patches, rests of infauna burrows, and so on, yielding a very irregular and even chaotic morphology (Fig. 6.4A). Relative relief and irregularities in the bottom and flanks diminish toward the mouth because the tidal flow tends to smooth them out by transporting and wearing out sediment blocks (Fig. 6.4B). Recently deposited sediments that have not achieved a certain level of compaction, preservation of small courses becomes difficult as tidal currents along creeks and channels resuspend and transport them and, as the courses are transversal to the flow, the potential to be filled up or eroded is high. In these cases, the flanks and bottom are smooth.

Tidal creeks and channels, when established, have relatively smooth flanks and bottoms, worn smooth by currents, and irregularities appear related to erosional processes at the outer bend of meanders, infauna activities, rotational slumps, and deformed microdeltas deposited at the mouth of tributaries. The presence of rills, grooves, and gullies (and creeks in channels) disturbs the smoothness of the flanks. Longitudinal profiles show a progressive higher relative relief toward the mouth. Tributary inlets often show sediment deposits that range in shape from typical deltas to extended banks parallel to the main channel. These deposits result from the sudden drop of the suspended and bedload material transported during the ebb, especially at or near low water slack. Depending on the strength of the currents, these deposits are either carried away or deformed, normally following the channel current dominance. Most examples found at the Bahía Blanca Estuary (Fig. 6.4C) and several others along the Argentina coast show a continuity of the inlet along the main channel flank bordered by a parallel shoal that, in most cases, is ebb-directed. Potentially, the displacement of the mouth induces an initial meander that may affect the circulation within the tributary. Although there are no studies that can assert it, this is clearly an instability on the course plan morphology that could displace headward, resulting in a meandering pattern developed inversely to accepted theories in fluvial systems (Leopold et al., 1964).

Courses in marshlands have an additional mechanism to deepen their channels: levee accretion. According to van der Wegen et al. (2012) levees in tidal environments are less developed than in fluvial counterparts and may be recognized because they emerge earlier than the surrounding flats as the tide recedes due to their higher elevation and coarser material. Restricted by surrounding vegetation, sediment deposition along the sides of the course increases the relative relief between channel bottom and marsh surface (Wells et al., 1990). This process also inhibits water exchange, resulting in water and sediment becoming trapped at high water and during surges. Both in marshes and mangroves, vegetation plays an important role in sediment trapping and stabilizing the wetland (Wolanski, 2007). Lateral and headward erosion activities also widen and lengthen the channel (Chapman, 1960). The type of wetland vegetation and the vertical distribution of plant species have significant influence on the degree and rate of stream channel migration and meandering (Garofalo, 1980), for instance, denser, well-rooted vegetation is more difficult to erode. Also, as water and sediment move from the channel to the platform, the typically denser vegetation located at the edge reduces flow velocity and allows higher sedimentation. Because of the increase in nutrient loading, plants at the edge tend to be taller resulting in more flow reduction and deposition, originating the levee. Similarly, these edge plants reduce the flow back from the platform to the channel, so sediment has less chance to be exported. The cyclicity of this process maintains itself as long as the equilibrium persists. If for some reason sediment input into the system is reduced, there is a tendency to the erosion of the levee as the plants no longer receive nutrients. Temmerman et al. (2007) summarized the influence of vegetation demonstrating that static vegetation patches reduce erosion, but dynamic vegetation patches (variable vegetation size and density) tend to reduce channel width, increasing current velocities, and, consequently, lateral erosion.

Although meandering flows have been extensively studied, especially in fluvial environments (Leopold and Wolman, 1960; Ikeda et al., 1981, among others), an adequate theory that fully explains meanders is still lacking. The leading process in meander development in fluvial rivers is the redistribution of momentum due to channel curvature (Ikeda et al.,

1981; Parker et al., 2011). Flow momentum concentrates along the outer bank originating its erosion. Secondary transversal flows driven by a superelevation transport the sediment, which deposits on the inside bank resulting in the development of point bars. Furthermore, alternate bars can lead to the formation of meander bends at initial stages of meander development (Blondeaux and Seminara, 1985; Seminara and Tubino, 1992). Although, Hood (2010) maintains that meander formation is predominantly depositional and is associated with delta progradation.

Beyond the specific mechanism of meander formation, the main hypothesis why a course meanders is based on the concept that meanders elongate the water path thus enabling it to contain more of the water discharge within the same valley distance. Although this is true for rivers and other contained flows, it is somewhat difficult to transfer this concept to tidal environments where the water is contained within the course only for a part of the time, even though it is the period with the highest velocities, whereas a large percentage of the flow spreads out over the contiguous wetland as the bankfull stage is overcome. The only correlation could be that once the water is concentrated in the course during the ebb, the total water volume may be larger than the actual volume that the course can discharge, thus deriving in a situation similar to that found in rivers.

Tidal meanders have a cross-sectional morphology similar to fluvial meanders. The outer bank tends toward a steeper flank, sometimes a cliff develops (Figs. 6.3C and 6.4D) and, eventually point bars evolve on the inner bank. Flow circulation on the meander appears, in general, similar to the circulation in river counterparts, although this is a simplification that cannot go beyond the cases in which the meander is symmetric. The fact that in coastal wetlands tides control circulation implying that the flow along courses is bidirectional. Both the bottom topography and overflow are causes for common flow asymmetry, which, eventually, could be enhanced by any river discharge. Therefore, flow around a bend in a tidal course is seldom symmetric and its consequence is the typical morphology observed in bends.

Meander shape in coastal wetlands, as also occurs in fluvial systems, is controlled by the sediment properties and erosion–sedimentation history of the environment. Wetlands that developed over older fluvial terraces or eroded deltas have layers of sediment of variable consistencies outcropping and, therefore, they can control the direction and planform shape of the whole course but most important that of the meanders. That is the case of both the Bahía Blanca Estuary and Anegada Bay (Argentina), both being part of the late Pleistocene–Early Holocene Colorado River Delta (Perillo and Piccolo, 1999; Melo et al., 2003, 2013), and it is common to find meanders with a "straight loop" (Fig. 6.4E) following the orientation of the original delta sets. In other cases, meanders also initiate by rotational slumps or "erosion cusps" (Ginsberg and Perillo, 1990).

Symmetrical currents erode different banks resulting in meanders that are narrower at the bend apices and wider between bends (Ahnert, 1960). Bend skewing, a normal feature in fluvial meanders due to their downflow asymmetry, produces "gooseneck loops" (Carson and La Pointe, 1983; Fagherazzi et al., 2004). However, because of the flow asymmetry, there are distinct planform morphologies depending on whether the course is ebb or flood dominated (Fagherazzi et al., 2004). As in rivers, wetland meanders also have point bars, but ebb-dominated tidal currents and the radius of curvature control their shape (Barwis, 1978).

The theoretical concept that equilibrium courses should have an exponential increase in their cross section from head to mouth could only apply to the large tidal channels and creeks. Model approaches have observed that this may be only acceptable for "equilibrium courses" (when sediment input and output is balanced) (Seminara et al., 2009). The degree of strong meandering paths in most small courses shows that equilibrium cannot be easily achieved in coastal wetlands. In these environments, the dynamical processes change constantly, flow seldom stays concentrated within the course and flow discharge vary, even with the same tidal amplitude, by a simple change in wind direction or speed. Sediment balance is rare, if ever occurs. Especially higher-order tidal channels, which conduct most of the tidal prism, may approach this equilibrium idea. Fagherazzi and Furbish (2001) presented a model for young marsh environments where sedimentation is active demonstrating, in contrast with terrestrial rivers, that salt marsh creeks experience a strong increase of width–depth ratio seaward due to the short duration time of the peak discharge. Course widening results from the imbalance between erosion versus deposition, then course cross section has little time to adjust. Furthermore, autoconsolidation of the bottom cohesive sediments and the consequent vertical gradients in resistance properties obstruct the formation of deep courses, yielding the formation of shallow wide courses (Fagherazzi and Furbish, 2001).

On the other extreme, course heads may have the same variability in shape as those observed in fluvial systems. Dietrich and Dunne (1993) defined course heads as the upslope boundary of concentrated water flow and sediment transport between definable banks. However, we follow the previous definition of course head given by Montgomery and Dietrich (1988) as the upstream limit of observable erosion and concentration of flow within definable banks. Minkoff and Perillo (2002) and Minkoff (2007) differentiated three types of course heads: (1) diffuse (with an undefined form, generally convergent with the surrounding land, Fig. 6.5A); (2) acute (ending in a point with a width < 15 cm, Fig. 6.5B), and (3) open (wide, mostly round shaped with a width ≥ 15 cm, Fig. 6.5C). In each of the head types, different processes act. In diffuse heads, the predominant process is the surface erosion of the sediment, whereas other factors as groundwater or headward erosion are of little consequence. In the case of acute heads, both subsurficial and headward retreat associated with water cascading are the most important factors, whereas subsurficial erosion and mass wasting are the major factors in the formation of open heads.

Studies of the time and spatial evolution of creek and gully heads indicate that their shapes varied as the courses were growing (Minkoff, 2007). For this particular case, the effects of the burrowing crab *Neohelice* (formerly *Chasgmanatus*) *granulata* play an important role. Landward advance of the heads is most significant (about 2–3 times faster) during the warmer (late spring, summer, and early autumn) than on colder months because of the high crab activity and population increase (Escapa et al., 2007). For instance, diffuse heads do not grow directly at the head, but there is subsurficial erosion (mostly due to a combination of crab burrowing and groundwater circulation) some distance from the border and, then, a sudden connection with the thalweg occurs by a collapsing of the surface. In all cases, Minkoff (2007) has demonstrated, after surveying more than 130 creek heads over a period of 4 years, that creek and gully landward growth is a pulsating process. There is a period of preparation after which the course retreats a certain distance in one single movement.

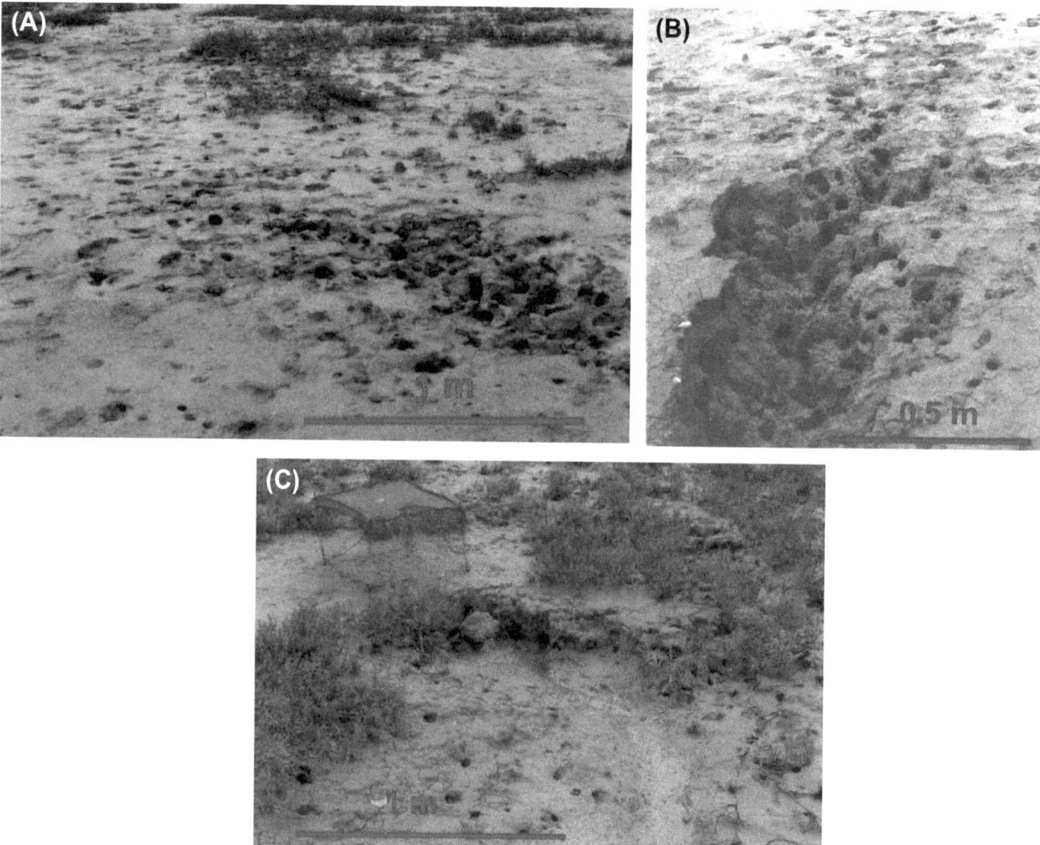

FIGURE 6.5 Examples of the three types of gully heads described by Minkoff (2007). (A) diffuse, (B) acute, (C) open.

Minkoff (2007) seems to be the first detailed analysis of the retreat processes in coastal wetlands, except for the specific study made by May (2002) in which she related headward creek retreat on the level of mass wasting of the terrace between the head and the frontal marsh area incised by the creek. She divided it into strong terrace wasting, weak terrace wasting (WTW), and no wasting (NW). The erosion rate observed on the terrace level defines the wasting. The former occurs in sectors where there are thick deposits of organic rich soil (at least 10 cm in depth). The transition zone of these creeks had a steep slope and a large drop in elevation of the marsh surface (>15 cm). The transition zone presented various erosional features such as large holes (at least 5 cm in diameter) in the sediment surface that were coalescing, as well as undercutting and slumping of sod. Those of WTW appear in areas with thin deposits of organic rich soil or overlying root mat with a smooth slope. The NW sectors have few erosive features. The biological induced head retreat can be complemented by other physical and chemical/mineralogic processes such as strong rainfall, freeze/thaw, groundwater erosion, or changes in the chemical composition of the mineral due to successive inundation of the sediments (Chen et al., 2011).

3.2 Course Networks and Drainage Systems

At first approach, drainage patterns on coastal wetlands resemble fluvial networks. But once they are analyzed in more detail, the first difference that appears is that the number of branches and channel hierarchy (in the Hortonian sense) is closer to mountain basins rather than lowlands to which coastal wetlands could be related. This factor by itself indicates that the network is draining large amounts of water in short time as opposed to flatland networks. As also occurs with their fluvial counterpart, for marshes (Ashley and Zeff, 1988) and tidal flats, courses at the head are close-ended channels. Close-ended courses are a mechanism to show local processes that control the network.

The second most striking difference is the lack of topographically defined basins. Although some geomorphologic control (likely developed after network inception) may exist, hydrodynamic processes control basin boundaries. Marani et al. (2003) found that the drainage density of a network, which Horton (1945) defined as proportional to the ratio of the basin's total channelized length divided by the watershed area, must, in fact, be defined by the statistical distribution and correlation structure of the lengths of unchanneled pathways. By modeling the current directions on the wetland surface, basin boundaries can be drawn and network structure defined. This situation arises when water on the wetland moves as overmarsh flow most of the time rather than in channelized flow.

Flow directions under these conditions may be strongly affected by winds especially in areas where wind is a major factor, for instance, by having dominant directions or by storm surges. Also, a wetland having different levels of inundation during the regular spring–neap cycle would be subject to different behavior for each sector resulting in modifications of the network. Therefore, drainage patterns at a high wetland could be rather different from that at the lower wetland.

Novakowski et al. (2004), on the other hand, have geomorphologically defined a marsh island as a section of marsh circumscribed by tidal channels that are deeper than 1 m at low tide. Although this definition is somewhat artificial, it provides an objective way for a first level estimation of hydraulic conditions in a wetland. Marsh course watershed was estimated by connecting end points of first-order courses.

Network density could also be affected by anthropic actions such as ditching. Ditched marshes (a common practice in Europe and North America in the late 19th and early 20th centuries) have both less course density and pool density (Lathrop et al., 2000; Adamowicz and Roman, 2005). Density of depressions is an important factor in course initiation and development because depression interconnection may lead to course formation. Furthermore, some basin control may occur due to morphological/artificial changes such as dikes or raised continental features. In some cases, the hydraulic control is based on the fact that tides, even during storm surges, cannot inundate the backmarsh. Changes in vegetation patterns can often be a good indicator of hydraulic patterns (Pethick, 1992).

Fig. 6.6 provides a series of examples of the various network patterns observed in coastal wetlands. A variety of linear, sinuous, rectangular, parallel, and dendritic patterns are common to all wetlands. Its diversity, unless artificially affected (Fig. 6.6E), is still a matter of discussion because there are no clear theories that can explain why a particular network pattern is present. Evidently, sediment characteristics, previous sediment history and wetland slope

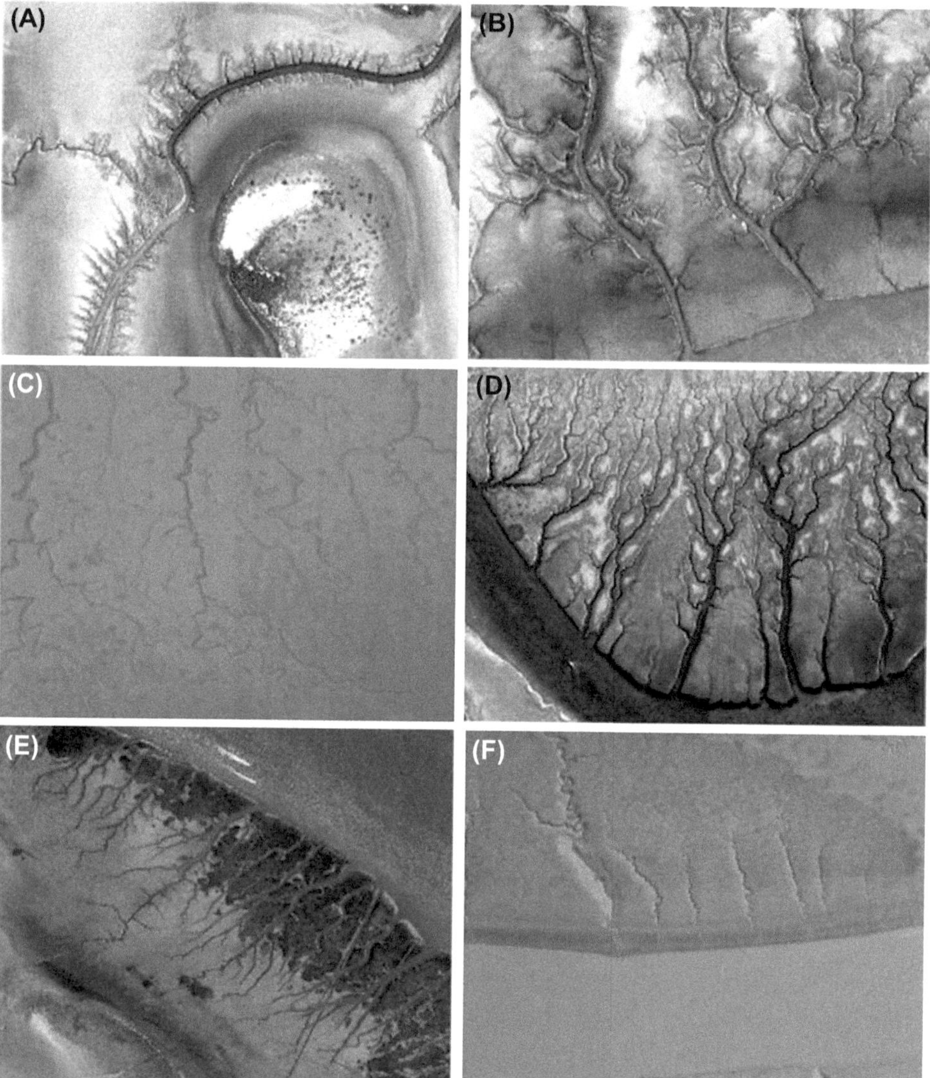

FIGURE 6.6 Examples of drainage patterns on a coastal wetland. (A) Rectangular, (B) linear dendritic, (C) sinuous, (D) sinuous dendritic, (E) parallel sinuous with some examples of ditching (straight creeks), and (F) linear parallel. *All photos from the Bahia Blanca Estuary. (A, B and D) Photos are a part of a Ikonos high resolution image. (E) Photo is a Google Earth image. All others are from the author.*

all play a role, but to what extent each of them is actually responsible for a particular pattern is presently unknown.

The case of Caleta Brightman (southernmost channel of BBE) (Fig. 6.6E) is an example of the degree of control that plants have in the development of tidal course networks. The main channel is bordered in most of its extension of relatively new developed *Spartina* marshes, which are from few hundred meters to 1 km wide, from there to the continent there are muddy tidal flats about 1–2 km wide. Channels and creeks crossing the marsh are mostly linear or mildly sinuous with very few, parallel tributaries connected at right angles with the courses. As these courses are followed into the unvegetated areas, the network suddenly develops in a well-defined dendritic pattern with sinuous to meandering courses. The number of tributaries increases at least threefold, bearing in mind that the flats and marshes are basically at the same level. Vandenbruwaene et al. (2012) obtained similar results when they found that the probability to have larger unchanneled flow lengths is much smaller in the vegetated tidal marsh than in a tidal flat.

Different from fluvial networks, the presence of a large number of nearly parallel runnels and ridges (Fagherazzi and Mariotti, 2012), as observed in Figs. 6.6F and 6.7A, characterize tidal environments. They are the initial stage in the formation of gullies and creeks. The high slope of the channel banks helps the erosional process although it does not explain the periodicity of the runnels. Several ideas were proposed and reviewed by

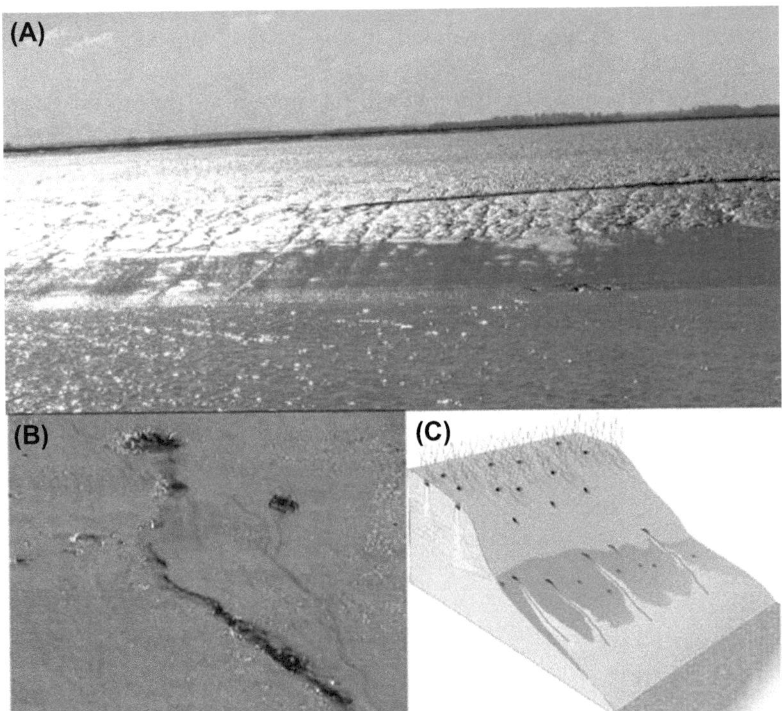

FIGURE 6.7 Rills and grooves at the flank of tidal channels. (A) Parallel rills and grooves showing periodic tendency; (B) groove formed by groundwater spillover from a crab burrow; (C) graphic model of formation of grooves from crab burrows *After Perillo, G.M.E., Minkoff, D.R., Piccolo, M.C., 2005. Novel mechanism of stream formation in coastal wetlands by crab–fish–groundwater interaction. Geo-Marine Letters 25, 124–220.*

Fagherazzi and Mariotti (2012), but none of them are conclusive. Nevertheless, they were able to measure tidal flows and suspended sediment concentrations in larger gullies concluding that they drive the higher currents and concentration of suspended sediments being at maximum during the ebb part of the flow.

The drainage network in coastal wetlands drives, below the bankfull level, a large percentage of the water controlled by the tides. Normally, because of the slight but natural channelward inclination, one expects that creek overflow during the flood will occur first at the mouth propagating inland. However, although the most common, this is not always the case as it is shown in one example in Fig. 6.7. In some sectors, particularly in the flats of the Bahia Blanca Estuary, overflow occurs at the head of the creek and inundation moves from the head to the mouth (Perillo, personal observation).

This condition also points to a geomorphologic characteristic of the tidal networks versus the fluvial ones, and that has not been discussed elsewhere: course interconnection. As can be observed in the various examples provided here (Figs. 6.6B,D,E and 6.8A), creeks and channels do interconnect with others from the same network or neighbor ones. To the best of my knowledge, there is no previous discussion about the processes giving rise to these connections, but they can be hypothesized to be the result of capturing by a headward retreat of one course into another adjacent. However, it can be further speculated that the lack of well-defined topographic basin boundaries controlling the circulation should be considered as a significant player in helping capture by adding more discharge to the capturing course.

In this sense, wind and different tidal stages at the mouth of the main channels may drive water parcels in different directions than the ones expected. When these conditions translate to major channels, interesting situations may arise in the circulation pattern. As an example occurring along the El Embudo Channel (Bahía Blanca Estuary, Angeles et al., 2004), its circulation changes direction up to four times during the same tidal cycle depending on the tidal stage at either of its mouths.

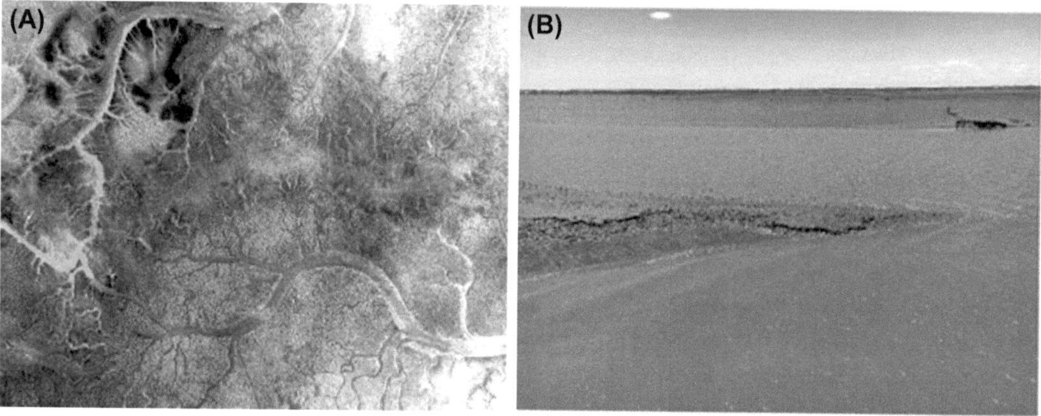

FIGURE 6.8 (A) Preliminary inundation process on the tidal flats in the Bahia Blanca Estuary. The Ikonos high-resolution image was taken 2 h after low tide showing that flooding of the tidal flat occurs first at the head of the creeks. Also the area shows that the creek heads are interconnected. (B) Example of the formation of a band of high suspended sediments along the channel flank due to the action of short, wind waves acting on it.

3.3 Origin of Tidal Courses

Courses are an integral part of the formation of a tidal flat. As sediment is being deposited water needs to move farther seaward, and courses follow depressions and irregularities of the surface topography. Therefore, the network must also be part of the original establishment of the tidal flat. This does not mean that the original network is sustained through time as the flat continues evolving. Most likely many of those original courses disappear to give place to the courses that form during a more mature stage, which in turn develop the observed drainage systems. Stefanon et al. (2010) and Vlaswinkel and Cantelli (2011) suggest that the network growth as the ebbing tides concentration produces an increase in bed shear stress, deepens the course. The deeper the incision, the higher flow concentration is observed.

After over 200 years of fluvial studies, a detailed description of course initiation is still lacking (Montgomery and Dietrich, 1988, 1992; Istanbulluoglu et al., 2002; Kirkbya et al., 2003; Tucker et al., 2006). Obviously, there are many problems not only in reaching consensus within the wetland community (Chapman, 1960; Pethick, 1969) but also in the well-documented fluvial literature. Based on Chapman (1960), Perillo et al. (1996) described the possible mechanism of tidal course formation as follows: during the ebb tide, sheet flows are guided by the topography, concentrating their discharge into depressions. These depressions become small courses by headward erosion. As the water velocity increases in the course during successive ebb tides due to the tendency to concentrate the retreating flow (Pestrong, 1965), these depressions become wider and deeper. Headward erosion is an important factor in further developing the course. As the course grows in size, meanders form, which is the final pattern a channel acquires when fully developed. Headward erosion lengthens the course in combination with the local annexation of nearby courses.

The main incognita still remains as for why a course form at the place it does and not somewhere else. For instance, a probabilistic approach was proposed associated with the flank slope and the erodability conditions of the sediment for continental situations (Istanbulluoglu et al., 2002). Following the original Horton (1945) idea that course inception occurs at a certain distance downslope of an incline, many rills and grooves appear at the flanks of tidal creeks and channels as well as the front edge of marshes during low tide. Horton's idea is that overland flows may require a certain distance to achieve enough bottom shear stress to overcome soil resistance. In other words, an erosion threshold controls the location of channel heads establishing that the same critical distance below a topographic divide is required for a sheet flow to erode sediment as for it to initiate a course. According to Horton's theory, such erosion features may expand rapidly upslope in response to changing climate and land use conditions and can even form during individual storm events.

Although this is true for thin overflow layers, typical of continental processes, in coastal wetlands during ebb, the water depth along the flanks is at maximum and diminishes toward low tide. Nevertheless, at the initial stage of the ebbing phase of the tide, most of the flow concentrates along the main course and the flow cascades from the surrounding flats or marshes as it is diverted toward the main course, and could not reach enough momentum to produce any erosion. Besides the experiment made by Fagherazzi and Mariotti (2012), there are no direct measurements of the flow dynamics very close to the surface along a channel margin during a tidal cycle; therefore, the following analysis is to be considered only as a hypothesis to be demonstrated. It is necessary to clarify that the study made by

Fagherazzi and Mariotti (2012) does not directly apply to the very thin layer (viscous sublayer) of the flow that we are referring to here.

Course inception on channel flanks is produced only at the final stages of sheet flow as it moves downslope. Likely this sheet is at that moment only a few centimeters deep; therefore, the flow is both supercritical and mostly turbulent as well. Supercritical flows induce stationary waves that act to dam the flow until enough depth is reached as to reduce the Froude number at or below the critical value. Then the waves disappear and the flow accelerates which, in itself, can be large enough to generate a bottom shear stress greater than the critical value for cohesive sediment erosion. As in most wetlands, this process occurs over fine sediments, which, as they are eroded, pass directly into suspension and thus increasing the flow density, which further augments the eroding capacity of the flow. In the waters along channel borders, at low tide, one can commonly observe a 10-cm to 1–2-m wide (depending on the channel breadth) band of water with high-suspended sediment concentration produced by this downslope erosion and wave action (Fig. 6.8B).

The incipient rills and grooves have a remarkable constant spacing of the order of 1–10 m (Figs. 6.1A and 6.7A,B), which may depend on the slope of the segment of the tidal flat where they develop (Perillo et al., 2005). Potentially, all the grooves may develop further into gullies, creeks, and channels, but only a few of them upgrade to the next level. However, Horton's general idea does not explain the presence of multiple, almost parallel rills and grooves formed over a smooth muddy surface. Evidently, there must be other factors that induce course inception at specific points on the surface. The presence of the periodic courses may be because of (1) periodic surface irregularities (i.e., along channel undulating surface); (2) presence of pebbles, infauna burrows, or mounds; (3) nonuniform sediment (i.e., size, degree of compaction, mineralogy, bioturbation, etc.); (4) presence of vascular plants or roots; and (5) variability of the layer of microphytobenthos and their excretions (i.e., EPS) (Cuadrado et al., 2014). Any or a sum of these items could be responsible for routing the flow, inducing hydraulic jumps.

Even though all those conditions can be found individually or in groups acting together helping in the development of grooves, there is an important issue along channel banks often overlooked: groundwater seepage. Most studies of groundwater exchange between the flats or marshes have been concentrated on the biogeochemical aspects (i.e., Findlay, 1995; Osgood, 2000; Hollins et al., 2000; Kelly and Moran, 2002; Duval and Hill, 2006). Gardner (2005) and Gardner and Wilson (2006) modeled the seepage from channel banks indicating that this process is a major mechanism for water and nutrient exchange between the course and the wetland. The process is driven mostly by the tide but affected by rain, plant roots, and infauna burrows, resulting in complex and highly transient variations in boundary conditions along creek banks and the marsh platform (Gardner, 2005), and the alternating development and disappearance of water tables, seepage faces, and zones of saturated and unsaturated flow. Tidal rills are one of the main consequences of seepage (Fig. 6.1A) as they mostly develop from the groundwater discharge through the saturated zone on channel and creek banks. Rill initiation associated with groundwater discharge, as also occurs on sandy beaches or drylands, is due to sediment dislodging at the inception point likely produced by the concentration of percolines resulting in an acceleration of the seepage at specific points (Perillo et al., 2005, Fig. 6.7).

In the case of rills, the longitudinal profile shows a maximum depth at the head, but the course soon becomes shallower and only a minimum channelization is observed, being at minimum at the water edge. The situation changes if several rills discharging groundwater converge into a single course, then course incision could be larger although discharge seldom is large enough to overcome a convex profile at the water edge.

Grooves are formed in channel margins of the Bahía Blanca Estuary (Figs. 6.1A and 6.7B) by groundwater discharge but enhanced by the percoline concentration due to crab burrows (Perillo et al., 2005). As groundwater fills the burrows, spills out over the burrow border, often producing an indentation on the sill that helps concentrate the flow and generate the groove. As these crabs are an important food source for the white croaker fish, occasionally the mouth of the burrow is enlarged (producing a crater) as the fish hits the surface when it traps a crab resulting in a greater groundwater concentration and larger spillover and, consequently, wider and deeper grooves. In this case, groundwater comes from the tidal filling of a dense concentration of crab burrows at the associated marsh platform.

Although the various methods described are probably the basic mechanisms for the initial development of courses on the lower tidal-flats (hydroperiod $= 360$ days year^{-1}), where tidal penetration occurs frequently, it is probably not the dominant mechanism of channel formation within salt marshes or mangroves at higher elevations. Higher (and older) areas may have reached an elevation that limits tidal penetration to only the highest tides (hydroperiod < 120 days year^{-1}). As a consequence, sheet flows are limited.

For the case of courses in marshes, it has been thought that they were inherited (and further modified) from the older tidal flats that they colonized. Some authors (e.g., Pestrong, 1972; Pethick, 1969, 1992; Pye, 1992) even suggested that channels were obliterated, rather than created in marshes. Yapp et al. (1917) describe the possible mechanism for the origin of courses in marshes based on the presence of hummocks formed by sediment retention due to the colonization of *Glyceria* (*Puccinellia*). Although Yapp et al. (1917) offered a valid explanation, it does not account for large gullies, creeks, and channels. Perillo et al. (1996) proposed the first recorded mechanism for creek formation on a middle salt marsh, which is located in southern Argentina. Creeks develop from the interconnection of series of ponds enlarged by wave erosion of the pond walls as a consequence of the intense, direction-concentrated winds prevailing on the Patagonia. Precreeks are finally connected to existing channels by cascading and seaward erosion as tides flood over the ponds. Further growth and maintenance of the creek depend on the intensity of the water exchange. If this exchange is poor, the reduction of soil salinity allows plant colonization, which further reduces the exchange, enhancing sedimentation, and finally attaining the complete obliteration of the creek.

A similar mechanism for gully and creek initiation was described for a freshwater marsh on the southern coast of the Río de la Plata Estuary (Perillo and Iribarne, 2003b). In this case, the ponds are much smaller (Fig. 6.9A and B) formed by soil marsh depression both by organic matter compaction and groundwater washing of interlayered sands. The interconnection between ponds and creek formation (Fig. 6.9C) is produced by wall erosion by waves, water cascading, and seaward retreat. Once the ponds are connected among themselves and with the estuary, water flows through the creek driven by the microtide dominant in the estuary and also by storm surges and their related waves frequent in the area.

Although there is no mention in the literature of processes similar to those described in the previous paragraphs, there are examples observed by the author in other marshes, i.e., Minas

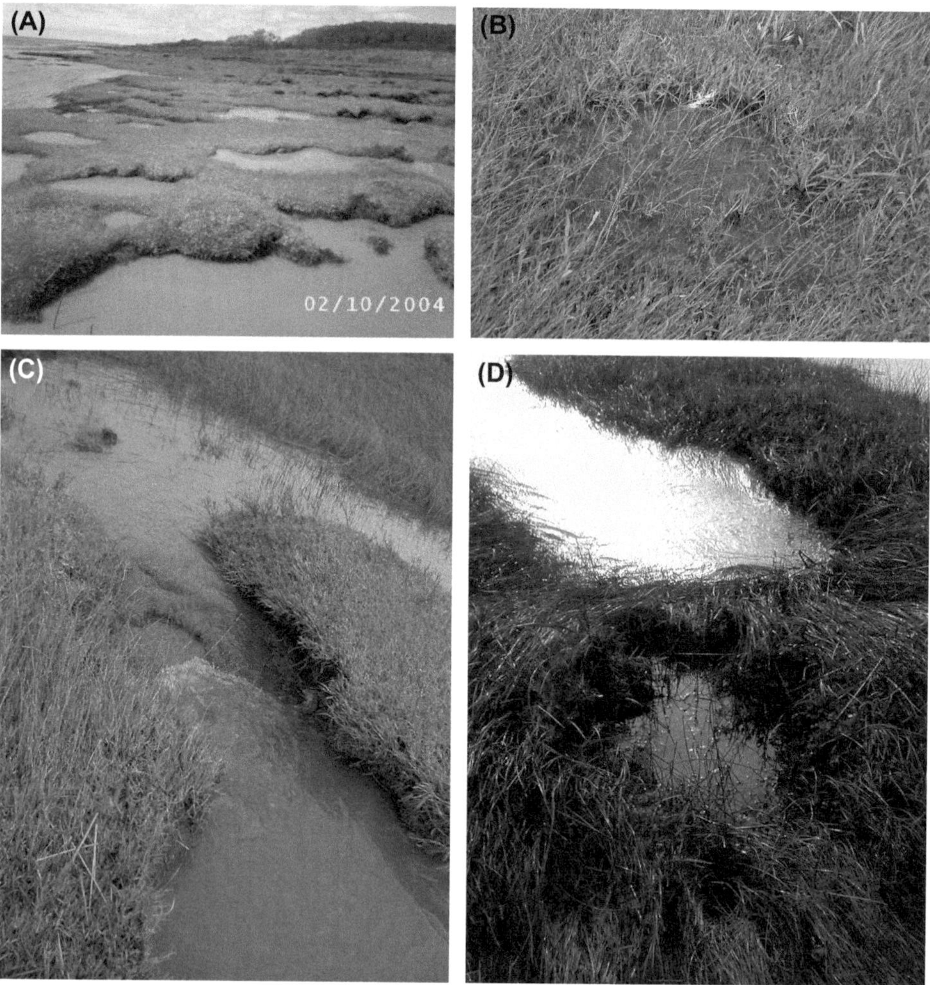

FIGURE 6.9 Formation of tidal creeks in a freshwater marsh. (A) View of the field of ponds along the southern coast of the Río de la Plata, (B) view of a pond showing the variety of plants inside and outside the pond, and (C) a threshold between two ponds. Cutting of the threshold interconnects the ponds as a step for the formation of a creek, (D) creek developed after the cutting of the threshold.

Basin (Fig. 6.10) that lead to the formation of creeks by pond connection. This idea is exactly the contrary to the development of channel pans proposed by Pethick (1974, 1992) where changes in the marsh conditions (i.e., lowering of the sea level, marsh disconnection, etc.) result in a smaller hydroperiod followed by vegetation growth and lateral erosion that may isolate parts of the course, especially those with meanders that become ponds.

A fourth mechanism of creek and gully formation was found and extensively studied because of the interesting physical–biological interaction observed in the Bahía Blanca Estuary (Perillo and Iribarne, 2003a,b). One of the most surprising and interesting findings is that creek formation can actually be a product of the intense action of crabs (*N. granulata*). In these

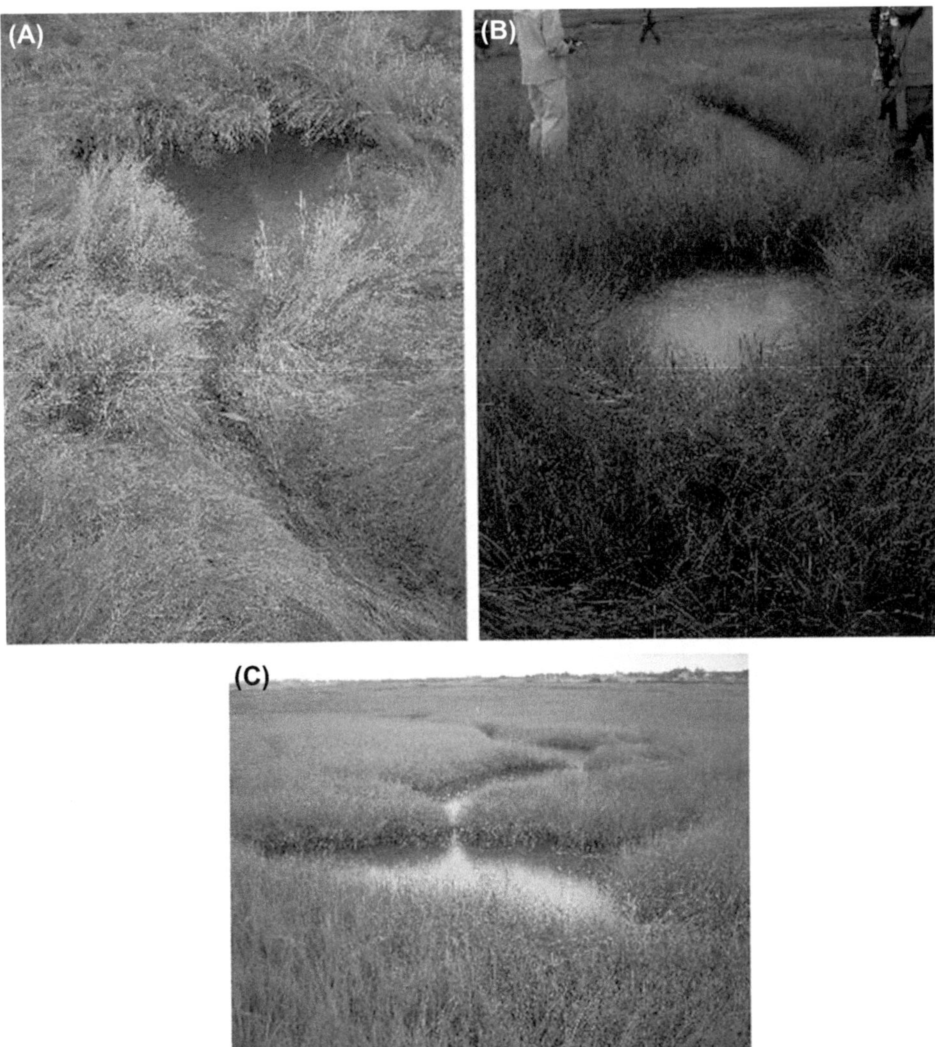

FIGURE 6.10 Example of the steps in creek formation observed at a Minas Basin (Canada) marsh. (A) Pond with a connected gully, (B) two independent ponds separated by a threshold, (C) creek formed by the interconnection of ponds.

settings, crabs first interact with a halophytic plant (*Sarcocornia perennis*) developing zones of high density of crab holes, which then are utilized by groundwater and tidal action to form tidal courses. When analyzing the spatial distribution of *Sarcocornia*, Perillo and Iribarne (2003a) discovered that the plants were mostly distributed in circles of up to 1.5–8 m in diameter, with the plants concentrated in a ring along the outer portion of the circle (Fig. 6.11A). These rings vary in width from 0.5 to 1.5 m. The central part of the circle is an unvegetated salt pan, which is densely excavated by the burrowing crab. Burrow density reaches

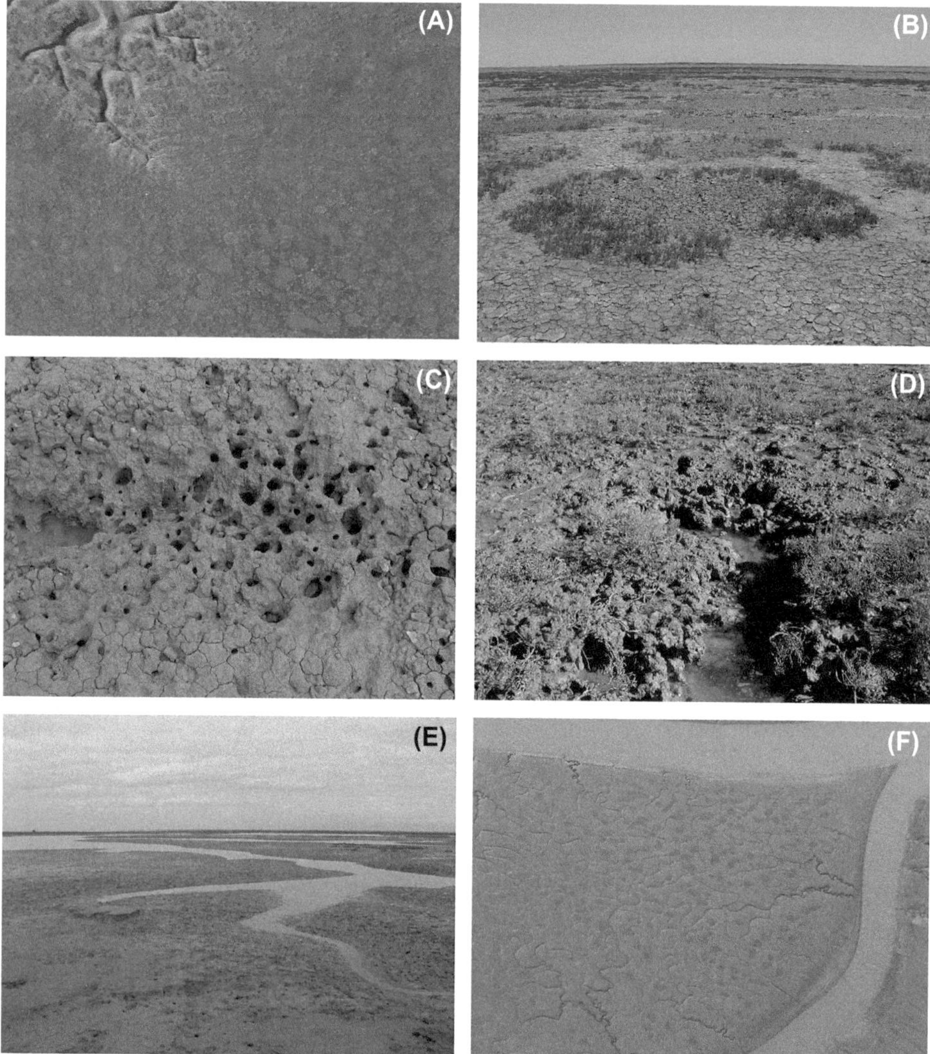

FIGURE 6.11 (A) Aerial view of the *Sarcocornia* rings, (B) an example of a *Sarcocornia* ring showing the dense distribution of crab burrows at the center, (C) head of a creek before the collapse of the surface soil where the large number of burrows mark the area that will be collapsed, (D) head of a creek after the collapse, (E) creek partially flooded, and (F) aerial view of the distribution of small ponds over a tidal flat of the Bahía Blanca Estuary.

40–60 holes per m^2 (Fig. 6.11B). The holes made by the crabs have a diameter of up to 12 cm (Fig. 6.11C), and they reach up to 70–100 cm into the sediment (Iribarne et al., 1997; Bortolus and Iribarne, 1999). An interesting feature of the plant rings and their interaction with the crabs is the effect on the erosive process of the salt marsh in an estuary, which is generally in an erosional stage. The formation of the *Sarcocornia* rings and the crab activity play a major role in the erosion of the marsh transferring 1380 m^3 of sediment from 270 ha marsh to the estuary

(Minkoff et al., 2005, 2006; Escapa et al., 2007). In 2004 alone, more than 13% (183 m^3) of the total sediment was exported indicating that the process keeps exporting more sediment every year in a pronounced exponential curve.

The fact that the crab holes are permanently inundated produces two effects. First, silty clay sediments (characteristic of this area) become partly loose and second, the water in the holes starts to migrate, breaking intercave walls and developing a groundwater stream that undermines the soil. Under the pan surface, caves develop where groundwater flows. A further stage in the development of the channel appears when soil surface fails. At this stage, the course has only a general structure filled with remnants of intercave walls (Fig. 6.11C and D), around which water circulates (Perillo and Iribarne, 2003b). The final stage is reached when the walls are eroded after a number of tidal inundations and the channel presents smooth, low-slope banks (Fig. 6.11D). Crabs can be found along the creek banks but no plants are observed. As a result, along the creeks, the original tidal flat moves landward at the expense of the salt marsh. This mechanism was described in the previous section. Application of a Cellular Automata model (Minkoff et al., 2006) that takes into account the various active interactions existing in the marsh predicted the changes in the geomorphology of the terrain and the formation and evolution of gullies and creeks. The effects of infauna activities on the growth of tidal creeks were also observed recently by Wilson et al. (2010, 2012) in US eastern coast marshes deeper near their heads than at their mouths.

Human intervention on coastal wetlands has been intensive in many sites of Europe and North America. One such intervention is ditching of channels to dry the wetlands, for instance, for mosquito control (Redfield, 1972) or salt hay farming (Adamowicz and Roman, 2005). On the other hand, courses can be developed by using vehicles (i.e., tractors) to transport goods and people along the wetlands (i.e., Jiangsu coast, China, Shu Gao, personal comm.). Either direct action by ditching or the effects of tires or people walking, as also occur in some places of the Bahía Blanca Estuary, disturb the surface of the wetland and provide the necessary conditions to initiate a course. In areas with high population density such as China, Bangladesh, or India, wetlands are employed often producing strong modifications, which may induce courses or depressions. On the other hand, in high latitude areas, ice scouring produces some form of ditching that could also initiate courses (Martini et al., this book).

3.4 Course Evolution

The morphology of large rivers depends on a great number of factors (i.e., tectonics, rock/sediment composition, slope, etc.). As size diminishes, local morphology plays an increasingly larger role in the valley and, specifically, in river channels. Tidal courses have similar influences. For instance, rills are controlled by surface sediment characteristics; the less compacted the material the deeper the rill. Also, slope is important because rills are more lineal with higher slopes.

In an evolving wetland, courses appear practically simultaneously with the deposition of sediments that forms a tidal flat (Wolanski, 2007). Rills, grooves, and gullies are the first to appear and their persistence is dependent mostly on local factors such as some particular irregularity on the sediment surface or the depth of the incision. A deeper incision tends to occur on spring tides as low water level induces more erosion than neap low water. Once a course has been preserved through several tidal cycles, the possibility to be erased by strong currents or wave activity diminishes proportionally with time. Then, further evolution is now controlled by ebb and flood currents, the former being particularly important as they convey the water discharge from the over-flat flooding.

Although tidal flats can be preserved, it is common that soil stability helps in establishing pioneer plants that, in a longer run, allow the formation of salt marshes and/or mangroves. Levees tend to form along courses where sedimentation is predominant even in the case of tidal flats, being most common on marsh courses. However, they seldom occur along courses in erosive environments or when the concentration of suspended sediments is very low (<100 mg L^{-1}). The formation of levees further stabilizes the courses and allows for their vertical growth. When they form, overflow onto the adjacent flat requires higher tides, which becomes a restriction in the dynamics and exchange between the course and the flatland behind the levee.

Levees are dissected by creeks and gullies but also by strong rain events and short-period, highly erosive waves. These processes help restore the exchange. Therefore, when considered as a whole, levees cannot grow indefinitely but must reach some kind of dynamic equilibrium based on tidal range, overflow velocities, suspended sediment concentration, plant species (involving all plant morphological characteristics affecting flow and sediment stability), and local geomorphology.

Further inland, courses are also elongated by the interaction with freshwater. Fine sediments react differently in freshwater than in saltwater. In the former, mud tends to remain in suspension longer than in saltwater conditions, thus allowing better conditions for sediment erosion (Wolanski, personal communication). Several examples in Australia (Mulrennan and Woodroffe, 1998; Cobb et al., 2007) show that freshwater affects mud deposited from saline water and, during continental floods, erosion of the salt marshes and flats is larger than during regular tidal inundation. Freshwater is not only introduced into the system by continental drainage but also in the form of rain which, when intense, also produces strong surface sediment erosion (Mwamba and Torres, 2002), which, when conducted by the geomorphology, can easily erode the sediment surface further and generate or prolong tidal courses. Then, further evolution of these features becomes a tidal process.

4. TIDAL DEPRESSIONS

Tidal depressions are seldom discussed in the analysis of the geomorphology of coastal wetlands even though they are common features. In some cases, depressions play a role in the concentration of benthic fauna (Molina, 2012). They were first described by Harshberger (1916) in salt marshes of United States. One year later, Yapp et al. (1917) proposed the first mechanisms for the formation of pans; later on, the term ponds was suggested by Shinn et al. (1969). In all these cases, but also all the authors that have analyzed these features (i.e., Yapp, 1922; Chapman, 1960; Pestrong, 1965; Redfield, 1972; Perillo et al., 1996; Adamowicz and Roman, 2005) so far concentrate their studies in salt marshes. Except for the recent study by Revollo Sarmiento et al. (2016), there is no previous work made in depressions in tidal flats except for a short comment by Yapp et al. (1917). I was not able to find any description in mangroves. Although seldom analyzed in detail, there are mentions of tidal depressions in carbonate platform in the Bahamas (Jarecki and Walkey, 2006; Rankey and Berkeley, 2012).

Even though depressions occur worldwide in present day wetlands, they are also rarely discussed in the geological analysis of ancient wetlands. One single example is the work of Suarez Gonzalez et al. (2015), although their study corresponds to a Cretaceous carbonate wetland. Wilson et al. (2010) also point out that there are very few analyses of the sediment structure found in the depressions; therefore, without present day analogues, it becomes difficult to discover them in the stratigraphic record.

4.1 Geomorphology of Tidal Depressions

Reviewing how different authors have described the geomorphology of tidal depression, seldom were they able to escape from relating them to their origin. Inclusive, some of the original geomorphologic descriptions (i.e., channel pans, secondary pan, etc.) have an integral part devoted to the initiation mechanism. In the present review, our classification was intended to move beyond this way because the form of the depression, although it can have some relationship with its origin, is not conclusive. This means that depressions with different origins and evolutions can have similar forms.

Although courses are elongated features observed on all coastal wetlands that may extend for several meters to many kilometers, depressions tend to cover a single surface with a generally elliptical shape. As indicated, they are concavities on the surface, which seldom are deeper than 75 cm. Therefore, pans and ponds are the most common depressions found. There is a wide variety of shapes (Fig. 6.12) as a result of the depending variables such as tidal range, geomorphology, and sedimentology of the wetland, waves, wind, biology (both plants and animals), and climate. In the present section, we only discuss the geomorphologic characteristics of these features without inferences as to their origin, which will be discussed in the next section.

Shapes range from circular (Fig. 6.12A and E) to elliptical (Fig. 6.12B and F) and complex (Fig. 6.12C and D). The complex forms can be subdivided into linked (Fig. 6.12C) and merged forms (Fig. 6.12D). The former can be assimilated to the channel pans proposed by Yapp et al. (1917). This simple classification is based on their general shape. When analyzed in detail, they may also have irregular borders. A typical characteristic of the ponds observed in the tidal flats of the Bahia Blanca Estuary (Figs. 6.11F and 6.12E,F) is the presence of a short groove that is seldom connected to another pond or creek. Because of their combination, historically, they were locally called "espermatoliths" for their similarity to spermatozoid cells. A worldwide search, using Google Earth, of ponds in tidal flats made by Revollo Sarmiento et al. (2016) looking for other fields was to test their classification methodology that indicated that these features do not appear elsewhere, at least on the numbers observed there.

A profile across a typical depression shows a smooth deepening toward the central part (Fig. 6.13A) similar to what one expects for a typical frying pan. In some cases, a deeper hole may develop due to a differential sinking of bottom sediments or subsurface erosion produced by groundwater flows or piping (Kesel and Smith, 2008). Asymmetrical shapes (Fig. 6.13C) are also common when some predominant dynamic conditions such as specific wind directions produce short erosive waves that erode one side more than the other of

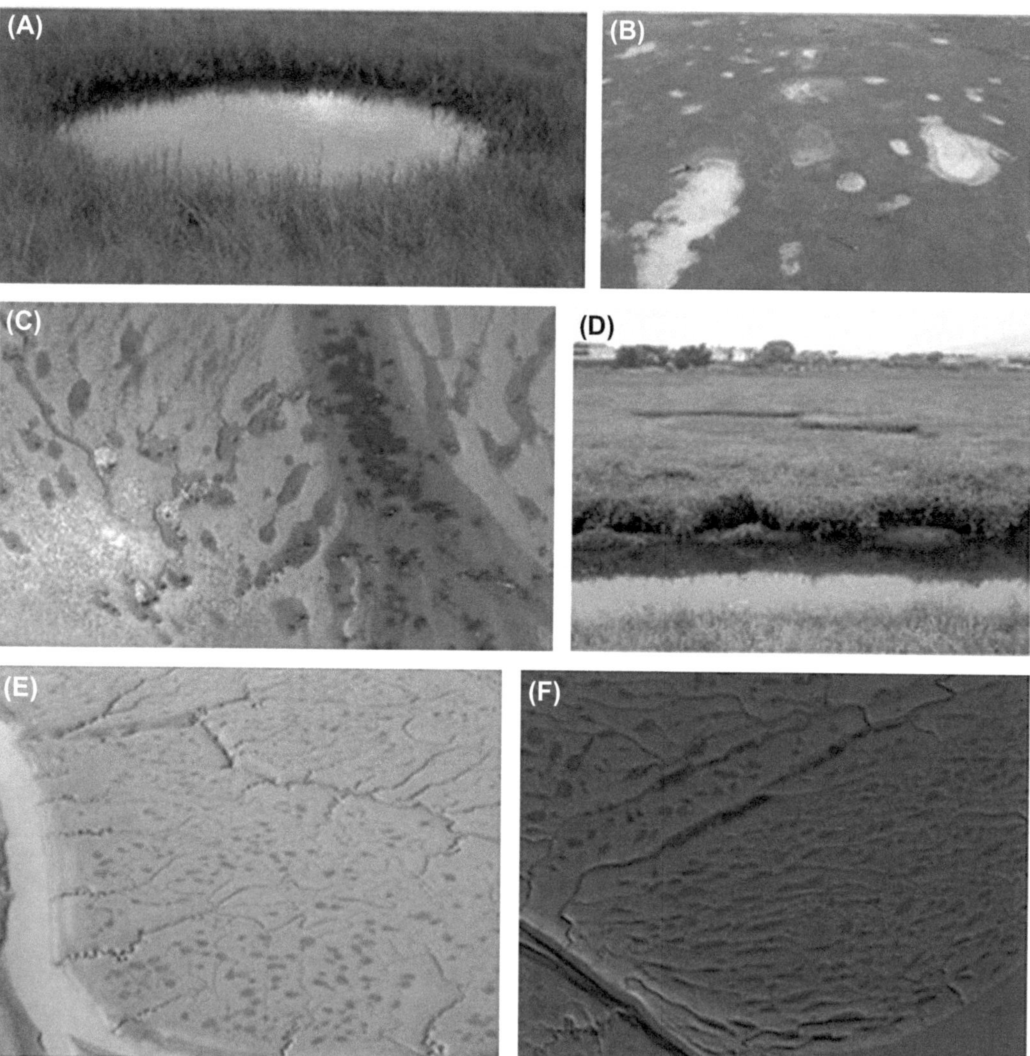

FIGURE 6.12 Common forms of tidal depressions found in tidal flats and marshes. (A) Circular (Minas Basin, Canada); (B) elliptical (San Blas, Argentina); (C) complex linked (Valdivia River, Chile); (D) complex merged (marsh at Santa Barbara, CA, USA); (E and F) fields of depressions circular, elongated, and channel pan types. Both images from Bahia Blanca Estuary, Argentina. *(A) Photo courtesy of Dr. Diana Cuadrado. (C) Photo courtesy of Mr. Mario Manzano.*

the depression inducing even its expansion and/or drift. These types of bottoms are more common in pans and shallower ponds. Other depressions show relatively flat bottoms with abrupt borders marked by a microcliff (Fig. 6.13C). Although possible, they are seldom asymmetrical.

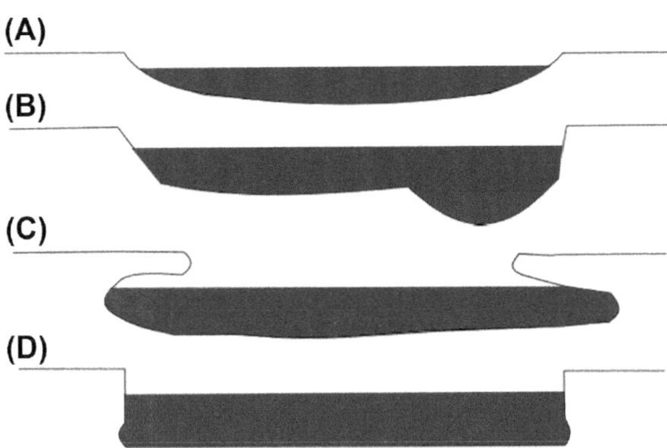

FIGURE 6.13 Some common examples of profiles across depressions as observed by the author. (A) Typical structure of a tidal pan or pool with some water retention; (B) a depression that includes a deeper hole; (C) depression with significant undercut along the border; (D) pool with cliffy borders that may or may not have notches. Vertical scale may vary in each case because we wanted to show form rather than dimensions. *Diagram based on field observations by the author.*

Regarding the sedimentology of the depressions, they normally do not differ from the material of the surrounding wetland. In areas where there is significant reworking of the interpan areas, water with high concentration of suspended sediments may get trapped in the depression and settle during low tide. Because they tend to be shallow, even the finest particles can have time to settle. Different is the case when the depression is flooded with water with high concentrations of organic matter (i.e., microalgae, marsh litter). Decay of vegetal material accumulated at the bottom of the depression represents a whole difference in sediment characteristics and marks a diverse way in the further evolution of the geoform.

In carbonate platforms sedimentation in depressions can be rather different from those observed in siliciclastic wetlands (Fig. 6.2C). For instance, Rankey and Berkeley (2012) describe pond sediments as yellow-brown peloidal mud with scattered foraminifera and gastropods. However, there are no descriptions of how a tidal depression can be filled by sediments. Suarez Gonzalez (personal communication) was able to detect ponds in their stratigraphic record based on a parallel deposition produced by the progressive infilling of the water bodies by black limestones and sandy black limestones (Suarez Gonzalez et al., 2015).

Depressions appear as the wetland emerges from the tidal inundation. Unless there is a rapid drainage, water remains in the depression. For instance, Revollo Sarmiento et al. (2016) used one of the segmentation mechanisms in detecting depressions in satellite images and the presence/absence of water in them. Depending on the level of infiltration and evaporation, water can be reduced or even totally eliminated from the depression. However, the latter seldom occur in a low or middle wetland (defined as a function of their hydroperiod) because they are rapidly covered by the following tide. Dry depressions are more common in high wetlands (those flooded few times a year) when local meteorological and hydrological conditions prevail, but also the depth of the depression: deeper ponds or pools have the chance to retain water for longer time.

The presence of water in the depression reduces the stability of the bottom sediments but also, because of the evaporation, may increase the salinity. Therefore, even if wetland plants are able to withstand significant changes in the salinity, the salinity values can be as large as 60–90, which are beyond the comfort zone for most plants. When the depression becomes totally dry, soil salinity (Fig. 6.2A) also prevents plant colonization (Perillo et al., 1996). However, in depressions located in freshwater marshes the presence of plants inside the pond is not uncommon (Fig. 6.9B); in some cases these plants are from different species than those surrounding the pond. Nevertheless, plants play an important role in the formation and evolution of the depression, as discussed in Sections 4.2. and 4.3.

Although depressions in a wetland may be an isolated feature (Fig. 6.12A and D), they appear more prevalent in what we call a "field of depressions" (Fig. 6.12B,C,E and F). Depression distribution has been a matter of concern since their earliest descriptions. Redfield (1972) described the pond distribution in a marsh in United Kingdom, where he observed that the higher density corresponded to the newest higher marsh diminishing toward the older sector of the same marsh; furthermore, he showed that ponds tend to disappear closer to the bordering land. The relationship between pond density and formation, which was also discussed by Pethick (1974) indicating that the original theory by Yapp et al. (1917) that depression density should be larger the younger the wetland cannot be proved totally at present.

On the other hand, Goudie (2013) made a study of depression distribution covering 67 marshes along the UK coast. She provides two clear conclusions: (1) depression density appears to have an inverse relationship with sea level change; (2) higher densities are found in meso and macrotidal marshes, with deltaic marshes having the lower densities. Moreover, Goudie (2013) also found an inverse relationship with course density. Adamowicz and Roman (2005) made similar observations for the case of ditched and unditched marshes from the NE, USA. Although they found that sizes were similar in both cases, ditched marshes have up to three times the depression density. Contrary to the results of Goudie (2013), Adamowicz and Roman (2005) indicated that there is no correlation with tidal range, but a weak correlation with latitude.

Another interesting but often-overlooked feature of tidal channels are the scour holes that form at the junction of two courses. Studies of scour holes have concentrated in fluvial junctions where their morphological features are related to the angle of the junction, tributary/main river discharge ratio, sediment erodability, etc. (Best, 1987). Scour holes in tidal environments (Figs. 6.2D and 6.14) were only mentioned by Shao (1977), Kjerfve et al. (1979), Ginsberg and Perillo (1999), Pierini et al. (2005), and Ginsberg et al. (2009). The latter results demonstrate that scour holes, having large relative relief in excess of 2 m and up to 17 m, at tidal environments differ significantly in morphology from those found in fluvial systems. Those observed in the Bahía Blanca Estuary, for instance, although elongated in shape as in the case of fluvial holes, they have the steeper side (3.5 degrees) at the mouth of the confluent channel and the gentler side (1.5 degrees) seaward. This structure is exactly the opposite to scour holes found in river environments. Based on current measurements and sediment transport estimations, Ginsberg and Perillo (1999) established flood dominance on the steep, inner face and ebb dominance over the gentler, outer flank (Fig. 6.5A). Another difference from holes in fluvial reaches is that tidal scour holes migrate headward. Scour holes that are not related to tidal channel junctions have been described for tidal inlets

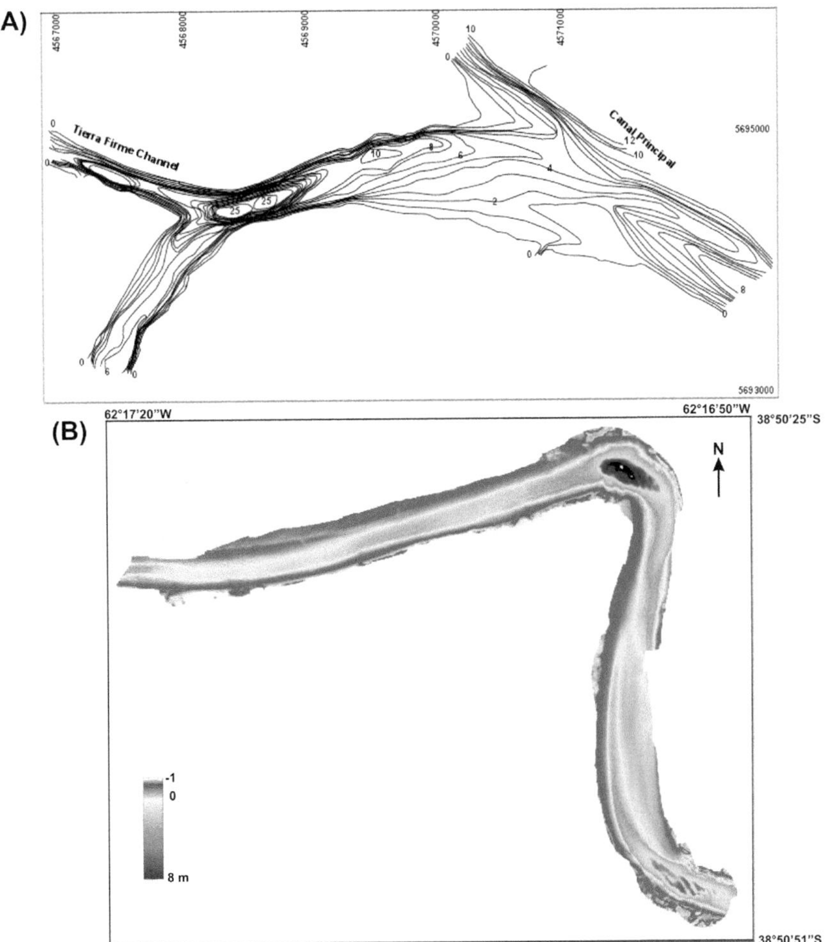

FIGURE 6.14 Examples of scour holes in tidal channels. (A) 26-m deep, scour hole at the intersection of the Tres Brazas and Tierra Firme channels; (B) scour hole at the center of a 90 degrees meander (unpublished data from the author). *(A) Modified from Ginsberg, S.S., Perillo, G.M.E., 2004. Characteristics of tidal channels in a mesotidal estuary of Argentina. Journal of Coastal Research 20, 489–497. Both examples are from the Bahía Blanca Estuary, Argentina.*

and, more recently, within specific channels (i.e., Shaw et al., 2012). In all these cases, the scouring process is due to flow acceleration product of a geomorphological constriction of the channel that increases bottom erosion. Scour holes may also appear at the center portion of very sharp (90 degrees) meanders as in the case of some of this type of meanders found in the Bahía Blanca Estuary (Fig. 6.14B). Although not studied in detail yet, we just found the presence of scour holes at the intersection of tidal creeks and gullies at intertidal flats and marshes (Fig. 6.2D). These depressions, by all accounts, have similar characteristics than their subtidal counterparts.

4.2 Origin of Tidal Depressions

Considering the previous descriptions, it becomes obvious that many depressions are an integral part of the geomorphology of the wetland (most likely a tidal flat) during its inception. As sediment was deposited, even the differential compaction results in uneven surface where the lower parts are more prone to retain water during low tide. As the flat evolves, depressions (as well as courses) develop further changing shapes and locations, converging or diverging and even disappearing altogether. Plant colonization marks a significant change in the wetland dynamics. Thus, surviving flat depressions must evolve as a function of the plant–geomorphology–dynamic interplay. Some depressions disappear, but others will be found adapted to the new conditions.

The present section concentrates on the mechanisms that give birth to depressions leaving their evolution to the next one. Some review of the original theories will be presented but directed to establish a modern view in which, as it will become clear, there is no single mechanism but a diversity that may result in similar geomorphologies.

Initial studies of depressions are concentrated in salt marshes of Europe (mostly United Kingdom) and North America where they are the dominant wetland. Tidal flats normally are reduced to a relatively narrow band along the tidal channels, which, as the wetland elevates trying to follow the relative sea level increase, is rapidly colonized by the plants. Therefore, it is no surprise that most of the pioneer work on the origin of depressions (Harshberger, 1916; Yapp et al., 1917; Pestrong, 1965, among others) comes from their analysis of salt marshes. Just a mere hint in Yapp et al. (1917) suggesting that the primary marsh (we understand that they were referring to the flat) relief is "... subjected to a continual struggle between the agents of erosion and retrogression on the one hand, and those of deposition and consolidation on the other." This means that depressions may result from the uneven relief of the flat surface.

Based on the suggested mechanisms for the formation of depressions proposed by the various authors on marshes (i.e., Harshberger, 1916; Yapp et al., 1917), depressions could be differentiated into primary and secondary. The former are those in which some conditions prevented the invasion of plants, whereas the latter are depressions that originated by processes such as plant die off, current or wave erosion, course segmentation, or formation of levees.

We consider that a better criterion to analyze their origin is based on the mechanism rather than if they were originally from the marsh or modified later on. Assuming that this approach is correct, we may divide the mechanism as (1) dynamical, (2) geomorphological, and (3) biological.

Dynamical processes refer to the action of waves and currents that rework the surface of the flat modifying the relief. Although sediment deposition may generate depressions by developing a levee, which dam water, most commonly, they produce the contrary effect of erosion that gives origin to a depression. Once a small pot hole appears on the surface, eddies may form that help enlarge the depression; normally enhancing the circular shape. In windy regions, locally generated waves, which tend to have large wave steepness, are highly erosive attacking the sides of the depression enlarging it but generating a more elongated feature as suggested by Perillo et al. (1996). In marshes, the dynamic mechanism is significantly affected by the presence of plants, which reduce the energy of the waves and currents. Nevertheless, this energy dissipation helps sediment deposition at the depression–plant interface developing a levee, which, in itself, is another way to deepen the depression and augmenting its chance of survival.

The classical "channel pans" proposed by Yapp et al. (1917) are the typical example of a geomorphological process. Undercut lateral blocks of sediment falls into the creek that is thus blocked, resulting in a series of elongated depressions that could follow the original creek extension. Undercutting course banks shows a particular characteristic of the vertical sediment structure and the influence of plants (when present). The upper levels of the sediment are more resistant, either by the sediment itself or the combination with the plants, than the level (normally without plants) under attack by the flow, allowing for the necessary formation of a significant hanging shelf, enough to block the course. In cases where the sediment is more homogeneous or plants are not protecting the upper levels, there is no course blockage even though sediment may be eroded in chunks as described by Perillo and Iribarne (2003a,b).

Other geomorphologic processes are related to levee formation, local relief conditions, and sediment subduction. When there is no levee breach along a course, after some extraordinary tides or rainfall, levees may act as dams allowing the formation of ponds even if there are no significant depressions, which could develop later by wave and current erosion. On the other hand, local relief may call for a lateral expansion of a groove or gully on the surface of tidal flat forming ponds as those observed in Fig. 6.11C and F. Sediment subduction due to underground erosion (i.e., water piping, groundwater erosion, etc.) is another geomorphic condition. For instance, many pools in carbonate platform are developed on sinkholes (Fig. 6.2C) as the platform surface falls into a hole excavated underneath. One last process was proposed by Dionne (1968) as the marsh can be eroded by ice blocks. The author describes the ice blocks in the St. Lawrence Estuary taking pieces of the marsh where they were established by spring tides leaving depressions of up to 25–35 cm deep and over 15 m in diameter. Martini et al. (this volume) provides another mechanism as marsh material frozen to the underside of the ice floes may be removed or scoured by ice push. When the tide recedes and the marsh is more vegetated, ice produce a pattern of pools with a "jigsaw" patterns (see their Fig. 4.8.)

Biological processes (including some anthropic ones) are the most active for the formation and further evolution of tidal depressions. The simplest biological mechanism is plants avoiding a particular spot. During plant colonization of a tidal flat, in places where some salinity, pH, or sediment instability conditions prevail, plants do not survive, and surfaces remain bare and exposed to further erosion processes by waves and currents (Fig. 6.2A). Perillo and Iribarne (2003a) and Escapa et al. (2007, 2015) have shown the effect that burrowing crabs have in the formation of pans in a *Saccocornia*-dominated high marsh. Crabs invaded *Saccocornia* hummocks forcing the plants to migrate outside the developing rings, while the center of the ring may contain up to 60 burrows m^{-2}. The crab activity results in a depression of the surface where water can be retained at low tide or during rainfall, but also due to surging of groundwater pushed by the tide. In some cases, the variation in relief between the original hummock to the final depression may be of the order of 10 cm, which results in an estimate of 0.38 $m^3\ ha^{-1}$ sediment lost only in 1 year of the study (Minkoff et al., 2005, 2006).

Although combined with tidal transport, plant litter accumulation in a particular spot of a marsh can produce plant die-off if the litter is thick and persists. However, the continuity of the depression is dependent on what happens after the litter and the rotten plants are either eliminated or incorporated in the depression sediment. If the chemical and physical conditions at the depressions return to the original ones, the surrounding plants may be able to recolonize the depression. If those conditions now become hostile to the plants, including some permanent flooding of the depression, it can survive further.

In summary, there are a significant number of processes that originate depressions into wetlands. Although we tried to separate them in three main categories, it is clear that combined processes are also very common. Considering that detailed studies of the formation of depressions are lacking, it is most likely that other mechanism may be discovered in the future.

4.3 Evolution of Tidal Depressions

The evolution of a tidal depression relates to its origin. How plants and animals interact with the depression, or what dynamic process acts, and how effective or in what particular direction impacts on the depression. The resulting changes primarily impact on the shape and the depth, but also their interaction with other depressions and courses. Therefore, the evolution of the depressions is naturally complex and, in most cases, depends on the local conditions.

The simplest process in the evolution requires the presence/absence of water permanently; that is moving from a pan to a pond. Lack of water due to evaporation can lead to increased soil salinization, which may spread out over the surrounding plants allowing minute growing of the pan. But the physical processes due to current eddies or short-period waves are much more effective in increasing and/or changing the size and form of the depression. As described by Perillo et al. (1996), creeks can be formed by the advance of ponds due to the erosion of wind waves and coalescence of adjacent ponds, a process exactly opposite to the formation of channel pans proposed by Yapp et al. (1917). On the other hand, depressions may be reduced in size or even completely obliterated when some of the conditions preventing plant colonization disappear. For instance, when an isolated pond connects with an active course, soil and water salinity may return to levels adequate for the plants to invade. Similar effects may occur if the bottom sediments become more stable for pioneer plants to grow. Of course, elimination of a depression may also occur by input of sediment filling.

During the evolution of depressions, one may imagine that modifications in plan view could be the most common way. However, it is becoming more and more common to find examples of vertical evolution. For the case of the ponds in Loyola Bay, Perillo et al. (1996) indicated that the formation of the microcliffs bordering the ponds was mostly due to bottom subsidence due to the compaction of organic matter because these ponds trap significant amount of macroalgae transported by the macrotides. There are no previous indications of this issue, but the presence of high concentrations of reduced organic matter in localized points of ancient wetland may be a potential recognition of these features in the stratigraphic record, in particular, when unconformable records are observed (Wilson et al., 2010).

The vertical growth of tidal depression has been considered previously by Shinn et al. (1969) based on the formation of natural levees formed around the depression. The increase of the levee results in deepening of the depression. Collins et al. (1987) suggest that even the formation of the depression (named *turf pans* by them) is a function of the upward building of the plants (turf) surrounding the pan.

With the concern of mean sea level rise in mind, some research has been made in the last decade regarding how these depressions may behave. Erwin et al. (2006) analyzed the evolution of several ponds on the east coast of United States in comparison of vertical changes of the surrounding marsh and sea level changes. They proposed three different conditions: (1) marsh and pond elevation rates were both significantly different than zero; (2) marsh and pond elevation rates were not significantly different from each other; and (3) sea level rate constant is less than both pond and marsh elevations. On the other hand, Wilson et al. (2014) found that both courses and depressions may have cyclic behavior to maintain a dynamic equilibrium with sea level. They propose that salt marsh elevation is reduced by degradation of organic matter and formation of depressions, but on short timescales (1–2 years) restoration of tidal exchange by creeks cutting into pools may induce a rapid increase of the marsh. Ultimately, comparisons made by Linhoss et al. (2016) show that elevation and topographic depressions are the primary environmental drivers for the formation and evolution of pans but affected by sea-level rise.

However, in some cases, bioturbation may play a role in the evolution of a depression. For instance, diatoms and worms tend to concentrate there, especially during warmer periods. Diatom oxygen productivity can be large enough to dislodge sediment and keep it floating. However, worms rework sediments and make them more unstable and easier to erode (Molina, 2012). Wind action is a major factor in enhancing ponds although this process has been rather overlooked in the literature. Typical short-period waves forming due to small fetch and water depth have high steepness, which makes them a very erosive process. In a place where some wind directions are very frequent (i.e., Argentine Patagonia) ponds could be enlarged along the wind path. Wind, in these cases, could play an important role in defining water circulation over the tidal flat (the effect on marshes and mangroves should be smaller) because of the relatively low water depth and velocities, wind shear transfer may easily change the pressure gradient induced by the tide or, at the beginning and end of the inundation time, by the topographic gradient.

5. SUMMARY

Tidal courses are an essential part of coastal wetlands as they play a major role in water and nutrient exchange. However, their origin and evolution are still a matter of discussion because of the complexities of the dynamic processes associated with their initiation. Some factors as the role of overland flows versus bankfull flow in the evolution of the courses are unknown. A major interrogation is what actually controls course meandering on tidal flats. Contradictory examples about the meandering distribution can be presented when the cases of the Anse d'Aiguillion (Eisma, 1997) and Bahía Blanca Estuary are compared. In the former, creeks in the higher flats are strongly meandering becoming straight and sinuous in the low flats, whereas in the latter occurs exactly the opposite; sinuosity increases significantly for creeks and gullies along the margins of tidal channels while on the flats they are much less sinuous.

A tidal course classification has been proposed leading to a unified description of these features and to avoid confusion. Although based on geomorphologic information, this classification may evolve further by integrating other descriptors. Therefore, this classification is open to consideration and discussion.

Although not new, tidal depressions have been absent from wetland geomorphologic research for many years. Those that worked in them from diverse points of view (mostly biological) have used a wide variety of names that can be, at least, confusing. We are introducing a specific definition and a pure geomorphologic classification that does not include origin; therefore, covering any possible form without restrictions.

We also analyzed the mechanisms of tidal depressions formation and evolution taking into account that in most cases there are more than one process intervening. Due also to their diversity and the areas they cover, there is a need to study them with high-resolution remote sensing techniques because most of the features considered are near the detection limit of most satellite products.

Acknowledgments

Partial funding for studies that led to the present review was provided by CONICET, Agencia de Promoción Científica y Tecnológica and Universidad Nacional del Sur from Argentina. We are grateful to the photos provided by various associates acknowledged in the text. Partial support was provided by the Inter-American Institute for Global Change Research (IAI) CRN3038 (under US NSF Award GEO-1128040). Special thanks are given to the two anonymous reviewers and Don Cahoon for their comments and suggestions that greatly improved the text.

References

Adam, P., 1997. Absence of creeks and pans in temperate Australian salt marshes. Mangroves and Salt Marshes 1, 239–241.

Adamowicz, S.C., Roman, C.T., 2005. New England salt marsh pools: a quantitative analysis of geomorphic and geographic features. Wetlands 25, 279–288.

Ahnert, F., 1960. Estuarine meanders in the Chesapeake Bay area. Geographical Review 50, 390–401.

Allen, J.R.L., 2000. Morphodynamics of Holocene salt marshes: a review sketch from the Atlantic and Southern North Sea coasts of Europe. Quaternary Science Reviews 19, 1155–1231.

Angeles, G.R., Perillo, G.M.E., Piccolo, M.C., Pierini, J.O., 2004. Fractal analysis of tidal channels in the Bahıa Blanca Estuary (Argentina). Geomorphology 57, 263–274.

Ashley, G.M., Zeff, M.L., 1988. Tidal channel classification for a low mesotidal salt marsh. Marine Geology 82, 17–32.

Barwis, J.H., 1978. Sedimentology of some South Carolina tidal-creek point bars, and a comparison with their fluvial counterparts. In: Miall, A.D. (Ed.), Fluvial Sedimentology. Canadian Society of Petroleum Geologists, pp. 129–160.

Best, J.L., 1987. Flow dynamics at river channel confluences: implications for sediment transport and bed morphology. In: Recent Developments in Fluvial Sedimentology. The Society of Economic Paleontologists and Mineralogists, Tulsa, OK. Special Publication 39.

Blondeaux, P., Seminara, G., 1985. A unified bar-bend theory of river meanders. Journal of Fluid Mechanics 157, 449–470.

Bortolus, A., Iribarne, O., 1999. Effects of the SW Atlantic burrowing crab Chasmagnathus granulata on a Spartina salt marsh. Marine Ecology Progress Series 178, 78–88.

Carson, M.A., La Pointe, M.F., 1983. The inherent asymmetry of river meander planform. The Journal of Geology 91, 41–55.

Chapman, V.J., 1960. Salt Marshes and Salt Deserts of the World. Interscience Publishers, London.

Chen, Y., Collins, M.B., Thompson, C.E.L., 2011. Creek enlargement in a low-energy degrading saltmarsh in southern England. Earth Surface Processes and Landforms 36, 767–778.

Cipolletti, M.P., Delrieux, C.A., Perillo, G.M.E., Piccolo, M.C., 2012. Super-resolution border segmentation and measurement in remote sensing images. Computers and Geosciences 40, 87–96.

Cipolletti, M.P., Delrieux, C.A., Perillo, G.M.E., Piccolo, M.C., 2014. Border extrapolation in remote sensing images. Computers and Geosciences 62, 25–34.

Cobb, S.M., Saynor, M.J., Eliot, M., Eliot, I., Hall, R., 2007. Saltwater Intrusion and Mangrove Encroachment of Coastal Wetlands in the Alligator Rivers Region, Northern Territory, Australia. Supervising Scientist Report 191. Supervising Scientist, Darwin NT.

Coco, G., Zhou, Z., van Maanen, B., Olabarrieta, M., Tinoco, R., Townend, I., 2013. Morphodynamics of tidal networks: advances and challenges. Marine Geology 346, 1–16.

Collins, L.B., Collins, J.N., Leopold, L.B., 1987. Geomorphic processes of an estuarine marsh. In: Gardiner, V. (Ed.), International Geomorphology 1986 Part I. John Willey & Sons, pp. 1049–1072.

Cuadrado, D.G., Perillo, G.M.E., Vitale, A.J., 2014. Modern microbial mats in siliciclastic tidal flats: evolution, structure and the role of hydrodynamics. Marine Geology 352, 367–380.

D'Alpaos, A., 2005. Tidal network ontogeny: channel initiation and early development. Journal of Geophysical Research 110, F02001. https://doi.org/10.1029/2004JF000182.

D'Alpaos, A., Lanzoni, S., Marani, M., Rinaldo, A., 2010. On the tidal prism–channel area relations. Journal of Geophysical Research 115, F01003. https://doi.org/10.1029/2008JF001243.

Dalrymple, R.W., Zaitlin, B.A., Boyd, R., 1992. A conceptual model of estuarine sedimentation. Journal of Sedimentary Petrology 62, 1130–1146.

Dietrich, W.E., Dunne, T., 1993. The channel head. In: Beven, K., Kirkby, M.J. (Eds.), Channel Network Hydrology. John Wiley & Sons Ltd., pp. 175–219

Dionne, J.-C., 1968. Action of shore ice on the tidal flats of the St. Lawrence estuary. Maritime Sediments 4, 113–115.

Duval, T.P., Hill, A.R., 2006. Influence of stream bank seepage during low-flow conditions on riparian zone hydrology. Water Resources Research 42, W10425. https://doi.org/10.1029/2006WR004861.

Dyer, K.R., 1998. Estuaries: A Physical Introduction. J. Wiley & Sons, Chichester, 195 pp.

Eisma, D., 1997. Intertidal Deposits: River Mouths, Tidal Flats, and Coastal Lagoons. CRC Press, Boca Raton, 525 pp.

Erwin, R.M., Cahoon, D.R., Prosser, D.J., Sanders, G.M., Hensel, P., 2006. Surface elevation dynamics in vegetated spartina marshes versus unvegetated tidal ponds along the Mid-Atlantic Coast, USA, with implications to waterbirds. Estuaries and Coasts 29, 96–106.

Escapa, C.M., Minkoff, D.R., Perillo, G.M.E., Iribarne, O.O., 2007. Direct and indirect effects of burrowing crab activities on erosion of Southwest Atlantic Sarcocornia-dominated marshes. Limnology and Oceanography 52, 2340–2349.

Escapa, C.M., Perillo, G.M.E., Iribarne, O.O., 2015. Biogeomorphically driven salt pan formation in Sarcocornia-dominated salt-marshes. Geomorphology 228, 147–157.

Fagherazzi, S., Furbish, D.J., 2001. On the shape and widening of salt marsh creeks. Journal of Geophysical Research Oceans 106, 991–1005.

Fagherazzi, S., Bortoluzzi, A., Dietrich, W.E., Adami, A., Lanzoni, S., Marani, M., Rinaldo, A., 1999. Tidal networks 1. Automatic network extraction and preliminary scaling features from DTMs. Water Resources Research 35, 3891–3904.

Fagherazzi, S., Gabet, E.J., Furbish, D.J., 2004. The effect of bidirectional flow on tidal channel planforms. Earth Surface Processes and Landforms 29, 295–309.

Fagherazzi, S., Kirwan, M.L., Mudd, S.M., Guntenspergen, G.R., Temmerman, S., D'Alpaos, A., van de Koppel, J., Rybczyk, J.M., Reyes, E., Craft, C., Clough, J., 2012. Numerical models of salt marsh evolution: ecological, geomorphic, and climatic factors. Reviews of Geophysics 50, RG1002.

Fagherazzi, S., Mariotti, G., 2012. Mudflat runnels: evidence and importance of very shallow flows in intertidal morphodynamics. Geophysical Research Letters 39, L14402. https://doi.org/10.1029/2012GL052542.

French, J.R., Stoddart, D.R., 1992. Hydrodynamics of salt marsh creek systems: implications for marsh morphological development and material exchange. Earth Surface Processes and Landforms 17, 235–252.

Findlay, S., 1995. Importance of surface-subsurface exchange in stream ecosystems: the hyporheic zone. Limnology and Oceanography 40, 159–164.

Gardner, L.R., 2005. A modeling study of the dynamics of pore water seepage from marsh sediments. Estuarine, Coastal and Shelf Science 62, 691–698.

Gardner, L.R., Wilson, A.M., 2006. Comparison of four numerical models for simulating seepage from salt marsh sediments. Estuarine, Coastal and Shelf Science 69, 427–437.

Garofalo, D., 1980. The influence of wetland vegetation on tidal stream channel migration and morphology. Estuaries 3, 258–270.

Ginsberg, S.S., Perillo, G.M.E., 1990. channel bank recession in the Bahía Blanca estuary, Argentina. Journal of Coastal Research 6, 999–1010.

Ginsberg, S.S., Perillo, G.M.E., 1999. Deep scour holes at the confluence of tidal channels in the Bahia Blanca Estuary, Argentina. Marine Geology 160, 171–182.

Ginsberg, S.S., Perillo, G.M.E., 2004. Characteristics of tidal channels in a mesotidal estuary of Argentina. Journal of Coastal Research 20, 489–497.

Ginsberg, S.S., Aliotta, S., Lizasoain, G.O., 2009. Morphodynamics and seismostratigraphy of a deep hole at tidal channel confluence. Geomorphology 104, 253–261.

Goudie, A., 2013. Characterising the distribution and morphology of creeks and pans on salt marshes in England and Wales using Google Earth. Estuarine, Coastal and Shelf Science 129, 112–123.

Harshberger, J.W., 1916. Origin and vegetation of salt marsh pools. Proceedings of the American Philosophical Society 55, 481–484.

Hollins, S.E., Ridd, P.V., Read, W.W., 2000. Measurement of the diffusion coefficient for salt in salt flat and mangrove soils. Wetlands Ecology and Management 8, 257–262.

Hood, W.G., 2010. Tidal channel meander formation by depositional rather than erosional processes: examples from the prograding Skagit River Delta (Washington, USA). Earth Surface Processes and Landforms 35, 319–330.

Horton, R.E., 1945. Erosional development of streams and their drainage basins; hydrophysical approach to quantitative morphology. Bulletin of the Geological Society of America 56, 275–370.

Hughes, Z.J., 2012. Tidal channels on tidal flats and marshes. In: Davis, R.A., Dalrymple, R.W. (Eds.), Principles of Tidal Sedimentology. Springer, Dordrecht, pp. 269–300.

Ikeda, S., Parker, G., Sawai, K., 1981. Bend theory of river meanders. part 1. Linear development. Journal of Fluid Mechanics 112, 363–377.

Iribarne, O., Bortolus, A., Botto, F., 1997. Between-habitat differences in burrow characteristics and trophic modes in the south western Atlantic burrowing crab Chasmagnathus granulata. Marine Ecology Progress Series 155, 137–145.

Istanbulluoglu, E., Tarboton, D.G., Pack, R.T., Luce, C., 2002. A probabilistic approach for channel initiation. Water Resources Research 38, 1325. https://doi.org/10.1029/2001WR000782.

Jarecki, L., Walkey, M., 2006. Variable hydrology and salinity of salt ponds in the British Virgin Islands. Saline Systems 2, 2. https://doi.org/10.1186/1746-1448-2-2.

Kelly, R.P., Moran, S.B., 2002. Seasonal changes in groundwater input to a well-mixed estuary estimated using radium isotopes and implications for coastal nutrient budgets. Limnology and Oceanography 47, 1796–1807.

Kesel, R.H., Smith, J.S., 2008. Tidal creek and pan formation in intertidal salt marshes, Nigg Bay, Scotland. Scottish Geographical Magazine 94, 159–168.

Kirkbya, M.J., Bullb, L.J., Poesenc, J., Nachtergaelec, J., Vandekerckhove, L., 2003. Observed and modelled distributions of channel and gully heads—with examples from SE Spain and Belgium. Catena 50, 415–434.

Kjerfve, B., 1978. Bathymetry as an indicator of net circulation in well mixed estuaries. Limnology and Oceanography 23, 816–821.

Kjerfve, B., Shao, C.-C., Stapor Jr., F.W., 1979. Formation of deep scour holes at the junction of tidal creeks: an hypothesis. Marine Geology 33, M9–M14.

Kleinhans, M.G., van Rosmalen, T.M., Roosendaal, C., van der Vegt, M., 2014. Turning the tide: mutually evasive ebb- and flood-dominant channels and bars in an experimental estuary. Advances in Geosciences 39, 21–26.

Lathrop, R.G., Cole, M.B., Showalter, R.D., 2000. Quantifying the habitat structure and spatial pattern of New Jersey (U.S.A.) salt marshes under different management regimes. Wetlands Ecology and Management 8, 163–172.

Linhoss, A.C., Underwood, W.V., 2016. Modeling salt panne land-cover suitability under sea-level rise. Journal of Coastal Research 32, 1116–1125.

Leopold, L.B., Wolman, M.G., Miller, J.P., 1964. Fluvial Processes in Geomorphology. W. H. Freeman, New York.

Leopold, L.B., Wolman, M.G., 1960. River meanders. The Geological Society of America Bulletin 71, 769–793.

Mallin, M.A., 2004. The importance of tidal creek ecosystems. Journal of Experimental Marine Biology and Ecology 298, 145–149.

Marani, M., Lanzoni, S., Zandolin, D., Seminara, G., Rinaldo, A., 2002. Tidal meanders. Water Resources Research 38, 1225. https://doi.org/10.1029/2001WR000404.

Marani, M., Belluco, E., D'Alpaos, A., Defina, A., Lanzoni, S., Rinaldo, A., 2003. On the drainage density of tidal networks. Water Resources Research 39, 1040. https://doi.org/10.1029/2001WR001051.

Melo, W.D., Schillizzi, R., Perillo, G.M.E., Piccolo, M.C., 2003. Influencia del área continental pampeana sobre el origen y la morfología del estuario de Bahía Blanca. Revista de la Asociación Argentina de Sedimentología 10, 65–72.

Melo, W.D., Perillo, G.M.E., Perillo, M.M., Schilizzi, R., Piccolo, M.C., 2013. Late Pleistocene-Holocene deltas in the southern Buenos Aires Province, Argentina. In: Young, G., Perillo, G.M.E. (Eds.), Deltas: Landforms, Ecosystems and Human Activities, vol. 358. IAHS Publ., pp. 187–195

Minkoff, D.R., 2007. Geomorfología y dinámica de canales de mareas en ambientes intermareales (Ph.D. dissertation). Departamento de Ingeniería, Universidad Nacional del Sur (unpublished).

Minkoff, D.R., Escapa, C.M., Ferramola, F.E., Perillo, G.M.E., 2005. Erosive processes due to physical – biological interactions based in a cellular automata model. Latin American Journal of Sedimentology and Basin Analysis 12, 25–34.

Minkoff, D.R., Escapa, C.M., Ferramola, F.E., Maraschin, S., Pierini, J.O., Perillo, G.M.E., Delrieux, C., 2006. A Cellular Automata model for study of the interaction between the crab Chasmagnatus granulatus and the halophyte plant Sarcocornia perennis in the evolution of tidal creeks in salt marshes. Estuarine, Coastal and Shelf Science 69, 403–413.

Minkoff, D.R., Perillo, G.M.E., 2002. Evolución de canales de marea en una marisma de Bahía Blanca. In: III Taller de Sedimentología y Medio Ambiente, Buenos Aires (abstract).

Molina, L.M., 2012. El rol de la biota en los procesos de estabilización-desestabilización de sedimentso estuariales (Ph.D. dissertation). Departamento de Biología, Bioquímica y Farmacia, Universidad Nacional del Sur (unpublished).

Montgomery, D.R., Dietrich, W.E., 1988. Where do channels begin? Nature 336, 232–234.

Montgomery, D.R., Dietrich, W.E., 1992. Channel initiation and the problem of landscape scale. Science 255, 826–830.

Mulrennan, M.E., Woodroffe, C.D., 1998. Holocene development of the lower Mary River plains, Northern Territory, Australia. The Holocene 8, 565–579.

Mwamba, M.J., Torres, R., 2002. Rainfall effects on marsh sediment redistribution, North Inlet, South Carolina, USA. Marine Geology 189, 267–289.

Nichols, M.H., Nearing, M., Hernandez, M., Polyakov, V.O., 2016. Monitoring channel head erosion processes in response to an artificially induced abrupt base level change using time-lapse photography. Geomorphology 265, 107–116.

Novakowski, K.I., Torres, R., Gardner, L.R., Voulgaris, G., 2004. Geomorphic analysis of tidal creek networks. Water Resources Research 40, W05401. https://doi.org/10.1029/2003WR002722.

Osgood, D.T., 2000. Subsurface hydrology and nutrient export from barrier island marshes at different tidal ranges. Wetlands Ecology and Management 8, 133–146.

Parker, G., Shimizu, Y., Wilkerson, G.V., Eke, E.C., Abad, J.D., Lauer, J.W., Paola, C., Dietrich, W.E., Voller, V.R., 2011. A new framework for modeling the migration of meandering rivers. Earth Surface Processes and Landforms 36, 70–86.

Perillo, G.M.E. (Ed.), 1995. Geomorphology and Sedimentology of Estuaries. Development in Sedimentology, vol. 53. Elsevier Science BV, Amsterdam.

Perillo, G.M.E., 2009. Tidal courses: classification, origin and functionality. In: Perillo, G.M.E., Wolanski, E., Cahoon, D.R., Brinson, M.M. (Eds.), Coastal Wetlands: An Integrated Ecosystem Approach. Elsevier, Amsterdam, pp. 185–210.

Perillo, G.M.E., Iribarne, O.O., 2003a. New mechanisms studied for creek formation in tidal flats: from crabs to tidal channels. EOS American Geophysical Union Transactions 84, 1–5.

Perillo, G.M.E., Iribarne, O.O., 2003b. Processes of tidal channels develop in salt and freshwater marshes. Earth Surface Processes and Landforms 28, 1473–1482.

Perillo, G.M.E., Minkoff, D.R., Piccolo, M.C., 2005. Novel mechanism of stream formation in coastal wetlands by crab–fish–groundwater interaction. Geo-Marine Letters 25, 124–220.

Perillo, G.M.E., Piccolo, M.C., 1999. Geomorphologic and physical characteristics of the Bahía Blanca Estuary. Argentina. In: Perillo, G.M.E., Piccolo, M.C., Pino Quivira, M. (Eds.), Estuaries of South America: Their Geomorphology and Dynamics. Springer-Verlag, Berlín, pp. 195–216.

Perillo, G.M.E., Piccolo, M.C., 2011. Global variability in estuaries and coastal settings. In: Wolanski, E., McLusky, D.S. (Eds.), Treatise on Estuarine and Coastal Science, vol. 1. Waltham, Academic Press, pp. 7–36.

Perillo, G.M.E., Ripley, M.D., Piccolo, M.C., Dyer, K.R., 1996. The formation of tidal creeks in a salt marsh: new evidence from the Loyola Bay Salt Marsh, Rio Gallegos Estuary, Argentina. In: Mangroves and Salt Marshes, vol. 1, pp. 37–46.

Pestrong, R., 1965. The Development of Drainage Patterns in Tidal Marshes, vol. 10. Stanford University Publications in Earth Science, pp. 1–87.

Pestrong, R., 1972. Tidal-flat-sedimentation at Cooley Landing, southwest San Francisco Bay. Sedimentary Geology 8, 251–288.

Pethick, J.S., 1969. Drainage in tidal marshes. In: Steers, J.R. (Ed.), The Coastline of England and Wales, 3rd. edition. Cambridge University Press, Cambridge, pp. 725–730.

Pethick, J.S., 1974. The distribution of salt pans on tidal salt marshes. Journal of Biogeography 1, 57–62.

Pethick, J.S., 1992. Saltmarsh geomorphology. In: Allen, J.R.L., Pye, K. (Eds.), Saltmarshes, Morphodynamics, Conservation and Engineering Significance. Cambridge University Press, Cambridge, pp. 41–62.

Pierini, J.O., Perillo, G.M.E., Carbone, M.E., Marini, M.F., 2005. Residual flow structure at a scour hole in Bahía Blanca Estuary, Argentina. Journal of Coastal Research 21, 784–796.

Poesen, J., Nachtergaele, J., Verstraeten, G., Valentin, C., 2003. Gully erosion and environmental change: importance and research needs. Catena 50, 91–133.

Pye, K., 1992. Saltmarshes on the barrier coastline of North Norfolk, eastern England. In: Allen, J.R.L., Pye, K. (Eds.), Saltmarshes: Morphodynamics, Conservation and Engineering Significance. Cambridge University Press, pp. 148–177.

Pye, K., French, P.W., 1993. Erosion and accretion processes on British Salt Marshes. In: Introduction: Saltmarsh Processes and Morphology, vol. 1. Cambridge Environmental Research Consultants, Cambridge. A final report to MAFF).

Rankey, E.C., Morgan, J., 2002. Quantified rates of geomorphic change on a modern carbonate tidal flat, Bahamas. Geology 30, 583–586.

Rankey, E.C., Berkeley, A., 2012. Holocene carbonate tidal flats. In: Davis, R.A., Dalrymple, R.W. (Eds.), Principles of Tidal Sedimentology. Springer Science, Dordrecht, pp. 507–535.

Redfield, A.C., 1972. Development of a New England salt marsh. Ecological Monographs 42, 201–237.

Revollo Sarmiento, G.N., Cipolletti, M.P., Perillo, M.M., Delrieux, C.A., Perillo, G.M.E., 2016. Methodology for classification of geographical features with remote sensing images: application to tidal flats. Geomorphology 257, 10–22.

Rinaldo, A., Belluco, E., D'Alpaos, A., Feola, A., Lanzonni, S., Marani, M., 2004. Tidal networks: form and function. In: Fagherazzi, S., Marani, M., Blum, L.K. (Eds.), The Ecogemorphology of Tidal Marshes. American Geophysical Union, Washington, DC, pp. 75–91.

Rinaldo, A., Fagherazzi, S., Lanzoni, S., Marani, M., Dietrich, W.E., 1999. Tidal networks 2. Watershed delineation and comparative network morphology. Water Resources Research 35, 3905–3917.

Seminara, G., Tubino, M., 1992. Weakly nonlinear theory of regular meanders. Journal of Fluid Mechanics 244, 257–288.

Seminara, G., Lanzoni, S., Tambroni, N., Toffolon, M., 2009. How long are tidal channels? Journal of Fluid Mechanics 643, 479–494.

Shao, C.-C., 1977. On the Existence of Deep Holes at Tidal Creek Junctions (MS thesis). Univ. of South Carolina, Columbia, SC, 31 p.

Shaw, J., Todd, B.J., Li, M.Z., Wu, Y., 2012. Anatomy of the tidal scour system at Minas Passage, Bay of Fundy, Canada. Marine Geology 323–325, 123–134.

Shinn, E.A., Lloyd, R.M., Ginsburg, R.N., 1969. Anatomy of a modern carbonate tidal-flat, Andros Island, Bahamas. Journal of Sedimentary Petrology 39, 1202–1228.

Stefanon, L., Carniello, L., D'Alpaos, A., Lanzoni, S., 2010. Experimental analysis of tidal network growth and development. Continental Shelf Research 30, 950–962.

Stribling, J.M., Cornwell, J.C., Glahn, O.A., 2007. Microtopography in tidal marshes: ecosystem engineering by vegetation? Estuaries and Coasts 30, 1007–1015.

Suarez-Gonzalez, P., Quijada, I.E., Benito, M.I., Mas, R., 2015. Sedimentology of ancient coastal wetlands: insights from a cretaceous multifaceted depositional system. Journal of Sedimentary Research 85, 95–117.

Temmerman, S., Bouma, T.J., Van de Koppel, J., Van der Wal, D., De Vries, M.B., Herman, P.M.J., 2007. Vegetation causes channel erosion in a tidal landscape. Geology 35, 631–634.

Tucker, G.E., Arnold, L., Bras, R.L., Flores, H., Istanbulluoglu, E., Solyom, P., 2006. Headwater channel dynamics in semiarid rangelands, Colorado high plains, USA. The Geological Society of America Bulletin 118, 959–974.

van der Wegen, M., Guo, J., Jaffe, B.E., van der Spek, A.J.F., Roelvink, J.A., 2012. Levee development along tidal channels. In: Jubilee Conference Proceedings, NCK-Days 2012, pp. 219–222.

Vandenbruwaene, W., Meire, P., Temmerman, S., 2012. Formation and evolution of a tidal channel network within a constructed tidal marsh. Geomorphology 151–152, 114–125.

Vincent, C.L., 1976. A Method for the Mathematical Analysis of the Cross-Sectional Geometry of Tidal Inlet Channels. Mathematical Geology 8, 635–647.

Vlaswinkel, B.M., Cantelli, A., 2011. Geometric characteristics and evolution of a tidal channel network in experimental setting. Earth Surface Processes and Landforms 36, 739–752.

Wells, J.T., Adams Jr., C.E., Park, Y.-A., Frankenberg, E.W., 1990. Morphology, sedimentology, and tidal channel processes on a high-tide-range mudflat, west coast of South Korea. Marine Geology 95, 11–130.

Wilson, C.A., Hughes, Z.J., FitzGerald, D.M., 2012. The effects of crab bioturbation on Mid-Atlantic saltmarsh tidal creek extension: geotechnical and geochemical changes. Estuarine, Coastal and Shelf Science 106, 33–44.

Wilson, C.A., Hughes, Z.J., FitzGerald, D.M., Hopkinson, C.S., Valentin, V., Kolker, A.S., 2014. Saltmarsh pool and tidal creek morphodynamics: dynamic equilibrium of northern latitude saltmarshes? Geomorphology 213, 99–115.

Wilson, K.R., Kelley, J.T., Croitoru, A., Dionne, M., Belknap, D.F., Steneck, R., 2009. Stratigraphic and ecophysical characterizations of salt pools: dynamic landforms of the Webhannet Salt Marsh, Wells, ME, USA. Estuaries and Coasts 32, 855–870.

Wilson, K.R., Kelley, J.T., Tanner, B.R., Belknap, D.F., 2010. Probing the Origins and Stratigraphic Signature of Salt Pools from North-Temperate Marshes in Maine, U.S.A. Journal of Coastal Research 26, 1007–1026.

Wolanski, E., 2007. Estuarine Ecohydrology. Elsevier, Amsterdam.

Yapp, R.H., 1922. The Dovey Salt Marshes in 1921. Journal of Ecology 10, 18–23.

Yapp, R.H., Johns, D., Jones, O.T., 1917. The salt marshes of the Dovey Estuary. The Journal of Ecology 5, 65–103.

Zhou, Z., Stefanon, L., Olabarrieta, M., D'Alpaos, A., Carniello, L., Coco, G., 2014. Analysis of the drainage density of experimental and modelled tidal networks. Earth Surface Dynamics 2, 105–116.

CHAPTER

7

Methods to Estimate Heat Balance in Coastal Wetlands

Maria Cintia Piccolo

Instituto Argentino de Oceanografia - Universidad Nacional del Sur, Bahia Blanca, Buenos Aires, Argentina

1. THE HEAT BALANCE EQUATION

By its unusually high specific and latent heat, water stands apart from other substances in its thermal properties for supporting the life cycles of a significant number of plant and animal species that depend on wetlands. Wetland plants alone exhibit a wide diversity of growth forms including emergent plants, submerged plants, floating-leafed plants, and a combination of these leaf forms within the same species (Sculthorpe, 1967). On the other hand, benthic animals of wetlands are significantly affected by the variability in the heat exchanged between the soil, water, and air. Given this variety of species and processes, it is useful to understand how is the heat balance in the different wetlands to learn in the future how these organisms have adapted to the various energy budgets.

Relatively few studies have been carried on coastal wetlands compared with those conducted in inland ones, where the tides and the marine microclimate do not affect their heat transfer. Yet, an understanding of how heat is transferred across air–sea–soil interfaces is fundamental for predicting, for example, how coastal wetlands will respond to global climate change and climate variability. Coastal wetlands develop particularly steep chemical and hydrological gradients as a result of their buffer position between continents and the ocean. Unlike vertical fluxes that dominate exchanges in upland ecosystems, the exchange of matter and energy in coastal wetlands is complicated by strong horizontal fluxes, particularly those driven by water movement. This chapter reviews studies that have contributed to measure or estimate the heat energy that is transferred across the different levels of these fluid and solid interfaces, and how this energy affects different biological and chemical processes.

Coastal Wetlands
https://doi.org/10.1016/B978-0-444-63893-9.00007-1

Studies of the energy budget over an earth surface elucidate how solar energy is locally redistributed to create a particular microclimate (Kjerfve, 1978). The following heat budget equation is the most used (Oke, 1978):

$$R_N = Q_S + Q_B + Q_E + Q_A \tag{7.1}$$

and it establishes that at any moment in time the available energy at the Earth's surface (net radiation, R_N) must be equivalent to a combination of convective exchange to or from the atmosphere (sensible [Q_S] and latent heat [$LE = Q_E$, where E is evaporation and L is the latent heat of vaporization]), conductive flux to or from the soil (Q_B), and incoming or outgoing advective flux (Q_A). Fig. 7.1 shows the components of the heat budget equation in an idealized bare tidal flat. The calculation of the heat balance involves the measurement of parameters that define terms of Eq. (7.1) and/or the use of bulk aerodynamic formulas with data obtained from the monitoring of meteorological and hydrological parameters. Different sensors, satellite images, and numerical models with different boundary conditions are used to estimate the heat balance.

The interface or boundary between the water and the air is dynamic. Matter and energy are continuously being transferred across the air–sea interface in both directions. Air either gains or loses heat from the water depending on the temperature difference between the water surface and the overlying air. Water evaporates to contribute to atmospheric moisture, and atmospheric water vapor condenses to form fog and clouds and, eventually, precipitation. Vegetated wetlands also receive and loose energy by radiation, conduction, convection, energy transfer into and out plant tissue, photosynthesis, respiration, and evaporation. However, water lost by plant transpiration must be separated from salts in seawater,

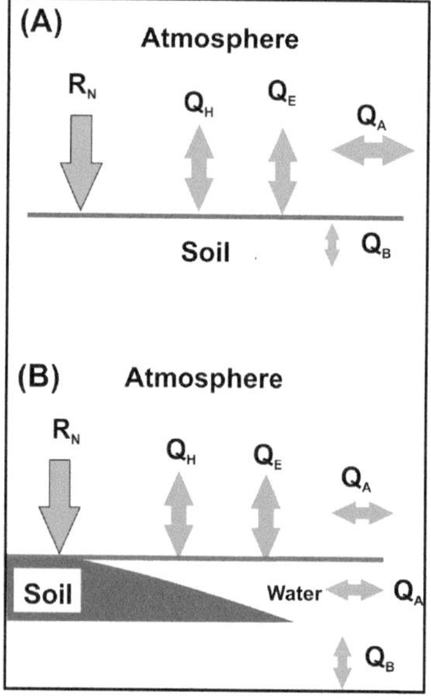

FIGURE 7.1 Components of the heat balance equation in a bare tidal flat. (A) Low tide and (B) high tide.

a process that comes at great cost to the plant (Teal and Kanwisher, 1970). Few species are well adapted to do this, which explains in part the low species richness of vascular plants in coastal wetlands. Eq. (7.2) is commonly used when the research involves vegetation (i.e., Drexler et al., 2004), although most investigations disregard the last two terms.

$$R_N = Q_S + Q_B + Q_E + Q_A + Q_M + Q_S \tag{7.2}$$

where Q_M is the energy flux for photosynthesis and respiration and Q_S is the energy transfer into and out of plant tissue.

Horizontal exchanges of matter and energy are affected by tides and wind. Of these, tides behave more predictably (Perillo and Piccolo, 1991), and extensive research is found in the literature using numerical models (i.e., Bai et al., 2016, Bakhtyar et al., 2016). They are responsible for the ebb and flow of water in all of the major groups of wetlands (sea grass meadows, mudflats, marshes, and mangroves). Winds, on the other hand, are less predictable (except for sea breezes), and few studies focus on how wind affects the energy budget of coastal areas (Leal and Lavín, 1998; Castro et al., 2003). In any case, advective fluxes may originate with tides and/or with wind (Fig. 7.1B). Furthermore, winds may serve to amplify or dampen the effects of astronomic tides.

Annual heat flux cycles in water bodies usually follow the seasonal fluctuations of incident solar radiation. Annual mean radiation is used to express the primary source of heat energy, and it has been measured in several studies to determine its magnitude and seasonality in coastal wetlands (i.e., Bianciotto et al., 2003; Jacobs et al., 2004; El-Metwally, 2004; Paulescu et al., 2006). Radiation on tidal flats is of particular interest because of its significance to ecological processes. In a study of sediment temperatures on the Bay of Fundy (Canada), Piccolo et al. (1993) confirmed that radiation and tides are the drivers of thermal behavior over and within tidal flats. Gould and Hess (2005) measured sediment exposure rate from environmental radiation on tidal flats using a high-pressure ion chamber so the shielding effects of the tidal cycle could be evaluated. They derived a theoretical model to predict the behavior of exposure rate as a function of time. In addition, they developed an empirical formula to calculate the total exposure on a tidal flat that requires measurements of only the slope of the tidal flat and the exposure rate when no shielding occurs (Gould and Hess, 2005). The experimental results were consistent with the model. The formula and the model can be applied to biological studies where radiation exposure is needed.

1.1 The Net Radiation

The Surface Radiation Balance (SRB) and its components in intertidal environments are of vital importance to any ecosystem ecology. Vitale et al. (2018) studied the behavior of the SRB in a temperate salt marsh in Villa del Mar (Bahía Blanca Estuary, Argentina) using bulk formulation and taking into consideration two conditions: nonflooded salt marsh (NFS) and flooded salt marsh (FS). The SRB equation is

$$R_N = K\downarrow - K\uparrow + L\downarrow - L\uparrow = K\downarrow(1-\beta) + L\downarrow - L\uparrow \tag{7.3}$$

where $K\downarrow$ and $K\uparrow$ are the incoming and outgoing shortwave radiations, respectively; $L\downarrow$ and $L\uparrow$ are the incoming and outgoing longwave radiations, respectively; and β is the albedo.

The average annual K↓ for the salt marsh was 280 W m^{-2}, whereas the average K↑ was 40 W m^{-2}, indicating a positive annual shortwave radiative flux (Vitale et al., 2018). Seasonally, the albedo varied according to the Fresnel's law. The annual albedo was 0.11 and 0.064 for NFS and FS conditions, respectively. The longwave radiation balance was negative in summer (−61 W m^{-2}) and winter (−87 W m^{-2}). The decrease in air and water surface temperature in winter generated an increase of the negative L↑ net flux. The results indicate that the tides slightly modified the annual SRB. In average terms, the tide led to an increase of 5% of R_N during summer (NFS = 3485 W m^{-2} day^{-1}; FS = 3663 W m^{-2} day^{-1}), whereas in winter, the resultant R_N under NFS and FS conditions was similar (NFS = −616 W m^{-2} day^{-1}; FS = −615 W m^{-2} day^{-1}), but with hourly differences along the day. Therefore, the role of the tide in the variability of radiative flux is important because the tide modifies both the surface temperature and albedo (Vitale et al., 2018).

Radiation data using satellite images can also be used to estimate parameters of the heat balance equation (i.e., Dutta, et al., 2016; Yang et al., 2017). Jacobs et al. (2004) estimated radiation derived from Geostationary Operational Environmental Satellite (GOES) to predict daily evapotranspiration (ET) in Florida wetlands (USA) with the Penman–Monteith, Turc, Hargreaves, and Makkink models. These estimates agreed well with ET measured with an eddy-correlation system. The authors demonstrate that the Penman–Monteith model provided the best estimates of ET (r^2 = 0.92). However, the empirical Makkink method showed similar agreement (r^2 = 0.90) using only the GOES solar radiation and the measured temperature. Then, using GOES-derived solar radiation, net radiation, and few additional surface measurements, we can generate spatially distributed daily potential ET (Jacobs et al., 2004). In addition, exploring wetlands in Europe and using various satellite data, Dabrowska-Zielinska et al. (2010) assessed soil moisture, heat fluxes, and ET values in areas of diverse soil-vegetation conditions. They based their research knowing that the incoming radiation plays a significant role in chlorophyll production in plants, in general, and that higher radiation induces a reduction in the chlorophyll production rate of different species.

In the literature, we found numerous formulations to calculate R_N and longwave radiation from solar radiation (K↓) measurements in different climates (i.e., Oke, 1978). An example is the linear relationships that have been found between daytime incoming radiation and both net ($R_N = 0.73\,K{\downarrow} - 13.45$, (W m^{-2})) and reflected radiation ($K{\uparrow} = 0.079\,K{\downarrow} + 3.3$, (W m^{-2})) over a *Spartina alterniflora* salt marsh during the summer (Crabtree and Kjerfve, 1978). They found that, on the average, net radiation was 70 ± 9% and reflected radiation was 9 ± 1% of incoming radiation. In spite of potentially damaging effects of K↓, Costa et al. (2006) found no evidence of differential sensitivity or resilience to UVB radiation between *Salicornia* species from low–mid latitudes and a high-latitude population in the Americas.

1.2 The Bottom Heat Flux

Soil properties play a significant role in plant composition, productivity, and zonation in coastal ecosystems because plant species are differentially tolerant to salinity and soil saturation (Pennings et al., 2005; Wang et al., 2007; Zhang et al., 2016; Liu et al., 2017). Variations in salinity and temperature in the intertidal zone restrict the distribution and abundance of marine organisms (Johnson, 1967). In tidal flats, species richness is higher where rates of salinity change are low (Sanders et al., 1965). In sandy intertidal areas where salinities fluctuate widely, the maximum number of species and individuals is small. This feature is not caused by sediment salinity alone because temperature, oxygen content, and other environmental factors also vary in the intertidal zone.

There is a strong relationship between evaporation and sediment salinity. Removal of plant cover, for example, induces higher soil temperature, pore water salinities increases, and water content decreases, most likely due to the greater sun exposure and higher evaporation (Whitcraft and Levin, 2007). On sandy beaches, Johnson (1967) found that an increase in evaporation caused an increase in the soil salinity in the first 20 cm of depth. Many of the effects of salt on plants occur as a result of water stress. Tolerance of plants to saline soils is due in part to biophysical, morphological, and biochemical adaptations. The narrow leaves characteristic of high marsh species may be an adaptation to help regulate leaf temperature in times of low latent cooling (Maricle et al., 2007).

Interaction processes at the sediment—water and sediment—atmosphere interfaces in a temperate tidal flat were analyzed by Beigt et al. (2003) seasonally using mass aerodynamic formulas in the Bahía Blanca estuary, Argentina, for a year. The soil temperature was measured every 10 min by a thermistor chain with three levels below the sediment surface (−0.05, −0.15, and −0.25 m). Water and air temperatures, solar radiation, and meteorological data were registered simultaneously, resulting in annual means of 14.07°C (soil at −0.15 m), 13.69°C (air), and 14.51°C (water). Atmospheric and tidal conditions were mechanisms that regulated the mudflat's thermal behavior. The soil temperature vertical profiles showed a diurnal and semidiurnal cycle, due to the influence of these factors. The diurnal thermal amplitude at −0.05 m reached 14.6°C on February 20th, whereas on July 1st there were only 2.8°C of amplitude. Most fluctuations of sediment temperatures were observed in the first 15 cm, with vertical gradients of 0.82°C cm^{-1} during summer (Beigt et al., 2003). When comparing soil heat fluxes during a typical summer day and a winter day, the authors found a net heat gain in the soil during the summer day (+840.6 KJ m^{-2}) and a heat loss in winter (−768.6 KJ m^{-2}).

1.3 The Sensible and Latent Heat Flux

For many wetlands, ET is the major component of water loss. When considered as its energy equivalent, latent heat flux, evaporation is a major energy sink (Wessel and Rouse, 1993; Souch et al., 1996). Despite numerous studies, evaporation from wetlands is little understood (Lafleur, 1990), and detailed studies of the physical processes involved are geographically restricted (Souch et al., 1996). Wetlands create local microclimates and differences in ET rates that are attributable to different biomasses and ground heat fluxes (Kelvin et al., 2017). The latent heat flux term in Eq. (7.1) is second only to the radiation flux. Evaporation uses most of the radiation energy. In spite of that, the hydrologic implications may be minor because water loss from coastal wetland sediment is typically and often replaced by precipitation or by infiltration of flooding estuarine water (Harvey and Nuttle, 1995).

Surface sensible and latent heat fluxes can be used to predict coastal storm development. Both affect wetland development. Surface latent flux provides a direct source of moisture needed for precipitation, whereas sensible heat flux can affect the stability of the storm environment, thereby modulating the timing and amount of rainfall (Persson et al., 1999). Data from any meteorological station placed almost anywhere across a wetland provide representative estimates of evaporation when atmospheric conditions are relatively homogeneous. Individual patches of vegetation in a wetland do not influence overlying atmospheric conditions significantly, and evaporation can be calculated using the well-known formula of Penman–Monteith (Gavin and Agnew, 2003). However, Acreman et al. (2003), using the eddy-correlation method to calculate evaporation from two types of wetlands, wet grassland and reed beds, in southwest England, demonstrated that the evaporation calculated by the Penman potential method did not represent actual evaporation. The best method of estimating evaporation on different coastal wetlands is still to be achieved (Kelvin et al., 2017).

ET is the dominant water loss from many different wetlands. The accuracy to estimate this parameter is critical to many investigations. A review of models and micrometeorological methods to determine wetland ET proved that there is no best technique to estimate this parameter (Drexler et al., 2004). All the bulk aerodynamic formulation depends on the reliability of the measurements of R_N and Q_B. Therefore, if the evaporation cannot be estimated with accuracy, the latent heat neither. A similar study performed by Amatya and Harrison (2016) shows significant variations of potential evapotranspiration (PET) at two sites of a wetland within 10 km of each other in the coastal South Carolina. They calculated the monthly and annual PET with five different bulk aerodynamic methods (Penman–Monteith (P-M), Turc, Thornthwaite (Thorn), Priestley–Taylor (P-T), and Hargreaves–Samani (H-S)) at two sites (grass and forest) using measured daily climatic data for the 2011–14 period. The results for the grass sites demonstrate that PET estimates are sensitive to the method used. Different estimates were obtained using a single method even for nearby sites because of differences in the complexity of the PET process, microclimatic factors, and interaction with site vegetation types. When compared with the P-M PET for the forest site, the P-T method was in the closest agreement with the highest R^2 of 0.96 and the least bias of 9.7% in mean monthly estimates, followed by the temperature-based H-S with an R^2 of 0.95 and a bias of 12.6% at the grass site (Amatya and Harrison, 2016).

Because different climates generate various types of coastal wetlands and, therefore, are the product of different sites, processes, etc., this chapter reviews heat balance studies on a diversity of wetland types, and with a variety of methods.

2. MID LATITUDES

The most common wetlands are tidal flats, and they are affected by tides, winds, and solar radiation exposure. Different methods are used to estimate the heat balance on those environments. Most of the methods are based on numerical models with different approaches and boundary conditions (i.e., Clulow et al., 2012; Smesrud et al., 2014). Most of the heat balance researches at temperate tidal flats are generally based on the bulk formulation (Smith and Kierspe, 1981; Vugts and Zimmerman, 1985; Harrison and Phizacklea, 1985; Piccolo et al., 1993, 1999; Piccolo and Dávila, 1993; Beigt and Piccolo, 2003; Kelleners et al., 2016). All of them highlight the significant influence of the tidal forces and the wind on heat fluxes.

The annual heat exchanges that occur at an estuarine tidal flat in the Bahía Blanca Estuary, Argentina, were analyzed across the water—atmosphere and the sediment—atmosphere interfaces at high and low tides, respectively (Beigt, 2007; Beigt et al., 2008). Different bulk aerodynamic formulas were employed to estimate the radiative and turbulent fluxes from available meteorological and oceanographic data. Net radiation (R_N) was determined from incident solar radiation and temperature data using (Evett, 2002)

$$R_N = K\downarrow(1 - \beta) - L\uparrow + L\downarrow \quad [W\,m^{-2}] \tag{7.4}$$

where $L\uparrow$ is terrestrial longwave radiation ($L\uparrow = \epsilon_s \sigma T_s^4$ [$W\,m^{-2}$]), ϵ_s is surface emissivity, σ is the Stefan–Boltzmann's constant [$W\,m^{-2}\,K^{-4}$], and T_s is surface temperature [°K] (water or sediment temperature, depending on tidal stage). Atmospheric longwave radiation ($L\downarrow$) is assessed using two different equations depending on temperature data. Swinbank (1963) method is employed when temperatures are over 0°C, whereas the Monteith (1973) equation is used for temperatures lower than 0°C and higher than −5°C.

Soil heat flux (Q_B) across the surface layer was determined from temperature data using the usual Fourier equation ($Q_B = -\lambda\,(\Delta T/\Delta z)$, Oke, 1978), where T is sediment temperature [°K], z is depth [m], λ is thermal conductivity ($\lambda = K_S\,C$, [$W\,m^{-1}\,K^{-1}$]), K_S is the thermal diffusivity [$m^2\,s^{-1}$], and C is the heat capacity [$J\,m^{-3}\,K^{-1}$]. Depending on the tidal stage, the sensible heat flux was estimated using two different equations. During tidal flat inundation, sensible heat flux is assessed using (Kantha and Clayson, 2000; Zaker, 2003)

$$Q_H = \rho\, c_\rho(U_a - U_s)C_H(T_w - T_a) \quad [W\,m^{-2}] \tag{7.5}$$

where ρ is air density [$kg\,m^{-3}$], c_ρ is specific heat of the air [$J\,kg^{-1}\,°C^{-1}$], U_a is wind speed at height z, U_s is wind speed at the water surface [$m\,s^{-1}$] (zero for a stationary surface), C_H is heat exchange coefficient [dimensionless], T_w is temperature at the water surface [°C], and T_a is air temperature [°C]. The heat exchange coefficient (C_H) for the water–atmosphere interface was taken from Friehe and Schmitt (1976).

During sediment exposure to atmospheric conditions, sensible heat flux was estimated (Evett et al., 1994; Evett, 2002) as

$$Q_H = \rho\, c_\rho(T_s - T_a)D_H \quad [W\,m^{-2}] \tag{7.6}$$

where D_H ($=k^2\,U\,[\ln(z/z_{oH})]^{-2}$) is the heat exchange coefficient (Kreith and Sellers, 1975; Ma et al., 2003) [$m\,s^{-1}$], k is the von Kármán's constant, and z_{oH} is the roughness length for sensible heat flux [m] taken from Kreith and Sellers (1975).

The authors calculated the latent heat flux across the sediment–atmosphere and water–atmosphere interfaces by the Penman–Monteith equation that is the potential evaporation rate, namely, the evaporation rate that occurs when water availability is not a limiting factor (Wallace and Holwill, 1997). It is usually applied in agronomical studies of vegetated and bare soils, although it has also been applied to salt marshes (Hughes et al., 2001) and tidal flats (Harrison and Phizacklea, 1985). Beigt (2007) estimated that annually, nearly 5978 MJ m^{-2} of heat entered the tidal flat as incident solar radiation. Of this, only 2954 MJ m^{-2} (49.4%) remained as available energy (R_N). Winds and tides helped to add heat to the ecosystem (1301 MJ m^{-2}). The annual budgets of sensible and soil heat fluxes showed that both processes provided heat energy to the tidal flat surface. Indeed, an amount of 947 MJ m^{-2} was transferred to the surface as sensible heat, whereas the annual budget of soil heat flux indicated an upward heat transfer of 25.2 MJ m^{-2}. The total energy that entered the tidal flat was balanced by an equal heat loss (5227 MJ $m^{-2} = 2127$ mm) as evaporation (Beigt, 2007; Beigt et al., 2008).

Surface heat fluxes through the air–sea interface for the coastal water of Kuwait was estimated by Sultan and Ahmad (1994) using the bulk formulas such as Beigt (2007) and showed similar annual behavior. Although Kuwait is representative of an arid region, both investigations indicate that a nocturnal inundation heats the tidal flat sediment (previously cooled by longwave emission), causing an upward circulation of sensible heat. On the contrary, a tidal inundation at midday or early afternoon usually cools the sediment, with the resultant flow of sensible heat from the air to the tidal flat.

The horizontal transport of heat or advection causes the addition (or subtraction) of energy to (from) an ecosystem. The most common agent of this process is wind; however, the tide must be taken into account when studying a tidal flat. Tidal energy acts as an "energy subsidy" to the coastal ecosystem, increasing its productivity as it enhances the amount of energy that is capable of being converted into production (Odum, 1975). Advective heat flux is estimated as the residual energy from the heat budget Eq. (7.1). The total advective flux was studied according to the tidal height, into two different fluxes ("advective flux at low tide" and "advective flux at high tide"). The former is developed by winds, whereas the tidal flow is considered to be the principal agent during tidal inundation. Atmospheric and tidal conditions regulate the heat exchanges. Tidal flooding affects the direction and magnitude of sensible and soil heat fluxes. The 2003 annual heat budget showed that net radiation, advective, sensible, and soil heat fluxes provided heat to the tidal flat surface, and the most significant heat fluxes were net radiation and latent heat. They are followed (in order of magnitude) by advective and sensible heat fluxes and finally by the soil heat flux (Beigt et al., 2008).

A three-dimensional numerical model of the water, sediment, and air temperature where the surface heat balance was the primary boundary condition was carried out on the tidal flats of the Bahía Blanca Estuary to complete previous studies (Vitale et al., 2014). Continuous and simultaneous meteorological and oceanographic high-frequency data from a buoy were used. The model considers three heat transfer processes: diffusion, convection, and radiation using bulk aerodynamic formulas as boundary conditions between the interfaces (water–sediment, water–air) and uses an open-source tool called Hemera 1.0 that has little complexity and low hardware requirements. To quantify thermal variations, Vitale et al. (2018) formulate the following surface boundary condition:

$$\lambda_{QG}\,\partial T/\partial z\big|_{z=Zsup} = Q_H - Q_E + K\downarrow(1-\beta) + L\downarrow - L\uparrow \qquad (7.7)$$

where λ_{QG} is the thermal conductivity [W m^{-1} K^{-1}] of the water, air, or sediment, Q_H is the sensible heat flux, Q_E is the latent heat flux, K is the shortwave radiation, and L is the longwave radiation. The modeled series adequately described the daily cycles, showing good agreement with the field measurements. The authors validated the model with measured data at 0.15 m water depth, where thermal variations are high, and they obtained an average relative error of 3.3% and a maximum relative error of 12.7%. The model calculated an adequate response to changes in the boundary conditions and the vegetation cover, and it reproduced the physical processes of the heat balance in coastal areas (Vitale et al., 2018) adequately.

Few heat balance studies were carried out on salt marshes. Teal and Kanwisher (1970) calculated the energy balance for the plants growing in a marsh on Cape Cod. They found that leaf temperature was well coupled to air temperature. If there were no evaporation of water from the leaves, they would have been from 3.6 to 9.2°C above air temperature when heat gain equaled heat loss. Table 7.1 shows some of their results where β is the Bowen ratio ($=Q_H/Q_E$). Latent heat fluxes were always greater than sensible heat ones. The same behavior is found in most of the heat balances in coastal wetlands (i.e., Vugts and Zimmerman, 1985; Rouse, 2000; Beigt et al., 2008).

The net balance between ET and rainfall infiltration is believed to be important in controlling soil salinity, particularly in the less frequently flooded high marsh zone. To elucidate the biophysical effects of drought and salinity on the interception and dissipation of solar energy in estuarine grasses, Maricle et al. (2007) studied leaf energy budget of 13 species. They found that latent heat loss decreased by as much as 65% under reducing water potential (a measure of the ability of a substance to absorb or release water relative to another element), causing an increase in leaf temperature of up to 4°C. Consequently, radiative and sensible heat losses increased under decreasing water potential. Sensible heat flux increased as much as 336% under reducing water potential. Latent heat loss appeared to be a primary mode of temperature regulation in all species, and sensible heat loss seems to be more important in high marsh species than in low marsh ones (Maricle et al., 2007). In general, heat budget estimates showed similar annual patterns (Sultan and Ahmad, 1994; Roads and Betts, 2000; Hughes et al., 2001; Rutgersson et al., 2001; Finch and Gash, 2002; etc.).

Another method to calculate the heat budget of a tidal flat area indirectly from downstream observations of temperature and horizontal velocity in a tidal course was presented by Onken et al. (2007). The advective heat flux (Qa) in tidal channels is monitored, and then the heat excess or deficit for the catchment area is calculated by integral methods. Instead of using the bulk formulation, the authors established a relationship between the velocity and the volume flux. The heat budget of the upstream region is then determined by integrating the heat flux over one tide (Qtide). The heat budget of a water volume fixed in space is defined as the sum of the heat input through its boundaries and the production of thermal energy inside the volume (Onken et al., 2007). The direction of the heat flux components between the tidal flat and the associated channels during all the different tidal stages is described in Fig. 7.2. At low tide, the bottom and the atmosphere exchange heat regarding the following vertical heat flux

$$Q_{at} = Q_{sw} + Q_{lw} + Q_E + Q_H \tag{7.8}$$

TABLE 7.1 Heat Fluxes (W m^{-2}) Measured in a *Spartina alterniflora* Salt Marsh on Cape Cod

Date	R_N	Q_E	Q_H	$\beta = Q_H/Q_E$
Aug. 30	425.353	599.678	104.595	0.17
Aug. 30	439.299	522.975	97.622	0.19
Sep. 28	320.758	278.920	83.676	0.30
Nov. 10	355.623	355.623	104.595	0.29

Modified from Teal, J.M., Kanwisher, J.W., 1970. Total energy balance in salt marsh grasses. Ecology 51, 690–695.

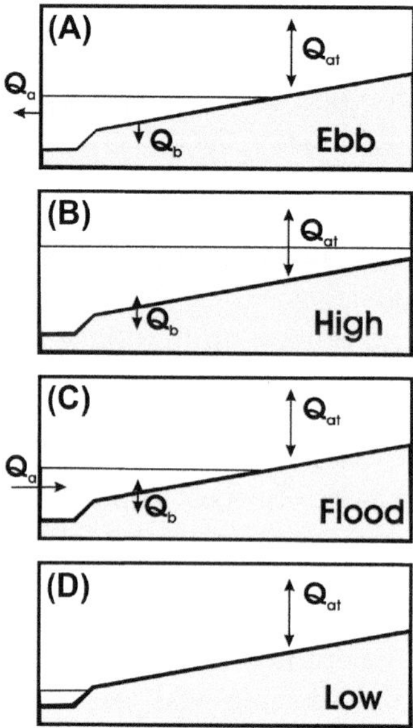

FIGURE 7.2 Heat flux components in different tide stages. (A) Ebb condition. (B) High tide. (C) Flood condition. (D) Low tide. *Modified from Onken, R., Callies, U., Vaessen, B., Riethmuller, R., 2007. Indirect determination of the heat budget of tidal flats. Continental Shelf Research 27, 1656–1676.*

where Q_{sw} is the shortwave radiative heat flux and Q_{lw} is the longwave radiative heat flux. The bottom heat content changes due to atmosphere–soil interaction (Fig. 7.2D). The advective flux in the channel is zero because there are no currents (slack water). During flood tide, there is a positive flux of heat toward the tidal flat by means of the advective flux Q_a (onshore flow). At the same time, the heat content of the water column is modified by the interaction with the atmospheric flux Q_{at}, and the heat exchange Q_b between the substrate and the water (Onken et al., 2007). The high water condition (Fig. 7.2B) is also characterized by zero heat advection, but the water temperature will change due to Q_{at} and Q_b fluxes. The heat gain (or loss) Q_{tide} of the water over one tidal cycle from low tide to low tide was determined by measuring the terms of Eq. (7.9),

$$Q_{tide} = \int_{t_{low}}^{t_{low}^{+}} Q_a dt = \int_{t_{low}}^{t_{low}^{+}} (Q_{at} + Q_b) dt \tag{7.9}$$

where t is time, t_{low} indicated the time of the low tide, and t_{low}^{+} the time of the subsequent one. Onken et al. (2007) developed, in addition, a simple model that can be used to determine the integral bottom heat flux of the tidal flats. An analytical estimate suggests that the tidal prism and the length of the drying period of the flats in the upstream region control the sign of the budget. This method presents another option for a good estimate of the heat balance equation without using the bulk formulation.

Other studies, such as heat budget investigations, performed in Pauatahanui Inlet, a small (3.5 km^2) New Zealand coastal inlet, indicate that the balance is essentially between solar and longwave radiation, evaporation, and advective heat exchange (Heath, 1977). The role of the mudflat in the heat balance is minor. The temperature at the entrance to the inlet exhibits high tidal fluctuations resulting from the exchange with the coastal waters and a diurnal inequality produced by the interaction of solar heating with the tidally controlled surface area and volume. The simulated temperature record for a month, calculated from a heat budget equation, exhibits the effect of the tidal/solar interaction in producing a 14.75-day pulse, variable diurnal inequality, and the generation of high-frequency components (Heath, 1977).

Several studies in different coastal environments related to heat balance equation focus on temperature fluctuations and/or applied numerical models. Piccolo et al. (1993) studied the thermal behavior of the air, water, and sediments over a tidal flat at Starr's Point, the Bay of Fundy, in summer. The temperature in the intertidal sediments showed rapid changes that occur principally during tidal inundation. Vertical gradients of 0.5×10^{-2} cm^{-1} were found in the upper 0.25 m layer. The presence of large populations of *Corophium volutator* increased thermal diffusivity because of their vertical migration. Therefore, the heat was distributed faster and through a greater depth. Vugts and Zimmerman (1985) predicted daily mean water temperatures with heat balance calculations of the tidal flat areas of the Dutch Wadden Sea. The daily heat balance interacts with the tidal cycle, resulting in a 15-day periodicity in the water temperature as well in the bottom temperature. They showed that with a simple model and some measured bulk parameters, it is possible to predict daily mean water temperatures from simple weather data measured at a nearby coastal station.

The traditional formulation of the SWIFT 2D model has been applied to numerous estuaries, bays, and harbors throughout the world. Swain (2005) made changes to expand SWIFT 2D for applicability to shallow coastal wetlands. These modifications included the representation of spatially and temporally varying rainfall and ET, wind sheltering owing to effects of emergent vegetation, and changes in frictional resistance with depth. Dietrich et al. (2004) presented a model-based method of determining the surface fluxes of heat and freshwater in the near shore coastal waters. The new method determines the fluxes as a residual within the framework of physically realistic and natural boundary conditions on the sea surface temperature and sea surface salinity.

By a balance model of the energy by surface waves in a coastal zone and experimental data about surface flows in shallow and deep-water zones, Panin et al. (2006) developed a model of the heat–mass exchange of a coastal zone reservoir with the atmosphere. The model allows the calculation of the values of the energy–mass exchange in the atmospheric boundary layer and the corresponding interaction characteristics based on standard micrometeorological information. For the construction of the model and its verifications, they used the direct measurement data (eddy-covariance) of the momentum, heat, and humidity turbulent fluxes, as well as the surface wave characteristics, and the main microcharacteristics of air and water. The model shows the intensification of the processes in a coastal zone in comparison with an open sea and allows the determination of the size intensification of flows at different distances from the coast.

The main methods used to study heat balance in midlatitudes coastal wetlands are the use of aerodynamic bulk formulas, satellite information, and different numerical models. The primary agents that affect these wetlands are the tides, the vegetation (or lack of it), and the wind. The magnitudes of the different terms of Eq. (7.1) depend on the particular microclimate of each ecosystem and the site conditions. In the last decade, the accuracy of the calculation of the heat balance has improved due to new and increasingly sophisticated methods. However, still much work is needed to find an accurate model or better measuring techniques to determine each term of Eq. (7.1).

3. LOW LATITUDES

Estimation of the heat budget in low latitudes (tropical and subtropical) is performed as it is in midlatitudes, although with a significant increase in radiation energy and dense vegetation. Salt marshes are ideal for studying patterns in plant community. Species interactions during colonization of bare patches are different than those found in dense vegetation. The metabolic process of the plant communities exhibits different rhythms of intensity. Variations in environmental factors such as light and temperature regulate them. Another important vegetation parameter such as photosynthesis is known to be temperature sensitive (e.g., Hargrave, 1969; Gallagher and Daiber, 1973). Because temperature affects respiration rates, an exogenous daily rhythm in respiration in salt marshes and mangroves would be expected in response to temperature cycles (Gallagher and Daiber, 1973).

The presence of plants affects ecosystem-level processes such as hydrology, sedimentation rate, and nutrient cycling (Bertness, 1988; Whitcraft and Levin, 2007). Plant cover is a fundamental feature of many coastal marine and terrestrial systems and controls the structure of associated animal communities. Studying the impact of shading in salt marshes, a relationship between temperature, salinity, water content, and macrofaunal density and diversity was determined by Whitcraft and Levin (2007). The authors found that increase in temperature and salinity and decrease in water content for *Salicornia virginica* was correlated with a decrease in macrofaunal density. Vegetation affects the distribution of incoming solar energy (reflection, ET, sensible heat, ground heat flux, and photosynthesis) and controls the daily dynamics of temperature on ecosystems (Huryna et al., 2014), therefore also controls the microclimate of them.

In recent years, Wu et al. (2016) tested whether the current understanding of the insignificant effect of drainage on ET in the temperate region wetlands applies to those in the subtropics. They identify the hydroclimatic drivers causing the changes in drained wetlands in the Florida's Everglades region with a subtropical climate. The authors compared the annual ET between a more intensively drained zone (836 mm) with another more humid (1271 mm) and found a significant variation (34%, $P = .001$). They concluded that the result was mainly due to drainage-driven differences in inundation and associated effects on net radiation (R_N) and local relative humidity (Wu et al., 2016). Similar results show the review of Huryna and Pokorny (2016). Based on bulk formulation they compared the magnitudes of the heat balance equation terms in drained and very humid landscapes (Fig. 7.3).

Estuaries in arid tropical regions differ significantly from their temperate and wet tropical counterparts. First, river discharge into the estuaries is often highly seasonal with very large flows in the wet season being followed by 5 to 10 months of negligible discharge. The second distinction is that large areas of salt marshes, mangrove swamps, and salt flats (where annual evaporation greatly exceeds annual precipitation) often fringe these tropical arid estuaries (Ridd and Stieglitz, 2002; de Silva Samarasinghe and Lennon, 2004) and they usually becomes hypersaline for most of the year. Evaporation plays an important role in concentrating salts and nutrients in soils and groundwater in estuarine wetlands. This is particularly true in zones of less frequent tidal inundation where the soil salinity depends on the balance between "evapoconcentration" of tidally supplied salts and rainfall or groundwater flushing (Hughes et al., 2001). The more arid the climate, the more extreme this effect becomes.

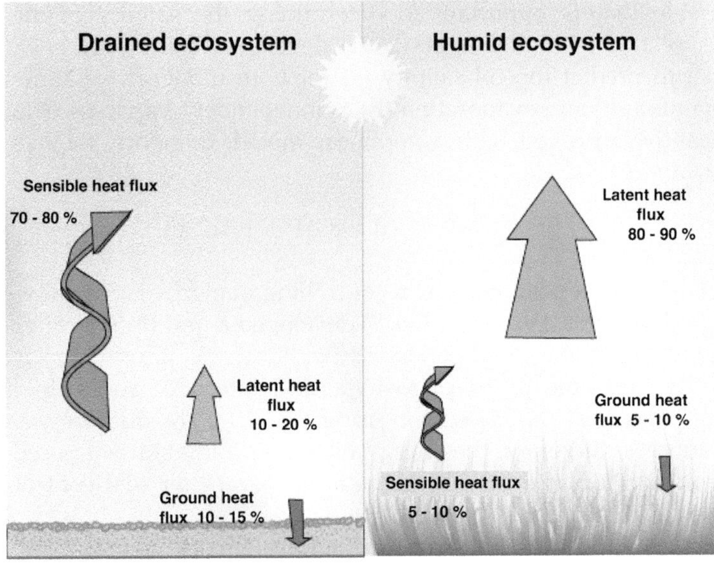

FIGURE 7.3 Redistribution of the Surface Radiation Balance in drained and humid wetlands. *Modified from Huryna, H., Pokorny, J., 2016. The role of water and vegetation in the distribution of solar energy and local climate: a review. Folia Geobotanica. https://doi.org/10.1007/s12224-016-9261-0.*

ET estimates are often the weak link in wetland water and solute balance modeling (Hughes et al., 2001). Therefore, latent heat flux is a major variable to calculate the salinity changes estimates over mangroves. The plant zonation in salt marshes is a consequence of local variation in soil patch salinities (Bertness, 1991). The different soil salinities are due to tidal flooding, annual variation in rainfall, ET, and small-scale topographic features that influence the drainage. Evaporation and ET increase the salinity of the swamp soils. A typical value of evaporation over open water in tropical areas is 5 mm day^{-1}. Wolanski et al. (1980) and Wolanski and Ridd (1986) calculated evaporation rate over mangroves at a rate of 2 mm day^{-1}. For salt flats, Hollins and Ridd (1997) estimated a monthly average evaporation rate of 2 mm day^{-1} with peak rates of 4–5 mm day^{-1} during spring tides (when the flats are saturated) falling to less than 1 mm day^{-1} when the salt flats form a hard surface crust. The salinity rate of change due to this effective evaporation rate E (Ridd and Stieglitz, 2002) is given by

$$\frac{\partial S}{\partial t} = \frac{ES}{h} \tag{7.10}$$

where S is the salinity, E the evaporation, and h is the depth. Field data from five arid estuaries fringed by mangroves and salt flats indicate that where a large area of salt flats and mangroves extends over the whole length of an estuary, the estuary becomes completely inverse with salinity rising up to 55 within a couple of months (Ridd and Stieglitz, 2002). The estuarine evaporation rates due to the presence of salt flats and mangroves cause a rapid increase in salinity. The persistence of freshwater in the upper reaches of this type of estuaries is likely to affect mangrove species assemblages (Ridd and Stieglitz, 2002).

Another variable that is important in determining the salinity of the soil in coastal wetlands is precipitation. Mondal et al. (2001) developed a multiple linear and nonlinear regression model to predict topsoil salinity (S) for both moderately saline and saline soils by using daily rainfall (P) and evaporation (E) as independent variables. The prediction level was not significantly improved with a nonlinear model; therefore, they suggest using the following linear model

$$S = 1.29077 - 0.49831P + 1.31230E \tag{7.11}$$

Salinity and tides have a primary role in controlling mangroves coastal wetlands communities of subtropical climates. Perri et al. (2017) developed a first-order ecohydrological model of the soil/plant atmosphere for *Avicennia marina*, known as gray mangrove, a highly salt-tolerant pioneer species able to adapt to hyperarid intertidal zones. In the model, they calculate the leaf balance of the vegetation using Eq. (7.1). The authors assumed that Q_B is negligible for canopies; therefore, the energy of the net radiation was used by Q_S and Q_E. The authors demonstrated that salinity is the main parameter that controls the mangrove transpiration in those ecosystems.

Another example of the use of a simplified surface energy budget is the contribution of Shoemaker et al. (2005) to estimate the factors that controls the budget in tropical coastal wetlands. They calculated the change in heat energy stored within a column of wetland surface water at seven sites in the wetland areas of southern Florida with the following equation

$$R_N - (W + G_{veg}) = Q_E + Q_H \tag{7.12}$$

where G_{veg} is biomass storage (heat energy storage in the vegetation) and W is the change in heat energy stored in wetland surface water. The difference between Eqs. (7.1) and (7.12) is that the advection and soil heat flux ($Q_A + Q_B$) terms were replaced by the energy stored term in wetland surface and also the inclusion of the biomass energy storage term to add the effect of the vegetation. This method estimate changes in stored heat energy that overcome a major data limitation, namely, the limited spatial and temporal availability of water temperature measurements because it is assumed that a change in surface water temperature reflects a change in stored heat energy (i.e., Edinger et al., 1968). The new method was based on readily available air temperature measurements and relies on the convolution of air temperature changes with a regression-defined transfer function to estimate changes in water temperature. These results can probably be adapted to other humid subtropical wetlands characterized by open water, sea grass, and diverse vegetation community types. It is important to point out that some authors incorporate not only the natural wetland heat energy stored in the vegetation but also that stored in the fauna. In the literature, several studies on the energy flow or trophic level production of coastal wetlands discuss the subject (i.e., Smalley, 1960; Teal, 1962; Wolff et al., 2000).

Most of the studies to predict the response of vegetated ecosystems to global and regional climatic changes are based on numerical models. They require accurate knowledge of the surface heat balance. Considering those coastal areas of low latitudes present dense vegetation not only ET and precipitation should be examined in the calculations but also the advective effect of the tides. Investigations of the heat flow in coastal zones affected by variable surface water levels and dense vegetation are few. Barr et al. (2013) analyzed the influence of the tidal advection on the summer heat balance in the mangrove forest of the Everglades, FL, USA, considering both (1) lateral advection of energy at the forest–estuary interface and (2) vertical exchanges of energy at the water–soil and water–air interfaces. The authors found that tides represented a significant sink of energy for the high net radiation loads that characterized the summer. Heat is transferred to the water when the cooler water inundates the warmer soil during flooding conditions. Then the ebb tides transport the dissipated heat to the river and estuary. They established that flood tides provided a microclimate buffer in the mangrove forest ecosystem. The same result was also presented in a heat budget analysis performed by Ohtsuki et al. (2015) in tidal flats and mangrove areas of Sumiyo Bay, Amami Ōshima Island. They describe how the sensible heat flux from the bottom of the tidal flat and mangrove areas significantly cooled the water column in the winter season.

Measuring the different terms of heat balance is expensive and challenging at some coastal sites, especially in those with close access. Therefore, the use of remote sensing has become a useful and accessible tool. R_N or solar radiation measurement is relatively accurate, but the estimation of the other terms of the heat balance equation is difficult. Two general classes are used to estimate ET by remote sensing-based methods: temperature- or energy-based methods and vegetation-based ones. Temperature-based methods typically compute latent heat as a residual of the surface energy balance and use radiometric surface temperature to estimate the sensible heat flux (Biggs et al., 2016). Therefore, latent heat flux and the corresponding water loss in wetlands are well correlated to radiation and temperature (Barraza et al., 2017). In arid and semiarid lands, Q_E is mostly driven by available water, and the vegetation exerts an active control over the rate of transpiration. Therefore, Q_E models that use optical vegetation indices to represent the plant component overestimate water fluxes in water-limited ecosystems (Barraza et al., 2017). Examples of estimations of Q_E are shown in Table 7.2 showing several field measurements and satellite modeled latent heat flux in various types of forested vegetation at low latitudes (Lagomarsino et al., 2015). The values differ according to the height and cover of the vegetation and also by inundation, severe storms, and tidal flushing that characterize each site.

To find the best remote sensing model of the energy surface balance (SEB) to be applied in the humid southeastern United States, Bhattarai et al. (2016) tested five models: the Surface Energy Balance Algorithm for Land (SEBAL), mapping ET at high resolution with Internalized Calibration (METRIC), Simplified Energy Balance Index (S-SEBI), Surface Energy Balance System (SEBS), and Simplified Surface Energy Balance (SSEBop). The authors compared the predictions of the ET using Landsat images for the period 2000–10 with measurements at four sites (marsh, grass, and citrus surfaces). SEBS model performed better than the others models in estimating daily ET from different land covers (the root mean squared method (RMSE) $= 0.74$ mm day^{-1}). For short grass conditions SEBAL, METRIC, and S-SEBI worked much better than SEBS. Therefore, for humid climates, SEB was the best model, although some modifications should be performed for short vegetation (Bhattarai et al., 2016).

The only simultaneous comparison is the work of Lagomarsino et al. (2015). The authors used satellite remote sensing (Landsat images) and data obtained from the tower described by Barr et al. (2013) on the mangroves coastal zone of the Everglades. They estimated net radiation, latent heat flux (ET), and other surface heat balance terms to model Eq. (7.1). Fig. 7.4 shows a comparison of their modeled and measured heat fluxes. The modeled

TABLE 7.2 Comparison of Satellite—Modeled and Field—Measured Latent Heat Flux (W m^{-2} day^{-1}) of Several Forested Vegetation

Forest Vegetation	Location	Satellite Models	References	Field Data	References
Mangrove forest	Florida (USA)	104–218	Lagomarsino et al. (2015)	115–231	Barr et al. (2014)
Mangrove forest	Thailand			200–425	Monji et al. (2002)
Mangrove forest	India	125–190	Ganguly et al. (2008)		
Mediterranean evergreen	Algeria	223–393	Zhaira et al. (2009)		
Tropical rain forest	Brazil			260–310	Malhi et al. (2002)
Coniferous forest	Sweden	100–180	Venalainen et al. (1999)		
Temperate deciduous forest	Tennessee (USA)			40–400	Baldocchi and Wilson (2001)

After Lagomarsino, D., Price, R.M., Whitman, D., Melesse, A., Oberbauer, S.F., 2015. Spatial and temporal variability in spectral-based surface energy evapotranspiration measured from Landsat 5TM across two mangrove ecotones. Agricultural and Forest Meteorology 213, 304–316.

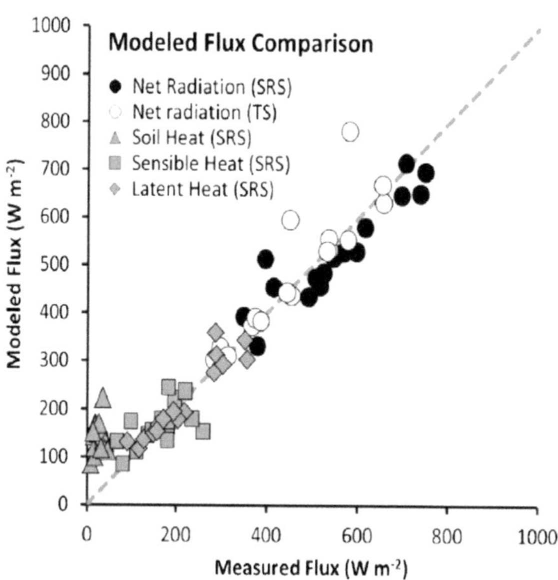

FIGURE 7.4 Comparison of the measured and modeled terms of the heat balance at the Everglades coastal zone (Lagomarsino et al., 2015).

data were in agreement with the field measurements despite some differences in specific dates. The results of Lagomarsino et al. (2015) show that the modeled values using satellite images constitute a good estimate of the surface energy balance of forest mangroves and it is one of the few examples of surface energy balance in mangrove forests using satellite images.

4. HIGH LATITUDES

High-latitude (subpolar and polar) wetlands are underlain by ice-rich permafrost, which helps maintain wetland systems and also imparts unique characteristics to their energy and water balances. In North America, components of the radiation balance decrease poleward ($-1.8\ W\ m^{-2}{}^{\circ}lat^{-1}$), whereas the poleward rate of decrease of temperature (i.e., between the latitudes of 50°–65°: $-1^{\circ}C\ {}^{\circ}lat^{-1}$) and precipitation lessens (i.e., between the latitudes of 50°–65°: $-22\ mm\ {}^{\circ}lat^{-1}$) (Rouse, 2000). Large annual changes in solar input characterize high latitudes. Albedo decreases sharply from winter when the surface is snow-covered to summer, especially in nonforested regions such as Arctic tundra and boreal wetlands. A primary characteristic of wetlands, in permafrost regions, is that the ice-rich frozen layer inhibits vertical water losses so that ponded water can persist through much of the summer. Components of the radiation balance tend to decrease with increasing latitude (Rouse, 2000; Rouse et al., 2004). During the 4-month summer of a high subarctic wetland, net radiation is large, and the latent heat flux dominates the energy cycle (Rouse, 2000). In winter, which typically lasts a minimum of 7 months, there is almost no evaporation, but there is sublimation loss. In winter, heat loss from the ground approximately balances negative net radiation. After the final departure of the snow in spring, there is a change in net radiation, evaporation, and ground heat flux. A primary requirement in high latitudes is documentation of the magnitudes of the forcing parameters, of which the most important is precipitation, in all its forms.

Globally annual Q_H on average is negative, and it constitutes a considerably smaller surface heat sink than does evaporation (Q_E). However, over the ice sheet of polar regions, the situation is different; the snow surface is highly reflective for shortwave radiation, whereas it effectively loses heat in the form of longwave radiation to the cold and clear overlying atmosphere. As a result, the surface is colder than the air, generating positive Q_H. Q_E is usually small because low temperatures limit the absolute moisture content. On the coastal areas of polar latitudes, cooling of the near-surface air originates strong and persistent katabatic winds. These winds provide the shear necessary to maintain turbulent exchange in the stably stratified surface layer, especially in winter (Van den Broeke et al., 2005a).

Synoptic weather systems play a significant role in day-to-day energy and water responses to climate forcing. Presented in Table 7.3 is an example of the influence of weather conditions on the energy balance of Hudson Bay coastal wetland. R_N is similar for both warm and cold overlying air masses. However, all other energy balance components are different. Under warm air mass conditions, latent heat flux is greater than sensible heat flux, as expected in low and midlatitudes, but in the cold air condition, the sensible heat flux is enhanced considerably (Rouse, 2000).

TABLE 7.3 Comparative Energy Balance Under a Warm and a Cold Air Mass Flow at Hudson Bay (Rouse, 2000)

Fluxes (W m^{-2})	Warm Air Mass	Cold Air Mass	Warm/Cold
R_N	161	161	1.00
Q_E	85	64	1.33
Q_H	47	80	0.59
Q_B	19	17	1.13

Cold and warm air advection actively controls the energy partitioning at coastal sites (Harazono et al., 1996; Lynch et al., 1999). A typical example in the Hudson Bay is the onshore winds that advect cold and humid air masses from the Arctic Ocean resulting in low air temperature, a large temperature gradient between the land surface and the air, and, therefore, a high sensible heat flux and low evaporative flux. Conversely, when offshore winds advected warm and dry continental air to the site, the temperature gradient between the land surface and the air was small, resulting in low sensible heat flux, but only slightly higher evaporative flux than during onshore wind conditions (Lynch et al., 1999). The authors employed a regional climate model in their study (ARCSyM) to simulated fluxes that were within the range of measured fluxes (Table 7.4), but overestimated both net radiation and latent heat fluxes. The regional model also captured site to site variations quite well, which appear to be more sensitive to mesoscale meteorological conditions than to site characteristics. The difference in energy partitioning was primarily due to larger heat gain of the open water ponds during the offshore conditions and to a minor increase in ground heat flux (Table 7.4). Yoshimoto et al. (1996) found a similar behavior in energy partitioning at Barrow, Alaska, during their 1993 field season.

The study of the effect of advection from the cold polar sea to a warmer terrestrial surface in the coastal area of Hudson Bay was carried out also by Weick and Rouse (1991). Three objectives were pursued: (1) to investigate the changes in the surface energy balance along a transect perpendicular to the coast line of the bay, (2) to identify local and regional effects on the energy balance along the transect, and (3) to document the use of a box model and the divergence and convergence of energy mass in the coastal boundary layer during onshore and offshore winds. Applying Eq. (7.1), the different fluxes and the Bowen ratio were estimated. The authors demonstrated that the largest advective influence on the turbulent fluxes

TABLE 7.4 Monthly Mean Surface Energy Balance (W m^{-2}) at Betty Pingo Coastal Site (70°18′N, 149°55′W) for the Summer (Lynch et al., 1999)

	R_N	Q_H	Q_E	Q_B
June 1995	128.1	39.6	61.0	27.5
July 1995	122.6	57.2	38.1	27.3
August 1995	75.7	14.7	48.6	12.4

TABLE 7.5 Seasonal Bowen Ratios (β) at Four Microclimate Stations Located at 0, 2.5, 9.4, and 12.4 km From the Coast for all Wind Conditions and for Onshore and Offshore Winds at Prudhoe Bay (Weick and Rouse, 1991)

Wind Conditions	Site 1 at the Coast	Site 2 (2.5 km)	Site 3 (9.4 km)	Site 4 (12.4 km)
All directions	0.95	0.79	0.58	0.58
Onshore winds	1.10	0.93	0.64	0.67
Offshore winds	0.71	0.58	0.48	0.43

occurs within 10 km of the coast, with a 2.7-fold downwind decrease in the Bowen ratio and a 1.8-fold decrease during offshore winds (Table 7.5). This decrease is due both to boundary layer adjustments to a new surface under onshore winds and to horizontal and vertical convergence and divergence in the atmosphere under all wind conditions.

Mendez et al. (1998) compared different ET models applied to an Arctic coastal wetland near Prudhoe Bay, Alaska, during the summers between 1994 and 1996. The objective was to gain a better understanding of ET in arctic wetlands. Evaporation after spring snowmelt averaged 3.11 mm day^{-1}. Latent heat flux was the dominant heat sink in wetlands, whereas sensible heat flux dominated in the drier upland area. Differences between the formulations were not significant.

The future increase in air temperature and precipitation due to climate change in Arctic wetlands could significantly affect ecosystem function; therefore, it is important to define controls on ET in high-latitude wetlands because it is the primary pathway of water loss from these systems. Liljedahl et al. (2011) studying the effects of the variations in near-surface soil moisture and atmospheric vapor pressure deficits on midday ET rates found that they have nonlinear effects. The midday ET was suppressed from both dry and wet soils but through different mechanisms. These parameters have the potential to a moderate inter-annual variation of total ET and reduce excessive water loss in a warmer climate. The authors emphasize that combined with the prevailing maritime winds and projected increases in precipitation these mechanisms will likely prevent extensive future soil drying and hence maintain the presence of coastal wetlands.

In the Southern Hemisphere using data of four automatic weather stations, Van den Broeke et al. (2005b) analyzed the turbulent exchange of sensible heat on the interior plateau in Dronning Maud Land, East Antarctica, snow surface for a 4-year period (1998–2001). Q_H was measured and also calculated using the aerodynamic bulk formulation. The authors found that at all the four sites, the surface layer was, on average, stably stratified, with a positive annual mean of Q_H. The annual mean was similar for the coastal ice shelf and the interior plateau (8 W m^{-2}). Clouds limit surface radiation loss on the flat coastal ice shelf; therefore, the vertical temperature gradient decreases, resulting in a small average Q_H. In contrast, clear sky conditions and high temperature gradients prevail on the interior plateau. The relatively weak winds, the aerodynamically smooth surface, and stability effects limit the average value of Q_H. The most favorable conditions for sensible heat exchange are found in the katabatic wind zone, where strong winds, in combination with clear skies and a relatively rough surface, result in annual mean Q_B values of 22–24 W m^{-2}.

In the McMurdo Dry Valleys, Victoria Land, Antarctica, the primary source of water to streams, lakes, and associated ecosystems is the melting of glacial ice. Hoffman et al. (2008) applied a one-dimensional energy balance model using 11 years of daily meteorological data and seasonal ablation measurements to analyze the geochemical fluxes and to understand the ecological responses of the area to past and future climates. Applying a daily time-step model, the surface energy balance was represented by

$$\chi(1-\beta)K\downarrow + L\downarrow + L\uparrow + Q_S + Q_E + Q_C = Q_m \quad \left[W\,m^{-2}\right] \tag{7.13}$$

where χ is the total solar radiation between the surface energy balance and the solar radiation source term in the heat transfer equation, β is albedo, Q_C is heat conduction in the ice, and Q_m is the energy available for melting, calculated as a residual. Their results showed good agreement between calculated and measured ablation and ice temperatures over the 11 years and also indicated that above freezing air temperatures did not necessarily result in melt and, in turn, melt occurred during subfreezing air temperatures under particular conditions. For air temperatures near freezing, low wind speed was essential for melt initiation. According to the model, subsurface melt occurred three times more frequently than surface melt occurred, no deeper than 50 cm below the glacier surface, and was small, never exceeding 8% by mass.

5. SUMMARY

The estimation of the heat balance depends on many variables. Some of them can be measured accurately, whereas others are calculated based on environmental variables. Although the latitude determines the amount of incoming radiation entering the ecosystem, the resultant net radiation also depends on the site conditions that characterize the ecosystem. The net radiation represents the available energy that different turbulent processes, which describe any energy budget, will have. Net radiation energy is employed by evaporation/ET (latent heat flux) and by heat conduction between the air–water and air–substrate (sensible heat flux). The typical vegetation of each ecosystem uses part of that energy in its particular biological processes. Therefore, the remaining heat is transported as advective heat flux either by wind or tides.

Temperate wetlands, especially tidal flats and salt marshes, are by far the sectors that have received more attention. Plant cover influences the microhabitat of the sediment by controlling the amount of light reaching the sediment surface. Significant differences might be found in heat balances between bare and vegetated marshes. These changes may induce variations in the sediment biotic community (Whitcraft and Levin, 2007). In low latitudes, besides the net radiation, the water balance and the evaporation are the most significant processes that define the heat balance and, then, the temperature variations in the soils, coastal waters, and lower layers of the atmosphere. Temperature not only affects respiration rates and photosynthesis, influences the distribution and movements of fish, but also affects many important biological processes (number of eggs laid, incubation time, etc.). Therefore, heat balance studies might help to understand temperature variations of wetlands and some of the processes that generate distribution patterns in coastal natural flora and fauna communities.

In the arid climate estuaries of low latitudes, the relationship between salinity and evaporation is significant. Although some formulas that relate both parameters are presented in this chapter; more measurements and experiments in diverse environments should be made. On the other hand, in high latitudes, because of the cold climate, numerical models are the most powerful tools to study the heat balance. However, in the last decade, an increase of field measurements to calculate the heat balance equation is found in the literature because measurements are needed to calibrate satellite images but most important because the changes in high-latitude regions (including wetlands) are affecting the whole planet. Remote sensing techniques are a powerful tool used by many scientists.

Many themes remain to be investigated on the heat budget, such as the high-frequency monitoring of the heat balance components in vegetated and bare coastal wetlands and its influence in the plant community. The effects of shading in temperature variation of wetlands soils, the effect of the heat balance in plant zonation, and animal interactions are also important subjects that should be studied. The results of such studies would provide some insight into flora and fauna distribution, biodiversity, and species behavior in diverse wetlands. Scientists are working with the bulk formulas in numerical models in several sites, time, and space scales. The interaction between biological and physical processes needs to be studied deeply. An exhaustive analysis of the underlying interactions among flora, fauna, and heat stored in coastal wetlands is still lacking. Future studies should investigate these interactions.

Different methods and techniques are used to study the heat balance in coastal wetlands. The reliability of the calculation of the heat balance has improved due to new and increasingly sophisticated methods. However, because of the variability and the different processes that characterize the wetlands, there is not a single approach that the researchers agree that is best to estimate the heat balance.

References

Acreman, M.C., Harding, R.J., Lloyd, C.R., McNeil, D.D., 2003. Evaporation characteristics of wetlands: experience from wet grassland and a reedbed using eddy correlation measurements. Hydrology and Earth System Sciences 7 (1), 11–21.

Amatya, D.M., Harrison, C.A., 2016. Grass and forest potential evapotranspiration comparison using five methods in the Atlantic coastal plain. Journal of Hydrologic Engineering. https://doi.org/10.1061/(ASCE)HE.1943-5584.0001341.

Bai, P., Gu, Y., Li, P., et al., 2016. Tidal energy budget in the Zhujiang (Pearl river) estuary. Acta Oceanologica Sinica 35–54. https://doi.org/10.1007/s13131-016-0850-9.

Bakhtyar, R., Dastgheib, A., Roelvink, D., Barry, D.A., 2016. Impacts of wave and tidal forcing on 3D nearshore processes on natural beaches. Part I: flow and turbulence fields. Ocean Systems Engineering 6 (1), 23–60.

Baldocchi, D.D., Wilson, K.B., 2001. Modeling CO2 and water vapor exchange of a temperate broadleaved forest across hourly to decadal time scales. Ecological Modelling 1–2, 155–184.

Barr, J.G., Fuentes, J.D., DeLonge, M.S., O'Halloran, T.L., Barr, D., Zieman, J.C., 2013. Summertime influences of tidal energy advection on the surface energy balance in a mangrove forest. Biogeosciences 10, 501–511.

Barr, J.G., DeLonge, M.S., Fuentes, J.D., 2014. Seasonal evapotranspiration patterns in mangrove forests. Journal of Geophysical Research - D: Atmospheres 119.

Barraza, V., Restrepo-Coupe, N., Huete, A., Grings, F., Beringer, J., Cleverly, J., Eamus, D., 2017. Estimation of latent heat flux over savannah vegetation across the North Australian Tropical Transect from multiple sensors and global meteorological data. Agricultural and Forest Meteorology 232, 689–703.

Beigt, D., 2007. Balance energético de las planicies de marea del estuario de Bahía Blanca y su relación con la productividad planctónica del estuario (Doctoral Thesis). Universidad Nacional del Sur, Bahia Blanca, Argentina.

Beigt, D., Piccolo, M.C., 2003. Estudio de la temperatura del agua en relación con la abundancia del microzooplancton en Puerto Cuatreros, estuario de Bahía Blanca. In: Contribuciones Científicas 2003 GAEA, Sociedad Argentina de Estudios Geográficos, pp. 49–55.

Beigt, D., Piccolo, M.C., Perillo, G.M.E., 2003. Soil heat exchange in Puerto Cuatreros tidal flats, Argentina. Ciencias Marinas 29 (4B), 595–602.

Beigt, D., Piccolo, M.C., Perillo, G.M.E., 2008. Surface heat budget of an estuarine tidal flat (Bahía Blanca Estuary, Argentina). Ciencias Marinas 34 (1), 1–15.

Bertness, M.D., 1988. Peat accumulation and success of marsh plants. Ecology 69, 703–713.

Bertness, M.D., 1991. Interspecific interactions among high marsh perennials in a New England salt marsh. Ecology 72, 125–137.

Bhattarai, N., Shaw, S.B., Quackenbush, L.J., Im, J., Niraula, N., 2016. Evaluating five remote sensing based single-source surface energybalance models for estimating daily evapotranspiration in a humidsubtropical climate. International Journal of Applied Earth Observation and Geoinformation 49, 75–86.

Bianciotto, O.A., Pinedo, L.B., San Roman, N.A., Blessio, A.Y., Collantes, M.B., 2003. The effect of natural UV-B radiation on a perennial *Salicornia* salt-marsh in Bahía San Sebastián, Tierra del Fuego, Argentina: a 3-year field study. Journal of Photochemistry and Photobiology B: Biology 70, 177–185.

Biggs, T.W., Marshall, M., Messina, A., 2016. Mapping daily and seasonal evapotranspiration from irrigated crops using global climate grids and satellite imagery: automation and methods comparison. Water Resources Research 52. https://doi.org/10.1002/2016WR019107.

Castro, R., Pares Sierra, A., Marinote, S.G., 2003. Evolución y extensión de los vientos Santa Ana de febrero de 2002 el océano frente a California y la peninsula de Baja California. Ciencias Marinas 29 (3), 275–281.

Clulow, A.D., Everson, C.S., Mengistu, M.G., Jarmain, C., Jewitt, G.P.W., Price, J.S., Grundling, P.L., 2012. Measurement and modelling of evaporation from a coastal wetland in Maputaland, South Africa. Hydrology and Earth System Sciences 16, 3233–3247.

Costa, C.S., Armstrong, R., Detres, Y., Koch, E.W., Bertiller, M., Beeskow, A., Neves, L.S., Tourn, G.M., Bianciotto, O.A., Pinedo, L.B., Blessio, A.Y., San Roman, N., 2006. Effect of ultraviolet-B radiation on salt marsh vegetation: trends of the genus *Salicornia* along the Americas. Photochemistry and Photobiology 82 (4), 878–886.

Crabtree Jr., S.J., Kjerfve, B., 1978. Radiation balance over a salt marsh. Boundary-Layer Meteorology 14, 59–66.

Dabrowska-Zielinska, K., Budzynska, M., Kowalik, W., Turlej, K., 2010. Soil moisture and evapotranspiration of wetlands vegetation habitats retrieved from satellite images. Hydrology and Earth System Sciences Discussions 7, 5929–5955. https://doi.org/10.5194/hessd-7-5929-2010.

de Silva Samarasinghe, J.R., Lennon, G.W., 2004. Hypersalinity, flushing and transient salt-wedges in a tidal gulf: an inverse estuary. Estuarine, Coastal and Shelf Science 24, 483–498.

Dietrich, D.E., Haney, R.L., Fernández, V., Josey, S.A., Tintore, J., 2004. Air–sea fluxes based on observed annual cycle surface climatology and ocean model internal dynamics: a non-damping zero-phase-lag approach applied to the Mediterranean Sea. Journal of Marine Systems 52, 145–165.

Drexler, J.Z., Snyder, R.L., Spano, D., Paw, U.K.T., 2004. A review of models and micrometeorological methods used to estimate wetland evapotranspiration. Hydrological Processes 18, 2071–2101.

Dutta, D., Mahalakshmi, D.V., Goel, P., Singh, M., Dadhwal, V.K., Reddy, R.S., Jha, C., 2016. Estimation of daily average net radiation and its variation over West Bengal, India using MODIS products. Geocarto Internacional 32 (3), 286–297.

Edinger, J.E., Duttweiler, D.W., Geyer, J.C., 1968. The response of water temperatures to meteorological conditions. Water Resources Research 4 (5), 1137–1143.

El-Metwally, M., 2004. Simple new methods to estimate global solar radiation based on meteorological data in Egypt. Atmospheric Research 69, 217–239.

Evett, S.R., 2002. Water and energy balances at soil-plant-atmosphere interfaces. In: Warrick, A. (Ed.), The Soil Physics Companion. CRC Press LLC, Florida, pp. 127–190.

Evett, S.R., Matthias, A.D., Warrick, A.W., 1994. Energy balance model of spatially variable evaporation from bare soil. Soil Science Society of America Journal 58 (6), 1604–1611.

Finch, J.W., Gash, J.H.C., 2002. Application of a simple finite difference model for estimating evaporation from open water. Technical Note. Journal of Hydrology 255, 253–259.

Friehe, C.A., Schmitt, K.F., 1976. Parameterization of air-sea interface fluxes of sensible heat and moisture by the bulk aerodynamic formulas. Journal of Physical Oceanography 6, 801–809.

Gallagher, J.L., Daiber, F.C., 1973. Diel rhythms in edaphic community metabolism in a Delaware salt marsh. Ecology 54, 1160–1163.

Ganguly, D., Dey, M., Mandai, S.K., De, T.K., Jana, T.K., 2008. Energy dynamics and its implication to biosphere-atmosphere exchange of CO_2, H_2O and CH_4 in a tropical mangrove forest canopy. Atmospheric Environment 42, 4172–4184.

Gavin, H., Agnew, C.T., 2003. Evaluating the reliability of point estimates of wetland reference evaporation. Hydrology and Earth System Sciences 7 (1), 3–10.

Gould, T.J., Hess, C.T., 2005. Measuring and modelling exposure from environmental radiation on tidal flats. Nuclear Instruments and Methods in Physics Research A 537, 658–665.

Harazono, Y., Yoshimoto, M., Vourlitis, G.L., Zulueta, R.C., Oechel, W.C., 1996. Heat, water and greenhouse gas fluxes over the arctic tundra ecosystems at Northslope in Alaska. In: Proc. IGBP/BAHC-LUCC, Kyoto, Japan, pp. 170–173.

Hargrave, B.T., 1969. Similarity of oxygen uptake by benthic communities. Limnology and Oceanography 14, 801–805.

Harrison, S.J., Phizacklea, A.P., 1985. Seasonal changes in heat flux and heat storage in the intertidal mudflats of the Forth Estuary, Scotland. Journal of Climatology 5, 473–485.

Harvey, J.W., Nuttle, W.K., 1995. Fluxes of water and solute in a coastal wetland sediment. 2. Effect of macropores on solute exchange with surface water. Journal of Hydrology 164, 109–125.

Heath, R.A., 1977. Heat balance in a small coastal inlet Pauatahanui inlet, North Island, New Zealand. Estuarine and Coastal Marine Science 5 (6), 783–792.

Hoffman, M.J., Fountain, A.G., Liston, G.E., 2008. Surface energy balance and melt thresholds over 11 years at Taylor Glacier, Antarctica. Journal of Geophysical Research 113, F04014. https://doi.org/10.1029/2008JF001029.

Hollins, S., Ridd, P.V., 1997. Evaporation over a tropical tidal saltflat. Mangroves and Salt Marshes 1, 95–102.

Hughes, C.E., Kalma, J.D., Binning, P., Willgoose, G.R., Vertzonis, M., 2001. Estimating evapotranspiration for a temperate salt marsh, Newcastle, Australia. Hydrological Processes 15, 957–975.

Huryna, H., Pokorny, J., 2016. The role of water and vegetation in the distribution of solar energy and local climate: a review. Folia Geobotanica. https://doi.org/10.1007/s12224-016-9261-0.

Huryna, H., Brom, J., Pokorny, J., 2014. The importance of wetlands in the energy balance of an agricultural landscape. Wetlands Ecology and Management 22, 363–381.

Jacobs, J.M., Anderson, M.C., Friess, L.C., Diak, G.R., 2004. Solar radiation, longwave radiation and emergent wetland evapotranspiration estimates from satellite data in Florida, USA. Hydrological Sciences Journal 49, 461–476.

Johnson, R.G., 1967. Salinity of interstitial water in a sandy beach. Limnology and Oceanography 12, 1–7.

Kantha, L.H., Clayson, C.A., 2000. Small scale processes in geophysical fluid flows. In: International Geophysics Series, vol. 67. Academic press, USA.

Kelleners, T.J., Koonce, J., Shillito, R., Dijkema, J., Berli, M., Young, M.H., Frank, J.M., Massman, W.J., 2016. Numerical modeling of coupled water flow and heat transport in soil and snow. Soil Science Society of America Journal 80, 247–263.

Kelvin, J., Acreman, M.C., Harding, R.J., Hess, T.M., 2017. Micro-climate influence on reference evapotranspiration estimates in wetlands. Hydrological Sciences Journal 62 (3), 378–388.

Kjerfve, b., 1978. Diurnal energy balance of a caribbean barrier reef environment. Bulletin of Marine Science 28 (1), 137–145.

Kreith, F., Sellers, W.D., 1975. General principles of natural evaporation. In: de Vries, D.A., Afgan, N.H. (Eds.), Heat and Mass Transfer in the Biosphere, Part 1. John Wiley and Sons, New York, pp. 207–227.

Lafleur, P.M., 1990. Evaporation from wetlands. The Canadian Geographer 34, 79–88.

Lagomarsino, D., Price, R.M., Whitman, D., Melesse, A., Oberbauer, S.F., 2015. Spatial and temporal variability in spectral-based surface energy evapotranspiration measured from Landsat 5TM across two mangrove ecotones. Agricultural and Forest Meteorology 213, 304–316.

Leal, J.C., Lavín, M.F., 1998. Comparación del viento costero y marino de la región norte del Golfo de California durante el invierno de 1994. GEOS 22, 12–17.

Liljedahl, A.K., Hinzman, L.D., Harazono, Y., Zona, D., Tweedie, C.E., Hollister, R.D., Engstrom, R., Oechel, W.C., 2011. Nonlinear controls on evapotranspiration in arctic coastal wetlands. Biogeosciences 8, 3375–3389.

Liu, X., Ruecker, A., Song, B., Xing, X., Conner, W.H., Chow, A.T., 2017. Effects of salinity and wet-dry treatments on C and N dynamics in coastal-forested wetland soils: implications of sea level rise. Soil Biology and Biochemistry 112, 56–67.

Lynch, A.H., Chapin III, F.S., Hinzman, L.D., Wu, W., Lilly, E., Vourlitis, G., Kim, E., 1999. Surface energy balance on the Arctic tundra: measurements and models. Journal of Climate 12, 2585–2606.

Ma, Y., Su, Z., Koike, T., Yao, T., Ishikawa, H., Ueno, K., Menenti, M., 2003. On measuring and remote sensing surface energy partitioning over the Tibetan Plateau – from GAME/Tibet to CAMP/Tibet. Physics and Chemistry of the Earth 28, 63–74.

Malhi, Y., Pegorano, E., Nobre, A.D., Pereira, M.G.P., Grace, J., Culf, A.D., Clement, R., 2002. The energy and water dynamics of a central Amazonia rain forest. Journal of Geophysical Research 107.

Maricle, B.R., Cobos, D.R., Campbell, C.S., 2007. Biophysical and morphological leaf adaptations to drought and salinity in salt marsh grasses. Environmental and Experimental Botany 60, 458–467.

Mendez, J., Hinzman, L.D., Kane, D.L., 1998. Evapotranspiration from a wetland complex on the arctic coastal plain of Alaska. Nordic Hydrology 29, 303–330.

Mondal, M.K., Bhuiyan, S.I., Franco, D.T., 2001. Soil salinity reduction and prediction of salt dynamics in the coastal ricelands of Bangladesh. Agricultural Water Management 47, 9–23.

Monji, N., Hamotaru, Hirano, T., Fukagawa, T., Yabuki, K., Jintana, V., 2002. CO_2 and heat exchange of mangrove forest in Thailand. Journal of Agricultural Meteorology 52, 489–492.

Monteith, J.L., 1973. Principles of Environmental Physics. Edward Arnold Publisher, London.

Odum, E.P., 1975. Ecology, second ed. Holt, Rinehart and Winston, New York.

Ohtsuki, K., Endo, R., Nihei, Y., Harada, W., Shimatani, Y., 2015. Thermal environment in tidal flat and mangrove areas based on field measurements of water and soil temperatures. In: E-proceedings of the 36th IAHR World Congress, the Hague, The Netherlands, p. 5.

Oke, T.R., 1978. Boundary Layer Climates. Methuen, London.

Onken, R., Callies, U., Vaessen, B., Riethmuller, R., 2007. Indirect determination of the heat budget of tidal flats. Continental Shelf Research 27, 1656–1676.

Panin, G.N., Nasonov, A.E., Foken, T., Lohse, H., 2006. On the parameterisation of evaporation and sensible heat exchange for shallow lakes. Theoretical and Applied Climatology 85, 123–129.

Paulescu, M., Fara, L., Tulcan-Paulescu, E., 2006. Models for obtaining daily global solar irradiation from air temperature data. Atmospheric Research 79, 227–240.

Pennings, S.C., Grant, M., Bertness, M.D., 2005. Plant zonation in low-latitude salt marshes: disentangling the roles of flooding, salinity and competition. Journal of Ecology 93, 159–167.

Perillo, G.M.E., Piccolo, M.C., 1991. Tidal response in the Bahía Blanca estuary. Journal of Coastal Research 7 (2), 437–449.

Perri, S., Viola, F., Noto, L.V., Annalisa Molini, A., 2017. Salinity and periodic inundation controls on themsoil-plant-atmosphere continuum of gray mangroves. Hydrological Processes 31, 1271–1282.

Persson, P.O., Walter, B., Neiman, G.P., Ralph, F.M., 1999. Contributions from California coastal-zone surface fluxes to heavy coastal precipitation: a case study from an El Niño year. In: Symposium on Observing and Understanding the Variability of Water in Weather and Climate, AMS, 9–13 February, Long Beach, Ca.

Piccolo, M.C., Dávila, P.M., 1993. El campo térmico de las planicies de marea del estuario de Bahía Blanca. In: Actas JNCMAR'91, pp. 11–15.

Piccolo, M.C., Perillo, G.M.E., Daborn, G.R., 1993. Soil temperature variations on a tidal flat in Mins basin, Bay of Fundy, Canada. Estuarine, Coastal and Shelf Science 35, 345–357.

Piccolo, M.C., Amador Rivero, J.D., Perillo, G.M.E., 1999. Aproximación al balance energético nocturno invernal de la playa del río Quequén Grande. GEOACTA 24, 77–86.

Ridd, P.V., Stieglitz, T., 2002. Dry season salinity changes in arid estuaries fringed by mangroves and saltflats. Estuarine, Coastal and Shelf Science 54, 1039–1049.

Roads, J., Betts, A., 2000. NCEP-NCAR and ECMWF reanalysis surface water and energy budgets for the Mississippi River Basin. Journal of Hydrometeorology 1, 88–94.

Rouse, W.R., 2000. The energy and water balance of high-latitude wetlands: controls and extrapolation. Global Change Biology 6 (Suppl. 1), 59–68.

Rouse, W.R., Carlson, D.W., Weick, E.J., 2004. Impacts of summer warming on the energy and water balance of wetland tundra. Climatic Change 22, 305–326.

Rutgersson, A., Smedman, A.S., Omstedt, A., 2001. Measured and simulated latent and sensible heat fluxes at two marine sites in the Baltic Sea. Boundary-Layer Meteorology 99, 53–84.

Sanders, H.L., Mangelsdorf, J.R., Hampson, G.R., 1965. Salinity and faunal distribution in the Pocasset River, Massachusetts. Limnology and Oceanography 10 (Suppl.), 216–229.

Sculthorpe, C.D., 1967. The Biology of Aquatic Vascular Plants. Edward Arnold, London.

Shoemaker, W.B., Sumner, D.M., Castillo, A., 2005. Estimating changes in heat energy stored within a column of wetland surface water and factors controlling their importance in the surface energy budget. Water Resources Research 41, W10411.

Smalley, A.E., 1960. Energy flow of a salt marsh grasshopper population. Ecology 41, 672–677.

Smesrud, J.K., Boyd, M.S., Cuenca, R.H., Eisner, S.L., 2014. A mechanistic energy balance model for predicting wáter temperature in surface flow wetlands. Ecological Engineering 67, 11–24.

Smith, N.P., Kierspe, G.H., 1981. Local energy exchanges in a shallow coastal lagoon: winter conditions. Estuarine, Coastal and Shelf Science 13, 159–167.

Souch, C., Wolfe, C.P., Grimmond, C., Susan, B., 1996. Wetland evaporation and energy partitioning: Indiana Dunes National Lakeshore. Journal of Hydrology 184, 189–208.

Sultan, S.A.R., Ahmad, F., 1994. Heat budget of the coastal water of Kuwait: a preliminary study. Estuarine, Coastal and Shelf Sciences 38, 319–325.

Swain, E.D., 2005. A Model for Simulation of Surface-Water Integrated Flow and Transport in Two Dimensions: Usr's Guide for Application to Coastal Wetlands: U.S. Geological Survey Open-file Report 2005-1033, 88 p.

Swinbank, W.C., 1963. Long – wave radiation from clear skies. Royal Meteorology Society 89, 339–348.

Teal, J.M., 1962. Energy flow in the salt marsh ecosystem of Georgia. Ecology 43 (4), 614–624.

Teal, J.M., Kanwisher, J.W., 1970. Total energy balance in salt marsh grasses. Ecology 51, 690–695.

Van den Broeke, M., Reijmer, C., Van As, D., van de Wal, R., Oerlemans, J., 2005a. Seasonal cycles of Antarctic surface energy balance from automatic weather stations. Annals of Glaciology 41, 131–139.

Van Den Broeke, M., Van As, D., Reijmer, C., Van De Wal, R., 2005b. Sensible heat exchange at the Antarctic snow surface: a study with automatic weather stations. International Journal of Climatology 25, 1081–1101.

Venalainen, A., Frech, M., Heikinheimo, M., Grelle, A., 1999. Comparison of latent and sensible heat fluxes over boreal lakes with concurrent fluxes over a forest: implications for regional averaging. Agricultural and Forest Meteorology 98–99, 535–546.

Vitale, A.J., Piccolo, M.C., Genchi, S.A., Delrieux, C., Perillo, G.M.E., 2014. 3D Numerical model of the thermal interaction between sediment-water-atmosphere. Environmental Modelling and Assessment 19, 467–485.

Vitale, A.J., Genchi, S.A., Piccolo, M.C., 2018. Assessing the surface radiation balance and its components in a coastal wetland environment. Journal of Coastal Research (in press).

Vugts, H.F., Zimmerman, J.T.F., 1985. The heat balance of a tidal flat area. Netherlands Journal of Sea Research 19 (1), 1–14.

Wallace, J.S., Holwill, C.J., 1997. Soil evaporation from tiger–bush in south-west Niger. Journal of Hydrology 188–189, 426–442.

Wang, H., Ping Hsieh, Y., Harwell, M.A., Huang, W., 2007. Modelling soil salinity distribution along topographic gradients in tidal salt marshes in Atlantic and Gulf coastal regions. Ecological Modelling 201, 429–439.

Weick, E.J., Rouse, W.R., 1991. Advection in the coastal Hudson Bay lowlands, Canada. I. Terrestrial surface energy balance. Artic and Alpine Research 23, 328–337.

Wessel, D.A., Rouse, W.R., 1993. Modelling evaporation from wetland tundra. Boundary-Layer Meteorology 68, 109–130.

Whitcraft, C.R., Levin, L.A., 2007. Regulation of benthic algal and animal communities by salt marsh plants: impact of shading. Ecology 88 (4), 904–917.

Wolanski, E., Ridd, P.V., 1986. Tidal mixing and trapping in mangrove swamps. Estuarine, Coastal and Shelf Science 23, 759–771.

Wolanski, E., Jones, M., Bunt, J.S., 1980. Hydrodynamics of a tidal creek-mangrove swamp system. Australian Journal of Marine and Freshwater Research 31, 431–450.

Wolff, M., Koch, V., Isaac, V., 2000. A trophic flow model of the Caeté mangrove estuary (North Brazil) with considerations for the sustainable use of its resources. Estuarine, Coastal and Shelf Science 50, 789–803.

Wu, C.-L., Shukla, S., Shrestha, N.K., 2016. Evapotranspiration from drained wetlands with different hydrologic regimes: drivers, modeling, and storage functions. Journal of Hydrology 538, 416–428.

Yang, Y., Anderson, M.C., Gao, F., Hain, C.R., Semmens, K.A., Kustas, W.P., Noormets, A., Wynne, R.H., Thomas, V.A., Sun, G., 2017. Daily Landsat-scale evapotranspiration estimation over a forested landscape in North Carolina, USA, using multi-satellite data fusion. Hydrology and Earth System Sciences 21, 1017–1037.

Yoshimoto, M., Harazono, Y., Miyata, A., Oechel, W.C., 1996. Micrometeorology and heat budget over the arctic tundra at Barrow, Alaska in the summer of 1993. Journal of Agricultural Meteorology 52, 11–20.

Zaker, N.H., 2003. In: Computation and Modelling of the Air–Sea Heat and Momentum Fluxes. International Centre for Theoretical Physics, Italy, 9 p.

Zhaira, S., Abderrahmane, H., Mederbal, K., Frederic, D., 2009. Mapping latent heat flux in western forest covered regions of Algeria using remote sensing data and a spatialized model. Remote Sensing 1, 795–817.

Zhang, G., Bai, J., Xi, M., Zhao, Q., Lu, Q., Jia, J., 2016. Soil quality assessment of coastal wetlands in the Yellow River Delta of China based on the minimum data set. Ecological Indicators 66, 458–466.

Further Reading

Hovel, K.A., Morgan, S.G., 1999. Susceptibility of estuarine crab larvae to ultraviolet radiation. Journal of Experimental Marine Biology and Ecology 237, 107–125.

National Research Council, 2000. Clean Coastal Waters: Understanding and Reducing the Effects of Nutrient Pollution. National Academy Press, Washington.

Talley, D.M., North, E.W., Juhl, A.R., Timothy, D.A., Conde, D., de Brouwer, J.F.C., Brown, C.A., Campbell, L.M., Garstecki, T., Hall, C.J., Meysman, F.J.R., Nemerson, D.M., Filho, P.W.S., Wood, R.J., 2003. Research challenges at the land-sea interface. Estuarine. Coastal and Shelf Sciences 58, 699–702.

Tsukamato, O., Ishida, H., Mitsuta, Y., 1995. Surface energy balance measurements around Ocean Weather Station-T during OMLET/WCRP. Journal of Meteorologic Society of Japan 73, 13–23.

CHAPTER

8

Hydrodynamics and Modeling of Water Flow in Coastal Wetlands

Iris Möller, Elizabeth Christie

Cambridge Coastal Research Unit, Department of Geography, University of Cambridge, Cambridge, United Kingdom

1. INTRODUCTION—IMPORTANCE OF UNDERSTANDING HYDRODYNAMICS AND WATER FLOWS IN COASTAL WETLANDS

The flow of water into and out of coastal wetland systems plays a key role in the formation, growth, and functioning of coastal wetland landforms and associated ecosystems. Water acts as a conveyor of materials (nutrients, sediment, pollutants) and energy (kinetic and gravitational), and the formation of a coastal wetland is critically dependent on both. For brackish or saline coastal wetlands to form, deposits either have to accumulate through a positive net transfer of sediment to shallow coastal regions or existing sediments have to be made accessible to salt water. Understanding the flow of water within and over coastal wetlands is thus critical to understanding their early evolution as landforms and ecosystems.

Once formed, the persistence of the wetland landform through time and the continued functioning of the associated ecosystem critically depend on the quantities of materials and energy transported via water movement. Both "too much" and "too little" input or output of water or hydrodynamic energy can lead to a change in ecosystem functioning and geomorphology, although what constitutes "too much" or "too little" is often not known. Variations in water exchange can lead to changes in the nutrient/pollutant budget of wetland systems, affecting water quality and potentially leading to the degradation of the ecosystem and adjacent waters through, for example, eutrophication. Furthermore, changes in salinity through changes in river versus tidal flow regimes or changes in surface sediment budgets (and surface elevation) can initiate the transitioning to a marine, permanently submerged, or a terrestrial, permanently emergent, system. The dynamics of water flow and its interaction with the coastal wetland surface thus arguably control all other geomorphological and ecological processes on and in the wetland.

Coastal Wetlands
https://doi.org/10.1016/B978-0-444-63893-9.00008-3

Although the flow of water plays a critical role in the functioning of coastal wetlands, the nature of this role changes over time as processes within the wetland begin to control how water and the material and energy it carries is transferred across the wetland. With respect to material budgets, the growing, maturing coastal wetland ecosystem contributes actively to its own nutrient and sediment budget through the growth and decay of microbial organisms, plant, and invertebrate biomass in situ. As far as hydrodynamic energy exposure is concerned, the wetland remains entirely dependent on the influx of water, but wetland geomorphology and ecology affects the transfer of hydrodynamic energy into, out of, and over the wetland. Although the system as a whole thus remains fundamentally at the mercy of external controls that determine the quantity, frequency, and speed of water, sediments and nutrients flowing through its channels and over its surfaces, what happens to this flow landward of its most seaward connection to the marine or estuarine waterbody is mediated heavily by the characteristics of the wetland itself.

A comprehensive level of understanding and informed management of coastal wetlands (e.g., for the purpose of maintaining the wetland's provisioning of ecosystem services) can only be achieved through an in-depth consideration of the wetland's interaction with water flow. Such an understanding is critically important to achieving the successful, integrative management of low-lying coasts and particularly for the maintenance of the coastal protection function (the reduction of flood and erosion risk landward of the wetland) provided by coastal wetlands.

Much progress has been made over recent decades in terms of the generation of knowledge on both the role of water as a conveyor of materials and on the transmission and dissipation of hydrodynamic energy by coastal wetlands. Technological improvements have transformed field and laboratory monitoring efforts, as well as numerical modeling capabilities (Luhar and Nepf, 2011). Increased storage capacity of automated data logging systems has allowed high-frequency measurements of flow velocity and waves over longer time periods, and data communication technology has revolutionized field monitoring in remote locations, facilitating the acquisition of quasi-continuous records of tidal, wind, and wave-driven water fluxes (Möller et al., 1999, 2011).

Although our understanding of hydrodynamic processes within coastal wetlands has rapidly grown, advances in numerical modeling have also been made. The simulation of tidal currents and waves over topographically complex surfaces with spatially varying surface roughness is now possible at a finer spatial and temporal resolution than ever before due to increased computational capacity (Loder et al., 2009). Advancements in laboratory and field studies have improved numerical model parameterization and validation through the provision of detailed hydrodynamic, ecological, and sedimentation observations, while bathymetry/topography data of increased resolution (e.g., through the use of airborne light detection and ranging (LiDAR) techniques) have allowed the creation of detailed digital terrain models for use in numerical grids.

Although these improvements in our ability to observe and numerically simulate water flow in and over wetland systems have been impressive, they have also led to the recognition of fundamentally important aspects of wetland interaction with water flows that have, arguably, so far been ignored. Such aspects include the importance of the physical properties of vegetation in its interaction with moving water (Luhar and Nepf, 2011) and particularly the dependency of flow-wetland interactions on the type of vegetation, water depth, and incident hydrodynamic energy (Jadhav and Chen, 2013; Zhang et al., 2012). Discussions around the validity and necessity of scaling our insights up/down from plant to canopy to landform scale within wetland systems are particularly interesting in this context (Friess et al., 2012).

In this chapter, we focus on the hydrodynamics of coastal wetlands at a range of scales and discuss, by way of illustrative examples, the current state of knowledge derived from both empirical observations (in the field and laboratory) and numerical modeling. Both allow us to understand and model the transfer of water into and out of coastal wetlands on a range of temporal and spatial scales, although some key knowledge gaps remain. The comparison between observational evidence and numerical modeling methods highlights key scaling issues which we address specifically at the end of the chapter.

2. TYPES OF WETLAND-FLOW INTERACTIONS

To understand the interaction between moving water and coastal wetlands, we must distinguish between different types of flow. The spatial and temporal scale at which a particular type of flow process is observed is critical as this determines the particular controls that are at work and affect the moving water. In other words, when considering water flows at kilometer scales compared with centimeter scales, different wetland attributes become important in controlling how the wetland interacts with the flow. This scaling is also important in terms of numerical models, as computational capacity sometimes limits the resolution of models; therefore, simplifications of hydrodynamics in complex wetland systems must be made.

As is often the case in environmental systems, time- and space scales are intricately connected, at least as far as continuous or (quasi-)periodic processes are concerned (Cowell and Thom, 1994). Fig. 8.1 schematically highlights those wetland properties that dominate the wetland-flow interaction for different types (duration) of flow "events" and over different spatial scales. Thus, residual flows over the wetland landform unit as a whole, for example, may be adequately understood over the duration of an entire tidal cycle or storm surge and interact most with "whole-wetland" or integrated landform properties, such as surface slope across the wetland width, degree of fragmentation by creeks, etc. Over shorter timescales, however, flows turn out to be highly variable and their explanation requires consideration of the geometry of individual creek sections or, as surface flows, vegetation canopy characteristics. At such smaller scales, too, flows over wetlands are affected by the presence of complex surface morphology, such as small cliffs, mud mounds, ridges, or runnels, as well as microtopographic variations (e.g., induced through the action of benthic invertebrates) and different types of vegetation canopy (or canopy layers) on the marsh surface. At the subcanopy scale, individual plant elements such as stems, branches, and leaves will interact with moving water (at the scale of wind waves or turbulent eddies). As the examples in this section will show, nonlinearities and high small-scale variability (both temporal and spatial) in flow behavior and wetland characteristics mean that variations in smaller-scale observations and insights may have to be regarded as "noise" and those at larger scales as fixed "boundary conditions" (see stippled lines in Fig. 8.1). When modeling the impact of wetland characteristics on flow at the larger scale, empirical parameterization of wetland characteristics is generally required at a scale-appropriate level. A particular challenge for

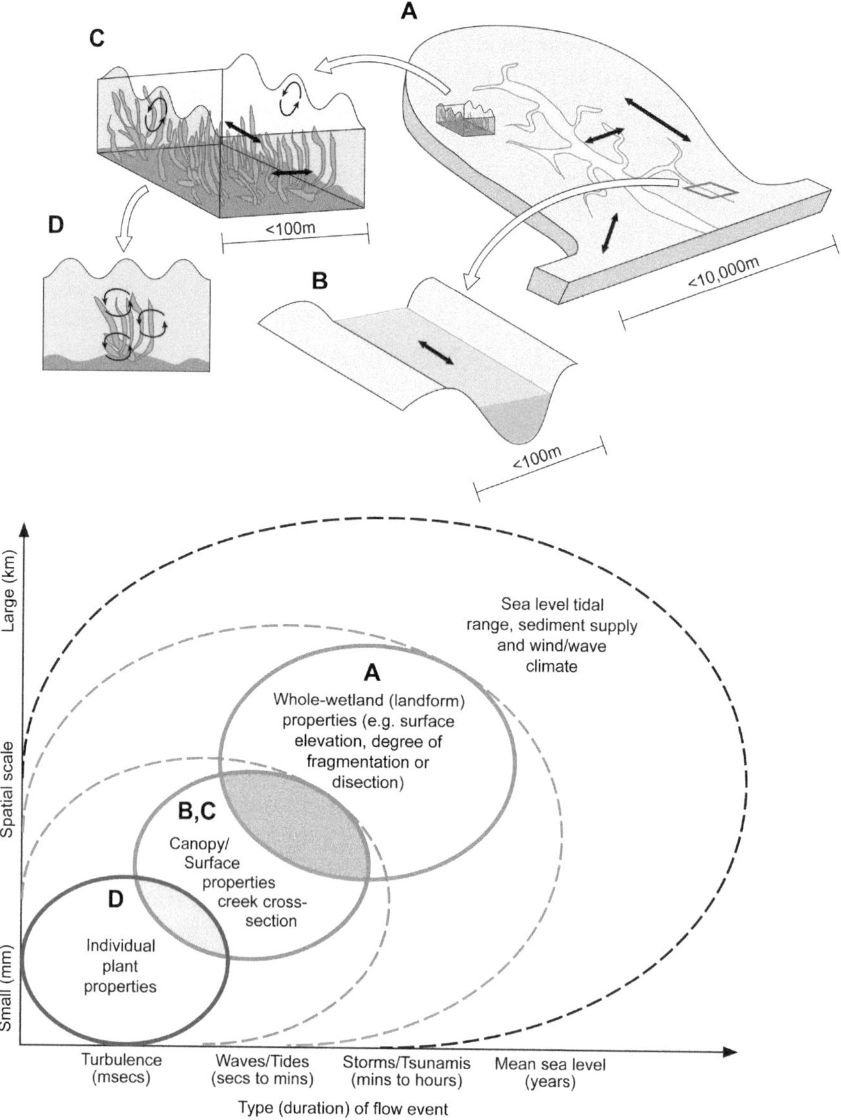

FIGURE 8.1 The interlinked spatial and temporal scales of hydrodynamic processes within coastal wetland systems (the different process scales (A–D) map onto a time-space continuum [lower graph] in which the, respectively, larger/longer process control acts as a boundary condition [*stippled lines*] to the smaller/shorter process, with clear regions of overlap [*shaded areas*] of scale-dependent controls).

such parameterization arises where flow events of different scales (duration) interact with wetland properties of different spatial scale levels (shaded areas in Fig. 8.1). Large, long-period, storm waves, for example, may require consideration of both large-scale (hundreds of meters) topographic variations over the wetland surface and small-scale (tens of meters) variations in canopy type/structure. We discuss the different scales separately, although there is a cascade of scales that link flow-wetland interactions from small to large scales.

Connected to the above considerations of scale are considerations of whether the wetland needs to be seen as a dynamic component or a passive mediator of flow. The timescale over which hydrodynamic interactions are observed largely determines whether it is permissible to think of the wetland as a passive feature (an "obstruction" to free flow) or whether it must be considered as mobile, actively changing its shape in response to water movement and thus affecting the movement in a classic morphodynamic feedback relationship (benthic vegetation, algal mats, and sediment can move and thus the wetland can change its shape and the way it interacts with flow). In the case of the latter, this active interaction may be noticeable over short (instantaneous to event-based) timescales (e.g., through scouring around plant elements (Bouma et al., 2009) under particular types of hydrodynamic conditions) or over long (landform evolution, engineering) timescales (where certain hydrodynamic conditions persist for long enough to alter the wetland landform itself, such as at the exposed Dengie Peninsula, Essex, UK (Reed, 1988)) (Fig. 8.2).

The above scale distinctions are critically important for modeling water flows as whether passive or active interactions at present determine the most appropriate temporal and spatial resolution of the model, how certain aspects of the wetland (e.g., vegetation) are parameterized, and whether or not those aspects feed back into the model.

Whether the wetland is passive or active in interacting with water flow also depends on (1) the relative **magnitude** of the forces induced by the flow compared with the forces that keep organisms and the sediment in place (e.g., storm vs. calm spring tide; flow over highly consolidated/cohesive vs. unconsolidated/mobile sediment; flexible vs. rigid biological elements); (2) their **duration** (e.g., flows acting over a few minutes may not noticeably change wetland morphology, but if they act over longer time periods, they may); and (3) their **direction** (e.g., flow reversal may cause plants to move in ways that cause erosion when unidirectional flow may not).

In addition, even when the wetland landform as a whole remains passive over the timescale of interest to us, the properties of the wetland surface at the smaller scale (i.e., microtopography and vegetation elements) may or may not alter under the influence of water

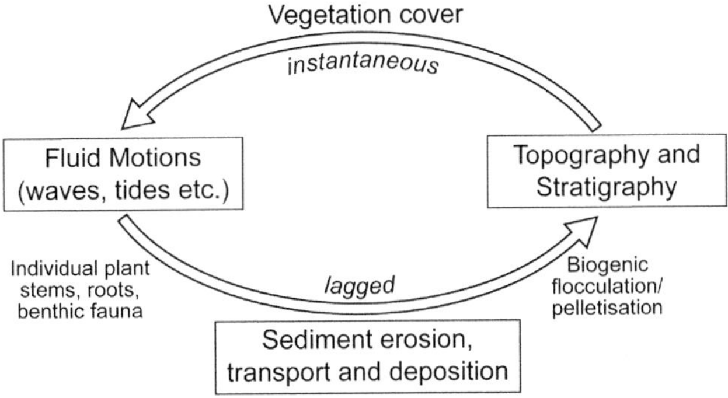

FIGURE 8.2 The biologically mediated morphodynamic feedback operating within coastal wetland systems. *Adapted from Cowell, P., Thom, B., 1994. Morphodynamics of coastal evolution. In Carter, R., Woodroffe, C., (Eds.), Coastal Evolution: Late Quaternary Shoreline Morphodynamics. Cambridge University Press, Cambridge, pp. 33–86.*

movement. On the wetland surface, plant interactions with flow may thus need to be thought of as active (the vegetation may move in the flow, e.g., Luhar and Nepf (2011)), whereas the landform as a whole remains passive (no sediment redistribution takes place (Spencer et al., 2016)), at least over the timescale over which an individual "hydrodynamic event," such as an individual tidal inundation, occurs. Indeed, the active interaction of the wetland surface with water flow at the smaller scale may be the very reason why the landform as a whole remains "stable" or "passive" over such timescales. As hydrodynamic energy is dissipated through small-scale plant movements in situ on the wetland surface and through meandering creeks and channels, for example, this dissipating function may protect the wetland landform itself from erosion.

3. LARGE-SCALE FLOWS

A range of what one might refer to as "large-scale flows," i.e., water movements on the scale of hundreds to thousands of meters, interacts with the wetland landform as a whole. The types of large-scale flows of greatest significance in terms of volume and energy exchanged depend on the setting of the wetland in question. Existing wetland classifications that are based on the relative exposure to tidal and river flow are useful in this context. Mangrove systems, e.g., may be seen as "basin," "fringe," or "riverine" systems based on three process controls: (1) frequency of flooding from tides, which increases between the former two categories, (2) decreasing salinity of floodwaters, which decreases between the latter two categories, and (3) increasing frequency of flooding from riverine sources, which increases between the first and last type (Ewel et al., 1998). Although these hydrological and hydrodynamic controls have been recognized as important controls on the functioning of other types of coastal wetlands (see, e.g., Childers et al. (2000) on the importance of hydrodynamic setting for nutrient flux studies), there are few comparable classification schemes for non-mangrove systems. One notable exception is that of Brooks et al. (2011), who argue for a hydromorphological classification of wetlands (fresh or saline) more broadly. Ultimately, any coastal wetland experiences a combination of different flow types and, as Allen (2000) recognizes, with respect to tidal flows and wave-driven flows in salt marshes, in any one wetland, different currents "are likely to act on a range of different paths during a tidal cycle" (p. 1183) or indeed a particular storm event.

In addition to river and tidal flows, wetland systems may interact with large-scale water currents resulting from elevated water levels (and thus water surface slopes) during storm surges and tsunamis, neither influence of which has been fully recognized in existing classification schemes (Spencer and Möller, 2013; Ewel et al., 1998). Although these types of flows occur infrequently, the interaction between the wetland and such rare but high-energy events can potentially determine (1) decadal- to century-scale wetland evolution and (2) wetland importance as a natural coastal protection feature than the daily or seasonal occurrence of tidal or river flooding (Cahoon, 2006).

As the magnitude and energy of the different types of large-scale flows varies markedly, we discuss evidence and insights from field and numerical modeling studies of wetland interaction with river and groundwater, tidal, storm surge, and tsunami-generated flows in turn below. Any given wetland may be exposed to any of these flows at different frequencies and to varying degrees such that their combined interaction with the wetland is very much location specific.

3.1 River and Groundwater Flow

Although most coastal wetlands are dominated by tidal exchange of water, freshwater discharge can play an important role in determining water depths over estuarine wetland surface, particularly in tidal freshwater wetlands, where freshwater inputs may dominate over tidal fluxes. Barendregt and Swarth (2013) use the ratio between the average maximum and minimum monthly river discharge as a criterion for identifying these freshwater-dominated systems. Extreme examples of such tidal freshwater wetlands are found on the Ivory Coast (Comoe estuary) and in Australia (Fitzroy estuary), both with ratios of minimum to maximum monthly discharge of <0.1.

The key hydrodynamic influence of varying river discharge is to alter tidally driven water levels at the margins of wetlands. Few studies have documented this effect in detail over timescales longer than individual flood events. Jay et al. (2015) found, for 21-year inundation hindcasts applied to the lower Columbia River estuary in the United States, freshwater discharge exerted a minimal control on tidal range seaward of Beaver, a distance of some 40 km from the estuarine mouth. Further landward, high discharge events suppressed tidal water level variations.

In terms of wetland hydrodynamics, low/high freshwater discharge events may lead to lower/higher than usual water surface slopes that then drive lower/higher flow speeds on the flood phase of the tidal cycle (Young et al., 2016). The interactions between these flows and the wetland are as described below for tidal forcing.

In addition to open-channel flow, groundwater discharge can have an effect on wetland hydrodynamics (Wolanski and Elliott, 2015), although evidence for this is sparse. Groundwater interaction with tidal influx of saline water, however, has been shown to lead to highly stratified flows in the outer Rio de la Plata estuary, Argentina, with notable impacts on wetland ecology (Carol et al., 2013). In arid regions, groundwater dynamics may be driven heavily by evaporation alongside precipitation and tidal fluctuation regimes (del Pilar Alvarez et al., 2015), and alongside freshwater fluxes, tidal groundwater forcing may be an important concern in wetland restoration projects (Glamore and Indraratna, 2009).

Within mangrove swamps, the salinity of groundwater exerts a key control on tree growth and distribution (Ridd and Sam, 1996). Simulating groundwater flow within the Bashita-Minato mangrove, Iriomote Island, Japan, Mazda et al. (1990) postulate three separate groundwater flow patterns: a flow toward the open sea due to a tidal mean pressure gradient in a seaward direction, a residual flow toward the mangrove swamp caused by damped tidal flow, and a tidally reversing flow (albeit damped and delayed in a landward direction). Evidence for the effect of soil burrows produced by crabs and other organisms on soil conductivity also exists and such processes may lead to large (several orders of magnitude) variations of hydraulic conductivity in specific locations (Mazda and Ikeda, 2006). Much less is known of groundwater drainage within salt marsh ecosystems, although Allen (2000) emphasizes that highly organic soils (peat) lead to less surface drainage and thus, consequently, a greater proportion of groundwater flow.

3.2 Tidal and Meteorologically Driven Flows

Coastal wetlands exist in a wide range of tidal contexts. Within the brackish water systems of enclosed embayments or regional seas, such as the Baltic Sea, water level fluctuations are largely driven by meteorological and hydrological factors with negligible tidal exchange. At the other end of the scale, within megatidal marshes, such as those of Nova Scotia, Canada, however, tidal fluxes are by far the most dominant physical control on all aspects of the coastal wetland landform. The degree to which the coastal wetland's interaction with tidal flows determines its functioning and future landform evolution is thus highly context dependent. For most coastal wetlands, however, hourly to daily fluctuations in water levels are driven by both tidal and meteorological factors acting together.

For those wetlands exposed to energetic tidal flow regimes, early studies of associated drainage systems such as tidal creeks in salt marshes have noted the unique channel morphology resulting from the bidirectional action of flood and ebb flow (Pethick, 1969; Redfield, 1972). The formation of such tidal drainage systems is covered elsewhere in this volume, and for salt marshes, the reader is referred to the excellent summary of salt marsh morphodynamics by Allen (2000). It is important here, however, to recognize the difference between tidal hydrodynamics in incipient, "young," coastal wetlands and older, more "mature" wetland systems, as the relative frequency with which within-channel to wetland surface water flow occurs alters as the channel system develops. The interactions between the wetland landform and water flow differ considerably between these two types of flow. Within mature vegetated coastal wetlands, the necessity to distinguish between flows that remain constrained within the channel system on the one hand and those that overtop the creek banks and flood the wetland surface on the other was recognized in the early studies of, e.g., Bayliss-Smith et al. (1979) and Reed (1987). More recent evidence has confirmed the importance of the transition from unvegetated to vegetated coastal wetlands in terms of flow routing efficiency of the tidal drainage system (Kearney and Fagherazzi, 2016).

The fundamental importance of vegetation in affecting flow dynamics further justifies the division of the discussion of tidal flow interactions into flows that remain constrained within tidal creeks and those that exceed creek banks and spread out over the intervening (vegetated) surfaces. We address each of these flows in turn.

3.2.1 *Flow Over Unvegetated Tidal Flats*

Water flows across unvegetated tidal wetland surfaces are due to both wave and tidal action, with the former generating higher current speeds than the latter. Thus, Shi et al. (2017) measure currents of between 0.08 and 0.52 m s^{-1} under waves with a significant wave height of between 0.02 and 0.77 m in water depths ranging from 0.8 to 2.0 m on a tidal mudflat on the Yangtze Delta, China, whereas tidal-averaged current velocities ranged from 0.05 to only 0.26 m s^{-1}. Modeling of tidal flat hydrodynamics suggests that the evolution of more permanent tidal creeks is dependent on the relative magnitude of tidally versus wave-induced currents on the unvegetated tidal flat (Mariotti and Fagherazzi, 2013). In addition to wave–tide interactions, however, the biologically mediated surface stability of tidal flats plays a further important role in the initiation of channel systems and thus the spatial variability in surface flow speeds (Widdows et al., 2004).

3.2.2 *Flow Within Tidal Creeks*

Tidal creeks are a common feature on coastal wetlands in which the substrate has reached a degree of surface stability, allowing flows to become channeled into surface depressions. Drainage of interchannel surfaces favors seedling establishment and vegetation growth such that positive feedback encourage channel evolution over time. Preferential deposition on channel margins, where the established vegetation slows down water currents, allows interchannel surface to become less convex and channel margins to steepen. Once vegetated, the higher surfaces become increasingly stable and the creek system increasingly "fixed," at least in the cross-channel dimension (Allen, 2000). As far as the interaction between tidal flows and wetlands is concerned, the wetland's drainage network of creeks or tidal channels, once established, constitutes the key mechanism by which tidal flow energy is dissipated such that bed shear stress is sufficiently reduced on the wetland surface to allow the settling of fine sediments and the germination of plant seeds. This dissipation is achieved on the flood tide through frictional resistance with the bed in shallow water depths and along creek side banks, as well as through an efficiently organized dendritic geometry of channels that distribute large volumes of water and effectively reduce flow speeds prior to the overtopping of channel margins (Pethick, 1969). Equally, on the ebb, water is efficiently transferred out of the wetland and water levels drop rapidly such that soil oxygen levels can be maintained to facilitate high wetland productivity. This efficient exchange of tidal water also serves to supply wetland surfaces with sediments and nutrients and flushes the system of toxins that negatively affect plant growth (Childers et al., 2000).

Evidence for the efficient conveyance of water through tidal creek systems can be found in numerous field studies that have monitored flow velocities throughout the tidal cycle within creek channels in both salt marsh and mangrove systems (Bayliss-Smith et al., 1979; Blanton et al., 2002; Horstman et al., 2013; Mazda et al., 2005; Murray and Spencer, 1997; Reed et al., 1985; Reed, 1987; Temmerman et al., 2005). Fig. 8.3 shows examples of typical velocity–stage curves for tidal channel flows within both types of system. Within-creek velocities are typically <1 m s^{-1}. On the flood part of the tidal cycle, flow velocities within the creeks are driven by the tidally induced water level rise at the creek mouth. Within-creek velocities during this part of the tidal phase thus depend on tidal amplitude and asymmetry within the adjacent estuary, embayment, or tidal flat, but they vary little throughout the flood phase, highlighting the fact that creek geometry and flow speeds coadjust over time to cause flow dissipation rates that scale with creek dimension (Knighton et al., 1992).

On the ebb tidal phase, the evidence from field observations (Fig. 8.3) suggests an equally efficient transfer of water out of the wetland system, albeit for a longer duration and lower speeds, as the greater bed friction reduces flow speeds within the channel network. Water levels outside of the marsh system fall more rapidly such that the driving force behind ebb tidal flows within the creek system is gravity. As creek dimensions reduce landward with distance from the main creek, however, water surface slopes on the flood part of the tidal cycle tend to exceed those on the ebb and lead to flood-dominant flow within the creek systems (i.e., average flow velocities on the flood part of the tidal cycle exceed those on the ebb part (Fig. 8.3)). Leopold et al. (1993) found that water surface slopes were greatest in the middle reaches of a tidal channel in Petaluma Marsh, California, and, over the 19,000 feet

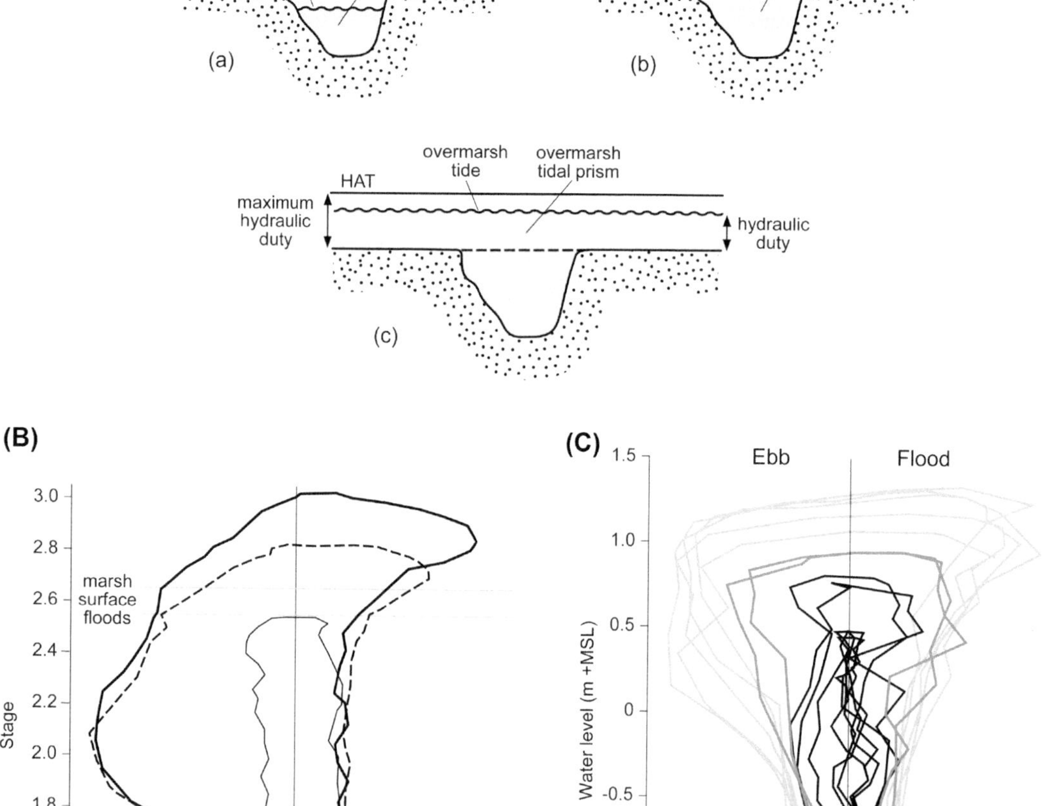

FIGURE 8.3 Tidal variation in flow velocity within a salt marsh and a mangrove creek system. Key water surface stages. (A) undermarsh, bankfull, and overmarsh, and representative velocity-stage curves in a salt marsh. (B) and mangrove. (C) creek location. *MSL*, mean sea level. *Adapted from Allen, J.R.L., 2000. Morphodynamics of Holocene salt marshes: a review sketch from the Atlantic and Southern North Sea coasts of Europe. Quaternary Science Reviews 19 (12), 1155–1231; Horstman, E.M., Dohmen-Janssen, C.M., Hulscher, S.J.M.H., 2013. Flow routing in mangrove forests: a field study in Trang province, Thailand. Continental Shelf Research 71, 52–67; Reed, D.J., Stoddart, D.R., Bayliss-Smith, T.P., 1985. Tidal flows and sediment budgets for a salt-marsh system, Essex, England. Vegetatio 62 (1–3), 375–380.*

(5800 m) long channel ranged from 1 in 10,000 to 5 in 10,000, with opposite slopes observed at either end (mouthward vs. headward) during some parts of the tidal cycle.

3.2.3 *Flows Over the Wetland Surface*

The frequency with which the vegetated wetland surface, rather than the tidal channel system that dissects it, interacts with tidal water flow is a function of surface elevation relative to the tidal water level. When inundated, the vegetated wetland surface acts as a hydraulically rough surface such that flow speeds within the vegetated boundary layer can be an order of magnitude lower than those at comparable vertical distances above the bed on the adjacent tidal flat (a phenomenon well illustrated by recorded flow speeds within the vegetated part of the intertidal shores of the Scheldt estuary in the southern Netherlands (Bouma et al., 2005)).

As the frictional resistance and the complex flow patterns around individual plants and groups of plants on the vegetated surface reduce flow speeds, water surface slopes develop over the wetland surface. Thus, alongside wetland margin water depths, the distance of any given location from the wetland margin determines flow speeds and depths above the wetland surface at any given time.

Vegetated wetland surfaces exist at a wide range of elevations relative to mean sea level. Mature coastal salt marsh may experience tidal inundation only during the highest spring tides of the year <10 times per month, whereas young, incipient marsh surfaces can experience tidal flow on over half the tidal cycles during the year, i.e., for more than 350 tides per year in semidiurnal tidal settings (Allen, 2000). Although tidal inundation frequency and duration is a key control on the establishment of coastal wetland plants, it is not the only one. Thus, a review of studies on the inundation frequency of *Spartina* spp. at their most seaward limit in the United Kingdom, Denmark, United States, and New Zealand suggests a range from 5800 to 7800 inundated hours per year (Friess et al., 2012). Although in mangroves, high salinity and permanent inundation limit plant growth and establishment, mangrove platforms may exist as either permanently or periodically (tidal or riverine) inundated surfaces (with hydroperiod controlling species composition).

Although flows over the vegetated wetland surfaces occur less frequently than within-channel flows, their capacity to convey water, nutrients, and sediments, as well as hydrodynamic energy onto and over the wetland surface gives them particular morphological and ecological significance. Thus, while creek flow has been observed to contribute far more to the tidal water flux into and out of mangrove systems than sheet flow (largely because of the high bed shear stresses over vegetated surfaces), sheet flow can take place over distances of several tens to hundreds of meters over those vegetated surfaces and is heavily affected by the properties of the mangrove trees (Mazda et al., 2005). Vertical differences in the arrangement and physical characteristics of mangrove tree constituents (roots, trunks, branches, and leaves) thus result in vertically differentiated interaction with tidal water flow.

In high macrotidal settings, such as those of the Bay of Fundy, water depths in excess of 1–2 m may occur regularly during high spring tides over the vegetated platform (Davidson-Arnott et al., 2002), but tidal water depths are much lower (<1 m) in mesotidal wetlands (Leonard and Croft, 2006; Leonard and Luther, 1995) and rarely above a few centimeters in-depth in microtidal systems (Fagherazzi et al., 2006). The relative importance of meteorologically, instead of tidally, induced water flows across wetland surfaces for wetland surface processes thus increases with decreasing tidal range (Friess et al., 2012; Cahoon et al., 2006).

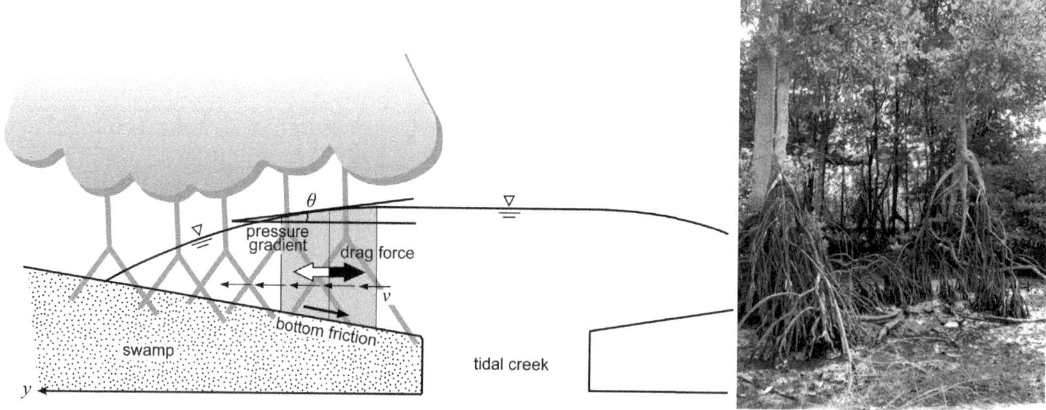

FIGURE 8.4 Water surface slopes within *Rhizophora*-type mangrove wetlands (left) and an example of *Rhizophora* mangrove species, Sungei Buloh Wetland Reserve, Singapore Island (right). *Adapted from Mazda, Y., Kobashi, D., Okada, S., 2005. Tidal-scale hydrodynamics within mangrove swamps. Wetlands Ecology and Management 13 (6), 647–655; Photograph: T. Spencer.*

When water flows across the wetland surface occur, their speeds are determined by water depth, the topography, and surface (vegetation) cover and are generally an order of magnitude less (at < 0.1 m s^{-1}) than those within the creeks (at < 1 m s^{-1}) (Bouma et al., 2005). As has been shown for flows within tidal channels, flows that extend onto the wetland platform induce water surface slopes of the order of 1:1000 at the landscape scale due to frictional drag induced by the hydraulically rough bed (Fig. 8.4). Thus, Young et al. (2016) measure water surface slopes of 0.0017 m per meter distance from main tidal channel onto the marsh surface in the Ogeechee River salt marshes, Georgia, USA. For low tidal inundation depths of several tens of centimeters in low-growing plant canopies, such as seagrass beds or salt marshes, such studies have thus shown that flow can be dissipated rapidly over short (1–10 s m) distances. Flows over the wetland surface in greater water depths that submerge such low-growing canopies may be much less affected by the presence of vegetation, as discussed below. The effect of taller vegetation, such as mangroves, affects water movement even during high inundation events (high tides, surges, or tsunamis), as has been documented by Zhang et al. (2012), who recorded a decrease in surge water levels by 23–48 cm km^{-1} distance into a South Florida mangrove area during the Category 3 hurricane Wilma in 2005 as compared with 15.8 cm km^{-1} of mangrove surface during hurricane Charley in 2004 (Krauss et al., 2009). Harvey et al. (2009) observe water surface slopes of less than 3:100,000 during the wet season, but higher slopes of up to 6:100,000 and 1:40,000 at other times within the mangrove systems in Florida. Mazda et al. (1999), however, provide evidence for even higher water surface slopes (of the order of 2:3000) within dense mangrove swamps.

Finally, it is important to note that water surface slopes are not necessarily entirely the result of the wetland's surface roughness: salt marsh restoration schemes have also been shown to have contributed significantly to water surface slopes on the ebb part of the tidal cycle, where breached embankments hold back water such that high flow velocities through constrained channel connections between restored and existing marsh can cause significant bank erosion (Friess et al., 2014; Symonds and Collins, 2007).

4. SMALL-SCALE FLOWS

In addition to evidence for the interaction between the coastal wetland and water flow through wetland creek systems and over wetland surfaces at the large (several hundreds to thousands of meter) scale discussed above, a number of smaller scale studies have attempted to more specifically investigate the vegetation characteristics that may help understanding how the interaction between coastal wetland surfaces and water movement takes place. Particularly for the more accessible shorter vegetation canopies of salt marshes and the root layer within mangrove canopies, it has been possible to acquire somewhat more detailed insights into the relationship between plant characteristics and unidirectional flow. It is worth considering the nature of these insights before discussing the modeling of such interaction and then moving on to a discussion of bidirectional (wave-driven) flows over wetland surfaces.

4.1 Field Evidence

Interactions between water flow and the plant canopy are affected not only by the amount of space available for water to travel through/around the plant structures but also by the shape and geometry of the structures themselves and their arrangement with respect to each other (e.g., vegetation clumps). Streamlined objects permit fast, efficient flow around them, whereas others may cause greater turbulence and eddy formation. When attempting to isolate those vegetation characteristics that determine the degree of streamlining, a major complication arises from the fact that plants and plant elements respond to forces exerted on them through flexing/bending/breaking and thus either realign themselves within the flow field or lose part of their obstructing mass.

Early studies using a tracer to record unidirectional flow speeds within a coastal reed bed by Hosokawa and Furukawa (1994) suggested that in low water depths of 30–50 cm and flow velocities $<10 \text{ cm s}^{-1}$, the presence of reed significantly affected the settling of fine particles through reduced flow velocities.

The presence of *Spartina alterniflora* and *Juncus roemerianus* salt marsh species, which, although smaller, are similar in overall structure to reed and often found in monospecific stands or patches on US east coast salt marshes, has also been shown to reduce mean flow velocity (Leonard and Luther, 1995). Such reduction in velocity, however, was shown to be vertically complex, as plant characteristics are vertically varied and so is their interference with flow. In addition to mean flow, the authors were also able to quantify canopy effects on flow turbulence through the deployment of hot-film anemometry sensor arrays. Compared with adjacent channel flows, turbulent kinetic energy (TKE) was reduced exponentially into the vegetation canopies such that less than 1% of TKE remained at a distance of 15 or more meters into the marsh.

Dense *Spartina maritima* vegetation (2340–3030 stems/m^2) has also been shown to increase the height of the boundary layer from 5 to 15 cm (Neumeier and Ciavola, 2004). This study also found, for similar flow speeds of around 5 cm s^{-1} over bare compared with submerged vegetated surfaces, that flow was diverted and increased in velocity above the vegetation canopy to >9 cm s^{-1}, whereas it decreased within the canopy to <3 cm s^{-1}. The presence of dense vegetation can thus clearly cause "skimming flow" at greater than initial velocities, as inflowing water is forced to accelerate where space for free flow is available.

Further field experiments conducted in a *Spartina alterniflora* marsh confirmed the importance of vertical biomass distribution to the dissipation of flow energy (Leonard and Croft, 2006). Importantly, this study's results showed that vertical components of the flow were reduced more than horizontal components of the flow and that a reduction of 50% of total TKE and the greatest reduction in flow velocities (from around 2.5 cm s^{-1} to less than 0.6 cm s^{-1}) occurred within 5 m horizontal distance into the vegetation canopy (Fig. 8.5).

Arguably the most marked vertical variability in coastal wetland plant structure and biomass can be found in mangrove vegetation. From the soil surface to several meters above the ground, water flowing at increasing depth encounters first exposed prop roots and pneumatophores alongside tree trunks, then the vertical main tree trunks, and finally the complex branches and leaves that constitute the tree canopy. It is thus to be expected that water flow through such a vegetation layer is affected in different ways at different water depths. Indeed, this phenomenon of vertically differentiated flow patterns has been documented

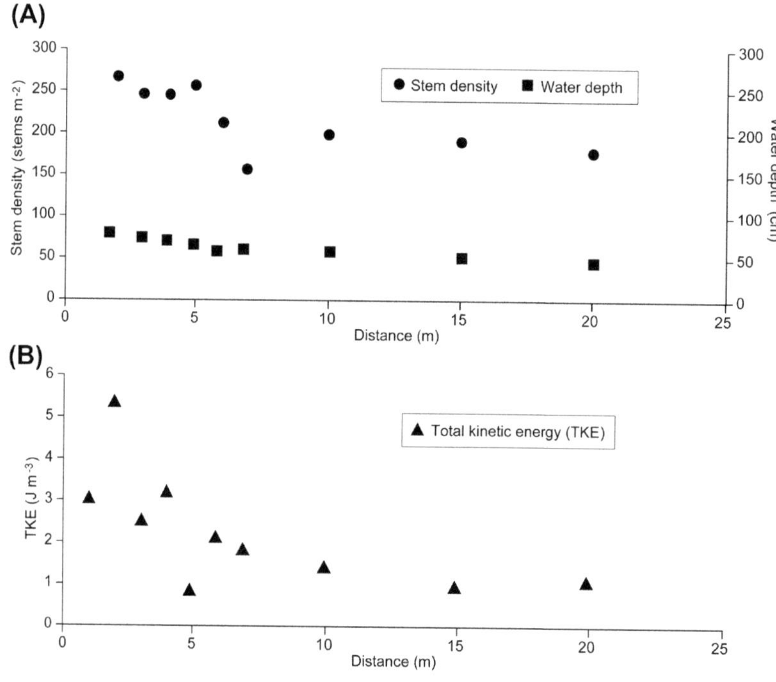

FIGURE 8.5 Variation of (A) stem density and water depth and (B) total turbulent kinetic energy (TKE) with distance into a *Spartina alterniflora* marsh canopy. *Adapted from Leonard, L.A., Croft, A.L., 2006. The effect of standing biomass on flow velocity and turbulence in* Spartina alterniflora *canopies. Estuarine, Coastal and Shelf Science 69 (3), 325–336.*

well by Horstman et al. (2013), who observe, in a tidally dominated mangrove system along the Andaman coast of Southern Thailand, that during spring tides, sheet flow across the vegetated zone between creeks contributes up to 20%–25% of the total inflow of water. Furthermore, they show that velocities of this flow increase as water levels rise above the zone of high bed roughness caused by the c. 50 cm thick exposed root layer. Even when water depths exceed the height of the root layer, however, flow velocities within the vegetated zone were observed to be an order of magnitude smaller than those within adjacent creeks (although even in the creeks, flow speeds were less than 0.3 m s^{-1}).

As in the salt marsh study of Leonard and Croft (2006), the observed vertical variability of flow was high such that velocities at 15 and 50 cm above the bed, for example, exceeded those recorded at 7 cm above the bed by a factor of 1.5 to 1.2 and 2.0 to 2.8 for the more and the less exposed sites. Horstman et al. (2013) attribute this vertical variability to the effects of micro-topographic and vegetation structure complexity on the localized flow field recorded with acoustic Doppler velocimeters (Fig. 8.6). Such localized variations in flow velocities were also picked up by field measurements of flow velocities within dense mangrove root systems, with Furukawa et al. (1997) recording jets and eddies with velocities up to three times the average flow velocity in the regions between root elements.

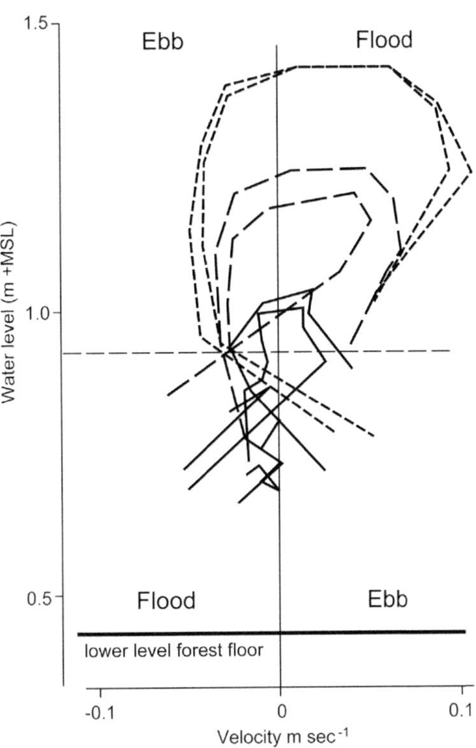

FIGURE 8.6 Flow velocities recorded within the mangrove root matrix, Trang Province, Southern Thailand. *Reproduced from Horstman, E.M., Dohmen-Janssen, C.M., Hulscher, S.J.M.H., 2013. Flow routing in mangrove forests: a field study in Trang province, Thailand. Continental Shelf Research 71, 52–67.*

4.2 Flume Studies and Dynamic Vegetation-Flow Feedback

While observations of flow velocities in field settings have contributed significantly to our present understanding of flow–canopy interactions within coastal wetlands, arguably the most detailed studies have emerged from the study of flow around plant stems and canopies within controlled laboratory flume environments. Such a controlled environment allows investigations into the specific effect of isolated parameters that are thought to be critical for determining the transmission of hydrodynamic (kinetic) energy through the system or the conversion of such energy to turbulent or heat energy and thus the loss of kinetic energy available for continued water movement or sediment erosion/transport. Here, we provide examples of key studies that have informed (1) the study of plant canopy drag and its dependence on incident flow conditions and vegetation properties and (2) the effect of plant canopy reconfiguration under flow.

Early flume studies used upright rigid circular cylinder arrays to represent wetland vegetation under varying flow regimes. Such studies showed that flow speeds could be significantly reduced when such cylinder arrays were present and highlighted clear relationships between cylinder geometry and density on the one hand and flow reduction on the other (see, e.g., Denny, 2014; Vogel, 1994; Kobayashi et al., 1993).

Indications that flow patterns within the vegetation canopy are more complex than can be captured through the simple representation of the vegetation by its projected area, however, emerged in one of the earliest systematic flume studies: Shi et al. (1996) exposed three different densities of *Spartina anglica* canopies cut to 30 cm height to low ($<0.05\ \mathrm{m\ s^{-1}}$) flow velocities in 30 cm deep water in a laboratory flume and found that at higher vegetation densities, a reversal of the flow velocity profile within the canopy can be achieved such that velocities in the central regions of the canopy exceed those near the bed and near the top of the canopy. Such flow variability ties in well with the reports on higher than average flow velocities within mangrove root systems discussed above (Furukawa et al., 1997).

Neumeier (2007) showed, in a flume study with *Spartina anglica*, that the submerged canopy characteristics, independent of water depth or flow velocity, determine where the zone of maximum vegetation roughness is located. The author was also able to show that, with an increased density of vegetation, the zone of highest turbulence moves upward (this zone exists some distance above the bed but within the vegetation layer and was also detected by Leonard and Croft (2006) and Neumeier and Ciavola (2004) in field conditions) (Fig. 8.7).

The plant canopy is evidently not necessarily rigid, and Neumeier (2007) also finds that the characteristics of the skimming flow that develops above the vegetation for submerged canopies depend on the depth of the unimpeded flow layer, i.e., water depth relative to reconfigured vegetation height. Thus, any flow-induced reconfiguration of the plant canopy can make as significant a difference to the routing of flow within/above the plant canopy as the vertical distribution of biomass (Neumeier, 2005). Flume studies have been able to highlight this phenomenon clearly through observations in controlled conditions of varying hydrodynamic energy and/or density and flexibility of wetland plants. Peralta et al. (2008), for example, exposed seagrass (*Zostera noltii*) and pioneer salt marsh (*Spartina anglica*) vegetation, as well as flexible plant mimics to uniform flow of between 0.045 and $0.30\ \mathrm{m\ s^{-1}}$ in 40 cm water depths within a recirculating flume. They found that canopy heights can reduce by more than 50% for low densities of 615 and 4989 structures $\mathrm{m^{-2}}$ of flexible mimics

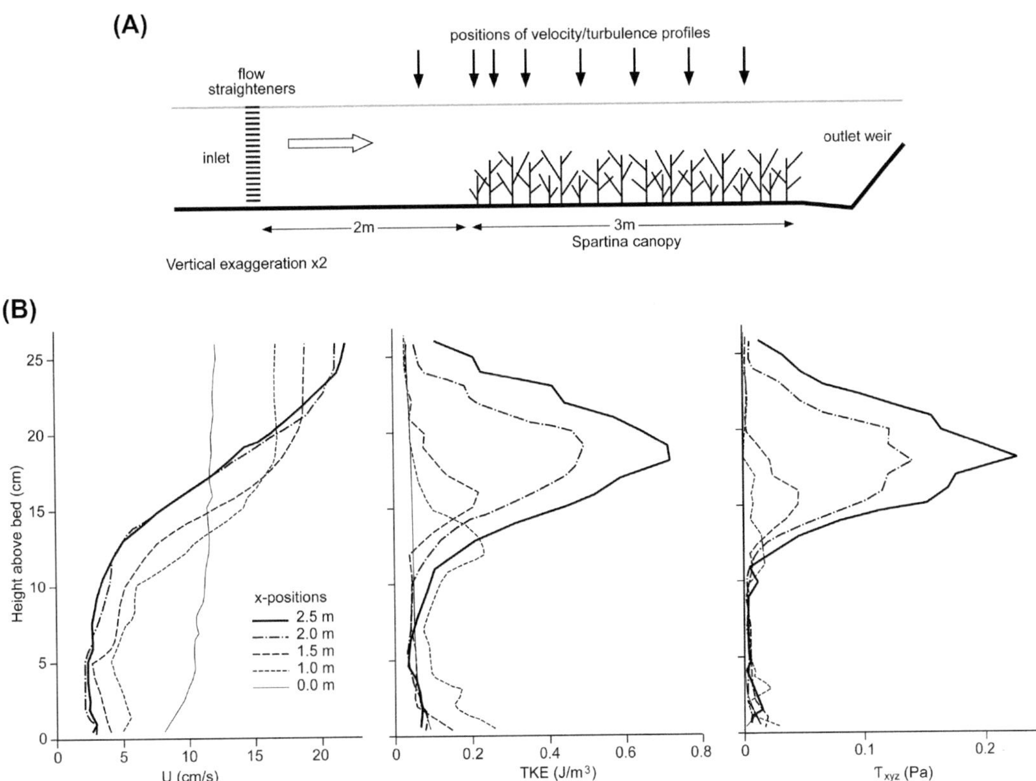

FIGURE 8.7 (A) Experimental setup and (B) recorded velocity, turbulent kinetic energy (TKE), and three-dimensional Reynolds stress (τ_{xyz}, a function of velocity in all three directions and water density) for different horizontal distances into a *Spartina* vegetation canopy (shoot density 1200 m^{-2}, canopy height 15.3 cm, water depth 32 cm, U = 11.3 cm s^{-1}). *Adapted from Neumeier, U., 2007. Velocity and turbulence variations at the edge of saltmarshes. Continental Shelf Research 27 (8), 1046–1059.*

and *Zostera noltii*, respectively. They found no flow effects on canopy heights for the stiffer mimics or marsh species, *Spartina anglica*. Mimics and species were selected to provide structures of contrasting flexibility without altering the dimensions of the structural elements. For flexible wheat plants, however, Järvelä (2005) found that the dynamic nature of the top of the vegetation canopy contributes significantly to locally generated levels of turbulence.

While many authors have thus now recognized the potentially important effect of plant flexibility on the interaction between flows and the vegetation canopy (e.g., Dijkstra, 2009; Loder et al., 2009; van Loon-Steensma, 2015), few have quantified how plant flexibility varies across different coastal wetland species. To the authors' knowledge, data in Table 8.1 are to date the only available data that allow such a direct comparison between flexibilities of coastal wetland species, as recorded using Young's modulus (E). Young's modulus provides a calculation of the ratio of stress (σ) to strain (ε) to determine elasticity of a material. It is

TABLE 8.1 Vegetation Flexibility Recorded in the Form of Young's Modulus Within a Range of Studies

Reference	Habitat	Species	Stem Part	Young's Modulus (MPa)	Stem Length (cm)	Stem Diameter (mm)
Zhang et al. (2013) and Zhang et al. (2015)	Mangrove	*Rhizophora* sp.	Prop roots	5800 ± 1700	20	36
Feagin et al. (2011)		*Spartina alterniflora*	Middle	1410 ± 710	51.67	4.3
Lara et al. (2016)	Salt marsh	*Puccinellia maritima*	Unspecified	13	47.29 (sd 8.65)	Not given
	Salt marsh	*Spartina anglica*	Unspecified	164	28.40 (sd 2.66)	Not given
Rupprecht et al. (2015)	Salt marsh	*Spartina anglica*	Bottom	118 ± 50	27.87 ± 4.66	4.5 ± 0.6
			Middle	123 ± 36		4.8 ± 0.5
			Top	311 ± 137		2.3 ± 0.7
	Salt marsh	*Puccinellia maritima*	Bottom	1995 ± 649	23.93 ± 6.94	1.4 ± 0.1
			Middle	1765 ± 354		1.5 ± 0.2
			Top	737 ± 281		1.5 ± 0.2
	Salt marsh	*Elymus athericus*	Bottom	4082 ± 1386	46.00 ± 12.30	1.6 ± 0.2
			Middle	2755 ± 694		1.7 ± 0.2
			Top	1952 ± 668		1.6 ± 0.2
Albert et al. (2013)	Fresh/brackish marsh	*Schoenoplectus pungens*	Unspecified	73 ± 27	54–118	5.7
			Unspecified	27 ± 7		12.1

calculated through beam loading experiments (Feagin et al., 2011). Young's modulus is calculated as

$$E = \frac{\sigma}{\varepsilon} = \frac{4s^3F}{3D\pi d^4} \tag{8.1}$$

where d is the stem diameter and s is the horizontal distance between two support bars on which the plant stem specimen rests while the force, F, is applied to the center of the stem section from above to produce the deflection, D, of the stem section. Young's modulus is thus a measure sensitive to stem damage. For healthy plant tissue, it must be calculated from the slope of the initial near-linear relationship between applied force and displacement of stem section; when excessive force is applied and/or plant stems are damaged, the relationship becomes nonlinear (see, e.g., Rupprecht et al., 2015).

The method by which flexural rigidity and Young's modulus is determined is thus sensitive to instrument setup and plant health, and this may explain some of the high variability in the values for Young's modulus that have been reported (Table 8.1). For individual coastal salt marsh plants in canopies of around 50 cm or less in height, reported Young's modulus values are highly variable, ranging from 13 MPa to over 4081 MPa. Even within species, such as *Puccinellia maritima*, different authors report very varied values, with Lara et al. (2016) reporting a value of 13 MPa but Rupprecht et al. (2015) reporting values of between 736 and 1995 MPa for the same species. Furthermore, at 27–73 MPa, the fresh- to brackish water marsh species *Schoenoplectus pungens* exhibits the lowest values, i.e., for these species little force is required to achieve the same displacement as for other species, given the diameter of the plants (although the triangular cross section of the stems of this species makes a direct comparison with circular stems of other species questionable). The greatest required force for any given displacement reported in the literature is required for the mangrove species, *Rhizophora*'s prop roots, for which recorded values of Young's modulus are of the order of 5800 MPa (Zhang et al., 2013).

5. MODELING FLOW OVER WETLANDS

Much of the above discussion around the interaction between water flow and wetlands revolves around an understanding of bed roughness and the complex flow patterns generated by topographic variability at a range of scales and the obstruction to flow provided by the vegetation canopy. Together, these bed topography and vegetation canopy effects can clearly cause hydrodynamic friction and drag effects that manifest themselves in water surface slopes over the wetland at scales of 10–1000 m and in reduced flow speeds detectable within wetland vegetation canopies even at small (tens of centimeter) scales. Any attempt at modeling unidirectional flow velocities and pathways over coastal wetlands thus has to parameterize the effect of such roughness, friction, and drag. As the examples below will highlight, the distinction between these three physical phenomena is not always clear or possible. The form drag due to vegetation structure, e.g., may be included in a model either by including a loss of momentum term that calculates the form drag or by increasing the bottom friction (e.g., Manning's n) determined empirically, from literature or through model calibration. Ultimately, all of the examples below attempt to achieve the same thing: a calculation of the energy lost due to the presence of vegetation and/or topographic roughness elements.

As it is difficult to resolve the exact flow structure over vegetated wetland surface areas of several hundreds to thousands of meters, energy or momentum loss due to vegetation has commonly been represented over vegetated wetland surface areas by a bottom friction coefficient.

As a simplification of the hydrodynamics, flow over a frictionally dominated tidal marsh platform can be assumed to be a balance between water surface slope and friction. Rinaldo et al. (1999) reduced the shallow water to a Poisson approximation, assuming that the tide propagates through the creek network instantaneously compared with the marsh platform, that friction is constant, and that spatial variations in water surface elevation are much smaller than water depth (Fagherazzi et al., 2012), as

$$\nabla^2 \eta_1(x,t) = \frac{\lambda}{D_0^2}\frac{\partial \eta_0}{\partial t} \tag{8.2}$$

where λ is the bottom friction coefficient; D_0 is the average water depth above the marsh platform; and η_1 is the local deviation of the water surface elevation from η_0, its average value; and t is time. The depth-averaged overmarsh flow velocity resulting from this water surface slope can thus be computed, assuming a linear frictionally dominated flow, as

$$U = -\frac{D_0 \nabla \eta_1}{\lambda} \tag{8.3}$$

As with all approaches addressing flow interactions with wetlands, the quantification of bottom friction is key to the accurate application of this formulation for the prediction of flow velocity. The bottom friction coefficient, λ, is a function of the Chézy friction coefficient, C, and the maximum tidal current U_0. The typically used bottom friction formula for flow in wetlands is the Manning's formula. The Manning's formula is based on the Chézy formula; it relates mean flow velocity, the hydraulic radius (R), the surface slope (S), and Manning's roughness coefficient (n) according to

$$U = \frac{1}{n}R^{2/3}S^{1/2} \tag{8.4}$$

Manning's n values are reported in the engineering literature for surfaces of varying sedimentological or surface roughness characteristics (e.g., vegetation type). Additionally, Manning's n values are used as calibration parameters to tune numerical models to empirical data.

In their early work on 2D hydrodynamic modeling of wetlands in Florida, Guardo and Tomasello (1995) used a fixed value Manning's coefficient to represent energy loss due to vegetation in the model. Wamsley et al. (2010) incorporate the additional energy losses due to the marsh vegetation in an ADCIRC model of the Louisiana wetlands using Manning's n values based on wetland type. In their study, categories of healthy wetlands include saline ($n = 0.035$), brackish ($n = 0.045$), intermediate ($n = 0.055$), fresh ($n = 0.055$), and open water ($n = 0.020$). Wamsley et al. (2010) note the importance of incorporating both the impact of bottom roughness and also the drag caused by the vegetation in their models. Although incorporating this drag as a bottom roughness term may not fully account for the momentum loss due to the plants over the full water column, it provides a good approximation.

Similarly, Loder et al. (2009) recognized the importance of incorporating both the bottom roughness and drag throughout the water column, as caused by wetland vegetation. They again approximated this drag through use of an addition to the bottom stress term. In an idealized ADCIRC model of storm surges over the Gulf of Mexico, the wetlands are represented by Manning's coefficient of $n = 0.020$ (sandy bed) through 0.035 (tall grass), 0.05 (scattered bush), 0.075 (dense bush), and 0.150 (dense woods) to 0.3 (maximum and, most probably, unrealistic). Additionally, as they used an idealized model grid with a uniform continental shelf slope and smooth bathymetry, the wetland platforms were incorporated into the model by applying an additional marsh elevation and uniform channels to account for marsh continuity. By modeling a range of combinations of bottom friction coefficient, marsh elevation heights, and marsh continuity, Loder et al. (2009) highlighted the sensitivity of storm surge modeling to these wetland characteristics.

As demonstrated through the empirical studies of water flows over wetland surfaces, the height of the roughness elements characterizing the wetland surface relative to the water depth is important in determining flow reduction, certainly for submerged vegetation structures: the greater the water depth, the less important bed friction becomes in slowing the progression of water flow over the wetland surface. For water flow that is driven by meteorological effects on water levels at the seaward wetland margin, this relationship may be less clear, as the progression and size of the storm may lead to complex water depth and wind-driven current variations over time that deviate from the rather more predictably sinusoidal rise and fall of tidally driven flow. Numerical models of surge progression over the Gulf of Mexico marshes thus suggest that bed friction effects on surge water levels may be greater during some relatively less energetic (but spatially larger) storm systems than during other more intense (but smaller) storms (Loder et al., 2009).

Where numerical models have been used to try to capture the effect of mangrove surfaces on flow propagation during storm surges, such models have also represented mangroves in terms of bed friction through Manning's n, i.e., as a bulk representation of bed friction rather than through direct empirical relationships between vegetation properties and hydrodynamic drag. Models showed a reasonable fit to observed water levels when a value of $0.14–0.15\ \mathrm{s\,m^{-1/3}}$ was used as Manning's n, representing dense tree canopies (Zhang et al., 2012; Xu et al., 2010), but they were also shown to be sensitive to small changes in this type of bed roughness parameterization. This suggests that differences in mangrove density or species may have nonnegligible effects on water flow and thus water level predictions.

Within models of unidirectional and wave-induced flow, accounting for energy/momentum loss due to the vegetation is commonly achieved through formulations based on the Morison equation (Morison et al., 1950), which describes the force on a cylinder. Vegetation is assumed to be rigid and of cylindrical structure such that it can be represented through stem density, diameter, and height. The Morison equation divides the force exerted on a cylinder into a form drag and skin friction component. Skin friction is neglected as it is of a much smaller magnitude than form drag.

The drag force, F_D, can be expressed as

$$F_D = \frac{1}{2}\rho C_D S_p U^2 \tag{8.5}$$

where ρ is the density, C_D is the drag coefficient, S_p refers to the area of the object projected into the direction of the flow, and U is the velocity. In their early laboratory studies, Denny (2014), Vogel (1994), and Kobayashi et al. (1993) all showed that, with rigid cylinders of varying density and diameter, the drag force (F_D) induced by such sets of objects can be approximated using this formula. The drag force is dependent on the shape of the object, roughness, and the flow characteristics. The drag coefficient is a function of the flow Reynolds number (Vogel, 1994), *Re*, which is the ratio between inertial and viscous forces:

$$Re = \frac{\rho UL}{\mu} \tag{8.6}$$

where μ is kinematic viscosity of water (in Nsm^{-2}) and L is the length scale, taken here as the water depth.

Mazda et al. (2005) express the effect of vegetation obstruction on water flow (perpendicular to the creek in the y direction) through the mangrove root layer by means of V_w, the volume of water within a given volume occupied jointly by water and plant (root) elements (Fig. 8.8):

$$g\frac{\partial \eta}{\partial y}V_w = F_{D,y} + F_{B,y} \tag{8.7}$$

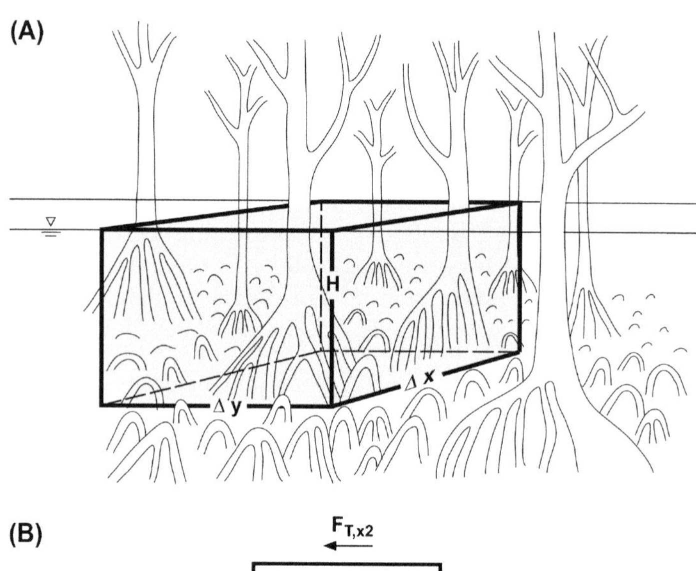

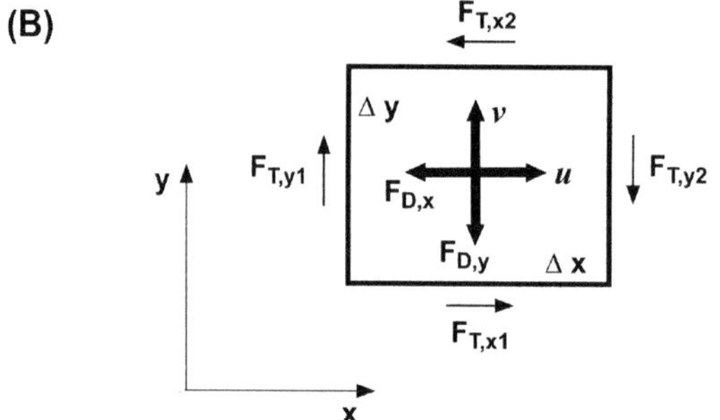

FIGURE 8.8 (A) Schematic illustration of *Rhizophora* stem and root elements within a given flow volume for computation of their effect on flow hydrodynamics (see text for explanation), and (B) $F_{T,x}$ and $F_{T,y}$ are the shear stresses in x and y, and $F_{D,x}$ and $F_{D,y}$ are the x and y components of the drag force. *Mazda, Y., Wolanski, E., King, B., Sase, A., Ohtsuka, D., Magi, M., 1997. Drag force due to vegetation in mangrove swamps. Mangroves and Salt Marshes 1 (3), 193–199.*

where η is the water surface elevation, F_D is the drag force due to the mangrove, and F_B is the bottom friction.

Mazda et al. (2005) propose that the bottom friction is too small to balance the water slope, and therefore that term is removed from Eq. (8.7). The drag force, F_D, is calculated using the Morison equation (Eq. 8.5) such that the drag coefficient is expressed as

$$C_D = -2g\frac{\partial \eta}{\partial y}\frac{D}{U|U|} \tag{8.8}$$

where D is the effective depth defined as V_w/A, with V_w, the volume of water present within the overall volume containing water and plant matter, and A, the total projected area of vegetation elements (obstacles) within the overall volume. This representation thus relies primarily on the a priori quantification of the obstruction provided by the plant elements themselves, assuming that their projected area is quantifiable for the wetland vegetation canopy. The latter is clearly difficult, however, when all mangrove canopy layers are to be considered, and it is thus perhaps not surprising that such an approach has thus far only been applied to the mangrove root layer. An improved understanding of the linkage between specific mangrove vegetation characteristics or canopy attributes and the roughness/drag parameterization within numerical models (and more validation data) is thus required to ensure better geographical transferability of models.

Temmerman et al. (2005) use the Delft 3D hydrodynamic model and incorporate the 3D effects of the vegetation with an additional source term in the momentum equations. This source term calculates momentum loss due to the vegetation canopy using the drag force formula (Eq. 8.4). This method models the impact of the vegetation over the water column, rather than just at the bed. The vegetation is parameterized in terms of its average diameter and density by measuring the dominant species at the marsh for inclusion in the model.

6. WAVE-INDUCED FLOWS AND INTERACTIONS WITH WETLAND SURFACES

Any statements around the interaction between waves and coastal wetlands may be considered to be somewhat of an oxymoron within wetland science. By definition, coastal wetlands form in sheltered locations, where low velocity flows allow the accumulation of fine substrates and the growth of, at least initially, physically sensitive plant seedlings. In water depths of <1 m, breaking waves of a few tens of centimeters in height can generate bed currents in excess of 0.7 m s^{-1}, strong enough to exceed the bed shear stress of fine particles and seeds.

Although low wave energy conditions are a prerequisite for wetland formation, mature wetlands are able to withstand significant wave impact (Spencer et al., 2016). This physical resilience of established coastal wetlands is a function of both their sedimentological strength and structure and the characteristics of their above- and belowground vegetation elements. With respect to the interaction of the wetland surface with wave-driven currents, aboveground vegetation characteristics have been shown to be critical in determining both (1) the likelihood of plant breakage/damage incurred and (2) the progression of wave energy across/through the vegetation canopy. Thus, flexible plants may be seen as "avoiding" physical stress due to wave action but contribute less toward the dissipation of wave energy, whereas rigid plants "tolerate" such stress and contribute more toward reducing such stress further along the direction of wave travel, as long as that stress does not lead to plant stem breaking (see, e.g., Schwarz et al., 2015).

Understanding and modeling wave hydrodynamics over coastal wetlands requires, initially, consideration of the fact that water depths over wetland surfaces are generally low (<2 m) and that waves are thus primarily depth-limited (Le Hir et al., 2007). Where water depths are less than twice the significant wave height (e.g., waves of 1 m height in 2 m depth), waves will be predominantly breaking with much wave-generated turbulence affecting the interaction with the marsh surface. For wave height to water depth ratios smaller than 0.5, wave-induced currents will be highly asymmetric, with stronger onshore than offshore currents. Thus, in water depths of <1 m, as are common on meso- to macrotidal coastal wetlands, the interaction between waves and the wetland surface is likely limited to waves of <30 cm in height (Moeller et al., 1996). For microtidal marshes, wave heights are typically much lower (Knutson et al., 1982).

It is thus not surprising that the first published field observations of waves over salt marshes concerned wind waves of relatively small magnitude (<5 cm). Initially, Wayne (1976) observed such waves over short (10 m) distances within a microtidal *Spartina alterniflora* marsh on the east coast of the United States. Waves lost an average of 77% and 66% of their energy over two consecutive 10 m long sections of the wetland surface. Knutson et al.'s (1982) experiment with boat-generated waves of 17 cm height in *Spartina alterniflora* marshes in Florida showed that wave heights reduced, on average, by 65% (wave energy by 88%) over a 10 m distance. In the United Kingdom, basic physical model experiments with assumed additional friction effects over plywood models of seawalls fronted by "marsh" (berms) suggested that an 80 m wide macrotidal salt marsh platform (with a 2 m high steep seaward "cliff" and inundation depths of 2 m above the platform) would reduce incident deep water wave heights of 1.75 m to less than 0.8 m at the landward end of the platform (Brampton, 1992). Field studies of Moeller et al. (1996) and Möller and Spencer, (2002) on the North Norfolk and Essex coast of the UK east coast show broad agreement with those modeled by Brampton (1992) for cliffed marshes (Fig. 8.9A). For more gradual transitions from tidal flat to vegetated marsh, however, wave height reduction observed in the field occurs more rapidly with distance than in Brampton's (1992) physical scale model (Fig. 8.9B). Those field studies also highlighted, however, that, as might be expected given their depth-limited nature, waves and their transformation with distance are highly dependent on water depth and incident wave conditions, as well as on vegetation characteristics (e.g., seasonal changes to the vegetation canopy) (Möller et al., 1999; Möller and Spencer, 2002).

7. MODELING WAVES OVER WETLANDS

In physical terms, the process of wave height change from an initial height (H_1) to a height some distance landward (H_2) over the coastal wetland surface can be expressed mathematically as

$$H_2 = H_1 K_s K_v K_f K_p \tag{8.9}$$

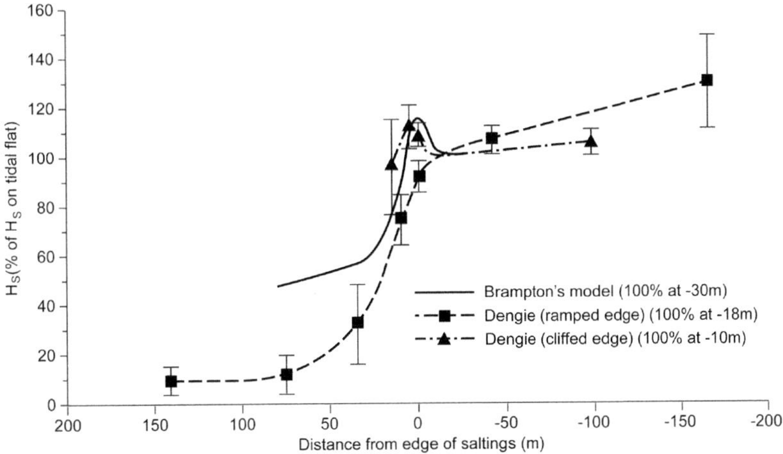

FIGURE 8.9 Observed and modeled wave dissipation over a cliffed and "ramped" open coast salt marsh, Essex, UK.

where the subscripts s, v, f, and p denote the shoaling coefficient, viscous friction, bottom friction, and percolation decay factors, respectively. Given the minimal slopes of wetland surfaces, fine-grained and compacted soils, shoaling, viscous friction, and percolation can be assumed to be minimal. As for the case of nonoscillatory water flows, questions around the effect of the wetland surface on wave-induced oscillatory water motion have thus focused on the effect of friction and drag caused by the presence of vegetation elements. Dalrymple et al. (1984) postulated that, some distance along the wave's direction of travel, the wave height (H_2) is a function of the initial wave height (H_1), the distance traveled (Δx), and a "damping factor" (a):

$$H_2 = H_1\left(\frac{1}{1 + a\Delta x}\right) \text{ or (approximated as)} H_2 = H_1 e^{-a\Delta x} \tag{8.10}$$

where $e^{-a\Delta x}$ is the "decay factor" (K_x). This relationship holds true for small values of $a\Delta x$ (<0.34). The inaccuracy in this relationship rises if $a\Delta x$ increases above 0.34, i.e., values of K_x fall below 0.71. As has been demonstrated by the field studies of Knutson et al. (1982) and Möller et al. (1999), however, K_x remains high for most conditions likely to be observed in the field.

Dalrymple's et al. (1984) model still underpins current modeling efforts for the dissipation of regular waves (see, e.g., Möller et al., 2014). For irregular waves, i.e., typically those generated by the action of wind on the water surface, however, alternative formulations are needed that represent dissipation/transformation of wave spectra, rather than monochromatic waves of a defined height and length/period. Mendez and Losada (2004) formulated such transformation with respect to the root mean square wave height over a distance x such that the relationship between incident wave height H_1 ($H_{rms,1}$) and damped wave height H_2 ($H_{rms,2}$) can be represented as

$$\frac{H_1 - H_2}{H_1} = \frac{\alpha x}{1 + \alpha x} \text{ (reg. waves)}, \quad \frac{H_{rms,1} - H_{rms,2}}{H_{rms,1}} = \frac{\alpha x}{1 + \alpha x} \text{ (irreg. waves)} \tag{8.11}$$

in which

$$\alpha = A \frac{S_D}{S_S^2} C_D k \left[\frac{\sinh^3 kS_H + 3 \sinh kS_H}{\sinh kh(\sinh 2\,kh + 2kh)} \right] \tag{8.12}$$

and $A = 4/(9\pi)H_1$ for regular waves and $A = 2/(3\sqrt{\pi})H_{rms,1}$ for irregular waves, $k = 2\pi/L$ (L = wave length of peak period T_p), and h = water depth (Möller et al., 2014). This model is based on the Morison equation, therefore, the representation of vegetation elements is through metrics that relate as in the case of Mazda et al.'s (2005) formulation for mangrove roots, to the projected area of the vegetation into the direction of wave approach; thus S_D = plant stem diameter, S_S = plant stem spacing, and S_H = plant stem height.

As for the case of steady unidirectional flow (Eqs. 8.5 and 8.6), the drag coefficient (C_D) can thus be empirically determined from the wave conditions and can be expressed in terms of the stem Reynolds number, Re, or the Keulegan–Carpenter number (Jadhav et al., 2013; Augustin et al., 2009). For their large-scale laboratory experiment in which regular and irregular waves were observed over a 40 m vegetated salt marsh platform, e.g., Möller et al. (2014) find that

$$C_D = -0.046 + \left(\frac{305.5}{Re_v}\right)^{0.977} \text{ (regular waves; } r^2 = 0.97) \tag{8.13}$$

$$C_D = 0.159 + \left(\frac{227.3}{Re_v}\right)^{1.615} \text{ (irregular waves; } r^2 = 0.99) \tag{8.14}$$

Although, in this case, the Reynolds number is computed differently to the unidirectional steady flow case above (Eqs. 8.4 and 8.5) to account for varying stem diameters and orbital wave currents (U_{max}):

$$Re_v = U_{max} \frac{S_D}{v_k} \tag{8.15}$$

where v_k is the kinematic viscosity of water ($1 \times 10^{-6}\ m^2s^{-1}$ for freshwater) and U_{max} is the maximum orbital near-bed velocity. The Re_v-dependent drag coefficients that are computed in this way, i.e., from an empirical fit between observed and modeled wave height dissipation, are more than two orders of magnitude lower than those modeled by others (Mendez and Losada, 2004) but compare favorably with those determined for seagrass beds (see Fig. 8.10).

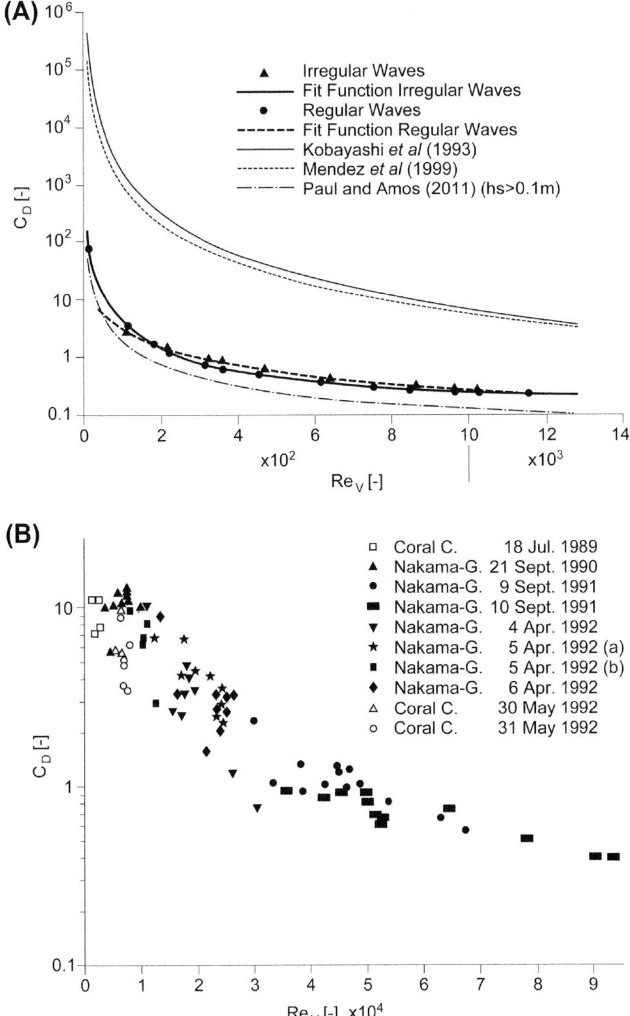

FIGURE 8.10 Flow-dependent drag as modeled empirically for different types of vegetation canopies. *(A) Adapted from Möller, I., Kudella, M., Rupprecht, F., Spencer, T., Paul, M., Van Wesenbeeck, B.K., Wolters, G., Jensen, K., Bouma, T.J., Miranda-Lange, M., Schimmels, S., 2014. Wave attenuation over coastal salt marshes under storm surge conditions. Nature Geoscience 7 (10), 727–731. (B) Adapted from Mazda, Y., Wolanski, E., King, B., Sase, A., Ohtsuka, D., Magi, M., 1997. Drag force due to vegetation in mangrove swamps. Mangroves and Salt Marshes 1 (3), 193–199.*

8. TSUNAMI WAVES

The experience of the Indian Ocean tsunami of 2004 led to much discussion around the role of coastal wetlands in reducing the run-up heights of these unusually large and fast-moving, tectonically forced waves. Unlike the meteorologically forced water level rise and wind-driven water currents discussed alongside tidal inundation of wetlands discussed above, tsunami waves represent hydrodynamic events of an altogether different character and magnitude. Although they are waves, their unusual length (typically of the order of several hundreds of kilometers) and period (10–60 min) means that they experience shoaling (the steepening and shortening of waves that occur when waves enter shallow water) much further offshore than wind-generated waves (Zhang et al., 2012; Alongi, 2008). With a deep water wave height of only around 0.5 m, their shallow water height at the seaward margin of coastal wetlands can thus exceed tens of meters. Much of the challenge of predicting the interaction between tsunami-generated water levels/currents and coastal wetlands arises from the fact that the prediction of the precise hydrodynamic characteristics (e.g., height and speed of the wave) at the wetland margin rely on knowledge of the precise propagation track and bathymetry with which the wave interacts *before* it arrives at the wetland's seaward edge. For tsunamis in the Indian Ocean with comparably high physical characteristics (intensity scores of XII), Suppasri et al. (2012) thus model the coastal height as varying between 23 and 42 m.

As far as the interaction with coastal wetlands is concerned, a series of studies reported, often anecdotally, reduced damage to people and property in the aftermath of the Asian tsunami on coasts on which mangrove wetlands were present (Alongi, 2008; Tanaka, 2009). Follow-on investigations confirmed, however, that the effect of mangroves on tsunami heights is dependent on the forest (tree, stem, and root) density and dimensions, as well as the adjacent bathymetry, exposure, and position of the wetland and the particular characteristics of the incident wave and tidal stage at the time of the occurrence of the tsunami (Alongi, 2008; Teo et al., 2009).

In terms of the representation of mangroves within numerical models of tsunami propagation, their effect on water flow has been parameterized in ways similar to that discussed above for flow through cylindrically shaped structures. Thus, Teo et al. (2009), for example, represent the drag forces induced by the presence of mangrove trees as

$$F = 0.5C_D D_t \rho_t \frac{q\sqrt{q_x^2 + q_y^2}}{H} \tag{8.16}$$

where C_D is, as discussed above, determined using previously established relationships with Reynolds conditions, D_t is the tree diameter, ρ_t is the number of trees per unit area, and q_x and q_y are the depth-integrated discharges per unit width in the x and y direction.

In addition to the drag force, however, models of such large-scale flow have been attempted in which the porosity of the full mangrove vegetation layer (θ) is parameterized as a function of D_t and ρ_t:

$$\theta = 1 - \frac{\pi}{4} D_t^2 \rho_t \tag{8.17}$$

Together, these terms have then been built into two-dimensional depth-integrated numerical models and used for the modeling of tidal flow propagation (Wu et al., 2001; Teo et al., 2009).

9. SCALING CHALLENGES

Much progress has been made in recent years on better understanding the biophysical linkages associated with coastal wetland hydrodynamics. The disciplinary boundary between ecological studies into coastal wetland functioning and geological and geomorphological studies into landform evolution has become less divisive in the context of the growing recognition of ecosystem service provisioning by wetlands. An increasing number of studies thus now appreciate the tight linkage between the presence of vegetation and benthic fauna and the water and sediment fluxes through and across the wetland system. The turn of the century has seen significant progress in isolating the effect of particular wetland characteristics on water flow speeds, as has been shown, e.g., in the studies of tidal flow through creek systems and over wetland surfaces. Innovative technologies for the high-frequency capture and storage of water depth, velocity, and flow direction observations have advanced this knowledge. Laboratory studies have become increasingly complex and moved away from simulating wetland vegetation as rigid cylinders and instead incorporating real plants with appropriate measurements of plant flexibility and structure. More detailed species-specific quantitative information on the latter is gradually becoming available, as plant samples from the field can be tested in engineering laboratories. Notably, the growing evidence for the stability of wetland substrates and for the effectiveness of vegetation in reducing flow energy even during extreme conditions gives full recognition of the natural flood and erosion protection provided by coastal wetlands.

In spite of these advances, however, two key challenges remain regarding our ability to accurately model hydrodynamic processes over coastal wetlands: Firstly, there is the issue of scale dependency of the varying marsh characteristics that determine the wetland's interaction with water. The particular scale at which such interaction occurs at any given time is a reflection of both the scale of the pattern of water movement and the scale of the wetland characteristic that most significantly affects it. Patterns of water movement vary depending on the force that causes such movement (e.g., gravitationally induced tides, atmospheric pressure–induced storm surges, wind-driven currents and waves, and tectonically induced tsunamis). Finding a particular, measurable, wetland characteristic (or set of characteristics) that most significantly affects the water movement associated with each of these different flow types is the fundamental challenge of the ongoing efforts to improve numerical modeling of water depths, flow speeds, and wave heights/periods over wetland surfaces.

At small spatial scales of 1–10 m, the effect of wetland surfaces on wave dissipation, for example, can be understood as a function of parameters that can be recorded at discrete and distinct locations along the cross-shore profile (notably vegetation type, structure, and density, as well as water depth and incident wave energy). How such parameterization is best achieved, however, is still unanswered. For vegetation canopies with complex structures, such as those of *Atriplex portulacoides* on salt marshes (Fig. 8.11), parameterizations based on stem metrics are meaningless because of the relatively large contribution of nonstem elements to the bulk of the canopy as a whole.

At larger spatial scales of 10–100 m, wave refraction and diffraction through surface topographical features (for example, creeks and salt pans) are likely to become equally important

FIGURE 8.11 The shrubby canopy of *Atriplex portulacoides* on a UK east coast salt marsh (the width of the photograph is approximately 70 cm). *Photograph: James Tempest.*

as effects of the vegetation canopy, but they are not always adequately resolved for the given scale of the hydrodynamic model. Landform characteristics such as fragmentation and drainage channel density may affect the interaction between the coastal wetland and water flow at the kilometer scale, with implications for wetland restoration and management (Wamsley et al., 2009). At those larger scales, however, it is still unclear how vegetation type, structure, and density might be most suitably parameterized and measured in practice (so as to link them to the wave dissipation effect), at least where vegetation cover is patchy and nonuniform. How best to aggregate those vegetation characteristics in patchy communities (such as those typical of mature, mixed European salt marsh communities) at the larger, marsh wide scale or whether such integration is even necessary or desirable for flow modeling remains a fundamental challenge for future research.

Secondly, growing evidence from extreme events such as hurricanes in the Gulf of Mexico (in the case of surge and wave dissipation) and the Asian tsunami illustrates the importance of the quantification of process thresholds or "tipping points." Such thresholds can be conceptualized as describing the switching from a system in which the biological elements act to mediate hydrodynamic energy (such as the reduction in surge elevations, flow velocities, or wind-generated waves) to one in which the presence of biological elements enhances the erosive impact of water flows (such as in sparsely vegetated marshes with high incident wave energy or where mangrove debris adds to the destructive impact of tsunami waves propagating inshore). Any serious incorporation of biophysical linkages into coastal management plans requires some assessment of the likelihood, frequency, and duration with which such process thresholds may be exceeded within a given time frame and further research is thus required to identify where those process thresholds lie.

Thus, while much progress has been made in understanding the interaction between coastal wetlands and hydrodynamic processes, for this knowledge to be used successfully in predictive models, e.g., for the purpose of its incorporation into coastal management approaches, further progress in these two areas is particularly needed.

Acknowledgments

Much of the content of this chapter is the result of discussions over many years between the authors and individuals within and beyond the Cambridge Coastal Research Unit. We thank them all for their invaluable indirect input. Particular thanks go to Professor Tom Spencer, without whose leadership as Director of the CCRU, much of the thinking that has informed this chapter would have been impossible. Dr Christie was funded through the "Physical and biological dynamic coastal processes and their role in coastal recovery (BLUE-coast)" project (NE/N015878/1, 2016-2021) by the Natural Environment Research Council (NERC) and Dr Möller conducted this work as input to the "Valuing the contribution which Coastal habitats make to human health and Wellbeing, with a focus on the alleviation of natural hazards (CoastWEB)" project (NE/N013573/1) and "Response of Ecologically-mediated Shallow Intertidal Shores and their Transitions to extreme hydrodynamic forcing in UK settings (RESIST-UK)" project (NE/R01082X/1), both also funded by NERC.

References

Albert, D.A., Cox, D.T., Lemein, T., Yoon, H.D., 2013. Characterization of Schoenoplectus pungens in a Great Lakes coastal wetland and a Pacific Northwestern estuary. Wetlands 33 (3), 445–458.

Allen, J.R.L., 2000. Morphodynamics of Holocene salt marshes: a review sketch from the Atlantic and Southern North Sea coasts of Europe. Quaternary Science Reviews 19 (12), 1155–1231.

Alongi, D.M., 2008. Mangrove forests: resilience, protection from tsunamis, and responses to global climate change. Estuarine, Coastal and Shelf Science 76 (1), 1–13.

Augustin, L.N., Irish, J.L., Lynett, P., 2009. Laboratory and numerical studies of wave damping by emergent and near-emergent wetland vegetation. Coastal Engineering 56 (3), 332–340.

Barendregt, A., Swarth, C.W., 2013. Tidal freshwater wetlands: variation and changes. Estuaries and Coasts 36 (3), 445–456.

Bayliss-Smith, T.P., Healey, R., Lailey, R., Spencer, T., Stoddart, D.R., 1979. Tidal flows in salt marsh creeks. Estuarine and Coastal Marine Science 9 (3), 235–255.

Blanton, J.O., Lin, G., Elston, S.A., 2002. Tidal current asymmetry in shallow estuaries and tidal creeks. Continental Shelf Research 22 (11–13), 1731–1743.

Bouma, T.J., Vries, M.D., Low, E., Kusters, L., Herman, P.M.J., Tanczos, I.C., Temmerman, S., Hesselink, A., Meire, P., Regenmortel, S.V., 2005. Flow hydrodynamics on a mudflat and in salt marsh vegetation: identifying general relationships for habitat characterisations. Hydrobiologia 540 (1–3), 259–274.

Bouma, T.J., Friedrichs, M., Klaassen, P., Van Wesenbeeck, B.K., Brun, F.G., Temmerman, S., Van Katwijk, M.M., Graf, G., Herman, P.M.J., 2009. Effects of shoot stiffness, shoot size and current velocity on scouring sediment from around seedlings and propagules. Marine Ecology Progress Series 388, 293–297.

Brampton, A.H., 1992. Engineering significance of British saltmarshes. In: Allen, J.R.L., Pye, K. (Eds.), Saltmarshes; Morphodynamics, Conservation and Engineering Significance. Cambridge University Press, Cambridge, pp. 115–122.

Brooks, R.P., Brinson, M.M., Havens, K.J., Hershner, C.S., Rheinhardt, R.D., Wardrop, D.H., Whigham, D.F., Jacobs, A.D., Rubbo, J.M., 2011. Proposed hydrogeomorphic classification for wetlands of the mid-atlantic region, USA. Wetlands 31 (2), 207–219.

Cahoon, D.R., Hensel, P.F., Spencer, T., Reed, D.J., McKee, K.L., Saintilan, N., 2006. Coastal wetland vulnerability to relative sea-level rise: wetland elevation trends and process controls. In: Verhoeven, J.T.A., et al. (Eds.), Wetlands and Natural Resource Management. Springer, Berlin, Heidelberg, pp. 271–292.

Cahoon, D.R., 2006. A review of major storm impacts on coastal wetland elevations. Estuaries and Coasts 29 (6), 889–898.

Carol, E., Kruse, E., Tejada, M., 2013. Surface water and groundwater response to the tide in coastal wetlands: assessment of a marsh in the outer Rio de la Plata estuary, Argentina. Journal of Coastal Research 65 (2), 1098–1103.

Childers, D.L., Day, J.W., McKellar, H.N., 2000. Twenty more years of marsh and estuarine flux studies: revisiting Nixon (1980). In: Weinstein, M.P., Kreeger, D.A. (Eds.), Concepts and Controversies in Tidal Marsh Ecology, pp. 391–423.

Cowell, P., Thom, B., 1994. Morphodynamics of coastal evolution. In: Carter, R., Woodroffe, C. (Eds.), Coastal Evolution: Late Quaternary Shoreline Morphodynamics. Cambridge University Press, Cambridge, pp. 33–86.

Dalrymple, R.A., Kirby, J.T., Hwang, P.A., 1984. Wave diffraction due to areas of energy dissipation. Journal of Waterway, Port, Coastal, and Ocean Engineering 110, 67–79.

Davidson-Arnott, R.G., van Proosdij, D., Ollerhead, J., Schostak, L., 2002. Hydrodynamics and sedimentation in salt marshes: examples from a macrotidal marsh, Bay of Fundy. Geomorphology 48 (1–3), 209–231.

del Pilar Alvarez, M., Carol, E., Hernández, M.A., Bouza, P.J., 2015. Groundwater dynamic, temperature and salinity response to the tide in Patagonian marshes: observations on a coastal wetland in San Jose Gulf, Argentina. Journal of South American Earth Sciences 62, 1–11.

Denny, M., 2014. Biology and the Mechanics of the Wave-Swept Environment. Princeton University Press.

Dijkstra, J.T., 2009. How to account for flexible aquatic vegetation in large-scale morphodyamic models. In: Smith, J.M. (Ed.), International Conference on Coastal Engineering 2008, pp. 2820–2831.

Ewel, K.C., Twilley, R.R., Ong, J.E., 1998. Different kinds of mangrove forests provide different goods and services. Global Ecology and Biogeography Letters 7 (1), 83–94.

Fagherazzi, S., Carniello, L., D'Alpaos, L., Defina, A., 2006. Critical bifurcation of shallow microtidal landforms in tidal flats and salt marshes. Proceedings of the National Academy of Sciences of the United States of America 103 (22), 8337–8341.

Fagherazzi, S., Kirwan, M.L., Mudd, S.M., Guntenspergen, G.R., Temmerman, S., D'Alpaos, A., Koppel, J., Rybczyk, J.M., Reyes, E., Craft, C., Clough, J., 2012. Numerical models of salt marsh evolution: ecological, geomorphic, and climatic factors. Reviews of Geophysics 50 (1).

Feagin, R.A., Irish, J.L., Möller, I., Williams, A.M., Colón-Rivera, R.J., Mousavi, M.E., 2011. Short communication: engineering properties of wetland plants with application to wave attenuation. Coastal Engineering 58 (3), 251–255.

Friess, D.A., Krauss, K.W., Horstman, E.M., Balke, T., Bouma, T.J., Galli, D., Webb, E.L., 2012. Are all intertidal wetlands naturally created equal? Bottlenecks, thresholds and knowledge gaps to mangrove and saltmarsh ecosystems. Biological Reviews 87 (2), 346–366.

Friess, D.A., Möller, I., Spencer, T., Smith, G.M., Thomson, A.G., Hill, R.A., 2014. Coastal saltmarsh managed realignment drives rapid breach inlet and external creek evolution, Freiston Shore (UK). Geomorphology 208, 22–33.

Furukawa, K., Wolanski, E., Mueller, H., 1997. Currents and sediment transport in mangrove forests. Estuarine, Coastal and Shelf Science 44 (3), 301–310.

Glamore, W.C., Indraratna, B., 2009. Tidal-forcing groundwater dynamics in a restored coastal wetland: implications of saline intrusion. Australian Journal of Earth Sciences 56 (1), 31–40.

Guardo, M., Tomasello, R.S., 1995. Hydrodynamic simulations of a constructed wetland in South Florida. JAWRA Journal of the American Water Resources Association 31 (4), 687–701.

Harvey, J.W., Schaffranek, R.W., Noe, G.B., Larsen, L.G., Nowacki, D.J., O'Connor, B.L., 2009. Hydroecological factors governing surface water flow on a low-gradient floodplain. Water Resources Research 45 (3).

Horstman, E.M., Dohmen-Janssen, C.M., Hulscher, S.J.M.H., 2013. Flow routing in mangrove forests: a field study in Trang province, Thailand. Continental Shelf Research 71, 52–67.

Hosokawa, Y., Furukawa, K., 1994. Surface flow and particle settling in a coastal reed field. Water Science and Technology 29 (4), 45–53.

Jadhav, R.S., Chen, Q., 2013. Probability distribution of wave heights attenuated by salt marsh vegetation during tropical cyclone. Coastal Engineering 82, 47–55.

Jadhav, R.S., Chen, Q., Smith, J.M., 2013. Spectral distribution of wave energy dissipation by salt marsh vegetation. Coastal Engineering 77, 99–107.

Järvelä, J., 2005. Effect of submerged flexible vegetation on flow structure and resistance. Journal of Hydrology 307 (1–4), 233–241.

Jay, D.A., Leffler, K., Diefenderfer, H.L., Borde, A.B., 2015. Tidal-fluvial and estuarine processes in the lower Columbia river: I. Along-channel water level variations, Pacific Ocean to Bonneville Dam. Estuaries and Coasts 38 (2), 415–433.

Kearney, W.S., Fagherazzi, S., 2016. Salt marsh vegetation promotes efficient tidal channel networks. Nature Communications 7, 12287.

Knighton, A.D., Woodroffe, C.D., Mills, K., 1992. The evolution of tidal creek networks, mary river, northern Australia. Earth Surface Processes and Landforms 17 (2), 167–190.

Knutson, P.L., Brochu, R.A., Seelig, W.N., Inskeep, M., 1982. Wave damping in *Spartina alterniflora* marshes. Wetlands 2 (1), 87–104.

Kobayashi, N., Raichle, A.W., Asano, T., 1993. Wave attenuation by vegetation. Journal of Waterway, Port, Coastal, and Ocean Engineering 119 (1), 30–48.

Krauss, K.W., Doyle, T.W., Doyle, T.J., Swarzenski, C.M., From, A.S., Day, R.H., Conner, W.H., 2009. Water level observations in mangrove swamps during two hurricanes in Florida. Wetlands 29 (1), 142–149.

Lara, J.L., Maza, M., Ondiviela, B., Trinogga, J., Losada, I.J., Bouma, T.J., Gordejuela, N., 2016. Large-scale 3-D experiments of wave and current interaction with real vegetation. Part 1: guidelines for physical modeling. Coastal Engineering 107, 70–83.

Le Hir, P., Monbet, Y., Orvain, F., 2007. Sediment erodability in sediment transport modelling: can we account for biota effects? Continental Shelf Research 27 (8), 1116–1142.

Leonard, L.A., Croft, A.L., 2006. The effect of standing biomass on flow velocity and turbulence in *Spartina alterniflora* canopies. Estuarine, Coastal and Shelf Science 69 (3), 325–336.

Leonard, L.A., Luther, M.E., 1995. Flow hydrodynamics in tidal marsh canopies. Limnology and Oceanography 40 (8), 1474–1484.

Leopold, L.B., Collins, J.N., Collins, L.M., 1993. Hydrology of some tidal channels in estuarine marshland near San Francisco. Catena 20 (5), 469–493.

Loder, N.M., Irish, J.L., Cialone, M.A., Wamsley, T.V., 2009. Sensitivity of hurricane surge to morphological parameters of coastal wetlands. Estuarine, Coastal and Shelf Science 84 (4), 625–636.

Luhar, M., Nepf, H.M., 2011. Flow-induced reconfiguration of buoyant and flexible aquatic vegetation. Limnology and Oceanography 56 (6), 2003–2017.

Mariotti, G., Fagherazzi, S., 2013. A two-point dynamic model for the coupled evolution of channels and tidal flats. Journal of Geophysical Research: Earth Surface 118 (3), 1387–1399.

Mazda, Y., Ikeda, Y., 2006. Behaviour of the groundwater in a riverine-type mangrove forest. Wetlands Ecology and Management 14 (6), 477–488.

Mazda, Y., Yokochi, H., Sato, Y., 1990. Groundwater flow in the Bashita-Minato mangrove area, and its influence on water and bottom mud properties. Estuarine, Coastal and Shelf Science 31 (5), 621–638.

Mazda, Y., Wolanski, E., King, B., Sase, A., Ohtsuka, D., Magi, M., 1997. Drag force due to vegetation in mangrove swamps. Mangroves and Salt Marshes 1 (3), 193–199.

Mazda, Y., Kanazawa, N., Kurokawa, T., 1999. Dependence of dispersion on vegetation density in a tidal creek-mangrove swamp system. Mangroves and Salt Marshes 3 (1), 59–66.

Mazda, Y., Kobashi, D., Okada, S., 2005. Tidal-scale hydrodynamics within mangrove swamps. Wetlands Ecology and Management 13 (6), 647–655.

Mendez, F.J., Losada, I.J., 2004. An empirical model to estimate the propagation of random breaking and nonbreaking waves over vegetation fields. Coastal Engineering 51 (2), 103–118.

Moeller, I., Spencer, T., French, J.R., 1996. Wind wave attenuation over saltmarsh surfaces: preliminary results from Norfolk, England. Journal of Coastal Research 12 (4), 1009–1016.

Möller, I., Spencer, T., 2002. Wave dissipation over macro-tidal saltmarshes: effects of marsh edge typology and vegetation change. Journal of Coastal Research 36 (1), 506–521.

Möller, I., Spencer, T., French, J.R., Leggett, D.J., Dixon, M., 1999. Wave transformation over salt marshes: a field and numerical modelling study from North Norfolk, England. Estuarine, Coastal and Shelf Science 49 (3), 411–426.

Möller, I., Mantilla-Contreras, J., Spencer, T., Hayes, A., 2011. Micro-tidal coastal reed beds: hydro-morphological insights and observations on wave transformation from the southern Baltic Sea. Estuarine, Coastal and Shelf Science 92 (3), 424–436.

Möller, I., Kudella, M., Rupprecht, F., Spencer, T., Paul, M., Van Wesenbeeck, B.K., Wolters, G., Jensen, K., Bouma, T.J., Miranda-Lange, M., Schimmels, S., 2014. Wave attenuation over coastal salt marshes under storm surge conditions. Nature Geoscience 7 (10), 727–731.

Morison, J., Johnson, J., Schaaf, S., 1950. The force exerted by surface waves on piles. Journal of Petroleum Technology 2 (5), 149–154.

Murray, A.L., Spencer, T., 1997. On the wisdom of calculating annual material budgets in tidal wetlands. Marine Ecology Progress Series 150 (1–3), 207–216.

Neumeier, U., Ciavola, P., 2004. Flow resistance and associated sedimentary processes in a *Spartina maritima* salt-marsh. Journal of Coastal Research 20 (2), 435–447.

Neumeier, U., 2005. Quantification of vertical density variations of salt-marsh vegetation. Estuarine, Coastal and Shelf Science 63 (4), 489–496.

Neumeier, U., 2007. Velocity and turbulence variations at the edge of saltmarshes. Continental Shelf Research 27 (8), 1046–1059.

Peralta, G., Van Duren, L.A., Morris, E.P., Bouma, T.J., 2008. Consequences of shoot density and stiffness for ecosystem engineering by benthic macrophytes in flow dominated areas: a hydrodynamic flume study. Marine Ecology Progress Series 368, 103–115.

Pethick, J.S., 1969. Drainage in tidal marshes. In: Steers, J.R. (Ed.), The Coastline of England and Wales. Cambridge University Press, Cambridge, pp. 725–730.

Redfield, A.C., 1972. Development of a New England salt marsh. Ecological Monographs 42 (2), 201–237.

Reed, D.J., Stoddart, D.R., Bayliss-Smith, T.P., 1985. Tidal flows and sediment budgets for a salt-marsh system, Essex, England. Vegetatio 62 (1–3), 375–380.

Reed, J., 1987. Temporal asymmetry sampling and discharge in salt marsh creeks. Estuarine, Coastal and Shelf Science 25, 459–466.

Reed, D.J., 1988. Sediment dynamics and deposition in a retreating coastal salt marsh. Estuarine, Coastal and Shelf Science 26, 67–79.

Ridd, P.V., Sam, R., 1996. Profiling groundwater salt concentrations in mangrove swamps and tropical salt flats. Estuarine, Coastal and Shelf Science 43, 627–635.

Rinaldo, A., Fagherazzi, S., Lanzoni, S., Marani, M., Dietrich, W.E., 1999. Tidal networks 2. Watershed delineation and comparative network morphology. Water Resources Research 35 (12), 3905–3917.

Rupprecht, F., Möller, I., Evans, B., Spencer, T., Jensen, K., 2015. Biophysical properties of salt marsh canopies - quantifying plant stem flexibility and above ground biomass. Coastal Engineering 100, 48–57.

Schwarz, C., Bouma, T.J., Zhang, L.Q., Temmerman, S., Ysebaert, T., Herman, P.M.J., 2015. Interactions between plant traits and sediment characteristics influencing species establishment and scale-dependent feedbacks in salt marsh ecosystems. Geomorphology 250, 298–307.

Shi, Z., Pethick, J.S., Burd, F., Murphy, B., 1996. Velocity profiles in a salt marsh canopy. Geo-Marine Letters 16 (4), 319–323.

Shi, B.W., Yang, S.L., Wang, Y.P., Li, G.C., Li, M.L., Li, P., Li, C., 2017. Role of wind in erosion-accretion cycles on an estuarine mudflat. Journal of Geophysical Research: Oceans 122, 193–206.

Spencer, T., Möller, I., 2013. Mangrove systems. In: Shroder, D.J., Sherman, J.F. (Eds.), Treatise on Geomorphology. Academic Press, San Diego, pp. 360–391.

Spencer, T., Möller, I., Rupprecht, F., Bouma, T.J., Wesenbeeck, B.K., Kudella, M., Paul, M., Jensen, K., Wolters, G., Miranda-Lange, M., Schimmels, S., 2016. Salt marsh surface survives true-to-scale simulated storm surges. Earth Surface Processes and Landforms 41 (4), 543–552.

Suppasri, A., Futami, T., Tabuchi, S., Imamura, F., 2012. Mapping of historical tsunamis in the Indian and Southwest Pacific Oceans. International Journal of Disaster Risk Reduction 1, 62–71.

Symonds, A.M., Collins, M.B., 2007. The development of artificially created breaches in an embankment as part of a managed realignment, Freiston Shore, UK. Journal of Coastal Research 130–134. Special Issue 50.

Tanaka, N., 2009. Vegetation bioshields for tsunami mitigation: review of effectiveness, limitations, construction, and sustainable management. Landscape and Ecological Engineering 5 (1), 71–79.

Temmerman, S., Bouma, T.J., Govers, G., Lauwaet, D., 2005. Flow paths of water and sediment in a tidal marsh: relations with marsh developmental stage and tidal inundation height. Estuaries 28 (3), 338–352.

Teo, F.Y., Falconer, R.A., Lin, B., 2009. Modelling effects of mangroves on tsunamis. Proceedings of the ICE - Water Management 162 (1), 3–12.

van Loon-Steensma, J.M., 2015. Salt marshes to adapt the flood defences along the Dutch Wadden Sea coast. Mitigation and Adaptation Strategies for Global Change 20 (6), 929–948.

Vogel, S., 1994. Life in Moving Fluids: The Physical Biology of Flow. Princeton University Press.

Wamsley, T.V., Cialone, M.A., Smith, J.M., Ebersole, B.A., Grzegorzewski, A.S., 2009. Influence of landscape restoration and degradation on storm surge and waves in southern Louisiana. Natural Hazards 51 (1), 207–224.

Wamsley, T.V., Cialone, M.A., Smith, J.M., Atkinson, J.H., Rosati, J.D., 2010. The potential of wetlands in reducing storm surge. Ocean Engineering 37 (1), 59–68.

Wayne, C.J., 1976. The effects of sea and marsh grass on wave energy. Coastal Research Notes 4 (7), 6–8.

Widdows, J., Blauw, A., Heip, C.H.R., Herman, P.M.J., Lucas, C.H., Middelburg, J.J., Schmidt, S., Brinsley, M.D., Twisk, F., Verbeek, H., 2004. Role of physical and biological processes in sediment dynamics of a tidal flat in Westerschelde Estuary, SW Netherlands. Marine Ecology Progress Series 274, 41–56.

Wolanski, E., Elliott, M., 2015. Estuarine Ecohydrology: An Introduction. Elsevier.

Wu, Y., Falconer, R., Struve, J., 2001. Mathematical modelling of tidal currents in mangrove forests. Environmental Modelling and Software 16 (1), 19–29.

Xu, H., Zhang, K., Shen, J., Li, Y., 2010. Storm surge simulation along the U.S. East and Gulf Coasts using a multi-scale numerical model approach. Ocean Dynamics 60 (6), 1597–1619.

Young, D.L., Bruder, B.L., Haas, K.A., Webster, D.R., 2016. The hydrodynamics of surface tidal flow exchange in saltmarshes. Estuarine, Coastal and Shelf Science 172, 128–137.

Zhang, K., Liu, H., Li, Y., Xu, H., Shen, J., Rhome, J., Smith, T.J., 2012. The role of mangroves in attenuating storm surges. Estuarine, Coastal and Shelf Science 102, 11–23.

Zhang, X., Chua, V.P., Cheong, H.F., 2013. A study on the properties of mangrove prop roots and their influence on current flow. In: Zhaoyin, W., Lee, J.H.W., Jizhang, G., Shuyou, C. (Eds.), Proceedings of the 35th IAHR World Congress, pp. 5151–5160.

Zhang, X., Chua, V.P., Cheong, H.F., 2015. Hydrodynamics in mangrove prop roots and their physical properties. Journal of Hydro-Environment Research 9 (2), 281–294.

CHAPTER

9

Mathematical Modeling of Tidal Flow Over Saltmarshes and Tidal Flats With Applications to the Venice Lagoon

L. D'Alpaos, L. Carniello, A. Defina

Department of Civil, Environmental and Architectural Engineering, University of Padua, Padova, Italy

1. INTRODUCTION

In this chapter attention is focused on the hydrodynamics and morphodynamics of saltmarshes and tidal flats in shallow tidal lagoons. Shallow tidal basins are often characterized by extensive tidal flats and marshes dissected by an intricate network of channels (Rinaldo et al., 1999a,b; Fagherazzi et al., 1999; Defina, 2000a; Marani et al., 2003b). Both tidal flats and saltmarshes are prevalently flat landforms located in the intertidal zone.

Saltmarshes have an elevation higher than the mean sea level and are periodically flooded by high tide. They are characterized by a very irregular surface and exhibit a dendritic and meandering structure of channels of varying sizes. These channels perform a drainage function, often continuing to flow long after the tide has receded and the marshes are exposed. They usually sustain a dense vegetation canopy of halophyte plants that withstand the relative infrequent flooding periods. Besides the biological processes related to plant colonization, such as biostabilization or soil production, vegetation sensitively affects the local hydrodynamics, reduces the bottom erosion, and increases sediment deposition.

Tidal flats are flat intertidal landforms occasionally dissected by wide and shallow meandering channels. For the sake of simplicity, the term "tidal flat" is herein used in its broadest sense to include the "shallow subtidal flats," i.e., the muddy platforms that do not emerge during ordinary low tide.

Saltmarsh elevation is controlled by mineral and organogenic sediment accumulation (Pethick, 1981), sea level variations, stabilizing effects of halophyte vegetation on its platform

Coastal Wetlands
https://doi.org/10.1016/B978-0-444-63893-9.00009-5

(Morris et al., 2002; Mudd et al., 2004; Silvestri et al., 2005; Marani et al., 2013), and the interaction between flora and fauna (Perillo et al., 2005; Minkoff et al., 2006).

Tidal flats stem from a delicate balance between sediment deposition and erosion by wind waves and tidal currents (Allen and Duffy, 1998; Carniello et al., 2005; Green and Coco, 2014). Indeed, biology plays a nonnegligible role on the morphology of tidal flats (for a thorough review, see Uncles, 2002). In particular, microphytobenthos and other organisms that colonize shallower areas affect the bottom shear stress threshold for sediment resuspension (Amos et al., 2004) and thus the morphological equilibrium of shallow tidal flats (Marani et al., 2007). However, besides the many studies that demonstrate the important interplay between biology and morphology on saltmarshes and tidal flats, mathematical models able to reliably predict the biological impact on flow dynamics and sediment processes are still lacking (Dietrich and Perron, 2006).

The distinctive characters of tidal flats and saltmarshes reflect on flow, wave field, transport and diffusion processes, and morphologic evolution as well. Therefore, different strategies must be followed when modeling the local hydrodynamics and morphology to maximize the accuracy and minimize the computational effort.

Although considerable progress has been made in the application of two-dimensional (2D) and three-dimensional (3D) models to simulate flow, waves, and sediment transport in estuaries and coastal lagoons, a number of problems still remain in this branch of computational fluid dynamics. These problems mainly stem from the need to model accurately the key physical processes when dealing with very shallow flows, time-dependent flow domains, and complex topography.

The use of well-structured numerical schemes, or extremely refined computational grids, is not the solution. An important effort must be addressed toward modeling the relevant physical phenomena, which are neglected or drastically filtered by the standard numerical models. This can be accomplished through the construction of suitable subgrid models, i.e., by setting up a phenomenological representation of the overall processes, which ensures a statistically equivalent description of the actual physics.

Among others, wetting and drying of saltmarshes and tidal flats and the hydrodynamics of the small-scale drainage networks dissecting saltmarshes are discussed in Section 1. Section 2 presents a simplified, computationally efficient wind wave model, which generates and propagates wind waves in shallow tidal environments by solving the wave action conservation equation on the same unstructured computational grid used by the hydrodynamic model. The adopted numerical scheme is a first-order finite volume explicit scheme. The problem of evaluating the bottom shear stress distribution due to the combined action of tidal currents and wind waves is also addressed in this section. Impact of saltmarsh vegetation on tidal currents and wind waves is shortly discussed in Section 3. Sediment transport and morphodynamics modeling of saltmarshes and tidal flats are discussed in Section 4, where long- and short-term approaches are distinguished. For long-term predictions, a conceptual model highlighting stable morphological states and evolutionary trends is presented, whereas a sediment transport and bed evolution model fully coupled with the hydrodynamic and wind wave model is developed for short-term analyses (storm event time scale). Finally, the main conclusions are summarized in Section 5.

The examples presented in this chapter use the Venice lagoon as typical irregular and shallow tidal basin. The Venice lagoon is a wide tidal basin crossed by a network of deep

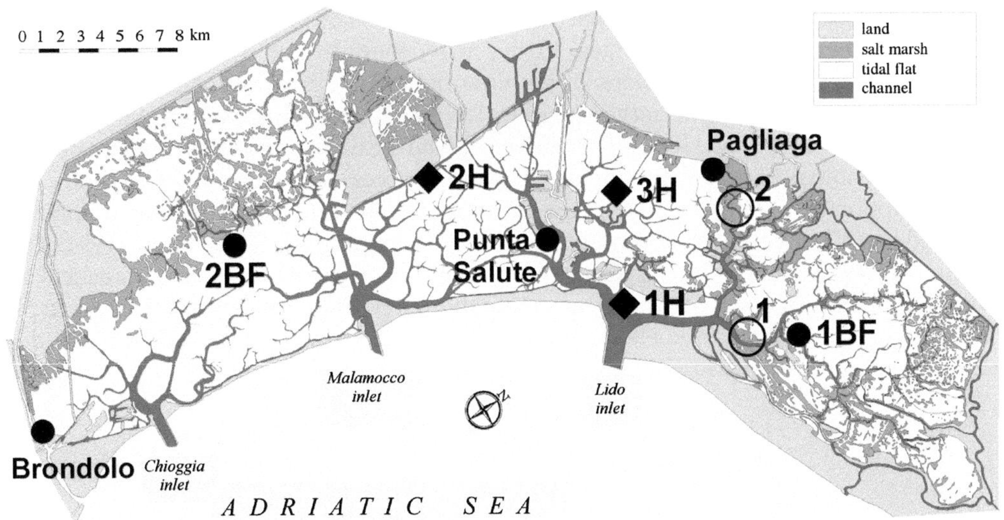

FIGURE 9.1 Map of the Venice lagoon. Symbols show the location of sites in the lagoon referred to in the text. *Adapted from Defina, A., 2000a. Two dimensional shallow flow equations for partially dry areas. Water Resources Research 36, 3251–3264.*

channels departing from three inlets, namely Lido, Malamocco, and Chioggia (Fig. 9.1). The lagoon is also characterized by the presence of wide tidal flats, small islands, and saltmarshes that exhibit a dendritic structure of channels of varying sizes (Rinaldo et al., 1999a,b; Defina, 2000a; Marani et al., 2003b).

2. WETTING AND DRYING AND THE DYNAMICS OF VERY SHALLOW FLOWS

The wetting and drying problem has received considerable attention in the recent past. Reviews, mainly concerned with numerical aspects of this problem, can be found in the works of Balzano (1998), Bates and Hervouet (1999), and Bates and Horritt (2005). The wetting and drying problem can be handled either by adapting the numerical grid at each time step to follow the deforming flow domain (Lynch and Gray, 1980; Kawahara and Umetsu, 1986; Akanbi and Katopodes, 1988) or by retaining a fixed computational grid and utilizing some additional algorithms to deal with the hydrodynamics of partially wet elements. Due to the great difficulty of developing efficient deformable grid techniques, the fixed grid approach is by far the most frequently used. In this case, a whole range of algorithms is available to identify wet elements and to control the flow over these elements (King and Roig, 1988; Leclerc et al., 1990; Falconer and Chen, 1991; Bates et al., 1992; Braschi et al., 1994; Defina et al., 1994; Hervouet and Janin, 1994; Defina and Zovatto, 1995; Ji et al., 2001; Oey, 2005). These algorithms are often related to a particular numerical scheme and their application to a different numerical model is not straightforward (Balzano, 1998). Moreover, when dealing with very small water depths and wetting/drying of large areas, the major source of inaccuracy comes from the fact that numerical models approximate the bottom with a piecewise homogeneous plane surface. In this way, they do not properly account for the effects because of the local variations of the flow field produced by small-scale topography (Defina et al., 1994; Bates and Hervouet, 1999; Defina, 2000a), thus yielding to approximate distributions of velocity and depth. The above problems can be partially overcome by setting up a phenomenological representation of the overall processes to supply a statistically equivalent description of the physics. To deal with partially wet and very irregular domains, we developed an effective subgrid model of ground topography (Defina et al., 1994; Defina and Zovatto, 1995; Defina, 2000a). Considering bottom irregularities from a statistical point of view and assuming the hydrostatic approximation, the 3D Reynolds equations were phase averaged over a representative elementary area (REA) and then integrated over the depth. The averaged equations read (Defina, 2000a):

$$\nabla h + \frac{1}{g}\frac{d}{dt}\left(\frac{\mathbf{q}}{Y}\right) + \mathbf{J} - \nabla\cdot\mathbf{R}_e = 0 \tag{9.1}$$

$$\eta(h)\frac{\partial h}{\partial t} + \nabla\cdot\mathbf{q} = 0 \tag{9.2}$$

where h is the free surface elevation, g is gravity, t is time, $q = (q_x, q_y)$ is the flow rate per unit width, Y is the equivalent water depth, defined as the volume of water per unit area actually ponding the bottom, η is the local fraction of wetted domain and accounts for the actual area that can be wetted or dried during the tidal cycle, R_e accounts for the horizontal turbulent stresses, and $\mathbf{J} = (J_x, J_y)$ is energy dissipation per unit length due to bottom shear stress and vegetation and energy gain due to wind shear stress acting on the free surface.

$$J = \frac{\tau_b + \tau_v - \tau_w}{g\rho Y} \tag{9.3}$$

where τ_b is bottom shear stress, τ_v is an equivalent shear stress accounting for vegetation resistance, τ_w is wind shear stress (see Section 2), and ρ is fluid density.

The bottom topography within an REA is assumed to be irregular with bottom elevations distributed according to a Gaussian probability density function. In this case, the functions η and Y are found to be (Defina, 2000a)

$$\eta = \frac{1}{2}\{1 + \mathrm{erf}[2D/a_r]\} \tag{9.4}$$

$$Y = a_r\left\{\eta(D/a_r) + \frac{1}{4\sqrt{\pi}}\exp\left[-4(D/a_r)^2\right]\right\} \tag{9.5}$$

where erf(•) is the error function, a_r is the typical height of bottom irregularities (i.e., the amplitude of bottom irregularities or approximately twice the standard deviation of bottom elevations), $D = h - z_b$ is the average water depth, and z_b being the average bottom elevation within an REA (Fig. 9.2).

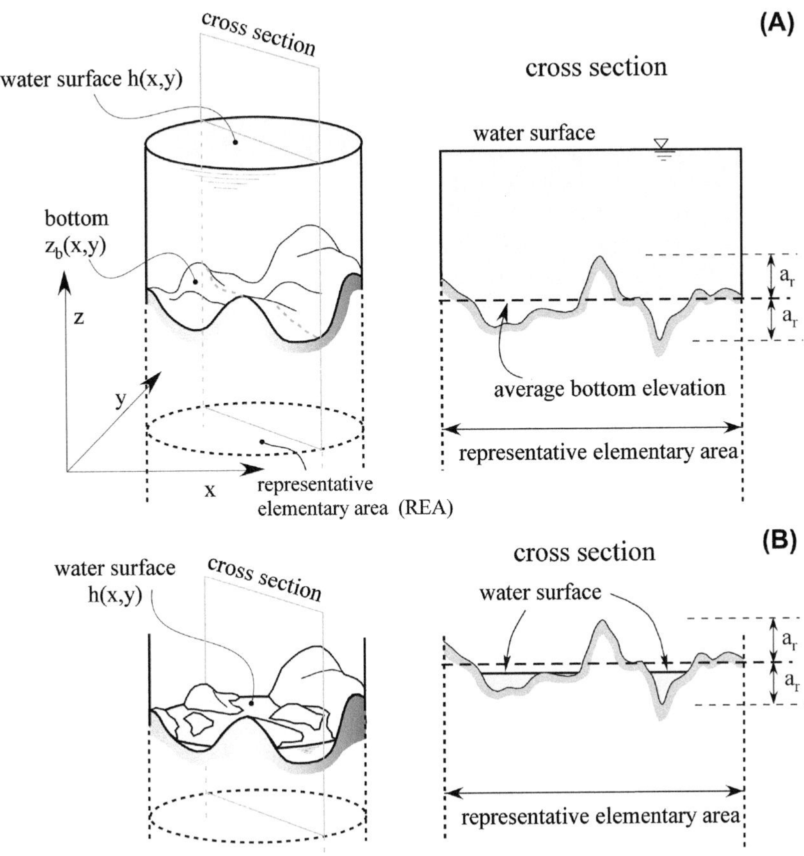

FIGURE 9.2 Representation of flow field and bottom topography within the representative elementary area (REA) when (A) the REA is completely wet and (A) the REA is partially wet. *Adapted from Defina, A., 2000a. Two dimensional shallow flow equations for partially dry areas. Water Resources Research 36, 3251–3264.*

For the case of turbulent flow over a rough wall, the energy dissipations due to bed shear stress can be written as (Defina, 2000a)

$$\frac{\tau_b}{g\rho Y} = \left(\frac{n^2|\mathbf{q}|}{H^{10/3}}\right)\mathbf{q} \tag{9.6}$$

where n is the Manning bed roughness coefficient, and H is an equivalent water depth, which can be approximated with the following interpolation formula (Defina, 2000a):

$$H/a_r \cong Y/a_r + 0.27\sqrt{Y/a_r}e^{-2Y/a_r} \tag{9.7}$$

The resulting subgrid model for ground irregularities requires the statistics of small-scale bottom topography. At present, remote sensing of topography (i.e., airborne laser altimetry, GPS-linked side scan sonar and wide swath bathymetry) is proving very effective in providing high-resolution terrain data capable of parameterizing the proposed approach, even at subgrid scale (Bates and Hervouet, 1999; Bates et al., 2003, 2005).

The above model proved very effective in the simulation of tide propagation in shallow lagoons, over saltmarshes and tidal flats. Examples can be found in the literature (D'Alpaos et al., 1994; Defina and Zovatto, 1995; Defina, 2000a; Lanzoni and Seminara, 2002; Carniello et al., 2005, 2011; Mariotti et al., 2010). All these examples clearly demonstrated the efficiency of the proposed equations. However, a number of open issues still need to be addressed: (1) the model does not account for water which may remain trapped within the REA during the drying phase. On the contrary, experimental evidence suggests that sometimes salt pans or ponds remain after the tidal wave recedes; (2) the model for bed shear stress (i.e., Eq. 9.6) neglects momentum exchange because of convective acceleration at the subgrid scale; (3) Eq. (9.6) for bed shear stress was derived on the assumption of an isotropic distribution of bottom irregularities. This is not always the case; often the creeks dissecting the marshes drive tidal flow along preferential directions (Fig. 9.3). The two latter issues are shortly discussed below.

The overall effects of momentum exchange because of convective acceleration at the subgrid scale can be accounted for by adjusting the friction coefficient. In fact, unresolved accelerations (i.e., subgrid accelerations) averaged over a sufficiently large area mostly produce extra dissipation, which is usually accounted for by suitably increasing the Manning friction coefficient. This is done during the model calibration step because, at present, no relationships are available relating the friction coefficient to filtered velocity distributions. It would clearly be useful to have relationships allowing the a priori estimation of the appropriate roughness corrections. Finding suitable solutions to this problem is a quite demanding task because of the large variety of mechanisms producing a spatially heterogeneous velocity field. Among the many, bottom topography in the presence of small water depths is possibly the easiest to handle. In this case, bottom topography generates small-scale momentum mixing, thus enhancing energy dissipation. The problem can be handled in a way similar to the "mixing length" approach in turbulence and a relationship relating the ratio of the equivalent

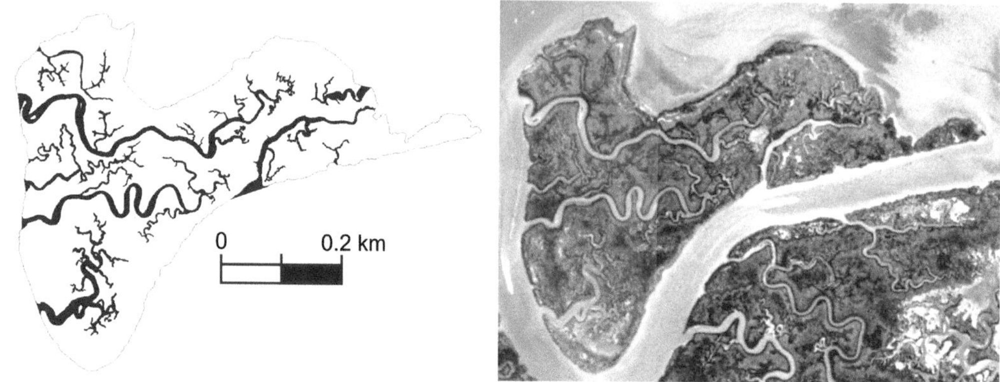

FIGURE 9.3 Sample of creeks aligned in the E–W direction, dissecting a saltmarsh of the Venice lagoon (site 1 in Fig. 9.1). The gray scale aerial photograph together with the extracted creek patterns is shown. *Adapted from Marani, M., Belluco, E., D'Alpaos, A., Defina, A., Lanzoni, S., Rinaldo, A., 2003b. On the drainage density of tidal networks. Water Resources Research 39, 1040. https://doi.org/10.1029/2001WR001051.*

(n_{eq}) to the actual (n) Manning coefficient to water depth and bottom topography can be established (Defina, 2000b; D'Alpaos and Defina, 2007)

$$\frac{n_{eq}}{n} = \frac{36(Y/a_r)^2 + 5\eta^3}{36(Y/a_r)^2 - \eta^3 + 6C_{bw}(\mathbf{r})\eta^3} \tag{9.8}$$

where $C_{bw}(\mathbf{r})$ is an autocorrelation function of bottom elevations, given as

$$C_{bw}(\mathbf{r}) = \frac{\int_{A_w} z(\mathbf{x})z(\mathbf{x}+\mathbf{r})dA - z_{wb}^2}{\sigma_{wb}^2} \tag{9.9}$$

where z_{wb} and σ_{bw} are the average and the root mean square of bottom elevations within the wetted part of the REA, $\mathbf{r}$ is a horizontal "mixing length," and z is local bottom elevation.

Eq. (9.8) for $C_{bw}(r) = 0$ and $C_{bw}(r) = 1$ is plotted in Fig. 9.4. The behavior of n_{eq}/n is not symmetric about $D/a_r = 0$. When $D/a_r \ll 0$, i.e., when the REA is nearly dry (Fig. 9.2B), the flow field is characterized by a braiding pattern with the flow in each branch being independent from the others. In this case, momentum mixing is negligibly small, and the equivalent Manning coefficient recovers its original value (i.e., $n_{eq}/n \approx 1$). When $D/a_r \gg 1$, i.e., when the water depth is nearly uniform within the REA (Fig. 9.2A), bottom irregularities have a minor impact on the velocity field and $n_{eq} \approx n$. Note that when the bottom is smooth, then $C_{bw}(\mathbf{r}) = 1$ and $n_{eq} = n$.

The results of two numerical experiments (Defina, 2000b; D'Alpaos and Defina, 2007) are also plotted in Fig. 9.4 for the sake of comparison. The numerical points lay within the region described by the two theoretical curves, and the behavior of the numerical solution is in quite good agreement with the theoretical curve for $C_{bw}(\mathbf{r}) = 0$. Importantly, the numerical solution approaches this limit curve as the Manning friction coefficient decreases. This is an expected

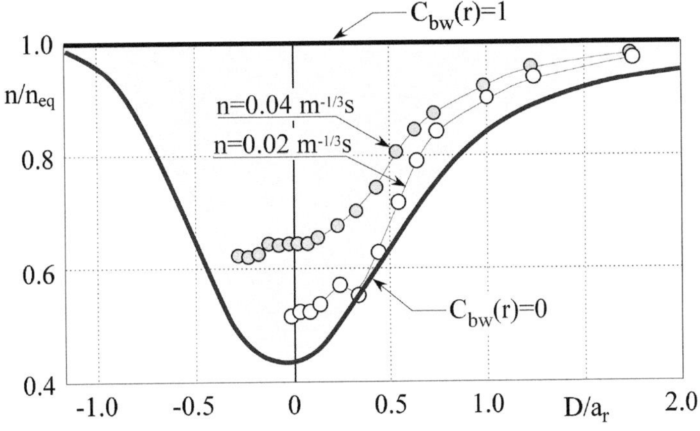

FIGURE 9.4 Computed equivalent Manning coefficient compared with theoretical prediction. *Adapted from Defina, A., 2000b. Alcune considerazioni sulla modellazione matematica di correnti bidimensionali caratterizzate da piccoli tiranti. In: Atti del XXVII Convegno di Idraulica e Costruzioni Idrauliche, vol. I, pp. 255–262. (in italian)*

result because increasing the Manning coefficient reduces friction and enhances large-scale momentum mixing. As a consequence, the mixing length, **r**, increases and $C_{bw}(\mathbf{r})$ decreases toward zero.

Eq. (9.8) must be considered as a first promising attempt at quantifying the effects produced by small-scale (i.e., subgrid) momentum mixing triggered by bottom irregularities. Indeed, further research is required to evaluate the mixing length **r** to be used in Eq. (9.9).

The second issue here shortly discussed focuses on the problem of modeling the creeks dissecting the marshes. A 2D model cannot resolve the small-scale drainage network, comprising channels having a very small width when the flow domain is comparably large, as this would require a large number of very small computational elements. An example is shown in Fig. 9.5 where the zoom of a refined grid of the Venice lagoon, comprising nearly 4×10^4 elements (Carniello et al., 2005), overlaps an aerial photograph of a marsh zone in the northern part of the lagoon. A tangle of small, highly meandering creeks with a width in range between 0.1 and 2 m covers most of the marsh surface. The mesh resolution required to describe all these channels is not acceptable for analysis at the scale of the entire tidal basin. In this case and as a first approximation, the smaller channels can be treated as "topographic irregularities." A clear distinction between actual ground irregularities, which are expected to behave quasi-isotropically, and small-scale channels and creeks cannot be identified, as it depends on the domain extension and on the required accuracy.

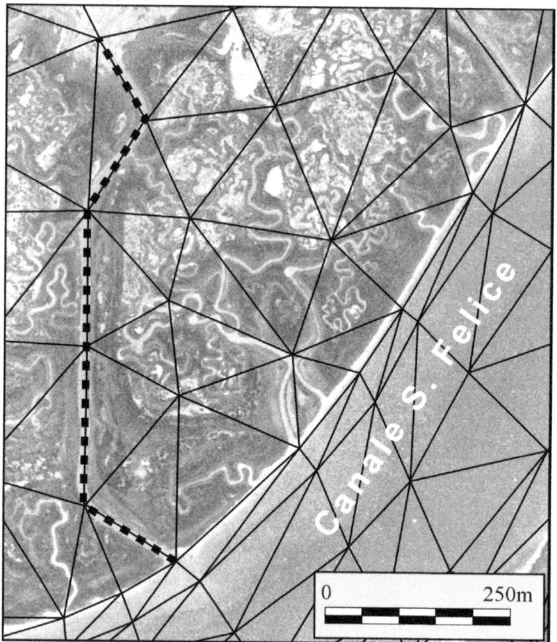

FIGURE 9.5 Zoom of the refined computational grid overlapping a marsh area in the northern Venice lagoon near Treporti (site 2 in Fig. 9.1). *Adapted from D'Alpaos, L. Defina, A., 2007. Mathematical modeling of tidal hydrodynamics in shallow lagoons: a review of open issues and applications to the Venice lagoon. Computers and Geosciences 33, 476–496. https://doi.org/10.1016/j.cageo.2006.07.009.*

Here, attention is focused on channels with a size that is small enough to prevent the use of 2D elements for their description and large enough to dissuade one from including them as bottom irregularities. These narrow channels are usually very numerous and their importance turns out to be comparable with that of large channels (Defina, 2004). The problem of accounting for this channel network can be tackled by observing that the flow within relatively deep and narrow channels flanked by shallow intertidal areas exhibits a distinctive one-dimensional (1D) character, thus suggesting the use of 1D elements to include them in the model.

The problem of coupling 1D and 2D elements has sometimes been found in the literature, e.g., SOBEK, developed by WL|Delft Hydraulics (http://www.sobek.nl), and LISFLOOD-FP (see also Bates and De Roo, 2000; Horritt and Bates, 2001a,b; Bates et al., 2005; Lin et al., 2006; Liang et al., 2007; Kuiry et al., 2010; Bladé et al., 2012). To keep a high accuracy and reduce the computational effort, a particular way of coupling a 2D model to describe the shallow water hydrodynamics with a 1D model to simulate the flow in the channels has been proposed by D'Alpaos and Defina (1993, 1995, 2007). In the model, channels are superimposed on the 2D domain. The effects due to momentum exchange between the channel and the 2D flow are neglected. In the model, the 2D and 1D flow equations are solved by a finite element scheme. The domain is divided into triangular and linear elements with each channel lying along the common side of two adjacent triangular elements. In this way each channel can be added to or removed from the domain without any change in the main 2D discretization. The number of nodes in the computational grid remains unchanged and the computational effort is only slightly increased because of inclusion of 1D elements. In the example shown in Fig. 9.5, Canale S. Felice, which is a very large channel, is described using 2D triangular elements, the very small creeks dissecting the tidal marsh are included in the model as bottom irregularities, whereas the large creek through the saltmarsh, departing from Canale S. Felice, is described with 1D elements (dotted segments) aligned along the edges of the 2D triangular elements. The model has been tested against the numerical solution computed with a 2D model for a number of test cases (D'Alpaos and Defina, 1993, 1995; 2007; D'Alpaos et al., 1995) and proved very effective.

3. WIND AND WIND WAVES

Surface gravity waves are one of the most important phenomena in shallow coastal lagoons and estuaries. Estimation of wave characteristics in estuaries, tidal basins, and coastal areas is essential to analyze sediment transport and local shoreline erosion processes (e.g., Anderson, 1972; Ward et al., 1984; Shoelhamer, 1995; Möller et al., 1999; Umgiesser et al., 2004). Swell waves approaching the coast from the open sea are relevant to study the shoreline morphodynamic in coastal areas. On the contrary, the locally generated wave field is of importance for lagoonal morphodynamics. Tidal currents alone are, in fact, unable to explain the erosion of saltmarshes and the flattening of the shallow lagoonal bottom as they produce shear stresses large enough to carry sediments into suspension only within the channel network where high velocities occur (see e.g., Fagherazzi and Overeem, 2007; de Swart and Zimmerman, 2009). Wind wave–induced bottom shear stress is, instead, the decisive process for mobilizing tidal flat sediments (Carniello et al., 2005, 2011; Fagherazzi and Wiberg, 2009) and influencing their equilibrium configuration (Fagherazzi et al., 2006; see also Section 4).

Two alternative methods are available to model wind wave generation and propagation, i.e., a phase-resolving approach, based on mass and momentum balance equations (for a review, see Dingemans, 1997), or a phase-averaged approach that solves the energy or wave action balance equation (e.g., Booij et al., 1999).

Phase-resolving models reproduce the sea surface in space and time and account for effects such as refraction and diffraction. Bottom friction and depth-induced wave breaking can be included in the model, but wind wave generation is usually absent or poorly reproduced. Phase-resolving models are thus unsuitable in lagoons where storm conditions and local wave generation are key processes. Furthermore, space and time resolutions required by phase-resolving models are of the order of a fraction of the wavelength and wave period, respectively, thus restricting their use to small domains and short duration events. For large-scale applications, phase-averaged models are by far more suitable. Since the pioneering work of Gelci et al. (1956), many models that use the phase-averaged approach have been developed. Among them are the GLERL model developed by Donelan (1977) and revised by Schwab et al. (1984); the HISWA model (hindcast shallow water waves model) (see Holthuijsen et al., 1989) and its successor the SWAN model (simulating wave nearshore) (Booij et al., 1999; Ris et al., 1999); the WAVAD model (Resio, 1987; Resio and Pierre, 1989), and the ACES model (Automated Coastal Engineering System) (Leenknecht et al., 1992). Lin et al. (1998) tested all the models mentioned above against a wind and wave data set collected in the northern Chesapeake Bay, US, during September 1992, when the tropical storm Danielle passed over the area. They found that no single model seems to be good at predicting all aspects of the surface wave field in that specific and morphologically irregular domain, but the GLERL and SWAN models were the most promising. Moreover, in shallow basins, the instantaneous local water depth is crucial to correctly predict the wave field because water depths strongly affect wave propagation. Wave prediction can, therefore, be accomplished only by coupling a wave model with a hydrodynamic model. Umgiesser et al. (2004) moved a preliminary step toward this direction. They combined a 2D finite elements model with the finite difference SWAN model run in stationary mode. For consistency, all the results produced by the hydrodynamic model were interpolated to the grid of the wave model, thus introducing significant numerical approximations. As shallow tidal basins have a very irregular morphology with large and sudden changes in bottom elevation, islands, and saltmarshes, which are periodically flooded and exposed, a specific framework must be adopted to model wind wave propagation in these environments.

A simplified, computationally efficient wind wave model was developed by Carniello et al. (2005) and coupled with the hydrodynamic model presented and discussed in the previous section (D'Alpaos and Defina, 2007), the two models sharing the same computational grid. More recently, the wind wave model has been improved (Carniello et al., 2011) by relaxing the original monochromatic assumption and using a parameterized approximation of the wave action conservation equation (Hasselmann et al., 1973). Wave action density, N, is defined as the ratio of wave energy, E, to the relative wave frequency, σ, and the zero-order moment, N_0, of the wave action spectrum in the frequency domain is chosen for the parameterization following the approach suggested by Holthuijsen et al. (1989). Based on this parameterization the equation solved by the model reads

$$\frac{\partial N_0}{\partial t} + \nabla \cdot (\mathbf{c}_g N_0) = S_{0w} - S_{0bf} - S_{0wc} - S_{0brk} \tag{9.10}$$

where $\mathbf{c_g} = (c_{gx}, c_{gy})$ is the wave group velocity. The parameterized source terms on the right-hand side of (Eq. 9.10) account for the wind energy input (S_{0w}), the energy dissipation by bottom friction (S_{0bf}) and by whitecapping (S_{0wc}), and the energy dissipation by depth-induced breaking (S_{0bf}). We refer the reader to Carniello et al. (2005, 2011) for a description of the source terms listed.

The model neglects the wave–current interaction because of the relatively low velocity that characterizes tidal currents on the tidal flats where wave-induced bottom shear stress is the main factor contributing to sediment resuspension. Furthermore, the model assumes that the direction of wave propagation instantaneously readjusts to match the wind direction, thus implicitly neglecting refraction.

To obtain the peak wave period, T_p, and the related mean frequency, $\sigma_0 \cong 1.1(2\pi/T_p)$, which are required to compute the source term and the wave group celerity, the model used an empirical equation relating the wave period to the local water depth and wind speed following the approach suggested by Young and Verhagen (1996) and Breugem and Holthuijsen (2007):

$$\widetilde{T} = a\widetilde{D}^{b} \tag{9.11}$$

where $\widetilde{T} = gT_p/U_{wind}$ and $\widetilde{D} = gY/U^2_{wind}$ are the dimensionless wave period and water depth, g is gravity, U_{wind} is wind speed, Y is water depth, and a and b are calibration parameters. Eq. (9.11) empirically accounts for the processes responsible for the spectrum modification in the frequency domain. The analysis of wave data collected in the Venice lagoon (1BF and 2BF stations) confirms that the relationship between the dimensionless wave period and the dimensionless water depth is actually well described by a power law with $a = 3.5$ and $b = 0.35$.

It has been demonstrated that wind speed and direction are generally not uniform over large basins (Brocchini et al., 1995) and wind nonuniformity may have a nonnegligible impact on the wave field prediction. Accordingly, the wind wave mode accounts for the spatial variability of the wind field that is determined adopting the interpolation technique proposed by Brocchini et al. (1995), which is an improvement of the standard and long-used technique in meteorological data interpolation proposed by Cressman (1959).

Examples showing the impact of the wind action on the hydrodynamics and the generated wave field inside the Venice lagoon under different wind and tidal conditions are briefly discussed. All simulations presented here are performed using a refined mesh reproducing the present topography of the Venice lagoon (Carniello et al., 2011). Numerical results are compared with field measurements (water levels and flow rates) from the Venice lagoon (see Fig. 9.1 for the location of the stations).

The comparison considers the storm event occurred in November 2004 characterized by the Bora wind blowing from the North-East with a maximum speed of $18\,m\,s^{-1}$.

The agreement between computed and measured water levels is very good (see Fig. 9.6B–D). The effect of the wind setup is particularly evident if we compare the numerical results obtained when considering or neglecting the effects of wind shear stress on the water surface (Fig. 9.6: solid lines and dashed lines, respectively). It clearly emerges that when intense

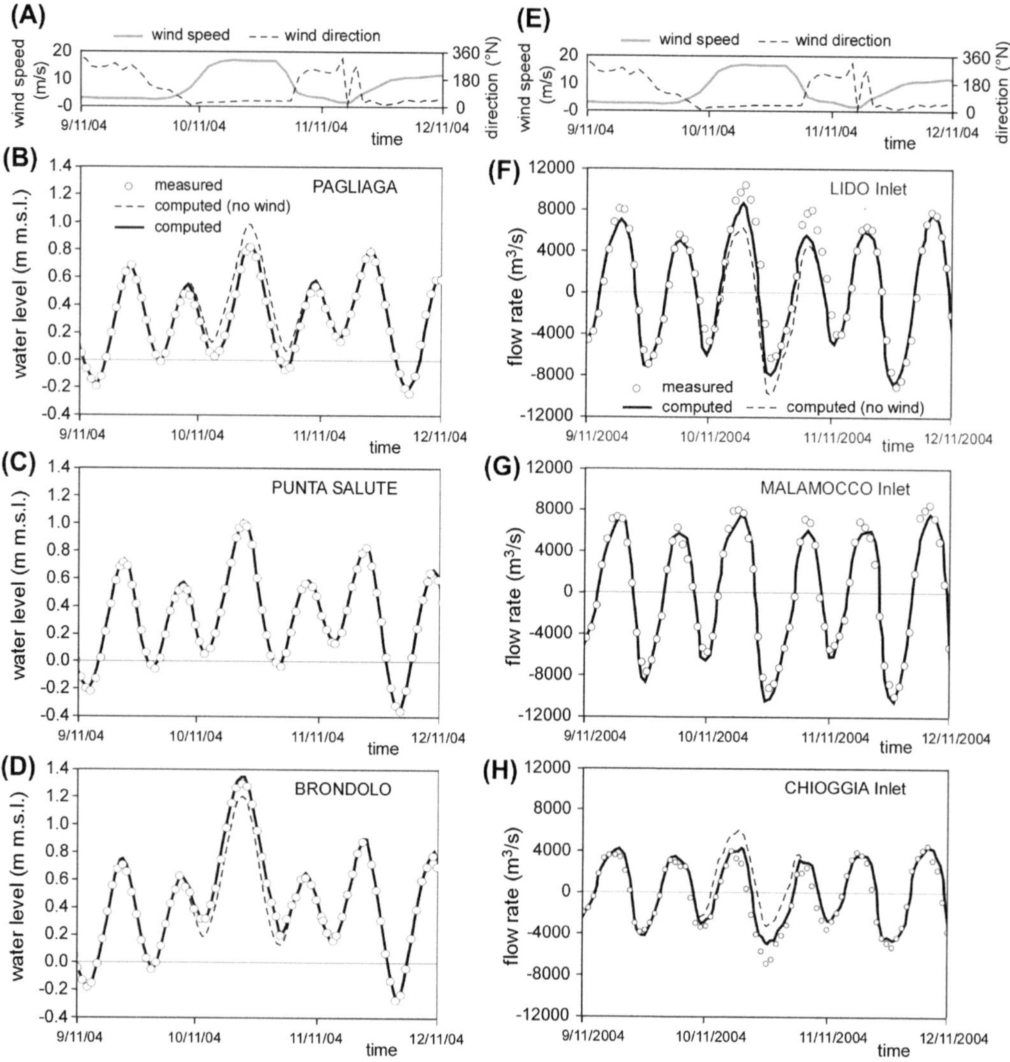

FIGURE 9.6 November 9–12, 2004. Measured wind speed and direction (A and B). Comparison of measured (*circles*) and computed water levels at the Pagliaga station (B), Punta della Salute station (C), and Brondolo station (D) and discharges at the Lido Inlet (F), Malamocco Inlet (G), and Chioggia Inlet (H). Numerical results were computed both considering (*solid lines*) and neglecting (*dashed lines*) the wind shear stress on the water surface. *Adapted from Carniello, L., D'Alpaos, A., Defina, A., 2011. Modeling wind waves and tidal flows in shallow micro-tidal basins, Estuarine, Coastal and Shelf Science 92, 263–276. https://doi.org/10.1016/j.ecss.2011.01.001.*

Bora wind blows over the lagoon, the wind setup leads to a reduction in the tidal levels leeward (at the Pagliaga station, Fig. 9.6B) and to an increase in tidal levels windward (at the Brondolo station, Fig. 9.6D) and the process is well reproduced by the model only if wind effect is accounted for. Considering the same storm event, Fig. 9.6F–H shows the quite good agreement between computed and measured flow rates at the three inlets. Interestingly, we observe that also the flow rates at the three inlets are affected by wind setup. The quite intense Bora wind blowing over the lagoon leads to an increase of the water volume flowing into the lagoon through the northern Lido Inlet (Fig. 9.6F) and to an increase of the water volume flowing outside the lagoon through the southern Chioggia Inlet (Fig. 9.6H), thus inducing a residual southward current within the lagoon. This process only marginally affects the Malamocco Inlet, located in the middle of the lagoon.

Fig. 9.7 compares the computed wave period and the computed significant wave height with that measured at stations 1BF and 2BF. The agreement for both storm events is quite good and confirms the capability of the coupled wind wave–tidal model to reproduce the modulation of the wave height induced by water level oscillations and wind speed variations. In particular, the detailed prediction of the sharp increase in wave height in response to the abrupt increase in the wind speed characterizing the reproduced storm event is noteworthy

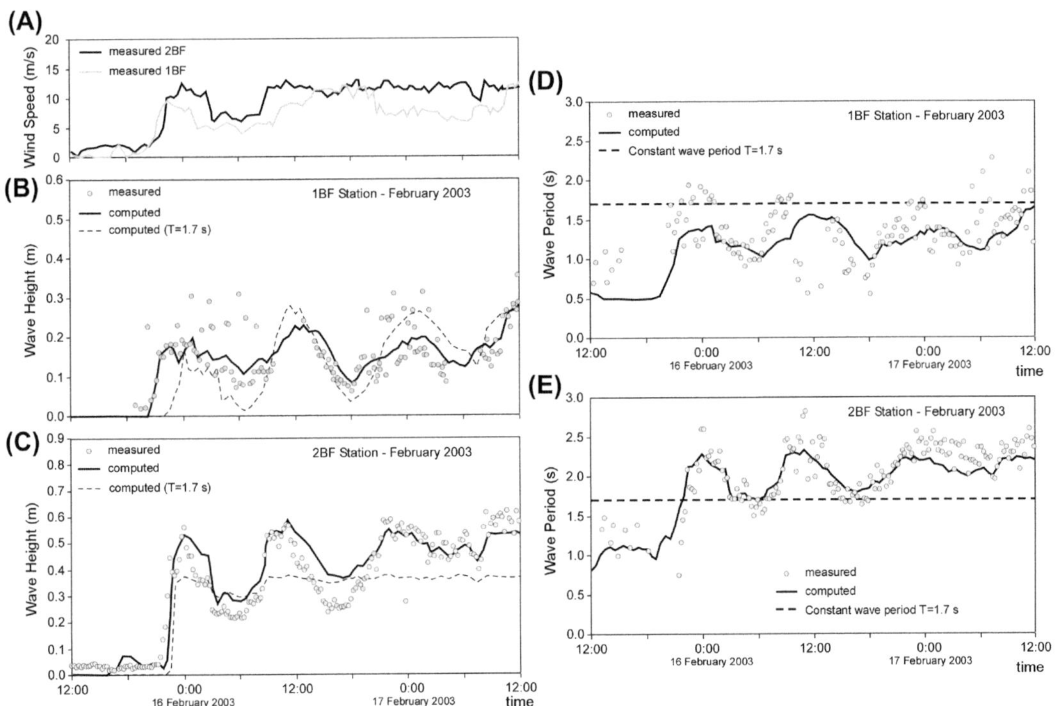

FIGURE 9.7 Storm event occurred the February 16–17, 2003: wind speed at the 1BF and 2BF stations (panel A); comparison of measured (*circles*) and computed wave height (panels B and C) and wave period (panels D and E) at the 1BF station and 2BF station. Results obtained considering a constant wave period (Carniello et al., 2005) are also reported (*dashed line*).

(see Fig. 9.7A–C). In the same figure, results obtained considering the constant value of the wave period, $T_p = 1.7$ s, assumed when using the previous version of the model, are also shown. The results obtained with the version of the model emphasize the improvement obtained by relaxing the original monochromatic assumption and introducing the empirical equation by Young and Verhagen (1996) for the estimation of the wave period.

The combined effect of tidal currents and wind waves on bottom shear stresses is crucial to predict the morphodynamic evolution of tidal flats in shallow tidal basins (see Section 4). Bottom shear stress due to waves ($\tau_{b,wave}$) is computed by the wind wave–tidal model as:

$$\tau_{b,wave} = \frac{1}{2} f_w \rho u_m^2 \text{ with } u_m = \frac{\pi H}{T \sinh(kY)} \text{ and } f_w = 1.39 \left[\frac{u_m T}{2\pi (D_{50}/12)} \right]^{-0.52} \tag{9.12}$$

where u_m is the maximum horizontal orbital velocity at the bottom according to the linear theory, f_w is the wave friction factor as given by Soulsby (1997), T is the wave period, k is the wave number, and D_{50} is the median grain diameter.

Because maximum shear stress τ_{max}, rather than average stress τ_m, is responsible for the bottom sediments mobilization, all the results presented and discussed herein after refer to the maximum total bottom shear stress, which is evaluated with the empirical formulation suggested by Soulsby (1997):

$$\tau_{max} = \left[(\tau_m + \tau_{b,wave} \cos\phi)^2 + (\tau_{b,wave} \sin\phi)^2 \right]^{1/2} \tag{9.13}$$

$$\tau_m = \tau_b \left[1 + 1.2 \left(\frac{\tau_{b,wave}}{\tau_b + \tau_{b,wave}} \right)^{3.2} \right] \tag{9.14}$$

where τ_b is given by Eq. (9.6), and ϕ is the angle between the current and the wave directions.

Fig. 9.8 shows the time evolution of the bottom shear stress at three different sites within the lagoon during the stormy event of February 16–17, 2003. The sites are chosen in a deep channel close to the Lido inlet (site 1H), on a tidal flat close to the Murano island (site 3H), and on a tidal flat next to the Casse di Colmata (site 2H). Each plot compares model results obtained with three different simulations, i.e., (1) the hydrodynamics is forced by the recorded tidal levels at the three inlets; wind shear stress and wind waves are neglected; (2) the wind shear stress is included, whereas wind waves are not; (3) both wind shear stress and wind waves are included in the model.

The results show that in deep channels, wind waves slightly affect bottom shear stresses (Fig. 9.8A), whereas no influence of wind stresses at the surface can be observed. On the contrary, bottom shear stresses on tidal flats are strongly enhanced when wind waves are included in the model (Fig. 9.8B and C). In this case, wind shear stress gives a minor contribution, feebly enhancing the bottom shear stresses produced by tidal currents.

In shallow areas, the bottom shear stresses exceed the critical value ($\tau_{cr} \cong 0.7$ Pa (Amos et al., 2004)) for sediment erosion only in the presence of waves. On the contrary, the bottom shear stresses are always smaller than the critical value when wind waves are not included in the model. This result is further supported by Fig. 9.8D and E mapping the regions where the

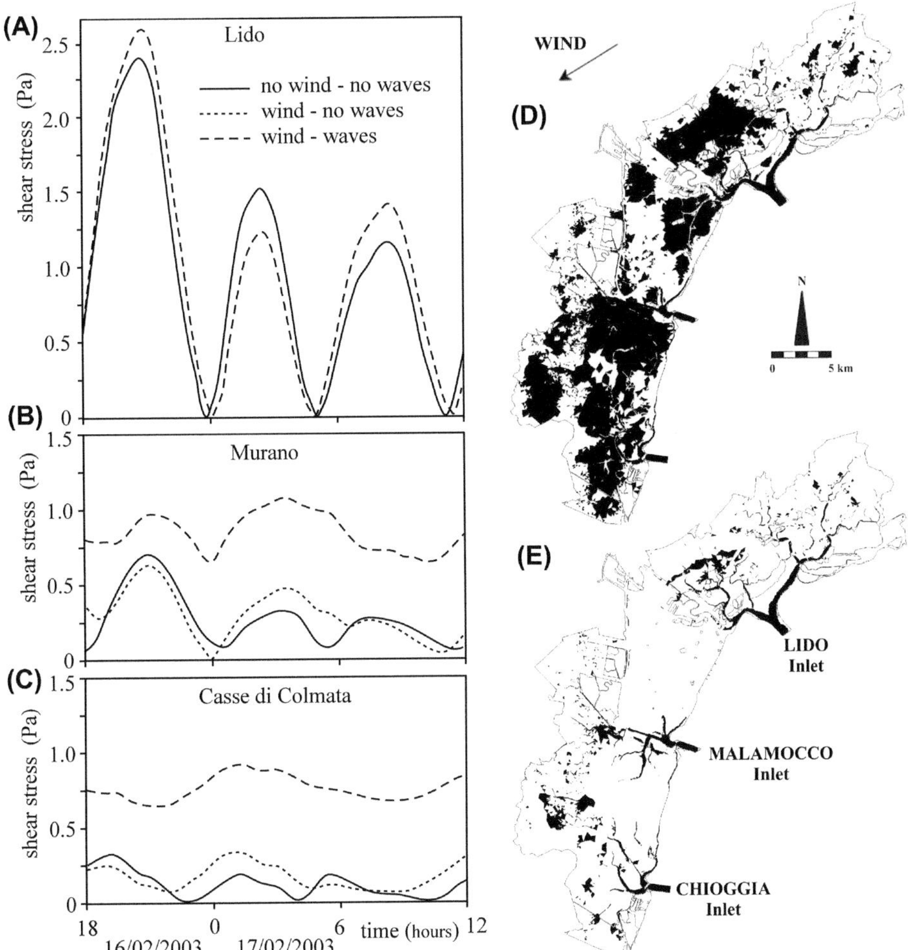

FIGURE 9.8 Storm events of February 16–17, 2003: comparison of the shear stress at the bottom produced by the combined effect of wind waves and tidal currents (— — — —) by tidal currents when the wind shear stress at the free surface is included (·······) or neglected (———). The comparison refers, respectively, to the Lido Inlet (site 1H) (A), a tidal flat close to Murano Island (site 3H) (B) and a tidal flat close to "Casse di Colmata" (site 2H) (C). Spatial distribution of the area experiencing a bottom shear stress greater than 0.7 Pa inside the Venice lagoon. The simultaneous effect of tidal currents and wind waves (D) is compared to the effect of tidal currents alone (E). The dotted line (wind-no waves) in panel (A) is hidden by the solid line (wind waves).

bottom shear stress exceeds τ_{cr}. No resuspension is possible on tidal flats and saltmarshes if wind waves are not considered.

Fig. 9.8D and E also confirms that wind wave resuspension is complementary to tidal current resuspension because waves are able to produce high bed shear stresses in shallower areas, whereas bed shear stresses due to tidal currents are high only for the deep channels where the tidal flow concentrates.

4. SALTMARSH VEGETATION

Tidal marshes are colonized by halophytic vegetation, i.e., macrophytes adapted to complete their life cycle in salty environments. Vegetation has a strong impact on the hydrodynamics over saltmarshes as it affects both tidal current (Burke and Stolzenbach, 1983; Kadlec, 1990; Leonard and Luther, 1995; Shi et al., 1995; Dunn et al., 1996; Nepf and Vivoni, 1999, 2000; Neumeier and Amos, 2006; Neumeier, 2007) and wind waves (Wayne, 1976; Knutson et al., 1982; Pethick, 1992; Koch and Gust, 1999; Möller et al., 1999; Möller and Spencer, 2002; Swales et al., 2004; Möller, 2006). Moreover, vegetation reduces bed shear stress, hence erosion, and strongly affects transport and diffusion processes (Lopez and Garcia, 1998; Nepf, 1999; Nepf and Koch, 1999; Leonard and Reed, 2002; Bouma et al., 2007; Defina and Peruzzo, 2010, 2012; Peruzzo et al., 2012).

Resistance to flow produced by vegetation can be included in the hydrodynamic model as an additional equivalent shear stress τ_V (see Eq. 9.3). To compute the equivalent shear stress τ_V any model (Shimizu and Tsujimoto, 1994; Klopstra et al., 1997; Lopez and Garcia, 2001; Righetti and Armanini, 2002; Defina and Bixio, 2005) able to predict the velocity profile in a uniform flow in the presence of vegetation can be used. Here, we focus on the case of rigid vegetation and consider a uniform flow in the x-direction. In this case, the velocity profile $u_x(z)$, z being the vertical direction, can be written as

$$u_x(z) = f(C_D, m, A_z, h_p, Y)\sqrt{S_{0x}} \tag{9.15}$$

where h_p is plant height, C_D the drag coefficient, A_z the frontal area of vegetation per unit depth, m the number of stems per unit area, S_{0x} the bottom slope, and f a function of water depth and vegetation characteristics. The flow rate per unit width is then given as

$$q_x = \sqrt{S_{0x}} \int_{z_z}^{h} f(C_D, m, A_z, h_p, Y)dz = \sqrt{S_{0x}} F(C_D, m, A_z, h_p, Y) \tag{9.16}$$

Recalling that $\tau_x = g\rho Y S_{0x}$, extension to 2D flow gives

$$\tau_V = \left(\frac{g\rho Y}{F^2}|\mathbf{q}|\right)\mathbf{q} \tag{9.17}$$

Once the velocity profile $u_x(z)$ is computed for a given slope S_{0x}, function F can be easily computed from (Eq. 9.15).

$$F(C_D, m, A_z, h_p, Y) = \int_{z_b}^{h} \frac{u_x(z)}{\sqrt{S_{0x}}} dz \tag{9.18}$$

Fig. 9.9 shows the behavior of the function F for three vegetation species that colonize the saltmarshes of the Venice lagoon. In this case, the velocity profiles under uniform flow condition have been computed using the model proposed by Defina and Bixio (2005).

Although the above results illustrate the effects of vegetation in producing additional flow resistance, the link between vegetation parameters and wave transformation remains qualitative in the absence of well-experimented quantitative relationships between vegetation structure and wave energy.

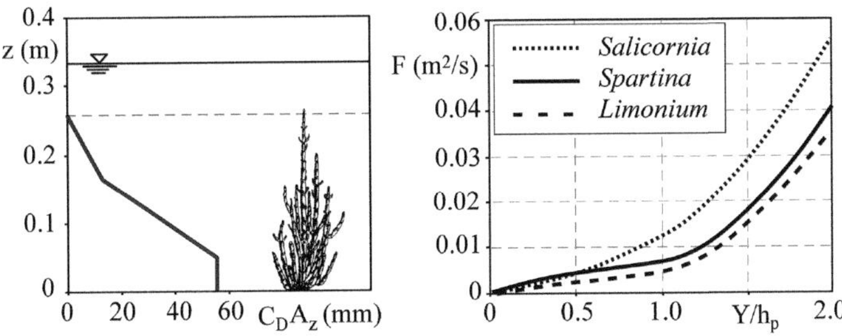

FIGURE 9.9 Behaviors of C_DA_z for Salicornia Veneta (left panel) and function F as given by Eq. (9.18) for three different species: Spartina (m = 350), Salicornia Veneta (m = 45), and Limonium (m = 100).

Wave–vegetation interaction has been investigated to predict wave attenuation produced by vegetation. Standard approaches to model wave attenuation by vegetation are based on the time-averaged conservation equation of wave energy and assume linear wave theory or linearized momentum equations to describe the local flow field (Dalrymple et al., 1984; Kobayashi et al., 1993). For the case of small-amplitude monochromatic waves, wave energy dissipation is commonly accounted for by introducing in Eq. (9.10) the additional source term (Kobayashi et al., 1993; Méndez and Losada, 2004).

$$S_{veg} = -\frac{4\sqrt{2g^3/\rho}}{3\pi} C_D m A_z \left(\frac{k}{\omega}\right)^3 \frac{\sin^3(k\,h_p) + 3\sinh(k\,h_p)}{3k\cosh^3(k\,Y)} E^{3/2} \qquad (9.19)$$

The drag coefficient C_{Dw} in Eq. (9.19) is different from the drag coefficient C_D used to estimate the resistance to flow produced by vegetation because it also accounts for the effects of periodic flow. In fact, C_{Dw} is likely to depend on the Keulegan–Carpenter number rather than on Reynolds number, as well as on the relative vegetation height h_p/Y (Méndez and Losada, 2004; Sánchez-González et al., 2011; Suzuki et al., 2011; Jadhav et al., 2013). However, the question of whether C_{Dw} correlates better with the Keulegan–Carpenter number or with the Reynolds number is still open (e.g., Anderson and Smith, 2014), and it deserves further investigations.

Eq. (9.19) can be extended to describe conditions of emergent vegetation by substituting h_p with Y. Wave number k in Eq. (9.19) depends not only on water depth and wave period but also on vegetation characteristics. However, in the limit of small wave energy damping, the standard dispersion equation based on linear wave theory can be used to compute k (Kobayashi et al., 1993; Méndez and Losada, 2004; Peruzzo et al., 2018).

Once the resistance to flow and the wave energy dissipation due to vegetation have been modeled, the problem of assessing the spatial distribution of vegetation must be addressed. Halophytic vegetation over saltmarshes is neither randomly distributed nor spatially uncorrelated but is, on the contrary, organized in characteristic patches (Pignatti, 1966; Chapman, 1976; Silvestri et al., 2000, 2005; Marani et al., 2004). Therefore, wide areas, which may extend over a few computational elements, are colonized by the same species, i.e., a set of parameters describing a single vegetation species can be associated to each computational element. Quantitative remote sensing, integrated with field observations, proved very effective for mapping the different species (Marani et al., 2003a, 2006; Silvestri and Marani, 2004; Belluco et al., 2006), thus allowing for a very accurate parameterization of vegetation in the numerical model.

5. SALTMARSHES AND TIDAL FLATS MORPHODYNAMICS

Different approaches are usually adopted to model short- and long-term morphologic evolution of coastal and lagoonal environments.

Long-term models were first introduced to investigate saltmarsh formation and evolution. In the pioneering point model suggested by Krone (1987), changes in marsh elevation are calculated as a function of sediment concentration, settling velocity of the suspended sediment flocs, and hydroperiod. When the marsh platform becomes emergent, the inundation period decreases so that less sediment has time to deposit leading to a reduction of marsh accretion. The modeling approach was then improved by considering the effect of sediment supply, bed composition, and sea level rise (Allen, 1990, 1995; French, 1993; Temmerman et al., 2004a,b; D'Alpaos et al., 2011). In recent years, a major development has been the inclusion in the marsh model of the vegetation effects on sediment dynamics, accumulation rates, and organic production by linking all these processes to the biomass of halophyte vegetation that colonizes the marsh surface (e.g., Morris et al., 2002; Mudd et al., 2004; D'Alpaos et al., 2006; Marani et al., 2013).

Besides the vertical accretion of the marsh platform, the recent research on the long-term morphodynamic evolution of saltmarshes considers the formation and the planimetric development of the tidal creek network (D'Alpaos et al., 2007; Belliard et al., 2015, 2016).

The above-mentioned complex biomorphodynamic evolution of saltmarshes, considering feedback between hydrodynamic, sediment transport, and vegetation, and the effect of marine transgression and regression are presented and discussed by D'Alpaos et al. (see Chapter 5).

However, these models disregard the incipient formation of saltmarshes. As a consequence, they can only be applied to locations in which the saltmarsh is already present but are ineffective in determining under what conditions the saltmarsh has evolved from tidal flats.

Besides the saltmarshes, the evolution of tidal flats is less considered in the literature and only recently it has been investigated more in detail accounting for the biostabilizing effect of vegetation and microphytobenthos (e.g., Marani et al., 2010; Carr et al., 2010, 2015). A first original contribution to tidal flat morphodynamics is given by Fagherazzi et al. (2006) and is based on the observation that the bathymetric data point out an abrupt transition between saltmarshes and tidal flats with very few areas lying at intermediate elevations. To describe this evidence, Fagherazzi et al. (2006) developed a conceptual model which indicates that this bimodal distribution of elevations strictly relates to wind wave shear stresses. It has been demonstrated (Carniello et al., 2005), in fact, that the role of sediment resuspension by wind waves is crucial in shallow tidal basins, whereas tidal fluxes alone are unable to produce the bottom shear stresses necessary to mobilize tidal flat sediments. The conceptual model follows from the wave model described in Section 2 of the present chapter. It mainly assumes that (1) wind waves are the main source of bottom shear stress (i.e., the model does not apply to tidal channels where bottom shear stress is mainly due to tidal current), (2) in shallow basins waves quickly adapt to external forcing, and (3) the fetch required to attain fully developed condition is short. Therefore, as a first approximation, the conservation Eq. (9.10) can be reduced to the local equilibrium between the source terms describing the unlimited fetch fully developed local wave field.

The conceptual model is based on the stability curve obtained by plotting the wind wave–induced bottom shear stress as a function of water depth (Fig. 9.10A). The model assumes that the rate of sediment erosion, E_S, is proportional to the difference between bottom shear stress (τ_b) and the critical shear stress for sediment erosion (τ_{cr}). Therefore, the curve is a proxy for bed erosion rate. The model further assumes some prescribed average annual sedimentation rate, D_S. Dynamic equilibrium, $E_S = D_S$, is achieved when $\tau_b = \tau_{eq}$ (points U and S). When $\tau_b < \tau_{eq}$, then deposition exceeds erosion and the bottom evolves toward higher elevations. On the contrary, when $\tau_b > \tau_{eq}$, erosion exceeds deposition and the bottom sinks toward lower elevations. Therefore, any point S along the right branch of the curve is a stable point, whereas any point U on the left branch of the curve is an unstable point. The conceptual model demonstrates that a stable morphodynamic equilibrium is possible only for saltmarshes (i.e., $Z_b > Z_{c1}$) and tidal flats (i.e., $Z_{c2} < Z_b < Z_{max}$).

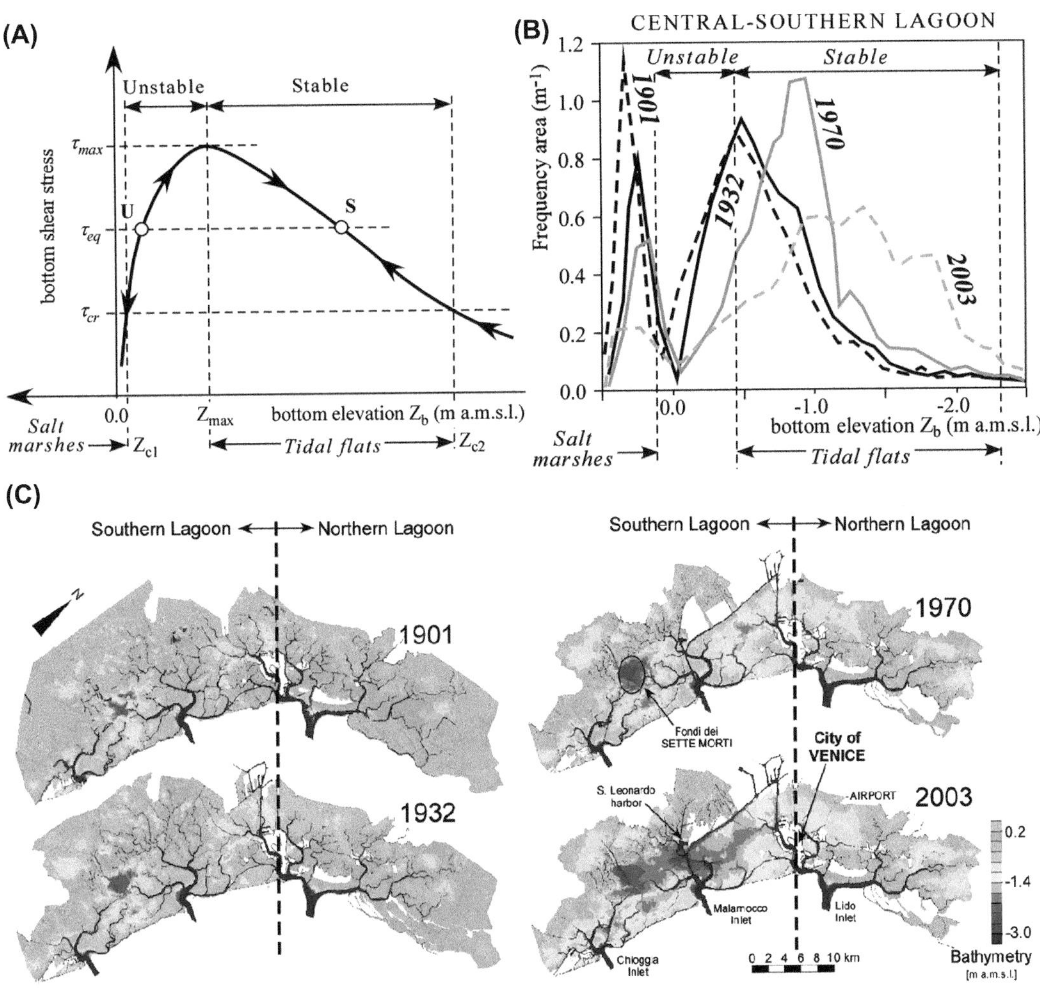

FIGURE 9.10 (A) Bed shear stress distribution as a function of bottom elevation; (B) frequency area distributions as a function of bottom elevation for the Southern Venice lagoon considering the four historical configurations shown in panel (C), namely 1901, 1932, 1970, and 2003.

We further tested the conceptual model through comparison with numerical results obtained with the 2D wind wave–tidal model described in Section 2 (Defina et al., 2007). Considering the Venice lagoon test case, we computed the total bottom shear stress during a storm event (occurred in February 2003) using the wind wave–tidal model. The analysis focuses on the Central-Southern part of the lagoon where the condition of fully developed wave field establishes over most of the domain.

The computed bottom shear stress plotted versus bottom elevation (Fig. 9.11) shows a remarkable concentration of points along the theoretical curve, whereas the points which do not cluster along the curve pertain to tidal channels or to fetch limited areas, i.e., regions that do not meet the main model hypotheses.

We further showed that all points falling along the stable branch of the curve are indeed tidal flats. The few points on the unstable branch of the curve are located on tidal flats close to saltmarsh edges where the lagoon morphology is likely far from equilibrium considered that, since the beginning of the last century, saltmarshes are progressively reducing their extension (Carniello et al., 2009; Molinaroli et al., 2009).

The bottom elevation density functions (Fig. 9.10B) obtained considering four historical configurations of the Venice lagoon (namely 1901, 1932, 1970, and 2003; Fig. 9.10C) clearly display a minimum corresponding to elevations in the unstable range thus confirming that just a small fraction of the basin is characterized by these intermediate elevations, in agreement with the stability curve.

Based on the stability model, Defina et al. (2007) suggested a conceptual model to describe the long-term evolution experienced by a tidal basin under erosion, such as the Venice lagoon during the last century. This model identifies an initial phase (phase 1) during which saltmarshes deteriorate and progressively shrink, feeding with sediments the adjacent tidal flats, which, therefore, are able to maintain their original elevation. In the following phase (phase 2), because marshes have largely reduced their extension, the amount of sediment they supply to tidal flats is no longer sufficient to balance the erosion affecting tidal flats. Accordingly, the average tidal flat elevation decreases. The two phases are clearly visible

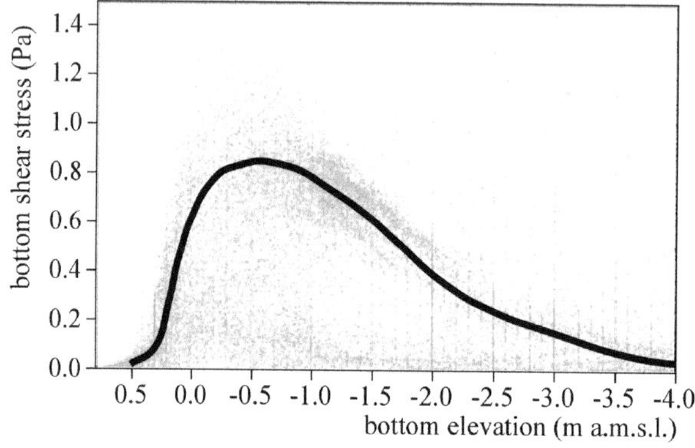

FIGURE 9.11 Computed bed shear stress distribution as a function of bottom elevation for the Venice lagoon test case. The solid line represents the theoretical curve on which the conceptual model is based. *Adapted from Defina, A., Carniello, L., Fagherazzi, S., D'Alpaos, L., 2007. Self organization of shallow basins in tidal flats and salt marshes. Journal of Geophysical Research – Earth Surface 112, F03001. https://doi.org/10.1029/2006JF000550.*

in Fig. 9.10B: phase 1 describes the evolution between 1901 and 1932, whereas phase 2 can be identified in the evolution from 1932 to 2003.

More consistent with the approach assumed in the present chapter is the short-term sediment dynamics and morphological evolution of tidal basins. Such analysis requires to take into account all the processes acting at the hourly/daily time scale, whereas processes such as organic soil production, soil compaction, eustatism, and sea level rise, extremely important in long-term evolution, are usually neglected.

Short-term sediment transport and bottom evolution of shallow tidal basins can be studied by coupling a hydrodynamic model that includes wave dynamics with a sediment transport and bed evolution model. In the following is a short description of the sediment transport model that has been implemented to study the short- and midterm morphodynamic evolution of the lagoon of Venice (Carniello et al., 2012).

The bed composition of the Venice lagoon is characterized by cohesive clayey silt with the exception of the bigger channels branching from the three inlets that are sandy and noncohesive (Amos et al., 2004). In the model we use two sedimentological classes: fine sand as a proxy of pure noncohesive sediments and mud (grain size less than 0.063 mm i.e., silt and clay) as a proxy of pure cohesive sediments, the local bed composition being expressed as mud percentage (p_m). We further assume an average grain size $d_{s50} = 150\ \mu m$ for the sand class and the grain size $d_{m50} = 20\ \mu m$ for the mud class. The model considers a 10% of mud content by dry weight to discriminate between noncohesive and cohesive behavior (Van Ledden, 2003; Van Ledden et al., 2004). Based on the analyses of the available bed composition data for the Venice Lagoon, we identified an empirical relationship between p_m and both the local bottom elevation and the distance from the inlets.

The sediment transport model neglects the horizontal diffusion which is small compared to advection (Pritchard and Hogg, 2003) and solves the advection equation for each sediment class:

$$\frac{\partial C_i Y}{\partial t} + \nabla_q C_i = E_{sand} + E_{mud} - D_{sand} - D_{mud} \tag{9.20}$$

where C ($m^3\ m^{-3}$) is the depth-averaged sediment concentration, E_{sand} and E_{mud} ($m\ s^{-1}$) are entrainment of sand and mud computed according to the equations suggested by Van Rijn (1993), Van Ledden (2003) and Van Ledden et al. (2004), which account for the different possible behaviors (i.e., noncohesive or cohesive) of the sand–mud mixture, D_{sand} ($m\ s^{-1}$) is the deposition rate for noncohesive sediments, which is proportional to the local sand concentration and still water settling velocity, D_{mud} ($m\ s^{-1}$) is the deposition rate for cohesive mud evaluated according to the Krone's formula.

It is worthwhile noting that our sediment transport model only considers sediment resuspension occurring when the bottom shear stress exceeds a critical threshold, thus neglecting processes such as mud fluidization associated to mud transport at subthreshold conditions in the presence of waves (e.g., Maa and Mehta, 1987; Ross and Mehta, 1991).

Eq. (9.20) is solved for each of the two sedimentological classes with a first-order finite volume explicit scheme to obtain the time and spatial evolution of suspended sand and mud concentrations.

A specific bed evolution module, based on the mixing layer concept (Hirano, 1971, 1972), has been developed to predict the time variation of bed elevation and bed composition as a consequence of sand/mud deposition and erosion. Bed elevation is governed by the following equation:

$$(1-n)\frac{\partial z_b}{\partial t} = D_{sand} + D_{mud} - E_{sand} - E_{mud} \tag{9.21}$$

where z_b is the bed elevation, and n is the bed porosity.

A peculiar feature of the sediment transport module is the stochastic approach adopted to reproduce the near-threshold conditions for sediment entrainment, which usually characterize shallow tidal basins where resuspension events occur periodically driven by bottom shear stresses that slightly exceed the erosion threshold. Both the total bottom shear stress (induced by the combined effect of waves and currents) and the critical shear stress for erosion are treated as random variables characterized by log-normal probability distributions (Carniello et al., 2012).

At each time step, the bed evolution model calculates the net variation of bed elevation distinguishing between sand and mud contribution. Based on the deposition and erosion rates of sand and mud, the model updates the composition of the active (mixing) layer. Importantly, the sediment transport and bed evolution model share the same computational grid with the hydrodynamic and wind wave model.

We calibrated and tested the sediment transport model against point turbidity measurements collected at different sites within the Venice lagoon, considering different tidal and meteorological conditions. To quantify the model's performance, we used conventional statistic parameters such as the Percentage Model Bias and the Scatter Index that testified the good agreement with field data. Fig. 9.12 shows the comparison of the measured and

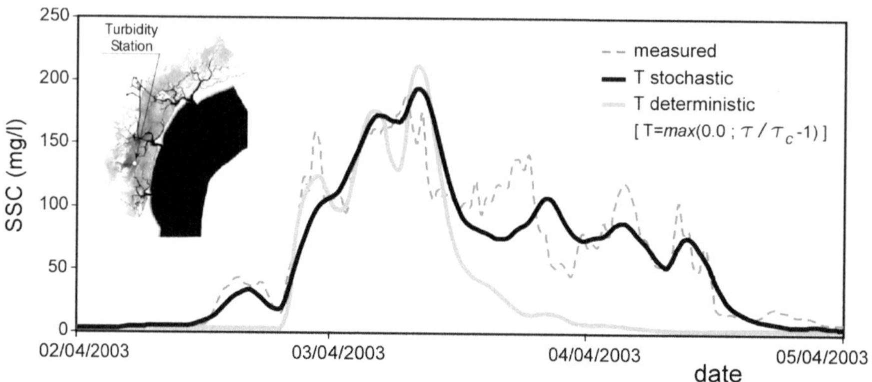

FIGURE 9.12 Comparison between measured (*dashed line*) and computed (*solid lines*) suspended sediment concentration (SSC) at one station (see inset). The computed SSC was obtained by considering both the probabilistic approach (*black line*) and the deterministic approach (*gray line*). *Adapted from Carniello, L., Defina, A., D'Alpaos, L., 2012. Modeling sand-mud transport induced by tidal currents and wind waves in shallow microtidal basins: application to the Venice Lagoon (Italy), Estuarine, Coastal and Shelf Science 102–103, 105–115. https://doi.org/10.1016/j.ecss.2012.03.016*

computed suspended sediment concentrations (SSCs) at one station during an intense storm event. Computed values obtained considering also the classic deterministic formulation to estimate sediment entrainment are shown in the figure to highlight the improvement obtained introducing the statistical approach. The sediment transport model correctly reproduced not only the magnitude of the SSC but also its modulation induced by tidal currents and wind wave variations.

More recently, the comparison with SSC retrievals from remote sensing data (see Fig. 9.13) further enabled us to assess the ability of the model to properly describe observed spatial patterns of SSC (Carniello et al., 2014). These studies highlighted, in particular, the crucial role

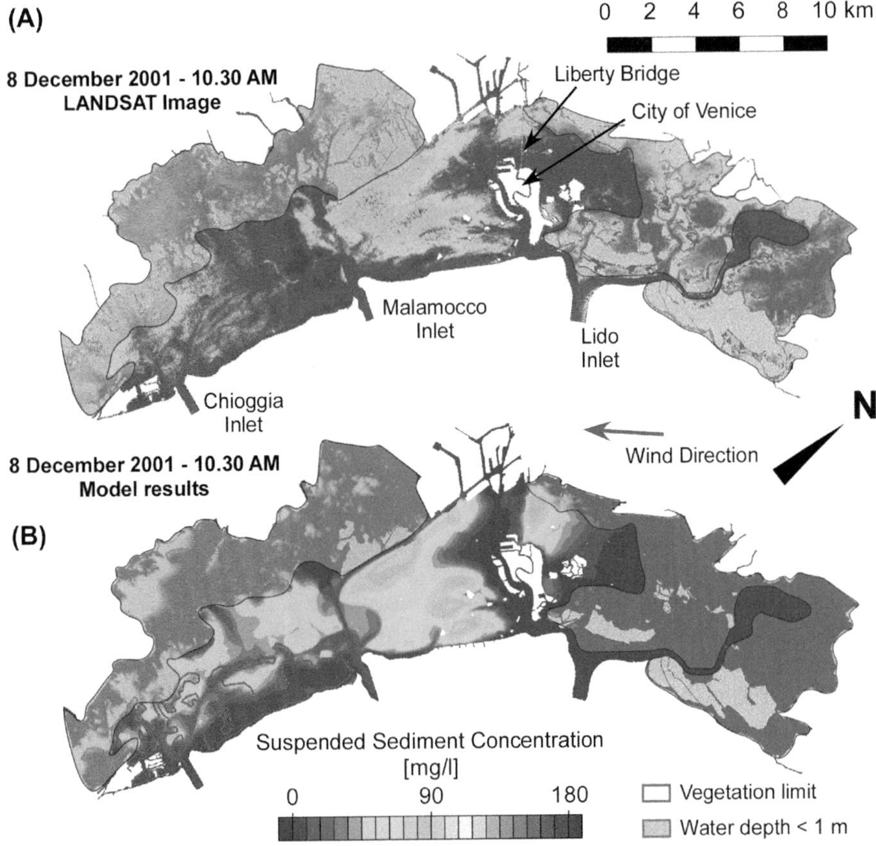

FIGURE 9.13 Resuspension event that occurred on December 8, 2001 at 10:30 a.m. Comparison between (A) the map of the suspended sediment concentration (SSC) obtained analyzing the Landsat satellite image and (B) the SSC map computed using the sediment transport model neglecting resuspension where benthic vegetation is present. Shaded areas refer to areas characterized by water depth lower than about 1 m where bed sediments may undermine SSC estimations from satellite. *Adapted from Carniello, L., Silvestri, S., Marani, M., D'Alpaos, A., Volpe, V., Defina, A., 2014. Sediment dynamics in shallow tidal basins: in situ observations, satellite retrievals, and numerical modeling in the Venice Lagoon, Journal of Geophysical Research – Earth Surface 119. https://doi.org/10.1002/2013JF003015.*

exerted by the sheltering effect associated to artificial structures and natural intertidal landforms on SSC patterns and clearly showed the stabilizing effect of benthic vegetation that, increasing the local critical shear stress for erosion because of the presence of roots, highly affects sediment dynamics at the system scale.

It is worthwhile noting that proper quantification of the stabilizing effect of benthic vegetation at the scale of the entire tidal basin is a difficult task. In fact, different vegetation species affect in different ways both local hydrodynamics and bed strength (e.g., Nepf and Vivoni, 2000; Venier et al., 2012) and their spatial distribution may be extremely variable in response to seasonal variations and vegetation–sediment–water flow interactions (e.g., Curiel et al., 2004; Carr et al., 2010). For the above reasons, in our model the stabilizing effect of vegetation is modeled by setting to zero the sediment resuspension flux in the areas of the computational domain colonized by vegetation (estimated on the basis of field surveys carried out by the Venice Water Authority in the period 2002–04), regardless of the species involved.

6. CONCLUSIONS

This review has attempted to make an examination of many aspects that must be considered when modeling the hydrodynamics and the morphodynamics of saltmarshes and tidal flats in shallow tidal lagoons. Although it is clear that considerable progress has been made in the development and application of shallow water models to simulate flow, waves, and sediment transport in estuaries and coastal lagoons it is also clear that, in this branch of computational fluid dynamics, many outstanding problems of physical process representation still remain for the future.

Among the many, a weak area is that of short-term morphodynamics modeling. Algorithms for calculating erosion, transport, and deposition of multiple sediment classes and evolution of sediment stratigraphy caused by wind waves and currents in tidal environments need improvement.

Feedback between hydrodynamics, biology, and morphology represents a further crucial aspect to be dealt with when modeling the tidal flow over saltmarshes and tidal flats. This is especially true with respect to the study of morphodynamic equilibrium. To this end, algorithms to describe each specific biological process, at tidal time scale, deserve to be developed, whereas long-term hydrobiological models still need to be improved.

In this work, it is shown that adequate solutions to many modeling problems can be accomplished through the construction of suitable subgrid models, i.e., by setting up a phenomenological representation of the overall processes, which provides a statistically equivalent description of the actual physics. This approach also simplifies the coupling of different models conceived to describe specific physical processes coming from different disciplines (e.g., hydraulics, hydrology, morphology, biology, ecology).

Finally, setup and validation procedures based on spatially distributed field data (e.g., wind and wave fields, spatial distribution of bottom sediments and sediment concentration, vegetation seasonal patterns, etc) are a key task to future model design in environmental science. To this end, remote sensing, which have become increasingly popular over the past few years owing to large advances in the technology sector, is expected to be a very helpful tool.

Acknowledgments

This research has been funded by Comune di Venezia "Modificazioni morfologiche della laguna, perdita e reintroduzione dei sedimenti." The authors wish to thank, among others, Bruno Matticchio, Sergio Fagherazzi, Sonia Silvestri, and Marco Marani for their contribution in conceiving and developing part of the presented models and analyses.

References

Akanbi, A.A., Katopodes, N.D., 1988. Model for flood propagation on initially dry land. Journal of Hydraulic Engineering 114, 689–706.

Allen, J.R.L., 1990. Salt-marsh growth and stratification: a numerical model with special reference to the Severn Estuary, Southwest Britain. Marine Geology 95, 77–96.

Allen, J.R.L., 1995. Salt-marsh growth and fluctuating sea level: implications of a simulation model for Flandrian coastal stratigraphy and peat-based sea-level curves. Sedimentary Geology 100, 21–45.

Allen, J.R.L., Duffy, M.J., 1998. Medium-term sedimentation on high intertidal mudflats and salt marshes in the Severn Estuary, SW Britain: the role of wind and tide. Marine Geology 150, 1–27.

Amos, C.L., Bergamasco, A., Umgiesser, G., Cappucci, S., Cloutier, D., DeNat, L., Flindt, M., Bonardi, M., Cristante, S., 2004. The stability of tidal flats in Venice Lagoon – the results of in-situ measurements using two benthic, annular flumes. Journal of Marine Systems 51, 211–241.

Anderson, F.E., 1972. Resuspension of estuarine sediments by small amplitude waves. Journal of Sedimentary Petrology 42, 602–607.

Anderson, M.E., Smith, J.M., 2014. Wave attenuation by flexible, idealized salt marsh vegetation. Coastal Engineering 83, 82–92.

Balzano, A., 1998. Evaluation of methods for numerical simulation of wetting and drying in shallow water flow models. Coastal Engineering 34, 83–107.

Bates, P.D., De Roo, A.P.J., 2000. A simple raster-based model for floodplain inundation. Journal of Hydrology 236, 54–77.

Bates, P.D., Hervouet, J.M., 1999. A new method for moving boundary hydrodynamic problems in shallow water. Proceedings of the Royal Society of London, Series A 455, 3107–3128.

Bates, P.D., Horritt, M.S., 2005. Modelling wetting and drying processes in hydraulic models. In: Bates, P.D., Lane, S.N., Ferguson, R.I. (Eds.), Computational Fluid Dynamics: Applications in Environmental Hydraulics. John Wiley & Sons, Ltd., pp. 121–146

Bates, P.D., Anderson, M.G., Baird, L., Walling, D.E., Simm, D.E., 1992. Modelling floodplain flows using a two dimensional finite element model. Earth Surface Processes and Landforms 17, 575–588.

Bates, P.D., Marks, K.J., Horritt, M.S., 2003. Optimal use of high-resolution topographic data in flood inundation models. Hydrological Processes 17, 537–557.

Bates, P.D., Dawson, R.J., Hall, J.W., Horritt, M.S., Nicholls, R.J., Wicks, J., Hassan, M.A.A.M., 2005. Simplified two-dimensional numerical modelling of coastal flooding and example applications. Coastal Engineering 52, 793–810.

Belliard, J.-P., Toffolon, M., Carniello, L., D'Alpaos, A., 2015. An ecogeomorphic model of tidal channel initiation and elaboration in progressive marsh accretional contexts. Journal of Geophysical Research – Earth Surface 120. https://doi.org/10.1002/2015JF003445.

Belliard, J.-P., Di Marco, N., Carniello, L., Toffolon, M., 2016. Sediment and vegetation spatial dynamics facing sea-level rise in microtidal salt marshes: insights from an ecogeomorphic model. Advances in Water Resources 93, 249–264. https://doi.org/10.1016/j.advwatres.2015.11.020.

Belluco, E., Camuffo, M., Ferrari, S., Modenese, L., Silvestri, S., Marani, A., Marani, M., 2006. Mapping salt-marsh vegetation by multispectral and hyperspectral remote sensing. Remote Sensing of Environment 105, 54–67.

Bladé, E., Gòmez-Valentìn, M., Dolz, J., Aragòn-Hernàndez, J.L., Corestein, G., Sànchez-Juny, M., 2012. Integration of 1D and 2D finite volume schemes for computations of water flow in natural channels. Advances in Water Resources 42, 17–29. https://doi.org/10.1016/j.advwatres.2012.03.021.

Booij, N., Ris, R.C., Holthuijsen, L.H., 1999. A third-generation wave model for coastal regions. 1. Model description and validation. Journal of Geophysical Research 104, 7649–7666.

Bouma, T.J., Van Duren, L.A., Temmerman, S., Claverie, T., Blanco-Garcia, A., Ysebaert, T., Herman, P.M.J., 2007. Spatial fl ow and sedimentation patterns within patches of epibenthic structures. Continental Shelf Research 27, 1020–1045. https://doi.org/10.1016/j.csr.2005.12.019.

Braschi, G., Dadone, F., Gallati, M., 1994. Plain flooding: near and far field simulations. In: Molinaro, P., Natale, L. (Eds.), Proceedings of the Specialty Conference on "Modelling of Flood Propagation Over Initially Dry Areas", Milan (Italy) 29 June -1 July 1994. ASCE, New York, pp. 45–59.

Breugem, W.A., Holthuijsen, L.H., 2007. Generalized shallow water wave growth from Lake George. Journal of Waterway, Port, Coastal, and Ocean Engineering 133 (3). https://doi.org/10.1061/(ASCE)0733-950X(2007)133:3(173).

Brocchini, M., Wurtele, M., Umgiesser, G., Zecchetto, S., 1995. Calculation of a mass-consistent two-dimensional wind-field with divergence control. Journal of Applied Meteorology 34, 2543–2555.

Burke, R.W., Stolzenbach, K.D., 1983. Free Surface Flow through Salt Marsh Grass. MIT-Sea Grant Report MITSG 83–16. Massachusset Institute of Technology, Cambridge, Massachusetts, 252 p.

Carniello, L., Defina, A., Fagherazzi, S., D'Alpaos, L., 2005. A combined wind wave-tidal model for the Venice lagoon, Italy. Journal of Geophysical Research – Earth Surface 110, F04007. https://doi.org/10.1029/2004JF000232.

Carniello, L., Defina, A., D'Alpaos, L., 2009. Morphological evolution of the Venice Lagoon: evidence from the past and trend for the future. Journal of Geophysical Research – Earth Surface 114, F04002. https://doi.org/10.1029/2008JF001157.

Carniello, L., D'Alpaos, A., Defina, A., 2011. Modeling wind waves and tidal flows in shallow micro-tidal basins. Estuarine, Coastal and Shelf Science 92, 263–276. https://doi.org/10.1016/j.ecss.2011.01.001.

Carniello, L., Defina, A., D'Alpaos, L., 2012. Modeling sand-mud transport induced by tidal currents and wind waves in shallow microtidal basins: application to the Venice Lagoon (Italy). Estuarine, Coastal and Shelf Science 102–103, 105–115. https://doi.org/10.1016/j.ecss.2012.03.016.

Carniello, L., Silvestri, S., Marani, M., D'Alpaos, A., Volpe, V., Defina, A., 2014. Sediment dynamics in shallow tidal basins: in situ observations, satellite retrievals, and numerical modeling in the Venice Lagoon. Journal of Geophysical Research – Earth Surface 119. https://doi.org/10.1002/2013JF003015.

Carr, J.A., D'Odorico, P., McGlathery, K., Wiberg, P., 2010. Stability and bistability of seagrass ecosystems in shallow coastal lagoons: role of feedbacks with sediment resuspension and light attenuation. Journal of Geophysical Research 115. https://doi.org/10.1029/2009JG001103.

Carr, J.A., D'Odorico, P., McGlathery, K., Wiberg, P., 2015. Spatially explicit feedbacks between seagrass meadow structure, sediment and light: habitat suitability for seagrass growth. Advances in Water Resources 93, 315–325. https://doi.org/10.1016/j.advwatres.2015.09.001.

Chapman, V.J., 1976. Coastal Vegetation, second ed. Pergamon Press, Oxford. 292 p.

Cressman, G.P., 1959. An operational objective analysis system. Monthly Weather Review 87, 367–374.

Curiel, D., Rismondo, A., Bellemo, G., Marzocchi, M., 2004. Macroalgal biomass and species variations in the lagoon of Venice (North Adriatic Sea, Italy). Scientia Marina 68, 57–67.

D'Alpaos, L., Defina, A., Matticchio, B., 1994. 2D Finite element modelling of flooding due to river bank collapse. In: Molinaro, P., Natale, L. (Eds.), Proceedings of the Specialty Conference on "Modelling of Flood Propagation over Initially Dry Areas", Milan (Italy) 29 June -1 July 1994, pp. 60–71.

D'Alpaos, L., Defina, A., Matticchio, B., 1995. A coupled 2D and 1D finite element model for simulating tidal flow in the Venice channel network. In: Proceedings of the 9th Int. Conf. on Finite Elements in Fluids, Venezia, 15–21 October, pp. 1397–1406.

D'Alpaos, A., Lanzoni, S., Marani, M., Rinaldo, A., 2007. Landscape evolution in tidal embayments: modelling the interplay of erosion, sedimentation, and vegetation dynamics. Journal of Geophysical Research – Earth Surface 112, F01008. https://doi.org/10.1029/2006JF000537.

D'Alpaos, A., Carniello, L., Mudd, S.M., 2011. Dynamic response of marshes to perturbations in suspended sediment concentrations and rates of relative sea level rise. Journal of Geophysical Research – Earth Surface 116, F04020. https://doi.org/10.1029/2011JF002093.

Dalrymple, R.A., Kirby, J.T., Hwang, P.A., 1984. Wave diffraction due to areas of energy dissipation. Journal of Waterway, Port, Coastal, and Ocean Engineering 10, 67–69.

de Swart, H., Zimmerman, J., 2009. Morphodynamics of tidal inlet systems. Annual Review of Fluid Mechanics 41, 203e229.

Defina, A., 2000a. Two dimensional shallow flow equations for partially dry areas. Water Resources Research 36, 3251–3264.

Defina, A., 2000b. Alcune considerazioni sulla modellazione matematica di correnti bidimensionali caratterizzate da piccoli tiranti. In: Atti del XXVII Convegno di Idraulica e Costruzioni Idrauliche, vol. I, pp. 255–262 (in italian).

Defina, A., 2004. Alcune considerazioni sulla stima delle dissipazioni di energia prodotte da opere fisse in una bocca lagunare. In: Atti dell'Istituto Veneto di SS.LL.AA., CLXII, II, pp. 441–478.

Defina, A., Bixio, A.C., 2005. Mean flow and turbulence in vegetated open channel flow. Water Resources Research 41. https://doi.org/10.1029/2004WR003475.

Defina, A., Peruzzo, P., 2010. Floating particle trapping and diffusion in vegetated open channel flow. Water Resources Research 46, W11525. https://doi.org/10.1029/2010WR009353.

Defina, A., Peruzzo, P., 2012. Diffusion of floating particles in flow through emergent vegetation: Further experimental investigation. Water Resources Research 48, W03501. https://doi.org/10.1029/2011WR011147.

Defina, A., Zovatto, L., 1995. Modellazione Matematica Delle Zone Soggette a Periodico Prosciugamento in Un Bacino a Marea. Istituto Veneto di SS.LL.AA., Rapporti e Studi. XII, pp. 337–351 (in italian).

Defina, A., D'Alpaos, L., Matticchio, B., 1994. A new set of equations for very shallow water and partially dry areas suitable to 2D numerical models. In: Molinaro, P., Natale, L. (Eds.), Proceedings of the Specialty Conference on "Modelling of Flood Propagation Over Initially Dry Areas", Milan (Italy) 29 June -1 July 1994, pp. 72–81.

Defina, A., Carniello, L., Fagherazzi, S., D'Alpaos, L., 2007. Self organization of shallow basins in tidal flats and salt marshes. Journal of Geophysical Research – Earth Surface 112, F03001. https://doi.org/10.1029/2006JF000550.

Dietrich, W.E., Perron, J.T., 2006. The search for a topographic signature of life. Nature 439, 411–418. https://doi.org/10.1038/nature04452.

Dingemans, M.W., 1997. Water wave propagation over uneven bottoms: part 1—linear wave propagation. In: Advanced Series on Ocean Engineering, vol. 13. World Scientific, pp. 248–398 (Chapter 3).

Donelan, M.A., 1977. A Simple Numerical Model for Wave and Wind Stress Prediction. Report. Nat., Water Res. Inst., Burlington, Ont. Canada, 28 p.

Dunn, C., Lopez, F., Garcia, M., 1996. Mean Flow and Turbulence Structure Induced by Vegetation: Experiments. In: Hydraulic Engineering Series, vol. 51. Dept. of Civ. Eng., Univ. of Illinois at Urbana-Champaign, Urbana, Ill.

D'Alpaos, L., Defina, A., 1993. Venice lagoon hydrodynamics simulation by coupling 2D and 1D finite element models. In: Proceedings of the 8th Conference on "Finite Elements in Fluids. New Trends and Applications", Barcelona (Spain), 20–24 September 1993, pp. 917–926.

D'Alpaos, L., Defina, A., 1995. Modellazione matematica del comportamento idrodinamico di zone di barena solcate da una rete di canali minori. Istituto Veneto di SS.LL.AA., Rapporti e Studi. XII, pp. 353–372 (in italian).

D'Alpaos, L., Defina, A., 2007. Mathematical modeling of tidal hydrodynamics in shallow lagoons: a review of open issues and applications to the Venice lagoon. Computers and Geosciences 33, 476–496. https://doi.org/10.1016/j.cageo.2006.07.009.

D'Alpaos, A., Lanzoni, S., Mudd, S.M., Fagherazzi, S., 2006. Modelling the influence of hydroperiod and vegetation on the cross-sectional formation of tidal channels. Estuarine, Coastal and Shelf Science 69, 311–324. https://doi.org/10.1016/j.ecss.2006.05.02.

Fagherazzi, S., Overeem, I., 2007. Models of deltaic and inner continental shelf landform evolution. Annual Review of Earth and Planetary Sciences 35, 685–715. https://doi.org/10.1146/annurev.earth.35.031306.140128.

Fagherazzi, S., Wiberg, P., 2009. Importance of wind conditions, fetch, and water levels on wave-generated shear stresses in shallow intertidal basins. Journal of Geophysical Research 114, F03022. https://doi.org/10.1029/2008JF001139.

Fagherazzi, S., Bortoluzzi, A., Dietrich, W.E., Adami, A., Lanzoni, S., Marani, M., Rinaldo, A., 1999. Tidal networks 1. Automatic network extraction and preliminary scaling features from digital elevation maps. Water Resources Research 35, 3891–3904.

Fagherazzi, S., Carniello, L., D'Alpaos, L., Defina, A., 2006. Critical bifurcation of shallow microtidal landforms in tidal flats and salt marshes. Proceedings of the National Academy of Sciences of the United States of America 103, 8337–8341. https://doi.org/10.1073/pnas.0508379103.

Falconer, R.A., Chen, Y., 1991. An improved representation of flooding and drying and wind stress effects in a two-dimensional tidal numerical model. Proceedings of the Institution of Civil Engineers 91, 659–687.

French, J.R., 1993. Numerical simulation of vertical marsh growth and adjustment to accelerated sea-level rise, North Norfolk, UK. Earth Surface Processes and Landforms 18, 63–81.

Gelci, R., Calazé, H., Vassal, J., 1956. Utilization des diagrammes de propagation à la prévision énergétique de la houle. In: Info. Bull., vol. 8, pp. 160–197. Com. Cent. d'Océanogr. et d'Etudes des Côtes, Paris. (in french).

Green, M.O., Coco, G., 2014. Review of wave-driven sediment resuspension and transport in estuaries. Reviews of Geophysics 52, 77–117. https://doi.org/10.1002/2013RG000437.

Hasselmann, K., Barnett, T.P., Bouws, E., Carlson, H., Cartwright, D.E., Enke, K., Ewing, J.A., Giennapp, H., Hasselmann, D.E., Kruseman, P., Meersburg, A., Muller, P., Olbers, D.J., Richter, K., Sell, W., Walden, H., 1973. Measurements of wind-wave growth and swell decay during the joint north sea wave project (JONSWAP). Deutsche Hydrograph Zeit Supplement 12, 1–95.

Hervouet, J.M., Janin, J.M., 1994. Finite element algorithms for modelling flood propagation. In: Molinaro, P., Natale, L. (Eds.), Proceedings of the Specialty Conference on "Modelling of Flood Propagation Over Initially Dry Areas", Milan (Italy) 29 June -1 July 1994, pp. 101–113.

Hirano, M., 1971. River bed degradation with armouring. Transactions Japan Society of Civil Engineers 3, 194–195.

Hirano, M., 1972. Studies on variation and equilibrium state of a river bed composed of non-uniform material. Transactions Japan Society of Civil Engineers 4, 128–129.

Holthuijsen, L.H., Booij, N., Grant, W.D., 1989. A prediction model for stationary, short-crested waves in shallow water with ambient currents. Coastal Engineering 13, 23–54.

Horritt, M.S., Bates, P.D., 2001a. Predicting floodplain inundation: raster-based modelling versus the finite element approach. Hydrological Processes 15, 825–842.

Horritt, M.S., Bates, P.D., 2001b. Effects of spatial resolution on a raster based model of flood flow. Journal of Hydrology 253, 239–249.

Jadhav, R.S., Chen, Q., Smith, J.M., 2013. Spectral distribution of wave energy dissipation by salt marsh vegetation. Coastal Engineering 77, 99–107.

Ji, Z.G., Morton, M.R., Hamrick, J.M., 2001. Wetting and drying simulation of estuarine processes. Estuarine, Coastal and Shelf Science 53, 683–700. https://doi.org/10.1006/ecss.2001.0818.

Kadlec, R., 1990. Overland flow in wetlands: vegetation resistance. Journal of Hydraulic Engineering 116, 691–707.

Kawahara, M., Umetsu, T., 1986. Finite element method for moving boundary problems in river flow. International Journal for Numerical Methods in Fluids 6, 365–386.

King, I.P., Roig, L., 1988. Two dimensional finite element models for floodplains and tidal flats. In: Niki, K., Kawahara, M. (Eds.), Proc. Int. Conf. On Computational Methods in Flow Analysis, Okajama, Japan, pp. 711–718.

Klopstra, D., Barneveld, H.J., Van Noortwijk, J.M., Van Velzen, E.H., 1997. Analytical model for hydraulic roughness of submerged vegetation. In: Proc. Managing Water: Coping with Scarcity and Abundance, IAHR, San Francisco, ASCE, Menphis, Tennessee, USA, pp. 775–780.

Knutson, P.L., Broch, R.A., Seelig, W.N., Inskeep, M., 1982. Wave damping in Spartina alterniyora marshes. Wetlands 2, 87–104.

Kobayashi, N., Raichle, A.W., Asano, T., 1993. Wave attenuation by vegetation. Journal of Waterway, Port, Coastal and Ocean Engineering (ASCE) 119, 30–48.

Koch, E.W., Gust, G., 1999. Water flow in tide- and wave-dominated beds of the seagrass Thalassia testudinum. Marine Ecology Progress Series 184, 63–72.

Krone, R.B., 1987. A method for simulating historic marsh elevations. In: Krause, N.C. (Ed.), Coastal Sediments '87. American Society of Civil Engineers, New York, pp. 316–323.

Kuiry, S., Sen, D., Bates, P.D., 2010. Coupled 1D-quasi-2D flood inundation model with unstructured grids. Journal of Hydraulic Engineering 136, 493–506.

Lanzoni, S., Seminara, G., 2002. Long term evolution and morphodynamic equilibrium of tidal channels. Journal of Geophysical Research 107, 1–13.

Leclerc, M., Bellemare, J.F., Dumas, G., Dhatt, G., 1990. A finite element model of estuarine and river flows with moving boundaries. Advances in Water Resources 13, 158–168.

Leenknecht, D.A., Szuwalski, A., Sherlock, A.R., 1992. Automated Coastal Engineering System. Coastal Engineering Research Center, U.S. Army Engineer Waterways Experiment Station, Vicksburg, MS, USA.

Leonard, L., Luther, M., 1995. Flow hydrodynamics in tidal marsh canopies. Limnology and Oceanography 40, 1474–1484.

Leonard, L.A., Reed, D.J., 2002. Hydrodynamics and sediment transport through tidal marsh canopies. Journal of Coastal Research 36, 459–469.

Liang, D., Falconer, R.A., Lin, B., 2007. Linking one- and two-dimensional models for free surface flows. Proceedings of the Institution of Civil Engineers, Water Management 160, 145–151.

Lin, W., Sanford, L.P., Alleva, B.J., Schwab, D.J., 1998. Surface wind wave modeling in Chesapeake Bay. In: Proceedings of the Third International Conference on Ocean Wave Measurements and Analysis – ASCE, Virginia Beach, VA, pp. 1048–1062.

Lin, B., Wicks, J.M., Falconer, R.A., Adams, K., 2006. Integrating 1D and 2D hydrodynamic models for flood simulation. Proceedings of the Institution of Civil Engineers, Water Management 159, 19–25.

Lopez, F., Garcia, M., 1998. Open-channel flow through simulated vegetation: suspended sediment transport modeling. Water Resources Research 34, 2341–2352.

Lopez, F., Garcia, M., 2001. Mean flow and turbulence structure of open-channel flow through non-emergent vegetation. Journal of Hydraulic Engineering 127, 392–402.

Lynch, D.R., Gray, W.G., 1980. Finite element simulation of flow in deforming regions. Journal of Computational Physics 36, 135–153.

Maa, Y.P., Mehta, A.J., 1987. Mud erosion by wave: a laboratory study. Continental Shelf Research 7, 1269–1284.

Marani, M., Silvestri, S., Belluco, E., Camuffo, M., D'Alpaos, A., Defina, A., Lanzoni, S., Marani, A., Tortato, M., Rinaldo, A., 2003a. Patterns in tidal environments: salt-marsh channel networks and vegetation. In: Proceedings of Geoscience and Remote Sensing Symposium, 2003. IGARSS '03, vol. 5. IEEE Press, Piscataway, NJ, pp. 3269–3271.

Marani, M., Belluco, E., D'Alpaos, A., Defina, A., Lanzoni, S., Rinaldo, A., 2003b. On the drainage density of tidal networks. Water Resources Research 39, 1040. https://doi.org/10.1029/2001WR001051.

Marani, M., Lanzoni, S., Silvestri, S., Rinaldo, A., 2004. Tidal landforms, patterns of halophytic vegetation and the fate of the lagoon of Venice. Journal of Marine Systems 51, 191–210.

Marani, M., Belluco, E., Ferrari, S., Silvestri, S., D'Alpaos, A., Lanzoni, S., Feola, A., Rinaldo, A., 2006. Analysis, synthesis and modelling of high-resolution observations of salt-marsh ecogeomorphological patterns in the Venice lagoon. Estuarine, Coastal and Shelf Science 69, 414–426. https://doi.org/10.1016/j.ecss.2006.05.021.

Marani, M., D'Alpaos, A., Lanzoni, S., Carniello, L., Rinaldo, A., 2007. Biologically-controlled multiple equilibria of tidal landforms and the fate of the Venice lagoon. Geophysical Research Letters 34, L11402. https://doi.org/10.1029/2007GL030178.

Marani, M., D'Alpaos, A., Lanzoni, S., Carniello, L., Rinaldo, A., 2010. The importance of being coupled: stable states and catastrophic shifts in tidal biomorphodynamics. Journal of Geophysical Research – Earth Surface 115, F04004. https://doi.org/10.1029/2009JF001600.

Marani, M., Da Lio, C., D'Alpaos, A., 2013. Vegetation engineers marsh morphology through multiple competing stable states. Proceedings of the National Academy of Sciences of the United States of America 110, 3259–3263. https://doi.org/10.1073/pnas.1218327110.

Mariotti, G., Fagherazzi, S., Wiberg, P.L., McGlathery, K.J., Carniello, L., Defina, A., 2010. Influence of storm surges and sea level on shallow tidal basin erosive processes. Journal of Geophysical Research – Oceans 115, C11012. https://doi.org/10.1029/2009JC005892.

Méndez, F.J., Losada, I.J., 2004. An empirical model to estimate the propagation of random breaking and nonbreaking waves over vegetation fields. Coastal Engineering 51, 103–118.

Minkoff, D.R., Escapa, M., Ferramola, F.E., Maraschn, S.D., Pierini, J.O., Perillo, G.M.E., Delrieux, C., 2006. Effects of crabhalophytic plant interactions on creek growth in a S.W. Atlantic salt marsh: a cellular automata model. Estuarine Coastal Shelf Science 69, 403–413. https://doi.org/10.1016/j.ecss.2006.05.008.

Molinaroli, E., Guerzoni, S., Sarretta, A., Masiol, M., Pistolato, M., 2009. Thirty-year changes (1970 to 2000) in bathymetry and sediment texture recorded in the lagoon of Venice sub-basins, Italy. Marine Geology 258, 115–125. https://doi.org/10.1016/j.margeo.2008.12.001.

Möller, I., 2006. Quantifying saltmarsh vegetation and its effect on wave height dissipation: results from a UK East coast saltmarsh. Estuarine, Coastal and Shelf Science 69, 337–351. https://doi.org/10.1016/j.ecss.2006.05.003.

Möller, I., Spencer, T., 2002. Wave dissipation over macro-tidal saltmarshes: effects of marsh edge typology and vegetation change. Journal of Coastal Research 36, 506–521.

Möller, I., Spencer, T., French, J.R., Leggett, D.J., Dixon, M., 1999. Wave transformation over salt marshes: a field and numerical modelling study from North Norfolk, England. Estuarine, Coastal and Shelf Science 49, 411–426.

Morris, J.T., Sundareshwar, P.V., Nietch, C.T., Kjerfve, B., Cahoon, D.R., 2002. Responses of coastal wetlands to rising sea level. Ecology 83, 2869–2877.

Mudd, S.M., Fagherazzi, S., Morris, J.T., Furbish, D.J., 2004. Flow, sedimentation, and biomass production on a vegetated salt marsh in South Carolina: toward a predictive model of marsh morphologic and ecologic evolution. In: Fagherazzi, S., Marani, M., Blum, L.K. (Eds.), The Ecogeomorphology of Tidal Marshes, American Geophysical Union, Coastal and Estuarine Studies, Washington DC, vol. 59, pp. 165–188.

Nepf, H.M., 1999. Drag, turbulence and diffusion in flow through emergent vegetation. Water Resources Research 35, 479–489.

Nepf, H.M., Koch, E.W., 1999. Vertical secondary flows in submersed plant-like arrays. Limnology and Oceanography 44, 1072–1080.

Nepf, H.M., Vivoni, E.R., 1999. Turbulence structure in depth-limited vegetated flows: transition between emergent and submerged regimes. In: Proc. XXVIII IAHR Congress, Graz, Austria, pp. 1–8.

Nepf, H.M., Vivoni, E.R., 2000. Flow structure in depth-limited, vegetated flow. Journal of Geophysical Research 105, 28547–28557.

Neumeier, U., 2007. Velocity and turbulence variations at the edge of saltmarshes. Continental Shelf Research 27, 1046–1059. https://doi.org/10.1016/j.csr.2005.07.009.

Neumeier, U., Amos, C.L., 2006. The influence of vegetation on turbulence and flow velocities in European saltmarshes. Sedimentology 53, 259–277.

Oey, L.-Y., 2005. A wetting and drying scheme for POM. Ocean Modelling 9, 133–150.

Perillo, G.M.E., Minkoff, D.R., Piccolo, M.C., 2005. Novel mechanism of stream formation in coastal wetlands by crab-fish-groundwater interaction. Geo-Marine Letters 25, 214220. https://doi.org/10.1007/s00367-005-0209-2.

Pethick, J.S., 1981. Long-term accretion rates on tidal salt marshes. Journal of Sedimentary Petrology 51, 571–577.

Pethick, J.S., 1992. Saltmarsh geomorphology. In: Allen, J.R.L., Pye, K. (Eds.), Saltmarshes, Morphodynamics, Conservation and Engineering Significance. Cambridge University Press, Cambridge, pp. 41–62.

Pignatti, S., 1966. La vegetazione alofila della laguna veneta, vol. XXXIII. Istituto Veneto di Scienze, Lettere ed Arti, Memorie. Fascicolo I, Venezia, 174 pp. (in italian).

Pritchard, D., Hogg, A.J., 2003. Cross-shore sediment transport and the equilibrium morphology of mudflats under tidal currents. Journal of Geophysical Research 108, 3313. https://doi.org/10.1029/2002JC001570.

Peruzzo, P., Defina, A., Nepf, H., 2012. Capillary trapping of buoyant particles within regions of emergent vegetation. Water Resources Research 48, W07512. https://doi.org/10.1029/2012WR011944.

Peruzzo, P., De Serio, F., Defina, A., Mossa, M., 2018. Wave Height Attenuation and Flow Resistance Due to Emergent or Near-Emergent Vegetation. Water 10, 402. https://doi.org/10.3390/w10040402.

Resio, D.T., 1987. Shallow-water waves. I: theory. Journal of Waterway, Port, Coastal, and Ocean Engineering 113, 264–281.

Resio, D.T., Pierre, W., 1989. Implication of an f-4 equilibrium range for wind generated waves. Journal of Physical Oceanography 19, 193–204.

Righetti, M., Armanini, A., 2002. Flow resistance in open channel flows with sparsely distributed bushes. Journal of Hydrology 269, 55–64.

Rinaldo, A., Fagherazzi, S., Lanzoni, S., Marani, M., Dietrich, W.E., 1999a. Tidal networks, 2, Watershed delineation and comparative network morphology. Water Resources Research 35, 3905–3917.

Rinaldo, A., Fagherazzi, S., Lanzoni, S., Marani, M., Dietrich, W.E., 1999b. Tidal networks, 3, Landscape-forming discharges and studies in empirical geomorphic relationships. Water Resources Research 35, 3919–3929.

Ris, R.C., Holthuijsen, L.H., Booij, N., 1999. A third-generation wave model for coastal regions. 2. Verification. Journal of Geophysical Research 104, 7667–7681.

Ross, M.A., Mehta, A.J., 1991. Fluidization of soft estuarine mud by waves. In: Bennett, R.H., et al. (Eds.), Microstructure of Fine-Grained Sediments. Frontiers in Sedimentary Geology. Springer, New York, NY.

Sánchez-González, J.F., Sánchez-Rojas, V., Memos, C.D., 2011. Wave attenuation due to Posidonia oceanica meadows. Journal of Hydraulic Research 49, 503–514.

Schwab, D.J., Benett, J.R., Liu, P.C., Donelan, M.A., 1984. Application of a simple numerical wave prediction model to Lake Erie. Journal of Geophysical Research 89, 3586–3592.

Shi, Z., Pethick, J., Pye, K., 1995. Flow structure in and above the various heights of a saltmarsh canopy: a laboratory flume study. Journal of Coastal Research 11, 1204–1209.

Shimizu, Y., Tsujimoto, T., 1994. Numerical analysis of turbulent open-channel flow over a vegetation layer using a $\kappa-\varepsilon$ turbulence model. Journal of Hydroscience and Hydraulic Engineering 11, 57–67.

Shoelhamer, D.H., 1995. Sediment resuspension mechanisms in Old Tampa Bay, Florida. Estuarine, Coastal and Shelf Science 40, 603–620.

Silvestri, S., Marani, M., 2004. Salt marsh vegetation and morphology, modelling and remote sensing observations. In: Fagherazzi, S., Marani, M., Blum, L. (Eds.), Ecogeomorphology of Tidal Marshes, American Geophysical Union, Coastal and Estuarine Monograph Series, Washington, vol. 59, 266 p.

Silvestri, S., Marani, M., Rinaldo, A., Marani, A., 2000. Vegetazione alofila e morfologia lagunare. In: Atti dell'Istituto Veneto di Scienze, Lettere ed Arti, Tomo CLVIII (1999-2000), Classe di scienze fisiche, matematiche e naturali, pp. 333–359 (in italian).

Silvestri, S., Defina, A., Marani, M., 2005. Tidal regime, salinity and salt-marsh plant zonation. Estuarine, Coastal and Shelf Science 62, 119–130.

Soulsby, R.L., 1997. Dynamics of marine sands. In: Telford, T. (Ed.), A Manual for Practical Applications, 248 p.

Suzuki, T., Zijlema, M., Burger, B., Meijer, M.C., Narayan, S., 2011. Wave dissipation by vegetation with layer schematization in SWAN. Coastal Engineering 59, 64–71.

Swales, A., MacDonald, I.T., Green, M.O., 2004. Influence of wave and sediment dynamics on cordgrass (*Spartina anglica*) growth and sediment accumulation on an exposed intertidal flat. Estuaries 27, 225–243.

Temmerman, S., Govers, G., Wartel, S., Meire, P., 2004a. Modelling estuarine variations in tidal marsh sedimentation: response to changing sea level and suspended sediment concentrations. Marine Geology 212, 1–19.

Temmerman, S., Govers, G., Meire, P., Wartel, S., 2004b. Simulating the long-term development of levee-basin topography on tidal marshes. Geomorphology 63, 39–55.

Umgiesser, G., Sclavo, M., Carniel, S., Bergamasco, A., 2004. Exploring the bottom stress variability in the Venice Lagoon. Journal of Marine Systems 51, 161–178.

Uncles, R.J., 2002. Estuarine physical processes research: some recent studies and progress. Estuarine, Coastal and Shelf Science 55, 829–856. https://doi.org/10.1006/ecss.2002.1032.

Van Ledden, M., 2003. Sand-Mud Segregation in Estuaries and Tidal Basins (Ph.D. thesis). T.U. Delft, Dep. of Civil Engineering and Geosciences. Report 03-2, ISSN 0169–6548.

Van Ledden, M., Wang, Z.B., Winterwerp, H., De Vriend, H., 2004. Sand-mud morphodynamics in a short tidal basin. Oceans Dynamics 54, 385–391. https://doi.org/10.1007/s10236-003-0050-y.

Van Rijn, L.C., 1993. Principles of Sediment Transport in Rivers Estuaries and Coastal Seas. Aqua Publications, Amsterdam, The Netherlands, ISBN 90-800356-2-9.

Venier, C., Figueiredo da Silva, J., McLelland, S.J., Duck, R.W., Lanzoni, S., 2012. Experimental investigation of the impact of macroalgal mats on flow dynamics and sediment stability in shallow tidal areas. Estuarine Coastal Shelf Science 112, 52–60. https://doi.org/10.1016/j.ecss.2011.12.

Ward, L.G., Kemp, W.M., Boynton, W.R., 1984. The influence of waves and seagrass communities on suspended particulates in an estuarine embayment. Marine Geology 59, 85–103.

Wayne, C.J., 1976. The effects of sea and marsh grass on wave energy. Coastal Research Notes 14, 6–8.

Young, I.R., Verhagen, L.A., 1996. The growth of fetch-limited waves in water of finite depth. Part 1: total energy and peak frequency. Coastal Engineering 29, 47–78.

PART III

TIDAL FLATS

CHAPTER

10

Geomorphology and Sedimentology of Tidal Flats

Shu Gao

State Key Laboratory for Estuarine and Coastal Research, East China Normal University, Shanghai, China

1. INTRODUCTION

Tidal flats are distributed widely along the world coastlines, representing an important part of coastal wetlands. A tidal flat can be divided into three parts, according to the characteristic tidal water levels (Amos, 1995): (1) supratidal zone, which is located above the high water on springs and is inundated only under extreme conditions (e.g., storm surge events); (2) intertidal zone, located between the high water and low water on springs and is inundated periodically during spring–neap tidal cycles; and (3) subtidal zone, which is below the low water on springs and is rarely exposed in air. In literature, the studies on tidal flats have been concentrated mainly on the intertidal part; some authors use the term "tidal flat" to represent the intertidal zone (which is adopted in the present study), whereas others prefer the term "intertidal flat." This chapter will concentrate mainly on the physical aspects of the intertidal flat; salt marshes will be described only briefly when necessary because they are treated in detail in a separate chapter. The importance of biological processes is evaluated in terms of the physical consequences, with an emphasis on future research directions.

The well studied tidal flats include those along the Dutch, German, and Danish coasts (Reineck, 1972; Reineck and Singh, 1980; Pejrup, 1988), in the Wash embayment of England (Evans and Collins, 1975; Collins et al., 1981), over the bay-head areas of the Bay of Fundy, eastern Canada (Amos and Mosher, 1985), and on the Jiangsu coast in eastern China (Wang, 1983; Ren, 1986) (Fig. 10.1). Generally, these tidal flat systems are characterized by accumulation of fine-grained sediments and gentle bed slopes. Tidal currents are relatively strong on the lower and middle parts of the tidal flat, resulting in high mobility of bed materials. Over the upper parts of the intertidal zone, salt marshes may be present, with water and nutrients being supplied by the tidal flow shortly before and after high water slack (Blanton et al., 2002; Zhang et al., 2004). In tropical areas, mangroves may develop on mudflats (Wells and Coleman, 1981).

Coastal Wetlands
https://doi.org/10.1016/B978-0-444-63893-9.00010-1

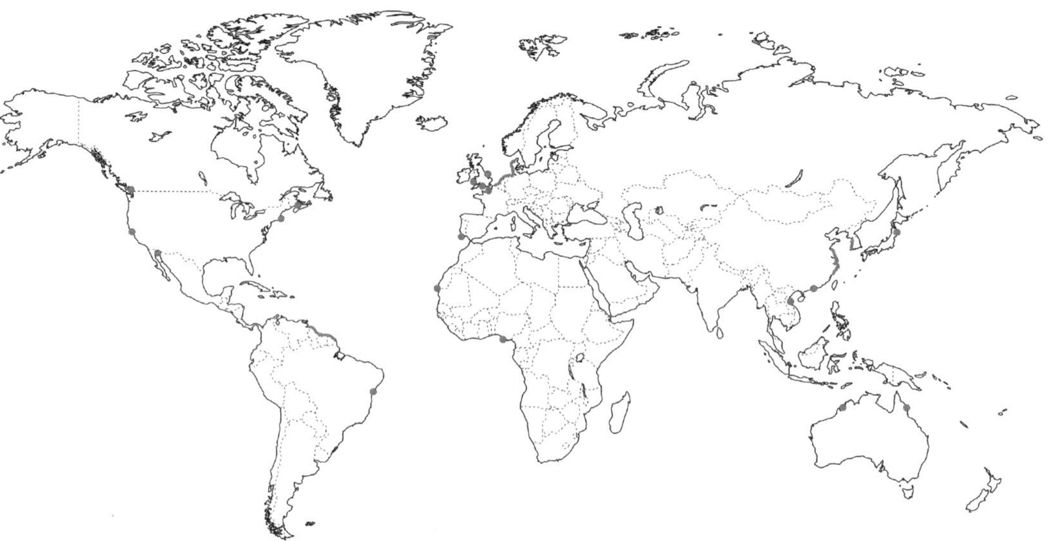

FIGURE 10.1 Coastal sections (*dashed lines*) and locations (*circles*) along the world coastlines where extensive or detailed studies on tidal flat sedimentology and geomorphology have been undertaken (Allen, 1965; Reineck and Singh, 1980; Klein, 1985; Ren, 1986; Isla et al., 1991; Daborn et al., 1993; Perillo et al., 1996; Netto and Lana, 1997; Perillo and Piccolo, 1999; Kjerfve et al., 2002; Lim and Park, 2003; Deloffre et al., 2005; Falcão et al., 2006; Quaresma et al., 2007; Sakamaki and Nishimura, 2007; Anthony et al., 2008; Proske et al., 2008; Talke and Stacey, 2008).

Progress has been made in the understanding of the characteristics, processes, and evolution of tidal flats, which is important for the purpose of coastal wetland protection and restoration. In early times, the unique morphological features, especially zonation in geomorphology, attracted the researchers. From high water to low water marks, there are systematic changes in sediment grain size, bedforms, sedimentary structures, and biological activities, which have been studied since the 1930s (e.g., Haentzschel, 1939; Linke, 1939). Then, from the 1950s, sediment dynamic and morphodynamic studies have been carried out, in an attempt to understand the mechanisms of sediment transport and accumulation (Postma, 1954; van Straaten and Kuenen, 1957, 1958). Extensive in situ measurements have been undertaken, on the flat surface and in tidal creeks, to obtain information on tidal current, suspended sediment concentration, and the benthic boundary layer (Evans and Collins, 1975; Letzsch and Frey, 1980; Collins et al., 1981, 1998; Stumpf, 1983; Bartholdy and Madsen, 1985; Pejrup, 1988; Wells et al., 1990; Alexander et al., 1991; Gouleau et al., 2000; Andersen and Pejrup, 2001; Davidson-Arnott et al., 2002). On such a basis, quantitative models for tidal flat sedimentation and morphodynamics have been proposed (Allen, 1989, 2000, 2003; French, 1993; Allen and Duffy, 1998; Roberts et al., 2000; Pritchard et al., 2002; Malvarez et al., 2004; Temmerman et al., 2004). At the same time, studies on sedimentary sequences and associated physically and biologically induced sedimentary structures were documented in detail, to obtain information on the environmental conditions under which the deposits were formed and on the environmental changes (Reineck and Singh, 1980; Klein, 1985).

Recently, research has been focused on the formation of tidal flat sediment systems and the information on climate, environmental, and ecosystem changes contained in the sedimentary record (Cundy and Croudace, 1995; Dellwig et al., 2000; Gerdes et al., 2003; Gao, 2009). The evolution of tidal flats in response to global climate change and intensified anthropogenic activities has become an important research area (Vos and van Kesteren, 2000; Wang et al., 2012). Furthermore, biogeomorphology of tidal flats has become an important research field, to understand the role played by organisms in morphodynamics and the coupling effects of physical, chemical, and biological processes (D'Alpaos et al., 2012; Da Lio et al., 2013; Gao et al., 2014; D'Alpaos and Marani, 2016).

The purpose of the present contribution is to provide a general description about the sediment distribution patterns and morphological features of tidal flats, together with an overview of the conditions and dynamic processes for the formation and evolution of tidal flats.

2. BASIC CONDITIONS FOR THE FORMATION OF TIDAL FLATS

Tidal flats are formed in areas where there is a sufficient supply of fine-grained sediment (i.e., clays, silts, and fine to very fine sands), and that tides and tidal currents dominate over other hydrodynamic forces (Klein, 1985). The first condition is satisfied for many coastal environments: fine-grained sediment is delivered by rivers and discharged into estuaries and adjacent coastal areas, erosion on the seabed and cliff recession provide additional sources of sedimentary materials, and organisms living in coastal waters and salt marshes produce shell debris and particulate organic matter. The second condition determines whether or not fine-grained sediment will be deposited on the tidal flat. Several factors influence this condition.

Firstly, the tidal action should be significant. The average tidal range (R) has been used in the classification of the coast: the coast can be microtidal ($R < 2$ m), mesotidal ($R = 2-4$ m), or macrotidal ($R > 4$ m) (Davies, 1964). On a microtidal coast, tidal currents are relatively weak unless the slope of the seabed is extremely small. In mesotidal and macrotidal environments, tidal currents tend to be relatively strong compared with those on the microtidal coast, which favors the formation of tidal flats. It should be noted that tidal flats can be formed in microtidal areas of sheltered coastal embayments or semi-enclosed seas; this is because in such environments the wave action is of only a secondary importance compared with tidal currents. In addition to tide range, the period of the tidal cycle (e.g., diurnal or semidiurnal) also influences the tidal currents; under the same conditions of tide range and seabed morphology, tidal currents will be stronger in semidiurnal areas than in diurnal areas (e.g., Gao and Collins, 1994; Archer, 1995).

Secondly, the dominance of tidal action should be understood in a relative sense: tidal flats cannot be formed where wave action dominates, even if the tidal range is large. On open coasts, waves may represent a dominant force, which break in the surf zone, causing transport of fine-grained sediment toward offshore areas (King, 1972). In this case, if the supply of fine-grained materials is not abundant, then sandy or gravelly beaches will be formed, rather than tidal flats. However, if the rate of fine-grained sediment supply is high, then the accumulation of fine-grained materials reduces the bed slope in the intertidal area. Eventually, wave breaking will less frequently occur on the bed, i.e., the wave energy is dissipated over the wide intertidal area due to bed friction, without causing breaking. The reduction of the bed slope will lead to enhanced tidal currents. Thus, in response to fine-grained sediment accumulation, wave action is weakened and the tidal action is enhanced. However, this observation does not imply that waves are unimportant on tidal flats; waves are important in the transport of sediment and the shaping of the tidal flat morphology (Christiansen et al., 2006; Friedrichs, 2011; Fan, 2012). In situ measurements have shown that combined wave–tide action (without wave breaking) can cause intense sediment movement on the flat (Fan et al., 2006; Wang et al., 2006; Fan, 2012). Likewise, a tidal flat profile has a convex-up morphology due to tidal action, but there is a tendency for the profile to become concave-up if wave action is added to the system (Friedrichs, 2011). On the other hand, numerical experiments indicate that for back-barrier tidal flats, where wave action is weak, such morphology can also be caused purely by tidal action: a large tidal basin with small tidal range will lead to a concave-up morphology, but a small basin with large tidal range favors a less concave-up feature (Yu et al., 2012).

Finally, although tidal action is a dominant agent, storm events can significantly and rapidly modify the tidal flat environment. For example, during a typhoon event, storm surges become temporally the dominant forcing for sediment erosion, transport, and accumulation on a tidal flat, causing erosion over the middle–lower parts of the intertidal zone and in tidal creeks, and accumulation of thick layers of sediment over the upper part (Ren et al., 1985; Ren, 1986; Gao and Zhu, 1988; Andersen and Pejrup, 2001).

The arguments outlined above imply that tidal flats may develop in sheltered tidal estuaries and coastal embayments where there is continuous supply of fined-grained sediment (although the rate of supply may be small) or on open coasts where tidal range is sufficiently large and sediment supply is abundant. A typical example of the former is the tidal flats in the Dutch Wadden Sea (van Straaten and Kuenen, 1957, 1958). Here, the tidal flat areas are sheltered by a series of barrier islands, the exchange of water and fine-grained sediment between the Wadden Sea and the open North Sea being via tidal inlets cutting through the barrier islands (Yu et al., 2012, 2014; Wang et al., 2014). The sediment source is provided mainly by the North Sea (Postma, 1961; Pejrup et al., 1997), whereas river input represents a secondary source. For the open coast tidal flats, a typical example is the tidal flats on the Jiangsu coast, eastern China (Ren, 1986; Gao and Zhu, 1988). During the Holocene period, because of the abundant sediment supply from two large rivers (i.e., the Changjiang and the Yellow rivers), extensive tidal flats have been formed (Liu et al., 2011, 2015).

3. ZONATION IN SEDIMENTATION AND FLAT SURFACE MORPHOLOGY

3.1 Vertical Sediment Sequences

Sediment cores taken from the upper part of the intertidal flat tend to show a "fining-upward" sequence (Klein, 1985). Although the sediment deposited is ultimately determined by the source characteristics, in most cases the sediment source for tidal flats contain materials from sand- to clay-sized materials. Sandy materials tend to accumulate in the lower and middle parts of the flat, whereas muddy materials deposit over the upper parts of the flat. The top part consists of clayey or muddy materials, often with high organic carbon content. Below this layer is a mud layer corresponding to the elevation near the high water, with very thin laminae as a major type of sedimentary structure. Then there is a mixed sand–mud layer, interbedded with sandy and muddy materials; such sedimentary structures are known as "tidal bedding" (Reineck and Singh, 1980). The lowest part of the core is a sand layer, corresponding to the lower part of the intertidal zone and the subtidal zone.

Such a vertical sequence reflects the spatial distributions of sediments on tidal flats. A general pattern is that along a transect from the supratidal zone to subtidal zone, there are salt marshes, mudflats, mixed sand–mud flats, and sand flats (Fig. 10.2A). On the salt marshes, the bed material is finest, with organic matter being derived from marsh plants and organisms. The mudflat is covered with clay and fine silts; this area is often located between the high water levels on springs and on neaps. In some regions, e.g., the Jiangsu coast in eastern China, salt marsh vegetation may extend from the supratidal zone into the mudflat (Ren, 1986). Mixed sand–mud flats are close to the mean sea level; here, sands are deposited during spring tides and muds are deposited on neaps (for details, see Section 4.2). Over the lower parts of tidal flats, well-sorted sands are present, with various types of bedforms (e.g., dunes and ripples). On the tidal flats, tidal creeks may be formed, consisting of small creeks and large tidal channels. Sedimentation in tidal creeks may be different from that on the adjacent flat surface, but the general "fining-upward" pattern is also maintained for the creek systems. Superimposed on the above-mentioned patterns, if examined in details, the zonation for some tidal flat environments can have slightly different patterns (Evans, 1965; Amos, 1995), as indicated by Fig. 10.2B. The reason for this is that, in addition to bed elevation, other factors such as bed slope and tidal water level also influence the transport and accumulation of sediments (see below).

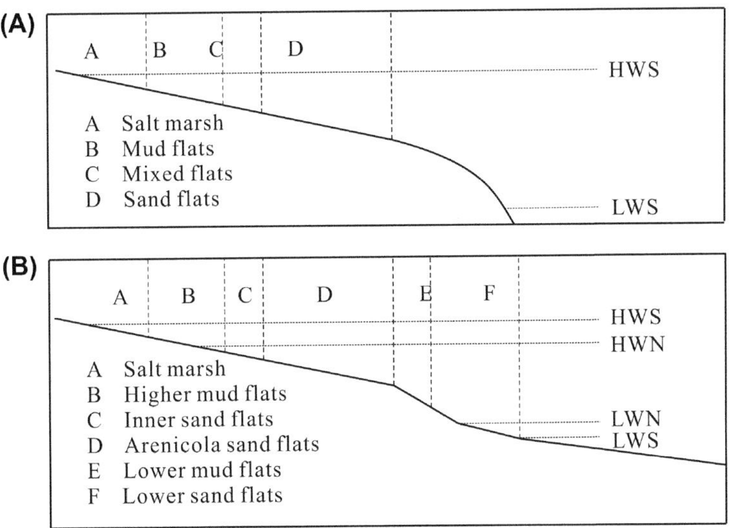

FIGURE 10.2 General patterns of (A) zonation of tidal flats (Reineck and Singh, 1980) and (B) the geomorphic and sedimentary features of the Wash, England (Evans, 1965). *HWN*, high water on neaps; *HWS*, high water on springs; *LWN*, low water on neaps; *LWS*, low water on springs.

TABLE 10.1 Thickness of the Sediment Layer Above the Mixed Sand–Mud Flat

Location	Tide Range (m)					
	<R>	R_S	R_N	H_m (m)	W (km)	References
Wadden Sea (The Netherlands)	1.3–2.8			2.0	7–10	van Straaten (1961)
The Wash (England)	5.0	6.5	3.5	1.5–2.0	1.0–6.5	Evans (1965)
Colorado River Estuary (USA)	4–5	6–8		>8		Ginsberg (1975)
Wadden Sea (Germany)		2.6–4.1	1.8–3.1	2.5	5	Ginsberg (1975)
Fraser River Estuary, Boundary Bay (Canada)	2.7	4.1	1.5	<0.5		Ginsberg (1975)
Central Jiangsu coast (China)				3	7–10	Ren (1986)
Salmon River Estuary, Bay of Fundy (Canada)	11.9	15.2	8.7	4	5	Dalrymple et al. (1990)
Haenam Bay (South Korea)	3.0	4.0	1.8	>4	2–2.5	Lim and Park (2003)
Newtownards (Northern Ireland)	3.0	3.5		0	0.5–1.2	Malvarez et al. (2004)
Baeksu Tidal Flat (South Korea)	3.9			<0.5	4–6	Yang et al. (2005)

<*R*>, average tide range; H_m, thickness of mud layer; R_N, neap tide range; R_S, spring tide range; *W*, width of intertidal zone.

The zonation of tidal flats implies that the finest materials tend to accumulate as the upper part of the sedimentary sequence. This observation raises a question of the thickness of the mud layer on the upper part of the tidal flat. Data sets from the different parts of the world (Table 10.1) show that the thickness varies considerably; in some places the entire profile is covered with muddy sediments, whereas on the other extreme mud is absent. Apparently, the correlation between the thickness and tidal range is poor (cf. data listed in Table 10.1). In Section 4, an explanation about the factors that control the thickness of the mud layer will be given on the basis of sediment dynamic analysis, for the tidal flats that are associated with a supply of different types of sediments ranging from clay to fine sand.

3.2 Sediment and Morphology on Intertidal Mudflats

Generally, mudflats are located on the upper part of the intertidal zone, with gentle bed slopes (Fig. 10.3). Here, the sediment is the finest for the entire tidal flat, and deposition takes places because of the settling of fine-grained materials from the water column. Therefore, although there are tidal cycle changes in grain size, mud (i.e., a mixture of clayey and silty sediments) is the major component of mudflat sediments. The deposition of coarser materials, which occasionally occur in the mudflat sediment layers, is due to extreme events (e.g., storm surges). The sedimentary record consists of laminated mud, with alternating silty and clayey layers of less than 1 mm in thickness. Because resuspension of the bed material is relatively weak here, the continuity of the mud deposit is high compared with the other parts of the tidal flat. However, in some places, intense bioturbation often causes destruction of the original sedimentary structure.

FIGURE 10.3 Bed features of the bare mudflat (with benthic algae at the sediment surface), Jiangsu coast, eastern China.

The landward part of the mudflat may be covered by pioneer plants extending from the salt marshes in the supratidal zone, or even covered with salt marshes. On the Jiangsu coast, eastern China, for example, extensive *Spartina* marshes are formed over the landward part of the mudflats, with an upper limit close to mean high water on springs and a lower limit slightly lower than high water on neaps (Zhang et al., 2004; Gao et al., 2014).

Under tidal action alone, accretion becomes progressively slower when the bed elevation is enhanced due to sediment accumulation (Pethick, 1981); thus, the high water level on springs represents the upper limit for the accretion (Amos and Mosher, 1985; Amos, 1995). However, the bed accretion continues beyond the high water level, to form the supratidal zone. There are several reasons for this phenomenon. First, during storm events the water level can become much higher than during normal tidal cycles, especially when the surge coincides with an astronomical spring tide. During such events, a large amount of sediment can be transported to the upper parts of the flat and deposits there. Thick storm deposits have been identified in the upper tidal flat sequences (Ren et al., 1985). Second, benthic fauna living in the flat environment will modify the bed morphology through bioturbation, generating unique sedimentary structures (Reineck and Singh, 1980). Such activities produce uneven bed morphology (e.g., mud mounds formed by crabs and mud skippers): the elevated features may become above the high water, and the depressions will receive an increased amount of sediment during high water. Finally, the deposition of aerosol forms an additional sediment source (Li et al., 1997), which is not constrained by the high water mark.

On the tidal flat, tidal creeks form a drainage network (Bayliss-Smith et al., 1979; Marani et al., 2003). As a part of this network, tidal creeks on the mudflat tend to have a small width to depth ratio, but they are different from a fluvial system in that the upper end of the creeks (often located in salt marshes) does not often represent an eroding phase. Initially, the formation of a creek may be associated with headward erosion (Perillo et al., 1996; Fagherazzi and

Furbish, 2001), but eventually the creek end reaches a final stage of tidal creek development. Here, the tidal current is weak due to reduced tidal prism, with accumulation of fine-grained sediment. On the bare mudflat, the tidal creeks may experience relatively strong tidal currents, which cause lateral migration of the channel and meandering pattern formation (Weimer et al., 1982; Eisma, 1998; Wang et al., 1999).

At the boundary between the mudflat and salt marshes, cliffs or scarps tend to occur, with a height of less than 1 m (Reineck and Singh, 1980). In some places, much higher scarps can be formed, for example, those found in the Severn Estuary, UK, which has been interpreted as periodic erosion and accretion cycles in response to changes in hydrodynamic and sediment supply conditions (Allen, 1989). However, disagreement exists with regard to the significance of the low cliffs that are more widely distributed. Some researchers believe that the cliffs represent an indication of coastal erosion, but others argue that they result from localized scour. Observations show that because the accretion on the marsh that is more rapid than the adjacent bare flat (the flat with plants traps more sediment than the bare flat, and the organic matter further adds to the sedimentary materials in the marsh), the bed elevation gradually becomes different (Amos, 1995). Then, the edge of the marsh becomes progressively steeper, causing concentration of wave energy (the small waves would otherwise not break at this location).

In coastal embayments, such low cliffs are often associated with low hydrodynamic forcing and low sediment supply, for example, in Christchurch Harbour, southern England, where the deposition rate has been higher on the marsh than on the bare flat. Subsequently, the small waves formed in the estuarine waters (with a fetch of only 1 km) started to erode part of the materials at the marsh edge, forming low cliffs of 0.4–0.6 m in height (Gao and Collins, 1997). Numerical model output (Gao and Collins, 1997) indicated that the action of small waves, which break at the marsh–bare flat boundary because of the enhanced bed slope at this location, is responsible for the localized scour and the cliffs can be stable or retreat slowly. In the German and Danish Wadden Sea, marsh cliff erosion was measured by Pejrup et al. (1997) and similar results were obtained. Elsewhere, measurements in Rehoboth Bay, USA, revealed that the rate of cliff recession is also small, ranging between 0.14 and 0.43 m year^{-1} (Schwimmer, 2001). Hence, it is the evolution of the tidal flat itself that creates the condition for the formation of the cliff.

3.3 Sediment and Morphology on Mixed Sand–Mud Flats

Mixed sand–mud flats are characterized by alternating deposition of muddy materials on neaps and sandy material on springs. The spatial distribution of this morphological unit on the tidal flat depends on a number of factors, including tidal regime, sediment supply, and suspended sediment concentration of seawater. On the Jiangsu coast, where the tidal currents during the flood and ebb maximum periods exceed the threshold for bedload transport over the entire spring–neap tidal cycle, the mixed flat is located between high water on neaps and mean sea level (Zhu and Xu, 1982).

In the sediment sequences, typical "tidal bedding" (i.e., interlayered relatively coarse- and fine-grained sediments) is found on the mixed sand–mud flats. These beddings contain information on the sedimentary processes during flood–ebb cycles and spring–neap cycles (Dalrymple et al., 1990). A well-preserved sediment record at this location will show that the thickness of the sand layer decreases with decreasing tidal range from the springs toward the neaps. Likewise, the mud layer increases its thickness toward the neap tides. However, the preservation potential for the sand–mud flat is usually not high, and most of the sedimentary record formed in tidal cycles is destroyed subsequently by reworking or resuspension of the bed materials (Fan et al., 2002; Deloffre et al., 2005, 2007; Gao, 2009). On the bed, scour features are present, resulting from sediment reworking (Fig. 10.4A and B). Generally, on the lower part of bare mudflats and mixed mud–sand flats, a part of the muddy material accumulated on neaps can survive until the next neap tidal phase. This is the reason for the formation of the scour features. Similar but more stable features (known as tidal depressions, including pans, ponds, and pools) may be developed on low-energy mudflats (see Chapter 6 of this book, for details). It should be noted that such scour does not indicate long-term net erosion of the bed; on the contrary, the accretion rate on the mixed sand–mud flat is actually higher than the other parts of the tidal flat system (Gao and Zhu, 1988). The scour represents only short-term effects within a long-term accretion trend.

Tidal creeks are highly dynamic on mixed sand–mud flats. For instance, on the central coast of Jiangsu Province, China, the strong currents in combination with the silty sediments on such flats result in rapid lateral migration of the tidal channels. Point-bar deposits are formed due to lateral migration associated with meanders. Channel migration generates new tidal creeks, which can extend toward the upper part of the intertidal zone rapidly, in the form of headward erosion. During storms, the tidal creeks are even more active; significant deepening of the channel bottom, development of many new channels, and rapid migration may be observed (Ren, 1986).

3.4 Sediment and Morphology on Sand Flats

In a typical tidal flat system with abundant sandy sediment supply and strong tidal currents, sand flats occupy the lower part of the intertidal zone. The sands are normally well sorted. Tidal creeks on the sand flat are relatively wide, with a small depth to width ratio (Fig. 10.4C). The migration of the tidal channels is active and is influenced by bedload transport and accumulation. Cross bedding is a common type of sedimentary structure in tidal creek sequences.

During the late stages of the ebb, before the bed is exposed to air, the tidal currents are too strong for the fine-grained, suspended sediment to settle onto the bed, and the material settled during the flood slack tends to be suspended during the ebb. Numerical calculations show that net accumulation of mud on the sand flat is possible only when the suspended sediment concentration is extremely high (Amos, 1995), in which case sand flats with pure sand deposition cannot be formed.

Where the bed slope is sufficiently large (i.e., 1.0×10^{-3} or greater) to allow rapid draining of water mass, the geometry of bedforms (dunes and ripples) is preserved and exposed during low tide (Klein, 1985; Dalrymple et al., 1990). The dunes have a wavelength (or spacing) of 0.6–6 m, whereas the ripples are small bedforms (wavelength <0.6 m). Compared with the dunes, ripples are more extensively distributed. Field observations show that during the high water slack ripples are formed by combined wave–current action; during the ebb the upper part of the sand flat will be exposed rapidly, and the morphology of wave ripples may be preserved, but the lower sand flat can be modified by ebb currents to form ebb-oriented,

FIGURE 10.4 Geomorphic and sedimentary features of the middle to lower parts of the intertidal zone on the Jiangsu coast, eastern China: (A) scour features on mudflats; (B) scour features on mixed mud–sand flats; (C) a tidal creek on mixed sand–mud flats; (D) sand flat bed, with current ripples; and (E) sand flat near the low water mark.

flow-generated ripples (Amos and Collins, 1978). If the bed has an extremely gentle slope, then these bedforms may be modified by water flow generated by seepage from the substrate; because the flow depth is very small (i.e., <10–1 m) and the Froude number is large (i.e., close to 1), plane bed without bedforms can be formed in a short period of time (i.e., 1–2 h). On the Jiangsu coast, eastern China, the central parts of the sand flat (with a mean grain size of 0.06–0.1 μm) have a bed slope of around 0.0005, and the slope increases to more than 0.001 toward either the low water mark or the boundary between the sand flat and the mixed sand–mud flat (Zhu and Xu, 1982). Therefore, during low water on springs, wave ripples are present near the boundary between the sand flat and the mixed sand–mud flat, plane beds are present over the central part of the sand flat, and current ripples are found over the lower sand flat (Fig. 10.4D and E).

4. FACTORS AND PROCESSES

4.1 Influences of Quantity and Composition of Sediment Supply

The evolution of present-day tidal flats began when sea level reached its highest elevation during the Holocene period, from a base that was left from the last glacial period. Thus, the accommodation space for the tidal flat deposit is confined by the high water mark, the original topography/bathymetry, and the tidal flat profile. Although the high water level and the topographic baseline are influenced by sea-level changes and crustal movements, the rate of shoreline advancement is determined by the amount of sediment supply, and the shape of the profile is influenced by the grain size composition of the materials supplied.

As shown in a simplified shore-normal transect for the geometry of a tidal flat sediment system (Fig. 10.5), the increase of the tidal flat sediment body through time is associated with upward accretion and the advancement of the tidal flat profile toward the sea. The vertical accretion rate of a tidal flat can be related to the portion of sediment supply that contributes to the tidal flat formation by

$$D = \frac{\Delta S}{\Delta L} \frac{\sin(\beta - \alpha)}{\sin\alpha \sin\beta} \tan\beta \qquad (10.1)$$

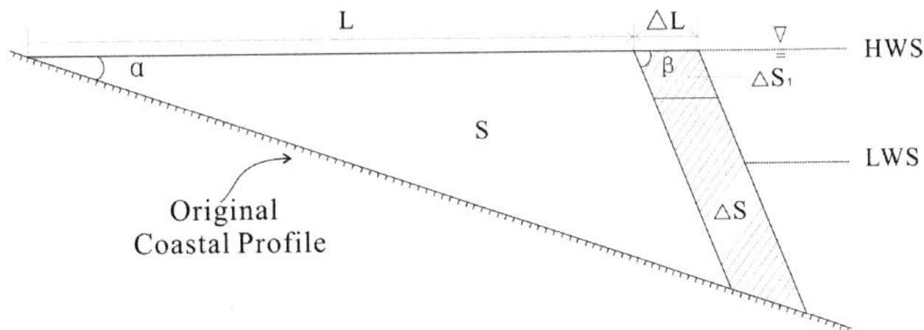

FIGURE 10.5 Schematic diagram showing a shore-normal cross section of tidal flat sedimentation (L, the width of coastal plain formed by tidal flat sedimentation; S, volume of sediment accumulated over unit length of the shoreline; α, the slope angle of the original topography; β, the average bed slope angle of the flat profile; ΔL, the distance of shoreline advancement during a unit time; ΔS, the amount of sediment supply per unit time over unit length of the shoreline). Note: the vertical scale is exaggerated.

where D is the vertical accretion rate averaged over the entire profile, ΔS is the proportion of sediment supply that is deposited on the flat per unit time over unit length of the shoreline, α is the slope angle of the original topography, β is the average bed slope angle of the flat profile, and ΔL is the distance of shoreline advancement during a unit period of time.

In addition to the accretion rate, the thickness of the mud layer of the upper part of the sequences is also an important parameter for the tidal flat. Here, the mud layer represents the deposits of the mud flat and the mixed sand–mud flat (see Sections 3.2 and 3.3). Assuming that most of the muddy materials are deposited on the upper part of the tidal flat, as is the case for the Jiangsu coast, this thickness can be expressed approximately as (cf. the geometric relationship shown in Fig. 10.5)

$$H_m \approx \frac{\Delta S_1}{\Delta S}(L + \Delta L)\tan\alpha \tag{10.2}$$

where H_m is the thickness of the mud layer, ΔS_1 is the mud fractions within ΔS, and L is the total width of coastal plain formed by tidal flat sedimentation. Eq. (10.2) implies that the thickness is related to the composition of sediment supply and the stage of the tidal flat evolution (since it is a function of ΔS_1 and L), but it is not directly related to tide range. Observations of the world tidal flats have provided support to this inference (cf. Table 10.1): the thickness of the mud layer does not increase with the increasing tide range. In an extreme case (Malvarez et al., 2004), the profile is covered entirely with mud because in this system the supply of sandy sediments is very small. On the Jiangsu coast, the sediment input from the Yellow River has relatively stable composition; as a result, there is a trend of increase in the thickness of the mud layer during the tidal flat development (Gao, 2009). Such an observation is also consistent with the prediction by Eq. (10.2).

The thickness of the mud layer determines the boundary on the tidal flat where sandy and muddy deposits are divided. At this boundary, the bed slope can be expressed as a function of the threshold for the initial motion of the sands, as demonstrated below. At any site over the intertidal flat, the instantaneous tidal current velocity is a vector, which consists of an onshore–offshore component, u, and a longshore component, v (Anderson, 1973; Perillo et al., 1993; Wang et al., 1999). The onshore–offshore component is controlled by the water level changes and the bed morphology, on the basis of the principle of mass conservation (Zhu and Gao, 1985; Wang et al., 1999):

$$u = \frac{1}{\tan\beta}\frac{dh}{dt} \tag{10.3}$$

where $\tan\beta$ is the average bed slope over the inundated section and h is the tidal water level.

At the sand–mud boundary, maximum current speeds during the flood should not exceed the threshold for initial bedload motion; otherwise, sandy materials will be transported across the boundary further toward the upper tidal flat. At the same time, it should not be smaller than the threshold; otherwise, sandy materials cannot be transported to this location. Hence, there is only one possibility: the sand–mud boundary is located where the maximum shore-normal current speed is equal to the threshold:

$$u_{max} = u_{cr}\sin\theta = \frac{1}{\tan\beta_b}\frac{dh_b}{dt} \tag{10.4}$$

where θ is the angle between the current direction and the longshore direction, the subscript b denotes the sand–mud boundary, and dh_b/dt denotes the rate of water level change when the flow reaches the boundary. Thus, the slope at the sand–mud boundary can be defined by

$$\tan\beta_b = \frac{1}{u_{cr}\sin\theta}\frac{dh_b}{dt} \tag{10.5}$$

The threshold current speed for initial sediment motion is a function of near-bed shear stress. Eq. (10.5) implies that a steep bed slope at the sand–mud boundary is associated with a large value of dh_b/dt, or a low elevation for the sand–mud boundary (because the rate of water level change cannot be large near the high water mark). Because the thickness of the mud layer is the vertical difference between the elevation at the boundary and the high water mark, the steep slope also means a lager thickness of the mud deposit in the tidal flat sequence. This observation may explain the phenomenon that on a tidal flat with a thick mud layer, maximum slope gradients occur on the mixed sand–mud flat (Zhu and Xu, 1982). The elevation associated with the critical current speed differs between the spring and neap conditions. As a result, there will be two critical elevation values, one for neap tides and the other for spring tides. Between the two critical elevations, mud is deposited on neaps and sand is deposited on springs. Therefore, the thickness of the mud layer (consisting of mud and mixed mud–sand deposits) is related to the lower critical elevation for the neap tides, and the mixed sand–mud flat itself is a result of spring–neap tidal cycles.

The relationship between sediment supply and the tidal flat morphology can explain the effect of land reclamation. Reclamation reduces the width of the intertidal zone and, therefore, the tidal current speed. As a result, landward transport of coarse-grained sediment (i.e., bedload) is weakened. Suspended sediment transport becomes a dominant mode. In this case, if there is a sufficient supply of fine-grained sediment, then accretion will continue to take place in the intertidal zone, with an increase in mud sediment thickness. Wang et al. (2012) have reported a case study which supports this prediction.

4.2 Sedimentation During Tidal Cycles

On a tidal flat, why are the sediments transported toward the land? Two mechanisms have been identified, one for suspended load and the other for bedload. The movement of the suspended load is related to a physical mechanism known as "settling- and scour-lag effects" (Postma, 1954; van Straaten and Kuenen, 1957, 1958). The basic condition for these effects is the particular patterns of current speed variations over the tidal flat. During a tidal cycle, the rate of water level changes and, according to Eq. (10.3), the current speed also changes. Minimum rates of water level change occur during high and low water periods, and maximum rates appear at the middle of flood or ebb phases. Consequently, only the middle and lower parts of the intertidal zone will experience large tidal currents, and the upper part is associated with weak currents. In such an environment, it will be difficult for the suspended material originated from subtidal areas to permanently stay on the middle and lower parts of the tidal flat, unless the concentration is sufficiently high (Amos, 1995).

The "settling- and scour-lag effects" explain why a suspended particle is carried by currents toward the upper tidal flat, as shown in Fig. 10.6. On the tidal flat, P represents the

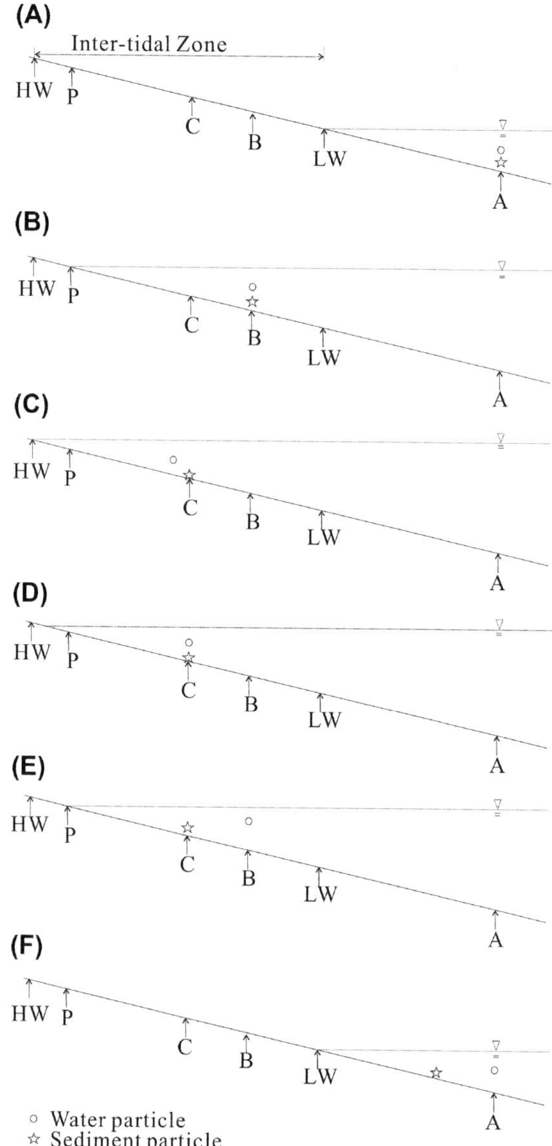

FIGURE 10.6 Diagram showing the settling and scour effects, by comparing the tidal cycle movement of a water parcel and a sediment particle at the stages of (A) low water (representing the beginning of the tidal cycle); (B) water level reaching the point P when the current speed is reduced to the critical value for bed erosion; (C) high water; (D) the water parcel reaching the site C where the sediment particle is settled to the bed; (E) water level reaching the point P; and (F) low water (the end of the tidal cycle).

location above which the tidal current speed never exceeds the critical value for resuspension and below which the speed is below the critical value only during the slack periods. Let us suppose that a water particle and a suspended sediment are located at the same site, A, at the beginning of a flood tide (Fig. 10.6A). During the flood, when the current speed decreases to the threshold at location B (at this time the water level reaches the elevation of site P), the sediment particle starts to settle, but it cannot reach the bed at location B (Fig. 10.6B). It may reach the bed at a site, C, which is further toward the high water mark (Fig. 10.6C). Such an effect is called "settling-lag." During the ebb, when the water particle reaches the location C, its speed is still below the threshold for resuspension because at this time the water is above site P (Fig. 10.6D). Resuspension for the sediment in consideration does not occur until the water particle arrives at B (Fig. 10.6E). At the slack water of the ebb, the positions for the water parcel and sediment particle become different (Fig. 10.6F). The process described in Fig. 10.6D–F is known as "scour-lag." Thus, during each tidal cycle, the net transport is directed toward the high water mark. Although some biological and biogeochemical processes also influence the transport of suspended materials (Neumeier and Amos, 2006), the lag effects represent a basic physical mechanism.

For bedload transport, the time–velocity asymmetry patterns caused by the deformation of tidal waves favor landward transport. Over the tidal flat, shallow water tides (known as "over tides") are generated by seabed friction. As a result, the flood duration becomes shorter and the ebb duration becomes longer on the flat surface (Zhang, 1992). At the same time, the peak current speed during the flood becomes larger than during the ebb. In addition, the transport rate for bedload has a nonlinear relationship with the current speed (Hardisty, 1983; Wang and Gao, 2001). Thus, it has been shown mathematically that the transport capacity during the flood is greater than during the ebb (Zhu and Gao, 1985). This analysis indicates that if sufficient bedload is available, then net transport will be directed to landward during a tidal cycle. However, such a pattern may not be observed if the sites for measurements are close to tidal creek systems (Wang et al., 2012; cf. Section 4.4).

The landward transport of bedload is important for the maintenance of accretion of tidal flats. Otherwise, the accumulation of fine-grained materials over the mudflat will lead to a narrower intertidal zone and reduce the average bed slope. This will, in turn, reduce the tidal current speed according to Eq. (10.3). If the tidal dominance disappears, then development of tidal flats cannot continue. Thus, it is crucial for sands to be deposited over the lower parts of the tidal flat; only in this way can the small bed slope of tidal flats be maintained.

On tidal flats, especially on the middle parts, tidal surges are often present Fig. 10.7. During the flood, the bed at this location will experience large tidal currents (Gao and Zhu, 1988). The water depth is so small that the near-bed shear stress becomes extremely high, resulting breaking at the water mass front to form the surge. According to sediment dynamic analysis, the surge height can be expressed as follows (Gao, 2010):

$$H_s = 4z_0 \tag{10.6}$$

where H_s is the surge height and z_0 is the bed roughness length. The surge is associated with an extremely high suspended sediment concentration, but soon after the passing through of

FIGURE 10.7 Tidal surges occurring on the middle parts of the tidal flat during the flood phase; the turbulent front has a height of around 5 cm, with a tidal current speed of 0.5–1.0 m s^{-1}. *Photograph taken on June 29, 2003. Courtesy of Professor Ya Ping Wang.*

the surge the concentration is dropped rapidly (Li et al., 2006). Hence, it does not appear that tidal surges will lead to intense sediment transport.

4.3 Long-Term Accretion–Erosion Cycles

Sediment supply is a necessary condition for the growth or accretion of tidal flats. For accreting tidal flats, the shore-normal profile will advance toward the sea, with a more or less stable morphology (Liu et al., 2011, 2015). It appears that an equilibrium status can be maintained. Such equilibrium may result from a negative feedback mechanism: the advance of the low water mark widens the intertidal zone, which intensifies the tidal action and landward transport of fine-grained sediment, and the accretion near the high water mark reduces the width of the intertidal zone.

When the sediment supply is cut off, or it becomes too small, erosion will occur. The coastline near the old Yellow River Delta, in northern Jiangsu Province, China, is a typical example which indicates the effect of sediment cut off. Before 1855, the sediment discharge of the Yellow River formed a large delta, and tidal flats were well developed. Then, in 1855 the Yellow River shifted its course to discharge into the Bohai Sea in northern China. Since then the shoreline in northern Jiangsu has been retreating. The original tidal flat has been modified in terms of sediment distribution and profile morphology. The width of the intertidal zone has been reduced to less than 2 km; sandy materials and mud pebbles are found near the high water mark. The overall profile shape is approaching to a wave-dominated beach profile (Gao and Zhu, 1988).

Such responses can be explained by sediment dynamics. Because the supply is reduced, the landward transport capacity due to tidal currents cannot be satisfied. However, at this time, the transport capacity during the ebb remains (i.e., the tidal flat surface is now transformed from a sediment sink into a sediment source). The removal of sediment from the lower part of the flat results in reduction in bed slope and in the strength of tidal currents on the flat. Eventually wave action becomes a dominant factor. Wave breaking takes place on the shore face, and the fine-grained material is transported toward deeper waters, just as observed on a sandy beach. During shoreline recession, if the sediment strata contain shells and other coarse-grained materials, then this debris may form cheniers (Augustinus, 1989; Wang and Ke, 1989). Thus, the presence of cheniers in tidal flats is indicative of coastal erosion periods. Often such cycles are related to sediment supply changes.

Even if the sediment supply is maintained, there may still be a limit of growth for the tidal flat sedimentary system. The growth of a river delta may represent an analogue: in response to reduction in the Sediment Retention Index (SRI, the ratio of the sediment permanently retained in the system to the total sediment input provided by fluvial, marine, atmospheric, and sources) when the delta progradates toward deeper water areas, the rate of delta growth will decrease (Gao, 2007). It is likely that similar processes are associated with the tidal flats, that is, they have a limited space for its development and after their growth over the Holocene period they may be already approaching the growth limit.

4.4 Tidal Creek Systems

Tidal creeks are a secondary morphology superimposed on the tidal flat. They have a different function compared with the flat surface. During the flood there exist several velocity maxima in a creek, in response to rapid enlargement of the inundated area on the flat adjacent to the creek, and during the ebb extra water enters the creek from the surface or through seepage from the bed (Bayliss-Smith et al., 1979; Pethick, 1980; Wang et al., 1999). As a result, water balance in the creek can be asymmetric: the water discharge during the ebb is larger than during the flood. Observations demonstrate that net sediment transport in creeks tends to be seaward, especially during spring tides (Yang et al., 2003). This pattern may be further enhanced when storm occurs. Therefore, it appears that tidal creeks reduce the overall accretion rate of the tidal flat.

Each tidal creek system occupies a "drainage basin" on the tidal flat (Zhang, 1992). In macrotidal environments, tidal creeks may migrate laterally intensively, forming various sizes of point bars. On the sand flat, the creek channel migration is affected by bedload transport in the longshore direction, as observed on the Jiangsu coast (Wang et al., 2006). However, the migration of the creek is confined with the drainage basin. Generally, within the drainage basin the stability or equilibrium status may be indicated by a constant ratio of total drainage area to total network length (Wu and Gao, 2012). An investigation from the Changjiang River estuary shows that such a ratio indeed exists for two different evolution patterns: the "concentration pattern" is characterized by a single, strong-developed network with larger drainage area, coupled with stronger hydrodynamic conditions, whereas the "scattering pattern" is associated with several, small-scale networks and smaller drainage area (Wu and Gao, 2012).

Both patterns have a similar channelization degree. In terms of spatial scales, there is a power law relationship between the tidal channel area and the tidal basin area (Yu et al., 2014):

$$A_c = KA_b^{1.5} \tag{10.7}$$

where A_c is tidal channel area and A_b is tidal basin area; the coefficient K is a constant when equilibrium is reached. This relationship implies that the influence of the tidal creeks on the tidal flat morphology will be different if the magnitude of the tidal flat differs.

4.5 Biological Processes

The importance of biological processes in shaping the tidal flat morphology and sedimentation has been known for a long time (e.g., Reineck and Singh, 1980). Benthic communities influence the bed roughness, salt marsh vegetation modifies the flow structure, and the original sedimentary structure is destructed by bioturbation. Apparently, the accretion rate on a salt marsh surface differs significantly from a bare flat (see Section 3.2). Between a tidal flat without vegetation and one with vegetation there is a system regime shift: the vegetation retards the tidal flow to enhance the vertical settling flux of suspended sediment, and the organic particles from the vegetation contribute to the deposit. Similarly, an environment associated with different marsh plants will have a different sedimentation pattern (Fei et al., 2014). On the Jiangsu coast, eastern China, because the artificially introduced *Spartina alterniflora* replaced the native marsh plants, the sedimentation rate is enhanced (Gao et al., 2014).

To understand such a regime shift, it would be necessary to analyze the interaction between the physical–chemical processes and the biological processes. There are basically two ways of such analyses. The first is to consider the feedbacks between the two types of processes (D'Alpaos et al., 2012; Da Lio et al., 2013; D'Alpaos and Marani, 2016). It should be noted that in this approach the effect of biological processes is expressed as a function of physical variables (Gao and Collins, 2014). The presence of marsh plants modifies the hydrodynamics and suspended sediment concentration variations, thus enabling the establishment of the interrelationships between the vegetation characteristics and the sedimentation patterns (Davidson-Arnott et al., 2002; D'Alpaos et al., 2011). The second approach is to quantify the biological processes independently from the physical processes. Although the presence can be translated into physical conditions that shape the geomorphology, the characteristics of biota cannot be explained. As pointed by Stallins (2006), biogeomorphic systems are "path dependent." Thus, there is a need to consider the biological processes independently from physical processes. However, so far there have been few efforts to quantify the biological processes in the form of "governing equations."

5. SUMMARY

The basic sedimentological and geomorphological characteristics of tidal flats may be summarized as follows:

1. Tidal flats are formed under the condition that tides dominate over other hydrodynamic processes. They have a significant pattern of zonation: from the supratidal zone to subtidal zone salt marshes, mudflats, mixed sand–mud flats and sand flats are distributed.

Such a pattern may be modified by other factors such as tidal creek formation and sediment supply. In a sediment core from an upper part of the tidal flat, the zonation is reflected by a "fining-upward" sequence.

2. Settling- and scour-lag effects are responsible for the transport and accumulation of muddy sediments over the upper parts of tidal flats, whereas the deformation of tidal waves causes landward transport of sandy materials, with the upper limit of sand accumulation being controlled by the tidal current speed during the flood. In addition to these physical mechanisms, biological effects on the suspended sediment transport are also important.
3. Some characteristics and/or parameters of tidal flats (e.g., the bed slope gradient, sedimentary structures, salt marsh cliffs, scour over the mixed sand–mud flat, the thickness of the mud layer, the tidal creeks, and the sediment retention index) contain important information on the system behavior and evolution of tidal flats. Such information may be obtained by means of sediment dynamic analysis. Furthermore, in the study of biogeomorphology, there is a need to quantify the biological/anthropogenic processes which are independent from physical processes.

Acknowledgments

This study has been supported by the SCOR-LOICZ-IAPSO Working Group 122. Financial support is also provided by the Natural Science Foundation of China (Grant Number 40476041), the Ministry of Science and Technology of China (Grant number 2006CB708410), and the Science-Technology Administration of Jiangsu Province (Grant number BK2011012). Mr. Niu Zhan-sheng is thanked for his help with the preparation of some of the figures, and Dr. Ya Ping Wang is thanked for providing the in situ photographs for Figs. 10.4C–E. Dr. Gerardo M.E. Perillo provided the definition of sediment retention index. The author wishes to thank the reviewers (Professor Carl Amos, Dr. Morton Pejrup, and Dr. Gerardo M.E. Perillo) for their constructive comments on an early version of the manuscript.

References

Alexander, C.R., Nittrouer, C.A., Demaster, D.J., Park, Y.A., Park, S.C., 1991. Macro-tidal mudflats of the southwestern Korean coast: a model for interpretation of inter-tidal deposits. Journal of Sedimentary Petrology 61, 805–824.

Allen, J.R.L., 1965. Coastal geomorphology of eastern Nigeria: beach-ridge barrier islands and vegetated tidal flats. Geologie en Mijnbouw 44 (1), 1–21.

Allen, J.R.L., 1989. Salt-marsh growth and stratification: a numerical model with special reference to the Severn Estuary, southwest Britain. Marine Geology 95, 77–96.

Allen, J.R.L., 2000. Morphodynamics of Holocene salt marshes: a review sketch from the Atlantic and southern North Sea coasts of Europe. Quaternary Science Reviews 19, 1155–1231.

Allen, J.R.L., 2003. An eclectic morphostraphic model for the sedimentary response to Holocene sea-level rise in northwest Europe. Sedimentary Geology 161, 31–54.

Allen, J.R.L., Duffy, M.J., 1998. Medium-term sedimentation on high intertidal mudflats and salt marshes in the Severn Estuary, SW Britain: the role of wind and tide. Marine Geology 150, 1–27.

Amos, C.L., 1995. Siliciclastic tidal flats. In: Perillo, G.M.E. (Ed.), Geomorphology and Sedimentology of Estuaries. Elsevier, Amsterdam, pp. 273–306.

Amos, C.L., Collins, M.B., 1978. The combined effects of wave motion and tidal currents on the morphology of intertidal ripple marks: the Wash, UK. Journal of Sedimentary Petrology 48, 849–856.

Amos, C.L., Mosher, D.C., 1985. Erosion and deposition of fine-grained sediments from the Bay of Fundy. Sedimentology 32, 815–832.

Andersen, T.J., Pejrup, M., 2001. Suspended sediment transport on a temperate, microtidal mudflat, the Danish Wadden Sea. Marine Geology 173, 69–85.

Anderson, F.E., 1973. Observations of some sedimentary processes acting on a tidal flat. Marine Geology 14, 101–116.

Anthony, E.J., Dolique, F., Gardel, A., Gratiot, N., Proisy, C., Polidori, L., 2008. Nearshore intertidal topography and topographic-forcing mechanisms of an Amazon-derived mud bank in French Guiana. Continental Shelf Research 28, 813–822.

Archer, A.W., 1995. Modeling of cyclic tidal rhythmites based on a range of diurnal to semidiurnal tidal-station data. Marine Geology 123, 1–10.

Augustinus, P.G.E.F., 1989. Cheniers and chenier plains: a general introduction. Marine Geology 90, 219–229.

Bartholdy, J., Madsen, P.P., 1985. Accumulation of fine-grained material in a Danish tidal area. Marine Geology 67, 121–137.

Bayliss-Smith, T.P., Healey, R., Lailey, R., Spencer, T., Stoddart, D.R., 1979. Tidal flows in salt marsh creeks. Estuarine and Coastal Marine Science 9, 235–255.

Blanton, J.O., Lin, G., Elston, S.A., 2002. Tidal current asymmetry in shallow estuaries and tidal creeks. Continental Shelf Research 22 (11), 1731–1743.

Christiansen, C., Vølund, G., Lund-Hansen, L.C., Bartholdy, J., 2006. Wind influence on tidal flat sediment dynamics: field investigations in the Ho Bugt, Danish Wadden Sea. Marine Geology 235 (1), 75–86.

Collins, M.B., Amos, C.L., Evans, G., 1981. Observation of some sediment-transport processes over intertidal flat. IAS Special Publications 5, 81–98.

Collins, M.B., Ke, X.K., Gao, S., 1998. Tidally-induced flow structure over sandy intertidal flats. Estuarine, Coastal and Shelf Science 46, 233–250.

Cundy, A.B., Croudace, I.W., 1995. Sedimentary and geochemical variations in a salt marsh/mud flat environment from the mesotidal Hamble estuary, southern England. Marine Chemistry 51, 115–132.

Da Lio, C., D'Alpaos, A., Marani, M., 2013. The secret gardener: vegetation and the emergence of biogeomorphic patterns in tidal environments. Philosophical Transactions of the Royal Society of London A: Mathematical, Physical and Engineering Sciences 371 (2004), 20120367.

Daborn, G.R., Amos, C.L., Brylinsky, M., Christian, H., Drapeau, G., Grant, J., Long, B., Paterson, D., Perillo, G.M.E., Piccolo, M.C., 1993. An ecological "cascade" effect: migratory birds affect stability of intertidal sediments. Limnology and Oceanography 38, 225–231.

D'Alpaos, A., Da Lio, C., Marani, M., 2012. Biogeomorphology of tidal landforms: physical and biological processes shaping the tidal landscape. Ecohydrology 5 (5), 550–562.

D'alpaos, A., Mudd, S.M., Carniello, L., 2011. Dynamic response of marshes to perturbations in suspended sediment concentrations and rates of relative sea level rise. Journal of Geophysical Research: Earth Surface 116 (F4).

D'Alpaos, A., Marani, M., 2016. Reading the signatures of biologic–geomorphic feedbacks in salt-marsh landscapes. Advances in Water Resources 93, 265–275.

Dalrymple, R.W., Knight, R.J., Zaitlin, B.A., Middleton, G.V., 1990. Dynamics and facies of macrotidal sand-bar complex, Cobequid Bay–Salmon River Estuary (Bay of Fundy). Sedimentology 37, 577–612.

Davidson-Arnott, R.G.D., van Proosdij, D., Ollerhead, J., Schostak, L., 2002. Hydrodynamics and sedimentation in salt-marshes: examples from a macrotidal marsh, Bay of Fundy. Geomorphology 48 (1), 209–231.

Davies, J.L., 1964. A morphogenic approach to world shoreline. Zeitschrift für Geomorphologie 8, 27–42.

Dellwig, O., Hinrichs, J., Hild, A., Brumsack, H.-J., 2000. Changing sedimentation in tidal flat sediments of the southern North Sea from the Holocene to the present: a geochemical approach. Journal of Sea Research 44, 195–208.

Deloffre, J., Lafite, R., Lesueur, P., Lesourd, S., Verney, R., Gu-ézennec, L., 2005. Sedimentary processes on an intertidal mudflat in the upper macrotidal Seine estuary, France. Estuarine, Coastal and Shelf Science 64, 710–720.

Deloffre, J., Verney, R., Lafite, R., Lesueur, P., Lesourd, S., Cundy, A.B., 2007. Sedimentation on intertidal flats in the lower part of macrotidal estuaries: sedimentation rhythms and their preservation. Marine Geology 241, 19–32.

Eisma, D., 1998. Intertidal Deposits: River Mouths, Tidal Flats, and Coastal Lagoons. CRC Press, Boca Raton, 525 p.

Evans, G., 1965. Intertidal flat sediments and their environments of deposition in the Wash. Quarterly Journal of the Geological Society 121, 209–245.

Evans, G., Collins, B.M., 1975. The transportation and deposition of suspended sediment over the intertidal flats of the Wash. In: Hails, J., Carr, A. (Eds.), Nearshore Sediment Dynamics and Sedimentation. John Wiley & Sons, Chichester, pp. 273–306.

Fagherazzi, S., Furbish, D.J., 2001. On the shape and widening of salt marsh creeks. Journal of Geophysical Research: Oceans 106 (C1), 991–1003.

Falcão, M., Caetano, M., Serpa, D., Gaspar, M., Vale, C., 2006. Effects of infauna harvesting on tidal flats of a coastal lagoon (Ria Formosa, Portugal): implications on phosphorus dynamics. Marine Environmental Research 61, 136–148.

Fan, D., 2012. Open-coast tidal flats. In: Principles of Tidal Sedimentology. Springer, Netherlands, pp. 187–229.

Fan, D.D., Guo, Y., Wng, P., Shi, Z., 2006. Cross-shore variations in morphodynamic processes of an open-coast mudflat in the Changjiang Delta, China: with an emphasis on storm impacts. Continental Shelf Research 26, 517–538.

Fan, D.D., Li, C.X., Archer, A.W., Wang, P., 2002. Temporal distribution of diastems in deposits of an open-coast tidal flat with high suspended sediment concentrations. Sedimentary Geology 152, 173–181.

Fei, S., Phillips, J., Shouse, M., 2014. Biogeomorphic impacts of invasive species. Annual Review of Ecology, Evolution, and Systematics 45 (1), 69–87.

French, J.R., 1993. Numerical simulation of vertical marsh growth and adjustment to accelerated sea-level rise, north Norfolk, UK. Earth Surface Processes and Landforms 18, 63–81.

Friedrichs, C.T., 2011. Tidal Flat Morphodynamics: A Synthesis. Treatise on Estuarine and Coastal Science. Academic Press, Waltham, pp. 137–170.

Gao, S., 2007. Modeling the growth limit of the Changjiang delta. Geomorphology 85, 225–236.

Gao, S., 2009. Modeling the preservation potential of tidal flat sedimentary records, Jiangsu coast, eastern China. Continental Shelf Research 29 (16), 1927–1936.

Gao, S., 2010. Extremely shallow water benthic boundary layer processes and the resultant sedimentological and geomorphological characteristics. Acta Sedimentologica Sinica 28 (5), 926–932 (in Chinese with English abstract).

Gao, S., Collins, M., 1994. Tidal inlet stability in response to hydrodynamic and sediment dynamic conditions. Coastal Engineering 23, 61–80.

Gao, S., Collins, M., 1997. Formation of salt-marsh cliffs in an accretional environment, Christchurch Harbour, southern England. In: Wang, P.X., Bergran, W. (Eds.), Proceedings of the 30th International Geological Congress (Vol. 13: Marine Geology and Palaeoceanography). VSP Press, Amsterdam, pp. 95–110.

Gao, S., Collins, M.B., 2014. Holocene sedimentary systems on continental shelves. Marine Geology 352, 268–294.

Gao, S., Du, Y.F., Xie, W.J., Gao, W.H., Wang, D.D., Wu, X.D., 2014. Environment-ecosystem dynamic processes of Spartina alterniflora salt-marshes along the eastern China coastlines. Science China (Earth Sciences) 57 (11), 2567–2586.

Gao, S., Zhu, D.K., 1988. The profile of Jiangsu's mud coast. Journal of Nanjing Forestry University (Natural Sciences Edition) 21 (1), 75–84 (in Chinese, with English abstract).

Gerdes, G., Petzelberger, B.E.M., Scholz-Bottcher, B.M., Streif, H., 2003. The record of climatic change in the geological archives of shallow marine, coastal, and adjacent lowland aareas of northern Germany. Quaternary Science Reviews 22, 101–124.

Ginsberg, R.N. (Ed.), 1975. Tidal Deposits. Springer-Verlag, New York, 428 p.

Gouleau, D., Jouanneau, J.M., Weber, O., Sauriau, P.G., 2000. Short- and long-term sedimentation on Montportail-Brouage intertidal mudflat, Marennes-Oleron Bay (France). Continental Shelf Research 20, 1513–1530.

Haentzschel, W., 1939. Tidal flat deposits. In: Trask, P.D. (Ed.), Recent Marine Sediments. AAPG, Tulsa, pp. 195–206.

Hardisty, J., 1983. An assessment and calibration of formulations for Bagnold's bedload equation. Journal of Sedimentary Petrology 53, 1007–1010.

Isla, F.I., Vilas, F.E., Bujalesky, G.G., Ferrero, M., Bonorino, G., Miralles, A.A., 1991. Gravel drift and wind effect on the macrotidal San Sebastian Bay, Tierra del Fuego, Argentina. Marine Geology 97, 211–224.

King, C.A.M., 1972. Beaches and Coasts, second ed. Edward Arnold, London. 570 p.

Kjerfve, B., Perillo, G.M.E., Gardner, L.R., Rine, J.M., Dias, G.T.M., Mochel, F.R., 2002. Morphodynamics of muddy environments along the Atlantic coasts of North and South America. In: Healy, T., Wang, Y., Healy, J.A. (Eds.), Muddy Coasts of the World: Processes, Deposits and Functions. Elsevier, Amsterdam, pp. 479–532.

Klein, G. de V., 1985. Intertidal flats and intertidal sand bodies. In: Davis, R.A. (Ed.), Coastal Sedimentary Environments, second ed. Springer-Verlag, New York, pp. 187–224.

Letzsch, W.S., Frey, R.W., 1980. Deposition and erosion in a Holocene salt marsh, Sapelo Island Georgia. Journal of Sedimentary Petrology 50, 529–542.

Li, A.C., Chen, L.R., Wang, P.G., 1997. An atmospheric aerosol transport event over the Qingdao region and the material source. Chinese Science Bulletin 43, 62–65.

Li, Z.H., Gao, S., Chen, S.L., Wang, Y.P., 2006. Grain size distribution patterns of suspended sediment in response to hydrodynamics on the Dafeng intertidal flats, Jiangsu, China. Acta Oceanologica Sinica 25 (6), 63–77.

Lim, D.I., Park, Y.A., 2003. Late quaternary stratigraphy and evolution of a Korean tidal flat, Haenam Bay, southeastern Yellow Sea, Korea. Marine Geology 193, 177–194.

Linke, O., 1939. Die biota des Jadebusens. Helgoländer Wissenschaftliche Meeresuntersuchungen 1, 201–348.

Liu, X.J., Gao, S., Wang, Y.P., 2011. Modeling profile shape evolution for accreting tidal flats composed of mud and sand: a case study of the central Jiangsu coast, China. Continental Shelf Research 31 (16), 1750–1760.

Liu, X.J., Gao, S., Wang, Y.P., 2015. Modeling the deposition system evolution of accreting tidal flats: a case study from the coastal plain of central Jiangsu, China. Journal of Coastal Research 31 (1), 107–118.

Malvarez, G., Navas, F., Jackson, D.W.T., 2004. Investigations on the morphodynamics of sandy tidal flats: a modeling application. Coastal Engineering 51, 731–747.

Marani, M., Belluco, E., D'Alpaos, A., Defina, A., Lanzoni, S., Rinaldo, A., 2003. On the drainage density of tidal networks. Water Resources Research 39 (2).

Netto, S.A., Lana, P.C., 1997. Influence of *Spartina alterniflora* on superficial sediment characteristics of tidal flats in Paranagua Bay (South-Eastern Brazil). Estuarine, Coastal and Shelf Science 44, 641–648.

Neumeier, U., Amos, C., 2006. The influence of vegetation on turbulence and flow velocities in European saltmarshes. Sedimentology 53, 259–277.

Pejrup, M., 1988. Suspended sediment transport across a tidal flat. Marine Geology 82, 187–198.

Pejrup, M., Larsen, M., Edelvang, K., 1997. A fine-grained sediment budget for the Sylt-Romo tidal basin. Helgoländer Meeresuntersuchungen 51, 253–268.

Perillo, G.M.E., Drapeau, G., Piccolo, M.C., Chaouq, N., 1993. Tidal circulation pattern on a tidal flat, Minas Basin, Canada. Marine Geology 112, 219–236.

Perillo, G.M.E., Piccolo, M.C., 1999. Geomorphologic and physical characteristics of the Bahía Blanca Estuary. Argentina. In: Perillo, G.M.E., Piccolo, M.C., Pino Quivira, M. (Eds.), Estuaries of South America: Their Geomorphology and Dynamics. Springer-Verlag, Berlín, pp. 195–216.

Perillo, G.M.E., Ripley, M.D., Piccolo, M.C., Dyer, K.R., 1996. The formation of tidal creeks in a salt marsh: new evidence from the Loyola Bay Salt Marsh, Rio Gallegos Estuary, Argentina. Mangroves and Salt Marshes 1, 37–46.

Pethick, J.S., 1980. Velocity surges and asymmetry in tidal channels. Estuarine and Coastal Marine Science 11, 331–345.

Pethick, J.S., 1981. Long-term accretion rates on tidal salt marshes. Journal of Sedimentary Petrology 51, 571–577.

Postma, H., 1954. Hydrography of the Dutch Wadden Sea. Archives Néerlandaises de Zoologie 10, 405–511.

Postma, H., 1961. Transport and accumulation of suspended matter in the Dutch Wadden Sea. Netherlands Journal of Sea Research 1, 148–190.

Pritchard, D., Hogg, A.J., Roberts, W., 2002. Morphological modeling of intertidal mudflats: the role of cross-shore tidal currents. Continental Shelf Research 22, 1887–1895.

Proske, U., Hanebuth, T.J.J., Meggers, H., Leroy, S.A.G., 2008. Tidal flat sedimentation during the last millennium in the northern area of Tidra Island, Banc d'Arguin, Mauritania. Journal of African Earth Sciences 50, 37–48.

Quaresma, V.da S., Bastos, A.C., Amos, C.L., 2007. Sedimentary processes over an intertidal flat: a field investigation at Hythe flats, Southampton Water (UK). Marine Geology 241, 117–136.

Reineck, H.E., 1972. Tidal flats. In: Rigby, J.K., Hamblin, W.K. (Eds.), Recognition of Ancient Sedimentary Environments. SEPM, Tulsa, pp. 146–159. SEPM Special Publication 16.

Reineck, H.-E., Singh, I.B., 1980. Depositional Sedimentary Environments, second ed. Springer- Verlag, Berlin. 549 p.

Ren, M.E., 1986. Tidal mud flat. In: Ren, M.E. (Ed.), Modern Sedimentation in the Coastal and Nearshore Zones of China. China Ocean Press, Beijing, pp. 78–127.

Ren, M.E., Zhang, R.S., Yang, J.H., 1985. Effect of typhoon no. 8114 on coastal morphology and sedimentation of Jiangsu Province, People's Republic of China. Journal of Coastal Research 1, 21–28.

Roberts, W., Le Hir, P., Whitehouse, R.J.S., 2000. Investigation using simple mathematical models of the effect of tidal currents and waves on the profile shape of intertidal mudflats. Continental Shelf Research 20, 1079–1097.

Sakamaki, T., Nishimura, O., 2007. Physical control of sediment carbon content in an estuarine tidal flat system (Nanakita River, Japan): a mechanistic case study. Estuarine, Coastal and Shelf Science 73, 781–791.

Schwimmer, R.A., 2001. Rates and processes of marsh shoreline erosion in Rehoboth Bay, Delaware, USA. Journal of Coastal Research 17, 672–683.

Stallins, J.A., 2006. Geomorphology and ecology: unifying themes for complex systems in biogeomorphology. Geomorphology 77 (3), 207–216.

Stumpf, R.P., 1983. The process of sedimentation on the surface of a salt marsh. Estuarine, Coastal and Shelf Science 17, 495–508.

Talke, S.A., Stacey, M.T., 2008. Suspended sediment fluxes at an intertidal flat: the shifting influence of wave, wind, tidal, and freshwater forcing. Continental Shelf Research 28, 710–725.

Temmerman, S., Govers, G., Meire, P., Wattel, S., 2004. Simulating the long-term development of levee-basin topography on tidal marshes. Geomorphology 63, 39–55.

van Straaten, L.M.J.U., 1961. Sedimentation in tidal flat areas. Journal of the Alberta Society of Petroleum Geologists 9, 203–226.

van Straaten, L.M.J.U., Kuenen, P.D., 1957. Accumulation of fine-grained sediments in the Dutch Wadden Sea. Geologie en Mijnbouw 19, 329–354.

van Straaten, L.M.J.U., Kuenen, P.D., 1958. Tidal action as a cause of clay accumulation. Journal of Sedimentary Petrology 28, 406–413.

Vos, P.C., van Kesteren, W.P., 2000. The long-term evolution of intertidal mudflats in the northern Netherlands during the Holocene: natural and anthropogenic processes. Continental Shelf Research 20, 1687–1710.

Wang, Y., 1983. The mudflat system of China. Canadian Journal of Fisheries and Aquatic Sciences 40 (Suppl. 1), 160–171.

Wang, Y.P., Gao, S., 2001. Modification to the Hardisty equation, regarding the relationship between sediment transport rate and grain size. Journal of Sedimentary Research 71, 118–121.

Wang, Y., Ke, X.K., 1989. Cheniers on the eastern coastal plain of China. Marine Geology 90, 321–335.

Wang, Y.P., Gao, S., Jia, J., Thompson, C.E., Gao, J., Yang, Y., 2012. Sediment transport over an accretional intertidal flat with influences of reclamation, Jiangsu coast, China. Marine Geology 291, 147–161.

Wang, Y.P., Gao, S., Jia, J.J., 2006. High-resolution data collection for analysis of sediment dynamic processes associated with combined current-wave action over intertidal flats. Chinese Science Bulletin 51, 866–877.

Wang, Y.P., Zhang, R.S., Gao, S., 1999. Velocity variations in salt marsh creeks, Jiangsu, China. Journal of Coastal Research 15, 471–477.

Wang, Y.W., Yu, Q., Gao, S., Flemming, B., 2014. Modeling the effect of progressive grain-size sorting on the scale dependence of back-barrier tidal basin morphology. Continental Shelf Research 91, 26–36.

Weimer, R.J., Howard, J.D., Lindsay, D.R., 1982. Tidal flats and associated tidal channels. Sandstone depositional environments. American Association of Petroleum Geologists, Memoir 31, 191–245.

Wells, J.T., Adams, C.E., Park, Y.A., Frankenberg, E.W., 1990. Morphology, sedimentology and tidal channel processes on a high-tide-range mudflat, west coast of South Korea. Marine Geology 95, 111–130.

Wells, J.T., Coleman, J.M., 1981. Periodic mudflat progradation, northeastern coast of South America: a hypothesis. Journal of Sedimentary Petrology 51, 1069–1075.

Wu, X.D., Gao, S., 2012. A morphological analysis of tidal creek network patterns on the Jiuduansha Shoal, Changjiang estuary. Acta Oceanologica Sinica (Chinese Version) 34 (6), 126–132.

Yang, B.C., Dalrymple, R.W., Chun, S.S., 2005. Sedimentation on a wave-dominated, open-coast tidal flat, southwestern Korea: summer tidal flat – winter shoreface. Sedimentology 52, 235–252.

Yang, Y., Wang, Y.P., Gao, S., 2003. Hydrodynamic processes and suspended sediment transport in salt-marsh creeks, Wanggang, Jiangsu, China. In: Choi, B.H., Park, Y.A. (Eds.), Sedimentation in the Yellow Sea Coasts. Hanrimwon Publishing Co., Seoul, pp. 79–87.

Yu, Q., Wang, Y.W., Flemming, B., Gao, S., 2012. Modeling the equilibrium hypsometry of back-barrier tidal flats in the German Wadden Sea (southern North Sea). Continental Shelf Research 49, 90–99.

Yu, Q., Wang, Y.W., Flemming, B., Gao, S., 2014. Scale-dependent characteristics of equilibrium morphology of tidal basins along the Dutch-German North Sea coast. Marine Geology 348, 63–72.

Zhang, R.S., 1992. Suspended sediment transport processes on tidal mud flat in Jiangsu Province, China. Estuarine, Coastal and Shelf Science 35, 225–233.

Zhang, R.S., Shen, Y.M., Lu, L.Y., Yan, S.G., Wang, Y.H., Li, J.L., Zhang, Z.L., 2004. Formation of *Spartina alterniflora* salt marshes on the coast ofJiangsu Province, China. Ecological Engineering 23, 95–105.

Zhu, D.K., Gao, S., 1985. A mathematical model for the geomorphic evolution and sedimentation of tidal flats. Marine Science Bulletin 4 (5), 15–21 (in Chinese with English abstract).

Zhu, D.K., Xu, T.G., 1982. Wetland evolution and management issues of the central Jiangsu coast. Journal of Nanjing Forestry University (Natural Sciences Edition) 18 (3), 799–818 (in Chinese with English abstract).

CHAPTER

11

Intertidal Flats: Form and Function

David M. Paterson[1], *Irene Fortune*[1], *Rebecca J. Aspden*[1], *Kevin S. Black*[2]

[1]Scottish Oceans Institute, School of Biology, University of St Andrews, Fife, Scotland; [2]Partrac, Glasgow, Scotland

1. INTRODUCTION

Ecology is a complex science that implies a strong understanding of the interaction between the physical and biological components of a habitat. The conceptual understanding that the delivery of ecosystem services and the benefits derived from any specific habitat is dependent on the "health" and biodiversity of a system is manifest in the widespread adoption of the "ecosystem approach" in legislation and management (e.g., EU Marine Strategy Framework Directive). However, this conceptual realization is often unsupported by the status of knowledge needed to implement relevant management practice. In addition, the natural variability of systems is enhanced by external drivers, such as the effects of anthropogenic climate change, which presents a further challenge to the ecosystem approach. In the next decades, the impacts of climate change will become clearer and the requirement to understand and manage this change will become a priority. Such changes will be more manifest in some ecosystems than others and the coastal zone is likely to show effects more rapidly than many. Unvegetated coastal depositional habitats, while important, have often been less considered than more charismatic vegetated systems such as salt marshes or sea grasses. We emphasize the importance of these less charismatic habitats and the services they provide.

1.1 Intertidal Systems

Intertidal soft sediments are important dynamic systems and sites of significant primary productivity, whereby autotrophic species convert light energy into biomass, absorbing carbon dioxide and releasing oxygen. The energy from primary productivity is usually very rapidly transferred through the food chain to macrofauna, meiofauna, and bacteria (Middelburg et al., 2000; Miyatake et al., 2014). This cycling of carbon is central to climate regulation. The majority of the terrestrial carbon pool is provided by wetlands, and tidal

Coastal Wetlands
https://doi.org/10.1016/B978-0-444-63893-9.00011-3

saline wetlands are estimated to store at least 44.6 Tg C year^{-1} (Chmura et al., 2003). Coastal wetlands provide further vital ecosystem services such food provision, waste remediation, and erosion control. However, coastal wetlands around the world are under increasing pressure from human impacts and are exposed to both land-based activities, such as dam building and nutrient and pollutant runoff, and hydrological processes, such as changing current patterns and storm events. Understanding the function of these key habitats is vital for sustainable development, where an ecosystem approach will allow the integration of ecological conservation with benefits to society.

1.2 Ecological Terminology

Early ecological research was largely descriptive of species and habitats, often concerned with changes in plant communities over time. This work formed the classical study of the successional progression of assemblages toward a supposed vegetative climax. The abiotic component of the habitat was considered to provide the overarching framework within which species might compete and fail or be successful, leading to Clements' original monoclimax theory of successional change (Townsend et al., 2008). The physical environment was therefore the stage on which the biotic actors played their parts. This paradigm is now rejected as too simplistic because it is now recognized that the organisms inhabiting an ecosystem have a range of effects on the physical structure and dynamics of the system and contribute as architects of their own habitat (Hansell, 2005; Malarkey et al., 2015). For example, on intertidal flats, benthic organisms commonly restructure and process the material of their surroundings in a process known as bioturbation (Reise, 2002). Where an organism is considered to have a significant impact on the surrounding habitat it can be classified as an "ecosystem engineer" (Jones et al., 1994) (Table 11.1). However, it might be considered that all organisms have some effect on their surroundings, and that the characterization as an ecosystem engineer has more to do with scale, magnitude, and our perception of effects than a clear classification. Further to the concept of ecosystem engineering is the theory of niche construction (Laland et al., 2004). Niche construction has been used to describe the selective pressures that modification of the environment by the metabolic, physiological, and behavioral activities of species place on the evolution of future generations of similar and different populations. Examples include not only burrowing activity and the excretion of compounds that alter conditions but also biochemical processes that change the redox potential of the bed. Niche construction includes ecosystem engineers (i.e., organisms that change their surroundings by nontrophic methods) and foundation and facultative species (i.e., habitat-creating species; Odling-Smee et al., 2013 and references therein).

1.3 Ecosystem Services

One of the strongest recent intellectual drivers is to understand the processes that occur in an ecosystem that are beneficial, or even essential, to humans (MEA, 2005). In trying to conceptualize these varied ecological dynamics across systems, a new terminology is now commonly applied. The transformations of matter or energy driven by biota within natural ecosystems are termed "ecosystem functions," whereas those functions or groups of functions that are, rather subjectively, deemed to be important to humans are termed "ecosystem services" (Chapin et al., 1997) (Table 11.1). The Millennium Ecosystem Assessment (MEA, 2005)

TABLE 11.1 Terminology Commonly Used in the Ecosystem Function Debate

Term	Definition
Ecosystem process	Any process of transformation that occurs in an ecosystem, theoretically whether measurable or not. This includes all metabolism, catabolism, and dynamic processes such as sediment bioturbation or active resuspension.
Ecosystem function	Largely used as a synonym for "ecosystem process" but often given a more "practical" role: a process that can be measured as an attribute of the system under study. Habitat or its inherent three-dimensional structure (architecture) may itself be regarded by some authors as "functional" such as in the statement "The seagrass meadows provide refuge for juvenile fish."
Ecosystem services	A very anthropocentric term in common usage. Ecosystem services are ecosystem functions that have implicit value to mankind. Common examples are carbon fixation, oxygen generation, and nutrient turnover. However, physical attributes can also be recognized as services such as the greater resilience of some habitats (e.g., mangroves) when faced with extreme conditions (tsunami, etc.) (Alongi, 2008).
Ecosystem engineer	An organism whose activity has a significant impact on its habitat. It has varied definitions under a variety of contexts and is widely used. In fact, most organisms could be argued to act as ecosystem engineers on some scale, but a common example is the lugworm, *Arenicola marina*. *A. marina* has a great impact on sediment turnover as a bioturbator. However, even diatoms or bacteria can be regarded as ecosystem engineers because, in concert, they can produce organic material which can increase the erosion resistance of sediments (Paterson and Black, 1999).
Ecosystem engineering	The activity of an ecosystem engineer.
Niche construction	Recently, a new evolutionary paradigm has been proposed, which suggests that organisms that engineer their environment create a selective pressure on their own future generations and those of other inhabitants of that environment. Thus, there is an evolutionary pressure inherent in ecosystem engineering. "Niche construction" is the name given to this form of evolutionary ecosystem influence (Laland et al., 2004).

Definitions are given with a perspective toward depositional habitats.

defines ecosystem services as "benefits people obtain from ecosystems." These benefits may be derived directly or indirectly (Costanza et al., 1997), for example, food, timber, genetic materials, and may include regulatory or supporting services such as waste remediation (Watson et al., 2016), water purification, oxygen production, and carbon storage. These definitions remain important continually and are being applied more widely (Luisetti et al., 2014; Beaumont et al., 2008). It is also more common to refer to "ecosystem health," a subjective term which encompasses the implicit suggestion that a "healthy" system will deliver more appropriate levels of goods and services to humanity than an unhealthy one. A central question arises, "how many species does an ecosystem need to remain healthy?" This debate is now being advanced on many levels (Davies et al., 2011; McCann, 2007). The authors have provided their own working understanding of these terms in a simplified form (Table 11.1), but there are many other interpretations possible. For example, the term "ecosystem engineer" has broadened with time to include a wider range of organism than latterly acknowledged (Boogert et al., 2006). Furthermore, reading across the field is advisable to fully understand the correct usage and lineage of these terms (Lawton, 1994; Loreau et al., 2002; Worm and Duffy, 2003; Passarelli et al., 2014).

The biotic component of the ecosystem is often reported as some measure of the variety of species that contribute to the functionality, under the general term of "biodiversity" (Magurran, 2004). Thus, the question becomes "How does biodiversity affect ecosystem function?" The answer to this would provide a clear scientific and political message concerning the value of biology to the economy and health of marine ecosystems. Although coastal systems have received less attention in this debate (Duarte et al., 2008), depositional environments have an advantage that sediment systems can be recreated in laboratory mesocosms (Ford et al., 2016; Benton et al., 2007), manipulated in the field (Kenworthy et al., 2016), and their properties and processes relatively easily characterized. The time scale of change and relatively small size of the dominant benthic organisms make this observation and manipulation more logistically amenable than for many terrestrial systems. The biodiversity–ecosystem function debate is therefore an active area of marine research and is shedding further light on the dynamics and functional role of depositional systems (Wohlgemuth et al., 2016; Solan et al., 2006). One approach is to develop mathematical models of system response. Again, there is a clear lag in the application of these modeling approaches to depositional systems and aquatic systems in general (Raffaelli et al., 2005; Hyder et al., 2015). This may be simply because there are fewer researchers concentrating on these areas but also probably because of the inherent complexity and "remoteness" of the activity. The tidal rise and fall makes boundary conditions difficult for modeling studies, and the natural variability of the system is extreme. There is, however, a new urgency in addressing coastal ecology driven by the challenge of managing marine systems under scenarios of the combined impacts of global climate change and human activities (Paterson et al., 2011).

2. ENGINEERING THE INTERTIDAL

Intertidal depositional systems are often considered as physically challenging and stressful environments (Kaiser et al., 2005) that are highly variable in terms of physicochemical conditions such as temperature and salinity. It is this constant variation, combined with the physical forcing, that organisms inhabiting depositional environments must tolerate to be successful (Fig. 11.1).

Fewer species occur in this harsh environment than found in more amenable and structurally complex systems (Kaiser et al., 2005), such as tropical rainforest and meadowlands. In relatively homogenous habitats, available niche space is restricted and fewer species must compete for resources (Ricklefs and Miller, 1999). Productive depositional habitats can only occur where the correct boundary conditions of flow energy exist. Too much energy and the sediments are swept away, too little and the system may stagnate. The extent of the physical forcing controls the type of depositional habitat formed. However, within these boundaries, great opportunities exist for ecosystem engineers to moderate the habitat and increase their own success and fitness. Some extreme forms may even create a depositional habitat where none existed before. The red alga *Audouinella* sp. (Dillwyn) is a perennial rhodophyte characterized by its ability to retain sediment particles, which then become an important structural component of the algal turf. The ability to trap and retain sediments is an excellent example of "ecosystem engineering" due to the mediation of the surrounding environment in a manner that changes the nature of resource availability and therefore fitness (Fig. 11.2).

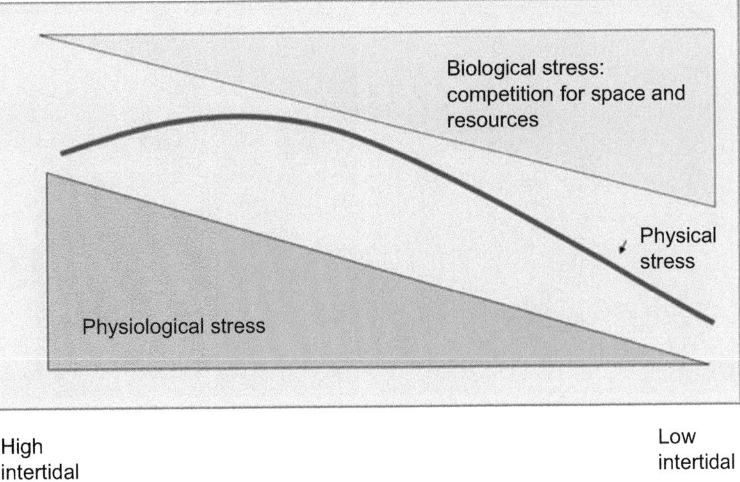

FIGURE 11.1 The balance between biological (competition for food and space) and physiological stress (desiccation) with position on the intertidal shore. Physical (wave) stress tends to peak at mid-shore level.

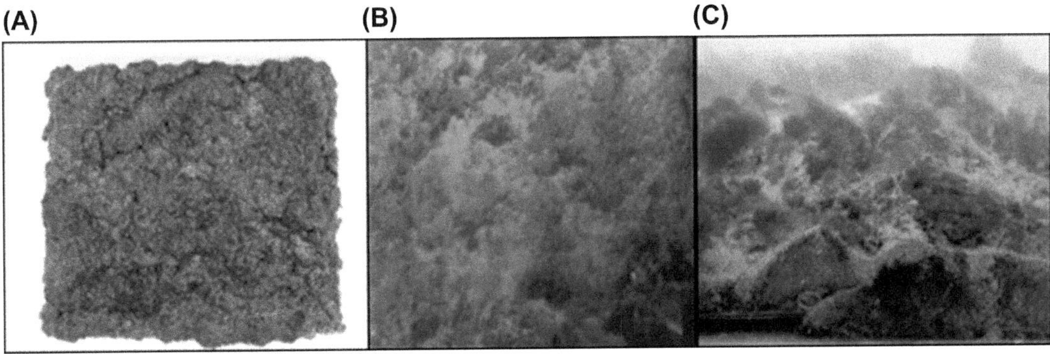

FIGURE 11.2 The red alga *Audouinella sp.* (Dillwyn) forms a matrix over the surface of intertidal boulders, trapping sediments and creating niche space for other organisms. (A) Algal mat. (B) Detail of surface. (C) Mat exposed to flow in a laboratory flume. *Image by the author, RJA.*

Flume studies show the presence of the algae reduces bed load transport and sediment was trapped within the filamentous matrix of the algae, creating a new biogenic "depositional" environment (Aspden, 2005). This is a modern example but the evolution of this mechanism has perhaps the longest traceable record on the surface of the planet (Reitner et al., 2011). Stromatolites are laminated patterns in ancient rock, created by bacterial action and first visible evidence of life on Earth dating from approximately 3.8 billion year bp. Early bacterial assemblages captured and retained sediments, stabilizing gradients at the sediment–water interface creating the diversity of conditions that helped generate diversification. Work with modern stromatolites, living relatives of the early forms, suggests that biogenic stabilization (Fig. 11.3) and niche construction was an important component of their ecology (Paterson et al., 2008). Some authors even suggest that sediment stabilization by

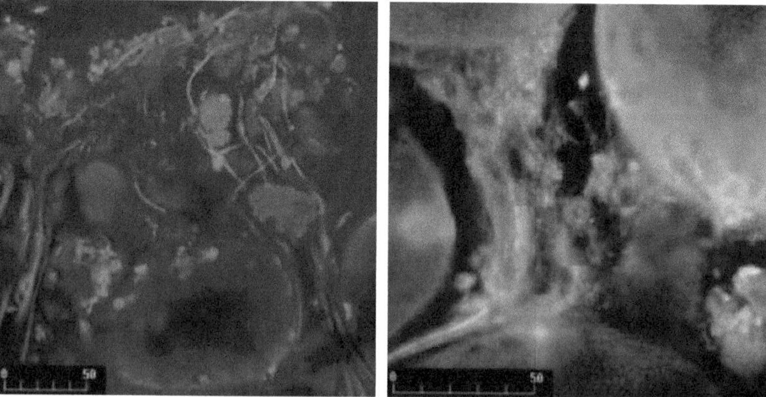

FIGURE 11.3 Confocal laser imagery of the microbial matrix of a living stromatolitic assemblage from the Bahamas. Left: Cyanobacterial filaments fluorescing red. Right: Organic material between stromatolite ooids fluorescing green. *Images courtesy of Prof Alan Decho.*

biological action may mediate the existence of alternate stable states (van de Koppel et al., 2001). This form of habitat mediation may be an adaptation to sediment redistribution where high sedimentation rates and particle movement can have detrimental effects on the diversity and overall richness of intertidal sediment communities (Airoldi and Cinelli, 1997). The relationship between sediment transport and biota is therefore varied and becoming increasingly studied (Black et al., 2002; Orvain et al., 2004 and Oravin, 2005; Aspden et al., 2004a) and modeled (Widdows et al., 2004). However, there is a balance because sediment movement may also be responsible for promoting diversity due to the creation of patches and maintenance of spatial and temporal heterogeneity, within an otherwise be a fairly homogenous environment (Littler et al., 1983; McQuaid and Dower, 1990; Ford et al., 1999).

Another major engineering impact on depositional systems is caused by the movement of organisms through the material, bioturbating the surrounding medium. The intertidal burrowing crab *Chasmagnathus granulatus* is an effective bioturbator and can modify the soft sediments in which it occurs to such a degree that the areas containing populations of *C. granulatus* are more humid, softer, and homogeneous than the areas without *C. granulatus*. As a result of the change in sediment properties, the vertical migrations of infaunal assemblages were observed to be greater in sediments with no *C. granulatus* compared with those with *C. granulatus* (Escapa et al., 2004). The functional role of bioturbation is now being studied extensively and used as an index for predicting the effect of species ecosystem function (Solan et al., 2004).

2.1 The Physical Context

Intertidal mudflats are largely found fringing the shorelines of estuaries although systems are also found on exposed coastal shores (e.g., as in Morecambe Bay, UK). The study of the sedimentology is a study of the distribution of sediments and for intertidal flats can be traced back to the classic work of Evans (1965) in the Wash, UK. In the simplest terms, sediment grains are defined by size and mineralogy and generally classed as cohesive or noncohesive sediments. Cohesive sediments, such as muddy systems, consist of smaller grains (<63 μm) where surface charges on the particles allow them to attract each other thus "cohering" together. As particles increase in size, surface charges have little or no effect and the larger grains respond to external forces independently and are hence normally noncohesive (usually >63 μm). This description is largely based on laboratory studies of clean and well-sorted sediments, but in nature, sediments may not always behave as expected, being redistributed by burrowing organisms or kept in place by biofilms or mucilage (Black et al., 2002).

Evans (1965) provided the first comprehensive description of the sedimentary composition and sediment dynamics of temperate intertidal flat environments. He outlined a sequence of shore-normal depositional sediment types (or "facies") which can be found between the low- and high-water marks and which can be related to the tidally driven hydrodynamic conditions across the flat. The occurrence of accelerating tidal currents on the lower flat area usually (assuming no sediment supply limitations) winnows any fine sediments and gives rise to a well-sorted sandy sediment bed. As the tide rises (and falls) the midflat area is exposed to the strongest currents found within the tidal frame and this also results in a coarser sediment texture although often there is a small secondary fine fraction. Shoreward of the mid-tide datum, current velocities monotonically decrease and this permits deposition of progressively finer sediments. The upper intertidal environment is, due to distortion of the tidal wave, dominantly depositional in nature. Although Evans' geological–physiographic description is set principally within a tidal context, broad intertidal regions are susceptible to other hydrodynamic forcing factors which can influence the sediment character (and thus also the ecosystem composition), and these include wave action and rainfall (Tolhurst et al., 2006), which can directly impact surface sediments during periods of receding and low tide. The transport of sediments within intertidal flats is a continuous process but can be subdivided into a number of contiguous processes: these are erosion (E), transport (T), deposition (D), and consolidation (C). This continuum of processes is often referred to as the erosion, transport, and deposition cycle (ETDC) (Tolhurst et al., 2006). The daily progression of the tidal wave across the intertidal zone is a fundamental factor governing sediment transport and hence system geomorphology. The tidal wave within estuaries but distant from the mouth is typified by currents that accelerate to a maximum at the mid-tide stage and then decelerate toward the high-water slack tide stage (Pethick, 1984). Furthermore, due to asymmetric distortion of the tidal wave, flood currents are typically stronger (but of shorter duration) than corresponding ebb currents. These two features indicate firstly that the elevation of the intertidal flat within the tidal frame can be related to the local current velocities with peak current velocities associated with the midtidal flat area. Second, there is greater potential for a flood-directed sediment transport than ebb-directed movement, a situation which gives rise to intertidal flat regions being net, medium- to long-term depositional areas (Christie and Dyer, 1998; Bassoullet et al., 2000).

All depositional systems are thus driven by the supply and movement of their constituent grains and share many similarities when viewed from a distance. They are generally relatively flat and largely devoid of obvious structure which lend terrestrial and subtidal systems architectural complexity. The ecology of both sandy and muddy systems is largely driven by visitors that feed from the surface and organisms that live below the surface of the sediments. However, on closer examination, the mud and sandy systems show considerable divergence. Physical difference in the grains can be manifest in surface topography. Ripple structure caused by waves are evident on sandy sediments, whereas muddy sediments are usually planar with the exception of structures such as casts and burrows created by infaunal organisms.

Abiotic attributes, such as the erosional and depositional events (ETDC cycle), are critical to the occurrence of species within a habitat (Paterson and Black, 1999) and may determine the type of community that occurs (Sousa, 1984; Dernie et al., 2003). The changing nature of the physical environment and the immersion/emersion cycles lead to a zonation of species. The evolution of strategies enabling organisms to adapt to life in sediments has led to different niches being exploited using a range of individual abilities. However, the outcome (fitness) of the species represents a successful balance that has been achieved between the driving pressures of nature and the possible niche spaces available. These pressures can be expressed as physical (wind and waves), biological (competition predation and grazing), and physiological (desiccation, salinity tolerance, temperature tolerance, etc.). The most obvious outcome of these pressures is the simple distributional variations of species across environmental gradients that inherently reflect these drivers. However, the classic extreme zonation of rocky shores is not found so readily within depositional systems (Nybakken and Bertness, 2005). This is because the sediments themselves act as an efficient buffer helping to reduce the effects of air exposure and minimize changes in factors such as salinity (Little, 2000). Salinity within the pore water varies much less than the salinity of the surface water (Lindberg and Harriss, 1973); however, there are always exceptional circumstances. On hot days at low tides, the sediments can be exposed to high evaporation rates and therefore the salinity of pore fluid can increase dramatically. These effects are more extreme on the high shore than the lower shore due to exposure time. Such variations in water content and evaporation help to drive patterns of change, and different life phases of intertidal organisms may show different tolerances and hence zonation patterns with age and development. It is considered that in muddy intertidal sediments, the upper limit of zonation is governed by the abiotic process such as immersion period, salinity, and desiccation, whereas the lower limits are controlled by biotic factors such as the presence of competitors or predators. In this way, soft sediments are very similar to rocky shore environments (Little and Kitching, 1996). For example in the Wadden Sea, the lower limits of the populations of *Corophium* spp. were controlled by the presence or absence of *Arenicola marina*. In muddy sediments, in which *A. marina* were absent, *Corophium* spp. were able to extend to a lower tidal level than those in sandy sediments with *A. marina*. The occurrence of communities of decapod crustaceans studied along the Italian coast of the Tyrrhenian Sea was observed to be vertically controlled by the dynamics of the water flow and the sedimentation rates within the habitat (Scipione et al., 2005).

2.2 Sandy Systems

Sandy systems provide a habitat that is highly dynamic, the smaller grains have been removed by hydrodynamic sorting (winnowing) because of the high energy of the system, and the sand bed is in almost constant motion. The grains are packed together but are sufficiently large that there is considerable void space between them allowing specialist organisms to inhabit the interstitial spaces (meiofauna). These voids also allow advective transport through the pore spaces and this allows oxygen penetration, driven by the pore water advection (Janssen et al., 2005). The depth of penetration of oxygen is a critical difference between sandy and muddy shores. When oxygen is available, the system is oxygenic, whereas without oxygen, the system becomes anoxic and is largely toxic to macrofauna. There is often a rapid transition zone between the oxic surface conditions and the deeper anoxic zone. This region is called the redox potential discontinuity zone and the depth of this layer varies from meters in sandy system to submillimeter levels in compact muddy systems. The availability of oxygen has a profound influence on the breakdown of organic material and nature and speed of the metabolic processes including carbon and nutrient recycling.

The traditional view of sandy habitats has been of relatively low organic content supporting an assemblage of filter-feeding organisms that largely rely on material from the water column. At a functional level, this leaves the sand as a rather inert passive substratum acting as housing for invertebrates. This view often incorporates the implicit idea that productivity, metabolic rates, and nutrient turnover are low in sandy environments as compared with the high organic content and active metabolism of mudflats. This has been recently, and effectively, challenged providing a different perspective (Zetsche et al., 2011). Light penetration in sandy sediment is much deeper than muddy systems, and microphytobenthic photosynthesis can continue during tidal submersion, which is rare for muddy systems because the overlying water is turbid and occludes too much light for this strategy to be energetically viable. The biomass of microphytobenthos is more dispersed with depth and can photosynthesize over much longer time periods. The integrated photosynthetic response can be much greater than previously considered (Billerbeck et al., 2007), thus the functional capacity of sandy flats to deliver carbon fixation (autochthonous production) may be a significant ecosystem service. Infauna may also "engineer" conditions to their advantage. The lugworm *Arenicola marina* lives in J-shaped burrows in which it ventilates in a peristaltic pumping action, this results in bioadvection whereby nutrients from lower levels are transported to surface pore waters. Controlled experiments have shown this activity promotes the growth of microphytobenthos (Chennu et al., 2015). Thus, the primary productivity of sandy systems may be greater than first supposed on the basis of absolute chlorophyll content.

In terms of metabolic performance, De Beer et al. (2005) carried out studies demonstrating that sandy sediment systems had high aerobic degradation rates despite a much lower organic content than fine cohesive sediments with similar decomposition rates. Similar conclusions have been reached on the processing of external (allochthonous) carbon. Huettel et al. (1996) propose that organic mixing into the pore waters can occur due to hydromechanical mechanisms (Zetsche et al., 2011) as well as bioturbatory and filtration activities of organisms. This transport of organic material into the sediment only occurs within sediments that have permeability above a given threshold. Rather than considering sandy systems as areas with relatively little activity, the same system can be regarded as one that relies on the episodic allochthonous input of carbon that precipitates a period of extremely high turnover (Buhring et al., 2006). After the organic supply is exhausted, the system is quiescent again until another pulse of activity is initiated. This is similar to the model proposed for the organic supply to deep oceanic benthic systems. Both models provide a view of a relatively low turnover system for most of the time but the reasoning is different. Can the organic processing capacity of the sandy system be overloaded? In certain circumstances, the build up of organic material becomes too great to process. This is the situation where oxygen is "used up" in the breakdown process and leads to the development of recognizable, characteristically dark (iron sulfide), anoxic regions in some tidal flats termed "black spots" (Rusch et al., 1998).

2.3 Muddy Systems

Muddy systems have characteristically higher organic loads than sandy deposits. The presence of organic material associated with mud is partly due to in situ productivity but also due to the steady supply of allochthonous organic material from the coastal and riverine catchments (Nybakken and Bertness, 2005) that is deposited and retained by the mud under conditions of relatively low energy. This supply is almost never completely exhausted but the close packing of the fine particles confines the oxygenic activity to the very surface layers (often <500 μm), and this establishes a system of extreme gradients and constant organic recycling (Aspden et al., 2004a). Muddy sediments support active assemblages of microphytobenthos (Paterson et al., 2003; Underwood et al., 2005) with high level of biomass (Chl a) concentrated at the sediment surface. This concentration at the sediment–air interface is a requirement for photosynthetic organisms because light penetration on cohesive sediments is extremely limited (Consalvey et al., 2004). Nutrients are also usually plentiful given the charged nature of the very fine particles and their ability to sequester nutrients from the water column (Nedwell et al., 1999). The primary productivity of such systems is well known (Underwood and Kromkamp, 1999; Billerbeck et al., 2007) and their functional role in contributing to the primary productivity of the coastal zone often cited (Macintyre et al., 1996). This productivity is often enhanced by coastal eutrophication.

2.4 Functional Characterization

Mudflats have been given more functional prominence than sand flats, but this is probably debatable especially given some of the recent evidence cited above. The public perceptions of the systems can also be quite different, but each provides ecosystem services that cannot easily be replaced. The appreciation and valuation of these services (Beaumont et al., 2007, 2008) is another matter and one that is now attracting considerable attention. The different habitats may harbor organisms that carry out very similar functionality and contribute similar ecosystem services. For example, the tidal flats of the Dutch Wadden Sea are composed of areas of varying sediment particle size. High densities of *Corophium volutator* occur in silty areas, and *C. arenarium* dominate in sandy areas (Beukema and Flach, 1995). Thus, the nature of the substratum is influential on the species that occur but different forms may have similar ecological strategies. On a broader scale, benthic assemblages can be placed in five categories according to body size and the type of substratum (Table 11.2) but can also be classified related to their functional attributes (Fig. 11.4).

2.5 Patterns of Life

The functions of ecosystems may be understood by examining the biological traits of the species living within them. Biological traits analysis (BTA) uses characteristics such as feeding method, movement, morphology, life span, etc., to build a profile of the functions and services of populations within an ecosystem. The greater the number of traits identified (Fig. 11.4), the clearer the description of the ecosystem functions (Bremner et al., 2006). This enables comparisons to be made between sites and for change to be monitored.

TABLE 11.2 Broad Classification of Benthic Organisms by Body Size and Lifestyle

Microbenthos	The microbenthos inhabit interstitial spaces between sediment particles and comprise unicellular organisms such as bacteria, diatoms, euglenoids, and ciliates. The photosynthetic, pigment-containing forms are known as microphytobenthos. When at high biomass, these organisms can be seen by the naked eye as a green or golden-brown biofilm in the sediment surface. These biofilms are adapted to survive depositional and highly dynamic environments (Paterson and Black 1999). Not only do microphytobenthic biofilms serve as primary producers in the ecosystem but the group also has a number of other ecosystem services (Chapin et al., 1997), including the stabilization of cohesive sediment. The microphytobenthos are important primary produces and provide a significant source of autochthonous carbon within the ecosystem. The microbenthos could soon be commonly divided into picoheterobenthos, picophytobenthos, microphytobenthos, and microheterobenthos.
Picoheterobenthos	Prokaryotes such as bacteria and viruses.
Picophytobenthos	Photosynthetic cells $<2\ \mu m$.
Microphytobenthos	Unicellular photosynthetic organisms $>2\ \mu m$.
Microheterobenthos	Unicellular heterotrophic organisms $>2\ \mu m$.
Meiobenthos	The meiobenthos also occur within the interstitial spaces of sediments but consist of any multicellular organism less than 1 mm in length. This group consists of a variety of invertebrates such as nematodes and copepods.
Hyperbenthos	The hyperbenthos occur in the water column just above the sediment surface. However, they are often found within the sediments due to the close proximity to the sediment bed and some may also burrow. Hyperbenthic organisms tend to be small, only a few millimeters in length.
Macrobenthos	The macrobenthos are over 1 mm in length and move freely through soft sediments. This group consists of species such polychaete worms, bivalves, and amphipods. This group has a major effect on their surrounding habitat and is important ecosystem engineer.
Epibenthos	The epibenthos are often predatory and consist of large, active organisms, such as crabs, lobsters, and bottom fish, that exploit the habitat and as such makes up a large proportion of the benthic biomass.

Taxa may be scored on how much they display a trait using "fuzzy coding." Some traits may vary in a species over time due to changes in life form, in response to competition, predation, or environmental change.

Thus BTA should be repeated at suitable time intervals. A limitation of the method is the time and level of information necessary to conduct analysis. The BIOTIC database facilitates scientific studies by allowing users to link survey data to 40 functional traits of benthic species (MarLIN, 2006). Recent studies describe the resilience of ecosystem functions in coastal and subtidal sites despite significant changes to species composition (Clare et al., 2015; Munari, 2013). This suggests that the functions of replacement species may compensate for changes to the original assemblages. However, functional stability may be disrupted following disturbances such as trawling if the species are substituted with dissimilar taxa (Clare et al., 2015). The strength of BTA analysis should improve with time as more biological information on species is recorded.

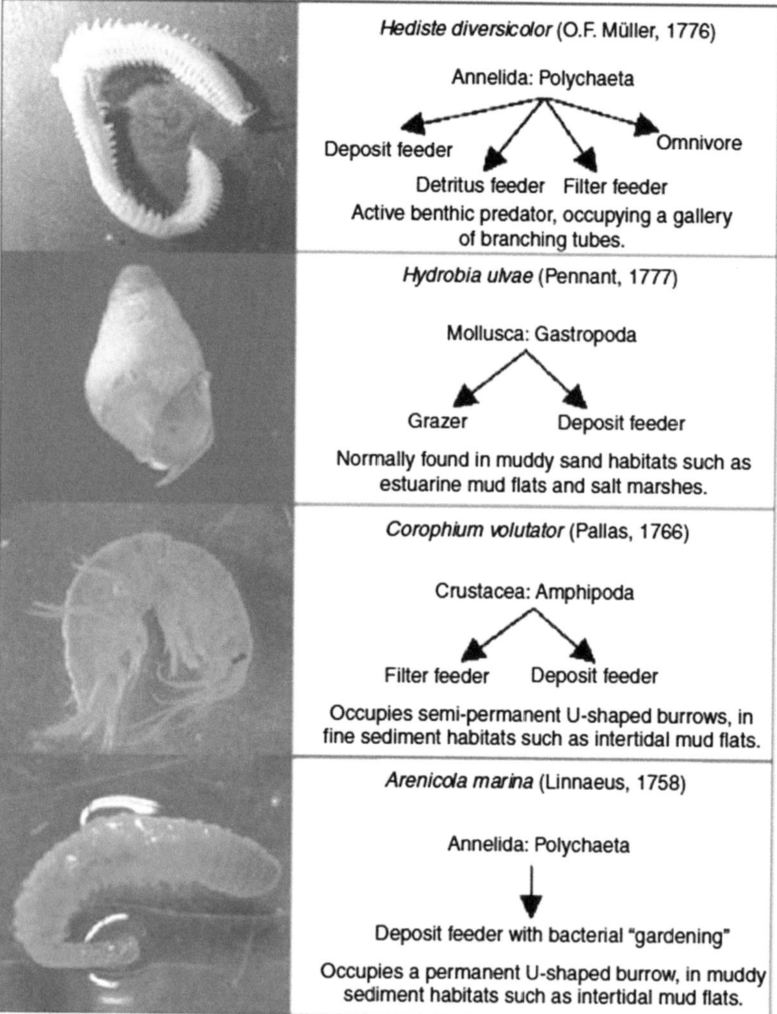

FIGURE 11.4 Examples of benthic infauna and their functional capabilities/traits.

3. BIODIVERSITY IMPACTS

Although physical attributes may have a strong influence in dictating the assemblages present, the biotic components play the primary role in habitat functioning. Faunal assemblages within the sediment carry out essential processes such as bioturbation, biodeposition, pelletization, biogenic stabilization, biogeochemical regulation, oxygenation of sediments, the distribution of solutes, and the transport of reactants and metabolites across the sediment–water interface (Aller, 1994; Grant et al., 1982; Norling et al., 2007). The species-specific traits and the biodiversity of functional processes of benthic macrofauna can affect key functions of a community, habitat, and ecosystem (Solan et al., 2006). Indeed, biodiversity is often deemed to be a measure of ecosystem health and because there has been a decline in global biodiversity, at rates exceeding species extinctions in the fossil record, efforts have been made to conserve and restore biodiversity through international accords such as the Convention on Biodiversity (2010). Thus, the impact of change in species diversity is strongly related not to the identity of the species per se but rather to the attributes of the species and the regulation of ecosystem functions and corresponding structural variables are influenced by species-specific traits rather than by species richness (Giller et al., 2004; Hooper et al., 2005). As a result of these earlier studies, it is becoming commonplace to consider the diversity of functional groups as opposed to species diversity or species richness when measuring ecosystem processes within coastal systems. Norling et al. (2007) defined functional biodiversity as the functional group richness of benthic fauna and classified dominant benthic invertebrates from the Baltic Sea and the Skagerrak as functional groups in accordance with their feeding behavior and sediment reworking activities. Great care must be taken when assigning organisms to a functional group, as many organisms can be defined as belonging to more than one group, showing functional plasticity (Boogert et al., 2006). It may be more appropriate to place organisms into a functional group based on an amalgamation of traits rather than one characteristic (Hooper et al., 2002). The dominant functions carried out by the organisms may depend on the community assemblage and how the assemblage functions as a whole or/and the stage in an individual's life cycle. A coastal management strategy, which focuses on maintaining biodiversity, reduces the risks of unknowns related to species loss and ecosystem function (Emmerson et al., 2001).

Many recent studies seek to examine the relationship between biodiversity and ecosystem function and outcomes for ecosystem services (Bulling et al., 2010). A positive relationship was observed in species richness (as a measure of biodiversity) and ecosystem function with regard to nutrient release (Emmerson et al., 2001). This relationship was noted to be negatively altered by increased temperatures and atmospheric CO_2 levels (Bulling et al., 2010). In addition, there is evidence for a positive but idiosyncratic relationship between coastal species diversity and greater complementarity in resource use (Emmerson et al., 2001). This may be due to the differing functional roles of individual species, for example, the crustacean *Corophium volutator* resuspends surface sediments while bivalves including the mussel (*Mytilius edulis*) filter feed from the water column, effectively performing the service of nutrient cycling. Furthermore, greater consistency of function was shown with increased species richness (Emmerson et al., 2001).

The outcome of a decline in biodiversity on ecosystem processes has been shown to be related to the dominance (i.e., biomass) of species within intertidal ecosystems (Davies et al., 2011), although this relationship can be complex (Wohlgemuth et al., 2016). Dominant species were found to conduct most of the ecosystem processes (community productivity, nutrient uptake, and grazing of microalgae) in a variety of coastal wetland habitats. This link between biomass and productivity may serve as a useful proxy for estimating the contribution of species to ecosystem processes and the impact of biodiversity decline on ecosystem function. Dominant species may be most resilient to extinction and may offset the loss of function of rarer species. Thus, the role of dominant organisms should not be underestimated. Ieno et al. (2006) noted that changes in the composition of coastal benthic macrofauna changed the biogeochemistry of the system. An increase in species diversity was positively related to nutrient regeneration, although this relationship was believed to be species specific.

Bioturbatory activity was believed to be the key to the relationship, suggesting that species identity and density must be considered during studies of this nature (Biles et al., 2003). Changes in biogeochemistry can be subtle, but there are many more obvious examples of the effects of organismal activity on habitat structure and function. *Lanice* sp. reefs may act as refuge for organisms such as bivalves, gastropods, polychaetes, amphipods, and isopods, and therefore these areas tend to be higher in individual numbers and diversity than areas of sediment with no reef structure present. Effects may be subtle interactions with other system drivers. The modification of bed roughness at the sediment surface due to biotic processes, such as the production of mounds or presence of tubes, can cause the flow to vary to such a degree that erosion thresholds change (Ziebis et al., 1996). Classic flume experiments, carried out by Eckman et al. (1981), studying the effects of *Owenia fusiformis* (a tube building polychaete) suggested that at higher densities the erosion threshold was lower than that at lower densities. Carey (1983) carried out similar experiments using the polychaete *Lanice conchilega*. The tubes of the polychaete caused a reduction in flow of up to 20%. Resuspension of sediments in front of the tubes occurred due to the alteration of the flow; however, in the wake of the tubes the flow was reduced, thereby allowing the deposition of sediments previously held in suspension. The presence of tubes has been suggested to encourage bacterial and microbial colonization within the surface sediments (Eckman, 1985), thereby stabilizing the sediments further due to microbial stabilization. These examples indicate organisms capable of mediating their habitats and surviving local conditions. In addition, the action of organisms may be complimentary in terms of the delivery of a functional services and this also applies to ecosystem engineers. When two different species both contribute to a particular functional outcome (sediment stability, for example), then the activity has been defined as cooperative ecosystem engineering (Passarelli et al., 2014). However, there is often a balance to be struck between the engineering activity and aspects of lifestyle and ecology. Particular organisms may depend on disturbance events to complete their life cycles, such as the sabellid worms which only spawn during periods of high storm activity (Barry, 1989). This dependence on a disturbance event is a response to the probability of high adult mortality due to the storm, therefore reducing competition for space and resources and providing optimum conditions for recruitment, so disturbance is useful and too strong an engineering activity might be harmful overall.

4. SYSTEM MONITORING

To understand the relationship between species diversity and ecosystem processes, it is important take into consideration that these relationships may vary with scale (i.e., local to global), and as the habitats/ecosystems get larger the effects of diversity change will change accordingly. Ninety-seven percent of the available living volume on the planet can be defined as marine environment (Raffaelli, 2006), providing a potentially huge volume to study. The examination of such large sample areas generally means that replication can rarely be adequate. In today's studies, many of the functional measurements, such as primary productivity and nutrient flux, are measured at a very small scale. However, when contemplating these measurements in terms of entire habitats or ecosystem processes, it is not as simple as scaling up the results, as an increase in spatial variability also increases the variation of biodiversity. Approaches to this problem include large-scale mapping, remote sensing (RS), and modeling.

4.1 Remote Sensing

RS is an indispensable tool for monitoring change in habitats over time and allows objective and comprehensive studies over larger geographic regions than is possible with fieldwork alone. In terms of RS, not all life forms are equally relevant (e.g., soil bacteria are poor candidates), and the focus is usually on plant cover although in specific instances other fairly well-defined elements of biodiversity can be assessed, for example, coral reefs (Hedley et al., 2016). The determination of biodiversity from RS depends on the optical properties of the target systems and the ability to differentiate between systems comprising different levels or types of biodiversity. The applicability of RS will therefore depend on the nature of the questions being asked. A significant change in land cover, from forest to grassland, for example, represents a significant change in biodiversity and consequent service delivery and is more easily described. Field sampling provides detailed, local biological information for small areas, but it can be expensive. Used together (Fig. 11.4), strategic ground sampling, expert knowledge, and interpretation of RS imagery can form a reliable, repeatable, and cost-effective analytical framework for accurately assessing the rate of change in biological diversity (Strand et al., 2007). The raw data from RS are more easily linked to some ecosystem services than others. For example, primary production may be estimated using remotely acquired chlorophyll measurements, whereas regulatory services may be only indirectly inferred. Therefore, other datasets may be used in conjunction with spectral data to assess the value of these ecosystem services (de Araujo Barbosa et al., 2015). The majority of RS studies on ecosystem services use the variables of land cover and Normalized Difference Vegetation Index (NDVI). NDVI was found to correlate strongly with sediment chlorophyll *a* and phaeopigments on four midlatitude estuaries, with the strength of the relationship being greater on muddy than on sandy sediments (Kromkamp et al., 2006).

Regularly captured data have been shown to be effective in measuring changes to coastal ecosystems and services. In addition, the impact of storm surges on wetlands ecosystems may be monitored. Both spatial area and surge height were captured on a regular and consistent pattern and when used with water classification models, produced maps of water boundaries. Such maps are crucial in recognizing the risks posed to coastal zones from erosion and rising seawater levels. RS data used in spatial value transfer models for the assessment of ecosystem services, is dependent on spatial resolution and the image classifications. However, improvements to RS technology, such as continental scale mapping, will provide more data to evaluate ecosystem services and adaptation measures for sea level rise (Murray et al., 2012; de Araujo Barbosa et al., 2015) (Fig. 11.5).

4.2 Effects of Disturbance

Coastal management often requires an assessment of the outcome of various intrusive coastal activities on benthic communities. The knowledge of the recovery process for the habitat and in turn the recovery of the faunal assemblage is essential, and it is therefore crucial to understand and predict the association between biodiversity and ecosystem functioning. The coastal disposal of dredged material is one such activity (Pilskaln et al., 1998). Dredging is essential to maintain safe navigation in estuaries and ports. Recent focus has been on disposing of this material where it may be most beneficial for the environment,

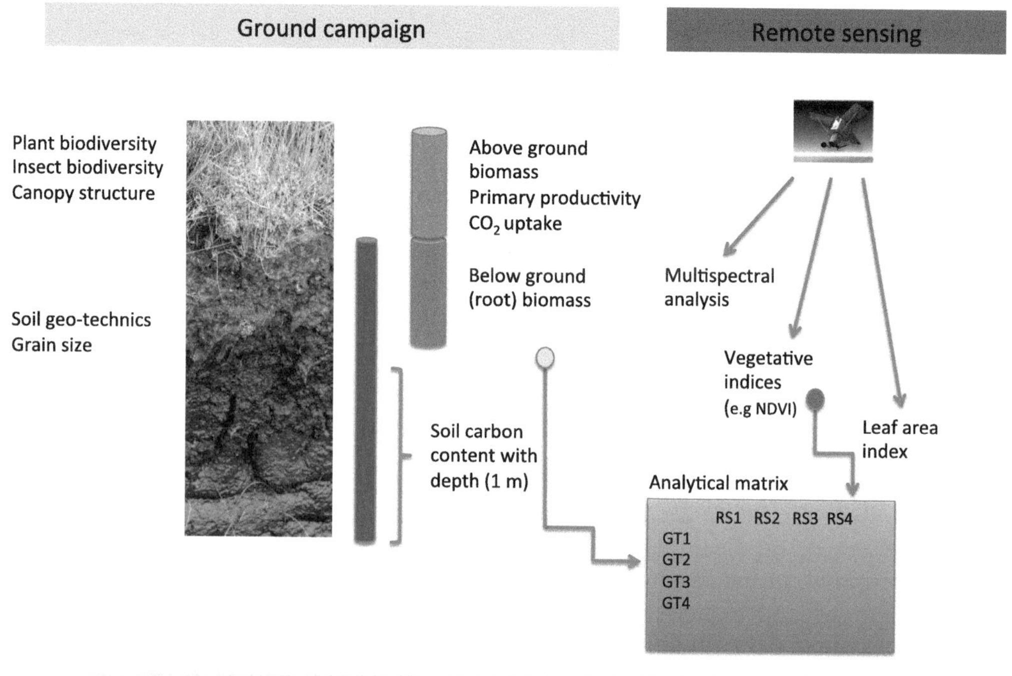

FIGURE 11.5 Combined ground truthing and remote sensing can be used to asses change in ecosystem functions using proxy measures of system performance such as Normalized Difference Vegetation Index (NDVI) and leaf area index.

i.e., by regarding the material as a potential resource to recharge existing mudflats or create new ones (Bolam, 2014). However, the relocation of large volumes of sediment alters the tidal height of receiving sites and buries existing benthic assemblages. A 4-year study compared the taxonomic structure and ecosystem function of two "recharged" mudflat sites with nearby reference sites to assess recovery. It found benthic faunal assemblages at the recharged sites to be less species rich and less densely populated than at reference sites. These taxonomic changes persisted throughout the study period (Bolam, 2014). However, ecosystem function was more resilient. Functional diversity and secondary productivity at the recharge sites were found to be similar to reference sites and could be considered to be fully recovered. A third functional measure, BTA contradicted these findings (Bolam, 2014). BTA analysis has shown that ecological function is more resilient than taxonomic structure in response to the impact of coastal defense barriers in subtidal environments (Munari, 2013). However, the lower biodiversity found on intertidal mudflats may explain altered trait composition. There is the greater chance for a taxonomic change in a species-poor community to change the functional traits as there are fewer species possessing each trait (Bolam, 2014). This supports the theory that functional redundancy is lower in species-poor assemblages. Identifying the traits most impacted by disturbance events and their importance to ecological function will be central to ensuring sustainable development. Therefore, it is necessary to

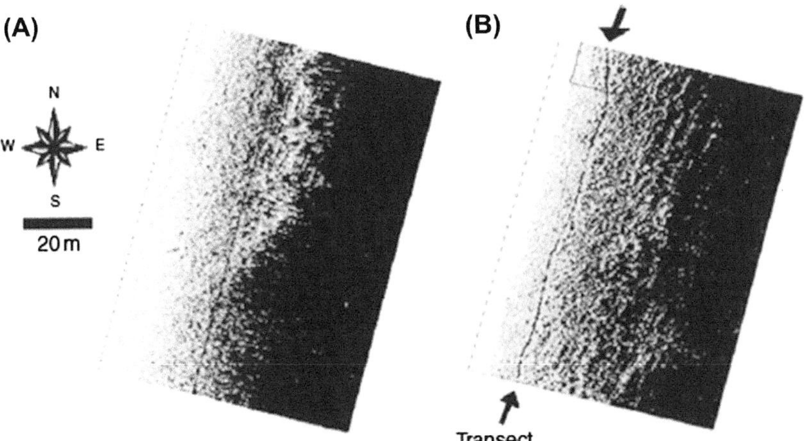

FIGURE 11.6 Side scan sonar images of the same area of sea bed taken prior to (A) and post (B) the impacts of a harvesting event. The images of the sediment disturbance (B) provide an excellent visual representation of the damage caused by a harvesting event when compared with the homogeneous nature of the sediment prior to the harvesting event. A furrow can be seen clearly after the harvesting activity had taken place (indicated by *arrows*).

consider the impact of disturbance on both structural taxonomy and functional indices (Bolam, 2014). Dernie et al. (2003) carried out a study on the effects of physical disturbance on both the habitat and fauna of a sheltered sand flat to ascertain whether the recovery of the biota could be predicted from physical attributes of the system. The outcome of this study suggested that benthic community recovery in a lower disturbance event took place within 64 days, whereas recovery after higher intensity disturbance event did not occur until 208 days postdisturbance. Aspden et al. (2004b) determined that clam (*Tapes philippinarum*) harvesting, in the Venice Lagoon (Italy), could be associated with decreases in sediment stability. A high frequency of harvesting prevented sediment consolidation, the normal succession of biological communities, and the development of stable microphytobenthic assemblages. The microphytobenthic assemblage composition did not significantly differ before and after harvesting at intensely harvested sites; however, community composition was affected at a site that had been subjected to negligible harvesting intensity prior to the study. Side scan sonar was used to visualize the physical disturbance to the bed before and after harvesting (Fig. 11.6).

5. FUTURE SHOCK: CURRENT CHALLENGES

5.1 Invasive Nonnative Species

An increasingly concerning biological perturbation to intertidal systems is the introduction of invasive species (Williams and Grosholz, 2008). Increased globalization of commerce has resulted in the redistribution of species, primarily via ship ballast water (Cloern and Jasby, 2012) but also by other means even deliberate introduction. Introduction of nonnative species represents a challenge to the integrity of natural assemblages. The debate about the relative harms, or sometimes benefits, of invasive species is still ongoing. Global climate changes will promote the expansion and reduction of species ranges and recognition of this and the mathematical modeling of effects is now being attempted for known invaders (Dunstan and Bax, 2007). Species have also often been introduced by man for some specific purpose such as the biocontrol of pests (the Rhodolia beetle) or for leisure activity (e.g., game fish). The aquatic system is no different, and invasive species may have a large impact on coastal zone systems. The classic example is probably that of the green alga *Caulerpa taxifolia* in the Mediterranean. *Caulerpa taxifolia* is a popular aquarium alga and was released into the Mediterranean from the Musée Oceanographique de Monaco in 1984. Since that time, it has spread and replaced the natural seagrass meadows with its own simple monophyletic stands. This leads to a reduction in species richness and functionality of the systems and is considered by some to be a major ecological threat (Longepierre et al., 2005). However, there is also evidence that *Caulerpa* may be benefiting as a result of anthropogenic pressures, and some aspects of the "invasion" may be beneficial and not as damaging as first considered. There is also an implication that mild winters and an increase in temperature is also favoring *Caulerpa* spp. (Aplikioti et al., 2016).

The scale of the invasive problem in some habitat is perhaps brought home by the work in the Mediterranean and in San Francisco Bay. Galil (2007) lists more than 500 alien species in the Mediterranean alone with many authors blaming the Suez Canal. In San Francisco Bay, introduced species form 97% of individuals and 99% of biomass in some communities (Cloern and Jasby, 2012). The arrival of a nonnative bivalve, *Corbula amurensis*, had far-reaching consequences including the restructure of a planktonic food web and a shift in zooplankton species from native to nonnative forms. Changes to zooplankton composition leads to dietary shifts by their predator fish. Fish whose diets changed most (the Northern anchovy, *Engraulis mordax*) suffered the greatest population declines (94% decrease in summer abundance (Cloern and Jassby, 2012)). Invasive species have been cited as the most important cause of bird extinction and second most important cause in fish (Clavero and García-Berthou, 2005). Invasive species can also be vectors of disease as shown by the infection caused by the nonnative parasite *Haplosporidium nelsoni* leading to huge losses in the natural populations of oysters (*Crassostrea virginica*) on the American Atlantic coast. Thus, invasive species can have serious economic consequences (Cloern and Jassby, 2012). Given, the impact on natural assemblages and economic species, management policies have been introduced to restrict the discharge of ballast water around coastlines. However, a more concerted international effort is required to support this. On a functional level, the debate is not so clear cut. If invasive species exploit unoccupied niche space, then can they not be seen as beneficial? This has certainly been argued (Oliverio and Taviani, 2003), and on a functional level it is not species identity that is important but the delivery of ecosystem services and the resistance and resilience of the system. For example, the invasive polychaete *Marenzelleria cf. viridis* (found in the Firth of Forth in 1982 and in the Firth of Tay in 1984) (Atkins et al., 1987) replaced the native species *Hediste diversicolor*. Although this invasive species may have changed the food web dynamics, there may have been little variation in trophic structure as both species are directly competitive and occupy the same functional group. In addition, if invasive species increase biodiversity, then they may also stabilize the system in its response to stress. However, most studies on invasive species have found negative impacts on the ecosystems involved (Williams and Grosholz, 2008). Some programs to control or eradicate invasive species have been successful, particularly in Australia and New Zealand where resources have been focused on this cause (Williams and Grosholz, 2008). Globally, however, it is inevitable that we must become used to changes in species populations in coastal habitats.

5.2 Climate Change

Anthropogenic global change is a phenomenon we have not experienced before, and as the climate alters in response to the unnaturally high levels of CO_2 (currently over 400 ppm from preindustrial levels of 280 ppm), the consequences of that change are difficult to predict. Around half of CO_2 emitted has been absorbed by oceans, changing the chemistry of seawater with the formation of carbonic acid and resulting in a warmer ocean, 26% more acidic than preindustrial times (NOAA). Sea level is also increasing leading to enhanced coastal erosion and terrestrial habitat loss. The real crisis, from an anthropocentric perspective, will be when ecosystems no longer supply the goods and services we require. Coastal intertidal and wetland systems may be the first to degrade in this way. Coastal zone management has traditionally focused on land-based human disturbances. However, ecosystem processes on intertidal systems are also influenced by large-scale climate shifts to oceanographic systems (Cloern and Jassby, 2012). A 1.7°C increase in the temperature of winter coastal waters since 1970 coincided with a 40%–50% reduction in primary production, decreased benthic metabolism and abundance of demersal fish, and a switch of N cycling from net denitrification to net N fixation in Narragansett Bay (Nixon et al., 2009). In the future, coastal systems will be experiencing the combined effects of multiple stressors from climate change, including temperature, sea level rise, increasing storm frequency, and ocean acidification, in addition to local effects such as nutrient runoff, fisheries impacts, pollution, dredging, and other more localized changes. This suggest that the management of coastal systems or at least a fuller understanding of the impacts of the pressures they are likely to experience is essential and will be a significant task for the future.

Acknowledgments

This chapter was written with support from the University of St. Andrews and Partrac Ltd. The authors acknowledge the MarBEF Network of Excellence "Marine Biodiversity and Ecosystem Functioning" funded by the European Community's Sixth Framework Program (contract no. GOCE-CT-2003-505446). This publication is contribution number MPS-08032 of MarBEF. DMP received funding from the NERC CBESS (NE/J015644/1) and Blue Coast (NE/N016009/1) projects and the MASTS pooling initiative (The Marine Alliance for Science and Technology for Scotland) and their support is gratefully acknowledged. MASTS is funded by the Scottish Funding Council (grant reference HR09011) and contributing institutions.

References

Airoldi, L., Cinelli, F., 1997. Effects of sedimentation on subtidal macroalgal assemblages: an experimental study from a Mediterranean rocky shore. Journal of Experimental Marine Biology and Ecology 215, 269–288.

Aller, R.C., 1994. Bioturbation and remineralization of sedimentary organic matter: effects of redox oscillation. Chemical Geology 114, 331–345.

Alongi, D.M., 2008. Mangrove forests: Resilience, protection from tsunamis, and responses to global climate change. Estuarine, Coastal and Shelf Science 76 (1), 1–13. https://doi.org/10.1016/j.ecss.2007.08.024.

Aplikioti, M., Louizidou, P., Mystikou, A., Marcou, M., Stavrou, P., Kalogirou, S., Tsiamis, K., Panayotidis, P., Kupper, F.C., 2016. Further expansion of the alien seaweed Caulerpa taxifolia var. distichophylla (Sonder) Verlaque, Huisman & Procacini (Ulvophyceae, Bryopsidales) in the Eastern mediterranean sea. Aquatic Invasions 11 (1), 11–20.

Aspden, R.J., Vardy, S., Paterson, D.M., 2004a. Salt marsh microbial ecology: microbes, benthic mats and sediment movement. In: Fagherazzi, S., Marani, M., Blum, L.K. (Eds.), The Ecogeomorphology of Tidal Marshes. Coastal and Estuarine Studies Series. (AGU), American Geophysical Union, Washington, DC, pp. 115–136.

Aspden, R.J., Vardy, S., Perkins, R.G., Davidson, I.R., Bates, R., Paterson, D.M., 2004b. The effects of clam fishing on the properties of surface sediments in the lagoon of Venice, Italy. Hydrology and Earth System Sciences 8, 160–169.

Aspden, R.J., 2005. Studies on the Biogenic Mediation of Sediment Dynamics in Coastal Systems. University St Andrews, UK. Ph.D. thesis.

Atkins, S.M., Jones, A.M., Garwood, P.R., 1987. The ecology and reproductive cycle of a population of *Marenzelleria viridis* (Annelida: Polychaeta: Spionidae) in the Tay estuary. In the natural environment of the Tay estuary. Proceedings of the Royal Society of Edinburgh, Section B: Biological Science 92, 311–322.

Barry, J.P., 1989. Reproductive response of a marine annelid to winter storms: an analogue to fire adaptation in plants? Marine Ecology Progress Series 54, 99–107.

Bassoullet, P., le Hir, P., Gouleau, D., Robert, D., 2000. Sediment transport over an intertidal mudflat: field investigations and estimated fluxes within baie de marennes Oleron (France). Continental Shelf Research 20, 1635–1654.

Beaumont, N.J., Austen, M.C., Atkins, J.P., Burdon, D., Degraer, S., Dentinho, T.P., Derous, S., Holm, P., Horton, T., van Ierland, E., Marboe, A.H., Starkey, D.J., Townsend, M., Zarzycki, T., 2007. Identification, definition and quantification of goods and services provided by marine biodiversity: implications for the ecosystem approach. Marine Pollution Bulletin 54, 253–265.

Beaumont, N.J., Austen, M.C., Mangi, S.C., Townsend, M., 2008. Economic valuation for the conservation of marine biodiversity. Marine Pollution Bulletin 56, 386–396.

Benton, T.G., Solan, M., Travis, J.M.J., Sait, S.M., 2007. Microcosm experiments can inform global ecological problems. Trends in Ecology and Evolution 22, 516–521.

Beukema, J.J., Flach, E.C., 1995. Factors controlling the upper and lower limits of the intertidal distribution of two *Corophium* species in the Wadden Sea. Marine Ecology Progress Series 125, 117–126.

Biles, C.L., Solan, M., Isaksson, I., Paterson, D.M., Emes, C., Raffaelli, D.G., 2003. Flow modifies the effect of biodiversity on ecosystem functioning: an in situ study of estuarine sediments. Journal of Experimental Marine Biology and Ecology 285–286, 165–177.

Billerbeck, M., Roy, H., Bosselmann, K., Huettel, M., 2007. Benthic photosynthesis in submerged Wadden Sea intertidal flats. Estuarine, Coastal and Shelf Science 71, 704–716.

Black, K.S., Tolhurst, T.J., Hagerthey, S.E., Paterson, D.M., 2002. Working with natural cohesive sediments. Journal of Hydraulic Engineering Forum 128, 1–7.

Bolam, S.G., 2014. Macrofaunal recovery following the intertidal recharge of dredged material: a comparison of structural and functional approaches. Marine Environmental Research 97, 15–29.

Boogert, N.J., Paterson, D.M., Laland, K.N., 2006. The implications of niche construction and ecosystem engineering for conservation biology. Biosciences 57, 570–578.

Bremner, J., Rogers, S.I., Frid, C.L.J., 2006. Methods for describing ecological functioning of marine benthic assemblages using biological traits analysis (BTA). Ecological Indicators 6, 609–622.

Buhring, S.I., Ehrenhauss, S., Kamp, A., 2006. Enhanced benthic activity in sandy sublittoral sediments: evidence from C-13 tracer experiments. Marine Biology Research 2, 120–129.

Bulling, M.T., Hicks, N., Murray, L., Paterson, D.M., Raffaelli, D., White, P.C., Solan, M., 2010. Marine biodiversity-ecosystem functions under uncertain environmental futures. Philosophical Transactions of the Royal Society of London B Biological Sciences 365 (1549), 2107–2116.

Carey, D.A., 1983. Particle resuspension in the benthic boundary layer induced by flow around polychete tubes. Canadian Journal of Fisheries and Aquatic Sciences 40, 301–308.

Chapin, F.S., Walker, B.H., Hobbs, R.J., Hooper, D.U., Lawton, J.H., Sala, O.E., Tilman, D., 1997. Biotic control over the functioning of ecosystems. Science 277, 277–500.

Chennu, A., Volkenborn, N., de Beer, D., Wethey, D.S., Woodin, S.A., Polerecky, L., 2015. Effects of bioadvection by Arenicola marina on microphytobenthos in permeable sediments. PLoS One 10 (7), e0134236.

Chmura, G.L., Anisfeld, S.C., Cahoon, D.R., Lynch, J.C., 2003. Global carbon sequestration in tidal, saline wetland soils. Global Biogeochemical Cycles 17 (4), 1111.

Christie, M.C., Dyer, K.R., 1998. Measurement of the turbid edge over the skeffling mudflats. In: Black, K.S., Paterson, D.M., Cramp, A. (Eds.), Sedimentary Processes in the Intertidal Zone. Geological Society, London, pp. 45–55. Special Publications 39.

Clare, D.S., Robinson, L.A., Frid, C.L., 2015. Community variability and ecological functioning: 40 years of change in the North Sea benthos. Marine Environmental Research 107, 24–34.

Clavero, M., García-Berthou, E., 2005. Invasive species are a leading cause of animal extinctions. Trends in Ecology and Evolution 20, 110.

Cloern, J.E., Jassby, A.D., 2012. Drivers of change in estuarine-coastal ecosystems: discoveries from four decades of study in San Francisco Bay. Reviews of Geophysics 50, RG4001.

Consalvey, M., Paterson, D.M., Underwood, G.J.C., 2004. The ups and downs of life in a benthic biofilm: migration of benthic diatoms. Diatom Research 19, 181–202.

Costanza, R., dArge, R., de Groot, R., Farber, S., Grasso, M., Hannon, B., Limburg, K., Naeem, S., Oneill, R.V., Paruelo, J., Raskin, R.G., Sutton, P., van den Belt, M., 1997. The value of the world's ecosystem services and natural capital. Nature 387, 253–260.

Davies, T.W., Jenkins, S.R., Kingham, R., Kenworthy, J., Hawkins, S.J., Hiddink, J.G., 2011. Dominance, Biomass and Extinction Resistance Determine the Consequences of Biodiversity Loss for Multiple Coastal Ecosystem Processes. PLoS ONE 6 (12), e28362.

de Araujo Barbosa, C.C., Atkinson, P.M., Dearing, J.A., 2015. Remote sensing of ecosystem services: a systematic review. Ecological Indicators 52, 430–443.

De Beer, D., Wenzhöfer, F., Ferdelman, T.G., Boehme, S.E., Huettel, M., van Beusekom, J.E.E., Böttcher, M.E., Musat, N., Dubilier, N., 2005. Transport and mineralization rates in North Sea sandy intertidal sediments, Sylt-Rømø Basin, Wadden Sea. Limnology and Oceanography 50, 113–127.

Dernie, K.M., Kaiser, M.J., Richardson, E.A., Warwick, R.M., 2003. Recovery of soft sediment communities and habitats following physical disturbance. Journal of Experimental Marine Biology and Ecology 285–286, 415–434.

Duarte, C.M., Dennison, W.C., Orth, R.J.W., Carruthers, T.J.B., 2008. The charisma of coastal ecosystems: addressing the imbalance. Estuaries and Coasts 31, 233–238.

Dunstan, P.K., Bax, N.J., 2007. How far can marine species go? Influence of population biology and larval movement on future range limits. Marine Ecology Progress Series 344, 15–28.

Eckman, J.E., Nowell, A.R.M., Jumars, P.A., 1981. Sediment eestabilisation by animal tubes. Journal of Marine Research 39, 361–374.

Eckman, J.E., 1985. Flow disruption by an animal-tube mimic affects sediment bacterial colonization. Journal of Marine Research 43, 419–435.

Emmerson, M.C., Solan, M., Emes, C., Paterson, D.M., Raffaelli, D., 2001. Consistent patterns and the idiosyncratic effects of biodiversity in marine ecosystems. Nature 411, 73–77.

Escapa, C.M., Iribarne, O., Navarro, D., 2004. Effects of the SW burrowing crab Chasmagnathus granulatus on infaunal zonation patterns, tidal behavior and risk of mortality. Estuaries 27, 120–131.

Evans, G., 1965. Intertidal flat sediments and their environments of deposition in the Wash. Quarterly Journal of the Geological Society of London 121, 209–241.

Ford, R.B., Thrush, S.F., Probert, P.K., 1999. Macrobenthic colonisation of disturbances on an intertidal sandflat: the influence of season and buried algae. Marine Ecology Progress Series 191, 163–174.

Ford, H., Garbutt, A., Ladd, C., Malarkey, J., Skov, M.W., 2016. Soil stabilization linked to plant diversity and environmental context in coastal wetlands. Journal of Vegetable Science 27 (2), 259–268.

Galil, B.S., 2007. Loss or gain? Invasive aliens and biodiversity in the Mediterranean Sea. Marine Pollution Bulletin 55, 314–322.

Giller, P.S., Hillebrand, H., Berninger, U.G., Gessner, M.O., Hawkins, S., Inchausti, P., Inglis, C., Leslie, H., Malmqvist, B., Monaghan, M., Morin, P., O'Mullan, G., 2004. Biodiversity effects on ecosystem functioning: emerging issues and their experimental test in aquatic environments. Oikos 104, 423–436.

Grant, W.D., Boyer, L.F., Sanford, L.P., 1982. The effects of bioturbation on the initiation of motion of interidal sands. Journal of Marine Research 40, 659–677.

Hansell, M.H., 2005. Animal Architecture. Oxford University Press, Oxford, p. 343.

Hedley, J.D., Roelfsema, C.M., Chollett, I., Harborne, A.R., Heron, S.F., Weeks, S., Skirving, W.J., Strong, A.E., Eakin, C.M., Christensen, T.R.L., Ticzon, V., Bejarano, S., Mumby, P.J., 2016. Remote sensing of coral reefs for monitoring and management: a review. Remote Sensing 8, 118. https://doi.org/10.3390/rs8020118. On line.

Hooper, D.U., Solan, M., Symstad, A., 2002. Species diversity, functional diversity and ecosystem functioning. In: Loreau, M., Naeem, S., Inchausti, P. (Eds.), Biodiversity and Ecosystem Functioning Synthesis and Perspectives. Oxford University Press, UK, pp. 195–208.

Hooper, D.U., Chapin, F.S., Ewel, J.J., Hector, A., Inchausti, P., Lavorel, S., Lawton, J.H., Lodge, D.M., Loreau, M., Naeem, S., Schmid, B., Setälä, H., Symstad, A.J., Vandermeer, J., Wardle, D.A., 2005. Effects on biodiversity on ecosystem functioning: a consensus of current knowledge. Ecological Monographs 75, 3–35.

Huettel, M., Ziebis, W., Forster, S., 1996. Flow-induced uptake of particulate matter in permeable sediments. Limnology and Oceanography 41, 309–322.

Hyder, K., Rossberg, A.G., Allen, J.I., Austen, M.C., Bannister, H.J., Barciela, R.M., Blackwell, P.G., Blanchard, J.L., Burrows, M.T., Defriez, E., Dorrington, T., Edwards, K.P., Garcia-Carreras, B., Heath, M.R., Hembury, D.J., Heymans, J.J., Holt, J., Houle, J.E., Jennings, S., Mackinson, S., Malcolm, S.J., McPike, R., Mee, L., Mills, D.K., Montgomery, C., Pearson, D., Pinnegar, J.K., Pollicino, M., Popova, E.E., Rae, L.R., Rogers, S.I., Speirs, D., Spence, M.A., Thorpe, R., Turner, R.K., van der Molen, J., Yool, A., Paterson, D.M., 2015. Making modelling count - increasing the contribution of shelf-seas community and ecosystem models to policy development and management. Marine Policy 61, 291–302.

Ieno, E.N., Solan, M., Batty, P., Pierce, G.J., 2006. How biodiversity affects ecosystem functioning: roles of infaunal species richness, identity and density in the marine benthos. Marine Ecology Progress Series 311, 263–271.

Janssen, F., Huettel, M., Witte, U., 2005. Pore-water advection and solute fluxes in permeable marine sediments (II): benthic respiration at three sandy sites with different permeabilities (German Bight, North Sea). Limnology and Oceanography 50, 779–792.

Jones, C.G., Lawton, J.H., Shachak, M., 1994. Organisms as ecosystem engineers. Oikos 69, 373–386.

Kaiser, M.J., Attrill, M.J., Jennings, S., Thomas, D.N., Barnes, D.K.A., Brierley, A.S., Polunin, N.V.C., Raffaelli, D.G., Williams, P.J., Le, B., 2005. Marine Ecology: Processes, Systems and Impacts. Oxford University Press, Oxford, p. 584.

Kenworthy, J., Paterson, D.M., Bishop, M.J., 2016. Response of benthic assemblages to multiple stressors: comparative effects of nutrient enrichment and physical disturbance. Marine Ecology Progress Series 562, 37–71.

Kromkamp, J.C., Morris, E.P., Forster, R.M., Honeywill, C., Hagerthey, S., Paterson, D.M., 2006. Relationship of intertidal surface sediment chlorophyll concentration to hyperspectral reflectance and chlorophyll fluorescence. Estuaries and Coasts 29 (No. 2), 183–196.

Laland, K.N., Odling-Smee, F.J., Feldman, M.W., 2004. Causing a commotion. Niche construction: do the changes that organisms make to their habitats transform evolution and influence natural selection? Nature 429, 609.

Lawton, J., 1994. What do species do in ecosystems? Oikos 71, 367–374.

Lindberg, S.E., Harriss, R.C., 1973. Mechanisms controlling pore water salinities in a salt marsh. Limnology and Oceanography 18, 788–791.

Little, C., Kitching, J.A., 1996. The Biology of Rocky Shores. Oxford University Press, Oxford, p. 240.

Little, C., 2000. The Biology of Soft Shores and Estuaries. Oxford University Press, Oxford, p. 264.

Littler, M.M., Martz, D.R., Littler, D.S., 1983. Effects of recurrent sand deposition on rocky intertidal organisms: importance of substrate heterogeneity in a fluctuating environment. Marine Ecology Progress Series 11, 129–139.

Longepierre, S., Robert, A., Levi, F., Francour, P., 2005. How an invasive alga species (*Caulerpa taxifolia*) induces changes in foraging strategies of the benthivorous fish *Mullus surmuletus* in coastal mediterranean ecosystems. Biodiversity and Conservation 14, 3609–3622.

Loreau, M., Naeem, S., Inchausti, P. (Eds.), 2002. Biodiversity and Ecosystem Functioning. Oxford University Press, Oxford.

Luisetti, T., Turner, R.K., Jickells, T., Andrews, J., Elliott, M., Schaafsma, M., Beaumont, N., Malcolm, S., Burdon, D., Adams, C., Watts, W., 2014. Coastal Zone Ecosystem Services: from science to values and decision making; a case study. The Science of the Total Environment 15 (493), 682–689.

MacIntyre, H.L., Geider, R.J., Miller, D.C., 1996. Microphytobenthos: the ecological role of the "Secret Garden" of unvegetated, shallow-water marine habitats. I. Distribution, abundance and primary production. Estuaries 19, 186–201.

Magurran, A.E., 2004. Measuring Biological Diversity. Blackwell Science, Oxford.

Malarkey, J., Baas, J.H., Hope, J.A., Aspden, R.J., Parsons, D.R., Peakall, J., Paterson, D.M., Schindler, R.J., Ye, L., Lichtman, I.D., Bass, S.J., Davies, A.G., Manning, A.J., Thorne, P.D., 2015. The pervasive role of biological cohesion in bedform development. Nature Communications 6, 6257.

MarLIN, 2006. BIOTIC - Biological Traits Information Catalogue. Marine Life Information Network. Marine Biological Association of the United Kingdom, Plymouth. www.marlin.ac.uk/biotic.

McCann, K., 2007. Protecting biostructure. Nature 446, 29.

McQuaid, C., Dower, K.M., 1990. Enhancement of habitat heterogeneity and species richness on rocky shores inundated by sand. Oceologia 84, 142–144.

MEA, 2005. Millennium Ecosystem Assessment, 2005. Ecosystems and Human Well-being: Synthesis. Island Press, Washington, DC. Copyright © 2005 World Resources Institute.

Middelburg, J.J., Barranguet, C., Boschker, H.T.S., Herman, P.M.J., Moens, T., Heip, C.H.R., 2000. The fate of intertidal microphytobenthos carbon: an in situ 13C-labeling study. Limnology and Oceanography 45 (6), 1224–1234.

Miyatake, T., et al., 2014. Tracing carbon flow from microphytobenthos to major bacterial groups in an intertidal marine sediment by using an in situ 13C pulse-chase method. Limnology and Oceanography 59 (4), 1275–1287.

Munari, C., 2013. Benthic community and biological trait composition in respect to artificial coastal defence structures: a study case in the northern Adriatic Sea. Marine Environmental Research 90, 47–54.

Murray, N.J., Phinn, S.R., Clemens, R.S., Roelfsema, C.M., Fuller, R.A., 2012. Continental scale mapping of tidal flats across east Asia using the landsat archive. Remote Sensing 4, 3417–3426.

Nedwell, D.B., Jickells, T.D., Trimmer, M., Sanders, R., 1999. Nutrients in estuaries. Advances in Ecological Research 29, 43–92.

Nixon, S.W., Fulweiler, R.W., Buckley, B.A., Granger, S.L., Nowicki, B.L., Henry, K.M., 2009. The impact of changing climate on phenology, productivity, and benthic-pelagic coupling in Narragansett Bay. Estuarine, Coastal and Shelf Science 82 (1), 1–18.

Norling, K., Rosenberg, R., Hulth, S., Grémare, A., Bonsdorff, E., 2007. Importance of functionalbiodiversity and species-specific traits of benthic fauna for ecosystem functions in marine sediment. Marine Ecology Progress Series 332, 11–23.

Nybakken, J.W., Bertness, M.D., 2005. Marine Biology: An Ecological Approach. Benjamin Cummings Publishers, San Francisco, p. 579.

Odling-Smee, J., Erwin, D.H., Palkovacs, E.P., Feldman, M.W., Laland, K.N., 2013. Niche construction theory: a practical guide for ecologists. The Quarterly Review of Biology 88 (1), 3–28.

Oliverio, M., Taviani, M., 2003. The Eastern Mediterranean Sea: tropical invasions and niche opportunities in a "Godot Basin". Biogeographia 24, 313–327.

Orvain, F., Sauriau, P.G., Sygut, A., Joassard, L., Le Hir, P., 2004. Interacting effects of *Hydrobia ulvae* bioturbation and microphytobenthos on the erodibility of mudflat sediments. Marine Ecology Progress Series 278, 205–223.

Orvain, F., 2005. A model of sediment transport under the influence of surface bioturbation: generalisation to the facultative suspension-feeder *Scrobicularia plana*. Marine Ecology Progress Series 286, 43–56.

Passarelli, C., Olivier, F., Paterson, D.M., Meziane, T., Hubas, C., 2014. Organisms as cooperative ecosystem engineers in intertidal flats. Journal of Sea Research 92, 92–101.

Paterson, D.M., Black, K.S., 1999. Water flow, sediment dynamics, and benthic biology. In: Raffaelii, D.B., Nedwell, D.G. (Eds.), Advances in Ecological Research. Oxford University Press, Oxford, pp. 155–193.

Paterson, D.M., Perkins, R., Consalvey, M., Underwood, G.J.C., 2003. Ecosystem function, cell micro-cycling and the structure of transient biofilms. In: Krumbein, W.E., Paterson, D.M., Zavarzin, G.A. (Eds.), Fossil and Recent Biofilms A. Natural History of Life on Earth.Kluwer, London, pp. 47–63.

Paterson, D.M., Aspden, R.J., Visscher, P.T., Consalvey, M., Andres, M.S., Decho, A.W., Stolz, J., Reid, R.P., 2008. Light-dependant biostabilisation of sediments by stromatolite assemblages. PLoS One 3 (9), e3176. https://doi.org/10.1371/journal.pone.000317.

Paterson, D.M., Hanley, N.D., Black, K., Defew, E.C., Solan, M., 2011. Biodiversity, ecosystems and coastal zone management: linking science and policy. Marine Ecology Progress Series 434, 201–202.

Pethick, J., 1984. An Introduction to Coastal Geomorphology. Edward Arnold, London.

Pilskaln, C.H., Churchill, J.H., Mayer, L.M., 1998. Resuspension of sediment by bottom trawling in the Gulf of Maine and potential geochemical consequences. Conservation Biology 12, 1523–1739.

Raffaelli, D., Solan, M., Webb, T.J., 2005. Do marine and terrestrial ecologists do it differently? Marine Ecology Progress Series 304, 283–289.

Raffaelli, D.G., 2006. Biodiversity and ecosystem functioning: issues of scale and trophic complexity. Marine Ecology Progress Series 311, 285–294.

Reise, K., 2002. Sediment mediated species interactions in coastal waters. Journal of Sea Research 48, 127–141.

Reitner, J., Quéric, N.-V., Arp, G., 2011. Advances in stromatolite geobiology. In: Lecture Notes in Earth Sciences, vol. 131. Springer.

Ricklefs, R.E., Miller, G.L., 1999. Ecology. W.H. Freeman and Company, New York.

Rusch, A., Topken, H., Bottcher, M.E., Hopner, T., 1998. Recovery from black spots: results of a loading experiment in the Wadden Sea. Journal of Sea Research 40, 205–219.

Scipione, M.B., Lattanzi, L., Tomassetti, P., Chimenz, G., Maggiore, F., Mariniello, L., Cironi, R., Taramelli, E., 2005. Biodiversity and zonation patterns of crustacean peracarids and decapods of coastal soft bottom assemblages (Central Tyrrhenian Sea, Italy). Vie et Milieu 55, 143–161.

Solan, M., Cardinale, B.J., Downing, A.L., Engelhardt, K.A., Ruesink, J.L., Srivastava, D.S., 2004. Extinction and ecosystem function in the marine benthos. Science 306 (5699), 1177–1180.

Solan, M., Raffaelli, D.G., Paterson, D.M., White, P.C.L., Pierce, G.J., 2006. Introduction to Marine biodiversity and ecosystem functioning: empirical approaches and future research. Marine Ecology Progress Series 311, 175–178.

Sousa, W.P., 1984. The role of disturbance in natural communities. Annual Review of Ecology and Systematics 15, 353–391.

Strand, H., Höft, R., Strittholt, J., Miles, L., Horning, N., Fosnight, E., Turner, W., 2007. Sourcebook on Remote Sensing and Biodiversity Indicators. Secretariat of the Convention on Biological Diversity, Montreal. Technical Series no. 32.

Tolhurst, T.J., Friend, P.L., Watts, C., Wakefield, R., Black, K.S., Paterson, D.M., 2006. The effects of rain on the erosion threshold of intertidal cohesive sediments. Aquatic Ecology 40, 533–541.

Townsend, C.R., Begon, M., Harper, J.L., 2008. Essentials of Ecology. Backwell, London, p. 532.

Underwood, G.J.C., Kromkamp, J., 1999. Primary production by phytoplankton and microphytobenthos in estuaries. In: Raffaelii, D.B., Nedwell, D.G. (Eds.), Advances in Ecological Research in Estuaries. Academic Press, New York, pp. 93–153.

Underwood, G.J.C., Perkins, R.G., Consalvey, M.C., Hanlon, A.R.M., Oxborough, K., Baker, N.R., Paterson, D.M., 2005. Patterns in microphytobenthic primary productivity: species-specific variation in migratory rhythms and photosynthesis in mixed-species biofilms. Limnology and Oceanography 50, 755–767.

van De Koppel, J., Herman, P.M.J., Thoolen, P., Heip, C., 2001. Do alternate stable states occur in natural ecosystems? Evidence from a tidal flat. Ecology 82, 3449–3461.

Watson, S.C.L., Paterson, D.M., Queiros, A.M., Rees, A.P., Stephens, N., Widdicombe, S., Beaumont, N.J., 2016. A conceptual framework for assessing the ecosystem service of waste remediation:In the marine environment. Ecosystem Services 20, 69–81.

Widdows, J., Blauw, A., Heip, C.H.R., Herman, P.M.J., Lucas, C.H., Middelburg, J.J., Schmidt, S., Brinsley, M.D., Twisk, F., Verbeek, H., 2004. Role of physical and biological processes in sediment dynamics of a tidal flat in Westerschelde Estuary, SW Netherlands. Marine Ecology Progress Series 274, 41–56.

Williams, S.L., Grosholz, E.D., 2008. The invasive species challenge in estuarine and coastal environments: marrying management and science. Estuaries and Coasts 31, 3–20.

Wohlgemuth, D., Solan, M., Godbold, J.A., 2016. Specific arrangements of species dominance can be more influential than evenness in maintaining ecosystem process and function. Scientific Reports 6, 39325.

Worm, B., Duffy, J.E., 2003. Biodiversity, productivity and stability in real food webs. Trends in Ecology and Evolution 18, 628–632.

Zetsche, E., Paterson, D.M., Lumsdon, D.G., Witte, U., 2011. Temporal variation in the sediment permeability of an intertidal sandflat. Marine Ecology Progress Series 441, 49–63.

Ziebis, W., Forster, S., Huettel, M., Jørgensen, B.B., 1996. Complex burrows of the mud Shrimp *Callianasa truncata* and their geochemical impact in the sea bed. Nature 382, 619–623.

CHAPTER

12

Biogeochemical Dynamics of Coastal Tidal Flats

C.A. Schutte[1], S. Ahmerkamp[2], C.S. Wu[2], M. Seidel[3], Dirk de Beer[2], P.L.M. Cook[4], S.B. Joye[5]

[1]Louisiana Universities Marine Consortium (LUMCON), Chauvin, LA, United States; [2]Max Planck Institute for Marine Microbiology, Bremen, Germany; [3]Institute for Chemistry and Biology of the Marine Environment (ICBM), University of Oldenburg, Oldenburg, Germany; [4]Water Studies Centre, Monash University, Clayton, VIC, Australia; [5]Department of Marine Sciences, University of Georgia, Athens, GA, United States

1. INTRODUCTION

Coastal sediments are well recognized for their importance in biogeochemical cycling of critical bioelements, such as carbon, nitrogen, and phosphorus (Boynton et al., 1980; Alongi, 1997; Epstein, 1997; Jassby et al., 2002; Yamamoto, 1997). The coastal ocean is a shallow region (<200 m) with an area of 26×10^6 km^2, which accounts for about 7% of total ocean area. These regions support dynamic interactions between the land, ocean, and atmosphere and account for about 15% of global ocean productivity, 90% of sediment-based mineralization, and 80% of organic matter (OM) burial (Gattuso et al., 2009). Although intertidal flats comprise a small fraction of the world's coastlines (<10%) (Stutz and Pilkey, 2002), they are among the most productive components of shelf ecosystems. Tidal flats are important sites of recycling for both terrestrially and marine-derived OM and nutrients and also support substantial rates of primary productivity. The best-studied tidal flats are in Europe and the United States. Large areas on the coasts of Africa (the Niger and Congo delta), South America (the Amazon delta), and South Asia (the Ganges delta) are much less explored (Alongi, 1997).

The largest intertidal flat area, the Wadden Sea, is located in Northern Europe and is to a great extent man-made. The original ecosystems of the Wadden Sea were salt marshes, mudflats, and seagrass beds that developed after the last ice age. The salt marshes disappeared due to peat harvesting and the flood plain areas decreased, primarily due to land reclamation. Thus, the sediments of the Wadden Sea are changing via a gradual displacement of silt with

Coastal Wetlands
https://doi.org/10.1016/B978-0-444-63893-9.00012-5

sands, especially toward the barrier islands and the large tidal channels where intertidal sand flats now dominate (Lotze, 2005). Similar coastal areas in the United States are still salt marshes. The observed and expected increases of sea level will affect intertidal areas worldwide. Where coastal regions are protected by dikes, the total area occupied by tidal flats will shrink substantially.

Intertidal flats are located between the spring high- and low-tide marks and lack rooted vegetation. They are highly heterogeneous and differ in sediment composition and microbial processes depending on location within the system. As intertidal flats are found in sheltered bays, estuaries, and before coasts more or less protected from exposure to oceans by barrier islands, they can range from mudbanks to coarse sand flats and receive in varying degrees terrestrial inputs from rivers and coastal seas. Intertidal flats consist of lithogenous or biogenous sediments or a mixture of both. Lithogenous sediments originate from eroded rock and are rich in silicates, whereas biogenous sediments are mainly carbonates, usually deriving from shells or corals. Biogenous sediments are more abundant in the tropics, especially in coral seas. We will focus on the temperate tidal flats characterized by lithogenous material, as such areas dominate most margins.

The grain size distribution of lithogenous sediments ranges from fine clay ($<2\ \mu m$) to coarse sand ($>250\ \mu m$) and is determined by the hydrodynamic regime (Dyer, 1979). Clay-rich and silty sediments exhibit porosities ranging from $\sim$50% to over 80% (Taylor Smith and Li, 1966). Sands, on the other hand, exhibit porosities ranging from 37% to 50%, with fine sands having a slightly higher porosity than medium-grained sands (Taylor Smith and Li, 1966). Typically, an inverse relationship is observed between grain size and nutritional content (often reported as percentage of organic carbon) (Dale, 1974; Longbottom, 1970; Panutrakul et al., 2001): Mudbanks and silty tidal flats are typically rich in organics (typically 0.5%–3%), nutrients, and trace metals that easily bind to the large surface area of the clay particles, whereas sandy sediments contain little OM (around 0.1%) and lower contents of metals and nutrients. However, some studies have failed to show a strong relationship between grain size and biological parameters such as bacterial abundance or chlorophyll *a* (Cammen, 1982), whereas others have illustrated good correlations between grain size, organic content, and microbial activity (sulfate reduction (Panutrakul et al., 2001)).

When a storm hits a tidal flat, large amounts of sediments can be suspended and redeposited (Christiansen et al., 2006; Graber et al., 1989), thereby mixing the top few centimeters to decimeters of sediments irregularly. Bioturbation provides a constant reworking of the top 10 cm (Boudreau, 1998). This, together with their daily cycles of exposure and inundation, makes tidal flat sediments highly dynamic. The inundation/exposure ratio strongly varies with distance from the low water line and is further dependent on the tidal range. Moreover, there is temporal variability, with biogeochemical processes in tidal flats varying significantly with season and weather. Due to the heterogeneity within and between flats, it is difficult to generalize the biogeochemistry of intertidal flats. In fact, their temporal and spatial heterogeneities and their extreme dynamics are typical characteristics that distinguish these habitats from permanently inundated coastal sediments.

The relative inaccessibility of intertidal flats has, to a large extent, limited their study. Being intermittently inundated and uncovered, as well as exposed to strong currents during rising and falling tide, a solid ship that can fall dry is needed to access the flats during complete tidal cycles. Explorations on foot during low tide must be planned carefully and can be difficult in soft muds and drift sands. Storms are interesting events, as this is when most sediment reposition occurs, but during such occasions the flats must be abandoned. As a consequence of the spatial and temporal heterogeneities and the difficulty of their in situ study, the phenomena reported here may be conceptually valid but may not accurately reflect the in situ situation.

2. TRANSPORT PROCESSES ON INTERTIDAL FLATS

Biogeochemical processes in tidal flat sediments are regulated by the dynamic interaction of microbial reactions and transport processes. The microbial reactions include primary production and respiration. Primary production is controlled by the availability of light, nutrients, and temperature. Respiration is mainly dependent on the availability of OM, oxygen (or other electron acceptors), and temperature. The dominant transport processes for solute exchange are advection driven by water currents and waves or bioirrigation, diffusion, and sediment transport. The important governing transport processes determine the exchange of solutes and suspended matter between the intertidal flat with the water column, as well as the internal transport of both solutes and sediment.

The resulting interaction of microbial reactions and transport processes can be best summarized by quantifying net fluxes of solutes (in $\mu mol\ m^{-2}\ d^{-1}$). While fluxes represent only transport, net fluxes represent the difference of outflow and inflow into the marine sediments and this also captures microbial processing. Solute and particulate exchange processes between seawater and sediments and within the pore water are regulated by diffusion and advection. Following Fick's law of diffusion, diffusive fluxes (J) are determined by the steepness of the solute concentration gradients (dC/dx) and the effective diffusion coefficient (D_{eff}) for the solute (Eq. 12.1):

$$J = -D_{eff}\frac{dC}{dx} \tag{12.1}$$

The effective diffusion coefficient in sediments is a function of the diffusion coefficient of the solute in water (D_w) corrected for the tortuosity (θ) and porosity (φ) of the sediments (Eq. 12.2):

$$D_{eff} = D_w\varphi\theta^2 \tag{12.2}$$

For fine clays with high porosity (>0.7), an empirical approximation is commonly used (Ullman and Aller, 1982):

$$D_{eff} = D_w\varphi^{-2} \tag{12.3}$$

In porous sands, water–sediment exchange and transport inside sediments is governed by advection, pore water flow (v) (Eq. 12.4):

$$J = -D_{eff}\frac{dC}{dx} + vC \tag{12.4}$$

The relative importance of diffusion or advection is determined by an intrinsic characteristic of the sediment: the permeability (k), which is the ability to allow for pore water flow. The permeability is mainly determined by the grain size distribution of the intertidal flats (Neumann et al., 2016). Grain sizes between 100 and 200 μm typically represent the transition between impermeable ($k < 10^{-12}$ m^2) and permeable ($k > 10^{-12}$ m^2) sediments (Huettel et al., 2014). In most cases tidal flats are dominated by one of the two transport processes, either diffusion, e.g., mudflats, or advection, e.g., sand banks.

In diffusion controlled systems, the concentration gradient, and therefore total transport, is controlled by the magnitude of the volumetric reaction rates. Therefore, the activity of the microbial community itself influences the solute flux. This hydrodynamic stability allows bacterial communities to colonize the entire pore space available. To a small extent, diffusive exchange might be altered by changes of currents, i.e., thinning or thickening of the diffusive boundary layer (Jørgensen and Des Marais, 1990). On larger spatial scales, diffusion is a slow process and oxygen only penetrates into the first few millimeters of tidal flats (Llobet-Brossa et al., 2002).

In advection controlled sediments, on the other hand, solute exchange by pore water advection is determined by pressure gradients, i.e., the physical forcing. Pressure gradients induce flow through sediments at all spatial scales. The interaction of currents with the rippled surface of the tidal flat induces **"skin" circulation**. Drainage of the sediment body induces "**body**" **circulation**. In case of advective flow, the microbial communities adapt by mainly colonizing sediment grains (Musat et al., 2006) and taking advantage of the advective supply of electron acceptors—such as oxygen—and OM. Although advectional transport seems simple in theory, the local solute concentration and an estimate for the pore water velocity must be known. In practice, these terms are difficult to quantify. Because pore water flow is driven by currents and waves, all factors that influence the local hydrodynamics change the advectional flow regime. Such factors include measuring equipment, vehicles and ships, and the observers themselves. Thus, minimally invasive techniques need to be used for in situ measurements in permeable sediments.

2.1 Skin Circulation

For decades, sandy sediments have been seen as biogeochemical deserts, but this picture has altered with an improved understanding of the "skin" circulation, i.e., ripple-induced flow (Boudreau et al., 2001). The principle was discovered on gravel beds in rivers (Thibodeaux and Boyle, 1987) and extended to finer sediments and oxygen transport in the 90s (Elliott and Brooks, 1997; Rutherford et al., 1993). Its major importance for global biogeochemical cycling in the coastal ocean was only recognized over the past few decades (Forster et al., 1996; Huettel and Gust, 1992; Huettel et al., 1996; Røy et al., 2002; Ziebis et al., 1996).

The concept can be understood from small pressure gradients that are induced by a current traveling over a bedform of centimeter scale, so-called ripples (Fig. 12.1). Following Bernoulli's principle, the flow accelerates at the stoss or oncoming current side of the ripple, with the maximum at the ripple crest, and decelerates at the lee side. This leads to a pressure gradient at the sediment–water interface, which typically exhibits a strength of 1–10 Pa equaling a water head of less than a millimeter. This apparently small pressure gradient constantly pumps bottom water through the pore space providing solutes and particles from the water column to the benthic microbial communities. The induced pore water velocities range from millimeter to

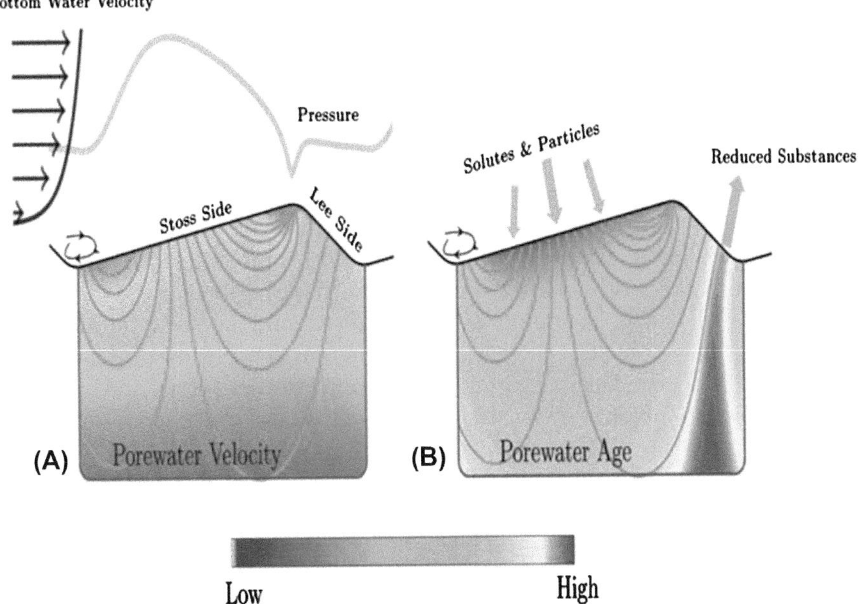

FIGURE 12.1 Schematic illustrations to emphasize the "skin" circulation, i.e., ripple-induced flow. The left panel (A) shows schematically the porewater velocity with the pressure distribution at the sediment–water interface. The right panel (B) shows the porewater age; at the lee side a narrow outlet becomes apparent, which releases reduced substances to the water column (Ahmerkamp, 2016)

centimeter per hour (Reimers et al., 2004), leading to a filtration capacity of 0.1 $m^3\ m^{-2}\ d^{-1}$ (Santos et al., 2012). Ripple-induced flow is, therefore, of major significance and leads to distinct two-dimensional to three-dimensional oxygen distributions.

Oxygen penetrates deep below the valleys and anoxic zones and may almost reach the sediment surface near the peak of the ripples. Such flow enhances the exchange of solutes, such as oxygen, nutrients, and metals (Mn^{2+} and Fe^{2+}) (Ahmerkamp et al., 2017; Berninger and Huettel, 1997; Huettel and Rusch, 2000; Huettel et al., 1998; Precht and Huettel, 2004; Rasheed et al., 2003; Røy et al., 2002; Rusch and Huettel, 2000; Ziebis et al., 1996). Bioirrigation, the pumping of pore water through worm burrows, might contribute significantly to this process (Meysman et al., 2006). However, as a result of the heterogeneity of this process, accurate quantifications on larger scales lag behind.

The mere presence of ripples and bedforms is indicative of a regularly mobilized sediment bed. Sediment transport is either induced by biological processes, such as bioturbation, or physical transport, such as sediment resuspension and redeposition during storm events and ripple migration. These processes lead to a mixing of the sediment matrix, which exposes nutrients from deeper sediment layers to the surface fueling primary production (Chennu et al., 2013). Furthermore, OM, which might be trapped in the pore space, is released and reworked (Ahmerkamp et al., 2017).

Biogeochemically, the most important mechanism of sediment transport is ripple migration. The combination of sediment erosion at the stoss side of ripples and deposition at the lee side leads to a train-like motion. This means that the pressure gradient at the sediment–water interface is traveling, which has several implications. At the onset of bedform migration, the sediment matrix is separated into a mobile sediment layer and a stationary layer underneath. When the speed of bedform migration exceeds the pore water advection, exchange processes become limited to the mobile sediment layer, i.e., the first 2–3 cm (Ahmerkamp et al., 2015). Under dynamic in situ conditions, the traveling pressure gradient in combination with changing current direction and magnitude leads to highly dynamic oxygen penetration patterns into intertidal flats (De Beer et al., 2004; Franke et al., 2006; Precht and Huettel, 2004; Werner et al., 2006). Ripple-induced flow and ripple migration govern the upper 5–10 cm of the sediments and generate a constantly changing mosaic of inflow and outflow areas and thus a constantly changing pattern of oxic and anoxic volumes in the upper sediments (Ahmerkamp et al., 2017). Despite the inherent complexity of the flow field within ripples, recent work succeeded in estimating potential nitrogen loss and oxygen fluxes by reducing the driving variables to grain size, current velocities, and reaction rates (Ahmerkamp et al., 2017; Marchant et al., 2016).

2.2 Body Circulation

During the transition from high tide to low tide, reduced permeability of the intertidal flats holds the water and the flats remain covered with a thin layer of water. Consequently, the existing pressure head drives a flow from the center of the flats toward the low water line, where water seeps out during low tide. The pore water near the seeps is highly enriched in remineralization by-products such as nutrients, dissolved inorganic carbon (DIC), sulfide, and methane (Billerbeck et al., 2006a; Seidel et al., 2015). At high tide when the flats are completely covered, the pressure differences are dissipated and this flow stops. Thus, an intermittent, unidirectional internal circulation pattern is common on tidal flats.

Based on this concept, pore water flow patterns were quantitatively modeled in a permeable flat in the German Wadden Sea (the Janssand) using Comsol (Røy et al., 2007). The calculated flow velocities were on the order of a millimeter to a centimeter per day, depending on the length of the pathway. The residence time of the pore water in the flats varied from months to centuries and the average residence time of the pore water leaving the flat near the low water line was estimated to be several decades. This age was confirmed by ^{14}C dating of methane and DIC in the seep, which still carried the so-called bomb signal, a peak in radioactivity caused by the hydrogen bomb testing in the early 1960s. The level of ^{14}C was equal to the atmospheric levels of about 1970 (Røy et al., 2007).

An even slower and deeper circulation was found in a biogeochemical study on a 20-m deep drill core in the center of the same intertidal flat, the Jansand. Most basic geochemical data and the lithostratigraphy were published previously. Sulfate levels decreased rapidly in the upper 3 m because of sulfate reduction and then increased again. Sulfate reduction rates, measured by the ^{35}S method (Kallmeyer et al., 2004), were highest in the upper 3 m and then decreased but remain detectable up to 20 m below the seafloor (bsf) (Fig. 12.2). In addition, the sulfate/salinity profile indicates active sulfate consumption, as dilution by freshwater would keep the ratio constant with depth, while a clear sulfate depletion occurs. The increase of this ratio 5 m below the seafloor indicates a sulfate resupply coming from the side.

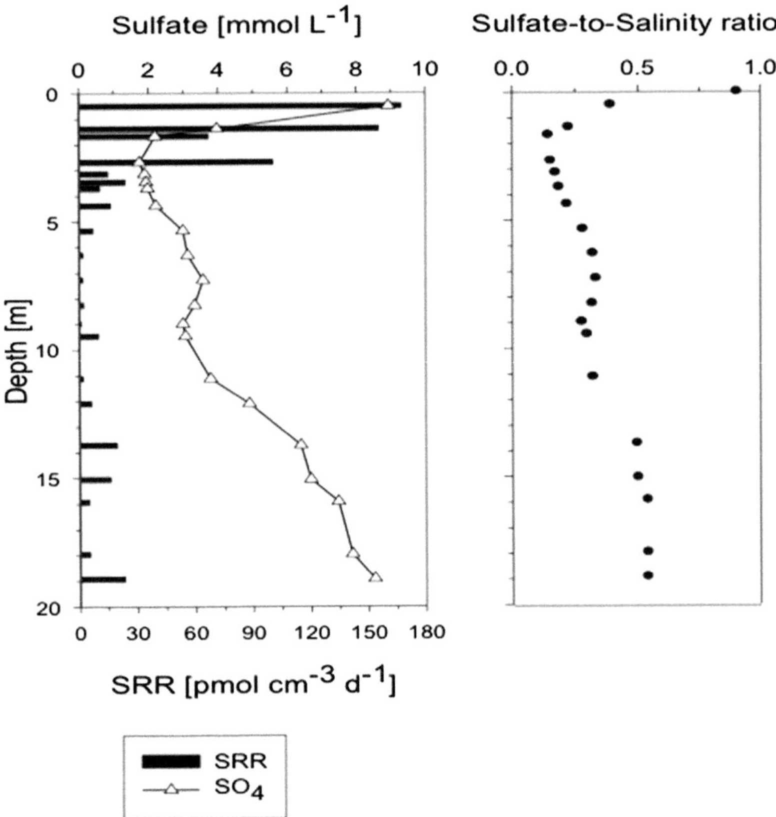

FIGURE 12.2 Sulfate reduction rates and sulfate profiles (left). Sulfate was depleted by consumption and not by dilution with low-saline water as the sulfate/salinity ratios strongly deviate from the seawater value of almost 1 (right).

Importantly, the sulfate turnover rates (the concentration/consumption rate) were much shorter than the sediment age (Fig. 12.3). Below 5 m depth sulfate turnover was in the order of 1000 years, while these sediments were deposited 10,000–400,000 years ago (Fig. 12.3). Thus, transport is needed from the surface to at least 20 m below the seafloor to resupply the sulfate pools. Without this "deep biosphere circulation," sulfate would have been depleted below 5 m depth, as is indeed the case in large parts of the flat (Beck et al., 2008; Røy et al., 2008). The term "deep biosphere circulation" is based on the observation that the microbial communities 3–5 m or greater below the seafloor in these intertidal flats are comparable with those in the deep biosphere (Engelen and Cypionke, 2009). The turnover time of organics is much longer than the age of their sediment layers (Fig. 12.3, right panel), which means that the buried pools are more than enough to sustain sulfate reduction for almost 1,000,000 years. Yet, the organics provide additional evidence for a deep exchange process.

The age of the mobile organic fraction in the sediments (humics) was determined by ^{14}C dating. There was an interesting discrepancy with the sediment age as determined by a variety of geological methods and from the ^{14}C-age of immobile organic fractions

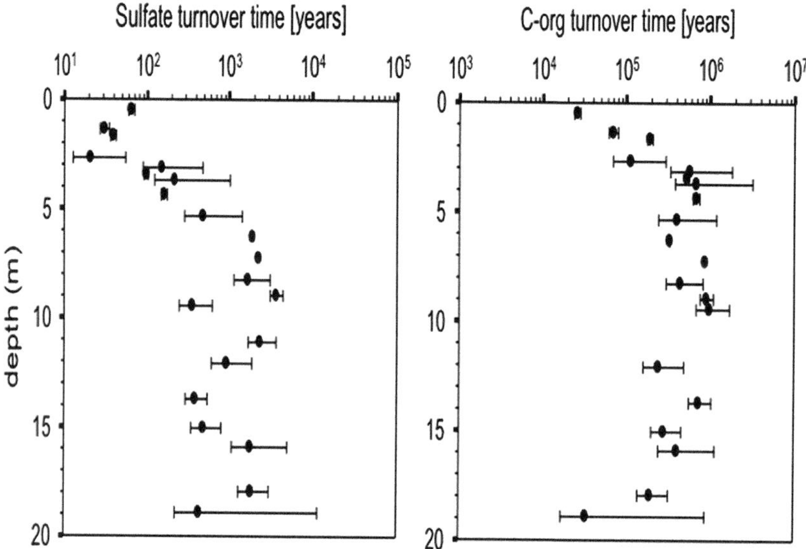

FIGURE 12.3 Local sulfate turnover time (left) calculated from the experimentally determined SRR, and the SO_4^{2-} profile, and organic carbon turnover time, calculated from the TOC profile and the local carbon degradation rates (obtained from the SRR, assuming a stoichiometry organic carbon: SO_4^{2-} of 2:1).

(including humins, wood, and peat). The mobile OM in the upper sediments, less than 7 m below the seafloor, was older than the sediments, and in the deeper sediments, the mobile organics were younger than the sediments (Fig. 12.4). This can only be explained by a gradual exchange of pore water with solutes and colloids between the seawater and deeper sediments, transporting "old" material upward and "young" material downward. This transport brings down sulfate and relatively fresh OM that drives sulfate reduction in very old sediment layers. This "deep biosphere circulation" is fundamentally different than the deep (but much less deep) transport from the center of the flat to the low water line. That flow is well understood and simple. The flow paths to the deep biosphere are probably convoluted and have changed during their typical life span. The forces driving this flow can only be speculated upon.

In summary, three transport phenomena control exchange between intertidal flats and seawater (Billerbeck et al., 2006a) (Fig. 12.5). Surficial "skin" exchange occurs during high tide and is driven by the interaction of currents and waves with ripples. This process interacts with sediment transport, i.e., migration of ripples. During low tide, "deep body" circulation occurs, leading to seepage of highly reduced pore water at the low water line. The "deep body" circulation might lead to a significant release of methane and nutrients (Middelburg et al., 2002; Røy et al., 2007). A third "deep biosphere" circulation brings down seawater components to a depth of at least 20 m, sustaining microbial degradation processes. The "skin" exchange occurs in minutes, the "deep body" circulation in decades, and the "deep biosphere" circulation in millennia. The interaction and dynamics of the transport phenomena on intertidal flats characterize a microbial realm that has large variations of chemical gradients

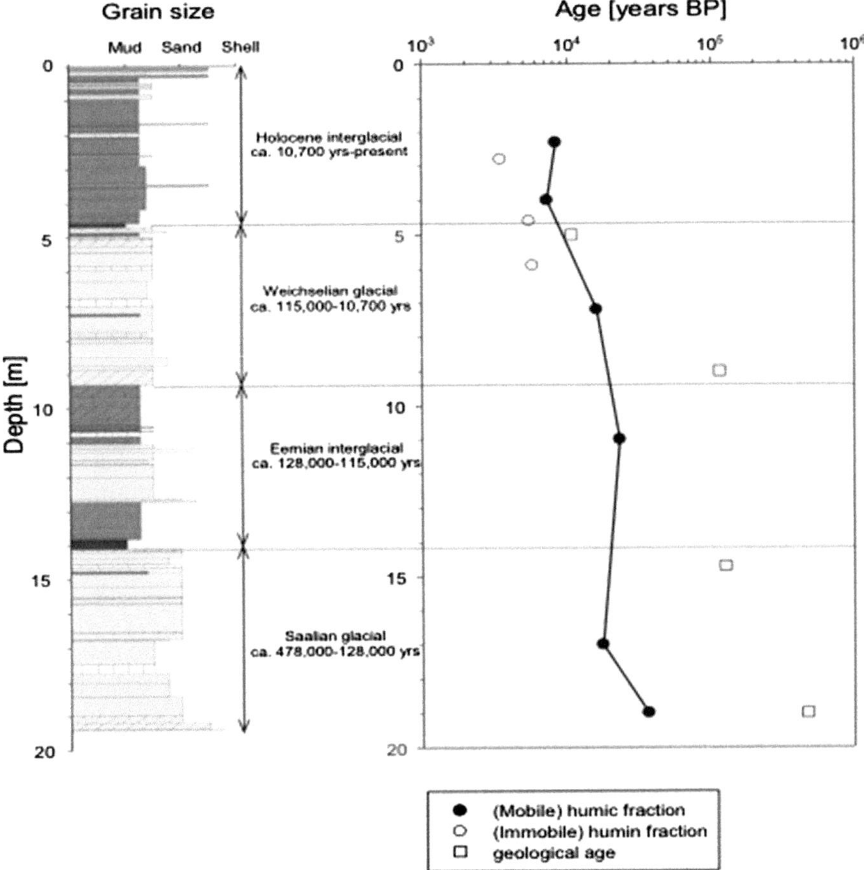

FIGURE 12.4 Left: lithological profile of the sediment core reproduced from Beck et al. (2011) (Orange: shell bed; yellow: sand; green: mud; red: peat; diagonally hatched: intervals stained by humous matter). Right: radiocarbon age of mobile humic fraction (*filled circles*) and immobile humin fraction (*open circles*), and geological age based on lithostratigraphic methods (*open squares*).

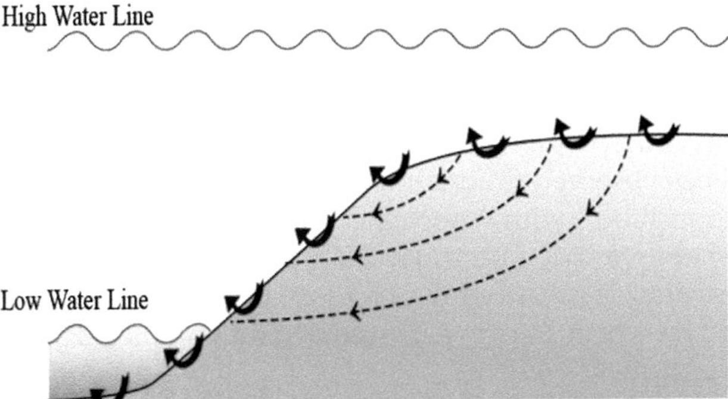

FIGURE 12.5 Cartoon illustrating the flow through intertidal sand banks at high and low tide. Bold arrows indicate "skin" circulation during high tide, whereas *dashed lines* indicate "body" circulation. The "deep biosphere" circulation proceeds at depths well below (penetrating depths of at least 20m) where body circulation occurs.

and fluctuating availabilities of electron donors and electron acceptors. This couples microbial activity and communities rather to physical parameters, such as permeability (Probandt et al., 2017), than to solute distributions and leads to specific metabolic adaptations (Marchant et al., 2017). The consequences of these processes and fluxes for global elemental budgets need to be further assessed.

3. MICROBIAL PROCESSES

The role of microbes in element cycling and biogeochemistry of tidal flats is easily underestimated. The original concept that life on tidal flats is dominated by herbivorous macrobenthos has been replaced by a view of microbial primary producers and heterotrophic bacteria regulating energy flow. For example, all benthic filter and deposit feeders (bivalves and polychaetes) were estimated to consume less than 25% of the yearly planktonic primary production; the remaining was either washed out to sea or microbially degraded (Warwick et al., 1979). Similarly, in tidal flats of the Wadden Sea, 75% of the annual organic input (the sum of primary production and deposition of detritus) was primarily degraded by microorganisms, with only 25% consumed by macrofauna (Kuipers et al., 1981).

As in other sediments, the biogeochemical processes in tidal flat sediments are more or less stratified. Primary production is limited to the photic zone of several mm (De Beer et al., 2004; Werner et al., 2006) during low tide or an even thinner zone during inundation (Billerbeck et al., 2007). Chemoautotrophic processes occur in the oxic zone, particularly the oxidation of ammonium, Fe^{2+}, and sulfur, but these CO_2 fixation rates are much lower than photosynthesis. A first step in the degradation of OM is hydrolysis of biopolymers, a complex process in which the Cytophaga–Flavobacterium consortium is essential (Kim et al., 2004; Moss et al., 2006; Musat et al., 2006). The main reductive processes are stratified in the classical sequence of electron acceptors: O_2, NO_3^-, Mn(IV)/Fe(III), SO_4^{2-}, and CO_2. Degradation of OM is intense in the oxic zone; denitrification and sulfate reduction rates are highest near the oxic–anoxic transition; and these processes do not appear to be entirely inhibited within the oxygen-rich zone (Dilling and Cypionka, 1990; Marschall et al., 1993; Wilms et al., 2006).

In marine sediments, approximately half of the OM is degraded by sulfate reduction (Fossing and Jørgensen 1989, 1990; Jørgensen, 1982; Sorensen et al., 1981). Methanogenesis was thought to be of minor importance in the majority of marine sediments because of competition with sulfate reduction (Reeburgh, 1983). In particular, on well-flushed tidal flats, frequent replenishing of sulfate pools was thought to limit the importance of methanogenesis. This concept might need revision, however, as a significant methane efflux from estuaries has been observed (Middelburg et al. 1996, 2002; Røy et al., 2007) and a clear sulfate–methane transition with associated distribution of sulfate reducers and methanotrophic archaea was found in tidal flat sediments (Wilms et al., 2007). The typical pattern of vertical stratification in redox processes is broken by bioventilating organisms, particularly around the worm burrows, resulting in a complex 3D geometry of oxic and anoxic zones.

4. ORGANIC MATTER SOURCES AND TRANSFORMATIONS

Tidal flats receive OM from a wide range of terrestrial and marine sources. Marine sources include autochthonous (on-site) production by benthic micro- and macroalgae, bacteria, and burrowing animals such as mussels (Meziane et al., 1997; Meziane and Tsuchiya 2000, 2002; Rohjans et al., 1998; Seidel et al., 2012; Volkman et al., 1980) and detritus from seagrasses and from pelagic algal blooms (Freese et al., 2008; Rohjans et al., 1998; Volkman et al., 1980). Allochthonous (external) sources include terrigenous material from adjacent mangroves and salt marshes (Meziane et al., 1997; Meziane and Tsuchiya, 2000), eroded peats (Volkman et al., 2000), and discharge of OM from anthropogenic sources (Meziane and Tsuchiya, 2002). Additionally, freshwater-impacted tidal flats receive high amounts of terrigenous OM via surface river runoff or when meteoric groundwater enters coastal aquifers in a subterranean estuary (Kim et al., 2012; Middelburg and Herman, 2007).

A major proportion of the deposited OM in tidal flats with limited riverine inputs originates from marine primary production (Freese et al., 2008). The dissolved OM (DOM) and particulate OM (POM) can be supplied to the sediments via rapid burial and sedimentation, bedform-generated convection, and tide-driven seawater circulation. This OM, in turn, fuels aerobic and anaerobic microbial remineralization in the sediments (Beck et al., 2009; Billerbeck et al., 2006a; Gao et al., 2009; Huettel and Rusch, 2000; Røy et al., 2008; Seidel et al., 2012). Consequently, most sediment respiration in tidal flats appears to be driven by autochthonous production by benthic microalgae (Cook et al., 2004b; Moens et al., 2002). The carbon isotopic composition of the terminal remineralization products of OM (methane and DIC) in the seeping pore water from an intertidal sand flat in the Wadden Sea pointed to less than 50-year-old carbon previously fixed by benthic photosynthesis (Røy et al., 2008). This indicates that in sandy tidal flats, the continuous advective transport of algal material contributes significantly to OM remineralization.

Oxygen depletion within the top few millimeters of diffusion-dominated muddy tidal flats documents intense microbial remineralization (Böttcher et al., 1998; Jansen et al., 2009). Pore water advection in permeable sands at tidal flat margins can further increase microbial activities because of the supply of electron acceptors and OM (Billerbeck et al., 2006a; Huettel and Rusch, 2000). This is confirmed by high microbial cell numbers in the sediments (10^8–10^9 cells per gram sediment), high sulfate reduction rates, and an enrichment of remineralization products (Beck et al., 2008; Gittel et al., 2008; Riedel et al., 2011). Deeper sediments (meters), central parts of tidal flats, and fine-grained diffusion-dominated mudflats are generally microbially less active than sandy tidal flat margins because of the lack of supply of fresh OM and nutrients by pore water advection (Beck et al. 2009, 2011; Røy et al., 2008).

Low molecular weight organic molecules such as monosaccharides, amino acids, and organic acids can be taken up directly by bacteria and are therefore turned over quickly. Heterotrophic bacteria in sediments are thus coupled closely to these substrates, and carbon from microphytobenthos is rapidly transferred to the heterotrophic bacterial biomass (Middelburg et al., 2000; Miyatake et al., 2014). Permeable tidal flat recharge zones with percolating seawater therefore act as effective biofilters for bioavailable marine DOM such as dissolved carbohydrates from benthic and planktonic primary production (Seidel et al., 2014).

The more easily degradable marine OM is largely consumed in the upper meter of tidal flat sediments, resulting in a relative increase of terrestrial POM in deeper, fine-grained sediments (Freese et al., 2008). The presence and shift of different types of organic substrates can, in turn, affect the composition of the bacterial communities in the sediments (Bohórquez et al., 2017; Seidel et al., 2012; Wilms et al., 2006).

While bioavailable DOM in shallow permeable sediments is degraded quickly, less bioavailable DOM can accumulate in the deeper sediments (Beck et al., 2008; Kim et al., 2012; Seidel et al., 2014). An increase of dissolved organic carbon (DOC) concentrations with depth generally documents the microbial breakdown of particulate sediment OM and the incomplete remineralization of DOM under anoxic conditions (Burdige and Gardner, 1998). The enzymatic hydrolysis of macromolecular POM by microbes leads to the formation of high molecular weight DOM, which is then further fermented or hydrolyzed to low molecular weight DOM and monomeric compounds. The latter serves as electron donors for terminal oxidizing microbes, such as methanogens or sulfate-reducing bacteria (Burdige and Gardner, 1998). In turn, the reduction of sulfate produces sulfide and the DOM is remineralized to metabolic end products such as DIC, ammonium, and phosphate. The water column enrichment of remineralization products from pore water discharge is documented by strong tidal cycles of biogeochemical constituents in the water column (Beck and Brumsack, 2012; Grunwald et al., 2009; Moore et al., 2011).

The detailed molecular-level characterization of tidal flat pore water DOM by ultrahigh-resolution mass spectrometry revealed a strong increase of nitrogen- and sulfur-containing molecules that increased with pore water age toward the seepage zone at the low-tide water line (Seidel et al., 2014). This was proposed to be the result of the incomplete microbial degradation of nitrogen- and sulfur-rich marine POM. At the same time, more oxidized and more aromatic compounds were found in the deeper and older pore waters, suggesting an accumulation of these compounds as part of the DOM pool that is less reactive under anoxic conditions (Seidel et al., 2014). There is, however, also evidence for the degradation of aromatic terrestrial DOM such as lignins under sulfate-reducing conditions in tidal sediments (Dittmar and Lara, 2001).

Sulfide can abiotically react with sedimentary OM, which has been proposed as a stabilizing mechanism that may render POM more biorefractory (Sinninghe Damsté and De Leeuw, 1990). There is accumulating evidence that DOM also reacts abiotically with sulfide producing dissolved organic sulfur compounds in a wide range of sulfidic environments including tidal flat sediments (Gomez-Saez et al., 2016; Schmidt et al., 2009; Seidel et al., 2014). Yet, whether the sulfurized DOM is also intrinsically more stable in the pore water or even in the water column on discharge remains to be answered.

When rivers pass through intertidal estuaries with fringing mangroves or salt marshes, their terrigenous DOM signature is significantly altered by in situ processes leaving specific DOM imprints (Bittar et al., 2016; Dittmar et al., 2001; Medeiros et al., 2015). Comparable with estuaries, the export of pore water DOM from tidal flats to the coastal ocean by submarine groundwater discharge can be traced as chromophoric DOM (Lübben et al., 2009) and it can include terrestrial (Kim et al., 2012) and refractory DOM, such as dissolved black carbon (Dittmar et al., 2012). Applying radium and radon isotope measurements, it was estimated that about 1% of the Wadden Sea low-tide water is derived from the adjacent tidal flats (Moore et al., 2011; Santos et al., 2015). The fact that pore water concentrations of nutrients and DOM can be orders of magnitude higher than seawater concentrations demonstrates that pore water drainage, exchange, and mixing from tidal flats exhibit an important control on carbon dynamics in coastal oceans.

5. NITROGEN CYCLE

Nitrogen is a critical nutrient element that is required by all organisms for cellular biosynthesis. Primary production and/or decomposition can be limited by the availability of fixed nitrogen (Ryther and Dunstan, 1971; Smith, 1984). In addition, oxidized nitrogen species, mainly nitrite or nitrite, can be used as a terminal electron acceptor during anaerobic respiration. The sediment nitrogen cycle has received considerable attention in recent years because of the key role of nitrogen in regulating coastal productivity (Canfield et al., 2005). Here we focus on the literature relating specifically to tidal flats.

5.1 Nitrogen Fixation

Nitrogen (N_2) fixation in aquatic sediments is carried out by bacteria and some members of the archaea (Canfield et al., 2005). Rates of N_2 fixation by benthic cyanobacteria are generally much higher than those supported by heterotrophic bacteria and chemoautotrophs; however, their importance to total system N_2 fixation is often constrained by their limited areal coverage (Howarth et al., 1988a). The exception to this appears to be in the tropics, where cyanobacteria-dominated microbial mats thrive in the sediments of shallow subtidal and intertidal areas and contribute high rates of N_2 fixation to the ecosystem (Joye and Lee, 2004; Lee and Joye, 2006; Paerl et al., 2000; Pinckney and Paerl, 1997). Tidal flat cyanobacteria are often associated high rates of N_2 fixation in temperate habitats as well (Bautista and Paerl, 1985; Cook et al., 2004a; Gotto et al., 1981; Howarth et al., 1988b; Stal, 1995).

Many factors identified as being conducive to cyanobacterial growth and N_2 fixation (Howarth et al., 1988a) exist on tidal flats. Relatively high light levels mean that energy is not a limiting factor for the ATP-demanding process of N_2 fixation. There is an abundant supply of the trace metals, e.g., molybdenum and iron, which are required for synthesis of nitrogenase, the enzyme that mediates N_2 fixation and their availability could stimulate sediment N_2 fixation. High concentrations of DOC within the sediment may also chelate trace metals, increasing their bioavailability. In fact, trace metal additions did not stimulate N_2 fixation in tidal flat sediments (Lee and Joye, 2006; Paerl et al., 1993), suggesting that benthic N_2 fixing microbes were not metal limited. Most of the terrestrially derived bioavailable phosphorus entering the coastal zone is adsorbed to particle surfaces. On deposition of these particles on tidal flats, reductive dissolution may lead to a release of phosphorus from the sediment, enriching the benthic flux of P relative to N, thus potentially stimulating N_2 fixation.

Strong diurnal variations in N_2 fixation are often observed depending on the type of cyanobacteria present. Where nonheterocystous cyanobacteria such as *Oscillatoria* spp. dominate tidal flat communities, maximum rates of N_2 fixation may be observed at sunrise and sunset or in the dark (Cook et al., 2004a; Stal et al., 1984; Villbrandt et al., 1991) because of oxygen inhibition of nitrogenase activity during the day. In contrast, heterocystous cyanobacteria may exhibit maximum N_2 fixation rates during illumination

(Carpenter et al., 1978; Currin and Paerl, 1998), although some heterocystous species are also known to temporally decouple N_2 fixation and photosynthesis (Paerl, 1990). Variations in N_2 fixation rates have also been observed depending on inundation and exposure, with reduced N_2 fixation rates being observed during submersion in epiphytic cyanobacterial communities (Currin and Paerl, 1998). No difference in N_2 fixation rates was, however, observed between inundation and exposure on a tidal flat sediment (Cook, 2003).

Inorganic N availability also plays an important role in regulating N_2 fixation rates (Carpenter et al., 1978), and rapid decreases in N_2 fixation rates have been observed on tidal flats in response to nitrate delivery by storm runoff events (Joye and Paerl, 1993). However, in some tropical microbial mats, high concentrations of ammonium (>500 μM) inhibited N_2 fixation (Lee and Joye, 2006). In temperate systems, strong seasonal variations in N_2 fixation on tidal flats are often observed with the highest rates generally occurring during spring and summer, when light levels are highest and nitrogen is most limiting (Carpenter et al., 1978; Cook et al., 2004a; Villbrandt et al., 1991). In tropical systems, rates of N_2 fixation vary less seasonally with observed activity patterns driven mainly by changes in precipitation or tidal inundation (Lee and Joye, 2006).

In comparison to temporal variations in N_2 fixation, much less is known about the spatial distribution of N_2 fixation on tidal flats because most published studies so far have focused on light and nutritional controls on N_2 fixation. Substantial gradients in algal biomass and N_2 fixation occur across tidal flats and activity varies by an order of magnitude over distances of meters (Cook et al., 2004a). It is likely these gradients are mediated by light levels, exposure to wave energy, and grazing.

Quantifying the exact rates of N_2 fixation is complicated by the fact that the generally used technique for measuring N_2 fixation, the acetylene reduction assay, is an indirect method. Thus, assumptions must be made to convert moles of acetylene reduced to N_2 equivalents. For cyanobacterial mats, a factor close to the theoretical range of 3–4 can be used based on calibrations of the acetylene reduction assay with $^{15}N_2$ (Cook et al., 2004a; Howarth et al., 1988b). For consistency with the existing literature, we use a ratio of 3 here where rates are reported as acetylene reduction rates. Another complicating factor is that rates are often reported relative to chlorophyll *a* rather than on an areal basis (Potts, 1979), making it difficult to compare rates between studies.

In general, rates of N_2 fixation in cyanobacterial mat communities are among the highest measured in aquatic systems, ranging from 0.09 to 5.4 mol N m^{-2} y^{-1}. These rates are much greater than those measured in unvegetated (0–0.11 mol N m^{-2} y^{-1}) or vegetated (0–3.64 mol N m^{-2} y^{-1}) subtidal sediments (Howarth et al., 1988b). These rates of nitrogen fixation (up to 75 g N m^{-2} y^{-1}) reflect a higher load than experienced by highly fertilized farm fields (20 m^{-2} y^{-1}). Therefore, one would expect tidal flats colonized by cyanobacteria to exhibit some of the highest N_2 fixation rates observed in aquatic ecosystems. Given the high spatiotemporal variation in N_2 fixation rates and the limited data available, we have represented the rates measured on mudflats on an hourly basis. Rates are often essentially 0 when there is low cyanobacterial biomass (during autumn and winter). During spring and summer, when cyanobacteria are abundant, rates generally fall in the range of ~10–100 μmol N m^{-2} h^{-1} with extreme highs of ~350 μmol N m^{-2} h^{-1} also noted (Abd. Aziz and Nedwell, 1986; Bautista and Paerl, 1985; Carpenter et al., 1978; Cook et al., 2004a; Gotto et al., 1981; Joye and Paerl, 1993; Stal et al., 1984).

5.2 Benthic Microalgal N Assimilation

Another important factor influencing nitrogen cycling on tidal flats is the ubiquitous benthic microalgae that coat the flats (Fig. 12.6). Nitrogen fluxes across the sediment–water interface of tidal flats often show a distinct diurnal variation as a consequence of nitrogen assimilation by benthic microalgae at the sediment surface. During the light periods, sediment nitrogen effluxes are greatly reduced or reversed and sediment uptake rates increase (Cabrita and Brotas, 2000). The result is efficient nitrogen retention and recycling within the sediment that can promote its productivity (Sundbäck and Miles, 2000; Veuger et al., 2007). Measuring the nutrient assimilation rate of benthic microalgae is much more complicated than for pelagic algae because they derive a large fraction of their nutritional requirement from nutrients produced within the sediment. Simply estimating the difference between light and dark fluxes will underestimate nutrient assimilation rates because benthic microalgae are capable of assimilating nutrients for long periods in the dark. Estimates of N assimilation by benthic microalgae are often made based on measured rates of photosynthesis and then applying a C:N ratio measured for algal cells (often ~8–10) (Dong et al., 2000). This stoichiometric approach is often used to estimate the relative amount of N assimilated by benthic microalgae so that this value can be compared to the amount of N that is denitrified. Within the Colne estuary, benthic microalgae assimilated a similar amount of N to that denitrified at a tidal flat site in the upper estuary, but that the amount of N assimilated was ~2–4-fold greater than that denitrified at a site in the lower estuary. Similarly, in the Tagus estuary,

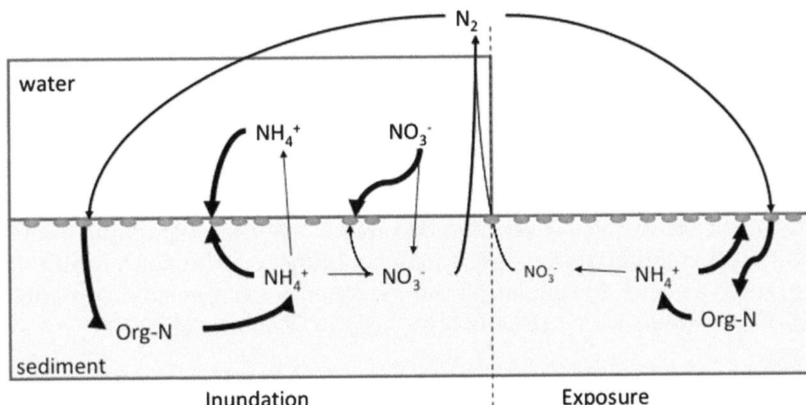

FIGURE 12.6 A conceptual diagram of aspects of the N cycle pertinent to tidal flats. Benthic microalgae (●) play a key role in the N cycle on tidal flats assimilating N from both the sediment and the water column, the flows of N through benthic microalgae are likely to dominate the N cycle. Organic N is mineralized in the sediment-releasing NH_4^+, a large fraction of which is assimilated by benthic microalgae. Nitrifying bacteria may be out competed by benthic microalgae for NH_4^+, and thus nitrification rates are likely to be low (thus limiting denitrification rates), particularly in oligotrophic and mesotrophic systems. In eutrophic systems, rates of denitrification may be high, driven to a large extent by NO_3^- from the water column and under these circumstances tidal flats mat be an important sink for fixed N. During exposure, exchange between the sediment and the water column is cut off resulting in a depletion of NO_3^- in the porewaters and an enrichment in NH_4^+, which may be released on inundation. Where the biomass of cyanobacteria is high, N_2 fixation rates may be extremely high, and in these circumstances tidal flats may be a net source of fixed N. The thickness of the arrows gives a rough indication of the relative magnitude of the N flows.

on average twice as much N was assimilated by microphytobenthos (MPB) than was denitrified (Cabrita and Brotas, 2000).

We note, however, that the stoichiometric approach used for these calculations neglects the fact that benthic algae exude a large fraction of C fixed as DOM (Porubsky et al., 2008). Therefore, applying the C:N ratio measured in cells may not be appropriate, particularly in systems with low dissolved inorganic N concentrations. A more direct estimate of nutrient assimilation by benthic microalgae on a tidal flat was made by Cook et al. (2004a), who estimated total nutrient assimilation from nutrient uptake in cores, the upward flux of ammonium (based on pore water profiles), and measured rates of N_2 fixation. When compared with rates of photosynthesis, the C:N uptake ratio was conservatively estimated at being between 17 and 52 during all the seasons except winter. Using nitrogen stable isotopes (^{15}N), it was possible to show that benthic microalgae were responsible for ~25% of nitrogen uptake in the light, mostly as ammonium (Dähnke et al., 2012). Nitrate uptake was much less important, accounting for only 6% of the total ^{15}N-nitrate added.

5.3 Benthic Microalgal N Uptake and Storage

Nitrogen assimilation is not the only mechanism by which benthic microalgae influence nitrogen availability in tidal flat sediments. Diatoms, an important component of the benthic microalgal community, at least some of the time (Longphuirt et al., 2009; Stief et al., 2013), take up and store nitrate at high concentrations within their cells (Kamp et al. 2011, 2015; Lomas and Glibert, 2000). They assimilate this stored nitrate when light and oxygen are available (Kamp et al., 2015) but reduce it to ammonium when grown under dark, anoxic conditions (Kamp et al., 2011). It is likely that they use the energy generated from nitrate reduction to enter a resting state that allows them to survive up to months of burial. This may be an important survival strategy for organisms inhabiting an environment where burial by sediment transport is a regular part of life. Intracellular nitrate can be an important component of the total nitrogen pool in tidal flat sediments and can be found in deep sediment layers where pore water nitrate is no longer available (Garcia-Robledo et al., 2016; Stief et al., 2013). The extent to which this nitrate "leaks" and becomes available to the sediment microbial community is still an open question, though it may be a function of diatom health. For example, Intracellular nitrate fueled extracellular nitrate reduction activity in sinking marine diatom aggregates in which diatoms may not have been healthy (Kamp et al., 2016).

5.4 Nitrification and Nitrate Reduction

Nitrification refers to the sequential oxidation of ammonia to nitrite and ultimately nitrate. Nitrification is regulated by substrate availability (oxygen and ammonium) and by hydrogen sulfide, which inhibits ammonia oxidation, the first step of the process (Joye and Anderson, 2008; Joye and Hollibaugh, 1995). Nitrification links the reduced and oxidized portions of the nitrogen cycle and is often closely coupled to nitrate reduction processes. The most well-studied pathway of nitrate reduction is denitrification, the process whereby nitrate is utilized as a terminal electron acceptor for an oxidation reaction and in the process is reduced to nitrous oxide or dinitrogen gas. Denitrification has received much interest over the past 30 years because it represents a significant sink for fixed nitrogen in aquatic ecosystems.

Denitrification is a facultatively anaerobic process that usually takes place below the oxic surface layer of sediment. At least in permeable sediments, denitrification is also active in the presence of oxygen (Gao et al., 2010; Rao et al., 2008), though its rate is typically lower than in deeper anoxic sediment (Gao et al., 2010). Aerobic respiration and denitrification have even been shown to occur simultaneously in sediment from an intertidal sand flat (Marchant et al., 2017). In spite of relatively low denitrification rates in the presence of oxygen, aerobic denitrification can be an important nitrogen sink at the ecosystem level due to the deep penetration of oxygen and nitrate (up to 5–6 cm) into permeable sediments (Billerbeck et al., 2006b; Werner et al., 2006).

Conceptually, denitrification is often partitioned into denitrification fueled by nitrate transported into the sediment from the water column, and that produced within the sediment by nitrification. Nitrate transport into the sediment, in turn, can be driven by three primary processes: diffusion, advection, and bioturbation. Denitrification and coupled nitrification–denitrification will be considered for each of these three scenarios in the following paragraphs. Most of the measurements of denitrification on tidal flats over the last 20 years have used the isotope pairing technique and for consistency we only consider the rates measured using this method here.

In cohesive, impermeable tidal flat sediments, denitrification is influenced strongly by the amount of nitrate available in the water column. If water column nitrate concentrations are low, denitrification will be driven mainly by nitrification within the sediment. As water column nitrate concentrations increase, the diffusive flux of nitrate into the sediment increases proportionally, as does the denitrification rate that it fuels. This effect may plateau at extremely high nitrate concentrations (Dong et al., 2000). On tidal flats, the highest rates of denitrification take place when the sediments are exposed to high nitrogen concentrations, such as those on the European north Atlantic coast where nitrate concentrations can be up to 1 mM (Dong et al., 2000; Nedwell et al., 1999). Because water column nitrate concentrations are generally highest during winter, the maximum rates of denitrification are often measured during the winter as well (Cabrita and Brotas, 2000; Ogilvie et al., 1997). Denitrification rates on tidal flats in eutrophic estuaries fall in the range of 10–100 μmol N m^{-2} h^{-1} (Cabrita and Brotas, 2000; Dong et al., 2000; Ogilvie et al., 1997). In more mesotrophic tidal flats, denitrification rates are much lower, ranging from undetectable to generally <10 μmol N m^{-2} h^{-1} (Cook et al., 2004a; Trimmer et al., 2000). In the oligotrophic tidal flats of coastal Georgia, in situ denitrification rates were low (1–10 μmol N m^{-2} h^{-1}) (Cook et al., 2004a; Trimmer et al., 2000) but rates increased by more than tenfold when the sediments were amended with nitrate to mimic a nutrient-rich runoff event (Porubsky et al., 2009).

During exposure, the nitrate source from the water column is eliminated, leaving only sediment nitrification to fuel denitrification. To date only one study has measured denitrification during exposure of intertidal flats, finding that denitrification rates were similar during exposure and inundation at night (Ottosen et al., 2001). Coupled nitrification–denitrification was much lower during the day, possibly because increased salinity and competition for NH_4^+ with benthic microalgae inhibited nitrification. These authors also noted that denitrification rates were three times higher in a nearby subtidal location, suggesting that intertidal conditions may have a generally inhibiting influence on nitrification and denitrification. A similar observation was made by Cook et al. (2004a) who found that primarily coupled nitrification–denitrification rates were consistently highest on the lowest parts of tidal flats.

This observation is not universal, however, and we note that Cabrita and Brotas (2000) observed no effect of elevation on denitrification rates. On a more localized scale, there have been observations of differences in sedimentary nitrification (and presumably denitrification rates) between ridge and runnel structures on tidal flats, with nitrification rates being sixfold higher in runnel structures (Laima et al., 1999).

In permeable sediments, such as those found on sandy tidal flats, the advective supply of nitrate increases denitrification rates by a factor of 7–10 over rates measured in the same sediment without advection (Canion et al., 2014; Gao et al., 2012; Rao et al., 2008). Potential denitrification rates under advective conditions are typically in the 200–300 μmol N m^{-2} h^{-1} and have been measured as high as 1000 μmol N m^{-2} h^{-1} (Gao et al., 2012; Marchant et al., 2014). Interestingly, these rates are independent of the water column nitrate concentration, at least over concentration ranges of tens to hundreds of micromolar typical of mesotrophic to eutrophic tidal flats. This is due to the high affinity of denitrifying microorganisms for nitrate; with apparent half-saturation constants as low as 1.5 μM (Evrard et al., 2013), the advection of seawater containing only a few micromolar nitrate through the sediment is sufficient to saturate denitrification. Coupled nitrification–denitrification is less important than water column–driven denitrification in this setting because (1) the nitrate flux from the water column dominates that derived from nitrification, (2) pore water advection flushes out ammonium generated from remineralization of OM in the sediment, making it less available for nitrification, and (3) flow dynamics beneath rippled sand beds limits interactions between oxic and anoxic sediment layers (Kessler et al., 2012), preventing overlap of ammonium and oxygen within the sediment. Furthermore, there is some evidence that nitrification rates are quite low in permeable intertidal flats relative to other coastal and benthic marine sediments (Marchant et al., 2016).

Coupled nitrification–denitrification is enhanced in sediment where bioirrigation is an important transport process (Heisterkamp et al., 2012; Henriksen et al., 1983; Na et al., 2008). Infauna such as worms consume particulate matter and excrete ammonium while pumping oxygenated water into the sediment, stimulating nitrification rates and increasing nitrate availability for denitrification. Worm activity also stimulates the growth and activity of benthic microalgae by transporting nutrients from deep remineralization sites to the surface (Chennu et al., 2015). The extreme spatial heterogeneity in sediment structure and chemistry induced by worm activity (Chennu et al., 2015) makes it difficult to quantify areal rates of processes without combining high-resolution imaging, rate measurements, and modeling.

The activity of benthic microalgae also influences rates of nitrification and denitrification in tidal flat sediments. There are two primary mechanisms behind this influence: alteration of redox regimes and direct assimilation or uptake and storage of inorganic nitrogen (De Beer 2000, 2002). This relationship has been extensively studied under the condition that the supply of nitrate from the water column was driven by diffusion. Overall, the impact of benthic microalgae on denitrification seems to be negative based on a metaanalysis of denitrification from 18 European estuaries: net autotrophic sediments had significantly lower rates of denitrification than did net heterotrophic sediments (Risgaard-Petersen, 2003). This results from direct competition for inorganic nitrogen between benthic microalgae and nitrifying microorganisms. Mesocosm studies have clearly demonstrated a negative impact of the presence of benthic algae on nitrification and denitrification rates (Risgaard-Petersen, 2003, Risgaard-Petersen et al., 2004), and it appears that the competition for nutrients arises from both benthic microalgae and from actively growing heterotrophic bacteria that are fed by algal exudates. These findings also accord with field studies of denitrification on tidal flats, which have reported negative correlations with benthic microalgae-associated parameters (such as chlorophyll *a* and photosynthesis) and denitrification (Cabrita and Brotas, 2000; Cook et al., 2004a).

There are also reports of benthic microalgae enhancing coupled nitrification–denitrification by increasing the depth of the oxic zone in the sediment through photosynthetic oxygen production (An and Joye, 2001; Risgaard-Petersen et al., 1994). Denitrification driven by nitrate supplied from the water column will be reduced, however, as the distance that nitrate has to travel through the sediment oxic zone to the zone of denitrification is increased (Risgaard-Petersen et al., 1994). Enhanced nitrification–denitrification is only possible if sufficient inorganic nitrogen is supplied to the sediment to overcome inhibition of nitrification by competition with the benthic microalgae. The influence of benthic microalgae on denitrification under the condition that advection drives nitrate supply to the sediment has not been well studied. One study showed that 20%–40% of nitrate supplied to sediment from a sandy tidal flat was taken up and stored in the intracellular nitrate pool (Marchant et al., 2014).

The importance of denitrification on tidal flats in attenuating N loads to the coastal zone is quite variable. In the Lower Great Ouse estuary, it was estimated that between 2% and 56% of the nitrate load was attenuated by denitrification on the tidal flats depending on season (Trimmer et al., 1998). This compares to an attenuation rate of 32%–44% of the nitrate load estimated for the Colne estuary (Ogilvie et al., 1997). In the Tagus estuary, the tidal flats in the lower estuary attenuated 35% of the incoming dissolved inorganic nitrogen (DIN) load compared with only 4% in the inner estuary (Cabrita and Brotas, 2000). Permeable intertidal and subtidal sediments in the Wadden Sea were estimated to attenuate ~30% of the total nitrogen loading to the system (Gao et al., 2012).

Other nitrate reduction pathways include the anammox reaction and dissimilatory nitrate reduction to ammonium (DNRA). DNRA tends to occur in reducing sulfidic sediments (Brunet and Garcia-Gil, 1996; Christensen et al., 2000), as well as at temperature extremes (Kelly-Gerreyn et al., 2001). Such conditions often occur on tidal flats, and substantial rates of DNRA have been inferred in these environments, with the proportion of nitrate reduced to ammonium varying between 1% and 10% for cohesive sediments with a steady-state diffusive flux of nitrate from the water column (Kelly-Gerreyn et al., 2001). However, this fraction was found to increase to 37%–77% when the same sediments were slurried with seawater (Behrendt et al., 2013). This was most likely an artifact of increased nitrate supply to DNRA microorganisms in the slurry incubations. In intact cores, maximum DNRA activity was observed slightly deeper than maximum denitrification activity, where more sulfide and less nitrate were available (Stief et al., 2010). Thus, slurry incubations may overestimate the importance of DNRA in many sediment types. The fraction of nitrate reduced to ammonium may be systematically higher in permeable sediment where nitrate is delivered to deeper sediment strata by advection. Indeed, DNRA made up ~20% of total nitrate reduction at a sandy tidal flat with an advective nitrate supply (Marchant et al., 2014). Anammox generally makes up only a small fraction of total nitrate reduction rates in nearshore sediments (Dalsgaard and Thamdrup, 2002). This is also generally true for tidal flat sediments. Maximum anammox contributions of 6%–14% of total nitrate reduction have been observed for arctic and temperate sites (Canion et al., 2014), with negligible contributions reported for subtropical sites (Canion et al., 2014) and another temperate location (Marchant et al., 2014).

The ratio of electron donor (i.e., labile organic carbon) to electron acceptor (i.e., nitrate) plays a key role in determining, whether heterotrophic, i.e., denitrification, DNRA, or autotrophic, i.e., anammox, processes (Algar and Vallino, 2014). Modeling results showed that anammox dominated when the ratio of CH_2O/nitrate was less than 1.5, denitrification dominated between CH_2O/nitrate ratios of 1.5–2.5 and DNRA dominated at CH_2O/nitrate ratios greater than 4. In marsh sediments, hydrogen sulfide, which can fuel autotrophic denitrification and DNRA, played an important role in regulating the pathway of nitrate reduction (Porubsky et al., 2009). Future studies in tidal flats should explore the role(s) of organic carbon and sulfide as potential mechanisms that regulate the fate of nitrate.

5.5 Exchange of Dissolved Nitrogen Between the Sediment and the Water Column

The exchange of nitrogen between the sediment and the overlying water is modulated by the periodic exposure to the atmosphere. During exposure, ammonium accumulated in the pore water and may be released in a pulse into the overlying water during inundation. In permeable sediments, this process is enhanced by buoyancy-driven pore water exchange (Falcao and Vale, 1995; Rocha, 1998). At the edge of tidal sand flats, extremely high fluxes of ammonium may be observed on the low/falling tide driven by tidally induced pore water circulation phenomena described earlier (Billerbeck et al., 2006a, Fig. 12.7). Such circulation patterns would also impact phosphorus and silicon cycling (discussed further later in the chapter). In situ flume experiments have also shown that ammonium exchange rates scale with current velocity, underscoring the importance of advective flushing in mediating fluxes on inundation of tidal flats (Asmus et al., 1998).

Impacts of exposure have also been observed in cohesive tidal flat sediments, with an increased efflux of ammonium observed after inundation following a 6.5 h exposure period in the dark. Interestingly, the same exposure period in the light lead to a large net uptake of ammonium on inundation, highlighting how interactions between exposure and illumination can influence benthic fluxes in tidal flats (Thornton et al., 1999). In areas with high pore water nitrate concentrations, nitrate becomes depleted in the pore water through nitrate reducing processes during exposure (Malcolm and Sivyer, 1997) and increased sediment uptake rates of nitrate have been observed after exposure (Feuillet-Girard et al., 1997).

5.6 N Cycle Summary

The N cycle on tidal flats is highly dynamic with large changes in rates occurring temporally and spatially driven by interactions between sediment permeability, mass transport mechanisms, tidal cycles, and benthic microalgal production. Perhaps of

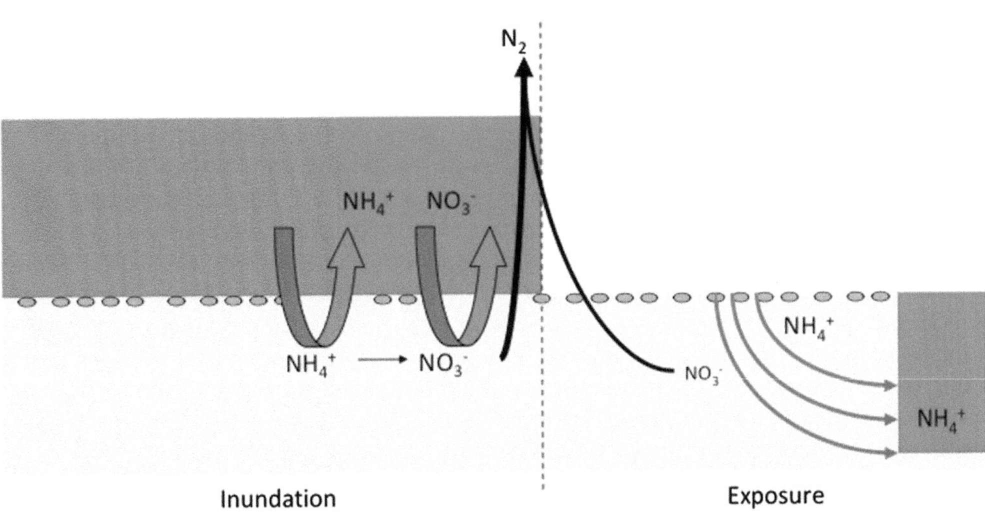

FIGURE 12.7 The same general features of tidal flats apply to those formed on a permeable sandy substrate. A key difference, however, is that during inundation, the transport of dissolved N species between the sediment and the water column will be enhanced by advective porewater exchange (water flow through the sediments) . This process may also enhance important reaction rates such as denitrification. During the falling tide and exposure, water drains out through the flat ( ) leading to locally extreme fluxes of solutes such as NH_4^+ at the edge of the flat.

greatest significance, tidal flats may be both substantial sources and sinks of fixed nitrogen. On the intertidal flats of Tomales Bay (California, USA), Joye and Paerl (1994) documented simultaneous N_2 fixation and denitrification and concluded that these sediments, inhabited by dense microbial mats of cyanobacteria, were a net source of nitrogen to the system over an annual cycle. However, given in general the lack of simultaneous measurements of nitrogen fixation and denitrification, it is hard to comment on the role of tidal flats as net sources or sink of nitrogen. We suggest that in highly eutrophic systems, tidal flats are likely to be an important sink for nitrate, whereas in more mesotrophic to oligotrophic systems, nitrogen fixation may be an important source of N for local ecosystems (Fulweiler et al., 2007; Joye and Paerl, 1994). Simultaneous measurements of nitrogen fixation and denitrification are rare, and future studies should consider both these processes of appropriate spatial and temporal scales to evaluate the nitrogen balance of tidal flats. The controls on denitrification rates in permeable tidal sand flats remain poorly characterized and it will require combined experimental and modeling approaches to better determine these controls. Understanding the nitrogen balance of tidal flats has important implications for other coastal ecosystems because they trap particulate nitrogen from the water column, remineralize it to ammonium, and release this ammonium back to the water column (Billerbeck et al., 2006a; Santos et al., 2014). Ammonium is often a key nutrient limiting the productivity of phytoplankton in coastal waters (Ryther and Dunstan, 1971; Howarth et al., 1998b), so its slow release from tidal flat pore water may help to sustain coastal primary productivity in spite of variability in terrestrially derived nutrient loading (Santos et al., 2014).

6. PHOSPHORUS CYCLE

Like nitrogen, phosphorus (P) is an essential nutrient element that is required for cellular biosynthesis and thus can limit rates of biological processes in the environment. Unlike nitrogen, sediment phosphorus is present in only one inorganic species/valence state, dissolved phosphate, which is typically present as HPO_4^{2-} at circumneutral pH and seawater salinity (Ruttenberg, 2003). Phosphorus is present in sediments as a variety of organic compounds, including monophosphate esters, nucleotides and their derivatives, vitamins, phosphonates, and humic acids (Ruttenberg, 2003). Phosphorus is present in a number of inorganic mineral forms. The microbially mediated component of phosphorus cycling is restricted to biological uptake and regeneration of inorganic phosphorus from organic forms because the phosphorus cycle has no dominant gaseous component.

Continental weathering and subsequent sedimentation of particulate organic and inorganic phosphorus is the primary input of phosphorus to coastal sediments. Ensuing regeneration/release of dissolved inorganic phosphorus from the solid phase (both organic and inorganic forms) regulates P availability in sediments (Paytan and Mclaughlin, 2007; Ruttenberg, 2003). A large portion of the solid-phase P entering coastal sediments is inorganic phosphorus sorbed onto iron oxyhydroxides (Ruttenberg, 2003). Iron oxyhydroxides undergo reductive dissolution in anaerobic sediments by biotic and abiotic mechanisms, thereby releasing dissolved inorganic phosphorus into the pore water where it is then available for biological consumption (Fig. 12.8). Several excellent reviews of the phosphorus cycle are available; for example, see Benitez-Nelson (2000), Ruttenberg (2003), and Paytan and Mclaughlin (2007).

Phosphorus dynamics on tidal flats has received much less attention than N dynamics, possibly because P is not often considered the limiting nutrient in marine systems

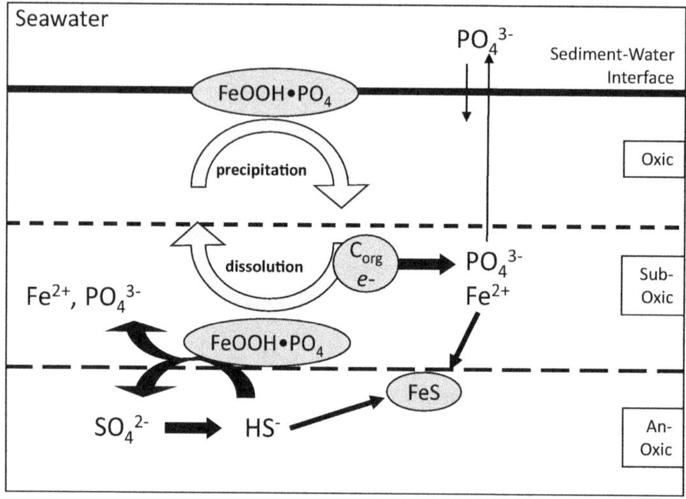

FIGURE 12.8 Simplified schematic of reductive dissolution of Fe–P oxides in sediments. Reactions on the right side illustrate biological iron reduction resulting in release of reduced iron (Fe^{2+}) and dissolved phosphate (PO_4^{3-}). Reactions on the left show the abiotic reduction of iron oxyhydroxides coupled to hydrogen sulfide oxidation. This process also leads to release of Fe^{2+} and PO_4^{3-}. In oxic sediment, iron oxyhydroxides may reprecipitate. This gives rise to a complex internal cycle of dissolution and (re)precipitation that helps sequester both Fe and P in sediments as long as the sediments are overlain by an oxic layer. In anoxic sediments without MPB, both Fe^{2+} and PO_4^{3-} are released from the sediment. *Adapted from Ruttenberg, 2003.*

(Ryther and Dunstan, 1971). However, recent data illustrate the previously proposed belief (Smith, 1984) that widespread P limitation of biological activity can occur during certain seasons or within certain components of the ecosystem (Nicholson et al., 2006; Sundareshwar et al., 2003; Thingstad et al., 1998). Phosphorus availability can also regulate nitrogen fixation in some environments (Ruttenberg, 2003; Tyrrell, 1999).

Variation in the position of the oxic–anoxic interface within sediments may result from daily variations in benthic primary production rates or from hydrodynamic forcing. These changes in sediment oxygen distribution strongly regulate benthic P cycling. Sediments experiencing daily cycles of inundation and exposure, such as those of tidal flats, have a higher binding capacity for inorganic P because Fe oxides are recycled efficiently during periods of exposure (Fig. 12.8; Baldwin and Mitchell, 2000; Coelho et al., 2004). Iron oxyhydroxides effectively scavenge and retain inorganic P in sediments, which may ultimately sustain or stimulate the activity of benthic primary producers. Seasonality in inputs of particulate phosphorus versus organic phosphorus can further influence the impact of oxic–anoxic fluctuations on benthic P dynamics (Coelho et al., 2004). During periods of time with high particulate Fe–P inputs, movement of P between dissolved and inorganic pools can be largely driven by diurnal variations in oxygen availability (Fig. 12.8). If, however, particulate Fe–P pools are limited, recycling of organic P may be a much more important mechanism for providing inorganic P to fuel biological processes.

Benthic microalgal activity suppresses P release from sediments, either by direct P assimilation into biomass or by increased oxygen availability, which may stimulate formation of Fe oxides and thus increase the inorganic P sorption capacity of the sediment (Figs. 12.8 and 12.9). Typically, during submerged light incubation, P uptake from the overlying water rather than P release to the overlying water is observed (Figs. 12.5 and 12.6; Joye et al., 2003, 1996;

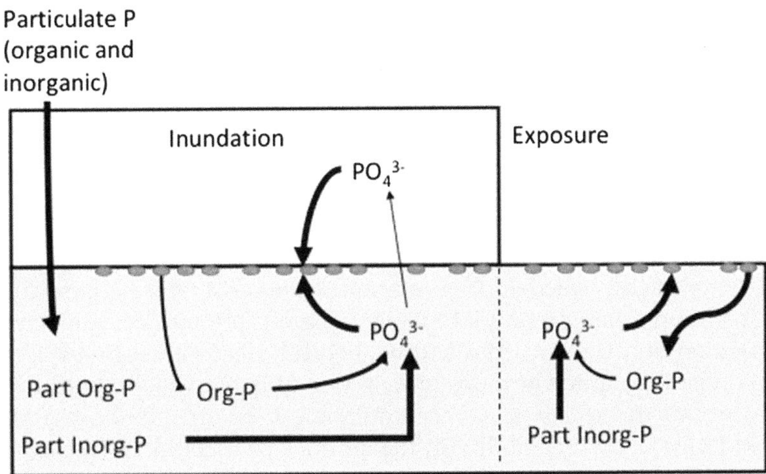

FIGURE 12.9 Phosphorus dynamics in tidal flat sediments. Benthic microalgal ( ) assimilation plays a significant role in limiting the P flux from the sediment and form sequestering water column P into sediments. As with N, P flows through benthic microalgae may dominate the P cycle. Organic P is mineralized releasing PO_4^{3-}, which can be assimilated by benthic microalgae. Dissolved inorganic P is released from particulate inorganic P in anoxic sediments, providing another P source to biota. During exposure, exchange between the sediment and the water column is eliminated and an enrichment in PO_4^{3-} may occur; this may drive P release from sediments on inundation. The thickness of the arrows gives a rough indication of the relative magnitude of the P flows.

Porubsky et al., 2008). Sediments inhabited by benthic microalgae appear to be able to take up P from water column and prevent sediment P release, even during nighttime (dark) incubation (Joye et al., 1996; Porubsky et al., 2008). Prolonged (>24 h) dark incubation in combination with anoxic conditions is required to initiate P release from sediments inhibited by benthic microalgae (Joye et al., 1996).

Anoxic conditions promote P release by stimulating either biological reduction of Fe oxyhydroxides or hydrogen sulfide–mediated reductive dissolution of Fe oxyhydroxides (Porubsky et al., 2008). Release of inorganic P from Fe oxyhydroxides was strongly correlated with sulfate reduction, and presumably hydrogen sulfide production, rates (Joye et al., 1996; Roden and Edmonds, 1997). Anoxic bottom water further stimulates P release from sediments, presumably by increasing both biological reduction of Fe oxyhydroxides and hydrogen sulfide–mediated Fe oxyhydroxides reductive dissolution (Roden and Edmonds, 1997). Additional research is needed to fully understand the mechanisms and controls of P dynamics on tidal flats.

7. SILICON CYCLE

Silicon dynamics on tidal flats is also poorly understood. Silicon (Si) is required by benthic and pelagic diatoms that take up dissolved silicate (silicic acid, $Si(OH)_4$) and produce elaborate biogenic silica (SiO_2) frustules (Joye et al., 1996). Silicon is supplied to coastal systems mainly via terrestrial runoff and sedimentation of lithogenic inorganic primary and secondary Si minerals and biogenic Si. Groundwater may also supply the Si required by diatoms to some coastal systems; such inputs are important in some lakes (Davis, 1976) but have not been documented so far in coastal waters. Dissolved silicate is released through weathering and/or dissolution reactions of primary minerals or via dissolution (regeneration) of biogenic silica phases (e.g., opal). Biogenic silica dissolves about five times faster than lithogenic (mineral) silica (Jurley et al., 1985). In sediments, dissolved silicate accumulates in the pore water as accumulated diatom frustules dissolve (Ragueneau et al., 2006) and then fluxes upward against the concentration gradient, where it may be consumed by benthic microalgae or may escape across the sediment–water interface (Fig. 12.10).

While benthic microalgae take advantage of the immediate sediment source of dissolved silicate, they also assimilate dissolved silicate from the water column (Demaster, 1981). Benthic microalgae often contain more silicon per chlorophyll *a* than their pelagic counterparts and are often highly silicified (Sigmon and Cahoon, 1997). Because diatoms cannot store silicon internally, they accumulate it immediately prior to cell division (Ragueneau et al., 2006; Sigmon and Cahoon, 1997); thus, benthic silicon fluxes may follow the cycle of diatom cell division as much or more so than that of light availability (Busby and Lewin, 1967). The reason for this excess silica accumulation in benthic diatoms is not known but may be related to low light availability in the benthos or it could assist benthic diatoms in resisting suspension by tidal currents or help them to return (sink) to sediments faster following a suspension event (Ragueneau et al., 2006).

The availability of dissolved silicate can regulate primary production in pelagic waters (Sigmon and Cahoon, 1997). Because benthic microalgae often regulate dissolved silicate fluxes in coastal systems, their activity can generate Si limitation in pelagic waters, leading

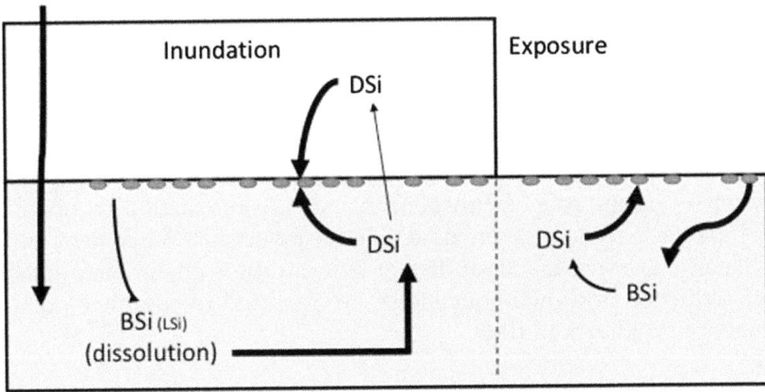

FIGURE 12.10 Silica dynamics in tidal flat sediments. Benthic microalgal (●) assimilation plays a significant role in regulating both the dissolved silicate (DSi) flux at the sediment–water interface and in sequestering Si as biogenic Si (BSi) in the sediments. As with N and P, benthic microalgal-mediated Si uptake may dominate the Si cycle. Biogenic Si dissolves releasing DSi, which can be assimilated or may flux from the sediment. Dissolved inorganic Si can also be released during the weathering of lithogenic Si, but rates of such processes are poorly constrained. During exposure, exchange between the sediment and the water column is eliminated and an enrichment of DSi in the pore water may occur; this may drive DSi release from sediments on inundation. The thickness of the *arrows* gives a rough indication of the relative magnitude of the Si flows.

to shifts in the dominant primary producer from silicified diatoms to nonsilicified phytoplankton (Officer and Ryther, 1980). Low water column dissolved silicate concentrations (<2 μM) drive silicate limitation of diatoms in some environments (Sigmon and Cahoon, 1997). Whether dissolved silicate availability limits production of benthic microalgae is unclear but the limited data available suggest that it is common during certain times of year (Billen et al., 1991). Sequential nutrient limitation of benthic diatoms has been documented in some coastal waters (Sigmon and Cahoon, 1997). In coastal Georgia (USA) intertidal flats occupied by benthic diatoms, nitrate pulses to the overlying water resulted in subsequent silicate and/or phosphorus uptake from the overlying water, suggesting that benthic diatoms are primarily nitrogen limited. Once nitrogen demands were met, the diatoms become limited by either silicate (more often) or phosphorus (Joye et al., 2003; Porubsky et al., 2008). The potential for nutrient limitation of benthic diatoms on tidal flats warrants further study.

8. CONCLUDING REMARKS

Temporal and spatial heterogeneity and strong forcing by tidal and wind dynamics generate extreme biogeochemical variability in intertidal flats and distinguish these habitats from subtidal coastal sediments. Because of this inherent variability and a limited number of interdisciplinary studies, it is difficult to generalize regarding the biogeochemistry of intertidal flats. Nonetheless, it is clear that benthic intertidal flats play important roles in cycling of carbon, nitrogen, phosphorus, and silicon in coastal ecosystems. Physical (tidal, storm) forcing in tidal flat sediments modulates benthic exchange and the position of redox gradients in the sediments. Enhanced fluid exchange at low tide may stimulate N, P, and Si exchange and return to the overlying water column. More detailed studies of the interactions between biogeochemical cycles, such as documenting feedback between benthic primary production (e.g., C fixation and DOC release) and associated heterotrophic processes, regulation of benthic primary production by nutrient and light availability, and the balance between nitrogen fixation and denitrification, are needed in a broad array of intertidal flat habitats. Understanding the role of phosphorus and silicon availability, relative to nitrogen, in regulating biological dynamics on tidal flats is potentially important but very poorly documented. Finally, more detailed studies of heterotrophic anaerobic microbial processes, such as sulfate reduction and methanogenesis, are required to complete our understanding of carbon dynamics on intertidal flats.

Acknowledgments

The original chapter was supported by grants from the National Science Foundation's Long-Term Ecological Research Program (OCE 99-82133 to SBJ) and National Oceanic and Atmospheric Administration Sea Grant Programs in Georgia (award numbers NA06RG0029-R/WQ11 and R/WQ12A to SBJ) and South Carolina (award number: NA960PO113 to SBJ). Support for revision and update of this chapter was provided by the University of Georgia Athletic Foundation Professorial Endowment.

References

Abd. Aziz, S.A., Nedwell, D.B., 1986. The nitrogen cycle of an East Coast U.K. saltmarsh: II. Nitrogen fixation, nitrification, denitrification, tidal exchange. Estuarine, Coastal and Shelf Science 22, 689–704.

Ahmerkamp, S., 2016. Regulation of Oxygen Dynamics by Transport Processes and Microbial Respiration in Sandy Sediments. Ph.D. thesis).

Ahmerkamp, S., Winter, C., Janssen, F., Kuypers, M.M., Holtappels, M., 2015. The impact of bedform migration on benthic oxygen fluxes. Journal of Geophysical Research: Biogeosciences 120, 2229–2242.

Ahmerkamp, S., et al., 2017. Regulation of benthic oxygen fluxes in permeable sediments of the coastal ocean. Limnology and Oceanography 62 (5), 1935–1954.

Algar, C.K., Vallino, J.J., 2014. Predicting microbial nitrate reduction pathways in coastal sediments. Aquatic Microbial Ecology 71 (3), 223–238. https://doi.org/10.3354/ame01678.

Alongi, D.M., 1997. Coastal Ecosystem Processes. CRC Press.

An, S., Joye, S.B., 2001. Enhancement of coupled nitrification-denitrification by benthic photosynthesis in shallow estuarine sediments. Limnology and Oceanography 46, 62–74.

Asmus, R.M., Jensen, M.H., Jensen, K.M., Kristensen, E., Asmus, H., Wille, A., 1998. The role of water movement and spatial scaling for measurement of dissolved inorganic nitrogen fluxes in intertidal sediments. Estuarine, Coastal and Shelf Science 46, 221–232.

Baldwin, D.S., Mitchell, A.M., 2000. The effects of drying and re-flooding on the sediment and soil nutrient dynamics of lowland river–floodplain systems: a synthesis. River Research and Applications 16, 457–467.

Bautista, M.F., Paerl, H.W., 1985. Diel N_2 fixation in an intertidal marine cyanobacterial mat community. Marine Chemistry 16, 369–377.

Beck, M., Brumsack, H.J., 2012. Biogeochemical cycles in sediment and water column of the Wadden Sea: the example Spiekeroog Island in a regional context. Ocean and Coastal Management 68, 102–113.

Beck, M., et al., 2008. Sulphate, dissolved organic carbon, nutrients and terminal metabolic products in deep pore waters of an intertidal flat. Biogeochemistry 89, 221–238.

Beck, M., et al., 2009. Deep pore water profiles reflect enhanced microbial activity towards tidal flat margins. Ocean Dynamics 59, 371–384.

Beck, M., et al., 2011. Imprint of past and present environmental conditions on microbiology and biogeochemistry of coastal Quaternary sediments. Biogeosciences 8, 55–68.

Behrendt, A., De Beer, D., Stief, P., 2013. Vertical activity distribution of dissimilatory nitrate reduction in coastal marine sediments. Biogeosciences 10, 7509–7523.

Benitez-Nelson, C.R., 2000. The biogeochemical cycling of phosphorus in marine systems. Earth-Science Reviews 51, 109–135.

Berninger, U., Huettel, M., 1997. Impact of flow on oxygen dynamics in photosynthetically active sediments. Aquatic Microbial Ecology 12, 291–302.

Billen, G., Lancelot, C., Meybeck, M., Mantoura, R.F.C., Martin, J.M., Wollast, R., 1991. N, P and Si retention along the aquatic continuum from land to ocean. In: Ocean Margin Processes in Global Change, first ed. John Wiley & Sons, pp. 19–44.

Billerbeck, M., Roy, H., Bosselmann, K., Huettel, M., 2007. Benthic photosynthesis in submerged Wadden Sea intertidal flats. Estuarine, Coastal and Shelf Science 71, 704–716.

Billerbeck, M., Werner, U., Bosselman, K., Walpersdorf, E., Huettel, M., 2006a. Nutrient release from an exposed intertidal sand flat. Marine Ecology Progress Series 316, 35–51.

Billerbeck, M., Werner, U., Polerecky, L., Walpersdorf, E., De Beer, D., Huettel, M., 2006b. Surficial and deep pore water circulation governs spatial and temporal scales of nutrient recycling in intertidal sand flat sediment. Marine Ecology Progress Series 326, 61–76.

Bittar, T.B., et al., 2016. Seasonal dynamics of dissolved, particulate and microbial components of a tidal saltmarsh-dominated estuary under contrasting levels of freshwater discharge. Estuarine, Coastal and Shelf Science 182, 72–85.

Bohórquez, J., Mcgenity, T.J., Papaspyrou, S., García-Robledo, E., Corzo, A., Underwood, G.J.C., 2017. Different types of diatom-derived extracellular polymeric substances drive changes in heterotrophic bacterial communities from intertidal sediments. Frontiers in Microbiology 8.

Böttcher, M.E., Oelschläger, B., Höpner, T., Brumsack, H.-J., Rullkötter, J., 1998. Sulfate reduction related to the early diagenetic degradation of organic matter and "black spot" formation in tidal sandflats of the German Wadden Sea (southern North Sea): stable isotope (^{13}C, ^{34}S, ^{18}O) and other geochemical results. Organic Geochemistry 29, 1517–1530.

Boudreau, B.P., 1998. Mean mixed depth of sediments: the wherefore and the why. Limnology and Oceanography 43, 524–526.

Boudreau, B.P., et al., 2001. Permeable marine sediments: overturning an old paradigm. EOS. Transactions American Geophysical Union 82, 133–140.

Boynton, W.R., Kemp, W.M., Osbourne, C.G., 1980. Nutrient fluxes across the sediment-water interface in the turbid zone of a coastal plain estuary. In: Kennedy, V.S. (Ed.), Estuarine Perspectives. Academic Press, New York, pp. 93–109.

Brunet, R.C., Garcia-Gil, L.J., 1996. Sulfide-induced dissimilatory nitrate reduction to ammonia in anaerobic freshwater sediments. FEMS Microbiology Ecology 21, 131–138.

Burdige, D.J., Gardner, K.G., 1998. Molecular weight distribution of dissolved organic carbon in marine sediment pore waters. Marine Chemistry 62, 45–64.

Busby, W.F., Lewin, J., 1967. Silicate uptake and silica shell formation by synchronously dividing cells of the diatom *Navicula pelliculosa* (Bréb.) Hilse. Journal of Phycology 3, 127–131.

Cabrita, M.T., Brotas, V., 2000. Seasonal variation in denitrification and dissolved nitrogen fluxes in intertidal sediments of the Tagus estuary, Portugal. Marine Ecology Progress Series 202, 51–65.

Cammen, L.M., 1982. Effect of particle size on organic content and microbial abundance within four marine sediments. Marine Ecology Progress Series 273–280.

Canfield, D.E., Thamdrup, B., Kristensen, E., 2005. Aquatic Geomicrobiology. Elsevier.

Canion, A., et al., 2014. Temperature response of denitrification and anammox reveals the adaptation of microbial communities to in situ temperatures in permeable marine sediments that span 50 degrees in latitude. Biogeosciences 11, 309–320.

Carpenter, E.J., Raalte, V., Valiela, I., 1978. Nitrogen fixation by algae in a Massachusetts salt marsh. Limnology and Oceanography 23, 318–327.

Chennu, A., et al., 2013. Hyperspectral imaging of the microscale distribution and dynamics of microphytobenthos in intertidal sediments. Limnology and Oceanography: Methods 11, 511–528.

Chennu, A., Volkenborn, N., De Beer, D., Wethey, D.S., Woodin, S.A., Polerecky, L., 2015. Effects of bioadvection by *Arenicola marina* on microphytobenthos in permeable sediments. PLoS One 10.

Christensen, P.B., Rysgaard, S., Sloth, N.P., Dalsgaard, T., Schwærter, S., 2000. Sediment mineralization, nutrient fluxes, denitrification and dissimilatory nitrate reduction to ammonium in an estuarine fjord with sea cage trout farms. Aquatic Microbial Ecology 21, 73–84.

Christiansen, C., Vølund, G., Lund-Hansen, L.C., Bartholdy, J., 2006. Wind influence on tidal flat sediment dynamics: field investigations in the Ho Bugt, Danish Wadden Sea. Marine Geology 235, 75–86.

Coelho, J.P., Flindt, M.R., Jensen, H.S., Lillebø, A.I., Pardal, M.A., 2004. Phosphorus speciation and availability in intertidal sediments of a temperate estuary: relation to eutrophication and annual P-fluxes. Estuarine, Coastal and Shelf Science 61, 583–590.

Cook, P.L.M., 2003. Carbon and Nitrogen Cycling on Intertidal Mudflats of a Temperate Australian Estuary. University of Tasmania.

Cook, P.L.M., Revill, A.T., Butler, E.C.V., Eyre, B.D., 2004a. Benthic carbon and nitrogen cycling on intertidal mudflats of a temperate Australian estuary II. Nitrogen cycling. Marine Ecology Progress Series 280, 39–54.

Cook, P.L.M., Revill, A.T., Clementson, L.A., Volkman, J.K., 2004b. Benthic carbon and nitrogen cycling on intertidal mudflats of a temperate Australian estuary III. Sources of organic matter. Marine Ecology Progress Series 280, 55–72.

Currin, C.A., Paerl, H.W., 1998. Environmental and physiological controls on diel patterns of N_2 fixation in epiphytic cyanobacterial communities. Microbial Ecology 35, 34–45.

Dähnke, K., Moneta, A., Veuger, B., Soetaert, K., 2012. Balance of assimilative and dissimilative nitrogen processes in a diatom-rich tidal flat sediment. Biogeosciences 9, 4059–4070.

Dale, N.G., 1974. Bacteria in intertidal sediments: factors related to their distribution. Limnology and Oceanography 19, 509–518.

Dalsgaard, T., Thamdrup, B., 2002. Factors controlling anaerobic ammonium oxidation with nitrite in marine sediments. Applied and Environmental Microbiology 68, 3802–3808.

Davis, C.O., 1976. Continuous culture of marine diatoms under silicate limitation. II. Effect of light intensity on growth and nutrient uptake of *Skeletonema costatum*. Journal of Phycology 12, 291–300.

De Beer, D., 2000. Potentiometric microsensors for *in situ* measurements in aquatic environments. In: Buffle, J., Horvai, G. (Eds.), In Situ Monitoring of Aquatic Systems: Chemical Analysis and Speciation. Wiley & Sons, pp. 161–194.

De Beer, D., 2002. Microsensor studies of oxygen, carbon, and nitrogen cycles in lake sediments and microbial mats. In: Taillefert, M., Rozan, T.F. (Eds.), Environmental Electrochemistry: Analyses of Trace Element Biogeochemistry. ACS Publications.

De Beer, D., et al., 2004. Transport and mineralization rates in North sea sandy intertidal sediments (Sylt-Rømø basin, Waddensea). Limnology and Oceanography 50, 113–127.

Demaster, D.J., 1981. The supply and accumulation of silica in the marine environment. Geochimica et Cosmochimica Acta 45, 1715–1732.

Dilling, W., Cypionka, H., 1990. Aerobic respiration in sulfate reducing bacteria. FEMS Microbiology Letters 71, 123–128.

Dittmar, T., Lara, R.J., 2001. Molecular evidence for lignin degradation in sulfate-reducing mangrove sediments (Amazônia, Brazil). Geochimica et Cosmochimica Acta 65, 1417–1428.

Dittmar, T., Lara, R.J., Kattner, G., 2001. River or mangrove? Tracing major organic matter sources in tropical Brazilian coastal waters. Marine Chemistry 73, 253–271.

Dittmar, T., Paeng, J., Gihring, T., Suryaputra, I.G.N.A., Huettel, M., 2012. Discharge of dissolved black carbon from a fire-affected intertidal system. Limnology and Oceanography 57, 1171–1181.

Dong, L.F., Thornton, D.C.O., Nedwell, D.B., Underwood, G.J.C., 2000. Denitrification in sediments of the river Colne estuary, England. Marine Ecology Progress Series 203, 109–122.

Dyer, K.R., 1979. Estuarine Hydrography and Sedimentation: A Handbook. CUP Archive.

Elliott, A.H., Brooks, N.H., 1997. Transfer of nonsorbing solutes to a streambed with bed forms: Theory. Water Resources Research 33, 123–136.

Engelen, B., Cypionke, H., 2009. The subsurface of tidal-flat sediments as a model for the deep biosphere. Ocean Dynamics 59, 385–391.

Epstein, S.S., 1997. Microbial food webs in marine sediments. II Seasonal changes in trophic interactions in a sandy tidal flat community. Microbial Ecology 34, 199–209.

Evrard, V., Glud, R.N., Cook, P.L.M., 2013. The kinetics of denitrification in permeable sediments. Biogeochemistry 113, 563–572.

Falcao, M., Vale, C., 1995. Tidal flushing of ammonium from intertidal sediments of Ria Formosa, Portugal. Netherlands Journal of Aquatic Ecology 29, 239–244.

Feuillet-Girard, M., Gouleau, D., Blanchard, G., Joassard, L., 1997. Nutrient fluxes on an intertidal mudflat in marennes-oleron bay, and influence of the emersion period. Aquatic Living Resources 10, 49–58.

Forster, S., Huettel, M., Ziebis, W., 1996. Impact of boundary layer flow velocity on oxygen utilisation in coastal sediments. Marine Ecology Progress Series 143, 173–185.

Fossing, H., Jørgensen, B.B., 1989. Measurement of bacterial sulfate reduction in sediments: evaluation of single-step chromium reduction method. Biogeochemistry 8, 205–222.

Fossing, H., Jørgensen, B.B., 1990. Oxidation and reduction of radiolabeled inorganic sulfur compounds in an estuarine sediment, Kysing Fjord, Denmark. Geochima et Cosmochimica Acta 54, 2731–2742.

Franke, U., Polerecky, L., Precht, E., Huettel, M., 2006. Wave tank study of particulate organic matter degradation in permeable sediments. Limnology and Oceanography 51, 1084–1096.

Freese, E., Köster, J., Rullkötter, J., 2008. Origin and composition of organic matter in tidal flat sediments from the German Wadden Sea. Organic Geochemistry 39, 820–829.

Fulweiler, R.W., Nixon, S.W., Buckley, B.A., Granger, S.L., 2007. Reversal of the net dinitrogen gas flux in coastal marine sediments. Nature 448, 180.

Gao, H., et al., 2012. Intensive and extensive nitrogen loss from intertidal permeable sediments of the Wadden Sea. Limnology and Oceanography 57, 185–198.

Gao, H., et al., 2010. Aerobic denitrification in permeable Wadden Sea sediments. The ISME Journal 4, 417–426.

Gao, H., et al., 2009. Aerobic denitrification in permeable Wadden Sea sediments. The ISME Journal 4, 417–426.

Garcia-Robledo, E., Bohorquez, J., Corzo, A., Jimenez-Arias, J.L., Papaspyrou, S., 2016. Dynamics of inorganic nutrients in intertidal sediments: porewater, exchangeable, and intracellular pools. Frontiers in Microbiology 7, 761.

Gattuso, J.-P., Smith, S.V., Hogan, C.M., Duffy, J.E., 2009. Coastal Zone. Environmental Information Coalition, National Council for Science and the Environment.

Gittel, A., Mußmann, M., Sass, H., Cypionka, H., Könneke, M., 2008. Identity and abundance of active sulfate-reducing bacteria in deep tidal flat sediments determined by directed cultivation and CARD-FISH analysis. Environmental Microbiology 10, 2645–2658.

Gomez-Saez, G.V., et al., 2016. Molecular evidence for abiotic sulfurization of dissolved organic matter in marine shallow hydrothermal systems. Geochimica et Cosmochimica Acta 190, 35–52.

Gotto, J.W., Tabita, F.R., Baalen, C.V., 1981. Nitrogen fixation in intertidal environments of the Texas Gulf coast. Estuarine, Coastal and Shelf Science 12, 231–235.

Graber, H.C., Beardsley, R.C., Grant, W.D., 1989. Storm-generated surface waves and sediment resuspension in the East China and Yellow Seas. Journal of Physical Oceanography 19, 1039–1059.

Grunwald, M., et al., 2009. Methane in the southern North Sea: sources, spatial distribution and budgets. Estuarine, Coastal and Shelf Science 81, 445–456.

Heisterkamp, I.M., Kamp, A., Schramm, A.T., De Beer, D., Stief, P., 2012. Indirect control of the intracellular nitrate pool of intertidal sediment by the polychaete *Hediste diversicolor*. Marine Ecology Progress Series 445, 181–192.

Henriksen, K., Rasmussen, M.B., Jensen, A., 1983. Effect of bioturbation on microbial nitrogen transformations in the sediment and fluxes of ammonium and nitrate to the overlaying water. Ecological Bulletins 193–205.

Howarth, R.W., Marino, R., Cole, J.J., 1988a. Nitrogen fixation in freshwater, estuarine, and marine ecosystems. 2. Biogeochemical controls. Limnology and Oceanography 33, 688–701.

Howarth, R.W., Marino, R., Lane, J., Cole, J.J., 1988b. Nitrogen fixation in in freshwater, estuarine and marine ecosystems. I. Rates and importance. Limnology and Oceanography 33, 669–687.

Huettel, M., Berg, P., Kostka, J.E., 2014. Benthic Exchange and Biogeochemical Cycling in Permeable Sediments.

Huettel, M., Gust, G., 1992. Impact of bioroughness on interfacial solute exchange in permeable sediments. Marine Ecology Progress Series 89, 253–267.

Huettel, M., Rusch, A., 2000. Transport and degradation of phytoplankton in permeable sediment. Limnology and Oceanography 45, 534–549.

Huettel, M., Ziebis, W., Forster, S., 1996. Flow-induced uptake of particulate matter in permeable sediments. Limnology and Oceanography 41, 309–322.

Huettel, M., Ziebis, W., Forster, S., Luther, G.W., 1998. Advective transport affecting metal and nutrient distributions and interfacial fluxes in permeable sediments. Geochimica et Cosmochimica Acta 62, 613–631.

Jansen, S., Walpersdorf, E., Werner, U., Billerbeck, M., Böttcher, M., De Beer, D., 2009. Functioning of intertidal flats inferred from temporal and spatial dynamics of O_2, H_2S and pH in their surface sediment. Ocean Dynamics 59, 317–332.

Jassby, A.D., Cloern, J.E., Cole, B.E., 2002. Annual primary production: patterns and mechanisms of change in a nutrient rich tidal ecosystem. Limnology and Oceanography 47, 698–712.

Jørgensen, B.B., 1982. Mineralization of organic matter in the sea bed - the role of sulphate reduction. Nature 296, 643–645.

Jørgensen, B.B., Des Marais, D.J., 1990. The diffusive boundary layer of sediments: oxygen microgradients over a microbial mat. Limnology and Oceanography 35, 1343–1355.

Joye, S., Porubsky, W., Weston, N., Lee, R., 2003. Benthic microalgal production and nutrient dynamics in intertidal sediments. Berichte - Forschungszentrum Terramare 12, 67–70.

Joye, S.B., Anderson, I.C., 2008. Nitrogen cycling in coastal sediments. In: Capone, D.G., Bronk, D.A., Mulholland, M.R., Carpenter, E.J. (Eds.), Nitrogen in the Marine Environment, second ed. Elsevier, pp. 686–902.

Joye, S.B., Hollibaugh, J.T., 1995. Influence of sulfide inhibition of nitrification on nitrogen regeneration in sediments. Science 270, 623–625.

Joye, S.B., Lee, R.Y., 2004. Benthic microbial mats: important sources of fixed nitrogen and carbon to the Twin Cays, Belize ecosystem. Atoll Research Bulletin 528, 1–24.

Joye, S.B., Mazzotta, M.L., Hollibaugh, J.T., 1996. Community metabolism in intertidal microbial mats: the importance of iron and manganese reduction. Estuarine, Coastal and Shelf Science 43, 747–766.

Joye, S.B., Paerl, H.W., 1993. Contemporaneous nitrogen fixation and denitrification in intertidal microbial mats: rapid response to runoff events. Marine Ecology Progress Series 94, 267–274.

Joye, S.B., Paerl, H.W., 1994. Nitrogen cycling in marine microbial mats: rates and patterns of nitrogen fixation and denitrification. Marine Biology 119, 285–295.

Jurley, J.P., Armstrong, D.E., Kenoyer, G.J., Bowser, C.J., 1985. Ground water as a silica source for diatom production in a precipitation-dominated lake. Science 227, 1576–1579.

Kamp, A., De Beer, D., Nitsch, J.L., Stief, P., 2011. Diatoms respire nitrate to survive dark and anoxic conditions. Proceedings of the National Academy of Sciences 108, 5649–5654.

Kamp, A., Høgslund, S., Risgaard-Petersen, N., Stief, P., 2015. Nitrate storage and dissimilatory nitrate reduction by eukaryotic microbes. Frontiers in Microbiology 6.

Kamp, A., Stief, P., Bristow, L.A., Thamdrup, B., Glud, R.N., 2016. Intracellular nitrate of marine diatoms as a driver of anaerobic nitrogen cycling in sinking aggregates. Frontiers in Microbiology 7.

Kelly-Gerreyn, B.A., Trimmer, M., Hydes, D.J., 2001. A diagenetic model discriminating denitrification and dissimilatory nitrate reduction to ammonium in a temperate estuarine sediment. Marine Ecology Progress Series 220, 33–46.

Kessler, A.J., Glud, R.N., Cardenas, M.B., Larsen, M., Bourke, M.F., Cook, P.L.M., 2012. Quantifying denitrification in rippled permeable sands through combined flume experiments and modeling. Limnology and Oceanography 57, 1217–1232.

Kim, B.S., Oh, H.M., Kang, H., Park, S.S., Chun, J., 2004. Remarkable bacterial diversity in the tidal flat sediment as revealed by 16S rDNA analysis. Journal of Microbiology and Biotechnology 14, 205–211.

Kim, T.-H., Waska, H., Kwon, E., Suryaputra, I.G.N., Kim, G., 2012. Production, degradation, and flux of dissolved organic matter in the subterranean estuary of a large tidal flat. Marine Chemistry 142–144, 1–10.

Kallmeyer, J., Ferdelman, T.G., Weber, A., Fossing, H., Jørgensen, B.B., 2004. A cold chromium distillation procedure for radiolabeled sulfide applied to sulfate reduction measurements. Limnology and Oceanography: Methods 2, 171–180.

Kuipers, B.R., De Wilde, P. a. W., Creutzberg, F., 1981. Energy flow in a tidal flat ecosystem. Marine Ecology Progress Series 5, 215–222.

Laima, M.J.C., Girard, M.F., Vouve, F., Richard, P., Blanchard, G., Gouleau, D., 1999. Nitrification rates related to sedimentary structures in an Atlantic intertidal mudflat, Marennes-Olèron Bay, France. Marine Ecology Progress Series 191, 33–41.

Lee, R., Joye, S., 2006. Seasonal patterns of nitrogen fixation and denitrification in oceanic mangrove habitats. Marine Ecology Progress Series 307, 127–141.

Llobet-Brossa, E., et al., 2002. Community structure and activity of sulfate-reducing bacteria in an intertidal surface sediment: a multi-method approach. Aquatic Microbial Ecology 29, 211–226.

Lomas, M.W., Glibert, P.M., 2000. Comparisons of nitrate uptake, storage, and reduction in marine diatoms and flagellates. Journal of Phycology 36, 903–913.

Longbottom, M.R., 1970. The distribution of *Arenicola marina* (L.) with particular reference to the effects of particle size and organic matter of the sediments. Journal of Experimental Marine Biology and Ecology 5, 138–157.

Longphuirt, S.N., et al., 2009. Dissolved inorganic nitrogen uptake by intertidal microphytobenthos: nutrient concentrations, light availability and migration. Marine Ecology Progress Series 379, 33–44.

Lotze, H.K., 2005. Radical changes in the Wadden Sea fauna and flora over the last 2000 years. Helogland Marine Research 59, 71–83.

Lübben, A., et al., 2009. Distributions and characteristics of dissolved organic matter in temperate coastal waters (Southern North Sea). Ocean Dynamics 59, 263–275.

Malcolm, S.J., Sivyer, D.B., 1997. Nutrient recycling in intertidal sediments. In: Jickells, T.D., Rae, J.E. (Eds.), Biogeochemistry of Intertidal Sediments. Cambridge University Press, pp. 85–98.

Marchant, H.K., et al., 2017. Denitrifying community in coastal sediments performs aerobic and anaerobic respiration simultaneously. The ISME Journal 11, 1799–1812.

Marchant, H.K., Holtappels, M., Lavik, G., Ahmerkamp, S., Winter, C., Kuypers, M.M.M., 2016. Coupled nitrification–denitrification leads to extensive N loss in subtidal permeable sediments. Limnology and Oceanography 61, 1033–1048.

Marchant, H.K., Lavik, G., Holtappels, M., Kuypers, M.M., 2014. The fate of nitrate in intertidal permeable sediments. PLoS One 9, e104517.

Marschall, C., Frenzel, P., Cypionka, H., 1993. Influence of oxygen on sulfate reduction and growth of sulfate-reducing bacteria. Archives of Microbiology 159, 168–173.

Medeiros, P.M., Seidel, M., Dittmar, T., Whitman, W.B., Moran, M.A., 2015. Drought-induced variability in dissolved organic matter composition in a marsh-dominated estuary. Geophysical Research Letters 42, 6446–6453.

Meysman, F.J., Galaktionov, O.S., Gribsholt, B., Middelburg, J.J., 2006. Bioirrigation in permeable sediments: advective pore-water transport induced by burrow ventilation. Limnology and Oceanography 51, 142–156.

Meziane, T., Bodineau, L., Retiere, C., Thoumelin, G., 1997. The use of lipid markers to define sources of organic matter in sediment and food web of the intertidal salt-marsh-flat ecosystem of Mont-Saint-Michel Bay, France. Journal of Sea Research 38, 47–58.

Meziane, T., Tsuchiya, M., 2000. Fatty acids as tracers of organic matter in the sediment and food web of a mangrove/intertidal flat ecosystem, Okinawa, Japan. Marine Ecology Progress Series 200, 49–57.

Meziane, T., Tsuchiya, M., 2002. Organic matter in a subtropical mangrove-estuary subjected to wastewater discharge: origin and utilisation by two macrozoobenthic species. Journal of Sea Research 47, 1–11.

Middelburg, J.J., et al., 1996. Organic matter mineralization in intertidal sediments along an estuarine gradient. Marine Ecology Progress Series 132, 157–168.

Middelburg, J.J., Barranguet, C., Boschker, H.T.S., Herman, P.M.J., Moens, T., Heip, C.H.R., 2000. The fate of intertidal microphytobenthos carbon: an in situ C-13-labeling study. Limnology and Oceanography 45, 1224–1234.

Middelburg, J.J., Herman, P.M.J., 2007. Organic matter processing in tidal estuaries. Marine Chemistry 106, 127–147.

Middelburg, J.J., et al., 2002. Methane distribution in European tidal estuaries. Biogeochemistry 59, 95–119.

Miyatake, T., Moerdijk-Poortvliet, T.C.W., Stal, L.J., Boschker, H.T.S., 2014. Tracing carbon flow from microphytobenthos to major bacterial groups in an intertidal marine sediment by using an in situ ^{13}C pulse-chase method. Limnology and Oceanography 59, 1275–1287.

Moens, T., Luyten, C., Middelburg, J.J., Herman, P.M., Vincx, M., 2002. Tracing organic matter sources of estuarine tidal flat nematodes with stable carbon isotopes. Marine Ecology Progress Series 234, 127–137.

Moore, W.S., et al., 2011. Radium-based pore water fluxes of silica, alkalinity, manganese, DOC, and uranium: a decade of studies in the German Wadden Sea. Geochimica et Cosmochimica Acta 75, 6535–6555.

Moss, J.A., Nocker, A., Lepo, J.E., Snyder, R.A., 2006. Stability and change in estuarine biofilm bacterial community diversity. Applied and Environmental Microbiology 72, 5679–5688.

Musat, N., et al., 2006. Microbial community structure of sandy intertidal sediments in the north sea, Sylt-Romo Basin, Wadden Sea. Systematic and Applied Microbiology 29, 333–348.

Na, T., Gribsholt, B., Galaktionov, O.S., Lee, T., Meysman, F.J.R., 2008. Influence of advective bio-irrigation on carbon and nitrogen cycling in sandy sediments. Journal of Marine Research 66, 691–722.

Nedwell, D.B., Jickells, T.D., Trimmer, M., Sanders, R., 1999. Nutrients in estuaries. In: Nedwell, D.B., Raffaelli, D.G. (Eds.), Advances in Ecological Research - Estuaries. Advance in Ecological Research. Academic Press, pp. 43–92.

Neumann, A., Möbius, J., Hass, H.C., Puls, W., Friedrich, J., 2017. Empirical model to estimate permeability of surface sediments in the German Bight (North Sea). Journal of Sea Research 127, 36–45.

Nicholson, D., Dyhrman, S., Chavez, F., Paytan, A., 2006. Alkaline phosphatase activity in the phytoplankton communities of Monterey bay and San Francisco bay. Limnology and Oceanography 51, 874–883.

Officer, C.B., Ryther, J.H., 1980. The possible importance of silicon in marine eutrophication. Marine Ecology Progress Series 83–91.

Ogilvie, B., Nedwell, D.B., Harrison, R.M., Robinson, A., Sage, A., 1997. High nitrate, muddy estuaries as nitrogen sinks - the nitrogen budget of the River Colne Estuary (United Kingdom). Marine Ecology Progress Series 150, 217–228.

Ottosen, L.D.M., Risgaard-Petersen, N., Nielsen, L.P., Dalsgaard, T., 2001. Denitrification in exposed intertidal mudflats, measured with a new ^{15}N-ammonium spray technique. Marine Ecology Progress Series 209, 35–42.

Paerl, H.W., 1990. Physiological ecology and regulation of N_2 fixation in natural waters. Advances in Microbial Ecology 11, 305–344.

Paerl, H.W., Joye, S.B., Fitzpatrick, M., 1993. Evaluation of nutrient limitation of CO_2 and N_2 fixation in marine microbial mats. Marine Ecology Progress Series 297–306.

Paerl, H.W., Pinckney, J.L., Steppe, T.F., 2000. Cyanobacterial–bacterial mat consortia: examining the functional unit of microbial survival and growth in extreme environments. Environmental Microbiology 2, 11–26.

Panutrakul, S., Monteny, F., Baeyens, W., 2001. Seasonal variations in sediment sulfur cycling in the Ballastplaat mudflat, Belgium. Estuaries 24, 257–265.

Paytan, A., Mclaughlin, K., 2007. The oceanic phosphorus cycle. Chemical Reviews 107, 563–576.

Pinckney, J.L., Paerl, H.W., 1997. Anoxygenic photosynthesis and nitrogen fixation by a microbial mat community in a bahamian hypersaline lagoon. Applied and Environmental Microbiology 63, 420–426.

Porubsky, W.P., Velasquez, L.E., Joye, S.B., 2008. Nutrient-replete benthic microalgae as a source of dissolved organic carbon to coastal waters. Estuaries and Coasts 31, 860–876.

Porubsky, W.P., Weston, N.B., Joye, S.B., 2009. Benthic metabolism and the fate of dissolved inorganic nitrogen in intertidal sediments. Estuarine, Coastal and Shelf Science 83, 392–402.

Potts, M., 1979. Nitrogen fixation (acetylene reduction) associated with communities of heterocystous and non-heterocystous blue-green algae in mangrove forests of Sinai. Oecologia 39, 359–373.

Precht, E., Huettel, M., 2004. Rapid wave-driven advective pore water exchange in a permeable coastal sediment. Journal of Sea Research 51, 93–107.

Probandt, D., Knittel, K., Tegetmeyer, H., Ahmerkamp, S., Holtappels, M., Amann, R., 2017. Permeability shapes bacterial communities in sublittoral surface sediments. Environmental Microbiology 19, 1584–1599.

Ragueneau, O., Conley, D.J., Leynaert, A., Longphuirt, S.N., Slomp, C.P., 2006. Role of diatoms in silicon cycling and coastal marine food webs. In: Ittekkot, V., Unger, D., Humborg, C., An, N.T. (Eds.), The Silicon Cycle: Human Perturbations and Impacts on Aquatic Systems. Island Press, pp. 163–195.

Rao, A.M.F., Mccarthy, M.J., Gardner, W.S., Jahnke, R.A., 2008. Respiration and denitrification in permeable continental shelf deposits on the South Atlantic Bight: N_2:Ar and isotope pairing measurements in sediment column experiments. Continental Shelf Research 28, 602–613.

Rasheed, M., Badran, M.I., Huettel, M., 2003. Particulate matter filtration and seasonal nutrient dynamics in permeable carbonate and silicate sands of the Gulf of Aqaba, Red Sea. Coral Reefs 22, 167–177.

Reeburgh, W.S., 1983. Rates of biogeochemical processes in anoxic sediments. Annual Review of Earth and Planetary Science 11, 269–298.

Reimers, C.E., et al., 2004. In situ measurements of advective solute transport in permeable shelf sands. Continental Shelf Research 24, 183–201.

Riedel, T., Lettmann, K., Schnetger, B., Beck, M., Brumsack, H.-J., 2011. Rates of trace metal and nutrient diagenesis in an intertidal creek bank. Geochimica et Cosmochimica Acta 75, 134–147.

Risgaard-Petersen, N., 2003. Coupled nitrification-denitrification in autotrophic and heterotrophic estuarine sediments: on the influence of benthic microalgae. Limnology and Oceanography 48, 93–105.

Risgaard-Petersen, N., Nicolaisen, M.H., Revsbech, N.P., Lomstein, B.A., 2004. Competition between ammonia-oxidizing bacteria and benthic microalgae. Applied and Environmental Microbiology 70, 5528–5537.

Risgaard-Petersen, N., Rysgaard, S., Nielsen, L.P., Revsbech, N.P., 1994. Diurnal variation of denitrification and nitrification in sediments colonized by benthic microphytes. Limnology and Oceanography 39, 573–579.

Rocha, C., 1998. Rhythmic ammonium regeneration and flushing in intertidal sediments of the Sado Estuary. Limnology and Oceanography 43, 823–831.

Roden, E.E., Edmonds, J.W., 1997. Phosphate mobilization in iron-rich anaerobic sediments: microbial Fe(III) oxide reduction versus iron-sulfide formation. Archiv für Hydrobiologie 139, 347–378.

Rohjans, D., Brocks, P., Scholz-Böttcher, B.M., Rullkötter, J., 1998. Lipid biogeochemistry of surface sediments in the Lower Saxonian Wadden Sea, northwest Germany, and the effect of the strong winter 1995-1996. Organic Geochemistry 29, 1507–1516.

Røy, H., Huettel, M., Jørgensen, B.B., 2002. The role of small-scale sediment topography for oxygen flux across the diffusive boundary layer. Limnology and Oceanography 47, 837–847.

Røy, H., Lee, J.S., Jansen, S., De Beer, D., 2007. Tide-driven deep pore-water flow in intertidal sand flats. Limnology and Oceanography 53, 1521–1530.

Røy, H., Lee, J.S., Jansen, S., De Beer, D., 2008. Tide-driven deep pore-water flow in intertidal sand flats. Limnology and Oceanography 53, 1521–1530.

Rusch, A., Huettel, M., 2000. Advective particle transport into permeable sediments-evidence from experiments in an intertidal sandflat. Limnology and Oceanography 45, 525–533.

Rutherford, J., Latimer, G., Smith, R., 1993. Bedform mobility and benthic oxygen uptake. Water Research 27, 1545–1558.

Ruttenberg, K.C., 2003. The global phosphorus cycle. Treatise on geochemistry 8, 682.

Ryther, J.H., Dunstan, W.M., 1971. Nitrogen, phosphorus, and eutrophication in the coastal marine environment. Science 171, 1008–1013.

Santos, I.R., et al., 2015. Porewater exchange as a driver of carbon dynamics across a terrestrial-marine transect: insights from coupled ^{222}Rn and pCO_2 observations in the German Wadden Sea. Marine Chemistry 171, 10–20.

Santos, I.R., Bryan, K.R., Pilditch, C.A., Tait, D.R., 2014. Influence of porewater exchange on nutrient dynamics in two New Zealand estuarine intertidal flats. Marine Chemistry 167, 57–70.

Santos, I.R., Eyre, B.D., Huettel, M., 2012. The driving forces of porewater and groundwater flow in permeable coastal sediments: a review. Estuarine, Coastal and Shelf Science 98, 1–15.

Schmidt, F., Elvert, M., Koch, B.P., Witt, M., Hinrichs, K.-U., 2009. Molecular characterization of dissolved organic matter in pore water of continental shelf sediments. Geochimica et Cosmochimica Acta 73, 3337–3358.

Seidel, M., et al., 2015. Benthic-pelagic coupling of nutrients and dissolved organic matter composition in an intertidal sandy beach. Marine Chemistry 176, 150–163.

Seidel, M., et al., 2014. Biogeochemistry of dissolved organic matter in an anoxic intertidal creek bank. Geochimica et Cosmochimica Acta 140, 418–434.

Seidel, M., Graue, J., Engelen, B., Köster, J., Sass, H., Rullkötter, J., 2012. Advection and diffusion determine vertical distribution of microbial communities in intertidal sediments as revealed by combined biogeochemical and molecular biological analysis. Organic Geochemistry 52, 114–129.

Sigmon, D.E., Cahoon, L.B., 1997. Comparative effects of benthic microalgae and phytoplankton on dissolved silica fluxes. Aquatic Microbial Ecology 13, 275–284.

Sinninghe Damsté, J.S., De Leeuw, J.W., 1990. Analysis, structure and geochemical significance of organically-bound sulphur in the geosphere: state of the art and future research. Organic Geochemistry 16, 1077–1101.

Smith, S.V., 1984. Phosphorus versus nitrogen limitation in the marine environment. Limnology and Oceanography 29, 1149–1160.

Sorensen, J., Jørgensen, B.B., Revsbech, N.P., 1981. A comparison of oxygen, nitrate, and sulfate respiration in coastal marine sediments. Microbial Ecology 5, 105–115.

Stal, L.J., 1995. Physiological ecology of cyanobacteria in microbial mats and other communities. New Phytologist 131, 1–32.

Stal, L.J., Grossberger, S., Krumbein, W.E., 1984. Nitrogen fixation associated with the cyanobacterial mat of a marine laminated microbial ecosystem. Marine Biology 82, 217–224.

Stief, P., Behrendt, A., Lavik, G., De Beer, D., 2010. Combined gel probe and isotope labeling technique for measuring dissimilatory nitrate reduction to ammonium in sediments at millimeter-level resolution. Applied and Environmental Microbiology 76, 6239–6247.

Stief, P., Kamp, A., De Beer, D., 2013. Role of diatoms in the spatial-temporal distribution of intracellular nitrate in intertidal sediment. PLoS One 8.

Stutz, M.L., Pilkey, O.H., 2002. Global distribution and morphology of deltaic barrier island systems. Journal of Coastal Research 36, 694–707.

Sundareshwar, P.V., Morris, J.T., Koepfler, E.K., Fornwalt, B., 2003. Phosphorus limitation of coastal ecosystem processes. Science 299, 563–565.

Sundbäck, K., Miles, A., 2000. Balance between denitrification and microalgal incorporation of nitrogen in microtidal sediments, NE Kattegat. Aquatic Microbial Ecology 22, 291–300.

Taylor Smith, D., Li, W.N., 1966. Echo-sounding and sea-floor sediments. Marine Geology 4, 353–364.

Thibodeaux, L.J., Boyle, J.D., 1987. Bedform-generated convective-transport in bottom sediment. Nature 325, 341–343.

Thingstad, T.F., Zweifel, U.L., Rassoulzadegan, F., 1998. P limitation of heterotrophic bacteria and phytoplankton in the northwest Mediterranean. Limnology and Oceanography 43, 88–94.

Thornton, D.C.O., Underwood, G.J.C., Nedwell, D.B., 1999. Effect of illumination and emersion period on the exchange of ammonium across the estuarine sediment-water interface. Marine Ecology Progress Series 184, 11–20.

Trimmer, M., Nedwell, D.B., Sivyer, D.B., Malcolm, S.J., 1998. Nitrogen fluxes through the lower estuary of the River Great Ouse, England - the role of the bottom sediments. Marine Ecology Progress Series 163, 109–124.

Trimmer, M., Nedwell, D.B., Sivyer, D.B., Malcolm, S.J., 2000. Seasonal organic mineralisation and denitrification in intertidal sediments and their relationship to the abundance of *Enteromorpha* sp. and *Ulva* sp. Marine Ecology Progress Series 203, 67–80.

Tyrrell, T., 1999. The relative influences of nitrogen and phosphorus on oceanic primary production. Nature 400, 525.

Ullman, W.J., Aller, R.C., 1982. Diffusion coefficients in nearshore marine sediments. Limnology and Oceanography 27, 552–556.

Veuger, B., Eyre, B.D., Maher, D., Middelburg, J.J., 2007. Nitrogen incorporation and retention by bacteria, algae, and fauna in a subtropical, intertidal sediment: an in situ ^{15}N-labeling study. Limnology and Oceanography 52, 1930–1942.

Villbrandt, M., Krumbein, W.E., Stal, L.J., 1991. Diurnal and seasonal variations of nitrogen fixation and photosynthesis in cyanobacterial mats. Plant and Soil 137, 13–16.

Volkman, J.K., Johns, R.B., Gillan, F.T., Perry, G.J., 1980. Microbial lipids of an intertidal sediment - I. Fatty acids and hydrocarbons. Geochimica et Cosmochimica Acta 44, 1133–1143.

Volkman, J.K., Rohjans, D., Rullkötter, J., Scholz-Böttcher, B.M., Liebezeit, G., 2000. Sources and diagenesis of organic matter in tidal flat sediments from the German Wadden Sea. Continental Shelf Research 20, 1139–1158.

Warwick, R.M., Joint, I.R., Radford, P.J., 1979. Secondary production of the bethos in an estuarine environment. In: Jeffries, R.L., Davey, A.L. (Eds.), Ecological Processes in Coastal Environments. Blackwell, pp. 429–467.

Werner, U., et al., 2006. Spatial and temporal patterns of mineralization rates and oxygen distribution in a permeable intertidal sand flat (Sylt, Germany). Limnology and Oceanography 51, 2549–2563.

Wilms, R., Sass, H., Koepke, B., Cypionka, H., Engelen, B., 2006. Specific bacterial, archeal, and eukaryotic communities in tidal flat sediments along a vertical profile of several meters. Applied and Environmental Microbiology 72, 2756–2764.

Wilms, R., Sass, H., Koepke, B., Cypionka, H., Engelen, B., 2007. Methane and sulfate profiles within the subsurface of a tidal flat are reflected by the distribution of sulfate-reducing bacteria and methanogenic archaea. FEMS Microbiology Ecology 59, 611–621.

Yamamoto, T., 1997. Importance of coastal seas on the absorption of carbon dioxide in the atmosphere. Bulletin of the Japanese Society of Fisheries Oceanography 61, 381–393.

Ziebis, W., Huettel, M., Forster, S., 1996. Impact of biogenic sediment topography on oxygen fluxes in permeable seabeds. Marine Ecology Progress Series 140, 227–237.

PART IV

MARSHES AND SEAGRASSES

CHAPTER

13

Productivity and Biogeochemical Cycling in Seagrass Ecosystems

Marianne Holmer

Department of Biology, University of Southern Denmark, Odense, Denmark

1. INTRODUCTION

Seagrass beds are widely distributed in subtidal coastal zones around the world, except in the high-Arctic and Antarctic, where ice cover limits their expansion (Green and Short, 2003; Larkum et al., 2006a). In the tropics high diversity is found with up to nine species present in the same meadow. Under temperate conditions fewer species or monospecific stands are more common (Green and Short, 2003; Larkum et al., 2006a), with some exceptions such as along the coasts of temperate Australia where high species diversity can be found (Carruthers et al., 2007). Compared with macroalgal communities, the diversity of seagrasses is relatively low, but among the more than 60 species found, there are large morphological differences extending from small species of a few centimeters in length to the meter-long leaves of *Zostera marina* and *Enhalus acoroides* (Fig. 13.1). Seagrasses are anchored in the sediments by rhizomes and roots extending from a few centimeters to several meters for *Halophila* sp. and *E. acoroides*, respectively (Duarte et al., 1998). Seagrass beds are some of the most productive ecosystems in the world, and they provide a range of ecosystem services which can be divided into provision, regulation, support, and cultural (Ruiz-Frau et al., 2017) and include food, raw materials, climate regulation, coastal protection, bioremediation of waste, life cycle maintenance, water conditions, research and education, recreation and tourism, cultural heritage, and identity (Jackson et al., 2001; Heck and Valentine, 2006; McGlathery et al., 2007; Fourqurean et al., 2012; Ruiz-Frau et al., 2017). The rhizomes and roots of seagrasses stabilize sediments and at the same time organic matter is buried in the meadows, which affects sediment biogeochemical conditions, including microbial activity and redox potential. Seagrass beds are threatened all over the world because of increasing human activity in coastal zones, where urbanization, tourism, and eutrophication are among the major threats (Waycott et al., 2009; Boström et al., 2014). This chapter reviews the importance of subtidal sediment biogeochemistry by (1) describing the role of seagrasses as primary

Coastal Wetlands
https://doi.org/10.1016/B978-0-444-63893-9.00013-7

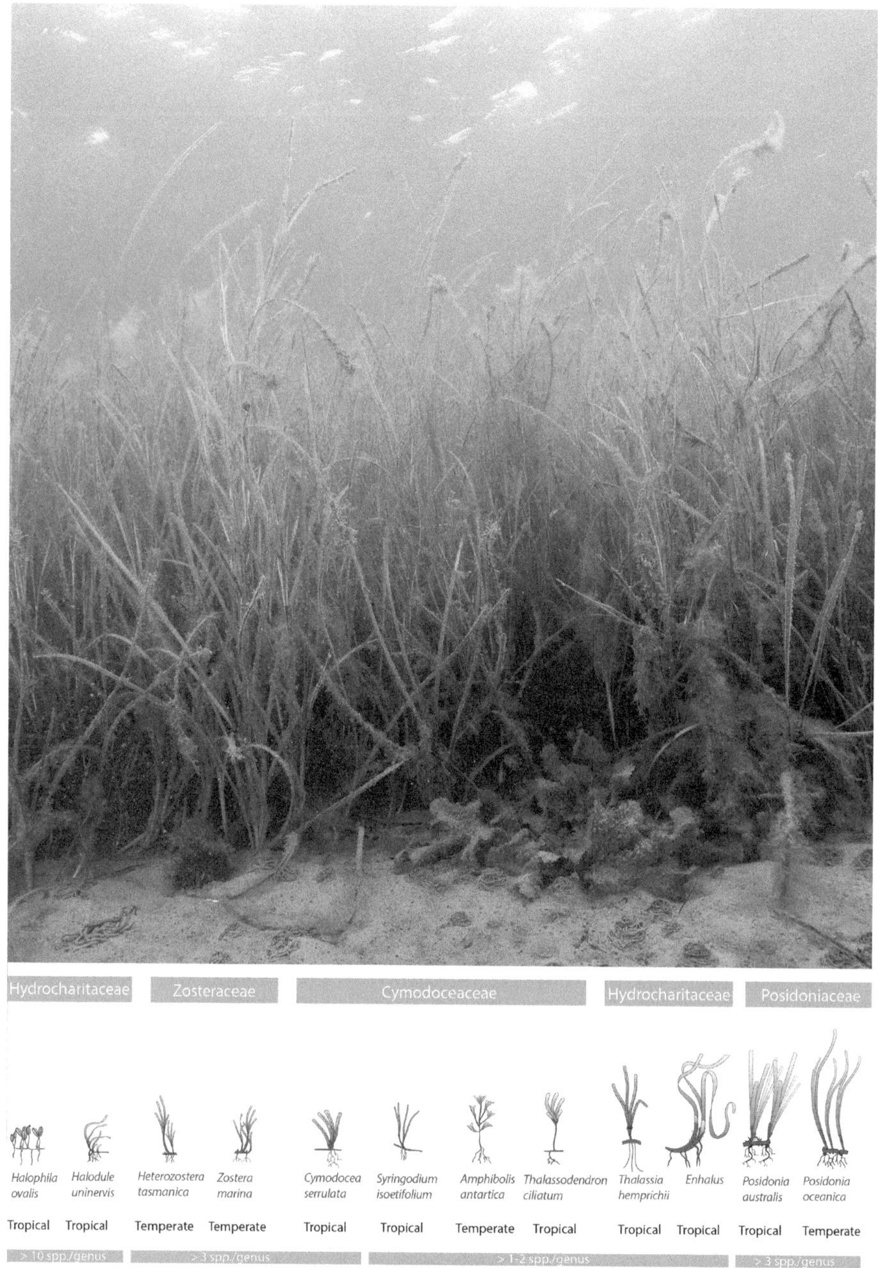

FIGURE 13.1 Top: Photo of a *Zostera marina* meadow in Denmark. Bottom: The relative sizes of average adult plants across the 12 genera and 4 families to which seagrass mostly belong (*green bars*). *Photo: Jonas Thomar.*

producers in the coastal zone, (2) describing the fate of produced organic material within the meadows including the role as Blue Carbon storage, (3) elucidating the effects of seagrasses on sediment biogeochemistry from micro- to macroscale, (4) describing seagrass–sediment interactions with focus on nutrients and sulfur cycling, and finally (5) discussing the consequences of human pressures on seagrass–sediment interactions.

1.1 Primary Productivity

Seagrass beds are one of the most productive marine ecosystems, along with coral reefs and tidal wetlands (Duarte et al., 2004). In comparison with macroalgae, seagrasses take advantage of their ability to acquire nutrients from the sediments and the water column. Their light requirements, however, are higher because seagrass tissues support a larger fraction of respiratory organs (rhizomes and roots) (Borum et al., 2006). Because of the large size variation in seagrasses, production (gDW $m^{-2}d^{-1}$) varies by a factor of 500 between the least (*Halophila ovalis*; 0.01 g dry weight (DW) $m^{-2}d^{-1}$) and the most productive species (*Phyllospadix torreyi*; 11.3 gDW $m^{-2}d^{-1}$) (Duarte and Chiscano, 1999). A recent estimate of net primary production for seagrasses ranges between 349 and 449 gC m^{-2} $year^{-1}$ similar to macroalgae communities and lower ranges of salt marshes and mangroves (Duarte, 2017). As for wetland and terrestrial systems, belowground production is seldom measured, but it is expected to be less than aboveground production. The studies available from subtropical and tropical seagrasses show that the belowground production (3.6–130 gDW m^{-2} $year^{-1}$) can account for half of the aboveground production (9.5–2323 gDW m^{-2} $year^{-1}$, Duarte et al., 1998). Recent studies suggest the use of belowground dynamics as a proxy to assess long-term effects of environmental stressors due to much longer turnover time compared with leaf biomass (Vonk et al., 2015). Differences in nutrient availability between the water column and sediments may stimulate relative growth of leaves versus roots, respectively, as reported by Hillman et al. (1995) for *Halophila ovalis* in an estuary in Western Australia.

Seagrasses are widely distributed from the equator to high latitudes with observations of *Z. marina* as far north as in Greenland (Clausen et al., 2014). The production of *Z. marina* has been measured as far north as at 63°35′65″N, where it was about 62.5 mmol C $m^{-2}d^{-1}$, only slightly lower compared with lower latitudes (Duarte, 2002). A study of *Z. marina* biomass and growth dynamics in the distribution area of *Z. marina* across gradients in temperature and latitude (29.1–66.2°N) showed limited effect of temperature and latitude, suggesting that other environmental factors are more important in controlling production (Clausen et al., 2014). In temperate areas, seagrasses show significant seasonal variation with maximum production and biomass during summer, as well as low biomass and low or insignificant net production during winter (*Z. marina* in Europe, USA, and Japan; Olesen and Sand-Jensen, 1994, *Z. marina* in Denmark; Risgaard-Petersen and Ottosen, 2000, *Z. marina* from 29.1 to 66.2°N Clausen et al., 2014). In the subtropics and the tropics, seasonal studies show less variation, although changes in light intensity, salinity, and temperature may significantly affect productivity (*Cymodocea rotundata* and *Enhalus acoroides* Sri Lanka; Johnson and Johnstone, 1995, *Enhalus acoroides* Thailand; Rattanachot and Prathep, 2011, *Thalassia hemprichii* Taiwan; Chiu et al., 2013).

Seagrasses are generally the dominant source of the total primary production in seagrass beds, although epiphytes may contribute up to 60% of the total (Hemminga and Duarte, 2000). There are many different types of epiphytes from small calcareous algae to large thread-like macroalgae. A growing problem due to increasing eutrophication in coastal zones is the invasion of seagrass beds by drifting macroalgae (Hauxwell et al., 2001; Kopecky and Dunton, 2006) or invasive species (Holmer et al., 2009a; Drouin et al., 2016). These may contribute significantly to system net primary production but negatively affect the seagrass ecosystem because of anoxia that results from decomposition of the excessive macroalgae as found in eutrophic coastal zones in Denmark (Krause-Jensen et al., 1999) and in experiments with *Z. marina* (Holmer and Nielsen, 2007). A worldwide metaanalysis of 59 experiments with seaweeds and seagrasses also showed negative effects of seaweeds on general seagrass performance (Thomsen et al., 2012).

The net production of seagrasses is most often assessed from leaf growth measurements, e.g., by puncturing leaves or by following changes in biomass over time (Zimmerman, 2006). Similarly, total seagrass ecosystem metabolism of seagrass beds is measured in situ by using benthic chambers or eddy covariance techniques. Benthic chambers are typically deployed over a 24-h period, and the fluxes of O_2 and/or CO_2 are measured to assess the total ecosystem production and respiration (e.g., Barrón et al., 2006; Castorani et al., 2015). The eddy covariance technique is noninvasive and integrates over a large area ($>100\ m^2$) relative to chambers ($<0.1\ m^2$), e.g., a seagrass meadow, but it is technically demanding (Berg et al., 2007) and relatively few studies are available from seagrass beds to date (Rheuban et al. 2014a, 2014b; Long et al. 2015a, 2015b). Common for both chamber and covariance techniques are large diurnal and seasonal variations in net ecosystem metabolism showing autotrophy during the summer and heterotrophy during winter (Barrón et al., 2006; Apostolaki et al., 2010; Long et al., 2015b). Light is the most important controlling parameter of seagrass productivity (Zimmerman, 2006), whereas nutrients are most important for tropical seagrasses growing under oligotrophic conditions (Romero et al., 2006). Furthermore, seagrasses are generally considered to be carbon limited because of the low availability of CO_2 at seawater pH (Larkum et al., 2006b). Recent studies simulating higher CO_2 in the atmosphere show variable response reflecting complex interactions between environmental factors and seagrass productivity and that it is far from likely that seagrasses will benefit from ocean acidification (Apostolaki et al., 2014; Ow et al. 2016a, 2016b; Hendriks et al., 2015). Experiments with three tropical seagrasses (*Cymodocea serrulata*, *Halodule uninervis*, and *Thalassia hemprichii*) showed higher growth at higher CO_2, but the response varied between species suggesting that some species may benefit more and potentially increase their competitiveness resulting in community changes (Ow et al., 2015). In a study of *Posidonia oceanica* in the Mediterranean at a CO_2 venting site, there was no change in *P. oceanica* photosynthesis but potentially a higher susceptibility toward herbivory, as the epibiont community increased in abundance and diversity possibly decreased the standing stock of the seagrass (Guilini et al., 2017). Several authors suggest that environmental factors, e.g., increased nutrients and reduced lights, will have stronger negative impact compared with the benefits of increased CO_2 in a future climatic setting based on laboratory experiments (*P. oceanica* Hendriks et al., 2015; *Cymodocea nodosa* de los Santos et al., 2017).

The release of oxygen from the roots may also affect the pools of inorganic carbon through enhanced aerobic respiration and oxidation of rhizosphere sediments as found in seagrass meadows dominated by *Thalassia testudinum* in the Bahamas (Burdige and Zimmerman, 2002). Aerobic respiration increases pools of CO_2 and may also decrease pH, which in carbonate sediments may result in significant carbonate dissolution (Hu and Burdige, 2007; Burdige et al., 2010). One advantage of the dissolution of carbonates is the associated release of phosphorus, which is considered a limiting nutrient for seagrasses in carbonate systems such as *T. testudinum* in Florida Bay (Nielsen et al., 2006). Recently new methodologies have documented the dynamics of pH in rhizosphere sediments of *Zostera marina* and *Zostera muelleri*, where the pH is typically higher compared with bulk sediment (Koren et al., 2015; Brodersen et al., 2016), affecting the biogeochemistry of sediments on short timescales and further studies are recommended for seagrasses growing in carbonate sediments.

In spite of an estimated cover of only <2% of the ocean surface, seagrass beds contribute significantly to the oceanic carbon cycle because of their locally high productivity. In particular, their capacity to bury allochthonous and autochthonous organic carbon has been highlighted in recent studies of carbon sequestration in the oceans (e.g., Duarte et al., 2004; Fourqurean et al., 2012; Duarte, 2017). This is also of major importance for the biogeochemical processes in sediments as they are generally limited by the availability of organic matter.

1.2 Fate of Primary Productivity—Export

Only a limited part of the primary production from seagrasses enters directly into the grazing food chain, and with the loss of large herbivore megafauna, such as dugong and turtles, grazing has declined even further (Valentine and Duffy, 2006). An average value for grazing in seagrass beds is estimated to be 10%, but large variability exists and depends on grazer and seagrass species, as well as environmental factors (Cebrián and Duarte, 1998; Duffy et al., 2015). The nongrazed biomass supports an assemblage of species in the detritus food chain (Heck and Valentine, 2006). Seagrass detritus is relatively recalcitrant compared with macroalgae and phytoplankton but less recalcitrant than wetland macrophytes as recently shown in a study of Blue Carbon potentials for different marine macrophytes in Australia (Trevathan-Tackett et al., 2015). As such, detritivores and microbial decomposers require the activity of specific enzymes such as cellulases and pectinases, which limit their rate of decomposition, especially under low nutrient availability (global study terrestrial and marine plants Enríquez et al., 1993, *Z. marina* Holmer and Bachmann Olsen, 2002, *Posidonia oceanica* Pedersen et al., 2011). Detritivores may contribute significantly to the decomposition process by fragmenting detritus, in particular in the tropics where detritivore diversity is high (Valentine and Duffy, 2006).

An important research question is to what extent is seagrass detritus decomposed within the meadows or exported? Seagrass detritus is exported from seagrass beds in dissolved and particulate forms. Seagrasses release dissolved organic matter (DOM) directly from the leaves as found in mixed subtropical seagrass meadows (Ziegler and Benner, 1999) and for *Posidonia oceanica* (Barrón and Duarte, 2009) and high release rates have also been found during decomposition of *Zostera marina* detritus (Vichkovitten and Holmer, 2004).

Seagrass beds appear to be net dissolved organic carbon (DOC) producers on an annual basis and export may stimulate bacterial growth beyond their boundaries (Barrón et al. 2004, 2006; Apostolaki et al., 2010; Hyndes et al., 2014; Wang et al., 2014). Net DOM release is highest during the daytime, and for a *P. oceanica* meadow, DOM represents up to 71% of the net community production (Barrón and Duarte, 2009). Large piles of *P. oceanica* and *Z. marina* aboveground biomass (beach wrack) accumulate on the beaches after autumn storms as found throughout the Mediterranean for *P. oceanica* (Simeone and de Falco, 2012) and in the distribution range of *Z. marina* (Sand-Jensen et al., 1994), indicating a significant export of seagrass detritus in particulate forms (Duarte et al., 2004; Heck et al., 2008; Duarte, 2017). Massive accumulations of beach wrack have been utilized for commercial purposes such as roofs and as fertilizers in Denmark (Sand-Jensen et al., 1994) and China (Photo 13.1) and have recently gained new interest as sustainable insulation. Beach wrack is, however, often perceived a nuisance by beach tourists, and the material is regularly removed and deposited on land (Simeone and de Falco, 2012). This removal from the aquatic ecosystem may significantly impact the biogeochemical cycles of the sea, as most of the nutrients bound in detritus are moved from the beach back to the sea, where it is again incorporated into aquatic system nutrient cycles (Mateo et al., 2003). Seagrass detritus is also important for the stability of beaches and hinders erosion during storms. Thus mechanical removal of seagrass detritus may increase erosion (Simeone and De Falco, 2012). The export of detritus from seagrass beds varies a lot, with estimates ranging from 1% to 80% of production being reported (Hemminga et al., 1991; Mateo et al., 2006; Lee et al., 2016) with a recent average global estimate of 24.3% of net primary production (NPP) (Duarte, 2017). The large range is due to many different factors such as local currents, seafloor characteristics, wind exposure, and leaf morphology. A considerable fraction of the detritus is considered to be exported to greater water depths where it contributes to the organic matter and nutrient pools in the sediments (Duarte et al., 2004; Duarte, 2017).

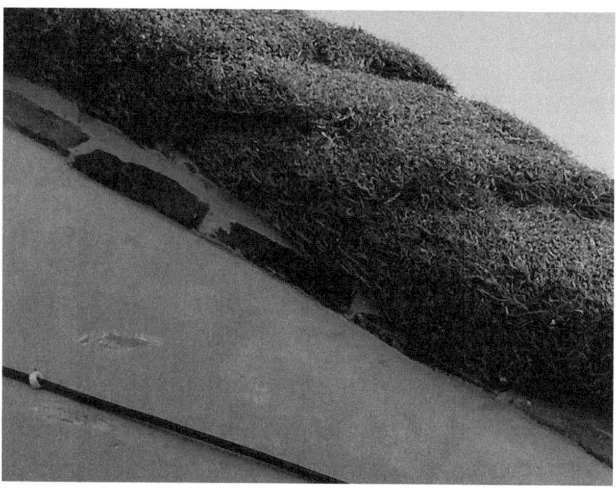

PHOTO 13.1 Houses with roofs made of seagrass wrack in Rongcheng, Weihai City, Shandong Province. China. By Xiao Xi.

PHOTO 13.2 Accumulation of rhizomes and roots in a Posidonia oceanica meadow at Formentera Island, Spain. By: Christoffer Boström.

1.3 Fate of Primary Productivity—Burial and Seagrass Blue Carbon

Dead seagrass roots and rhizomes often make a significant contribution to the pool of detritus in the sediments (Duarte et al., 1998; Mateo et al., 2003). This is particularly the case for *Posidonia oceanica* in the Mediterranean, which forms meter deep mats (Photo 13.2), but most other seagrasses with large roots and rhizomes accumulate seagrass detritus as found for *Z. marina* in the Baltic Sea (Rohr et al., 2016), as well as *Thalassia testudinum* (Mexican Caribbean, Costa Rica, USA), *T. hemprichii* (Philippines), *Enhalus acoroides* (Philippines), *Cymodocea nodosa* (Spain), *Zostera noltii* (Spain), and *Thalassodendron ciliatum* (Kenya) (Duarte et al., 1998; van Tussenbroek, 1998; Kaldy and Dunton, 2000; Paynter et al., 2001). In addition, organic matter imported from other ecosystems and production by epiphytes within beds contributes to sediment detritus (Gacia et al., 2002; Kennedy et al., 2010; Rohr et al., 2016; Trevathan-Tackett et al., 2015). Organic enrichment from allochthonous sources (e.g., macroalgae and phytoplankton) stimulates microbial activity and enhances the turnover and regeneration of nutrients because of their less recalcitrant composition compared with seagrasses (Holmer et al., 2003a, Holmer et al., 2003b, Duarte et al., 2005, Marbà et al., 2006, Mazarrasa et al., 2017a,b).

Burial of organic matter in seagrass beds depends on several factors such as the flux of organic matter to the sediment, sediment accumulation rate, sediment grain size and surface area, availability of oxygen, and the lability of various organic components (Gacia et al., 2002; Marbà et al., 2006; Mateo et al., 2006). Current knowledge on burial is rapidly increasing because of interest in seagrass meadows as Blue Carbon hot spots (see below), but until recently most knowledge on burial has been obtained from the peat-like accumulations as observed for *P. oceanica* where mats several meters thick and thousands of years old can be found (Mateo et al., 2003). For *Posidonia australis, Posidonia sinuosa,* and *Cymodocea nodosa,* it is common to find intact fractions of rhizomes in deeper parts of the sediments (Pérez et al., 2001).

The use of bacterial markers such as $\delta^{13}C$ phospholipids has led to interesting observations of the fate of organic matter in seagrass beds (Boschker et al. 1999, 2000; Holmer et al., 2003a; Bouillon et al., 2004). These techniques are useful because seagrass is typically $\delta^{13}C$ heavy ($\sim -10‰$) compared with algal, phytoplankton, mangrove, and terrestrial C sources. It is thus possible to identify the origin of carbon for bacteria from such sources as seagrass detritus, macroalgae, phytoplankton, or seston through the use of mixing models. Seagrass detritus plays an important role in oligotrophic conditions in the tropics (Holmer et al., 2001; Bouillon et al., 2004; Bouillon and Boschker, 2006), whereas in nutrient-rich seagrass ecosystems, detritus in the sediments is dominated by allochthonous sources such as macroalgae and phytoplankton (Rohr et al., 2016; Mazarrasa et al., 2017a). These substrates show higher degradability compared with seagrass detritus, which increases the microbial activity compared with less nutrient-enriched seagrass beds (Boschker et al., 2000; Holmer et al., 2003a). A similar pattern occurs for the $\delta^{13}C$ signal of bulk carbon, where the contribution of allochthonous sources is higher in eutrophic than oligotrophic sediments (Kennedy et al., 2010; Watanabe and Kuwae, 2015).

Global climate change research during the last two decades has increased attention on seagrass meadows because of their high organic carbon content (Duarte et al., 2005; Duarte, 2017). Seagrasses are considered key ecosystems for carbon cycling and burial because of their broad distribution in coastal zones, high productivity, capacity to trap particles, and high sediment organic content. Seagrass sediments accumulate organic matter, and recent research has focused on their significance as a global sink for atmospheric CO_2. Although seagrass meadows only occupy $<0.2\%$ of the ocean floor, they are estimated to be responsible for $\sim 20\%$ of the global organic carbon burial in marine sediments and are thus important carbon sinks and contribute to mitigation of climate change through the sequestration of atmospheric CO_2 (Kennedy et al., 2010). Their carbon sink capacity varies significantly between species and along environmental gradients. The Mediterranean *P. oceanica* is considered the most significant carbon sink, storing the largest carbon stocks recorded so far (Fourqurean et al., 2012; Mazarrasa et al., 2017a,b). *P. oceanica* develops a "matte" of roots and rhizomes due to slow decomposition of this material under the anoxic conditions in the sediments (Pedersen et al., 2011) and the matte may build up to several meters in thickness (Mateo et al., 2003). Other species of *Posidonia,* the widespread *Z. marina* (Röhr et al., 2016), and the small-sized *Halophila* species (Campbell et al., 2015) also show significant carbon burial (Lavery et al., 2013). The effects of environmental factors such as hydrodynamic forcing, water depth, temperature, and latitude on carbon sink capacity are currently being investigated. There is often a lack of direct correlation between the productivity of seagrasses and carbon sink capacity as local factors, such as hydrodynamic forcing, are more important (Kennedy et al., 2010; Miyajima et al., 2015; Röhr et al., 2016). Sink capacity appears to vary with sediment grain size, which to some extent reflects local hydrodynamic regimes. Fine-grained sediments typically have higher sink capacity than coarse, sandy sediments (Miyajima et al., 2015, Röhr et al., 2016). In addition, the available sources of carbon at the site of the seagrass meadows play a role, where allochthonous sources can augment that contributed by local seagrass (Kennedy et al., 2010; Trevathan-Tackett et al., 2015; Watanabe and Kuwae, 2015). Nutrient enrichment stimulates seagrass productivity and carbon burial in oligotrophic systems but contributes to overproduction of phytoplankton and epiphytic algae in eutrophic systems. The effect on C burial in eutrophic sites is nonconclusive, where both higher and lower burials have been found (Holmer et al., 2004; Armitage and Fourqurean, 2016; Howard et al., 2016; Mazarrasa et al., 2017a,b).

While most studies have examined burial of organic carbon, in carbonate-rich environments, inorganic carbon and the effect of pH should also be considered (Howard et al., 2018). Inorganic carbon is derived from carbonate skeletons and shells of organisms inhabiting seagrass meadows or from sedimentation of particles from the water column. The in situ production of carbonate has important implications for the carbon sink capacity of seagrass sediments (Mateo and Serrano, 2012; Macreadie et al., 2014). Precipitation of $CaCO_3$ releases CO_2 to the atmosphere, effectively reducing CO_2 sequestration. A global study of 403 seagrass and 34 adjacent bare sites showed an average ratio of POC:PIC (organic:inorganic particulate carbon) of 0.74, indicating higher accumulation of inorganic carbon compared with organic carbon, but the organic carbon burial was still much higher in vegetated compared with bare sites demonstrating the CO_2 sink capacity of seagrass meadows (Mazarrasa et al., 2015).

Degradation or loss of seagrass meadows worldwide reduces their sequestration potential, especially when sediment organic matter in the lost seagrass meadows is remineralized and CO_2 returned to the atmosphere (Marbà et al., 2015; Rozaimi et al., 2016). The loss of *Z. marina* in Danish coastal waters over the past century corresponds to a reduction in carbon burial capacity of 23–27 Gg C (Rohr et al., 2016), and similarly, the loss of *P. australis* in Oyster Harbour during eutrophication in the 1980s was estimated to reduce burial by 37–41 Gg C over 40 years. In both cases the lost capacity is greater than the remaining carbon burial. Restoration of the seagrass bed presumably could reverse these anthropogenic losses (Greiner et al., 2013; Marbà et al., 2015).

2. SEDIMENT BIOGEOCHEMISTRY—MODIFIED BY SEAGRASSES

High organic carbon content and high loading rates contribute to high rates of microbial activity in seagrass sediments (Marbà et al., 2006; Koren et al., 2015). The presence of live and dead roots and rhizomes with their large surface area (Duarte et al., 1998) increases the complexity of the sediments considerably (Fig. 13.2). Roots have a major effect on sediment biogeochemistry because of their fast elongation rates, where, for example, *Zostera marina* and *Z. muelleri* elongate several millimeters per day (Frederiksen and Glud, 2006; Koren et al., 2015) and they exchange oxygen and DOM across young root tips as found for several species including *Zostera marina* and *Z. muelleri* (Isaksen and Finster, 1996; Jovanovic et al., 2015; Koren et al., 2015). Most seagrasses also show fast vegetative growth with the ability to spread into bare sediments. As seagrass systems mature, the biogeochemical conditions change as well (Pedersen et al., 1997; Pérez et al., 2001; Barrón et al., 2004; Marbà et al., 2015).

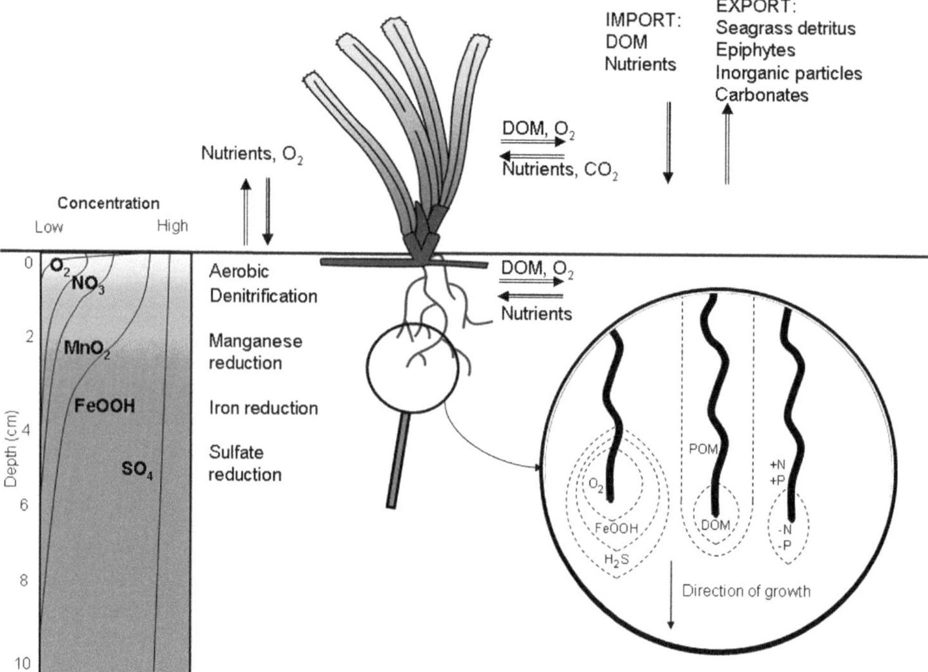

FIGURE 13.2 Conceptual model of macro- and microscale biogeochemical gradients and biogeochemical processes that create them in seagrass ecosystems. IAN Symbol Libraries is acknowledged for the seagrass symbols. *DOM*, dissolved organic matter.

Numerous important microbial processes occur throughout the sediment profile in marine sediments (Canfield et al., 1993). In the upper sediment layers, where oxygen penetrates, aerobic respiratory processes dominate the decomposition of organic matter. In unvegetated sediments, aerobic respiration often dominates decomposition, as organic matter is supplied from the water column and mineralized rapidly in the oxic layer (Fig. 13.2, Rasmussen and Jorgensen, 1992). The residual organic matter that accumulates in deeper anoxic layers is often relatively refractory resulting in much lower rates of decomposition (Canfield et al., 1993). In seagrass sediments, labile organic matter supplied to deeper layers through root and rhizome death and live root exudates stimulates microbial activity in these layers (Moriarty et al., 1986; Blaabjerg et al., 1998; Nielsen et al., 2001).

The general depth gradient, but with patchy microzones of electron donors and acceptors, results in a redox gradient where there is an overall decrease in free energy yield of microbial metabolism. In the oxic layers and microzones, nitrification is often an important process resulting in the oxidation of ammonium to nitrate, the latter which can either diffuse to the water column or to more reducing layers, where it is denitrified (Canfield et al., 1993). The denitrification process is succeeded by iron (Fe) and manganese (Mn) reduction, and the importance of both these processes for organic matter decomposition is controlled by their availability in oxidized forms (Thamdrup, 2000). The supply of oxygen from roots is expected to increase the reoxidation of reduced manganese and iron, thereby increasing the potential importance of Fe and Mn recycling and reduction (Jovanovic et al., 2015; Brodersen et al., 2015a,b). In carbonate-rich sediments, which are important seagrass substrates in the subtropics and tropics, iron and manganese pools are often low because of the low terrigenous influence (Berner, 1984). Here the terminal mineralization step is dominated by sulfate reduction. Methanogenesis is considered of limited importance because of high concentration of sulfate in the relatively shallow seagrass sediments (Holmer et al. 2001, 2003a). But even in Fe- and Mn-rich sediments, sulfate reduction is typically the dominant process in organic carbon oxidation after aerobic respiration. In organic-enriched seagrass sediments where the oxygen penetration is limited, sulfate reduction can dominate the mineralization (Holmer et al., 2003a).

The belowground system of rhizomes and roots has several functions, in addition to anchoring seagrasses in the sediments. Carbohydrate reserves in rhizomes can be utilized when light availability is reduced, for example, in *Zostera marina* during winter in temperate areas (Burke et al., 1996; Vichkovitten et al., 2007). Rhizomes also translocate nutrients between shoots, thus allowing seagrasses to maintain optimal growth conditions, despite a heterogenous nutrient distribution in the sediments as shown for five tropical and three temperate species (Marbà et al., 2002). Studies on rhizome and root oxygen leakage are rapidly growing as the technology improves, and with this, new knowledge on the plant—sediment interactions is expanding. Recent studies of *Zostera muelleri* show that rhizomes leak oxygen in general (Brodersen et al., 2015b), while oxygen leakage from root tips may be restricted to young roots as demonstrated for *Zostera marina* and *Z. muelleri* (Brodersen et al., 2015b; Jovanovic et al., 2015). Exchanges of substrates across the sediment—plant interface are often deduced from bulk analysis of sediments or observations of the distribution of bacteria in, on, and around the roots (Blaabjerg and Finster, 1998; Marbà et al., 2006; Trevathan-Tackett et al., 2014; Garcias-Bonet et al., 2016). Bulk analyses of sediments show depletion of nutrient pools such as ammonium and phosphate in *Z. marina* meadows (Risgaard-Petersen and Ottosen, 2000; Holmer et al., 2006), as well as oxidation of sulfide and iron pools in *Posidonia oceanica* and *Z. muelleri* meadows (Holmer et al., 2003a; Pagès et al., 2012). Stress studies, such as from shading, suggest a release of organic compounds (sugars, ethanol, and amino acids) probably because of root anoxia and associated anaerobic respiration in roots (*Z. marina* Smith et al., 1988, *Posidonia oceanica* Pérez et al., 2007, *Z. marina* Hasler-Sheetal et al., 2015). Enhanced activity of sediment microbes during the day suggests a release of photosynthetic compounds to the sediment (*Z. noltii* Isaksen and Finster, 1996, *Z. marina* Blaabjerg et al., 1998). Microcosm experiments have shown significant release of DOC through the root system, which was rapidly taken up by bacteria (*Thalassia testudinum* Kaldy et al., 2006). There is thus significant evidence that the release and uptake of compounds from belowground plant tissues affect the biogeochemistry of the seagrass rhizosphere.

Despite the fact that many seagrass meadows are autotrophic, the sediment compartment is net heterotrophic with relatively high oxygen consumption rates and low oxygen penetration depth as demonstrated for temperate (*Zostera marina*) and subtropical (*Thalassia testudinum*) species (Castorani et al., 2015; Jovanovic et al., 2015; Long et al., 2015b). There is little consensus on the magnitude of nitrification in seagrass sediments: the earliest studies showed enhanced rates in *Z. marina* meadows compared with unvegetated sites (Caffrey and Kemp, 1990), whereas the later studies of *Z. marina* and *Z. noltii* showed low rates due to competition between plants and microbes for ammonium (Risgaard-Petersen et al., 1998; Ottosen et al., 1999; Welsh et al., 2000; Table 13.1). Most studies show low rates of denitrification for temperate species (Table 13.1), whereas Eyre et al. (2011) found high rates in the subtropics. So far there are no reports on the importance of microbial iron reduction in seagrass sediments, but iron reduction may play an important role in sediments with high Fe content, such as terrigenous sediments or in sediments where seagrass species release oxygen at high rates, thus speeding up Fe reoxidation (Brodersen et al., 2015a,b). Dissolved pools of iron show diurnal dynamics in seagrass sediments (Pagès et al., 2012), and Fe additions can increase seagrass growth as found for *Posidonia oceanica* and *Thalassia testudinum* (Marbà et al., 2008; Ruiz-Halpern et al., 2008). The mechanisms behind these effects are thought to relate to sulfide dynamics and sulfide toxicity (Holmer and Hasler-Sheetal, 2014). Sulfate reduction has been examined in increasing detail in recent years, stimulated by observations of high sulfide pools in dieback areas of *T. testudinum* in Florida Bay (Carlson et al., 1994).

Sulfate reduction is often the dominant electron acceptor in seagrass sediments, especially in the deeper layers where other electron acceptors are exhausted (Holmer and Nielsen, 2007). There are, however, large variations within and between species (Table 13.2). Seasonal variations in water temperature and inputs of organic matter are important factors controlling sulfate reduction rates (Holmer and Laursen, 2002; Holmer et al., 2003a; Vichkovitten and Holmer, 2005). In a study of *P. oceanica* at two locations, sulfate reduction rates were four times higher in the late spring compared with winter when water *temperature* was 8°C lower (Fig. 13.3). Microbial activity is controlled by temperature, and sulfate reduction rates typically increase by a factor of 2–3 for every 10°C increase. The temperature effect is somewhat higher than observed for bare sediments (Moeslund et al., 1994) but consistent with observations of enriched fish farm sediments, where the input of organic matter in combination with increasing temperatures resulted in very high sulfate reduction rates, suggesting synergistic effects of temperature and organic matter loading (Holmer and Kristensen, 1996).

TABLE 13.1 Rates of Nitrification and Denitrification in Seagrass Sediments

Species	Location	Nitrification (μmol N $m^{-2}d^{-1}$)	Denitrification (μmol N $m^{-2}d^{-1}$)	References
Halophila ovalis and *Halophila spinulosa*	Australia	–	1920	Eyre et al. (2011)
Zostera capricorni	Australia	–	7680	Eyre et al. (2011)
Zostera marina	USA	410–2481	71–209	Caffrey and Kemp (1990)
	Denmark	–	17	Risgaard-Petersen et al. (1998)
		–	1.5	Ottosen et al. (1999)
Zostera noltii	France	–	3–12	de Wit et al. (2001)
		–	2–6	Welsh et al. (1996, 2000)

TABLE 13.2 Sulfate Reduction Rates (SRR) and Pools of Acid-Volatile (AVS) and Chromium-Reducible Sulfide (CRS) in Seagrass Sediments

Species	Location	SRR (mmol $m^{-2}d^{-1}$)	AVS (mol m^{-2})	CRS (mol m^{-2})	References
Amphibolis australis	Australia	2–8	–	0.07–3.28[a]	Holmer and Kendrick (2013)
Cymodocea nodosa	Spain	21	–	12.80[a]	Holmer et al. (2004)
Cymodocea rotundata	Thailand	6.4	–	9.50[a]	Holmer et al. (2001)
	Thailand	110	0.3	3.0	Holmer et al. (2006)
	Thailand	–	<0.5	1.3–7.4	Delefosse (2007)
Cymodocea serrulata	Thailand	–	<0.5	2.2–6.0	Delefosse (2007)
Enhalus acoroides	Australia	90	–	–	Pollard and Moriarty (1991)
	Thailand	80	0.5	11.5	Holmer et al. (2006)
	Thailand	–	<0.5	1.7–8.2	Delefosse (2007)
Halodule beaudetti	Jamaica	34	–	–	Blackburn et al. (1994)
Halophila ovalis	Thailand	80–120	0.1	3.5–4.0	Holmer et al. (2006)
	Thailand	–	<0.5	1.9–7.8	Delefosse (2007)
Halophila ovalis	Australia	4–7	–	0.07–3.28[a]	Holmer and Kendrick (2013)
Posidonia australis	Australia	1–6	–	0.07–3.28[a]	Holmer and Kendrick (2013)
Posidonia oceanica	Spain	3–12	–	0.16–3.82[a]	Holmer et al. (2004)
	Spain	–	–	0.05–1.81[a]	Calleja et al. (2006)
	Spain	10–15	0–0.16	0.86–11.98	Holmer (unpublished)
	Spain	8–24	0.04–0.12	15.57–23.83	Frederiksen et al. (2007)
	Italy	19–42	0–0.40	11.72–17.54	Frederiksen et al. (2007)
	Greece	6–19	0.04–0.08	0.70–3.21	Frederiksen et al. (2007)
	Cyprus	17–27	0.05–0.07	9.23–20.59	Frederiksen et al. (2007)
	Spain	–	0.37–2.29	12.24–17.38	Pérez et al. (2007)
	Spain	2	0.005	–	Gacia et al. (2012)
Posidonia sinuosa	Australia	1–4	–	0.07–3.28[a]	Holmer and Kendrick (2013)
Thalassia hemprichii	Thailand	2.0	–	7.50[a]	Holmer et al. (2001)
	Thailand	40–130	0.1	3.2–4.2	Holmer et al. (2006)

(Continued)

TABLE 13.2 Sulfate Reduction Rates (SRR) and Pools of Acid-Volatile (AVS) and Chromium-Reducible Sulfide (CRS) in Seagrass Sediments—cont'd

Species	Location	SRR (mmol $m^{-2}d^{-1}$)	AVS (mol m^{-2})	CRS (mol m^{-2})	References
Thalassia testudinum	Florida	–	–	0–40[b]	Chambers et al. (2001)
	Florida	2–20 10–55	0.27	2.25	[c]Devereux et al. (2011)
	US Virgin Island	44–83	0.1–1.9	1.9–23.6	Holmer et al. (2009a,b)
Zostera marina	Denmark	25–59.1	–	0.07–0.20	Holmer and Nielsen (1997)
	Denmark	13.2–29.6	–	–	Boschker et al. (2000)
	Netherlands	7.4	–	–	Boschker et al. (2000)
	Denmark	12–70	–	–	Blaabjerg et al. (1998)
	Denmark	19–41	–	0.07–7.0[a]	Holmer and Laursen (2002)
	Denmark	–	0.00–0.63	4.56–8.19	Frederiksen (2005)
	Denmark	16–40	0–1.84	0.76–40.3	Frederiksen et al. (2006)
	Denmark	38–110	0.03–0.20	0.46–0.59	Vinther et al. (2008)
	Denmark	–	0.29–0.42	0.62–1.17	Borum et al. (2014)
Zostera nigraulis	Australia	3–10	–	0.07–3.28[a]	Holmer and Kendrick (2013)
Zostera noltii	France	13.7	–	–	Welsh et al. (1996, 2000)
		29	–	–	Isaksen and Finster (1996)

Values are depth-integrated over the upper 10 cm.
[a]*Given as total reducible sulfur (TRS = AVS + CRS).*
[b]*Given in* $\mu mol\ g^{-1}$.
[c]*Stations with lowest and highest SRR are presented; AVS and CRS estimated from depth profiles.*

For *Zostera* species, positive relationships with sulfate reduction rates have been found for shoot density and sulfate reduction rate (Holmer and Nielsen, 1997) and root biomass (Vichkovitten and Holmer, 2004). The root biomass effect may be due to root exudation of DOM and the shoot density effect may be due to the seagrass beds ability to trap allochthonous organic matter. For species like *P. oceanica*, a negative correlation between the root biomass and sulfate reduction rate has been found, suggesting an inhibition of microbial rates. In contrast, no correlations have been reported between *Thalassia testudinum* root mass and sulfate reduction, showing that the interactions between seagrass and sediment biogeochemistry are complex (Holmer et al., 2009b). These interactions influence the depth profile of sulfate reduction rates. Comparing the depth profiles in *P. oceanica* between pristine and impacted

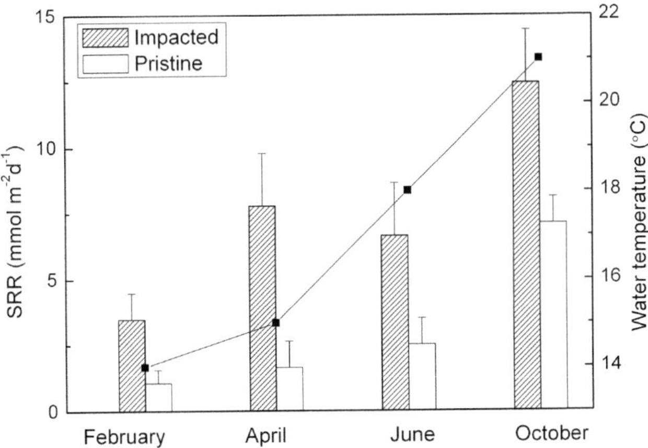

FIGURE 13.3 Seasonal variation in depth-integrated sulfate reduction rates in two different *Posidonia oceanica* meadows, one located at a pristine site (Pristine) and the other in a human impacted site (Impacted). Water temperature is shown by the line with *black squares. Modified from Holmer, M., Duarte, C.M., Marbá, N., 2003a. Sulfur cycling and seagrass* (Posidonia oceanica) *status in carbonate sediments. Biogeochemistry 66, 223–239.*

sites showed low rates throughout the depth profile at the pristine site, only stimulated in the surface layers in late summer probably due to input of organic matter from the water column (Holmer et al., 2003a). At the impacted site, rates were higher throughout the entire depth profile at all times and increased most in the upper layers as temperature and organic matter loading increased (Holmer et al., 2003a). In *Z. marina* meadows, sulfate reduction rates typically increase with depth with a peak in the live root zone probably stimulated by the release of DOM from the root tips (Holmer and Nielsen, 1997). In addition, the sediment composition affects sulfate reduction rates with higher sulfate reduction rates in fine-grained sediments (Holmer et al., 2009a). The sediment composition typically reflects the hydrodynamics at sites with fine-grained sediments confined to sheltered sites with higher organic matter loading and higher sulfate reduction rates compared with exposed sites and in particular in the surface sediments due to the loading from the water column. The profile of the belowground biomass further affects the depth profiles as shallow species, such as *Halophila* sp. and *Halodule* sp., primarily affect the surface layers, whereas deep-rooted seagrasses such as *Enhalus acoroides* affect the deeper layers (Holmer et al., 2006, 2009a,b).

High concentrations of reduced sulfides have been reported for many seagrass sediments (Table 13.2) with acid-volatile sulfides (AVSs) and chromium-reducible sulfur pools being higher than in unvegetated sediments. The dissolved sulfide pools do not necessarily reflect high sulfate reduction rates, as dissolved sulfide pools range from below detection limits (<1 μM) up to 50–100 μM in pristine seagrass sediments with large diurnal and seasonal variation (Holmer and Nielsen, 1997; Lee and Dunton, 2000; Pedersen et al., 2004; Borum et al., 2005; Frederiksen et al., 2006; Calleja et al., 2007; Pagès et al., 2012). In dieback areas of *T. testudinum* in Florida Bay, sulfide concentrations up to several millimolars can be found (Carlson et al., 1994; Borum et al., 2005). Seagrass growth can be inhibited by high sediment sulfide concentrations and for *P. oceanica* meadows around the Balearic Islands net population growth began to be inhibited at concentrations as low as 10 μM (Calleja et al., 2007). Frederiksen et al. (2006) found in the eutrophic area of Roskilde Fjord, Denmark, high pools of dissolved sulfide in the summer months for *Z. marina*, indicating that sediment reoxidation of sulfide does not keep up with sulfate reduction, despite a high capacity for oxygen release from roots. Accumulation of sulfide was coincident with a decline in carbohydrate content of the belowground biomass. This suggests that plants were not able to maintain their carbohydrate reserves during the summer period but instead utilized these reserves to cope with root anoxia (Vichkovitten et al., 2007; Hasler-Sheetal et al., 2015). Similarly, *P. oceanica* showed a decline in nonstructural carbohydrate pools as a result of sediment organic matter enrichment followed by the development of high sedimentary sulfide pools (Pérez et al., 2007).

2.1 Sulfide Intrusion in Seagrasses

Stable sulfur isotopes have been used to track intrusion of sulfides into seagrasses from sediments (reviewed in Holmer and Hasler-Sheetal, 2014). ^{34}S is a sensitive indicator of sulfide intrusion because of the difference in the $\delta^{34}S$ signal between seawater sulfate ($\sim +20‰$) and sediment sulfides (-15 to $-30‰$). Sulfate-reducing bacteria discriminate toward the lighter isotope and sediment sulfides can be up to 50‰ lighter than seawater sulfate.

Sulfate and sulfide isotope values remain unchanged in plant uptake and assimilation such that the isotopic composition of plant tissues reflects the sulfur source. Intrusion of sulfides leads to lighter isotopic signals in plant tissues. The $\delta^{34}S$ composition is now known for about half of the seagrass species (Table 13.3), from all climate zones. All seagrasses species show intrusion of sulfides into above- and belowground tissues, with a large variation within the same species and between different species (Table 13.3). *Posidonia* sp. shows the lowest intrusion, whereas *Zostera* sp., *Thalassia* sp., and *Halophila* sp. show high sulfide intrusion and to some extent also the highest within species variability (Holmer and Hasler-Sheetal, 2014). The $\delta^{34}S$ of the belowground tissues is significantly lower than that of the leaves, indicating that the intrusion of sulfides is through the roots (Borum et al., 2006; Hasler-Sheetal and Holmer, 2015) consistent with observations of root tips open for active uptake or passive diffusion of compounds, such as gaseous sulfides. Gaseous sulfide is considered to move freely in plants, controlled by the gradient in partial pressures between the plant and the sediment, but so far gaseous sulfides have not been observed in leaves (Pedersen et al., 2004). Sulfides may also enter the plant as sulfate after reoxidation of sulfides in the rhizosphere (Hasler-Sheetal and Holmer, 2015). Inside the plant, there are several pathways for sulfides. Sulfides typically enter the plants during nighttime, when the oxygen pressure is low in the plant, and on arrival of daylight and initiation of photosynthesis, sulfides are reoxidized (Pedersen et al., 2004). Sulfides are reoxidized to sulfate and stored in vacuoles or to elemental sulfur and precipitating on the cell surfaces in the plant (Hasler-Sheetal and Holmer, 2015). Large accumulations of elemental sulfur have been found in several seagrass species, including *Posidonia oceanica*, *Thalassia testudinum*, and *Zostera marina*, particularly in more reducing, organic-rich sediments with high sulfide concentrations (Holmer et al., 2005b; Frederiksen et al., 2006; Holmer and Nielsen, 2007; Koch et al., 2007a,b; Mascaro et al., 2009; Hasler-Sheetal and Holmer, 2015). $\delta^{34}S$ light organic sulfur compounds have also been found in plant tissues, with thiols as the most likely precursor (Fig. 13.4, Hasler-Sheetal and Holmer, 2015). All these

TABLE 13.3 $\delta^{34}S$ of Leaves, Rhizomes, and Roots of Several Seagrass Species. References Were Reported in Holmer and Hasler-Sheetal (2014)

	Leaf		Rhizome		Root	
Species	N	$\delta^{34}S$ (‰)	N	$\delta^{34}S$ (‰)	N	$\delta^{34}S$ (‰)
Amphibolis antarctica	5	+19.3 ± 0.8	5	−2 ± 13.1	5	−15.7 ± 3.1
Amphibolis australis	6	+16.7 ± 3.2	6	+2.5 ± 3.9	6	−3.8 ± 12.9
Amphibolis griffithii	1	+16.5	1	+7.6	1	−6.5
Cymodocea angustata	1	+10.1	1	−11.7	1	+1.8
Cymodocea nodosa	1	+17	1	+8.5	1	+6.9
Cymodocea rotundata	1	+17.7	1	+11.5	1	+1.2
Cymodocea serrulata	3	+15.1 ± 1.5	3	+8.2 ± 6.1	3	+5.7 ± 6.3
Halodule sp.	1	+19.1	nd	nd	1	+12.6
Halodule uninervis	2	+14.6 ± 1.9	2	−6.5 ± 4.6	2	+1.6 ± 5.1
Halodule wrightii	9	+9.3 ± 6.9	nd	nd	3	−7.4 ± 5.6
Halophila ovalis	6	+16.2 ± 2.3	6	+2.7 ± 14.6	6	−5.5 ± 7.5
Halophila engelmannii	1	+11.2	nd	nd	1	+11.5
Lepilaena sp.	1	+14.3	nd	nd	nd	nd
Posidonia angustifolia	1	+15.5	1	−3.2	1	−4.7
Posidonia australis	6	+17.2 ± 3.8	6	+4.7 ± 6	6	+2.6 ± 4.5
Posidonia coriacea	1	+14.7	1	−0.4	1	+0.6
Posidonia oceanica	34	+20.9 ± 1.7	32	+13.2 ± 2.8	32	+11.3 ± 5.5
Posidonia sinuosa	4	+18.3 ± 2.7	4	+4.6 ± 1.6	4	+0.9 ± 4
Ruppia maritima	2	−4.5 ± 2.5	nd	nd	2	−9.6 ± 6.4
Ruppia megacarpa	3	+11.1 ± 5.2	2	+11.7 ± 5.8	2	−2.1 ± 2.1
Syringodium filiforme	1	+11.5	nd	nd	1	−4
Syringodium isoetifolium	2	+16.2 ± 2.5	2	+2.4 ± 13.9	2	+10.3 ± 2.1
Syringodium sp.	3	+11.6 ± 3.2	2	−7.3 ± 4.2	3	+0.4 ± 8.8
Thalassia testudinum	13	+11.7 ± 5.6	4	−2.7 ± 5	6	−11.8 ± 6.8
Zostera capricorni	2	+15.2 ± 2.5	nd	nd	nd	nd
Zostera marina	44	+4 ± 6.9	44	+1.2 ± 7	52	−5.2 ± 7.8
Zostera mucronata	1	+13.5	1	+9.8	1	+1.1
Zostera nigraulis	3	+14.9 ± 2.7	3	+11.8 ± 5.6	3	+3 ± 5.2
Zostera noltii	1	+15.6	nd	nd	nd	nd
Grand total	159	+12.4 ± 8.2	123	+5.1 ± 8.6	142	+0.1 ± 10.1

Values are given as average plus or minus the standard deviation. N, number of observations included = 159.

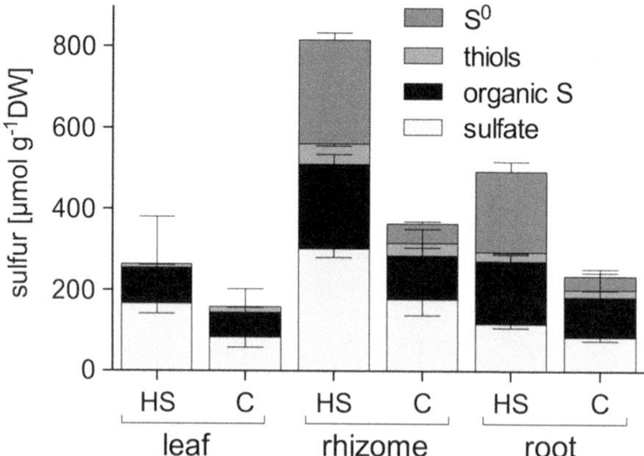

FIGURE 13.4 Distribution of sulfur compounds in leaves, rhizomes, and roots of *Zostera marina* in an experiment with elevated sulfide levels in the sediments. *C*, control; *DW*, dry weight; *HS*, high sulfide. *Modified from Hasler-Sheetal, H., Holmer, M., 2015. Sulfide intrusion and detoxification in the seagrass* Zostera marina. *PLoS One 10, e0129136.*

pathways can be considered as mechanisms to detoxify sulfides. The reduction of negative effects of sulfide intrusion in seagrasses can be considered a central adaptation to the successful evolution of seagrasses in anoxic, reduced sediments (Fig. 13.5).

It is possible to estimate the contribution of sediment sulfides to the total sulfur content of the seagrasses by calculating $F_{sulfide}$, which is a ratio between the $\delta^{34}S$ signal in plant tissue and the $\delta^{34}S$ signal of the potential sulfur sources:

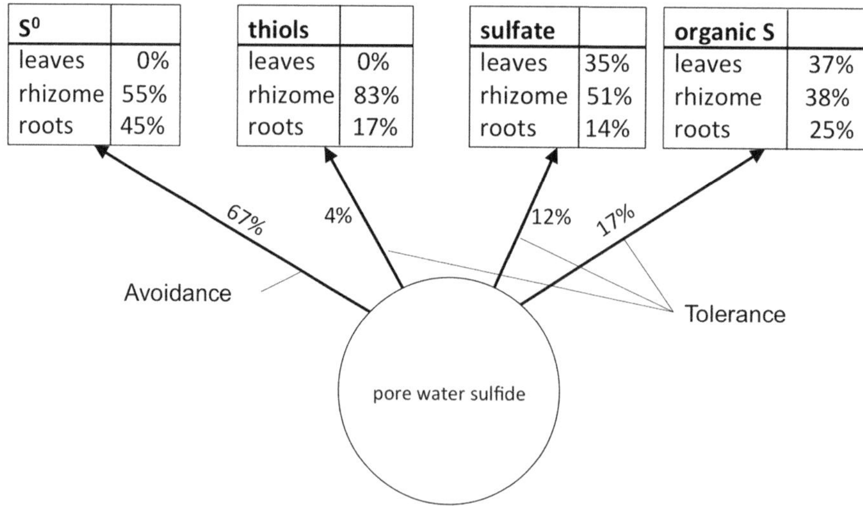

FIGURE 13.5 Contribution of avoidance and tolerance mechanisms in *Zostera marina* on intrusion of pore water sulfide. The percentage distribution of the different sulfur compounds between leaves, rhizomes, and roots is given in boxes. *Modified from Hasler-Sheetal, H., Holmer, M., 2015. Sulfide intrusion and detoxification in the seagrass* Zostera marina. *PLoS One 10, e0129136.*

$$F_{\text{sulfide}} = \frac{\delta^{34}S_{\text{tissue}} - \delta^{34}S_{\text{sulphate}}}{\delta^{34}S_{\text{sulfide}} - \delta^{34}S_{\text{sulphate}}}$$

where $\delta^{34}S_{tissue}$ is the value measured in the leaf, rhizome, or root; $\delta^{34}S_{sulfate}$ is the seawater value; and $\delta^{34}S_{sulfide}$ is the sediment sulfide value (derived from pore water sulfides or AVS) (Frederiksen et al., 2006).

$F_{sulfide}$ has been reported from a number of species and ranges from 0% in *Posidonia* sp. leaves to 100% in *Halophila ovalis* roots. $F_{sulfide}$ is typically highest in roots, decreasing in rhizomes, and lowest in leaves reflecting the intrusion path of sulfides. For *Z. marina* and *T. testudinum*, species showing high sulfide intrusion, highest values were 68%/86% and 21%/96% for leaves and roots, respectively. Increased $F_{sulfide}$ is typically observed in sediment organic enrichment experiments (Frederiksen et al., 2008) and with exposure to hypoxia (Hasler-Sheetal and Holmer, 2015), suggesting that $F_{sulfide}$ is an early indicator of sulfide stress. The indicator should be examined further perhaps to better understand the impacts of multiple stressors common to seagrass meadows.

2.2 Microscale Effects

In recent years, studies of the rhizosphere have contributed high-resolution temporal and spatial information on rhizosphere biogeochemistry. These new studies complement the above-mentioned studies on relations between seagrass roots and bulk sediment biogeochemistry. The use of microelectrode, optode, nanoparticle-based sensors, DGT (diffusive gradients in a thin film), and DET (diffusive equilibration in a thin film) techniques has provided new information on the rhizosphere biogeochemistry. Most recent studies have focused on oxygen dynamics and oxygen leakage from root tips, which is referred to as radial oxygen loss (Jensen et al., 2005; Frederiksen and Glud, 2006; Brodersen et al. 2014, 2015a, 2015b; Koren et al., 2015). These techniques provide detailed information on sulfides, pH, and nutrients in the rhizosphere as well (Pagès et al., 2012; Koren et al., 2015). Root tips are the most dynamic seagrass belowground tissues for oxygen exchange (Fig. 13.6), but oxygen is also released from the basal leaf meristems and the root–shoot junction in *Z. marina* (Brodersen et al., 2016) and *Z. muelleri* (Koren et al., 2015). Mature seagrass tissue has high resistance to oxygen leakage because of the presence of Casparian bands along the longitudinal interior, leaving only the root tips and meristematic tissues without a barrier (Barnabas, 1994; Koren et al., 2015). Biogeochemical cycling of elements in the rhizosphere is especially dynamic during times of fast root growth (e.g., 8.7 ± 1.7 mm day^{-1} in *Z. marina*, Frederiksen and Glud, 2006). Microelectrode and optode studies have shown that oxygen diffuses 1 cm from the root tip of *Z. marina* (Fig. 13.5) when no sulfide is present in this zone. In contrast, root tips of *Z. muelleri* showed relatively little oxygen leakage and it was confined to the meristems of leaf and rhizomes where it resulted in an oxic microshield out to 300 μm from these tissues. Within the microshield, oxygen leakage caused a decrease in sulfide concentration from 6 μM to 0 at the plant surface, thereby shielding against sulfide intrusion. Concomitant decreases in pH (from ~7.00 to ~6.50) suggested that sulfides were reoxidized to sulfate, a process generating protons and lowering the pH (Brodersen et al., 2015b). Similarly, Brodersen et al. (2016) found microniches of low and high pH in the rhizosphere of

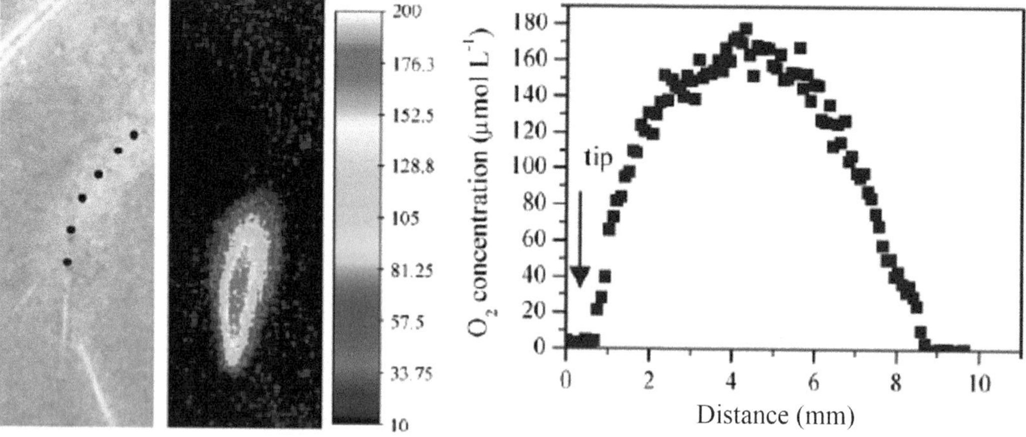

FIGURE 13.6 Oxygen distribution around roots of *Zostera marina* at a light intensity above photosynthetic saturation (500 μmol photons $m^{-2}s^{-1}$). The root distribution is shown in the black-and-white image (left), where roots appear as *white lines* (marked by *black dots*). Planar optode images of oxygen distribution are shown in the middle image and peak oxygen concentrations along the root length axis starting from the root tip (0 cm) are shown in the right image. *Modified from Frederiksen, M.S., Glud, R.N., 2006. Oxygen dynamics in the rhizosphere of* Zostera marina*: a two-dimensional planar optode study. Limnology and Oceanography 51, 1072–1083.*

Z. marina by using nanoparticle-based pH sensors in an experimental setup. Microniches of low pH were found in areas of root oxygen leakage not only at root tips but also around prophyllums (single leaves originating from the root/shoot junctions), as well as at the base of leaf sheaths. In this setup, the decrease in pH was also attributed to sulfide reoxidation. Interestingly, in light treatments, elevated pH was found at rhizome and root surfaces, and pH was higher near the plant tissues than in bulk sediment. The increase in pH was considered to be due to secretion of allelochemicals such as amines, CO_2 uptake by the belowground tissue, which changes the carbonate equilibrium, or stimulation of sulfate reduction by root exudates that consumed protons (Brodersen et al., 2016). These new nanoparticle-based sensor techniques have great potential for furthering studies of the seagrass rhizosphere. Unfortunately, the technique is limited to the use of artificial, transparent sediments and at the moment not applicable to natural sediments under in situ conditions (Koren et al., 2015).

The microshield of oxygen around seagrass belowground tissues provides important protection against sulfide intrusion and potential sulfide toxicity (Lamers et al., 2013; Holmer and Hasler-Sheetal, 2014). The extent of the microshield is highly dynamic and varies in response to changes in environmental conditions, such as light, water column oxygen concentration, and temperature (Borum et al., 2005; Frederiksen and Glud, 2006; Brodersen et al., 2015b). Seagrass sediments are generally highly reduced and oxygen is consumed quickly at the sediment surface by aerobic respiration and in the deeper layers by reoxidation of reduced iron and sulfides. Seagrass sediments are typically anoxic within the first millimeters (Frederiksen and Glud, 2006) and oxygen in the deeper sediments has to be supplied from the seagrasses. The storage of oxygen within seagrass aerenchyma is consumed within minutes of the cessation of photosynthesis, so maintenance of the microshield in the dark is dependent on oxygen in the water column (Sand-Jensen et al., 2005). This is possible as long as the oxygen concentration in the overlying water is >20%–50% of air saturation and plants are not stressed by other factors, such as high water temperature, substantial epiphyte cover, deposition of resuspended material on leaves, or high sulfide pressure from the sediments (Borum et al., 2005; Mascaro et al., 2009; Brodersen et al. 2015a, 2015b, 2017). Experimental studies in the laboratory with reduced oxygen concentrations in the overlying water column have shown a threshold of 20%–50% of air saturation for sulfide intrusion into plants (Pedersen et al., 2004, Brodersen et al., 2015b). Similarly, in the field, sulfide has been found intruding during the night at low oxygen saturation in the plant, e.g., in areas of high *Thalassia testudinum* mortality in Florida Bay (Borum et al., 2005). Although we know the microshield helps to prevent sulfide toxicity in seagrass, further work is needed to understand the full complexity of microshield dynamics for more seagrass species under a wider range of environmental conditions. The oxygen release from roots is estimated to contribute up to 12% of the total sediment–water oxygen flux in *Z. marina* sediments (Jensen et al., 2005; Frederiksen and Glud, 2006). This indicates a limited effect of root oxygen leakage and supports findings of limited stimulation of nitrification and coupled nitrification–denitrification in rhizosphere sediments of *Z. marina* (Risgaard-Petersen and Ottosen, 2000). Sediment oxygen uptake in typical seagrass beds is mostly attributable to abiotic sulfide reoxidation rather than to bacterial respiration. Despite the unfavorable conditions for nitrifiers (limited oxygen supply and competition with plants for NH_4^+) and denitrifiers (limited nitrate availability), anaerobic bacterial activity is stimulated in the area around the root tips because of release of DOM from the roots (Nielsen et al., 2001; Long et al., 2008), especially bacterial sulfate reduction (Marbà et al., 2006). These hot spots of bacterial activity and strong redox gradients may be crucial to plant survival and nutrient uptake. Recent microscale studies have advanced our knowledge substantially on the complexity of oxygen, sulfide, and pH biogeochemistry in the very active rhizosphere.

Recently it has been suggested that lucinid bivalves may be important in the reoxidation of sulfides in seagrass sediments (van der Heide et al., 2012). Lucinid bivalves host sulfide-oxidizing bacteria that can reduce the sulfide levels in seagrass sediments. Where the seagrasses benefit from sulfide reoxidation, the bivalves and their endosymbionts profit from the organic matter accumulation and radial oxygen release from the seagrass belowground tissues. The global distribution of lucinids in seagrass meadows and the interaction are currently being explored and considered for seagrass restoration efforts (de Fouw et al., 2016). Other bivalves, such as *Mytilus edulis*, have shown both positive and negative effects on performance of *Zostera marina*: positive from their nutrient remineralization (Reusch and Williams, 1998) but negative from their contribution to increased anoxia and resultant sulfide intrusion (Vinther et al., 2008; Castorani et al., 2015).

2.3 Nutrient Cycling—Importance of Root Uptake

Seagrasses contribute in many ways to the nutrient cycles in coastal zones, not least because of their uptake and retention of nutrients during the growth season (Risgaard-Petersen and Ottosen, 2000). Enhanced sedimentation of organic matter from allochthonous and autochthonous sources in seagrass beds also contributes to nutrient retention. Access to sediment pools of nutrients ensures a wide distribution of seagrasses in oligotrophic waters (Hemminga, 1998). Nutrient uptake from sediments is to a large extent determined by rates of organic matter mineralization, which in turn is determined from an array of factors, such as organic matter pool size and lability, temperature, and physical conditions of the sediments (Canfield et al., 1993). Nutrient uptake kinetics in seagrasses has been explored by use of split chambers (e.g., *Thalassia testudinum*) or in sediments with flow-through of nutrient solutions (e.g., *Zostera marina*) (Gras et al., 2003; Nielsen et al., 2006; Villazán et al., 2013). Roots of *Z. marina* show a high affinity for ammonium uptake (Villazán et al., 2013), but both V_{max} and affinity for ammonium may not be as high as in the leaves probably because of the lower ammonium concentrations in the water column compared with sediments (Rubio et al., 2007; Alexandre et al., 2011). Uptake of nitrate is considered negligible in roots, whereas it can be important in leaves (Nayar et al., 2010). Small species, such as *Halophila stipulacea*, show similar uptake capacity and efficiency in leaves and roots, which may favor their rapid colonization under variable nutrient conditions (Alexandre et al., 2014). Seagrasses can take up not only low molecular weight organic N compounds with urea as the most important but also amino acids and other forms of organic N under oligotrophic conditions (Vonk et al., 2008; Van Engeland et al., 2011; La Nafie et al., 2014). N acquisition by roots of *Z. marina* has been found to contribute more to N uptake than leaves (Park et al., 2013) but will most likely be dependent on the nutrient status of the environment as shown for three tropical seagrass species (Holmer et al., 2001).

The affinity for phosphate uptake by roots of *Z. marina* is lower than that reported for ammonium (Villazán et al., 2013), but phosphate can be taken up at very low concentrations, which is an advantage particularly in carbonate-rich sediments, where P is carbonate bound (Nielsen et al., 2006). P limitation is common in carbonate-rich sediments as observed from *Thalassia testudinum* in several locations (Jensen et al., 1998; Gras et al., 2003; Nielsen et al., 2006; Armitage et al., 2011), but both N and P can be limiting under oligotrophic conditions as found in Shark Bay, Australia (Fraser et al., 2012; Burkholder et al., 2013). While oligotrophic water columns have very low P concentrations, large pools of P in the sediments can be sufficient to support growth for decades as suggested for *T. testudinum* in Florida Bay (Jensen et al., 1998; Nielsen et al., 2006). Dissolution of carbonates in conjunction with sulfide reoxidation has been proposed as a mechanism that increases P availability in carbonate-rich sediments of *T. testudinum* (Jensen et al., 2009). Similarly, exudation of organic acids from *T. testudinum* roots has been found to increase P availability in carbonate-rich seagrass sediments, where increased mineralization and acidification are possible mechanisms contributing to P release (Long et al., 2008). In seagrasses with terrigenous, mineral sediments, P bound in the oxidized iron pool is an important source of P when it is released following Fe reduction and coupling with sulfide (Holmer et al., 2006; Deborde et al., 2008).

3. MULTIPLE STRESSORS AND BIOGEOCHEMICAL CONDITIONS IN SEAGRASS MEADOWS

The effects of multiple stressors on seagrass biogeochemistry have been a hot topic in the past decade. Ammonium toxicity in *Zostera marina* has been found to be most severe when combined with elevated temperature and low light (van Katwijk et al., 1997). High concentrations of ammonium are typically found in eutrophic ecosystems and the presence of drifting macroalgae (Hauxwell et al., 2001) or blue mussels (Vinther et al., 2008) has resulted in high ammonium concentrations, and ammonium toxicity has been proposed as a contributing factor to negative effects on *Z. marina* performance in the United States and Denmark, as well as for *Z. noltii* in Spain (Brun et al., 2002, 2008). More recent studies of *Z. noltii* suggest that the effect of ammonium is strongly dependent on the environmental conditions, where ammonium toxicity can be alleviated by the presence of the macroalgae *Ulva* sp. or benthic microalgae because of their uptake of ammonium (Moreno-Marin et al., 2016) and combinations of flow (from low to high) and light (from low to high) leads to a nonlinear response to ammonium enrichment (Villazán et al., 2016). In contrast to temperate seagrasses, few studies have addressed multiple stressors in combination with ammonium enrichment. *Halodule uninervis* and *Thalassia hemprichii* can cope with high ammonium levels better than *Z. marina* (Christianen et al., 2011).

Dieback events of *Thalassia testudinum* have also been proposed to occur when multiple stressors interact (Zieman et al., 1999; Koch et al., 2007a). The overacting factor of the dieback is most likely plant O_2 imbalance (Borum et al., 2005) as a result of combinations of hypersalinity (40–60), high temperatures (33–38°C), and high pore water sulfides (2–5 mmol L^{-1}). A conceptual model has been developed based on field observations and laboratory experiments and involves both changes in sediment biogeochemistry and plant performance (Koch et al., 2007b). In Florida Bay seagrasses, where low Fe availability limits pyrite formation, diebacks are initialized by inputs of nutrients (phosphorus is limiting in Florida Bay) from the Gulf of Mexico to Florida Bay, whereby the productivity of nutrient-limited *T. testudinum* is enhanced. The plants release DOM into the rhizosphere, and sulfate reduction rates are stimulated, increasing the consumption of oxygen in the rhizosphere due to reoxidation of sulfides (Chambers et al., 2001; Zhang et al., 2004). Sediment oxygen consumption is further enhanced because of high water temperatures and high salinity, which particularly affects areas with high water residence times (Nuttle et al., 2000). Plant O_2 imbalance is most significant at the end of the growing season, when plants are downregulating and the water column is exposed to nighttime hypoxia due to large pools of detritus (phytoplankton and seagrass). Seagrass mortality escalates due to sulfide intrusion into the plants leading to dieback events (Borum et al., 2005). The large dieback events on the scale of 40 km^2 were abundant in the late 1980s and then dropped off until 2015, when similar driving conditions were found (Hall et al., 2016).

4. HUMAN PRESSURES AND EFFECTS ON BIOGEOCHEMISTRY

Seagrasses are increasingly impacted by human activities, especially because of their proximity to human populations in the coastal zone (Duarte, 2002; Orth et al., 2006; Waycott et al., 2009). Impacts stem from agricultural nutrient runoff, wastewater discharge, siltation due to watershed soil erosion, tourist development, beach restoration, desalinization plants, extraction of sand, expansion of harbors, etc. Regardless of the specific impact, a common factor stemming from many of these pressures is reduced light availability, as a result of higher phytoplankton biomass, increased sediment turbidity, increased epiphytization, or blooms of macroalgae. The inner Danish coastal waters and the west coast of Sweden have experienced drastically reduced depth penetration of seagrass over the past 100 years, especially in the 1970s because of increased agricultural fertilizer runoff (Nielsen et al., 2002; Frederiksen et al., 2004; Boström et al., 2014) that led to light reductions from excessive growth of phytoplankton, epiphytization of seagrasses, blooms of macroalgae, or presence of drifting macroalgae (Nielsen et al., 2002; Baden et al., 2003). Phytoplankton production is stimulated directly by increased loading of nutrients, whereas the growth of epiphytes and macroalgae may be an indirect effect because of altered trophic cascades due to overfishing and reduced grazing in seagrass meadows (Moksnes et al., 2008; Baden et al., 2012). Many seagrasses require on the order of 11% to more than 20% of incident light to support the maintenance of belowground nonphotosynthetic biomass, more than that required by phytoplankton and macroalgae (Steward et al., 2005; Ralph et al., 2007). Shading experiments for a wide selection of seagrasses have shown a sequence of events that ultimately leads to declining redox potentials and increasing pools of sulfides in the sediments, suggesting that low light not only affects plant growth but also contributes to altered sediment biogeochemical conditions (Enríquez et al., 2001; Holmer and Laursen, 2002; Calleja et al., 2006; Gacia et al., 2012). Increased root anoxia and sulfide toxicity likely contribute to the observed declines under light reduction due to plant O_2 imbalance and sulfide intrusion (Brodersen et al., 2015b, 2016; Hasler-Sheetal and Holmer, 2015).

Eutrophication and increased loading of particulates from land can contribute to organic enrichment of sediments, increasing sediment oxygen demand and affecting the internal plant O_2 balance due to oxygen drainage by rhizosphere sediments (Borum et al., 2005; Brodersen et al., 2017). While low levels of enrichment may stimulate the growth of seagrasses, as for *Thalassia testudinum* in Florida Bay (Armitage et al., 2011), high rates of loading can enhance sulfate reduction and the accumulation of sulfides in pore waters (e.g., Holmer et al., 2003a). Several experiments have added organic matter to the sediment or have injected sulfide solutions, and seagrasses generally show large variation in response. Some species are tolerant to pore water sulfides, such as *Z. marina* and *T. testudinum*, probably because of their large capacity of sulfide avoidance by reoxidating sulfides in the rhizosphere or reoxidation inside the plant (Hasler-Sheetal and Holmer, 2015). Others, such as *P. oceanica*, are highly sensitive and low accumulations of sulfides in pore water lead to increased mortality (Calleja et al., 2007; Frederiksen et al., 2007; García et al., 2013). However, combining multiple stressors, such as organic matter additions with light reductions, water column hypoxia, or elevated water temperature, significantly impedes growth and increases mortality even in tolerant species such as *T. testudinum* and *Z. marina* (Koch et al., 2007a,b; Mascaro et al., 2009; Holmer et al., 2011; Höffle et al., 2012).

In carbonate-rich sediments, an organic enrichment may show even more dramatic impacts compared with terrigenous sediments, as the buffer capacity toward sulfide is reduced because of the low iron content (Koch et al., 2007b). In terrigenous sediments, iron buffers sulfide by precipitation of iron sulfides, which reduces the pool of dissolved sulfides in pore water. In carbonate sediments, iron pools are rapidly exhausted, as the degree of pyritization already is close to maximum (Berner, 1984), and the sediments hence have limited capacity for binding sulfides (Zhang et al., 2004). Sulfide may accumulate in the pore water, provided that it is not reoxidized by the release of oxygen from belowground tissues. As sediment oxygen demand increases, the risk of root anoxia and sulfide intrusion increases as well (Calleja et al., 2006; Borum et al., 2014; Holmer and Kendrick, 2013). A second important aspect is that iron availability for plant growth can decrease if it is bound irreversibly to iron sulfides, possibly resulting in Fe limitation of plant growth (Duarte et al., 1995). Experiments where iron is added to organic-enriched sediments of *Posidonia oceanica* and *Thalassia testudinum* thus showed that iron availability was enhanced and resulted in increased seagrass growth (Holmer et al., 2005a; Ruiz-Halpern et al., 2008). In a 2-year Fe addition experiment in organic-enriched sediments of *P. oceanica*, a decline in seagrass growth was reversed as a result of reduced sulfide stress and increased iron availability (Marbà et al., 2008).

5. FUTURE PERSPECTIVES AND CONCLUSIONS

Seagrass ecosystems are facing a global crisis because of human pressures in the coastal zone (Orth et al., 2006; Waycott et al., 2009; Short et al., 2011). This chapter examines several examples of how altered sediment biogeochemistry is linked to observed seagrass decline. The most pronounced impact of human activity on sediment biogeochemistry is the reduction of sediment redox potential. Redox potential decreases when sediment oxidation from root exudates declines as a result of decreased seagrass production. Seagrass production often declines as a result of decreased levels of dissolved oxygen in the water column, which is often the result of organic enrichment associated with eutrophication. Cumulative impacts of multiple stressors also lead to seagrass decline and this is an active area of current research. Seagrass outcomes under increasing human pressures are challenging to predict, considering the limited knowledge of multiple stressors (Koch and Erskine, 2001; Koch et al., 2007b). Another important aspect of seagrass decline and areal loss is the impact of their loss on the rest of the coastal system. Some examples can be seen in systems that were subject to wasting disease in *Z. marina* in the northern hemisphere during the 1930s. In these places there were major increases in coastal erosion, loss of fisheries habitat, and declining fisheries production. Many of the systems experiencing wasting disease had not fully recovered before new seagrass losses started to occur in response to coastal eutrophication (Frederiksen et al., 2004; Boström et al., 2014). Losses of slow-growing species, such as *P. oceanica* in the Mediterranean (Marbà et al., 2014), may be even more dramatic because of their slow regrowth rates. It is hard to predict whether better coastal zone land management and reduced nutrient loading will allow *Z. marina* recolonization of Danish coastal waters because of many other concurrent pressures (Kuusemäe et al., 2016). In some areas the organic enrichment of the sediments during eutrophication in the 1970 and 1980s led to repeated oxygen depletion events that prevented recolonization (Frederiksen et al., 2004). Full recovery may take decades, if at all (Valdemarsen et al., 2014). In some areas, a faster colonization by the blue mussel *Mytilus edulis* into degraded areas has hampered seagrass recolonization (Vinther et al., 2008). To be able to understand, manage, and conserve seagrass meadows, it will be necessary to expand biogeochemical studies to a wider geographical extent and for more seagrass species. One important objective should be to determine ecological thresholds for seagrass performance in ecosystems under human pressures so that further losses of seagrasses can be avoided through wiser and better management.

Acknowledgments

Thanks to Harald Hasler-Sheetal for preparing the artwork and list of references.

References

Alexandre, A., Silva, J., Bouma, T.J., Santos, R., 2011. Inorganic nitrogen uptake kinetics and whole-plant nitrogen budget in the seagrass *Zostera noltii*. Journal of Experimental Marine Biology and Ecology 401, 7–12.

Alexandre, A., Georgiou, D., Santos, R., 2014. Inorganic nitrogen acquisition by the tropical seagrass *Halophila stipulacea*. Marine Ecology 35, 387–394.

Apostolaki, E.T., Holmer, M., Marbà, N., Karakassis, I., 2010. Metabolic imbalance in coastal vegetated (*Posidonia oceanica*) and unvegetated benthic ecosystems. Ecosystems 13, 459–471.

Apostolaki, E.T., Vizzini, S., Hendriks, I.E., Olsen, Y.S., 2014. Seagrass ecosystem response to long-term high CO_2 in a Mediterranean volcanic vent. Marine Environmental Research 99, 9–15.

Armitage, A.R., Fourqurean, J.W., 2016. Carbon storage in seagrass soils: long-term nutrient history exceeds the effects of near-term nutrient enrichment. Biogeosciences 13, 313321.

Armitage, A.R., Frankovich, T.A., Fourqurean, J.W., 2011. Long-term effects of adding nutrients to an oligotrophic coastal environment. Ecosystems 14, 430–444.

Baden, S., Gullström, M., Lunden, B., Pihl, L., Rosenberg, R., 2003. Vanishing seagrass (*Zostera marina*, L.) in Swedish coastal waters. Ambio 32, 374–377.

Baden, S., Emanuelsson, A., Pihl, L., Svensson, C.J., Aberg, P., 2012. Shift in seagrass food web structure over decades is linked to overfishing. Marine Ecology Progress Series 451, 61–73.

Barnabas, A.D., 1994. Apoplastic and symplastic pathways in leaves and roots of the seagrass *Halodule uninervis* (Forssk.) Aschers. Aquatic Botany 47, 155–174.

Barrón, C., Duarte, C.M., 2009. Dissolved organic matter release in a *Posidonia oceanica* meadow. Marine Ecology Progress Series 374, 75–84.

Barrón, C., Marbé, N., Terrados, J., Kennedy, H., Duarte, C.M., 2004. Community metabolism and carbon budget along a gradient of seagrass (*Cymodocea nodosa*) colonization. Limnology and Oceanography 49, 1642–1651.

Barrón, C., Duarte, C.M., Frankignoulle, M., Borges, A.V., 2006. Organic carbon metabolism and carbonate dynamics in a Mediterranean seagrass (*Posidonia oceanica*), meadow. Estuaries and Coasts 29, 417–426.

Berg, P., Røy, H., Wiberg, P.L., 2007. Eddy correlation flux measurements: the sediment surface area that contributes to the flux. Limnology and Oceanography 52, 1672–1684.

Berner, R.A., 1984. Sedimentary pyrite formation: an update. Geochimica et Cosmochimica Acta 48, 605–615.

Blaabjerg, V., Finster, K., 1998. Sulphate reduction associated with roots and rhizomes of the marine macrophyte *Zostera marina*. Aquatic Microbial Ecology 15, 311–314.

Blaabjerg, V., Mouritsen, K.N., Finster, K., 1998. Diel cycles of sulphate reduction rates in sediments of a *Zostera marina* bed (Denmark). Aquatic Microbial Ecology 15, 97–102.

Blackburn, T.H., Nedwell, D.B., Wiebe, W.J., 1994. Active mineral cycling in a Jamaican seagrass sediment. Marine Ecology Progress Series 110, 233–239.

Borum, J., Pedersen, O., Greve, T.M., Frankovich, T.A., Zieman, J.C., Fourqurean, J.W., Madden, C.J., 2005. The potential role of plant oxygen and sulphide dynamics in die-off events of the tropical seagrass, *Thalassia testudinum*. Journal of Ecology 93, 148–158.

Borum, J., Sand-Jensen, K., Binzer, T., Pedersen, O., Greve, T.M., 2006. Oxygen movement in seagrasses. In: Larkum, A.W.D., Orth, R.J., Duarte, C.M. (Eds.), Seagrasses: Biology, Ecology and Conservation. Springer Netherlands, Dordrecht, Nederlands.

Borum, J., Raun, A.L., Hasler-Sheetal, H., Pedersen, M.Ø., Pedersen, O., Holmer, M., 2014. Eelgrass fairy rings: sulfide as inhibiting agent. Marine Biology 161, 351–358.

Boschker, H.T.S., de Brouwer, J.F.C., Cappenberg, T.E., 1999. The contribution of macrophytederived organic matter to microbial biomass in salt-marsh sediments: stable carbon isotope analysis of microbial biomarkers. Limnology and Oceanography 44, 309–319.

Boschker, H.T.S., Wielemaker, A., Schaub, B.E.M., Holmer, M., 2000. Limited coupling of macrophyte production and bacterial carbon cycling in the sediments of *Zostera* spp. meadows. Marine Ecology Progress Series 203, 181–189.

Boström, C., Baden, S., Bockelmann, A.-C., Dromph, K., Fredriksen, S., Gustafsson, C., Krause- Jensen, D., Möller, T., Nielsen, S.L., Olesen, B., Olsen, J., Pihl, L., Rinde, E., 2014. Distribution, structure and function of Nordic eelgrass (*Zostera marina*) ecosystems: implications for coastal management and conservation. Aquatic Conservation: Marine and Freshwater Ecosystems 24, 410–434.

Bouillon, S., Boschker, H.T.S., 2006. Bacterial carbon sources in coastal sediments: a crosssystem analysis based on stable isotope data of biomarkers. Biogeosciences 3, 175–185.

Bouillon, S., Moens, T., Dehairs, F., 2004. Carbon sources supporting benthic mineralization in mangrove and adjacent seagrass sediments (Gazi Bay, Kenya). Biogeosciences 1, 7178.

Brodersen, K., Nielsen, D., Ralph, P., Kühl, M., 2014. A split flow chamber with artificial sediment to examine the below-ground microenvironment of aquatic macrophytes. Marine Biology 1–10.

Brodersen, K., Lichtenberg, M., Paz, L.-C., Kühl, M., 2015a. Epiphyte-cover on seagrass (*Zostera marina* L.) leaves impedes plant performance and radial O_2 loss from the belowground tissue. Frontiers in Marine Science 2.

Brodersen, K.E., Nielsen, D.A., Ralph, P.J., Kühl, M., 2015b. Oxic microshield and local pH enhancement protects *Zostera muelleri* from sediment derived hydrogen sulphide. New Phytologist 205, 1264–1276.

Brodersen, K.E., Koren, K., Lichtenberg, M., Kühl, M., 2016. Nanoparticle-based measurements of pH and O_2 dynamics in the rhizosphere of *Zostera marina* L.: effects of temperature elevation and light-dark transitions. Plant, Cell and Environment 39, 16191630.

Brodersen, K., Hammer, K.J., Schrameyer, V., Floytup, A., Rasheed, M.A., Ralph, P.J., Kuhl, M., Pedersen, O., 2017. Sediment resuspension and deposition on seagrass leaves impedes internal plant aeration and promotes phytotoxic H_2S intrusion. Frontiers of Plant Science 8, 657. https://doi.org/10.3389/fpls.2017.00657.

Burdige, D.J., Zimmerman, R.C., 2002. Impact of sea grass density on carbonate dissolution in Bahamian sediments. Limnology and Oceanography 47, 1751–1763.

Burdige, D.J., Hu, X., Zimmerman, R.C., 2010. The widespread occurrence of coupled carbonate dissolution/reprecipitation in surface sediments on the Bahamas Bank. American Journal of Science 310, 492–521.

Burke, M.K., Dennison, W.C., Moore, K.A., 1996. Non-structural carbohydrate reserves of eelgrass *Zostera marina*. Marine Ecology Progress Series 137, 195–201.

Burkholder, D.A., Fourqurean, J.W., Heithaus, M.R., 2013. Spatial pattern in seagrass stoichiometry indicates both N-limited and P-limited regions of an iconic P-limited subtropical bay. Marine Ecology Progress Series 472, 101–115.

Caffrey, J.M., Kemp, W.M., 1990. Nitrogen cycling in sediments with estuarine populations of *Potamogeton perfoliatus* and *Zostera marina*. Marine Ecology Progress Series 66, 147–160.

Calleja, M.L., Barrón, C., Hale, J.A., Frazer, T.K., Duarte, C.M., 2006. Light regulation of benthic sulfate reduction rates mediated by seagrass (*Thalassia testudinum*) metabolism. Estuaries and Coasts 29, 1255–1264.

Calleja, M.L., Marbà, N., Duarte, C.M., 2007. The relationship between seagrass (*Posidonia oceanica*) decline and sulfide porewater concentration in carbonate sediments. Estuarine, Coastal and Shelf Science 73, 583–588.

Campbell, J.E., Lacey, E.A., Decker, R.A., Crooks, S., Fourqurean, J.W., 2015. Carbon storage in seagrass beds of Abu Dhabi, United Arab Emirates. Estuaries and Coasts 38, 242–251.

Canfield, D.E., Jørgensen, B.B., Fossing, H., Glud, R., Gundersen, J., Ramsing, N.B., Thamdrup, B., Hansen, J.W., Nielsen, L.P., Hall, P.O.J., 1993. Pathways of organic carbon oxidation in three continental margin sediments. Marine Geology 113, 27–40.

Carlson, P.R.J., Yarbro, L.A., Barber, T.R., 1994. Relationship of sediment sulfide to mortality of *Thalassia testudinum* in Florida Bay. Bulletin of Marine Science 54, 733–746.

Carruthers, T.J.B., Dennison, W.C., Kendrick, G.A., Waycott, M., Walker, D.I., Cambridge, M.L., 2007. Seagrasses of south–west Australia: a conceptual synthesis of the world's most diverse and extensive seagrass meadows. Journal of Experimental Marine Biology and Ecology 350, 21–45.

Castorani, M.C.N., Glud, R.N., Hasler-Sheetal, H., Holmer, M., 2015. Light indirectly mediates bivalve habitat modification and impacts on seagrass. Journal of Experimental Marine Biology and Ecology 472, 41–53.

Cebrián, J., Duarte, C.M., 1998. Patterns in leaf herbivory on seagrasses. Aquatic Botany 60, 67–82.

Chambers, R.A., Fourqurean, J.W., Macko, S.A., Hoppenot, R., 2001. Biogeochemical effects of iron availability on primary producers in a shallow marine carbonate environment. Limnology and Oceanography 46, 1278–1286.

Chiu, S.-H., Huang, Y.-H., Lin, H.-J., 2013. Carbon budget of leaves of the tropical intertidal seagrass *Thalassia hemprichii*. Estuarine, Coastal and Shelf Science 125, 27–35.

Christianen, M.J.A., van der Heide, T., Bouma, T.J., Roelofs, J.G.M., van Katwijk, M.M., Lamers, L.P.M., 2011. Limited toxicity of NHx pulses on an early and late successional tropical seagrass species: interactions with pH and light level. Aquatic Toxicology 104, 73–79.

Clausen, K.K., Krause-Jensen, D., Olesen, B., Marbà, N., 2014. Seasonality of eelgrass biomass across gradients in temperature and latitude. Marine Ecology Progress Series 506, 71–85.

De Fouw, J., Govers, L.L., van de Koppel, J., van Belzen, J., Dorigo, W., Cheikh, M.A.S., Christianen, M.J.A., van der Reijden, K.J., van der Geest, M., Piersma, T., Smolders, A.J.P., Olff, H., Lamers, L.P.M., van Gils, J.A., van der Heide, T., 2016. Drought, mutualism breakdown, and landscape scale degradation of seagrass beds. Current Biology 26, 1051–1056.

de los Santos, C.B., Godbold, J.A., Solan, M., 2017. Short-term growth and biomechanical responses of the temperate seagrass Cymodocea nodosa to CO_2 enrichment. Marine Ecology Progress Series 572, 91–102.

de Wit, R., Stal, L.J., Lomstein, B.A., Herbert, R.A., van Gemerden, H., Viaroli, P., Cecherelli, V.U., Rodriguez-Valera, F., Bartoli, M., Giordani, G., Azzoni, R., Schaub, B., Welsh, D.T., Donnelly, A., Cifuentes, A., Anton, J., Finster, K., Nielsen, L.B., Pedersen, A.G.U., Neubauer, A.T., Colangelo, M.A., Heijs, S.K., 2001. ROBUST: The ROle of BUffering capacities in STabilising coastal lagoon ecosystems. Continential Shelf Research 21, 2021–2041.

Deborde, J., Abril, G., Mouret, A., Jézéquel, D., Thouzeau, G., Clavier, J., Bachelet, G., Anschutz, P., 2008. Effects of seasonal dynamics in a *Zostera noltii* meadow on phosphorus and iron cycles in a tidal mudflat (Arcachon Bay, France). Marine Ecology Progress Series 355, 59–71.

Delefosse, M., 2007. The effects of seagrasses on the sediment biogeochemistry with emphasis on the interactions between sulphur and iron in Trang Province. Thailand, Master thesis, Odense 47.

Devereux, R., Yates, D.F., Aukamp, J., Quarles, R.L., Jordan, S.J., Stanley, R.S., Eldridge, P.M., 2011. Interactions of Thalassia testudinum and sediment biogeochemistry in Santa Rosa Sound, NW Florida. Marine Biology Research 7, 317–331.

Drouin, A., McKindsey, C.W., Johnson, L.E., 2016. Dynamics of recruitment and establishment of the invasive seaweed *Codium fragile* within an eelgrass habitat. Marine Biology 163, 1–12.

Duarte, C.M., 2002. The future of seagrass meadows. Environmental Conservation 29.

Duarte, C.M., 2017. Reviews and syntheses: hidden forests, the role of vegetated coastal habitats in the ocean carbon budget. Biogeosciences 14, 301–310.

Duarte, C.M., Chiscano, C.L., 1999. Seagrass biomass and production: a reassessment. Aquatic Botany 65, 159–174.

Duarte, C.M., Martín, M., Margarita, G., 1995. Evidence of iron deficiency in seagrasses growing above carbonate sediments. Limnology and Oceanography 40, 1153–1158.

Duarte, C.M., Merino, M., Agawin, N.S.R., Uri, J., Fortes, M.D., Gallegos, M.E., Marbá, N., Hemminga, M.A., 1998. Root production and belowground seagrass biomass. Marine Ecology Progress Series 171, 97–108.

Duarte, C.M., Middelburg, J.J., Caraco, N., 2004. Major role of marine vegetation on the oceanic carbon cycle. Biogeosciences Discussions 1, 659–679.

Duarte, C.M., Holmer, M., Marbà, N., 2005. Plant-Microbe Interactions in Seagrass Meadows. Coastal and Estuarine Studies. Wiley-Blackwell.

Duffy, J.E., Reynolds, P.L., Boström, C., Coyer, J.A., Cusson, M., Donadi, S., Douglass, J.G., Eklöf, J.S., Engelen, A.H., Eriksson, B.K., Fredriksen, S., Gamfeldt, L., Gustafsson, C., Hoarau, G., Hori, M., Hovel, K., Iken, K., Lefcheck, J.S., Moksnes, P.-O., Nakaoka, M., O'Connor, M.I., Olsen, J.L., Richardson, J.P., Ruesink, J.L., Sotka, E.E., Thormar, J., Whalen, M.A., Stachowicz, J.J., 2015. Biodiversity mediates top–down control in eelgrass ecosystems: a global comparative-experimental approach. Ecology Letters 18, 696–705.

Enríquez, S., Duarte, C.M., Sand-Jensen, K., 1993. Patterns in decomposition rates among photosynthetic organisms: the importance of detritus C: N:P content. Oecologia 94, 457–471.

Enríquez, S., Marbà, N., Duarte, C.M., van Tussenbroek, B.I., Reyes-Zavala, G., 2001. Effects of seagrass *Thalassia testudinum* on sediment redox. Marine Ecology Progress Series 219, 149–158.

Eyre, B.D., Ferguson, A.J.P., Webb, A., Maher, D., Oakes, J.M., 2011. Denitrification, N-fixation and nitrogen and phosphorus fluxes in different benthic habitats and their contribution to the nitrogen and phosphorus budgets of a shallow oligotrophic sub-tropical coastal system (southern Moreton Bay, Australia). Biogeochemistry 102, 111–133.

Fourqurean, J.W., Duarte, C.M., Kennedy, H., Marba, N., Holmer, M., Mateo, M.A., Apostolaki, E.T., Kendrick, G.A., Krause-Jensen, D., McGlathery, K.J., Serrano, O., 2012. Seagrass ecosystems as a globally significant carbon stock. Nature Geoscience 5, 505–509.

Fraser, M.W., Kendrick, G.A., Grierson, P.F., Fourqurean, J.W., Vanderklift, M.A., Walker, D.I., 2012. Nutrient status of seagrasses cannot be inferred from system-scale distribution of phosphorus in Shark Bay, Western Australia. Marine and Freshwater Research 63, 1015–1026.

Frederiksen, M.S., 2005. Seagrass response to organic loading of meadows caused by fish farming or eutrophication. Ph.d. thesis. University of Southern Denmark, Odense, p. 209.

Frederiksen, M.S., Glud, R.N., 2006. Oxygen dynamics in the rhizosphere of *Zostera marina*: a two-dimensional planar optode study. Limnology and Oceanography 51, 1072–1083.

Frederiksen, M., Krause-Jensen, D., Holmer, M., Laursen, J.S., 2004. Long-term changes in area distribution of eelgrass (*Zostera marina*) in Danish coastal waters. Aquatic Botany 78, 167181.

Frederiksen, M.S., Holmer, M., Borum, J., Kennedy, H., 2006. Temporal and spatial variation of sulfide invasion in eelgrass (*Zostera marina*) as reflected by its sulfur isotopic composition. Limnology and Oceanography 51, 2308–2318.

Frederiksen, M.S., Holmer, M., Díaz-Almela, E., Marba, N., Duarte, C., 2007. Sulfide invasion in the seagrass *Posidonia oceanica* at Mediterranean fish farms: assessment using stable sulfur isotopes. Marine Ecology Progress Series 345, 93–104.

Frederiksen, M.S., Holmer, M., Perez, M., Invers, O., Ruiz, J.M., Knudsen, B.B., 2008. Effect of increased sediment sulfide concentrations on the composition of stable sulfur isotopes (delta S-34) and sulfur accumulation in the seagrasses *Zostera marina* and *Posidonia oceanica*. Journal of Experimental Marine Biology and Ecology 358, 98–109.

Gacia, E., Duarte, C.M., Middelburg, J.J., 2002. Carbon and nutrient deposition in a Mediterranean seagrass (*Posidonia oceanica*) meadow. Limnology and Oceanography 47, 23–32.

Gacia, E., Marbà, N., Cebrián, J., Vaquer-Sunyer, R., Garcias-Bonet, N., Duarte, C.M., 2012. Thresholds of irradiance for seagrass *Posidonia oceanica* meadow metabolism. Marine Ecology Progress Series 466, 69–79.

García, R., Holmer, M., Duarte, C.M., Marbà, N., 2013. Global warming enhances sulphide stress in a key seagrass species (NW Mediterranean). Global Change Biology 19, 3629–3639.

Garcias-Bonet, N., Arrieta, J.M., Duarte, C.M., Marbà, N., 2016. Nitrogen-fixing bacteria in Mediterranean seagrass (*Posidonia oceanica*) roots. Aquatic Botany 131, 57–60.

Gras, A.F., Koch, M.S., Madden, C.J., 2003. Phosphorus uptake kinetics of a dominant tropical seagrass *Thalassia testudinum*. Aquatic Botany 76, 299–315.

Green, E.P., Short, F.T., 2003. World Atlas of Seagrasses. Univ of California Press, Berkeley, California, USA.

Greiner, J.T., McGlathery, K.J., Gunnell, J., McKee, B.A., 2013. Seagrass restoration enhances "blue carbon" sequestration in coastal waters. PLoS One 8, e72469.

Guilini, K., Weber, M., de Beer, D., Schneider, M., Molari, M., Lott, C., Bodnar, W., Mascart, T., De Troch, M., Vanreusel, A., 2017. Response of *Posidonia oceanica* seagrass and its epibiont communities to ocean acidification. PLoS One 12, e0181531.

Hall, M.O., Furman, B.T., Merello, M., Durako, M.J., 2016. Recurrence of *Thalassia testudinum* seagrass die-off in Florida Bay, USA: initial observations. Marine Ecology Progress Series 560, 243–249.

Hasler-Sheetal, H., Holmer, M., 2015. Sulfide intrusion and detoxification in the seagrass *Zostera marina*. PLoS One 10, e0129136.

Hasler-Sheetal, H., Fragner, L., Holmer, M., Weckwerth, W., 2015. Diurnal effects of anoxia on the metabolome of the seagrass *Zostera marina*. Metabolomics 11, 1208–1218.

Hauxwell, J., Cebrian, J., Furlong, C., Valiela, I., 2001. Macroalgal canopies contribute to eelgrass (*Zostera marina*) decline in temperate estuarine ecosystems. Ecology 82, 1007–1022.

Heck, K.L., Valentine, J.F., 2006. Plant–herbivore interactions in seagrass meadows. Journal of Experimental Marine Biology and Ecology 330, 420–436.

Heck, L.K., Carruthers, B.T.J., Duarte, M.C., Hughes, R.A., Kendrick, G., Orth, J.R., Williams, W.S., 2008. Trophic transfers from seagrass meadows subsidize diverse marine and terrestrial consumers. Ecosystems 11, 1198–1210.

Hemminga, M.A., 1998. The root/rhizome system of seagrasses: an asset and a burden. Journal of Sea Research 39, 183–196.

Hemminga, M.A., Duarte, C.M., 2000. Seagrass Ecology. Cambridge University Press.

Hemminga, M.A., Harrison, P.G., van Lent, F., 1991. The balance of nutrient losses and gains in seagrass meadows. Marine Ecology Progress Series 71, 85–96.

Hendriks, I.E., Olsen, Y.S., Duarte, C.M., 2015. Light availability and temperature, not increased CO_2, will structure future meadows of *Posidonia oceanica*. Aquatic Botany 139, 32–36.

Hillman, K., McComb, A.J., Walker, D.I., 1995. The distribution, biomass and primary production of the seagrass *Halophila ovalis* in the Swan/Canning Estuary, Western Australia. Aquatic Botany 51, 1–54.

Höffle, H., Wernberg, T., Thomsen, M.S., Holmer, M., 2012. Drift algae, an invasive snail and elevated temperature reduce ecological performance of a warm-temperate seagrass, through additive effects. Marine Ecology Progress Series 450, 67–80.

Holmer, M., Bachmann Olsen, A., 2002. Role of decomposition of mangrove and seagrass detritus in sediment carbon and nitrogen cycling in a tropical mangrove forest. Marine Ecology Progress Series 230, 87–101.

Holmer, M., Hasler-Sheetal, H., 2014. Sulfide intrusion in seagrasses assessed by stable sulfur isotopes – a synthesis of current results. Frontiers in Marine Science 1, 64.

Holmer, M., Kendrick, G., 2013. High sulfide intrusion in five temperate seagrasses growing under contrasting sediment conditions. Estuaries and Coasts 36, 116–126.

Holmer, M., Kristensen, E., 1996. Seasonality of sulfate reduction and pore water solutes in a marine fish farm sediment: the importance of temperature and sedimentary organic matter. Biogeochemistry 32, 15–39.

Holmer, M., Laursen, L., 2002. Effect of shading of *Zostera marina* (eelgrass) on sulfur cycling in sediments with contrasting organic matter and sulfide pools. Journal of Experimental Marine Biology and Ecology 270, 25–37.

Holmer, M., Nielsen, S.L., 1997. Sediment sulfur dynamics related to biomass-density patterns in *Zostera marina* (eelgrass) beds. Marine Ecology Progress Series 146, 163–171.

Holmer, M., Nielsen, R.M., 2007. Effects of filamentous algal mats on sulfide invasion in eelgrass (*Zostera marina*). Journal of Experimental Marine Biology and Ecology 353, 245–252.

Holmer, M., Andersen, F.Ø., Nielsen, S.L., Boschker, H.T.S., 2001. The importance of mineralization based on sulfate reduction for nutrient regeneration in tropical seagrass sediments. Aquatic Botany 71, 1–17.

Holmer, M., Duarte, C.M., Marbá, N., 2003a. Sulfur cycling and seagrass (*Posidonia oceanica*) status in carbonate sediments. Biogeochemistry 66, 223–239.

Holmer, M., Pérez, M., Duarte, C.M., 2003b. Benthic primary producers––a neglected environmental problem in Mediterranean maricultures? Marine Pollution Bulletin 46, 1372–1376.

Holmer, M., Duarte, C.M., Boschker, H.T.S., Barrón, C., 2004. Carbon cycling and bacterial carbon sources in pristine and impacted Mediterranean seagrass sediments. Aquatic Microbial Ecology 36, 227–237.

Holmer, M., Duarte, C.M., Marbá, N., 2005a. Iron additions reduce sulfate reduction rates and improve seagrass growth on organic-enriched carbonate sediments. Ecosystems 8, 721–730.

Holmer, M., Frederiksen, M.S., Møllegaard, H., 2005b. Sulfur accumulation in eelgrass (*Zostera marina*) and effect of sulfur on eelgrass growth. Aquatic Botany 81, 367–379.

Holmer, M., Carta, C., Andersen, F.Ø., 2006. Biogeochemical implications for phosphorus cycling in sandy and muddy rhizosphere sediments of *Zostera marina* meadows (Denmark). Marine Ecology Progress Series 320, 141–151.

Holmer, M., Marbà, N., Lamote, M., Duarte, C.M., 2009a. Deterioration of sediment quality in seagrass meadows (*Posidonia oceanica*) invaded by macroalgae (*Caulerpa* sp.). Estuaries and Coasts 32, 456–466.

Holmer, M., Pedersen, O., Krause-Jensen, D., Olesen, B., Hedegård Petersen, M., Schopmeyer, S., Koch, M., Lomstein, B.A., Jensen, H.S., 2009b. Sulfide intrusion in the tropical seagrasses *Thalassia testudinum* and *Syringodium filiforme*. Estuarine, Coastal and Shelf Science 85, 319–326.

Holmer, M., Wirachwong, P., Thomsen, M., 2011. Negative effects of stress-resistant drift algae and high temperature on a small ephemeral seagrass species. Marine Biology 158, 297–309.

Howard, J.L., Perez, A., Lopes, C.C., Fourqurean, J.W., 2016. Fertilization changes seagrass community structure but not Blue Carbon storage: results from a 30-year field experiment. Estuaries and Coasts 39, 1422–1434.

Howard, J.L., Creed, J.C., Aguiar, M.V.P., Fourqurean, J.W., 2018. CO_2 released by carbonate sediment production in some coastal areas may offset the benefits of seagrass "Blue Carbon" storage. Limnology and Oceanography 63, 160–172.

Hu, X., Burdige, D.J., 2007. Enriched stable carbon isotopes in the pore waters of carbonate sediments dominated by seagrasses: evidence for coupled carbonate dissolution and reprecipitation. Geochimica et Cosmochimica Acta 71, 129–144.

Hyndes, G.A., Nagelkerken, I., McLeod, R.J., Connolly, R.M., Lavery, P.S., Vanderklift, M.A., 2014. Mechanisms and ecological role of carbon transfer within coastal seascapes. Biological Reviews 89, 232–254.

Isaksen, M.F., Finster, K., 1996. Sulphate reduction in the root zone of the seagrass Zostera noltii on the intertidal flats of a coastal lagoon (Arcachon, France). Marine Ecology Progress Series 137, 187–194.

Jackson, E.L., Rowden, A.A., Attrill, M.J., Bossey, S.J., Jones, M.B., 2001. The importance of seagrass beds as a habitat for fishery species. Oceanography and Marine Biology 39, 269–304.

Jensen, H.S., McGlathery, K.J., Marino, R., Howarth, R.W., 1998. Forms and availability of sediment phosphorus in carbonate sand of Bermuda seagrass beds. Limnology and Oceanography 43, 799–810.

Jensen, S.I., Kühl, M., Glud, R.N., Jørgensen, L.B., Priemé, A., 2005. Oxic microzones and radial oxygen loss from roots of *Zostera marina*. Marine Ecology Progress Series 293, 49–58.

Jensen, H.S., Nielsen, O.I., Koch, M.S., de Vicentea, I., 2009. Phosphorus release with carbonate dissolution coupled to sulfide oxidation in Florida Bay seagrass sediments. Limnology and Oceanography 54, 1753–1764.

Johnson, P., Johnstone, R., 1995. Productivity and nutrient dynamics of tropical seagrass communities in Puttalam Lagoon, Sri Lanka. Ambio 24, 411–417.

Jovanovic, Z., Pedersen, M., Larsen, M., Kristensen, E., Glud, R.N., 2015. Rhizosphere O2 dynamics in young *Zostera marina* and *Ruppia maritima*. Marine Ecology Progress Series 518, 95105.

Kaldy, J.E., Dunton, K.H., 2000. Above- and below-ground production, biomass and reproductive ecology of *Thalassia testudinum* (turtle grass) in a subtropical coastal lagoon. Marine Ecology Progress Series 193, 271–283.

Kaldy, J.E., Eldridge, P.M., Cifuentes, L.A., Jones, W.B., 2006. Utilization of DOC from seagrass rhizomes by sediment bacteria: 13C-tracer experiments and modeling. Marine Ecology Progress Series 317, 41–55.

Kennedy, H., Beggins, J., Duarte, C.M., Fourqurean, J.W., Holmer, M., Marbà, N., Middelburg, J.J., 2010. Seagrass sediments as a global carbon sink: isotopic constraints. Global Biogeochemical Cycles 24.

Koch, M.S., Erskine, J.M., 2001. Sulfide as a phytotoxin to the tropical seagrass *Thalassia testudinum*: interactions with light, salinity and temperature. Journal of Experimental Marine Biology and Ecology 266, 81–95.

Koch, M.S., Schopmeyer, S., Kyhn-Hansen, C., Madden, C.J., 2007a. Synergistic effects of high temperature and sulfide on tropical seagrass. Journal of Experimental Marine Biology and Ecology 341, 91–101.

Koch, M.S., Schopmeyer, S., Nielsen, O.I., Kyhn-Hansen, C., Madden, C.J., 2007b. Conceptual model of seagrass die-off in Florida Bay: links to biogeochemical processes. Journal of Experimental Marine Biology and Ecology 350, 73–88.

Kopecky, A.L., Dunton, K.H., 2006. Variability in drift macroalgal abundance in relation to biotic and abiotic factors in two seagrass dominated estuaries in the Western Gulf of Mexico. Estuaries and Coasts 29, 617–629.

Koren, K., Brodersen, K.E., Jakobsen, S.L., Kühl, M., 2015. Optical sensor nanoparticles in artificial sediments–a new tool to visualize O_2 dynamics around the rhizome and roots of seagrasses. Environmental Science and Technology 49, 2286–2292.

Krause-Jensen, D., Christensen, P.B., Rysgaard, S., Rysgaard, S., 1999. Oxygen and nutrient dynamics within mats of the filamentous macroalga *Chaetomorpha linum*. Estuaries 22, 31.

Kuusemäe, K., Rasmussen, E.K., Canal-Vergés, P., Flindt, M.R., 2016. Modelling stressors on the eelgrass recovery process in two Danish estuaries. Ecological Modelling 333, 11–42.

La Nafie, Y.A., Van Engeland, T., van Katwijk, M.M., Bouma, T.J., 2014. Uptake of nitrogen from compound pools by the seagrass *Zostera noltii*. Journal of Experimental Marine Biology and Ecology 460, 47–52.

Lamers, L.P.M., Govers, L.L., Janssen, I.C.J.M., Geurts, J.J.M., Van der Welle, M.E.W., Van Katwijk, M.M., Van der Heide, T., Roelofs, J.G.M., Smolders, A.J.P., 2013. Sulfide as a soil phytotoxin - a review. Frontiers of Plant Science 4, 1–14.

Larkum, A.W., Orth, R.J., Duarte, C.M., 2006a. Seagrasses: Biology, Ecology and Conservation. Springer, Dordrecht, NL.

Larkum, A.W.D., Drew, E.A., Ralph, P.J., 2006b. Photosynthesis and metabolism in seagrasses at the cellular level. In: Larkum, A.W.D., Orth, R.J., Duarte, C.M. (Eds.), Seagrasses: Biology, Ecology and Conservation. Springer Netherland, DordrecCht, Netherland, pp. 323–345.

Lavery, P.S., Mateo, M.-Á., Serrano, O., Rozaimi, M., 2013. Variability in the carbon storage of seagrass habitats and its implications for global estimates of blue carbon ecosystem service. PLoS One 8, e73748.

Lee, K.-S., Dunton, K.H., 2000. Diurnal changes in pore water sulfide concentrations in the seagrass *Thalassia testudinum* beds: the effects of seagrasses on sulfide dynamics. Journal of Experimental Marine Biology and Ecology 255, 201–214.

Lee, C.-L., Huang, Y.-H., Chen, C.-H., Lin, H.-J., 2016. Remote underwater video reveals grazing preferences and drift export in multispecies seagrass beds. Journal of Experimental Marine Biology and Ecology 476, 17.

Long, M.H., McGlathery, K.J., Zieman, J.C., Berg, P., 2008. The role of organic acid exudates in liberating phosphorus from seagrass-vegetated carbonate sediments. Limnology and Oceanography 53, 2616–2626.

Long, M.H., Berg, P., Falter, J.L., 2015a. Seagrass metabolism across a productivity gradient using the eddy covariance, Eulerian control volume, and biomass addition techniques. Journal of Geophysical Research: Oceans 120, 3624–3639.

Long, M.H., Berg, P., McGlathery, K.J., Zieman, J.C., 2015b. Sub-tropical seagrass ecosystem metabolism measured by eddy covariance. Marine Ecology Progress Series 529, 75–90.

Macreadie, P.I., Baird, M.E., Trevathan-Tackett, S.M., Larkum, A.W., Ralph, P.J., 2014. Quantifying and modelling the carbon sequestration capacity of seagrass meadows—a critical assessment. Marine Pollution Bulletin 83, 430–439.

Marbà, N., Hemminga, M.A., Mateo, M.A., Duarte, M.D., Mass, Y.E.M., Terrados, J., Gacia, E., 2002. Carbon and nitrogen translocation between seagrass ramets. Marine Ecology Progress Series 226, 287–300.

Marbà, N., Holmer, M., Gacia, E., Barrón, C., 2006. Seagrass beds and coastal biogeochemistry. In: Larkum, A.W.D., Orth, R.J., Duarte, C.M. (Eds.), Seagrasses: Biology, Ecology and Conservation. Springer Netherlands.

Marbà, N., Duarte, C.M., Holmer, M., Calleja, M.L., Álvarez, E., Díaz-Almela, E., Garcias-Bonet, N., 2008. Sedimentary iron inputs stimulate seagrass (*Posidonia oceanica*) population growth in carbonate sediments. Estuarine, Coastal and Shelf Science 76, 710–713.

Marbà, N., Díaz-Almela, E., Duarte, C.M., 2014. Mediterranean seagrass (*Posidonia oceanica*) loss between 1842 and 2009. Biological Conservation 176, 183–190.

Marbà, N., Arias-Ortiz, A., Masqué, P., Kendrick, G.A., Mazarrasa, I., Bastyan, G.R., Garcia-Orellana, J., Duarte, C.M., 2015. Impact of seagrass loss and subsequent revegetation on carbon sequestration and stocks. Journal of Ecology 103, 296–302.

Mascaro, O., Valdemarsen, T., Holmer, M., Perez, M., Romero, J., 2009. Experimental manipulation of sediment organic content and water column aeration reduces *Zostera marina* (eelgrass) growth and survival. Journal of Experimental Marine Biology and Ecology 373, 26–34.

Mateo, M.A., Serrano, O., 2012. The carbon sink associated to *Posidonia oceanica*. In: Pergent, G., et al. (Eds.), Mediterranean Seagrass Meadows: Resilience and Contribution to Climate Change Mitigation. IUCN, Gland, Switzerland and Málaga, Spain.

Mateo, M.Á., Sánchez-Lizaso, J.L., Romero, J., 2003. *Posidonia oceanica* 'banquettes': a preliminary assessment of the relevance for meadow carbon and nutrients budget. Estuarine, Coastal and Shelf Science 56, 85–90.

Mateo, M.A., Cebrián, J., Dunton, K., Mutchler, T., 2006. Carbon Flux in Seagrass Ecosystems. Seagrasses: Biology, Ecology and Conservation. Springer Science+Business Media.

Mazarrasa, I., Marbà, N., Lovelock, C.E., Serrano, O., Lavery, P.S., Fourqurean, J.W., Kennedy, H., Mateo, M.A., Krause-Jensen, D., Steven, A.D.L., Duarte, C.M., 2015. Seagrass meadows as a globally significant carbonate reservoir. Biogeosciences 12, 4993–5003.

Mazarrasa, I., Marbà, N., Garcia-Orellana, J., Masque, P., Arias-Ortiz, A., Duarte, C.M., 2017a. Dynamics of carbon sources supporting burial in seagrass sediments under increasing anthropogenic pressure. Limnology and Oceanography 62, 1451–1465.

Mazarrasa, I., Marbà, N., Garcia-Orellana, J., Masque, P., Arias-Ortiz, A., Duarte, C.M., 2017b. Effect of environmental factors (wave exposure and depth) and anthropogenic pressure in the C sink capacity of *Posidonia oceanica* meadows. Limnology and Oceanography 62, 1436–1450.

McGlathery, K.J., Sundbäck, K., Anderson, I.C., 2007. Eutrophication in shallow coastal bays and lagoons: the role of plants in the coastal filter. Marine Ecology Progress Series 348, 1–18.

Miyajima, T., Hori, M., Hamaguchi, M., Shimabukuro, H., Adachi, H., Yamano, H., Nakaoka, M., 2015. Geographic variability in organic carbon stock and accumulation rate in sediments of East and Southeast Asian seagrass meadows. Global Biogeochemical Cycles 29, 397–415.

Moeslund, L., Thamdrup, B., Barker Jørgensen, B., 1994. Sulfur and iron cycling in a coastal sediment: radiotracer studies and seasonal dynamics. Biogeochemistry 27.

Moksnes, P.O., Gullström, M., Tryman, K., Baden, S., 2008. Trophic cascades in a temperate seagrass community Oikos, 117, 763–777.

Moreno-Marin, F., Vergara, J.J., Perez-Llorens, J.L., Pedersen, M.F., Brun, F.G., 2016. Interaction between ammonium toxicity and green tide development over seagrass meadows: a laboratory study. PLoS One 11, e0152971. https://doi.org/10.1371/journal.pone.0152971.

Moriarty, D.J.W., Iverson, R.L., Pollard, P.C., 1986. Exudation of organic carbon by the seagrass *Halodule wrightii* Aschers. And its effect on bacterial growth in the sediment. Journal of Experimental Marine Biology and Ecology 96, 115–126.

Nayar, S., Collings, G.J., Miller, D.J., Bryars, S., Cheshire, A.C., 2010. Uptake and resource allocation of ammonium and nitrate in temperate seagrasses Posidonia and Amphibolis. Marine Pollution Bulletin 60, 1502–1511.

Nielsen, L.B., Finster, K., Welsh, D.T., Donelly, A., Herbert, R.A., de Wit, R., Lomstein, B.A.A., 2001. Sulphate reduction and nitrogen fixation rates associated with roots, rhizomes and sediments from *Zostera noltii* and *Spartina maritima* meadows. Environmental Microbiology 3, 63–71.

Nielsen, S.L., Sand-Jensen, K., Borum, J., Geertz-Hansen, O., 2002. Depth colonization of eelgrass (*Zostera marina*) and macroalgae as determined by water transparency in Danish coastal waters. Estuaries 25, 1025–1032.

Nielsen, O.I., Koch, M.S., Jensen, H.S., Madden, C.J., 2006. *Thalassia testudinum* phosphate uptake kinetics at low in situ concentrations using a 33P radioisotope technique. Limnology and Oceanography 51, 208–217.

Nuttle, W.K., Fourqurean, J.W., Cosby, B.J., Zieman, J.C., Robblee, M.B., 2000. Influence of net freshwater supply on salinity in Florida Bay. Water Resources Research 36, 1805–1822.

Olesen, B., Sand-Jensen, K., 1994. Biomass-density patterns in the temperate seagrass *Zostera marina*. Marine Ecology Progress Series 109, 283–291.

Orth, R.J., Carruthers, T.J.B., Dennison, W.C., Duarte, C.M., Fourqurean, J.W., Heck, K.L., Hughes, A.R., Kendrick, G.A., Kenworthy, W.J., Olyarnik, S., Short, F.T., Waycott, M., Williams, S.L., 2006. A global crisis for seagrass ecosystems. BioScience 56, 987.

Ottosen, L.D.M., Risgaard-Petersen, N., Nielsen, L.P., 1999. Direct and indirect measurements of nitrification and denitrification in the rhizosphere of aquatic macrophytes. Aquatic Microbial Ecology 19, 81–91.

Ow, Y.X., Collier, C.J., Uthicke, S., 2015. Responses of three tropical seagrass species to CO_2 enrichment. Marine Biology 162, 1005–1017.

Ow, Y.X., Uthicke, S., Collier, C.J., 2016a. Light levels affect carbon utilisation in tropical seagrass under ocean acidification. PLoS One 11, e0150352.

Ow, Y.X., Vogel, N., Collier, C.J., Holtum, J.A.M., Flores, F., Uthicke, S., 2016b. Nitrate fertilisation does not enhance CO2 responses in two tropical seagrass species. Scientific Reports 6, 23093.

Pagès, A., Welsh, D.T., Robertson, D., Panther, J.G., Schäfer, J., Tomlinson, R.B., Teasdale, P.R., 2012. Diurnal shifts in co-distributions of sulfide and iron(II) and profiles of phosphate and ammonium in the rhizosphere of *Zostera capricorni*. Estuarine, Coastal and Shelf Science 115, 282290.

Park, S.R., Kim, Y.K., Kim, S.H., Lee, K.-S., 2013. Nitrogen budget of the eelgrass, *Zostera marina* in a bay system on the south coast of Korea. Ocean Science Journal 48, 301–310.

Paynter, C.K., Cortés, J., Engels, M., 2001. Biomass, productivity and density of the seagrass *Thalassia testudi*num at three sites in Cahuita National Park, Costa Rica. Revista de Biología Tropical 49, 265–272.

Pedersen, M.F., Duarte, C.M., Cebrián, J., 1997. Rate of changes in organic matter and nutrient stocks during seagrass *Cymodocea nodosa* colonization and stand development. Marine Ecology Progress Series 159, 29–36.

Pedersen, O., Binzer, T., Borum, J., 2004. Sulphide intrusion in eelgrass (*Zostera marina* L.). Plant, Cell and Environment 27, 595–602.

Pedersen, M., Serrano, O., Mateo, M., Holmer, M., 2011. Temperature effects on decomposition of a *Posidonia oceanica* mat. Aquatic Microbial Ecology 65, 169–182.

Pérez, M., Mateo, M.A., Alcoverro, T., Romero, J., 2001. Variability in detritus stocks in beds of the seagrass *Cymodocea nodosa*. Botanica Marina 44.

Pérez, M., Invers, O., Ruiz, J.M., Frederiksen, M.S., Holmer, M., 2007. Physiological responses of the seagrass *Posidonia oceanica* to elevated organic matter content in sediments: an experimental assessment. Journal of Experimental Marine Biology and Ecology 344, 149–160.

Pollard, P.C., Moriarty, D.J.W., 1991. Organic-carbon decomposition, primary and bacterial productivity, and sulfate reduction, in tropical seagrass beds of the Gulf of Carpentaria, Australia. Marine Ecology Progress Series 69, 149–159.

Ralph, P.J., Durako, M.J., Enríquez, S., Collier, C.J., Doblin, M.A., 2007. Impact of light limitation on seagrasses. Journal of Experimental Marine Biology and Ecology 350, 176–193.

Rasmussen, H., Jorgensen, B.B., 1992. Microelectrode studies of seasonal oxygen uptake in a coastal sediment: role of molecular diffusion. Marine Ecology Progress Series 81, 289–303.

Rattanachot, E., Prathep, A., 2011. Temporal variation in growth and reproduction of *Enhalus acoroides* (L.f.) Royle in a monospecific meadow in Haad Chao Mai National Park, Trang Province, Thailand. Botanica Marina. Book 54.

Reusch, T.B.H., Williams, S.L., 1998. Variable responses of native eelgrass *Zostera marina* to a non-indigenous bivalve *Musculista senhousia*. Oecologia 113, 428–441.

Rheuban, J.E., Berg, P., McGlathery, K.J., 2014a. Ecosystem metabolism along a colonization gradient of eelgrass (*Zostera marina*) measured by eddy correlation. Limnology and Oceanography 59, 1376–1387.

Rheuban, J.E., Berg, P., McGlathery, K.J., 2014b. Multiple timescale processes drive ecosystem metabolism in eelgrass (*Zostera marina*) meadows. Marine Ecology Progress Series 507, 1–13.

Risgaard-Petersen, N., Ottosen, L.D.M., 2000. Nitrogen cycling in two temperate *Zostera marina* beds: seasonal variation. Marine Ecology Progress Series 198, 93–107.

Risgaard-Petersen, N., Dalsgaard, T., Rysgaard, S., Christensen, P.B., Borum, J., McGlathery, K., Nielsen, L.P., 1998. Nitrogen balance of a temperate eelgrass *Zostera marina* bed. Marine Ecology Progress Series 174, 281–291.

Röhr, M.E., Boström, C., Canal-Vergés, P., Holmer, M., 2016. Blue carbon stocks in Baltic Sea eelgrass (*Zostera marina*) meadows. Biogeosciences 13, 6139–6153.

Romero, J., Lee, K.-S., Peréz, M., Mateo, M.A., Alcoverro, T., 2006. Nutirent dynamics in seagrass ecosystems. In: Larkum, A.W.D., Orth, R.J., Duarte, C.M. (Eds.), Seagrasses: Biology, Ecology and Conservation. Springer Netherland, Dordrecht, Netherland, pp. 227–254.

Rozaimi, M., Lavery, P.S., Serrano, O., Kyrwood, D., 2016. Long-term carbon storage and its recent loss in an estuarine *Posidonia australis* meadow (Albany, Western Australia). Estuarine, Coastal and Shelf Science 171, 58–65.

Rubio, L., Linares-Rueda, A., García-Sánchez, M.J., Fernández, J.A., 2007. Ammonium uptake kinetics in root and leaf cells of *Zostera marina L.* Journal of Experimental Marine Biology and Ecology 352, 271–279.

Ruiz-Frau, A., Gelcich, S., Hendriks, I.E., Duarte, C.M., Marba, N., 2017. Current state of seagrass ecosystem services: research and policy integration. Ocean and Coastal Management 149, 107–115.

Ruiz-Halpern, S., Macko, S.A., Fourqurean, J.W., 2008. The effects of manipulation of sedimentary iron and organic matter on sediment biogeochemistry and seagrasses in a subtropical carbonate environment. Biogeochemistry 87, 113–126.

Sand-Jensen, K., Borum, J., Duarte, C.M., 1994. Havgræsserne i verdenshavene. Naturens Verden 27–40.

Sand-Jensen, K., Pedersen, O., Binzer, T., Borum, J., 2005. Contrasting oxygen dynamics in the freshwater isoetid *Lobelia dortmanna* and the marine seagrass *Zostera marina*. Annals of Botany 96, 613–623.

Short, F.T., Polidoro, B., Livingstone, S.R., Carpenter, K.E., Bandeira, S., Bujang, J.S., Calumpong, H.P., Carruthers, T.J.B., Coles, R.G., Dennison, W.C., Erftemeijer, P.L.A., Fortes, M.D., Freeman, A.S., Jagtap, T.G., Kamal, A.H.M., Kendrick, G.A., Judson Kenworthy, W., La Nafie, Y.A., Nasution, I.M., Orth, R.J., Prathep, A., Sanciangco, J.C., Tussenbroek, B., Vergara, S.G., Waycott, M., Zieman, J.C., 2011. Extinction risk assessment of the world's seagrass species. Biological Conservation 144, 1961–1971.

Simeone, S., De Falco, G., 2012. Morphology and composition of beach-cast *Posidonia oceanica* litter on beaches with different exposures. Geomorphology 151–152, 224–233.

Smith, R.D., Pregnall, A.M., Alberte, R.S., 1988. Effects of anaerobiosis on root metabolism of *Zostera marina* (eelgrass): implications for survival in reducing sediments. Marine Biology 98, 131–141.

Steward, J.S., Virnstein, R.W., Morris, L.J., Lowe, E.F., 2005. Setting seagrass depth, coverage, and light targets for the Indian River Lagoon system, Florida. Estuaries 28, 923–935.

Thamdrup, B., 2000. Bacterial manganese and iron reduction in aquatic sediments. Advances in Microbial Ecology. Springer Science + Business Media).

Thomsen, M.S., Wernberg, T., Engelen, A.H., Tuya, F., Vanderklift, M.A., Holmer, M., McGlathery, K.J., Arenas, F., Kotta, J., Silliman, B.R., 2012. A meta-analysis of seaweed impacts on seagrasses: generalities and knowledge gaps. PLoS One 7, e28595.

Trevathan-Tackett, S., Macreadie, P., Ralph, P., Seymour, J., 2014. Detachment and flow cytometric quantification of seagrass-associated bacteria. Journal of Microbiological Methods 102, 23–25.

Trevathan-Tackett, S.M., Kelleway, J., Macreadie, P.I., Beardall, J., Ralph, P., Bellgrove, A., 2015. Comparison of marine macrophytes for their contributions to blue carbon sequestration. Ecology 96, 3043–3057.

Valdemarsen, T., Quintana, C.O., Kristensen, E., Flindt, M.R., 2014. Recovery of organic-enriched sediments through microbial degradation: implications for eutrophic estuaries. Marine Ecology Progress Series 503, 41–58.

Valentine, J.F., Duffy, J.E., 2006. The Central Role of Grazing in Seagrass Ecology. Seagrasses: Biology, Ecology and Conservation. Springer Science + Business Media.

van der Heide, T., Govers, L.L., De Fouw, J., Olff, H., van der Geest, M., Van Katwijk, M.M., Piersma, T., van de Koppel, J., Silliman, B.R., Smolders, A.J., van Gils, J.A., 2012. A three-stage symbiosis forms the foundation of seagrass ecosystems. Science 336, 1432–1434.

Van Engeland, T., Bouma, T.J., Morris, E.P., Brun Murillo, F.G., Peralta, G., Lara, M., Hendriks, I.E., Soetaert, K.E.R., Middelburg, J.J., 2011. Potential uptake of dissolved organic matter by seagrasses and macroalgae. Marine Ecology Progress Series 427, 71–81.

Van Katwijk, M.M., Vergeer, L.H.T., Schmitz, G.H.W., Roelofs, J.G.M., 1997. Ammonium toxicity in eelgrass *Zostera marina*. Marine Ecology Progress Series 157, 159–173.

Van Tussenbroek, B.I., 1998. Above- and below-ground biomass and production by *Thalassia testudinum* in a tropical reef lagoon. Aquatic Botany 61, 69–82.

Vichkovitten, T., Holmer, M., 2004. Contribution of plant carbohydrates to sedimentary carbon mineralization. Organic Geochemistry 35, 1053–1066.

Vichkovitten, T., Holmer, M., 2005. Dissolved and particulate organic matter in contrasting (eelgrass) sediments. Journal of Experimental Marine Biology and Ecology 316, 183–201.

Vichkovitten, T., Holmer, M., Frederiksen, M.S., 2007. Spatial and temporal changes in nonstructural carbohydrate reserves in eelgrass (*Zostera marina* L.) in Danish coastal waters. Botanica Marina 50.

Villazán, B., Pedersen, M.F., Brun, F.G., Vergara, J.J., 2013. Elevated ammonium concentrations and low light form a dangerous synergy for eelgrass *Zostera marina*. Marine Ecology Progress Series 493, 141–154.

Villazán, B., Brun, F.G., Gonzalez-Ortiz, V., Moreno-Marin, F., Bouma, T.J., Vergara, J.J., 2016. Flow velocity and light level drive non-linear response of seagrass *Zostera noltei* to ammonium enrichment. Marine Ecology Progress Series 545, 109–121.

Vinther, H.F., Laursen, J.S., Holmer, M., 2008. Negative effects of blue mussel (*Mytilus edulis*) presence in eelgrass (*Zostera marina*) beds in Flensborg fjord, Denmark. Estuarine, Coastal and Shelf Science 77, 91–103.

Vonk, J.A., Middelburg, J.J., Stapel, J., Bouma, T.J., 2008. Dissolved organic nitrogen uptake by seagrasses. Limnology and Oceanography 53, 542–548.

Vonk, J.A., Christianen, M.J.A., Stapel, J., O'Brien, K.R., 2015. What lies beneath: why knowledge of belowground biomass dynamics is crucial to effective seagrass management. Ecological Indicators 57, 259–267.

Wang, X., Chen, R.F., Cable, J.E., Cherrier, J., 2014. Leaching and microbial degradation of dissolved organic matter from salt marsh plants and seagrasses. Aquatic Sciences 76, 595–609.

Watanabe, K., Kuwae, T., 2015. How organic carbon derived from multiple sources contributes to carbon sequestration processes in a shallow coastal system? Global Change Biology 21, 2612–2623.

Waycott, M., Duarte, C.M., Carruthers, T.J., Orth, R.J., Dennison, W.C., Olyarnik, S., Calladine, A., Fourqurean, J.W., Heck Jr., K.L., Hughes, A.R., Kendrick, G.A., Kenworthy, W.J., Short, F.T., Williams, S.L., 2009. Accelerating loss of seagrasses across the globe threatens coastal ecosystems. Proceedings of the National Academy of Sciences of the United States of America 106, 12377–12381.

Welsh, D.T., Bourgues, S., de Wit, R., Herbert, R.A., 1996. Seasonal variation in rates of heterotrophic nitrogen fixation (acetylene reduction) in Zostera noltii meadows and uncolonised sediments of the Bassin d'Arcachon, south-west France. Hydrobiologia 329, 161–174.

Welsh, D.T., Bartoli, M., Nizzoli, D., Castaldelli, G., Riou, S.A., Viaroli, P., 2000. Denitrification, nitrogen fixation, community primary productivity and inorganic-N and oxygen fluxes in an intertidal *Zostera noltii* meadow. Marine Ecology Progress Series 208, 65–77.

Zhang, J.Z., Fischer, C.J., Ortner, P.B., 2004. Potential availability of sedimentary phosphorus to sediment resuspension in Florida Bay. Global Biogeochemical Cycle 18, GB4008.

Ziegler, S., Benner, R., 1999. Dissolved organic carbon cycling in a subtropical seagrass dominated lagoon. Marine Ecology Progress Series 180, 149–160.

Zieman, J.C., Fourqurean, J.W., Frankovich, T.A., 1999. Seagrass die-off in Florida Bay: long-term trends in abundance and growth of turtle grass, *Thalassia testudinum*. Estuaries 22, 460–470.

Zimmerman, R.C., 2006. Light and photosynthesis in seagasses. In: Larkum, A.W.D., Orth, R.J., Duarte, C.M. (Eds.), Seagrasses: Biology, Ecology and Conservation. Springer Netherland, Dordrecht, Netherland, pp. 303–321.

CHAPTER

14

Tidal Salt Marshes: Sedimentology and Geomorphology

Jon French

Coastal and Estuarine Research Unit, UCL Department of Geography, University College London, Gower Street, London, United Kingdom

1. INTRODUCTION

Salt marsh is a distinctive wetland type found within the upper part of the intertidal zone in a variety of coastal and estuarine geomorphological contexts. Their high elevation and the presence of halophytic macrophytes distinguish salt marshes from unvegetated tidal flats and also from mangroves, which occupy a similar intertidal niche within the tropics (Chapman, 1976; Woodroffe, 1992; Ellison, 2018). Both globally and regionally, marked differences in salt marsh ecology and function arise from the interaction between the character and diversity of the halophytic flora with climatic, edaphic, and hydrographic influences (Frey and Basan, 1985; Isacch et al., 2006; Porter et al., 2015; Suchrow et al., 2015). Geographical variation in plant zonation has been used to infer differences in succession and provides a basis for the identification of discrete marsh types (Beeftink, 1966; Chapman, 1974; Adam, 1990). Halophytic vegetation is not only fundamental to the definition of salt marsh but also widely considered to be instrumental in stabilizing the upper tidal flat and effecting an increase in the sediment trapping efficiency that drives the emergence of an elevated salt marsh platform (Morris et al., 2002; Li and Yang, 2009; Mudd et al., 2010; Vandenbruwaene et al., 2015). As the marsh builds and its elevation diverges from that of the tidal flat, plant productivity also becomes a contributor to the sedimentary fabric through the accumulation of refractory root and rhizome tissue and refractory organic particles from the aboveground biomass (Nyman et al., 2006; Neubauer, 2008; Morris et al., 2016). Long-term carbon storage, however, is influenced not only by plant productivity but also by sedimentology (Kelleway et al., 2016).

Where the tidal influence is weak and/or external inputs of mineral sediments are limited, it may be appropriate to view salt marshes first and foremost as ecosystems and to relegate their geomorphology to a set of subsidiary influences. On the other hand, where the tidal influence is stronger and sediment supply more abundant, it can be argued that ecosystem

Coastal Wetlands
https://doi.org/10.1016/B978-0-444-63893-9.00014-9

functions and values become more fundamentally underpinned by sedimentary processes that give rise to the accumulation of the fine sediments and the maintenance of marsh platform elevations within the tidal frame (Stevenson et al., 1986; Allen, 1990; French, 1994, 2006; Silva et al., 2009; Kirwan and Guntenspergen, 2010). Meso- and macrotidal salt marshes typically have larger inputs of inorganic (mineral) sediment and the contribution of organic material to marsh substrates may be rather small (French and Spencer, 1993; Allen, 2000; Bartholdy et al., 2014). It is reasonable, in such systems, to ascribe greater importance to the physical processes that drive morphological evolution of the salt marsh as a sedimentary landform. This landform is a composite feature that comprises two main components: a gently sloping vegetated platform and a channel system that accounts for a large proportion of the water and sediment exchange with adjacent coastal or estuarine environments (Fig. 14.1). From a geomorphological perspective, salt marshes provide one of the best examples or morphodynamic behavior (i.e., two-way interaction between morphology and processes) and extremely rapid landform evolution toward dynamic equilibrium with continually varying boundary conditions (e.g., tidal range, sea level, sediment supply). At the same time, more complex interdependencies between physical, biological, and geochemical processes influence the landform morphodynamics within the intertidal zone more generally (e.g., Friedrichs and Perry, 2001; van de Koppel et al., 2005; Marani et al., 2010; Wang and Temmerman, 2013).

This chapter outlines the physical contexts for the formation and persistence of tidal salt marshes, including factors that favor the accumulation of fine sediment within the upper intertidal zone and its colonization by halophytic vegetation. It then describes the essential characteristics of salt marsh sediments and the morphology and development of the various components of the salt marsh landform complex. Lastly, conceptual frameworks for longer-term salt marsh evolution are considered together with some insights into the resilience of salt marshes to sea level rise. The emphasis is on tidally dominated systems that are primarily formed by the accumulation of mineral sediments, although the mediation of physical processes and landform morphodynamics by biological influences is also considered.

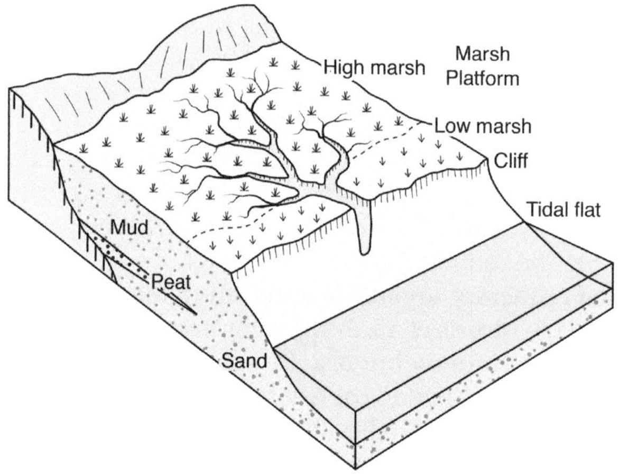

FIGURE 14.1 Definition sketch of the tidal salt marsh landform and its relation to the tidal flat environment. *Modified from Woodroffe, C.D., 2003. Coasts. Form, Process and Evolution. Cambridge University Press, Cambridge, 623p.*

2. CONTEXTS FOR TIDAL SALT MARSH FORMATION

Although salt marsh is sometimes considered to constitute a single class of coastal wetland, the range of settings in which they are found and the interplay of regionally specific geological, coastal oceanographic, hydrological, and biogeographical contexts give rise to considerable diversity in morphology and function. First-order controls on the occurrence and persistence of tidal salt marshes include not only the tidal regime but also the coastal configuration, fine sediment regime, and relative sea level history. Each of these controls is considered in turn below.

2.1 Tidal Regime

The common depiction of salt marshes as low-energy depositional systems is somewhat misleading because many of the best-developed examples occur at the highest tidal ranges, where exchanges of water and materials with adjacent coastal and estuarine environments are extremely energetic (Gordon et al., 1985; van Proosdij et al., 2006; Détriché et al., 2011). Mean tidal range is widely used as a first-order classification of the coast, and the Davies (1964) division into micro- (mean tidal range < 2 m), meso- (2 m $<$ mean tidal range < 4 m), and macrotidal (mean tidal range > 4 m) is still commonly used. Tidal range defines a vertical and lateral envelope, or "accommodation space," within which sedimentary infilling can occur (Allen, 1990). The width of the intertidal zone tends to be smaller in microtidal settings (Fan, 2012) and this limits the potential extent of salt marsh in the upper part of that zone. Exchanges of water and materials with adjacent coastal and estuarine waters can be highly energetic in macrotidal salt marshes and there is empirical evidence to support a correlation between tidal range and the potential for sediment transport, as indicated by sedimentation rate (Harrison and Bloom, 1977; Stevenson et al., 1986; French, 2006) or tidal current velocities (Friedrichs, 1995).

These generalizations are complicated by interactions with other factors. For example, even on microtidal coasts, tidal currents may be strong within constricted inlets and these may be effective in sustaining locally high suspended sediment concentrations and correspondingly high rates of sedimentation within salt marshes adjacent to estuary channels (e.g., French and Burningham, 2003; French et al., 2008). In addition, nontidal storm inundation events are often of greater relative importance in microtidal situations (French, 2006) and can sustain inorganic sedimentation within salt marshes where normal tidal exchange is limited (Stumpf, 1983; Cahoon et al., 1995; Lagomasino et al., 2013).

Local tidal characteristics are important in partitioning the intertidal zone and give rise to a set of datum levels that are known to influence physical and biological processes on tidal flats and salt marshes, not least through variation in the frequency and duration (hydroperiod) of tidal inundation. Of particular importance is the threshold elevation above which pioneer halophytes are able to colonize and thrive on the upper tidal flat. Various studies have attempted to approximate or generalize this with respect to specific tidal datum levels. A classic study by Hinde (1954) suggested that the lower limit of salt marsh can be approximated by Mean Low Water (MLW), although this is based on observations from microtidal marshes on Long Island, USA. McKee and Patrick (1988) showed that growth limits of *Spartina alterniflora* usually lie between MLW and Mean High Water (MHW) within a broader sample of US Atlantic and Gulf Coast salt marshes. However, systematic differences in actual growth limits were apparent, and these appeared to be correlated with mean tidal range, with additional variation due to factors such as salinity, drainage, soil chemistry, and interspecific competition. Many of these secondary influences are in turn partly controlled by inundation and therefore elevation (e.g., Pennings and Callaway, 1992; Moffett et al., 2010). This is consistent with a large body of work on the factors influencing halophyte establishment in both the northern (e.g., Gray, 1972; Nestler, 1977) and southern hemispheres (e.g., Veldkornet et al., 2016). However, detailed studies show that distributions of plants within the tidal frame may be quite heterogeneous, as demonstrated for a Dutch barrier island salt marsh dominated by *Elymus athericus*, *Artemisia maritima*, and *Festuca rubra* by Bockelmann et al. (2002). In northwest Europe, where most salt marshes are meso- to macrotidal, Mean High Water Neap Tides is often assumed to coincide roughly with the lower limit of the salt marsh vegetation (e.g., Adam, 2002; Bakker et al., 2002; French, 2006). In the microtidal environment of southern Europe, mean sea level tends to be the lower limit of salt marsh vegetation (e.g., Silvestri et al., 2005). There is some evidence that inundation frequency exerts a critical control on the establishment of certain pioneers (notably *Salicornia* spp.; Wiehe, 1935). A comprehensive regional to global analysis of the vertical limits to salt marsh across a wide spectrum of tidal ranges by Balke et al. (2016) also suggests that the frequency of inundation is more important than its cumulative duration.

The elevations associated with MHW or Mean High Water Spring Tides have been variously associated with approximate equilibrium marsh elevations (Kestner, 1975; Krone, 1987; Pethick, 1981). Elevation relative to MHW has also been associated with transitions between key habitat types (e.g., Suchrow and Jensen, 2010).

2.2 Coastal Configuration and Geomorphological Setting

At a continental scale, the configuration and extent of coastal margins exert a first-order control on the potential "accommodation space" for intertidal sediment accumulation within the intertidal zone. Variation in the width of the continental shelf also exerts a control on tidal range (Davies, 1964; Hayes, 1979). Frey and Basan (1985), for example, highlight the contrasts in the space available for salt marsh development between the Pacific and Atlantic coasts of North America. On the tectonically active Pacific coast, tidal salt marsh is fragmented and is restricted to relatively narrow fringes around protected embayments and estuaries. In comparison, Atlantic coastal plain marshes are much more extensive and are continuous over large areas. Wide shallow coastal shelves favor a flood-dominated hydrodynamic regime, leading to a net landward transport of sediments, especially within the intertidal (Postma, 1961; Amos, 1995), and the retention of sediments within estuaries and embayments (Dalrymple et al., 2012).

At smaller regional and local scales, it is useful to consider the geomorphological setting in terms of the relative significance of riverine, tidal, and wave-related processes. This ternary classification has been widely used for deltas, estuaries, and coastal environments in general (Boyd et al., 1992). The distribution of salt marsh is determined by the occurrence of intertidal accommodation space that is sufficiently sheltered from ocean waves and higher-energy wind waves generated over estuarine or lagoonal fetches. Salt marsh thus tends to be extensive not only at higher tidal ranges (where riverine influences are diminished), but also where wave energy is low. Even within restricted fetch settings, locally generated wind waves can generate intertidal bottom stresses that greatly exceed those due to tidal currents and can exert a major influence on the geomorphological evolution of the tidal flats (Amos, 1995; Fan, 2012; Zhou et al., 2016). Wave energy also influences the viability of halophytic vegetation establishment (Eisma and Dijkema, 1997; Silinski et al., 2016), as well the stability of the salt marsh–tidal flat boundary (van de Koppel et al., 2005; Pedersen and Bartholdy, 2007; Callaghan et al., 2010; Bouma et al., 2016). As illustrated in Fig. 14.2, the range of wave-sheltered geomorphological settings is quite diverse and provides a basis for classifying marsh types.

Deltaic settings are purposely omitted in Fig. 14.2 given that the focus here is on tide-dominated salt marshes. These tend to have more limited distribution in deltas, where more brackish and microtidal wetlands predominate. Most of the major tide-dominated deltas lie in the tropics, where mangrove dominates the upper intertidal (Goodbred and Saito, 2012). Tidal salt marshes are much more widely distributed in estuaries, tidal inlets, and embayments, although here they are often fragmented and comprise numerous units that are individually small in area. Macrotidal embayment and estuarine marshes in the Bay of Fundy, Canada (Shaw and Ceman, 1999; Davidson-Arnott et al., 2002; van Proosdij et al., 2006), and the Severn Estuary, UK (Allen, 1990; Allen and Duffy, 1998; Allen and Haslett, 2014), have been especially well studied. In South America, the temperate estuarine marshes of Argentina have also been extensively studied (Bortolus et al., 2009; Pratolongo et al., 2010, 2013). The extensive salt marshes in the major estuaries of China are also well

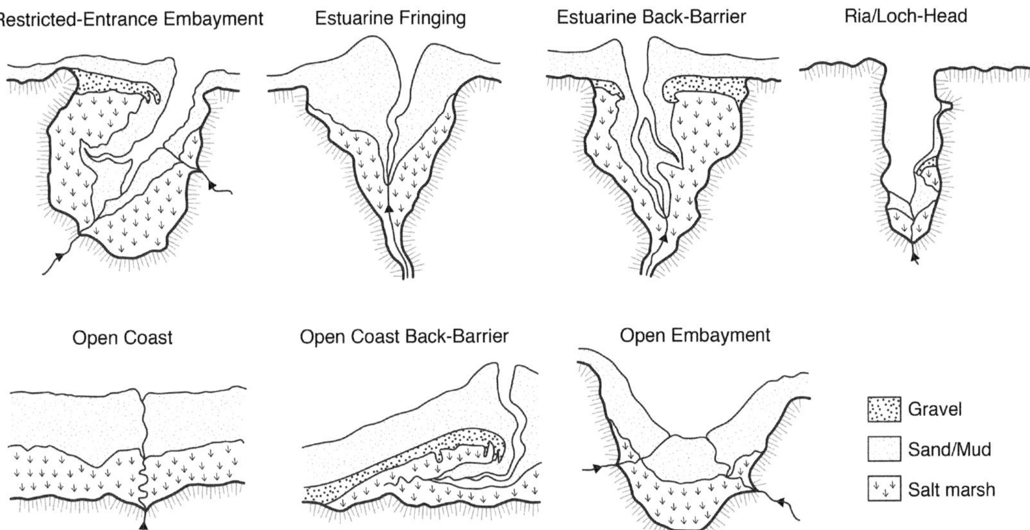

FIGURE 14.2 Schematic illustration of coastal and estuarine contexts for the occurrence of tidal salt marsh. *Modified from similar figures in Dijkema, K.S., 1987. The geography of salt marshes in Europe. Zeitschrift für Geomorphologie 31, 489–499, Allen, J.R.L., 2000. Morphodynamics of Holocene saltmarshes: a review sketch from the Atlantic and southern north sea coasts of Europe. Quaternary Science Reviews 19, 1155–1231 and Woodroffe, C.D., 2003. Coasts. Form, Process and Evolution. Cambridge University Press, Cambridge, 623p.*

known on account of the widespread impacts of human activity (Yang and Chen, 1995) in conjunction with the progressive displacement of the native flora by invasives such as the cordgrass *Spartina alterniflora* (Zuo et al., 2012). On the open coast, back-barrier environments often contain extensive salt marshes. Those of the Dutch and German Wadden Sea (Reise, 2005; de Groot et al., 2011a; Suchrow et al., 2012; Schuerch et al., 2014) are especially well developed, and sites on the macrotidal North Norfolk coast of the United Kingdom (Pethick, 1981; French and Stoddart, 1992; French and Spencer, 1993) and the microtidal Skallingen peninsula in Denmark (Bartholdy et al., 2004, 2010) are also exceptionally well-documented. True open coast salt marsh is restricted to a few localities. The Dengie Marshes (Essex, UK) provide an example of muddy open coast salt marsh in a mesotidal setting; even here, the salt marsh area has been significantly reduced by erosion over the last few decades (Harmsworth and Long, 1986).

2.3 Fine Sediment Regime

Stratigraphic studies demonstrate the effectiveness of tidal salt marshes as sediment sinks at centennial to millennial timescales (Redfield, 1972; Shaw and Ceman, 1999; Allen, 2003). However, the nature of this sediment sink function varies markedly between systems driven primarily by the accumulation of internally produced organic material and those characterized by the deposition of externally derived inorganic sediments. Dijkema (1987) refers to these as autochthonous and allochthonous, respectively; Allen (2000) uses the terms organogenic and mineralogenic. Autochthonous marshes tend to be more characteristic of microtidal regimes, where the effective supply of inorganic sediment is limited by generally lower background suspended sediment concentrations and by less energetic tidal exchanges (French, 1994; Friedrichs and Perry, 2001). Plant productivity is a primary driver of marsh elevation adjustment within these systems (Morris et al., 2002, 2016). Under accelerated sea level rise, especially in deltaic systems, where relative sea level rise is enhanced by high rates of subsidence, marsh collapse may occur (e.g., Day et al., 2000).

The relative magnitude of inorganic and organic matter accumulation exerts a strong influence on marsh geomorphology and ecology and has been used as a starting point for some classification schemes—notably, that for northwest Europe by Dijkema (1987). The supply of fine sediment is a key factor in both the formation and persistence of marshes. At decadal to centennial scales, magnitude of allochthonous sediment input determines the ability of a marsh to adjust to changing boundary conditions (especially sea level rise) along regional gradients in tidal range and sediment supply (Stevenson et al., 1986; French, 1994, 2006). Sediment supply is not a constant but exhibits variability at various timescales. Bartholdy et al. (2004) and Kim et al. (2013) present convincing evidence that time variation in sedimentation within the salt marshes of the Skallingen peninsula, Denmark, tracks the North Atlantic Oscillation (NAO) over a 60-year period (Fig. 14.3). Sedimentation depends on the frequency and magnitude of overmarsh tides, which are influenced, in turn, by the strength and persistence of the prevailing westerlies for which the NAO is considered to be a proxy (see also Burningham and French, 2013). French and Burningham (2003) also showed an association between the NAO winter index and coherent variations in elevation changes for multiple sites in a UK east coast estuary (although the measurement record here is much shorter). The implication here is that the NAO is a proxy for variation in

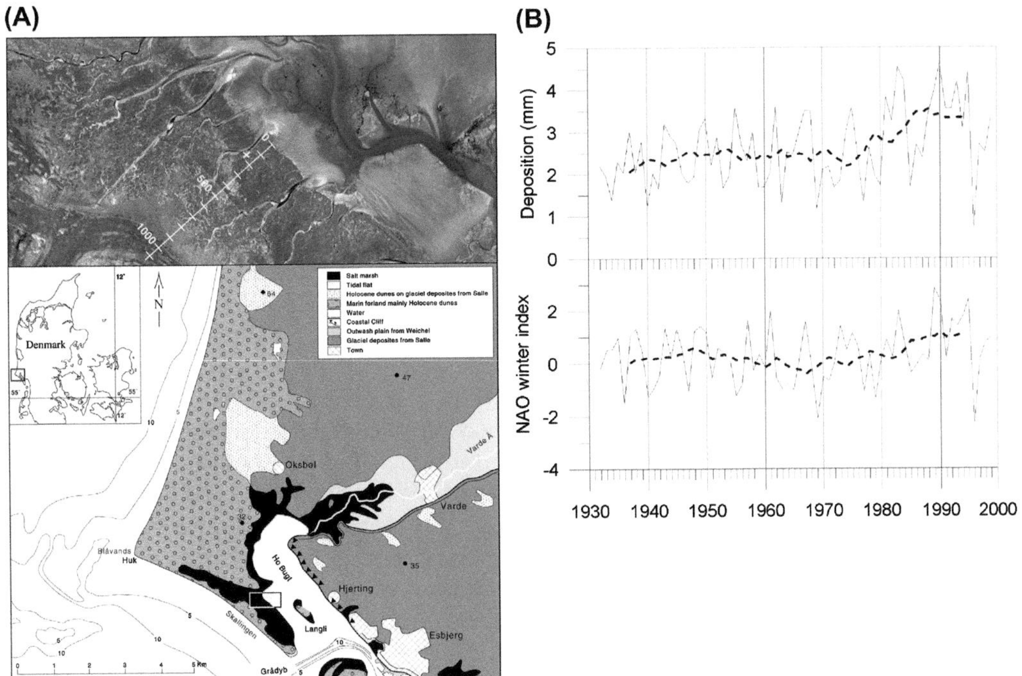

FIGURE 14.3 (A) The salt marsh at Skallingen, Denmark, one of the largest undiked salt marshes in Europe. Line shows transect along which annual sedimentation was measured and modeled; (B) time variation in empirically modeled annual sedimentation rate (upper panel, with moving average fitted) and North Atlantic Oscillation (NAO) winter index (lower panel). *Reproduced from Bartholdy, J., Christiansen, C., Kunzendorf, H., 2004. Long term variations in backbarrier salt marsh deposition on the Skallingen peninsula – the Danish Wadden Sea. Marine Geology 203, 1–21.*

wind wave action on the adjacent tidal mud flats, which modulates the local sediment supply to the salt marshes. In the marshes of the Rio del Plata (Argentina and Uruguay), Schuerch et al. (2016) have documented the importance of fluvial sediment inputs and their modulation by ENSO-related climate variation.

2.4 Sea Level History

Stratigraphic studies have shown the formation and vertical buildup of salt marshes to be intimately related to movements in mean sea level. Holocene sea level changes have been especially complex in northern hemisphere locations influenced by glacio- and hydroisostatic effects. As Allen (2000) observes, these force salt marsh dynamics in two main ways. First, sea level rise creates new accommodation space for the accumulation of intertidal sediments. Second, time variation in the rate (and sometimes the direction) of sea level change exerts a subtle influence over the nature of intertidal sedimentation, including the balance between autochthonous and allochthonous sedimentation. Many salt marshes are known to have been initiated and to have maintained their integrity through vertical and lateral accretion during long periods of sustained sea level rise (Redfield, 1972; Gunnell et al., 2013). Oscillations in sea level tendency, both positive and negative, also appear to have triggered expansion of salt marsh within settings conducive to fine sediment accumulation (Pethick, 1980; Orson et al., 1987; Shaw and Ceman, 1999). Within an estuarine salt marsh in the Danish Wadden Sea, Madsen et al. (2007) used optically stimulated luminescence dating of sediment cores to examine time variation in the rate of sedimentation over the last 2000 years. They attribute distinct phases of rapid vertical marsh growth (at up to 9 mm yr^{-1}) to accelerations in the rate of sea level rise. These created additional accommodation space and may also have increased sediment availability.

Sea level rise thus provides a moving boundary condition under which the sediment accumulating potential (both organic and inorganic) of the salt marsh environment can be most fully realized. This "depositional paradigm" (Stevenson et al., 1988) was challenged by the discovery of sedimentary deficits within subsiding deltaic marshes, some of which accumulate insufficient material to keep pace with high rates of relative sea level rise (Baumann et al., 1984; Rybczyk and Cahoon, 2002). The likelihood of accelerated and more widespread sea level rise has given rise to a "submergence paradigm," under which it is envisaged that more frequent inundation will force a reversion of the vegetation succession (e.g., Warren and Niering, 1993), possibly culminating in physical degradation and reversion to lower intertidal or subtidal environments. However, the concept of "submergence" oversimplifies the complex linkages between sea level, sedimentation accumulation, topography, and biotic response. These are only just being understood through a new generation of coupled morphodynamic and ecosystem models (e.g., Morris et al., 2002; Kirwan and Murray, 2007; Marani et al., 2007; see also Section 5).

2.5 Anthropogenic Influences

Although the morphology and dynamics of coastal landforms are usually conceptualized in terms of an interplay of geological context, sediment supply, and coastal oceanographic processes (tidal regime and wave climate), the effects of anthropogenic activities are pervasive at the coast (Hapke et al., 2013). A combination of simple topography, ecosystem resources, and potentially fertile soils mean that humans have long exploited tidal salt marshes. The high primary productivity of salt halophytes has been harvested for animal feed and thatch, and livestock grazing remains widespread (Gedan et al., 2009). Drainage (including ditching for mosquito control; Kirby and Widjeskog, 2013) and reclamation have been more destructive and a large proportion of the natural salt marsh resource has been lost since medieval times, especially in Europe (Allen, 2000; Reise, 2005), Asia (Koh and de Jonge, 2014; Du et al., 2016), Australia (Zann, 2000), and North America (Atwater et al., 1979; Kennish, 2001).

There are thus relatively few instances of tidal salt marshes formed in settings outlined in Fig. 14.1 that do not bear at least some imprint of anthropogenic activities past or present. In the case of predominantly allochthonous tidal salt marshes, human-induced perturbations of the sediment budget are implicated in significant phases of expansion and contraction over at the last 1000 years or so. Major periods of tidal salt marsh building have been attributed to pulses of sediment generated by post-9th century land use changes that affected sediment discharge of the Yangtze River, China (Yang et al., 2001), by post-17th century European settlement of the Northeastern United States (Kirwan et al., 2011), and by 19th century hydraulic mining in the catchments feeding into San Francisco Bay, California (Atwater et al., 1979). More recent historic episodes of marsh degradation and loss have been attributed to declining sediment supply, for example, in the Yangtze Delta region because of dam construction (Du et al., 2016). In the United States, the sheer scale of catchment impoundment for water supply purposes has led to significant reductions in sediment delivery via estuaries and is increasingly highlighted as a constraint on the ability of the remaining tidal salt marshes to track future sea level rise (e.g., Weston, 2014).

The indicative effects of this anthropogenic perturbation of estuarine and coastal sediment budgets on marsh expansion and persistence have been conceptualized by Mudd (2011) and are summarized here in Fig. 14.4. A key aspect of this diagram is the hysteresis of the relationship between sediment supply and the various state variables of the salt marsh system. Thus, even if the sediment supply returns to its prepulse level, the salt marsh may persist in an altered state. One reason for this is that salt marsh is so effective as a sediment trap that, once established, it may alter both the hydrodynamics and sediment budget of an entire estuarine system.

A less common anthropogenic impact in an age when many salt marsh systems are faced with increasing inundation because of accelerated sea level rise is the reduction of tidal range due to the construction of tidal barrages. In the Netherlands, de Jong et al. (1994) document loss and degradation of salt marsh following completion of the Oosterschelde storm surge barrier in 1987. Tidal range and MHW levels in the estuary have been reduced by over 10%, leading to desiccation of the fringing salt marsh deposits and enhanced erosion of microcliffs that mark the transition to the tidal flat.

3. SEDIMENTOLOGY

3.1 Characteristic Sediments

Much of the coastal wetlands literature uses the term soil to refer to the salt marsh deposit, even in tide-dominated allochthonous systems. In the geomorphological literature, sediment appears to be the preferred term. As Frey and Basan (1985) argue, the term sediment is preferable as it does not imply any genetic relationship with underlying bedrock or regolith. The fundamental distinction between autochthonous deposits, formed primarily from in situ accumulation of organic matter, and allochthonous deposits comprised mainly of externally sourced inorganic mineral sediments has already been noted. Tide-dominated salt marshes are primarily allochthonous in nature and the discussion that follows is largely restricted to inorganic sediments.

Allochthonous inorganic sediments can be derived from fluvial sources usually via estuarine inflows (e.g., Mattheus et al., 2010; Weston, 2014; Schuerch et al., 2016); coastal cliff erosion (Amos and Tee, 1989); and/or local marine sources (e.g., Holden et al., 2011). Coastal cliff sources often involve long-range transport of fine suspended muds (e.g., Dyer and Moffat, 1998) and it can be difficult to determine source–sink pathways with any certainty, let alone quantify their magnitude. Even in sandier marshes, finding definitive evidence to support more proximal intertidal or subtidal sediment sources can be difficult. The attempt by Holden et al. (2011) to use textural and mineral magnetic properties to demonstrate a local

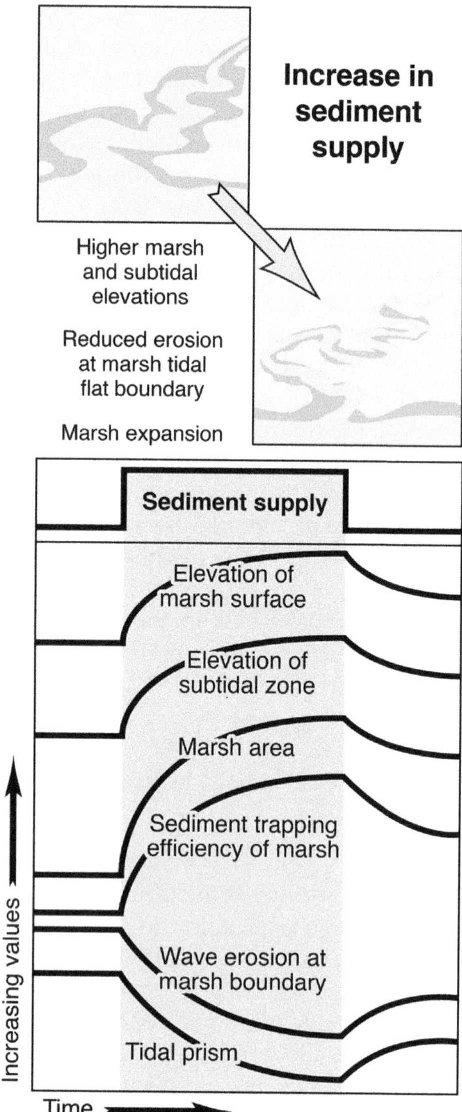

FIGURE 14.4 Schematic diagram illustrating the sensitivity of tidal salt marsh extent and characteristics to variation in sediment supply, such as might result from anthropogenic disturbance. *Adapted from Mudd, S.M., 2011. The life and death of salt marshes in response to anthropogenic disturbance of sediment supply. Geology 39, 511–512.*

provenance for salt marsh sediments on the Sefton Coast, Northwest England, illustrates this well. Tidal salt marsh deposits also contain a small biogenic sediment fraction (including carbonate tests and shells and siliceous diatom and foraminiferal remnants). The biogenic fraction is especially diagnostic of paleoenvironmental factors such as tidal range or salinity (Strotz, 2015; Strachan et al., 2016) and can provide invaluable insights into local sea level history and the adjustment of the salt marsh platform (Edwards et al., 2004; Stephan et al., 2015; Kemp et al., 2017), as well as evidence for episodic land movements (Hayward et al., 2016) or storm events (Hippensteel and Garcia, 2014). However, studies in eroding New Zealand salt marshes have shown that in situ foraminiferal faunas may be mixed with material reworked from marsh cliff erosion, compromising their use in paleoenvironmental reconstruction (Figueira and Hayward, 2014).

Salt marsh sediments typically show a continuation of the landward-fining sequence that is observed on tidal flats (e.g., Amos, 1995; Yang et al., 2008). The proximal tidal flat is typically muddy, although some salt marshes are fronted by sand flats, especially in back-barrier or open coastal settings. Suspension is the dominant mode of sediment transport within the salt marsh, with a variable proportion of fine sands and cohesive muds being transported within the tidal channels but predominantly muds in the shallower flows that inundate the marsh platform. Bartholdy et al. (2004) report a typical proportion of 40%–45% clay (<2 μm), 40%–55% silt (2–63 μm), and 5%–15% sand (>63 μm) within the back-barrier salt marsh at Skallingen, Denmark. In the high macrotidal salt marshes of the Cumberland Basin, Bay of Fundy, more than 95% of the sediments are coarse silt with 2.5% clay and only 1.5% sand (Davidson-Arnott et al., 2002). The suspended material within an instrumented marsh creek here had a median grain size of 36 μm (van Proosdij et al., 2006). Most salt marsh deposits have a detectable sand fraction and this is typically associated with discrete higher-energy tidal or storm surge events (Ehlers et al., 1993; de Groot et al., 2011b) or more diffuse aeolian input from adjacent sand flats or dune systems (de Groot et al., 2011a; Rodriguez et al., 2013).

In muddy environments, the cohesive nature of the suspended material means that the effective grain size (and therefore settling behavior) is strongly influenced by various processes that drive the formation and breakup of composite particles within the water column (Eisma, 1986; Manning and Dyer, 2007). Flocculation occurs under the influence of salinity and clay mineralogy (Gibbs, 1985) and leads to the formation of floc (Fig. 14.5) that are larger and have significantly higher settling velocities compared with the constitute mineral particles. The process is influenced by suspended sediment concentration and settling velocity is often represented as a function of concentration in models of salt marsh deposition (e.g., Temmerman et al., 2003). However, flocculation dynamics are also strongly influenced by turbulent shear within the flow. Measurements from tidal creek in South Carolina, USA, by Voulgaris and Meyers (2004) show considerable differences in the characteristics of suspended sediment between neap and spring tides. On neap tides, the mean floc size was around 50 μm, with an implied settling velocity of about 0.1 mm s^{-1}. On higher spring tides, the mean floc size is up to 150 μm, with settling velocity as high as 1 mm s^{-1}. This system is microtidal (mean tidal range about 1.4 m) and the tidal current velocities are quite low. In situ observations by Benson and French (2007) from a narrow salt marsh–fringed estuarine channel (tidal range approximately 1.9 m) with stronger flood and ebb tidal currents (up to 1.0 and 1.2 m s^{-1}, respectively) show a wider floc size range both within and between tidal cycles (Fig. 14.6). There are very few comparable studies of flocculation effects within shallower overmarsh flows. However, Wang et al. (2010) present data for a *Spartina alterniflora* marsh in Luoyuan Bay, Fujian Province, China, where the mean tidal range is 5 m. Peak current velocities decline from about 0.17 m s^{-1} at the marsh edge to 0.06 m s^{-1} in the marsh interior. The mean grain size of the primary mineral grains varies from 7.0 to 9.6 μm, while

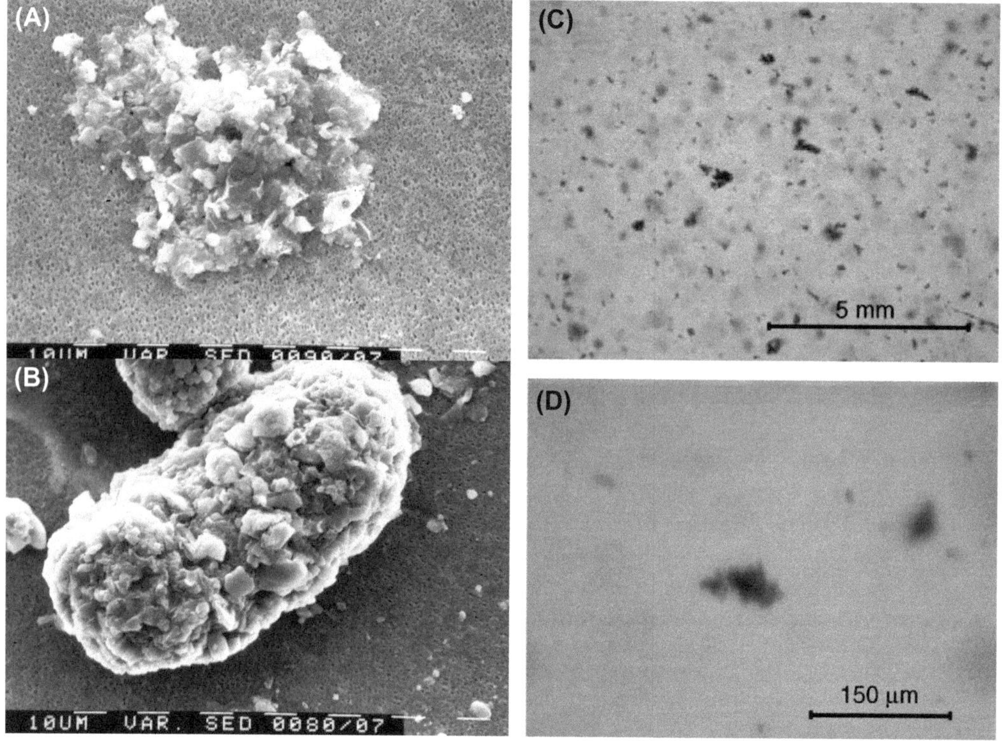

FIGURE 14.5 (A and B) Scanning electron microscope image of composite particles and a fecal pellet from a macrotidal salt marsh (Norfolk, E England); (C and D) in situ optical images of suspended sediment from a salt marsh–fringed estuarine channel (Suffolk, E England). *Reproduced from Benson, T., French, J.R., 2007. InSiPID: a new low cost instrument for in situ particle size measurements in estuaries. Journal of Sea Research 58, 167–188.*

flocculation sizes are 30–69 μm and settling velocities 0.17–0.32 mm s^{-1}. Again, current velocity and the suspended sediment concentration are the main factors determining the flocculation process and the settling velocity.

In addition to physiochemical processes of flocculation, various bioprocessing mechanisms are also important. Most of the observations of flocculation within the salt marsh environment actually relates to true electrochemical flocs in combination with a variety of particles composites that involves a mixture of inorganic and organic material. A significant proportion of estuarine intertidal fine sediments are processed by filter-feeding organisms into fecal pellets, which may be more or less compacted, and are abundant within the salt marsh environment (e.g., French et al., 1993, Fig. 14.5B). Of particular importance within muddy intertidal areas is the effect of algal mats and thin surface films produced by diatoms and bacterial action (Tolhurst et al., 2002, 2003) in stabilizing the sediment surface and increasing the critical shear stress for erosion (Widdows et al., 2000). In contrast, various bioturbating organisms are implicated in reduced stability of cohesive marsh sediments. Schultz et al. (2016), for example, highlight the role played by the Purple Marsh Crab (*Sesarma reticulatum*) as a factor that combines with submergence to exacerbate the

FIGURE 14.6 Time variation in suspended size distribution within a salt marsh–fringed estuarine channel (Suffolk, E England) for neap and spring tidal conditions obtained from in situ optical imaging. *Modified from Benson, T., French, J.R., 2007. InSiPID: a new low cost instrument for in situ particle size measurements in estuaries. Journal of Sea Research 58, 167–188.*

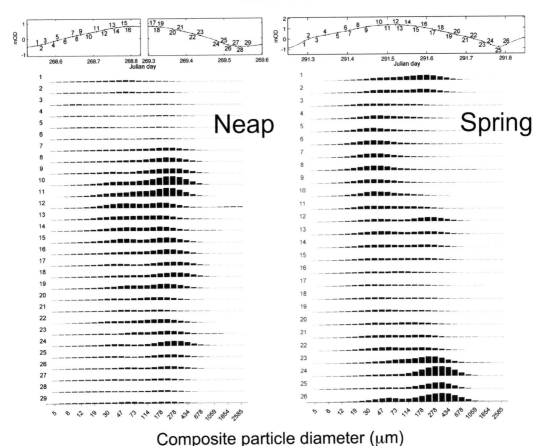

degradation and loss of tidal salt marshes in Long Island Sound, USA. As Escapa et al. (2008) have shown, however, the interplay between sediment dynamics and biota is extremely complex and the same organism can act to enhance deposition at certain times and locations and promote erosion elsewhere.

3.2 Sedimentary Structures

In addition to a general tendency toward landwards fining, salt marsh deposits also tend to exhibit an upward fining (Reineck and Gerdes, 1996; Yang et al., 2008). Like most generalizations, this has its exceptions: Allen (1996) presents a conceptual model that predicts an upward coarsening in some situations when a salt marsh is actively retreating (and therefore a given location becomes progressively closer to the marsh edge and the local sediment source).

In general, however, an upward fining is consistent with a conceptual model in which the frequency of tidal inundation, competence of overmarsh currents to transport sediment, and also the rate of deposition all decline as the marsh platform builds vertically in the tidal frame (Pethick, 1981; Krone, 1987; Allen, 1990; French, 1993; see also Section 4). Upward fining is often accompanied by thin near-parallel laminations that are perhaps the most prominent sedimentary structure within salt marsh deposits (Fig. 14.7A–C). These have been variously interpreted attributed to variations in sediment size (e.g., Pestrong, 1972), various tidal periodicities (e.g., Allen, 1996), or annual layers associated with algal mats (Redfield, 1972). Laminations are often absent in deposits that lack any obvious physical stratification.

High-energy inundation events associated with storms or tsunamis (Wheeler et al., 1999; Haslett and Bryant, 2007) can also be preserved within salt marsh deposits as layers of coarser sediment that thin away from the sediment source, thus contributing to both vertical and horizontal variations in sedimentology (Ehlers et al., 1993; de Groot et al., 2011b; Tsompanoglou et al., 2011). Wheeler et al. (1999) studied laminated salt marsh deposits from two sandy

FIGURE 14.7 Salt marsh edge exposures showing characteristic laminations: (A) macrotidal marsh in Morecambe Bay, NW England (cliff about 1 m high); (B) high macrotidal marsh in Cumberland Basin, Nova Scotia, Canada, showing less regular laminae and secondary marsh establishment at base of cliff; and (C) microtidal sand-rich marsh in Donegal, NW Ireland, showing transition from sandy tidal flat.

estuaries on the northwest coast of Ireland and found that storm surge frequency and magnitude accounted for about half the variance in deposition over a centennial timescale constrained by ^{210}Pb dating. One of the most detailed studies is that of de Groot et al. (2011b) on the barrier island of Schiermonnikoog and in the Netherlands. They show that a considerable part of a salt marsh can initiate under conditions dynamic enough for the transport of sand, such that earlier storm layers are more extensive. As the marsh builds vertically, subsequent sand layers result from less frequent events and are less spatially extensive. At Schiermonnikoog, most sand layers extend over about 20% of the salt marsh area (Fig. 14.8) but account for less than 10% of the total marsh deposit.

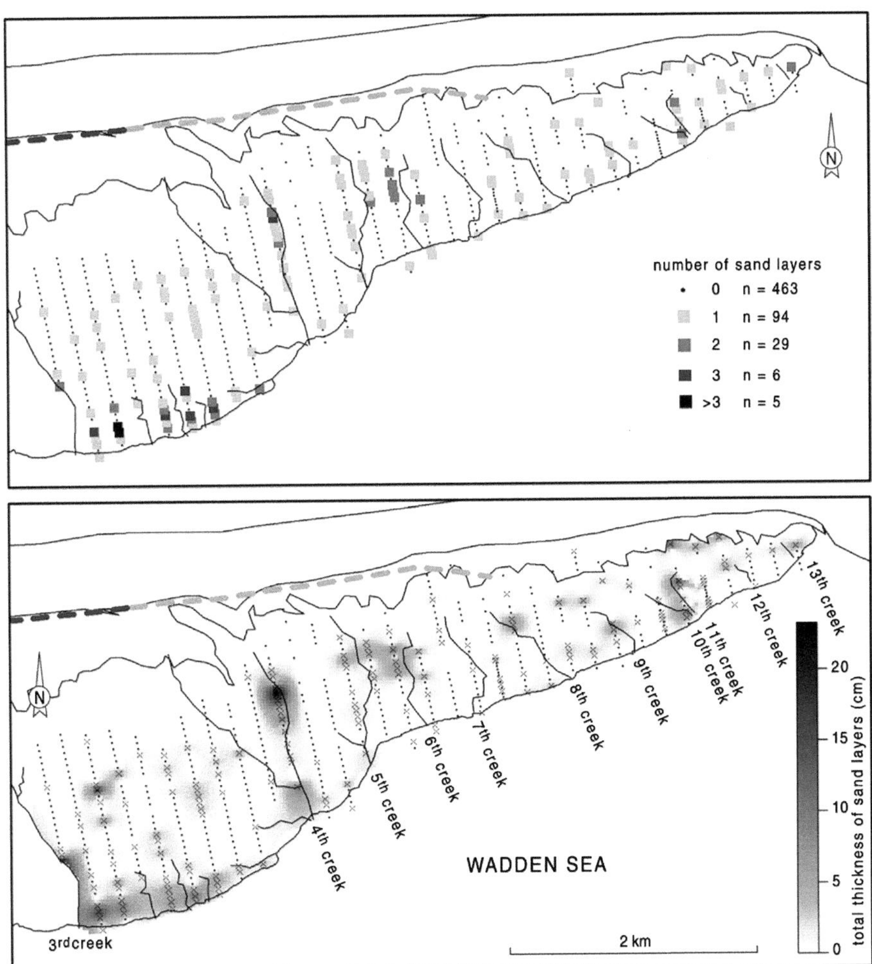

FIGURE 14.8 Sand layers within back-barrier salt marsh at Schiermonnikoog, Dutch Wadden Sea. Upper panel: average number of sand layers; lower panel: thickness of sand layers. *Reproduced from de Groot, A.V., Veeneklaas, R.M., Bakker, J.P., 2011a. Sand in the salt marsh: contribution of high-energy conditions to salt-marsh accretion. Marine Geology 282, 240–254.*

Bioturbation can be locally important over the top few decimeters of the sediment profile. Burrowers include Polychaete worms (Paramor and Hughes, 2004) and crabs (Gribsholt et al., 2003), and plant roots may also blur sediment stratification (Phleger, 1977; Roychoudhury, 2007). To some extent, stratification and biological disturbance are mutually exclusive; Frey and Basan (1985) note that marsh deposits tend to be either well-stratified, with relatively little evidence of burrowing, bioturbation, and root disturbance (see also Bouma, 1963), or else lack stratification on account of more active bioturbation and disturbance by root penetrations.

3.3 Stratigraphy and Facies Models

The classic facies model is that of Bouma (1963), which was developed based on observations in the Dutch Wadden Sea. Bouma's model emphasizes the generally thin nature of the salt marsh deposit, which rarely exceeds a few meters in thickness, the prevalence of wavy to parallel laminae, mud cracks associated with desiccation of high intertidal surfaces, and a general absence of plant and animal fossils. The poor preservation potential for animal fossils in particular has also been noted by Frey and Basan (1985). Despite the broadly similar physical processes of sedimentation within most tidal salt marshes, stratigraphic sequences vary considerably depending on the geological and geographical context (Fig. 14.1). Ward et al. (2008), for example, show that five tidal marshes developed within the Great Bay Estuary (New Hampshire, USA) differ considerably in their stratigraphy because of local variations in the geological context. Estuarine marshes are typically characterized by intercalations of peats and muds that reflect Holocene variability in the rate of sea level rise and the degree of marine or fluvial influence (e.g., Wang et al., 2006). Back-barrier salt marshes often sit on top of much coarser sand sheets, possibly with embedded storm ridges and gravel bodies (e.g., Bartholdy et al., 2004). Dashtgard and Gingras (2005) present a model for a high macrotidal transition from open coast to embayment. This distinguishes between salt marsh deposits comprising the marsh platform and tidal creek deposits, with some associated beach deposits also identified.

4. TIDAL SALT MARSH LANDFORM DEVELOPMENT

4.1 Overview of Landform Components

The salt marsh landform arises through the emergence of an elevated sedimentary platform above some critical elevation at which colonization of the upper tidal flat by halophytic vegetation can occur. As the platform builds vertically within the tidal frame, its gradient tends to decline relative to that of the original tidal flat and may be nearly horizontal in old established systems. Morphological diversity comes from differences in the nature of the salt marsh–tidal flat transition. This often takes the form of a gentle ramp, usually with a fairly distinct vegetation boundary (Fig. 14.9A). However, many marshes have a cliffed seaward edge, which may show active retreat, or else be embedded within a subsequent stage of marsh buildup (Fig. 14.9B). The subtle topography of the marsh platform often includes low levee features along channel margins (Fig. 14.10A) and many marshes are

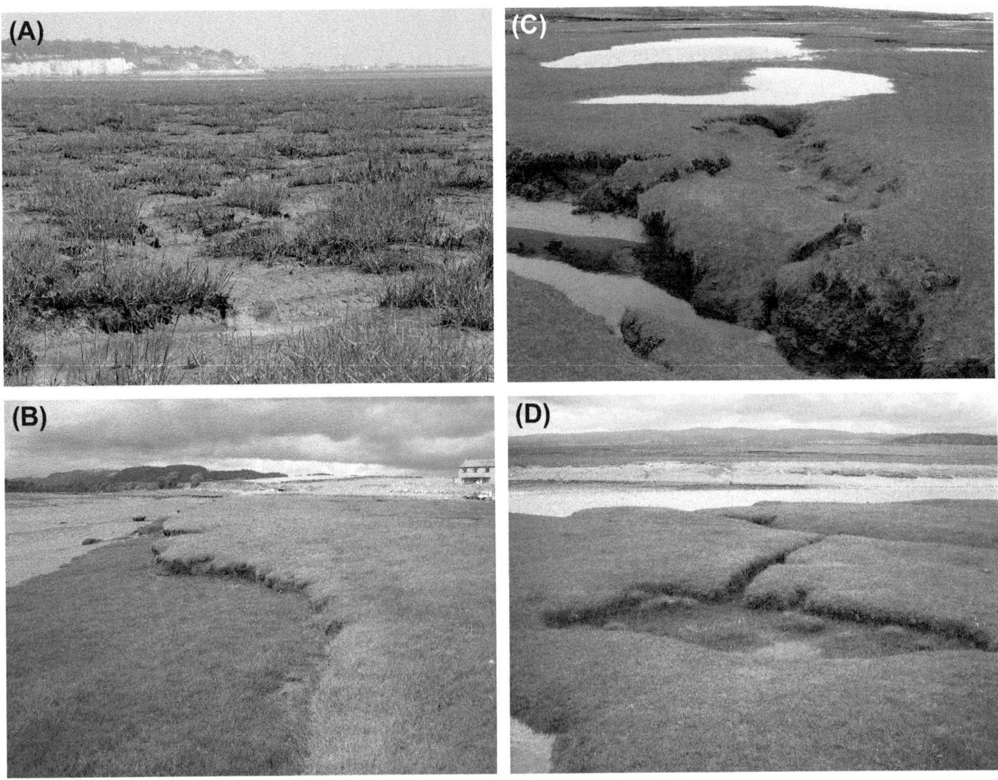

FIGURE 14.9 (A) Ramped transition from tidal flat to salt marsh, Kent, SE England; (B) cliffed tidal flat transition, with secondary marsh formation, Morecambe Bay, NW England; (C) salt pans, Donegal, NW Ireland; (D) channel-connected salt pan, Morecambe Bay, NW England. *(A) Photo: Dr. H. Burningham.*

characterized by numerous shallow depressions (or "pans"), which may or may not connect to the channel system (Fig. 14.9C and D).

From a systems perspective, the elevation of the marsh platform is a key state variable in that it determines, and is in turn determined by, the rate of sedimentation and also exerts a major control on the depth and frequency of tidal inundation and on the spatial distribution of the marsh vegetation. Measurement and modeling of the rate and pattern of sedimentation and its translation into elevation change thus account for a large proportion of geomorphological studies of tidal salt marshes. However, considerable attention has also been devoted to the morphology and evolution of the channel networks, often referred to as "creeks" (or "sloughs" in the United States), which are usually incised into the marsh platform (Fig. 14.10). These have attracted scientific interest on account of their similarities and differences to fluvial drainage networks. Channels are also important as conduits for the exchange of water, sediment, and nutrients between salt marshes and adjacent tidal waters. This means that an understanding of tidal channel formation and morphodynamics is important to inform marsh creation or restoration schemes.

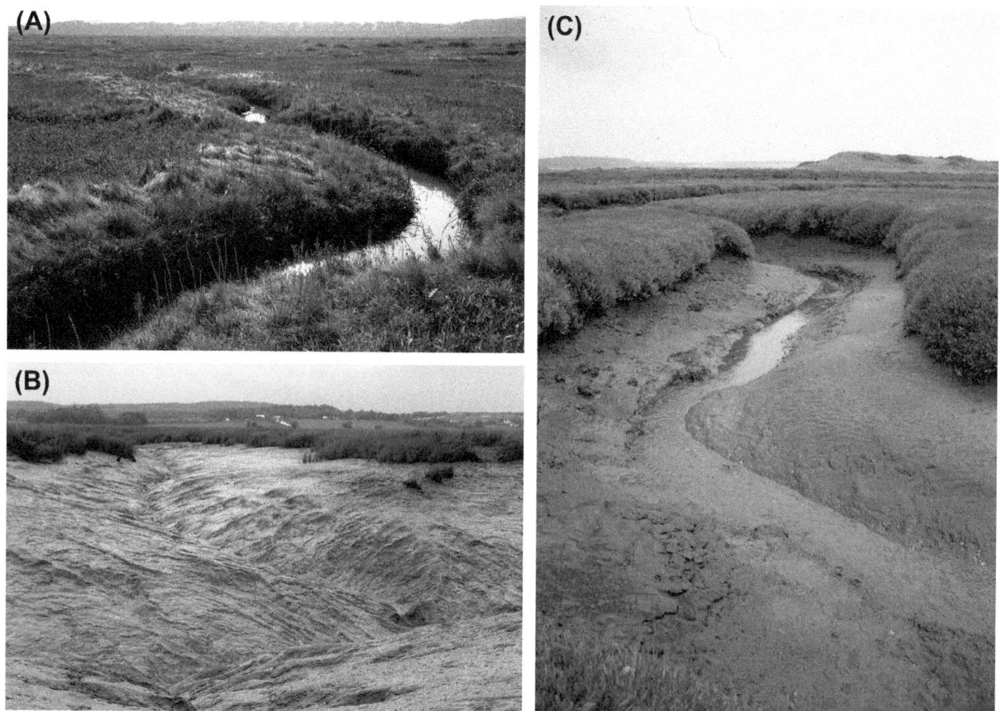

FIGURE 14.10 (A) Tidal creek in a "mature" salt marsh with channel-side levee topography reflected in vegetation zonation, Norfolk, E England; (B) high macrotidal marsh creeks with characteristic vee-shaped profile, Cumberland Basin, Nova Scotia, Canada; (C) salt marsh channel incised into underlying back-barrier sand sheet, Norfolk, E England.

4.2 Sedimentation on the Salt Marsh Platform

Early studies of tidal marshes focused on the marsh platform as evidence of the "land-building" capabilities of halophytic vegetation (Ganong, 1903; Yapp et al., 1917). Sedimentation, via its effect on elevation, was also seen as a fundamental driver of vertical and horizontal zonations of the various plant species (e.g., Johnson and York, 1915). From a geomorphological perspective, it is interesting to note that some of the earliest direct measurements of landform change were made on salt marshes. In Denmark, Nielsen applied marker layers of colored sand within the salt marsh at Skallingen in 1931 and these provided a reference horizon for the determination of subsequent deposition of marsh mud (Nielsen, 1935; Nielsen and Nielsen, 2002; see also Bartholdy et al., 2004). In the United Kingdom, Steers (1935, 1946) undertook similar work in the macrotidal back-barrier salt marshes at Scolt Head Island on the North Norfolk coast. This work influenced the parallel development of ideas in salt marsh successional ecology, especially the work of Chapman (1938, 1976). These early studies quickly established the essential morphodynamic feedback between sedimentation, the elevation of the marsh surface, and the frequency and depth of tidal inundation.

Subsequent work has quantified annual rates of sedimentation in a wide variety of tidal marsh settings using a variety of artificial marker horizons (e.g., Guilcher and Berthois, 1957; Harrison and Bloom, 1977; Stoddart et al., 1989; Goodman et al., 2007) or comparative topographic surveys (e.g., Esselink et al., 1998; Suchrow et al., 2012). Rates of sedimentation are typically of the order of 1–10 mm yr^{-1}, with higher rates in younger marshes. A key motivation for the earlier studies was to understand the relative importance of external forcing factors (e.g., tidal regime, storms, variation in sediment supply) and intrinsic morphodynamic feedback within the salt marsh system itself. Later work has inevitably used observations of the rate of sediment accumulation or elevation change to investigate the extent to which surfaces are able to keep pace with sea level rise (e.g., Goodman et al., 2007; Suchrow et al., 2012).

Back-barrier settings that have experienced relatively little direct human impact provide excellent natural laboratories within which to study the processes of salt marsh sedimentation. Scolt Head Island (Fig. 14.11A) is the site of some of the most intensive sedimentation

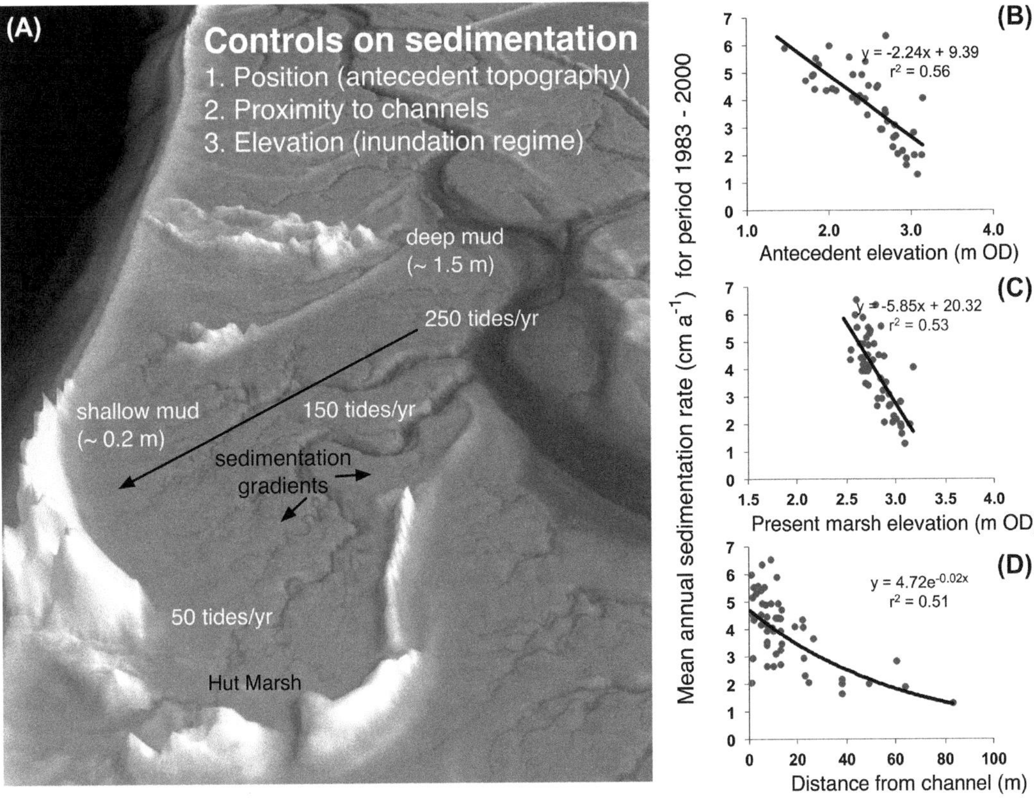

FIGURE 14.11 (A) Back-barrier salt marshes at Scolt Head Island, Norfolk, UK (visualization based on airborne LiDAR data provided by the Environment Agency under an Open Government License), illustrating various controls on the contemporary pattern of sedimentation on Hut Marsh (see also French and Spencer, 1993); (B) 17-year mean annual sedimentation versus underlying tidal flat elevation at Hut Marsh; (C) sedimentation as a function of present marsh platform elevation; (D) sedimentation as a function of proximity to channel system.

studies (Stoddart et al., 1989; French and Spencer, 1993; French et al., 1995; Reed et al., 1999). Here, multiannual sedimentation measurements at a large number of sample locations provide interesting insights into the relative importance of three key drivers of sedimentation: (1) antecedent topography of the tidal flat; (2) contemporary marsh platform elevation; and (3) proximity to tidal channels and/or the seaward edge of the salt marsh. Tidal flat gradients are steeper than those of marsh platform: sedimentary infilling within the tidal frame is thus a leveling processes, in which the antecedent topography continues to influence contemporary rates of sedimentation (Fig. 14.11B). This reflects topographic control on the equilibrium relationship between hydroperiod and sedimentation (Fig. 14.11C). The narrower range of elevations on the modern marsh surface still experiences very different frequencies and cumulative depths of inundation due to tides and surges. The morphodynamic feedback linking elevation, inundation, and sedimentation is thus spatially distributed and influenced by both inherited and acquired topographies. A third control results from the interaction of marsh platform and tidal creek hydrodynamics and is responsible for significant local variation in sedimentation (Fig. 14.11D). French and Spencer (1993) recorded as much variability in vertical accretion within 25 m of a major channel as over the whole of this salt marsh (see also Reed et al., 1999).

Detailed studies of hydrodynamics and sedimentation along pathways of water movement provide further insight into the fundamental processes that drive the morphological development of the marsh platform. Neumeier and Amos (2006) have carried out in situ measurements of turbulence and flow velocities within a sample of Portuguese and English salt marshes. These show that in emergent *Spartina* and short *Salicornia/Suaeda* vegetation canopies, the position of the maximum velocity gradient is shifted upward compared with a boundary layer over bare sediment. Turbulence is also attenuated near the bed, an effect which is greater in fully submerged *Spartina* canopies and which can be expected to enhance deposition and protect the bed against subsequent erosion. In contrast to the widespread reliance of accretion markers for longer-term studies, short-term sedimentation studies generally use artificial sediment traps to record the vertical mass flux between the water column and the marsh surface (e.g., French et al., 1995; Leonard, 1997; Marion et al., 2009). This is advantageous for sediment budget studies (e.g., van Proosdij et al., 2006), although it does complicate the translation of measured sedimentation into equivalent rates of elevation change (French, 2006). Short-term measurements of both suspended sediment concentration and mass deposition have demonstrated progressive sedimentation along inundation pathways within range of tidally influenced marshes (Leonard et al., 1995; Reed et al., 1999), as well as some dependence of the deposition rate on the local source concentration (e.g., Leonard, 1997) and the vegetation characteristics (e.g., Li and Yang, 2009). Within tidal, brackish, and freshwater marshes in the Elbe Estuary, Germany, sediment trap measurements by Butzeck et al. (2015) showed decreasing sediment deposition rates with increasing distances from the estuarine sediment source. However, the background suspended sediment concentration accounted for more of the overall variance in a multivariate statistical model.

A partial analogy can be drawn between tidal marsh platforms and river flood plains because both environments tend to show a decline in both sedimentation rate and grain size away from a major channel (Allen, 1992). In the case of river flood plains, however, inundation events tend to be longer in duration and diffusive processes tend to dominate the transfer of sediment away from main river channel (e.g., Pizzuto, 1987). Salt marshes, in contrast, experience more rapid (and more frequent) inundation events in which advective transfer by currents flowing over the marsh tend to be important. Even so, current velocity diminishes away from the creeks and marsh edge and rapid settling occurs on account of the predominance of flocculations or other particle aggregates within the water column (Wang et al., 2010). An exception to the general decline in sedimentation away from the marsh edge has been documented in high macrotidal marshes in the Cumberland Basin, Nova Scotia. Here, van Proosdij et al. (2006) show that sedimentation rates are lower along the marsh edge and attribute this to higher inundation depths that allow waves to propagate further into the marsh than is the case at lower tidal ranges.

Very few studies have attempted to quantify the efficiency of this sink. French (2006) estimated this based on a simple box model (MARSH-0D) and showed that the efficiency of the vegetated marsh platform in sequestering sediment can be over 90% for microtidal systems but decreases with tidal range and also sediment settling velocity to about 50%–70% for the tidal ranges modeled (Fig. 14.12). These findings are in broad agreement with the limited observational data. In a more recent sediment trap study carried out in a microtidal Delaware salt marsh, Moskalski and Sommerfield (2010) found that trapping efficiency (defined as the ratio of measured deposition and the amount of sediment available

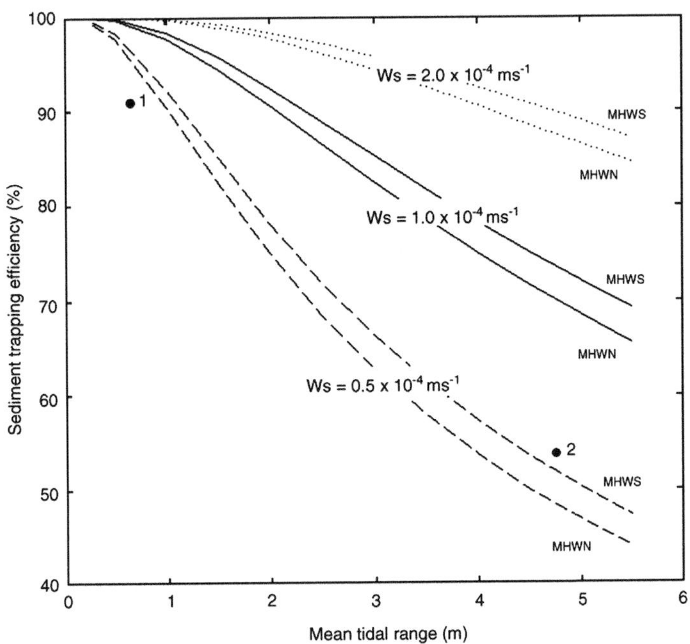

FIGURE 14.12 Modeled sediment trapping efficiency (percentage of sediment introduced on flood tide retained on the marsh surface) for marshes at Mean High Water Neap (MHWN) and Mean High Water Spring (MHWS) tides as a function of mean tidal range and particle settling velocity, w_s. Equivalent field data points for (1) Holland Glade Marsh, Delaware, USA (Stumpf, 1983) and (2) Hut Marsh, Norfolk, England, are also plotted. *Reproduced from French, J.R., 2006. Tidal marsh sediment trapping efficiency and resilience to environmental change: exploratory modelling of tidal, sea-level and sediment supply forcing in predominantly allochthonous systems. Marine Geology 235, 119–136.*

for deposition within the water column) declined with distance from the source channel. They attributed this to exhaustion of the sediment supply as the larger particle aggregates settled, leaving only small particles with low settling velocities within the marsh interior.

Shallow, water-filled depressions, or "salt pans," are common features of vegetated marsh platforms. An early hypothesis for their formation was proposed by Yapp et al. (1917), who distinguished between primary pans (more circular and generally flat bottomed) and channel pans (elongated, sinuous, and even branching). Primary pans were attributed to the uneven colonization of the tidal flat. Channel pans form during the morphological development of the marsh through blockage and abandonment of channel reaches. The interpretation of primary pans was challenged by Pethick (1974), who noted a relationship between pan density and platform elevation (and therefore marsh age). This implies that these features may continue to form as the marsh builds. Pethick suggested that accumulations of rafted plant debris might lead to localized vegetation dieback and pool formation, although impeded drainage, possibly exacerbated by sea level rise, has also been implicated (Warren and Niering, 1993; Millette et al., 2010). Faunal activity may also be important in maintaining and enlarging surface depressions. In Argentina, Perillo and Iribarne (2003) and Escapa et al. (2015) have demonstrated the role of burrowing crabs in the formation of pans in high *Saccocornia*-dominated marshes. Crabs invade *Saccocornia* hummocks, causing these to evolve into enlarged circular rings with a central depression with a high density of crab burrows. Seemingly minor aspects of salt marsh platform geomorphology such as this have been shown to be potentially important in influencing the stability of marsh platform elevations in response to sea level rise (Kirwan et al., 2008; Mariotti, 2016; see also Section 5). Wilson et al. (2014) present evidence from salt marshes in Massachusetts, which suggests that a recent increase in salt pan area coincides with changes in drainage density following reversion from ditched to more natural drainage conditions. Both salt pan and channel dynamics were found to be cyclic in that salt marsh elevation may be lowered because of degradation of organic matter and formation of a pool, with recovery following with channel incision into the pan and restoration of tidal exchange.

4.3 Tidal Channel Systems

Channels are virtually ubiquitous in natural tidal salt marshes. They occasionally occur as individual channels but more commonly as networks of varying pattern and complexity (Fig. 14.13). Geomorphological interest has focused on their formation and the subsequent evolution of their cross-sectional, planform, and network characteristics. Channel systems provide important conduits for the exchange of water and materials (both particulate and dissolved) between salt marshes and adjacent coastal and estuarine environments and have thus been the focus of numerous sediment flux studies (Stevenson et al., 1988; French and Stoddart, 1992; Ganju et al., 2013). The channelized nature of these exchanges has implications for the spatial pattern of sedimentation and the morphological development of the marsh as a whole. A further strand of work has examined this at a network level and sought to formulate morphometric indices and mechanistic models that provide insights into the emergence of organized landscapes through the interplay of hydrodynamic, sedimentary, and biological processes (e.g., D'Alpaos et al., 2012).

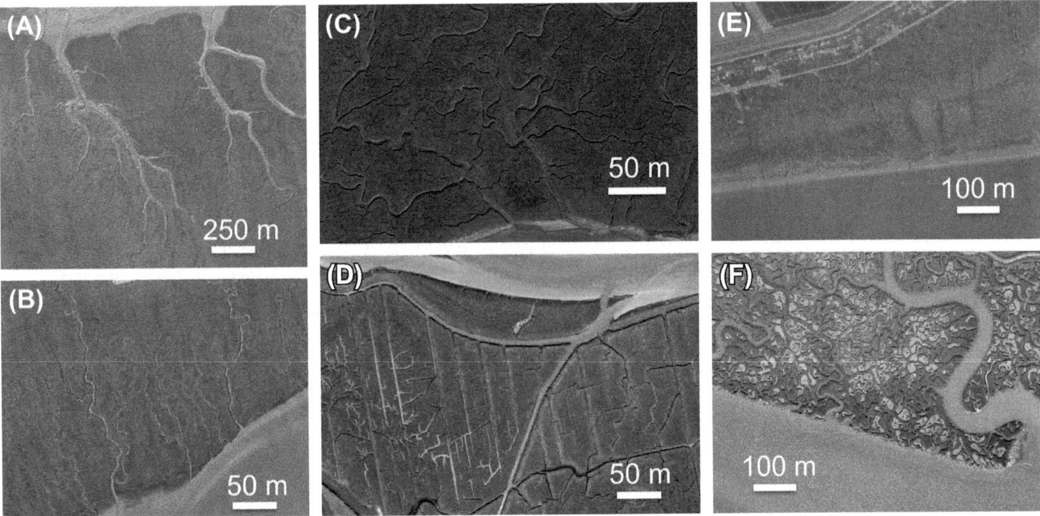

FIGURE 14.13 Tidal channel networks of varying pattern and complexity: (A) dendritic marshes, Ribble Estuary, NW England; (B) linear dendritic network at China Camp, San Francisco Bay, USA; (C) meandering dendritic network, Norfolk, E England; (D) superimposed network, Essex, SE England; (E) linear channels developed on estuarine fringing marshes, Ouse Estuary, Lincolnshire, E England; (F) complex degradational network, Blackwater Estuary, Essex, SE England.

There is some evidence that the principal salt marsh channel alignments are inherited from tidal flat channels and that drainage density increases as the marsh platform builds vertically (Redfield, 1972; Shi et al., 1995). Salt marsh channels appear to be less mobile and also more sinuous than their tidal flat counterparts (Ashley and Zeff, 1988). The generally cohesive nature of salt marsh sediments allows the maintenance of narrow and deep sections with near vertical bank sections. These microcliffs often give the impression of active lateral migration, although measured rates of bank retreat are generally low (Gabet, 1998). Within the salt marsh, channels are typically incised into the underlying tidal flat and may be wholly intertidal (e.g., French and Spencer, 1992) or else partly subtidal. Channel widths usually decline sharply landward, which is a characteristic of tidal channels in general (Hughes, 2012).

As Allen (2000) notes, the network type reflects the combined influence of natural factors such as the slope of the antecedent tidal flat, sediment texture, and the developmental stage of the marsh, as well as human interference. The more intricate dendritic networks (Fig. 14.13A–C) are often associated with more "mature" marshes, although simple linear channels can be found on well-established estuarine fringing marshes, which tend to be of limited area (Fig. 14.13E). The complex type of networks is especially characteristic of certain locations in south east England, where erosional degradation is widespread (Fig. 14.13F; van der Wal and Pye, 2004). Here, tidal channel widening has been accompanied by the development of shallow basins at the extremities of the low-order channels. Superimposed networks (Fig. 14.13D) bear the imprint of past efforts to improve drainage by ditching or enhance siltation prior to reclamation. It is by no means clear that superficially similar networks in different parts of the world have common origins and the diversity in channel system characteristics, both at the network scale and in detail, makes it hard to generalize.

Despite the relative stability of channel planform alignments, it is clear that the network evolves significantly as the marsh platform builds. As marsh surface elevation increases, the effective gradient at the channel headwaters increases, and this drives network extension through headwater erosion, although this has also been viewed as a response to sea level rise (e.g., Hughes et al., 2009; Hughes, 2012). Allen (1997) has also emphasized the role of tidal prism changes, which are accommodated by rationalization of the network and enlargement or infilling of individual sections. In most marshes, the total tidal prism at high spring tides is dominated by the water overlying the marsh surface, rather than the comparatively small volume contributed by the deeper, but much less extensive channels. However, because much of this flow will be routed via the major channels (French and Stoddart, 1992), the tidal prism provides a crude surrogate for the discharge that they must convey (see also Friedrichs, 1995).

In established marshes, channelization of a major proportion of the tidal exchange leads to spatial contrasts in flow intensity such that suspended sediment is more readily transported within the channels but readily deposited once it is transported over bank and across the marsh surface. This explains why channel-related processes appear to cause much of the spatial variation in sedimentation on the marsh surface (Reed et al., 1999). Furthermore, the interaction of shallow water tides with marshes comprised of deep channels and shallow surfaces generates highly distinctive flow regimes. Well-defined flow peaks are typically observed during both the flood and ebb, around the time at which inundation or drying of the marsh surface occurs (Pestrong, 1965; Bayliss-Smith et al., 1979; Dankers et al., 1984). The mechanics of these transients have been investigated by French and Stoddart (1992), who argued that the marsh surface acts as a topographic threshold separating two distinctive flow regimes: a flood-dominated, "belowmarsh" regime and an ebb-dominated, "overmarsh" regime (Fig. 14.14). The former is associated with neap tides and sediment deposition within the channel system; the latter is associated with spring tides, scouring of the channel system and the introduction of suspended sediments onto the marsh surface. The flood transient occurs just above channel bankfull, due to the sudden increase in tidal prism at a time when overall tidal exchange is still largely routed via the channel system. High turbulent stresses generated at this time (French and Clifford, 1992) appear to be important in maintaining the suspension of sediment within the upper part of the water column such that a proportion of this material can be advected laterally over the adjacent marsh surfaces, where it tends to deposit within a very short distance (French et al., 1993; Leonard et al., 1995; Reed et al., 1999). On the ebb, maximum channel velocities and stresses occur just below bankfull, associated with steep water surface slopes generated as water retained on the vegetated marsh provides a delayed inflow to the channels. These ebb flows are often (though not always) greater than the flood maxima and are clearly important in scouring the channel system, as well as in extending it through erosion in headwater regions.

It should be noted that salt marshes with strongly ebb-dominated channel systems are usually highly efficient sediment sinks because a large proportion of the sediment introduced on the flood is retained on the surface (Dankers et al., 1984; French and Stoddart, 1992; French, 2006). Ebb dominance within a marsh does not, therefore, carry the same connotation as it

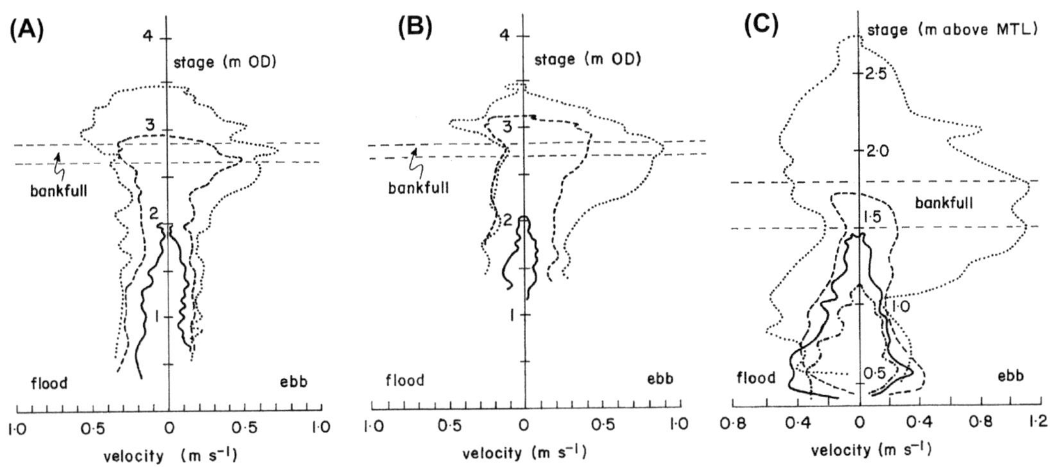

FIGURE 14.14 Representative tidal current velocity versus water level plots for belowmarsh, bankfull, and overmarsh tides at (A) and (B) mid-network and creek mouth locations in Hut Creek, Scolt Head Island, Norfolk, England (French and Stoddart, 1992); (C) Ems-Dollard marshes, Dutch–German border (Dankers et al., 1984). *Simplified from Allen, J.R.L., 2000. Morphodynamics of Holocene saltmarshes: a review sketch from the Atlantic and southern north sea coasts of Europe. Quaternary Science Reviews 19, 1155–1231.*

does in a wider estuarine or embayment context, where it may be associated with a tendency for sediment export. Flood tidal dominance occurs in some salt marshes (e.g., Leonard et al., 1995), where it is possibly a result of lower tidal range or diurnal rather than semidiurnal tides (making the momentum contrasts between channel and surface less marked) or because strongly flood-dominant characteristics of the broader coastal setting override any internal form–process interaction (Friedrichs and Aubrey, 1988).

The hydrodynamic and sedimentary consequences of channel–surface interactions in salt marshes are only partly dependent on vegetation characteristics. Extremely steep sedimentation gradients relative to channel margins have been observed within uniform vegetation (French and Spencer, 1993; see also above), implying that channel surface hydraulic gradients are more important than local variation in plant-induced roughness. In addition, gross hydrodynamic characteristics (i.e., flood and ebb velocity and stress transients, and the resulting tidal asymmetry), while modified by the effects of vegetation-induced roughness, are not necessarily dependent on them (French and Stoddart, 1992; see also Dronkers, 1986; Friedrichs, 1995). However, numerical hydrodynamic model experiments by Temmerman et al. (2005) showed vegetation structure to be a key determinant for both flow routing and the pattern of sedimentation, especially when the vegetation is emergent. The contrast in hydraulic roughness between vegetated and unvegetated surfaces concentrates flow into unvegetated areas, with water movements into the salt marsh being more or less perpendicular to the vegetation edge. Rapid settling occurs within the vegetation. More homogeneous patterns of deposition occur when the vegetation is fully submerged and a simpler sheet flow occurs on or off the marsh. This conclusion is further supported by the later modeling work of Vandenbruwaene et al. (2015).

4.4 Morphodynamics of the Salt Marsh—Tidal Flat Transition

Although the pioneer zone for marsh halophytes can sometimes have very little topographic expression, the transition between tidal flat and salt marshes is very often marked by a microcliff (Fig. 14.9; van Eerdt, 1985; Allen, 1989; van der Waal et al., 2008). By definition, these are erosional and the seaward boundary of the salt marsh can sometimes be quite dynamic at decadal timescales. This has attracted interest on account of the potentially high rates of marsh area loss (e.g., Harmsworth and Long, 1986; de Jong et al., 1994; Pringle, 1995), although erosion of the marsh edge can also provide a local source of sediment that can be reworked to contribute to the adjustment of the marsh platform to rising sea level (e.g., Reed, 1988). Seaward progradation is also possible, and Pedersen and Bartholdy (2007) present a conceptual model for the incremental progradation, erosion, and extended progradation of exposed open coastal salt marsh in the Juvre Dyb area of the Danish Wadden Sea.

The various mechanisms and modes of marsh cliff retreat are also of interest from a geotechnical perspective and have been the subject of thorough treatments by Allen (1989) and Francalanci et al. (2013). It is useful to distinguish between small-scale surface erosion (chiefly under the influence of waves) and larger scale discrete mass failures. The latter include rotational slips, cantilever failures, toppling, and other failure types that depend largely on the configuration and composition of the bank. Toppling failures are especially common in cohesive salt marsh cliffs (Allen, 1989) and are often triggered by the formation of tension cracks, as well as basal scour by waves (Francalanci et al., 2013).

Sequential phases of marsh growth, erosional retreat, and marsh reestablishment occur at many locations (e.g., Figs. 14.7B and 14.9B) and were noted in some of the earliest salt marsh studies (e.g., Yapp et al., 1917). While some salt marshes clearly respond to wider coastal or estuarine system dynamics (e.g., the shifting of tidal channels in Morecambe Bay, northwest England; Pringle, 1995), changes elsewhere appear to reflect an intrinsic dynamic of the salt marsh geomorphic system rather than any systemic variability in extrinsic factors such as sea level, wave climate, or sediment supply. Chauhan (2009) presents an example of such "autocyclic" behavior from the Solway Firth, also in northwest England, where the salt marsh profile is characterized by a set of embedded relict clifflets. This kind of self-organizing behavior is a feature of many geomorphic systems, although, as van de Koppel et al. (2005) note, it generally promotes stability, whereas in the case of salt marsh morphological development, it leads to instability and erosion. In their conceptual model, self-organization leads to the steepening of the marsh front, the focusing of wave energy, and a cascade of erosional cliff formation and vegetation collapse. The key point here is that retreat of salt marsh cliffs, which is widely observed in mature salt marshes, very often arises due to intrinsic landform morphodynamics rather than any external forcing.

5. SALT MARSH MORPHODYNAMICS AND RESILIENCE TO SEA LEVEL RISE

It is evident from the preceding discussion that the geomorphology of tidal salt marshes reflects a complex web of interactions between hydrodynamic, sedimentary, and biological processes and the morphodynamic feedback that mediate these at the landform scale. Some of this complexity can be captured by relatively simple zero-dimensional "box" models that simulate the evolution of a small set of system state variables (Fig. 14.15). These models generally focus on the adjustment of marsh platform elevation and the nature of its equilibrium with relative sea level rise (Rybzyck and Callaway, 2018). The strong negative feedback loop that involves platform elevation (taken as an average for the marsh as a whole in a zero-dimensional model), the frequency and magnitude of inundation, and the rate of sedimentation, is of particular importance, as it favors an asymptotic evolution toward a platform elevation at which the rate of sedimentation is just sufficient to balance the effect of sea level rise (Allen, 1990; French, 1993). Under conditions of stable sea level, the additional contribution of plant productivity to the marsh deposit means that emergence and transition to a terrestrial habitat is possible. Plant productivity also changes the asymptotic equilibrium level even when sea level is not stable (see, for example, Morris et al., 2002 and Marani et al., 2007, 2010).

Vertical (and possibly lateral) accretion of the marsh platform is accompanied by a decline in tidal prism and by adjustment of the tidal channel system as the network extends and then rationalizes (Fig. 14.16A). The morphological adjustment of the tidal channels varies somewhat according to the geomorphological setting but, for back-barrier environments, accretion of the marsh platform is accompanied by scouring of the channels and their incision into the underlying tidal flat deposits (Fig. 14.16B).

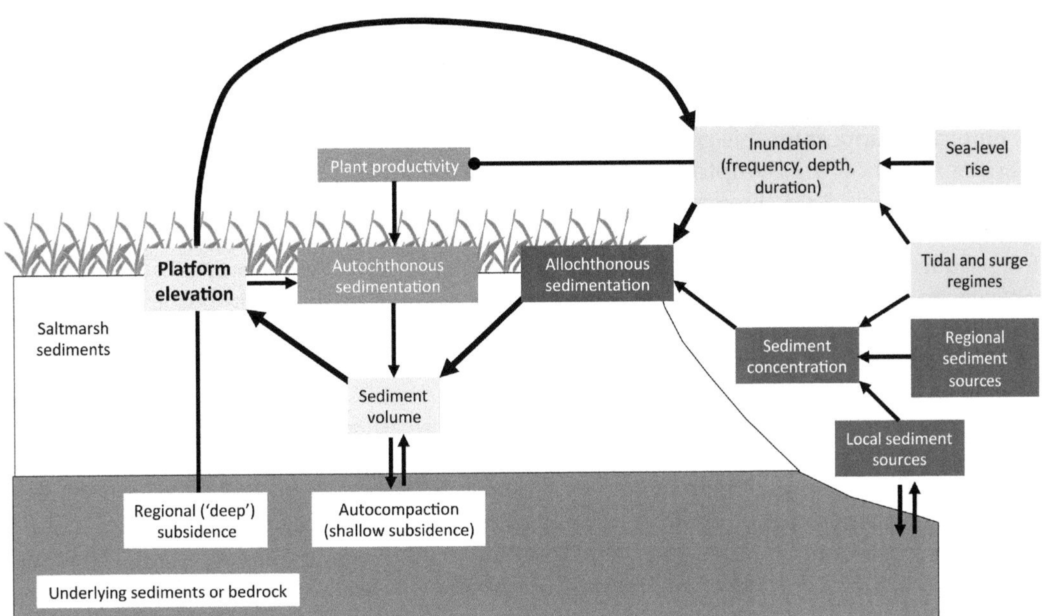

FIGURE 14.15 Schematic representation of principal geomorphic system state variables relevant to understanding the resilience of the tidal salt marsh landform to sea level rise. *Modified from French, J.R., 2006. Tidal marsh sediment trapping efficiency and resilience to environmental change: exploratory modelling of tidal, sea-level and sediment supply forcing in predominantly allochthonous systems. Marine Geology 235, 119–136.*

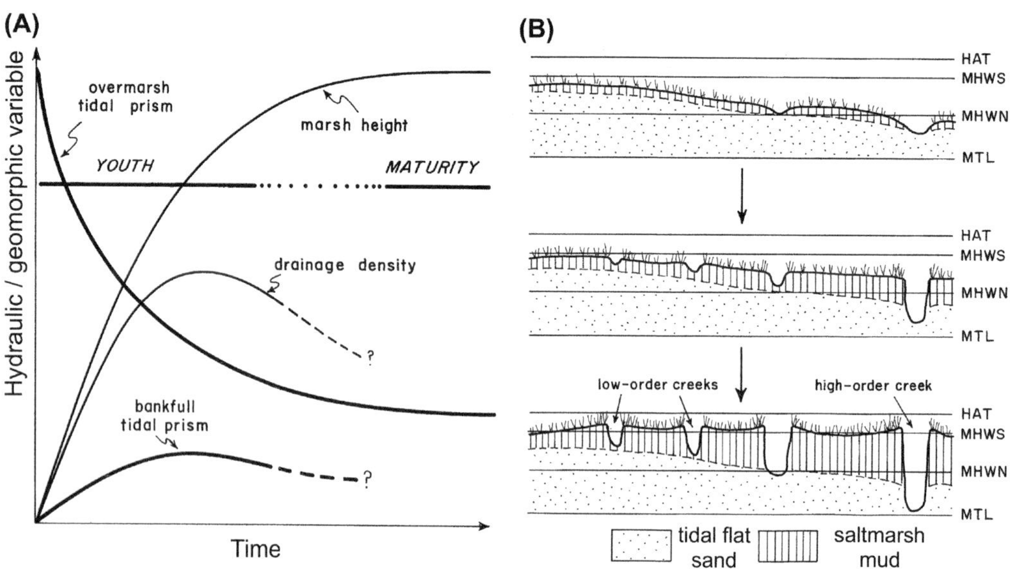

FIGURE 14.16 (A) Illustrative time evolution of salt marsh morphological variables; and (B) conceptual model of coevolution of marsh platform and tidal channel system based on observations from back-barrier salt marshes. *HAT*, Highest Astronomical Tide; *MHWN*, Mean High Water Neap; *MHWS*, Mean High Water Spring; *MTL*, Mean Tide Level. *(A) Modified from Allen, J.R.L., 2000. Morphodynamics of Holocene saltmarshes: a review sketch from the Atlantic and southern north sea coasts of Europe. Quaternary Science Reviews 19, 1155–1231. (B) After French, J.R., Stoddart, D.R., 1992. Hydrodynamics of salt-marsh creek systems – implications for marsh morphological development and material exchange. Earth Surface Processes and Landforms 17, 235–252.*

Empirical support for this general model comes from geographical chronosequences of salt marshes that show an asymptotic increase in marsh platform elevation with increasing marsh age (e.g., Pethick, 1981; French, 1993; Temmerman et al., 2004). Global metaanalyses of the rate of the balance between sedimentation and relative sea level rise also highlight a tendency for sedimentation to evolve toward a dynamic equilibrium with the rate of sea level rise (Stevenson et al., 1986; French, 1994, 2006; Kirwan et al., 2010). An important caveat here is that many studies make the assumption that sedimentation, as measured by the thickness of new sediment deposition above a marker of some kind, translates into an equivalent change in the elevation of the marsh surface. As Cahoon et al. (1995) demonstrate, this assumption is by no means justified, especially in more autochthonous systems where rates of autocompaction (or "shallow subsidence" to distinguish the effect from deeper geological subsidence) may be quite high. Even in more allochthonous systems, autocompaction effects are important at Holocene timescales in estuarine settings characterized by intercalated marine muds freshwater peats and probably continue to increase the accommodation space and therefore drive higher rates of sediment accumulation (Allen, 1999).

Determination of net elevation change as opposed to gross sedimentation is important when observations of marsh platform change are compared with past or likely future sea level rise. Instrumentation such as the sedimentation-erosion table (Cahoon et al., 1995) permits the generation of high-precision datasets on elevation change. Even so, direct comparisons with sea level trends are not always appropriate (Cahoon, 2015). For a start, the time integration of sedimentation or elevation change is typically performed over a few years, whereas sea level rise usually requires several decades of tide gage data to obtain a robust trend. Important variation in tidal amplitude occurs with a periodicity of 18.6 years (French, 2006). Very few direct observations of salt marsh sedimentation achieve this duration (see Bartholdy et al., 2004; Goodman et al., 2007), although useful insights can be gained from sediment core analysis at historical timescales (e.g., Hill and Anisfeld, 2015). Shorter-term comparisons that incorporate on a fraction of this nodal tidal variation are of dubious value, however. More important, the nonlinearity in vertical marsh growth (Fig. 14.16) and the involvement of multiple factors (antecedent morphology, tidal and nontidal flooding, sediment supply, etc.) mean that mechanistic modeling provides the best insights into the nature of present and future marsh platform adjustment within the tidal frame. Most of the later marsh morphodynamic models include explicit treatment of autocompaction as well as a broader suite of biological factors that feedback into landform behavior (e.g., Morris et al., 2002; French, 2006; Morris et al., 2009; Kirwan et al., 2010).

A general consensus from modeling studies is that tidal salt marshes dominated by allochthonous sediments exhibit far greater resilience to sea level rise than their microtidal autochthonous counterparts. The effect of sea level rise on inundation is countered by increased sedimentation and this buffering capability increases with both tidal range (French, 1994, Fig. 14.17A) and sediment supply (French, 2006). Kirwan et al. (2010) use an ensemble

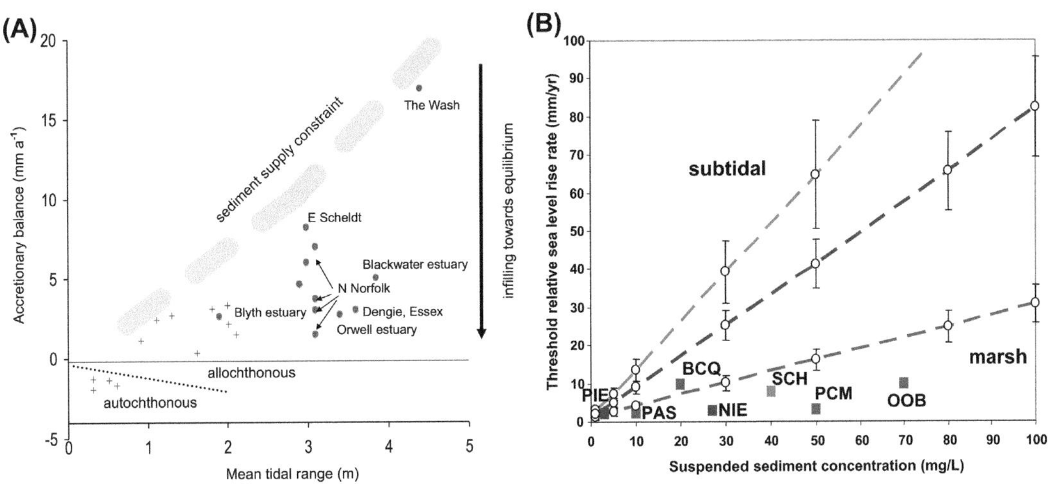

FIGURE 14.17 (A) Conceptualization of salt marsh infilling along a gradient of tidal energy; (B) threshold rates of sea level rise above which salt marshes become replaced by subtidal environments as the stable ecosystem. Each line represents the mean threshold rate (±1 standard error) predicted by five models as a function of suspended sediment concentration and spring tidal range. Pink, blue, and green lines denote thresholds for tidal ranges of 1, 3, and 5 m, respectively. Observational data overlay: PIE, Plum Island Estuary, Massachusetts; PAS, Pamlico Sound, North Carolina; BCQ, Bayou Chitique, Louisiana; NIE, North Inlet Estuary, South Carolina; SCH, Scheldt Estuary, Netherlands; PCM, Phillips Creek Marsh, Virginia; OOB, Old Oyster Bayou, Louisiana. *(A) Modified after French, J.R., 1994. Tide-dominated coastal wetlands and accelerated sea level rise: a Northwestern European perspective. Journal of Coastal Research Special Issue 12, 91–101. Kirwan, M.L., Guntenspergen, G.R., D'Alpaos, A., Morris, J.T., Mudd, S.M., Tmmermann, S., 2010. Limits on the adaptability of coastal marshes to rising sea level. Geophysical Research Letters 37, L23401. https://doi.org/10.1029/2010GL045489.*

of marsh models to define more precisely the portion of the tidal range, sediment supply, and sea level rise parameter space within which salt marsh can persist (Fig. 14.17B).

As already noted, the continuing emphasis on vertical adjustment neglects importance aspects of the marsh landform evolution. Erosion of the marsh edge is known to be an important driver of marsh area changes, and the existence of autocyclical behavior can complicate the interpretation of observed changes and their relationship to wider environmental changes (e.g., van der Koppel et al., 2005). A new generation of spatial models can provide further insights into some of these points. These are reviewed by Fagherazzi et al. (2012), who highlight a spectrum of model complexity that ranges from empirical prediction of sedimentation as a function of topographic variables (Temmerman et al., 2003) to simplified hydrodynamic schemes (e.g., D'Alpaos et al., 2007) or full numerical hydrodynamic treatments (Temmerman et al., 2005).

Acknowledgments

This chapter is dedicated to the memory of Professor David Stoddart, who inducted me into the muddy waters of salt marsh geomorphology some 35 years ago. Miles Irving, of the Drawing Office in the UCL Department of Geography, is thanked for his invaluable assistance with the drafting of several of the figures. Dr. Helene Burningham provided helpful comments on an earlier draft and for the photograph in Fig. 14.9A.

References

Adam, P., 1990. Saltmarsh Ecology. Cambridge University Press, Cambridge, 461p.

Adam, P., 2002. Saltmarshes in a time of change. Environmental Conservation 29, 39–61. https://doi.org/10.1017/S0376892902000048.

Allen, J.R.L., Duffy, M.J., 1998. Medium-term sedimentation in high intertidal mudflats and salt marshes in the Severn Estuary, SW Britain: the role of wind and tide. Marine Geology 150, 1–27.

Allen, J.R.L., Haslett, S.K., 2014. Salt-marsh evolution at Northwick and Austwarths, severn estuary, UK: a case of constrained autocyclicity. Atlantic Geology 50, 1–17.

Allen, J.R.L., 1989. Evolution of salt-marsh cliffs in muddy and sandy systems: a qualitative comparison of British west-coast estuaries. Earth Surface Processes and Landforms 14, 85–92.

Allen, J.R.L., 1990. Salt-marsh growth and stratification: a numerical model with special reference to the Severn Estuary, southwest Britain. Marine Geology 95, 77–96.

Allen, J.R.L., 1992. Large-scale textural patterns and sedimentary processes on salt marshes in the Severn Estuary, southwest Britain. Sedimentary Geology 81, 299–318.

Allen, J.R.L., 1996. Shoreline movement and vertical textural patterns in salt marsh deposits: implications of a simple model for flow and sedimentation over tidal marshes. Proceedings of the Geologists Association 107, 15–23.

Allen, J.R.L., 1997. Simulation models of salt-marsh morphodynamics: some implications for high-intertidal sediment couplets related to sea-level change. Sedimentary Geology 113, 211–223.

Allen, J.R.L., 1999. Geological impacts on coastal wetland landscapes: some general effects of sediment autocompaction in the Holocene of Northwest Europe. The Holocene 9, 1–12.

Allen, J.R.L., 2000. Morphodynamics of Holocene saltmarshes: a review sketch from the Atlantic and southern north sea coasts of Europe. Quaternary Science Reviews 19, 1155–1231.

Allen, J.R.L., 2003. An eclectic morphostratigraphic model for the sedimentary response to Holocene sea-level rise in northwest Europe. Sedimentary Geology 161, 31–54.

Amos, C.L., Tee, K.T., 1989. Suspended sediment transport processes in Cumberland Basin, Bay of Fundy. Journal of Geophysical Research 94C, 14407–14417.

Amos, C., 1995. Siliclastic tidal flats. In: Perillo, G.M.E. (Ed.), Geomorphology and Sedimentology of Estuaries. Advances in Sedimentology, vol. 53. Elsevier, Amsterdam, pp. 273–306.

Ashley, G.M., Zeff, M.L., 1988. Tidal channel classification for a low-mesotidal salt marsh. Marine Geology 82, 17–32.
Atwater, B.F., Conard, S.G., Dowden, J.N., Hedel, C.W., MacDonald, R.L., Savage, W., 1979. History, landforms, and vegetation of the estuary's tidal marshes. In: Conomos, T.J., Leviton, A.E., Berson, M. (Eds.), San Francisco Bay: The Urbanized Estuary, Investigations into the Natural History of San Francisco Bay and Delta with Reference to the Influence of Man. Pacific Division of the American Association for the Advancement of Science, San Francisco, pp. 347–385.
Bakker, J.P., Esselink, P., Dijkema, K.S., Van Duin, W.S., De Jong, D.J., 2002. Restoration of salt marshes in The Netherlands. In: Nienhuis, P.H., Gulati, R.D. (Eds.), Ecological Restoration of Aquatic and Semi-aquatic Ecosystems in the Netherlands (NW Europe). Springer, Dordrecht, pp. 29–51.
Balke, T., Stock, M., Jensen, K., Bouma, T.J., Kleyer, M., 2016. A global analysis of the seaward salt marsh extent: the importance of tidal range. Water Resources Research 52, 3775–3786.
Bartholdy, J., Christiansen, C., Kunzendorf, H., 2004. Long term variations in backbarrier salt marsh deposition on the Skallingen peninsula - the Danish Wadden Sea. Marine Geology 203, 1–21.
Bartholdy, A.T., Bartholdy, J., Kroon, A., 2010. Salt marsh stability and patterns of sedimentation across a backbarrier platform. Marine Geology 278, 31–42.
Bartholdy, J., Bartholdy, A.T., Kim, D., Pedersen, J.B.T., 2014. On autochthonous organic production and its implication for the consolidation of temperate salt marshes. Marine Geology 351, 53–57.
Baumann, R.H., Day, J.W., Miller, C.A., 1984. Mississippi deltaic wetland survival: sedimentation versus submergence. Science 224, 1093–1095.
Bayliss-Smith, T.P., Healey, R., Lailey, R., Spencer, T., Stoddart, D.R., 1979. Tidal flows in salt marsh creeks. Estuarine and Coastal Marine Science 9, 235–255.
Beeftink, W.G., 1966. Vegetation and habitat of the salt marshes and beach plains in the south-western part of The Netherlands. Wentia 15, 83–108.
Benson, T., French, J.R., 2007. InSiPID: a new low cost instrument for in situ particle size measurements in estuaries. Journal of Sea Research 58, 167–188.
Bockelmann, A., Bakker, J.P., Neuhaus, R., Lage, J., 2002. The relation between vegetation zonation, elevation and inundation frequency in a Wadden Sea salt marsh. Aquatic Botany 73, 211–221.
Bortolus, A., Schwindt, E., Bouza, P.J., Idaszkin, Y.L., 2009. A characterization of Patagonian salt marshes. Wetlands 29, 772–780.
Bouma, T.J., van Belzen, J., Balke, T., et al., 2016. Short-term mudflat dynamics drive long-term cyclic saltmarsh dynamics. Limnology and Oceanography 61, 2261–2275.
Bouma, A.H., 1963. A graphic presentation of the facies model of saltmarsh deposits. Sedimentology 2, 122–129.
Boyd, R., Dalrymple, R., Zaitlin, B.A., 1992. Classification of clastic coastal depositional environments. Sedimentary Geology 80, 139–150.
Burningham, H., French, J.R., 2013. Is the NAO winter index a reliable proxy for wind climate and storminess in northwest Europe? International Journal of Climatology 33, 2036–2049.
Butzeck, C., Eschenbach, A., Grongroft, A., Hansen, K., Nolte, S., Jensen, K., 2015. Sediment deposition and accretion rates in tidal marshes are highly variable along estuarine salinity and flooding gradients. Estuaries and Coasts 38, 434–450.
Cahoon, D.R., Reed, D.J., Day, J.W., 1995. Estimating shallow subsidence in microtidal saltmarshes of the southeastern United States: Kaye and Barghoorn revisited. Marine Geology 128, 1–9.
Cahoon, D.R., 2015. Estimating relative sea-level rise and submergence potential at a coastal wetland. Estuaries and Coasts 38, 1077–1084.
Callaghan, D.P., Bouma, T.J., Klaassen, P., van der Wal, D., Stive, M.J.F., Herman, P.M.J., 2010. Hydrodynamic forcing on salt marsh development: distinguishing the relative importance of waves and tidal flows. Estuarine, Coastal and Shelf Science 89, 73–88. https://doi.org/10.1016/j.ecss.2010.05.013.
Chapman, V.J., 1938. Studies in saltmarsh ecology I-III. Journal of Ecology 26, 144–176.
Chapman, V.J., 1974. Salt Marshes and Salt Deserts of the World, second ed. Cramer, Lehere. 392p.
Chapman, V.J., 1976. Mangrove Vegetation. Cramer, Lehere, 447p.
Chauhan, P.P.S., 2009. Autocyclic erosion in tidal marshes. Marine Geology 110, 45–57.
D'Alpaos, A., Lanzoni, S., Marani, M., Rinaldo, A., 2007. Landscape evolution in tidal embayments: modeling the interplay of erosion sedimentation and vegetation dynamics. Journal of Geophysical Research 112, F01008. https://doi.org/10.1029/2006JF000537.

D'Alpaos, A., Da Lio, C., Marani, M., 2012. Biogeomorphology of tidal landforms: physical and biological processes shaping the tidal landscape. Ecohydrology 5, 550–562.

Dalrymple, R.W., Mackay, D.A., Ichaso, A.A., Choi, K.S., 2012. Processes, morphodynamics and facies of tide-dominated estuaries. In: Davis, R.A., Dalrymple, R.W. (Eds.), Principles of Tidal Sedimentology. Springer, New York, pp. 79–107.

Dankers, N., Binsberger, M., Zegers, K., Laane, R., van de Loeff, M.R., 1984. Transport of water, particulate and dissolved organic and inorganic matter between a salt marsh and the Ems-Dollard Estuary, The Netherlands. Estuarine, Coastal and Shelf Science 19, 143–165.

Dashtgard, S.E., Gingras, M.K., 2005. Facies architecture and ichnology of recent salt-marsh deposits: Waterside Marsh, New Brunswick, Canada. Journal of Sedimentary Research 75, 596–607.

Davidson-Arnott, R.G.D., van Proosdij, D., Ollerhead, J., Schostak, L., 2002. Hydrodynamics and sedimentation in salt marshes: examples from a macrotidal marsh, Bay of Fundy. Geomorphology 48, 209–231.

Davies, J.L., 1964. A morphogenic to world shorelines. Zeitschrift für Geomorphology 8, 127–142.

Day, J.W., Kemp, G.P., Reed, D.J., Cahoon, D.R., Boumans, R.M., Suhayda, J.M., Gambrell, R., 2000. Vegetation death and rapid loss of surface elevation in two contrasting Mississippi delta salt marshes: the role of sedimentation, autocompaction and sea-level rise. Ecological Engineering 37, 229–240.

de Groot, A.V., Veeneklaas, R.M., Bakker, J.P., 2011a. Sand in the salt marsh: contribution of high-energy conditions to salt-marsh accretion. Marine Geology 282, 240–254.

de Groot, A.V., Veeneklaas, R.M., Kuijper, D.P.J., Bakker, J.P., 2011b. Spatial patterns in accretion on barrier-island salt marshes. Geomorphology 134, 280–296.

de Jong, D.J., Dejong, Z., Mulder, J.P.M., 1994. Changes in area, geomorphology and sediment nature of salt marshes in the Oosterschelde estuary (SW Netherlands) due to tidal changes. Hydrobiologia 283, 303–316.

Détriché, S., Susperregui, A.S., Feunteun, E., Lefeuvre, J.C., Jigorel, A., 2011. Interannual (1999–2005) morphodynamic evolution of macro-tidal salt marshes in Mont-Saint-Michel Bay (France). Continental Shelf Research 31, 611–630.

Dijkema, K.S., 1987. The geography of salt marshes in Europe. Zeitschrift für Geomorphologie 31, 489–499.

Dronkers, J., 1986. Tidal asymmetry and estuarine morphology. Netherlands Journal of Sea Research 20, 117–131.

Du, J.L., Yang, S.L., Feng, H., 2016. Recent human impacts on the morphological evolution of the Yangtze River delta foreland: a review and new perspectives. Estuarine, Coastal and Shelf Science 181, 160–169.

Dyer, K.R., Moffat, T.J., 1998. Fluxes of suspended matter in the East Anglian plume, southern North Sea. Continental Shelf Research 19, 1311–1331.

Edwards, R.J., Wright, A., Van de Plassche, O., 2004. Surface distributions of salt-marsh foraminifera from Connecticut, USA: modem analogues for high-resolution sea level studies. Marine Micropaleontology 51, 1–21.

Ehlers, J., Nagorny, K., Schmidt, P., Stieve, B., Zietlow, K., 1993. Storm-surge deposits in North-Sea salt marshes dated by Cs-134 and Cs-137 determination. Journal of Coastal Research 9, 698–701.

Eisma, D., Dijkema, K.S., 1997. The influence of salt marsh vegetation on sedimentation. In: Eisma, D. (Ed.), Intertidal Deposits. CRC Press, Boca Raton, pp. 403–414.

Eisma, D., 1986. Flocculation and deflocculation of suspended matter in estuaries. Netherlands Journal of Sea Research 20, 183–199.

Ellison, J.C., 2018. Chapter 20: Geomorphology and Sedimentology of Mangroves.

Escapa, M., Perillo, G.M.E., Iribarne, O., 2008. Sediment dynamics modulated by burrowing crab activities in contrasting SW Atlantic intertidal habitats. Estuarine, Coastal and Shelf Science 80, 365–373.

Escapa, C.M., Perillo, G.M.E., Iribarne, O.O., 2015. Biogeomorphically driven salt pan formation in Sarcocornia-dominated salt-marshes. Geomorphology 228, 147–157.

Esselink, P., Dijkema, K.S., Reents, S., Hageman, G., 1998. Vertical accretion and profile changes in abandoned man-made tidal marshes in the Dolland estuary, The Netherlands. Journal of Coastal Research 14, 570–582.

Fagherazzi, S., Kirwan, M.L., Mudd, S.M., et al., 2012. Numerical models of salt marsh evolution: ecological, geomorphic, and climatic factors. Reviews of Geophysics 50.

Fan, D., 2012. Open-coast tidal flats. In: Davis, R.A., Dalrymple, R.W. (Eds.), Principles of Tidal Sedimentology. Springer, New York, pp. 187–229.

Figueira, B., Hayward, B.W., 2014. Impact of reworked foraminifera from an eroding salt marsh on sea-level studies. New Zealand Journal of Geology and Geophysics 57, 378–389.

Francalanci, S., Bendoni, M., Rinaldi, M., Solari, L., 2013. Ecomorphodynamic evolution of salt marshes: experimental observations of bank retreat processes. Geomorphology 195, 53–65. https://doi.org/10.1016/j.geomorph.2013.04.026.

French, J.R., Burningham, H., 2003. Tidal marsh sedimentation versus sea-level rise: a southeast England estuarine perspective. In: Kraus, N.C., McDougal, W.G. (Eds.), Coastal Sediments 03. American Society of Civil Engineers, New York, pp. 1–14.

French, J.R., Clifford, N.J., 1992. Characteristics and 'event-structure' of near-bed turbulence in a macrotidal saltwater channel. Estuarine, Coastal and Shelf Science 34, 49–69.

French, J.R., Spencer, T., 1993. Dynamics of sedimentation in a tide-dominated backbarrier salt marsh, Norfolk, UK. Marine Geology 110, 315–331.

French, J.R., Stoddart, D.R., 1992. Hydrodynamics of salt-marsh creek systems - implications for marsh morphological development and material exchange. Earth Surface Processes and Landforms 17, 235–252.

French, J.R., Clifford, N.J., Spencer, T., 1993. High frequency flow and suspended sediment measurements in a tidal wetland channel. In: Clifford, N.J., French, J.R., Hardisty, J. (Eds.), Turbulence: Perspectives on Flow and Sediment Transport. John Wiley, Chichester, pp. 93–120.

French, J.R., Spencer, T., Murray, A.L., Arnold, N.A., 1995. Geostatistical analysis of sediment deposition in two small tidal wetlands, Norfolk, UK. Journal of Coastal Research 10, 308–321.

French, J.R., Burningham, H., Benson, T., 2008. Tidal and meteorological forcing of suspended sediment flux in a muddy mesotidal estuary. Estuaries and Coasts 31, 843–859.

French, J.R., 1993. Numerical modelling of vertical marsh growth and response to rising sea-level, Norfolk, UK. Earth Surface Processes and Landforms 18, 63–81.

French, J.R., 1994. Tide-dominated coastal wetlands and accelerated sea level rise: a Northwestern European perspective. Journal of Coastal Research Special Issue 12, 91–101.

French, J.R., 2006. Tidal marsh sediment trapping efficiency and resilience to environmental change: exploratory modelling of tidal, sea-level and sediment supply forcing in predominantly allochthonous systems. Marine Geology 235, 119–136.

Frey, R.W., Basan, P.B., 1985. Coastal salt marshes. In: Davis, R.A. (Ed.), Coastal Sedimentary Environments, second ed. Springer-Verlag, New York, pp. 225–301.

Friedrichs, C.T., Aubrey, D.G., 1988. Non-linear tidal distortion in shallow well-mixed estuaries: a synthesis. Estuarine, Coastal and Shelf Science 27, 521–545.

Friedrichs, C.T., Perry, J.E., 2001. Tidal salt marsh morphodynamics: a synthesis. Journal of Coastal Research Special Issue 27, 7–37.

Friedrichs, C.T., 1995. Stability shear stress and equilibrium cross-sectional geometry of sheltered tidal channels. Journal of Coastal Research 11, 1062–1074.

Gabet, E.J., 1998. Lateral migration and bank erosion in a saltmarsh tidal channel in San Francisco Bay, California. Estuaries 21, 745–753.

Ganju, N.K., Nidzieko, N.J., Kirwan, M.L., 2013. Inferring tidal wetland stability from channel sediment fluxes: observations and a conceptual model. Journal of Geophysical Research – Earth Surface 118, 2045–2058.

Ganong, W.F., 1903. The vegetation of the Bay of Fundy salt and diked marshes: an ecological study. Botanical Gazette 36, 161-186, 280-302, 349-367, 429–455.

Gedan, K.B., Silliman, B.R., Bertness, M.D., 2009. Centuries of human-driven change in salt marsh ecosystems. Annual Review of Marine Science 1, 117–141.

Gibbs, R.J., 1985. Settling velocity, diameter and density for flocs of illite, kaolinite and montmorillonite. Journal of Sedimentary Petrology 55, 65–68.

Goodbred, S.L., Saito, Y., 2012. Tide-dominated deltas. In: Davis, R.A., Dalrymple, R.W. (Eds.), Principles of Tidal Sedimentology. Springer, New York, pp. 79–107.

Goodman, J., Wood, M.E., Gehrels, W.R., 2007. A 17-yr record of sediment accretion in the salt marshes of Maine (USA). Marine Geology 242, 109–121.

Gordon, D.C., Cranford, P., Despanque, C., 1985. Observations on the ecological importance of salt marshes in the Cumberland Basin, a macrotidal estuary in the Bay of Fundy. Estuarine, Coastal and Shelf Science 20, 205–227.

Gray, A.J., 1972. The ecology of Morecambe Bay V: the saltmarshes. Journal of Applied Ecology 9, 207–220.

Gribsholt, B., Kostka, J.E., Kristensen, E., 2003. Impact of fiddler crabs and plant roots on sediment biogeochemistry in a Georgia saltmarsh. Marine Ecology Progress Series 259, 237–251.

Guilcher, A., Berthois, L., 1957. Cinq annees d'observations sedimentologiques dans quatre estuaires-tremoins de l'ouest de la Bretagne. Revue de Geomorphol. Dynamique 8, 67–76.

Gunnell, J.R., Rodriguez, A.B., McKee, B.A., 2013. How a marsh is built from the bottom up. Geology 41, 859–862.

Hapke, C.J., Kratzmann, M.G., Himmelstoss, E.A., 2013. Geomorphic and human influence on large-scale coastal change. Geomorphology 199, 160–170.

Harmsworth, G.C., Long, S.P., 1986. An assessment of salt-marsh erosion in Essex, England, with reference to the Dengie Peninsula. Biological Conservation 35, 377–387.

Harrison, E.Z., Bloom, A.L., 1977. Sedimentation rates on tidal salt marshes in Connecticut. Journal of Sedimentary Petrology 47, 1484–1490.

Haslett, S.K., Bryant, E.A., 2007. Reconnaissance of historic (post-AD 1000) high-energy deposits along the Atlantic coast of southwest Britain, Ireland and Brittany, France. Marine Geology 242, 207–220.

Hayes, M.O., 1979. Barrier island morphology as a function of tidal and wave regime. In: Leatherman, S.P. (Ed.), Barrier Islands. Academic Press, Orlando, pp. 1–27.

Hayward, B.W., Grenfell, H.R., Sabaa, A.T., Cochran, U.A., Clark, K.J., Wallace, L., Palmer, A.S., 2016. Salt-marsh foraminiferal record of 10 large Holocene (last 7500 yr) earthquakes on a subducting plate margin, Hawkes Bay, New Zealand. The Geological Society of America Bulletin 128, 896–915.

Hill, T.D., Anisfeld, S.C., 2015. Coastal wetland response to sea level rise in Connecticut and New York. Estuarine, Coastal and Shelf Science 163, 185–193.

Hinde, H.P., 1954. The vertical distribution of salt marsh phanerogams in relation to tide levels. Ecological Monographs 24, 209–225. https://doi.org/10.2307/1948621.

Hippensteel, S.P., Garcia, W.J., 2014. Micropaleontological evidence of prehistoric hurricane strikes from southeastern North Carolina. Journal of Coastal Research 30, 1157–1172.

Holden, V.J.C., Worsley, A.T., Booth, C.A., Lymbery, G., 2011. Characterisation and sediment-source linkages of intertidal sediment of the UK's north Sefton Coast using magnetic and textural properties: findings and limitations. Ocean Dynamics 61, 2157–2179.

Hughes, Z.J., Fitzgerald, D.M., Wilson, C.A., Pennings, S.C., Wieiski, K., Mahadevan, A., 2009. Rapid headward erosion of marsh creeks in response to relative sea-level rise. Geophysical Research Letters. https://doi.org/10.1029/2008GL036000.

Hughes, Z.J., 2012. Tidal channels on tidal flats and marshes. In: Davis, R.A., Dalrymple, R.W. (Eds.), Principles of Tidal Sedimentology. Springer, New York, pp. 269–300.

Isacch, J.P., Costa, C.S.B., Rodriguez-Gallego, L., Conde, D., Escapa, M., Gagliardini, D.A., Iribarne, O.O., 2006. Distribution of saltmarsh plant communities associated with environmental factors along a latitudinal gradient on the south-west Atlantic coast. Journal of Biogeography 33, 888–900.

Johnson, D.S., York, H.H., 1915. The Relations of Plants to Tide Levels, a Study of Factors Affecting the Distribution of Marine Plants at Cold Spring Harbor, vol. 206. Carnegie Institution Washington Publication, Long Island, New York, 161 p.

Kelleway, J.J., Saintilan, N., Macreadie, P.I., Ralph, P.J., 2016. Sedimentary factors are key predictors of carbon storage in SE Australian saltmarshes. Ecosystems 19, 865–880.

Kemp, A.C., Wright, A.J., Barnett, R.L., et al., 2017. Utility of salt-marsh foraminifera, testate amoebae and bulk-sediment delta C-13 values as sea-level indicators in Newfoundland, Canada. Marine Micropaleontogy 130, 43–59.

Kennish, M.J., 2001. Coastal salt marsh systems in the US: a review of anthropogenic impacts. Journal of Coastal Research 17, 731–748.

Kestner, J.T., 1975. The loose boundary regime of the Wash. Geographical Journal 141, 389–414.

Kim, D., Grant, W.E., Cairns, D.M., Bartholdy, J., 2013. Effects of the North Atlantic Oscillation and wind waves on salt marsh dynamics in the Danish Wadden Sea: a quantitative model as proof of concept. Geo-Marine Letters 33, 253–261.

Kirby, R.E., Widjeskog, L.E., 2013. Sediment redistributed by coastal marsh mosquito ditching in Cape May County, New Jersey, USA. Journal of Coastal Research 29, 86–93.

Kirwan, M.L., Guntenspergen, G.R., 2010. Influence of tidal range on the stability of coastal marshland. Journal of Geophysical Research 115, F02009. https://doi.org/10.1029/2009JF001400.

Kirwan, M.L., Murray, A.B., 2007. A coupled geomorphic and ecological model of tidal marsh evolution. Proceedings of the National Academy of Sciences of the United States of America 104, 6118–6122.

Kirwan, M.L., Murray, A.B., Boyd, W.S., 2008. Temporary vegetation disturbance as an explanation for permanent loss of tidal wetlands. Geophysical Research Letters 35, L05403.

Kirwan, M.L., Guntenspergen, G.R., D'Alpaos, A., Morris, J.T., Mudd, S.M., Tmmermann, S., 2010. Limits on the adaptability of coastal marshes to rising sea level. Geophysical Research Letters 37, L23401,. https://doi.org/10.1029/2010GL045489.

Kirwan, M.L., Murray, A.B., Donnelly, J.P., Corbett, D.R., 2011. Rapid wetland expansion during European settlement and its implication for marsh survival under modern sediment delivery rates. Geology 39, 507–510.

Koh, C.H., de Jonge, V.N., 2014. Stopping the disastrous embankments of coastal wetlands by implementing effective management principles: Yellow Sea and Korea compared to the European Wadden Sea. Ocean and Coastal Management 102, 604–621.

Krone, R.B., 1987. A method for simulating historic marsh elevations. In: Kraus, N.C. (Ed.), Proceedings Coastal Sediments '87. American Society of Civil Engineers, New York, pp. 316–323.

Lagomasino, D., Corbett, D.R., Walsh, J.P., 2013. Influence of wind-driven inundation and coastal geomorphology on sedimentation in two microtidal marshes, Pamlico River Estuary, NC. Estuaries and Coasts 36, 1165–1180.

Leonard, L.A., Hine, A.C., Luther, M.E., 1995. Surficial sediment transport and deposition processes in a *Juncus roemerianus* marsh, west-central Florida. Journal of Coastal Research 11, 322–336.

Leonard, L.A., 1997. Controls of sediment transport and deposition in an incised mainland marsh basin, southeastern North Carolina. Wetlands 17, 263–274.

Li, H., Yang, S.L., 2009. Trapping effect of tidal marsh vegetation on suspended sediment, Yangtze Delta. Journal of Coastal Research 25, 915–936.

Madsen, A.T., Murray, A.S., Andersen, T.J., Pejrup, M., 2007. Temporal changes of accretion rates on an estuarine salt marsh during the late Holocene – reflection of local sea level changes? The Wadden Sea, Denmark. Marine Geology 242, 221–233.

Manning, A.J., Dyer, K.R., 2007. Mass settling flux of fine sediments in Northern European estuaries: measurements and predictions. Marine Geology 245, 107–122.

Marani, M., D'Alpaos, A., Lanzoni, S., Carniello, L., Rinaldo, A., 2007. Biologically-controlled multiple equilibria of tidal landforms and the fate of the Venice lagoon. Geophysical Research Letters 34, L11402.

Marani, M., D'Alpaos, A., Lanzoni, S., Carniello, L., Rinaldo, A., 2010. The importance of being coupled: stable states and catastrophic shifts in tidal biomorphodynamics. Journal of Geophysical Research 115, F04004.

Marion, C., Anthony, E.J., Trentesaux, A., 2009. Short-term (<=2 yrs) estuarine mudflat and saltmarsh sedimentation: high-resolution data from ultrasonic altimetery, rod surface-elevation table, and filter traps. Estuarine, Coastal and Shelf Science 83, 475–484.

Mariotti, G., 2016. Revisiting salt marsh resilience to sea level rise: are ponds responsible for permanent land loss? Journal of Geophysical Research – Earth Surface 121, 1391–1407.

Mattheus, C.R., Rodriguez, A.B., McKee, B.A., Currin, C.A., 2010. Impact of land-use change and hard structures on the evolution of fringing marsh shorelines. Estuarine, Coastal and Shelf Science 88, 365–376.

McKee, K., Patrick, W.H., 1988. The relationship of Smooth Cordgrass (*Spartina alterniflora*) to tidal datums: a review. Estuaries 11, 143–151.

Millette, T.L., Argow, B.A., Marcano, E., Hayward, C., Hopkinson, C.S., Valentine, V., 2010. Salt marsh geomorphological analyses via integration of multitemporal multispectral remote sensing with LIDAR and GIS. Journal of Coastal Research 26, 809–816.

Moffett, K.B., Robinson, D.A., Gorelick, S.M., 2010. Relationship of salt marsh vegetation zonation to spatial patterns in soil moisture, salinity, and topography. Ecosystems 13, 1287–1302.

Morris, J.T., Sundareshwar, P.V., Nietch, C.T., Kjervfe, B.J., Cahoon, D.R., 2002. Responses of coastal wetlands to rising sea level. Ecology 83, 2869–2877.

Morris, J.T., Barber, D.C., Callaway, J.C., et al., 2016. Contributions of organic and inorganic matter to sediment volume and accretion in tidal wetlands at steady state. Earth's Future 4. https://doi.org/10.1002/2015EF000334.

Moskalski, S.M., Sommerfield, C.K., 2010. Suspended sediment deposition and trapping efficiency in a Delaware salt marsh. Geomorphology 139–140, 195–204.

Mudd, S.M., D'Alpaos, A., Morris, J.T., 2010. How does vegetation affect sedimentation on tidal marshes? Investigating particle capture and hydrodynamic controls on biologically mediated sedimentation. Journal of Geophysical Research – Earth Surface 115, F03029. https://doi.org/10.1029/2009JF001566.

Mudd, S.M., 2011. The life and death of salt marshes in response to anthropogenic disturbance of sediment supply. Geology 39, 511–512.

Nestler, J., 1977. Interstitial salinity as a cause of ecophenic variation in *Spartina alternifora*. Estuarine and Coastal Marine Science 5, 707–714.

Neubauer, S.C., 2008. Contributions of mineral and organic components to tidal freshwater marsh accretion. Estuarine, Coastal and Shelf Science 78, 78–88.

Neumeier, U., Amos, C.L., 2006. The influence of vegetation on turbulence and flow velocities in European saltmarshes. Sedimentology 53, 259–277.

Nielsen, N., Nielsen, J., 2002. Vertical growth of a young backbarrier salt marsh, Skallingen, SW Denmark. Journal of Coastal Research 18, 287–299.

Nielsen, N., 1935. Eine methode zur exakten sedimentations messung. Meddelelseries fra Skalling Laboratoriet 1, 97 p.

Nyman, J.A., Walters, R.J., Delaune, R.D., Patrick Jr., W.H., 2006. Marsh vertical accretion via vegetative growth. Estuarine, Coastal and Shelf Science 69, 370–380.

Orson, R.A., Warren, R.S., Niering, W.A., 1987. Development of a tidal marsh in a new England river valley. Estuaries 10, 20–27.

Paramor, O.A.L., Hughes, R.G., 2004. The effects of bioturbation and herbivory by the polychaete *Nereis diversicolor* on loss of saltmarsh in south-east England. Journal of Applied Ecology 41, 449–463.

Pedersen, J.B.T., Bartholdy, J., 2007. Exposed salt marsh morphodynamics: an example from the Danish Wadden Sea. Geomorphology 90, 115–125.

Pennings, S.C., Callaway, R.M., 1992. Salt marsh plant zonation: the relative importance of competition and physical factors. Ecology 73, 681–690.

Perillo, G.M.E., Iribarne, O.O., 2003. Processes of tidal channel development in salt and freshwater marshes. Earth Surface Processes and Landforms 28, 1473–1482.

Pestrong, R., 1965. The Development of Drainage Patterns in Tidal Marshes. Stanford University Publications in Earth Science 10, pp. 1–87.

Pestrong, R., 1972. Tidal flat sedimentation at Cooley Landing, southwest san Francisco Bay. Sedimentary Geology 8, 251–288.

Pethick, J.R., 1974. The distribution of salt pans on tidal salt marshes. Journal of Biogeography 1, 57–62.

Pethick, J.S., 1980. Salt marsh initiation during the Holocene transgression: the example of the north Norfolk marshes. Journal of Biogeography 7, 1–9.

Pethick, J.S., 1981. Long-term accretion rates on tidal marshes. Journal of Sedimentary Petrology 61, 571–577.

Phleger, F.B., 1977. Soils of marine marshes. In: Chapman, V.J. (Ed.), Wet Coastal Ecosystems. Elsevier, Amsterdam, pp. 69–77.

Pizzuto, J.E., 1987. Sediment diffusion during overbank flows. Sedimentology 34, 301–317.

Porter, C., Lundholm, J., Bowron, T., Lemieux, B., van Proosdij, D., Neatt, N., Graham, J., 2015. Classification and environmental correlates of tidal wetland vegetation in Nova Scotia, Canada. Botany 93, 825–841.

Postma, H., 1961. Transport and accumulation of suspended matter in the Dutch Wadden Sea. Netherlands Journal of Sea Research 1, 148–190.

Pratolongo, P., Perillo, G.M.E., Piccolo, M.C., 2010. Combined effects of waves and plants on a mud deposition event at a mudflat-saltmarsh edge in the Bahía Blanca estuary. Estuarine, Coastal and Shelf Science 87, 207–212.

Pratolongo, P., Mazzon, C., Zapperi, G., Piovan, M.J., Brinson, M.M., 2013. Land cover changes in tidal salt marshes of the Bahía Blanca estuary (Argentina) during the past 40 years. Estuarine, Coastal and Shelf Science 133, 23–31.

Pringle, A.W., 1995. Erosion of a cyclic saltmarsh in Morecambe Bay, north-west England. Earth Surface Processes and Landforms 20, 387–405.

Redfield, A.C., 1972. Development of a new England salt marsh. Ecological Monographs 41, 201–237.

Reed, D.J., Spencer, T., Murray, A., French, J.R., Leonard, L., 1999. Marsh surface sediment deposition and the role of tidal creeks: implications for created and managed coastal marshes. Journal of Coastal Conservation 5, 81–90.

Reed, D.J., 1988. Sediment dynamics and deposition in a retreating coastal salt marsh. Estuarine, Coastal and Shelf Science 26, 67–79.

Reineck, H.E., Gerdes, G., 1996. A seaward prograding siliciclastic sequence from upper tidal flats to salt marsh facies (Southern North Sea). Facies 34, 209–218.

Reise, K., 2005. Coast of change: habitat loss and transformations in the Wadden Sea. Helgoland Marine Research 59, 9–12.

Rodriguez, A.B., Fegley, S.R., Ridge, J.T., VanDusen, B.M., Anderson, N., 2013. Contribution of aeolian sand to back-barrier marsh sedimentation. Estuarine, Coastal and Shelf Science 117, 248–259.

Roychoudhury, A.N., 2007. Spatial and seasonal variations in depth profile of trace metals in saltmarsh sediments from Sapelo Island, Georgia, USA. Estuarine, Coastal and Shelf Science 72, 675–689.

Rybczyk, J.M., Cahoon, D.R., 2002. Estimating the potential for submergence for two wetlands in the Mississippi River Delta. Estuaries 25, 985–998.

Rybzyck, J.M., Callaway, J.C., 2018. Chapter 30: Surface Elevation Models.

Schuerch, M., Dolch, T., Reise, K., Vafeidis, A.T., 2014. Unravelling interactions between salt marsh evolution and sedimentary processes in the Wadden Sea (southeastern North Sea). Progress in Physical Geography 38, 691–715.

Schuerch, M., Scholten, J., Carretero, S., Garcia-Rodriguez, F., Kumbier, K., Baechtiger, M., Liebetrau, V., 2016. The effect of long-term and decadal climate and hydrology variations on estuarine marsh dynamics: an identifying case study from the Rio de la Plata. Geomorphology 269, 122–132.

Schultz, R.A., Anisfeld, S.C., Hill, T.D., 2016. Submergence and herbivory as divergent causes of marsh loss in Long Island Sound. Estuaries and Coasts 39, 1367–1375.

Shaw, J., Ceman, J., 1999. Salt-marsh aggradation in response to late-Holocene sea-level rise at Amherst Point, Nova Scotia, Canada. The Holocene 9, 439–451.

Shi, Z., Lamb, H.F., Collin, R.L., 1995. Geomorphic change of saltmarsh tidal creek networks in the Dyfi Estuary, Wales. Marine Geology 128, 73–83.

Silinski, A., Heuner, M., Troch, P., Puijalon, S., Bouma, T.J., Schoelynck, J., Schroder, U., Fuchs, E., Meire, P., Temmerman, S., 2016. Effects of contrasting wave conditions on scour and drag on pioneer tidal marsh plants. Geomorphology 255, 49–62.

Silva, H., Dias, J.M., Cacador, I., 2009. Is the salt marsh vegetation a determining factor in the sedimentation processes? Hydrobiologia 621, 33–47.

Silvestri, S., Defina, A., Marani, M., 2005. Tidal regime, salinity and salt marsh plant zonation. Estuarine, Coastal and Shelf Science 62, 119–130.

Steers, J.A., 1935. A note on the rate of sedimentation on a salt marsh on Scolt Head Island. Geological Magazine 72, 443–435.

Steers, J.A., 1946. Twelve years measurement of accretion on North Norfolk salt marshes. Geological Magazine 85, 163–166.

Stephan, P., Goslin, J., Pailler, Y., Manceau, R., Suanez, S., Van Vliet-Lanoe, B., Henaff, A., Delacourt, C., 2015. Holocene salt-marsh sedimentary infilling and relative sea-level changes in West Brittany (France) using foraminifera-based transfer functions. Boreas 44, 153–177.

Stevenson, J.C., Ward, L.G., Kearney, M.S., 1986. Vertical accretion in marshes with varying rates of sea-level rise. In: Wolfe, D.A. (Ed.), Estuarine Variability. Academic Press, Orlando, pp. 241–260.

Stevenson, J.C., Ward, L.G., Kearney, M.S., 1988. Sediment transport and trapping in marsh systems: implications of tidal flux studies. Marine Geology 80, 37–59.

Stoddart, D.R., Reed, D.J., French, J.R., 1989. Understanding salt marsh accretion, Scolt Head island, North Norfolk, England. Estuaries 12, 228–236.

Strachan, K.L., Finch, J.M., Hill, T.R., Barnett, R.L., Morris, C.D., Frenzel, P., 2016. Environmental controls on the distribution of salt-marsh foraminifera from the southern coastline of South Africa. Journal of Biogeography 43, 887–898.

Strotz, L.C., 2015. Spatial patterns and diversity of foraminifera from an intermittently closed and open lagoon, Smiths Lake, Australia. Estuarine, Coastal and Shelf Science 164, 340–352.

Stumpf, R.P., 1983. The processes of sedimentation on the surface of a saltmarsh. Estuarine, Coastal and Shelf Science 17, 495–508.

Suchrow, S., Jensen, K., 2010. Plant species responses to an elevational gradient in German North Sea salt marshes. Wetlands 30, 735–746. https://doi.org/10.1007/s13157-010-0073-3.

Suchrow, S., Pohlmann, N., Stock, M., Jensen, K., 2012. Long-term surface elevation changes in German North Sea salt marshes. Estuarine, Coastal and Shelf Science 98, 71–83.

Suchrow, S., Stock, M., Jensen, K., 2015. Patterns of plant species richness along environmental gradients in German North Sea salt marshes. Estuaries and Coasts 38, 296–309.

Temmerman, S., Govers, G., Meire, P., Wartel, S., 2003. Modeling long-term tidal marsh growth under changing tidal conditions and suspended sediment concentrations, Scheldt estuary, Belgium. Marine Geology 193, 151–169.

Temmerman, S., Govers, G., Wartel, S., Meire, P., 2004. Modelling estuarine variations in tidal marsh sedimentation: response to changing sea level and suspended sediment concentrations. Marine Geology 212, 1–19.

Temmerman, S., Bouma, T.J., Govers, G., Wang, Z.B., De Vries, M.B., Herman, P.M.J., 2005. Impact of vegetation on flow routing and sedimentation patterns: three-dimensional modeling for a tidal marsh. Journal of Geophysical Research 110, F04019. https://doi.org/10.1029/2005JF000301.

Tolhurst, T.J., Güst, G., Paterson, D.M., 2002. The influence of an extracellular polymeric substance (EPS) on cohesive sediment stability. In: Winterwerp, J.C., Kranenburg, C. (Eds.), Fine Sediment Dynamics in the Marine Environment. Proceedings in Marine Science, vol. 5, pp. 409–425.

Tolhurst, T.J., Jesus, B., Brotas, V., Paterson, D.M., 2003. Diatom migration and sediment armoring: an example from the Tagus Estuary, Portugal. Hydrobiologia 503, 183–193.

Tsompanoglou, K., Croudace, I.W., Birch, H., Collins, M., 2011. Geochemical and radiochronological evidence of North Sea storm surges in salt marsh cores from the Wash embayment (UK). The Holocene 21, 225–236.

van de Koppel, J., van der Wall, D., Bakker, J.P., Herman, P.M.J., 2005. Self-organization and vegetation collapse in salt marsh ecosystems. American Naturalist 165, E1–E12.

van der Waal, D., Wielemaker-Van den Dool, A., Herman, P.M.J., 2008. Spatial patterns, rates and mechanisms of saltmarsh cycles (Westerschelde, The Netherlands). Estuarine, Coastal and Shelf Science 76, 357–368.

van der Wal, D., Pye, K., 2004. Patterns, rates and possible causes of saltmarsh erosion in the Greater Thames area (UK). Geomorphology 61, 373–391.

van Eerdt, M.M., 1985. Salt marsh cliff stability in the Oosterschelde. Earth Surface Processes and Landforms 10, 95–106.

van Proosdij, D., Davidson-Arnott, R.G.D., Ollerhead, J., 2006. Controls on spatial patterns of sediment deposition across a macro-tidal salt marsh surface over single tidal cycles. Estuarine, Coastal and Shelf Science 69, 64–86.

Vandenbruwaene, W., Schwarz, C., Bouma, T.J., Meire, P.J., Temmerman, S., 2015. Landscape-scale flow patterns over a vegetated tidal marsh and an unvegetated tidal flat: implications for the landform properties of the intertidal floodplain. Geomorphology 231, 40–52.

Veldkornet, D.A., Potts, A.J., Adams, J.B., 2016. The distribution of salt marsh macrophyte species in relation to physiochemical variables. South African Journal of Botany 107, 84–90.

Voulgaris, G., Meyers, S.T., 2004. Temporal variability of hydrodynamics, sediment concentration and sediment settling velocity in a tidal creek. Continental Shelf Research 24, 1659–1683.

Wang, C., Temmerman, S., 2013. Does biogeomorphic feedback lead to abrupt shifts between alternative landscape states? An empirical study on intertidal flats and marshes. Journal of Geophysical Research 118, 229–240.

Wang, J.H., Masse, L., Tastet, J.P., 2006. Sedimentary facies and paleoenvironmental interpretation of a Holocene marsh in the Gironde Estuary in France. Acta Oceanologica Sinica 25, 52–62.

Wang, A.J., Ye, X.A., Chen, J.A., 2010. Observations and analyses of floc size and floc settling velocity in coastal salt marsh of Luoyuan Bay, Fujian Province, China. Acta Oceanologica Sinica 29, 116–126.

Ward, L.G., Zaprowski, B.J., Trainer, K.D., Davis, P.T., 2008. Stratigraphy, pollen history and geochronology of tidal marshes in a Gulf of Maine estuarine system: climatic and relative sea level impacts. Marine Geology 256, 1–17.

Warren, R.S., Niering, W.A., 1993. Vegetation change on a northeast tidal marsh: interactions of sea-level rise and marsh accretion. Ecology 74, 96–103.

Weston, N.B., 2014. Declining sediments and rising seas: an unfortunate convergence for tidal wetlands. Estuaries and Coasts 37, 1–23.

Wheeler, A.J., Orford, J.D., Dardis, O., 1999. Saltmarsh deposition and its relationship to coastal forcing over the last century on the north-west coast of Ireland. Netherlands Journal of Geosciences – Geologie en Mijnbouw 77, 295–310.

Widdows, J., Brown, S., Brinsley, M., Salkield, P.N., Elliot, M., 2000. Temporal changes in intertidal sediment erodability: influence of biology and climactic factors. Continental Shelf Research 20, 1275–1290.

Wiehe, P.O., 1935. A Quantitative study of the influence of tide upon populations of *Salicornia europea*. Journal of Ecology 23, 323–333. https://doi.org/10.2307/2256124.

Wilson, C.A., Hughes, Z.J., FitzGerald, D.M., Hopkinson, C.S., Valentine, V., Kolker, A.S., 2014. Saltmarsh pool and tidal creek morphodynamics: dynamic equilibrium of northern latitude saltmarshes? Geomorphology 213, 99–115.

Woodroffe, C.D., 1992. Mangrove sediments and geomorphology. In: Alongi, D., Robertson, A. (Eds.), Tropical Mangrove Ecosystems. Coastal and Estuarine Studies. American Geophysical Union, Washington DC, pp. 7–41.

Woodroffe, C.D., 2003. Coasts. Form, Process and Evolution. Cambridge University Press, Cambridge, 623p.

Yang, S.L., Chen, J.Y., 1995. Coastal salt marshes and mangrove swamps in China. Chinese Journal of Oceanology and Limnology 13, 318–324.

Yang, S.L., Ding, P.X., Chen, S.L., 2001. Changes in progradation rate of the tidal flats at the mouth of the Changjiang (Yangtze) River, China. Geomorphology 38, 167–180.

Yang, S.L., Li, H., Ysebaert, T., Bouma, T.J., Zhang, W.X., Wang, Y., Li, P., Li, M., Ding, P., 2008. Spatial and temporal variations in sediment grain size in tidal wetlands, Yangtze Delta: on the role of physical and biotic controls. Estuarine, Coastal and Shelf Science 77, 657–671.

Yapp, R.H., Johns, D., Jones, O.T., 1917. The saltmarshes of the Dovey Estuary: part II. The salt marshes. Journal of Ecology 5, 65–103.

Zann, L.P., 2000. The East Australian Region: a dynamic tropical/temperate biotone. Marine Pollution Bulletin 41, 188–203.

Zhou, Z., Ye, Q., Coco, G., 2016. A one-dimensional biomorphodynamic model of tidal flats: sediment sorting, marsh distribution, and carbon accumulation under sea-level rise. Advances in Water Resources 93, 288–302.

Zuo, P., Zhao, S.H., Liu, C.A., Wang, C.H., Liang, Y.B., 2012. Distribution of *Spartina* spp. along China's coast. Ecological Engineering 40, 160–166.

CHAPTER

15

Ecosystem Structure of Tidal Saline Marshes

Jenneke M. Visser[1], *Stephen Midway*[2], *Donald M. Baltz*[2], *Charles E. Sasser*[2]

[1]Institute for Coastal and Water Research, and School of Geosciences, University of Louisiana at Lafayette, Lafayette, LA, United States; [2]Department of Oceanography and Coastal Sciences, Louisiana State University, Baton Rouge, LA, United States

1. INTRODUCTION

Saline marshes occur throughout the world as coastal features that often fringe shorelines and can dominate vegetated areas of estuarine environments. They vary from small and discrete marshes to broad and expansive areas and are typically located in sheltered, low-energy shoreline areas forming the interface between marine and terrestrial environments. Salt marshes are valuable ecosystems that are known to be highly productive, ranking with the most productive ecosystems of the world.

From a global perspective, tidal saline marshes are found in middle and high latitudes along shores throughout the world (Chapman, 1977). The largest concentrations of tidal marshes are found along the South Atlantic and Gulf coasts of North America followed by China (Greenberg et al., 2006). The physical features of tides, sediments, freshwater inputs, and shoreline geomorphology determine the development and extent of tidal saline wetlands, and their regional differences in productivity are likely related to the available solar energy (Mitsch and Gosselink, 2000). Tidal saline marshes are generally found in sedimentary environments and can be broadly classified into those that originate on reworked marine sediments and those that are formed at the margins of river deltas on riverine sediments. Tidal regimes vary from microtidal (<2 m) to macrotidal (>6 m) and can also be diurnal, semidiurnal, or mixed.

The impacts of human activities on coastal systems are not included in this chapter (but see Chapter 23), although we recognize that human activities and development have had

Coastal Wetlands
https://doi.org/10.1016/B978-0-444-63893-9.00015-0

substantial impacts, including salt marshes and associated estuaries (Baltz and Yáñez-Arancibia, 2009). Over several centuries, expanding human populations in the coastal zone and expanding fishing pressure on finite resources have led to altered estuarine and coastal ecosystems through a series of anthropogenic effects and can be characterized as the shifting baseline syndrome (Pauly, 1995). Fishing has generally been first and foremost, coming before pollution, habitat destruction, introductions of exotic species, and climatic change in the timing and degree of impact (Jackson et al., 2001). Historic alterations have significantly modified saltmarsh ecosystems to the point of dramatic functional impairment or outright marsh destruction (Bertness et al., 2002). To fully appreciate the current condition of saline marshes and foresee the direction of future changes, we need a better understanding of the history of interactions between nature and society in our coastal systems (Kates et al., 2001).

2. SALINE MARSH COMMUNITIES

In describing saltmarsh communities, we have attempted to not only cover the species that are best represented in the literature but also recognize that many important salt marsh species may not be included in our overview. Our coverage includes emergent vegetation and algae, up through fish, birds, and mammals.

2.1 Emergent Vegetation

Tidal saline marshes are defined as natural or seminatural halophytic emergent vegetation on alluvial sediments, with a connection to saline waterbodies (Beeftink, 1977). Typical hydrology involves periodic flooding of the marsh surface as water moves on and off the marsh, controlled by the local tidal regime. The emergent vegetation of tidal saline marshes changes in species composition depending on geographical location (Adam, 1990; Table 15.1). Salt marshes are generally replaced by mangrove swamps in coastal regions of the tropics and subtropics—between 25°N and 25°S latitude (Mitsch and Gosselink, 2000).

Within geographical locations, species composition varies along inundation gradients. A low (submergence) and a high (emergence) marsh zone are generally recognized. The marsh below mean high water is regularly flooded and has reduced soils for most of the time except along creek banks with good drainage (Armstrong et al., 1985). This zone is generally species poor and is dominated by species that are both salt and flood tolerant. The high marsh is found above mean high tide and is less frequently flooded and has oxidized soils that may briefly become reduced during flood events (Armstrong et al., 1985). In some areas, an intermediate middle marsh zone may be distinguished by different plant species composition. Salinity often decreases inland, but salinity inversions are relatively common, particularly when evapotranspiration exceeds precipitation (Mahall and Park, 1976; Callaway et al., 1990). In areas that are irregularly flooded, zonation may be absent or zonation may be related to distance from tidal creeks (Zedler et al., 1999; Costa et al., 2003). Flooding effects on biogeochemical cycling are described in detail in Chapter 16.

TABLE 15.1 Dominant Emergent Plant Species of Tidal Saline Marshes

Family	Species	Marsh Zone	Geographic Extent	Source
Amaranthaceae	*Salkomia virginica*	Low	Western North America	Pennings and Callaway (1992)
Amaranthaceae	*Sarcocornia quinqueflora*	Low	South Pacific	Thannheiser and Holland (1994)
Chenopodiaceae	*Arthrocnemum perenne*	Middle	Europe	Castellanos et al. (1994)
Chenopodiaceae	*Arthrocnemum subterminale*	Middle	Western North America	Pennings and Callaway (1992)
Chenopodiaceae	*Suaeda maritima*	Low	Japan	Adam (1990)
Cyperaceae	*Carex glareosa*	High	Arctic	Adam (1990)
Goodeniaceae	*Seleria radicans*	Middle	South Pacific	Thannheiser and Holland (1994)
Juncaceae	*Juncus kraussii*	High	South Pacific	Adam (1990)
Juncaceae	*Juncus roemerianus*	Middle to high	Eastern North America	Mitsch and Gosselink (2000)
Juncaginaceae	*Triglochin maritima*	Middle to low	Western Europe, Arctic, Asia	Bakker (1985) and Davy and Bishop (1991)
Plumbaginaceae	*Limonium vulgare*	Middle to Low	Europe and North Africa	Boorman (1967) and Bakker (1985)
Poaceae	*Distichlis scoparia*	High	South America	Cantero et al. (1998)
Poaceae	*Distichlis spicata*	Middle	Western North America, South America	Adam (1990) and Cantero et al. (1998)
Poaceae	*Festuca rubra*	High	Western Europe	Gray and Mogg (2001)
Poaceae	*Puccinellia maritima*	Middle to low	Western Europe	Gray and Mogg (2001)
Poaceae	*Puccinellia phryganodes*	Low	Arctic	Beaulieu and Allard (2003)
Poaceae	*Spartina alterniflora*	Low	Eastern North America	Mitsch and Gosselink (2000)
Poaceae	*Spartina anglica*	Low	Western Europe	Gray and Mogg (2001)
Poaceae	*Spartina densiflora*	Low	South America	Cantero et al. (1998)
Poaceae	*Spartina foliosa*	Low	Western North America	Adam (1990)
Poaceae	*Spartina maritima*	Low	Europe	Castellanos et al. (1994)
Poaceae	*Spartina patens*	High	Eastern North America	Mitsch and Gosselink (2000)
Poaceae	*Sporobolus virginicus*	High	Tropics	Adam (1990)
Poaceae	*Zoysia sinica*	High	Japan	Adam (1990)
Primulaceae	*Samolus repens*	Low	South Pacific	Thannheiser and Holland (1994)

The distribution of plant species in tidal saline marshes is determined by the physical, chemical, and biological environment. Physical and geochemical factors that affect the distribution of tidal saline marsh species include flooding, salinity, and the ratio of sodium to potassium, as well as the ratio of calcium to magnesium (Clarke and Hannon, 1970; Olff et al., 1988; Partridge and Wilson, 1989; Cantero et al., 1998; Alvarez Rogel et al., 2000; Huckle et al., 2000). Over a stress gradient, the distribution of a species is determined by physical constraints at higher stress levels and competition with other species at lower stress levels (Pennings and Callaway, 1992). Plant establishment where physical stress is high may decrease stress levels by increasing elevation, oxygenating the rhizosphere, or reducing soil salinity, thereby facilitating the establishment of higher marsh species (Bertness and Shumway, 1993; Castellanos et al., 1994; Figueroa et al., 2003). However, the relative effects of facilitation and competition are highly dependent on the stress tolerance of the local species (Pennings et al., 2003). Increasing nitrogen availability allows species to better compete and expand (van Wijnen and Bakker, 1999; Bertness and Pennings, 2000).

Sometimes parasitic plants can alter competitive outcomes. For example, in tidal saline marshes of the western United States, the parasitic saltmarsh dodder (*Cuscuta salina*) facilitates two relatively uncommon plant species, sea lavender (*Limonium californicum*) and sea heath (*Frankenia salina*), by selectively infecting and suppressing the competitive dominant glasswort (*Salicornia virginica*) (Pennings and Callaway, 1996). Interaction with animals also plays an important role. In many tidal saline marshes across the globe, invasive plant species have started to replace the historic/native vegetation due to introductions of nonnative plant and animal species, human alterations to the local hydrology, and increasing nutrient levels in estuarine waters (Thannheiser and Holland, 1994; Moyle, 1996; Castillo et al., 2000; Talley and Levin, 2001; Bertness et al., 2002; Valéry et al., 2004). In China, *Spartina alterniflora* (introduced in 1979) has spread over much of the coast and has proven to be competitive interacting with the native mangroves (Zhang et al., 2012).

2.2 Benthic Algae

Benthic algae occur on the sediments below and adjacent to the emergent vegetation of tidal saline marshes, as well as on the culms of the emergent vegetation (epiphytic algae). Diatoms are universally present (Sullivan and Currin, 2000). Extensive cyanobacterial populations develop during the summer in Europe (Birkemoe and Liengen, 2000; Quintana and Moreno-Amich, 2002), as well as northeast and southwest coasts of the United States (Blum, 1968; Zedler, 1982). Green and brown algae reach the largest population sizes during seasons in which emergent vegetation is not dominant (Brinkhuis, 1977; Houghton and Woodwell, 1980; Sullivan and Currin, 2000). Distinct benthic algal communities are associated with different emergent vegetation communities and are related to differences in elevation, soil temperature, soil moisture, interstitial ammonium concentration, and canopy height (Sullivan and Currin, 2000).

2.3 Nekton

Numerous factors and classifications have been proposed to characterize marsh habitat and the nekton that inhabit them. Marsh habitat (e.g., zonation) may drive structuring of nekton, and the collection of conditions has been referred to as the "marsh gradient" (Rountree and Able, 2007). From a species perspective, use of marshes may be driven by life history traits that include spawning, feeding, migration, and nursery (growth and refuge) functions, among others. Furthermore, marsh use within a species may vary. In South Carolina, resident and transient species entered during different phases of the tide (Bretsch and Allen, 2006). Species-specific differences in marsh use were also found to be size-dependent in a North Carolina marsh (Meyer and Posey, 2009).

The broadest division of marsh nekton might separate resident and transient species (Cowan et al., 2012). Transient species may be found in marshes (based on tide, season, or numerous other environmental variables) and include both fresh- and saltwater visitors, as well as diadromous forms such as salmonids, clupeids, and anquillids that are present or migratory for one or more life history stages (Baltz et al., 1993; Dionne et al., 1999). The resident and transient groups are still coarse, however, and subsequent classifications have been proposed. Peterson and Turner (1994), for example, suggest four categories based on marsh use: (1) residents on the marsh surface that generally remain at low tide in pools and puddles, (2) regular visitors at high tide that retreat to fringing vegetation along the marsh edge at low tide, (3) individuals of larger species that associate strongly with the marsh edge as juveniles and penetrate only a few meters into the marsh at high tide, and (4) other subtidal species that rarely penetrate far onto the flooded marsh but may be associated with tidal creeks. Although this may be regarded as a useful classification, others groupings may have value.

In Louisiana marshes, tidal amplitude (~30 cm) is at the low end of the microtidal range and it is easily dominated by winds making marsh flooding less predictable. Thus, interannual climatological variation influences flooding duration and frequency of saltmarsh habitats (Childers et al., 1990). In marshes with greater micro- and mesotidal ranges (>1 m), inundation is more predictable and some fishes may spend as much as one-third of their time in flooded smooth cordgrass (*Spartina alterniflora*) (Hettler, 1989). For species that use intertidal zones as nurseries, interannual variation in habitat availability may have a strong influence on recruitment, particularly in microtidal systems (Childers et al., 1990; Baltz et al., 1993, 1998). For example, climatic conditions affect marsh accessibility for juvenile shrimp and are related to interannual variation in shrimp landings (Childers et al., 1990).

Connolly (1999) reviewed study limitations that have hampered our understanding of the direct use of tidal saline marshes by nekton, primarily fishes, shrimps, and crabs, and suggested improvements to standardize sampling methods, overcome poor sampling designs, and improve assessment of flooding regimes and landscape structure. He also noted the uneven distribution of studies, largely limited to North America (90%), Europe (7%), and Australia (3%), although a disproportionate amount of the world's salt marshes occurs on these continents. Based on a review of 20 North American salt marshes along the Atlantic Ocean and Gulf of Mexico, 237 species of fish were detected (with the highest richness at a site being 86 species; Nordlie, 2003). Marine transients were the most represented group (>50%), whereas other groupings were much less represented, and interestingly, the proportion of marine transients was variable among sites, but did not change along a latitudinal gradient. Cattrijsse and Hampel (2006) reviewed intertidal marsh studies in Europe and contrasted floral, faunal, and physical patterns with North American coastal systems. In contrast to North American salt marshes, European salt marshes differ in having the lower limit of marsh vegetation defined by mean high water neap tides rather than mean tide level, vegetation typically dominated by sea purslane (*Halimione* spp.) rather than smooth cordgrass and a much higher stem density that inhibits nekton movement on the marsh surface. Laffaille et al. (2000) examined fish use of saltmarsh vegetation in the macrotidal system of Mont Saint-Michel Bay, France, which is accessible to fishes for only a few minutes or hours during high spring tides (5%–10% of tides). Due to this brief flood duration, no fishes are considered residents, but the annual pattern and three seasonal patterns of community structure are stable. Thirty-one fishes, netted in creeks as tides ebbed off of flooded marsh, included 7 marine stragglers, 13 estuarine-dependent marine species, 3 catadromous species, and 8 estuarine species. Most of the species can be characterized as euryhaline and eurythermal migrants.

In Australia, Connolly et al. (1997) examined fish use of flooded saline marsh in the high intertidal zone that is generally separated from open water by fringing mangroves. Compared with tidal creeks draining the same marsh flats, the number of species and individuals caught in the high intertidal zone was lower; however, the density of fishes on marsh flats (1 per 23 m^2) was higher than expected when creek numbers were parsed over drainage areas (1 per 134 m^2). Only the two most abundant species found in creeks were captured on high marsh flats, glass goby (*Gobiopterus semivestitus*) and smallmouth hardyhead (*Atherinosoma microstoma*).

2.4 Reptiles and Amphibians

Only a few species of snakes, turtles, and crocodilians can be considered common residents of tidal saline marshes (Neill, 1958; Greenberg et al., 2006). The saltmarsh snake (*Nerodia clarkii*, but comprised of three subspecies) is restricted to saline and brackish tidal wetlands along the Atlantic and Gulf coasts of North America. As a way to avoid predators, salt marsh snakes are nocturnal. The saltmarsh snake does not drink saline water and obtains water primarily from the food it eats (Pettus, 1958). Diamondback terrapin (*Malaclemys terrapin*) is the only North American turtle restricted to tidal saline marshes (Greenberg et al., 2006). Special glands in the turtle's eye region excrete excess sodium (Bentley et al., 1967). Juvenile and smaller male terrapins rely on the nearshore area where they forage on readily available prey such as clams, crabs, and small crustaceans. Interestingly, diamondback terrapins show high site fidelity, yet have less genetic structuring than might be expected because of historic mixing and stocking among populations (Hauswaldt and Glenn, 2005). A recent study determined an increase in contemporary gene flow into Chesapeake Bay populations (Converse et al., 2015). Another recent study in the Chesapeake Bay area highlighted vulnerability of species such as the Diamondback terrapin to potential loss of habitat and sea level rise (Woodland et al., 2017). A large number of amphibian and reptile species are occasional visitors to tidal saline marshes but are generally found in fresh to slightly brackish water (Greenberg et al., 2006).

2.5 Birds

Typically, saline marsh avifauna is dominated, at least numerically, by large numbers of Anseriformes (waterfowl), Ciconiiformes (long-legged wading birds), and Charadriiformes (shorebirds, gulls, and terns) (Goss-Custard et al., 1977; Custer and Osborn, 1978; Bildstein et al., 1982; Erwin, 1996). In addition, Passeriformes (songbirds) feed and breed in saline marshes (Brown and Atkinson, 1996; Dierschke and Bairlein, 2004). In many marshes, avian populations increase considerably seasonally, not only during migratory periods, when large numbers of waterfowl and shorebirds congregate to feed and rest, but also during the breeding season, when wading birds congregate at traditional coastal colonies to nest. The total number of breeding bird species whose habitat primarily consists of tidal marshes has been estimated to be between 11 and 21, with only 2 songbird species that are entirely restricted to tidal saline marsh in North America: the seaside sparrow (*Ammodramus maritimus*) and the saltmarsh sharp-tailed sparrow (*Ammodramus caudacutus*) (Greenberg et al., 2006). These species are adapted to nesting in cordgrass-dominated tidal marshes (Benoit and Askins, 1999). Several subspecies of birds restricted to tidal marshes have wider distributions (e.g., a subspecies of the slender-billed thornbill, *Acanthiza iredalei rosinae*) (Greenberg et al., 2006).

2.6 Mammals

The total number of mammalian species whose habitat primarily consists of tidal marshes has been estimated to range from 13 to 26, with rodents predominant (Greenberg et al., 2006). Of these, only the saltmarsh harvest mouse (*Reithrodontomys raviventris*) is restricted to coastal marshes. In addition to wild mammals, domestic or feral mammals (primarily cattle, sheep, and horses) graze in some tidal saline marshes (Bakker, 1985; Turner, 1987). Although saltmarsh mammals represent a relatively small portion of the potential species richness, they are notable for their ability to cause a dieback event through an "eat out" (Alber et al., 2008). "Eat out" events occur when overgrazing by mammals like the invasive nutria (*Myocastor coypus*; as well as by cattle, horses, and other mammals) destroy more vegetation than they consume (Mitsch and Gosselink, 2000), typically resulting in patches of bare mud. In some areas such as coastal Louisiana, the damaged marsh areas might not recover because of rising sea level and high marsh subsidence rates (Mitsch and Gosselink, 2007).

3. INTERACTIONS AMONG COMMUNITIES

3.1 Effects of Animals on Emergent Vegetation Distribution

Burrowing organisms such as fiddler crabs (*Uca* spp.) increase tidal saline marsh plant production through moderation of soil stresses; for example, they increase soil aeration, oxidation–reduction potential, and in situ decomposition of belowground plant debris (Bertness, 1985), although the increase in production may depend on the severity of the stresses (Nomann and Pennings, 1998). In return, the crabs benefit from the shade and the shelter that the increased plant cover provides (Bortolus et al., 2002).

Grazing by herbivorous waterfowl or cattle can significantly change the vegetation community in a tidal saline wetland (Ranwell, 1961; Bakker, 1985; Pehrsson, 1988). Grazing by breeding waterfowl in the Arctic can change the trajectory of plant succession by physically changing the environment. Moderate grazing preserves the dominance of grasses (Bazely and Jefferies, 1986), whereas heavy grazing can convert marsh to unvegetated mudflats (Handa et al., 2002). Intense grazing can also limit the distribution of plants along a maturation (time since plant establishment) gradient that is not related to elevation. Van der Wal et al. (2000) showed that the distribution of seaside arrowgrass (*Triglochin maritima*) in younger marshes is limited because of intense grazing by geese, hares, and rabbits. The distribution in more mature marshes is limited because of competition for light by taller species that are slower colonizers such as saltbush (*Atriplex portulacoides*). Field experiments indicate that the periwinkle (*Littoraria irrorata*) can overgraze otherwise healthy stands of smooth cordgrass and reduce them to bare mudflats when periwinkle predator density is low. Thus, periwinkle predators including blue crabs (*Callinectes sapidus*) and terrapins are capable of exerting top-down control of smooth cordgrass production (Silliman and Bertness, 2002). The density of grazers can be positively correlated with nitrogen availability (Bowdish and Stiling, 1998; Visser et al., 2006). The more stressful conditions in the low marsh may make plants at this elevation more palatable (Dormann et al., 2000) and grazing effects may be more pronounced in these more stressful environments (Kuijper and Bakker, 2005). Neighboring plant species may have both positive and negative effects on the level of grazing of a palatable species. Grazing on smooth cordgrass by an herbivorous crab (*Chasmagnathus granulata*) is more intense when the plant grows with alkali bulrush (*Scirpus maritimus*) than when it grows in monotypic cordgrass stands (Costa et al., 2003).

3.2 Emergent Vegetation as Animal Habitat

In some tidal saline marshes several species of wintering songbirds forage on the seeds of Limonium, Sueda, and Salicornia (Brown and Atkinson, 1996). In addition, bulrush and cordgrass seeds are eaten by waterfowl (Mendall, 1949; Hartman, 1963; Landers et al., 1976; Gordon et al., 1998). Birds feeding on small aquatic organisms along the marsh edge occur more frequently in areas that have more open water in the form of marsh creeks and ponds (Craig and Beal, 1992). Both rodents and birds nesting in tidal saline marshes use the available vegetation as nesting material (Shanholtzer, 1974). Nests are generally constructed from blades of grass. Similar to forest species, breeding birds in tidal marshes are often associated with larger marsh tracts (Craig and Beal, 1992). Although tidal saline marshes probably function in a variety of ways to enhance growth and survival of a particular nekton species, the relative importance of food versus refuge from predation is poorly understood (Boesch and Turner, 1984) and probably varies across species and life history stages. Both shallow water and vegetation in the marsh provide protection for small nekton from large predators, particularly from large piscivorous birds and fishes (Kneib, 1982a,b), and both provide a food-rich environment (Van Dolah, 1978; Gleason, 1986; Cyrus and Blaber, 1987; Gleason and Wellington, 1988). In experimental tests of predation as a factor determining the size-specific habitat difference between killifish (*Fundulus heteroclitus*) age classes, Kneib (1987) found that young killifish remained in high intertidal cordgrass habitat avoiding concentrations of larger piscivorous fishes in subtidal habitats, whereas habitat use by larger killifish is influenced by avian predators. Other field experiments indicate that predation pressure is lower and food availability is higher in vegetated than in unvegetated habitats (Rozas and Odum, 1988).

3.3 Nursery Function

Nurseries foster the growth and/or survival of early life history stages of fishes and macroinvertebrates (Beck et al., 2001). The particular environment used by a species may be characterized as nursery habitat if it can be shown that individuals are found at higher densities and experience enhanced survival and/or growth compared with nearby habitat types (Pearcy and Myers, 1974; Weinstein, 1979). It is difficult to separate the nursery function of flooded cordgrass marsh from that of adjacent habitat types. Baltz et al. (1993, 1998) examined the use of shallow open water and flooded smooth cordgrass as nursery habitat for a variety of resident and transient fishes in Louisiana estuaries. Small fishes and early life history stages of larger species often use shallow turbid water along the marsh edge at low tides and move onto flooded marsh at higher tides. Evidence suggests that the magnitude of fishery landings is correlated with the spatial extent of estuarine vegetation (Turner, 1977, 1992; Pauly and Ingles, 1988); therefore, extensive marsh loss in the northern Gulf of Mexico is a major concern for the sustainability of fisheries. Construction of dams and weirs in the upper Mississippi River Basin in the last century and downstream channelization have resulted in reduced sediment available for overland flow into the deltaic plain wetlands in Louisiana (Kesel, 1988), resulting in significant marsh degradation. However, the connection between fishery landings and marsh habitat loss in the northern Gulf of Mexico is not clear. Moreover, landings have increased for many species in spite of accumulating habitat alterations (Zimmerman et al., 1991). One hypothesis is that the marsh edge (i.e., the perimeter at the marsh–open water interface) is the essential habitat for many species and that the nursery function and value will not decline and result in reduced landings until the quantity of marsh edge perimeter begins to decline. During the process of marsh deterioration, the amount of marshedge initially increases as solid marsh is converted to broken marsh and then it declines as broken marsh is converted to open water (Chesney et al., 2000). A temporary increase in marshedge perimeter, which occurs in the broken marsh phase, may be masking the ultimate effect of habitat loss on landings (Browder et al., 1985, 1989). An alternative hypothesis is that the marshedge is not the essential habitat per se but serves as access to flooded marsh, which is essential. However, neither hypothesis may be appropriate for all species because a variety of species, whose microhabitat use patterns often differ, occur in high densities in marshedge and flooded cordgrass habitat types (Zimmerman and Minello, 1984; Rakocinski et al., 1992; Baltz et al., 1993; Minello et al., 1994). These alternative hypotheses are testable in experiments that examine growth and/or survival along the marsh edge by contrasting sites with and without access to flooded marsh.

3.4 Saline Marsh Food Webs

Saline marsh food webs are probably far more complex than we recognize. Certainly, the fishes and macroinvertebrates that form upper trophic levels add to this complexity, particularly if we take their large variation in size into account. Because many species of fish and macroinvertebrates continue to grow throughout their lives, they can individually function as multiple "species," as their predator–prey relationships and habitat utilization patterns change with ontogeny (Livingston, 1988). The food webs are also complicated by interactions with subtidal ecosystem components. The traditional view of bottom-up control in saltmarsh food webs continues being supported in some work, yet it is clear that top-down effects play a larger role than previously thought (Sala et al., 2008).

3.4.1 Species Interactions

Kneib (1984) highlighted our lack of knowledge about the importance of complex species interactions in saline marsh communities that involve more than two trophic levels. Besides predation, there are many factors acting alone or in concert that may influence the distribution and abundance of invertebrates and their predators, primarily crustaceans, fishes, and birds. These include density-dependent processes, selective larval settlement or mortality, physical gradients that influence habitat selection, and both unpredictable and cyclical physical disturbances.

Pennings et al. (2001) compared the palatability of northern and southern populations of smooth cordgrass and saltmeadow cordgrass (*Spartina patens*) to a variety of grazing insects. In 28 of 32 trials, the insects showed significant preferences for the northern plants and supported the biogeographic hypothesis that lower latitude plants are better defended from herbivory. Whether the preferences are based on toughness, nutrient and mineral content, or secondary metabolites remains unclear and is likely to vary between plant-herbivore species pairs, although more recent experimental work (Ho and Pennings, 2013) also supports latitudinal trends. Somewhat counter to these findings, Marczak et al. (2011) demonstrated that latitudinal variation in plant quality (*Iva frutescens*) was less important than latitudinal variation in top consumers. Although climate may structure herbivores (e.g., through generation time), saltmarsh consumers still likely exert some top-down control. Although results from individual studies prove interesting, a large review (443 studies) of consumer control on vegetation found that saltmarsh herbivores often strongly suppress plant survival (He and Silliman, 2016).

Kneib (1991) explored the importance of indirect effects, particularly involving chains of predator–prey interactions, in soft-sediment communities. Indirect effects can influence primary producers, macrofauna, and meiofauna in marsh communities and are often implicated in counterintuitive outcomes of experiments intended to examine direct effects. Soft sediments reduce the feeding efficiency of predators on epibenthic meiofauna (Gregg and Fleeger, 1998). Large crustaceans are important predators in marshes. Grass shrimp have been well studied and are implicated as a connecting link between meiofaunal and nekton communities, especially as effective predators that focus their activities around the bases of smooth cordgrass stems (Gregg and Fleeger, 1998).

In an elegant field-caging experiment, Silliman and Zieman (2001) demonstrated that periwinkle (*Littoraria irrorata*) grazing could exert top-down control of smooth cordgrass annual net primary production. The periwinkle is an important grazer in marshes, but it is also an important prey of a large number of predators that forage on marshes, so its ability to control smooth cordgrass is questionable where it suffers normal predation. Sala et al. (2008) reported similar evidence of potential top-down control. When New England salt marshes became eutrophic, it was found that grazing increased such that dual control was likely structuring the community. In perhaps the most extreme investigation of top-down control, Nifong and Silliman (2013) manipulated the American alligator (*Alligator mississippiensis*) to demonstrate trophic cascades, behavioral changes on mesopredators, and indirect effects on grazing.

In addition to species interactions having to do with predation, other relationships are important for structure and function of salt marsh ecosystems. For example, Angelini et al. (2016) found that mussel mounds greatly enhanced cordgrass survival during drought conditions.

3.4.2 *Primary Producers*

Stable isotopes are useful for tracing the flow of primary production and nutrients through food webs (Fry, 2006). Stable isotopes of nitrogen, sulfur, and carbon, when used in combination, can greatly increase the power of the isotopic tracer approach in coastal food webs (Peterson et al., 1985) and help address questions about the role of cordgrass marshes in supporting marsh and estuarine consumers. Benthic microalgae and standing dead material may overshadow live cordgrass, macroalgae, and phytoplankton as sources of carbon (Sullivan and Moncreiff, 1990; Currin et al., 1995). Peterson et al. (1986) found that cordgrass detritus and phytoplankton were much more important than upland vegetation and sulfur-oxidizing bacteria as carbon sources for marsh macroconsumers. Killifish and mud snails relied more on cordgrass, while filter feeders typified by oysters and mussels relied on a combination of cordgrass and plankton. In addition to marsh and estuarine consumers, organic matter from marsh vegetation and estuarine plankton is exported to the offshore environment where it is consumed by marine organisms (Teal, 1962; Odum, 2000). Alber and Valiela (1994) provide evidence that microbial organic aggregates are more important in the nutrition of two marine mussels than particulate detritus or dissolved organic matter. Despite the longstanding support of cordgrass detritus as the base of the saltmarsh food web, Galván et al. (2008) used stable isotopes to suggest limited importance of cordgrass detritus in a New England salt marsh. Benthic algae and phytoplankton were found to be the dominant food source. In another study in southern California, cyanobacteria (microalgae) was found to play an important role in saltmarsh food webs and young salt marshes in particular (Currin et al., 2011). It is likely that the importance of primary producers may be more spatially and temporally variable than traditionally thought.

Nitrogen fixation by the community of epiphytes growing on standing dead stems contributes significantly to total nitrogen fixation in marshes where senescent plants are not flattened by ice (Currin and Paerl, 1998). The rate for natural salt marsh is 2.6 $g\,N\,m^{-2}$ stem surface per year and is comparable with sediment rates and about half of rhizospheric nitrogen fixation. Rates are comparable to cyanobacterial mats 2–8 $g\,N\,m^{-2}\,year^{-1}$. While of little direct benefit to cordgrass, nitrogen fixation by epiphytes is important to animals that graze on them and to the nutrient cycle as a source of new biologically available nitrogen. Epiphytes are an important food resource for consumers (Currin et al., 1995). Diverse meiofaunal communities associated with algal epiphytes on smooth cordgrass stems are utilized by shrimp and fish (Rutledge and Fleeger, 1993; Gregg and Fleeger, 1998). Three saltmarsh macroinvertebrates, periwinkles (*Littoraria irrorata*), saltmarsh coffee bean snails (*Melampus bidentatus*), and talitrid amphipods (*Uhlorchestia spartinophila*), which feed by shredding dead and senescing smooth cordgrass leaves, also benefit from ingesting fungi; snails and amphipods also stimulate fungal growth (Graça et al., 2000). These grazing macroinvertebrates contribute to nutrient cycling and connect microbial decomposers to higher-order consumers such as blue crab and *Fundulus* species that prey on them (Graça et al., 2000).

3.4.3 Indirect Interactions Among Species

A general and realistic qualitative loop model (Levins, 1966; Puccia and Levins, 1991) of a simplified saltmarsh food web (Fig. 15.1A) was compiled based in part on the literature reviewed above to illustrate the importance of indirect interactions. The model includes 20 nodes and covers several trophic levels. In this graphical model, direct positive effects of one node on another are indicated by a link terminating in an arrowhead (→), and direct negative effects are indicated by a link terminating in a filled circle (•). A loop is a path of interactions through nodes (variables) that return to the node of origin without retracing itself through any node previously encountered. Model input is a community interaction matrix of direct interaction (i.e., −1, 0, or +1), and output (i.e., the adjoint matrix) is an evaluation of the net number of positive or negative feedback loops influencing each node of the community (Dambacher et al., 2002a,b). Loop models can predict the outcomes of one or more perturbations on a system providing that the community structure is stable (Bender et al., 1984; Dambacher et al., 2002a). A press perturbation (*sensu* Bender et al., 1984), which may be positive or negative, is a sustained alteration of species densities (or environmental variables), and if the press is maintained, the unperturbed species (or variables) reach a new equilibrium. In essence, the elements of an adjoint matrix are the algebraic summation of the number of positive or negative feedback loops that contribute to the direction of change of a given variable. We have simplified the model output to a graphical representation of a positive press on the variables (Fig. 15.1B).

While the analysis of many similar models is necessary to reveal general truths (Levins, 1966) about marsh food webs, we can gain some insights from a single model. Our greatly simplified model with only 20 nodes has several million feedback loops and supports Kneib's (1991) contention that indirect effects can influence primary producers, macrofauna, and meiofauna in marsh communities and are often implicated in counterintuitive outcomes of experiments intended to examine direct effects. As an illustration, we can revisit Silliman and Zieman's (2001) cage experiment in which nitrogen and periwinkles (gastropods) were manipulated. In our model, a positive press on nitrogen or periwinkles predicts an enhancement and reduction of *Spartina*, respectively (Fig. 15.1B). However, the experimental design also excluded major predators on periwinkles. In effect, the cage experiment can be interpreted as simultaneous perturbations of nitrogen, periwinkles, and their predators, most importantly blue crabs (Graça et al., 2000). A negative press on periwinkle predators, including blue crabs (−15,157 feedback loops) and raccoons (−13,493), greatly enhances the direct effect of augmenting periwinkles (−26,103) and offsets the enhancement due to nutrients (21,244). The overall effect on *Spartina* (−33,509) is strongly negative (Table 15.2). Moreover, the model indicates that virtually any positive or negative press, either alone or in concert, can strongly influence *Spartina* (Fig. 15.1B). The often-overlooked influence of indirect effects can be manifested by perturbations on all nodes in the model and should be more generally evaluated in nature before assuming bottom-up or top-down control of ecosystems.

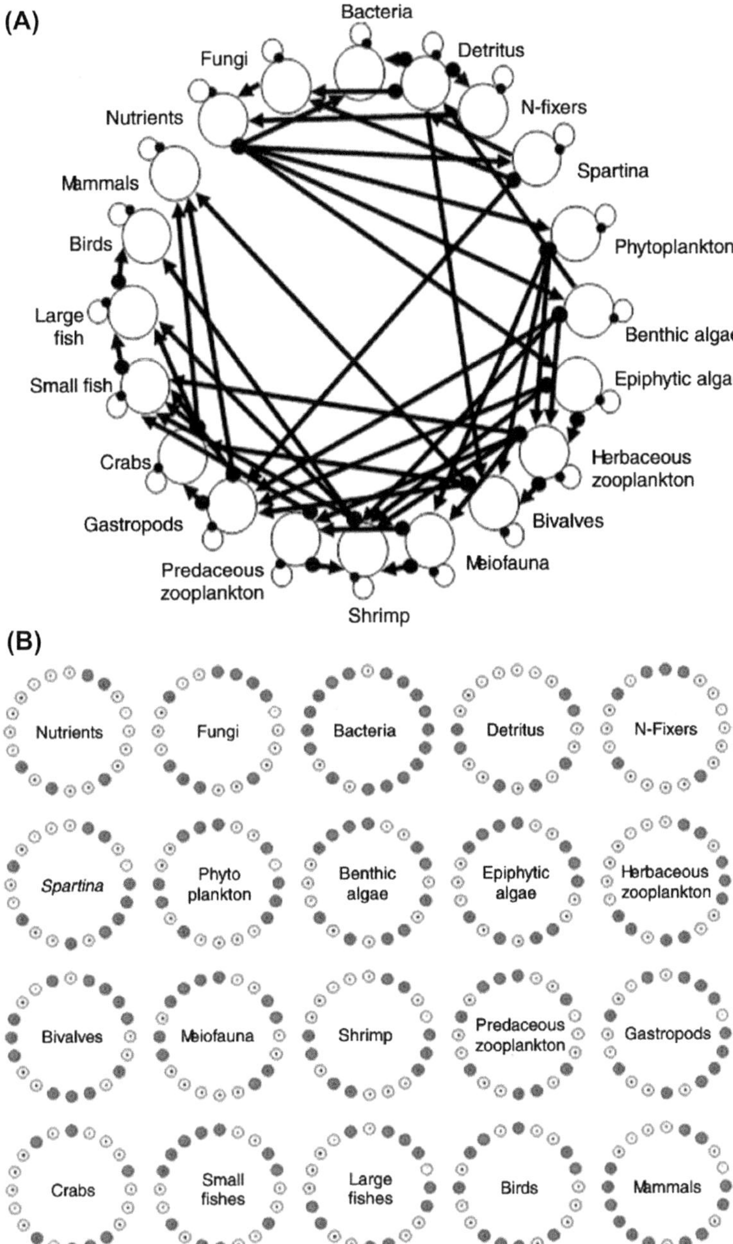

FIGURE 15.1 (A) A pictorial model of a *Spartina* marsh food web used to generate a qualitative loop analysis. Connections indicate positive (→) and negative (•) direct interactions between nodes and result in many positive and negative feedback loops that amount to indirect interactions. (B) A simplified graphical representation of a positive press on the named variables. Named variables occur inside a *circle* of 20 points, which correspond to the 20 *circles* in Fig. 15.1A. For example, in the first *circle*, nutrients are enhanced, which results in negative responses in the small red nodes (*circles*) corresponding to detritus, N-fixers, bivalves, predaceous zooplankton, and crabs. All other (white) nodes are enhanced. A negative press on a variable of interest can be seen by simply reversing the signs of all other elements.

TABLE 15.2 Individual Positive Presses on Nutrients and Gastropods and Negative Presses on Crabs and Mammals Result in a Mix of Responses by *Spartina*

	Nutrients (+)	Gastropods (+)	Crabs (−)	Mammals (−)	Summation
Nutrients	54,740	4856	10,556	−3832	66,320
Fungi	7559	−27,317	−17,796	−12,535	−50,089
Bacteria	41,055	3642	7917	−2874	49,740
Detritus	−13,685	−1214	−2639	958	−16,580
N-fixers	−13,685	−1214	−2639	958	−16,580
Spartina	21,244	−26,103	−15,157	−13,493	−33,509
Phytoplankton	28,852	29,074	14,415	−9424	62,917
Benthic algae	13,188	−20,000	−19,083	9023	−16,872
Epiphytic algae	13,188	−20,000	−19,083	9023	−16,872
Herbaceous zooplankton	4279	−11,900	−21,166	−2659	−31,446
Bivalves	−3452	−37,510	16,033	2562	−22,367
Meiofauna	25,061	25,192	1274	5689	57,216
Shrimp	11,336	−21,520	7296	−32,392	−35,280
Predaceous zooplankton	−3266	13,502	−15,321	14,620	9535
Gastropods	25,937	58,276	43,509	22,196	149,918
Crabs	−12,762	−12,786	−64,237	58,876	−30,909
Small fishes	21,270	21,310	−11,867	20,802	51,515
Large fishes	4254	4262	−38,052	39,839	10,303
Birds	15,590	−17,258	−30,756	7447	−24,977
Mammals	9723	7980	−4695	−94,759	−81,751

The summation of all four presses, two positive and two negative, results in an overall reduction of *Spartina*, and all other variables in the model are also influenced to some degree. This is an extract of the adjoint matrix and negative presses for crabs and mammals were generated by reversing the signs on all variables to reflect the effects of reducing crab and mammal abundances.

References

Adam, P., 1990. Saltmarsh Ecology. Cambridge University Press, Cambridge, Great Britain, 461 p.

Alber, M., Valiela, I., 1994. Incorporation of organic aggregates by marine mussels. Marine Biology 121, 259–265.

Alber, M., Swenson, E.M., Adamowicz, S.C., Mendelssohn, I.A., 2008. Salt marsh dieback: an overview of recent events in the US. Estuarine, Coastal and Shelf Science 80 (1), 1–11.

Alvarez Rogel, J., Alcaraz Ariza, F., Ortiz Silla, R., 2000. Soil salinity and moisture gradients and plant zonation in Mediterranean salt marshes of southeast Spain. Wetlands 20, 357–372.

Angelini, C., Griffin, J., van de Koppel, J., Derksen-Hooijberg, M., Lamers, L., Smolders, A.J.P., van der Heide, T., Silliman, B.R., 2016. A keystone mutualism underpins resilience of a coastal ecosystem to drought. Nature Communications 7, 12473.

Armstrong, W., Beckett, P.M., Lythe, S., Gaynard, T.J., 1985. Plant zonation and the effects of the spring-neap tidal cycle on soil aeration in a Humber salt marsh. Journal of Ecology 73, 323–339.

Bakker, J.P., 1985. The impact of grazing on plant communities, plant populations and soil conditions on salt marshes. Vegetation 62, 391–398.

Baltz, D.M., Rakocinski, C., Fleeger, J.W., 1993. Microhabitat use by marsh-edge fishes in a Louisiana estuary. Environmental Biology of Fishes 36, 109–126.

Baltz, D.M., Fleeger, J.W., Rakocinski, C.F., McCall, J.N., 1998. Food, density, and microhabitat: factors affecting growth and recruitment potential of juvenile saltmarsh fishes. Environmental Biology of Fishes 53, 89–103.

Baltz, D.M., Yáñez-Arancibia, A., 2009. Ecosystem-based management of coastal fisheries in the Gulf of Mexico: environmental and anthropogenic impacts and essential habitat protection. In: Day, J.W., Yáñez-Arancibia, A. (Eds.), The Gulf of Mexico: Ecosystem Based Management. The Gulf of Mexico: Its Origin, Waters, Biota, Economics & Human Impacts (Series). Harte Research Institute for Gulf of Mexico Studies, Texas A&M University – Corpus Christi, Texas A&M University Press, College Station, TX.

Bazely, D.R., Jefferies, R.L., 1986. Changes in the composition and standing crop of salt-marsh communities in response to the removal of a grazer. Journal of Ecology 74, 693–706.

Beaulieu, N., Allard, M., 2003. The impact of climate change on an emerging coastline affected by discontinuous permafrost: Manitounuk Strait, northern Quebec. Canadian Journal of Earth Sciences 40, 1393–1404.

Beeftink, W.G., 1977. Salt marshes. In: Barnes, R.S.K. (Ed.), The Coastline. John Wiley & Sons, New York, USA, pp. 109–149.

Beck, M.W., Heck Jr., K.L., Able, K.W., Childers, D.L., Eggleston, D.B., Gillanders, B.M., Halpern, B., Hays, C.G., Hosino, K., Minello, T.J., Orth, R.J., Sheridan, P.F., Weinstein, M.P., 2001. The identification, conservation, and management of estuarine and marine nurseries for fish and invertebrates. BioScience 51, 633–641.

Bender, E.A., Case, T.J., Gilpin, M.E., 1984. Perturbation experiments in community ecology: theory and practice. Ecology 65, 1–13.

Benoit, L.K., Askins, R.A., 1999. Impact of the spread of phragmites on the distribution of birds in Connecticut tidal marshes. Wetlands 19, 194–208.

Bentley, P.J., Bretz, W.L., Schmidt-Nielsen, K., 1967. Osmoregulation in the diamondback terrapin, *Malaclemys terrapin centrata*. Journal of Experimental Biology 46, 161–167.

Bertness, M.D., 1985. Fiddler crab regulation of *Spartina alterniflora* production on a New England salt marsh. Ecology 66, 1042–1055.

Bertness, M.D., Ewanchuk, P.J., Silliman, B.R., 2002. Anthropogenic modification of New England salt marsh landscapes. Proceedings of the National Academy of Sciences USA 99, 1395–1398.

Bertness, M.D., Pennings, S.C., 2000. Spatial variation in process and pattern in salt marsh plant communities in eastern North America. In: Weinstein, M.P., Kreeger, D.A. (Eds.), Concepts and Controversies in Tidal Marsh Ecology. Kluwer Academic Publishers, Dordrecht, The Netherlands, pp. 39–58.

Bertness, M.D., Shumway, S.W., 1993. Competition and facilitation in marsh plants. The American Naturalist 142, 718–724.

Bildstein, K.L., Christy, R., Decoursey, P., 1982. Size and structure of a South Carolina salt marsh avian community. Wetlands 2, 118–137.

Birkemoe, T., Liengen, T., 2000. Does collembolan grazing influence nitrogen fixation by cyanobacteria in the high Arctic? Polar Biology 23, 589–592.

Blum, J.L., 1968. Salt marsh *Spartinas* and associated algae. Ecological Monographs 38, 199–221.

Boesch, D.F., Turner, R.E., 1984. Dependence of fishery species on salt marshes: the role of food and refuge. Estuaries 7 (4A), 460–468.

Boorman, L.A., 1967. *Limonium vulgare* Mill. And L. Humile Mill. (in biological Flora of the British Isles). Journal of Ecology 55, 221–232.

Bortolus, A., Schwindt, E., Iribarne, O., 2002. Positive plant–animal interactions in the high marsh of an Argentinean coastal lagoon. Ecology 83, 733–742.

Bowdish, T.I., Stiling, P., 1998. The influence of salt and nitrogen on herbivore abundance: direct and indirect effects. Oecologia 113, 400–405.

Brinkhuis, P.H., 1977. Seasonal variations in salt-march macroalgae photosynthesis. II. *Fucus vesiculosus* and *Ulva lactuca*. Marine Biology 44, 177–186.

Bretsch, K., Allen, D.M., 2006. Tidal migrations of nekton in salt marsh intertidal creeks. Estuaries and Coasts 29 (3), 474–486.

Browder, J.A., Bartley, H.A., Davis, K.S., 1985. A probabilistic model of the relationship between marshland-water interface and marsh disintegration. Ecological Modeling 29, 245–260.

Browder, J.A., May Jr., J.L.N., Rosenthal, A., Gosselink, J.G., Baumann, R.H., 1989. Modeling future trends in wetland loss and brown shrimp production in Louisiana using thematic mapper imagery. Remote Sensing of Environment 28, 45–59.

Brown, A.F., Atkinson, P.W., 1996. Habitat associations of coastal wintering passerines. Bird Study 43, 188–200.

Callaway, R.M., Jones, S., Ferren, W.R., Parikh, A., 1990. Ecology of a Mediterranean-climate estuarine wetland at Carpinteria, California: plant distributions and soil salinity in the upper marsh. Canadian Journal of Botany 68, 1139–1146.

Cantero, J.J., Leon, R., Cisneros, J.M., Cantero, A., 1998. Habitat structure and vegetation relationships in central Argentina salt marsh landscapes. Plant Ecology 137, 79–100.

Castellanos, E.M., Figueroa, M.E., Davy, A.J., 1994. Nucleation and fascilitation in saltmarsh succession: interactions between *Spartina maritima* and *Arthrocnemum perenne*. Journal of Ecology 82, 239–248.

Castillo, J.M., Fernández-Baco, L., Castellanos, E.M., Luque, C.J., Figueroa, M.E., Davy, A.J., 2000. Lower limits of *Spartina densiflora* and *S. maritima* in a Mediterranean salt marsh determined by different ecophysiological tolerances. Journal of Ecology 88, 801–812.

Cattrijsse, A., Hampel, H., 2006. European intertidal marshes: a review of their habitat functioning and value for aquatic organisms. Marine Ecology Progress Series 324, 293–307.

Chapman, V.J., 1977. Wet Coastal Ecosystems. Elsevier, Amsterdam, 428 p.

Chesney, E.J., Baltz, D.M., Thomas, R.G., 2000. Louisiana estuarine and coastal fisheries and habitats: perspectives from a fish's eye view. Ecological Applications 10, 350–366.

Childers, D.L., Day Jr., J.W., Muller, R.A., 1990. Relating climatological forcing to coastal water levels in Louisiana estuaries and the potential importance of El Nino-Southern Oscillation events. Climate Research 1, 31–42.

Clarke, L.D., Hannon, N.J., 1970. The mangrove swamp and salt marsh communities of the Sydney District: III. Plant growth in relation to salinity and waterlogging. Journal of Ecology 58, 351–369.

Connolly, R.M., Dalton, A., Bass, D.A., 1997. Fish use of an inundated saltmarsh flat in a temperate Australian estuary. Australian Journal of Ecology 22, 222–226.

Connolly, R.M., 1999. Saltmarsh as habitat for fish and nektonic crustaceans: challenges in sampling designs and methods. Australian Journal of Ecology 24, 422–430.

Converse, P.E., Kuchta, S.R., Roosenburg, W.M., Henry, P.F.P., Haramis, G.M., King, T.L., 2015. Spatiotemporal analysis of gene flow in Chesapeake Bay Diamondback terrapins (*Malaclemys terrapin*). Molecular Ecology 24, 5864–5876.

Costa, C.S.B., Marangoni, J.C., Azevedo, A.M.G., 2003. Plant zonation in irregularly flooded salt marshes: relative importance of stress tolerance and biological interactions. Journal of Ecology 91, 951–965.

Cowan, J.H., Yáñez-Arancibia, A., Sánchez-Gil, P., Deegan, L.A., 2012. Estuarine nekton. In: Day, J.W., Crump, B.C., Kemp, W.M., Yáñez-Arancibia, A. (Eds.), Estuarine Ecology, second ed. John Wiley & Sons, Inc., Hoboken, NJ, USA.

Craig, R.J., Beal, K.G., 1992. The influence of habitat variables on marsh bird communities of the Connecticut River Estuary. Wilson Bulletin 104, 295–311.

Currin, C.A., Newell, S.Y., Paerl, H.W., 1995. The role of standing dead *Spartina alterniflora* and benthic microalgae in salt marsh food webs: considerations based on multiple stable isotope analysis. Marine Ecology Progress Series 121, 99–116.

Currin, C.A., Paerl, H.W., 1998. Epiphytic nitrogen fixation associated with standing dead shoots of smooth cordgrass, *Spartina alterniflora*. Estuaries 21, 108–117.

Currin, C.A., Levin, L.A., Talley, T.S., Michener, R., Talley, D., 2011. The role of cyanobacteria in Southern California salt marsh food webs. Marine Ecology 32, 346–363.

Custer, T.W., Osborn, R.G., 1978. Feeding habitat use by colonially-breeding herons, egrets, and ibises in North Carolina. The Auk 95, 733–743.

Cyrus, D.P., Blaber, S.J.M., 1987. The influence of turbidity on juvenile marine fish in the estuaries of Natal, South Africa. Continental Shelf Research 7, 1411–1416.

Dambacher, J.M., Li, H.W., Rossignol, P.A., 2002a. Relevance of community structure in assessing indeterminacy of ecological predictions. Ecology 83, 1372–1385.

Dambacher, J.M., Li, H.W., Rossignol, P.A., 2002b. Maple V program commands for qualitative and symbolic analysis of the community matrix. Ecological Archives (Ecology). E083022–S1 (2005 update).
Davy, A.J., Bishop, G.F., 1991. *Triglochin maritima* L. (in biological Flora of the British Isles). Journal of Ecology 79, 531–555.
Dierschke, J., Bairlein, F., 2004. Habitat selection of wintering passerines in salt marshes of the German Wadden Sea. Journal of Ornithology 145, 48–58.
Dionne, M., Short, F.T., Burdick, D.M., 1999. Fish utilization of restored, created, and reference salt-marsh habitat in the Gulf of Maine. American Fisheries Society Symposium 22, 384–404.
Dormann, C.F., van Der Wal, R., Bakker, J.P., 2000. Competition and herbivory during salt marsh succession: the importance of for growth strategy. Journal of Ecology 88, 571–583.
Erwin, R.M., 1996. Dependence of waterbirds and shorebirds on shallow-water habitats in the Mid-Atlantic Coastal Region: an ecological profile and management recommendations. Estuaries 19, 213–219.
Figueroa, M.E., Castillo, J.M., Redondo, S., Luque, T., Castellanos, E.M., Nieva, F.J., Luque, C.J., Rubio-Casal, A.E., Davy, A.J., 2003. Facilitated invasion by hybridization of *Sarcocornia* species in a salt-marsh succession. Journal of Ecology 91, 616–626.
Fry, B., 2006. Stable Isotope Ecology. Springer, New York, p. 300.
Galván, K., Fleeger, J.W., Fry, B., 2008. Stable isotope addition reveals dietary importance of phytoplankton and microphytobenthos to saltmarsh infauna. Marine Ecology Progress Series 359, 37–49.
Gleason, D.F., 1986. Utilization of salt marsh plants by postlarval brown shrimp: carbon assimilation rates and food preferences. Marine Ecology Progress Series 31, 151–158.
Gleason, D.F., Wellington, G.M., 1988. Food resources of postlarval brown shrimp *(Penaeus aztecus)* in a Texas salt marsh. Marine Biology 97, 329–337.
Gordon, D.H., Gray, B.T., Kaminski, R.M., 1998. Dabbling duck-habitat associations during winter in coastal South Carolina. Journal of Wildlife Management 62, 569–580.
Goss-Custard, J.D., Jones, R.E., Newbery, P.E., 1977. The ecology of the Wash I: distribution and diet of wading birds (Charadrii). Journal of Applied Ecology 14, 681–700.
Graça, M.A., Newell, S.Y., Kneib, R.T., 2000. Grazing rates of organic matter and living fungal biomass of decaying *Spartina alterniflora* by three species of salt-marsh invertebrates. Marine Biology 136, 281–289.
Gray, A.J., Mogg, R.J., 2001. Climate impacts on pioneer salt marsh plants. Climate Research 18, 105–112.
Greenberg, R., Maldonado, J.E., Droege, S., McDonald, M.V., 2006. Tidal marshes: a global perspective on the evolution and conservation of their terrestrial vertebrates. BioScience 56, 675–685.
Gregg, C.S., Fleeger, J.W., 1998. Grass shrimp *Palaemonetes pugio* predation on sediment- and stem-dwelling meiofauna: field and laboratory experiments. Marine Ecology Progress Series 175, 77–86.
Handa, I.T., Harmsen, R., Jefferies, R.L., 2002. Patterns of vegetation change and the recovery potential of degraded areas in a coastal marsh system of the Hudson Bay lowlands. Journal of Ecology 90, 86–99.
Hartman, F.E., 1963. Estuarine wintering habitat for black ducks. Journal of Wildlife Management 27, 339–347.
Hauswaldt, J.S., Glenn, T.C., 2005. Population genetics of the diamondback terrapin (*Malaclemys terrapin*). Molecular Ecology 14, 723–732.
He, Q., Silliman, B.R., 2016. Consumer control as a common driver of coastal vegetation worldwide. Ecological Monographs 86 (3), 278–294.
Hettler Jr., W.F., 1989. Nekton use of regularly-flooded saltmarsh cordgrass habitat in North Carolina, USA. Marine Ecology Progress Series 56, 111–118.
Ho, C.-K., Pennings, S.C., 2013. Preference and performance in plant-herbivore interactions across latitude – a study in U.S. Atlantic salt marshes. PLoS One 8 (3).
Houghton, R.A., Woodwell, G.M., 1980. The Flax Pond ecosystem study: exchanges of CO_2 between a salt marsh and the atmosphere. Ecology 61, 1434–1445.
Huckle, J.M., Potter, J.A., Marrs, R.H., 2000. Influence of environmental factors on the growth and interactions between salt marsh plants: effects of salinity, sediment and waterlogging. Journal of Ecology 88, 492–505.
Jackson, J.B.C., Kirby, M.X., Berger, W.H., Bjorndal, K.A., Botsford, L.W., Bourque, B.J., Bradbury, R.H., Cooke, R., Erlandson, J., Estes, J.A., Hughes, T.P., Kidwell, S., Lange, C.B., Lenihan, H.S., Pandolfi, J.M., Peterson, C., Steneck, R.S., Tegner, M.J., Warner, R.R., 2001. Historical overfishing and the recent collapse of coastal ecosystems. Science 293, 629–638.

Kates, R.W., Clark, W.C., Corell, R., Hall, J.M., Jaeger, C.C., Lowe, I., McCarthy, J.J., Schellnhuber, H.J., Bolin, B., Dickson, N.M., Faucheux, S., Gallopin, G.C., Grübler, A., Huntley, B., Jäger, J., Jodha, N.S., Kasperson, R.E., Mabogunje, A., Matson, P., Mooney, H., Moore III, B., O'Riordan, T., Svedin, U., 2001. Sustainability science. Science 292 (5517), 641–642.

Kesel, R.H., 1988. The decline in the suspended load of the lower Mississippi River and its influence on adjacent wetlands. Environmental Geology and Water Sciences 11 (3), 271–281.

Kneib, R.T., 1982a. The effects of predation by wading birds (Ardeidae) and blue crabs *(Callinectes sapidus)* on the population size structure of the common mummichog, *Fundulus heteroclitus*. Estuarine, Coastal, and Shelf Science 14, 159–165.

Kneib, R.T., 1982b. Habitat preference, predation, and the intertidal distribution of gammaridean amphipods in a North Carolina salt marsh. Journal of Experimental Marine Biology and Ecology 59, 219–230.

Kneib, R.T., 1984. Patterns of invertebrate distribution and abundance in the intertidal salt marsh: causes and questions. Estuaries 7, 392–412.

Kneib, R.T., 1987. Predation risk and use of intertidal habitats by young fishes and shrimp. Ecology 68, 379–386.

Kneib, R.T., 1991. Indirect effects in experimental studies of marine soft-sediment communities. American Zoologist 31, 874–885.

Kuijper, D.P.J., Bakker, J.P., 2005. Top-down control of small herbivores on salt-marsh vegetation along a productivity gradient. Ecology 86, 914–923.

Laffaille, P., Feunteun, E., Lefeuvre, J.C., 2000. Composition of fish communities in a European macrotidal salt marsh (the Mont Saint-Michel Bay, France). Estuarine, Coastal and Shelf Science 51, 429–438.

Landers, L., Johnson, A.S., Morgan, P.H., Baldwin, W.P., 1976. Duck foods in managed tidal impoundments in South Carolina. Journal of Wildlife Management 40, 721–728.

Levins, R., 1966. The strategy of model building in population biology. American Scientist 54, 421–431.

Livingston, R.J., 1988. Inadequacy of species-level designations for ecological studies of coastal migratory fishes. Environmental Biology of Fishes 22, 225–234.

Mahall, B.E., Park, R.B., 1976. The ecotone between *Spartina foliosa* Trin. and *Salicornia virginica* L. in salt marshes of northern San Francisco Bay: II. Soil water and salinity. Journal of Ecology 64, 793–809.

Marczak, L.B., Ho, C.K., Wieski, K., Vu, H., Denno, R.F., Pennings, S.C., 2011. Latitudinal variation in top-down and bottom-up control of a salt marsh food web. Ecology 92, 276–281.

Mendall, H.L., 1949. Food habits in relation to black duck management in Maine. Journal of Wildlife Management 13, 64–101.

Meyer, D.L., Posey, M.H., 2009. Effects of life history strategy on fish distribution and use of estuarine salt marsh and shallow-water flat habitats. Estuaries and Coasts 32, 797–812.

Minello, T.J., Zimmerman, R.J., Medina, R., 1994. The importance of edge for natant macrofauna in a created salt marsh. Wetlands 14, 184–198.

Mitsch, W.J., Gosselink, J.G., 2000. Wetlands. John Wiley & Sons, New York, p. 920.

Mitsch, W.J., Gosselink, J.G., 2007. Wetlands 4th Edition. John Wiley & Sons, New York, p. 582.

Moyle, P.B., 1996. Aliens: let's eliminate the term "Accidental Introduction". Aliens 4, 1–2.

Neill, W.T., 1958. The occurrence of amphibians and reptiles in saltwater areas and a bibliography. Bulletin of Marine Science 8, 1–95.

Nifong, J.C., Silliman, B.R., 2013. Impacts of a large-bodied, apex predator (*Alligator mississippiensis* Daudin 1801) on salt marsh food webs. Journal of Experimental Marine Biology and Ecology 440, 185–191.

Nomann, B.E., Pennings, S.C., 1998. Fiddler crab–vegetation interactions in hypersaline habitats. Journal of Experimental Biology and Ecology 225, 53–68.

Nordlie, F.G., 2003. Fish communities of estuarine salt marshes of eastern North America, and comparisons with temperate estuaries of other continents. Reviews in Fish Biology and Fisheries 13, 281–325.

Odum, E.P., 2000. Tidal marshes as outwelling/pulsing systems. In: Weinstein, W.P., Kreeger, D.A. (Eds.), Concepts and Controversies in Tidal Marsh Ecology. Kluwer Academic Publishers, Dordrecht, pp. 3–7.

Olff, H., Bakker, J.P., Fresco, L.F.M., 1988. The effect of fluctuations in tidal inundation frequency on a salt-marsh vegetation. Plant Ecology 78, 13–19.

Partridge, T.R., Wilson, J.B., 1989. Methods for investigating vegetation/environment relations a test using the salt marsh vegetation of Otago, New Zealand. New Zealand Journal of Botany 27, 35–47.

Pauly, D., 1995. Anecdotes and the shifting baseline syndrome of fisheries. Tree 10 (10).

Pauly, D., Ingles, J., 1988. The relationship between shrimp yields and intertidal vegetation (mangrove) areas: a reassessment. In: Yanez-Arancibia, A., Pauly, D. (Eds.), IOC/FAO Workshop on Recruitment in Tropical Coastal Demersal Communities. UNESCO Intergovernmental Oceanographic Commission Workshop Report Supplement 44, pp. 277–283.
Pearcy, W.G., Myers, S.S., 1974. Larval fishes of Yaquina Bay Oregon: a nursery ground for marine fishes? Fishery Bulletin 72, 201–213.
Pehrsson, O., 1988. Effects of grazing and inundation on pasture quality and seed production in a salt marsh. Plant Ecology 74, 113–124.
Pennings, S.C., Callaway, R.M., 1992. Salt marsh plant zonation: the relative importance of competition and physical factors. Ecology 73, 681–690.
Pennings, S.C., Callaway, R.M., 1996. Impact of a parasitic plant on the structure and dynamics of salt marsh vegetation. Ecology 77, 1410–1419.
Pennings, S.C., Siska, E.L., Bertness, M.D., 2001. Latitudinal differences in plant palatability in Atlantic coast salt marshes. Ecology 82, 1344–1359.
Pennings, S.C., Selig, E.R., Houser, L.T., Bertness, M.D., 2003. Geographic variation in positive and negative interactions among salt marsh plants. Ecology 84, 1527–1538.
Peterson, B.J., Howarth, R.W., Garritt, R.H., 1985. Multiple stable isotopes used to trace the flow of organic matter in estuarine food webs. Science 227, 1361–1363.
Peterson, B.J., Howarth, R.W., Garritt, R.H., 1986. Sulfur and carbon isotopes as tracers of salt marsh organic matter flow. Ecology 67, 865–874.
Peterson, G.W., Turner, R.E., 1994. The value of salt marsh edge vs interior as a habitat for fish and decapod crustaceans in a Louisiana salt marsh. Estuaries 17, 235–262.
Pettus, D., 1958. Water relationships in *Natrix sipedon*. Copeia 1958, 207–211.
Puccia, C.J., Levins, R., 1991. Qualitative modeling in ecology: loop analysis, signed diagraphs, and time averaging. In: Fishwick, P.A., Lucker, P.A. (Eds.), Qualitative Simulation Modeling and Analysis. Springer-Verlag, New York, pp. 119–143.
Quintana, X.D., Moreno-Amich, R., 2002. Phytoplankton composition of Emporda salt marshes, Spain and its response to freshwater flux regulation. Journal of Coastal Research 36, 581–590.
Rakocinski, C.F., Baltz, D.M., Fleeger, J.W., 1992. Correspondence between environmental gradients and the community structure of marsh-edge fishes in a Louisiana estuary. Marine Ecology Progress Series 80, 135–148.
Ranwell, D.S., 1961. Spartina salt marshes in southern England: I. The effects of sheep grazing at the upper limits of Spartina marsh in Bridgwater Bay. Journal of Ecology 49, 325–340.
Rozas, L.P., Odum, W.E., 1988. Occupation of submerged aquatic vegetation by fishes: testing the roles of food and refuge. Oecologia 77, 101–106.
Rountree, R.A., Able, K.W., 2007. Spatial and temporal habitat use patterns for salt marsh nekton: implications for ecological functions. Aquatic Ecology 41, 25–45.
Rutledge, P.A., Fleeger, J.W., 1993. Abundance and seasonality of meiofauna, including harpacticoid copepod species, associated with stems of the salt-marsh cord grass, *Spartina alterniflora*. Estuaries 16, 760–768.
Sala, N.M., Bertness, M.D., Silliman, B.R., 2008. The dynamics of bottom–up and top–down control in a New England salt marsh. Oikos 117, 1050–1056.
Shanholtzer, G.F., 1974. Relationship of vertebrates to salt marsh plants. In: Reimold, R.J., Queen, W.H. (Eds.), Ecology of Halophytes. Academic Press, New York, pp. 463–474.
Silliman, B.R., Zieman, J.C., 2001. Top-down control of *Spartina alterniflora* production by periwinkle grazing in a Virginia salt marsh. Ecology 82, 2830–2845.
Silliman, B.R., Bertness, M.D., 2002. A trophic cascade regulates salt marsh primary production. Proceedings of the National Academy of Sciences USA 99, 10500–10505.
Sullivan, M.J., Moncreiff, C.A., 1990. Edaphic algae are an important component of salt marsh foodwebs: evidence from multiple stable isotope analyses. Marine Ecology Progress Series 62, 149–159.
Sullivan, M.J., Currin, C.A., 2000. Community structure and functional dynamics of benthic microalgae in salt marshes. In: Weinstein, M.P., Kreeger, D.A. (Eds.), Concepts and Controversies in Tidal Marsh Ecology. Kluwer Academic Publishers, Dordrecht, pp. 81–106.
Talley, T.S., Levin, L.A., 2001. Modification of sediments and macrofauna by an invasive marsh plant. Biological Invasions 3, 51–68.

Teal, J.M., 1962. Energy flow in the salt marsh ecosystem of Georgia. Ecology 43, 614–624.

Thannheiser, D., Holland, P., 1994. The plant communities of New Zealand salt meadows. Global Ecology and Biogeography Letters 4, 107–115.

Turner, R.E., 1977. Intertidal vegetation and commercial yields of penaeid shrimp. Transactions of the American Fisheries Society 106, 411–416.

Turner, M.G., 1987. Effects of grazing by feral horses, clipping, trampling, and burning on a Georgia salt marsh. Estuaries 10, 54–60.

Turner, R.E., 1992. Coastal wetlands and penaeid shrimp habitat. In: Stroud, R.H. (Ed.), Stemming the Tide of Coastal Fish Habitat Loss. Proceedings of the 14th Annual Marine Recreational Fisheries Symposium Baltimore, Maryland. Mar. Recreational Fish, vol. 14, pp. 97–104.

Valéry, L., Bouchard, V., Lefeuvre, J.C., 2004. Impact of the invasive native species *Elymus athericus* on carbon pools in a salt marsh. Wetlands 24, 268–276.

Van der Wal, R., Egas, M., van der Veen, A., Bakker, J., 2000. Effects of resource competition and herbivory on plant performance along a natural productivity gradient. Journal of Ecology 88, 317–330.

Van Dolah, R.F., 1978. Factors regulating the distribution and population dynamics of the amphipod *Gammarus palustris* in an intertidal salt marsh. Ecological Monographs 48, 191–217.

van Wijnen, H.J., Bakker, J.P., 1999. Nitrogen and phosphorus limitation in a coastal barrier salt marsh: the implications for vegetation succession. Journal of Ecology 87, 265–272.

Visser, J.M., Sasser, C.E., Cade, B.S., 2006. The effect of multiple stressors on salt marsh end- of-season biomass. Estuaries and Coasts 29, 331–342.

Weinstein, M.P., 1979. Shallow marsh habitats as primary nurseries for fishes and shellfish, Cape Fear River, North Carolina. Fishery Bulletin 77, 339–357.

Woodland, R.J., Rowe, C.L., Henry, P.F.P., 2017. Changes in habitat availability for multiple life stages of Diamondback Terrapins (Malaclemys terrapin) in Chesapeake Bay in response to sea level rise. Estuaries and Coasts. https://doi.org/10.1007/s12237-017-0209-2.

Zedler, J.B., 1982. Salt marsh algal mat composition: spatial and temporal comparisons. Bulletin of the Southern California Academy of Sciences 81, 41–50.

Zedler, J.B., Callaway, J.C., Desmond, J.S., Vivian-Smith, G., Williams, G.D., Sullivan, G., Brewster, A.E., Bradshaw, B.K., 1999. Californian salt-marsh vegetation: an improved model of spatial pattern. Ecosystems 2, 19–35.

Zhang, Y., Huang, G., Wang, W., Chen, L., Lin, G., 2012. Interactions between mangroves and exotic Spartina in an anthropogenically disturbed estuary in southern China. Ecology 93 (3), 588–597.

Zimmerman, R.J., Minello, T.J., 1984. Densities of *Penaeus aztecus, Penaeus setiferus*, and other natant macrofauna in a Texas salt marsh. Estuaries 7, 421–433.

Zimmerman, R., Minello, T., Klima, E., Nance, J., 1991. Effects of accelerated sea level rise on coastal secondary production. In: Bolten, H.S., Magoon, O.T. (Eds.), Coastal Wetlands, Coastal Zone '91 Conference. American Society of Civil Engineers, New York, pp. 110–124.

CHAPTER

16

Salt Marsh Biogeochemistry—An Overview

Craig Tobias[1], *Scott C. Neubauer*[2]

[1]University of Connecticut, Groton, CT, United States; [2]Department of Biology, Virginia Commonwealth University, Richmond, VA, United States

1. INTRODUCTION

Salt marshes have long been considered important sources, sinks, and/or transformers of biologically important compounds in the coastal landscape. Like other vegetated wetlands, the high rates of macrophyte primary production in salt marshes provide large reservoirs of organic matter and nutrients in both the particulate and dissolved phases. Marsh geomorphology leads to trapping of allochthonous organic and mineral particulates. The high rates of respiration in marsh substrates create an electron-rich, chemically reduced environment proximal to oxidized surface waters. The subsurface redox environment is further modified by root-mediated release of O_2 into the rhizosphere, physical mixing by macroorganisms, and water movements caused by tidal infiltration and drainage, factors that vary over tidal to seasonal and interannual scales. This juxtaposition of anaerobic and aerobic environments creates conditions that favor rapid elemental cycling between reduced and oxidized molecules. As such marshes provide highly reactive surfaces that modify water quality. High organic matter production rates, variable redox environments, and physically dynamic ecosystems that are regularly pulsed by tides are characteristic features of salt marshes, mangrove swamps, tidal freshwater wetlands, and seagrass systems. Salt marshes differ considerably from nonvegetated mudflats in terms of overall autotrophic productivity and impacts of macrophyte root production on the cycling of macroelements. Although there are many commonalities between saline and tidal freshwater wetlands, the chemistry of seawater (notably the accumulation of salts in porewater and the presence of SO_4^{2-}) leads to significant biogeochemical differences in processes such as organic matter oxidation and the cycling of N, P, S, and Fe.

Coastal Wetlands
https://doi.org/10.1016/B978-0-444-63893-9.00016-2

2. CARBON

2.1 Inputs

2.1.1 Photoautotrophy

Marsh macrophytes serve as a dominant source of new C to marshes, play a key role in stabilizing the marsh platform, trap sediments, aerate the soil through root O_2 loss (ROL), and influence the biogeochemical cycling of C, N, Fe, and S (Bodelier, 2003; Hines, 2006). Rates of net primary production (NPP) in *Spartina alterniflora* marshes range from 100 to >2500 g C m^{-2} $year^{-1}$ (Mitsch and Gosselink, 1993; Dame et al., 2000), with highest values in south Atlantic and Gulf Coast marshes (Mendelssohn and Morris, 2000). Belowground production often equals or exceeds that of aboveground tissues (Valiela et al., 1976; Schubauer and Hopkinson, 1984; Darby and Turner, 2008a). The productivity of other marsh plants including *Spartina patens*, *Distichlis spicata*, and *Juncus roemerianus* can be comparable with that of *S. alterniflora* (Mitsch and Gosselink, 1993). NPP can vary within a single marsh as a function of anoxia, sulfide, and salinity stresses that affect nitrogen uptake and assimilation (King et al., 1982; Mendelssohn and Morris, 2000). Furthermore, sea level anomalies lead to high interannual variability by affecting flooding frequency and thus the salinity of marsh porewaters; for each plant species, there is an optimal elevation that maximizes plant productivity (Morris, 2000; Morris et al., 2002). The majority of C fixed by plants is atmospheric CO_2, although small amounts (<10% of atmospheric fixation) can come from porewater dissolved inorganic C (DIC) or CO_2 that is recycled in lacunar spaces (Hwang and Morris, 1992).

Relative to rates of vascular plant productivity, less is known about primary productivity by salt marsh benthic microalgae and macroalgae. Benthic microalgal productivity ranges from ~30 to 300 g C m^{-2} $year^{-1}$ (Table 16.1) and shows a similar latitudinal pattern as for macrophytes. The highest rates of productivity are observed along the southeast Atlantic and California coasts (Zedler, 1980). Benthic microalgal production generally accounts for ~10%–25% of total ecosystem (i.e., plant + algal) productivity but sometimes exceeds plant productivity (Table 16.1). Microalgal productivity is often higher during winter when the plant cover is at a minimum. Despite lower rates of productivity, benthic microalgal biomass is generally more labile than that of macrophytes and is preferentially assimilated by secondary consumers (Sullivan and Moncreiff, 1990; Currin et al., 1995; Sullivan and Currin, 2000). Younger marshes possess a higher proportion of trapped allochthonous C relative to more mature systems that have built rhizosphere (Hansen et al., 2017).

For decades, estimates of marsh primary production have been based on measurements of plant/algal biomass or small-scale (<1 m^2) chamber-based measures of CO_2 uptake. In the past decade, eddy correlation techniques have been applied in coastal wetlands to quantify the net ecosystem exchange (NEE) of CO_2 between marshes and the atmosphere over a broad spatial scale (hundreds of m^2) and at high frequency (fluxes typically calculated every 30 min), with NEE serving as an analog for net ecosystem productivity (NEP; Kathilankal et al., 2008; Guo et al., 2009, Moffett et al., 2010; Artigas et al., 2015, Forbrich and Giblin, 2015). Eddy correlation measurements provide better temporal resolution and

TABLE 16.1 Net Benthic Microalgal (BMP) and Vascular Plant Production (VPP) in Tidal Salt Marshes, and the Ratio of BMP to Net Primary Production (NPP), Where NPP = BMP + VPP

	Productivity (g C m^{-2} $year^{-1}$)		(%)	
State	**Microalgae**	**Vascular Plants**	**BMP/NPP**	**References**
Massachusetts	53[a]	212[a]	20	Van Raalte et al. (1976)
New York	50	292	15	Woodwell et al. (1979)
Delaware	61–99	NR	25	Gallagher and Daiber (1974)
Virginia	27.8	254	10	Anderson et al. (1997)
Virginia	67[b]	161	29	Miller et al. (2001)
Virginia	235	831	22	Buzzelli et al. (1999)
South Carolina	98, 234[c]	534, 296[c]	16, 44[c]	Pinckney and Zingmark (1993)
Georgia	180	732	20	Pomeroy (1959) and Teal (1962)
Georgia	208	1127	16	Pomeroy et al. (1981)
Mississippi	28–151	248–742[a]	9–38	Sullivan and Moncreiff (1988)
Texas	71	550–900	9–11	Hall and Fisher (1985)
California	185–341	243–340	43–58	Zedler (1980)

Note that this calculation differs from the oft-reported ratio that expresses BMP as a fraction of plant production (i.e., BMP/VPP). Plant production was estimated based on aboveground biomass in all studies except Miller et al. (2001) where a CO_2-based gas flux model was utilized. "NR", not reported.

[a]*Converted from original estimates, assuming that algal and/or plant biomass is 50% C.*

[b]*BMP was measured in cleared zones within a dense* Spartina patens/Distichlis spicata *canopy; rates in these cleared zones likely overestimate rates beneath the canopy (e.g., Sullivan and Daiber, 1975).*

[c]*Averages for tall and short* Spartina alterniflora *zones, respectively.*

broader spatial integration than chamber methods and may improve scaling of NEP fluxes. Remaining challenges with eddy correlation techniques include deciphering physical and biological controls on NEE during tidal inundation, which could affect fidelity between NEE and NEP (Kathilankal et al., 2008), and refining modeling approaches for deriving gross primary production and ecosystem respiration from NEE under changing periods of wetting/drying (Wang et al., 2017). Eddy correlation approaches can also be used for CH_4 fluxes in coastal wetlands, but the deployment of these systems primarily has been limited to fresh to brackish marshes (Holm et al., 2016; Krauss et al., 2016; and see Tobias and Neubauer, this volume) because CH_4 fluxes are typically low at high salinities (Poffenbarger et al., 2011).

2.1.2 *Chemoautotrophy*

Reduced compounds such as Fe(II) and H_2S contain considerable energy that can be released on oxidation. When this oxidation occurs through a chemoautotrophic microbial process, the energy in the chemical bonds can be utilized to fix inorganic C. Thus, from a

C-cycling perspective, this C-fixation represents primary production. However, from an energetic perspective, chemoautotrophy is a form of secondary production (after Howarth, 1993, Chapin et al., 2006) because the chemical energy in the inorganic compounds was originally released from organic matter (or H_2).

Sulfides produced through SO_4^{2-} reduction are rapidly incorporated into minerals such as iron monosulfides (FeS) and pyrite (FeS_2). Given that >70% of the energy in organic compounds utilized by SO_4^{2-} reducers ends up in these reduced sulfur compounds, it is apparent that FeS, FeS_2, and H_2S in marsh substrates represent a considerable energy source. Mass balance calculations indicate that >90% of the sulfides are ultimately reoxidized or exported from the marsh (Howes et al., 1984; Gardner, 1990). Rates of S-mediated chemoautotrophy may range from ~20 to 480 g C m^{-2} $year^{-1}$ (Howarth and Teal, 1980, Howarth, 1984, 1993), a rate that is comparable to benthic microalgal production. However, if these sulfides are oxidized chemically, the energy is dissipated in the environment and not used for chemoautotrophic production.

Lithoautotrophic Fe(II)-oxidizing bacteria (FeOB) also exist in salt marshes (Weiss et al., 2003) and subtidal marine environments (Emerson and Moyer, 2002; Edwards et al., 2003). Based on estimates of Fe(II) supply rates through the turnover of Fe–S minerals and oxidized Fe(III), the rate of chemoautotrophic production associated with the activity of FeOB could be as much as 2–185 g C m^{-2} $year^{-1}$ (Tobias and Neubauer, 2009). However, the actual contribution of FeOB to chemoautotrophic production is likely to be considerably lower because FeOB are a very small component (<0.01%) of the total salt marsh microbial community and account for 50% or less of total Fe(II) oxidation (Neubauer et al., 2002b, 2007).

2.1.3 *Tidal Inputs—Sediment Deposition*

Tidal flooding provides a mechanism for the delivery and deposition of water column–suspended sediments. Rates of sedimentation depend on suspended sediment concentrations, tidal range, vegetation, creek proximity, and hydroperiod (Friedrichs and Perry, 2001). Thus, sedimentation and organic C deposition varies widely from marsh to marsh, laterally within a single site and temporally within the year. Paulina Marsh, in the Netherlands, provides an example of varied deposition patterns (Fig. 16.1, Temmerman et al., 2003). The lowest deposition (<100 g m^{-2} spring–neap $cycle^{-1}$) occurred near the marsh-upland border. Highest rates (2000 g m^{-2} $cycle^{-1}$) were measured at the marsh–mudflat boundary. Average deposition was ~200 g m^{-2} $cycle^{-1}$ (Fig. 16.1). Assuming an organic C content of ~5% for suspended sediments (Middelburg and Herman, 2007), deposition across this marsh delivers on the order of 250 g C m^{-2} $year^{-1}$. This rate falls within the range of deposition reported for other salt marshes (e.g., Cahoon and Reed, 1995; Salgueiro and Caçador, 2007) and tidal freshwater marshes (Megonigal and Neubauer, this volume). Thus, sedimentation represents a source of C to salt marshes comparable with C-fixation by benthic microalgae and chemoautotrophs. Such inputs of allochthonous C have been suggested as a mechanism to explain the 9–12‰ depletion in soil organic matter (SOM) $\delta^{13}C$ values (relative to *S. alterniflora* biomass), which is often observed in mineral-dominated marshes (Middelburg et al., 1997). Marshes in sediment-starved regions or those that are

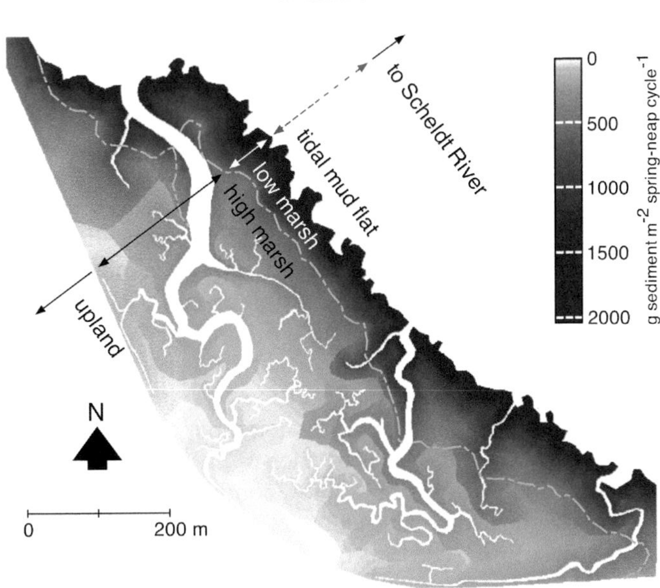

FIGURE 16.1 Spatial patterns of sedimentation in the Paulina salt marsh, the Netherlands. Deposition rates were calculated using a GIS model linked to marsh surface elevation, distance from the nearest creek or marsh edge, and distance from the marsh edge as measured along the nearest creek (figure slightly modified from Temmerman et al., 2003). Deposition was not calculated in the tidal creeks. The model successfully predicted measured deposition rates ($r^2 = 0.96$ when averaged across the entire year). The dominant plant in the low marsh was *Spartina townsendii*. The high marsh contained *Puccinellia maritima*, *Aster tripolium*, *Atriplex portulacoides*, and *Elytrigia pungens*. On the figure, the *dotted gray line* shows the approximate location of the low marsh—high marsh border.

infrequently flooded will have lower inputs of allochthonous C and will therefore be more dependent on autochthonous production (Kirwan and Megonigal, 2013).

Considerable quantities of sediment-associated C can be delivered by hurricanes and other large storms (Parsons, 1998; Turner et al., 2006a). Hurricanes Katrina and Rita (US Gulf of Mexico Coast, Aug/Sept 2005) deposited 22.3 kg sediment m^{-2} (Turner et al., 2006a). Assuming that these sediments originated from offshore (as hypothesized by Turner et al., 2006a) with an average organic C content of 2.2% (Mayer et al., 2007), this deposition equates with 490 g C m^{-2} to the marsh surface. The inorganic fraction can contribute to significant vertical marsh growth (Turner et al., 2006a), but the long-term fate of the organic fraction is currently unknown. Although large storms can deliver sediments and allochthonous carbon to the marsh platform, storms can also destroy large areas of marsh, causing a geologically instantaneous loss of sequestered soil carbon (DeLaune and White, 2012).

2.2 Transformations

The mineralization of SOM, an important process for recycling inorganic nutrients in marsh soils and influencing the preservation versus loss of SOM, is a multistage process that is carried out by a consortium of soil bacteria, archaea, and fungi. The depolymerization of complex substrates (e.g., cellulose, lignin, proteins) to simpler organic molecules (e.g., monosaccharides, amino acids) by extracellular enzymes is typically the rate-limiting step of decomposition (Megonigal et al., 2004). Fermentation is an exergonic (energy-yielding) microbial process that does not require an external electron acceptor—organic molecules such as glucose serve as both electron acceptor and electron donor—and produces acetate and simple organic acids, plus CO_2 and H_2. Under anaerobic conditions, fermentation is often a key step of the mineralization process because most nonfermentative anaerobes cannot use the primary products of enzyme-mediated depolymerization (Megonigal et al., 2004). However, wetland plant roots can release simple organic acids and alcohols into anaerobic soils, making fermenters somewhat less important in vegetated salt marsh soils (Hines et al., 1994, 1999). The simple organic molecules produced by fermenters can undergo terminal microbial respiration under aerobic or anaerobic conditions, yielding CO_2 and/or CH_4. Rates of mineralization are influenced by the supply of electron acceptors (see Sections 2.2.1 and 2.2.2 for more detail), the abundance/composition of organic matter (electron donors), and the physicochemical environment (e.g., redox, temperature, pH, salinity; Benner et al., 1985, Reddy and DeLaune, 2008, Sutton-Grier et al., 2011, Ouyang et al., 2017).

Soils with a higher organic matter content tend to have higher mineralization rates, as expressed on a per gram of soil basis (e.g., Sutton-Grier et al., 2011), reflecting the greater availability of organic matter that could be mineralized. Expressing rates per gram of SOM (or per gram of soil carbon) provides information on the relative lability/recalcitrance of different SOM sources, which will often reflect the quality (e.g., C:N ratio or lignin content) of the dominant primary producers in the system. As labile organic matter sources are mineralized, the remaining organic matter generally becomes more recalcitrant and less susceptible to mineralization (Neubauer et al., 2005a; Sutton-Grier et al., 2011). There is emerging evidence that further decomposition can be induced, when old organic matter is exposed to different environmental conditions (e.g., during erosion) or primed by additions of new organic carbon sources (Bernal et al., 2016, 2017).

2.2.1 Aerobic Mineralization

Molecular oxygen is rapidly used in salt marsh soils as an electron acceptor for microbial respiration and as an oxidant for reduced chemical species. Marsh O_2 uptake (maximal at low tide) ranges from 5 to 65 mol O_2 m^{-2} $year^{-1}$ (Howes et al., 1984; Howarth, 1993; Cai et al., 1999). Many plants including *S. alterniflora*, *Spartina anglica*, and *J. roemerianus* can oxidize subsurface soils through ROL (Mendelssohn et al., 1995; Holmer et al., 2002; Maricle and Lee, 2002; Colmer, 2003; Koretsky and Miller, 2008). The amount of O_2 uptake because of air entry into soil pore spaces can be ~50% of the diffusive O_2 uptake by marsh soils (Morris and Whiting, 1985). Filling pore spaces with fully aerated tidal water (rather than air) contributes a much smaller amount of O_2 because of the lower concentration of O_2 in water. Infauna can increase soil oxidation through burrowing, bioirrigation, and physical mixing of the substrate (Meile et al., 2001; Gribsholt and Kristensen, 2002; Kristensen and Kostka, 2005), leading to accelerated organic matter turnover and decreases in SOM concentrations (Thomas and Blum, 2010; Wang et al., 2010). Because there are no in situ methods for quantifying O_2 added during ROL or faunal activity, estimates of aerobic respiration are likely to be seriously underestimated if only O_2 fluxes across the marsh surface are considered. Howarth (1993) estimated that aerobic respiration accounted for 18%–30% of total soil respiration but diagenetic simulation results suggest a much smaller contribution (1%–5% of the total organic matter decomposition; Furukawa et al., 2005).

Because organic matter can persist above the soil surface as standing dead stems and wrack, measurements of soil processes alone will underestimate the total role of aerobic decomposition. Fungi and bacteria carry out a significant amount of C mineralization subaerially and under aerobic conditions. For example, peak microbial respiration rates on senescent leaves of tall *S. alterniflora* coincided with peak fungal biomass (Buchan et al., 2003). Fungal degradation can remove up to 60% of the original aboveground organic matter (Newell and Porter, 2000) with aerobic degradation continuing to occur, as the bacterial standing crop increases during later stages of decomposition (Benner et al., 1986; Newell et al., 1989).

2.2.2 Anaerobic Mineralization

Thermodynamic theory indicates that the availability of electron acceptors and the competition between microbes for electron donors will govern the relative importance of different catabolic processes (Megonigal et al., 2004). Respiration using O_2 as the electron acceptor has the highest energy yield. Following the depletion of O_2, a predictable sequence of anaerobic processes follows: denitrification, Mn reduction, Fe(III) reduction, SO_4^{2-} reduction, and methanogenesis (Ponnamperuma, 1972). There is some recent evidence that humic acids can also be significant electron acceptors. One result of the competition between microbes for electron donors can be vertical redox stratification that reveals itself as gradients in solid phase or porewater geochemistry (Griffin et al., 1989; Taillefert et al., 2007). However, due to fine-scale heterogeneity in distributions of electron donors and electron acceptors, multiple pathways can coexist within the same volume of soil (Højberg et al., 1994).

Despite the importance of denitrification for NO_3^- removal (Section 3.3.1), denitrification is not significant for organic C turnover in marshes because of relatively low NO_3^- supply. It accounts for $\leq$1% of C mineralization in salt marsh soils (Table 16.2). Additions of NO_3^- to Georgia marsh soils did not increase total C mineralization rates (Hyun et al., 2007).

To our knowledge, rates of Mn(III, IV) reduction have not been measured in salt marsh substrates. However, voltametric (microelectrode) studies have identified dissolved Mn^{2+} in vegetated salt marsh soils (Brendel and Luther, 1995) and in nonvegetated intertidal salt marsh creekbank sediments (Taillefert et al., 2007), indicating that Mn(III, IV) reduction was occurring. Based only on the presence/absence of reduced Mn, it is not possible to determine if the Mn reduction was driven by biological (enzymatic) or strictly chemical reactions. Regardless, M(III, IV) reduction will probably be less important than Fe(III) reduction with respect to C mineralization because Mn is generally present at lower concentrations than Fe (e.g., Luther et al., 1992). However, Mn reduction can be a dominant pathway in near-surface subtidal marine sediments (Canfield et al., 1993; Thamdrup, 2000).

Fe(III) reduction can be an important pathway for organic matter turnover in salt marsh soils, accounting for 50%–100% of anaerobic metabolism (Table 16.2). The relatively recent recognition that metal reduction can be a significant biological process contrasts with the historical view that SO_4^{2-} reduction dominates anaerobic metabolism in salt marshes (Howarth and Hobbie, 1982; Howarth, 1993). Many early studies were conducted in short *S. alterniflora* or mid-marsh habitats where H_2S accumulates because of low porewater turnover rates.

TABLE 16.2 Importance of Anaerobic to Total (Aerobic + Anaerobic) Metabolism in Several Salt Marshes and the Partitioning of Anaerobic C Decomposition Through Various Pathways

		Percentage of Anaerobic Respiration Due to				
	Anaerobic/total	NO_3^- Reduction	Metal Reduction	SO_4^{2-} Reduction	Methanogenesis	Citations
Sippewissett (Massachusetts)	0.82	0.2	Negligible	99.4	0.4	Howarth (1993)
Jack Bay (Maryland)	Nd	Nd	2.7–51.4[a]	48.6–95.2[a]	0–2.0[a]	Neubauer et al. (2005b)
Skidaway Island (Georgia)	Nd	Nd	0–109[b]	6–82[b]	Nd	Kostka et al. (2002a)
	Nd	Nd	28–96[c]	4–72[c]	Nd	Gribsholt et al. (2003)
	0.95–0.99[d]	0.9–1.0[d]	1.7–61.6[d]	36.8–97.5[d]	Nd	Furukawa et al. (2004)
	Nd	Nd	0–71.8[e]	22.1–95[e]	Nd	Hyun et al. (2007)
Sapelo Island (Georgia)	0.70	1.1	Negligible	94.4	4.4	Howarth (1993)
	Nd	Nd	Nd[f]	0–106[f]	Nd	Kostka et al. (2002b)

The relative importance of each pathway is taken directly from each citation, or calculated after summing the measured pathways and assuming that unmeasured pathways did not significantly contribute to total anaerobic metabolism. As discussed in Neubauer et al., (2005b) and Hyun et al. (2007), this is not always a valid assumption since the sum of individual metabolic rates does not always equal total $CO_2 + CH_4$ production. Mn reduction rates have not been measured in salt marsh soils and sediments, so "metal reduction" refers to biological Fe(III) reduction only. "Negligible" is a quote from Howarth (1993). "Nd" indicates that aerobic respiration was not reported or that rates for a specific metabolic pathway were not measured.

[a]*Range due to monthly variability in rates measured in June, July, and August.*

[b]*Range due to spatial variability between a bioturbated, vegetated (*Spartina alterniflora*) levee site, and an unvegetated, nonbioturbated creekbank.*

[c]*Range due to spatial variability with increasing distance from* Uca pugnax *burrows (0–35 cm from burrow wall), as well as variations between rhizosphere and levee bulk soils.*

[d]*Range due to spatial variability in modeled rates (integrated to 9 cm depth) between locations that are vegetated with* S. alterniflora*, contain abundant bioturbating fiddler crabs (*U. pugnax*) but no plants, or are unvegetated and lack* U. pugnax.

[e]*Range due to lateral variability between tall* S. alterniflora*, short* S. alterniflora*, and creekbank zones, as well as depth-related differences (0–3 cm and 3–6 cm) within each zone.*

[f]*Range due to lateral variability between a bioturbated unvegetated creekbank, a bioturbated vegetated (*S. alterniflora*) levee, and* S. alterniflora *mid-marsh zones, as well as depth-related differences (0–5 cm and 10–15 cm) within each zone. In this study, Fe(III) reduction was measured. Although no distinction was made between chemical and biological Fe(III) reduction, Kostka et al. (2002b) suggested that abiotic Fe(III) reduction dominated.*

Under these conditions, SO_4^{2-} reduction should dominate metabolism, and Fe(III) reduction will be primarily an abiotic process (Eq. 16.1), where Fe oxides serve as an oxidant for sulfides:

$$8FeOOH + H_2S + 14H^+ \rightarrow 8Fe^{2+} + SO_4^{2-} + 12H_2O \tag{16.1}$$

Biological Fe(III) reduction will be more important than chemical reduction when amorphous Fe(III) oxides are plentiful and continually regenerated, or H_2S production is low relative to the Fe(III) concentration (Jacobson, 1994). Indeed, experimental evidence indicates that marsh zones with heavy bioturbation activity (especially by fiddler crabs, *Uca* spp.) can have anaerobic metabolism dominated by biotic Fe(III) reduction (Gribsholt et al., 2003; Furukawa et al., 2004; Hyun et al., 2007). Plant-driven inputs of O_2 (via ROL) and organic C can also lead to significant rates of C turnover coupled to Fe(III) reduction (Neubauer et al., 2005b), although this mechanism appears to be less important than bioturbation in some salt marsh soils (Furukawa et al., 2004). Furthermore, Fe(III) oxides formed in the rhizosphere can be more amorphous (a factor that favors enzymatic Fe(III) reduction) than those in bulk soil (Weiss et al., 2004). Gribsholt and Kristensen (2002) suggested that the highest rates of organic matter decomposition and Fe(III) reduction will occur in substrates that are bioturbated (to regenerate Fe(III) oxides) and vegetated (with plants acting as an organic C source). Hydrology, including diurnal tidal cycles (Taillefert et al., 2007) and seasonal changes in water level (Neubauer et al., 2005b), may also be important in driving O_2 penetration into marsh soils and sediments, regulating Fe(III) regeneration, and leading to significant rates of biotic Fe(III) reduction.

Humic acids can be used as terminal electron acceptors for anaerobic metabolism with reasonable thermodynamic efficiency (Cervantes et al., 2000), in addition to their role as electron shuttles for Mn(III, IV) and Fe(III) reduction (Lovley et al., 1996). In salt marshes, the role of humic acid reduction as a pathway for organic C oxidation has been inferred from differences between total rates of microbial respiration ($\sum CO_2 + CH_4$ production) and anaerobic N, S, Fe stoichiometries. For example, the sum of individually measured anaerobic pathways accounted for only 43% of total metabolism in a Georgia salt marsh (Hyun et al., 2007) and 20%–30% in a Maryland brackish marsh (Neubauer et al., 2005b). In both tidal and nontidal freshwater wetlands, humic acids have been shown as important electron acceptors (e.g., Keller et al., 2009a, Minderlein and Blodau, 2010, Klüpfel et al., 2014).

SO_4^{2-} reduction is often a dominant pathway for anaerobic C metabolism (Table 16.2; Howarth and Hobbie, 1982, Howarth, 1984, Hines et al., 1989, Kostka et al., 2002b, Weston et al., 2014). SO_4^{2-} concentrations typically exceed several millimolars in flooding water, so SO_4^{2-} availability will not limit reduction (Boudreau and Westrich, 1984; Roychoudhury et al., 2003b). Instead, competition from other electron-accepting processes and the availability of labile organic C regulates the relative importance of SO_4^{2-} reduction. Based on considerations presented earlier, SO_4^{2-} reduction should have the largest contribution to total anaerobic C metabolism in soils that are unvegetated, largely undisturbed (i.e., low bioturbation), and/or poorly flushed (allowing H_2S to accumulate and minimizing advective O_2 delivery)—this has been demonstrated in several studies (Kostka et al., 2002a, 2002b; Furukawa et al., 2004). Sulfate reducers depend on a suite of relatively simple organic molecules including volatile fatty acids and acetate (Hines et al., 1994), so the rate at which these molecules are generated and/or the degree of competition for these electron donors will affect SO_4^{2-} reduction rates. High rates of SO_4^{2-} reduction occur during summer when plant growth and releases of dissolved organic C (DOC) from the rhizosphere are greatest (Hines et al., 1989), a pattern only partially explained by changing temperatures (Howarth and Teal, 1979; Hines et al., 1989). Temperature does play a significant role because there is an imbalance between the temperature responses of microbes that hydrolyze and ferment complex organic substrates and SO_4^{2-} reducers that utilize the resulting simple organic molecules (Weston and Joye, 2005). At low temperatures, the production of low molecular weight

DOC exceeds its consumption, resulting in accumulation. At high temperatures, however, SO_4^{2-} reduction becomes limited by the rate at which low molecular weight electron donors are generated (Weston and Joye, 2005).

There has been little work on rates of CH_4 formation in salt marshes, and existing evidence suggests that methanogenesis accounts for less than 5% of anaerobic respiration (Table 16.2 and references therein). The paradigm is that methanogenesis in salt marshes is limited because methanogens are outcompeted for electron donors (King and Wiebe, 1980a), although CH_4 production can be substantial in some humus-rich salt marsh soils (Giani et al., 1996). In a comparison of tall and short *S. alterniflora* zones, King and Wiebe (1980b) found that rates of CH_4 production were considerably greater in the short *Spartina* zone. At the same time, SO_4^{2-} was relatively more depleted in the short *Spartina* zone, which allowed methanogenesis to increase in importance (King and Wiebe, 1980b). Given higher Fe(III) reduction creekbank *Spartina* zones (Furukawa et al., 2004; Hyun et al., 2007), electron donor availability to methanogens may be limited by metal reducers along creekbanks and SO_4^{2-} reducers in the marsh interior. Despite the thermodynamic constraints against methanogenesis, some C is mineralized through this process (e.g., Parkes et al., 2012), an indication of microzones in the soil matrix that are depleted in other electron acceptors and/or regions where substrates exist that can be converted to CH_4 but cannot be used by SO_4^{2-} reducers (e.g., methanol and methylated amines; Oremland et al., 1982).

2.2.3 *Carbonate Mineral Formation*

In most marine environments, Fe^{2+} rapidly precipitates with sulfides (Section 4.2.2). However, large concretions containing siderite ($FeCO_3$) have been found in some salt marshes (Pye, 1981). The $\delta^{13}C$ evidence suggests that these carbonates are partially derived from the degradation of marsh organic matter (Pye et al., 1990). Adams et al. (2006) suggested that siderite is more likely to form in saline/brackish systems when rates of Fe(III) reduction are high relative to SO_4^{2-} reduction due to low organic C availability. Under such conditions or when biological Fe(III) reduction is favored over chemical Fe(III) reduction (see above; Jacobson, 1994), low sulfide concentrations mean that Fe^{2+} and Mn^{2+} are more likely to complex with CO_3^{2-} than with sulfides (Berner, 1969; Pye et al., 1990; Adams et al., 2006). However, Giblin and Howarth (1984) reported that porewaters of Great Sippewissett Marsh (Massachusetts) were highly undersaturated with respect to both siderite and rhodochrosite ($MnCO_3$), providing strong evidence that the formation of these minerals is limited.

2.3 Losses

2.3.1 *Organic C Export*

The exchanges of C and nutrients between marshes and tidal waters have been studied for decades, often in the context of the outwelling hypothesis (Kalber, 1959; Odum, 1968, see reviews by Nixon, 1980; Dame, 1994; Childers et al., 2000). High variability in terms of hydrology, basin age, and geomorphological setting, among other features, makes it a futile task to draw broad conclusions with respect to the direction and magnitude of particulate organic C (POC) and DOC fluxes that will apply to all marshes at all times. Childers et al. (2000) summarized the salt marsh flux literature that has appeared since Nixon's (1980) influential review of marsh–estuarine exchanges. In their compilation, three of eight studies showed a net annual POC export, with rates of 11–128 g C m^{-2} $year^{-1}$ (the other five studies had net POC imports of 3–140 g C m^{-2} $year^{-1}$). In the North Inlet (SC) system, the entire 3200 ha marsh basin exported POC at a rate of 128 g C m^{-2} $year^{-1}$, whereas the geologically young Bly Creek subbasin (66 ha) imported POC at 31 g C m^{-2} $year^{-1}$ (Dame et al., 1986, 1991). In contrast to the high variability in the rates and direction of POC exchange, 11 of the 13 studies reporting DOC fluxes showed net DOC export (15–328 g C m^{-2} $year^{-1}$ vs. 7–13 g C m^{-2} $year^{-1}$ for the two studies that showed DOC import). In the Plum Island Estuary (Massachusetts), the largest amount of DOC is exported during midtides and that this material reflects old/refractory sources, suggesting that the signal strength of salt marsh DOC export varies throughout the tidal cycle (Fagherazzi et al., 2013). A similarly large export of high molecular weight DOC was reported for Rhode River, Maryland, marshes (Tzortziou et al., 2008). Both POC and DOC export varies seasonally with respect to sources, magnitudes, and lability (Osburn et al., 2015).

2.3.2 *Inorganic C Export*

Mass balance calculations indicate that a significant fraction of marsh primary production (>90%) must be decomposed or otherwise exported from salt marshes to explain long-term C accumulation rates (Howes et al., 1985; Gardner, 1990; Middelburg et al., 1997). Gaseous CO_2 emissions from salt marshes range from 240 to 720 g C m^{-2} $year^{-1}$ (Blum et al., 1978; Howes et al., 1985; Morris and Whiting, 1986; Morris and Jensen, 1998; Miller et al., 2001) with an additional loss of DIC of 120–240 g C m^{-2} $year^{-1}$ (Howes et al., 1985; Morris and Whiting, 1986; Nietch, 2000; Wang and Cai, 2004). Across saline and brackish marshes, the export of DIC accounts for ~20%–30% of total inorganic C losses (i.e., CO_2 + DIC). Tidal marshes can export significant amounts of DIC to adjacent coastal waters and affect the magnitude and direction of CO_2 exchange between coastal waters and the atmosphere (Wang et al., 2017, Wang et al., 2016, Cai and Wang, 1998; Cai et al. 1999, 2000, Raymond et al., 2000, Neubauer and Anderson, 2003, Wang and Cai, 2004, Borges, 2005). The CO_2 and DIC that accumulate (and can be lost) from salt marsh soils can be derived from the decomposition of plant roots or bulk SOM. Although SOM is less labile than fresh plant-derived material, the mineralization of the large pool of SOM can proceed at appreciable rates, even many years after in situ plant primary production has stopped (Morris and Whiting, 1986). Decomposition rates of "old" SOM once eroded and released to adjacent waters are currently unknown.

Emissions of CH_4 from salt marshes are generally lower than fluxes from freshwater wetlands (Bartlett et al., 1987, Bridgham et al., 2006, Poffenbarger et al., 2011). In a compilation of flux data from North American wetlands, Bridgham et al. (2006) reported that the average CH_4 flux from freshwater wetlands was 36.0 ± 5.0 g C m^{-2} $year^{-1}$ (average ± SE), whereas that from salt marshes was 3.6 ± 2.3 g C m^{-2} $year^{-1}$. In vegetated soils, CH_4 emission rates are largely described by temperature because plant phenology is largely driven by temperature, and temperature directly affects microbial activity, but there is a lag between periods of CH_4 production and emission in unvegetated sediments (Reid et al., 2013). Salt marsh methane emission rates are also affected by marsh invertebrates (e.g., mussels), possibly related to the biodeposition of feces that act as a carbon/nutrient source and also lead to localized O_2 depletion, both of which could enhance CH_4 production rates (Rietl et al., 2017). Sulfate reduction coupled to anaerobic CH_4 oxidation (e.g., Martens and Berner, 1977;

Boetius et al., 2000) would further reduce CH_4 emissions, but this process has not been explored in salt marsh soils.

2.3.3 Burial

Buried carbon in salt marshes has been ascribed importance as a repository of "blue carbon" (Chmura et al., 2003; Hopkinson et al., 2012; Mcleod et al., 2011; Duarte et al., 2005; Nellemann et al., 2009). As determined using ^{137}Cs and ^{210}Pb radiotracers and shorter-term accretion methods, rates of salt marsh C accumulation ranged from ~10,057 to 218 g C m^{-2} $year^{-1}$ over the last 50–100 year (e.g., Turner et al., 2000; Chmura et al., 2003; Bridgham et al., 2006; Craft, 2007). This range is comparable with that for tidal freshwater marshes (Craft, 2007, Neubauer, 2008, Megonigal and Neubauer, this volume), although there is some evidence that rates of C accumulation are lower in salt marshes than in tidal freshwater and brackish marshes (Craft, 2007). The range of carbon burial estimates includes relatively few estimates of C accumulation in the subsurface beyond a 100-year time horizon and/or derived from multidecadal- to century-scale geochronology. Average accretion and accumulation rates decrease over the first ~1000 year following burial because of decomposition and compaction of deeper soil horizons (Neubauer et al., 2002a; Turner et al., 2006b). It is not wholly clear whether such lower rates of C storage measured over longer timescales also reflect lower rates of sea level rise (SLR) in the past (Choi and Wang, 2004). Regardless of the mechanism, the differences in accretion/burial rates as a function of timescale imply that geochronology-derived estimates of C storage/burial yield a more conservative estimate of total C buried in marshes and extant blue carbon benefits than do measurements over shorter timescales (Turner et al., 2006b; Bridgham et al., 2014). Standardization of timescales and methodologies for defining carbon accretion, burial, and blue carbon are needed.

Mean rates when extrapolated to globally estimated salt marsh areas indicate marsh C burial of 11–87 Tg C $year^{-1}$ (Hopkinson et al., 2012). Due to high rates of C burial/sequestration and low rates of CH_4 emissions (Section 2.3.2), tidal marshes with a salinity of 18–23 or greater tend to have a negative radiative balance (that is, they have a net cooling effect on global climate over a typical 100-year period) (Poffenbarger et al., 2011; Neubauer and Megonigal, 2015). Elevated CO_2 levels enhance C accumulation on decadal scales but can come with increased emissions of CH_4 that offset some of the cooling due to salt marsh carbon sequestration (Pastore et al., 2017). The organic matter that is buried long-term likely represents a combination of in situ production (litter, detritus, aquatic, and subsurface roots) and allochthonous organic materials associated with mineral sediments (Wolaver et al., 1988; Craft et al., 1993; Rooth et al., 2003; Nyman et al., 2006; Neubauer, 2008).

The variance in long-term C burial estimates compounded with projected changes in organic carbon density, sediment supply/source, and overall marsh sustainability in the face of SLR confounds clear determination of how much C will continue to be buried globally in the coming decades to centuries (Morris et al. 2002, 2016; Kirwan and Megonigal, 2013). Marshes with higher tidal ranges and suspended sediment concentrations are more likely to keep pace with SLR and thus continue to bury carbon (Kirwan et al., 2010). Drowned marshes represent a repository of C potentially available for mineralization and return to the atmosphere as CO_2. The lability of this formerly buried C and the timescale of its possible mineralization are unknown.

3. NITROGEN

The total salt marsh N pool varies with marsh age and can be 5–30 times larger than the sum of all N transformations within a year, leading to long turnover times for marsh N (Anderson et al., 1997; Rozema et al., 2000). The largest marsh N pool is in bulk soils (200–1000 $g\,N\,m^{-2}$), with plant biomass containing 1–22 $g\,N\,m^{-2}$ (Hopkinson and Schubauer, 1984; Morris, 1991; Anderson et al., 1997; Rozema et al., 2000; Hopkinson and Giblin, 2018). In porewaters, dissolved organic N (DON) and NH_4^+ are the dominant species (concentrations often ≥100 s μM), while NO_3^- concentrations are generally low (≤10 μM). Despite the smaller overall inventory of porewater dissolved inorganic N (DIN), it is the currency for rapid internal N cycling.

3.1 Inputs

3.1.1 N-Fixation

N-fixation may be important for building N stocks in young marshes, but its importance generally lessens as marshes trap more exogenous N and increase internal recycling. In mature marshes, N-fixation ranges from <0.5 to 6.8 $g\,N\,m^{-2}\,year^{-1}$ (e.g., Teal et al., 1979; Rozema et al., 2000; Moisander et al., 2005), is usually a small part (~10%) of the total N inputs, and is an order of magnitude smaller that internal N recycling rates (Valiela and Teal, 1979; Anderson et al., 1997; Rozema et al., 2000). N-fixation rates and community assemblages vary seasonally and with marsh elevation (Davis et al., 2011). In young or restored marshes, rates can exceed 35 $g\,N\,m^{-2}\,year^{-1}$ and be of sufficient magnitude to support macrophyte N demand (Currin et al., 1996; Tyler et al., 2003). N-fixation is concentrated in the marsh surface and performed principally by hetero- and nonheterocystous cyanobacteria (Currin and Paerl, 1998a,b; Piehler et al., 1998). N-fixation patterns represent the sum total effects of irradiance, cyanobacterial diversity, tidal inundation, and O_2 inhibition of nitrogenase (Ubben and Hanson, 1980; Joye and Paerl, 1994; Currin et al., 1996; Moseman, 2007). There is a potential for additional N-fixation to occur in deeper sediments via sulfate reducers (Andersson et al., 2014). The importance of this pathway on total marsh N-fixation is unknown.

3.1.2 Atmospheric Deposition

Atmospheric deposition, which is typically dominated by NO_3^- and DON, can also occur as NH_4^+ and particulate N (PN) (Russell et al., 1998; Paerl et al., 2002). The DIN concentrations in precipitation are comparable with those in coastal tidal waters (Paerl et al., 2002; Russell et al., 1998). However, because precipitation comprises only a small part of the marsh water budget (Morris, 1991; Lent et al., 1997), atmospheric deposition rates are a small part of the marsh N budget. Along the Atlantic coast (USA), atmospheric deposition (<0.5–1.2 $g\,N\,m^{-2}\,year^{-1}$) is ~15% of the water column inputs (e.g., Valiela et al., 1978; Morris, 1991; Paerl et al., 2002; Seitzinger et al., 2002; Anderson et al., 1997; Buzzelli, 2008); higher loadings (~3g $N\,m^{-2}\,year^{-1}$) are seen in the marshes of northwestern Europe (Rozema et al., 2000). Timing of N delivery via precipitation and marsh elevation are variables that affect the importance of atmospheric N to macrophytes. Plants located higher in the tidal frame, which are less frequently flooded, derive more N nutrition from atmospheric sources (Oczkowski et al., 2016). There is a need for more studies in areas such as East Asia that can have even higher N-deposition rates.

3.1.3 Groundwater Inputs

Groundwater (GW) can be an important route of N delivery to salt marshes that are adjacent to N-loaded watersheds with low evapotranspiration, conductive soils, and short GW marsh flow paths. Discharge occurs at either the marsh-upland border, directly into tidal creekbeds, or can flow under the marsh and discharge subtidally offshore (Howes et al., 1996; Portnoy et al., 1997). GW N fluxes to salt marshes (0.2–100 g N m^{-2} $year^{-1}$; Table 16.3) reflect the net result of discharge, N concentration in GW, marsh area, and some influence of porewater drainage. Great Sippewissett Marsh overlies an N-rich, highly conductive aquifer, and >50% of the new N to the marsh is supplied by GW (Valiela and Teal, 1979). However, Anderson et al. (1997) presented a marsh N budget where inputs of GW N were trivial, despite close proximity to an N-rich aquifer. In other cases, GW N delivery is low because of N-poor GW (despite high flow rates) or because GW bypasses (i.e., flows under) the marsh (Portnoy et al., 1997; Tobias et al., 2001a,b,c). Because discharge typically decreases with distance from the upland-marsh border (Howes et al., 1996), the GW N flux can be less important in large marshes with short stretches of marsh-upland border than in fringing marshes. Flux estimates provided by watershed-scale numerical models (e.g., Colman and Masterson, 2008) may provide a better-constrained estimate of GW N flux when compared with small-scale direct measurements extrapolated to the whole ecosystem. The direct measure flux estimates, or those based on geochemical tracers (e.g., radium budgets;

TABLE 16.3 Selected Groundwater (GW) Nitrogen Fluxes Estimated for a Range of Salt Marshes

Marsh	GW N Flux (gN m^{-2} $year^{-1}$)	Dominant N Species	Method	References
Ringfield – VA, USA	0.2	NO_3^-	Salt balance, hydraulic head	Tobias et al. (2001a)
Philips Creek, VA, USA	<1.7	NO_3^-	Spring discharge	Anderson et al. (1997)
Bly Creek, SC, USA	0.4	Dissolved organic N	Hydraulic head	Dame et al. (1991) and Wolaver et al. (1988)
Nauset Marsh, MA USA	0.4	NO_3^-	Simulation model	Colman and Masterson (2008) and Nowicki et al. (1999)
Nauset Marsh	2–28	NO_3^-	Multiple methods	Giblin and Gaines (1990)
Great Sippewissett, MA	26	NO_3^-	Flood–ebb salt balance	Valiela et al. (1978)
Pamet River Estuary, MA	23.8–99.4	Dissolved inorganic N	Radium budget	Charette (2007)
North Inlet, SC	12	NH_4^+	Radium budget	Krest et al. (2000)

Method Denotes the Technique Used to Calculate the Water flux Component of the GW N flux.

Krest et al., 2000), include some component of recycled porewater N that can lead to an overestimate of GW N delivery but may instead provide estimates of marsh-derived export (Bollinger and Moore, 1993). More recent studies have used radium isotopes to parse out the contribution of "fresh" GW from that of recycled porewaters (Charette, 2007). Despite some of the uncertainties associated with radium-based estimates of GW N–marsh interaction, this approach operates at expanded scales not afforded by other techniques (Charette et al., 2003; Krest et al., 2000; Charette, 2007).

3.1.4 Tidal Inputs/Outputs

Tidal exchange usually dominates the mass fluxes of N between salt marshes and adjacent ecosystems (Morris, 1991; Rozema et al., 2000). Import/export has been calculated from scaling up porewater drainage measurements or from flume/whole creek estimates of net tidal exchange. Porewater drainage techniques use either by direct collection of draining porewaters or more by estimating the drainage flux using hydraulic gradient methods (e.g., Darcy's Law) and porewater concentrations. Whole creek methods are challenged by accounting for asymmetries in flood and ebb water volumes and continuous measurements of dissolved constituents (Ganju, 2011; Etheridge et al., 2014). Hydraulic gradient approaches are challenged by uncertainty in hydraulic conductivity. Compared with the porewater in the interior of the marsh, the creekbank porewaters have a shorter residence time and lower concentrations of dissolved N, C, and P. Recharge of the creekbank zone by the marsh interior is small, so dissolved N, C, and P exported via drainage is largely produced in the near creekbank zone with only small contributions from the marsh interior. Because drainage is driven by tidally driven hydraulic gradients, marshes with higher tidal amplitudes yield higher drainage fluxes of dissolved constituents (Lettrich, 2011). Scaling drainage fluxes to the whole marsh can be geomorphology dependent. Drainage fluxes are reported in units of mass per linear shoreline distance, so scaling to the whole marsh depends on the scale of measurement of the shoreline. Marshes with highly sinuous creek networks will yield a higher total system flux via drainage because of the longer shoreline per marsh area. Similarly, the total drainage flux increases nonlinearly, as smaller and smaller tidal creeks are considered in the scaling of the drainage flux.

Thirty years of tidal flux studies show that marshes transform N, sometimes acting as a net importer or exporter of dissolved or particulate N. The magnitude, and in some cases the direction, of such exchanges can switch on short timescales within a single marsh (e.g., Dankers et al., 1984; Wolaver et al., 1988; Anderson et al., 1997; Whiting et al., 1989), whereas others consistently import NO_3^- and export NH_4^+ (Poulin et al., 2009). It remains difficult to predict if a particular marsh will import or export various N species at any given time.

Over 20 salt marsh N studies provide an accounting of N exchanges as DIN, DON, and particulate organic N (PON)/PN (Table 16.4). In this dataset, 40% of the marshes were net annual importers of NH_4^+, 35% imported NO_3^-, 27% imported DON, and 26% imported PON/PN. The exchanges (either import or export) of NH_4^+ tended to be larger than those of NO_3^-. The magnitudes of import/export of DON and PON were higher than those of the DIN fractions but were highly variable. When tidal exchanges are broken down as a function of marsh age (Dame, 1994; Valiela et al., 2000), younger marshes tend to import total N (although DON is often exported), whereas older marshes often export N as both DIN and DON. Some of the NO_3^- export from these mature marshes may be explained by

TABLE 16.4 A Representative Subset of Reported Tidal Exchange Fluxes (gN m^{-2} $Year^{-1}$)

	Net Import				Net Export			
N fraction	**Range**	**Mean**	**Std**	**n**	**Range**	**Mean**	**Std**	**n**
NO_3^-	0.6–2.7	1.6	0.8	8	0.2–6.3	2.1	2.1	11
NH_4^+	0.4–4.8	2.9	1.9	6	1.6–8.7	4.2	2.8	7
Dissolved organic N	0.9–24.1	7.4	11.2	4	0.3–9.2	3.3	17.3	4
Particulate N	0.9–31	8.7	15.2	4	4.5–42	10.7	16.1	5

The range, mean flux, number of studies, and standard deviation (std) of the flux magnitudes are reported. Exchange data from Anderson et al. (1997), Chambers et al. (1994), and compilations of previous work presented in Dame (1994), Childers (1994), Childers et al. (2000), and Valiela et al. (2000).

NO_3^--rich GW discharge (Howes et al., 1996) or rapid oxidation (i.e., nitrification) of porewater that drain into exposed creekbeds during ebb tide. The latter is consistent with very high estimates of system-scale nitrification derived from whole marsh isotope additions in marshes (Gribsholt et al., 2005, 2006; Drake et al., 2009). Mid-aged marshes generally import particulate N and export dissolved N (Dame, 1994). Tidal amplitude is superimposed on marsh age as an added control on tidal fluxes—younger marshes are more sensitive to changes in tidal amplitude—a response that reflects changes in hydraulic conductivity, porewater chemistry, and rates of porewater drainage as marshes age (Whiting and Childers, 1989; Childers, 1994). Using data exclusive of flume studies, Childers et al. (2000) showed that tidal amplitude alone accounted for 40% of the variance in NO_3^- exchanges. At tidal amplitudes above 1.2 m, marshes tended to switch from NO_3^- import to export. This switch is consistent with higher tidal ranges delivering more direct drainage of high NH_4^+ porewater that can be rapidly oxidized and exported as NO_3^-.

The largest PON (and POC) import is observed in younger marshes (Childers, 1994; Dame, 1994; Osgood, 2000; Valiela et al., 2000). Tidal range positively influences PON import in younger systems but does not affect older marshes, which tend to be net exporters (Childers, 1994; Dame, 1994). Across a range of marsh types, net annual import PON is observed about as often as net export of PON (Table 16.4). Stoichiometric PON flux estimates (POM flux × N:C_{POM}) agree reasonably well with other direct estimates of PON import and range from 0.9 to 42 g N m^{-2} $year^{-1}$.

3.2 Transformations

3.2.1 *Autotrophic Uptake*

Except for some very young marshes, autotroph N demand exceeds inputs and primary production is fueled by recycled N. Macrophyte N uptake is one of the larger N flux terms ranging from 1 to 33 g N m^{-2} $year^{-1}$ (Gallagher et al., 1980, Hopkinson and Schubauer, 1984, Morris, 1991, Blum, 1993, Dai and Wiegert, 1996, Anderson et al., 1997, Rozema et al., 2000, Hopkinson and Giblin, 2018). Microalgal uptake in marshes is ~10%–15% of emergent macrophyte demand (Anderson et al., 1997; Rozema et al., 2000; Tyler et al., 2003).

Plant N content ranges from <0.2% to 3% (Anderson et al., 1997, Buresh et al., 1980). Nitrogen frequently limits plant production; plants respond to nitrogen additions with increased N content and enhanced biomass production (Drake et al., 2008, Deegan et al., 2007, and others). Marsh N turnover is on the order of a few years up to about 20 years. Anthropogenic additions of N to marshes enhance aboveground production (Fox et al., 2012; Anisfeld and Hill, 2012). The effect on belowground biomass and extant long-term marsh stability is less clear and may be dependent of several factors including marsh type (peat forming vs. mineral accreting; Deegan et al., 2012, Morris et al., 2013, Wigand et al., 2015, Kirwan and Mudd, 2012, Turner et al., 2009, Wigand et al., 2015)

3.2.2 *Mineralization and Immobilization*

The sum of new N inputs to salt marshes is on the order of 0.5% to <5% of the total N necessary to support macrophyte production (Hopkinson and Schubauer, 1984; Anderson et al., 1997; DeLaune et al., 1989; Dame et al., 1991). The difference is made up by recycling which is dominated by the mineralization of organic N into the quick turnover NH_4^+ pool (Fig. 16.2). Both aerobic and anaerobic (e.g., sulfate reduction) mineralization pathways yield NH_4^+.

Aerobic mineralization (y/z = C:N ratio of the organic matter)

$$(CH_2O)_y(\mathbf{N}H_3)_zH_3PO_4 + 106O_2 \rightarrow yCO_2 + z\mathbf{N}H_3 + H_3PO_4 + yH_2O$$

Sulfate reduction

$$(CH_2O)_y(\mathbf{N}H_3)_zH_3PO_4 + 53SO_4^{2-} \rightarrow yCO_2 + z\mathbf{N}H_3 + H_3PO_4 + 0.8yS^{2-} + yH_2O$$

Consequently porewater NH_4^+ concentrations in well-developed marshes are high. Mineralization rates vary widely between marshes (3.0–122 g N m^{-2} $year^{-1}$; Morris, 1991,

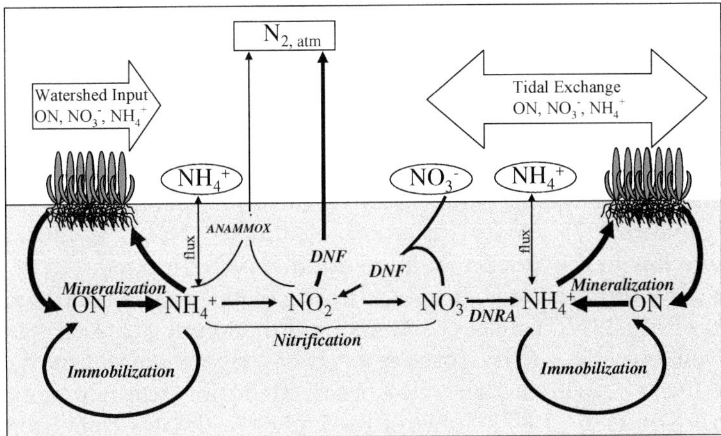

FIGURE 16.2 Summary of the major N transformation pathways in salt marshes. The sizes of the *arrows* denote the relative magnitudes of the processes. *ANAMMOX*, anaerobic ammonium oxidation; *DNF*, denitrification; *DNRA*, dissimilatory nitrate reduction to ammonium.

Anderson et al., 1997, Rozema et al., 2000, Thomas and Christian, 2001, Tobias et al., 2001b). Mineralization rates decrease exponentially with depth due to decreasing organic matter quantity and lability (Howes et al., 1985; Tobias et al., 2001b; Bowden, 1984).

For a variety of marshes, mineralization meets between 50% and 200% of autotrophic N requirements. Even in systems that appear to have a large mineralization subsidy relative to N sinks, no large exports of N are observed (e.g., Anderson et al., 1997). Several lines of evidence indicate that microbial immobilization plays an important role in retaining a significant fraction (15%–50%) of mineralized N (Anderson et al., 1997; Hopkinson and Giblin, 2018). Increases in the N content of macrophyte detritus during diagenesis are consistent with the immobilization of porewater NH_4^+ (Benner et al., 1991). Significant retention of ^{15}N tracers within marsh soils over 100 d (93%; DeLaune et al., 1983) to 7 y (40%; White and Howes, 1994a, 1994b) indicate that the NH_4^+ pool acts as the intermediary for exchanges between plant assimilation, microbial recycling, and removal via denitrification coupled to nitrification and burial.

3.2.3 *Nitrification*

Nitrification oxidizes porewater ammonium to nitrate, consumes molecular O_2, and links mineralization to N export via denitrification. Nitrification can proceed via two oxidation steps carried out in different microbial taxa or completely within a single taxa (Daims et al., 2015). Nitrification is typically the rate-limiting step for denitrifying mineralized N out of the marsh.

$$NH_4^+ + 1.5O_2NO_2^- \rightarrow 2H^+ + H_2O$$

$$NO_2^- + 0.5O_2 \rightarrow NO_3^-$$

Nitrification occurs in the presence of adequate supplies of O_2 and NH_4^+—because porewaters are typically rich in NH_4^+, nitrification is generally limited by O_2 availability. It is restricted either to surface soils or oxic microzones, declines with depth, and is influenced by tidal wetting and drying in close proximity to root channels or burrows (Howes et al., 1981; Tobias et al., 2001a,b,c; Eriksson et al., 2003; Dollhopf et al., 2005; Costa et al., 2007). Nitrification activity is inhibited by elevated sulfide levels (Joye and Hollibaugh, 1995), high salinities (Seitzinger et al., 1991; Rysgaard et al., 1999), and low pH (<4.50) characteristic of drained marshes undergoing acidification because of high rates of sulfur oxidation (Portnoy and Giblin, 1997). Under anaerobic conditions, NO_3^- is quickly reduced, so concentrations are low in the absence of large external NO_3^- inputs.

Annual rates of nitrification range from 0.26 to 52 g N m^{-2} $year^{-1}$ (Abd. Aziz and Nedwell 1986; Anderson et al., 1997; Tobias et al., 2001b, Eriksson et al., 2003; Hammersley and Howes, 2003; Dollhopf et al., 2005; Costa et al., 2007). Eighty percent of the reported rates are <10 g N m^{-2} $year^{-1}$, with higher rates observed during warmer months (Thompson et al., 1995; Anderson et al., 1997). Some of the highest estimates come from marshes that are heavily N loaded (Eriksson et al., 2003; Hammersley and Howes, 2005), suggesting that competition for N can occur between nitrifiers and autotrophs. In marsh N budgets, nitrification is 4–20-fold smaller than mineralization (Abd. Aziz and Nedwell 1986;

Anderson et al., 1997; Neubauer et al., 2005a) and is roughly equivalent to coupled denitrification. In situ estimates of nitrification following a whole ecosystem $^{15}NH_4^+$ release showed that rapid nitrification on the marsh surface accounted for 30% of the NH_4^+ transformations (Gribsholt et al., 2005, 2006). This in situ technique yielded nitrification rates 4–8-fold higher than those measured in the laboratory and illustrated the importance of natural gradients of O_2 and NH_4^+ in maintaining this reaction. Two whole-system $^{15}NO_3^-$ isotope dilution studies estimated nitrification at ecosystem scales. Rates of 11 gN m^{-2} $year^{-1}$ was calculated for a southeast coastal marsh (Duernberger et al., 2018). A New England macrotidal marsh yielded rates that were 30-fold higher, suggesting that large-scale wetting and drying significantly influences marsh nitrification at the whole marsh scale. Marsh nitrification based on small-scale incubations may therefore underestimate the system nitrification rate, and further nitrification may be enhanced at higher tidal amplitudes (Drake et al., 2009; Koop-Jakobsen and Giblin, 2010). These types of in situ approaches show promise in addressing nitrification (and other processes) at the ecosystem scale.

3.2.4 *Dissimilatory Nitrate Reduction to Ammonium*

Dissimilatory nitrate reduction to ammonium (DNRA) retains N in the marsh.

Fermentative DNRA from Giblin et al. (2013) is given as

$$CH_2O + 0.5NO_3^- + 5H_2O \rightarrow NH_4^+ + HCO_3^-$$

Autotrophic DNRA from Giblin et al. (2013) is given as

$$HS^- + NO_3^- + H^+ \rightarrow SO_4^{2-} + NH_4^+$$

DNRA can proceed through a fermentative pathway (Megonigal et al., 2004; Nijburg and Laanbroek, 1997; Tiedje, 1988), or through chemoautrophy that couples HS^- oxidation to NO_3^- reduction (Giblin et al., 2013; Brunet and Garcia-Gil, 1996). Neither pathway is currently thought to evolve N_2O. Enhanced carbon availability can support DNRA through either pathways and the autotrophic mechanisms suggest a link between DNRA and sulfate reduction. Temperature, salinity, sulfide, redox, labile organic carbon, and $[NO_3^-]$ have all been suggested as potential controls on DNRA (King and Nedwell, 1985; An and Gardner, 2002; Laverman et al., 2007). DNRA appears to be favored at high and low temperature extremes (Kelly-Gerreyn et al., 2001) and seems to be enhanced by a high organic carbon:NO_3^- ratio or when ample sulfide provides an alternate electron source (Gardner et al., 2006). DNRA can be further coupled to sulfur cycling through sulfide inhibition of denitrification, which allows DNRA to dominate (Brunet and Garcia Gil, 1996; An and Gardner, 2002; Christensen et al., 2000; Senga et al., 2006), or by inducing some sulfate-reducing bacteria to switch to NO_3^- as the terminal electron acceptor and produce NH_4^+ (Dalsgaard and Bak, 1994; An and Gardner, 2002). Different sulfide and carbon loads may also influence microbial community controls on DNRA and contribute to variations in observed rates (Scott et al., 2008; Burgin and Hamilton, 2007). The ratio of organic carbon (OC) to NO_3^- is thought to exert control on the end products of nitrate reduction favoring DNRA at high OC:NO_3 ratios (Algar and Vallino, 2014). Both carbon quality and quantity play a role in DNRA rates and the ratio of DNF to DNRA as a dominant fate for NO_3^- in the marsh (Plummer et al., 2015; Tobias et al., 2001a,b,c).

DNRA effectively competes with denitrification for NO_3^- across a range of "wet" ecosystems where it often accounts for >50%–90% of the NO_3^- reduction (Kelly-Gerreyn et al., 2001; Megonigal et al., 2004). DNRA is currently believed to be most prevalent in marine sediments where rates range from 0 to 40 g N m^{-2} $year^{-1}$ (Gilbert et al., 1997; Bonin et al., 1998; Christensen et al., 2000; Kelly-Gerreyn et al., 2001; An and Gardner, 2002; Gardner et al., 2006; Thornton et al., 2007). Approximately two dozen studies have considered DNRA in salt marshes (see review by Giblin et al., 2013; Peng et al., 2016). It has also been considered in freshwater wetlands (Bowden, 1986; Neubauer et al., 2005a; Scott et al., 2008). DNRA in mesohaline marsh soils measured in the laboratory and in situ using a $^{15}NO_3^-$ tracer yielded "annualized" rates ranging from 1.2 to 92 g N m^{-2} $year^{-1}$. These rates were from 0.3 to 2 times that of denitrification (Tobias et al., 2001b, 2001c). DNRA appears to be an important fate for NO_3^- in salty, high sulfide, organic-rich sediments. The positive relationship between DNRA and increasing salinity/sulfide suggests increased N retention via DNRA in tidal fresh and mesohaline marshes as seawater encroaches up estuaries due to SLR (Giblin et al., 2010). Responses in DNRA rates and the DNF:DNRA ratio at different N loads are inconsistent between studies. Direct measures of DNRA rates in whole-system NO_3^- additions were enhanced with NO_3^- load, but higher NO_3^- showed little affect on the ratio of DNF to DNRA because direct denitrification was similarly enhanced by the NO_3^- addition (Drake et al., 2009; Koop-Jakobsen and Giblin, 2010). In contrast, other studies show a clear proportional decrease in DNRA relative to denitrification at high NO_3^- loads (Peng et al., 2016).

The increasing number of DNRA measurements in salt marshes over the last decade provides good progress toward understanding this aspect of net N removal versus retention services provided by salt marshes. The positive relationship between DNRA and higher salinity/sulfide suggests increasing N retention in tidal fresh and mesohaline marshes as seawater encroaches up estuaries due to SLR.

3.3 Losses

3.3.1 Gaseous Losses of N—Denitrification and Anammox

Denitrification is the primary route for losses of gaseous N from salt marshes. In addition to N_2, several intermediate N gases are produced during denitrification (e.g., N_2O). N_2 is typically by far the dominant end product of denitrification relative to N_2O (e.g., 5%), particularly when O_2 is depleted under flooding conditions, although fungal and/or chemodenitrification produce N_2O exclusively (Wankel et al., 2017). Denitrification in saturated substrates usually yields <10% N_2O as an end product (Seitzinger and Kroeze, 1998; Smith et al., 1983). Under exceptionally high NO_3^- loads, Tobias et al. (2001c) reported a N_2O:N_2 ratio closer to 0.40 and suggested denitrification as a source of N_2O in heavily N-impacted marshes. N_2O emissions derived from incomplete denitrification increase at higher NO_3^- loads and are accompanied by nitrification-derived N_2O under lower N conditions in the presence of trace O_2. These conditions are most encountered during marsh emersion, particularly nearest creekbanks which exhibit that largest drawdown of the porewater elevation at low tide

(Ji et al., 2015; Harvey et al., 1987). N_2O emission from pristine salt marshes is typically a small constituent of the marsh N budget. These systems are a net sink or small source of N_2O under low N loads on the order of +1 to −10s mol N m^{-2} d^{-1} but switch to a larger source of N_2O to the atmosphere at higher NO_3^- loads particularly under conditions of low plant competition for N (Smith et al., 1983; Moseman-Valtierra et al., 2011). N_2O contribution to the greenhouse gas balance in salt marshes under high N load can contribute significantly to the total greenhouse gas carbon equivalent balance.

Most N_2O fluxes are too small to be measured using eddy covariance methods, and estimates in the literature are derived from in situ, short duration, closed chamber techniques (Moseman-Valtierra et al., 2011; Adams et al., 2012).

Although NH_4^+ is abundant in porewaters, soil pHs are typically low enough (NH_3 $pK_a = 9.7$) to prevent significant volatilization of NH_4^+ to $NH_{3(g)}$ (Morris, 1991).

Denitrification reduces 2 mol of N from NO_3 to N_2 gas. Organic carbon is the principal electron donor; the importance of alternate electron donors (e.g., reduced sulfur or iron) to denitrification remains an open question.

Denitrification coupled to organic matter oxidation

$$\left[(CH_2O)_y(NH_3)_z\right] + 0.8yNO_3^- \rightarrow 0.4yN_2 + (0.2y - z)CO_2 + (0.8y + z)HCO_3^- + (0.6y - z)H_2O$$

Denitrification rates range from 0 to 60 g N m^{-2} $year^{-1}$ (Table 16.5) and between-marsh comparisons are hampered by methodological differences between different studies (see discussion in Seitzinger et al., 2006). Having said that, differences in rates (Table 16.5) cannot be explained solely by different methods. Total denitrification depends on O_2 concentrations in the subsurface, NO_3^- supply, quantity and quality of organic carbon, and the presence of inhibitors (e.g., sulfide). The source of NO_3^- for denitrification comes from allochthonous sources such as tidal water, GW, and precipitation (direct denitrification) or is supplied from mineralization–nitrification (coupled denitrification). Seitzinger et al. (2006) suggested a breakpoint around 20 μM NO_3^- where coupled denitrification dominates at lower NO_3^- concentrations and direct denitrification dominates when NO_3^- exceeds 20 μM. Direct denitrification capacity typically increases with NO_3^- concentration due to high organic matter availability (Huntington et al., 2012) but may not scale proportionally to increasing NO_3^- loads (Davis et al., 2004). While O_2 in high abundance inhibits direct denitrification, its delivery into the NH_4^+-rich subsurface promotes the high rates of coupled nitrification–denitrification observed in close proximity to root channels and infaunal burrows (Howes et al., 1981; Dollhopf et al., 2005). Hammersley and Howes (2005) attribute several-fold higher coupled denitrification rates measured in situ versus laboratory rates to the enhancement of denitrification by plants. Higher coupled and total rates of denitrification in creekbank zones reflect more frequent oxygenation of the subsurface fueling coupled nitrification/denitrification and enhanced flushing of H_2S that inhibits denitrification (Lettrich, 2011). Seasonal variations in direct denitrification reflect changes in the external NO_3^- loading (Koch et al., 1992). Coupled denitrification is enhanced by external inputs of reduced N (Hammersley and Howes, 2005) and responds to increased NH_4^+ supply from mineralization. Positive relationships have been observed between mineralization and denitrification in freshwater

TABLE 16.5 Select Salt Marsh Denitrification Rates. Rates Are Either Reported by Authors as Annual Rates or "Annualized" for Comparison

References	$gN\ m^{-2}\ year^{-1}$	Geographic Region	Marsh	Ambient NO_3^-
Anderson et al. (1997)	0.6	Mid-Atlantic VA	Philips Creek	Low
Lettrich (2011)	0.6–3.6	New River, NC	Multiple	Low
Smith et al. (1985)	1.7	Gulf Coast, LA	Four League Bay	Moderate to high
Hinshaw et al. (2017)	0.9uv–3.6uv	Gulf Coast, MS	Chandeleur Islands	Low
Kaplan et al. (1979) and Valiela and Teal (1979)	0–44	New England, MA	Great Sippewissett	High
Abd. Aziz and Nedwell (1986)	0.3–0.8	UK	Colne Pt	Moderate to high
Koch et al. (1992)	0.3–0.7	UK	Torridge Marsh	Moderate to high
Thompson et al. (1995)	0–1.8	Southeastern NC	Newport River	Low
Eriksson et al. (2003)	0–19.6	Italy	Venice Lagoon	High
Piehler and Smyth (2011)	5.6–13.7	Southeastern NC	Multiple	Low
Poulin et al. (2007)	2.3–5.5	Canada	St. Lawrence	Low to moderate
Tobias et al. (2001b)	102	Mid-Atlantic, VA	Ringfield	Very high
Hammersley and Howes (2003)	30	New England, MA	Mashapaquit	High
Hammersley and Howes (2005)	2–60	New England, MA	Great Sippewissett	High
Compilation by Morris (1991)	0.4–14.3	East Coast, USA, Europe	Several	Variable
Smith et al. (1985)	7	Gulf Coast	Barataria Bay	Moderate

Rates encompass various laboratory and in situ techniques including N_2 accumulation, $^{15}N_2$, $^{15}N_2O$ dilution, isotope pairing, acetylene block, and mass balance methods. *uv*, unvegetated.

systems (Seitzinger, 1994; Mulholland et al., 2008). This pattern is mirrored in some salt marshes where higher denitrification is coincident with seasonally high mineralization rates (Anderson et al., 1997). Piehler and Smyth (2011) report that marsh denitrification rates are linearly correlated with mineralization as measured by sediment oxygen demand, although the denitrification rate per unit loss of O_2 is lower than that observed in other coastal environments where organic matter is more phytoplankton derived. For marshes dominated by either type of denitrification, the highest rates are encountered in surface soils which are in closest contact with NO_3^- in overlying water, nearest the zone of nitrification, and closest to near-surface labile organic matter supplies (Koop-Jakobsen and Giblin, 2009a,b; Koch et al., 1992, Tobias et al., 2001b). Denitrification rates in vegetated soils are several folds higher than nonvegetated soils due to enhanced nitrification and/or labile carbon availability (Hinshaw et al., 2017).

Because denitrification exports N, it plays an important role in overall N residence time in the marsh. Competition for N between plants and denitrification is a regulator of long-term N retention, with denitrification reported at roughly 20% of plant uptake (White and Howes, 1994b). There are examples of marshes where plant growth far outcompetes denitrification for N (Anderson et al., 1997) and those where denitrification and plant N demand are roughly equal (Morris, 1991). Studies in several mature marshes put losses of N through denitrification approximately equivalent to inputs via N-fixation suggesting that the balance between these two pathways is important for overall marsh N balance (Valiela and Teal, 1979; Morris, 1991; White and Howes, 1994b; Anderson et al., 1997; Rozema et al., 2000). Denitrification ranges between 15% and 100% of N burial, with most marshes showing similar magnitudes for each pathway. There are only a few examples of systems where denitrification vastly exceeds burial and are likely explained by high NO_3^- inputs to these marshes fueling high rates of direct denitrification (e.g., Kaplan et al., 1979). As with all other input and removal pathways, the magnitude of denitrification is small relative to internal N cycling.

Anaerobic ammonium oxidation (ANAMMOX) is an alternate pathway by which N can be exported from ecosystems. ANAMMOX uses ammonium to reduce nitrite to produce N_2. Nitrite is combined with ammonium through NO and hydrazine intermediates to yield N_2 gas and NO_3^- in an molar N ratio of 6 N_2–N per NO_3–N (Brunner et al., 2013).

$$1.3\ NO_2^- + NH_4^+ \rightarrow N_2 + 0.3\ NO_3^- + 2\ H_2O$$

It is chemoautotrophic and unlike denitrification does not directly require organic carbon. ANAMMOX can rival rates of denitrification in a variety of marine sediments. The importance of ANAMMOX relative to denitrification is likely dependent on the ratio of organic carbon to NO_3^- (Algar and Vallino, 2014; Hardison et al., 2015). The high organic C in marshes is not considered favorable for ANAMMOX. Correlations between ANAMMOX and denitrification rates suggest that the ANAMMOX nitrite requirement is linked to the nitrate reduction step of denitrification (Hou et al., 2013; Song and Tobias, 2011). ANAMMOX rates in marshes are typically small relative to denitrification and comprise from <1% to 12% of the total N_2 evolved from both reactions (Naeher et al., 2015; Dalsgaard et al., 2005; Hou et al., 2013; Hou et al., 2015; Koop-Jakobsen and Giblin, 2009a,b). Most reported rates are less than 4% with slightly higher values in brackish relative to full strength salinity wetlands. There is no apparent effect of NO_3^- availability on the ANAMMOX rate (Koop-Jakobsen and Giblin, 2009a,b), despite the presumed shift toward a more favorable organic carbon to nitrate ratio at high N loads. Salinity influences ANAMMOX bacterial diversity in marsh sediments but not the rates (Hou et al., 2013). The ANAMMOX: denitrification ratio increases at lower temperatures (Rysgaard et al., 2004) and thus figures more prominently in marsh N_2 production in the winter (Hou et al., 2015). But this effect is small, given that ANAMMOX rates remain about one-tenth that of denitrification. Nevertheless, ANAMMOX plays a secondary role, after denitrification, in removal of reactive N from the coastal landscape but is less important in high organic carbon systems (such as salt marshes), where it accounts for less than 10% of the N_2 production

(Dalsgaard et al., 2005). It is not currently considered an important route of N loss from salt marshes. ANAMMOX coupled to iron reduction (Feammox) accounted for 3%–5% of N losses from a Yangtze Estuary marsh, but the broader importance of Feammox in marsh N cycling is not known (Li et al., 2015).

3.3.2 *Burial*

Long-term nitrogen burial rates in salt marshes have been estimated directly using sediment traps, horizon markers, radioisotope profiles (^{210}Pb, ^{137}Cs), and extrapolated from SLR estimates (Hutchinson et al., 1995; Merrill and Cornwell, 2000; Craft, 2007; Goodman et al., 2007). N (and P) burial combines contributions from allochthonous suspended sediments and autochthonous organic matter. N is recycled multiple times within marsh soils, porewater, and biota before it ultimately becomes sequestered primarily in belowground organic matter (White and Howes, 1994b). Nitrogen burial rates along the Atlantic and Gulf coasts (USA) range from 1 to 23 g N m^{-2} $year^{-1}$ (e.g., Craft, 2007 and references therein). For Wadden Sea salt marshes, net sedimentation of N is on the order of 10–50 g N m^{-2} $year^{-1}$ (Rozema et al., 2000). Estuarine marshes in the United Kingdom have a similar range with the highest rates reported for "newly" created managed marshes that have high but transient rates of sedimentation (Adams et al., 2012). For systems that lack direct measurements, an SLR proxy can be used to estimate N burial. With some exceptions (Chmura and Hung, 2004), there is good agreement between marsh accretion and SLR (Donnelly et al., 2004). By combining typical bulk density and N content estimates for marsh soils with local measures of SLR, one can broadly generalize that marshes keeping pace with SLR are burying N on the order of 2–6 g N m^{-2} $(mm\ SLR)^{-1}$. In rapidly accreting marshes (>5 mm/year), N burial can be comparable with rates of N turnover in soils. For most other systems, N burial is on the order of 50%–60% of the total N inputs (Anderson et al., 1997; Neubauer et al., 2005a). Network analysis of three salt marshes (New England, Georgia, and Mid-Atlantic, USA) indicated that burial is second only to tidal export as the fate of imported PN across these marsh types (Thomas and Christian, 2001). With the exception of high nitrate systems that drive high rates of direct denitrification, burial is typically the dominant N retention mechanism in salt marshes. Across salinity gradients, N burial is inversely related to salinity due to a lower soil N content at euryhaline sites and enhanced decomposition of belowground organic matter at higher salinity marshes (Craft, 2007).

4. IRON AND SULFUR

The biogeochemical cycles of iron and sulfur are intimately linked in salt marsh soils (Fig. 16.3), primarily through the abiotic reduction of Fe(III) by sulfides and via the reactions that control the formation and dissolution of Fe–S minerals such as iron monosulfides (FeS) and pyrite (FeS_2). Furthermore, both elements undergo active redox cycling, play a role in organic C catabolism (Section 2.2.2), and may also contribute to significant chemoautotrophic production (Section 2.1.2). There is tight coupling between oxidation and reduction reactions of both Fe and S at aerobic/anaerobic interfaces (Megonigal et al., 2004; Wilbanks et al., 2014; Ionescu et al., 2015), as is also seen with the N cycle (e.g., coupled nitrification–denitrification; Section 3.3.1)

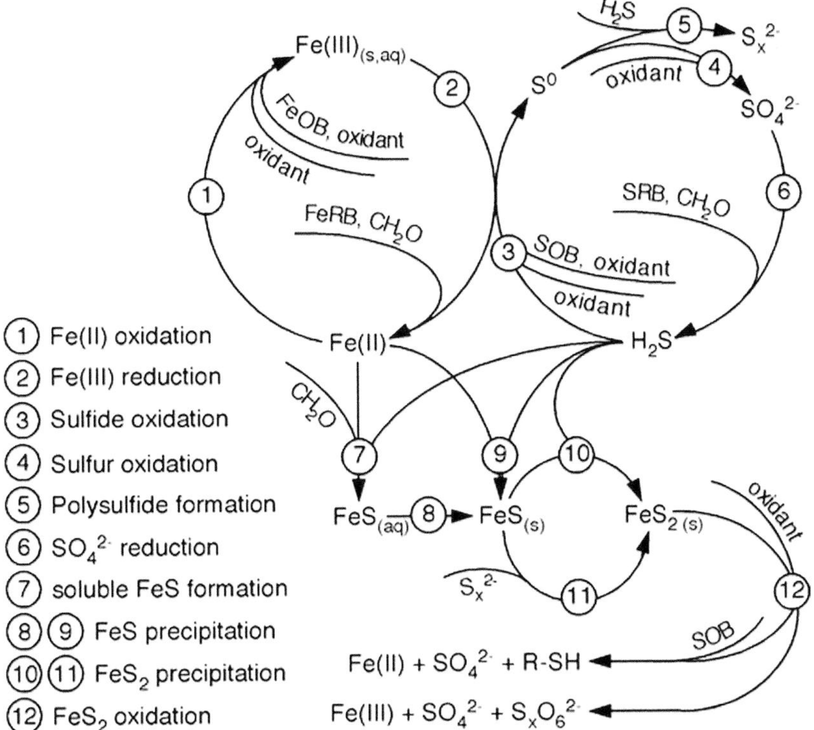

FIGURE 16.3 The Fe and S cycles are linked in salt marsh soils through a series of biologically mediated and abiotic reactions. This figure focuses on inorganic forms of Fe and S; with the exception of thiol (R–SH) formation during biological pyrite oxidation, the complexities of the organic Fe or S cycles are not considered here. For simplicity, the oxidation of FeS is not shown, although the products of this reaction are similar to those of pyrite oxidation (i.e., SO_4^{2-}, Fe(II), and/or Fe(III)). FeOB and SOB = Fe(II)- and sulfide-oxidizing bacteria, respectively; FeRB and SRB = Fe(III)- and SO_4^{2-}-reducing bacteria, respectively. Oxidants are compounds including O_2, NO_3^-, MnO_2, and Fe(III) oxides that can oxidize reduced compounds. CH_2O generically refers to organic C, which can be in the form of organic molecules that are catabolized by heterotrophic microbes such as FeRB and SRB or as ligands that play a key role in solubilizing and/or reducing Fe minerals (e.g., Luther et al., 1992; Carey and Taillefert, 2005).

4.1 Inputs

4.1.1 Dissolved and Atmospheric Inputs

Delivery of seawater SO_4^{2-} (28 mmol L^{-1}) occurs via tidal infiltration and diffusion. In some cases, tidal inputs of SO_4^{2-} can exceed rates of sulfate reduction (Morris, 1995; Kostka et al., 2002b). SO_4^{2-} delivery rates vary with soil depth, distance from tidal creeks, and location within the estuary. Sulfate concentrations in marsh porewaters are typically depleted (relative to a conservative tracer), indicating net SO_4^{2-} utilization (Gardner et al., 1988; Hines et al., 1989; Hseih and Yang, 1997).

Atmospheric SO_4^{2-} deposition in the coastal region is relatively minor at <1 g SO_4^{2-}-S m^{-2} $year^{-1}$ (NADP, 2018) and Fe deposition (~0.1 g Fe m^{-2} $year^{-1}$) is orders of magnitude lower than Fe deposition via sedimentation. Both fluxes may be locally elevated in close proximity to industrial emissions (Lecoanet et al., 2001).

4.1.2 *Sediment Deposition*

Recently deposited sediments and near-surface marsh soils often have Fe contents in the range of 1%–2% (weight basis; DeLaune et al., 1981, Gardner et al., 1988, Cornwell and Zelenke, 1998) and S contents of 0.5%–2% (although values up to 5% can occur; Cutter and Velinsky, 1988, Gardner et al., 1988, Giblin, 1988). Using these Fe and S concentrations and mass deposition rates for Paulina marsh (Fig. 16.1), annual deposition of these elements is ~50–100 g Fe m^{-2} $year^{-1}$ and ~25–100 g S m^{-2} $year^{-1}$. These deposition rates roughly overlap with long-term accumulation (burial) rates of these elements (Section 4.3.3)

4.2 Transformations

4.2.1 *Iron and Sulfur Reduction*

Iron(III) reduction can be biotic or abiotic (Fig. 16.3). Microbial and geochemical analyses have suggested a significant, if not dominant, role for microbial Fe(III) reduction, although there is spatial variability in the importance of each pathway (Section 2.2.2). Large populations of Fe(III)-reducing bacteria (FeRB) can be found in vegetated soils and around roots (Lowe et al., 2000; Kostka et al., 2002a; Koretsky et al., 2003; Weiss et al., 2003), proximal to SO_4^{2-}-reducing bacteria (SRB) and aerobic microbes (Koretsky et al., 2005). Iron reduction is further linked to the N-cycle through Feammox and chemodenitrifcation (Li et al., 2015, Wankel et al. 207). Microbial Fe(III) reduction accounts for 80%–100% of iron reduction and 39%–53% of total reduction in mineral and organic-rich marsh soils, respectively (Kostka et al., 2002a; Gribsholt et al., 2003; Hyun et al., 2007; Neubauer et al., 2005b). The importance of microbial iron reduction as a catabolic pathway is influenced by the availability of poorly crystalline Fe(III), which varies as a function of tidal inundation (Luo et al., 2016). Koretsky et al. (2003) proposed that seasonal changes in rates of H_2S production cause oscillation between chemically dominated Fe(III) reduction in summer and biologically dominated Fe(III) reduction in other seasons. When burial rates are high, subsurface soils can contain high concentrations of reactive Fe(III) oxides that serve as an effective chemical sink for sulfides produced via sulfate reduction, producing FeS minerals (Pye, 1981). The low solubility of FeS creates a diffusion gradient that drives Fe(II) movement to the site, creating conditions that are oversaturated with respect to siderite (i.e., high in Fe(II) with high OH^- alkalinity) and leading to the precipitation of $FeCO_3$ in addition to the FeS minerals that are typical of salt marshes (Mortimer et al., 2011).

The reduction of SO_4^{2-} is primarily a biological process coupled to the oxidation of simple organic compounds (Hines et al., 1999; Boschker et al., 2001). The H_2S produced tends to rapidly convert to pyrite, although some fraction (<5%–23%) appears as organosulfur compounds (Howarth, 1979; Neubauer et al., 2005b). Strictly anaerobic periods favor sulfate reduction, while suboxic conditions associated with lower porewater elevation favor iron reduction and denitrification (Palomo et al., 2013). Seasonal patterns of SO_4^{2-} reduction tend to follow temperature patterns and are also influenced by primary production within the marsh (Howarth and Teal, 1979; Hines et al., 1999; Koretsky et al., 2003). During flowering, SO_4^{2-} reduction decreases as plants allocate more C to flowering/reproductive structures and less C leaks into the soil to fuel SO_4^{2-} reducers (Lytle and Hull, 1980; Hines et al., 1999). The relative abundance of SRB rRNA associated with plant roots increased in parallel with changes in SO_4^{2-} reduction activity (Hines et al., 1999). Similarly, the SRBs *Desulfobacter* and *Desulfovibrio* were more prevalent in summer than winter (Koretsky et al., 2005). However, Hines et al. (1999) reported that the rRNA that could be ascribed to SRB did not change significantly over the year (relative to bacterial rRNA), even though rates of SO_4^{2-} reduction did show seasonal variability.

4.2.2 Formation and Oxidation of Fe–S Minerals

Solid-phase sulfides are operationally divided into acid-volatile sulfides (consisting of H_2S and FeS) and chromium (II)-reducible sulfides (CRS: FeS_2 and S^0). Early work measuring SO_4^{2-} reduction rates demonstrated that a large fraction of the added $^{15}SO_4^{2-}$ tracer appeared as solid-phase sulfides, indicating that Fe–S minerals form rapidly in salt marsh soils (Howarth, 1979). Pyrite is a tenfold more important end product than iron monosulfides, although the mechanisms of pyrite formation involve an FeS intermediary (Fig. 16.3; Giblin and Howarth, 1984, Cutter and Velinsky, 1988, Gardner et al., 1988). The reaction of Fe(II) and H_2S leads to the formation of a soluble FeS phase when suitable organic ligands are present (Taillefert et al., 2000; Carey and Taillefert, 2005) or to direct precipitation of FeS minerals. Reaction with additional reduced S (as H_2S or polysulfides, S_x^{2-}) then leads to FeS_2 precipitation (Rickard and Luther, 2007; Canfield et al., 1998).

Inorganic sulfide minerals are not always the dominant form of S in salt marsh soils (Fig. 16.4A)—across a range of salt marshes, CRS accounts for 20%–100% of total soil S (average CRS:TS [total sulfur] ratio = 0.57 ± 0.25 SD). The nonpyritic sulfur is likely dominated by organic S forms (Giblin, 1988). Cutter and Velinsky (1988) showed that organic S accounted for 45 ± 19% of the total S pool in the top 30–50 cm of Great Marsh, Delaware. Similarly, using depth profile data from marshes in Delaware (Great Marsh, Cutter and

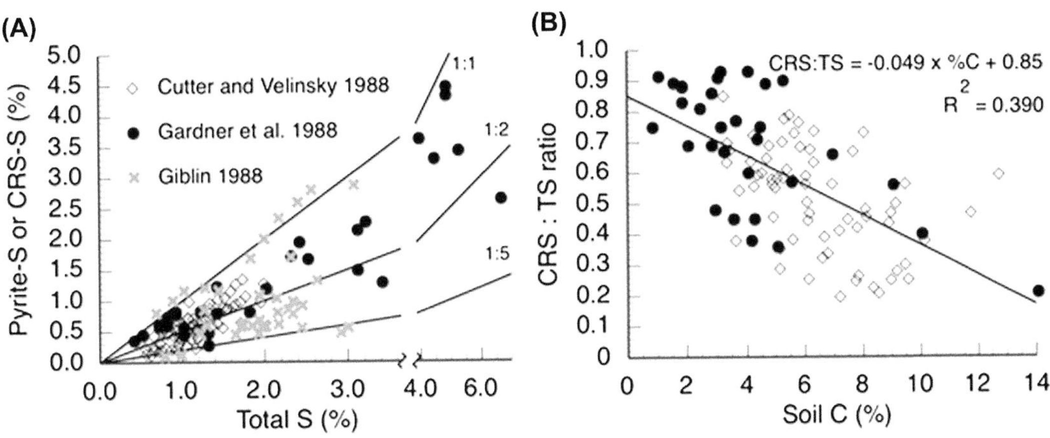

FIGURE 16.4 Soil pyrite relationships in salt marsh soils. Most data are from *Spartina alterniflora* marshes, although the Giblin (1988) study also included data from sites vegetated with *Spartina patens*, *Scirpus*, and *Typha*. Data points are individual samples and reflect temporal and spatial (lateral and depth-related) variability in solid-phase soil chemistry. (A) Chromium(II)-reducible sulfur (CRS) versus total sulfur (TS) in marsh soils. *Solid lines* indicate contours where CRS = TS, CRS = 0.5 TS, and CRS = 0.2 TS. (B) Relationship between the CRS:TS ratio and the soil C content. The regression line is fit through all data.

Velinsky, 1988) and South Carolina (North Inlet, Gardner et al., 1988), we demonstrate that the soil C content is inversely correlated with the fraction of CRS-S (Fig. 16.4B). In this same dataset, there was a positive relationship between the degree of pyritization (DOP) and the CRS:TS ratio for Great Marsh but not for North Inlet (data not shown).

The DOP (Berner, 1970), an index of how much of the available Fe has reacted with sulfides to form pyrite, is calculated as

$$\mathrm{DOP} = \frac{\mathrm{Fe_{pyrite}}}{\mathrm{Fe_{pyrite}} + \mathrm{Fe_{reactive}}}$$

where Fe_{pyrite} is the amount of Fe in pyrite and $Fe_{reactive}$ is the amount of reactive or amorphous Fe in the sample. In salt marshes, DOP values for interior/high marsh sites are generally higher than for creekbank/levee sites, (King et al., 1982; Gardner et al., 1988; Giblin, 1988; Otero and Macias, 2003; Roychoudhury et al., 2003a; Roychoudhury, 2007). Sulfide production in interior marsh sites appears ~1–2 orders of magnitude greater than the rate of Fe delivery (Gardner et al., 1988) and DOP values can approach 1.0 at these sites (DOP ~.6–1.0). The lower DOP values in creekbank marshes (~0.2–0.7) indicate an enhanced supply of reactive Fe (Section 2.2.2), decreased sulfide availability (King et al., 1982), and/or processes that limit the formation/accumulation of pyrite that are likely related to higher flushing rates relative to the marsh interior.

Net rates of pyrite accumulation are considerably lower than gross rates of formation (Gardner, 1990). Net pyrite oxidation exists during the spring/summer growing season, whereas net pyrite formation occurs in winter (Giblin and Howarth, 1984; Cutter and Velinsky, 1988). Variations in the timing and frequency of flooding of the marsh surface can override some of this seasonal generality (see discussion in Giblin, 1988). Hseih and Yang (1997) speculated that roots play a major role in rates of pyrite accumulation by oxidizing the soil and providing labile organic C to fuel Fe(III) reduction and SO_4^{2-} reduction. Below the rooting zone, pyrite concentrations do not show large seasonal or spatial variability, again emphasizing the role of plants in both FeS_2 formation and oxidation (Cutter and Velinsky, 1988; Hseih and Yang, 1997).

Diagenetic model results suggest that rhizosphere oxidation (with either O_2 or organic compounds as oxidants) was responsible for the majority of FeS_2 oxidation (Gardner, 1990). Gardner et al. (1988) suggested that bioturbation near creekbanks could remove 18% of subsurface pyrite by bringing it to the surface where it could be oxidized and/or washed away. Advection of O_2-rich tidal water through soils contributes to subsurface FeS_2 oxidation, and the seepage of water out of creekbanks provides advective removal for H_2S. Roychoudhury et al. (2003a) suggested that this tidal flushing, rather than effects of macroorganisms and plants, was most significant in affecting pyritization in creekbank soils. Away from the creekbanks, where infaunal densities and porewater flushing are low, rhizosphere oxidation is likely to be the most significant mechanism for pyrite turnover. Thermodynamically, FeS_2 can be oxidized anaerobically with NO_3^-, MnO_2, or Fe(III) as oxidants. With some exceptions (Schippers and Jørgensen, 2002; Carey and Taillefert, 2005), the significance of these reactions has not been fully assessed in salt marshes.

4.2.3 *Iron Oxidation*

Oxidized Fe, in the form of root plaques or otherwise (Luther et al., 1982a; Mendelssohn et al., 1995; Weiss et al., 2003), can serve as electron acceptors for anaerobic metabolism (Lowe et al., 2000; Kostka et al., 2002a; Neubauer et al., 2005b), as a sink for trace metals and PO_4^{3-} (Weis and Weis, 2004; Neubauer et al., 2008; Scudlark and Church, 1989; Chambers and Odum, 1990). In tidal wetlands, the solid-phase Fe in the upper portions of the soil profile is generally rich in Fe oxides, whereas iron sulfide minerals dominate at lower depths (Griffin et al., 1989). In shallow soils, O_2 is the likely oxidant that drives both chemical and biologically mediated Fe(II) oxidation. Microbial Fe(II) oxidation linked to anaerobic photosynthesis (Widdel et al., 1993), NO_3^- reduction (Straub et al., 1996), and perchlorate reduction (Chaudhuri et al., 2001) may also play a role. Plant roots are hot spots for oxidation, forming Fe oxide–rich plaques tens of micrometers up to 0.4 cm in thickness (Taylor et al., 1984; Vale et al., 1990). Sundby et al. (2003) proposed that rapid rates of biological and/or chemical Fe(II) oxidation (and therefore plaque formation) in a Tagus Estuary (Portugal) salt marsh occurred in spring as new highly active roots invaded anoxic sediments with high Fe^{2+} concentrations. Over multiple years, the oxidizing activity of plant roots can increase extractable Fe(III) pools relative to unvegetated sites (Kostka and Luther, 1995; Roden and Wetzel, 1996; Weiss et al., 2004). Tidal delivery of O_2 also influences temporal patterns of Fe(II) oxidation (Neubauer et al., 2005b). Because Fe(II) availability is low in mid- and high marsh areas (i.e., DOP is high, see above), overall rates of Fe(II) oxidation are likely to be lowest in these parts of the marsh. However, lower pH (Gardner et al., 1988) may lead to biological Fe(II) oxidation being relatively more important in these areas because chemical Fe(II) oxidation rates decrease as the pH decreases (Singer and Stumm, 1970).

4.3 Losses

4.3.1 *Tidal Export*

The turnover of marsh porewater and diffusive fluxes each play a role in exporting dissolved S and Fe. Tidal export of 20 g H_2S–S m^{-2} $year^{-1}$ and 1504 g S_2O_3–S m^{-2} $year^{-1}$ was estimated for Great Sippewissett Marsh (Howarth and Teal, 1980). The energy contained in these reduced S effluxes was equivalent to 23% of total NPP. Using an H_2S concentration of 100 μmol L^{-1} for tall *S. alterniflora* soils in Georgia (King et al., 1982) and applying a porewater turnover of 5–20 L m^{-2} d^{-1} (as above), we estimate an export of 6.4–22 g H_2S–S m^{-2} $year^{-1}$, a rate that overlaps with the value calculated by Howarth and Teal (1980). Luther et al. (1982b) observed that the SO_4^{2-}:Cl^- ratio over the tidal cycle decreased during ebb tide in two marsh creeks but increased dramatically in a third creek, indicating that total S import/export depends on spatially distinct balances between sulfate reduction and oxidation of metal sulfides. Guo et al. (2000) reported a Fe^{2+} flux of 2.6–5.2 mg Fe m^{-2} d^{-1} (0.9–1.9 g Fe m^{-2} $year^{-1}$) from saline and brackish marsh microcosms. The loss of Fe(II) due to porewater drainage will be greater at creekbanks than in mid- and high marsh locations due to a higher Fe^{2+} supply (more Fe(III) reduction and a lower DOP) and more rapid rates of porewater turnover in creekbank marshes. Given the kinetics of Fe(II) oxidation, it is likely that a significant fraction of Fe^{2+} in tidal waters will be rapidly reoxidized and may settle back onto the marsh surface. However, reactions with organic ligands may keep the resulting Fe(III) in a soluble form (Luther et al., 1996) and allow $Fe_{(aq)}$ to be exported from the marsh.

4.3.2 *Atmospheric Losses*

Saline and brackish marshes emit H_2S, dimethyl sulfide ($(CH_3)_2S$), carbonyl sulfide (COS), methanethiol (CH_3SH), carbon disulfide (CS_2), and dimethyl disulfide ($(CH_3S)_2$) (Morrison and Hines, 1990; DeLaune et al., 2002b). Most salt marshes are net sources of S gases to the atmosphere (Cooper et al., 1987; Dacey et al., 1987; Morrison and Hines, 1990; Crozier et al., 1995; DeLaune et al., 2002b; Whelan et al., 2013), although there are reports of uptake of specific gases (e.g., COS; Morrison and Hines, 1990). DeLaune et al. (2002b) reported an average total gaseous emission rate of 0.6 g S m^{-2} $year^{-1}$ for a salt marsh and 0.3 g S m^{-2} $year^{-1}$ for a brackish marsh.

The production of sulfur-containing gases is positively correlated with SO_4^{2-} concentration supply (Crozier et al., 1995), a pattern that was observed for total gaseous S emissions along a salinity gradient in Louisiana (DeLaune et al., 2002b). Because the production of these gases is biologically driven, emission rates are generally greater with higher plant biomass and growth (e.g., DeLaune et al., 2002b; Morrison and Hines, 1990). Dimethyl sulfide dominates emissions (>50%) in *S. alterniflora* marshes with low emissions as H_2S (<10%; Cooper et al., 1987, DeLaune et al., 2002b). A *S. patens* brackish marsh emits primarily H_2S and COS (DeLaune et al., 2002b). High dimethyl sulfide emission rates from *S. alterniflora* versus *S. patens* brackish marshes reflect some plant physiology; *Spartina alterniflora* contains high concentrations of dimethylsulfoniopropionate, the precursor to dimethyl sulfide, whereas *S. patens* does not (Dacey et al., 1987). The differential emissions along the salinity gradient also show the effect of sulfide sequestration in metal S complexes in the euryhaline marsh. Despite emission of several S gases, it is not an important loss term for individual marshes relative to burial and tidal export. Although S gas emissions are important for atmospheric chemistry and radiative balance (Shaw, 1983; Kelly and Smith, 1990), salt marshes contribute little as a global source/sink.

4.3.3 *Burial*

Howarth (1984) reported a net rate of S accretion of 13 g S m^{-2} $year^{-1}$ in the Great Sippewissett and Sapelo Island marshes, although this estimate does not include the burial of organic-bound S. Assuming that the S is primarily buried as pyrite (but see below), we can then estimate that Fe burial is ~11 g Fe m^{-2} $year^{-1}$. Data from more organic-rich brackish and salt marshes show higher rates of total S accumulation (24–46 g S m^{-2} $year^{-1}$; Krairapanond et al., 1992). Furthermore, in these marshes, the majority of the S accumulated as C-bonded S (56%–65%) and ester sulfates (21%–23%) with <3% because of the preservation of pyrite. Using soil data from Cutter and Velinsky (1988) and an accretion rate of 0.47 cm $year^{-1}$ (Church et al., 1981), we calculate that the S accumulation rate has averaged 29 g S m^{-2} $year^{-1}$ over the last 75–100 year at Delaware's Great Marsh, with the accumulation split roughly evenly between pyrite and organic S. Using the same dataset, we estimate a long-term accretion rate of 21 g Fe m^{-2} $year^{-1}$ that was driven by pyrite (62% of total Fe accumulation) and oxidized Fe (27%). The accumulation of pyrite increased with depth, whereas that of reactive oxides generally decreased with depth. Data from other salt marshes are similar in terms of the magnitude of Fe burial, with rates ranging from ~11 to 60 g Fe m^{-2} $year^{-1}$ (e.g., DeLaune et al., 1981; Gardner et al., 1988).

5. PHOSPHORUS

Phosphorus (P) in salt marshes is diversely speciated in the solid and dissolved phases. P cycling is affected by iron and sulfur reactions and thus differs considerably in salt marshes versus tidal freshwater marshes. In salt marshes, less P is adsorbed to soils, due in part to higher ionic strength, and porewaters contain more dissolved inorganic P (DIP) (Paludan and Morris, 1999). N-limitation of primary production in salt marshes versus P-limitation in freshwaters arises in part from this P speciation (Hartzell and Jordan, 2012). The resultant 4–5-fold drop in the DIN:DIP ratio of salt marsh porewater typically prevents P-limitation of autotrophs, although marsh bacteria can remain P-limited because of scarcity of organic P substrates (Patrick and Delaune, 1976; Sundareshwar and Morris 1999). P fertilization alone does not typically affect aboveground production but can lead to a decrease in belowground biomass (Darby and Turner, 2008a). There are some reports of P-limitation of macrophytes production in young salt marshes, where P recycling from organic sources is low, and in marshes that receive high external N loads (Van Wijnen and Bakker, 1999).

Solid phases (including sorbed PO_4^{3-}) dominate P inventories. Soil-bound P (0.05%–0.08%) plus that in dead plant biomass is the largest reservoir ($\sim$30–100 g P m^{-2}), exceeding that of live plant biomass ($\sim$10–20 g P m^{-2}; 0.05%–2% by weight) and porewater PO_4^{3-} (<0.5 g P m^{-2}; 0–100 s μM; Buresh et al., 1980, Sundareshwar and Morris, 1999, Zhou et al., 2007). Large fractions of total phosphorus (TP) are associated with organic matter. Marshes low in organic content tend to show lower overall TP and porewater PO_4^{3-} (Zhou et al., 2007). The various P fractions in soils are defined as (1) loosely exchangeable (salt extractable), which includes both the dissolved PO_4^{3-} pool and the dissolved nonreactive organic P (DNRP); (2) PO_4^{3-} bound to oxidized iron and aluminum (of which the iron fraction is redox sensitive); (3) soil organic (humic) P, which is removable following sequential NaOH/HCl treatment and which tends to be low in young or newly constructed marshes; (4) calcium (carbonate)-bound P, which is liberated following acidification; and (5) residual refractory P (Sundby et al., 1998; Paludan and Morris, 1999; Coelho et al., 2004; Zhou et al., 2007; Xie et al., 2015). The amount of TP through the soil profile can either be relatively constant or decline with depth, but TP inventories change little throughout the year (Du et al., 2016; Stribling and Cornwell, 2001; Coelho et al., 2004; Weston et al., 2006). Increases in the refractory P content with depth are consistent with organic matter diagenesis during burial (Lillebø et al., 2007). Approximately 20% to >40% of the bound P is organic (humic + nonreactive + refractory fractions; Coelho et al., 2004, Álverez-Rogel et al., 2007). Approximately 40% of the total P in soils is readily exchangeable and available for equilibrium maintenance of porewater PO_4^{3-} (Lillebø et al., 2007). Spatial and temporal variability in porewater PO_4^{3-} coincides with changes in biological activity (e.g., mineralization, plant demand) and physicochemical conditions that affect (de)sorption to/from mineral phases. The highest PO_4^{3-} concentrations are seen in soils where mineralization is high and low Eh reduces PO_4^{3-} sorption to iron oxides (Chambers et al., 1992; Chambers, 1997; Sundareshwar and Morris, 1999; Stribling and Cornwell, 2001; Lillebø et al., 2007). Exchanges between solid and aqueous phases maintain porewater PO_4^{3-} concentrations, and spatial/temporal differences represent adjustments to that dynamic equilibrium rather than large changes in net P import/export.

5.1 Inputs and Tidal Exchanges

Aside from some instances of GW inputs in areas of P-rich geology (Weston et al., 2006), tides are the principal routes for delivering and removing P. These exchanges are small, however, relative to the size of the marsh P reservoir (Paludan and Morris, 1999; Coelho et al., 2004). While marshes accumulate P over their life span, they can act at any given time as sinks or sources for particulate and/or DIP.

5.1.1 *Tidal Exchanges—Particulate P*

Settling of particulate P (organic and inorganic bound) represents a major source of P for most salt marshes. There are robust examples of both net import and export of particulate P (Dankers et al., 1984; Dame, 1994; Childers, 1994). The magnitude and distribution of particulate P delivery is controlled by settling patterns of suspended P as a function of water column load, partitioning of P between organic and mineral phases, hydroperiod, and flooding frequency (Friedrichs and Perry, 2001). The mineral-attached fraction of particulate P depends on the geochemical composition of suspended sediments. Changes in ionic strength, pH, and availability of different sorptive minerals along salinity gradients cause difference in the magnitude and form of P delivery to mesohaline and euryhaline marshes (Sundby et al., 1992; Paludan and Morris, 1999; Jordan et al., 2008). Contributions of P associated with Fe and Al oxides are greatest in low salinity environments where Fe and Al minerals are in ample supply (Fox et al., 1986; Froehlich, 1988; Sundareshwar and Morris, 1999). At euryhaline salinities with higher pHs, the Fe- and Al-bound fractions decrease as does the total P adsorbed per unit soil unless there are external sources of Fe (Jiménez-Cárceles and Álvarez-Rogel, 2007). The calcite-bound/sorbed fraction of P becomes a more important source of settling P to euryhaline marshes and its distribution can be affected by variation in pH (Coelho et al., 2004; Álverez-Rogel et al., 2007, Xie et al., 2015, Fig. 16.5).

Relative to tidal flux studies of POC or PON, there are fewer estimates of organic particulate P fluxes to the marsh. However, one can derive estimates of particulate organic P (POP) fluxes from previously measured PON and POC fluxes. Using existing compilations of POC/PON flux data (Dame, 1994; Rozema et al., 2000), we assigned a conservative C:P ratio of 120 to the POC fluxes and an intermediate N:P ratio of 20 (Craft, 2007) to reported PON fluxes. Using these stoichiometries, net POP fluxes ranged from -0.2 to $0.79\ \mathrm{g\ P\ m^{-2}\ year^{-1}}$ (from POC data, negative sign indicates export; Dame, 1994) and -3.0 to $2.0\ \mathrm{g\ P\ m^{-2}\ year^{-1}}$ (from PON data; Rozema et al., 2000). For studies that showed net POP import, average rates were $\sim 0.6\ \mathrm{g\ P\ m^{-2}\ year^{-1}}$ (from both POC and PON stoichiometry). The export of POP averaged 0.2 and $1.0\ \mathrm{g\ P\ m^{-2}\ year^{-1}}$ from POC and PON data, respectively. Tidal flux estimates of net P import tend to be at the low end of measured long-term P accumulation rates (Section 5.3.1), probably because buried P contains mineral-bound P that is not included in the POP flux calculations.

Younger to middle age marshes are more likely to show net import of particulate P (Dame, 1994). Marsh flume studies in systems with tidal amplitudes ≤ 1 m showed a close balance between P import and export (i.e., insignificant low net particulate P exchanges). There was a small but significant net export of sediment-bound P at higher tidal ranges (Childers, 1994). An opposite tidal amplitude effect was reported for organic particulate P.

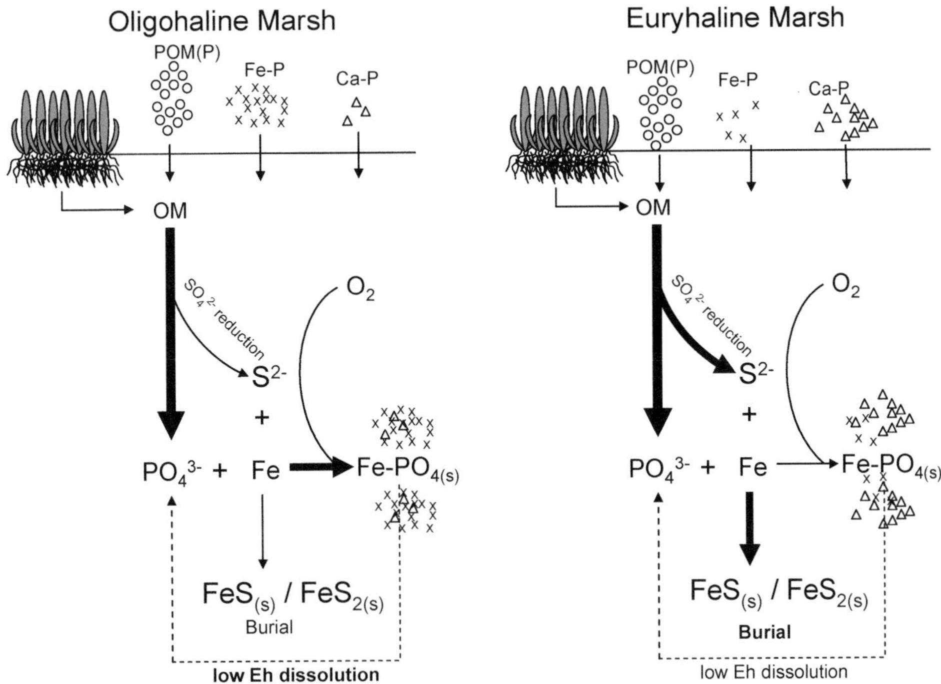

FIGURE 16.5 Summary of P dynamics in marsh sediments contrasting freshwater and salt marsh ecosystems. Salt marshes receive and contain more Ca–P relative to Fe–P. Less PO_4^{3-} is sorbed to iron in salt marshes due to sulfur cycling.

This disparity between inorganic bound and organic particulate P likely reflects asymmetry in flood versus ebb energy and/or density differences between organic and inorganic particles. Export over mean tidal conditions probably underestimates the total annual particulate P efflux. Large export events of sediment bound P have been observed when storms scour exposed marsh sediments during low tide (Dankers et al., 1984). These extreme events remain difficult to adequately quantify.

5.1.2 *Tidal Exchanges of Dissolved P*

Net tidal exchanges of DIP tend to be 2–3-fold smaller than particulate P fluxes. The majority of tidal marsh flux studies, particularly when they include measures of porewater drainage, indicate that salt marshes are small net exporters of PO_4^{3-} (0.03–2.25 g P m^{-2} $year^{-1}$; Chambers et al., 1992, Dame, 1994, Childers, 1994, Lillebø et al., 2007). Whole creek estimates of DIP flux in salt marshes of the St. Lawrence Estuary showed variable import/export ranging from −0.003 to 0.006 g m^{-2} d^{-1}. DIP export relative to that of DIN in marshes of the Yangtze and Great Bay Estuaries exceeded the Redfield ratio suggesting that marsh drainage contributes to N limitation of adjacent coastal waters (Wang et al., 2011; Flynn, 2008). Of the 11 whole tidal creek marsh exchange studies reviewed by Dame (1994), only two document a net import of PO_4^{3-}, although there can be seasonal reversals of import/export at individual sites (Jordan and Correll, 1991). Both marshes showing annual

PO_4^{3-} import were geologically/ecologically young systems. Even young marshes will export PO_4^{3-} when porewater PO_4^{3-} is high and its exchange is dominated by porewater drainage (Osgood, 2000). Phosphate effluxes are most prevalent at the early stages of the flooding tide when low Eh porewater pools are flushed of PO_4^{3-} recently produced from organic matter mineralization (Lillebø et al., 2004). At larger tidal ranges, some marshes respond with enhanced PO_4^{3-} export (Childers, 1994), whereas others show a switch from export to import (Childers et al., 2000). These opposite patterns probably represent trade-offs at higher tides between enhanced tidal drainage and an expanded oxidized unsaturated zone favorable to PO_4^{3-} sorption to iron oxides.

Several factors contribute to seasonality in PO_4^{3-} export. Mineralization is the ultimate source of PO_4^{3-}, and higher PO_4^{3-} effluxes from the marsh occur during warmer months with up to fourfold increases reported in the summer (Lillebø et al., 2004). Seasonal plant dynamics also influence PO_4^{3-} exchange with lower PO_4^{3-} effluxes from vegetated soils (Lillebø et al., 2004). Both vegetated and unvegetated regions of a Portuguese marsh were net sources of PO_4^{3-} to the water column through the year except when autotrophic demand by plants and epiphytes was maximal (Lillebø et al., 2004, 2007). The plant modulation of PO_4^{3-} efflux over diel and seasonal scales (Lillebø et al., 2004) is attributed to both plant assimilation and to O_2 pumping into the soils that induces changes in Fe and S cycling and promotes mineral (Fe and carbonate) scavenging of P. The 2–4-fold greater PO_4^{3-} adsorption capacity observed in vegetated soils translates to 50% less PO_4^{3-} export compared with mudflats. Plant effects are seen on seasonal and diel timescales.

Although the Ca-(and Mg)-bound P pool dominates the mineral-bound P reservoir, water column–sediment PO_4^{3-} fluxes may be controlled in just a thin layer at the surface through redox-sensitive Fe oxide–P sorption dynamics (Chambers et al., 1992; Lillebø et al., 2007). It is not clear if/how the Ca–P fraction is involved in PO_4^{3-} fluxes. Theoretically, localized pH decreases would release this Ca-bound fraction; sulfide oxidation is capable of dropping pH low enough (e.g., <6.0) to mobilize P bound to Ca/carbonates (Giblin and Howarth, 1984; Kostka and Luther, 1995). The extent to which this mechanism could control surface P sorption dynamics and subsequent PO_4^{3-} fluxes to overlying water may be important but is not well characterized.

5.2 Transformations

5.2.1 Autotrophic Uptake and Mineralization of P

Is the import of P sufficient to support macrophyte production in mature salt marshes? Using the range of macrophyte N uptake presented earlier in this chapter and a plant N:P of 10–20, we estimate macrophyte P demand of 0.1–6.0 $g\,P\,m^{-2}\,year^{-1}$. Direct PO_4^{3-} uptake is small, as most marshes export DIP, and particle settling supplies $\sim 0.6 \pm 0.8\,g\,P\,m^{-2}\,year^{-1}$ (see Sections 5.1.1 and 5.1.2). Comparing this supply rate with a median estimate of plant demand ($3\,g\,P\,m^{-2}\,year^{-1}$) shows that incoming particulate P is about 20% of plant P demand. Although some marshes can import P at rates nearly sufficient to support macrophyte production (Wolaver and Zieman, 1984), autotrophy in many marshes must be supported by internal recycling of P. P storage in macrophyte biomass is an important mechanism for P retention, particularly in marshes where sediment import is low (Negrin et al., 2016). Faster decomposition of leaf litter at higher temperatures and salinity is expected to accelerate turnover of P (and N) in marshes in future climate and sea level scenarios (Wei et al., 2017). Based on the size of the bioavailable P pool and macrophyte P demand, useable P turns over on the order of 15 years (Paludan and Morris, 1999). Mineralization of organic matter is an important source of PO_4^{3-}, but its subsequent availability for plant uptake is controlled by geochemical speciation reactions. The porewater DIP pool, while spatially variable, can be relatively invariant seasonally (Negrin et al., 2011).

P release during mineralization is consistent with the rates of organic matter respiration (aerobic and anaerobic) and C:N:P stoichiometry presented earlier in this chapter. Using the range of reported N mineralization rates and N:P ratios for plants and marsh soils of 10–25 (Buresh et al., 1980; Zhou et al., 2007; Craft, 2007), P mineralization ranges from 0.3 to 12 g P m^{-2} $year^{-1}$ with the majority of mature marshes in the range of 1.0–8.0 g P m^{-2} $year^{-1}$. These values are on the order of 5–10-fold higher than P import rates and are comparable with macrophyte demand. Bacteria strongly prefer organic P (Sundareshwar et al., 2003), so only a small fraction of any excess PO_4^{3-} created during mineralization is microbially immobilized. Instead, it is geochemically sorbed and re-speciated. These processes represent exchanges between sorbed P and porewater PO_4^{3-}, which help to regulate bioavailability, and affect the resultant exchanges of DIP with tidal waters.

5.2.2 *Geochemical Cycling of P*

Bacteria perform the initial mineralization of organic matter to release PO_4^{3-}. This PO_4^{3-} sorbs readily to humics and oxidized mineral phases and gets incorporated into solid organic fractions of variable lability. Ranges of sorbed P are 0.15–1 mg P gdw^{-1} in both vegetated and nonvegetated marsh soils. Exchangeable fractions of sorbed P are on the order of 20%–50% depending on mineral composition. Based on radio-labeled ^{32}P experiments, the dissolved PO_4^{3-}, DNRP, Al-, and Fe-bound P pools are potentially available for maintaining equilibrium with porewater PO_4^{3-} and thus available for biological uptake, but the Ca–P fraction shows little exchangeability (Jensen and Thamdrup, 1993).

Salt marsh soils (relative to oligohaline systems) have lower organic N:P, higher porewater PO_4^{3-}, less Fe oxide–bound P, and more Ca-bound P (Paludan and Morris, 1999; Coelho et al., 2004; Zhou et al., 2007). Despite the smaller contribution of Fe oxide–bound P, this fraction remains an important mediator of P dynamics in salt marshes. Soil Fe largely accounted for the distribution of sorbed P among Gulf of Mexico and Hawaiian polyhaline marshes (Marton and Roberts, 2014; Bruland and DeMent, 2009). Zones of Fe-bound P occur in the uppermost soils in contact with the atmosphere and around oxidized root channels (Krom and Berner, 1980). These zones limit upward diffusion of PO_4^{3-} (Chambers and Odum, 1990; Coelho et al., 2004). The Fe-bound fraction is the most bioavailable of the sorbed P fractions because of its redox sensitivity and reactivity with sulfur (Fig. 16.5). Salt marshes retain more P in vegetated soils where high rates of photosynthesis enhance soil oxidation. Increased Fe oxide formation sorbs more P and draws down porewater PO_4^{3-}, leading to greater retention of PO_4^{3-} (Mendelssohn and Postek, 1982; Sundby et al., 1998; Lillebø et al., 2004). Maximum Fe oxide formation and P sorption was observed in a Delaware,

USA, marsh in summer when photosynthesis was highest (Kostka and Luther, 1995). Coelho et al. (2004) measured decreased daytime PO_4^{3-} export in excess of that attributable to direct plant assimilation, and the formation of P-rich root plaques associated with Ca and Fe oxides in vegetated marsh soils has been widely observed (Caetano and Vale, 2002; Lillebø et al., 2007). Some P-rich plaques are retained on long timescales, whereas others appear to be more transient. The reduction of Fe(III) oxides due to biological processes or chemical reactions with sulfides (Sections 2.2.2 and 4.2.1) can liberate PO_4^{3-} and thus reduce the amount of potential P burial (Anschutz et al., 1998; Rozan et al., 2002).

PO_4^{3-}—Fe interactions are heavily modified by S. Sulfides from SO_4^{2-} reduction bind tightly to Fe-forming pyrite and/or iron monosulfides (Fig. 16.3), which inefficiently bind PO_4^{3-}. Additionally, SO_4^{2-} competes for P-binding sites on Fe oxides. Iron sulfide minerals limit Fe from further redox cycling and constrain future binding with PO_4^{3-} (Sundby et al., 1992). The net result of a high S environment is that less PO_4^{3-} sorbs to solid mineral phases and more remains in porewaters. Redox variations with seasons, tides, and daily photosynthesis generate patterns of Fe and S cycling, which modify the P speciation and subsequent release of PO_4^{3-} on the ecosystem scale (Giblin and Howarth, 1984; Scudlark and Church, 1989; Kostka and Luther, 1995).

Ca-bound P represents a large fraction (40%—80%) of the mineral-bound soil P, yet it is less well-characterized than other pools (Paludan and Morris, 1999; Coelho et al., 2004; Álverez-Rogel et al., 2007; Zhou et al., 2007). Sedimentation of Ca—P particles at the seaward end of estuaries traps P that might otherwise migrate upstream to more P-limited ecosystems. High Ca^{2+} (e.g., seawater) competes for P with organics, P-rich root plaques are also carbonate rich, and Ca—P is soluble at low pH (Van Dijk et al., 2015). However, Ca—P is not considered readily exchangeable and is not thought to participate in exchanges with porewater PO_4^{3-}. On one hand, the Ca—P fraction appears to be dynamic, whereas on the other it may represent an essential step toward P burial. Given the magnitude of P bound to Ca in salt marshes, its dynamics may represent a relatively unknown but potentially important component of salt marsh P dynamics.

5.3 Losses

5.3.1 *Burial of P*

Burial is the dominant mechanism of P loss for all salt marshes. The amount of residual bound P (0—0.26 μg P gdw^{-1}) provides a lower limit on the extent of P burial (Paludan and Morris, 1999; Coelho et al., 2004; Zhou et al., 2007). Assuming that marsh accretion reflects SLR, a conservative P burial rate of ~0.2 g P m^{-2} (mm SLR)$^{-1}$ can be estimated. If the Ca—P bound fraction (assumed to be relatively bio-unavailable) is added to the burial inventory, the rate of P sequestration by salt marshes increases by a factor of 2—3. Measured rates of P burial for a variety of salt marshes range between 0.36 and 24.75 g P m^{-2} $year^{-1}$ (Craft, 2007 and references therein; Adams et al., 2012) with highest rates measured in areas of rapid SLR and high accretion rates. These P burial rates in salt marshes are generally lower than P burial in freshwater and oligohaline marshes due to higher sediment availability and a greater sedimentary P content in up-estuary environments, as well as increased decomposition in higher salinity systems (Loomis and Craft, 2010; Merrill and Cornwell, 2000; Craft, 2007; Jordan et al., 2008).

6. MARSHES IN TRANSITION

Salt marshes are fluid in terms of geomorphology, biogeochemistry, and their role in the coastal landscape. They are subject to the synergistic effects of global scale climate forcings and local-scale human impacts (e.g., Gedan et al., 2009; McKee et al., 2012; Morris et al., 2013; Megonigal et al., 2016). Since the first version of this chapter was published, there has been a large increase in the number of published papers that have examined biogeochemical responses to environmental changes such as increased atmospheric CO_2 concentrations, eutrophication of coastal waters, and climate change/global warming. Salt marsh responses to these and other stressors occur at the scale of soil biogeochemistry, plant productivity, and geomorphology and may likely alter the role of marshes in the coastal landscape. One could easily write an entire chapter on these topics; here we summarize some of the major findings thus far.

Both elevated CO_2 and increased nitrogen tend to increase salt marsh plant productivity, either at the level of individual plants or at the ecosystem scale (e.g., Langley et al., 2009; McKee et al., 2012; Johnson et al., 2016). However, there are broad differences between C3 and C4 plants in terms of their responses to elevated CO_2 (Curtis et al., 1989) plus species-specific differences in terms of if/how increased nitrogen affects the ability of plants to withstand SLR (Langley et al., 2013). With increases in N availability, plants shift resource allocation from belowground roots/rhizomes to aboveground tissues (Darby and Turner, 2008b; White et al., 2012). But this increase in plant productivity has resulted in both positive and negative effects on marsh elevation and/or the stability of the marsh platform (Deegan et al., 2012, Turner et al., 2009, Turner, 2011, Langley et al., 2009, Graham and Mendelssohn, 2014, Anisfeld and Hill, 2012). There are also cascading effects of nutrient loading on porewater chemistry and rates/pathways of metabolism (e.g., Wolf et al., 2007; Keller et al., 2009b; Koop-Jakobsen and Giblin, 2010; Vieillard and Fulweiler, 2012). Changing soil water content, as might be expected with increased flooding due to sea level, can alter SOM mineralization rates (Palomo et al., 2013; Lewis et al., 2014), although larger inundation effects on decomposition are driven indirectly through plant responses to changing hydrology (Mueller et al., 2016). In contrast, CH_4 emissions may be more sensitive to the direct effects of flooding rather than through flooding effects on plant-mediated forcings (Ding et al., 2010). Warming can accelerate rates of biogeochemical transformations (Charles and Dukes, 2009; Kirwan and Blum, 2011; Lewis et al., 2014), drive species transitions (e.g., marsh to mangrove; Perry and Mendelssohn, 2009, Cavanaugh et al., 2014, Saintilan et al., 2014), and influence soil development and carbon accumulation rates (Kirwan and Mudd, 2012; Henry and Twilley, 2013). The marsh responses to environmental changes can be driven by changes in the physicochemical environment, but these responses may be mediated through the responses of the microbial community (e.g., Nie et al., 2009; Lage et al., 2010; Bowen et al., 2011; Lee et al., 2017), which plays a key role in many biogeochemical processes.

In the simplest of terms, salt marshes trap C, N, and P through marsh accretion, and this term typically dominates net system mass balances. This function in the marsh landscape, along with that of denitrification, scales with marsh area and is contingent on marsh sustainability in the face of accelerated SLR and changing anthropogenic stressors (Deegan et al., 2012; Morris et al. 2002, 2016; Kirwan and Megonigal, 2013; DeLaune and White, 2012). Setting all factors aside governing marsh sustainability (e.g., SLR rate, sediment supply, eutrophication response), some marshes will keep pace, others won't. If a marsh survives, its net C, N, and P trapping function remains intact over the long term and it stores "blue" N,C,P at the rate that is a function of SLR and the soil density of each of these elements. If the marsh is not sustainable, then this primary biogeochemical function ceases, but there are other ramifications beyond lack of element storage. First, for systems that are in decline, there will be a shift in geomorphology as the system declines. Fragmentation of the marsh will increase the shoreline distance relative to marsh area, which will increase the marsh-scale lateral porewater flux export and decrease functions such as retention through accretion or denitrification that scales with marsh area. It is possible that this change in geomorphology will influence the net biogeochemical balance between dissolved and particulate fluxes, net C, N, and P retention via accretion and alter adjacent water quality. The fragmentation is analogous to the effect of marsh ditching to enhance drainage, and effects on adjacent waters may be similar (Koch and Gobler, 2009). Declining marshes are likely to function biogeochemically differently than their stable counterparts, and resolving these differences is important. Secondarily, once marshes are gone, what happens to the stored C, N, and P? Current estimates of blue carbon loss assume that stored carbon is respirable on timescales relevant to the atmospheric residence time of CO_2 and hence represents a significant source of C to the atmosphere. Yet this carbon has in some cases survived hundreds to thousands of years of mineralization. Its lability once delivered to a photic and O_2-rich environments following erosion is unclear, particularly on a timescale that is relevant for greenhouse gas consideration (Pendleton et al., 2012). Future efforts should therefore focus on the shifting biogeochemistry for marshes in transition and on the fate of previously "buried" N,C,P once marshes succumb to SLR. These are keystone pieces of information for considering shifts in ecosystem service at the local coastal landscape scale and global implications of marsh sustainability.

Our understanding of how salt marshes respond to climate forcings and human impacts needs improvement. Understanding the interactions between multiple stressors at a variety of spatial and temporal scales is a necessary challenge before we can make robust projections of marsh responses under future climate. This is perhaps the biggest challenge for future salt marsh biogeochemistry work.

7. DIRECTIONS FOR FUTURE WORK

In the previous version of this chapter, we proposed a list of salt marsh biogeochemistry questions that need further study (Tobias and Neubauer, 2009). In the intervening years, some progress has been made in understanding those topics, which ranged from characterizing the pathways of organic carbon catabolism to describing the DNRA linkages between the N and S cycles to advancing a methodology for quantifying hydrologic and nutrient fluxes between salt marshes and coastal/estuarine waters. Additional research into these topics, and the others mentioned in that chapter, is still needed. Here, we offer several more specific directions for future research.

- Tighten estimates of C, N, and P burial on decadal to century timescales and assess the fate of "old" eroded C as a source of CO_2 to the atmosphere. An increased focus on blue carbon burial rates derived from geochronology better matches the timescale of atmospheric CO_2 dynamics and accounts for postdepositional diagenesis after short-term accretion. A better understanding of how likely previously buried C is to be respired and returned to the atmosphere will better constrain current and future salt marsh blue carbon contributions to the global C cycle.

- Standardization of timescales, and methodologies for defining carbon accretion, burial, and blue carbon are needed.
- Determine sources of buried C, specifically the extent to which further marsh accretion is dependent on local erosional sources versus newly fixed C and how much does marsh sustainability depend on cannibalizing other marshes.
- Develop a better understanding of anaerobic oxidation pathways (e.g., of ferrous iron or CH_4) in salt marsh soils. Active biogeochemical cycles regularly occur at aerobic–anaerobic interfaces, with reduction reactions occurring in anaerobic regions and oxidation reactions that serve to regenerate electron acceptors and other oxidized substrates occurring in close proximity under aerobic conditions. This type of cycling could also be important in completely anaerobic environments, where oxidants such as NO_3^- or SO_4^{2-} are used instead of molecular O_2.
- Further evaluate the magnitude, controls, and importance of N-fixation by sulfate reducers.
- Improve/standardize techniques for evaluating lateral fluxes of C, N, and P and how those fluxes vary as a function of physical drivers, geomorphology, and ecosystem productivity (e.g., tidal amplitude, temperature, episodic events, sinuosity, edge:open water ratio, gross primary production, etc.).
- Assess how salt marsh biogeochemical cycles and accretion rates respond to multiple, simultaneous, and sustained environmental changes. This is likely to involve a combination of multifactor field experiments, focused lab studies to address mechanisms, and simulation modeling.

LIST OF ABBREVIATIONS

AVS Acid-volatile sulfides
BMP Benthic microalgal production
CRS Chromium(II)-reducible sulfides
DIC Dissolved inorganic carbon
DNRA Dissimilatory NO_3^- reduction to NH_4^+
DOC Dissolved organic carbon
DOP Degree of pyritization
FeOB Fe(II)-oxidizing bacteria.
FeRB Fe(III)-reducing bacteria
NPP Net primary production
POC Particulate organic carbon
ROL Root O_2 loss
rRNA ribosomal ribonucleic acid
SOB Sulfide-oxidizing bacteria
SOM Soil organic matter
SRB SO_4^{2-}-reducing bacteria
TS Total sulfur
US United States
VPP Vascular plant production

References

Abd. Aziz, A.A.B., Nedwell, 1986. The nitrogen cycle of an East Coast, U.K. salt marsh: II nitrogen fixation, nitrification, denitrification, tidal exchange. Estuarine, Coastal and Shelf Science 22, 689–704.

Adams, L.K., Macquaker, J.H.S., Marshall, J.D., 2006. Iron(III)-reduction in a low-organic-carbon brackish-marine system. Journal of Sedimentary Research 76, 919–925.

Adams, C.A., Andrews, J.E., Jickells, T., 2012. Nitrous oxide and methane fluxes vs. carbon, nitrogen and phosphorous burial in new intertidal and saltmarsh sediments. The Science of the Total Environment 434, 240–251.

Algar, C.K., Vallino, J.J., 2014. Predicting microbial nitrate reduction pathways in coastal sediments. Aquatic Microbial Ecology 71, 223–238. https://doi.org/10.3354/ame01678.

Álverez-Rogel, J., Jiménez-Cárceles, F.J., Egea-Nicolás, C., 2007. Phosphorus retention in a coastal salt marsh in SE Spain. The Science of the Total Environment 378, 71–74.

An, S., Gardner, W.S., 2002. Dissimilatory nitrate reduction to ammonium (DNRA) as a nitrogen link, versus denitrification as a sink in a shallow estuary (Laguna Madre/Baffin Bay, Texas). Marine Ecology Progress Series 237, 41–50.

Anderson, I.C., Tobias, C.R., Neikirk, B.B., Wetzel, R.L., 1997. Development of a process-based nitrogen mass balance model for a Virginia (USA) *Spartina alterniflora* salt marsh: implications for net DIN flux. Marine Ecology Progress Series 159, 13–27.

Andersson, B., Sundback, K., Hellman, M., Hallin, S., Alsterberg, C., 2014. Nitrogen fixation in shallow-water sediments: spatial distribution and controlling factors.

Anisfeld, S.C., Hill, T.D., 2012. Fertilization effects on elevation change and belowground carbon balance in a Long Island Sound tidal marsh. Estuaries and Coasts 35, 201–211.

Anschutz, P., Zhong, S., Sundby, B., Mucci, A., Gobeil, C., 1998. Burial efficiency of phosphorus and the geochemistry of iron in continental margin sediments. Limnology and Oceanography 43, 53–64.

Artigas, F., Shin, J.Y., Hobble, C., Marti-Donati, A., Schäfer, K.V., Pechmann, I., 2015. Long term carbon storage potential and CO_2 sink strength of a restored salt marsh in New Jersey. Agricultural and Forest Meteorology 200, 313–321.

Bartlett, K.B., Bartlett, D.S., Harriss, R.C., Sebacher, D.I., 1987. Methane emissions along a salt marsh salinity gradient. Biogeochemistry 4, 183–202.

Benner, R., Moran, M.A., Hodson, R.E., 1985. Effects of pH and plant source on lignocellulose biodegradation rates in two wetland ecosystems, the Okefenokee Swamp and a Georgia salt marsh. Limnology and Oceanography 30 (3), 489–499.

Benner, R., Moran, M.A., Hodson, R.E., 1986. Biogeochemical cycling of lignocellulosic carbon in marine and freshwater ecosystems: relative contributions of procaryotes and eucaryotes. Limnology and Oceanography 31, 89–100.

Benner, R., Fogel, M.L., Sprague, E.K., 1991. Diagenesis of belowground biomass of *Spartina alterniflora* in salt-marsh sediments. Limnology and Oceanography 36, 1358–1374.

Bernal, B., McKInley, D.C., Hungate, B.A., White, P.M., Mozdzer, T.J., Megonigal, J.P., 2016. Limits to soil carbon stability; deep, ancient soil carbon decomposition stimulated by new labile organic inputs. Soil Biology and Biochemistry 98, 85–94.

Bernal, B., Megonigal, J.P., Mozdzer, T.J., 2017. An invasive wetland grass primes deep soil carbon pools. Global Change Biology 23 (5), 2104–2116.

Berner, R.A., 1969. Migration of iron and sulfur within anaerobic sediments during early diagenesis. American Journal of Science 267, 19–42.

Berner, R.A., 1970. Sedimentary pyrite formation. American Journal of Science 268, 1–23.

Blum, U., Seneca, E.D., Stroud, L.M., 1978. Photosynthesis and respiration of *Spartina* and *Juncus* salt marshes in North Carolina: some models. Estuaries 1, 228–238.

Blum, L.K., 1993. *Spartina alterniflora* root dynamics in a Virginia Marsh. Marine Ecology Progress Series 102, 169–178.

Bodelier, P.L.E., 2003. Interactions between oxygen-releasing roots and microbial processes in flooded soils and sediments. In: de Kroon, H., Visser, E.J.W. (Eds.), Ecological Studies: Root Ecology. Springer-Verlag, Berlin, pp. 331–362.

Boetius, A., Ravenschlag, K., Schubert, C.J., Rickert, D., Widdel, F., Gieseke, A., Amann, R., Jorgensen, B.B., Witte, U., Pfannkuche, O., 2000. A marine microbial consortium apparently mediating the anaerobic oxidation of methane. Nature 407, 623–626.

Bollinger, M.S., Moore, W., 1993. Evaluation of salt marsh hydrology using radium as a tracer. Geochimica et Cosmochimica Acta 57, 2203–2212.

Bonin, P., Omnes, P., Chalamet, A., 1998. Simultaneous occurrence of denitrification and nitrate ammonification in sediments of the French Mediterranean Coast. Hydrobiologia 389, 169–182.

Borges, A., 2005. Do we have enough pieces of the jigsaw to integrate CO_2 fluxes in the coastal ocean. Estuaries 28, 3–27.

Boschker, H.T.S., deGraff, W., Koster, M., MeyerReil, L.A., Cappenberg, T.E., 2001. Bacterial populations and processes involved in acetate and propionate consumption in anoxic brackish sediment. FEMS Microbiology Ecology 35, 97–103.

Boudreau, B.P., Westrich, J.T., 1984. The dependence of bacterial sulfate reduction on sulfate concentration in marine sediments. Geochimica et Cosmochimica Acta 48, 2503–2516.

Bowden, W.B., 1984. A nitrogen-15 isotope dilution study of ammonium production and consumption in a marsh sediment. Limnology and Oceanography 29, 1004–1015.

Bowden, W.B., 1986. Nitrification, nitrate reduction, and nitrogen immobilization in a tidal freshwater marsh sediment. Ecology 67, 88–99.

Bowen, J.L., Ward, B.B., Morrison, H.G., Hobbie, J.E., Valiela, I., Deegan, L.A., Sogin, M.L., 2011. Microbial community composition in sediments resists perturbation by nutrient enrichment. The ISME Journal 5 (9), 1540.

Brendel, P.J., Luther III, G.W., 1995. Development of a gold amalgam voltametric microelectrode for the determination of dissolved Fe, Mn, O_2, and S(-II) in porewaters of marine and freshwater sediments. Environmental Science and Technology 29, 751–761.

Bridgham, S.D., Megonigal, J.P., Keller, J.K., Bliss, N.B., Trettin, C.C., 2006. The carbon balance of North American wetlands. Wetlands 26, 889–916.

Bridgham, S.D., Moore, T.R., Richardson, C.J., Roulet, N.T., 2014. Errors in greenhouse forcing and soil carbon sequestration estimates in freshwater wetlands: a comment on Mitsch et al. 2013. Landscape Ecology 29 (9), 1481–1485.

Bruland, G.L., DeMent, G., 2009. Phosphorus sorption dynamics of Hawaii's coastal wetlands. Estuaries and Coasts 32, 844–854.

Brunet, R.C., Garcia-Gil, L.J., 1996. Sulfide-induced dissimilatory nitrate reduction to ammonia in anaerobic freshwater sediments. FEMS Microbiology Ecology 21, 131–138.

Brunner, B., Contreras, S., Lehman, M.F., Matantseva, O., Rollog, M., Kalvelage, T., Klockgether, G., Lavik, G., Jetten, M.S.M., Kartal, B., Kuypers, M.M., 2013. Nitrogen isotope effects induced by anammox bacteria. Proceeding of the National Academy of Science 110, 18994–18999.

Buchan, A., Newell, S.Y., Butler, M., Biers, E.J., Hollibaugh, J.T., Moran, M.A., 2003. Dynamics of bacterial and fungal communities on decaying salt marsh grass. Applied and Environmental Microbiology 69, 6676–6687.

Buresh, R.J., DeLaune, R.D., Patrick, W.H., 1980. Nitrogen and phosphorous distribution and utilization by *Spartina alterniflora* in a Lousiana Gulf Coast Marsh. Estuaries 3, 111–121.

Burgin, A.J., Hamilton, S.K., 2007. Have we overemphasized denitrification in aquatic ecosystems: a review of nitrate removal pathways. Frontiers in Ecology and the Environment 5, 89–96.

Buzzelli, C.P., Wetzel, R.L., Meyers, M.B., 1999. A linked physical and biological framework to assess biogeochemical dynamics in a shallow estuarine ecosystem. Estuarine, Coastal and Shelf Science 49, 829–851.

Buzzelli, C., 2008. Development and application of tidal creek ecosystem models. Ecological Modelling 210, 127–143.

Caetano, M., Vale, C., 2002. Retention of arsenic and phosphorus in iron-rich concretions of Tagus salt marshes. Marine Chemistry 79, 261–271.

Cahoon, D.R., Reed, D.J., 1995. Relationships among marsh surface topography, hydroperiod, and soil accretion in a deteriorating Louisiana salt marsh. Journal of Coastal Research 11, 357–369.

Cai, W.-J., Wang, Y., 1998. The chemistry, fluxes, and sources of carbon dioxide in the estuarine waters of the Satilla and Altamaha Rivers, Georgia. Limnology and Oceanography 43, 657–668.

Cai, W.-J., Pomeroy, L.R., Moran, M.A., Wang, Y., 1999. Oxygen and carbon dioxide mass balance for the estuarine-intertidal marsh complex of five rivers in the southeastern U.S. Limnology and Oceanography 44, 639–649.

Cai, W.-J., Wiebe, W.J., Wang, Y., Sheldon, J.E., 2000. Intertidal marsh as a source of dissolved inorganic carbon and a sink of nitrate in the Satilla River-estuarine complex in the southeastern U.S. Limnology and Oceanography 45, 1743–1752.

Canfield, D.E., Jørgensen, B.B., Fossing, H., Glud, R., Gundersen, J., Ramsing, N.B., Thamdrup, B., Hansen, J.W., Nielsen, L.P., Hall, P.O.J., 1993. Pathways of organic carbon oxidation in three continental margin sediments. Marine Geology 113, 27–40.

Canfield, D.E., Thamdrup, B., Fleischer, S., 1998. Isotope fractionation and sulfur metabolism by pure and enrichment cultures of elemental sulfur-disproportionating bacteria. Limnology and Oceanography 43, 253–264.

Carey, E., Taillefert, M., 2005. The role of soluble Fe(III) in the cycling of iron and sulfur in coastal marine sediments. Limnology and Oceanography 50, 1129–1141.

Cavanaugh, K.C., Kellner, J.R., Forde, A.J., Gruner, D.S., Parker, J.D., Rodriguez, W., Feller, I.C., 2014. Poleward expansion of mangroves is a threshold response to decreased frequency of extreme cold events. Proceedings of the National Academy of Sciences 111 (2), 723–727.

Cervantes, F.J., van der Velde, S., Lettinga, G., Field, J.A., 2000. Competition between methanogenesis and quinone respiration for ecologically important substrates in anaerobic consortia. FEMS Microbiology Ecology 34, 161–171.

Chambers, R.M., Odum, W.E., 1990. Porewater oxidation, dissolved phosphate and the iron curtain: iron-phosphorus relations in tidal freshwater marshes. Biogeochemistry 10, 37–52.

Chambers, R.M., Harvey, J.W., Odum, W.E., 1992. Ammonium and phosphate dynamics in a Virginia salt marsh. Estuaries 15, 349–359.

Chambers, R.M., Smith, S.V., Hollibaugh, J.T., 1994. An ecosystem-level context for tidal exchange studies in salt marshes of Tomales Bay, California, USA. In: Mitsch, W.J. (Ed.), Global Wetlands: Old World and New. Elsevier Science, New York, pp. 265–276.

Chambers, R.M., 1997. Porewater chemistry associated with *Phragmites* and *Spartina* in a Connecticut tidal marsh. Wetlands 17, 360–367.

Chapin III, F.S., Woodwell, G.M., Randerson, J.T., Rastetter, E.B., Lovett, G.M., Baldocchi, D.D., Clark, D.A., Harmon, M.E., Schimel, D.S., Valentini, R., Wirth, C., Aber, J.D., Cole, J.J., Goulden, M.L., Harden, J.W., Heimann, M., Howarth, R.W., Matson, P.A., McGuire, A.D., Melillo, J.M., Mooney, H.A., Neff, J.C., Houghton, R.A., Pace, M.L., Ryan, M.G., Running, S.W., Sala, O.E., Schlesinger, W.H., Schulze, E.-D., 2006. Reconciling carbon-cycle concepts, terminology, and methods. Ecosystems 9, 1041–1050.

Charette, M.A., Splivallo, R., Herbold, C., Bollinger, M.A., Moore, W.S., 2003. Salt marsh submarine groundwater discharge as traced by radium isotopes. Marine Chemistry 84, 113–121.

Charette, M.A., 2007. Hydrologic forcing of submarine groundwater discharge: insight from a seasonal study of radium isotopes in a groundwater-dominated salt marsh estuary. Limnology and Oceanography 52, 230–239.

Charles, H., Dukes, J.S., 2009. Effects of warming and altered precipitation on plant and nutrient dynamics of a New England salt marsh. Ecological Applications 19 (7), 1758–1773.

Chaudhuri, S.K., Lack, J.G., Coates, J.D., 2001. Biogenic magnetite formation through anaerobic biooxidation of Fe(II). Applied and Environmental Microbiology 67, 2844–2848.

Childers, D.L., Day Jr., J.W., McKellar Jr., H.N., 2000. Twenty more years of marsh and estuarine flux studies: revisiting Nixon (1980). In: Weinstein, M., Kreeger, D.A. (Eds.), Concepts and Controversies in Tidal Marsh Ecology. Kluwer Academic Publishing, Dordrecht, Netherlands, pp. 391–423.

Childers, D.L., 1994. Fifteen years of marsh flumes: a review of marsh-water column interactions in Southeastern USA estuaries. In: Mitsch, W.J. (Ed.), Global Wetlands: Old World and New. Elsevier Science, New York, pp. 277–293.

Chmura, G.L., Hung, G.A., 2004. Controls on salt marsh accretion: a test in salt marshes of eastern Canada. Estuaries 27, 70–81.

Chmura, G.L., Anisfeld, S.C., Cahoon, D.R., Lynch, J.C., 2003. Global carbon sequestration in tidal, saline wetland soils. Global Biogeochemical Cycles 17. Article 1111.

Choi, Y., Wang, Y., 2004. Dynamics of carbon sequestration in a coastal wetland using radiocarbon measurements. Global Biogeochemical Cycles 18, GB4016. https://doi.org/10.1029/2004GB002261.

Christensen, P.B., Rysgaard, S., Sloth, N.P., Dalsgaard, T., Schwaerter, S., 2000. Sediment mineralization, nutrient fluxes, denitrification, and dissimilatory nitrate reduction to ammonium in an estuarine fjord with sea cage trout farms. Aquatic Microbial Ecology 21, 73–84.

Church, T.M., Lord III, C.J., Somayajulu, B.L.K., 1981. Uranium, thorium, and lead nuclides in a Delaware salt marsh sediment. Estuarine, Coastal and Shelf Science 13, 267–275.

Coelho, J.P., Flindt, R.R., Jensen, H.S., Lillebø, A.I., Pardal, M.A., 2004. Phosphorus speciation and availability in intertidal sediments of a temperate estuary: relation to eutrophication and annual P-fluxes. Estuarine, Coastal and Shelf Science 61, 583–590.

Colman, J.A., Masterson, J.P., 2008. Transient simulations of nitrogen load for a coastal aquifer and embayment, Cape Cod, MA. Environmental Science and Technology 42, 207–213.

Colmer, T.D., 2003. Long-distance transport of gases in plants: a perspective on internal aeration and radial oxygen loss from roots. Plant, Cell and Environment 26, 17–36.

Cooper, D.J., DeMello, W.Z., Cooper, W.J., Zika, R.G., Saltzman, E.S., Prospero, J.M., Savoie, D.L., 1987. Short-term variability in biogenic sulphur emissions from a Florida *Spartina alterniflora* marsh. Atmospheric Environment 21, 7–12.

Costa, A.L., Carolino, M., Caçador, I., 2007. Microbial activity profiles in Tagus estuary salt marsh sediments. Hydrobiologia 587, 169–175.

Craft, C.B., Seneca, E.D., Broome, S.W., 1993. Vertical accretion in microtidal regularly and irregularly flooded estuarine marshes. Estuarine, Coastal and Shelf Science 37, 371–386.

Craft, C.B., 2007. Freshwater input structures soil properties, vertical accretion, and nutrient accumulation of Georgia and U.S. tidal marshes. Limnology and Oceanography 52, 1220–1230.

Crozier, C.R., Devai, I., DeLaune, R.D., 1995. Methane and reduced sulfur gas production by fresh and dried wetland soils. Soil Science Society of America Journal 59, 277–284.

Currin, C.A., Paerl, H.W., 1998a. Environmental and physiological controls on diel patterns of N_2 fixation in epiphytic cyanobacterial communities. Microbial Ecology 35, 34–35.

Currin, C.A., Paerl, H.W., 1998b. Epiphytic nitrogen fixation associated with standing dead shoots of smooth cordgrass, *Spartina alterniflora*. Estuaries 21, 108–117.

Currin, C.A., Newell, S.Y., Paerl, H.W., 1995. The role of benthic microalgae and standing dead *Spartina alterniflora* in salt marsh food webs: implications based on multiple stable isotope analysis. Marine Ecology Progress Series 21, 99–116.

Currin, C.A., Joye, S.B., Paerl, H.W., 1996. Diel rates of N_2-fixation and denitrification in a transplanted *Spartina alterniflora* marsh: implications for N-flux dynamics. Estuarine, Coastal, and Shelf Science 42, 597–616.

Curtis, P.S., Drake, B.G., Leadley, P.W., Arp, W.J., Whigham, D.F., 1989. Growth and senescence in plant communities exposed to elevated CO_2 concentrations on an estuarine marsh. Oecologia 78 (1), 20–26.

Cutter, G.A., Velinsky, D.J., 1988. Temporal variations of sedimentary sulfur in a Delaware salt marsh. Marine Chemistry 23, 311–327.

Dacey, J.W.H., King, G.M., Wakeham, S.G., 1987. Factors controlling emission of dimethylsulfide from salt marshes. Nature 330, 643–645.

Dai, T., Wiegert, R.G., 1996. Ramet population dynamics and net aerial primary productivity of *Spartina alterniflora*. Ecology 77, 276–288.

Daims, H., Lebedeva, E.V., Pjevac, P., Han, P., Herbold, C., Albertsen, M., Jehmlich, N., Palatinszky, M., Vierheilig, J., Bulaev, A., Kirkegaard, R.H., 2015. Complete nitrification by *Nitrospira* bacteria. Nature 528 (7583), 504–509.

Dalsgaard, T., Bak, F., 1994. Nitrate reduction in a sulfate-reducing bacterium, *Desulfovibrio desulfuricans*, isolated from rice paddy soil: sulfide inhibition, kinetics, and regulation. Applied and Environmental Microbiology 60, 291–297.

Dalsgaard, T., Thamdrup, B., Canfield, D.E., 2005. Anaerobic ammonium oxidation (anammox) in the marine environment. Research in Microbiology 156, 457–464.

Dame, R.F., Chrzanowski, T.H., Bildstein, K.L., Kjerfve, B., McKellar Jr., H.N., Nelson, D.C., Spurrier, J., Stancyk, S., Stevenson, H., Vernberg, J., Zingmark, R.G., 1986. The outwelling hypothesis and North Inlet, South Carolina. Marine Ecology Progress Series 33, 217–229.

Dame, R.F., Spurrier, J.D., Williams, T.M., Kjerfve, B., Zingmark, R.G., Wolaver, T.G., Chrzanowski, T.H., McKellar Jr., H.N., Vernberg, F.J., 1991. Annual material processing by a salt marsh-estuarine basin in South Carolina. Marine Ecology Progress Series 72, 153–166.

Dame, R.F., Alber, M., Allen, D.M., Mallin, M., Montague, C.L., Lewitus, A., Chamlers, A., Gardner, L.R., Gilman, C., Kjerfve, B., Pinckney, J.L., Smith, N., 2000. Estuaries of the south Atlantic coast of north America: their geographical signatures. Estuaries 23, 793–819.

Dame, R.F., 1994. The net flux of materials between marsh-estuarine systems and the sea: the Atlantic coast of the United States. In: Mitsch, W.J. (Ed.), Global Wetlands: Old World and New. Elsevier, Amsterdam, pp. 295–305.

Dankers, N., Binsbergen, M., Zegers, K., Laane, R., van der Loeff, M.R., 1984. Transportation of water, particulate and dissolved organic and inorganic matter between a salt marsh and the Ems-Dollard estuary, The Netherlands. Estuarine, Coastal, and Shelf Science 19, 143–165.

Darby, F.A., Turner, R.E., 2008a. Below- and aboveground *Spartina alterniflora* production in a Louisiana salt marsh. Estuaries and Coasts 31, 223–231.

Darby, F.A., Turner, R.E., 2008b. Effects of eutrophication on salt marsh root and rhizome biomass accumulation. Marine Ecology Progress Series 363, 63–70.

Davis, J.L., Nowicki, B., Wigand, C., 2004. Denitrification in fringing salt marshes of Narragansett Bay, Rhode Island, USA. Wetlands 24, 870–878.

Davis, D.A., Gamble, M.D., Bagwell, C.E., Bergholz, P.W., Lovell, C.R., 2011. Responses of salt marsh plant rhizosphere diazotroph assemblages to changes in marsh elevation, edaphic conditions and plant host species. Microbial Ecology 61, 386–398.

Deegan, L.A., et al., 2007. Susceptibility of salt marshes to nutrient enrichment and predatory removal. Ecological Applications 17 (Suppl.), s42–s63.

Deegan, L.A., Johnson, D.S., Warren, R.S., Peterson, B.J., Fleeger, J.W., Fagherazzi, S., Wollheim, W.M., 2012. Coastal eutrophication as a driver of salt marsh loss. Nature 490, 380–392.

DeLaune, R.D., White, J.R., 2012. Will coastal wetlands continue to sequester carbon in response to an increase in global sea level?: A case study of the rapidly subsiding Mississippi River deltaic plain. Climatic Change 110, 297–314.

DeLaune, R.D., Reddy, C.N., Patrick Jr., W.H., 1981. Accumulation of plant nutrients and heavy metals through sedimentation processes and accrtion in a Louisiana salt marsh. Estuaries 4, 328–334.

DeLaune, R.D., Smith, C.J., Patrick III, W.H., 1983. Nitrogen losses from a Louisiana Gulf Coast salt marsh. Estuarine, Coastal and Shelf Science 17, 133–141.

DeLaune, R.D., Feijtel, T.C., Patrick Jr., W.H., 1989. Nitrogen flows in Lousiana Gulf Coast salt marsh: spatial considerations. Biogeochemistry 8, 25–37.

DeLaune, R.D., Devai, I., Lindau, C.W., 2002b. Flux of reduced sulfur gases along a salinity gradient in Louisiana coastal marshes. Estuarine, Coastal and Shelf Science 54, 1003–1011.

Ding, W., Zhang, Y., Cai, Z., 2010. Impact of permanent inundation on methane emissions from a *Spartina alterniflora* coastal salt marsh. Atmospheric Environment 44 (32), 3894–3900.

Dollhopf, S.L., Hyun, J., Smith, A.C., Adams, H.J., O'Brien, S., Kostka, J., 2005. Quantification of ammonia-oxidizing bacteria and factors controlling nitrification in salt marsh sediments. Applied and Environmental Microbiology 71, 240–246.

Donnelly, J.P., Cleary, P., Newby, P., Ettinger, R., 2004. Coupling instrumental and geologic records of sea-level change: evidence from southern New England of an increase in the rate of sea-level rise in the late 19th century. Geophysical Research Letters 31, L05203.

Drake, D.C., Peterson, B.J., Deegan, L.A., Harris, L.A., Miller, E.E., Warren, R.S., 2008. Plant nitrogen dynamics in fertilized and natural New England salt marshes: a paired ^{15}N tracer study. Marine Ecology Progress Series 354, 35–46.

Drake, D.C., Peterson, B.J., Galvan, K.A., Deegan, L.A., Hopkinson, C., Johnson, J.M., Koop-Jakobsen, K., Lemay, L.E., Picard, C., 2009. Salt marsh ecosystem biogeochemical responses to nutrient enrichment: a paired ^{15}N tracer study. Ecology 90 (9), 2535–2546. https://doi.org/10.1890/08-1051.1.

Du, Y., Zhen, X., Wenxia, Z., Zhang, Y., 2016. Distribution characteristics of phosphorus under different vegetation communities in salt marshes of Jiaozhou Bay. Wetland Science 14, 3.

Duarte, C.M., Middelburg, J.J., Caraco, N., 2005. Major role of marine vegetation on the oceanic carbon cycle. Biogeosciences 2, 1–8.

Duernberger, K., Tobias, C., Mallin, M., 2018. Ecosystem scale nitrification and watershed support of tidal creek productivity revealed using whole system isotope tracer labeling. Limnology and Oceanography. https://doi.org/10.1002/lno.10927.

Edwards, K.J., Rogers, D.R., Wirsen, C.O., McCollom, T.M., 2003. Isolation and characterization of novel psychrophilic, neutrophilic, Fe-oxidizing, chemolithoautotrophic alpha and gamma-*Proteobacteria* from the deep sea. Applied and Environmental Microbiology 69, 2906–2913.

Emerson, D., Moyer, C.L., 2002. Neutrophilic Fe-oxidizing bacteria are abundant at Loihi Seamount hydrothermal vents and play a major role in Fe oxide deposition. Applied and Environmental Microbiology 68, 3085–3093.

Eriksson, P.G., Svensson, J.M., Carrer, G.M., 2003. Temporal changes and spatial variation of soil oxygen consumption, nitrification, and denitrification rates in a tidal salt marsh of the Lagoon of Venice, Italy. Estuarine, Coastal and Shelf Science 58, 861–871.

Etheridge, J.R., Birgand, F., Osborne, J.A., Osburn, C.L., Burchell, M.R., Irving, J., 2014. Using in situ ultraviolet-visual spectroscopy to measure nitrogen, carbon, phospsphorus, and suspended solids concentrations at a high frequency in a brackish tidal marsh. Limnology and Oceanography 12, 10–22.

Fagherazzi, S., Wiberg, P.L., Temmerman, S., Struyf, E., Zhao, Y., Raymond, P.A., 2013. Fluxes of water, sediments, and biogeochemical compounds in salt marshes. Ecological Processes 2 (1), 3.

Flynn, A.M., 2008. Organi matter and nutrient cycling in a coastal plain estuary: carbon, nitrogen, and phosphorus distributions, budgets, and fluxes. Journal of Coastal Research 55, 76–94.

Forbrich, I., Giblin, A.E., 2015. Marsh-atmosphere CO_2 exchange in a New England salt marsh. Journal of Geophysical Research: Biogeosciences 120 (9), 1825–1838. https://doi.org/10.1002/2015JG003044.

Fox, L.E., Sager, S.L., Wofsky, S.T., 1986. The chemical control of soluble phosphorus in the Amazon estuary. Geochimica et Cosmochimica Acta 50, 783–794.

Fox, L.I., Valiela, Kinney, E., 2012. Vegetation cover and elevation in long-term experimental nutrient enrichment plots in Great Sippewissett Salt Marsh, Cape Cod Massachusetts: implications for eutrophication and sea level rise. Estuaries and Coasts 35, 445–458.

Friedrichs, C.T., Perry III, J.E., 2001. Tidal salt marsh morphodynamics. Journal of Coastal Research 27 (special issue), 7–37.

Froehlich, P.N., 1988. Kinetic control of dissolved phosphate in natural rivers and estuaries: a primer on the phosphate buffer mechanism. Limnology and Oceanography 33, 649–668.

Furukawa, Y., Smith, A.C., Kostka, J.E., Watkins, J., Alexander, C.R., 2004. Quantification of macrobenthic effects on diagenesis using a multicomponent inverse model in salt marsh sediments. Limnology and Oceanography 49, 2058–2072.

Furukawa, Y., Inubushi, K., Ali, M., Itang, A.M., Tsuruta, H., 2005. Effect of changing groundwater levels caused by land-use changes on greenhouse gas fluxes from tropical peat lands. Nutrient Cycling in Agroecosystems 71, 81–91.

Gallagher, J.L., Daiber, F.C., 1974. Primary production of edaphic algal communities in a Delaware salt marsh. Limnology and Oceanography 19, 390–395.

Gallagher, J., Reimold, R., Linthurst, R., Pfeiffer, W.J., 1980. Aerial Production, Mortality, and Mineral Accumulation-Export Dynamics in *Spartina alterniflora* and *Juncus roemerianus* Plant Stands in a Georgia Salt Marsh. Ecology 61, 303–312.

Ganju, N.K., 2011. A novel approach for direct estimation of fresh groundwater discharge to an estuary. Geophysical Research Letters 38, L11402. https://doi.org/10.1029/2011GL047718.

Gardner, L.R., Wolaver, T.G., Mitchell, M., 1988. Spatial variations in the sulfur chemistry of salt marsh sediments at North Inlet, South Carolina. Journal of Marine Research 46, 815–836.

Gardner, W.S., McCarthy, M.J., An, S., Sobolev, D., Sell, K.S., Brock, D., 2006. Nitrogen fixation and dissimilatory nitrate reduction to ammonium (DNRA) support nitrogen dynamics in Texas estuaries. Limnology and Oceanography 51, 558–568.

Gardner, L.R., 1990. Simulation of the diagenesis of carbon, sulfur, and dissolved oxygen in salt marsh sediments. Ecological Monographs 60, 91–111.

Gedan, K.B., Silliman, B.R., Bertness, M.D., 2009. Centuries of human-driven change in salt marsh ecosystems. Annual Review of Marine Science 2009 (1), 117–141. https://doi.org/10.1146/annurev.marine.010908.163930.

Giani, L., Dittrich, K., Martsfeld Hartmann, A., Peters, G., 1996. Methanogenesis in saltmarsh soils of the North Sea coast of Germany. European Journal of Soil Science 47, 175–182.

Giblin, A.E., Gaines, A.G., 1990. Nitrogen inputs to a marine embayment: the importance of groundwater. Biogeochemistry 10, 309–328.

Giblin, A.E., Howarth, R.W., 1984. Porewater evidence for a dynamic sedimentary iron cycle in salt marshes. Limnology and Oceanography 29, 47–63.

Giblin, A.E., Weston, N.B., Banta, G.T., Tucker, J., Hopkinson, C.S., 2010. The effects of salinity on nitrogen losses from an oligohaline estuarine sediment. Estuaries and Coasts 33, 1054–1068.

Giblin, A., Tobias, C., Song, B., Weston, N., Banta, G., Rivera-Monroy, V., 2013. The importance of dissimilatory nitrate reduction to ammonium (DNRA) in the nitrogen cycle of coastal ecosystems. Oceanography 26 (3), 124–131. https://doi.org/10.5670/oceanog.2013.54.

Giblin, A.E., 1988. Pyrite formation in marshes during early diagenesis. Geomicrobiology Journal 6, 77–97.

Gilbert, F., Souchu, P., Bianchi, M., Bonin, P., 1997. Influence of shellfish farming activities on nitrification, nitrate reduction to ammonium and denitrification a the water-sediment interface of the Thau lagoon, France. Marine Ecology Progress Series 151, 143–153.

Goodman, J.E., Wood, M.E., Gehrels, W.R., 2007. A 17-yr record of sediment accretion in the salt marshes of Maine (USA). Marine Geology 242, 109–121.

Graham, S.A., Mendelssohn, I.A., 2014. Coastal wetland stability maintained through counterbalancing accretionary responses to chronic nutrient enrichment. Ecology 95 (12), 3271–3283.

Gribsholt, B., Kristensen, E., 2002. Effects of bioturbation and plant roots on salt marsh biogeochemistry: a mesocosm study. Marine Ecology Progress Series 241, 71–87.

Gribsholt, B., Kostka, J.E., Kristensen, E., 2003. Impact of fiddler crabs and plant roots on sediment biogeochemistry in a Georgia saltmarsh. Marine Ecology Progress Series 259, 237–251.

Gribsholt, B., et al., 2005. Nitrogen processing in a tidal freshwater marsh: a whole-ecosystem ^{15}N labeling study. Limnology and Oceanography 50, 1945–1959.

Gribsholt, B., et al., 2006. Ammonium transformations in a nitrogen-rich tidal freshwater marsh. Biogeochemistry 80, 289–298.

Griffin, T.M., Rabenhorst, M.C., Fanning, D.S., 1989. Iron and trace metals in some tidal marsh soils of the Chesapeake Bay. Soil Science Society of America Journal 53, 1010–1019.

Guo, T., DeLaune, R.D., Patrick Jr., W.H., 2000. Iron and manganese transformation in Louisiana salt and brackish marsh sediments. Communications in Soil Science and Plant Analysis 31, 2997–3009.

Guo, H., Noormets, A., Zhao, B., Chen, J., Sun, G., Gu, Y., Li, B., Chen, J., 2009. Tidal effects on net ecosystem exchange of carbon in an estuarine wetland. Agricultural and Forest Meteorology 149 (11), 1820–1828.

Hall, S.L., Fisher Jr., F.M., 1985. Annual productivity and extracellular release of dissolved organic compounds by the epibenthic algal community of a brackish marsh. Journal of Phycology 21, 277–281.

Hammersley, M.R., Howes, B.L., 2003. Contribution of denitrification to nitrogen, carbon, and oxygen cycling in tidal creek sediments of a New England salt marsh.

Hammersley, M.R., Howes, B.L., 2005. Coupled nitrification-denitrification measured in situ in a *Spartina alterniflora* marsh with a $^{15}NH_4^+$ tracer. Marine Ecology Progress Series 299, 123–135.

Hansen, K., Butzeck, C., Eachenbach, A., Grongroft, A., Jensen, K., et al., 2017. Factors influencing the organic carbon pools in tidal marsh soils of the Elbe estuary (Germany). Journal of Soils and Sediments 17, 47–60.

Hardison, A.K., Algar, C.K., Giblin, A.E., Rich, J.E., 2015. Influence of organic carbon and nitrate loading on partitioning between dissimilatory nitrate reduction to ammonium (DNRA) and N_2 production. Geochimica et Cosmochimica Acta 164, 146–160.

Hartzell, J., Jordan, T.E., 2012. Shifts in the relative availability of phosphorus and nitrogen along estuarine salinity gradients. Biogeochemistry 107, 489–500.

Harvey, J.W., Germann, P.F., Odum, W.E., 1987. Geomorphological control of subsurface hydrology in the creekbank zone of tidal marshes. Estuarine, Coastal and Shelf Science 25, 677–691.

Henry, K.M., Twilley, R.R., 2013. Soil development in a coastal Louisiana wetland during a climate-induced vegetation shift from salt marsh to mangrove. Journal of Coastal Research 29 (6), 1273–1283.

Hines, M.E., Knollmeyer, S.L., Tugel, J.B., 1989. Sulfate reduction and other sedimentary biogeochemistry in a northern new England salt marsh. Limnology and Oceanography 34, 578–590.

Hines, M.E., Banta, G.T., Giblin, A.E., Hobbie, J.E., Tugel, J.B., 1994. Acetate concentrations and oxidation in salt-marsh sediments. Limnology and Oceanography 39, 140–148.

Hines, M.E., Evans, R.S., Sharak Genthner, B.R., Willis, S.G., Friedman, S., Rooney-Varga, J.N., Devereux, R., 1999. Molecular phylogenetic and biogeochemical studies of sulfate-reducing bacteria in the rhizosphere of *Spartina alterniflora*. Applied and Environmental Microbiology 65, 2209–2216.

Hines, M.E., 2006. Microbially mediated redox cycling at the oxic-anoxic boundary in sediments: comparison of animal and plant habitats. Water, Air, and Soil Pollution: Focus 6, 523–536.

Hinshaw, S.E., Tatariw, C., Flournoy, N., Kleinhuizen, A., Taylor, C., Sobecky, P., Mortazavi, B., 2017. Vegetation loss decreases salt marsh denitrification capacity: implications for marsh erosion. Environmental Science and Technology. https://doi.org/10.1021/acs.est.7b00618.

Højberg, O., Revsbech, N.P., Tiedje, J.M., 1994. Denitrification in soil aggregates analyzed with microsensors for nitrous oxide and oxygen. Soil Science Society of America Journal 58, 1691–1698.

Holm, G.O., Perez, B.C., McWhorter, D.E., Krauss, K.W., Johnson, D.J., Raynie, R.C., Killebrew, C.J., 2016. Ecosystem level methane fluxes from tidal freshwater and brackish marshes of the Mississippi River Delta: Implications for coastal wetland carbon projects.

Holmer, M., Gribsholt, B., Kristensen, E., 2002. Effects of sea level rise on growth of *Spartina anglica* and oxygen dynamics in rhizosphere and salt marsh sediments. Marine Ecology Progress Series 225, 197–204.

Hopkinson, C.S., Giblin, A.E., 2008. Nitrogen dynamics of coastal salt marsh ecosystems. In: Capone, D.G., Bronk, D.A., Carpenter, E.J. (Eds.), Nitrogen in the Marine Environment.

Hopkinson, C.S., Schubauer, J.P., 1984. Static and dynamic aspects of nitrogen cycling in the salt marsh graminoid *Spartina alterniflora*. Ecology 65, 961–969.

Hopkinson, C.S., Cai, W., Hu, X., 2012. Carbon sequestration in wetland dominated coastal systems – a global sink of rapidly diminishing magnitude. Current Opinions in Environmental Sustainability 4, 186–194.

Hou, L., Zheng, Y., Liu, M., Gong, J., Zhang, X., Yin, G., You, L., 2013. Anaerobic ammonium oxidation (anammox) bacterial diversity, abundance, and activity in marsh sediments. Journal of Geophysical Research Biogeosciences 118, 1–10.

Hou, L., Zheng, Y., Liu, M., Li, X., Lin, X., Yin, G., Gao, J., Deng, F., Chen, F., Jiang, X., 2015. Anaerobic ammonium oxidation and its contribution to nitrogen removal in China's coastal wetlands. Scientific Reports 5, 15621. https://doi.org/10.1038/srep15621.

Howarth, R.W., Hobbie, J.E., 1982. The regulation of decomposition and heterotrophic microbial activity in salt marsh soils: a review. In: Kennedy, V.S. (Ed.), Estuarine Comparisons. Academic, New York, pp. 183–207.

Howarth, R.W., Teal, J.M., 1979. Sulfate reduction in a New England salt marsh. Limnology and Oceanography 24, 999–1013.

Howarth, R.W., Teal, J.M., 1980. Energy flow in a salt marsh ecosystem: the role of reduced inorganic sulfur compounds. The American Naturalist 116, 862–872.

Howarth, R.W., 1979. Pyrite: its formation in a salt marsh and its importance in ecosystem metabolism. Science 203, 49–51.

Howarth, R.W., 1984. The ecological significance of sulfur in the energy dynamics of salt marsh and coastal marine sediments. Biogeochemistry 1, 5–27.

Howarth, R.W., 1993. Microbial processes in salt-marsh sediments. In: Ford, T.E. (Ed.), Aquatic Microbiology: An Ecological Approach. Blackwell Scientific, London, pp. 239–258.

Howes, B., Howarth, R., Teal, J., Valiela, I., 1981. Oxidation reduction potential in a salt marsh: spatial patterns and interactions with primary production. Limnology and Oceanography 22, 350–360.

Howes, B.L., Dacey, J.W.H., King, G.M., 1984. Carbon flow through oxygen and sulfate reduction pathways in salt marsh sediments. Limnology and Oceanography 29, 1037–1051.

Howes, B.L., Dacey, J.W.H., Teal, J.M., 1985. Annual carbon mineralization and belowground production of *Spartina alterniflora* in a New England salt marsh. Ecology 66, 595–605.

Howes, B.L., Weiskel, P.K., Goehringer, D.D., Teal, J.M., 1996. Interception of freshwater and nitrogen transport from uplands to coastal waters: the role of saltmarshes. In: Nordstrom, K.F., Roman, C.T. (Eds.), Estuarine Shores: Evolution, Environments and Human Alterations. John Wile and Sons, New York, New York, USA, pp. 287–310.

Hseih, Y.P., Yang, C.H., 1997. Pyrite accumulation and sulfate depletion as affected by root distribution in a *Juncus* (needle rush) salt marsh. Estuaries 20, 640–645.

Huntington, T.G., Culbertson, C.W., Duff, J.H., 2012. Ambient and Potential Denitrification Rates in Marsh Soils of Northeast Creek and Bass Harbor Marsh Watersheds, Mount Desert Island, Maine. USGS Scientific Investigations Report 2012–5166.

Hutchinson, S.E., Sklar, F.H., Roberts, C., 1995. Short term sediment dynamics in a Southeastern USA Spartina marsh. Journal of Coastal Research 11, 370–380.

Hwang, Y.-H., Morris, J.T., 1992. Fixation of inorganic carbon from different sources and its translocation in *Spartina alterniflora* Loisel. Aquatic Botany 43, 137–147.

Hyun, J.-H., Smith, A.C., Kostka, J.E., 2007. Relative contributions of sulfate- and iron(III) reduction to organic matter mineralization and process controls in contrasting habitats of the Georgia saltmarsh. Applied Geochemistry 22, 2637–2651.

Ionescu, D., Heim, C., Polerecky, L., Thiel, V., De Beer, D., 2015. Biotic and abiotic oxidation and reduction of iron at circumneutral pH are inseparable processes under natural conditions. Geomicrobiology Journal 32 (3–4), 221–230.

Jacobson, M.E., 1994. Chemical and biological mobilization of Fe(III) in marsh sediments. Biogeochemistry 25, 41–60.

Jensen, H.S., Thamdrup, B., 1993. Iron-bound phosphorus in marine sediments as measured by bicarbonate-dithionite extraction. Hydrobiologia 253, 47–59.

Ji, Q., Babbin, A.R., Peng, X., Bowen, J.L., Ward, B.B., 2015. Nitrogen substrate-dependent nitrous oxide cycling in salt marsh sediments. Journal of Marine Research 73, 71–92.

Jiménez-Cárceles, F.J., Álvarez-Rogel, J., 2007. Phosphorus fractionation and distribution in salt marsh soils affected by mine wastes and eutrophicated water: a case study in SE Spain. Geoderma 144, 299–309.

Johnson, D.S., Warren, R.S., Deegan, L.A., Mozdzer, T.J., 2016. Saltmarsh plant responses to eutrophication. Ecological Applications 26 (8), 2647–2659.

Jordan, T., Correll, D., 1991. Continuous automated sampling of tidal exchanges of nutrients by brackish marshes of Chesapeake Bay. Estuarine, Coastal and Shelf Science 32, 527–545.

Jordan, T.E., Cornwell, J.C., Boynton, W.R., Anderson, J.T., 2008. Changes in phosphorus biochemistry along an estuarine salinity gradient: the iron conveyor belt. Limnology and Oceanography 53, 172–184.

Joye, S.B., Hollibaugh, J.T., 1995. Influence of sulfide inhibition of nitrification on nitrogen regeneration in sediments. Science 270, 623–625.

Joye, S.B., Paerl, H.W., 1994. Nitrogen cycling in microbial mats: rates and patterns of denitrification and nitrogen fixation. Marine Biology 119, 285–295.

Kalber Jr., F.A., 1959. A hypothesis on the role of tide-marshes in estuarine productivity. Estuarine Bulletin 4, 3.

Kaplan, W., Valiela, I., Teal, J.M., 1979. Denitrification in a salt marsh ecosystem. Limnology and Oceanography 22, 726–734.

Kathilankal, J.C., Mozdzer, T.J., Fuentes, J.D., D'Odorico, P., McGlathery, K.J., Zieman, J.C., 2008. Tidal influences on carbon assimilation by a salt marsh. Environmental Research Letters 3 (4), 044010. https://doi.org/10.1088/1748-9326/3/4/044010.

Keller, J.K., Weisenhorn, P.B., Megonigal, J.P., 2009a. Humic acids as electron acceptors in wetland decomposition. Soil Biology and Biochemistry 41 (7), 1518–1522.

Keller, J.K., Wolf, A.A., Weisenhorn, P.B., Drake, B.G., Megonigal, J.P., 2009b. Elevated CO_2 affects porewater chemistry in a brackish marsh. Biogeochemistry 96 (1–3), 101–117.

Kelly, D.P., Smith, N.A., 1990. Organic sulfur compounds in the environment: biogeochemistry, microbiology, and ecological aspects. Advances in Microbial Ecology 11, 345–385.

Kelly-Gerreyn, B.A., Trimmer, M., Hydes, D.J., 2001. A diagenetic model discriminating denitrification and dissimilatory nitrate reduction to ammonium in a temperate estuarine sediment. Marine Ecology Progress Series 220, 33–46.

King, D., Nedwell, D.B., 1985. The influence of nitrate concentration upon end-products of nitrate dissimilation by bacteria in anaerobic salt marsh sediment. FEMS Microbiology Ecology 31, 23–28.

King, G.M., Wiebe, W.J., 1980a. Regulation of sulfate concentrations and methanogenesis in salt-marsh soils. Estuarine and Coastal Marine Science 10, 215–223.

King, G.M., Wiebe, W.J., 1980b. Tracer analysis of methanogenesis in salt marsh soils. Applied and Environmental Microbiology 39, 877–881.

King, G.M., Klug, M.J., Wiegert, R.G., Chalmers, A.G., 1982. Relation of soil water movement and sulfide concentration to *Spartina alterniflora* production in a Georgia salt marsh. Science 218, 61–63.

Kirwan, M.L., Blum, L.K., 2011. Enhanced decomposition offsets enhanced productivity and soil carbon accumulation in coastal wetlands responding to climate change. Biogeosciences 8 (4), 987.

Kirwan, M.L., Megonigal, J.P., 2013. Tidal wetland stability in the face of human impacts and sea-level rise. Nature 504, 53–60.

Kirwan, M.L., Mudd, S.M., 2012. Response of salt-marsh carbon accumulation to climate change. Nature 489, 550–553.

Kirwan, M.L., Guntenspergen, G.R., D'Alpaos, A., Morris, J.T., Mudd, S.M., Temmerman, S., 2010. Limits on the adaptability of coastal marshes to rising sea level. Geophysical Research Letters 37, L23401. https://doi.org/10.1029/2010GL045489.

Klüpfel, L., Piepenbrock, A., Kappler, A., Sander, M., 2014. Humic substances as fully regenerable electron acceptors in recurrently anoxic environments. Nature Geoscience 7 (3), 195.

Koch, F., Gobler, C.J., 2009. The effects of tidal export from salt marsh ditches on estuarine water quality and plankton communities. Estuaries and Coasts 32, 261–275.

Koch, M.S., Maltby, E., Oliver, G.A., Bakker, S.A., 1992. Factors controlling denitrification rates of tidal mudflats and fringing salt marshes in South-west England. Estuarine, Coastal and Shelf Science 34, 471–485.

Koop-Jakobsen, K., Giblin, A., 2009a. Anammox in tidal marsh sediments: the role of salinity, nitrogen loading, and marsh vegetation. Estuaries and Coasts 32, 238–245.

Koop-Jakobsen, K., Giblin, A., 2009b. New approach for measuring denitrification in the rhizosphere of vegetated marsh sediments. Limnology and Oceanography 7, 626–637.

Koop-Jakobsen, K., Giblin, A.E., 2010. The effect of increased nitrate loading on nitrate reduction via denitrification and DNRA in salt marsh sediments. Limnology and Oceanography 55 (2), 789–802.

Koretsky, C.M., Miller, D., 2008. Seasonal influence of the needle rush *Juncus roemerianus* on saltmarsh porewater geochemistry. Estuaries and Coasts 31, 70–84.

Koretsky, C.M., Moore, C.M., Lowe, K.L., Meile, C., Dichristina, T.J., van Cappellen, P., 2003. Seasonal oscillation of microbial iron and sulfate reduction in saltmarsh sediments (Sapelo Island, GA, USA). Biogeochemistry 64, 179–203.

Koretsky, C.M., van Cappellen, P., DiChristina, T.J., Kostka, J.E., Lowe, K.L., Moore, C.M., Roychoudhury, A.N., Viollier, E., 2005. Salt marsh porewater geochemistry does not correlate with microbial community structure. Estuarine, Coastal and Shelf Science 62, 233–251.

Kostka, J.E., Luther III, G.W., 1995. Seasonal cycling of Fe in salt marsh sediments. Biogeochemistry 29, 159–181.

Kostka, J.E., Gribsholt, B., Petrie, E., Dalton, D., Skelton, H., Kristensen, E., 2002a. The rates and pathways of carbon oxidation in bioturbated saltmarsh sediments. Limnology and Oceanography 47, 230–240.

Kostka, J.E., Roychoudhury, A., van Cappellen, P., 2002b. Rates and controls of anaerobic microbial respiration across spatial and temporal gradients in saltmarsh sediments. Biogeochemistry 60, 49–76.

Krairapanond, N., DeLaune, R.D., Patrick Jr., W.H., 1992. Distribution of organic and reduced sulfur forms in marsh soils of coastal Louisiana. Organic Geochemistry 18, 489–500.

Krauss, K.W., Holm Jr., G.O., Perez, B.C., McWhorter, D.E., Cormier, N., Moss, R.F., Johnson, D.J., Neubauer, S.C., Raynie, R.C., 2016. Component greenhouse gas fluxes and radiative balance from two deltaic marshes in Louisiana: Pairing chamber techniques and eddy covariance. Journal of Geophysical Research: Biogeosciences 121 (6), 1503–1521. https://doi.org/10.1002/2015JG003224.

Krest, J.M., Moore, W.S., Gardner, L.R., 2000. Marsh nutrient export supplied by groundwater discharge: evidence from radium measurements. Global Biogeochemical Cycles 14, 167–176.

Kristensen, E., Kostka, J.E., 2005. Macrofaunal burrows and irrigation in marine sediment: microbiological and biogeochemical interactions. In: Kristensen, E., Haese, R.R., Kostka, J.E. (Eds.), Interactions between Macro- and Microorganisms in Marine Sediments. American Geophysical Union, Washington, DC, pp. 125–158.

Krom, M.D., Berner, R.A., 1980. Adsorption of phosphate in anoxic marine sediments. Limnology and Oceanography 25, 797–806.

Lage, M.D., Reed, H.E., Weihe, C., Crain, C.M., Martiny, J.B., 2010. Nitrogen and phosphorus enrichment alter the composition of ammonia-oxidizing bacteria in salt marsh sediments. The ISME Journal 4 (7), 933.

Langley, J.A., McKee, K.L., Cahoon, D.R., Cherry, J.A., Megonigal, J.P., 2009. Elevated CO_2 stimulates marsh elevation gain, counterbalancing sea-level rise. Proceedings of the National Academy of Sciences 106 (15), 6182–6186.

Langley, J.A., Mozdzer, T.J., Shepard, K.A., Hagerty, S.B., Patrick Megonigal, J., 2013. Tidal marsh plant responses to elevated CO2, nitrogen fertilization, and sea level rise. Global Change Biology 19 (5), 1495–1503.

Laverman, A.M., Canavan, R.W., Slomp, C.P., Van Cappellen, P., 2007. Potential nitrate removal in a coastal freshwater sediment (Haringvliet Lake, The Netherlands) and response to salinization. Water Research 41, 3061–3068.

Lecoanet, H., Leveque, F., Arnbrosi, J.P., 2001. Magnetic properties of salt-marsh soils contaminated by iron industry emissions (southeast France). Journal of Applied Geophysics 48, 67–81.

Lee, S.H., Megonigal, P.J., Langley, A.J., Kang, H., 2017. Elevated CO_2 and nitrogen addition affect the microbial abundance but not the community structure in salt marsh ecosystem. Applied Soil Ecology 117, 129–136.

Lent, R.M., Weiskel, P.K., Lyford, F.P., Armstrong, D.S., 1997. Hydrologic indicies for nontidal wetlands. Wetlands 17, 19–30.

Lettrich, M., 2011. Nitrogen Advection and Denitrification Loss in Southeastern North Carolina Salt Marshes. M.Sc. thesis. University of North Carolina Wilmington, 109 p.

Lewis, D.B., Brown, J.A., Jimenez, K.L., 2014. Effects of flooding and warming on soil organic matter mineralization in *Avicennia germinans* mangrove forests and *Juncus roemerianus* salt marshes. Estuarine, Coastal and Shelf Science 139, 11–19.

Li, X., Hou, L., Liu, M., Zheng, Y., Yin, G., Lin, X., Cheng, L., Li, Y., Hu, X., 2015. Evidence of nitrogen loss from anaerobic ammonium oxidation coupled with ferric iron reduction in an intertidal wetland. Environmental Science and Technology 49, 11560–11568.

Lillebø, A.I., Neto, J.M., Flindt, M.R., Marques, J.C., Pardal, M.A., 2004. Phosphorous dynamics in a temperate intertidal estuary. Estuarine, Coastal and Shelf Science 61, 101–109.

Lillebø, A.I., Coelho, J.P., Flindt, M.R., Jensen, H.S., Marques, J.C., Pedersen, C.B., Pardal, M.A., 2007. *Spartina maritima* influence on the dynamics of the phosphorus sedimentary cycle in a warm temperate estuary (Mondego estuary, Portugal). Hydrobiologia 587, 195–204.

Loomis, M.J., Craft, C.B., 2010. Carbon sequestration and nutrient (nitrogen and phosphorus) accumulation in river-dominated tidal marshes. Soil Science Society of America Journal 74, 1028–1036.

Lovley, D.R., Coates, J.D., Blunt-Harris, E.L., Phillips, E.J.P., Woodward, J.C., 1996. Humic substances as electron acceptors for microbial respiration. Nature 382, 445–448.

Lowe, K.L., DiChristina, T.J., Roychoudhury, A.N., van Cappellen, P., 2000. Microbiological and geochemical characterization of microbial Fe(III) reduction in salt marsh sediments. Geomicrobiology Journal 17, 163–178.

Luo, M., Zeng, C.S., Tong, C., Huang, J.F., Chen, K., Liu, F.Q., 2016. Iron reduction along an inundation gradient in a tidal sedge (*Cyperus malaccensis*) marsh: the rates, pathways, and contributions to anaerobic organic matter mineralization. Estuaries and Coasts 39 (6), 1679–1693.

Luther, G.W., Giblin, A.E., Howarth, R.W., Ryans, R.A., 1982a. Pyrite and oxidized iron mineral phases formed from pyrite oxidation in salt marsh and estuarine sediments. Geochimica et Cosmochimica Acta 46, 2665–2669.

Luther III, G.W., Meyerson, A.L., Rogers, K., Hall, F., 1982b. Tidal and seasonal variations of sulfate ion in a New Jersey marsh system. Estuaries 5, 189–196.

Luther III, G.W., Kostka, J.E., Church, T.M., Sulzberger, B., Stumm, W., 1992. Seasonal iron cycling in the salt-marsh sedimentary environment: the importance of Fe(II) and Fe(III) in the dissolution of Fe(III) minerals and pyrite, respectively. Marine Chemistry 40, 81–103.

Luther III, G.W., Schellenbarger, P.A., Brendel, P.J., 1996. Dissolved organic Fe(III) and Fe(II) complexes in salt marsh porewaters. Geochimica et Cosmochimica Acta 60, 951–960.

Lytle Jr., R.W., Hull, R.J., 1980. Photoassimilate distribution in *Spartina alterniflora* Loisel. I. Vegetative and floral development. Agronomy Journal 72, 933–938.

Maricle, B.R., Lee, R.W., 2002. Aerenchyma development and oxygen transport in the estuarine cordgrasses *Spartina alterniflora* and *S. anglica*. Aquatic Botany 74, 109–120.

Martens, C.S., Berner, R.A., 1977. Interstitial water chemistry of anoxic Long Island Sound sediments: I. Dissolved gases. Limnology and Oceanography 22, 10–25.

Marton, J.M., Roberts, B.J., 2014. Spatial variability of phosphorous sorption dynamics in Lousiana salt marshes. Journal of Geophysical Research: Biogeosciences 119, 451–465.

Mayer, L.M., Schick, L.L., Allison, M.A., Ruttenberg, K.C., Bentley, S.J., 2007. Marine vs. terrigenous organic matter in Louisiana coastal sediments: the uses of bromine:organic carbon ratios. Marine Chemistry 107, 244–254.

McKee, K., Rogers, K., Saintilan, N., 2012. Response of salt marsh and mangrove wetlands to changes in atmospheric CO2, climate, and sea level. In: Global Change and the Function and Distribution of Wetlands. Springer, Netherlands, pp. 63–96.

Mcleod, E., Chmura, G.L., Bouillon, S., Salm, R., Björk, M., Duarte, C.M., Lovelock, C.E., Schlesinger, W.H., Silliman, B., 2011. A blueprint for blue carbon: toward an improved understanding of the role of vegetated coastal habitats in sequestration. Frontiers in Ecology and the Environment 9, 552–560.

Megonigal, J.P., Hines, M.E., Visscher, P.T., 2004. Anaerobic metabolism: linkages to trace gases and aerobic metabolism. In: Schlesinger, W.H. (Ed.), Biogeochemistry. Elsevier-Pergamon, Oxford, United Kingdom, pp. 317–424.

Megonigal, J.P., Chapman, S., Crooks, S., Dijkstra, P., Kirwan, M., Langley, A., 2016. 3.4 Impacts and effects of ocean warming on tidal marsh and tidal freshwater forest ecosystems. In: Laffoley, D., Baxter, J.M. (Eds.), Explaining Ocean Warming: Causes, Scale, Effects and Consequences. IUCN, Gland, Switzerland, pp. 105–120. Full Report.

Meile, C., Koretsky, C.M., van Cappellen, P., 2001. Quantifying bioirrigation in aquatic sediments: an inverse modeling approach. Limnology and Oceanography 46, 164–177.

Mendelssohn, I.A., Morris, J.T., 2000. Eco-physiological controls on the productivity of *Spartina alterniflora* Loisel. In: Weinstein, M., Kreeger, D.A. (Eds.), Concepts and Controversies in Tidal Marsh Ecology. Kluwer Academic Publishing, Dordrecht, Netherlands, pp. 59–80.

Mendelssohn, I.A., Postek, M.T., 1982. Elemental analysis of deposits on the roots of *Spartina alterniflora* Loisel. American Journal of Botany 69, 904–912.

Mendelssohn, I.A., Kleiss, B.A., Wakely, J.S., 1995. Factors controlling the formation of oxidized root channels: a review. Wetlands 15, 37–46.

Merrill, J.Z., Cornwell, J.C., 2000. The role of oligohaline marshes in estuarine nutrient cycling. In: Weinstein, M.P., Kreeger, D.A. (Eds.), Concepts and Controversy in Tidal Marsh Ecology. Kluwer Academic Publishers, Dordrecht, The Netherlands, pp. 425–442.

Middelburg, J.J., Herman, P.M.J., 2007. Organic matter processing in tidal estuaries. Marine Chemistry 106, 127–147.

Middelburg, J.J., Nieuwenhuize, J., Lubberts, R.K., van de Plassche, O., 1997. Organic carbon isotope systematics of coastal marshes. Estuarine, Coastal and Shelf Science 45, 681–687.

Miller, W.D., Neubauer, S.C., Anderson, I.C., 2001. Effects of sea level induced disturbances on high salt marsh metabolism. Estuaries 24, 357–367.

Minderlein, S., Blodau, C., 2010. Humic-rich peat extracts inhibit sulfate reduction, methanogenesis, and anaerobic respiration but not acetogenesis in peat soils of a temperate bog. Soil Biology and Biochemistry 42 (12), 2078–2086.

Mitsch, W.J., Gosselink, J.G., 1993. Wetlands, second ed. Van Nostrand Reinhold, New York, NY, USA.

Moffett, K.B., Wolf, A., Berry, J.A., Gorelick, S.M., 2010. Salt marsh-atmosphere exchange of energy, water vapor, and carbon dioxide: Effects of tidal flooding and biophysical controls. Water Resources Research 46 (10), W10525. https://doi.org/10.1029/2009WR009041.

Moisander, P.H., Piehler, M.F., Paerl, H.W., 2005. Diversity and activity of epiphytic nitrogen-fixers on standing dead stems of the salt marsh grass *Spartina alterniflora*. Aquatic Microbial Ecology 39, 271–279.

Morris, J.T., Jensen, A., 1998. The carbon balance of grazed and non-grazed *Spartina alterniflora* saltmarshes at Skallingen, Denmark. Journal of Ecology 86, 229–242.

Morris, J.T., Whiting, G.J., 1985. Gas advection in sediments of a South Carolina salt marsh. Marine Ecology Progress Series 27, 187–194.

Morris, J.T., Whiting, G.J., 1986. Emission of gaseous carbon dioxide from salt-marsh sediments and its relation to other carbon losses. Estuaries 9, 9–19.

Morris, J.T., Sundareshwar, P.V., Nietch, C.T., Kjerve, B., Cahoon, D.R., 2002. Responses of coastal wetlands to rising sea level. Ecology 83, 2869–2877.

Morris, J.T., Shaffer, G.P., Nyman, J.A., 2013. Brinson review: perspectives on the influence of nutrients on the sustainability of coastal wetlands. Wetlands 33, 975–988. https://doi.org/10.1007/s13157-013-0480-3.

Morris, J.T., Barber, D.C., Callaway, J.C., Chambers, R., Hagen, S.C., Hopkinson, C.S., Johnson, B.J., Megonigal, P., Neubauer, S.C., Troxler, T., Wigand, C., 2016. Contributions of organic and inorganic matter to sediment volume and accretion in tidal wetlands at steady state. Earths Future 4, 110–121. https://doi.org/10.1002/2015EF000334.

Morris, J.T., 1991. Effects of nitrogen loading on wetland ecosystems with particular reference to atmospheric deposition. Annual Reviews in Ecological Sytematics 22, 257–279.

Morris, J.T., 1995. The salt and water balance of intertidal sediments: Results from North Inlet, South Carolina. Estuaries 18, 556–567.

Morris, J.T., 2000. Effects of sea-level anomalies on estuarine processes. In: Hobbie, J. (Ed.), Estuarine Science: A Synthetic Approach to Research and Practice. Island Press, pp. 107–127.

Morrison, M.C., Hines, M.E., 1990. The variability of biogenic sulfur flux from a temperate salt marsh on short time and space scales. Atmospheric Environment 24, 1771–1779.

Mortimer, R.J., Galsworthy, A.M., Bottrell, S.H., Wilmot, L.E., Newton, R.J., 2011. Experimental evidence for rapid biotic and abiotic reduction of Fe(III) at low temperatures in salt marsh sediments: a possible mechanism for formation of modern sedimentary siderite concretions. Sedimentology 58 (6), 1514–1529.

Moseman, S.W., 2007. Opposite diel patterns of nitrogen fixation associated with salt marsh plant species (*Spartina foliosa* and *Salicornia virginica*) in southern California. Marine Ecology Progress Series 28, 276–287.

Moseman-Valtierra, S., Gonzalez, R., Kroeger, K.D., Tang, J., Chun Chao, W., Crusius, J., Bratton, J., Green, A., Shelton, J., 2011. Short-term nitrogen additions can shift a coastal wetland from a sink to a source of N_2O. Atmospheric Environment 45, 4390–4397.

Mueller, P., Jensen, K., Megonigal, J.P., 2016. Plants mediate soil organic matter decomposition in response to sea level rise. Global Change Biology 22 (1), 404–414.

Mulholland, P.J., et al., 2008. Stream denitrification across biomes and its response to anthropogenic nitrate loading. Nature 452, 202–206.

NADP, 2018. National Atmospheric Deposition Program. Illinois State Water Survey, Champaign, IL. http://nadp.sws.uiuc.edu. accessed 06 September 2018.

Naeher, S., Huguet, A., Roose-amsaleg, C.L., Laverman, A.M., Fosse, C., et al., 2015. Molecular and geochemical constraints on anaerobic ammonium oxidation (anammox) in a riparian zone of the Seine Estuary (France). Biogeochemistry 123, 237–250.

Negrin, V.L., Spetter, C.V., Asteasuain, R.O., Perillo, M.E., Marcovecchio, J.E., 2011. Influence of flooding and vegetation on carbon, nitrogen, and phosphorus dynamics in the porewater of a *Spartina alterniflora* salt marsh. Journal of Environmental Sciences 23, 212–221.

Negrin, V.L., Botte, S.E., Pratolongo, P.D., Gonzalez, T.G., Marcovecchio, J.E., 2016. Ecological processes and biogeochemical cycling in salt marshes: synthesis of studies in the Bahia Blanca estuary. Hydrobiologia 774, 217–235.

Nellemann, C., Corcoran, E., Duarte, C.M., Valdes, L., De Young, C., Fonseca, L., Grimsditch, G., 2009. Blue carbon. A rapid response assessment. In: United Nations Environment Programme. GRID-Arendal, ISBN 978-82-7701-060-1. www.grida.no.

Neubauer, S.C., Anderson, I.C., 2003. Transport of dissolved inorganic carbon from a tidal freshwater marsh to the York River estuary. Limnology and Oceanography 48, 299–307.

Neubauer, S.C., Megonigal, J.P., 2015. Moving beyond global warming potentials to quantify the climatic role of ecosystems. Ecosystems 18 (6), 1000–1013.

Neubauer, S.C., Anderson, I.C., Constantine, J.A., Kuehl, S.A., 2002a. Sediment deposition and accretion in a mid-Atlantic (U.S.A.) tidal freshwater marsh. Estuarine, Coastal and Shelf Science 54, 713–727.

Neubauer, S.C., Emerson, D., Megonigal, J.P., 2002b. Life at the energetic edge: Kinetics of circumneutral iron oxidation by lithotrophic iron-oxidizing bacteria isolated from the wetland-plant rhizosphere. Applied and Environmental Microbiology 68, 3988–3995.

Neubauer, S.C., Anderson, I.C., Neikirk, B.B., 2005a. Nitrogen cycling and ecosystem exchanges in a Virginia tidal freshwater marsh. Estuaries 28, 909–922.

Neubauer, S.C., Givler, K., Valentine, S., Megonigal, J.P., 2005b. Seasonal patterns and plant-mediated controls of subsurface wetland biogeochemistry. Ecology 86, 3334–3344.

Neubauer, S.C., Toledo-Durán, G.E., Emerson, D., Megonigal, J.P., 2007. Returning to their roots: Iron-oxidizing bacteria enhance short-term plaque formation in the wetland-plant rhizosphere. Geomicrobiology Journal 24, 65–73.

Neubauer, S.C., Emerson, D., Megonigal, J.P., 2008. Microbial oxidation and reduction of iron in the root zone and influences on metal mobility. In: Violante, A., Huang, P.M., Gadd, G.M. (Eds.), Biophysico-chemical Processes of Heavy Metals and Metalloids in Soil Environments. John Wiley and Sons, Hoboken, New Jersey, pp. 339–371.

Neubauer, S.C., 2008. Contributions of mineral and organic components to tidal freshwater marsh accretion. Estuarine, Coastal and Shelf Science 78, 78–88.

Newell, S.Y., Porter, D., 2000. Microbial secondary production from saltmarsh grass shoots and its known potential fates. In: Weinstein, M., Kreeger, D.A. (Eds.), Concepts and Controversies in Tidal Marsh Ecology. Kluwer Academic Publishing, Dordrecht, Netherlands, pp. 159–185.

Newell, S.Y., Porter, D., Lingle, W.L., 1989. Decomposition and microbial dynamics for standing, naturally positioned leaves of the salt marsh grass *Spartina alterniflora*. Marine Biology 101, 471–481.

Nie, M., Wang, M., Li, B., 2009. Effects of salt marsh invasion by *Spartina alterniflora* on sulfate-reducing bacteria in the Yangtze River estuary, China. Ecological Engineering 35 (12), 1804–1808.

Nietch, C.T., 2000. Carbon Biogeochemistry in Tidal Marshes of South Carolina: The Effect of Salinity and Nutrient Availability on Marsh Metabolism in Estuaries with Contrasting Histories of Disturbance and River Influence. Ph.D. dissertation. University of South Carolina, Columbia, South Carolina.

Nijburg, J., Laanbroek, H.J., 1997. The influence of *Glyceria maxima* and nitratae inpupt on the composition of nitrate-reducing bacteria community. Applied and Environmental Microbiology 63, 931–937.

Nixon, S.W., 1980. Between coastal marshes and coastal waters – a review of twenty years of speculation and research on the role of salt marshes in estuarine productivity and water chemistry. In: Hamilton, P., MacDonald, K.B. (Eds.), Estuarine and Wetland Processes. Plenum Press, New York, pp. 437–525.

Nowicki, B.L., Requintina, E., Van Keuren, D., Portnoy, J., 1999. The role of sediment denitrification in reducing groundwater-derived nitrate inputs to Nauset Marsh estuary, Cape Cod, Massachusetts. Estuaries 22, 245–259.

Nyman, J.A., Walters, R.J., DeLaune, R.D., Patrick Jr., W.H., 2006. Marsh vertical accretion via vegetative growth. Estuarine, Coastal and Shelf Science 69, 370–380.

Oczkowski, A., Wigand, C., Hanson, A., Markham, E., Miller, M., Johnson, R., 2016. Nitrogen retention in salt marsh systems across nutrient-enrichment, elevation, and precipitation regimes: a multiple stressor experiment. Estuaries and Coasts 39, 68–81.

Odum, E.P., 1968. A research challenge: evaluating the productivity of coastal and estuarine waters. In: Second Sea Grant Conference. Rhode Island, Newport, pp. 63–64.

Oremland, R.S., Marsh, L.M., Polcin, S., 1982. Methane production and simultaneous sulphate reduction in anoxic, salt marsh sediments. Nature 296, 143–145.

Osburn, C.L., Mikan, M.P., Etheridge, J.R., Burchell, M.R., Birgand, F., 2015. Seasonal variation in the quality of dissolved and particulate organic matter exchanged between a salt marsh and its adjacent estuary. Journal of Geophysical Research Biogeosciences 120, 1430–1449. https://doi.org/10.1002/2014JG002897.

Osgood, D.T., 2000. Subsurface hydrology and nutrient export from barrier island marshes at different tidal ranges. Wetlands Ecology and Management 8, 133–146.

Otero, X.L., Macias, F., 2003. Spatial variation in pyritization of trace metals in salt-marsh soils. Biogeochemistry 62, 59–86.

Ouyang, X., Lee, S.Y., Connolly, R.M., 2017. The role of root decomposition in global mangrove and saltmarsh carbon budgets. Earth Science Reviews 166, 53–63.

Paerl, H.W., Dennis, R.L., Whitall, D.R., 2002. Atmospheric deposition of nitrogen: implications for nutrient over-enrichment of coastal waters. Estuaries 25, 677–693.

Palomo, L., Meile, C., Joye, S.B., 2013. Drought impacts on biogeochemistry and microbial processes in salt marsh sediments: a flow-through reactor approach. Biogeochemistry 1121, 389–407.

Paludan, C., Morris, J.T., 1999. Distribution and speciation of phosphorous along a salinity gradient in intertidal marsh sediments. Biogeochemistry 45, 197–221.

Parkes, R.J., Brock, F., Banning, N., Hornibrook, E.R., Roussel, E.G., Weightman, A.J., Fry, J.C., 2012. Changes in methanogenic substrate utilization and communities with depth in a salt-marsh, creek sediment in southern England. Estuarine, Coastal and Shelf Science 96, 170–178.

Parsons, M.L., 1998. Salt marsh sedimentary record of the landfall of Hurricane Andrew on the Louisiana coast: diatoms and other paleoindicators. Journal of Coastal Research 14, 939–950.

Pastore, M.A., Megonigal, P.J., Langley, A.J., 2017. Elevated CO_2 and nitrogen additions accelerate net carbon gain in a brackish marsh. Biogeochemistry 133, 73–87.

Patrick, W.H., Delaune, R.D., 1976. Nitrogen and phosphorous utilization by *Spartina alterniflora* in a salt marsh in Barataria Bay, Lousiana. Estuarine and Coastal Marine Science 4, 59–64.

Pendleton, L., et al., 2012. Estimating global 'blue carbon' emissions from conversion and degradation of vegetated coastal ecosystems. PLoS One 7 (9), e43542.

Peng, X., Ji, Q., Angell, J.H., Kearns, P.J., Yang, H.J., Bowen, J.L., Ward, B.B., 2016. Long-term fertilization alters the relative importance of nitrate reduction pathways in salt marsh sediments. Journal of Geophysical Research Biogeosciences 121, 2082–2095.

Perry, C.L., Mendelssohn, I.A., 2009. Ecosystem effects of expanding populations of *Avicennia germinans* in a Louisiana salt marsh. Wetlands 29 (1), 396–406.

Piehler, M.F., Smyth, A.R., 2011. Habitat-specific distinctions in estuarine denitrification affect both ecosystem function and services. Ecosphere 2, 1–16.

Piehler, M.F., Currin, C.A., Cassanova, R., Paerl, H.W., 1998. Development and N_2-fixing activity of the benthic microbial community in transplanted *Spartina alterniflora* marshes in North Carolina. Restoration Ecology 6, 290–296.

Pinckney, J.L., Zingmark, R.G., 1993. Modeling the annual production of intertidal benthic microalgae in Estuarine ecosystems. Journal of Phycology 29, 396–407.

Plummer, P., Tobias, C., Cady, D., 2015. Nitrogen reduction pathways in estuarine sediments: Influences of Organic Carbon and Sulfide. Journal of Geophysical Research: Biogeosciences 120. https://doi.org/10.1002/2015JG003057.

Poffenbarger, H., Needelman, B., Megonigal, J.P., 2011. Salinity influence on methane emissions from tidal marshes. Wetlands 31, 831–842.

Pomeroy, L.R., Darley, W., Dunn, E.L., Gallagher, J.L., Haines, E.B., Whitney, D., 1981. Primary production. In: Pomeroy, L.R., Wiegert, R.G. (Eds.), The Ecology of a Salt Marsh. Springer-Verlag, New York, New York, pp. 39–67.

Pomeroy, L.R., 1959. Algal Productivity in Salt Marshes of Georgia. Limnology and Oceanography 4, 386–398.

Ponnamperuma, F.N., 1972. The chemistry of submerged soils. In: Brady, N.C. (Ed.), Advances in Agronomy. Academic Press, New York, NY, pp. 29–96.

Portnoy, J.W., Giblin, A.E., 1997. Effects of historic tidal restrictions on salt marsh sediment chemistry. Biogeochemistry 36, 275–303.

Portnoy, J.W., Nowicki, B.L., Roman, C.T., Urish, D.W., 1997. The discharge of nitrate-contaminated groundwater from a developed shoreline to a marsh-fringed estuary. Water Resources Research 34, 3095–3104.

Poulin, P., Pelletier, E., Saint-Louis, R., 2007. Seasonal variability of denitrification efficiency in northern salt marshes; an example from the St. Lawrence Estuary. Marine Environmental Research 63, 490–505.

Poulin, P., Pelletier, E., Koutitonski, V.G., Neumeier, U., 2009. Seasonal nutrient fluxes variability of northern salt marshes: examples from the lower St. Lawrence Estuary. Wetlands Ecology and Management 17, 655–673.

Pye, K., Dickson, J.A.D., Schiavon, N., Coleman, M.L., Cox, M., 1990. Formation of siderite-Mg-calcite-iron sulphide concretions in intertidal marsh and sandflat sediments, north Norfolk, England. Sedimentology 37, 325–343.

Pye, K., 1981. Marshrock formed by iron sulphide and siderite cementation in saltmarsh sediments. Nature 294, 650–652.

Raymond, P.A., Bauer, J.E., Cole, J.J., 2000. Atmospheric CO_2 evasion, dissolved inorganic carbon production, and net heterotrophy in the York River Estuary. Limnology and Oceanography 45, 1707–1717.

Reddy, K.R., DeLaune, R.D., 2008. Biogeochemistry of Wetlands: Science and Applications. CRC press.

Reid, M.C., Tripathee, R., Schäfer, K.V.R., Jaffé, P.R., 2013. Tidal Marsh Methane dynamics: Difference in seasonal lags in emissions driven by storage in vegetated versus unvegetated sediments. Journal of Geophysical Research Biogeosciences 118, 1802–1813. https://doi.org/10.1002/jgrg.20152.

Rickard, D., Luther III, G.W., 2007. Chemistry of iron sulfides. Chemical Reviews 107, 514–562.

Rietl, A.J., Nyman, J.A., Lindau, C.W., Jackson, C.R., 2017. Gulf ribbed mussels (*Geukensia granosissima*) increase methane emissions from a coastal *Spartina alterniflora* marsh. Estuaries and Coasts 40 (3), 832–841.

Roden, E.E., Wetzel, R.G., 1996. Organic carbon oxidation and suppression of methane production by microbial Fe(III) oxide reduction in vegetated and unvegetated freshwater wetland sediments. Limnology and Oceanography 41, 1733–1748.

Rooth, J.E., Stevenson, J.C., Cornwell, J.C., 2003. Increased sediment accretion rates following invasion by *Phragmites australis*: the role of litter. Estuaries 26, 475–483.

Roychoudhury, A.N., Kostka, J.E., van Cappellen, P., 2003a. Pyritization: a paleoenvironmental and redox proxy reevaluated. Estuarine, Coastal and Shelf Science 57, 1183–1193.

Roychoudhury, A.N., van Cappellen, P., Kostka, J.E., Viollier, E., 2003b. Kinetics of microbially mediated reactions: dissimilatory sulfate reduction in saltmarsh sediments (Sapelo Island, Georgia, USA). Estuarine, Coastal and Shelf Science 56, 1001–1010.

Roychoudhury, A.N., 2007. Spatial and seasonal variations in depth profile of trace metals in saltmarsh sediments from Sapelo Island, Georgia, USA. Estuarine, Coastal and Shelf Science 72, 675–689.

Rozan, T.F., Tailleferet, M., Trouwborst, R.E., Glazer, B.T., Ma, S., Herszage, J., Valdes, L.M., Price, K.S., Luther III, G.W., 2002. Iron-sulfur-phosphorus cycling in the sediments of a shallow coastal bay: implications for sediment nutrient release and benthic macroalgal blooms. Limnology and Oceanography 47, 1346–1354.

Rozema, J., Leendertse, P., Bakker, J., van Wijnen, H., 2000. Nitrogen and vegetation dynamics in European salt marshes. In: Weinstein, M.P., Kreeger, D.A. (Eds.), Concepts and Controversy in Tidal Marsh Ecology. Kluwer Academic Publishers, Dordrecht, The Netherlands, pp. 469–494.

Russell, K., Galloway, J., Macko, S., Moody, J., Scudlark, J., 1998. Sources of nitrogen in wet deposition to the Chesapeake Bay region. Atmospheric Environment 32, 2453–2465.

Rysgaard, S., Thastum, P., Dalsgaard, T., Christensen, P.B., Sloth, N.P., 1999. Effects of salinity on NH_4^+ adsorption capacity, nitrification, and denitrification in Danish estuarine sediments. Estuaries 22, 21–30.

Rysgaard, S., Glud, R.N., Risgaard-Petersen, N., Dalsgaard, T., 2004. Denitrification and anammox activity in Arctic marine sediments. Limnology and Oceanography 49 (1493), 1502.

Saintilan, N., Wilson, N.C., Rogers, K., Rajkaran, A., Krauss, K.W., 2014. Mangrove expansion and salt marsh decline at mangrove poleward limits. Global Change Biology 20 (1), 147–157.

Salgueiro, N., Caçador, I., 2007. Short-term sedimentation in Tagus estuary, Portugal: The influence of salt marsh plants. Hydrobiologia 587, 185–193.

Schippers, A., Jørgensen, B.B., 2002. Biogeochemistry of pyrite and iron sulfide oxidation in marine sediments. Geochimica et Cosmochimica Acta 66, 85–92.

Schubauer, J.P., Hopkinson, C.S., 1984. Above- and belowground emergent macrophyte production and turnover in a coastal marsh ecosystem. Limnology and Oceanography 29, 1052–1065.

Scott, J.T., McCarthy, M.J., Gardner, W.S., Doyle, R.D., 2008. Denitrification, dissimilatory nitrate reduction to ammonium, and nitrogen fixation along a nitrate concentration gradient in a created freshwater wetland. Biogeochemistry 87, 99–111.

Scudlark, J.R., Church, T.M., 1989. The sedimentary flux of nutrients at a Delaware salt marsh site: a geochemical perspective. Biogeochemistry 7, 55–75.

Seitzinger, S.P., Kroeze, C., 1998. Global distribution of nitrous oxide production and N inputs in freshwater and coastal marine ecosystems. Global Biogeochemical Cycles 12, 93–113.

Seitzinger, S.P., Gardner, W.S., Spratt, A.K., 1991. The effect of salinity on ammonium sorption in aquatic sediments: implications for benthic nutrient cycling. Estuaries 14, 167–174.

Seitzinger, S.P., Kroeze, C., Bouman, A.F., Caraco, N., Dentener, F., Styles, R.V., 2002. Global pattern of dissolved inorganic and particulate nitrogen inputs to coastal systems: recent conditions and future projections. Estuaries 25, 640–655.

Seitzinger, S., Harrison, J.A., Bölke, J.K., Bouwman, A.F., Lowrance, R., Peterson, B., Tobias, C., Van Drecht, G., 2006. Denitrification across landscapes and waterscapes: a synthesis. Ecological Applications 16, 2064–2090.

Seitzinger, S.P., 1994. Linkages between organic matter mineralization and denitrification in eight riparian wetlands. Biogeochemistry 25, 19–39.

Senga, Y., Mochida, K., Fukumori, R., Okamoto, N., Seike, Y., 2006. N_2O accumulation in estuarine and coastal sediments: the influence of H_2S on dissimilatory nitrate reduction. Estuarine, Coastal and Shelf Science 67, 231–238.

Shaw, G.E., 1983. Bio-controlled thermostasis involving the sulfur cycle. Climatic Change 5, 297–303.

Singer, P.C., Stumm, W., 1970. Acidic mine drainage: the rate-determining step. Science 167, 1121–1123.

Smith, C.J., Delaune, R.D., Patrick Jr., W.H., 1983. Nitrous oxide emission from Gulf Coast wetlands. Geochimica et Cosmochimica Acta 47, 1805–1814.

Smith, C., Delaune, Patrick, W., 1985. Fate of riverine nitrate entering an estuary: I Denitrification and nitrogen burial. Estuaries 8, 15–21.

Song, B.K., Tobias, C.R., 2011. Molecular and stable isotope methods to detect and measure anaerobic ammonium oxidation (anammox) in aquatic ecosystems. In: Martin, G. Klotz, Stein, L.Y. (Eds.), Methods in Enzymology, vol. 496. Academic Press, Burlington, pp. 63–89.

Straub, K.L., Benz, M., Schink, B., Widdel, F., 1996. Anaerobic, nitrate-dependent microbial oxidation of ferrous iron. Applied and Environmental Microbiology 62, 1458–1460.

Stribling, J.M., Cornwell, J.C., 2001. Nitrogen, phosphorus, and sulfur dynamics in a low salinity marsh system dominated by *Spartina alterniflora*. Wetlands 21, 629–638.

Sullivan, M.J., Currin, C.A., 2000. Community structure and functional dynamics of benthic microalgae in salt marshes. In: Weinstein, M., Kreeger, D.A. (Eds.), Concepts and Controversies in Tidal Marsh Ecology. Kluwer Academic Publishing, Dordrecht, Netherlands, pp. 81–106.

Sullivan, M.J., Daiber, F.C., 1975. Light, nitrogen and phosphorus limitation of edaphic algae in a Delaware salt marsh. Journal of Experimental Marine Biology and Ecology 18, 79–88.

Sullivan, M.J., Moncreiff, C.A., 1988. Primary production of edaphic algal communities in a Mississippi salt marsh. Journal of Phycology 24, 49–58.

Sullivan, M.J., Moncreiff, C.A., 1990. Edaphic algae are an important component of salt marsh food-webs: evidence from multiple stable isotope analyses. Marine Ecology Progress Series 62, 149–159.

Sundareshwar, Morris, J.T., 1999. Phosphorous sorption characteristics of intertidal marsh sediments along an estuarine salinity gradient. Limnology and Oceanography 44, 1693–1701.

Sundareshwar, P.V., Morris, J.T., Koepfler, E.K., Fornwalt, B., 2003. Phosphorous limitation of coastal ecosystem processes. Science 299, 563–565.

Sundby, B., Gobeil, C., Silverberg, N., Mucci, A., 1992. The phosphorus cycle in coastal marine sediments. Limnology and Oceanography 37, 1129–1145.

Sundby, B., Vale, C., Cacador, I., Catarino, F., Madureira, M., Caetano, M., 1998. Metal-rich concretions on the roots of salt marsh plants: mechanisms and rate of formation.

Sundby, B., Vale, C., Caetano, M., Luther III, G.W., 2003. Redox chemistry in the root zone of a salt marsh sediment in the Tagus Estuary, Portugal. Aquatic Geochemistry 9, 257–271.

Sutton-Grier, A.E., Keller, J.K., Koch, R., Gilmour, C., Megonigal, J.P., 2011. Electron donors and acceptors influence anaerobic soil organic matter mineralization in tidal marshes. Soil Biology and Biochemistry 43 (7), 1576–1583.

Taillefert, M., Bono, A.B., Luther III, G.W., 2000. Reactivity of freshly formed Fe(III) in synthetic solutions and (pore) waters: voltametric evidence of an aging process. Environmental Science and Technology 34, 2169–2177.

Taillefert, M., Neubhuber, S., Bristow, G., 2007. The effect of tidal forcing on biogeochemical processes in intertidal salt marsh sediments. Geochemical Transactions 8, 6.

Taylor, G.J., Crowder, A.A., Rodden, R., 1984. Formation and morphology of an iron plaque on the roots of *Typha latifolia* L. grown in solution culture. American Journal of Botany 71, 666–675.

Teal, J.M., Valiela, I., Berlo, D., 1979. Nitrogen fixation by rhizosphere and free-living bacteria in salt marsh sediments. Limnology and Oceanography 24, 126–132.

Teal, J.M., 1962. Energy flow in the salt marsh ecosystem of Georgia. Ecology 43, 614–624.

Temmerman, S., Govers, G., Wartel, S., Meire, P., 2003. Spatial and temporal factors controlling short-term sedimentation in a salt and freshwater tidal marsh, Scheldt Estuary, Belgium, SW Netherlands. Earth Surface Processes and Landforms 28, 739–755.

Thamdrup, B., 2000. Bacterial manganese and iron reduction in aquatic sediments. In: Schink, B. (Ed.), Advances in Microbial Ecology. Kluwer Academic/Plenum Publishers, New York, pp. 41–84.

Thomas, C.R., Blum, L.K., 2010. Importance of the fiddler crab *Uca pugnax* to salt marsh soil organic matter accumulation. Marine Ecology Progress Series 414, 167–177.

Thomas, C.R., Christian, R.R., 2001. Comparison of nitrogen cycling in salt marsh zones related to sea-level rise. Marine Ecology Progress Series 221, 1–16.

Thompson, S.P., Paerl, H.W., Malia, C.G., 1995. Seasonal patterns of nitrification and denitrification in a natural and a restored salt marsh. Estuaries 18, 399–408.

Thornton, D.C.O., Dong, L.F., Underwood, J.C., Nedwell, D.B., 2007. Sediment-wataer inorganic nutrient exchange and nitrogen budgets in the Colne Estuary, UK. Marine Ecology Progress Series 337, 63–77.

Tiedje, J.M., 1988. Ecology of denitrification and dissimilatory nitrate reduction to ammonium. In: Zehnder, A.J.B. (Ed.), Biology of Anaerobic Microorganisms. Wiley, New York.

Tobias, C.R., Neubauer, S.C., 2009. Salt marsh biogeochemistry: an overview. In: Perillo, G., Wolanski, E., Cahoon, D., Brinson, M. (Eds.), Coastal Wetlands: An Integrated Ecosystem Approach. Elsevier, pp. 445–492.

Tobias, C.R., Anderson, I.C., Canuel, E.A., Macko, S.A., 2001a. Nitrogen cycling through a fringing marsh-aquifer ecotone. Marine Ecology Progress Series 210, 25–39.

Tobias, C.R., Harvey, J.W., Anderson, I.C., 2001b. Quantifying groundwater discharge through fringing wetlands to estuaries: seasonal variability, methods comparison, and implications for wetland-estuary exchange. Limnology and Oceanography 46, 604–615.

Tobias, C.R., Macko, S.A., Anderson, I.C., Canuel, E.A., Harvey, J.W., 2001c. Tracking the fate of a high concentration groundwater plume through a fringing marsh: a combined groundwater tracer and in situ isotope enrichment study. Limnology and Oceanography 46, 1977–1989.

Turner, R.E., Swenson, E.M., Milan, C.S., 2000. Organic and inorganic contributions to vertical accretion in salt marsh sediments. In: Weinstein, M., Kreeger, D.A. (Eds.), Concepts and Controversies in Tidal Marsh Ecology. Kluwer Academic Publishing, Dordrecht, Netherlands, pp. 583–595.

Turner, R.E., Baustian, J.J., Swenson, E.M., Spicer, J.S., 2006a. Wetland sedimentation from Hurricanes Katrina and Rita. Science 314, 449–452.

Turner, R.E., Milan, C.S., Swenson, E.M., 2006b. Recent volumetric changes in salt marsh soils. Estuarine, Coastal and Shelf Science 69, 352–359.

Turner, R.E., Howes, B.L., Teal, J.M., Milan, C.S., Swenson, E.M., Goehringer-Toner, D., 2009. Salt marshes and eutrophication: An unsustainable outcome. Limnology and Oceanography 54, 1634–1642.
Turner, R.E., 2011. Beneath the salt marsh canopy: loss of soil strength with increasing nutrient loads. Estuaries and Coasts 34, 1084–1093.
Tyler, A.C., Mastronicola, T.A., McGlathery, K.J., 2003. Nitrogen fixation and nitrogen limitation of primary production along a natural marsh chronosequence. Oecologia 136, 431–438.
Tzortziou, M., et al., 2008. Tidal marshes as a source of optically and chemically distinctive colored dissolved organic matter in the Chesapeake Bay. Limnology and Oceanography 53 (1), 148–159.
Ubben, M.S., Hanson, R.B., 1980. Tidal induced regulation of nitrogen fixation activity (C_2H_4 production) in a Georgia salt marsh. Estuarine and Coastal Marine Science 10, 445–453.
Vale, C., Caterino, F., Cortesão, C., Caçador, I., 1990. Presence of metal-rich rhizoconcretions on the roots of *Spartina maritima* from the salt marshes of the Tagus Estuary, Portugal. The Science of the Total Environment 97/98, 617–626.
Valiela, I., Teal, J.M., 1979. The nitrogen budget of a salt marsh ecosystem. Nature 280, 652–656.
Valiela, I., Teal, J.M., Persson, N.Y., 1976. Production and dynamics of experimentally enriched salt marsh vegetation: belowground biomass. Limnology and Oceanography 21, 245–252.
Valiela, I., Teal, J.M., Volkmann, S., Shafer, D., Carpenter, E.J., 1978. Nutrient and particulate fluxes in a salt marsh ecosystem: tidal exchanges and inputs by precipitation and groundwater. Limnology and Oceanography 23, 798–812.
Valiela, I., Cole, M.L., McClelland, J., Hauxwell, J., Cebrian, J., Joye, S.B., 2000. Role of salt marshes as part of coastal landscapes. In: Weinstein, M.P., Kreeger, D.A. (Eds.), Concepts and Controversy in Tidal Marsh Ecology. Kluwer Academic Publishers, Dordrecht, The Netherlands, pp. 23–38.
Van Dijk, G., Smolders, A.J., Loeb, R., Bout, A., Roelofa, J.G., et al., 2015. Salinization of coastal freshwater wetlands; effects of constant versus fluctuating salinity on sediment biogeochemistry. Biogeochemistry 126, 71–84.
Van Raalte, C.D., Valiela, I., Teal, J.M., 1976. Production of epibenthic salt marsh algae: light and nutrient limitation. Limnology and Oceanography 21, 862–872.
Van Wijnen, H.J., Bakker, J.P., 1999. Nitrogen and phosphorous limitation in a coastal barrier salt marsh: the implications for vegetation succession. Journal of Ecology 87, 265–272.
Vieillard, A.M., Fulweiler, R.W., 2012. Impacts of long-term fertilization on salt marsh tidal creek benthic nutrient and N2 gas fluxes. Marine Ecology Progress Series 471, 11–22.
Wang, S.R., Di lorio, D., Cai, W., Hopkinson, C.S., 2017. Inorganic carbon and oxygen dynamics in a marsh-dominated estuary. Limnology and Oceanography. https://doi.org/10.1002/lno.10614.
Wang, Z.A., Cai, W.-J., 2004. Carbon dioxide degassing and inorganic carbon export from a marsh-dominated estuary (the Duplin River): A marsh CO_2 pump. Limnology and Oceanography 49, 341–354.
Wang, J.Q., Zhang, X.D., Jiang, L.F., Bertness, M.D., Fang, C.M., Chen, J.K., Hara, T., Li, B., 2010. Bioturbation of burrowing crabs promotes sediment turnover and carbon and nitrogen movements in an estuarine salt marsh. Ecosystems 13 (4), 586–599.
Wang, W., Li, D.J., Zhou, J.L., Gao, L., 2011. Nutrient dynamics in pore water of tidal marshes near the Yangtze Estuary and Hangzhou Bay, China. Environmental Earth Sciences 63, 1067–1077.
Wang, Z.A., Kroeger, K.D., Ganju, N.K., Gonneea, M.E., Chu, S.N., 2016. Intertidal salt marshes as an important source of inorganic carbon to the coastal ocean. Limnology and Oceanography 61 (5), 1916–1931.
Wankel, S.D., Ziebis, W., Buchwald, C., Chawalit, N., Charoenpong, de Beer, Dirk, Dentinger, J., Xu, Z., Zengler, K., 2017. Evidence for fungal and chemodenitrification based N_2O flux from nitrogen impacted coastal sediments. Nature Communications. https://doi.org/10.1038/ncomms15595.
Wei, W., Huang, H., Biber, P., Bethel, M., 2017. Litter decomposition of *Spartina alterniflora* and *Juncus Roemerianus*: implications of climate change in salt marshes. Journal of Coastal Research 33, 372–384.
Weis, J.S., Weis, P., 2004. Metal uptake, transport and release by wetland plants: Implications for phytoremediation and restoration. Environment International 30, 685–700.
Weiss, J.V., Emerson, D., Backer, S.M., Megonigal, J.P., 2003. Enumeration of Fe(II)-oxidizing and Fe(III)-reducing bacteria in the root zone of wetland plants: Implications for a rhizosphere iron cycle. Biogeochemistry 64, 77–96.
Weiss, J.V., Emerson, D., Megonigal, J.P., 2004. Geochemical control of microbial Fe(III) reduction potential in wetlands: comparison of the rhizosphere to non-rhizosphere soil. FEMS Microbiology Ecology 48, 89–100.
Weston, N.B., Joye, S.B., 2005. Temperature-driven decoupling of key phases of organic matter degradation in marine sediments. Proceedings of the National Academy of Sciences 102, 17036–17040.

Weston, N.B., Porubsky, W.P., Samarkin, V.A., Erickson, M., Macavoy, S.E., Joye, S.B., 2006. Porewater stoichiometry of terminal metabolic products, sulfate, and dissolved organic carbon and nitrogen in estuarine intertidal creek-bank sediments. Biogeochemistry 77, 375–408.

Weston, N.B., Neubauer, S.C., Velinsky, D.J., Vile, M.A., 2014. Net ecosystem carbon exchange and the greenhouse gas balance of tidal marshes along an estuarine salinity gradient. Biogeochemistry 120 (1–3), 163–189.

Whelan, M.E., Min, D.H., Rhew, R.C., 2013. Salt marsh vegetation as a carbonyl sulfide (COS) source to the atmosphere. Atmospheric Environment 73, 131–137.

White, D.S., Howes, B.L., 1994a. Nitrogen incorporation into decomposing litter of *Spartina alterniflora*. Limnology and Oceanography 39, 133–140.

White, D.S., Howes, B.L., 1994b. Long-term ^{15}N-nitrogen retention in the vegetated sediments of a New England salt marsh. Limnology and Oceanography 39, 1878–1892.

White, K.P., Langley, J.A., Cahoon, D.R., Megonigal, J.P., 2012. C-3 and C-4 biomass allocation responses to elevated CO_2 and nitrogen: contrasting resource capture strategies. Estuaries and Coasts 35, 1028–1035.

Whiting, G., Childers, D., 1989. Subtidal advective water as a potentially important nutrient input to southeastern USA salt marsh estuaries. Estuarine, Coastal and Shelf Science 28, 417–431.

Whiting, G., McKellar, H., Spurrie, J., Wolaver, T., 1989. Nitrogen exchange between a portion of vegetated salt marsh and the adjoining creek. Limnology and Oceanography 34, 463–473.

Widdel, F., Schnell, S., Heising, S., Ehrenreich, A., Assmus, B., Schink, B., 1993. Ferrous iron oxidation by anoxygenic phototrophic bacteria. Nature 362, 834–835.

Wigand, C., Davey, E., Johnson, R., Sundberg, K., Morris, J., Kenny, P., Smith, E., Holt, M., 2015. Nutrient effects on belowground organic matter in a minerogenic salt marsh, North Inlet, SC. Estuaries and Coasts 38 (1838), 1853.

Wilbanks, E.G., Jaekel, U., Salman, V., Humphrey, P.T., Eisen, J.A., Facciotti, M.T., Orphan, V.J., 2014. Microscale sulfur cycling in the phototrophic pink berry consortia of the Sippewissett Salt Marsh. Environmental Microbiology 16 (11), 3398–3415.

Wolaver, T.G., Zieman, J., 1984. The role of tall and medium *Spartina alterniflora* zones in the processin of nutrients in tidal waters. Estuarine, Coastal and Shelf Science 19, 1–13.

Wolaver, T.G., Dame, R.F., Spurrier, J.D., Miller, A.B., 1988. Sediment exchange between a euhaline salt marsh in South Carolina and the adjacent tidal creek. Journal of Coastal Research 4, 17–26.

Wolf, A.A., Drake, B.G., Erickson, J.E., Megonigal, J.P., 2007. An oxygen-mediated positive feedback between elevated carbon dioxide and soil organic matter decomposition in a simulated anaerobic wetland. Global Change Biology 13 (9), 2036–2044.

Woodwell, G.M., Houghton, R.A., Hall, C.A.S., Whitney, D.E., Moll, R.A., Juers, D.W., 1979. The Flax Pond ecosystem study: the annual metabolism and nutrietn budgets of a salt marsh. In: Jeffries, R.L., Davy, A.J. (Eds.), Ecological Processes in Coastal Environments. Blackwell Scientific, Oxford, England, pp. 491–511.

Xie, Y., Xiao, R., Cui, Y., Chen, J., Zhang, M., et al., 2015. Distributions characteristics of phosphorus in soils of natural and restored salt marshes in the Yellow River Delta. Wetland Science 13, 6.

Zedler, J.B., 1980. Algal mat productivity: comparisons in a salt marsh. Estuaries 3, 122–131.

Zhou, J., Wu, Y., Kang, Q., Zhang, J., 2007. Spatial variations of carbon, nitrogen, phosphorous and sulfur in the salt marsh sediments of the Yangtze Estuary in China. Estuarine, Coastal and Shelf Science 71, 47–59.

CHAPTER

17

The Role of Freshwater Flows on Salt Marsh Growth and Development

Laurence A. Boorman

L A B Coastal, Cambridgeshire, United Kingdom

1. INTRODUCTION

Salt marshes are characterized by the presence of plants tolerant of both immersion in water for varying periods and some degree of salinity, although freshwater inputs can also be a significant factor in many marshes (Boorman, 2003).

Salt marshes have been shown to play a major role in many processes within the estuarine and coastal ecosystems (Boorman, 1999). There are various sources and routes of freshwater into a salt marsh. These can include river flow into the estuary, groundwater flow along a defined aquifer or channel or diffuse seepage, and also directly as a result of rainfall on the marsh and through surface flow from adjacent slopes. However, in general terms, little attention has been paid to freshwater inputs and impacts on salt marshes except from the point of view of the effects of rainfall on the acceleration of seed germination of many plant species and the effect of river flow on the overall salinity of the waterbody at particular points in an estuary (Boorman and Hazelden, 2004).

It has been shown that the tidally driven flow of saltwater can be the agent of transport to and from the marsh itself of sediment, mineral nutrients, pollutants and particulate, or dissolved organic carbon (Hazelden and Boorman, 1999). It would seem probable that where there are significant flows of freshwater in a salt marsh, a similar effect may be expected. Additionally, as excessive nutrient levels in an estuary (White et al., 2004) can affect the marsh plant communities, it would seem likely that nutrients brought in by freshwater could have a similar impact.

Salt marshes are commonly developed on fine-textured sediments with the particle size in the clay and fine silt range, and consequently the permeability of such soil might be expected to be low. An inspection of many salt marshes shows that this is a simplification of the true

Coastal Wetlands
https://doi.org/10.1016/B978-0-444-63893-9.00017-4

picture. At low tide, water is seen to seep from the sides of marsh creeks from a variety of fissures and holes in the otherwise slowly permeable marsh clays and silts. These more permeable layers can be of either physical or biological in origin. Physically, cracks and fissures develop when the soil dries out and the clay shrinks. Although these will close up on rewetting, they remain a permanent feature of the soil structure. In addition, coarse-textured horizons within the soil (perhaps sand and shell debris deposited during a storm) can give rise to more permeable layers within the soil (Boorman and Hazelden, 2005).

Permeable layers, of biological origin, can result from the residual channels left after the death and decay of roots and other underground plant material (Boorman and Hazelden, 2005). They also result from the burrowing activities of varied intertidal fauna, from crabs and mollusks through to the many different groups in the meso- and microfauna. Water movement paths are also created in the marsh soil by the burial of layers of organic matter such as are created when the autumn fall of leaf material is buried by high rates of accretion during equinoctial tides. These layers are quite persistence as they are sometimes visible a 100 mm below the surface which, in a marsh with a mean annual accretion rate of around 3 mm, represents an age of the order of 30 years.

These changing processes are all interrelated and they underline the need for a better understanding of an ever-changing coastal zone. Salt marshes are currently under increasing threats both directly from rising sea levels and indirectly from a variety of coastal developments (Goudie and Viles, 2016; Boorman and Hazelden, 2017).

2. FRESHWATER ROUTES IN SALT MARSHES

The dominant role of saline water (usually seawater) can be modified by the addition of freshwater in various ways, and the magnitudes of the extent and impact of freshwater depend largely on the routes and quantities involved (Fig. 17.1) (Boorman and Hazelden, 2005). The hydrodynamics of a marsh system will be determined by the topography of the marsh and its setting, by the variations in the porosity of the soil and the underlying strata, and by the tidal regimes of the area. However, the overall freshwater input to an area will be determined by meteorological conditions, principally the precipitation/evaporation balance. It should also be noted that elevated temperatures and low humidity will not only reduce the effective rainfall but also tend to elevate the natural salinity by increasing evaporation. Furthermore, the impact of any salinity changes will be at their greatest under favorable growing conditions, whereas they will be reduced when soil and/or atmospheric temperatures fall to a level at which seed germination and plant growth is inhibited. The details of any freshwater input to a salt marsh will, however, depend on the route involved, including direct static effects and dynamic effects through modified fluxes and exchanges. The biggest effects are generally through the long-term semipermanent flows above- or belowground level in the form of river/stream flow or the flow of groundwater.

2.1 Stream Flow

Stream flow in this section include any discrete aboveground flow of freshwater into a salt marsh; although it is very often in the form of a stream, this can range from the merest of

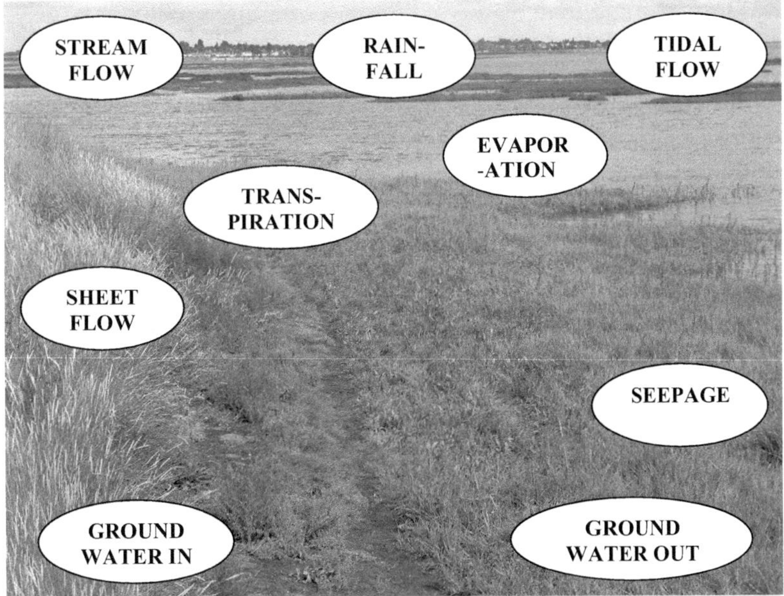

FIGURE 17.1 Water exchange routes associated with salt marshes. The background picture is of the marshes at Tollesbury, Essex, England. These are lowland estuarine marshes with a macrotidal regime (tidal range up to 5 m). The main estuary (River Blackwater) and the open sea are to the right of the picture, whereas the gradients to the terrestrial ecosystems are to the left.

trickles through to that of a major river flow into an estuary fringed with salt marshes. Whatever the size of the freshwater inflow, there are two distinct situations; the first being where the freshwater flows directly into a salt marsh system and the second where the freshwater input is less direct and through admixture with the seawater of an estuary. The distinction is thus between the situation where the freshwater impacts on the salt marsh directly and where the freshwater reduces the salinity of the inshore waters adjacent to the marsh system. In practice the distinction may be blurred not only because in many salt marshes both routes can occur but also the definition of an individual salt marsh system is variable and flexible. Even where the bank of an estuary is fringed with salt marshes, these may collectively be regarded as part of the whole or as separate discrete systems, particularly when the marshes have their own distinct outflows to the estuary (Boorman, 2003).

The impact of stream flow on a salt marsh will depend on the proportional contribution of freshwater and the resultant salinity at various points across the marsh. Particularly where the flow of freshwater is directly to the marsh, any seasonality in flow rates will affect the resultant impact with short-term high flow rates having a much smaller impact on the vegetation than the same total volume spread out over a long period. The setting of a salt marsh is the major factor determining the stream flow. Salt marshes in lowland situations, such as is found in the east and south of the Britain, tend to have limited inflows of freshwater; in estuarine situations, river flows are relatively constant and thus there is limited variability in the extent of the penetration of seawater. Very often the adjoining land is reclaimed land which was formerly salt marsh and in these cases the stream flow is largely as a result of land drainage systems in the adjoining farm land. The small flow rates involved results in there being only a limited impact, although nutrient and pollutant runoff from agricultural land can have serious environmental effects on salt marshes (Section 3). In mountainous upland situations such as those found in the west of Scotland, the salt marshes tend to be found at the head of the sea lochs and there can often be major stream flows into and across these marshes (Boorman, 2002). The flows involved can be relatively small in daily volume but with a high degree of seasonal variability. Although the catchment areas may be relatively small in area, the enhanced rainfall over high ground can result in these high short-term flow rates with their own special influence on salt marsh processes (Fig. 17.2).

The nature of a freshwater flow into and through a salt marsh can also be affected by the subsurface structure with a rather different range of impacts when there is a porous substrate as compared with marshes underlain by solid impervious rock or other impermeable substrates. The water routes created when there is a porous subsurface layer will be considered more fully in the next section. When stream flows through salt marshes have high short-term flow rates, the bed material of the stream itself will be bedrock rather than marine sediments or terrestrial alluvium as is the case when there are consistently low flow rates, which is the situation in the majority of lowland salt marshes. The pattern of exchanges in these upland marshes is markedly different from that in lowland marshes where creek water velocities are relatively low and stable (Boorman, 2002). The pattern of water flow can be further modified by the magnitude of the tidal range interacting with the relative volume of freshwater discharge (Fitzgerald et al., 2002).

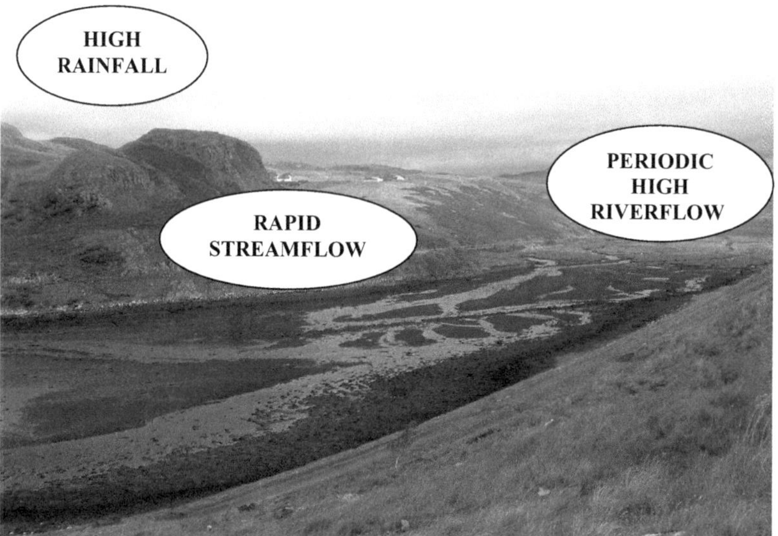

FIGURE 17.2 Fluxes associated with upland salt marshes, Loch Beag, Isle of Skye, Scotland. The Isle of Skye has extensive sea lochs with limited deposits of often rather coarse sediment in rock basins extending up to 15 km from the open sea and there are tidal ranges of around 4 m. The whole area has a high rainfall (up to 1.0 m year^{-1}), and with largely impermeable substrates, this results in high runoff with periodic flushing of the salt marsh by high volumes of freshwater.

2.2 Groundwater Flow

While stream flow over the land surface is the visible input of freshwater into salt marshes, the flow of groundwater below the soil surface may in fact have greater impact at least in some salt marshes. Two conditions need to be fulfilled before groundwater flow into salt marshes can occur: the existence of permeable layer or layers within the substrate and a sufficient water pressure to initiate the flow. The critical factor for the latter is the presence of higher ground adjacent to the salt marsh (Boorman and Hazelden, 2005). Typically this situation is found where marsh-fringed estuaries are bordered by higher ground such as that of the River Stour in north Essex, England, where local reductions in soil salinity marked by the invasion of *Phragmites australis* reveal the local upwelling of freshwater (Fig. 17.3).

Not all underground flows will be visible anywhere on the marsh surface or will indeed have any significant direct effect on the marsh. If the permeable soil layers are continuous under the marsh, freshwater may partially or totally pass under the marsh and discharge directly into the sea. In this case, the impact of nonsaline groundwater will be virtually inseparable from that of the stream flow contributing only through a limited reduction in estuarine salinity. The situation described above of estuaries being bordered by higher ground relates to the lowland situation. In upland situations the occurrence of higher ground adjacent to a salt marsh is more or less universal; however, the underlying material is very often impermeable bedrock. Even when there is a degree of subsoil permeability, the relatively high rainfall tends to result in overground stream or sheet flow being the dominant route of drainage water of terrestrial origin.

FIGURE 17.3 Scattered clumps of *Phragmites communis* are found along the upper margins of the macrotidal salt marshes on the estuary of the River Stour, Essex, England, marking the occurrence of local upwelling of fresh groundwater moving down from higher ground to the south (right-hand side).

Wilson and Gardner (2006) demonstrated the importance of tidal activity in maintaining the flow of groundwater through salt marsh sediments. They demonstrated that groundwater flow principally occurred in the creek bank and even when the marsh surface was covered by the tide. However, the porosity of the marsh sediments was an important controlling factor for groundwater fluxes. A sandy layer or layers or underlying sandy strata were shown to have a big influence in groundwater flow (Fig. 17.4). Groundwater movements were less in muddy sediments particularly with decreasing particle size but the sediment porosity was also affected by the degree of soil compression.

While the major movements of groundwater in a salt marsh are in the form of discrete groundwater flows, there are also the diffuse movements of pore water through the soil. As with major groundwater movements, the seepage of pore water is driven by pore water pressure gradients but relatively small quantities of water are involved and then over shorter distances. While the generally small particle size found in salt marsh sediments results in limited soil permeability, the occurrence of pore water movement can be seen through the way that a pit dug in the salt marsh can collect water long before it is reached by the rising tide and by the seepage visible along the creek side during low tide.

Subsurface water flow can induce complex water table dynamics particularly in relation to the distance from the nearest creek (Ursino et al., 2004). They also showed that, often, there was a persistent unsaturated zone below the soil surface, which remained aerobic, facilitating root respiration and growth. This persistent unsaturated zone was enhanced by the presence

FIGURE 17.4 Seepages of pore water during low tide from the creek bank of a salt marsh at Mill Bay, County Down, Northern Ireland. The main seepage is associated with a layer of slightly coarser sediment with a higher porosity over a more compact and less porous layer of fine sediment, but usually some seepage occurs right down the creek bank soil profile.

of plants, and moreover, the presence of pioneer plants along the marsh edges increased the availability of soil oxygen facilitating the development and growth of subsequent plant communities.

The dynamics of pore water seepage in marsh sediments has been the subject of a recent paper by Gardner (2005). He developed a mathematical model in which he showed that pore water discharge from the marsh occurred predominantly through the creek face rather than the channel bottom. Furthermore, he suggested that these exchanges took place mainly within a few meters of the creek itself and that the pore water dynamics of the creek bank environment differs markedly from that of the distal areas of the marsh. The situation is, however, likely to be different in marshes based on sandy substrates or where permeable sandy layers underlie the marsh. Less information is available regarding the movement of fresh or brackish as opposed to saline pore water in salt marshes, but this has been studied by Gardner et al. (2002) who showed that these groundwater movements were subject to change through evapotranspiration, seepage and tidal movements, as well as rainfall and surface flow (*q.v.*).

2.3 Rainfall

Atmospheric precipitation provides the input of freshwater into virtually all salt marshes, and although the water volumes are generally small compared with the saline water reserve in the marsh soil, the effect of the rainwater is emphasized through its effect being mainly limited to the surface layers of the marsh soil. The effects of rainfall on the marsh are twofold. Firstly, there is the reduction of the soil salinity, and secondly, there are the effects of the water flow itself across the surface of the marsh. The effects of salinity are mainly on seedling germination and plant growth and this will be considered under Section 3.5. While these are the most studied effects of rainfall on a salt marsh, the appearance of colored sediment-laden water running off the marsh surface into marsh creeks during periods of heavy rainfall suggests that rainfall and the resultant sheet flow can entrain and transport sediment. Mwamba and Torres (2002) have shown that the impact of water droplets rapidly mobilizes recently deposited sediments during low-tide rainstorm events facilitating subsequent transport by sheet flow of the released sediment. They concluded that although the freshwater flux over salt marshes is negligible compared with that of the tidal prism, the rainfall may facilitate the redistribution of disproportionately large volumes of sediment. The effect of rainfall appears to be indirect and direct. Tolhurst et al. (2006) showed that rain showers during low tide were correlated with a general reduction in the erosive threshold of the intertidal cohesive sediments, thus facilitating the tidal recycling of sediment. **Rainfall can also have a negative effect on salt marshes. Charles and Dukes (2009) showed that a drought could have a positive effect, markedly increasing the biomass of *Spartina* through a reduction in the waterlogging of salt marsh sediments.**

2.4 Surface Flow

The flow of seawater to and fro across a salt marsh is perhaps the defining feature of the salt marsh. Not only does the incoming tide bring the sediment necessary for marsh building but also the salinity of the seawater controls the composition of both the flora and the fauna, which together characterize the marsh itself. Section 2.3 described the flow of freshwater resulting from rainfall over the marsh surface. While both freshwater and saltwater flows across the surface of the marsh start as sheet flow, this is broken down by the marsh vegetation into discrete microchannels, which coalesce ultimately into the complexities of the system of salt marsh creeks. The wide-scale deposition of sediment across the whole salt marsh is modified by the development of the creek system which is itself the result of the role of salt marsh vegetation in determining the patterns of accretion within the marsh. Much of the sediment initially deposited is subsequently reworked by the surface flow and the channeled flow of water. Together, these are modified by the overriding influence of the salt marsh vegetation in determining patterns of accretion (Boorman et al., 1998)

3. ASSOCIATED PROCESSES

Both the runoff of freshwater from inland to the salt marsh and the flow of groundwater provide a direct route and mechanism for the transport of various materials to the salt marsh and also through the marsh to the sea. The material can be transported both in the dissolved and the suspended solid states, although unless the groundwater flows are through distinct channels, much of the suspended material is likely to be filtered off. The substances involved can include plant nutrients, dissolved organic matter, potentially toxic pollutants, and suspended sediments; in addition, the freshwater itself can have an important effect in diluting dissolved substances in the marsh pore water.

3.1 Nutrient Transport

The freshwater runoff, direct or through groundwater, from land which has been subjected to urban or agricultural development often contains elevated levels of nitrate and particulate organic nitrogen relative to marsh surface waters (Page et al., 1995). The nitrate imports from the land are utilized both for the primary production on the marsh and for being exported to the estuarine waters either as nitrate or ammonium nitrogen (Fig. 17.5).

The dynamics of pore water nutrients has not so far been studied in great detail in salt marshes, but it is clear from work on the pore water of an intertidal sand flat that nutrient concentration gradients can generate diffusive fluxes to and from the deeper sediments and that the increased oxygenation during emersion affected nitrification and nitrate reduction rates (Kuwae et al., 2003). Microbial nitrate reduction occurred in the deeper subsurface sediments and this process was supported by the downward diffusive flux of nitrate from the surface sediment. It might be expected that in the less porous salt marsh soils, similar processes might occur at a slower rate; however, even in the more porous sandy sediments both the soil water content and the levels of the water table change little during immersion suggesting that porosity was not a particularly important controlling factor.

To understand the effect of freshwater flows on the nutrient economy of salt marshes, it is necessary to consider the effects of the normal tidal fluxes affected by seawater. The fluxes of mineral nutrients between salt marsh and estuary are dependant on the external nutrient loading of the estuary, on the degree of eutrophication, and on the release of nutrients by the decay of organic matter within the salt marsh. Overall, in studies on a range of Western

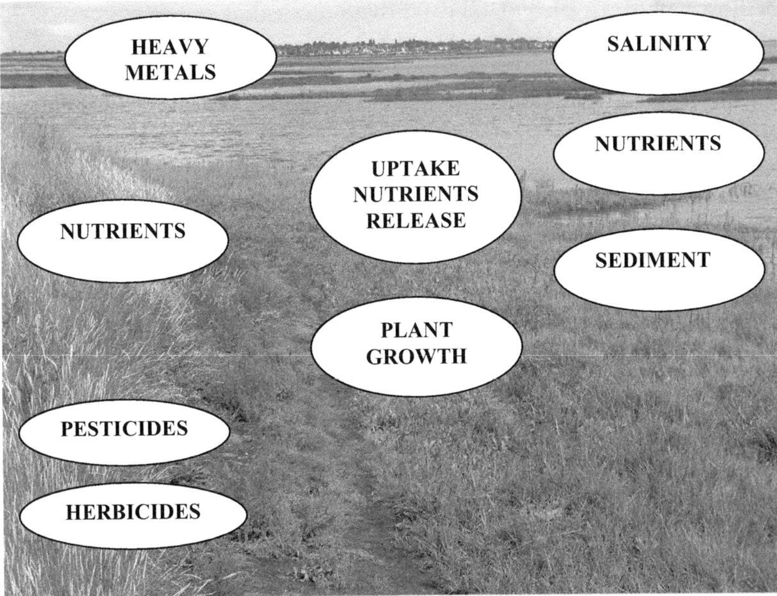

FIGURE 17.5 Exchanges of plant nutrients, pollutants, and sediments in salt marshes. The background picture is of the marshes at Tollesbury, Essex, England. The main estuary (River Blackwater) is to the right with tidal inputs of mineral nutrients and fine sediments. To the left are gradients to the terrestrial ecosystems, mainly agricultural land, where fertilizers, pesticides, and herbicides are used, whereas in the background there are some boatyards, formerly a significant source of heavy metals.

European estuaries, there appeared to be a net export of dissolved nitrogen out of the salt marshes (Boorman, 2000). The picture regarding available phosphorus was less clear, although it did appear that, at least at certain times of the year, there were net exports of phosphate. The concentrations of both nitrogen and phosphorous in salt marsh creeks will depend on the balance between the supply of that component from inside and outside the marsh and the rate of uptake of the component by the growth of salt marsh vegetation. Nitrogen is generally considered to be the critical factor for algal production in saltwater (Doering et al., 1995), but this is not always the case with higher plants. It appears that, at times, low concentrations of phosphorus can be the main limiting factor in determining the productivity of salt marsh vegetation with eutrophic estuaries enhancing vegetation growth (Boorman, 2000). While generally the release of nitrogen and phosphorus from the salt marsh occurs during the processes of the decomposition of organic matter, direct losses by the leaching of nitrogen and phosphorus, as well as carbon, from live plant tissues can also occur (Turner, 1993). The amounts released are high enough to account for significant increases in the activity of the estuarine plankton community and are thus are of potential significance for many other estuarine communities. A discharge of nutrient-rich groundwater not only affects the salt marsh but also has a significant role in the primary production of adjacent coastal waters. It is very likely that a proportion of these nutrients could feed back and enhance the productivity of the salt marsh itself (Slomp and van Cappellen, 2004).

Freshwater flows with a nutrient load equivalent to that of seawater would not significantly affect nutrient fluxes except in situations where enhanced total water volumes were involved, thus augmenting flux opportunities. However, in many cases the runoff or discharge from urban or agricultural areas is eutrophic with enhanced levels of nitrogen or phosphorus. The concentrations of these macronutrients in the groundwater flowing into a salt marsh influence the growth and development of the marsh vegetation (Wolanski et al., 2004). There is a fine line between the marginally enhanced growth of salt marsh plants in this way and the gross changes that result from the input of highly eutrophic freshwater. This can significantly affect the composition of the salt marsh vegetation through the enhanced growth of the less salt-tolerant species and the competitive exclusion of the less vigorous halophytes (Alexander and Dunton, 2006). Not all nitrogen coming into the marsh is taken up there; a proportion is exported to the estuarine waters, although often nitrogen coming into a marsh as nitrate nitrogen via the groundwater is converted to ammonium nitrogen before being exported to the estuary (Page et al., 1995). Thus the salt marsh is, in effect, a processor facilitating and transforming the exchange of nutrients between land and sea (Boorman, 2000). The changing nature of this role has been illustrated by Jickells and Andrews (2000) who showed that as the extent of salt marsh in the Humber estuary, England, was reduced by reclamation, the trapping of nitrogen and phosphorus was also reduced, and they suggest that this process could be reversed by large-scale salt marsh creation.

Mention has already been made of the serious effects groundwater pollution can have on salt marshes, but while efforts are being made to reduce the inputs of excessive nutrient levels as yet there has been little work on controlling groundwater pathways. However, in a situation where the reverse problem has occurred, that of saline intrusions in fresh groundwater, a degree of groundwater control has been achieved by modeling the discharge matrix and then by the selective drawdown through controlled pumping (Zhou et al., 2003).

3.2 Sediment Transport

Sediment can be transported equally well by freshwater as by salt, but for many marshes the freshwater inputs are small in comparison with the daily saline tidal flow and thus only limited sediment inputs are possible. Higher relative freshwater flows can be found in the loch head tidal marshes in upland areas, but the peak stream flows occur with high water velocities and most bedload and suspended sediment go straight through to the sea. The major input of freshwater sediment occurs in salt marshes associated with major river systems such as the Mississippi, United States, or the Humber, England, where the river flow is of very turbid water with a high sediment load. This is in marked contrast to other estuaries where the sediment load is mainly of marine origin. **The supply of sediment to salt marshes is a key to wetland stability and salt marsh growth. This is shown in recent modeling of sediment fluxes (Ganju et al., 2013)** It has been shown that even in cases where marsh accretion from marine sources is insufficient to compensate for rising sea levels, the introduction of river water with the associated suspended sediment load can enhance marsh accretion and stability, thus reversing the rate of wetland loss (DeLaune et al., 2003). The benefits appear to result from a combination of sediment input, lowered salinity, and enhanced levels of extractable phosphorus.

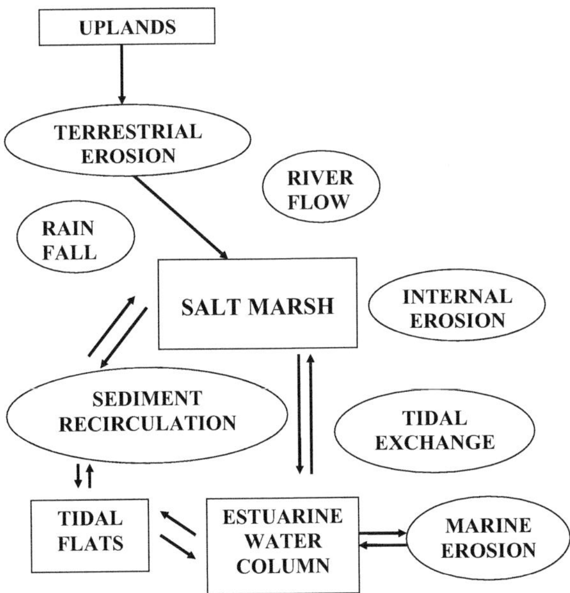

FIGURE 17.6 Sediment sources and sediment transport routes in a salt marsh. Boxes indicate locations, whereas ellipses indicate the processes involved. *Arrows* indicate the major routes for sediment exchanges. "Uplands" is a relative term implying both hill country and lowlands draining to marshes and estuaries.

Sediment transport in salt marshes is mainly thought of in terms of tidally driven sedimentary processes; however, freshwater in the form of rainfall can also make a significant contribution. It has been shown that rainfall can decrease the erosive threshold and thus remobilize recently deposited sediments (Tolhurst et al., 2006). This will further add to the recirculation of sediment brought in by the tide and initially trapped by the salt marsh vegetation (Boorman et al., 1998). The accretion of the salt marsh surface is thus affected by the combination of sediment input from the river and from the sea, by sediment recirculated from the low-level intertidal mud flats, and by internal processes in the marsh itself (Fig. 17.6).

3.3 Organic Matter Transport

In addition to the ionic transport of plant mineral nutrients, there can be significant fluxes of dissolved inorganic carbon with significant contributions from the degradation of organic carbon (Cai et al., 2003). These studies showed that the groundwater in the marsh in South Carolina is a mixture of seawater and freshwater and that the end-members are modified by the input of CO_2 from the degradation of organic matter. Furthermore, the work demonstrated that there were significant groundwater fluxes of dissolved inorganic carbon from the land to the sea via the salt marshes.

While much attention has been paid to fluxes of carbon associated with dissolved and fine particulate organic matter in the water column, at times there can also be significant fluxes of coarse floating organic matter comprising plant remains, including the seeds of salt marsh plants, lifted from the surface of the marsh by spring tides. Depending on the wind direction,

FIGURE 17.7 The wind-driven transport of floating organic matter can lead to significant accumulation of coarse organic matter along the high tide line, particularly when there are spring tides and onshore winds. This drift line of plant debris is primarily composed of dead leaves and partially decayed remains of vascular plant together with macroalgae, and it often includes the seeds of salt marsh plants. This example of the drift line was photographed at Tollesbury, Essex, England.

this material may be carried out to sea when it is dispersed with little visible effect or, with an onshore wind, it can be blown ashore at the upper margin of the marsh in the form of a visible drift line of plant debris (Fig. 17.7).

3.4 Pollutants

In addition to the loading of plant macronutrients, freshwater flowing into a salt marsh can also contain significant levels of various agricultural chemicals, including herbicides such as atrazine and organochlorine insecticides (Boorman, 2003). It is difficult to assess the impact of these chemicals on the biota as they only occur at relatively low levels and most of the available data on their specific effects on plant and animal species relate to much higher levels, but given their persistence there is at least a need for some caution.

Various industrial chemicals such as polychlorinated biphenyls also occur widely in the environment and have been detected in salt marsh sediments. It is likely that water flow of terrestrial origin is the main source of input, and given the possibility for the redistribution of these pollutants following the reworking of sediments, there would also appear to be a potential for damaging effects. This reworking of sediments can result from cyclical changes in the patterns of salt marsh growth and this process is likely to be enhanced by the impact of rising sea level with the accompanying risks of dangerous pollutants being brought back into circulation (Boorman 1999, 2003).

Various heavy metals have also been shown to occur in salt marsh sediments, including cadmium, lead, chromium, and mercury (Windham et al., 2001). Although there is no evidence that the levels recorded has any effect of these metals on the salt marsh plants, which can absorb and accumulate quite high concentrations, it is considered that plant feeders, including grazing stock, could well be deleteriously affected. Studies at the managed realignment site at Tollesbury, Essex, England, show, however, that vegetated salt marsh can develop despite the presence of metallic contaminants (Chang et al., 2001).

In climatically warmer and drier areas, there is a tendency for marshes to become hypersaline through enhanced evaporation (Hughes et al., 1998). This process can be reduced by the inflow of freshwater; however, river flows are frequently dramatically reduced to provide water supplied to urban developments and for agriculture. When attempts are made to restore river flow levels using treated waste water effluent, the degree of eutrophication can significantly affect the composition of the salt marsh vegetation through the enhanced growth of the less salt-tolerant species and the competitive exclusion of the less vigorous halophytes (Alexander and Dunton, 2006).

3.5 Salinity Changes

3.5.1 Seed Germination

While the flora of the older and higher salt marsh is composed mainly of perennial species, which set seed infrequently, the colonizing margin and the lower areas of marsh are dominated by annual species replaced by the germination and establishment of seed each year (Adam, 1990). The dispersal of the seeds of salt marsh species is primarily by the sea; nevertheless, the germination of many of the salt marsh species is facilitated by a reduction in salinity. It has been suggested that this may be of adaptive advantage to reduce losses through seeds germinating while they are still floating in seawater. The lowest areas colonized by salt marsh species are at a level not reached by the smallest of neap tides over a period of a few days; thus the surface salinity can fall if there is rainfall during this period and soil temperatures are high enough. This typically occurs in the early spring, although increasingly seed germination can occur in salt marshes during mild periods in the winter months.

The germination of the seeds of salt marsh plants is at a minimum either during high spring tides, which cover the whole marsh with fully saline seawater, or during hot dry periods in the summer, even in temperate areas. Under these circumstances, evaporation from shallow pools in the upper marsh left by the retreating tide can raise the salinity to levels considerably in excess of those in seawater resulting in the inhibition of seed germination and the formation and spread of areas devoid of vegetation through the inhibition of seed germination (Adam, 1990). This effect is even more marked in tropical and subtropical areas. However, although the reduction in soil salinity is essential for the germination and growth of most salt marsh plants, it has been shown that exposure to sea can be a prerequisite for the subsequent freshwater-enhanced germination of certain salt marsh species, notably, *Limonium* spp (Boorman, 1971).

3.5.2 Seedling Growth

Like seed germination, the growth of the seedlings themselves is inhibited by high salinities and there may even be significant seedling mortality before they become better adapted

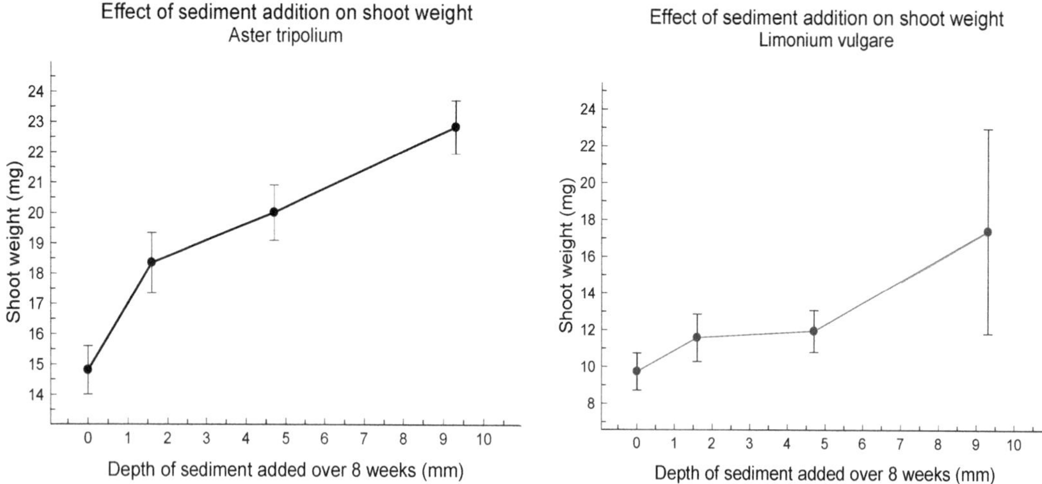

FIGURE 17.8 The effect of depth of sediment addition on the shoot weight of the pioneer species *Aster tripolium* and the mature salt marsh species *Limonium vulgare* grown in a mesocosm under simulated tidal conditions. *Partly adapted from Boorman, L.A., Hazelden, J., Boorman, M., 2001. The effect of rates of sedimentation and tidal submersion regimes on the growth of salt marsh plants. Continental Shelf Research, 21, 2155–2165.*

to withstand full exposure to seawater (Adam, 1990). It is also worth noting that the duration of immersion itself can reduce the growth rates of pioneer salt marsh species tolerant of the salinity of seawater. However, other factors are also involved. For example, the burial of the root system by the deposition of fresh sediment can enhance seedling growth of both pioneer and upper marsh species (Fig. 17.8). Clearly, there is a complex of other factors and the level of salinity, which determines the rate of seedling growth in respect of both root and shoot parameters (Boorman et al., 2001). For seedlings growing in a mobile habitat such as a pioneer salt marsh, the rate of growth, particularly root growth, can be critical to ensure that the plant can withstand tidal water movements and not be washed away (Adam, 1990).

3.5.3 *Plant Growth and Reproduction*

While the germination and seedling growth of many salt marsh plants are enhanced by a reduction in salinity, the majority of the perennial species of the more mature higher level salt marshes are little affected by fluctuations in salinity and are well able to withstand regular immersion in seawater. Many salt marsh species can grow and thrive in nonsaline soils but their ability to withstand saline conditions appears to be at the expense of reduced competitive vigor. Salt marsh plants do not occur naturally in nonsaline soils because they are unable to withstand competition from the more aggressive non–salt-tolerant plant species. It is sometimes possible to find individual salt marsh plant species growing on the landward side of sea banks, but this almost invariably indicates points of seepages of seawater through the sea bank. There is, however, a small group of perennial plant species that have a degree of salt tolerance, although less than that of the majority of salt marsh plant species. Typical of these are species such as *Phragmites australis*, *Bolboschoenus* spp., and *Schoenoplectus* spp. and these species characteristically provide a fringe between salt marsh and adjacent freshwater marshes on the landward side (Adam, 1990). Perhaps more common than this situation is one where there are visible clumps of, for example, *Phragmites australis* along the edge of the upper marshes marking localized freshwater seepages generally associated with groundwater movement from adjacent high ground (Boorman and Hazelden, 2005). Although generally such invasions are kept in check by full salinity in other parts of the marsh, there have been examples reported where the invasion of *Phragmites australis* has become a serious problem in the management of salt marshes, particularly where human activities have resulted in salinity reduction over wide areas (Bart et al., 2006).

In salt marshes in temperate, relatively damp and cool, areas, the majority of plant species are little affected by the general level of salinity. In drier and warmer areas of the world, the input of freshwater becomes of increasing importance to the salt marsh. In South Africa it has been shown that the salt marsh plants are only in active growth during the winter rainfall period (Bornman et al., 2002). During the dry season, plants are dependant for their survival on access to saline groundwater. The occurrence of winter rainfall ensures the replenishment of the saline groundwater with freshwater both decreasing the depth of the water table and reducing its salinity, thus facilitating plant growth.

4. HYDROLOGICAL IMPACTS IN SALT MARSHES

Although surface and groundwater flow can provide necessary plant nutrients, excessive nutrient loading can result in hypereutrophic conditions with major effects on the biodiversity. It has been shown that groundwater flows can cause the transport of these nutrients over considerable distances (Mayer et al., 2000). Groundwater dynamics along forest-marsh transects was studied by Gardner et al. (2002) using long-term piezometric transects coupled with salinity measurements. This study also showed that there were mechanisms by which excessive nutrient levels could be transported through to nearshore sediments with possible effects on marine habitats. This is an extreme situation and more generally salt marshes can be regarded as sinks that control the eutrophication of coastal waters by removing excessive nutrients from the system (Teal and Howes, 2000).

In addition to affecting the concentrations and fluxes of nutrients, organic matter, and sediment associated with a salt marsh, the hydrology of the marsh can also affect the physical conditions within a marsh. It has been shown that variability in evapotranspiration and tidal flooding can affect the soil volume and consequently the precise level of the surface of the marsh (Paquette et al., 2004). This effect is of primary importance in making accurate measurements of accretion/erosion in marsh development. Such measurements are crucial both in the study of salt marsh processes and in the monitoring of success in salt marsh creation. Undetected changes in marsh levels could also have significant consequences for physical and biological processes on the surface of the marsh, in particular on the patterns of seed dispersal and germination and thus the subsequent resultant patterns of plant colonization.

5. TECHNIQUES FOR THE STUDY OF MARSH HYDROLOGY

Groundwater flows, with the possibilities of their transporting nutrients over considerable distances, necessitate the use of special techniques to determine their source. In one study, involving the leakage of partially treated sewage, the molecular marker coprostanol was used to assess nutrient inputs to a marsh (Mayer et al., 2000). Radioisotopes have also been used to trace groundwater pathways. Routes and flux rates of submarine groundwater discharge in a Massachusetts salt marsh were determined using four radium isotopes (Charette et al., 2003). These workers also showed that under drought conditions seawater–sediment interactions were important in delivery of certain dissolved substances to coastal waters. In another study in North Carolina, the isotopic composition of dissolved inorganic carbon was used to define a component of the surface water–groundwater system (Gramling et al., 2003). The work demonstrated that, when precipitation was low, artesian groundwater discharge accounted for virtually all the freshwater input to the marsh while in wet periods there was a negligible groundwater contribution.

Studies continue to collect long-term real-time data on the ecohydrology of salt marshes and to develop mathematical models to interpret the various processes involved (Crowe et al., 2004). More is known about groundwater dynamics in wet coastal grasslands, enabling the prediction of changes (Mohrlok, 2002). Reeves and Fairborn (1996) installed extensive instrumentation to enable the development of a numerical model to study the groundwater dynamics of the forest-marsh interface. Crucius et al. (2005) studied groundwater flow and discharge in a small estuary using radon and salinity measurements as markers and constructed a box model. The model was used to aid the understanding of both rates and locations of discharge, and the best fit was achieved on the assumption that groundwater discharge is fresh and it occurs in the channel or adjacent areas of the marsh. The authors concluded that more data were needed for the results to be representative of long-term averages. Wilson and Gardner (2006) used numerical simulations to show the effect of tidal activity as the driver of groundwater flow and solute exchanges and underlined the importance of adequate data on sediment porosity in controlling the magnitude of these flows and exchanges. The next major step will be to integrate these various models, possibly through the use of a decision-based support system, in such a way that for any given salt marsh, the underlying ecological processes, including the magnitude and direction of the various fluxes, can be understood sufficiently to develop effective management techniques.

6. IMPLICATIONS OF FRESHWATER FLOWS FOR SALT MARSH MANAGEMENT

The most direct effect of groundwater on salt marshes is the opportunity it offers for the transport of pollutants into the salt marsh ecosystem. Salt marshes adjacent to intensively used farm land can have significant concentrations of selective herbicides (Fletcher et al., 2004). The transport was by both surface and subsurface routes. While it was not possible to demonstrate a detectable effect on the vegetation, the residual herbicide concentrations measured in this study were above the United Kingdom environmental safety guidelines.

The implications of groundwater quality for the management of salt marshes can also be inferred indirectly. Studies in Japan showed that the use of excessive fertilizer could affect the use of the water for irrigation (Fujiwara et al., 2002). Seawater intrusion into the aquifer was also shown to be having an impact on water quality, but the situation was complicated by the activity of cation exchange phenomena.

The relation of the salt marsh and freshwater flows is often seen simply in terms of a stream or river flowing to the sea, through an area of salt marsh, and measurement of the incoming river flow will thus be considered to characterize the freshwater input. However, in practice salt marsh areas often have many freshwater inputs from a number of distinct areas with very different types of land cover and land use. Such was the case in a study of salt marshes in South Carolina, USA, where, through the development of a conceptual model, it was shown that the monitoring of creek headwaters could give early warning of possible harmful effects on tidal areas with serious implications both for conservation and economically important activities (Holland et al., 2004).

7. IMPLICATIONS OF FRESHWATER FLOWS FOR SALT MARSH CREATION

The recreation of salt marshes on land, which was originally salt marsh, would in the hydrological sense seem fairly straightforward. However, there can be problems caused by the changes that will have taken place to the sediment and soils while the land has been used for agriculture or grazing (Hazelden and Boorman, 2001). The most obvious physical change is the "ripening" of the soil; this is the irreversible drying of the sediment by evapotranspiration during which the bulk density increases and porosity decreases. Soil structure, a semipermanent network of cracks throughout the soils delineating soil "peds," also develops, and the salt (NaCl) will have been leached from at least the upper layers of the soil. On some newly created salt marshes, the old agricultural soil is rapidly buried by the accumulation of new sediment, which provides a good medium for the germination and growth of salt marsh plants. However, where this does not happen, the establishment of salt marsh vegetation may be hindered in these dry, dense soils.

The physical properties of reclaimed marsh soils are little altered by the reversion of the land to salt marsh and their burial by new sediment. However, the relatively dense subsurface layer can affect subsequent creek development. Drainage patterns established on a site prior to its reversion to salt marsh will, to a great extent, control those that subsequently become established (Fig. 17.9).

It might be thought that the salinity of a salt marsh depended only on the balance between saline inundation and local rainfall. Recent studies on the northwest coasts of the Gulf of Mexico have shown, however, that the freshwater inputs also depended on the balance between local rainfall and watershed-based freshwater flows (Osland et al., 2014).

In some sites salt marsh recreation may be complicated by other factors. It has been shown that some grassland communities of saline areas are very much dependant on the upwelling of groundwater through a saline peat layer (Beyen and Meire, 2003). To compensate for the loss of such areas, it was necessary to make detailed hydrological studies, albeit on a fairly local scale, to locate the relative rare occurrence of sites suitable for this type of habitat creation.

Even when there are no such special conditions for the recreation of salt marsh, the changes in the soil hydrological regime, which occurred while the marsh was under agricultural use and no longer subject to regular tidal flooding, are considerable. The effects of the changes in tidal level were limited to small changes in the level of the underlying water table

FIGURE 17.9 A linear creek in the salt marshes at North Fambridge, Essex, England. These marshes were recreated naturally following a break in the sea wall during a major storm over a 100 years ago. It is generally considered that the unnaturally straight form of most of the creeks reflects the lines of the man-made drainage system that had been constructed while the land was under agricultural use. This pattern is still visible today, despite the intervening accumulation of more than a meter of new sediment on the former land surface.

(Blackwell et al., 2004). Consequently there were major adjustments following the return of tidal flooding. Not only was there the direct effect of the immersion in saltwater but there was also a wide range of changes in both physical and chemical soil properties. Changes in soil water table resulted in the soil environment changes from an oxidizing to a reducing environment. In the short term, there were changes in soil pH, with the topsoil water becoming markedly acid. There were also large decreases in the rates of decomposition of organic matter. All of these effects have serious implications for the establishment of salt marsh vegetation and subsequent salt marsh management.

The sustainable long-term management of created salt marshes must be a key part of any such program and there is a range of issues involved (Boorman and Hazelden, 2004). While the successful establishment of vegetation cover may only take a few years, much longer time periods are needed before anything like full ecosystem function is achieved. A recent study of the rate of ecosystem development in created *Spartina alterniflora* marshes (Craft et al., 2003) showed that while most of the functional ecological attributes have achieved equivalence to those in nearby natural marshes in 5–15 years, the levels of pools of organic carbon and nitrogen are still lower than in the natural marshes even 28 years after the marsh creation. This work involved the study of a wide range of ecological processes and this may not always be possible to achieve on economic grounds when there is extensive and widespread marsh creation.

It is important, however, to note that simpler methods of assessment may give misleading results. Studies in a range of healthy and impaired salt marshes in Louisiana showed that the state of the aboveground biomass was not a good indicator of marsh health (Turner et al.,

2004). However, the work did show that marshes under stress have a reduced belowground biomass which could be detected long before there was any detectable effect on the vegetation aboveground, thus giving the possibility of applying appropriate management techniques.

8. THE ECOHYDROLOGICAL APPROACH IN SALT MARSH STUDIES

The case for salt marsh creation to compensate for lost or degraded marshes has been well made, but at present the remedial measures suggested are considered to be inadequate to restore fully the ecological processes of a healthy robust estuary or to reinstate the full beneficial functions of the estuarine ecosystem (Wolanski et al., 2004; Boorman and Hazelden 2017). These authors consider that the successful management of estuaries and coastal waters requires an ecohydrology-based catchment-wide approach to ensure the survival of the full range of marine coastal, brackish and freshwater habitats. This will necessitate changing present practices which are based on local administrative units and on the narrowly focused approaches of managers of specific activities (including fisheries, water resources, and urban development). Without this change in thinking and in management concepts, estuaries and coastal waters will continue to degrade whatever management plans are put in place.

9. THE WAY AHEAD—PROBLEMS AND CHALLENGES

It should be clear from the preceding text that flows of freshwater can and frequently do have a distinctive role in many of the components of the ecological functioning of a salt marsh. The precise details of this do seem to vary considerably with the geological, chemical, and biological parameters of individual salt marshes. Because of this, it is often difficult to extrapolate from the results of detailed studies on specific marshes to the general situation. The implications of the various freshwater-induced effects have also to be seen both from the standpoint of a fundamental understanding of salt marsh function and from the more practical aspects of salt marsh creation and management (see Section 7 and also the Chapter on the Creation and Management of Coastal Wetlands).

The advancement of the understanding of salt marsh function depends both on the collection of good long-term sets of real-time data on the levels and fluxes of each of the significant plant nutrients and on the development of functional models of the magnitude and direction of all major marsh fluxes. The achievement of these aims also depends on the availability of adequate data on sediment porosity and potential water pathways. Quite a lot has already been revealed by the various models and simulation of salt marsh fluxes and logically the next step will be to try and integrate the various models and resolve any anomalies that may appear.

Reference has been made to the importance of acquiring adequate data sets to represent the annual variations in the levels and fluxes of important plant nutrients. However, in the times of global climatic change, data sets that were previously regarded as adequate for modeling purposes may need to be reexamined in relation to setting values for climatic extremes.

References

Adam, P., 1990. *Saltmarsh* Ecology. The University Press, Cambridge.

Alexander, H.D., Dunton, K.H., 2006. Treated wastewater effluent as an alternative freshwater source in a hyper saline salt marsh: impacts on salinity, inorganic nitrogen, and emergent vegetation. Journal of Coastal Research 22, 377–392.

Bart, D., Burdick, D., Chambers, R., Hartman, J.M., 2006. Human facilitation of *Phragmites australis* invasions in tidal marshes: a review and synthesis. Wetlands Ecology and Management 14, 53–65.

Beyen, W., Meire, P., 2003. Ecohydrology of saline grasslands: consequences for their restoration. Applied Vegetation Science 6, 153–160.

Blackwell, M.S.A., Hogan, D.V., Maltby, E., 2004. The short-term impact of managed realignment on soil environmental variables and hydrology. Estuarine, Coastal and Shelf Science 59, 678–701.

Boorman, L.A., 1971. Studies in salt marsh ecology with special reference to the genus Limonium. Journal of Ecology 59, 103–120.

Boorman, L.A., 1999. Salt marshes – present functioning and future change. Mangroves and Salt Marshes 3, 227–241.

Boorman, L.A., 2000. The functional role of salt marshes in linking land and sea. In: Sherwood, B.R., Gardiner, B.G., Harris, T. (Eds.), British Saltmarshes. Forrest Text for the Linnean Society, pp. 1–24.

Boorman, L.A., 2002. Feeding the fish - a Celtic perspective. Changing potentials for land-sea exchanges of organic matter and other materials in selected sea lochs. In: Celtic Water in a European Framework. Third Inter-celtic Colloquium on Hydrology and the Management of Water Resources. National University of Ireland, Galway, pp. 241–256.

Boorman, L.A., 2003. Salt marsh review. An overview of coastal saltmarshes, their dynamic and sensitivity characteristics for conservation and management. In: Joint Nature Conservation Committee Report No. 334. JNCC, Peterborough.

Boorman, L.A., Hazelden, J., 2004. The sustainable management of biodiversity in natural and created salt marshes. In: Green, D.R. (Ed.), *Developing Sustainable Coasts: Connecting Science and* Policy, pp. 508–513. Proceedings of Littoral 2004, Aberdeen, Scotland.

Boorman, L.A., Hazelden, J., 2017. Managed realignment: a salt marsh dilemma? Wetlands Ecology and Management 25, 387403. https://doi.org/10.1007/s11273-0174-9556-9.

Boorman, L.A., Hazelden, J., 2005. In: Herrier, J.-L., et al. (Eds.), Proceedings *'Dunes and Estuaries'* – International Conference on Nature Restoration. Practices in European Coastal Habitats, Kokzijde, Belgium, 19-23 November 2005, pp. 335–343. VLIZ Special Publication 19.

Boorman, L.A., Garbutt, A., Barratt, D., 1998. The role of vegetation in determining patterns of the accretion of salt marsh sediments. In: Black, K.S., Patterson, D.M., Cramp, A. (Eds.), Sedimentary Processes in the Intertidal Zone, vol. 139. Geological Society, London, pp. 389–399. Special Publications.

Boorman, L.A., Hazelden, J., Boorman, M., 2001. The effect of rates of sedimentation and tidal submersion regimes on the growth of salt marsh plants. Continental Shelf Research 21, 2155–2165.

Bornman, T.G., Adams, J.B., Bate, G.C., 2002. Freshwater requirements of a semi-arid supratidal and floodplain salt marsh. Estuaries 25, 1394–1405.

Cai, W.J., Wang, Y.C., Krest, J., Moore, W.S., 2003. The geochemistry of dissolved inorganic carbon in a surficial groundwater aquifer in North Inlet, South Carolina, and the carbon fluxes to the coastal ocean. Geochimica et Cosmochimica Acta 67, 631–639.

Chang, Y.H., Scrimshaw, M.D., MacLeod, C.L., Lester, J.N., 2001. Flood defence in the Blackwater Estuary, Essex, UK: the impact of sedimentological and geochemical changes on salt marsh development in the Tollesbury Managed Realignment Site. Marine Pollution Bulletin 42, 470–481.

Charles, H., Dukes, J.S., 2009. Effects of warming and altered precipitation on plant and nutrient dynamics of a New England salt marsh. Ecological Applications 19, 1758–1773.

Charette, M.A., Splivallo, R., Herbold, C., Bollinger, M.S., Moore, W.S., 2003. Salt marsh submarine groundwater discharge as traced by radium isotopes. Marine Chemistry 84, 113–121.

Craft, C.P., Megonigal, S., Broome, J., Stevenson, R., Freese, J., Cornell, L., Zheng, I., Sacco, J., 2003. The pace of ecosystem development of constructed *Spartina alterniflora* marshes. Ecological Applications 13, 1417–1432.

Crucius, J.D., Koopmans, J.F., Bratton, M.A., Charette, K., Kroeger, P., Henderson, L., Ryckman, L., Halloran, K., Colman, J.A., 2005. Submarine groundwater discharge to a small estuary estimated from radon and salinity measurements and a box model. Biogeosciences 2, 141–157.

Crowe, A.S., Shikaze, S.G., Ptacek, C.J., 2004. Numerical modeling of groundwater flow and contaminant transport to Point Pelee Marsh, Ontario, Canada. Hydrological Processes 18, 293–314.

DeLaune, R.D., Jugsujnda, A., Peterson, G.W., Patrick Jr., W.H., 2003. Impact of Mississippi River freshwater reintroduction on enhancing marsh accretionary processes in a Louisiana estuary. Estuarine, Coastal and Shelf Science 58, 6531–6662.

Doering, P.H., Oviatt, C.A., Nowicki, B.L., Klos, E.G., Reed, L.W., 1995. Phosphorus and nitrogen limitation of primary production on a simulated estuarine gradient. Marine Ecology Progress Series 124, 271–287.

Fitzgerald, D.M., Buynevich, I.V., Davis Jr., R.A., Fenster, M.S., 2002. New England tidal inlets with special reference to riverine-associated systems. Geomorphology 48, 179–208.

Fletcher, C.A., Scrimshaw, M.D., Lester, N., 2004. Transport of mecoprop from agricultural soils to an adjacent salt marsh. Marine Pollution Bulletin 48, 313–320.

Fujiwara, T., Ohtoshi, K., Tang, X., Yamabe, K., 2002. Sequential variation of groundwater quality in an agricultural area with greenhouses near the coast. Water Science and Technology 45, 53–61.

Gardner, L.R., 2005. A modeling study of the dynamics of pore water seepage from intertidal marsh sediments. Estuarine. Coastal and Shelf Science 62, 691–698.

Gardner, L.R., Reeves, H.W., Thibodeau, P.M., 2002. Groundwater dynamics along forest-marsh transects in a southeastern salt marsh, USA, Description, interpretation and challenges for numerical modeling. Wetlands Ecology and Management 10, 145–149.

Ganju, N.K., Nidzieko, N.J., Kirwan, M.L., 2013. Inferring tidal wetland stability from channel sediment fluxes, observations and a conceptual model. Journal of Geophysical Research: Earth Surface 118 (4), 2045–2058.

Goudie, A.S., Viles, H.A., 2016. Geomorphology in the Anthropocene. University Press, Cambridge.

Gramling, C.M., McCorkle, D.C., Mulligan, A.E., Woods, T.L., 2003. A carbon isotope method to quantify groundwater discharge at the land-sea interface. Limnology and Oceanography 48, 957–970.

Hazelden, J., Boorman, L.A., 1999. The role of soil and vegetation processes in the control of organic and mineral fluxes in some western European salt marshes. Journal of Coastal Research 15, 15–31.

Hazelden, J., Boorman, L.A., 2001. Soils and 'managed retreat' in south-east England. Soil Use and Management 17, 150–154.

Holland, A.F., Sanger, D.M., Gawle, C.P., Ledberg, S.B., Santiago, M.S., Riekerk, G.H.M., Zimmerman, L.E., 2004. Linkages between tidal creek ecosystems and the landscape and demographic attributes of their watersheds. Experimental Marine Biology and Ecology 28, 151–178.

Hughes, C.E., Binning, P., Willgoose, G.R., 1998. Chacterisation of the hydrology of an estuarine wetland. Journal of Hydrology 211, 34–49.

Jickells, T.J., Andrews, J.E., 2000. Nutrient fluxes through the Humber, UK estuary – past, present and future. Ambio 29, 130–135.

Kuwae, T., Kibe, E., Nakamura, Y., 2003. Effects of emersion and immersion on the pore water nutrient dynamics of an intertidal sand flat in Tokyo Bay. Estuarine, Coastal and Shelf Science 57, 929–940.

Mayer, T., Bourbonniere, R.A., Crowe, A.S., 2000. Assessment of sewage derived phosphorous input to Point Pelee marsh. International Wetlands and Remediation Conference 205–214.

Mohrlok, U., 2002. Prediction of changes in groundwater dynamics caused by relocation of river embankments. Hydrology, Earth System Science 7, 67–74.

Mwamba, M.J., Torres, R., 2002. Rainfall effects on marsh sediment redistribution, North Inlet, South Carolina, USA. Marine Geology 189, 267–287.

Osland, M.J., Enwright, N., Stagg, C.L., 2014. Freshwater availability and coastal wetland foundation species, ecological transitions along a rainfall gradient. Ecology 95, 2789–2802.

Page, H.M., Petty, R.L., Meade, D.E., 1995. Influence of watershed run-off on nutrient dynamics in a southern California salt marsh. Estuarine, Coastal and Shelf Science 41, 163–180.

Paquette, C.H., Sundberg, K.L., Boumans, R.M.J., Chmura, G.L., 2004. Changes in salt marsh surface elevation due to variability in evapo-transpiration and tidal flooding. Estuaries 27, 82–89.

Reeves, H.W., Fairborn, L.W., 1996. Application of an inverse model, SUTRAP, to a tidally driven groundwater system. IAHS Publications 237, 115–123.

Slomp, C.P., van Cappellen, P., 2004. Nutrient inputs to the coastal ocean through groundwater discharge: controls and potential impact. Journal of Hydrology 295, 64–86.

Teal, J.M., Howes, B.L., 2000. Salt marsh values: retrospection from the end of the century. In: Weinstein, M.P., Kreeger, D.A. (Eds.), Concepts and Controversies in Tidal Marsh Ecology, pp. 9–19.

Tolhurst, T.J., Friend, P.L., Watts, C., Wakefield, R., Black, K.S., Paterson, D.M., 2006. The effect of rain on the erosion threshold of cohesive intertidal sediments. Aquatic Ecology 40, 533–541.

Turner, R.E., 1993. Carbon, nitrogen and phosphorus leaching rates from *Spartina alterniflora* salt marshes. Marine Ecology Progress Series 92, 135–140.

Turner, R.E., Swenson, E.M., Milan, C.S., Lee, J.M., Oswald, T.A., 2004. Below-ground biomass in healthy and impaired salt marshes. Ecological Research 19, 29–35.

Ursino, N., Silvestri, S., Marini, M., 2004. Subsurface water flow and vegetation patterns in tidal environments. Water Resources Research 40, W05115.

White, D.L., Porter, D.E., Lewitus, A.J., 2004. Spatial and temporal analyses of water quality and phytoplankton biomass in an urbanized versus a relatively pristine salt marsh estuary. Journal of Experimental Marine Biology and Ecology 298, 255–273.

Wilson, A.M., Gardner, L.R., 2006. Tidally driven groundwater flow and solute exchange in a marsh: numerical simulations. Water Resources Research 42, W01405.

Windham, L., Weis, J.S., Weis, P., 2001. Patterns and process of mercury release from the leaves of two dominant salt marsh macrophytes, *Phragmites australis* and *Spartina alterniflora*. Estuaries 24, 787–795.

Wolanski, E., Boorman, L.A., Chicaro, L., Langlois-Saliou, E., Lara, R., Plater, A.J., Uncles, R.J., Zalewski, M., 2004. Ecohydrology as a new tool for sustainable management of estuaries and coastal waters. Wetlands Ecology and Management 12, 235–276.

Zhou, X., Chen, M., Liang, C., 2003. Optimal schemes of groundwater exploitation for the prevention of seawater intrusion in the Leizhou Peninsular in southern China. Environmental Geology 43, 978–985.

CHAPTER

18

Tidal Freshwater Wetlands

Dennis F. Whigham[1], *Andrew H. Baldwin*[2], *Aat Barendregt*[3]

[1]Smithsonian Environmental Research Center, Edgewater, MD, United States; [2]Department of Environmental Science and Technology, University of Maryland, College Park, MD, United States; [3]Utrecht University, Utrecht, The Netherlands

1. INTRODUCTION

Tidal freshwater wetlands (hereafter referred to as TFW) are the orphans of coastal wetland ecosystems. In many places, they are not recognized as a distinct type of coastal wetland, and in most developed parts of the world they have been historically heavily impacted by human activities, resulting in their destruction or degradation. One consequence of their orphan status is that the literature on the distribution and ecology of TFW is relatively scant compared to the large number of publications on saline and brackish coastal wetlands. There are, however, recent efforts to remedy this situation by compiling and summarizing the literature on TFW. First, Barendregt et al. (2006) summarized information on TFW in Europe and North America. In North America, many publications appeared in the 1980s, particularly from studies of TFW along the Atlantic coast (e.g., Odum, 1988; Odum et al., 1984 and references in Yozzo and Steineck, 1994). Another synthesis, relying heavily on the material compiled in Odum et al. (1984), for North American TFW appeared in Mitsch and Gosselink (2015). Prior to the Barendregt et al. (2006) summary, only one European review (Meire and Vincx, 1993) was available. The most recent comprehensive effort to summarize the work on TFW is an edited volume by Barendregt et al. (2009a). A book that focuses on tidal swamps of the southeastern United States, including forested TFW, was published in 2007 (Conner et al., 2007) and a comparison plant community patterns in forested tidal wetlands in the United States was published in 2014 (Duberstein et al., 2014). The overview that we present in this chapter is based on the earlier reviews cited above, information summarized in Barendregt et al. (2006) and Barendregt and Swarth (2013).

This chapter has five sections. We begin with an overview of the hydrogeomorphic settings in which TFW occur. We describe where TFW are known to occur, with the knowledge that a description of their global distribution is incomplete because a global inventory is lacking. In Section 3, we describe elements of biodiversity. TFW often have high plant species

https://doi.org/10.1016/B978-0-444-63893-9.00018-6

biodiversity and high community diversity, but few species are known to be restricted to TFW. Because of the importance of annual species, vegetation is also very often dynamic and we consider the relationship between the diversity of the seed bank and the diversity of extant vegetation in one well-studied TFW. Compared with plants, fewer studies have focused on animals in TFW. In this review, we focus on fish, birds, and mammals. Ecological processes are the theme of Section 4. Some of the highest levels of net annual primary production in temperate zone wetlands have been measured in TFW, but the level of productivity varies widely depending on geographic location and within-wetland habitat variation. TFW have also been shown to provide important water quality functions. In Section 5, we examine threats to TFW. Given their location near the upper limit of tide estuaries (i.e., also historically often the upper limit of navigation), TFW have been almost completely destroyed in some countries and the remaining areas are now viewed as being important and worthy of intervention to assure their survival or restoration. In other parts of the world, human impacts have been minimal and the major threats to TFW are increasing stresses associated with global environmental changes, such as sea level rise and intrusion of brackish water into areas that are currently tidal freshwater habitats. In Section 5, we also consider approaches that have been and are being used to conserve and restore TFW, a topic treated in detail in Baldwin et al. (2009) and in Chapter 29. We end with a prospectus.

2. HYDROGEOMORPHIC SETTING

TFW are almost always restricted to the upper limit of tide where coastal brackish water meets freshwater flow from nontidal rivers (Fig. 18.1), resulting in a tidal freshwater zone where there is bidirectional flow of freshwater. These conditions primarily exist when there is sufficient freshwater flow from a river, where there is a relatively flat and long gradient from the ocean inland, and where there is a tidal range of 0.5 m or more (Odum et al., 1984; Mitsch and Gosselink, 2015; Barendregt et al., 2006). The tidal freshwater zone probably occurs in most rivers with an appropriate geomorphic setting, but the extent of the zone would vary seasonally in response to annual rainfall patterns. For example, the tidal freshwater zone varies seasonally in estuarine systems in arid Mediterranean climates. During dry periods, freshwater flows are so low that brackish or saline water extends to the upper limit of tide. In wet periods, freshwater flows are large enough to create a tidal freshwater zone within the tidal portion of the river.

Patterns of sedimentation within tidal freshwater zones also influence the development and dynamics of TFW (Pasternack, 2009). Pasternack described two landscape positions in which TFW form: deltaic and fringe. TFW develop on dynamic deltaic deposits that form at the mouth of tidal basins where the sediment-carrying capacity of river has been exceeded. Fringing TFW occur at any location in the tidal freshwater zone where the local supply of sediment is greater than the transport capacity of the water. Studies of sediment cores from TFW demonstrate that they are a recent landform, ranging in age from a little more than 100 years to almost 4000 years (Pasternack, 2009). The influences that humans have had on sediment dynamics in the tidal freshwater zone have been especially important in recent history. Khan and Brush (1994) analyzed sediment cores from a TFW on the Patuxent River (Maryland, USA) and found that sedimentation rates between 630 and 1603 years AD

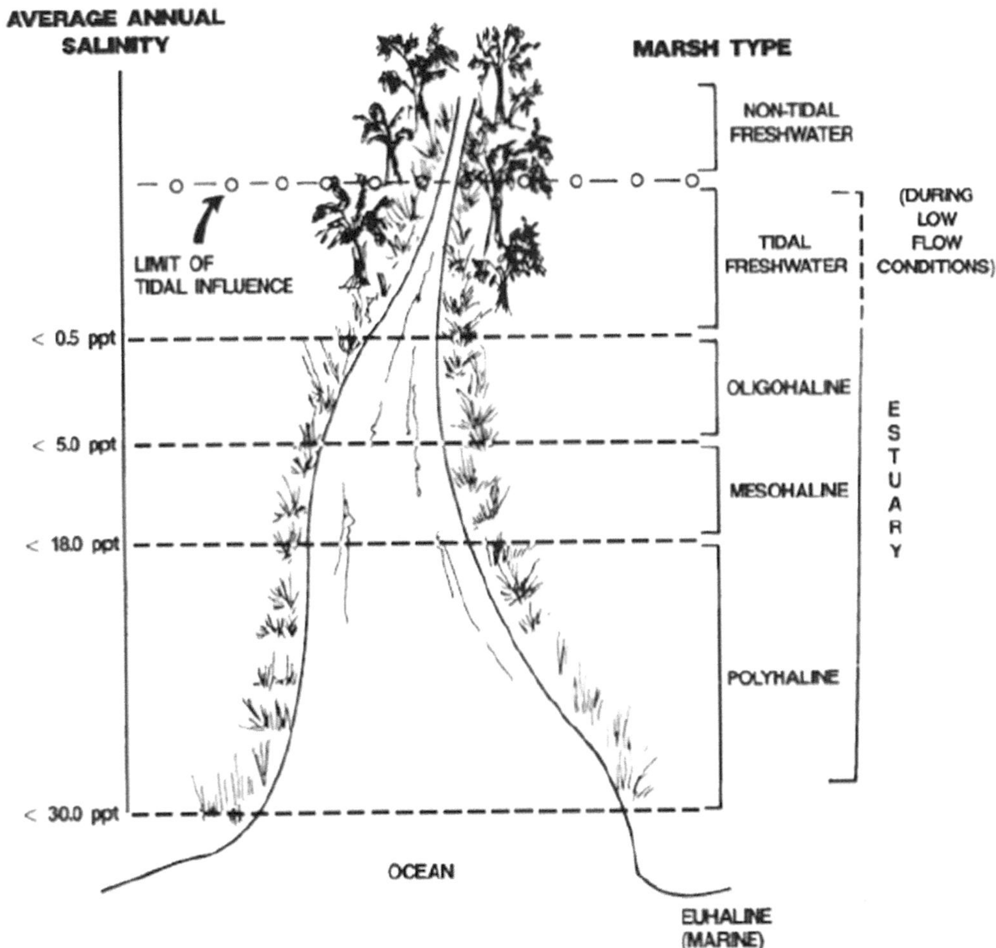

FIGURE 18.1 Distribution of wetlands along a salinity gradient from the open ocean to a nontidal river. Tidal freshwater wetlands occur in the tidal freshwater zone. *Reproduced from Odum, W.E., Smith III, T.J., Hoover, J.K., McIvor, C.C., 1984. The Ecology of Tidal Freshwater Marshes of the United States East Coast: A Community Profile. FWS OBS-83-17, US Fish and Wildlife Service, Washington, DC, 177 p.*

ranged from 0.05 to 0.08 cm year^{-1}. Rates of sedimentation increased dramatically after European settlement, which signaled the onset of land clearing. The range of sedimentation rates for periods of time between AD 1690 and 1990 varied from 0.13 cm year^{-1} between 1686 and 1694 to a high of 0.77 cm year^{-1} in 1972–73. In only one time period (1924–27) did Khan and Brush measure sedimentation rates lower than 0.40 cm year^{-1}.

There has never been a global inventory of TFW and consequently there are no estimates of their worldwide extent. In North America, TFW are abundant along the mid-Atlantic Coast from southern New England to Florida (Odum et al., 1984). With the exception of the St. Lawrence Estuary (Glooschenko et al., 1993), TFW along the New England coast are fewer and smaller in extent because there are few geomorphic settings that are appropriate for their development (Leck and Crain, 2009). TFW occur along the Gulf Coast of the United States, but they are hydrologically and geomorphically distinct from the definition for TFW given earlier. TFW on the Gulf Coast typically have a tidal amplitude that is less than 0.5 m, they occur far inland in areas where there is little slope to the land, and they are not associated with specific river systems (Sasser et al., 2009). Along the west coast of the United States, TFW are not abundant in areas where a Mediterranean climate dominates river hydrology (Leck et al., 2009). In Mediterranean climates, the tidal freshwater zone in rivers can be extensive during the rainy season, but it disappears or is very narrow during the dry season. As a result, saline and brackish waters intrude far into river systems during the dry season, and the wetlands are typically dominated by species that are associated with brackish tidal wetlands. TFW are more extensive along the larger rivers in the Pacific Northwest (e.g., Columbia) and British Columbia (Fraser), but there have been few detailed studies of TFW in those areas (Leck et al., 2009). However, these often-forested systems are spatially complex and support diverse floral and faunal communities (Johnson and Simenstad, 2015). The largest extent of tidal freshwater habitat in the United States occurs in Alaska (Hall, 2009), where acreage is probably greater than the estimate for all TFW in the lower 48 states. We assume that extensive TFW also exist in northeastern Russia where the landscape is very similar to Alaska, but we know of no assessment of TFW for Russia or any other part of eastern Asia. One of us (D.F.W.) has seen TFW dominated by herbaceous plant species on Hokkaido Island in northern Japan and forested TFW on Iriomote Island in southern (subtropical) Japan, but we are unaware of any assessments of either their extent or ecology in Japan.

TFW were historically common in northwestern Europe, but many have been destroyed during centuries of human activities (Barendregt et al., 2006) and some of the remaining areas continue to be used for cultural activities (Fig. 18.2). In The Netherlands, TFW were diked

FIGURE 18.2 Tidal freshwater wetland on the Oude Maas (The Netherlands) in winter. On the right side of the tidal stream is a coppiced stand of osier (*Salix*). In historical and modern times, managed osier beds are sources of stems used for a variety of purposes (e.g., basketry, mats used in dike construction and maintenance). *Source: A. Barendregt.*

and drained in ancient times, using some of the first techniques to control water movement (Barendregt et al., 2006). Port development and diking eliminated most of the original TFW habitats in Germany, Belgium, and England. In The Netherlands, the massive Delta project resulted in the elimination or deterioration of most of the remaining TFW, and they only remain in Belgium because the Scheldt estuary was not closed as part of the Delta project. Along the river Elbe, remnants of many TFW still occur, although they suffer from the deepening of the channel for shipping (Garniel and Mierwald, 1996). Other European estuaries, likewise, contain fragments of TFW, generally associated with depositional areas along primary river channels (Fig. 18.3). The presence and abundance of TFW in other parts of the world are poorly documented. Junk (1983) briefly described the presence of TFW on the Atlantic Coast of South America, but he offered no details on their locations or extent. Characteristics of TFW in the Río de la Plata Estuary in Argentina have been described (Kandus and Malvárez, 2004; Pratolongo et al., 2007).

FIGURE 18.3 Tidal freshwater wetlands of three European Atlantic coast estuaries: A. Elbe (Germany), B. Loire (France), and C. Minho (Spain/Portugal). Common species in these wetlands include *Phalaris arundinacea* L., *Phragmites australis*, and *Typha* sp. *Source: A.H. Baldwin.*

3. BIODIVERSITY

3.1 Plants

Plant diversity was often a topic for study by researchers of TFW on the Atlantic Coast of eastern North America, where factors associated with tides, such as increased soil aeration, combined with lack of salt water stress result in high species diversity and high primary production (the latter discussed in Section 4) (Odum et al., 1995). TFW almost always have a higher diversity of plants than brackish or saline tidal wetlands (Odum et al., 1984; Sharpe and Baldwin, 2009). TFW in Europe seem to have lower diversity than their counterparts in North America, most likely due to high rates of sedimentation and the highly eutrophic conditions in most European TFW, a condition that often results in dominance by fewer species (Barendregt et al., 2006). Although overall plant species diversity is high in TFW, diversity varies from one habitat to another, and the variation can be explained by differences

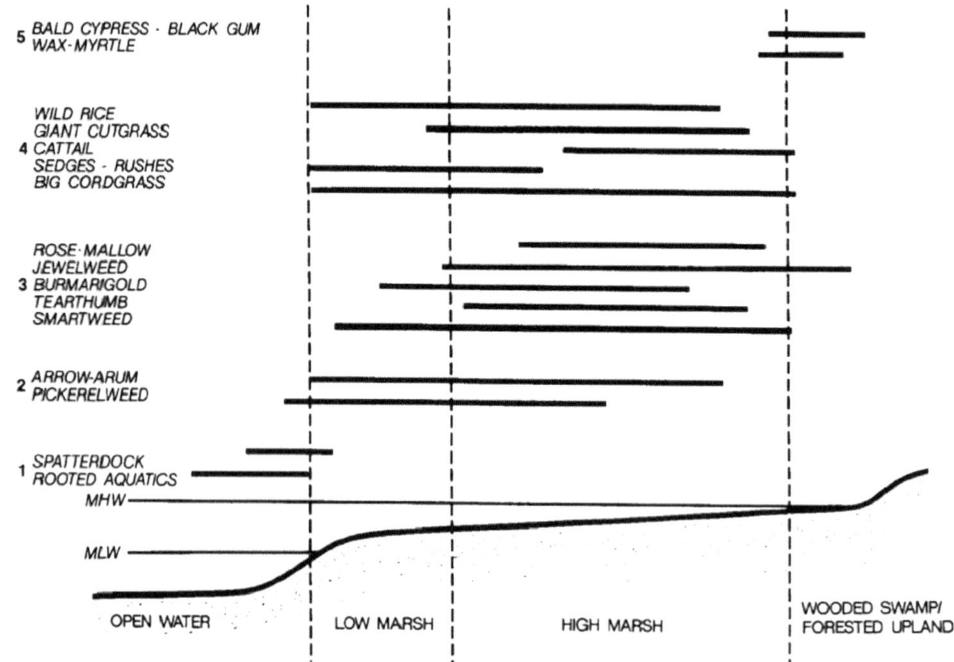

FIGURE 18.4 Cross section of a typical tidal freshwater wetland showing major habitats and distributions of species. Scientific names are as follows: bald cypress = *Taxodium distichum,* black gum = *Nyssa sylvatica,* wax myrtle = *Morella (Myrica) cerifera,* wild rice = *Zizania aquatica,* giant cutgrass = *Leersia oryzoides,* cattail = *Typha spp.,* sedges and rushes = *Carex* spp. and *Juncus* spp., big cordgrass = *Spartina cynosuroides,* rose mallow = *Hibiscus moscheutos,* jewelweed = *Impatiens capensis,* bur marigold = *Bidens laevis,* tearthumb = *Polygonum arifolium* and *Polygonum sagittatum,* smartweed = *Polygonum punctatum,* arrow arum = *Peltandra virginica,* pickerelweed = *Pontederia cordata,* spatterdock = *Nuphar lutea,* rooted aquatics = for example, *Myriophyllum spicatum, Vallisneria americana, MLW,* mean low water; *MHW,* mean high water. *Reproduced from Odum, W.E., Smith III, T.J., Hoover, J.K., McIvor, C.C., 1984. The Ecology of Tidal Freshwater Marshes of the United States East Coast: A Community Profile. FWS OBS-83-17, US Fish and Wildlife Service, Washington, DC, 177 p.*

in the relationship between habitat setting and hydrology. A typical cross section through a TFW in the United States is shown in Fig. 18.4, and similar zonation patterns occur in European TFW (Barendregt, 2005; Barendregt et al., 2006). Vegetation in the open water, low marsh, and high marsh habitats is dominated by herbaceous species with diversity increasing from the open water to high marsh habitats (Simpson et al., 1983a; Engels and Jensen, 2009). At the upper extreme of tide, TFW are often dominated by woody species with areas being dominated by shrubs or by trees (Barendregt et al., 2006; Conner et al., 2007; Leck et al., 2009). Extensive lists of species for TFW vegetation can be found in Leck et al. (2009) and Odum et al. (1984). In general, subtidal habitats and low marsh areas that are exposed briefly at low tide (Fig. 18.5) are dominated by species with relatively large leaves that are held above the water (e.g., *Nuphar lutea* (L.) Sm., *Peltandra virginica* (L.) Schott, and *Pontederia cordata* L. in North America). In northwestern Europe, most open water systems have no aquatic plants due to the high sedimentation rates and eutrophication. Low marshes in northwestern Europe with extensive mudflats that are flooded twice a day are dominated at the upper border by *Schoenoplectus triqueter* (L.), *Schoenoplectus lacustris* (L.) Palla, and *Bolboschoenus maritimus* (L.) Palla. The low marsh in the United States has many of the same species that occur in open water areas but it is also the habitat in which *Zizania aquatica* L. and *Polygonum punctatum* Elliott are often abundant. The creek bank associated with the low marsh commonly has several low-growing species (*Callitriche heterophylla* Pursh, *Gratiola neglecta* Torrey, *Lindernia dubia* (L.) Pennell, *Ludwigia palustris* (L.) Ell.) that form ground covers. Fig. 18.6 shows the transition zone between a low marsh and a high marsh habitat along the Nanticoke River (Maryland, USA). The high marsh habitat has the highest species diversity in both the United States and Europe. In the United States, high marsh habitats consist of a diversity of annual (e.g., *Ambrosia trifida* L., *Bidens laevis* (L.) BSP, *Impatiens capensis* Meerb., *Pilea pumila* (L.) A. Gray, *Polygonum arifolium* L., *Polygonum sagittatum* L.) and perennial

FIGURE 18.5 Mudflat with *Schoenoplectus lacustris* in the Elbe river (Germany). *Source: A. Barendregt.*

FIGURE 18.6 Low marsh to high marsh transition on the Nanticoke River, Delaware (USA). The dominant species in the low marsh (left side of photograph) is *Nuphar lutea*. The dominant species in the high marsh (right side of photograph) is *Acorus calamus*. In the high marsh annual species become dominant toward the end of the growing season. *Source: A.H. Baldwin.*

(*Acorus calamus, Leersia oryzoides* (L), Swartz, *Peltandra virginica, Typha* spp.) species. The lowest diversity on the high marsh occurs when clonal perennials form dense patches in which few other species become established. Examples of patch-forming perennials are species of *Typha* and *Phragmites australis* (Cav.) Trin. ex Steudel. In Europe, perennials (e.g., *Lythrum, Phalaris, Epilobium, Typha, Symphytum, Valeriana, Sparganium*) dominate high marsh habitats (Barendregt et al., 2006).

TFW habitats dominated by shrubs and trees also can also have high species diversity (Peterson and Baldwin, 2004; Baldwin, 2007) because, in addition to trees and shrubs and a few herbs that rarely occur in more open habitats (e.g., *Osmunda regalis* var. *spectabilis* (Willd.) Gray), they contain many of the herbaceous species that occur on the high marsh. Examples of shrub and tree species (e.g., *Acer rubrum* L., *Viburnum dentatum* L., *Fraxinus pennsylvanica* Marsh, *Picea sitchensis* (Bong.) Carrière, *Alnus rubra* Bong.) in the United States (Figs. 18.7 and 18.8) and Europe can be found in Barendregt et al. (2006), Odum et al. (1984), Rheinhardt (1992), and Barendregt (2005). Living and fallen trees and shrubs are often the focal points for the development of mounds or hummocks (Duberstein and Conner, 2009) that are the preferred habitat for a variety of herbs that are less tolerant of flooding, including species of *Carex*, grasses (*Cinna arundinacea* L.), ferns (*Osmunda cinnamomea* L., *O. regalis, Thelypteris palustris* Schott), and *Viola cucullata* Aiton (Rheinhardt, 1992; Leck et al., 2009). Species richness and the diversity of wetland types associated with tidal forests can also vary along estuarine salinity gradients (Johnson and Simenstad, 2015).

FIGURE 18.7 Forested tidal freshwater wetland on the Nanticoke River, Maryland (USA). Note relatively open forest canopy and diverse assemblage of herbaceous plants in the understory. *Source: A.H. Baldwin.*

FIGURE 18.8 Tidal freshwater forested wetlands dominated by Sitka spruce, *Picea sitchensis*, on the Columbia River, Oregon, USA. *Source: A.H. Baldwin.*

Almost all plant species in TFW also occur in non-TFW. In the United States, only one TFW plant species (*Aeschynomene virginica* (L.) BSP) has been listed as endangered (Griffith and Forseth, 2003). In Europe, almost all TFW plant species also occur in other types of freshwater wetlands, and there are few species that have been identified as threatened or endangered. In TFW of the Elbe estuary close to Hamburg, there are two endemic species, *Oenanthe conioides* Lange and *Deschampsia wibeliana* (Sond.) Parl (Burkart, 2001). Both species are listed in Germany and incorporated into the EU Habitats Directive to preserve the species. In TFW in The Netherlands and Belgium, a variety of *Caltha palustris* L. (var. *araneosa*) occurs that produces roots on the nodes below the flower that after the breaking of the stem can be transported by the tides permitting dispersal to almost all TFW in the region (van Steenis, 1971). An endangered European species that occurs in low marsh habitats that experience some erosion is *Schoenoplectus triqueter* (Deegan and Harrington, 2004). This species is distributed from the Elbe in Germany to the Gironde in the south of France.

There have been few studies of rarity in TFW. However, long-term studies in New Jersey (USA), based on monitoring of seed bank and vegetation in a created TFW in the Delaware River that is adjacent to a natural TFW, provide insights into the dynamics of TFW vegetation including rare species (Leck et al., 1988; Leck and Leck, 2005 and references cited therein). In 1988, 426 species were reported in the study area (Leck et al., 1988); by 2005, the number had increased to 875 with a number of rare (29) and endangered (8) species mostly from the created wetland (Leck and Leck, 2005). The increased number of taxa was the result of continued exploration of the study area, disturbances of the natural wetlands due to road construction, and inclusion of vegetation in upland habitats within the marsh complex. From a wetland perspective, one of the most interesting results was the number of rare species that appeared in the constructed wetlands attributed to the availability of new substrates for colonization. Over a 5-year period, 177 species emerged from soil seed bank samples from the constructed wetland, compared with 96 species from soils in the natural wetland over more than 15 years. Eighty-three of these species only occurred in soils in the constructed wetland, an indication of the potential for dispersion of rare species within the tidal freshwater zone of the river. In both the constructed and natural wetlands, the number of established plant species was much lower than the number of species that emerged as seedlings from the soil samples. Leck and Leck (2005) suggested that the differences were due, in part, to the absence of suitable field germination sites in both types of wetlands. The presence of a relatively high number of rare and endangered species at the constructed wetland, which was only one small portion of a larger tidal freshwater zone in the Delaware River, suggests the importance of maintaining a diversity of TFW habitats to assure the persistence of a diverse flora, especially species requiring open habitats with limited competition.

3.2 Animals

Animals associated with TFW have received less attention than flowering plants, as have other groups of plants (e.g., algae, bryophytes, ferns) as well as fungi and microorganisms. Much of the information on animals is adapted from Barendregt et al. (2009b), Odum et al. (1984), Mitsch and Gosselink (2015), and Swarth and Kiviat (2009), and we focus on three groups of animals: fish, mammals, and birds. In general, benthic invertebrates may be less diverse in TFW compared with brackish and saline tidal wetlands, but the diversity of terrestrial invertebrates is higher (Barbour and Kiviat, 1986; Ysebaert et al., 1998, 2003; Barendregt et al., 2006).

Similar to plants, few animals are restricted to TFW, but beyond species identification, few animal groups have been examined in detail. Many animals that occur in TFW are wide ranging and also are common in brackish and saline wetlands or in non-TFW (Odum et al., 1984; Mitsch and Gosselink, 2015). Examples of wide-ranging fish, mammals, and birds in TFW in the United States are the yellow perch (*Perca flavescens* Mitchill), the predaceous river otter (*Lutra canadensis* Schreber), and herbivorous mammals such as the common muskrat (*Ondatra zibethicus* L.) and beaver (*Castor canadensis* Kuhl). Examples of widespread bird species are the great blue heron (*Ardea herodias* L.) and osprey (*Pandion haliaetus* L.).

Odum et al. (1984) listed 125 fish species for TFW, but only 59 were regular components of the fish community. The families with the greatest number of species were the Cyprinidae, Centrarchidae, and Ictaluridae. The fish fauna of TFW includes nonnative species such as the common carp (*Cyprinus carpio* L.). Cyprinid species of killifish (e.g., *Fundulus heteroclitus* L., *F. diaphanous* Lesueur) are examples of abundant forage fish.

Several fish species that are commercially important spawn in the tidal freshwater zone or as juveniles forage in that zone. Striped bass (*Morone saxatilis* Walbaum), yellow perch (*Perca flavescens* Mitchill), and American shad (*Alosa sapidissima* Wilson) are all abundant at one or more life history stages. Striped bass and American shad spawn in tidal freshwater zone. Yellow perch are protamodromous, migrating only within coastal rivers. They spawn in nontidal freshwater portions of rivers but larvae, juveniles, and adults forage in the tidal freshwater zone (Piavis, 1991).

The fish community has been described for many estuaries in Europe (Elliot and Dewailly, 1995), including the Minho, Lima, and Gironde (Lobry et al., 2003), Loire, Scheldt (Maes et al., 1998), Rhine, Meuse, and Elbe (Thiel and Potter, 2001), and Forth and Tyne (Pomfret et al., 1991). Similar to North America, some marine species that enter the estuary migrate through the tidal freshwater zone on a seasonal basis, either as adults or juveniles. Freshwater fish that occur in European TFW habitats also occur in nontidal freshwater habitats. Diadromous fish (anadromous and catadromous) that spend part of their life cycle at sea and part in nontidal portions of rivers use TFW habitats during migrations, and a few species, for example, Allis shad (*Alosa alosa* L.) and Twait shad (*Alosa fallax* Lace'pe`de), are protected at a European level, since they are listed in the EU Habitats Directive.

Odum et al. (1984) listed 10 mammals that are common in TFW. The most obvious mammals are the species that have visual impacts on the vegetation. The common muskrat builds lodges (up to 2 m high and 1–3 m wide) that are composed mostly of mounds of vegetation (Fig. 18.9). They also construct feeding stations (Fig. 18.9) that are not as large as lodges but are also distinct features within the vegetation mosaic. One consequence of lodge and feeding station construction is that muskrats apparently harvest more aboveground biomass than belowground biomass even though rhizomes of several species are preferred food (Lynch et al., 1947). Muskrats, however, appear to have little impact on plant diversity, but feeding activities alter soil nitrogen dynamics (Connors et al., 1999).

Beavers, eradicated throughout much of the Atlantic coast of the United States, have made a remarkable recovery in recent decades and are now common in TFW where they build lodges and consume large amounts of woody biomass. In some situations, beaver lodges occur on the high marsh, but they are most often found in areas where dams have been placed across shallow tidal areas, often located near food sources (Fig. 18.10). In addition to the larger mammals, some smaller species (e.g., marsh rice rat (*Oryzomys palustris* Harlan))

FIGURE 18.9 Muskrat lodges at the Jug Bay National Estuarine Reserve on the Patuxent River, Maryland (USA). Several lodges can be seen in the photograph (dark mounds), as well as a lower feeding station to the left of the lodge in the foreground. Two people standing in the marsh provide scale. *Source: A.H. Baldwin.*

FIGURE 18.10 Beaver dam across a tidal freshwater creek at the Jug Bay Wetlands Sanctuary on the Patuxent River in Maryland (USA). The dam can be seen running diagonally across the photograph from right to left in the foreground; the low marsh plant *Nuphar lutea* is visible in the background on the far side of the small pond created by the dam. In the foreground, to the right of the dam is a tidal creek and associated tidal freshwater wetland. To the left of the dam, the wetlands no longer experience any significant tidal influence. *Source: A. Barendregt.*

can impact vegetation through their feeding activities. One of the authors (D.F.W.) has observed marsh rice rats consuming seedlings and juveniles of wild rice on numerous occasions, to the point where population size was reduced on a small scale.

Birds are also conspicuous components of TFW. Species may nest in TFW vegetation, forage on vegetation, or hunt animal prey. The most common birds that nest in TFW vegetation on the Atlantic Coast are the least bittern (*Ixobrychus exilis* Gmelin), Canada goose (*Branta canadensis* L.), Virginia rail (*Rallus limicola* Vieillot), king rail (*Rallus elegans* Audubon), marsh wren (*Cistothorus palustris* Wilson), common yellowthroat (*Geothlypis trichas* L.), and red-winged blackbird (*Agelaius phoeniceus* L.).

In North America, the American black duck (*Anas rubripes* Brewster) is the most abundant waterfowl species, especially in the winter (Swarth and Burke, 2000), but many other species of waterfowl use TFW for resting and feeding during migration. Gulls are often abundant throughout the year. Large numbers of gulls, *Larus argentatus* (Pontoppidan, 1763, Denmark), *Larus atricilla* (Linnaeus, 1758, Bahamas), *Larus delawarensis* (Ord, 1815; Philadelphia) congregate in TFW at low tide (Wondolowski, 2001), and Chris Swarth (personal communication) has observed up to 12,000 *Larus atricilla* resting in Maryland (USA) TFW prior to continuing on to summer breeding grounds.

TFW are used by a large number of wintering songbirds that roost individually or in small-to-large flocks and forage activity in all types of habitats. Red-winged blackbirds and common grackles (*Quiscalus quiscula* L.) are two species that congregate in enormous flocks, often roosting in tall emergent vegetation (Meanley, 1965 as cited in Swarth and Kiviat, 2009). Red-winged blackbirds and bobolink (*Dolichonyx oryzivorus* L.) specialize on wild rice seeds (Fig. 18.11) in the late summer (Meanley, 1965 as cited in Swarth and Kiviat, 2009).

The European TFW are rich in bird species, with nesting birds in the reed beds, marshlands, and tidal forests during summer and additional migrating birds in the winter season (Ysebaert et al., 2000; Barendregt et al., 2006). Ducks and waders especially in the winter period are important. In some locations, their numbers are so great that TFW are of extreme conservation value, supporting 1% of the world population for many species.

4. PRIMARY PRODUCTION AND NUTRIENT CYCLING

TFW are one of the most productive types of wetlands in the temperate zone, but the level of biomass production varies among species and habitats, with a range of approximately 400–2500 g m^{-2} for aboveground biomass (Whigham et al., 1978; Odum et al., 1984; Mitsch and Gosselink, 2015; Barendregt et al., 2006). Open water and low marsh habitats are also less productive because those sites are inundated for longer periods compared with high marsh and shrub- and tree-dominated habitats. The most productive habitat appears to be the high marsh (Neubauer et al., 2000) where annual net biomass production of more than 3,000 g m^{-2} has been measured for individual species (e.g., Sickels and Simpson, 1985).

An interesting feature of many high marsh habitats is that there is less annual variation in aboveground production compared to brackish and saline tidal wetlands. Whigham and Simpson (1992) reported results from an 11-year study of a TFW in the Delaware River estuary. TFW had a lower coefficient of variation in annual production compared with brackish

FIGURE 18.11 The annual grass wild rice, Zizania aquatica, with inflorescences. *Source: A.H. Baldwin (shown in the picture).*

and saline tidal wetlands. They suggested that the low annual variation was due to the three factors. Plants in TFW are not stressed by salinity, nutrient levels are high in TFW because most of them are located near urban and suburban areas with high nutrient loading rates, and they have a high diversity of annual species. The high diversity of annual species allows for compensation among species resulting in fairly constant levels of biomass production even though the abundance and growth of one or more species may vary considerably from year to year. This feature of TFW appears to be unique among tidal wetland ecosystems.

Most of the in situ organic matter produced by plants in TFW flows through the detritus food chain, and leaves of most species have high decomposition rates (Odum and Heywood, 1978; Findlay et al., 1990). Internal cycling of nutrients seems to be sufficient to support the high rates of primary production (Morris and Lajtha, 1986; Bowden et al., 1991) and experiments to test the hypothesis that production is nitrogen limited did not result in an increase in aboveground biomass, an indication of the relatively high N status of many TFW (Chambers and Fourqurean, 1990; Bowden et al., 1991; Morse et al., 2004). The response of annual and perennial species appeared to be different for nitrogen and phosphorus; especially the annual species are susceptible to P-limitation (Baldwin, 2013).

Sediment deposition is also an important source of nutrients in TFW (Orson et al., 1990; Darke and Megonigal, 2003; Morse et al., 2004; Pasternack, 2009), and sediment inputs may enable surface elevations in TFW to keep pace with an accelerated rate of sea level rise. Because they are accreting environments, TFW substrates also accumulate heavy metals (Khan and Brush, 1994), resulting in elevated concentrations in plant tissues (Simpson et al., 1983b).

High rates of primary production in many habitats and high rates of sedimentation both indicate that TFW would be net sinks for nutrients. The few nutrient-budget studies that have been conducted on TFW suggest that there is a net accumulation of nutrients during the growing season and a net release of nutrients during the fall and winter months (Simpson et al., 1978, 1983b). Their primary contribution to coastal estuarine systems seems to be as sites for nutrient transformation and export, with particulate forms of nutrients dominating flood tides and dissolved nutrients dominating ebb tides (Odum et al., 1984; Bowden et al., 1991; Zhang et al., 2014; Lehman et al., 2015). Bowden et al. (1991) concluded that the nitrogen budget of a TFW in Massachusetts (USA) was "largely independent of the nitrogen budget of the river." Nitrate retention appeared to be a function of the water volume. It is controlled by the hydrological transport and not limited by the nitrogen input (Seldomridge and Prestegaard, 2014). In Europe, the TFW appeared to be the essential link between the rivers and the estuaries, where nutrients and suspended matter are transformed into detritus. The silica cycle appeared to be especially important in TFW (Barendregt et al., 2006).

5. THREATS AND FUTURE PROSPECTUS

Barendregt et al. (2006) described the fate of many TFW in northwest Europe and the Atlantic coast of the United States. The location of TFW near the upper limit of tide in major river systems resulted in their destruction, especially in European estuaries, as cities and associated port facilities developed. A small-scale example of the long-term effects of human activities in the United States can be observed in the Anacostia River, a tributary of the Potomac River within the city of Washington, DC. Most of the original 1000 ha of TFW have been destroyed by dredging and filling, and the sites that were not destroyed are now highly degraded (Baldwin, 2004). Ongoing efforts are currently directed toward protection and restoration of TFW on the Anacostia (Baldwin, 2004, Chapter 29). An important component of the restoration activity is a watershed-level effort to improve water quality. The responses of existing TFW to improvements in water quality will be interesting to document because in recent history, TFW typically occur in areas that are rich in nutrients and sediments. Improvement in water quality will result in a decrease in nutrients and a reduction in sediment inputs. These changes are likely to result in shifts in species abundances within TFW habitats (Baldwin, 2013).

In other parts of the United States, different activities were responsible for the historical losses of TFW. In New England, the placement of dams near the upper limit of tide was responsible for the losses of TFW (Leck and Crain, 2009). In South Carolina, large areas of TFW were diked and converted into rice fields, and only recently have there been efforts to restore them to their original condition (Whigham et al., 2009). Diking and filling were also responsible for losses in the Sacramento–San Joaquin river Delta (California, USA), and restoration efforts are underway to restore the important ecological functions associated with the TFWs (Jassby and Cloern, 2000; Hammersmark et al., 2005); however, loss of organic matter due to aeration has lowered substrate levels exacerbating flooding and negatively affecting restoration efforts.

In the United States, national and state regulations have resulted in the protection of most coastal wetlands, and wetland losses in the coastal zone have been reduced dramatically (Dahl, 2006), but degradation continues. Restoration of TFW in The Netherlands and Belgium, where there had been significant historical losses, is currently under consideration (Barendregt et al., 2006). Technical procedures for restoration of TFW in Europe are well established, and the ecosystems become well established within a few years when the conditions are optimal (e.g., Zonneveld, 1999; Beauchard et al., 2013). A range of restoration projects are planned or even in the implementation phase (Storm et al., 2005; Van den Bergh et al., 2005). However, in North America a suite of restoration techniques have been attempted, with varying degrees of success in establishing ecosystem structure and function comparable to undisturbed TFW (Baldwin et al., 2009; Sloey and Hester, 2016).

On a global scale, as indicated in Section 2, there are undoubtedly large TFW areas that have not been heavily impacted by human activities. The extensive TFW that exist in Alaska, for example, do not face any immediate threat. Similar conditions probably prevail in other northern areas (e.g., Siberia) where human impacts have been minimal. In those areas, the greatest threats are undoubtedly associated with the consequences of global climate change. In Alaska, increasing temperatures are causing glaciers to melt at a faster rate, resulting in increased sediment input to coastal estuaries. The long-term impacts of increased sediment loading are unknown, and the consequences can be either positive or negative. Increased sediment input will enable TFW to increase their relative surface elevation and thus keep pace with rising sea levels. Too much sediment, however, can result in negative impacts of vegetation. In the Kenilworth Marsh in Washington, DC, for example, sediment was placed at a higher elevation in one of the cells that was constructed for restoration purposes. The cell became dominated by invasive species as a result of the higher surface elevation in the cell.

The consequences from global climate change will affect TFW in many ways. Sea water rise will impact hydrology, and salinity will increase upstream. The major effects are described by Neubauer and Craft (2009) and Craft et al. (2009). Accretion will change, but other studies indicate that the accretion rates will locally counteract the sea level rise (Butzeck et al., 2015). However, marshes across tidal freshwater as well as oligohaline and brackish zones of a Chesapeake Bay sub-estuary are not building elevation at a sufficient rate to keep pace with rising relative sea level, despite high levels of surface accretion (Beckett et al., 2016). Some tidal freshwater forested wetlands are also not keeping up with sea level rise, although some of these may convert into marshes (Craft, 2012). Carbon pools will be impacted by higher salinity for the oxidation of sulfate that will reduce the carbon stocks (Weston et al., 2006, 2011, 2014; Loomis and Craft, 2011), but Neubauer et al. (2013) indicate that different contrary processes with CO_2 and NH_4 production might be present. Hansen et al. (2016) indicate differences in carbon storages with salinity levels and introduce the suggestion that the decreased aboveground biomass in TFW will affect the balance. Connected with the storage and mineralization of organic matter is the availability of nitrogen and phosphorus that can be stimulated with increasing salinity (Noe et al., 2013). Another effect of global change is that the temperature will rise; this appeared to have negative effect on the species numbers (Baldwin et al., 2014).

Threats from invasive species also have the potential to alter the dynamics of TFW. One example is the expansion of a European haplotype of common reed (*Phragmites australis*). The primary threat is that the nonnative haplotype has the potential to dominate sites and convert them from species-rich assemblages into monocultures (Hazelton et al., 2015). In addition to altering diversity patterns, there may also be impacts of nutrient dynamics that result from efforts to remove the invasive form of *Phragmites* (Alldred et al., 2016). Another example is the devastation of forested TFW dominated by ashes (*Fraxinus* spp.) on the US Atlantic coast by the emerald ash borer (*Agrilus planipennis* Fairmaire), an Asian phloem-feeding beetle introduced inadvertently in the 1990s (Fig. 18.12; Flower et al., 2013). First detected in Michigan in 2002, emerald ash borer has now spread to all US Atlantic Coast states (Emerald Ash Borer Information Network, www.emeraldashborer.info, accessed 26 November 2016), where it is likely to soon kill trees in vast areas of TFWs containing *Fraxinus pennsylvanica* Marshall (green ash) and *Fraxinus profunda* (Bush) Bush (pumpkin ash), altering subcanopy species composition and ecosystem function (Flower et al., 2013).

FIGURE 18.12 Tidal freshwater forested wetland on the Patuxent River, Maryland, USA where the dominant canopy species, *Fraxinus pennsylvanica* (green ash), succumbed to infestation by emerald ash borer in 2016. With the tree canopy almost entirely gone, the understory was colonized rapidly by *Bidens* spp. (yellow flowers visible). *Source: A.H. Baldwin.*

Ongoing efforts to protect and restore TFW present a paradox against the backdrop of the potential effects associated with global climate change (e.g., Tabak et al., 2016). Salinity encroachment into TFW not only threatens species composition but also influences patterns of nutrient cycling (e.g., Gao et al., 2014). We strongly recommend that these efforts around the world be undertaken in the context of the dynamic location of TFW within the coastal zone. Effective conservation, restoration, and management will require vigilance and commitment by governmental and nongovernmental organizations.

References

Alldred, M., Baines, S.B., Findlay, S., 2016. Effects of invasive-plant management on nitrogen-removal services in freshwater tidal marshes. PLoS One 11 (2), e0149813. https://doi.org/10.1371/journal.pone.0149813.

Baldwin, A.H., 2004. Restoring complex vegetation in urban settings: the case of tidal freshwater marshes. Urban Ecosystems 7, 125–137.

Baldwin, A.H., 2007. Vegetation and seed bank studies of salt-pulsed swamps of the Nanticoke River, Chesapeake Bay. In: Conner, W.H., Doyle, T.W., Krauss, K.W. (Eds.), Ecology of Tidal Freshwater Forested Wetlands of the Southeastern United States. Springer, Dordrecht, The Netherlands.

Baldwin, A.H., 2013. Nitrogen and phosphorus differentially affect annual and perennial plants in tidal freshwater and oligohaline wetlands. Estuaries and Coasts 36, 547–558.

Baldwin, A.H., Hammerschlag, R.S., Cahoon, D.R., 2009. Evaluation of restored tidal freshwater wetlands. In: Perillo, G.M.E., Wolanski, E., Cahoon, D.R., Brinson, M.M. (Eds.), Coastal Wetlands: An Integrated Ecosystem Approach. Elsevier Science, Amsterdam, pp. 801–832.

Baldwin, A.H., Jensen, K., Schönfeld, M., 2014. Warming influences plant biomass and reduces diversity across continents, latitudes, and species migration scenarios in experimental wetland communities. Global Change Biology 20, 835–850.

Barbour, S., Kiviat, E., 1986. A survey of Lepidoptera in Tivoli North Bay (Hudson River estuary). In: Cooper, J.D. (Ed.), Polgar Fellowship Reports of the Hudson River National Estuarine Research Reserve Program, 1985. Hudson River Foundation, New York, NY, USA, pp. IV.1–IV.26.

Barendregt, A., 2005. The impact of flooding regime on ecosystems in a tidal freshwater area. International Journal of Ecohydrology and Hydrobiology 5, 95–102.

Barendregt, A., Swarth, C.W., 2013. Tidal freshwater wetlands: variation and changes. Estuaries and Coasts 36, 445–456.

Barendregt, A., Whigham, D.F., Baldwin, A.H. (Eds.), 2009a. Tidal Freshwater Wetlands. Backhuys Publishers, Leiden.

Barendregt, A., Ysebaert, T., Wolff, W.J., 2009b. Animal communities in European tidal freshwater wetlands. In: Barendregt, A., Whigham, D.F., Baldwin, A.H. (Eds.), Tidal Freshwater Wetlands, Chapter 8. Backhuys Publishers, Leiden, pp. 89–104.

Barendregt, A., Whigham, D.F., Baldwin, A.H., van Damme, S., 2006. Wetlands in the tidal freshwater zone. In: Bobbink, R., Beltman, B., Verhoeven, J.T.A., Whigham, D.F. (Eds.), Wetlands: Functioning, Biodiversity Conservation, and Restoration. Springer-Verlag, Berlin, Germany, pp. 117–148.

Beauchard, O., Jacobs, S., Ysebaert, T., Meire, P., 2013. Avian response to tidal freshwater habitat creation by controlled reduced tide system. Estuarine, Coastal and Shelf Science 131, 12–23.

Beckett, L.H., Baldwin, A.H., Kearney, M.S., 2016. Tidal marshes across a Chesapeake Bay subestuary are not keeping up with sea-level rise. PLoS One. https://doi.org/10.1371/journal.pone.0159753.

Bowden, W.B., Vörösmarty, C.J., Morris, J.T., Peterson, B.J., Hobbie, J.E., Steudler, P.A., Moore III, B., 1991. Transport and processing of nitrogen in a tidal freshwater wetland. Water Resources Research 27, 389–408.

Burkart, M., 2001. River corridor plants (Stromtalpflanzen) in Central European Lowland: a review of a poorly understood plant distribution pattern. Global Ecology and Biogeography 10, 449–468.

Butzeck, C., Eschenbach, A., Gröngröft, A., Hansen, K., Nolte, S., Jensen, K., 2015. Sediment deposition and accretion rates in tidal marshes are highly variable along estuarine salinity and flooding gradients. Estuaries and Coasts 38, 434–450.

Chambers, R.M., Fourqurean, J.W., 1990. Alternative criteria for assessing nutrient limitation of a wetland macrophyte (*Peltandra virginica* (L.) Kunth). Aquatic Botany 40, 305–320.

Conner, W., Doyle, T., Krauss, K., 2007. Ecology of Tidal Freshwater Swamps of the Southeastern United States. Springer, Dordrecht.

Connors, L.M., Kiviat, E., Groffman, P.M., Ostfeld, R.S., 1999. Muskrat (*Ondatra zibethicus*) disturbance to vegetation and potential net nitrogen mineralization and nitrification rates in a freshwater tidal marsh. The American Midland Naturalist 143, 53–63.

Craft, C.B., 2012. Tidal freshwater forest accretion does not keep pace with sea level rise. Global Change Biology 18, 3615–3623.

Craft, C., Clough, J., Ehman, J., Joye, S., Park, R., Pennings, S., Guo, S., Machmuller, M., 2009. Forecasting the effects of accelerated sea level rise on tidal marsh ecosystem services. Frontiers in Ecology and the Environment 7, 73–78.

Dahl, T.E., 2006. Status and Trends of Wetlands in the Conterminous United States 1998 to 2004. US Department of the Interior, Fish and Wildlife Service, Washington, DC, USA, 112 p.

Darke, A.K., Megonigal, J.P., 2003. Control of sediment deposition rates in two mid-Atlantic coast tidal freshwater wetlands. Estuarine, Coastal and Shelf Science 57, 259–272.

Deegan, B.M., Harrington, T.J., 2004. The distribution and ecology of *Schoenoplectus triqueter* in the Shannon estuary. Proceedings of the Royal Irish Academy 104, 107–117.

Duberstein, J.A., Conner, W.H., 2009. Use of hummocks and hollows by trees in tidal freshwater forested wetlands along the Savannah River. Forest Ecology and Management 258, 1613–1618.

Duberstein, J.A., Conner, W.H., Krauss, K.W., 2014. Woody vegetation communities of tidal freshwater swamps in South Carolina, Georgia and Florida (US) with comparison to similar systems in the US and South America. Journal of Vegetation Science 25, 848–862.

Elliot, M., Dewailly, F., 1995. The structure and components of European estuarine fish assemblages. Netherlands Journal of Aquatic Ecology 29, 397–417.

Engels, J.G., Jensen, K., 2009. Patterns of wetland plant diversity along estuarine stress gradients of the Elbe (Germany) and Connecticut (USA) Rivers. Plant Ecology and Diversity 2, 301–311.

Findlay, S., Howe, K., Austin, H.K., 1990. Comparison of detritus dynamics in two tidal freshwater wetlands. Ecology 71, 288–295.

Flower, C.E., Knight, K.S., Gonzalez-Meler, M.A., 2013. Impacts of the emerald ash borer (*Agrilus planipennis* Fairmaire) induced ash (*Fraxinus* spp.) mortality on forest carbon cycling and successional dynamics in the eastern United States. Biological Invasions 15, 931–944.

Gao, H., Bai, J., He, X., Zhao, Q., Lu, Q., Want, J., 2014. High temperature and salinity enhance soil nitrogen mineralization in a tidal freshwater marsh. PLoS One 9 (4), e9501. https://doi.org/10.1371/journal.pone.0095011.

Garniel, A., Mierwald, U., 1996. Changes in the morphology and vegetation along the human-altered shoreline of the Lower Elbe. In: Nordstrom, K.F., Roman, C.T. (Eds.), EstuarineShores – Evolution, Environments and Human Alterations. John Wiley & Sons, Chichester, pp. 375–396.

Glooschenko, W.A., Tarnocai, C., Zoltai, S., Glooschenko, V., 1993. Wetlands of Canada and Greenland. In: Whigham, D.F., Dykyjová, D., Hejný, S. (Eds.), Wetlands of the World: Inventory, Ecology and Management, vol. 1. Kluwer Academic Publishers, Dordrecht, pp. 415–515. Africa, Australia, Canada and Greenland, Mediterranean, Mexico, Papua New Guinea, South Asia, Tropical South America, United States.

Griffith, A.B., Forseth, I.N., 2003. Establishment and reproduction of *Aeschynomene virginica* (L.) (Fabaceae) a rare, annual, wetlands species in relation to vegetation removal and water level. Plant Ecology 167, 117–125.

Hall, J.V., 2009. Freshwater tidal wetlands of Alaska. In: Barendregt, A., Whigham, D.F., Baldwin, A.H. (Eds.), Tidal Freshwater Wetlands. Backhuys Publishers, Leiden.

Hammersmark, C.T., Fleenor, W.E., Schaldow, S.G., 2005. Simulation of flood impact and habitat extent for a tidal freshwater marsh restoration. Ecological Engineering 25, 137–152.

Hansen, K., Butzeck, C., Eschenbach, A., Gröngröft, A., Jensen, K., Pfeiffer, E.-M., 2017. Factors influencing the organic carbon pools in tidal marsh soils of the Elbe estuary (Germany). Journal of Soils and Sediments 17, 47–60.

Hazelton, E.L.G., McCormick, M.K., Sievers, M., Kettenring, K.M., Whigham, D.F., 2015. Stand age is associated with clonal diversity, byt not vigor, community structure, or insect herbivory in Chesapeake Bay Phragmites australis. Wetlands 35, 877–888.

Jassby, A.D., Cloern, J.E., 2000. Organic matter sources and rehabilitation of the Sacramento-San Joaquin delta (California, USA). Aquatic Conservation: Marine and Freshwater Ecosystems 10, 323–352.

Johnson, L.K., Simenstad, C.A., 2015. Variation in the flora and fauna of tidal freshwater forest ecosystems along the Columbia River estuary gradient: controlling factors in the context of river flow regulation. Estuaries and Coasts 38, 679–698.

Junk, W.J., 1983. Wetlands of tropical south America. In: Whigham, D.F., Dykyjová, D., Hejný, S. (Eds.), Wetlands of the World: Inventory, Ecology and Management, vol. 1. Kluwer Academic Publishers, Dordrecht, pp. 679–739. Africa, Australia, Canada and Greenland, Mediterranean, Mexico, Papua New Guinea, South Asia, Tropical South America, United States.

Kandus, P., Malvárez, A.I., 2004. Vegetation patterns and change analysis in the lower delta islands of the Parana River (Argentina). Wetlands 24, 620–632.

Khan, H., Brush, G.S., 1994. Nutrient and metal accumulation in a freshwater tidal marsh. Estuaries 17, 345–360.

Leck, M.A., Baldwin, A.H., Parker, V.T., Schile, L., Whigham, D.F., 2009. Plant communities of North American tidal freshwater wetlands. In: Barendregt, A., Whigham, D.F., Baldwin, A.H. (Eds.), Tidal Freshwater Wetlands. Backhuys Publishers, Leiden.

Leck, M.A., Crain, C.M., 2009. Northeastern North American case studies - New Jersey and new England. In: Barendregt, A., Whigham, D.F., Baldwin, A.H. (Eds.), Tidal Freshwater Wetlands. Backhuys Publishers, Leiden, The Netherlands.

Leck, M.A., Leck, C.F., 2005. Vascular plants of a Delaware River tidal freshwater wetland and adjacent terrestrial areas: seed bank and vegetation comparisons of reference and constructed marshes and annotated species list. Journal of the Torrey Botanical Society 132, 323–354.

Leck, M.A., Simpson, R.L., Whigham, D.F., Leck, C.F., 1988. Plants of the Hamilton marshes: a Delaware river freshwater tidal wetlands. Bartonia 54, 1–17.

Lehman, P.W., Mayr, S., Liu, L., Tang, A., 2015. Tidal day organic and inorganic material flux of points in the Liberty Island freshwater tidal wetlands. SpringerPlus 4, 273. https://doi.org/10.1186/s40064-015-1068-6.

Lobry, J., Mourand, L., Rochard, E., Elie, P., 2003. Structure of the Gironde estuarine fish assemblages: a comparison of European estuaries perspective. Aquatic Living Resources 16, 47–58.

Loomis, M.J., Craft, C.B., 2011. Carbon sequestration and nutrient (nitrogen, phosphorus) accumulation in river-dominated tidal marshes, Georgia, USA. Soil Science Society of America Journal 74, 1028–1036.

Lynch, J.J., O'Neal, T., Lay, D.W., 1947. Management significance of damage by geese and muskrats to gulf coast marshes. Journal of Wildlife Management 11, 50–76.

Maes, J., Taillieu, A., van Damme, P.A., Cottenie, K., Ollevier, F., 1998. Seasonal patterns in the fish and crustacean community of a turbid temperate estuary (Zeeschelde estuary, Belgium). Estuarine, Coastal and Shelf Science 47, 143–151.

Meanley, B., 1965. Early-fall food and habitat of the sora in the Patuxent River marsh, Maryland. Chesapeake Science 6, 235–237.

Meire, P., Vincx, M., 1993. Marine and estuarine gradients. Netherlands Journal of Aquatic Ecology 27, 75–493.

Mitsch, W.J., Gosselink, J.G., 2015. Wetlands, fifth ed. Wiley, New York.

Morris, J.T., Lajtha, K., 1986. Decomposition and nutrient dynamics of litter from four species of freshwater emergent macrophytes. Hydrobiologia 11, 215–223.

Morse, J.L., Megonigal, J.P., Walbridge, M.R., 2004. Sediment nutrient accumulation and nutrient availability in two tidal freshwater marshes along the Mattaponi River, Virginia, USA. Biogeochemistry 69, 175–206.

Neubauer, S.C., Craft, C.C., 2009. Global change and tidal freshwater wetlands: scenarios and impacts. In: Barendregt, A., Whigham, D.F., Baldwin, A.H. (Eds.), Tidal Freshwater Wetlands. Backhuys Publishers, Leiden.

Neubauer, S.C., Miller, W.D., Anderson, I.C., 2000. Carbon cycling in a tidal freshwater marsh ecosystem: a carbon gas flux study. Marine Ecology Progress Series 199, 13–30.

Neubauer, S.C., Franklin, R.B., Berrier, D.J., 2013. Saltwater intrusion into tidal freshwater marshes alter the biogeochemical processing of organic carbon. Biogeosciences 10, 8171–8181-83.

Noe, G.B., Krauss, K.W., Lockaby, B.G., Conner, W.H., Hupp, C.R., 2013. The effect of increasing salinity and forest mortality on soil nitrogen and phosphorus mineralization in tidal freshwater forested wetlands. Biogeochemistry 114, 225–244.

Odum, W.E., 1988. Comparative ecology of tidal freshwater and salt marshes. Annual Review of Ecology and Systematics 19, 147–176.

Odum, W.E., Heywood, M.A., 1978. Decomposition of intertidal freshwater marsh plants. In: Good, R.E., Whigham, D.F., Simpson, R.L. (Eds.), Freshwater Wetlands. Ecological Processes and Management Potential. Academic Press, New York, NY, pp. 89–97.

Odum, W.E., Odum, E.P., Odum, H.T., 1995. Nature's pulsing paradigm. Estuaries 18, 547–555.

Odum, W.E., Smith III, T.J., Hoover, J.K., McIvor, C.C., 1984. The Ecology of Tidal Freshwater Marshes of the United States East Coast: A Community Profile. FWS OBS-83-17. US Fish and Wildlife Service, Washington, DC, 177 p.

Orson, R.A., Simpson, R.L., Good, R.E., 1990. Rates of sediment accumulation in a tidal freshwater marsh. Journal of Sedimentology Petroleum 60, 859–869.

Pasternack, G.B., 2009. Hydrogeomorphology and sedimentation. In: Barendregt, A., Whigham, D.F., Baldwin, A.H. (Eds.), Tidal Freshwater Wetlands. Backhuys Publishers, Leiden.

Peterson, J.E., Baldwin, A.H., 2004. Variation in wetland seed banks across a tidal freshwater landscape. American Journal of Botany 91, 1251–1259.

Piavis, P.G., 1991. Yellow perch (*Percaflavescens*). In: Funderburk, S.L., Mihursky, J.A., Jordan, S.J., Riley, D. (Eds.), Habitat Requirements for Chesapeake Bay Living Resources. National Oceanic and Atmospheric Administration, Annapolis, MD, 14–1–14–15.

Pomfret, J.R., Elliott, M., O'Reilly, M.G., Phillips, S., 1991. Spatial and temporal patterns in the fish communities in two UK North Sea estuaries. In: Elliott, M., Ducrotoy, J.P. (Eds.), Estuaries and Coasts: Spatial and Temporal Intercomparisons. Olson & Olson, Fredensborg, pp. 277–284.

Pratolongo, P., Kandus, P., Brinson, M.M., 2007. Net aboveground primary production and soil properties of floating and attached freshwater tidal marshes in the Río de la Plata estuary, Argentina. Estuaries and Coasts 30, 618–626.

Rheinhardt, R., 1992. A multivariate analysis of vegetation patterns in tidal freshwater swamps of lower Chesapeake Bay, U.S.A. Bulletin of the Torrey Botanical Club 119, 192–207.

Sasser, C.E., Gosselink, J.G., Holm, G.O., Visser, J.M., 2009. Freshwater tidal wetlands of the Mississippi River delta. In: Barendregt, A., Whigham, D.F., Baldwin, A.H. (Eds.), Tidal Fresh-water Wetlands. Backhuys Publishers, Leiden.

Seldomridge, E., Prestegaard, K., 2014. Geochemical, temperature, and hydrological transport limitations on nitrate retention in tidal freshwater wetlands, Patuxent river, Maryland. Wetlands 34, 641–651.

Sharpe, P.J., Baldwin, A.H., 2009. Patterns of wetland plant species richness across estuarine gradients of Chesapeake Bay. Wetlands 29, 225–235.

Sickels, F.A., Simpson, R.L., 1985. Growth and survival of giant ragweed (*Ambrosia trifida* L.) in a Delaware River freshwater tidal wetland. Bulletin of the Torrey Botanical Club 112, 368–375.

Simpson, R.L., Good, R.E., Leck, M.A., Whigham, D.F., 1983a. The ecology of freshwater tidal wetlands. BioScience 33, 255–259.

Simpson, R.L., Good, R.E., Walker, R., Frasco, B.R., 1983b. The role of Delaware River freshwater tidal wetlands in the retention of nutrients and heavy metals. Journal of Environmental Quality 12, 41–48.

Simpson, R.L., Whigham, D.F., Walker, R., 1978. Seasonal patterns of nutrient movement in a freshwater tidal marsh. In: Good, R.E., Whigham, D.F., Simpson, R.L. (Eds.), Freshwater Wetlands. Ecological Processes and Management Potential. Academic Press, New York, NY, pp. 243–258.

Sloey, T.M., Hester, M.W., 2016. Interactions between soil physicochemistry and belowground biomass production in a freshwater tidal marsh. Plant and Soil 401, 397–408.

Storm, C., Van der Velden, J.A., Kuijpers, J.W.M., 2005. From nature conservation towards restoration of estuarine dynamics in the heavily modified Rhine-Meuse estuary, The Netherlands. Archiv für Hydrobiologie 15 (Supplement 155), 305–318 (Large Rivers).

Swarth, C.W., Burke, J., 2000. Waterbirds in Freshwater Tidal Wetlands: Population Trends and Habitat Use in the Non-breeding Season. Technical Report of the Jug Bay Wetlands Sanctuary, Lothian, MD, 37 p.

Swarth, C.W., Kiviat, E., 2009. Animal communities – North America. In: Barendregt, A., Whigham, D.F., Baldwin, A.H. (Eds.), Tidal Freshwater Wetlands. Backhuys Publishers, Leiden.

Tabak, N.M., Laba, M., Spector, S., 2016. Simulating the effects of sea level rise on the resilience and migration of tidal wetlands along the Hudson River. PLoS One 11 (4), e0152437. https://doi.org/10.1371/journal.pone.0152437.

Thiel, R., Potter, I.C., 2001. The ichthyofaunal composition of the Elbe Estuary: an analysis in space and time. Marine Biology 138, 603–616.

Van den Bergh, E., Van Damme, S., Graveland, J., De Jong, D., Baten, I., Meire, P., 2005. Ecological rehabilitation of the Schelde Estuary (The Netherlands-Belgium; Northwest Europe): linking ecology, safety against floods and accessibility for port development. Restoration Ecology 13, 204–214.

van Steenis, C.G.G.J., 1971. De zoetwatergetijdedotter van de Biesbosch en de Oude Maas, *Caltha palustris* L. var. *araneosa*, var.nov. Gorteria 5, 213–219.

Weston, N.B., Dixon, R.E., Joye, S.B., February 7, 2006. Ramifications of increased salinity in tidal freshwater sediments: geochemistry and microbial pathways of organic matter mineralization. Journal of Geophysical Research-Biogeosciences 111, G01009.

Weston, N.B., Vile, M.A., Neubauer, S.C., Velinsky, D.J., 2011. Accelerated microbial organic matter mineralization following salt-water intrusion into tidal freshwater marsh soils. Biogeochemistry 102, 135–151.

Weston, N.B., Neubauer, S.C., Velinsky, D.J., Vile, M.A., 2014. Net ecosystem carbon exchange and the greenhouse gas balance of tidal marshes along an estuarine salinity gradient. Biogeochemistry 120, 163–189.

Whigham, D.F., Baldwin, A.H., Swarth, C., 2009. Conservation of tidal freshwater wetlands in North America. In: Barendregt, A., Whigham, D.F., Baldwin, A.H. (Eds.), Tidal Freshwater Wetlands. Backhuys Publishers, Leiden.

Whigham, D.F., McCormick, J., Good, R.E., Simpson, R.L., 1978. Biomass and primary production in freshwater tidal wetlands of the middle Atlantic coast. In: Good, R.E., Whigham, D.F., Simpson, R.L. (Eds.), Freshwater Wetlands. Ecological Processes and Management Potential. Academic Press, New York, pp. 3–20.

Whigham, D.F., Simpson, R.L., 1992. Annual variation in biomass and production of a tidal freshwater wetland and comparisons with other wetland systems. Virginia Journal of Science 43, 5–14.

Wondolowski, L., 2001. Diurnal Activity Patterns of Wintering Gulls at Jug Bay Wetlands Sanctuary, Maryland. M.Sc. thesis. Bard College, Annandale-on-Hudson, New York, 92pp.

Yozzo, D.L., Steineck, P.L., 1994. Ostracoda from tidal freshwater wetlands at Stockport, Hudson River estuary: abundance, distribution, and composition. Estuaries 17, 680–684.

Ysebaert, T., Herman, P.M.J., Meire, P., Craeymeersch, J., Verbeek, H., Heip, C.H.R., 2003. Large-scale spatial patterns in estuaries: estuarine macrobenthic communities in the Schelde estuary,NW-Europe. Estuarine, Coastal and Shelf Science 57, 335–355.

Ysebaert, T., Meininger, P.L., Meire, P., Devos, K., Berrevoets, C.M., Strucker, R.C.W., Kuijken, E., 2000. Waterbird communities along the estuarine salinity gradient of the Schelde estuary, NW-Europe. Biodiversity and Conservation 9, 1275–1296.

Ysebaert, T., Meire, P., Coosen, J., Essink, K., 1998. Zonation of intertidal macrobenthos in the estuaries of Schelde and Ems. Aquatic Ecology 32, 53–71.

Zhang, J., Jørgensen, S.E., Lu, J., Nielsen, S.N., Wang, Q., 2014. A model for the contribution of macrophyte-derived organic carbon in harvested tidal freshwater marshes to surrounding estuarine and oceanic ecosystems and its response to global warming. Ecological Modelling 294, 105–116.

Zonneveld, I.S., 1999. De Biesbosch Een Halve Eeuw Gevolgd. Uniepers, Abcoude, 223 p.

Further Reading

Baldwin, A.H., 2009. Restoration of tidal freshwater wetlands in North America. In: Barendregt, A., Whigham, D.F., Baldwin, A.H. (Eds.), Tidal Freshwater Wetlands. Backhuys Publishers, Leiden.

Lippson, A.J., Lippson, R.L., 1997. Life in the Chesapeake Bay. The Johns Hopkins University Press, Baltimore, MD, 344 p.

Mitsch, W.J., Gosselink, J.G., 2000. Wetlands. Van Nostrand Reinhold, New York, NY, 722 p.

CHAPTER

19

Biogeochemistry of Tidal Freshwater Wetlands

J. Patrick Megonigal[1], *Scott C. Neubauer*[2]

[1]Smithsonian Environmental Research Center, Edgewater, MD, United States; [2]Department of Biology, Virginia Commonwealth University, Richmond, VA, United States

1. INTRODUCTION

Biogeochemical cycles in tidal freshwater wetlands are regulated by many of the same factors that operate in saline tidal wetlands, yet the interplay among element cycles in tidal freshwater wetlands is unique because of their position at the interface of nontidal rivers and brackish estuaries. Here we focus on the exchanges, transformations, and storage of the major elements, recognizing that these processes govern the contribution of tidal freshwater wetlands to the metabolism of coastal landscapes.

Tidal freshwater and tidal saline wetlands share ecological characteristics that ultimately regulate element cycles. Tidal freshwater wetlands support the full range of plant functional types that occur in saline tidal wetlands, including herbaceous species, trees, and shrubs, which differ with respect to primary productivity, root–leaf–wood C allocation, and C quality. As with tidal saline wetlands, tidal freshwater wetlands occur in geomorphic settings that promote not only high rates of element sequestration in biomass and soils but also high exchange rates of water, solutes, solids, and gases with terrestrial and aquatic ecosystems, groundwater, and the atmosphere. Tidal freshwater and tidal saline systems both develop on mineral and organic soils, the chemical composition of which is expected to affect a variety of ecosystem processes, including the contribution of Fe(III) to anaerobic microbial respiration, and both are found on eutrophic and oligotrophic rivers that place limits on nutrient availability and productivity.

Tidal freshwater wetland biogeochemistry differs fundamentally from wetlands that are either freshwater and nontidal or tidal and saline. Compared with nontidal freshwater wetlands, tidal freshwater hydrology is predictable and aseasonal. Also, tidal hydrology drives more open element cycles in tidal freshwater wetlands than nontidal freshwater wetlands, an

Coastal Wetlands
https://doi.org/10.1016/B978-0-444-63893-9.00019-8

observation that inspired the Outwelling hypothesis (Kalber, 1959; Odum, 1968) and contemporary research on import, export, and chemical transformation in tidal wetlands.

Compared with tidal saline wetlands, tidal freshwater plant and microbial activity is less influenced by salts and sulfate. Low concentrations of sulfate in freshwater tend to make tidal freshwater wetlands stronger CH_4 sources than saline tidal wetlands. The influence of alkalinity exported from tidal freshwater wetlands on adjacent estuarine waters is relatively dramatic in tidal freshwater rivers because they are poorly buffered compared with saline waters closer to oceans (Stets et al., 2017). Thus, the biogeochemistry of tidal freshwater wetlands is unique in the coastal landscape because of a combination of flushing by tides, the chemical milieu of freshwater, and position at the limit of tidal influence.

2. CARBON BIOGEOCHEMISTRY

Carbon cycling is a fundamental driver of biogeochemical transformations in ecosystems. Gross primary production (GPP) largely establishes the upper limit of heterotrophic activity, including the secondary productivity of consumers. Decomposition and microbial respiration (R) releases or sequesters nitrogen and other nutrients, depending on the chemical characteristics of the detritus. The chemical composition of organic carbon compounds and interactions with mineral surfaces regulate microbial transformations such as denitrification. The capacity of ecosystems to sequester CO_2 in biomass or soil organic matter represents an imbalance between GPP and R (i.e., net ecosystem production, NEP), as well as the net exchange of particulate and dissolved organic carbon (DOC) with adjacent ecosystems.

The most complete C budget of a tidal freshwater wetland completed to date concerned Sweet Hall marsh, located on the Pamunkey River in Virginia, USA (Neubauer et al., 2000, 2002; Neubauer and Anderson, 2003, Fig. 19.1). The budget was developed from a variety of measurements, including repeated measurements of whole-system CO_2 and CH_4 exchange. This approach avoids several problems with estimating C input from biomass harvests, such as accounting for biomass turnover and translocation (see discussion in Neubauer et al., 2000), and is especially useful for understanding ecosystem-level C cycling. We present the Sweet Hall C budget as a case study to illustrate the processes that distribute C among pools and fluxes in a typical tidal freshwater wetland and review studies of other tidal freshwater wetland sites that focused on one or more components of the C cycle.

2.1 Carbon Inputs

GPP is the photosynthetic assimilation of CO_2 by macrophytes and microalgae and the dominant source of metabolic energy driving biogeochemical cycles in tidal wetlands. GPP contributes two-thirds of annual C inputs to Sweet Hall marsh (Fig. 19.1), with the balance from allochthonous sources associated with sediment deposition. About 37% of macrophyte GPP is consumed for plant growth and maintenance respiration and thus quickly returned to the atmosphere as CO_2. The remainder is net primary production (NPP; 625 g C m^{-2} $year^{-1}$), some of which may be allocated to root exudates or mycorrhizae. NPP is supported both by photosynthates produced in a given year and by C from previous years that is translocated from storage organs such as rhizomes. Accounting for both photosynthesis and translocation,

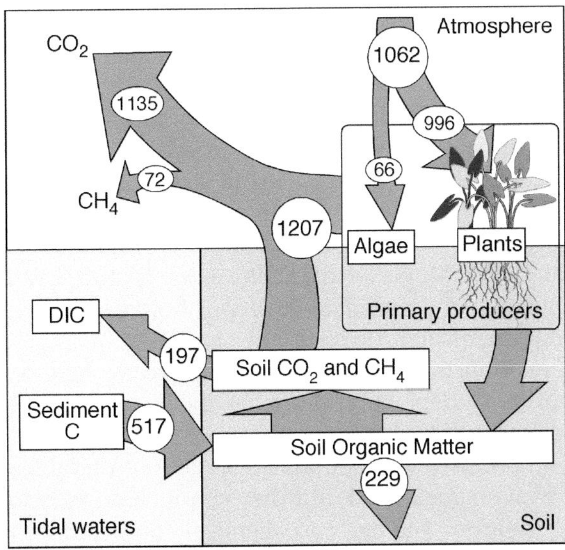

FIGURE 19.1 Carbon budget for Sweet Hall marsh, USA, showing the major pools and fluxes. All fluxes are in units of $g\ C\ m^{-2}\ year^{-1}$. Ecosystem boundaries for the purpose of this budget are the soil–atmosphere interface, the soil–tidal water interface, and the 30 cm depth contour (which is based on the 1963 ^{137}Cs peak). Most of the belowground biomass and the most active zone of biological activity lie within the top 30 cm of the marsh. *DIC*, dissolved inorganic carbon.

macrophyte NPP was 557–736 $g\ C\ m^{-2}\ year^{-1}$ (1150–1500 g biomass $m^{-2}\ year^{-1}$), which is about double the peak aboveground biomass at Sweet Hall marsh (Neubauer et al., 2000). It is difficult to generalize about the relative importance of below- and aboveground NPP because belowground production is difficult to measure and estimates vary widely from NPP ratios $\ll 1$ to $\gg 1$ (Whigham and Simpson, 1978; Schubauer and Hopkinson, 1984; Bellis and Gaither, 1985). Tidal freshwater marshes dominated by annual species are expected to have higher above: below NPP ratios than tidal saline marshes because annuals lack permanent storage organs such as rhizomes (Whigham, 2009).

Weston et al. (2014) applied similar techniques to three sites arranged along a salinity gradient in Delaware Bay, USA, and concluded that tidal freshwater wetland marsh productivity is equivalent to more saline marshes on average, but that tidal freshwater marshes exhibit far more interannual variation. Salinity explained 50%–60% of the interannual variation in oligohaline and mesohaline marshes but not in a tidal freshwater marsh where factors such as spring inundation may regulate production indirectly through the recruitment success of annual plants (Weston et al., 2014). In general, the aboveground productivity of tidal freshwater wetlands is higher (>1000 $g\ m^{-2}\ year^{-1}$) on sites dominated by annual compared with perennial species, and this appears to hold for tidal freshwater wetland forests (Whigham, 2009).

Sediment deposition is an important vehicle for importing allochthonous particulate organic C into tidal freshwater wetland soils, and it enhances organic C preservation and nutrient removal through burial. Tidal freshwater wetlands that can maintain vertical accretion and areal extent with sea level rise are net C sinks as organic matter is buried in accreting sediments (Stevenson et al., 1988). The amount of C added via sediment deposition at Sweet Hall marsh (517 g C m^{-2} $year^{-1}$; Neubauer et al., 2002) is one-third of all organic C inputs to the site (Fig. 19.1). This rate may not be typical because it exceeds estimates from other tidal freshwater wetlands (Craft, 2007; Neubauer, 2008; Weston et al., 2014; Noe et al., 2016). Sediment carbon fluxes vary dramatically with extreme events such as hurricanes (Noe et al., 2016) and with geomorphic setting (Pasternack, 2009). Of particular importance is the location of the site in relation to the estuarine turbidity maximum, a feature that forms at the freshwater–saltwater interface of tidal rivers and has a profound effect on C deposition rates (Darke and Megonigal, 2003; Morse et al., 2004; Loomis and Craft, 2010; Ensign et al., 2014a,b). Sediment deposition and associated C deposition are generally faster in tidal freshwater marshes than tidal freshwater forests in the short term (Ensign et al., 2014a and references therein), but this relationship is less prognostic in areas where sea level rise is changing the factors that control local deposition, such as the position of the estuarine turbidity maximum, plant biomass, shallow subsidence rates, or soil elevation (Noe et al., 2016).

The sources of allochthonous C compounds imported with tidal freshwater wetland sediment deposits are not well characterized, but they can include upland soils, DOC sorbed to mineral particles, and plankton. The age and chemical composition of these sources varies, affecting the extent to which they are ultimately preserved as soil organic matter or exported. A metaanalysis of >300 samples from salt marshes, mangroves, and seagrass beds suggested that the organic matter preserved in coastal sediments is dominated by allochthonous sources at sites with ≤10% organic soil C (Boullion and Boschker, 2006). At Sweet Hall marsh, allochthonous C is ~33% of C inputs but only ~10% of microbial soil respiration (54–71 g C m^{-2} $year^{-1}$; Neubauer et al., 2002), suggesting that allochthonous C sources are better protected against microbial decomposition than autochthonous C (mainly fresh plant) sources. Sequestered organic matter is a sink for N, P, and other elements in organic tissues (see Sections 4 and 5).

2.2 Carbon Outputs

The collective respiration of plants, microbes, and animals at Sweet Hall marsh is supported by total organic C inputs (GPP + sediment-associated C) of 1579 g C m^{-2} $year^{-1}$ (Fig. 19.1). One-third of GPP is consumed by plant respiration that releases CO_2 directly to the atmosphere, soil atmosphere, or soil solution. Most of the remaining GPP takes the form of plant biomass that supports the heterotrophic respiration of bacteria, fungi, insects, grazing snails, and a variety of other organisms (Hines et al., 2006). An uncertain fraction of GPP is lost from plants as root exudates.

Organic C inputs from plants and sediments are eventually subjected to decomposition and microbial degradation, producing soil organic matter, dissolved inorganic carbon (DIC), DOC, and CH_4. At Sweet Hall marsh, 15% of organic C inputs (229 g C m^{-2} $year^{-1}$) are buried by accreting soil and enter a very slowly decomposing soil organic matter pool. The remaining 85% is cycled in time frames of hours to months via plant, animal, and microbial respiration, as well as other microbial degradation processes such as fermentation. Rates of organic C burial in tidal freshwater wetlands along the Atlantic and Gulf coasts of North America, and on the Scheldt River of Europe, ranged from 10 to 930 g C m^{-2} $year^{-1}$ (Table 19.1; Neubauer, 2008 and references therein). In addition to sequestering organic C,

TABLE 19.1 Vertical Accretion and Nutrient Burial Rates in Tidal Freshwater and Oligohaline Wetlands

		(mm/year)	(g C, N, or P m^{-2} $year^{-1}$)			
Location	**Method**	**Accretion**	**C Burial**[a]	**N Burial**	**P Burial**	**References**
North River, MA	model[b]	(1.3–3.6)	nd	(3.1–8.6)	(0.3–0.8)	Bowden et al. (1991)
Tivoli Bays, Hudson River, NY	^{210}Pb	4.5 ± 1.4 (3.6–6.9)	142 ± 90 (33–285)	11.6 ± 3.6 (8.0–15.9)	2.3 ± 1.0 (1.1–3.6)	Merrill (1999)
Two sites, Delaware River, NJ	^{137}Cs, ^{210}Pb	(6–12)[c]	(139–204)	(7.1–8.8)	(3.2–6.1)	Church et al. (2006)
Otter Point Creek, MD	^{210}Pb	5.0 ± 4.5 (2.1–10.2)	163 ± 153 (65–339)	(2.7–11.7)	(0.5–2.1)	Merrill (1999) and Merrill and Cornwell (2000)
Three sites, Choptank River, MD	^{210}Pb	9.2 ± 2.7 (6.1–10.9)	366 ± 103 (265–470)	(19.2–27.1)	(0.2–2.0)	Merrill (1999) and Merrill and Cornwell (2000)
Eight sites, Choptank River, MD	^{210}Pb	8.4 ± 4.7 (3.2–21.5)	nd	21 ± 11 (7–44)	1.7 ± 1.1 (0.5–4.3)	Malone et al. (2003)
25 sites[d], Patuxent River, MD	^{210}Pb	8.5 ± 6.0 (1.1–21.9)	358 ± 258 (9–930)	16.9 ± 11.9 (0.5–36.1)	4.3 ± 3.9 (0.1–12.6)	Merrill (1999)
Jug Bay marsh, Patuxent River, MD	Pollen[e]	(3.7–8.9)	(70–249)	(8–26)	(0.6–2.3)	Khan and Brush (1994)
Sweet Hall marsh, Pamunkey River, VA	^{137}Cs	8.5	229 ± 45	18.1 ± 3.1	nd	Neubauer et al. (2002; 2005a)
Carr's Island, Altamaha River, GA	^{137}Cs	(3.5–4.6)	(103–122)	(7–8)	(0.4–0.8)	Craft (2007)
Barataria Basin, Mississippi River, LA	^{137}Cs	(6.5–10.6)	(153–239)	(9–16)	(0.5–1.0)	Hatton et al. (1983)
Barataria Basin, Mississippi River, LA	^{137}Cs	7.5	198	12	nd	DeLaune et al. (1986)

Values are presented as means ± 1 standard deviation, with reported ranges in parentheses. nd = no data. Studies that did not contain burial rates for at least two of the three elements (C, N, P) are not shown.

[a] *As needed, organic matter accumulation rates were converted to C assuming percentage of C = 0.5 (% OM).*

[b] *Ranges calculated from a mechanistic model of sediment decomposition. P burial determined from soil P content in Bowden (1984).*

[c] *Cores (1 per site) were dated using both ^{137}Cs and ^{210}Pb. At one site, accretion rates from the two methods were identical. At the other site, this table reports the midpoint for the ^{137}Cs (14 mm/year) and ^{210}Pb methods (9 mm/year), as was done by Church et al. (2006).*

[d] *17 sites for N burial.*

[e] *Range in rates estimated since 1900 for high and low marsh cores.*

sediment accumulation adds elevation to wetland soils at a rate that is regulated by relative sea level rise (Morris et al., 2002).

2.2.1 Exports of CO_2, DIC, DOC, and POC

C compounds exported from tidal wetlands influence the chemical composition of adjacent estuaries and the atmosphere. The plant and microbial respiration that takes place in saturated or flooded soils generates DIC, which partitions into CO_2, HCO_3^-, and CO_3^{2-} according to pH. At Sweet Hall marsh, 12% of all organic C added to the marsh is exported as DIC. A portion of the DIC pool is emitted directly to the atmosphere as CO_2, while the remainder is exported to the York River (Fig. 19.1; Neubauer and Anderson, 2003).

Evidence is mounting that tidal wetlands are a significant source of DIC and alkalinity to estuaries and coastal shelves (Cai, 2011; Bauer et al., 2013; Wang et al., 2016). Extrapolating seasonal estimates of DIC export from Sweet Hall marsh (197 g C m^{-2} $year^{-1}$) to all tidal marshes of the York River estuary suggested that 47% of excess water column DIC (DIC unexplained by conservative mixing of fresh and marine end-members) is imported from wetlands (Neubauer and Anderson, 2003). High temporal resolution measurements in a polyhaline tidal marsh show that DIC exports can be larger than previously estimated and comparable with major fluxes such as riverine DOC and DIC fluxes to continental shelves (Wang et al., 2016). It is presently unclear whether there are meaningful differences between tidal freshwater wetlands and more saline tidal wetlands in the total or area-based rate of DIC export. However, several studies suggest that freshwater and low-salinity tidal wetlands are significant DIC sources to estuaries and coastal oceans (Smith and Hollibaugh, 1993; Frankignoulle et al., 1996, 1998; Nietch, 2000).

Far less DOC is exported from tidal wetlands to estuaries compared with DIC, yet DOC is arguably the most important form of C exported from these systems. Saline tidal marshes generally export DOC to estuaries (Nixon, 1980 and references therein; Tobias and Neubauer, this volume; Tzortziou et al., 2011), where it influences estuarine microbial metabolism, nutrient cycling, and ultraviolet (UV) light penetration of the water column (Epp et al., 2007). Although there is no evidence of DOC export from Sweet Hall marsh (Neubauer, 2000), DOC is exported from tidal freshwater wetlands on the Patuxent River, USA (Fig. 19.2), the Hudson River, USA (Findlay et al., 1998), and the Dovey River, Wales, UK (Dausse et al., 2012). Raymond and Bauer (2001) proposed that tidal freshwater wetlands are 30% of DOC sources to the comparatively pristine York River, USA.

The chemical composition of DOC influences the biogeochemical effects it will have on receiving estuaries. Features such as the aromatic ring content affect UV radiation adsorption (Tzortziou et al., 2007) and perhaps the ability of DOC to support microbial respiration. Chromophoric dissolved organic matter (CDOM) is the light adsorbing component of DOC and a dominant fraction of the DOC pool in estuaries. The spectral slope (S_{CDOM}) of CDOM exported from a tidal freshwater marsh on the Patuxent River, USA, indicated that the wetland is a source of relatively complex, high molecular weight, and aromatic-rich DOC (Fig. 19.2). Presumably the S_{CDOM} of this tidal freshwater marsh reflects the relatively high lignin content and complexity of organic C compounds in emergent wetland plants compared with phytoplankton (Enriquez et al., 1993; Tzortziou et al., 2008).

Particulate organic carbon (POC) in rivers is derived from aquatic (e.g., phytoplankton) and terrestrial (e.g., emergent plant) sources. Although the fraction of POC from terrestrial

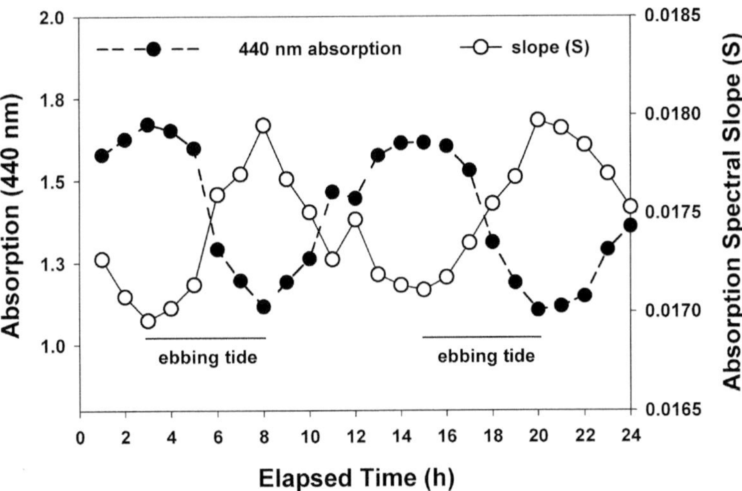

FIGURE 19.2 Tidal cycle variation in chromophoric dissolved organic matter (CDOM) optical properties as a function of elapsed time since the first measurement (h) for the tidal cycle sampled at Jug Bay, USA, on 7–8 September 2005. Left axis: CDOM absorption magnitude at 440 nm, aCDOM(440). Right axis: CDOM absorption spectral slope (S). The data were collected by M. Tzortziou, unpublished. Two *horizontal lines* mark the period of flooding tide when waters are dominated by estuarine CDOM sources.

sources increases toward freshwater end of tidal rivers (Hoffman and Bronk, 2006), studies of POC dynamics in tidal freshwater wetlands indicate that they can vary from net sinks to net sources (Findlay et al., 1990; Neubauer et al., 2002; Hunsinger et al., 2010, 2012). Perhaps the most detailed study of tidal freshwater wetland export of POC to date reported that river POC concentrations were positively correlated with wetland area in tidal freshwater reaches of the Hudson River, USA (Hunsinger et al., 2012). Based on temporal and spatial variation in lignin phenols and stable isotopes, Hudson River tidal freshwater wetlands are sources of aromatic-rich compounds derived from emergent and submerged vascular plants (Hunsinger et al., 2010, 2012). Generalizations about POC dynamics require long-term studies that quantify the influence of seasonal and episodic events such as storms on POC fluxes.

2.2.2 *Methane Emissions and Export*

A portion of organic matter decomposition in saturated soils yields methane (CH_4) rather than CO_2. Emissions from 13 tidal freshwater wetlands averaged 32 ± 37 g CH_4 m^2 $year^{-1}$ (Table 19.2), which is similar to 36 ± 47 g CH_4 m^{-2} $year^{-1}$ from all freshwater wetlands in the conterminous United States (mean $\pm$ sd) (Bridgham et al., 2006). It is noteworthy that the lowest CH_4 emissions in this compilation are from six sites located along a 2.5 km reach of a single tidal freshwater river (Kelley et al., 1995; Megonigal and Schlesinger, 2002). The average of the remainder of the sites is 58 ± 36 g CH_4 m^{-2} $year^{-1}$, which is perhaps a more representative sample of tidal freshwater wetlands across the Atlantic and Gulf coasts of the USA. Indeed, annual emissions quantified with high-frequency eddy flux (Holm et al., 2016) agree better with a log-linear fit of tidal wetland CH_4 emissions when the data from this river (White Oak River, NC) are excluded (Fig. 19.3). This insight suggests that increasing

TABLE 19.2 Methane Emissions From Tidal Freshwater Wetlands

Location	Site Name	Method	Data Points	Period	Total Flux	Flux Used	References
					g CH_4 m^{-2} $year^{-1}$		
White Oak River, NC[a]	UF-FB	Chamber	8	7 Months (Mar–Nov)	3.2	3.2	Kelley et al. (1995)
White Oak River, NC[a]	GI-FB	Chamber	8	7 Months (Mar–Nov)	3.8	3.8	
White Oak River, NC[a]	UF-NB	Chamber	8	7 Months (Mar–Nov)	4.6	4.6	
White Oak River, NC[a]	GI-NB	Chamber	8	7 Months (Mar–Nov)	5.3	5.3	
White Oak River, NC[b]	Upper site	Chamber	20	24 Months	1.4	1.4	Megonigal and Schlesinger (2002)
White Oak River, NC[b]	Lower site	Chamber	18	24 Months	1.8	1.8	
York River, VA	TFW marsh	Chamber	8	16 Months	96.0	96.0	Neubauer et al. (2000)
Waccamaw River, SC	TFW control	Chamber	24	21 Months	56.0	56.0	Neubauer (2013b)
Patuxent River, VA[c]	Plant	Chamber	6	4 Months (Jul–Oct)	18.0	36.0	Keller et al. (2013)
Delaware River, DE	TFW site	Chamber	20	36 Months	29.3	29.3	Weston et al. (2014)
Mississippi River, LA	TFW site	Eddy flux	Continuous	24 Months	62.3	62.3	Holm et al. (2016)
Mississippi River, LA	TFW site	Chamber	11	21 Months	122.5	122.5	Krauss et al. (2016)
Pamunkey River, VA					77.3	77.3	Neubauer and Lee (in preparation)
Mean ± SD all studies					38 ± 40	38 ± 40	
Mean ± SD White Oak River excluded					64 ± 34	66 ± 33	

All studies are in the United States. Means and standard deviations are given for all studies, excluding those from the White Oak River, NC, which are all considerably lower than from other sites. *GI-FB*, Goldhaber's Island-far bank; *GI-NB*, Goldhaber's Island-near bank; *TFW*, tidal freshwater wetland; *UF-FB*, Upper Fork- far bank; *UF-NB*, Upper Fork- near bank.

[a]*Calculated from Kelley et al. (1995) using linear interpolation between successive sample dates, starting with Mar 1991. Fluxes were not extrapolated to 12 months because winter fluxes were not measured, and winter fluxes are 4% of total annual flux based on the winter flux data reported from the same ecosystem by Megonigal and Schlesinger (2002).*

[b]*Calculated from mean flux data reported in Megonigal and Schlesinger (2002) using linear interpolation between successive sample dates. The cumulative flux over the 746 day period was corrected down to a 720 day period (2 years) and divided by 2 for an average annual flux. These figures differ from those in a review by Bridgham et al. (2006) because of difference in the extrapolation method.*

[c]*Cumulative CH_4 emissions over the period were 18 g C–CH_4 m^{-2}. This figure was multiplied by 2 for an estimated annual rate of 36 g C–CH_4 m^{-2}. The rationale was that missing fluxes from April to June were comparable measured fluxes from July to Octpber, and that winter fluxes are relatively small.*

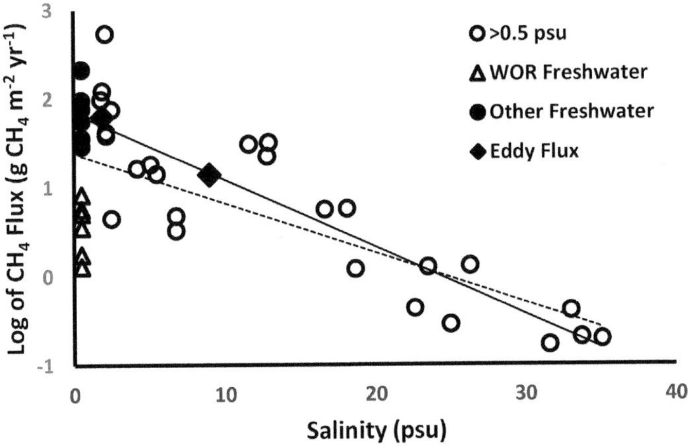

FIGURE 19.3 Relationship of tidal marsh CH_4 emissions with floodwater salinity. Tidal freshwater wetlands are divided into sites located on the White Oak River (WOR) in North Carolina, USA (*open triangles*), and all other sites in Table 19.2 (*closed circles*). Fluxes were measured with static chambers, with the exception of one tidal freshwater site and one brackish site where fluxes were determined by eddy covariance flux (*filled diamonds*). *Open circles* are the saline (>0.5 psu) sites reported by Poffenbarger et al. (2011). A linear regression fit to all data except the WOR (*solid line*; $Y = -0.076X + 1.843$, $R^2 = 0.805$) is a better fit to the eddy flux data than the original model reported by Poffenbarger et al. (2011); *dashed line*.

the sample size will effectively reduce the high variation in current estimates of tidal freshwater wetland CH_4 emissions and that there are specific sites where new studies can be focused to understand the mechanisms that cause low emissions.

Salinity is currently the most useful predictor of spatial variation in CH_4 emissions across tidal wetlands (Bartlett et al., 1987; Poffenbarger et al., 2011, Fig. 19.3). Salinity is a proxy for SO_4^{2-} supply, which regulates CH_4 production indirectly through the activity of SO_4^{2-}-reducing bacteria that compete with methanogens for electron donor compounds (see Section 3.4). However, salinity is also an imperfect proxy because it fails to explain a great deal of spatial and temporal variation in emissions. Across tidal salinity gradients, both rates and variation in CH_4 emissions peak in oligohaline wetlands (salinity 0.5–5) compared with wetlands in either fresher or saltier estuaries (Poffenbarger et al., 2011). This pattern has been reported across transects within single estuaries, including Delaware Bay, USA, where CH_4 emissions were nearly an order of magnitude higher in an oligohaline marsh than a tidal freshwater marsh (Weston et al., 2014). However, the pattern was absent in Mobile Bay, USA, where emissions were similar across sites ranging from oligohaline to polyhaline (Wilson et al., 2015). Plant community composition may explain some of the oligohaline site variation. For example, two oligohaline marshes in the Scheldt Estuary, EU, which differed by an order of magnitude, were dominated by species with distinct morphology and physiology, both of which influence CH_4 emissions (van der Nat and Middelburg, 2000). In particular, emissions from a *Phragmites australis* marsh were higher than from a *Scirpus lacustris* marsh due to a combination of higher carbon additions and lower rhizosphere CH_4 oxidation (van der Nat and Middelburg, 2000). Flooding frequency is another important variable that influences CH_4 emissions (van der Nat and Middelburg, 2000; Megonigal and

Schlesinger, 2002). However, it is clear that part of the variability arises from biogeochemical interactions that do not conform to our current understanding of anaerobic metabolism (Section 3.3).

Rates of CH_4 emission underestimate the contribution of methanogens to overall microbial respiration because they do not account for export of dissolved CH_4 in groundwater, microbial CH_4 oxidation to CO_2, or ebullition (bubble export). Lateral export of dissolved CH_4 from a tidal freshwater swamp on the White Oak River, USA, amounted to 30% of CH_4 production (Kelley et al., 1995). At two nearby tidal freshwater swamps (i.e., forested wetlands) on the same river, CH_4 oxidation reduced CH_4 emissions 50%–80% (Megonigal and Schlesinger, 2002, Fig. 19.5). Methane export by ebullition is rarely measured, but it is expected to be a relatively minor pathway in emergent wetlands where plants effectively vent CH_4 through aerenchyma tissue (van der Nat and Middelburg, 1998a, 2000). Ebullition is an important pathway of CH_4 export in the absence of emergent vegetation, occurring episodically in response to changes in hydrostatic pressure or temperature (Chanton et al., 1989). Ebullition accounted for 44% of CH_4 efflux from tidal freshwater river sediment (Chanton et al., 1989) and 90% of emissions from a plant-free tidal freshwater wetland soil (van der Nat and Middelburg, 1998a). Ebullition can increase in importance during seasonal transitions in plant biomass, such as plant emergence or senescence (van der Nat and Middelburg, 1998a).

3. ORGANIC CARBON PRESERVATION AND METABOLISM

A small fraction of the organic carbon produced by plants is ultimately preserved in soils. The fate of organic matter in wetland ecosystems is regulated by complex interactions between plant processes that affect electron donor and electron acceptor availability and microbial processes that degrade and modify organic matter (Megonigal et al., 2004). Here we review the tidal freshwater wetland literature on C cycling processes, focusing on pathways of microbial respiration that are ultimately responsible for key wetland functions such as CO_2 sequestration, organic carbon burial, and greenhouse gas emissions.

3.1 Carbon Quality

A major initial constraint on organic carbon metabolism in wetlands is the chemical composition of plant material. Mass loss from freshly senescent litter follows an exponential decay curve, with an initial period of rapid loss as the litter is physically fractured and subjected to leaching (Day, 1983), and progressively slower mass loss as organisms depolymerize macromolecules to relatively simple monomers or as molecules are transformed by microbial and chemical reactions into new, decay-resistant compounds (Arndt et al., 2013). Proteins, carbohydrates, and lipids tend to be more susceptible to degradation than lignin, alkyl-C, or S-rich compounds (Baldock et al., 2004; Arndt et al., 2013). Mechanistic explanations for differences in the degradation rate of organic compounds have been proposed, but a cohesive conceptual model that integrates organic transformations with microbial and environmental agents remains elusive (Arndt et al., 2013).

A fundamental constraint on organic matter mineralization rates is the amount of energy produced to support microbial respiration and growth, a characteristic that can be quantified by the free energy yield of coupled reduction–oxidation reactions. Differences in free energy yield are commonly invoked to explain the outcome of competition for organic carbon among microorganisms using different terminal electron acceptors, such as the suppression of methanogenesis by sulfate reduction. Of equal importance is the thermodynamic yield of the electron donor (i.e., organic carbon), a characteristic that varies with the elemental stoichiometry of the compound and is conveniently expressed as the nominal oxidation state of carbon (NOSC; LaRowe and Van Cappellan, 2011). NOSC is inversely related to thermodynamic yield and appears promising for understanding the influence of carbon quality on anaerobic carbon degradation rates (Boye et al., 2017). This quantity may prove particularly useful in tidal wetlands where organic matter comes from disparate sources such as herbaceous plants, woody plants, algae, phytoplankton, and eroded terrestrial soil organic matter.

Because tidal freshwater wetland plant communities are species-rich compared with tidal saline wetlands (Odum, 1988), organic matter quality may be particularly sensitive to shifts in plant species (Kögel-Knabner, 2002). Plant community composition indirectly controlled root and rhizome decomposition rates across modest salinity gradients (salinity range 0–5) in two tidal freshwater rivers (Stagg et al., 2017). In this case, the dominant plants changed from tidal freshwater tree species with high lignin content to oligohaline herbaceous species with low lignin content. Because salt marshes generally lack genera with very low lignin content such as *Nuphar*, *Peltandra*, and *Pontederia*, it can be hypothesized that litter decomposition rates should be slower in salt marshes than tidal freshwater wetlands. For example, Williams and Rosenheim (2015) concluded that soil organic matter stability in three tidal marshes ranging from oligohaline to saline was explained by the lignin content of the dominant plant species. However, Craft (2007) and Stagg et al. (2017) reported higher rates of root decomposition in relatively high-salinity sites across five USA estuaries, suggesting that factors other than tissue quality also regulate tidal freshwater plant decomposition rates.

Sulfate availability is one of the most commonly cited factors affecting decomposition along salinity gradients (Craft, 2007). Sutton-Grier et al. (2011) investigated the interaction of carbon quality and sulfate availability by performing a reciprocal transplant of soils between a brackish (mean salinity 14) and a tidal freshwater marsh over a 24–31-month period. They found that soil organic matter from the tidal freshwater marsh decomposed faster than the brackish marsh regardless of salinity regime. However, soil organic matter from both sites decomposed more rapidly at the brackish site, and SO_4^{2-} additions tended to increase mineralization rates. Thus, the chemical nature of electron donors (organic matter) interacts with the availability of electron acceptors $\left(SO_4^{2-}\right)$ to regulate decomposition rates in tidal freshwater wetlands.

3.2 Aerobic Respiration and Carbon Preservation

Organic matter preservation in saturated soils and sediments is governed by interactions between the chemical composition of tissues and environmental factors (Day, 1982; Benner et al., 1985). Factors that regulate O_2 supply are particularly important because aerobic respiration yields far more free energy than anaerobic respiration and does not require a full consortium of microorganisms to complete (Megonigal et al., 2004). In addition, O_2 supply can indirectly affect extracellular enzyme activity by regulating phenol oxidase and therefore concentrations of enzyme-inhibiting phenolic compounds (Freeman et al., 2001). Belowground plant biomass is presumably preserved more efficiently than aboveground biomass in tidal wetlands because it is deposited directly into hypoxic or anaerobic environments. Similarly, organic matter deposited at the soil surface is preserved better in sites with faster rates of burial, due in part to the decrease in O_2 supply (Hedges and Keil, 1995). For example, long-term (^{210}Pb-determined) C accumulation rates increased with sediment accretion rates in tidal freshwater and oligohaline wetlands (Noe et al., 2016, Table 19.1). Despite the importance of aerobic respiration for carbon preservation, there are virtually no estimates of aerobic respiration in wetlands (Howes et al., 1984) and no methods for measuring in situ O_2 flux that accounts for root O_2 loss (ROL).

Extracellular enzyme activity is an important control of organic matter degradation rates. Morrissey et al. (2014) found that small differences in salinity (salinity range 0–2) had a direct positive effect on the activity of key carbon-degrading extracellular enzymes across eight tidal freshwater and oligohaline marshes. They attribute this pattern to salinity-induced increases in enzyme activity, organic matter bioavailability, and microbial community structure. If confirmed by other studies, stimulation of hydrolytic enzyme activity by salinity may contribute to the widespread observation that soil organic matter content is lower in tidal saline than tidal freshwater wetlands (Craft, 2007). However, other studies on extracellular enzyme responses to salinity show positive, negative, and null responses (Morrissey et al., 2014 and citations therein). Organic compounds are protected from extracellular enzymes through interactions with mineral surfaces or sequestration in mineral microaggregates (Blair and Aller, 2012), which are mediated by reactive Fe compounds in tidal freshwater wetlands (Shields et al., 2016).

3.3 Pathways of Anaerobic Respiration

Microorganisms derive energy by transferring electrons from an external electron donor to an external electron acceptor. Most respiration in wetland soils depends directly or indirectly (in the case of H_2) on organic C as the electron donor, and carbon supply tends to limit microbial respiration. This is true of even organic soils where carbon accumulates because it is protected from microbial activity by a variety of factors that ultimately relate to anaerobiosis (Keiluweit et al., 2016). Competition for electron donors favors the respiration pathway that yields the most free energy in the order: aerobic respiration > denitrification > manganese reduction > iron reduction > sulfate reduction > methanogenesis (Megonigal et al., 2004). Humic substances act as electron acceptors under circumstances that are poorly understood and appear to yield more free energy than methanogenesis (Megonigal et al., 2004).

Many tidal freshwater wetlands occur in urbanized watersheds and are exposed to high NO_3^- concentrations in floodwater, but the contribution of the denitrification pathway to organic C mineralization has not been quantified in tidal freshwater wetland soils to our knowledge. In tidal freshwater river sediments from the Altamaha River, USA, denitrification supported 10% of anaerobic C mineralization (Weston et al., 2006). The NO_3^- concentrations in this study were 20 μM, which is similar to NO_3^- concentrations in the Hudson River and many other tidal freshwater wetland systems. However, Krauss et al. (2016) observed net N_2O uptake in a tidal freshwater wetland, suggesting very low rates of denitrification. Although it seems unlikely that denitrification is ever a dominant pathway of microbial respiration in tidal freshwater wetland soils, it can be an important NO_3^- sink in tidal freshwater wetland-dominated estuaries (see Section 4.2).

Iron oxide minerals can be the dominant electron acceptor in tidal freshwater wetland mineral soils (Roden and Wetzel, 1996; Megonigal et al., 2004), and tidal freshwater wetlands have been used extensively for studies of Fe(III) reduction. Tidal freshwater river sediments were used in the first studies to conclusively establish that Fe(III) reduction supports microbial growth (Lovley and Phillips, 1986, 1987). A decade later, van der Nat and Middelburg (1998a) concluded that Fe(III) reduction explained up to 80% of anaerobic respiration in tidal freshwater wetland mesocosms and that the contribution was higher in mesocosms planted with *S. lacustris* than *P. australis*, suggesting that species-specific plant characteristics influence Fe(III) reduction rates. A field study in a *Peltandra virginica*–dominated tidal freshwater wetland showed that Fe(III) reduction mediated 20%–98% of anaerobic C metabolism (Neubauer et al., 2005b). The importance of Fe(III) reduction declined during the growing season in parallel to plant activity, again suggesting that plants indirectly regulate this microbial process (Fig. 19.4). However, this seasonal pattern can also be explained by direct

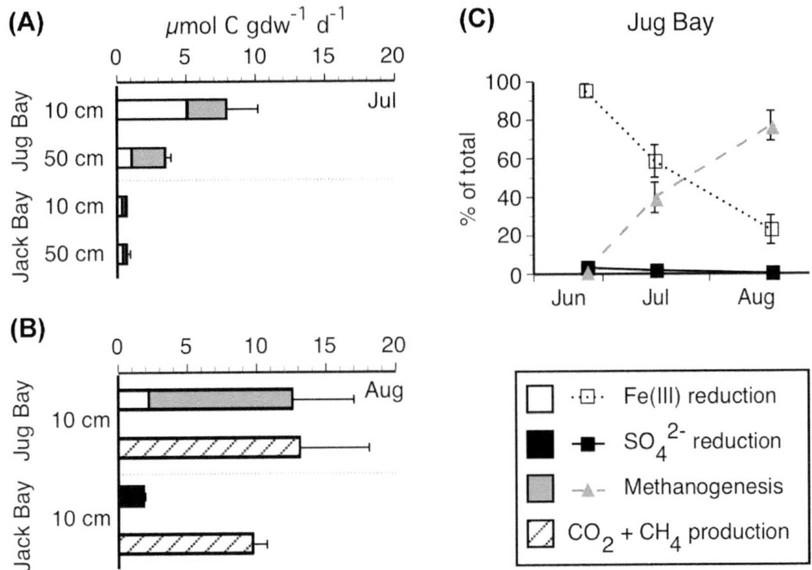

FIGURE 19.4 (A and B) Soil organic carbon mineralization rates in a tidal freshwater marsh (Jug Bay) and a brackish marsh (Jack Bay) in July and August 2002. The July data (A) provide a comparison of rates at 10 and 50 cm depth. The August data (B) provide a comparison of the total rate of anaerobic carbon decomposition as determined by the sum of CO_2 and CH_4 production (striped bars) versus the sum of carbon mineralization from three possible anaerobic pathways (Fe(III) reduction, SO_4^{2-} reduction, and CH_4 production). (C) Seasonal changes in the relative importance of Fe(III) reduction (*open squares*), SO_4^{2-} reduction (*filled squares*), and methanogenesis (*filled triangles*) from June to August. *Error bars* show ±SE, n = 3–5 replicate cores. *From Neubauer, S.C., Givler, K., Valentine, S., Megonigal, J.P., 2005b. Seasonal patterns and plant-mediated controls of subsurface wetland biogeochemistry. Ecology 86, 3334–3344.*

temperature effects on microbial processes. Bullock et al. (2013) subjected tidal freshwater wetland sediments to a range of temperatures and found that rates of Fe(III) reduction were 50% more sensitive to temperature than Fe(II) oxidation. The result was that Fe(III) oxide pools declined with warming temperatures when the two processes are coupled, similar to patterns observed in the field (Neubauer et al., 2005b; Keller et al., 2013).

Manganese respiration has received very little attention because concentrations of Mn(III, IV) are usually far lower than Fe(III) in soils (Neubauer et al., 2005b). In theory, this limitation could be overcome by differences in Fe and Mn chemistry, such as the fact that Mn(III,IV) reduction is favored thermodynamically over Fe(III) reduction. Indeed, solid-state Au/Hg voltammetric microelectrode profiles in a tidal freshwater wetland suggested that Mn(III, IV) reduction is more important than Fe(III) reduction in some locations (Ma et al., 2008).

Humic substances are the most recent class of terminal electron acceptor identified in anaerobic substrates (Nevin and Lovley, 2000), and the process remains virtually unstudied in any ecosystem. Keller et al. (2009) amended tidal freshwater wetland soil with humic substances extracted from tidal freshwater plant species (*Nuphar advena*, *P. australis*, *Salix nigra*, and *Typha latifolia*). The extract amendments inhibited CH_4 production, supporting the notion that microbial reduction of humic substances yields more free energy than methanogenesis and that humic substance reduction can suppress CH_4 production. Indeed, humic substance respiration contributed 33%–61% of anaerobic respiration in bog soils (Keller and Takagi, 2013). Humic substance respiration may explain why the amount of CO_2 and CH_4 produced in root-free, anaerobic soil incubations often far exceeds the summed contributions of denitrification, metal reduction, and sulfate reduction (Neubauer et al., 2005b; Keller and Bridgham, 2007).

Sulfate reduction is often assumed to be unimportant in tidal freshwater wetlands because of limitation by SO_4^{2-} at concentrations <1 mM (Weston et al., 2006), but the sparse literature on the process in tidal freshwater wetlands suggests there is a need for more research. For example, SO_4^{2-} reduction rates were an order of magnitude higher in a tidal freshwater marsh than a polyhaline marsh (7 vs. 144 mmol m^{-2} d^{-1}, respectively), despite lower SO_4^{2-} concentrations at the tidal freshwater wetland (Segarra et al., 2013). A possible explanation for this unexpected pattern is that SO_4^{2-}-reducing bacteria had a greater affinity for SO_4^{2-} in the tidal freshwater wetland (Ingvorsen and Jørgensen, 1984). Similarly, Weston et al. (2014) reported that SO_4^{2-} reduction mineralized 60% as much organic C as methanogenesis. By contrast, Neubauer et al. (2005b) reported that SO_4^{2-} reduction in a tidal freshwater wetland was <2% of anaerobic C metabolism, while methanogenesis was 40%–70% (Fe reduction was the remainder). It would be profitable to resolve the causes of spatial and temporal variation in tidal freshwater wetland SO_4^{2-} reduction rates, in part because of the important implications for CH_4 emission rates.

3.4 Methane Regulation by Other Anaerobic and Aerobic Microbial Processes

Plants enhance CH_4 emissions as the primary source of organic C that supports methanogenesis and simultaneously depresses CH_4 emissions as a source of O_2 that either inhibits CH_4 production or enhances CH_4 oxidation. Several lines of evidence suggest that methanogenesis is tightly coupled to photosynthesis and NPP (Megonigal et al., 2004). In tidal freshwater wetlands, the evidence includes plant removal experiments (van der Nat and Middelburg, 1998a), a $^{14}CO_2$ tracing experiment (Megonigal et al., 1999), and relationships between CH_4 emissions and photosynthetic rates (Vann and Megonigal, 2003). Plants reduce potential CH_4 emissions by releasing O_2 into the rhizosphere that generates competing terminal electron acceptors such as Fe(III) and supports microbial CH_4 oxidation. The combined effects of plants on CH_4 production, oxidation, and transport generally favor higher net CH_4 emissions. Evidence of this includes higher CH_4 emissions from tidal freshwater wetlands exposed to elevated versus ambient CO_2 (Megonigal and Schlesinger, 1997; Vann and Megonigal, 2003) and higher emissions in the presence versus absence of plants (Kelley et al., 1995; van der Nat and Middelburg, 2000). However, when Keller et al. (2013) tested this hypothesis by removing plants in a tidal freshwater marsh field experiment, there was no difference in potential CH_4 production between treatments. One explanation for this result is that plant effects on CH_4 production and oxidation approximately balanced in this case.

Methanogenesis is suppressed when there is an adequate supply of competing electron acceptors. In mineral-rich tidal freshwater wetland soils, the dominant competing electron acceptor is Fe(III), which plants regenerate as poorly crystalline Fe(III) oxides in the rhizosphere (Weiss et al., 2004). Suppression of methanogenesis by Fe(III) can range from complete to negligible depending on several factors, including plant activity (van der Nat and Middelburg, 1998a; Neubauer et al., 2005b; Keller et al., 2013). Two studies at the same tidal freshwater marsh arrived at opposite conclusions about the importance of plants as sources of Fe(III) oxides for Fe-reducing bacteria. Neubauer et al. (2005b) found that Fe(III) reduction was strongly related to plant activity, whereas Keller et al. (2013) reported Fe(III) reduction rates that were 75% lower and largely unaffected by plants. The two study sites were separated by 440 m but differed in soil mineral content and plant species (Keller et al., 2013), factors that influence ROL and Fe(III) oxide availability.

Species-specific plant traits can indirectly regulate CH_4 production via rhizosphere Fe cycling. Sutton-Grier and Megonigal (2011) planted mesocosms with the tidal freshwater wetland species *Peltandra virginica*, *T. latifolia*, *Juncus effusus*, and *P. australis* and found that plant traits such as biomass are related to ROL, rhizosphere generation of Fe(III) oxides, and CH_4 emission rates. Tidal freshwater wetland sites and plant species have been used to elucidate the mechanisms that create rhizosphere hotspots of Fe-cycling compared with nonrhizosphere soil. Fe plaque deposits on wetland plant roots are enriched in poorly crystalline Fe(III) oxide minerals and Fe-reducing bacteria and therefore support relatively high rates of Fe(III) reduction under anaerobic conditions (Weiss et al., 2003, 2004, 2005). Under aerobic rhizosphere conditions, Fe(II)-oxidizing bacteria accelerate Fe(II) oxidation by 18%–83% (Neubauer et al., 2007, 2008).

Sulfate reduction inhibits methanogenesis, but the two processes nevertheless coexist even in highly saline soils because of spatial and temporal variation in the supply of electron donors and acceptors (Megonigal et al., 2004). Sulfate concentrations typically decrease with depth in tidal freshwater wetland soils (Segarra et al., 2013), but this does not necessarily translate into lower SO_4^{2-} reduction rates. Sulfate reduction in tidal freshwater wetlands sometimes maintains high rates at depth and appears to occur simultaneously with CH_4 production (Segarra et al., 2013; Weston et al., 2011). One mechanism that can explain the coexistence of these processes is that the two groups of microorganisms use different electron donors and therefore are not competing (Oremland et al., 1982; Segarra et al., 2014).

Aerobic CH_4-oxidizing bacteria are abundant in microaerobic zones of tidal freshwater wetland rhizospheres and have the potential to respond to variations in plant physiology and morphology that influence ROL. In tidal freshwater wetland mesocosms planted with *P. australis* and *S. lacustris*, both absolute CH_4 oxidation rate and CH_4 oxidation efficiency (i.e., as a percentage of CH_4 production) were significantly greater when the plants were actively growing than after they matured (van der Nat and Middelburg, 1998a). CH_4 oxidation apparently became O_2-limited as ROL declined. Because the capacity to transport O_2 was similar in growing versus mature plants, an increase in O_2 demand by roots or aerobic rhizosphere bacteria was the most likely cause of seasonality in CH_4 oxidation (van der Nat and Middelburg, 1998b).

Aerobic CH_4 oxidation in wetland soils can be O_2-limited or CH_4-limited (Lombardi et al., 1997; Bosse and Frenzel, 1998). CH_4 oxidation in a tidal freshwater swamp was linearly related to CH_4 production (Fig. 19.5), indicating that the process was CH_4-limited (Megonigal and Schlesinger, 2002). By contrast, CH_4 oxidation was O_2-limited in tidal freshwater marsh mesocosms (van der Nat and Middelburg, 1998b). These different conclusions may be explained by the observation that CH_4 transport in the continuously flooded marsh mesocosms occurred across the thin, oxidized zone around roots, whereas transport in the tidal freshwater swamp also occurred across a relatively thick (5 cm) oxidized zone at the soil surface. The high efficiency of CH_4 oxidation at the soil surface may have made the process less dependent on O_2 availability at the swamp site.

Anaerobic CH_4 oxidation rates were quantified in great detail at a tidal freshwater marsh site on the Altamaha River, USA. Rates of anaerobic CH_4 oxidation were comparable at a

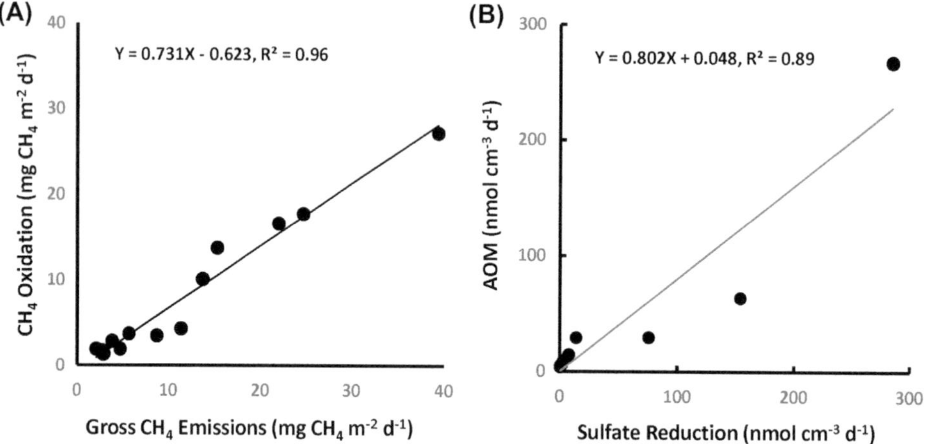

FIGURE 19.5 (A) Relationship between rates of gross CH_4 emission and aerobic CH_4 oxidation measured over a 13-month period in two tidal freshwater wetlands, suggesting CH_4 limitation of aerobic oxidation rates. (B) Relationship between rates of anaerobic oxidation of methane (AOM) and sulfate reduction, suggesting SO_4 limitation of AOM. *(A) From Megonigal, J.P., Schlesinger, W.H., 2002. Methane-limited methanotrophy in tidal freshwater swamps. Global Biogeochemical Cycles 16, 1062. https://doi.org/10.1029/2001GB001594. (B) Redrawn with permission from Segarra, K.E.A., Schubotz, F., Samarkin, V., Yoshinaga, M.Y., Hinrichs, K-U., Joye, S.B., 2014. High rates of anaerobic methane oxidation in freshwater wetlands reduce potential atmospheric methane emissions. Nature Communications 6, 7477. https://doi.org/10.1038/ncomms8477.*

tidal freshwater marsh and a polyhaline marsh (Fig. 19.5; Segarra et al., 2013). Rates of anaerobic CH_4 oxidation at the tidal freshwater marsh exceeded rates of methanogenesis, suggesting that the process has the capacity to significantly suppress CH_4 emissions. The electron acceptor coupled to anaerobic CH_4 oxidation is uncertain but includes SO_4^{2-}, NO_3^-, NO_2^-, Fe and Mn oxides, and humic substances. A strong correlation between SO_4^{2-} reduction and anaerobic CH_4 oxidation rates (Segarra et al., 2014) suggests that anaerobic CH_4 oxidation is a process that allows methanogenesis and SO_4^{2-} reduction to coexist.

4. NITROGEN BIOGEOCHEMISTRY

The biogeochemistry of N, like that of C, is strongly influenced by the supply and availability of electron acceptors and electron donors, as well as the degree of soil oxidation, leading to considerable variability in the importance of individual processes across the marsh surface and over time. For example, sediment deposition and long-term burial of N are often higher close to creekbanks than in the marsh interior (Merrill, 1999; Neubauer et al., 2005a). Porewater concentrations of NH_4^+ can be higher in topographically low hollows versus elevated hummock areas (Courtwright and Findlay, 2011), a pattern that may be driven by higher rates of nitrification of NH_4^+ to NO_3^- in hummocks (Noe et al., 2013). Daily rates of denitrification depend on multiple factors including the duration of inundation (Ensign et al., 2013), which will vary between high marsh and low marsh zones and between hummocks and hollows. Determining the existence and persistence of "hotspots and hot moments" (McClain et al., 2003) and "control points" (Bernhardt et al., 2017) within tidal freshwater wetlands remains an enduring challenge for determining ecosystem rates of N cycling.

Tidal freshwater marsh nutrient studies have historically focused on understanding marsh effects on estuarine water quality and were designed to quantify exchanges of dissolved inorganic nitrogen (DIN) between marshes and tidal waters (i.e., "flux studies") (Grant and Patrick, 1970; Heinle and Flemer, 1976; Simpson et al., 1978; Bowden, 1986; Chambers, 1992; Campana, 1998; Ziegler et al., 1999). This approach provides valuable information and can average across small-scale spatial variability within an individual basin, but it is limited by high variability between marsh basins and across time and difficulties in obtaining accurate hydrologic budgets to scale up the measurements. Flux studies do not provide detailed information on internal transformations that are occurring within marsh soils and sediments. Numerical simulation models allow process rates to be calculated from measurements of soil organic and inorganic nutrients (Morris and Bowden, 1986) and, if robustly designed, can be used to explore how the system might respond to future environmental changes (e.g., level of watershed nutrient loading). Recently, isotope tracers have been used to determine both the fate of water column N and the processes by which the N is transformed or removed from the water column (Gribsholt et al., 2005, 2006; 2007; Drake et al., 2009). This approach provides an elegant means of quantifying N transformations, which eliminates many of the issues associated with flux studies and process rate measurements. Isotope tracer studies can be used to quantify fluxes and transformations in the water column (Tobias et al., 2003; Gribsholt et al., 2005, 2006, 2007) and processes occurring in the root zone across a range of temporal scales (White and Howes, 1994; Tobias et al., 2001b; Drake et al., 2009).

We only know of two comprehensive N models for tidal freshwater marshes that consider exchanges of N between the marsh, estuary, and atmosphere, as well as internal N transformations in soils (Fig. 19.6A and B; Bowden et al., 1991; Neubauer et al., 2005a). Both models

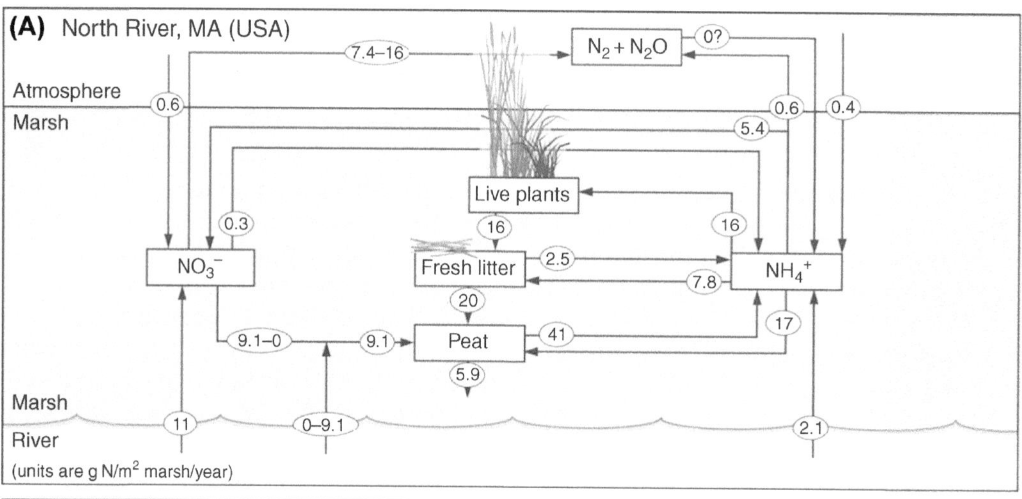

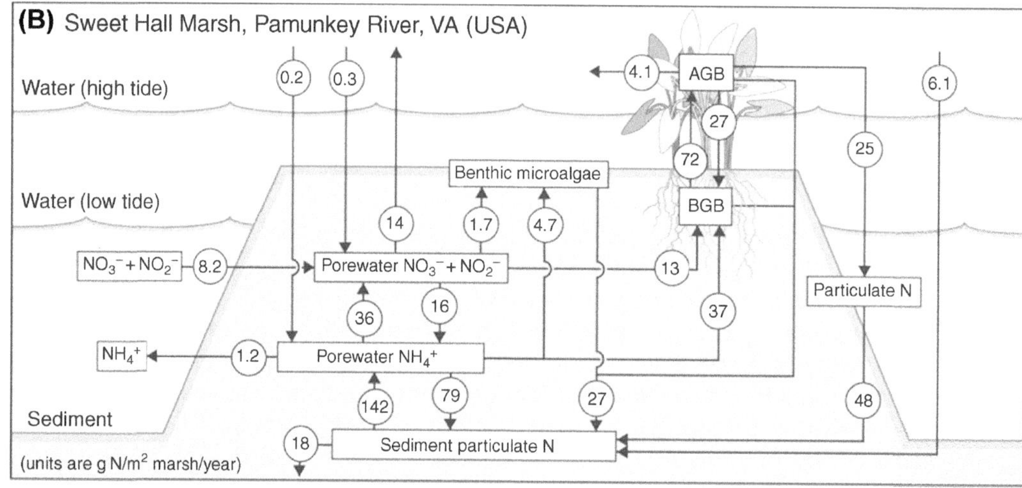

FIGURE 19.6 Ecosystem-scale tidal freshwater marsh N models. (A) Process-oriented N budget for a 15.8 ha marsh based on field, laboratory, and modeling efforts. (B) Process-based nitrogen mass balance model for a 401 ha marsh based on measured field fluxes, with unmeasured fluxes calculated from literature values or to force the model to steady-state conditions. Fluxes in each figure are g N m^{-2} $year^{-1}$. *AGB*, aboveground biomass; *BGB*, belowground biomass. *(A) Figure is redrawn from Bowden, W.B., Vorosmarty, C.J., Morris, J.T., Peterson, B.J., Hobbie, J.E., Steudler, P.A., Moore, B., 1991. Transport and processing of nitrogen in a tidal freshwater wetland. Water Resources Series 27, 389–408. (B) Figure is reproduced, with slight modifications, from Neubauer, S.C., Anderson, I.C., Neikirk, B.B., 2005a. Nitrogen cycling and ecosystem exchanges in a Virginia tidal freshwater marsh. Estuaries 28, 909–922.*

were built with seasonal data from different locations. Although neither is seasonally or spatially explicit, both integrate measurements made at multiple times and in multiple locations within a wetland. The model by Bowden and colleagues (Fig. 19.6A) describes N cycling at a North River, Massachusetts, USA, marsh that had organic-rich soils (40%–63% organic matter) and a well-developed, persistent plant litter layer. Neubauer and collaborators (Fig. 19.6B) studied Sweet Hall marsh on the Pamunkey River, USA, which had relatively little plant litter and more mineral soils (16%–21% organic). Average nutrient concentrations (NH_4^+ and NO_3^-) in the North River were ~4 times greater than those in the Pamunkey River (Neubauer et al., 2005a). Despite the limitations of mass balance modeling and differences between these marshes in plant community, soil type, marsh elevation, nutrient loading, and climate, several features of the N cycle were similar and may be common to tidal freshwater wetlands generally:

1. Exchanges of NH_4^+ and NO_3^- between the tidal freshwater wetlands and tidal waters were small compared with rates of internal N cycling and transformations in soils. Nitrogen in sediments deposited on the soil surface can be a significant source of new N to marshes, although the importance of this N source varies with flooding frequency and a suite of factors that influence sediment deposition rates (Darke and Megonigal, 2003).
2. Marsh-estuary exchanges of NO_3^- are generally directed into the marsh (i.e., net uptake by the marsh) and are similar in magnitude to rates of denitrification, suggesting the two processes are coupled.
3. Tidal freshwater wetlands are efficient at recycling and retaining nutrients within the soil profile. The efficiency of nutrient recycling may be greater in older wetlands with deep soils than in younger wetlands (Morris and Bowden, 1986) but less efficient than more closed, nontidal wetlands (Hopkinson, 1992).
4. The generation of inorganic N via organic matter mineralization can provide more than enough N to support primary production. This suggests that plant production may be largely uncoupled from nutrient loading in the adjacent tidal waters over relatively short periods of time (ca. one to several years). Over longer periods of time, the progressive assimilation and accumulation of water column N by the tidal freshwater wetland offsets N losses to denitrification and helps build the soil N pool, which can then be mineralized to support plant demands.

4.1 Nitrogen Exchanges

Exchanges of NH_4^+ between tidal freshwater wetlands and floodwaters are controlled by the diffusive gradient between soil porewaters and tidal waters, which is influenced strongly by microbial NH_4^+ assimilation and by advection in creekbank wetlands with significant tidal drainage. Tidal freshwater wetlands with an extensive litter layer often show net NH_4^+ uptake from the water column (Heinle and Flemer, 1976; Bowden, 1986), despite high porewater concentrations because the litter layer is acting as a sink for both porewater and water column NH_4^+. Other wetlands are sources of NH_4^+ to tidal waters (Campana, 1998; Ziegler et al., 1999; Neubauer et al., 2005a). Although wetland-estuary exchanges of NH_4^+ (and other N forms) may be significant on a whole-estuary basis, the magnitude of these fluxes is generally small relative to internal N transformations occurring in soils (e.g., Fig. 19.6A and B). In a pair of elegant $^{15}NH_4^+$ labeling experiments conducted in May (early growing season with active plant growth) and September (late growing season with senescent, flowering plants), the fate of water column NH_4^+ was tracked over about 15 days in an N-rich tidal freshwater marsh in Belgium, EU (Gribsholt et al., 2005, 2006, 2007). Despite the temporal separation between the experiments, the fates of NH_4^+ were remarkably similar between the months. In each experiment, the majority of the water column NH_4^+ was exported from the system without being transformed by the marsh (Fig. 19.7). Approximately 4% of the NH_4^+ was sequestered in plant biomass, litter, or soil (either via physical sorption or microbial assimilation). Overall, microbial pathways of N uptake were more important than the direct assimilation of tidal water NH_4^+ by plants (Gribsholt et al., 2006). However, plants are likely to play indirect roles in modifying water column N loads by providing both O_2 and labile organic C to soil microbes. The ^{15}N label was also found in other N pools within the water column, indicating that active N transformations were occurring within the water column and/or in flooded marsh soils. Of these N transformations, nitrification accounted for the largest fraction of the added NH_4^+, with smaller amounts in the suspended particulate N, N_2, and N_2O pools. The marsh soils appeared to be a significant site for nitrification in May (Gribsholt et al., 2005), whereas soil denitrification rates were highest in September (Gribsholt et al., 2006).

In contrast to the high variability in the direction and rate of NH_4^+ exchanges between tidal freshwater wetland soils and tidal waters, tidal freshwater wetlands are generally strong sinks for water column NO_3^- (Arrigoni et al., 2008; Bowden et al., 1991; Findlay and Fischer, 2013; McKellar et al., 2007; Neubauer et al., 2005a; Van Damme et al., 2009). This NO_3^- is

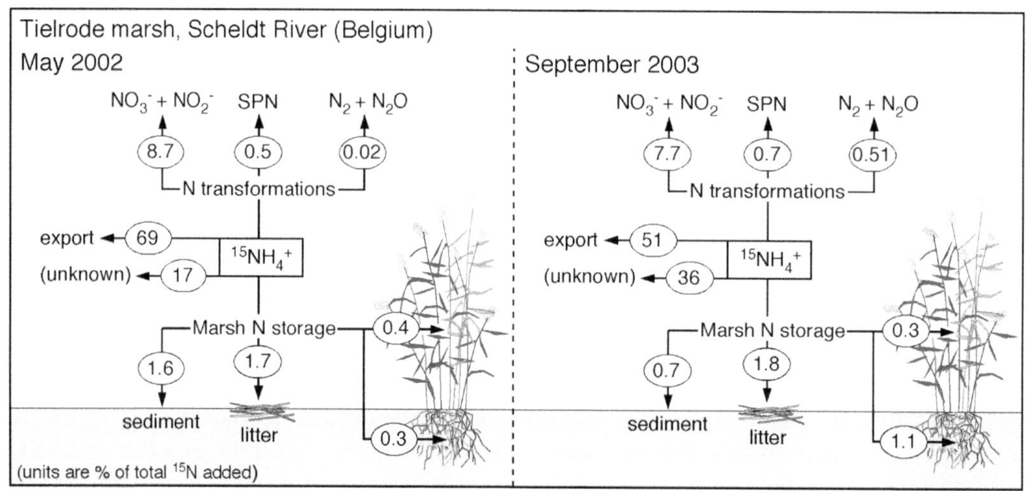

FIGURE 19.7 Transformations and uptake of NH_4^+, as inferred by whole-ecosystem $^{15}NH_4^+$ labeling of a 3477 m^2 marsh area (Gribsholt et al., 2006). Fluxes indicate the percentage of the added $^{15}NH_4^+$ label that was recovered in each N pool after label addition. The total input of ^{15}N was 1.97 and 1.41 mol $^{15}N{-}NH_4^+$ in May and September, respectively; the label addition increased the total tidal water NH_4^+ concentration by 14% (May) to 73% (September). *SPN*, suspended particulate N.

often used to support denitrification in tidal freshwater wetland soils (see Section 4.2). Although there is generally a consistent trend of NO_3^- uptake by tidal freshwater wetland soils, it has proven difficult to develop predictive relationships to quantify the amount of NO_3^- uptake. Findlay and Fischer (2013) reported that 40% of the variation in the amount of NO_3^- decline between flood and ebb tide could be explained by the coverage of graminoid-dominated high intertidal vegetation within a wetland basin. However, Arrigoni et al. (2008), who worked with a smaller subset of tidal freshwater wetlands in the same Hudson River (NY, USA) region studied by Findlay and Fischer (2013), reported no ability to predict NO_3^- fluxes on the basis of vegetation type or geomorphic factors. Flux studies in tidal freshwater wetlands on the Patuxent River, USA, have shown that the amount of NO_3^- removal is strongly correlated with the incoming load of NO_3^-, but only weakly with the concentration of incoming NO_3^- (Seldomridge and Prestegaard, 2014), suggesting that nitrate removal is limited by the hydrological delivery of NO_3^- to tidal freshwater wetlands and not to kinetic factors that control rates of denitrification (Seldomridge and Prestegaard, 2011). Structural equation modeling pointed toward tidal exchange volume, ambient NO_x concentrations (which determine the NO_3^- load), and ecosystem respiration as the primary controls on NO_3^- removal in a recently restored tidal freshwater wetland in Virginia, USA (Bukaveckas and Wood, 2014). Despite lower fluvial NO_3^- loading, total NO_3^- removal in tidal freshwater rivers is greatest in warmer months because denitrification is a biological process that increases with temperature (Bukaveckas et al., 2018). Denitrification is often the ultimate sink for water column NO_3^-, converting it to N_2 gas.

Tidal freshwater wetlands are often significant sinks for water column particulate N deposited on the soil during tidal flooding. The N content of accumulated sediments ranged from 4 to 16 mg N g^{-1} sediment in several tidal freshwater marshes in Virginia (Morse et al., 2004; Neubauer et al., 2005a) and is significantly correlated with the soil N content (Morse et al., 2004). The sediment-associated N is presumably a combination of detrital material, microbial biomass, and NH_4^+ sorbed to mineral surfaces. On an annual basis, inputs of allochthonous particulate N can be large with respect to the marsh N budget, contributing up to 20 g N m^{-2} $year^{-1}$ (Bowden et al., 1991; Morse et al., 2004; Neubauer et al., 2005a). There is significant spatial variation in deposition rates driven by marsh elevation and flooding frequency (Morse et al., 2004). Over decadal scales, the burial of N sequesters significant amounts of N in tidal freshwater marsh soils, on the order of 10–30 g N m^{-2} $year^{-1}$ (Table 19.1 and references therein). Deposition and burial rates are much higher in Sweet Hall marsh than the North River marsh (Fig. 19.6A and B), a difference reflected in the lower organic content of Sweet Hall marsh soils. Indeed, regional patterns of sedimentation may explain why soil accretion rates in tidal freshwater wetlands of the Northeast United States are correlated only with organic accumulation, whereas those in the Southeastern United States are correlated with both mineral and organic accumulation (Neubauer, 2008).

Over each tidal cycle, a large volume of water floods and ebbs from the surface of tidal freshwater wetlands. Because of the high rate of surface water exchange, wetland uptake of DIN from floodwaters is inefficient (Hopkinson, 1992) and can meet only a small fraction of plant N demand. This leads to the "requirement" that existing nutrients are retained within the wetland. Indeed, only ~1% of the total N supplied to the NH_4^+ pool from external and internal sources is lost; the remainder is recycled to other N reservoirs in the marsh (Fig. 19.6A and B). It is likely that the slow turnover of marsh porewater (67 to >800 days in the upper 30 cm at Sweet Hall marsh) drives this efficient N retention. Microbial immobilization of NH_4^+ into particulate matter is a primary mechanism by which N is retained in the marsh; this mechanism can retain >50% of mineralized N in tidal freshwater (Bowden et al., 1991; Neubauer et al., 2005a) and salt marshes (Anderson et al., 1997). In contrast to the porewater NH_4^+ pool, plant biomass N in tidal freshwater wetlands is directly exposed to the tides and less efficiently retained, with ~50% exported as dissolved or particulate organic matter (Hopkinson, 1992; Neubauer et al., 2005a). In support of the link between water turnover and system closure, there was little evidence of N export from a "periodically flooded" high marsh (Bowden et al., 1991).

4.2 Nitrogen Transformations

Tidal freshwater marshes permanently remove DIN from riverine and estuarine waters via burial and denitrification (Section 4.1), the reduction of NO_3^- to gaseous N_2. Sources of NO_3^- include nitrification within the marsh and the uptake of external (water column) NO_3^-. Mass balance calculations indicate high rates of N removal in upper estuaries (Howarth et al., 1996), and many characteristics of tidal freshwater wetlands appear to favor denitrification, such as high active surface area, shallow depth to anaerobic zone, and high organic matter availability. High denitrification rates have been confirmed in multiple tidal freshwater wetland studies (Groszkowski, 1995; Merrill, 1999; Merrill and Cornwell, 2000; Elsey-Quirk et al., 2013; Ensign et al., 2008, 2013; Von Korff et al., 2014). Greene (2005) reported that median denitrification rates for a tidal freshwater marsh (~120 $\mu mol\ N\ m^{-2}\ h^{-1}$) were slightly larger than the median rate for a wide range of intertidal and aquatic systems (~75 $\mu mol\ N\ m^{-2}\ h^{-1}$). However, there is considerable spatial variability between and within tidal freshwater wetlands (Merrill, 1999; Greene, 2005).

The environmental controls on denitrification have been extensively reviewed (Seitzinger, 1988; Cornwell et al., 1999; Wallenstein et al., 2006) and will not be covered in great detail here. In tidal freshwater marshes, denitrification rates are correlated with benthic sediment O_2 demand in a New York marsh, but not in a tidal freshwater wetland in Maryland, USA (Merrill, 1999). Based on laboratory manipulations, denitrification rates increase with increases in water column NO_3^- (Merrill, 1999; Greene, 2005). Similarly, the N models of Bowden et al. (1991) and Neubauer et al. (2005a) suggest that denitrification is supported primarily by water column NO_3^-. Because O_2 diffusing from roots and across the soil surface can support oxidation in tidal freshwater marsh soils (Neubauer et al., 2005b), denitrification is likely to be coupled to both in situ nitrification and water column NO_3^- uptake. Gribsholt et al. (2005, 2006, 2007) presented evidence for coupled nitrification–denitrification in tidal freshwater wetlands; following the addition of a $^{15}NH_4^+$ to tidal floodwaters, some of the ^{15}N label appeared in the dissolved N_2 and N_2O pools (Fig. 19.7). Much of this nitrification takes place in marsh soils—possibly associated with plant roots—rather than in the water column (Gribsholt et al., 2005). The importance of the soil as a site for nitrification can vary seasonally (Neubauer et al., 2005a; Gribsholt et al., 2006) and spatially, with higher-elevation hummocks having roughly three times higher rates of nitrification than lower-elevation hollows (Noe et al., 2013). Most laboratory measurements of denitrification give potential rates in that incubations are done under anaerobic conditions. Scaling these measurements across space and time requires considering the duration that a particular location in a tidal freshwater wetland is sufficiently reduced (i.e., has a low-enough redox potential) to support denitrification, which itself is a function of soil characteristics and marsh topography (Elsey-Quirk et al., 2013; Ensign et al., 2008, 2013; Von Korff et al., 2014).

When integrated over the entire network of tidal freshwater wetlands within an estuary, nutrient removal may be substantial because the small contributions of individual marshes can have a large cumulative impact on water quality. This is especially true in systems with large areas of tidal marsh relative to open water. For example, in the Patuxent and Choptank rivers, USA, slightly more than 30% of the total N input at the fall line is permanently removed by low-salinity tidal marshes via burial and denitrification (Merrill, 1999; Malone et al., 2003). In contrast, N removal by tidal freshwater marshes in larger systems such as the Hudson and Delaware Rivers, USA, is less efficient, with only ~2%–5% of the N sequestered or denitrified (Academy, 1998; Merrill, 1999; Elsey-Quirk et al., 2013).

Dissimilatory NO_3^- reduction to NH_4^+ (DNRA) is a mechanism by which NO_3^- can be retained in the marsh, rather than lost to the atmosphere as N_2 as in denitrification. There is very little evidence from tidal freshwater wetlands on the relative importance of DNRA and denitrification as alternative fates for soil NO_3^-. Bowden (1986) determined that DNRA rates were 5% of NO_3^- supply (i.e., nitrification) rates. In contrast, Neubauer et al. (2005a) calculated that DNRA was about 40% of nitrification. Based on work in other systems, the availability of labile C relative to NO_3^- (i.e., electron donor: electron acceptor ratio) is important in determining the fate of NO_3^-, with high organic C availability favoring DNRA (Fazzolari et al., 1998; Christensen et al., 2000; Giblin et al., 2013; Algar and Vallino, 2014) and denitrification increasing in importance at higher NO_3^- concentrations (Nijburg et al., 1997; Tobias et al., 2001a,b). Thus, relatively low NO_3^- in the Pamunkey River (Neubauer et al., 2005a) may explain the higher importance of DNRA in that system. Across the estuarine gradient, DNRA is generally more important (relative to denitrification) in estuarine and marine systems, whereas denitrification is more important in freshwater systems (Tobias et al., 2001b). This pattern may be related to sulfide inhibition of denitrification (Brunet and Garcia-Gil, 1996; An and Gardner, 2002).

4.3 Nutrient Regulation of Plant Production

It has proven difficult to unambiguously determine the nature (or existence) of nutrient limitations in tidal freshwater wetlands (Chambers and Fourqurean, 1991). Inconsistent responses to allochthonous nutrient inputs suggest that rates of plant production in tidal freshwater marshes are largely uncoupled from allochthonous nutrient inputs. Indeed, rates of N mineralization in tidal freshwater wetlands are considerably greater than plant N demand and rates of diffusive DIN uptake from tidal waters (Fig. 19.6A and B; Bowden et al., 1991; Neubauer et al., 2005a). In these studies, gross N mineralization provided almost three times more NH_4^+ than was needed to support annual plant N requirements. Furthermore, there are multiple examples in which direct fertilization of tidal freshwater wetlands with N, P, or N + P did not increase either aboveground biomass or biomass nutrient content (Whigham and Simpson, 1978; Walker, 1981; Booth, 1989; Chambers and Fourqurean, 1991; Morse et al., 2004).

There are some examples in which fertilization increased plant growth (Booth, 1989; DeLaune et al., 1986; Frost et al., 2009). Instances when tidal freshwater wetland plant productivity responds to fertilization may reflect differences in plant species composition (e.g., annuals vs. perennials, or graminoid vs. broadleaf vs. woody) and/or the level of applied fertilizer. A gradient fertilization study of a Virginia, USA, tidal freshwater wetland applied N and P fertilizers in four annual doses ranging from 18 to 73 $g\,N\,m^{-2}$ and 0.5–1.7 $g\,P\,m^{-2}$ (for comparison, annual tidal nutrient inputs to the site were roughly 60 $g\,N\,m^{-2}$ and 2.9 $g\,P\,m^{-2}$). Compared to control plots, there were few significant plant responses to fertilization at lower fertilization levels, but several (not all) species responded with higher growth at the highest levels of fertilization (Burton and Neubauer, 2018).

There is a generalization that aquatic primary production is limited by phosphorus (P) availability in freshwaters and by N in brackish and saline waters, but this does not appear to hold in tidal wetlands, where primary production is often limited by N in both tidal freshwater wetlands and salt marshes. This is illustrated by a factorial nutrient addition study in a *Zizaniopsis miliacea*–dominated tidal freshwater wetland in Georgia, USA. Aboveground plant productivity significantly increased following N additions but did not change (relative to controls) with P fertilization. Furthermore, plant production was no higher in experimental plots that received both N and P, compared with N-only plots (Gautam, 2015). Similarly, P additions did not affect primary productivity in tidal freshwater wetlands in Virginia and Georgia, USA (Morse et al., 2004; Frost et al., 2009), although Baldwin (2013) observed that N and P had contrasting effects on annuals versus perennials.

5. PHOSPHORUS BIOGEOCHEMISTRY

Because tidal freshwater wetlands are located in upper estuaries where watershed-derived inputs of P are concentrated, these intertidal systems may be key sites in landscapes for P sequestration and transformation. In a pair of tidal freshwater marshes, sediments deposited on the soil surface contained 0.3–1.7 $mg\,P\,g^{-1}$ sediment and contributed inputs of 0.6–2.3 $g\,P\,m^{-2}\,year^{-1}$ (Morse et al., 2004). Uptake of inorganic P by organisms results in a relative enrichment of organic P in surface soils (Morse et al., 2004). In a South Carolina tidal freshwater wetland, much of the soil organic P was bound to humic acids, whereas the inorganic P was primarily associated with Fe or Al (Paludan and Morris, 1999).

Phosphatase is an enzyme that liberates organic-bound P and is secreted by plants, algae, and bacteria under conditions of PO_4^{3-} limitation. Phosphatase activity is expected to be greatest where the biological demand for inorganic P is high, most soil P is in organic forms, and soil sorption limits porewater PO_4^{3-} concentrations. The activities of three phosphatase enzymes were highly positively correlated with aboveground plant biomass and soil organic content in a successional sequence of tidal freshwater wetlands (Huang and Morris, 2003). Similar correlations between phosphatase activity, soil organic matter, and soil organic P were observed along a salinity gradient, with the highest activity in tidal freshwater wetlands

(Huang and Morris, 2005). By comparison, Morrissey et al. (2014) reported lower phosphatase activity in a tidal freshwater wetland than in oligohaline wetlands.

Organic P made up a greater fraction of total P in the late (intertidal marsh) versus early (open water) successional stages. The interactions between P availability and demand may lead to a positive feedback that drives ecosystem succession, whereby high demand for inorganic P reduces available PO_4^{3-} concentrations, leading to an increase in phosphatase activity and increased organic P mineralization, further increasing plant growth (Huang and Morris, 2003). Over years to decades, P mineralization results in a significant decrease in organic P concentrations with increasing soil depth (Paludan and Morris, 1999). Despite high phosphatase enzyme activities in tidal freshwater wetland soils, concentrations of dissolved PO_4^{3-} are often low because of the combined effects of biological demand and chemical sorption processes that remove free PO_4^{3-} from marsh porewaters. Sundareshwar and Morris (1999) showed that P sorption rates were lower in systems where sediments had lower surface areas and lower Fe/Al mineral contents. The lack of significant accumulation of inorganic P in deeper soils (Paludan and Morris, 1999) implies that organic P is not simply mineralized to inorganic P but is instead removed from the soil via plant uptake or hydrological export. The amount of extractable total P and extractable inorganic P is related to soil type, where soils that were higher in clay content and bulk density had greater P concentrations, compared with soils that were higher in organic matter and sand content (Noe et al., 2013). Higher total P content was associated with higher rates of P mineralization and a greater P turnover rate (Noe et al., 2013).

Although tidal freshwater wetlands are sinks for sediment-associated particulate P, PO_4^{3-} fluxes are highly variable and there can be net PO_4^{3-} uptake (Simpson et al., 1978; Gilbert, 1990), seasonal variability (Simpson et al., 1983; Campana, 1998), or negligible PO_4^{3-} fluxes (Anderson et al., 1998) between tidal freshwater wetlands and tidal waters. One factor that may affect spatial and temporal variations in marsh-estuary tidal fluxes of PO_4^{3-} is the interplay between P, Fe, and O_2 dynamics. In soils that are regularly exposed to O_2 during low tides (e.g., the marsh surface and creekbank edges), Fe(II) can oxidize to Fe(III) and lead to the formation of an "iron curtain" of iron oxyhydroxide minerals that efficiently sorb PO_4^{3-}, causing P retention in the marsh (Chambers and Odum, 1990). In combination with diagenetic effects, this mechanism may contribute to decreases in total soil P content with increasing depth, which has been observed in some tidal freshwater wetlands (Bowden, 1984; Chambers and Odum, 1990; Merrill, 1999; Paludan and Morris, 1999), though not in all (Simpson et al., 1983; Greiner and Hershner, 1998). The mineral content of the soils is likely to play a role in the efficiency of such an iron curtain because minerals are rich in Fe. The ecological implications of the iron curtain on ecosystem P dynamics are unclear because it retains a potentially limiting nutrient within the marsh, but not necessarily in a bioavailable form. That said, P mineralization rates in tidal wetlands along an alluvial river were positively correlated with total soil Fe, suggesting that reduction of Fe—P minerals was contributing to P release (Noe et al., 2013). Similarly, the concentration of porewater PO_4^{3-} was inversely correlated with the concentration of Fe—P minerals in subtidal estuarine sediments (Jordan et al., 2008). Another implication is that high marshes, which are flooded less frequently and therefore exposed to air for longer periods of time, may have an extensive iron curtain that allows for more efficient P retention and recycling than low marshes, leading to increased P accumulation in high versus low marsh habitats (Khan and Brush, 1994).

The storage (burial) of P in marsh soils is an important mechanism by which P can be removed and sequestered from estuarine waters. Over timescales ranging from decades to centuries, tidal freshwater wetlands sequester significant amounts of P, with burial rates ranging from <0.5 to >4 g P m^{-2} $year^{-1}$ (Table 19.1 and references therein). These burial rates are roughly comparable with rates of sediment-associated P deposition onto the marsh surface (Morse et al., 2004). There is often significant variability in P burial rates within an individual marsh. This is correlated with plant community composition, which is a function of marsh elevation and proximity to creekbanks, and differs strongly between watersheds (Merrill, 1999; Merrill and Cornwell, 2000). For example, in the Patuxent River, USA, total P burial rates were significantly higher in the *Nuphar luteum* (spatterdock)–dominated low marsh (12 g P m^{-2} $year^{-1}$) relative to high marsh areas dominated by *Hibiscus moscheutos* (marsh hibiscus), *Typha* spp., or *Zizania aquatica* (northern wild rice) (1.2–2.5 g P m^{-2} $year^{-1}$) (Merrill, 1999). Extrapolating marsh P burial rates to a landscape scale shows that low-salinity tidal marshes can sequester a significant fraction (>60%) of watershed-derived P in relatively small, marsh-dominated estuarine systems (Merrill, 1999; Malone et al., 2003). In contrast, about 12% of the combined sewage and riverine-derived P to the upper Hudson River, USA, is buried in freshwater tidal and nontidal marshes (Limburg et al., 1986; Phillips and Hanchar, 1996; Merrill, 1999). Based on average literature values for marsh P burial, only ~7% of the P entering the upper Delaware River estuary, USA, is permanently buried in tidal marshes (Academy, 1998). Thus, tidal freshwater wetlands can be long-term sinks for significant amounts of watershed-derived P, but the extent of wetlands within the estuary (relative to the size of the estuary or watershed) appears to be important in determining how efficiently tidal freshwater wetlands perform this function.

6. SILICON BIOGEOCHEMISTRY

The weathering of terrestrial silicate (Si) minerals ultimately leads to inputs of dissolved silica (DSi) to estuaries and the coastal ocean where diatom production can be limited by low DSi availability. Evidence suggests that silica transformations in tidal freshwater wetlands play a key role in transforming silica from biogenic (BSi) to DSi forms. For example, under low discharge conditions (summer) in the Scheldt estuary, EU, the input of DSi from fluvial sources was as low as 10,000 kg $month^{-1}$, an amount that can be exported from the 450 ha of tidal freshwater wetlands in the system in six tidal cycles (Struyf et al., 2006). Because rates of DSi export increased with decreasing concentrations of DSi in estuarine waters, marsh-mediated recycling of Si may be especially important when low, ambient DSi concentrations otherwise limit aquatic primary productivity (Struyf et al., 2006). Indeed, marsh DSi export has the greatest effect on estuarine N:DSi ratios during summer when the concentrations of both DSi and inorganic N are lowest (van Damme et al., 2009). Understanding the factors that regulate DSi export from tidal freshwater wetlands requires additional research on biogeochemical transformations in tidal freshwater wetland soils (Struyf et al., 2005a, 2005b, 2007).

7. RESPONSES AND CONTRIBUTIONS TO GLOBAL CHANGE

In their landscape position between uplands and estuaries, tidal freshwater wetlands are subject to changes in the terrestrial landscape (freshwater, nutrient, and sediment delivery), the estuarine/oceanic seascape (sea level rise, saltwater intrusion), and the atmosphere (global warming, rising CO_2) (Neubauer and Craft, 2009). In this section, we briefly discuss the effects of saltwater intrusion on tidal freshwater wetland biogeochemistry, both because saltwater intrusion changes a defining characteristics of tidal freshwater wetlands—the existence of freshwater versus brackish conditions—and because saltwater intrusion has impacts on the biogeochemical cycling of multiple elements including C, N, P, Fe, and S. We also consider tidal freshwater wetlands as sinks for CO_2 and sources for CH_4 and the potential for tidal freshwater wetlands to influence global radiative forcing.

7.1 Biogeochemical Effects of Saltwater Intrusion

Saltwater intrusion is the upstream movement of brackish or saline water that causes concentrations of salt and other seawater-derived ions to increase above natural background levels (Herbert et al., 2015). The increase in global sea level has the potential to affect tidal freshwater wetlands by modifying their hydroperiod and by pushing the salt front up-estuary so that these systems are exposed to more saline waters. Similarly, decreases in river discharge that are caused by changes in watershed precipitation and/or anthropogenic activities (e.g., construction of dams) that reduce river flow can lead to increased salinity in the tidal freshwater zone. Future changes in salinity in a given tidal freshwater wetland will depend on the wetland's position on the current salinity gradient, rates of sea level rise, and changes in river discharge. Both sea level rise and river discharge are influenced by global warming (Burkett et al., 2001). We briefly discuss the impacts of saltwater intrusion on the biogeochemical cycling of C, N, P, Fe, and S in tidal freshwater wetlands, but this global change stressor affects tidal freshwater wetlands from physiological to ecosystem to landscape scales (Fig. 19.8).

Changes in salinity can affect a number of nutrient cycling processes including N and P sorption, denitrification, and nitrification (Howarth et al., 1988; Caraco et al., 1989; Rysgaard et al., 1999). The intrusion of saltwater and associated SO_4^{2-} can lead to the breakdown of the "iron curtain" described by Chambers and Odum (1990). For example, Meiggs and Taillefert (2011) showed enhanced Fe(III) reduction rates following saltwater intrusion. Along estuarine gradients, more sediment P is bound to Fe(III) oxides in freshwater, whereas the abundances of organic P and free PO_4^{3-} are greater in brackish water (Jordan et al., 2008). As Fe oxides are reduced either biologically (by Fe(III)-reducing bacteria) or chemically (via H_2S), sediment-bound P is released and the Fe(II) can be sequestered in Fe–S compounds (e.g., pyrite) (Caraco et al., 1989; Lamers et al., 2001). The release of soil P and increases in soil sulfides can lead to internal eutrophication (Smolders et al., 2010), reductions in plant productivity and diversity (Lamers et al., 2013; Sutter et al., 2014), and alterations to soil N cycling.

Production of the greenhouse gas N_2O can increase if high H_2S concentrations inhibit nitrification and denitrification, as has been shown for unvegetated sediments (Joye and Hollibaugh, 1995; Brunet and Garcia-Gil, 1996; An and Gardner, 2002). However, these

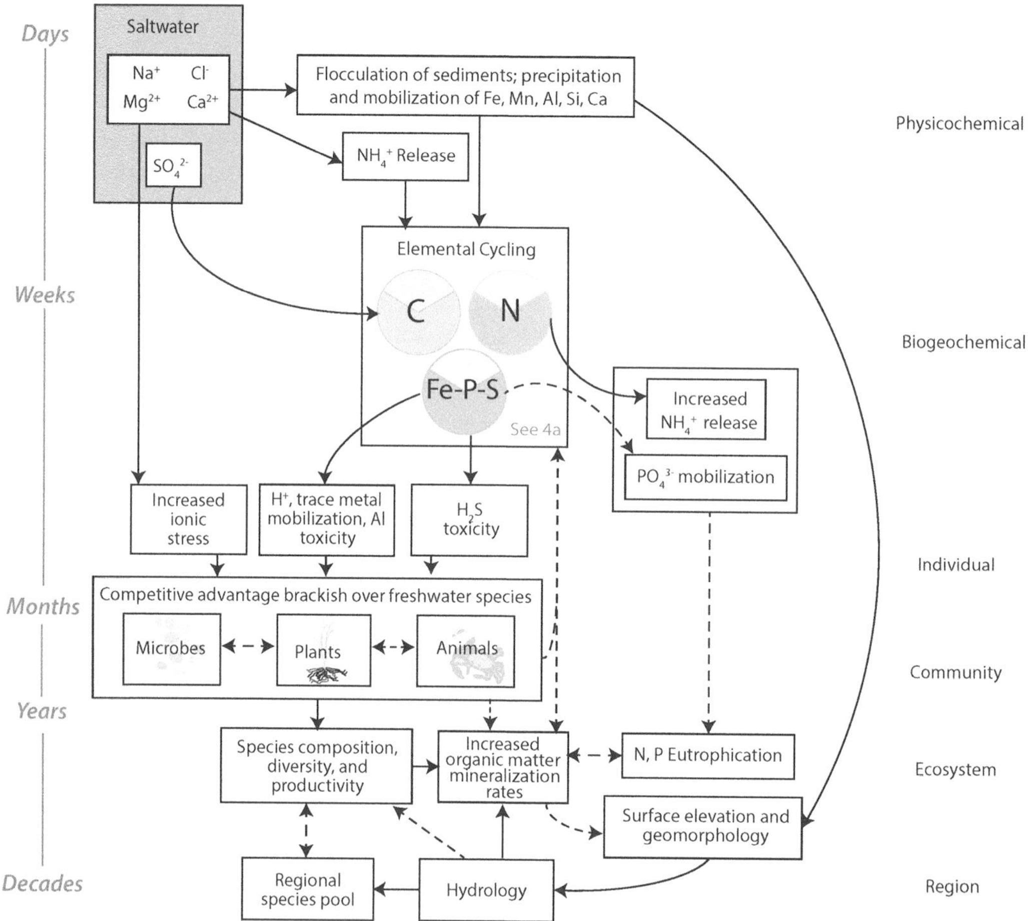

FIGURE 19.8 The predicted effects of saltwater intrusion span spatial scales from the microbial to the landscape and temporal scales from days to decades. Saltwater intrusion leads to increases in the concentrations of cations and anions (grey box in upper left), which subsequently affects wetland biogeochemistry and leads to effects than span multiple scales. Pathways for which there is a high degree of consensus on the expected direction of change are indicated by solid lines. The dashed lines highlight pathways for which the direction of change is unknown, either because existing research offers conflicting expectations or there is insufficient research to make confident predictions. *From Herbert, E.R., Boon, P., Burgin, A.J., Neubauer, S.C., Franklin, R.B., Ardón, M., Hopfensperger, K.N., Lamers, L.P.M., Gell, P., 2015. A global perspective on wetland salinization: ecological consequences of a growing threat to freshwater wetlands. Ecosphere 6 (10), 1–43. https://doi.org/10.1890/ES14-00534.1.*

effects may be minimized in tidal freshwater wetlands with large quantities of soil Fe because Fe(III) can scavenge sulfides that are produced by SO_4^{2-} reduction during saltwater intrusion episodes (Schoepfer et al., 2014). Elevated salinity can also lead to decreased sorption of NH_4^+ to soil particles as NH_4^+ is displaced by positively charged cations such as Na^+ and Mg^{2+}. Increased NH_4^+ concentrations may suppress N_2 fixation because nitrogenase is inhibited by free NH_4^+ (Howarth et al., 1988). The physiological effects of salinity on nitrifying and denitrifying microbes reduce the activity of these organisms (MacFarlane and Hebert, 1984; Furumai et al., 1988; Stehr et al., 1995; Rysgaard et al., 1999) and can lead to changes in denitrification rates (Craft et al., 2009; Marton et al., 2012). If increased SO_4^{2-} concentrations accelerate soil organic matter decomposition (see Section 3.1), rates of nutrient mineralization would also increase.

Over short time periods (months to years), saltwater intrusion acts as a stressor to tidal freshwater wetland plants and consequently reduces ecosystem-scale primary productivity. Over periods of several days (Chambers et al., 2011; Neubauer et al., 2013) to a year (Weston et al., 2011), saltwater intrusion also tends to increase soil carbon mineralization rates. Overall, salinity tends to lower marsh-atmosphere CO_2 emissions in the field due to a combination of lower plant respiration and lower inputs of labile organic matter to soils (Neubauer, 2013b; Lee et al., 2016; Neubauer et al., 2013).

Thermodynamics dictates that microbial SO_4^{2-} reducers will outcompete methanogens for substrates such as organic C. The contribution of SO_4^{2-} reduction to C mineralization is expected to increase with tidal freshwater wetland proximity to the saltwater boundary, which makes tidal freshwater wetlands and oligohaline wetlands susceptible to incursions of SO_4^{2-} due to drought, storm surge, and sea level rise. Saltwater intrusion favors the SO_4^{2-} reducers that compete against methanogens, leading to lower emissions of the greenhouse gas CH_4 (Megonigal et al., 2004; Neubauer and Craft, 2009). Measurements consistent with this prediction have been reported in multiple laboratory (Chambers et al., 2011; Neubauer et al., 2013) and field studies (Neubauer, 2013b; Lee et al., 2016; Herbert et al., 2018), although a smaller number of studies have reported unexpected increases in CH_4 production and/or emission following saltwater intrusion (Weston et al., 2011, 2014). Furthermore, there is evidence that increased SO_4^{2-} reduction rates will stimulate organic matter mineralization (Portnoy and Giblin, 1997; Weston et al., 2006; Craft, 2007; Neubauer et al., 2013). However, other studies reported no differences in rates of anaerobic organic matter mineralization in comparisons of brackish versus tidal freshwater marsh soils (Neubauer et al., 2005b, Fig. 19.4) and sediments (Kelley et al., 1990). Over days to months, saltwater intrusion may increase mineralization as existing C pools are degraded more rapidly because of increased rates of SO_4^{2-} reduction (Sutton-Grier and Megonigal, 2011), but longer-term (multiyear) saltwater intrusion can reduce pools of soil organic matter and lead to lower overall rates of mineralization (Neubauer et al., 2013). This is an important question to understand with respect to climate change because of the implications for soil C pools and nutrient turnover in former tidal freshwater wetlands (Neubauer and Craft, 2009). In tidal freshwater wetlands, an average of 62% of vertical accretion is because of the accumulation of organic matter (Neubauer, 2008). If saltwater intrusion reduces rates of organic matter accumulation by reducing rates of primary production and/or by increasing rates of decomposition, tidal freshwater wetlands will be less able to accrete vertically and keep pace with ever-increasing rates of sea level rise.

Thus far, there are no consistent trends in the impact of saltwater intrusion on NEP. NEP is the balance between GPP and ecosystem respiration and is used as a proxy for ecosystem carbon storage or loss. In a multiyear in situ salinity manipulation in a South Carolina, USA, tidal freshwater marsh, saltwater intrusion decreased annual NEP by 55%–63% during

2 years but had no effect in the third year (Neubauer, 2013b; Neubauer unpublished). Similarly, experimental in situ saltwater intrusion led to lower NEP in a Georgia, USA, tidal freshwater wetland (Herbert et al., 2018) but had no effect on NEP in a Virginia, USA, marsh (Neubauer and Lee, in preparation). Over longer time periods, salt-sensitive plants, animals, and microbes will likely be replaced by salt-tolerant species (Magalhães et al., 2005). There is relatively little known from direct manipulations of salinity about the direction of these longer-term effects or how they will affect ecosystem fluxes and carbon storage.

7.2 Radiative Forcing

High rates of primary production and carbon preservation (i.e., the inverse of decomposition) lead to high rates of C sequestration in tidal freshwater wetlands (Neubauer, 2013a, Table 19.1). However, tidal freshwater wetlands are also sources of CH_4 to the atmosphere (see Section 2.2.2) and may be either sources or sinks of N_2O (e.g., Krauss and Whitbeck, 2012; Krauss et al., 2016). Because tidal freshwater wetlands are both sources and sinks of greenhouse gases, there is interest in determining the overall climatic effect of these ecosystems; that is, to ask whether they have a net warming or cooling effect. Neubauer and Megonigal (2015) developed the sustained-flux global warming potential (SGWP) as a straightforward metric of the radiative balance in ecosystems that emit and remove multiple greenhouse gases from the atmosphere. The SGWP avoids some inherent limitations of the global warming potential (GWP) when applied to ecosystems, especially the GWP assumption that greenhouse gas exchanges occur as a single discrete pulse rather than as a continuous flux. Over the commonly used 100-year time frame, tidal freshwater wetlands consistently have a net warming effect on the climate (Table 19.3), meaning that the warming due to the emission of CH_4 exceeds the cooling due to the uptake of CO_2 (annual N_2O fluxes have rarely been measured in tidal freshwater wetlands, but typical wetland N_2O exchange rates would have only a small effect on the radiative balance) (Neubauer, 2014). This is true whether one uses NEP as a proxy for the rate of ecosystem C storage (Chapin et al., 2006) or radionuclide-based (^{137}Cs- or ^{210}Pb) estimates of organic C burial and CO_2 sequestration (Bridgham et al., 2014, Table 19.3). However, a warming effect on climate does not mean that tidal freshwater wetlands contribute to anthropogenic climate change; such an effect (i.e., radiative forcing) only occurs if the radiative balance of tidal freshwater wetlands has changed since the pre-Industrial era baseline (pre-1750). This would occur if the balance of CO_2 sequestration and CH_4 emission rates had changed, such as in response to saltwater intrusion, human modifications, or another global change, or if the wetland had formed since 1750.

As an ecosystem develops, there is a time when the cumulative lifetime warming due to CH_4 emissions is exceeded by the cumulative lifetime cooling due to CO_2 sequestration (Frolking et al., 2006; Neubauer, 2014). This point is known as the radiative forcing switchover time. Using the atmospheric perturbation model of Neubauer and Megonigal (2015), we calculate that the radiative forcing switchover time of tidal freshwater wetlands is on the order of 400–1000 years (Table 19.3), although there are some tidal freshwater wetlands that emit both CO_2 and CH_4 and therefore may not ever have a lifetime cooling effect. There are few tidal freshwater wetlands where CH_4 emissions, CO_2 sequestration rates, and wetland ages have been determined; Sweet Hall marsh, Virginia, USA (Fig. 19.1), is one

TABLE 19.3 Radiative Balance of Six Tidal Freshwater Wetlands and One Oligohaline Wetland Calculated From Annual Net Ecosystem Production (NEP) and CH_4 Emissions Expressed as CO_2 Equivalents

Site	Annual Fluxes ($g\ C\ m^{-2}\ year^{-1}$)			Radiative Balance ($g\ CO_2$-eq $m^{-2}\ year^{-1}$)		Switchover Time (year)		References
	CH_4 Emission	NEP	Soil Seq.	NEP	Soil Seq.	NEP	Soil Seq.	
Racoon, Delaware River, NJ	20–25	98–257	75	990	1075	618	879	Weston et al. (2014)
Cumberland, Pamunkey River, VA	58	301	nd	2381	nd	490	nd	Neubauer and Lee (in preparation)
Sweet Hall, Pamunkey River, VA	72	−145	229	4852	3480	n/a	756	This chapter, Fig. 19.1
Brookgreen, Waccamaw River, SC	42–70	176–471	nd	2137	nd	429	nd	Neubauer (2013a,b)
SALT-Ex, Altamaha River, GA	114	368	nd	5491	nd	924	nd	Herbert et al. (2018)
Week's Bay, Mobile Bay, AL	11	−893	nd	3907	nd	n/a	nd	Wilson et al. (2015)[a]
Salvador, Mississippi River, LA	47	290	nd	1752[b]	nd	376[a]	nd	Krauss et al. (2016)

Switchover time represents the number of years required for the cumulative lifetime warming due to CH_4 emissions to be exceeded by the cumulative lifetime cooling due to CO_2 sequestration during ecosystem development.
[a]*This Wilson et al. (2015) study does not appear in Table 19.1 because the site is not a true tidal freshwater marsh (mean salinity = 2.3).*
[b]*The radiative balance and switchover times also consider the 0.02 g N m^{-2} $year^{-1}$ uptake of N_2O.*

such example. Because the switchover time for Sweet Hall (756 years; Table 19.3) is less than the age of the site (3640 years BP; Neubauer et al., 2002), this tidal freshwater wetland has had a lifetime net cooling effect on the climate, although it has a positive radiative balance over a 100-year period. This suggests that tidal freshwater wetlands created today will contribute to climate change for several centuries; however, the global area of tidal freshwater wetlands is insignificant in the context of anthropogenic climate change, while it is highly significant for the many nonclimate ecosystem services that tidal freshwater wetland provide.

8. CONCLUDING COMMENTS

It is perhaps appropriate that tidal freshwater wetland biogeochemistry has not been well studied, given the fact tidal freshwater wetlands occupy less area than many other wetland ecosystems. However, tidal freshwater wetlands are grossly underrepresented even in the tidal wetland literature. A Web of Science search (1980–2017) shows that <6% of all tidal wetland literature concerns tidal freshwater wetlands, whereas tidal freshwater wetlands can represent ~20% of the total tidal wetland area in some regions (Stevenson et al., 1988; Dahl, 1999). Tidal wetlands deserve increased attention for many reasons. They are species-rich ecosystems that support waterfowl, fish, and terrestrial wildlife. They influence the chemistry of adjacent estuarine waters through exchange and transformation of organic C and nutrients. Because of their location at the head of estuaries, tidal freshwater wetlands are important sites for the deposition of nutrient-laden sediments, and they can rapidly sequester C, N, and P through burial. In short, tidal freshwater wetlands are important features of the landscapes in which they occur.

Tidal freshwater wetlands are sentinel ecosystems for monitoring the influence of global climate change on coastal ecosystems. Poised at the interface of nontidal rivers and saline estuarine waters, they are influenced by river discharge and sea level rise. River discharge is sensitive to precipitation and evapotranspiration, which are in turn sensitive to global warming (Palmer et al., 2008). Increasingly frequent incursions of saline water into tidal freshwater wetland ecosystems can be expected as sea levels rise and droughts become more common (Herbert et al., 2015). It is uncertain what the long-term effects of such episodic events will have on element cycling and plant community composition in tidal freshwater wetlands, but based on current distributions of plant species, even small increases in salinity will elicit dramatic changes in plant and microbial community composition and fundamentally alter the characteristics of tidal freshwater wetland biogeochemical cycles (Neubauer and Craft, 2009). At present, there are many hypotheses about the response of tidal freshwater wetlands to these perturbations. For example, we could expect tidal freshwater wetlands to be less sensitive to sea level rise than tidal saline wetlands because they are located at the head of estuaries near riverine sediment sources. However, their steep slopes and human infrastructure present barriers for tidal freshwater wetland transgression inland. We do not understand tidal freshwater wetlands well enough to fully predict how they will respond to climate and land use change.

Tidal freshwater wetlands have not received the level of biogeochemical scrutiny that has been directed toward nontidal freshwater and tidal saline wetlands. In the absence of more complete knowledge, it is often assumed that tidal freshwater wetland processes adhere to generalizations drawn from better-studied ecosystems. This approach has proved fruitful, but limited. For example, there is now doubt about the assumption that tidal freshwater wetland plants decompose more rapidly than saline tidal wetland plants (Craft, 2007). Based on hydrology, it seems reasonable to assume that tidal freshwater wetlands have relatively open nutrient cycles, but that was not the case in two tidal freshwater wetland systems that have been fully studied (Fig. 19.6). Clearly, understanding biogeochemical processes in tidal freshwater wetlands will require more direct observations of tidal freshwater wetlands in relation to their tidal saline and nontidal analogs.

LIST OF ABBREVIATIONS

BSi Biogenic silica
CDOM Chromophoric dissolved organic matter
DIC Dissolved inorganic carbon
DIN Dissolved inorganic nitrogen
DNRA Dissimilatory nitrate reduction to ammonium
DOC Dissolved organic carbon
DSi Dissolved silica
EU European Union
GPP Gross primary production
NPP Net primary production
POC Particulate organic carbon
ROL Root oxygen loss
TFW Tidal freshwater wetland
USA United States of America

References

Academy, 1998. Impact of Aquatic Vegetation on Water Quality of the Delaware River Estuary. 98–5F. By the Academy of Natural Sciences, Patrick Center for Environmental Research for the Delaware River Basin Commission, Philadelphia, PA, 109 p.

Algar, C., Vallino, J., 2014. Predicting microbial nitrate reduction pathways in coastal sediments. AME 71, 223–238.

An, S., Gardner, W.S., 2002. Dissimilatory nitrate reduction to ammonium (DNRA) as a nitrogen link, versus denitrification as a sink in a shallow estuary (Laguna Madre/Baffin Bay, Texas). Marine Ecology Progress Series 237, 41–50.

Anderson, I.C., Tobias, C.R., Neikirk, B.B., Wetzel, R.L., 1997. Development of a process-based nitrogen mass balance model for a Virginia (USA) *Spartina alterniflora* Salt Marsh: implications for net DIN flux. Marine Ecology Progress Series 159, 3–27.

Anderson, I.C., Neubauer, S.C., Neikirk, B.B., Wetzel, R.L., 1998. Exchanges of carbon and nitrogen between tidal freshwater wetlands and adjacent tributaries. In: Final Report for the Virginia Coastal Resources Management Program. Virginia Department of Environmental Quality, Richmond, VA, 56 p.

Arndt, S., Jørgensen, B.B., LaRowe, D.E., Middelburg, J.J., Pancost, R.D., Regnier, P., 2013. Quantifying the degradation of organic matter in marine sediments: a review and synthesis. Earth Science Reviews 123, 53–86. https://doi.org/10.1016/j.earscirev.2013.02.008.

Arrigoni, A., Findlay, S., Fischer, D., Tockner, K., 2008. Predicting carbon and nutrient transformations in tidal freshwater wetlands of the Hudson River. Ecosystems 11 (5), 790–802.

Baldock, J.A., Masiello, C.A., Gelinas, Y., Hedges, J.I., 2004. Cycling and composition of organic matter in terrestrial and marine ecosystems. Marine Chemistry 92, 39–64. https://doi.org/10.1016/j.marchem.2004.06.016.

Baldwin, A.H., 2013. Nitrogen and phosphorus differentially affect annual and perennial plants in tidal freshwater and oligohaline wetlands. Estuaries and Coasts 36 (3), 547–558.

Bartlett, K.B., Bartlett, D.S., Harriss, R.C., Sebacher, D.I., 1987. Methane emissions along a salt marsh salinity gradient. Biogeochemistry 4, 183–202.

Bauer, J.E., Cai, W.J., Raymond, P.A., Bianchi, T.S., Hopkinson, C.S., Regnier, P.A.G., 2013. The changing carbon cycle of the coastal ocean. Nature 504, 61–70. https://doi.org/10.1038/nature12857.

Bellis, V.J., Gaither, A.C., 1985. Seasonality of aboveground and belowground biomass for six salt marsh plant species. Journal of the Elisha Mitchell Scientific Society 101 (2), 95–109.

Benner, M.A., Moran, M.A., Hodson, R.E., 1985. Effects of pH and plant source on lignocellulose biodegredation rates in two wetland ecosystems, the Okefenokee Swamp and a Georgia salt marsh. Limnology and Oceanography 30, 489–499.

Bernhardt, E.S., Blaszczak, J.R., Ficken, C.D., Fork, M.L., Kaiser, K.E., Seybold, E.C., 2017. Control points in ecosystems: moving beyond the hot spot hot moment concept. Ecosystems 20 (4), 665–682.

Blair, N.E., Aller, R.C., 2012. The fate of terrestrial organic carbon in the marine environment. Annual Review of Marine Science 4, 401–423.

Booth, P.M., 1989. Nitrogen and Phosphorus Cycling Strategies in Two Tidal Freshwater Macrophytes, Peltandra Virginica and *Spartina cynosuroides*. Ph.D. dissertation. The College of William and Mary, Gloucester Point, VA, 264 p.

Bosse, U., Frenzel, P., 1998. Methane emissions from rice microcosms: the balance of production, accumulation and oxidation,. Biogeochemistry 41, 199–214.

Boullion, S., Boschker, H.T.S., 2006. Bacterial carbon sources in coastal sediments: a cross-system analysis based on stable isotope data of biomarkers. Biogeosciences 3, 175–185.

Bowden, W.B., Vorosmarty, C.J., Morris, J.T., Peterson, B.J., Hobbie, J.E., Steudler, P.A., Moore, B., 1991. Transport and processing of nitrogen in a tidal freshwater wetland. Water Resources Series 27, 389–408.

Bowden, W.B., 1984. Nitrogen and phosphorus in the sediments of a tidal, freshwater marsh in Massachusetts. Estuaries 7, 108–118.

Bowden, W.B., 1986. Nitrification, nitrate reduction, and nitrogen immobilization in a tidal fresh-water marsh sediment. Ecology 67, 88–99.

Boye, K., Noël, V., Tfaily, M.M., Bone, S.E., Williams, K.H., Bargar, J.R., Fendorf, S., 2017. Thermodynamically controlled preservation of organic carbon in floodplains. Nature Geoscience 10, 415–419.

Bridgham, S.D., Megonigal, J.P., Keller, J.K., Bliss, N.B., Trettin, C., 2006. The carbon storage of North American wetlands. Wetlands 26, 889–916.

Bridgham, S.D., Moore, T.R., Richardson, C.J., Roulet, N.T., 2014. Errors in greenhouse forcing and soil carbon sequestration estimates in freshwater wetlands: a comment on Mitsch et al. (2013). Landscape Ecology 29 (9), 1481–1485.

Brunet, R.C., Garcia-Gil, L.J., 1996. Sulfide-induced dissimilatory nitrate reduction to ammonia in anaerobic freshwater sediments. FEMS Microbiology Ecology 21, 131–138.

Bukaveckas, P.A., Wood, J., 2014. Nitrogen retention in a restored tidal stream (Kimages Creek, VA) assessed by mass balance and tracer approaches. Journal of Environmental Quality 43 (5), 1614–1623.

Bukaveckas, P.A., Beck, M., Devore, D., Lee, W.M., 2018. Climatic variability and its role in regulating C, N and P retention in the James River Estuary. Estuarine, Coastal and Shelf Science 205, 161–173.

Bullock, A., Sutton-Grier, A., Megonigal, J., 2013. Anaerobic metabolism in tidal freshwater wetlands: III. Temperature regulation of iron cycling. Estuaries and Coasts 36 (3), 482–490. https://doi.org/10.1007/s12237-012-9536-5.

Burkett, V., Ritschard, R., McNulty, S., O'Brien, J.J., Abt, R., Jones, J., Hatch, U., Murray, B., Jagtop, S., Cruise, J., 2001. Potential consequences of climate change variability and change for the southeastern United States. In: Climate Change Impacts on the United States: The Potential Consequences of Climate Variability and Change. Cambridge University Press, Cambridge, pp. 137–164.

Burton, K., Neubauer, S.C., 2018. Species-Specific and Dose-dependent Responses of Tidal Freshwater Marsh Vegetation to Nitrogen and Phosphorus Enrichment. for submission to Wetlands.

Cai, W.J., 2011. Estuarine and coastal ocean carbon paradox: CO_2 sinks or sites of terrestrial carbon incineration? Annual Review of Marine Science 3, 123–145. https://doi.org/10.1146/annurev-marine-120709-142723.

Campana, M.L., 1998. The Effect of Phragmites Australis Invasion on Community Processes in a Tidal Freshwater Marsh. M.Sc. thesis. The College of William and Mary, Gloucester Point, VA, 88 p.

Caraco, N.F., Cole, J.J., Likens, G.E., 1989. Evidence for sulphate-controlled phosphorus release from sediments of aquatic systems. Nature 341, 395–413.

Chambers, R.M., Fourqurean, J.W., 1991. Alternative criteria for assessing nutrient limitation of a wetland macrophyte (*Peltandra-virginica* (L) Kunth). Aquatic Botany 40, 305–320.

Chambers, R.M., Odum, W.E., 1990. Porewater oxidation, dissolved phosphate and the iron curtain: iron-phosphorus relations in tidal freshwater marshes. Biogeochemistry 10, 37–52.

Chambers, L.G., Reddy, K.R., Osborne, T.Z., 2011. Short-term response of carbon cycling to salinity pulses in a freshwater wetland. Soil Science Society of America Journal 75 (5), 2000–2007.

Chambers, R.M., 1992. A fluctuating water-level chamber for biogeochemical experiments in tidal marshes. Estuaries 15, 53–58.

Chanton, J.P., Martens, C.S., Kelley, C.A., 1989. Gas transport from methane-saturated, tidal freshwater and wetland sediments. Limnology and Oceanography 34, 807–819.

Chapin, F.S., Woodwell, G.M., Randerson, J.T., Rastetter, E.B., Lovett, G.M., Baldocchi, D.D., Clark, D.A., Harmon, M.E., Schimel, D.S., Valentini, R., Wirth, C., et al., 2006. Reconciling carbon-cycle concepts, terminology, and methods. Ecosystems 9 (7), 1041–1050.

Christensen, P.B., Rysgaard, S., Sloth, N.P., Dalsgaard, T., Scheaerter, S., 2000. Sediment mineralization, nutrient fluxes, denitrification, and dissimilatory nitrate reduction to ammonium in an estuarine fjord with sea cage trout farms. Aquatic Microbial Ecology 21, 73–84.

Church, T.M., Sommerfield, C.K., Velinsky, D.J., Point, D., Benoit, C., Amouroux, D., Plaa, D., Donard, O.F.X., 2006. Marsh sediments as records of sedimentation, eutrophication and urban pollution in the urban Delaware Estuary. Marine Chemistry 102, 72–95.

Cornwell, J.C., Kemp, W.M., Kana, T.M., 1999. Denitrification on coastal ecosystems: methods, environmental controls, and ecosystem level controls, a review. Aquatic Ecology 33, 41–54.

Courtwright, J., Findlay, S.E.G., 2011. Effects of microtopography on hydrology, physicochemistry, and vegetation in a tidal swamp of the Hudson River. Wetlands 31 (2), 239–249.

Craft, C., Clough, J., Ehman, J., Joye, S., Park, R., Pennings, S., Guo, H., Machmuller, M., 2009. Forecasting the effects of accelerated sea-level rise on tidal marsh ecosystem services. Frontiers in Ecology and the Environment 7 (2), 73–78.

Craft, C.B., 2007. Freshwater input structures soil properties, vertical accretion, and nutrient accumulation of Georgia and U.S. tidal marshes. Limnology and Oceanography 52, 1220–1230.

Dahl, T.E., 1999. South Carolina's Wetlands—status and Trends 1982–1989. Department of the Interior, US Fish and Wildlife Service, Washington, D.C, 58 p.

Darke, A.K., Megonigal, J.P., 2003. Control of sediment deposition rates in two mid-Atlantic Coast tidal freshwater wetlands. Estuarine, Coastal and Shelf Science 57, 255–268.

Dausse, A., Garbutt, A., Norman, L., Papadimitriou, S., Jones, L.M., Robins, P.E., Thomas, D.N., 2012. Biogeochemical functioning of grazed estuarine tidal marshes along a salinity gradient. Estuarine, Coastal and Shelf Science 100, 83–92. https://doi.org/10.1016/j.ecss.2011.12.037.

Day, F.P., 1982. Litter decomposition rates in the seasonally flooded Great Dismal Swamp. Ecology 63, 670–678.

Day, Frank P., 1983. Effects of flooding on leaf litter decomposition in microcosms. Oecologia 56 (2–3), 180–184.

DeLaune, R.D., Smith, C.J., Sarafyan, M.N., 1986. Nitrogen cycling in a freshwater marsh of *Panicum hemitomon* on the deltaic plain of the Mississippi River. Journal of Ecology 74, 249–256.

Drake, D., Peterson, B., Galvan, K., Deegan, L., Hopkinson, C., Johnson, J., Koop-Jakobsen, K., LeMay, L., Picard, C., 2009. Salt marsh ecosystem biogeochemical responses to nutrient enrichment: a paired 15N tracer study. Ecology 90, 2535–2546.

Elsey-Quirk, T., Smyth, A., Piehler, M., Mead, J.V., Velinsky, D.J., 2013. Exchange of nitrogen through an urban tidal freshwater wetland in Philadelphia, Pennsylvania. Journal of Environmental Quality 42 (2), 584–595.

Enriquez, S., Duarte, C.M., Sand-Jensen, K., 1993. Patterns in decomposition rates among photo- synthetic organisms: the importance of detritus C: N:P content. Oceologia 94, 457–471.

Ensign, S.H., Piehler, M.F., Doyle, M.W., 2008. Riparian zone denitrification affects nitrogen flux through a tidal freshwater river. Biogeochemistry 91 (2–3), 133–150.

Ensign, S.H., Siporin, K., Piehler, M., Doyle, M., Leonard, L., 2013. Hydrologic versus biogeochemical controls of denitrification in tidal freshwater wetlands. Estuaries and Coasts 36 (3), 519–532.

Ensign, S.H., Hupp, C.R., Noe, G.B., Krauss, K.W., Stagg, C.L., 2014a. Sediment accretion in tidal freshwater forests and oligohaline marshes of the Waccamaw and Savannah Rivers, USA. Estuaries and Coasts 37, 1107–1119. https://doi.org/10.1007/s12237-013-9744-7.

Ensign, S.H., Noe, G.B., Hupp, C.R., 2014b. Linking channel hydrology with riparian wetland accretion in tidal rivers. Journal of Geophysical Research, Earth Surface 119, 28–44. https://doi.org/10.1002/2013JF002737.

Epp, R.G., Erickson, D.J., Paul, N.D., Sulzberger, B., 2007. Interactive effects of solar UV radiation and climate change on biogeochemical cycling. Photochemical and Photobiological Sciences 6, 286–300.

Fazzolari, É., Nicolardot, B., Germon, J.C., 1998. Simultaneous effects of increasing levels of glucose and oxygen partial pressures on denitrification and dissimilatory nitrate reduction to ammonium in repacked soil cores. European Journal of Soil Biology 34, 47–52.

Findlay, S., Fischer, D., 2013. Ecosystem attributes related to tidal wetland effects on water quality. Ecology 94 (1), 117–125.

Findlay, S., Howe, K., Austin, H.K., 1990. Comparison of detritus dynamics in two tidal freshwater wetlands. Ecology 71, 288–295.

Findlay, S., Sinsabaugh, R.L., Fischer, D.T., Franchini, P., 1998. Sources of dissolved organic carbon supporting planktonic bacterial production in the tidal freshwater Hudson River. Ecosystems 1, 227–239.

Frankignoulle, M., Bourge, I., Wollast, R., 1996. Atmospheric CO_2 fluxes in a highly polluted estuary (the Scheldt. Limnology and Oceanography 41. https://doi.org/10.4319/lo.1996.41.2.0365.

Frankignoulle, M., Abril, G., Borges, A., Bourge, I., Canon, C., Delille, B., Libert, E., Théate, J., 1998. Carbon dioxide emission from European estuaries. Science 282, 434–436.

Freeman, C., Ostle, N., Kang, H., 2001. An enzymatic 'latch' on a global carbon store. Nature 409, 149.

Frolking, S., Roulet, N., Fuglestvedt, J., 2006. How northern peatlands influence the Earth's radiative budget: sustained methane emission versus sustained carbon sequestration. Journal of Geophysical Research 111, G01008. https://doi.org/10.1029/2005JG000091.

Frost, J.W., Schleicher, T., Craft, C.B., 2009. Nitrogen limits primary and secondary production in a Georgia (USA) tidal freshwater marsh. Wetlands 29, 196–203.

Furumai, H., Kawasaki, T., Futuwatari, T., Kusuda, T., 1988. Effect of salinity on nitrification in a tidal river. Water Science and Technology 20, 165–174.

Gautam, S., 2015. Controls of Organic Matter Decomposition in Coastal Wetlands. Ph.D. Dissertation, South Dakota School of Mines and Technology, Rapid City, South Dakota.

Giblin, A.E., Tobias, C.R., Song, B., Weston, N., Banta, G.T., Rivera-Monroy, V.H., 2013. The importance of dissimilatory nitrate reduction to ammonium (DNRA) in the nitrogen cycle of coastal ecosystems. Oceanography 26 (3), 124–131. https://doi.org/10.5670/oceanog.2013.54.

Gilbert, H., 1990. Éléments nutritifs (N et P), métaux lourds (Zn, Cu, Pb et Hg) et productivité végétale dans un marias intertidal d'eau douce, Québec (Québec). Canadian Journal of Botany 68, 857–863.

Grant Jr., R.R., Patrick, R., 1970. Tinicum Marsh as a water purifier. In: Two Studies of Tinicum Marsh. The Conservation Foundation, Washington, D.C., USA, pp. 105–123.

Greene, S.E., 2005. Nutrient Removal by Tidal Fresh and Oligohaline Marshes in a Chesapeake Bay Tributary. M.Sc. thesis. University of Maryland, Solomons, MD, USA, 149 p.

Greiner, M., Hershner, C., 1998. Analysis of wetland total phosphorus retention and watershed structure. Wetlands 18, 142–149.

Gribsholt, B., Boschker, H.T.S., Struyf, E., Andersson, M., Tramper, A., De Brabandere, L., van Damme, S., Brion, N., Meire, P., Dehairs, F., Middelburg, J.J., Heip, C.H.R., 2005. Nitrogen processing in a tidal freshwater marsh: a whole-ecosystem N-15 labeling study. Limnology and Oceanography 50, 1945–1959.

Gribsholt, B., Struyf, E., Tramper, A., Andersson, M.G.I., Brion, N., De Brabandere, L., van Damme, S., Meire, P., Middelburg, J.J., Dehairs, F., Boschker, H.T.S., 2006. Ammonium transformation in a nitrogen-rich tidal freshwater marsh. Biogeochemistry 80, 289–298.

Gribsholt, B., Struyf, E., Tramper, A., De Brabandere, L., Brion, N., van Damme, S., Meire, P., Dehairs, F., Middelburg, J.J., Boschker, H.T.S., 2007. Nitrogen assimilation and short term retention in a nutrient-rich tidal freshwater marsh – a whole ecosystem ^{15}N enrichment study. Biogeosciences 4, 11–26.

Groszkowski, K.M., 1995. Denitrification in a Tidal Freshwater Marsh. Senior Thesis. Harvard College, Cambridge, MA, USA, 85 p.

Hatton, R.S., DeLaune, R.D., Patrick Jr., W.H., 1983. Sedimentation, accretion and subsidence in marshes of Barataria Basin, Louisiana. Limnol. Oceanogr 28, 494–502.

Hedges, J.I., Keil, R.G., 1995. Sedimentary organic matter preservation: an assessment and speculative synthesis. Marine Chemistry 49, 81–115.

Heinle, D.R., Flemer, D.A., 1976. Flows of materials between poorly flooded tidal marshes and an estuary. Marine Biology 35, 359–373.

Herbert, E.R., Boon, P., Burgin, A.J., Neubauer, S.C., Franklin, R.B., Ardón, M., Hopfensperger, K.N., Lamers, L.P.M., Gell, P., 2015. A global perspective on wetland salinization: ecological consequences of a growing threat to freshwater wetlands. Ecosphere 6 (10), 1–43. https://doi.org/10.1890/ES14-00534.1.

Herbert, E.R., Schubauer-Berigan, J., Craft, C.B., 2018. Differential effects of chronic and acute simulated seawater intrusion on tidal freshwater marsh carbon cycling. Biogeochemistry 138 (2), 137–154.

Hines, J., Megonigal, J.P., Denno, R.F., 2006. Nutrient subsidies to belowground microbes impact aboveground food web interactions. Ecology 87, 1542–1555.

Hoffman, J.C., Bronk, D.A., 2006. Interannual variation in stable carbon and nitrogen isotope biogeochemistry of the Mattaponi River, Virginia. Limnology and Oceanography 51, 2319–2332.

Holm, G.O., Perez, B.C., McWhorter, D.E., Krauss, K.W., Johnson, D.J., Raynie, R.C., Killebrew, C.J., 2016. Ecosystem level methane fluxes from tidal freshwater and brackish marshes of the Mississippi river delta: implications for coastal wetland carbon projects. Wetlands 36, 401. https://doi.org/10.1007/s13157-016-0746-7.

Hopkinson, C.S., 1992. A comparison of ecosystem dynamics in freshwater wetlands. Estuaries 15, 549–562.

Howarth, R.W., Marine, R., Cole, J.J., 1988. Nitrogen-fixation in fresh-water, estuarine, and marine ecosystems. 2. Biogeochemical controls. Limnology and Oceanography 33, 688–701.

Howarth, R.W., Schneider, R., Swaney, D., 1996. Metabolism and organic carbon fluxes in the tidal freshwater Hudson River. Estuaries 19, 848–865.

Howes, B.L., Dacey, J.W.H., King, G.M., 1984. Carbon flow through oxygen and sulfate reduction pathways in salt marsh sediments. Limnology and Oceanography 29, 1037–1051.

Huang, X.Q., Morris, J.T., 2003. Trends in phosphatase activity along a successional gradient of tidal freshwater marshes on the Cooper River, South Carolina. Estuaries 26, 1281–1290.

Huang, X.Q., Morris, J.T., 2005. Distribution of phosphatase activity in marsh sediments along an estuarine salinity gradient. Marine Ecology Progress Series 292, 75–83.

Hunsinger, G.B., Mitra, S., Findlay, S.E.G., Fischer, D.T., 2010. Wetland-driven shifts in suspended particulate organic matter composition of the Hudson River estuary, New York. Limnology and Oceanography 55, 1653–1667. https://doi.org/10.4319/lo.2010.55.4.1653.

Hunsinger, G.B., Mitra, S., Findlay, S.E.G., Fischer, D.T., 2012. Littoral-zone influences on particulate organic matter composition along the freshwater-tidal Hudson River, New York. Limnology and Oceanography 57 (5), 1303–1316. https://doi.org/10.4319/lo.2012.57.5.1303.

Ingvorsen, K., Jørgensen, B., 1984. Kinetics of sulfate uptake by freshwater and marine species of *Desulfovibrio*. Archives of Microbiology 139, 61–66.

Jordan, T.E., Cornwell, J.C., Boynton, W.R., Anderson, J.T., 2008. Changes in phosphorus biogeochemistry along an estuarine salinity gradient: the iron conveyer belt. Limnology and Oceanography 53 (1), 172–184.

Joye, S.B., Hollibaugh, J.T., 1995. Influence of sulfide inhibition of nitrification on nitrogen regeneration in sediments. Science 270, 623–625.

Kalber Jr., F.A., 1959. A hypothesis on the role of tide-marshes in estuarine productivity. Estuarine Bulletin 4, 3.

Keiluweit, M., Nico, P.S., Kleber, M., Fendorf, S., 2016. Are oxygen limitations under recognized regulators of organic carbon turnover in upland soils? Biogeochemistry. https://doi.org/10.1007/s10533-015-0180-6.

Keller, J.K., Bridgham, S.D., 2007. Pathways of anaerobic carbon cycling across an ombrotrophic–minerotrophic peatland gradient. Limnology and Oceanography 52, 96–107.

Keller, J.K., Takagi, K.K., 2013. Solid-phase organic matter reduction regulates anaerobic decomposition in bog soil. Ecosphere 4 (5), 54.

Keller, J.K., Weisenhorn, P.B., Megonigal, J.P., 2009. Humic acids as electron acceptors in wetland decomposition. Soil Biology and Biochemistry 41 (7), 1518–1522. https://doi.org/10.1016/j.soilbio.2009.04.008.

Keller, J., Sutton-Grier, A., Bullock, A., Megonigal, J.P., 2013. Anaerobic metabolism in tidal freshwater wetlands: I. Plant removal effects on iron reduction and methanogenesis. Estuaries and Coasts 36 (3), 457–470. https://doi.org/10.1007/s12237-012-9527-6.

Kelley, C.A., Martens, C.S., Chanton, J.P., 1990. Variations in sedimentary carbon remineralization rates in the White Oak River estuary, North Carolina. Limnology and Oceanography 35, 372–383.

Kelley, C.A., Martens, C.S., Ussler, W.I., 1995. Methane dynamics across a tidally flooded riverbank margin. Limnology and Oceanography 40, 1112–1129.

Khan, H., Brush, G.S., 1994. Nutrient and metal accumulation in a fresh-water tidal marsh. Estuaries 17, 345–360.

Kögel-Knabner, I., 2002. The macromolecular organic composition of plant and microbial residues as inputs to soil organic matter. Soil Biology and Biochemistry 34, 139–162.

Krauss, K.W., Whitbeck, J.L., 2012. Soil greenhouse gas fluxes during wetland forest retreat along the lower Savannah River, Georgia (USA). Wetlands 32 (1), 73–81.

Krauss, K.W., Perez, B.C., Holm Jr., G.O., McWhorter, D.E., Cormier, N., Moss, R.F., Johnson, D.J., Neubauer, S.C., Raynie, R.C., 2016. Component greenhouse gas fluxes and radiative forcing from degrading and healthy coastal deltaic marshes: pairing chamber techniques and eddy covariance. Journal of Geophysical Research: Biogeosciences 121, 1503–1521. https://doi.org/10.1002/2015JG003224.

Lamers, L.P.M., Dolle, G., Van Den Berg, S.T.G., Van Delft, S.P.J., Sebastiaan, P.J., Roelofs, J.M., 2001. Differential responses of freshwater wetland soils to sulphate pollution. Biogeochemistry 55, 87–102.

Lamers, L.P., Govers, L.L., Janssen, I.C., Geurts, J.J., Van der Welle, M.E., Van Katwijk, M.M., Van der Heide, T., Roelofs, J.G., Smolders, A.J., 2013. Sulfide as a soil phytotoxin—a review. Frontiers of Plant Science 4, 268. https://doi.org/10.3389/fpls.2013.00268.

LaRowe, D.E., Van Cappellen, P., 2011. Degradation of natural organic matter: a thermodynamic analysis. Geochimica et Cosmochimica Acta 75, 2030–2042.

Lee, D.Y., De Meo, O.A., Tillett, A.L., Thomas, R.B., Neubauer, S.C., 2016. Design and construction of an automated irrigation system for simulating saltwater intrusion in a tidal freshwater wetland. Wetlands 36, 889–898. https://doi.org/10.1007/s13157-016-0801-4.

Limburg, K.E., Moran, M.A., McDowell, W. (Eds.), 1986. The Hudson River Ecosystem. Springer-Verlag, New York, NY, USA, 331 p.

Lombardi, J.E., Epp, M.A., Chanton, J.P., 1997. Investigation of the methylfluoride technique for determining rhizospheric methane oxidation. Biogeochemistry 36, 153–172.

Loomis, M.J., Craft, C.B., 2010. Carbon sequestration and nutrient (nitrogen, phosphorus) accumulation in river-Dominated tidal marshes, Georgia, USA. Soil Science Society of America Journal 74, 1028–1036. https://doi.org/10.2136/sssaj2009.0171.

Lovley, D.R., Phillips, E.J.P., 1986. Availability of ferric iron for microbial reduction in bottom sediments of the freshwater tidal Potomac River. Applied and Environmental Microbiology 52, 751–757.

Lovley, D.R., Phillips, E.J.P., 1987. Rapid assay for microbially reducible ferric iron in aquatic sediments. Applied and Environmental Microbiology 53, 1536–1540.

Ma, S., Luther, G.W.I., Keller, J., Madison, A.S., Metzger, E., Emerson, D., Megonigal, J.P., 2008. Solid-state Au/Hg microelectrode for the investigation of Fe and Mn cycling in a freshwater wetland: implications for methane production. Electroanalysis 20, 233–239.

MacFarlane, G.T., Hebert, R.A., 1984. Effect of oxygen tension, salinity, temperature, and organic matter concentration on the growth and nitrifying activity of an estuarine strain of Nitrosomonas. FEMS Microbiology Letters 23, 107–111.

Magalhães, C.M., Joye, S.B., Moreira, R.M., Wiebe, W.J., Bordalo, A.A., 2005. Effect of salinity and inorganic nitrogen concentrations on nitrification and denitrification rates in intertidal sediments and rocky biofilms of the Douro River estuary, Portugal. Water Research 39, 1783–1794.

Malone, T.C., Boicourt, W.C., Cornwell, J.C., Harding Jr., L.W., Stevenson, J.C., 2003. The Choptank River: a mid-Chesapeake Bay index site for evaluating ecosystem responses to nutrient management. In: Final Report to the Coastal Intensive Site Network (CISNet). Horn Point Environmental Laboratory, University of Maryland, Cambridge, MD, USA, 64 p.

Marton, J.M., Herbert, E.R., Craft, C.B., 2012. Effects of salinity on denitrification and greenhouse gas production from laboratory-incubated tidal forest soils. Wetlands 32 (2), 347–357.

McClain, M.E., Boyer, E.W., Dent, C.L., Gergel, S.E., Grimm, N.B., Groffman, P.M., Hart, S.C., Harvey, J.W., Johnston, C.A., Mayorga, E., McDowell, W.H., 2003. Biogeochemical hot spots and hot moments at the interface of terrestrial and aquatic ecosystems. Ecosystems 6 (4), 301–312.

McKellar, H.N., Tufford, D.L., Alford, M.C., Saroprayogi, P., Kelley, B.J., Morris, J.T., 2007. Tidal nitrogen exchanges across a freshwater wetland succession gradient in the upper Cooper River, South Carolina. Estuaries and Coasts 30, 989.

Megonigal, J.P., Schlesinger, W.H., 1997. Enhanced CH4 emissions from a wetland soil exposed to elevated CO_2. Biogeochemistry 37, 77–88.

Megonigal, J.P., Schlesinger, W.H., 2002. Methane-limited methanotrophy in tidal freshwater swamps. Global Biogeochemical Cycles 16, 1062. https://doi.org/10.1029/2001GB001594.

Megonigal, J.P., Whalen, S.C., Tissue, D.T., Bovard, B.D., Albert, D.B., Allen, A.S., 1999. A plant–soil–atmosphere microcosm for tracing radiocarbon from photosynthesis through methanogenesis. Soil Science Society of America Journal 63, 665–671.

Megonigal, J.P., Hines, M.E., Visscher, P.T., 2004. Anaerobic metabolism: linkages to trace gases and aerobic metabolism. In: Schlesinger, W.H. (Ed.), Biogeochemistry. Elsevier-Pergamon, Oxford, UK, pp. 317–424.

Meiggs, D., Taillefert, M., 2011. The effect of riverine discharge on biogeochemical processes in estuarine sediments. Limnology and Oceanography 56, 1797–1810.

Merrill, J.Z., Cornwell, J.C., 2000. The role of oligohaline marshes in estuarine nutrient cycling. In: Weinstein, M., Kreeger, D.A. (Eds.), Concepts and Controversies in Tidal Marsh Ecology. Kluwer Press, Dordrecht, Netherlands, pp. 425–441.

Merrill, J.Z., 1999. Tidal Freshwater Marshes as Nutrient Sinks: Particulate Nutrient Burial and Denitrification. Ph.D. dissertation. University of Maryland, College Park, MD, USA, p. 342.

Morris, J.T., Bowden, W.B., 1986. A mechanistic, numerical model of sedimentation, mineralization, and decomposition for marsh sediments. Soil Science Society of America Journal 50, 96–105.

Morris, J.T., Sundareshwar, P.V., Nietch, C.T., Kjerfve, B., Cahoon, D.R., 2002. Responses of coastal wetlands to rising sea level. Ecology 83, 2869–2877.

Morrissey, E.M., Gillespie, J.L., Morina, J.C., Franklin, R.B., 2014. Salinity affects microbial activity and soil organic matter content in tidal wetlands. Global Change Biology 20, 1351–1362. https://doi.org/10.1111/gcb.12431.

Morse, J.L., Megonigal, J.P., Walbridge, M.R., 2004. Sediment nutrient accumulation and nutrient availability in two tidal freshwater marshes along the Mattaponi River, Virginia, USA. Biogeochemistry 69, 175–206.

Neubauer, S.C., Anderson, I.C., 2003. Transport of dissolved inorganic carbon from a tidal freshwater marsh to the York River estuary. Limnology and Oceanography 48, 299–307.

Neubauer, S.C., Craft, C.B., 2009. Global change and tidal freshwater wetlands: scenarios and impacts. In: Barendregt, A., Whigham, D.F., Baldwin, A.H. (Eds.), Tidal Freshwater Wetlands. Backhuys Publishers, Leiden, The Netherlands, pp. 253–266.

Neubauer, S.C., Lee, D.Y., Saltwater intrusion suppresses ecosystem primary production and respiration in tidal freshwater wetlands (in preparation).

Neubauer, S.C., Megonigal, J.P., 2015. Moving beyond global warming potentials to quantify the climatic role of ecosystems. Ecosystems 18, 1000–1013. https://doi.org/10.1007/s10021-015-9879-4.

Neubauer, S.C., Miller, W.D., Anderson, I.C., 2000. Carbon cycling in a tidal freshwater marsh ecosystem: a carbon gas flux study. Marine Ecology Progress Series 199, 13–30.

Neubauer, S.C., Anderson, I.C., Constantine, J.A., Kuehl, S.A., 2002. Sediment deposition and accretion in a mid-Atlantic (USA) tidal freshwater marsh. Estuarine, Coastal and Shelf Science 54, 713–727.

Neubauer, S.C., Anderson, I.C., Neikirk, B.B., 2005a. Nitrogen cycling and ecosystem exchanges in a Virginia tidal freshwater marsh. Estuaries 28, 909–922.

Neubauer, S.C., Givler, K., Valentine, S., Megonigal, J.P., 2005b. Seasonal patterns and plant-mediated controls of subsurface wetland biogeochemistry. Ecology 86, 3334–3344.

Neubauer, S.C., Toledo-Durán, G.E., Emerson, D., Megonigal, J.P., 2007. Returning to their roots: iron-oxidizing bacteria enhance short-term plaque formation in the wetland-plant rhizosphere. Geomicrobiology Journal 24, 65–73.

Neubauer, S.C., Emerson, D., Megonigal, J.P., 2008. Microbial oxidation and reduction of iron in the root zone and influences on metal mobility. In: Violante, A., Huang, P.M., Gadd, G.M. (Eds.), Biophysico-Chemical Processes of Heavy Metals and Metalloids in Soil Environments. John Wiley & Sons, Hoboken, NJ, USA, pp. 339–371.

Neubauer, S.C., Franklin, R.B., Berrier, D.J., 2013. Saltwater intrusion into tidal freshwater marshes alters the biogeochemical processing of organic carbon. Biogeosciences 10, 8171–8183. https://doi.org/10.5194/bg-10-8171-2013.

Neubauer, S.C., 2000. Carbon Dynamics in a Tidal Freshwater Marsh. Ph.D. dissertation. The College of William and Mary, Gloucester Point, VA, USA, 221 p.

Neubauer, S.C., 2008. Contributions of mineral and organic components to tidal freshwater marsh accretion. Estuarine, Coastal and Shelf Science 78, 78–88.

Neubauer, S.C., 2013a. Carbon sequestration in wetland soils: importance, mechanisms, and future prospects. In: Society of Wetland Scientists Research Brief. Oct 2013-0001, 4 p.

Neubauer, S.C., 2013b. Ecosystem responses of a tidal freshwater marsh experiencing saltwater intrusion and altered hydrology. Estuaries and Coasts 36, 491–507. https://doi.org/10.1007/s12237-011-9455-x.

Neubauer, S.C., 2014. On the challenges of modeling the net radiative forcing of wetlands: reconsidering Mitsch et al. (2013). Landscape Ecology 29, 571–577. https://doi.org/10.1007/s10980-014-9986-1.

Nevin, K.P., Lovley, D.R., 2000. Potential for nonenzymatic reduction of Fe(III) via electron shuttling in subsurface sediments. Environmental Science and Technology 34, 2472–2478.

Nietch, C.T., 2000. Carbon Biogeochemistry in Tidal Marshes of South Carolina: The Effect of Salinity and Nutrient Availability on Marsh Metabolism in Estuaries with Contrasting Histories of Disturbance and River Influence. Ph.D. dissertation. University of South Carolina, Columbia, SC, USA, 156 p.

Nijburg, J.W., Coolen, M.J.L., Gerards, S., Gunnewiek, P.J.A.K., Laanbroek, H.J., 1997. Effects of nitrate availability and the presence of *Glyceria maxima* on the composition and activity of the dissimilatory nitrate-reducing bacterial community. Applied and Environmental Microbiology 63, 931–937.

Nixon, S.W., 1980. Between coastal marshes and coastal waters: a review of twenty years of speculation and research on the role of salt marshes in estuarine productivity and water chemistry. In: Hamilton, P., MacDonald, K. (Eds.), Estuarine and Wetland Processes. Plenum Press, New York, NY, USA, pp. 438–525.

Noe, G.B., Krauss, K.W., Lockaby, B.G., Conner, W.H., Hupp, C.R., 2013. The effect of increasing salinity and forest mortality on soil nitrogen and phosphorus mineralization in tidal freshwater forested wetlands. Biogeochemistry 114 (1–3), 225–244.

Noe, G.B., Hupp, C.R., Bernhardt, C.E., Krauss, K.W., 2016. Contemporary deposition and long-term accumulation of sediment and nutrients by tidal freshwater forested wetlands impacted by sea level rise. Estuaries and Coasts 39, 1006–1019. https://doi.org/10.1007/s12237-016-0066-4.

Odum, E.P., 1968. A research challenge: evaluating the productivity of coastal and estuarine water. In: Proceedings of the Second Sea Grant Conference. University of Rhode Island, Kingston, RI, USA, pp. 63–64.

Odum, W.E., 1988. Comparative ecology of tidal freshwater and salt marshes. Annual Review of Ecology and Systematics 19 (1), 147–176.

Oremland, R.S., Marsh, L.M., Polcin, S., 1982. Methane production and simultaneous sulfate reduction in anoxic salt marsh sediments. Nature 296, 143–145.

Palmer, M.A., Reidy Liermann, C.A., Nilsson, C., Flörke, M., Alcamo, J., Lake, P.S., Bond, N., 2008. Climate change and the world's river basins: anticipating management options. Frontiers in Ecology and the Environment 6, 81–89.

Paludan, C., Morris, J.T., 1999. Distribution and speciation of phosphorus along a salinity gradient in intertidal marsh sediments. Biogeochemistry 45, 197–221.

Pasternack, G.B., 2009. Hydrogeomorphology and sedimentation in tidal freshwater wetlands. In: Barendregt, A., Whigham, D., Baldwin, A. (Eds.), Tidal Freshwater Wetlands. Backhuys Publishers, The Netherlands.

Phillips, P.J., Hanchar, D.W., 1996. Water-Auality Assessment of the Hudson River Basin in New York and Adjacent States. Water-Resources Investigations Report 96-4065. U.S. Geological Survey, Troy, NY, USA, 76 p.

Poffenbarger, H.J., Needelman, B.A., Megonigal, J.P., 2011. Salinity influence on methane emissions from tidal marshes. Wetlands 31, 831–842. https://doi.org/10.1007/s13157-011-0197-0.

Portnoy, J.W., Giblin, A.E., 1997. Biogeochemical effects of seawater restoration to diked salt marshes. Ecological Applications 7, 1054–1063.

Raymond, P.A., Bauer, J.E., 2001. DOC cycling in a temperate estuary: a mass balance approach using natural ^{14}C and ^{13}C isotopes. Limnology and Oceanography 46, 655–667.

Roden, E.E., Wetzel, R.G., 1996. Organic carbon oxidation and suppression of methane production by microbial Fe(III) oxide reduction in vegetated and unvegetated freshwater wetland sediments. Limnology and Oceanography 41, 1733–1748.

Rysgaard, S., Thastum, P., Dalsgaard, T., Christensen, P.B., Sloth, N.P., 1999. Effects of salinity on NH_4^+ adsorption capacity, nitrification, and denitrification in Danish estuarine sediments. Estuaries 22, 21–31.

Schoepfer, V., Bernhardt, E.S., Burgin, A.J., 2014. Iron clad wetlands: soil iron–sulfur buffering determines coastal wetland response to salt water incursion. Journal of Geophysical Research: Biogeosciences 119, 2209–2219.

Schubauer, J.P., Hopkinson, C.S., 1984. Above- and belowground emergent macrophyte production and turnover in a coastal marsh ecosystem, Georgia. Limnology and Oceanography 29, 1052–1065.

Segarra, K., Comerford, C., Slaughter, J.B., Joye, S.B., 2013. Impact of electron acceptor availability on the anaerobic oxidation of methane in coastal freshwater and brackish wetland sediments. Geochimica et Cosmochimica Acta 115, 15–30. https://doi.org/10.1016/j.gca.2013.03.029.

Segarra, K.E.A., Schubotz, F., Samarkin, V., Yoshinaga, M.Y., Hinrichs, K.-U., Joye, S.B., 2014. High rates of anaerobic methane oxidation in freshwater wetlands reduce potential atmospheric methane emissions. Nature Communications 6, 7477. https://doi.org/10.1038/ncomms8477.

Seitzinger, S.P., 1988. Denitrification in freshwater and coastal marine ecosystems: ecological and geochemical significance. Limnology and Oceanography 33, 702–724.

Seldomridge, E., Prestegaard, K., 2011. Is denitrification kinetically-limited or transport-limited in tidal freshwater marshes? Applied Geochemistry 26, S256–S258.

Seldomridge, E., Prestegaard, K., 2014. Geochemical, temperature, and hydrologic transport limitations on nitrate retention in tidal freshwater wetlands, Patuxent River, Maryland. Wetlands 34 (4), 641–651.

Shields, M.R., Bianchi, T.S., Gélinas, Y., Allison, M.A., Twilley, R.R., 2016. Enhanced terrestrial carbon preservation promoted by reactive iron in deltaic sediments. Geophysical Research Letters 43, 1149–1157. https://doi.org/10.1002/2015GL067388.

Simpson, R.L., Whigham, D.F., Walker, R., 1978. Seasonal patterns of nutrient movement in a freshwater tidal marsh. In: Good, R.E., Whigham, D.F., Simpson, R.L. (Eds.), Freshwater Wetlands: Ecological Processes and Management Potential. Academic Press, New York, NY, USA, pp. 243–257.

Simpson, R.L., Good, R.E., Walker, R., Frasco, B.R., 1983. The role of Delaware River freshwater tidal wetlands in the retention of nutrients and heavy metals. Journal of Environmental Quality 12, 41–48.

Smith, S.V., Hollibaugh, J.T., 1993. Coastal metabolism and the oceanic organic carbon balance. Reviews of Geophysics 31, 75–89.

Smolders, A.J., Lucassen, E.C., Bobbink, R., Roelofs, J.G., Lamers, L.P., 2010. How nitrate leaching from agricultural lands provokes phosphate eutrophication in groundwater fed wetlands: the sulphur bridge. Biogeochemistry 98 (1–3), 1–7.

Stagg, C.L., Schoolmaster, D.R., Krauss, K.W., Cormier, N., Conner, W.H., 2017. Causal mechanisms of soil organic matter decomposition: deconstructing salinity and flooding impacts in coastal wetlands. Ecology 98, 2003–2018. https://doi.org/10.1002/ecy.1890.

Stehr, G., Böttcher, B., Dittberner, P., Rath, G., Koops, H.-P., 1995. The ammonia-oxidizing, nitrifying population of the River Elbe estuary. FEMS Microbiology Ecology 17, 177–186.

Stets, E.G., Butman, D., McDonald, C.P., Stackpoole, S.M., DeGrandpre, M.D., Striegl, R.G., 2017. Carbonate buffering and metabolic controls on carbon dioxide in rivers. Global Biogeochemical Cycles 31, 663–677. https://doi.org/10.1002/2016GB005578.

Stevenson, J.C., Ward, L.G., Kearney, M.S., 1988. Sediment transport and trapping in marsh systems: implications of tidal flux studies. Marine Geology 80, 37–59.

Struyf, E., Van Damme, S., Gribsholt, B., Meire, P., 2005a. Freshwater marshes as dissolved silica recyclers in an estuarine environment (Schelde estuary, Belgium). Hydrobiologia 540, 69–77.

Struyf, E., Van Damme, S., Gribsholt, B., Middelburg, J.J., Meire, P., 2005b. Biogenic silica in tidal freshwater marsh sediments and vegetation (Schelde estuary, Belgium). Marine Ecology Progress Series 303, 51–60.

Struyf, E., Dausse, A., Van Damme, S., Bal, K., Gribsholt, B., Boschker, H.T.S., Middelburg, J.J., Meire, P., 2006. Tidal marshes and biogenic silica recycling at the land-sea interface. Limnology and Oceanography 51, 838–846.

Struyf, E., Temmerman, S., Meire, P., 2007. Dynamics of biogenic Si in freshwater tidal marshes: Si regeneration and retention in marsh sediments (Scheldt estuary). Biogeochemistry 82, 41–53.

Sundareshwar, P.V., Morris, J.T., 1999. Phosphorus sorption characteristics of intertidal marsh sediments along an estuarine salinity gradient. Limnology and Oceanography 44, 1693–1701.

Sutter, L.A., Perry, J.E., Chambers, R.M., 2014. Tidal freshwater marsh plant responses to low level salinity increases. Wetlands 34 (1), 167–175.

Sutton-Grier, A.S., Megonigal, J.P., 2011. Plant species traits regulate methane production in freshwater wetland soils. Soil Biology and Biochemistry 43 (2), 413–420. https://doi.org/10.1016/j.soilbio.2010.11.009.

Sutton-Grier, A.E., Keller, J.K., Koch, R., Gilmour, C., Megonigal, J.P., 2011. Electron donors and acceptors influence anaerobic soil organic matter mineralization in tidal marshes. Soil Biology and Biochemistry 43 (7), 1576–1583. https://doi.org/10.1016/j.soilbio.2011.04.008.

Tobias, C.R., Anderson, I.C., Canuel, E.A., Macko, S.A., 2001a. Nitrogen cycling through a fringing marsh-aquifer ecotone. Marine Ecology Progress Series 210, 25–39.

Tobias, C.R., Macko, S.A., Anderson, I.C., Canuel, E.A., Harvey, J.W., 2001b. Tracking the fate of a high concentration groundwater nitrate plume through a fringing marsh: a combined groundwater tracer and in situ isotope enrichment study. Limnology and Oceanography 46, 1977–1989.

Tobias, C.R., Cieri, M., Peterson, B.J., Deegan, L.A., Vallino, J., Hughes, J., 2003. Processing watershed-derived nitrogen in a well-flushed New England estuary. Limnology and Oceanography 48, 1766–1778.

Tzortziou, M., Osburn, C.L., Neale, P.J., 2007. Photobleaching of dissolved organic material from a tidal marsh-estuarine system of the Chesapeake Bay. Photochemistry and Photobiology 83, 782–792.

Tzortziou, M., Neale, P.J., Osburn, C.L., Megonigal, J.P., Maie, N., Jaffe, R., 2008. Tidal marshes as a source of optically and chemically distinctive colored dissolved organic matter in the Chesapeake Bay. Limnology and Oceanography 53, 148–159.

Tzortziou, M., Neale, P.J., Megonigal, J.P., Pow, C.L., Butterworth, M., 2011. Spatial gradients in dissolved carbon due to tidal marsh outwelling into a Chesapeake Bay estuary. Marine Ecology Progress Series 426, 41–56. https://doi.org/10.3354/meps09017.

van Damme, S., Dehairs, F., Tackx, M., Beauchard, O., Struyf, E., Gribsholt, B., Cleemput, O., Meire, P., 2009. Tidal exchange between a freshwater tidal marsh and an impacted estuary: the Scheldt estuary, Belgium. Estuarine, Coastal and Shelf Science 85 (2), 197–207.

van der Nat, F.J.W.A., Middelburg, J.J., 1998a. Effects of two common macrophytes on methane dynamics in freshwater sediments. Biogeochemistry 43, 79–104.

van der Nat, F.J.W.A., Middelburg, J.J., 1998b. Seasonal variation in methane oxidation by the rhizosphere of *Phragmites australis* and *Scirpus lacustris*. Aquatic Botany 61, 95–110.

van der Nat, F.J.W.A., Middelburg, J.J., 2000. Methane emission from tidal freshwater marshes. Biogeochemistry 49, 103–121.

Vann, C.D., Megonigal, J.P., 2003. Elevated CO_2 and water depth regulation of methane emissions: comparison of woody and non-woody wetland plant species. Biogeochemistry 63, 117–134.

Von Korff, B.H., Piehler, M.F., Ensign, S.H., 2014. Comparison of denitrification between river channels and their adjoining tidal freshwater wetlands. Wetlands 34 (6), 1047–1060.

Walker, R., 1981. Nitrogen, Phosphorus and Production Dynamics for *Peltandra virginica* (L.) Kunth in a Southern New Jersey Freshwater Tidal Marsh. Ph.D. dissertation. Rutgers University, New Brunswick, NJ, USA, 180 p.

Wallenstein, M.D., Myrold, D.D., Firestone, M.K., Voytek, M.A., 2006. Environmental controls on denitrifying communities and denitrification rates: insights from molecular methods. Ecological Applications 16, 2143–2152.

Wang, Z.A., Kroeger, K.D., Ganju, N.K., Gonneea, M.E., Chu, S.N., 2016. Intertidal salt marshes as an important source of inorganic carbon to the coastal ocean. Limnology and Oceanography 61, 1916–1931. https://doi.org/10.1002/lno.10347.

Weiss, J.V., Emerson, D., Backer, S.M., Megonigal, J.P., 2003. Enumeration of Fe(II)-oxidizing and Fe(III)-reducing bacteria in the root zone of wetland plants: implications for a rhizosphere iron cycle. Biogeochemistry 64, 77–96.

Weiss, J.V., Emerson, D., Megonigal, J.P., 2004. Geochemical control of microbial Fe(III) reduction potential in wetlands: comparison of the rhizosphere to non-rhizosphere soil. FEMS Microbiology Ecology 48, 89–100.

Weiss, J.V., Emerson, D., Megonigal, J.P., 2005. Rhizosphere iron(III) deposition and reduction in a *Juncus effusus* *L.*-dominated wetland. Soil Science Society of America Journal 69, 1861–1870.

Weston, N.B., Dixon, R.E., Joye, S.B., 2006. Ramifications of increased salinity in tidal freshwater sediments: geochemistry and microbial pathways of organic mineralization. Journal of Geophysical Research 111, G01009.

Weston, N.B., Vile, M.A., Neubauer, S.C., Velinsky, D.J., 2011. Accelerated microbial organic matter mineralization following salt-water intrusion into tidal freshwater marsh soils. Biogeochemistry 102, 135–151.

Weston, N.B., Neubauer, S.C., Velinsky, D.J., Vile, M.A., 2014. Net ecosystem carbon exchange and the greenhouse gas balance of tidal marshes along an estuarine salinity gradient. Biogeochemistry 120, 163–189. https://doi.org/10.1007/s10533-014-9989-7.

Whigham, D.F., Simpson, R.L., 1978. Nitrogen and phosphorus movement in a freshwater tidal wetland receiving sewage effluent. In: Proceedings of Symposium on Technological, Environmental, Socioeconomic, and Regulatory Aspects of Coastal Zone Management. ASCE, San Fransisco, CA, USA, pp. 2189–2203.

Whigham, D.F., 2009. Primary production in tidal freshwater wetlands. In: Barendregt, A., Whigham, D.F., Baldwin, A.H. (Eds.), Tidal Freshwater Wetlands. Backhuys Publishers B.V., Leiden, The Netherlands.

White, D.S., Howes, B.L., 1994. Long-term ^{15}N-nitrogen retention in the vegetated sediments of a New England salt marsh. Limnology and Oceanography 39, 1878–1892.

Williams, E.K., Rosenheim, B.E., 2015. What happens to soil organic carbon as coastal marsh ecosystems change in response to increasing salinity? An exploration using ramped pyrolysis. Geochemistry, Geophysics, Geosystems 16, 2322–2335. https://doi.org/10.1002/2015GC005839.

Wilson, B.J., Mortazavi, B., Kiene, R.P., 2015. Spatial and temporal variability in carbon dioxide and methane exchange at three coastal marshes along a salinity gradient in a northern Gulf of Mexico estuary. Biogeochemistry 123, 329–347. https://doi.org/10.1007/s10533-015-0085-4.

Ziegler, S., Velinsky, D.J., Swarth, C.W., Fogel, M.L., 1999. Sediment-water Exchange of Dissolved Inorganic Nitrogen in a Freshwater Tidal Wetland. Jug Bay Wetlands Sanctuary, Lothian, MD, USA, 25 p.

PART V

MANGROVES

CHAPTER

20

Biogeomorphology of Mangroves

Joanna C. Ellison

Discipline of Geography and Spatial Sciences, School of Technology, Environments and Design, University of Tasmania, Launceston, TAS, Australia

1. INTRODUCTION

Mangroves are a taxonomically diverse group of tropical tree species, which by parallel evolution have developed special physiological and morphological adaptations to grow in intertidal environments (Polidoro et al., 2014), such as aerial roots and salt regulation strategies. Although mangroves are by definition a biogenic community primarily of trees and associated fauna, "mangrove" also refers to the saltwater wetland habitat in which they occur. Mangrove forest structure is characterized by "zones" of tree species, in patterns that run perpendicular to the shore or river mouth seaward margins, which reflect different species preferences to inundation frequency (Fig. 20.1). Inundation frequency processes are dependent on the microtopography and elevations (Watson, 1928; Ellison, 2009; Friess, 2016) in relation to sea level position and tidal range.

Owing to intertidal habitat preferences (Fig. 20.1), mangroves occur most extensively on low-gradient tropical sedimentary shorelines, between mean sea level (MSL) and high tide elevations. Topographic surveys show finer-grained shorelines to have a mangrove seaward edge elevation at about MSL, whereas sandy shorelines can extend marginally lower such as 0.36 m below MSL on offshore islands of the Great Barrier Reef (Ellison, 2009), which could be due to differences in oxygen availability in sediments at low tide.

Zonation creates variety in community structure between seaward, central, and landward mangrove zones (Fig. 20.1), which contributes to mangrove biodiversity. There are approximately 70 species of true mangroves, which are taxonomically diverse across 17 families (Polidoro et al., 2014), with the majority found in Southeast Asia (Fig. 20.2). The Atlantic East Pacific and Indo West Pacific groups have different species (Tomlinson, 1996; Lee et al., 2014), and of significance for an ecosystem that is co-located with sea level position, also have experienced differing relative sea level trends over the Holocene (Fig. 20.2).

Coastal Wetlands
https://doi.org/10.1016/B978-0-444-63893-9.00020-4

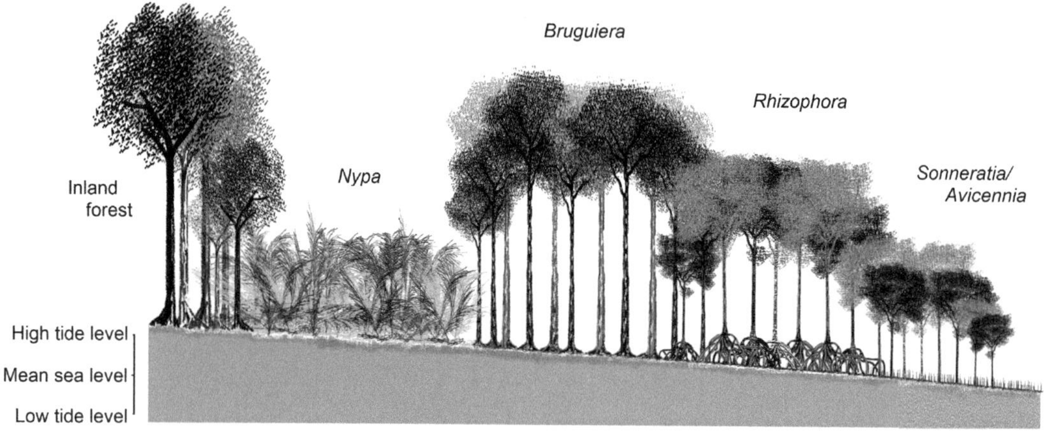

FIGURE 20.1 Typical gradient of mangrove zonation of the Southeast Asian region, indicating species preferences for elevation ranges within the intertidal limits.

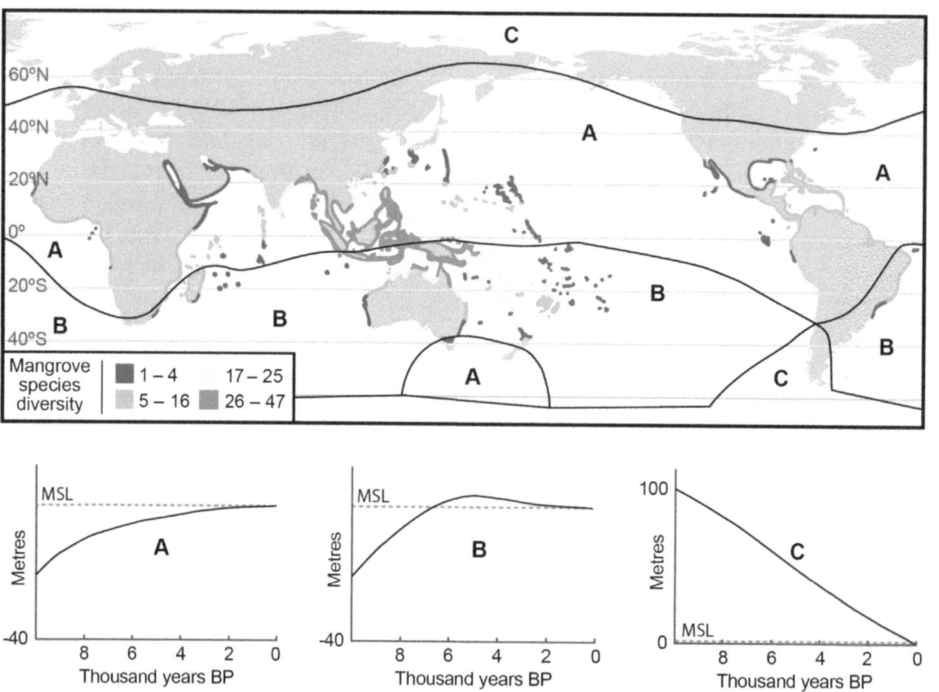

FIGURE 20.2 World distributions of mangroves and indicative Holocene relative sea level curve types. *Adapted from Spalding, M., Kainuma, M., Collins, L., 2010. World Atlas of Mangroves. Earthscan, London, Washington, DC.; Pirazzoli, P.A., 1996. Sea Level Changes: The Last 20,000 Years. J Wiley & Sons, Chichester; Ellison, J.C., 2009. Geomorphology and sedimentology of mangrove swamps. In: Perillo, G.M.E., Wolanski, E., Cahoon, D., Brinson, M. (Eds.), Coastal Wetlands: An Ecosystem Integrated Approach. Elsevier Science, Amsterdam, pp. 564–591; Lambeck, K., Woodroffe, C.D., Antonioli, F., Anzidei, M., Gehrels, W.R., Laborel, J., Wright, A.J., 2010. Paleoenvironmental records, geophysical modelling, and reconstruction of sea level trends and variability on centennial and longer timescales. In: Church, J.A., Woodworth, P.L., Aarup, T., Wilson, W.S. (Eds.), Understanding Sea Level Rise and Variability. Wiley-Blackwell, Chichester, pp. 61–121.*

2. BIOGEOMORPHOLOGY

Biogeomorphology considers the multiple interactions between ecological and geomorphological processes (Viles, 1988; Viles et al., 2008). In the longer term of geological timescales, landform position has influenced the distribution and development of species, and geomorphic-biological feedback mechanisms may have co-adjusted at the scales of ecological succession and organisms' evolution (Viles et al., 2008; Corenblit and Steiger, 2009). In the shorter term of ecological timescales, plants and animals influence geomorphic processes, such as through bioprotection of substrate (Naylor, 2005) and their ability to entrap, collect, and stabilize sediment (Ostling et al., 2009). Research techniques in biogeomorphology require a sharing of methods and understanding between the fields of geomorphology and ecology (Reinhardt et al., 2010; Nolte et al., 2013).

As concluded in their authoritative review of the ecology of mangroves, Lugo and Snedaker (1974) described mangroves as self-maintaining coastal landscape units that are responsive to long-term geomorphological processes and to continuous interactions with contiguous ecosystems in the regional mosaic. They are open systems with exchange of both energy and matter with upland terrestrial and offshore coastal ecosystems. This chapter reviews mangrove biogeomorphic values and processes, as well as resulting forms, as demonstrated from different mangrove settings and relative sea level trends.

2.1 Mangrove Biogeomorphic Values

Mangrove values have been widely appreciated in the literature (Spalding et al., 2010), and values related to biogeomorphology include protection and accretion. Widespread evidence exists for the paradigm that mangrove wetlands protect shorelines from wave damage and erosion, the vegetation slowing water flow, facilitating deposition, increasing shoreline cohesion, and building peat (Gedan et al., 2011), with significant values in natural hazards mitigation (Ostling et al., 2009). Mangrove vegetation density and low-gradient sedimentary expanses protect coastal communities from wave erosion, tropical cyclones, storm surges, and even moderate tsunami waves (Cochard et al., 2008; Gedan et al., 2011) (Fig. 20.3). Organic-rich soils tend to erode more slowly than mineral soils in wetlands (Gedan et al., 2011), with roots providing a physical barrier between the moving water and sediments, therefore reducing entrainment by hydraulic action.

Catchment runoff, as river discharge and through flow, is filtered by dense roots and root mats to reduce suspended sediments and excess nutrients, protecting offshore seagrass beds and coral reefs from turbidity (Vo et al., 2012). Ample evidence exists for the direct contribution of established mangroves to accretion in the vertical plane through peat formation (Lee et al., 2014), which contributes both to the mitigation of sea level rise and to carbon sequestration. Mangrove sediment carbon stores are far greater than those present in temperate, boreal, and tropical terrestrial forests (Donato et al., 2011; Bouillon, 2011; Alongi, 2014). Soil organic carbon content is variable between sites, however, extending to depths lower than terrestrial forests owing to low decomposition rates and high organic input from the mangroves and sedimentation (Kristensen et al., 2008).

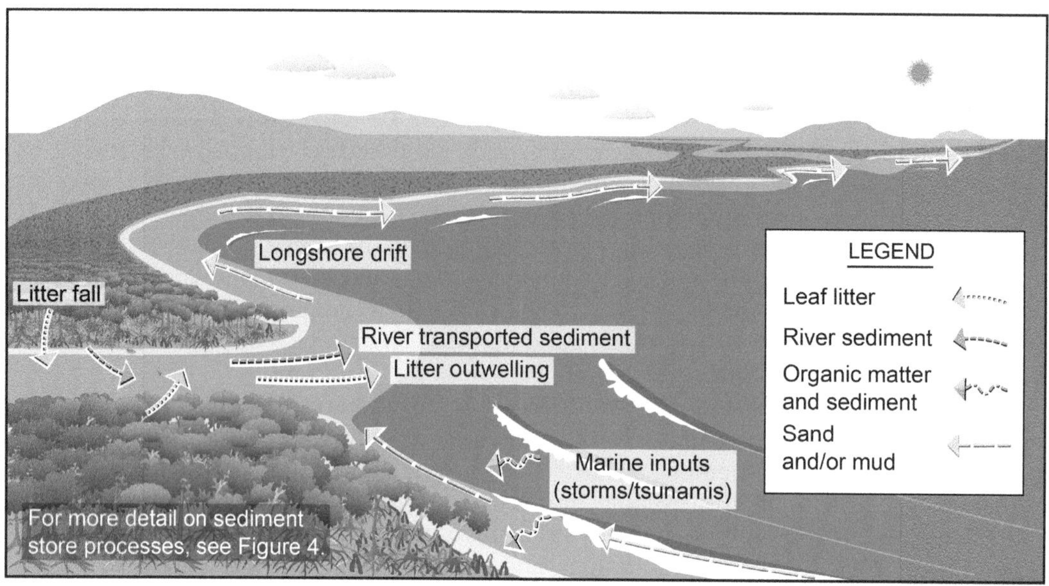

FIGURE 20.3 Mangrove ecosystem sediment sources and losses.

2.2 Mangrove Biogeomorphic Processes

Geomorphology recognizes a strong interdependence between processes and form (Brunsden and Thornes, 1979), where processes are mechanisms of change that can be inferred from landscape observations. This section reviews short-term biogeomorphic processes.

Sediment supply to mangroves derives from external and internal processes, involving interactions between mangrove plants and geomorphic settings. Processes of tidal exchange, wave action, and catchment runoff, including floods, are direct influences on the sediment budget of the mangrove area. Sediment sources include input from river catchments at estuaries and deltas, longshore drift particularly at wave-dominant coastlines, input from further offshore especially during high magnitude but low frequency storms or tsunamis, and autochthonous input from mangrove productivity (Fig. 20.3). Major losses include fluvial sediment not trapped in the mangroves, longshore transport down the coast, mangrove carbon export, and erosion caused by storms or tsunamis. The sediment budget is a balance of volumes of sediment entering or leaving the mangrove environment, influencing whether the mangrove environment accretes, progrades, or erodes over a number of timescales.

Mangrove sediment supply is important for mangrove resilience to the sea level rise scenarios that are predicted, and depends on the physiographic settings of mangroves (Di Nitto et al., 2014). Long-term ecosystem survival is primarily a balance between rates of sea level rise and positive vertical elevation change (Friess et al., 2012). A critical threshold is crossed if wetland substrate accretion fails to keep pace with accelerated sea level rise, thus compromising mangrove functions of protection (Gedan et al., 2011). Mangrove accretion rates are essential for mangrove survival with current eustatic sea level rise, and measurements by use of surface elevation tables (SET) with a sediment surface marker horizon can monitor

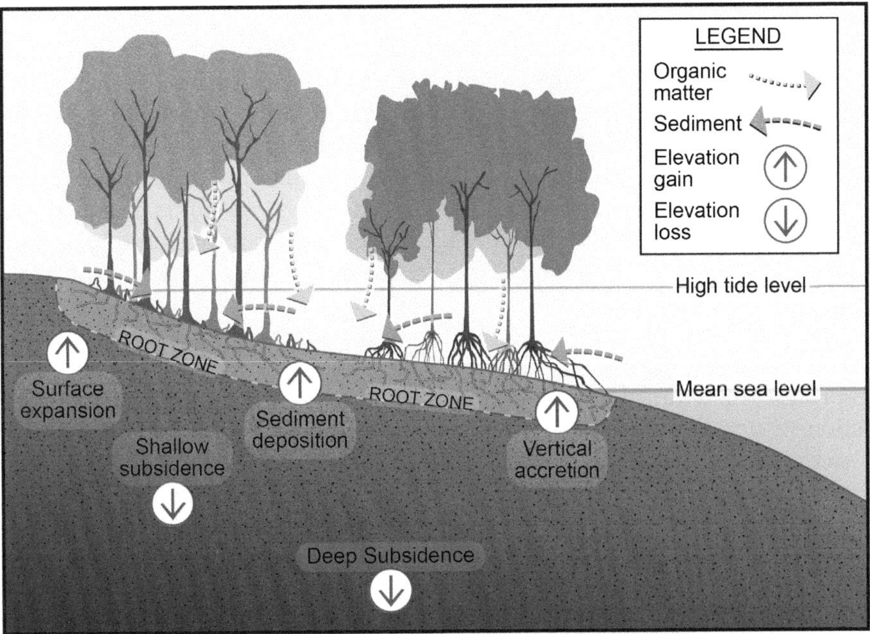

FIGURE 20.4 Processes influencing surface elevation change in mangroves.

intertidal surface elevation trajectories in coastal wetlands (Rybczyk and Callaway, 2009; Webb et al., 2013; Lovelock et al., 2015). Contributions to net mangrove sediment accretion (Fig. 20.4) can be quantified, such as organic detritus from the mangroves, mineral sediment from river discharge, and soil volume increase or compaction (McKee et al., 2007; Krauss et al., 2003, 2010; McKee, 2011). Such studies, however, are limited in scope and geographic extent and need to be repeated in a range of geomorphic settings to assess the broader role of mangroves in accretion and land development (Lee et al., 2014) and can have some potential limitations (Nolte et al., 2013).

Sediment deposition is the process of particles settling out of the water column onto the surface (Nolte et al., 2013) and derives from allochthonous sources external to the mangroves (Fig. 20.3), particularly from catchment erosion and river discharge, which can be higher during periods of catchment land use change. Fluvial suspended sediment tends to be minerogenic, and mangrove systems with such sediment supply have higher net sedimentation rates, which makes them less vulnerable to sea level rise impacts (Lovelock et al., 2015). Vertical accretion (Fig. 20.4) is measured by the accumulation above an exotic surface such as feldspar (Krauss et al., 2003; Cahoon, 2015) and acts to increase surface elevation. While in fluvial-dominant settings this is from mineral sediment deposition, at mangrove peat islands in Belize, McKee et al. (2007) found that surface deposition derives from algal/microbial mats and organic detritus. Elevation gain is the result of net accretion and affects the hydroperiod or the frequency, duration, and depth of flooding, which directly controls the sediment delivery to the wetland surface (Cahoon et al., 1999; Nolte et al., 2013), particularly in tide-dominated settings.

Root growth processes (Cahoon et al., 1999) contribute to surface expansion (Fig. 20.4), and in situ plant production contributes autochthonous sediment. In the anaerobic conditions of mangrove substrates, this organic sediment accumulates rather than decays, providing mangroves with the value of carbon accumulation and storage on a long-term basis. For root matter to accumulate, the rate of root decomposition must be slower than the rate of root production (McKee et al., 2007), and in mangrove peat, the inputs of fine roots contribute substantially to soil volume. The oxidative state of the mangrove soil influences the growth of plant shoots and roots, as well as decomposition of organic matter in the sediment, and is controlled by the frequency, duration, and depth of flooding (Cahoon et al., 1999).

Mangroves provide bioprotection values through the friction caused by aerial roots and dense stems to water movement (Furukawa et al., 1997), exerting a drag coefficient on tidal waters to protect the sediment from erosion (Mazda et al., 1995). Seedling density also provides friction to tidal water movement to promote sediment flocculation and settling and enhance accretion rates, especially at low tidal zones (Huxham et al., 2010; Kumara et al., 2010). Mangroves provide bioconstruction values through their enhancement of allochthonous sediment deposition. *Rhizophora* aerial roots assist in the settling of suspended sediment from estuarine waters (Krauss et al., 2003), whereas *Sonneratia* and *Avicennia* pneumatophores are successful at binding and stabilizing sediment (Spenceley, 1977) and promoting sediment accretion (Young and Harvey, 1996; Krauss et al., 2003).

Elevation gain can be negatively affected by shallow subsidence (Fig. 20.4), which is a decrease of the wetland surface due to sediment compaction, shrinking, and decomposition (McKee et al., 2007; Nolte et al., 2013). Autocompaction of the sediment column (Kaye and Barghoorn, 1964; Cahoon et al., 1999) occurs to some degree at most mangrove settings, and periodic soil compaction can be associated with hurricanes. The bottom zone of stratigraphy can be important to overall expansion and contraction of the entire profile, with groundwater increase having a direct positive effect on soil surface elevation (Cahoon et al., 1999). Groundwater head pressure leads to deep soil profile swelling (Whelan et al., 2005), linked to season rainfall patterns, river stage, and tidal stage.

In summary, mangrove biogeomorphic contribution to sediment accretion is through autochthonous processes and plant facilitation of allochthonous sediment trapping through slowing water flow (Fig. 20.4). Allochthonous sediment supply is the most variable component of mangrove environmental settings, dependent on differences in the prevailing processes operating and geomorphic settings.

3. GEOMORPHIC SETTINGS

Mangroves are facilitated in their expansive development by sheltered geomorphic settings (Ellison, 2009). Geomorphic settings characterize influences of the physical environment, prevailing processes, and their variability. Hydrodynamic processes include waves and turbulence, tides, seasonal variability, wind and wave climates, return periods of extreme events, and sea level rise (Friess et al., 2012). The geomorphology and settings of mangroves have been reviewed by several authors (Thom, 1982, 1984; Woodroffe, 1992; Augustinus, 1995;

Ellison, 2009). Supply and delivery of inorganic sediment depends strongly on local geomorphology and conditions of rainfall, tidal, wind, and storm regimes (Lee et al., 2014), and the below section reviews the differences in sediment supply in each setting, in the context of mangrove biogeomorphic processes.

3.1 Estuaries and Deltas

The largest areas of the world's mangrove forests are established in deltas or muddy coasts in the tropics, as well as estuaries where mangroves extend to higher temperate latitudes (Morrisey et al., 2010; Giri et al., 2015). Estuarine processes are influenced by tidal range, topography, morphology, and salinity structure (Dyer, 1997), receiving sediment from both fluvial and marine sources, these influenced by tide, wave, and fluvial processes (Dalrymple et al., 1992). Estuaries are morphologically recent, formed as river valleys submerged with eustatic sea level rise, and during the slowing of rise at around 6000 years ago (Fig. 20.2). Estuarine geomorphological evolution involves changing relative intensity of river, wave, and tidal influences (Dalrymple et al., 1992) (Table 20.1), and facies infill records differences in these sea level trends.

In wave-dominated estuaries, sediment is moved along the shore and into the estuary to build shore-parallel bars that tend to block the estuary mouth (Dalrymple et al., 1992), creating sheltered mangrove habitats in the central estuary (Table 20.1). Microtidal estuaries (<2 m range) are likely to have narrower mangrove areas between MSL and high tide levels and are less likely to have tidal creeks. By contrast, at the mouth of tide-dominated estuaries and deltas, tidal/current energy exceeds wave energy to cause elongated sand bars (Dalrymple et al., 1992), which are extensively colonized by mangroves such as in the Fly delta (Ellison, 2009). Currents and tidal ranges are generally larger, as well as more dominant relative to wave energy and river discharge. In meso- and macrotidal estuaries, turbulent flows, density-driven circulation, and tidal pumping maintain high suspended sediment concentrations in their upper reaches (Morrisey et al., 2010), enhancing sediment supply.

River-dominant estuaries (Table 20.1) are more likely to be microtidal, associated with high catchment sediment supply and nearshore conditions that do not favor coastal progradation (Cooper, 1993). They are also more likely to be at higher latitudes, whereas deltas are more common at lower latitudes (Morrisey et al., 2010). Sediment sources in temperate estuarine settings are dominated by allochthonous sources derived from the catchment soils and coastal margins, accelerated in New Zealand and Australia by catchment deforestation in the last 200 years (Morrisey et al., 2010).

Deltas occur at river mouths where progradation is far greater than is the case for estuaries (Dalrymple et al., 1992), as a result of a large fluvial sediment source and where low gradients approach the coast (Selvam, 2003). As with estuaries, deltas form and develop when sea level is relatively stable. Relative to estuaries, accumulation of catchment-derived sediment at the coast overwhelms any bedrock confinement to form large aprons of deposition through which channels distribute or split. Sediment contribution to the coast is variable with catchment size, river discharge, and sediment yield, as a function of geology, glacial history, and tectonic activity, as well as land use changes.

TABLE 20.1 Terrigenous Geomorphic Settings of Mangroves and Their Controlling Attributes

Attributes	Wave Dominated	Tide Dominated	River Dominated
Dominant process	Wave energy, sand bars enclosing lagoons with internal autochthonous sedimentation.	Tidal currents, allochthonous sedimentation.	River discharge, allochthonous sedimentation.
Landward morphology	River-dominated zone. Sand/gravel in bayhead delta(s).	River-dominated zone extending from an alluvial valley. Main river channels become funnel-shaped to seaward.	River floodplains above the tidal limit, low gradient. Rapid deposition of terrigenous material.
Central morphology	Low energy zone, open water with fine-grained organic muds, extensive mangroves.	Mixed energy zone, sands in tidal channels throughout estuary with strong bidirectional currents. Extensive intertidal areas of mangroves in back-barrier lagoons.	Broad alluvial basin with flood reworked fluvial sediment infill. Flow parallel bars
Seaward morphology	Barrier spit or bar across mouth or parallel to the shoreline, sand dominated.	Elongated depositional shoals. Extensive tidal flats colonized by mangroves.	Riverine facies extend to a delta, which expands seaward.
Examples	NSW barrier estuaries, Burdekin delta, Australia; Sao Francisco delta, Brazil; Niger delta, Nigeria; Douala Estuary, Cameroon.	S. Alligator River, Ord River delta complex, Australia; Fly River delta, Papua New Guinea; Sunderbans delta, India and Bangladesh. Klang-Langat delta, Malaysia.	Krishna and Godavari deltas, Andhra Pradesh, India; Rufiji delta, Tanzania. Ajkwa River delta, West Papua.
Sources	Allen (1970), Dalrymple et al. (1992), Roy (1984) and Ellison and Zouh (2012).	Thom et al. (1975), Coleman and Wright (1978), Selvam (2003), Dalrymple et al. (1992) and Ellison (2009).	Cooper (1993), Selvam (2003), Ganguly et al. (2006) and Ellison (2009).

While river processes are important, some deltas have greater influence from tidal processes, such as the macrotidal Sunderbans (Selvam, 2003), where tidal water penetrates c. 110 km toward land in the deltaic complex of the Ganges, Brahmaputra, and Meghna rivers (Ghosh et al., 2015). Hundreds of islands crisscrossed by a maze of anastomotic tidal creeks provide habitat for the largest continuous mangrove ecosystem in the world (Raha et al., 2014). This delta has existed through major Pleistocene sea level changes, with the modern delta building about 60 m of deposited layers on top of older delta substrates. Other deltas are wave-influenced (Bhattacharya and Giosan, 2003), with shore-parallel beach ridge barriers. Over 30% of the world's barrier islands occur on deltas, with one of the longest on the Niger Delta (Stutz and Pilkey, 2011). The Niger Delta has an offshore powerful wave swell from the southwest, and mangroves occur inland of the sand barriers (Allen, 1970).

Deltaic sedimentation causes both land subsidence and compaction of loose deltaic deposits (Raha et al., 2014), with relative sea level rise (RSLR) of up to 5 mm per year in the Sunderbans (Raha et al., 2014; Ghosh et al., 2015). Erosional features are common in the western sector, to which sediment and water flows in association with the Farakka Barrage may be contributing. Spatial analysis shows that island erosion and mangrove loss in this area is dominant (Ganguly et al., 2006; Raha et al., 2014; Ghosh et al., 2015), with a net loss of 200 km^2 in the last 50 years, of which RSLR is a major driver.

3.2 Lagoon Settings

A lagoon is a shallow coastal waterbody separated from the ocean by a barrier, connected intermittently to the ocean exchange by one or more restricted inlets, and usually orientated shore-parallel (Kjerfve, 1994). Where deltas or estuaries in wave-dominated settings develop shallow basins behind sand barriers, which are wide in a shore-parallel direction, these can be called lagoons (Kjerfve, 1994). The barriers are depositional, usually of sand or shingle (Bird, 1994), and lagoons are protected from coastal processes of wave action. Sediment supply is from coastal rivers, tidal, and wind processes, as well as organic autochthonous sources, and over time lagoons infill with inwashed sediment and organic deposits (Bird, 1994). On geologic timescales, lagoons are short-lived features, occupying about 13% of global coastlines, with variety in size, salinity, and morphology, and generally feature shallowness and propensity for trophic stresses (Pérez-Ruzafa et al., 2012).

In the context of mangrove settings, the traditional definition of lagoons (Kjerfve, 1994) is best broadened to include the terminology commonly used. In the mangrove literature, lagoons include marine semienclosed waters, such as the "Great Barrier Reef lagoon," as well as large waterbodies in terrigenous-dominated settings. Inside barrier coral reefs, shallow nearshore areas with patch reefs, seagrass, and fringing mangroves are commonly called lagoons, although the more classic use of the word does not refer to marine waters (Kjerfve, 1994; Bird, 1994). The adjacent reef and seagrass ecosystems provide co-benefits. Sediments and nutrients, carried by any freshwater runoff, are first filtered by coastal forests, then by mangrove wetlands, and finally by seagrass beds. The existence of coral reefs is directly dependant on the buffering capacity of the shoreward ecosystems, which help create the oligotrophic conditions under which coral reefs flourish, therefore limiting the algal growth that can threaten coral reef health (Golbuu et al., 2003). At locations where poor land management leads to high sediment discharge from rivers, this role can be critical (Victor et al., 2006). In turn, coral reefs and associated shingle ridges buffer the soft habitats of mangroves and seagrass from wave action (Frank and Jell, 2006; Villanoy et al., 2012).

3.3 Low Island Settings

Low islands differ from the terrigenous settings of deltas and estuaries primarily owing to lack of fluvial influences, with no point sources of freshwater and catchment-derived sediment supply. As a result, they have dominance by organic matter in sediment with autochthonous accumulation, derived primarily from in situ breakdown of mangrove production and root mat growth (Fig. 20.4). Low islands demonstrate biogeomorphic associations with coral reefs through evolved co-benefits, with a closer associations between mangroves, seagrasses, and coral reefs in island settings relative to continental margins (Linton and Warner, 2003). Reefs protect mangrove shores from wave action and also supply carbonate sediment to fringing mangrove habitats (Fig. 20.3), while mangroves provide habitats for reef fauna at stages of their life cycles, as well as outwelling dissolved organic matter protection to corals (Shank et al., 2010). The karst geomorphology of some low islands dominates the occurrences of types of inland mangroves, which provide a longer-term example of biogeomorphic relationships reviewed in the section on long-term mangrove biogeomorphic influences.

4. MANGROVE BIOGEOMORPHIC RESPONSE TO SEA LEVEL CHANGE

Mangrove geomorphic settings, with varying dominant processes and sediment sources (Table 20.1), show different biogeomorphic responses to medium-term processes of relative sea level (RSL) change. Over the last 10,000 years of the Holocene through which mangrove stratigraphic records are available, global eustatic sea level rise occurred owing to ice melt and ocean thermal expansion, combined with spatial variability in response to ice volume changes (Lambeck et al., 2010). Regional differences and generalized RSL trends are summarized in Fig. 20.2, and mangrove biogeomorphic response to each is reviewed below. RSL change is the sum of the regional and global components of sea level change acting at a particular site, but it may also include more local influences of subsidence due to sediment compaction and tectonic uplift or subsidence.

4.1 Rising Sea Level Sequences

The vast majority of modern coastlines of mangrove-occupied latitudes formed during RSL rise of the earlier Holocene, as shown by both types A and B of the sea level curves in Fig. 20.2. Conceptual models of lithofacies response in estuaries to Holocene RSL rise are provided by Dalrymple et al. (1992) for both wave-dominated and tide-dominated settings. Mangrove habitats, as represented in higher latitude estuaries by marshes, and mangrove paleoenvironmental records within these sequences indicate how mangroves adjusted. Mangrove strata can be identified by organic stratigraphy and mangrove domination of pollen proportions in sediments (Ellison, 2008).

River-dominated mangrove responses to a rising sea level sequence are demonstrated by the Tipoeka and Ajkwa river deltas of West Papua (Ellison, 2005, 2009), with mangrove pollen-dominated strata of >80% to depths far below the current mangrove intertidal elevation range. Within this Holocene stratigraphy, evidence of landward mangrove communities occurs from 6.0 to 1.5 m below present, above this replaced by a seaward mangrove community of *Rhizophora*, demonstrating landward movement of mangrove zones (Fig. 20.1) with slowly rising sea level. Pollen concentrations are the density of microfossils per volume of sediment (Ellison, 2008), indicative of the mangrove contribution to the sediment relative to the dilution by other processes, and cores from this site showed very high concentrations of >50,000 grains cm^{-3}. High pollen concentrations can also indicate a large mangrove area with healthy reproductive status of mangroves.

Similar pollen results are shown for shallower sequences from southern Papua New Guinea, with >90% mangrove dominance of pollen sums and 20,000–40,000 grains cm^{-3} to nearly 2 m of depth (Rowe et al., 2013), in a tidal range of <3 m (Paijmans and Rollet, 1976). In the central Torres Strait, Rowe (2007) described extensive mangrove records up to 4 m below present sea level between 6000 years BP and 3000 years BP, interpreting transgressive sea level from a transition from lower tidal *Rhizophora* forest to an upper intertidal community, with lower pollen concentrations of c. 3000 grains cm^{-3}. To the south, rising sea level sequences were described from the South Alligator River, Northern Territory of Australia, with mangrove fossil pollen dominance of stratigraphy up to 5 m below present MSL (Woodroffe et al., 1985; Grindrod, 1988). Sea level was transgressive earlier in the record, with increasing influence of marine processes and corresponding changes in mangrove communities. Extensive mangroves occurred at 7000–5500 years BP, around the time of sea level stabilization, then fluvial sedimentation and progradation caused conversion to freshwater habitats (Woodroffe et al., 1985; Grindrod, 1988).

Lagoonal setting mangrove response to a rising sea level sequence is demonstrated from the west coast of Viti Levu, Fiji (Ellison and Strickland, 2015). Mangroves are less extensive in lagoon settings relative to continental terrigenous settings (Table 20.1), though there is fluvial sediment supply from rivers of relatively small catchment size. Being a microtidal location, mangroves at Tikina Wai currently grow in the narrow elevation range shown by the red arrow on Fig. 20.5, yet mangrove pollen dominates sediment deposits down to >3 m with c. 80% presence. This shows that mangrove habitats earlier occurred at lower elevations than they do today, and have by accretion mostly retained spatial position with slowly rising sea level of 2.1 mm per year. Over time, from the base up (Fig. 20.5), a landward community of *Bruguiera* from 3 to 1 m depth is replaced by a seaward community of *Rhizophora* from 1 m to the surface, showing slow landward retreat of mangrove zones (Fig. 20.1). Organic content is useful in interpretation of biogeomorphic processes in different settings (Thorne et al., 2014) and was about 10% at depth, increasing to around 20% towards the surface (Ellison and Strickland, 2015). This record has strong similarities that described above from West Papua, demonstrating biogeomorphic processes of long-term bioaccretion with rising sea level. Both have high mangrove pollen proportions relative to the total pollen sum to depths well below the current intertidal range, as well as high fossil pollen concentrations, low organic content in sediment, accretion mostly keeping up with RSLR, and pollen diagrams showinga landward to seaward mangrove zone transition over time.

Low island mangrove biogeomorphic response to RSLR is demonstrated from the Caribbean region, which experienced slow RSLR for much of the Holocene (Fig. 20.2), reflecting ongoing isostatic adjustment to the melting of the Laurentide Ice Sheet (Lambeck et al., 2010). There are many mangrove stratigraphic studies from the region showing a similar record of mangrove response (Toscano and Macintyre, 2003; McKee et al., 2007). On the Caribbean coast of Belize at Twin Cays, a mangrove-dominated island with just a 0.2 m tidal range records 10 m depth of peat, with average loss on ignition of 65% showing that accumulation of mangrove organic matter through the Holocene allowed the low mangrove island to accrete vertically (Macintyre et al., 2004; McKee et al., 2007). Pollen records elsewhere give similar results; on the north coast of Trinidad, *Rhizophora*-dominated pollen assemblages with high levels of organic matter are described to 9 m depths dating back to 7000 years

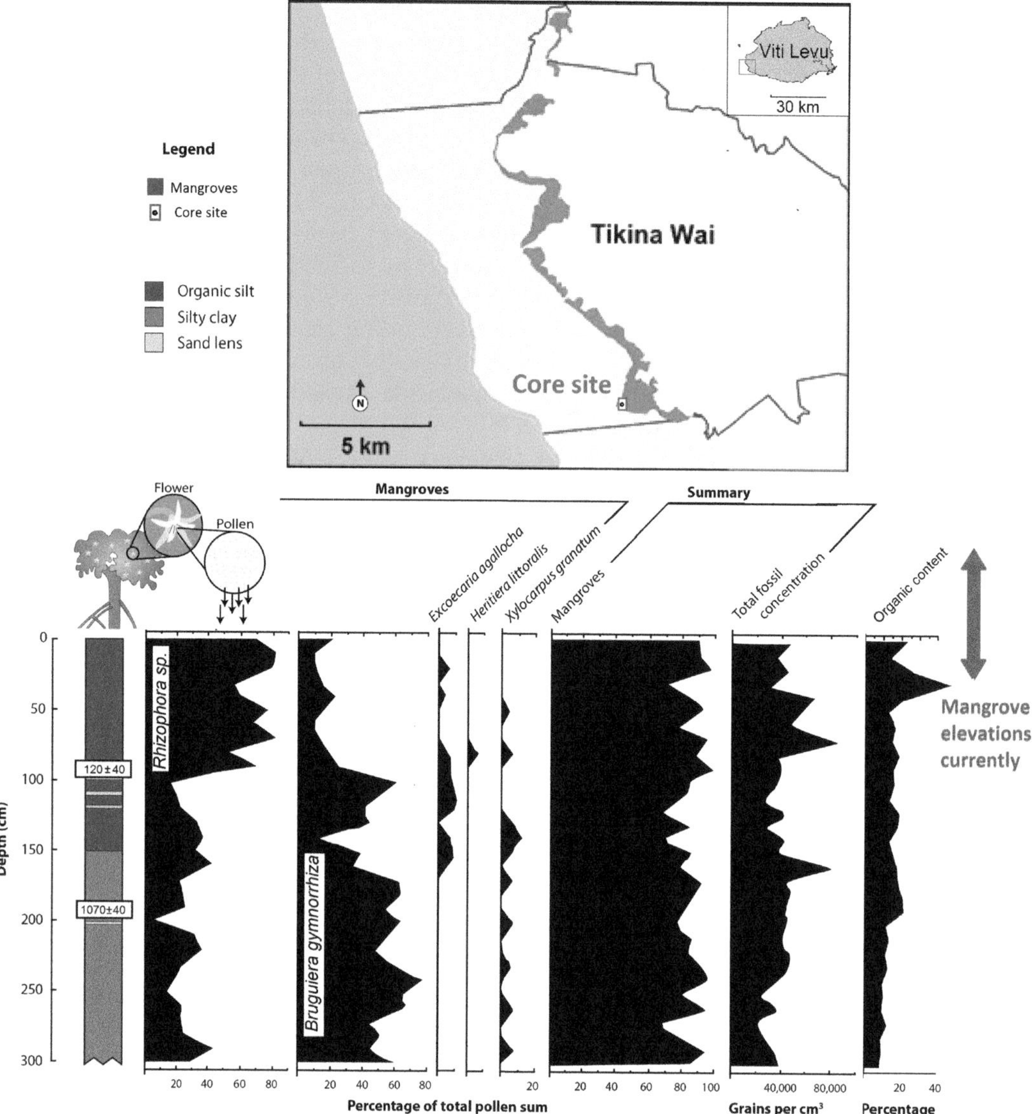

FIGURE 20.5 Mangrove paleohistory at Tikina Wai, Fiji, during long-term relative sea level rise of over 2 mm $year^{-1}$. *Adapted from Ellison, J.C., Strickland, P., 2015. Establishing relative sea level trends where a coast lacks a long term tide gauge. Mitigation and Adaptation Strategies for Global Change 20, 1211–1227.*

BP, formed during slow RSLR (Ramcharan, 2004). Similar concurring records are from the Caribbean coast of Mexico with mangrove organic sediment up to 2.5 m in depth (Islebe and Sánchez, 2002), and 2.5 m of mangrove peat developed since 2100 years BP further north at Bermuda (Ellison, 1996) in a tidal range of 0.75 m.

The low limestone island of Grand Cayman has extensive mangroves on the North Sound, where 20 km^2 of mangroves retreated between 4080 and 3230 years BP, during RSLR of 2.8–3.3 mm per year (Ellison, 2009). A new pollen record is provided here from a core on the current seaward edge of mangroves, collected and analyzed using standard techniques (Ellison, 2008), to show mangrove stratigraphy and pollen records of mangrove community change (Fig. 20.6). Tidal range in the North Sound is also very low, with a

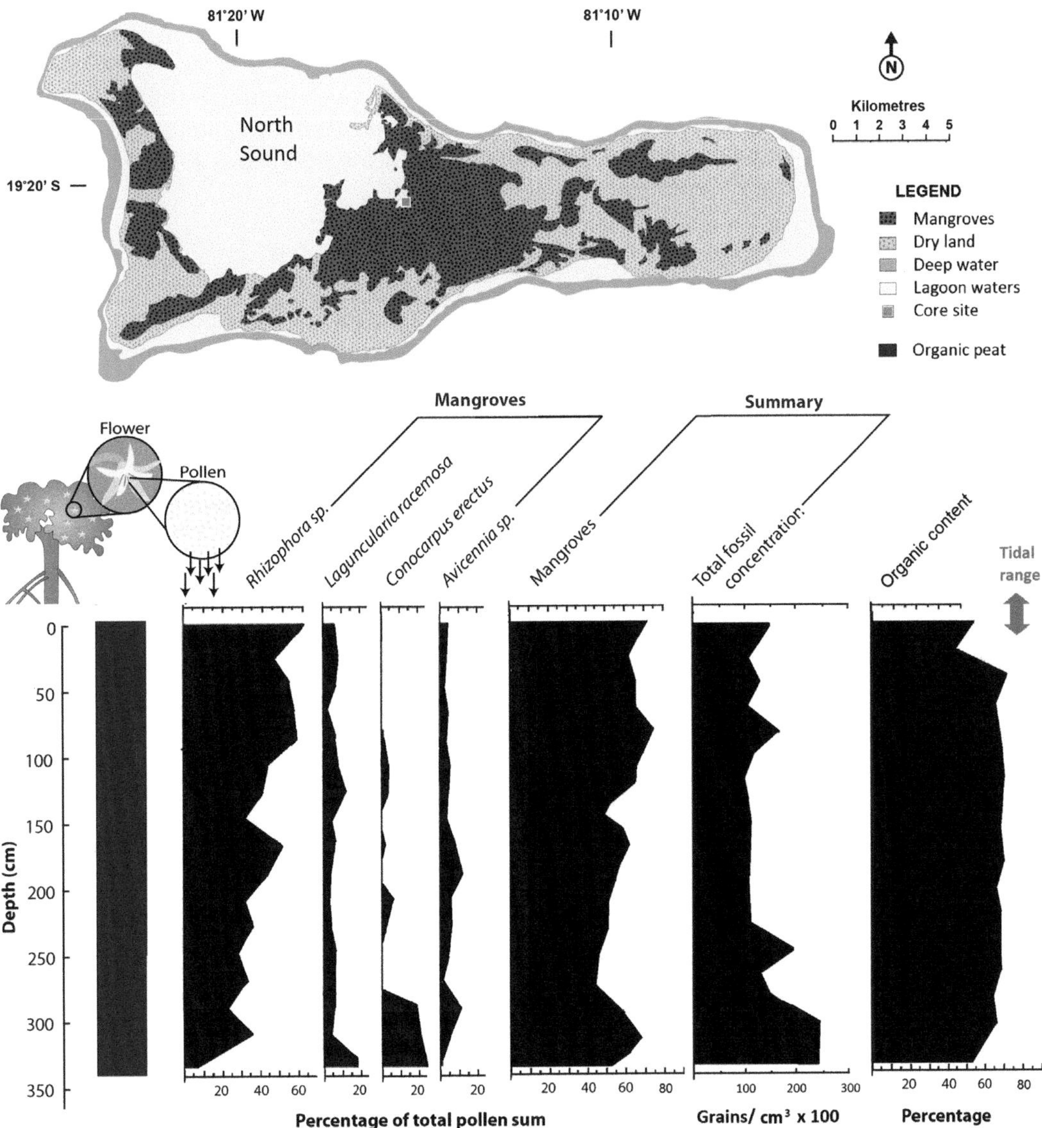

FIGURE 20.6 Map of mangroves of Grand Cayman, showing location of the core site, and pollen diagram from the seaward edge of mangroves, North Sound.

mean tidal amplitude of 28 cm (Brunt and Burton, 1994). Cayman has four mangrove tree species, with *Rhizophora mangle* growing at lower elevations and *Avicennia germinans*, *Laguncularia racemosa*, and *Conocarpus erectus* at higher elevations.

Pollen proportions of c. 60% occur throughout the core, with dominance by landward species at lower levels and increasing *Rhizophora* dominance in the upper 1 m (Fig. 20.6). Mangroves communities flourished at lower elevations up to 10× below the present tidal range, with presence of all four species, 60% organic content in peat, and 10,000–20,000 pollen grains cm^{-3} concentration. Although there has been some dieback to seaward (Ellison, 2009), mangroves by biogenic accumulation achieved net accretion with regional RSLR.

Both Figs. 20.5 and 20.6 show a more landward mangrove community at the base of the record, as shown in Fig. 20.6 by higher proportions of *Laguncularia*, *Conocarpus*, and *Avicennia* at lower levels, as well as increasing proportions of the seaward species *Rhizophora* toward the surface. This is indicative of a transgressional sequence.

4.2 Rising-Highstand-Falling Sequences

Holocene sequences of sedimentary infill and progradation are described from coastlines that experienced RSL rise to a highstand, followed by a sea level fall (Morrison and Ellison, 2017). As shown by RSL curve type B (Fig. 20.2), RSLs in the Indo-Pacific located far from major Pleistocene ice sheets reached close to present at least 7000 years ago to slightly higher (Lambeck et al., 2010), and then fell slowly to the present position with hydroisostatic adjustment. Rising sea level sequences of mangrove biogeomorphic response are described in the previous section, and mangrove records of a highstand and later RSL fall are available from the Eastern Australian coast and Tonga.

Transgressive tide-dominated estuaries show tidal-fluvial deposits flanked by brackish and freshwater marsh sediments accreting vertically above alluvial sediments, that move inland during Holocene sea level rise (Dalrymple et al., 1992). Mangrove stratigraphy of Missionary Bay, Queensland, shows similar brackish swamp sediments and intertidal organic muds from 15 to 5 m depth below current MSL, occurring before 6000 years BP on the RSL curve (Fig. 20.7). On low-gradient shorelines, stratigraphic diagrams can indicate the contemporary environments by considering a level line relative to the sea level position on the left hand RSL curve, showing that mangroves, as represented by the extent of intertidal organic muds, were <300 m wide.

Stratigraphy at Missionary Bay, Queensland, described by Grindrod and Rhodes (1984) includes later sequences of RSL fall as shown by the RSL curve (Fig. 20.7). During RSL fall, mangroves prograded to over 10× their previous area, with upper levels of intertidal organic muds showing mangroves colonizing the low tide muds at around MSL. RSL fall sequences are described in the next section, with a pollen record from Tonga.

4.3 Falling Sea Level Trends

RSL fall occurs at higher latitudes (type C on Fig. 20.2) on coastlines located near to Pleistocene major ice sheets, as a result of glacioisostatic rebound (Lambeck et al., 2010). These are

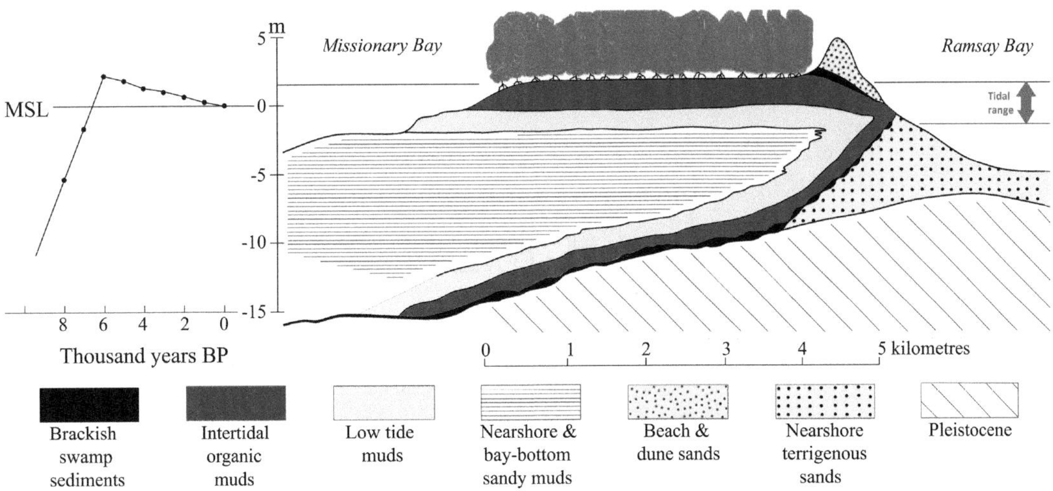

FIGURE 20.7 Holocene stratigraphy of Missionary Bay, Hinchinbrook Island (18° 17′S; 146° 14′E), (right), and the Townsville relative sea level curve of Chappell et al. (1983) redrawn to the same vertical scale (left). *MSL*, mean sea level. *Adapted from (right) Grindrod, J., Rhodes, E.G., 1984. Holocene sea-level history of a tropical estuary: missionary bay, North Queensland. In: Thom. B.G. (Ed.), Coastal Geomorphology in Australia. New York, Academic Press, pp. 151–178.*

not at latitudes where mangroves occur; however, RSL fall mangrove biogeomorphic response is recorded at mangrove areas where the coastal land is uplifting and is also observed late in the rising-highstand-falling sequences described above for Missionary Bay, which also occurred in Tonga (Ellison, 1989).

A similar Holocene record to Missionary Bay (Fig. 20.7) is found in the lagoon of Tongatapu, Tonga (Ellison, 2009), with lagoon calcareous sands and silts from −1.5 to +0.5 m relative to present sea level datum representing open lagoon waters during the sea level highstand. Above this developed a near-surface mangrove peat, which occurred during subsequent RSL fall, which is focused on in this section. Pollen analysis shows how the mangroves responded (Fig. 20.8), with this upper peat having c. 70% domination by mangrove pollen, with greater presence of *Rhizophora* at depth and increasing domination by mangroves of landward zones (*Bruguiera, Excoecaria*) towards the surface (Ellison, 1989). Pollen concentrations are high, of 30–50,000 grains cm^{-3}, increasing toward the surface. As sea level fell, this made areas of lagoon bed available for colonization, first by *Rhizophora* and then as peat accumulation raised ground levels, facilitating replacement by *Bruguiera*. Mangroves were initially opportunistic of new habitat availability and then became deterministic as sea level stabilized and accreted peat.

Normal regression or seaward migration of the shoreline occurs when progradation is driven by sediment supply, whereas forced regression is where progradation is driven by sea level fall (Catuneanu et al., 2009). In the latter case, progradational facies are still present but vertical aggradation is not possible and is replaced with downstepping or stacking patterns as described by Catuneanu et al. (2011), with younger layers forming at lower elevations down the slope.

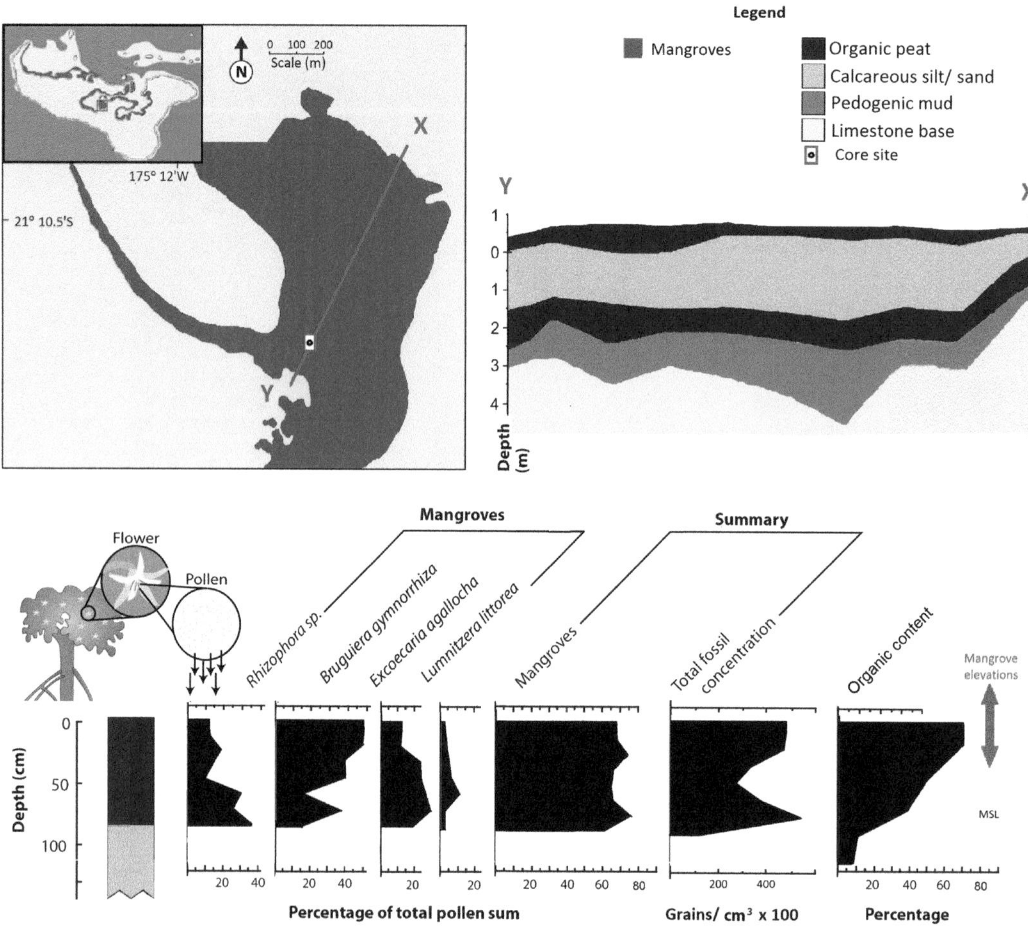

FIGURE 20.8 Folaha/Nukuhetulu mangrove area of Tongatapu, showing stratigraphy and pollen records in the upper peat unit. *Adapted from Ellison, J.C., 1989. Pollen analysis of mangrove sediments as a sea-level indicator: assessment from Tongatapu, Tonga. Palaeogeography, Palaeoclimatology, Palaeoecology 74, 327–341; Ellison, J.C., 2009. Geomorphology and sedimentology of mangrove swamps. In: Perillo, G.M.E., Wolanski, E., Cahoon, D., Brinson, M. (Eds.), Coastal Wetlands: An Ecosystem Integrated Approach. Elsevier Science, Amsterdam, pp. 564–591.*

A continental area experiencing RSL fall owing to land uplift is the central eastern coast of East Africa (Åse, 1981; Reuter et al., 2010). The Rufiji River is the largest in Tanzania and the delta includes about 43 islands and supports the largest mangrove area in Tanzania (Fig. 20.9), with half of the national mangroves (Wagner and Sallema-Mtui, 2016). The mangrove area changes over time with natural processes, such as a flood in the 1960s, which diverted the predominant flow from the southern delta to the north, with changes in salinity and inundation. From analysis of a 20 km landward to seaward transect of cores crossing the northern delta (Fig. 20.9), Punwong et al. (2013) used pollen analysis of stratigraphy to describe mangroves responding to wetter and drier conditions, as well as regional RSL

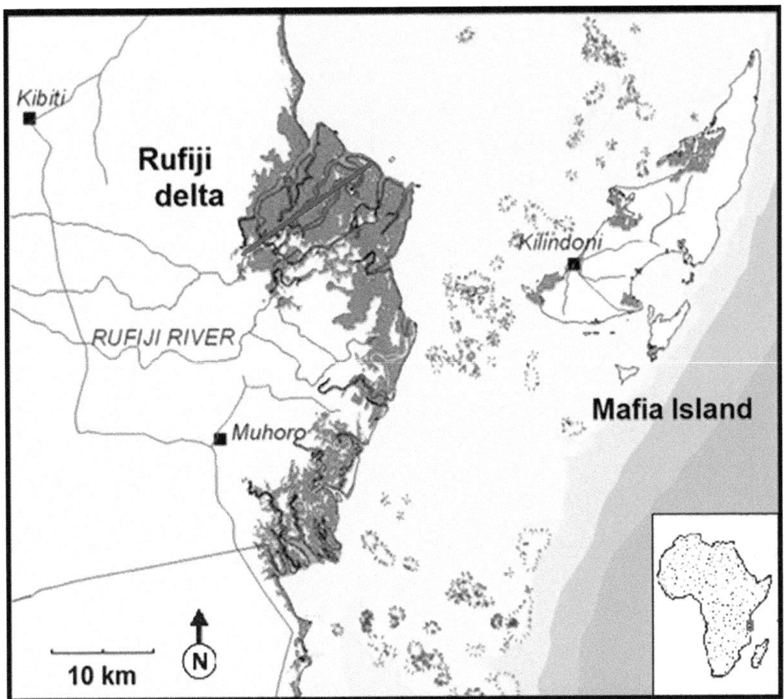

FIGURE 20.9 Rufiji river delta, Tanzania, showing mangroves in dark green and the core transect of Punwong et al. (2013) in red.

fall, with shallow marine substrate relative to the mesotidal range retreating seaward over the last several thousand years. Mangrove stratigraphy was dominated by inorganic sources, with <20% organic matter. Higher RSL was indicated earlier in core records, when there was greater inundation frequency in the central mangrove area than currently, and upper levels of the landward core showed reduction in mangrove species and an increase in back-mangrove and terrestrial grasses, indicative of a falling sea level.

4.4 Stable Sea Level

The most recent millennia of both RSL types A and B in Fig. 20.2 show a relatively stable sea level, up until the commencement of RSL rise of the Anthropocene. Douala Estuary, Cameroon (Fig. 20.10), shows records of development of mangrove lithofacies in a wave-dominated estuarine setting in the last few hundred years (Ellison and Zouh, 2012; Ellison, 2017). The mangroves extend landward of a >10 km sand barrier, which provides protection from offshore wave swell. Stratigraphy shows lower levels of inorganic sand (Fig. 20.10), fining upward to inorganic silty clay, with shallow levels of more organic silty clay near the surface. Organic matter content is overall low at <20%, indicative of significant allochthonous inorganic sediment input. Pollen concentrations are also very low at just c. 1000 grains cm^{-3}, further showing inorganic terrigenous sediment dilution of pollen deposition (Fig. 20.10).

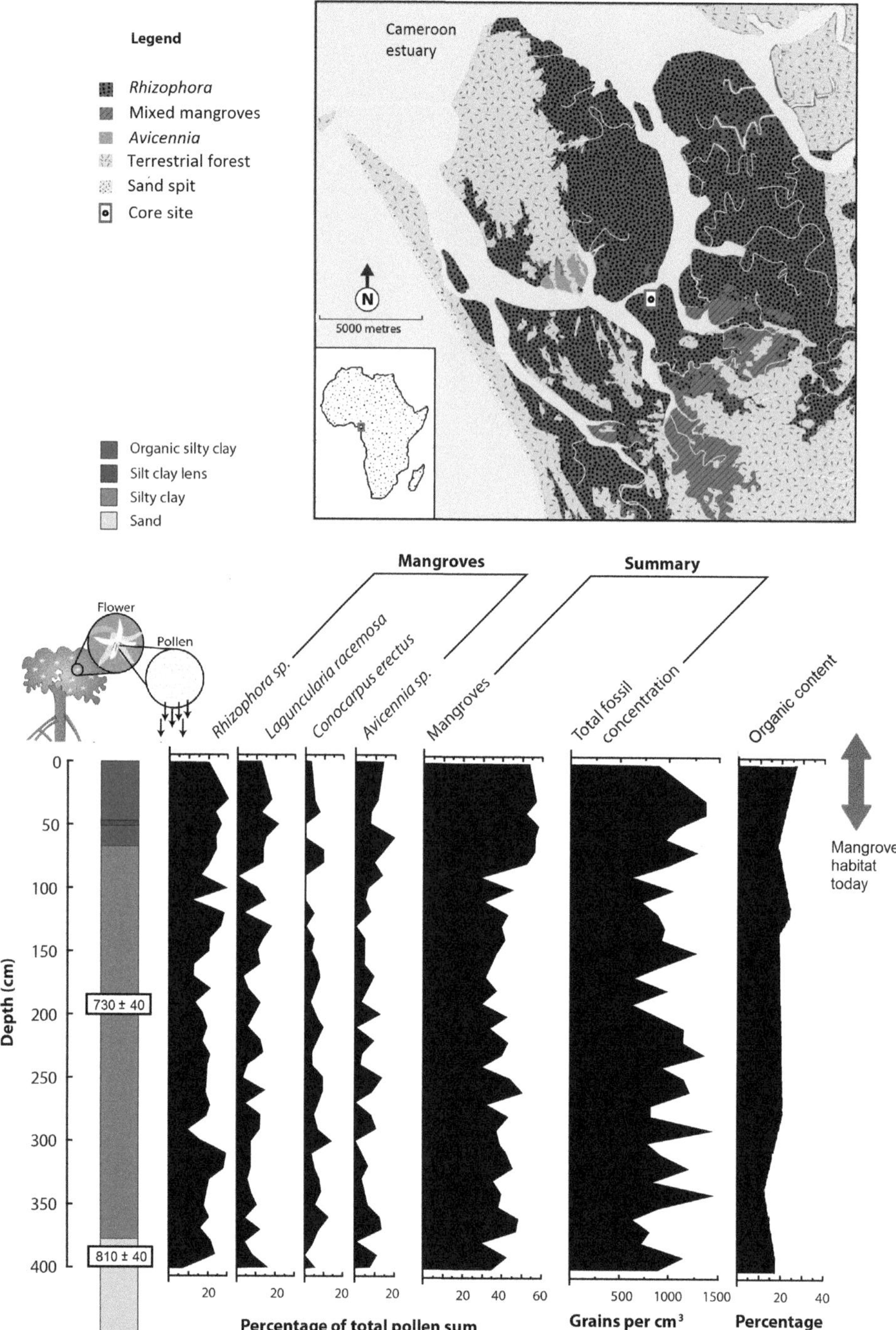

FIGURE 20.10 Mangroves of Douala, Cameroon, and pollen records from a 4-m core. *Adapted from Ellison, J.C., 2017. Applications of pollen analysis in estuarine systems. In: Weckström, K., Saunders, K.M., Gell, P.A., Skilbeck, C.G. (Eds.), Applications of Paleoenvironmental Techniques in Estuarine Studies. Springer, Netherlands, pp. 441–468.*

Higher proportions of mangrove pollen at 60% of the pollen sum occur at shallow levels under established mangroves. Where mangrove pollen declines to around 40%, this represents mudflats offshore (Fredoux, 1994; Van Campo and Bengo, 2004) and sediment becomes increasingly coarse grained, inorganic, and calcareous with depth. The deeper radiocarbon date (Fig. 20.10) indicates rapid sedimentation and mixing. Lacking is mangrove stratigraphy lower than present mangrove elevations, as found elsewhere on Atlantic coastlines (Bartlett and Barghoorn, 1973; Behling et al., 2001; Tossou et al., 2008; Cohen et al., 2009), this being indicative of a rising sea level trend described above for Tikina Wai (Fig. 20.5). This comparison indicates that mangroves of Cameroon at Douala are prograding seaward with high levels of terrigenous sediment supply and a stable sea level for the period of the record, as shown by low pollen concentrations and low organic fraction in all stratigraphy. Mangroves are occupying the advancing margin of the bayhead delta described in the wave-dominated conceptual model of Dalrymple et al. (1992), in conditions of sedimentation surplus, colonizing seaward as deposition occurs in an opportunistic manner.

5. OPPORTUNISTIC AND DETERMINISTIC BIOGEOMORPHOLOGY

Within pollen records, mangrove accretion with RSLR can be interpreted from the difference between the tidal range and the depth of the mangrove-dominated sequences, requiring accurate determinations of elevations (Ellison and Strickland, 2015). If mangrove-dominated pollen assemblages extend deeper than the contemporary tidal range (Figs. 20.5 and 20.6), this shows net mangrove accretion during rising sea level. If mangrove-dominated pollen assemblages occur only within the contemporary mangrove elevation range, then a prograding shore is the interpretation (Fig. 20.10). While mangroves contribute some biogenic matter, other sediment supply processes are more dominant. If sea level is falling, then progradation is exacerbated, with mangroves following the available habitat seaward (Fig. 20.8).

Mangrove sediment percent organic data indicates the net biogenic contribution by mangroves, which is also reflected in pollen concentrations. Mangrove pollen diagrams are traditionally displayed in percentages of the pollen sum, but pollen concentration provides information on mangrove biogeomorphic regimes. High percent organic matter along with high mangrove pollen concentrations in sediments show that mangroves are creating habitats in a biogenic environment as the coast accretes, as shown by the Tipoeka, Tikina Wai, and Cayman pollen diagrams (Ellison, 2005, Figs. 20.5 and 20.6). Low percent organic matter along with low mangrove pollen concentrations in sediments show that mangroves are dominated by external processes, following available habitats in a minerogenic environment, as shown by the Rufiji delta and Douala Estuary pollen diagrams (Ellison and Zouh, 2012; Punwong et al., 2013, Fig. 20.10). Minerogenic sediment supply or sea level fall both force mangroves to be opportunistic of suitable settings as they become available with habitat relocation as conditions change. The influence of mangroves on sedimentation processes and geomorphic evolution of estuaries and coasts has been researched for 130 years (Curtis, 1888; Hitchcock, 1891; Philipps, 1903; Pollard, 1903; Waming, 1925; Davis, 1940; Chapman, 1944; Egler, 1952; Thom, 1967; Carlton, 1974; Thom et al., 1975) and continues to be deliberated (Morrisey et al., 2010; Lee et al., 2014; Swales et al., 2015; Friess, 2016).

The ability of plants and animals to directly or indirectly influence their physical environment is long recognized (Viles, 1988; Naylor, 2005; Viles et al., 2008; Morrisey et al., 2010; Nolte et al., 2013), and recognition of the biogeomorphic role of mangroves as ecosystem engineers has an early history. Clements (1916) described concepts of succession after colonization of bare substrate and interpreted zonation of the mangroves in Florida as a successional sequence, generalized by Davis (1940) as typical of the tropical American mangroves. Seaward zones were interpreted as a stage toward a climax seral stage, facilitated by sediment accumulation, with biotic factors of mangrove root trapping of sediment (Davis, 1940) supporting the deterministic concept of mangroves as "land builders" (Curtis, 1888; Hitchcock, 1891; Philipps, 1903; Pollard, 1903; Waming, 1925; Chapman, 1944). The alternative view is that mangroves follow silting rather than cause it (Watson, 1928; Friess, 2016). Mangrove geomorphologists working in active deltaic settings supported this opportunistic interpretation of mangrove habitats, such as the Tabasco delta, Mexico (Thom, 1967), and the Ord delta (Thom et al., 1975). Mangrove distributions there follow habitat availability, dominated by processes of sedimentation and freshwater discharge (Thom, 1967). However, the relative contribution of mangroves to land formation probably varies with geomorphic settings (Lee et al., 2014; Friess, 2016) and RSL trends. Pollen analysis of long-term changes in mangrove stratigraphy, with elevation determination, has not been applied to this question before.

Where mangrove pollen-dominated stratigraphy occurs well below the present intertidal range, and during slowly rising sea level mangrove accretion raises the elevation of the habitat, this demonstrates the deterministic biogenic functions of mangroves (Figs. 20.5 and 20.6). Many mangrove researchers in the Caribbean region advocated this interpretation (Davis, 1940; Chapman, 1944; and all cited above in supporting the land-building concept), this region having a type A RSL trend (Fig. 20.2). Domination by minerogenic sediment and RSL fall causes mangroves to follow habitat migration (Fig. 20.8). As disturbance is reduced, mangroves may facilitate accretion and the maintenance of their habitat through biogenic processes of vertical accretion (Fig. 20.4). These alternative scenarios indicate that each view of opportunistic or deterministic mangroves is situational rather than alternative. Mangroves opportunistically colonize suitable habitats, and then in conditions of slow RSL rise, with reduced disturbances they can deterministically maintain their own habitat by net accretion.

Interpreting mangroves as land builders requires substantiation of their role in promoting lateral expansion, for which there is little evidence (Lee et al., 2014). Low tidal flats offshore of mangroves have been shown to contain organic matter derived from the adjacent mangroves (Coleman and Wright, 1978), which contribute to vertical accretion. About 50% of mangrove litter is exported to adjacent offshore waters (Alongi, 2014), and mangrove pollen deposition occurs offshore of mangroves (Fredoux, 1994; Van Campo and Bengo, 2004), demonstrating potential influences on offshore biogenic accretion. Mangrove influence extends at ground levels beyond their seaward edge canopy in several ways. Dense pneumatophores of seaward edge species in the *Avicennia* and *Sonneratia* genera extend offshore (Fig. 20.11A), which can promote sediment binding and settling (Spenceley, 1977; Young and Harvey, 1996), as well as sediment accretion (Krauss et al., 2003). *Rhizophora* roots also extend seaward of the mangrove canopy (Fig. 20.11B), likely also exacerbating accretion. Mangrove seedlings commonly sprout offshore of the mangrove canopy (Fig. 20.11C), and mangrove seedlings have been shown to enhance accretion rates by providing friction to tidal water movement

FIGURE 20.11 Mangrove influences extending seaward of their canopy. (A) *Avicennia* at Magnetic Island, Queensland. (B) *Rhizophora* at Exmouth Gulf, Western Australia. (C) *Rhizophora* seedlings in the Great Barrier Reef lagoon.

to promote sediment flocculation and settling (Huxham et al., 2010; Kumara et al., 2010). Mangrove contribution to lateral mudflat accretion could be confirmed by monitoring by surface elevation tables and marker horizons placed offshore of mangroves in different settings, including those that are shown by spatial analysis to be prograding.

6. LONG-TERM MANGROVE BIOGEOMORPHIC INFLUENCES

Biogeomorphology also considers the interactions between ecological and geomorphological processes at longer geological timescales (Naylor, 2005). Mangroves provide a strong record largely from pollen records of how landform position has influenced the distribution of species.

The evolution and global dispersal of mangroves has been reconstructed from palynological and macrofossil paleoecology records (Saenger, 1998; Ellison et al., 1999; Plaziat et al., 2001; Ellison, 2008). More recently, analyses of *Rhizophora* chloroplast and nuclear DNA sequences (Lo et al., 2014) support these interpretations from the fossil record that early *Rhizophora* evolved and dispersed along the Tethys seaway, migrating westward into the Atlantic through the closing Mediterranean Tethys Seaway and eastward to Southeast Asia and Australia. The pollen record while showing locations, arrival, and ecosystem development does not show clear trends of mangrove migration with climate change, rather that mangrove distributions have migrated over time due to biogeographic factors and habitat availability (Ellison, 2008). With the further application of mangrove DNA technology in combination with palynological and macrofossil records, there are opportunities to explore the resilience of mangroves during past climate and sea level changes.

One example of long-term influences of geomorphology on mangrove distributions is inland or landlocked mangrove settings.

6.1 Inland Settings

Mangroves that occur inland are rare and of small scale (Ellison, 2009); however, they have particular geomorphic interest as they can be relics of a former sea level and/or facilitated by karst geology. They are of intrinsic scientific interest being small mangrove populations, genetically isolated from others of the same species for centuries or more. Inland mangroves are described from several locations (Table 20.2). These inland settings have generally narrow mangrove margins because of lack of tides. Owing to pond or lake-like settings, sediment sources are biogenic and autochthonous, similar to coastal mangroves of low islands where many inland mangroves that have been described also occur.

With increasing fragmentation of coastal mangrove habitats owing to human impacts, more research is needed regarding constraints to gene flow at the landscape level (Friess et al., 2012). These isolated populations of inland mangroves have potential to provide an evolutionary record of the long-term impacts of population isolation.

While presenting suitable wetland habitats for stratigraphic preservation, pollen analysis, and dating, these inland mangrove sites have surprisingly been little used for paleoecological reconstruction. Basal deposits could indicate the dates and causes of their inland formation. This is an opportunity for future researchers to investigate the geomorphic causes of these unique mangroves, and habitat resilience over time.

TABLE 20.2 Locations of Inland Mangroves, Site Elevations, and Recorded Mangrove Species

Location	Description	Distance Inland	Elevation	Species	Sources
Bermuda	Saline, nontidal ponds, anchialine (tidal) ponds, or without pond	<1 km	Sea level	*Rhizophora mangle*, *Avicennia germinans*, and *Conocarpus erectus*	Thomas et al. (1992) and Thomas (1993)
Inagua, Bahamas	Large shallow lakes, salinity to 20 ppt	50 km		*Rhizophora mangle*	Lugo (1981)
Barbuda, Bahamas	Brackish ponds of 2 km^2 inside lithified beach ridges	2–4 km	<7 m	*Conocarpus erectus*, *Laguncularia racemosa*, and *Rhizophora mangle*	Stoddart et al. (1973)
Shravan Kavadia, Gujarat, India	Saline ponds, 0.7 ha	80 km	2–5 m	*Avicennia marina*	Patel and Agormoorthy (2012) and Tripathi et al. (2013)

TABLE 20.2 Locations of Inland Mangroves, Site Elevations, and Recorded Mangrove Species—cont'd

Location	Description	Distance Inland	Elevation	Species	Sources
West Papua (North)	Freshwater lake		75 m	*Sonneratia caseolaris*	Van Steenis (1963)
West Papua (South)	Tipoeka River bank	10 km	10 m	*Sonneratia caseolaris*	Ellison and Simmonds (2003)
Christmas Island, Australia	0.33 ha stand	1.2 km	24–37 m	*Bruguiera gymnorrhiza*	Van Steenis (1984) and Woodroffe (1988)
Mandora, Western Australia	Salt creek in limestone	25 km		*Avicennia marina*	Beard (1967) and Halse et al. (2011)
Lake MacLeod, Western Australia	22.5 ha fringing stands around a saline lake	18 km	Sea level	*Avicennia marina*	Ellison and Simmonds (2003) and Ellison (2012)
Tarawa and Abaiang, Kiribati	Ponds	<1 km	Sea level	*Bruguiera gymnorrhiza*	J.C. Ellison
Marshall Islands	Ponds	<1 km	Sea level	*Bruguiera gymnorrhiza*	Fosberg (1975) and J.C. Ellison unpublished data
Niutao and Nanumanga, Tuvalu	Landlocked lagoons	<1 km	Sea level	*Rhizophora stylosa*, *Lumnitzera littorea*	Woodroffe (1987)

7. CONCLUSION

Biogeomorphological processes operate on the different timescales, of short-term functions of trees in promoting accretion, medium-term processes of responses to RSL changes, and long-term processes of migration and the consequent distribution of genetic matter. Within different geomorphic settings, the role of mangrove biogeomorphic processes differs. Short-term biogeomorphic processes include biogenic substrate accretion owing to mangrove productivity and plant facilitation of minerogenic deposition. Allochthonous sediment supply is the most variable component of mangrove environmental settings, dependent on differences in prevailing processes operating and geomorphic settings.

For medium term biogeomorphic processes, pollen analysis of mangrove sediments with elevation determination can show the vertical biogeomorphological achievement of different mangroves, and clarify situations where mangroves are opportunistic within a changing environment or deterministic in fostering vertical accretion. Mangroves can be deterministic during slowly rising sea level, especially in lagoonal and low island settings. Mangroves are opportunistic during more rapid RSLR in all of terrigenous, lagoonal, and low island settings, in minerogenic-dominated settings with either wave or fluvial sediment supply, and with coastal progradation as sea level falls.

Use of pollen concentration and percent organic matter data in interpretation of mangrove changes over time is uncommon in the literature, and further research in different settings is needed, it having become common practice to only use percentage abundance of pollen in results. Pollen analysis is a low cost but time-consuming technique, where one diagram requires some months of careful laboratory analyses, hence existing reports tend to be site-focused. The potential for automation of the methods are so far limited. However, the results can add insight and long-term context into critical questions of mangrove response to sea level rise, beyond the timeframes available from site monitoring. Hence, multidisciplinary approaches that include pollen analysis along with elevation determination have potential to provide insight into mangrove long term responses to sea level change.

Monitoring of mangrove surface elevation change in real time as allowed by SET has potential to connect better with longer-term paleoecological records as their time-series increases. Installations offshore of mangroves could resolve questions as to whether mangroves contribute to their progradation through promoting accretion to create habitat that they may then colonize, or not. Expanded use of percent organic analysis of sediment techniques in association with SET research could clarify the sources of the sediment accretion, and associate better with research results available from longer-term sedimentology and paleoecology.

Application of pollen analysis to questions of long-term mangrove distribution and evolution as preserved in inland wetland habitats is a large and unexplored research area. Corenblit and Steiger (2009) consider an evolutionary relationship between plants and geomorphology through co-adjusting feedback, and further research could confirm these long-term links in mangrove development of aerial root systems and their promotion of vertical sediment accretion.

The umbrella concept of biogeomorphology has not recently been considered in mangrove research by many. It incorporates important areas of research and mangrove understanding from different disciplines, which could be integrated and expanded more widely, to increase our understanding of this increasingly threatened ecosystem.

References

Allen, J.R.L., 1970. Sediments of the modern Niger Delta: a summary and review. In: Morgan, J.P. (Ed.), Deltaic Sedimentation: Modern and Ancient. Society of Economic Paleontologists and Mineralogists, Tulsa, pp. 138–151.

Alongi, D.M., 2014. Carbon cycling and storage in mangrove forests. Annual Review of Marine Science 6, 195–219.

Åse, L.E., 1981. Studies of shores and shore displacement on the southern coast of Kenya. Especially in Kilifi District. Geografiska Annaler: Series A, Physical Geography 63, 303–310.

Augustinus, P.G.E.F., 1995. Geomorphology and sedimentology of mangroves. In: Perillo, G.M.E. (Ed.), Geomorphology and Sedimentology of Estuaries. Elsevier Science, Netherlands, pp. 333–357.

Beard, J.S., 1967. An inland occurrence of mangrove. Western Australian Naturalist 10, 112–115.

Bartlett, A.S., Barghoorn, E.S., 1973. Phytogeographic history of the isthmus of Panama during the past 12 000 years (A history of vegetation, climate and sea level change). In: Graham, A. (Ed.), Vegetation and Vegetational History of Northern Latin America. Elsevier, Amsterdam, pp. 203–299.

Behling, H., Cohen, M.C.L., Lara, R.J., 2001. Studies on Holocene mangrove ecosystem dynamics of the Bragança Peninsula in north-eastern Pará, Brazil. Palaeogeography, Palaeoclimatology, Palaeoecology 167, 225–242.

Bhattacharya, J.P., Giosan, L., 2003. Wave-influenced deltas: geomorphological implications for facies reconstruction. Sedimentology 50, 187–210.

Bird, E.C.F., 1994. Physical setting and geomorphology of coastal lagoons. In: Elsevier Oceanography Series, vol. 60, pp. 9–39.

Bouillon, S., 2011. Storage beneath mangroves. Nature Geoscience 4, 282–283.
Brunsden, D., Thornes, J.B., 1979. Landscape sensitivity and change. Transactions of the Institute of British Geographers NS 4, 463–484.
Brunt, M.A., Burton, F.J., 1994. Mangrove swamps of the Cayman Islands. In: Brunt, M.A., Davies, J.E. (Eds.), The Cayman Islands: Natural History and Biogeography. Springer, Netherlands, pp. 283–305.
Cahoon, D.R., 2015. Estimating relative sea-level rise and submergence potential at a coastal wetland. Estuaries and Coasts 38, 1077–1084.
Cahoon, D.R., Day Jr., J.W., Reed, D.J., 1999. The influence of surface and shallow subsurface soil processes on wetland elevation: a synthesis. Current Topics in Wetland Biogeochemistry 3, 72–88.
Carlton, J.M., 1974. Land-building and stabilization by mangroves. Environmental Conservation 1, 285–294.
Catuneanu, O., Abreu, V., Bhattacharya, J.P., Blum, M.D., Dalrymple, R.W., Eriksson, P.G., Fielding, C.G., Fisher, W.L., Galloway, W.E., Gibling, M.R., Giles, K.A., Holbrook, J.M., Jordan, R., Kendall, C.G.St.C., Macurda, B., Martinsen, O.J., Miall, A.D., Neal, J.E., Nummedal, D., Pomar, L., Posamentier, H.W., Pratt, B.R., Sarg, J.F., Shanley, K.W., Steel, R.J., Strasser, A., Tucker, M.E., Winker, C., 2009. Towards the standardization of sequence stratigraphy. Earth-Science Reviews 92, 1–33.
Catuneanu, O., Galloway, W.E., Kendall, C.G.St.C., Miall, A.D., Posamentier, H.W., Strasser, A., Tucker, M.E., 2011. Sequence stratigraphy: methodology and nomenclature. Newsletters on Stratigraphy 44, 173–245.
Chapman, V.J., 1944. 1939 Cambridge expedition to Jamaica Part 1. A study of the botanical processes concerned in the development of the Jamaican shoreline. Botanical Journal of the Linnean Society 52, 407–447.
Chappell, J., Chivas, A., Wallensky, E., Polach, H.A., Aharon, P., 1983. Hydro-isostasy and the sea level isobase of 5500 BP in north Queensland, Australia. Marine Geology 49, 81–90.
Clements, F.E., 1916. Plant Succession: An Analysis of the Development of Vegetation. Publications of the Carnegie Institute, p. 242.
Cochard, R., Ranamukhaarachchi, S.L., Shivakoti, G.P., Shipin, O.V., Edwards, P.J., Seeland, K.T., 2008. The 2004 tsunami in Aceh and Southern Thailand: a review on coastal ecosystems, wave hazards and vulnerability. Perspectives in Plant Ecology, Evolution and Systematics 10, 3–40.
Cohen, M.C.L., Behling, H., Lara, R.J., Smith, C.B., Matos, H.R.S., Vedel, V., 2009. Impact of sea-level and climatic changes on the Amazon coastal wetlands during the late Holocene. Vegetation History and Archaeobotany 18, 425–439.
Coleman, J.M., Wright, L.D., 1978. Sedimentation in an arid macrotidal alluvial river system: Ord River, Western Australia. The Journal of Geology 81, 621–642.
Cooper, J.A.G., 1993. Sedimentation in a river-dominated estuary. Sedimentology 40, 979–1017.
Corenblit, D., Steiger, J., 2009. Vegetation as a major conductor of geomorphic changes on the Earth surface: toward evolutional geomorphology. Earth Surface Processes and Landforms 34, 891–896.
Curtis, A.H., 1888. How the mangroves form islands. Garden and Forest 1, 100.
Dalrymple, R.W., Zaitlin, B.A., Boyd, R., 1992. Estuarine facies models: conceptual basis and stratigraphic implications. Journal of Sedimentary Petrology 62, 1130–1146.
Davis, J.H., 1940. The ecology and geologic role of mangroves in Florida. Papers from the Tortugas Laboratory 32, 303–412.
Di Nitto, D., Neukermans, G., Koedam, N., Defever, H., Pattyn, F., Kairo, J.G., Dahdouh-Guebas, F., 2014. Mangroves facing climate change: landward migration potential in response to projected scenarios of sea level rise. Biogeosciences 11, 857–871.
Donato, D.C., Kauffman, J.B., Murdiyarso, D., Kurnianto, S., Stidham, M., Kanninen, M., 2011. Mangroves among the most carbon-rich forests in the tropics. Nature Geoscience 4, 293–297.
Dyer, K.R., 1997. Estuaries: A Physical Introduction, second ed. John Wiley and Sons, Chichester.
Egler, F.E., 1952. Southeast Saline Everglades vegetation, Florida, and its management. Vegetatio 3, 213–265.
Ellison, A.M., Farnsworth, E.J., Merkt, R.E., 1999. Origins of mangrove ecosystems and the mangrove biodiversity anomaly. Global Ecology and Biogeography 8, 95–115.
Ellison, J.C., 1989. Pollen analysis of mangrove sediments as a sea-level indicator: assessment from Tongatapu, Tonga. Palaeogeography, Palaeoclimatology, Palaeoecology 74, 327–341.
Ellison, J.C., 1996. Pollen evidence of Late Holocene mangrove development in Bermuda. Global Ecology and Biogeography 5, 315–326.

Ellison, J.C., 2005. Holocene palynology and sea-level change in two estuaries in Southern Irian Jaya. Palaeogeography, Palaeoclimatology, Palaeoecology 220, 291–309.

Ellison, J.C., 2008. Long-term retrospection on mangrove development using sediment cores and pollen analysis: a review. Aquatic Botany 89, 93–104.

Ellison, J.C., 2009. Geomorphology and sedimentology of mangrove swamps. In: Perillo, G.M.E., Wolanski, E., Cahoon, D., Brinson, M. (Eds.), Coastal Wetlands: An Ecosystem Integrated Approach. Elsevier Science, Amsterdam, pp. 564–591.

Ellison, J.C., 2012. Mangrove swamps: causes of decline and mortality. In: Jenkins, J.A. (Ed.), Forest Decline: Causes and Impacts. Nova Science, New York, pp. 39–68.

Ellison, J.C., Vulnerability assessment of mangroves to climate change and sea-level rise impacts. Wetlands Ecology and Management 23, 115–137

Ellison, J.C., 2017. Applications of pollen analysis in estuarine systems. In: Weckström, K., Saunders, K.M., Gell, P.A., Skilbeck, C.G. (Eds.), Applications of Paleoenvironmental Techniques in Estuarine Studies. Springer, Netherlands, pp. 441–468.

Ellison, J.C., Simmonds, S., 2003. Structure and productivity of inland mangrove stands at Lake MacLeod, Western Australia. Journal of the Royal Society of Western Australia 86, 21–26.

Ellison, J.C., Strickland, P., 2015. Establishing relative sea level trends where a coast lacks a long term tide gauge. Mitigation and Adaptation Strategies for Global Change 20, 1211–1227.

Ellison, J.C., Zouh, I., 2012. Vulnerability to climate change of mangroves: assessment from Cameroon, Central Africa. Biology 1, 617–638.

Frank, T.D., Jell, J.S., 2006. Recent developments on a nearshore, terrigenous-influenced reef: low Isles Reef, Australia. Journal of Coastal Research 22, 474–486.

Fredoux, A., 1994. Pollen analysis of a deep sea core in the Gulf of Guinea: vegetation and climatic changes during the last 225,000 years B.P. Palaeogeography, Palaeoclimatology, Palaeoecology 109, 317–330.

Friess, D.A., 2016. JG Watson, Inundation classes, and their influence on paradigms in mangrove forest ecology. Wetlands. https://doi.org/10.1007/s13157-016-0747-6.

Friess, D.A., Krauss, K.W., Horstman, E.M., Balke, T., Bouma, T.J., Galli, D., Webb, E.L., 2012. Are all intertidal wetlands naturally created equal? Bottlenecks, thresholds and knowledge gaps to mangrove and saltmarsh ecosystems. Biological Reviews 87, 346–366.

Fosberg, F.R., 1975. Phytogeography of micronesian mangroves. In: Walsh, G., Snedaker, S., Teas, H. (Eds.), Proceedings of an International Symposium on the Biology and Management of Mangroves, vol. 1. University of Florida, Gainesville, pp. 23–42.

Furukawa, K., Wolanski, E., Mueller, H., 1997. Currents and sediment transport in mangrove forests. Estuarine, Coastal and Shelf Science 44, 301–310.

Ganguly, D., Mukhopadhyay, A., Pandey, R.K., Mitra, D., 2006. Geomorphological study of Sundarban deltaic estuary. Journal of the Indian Society of Remote Sensing 34, 431–435.

Gedan, K.B., Kirwan, M.L., Wolanski, E., Barbier, E.B., Silliman, B.R., 2011. The present and future role of coastal wetland vegetation in protecting shorelines: answering recent challenges to the paradigm. Climatic Change 106, 7–29.

Ghosh, A., Schmidt, S., Fickert, T., Nüsser, M., 2015. The Indian Sundarban mangrove forests: history, utilization, conservation strategies and local perception. Diversity 7, 149–169.

Giri, C., Long, J., Abbas, S., Murali, R.M., Qamer, F.M., Pengra, B., Thau, D., 2015. Distribution and dynamics of mangrove forests of South Asia. Journal of Environmental Management 148, 101–111.

Golbuu, Y., Victor, S., Wolanski, E., Richmond, R.H., 2003. Trapping of fine sediment in a semi-enclosed bay, Palau, Micronesia. Estuarine, Coastal and Shelf Science 57, 941–949.

Grindrod, J., 1988. The palynology of mangrove and saltmarsh sediments, particularly in Northern Australia. Review of Palaeobotany and Palynology 55, 229–245.

Grindrod, J., Rhodes, E.G., 1984. Holocene sea-level history of a tropical estuary: missionary bay, North Queensland. In: Thom, B.G. (Ed.), Coastal Geomorphology in Australia. Academic Press, New York, pp. 151–178.

Halse, A.S.S., Shiel, R.J., Creagh, S., 2011. Aquatic fauna and water chemistry of the mound springs and wetlands of Mandora Marsh, north-western Australia. Journal of the Royal Society of Western Australia 94, 419–437.

Hitchcock, A.S., 1891. A visit to the West Indies. Botanical Gazette 16, 130–141.

Huxham, M., Kumara, M.P., Jayatissa, L.P., Jayatissa, J.P., Krauss, K.W., Kairo, J., Langat, J., Mencuccini, M., Skov, M.W., Kirui, B., 2010. Intra- and interspecific facilitation in mangroves may increase resilience to climate change threats. Philosophical Transactions of the Royal Society of London B Biological Sciences 365, 2127–2135.

Islebe, G., Sánchez, O., 2002. History of late Holocene vegetation at Quintana Roo, Caribbean coast of Mexico. Plant Ecology 160, 187–192.

Kaye, C.A., Barghoorn, E.S., 1964. Late Quaternary sea-level change and crustal rise at Boston, Massachusetts, with notes on the autocompaction of peat. The Geological Society of America Bulletin 75, 63–80.

Kjerfve, B., 1994. Coastal lagoons. In: Kjerfve, B. (Ed.), Coastal Lagoon Processes. Elesevier Oceanography Series, vol. 60. Elsevier Science, Amsterdam, pp. 1–8.

Krauss, K.W., Allen, J.A., Cahoon, D.R., 2003. Differential rates of vertical accretion and elevation change among aerial root types in Micronesian mangrove forests. Estuarine, Coastal and Shelf Science 56, 251–259.

Krauss, K.W., Cahoon, D.R., Allen, J.A., Ewel, K.C., Lynch, J.C., Cormier, N., 2010. Surface elevation change and susceptibility of different mangrove zones to sea-level rise on Pacific high islands of Micronesia. Ecosystems 13, 129–143.

Kristensen, E., Bouillon, S., Dittmar, T., Marchand, C., 2008. Organic carbon dynamics in mangrove ecosystems: a review. Aquatic Botany 89, 201–219.

Kumara, M.P., Jayatissa, L.P., Krauss, K.W., Phillips, D.H., Huxham, M., 2010. High mangrove density enhances surface accretion, surface elevation change, and tree survival in coastal areas susceptible to sea-level rise. Oecologia 164, 545–553.

Lambeck, K., Woodroffe, C.D., Antonioli, F., Anzidei, M., Gehrels, W.R., Laborel, J., Wright, A.J., 2010. Paleoenvironmental records, geophysical modelling, and reconstruction of sea level trends and variability on centennial and longer timescales. In: Church, J.A., Woodworth, P.L., Aarup, T., Wilson, W.S. (Eds.), Understanding Sea Level Rise and Variability. Wiley-Blackwell, Chichester, pp. 61–121.

Lee, S.Y., Primavera, J.H., Dahdouh-Guebas, F., McKee, K., Bosire, J.O., Cannicci, S., Diele, K., Fromard, F., Koedam, N., Marchand, C., Mendelssohn, I., Mukherjee, N., Record, S., 2014. Ecological role and services of tropical mangrove ecosystems: a reassessment. Global Ecology and Biogeography 23, 726–743.

Linton, D.M., Warner, G.F., 2003. Biological indicators in the Caribbean coastal zone and their role in integrated coastal management. Ocean and Coastal Management 46, 261–276.

Lo, E.Y.Y., Duke, N.C., Sun, M., 2014. Phylogeographic pattern of *Rhizophora* (Rhizophoraceae) reveals the importance of both vicariance and long-distance oceanic dispersal to modern mangrove distribution. BMC Evolutionary Biology 14, 83.

Lovelock, C.E., Cahoon, D.R., Friess, D.A., Guntenspergen, G.R., Krauss, K.W., Reef, R., Rogers, K., Saunders, M.L., Sidik, F., Swales, A., Saintilan, N., 2015. The vulnerability of Indo-Pacific mangrove forests to sea-level rise. Nature 526, 559–565.

Lugo, A.E., 1981. The inland mangroves of Inagua. Journal of Natural History 15, 845–852.

Lugo, A.E., Snedaker, S.C., 1974. The ecology of mangroves. Annual Review of Ecology, Evolution, and Systematics 5, 39–64.

Macintyre, I.G., Toscano, M.A., Bond, G.B., 2004. Modern sedimentary environments, Twin Cays, Belize, Central America. Atoll Research Bulletin 509, 1–12.

Mazda, Y., Kanazawa, N., Wolanski, E., 1995. Tidal asymmetry in mangrove creeks. Hydrobiologia 295, 51–58.

McKee, K.L., Cahoon, D.R., Feller, I.C., 2007. Caribbean mangroves adjust to rising sea levels through biotic controls on change in soil elevation. Global Ecology and Biogeography 16, 545–556.

McKee, K.L., 2011. Biophysical controls on accretion and elevation change in Caribbean mangrove ecosystems. Estuarine, Coastal and Shelf Science 91, 475–483.

Morrisey, D.J., Swales, A., Dittmann, S., Morrison, M.A., Lovelock, C.E., Beard, C.M., 2010. The ecology and management of temperate mangroves. In: Gibson, R.N. (Ed.), Oceanography and Marine Biology: An Annual Review, vol. 48. CRC Press, Boca Raton, pp. 43–160.

Morrison, B.V., Ellison, J.C., 2017. Palaeo-environmental approaches to reconstructing sea level changes in estuaries. In: Weckström, K., Saunders, K.M., Gell, P.A., Skilbeck, C.G. (Eds.), Applications of Paleoenvironmental Techniques in Estuarine Studies. Springer, Netherlands, pp. 471–494.

Naylor, L.A., 2005. The contribution of biogeomorphology to the emerging field of geobiology. Palaeogeography, Palaeoclimatology, Palaeoecology 219, 35–51.

Nolte, S., Koppenaal, E.C., Esselink, P., Dijkema, K.S., Schuerch, M., De Groot, A.V., Bakker, J.P., Temmerman, S., 2013. Measuring sedimentation in tidal marshes: a review on methods and their applicability in biogeomorphological studies. Journal of Coastal Conservation 17, 301–325.

Ostling, J.L., Butler, D.R., Dixon, R.W., 2009. The biogeomorphology of mangroves and their role in natural hazards mitigation. Geography Compass 3, 1607–1624.

Paijmans, K., Rollet, B., 1976. The mangroves of Galley Reach, Papua New Guinea. Forest Ecology and Management 1, 119–140.

Patel, P., Agormoorthy, G., 2012. India's rare inland mangroves deserve protection. Environmental Science and Technology 46, 4261–4262.

Pérez-Ruzafa, Á.C., Marcos, C., Pérez-Ruzafa, I.M., 2012. Recent advances in coastal lagoons ecology: evolving old ideas and assumptions. Transitional Waters Bulletin 5, 50–74.

Philipps, O.P., 1903. How the mangrove tree adds new land to Florida. Journal of Geography 2, 10–21.

Pirazzoli, P.A., 1996. Sea Level Changes: The Last 20,000 Years. J Wiley & Sons, Chichester.

Plaziat, J.C., Cavagnetto, C., Koeniguer, J.C., Baltzer, F., 2001. History and biogeography of the mangrove ecosystem, based on a critical reassessment of the paleontological record. Wetlands Ecology and Management 9, 161–180.

Polidoro, B.A., Carpenter, K.E., Dahdouh-Guebas, F., Ellison, J.C., Koedam, N.E., Yong, J.W., 2014. Global patterns of mangrove extinction risk: implications for ecosystem services and biodiversity loss. In: Maslo, B., Lockwood, J.L. (Eds.), Coastal Conservation. Cambridge University Press, Cambridge, pp. 15–36.

Pollard, C.L., 1903. Plant agencies in the formation of the Florida Keys. Plant World 5, 8–10.

Punwong, P., Marchant, R., Selby, K., 2013. Holocene mangrove dynamics and environmental change in the Rufiji Delta, Tanzania. Vegetation History and Archaeobotany 22, 381–396.

Raha, A.K., Mishra, A., Bhattacharya, S., Ghatak, S., Pramanick, P., Dey, S., Sarkar, I., Jha, C., 2014. Sea level rise and submergence of Sundarban Islands: a time series study of estuarine dynamics. Journal of Ecology and Environmental Sciences 5, 114–123.

Ramcharan, E.K., 2004. Mid-to-late Holocene sea level influence on coastal wetland development in Trinidad. Quaternary International 120, 145–151.

Reinhardt, L., Jerolmack, D., Cardinale, B.J., Vanacker, V., Wright, J., 2010. Dynamic interactions of life and its landscape: feedbacks at the interface of geomorphology and ecology. Earth Surface Processes and Landforms 3, 78–101.

Reuter, M., Piller, W.E., Harzhauser, M., Berning, B., Kroh, A., 2010. Sedimentary evolution of a late Pleistocene wetland indicating extreme coastal uplift in southern Tanzania. Quaternary Research 73, 136–142.

Rowe, C., McNiven, I.J., David, B., Richards, T., Leavesley, M., 2013. Holocene pollen records from Caution Bay, southern mainland Papua New Guinea. The Holocene 23, 1130–1142.

Rowe, C., 2007. Vegetation change following mid-Holocene marine transgression of the Torres Strait shelf: a record from the island of Mua, northern Australia. The Holocene 17, 927–937.

Roy, P.S., 1984. New South Wales estuaries: their origin and evolution. In: Thom, B.G. (Ed.), Coastal Geomorphology in Australia. Academic Press, New York, pp. 69–121.

Rybczyk, R.J., Callaway, C., 2009. Surface elevation models. In: Perillo, G.M.E., Wolanski, E., Cahoon, D., Brinson, M. (Eds.), Coastal Wetlands: An Ecosystem Integrated Approach. Elsevier Science, Amsterdam, pp. 835–854.

Saenger, P., 1998. Mangrove vegetation: an evolutionary perspective. Marine and Freshwater Research 49, 277–286.

Selvam, V., 2003. Environmental classification of mangrove wetlands of India. Current Science 84, 757–765.

Shank, G.C., Lee, R., Vähätalo, A., Zepp, R.G., Bartels, E., 2010. Production of chromophoric dissolved organic matter from mangrove leaf litter and floating Sargassum colonies. Marine Chemistry 119, 172–181.

Spalding, M., Kainuma, M., Collins, L., 2010. World Atlas of Mangroves. Earthscan, London, Washington, DC.

Spenceley, A.P., 1977. The role on pneumatophores in sedimentary processes. Marine Geology 24, M31–M37.

Stoddart, D.R., Bryan, G.W., Gibbs, P.E., 1973. Inland mangroves and water chemistry, Barbuda, West Indies. Journal of Natural History 7, 33–46.

Stutz, M.L., Pilkey, O.H., 2011. Open-ocean barrier islands: global influence of climatic, oceanographic, and depositional settings. Journal of Coastal Research 27, 207–222.

Swales, A., Bentley, S.J., Lovelock, C.E., 2015. Mangrove-forest evolution in a sediment-rich estuarine system: opportunists or agents of geomorphic change? Earth Surface Processes and Landforms 40, 1672–1687.

Thom, B.G., 1967. Mangrove ecology and deltaic geomorphology, Tabasco, Mexico. Journal of Ecology 55, 301–343.

Thom, B.G., 1982. Coastal landforms and geomorphic processes. In: Snedaker, S.C., Snedaker, J.G. (Eds.), The Mangrove Ecosystem: Research Methods. UNESCO, Paris, pp. 3–15.

Thom, B.G., 1984. Mangrove ecology- A geomorphological perspective. In: Clough, B.F. (Ed.), Mangrove Ecosystems in Australia: Structure, Function and Management. Australian Institute of Marine Science, Townsville, pp. 3–18.

Thom, B.G., Wright, L.D., Coleman, J.M., 1975. Mangrove ecology and deltaic-estuarine geomorphology: Cambridge Gulf-Ord River, Western Australia. Journal of Ecology 63, 203–232.

Thomas, M.L.H., 1993. Mangrove swamps in Bermuda. Atoll Research Bulletin 386, 1–17.

Thomas, M.L.H., Logan, A., Eakins, K.E., Mathers, S.M., 1992. Biotic characteristics of the anchialine ponds of Bermuda. Bulletin of Marine Science 50, 133–157.

Thorne, K.M., Elliott-Fisk, D.L., Wylie, G.D., Perry, W.M., Takekawa, J.Y., 2014. Importance of biogeomorphic and spatial properties in assessing a tidal salt marsh vulnerability to sea-level rise. Estuaries and Coasts 37, 941–951.

Tomlinson, P.B., 1996. The Botany of Mangroves. Cambridge University Press, Cambridge.

Toscano, M.A., Macintyre, I.G., 2003. Corrected western Atlantic sea-level curve for the last 11,000 years based on calibrated 14C dates from *Acropora palmata* framework and intertidal mangrove peat. Coral Reefs 22, 257–270.

Tossou, M.G., Akoègninou, A., Ballouche, A., Sowunmi, M.A., Akpagana, K., 2008. The history of the mangrove vegetation in Bénin during the Holocene: a palynological study. Journal of African Earth Sciences 52, 167–174.

Tripathi, N., Singh, R.S., Bakhori, B., Dalal, C., Parmar, D., Mishra, B., 2013. The world's only inland mangrove in sacred grove of Kachchh, India, is at risk. Current Science 105, 1053–1055.

Van Campo, E., Bengo, D.M., 2004. Marine palynology in recent sediments off Cameroon. Marine Geology 208, 315–330.

Van Steenis, C.G.G.J., 1963. Miscellaneous notes on New Guinea plants VII. Nova Guinea (Botany) 12, 189.

Van Steenis, C.G.G.J., 1984. Three more mangrove trees growing locally in nature in freshwater. Blumea 29, 395–397.

Victor, S., Neth, L., Golbuu, Y., Wolanski, E., Richmond, R.H., 2006. Sedimentation in mangroves and coral reefs in a wet tropical island, Pohnpei, Micronesia. Estuarine, Coastal and Shelf Science 66, 409–416.

Viles, H.A. (Ed.), 1988. Biogeomorphology. Basil Blackwell, Oxford.

Viles, H.A., Naylor, L.A., Carter, N.E.A., Chaput, D., 2008. Biogeomorphological disturbance regimes: progress in linking ecological and geomorphological systems. Earth Surface Processes and Landforms 33, 1419–1435.

Villanoy, C., David, L., Cabrera, O., Atrigenio, M., Siringan, F., Aliño, P., Villaluz, M., 2012. Coral reef ecosystems protect shore from high-energy waves under climate change scenarios. Climatic Change 112, 493–505.

Vo, Q.T., Kuenzer, C., Vo, Q.M., Moder, F., Oppelt, N., 2012. Review of valuation methods for mangrove ecosystem services. Ecological Indicators 23, 431–446.

Wagner, G.M., Sallema-Mtui, R., 2016. The Rufiji estuary: climate change, anthropogenic pressures, vulnerability assessment and adaptive management strategies. In: Diop, S., Scheren, P., Machiwa, J. (Eds.), Estuaries: A Lifeline of Ecosystem Services in the Western Indian Ocean. Springer International Publishing, Switzerland, pp. 183–207.

Waming, E., 1925. Oecology of Plants. Oxford University Press, Oxford.

Watson, J.G., 1928. Mangrove forests of the Malay Peninsula. Malayan Forest Records 6, 1–275.

Webb, E.L., Friess, D.A., Krauss, K.W., Cahoon, D.R., Guntenspergen, G.R., Phelps, J., 2013. A global standard for monitoring coastal wetland vulnerability to accelerated sea-level rise. Nature Climate Change 3, 458–465.

Whelan, K.R., Smith, T.J., Cahoon, D.R., Lynch, J.C., Anderson, G.H., 2005. Groundwater control of mangrove surface elevation: shrink and swell varies with soil depth. Estuaries 28, 833–843.

Woodroffe, C.D., 1987. Pacific island mangroves: distributions and environmental settings. Pacific Science 41, 166–185.

Woodroffe, C.D., 1988. Relict mangrove stand on last Interglacial terrace, Christmas Island, Indian Ocean. Journal of Tropical Ecology 4, 1–17.

Woodroffe, C.D., 1992. Mangrove sediments and geomorphology. In: Robertson, A.I., Alongi, D.M. (Eds.), Tropical Mangrove Ecosystems. American Geophysical Union, Washington, DC, pp. 7–42.

Woodroffe, C.D., Thom, B.G., Chappell, J., 1985. Development of widespread mangrove swamps in mid- Holocene times in northern Australia. Nature 317, 711–713.

Young, B.M., Harvey, L.E., 1996. A spatial analysis of the relationship between mangrove (*Avicennia marina* var. *australasica*) physiognomy and sediment accretion in Hauraki Plains, New Zealand. Estuarine, Coastal and Shelf Science 42, 231–246.

CHAPTER

21

Mangrove Biogeochemistry at Local to Global Scales Using Ecogeomorphic Approaches

Robert R. Twilley[1], *Victor H. Rivera-Monroy*[1], *Andre S. Rovai*[1,2], *Edward Castañeda-Moya*[3], *Stephen Davis*[4]

[1]**Department of Oceanography and Coastal Sciences, College of the Coast and the Environment, Louisiana State University, Baton Rouge, LA, United States;** [2]**Programa de Pós-Graduação em Oceanografia, Universidade Federal de Santa Catarina, Florianópolis, Brazil;** [3]**Southeast Environmental Research Center, Florida International University, Miami, FL, United States;** [4]**Everglades Foundation, Palmetto Bay, FL, United States**

1. INTRODUCTION

Mangroves are forested wetlands found along 350,000 km of shoreline representing a diversity of geomorphological settings that can be subdivided into a continuum of landforms based on the relative processes of river input, tides, and waves (Boyd et al., 1992; Woodroffe, 1992). Mangroves cover from 82 to 240 × 10^3 km^2 (Table 21.1) of sheltered subtropical and tropical coastlines (Hamilton and Casey, 2016), comprising a relatively small portion of the world's total coastal ocean (shelf area = 32,242 × 10^3 km^2) and a similar small area of the global forested landscape (forested area = 40,000 × 10^3 km^2). Yet, there is some indication that these diverse geomorphological habitats, each with specific characteristics of mangrove wetlands structure and function (Thom, 1967, 1982; Twilley, 1995; Woodroffe, 2002), are considered important to the productivity of tropical estuaries and shelf environments (Bouillon and Connolly, 2009; Bouillon et al., 2008b; Lugo, 1990; Sousa and Dangremond, 2011; Twilley et al., 1996). In addition, these coastal zone regions, including continental and oceanic islands, are significant contributors to global biogeochemical processes (Bouillon et al., 2008a; Twilley et al., 1992). But there have been few approaches to account for how mangrove-dominated

Coastal Wetlands
https://doi.org/10.1016/B978-0-444-63893-9.00021-6

TABLE 21.1 Global Mangrove Area Estimates in km^2 by Year and Author. The Mangrove Area Estimates Within Each Decade are Highly Variable

ID No.	Source	Reference Year	No. of Countries	Mangrove Area (km^2)
1	FAO (2007, p. 9)	1980	Global	187,940
2	Lanly (1982, p. 43)	1980	76	154,620
3	Saenger et al. (1983, pp. 11–12)	1983	66	162,210
4	FAO (2004, Table 2.3)	1980–85	56	165,300
	1980s mean (sources 1–4)			167,518
5	FAO (2007, p. 9)	1990	Global	169,250
6	Groombridge (1992, pp. 325–6)	1992	87	198,478
7	ITTO & ISME (1993, p. 6)	1993	Global	141,973
8	Fisher and Spalding (1993, p. 11)	1993	91	198,817
9	Spalding et al. (1997, p. 23)	1997	112	181,077
	1990s mean (sources 5–9)			177,919
10	Spalding (2010, p. 6)	2000–01	123	152,361
11	FAO (2007, p. 9)	2000	Global	157,400
12	Aizpuru et al. (2000; secondary source)	2000	112	170,756
13	Giri et al. (2011, p. 156) (2010, p. 1, 3)	2000	Global	137,760
14	FAO (2007, p. 9)	2005	Global	152,310
	2000s mean (sources 10–14)			154,085
15	Hamilton and Casey 2016	2012	Global	81,849

Table is from Hamilton and Casey (2016), with their estimates of mangrove area included.

estuaries and coastal ecosystems contribute significantly to secondary productivity and biogeochemical processes in diverse coastal settings from river deltas, muddy coasts, estuaries (from rias to lagoons), and carbonate settings of continents and oceanic islands. Ecological processes may vary across these settings resulting in different perspectives as to the significance of mangrove contributions, particularly considering wet tropics where catchments deliver significant materials in runoff to the coastal ocean that includes exchange with mangrove wetlands. This correlation between coastal landform processes and ecological function has been documented relative to net primary productivity (NPP), net ecosystem production (NEP, e.g., wood and soil organic carbon [SOC] storage), and detritus exchange across a variety of coastal settings. This review will utilize several conceptual models and published data to test how the diverse geomorphological settings may explain patterns of nutrient biogeochemistry among different ecological types of mangrove wetlands (Twilley, 1997).

Diverse coastal settings in this review are defined by the source of sediment that represents a combination of geophysical processes and local geology, which may also influence mangrove ecology (Woodroffe, 2002). Clastic sediments that are delivered from upland catchments and deposited along the coastal zone form subaerial intertidal platforms that mangrove wetlands colonize forming unique forested habitats. Such terrigenous sediment systems can be classified into two broad subgroups based on the relative effects of river, tides, and waves controlling sediment transport. Muddy coasts represent those continental margins influenced by substantial river inputs characteristic of prograding deltaic coasts (see Chapter 8 in Woodroffe, 2002, Fig. 21.1). Such settings are formed by sediment derived from river basins (e.g., Ganges-Brahmaputra Delta) or from massive fluid muds transported by nearshore currents and accumulating downshore from deltaic environmental settings that create muddy shorelines (Woodroffe, 2002; Marchand et al., 2003). This group of river-dominated deltaic coasts has landforms shaped by a combination of tidal and wave processes forming a continuum of geomorphological settings (Boyd et al., 1992). The second group of clastic systems is represented by estuarine coasts that have a combination of freshwater and marine sources of sediment creating diverse landforms from rias (e.g., Guayas River estuary, Ecuador) to tide and wave-dominated lagoons (e.g., Terminos Lagoon, Mexico) (Fig. 21.1). These coastal landforms do not have the extensive muddy shorelines of deltaic systems, and mangroves colonize platforms formed by intertidal sediment transport. As a result, mangroves in such settings can comprise a significant component of the total estuarine area and can occur along both continental margins and oceanic islands. For example, high elevation islands in the Pacific Ocean in wet precipitation zones (>3000 mm per year) colonize an intertidal platform dominated by sediments eroded from upland watersheds (e.g., Kosrae, Micronesia). In contrast to these coastal landforms dominated

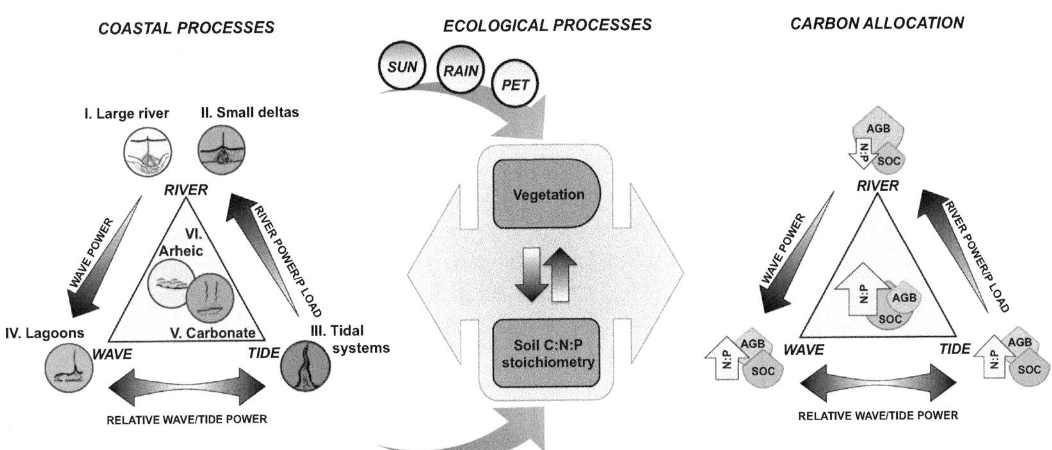

FIGURE 21.1 Coastal ecogeomorphology conceptual framework showing how bidirectional fluxes between abiotic and biotic components control nutrient stoichiometry and carbon (C) storage in mangroves. Distinct coastal environmental settings (e.g., deltas, estuaries, lagoons, carbonate) are formed by the relative contribution of geophysical variables (e.g., river discharge, tidal amplitude, wave energy). Along with regional climatic drivers, these geophysical forcings constrain C partitioning among ecosystem compartments (soil, above- and belowground biomass). *AGB*, aboveground biomass; *C:N:P*, carbon-to-nitrogen-to-phosphorus ratio; *PET*, potential evapotranspiration; *SOC*, soil organic carbon. Diagrams of the coastal environmental settings adapted from Boyd et al. (1992) and Dürr et al. (2011). *From Twilley et al. 2018.*

by clastic sediments, a third group of coastal depositional environments results from in situ processes such as biogenic carbonate formation along reef coasts (Gattuso et al., 1998). Mangroves in these continental or island landforms are dominated by peaty or calcareous soils that are a combination of carbonate sediment and mangrove organic matter, both of which originate largely through carbon fixation (e.g., production of mangrove roots and litter, as well as formation of reef structures and invertebrate shells).

A challenge has been to account for the ecological variation of mangrove function to food webs and biogeochemical processes across these diverse geomorphological settings coupled to processes that form either clastic or carbonate intertidal platforms. More recently, a spatially explicit global typology for nearshore coastal landforms based on numerical analyses of river basin runoff, tidal range, and coastal lithology has provided a robust spatial representation of dominant coastal environmental settings worldwide (Dürr et al., 2011). Similar to the early coastal environmental settings classification (Thom, 1982; Woodroffe, 2002), this global typology of coastlines recognizes muddy coasts (including large rivers and small deltas), estuarine coasts (estuaries, rias, bedrock, lagoons), and carbonate coasts, in addition to an arid, climate-dependent arheic coastline type. We will use these distinct types of coastal environmental settings to link geomorphological settings and geophysical processes (Fig. 21.1) to patterns of nutrient biogeochemistry in mangrove ecosystems. Moreover, we will illustrate a modeling opportunity linking the global distribution and diversity of coastal environmental settings to estimate large-scale biogeochemical analyses at the coastal zone. This is particularly important given the discussion of how the biogeochemistry of mangroves and other coastal habitats may respond to a changing climate (Day et al., 2008).

2. ECOGEOMORPHOLOGY OF MANGROVES

The ecogeomorphology model of mangroves is a hierarchical classification scheme of geomorphological settings and corresponding ecological types that are based on hypotheses concerning mangrove community structure and ecosystem processes (Fig. 21.2). This conceptual model of mangroves is dependent on gradients of geophysical and climate processes among coastal environmental settings (Thom, 1984; Woodroffe, 1992, 2002; Twilley, 1995; Twilley et al., 1996; see Fig. 21.1). At the local level, this conceptual model describes how multigradient patterns of resources, regulators, and hydroperiod control mangrove community structure and ecosystem function (Fig. 21.3; Twilley and Rivera-Monroy, 2005). As suggested in Fig. 21.2, these three gradients can describe patterns in mangrove forest structure and productivity. When considered along hydroperiod gradients, this can also determine both the levels of some regulators and how these regulators affect nutrient uptake. In some respects, hydroperiod may be considered a regulator, but here we distinguish its effect from other regulators given its overriding influence on mangrove soil physicochemical conditions and exchanges of materials.

Global climate constrains the latitudinal expanse of mangroves, and biogeographic provinces define which mangrove species will colonize intertidal platforms. These coastal geomorphic settings can be found in a variety of humidity zones (Blasco, 1984) that depend on regional climate and oceanographic processes, along with river input from catchments

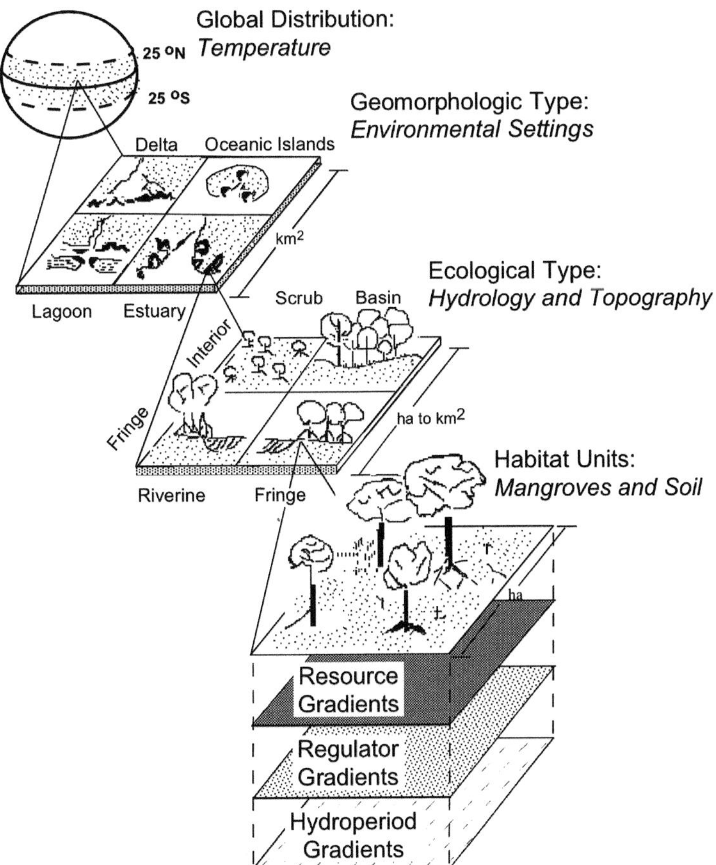

FIGURE 21.2 Hierarchical classification system to describe patterns of mangrove structure and function based on global, geomorphological (regional), and ecological (local) factors that control the concentration of nutrients resources and regulators in soil along gradients from fringe to more interior locations from shore. *Modified from Twilley et al. (1998) and Twilley and Rivera-Monroy (2005).*

that are far inland of coastal climates. Oceanic processes (tides and waves) and river input will also determine the source of sediments as either clastic or biogenic as discussed above (Woodroffe, 2002). However, these global and geomorphologic types are not easily delineated but instead reflect a continuum in the relative effects of climate, river, waves, and tides on coastal processes that determine mangrove distribution. Here, we use muddy, estuarine, and carbonate as different coastal environmental settings of mangroves that affect nutrient biogeochemistry. Given the dynamic nature of these landforms, it is important not to consider the association between landform and ecological function as a static model, particularly given climate change projections for the coastal zone over the next century.

These landform conditions constrain more granular ecological processes that explain the variety of habitats within a geomorphological setting. In each of the geomorphological settings we consider (muddy, estuarine, and carbonate coasts), local variations in topography and

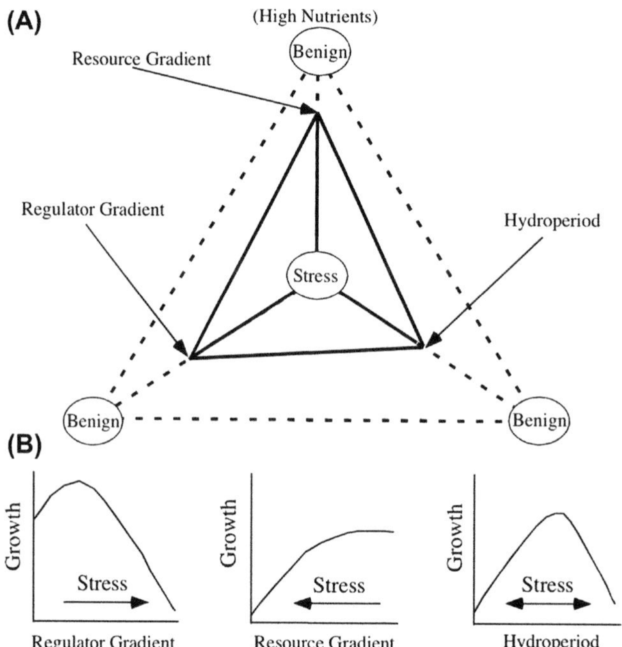

FIGURE 21.3 Factorial interactions controlling the productivity of coastal wetlands including regulator, resource, and hydroperiod gradients. (A) The production envelope (solid line) associated with levels of each factor interaction to demonstrate responding levels of net primary productivity. (B) The definition of stress associated with how gradients in each factor control growth of wetland vegetation. *From Twilley and Rivera-Monroy (2005).*

hydrology result in the development of distinct ecological types of mangroves such as riverine, fringe, and interior forests (Figs. 21.1 and 21.2; Lugo and Snedaker, 1974; Lugo, 1990; Woodroffe, 1992, 2002). Several conceptual models have described inundation classes (i.e., the number of tidal inundations per year) as a way to explain variation in community composition of mangrove forests (Watson, 1928; Chapman, 1976; see review by Friess, 2016). These inundation classes may also explain distinct patterns in nutrient dynamics of mangrove soils from fringe to interior locations that are particularly sensitive to local climate conditions affecting water budget and soil salinity (Twilley and Chen, 1998). Below, we describe how local patterns of hydroperiod (frequency and duration of inundation) serve to regulate patterns of mangrove nutrient biogeochemistry.

Ecological processes are constrained by a variety of stress conditions in mangrove soils (Fig. 21.3) that exist along gradients of resources, regulators, and hydroperiod (Huston, 1994; Twilley, 1997; Twilley and Rivera-Monroy, 2005). Resource gradients such as nutrients, light, and space are variables that are required for growth and are also consumed during mangrove productivity, compared with regulator gradients such as salinity, sulfide, pH, and metals that also control growth but are not consumed (Tilman, 1985). The third gradient, hydroperiod, is the frequency, duration, and depth of inundation that governs soil conditions, such as redox and chemical transformations, which can control wetland productivity (Gosselink and Turner, 1978). At low levels of stress along these environmental gradients (e.g., low salinity, high nutrients, and intermediate flooding in reference to Fig. 21.3), mangrove wetlands will reach their maximum levels of biomass and net ecosystem productivity. Coastal settings that result in increased stress conditions for one or more of these gradients should exhibit mangroves with lower total NPP. The combination of these three gradients have been proposed as a constraint envelope for defining the structure and productivity of mangrove wetlands based on the cumulative effects of stress conditions (Fig. 21.3; Twilley and Rivera-Monroy, 2005). And these stress gradients are associated with various combinations of geophysical processes and geomorphological settings. Using this model, we will describe critical feedback effects of resources, regulators, and hydroperiod that describe how biogeochemical processes may control community structure and ecosystem function of mangroves.

Experimental studies of nutrient limitation on mangrove NPP in carbonate settings have provided strong evidence of linkages between coastal environmental settings and ecological processes. Field experiments have shown that mangrove forests growing under nutrient-limited conditions respond quickly to fertilization (Feller, 1995). More specifically, studies in carbonate-dominated environments in the Caribbean show that phosphorus (P) enrichment significantly increases productivity of interior forests along a microtidal gradient, whereas trees were generally nitrogen (N) limited in the fringe zone (Feller et al., 1999, 2003a). Furthermore, P addition in scrub mangroves growing in marl sediments east of the Florida Coastal Everglades (FCE) sites also stimulated mangrove productivity (Koch, 1997). Phosphorus-rich sediment input from the Gulf of Mexico during hurricanes has been demonstrated as controlling the enhanced mangrove growth near the mouth of Shark River estuary, which is located in the carbonate setting of the FCE. The combined effect of P-rich sediment, P-rich Gulf of Mexico water (relative to Everglades freshwater), and relatively strong semidiurnal tides support optimum mangrove growth in this region (Castañeda-Moya et al., 2010; Chen and Twilley, 1999a).

Evidence of nutrient limitation in mangroves can also be evaluated by summarizing existing information on soil nutrient density among mangrove forests (about 125 sites summarized in Figs. 21.4 and 21.5). Over one-third ($n = 57$) of the total number of sites have soil total phosphorus (TP) density <0.20 mg cm^{-3}. More than half of these sites with low soil TP density represent mangroves in either carbonate or lagoon settings with limited clastic sediment input. In fact, 90% of these mangrove sites have N:P ratios >15, indicating that P is potentially limiting in these types of coastal environmental settings (Fig. 21.4). However, both of these surveys demonstrate that N is potentially limiting in many locations where sediment deposition in estuarine and muddy coasts provides soil TP density sufficient to meet growth requirements. The evidence that soil N:P ratio may be indicative of mangrove NPP is a significant relation between soil ratios and basal area of 60 forests in this survey (Fig. 21.5B). However, as N:P ratios increase, there is an increase in fine root productivity (Fig. 21.5B, see discussion below). A research study designed to compare soil characteristics across coastal environmental settings experimentally defined higher N:P in sediment-poor coasts compared with lower N:P in muddy coasts in the neotropics (Fig. 21.6A). This trend in the relative content of N and P among mangrove sites in muddy, estuarine, and carbonate settings is probably one of the strongest linkages between geomorphological settings and ecology of mangrove wetlands (Twilley et al., 2018).

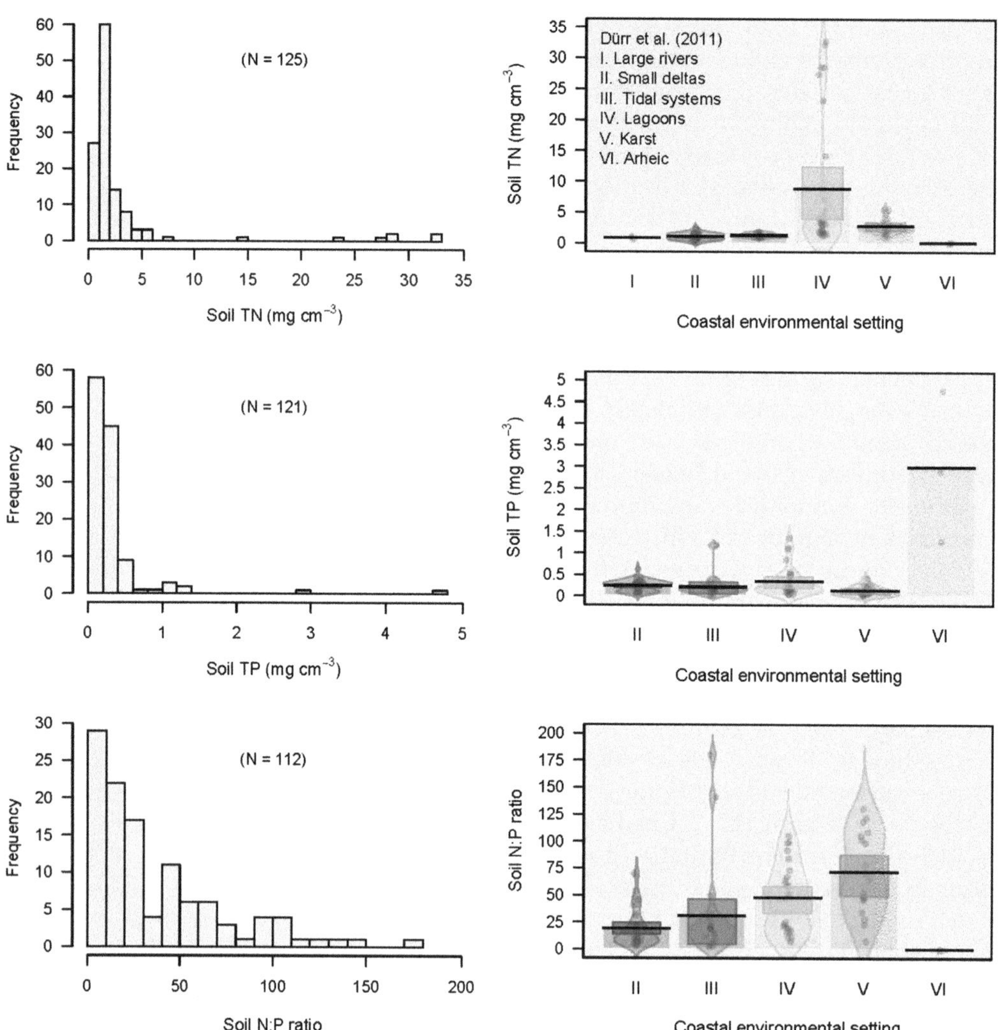

FIGURE 21.4 Left panels: Nutrient density (upper panel is total nitrogen and middle panel is total phosphorus) from about 125 mangrove sites around the world (based on mg per volume of sediment). Lower panel is of soil N:P ratios (atomic) for each respective site. Right panels: The nutrient density (TN and TP) and N:P ratios for soils of mangrove sites classified into coastal environmental settings defined by Dürr et al. (2011). The centered bars/lines and the bands display the mean and the 95% confidence intervals, respectively, while the beans and points represent the density and spread of the raw data.

Nutrient density across geomorphological settings that control resource limitation of mangroves may be significant in soil formation because plants respond differently to short- and long-term changes in soil nutrient resources according to species-specific evolutionary life history traits (Chapin et al., 1987). This is particularly significant in how mangroves allocate biomass to roots, as this contributes to soil formation and elevation gain in mangroves wetlands. Mangrove species respond to low nutrient availability with

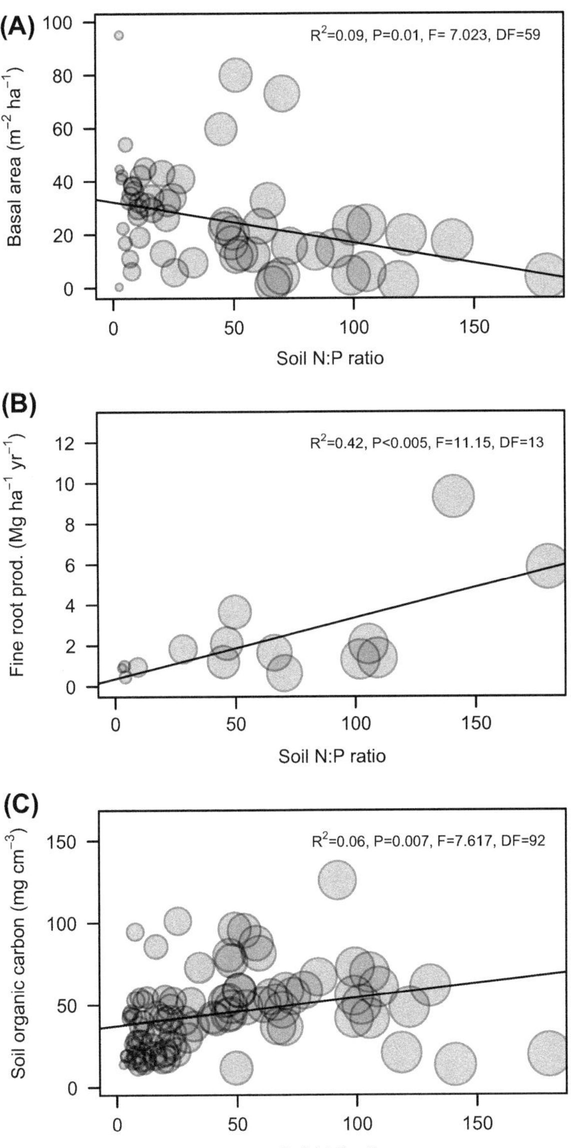

FIGURE 21.5 Relationships between structural and functional attributes: (A) basal area, (B) fine root productivity, and (C) soil organic carbon density, and soil N:P atomic ratios from mangroves around the world. The size of the dots is proportional to the N:P ratios observed for each site.

morphological and physiological plasticity (Feller et al., 2003a,b; Lovelock et al., 2004, 2006), which demonstrates the diversity of carbon allocation strategies across coastal environmental settings. Mangroves across the nutrient and hydroperiod gradients of oceanic islands of Belize and in the FCE have a significant negative relationship between fine root biomass

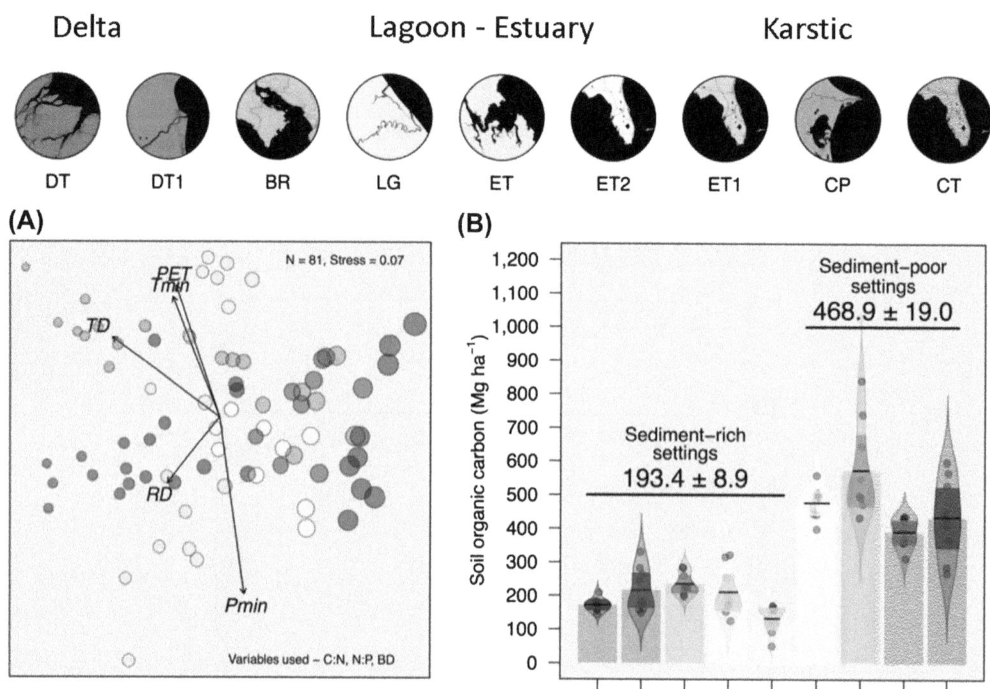

FIGURE 21.6 (B) Nonmetric multidimensional scaling analysis (nMDS) depicting the grouping of nine mangrove sites into distinct coastal environmental settings (CES) in the neotropics and the environmental correlates (vectors) that explained ($\alpha < 0.05$) the gradient formed in the ordination. The size of the dots is proportional to the N:P ratios observed for each site, spanning from sites that receive river input and have low N:P ratios (tide- and wave-dominated deltaic settings; to the left of the panel) to P-limited sites with negligible or no river input and high N:P ratios (carbonate and composite CES; to the right). The orientation and length of vectors represent the direction and the strength (e.g., correlation between ordination and environmental variable) of the gradient, respectively. Tmin (°C)—minimum temperature of the coldest month, Pmin (mm per year)—precipitation of the driest month, PET (mm per year)—average annual potential evapotranspiration, TD (m)—tidal range, and RD (m^3 per s)—river discharge. (A) Values of mangrove soil organic carbon density from sediment-rich to sediment-poor CES. The centered bars/lines and the bands display the mean and the 95% confidence intervals, respectively, while the beans and points represent the density and spread of the raw data.

and soil TP density, doubling the amount of fine root biomass in sites with N:P ratios >50 (Castañeda-Moya et al., 2011; Feller et al., 2003a, 2003b; Lovelock et al., 2004). Similar responses of root biomass along a soil fertility gradient have been found in Micronesia mangrove forests (Cormier et al., 2015) and with a long-term fertilization study in an *Avicennia marina* scrub mangrove in South Africa (Naidoo, 2009). Mangroves are highly adapted plants that can allocate a large proportion (up to 40%–60%) of their total biomass to belowground structures in response to nutrient limitation (Khan and Suwa, 2007; Komiyama et al., 2000), which is higher than the range of 10%–30% of fine roots that account for total forest tree biomass (Santantonio et al., 1977). Recent mass balance studies of belowground carbon allocation in mangrove forests around tropical and subtropical latitudes suggest that scrub mangroves allocate relatively more carbon belowground than do taller mangrove forests in response to low nutrient availability and anaerobic conditions (Castañeda-Moya et al., 2011; Lovelock, 2008). These patterns in biomass allocation associated with P limitation can explain the correlation of higher soil organic matter with higher N:P ratios (Fig. 21.5c) and increased fine root productivity (Fig. 21.5b) across 125 mangrove sites. Sediment-poor coastal settings such as karstic sediments have on the average 469 Mg ha^{-1} of SOC compared with 193 Mg ha^{-1} for sediment-rich sites such as deltas and estuaries (Fig. 21.6).

There are few studies that have addressed the influence of hydroperiod (duration, frequency, and depth of flooding) on mangrove root dynamics (Cardona-Olarte et al., 2006; Krauss et al., 2006). Greenhouse studies have documented significant shifts in biomass allocation between roots and shoots in neotropical mangrove seedlings under different hydroperiod conditions (Krauss et al., 2006). Patterns in root biomass allocation can be associated with the competitive ability of mangrove species and their tolerance limit to flooding (Cardona-Olarte et al., 2006; Krauss et al., 2006). A field study of mangrove forests in Belize showed that root production responds to different hydroperiod conditions from tall fringe to scrub interior mangroves and parallels the aboveground productivity gradient (McKee et al., 2007). Mangroves in nutrient-poor sites with long flooding duration and restricted tidal influence produce roots with greater longevity and low morphological plasticity as a mechanism to enhance nutrient conservation. In contrast, sites with higher nutrient stocks and tidal hydroperiods with more frequent and shorter durations of inundation produce short-lived roots with high nutrient uptake, rapid growth rates, and higher turnover (Castañeda-Moya et al., 2011; Cormier et al., 2015; Poret et al., 2007). These trade-offs are indicative of the strong link between belowground processes and the phenotypic plasticity of mangrove roots in response to the interactions among nutrient resource and hydroperiod gradients. These belowground allocation patterns are ecologically significant given that fine root production and organic matter accumulation are the primary processes controlling soil formation in mangrove forests (e.g., Belize, McKee et al., 2007; model simulations in Chen and Twilley, 1999b).

3. GEOCHEMICAL MODEL

The mechanisms that establish resource and regulator gradients, and how hydroperiod influences both of these gradients, are described by a geochemical model of mangrove soils developed by Clark et al. (1998; Fig. 21.7). Many of these gradients are associated with multiple transitions of oxidation and reduction zones that can occur with sediment depth and also with distance inland from shore (interior of either tidal or riverine fringe zones) that will change seasonally depending on presence and absence of dissolved oxygen. There are four redox zones that can be identified with depth in mangrove soils (Fig. 21.7; Clark et al., 1998). The upper oxidation zone is established when the supply of oxygen is sufficient to balance aerobic respiration, oxygen supply being largely a function of physical exchange with surface mangrove sediments. Below the upper oxidation zone is the upper reduction zone, where sulfate-reducing bacteria dominate respiration resulting in redox values well below −100 mV. Biological exchange of oxygen into deeper sediments, usually below the upper reduction zone, results in a lower oxidation zone, as a function of biological processes that reintroduce oxygen to otherwise anaerobic sediments. Below that is the lower

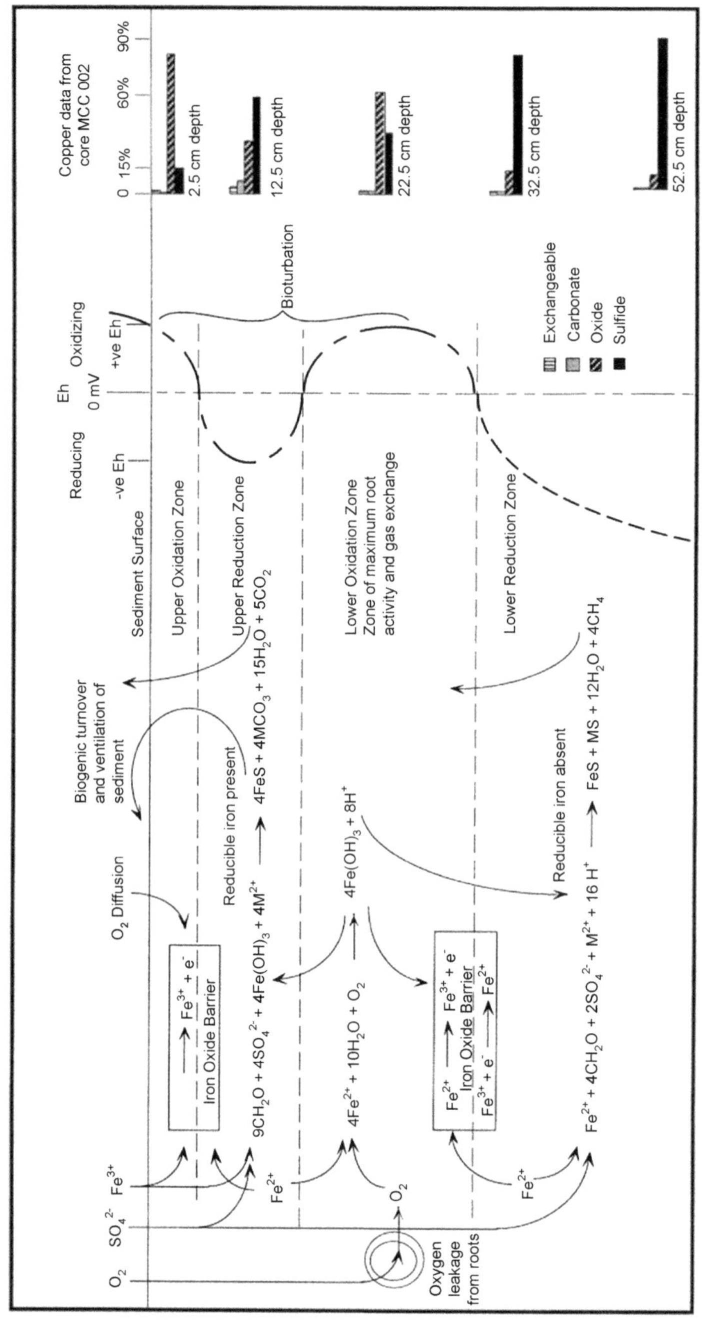

FIGURE 21.7 The geochemical model of mangrove biogeochemistry (Clark et al., 1998).

reduction zone, where anaerobic conditions prevail in the absence of oxygen supply (Fig. 21.7). How these zones vary in response to physical and biological processes across muddy, estuarine, and carbonate coastal settings is one mechanism that links coastal environmental settings to patterns in biogeochemistry.

Most mangrove soils lack oxygen and have redox potentials <200 mV (values above this potential usually indicate oxygen as terminal point of electron flow during respiration). Several mechanisms can enhance oxygen supply by increasing diffusion into otherwise oxygen-deficient soils (Clark et al., 1998), including (1) sediment texture, higher porosity, and permeability increasing diffusive flux (Hutchings and Saenger, 1987); (2) the position of the water table to soil surface elevation, which may vary seasonally with hydroperiod (balance of tides and rain events; Twilley and Chen, 1998); and (3) the influence of biota (directly and indirectly) on water and oxygen exchange with sediments (Kristensen, 2008; Thongtham and Kristensen, 2005). This supply of dissolved oxygen is balanced by soil oxygen consumption that is controlled by quality of decomposing plant tissue, soil organic matter, temperature, and nutrient content (Chambers et al., 2014, 2016; Kristensen et al., 2008a,b; Lovelock, 2008; Troxler et al., 2015).

When dissolved oxygen is absent in mangrove sediments, decomposition is controlled by a series of alternate electron acceptors used by facultative and obligate anaerobes during respiration. Each alternate electron acceptor is associated with decreasing free energy yield per mol of organic matter (CH_2O) respired. Collectively, these respiration pathways—in the absence of oxygen—represent the upper reduction zone that is just below the oxidation zone (Fig. 21.7). The collection of electrons during the process of respiration converts the oxidized form of reactants to reduced forms as follows: manganese (Mn^{3+}) and iron (Fe^{3+}) are reduced to Mn^{2+} and Fe^{2+}, NO_3^- is reduced to N_2 or N_2O (see discussion of denitrification below), SO_4^{2-} is reduced to sulfide (S^{2-}), and CO_2 is reduced to methane (CH_4). As SO_4^{2-} is normally the most abundant electron acceptor in most anoxic mangrove soils (because of presence of salinity), the "upper reduction zones" are characterized by the presence of decomposing organic matter and substantial populations of sulfate-reducing bacteria that produce sulfides (Fig. 21.7). The presence of metals in these sulfide-rich reduction zones produces a variety of reduced metal sulfides, dominated by pyrite (FeS_2) and mackinawite (FeS) (Berner, 1984; Jørgensen, 1982; Pons et al., 1982).

Exposure of anaerobic soils to oxygen may cause sulfides in the sediment to oxidize (Bloomfleld and Coulter, 1974), following a couple of possible reactions producing by-products that are chemically and ecologically significant (Singer and Stumm, 1970):

$$FeS_2 + 7/2O_2 + H_2O \rightarrow Fe^{2+} + 2SO_4^{2-} + 2H^+ \quad (21.1)$$

$$FeS_2 + 14Fe^{3+} + 8H_2O \rightarrow 15Fe^{2+} + 2SO_4^{2-} + 16H^+ \quad (21.2)$$

Large amounts of sulfuric acid are produced by these reactions, thereby lowering pore water pH and potentially resulting in mobilization of metals in mangrove sediment (Evangelou, 1995; Rimstidt and Vaughan, 2003). Nonferrous metal sulfides can also decompose by a similar sequence of reactions. Both the pyrite oxidation by either oxygen or oxidized Fe and subsequent jarosite formation yield large quantities of H^+ that can reduce pH of mangrove soils. If this acidity is not buffered in some way within the soil system, then pH values <4 can occur, which are typical of acid sulfate soils (Nath et al., 2013). At these low pH values, sulfur chemistry in soils can be considered a regulator of mangrove productivity and lethal to the biota of mangroves soils (Charoeuchamratcheep et al., 1987; Hart, 1963; Macdonald et al., 2007; Sammut et al., 1996; Webb et al., 1995).

These processes associated with decomposition and oxygen supply explain how the geochemical model explains variation in nutrient biogeochemistry among carbonate, estuarine, and muddy coasts. Rates of sulfate reduction in anaerobic environments and different forms of TP in mangrove soils represent regulators and resources, respectively, to mangrove productivity. Sulfate reduction forms hydrogen sulfide, which will react to form pyrite in the presence of iron, elemental sulfur, or iron monosulfides (Holmer et al., 1994, 2001). About 50% of total organic matter respiration on mangrove soils is aerobic, with the other 50% anaerobic respiration dominated by sulfate reduction. In iron-rich muddy coasts, iron respiration can reach rates of 35 μmol cm^{-3} with about 80% of carbon oxidation attributable to iron reduction (Bouillon et al., 2008a). In these types of environments, CH_4 flux and CO_2 respiration is a very small component of total respiration at about 0–5 mmol m^{-2} per day. In mature estuarine forests low in iron with highly reduced muds, about 80% of total organic carbon (TOC) oxidation is attributable to sulfate reduction. In similar coasts but with sandy sediments and more tidally inundated zones, sediments may be more oxidized resulting in less sulfate and iron reduction (Alongi and McKinnon, 2005).

The relative dominance of the oxidized and reduced conditions in the upper mangrove sediment zone is controlled by physical flushing and dissolved oxygen dynamics over the tidal cycle, compared with direct soil-atmosphere exchanges at low tide. Thus, the expansion and contraction of the upper oxidized zone is defined by the relative supply of dissolved oxygen as a function of hydroperiod, with the expansion of the reduced zone occurring when oxygen supply to sediments is limited. The depth and duration of the upper oxidation zone result from exposing pore space of mangrove soils to dissolved oxygen in water or exposing soils directly to the atmosphere. With semidiurnal tides (i.e., frequent tides of short duration), dissolved oxygen concentrations are more replenished, sufficient to support sediment oxygen demand. The upper oxidation zone (2–5 cm) in fringe mangroves is more consistently oxidized because of more frequent inundation with oxygenated waters that maintain elevated redox potentials. Near the mouth of muddy coasts such as French Guiana and the Fly River, mangrove soils exhibit high redox potentials to soil depths of 50 cm and generally lack vertical profiles in dissolved and particulate nutrients, bacterial numbers, and dissolved metal concentrations (Fabre et al., 1999; Marchand et al., 2003, 2006).

The geochemical model of redox zones also explains the distribution of resources and regulators in mangrove soils, which is controlled by the hydroperiod of a local site. Increased hydroperiod of mangrove soils will allow reduction zones to migrate toward the soil surface, with accumulation of anoxic respiration by-products including HS^- and H^+ ions that approach stress levels of regulator gradients. The effects of these regulators on mangrove NPP may be direct because HS^- and H^+ ions can be toxic to plants; it can also determine plant productivity by reducing nutrient uptake rates (Koch and Mendelssohn, 1989; Koch et al., 1990). This affects mangrove zonation patterns, as observed in the relative distribution of *Avicennia* and *Rhizophora* in the neotropics (Castañeda-Moya et al., 2006; McKee, 1993; Sherman et al., 1998) and species in Old World tropics (Matthijs et al., 1999). There are feedback mechanisms that can limit the negative effects of these reduction zones through rhizosphere oxidation. Thus, mangroves can develop in intertidal zones with long durations of flooding if lower oxidation zones can occur where upper reduction zones would otherwise predominate.

These patterns in biochemical pathways that contribute to formation of redox zones can also strongly influence phosphate availability along inundation gradients. Under oxidized conditions and in the presence of iron, phosphate adsorbs onto iron oxyhydroxides forming occluded forms of P reducing the concentration of soluble reactive phosphorus (SRP) in mangrove pore waters (Holmer et al., 1994; Sherman et al., 1998). Under anoxic conditions and in the presence of sulfide and ferrous iron, phosphate is released into the pore water solution increasing SRP (Krom and Berner, 1980). This process may vary from fringe to interior forests, but when oxygen is absent and both sulfate and iron are present, the pyrite-forming potential is high, increasing P release and solubility potential (Sherman et al., 1998; Gleason et al., 2003). However, even in the absence of oxygen, the supply of either sulfate or iron can influence the relative P solubility.

When pyrite is oxidized during the transition from reduced to oxidized zones in mangrove soils, sulfate and acidity are by-products as is the formation of iron oxyhydroxides that reduce the solubility and limit migration of P in the soil column (such as across the sediment–water interface). Pyrite formation is typically Fe-limited in marine carbonate environments (Berner, 1984; Chambers et al., 2001), but Fe is plentiful in coastal systems with clastic sediment inputs (Duarte et al., 1995). Analysis of sediment properties along four gradients of siltation in Southeast Asia demonstrated that silty mangrove habitats have higher contents of organic matter and nutrients (particularly N, P, and Fe) compared with more carbonate systems such as coral reefs and seagrass beds with higher calcium and lower organic matter (Kamp-Nielsen et al., 2002). Under these Fe-limited conditions, hydrogen sulfide concentrations accumulate (Sherman et al., 1998). When Fe is in low supply, P-sorption onto calcium carbonate minerals maintains low P concentrations in solution (Chambers et al., 2001; Morse et al., 1985; Rosenfeld, 1979). Under these circumstances primary producers are generally P-limited, but they are also susceptible to high regulator concentrations of hydrogen sulfide.

In anaerobic respiration, Fe undergoes two important reactions that affect regulator and resource gradients in mangrove soils: (1) Fe reacts with sulfide to reduce hydrogen sulfide toxicity and P is released into solution of mangrove pore waters; and (2) in absence of Fe, sulfate reduction forms hydrogen sulfide and P becomes bound to calcium carbonate minerals that remain insoluble under these conditions but are more vulnerable to low pH. Thus, the presence of Fe is a key nutrient that will determine the level of regulators that control mangrove productivity (Alongi, 2010; Attri et al., 2011). This interaction between resources (Fe and P) and regulator (HS^-) is a function of hydroperiod from fringe to interior topographic gradients because it controls redox zones in the geochemical model. And these complex mechanisms could explain the elevated biomass and net productivity of mangroves along muddy coasts with sufficient Fe and reduced regulator concentrations of salinity and sulfide in contrast to carbonate coastal settings.

Below the upper reduction zone is a lower oxidation zone that can develop because of the release of oxygen from mangrove roots (Armstrong and Armstrong, 1990; Armstrong, 1978; Armstrong and Wright, 1975; McKee et al., 1988) or animal burrows (Kristensen, 2008).

The resulting oxidation is accompanied by a reduction of pore water pH as iron sulfides are converted to oxyhydroxides and sulfuric acid (Eqs. 21.2 and 21.3). Radial release of oxygen from mangrove roots affects a microzone surrounding the root that may only be a few millimeters in depth (i.e., the rhizosphere); and sediments at a greater distance from the roots can remain reducing and sulfidic. The duration and depth of this lower oxidation zone can depend on the type of mangroves (radial roots vs. prop roots) and the density of mangrove roots in response to soil conditions (McKee et al., 1988; Clark et al., 1998; Gleason et al., 2003; Fig. 21.7). Oxygen that is transported with water that exchanges in animal burrows can modify respiratory processes from sulfate reduction to more aerobic processes in otherwise anaerobic conditions (Araújo et al., 2012; Ferreira et al., 2007; Kristensen et al., 2011; Nielsen et al., 2003; Ólafsson et al., 2002; Penha-Lopes et al., 2010). In mangrove forests, mud lobsters *Thalassina anomala* commonly provide obvious examples of sediment turnover (Bennett, 1968; Macnae, 1966). The resulting ventilation of anoxic sediments by oxygenated waters can increase the Cd, Cu, and Ni concentrations in oxidized surface sediments and enhance the fluxes of metals and other dissolved materials, for example, CO_2, HS^-, and CH_4, between the sediment and water (Aller, 1978; Morris et al., 1982; Emerson et al., 1984). The movement of water through burrows and mangrove root and pneumatophore casts also provides a means whereby sulfate-reducing bacteria in the sediment are supplied with sulfate (Bloch and Krouse, 1992) and the by-products of bacterial metabolism (e.g., HS^-) can be flushed from the sediment (Aller, 1980; Emerson et al., 1984; Kristensen, 2008; Kristensen and Alongi, 2006; Kristensen et al., 2011).

4. NET ECOSYSTEM NUTRIENT EXCHANGE

This section will focus on the net exchange of nutrients across the ecosystem boundaries of mangroves, which defines their function as a nutrient source, sink, or transformer depending on the net flux with atmosphere and coastal waters (Ewel et al., 1998; Jennerjahn, 2012). There is a particular interest in understanding the net exchange of nutrients across atmosphere and tidal boundaries with mangrove wetlands as a means of mitigating enrichment of greenhouse gases in the atmosphere and nutrients in coastal waters (Duarte et al., 2013; Heckbert et al., 2012; Palmer et al., 2014). If the net exchange of a nutrient across the atmosphere and tidal boundaries of a mangrove wetland is positive, the result is nutrient accumulation and mangroves are considered a nutrient sink. This net balance of nutrients, including carbon, nitrogen, and phosphorus, is considered an important ecosystem service that can mitigate environmental problems and make restoration more attractive (Barbier et al., 2011; Jerath et al., 2016; Murray et al., 2011). The challenge of defining the regional and global significance of such ecosystem services, and thus the benefits of investments in large-scale designs in ecosystem restoration, starts with the basic estimates of net nutrient sinks at local studies that can then be extrapolated to larger-scale estimates (Fig. 21.8). We introduce the net ecosystem nutrient exchange (NENE) concept to help define the function of mangroves as a nutrient sink, the net balance of fluxes across atmosphere and tidal boundaries. Given that exchanges across these boundaries may vary within the ecogeomorphology framework, we use this concept to test whether NENE will vary with coastal environmental settings. We review if there is evidence of how mangroves in different coastal settings can mitigate carbon and nutrient (N and P) enrichment in the atmosphere and/or coastal waters.

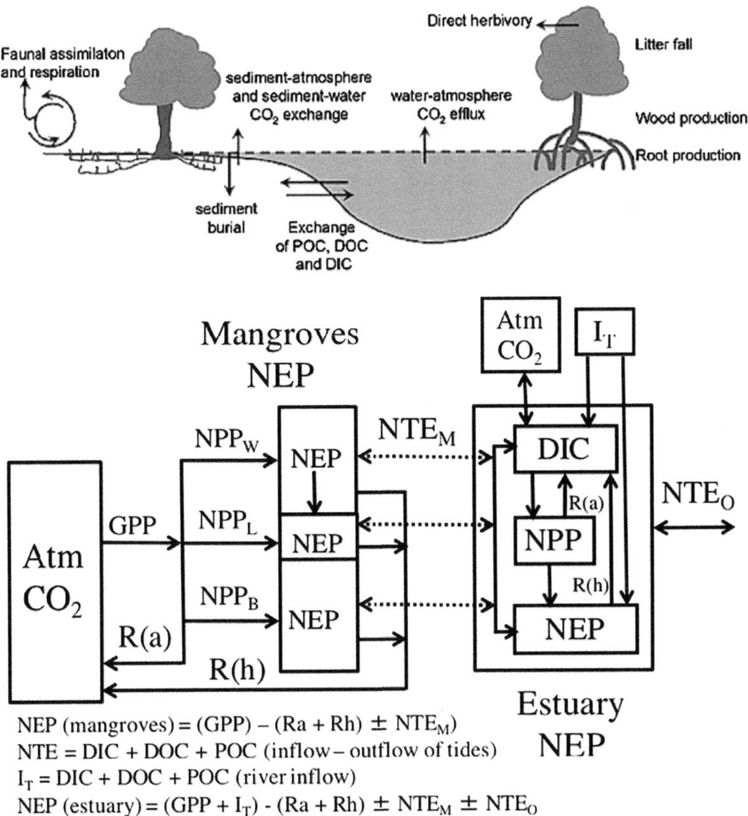

FIGURE 21.8 Top panel: summary of the major components in the mangrove carbon budgets normally measured in field observations including production (litter fall, wood, and root production) and various sinks (sediment burial) along with exchange with coastal waters (carbon forms) including particulate organic (POC), dissolved organic (DOC), and dissolved inorganic carbon (DIC) from Bouillon et al (2008a,b). Lower panel: mass balance of organic carbon using concepts of net ecosystem production (NEP) used in the text for mangroves (GPP is the gross primary production, Ra is the respiration autotrophs, Rh is the respiration heterotrophs, and NTE_M is the net exchange of tidal inflow-outflow). For estuary NEP, additional terms are IT for input from rivers; NTE_O is exchange with coastal ocean.

NENE of a mangrove wetland can be described by the following:

$$NENE = \Delta NU/dt \tag{21.3}$$

where ΔNU is the accumulation of a nutrient (NU) per unit time (dt) within an ecosystem boundary. This simple formula describes net ecosystem exchange as the residual of nutrient fluxes in mangrove wetlands after all the net exchanges with the atmosphere and tidal waters have been summed (Twilley et al., 2017). A positive NENE indicates an increase in the nutrient pool of mangrove wetlands (i.e., nutrient sink) and a negative value represents a loss from the nutrient pool of mangrove wetlands (i.e., nutrient source). Defining mangrove wetlands as a transformer of a given nutrient would mean that there is no net exchange of that nutrient across the aforementioned boundaries, but there is a change in the form of that nutrient between species of inorganic nutrients and/or between inorganic relative to organic concentrations. As an example, a mangrove wetland may exhibit no net exchange of N with nearshore tidal waters. However, there may be a net transformation of N from inorganic forms coming in (e.g., NO_3^- and NH_4^+) on a flooding tide to organic forms (e.g., dissolved organic nitrogen; DON) exported on an ebbing tide.

Open ecosystems such as mangrove wetlands located in the intertidal zones of continents or oceanic islands are associated with large gas exchanges with the atmosphere (e.g., carbon, nitrogen, sulfur) and net tidal exchanges of C, N, and P associated with a coupled estuary (Bouillon et al., 2008a; Twilley et al., 1992). Given the bidirectional flux of tides in mangroves, the net flux or net tidal exchange (NTE_M) results in negative values indicating a loss of nutrients from mangrove wetlands to coastal waters (Fig. 21.8) and positive values indicating net flux into mangrove ecosystem. The net sink of organic carbon in mangrove soils (ΔS_{org}) may integrate several of the processes that contribute to NEP. For example, the net amount of organic carbon in mangrove soils integrates net boundary exchanges with atmosphere and tidal waters by including net carbon balance of root production (NPP_B) and litter fall (NPP_L), losses due to sediment respiration (Rs), including sediment CO_2 efflux linked to leaching of organic carbon from canopy. ΔS_{org} also integrates the net effect of NTE_M. Thus, NEP can largely be determined by focusing on the sum of ΔS_{org} and wood production (NPP_W) based on the following substitutions described in Twilley et al. (2017) (Fig. 21.8):

$$NEP = (NPP_L + NPP_W + NPP_B) - (Rs \pm NTE_M) \tag{21.4}$$

$$\Delta S_{org} = (NPP_L + NPP_B) - (Rs \pm NTE_M) \tag{21.5}$$

Only wood production (NPP_W) remains after the substitution for ΔS_{org} into Eq. (21.6).

Thus,

$$NEP = NPP_W + \Delta S_{org} \tag{21.6}$$

Using this formulation of NEP to estimate NENE focuses on the fate of organic nutrients because the accounting is based on biological processes resulting in the accumulation of biomass and detritus in mangrove soils and wood accumulation. More thorough estimates of net nutrient budgets need to include all forms of inorganic exchange between atmosphere and tidal waters for those nutrients with a gas cycle, such as carbon, using net ecosystem carbon exchange (NECE) as follows (Twilley et al., 2017):

$$NECE = \sum NEE + F_{CO} + F_{CH4} + F_{VOC} + F_{DIC} + F_{DOC} + F_{PC} \tag{21.7}$$

where NEE is net ecosystem exchange of CO_2 (i.e., the net CO_2 flux from the atmosphere to the ecosystem, or net CO_2 uptake in mangrove wetlands = positive sign); F_{CO} is net carbon monoxide (CO) influx (or efflux = negative sign); F_{CH4} is net methane (CH_4) influx (or efflux = negative sign); F_{VOC} is net volatile organic C (VOC) influx (or efflux = negative sign); F_{DIC} is net dissolved inorganic carbon (DIC) input to the ecosystem (or net DIC export = negative sign); F_{DOC} is net dissolved organic carbon (DOC) input (or net DOC export = negative sign); and F_{PC} is the net lateral transfer of particulate (nondissolved, nongaseous) carbon into the ecosystem (or out of = negative sign, by fluxes represented by soot emission during fires, water, and wind deposition and erosion, animal movement, and anthropogenic transport or harvest). For other nutrients with a gas cycle, such as nitrogen and sulfur, there are some substitutions that account for NENE such as nitrogen fixation, denitrification, and sulfate reduction. For nutrients with a sedimentary cycle (e.g., P), gas exchange with the atmosphere may be much less important and therefore the focus of NENE should be on net exchanges with upland and tidal end-members.

A recent review by Twilley et al. (2017) reduced the complexity of NECE by focusing on what fluxes have actually been measured for mangrove wetlands. Because very little data are available to estimate F_{CO} and F_{VOC} for most mangrove wetlands, and methane fluxes in mangroves are minor to carbon balance (F_{CH_4}) (Alongi et al., 2005; Bartlett et al., 1989; Kreuzwieser et al., 2003; Purvaja and Ramesh, 2001), with the exception of Sundarbans (Dutta et al., 2013), these flux terms will not be included in our calculations in net ecosystem nutrient budgets. Thus, a modified NECE uses $\sum$NEE together with tidal exchange of inorganic and organic forms of carbon with coastal waters denoted as follows:

$$F_W = F_{DIC} + F_{DOC} + F_{POC} \tag{21.8}$$

Note that F_{POC} is used here assuming that all the particulate C exchange in water is organic.

We can compare NECE with NEP and include appropriate estimates with biometric methods:

$$\text{NECE} = \sum \text{NEE} + (F_W) = \text{NEP} = \text{NPP}_W + (\Delta S_{org}) \tag{21.9}$$

And what is Fw? Flux of all forms of C in water?

The problem is that NEP neither accounts for some of the inorganic fluxes in F_W, such as F_{CH4} with atmosphere or F_{DIC} across the tidal boundary, nor include relative rates of nitrogen fixation and denitrification for nitrogen. Again, NENE for sedimentary nutrients such as phosphorus may be adequately represented by knowledge of net annual storage in wood and sediments. We will review estimates for carbon and nitrogen exchange with the atmosphere, and we will consider the net exchange for C, N, and P associated with net tidal exchange, sediment accumulation, and incremental change in wood biomass to try and resolve estimates of NENE in mangroves across different coastal environmental settings.

5. CARBON EXCHANGE

5.1 Exchange With the Atmosphere

Sediment CO_2 efflux data are available for a wide range of mangrove systems (Bouillon et al., 2008a). Mangrove creek waters have consistently been found to show high CO_2 oversaturation and hence are a net source of CO_2 to the atmosphere, with an overall average of 258 g C m^{-2} per year (59 ± 52 mmol m^{-2} day, $n = 21$) (references from Bouillon et al., 2008a and include Ghosh et al., 1987; Ovalle et al., 1990; Millero et al., 2001; Borges et al., 2003; Bouillon et al., 2003, 2007a–c; Biswas et al., 2004). Dark fluxes from sediments average 267 g C m^{-2} per year (61 ± 46 mmol m^{-2} per day, $n = 82$), whereas about half of the available flux data under light conditions show a net CO_2 uptake of −66 g C m^{-2} per year (−15 ± 54 mmol m^{-2} per day, $n = 14$). These sediment and water column estimates relate only to net CO_2 fluxes and not to overall mineralization rates (see discussion below). Upscaling CO_2 fluxes for sediments and the water column separately is somewhat problematic because the relative surface areas in these intertidal systems shift during the tidal cycle. Bouillon et al. (2008a) showed a similar magnitude of CO_2 efflux from mangrove tidal creek waters and mangrove sediments of about 263 g C m^{-2} per year (60 ± 45 mmol m^{-2} per day) (ignoring the CO_2 efflux under light conditions). However, these CO_2 emission rates do not account for the lateral transport of DIC resulting from mineralization of organic-rich pore waters that drain from mangrove wetlands (Bouillon et al., 2008b). Work by Bouillon et al. (2008a) shows that DIC in mangrove creeks far exceeds DOC by a factor of ~3–10. This implies that DIC export exceeds DOC export by a similar range, assuming that both DOC and DIC originate mainly from the tidal exchange and therefore follow the same tidal variations (Bouillon et al., 2007c).

As discussed earlier in this chapter, CH_4 production in mangroves is negligible when compared with other types of wetlands (Matthews and Fung, 1987). On a global scale, the annual CH_4 emission from all freshwater wetlands is about 110 Tg (Matthews and Fung, 1987) or 17% of the total global CH_4 emissions (Carmichael et al., 2014). Mangroves are estimated to contribute only about 1.95 Tg CH_4/year (about 12 g C m^{-2} per year) to the atmosphere (Chauhan et al., 2008), which represents 1.7% and 0.3% of the total freshwater wetlands and the global emissions, respectively. In mangroves, soil-atmosphere C dynamics are mostly mediated by sulfate reduction, where carbon dioxide (CO_2) is the main by-product of soil microbial respiration (Twilley et al., 2017). However, CH_4 production can occur under strong reduction conditions through two major pathways: decarboxylation of acetic acid and reduction of CO_2 with hydrogen (i.e., H_2 released from organic matter) (Kumar and Ramanathan, 2014). These two main biogeochemical routes of CH_4 production are associated with freshwater residence time and organic matter flux into the soil (Kumar and Ramanathan, 2014). In addition, anthropogenic inputs of labile organic matter (i.e., eutrophication) can also boost soil metabolic activity leading to sulfate depletion and promoting a shift to methanogenesis (Chauhan et al., 2008). Temperature is also an important driver of microbial activity and it is expected to stimulate the production of both CO_2 and CH_4 by sulfate-reducing bacteria and methanogens, respectively (Hoehler and Alperin, 2014; Livesley and Andrusiak, 2012).

Under natural conditions, hydroperiod seems to be the most important driver of microbial-mediated process in mangroves soils. The frequency, duration, and depth of flooding are directly related to soil organic matter production and decomposition (Twilley et al., 2017). In contrast to sites regularly flushed by tides, seasonally or permanently flooded sites exhibit lower organic matter production and extreme anoxic conditions (Castañeda-Moya et al., 2013), which can favor methanogenesis. River discharge also influences the hydroperiod, regulates nutrient input, and reduces soil interstitial salinity (Lugo and Snedaker, 1974; Chen and Twilley, 1999a,b). River-born nutrient load controls C:N:P stoichiometry and resultant organic C stocks in mangrove soils (Rovai et al., 2018; Twilley et al., 2017). Reduced interstitial salinity has been linked to higher mangrove above- (i.e., litter fall) and belowground (i.e., roots) production (Twilley et al., 2017), two major sources of organic matter that fuels soil microbial activity. Methanogens are not tolerant to higher salinities and are outcompeted by sulfate-reducing bacteria; it has been shown that minor increases in salinity inhibit CH_4 production in mangrove soils (Kumar and Ramanathan, 2014). In addition to constraints imposed by sulfate-rich seawater and salinity as discussed above, only a small part of CH_4 produced is actually emitted to the atmosphere due to methane consumption by methanotrophic bacteria before it reaches the soil surface (Call et al., 2015).

Unfortunately, limited data on mangrove CH_4 production (but see Chauhan et al., 2008 and Linto et al., 2014) prevents one from delivering a robust quantitative assessment of the environmental drivers and spatial variability of CH_4 emissions at regional and global scales.

However, both CH_4 and CO_2 productions in mangroves are linked to hydrodynamic (i.e., fringe vs. interior) and geomorphic (i.e., coastal environmental settings) patterns (Twilley et al., 2017). We argue that the coastal environmental settings framework discussed earlier in this chapter represents a process-based approach that integrates the mechanisms responsible for organic matter production and decomposition in mangrove soils. We anticipate that, on the availability of new data, the coastal environmental settings framework can help elucidate questions related to the spatial variability and partitioning of soil CO_2 and CH_4 efflux and improve global coastal wetlands greenhouse gas inventories.

5.2 Net Tidal Exchange of Detritus

The fate of mangrove primary production across the tidal boundary of mangroves has been a major topic of debate in the literature during the past few decades (Bouillon et al., 2008b). In particular, the outwelling hypothesis described by Odum and Heald (1975) suggested that a large fraction of organic matter produced by mangrove trees was exported to coastal waters, where it formed the basis of a detritus food chain that supported economically important fisheries. Despite the large number of case studies dealing with various aspects of organic matter cycling in mangrove systems (Kristensen et al., 2008a), there is still no consensus on the ecological fate of mangrove-derived organic matter on the metabolism and food webs of the nearshore coastal margin (Bouillon et al., 2008b). There is a wide range in estimates of organic carbon export (Table 21.2) determined from a variety of techniques (i.e., DOC, particulate organic carbon [POC]). Organic carbon export from mangrove wetlands ranges from 0.5 to 138 g C m^{-2} per year, with an average rate of about 92 g C m^{-2} per year (Adame and Lovelock, 2011). When estimates include litter exchange (i.e., larger particulate fractions) in tidal creeks, the range increases to 401 g C m^{-2} per year for a high value, with an adjusted average of 151 g C m^{-2} per year. These estimates of organic carbon export from mangrove wetlands are similar to the average organic carbon export from salt marshes (Nixon, 1980), about 100 g C m^{-2} per year, but much lower than previously published (Twilley, 1988).

Isotope signatures in diets of consumers have been used to test the outwelling hypothesis of mangrove wetlands by describing a connection between mangrove net tidal exchange of organic matter and consumption by higher trophic level consumers (Sousa and Dangremond, 2011). Such evidence from isotopes may also provide some insights into the carbon biogeochemistry of mangroves by indicating the fate of carbon as part of the linkage to higher tropic levels, which also reflects the contribution of mangrove carbon to the metabolism of estuarine ecosystem coupled to mangrove production (Lee, 2004, 2005; Sousa and Dangremond, 2011). The assumption is that the strength of the mangrove carbon signal in tissues of consumers in estuaries and coastal waters is relative to the dominance of mangrove detritus production relative to estuary area (or volume). For example, sharp declines in mangrove detrital signature with distance from the mouth of a mangrove-dominated creek compared with lower signals in adjacent bays and no signal >2–4 km from mangroves are very common (Bouillon et al., 2002a,b, 2004a,b; Fleming et al., 1990; Fry and Smith, 2002; Hemminga et al., 1994; Mancera-Pineda, 2003; Rodelli et al., 1984). In Bocas del Toro, Panama, organic matter from mangroves declined by about 40%–50% over the first 250–300 m from the forest edge but was found in reef organisms living >10 km from a mangrove forest (Granek et al., 2009).

TABLE 21.2 Carbon Exchange (g Cm^{-2} per year) During Tidal Inundation in the Form of Particulate Organic Carbon (POC), Dissolved Organic Carbon (DOC), and Total Organic Carbon (TOC). Estimates of Carbon Exchange Are Also Determined by Litter Turnover. The Spatial Scale Considered in the Study and Methodology Used in the Field Are Included

Spatial scale	Methodology	Litter (g C m^{-2} per year)	POC (g C m^{-2} per year)	DOC (g C m^{-2} per year)	TOC (g C m^{-2} per year)	References
Estuarine basin and coastal ocean	Transects			~−138.1		Dittmar et al. (2006)
				−48		Dittmar et al. (2001)
			−9.1	−108.1	−117.1	Young et al. (2005)
Mangrove forest and adjacent creek	Detritus sampled with net at creek		−0.6	~18.7	~19.3	Ayukai et al. (1998)
			−56.9	~20.9	~77.8	Ayukai et al. (1998)
			−54.8	7.3	−62.1	Boto and Bunt (1981); Boto and Wellington (1988)
			0.1	0.4	0.5	Sanchez-Carrillo et al. (2009)
			−0.003	0	−0.003	Rajkaran and Adams (2007)
				~13.1		Rezende et al. (2007)
			−109.5			Woodroffe (1985)
	Flumes		0	67.3	67.3	Davis et al. (2001a,b)
				−56		Romigh et al. (2006)
			−15.1	−44.3	−59.4	Twilley (1985)
			−13.4	−36.7	−50.1	Twilley (1985)
	Based on litter decomposition			−0.1		Wafar et al. (1997)
Estuarine basin and coastal ocean	Litter sampled with net at creek	−198				Lugo and Snedaker (1974)
		−131.6				Rajkaran and Adams (2007)
		~−2.2				Woodroffe (1985)
		−1.0				Wattayarkon et al. (1990)
		−401.5				Golley et al. (1962)
Mangrove forest and adjacent creek	Litter production versus standing litter	−340.1				Twilley et al. (1997)
		−122.6				Mahmood et al. (2005)
		−176.0				Gong and Ong (1990)

TABLE 21.2 Carbon Exchange (g Cm^{-2} per year) During Tidal Inundation in the Form of Particulate Organic Carbon (POC), Dissolved Organic Carbon (DOC), and Total Organic Carbon (TOC). Estimates of Carbon Exchange Are Also Determined by Litter Turnover. The Spatial Scale Considered in the Study and Methodology Used in the Field Are Included—cont'd

Spatial scale	Methodology	Litter (g C m^{-2} per year)	POC (g C m^{-2} per year)	DOC (g C m^{-2} per year)	TOC (g C m^{-2} per year)	References
		−332.0				Woodroffe et al. (1988)
		−0.1				Wafar et al. (1997)
		−194.0				Robertson and Daniel (1989)
		−252.0				Robertson and Daniel (1989)
		−107.0				Robertson and Daniel (1989)
		−3.7				Lee (1989)
		~−3.9				Silva et al. (1998)
		−492.8				Flores-Verdugo et al. (1987)
		−365.0				Boto and Bunt (1981)
		−492.8				Flores-Verdugo et al. (1987)

Modified from the summary by Adame and Lovelock (2011).

These isotope observations support the dual gradient hypothesis of organic matter in estuaries proposed by Odum (1984), which explains the diverse patterns of organic detritus versus phytoplankton and seagrasses in the diets of estuarine food webs. In lower stream orders of tidal creeks with high amounts of mangrove area relative to water volume, along with sites in lower salinity regions of estuary, there is a significant quantity of mangrove detritus in diets of nekton and sessile organisms (Odum, 1984). As tidal creek stream size increases and salinity increases, phytoplankton and seagrasses typically dominate the food webs. Such trends support the claim that the significance of mangrove wetlands to fisheries depends on the total mangrove area compared with the area and/or volume of water habitats in the region (Cifuentes et al., 1996; Twilley, 1985, 1988) similar to the idea for salt marshes (Nixon, 1980). In regions with low ratio of wetland to water area, mangrove carbon contribution can range from 2% to 52% of the total available carbon pool for secondary productivity (Saintilan, 2004; Sousa and Dangremond, 2011; Twilley, 1988; Wafar et al., 1997). More expansive mangrove platforms along subtropical and tropical muddy coasts have a higher wetland to water area ratio, compared with lagoon and karstic coastal environmental settings, and this pattern may explain variations in contributions of mangrove detritus to food webs. Comparisons among coastal environmental settings may demonstrate the presence of mangrove carbon isotope signals in food webs relative to distance from mangrove sources (Mancera-Pineda, 2003).

5.3 Soil Nutrient Accumulation and Wood Production

Nutrient accumulation in mangrove soils and wood production can be used to estimate NENE and define the magnitude of nutrient sinks in mangrove wetlands (Eqs. 21.4 and 21.7). The net accumulation of nutrients in the aboveground compartment (as conceptualized in Fig. 21.9) represents long-term storage in wood, and the other losses from this compartment represent organic nutrient input to the soil compartment. The accumulation of nutrients in mangrove soils can be measured directly, using a variety of techniques using sediment vertical accretion multiplied by the nutrient density of soils (Breithaupt et al., 2012; Chmura et al., 2003; Lynch et al., 1989; Rivera-Monroy et al., 2013). Soil formation in mangrove wetlands, as in other intertidal wetlands, is the combination of several ecological processes including organic matter production (above- and belowground components), export, decomposition, and burial, as well as sedimentation of allochthonous inorganic matter (Fig. 21.9; Chen and Twilley, 1999b). Biogeochemical models that reconstruct sediment profiles can provide insights into the relative significance of these processes to the accumulation of organic matter and nutrients in wetland soils (Morris and Bowden, 1986).

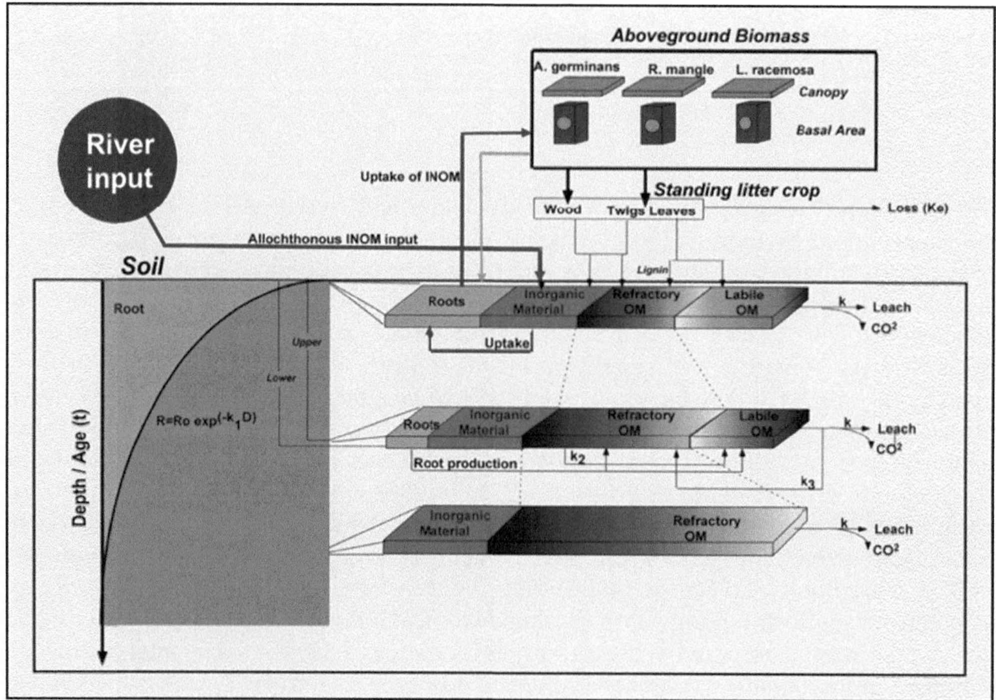

FIGURE 21.9 Conceptual model of variables included in the NUMAN model (Chen and Twilley, 1999b).

The NUMAN model is one of the only published mangrove biogeochemistry models that simulate concentrations of organic matter, N, and P with soil depth (Fig. 21.9). Another model with some structural similarities to the NUMAN model is the "Relative Elevation Model" developed by Rybczyk et al. (1998; Chapter 30) for marshes in coastal Louisiana, which has been adapted by Cahoon et al. (2003) to estimate changes in total sediment accretion after major hurricane disturbances in mangrove forest of the Caribbean region.

Sediments suspended in the water column are deposited in mangroves during flooding, and this allochthonous material enriches mangrove soils. Accumulation of organic matter was similar among all sites (Table 21.3), whereas accumulation of inorganic matter was higher in the forests influenced by river discharge. The contribution of inorganic material ranges from 133 to 5151 $g\,m^{-2}$ per year with higher values occurring in riverine mangrove forests such as the Guayas River estuary (Table 21.3). Sites with sedimentation rates $<500\,g\,m^{-2}$ per year include the fringe and basin sites of a lagoon in Rookery Bay, and the lowest rate of 94 $g\,m^{-2}$ per year of sediment input is to a fringe carbonate site in Belize (Table 21.2). The fringe and basin sites at Terminos Lagoon have sedimentation rates of 1094 and 618, respectively, whereas all of the riverine sites had sedimentation rates $>1{,}200\,g\,m^{-2}$ per year (Table 21.3). All of the riverine sites in the Guayas River estuary had sedimentation rates $>4000\,g\,m^{-2}$ per year, more than double the riverine sites in Belize or Terminos Lagoon and 10-fold higher than the basin mangrove sites in Rookery Bay. The IS:OM ratio of inorganic (IS) and organic (OM) sedimentation explains the effects of geophysical processes on sedimentation patterns in mangroves. The ratios are <1 in the basin sites at Rookery Bay and the carbonate site in Belize. The overwash site at Rookery Bay and fringe and basin sites in Terminos Lagoon have ratios that range from 1.7 to 2.1, while the subtidal fringe site has a ratio of 5.6. All the riverine sites have ratios that range from 3.6 to 7.8 (Table 21.3).

Root productivity and belowground biomass accumulation contribute to soil volume and consequently soil elevation in mangrove wetlands (Cahoon et al., 2003; McKee, 2011; McKee et al., 2007), particularly in carbonate settings (Lynch et al., 1989; Parkinson et al., 1994). Mangroves are touted as highly efficient carbon sinks in the wet tropics because of relatively high primary productivity and low rates of decomposition (Komiyama et al., 2008). Sedimentation and nutrient burial, particularly carbon and nitrogen, in mangrove wetlands not only include allochthonous inorganic matter input (sedimentation) but also net organic matter input resulting from high rates of root production relative to decomposition (Chen and Twilley, 1999b). Several studies have found that decay rates of belowground material are slower than leaf litter (McKee and Faulkner, 2000; Middleton and McKee, 2001; Poret et al., 2007; Van Der Valk and Attiwill, 1984a). A large part of sedimentary organic matter in mangroves is derived from root organic matter (Alongi et al., 2001) and, in many forest systems, can be the principle source of organic matter in deeper soil layers (Ludovici et al., 2002). For example, production and slow degradation of mangrove roots may contribute more to organic matter accumulation and vertical building of mangrove islands in Belize than total litter fall (McKee et al., 2007; Middleton and McKee, 2001). One of the most comprehensive reviews of carbon sequestration of mangrove soils by Breithaupt et al. (2012) demonstrates that global rates can range from 4 to 1100 $g\,C\,m^{-2}$ per year (Fig. 21.10). However, the largest proportion of observations, 41 of the 65 sites, exhibit

TABLE 21.3 Estimates of Nutrient Accumulation in Different Types of Mangrove Forests Around the World

Site Description		Ecological Type	Inorganic Sedimentation	Organic Sedimentation	Total Carbon	Total Nitrogen	Total Phosphorus	C:N:P	IS:OM
			Accumulation (g m^2 per year)	Accumulation (g m^2 per year)	Accumulation (g m^2 per year)	Accumulation (g m^2 per year)	Accumulation (g m^2 per year)	IS:OM	IS:OM
Southwest	Rookery Bay	Interior	239.19	246.83	114.50	6.64	0.23	1311:65:1	0.97
Florida		Interior	184.57	227.48	106.50	6.32	0.25	1116:57:1	0.81
		Interior	141.59	280.17	132.99	7.22	0.24	1460:68:1	0.51
		Interior	172.26	272.05	126.15	7.60	0.21	1518:79:1	0.63
		Fringe	307.82	54.78	32.27	1.34	0.27	307:11:1	5.62
		Fringe	173.16	100.29	51.59	1.97	0.62	214:7:1	1.73
Belize	West Pond	Interior	93.64	276.19	122.41	5.77	0.16	2024:81:1	0.34
	Sittee River	Riverine	1286.43	307.92	129.68	7.95	1.06	317:17:1	4.18
Terminos	Estero Pargo	Interior	744.80	348.93	157.03	11.01	0.63	647:39:1	2.13
Lagoon	Boca Chica	Interior	386.77	231.66	104.25	4.83	0.21	1304:52:1	1.67
		Riverine	1440.93	400.76	189.83	8.44	0.82	597:23:1	3.60
		Interior	1136.54	145.51	65.46	2.46	0.73	233:8:1	7.81
Ecuador	M1	Riverine	3581.60	803.35	287.84	10.07	3.01	247:7:1	4.46
		Riverine	4318.88	813.94	265.51	10.41	3.62	189:6:1	5.31
	M3	Riverine	3076.41	997.47	387.57	13.37	0.92	1087:32:1	3.08
Gulf of Thailand	Sawi Bay	S1	7500		225.6	8.18			
Matang Forest		MVR	2450		127	13.29			
		M18	3800		109.5	14.31			
		M5			100.7	17.89			

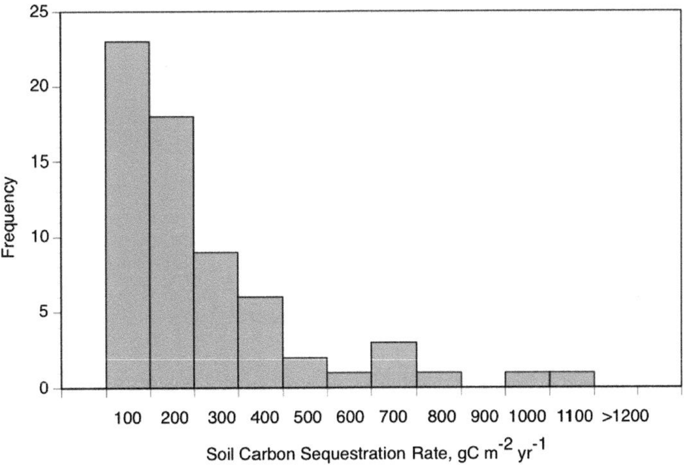

FIGURE 21.10 Distribution of soil carbon sequestration rates (g C m^2 per year) for mangrove forests based on review by Breithaupt et al. (2012).

carbon sequestration in the range of 4–200 g C m^{-2} per year, with a global mean burial rate of 163 g C m^{-2} per year for all sites. The importance of the contributions of mangrove wetlands to global carbon sequestration reinforces the need for a better understanding of biomass allocation patterns in different mangrove ecotypes. Coastal environmental settings framework can substantially improve our capacity to deliver more robust estimates of mangrove SOC stocks across hemispheric regions. For extensive and geomorphologic diverse coastlines such as Brazil, Australia, and Mexico, the use of ecogeomorphic diversity of shoreline allows for significant improvements in contemporaneous SOC stock assessments. The coastal environmental setting framework provides a powerful approach to address regional to macroscale variability in mangrove SOC (Rovai et al., 2018).

6. NITROGEN AND PHOSPHORUS EXCHANGE

6.1 Exchange With the Atmosphere

The net atmospheric exchange of nitrogen gas in mangrove ecosystems depends on the inputs of nitrogen via fixation relative to the loss via denitrification (Fig. 21.11). Nitrogen fixation represents a new source of nitrogen to the system and can compensate for nitrogen removal by denitrification (Howarth et al., 1988). Nitrogenase activity has been observed in decomposing leaves, root surfaces (prop roots and pneumatophores), and sediment, but few of these studies have interpreted nitrogen fixation rates relative to the net ecosystem nitrogen exchange of mangrove wetlands (Tables 21.4 and 21.5). As such, net nitrogen ecosystem exchange estimates are very seldom estimated, in contrast to increased awareness of carbon dynamics in mangrove ecosystems. The review below describes nitrogen inputs, storages,

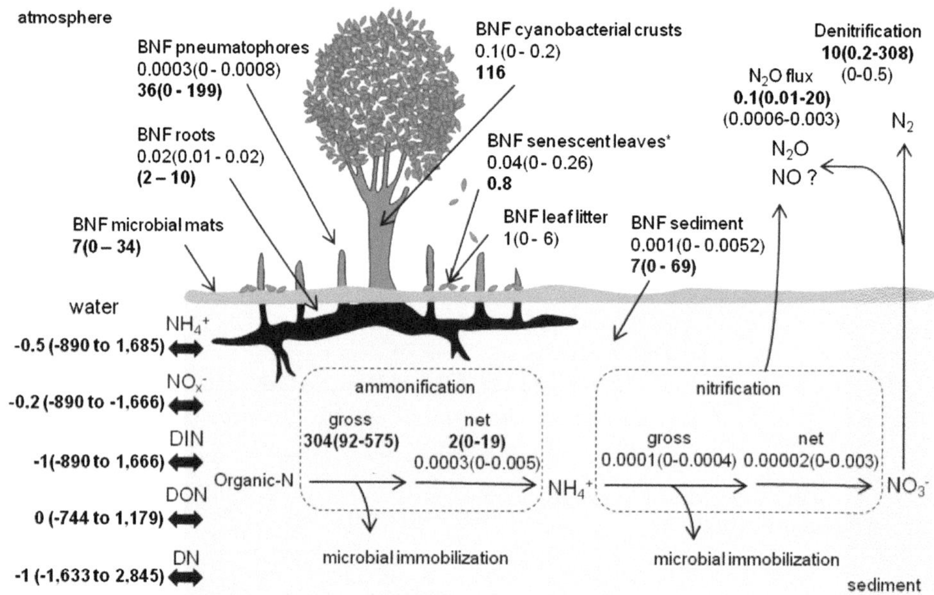

FIGURE 21.11 Nitrogen cycling rates in mangroves from Reis et al. (2017a). Median rates (and ranges) in mg N g^{-1} per day (unbold values) and mg N g^{-1} per day (bold values) are presented, including biological nitrogen fixation (BNF), nitrous oxide (N_2O) flux, and fluxes of ammonium (NH_4^+), nitrite plus nitrate (NO_X^-), dissolved inorganic nitrogen (DIN), dissolved organic nitrogen (DON), and total dissolved nitrogen (DN) between mangrove sediment and overlying water. Negative values indicate net flux into the sediment. For BNF, mean rates are presented. * incubated on the forest floor. *From Hesse (1961); Kimball and Teas (1975); Alongi et al. (1992); Gotto and Taylor (1976); Viner (1982); Howarth et al. (1988); Van Der Valk and Attiwill (1984a,b); Hicks and Silvester (1985); Iizumi (1986); Myint (1986); Shaiful (1987); Kristensen et al. (1988); Mann and Steinke (1992); Boto and Robertson (1990); Sengupta and Chaudhuri (1991); Sheridan (1991, 1992); Kristensen et al. (1992); Alongi et al. (1993); Nedwell et al. (1994); Rivera-Monroy et al. (1995a); Alongi (1996); Pelegri et al. (1997); Woitchik et al. (1997); Kristensen et al. (1998); Pelegri and Twilley (1998); Chen and Twilley (1999b); Corredor et al. (1999); Alongi et al. (1999, 2000); Kristensen et al. (2000); Davis et al. (2001a, b); Bauza et al. (2002); Lugomela and Bergman (2002); Mohammed and Johnstone (2002); Muñoz-Hincapié et al. (2002); Kyaruzi et al. (2003); Kreuzwieser et al. (2003); Alongi et al. (2004); Joye and Lee (2004); Lee and Joye (2006); Chauhan et al. (2008); Purvaja et al. (2008); Krishnan and Bharathi (2009); Chen et al. (2010); Fernandes et al. (2010, 2012a,b); Romero et al. (2012); Reis et al. (2017a).*

and exchange using a variety of units, with few integrated across a spatial scale of mangroves.

Decaying mangrove leaves are sites of relatively high rates of N fixation and thus may account for 15%–64% of nitrogen immobilization during decomposition of mangrove leaf litter on the forest floor (Gotto et al., 1981; Van Der Valk and Attiwill, 1984a,b; Woitchik et al., 1997). Sediment and roots fix nitrogen at rates from nondetectable to about 15 nmol N g^{-1} of substrate/h, compared with rates for leaf litter at >250 nmol N g^{-1} of substrate/h (Table 21.4). Nitrogen fixation could account for 45%–100% of the total nitrogen immobilized in leaf litter (1–8 mg N/dry wt nitrogen enrichment) in mangroves along the Shark River estuary in the FCE (USA, Pelegri et al., 1997). However, N fixation only supplied 7% (8.3 mg N m^{-2} per day) of the nitrogen required (53 mg N m^{-2} per day) for mangrove growth. Most of the studies of nitrogen fixation in mangroves average around 50 μmol m^{-2}

TABLE 21.4 Nitrogen Fixation Rates for a Variety of Locations Within Mangrove Systems. Original Rates Were Converted to Nmol of N Based on 3:1 Ratio of Ethylene to Mole of N_2

Site (Country)	Substrate	Rate (nmol N/g substrate/h)	Source
Tampa Bay (USA)	Sediment (*Avicennia germinans, Laguncularia racemosa, Rhizophora mangle*)	0.14[a] to 0.61[a]	Zuberer and Silver (1978)
West Port Bay (Australia)	Sediment (*Avicennia marina*)	5.3 + 4.1[bc]	Van Der Valk and Attiwill (1984a,b)
Florida Coastal Everglades (USA)	Sediment (*Avicennia germinans, Laguncularia racemosa, Rhizophora mangle*)	0[ac] to 20.9[ac]	Pelegri et al. (1997)
Florida Coastal Everglades (USA)	Sediment (*Avicennia germinans, Laguncularia racemosa, Rhizophora mangle*)	0.07[ac] to 0.27[ac]	Pelegri and Twilley (1998)
Tamsui estuary (Taiwan)	Soils (Fringe tall *Kandelia obovata*)	11.4[ac]	Shiau et al. (2017)
Tamsui estuary (Taiwan)	Soils (Scrub *Kandelia obovata*)	5.9[ac]	Shiau et al. (2017)
Tamsui estuary (Taiwan)	Bare soils tidal creek	0.9[ac]	Shiau et al. (2017)
Tampa Bay (USA)	Roots (*Rhizophora mangle*)	3[cd] to 10[cd]	Zuberer and Silver (1978)
Tampa Bay (USA)	Roots (*Laguncularia racemosa*)	7[cd] to 35[cd]	Zuberer and Silver (1978)
Tampa Bay (USA)	Roots (*Avicennia germinans*)	9[cd] to 11[cd]	Zuberer and Silver (1978)
West Port Bay (Australia)	Roots (*Avicennia marina*)	2.3[bcej] and 7.5 + 3.9[bcekl]	Van Der Valk and Attiwill (1984a,b)
Florida Coastal Everglades (USA)	Pneumatophores (*Avicennia germinans*)	0[ac] to 3.2[ac]	Pelegri et al. (1997)
Makham Bay (Thailand)	Roots (*Rhizophora apiculata*)	11 + 4[ace]	Kristensen et al. (1998)
Key Biscayne (USA)	Leaf litter (*Rhizophora mangle*)	50.0 + 50.0[bce] to 900.0 + 242.9[ade]	Gotto and Taylor 1976
Tampa Bay (USA)	Leaf litter (*Rhizophora mangle*)	23[b] and 25[a]	Zuberer and Silver (1978)
West Port Bay (Australia)	Leaf litter (*Avicennia marina*)	235 + 37[bcel] and 313 + 142[bcl]	Van Der Valk and Attiwill (1984a,b)
Auckland Province (New Zealand)	Leaf litter (*Avicennia marina*)	7.4 + 4.1 to 27.6 + 5.5	Hicks and Silvester (1985)
Gazi Bay (Kenya)	Leaf litter (*Rhizophora mucronata*)	250[bhl] and 780[bil]	Woitchik et al. (1997)

(*Continued*)

TABLE 21.4 Nitrogen Fixation Rates for a Variety of Locations Within Mangrove Systems. Original Rates Were Converted to Nmol of N Based on 3:1 Ratio of Ethylene to Mole of N_2—cont'd

Site (Country)	Substrate	Rate (nmol N/g substrate/h)	Source
Gazi Bay (Kenya)	Leaf litter (*Ceriops tagal*)	156[bhl] and 760[bil]	Woitchik et al. (1997)
Florida Coastal Everglades (USA)	Leaf litter (*Rhizophora mangle*)	44.1[acl]	Pelegri and Twilley (1998)
Florida Coastal Everglades (USA)	Leaf litter (*Avicennia germinans*)	43.1[acl]	Pelegri and Twilley (1998)
Florida Coastal Everglades (USA)	Leaf litter (*Rhizophora mangle*)	4.9[ac] to 149.0[ac]	Pelegri et al. (1997)
Florida Coastal Everglades (USA)	Leaf litter (*Laguncularia racemosa*)	22.5[ac] to 222.9[ac]	Pelegri et al. (1997)
Florida Coastal Everglades (USA)	Leaf litter (*Avicennia germinans*)	17[ac] to 359.9[ac]	Pelegri et al. (1997)
Iriomote Island (Okinawa)	Warty lenticellated bark (*Bruguiera gymnorrhiza*)	5[bc] to 8[bc]	Uchino et al. (1984)
Auckland Province (New Zealand)	CPOM[g] (*Avicennia marina*)	20.5	Hicks and Silvester (1985)
Missionary Bay (Australia)	Decomposing logs	0.63 + 0.30[f] to 4.46[f]	Boto and Robertson (1990)

[a]Anaerobic; [b]Aerobic; [c]Dark; [d]Light; eInundated; [f]24 h; [g]Course particulate organic matter; [h]Dry season; [i]Rainy season; [j]Live; [k]Dead; [l]Maximum.

All rates were obtained using the acetylene reduction technique. Rates originally reported as moles C2H4 produced were converted to moles of N2 using the theoretical 3:1 ratio. An effort was made to present the range of rates measured in each study. If study examined the rate of N fixation during decomposition of a substrate (i.e., leaves), the maximum rate was presented. If study measured a rate of 0, this was reported. The exception was if the study looked at rates of N fixation during decomposition (see note above). An effort was also made to keep track of the conditions of the incubations (i.e., light vs. dark, aerobic vs. anaerobic). If the conditions were unclear, then they were not specified.

per h, which is about 0.4% of the NPP of mangroves (Bouillon et al., 2008a). Results from mangrove sediments in south Florida indicate that nitrogen fixation rates range from 0.4 to 3.2 g N m^{-2} per year (3–26 μmol m^{-2} per h; Zuberer and Silver, 1978), similar to the natural rates of denitrification.

Sponges, tunicates, and a variety of other forms of epibiont communities on prop roots of mangroves are highly diverse (Ellison and Farnsworth, 1992; Rützler and Feller, 1988; Sutherland, 1980) and can have high rates of nitrogen fixation. The diversity and biomass of these communities may be limited to specific geomorphologic types, especially along carbonate shorelines with little terrigenous input. In such oligotrophic systems, these communities are sites of nitrogen fixation, which may be a source of nutrition for higher-level predators as well as influencing various processes in mangrove fringe forests. These processes of nutrient regeneration associated with sponge communities that colonize aerial root systems of mangroves have received comparatively little attention; but they may influence the productivity of fringe mangrove forests and enhance the exchange of nutrients

TABLE 21.5 Nitrogen Fixation Rates per Unit Area of Mangrove Forest Based on Experimental Conditions in Mangrove Forests Around the World. Original Rates Were Converted to Nmol of N Based on 3:1 Ratio of Ethylene to Mole of N_2

Site (Country)	Substrate	Rate (μmol Nm^{-2} per h)	Source
Auckland Province (New Zealand)	Sediment (*Avicennia marina*)	0.3[b] to 33[b]	Hicks and Silvester (1985)
Oyster Bay (Jamaica)	Sediment (*Rhizophora mangle, Avicennia germinans*)	0[e] to 100.0[e]	Nedwell et al. (1994)
Missionary Bay (Australia)	Sediments (*Rhizophora* spp., *Ceriops* spp.)	0[f] to 9.9 + 1.8[f]	Boto and Robertson (1990)
Joyuda Lagoon (Puerto Rico)	Sediment (*not specified*)	13.3[a] to 31.5[a]	Morell and Corredor (1993)
Makham Bay (Thailand)	Sediment (*Rhizophora apiculata*)	12 + 0.4[ace]	Kristensen et al. (1998)
Mekong Delta (Vietnam)	Sediment (*Rhizophora apiculata*)	20 + 11[bc] to 119 + 39[bc]	Alongi et al. (2000)
Mazizini (Tanzania)	Sediment (*Avicennia marina, Sonneratia alba*)	8.7[b] to 112.0[b]	Lugomela and Bergman (2002)
Sawi Bay (Thailand)	Sediment (*Avicennia alba, Ceriops decandra, Rhizophora apiculata*)	0[bcd] to 24 + 24[bcd]	Alongi et al. (2002)
Matang Mangrove Forest Reserve (Malaysia)	Sediment (*Rhizophora apiculata*)	0 + 0[bcd] to 250.0 + 108.3[bcd]	Alongi et al. (2004a)
Kisakasaka (Tanzania)	Sediment (*Avicennia marina, Bruguiera gymnorrhiza, Ceriops tagal, Rhizophora mucronata, Sonneratia alba*)	1.3[bf] to 41.3[bf]	Sjöling et al. (2005)
Jiulongjiang Estuary (China)	Sediment (*Kandelia candel, Aegiceras corniculatum*)	0[bcd] and 4 + 4[bcd]	Alongi et al. (2005)
Twin Cays (Belize)	Sediment (*Avicennia germinans, Laguncularia racemosa, Rhizophora mangle*)	0[b] to 100.8[b]	Lee and Joye (2006)
Mazizini (Tanzania)	Pneumatophores (*Avicennia marina, Sonneratia alba*)	40[b] to 790.0[b]	Lugomela and Bergman (2002)
Indian River Lagoon (Florida)	Soil slurries (scrub *Avicennia germinans*) control plots	2.92[ad]	Whigham et al. (2009)
Indian River Lagoon (Florida)	Soil slurries (scrub *Avicennia germinans*) fertilized plots	23.6[ad]	Whigham et al. (2009)
Lagunas Balandra and Enfermeria (Baja California, Mexico)	Soil slurries (*Avicennia germinans*)	0.32 to 0.38[ac]	Vovides et al. (2011)
Mazizini (Tanzania)	Pneumatophores (*Avicennia marina, Sonneratia alba*)	40[b] to 790.0[b]	Lugomela and Bergman (2002)
Mandovi and Chapora rivers (Goa, India)	Soil slurries (*A. officinalis, Excoecaria agallocha, Caesalpinia* spp.)	24.9[ac]	Fernandes et al. (2012a,b)

(*Continued*)

TABLE 21.5 Nitrogen Fixation Rates per Unit Area of Mangrove Forest Based on Experimental Conditions in Mangrove Forests Around the World. Original Rates Were Converted to Nmol of N Based on 3:1 Ratio of Ethylene to Mole of N_2—cont'd

Site (Country)	Substrate	Rate (μmol Nm^{-2} per h)	Source
Tamsui estuary (Taiwan)	Soils (fringe tall *Kandelia obovata*)	0.58[ac]	Shiau et al. (2017)
Tamsui estuary (Taiwan)	Soils (scrub *Kandelia obovata*)	0.13[ac]	Shiau et al. (2017)
Tamsui estuary (Taiwan)	Bare soils tidal creek	0.15[ac]	Shiau et al. (2017)

[a]Anaerobic; [b]Aerobic; [c]Dark; [d]Light; eInundated; [f]24 h; [g]Course particulate organic matter; [h]Dry season; [i]Rainy season; [j]Live; [k]Dead; [l]Maximum.
All rates were obtained using the acetylene reduction technique. Rates originally reported as moles C_2H_4 produced were converted to moles of N2 using the theoretical 3:1 ratio. An effort was made to present the range of rates measured in each study. If study examined the rate of N fixation during decomposition of a substrate (i.e., leaves), the maximum rate was presented. If study measured a rate of 0, this was reported. The exception was if the study looked at rates of N fixation during decomposition (see note above). An effort was also made to keep track of the conditions of the incubations (i.e., light vs. dark, aerobic vs. anaerobic). If the conditions were unclear, then they were not specified.

with coastal waters (Ellison et al., 1996). The specific contribution of these productive and diverse epibiont communities in predominately carbonate environments may demonstrate an important linkage between biodiversity and ecosystem function.

However, the spatial analysis of nitrogen fixation is still inadequate to provide a clear estimate of this contribution to the nitrogen budget of mangrove wetlands (Rivera-Monroy et al., 1999; Twilley, 1997). Comparisons to associative biological nitrogen fixation rates in mangroves, from which plants can also directly benefit, are difficult because of the reduced number of estimates reported on an aerial basis for mangroves. Nevertheless, these estimates range from 2 to 10 mg N m^{-2} per day (Fig. 21.11), falling within the global estimates range of symbiotic nitrogen fixation in tropical terrestrial forests (Reis et al., 2016). These rates are equivalent to 0.73–3.65 g N m^{-2} per year, and using 4000 g C m^{-2} per year for average gross primary production (GPP), then expected 292 g N m^{-2} per year is needed to supply given a stoichiometry of 16:1. Based on these estimates, nitrogen fixation provides only 1% of N required to support mangrove GPP.

Denitrification is primarily dependent on oxygen availability, an energy source, and availability of NO_3^- as an electron acceptor of microbial respiration. Based on the sources of NO_3^-, there are two types of denitrification: direct denitrification, which is fueled by NO_3^- diffusing from the overlying water column into sediments, and coupled nitrification, denitrification that is supported by NO_3^- being supplied by nitrification in the sediments (Nielsen, 1992). The contributions of these two NO_3^- sources are regulated by different mechanisms (Fig. 21.11). Direct denitrification is typically a linear function of NO_3^- concentrations in the overlying water and an inverse linear function of the oxygen penetration depth in sediments (Christensen et al., 1990; Nielsen et al., 1990). Coupled nitrification–denitrification is regulated by the rate and position of nitrification activity, which in turn is regulated by the nitrification capacity and the availability of ammonium and oxygen (Blackburn et al., 1994;

Henriksen and Kemp, 1988). An expansion of the reduction zones may either stimulate denitrification if the water column is the major source of NO_3^- or inhibit denitrification if NO_3^- produced by nitrification is the major source of NO_3^- (Rysgaard et al., 1994). Coupled nitrification–denitrification is thought to be quantitatively more important than influx of NO_3^- in most aquatic sediments (Christensen et al., 1990), yet this is not established for mangrove sediments. The seasonal fluctuation of oxidation and reduction zones in the upper sediments of mangroves could certainly stimulate these coupled reactions. And the strong presence of a lower oxidation zone associated with either mangrove roots or sediment bioturbation would enhance these coupled nitrogen processes.

There is a large range in reported rates of denitrification in mangrove sediments and waters (Table 21.6). A general average rate of denitrification among mangrove sediments is about 180 μmol m^{-2} per h, about four times that of nitrogen fixation (cf. Tables 21.4 and 21.5). This is similar to the medium range of rates measured in mangrove sediments described by Nedwell (1975). Estimates of denitrification based on NO_3 uptake range from a low of 0.53 μmol m^{-2} per h at Hinchinbrook Island (Iizumi, 1986) where concentrations were <10 μM compared with ranges of 9.7–261 mmol N m^{-2} per h in mangrove forests receiving effluents from sewage treatment plants with concentrations >1000 μM (Corredor and Morell, 1994; Nedwell, 1975). Measures of denitrification using small amendments of $^{15}NO_3$ followed by direct measures of $^{15}N_2$ production have shown that denitrification accounts for <10% of the applied isotope, suggesting that NO_3 uptake may not represent a significant transfer of nitrogen to the atmosphere (Rivera-Monroy et al., 1995b). Rates of denitrification linked to nitrification in mangrove sediments are also low, and thus little exchange of nitrogen gas can be associated with dissolved inorganic nitrogen uptake (Rivera-Monroy and Twilley, 1996). The lack of $^{15}N_2$ gas production indicates that much of the NO_3 uptake may not be loss to the atmosphere via denitrification but immobilized in litter on the forest floor. However, recent measurements of direct N_2 release from mangrove sediments suggest that more of this NO_3 may be lost to the atmosphere than suggested in the Terminos Lagoon studies. The N_2 gas flux estimates are commonly above 180 μmol m^{-2} per h, with some rates >900 μmol m^{-2} per h (Table 21.5). A more comprehensive comparison on fixation and denitrification is needed to balance the nitrogen budget relative to net flux with the atmosphere. Given the general observation that denitrification is greater than N fixation, nitrogen recycling or input with net tidal exchange must meet most of the nitrogen demand of NPP in mangrove forests.

Another potentially important nitrogen sink in mangrove sediments is anaerobic ammonium oxidation (anammox) (Francis et al., 2007; Hayatsu et al., 2008) (Fig. 21.10). Anammox is the anaerobic conversion of NO_2^- and NH_4^+ to N_2, and although its presence was suggested as early as 1965, direct evidence for this reaction was documented just over a decade ago (Kartal et al., 2007). Several studies have reported the presence of anammox in estuarine and offshore sediments, in permanently anoxic bodies of water, and in multiyear sea ice (Dalsgaard et al., 2005; Jensen et al., 2007; Jetten et al., 2003; Thamdrup et al., 2004, 2006). Anammox could be an important pathway in global nitrogen cycling because it can account for as much as 67% of benthic N_2 production, with the remainder being produced by denitrification as determined in oceanic and coastal nitrogen budgets (Meyer et al., 2005; Schmid et al., 2007). However, there are few estimates of anammox rates in mangrove sediments to determine any particular pattern of how this process impacts nitrogen fluxes in different sediment zones. Observations in a subtropical mangrove forest in Australia indicate

TABLE 21.6 Denitrification Rates Using Different Techniques and Various Experimental Conditions in Mangrove Forests Around the World

Site (Country)	Mangrove Type	Tech	Incubations	Rate (μmol N m^{-2} per h)	Source
Joyuda Lagoon (Puerto Rico)	Fringe	A	Dark, flooded, anaerobic	0.74 to 160.9	Morell and Corredor (1993)
Oyster Bay (Jamaica)	Fringe, center, rear[ab]	A	Flooded	0 to 83.3	Nedwell et al. (1994)
La Parguera (Puerto Rico)	Fringe receiving sewage effluent[abc]	A	Dark, flooded, anaerobic	9.7 to 183.0	Morell and Corredor (1993)
La Parguera (Puerto Rico)	Fringe receiving sewage effluent[abc]	A	Dark, flooded, anaerobic	49.9 + 0.3 to 89.9 + 83	Corredor et al. (1999)
Twin Cays (Belize)	Fringe, transition, dwarf[abc]	A	Diel, aerobic	0 to 92.5	Lee and Joye (2006)
Terminos Lagoon (Mexico)	Fringe, basin, riverine[abc]	B	Dark, flooded, aerobic	0.08 to 250	Rivera-Monroy and Twilley (1996)
Makham Bay (Thailand)	Mid-intertidal[d]	B	Dark, flooded, aerobic	1.9 + 0.4	Kristensen et al. (1998)
Hinchinbrook Channel (Australia)	Mature[dhi]	C	Flooded, aerobic	342 + 264 and 460 + 266	Alongi et al. (1999)
Mekong Delta (Vietnam)	Managed, high-intertidal[d]	C	Flooded, aerobic	0 and 183.3 + 41.7	Alongi et al. (2000)
Sawi Bay (Thailand)	Managed, mid and high-intertidal[def]	C	Dark, flooded, aerobic	0 to 160 + 300	Alongi et al. (2002)
Matang Mangrove Forest Reserve (Malaysia)	Managed[dikl]	C	Dark, flooded, aerobic	33.3 + 50.0 to 916.7 + 1066.7	Alongi et al. (2004)
Jiulongjiang Estuary (China)	Low-, mid-, high-intertidal[gj]	C	Dark, flooded, aerobic	92 + 53 to 315 + 186	Alongi et al. (2005)
Mandovi and Chapora rivers (Goa, India)	Fringe	B	Dark, flooded, aerobic	4 to 2080	Fernandes et al. (2012a,b)
Jiulongjiang Estuary (China)	Low-, mid-, high-intertidal[gj]	C	Dark, flooded, aerobic	92 + 53 to 315 + 186	Alongi et al. (2005)
Indian River Lagoon (Florida)	Scrub *Avicennia germinans* forest (control vs. fertilized plots)	A	Flooded, anaerobic	11.4 to 175.8	Whigham et al. (2009)

Techniques: *A*, acetylene blockage; *B*, isotope enrichment; *C*, N_2-gas flux.

that anammox to sediment N_2 production in subtidal sediments was relatively low (0%–9%) and comparable with other temperate estuaries (Meyer et al., 2005) and tidal marsh sediments (Koop-Jakobsen and Giblin, 2009). This is consistent with other studies of anammox in mangroves sediments (Balk et al., 2015; Chen and Gu, 2017; Fernandes et al., 2012b; Cao et al., 2016, 2017). This confirms the relative minor importance of anammox in shallow coastal sediments, as has been observed in most nearshore anammox studies. Yet, more studies are needed in tropical latitudes to further understand the significance of anammox in the nitrogen cycle of mangrove-dominated ecosystems. The leaching of organic matter from leaf litter that is produced daily in mangrove forests may provide an organic source that influences the relative balance of anammox, denitrification, or dissimilatory nitrate reduction as a fate of nitrate in tropical systems, based on model described for salt marshes (Algar and Vallino, 2014).

6.2 Net Tidal Exchange

Surveys of nitrogen and phosphorus tidal exchange demonstrate some of the principles of determining the function of mangrove wetlands as a nutrient sink and transformer of some nutrient species (Table 21.7). The average nitrogen tidal exchange for mangrove wetlands is about -1 g N m^{-2} per year (Adame and Lovelock, 2011). As observed for organic carbon, there is nearly a 100% coefficient of variation for N export, with values ranging from -5.0 to $+1.45$ g N m^{-2} per year (Table 21.7, excluding the indirect estimates). The average net tidal exchange of phosphorus is actually an import of $+1.7$ g P m^{-2} per year (Table 21.7, Adame and Lovelock, 2011). The coefficient of variation for these estimates is about 200%, demonstrating a large variation in estimates of exchange among mangrove sites. The C:N of net tidal exchange, using an average of 100 g C m^{-2} per year and 1 gN m^{-2} per year, respectively, is estimated at 117. This is consistent with the C:N of leaf litter and is not surprising given the chemical nature of dissolved organics that are exported by mangroves (Cawley et al., 2014).

Export and import flux studies of estuarine coasts in Mexico (Rivera-Monroy et al., 1995a) and Australia (Boto and Wellington, 1988) reported an import of NH_4 compared with slight release of NO_3 in Australia and import in Mexico. One of the main differences between the two sites is the flux of DON, which is imported into the wetlands in Australia and exported from the site in Mexico. The largest nitrogen flux at both sites was the export of particulate nitrogen, which coincided with net flux of organic carbon from most mangroves (described above). Compared with other flux studies of mangroves, there seems to be a pattern of net inorganic fluxes into the wetlands and corresponding flux of organic nutrients out (Rivera-Monroy et al., 1995a,b). Given large statistical variation around these flux estimates, the best conclusion may be that mangrove wetlands transform the tidal import of inorganic nutrients into organic nutrients that are then exported to coastal waters. Small-scale estimates of nutrient fluxes within the (<20 m^2) carbonate zone of Taylor Slough in the FCE (Davis et al., 2001a,b, 2003, 2004) indicated consistent uptake of ammonium (6.6–31.4 μmol m^{-2} per h) and of total nitrogen (98–502 μmol m^{-2} per h), whereas oxidized forms of inorganic N (7.1–139.5 μmol m^{-2} per h) were released to the water column (Davis et al., 2001a). Mangroves in both the sediment-rich (Terminos Lagoon) and sediment-poor (Taylor River) sites export organic nutrients to adjacent coastal waters. The sediment-poor, oligotrophic site with diverse prop root communities released nitrate, whereas the sediment-rich site removed nitrate from overlying water.

TABLE 21.7 Phosphorus and Nitrogen Exchange (gm^{-2} per year) and Chemical Form (Dissolved or Particulate) During Tidal Inundation for a Range of Mangrove Forests

Spatial scale	Methodology	Net N flux ($g\ m^{-2}$ per year)	Net P flux ($g\ m^{-2}$ per year)	Nutrient Form	References
Estuarine system and coastal ocean	Transects	+	+	D	Lin and Dushoff (2004)
		−1.57		D	Young et al. (2005)
		−4.21		P	
Estuarine basin and coastal ocean	Water sampled in creeks	−5.0	−0.19	D	Dittmar and Lara (2001)
		~1.60	~0.34	D	Ayukai et al. (1998)
		~−1.20	~−0.61	D	Ayukai et al. (1998)
		~−0.49	~−0.006	D	Wattayakorn et al. (1990)
		−1.79	0.18	D	Sanchez-Carrillo et al. (2009)
		1.45	0.5	D	Boto and Wellington (1988)
		−0.53	−0.007	P	Sutula et al. (2003)
		−1.46	19.71	P	Sanchez-Carrillo et al. (2009)
	Nutrients exported as litter based on export estimations	−0.002	−0.0002	D	Wafar et al. (1997)
		−1.60		P	Boto and Bunt (1981)
		−0.02		P	Silva et al. (1998)
		−0.002	−0.0003	P	Wafar et al. (1997)
		−2.66	−0.28	P	Gong and Ong (1990)
Mangrove forest and adjacent creek	Water sampled from island enclosures, flumes, and in the forest–creek interface	−	−	D	Wösten et al. (2003)
		0.61		D	Rivera-Monroy et al. (1995a,b)
		−0.52		P	Rivera-Monroy et al. (1995a,b)
		+	+	D	Adame et al. (2010)

TABLE 21.7 Phosphorus and Nitrogen Exchange (gm^{-2} per year) and Chemical Form (Dissolved or Particulate) During Tidal Inundation for a Range of Mangrove Forests—cont'd

Spatial scale	Methodology	Net N flux ($g\ m^{-2}$ per year)	Net P flux ($g\ m^{-2}$ per year)	Nutrient Form	References
		+	+	D	Adame et al. (2010)
Mangrove forest	Indirect estimate of exchange based on primary production and consumption.	0.02	~−0.12	P + D	Davis et al. (2001a, b)
		50.3	1.86		Simpson et al. (1997)
		112.98[a]	13.47[a]		Wösten et al. (2003)
		0.66[a]			Davie (1984)

The scale considered in the study and methodology used in the field is included. *D*, dissolved; *P*, particulate.
[a]*Data obtained from Valiela and Cole (2002).*

6.3 Soil Nutrient Accumulation

Total nitrogen accumulation was reported to range from 1.34 g m^{-2} per year in the subtidal fringe in Rookery Bay to 11.01 g m^{-1} per year in the fringe forest at Terminos Lagoon (Table 21.3). There is no clear pattern among ecological types or geographical locations. The accumulation of nitrogen among the riverine and basin forests is similar at about 5.5 g m^{-2} per year, which is higher than the range of denitrification and nitrogen fixation described below. Thus, nitrogen storage in sediment or soils is an important fate of nitrogen in mangrove ecosystems. The accumulation of carbon (59–185 g m^{-2} per year) and nitrogen (1.55–5.80 g m^{-2} per year) is associated with deposition of organic matter, and relative rates are similar among sites (Table 21.3). Atomic C:N ratios of accumulated material at riverine sites is >30, whereas sites with less riverine input have C:N ratios <20. Accumulation of phosphorus ranged from 0.11 to 0.78 g m^{-2} per year and the higher rates occurred at sites with high inorganic matter loading. The riverine forest in Terminos Lagoon receives increased inputs of inorganic sediment and phosphorus from the Palizada River, whereas sites dominated by tides accumulate a greater proportion of organic matter and nitrogen. The elevated phosphate input rates into riverine mangrove sites are associated with higher levels of litter productivity compared with tidal mangroves that have less phosphorus input and lower productivity. TP accumulation rates for riverine sites are >0.7 g m^{-2} per year and range from 0.73 to 3.62 g m^{-2} per year (Table 21.3). All of the higher rates >1.0 g m^{-2} per year were at the riverine sites in Guayas River estuary and Belize, whereas the rates in the riverine sites at Terminos Lagoon were 0.82 and 0.73 g m^{-2} per year for the fringe and interior site, respectively.

Tropical storm disturbance can result in substantial pulses of material related to wind-driven storm surge and high rainfall-related runoff (Davis et al., 2004). The passage of

Hurricane Wilma across the FCE in 2005 had sediment and phosphorus inputs only along the mouth of Shark River estuary; in contrast, there was very little sediment input to mangrove sites along Taylor River Slough (eastern FCE, Castañeda et al., 2010). Total P inputs from storm-derived sediments were twice the average surface soil P density (0.19 mg cm^{-3}), and vertical accretion resulting from this hurricane event was 8–17 times greater than the annual accretion rate (0.30 ± 0.03 cm per year) averaged over the last 50 years. Thus, an actual hurricane event corroborated historical evidence of this fertilization mechanism in soil cores over the last 100 years. The presence of sediments in Shark River estuary from the shelf is a key process to describe the landscape pattern of mangrove forest productivity and nutrient biogeochemistry (Castañeda-Moya et al., 2010; Chen and Twilley, 1999a). However, sediment-rich sites at the mouth of Shark River estuary in the FCE are very different from mangroves found along the muddy coast of French Guiana (Fabre et al., 1999). Phosphorus concentrations in muddy coast mangroves are high and mostly in an organic form (52%–58%) in fringe and interior mangroves. Soil TP concentrations range from 0.638 to 0.804 mg g^{-1} dry mass in pioneer and mixed mangroves, respectively: higher than the concentrations in Shark River estuary (0.5–1.1 mg g^{-1} dry mass) and much higher than Taylor and Belize carbonate systems. Extractable iron concentrations vary from 10.3 to 15 mg g^{-1} dry mass: much higher than FCE mangroves.

The nutrient biogeochemistry of carbonate settings that are biogenic (Taylor Slough and sites in Twin Cays, Belize) compared with sites from a muddy coast demonstrates the complex patterns that are associated with the delta–estuarine–carbonate continuum of geomorphological settings. The iron and phosphorus associated with clastic sediment systems along deltaic coasts, together with high-energy mixing of soils in fringe versus interior forests, results in much higher levels of mangrove net productivity and enhanced forest structure. The interior sites of mangroves along high-energy deltaic coasts may have more anoxic conditions that can promote phosphorus solubility and reduce sulfide toxicity with formation of pyrites. This interaction of sediment source and redox zones on resource and regulator availability needs more clarity across geomorphological settings, and the zonation patterns from fringe to interior sites can be obscured in wet tropical climates, compared with dry climates that have strong seasonal cycles of water levels and thus redox zones (Gleason et al., 2003). These differences are highlighted in the conditions presented in this review, which promote mangrove ecological productivity and thus help define the links between geomorphological setting and mangrove ecosystem function. Thus, the hierarchy of ecogeomorphic settings in Fig. 21.2 and the mechanisms described in this chapter explain some of the diverse patterns in nutrient biogeochemistry in mangrove ecosystems. But more comparisons of sites are needed to build a more comprehensive model of biogeochemical processes.

7. GLOBAL CARBON ESTIMATES

7.1 Carbon Stocks

Few studies have attempted to develop global mangrove C budgets (see recent review by Twilley et al., 2017, and others such as Bouillon et al., 2008a; Kristensen et al., 2008a; Alongi, 2014). Only recently have models been developed to account for global variation in mangrove

SOC stocks (Jardine and Siikamäki, 2014; Atwood et al., 2017). Despite methodological peculiarities, these past studies converge in their approach to use latitudinal (or climatic) gradients or mean reference values to extrapolate global mangrove SOC budgets. Twilley et al. (submitted) provide an overview of concurrent global mangrove SOC estimates, focusing on how ecogeomorphology models using the coastal environmental setting framework (see Section 2) may improve global estimates, in contrast to the more popular latitude-based hypotheses largely believed to explain hemispheric variation in mangrove ecosystem properties such as global C sinks. The coastal environmental setting approach may be used as a framework to perform macroecological analysis of coastal ecosystem processes to improve explanations as to why there are such striking hemispheric variations (Ellison, 2002). Local and regional estimates of SOC linked to coastal environmental settings can render a more realistic spatial representation of global mangrove SOC stocks, as most of the C in mangrove ecosystems is stored as SOC, where it remains stable for much longer (i.e., compared to aboveground biomass). Additionally, the scarcity of data that represent the heterogeneity of coastal environmental conditions where mangroves can possibly thrive, particularly in the Indo-West-Pacific (IWP) region (Atwood et al. 2017), poses a commensurate challenge for mangrove ecologists to robustly model the macrosystems variability of SOC stocks in these coastal forested wetlands.

In contrast to previous studies (Jardine and Siikamäki, 2014; Atwood et al., 2017), the dataset used by Twilley et al. (submitted) only includes soil profiles that were at least 0.3 m. This is noteworthy considering that SOC results mostly from the decay of roots (Saintilan et al., 2013), which are mostly distributed within the top meter of the soil profile (Chen and Twilley, 1999b; Castañeda-Moya et al., 2013), and that near-surface SOC density is subjected to seasonal variability (Suárez-Abelenda et al., 2014). In addition, only mangrove SOC density values obtained from elemental analyses or chemical determination (i.e., wet oxidation) were included in the Twilley et al. (submitted) dataset. After applying these criteria, the dataset consisted of 107 studies, reporting on 551 sites from 43 countries. Each site in the dataset was then classified into distinct coastal environmental settings according to two independent coastal system classifications (Thom, 1982; Dürr et al., 2011). The global extrapolations provided by Twilley et al. (submitted) are based on (1) independent published mangrove SOC values, (2) a high-resolution global mangrove forest cover map (Hamilton and Casey, 2016), and (3) a worldwide classification of nearshore coastal systems (Dürr et al., 2011). To calculate the total mangrove area per coastline type, they cropped a mangrove forest cover map to the global typology of nearshore coastal systems (Dürr et al., 2011). Mangrove SOC stock estimates per coastline type were then obtained by multiplying the mangrove area determined for each type of coastline by the mean SOC stock (Mg ha^{-1}) value computed from all the sites located in that same coastline type.

By coupling mangrove forest area coverage with coastal systems typology, Twilley et al. (submitted) show that nearly half of the global mangrove area is located along estuarine coasts, followed by deltaic (small deltas and large rivers combined), lagoonal, karstic, and arheic coastal settings (Fig. 21.12a). Mangrove SOC density varies substantially across these coastal environmental settings, ranging from 14.9 ± 0.8 in muddy coasts (deltaic, river-dominated) soils to 53.9 ± 1.6 mg cm^{-3} (mean $\pm$ SE) in carbonate coasts to 60.1 ± 11.3 mg cm^{-3} in arheic coasts (Fig. 21.12b). The global mangrove SOC budget is 2.3 Pg C (Fig. 21.13). This analysis shows that although total mangrove SOC storage

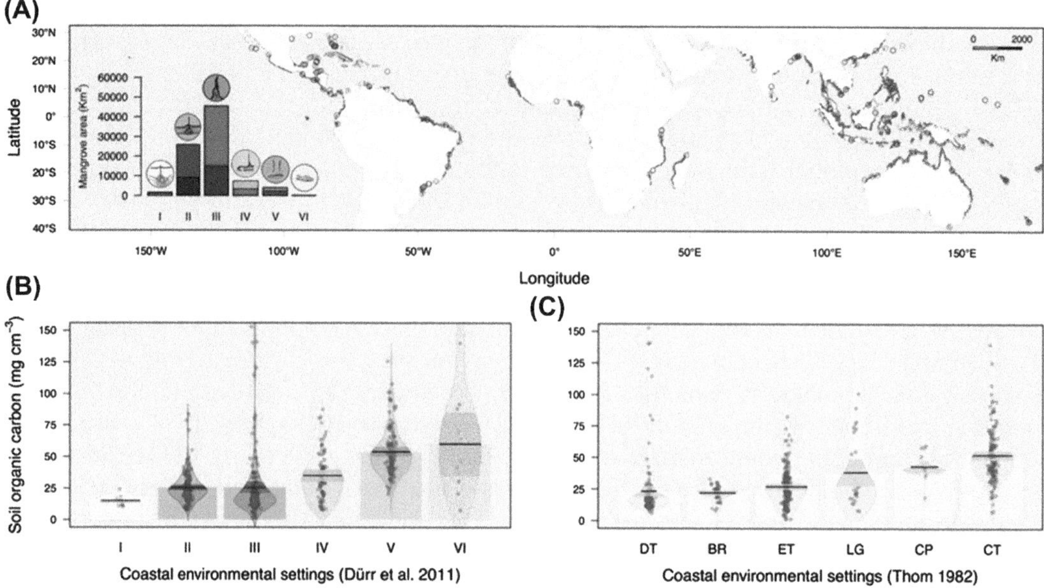

FIGURE 21.12 (A) Global mangrove forest cover (by area) distributed along distinct types of coastlines in the Atlantic East Pacific (hashed pattern in bar graph) and Indo West Pacific (solid color) regions. Blue circles represent mangrove soil organic carbon (SOC) density (mg cm^{-3}) values. (B) Mean mangrove SOC stocks increase from river-dominated (deltaic) to karstic coastal settings for both Dürr et al. (2011) and (C) Thom (1982) coastal environmental settings classification systems. The centered bars/lines and the bands display the mean and the 95% confidence intervals, respectively, while the beans and points represent the density and spread of the raw data. In b colors match the coastline classification shown in the map. Muddy coasts: I—deltas (large rivers), II—small deltas; estuarine coasts: III—tidal systems, IV—lagoons; carbonate and arheic coasts: V—karstic, VII—arheic. In **c** muddy coasts: DT—deltas; estuarine coasts: BR—bedrock, ET—estuaries, LG—lagoon, CP—composite; carbonate coasts CT—karstic. *Modified from Twilley et al., 2018.*

(i.e., in Pg) per coastline type increased with area, as expected, on a per unit area basis (i.e., Mg C ha^{-1}), these stocks are about twice as large in carbonate and arheic coastal settings compared with other coastline types. Despite mangroves in carbonate coastal settings accounting for less than 5% of the world's mangrove area, they have relatively high SOC content and represent almost 10% (213 Tg) of the global SOC stocks (Fig. 21.12a,b). Similarly, mangroves in arheic coastlines cover only about 0.2% of the global mangrove surface area, but the ratio of SOC stocks to mangrove area is above two, underscoring the potential of these coastal environmental settings as significant C stores despite the stunted physiognomy of trees. For other types of coastlines, the ratio of SOC to mangrove area decreases from 1.3 in lagoons to 0.9 for both estuaries (tidal systems) and small deltas to 0.6 in large rivers.

Twilley et al. (submitted; Fig. 21.12b,c) showed that mangrove SOC stocks increased from river-dominated to karstic coastal landforms, highlighting the role of geophysical controls (e.g., tides, river discharge) on the spatial variability of mangrove SOC stocks. This ecogeomorphology approach has not been taken into account on current global mangrove SOC stock estimates. The recent work by Twilley et al. (submitted) substantially improves recent global mangrove SOC stocks estimates by breaking down the within-country variability in

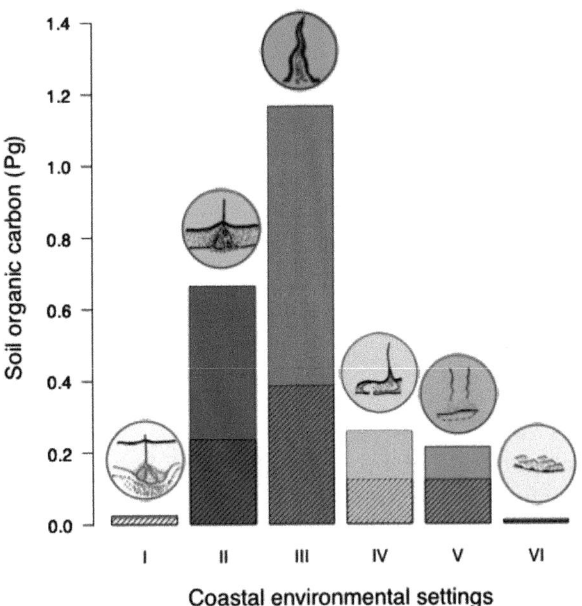

FIGURE 21.13 Global mangrove soil organic carbon stocks distributed along distinct types of coastlines in the Atlantic East Pacific (hashed pattern) and Indo-West-Pacific (solid color) regions. I—deltas (large rivers), II—small deltas, III—tidal systems, IV—lagoons, V—karstic, VII—arheic.

mangrove SOC stocks. In this respect, for extensive and geomorphologic diverse coastlines with abundant data such as Brazil (second largest mangrove area in the world), Australia (fifth), and Mexico (sixth), the ecogeomorphology model allows for significant improvements in contemporaneous SOC stock assessments. At the same time, for equally diverse but underrepresented coastlines—in terms of data availability—such as west Africa and the IWP region, the coastal environmental setting framework provides a powerful approach to address the macrosystem variability in mangrove SOC.

In contrast to early investigations, the analysis by Twilley et al. (submitted) highlights the role of carbonate settings as global blue carbon hotspots, particularly in the neotropics, which shelter 57% of the tropical karstic coastlines (Dürr et al., 2011). Unfortunately, the reduced occurrence of karstic coastlines in the Old World tropics (c. 25% of global karstic coastlines), in addition to the general poor quality of mangrove SOC data in this region (see Atwood et al., 2017), prevents one from delivering robust comparisons regarding mangrove SOC stocks between karstic settings from the neotropical and Old World tropical regions. In fact, this bias both in terms of spatial representativeness and data quality raises the concerns regarding the accuracy of the recent global extrapolations presented by Atwood et al. (2017). Their global estimates are based on country-level mean mangrove SOC stocks from 48 countries, out of a total of 105 nations mangroves can be found. For the remaining 57 countries that lack data on mangrove SOC stocks, a mean reference value of 283 $Mg\ C\ ha^{-1}$ was used to substitute missing data. Twilley et al. (submitted) showed that for those countries, many of which comprise mostly karstic coastal environmental settings, this global mean reference value advocated by Atwood et al. (2017) represents on average c. 50% lower than mangrove SOC stocks values reported in the literature (Fig. 21.12b,c).

The earlier work by Jardine and Siikamäki (2014) presented the variability in mangrove SOC stocks both within and across countries in more detail. However, there are considerable differences between their estimates and values reported in the literature. Their climate-based model projected considerably higher SOC density values (27–70 mg C cm^{-3}) for mangroves located in river-dominated coastlines (e.g., Louisiana, Amazon, Sundarbans) compared with much lower ranges reported in the literature (11–24 mg C cm^{-3}; Twilley et al. submitted). Their model also predicted lower mangrove SOC density values (32–70 mg C cm^{-3}) for mangroves in many karstic environments (i.e., South Florida, Bahamas, Malaysia, Papua), differing from published (21–126 mg C cm^{-3}) and unpublished (47–58 mg C cm^{-3}) values (Twilley et al. submitted). The coastal environmental setting framework has the potential to resolve these unexpected patterns observed between karstic and river-dominated coastal landforms identified in former global mangrove SOC budgets.

Some of the limitations on contemporary global SOC budgets represent a significant impact for blue carbon initiatives in countries with extensive and diverse coastlines, affecting the feasibility of national-level policy-making. Also, for several smaller countries that are almost entirely, if not entirely, dominated by karstic coastal landforms, the gaps discussed here compromise the validity of their national greenhouse gas inventories and underestimate their role as blue carbon hotspots. Here, we show how many conceptual and methodological caveats can be addressed by including ecogeomorphological key concepts as part of macroecological analyses. The use of coastal ecogeomorphology models improves our capacity to map the spatial variability of mangrove SOC stock estimates at global scale. The trends in mangrove SOC stocks among different coastline types using a spatially explicit ecogeomorphology model were consistent with a refined (i.e., site-specific) coastal environmental settings classification. In both classification systems, higher SOC stocks were found in karstic coastal settings and lower values in river-dominated coastlines. This new ecogeomorphology-based perspective should advance the macroecology of mangroves beyond the limitations inherent to the classic latitude-based paradigm.

7.2 Reassessment of Inorganic Carbon Exchange With Estuaries

Bouillon et al. (2008b) estimate that more than 50% of the carbon fixed by mangrove vegetation, estimated ~217 ± 72 Tg C per year, appears to be unaccounted for based on estimates of various carbon sinks (DIC export and mineralization; Table 21.8). This missing carbon sink is conservatively estimated at ~112 ± 85 Tg C per year, equivalent in magnitude to ~30%–40% of the global riverine organic carbon input to the coastal zone. This missing carbon is about 700 g C m^{-2} per year (based on global mangrove area of 150,000 km^2), which is about three times the amount thought to be buried annually in mangrove soils as a carbon sink (see previous section). However, estimates of NEP to balance carbon accounting in mangrove wetlands are very complex, especially when measuring the exchange of inorganic and organic forms of carbon with the atmosphere and across mangrove–estuarine boundaries (Fig. 21.8). Of the total NPP estimated at 204 Tg C per year (Bouillon et al., 2008b; 290 Tg C per year by Alongi, 2009, Table 21.8), it is well documented that litter fall (NPP_L)

TABLE 21.8 Global Mass Balance Estimates of Carbon Flow in Mangrove Ecosystems

	Bouillon et al. (2008a,b) (Tg C per year)	**Alongi (2009) (Tg C per year)**	**Bouillon et al. (2008a,b) ($g\ C\ m^{-2}$ per year)**	**Alongi (2009) ($g\ C\ m^{-2}$ per year)**
Inputs				
GPP	NA	690 ± 264		4313
NPP	204 ± 68	290 ± 107	1275	1813
Wood	63 ± 40	63 ± 42	394	394
Litter	64 ± 20	64 ± 20	400	400
Roots	77 ± 56	163	481	1019
Outputs	82	508	512	3175
POC Export	20 ± 22	27 ± 25	125	169
DOC Export	23 ± 21	13 ± 12	144	81
Tree respiration	NA	396 ± 151		2475
Soil + water respiration	39	72 ± 50	244	450
NEP (inputs – outputs)	122	182	763	1138
Soil burial	17 ± 20	27 ± 20	106	169
NEP (soil burial + wood production)	80	90	500	563
Missing DIC (NEP – soil burial)	105	155	656	969
Missing DIC (NPP – [NEP + outputs])	42	92	263	575

The values in Tg C per year are organized around estimates of net ecosystem production scaled to common total global area of 150,000 km^2. Estimates are also given on a per meter square basis of mangrove wetland area in intertidal zone ($g\ C\ m^{-2}$ per year). Values are ±1 SD. *DIC*, dissolved inorganic carbon; *DOC*, dissolved organic carbon; *POC*, particulate organic carbon.
Modified from Bouillon et al. (2010); Alongi, D.M., 2009. The Energetics of Mangrove Forests. Springer Science & Business Media.

does not accumulate in mangroves (typically decaying within 1 year), and soil carbon burial (ΔS_{org}) accounts for net belowground production (NPP_B) minus the loss from sediment respiration. Another way to estimate global NEP is to sum wood production and soil carbon sequestration ($NPP_W + \Delta S_{org}$, Eq. 21.6). Based on global reviews, wood production is about 600 g dry mass m^{-2} per year or about 240 g C m^{-2} per year. The average carbon sequestration in mangrove soils is about 163 g C m^{-2} per year (Breithaupt et al., 2012). The sum of these two measures, $NPP_W + \Delta S_{org}$, is an estimate of NEP at about 400 g C m^{-2} per year or about 84 Tg C per year (see Twilley et al., 2017). Another way to calculate the missing carbon is to account for NPP of 204 Tg C per year (290 Tg C per year from Alongi, 2009) by subtracting the NEP of 80 Tg C per year (90 Tg C per year from Alongi, 2009), the POC and DOC export of 43 Tg C per year (40 Tg C per year from Alongi, 2009), and assume that the soil and water respiration is Rh (heterotrophic respiration of the ecosystem) that is not accounted for in NEP, which is 39 Tg C per year (72 Tg C per year from Alongi, 2009). Based on these assumptions, the missing carbon is 42 Tg C per year (88 Tg C per year from Alongi, 2009), which is about 263 g C m^{-2} per year (550 g C m^{-2} per year from Alongi, 2009) (Table 21.8). Based on total organic export of about 250 g C m^{-2} per year (DOC + POC), DIC export would have to be 250–500 g C m^{-2} per year from mangrove forests to balance NEP estimates (Table 21.8). These fluxes of DIC are high compared with other coastal wetlands (Cai, 2011; Hopkinson et al., 2012) and may only be represented of systems in wet tropical environments where substantial DOC export rates have been observed (Dittmar et al., 2006).

The analysis above suggests that inorganic carbon flux from sediments and mangrove waters is severely underestimated and that DIC export from mangrove wetlands to adjacent waters must be nearly equal or double those of TOC, based on estimates by Bouillon et al. (2008b) or by Alongi (2009) (with refinements in estimates of NEP suggested above). The magnitude of this flux (sometimes referred to as the missing carbon sink) may vary among different mangrove systems (e.g., from muddy coasts to carbonate settings). Recent investigations have shown clearly that mangrove tidal creeks emit large amounts of CO_2 to the atmosphere (Table 21.9) (Barnes et al., 2006; Borges et al., 2003; Bouillon et al., 2003, 2007a,b,c; Hemminga et al., 1994; Koné and Borges, 2008; Kristensen et al., 2008b; Linto et al., 2014; Middelburg et al., 1996; Miyajima et al., 2009). A comparison of DIC and DOC concentrations has suggested that the DIC export from mangroves might be 3–10 times larger than the DOC export (Bouillon et al., 2008b; Miyajima et al., 2009). It remains uncertain as to how much of total ecosystem heterotrophic respiration is exported as DIC as opposed to released directly to the atmosphere as CO_2 (Fig. 21.8). CO_2 supersaturated within tidal waters may not be exchanged with the atmosphere while the tidal water remains in the mangrove wetland and thus not measured as Rh in sediment CO_2 efflux rates. The same holds for anaerobic respiration and CO_2 production that does not escape to the atmosphere. These two mechanisms result in transport of DIC in tidal waters as part of NTE_M to tidal creeks that enrich the DIC in estuarine waters. In either case, this DIC is a product of ecosystem respiration and should be incorporated into NEP estimates for mangrove wetlands (Fig. 21.8).

During ebb tide and at low tide, there is a strong influx of pore water from the mangrove intertidal platform that mixes with creek water, substantially affecting the chemical properties of the latter. This drainage of mangrove pore waters continues during low tide but decreases during flood high tide and stops when the mangrove is inundated at slack flood tide (Ovalle et al., 1990; Middelburg et al., 1996). The high concentrations of HCO_3 are a result of all anaerobic respiration pathways, including sulfate, nitrate, and iron reduction (Alongi et al., 1999; Kristensen et al., 2000; Ovalle et al., 1990). In some cases, aerobic degradation predominates (Alongi et al., 2001), demonstrating the importance of the redox zones discussed in the geochemical model above to the fate of DIC in the pore water exchanged during tidal inundation. The issues associated with estimating the flux of DIC during tidal inundation are somewhat similar to the methodological issues described for DOC and POC above. First, there must be frequent measures of DIC concentrations throughout a tidal cycle, and the flux of water between the tidal creek and the intertidal platform of mangrove wetlands has to be quantified. As this is a bidirectional flux, these measures must account for fluxes during both

TABLE 21.9 CH_4 and CO_2 Concentrations and Water to Air Fluxes in Mangrove Systems

Location	Country	CH_4 (nM)	CH_4 Flux (mmol C $m^{-2}\,d^{-1}$)	pCO2 (atm)	CO_2 Flux (mmol C $m^{-2}\,d^{-1}$)	References
Kalighat (dry season)	Andaman Is.	154–603	0.17–0.38*	1704–3392	22.7–49.3*	Linto et al. (2014)
Kalighat (wet season)	Andaman Is.	30–577	0.14–0.47*	2970–6140	51.9–172.8*	Linto et al. (2014)
Wright Myo (dry season)	Andaman Is.	146–585	0.12–0.29*	1246–6004	30.2–59.5*	Linto et al. (2014)
Wright Myo (wet season)	Andaman Is.	147–583	0.11–0.44*	1500–5490	38.7–113.8*	Linto et al. (2014)
Norman's Pond	Bahamas			385–750	13.8	Borges et al. (2003)
Itacuruca Creek	Brazil			660–7700	113.5	Borges et al. (2003)
Adyar	India				17.8	Ramesh et al. (2007)
Gaderu Creek	India			1380–4770	56.0	Borges et al. (2003)
Godavari	India				21.9–70.2	Bouillon et al. (2003)
Kakinada Bay	India				8.3	Bouillon et al. (2003)
Mooringanga Creek	India			800–1530	23.2	Ghosh et al. (1987)
Muthupet	India		0.04		31.8	Ramesh et al. (2007)
Pichavaram	India				6.1	Ramesh et al. (2007)
Saptamukhi Creek	India			1080–4000	56.7	Ghosh et al. (1987)
Sundarbans	India	11–129	0.02–0.4		0.4	Biswas et al. (2004, 2007)
Kidogoweni (Gazi)	Kenya			1480–6435	71.0	Bouillon et al. (2007c)
Kinondo Creek (Gazi)	Kenya				52.0	Bouillon et al. (2007c)
Tana River Delta	Kenya			490–10,035	58–131.2	Bouillon et al. (2007b)
Betsiboka Estuary	Madagascar				9.1	Ralison et al. (2008)
Nagada Creek	Papua New Guinea			540–1680	43.6	Borges et al. (2003)
Mtoni	Tanzania	105–190	0.07–0.35			Kristensen et al. (2008b)
Ras Dege	Tanzania	13–142	0.01–0.07			Kristensen et al. (2008b)
Ras Dege	Tanzania			430–5050	34.2	Bouillon et al. (2007a)
Florida Bay	USA			1533–6016	4.6	Millero et al. (2001)
Shark River, Florida	USA				20–302*	Ho et al. (2014)
Shark River, Florida	USA			975–5273	32–256*	Ho et al. (2014)

(Continued)

TABLE 21.9 CH_4 and CO_2 Concentrations and Water to Air Fluxes in Mangrove Systems—cont'd

Location	Country	CH_4 (nM)	CH_4 Flux (mmol C m^{-2} d^{-1})	pCO2 (atm)	CO_2 Flux (mmol C m^{-2} d^{-1})	References
Shark River, Florida	USA			920–2910	43.8	Koné and Borges (2008)
Tam Giang (dry season)	Vietnam			770–11,480	141.5	Koné and Borges (2008)
Tam Giang (wet season)	Vietnam			1210–7150	128.5	Koné and Borges (2008)
Ca Mau mangrove creeks	Vietnam				27.1	Koné and Borges (2008)
Ca Mau mangrove creeks	Vietnam				81.3	Koné and Borges (2008)
Kiến Va'ng (dry season)	Vietnam			705–4605	32.2	Koné and Borges (2008)
Kiến Va'ng (wet season)	Vietnam			1435–8140	154.7	Koné and Borges (2008)
Creek mouth—Moreton Bay	Australia	1.75–202	0.01–0.15*	409–9505	9.4–114.9*	Call et al., 2015
Upstream—Moreton Bay	Australia	1.82–889	0.05–0.63*	385–26,106	47–629.23*	Call et al., 2015
Sundarbans mangrove biosphere	India		0.41 ± 0.11			Dutta et al., 2013
Naples Bay	USA		0.02 to 0.14			Cabezas et al., 2017
Everglades	USA		0.62 to 6.82			Bartlett et al., 1989

Flux rates marked with * are from studies that used a range of empirical k models.

Table amended from Bouillon et al. (2008a,b); Chen and Borges (2009); Borges and Abril (2011); Linto et al. (2014). From Call et al. (2015).

flood and ebb tides. A complicating factor is that many DIC flux estimates also depend on assumptions associated with gas transfer velocities, assuming either constant estimates based on floating domes or using wind speed/gas exchange parameterizations (Borges et al., 2003, 2004, 2005; Ho et al., 2014). As for issues associated with changes in concentration with environmental gradients during a tidal exchange or seasonally, assuming a constant gas transfer velocity a priori has obvious drawbacks that might influence gas exchange during flux studies (Ho et al., 2014).

Borges et al. (2003) summarized all available pCO_2 (partial pressure of CO_2) data from mangrove-surrounded waters (including sites in India, Brazil, Bahamas, Papua New Guinea, and Florida) and calculated CO_2 fluxes that range from 20.1 to 544.8 g C m^{-2} per year, with an average for all those sites of 240 g C m^{-2} per year, which is within our range suggested for the missing carbon described above (250–500 g C m^{-2} per year) (see Table 21.9). Estimates for mangrove creeks in Vietnam show that the air-water CO_2 fluxes were five times higher during the rainy season than during the dry season in one site, whereas the air-water CO_2

fluxes were similar during both seasons at another location (Koné and Borges, 2008). The air-water CO_2 fluxes ranged from 130 g C m^{-2} per year to 620 g C m^2 per year during the dry season and from 356 g C m^{-2} per year to 678 g C m^{-2} per year during the rainy season. These values are within the range of values previously reported in other mangrove creeks and confirm that the efflux of CO_2 from waters surrounding mangrove forests is within the range of missing carbon values to balance NEP of mangroves globally. There is some consensus in the literature that an overall value of 219 g C m^2 per year given by Borges et al. (2003) was a reasonable first-order general estimate of DIC exchange from mangrove wetlands (Koné and Borges, 2008).

Crab burrows can greatly enhance the surface area of the sediment–air or sediment–water interface where exchange of CO_2 or DIC can take place, serving as significant conduits for enhancing CO_2 exchange between mangrove sediments and atmosphere (Kristensen, 2008). The process of tidal exchange enhanced by bulk hydraulic permeability associated with burrows built by animals such as crabs and other crustaceans is known as tidal pumping of DIC (Fig. 21.14). This process is also well established in mangrove creeks (Borges et al., 2003; Kristensen et al., 2008b; Maher et al., 2013; Stieglitz et al., 2013; Zablocki et al., 2011) and has been shown to dominate resulting pCO_2 distributions and CO_2 emissions to air (Borges et al., 2003; Bouillon et al., 2007c; Koné and Borges, 2008; Linto et al., 2014). The impact of these burrows, and possibly mangrove root formations, on the biochemical pathways of anaerobic and aerobic processes is defined by the relative distribution of lower redox zones described in the geochemical model above. What is critical is to define the relative volumes of water via subsurface advective exchange with tidal pumping that transports DIC produced within these redox zones (Maher et al., 2013).

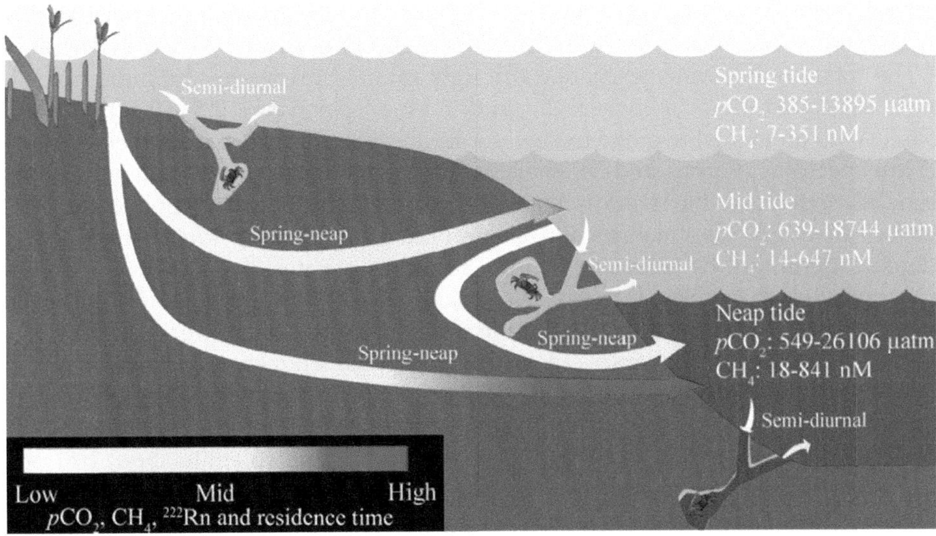

FIGURE 21.14 The "lunar mangrove pump," a model of the two-sediment pore water–surface water exchange mechanisms that operate over different time scales in a mangrove creek. The flushing of crab burrows occurs semidiurnally. At spring–neap time scales, infiltration of mangrove sediments by spring-tidal water seeps back into creek waters as tidal amplitude decreases due to decreasing hydrostatic pressure. It is hypothesized that the increased residence time in mangrove sediments results in elevated surface water pCO_2 and CH_4. *Arrows* represent the major flow pathways and the corresponding enrichment of ^{222}Rn, pCO_2, CH_4, and residence time. *From Call et al. (2015).*

The recent emphasis on DIC exchange associated with the tidal pump in mangrove wetlands has had limited impact on our understanding of the total carbon exchange with mangroves. There are few studies that have compared the additional exchange of DIC with the organic components of carbon exchange by comparing with DOC and POC export. In the subtropical southern Moreton Bay, on the east coast of Australia, a series of studies on C, N, and P exchange among habitats has carefully compared the contributions of mangroves with the estuary (Eyre et al., 2011; Maher et al., 2013). DIC and DOC were exported from the mangroves, with an average rate of 1095 g C m^{-2} per year for DIC compared with lower rates of 110 g C m^{-2} per year for DOC. POC was actually imported to the mangroves, mainly from seagrass wrack, with the largest import rate occurring during the summer large tide (726 g C m^{-2} per year). DIC dominated export of carbon, with ratios of DOC:POC in export varying from 8 to 15. The significance of tidal pumping was estimated for this mangrove wetland where groundwater accounted for 89%–92% of the DOC export and 93%–99% of the DIC export (Maher et al., 2013). Total fluxes of gaseous carbon for this system were estimated by integrating the creek mouth and upstream data based on gas transfer models that include the contribution of bottom-generated turbulence, generating a range of 429–538 g C m^{-2} per year, scaled to the entire catchment. This is similar to the globally averaged mangrove missing carbon of 250–500 g C m^{-2} per year described above. Maher et al. (2013) suggested that DIC export may constitute the missing carbon sink based on export rates of 183–341 g C m^{-2} per year (Call et al., 2015). This study suggests that current estimates of CO_2 efflux from mangrove creek waters in Moreton Bay are within the range of 250 g C m^{-2} per year described above for missing carbon of NEP.

One of the more comprehensive budgets of this tidal pumping process was calibrated from mangroves in the central Barrier Reef and extrapolated to the Australian wet tropics assuming a daily water flux value of 30.4 ± 4.7 L m^{-2} in the mangrove intertidal platform, a total forest area of 390 km^2, and an inundation-effective area of 252 km^2 (Stieglitz et al., 2013). The total annual water exchange through animal burrows in the mangroves of this region is estimated to be between 2.3 and 3.3 km^3 per year (2.8 ± 0.5 km^3 per year). Compared with an annual average freshwater influx (I_T in Fig. 21.8) from the region's rivers of 14.5 km^3 per year (Furnas and Mitchell, 2001), the mangrove pump thus may constitute between 16% and 22% of the river discharge on a coastal scale of many hundreds of kilometer (Stieglitz et al., 2013). The mangrove tidal pump across the entire Wet Tropics coastline is considered a significant hydrological process supporting water fluxes and associated solute fluxes, especially when the enhanced DIC concentration of burrow water is considered. This signal of DIC in tidal creek and estuarine waters may be even more significant in coastal environmental settings that have less river input, where this tidal pump could exceed annual river flows such as lagoons and carbonate environments. In these cases, the NTE_M source of DIC from mangroves may be much more significant than I_T (Fig. 21.8).

Organic carbon (OC) export from mangrove wetlands is equivalent to about 10% of terrestrial OC export to the coastal zone via rivers (Dittmar et al., 2006; Jennerjahn and Ittekkot, 2002). OC burial in mangrove sediments is comparable with that of salt marshes and seagrass meadows (Duarte et al., 2005). The fate of OC exported to the global ocean is a critical (Fig. 21.8) component of the ocean carbon budget (Twilley et al., 1992; Bouillon et al., 2008b; Alongi and Mukhopadhyay, 2015; Bauer et al., 2013). This may be particularly evident in river-dominated mangrove systems such as muddy coasts and deltas where organic material exchanged is greater at the boundary of mangrove wetlands (NTE_M) compared with other coastal settings (Twilley, 1985), as observed for major river systems such as Fly River, Papua New Guinea, or systems with large tidal exchanges in Australia (Alongi, 2009). The location of major mangrove regions globally in the wet tropics where most of the rainfall occurs is evident in Fig. 21.15. This may explain the strong coupling of mangrove carbon in coastal seas in the wet tropical regions (Alongi and Mukhopadhyay, 2015).

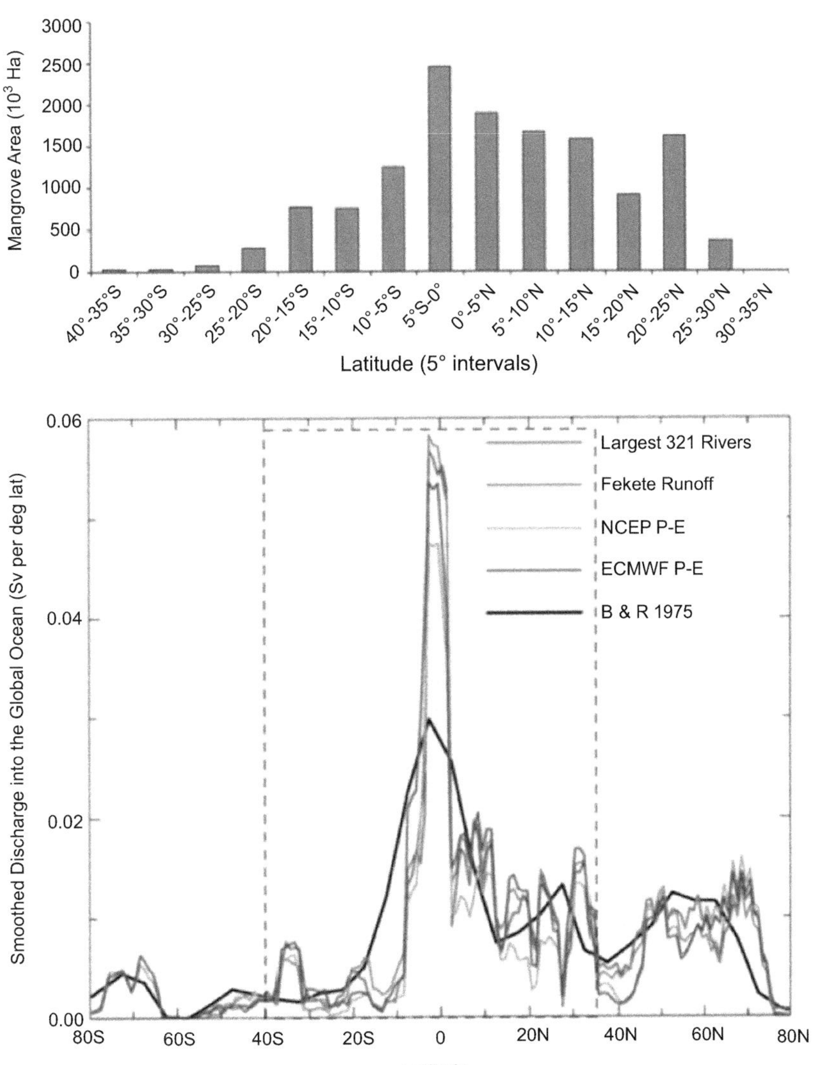

FIGURE 21.15 Upper panel shows the distribution of mangrove area estimated by latitude (Giri et al., 2011). Lower panel shows the annual river discharge into the global ocean by latitude based on several models (Dai and Trenberth, 2002). The *dotted box* on the river discharge pattern covers the latitudinal distribution of mangrove cover as defined by the upper panel.

Recent reviews of carbon dynamics in these coupled wetland/estuarine ecosystems describe organic matter from wetlands as contributing to the heterotrophic nature of estuaries and nearshore waters (Cai, 2011; Hopkinson et al., 2012) (Fig. 21.8). The assumption is that organic carbon exported from mangrove wetlands may contribute to the NEP of estuaries, depending on the extent to which heterotrophic respiration is stimulated by the allochthonous inputs. River DIC and TOC can also contribute to these two pathways of NEP in an estuary, again depending on how much Rh is stimulated by additional organics in the estuary. The degassing of CO_2 that has been observed in estuarine waters may result from release of DIC, which is the heterotrophic respiration of mangrove organic carbon that occurred on the mangrove platform, or TOC, which is heterotrophic respiration of mangrove organic carbon in the estuary. NEP of mangrove wetlands is based on the balance of these processes as they occur within the intertidal platform, thus defining this area of the coastal zone as a carbon sink. Whether TOC in the NTE_M becomes part of the sink in the estuary depends on the fate of this carbon as ecosystem respiration or NTE_O (Fig. 21.8). This is still a major question regarding the carbon budget of coastal ecosystems, but there have been some very interesting discoveries in the last decade.

Using estimates of NEE from flux towers, biometric estimates of NEP, and net tidal exchange measurements of organic carbon, preliminary estimate of F_{DIC} for mangroves at Shark River estuary was about 360 g C m^{-2} per year (Ho et al., 2014), within the range from 250 to 500 g C m^{-2} per year described above. An upper bound estimate of the aquatic C flux (including DIC, DOC, and POC) from this tidal forest is 784 g C m^{-2} per year (Troxler et al., 2013, 2015). As the area of mangroves around Shark River is c. 111 km^2, the maximum total carbon export from the forest is estimated to be 7.3×10^9 mol per year. The area of water is c. 17.5 km^2, so air-water CO_2 flux is $1.1-1.5 \times 10^9$ mol per year, representing 15%–20% of the estimated aquatic C sink. DOC flux is c.3×10^8 mol per year or about 5% of the aquatic C sink. Because mangrove-derived POC is of the same magnitude as DOC (Twilley, 1985), the organic C flux represents c. 10% of the aquatic C sink. The remainder of the aquatic C flux, between 70% and 75% of the total, is presumably lost as DIC via lateral transport to the coastal ocean. The calculation here represents a rough estimate as some of the DIC could be derived from remineralization or photodegradation of DOC that is not mangrove derived (Ho et al., 2014).

8. DISTURBANCE MODEL

The recent estimate of mangrove global extent (81,849 km^2; Hamilton and Casey, 2016) is only 55% of the value, 150,000 km^2, commonly used to extrapolate the global significance of mangrove biogeochemistry. The variation in global estimates of mangrove area (Table 21.1) has caused confusion about the global rate of mangrove loss, with some overestimates reaching several percentages per year and over half of total mangrove cover lost in the last several decades. More realistic estimates of mangrove loss globally have been based on an understanding of methodological challenges and with caution as to how such estimates are calculated (Friess and Webb, 2014). The new techniques developed by Hamilton and Casey (2016) estimate that the mangrove forests that existed in 2000 have decreased by 1646 km^2 between 2000 and 2012. This corresponds to a total loss over the analysis period of 137 km^2 from the year 2000 baseline or a linear global loss of 0.16% per year during the period analyzed. The authors contend that this consistent trend with little deviation allows future trends to be reliably extrapolated from the dataset with a high amount of certainty.

The significance of nonanthropogenic and anthropogenic disturbances on ecosystems associated with land use, and pulses such as floods, fires, and cyclones have to be assessed when estimating how the biogeochemistry of mangroves may mitigate global nutrient cycles, particularly management strategies stemming from increased atmospheric carbon. The NEP of wetlands changes with ecosystem development following disturbance and can be modified by anthropogenic effects such as drainage of organic soils (Armentano and Menges, 1986; Twilley et al., 2017). Such changes in NEP will alter the extent to which mangroves are sources or sinks of atmospheric carbon and nitrogen (Fig. 21.16; Twilley et al., 2017). For carbon, calculating these changes requires estimates of how shifts in the magnitude of sinks and sources in NEP may contribute to the mass balance of carbon in the atmosphere (Houghton, 1994, 2007). It is not certain if the net flux of carbon in coastal ecosystems such as mangrove wetlands may contribute to the unidentified sink of carbon that cannot be accounted for based on actual carbon increases over the last century (Houghton, 2007). There has been a focus on whether disturbances to terrestrial ecosystems

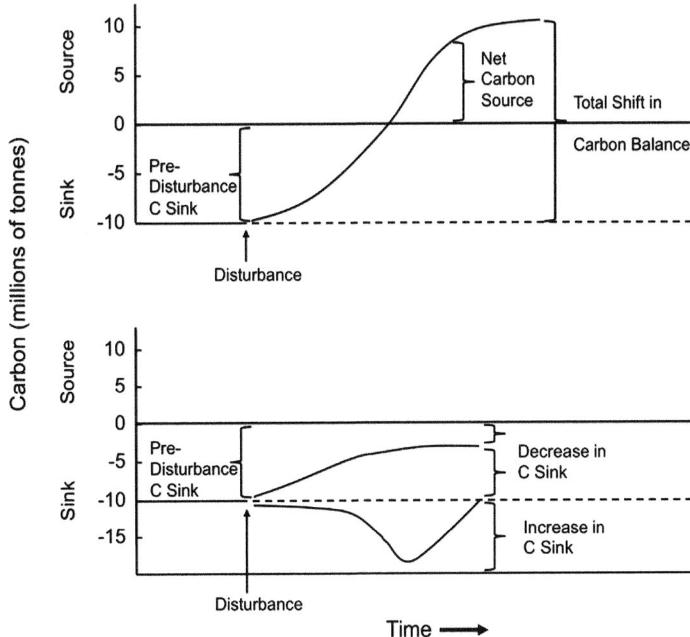

FIGURE 21.16 Definitions of carbon sources and sinks as exchanged with the atmosphere using two hypothetical wetland regions subject to disturbance. Top panel: the original net carbon sink is totally lost by the changes in ecosystem processes in response to disturbance, ultimately to a condition where the wetland is a net carbon source to the atmosphere. Bottom panel: the wetland currently functions as a diminished carbon sink because CO_2 release in disturbed wetlands is lower than net CO_2 fixation in undisturbed wetlands; with a modification to demonstrate that carbon sinks can actually increase in some scenarios following a disturbance by increasing net ecosystem production. *After Armentano and Menges (1986).*

may enhance carbon sinks or sources of NEP, but no global analysis has been done to account for role of tropical or subtropical coastal zones to this unidentified atmospheric carbon (Fig. 21.16).

Recovery of the mangrove wetlands following hurricanes shows how metabolism changes following disturbance. Within 5 years of disturbance, NPP and NEP were stimulated, including both enhanced litter fall and root production (Castañeda-Moya, 2010; Danielson et al., 2017). Root carbon storage after hurricane Wilma in FCE was 3.3 times higher compared with values reported before Wilma (Castañeda-Moya et al., 2011), with fine (<2 mm diameter) roots accounted for 51%–85% of the total carbon storage. These results demonstrate the resiliency of mangrove wetlands to recover to predisturbance NPP rates within a short (<5 years) period as a result of disturbances that caused significant changes in mangrove community structure and function. Similar observations have been made with carbon budgets in mangrove plantations following disturbance (Bosire et al., 2006; Fontalvo-Herazo et al., 2011; Jin-Eong et al., 1995; Kairo et al., 2008, 2009; Ren et al., 2008).

These estimates associated with tropical cyclones are very important to any global carbon exchange estimate of mangroves given the frequency of this disturbance in the subtropics. Mangrove wetlands are known as frequently disturbed ecosystems, particularly associated with tropical cyclones, and this forcing may have a significant impact on NEP mitigating carbon exchange with the atmosphere. This is similar to the debate for land use of forest ecosystems that may explain the unidentified carbon sink in the biosphere (Houghton, 1994). In addition, climate change is expanding mangroves into the warm temperate (Osland et al., 2012; Saintilan et al., 2009, 2014), potentially expanding the effects of mangrove NEP on mitigating atmospheric carbon. Finally, the continued contribution of mangrove wetlands to carbon sinks of the coastal zone is enhanced by accelerated rates of sea level rise, which can be considered an anthropogenic disturbance to coastal zone. Woodroffe (2002) summarized how the organic carbon storage (peat formation) of salt marshes changes with different responses of morphodynamics to sea level rise scenarios. If coastal wetlands accrete with sea level rise (and subsidence), then the NEP of wetlands is nonsteady state and there is constant annual net carbon sequestration. Accelerated sea level rise will potentially accelerate carbon sequestration as soil accretion also accelerates. Yet the latter scenario of accelerated rates depends on adequate source of sediment that can enhance sedimentation along with the contribution of NPP_B to soil accretion (Kirwan and Temmerman, 2009; Kirwan and Guntenspergen, 2012; Kirwan and Mudd, 2012). There are critical rates of relative sea level rise (eustatic sea level plus subsidence rates, RSLR) above which the drowning of the coastal wetlands will occur. Thus, ecogeomorphic types of coastal settings from carbonate lagoons to estuaries and river-dominated deltas will determine the capacity of coastal wetlands to adapt to increase RSLR in the next century and continue to sequester carbon in mangrove soils. This has important implications as to how mangrove wetlands may be able to mitigate an anthropogenic increase of CO_2 in the atmosphere with increased carbon sequestration with increased RSLR.

Geographic trends of disturbance in mangrove ecosystems are germane to understanding the potential feedback of mangroves in mitigating carbon dynamics with the atmosphere. An extensive review of global changes in mangrove area concluded that estimates exhibited high variability depending on which data points in each respective review were modeled (Friess and Webb, 2014). Myanmar appears to represent the current hotspot for mangrove deforestation, with a deforestation rate more than four times higher than the global average (Hamilton and Casey, 2016). Yet Indonesia has by far the largest area of loss at 749 km^2 from 2000 to 2012 (3.11% mangrove loss) and constitutes almost half of all global mangrove deforestation. Southeast Asia has experienced relatively high amounts of loss, which is concerning given that this region contains some of the most biodiverse mangrove forests and almost half (33.8%, Thomas et al., 2017) of the global mangrove area. A recent study shows that globally (1996–2010) the primary driver of anthropogenic mangrove loss is attributed to the conversion of mangroves to aquaculture/agriculture uses, with the greatest proportion of loss (50%) observed in Southeast Asia (Thomas et al., 2017). Other countries outside Southeast Asia have sustained significant mangrove losses from 2000 to 2012, including India and Guatemala. Within the Americas, Africa, and Australia, the deforestation of mangrove is approaching zero, with nominal rates in many countries (Hamilton and Casey, 2016).

There is a great need to improve estimates of mangrove areal extent to calculate their potential to mitigate atmospheric CO_2 levels. Many of the modeling efforts to improve global estimates of carbon stocks and NEP are really important to improve these types of assessments. Recently advances in a wealth of remote sensing studies assessing mangrove wetlands spatial distribution (see review in Kuenzer et al., 2011) have improved regional and local scale estimates of global carbon storage (e.g., Simard et al., 2006, 2008; Fatoyinbo et al., 2008; Fatoyinbo and Simard, 2013).

Acknowledgments

The authors were supported by the National Science Foundation under Grant No. DBI-0620409 and Grant No. DEB-9910514 (Florida Coastal Everglades, Long-Term Ecological Research). The development of ecogeomorphology concept was supported by the STC program of the National Science Foundation via the National Center for Earth-surface Dynamics under the agreement Number EAR-0120914. Any opinions, findings, conclusions, or recommendations expressed in the material are those of the authors and do not necessarily reflect the views of the National Science Foundation.

References

Adame, M.F., Lovelock, C.E., 2011. Carbon and nutrient exchange of mangrove forests with the coastal ocean. Hydrobiologia 663, 23–50.

Adame, M.F., Neil, D., Wright, S.F., Lovelock, C.E., 2010. Sedimentation within and among mangrove forests along a gradient of geomorphological settings. Estuarine, Coastal and Shelf Science 86, 21–30.

Algar, C.K., Vallino, J.J., 2014. Predicting microbial nitrate reduction pathways in coastal sediments. Aquatic Microbial Ecology 71, 223–238.

Aller, R.C., 1980. Quantifying solute distribution in the bioturbated zone of marine sediments by defining an average microenvironment. Geochimica et Cosmochimica Acta 44, 1955–1965.

Aller, R.C., 1978. The effect of animal-sediment interactions on geochemical processes near the sediment water interface. In: Wiley, M.L. (Ed.), Estuarine Interactions. Academic Press, New York.

Alongi, D.M., 1996. The dynamics of benthic nutrient pools and fluxes in tropical mangrove forests. Journal of Marine Research 54, 123–148.

Alongi, D.M., 2009. The Energetics of Mangrove Forests. Springer Science & Business Media.

Alongi, D.M., 2010. Dissolved iron supply limits early growth of estuarine mangroves. Ecology 91, 3229–3241.

Alongi, D.M., McKinnon, A.D., 2005. The cycling and fate of terrestrially-derived sediments and nutrients in the coastal zone of the Great Barrier Reef shelf. Marine Pollution Bulletin 51, 239–252.

Alongi, D.M., Mukhopadhyay, S.K., 2015. Contribution of mangroves to coastal carbon cycling in low latitude seas. Agricultural and Forest Meteorology 213, 266–272.

Alongi, D.M., Christoffersen, P., Tirendi, F., Robertson, A.I., 1992. The influence of freshwater and material export on sedimentary facies and benthic processes within the Fly Delta and adjacent Gulf of Papua (Papua New Guinea). Continental Shelf Research 12, 287–326.

Alongi, D.M., Christoffersen, P., Tirendi, F., 1993. The influence of forest type on microbial-nutrient relationships in tropical mangrove sediments. Journal of Experimental Marine Biology and Ecology 171, 201–223.

Alongi, D., Tirendi, F., Dixon, P., Trott, L., Brunskill, G., 1999. Mineralization of organic matter in intertidal sediments of a tropical semi-enclosed delta. Estuarine, Coastal and Shelf Science 48, 451–467.

Alongi, D.M., Tirendi, E., Clough, B.F., Tirendi, F., Clough, B.F., 2000. Below-ground decomposition of organic matter in forests of the mangroves Rhizophora stylosa and *Avicennia marina* along the arid coast of Western Australia. Aquatic Botany 68, 97–122.

Alongi, D., Wattayakorn, G., Pfitzner, J., Tirendi, F., Zagorskis, I., Brunskill, G., Davidson, A., Clough, B., 2001. Organic carbon accumulation and metabolic pathways in sediments of mangrove forests in southern Thailand. Marine Geology 179, 85–103.

Alongi, D.M., Trott, L., Wattayakorn, G., Clough, B.F., 2002. Below-ground nitrogen cycling in relation to net canopy production in mangrove forests of southern Thailand. Marine Biology 140, 855–864.

Alongi, D., Sasekumar, A., Chong, V., Pfitzner, J., Trott, L., Tirendi, F., Dixon, P., Brunskill, G., 2004. Sediment accumulation and organic material flux in a managed mangrove ecosystem: estimates of land–ocean–atmosphere exchange in peninsular Malaysia. Marine Geology 208, 383–402.

Alongi, D., Pfitzner, J., Trott, L., Tirendi, F., Dixon, P., Klumpp, D., 2005. Rapid sediment accumulation and microbial mineralization in forests of the mangrove *Kandelia candel* in the Jiulongjiang Estuary, China. Estuarine, Coastal and Shelf Science 63, 605–618.

Alongi, D.M., 2014. Carbon cycling and storage in mangrove forests. Annual Review of Marine Science 6 (1), 195–219. https://doi.org/10.1146/annurev-marine-010213-135020.

Araújo Jr., J., Otero, X., Marques, A., Nóbrega, G., Silva, J., Ferreira, T., 2012. Selective geochemistry of iron in mangrove soils in a semiarid tropical climate: effects of the burrowing activity of the crabs *Ucides cordatus* and *Uca maracoani*. Geo-Marine Letters 32, 289–300.

Armentano, T., Menges, E., 1986. Patterns of change in the carbon balance of organic soil-wetlands of the temperate zone. The Journal of Ecology 755–774.

Armstrong, W., 1978. Root aeration in the wetland condition. In: Hook, D.D., Crawford, R.M.M. (Eds.), Plant Life in Anaerobic Environments. Science Publications, Inc., Ann Arbor, pp. 269–297.

Armstrong, J., Armstrong, W., 1990. Light-enhanced convective throughflow increases oxygenation in rhizomes and rhizosphere of *Phragmites australis* (Cav.) Trin. ex Steud. New Phytologist 114, 121–128.

Armstrong, W., Wright, E.J., 1975. Radial oxygen loss from roots: the theoretical basis for the manipulation of flux data obtained by the cylindrical platinum electrode technique. Physiologia Plantarum 35, 21.

Attri, K., Kerkar, S., Lokabharathi, P., 2011. Ambient iron concentration regulates the sulfate reducing activity in the mangrove swamps of Diwar, Goa, India. Estuarine, Coastal and Shelf Science 95, 156–164.

Atwood, T.B., Connolly, R.M., Almahasheer, H., Carnell, P.E., Duarte, C.M., Ewers Lewis, C.J., Lovelock, C.E., 2017. Global patterns in mangrove soil carbon stocks and losses. Nature Climate Change 7 (7), 523–528. http://doi.org/10.1038/nclimate3326.

Ayukai, T., Miller, D., Wolanski, E., Spagnol, S., 1998. Fluxes of nutrients and dissolved and particulate organic carbon in two mangrove creeks in northeastern Australia. Mangroves and Salt Marshes 2, 223–230.

Balk, M., Laverman, A.M., Keuskamp, J.A., Laanbroek, H.J., 2015. Nitrate ammonification in mangrove soils: a hidden source of nitrite? Frontiers in Microbiology 6.

Barbier, E.B., Hacker, S.D., Kennedy, C., Koch, E.W., Stier, A.C., Silliman, B.R., 2011. The value of estuarine and coastal ecosystem services. Ecological Monographs 81, 169–193.

Barnes, J.R.R., Purvaja, R., Rajkumar, A., Kumar, B., Krithika, K., Ravichandran, K., Uher, G., Upstill-Goddard, R., 2006. Tidal dynamics and rainfall control N_2O ad CH_4 emissions from a pristine mangrove creek. Geophysical Research Letters 33.

Bartlett, D.S., Bartlett, K.B., Hartman, J.M., Harriss, R.C., Sebacher, D.I., Pelletier-Travis, R., Dow, D.D., Brannon, D.P., 1989. Methane emissions from the Florida Everglades: patterns of variability in a regional wetland ecosystem. Global Biogeochemical Cycles 3, 363–374.

Bauer, J.E., Cai, W.-J., Raymond, P.A., Bianchi, T.S., Hopkinson, C.S., Regnier, P.A., 2013. The changing carbon cycle of the coastal ocean. Nature 504, 61.

Bauza, J.F., Morell, J.M., Corredor, J.E., 2002. Biogeochemistry of nitrous oxide production in the red mangrove (Rhizophora mangle) forest sediments. Estuarine, Coastal and Shelf Science 55, 697–704.

Bennett, I., 1968. The mud-lobster. Australian Natural History 16, 22–25.

Berner, R.A., 1984. Sedimentary pyrite formation: an update. Geochimica et Cosmochimica Acta 48, 605–615.

Biswas, H., Mukhopadhyay, S., De, T., Sen, S., Jana, T., 2004. Biogenic controls on the air—water carbon dioxide exchange in the Sundarban mangrove environment, northeast coast of Bay of Bengal, India. Limnology and Oceanography 49, 95–101.

Biswas, H., Mukhopadhyay, S.K., Sen, S., Jana, T.K., 2007. Spatial and temporal patterns of methane dynamics in the tropical mangrove dominated estuary, NE coast of Bay of Bengal, India. Journal of Marine Systems 68 (1–2), 55–64. http://doi.org/10.1016/j.jmarsys.2006.11.001.

Blackburn, T., Nedwell, D., Wiebe, W., 1994. Active mineral cycling in a Jamaican seagrass sediment. Marine Ecology Progress Series 110, 233–239.

Blasco, F., 1984. Climatic factors and the biology of mangrove plants. In: Snedaker, S.C., Snedaker, J.G. (Eds.), The Mangrove Ecosystem: Research Methods. UNESCO, United Kingdom, pp. 18–35.

Bloch, J., Krouse, H.R., 1992. Sulfide diagenesis and sedimentation in the Albian Harmon Member, western Canada. Journal of Sedimentary Research 62 (2).

Bloomfleld, C., Coulter, J., 1974. Genesis and management of acid sulfate soils. Advances in Agronomy 25, 265–326.

Borges, A.V., Abril, G., 2011. 5.04-Carbon Dioxide and Methane Dynamics in Estuaries. Treatise on Estuarine and Coastal Science. Academic Press, Waltham, pp. 119–161.

Borges, A.V., Djenidi, S., Lacroix, G., Theate, J., Delille, B., Frankignoulle, M., 2003. Atmospheric CO_2 flux from mangrove surrounding waters. Geophysical Research Letters 30 (12-11)-(12-14).

Borges, A.V., Vanderborght, J.-P., Schiettecatte, L.-S., Gazeau, F., Ferrón-Smith, S., Delille, B., Frankignoulle, M., 2004. Variability of the gas transfer velocity of CO_2 in a macrotidal estuary (the Scheldt). Estuaries 27, 593–603.

Borges, A.V., Delille, B., Frankignoulle, M., 2005. Budgeting sinks and sources of CO_2 in the coastal ocean: diversity of ecosystem counts. Geophysical Research Letters 32 (14), 1–4. http://doi.org/10.1029/2005GL023053.

Bosire, J.O., Dahdouh-Guebas, F., Kairo, J.G., Wartel, S., Kazungu, J., Koedam, N., 2006. Success rates of recruited tree species and their contribution to the structural development of reforested mangrove stands. Marine Ecology Progress Series 325, 85–91.

Boto, K.G., Bunt, J.S., 1981. Tidal export of particulate organic matter from a northern Australian mangrove system. Estuarine, Coastal and Shelf Science 13, 247–255.

Boto, K.G., Robertson, A.I., 1990. The relationship between nitrogen fixation and tidal exports of nitrogen in a tropical mangrove system. Estuarine, Coastal and Shelf Science 31, 531–540.

Boto, K.G., Wellington, J.T., 1988. Seasonal variations in concentrations and fluxes of dissolved organic and inorganic materials in a tropical, tidally-dominated, mangrove waterway. Marine Ecology Progress Series 50, 151–160.

Bouillon, S., Connolly, R.M., 2009. Carbon exchange among tropical coastal ecosystems. In: Ecological Connectivity Among Tropical Coastal Ecosystems. Springer, pp. 45–70.

Bouillon, S., Koedam, N., Raman, A.V., Dehairs, F., 2002a. Primary producers sustaining macro-invertebrate communities in intertidal mangrove forests. Oecologia 130, 441–448.

Bouillon, S., Raman, A., Dauby, P., Dehairs, F., 2002b. Carbon and nitrogen stable isotope ratios of subtidal benthic invertebrates in an estuarine mangrove ecosystem (Andhra Pradesh, India). Estuarine, Coastal and Shelf Science 54, 901–913.

Bouillon, S., Frankignoulle, M., Dehairs, F., Velimirov, B., Eiler, A., Abril, G., Etcheber, H., Borges, A.V., 2003a. Inorganic and organic carbon biogeochemistry in the Gautami Godavari estuary (Andhra Pradesh, India) during pre-monsoon: the local impact of extensive mangrove forests. Global Biogeochemical Cycles 17 (25-21)-(25-12).

Bouillon, S., Koedam, N., Borges, A.V., Frankignoulle, M., Dehairs, F., 2003. Carbon Dynamics in Mangrove Ecosystmes: Interactions between Intertidal and Adjacent Auatic Habitats. Geophysical Research Abstracts 5.

Bouillon, S., Koedam, N., Baeyens, W., Satyanarayana, B., Dehairs, F., 2004a. Selectivity of subtidal benthis invertebrate communities for local microalgal production in an estuarine mangrove ecosystem during the post-monsoon period. Journal of Sea Research 51, 133–144.

Bouillon, S., Moens, T., Overmeer, I., Koedam, N., Dehairs, F., 2004b. Resource utilization patterns of epifauna from mangrove forests with contrasting inputs of local versus imported organic matter. Marine Ecology Progress Series 278, 77–88.

Bouillon, S., Dehairs, F., Schiettecatte, L.-S., Vieira Borges, A., 2007a. Biogeochemistry of the Tana estuary and delta (northern Kenya). Limnology and Oceanography 52, 46–59.

Bouillon, S., Dehairs, F., Velimirov, B., Abril, G., Borges, A.V., 2007b. Dynamics of organic and inorganic carbon across contiguous mangrove and seagrass systems (Gazi Bay, Kenya). Journal of Geophysical Research: Biogeosciences 112.

Bouillon, S., Middelburg, J.J., Dehairs, F., Borges, A.V., Abril, G., Flindt, M.R., Ulomi, S., Kristensen, E., 2007c. Importance of intertidal sediment processes and porewater exchange on the water column biogeochemistry in a pristine mangrove creek (Ras Dege, Tanzania). Biogeosciences Discussions 4, 317–348.

Bouillon, S., Borges, A.V., Castaneda-Moya, E., Diele, K., Dittmar, T., Duke, N.C., Kristensen, E., Lee, S.Y., Marchand, C., Middelburg, J.J., Rivera-Monroy, V.H., Smith, T.J., Twilley, R.R., 2008a. Mangrove production and carbon sinks: a revision of global budget estimates. Global Biogeochemical Cycles 22. https://doi.org/10.1029/2007GB003052.

Bouillon, S., Connolly, R.M., Lee, S.Y., 2008b. Organic matter exchange and cycling in mangrove ecosystems: recent insights from stable isotope studies. Journal of Sea Research 59, 44–58.

Boyd, R., Dalrymple, R., Zaitlin, B., 1992. Classification of clastic coastal depositional environments. Sedimentary Geology 80, 139–150.

Breithaupt, J.L., Smoak, J.M., Smith, T.J., Sanders, C.J., Hoare, A., 2012. Organic carbon burial rates in mangrove sediments: strengthening the global budget. Global Biogeochemical Cycles 26.

Cabezas, A., Mitsch, W., Macdonnell, C., Zhang, L., Bydałek, F., Lasso, A., 2017. Methane emissions from mangrove soils in hydrologically disturbed and reference mangrove tidal creeks in southwest Florida. Ecological Engineering 114, 57–65. 15 April 2018.

Cahoon, D.R., Hensel, P., Rybczyk, J., Mckee, K.L., Proffitt, C.E., Perez, B.C., 2003. Mass tree mortality leads to mangrove peat collapse at Bay Islands, Honduras after Hurricane Mitch. Journal of Ecology 91, 1093–1105.

Cai, W.-J., 2011. Estuarine and coastal ocean carbon paradox: CO_2 sinks or sites of terrestrial carbon incineration? Annual Review of Marine Science 3, 123–145.

Call, M., Maher, D.T., Santos, I.R., Ruiz-Halpern, S., Mangion, P., Sanders, C.J., Erler, D.V., Oakes, J.M., Rosentreter, J., Murray, R., 2015. Spatial and temporal variability of carbon dioxide and methane fluxes over semi-diurnal and spring–neap–spring timescales in a mangrove creek. Geochimica et Cosmochimica Acta 150, 211–225.

Cao, W., Yang, J., Li, Y., Liu, B., Wang, F., Chang, C., 2016. Dissimilatory nitrate reduction to ammonium conserves nitrogen in anthropogenically affected subtropical mangrove sediments in Southeast China. Marine Pollution Bulletin 110, 155–161.

Cao, W., Guan, Q., Li, Y., Wang, M., Liu, B., 2017. The contribution of denitrification and anaerobic ammonium oxidation to N_2 production in mangrove sediments in Southeast China. Journal of Soils and Sediments 1–10.

Cardona-Olarte, P., Twilley, R.R., Krauss, K.W., Rivera-Monroy, V., 2006. Responses of neotropical mangrove seedlings grown in monoculture and mixed culture under treatments of hydroperiod and salinity. Hydrobiologia 569, 325–341.

Carmichael, M.J., Bernhardt, E.S., Bräuer, S.L., Smith, W.K., 2014. The role of vegetation in methane flux to the atmosphere: should vegetation be included as a distinct category in the global methane budget? Biogeochemistry 119, 1–24.

Castañeda-Moya, E., Rivera-Monroy, V.H., Twilley, R.R., 2006. Mangrove zonation in the dry life zone of the Gulf of Fonseca, Honduras. Estuaries and Coasts 29, 751–764.

Castañeda-Moya, E., 2010. Landscape patterns of community structure, biomass and net primary productivity of mangrove forests in the Florida Coastal Everglades as a function of resources, regulators, hydroperiod, and hurricane disturbance Ph.D. Dissertation. Louisiana State University, Baton Rouge, LA.

Castañeda-Moya, E., Twilley, R.R., Rivera-Monroy, V.H., Zhang, K.Q., Davis, S.E., Ross, M., 2010. Sediment and nutrient deposition associated with hurricane Wilma in mangroves of the Florida coastal Everglades. Estuaries and Coasts 33, 45–58.

Castañeda-Moya, E., Twilley, R.R., Rivera-Monroy, V.H., Marx, B.D., Coronado-Molina, C., Ewe, S.M.L., 2011. Patterns of root dynamics in mangrove forests along environmental gradients in the Florida Coastal Everglades, USA. Ecosystems 14, 1178–1195.

Castañeda-Moya, E., Twilley, R.R.R.R., Rivera-Monroy, V.H.V.H., 2013. Allocation of biomass and net primary productivity of mangrove forests along environmental gradients in the Florida Coastal Everglades, USA. Forest Ecology and Management 307, 226–241.

Cawley, K.M., Yamashita, Y., Maie, N., Jaffé, R., 2014. Using optical properties to quantify fringe mangrove inputs to the dissolved organic matter (DOM) pool in a subtropical estuary. Estuaries and Coasts 37, 399–410.

Chambers, R.M., Fourqurean, J.W., Macko, S.A., Hoppenot, R., 2001. Biogeochemical effects of iron availability on primary producers in a shallow marine carbonate environment. Limnology and Oceanography 46, 1278–1286.

Chambers, L.G., Davis, S.E., Troxler, T., Boyer, J.N., Downey-Wall, A., Scinto, L.J., 2014. Biogeochemical effects of simulated sea-level rise on carbon loss in an Everglades mangrove peat soil. Hydrobiologia 726, 195–211.

Chambers, L.G., Guevara, R., Boyer, J.N., Troxler, T.G., Davis, S.E., 2016. Effects of salinity and inundation on microbial community structure and function in a mangrove peat soil. Wetlands 36, 361–371.

Chapin III, F.S., Bloom, A.J., Field, C.B., Waring, R.H., 1987. Plant responses to multiple environmental factors. BioScience 37, 49–57.

Chapman, V.J., 1976. Mangrove Vegetation. Liechtenstein: J. Cramer, Vaduz.

Charoeuchamratcheep, C., Smith, C.J., Satawathananout, S., Patrick Jr., W.H., 1987. Reduction and oxidation of acid sulfate soils of Thailand. Soil Science Society of America Journal 51, 630–634.

Chauhan, R., Ramanathan, A., Adhya, T., 2008. Assessment of methane and nitrous oxide flux from mangroves along Eastern coast of India. Geofluids 8, 321–332.

Chen, C.-T.A., Borges, A.V., 2009. Reconciling opposing views on carbon cycling in the coastal ocean: continental shelves as sinks and near-shore ecosystems as sources of atmospheric CO_2. Deep Sea Research Part II: Topical Studies in Oceanography 56, 578–590.

Chen, J., Gu, J.-D., 2017. Faunal burrows alter the diversity, abundance, and structure of AOA, AOB, anammox and n-damo communities in coastal mangrove sediments. Microbial Ecology 1–17.

Chen, R.H., Twilley, R.R., 1999a. Patterns of mangrove forest structure and soil nutrient dynamics along the Shark River estuary, Florida. Estuaries 22, 955–970.

Chen, R.H., Twilley, R.R., 1999b. A simulation model of organic matter and nutrient accumulation in mangrove wetland soils. Biogeochemistry 44, 93–118.

Chen, G.C., Tam, N.F.Y., Ye, Y., 2010. Summer fluxes of atmospheric greenhouse gases N_2O, CH_4 and CO_2 from mangrove soil in South China. The Science of the Total Environment 408, 2761–2767.

Chmura, G.L., Anisfeld, S.C., Cahoon, D.R., Lynch, J.C., 2003. Global carbon sequestration in tidal, saline wetland soils. Global Biogeochemical Cycles 17, 12.

Christensen, P.B., Nielsen, L.P., Sørensen, J., Revsbech, N.P., 1990. Denitrification in nitrate-rich streams: diurnal and seasonal variation related to benthic oxygen metabolism. Limnology and Oceanography 35, 640–651.

Cifuentes, L.A., Coffin, R.B., Solorzano, L., Cardenas, W., Espinoza, J., Twilley, R.R., 1996. Isotopic and elemental variations of carbon and nitrogen in a mangrove estuary. Estuarine, Coastal and Shelf Science 43, 781–800.

Clark, M.W., Mcconchie, D., Lewis, D.W., Saenger, P., 1998. Redox stratification and heavy metal partitioning in Avicennia-dominated mangrove sediments: a geochemical model. Chemical Geology 149, 147–171.

Cormier, N., Twilley, R.R., Ewel, K.C., Krauss, K.W., 2015. Fine root productivity varies along nitrogen and phosphorus gradients in high-rainfall mangrove forests of Micronesia. Hydrobiologia 750, 69–87.

Corredor, J.E., Morell, J.M., 1994. Nitrate depuration of secondary sewage effluents in mangrove sediments. Estuaries 17, 295–300.

Corredor, J.E., Morell, J.M., Bauza, J., 1999. Atmospheric nitrous oxide fluxes from mangrove sediments. Marine Pollution Bulletin 38, 473–478.

Dai, A., Trenberth, K.E., 2002. Estimates of freshwater discharge from continents: latitudinal and seasonal variations. Journal of Hydrometeorology 3, 660–687.

Dalsgaard, T., Thamdrup, B., Canfield, D.E., 2005. Anaerobic ammonium oxidation (anammox) in the marine environment. Research in Microbiology 156, 457–464.

Danielson, T.M., Rivera-Monroy, V.H., Castañeda-Moya, E., Briceño, H., Travieso, R., Marx, B.D., Farfán, L.M., 2017. Assessment of Everglades mangrove forest resilience: implications for above-ground net primary productivity and carbon dynamics. Forest Ecology and Management 404, 115–125. http://doi.org/10.1016/j.foreco.2017.08.009.

Day, J.W., Christian, R.R., Boesch, D.M., Yáñez-Arancibia, A., Morris, J., Twilley, R.R., Stevenson, C., 2008. Consequences of climate change on the ecogeomorphology of coastal wetlands. Estuaries and Coasts 31 (3), 477–491. http://doi.org/10.1007/s12237-008-9047-6.

Davie, J., 1984. Structural variation, litter production, and nutrient status of mangrove vegetation in Moreton Bay. In: Coleman, R.J., Covacevich, J., Davie., P. (Eds.), Focus on Stradbroke. Boolarong Press, Brisbane, pp. 208–223.

Davis, S.E., Childers, D.L., Day Jr., J.W., Rudnick, D.T., Sklar, F.H., 2001a. Nutrient dynamics in vegetated and unvegetated areas of a southern Everglades mangrove creek. Estuarine, Coastal and Shelf Science 52, 753–768.

Davis, S.E., Childers, D.L., Day, J.W., Rudnick, D.T., Sklar, F.H., 2001b. Wetland-water column exchanges of carbon, nitrogen, and phosphorus in a southern Everglades dwarf mangrove. Estuaries 24, 610–622.

Davis, S.E., Childers, D.L., Day Jr., J.W., Rudnick, D.T., Sklar, F.H., 2003. Factors affecting the concentration and flux of materials in two southern Everglades mangrove wetlands. Marine Ecology Progress Series 253, 85–96.

Davis, S.E., Cable, J.E., Childers, D.L., Coronado-Molina, C., Day Jr., J.W., Hittle, C.D., Madden, C.J., Reyes, E., Rudnick, D., Sklar, F., 2004. Importance of storm events in controlling ecosystem structure and function in a Florida gulf coast estuary. Journal of Coastal Research 1198–1208.

Dittmar, T., Hertkorn, N., Kattner, G., Lara, R.J., 2006. Mangroves, a major source of dissolved organic carbon to the oceans. Global Biogeochemical Cycles 20.

Dittmar, T., Lara, R.J., 2001. Do mangroves rather than rivers provide nutrients to coastal environments south of the Amazon River? Evidence from long-term flux measurements. Marine Ecology Progress Series 213, 67–77. http://doi.org/10.3354/meps213067.

Duarte, C.M., Merino, M., Gallegos, M., 1995. Evidence of iron deficiency in seagrasses growing above carbonate sediments. Limnology and Oceanography 40, 1153–1158.

Duarte, C.M., Middelburg, J.J., Caraco, N., 2005. Major role of marine vegetation on the oceanic carbon cycle. Biogeosciences 2, 1–8.

Duarte, C.M., Losada, I.J., Hendriks, I.E., Mazarrasa, I., Marbà, N., 2013. The role of coastal plant communities for climate change mitigation and adaptation. Nature Climate Change 3, 961–968.

Dürr, H.H., Laruelle, G.G., Van Kempen, C.M., Slomp, C.P., Meybeck, M., Middelkoop, H., 2011. Worldwide typology of nearshore coastal systems: defining the estuarine filter of river inputs to the oceans. Estuaries and Coasts 34, 441–458.

Dutta, M.K., Chowdhury, C., Jana, T.K., Mukhopadhyay, S.K., 2013. Dynamics and exchange fluxes of methane in the estuarine mangrove environment of the Sundarbans, NE coast of India. Atmospheric Environment 77, 631–639.

Ellison, A.M., Farnsworth, E.J., 1992. The ecology of Belizean mangrove-root fouling communities: patterns of epibiont distribution and abundance, and effects on root growth. Hydrobiologia 20, 1–12.

Ellison, A.M., Farnsworth, E.J., Twilley, R.R., 1996. Facultative mutualism between red mangroves and root-fouling sponges in Belizean mangal. Ecology 77, 2431–2444.

Ellison, A.M., 2002. Macroecology of mangroves: large-scale patterns and processes in tropical coastal forests. Trees 16 (2), 181–194. http://doi.org/10.1007/s00468-001-0133-7.

Emerson, S., Tahnke, R., Heggie, D., 1984. Sediment-water exchange in shallow water estuarine sediment. Journal of Marine Research 42, 709–730.

Evangelou, V., 1995. Pyrite Oxidation and its Control. CRC press.

Ewel, K.C., Twilley, R.R., Ong, J.E., 1998. Different kinds of mangrove forests provide different goods and services. Global Ecology and Biogeography Letters 7, 83.

Eyre, B.D., Ferguson, A.J., Webb, A., Maher, D., Oakes, J.M., 2011. Denitrification, N-fixation and nitrogen and phosphorus fluxes in different benthic habitats and their contribution to the nitrogen and phosphorus budgets of a shallow oligotrophic sub-tropical coastal system (southern Moreton Bay, Australia). Biogeochemistry 102, 111–133.

Fabre, A., Fromard, F., Trichon, V., 1999. Fractionation of phosphate in sediments of four representative mangrove stages (French Guiana). Hydrobiologia 392, 13–19.

FAO, 2004. Mangrove forest management guidelines. In: Department, F.a.A. (Ed.), FAO Forestry Paper. FAO, Rome, Italy.

FAO, 2007. The world's mangroves 1980–2005. In: FAO Forestry Paper. FAO, Rome, Italy.

Fatoyinbo, T.E., Simard, M., 2013. Height and biomass of mangroves in Africa from ICESat/GLAS and SRTM. International Journal of Remote Sensing 34, 668–681.

Fatoyinbo, T.E., Simard, M., Washington-Allen, R.A., Shugart, H.H., 2008. Landscape-scale extent, height, biomass, and carbon estimation of Mozambique's mangrove forests with Landsat ETM+ and Shuttle Radar Topography Mission elevation data. Journal of Geophysical Research: Biogeosciences 113.

Feller, I.C., Whigham, D.F., O'neil, J.P., Mckee, K.L., 1999. Effects of nutrient enrichment on within-stand cycling in a mangrove forest. Ecology 80, 2193–2205.

Feller, I.C., Mckee, K.L., Whigham, D.F., O'neill, J.P., 2002. Nitrogen vs. phosphorous limitation across an ecotonal gradient in a mangrove forest. Biogeochemistry 62, 145–175.

Feller, I.C., Mckee, K.L., Whigham, D.F., O'neill, J.P., 2003a. Nitrogen vs. phosphorus limitation across an ecotonal gradient in a mangrove forest. Biogeochemistry 62 (2), 145–175.

Feller, I.C., Whigham, D.F., Mckee, K.L., Lovelock, C.E., 2003b. Nitrogen limitation of growth and nutrient dynamics in a disturbed mangrove forest, Indian River Lagoon, Florida. Oecologia 134, 405–414.

Feller, I.C., 1995. Effects of nutrient enrichment on growth and herbivory of dwarf red mangrove (Rhizophora Mangle). Ecological Monographs 65 (4), 477–505. http://doi.org/10.2307/2963499.

Fernandes, S.O., Bharathi, P.L., Bonin, P.C., Michotey, V.D., 2010. Denitrification: an important pathway for nitrous oxide production in tropical mangrove sediments (Goa, India). Journal of Environmental Quality 39, 1507.

Fernandes, S.O., Bonin, P.C., Michotey, V.D., Garcia, N., Lokabharathi, P., 2012a. Nitrogen-limited mangrove ecosystems conserve N through dissimilatory nitrate reduction to ammonium. Scientific Reports 2, 419.

Fernandes, S.O., Michotey, V.D., Guasco, S., Bonin, P.C., Bharathi, P.a.L., Loka Bharathi, P.A., 2012b. Denitrification prevails over anammox in tropical mangrove sediments (Goa, India). Marine Environmental Research 74, 9–19.

Ferreira, T., Otero, X., Vidal-Torrado, P., Macías, F., 2007. Effects of bioturbation by root and crab activity on iron and sulfur biogeochemistry in mangrove substrate. Geoderma 142, 36–46.

Fisher, P., Spalding, M., 1993. Protected Areas with Mangrove Habitat. Unpublished report to WCMC, Cambridge, UK.

Fleming, M., Lin, G., Sternberg, L.S.L., 1990. Influence of mangrove detritus in an estuarine ecosystem. Bulletin of Marine Science 47, 663–669.

Flores-Verdugo, F.J., Day Jr., J.W., Briseño-Dueñas, R., 1987. Structure, litter fall, decomposition and detritus dynamics of mangroves in a Mexican Coastal Lagoon with an ephemeral inlet. Marine Ecology Progress Series 35, 83–90.

Fontalvo-Herazo, M.L., Piou, C., Vogt, J., Saint-Paul, U., Berger, U., 2011. Simulating harvesting scenarios towards the sustainable use of mangrove forest plantations. Wetlands Ecology and Management 19, 397–407.

Francis, C.A., Beman, J.M., Kuypers, M.M., 2007. New processes and players in the nitrogen cycle: the microbial ecology of anaerobic and archaeal ammonia oxidation. The ISME Journal 1, 19–27.

Friess, D., 2016. JG Watson, inundation classes, and their influence on paradigms in mangrove forest ecology. Wetlands 1–11.

Friess, D.A., Webb, E.L., 2014. Variability in mangrove change estimates and implications for the assessment of ecosystem service provision. Global Ecology and Biogeography 23, 715–725.

Fry, B., Smith, T.J., 2002. Stable isotope studies of red mangroves and filter feeders from the Shark River estuary, Florida. Bulletin of Marine Science 70, 871–890.

Furnas, M., Mitchell, A., 2001. Runoff of Terrestrial Sediment and Nutrients into the Great Barrier Reef World Heritage Area. CRC Press, Boca Raton.

Gattuso, J.P., Frankignoulle, M., Wollast, R., 1998. Carbon and carbonate metabolism in coastal aquatic ecosystems. Annual Reviews 29, 405–434.

Ghosh, S., Jana, T., Singh, B., Choudhury, A., 1987. Comparative study of carbon dioxide system in virgin and reclaimed mangrove waters of Sundarbans. Mahasagar 20, 155–161.

Giri, C., Ochieng, E., Tieszen, L.L., Zhu, Z., Singh, A., Loveland, T., Masek, J., Duke, N., 2011. Status and distribution of mangrove forests of the world using earth observation satellite data. Global Ecology and Biogeography 20, 154–159.

Gleason, S., Ewel, K., Hue, N., 2003. Soil redox conditions and plant–soil relationships in a Micronesian mangrove forest. Estuarine, Coastal and Shelf Science 56, 1065–1074.

Golley, F.B., Odum, H.T., Wilson, R.F., 1962. The structure and metabolism of a Puerto Rican red mangrove forest in May. Ecology 43, 9–19.

Gong, W.-K., Ong, J.-E., 1990. Plant biomass and nutrient flux in a managed mangrove forest in Malaysia. Estuarine, Coastal and Shelf Science 31, 519–530.

Gosselink, J.G., Turner, R.E., 1978. The role of hydrology in freshwater wetland ecosystems. In: Good, R.E., Whigham, D.F., Simpson, R.L. (Eds.), Freshwater Wetlands: Ecological Processes and Management Potential. Academic Press, New York, pp. 63–78.

Gotto, J.W., Taylor, B.F., 1976. N_2 fixation associated with decaying leaves of the red mangrove (*Rhizophora mangle*). Applied and Environmental Microbiology 31, 781–783.

Gotto, J.W., Tabita, F.R., Baalen, C.V., 1981. Nitrogen fixation in intertidal environments of the Texas gulf coast. Estuarine, Coastal and Shelf Science 12, 231–235.

Granek, E.F., Compton, J.E., Phillips, D.L., 2009. Mangrove-exported nutrient incorporation by sessile coral reef invertebrates. Ecosystems 12, 462–472.

Groombridge, B., 1992. Global Biodiversity: Status of the Earth's Living Resources. Chapman & Hall.

Hamilton, S.E., Casey, D., 2016. Creation of a high spatio-temporal resolution global database of continuous mangrove forest cover for the 21st century (CGMFC-21). Global Ecology and Biogeography 25, 729–738.

Hart, M.G.R., 1963. Observation on the source of acid in emplodered mangrove soils. II. Oxidation of soil polysulphides. Plant and Soil 19, 106–114.

Hayatsu, M., Tago, K., Saito, M., 2008. Various players in the nitrogen cycle: diversity and functions of the microorganisms involved in nitrification and denitrification. Soil Science and Plant Nutrition 54, 33–45.

Heckbert, S., Costanza, R., Poloczanska, E.S., Richardson, A.J., 2012. Climate regulation as a service from estuarine and coastal ecosystems. Treatise on Estuarine and Coastal Science 12, 199–216.

Hemminga, M.A., Slim, F.J., Kazungu, J., Ganssen, G.M., Nieuwenhuize, J., Kruyt, N.M., 1994. Carbon outwelling from a mangrove forest with adjacent seagrass beds and coral reefs (Gazi Bay, Kenya). Marine Ecology Progress Series 106, 291–301.

Henriksen, K., Kemp, W.M., 1988. Nitrification in estuarine and coastal marine sediments, methods, patterns and regulating factors. In: Sørensen, J., Blackburn, T.H., Rosswall, T. (Eds.), Nitrogen Cycling in Coastal Marine Sediments. J. Wiley and Sons, New York, pp. 207–249.

Hesse, P.R., 1961. Some differences between the soils of *Rhizophora* and *Avicennia* mangrove swamps in Sierra Leone. Plant and Soil 14, 335–346.

Hicks, B.J., Silvester, W.B., 1985. Nitrogen fixation associated with the New Zealand mangrove (*Avicennia marina* (Forsk.) Vierh. var. *resinifera* (Forst. f.) Bakh.). Applied and Environmental Microbiology 49, 955–959.

Ho, D.T., Ferrón, S., Engel, V.C., Larsen, L.G., Barr, J.G., 2014. Air-water gas exchange and CO_2 flux in a mangrove-dominated estuary. Geophysical Research Letters 41, 108–113.

Hoehler, T.M., Alperin, M.J., 2014. Methane minimalism. Nature 507, 436.

Holmer, M., Kristensen, E., Banta, G., Hansen, K., Jensen, M.H., Bussawarit, N., 1994. Biogeochemical cycling of sulfur and iron in sediments of a south-east Asian mangrove, Phuket Island, Thailand. Biogeochemistry 26, 145–161.

Holmer, M., Andersen, F.O., Nielsen, S.L., Boschker, H.T.S., 2001. The importance of mineralization based on sulfate reduction for nutrient regeneration in tropical seagrass sediments. Aquatic Botany 71, 1–17.

Hopkinson, C.S., Cai, W.J., Hu, X.P., 2012. Carbon sequestration in wetland dominated coastal systems - a global sink of rapidly diminishing magnitude. Current Opinion in Environmental Sustainability 4, 186–194.

Houghton, R.A., 1994. The worldwide extent of land-use change. BioScience 44, 305–313.

Houghton, R., 2007. Balancing the global carbon budget. Annual Review of Earth and Planetary Sciences 35, 313–347.

Howarth, R.W., Marino, R., Lane, J., Cole, J.J., 1988. Nitrogen fixation in freshwater, estuarine, and marine ecosystems. 1. Rates and importance. Limnology and Oceanography 33, 669–687.

Huston, M.A., 1994. Biological Diversity: The Coexistence of Species. Cambridge University Press.

Hutchings, P., Saenger, P., 1987. Ecology of Mangroves. Quensland University Press.

Iizumi, H., 1986. Soil nutrient dynamics. Workshop on mangrove ecosystem dynamics. In: UNDP/UNESCO Regional Project (RAS/79/002), New Delhi, pp. 171–180.

ITTO & ISME, 1993. Mangrove Ecosystems: Technical Reports. International Society for Mangrove Ecosystems (ISME), International Tropical Timber Organization (ITTO) and Japan International Association for Mangroves (JIAM), Nishihara, Japan.

Jardine, S.L., Siikamäki, J.V., 2014. A global predictive model of carbon in mangrove soils. Environmental Research Letters 9 (10), 104013. http://doi.org/10.1088/1748-9326/9/10/104013.

Jennerjahn, T.C., 2012. Biogeochemical response of tropical coastal systems to present and past environmental change. Earth-Science Reviews 114, 19–41.

Jennerjahn, T.C., Ittekkot, V., 2002. Relevance of mangroves for the production and deposition of organic matter along tropical continental margins. Naturwissenschaften 89, 23–30.

Jensen, M.M., Thamdrup, B., Dalsgaard, T., 2007. Effects of specific inhibitors on anammox and denitrification in marine sediments. Applied and Environmental Microbiology 73, 3151–3158.

Jerath, M., Bhat, M., Rivera-Monroy, V.H., Castañeda-Moya, E., Simard, M., Twilley, R.R., 2016. The role of economic, policy, and ecological factors in estimating the value of carbon stocks in Everglades mangrove forests, South Florida, USA. Environmental Science and Policy 66, 160–169.

Jetten, M.S., Sliekers, O., Kuypers, M., Dalsgaard, T., Van Niftrik, L., Cirpus, I., Van De Pas-Schoonen, K., Lavik, G., Thamdrup, B., Le Paslier, D., 2003. Anaerobic ammonium oxidation by marine and freshwater planctomycete-like bacteria. Applied Microbiology and Biotechnology 63, 107–114.

Jin-Eong, O., Khoon, G.W., Clough, B.F., 1995. Structure and productivity of a 20-year-old stand of *Rhizophora apiculata* Bl. Mangrove forest. Journal of Biogeography 22, 417–424.

Jørgensen, B.B., 1982. Mineralization of organic matter in the sea bed- the role of sulfate reduction. Nature 296, 643–645.

Joye, S.B., Lee, R.Y., 2004. Benthic microbial mats: important sources of fixed nitrogen and carbon to the Twin Cays, Belize ecosystem. Atoll Research Bulletin 528, 1–26.

Kairo, J.G., Lang'at, J.K.S., Dahdouh-Guebas, F., Bosire, J., Karachi, M., 2008. Structural development and productivity of replanted mangrove plantations in Kenya. Forest Ecology and Management 255, 2670–2677.

Kairo, J., Bosire, J., Langat, J., Kirui, B., Koedam, N., 2009. Allometry and biomass distribution in replanted mangrove plantations at Gazi Bay, Kenya. Aquatic Conservation: Marine and Freshwater Ecosystems 19, S63–S69.

Kamp-Nielsen, L., Vermaat, J.E., Wesseling, I., Borum, J., Geertz-Hansen, O., 2002. Sediment propterties along gradients of siltation in south-east Asia. Estuarine, Coastal and Shelf Science 54, 127–137.

Kartal, B., Kuypers, M.M., Lavik, G., Schalk, J., Op Den Camp, H.J., Jetten, M.S., Strous, M., 2007. Anammox bacteria disguised as denitrifiers: nitrate reduction to dinitrogen gas via nitrite and ammonium. Environmental Microbiology 9, 635–642.

Khan, N.I., Suwa, R., 2007. Carbon and nitrogen pools in a mangrove stand of Kandelia obovata (S., L.) Yong: vertical distribution in the soil-vegetation system. Wetlands Ecology and Management 15, 141–153.

Kimball, M.C., Teas, H.J., 1975. Nitrogen fixation in mangrove areas of southern Florida. In: Walsh, G., Snedaker, S., Teas, H. (Eds.), Proceedings of the International Symposium on the Biology and Management of Mangroves. Institute of Food and Agricultural Sciences, University of Florida, Gainesville, Florida, pp. 654–660.

Kirwan, M.L., Guntenspergen, G.R., 2012. Feedbacks between inundation, root production, and shoot growth in a rapidly submerging brackish marsh. Journal of Ecology 100, 764–770.

Kirwan, M.L., Mudd, S.M., 2012. Response of salt-marsh carbon accumulation to climate change. Nature 489, 550.

Kirwan, M., Temmerman, S., 2009. Coastal marsh response to historical and future sea-level acceleration. Quaternary Science Reviews 28, 1801–1808.

Koch, M.S., 1997. Rhizophora mangle L. seedling development into the sapling stage across resource and stress gradients in subtropical Florida. Biotropica 29, 427–439.

Koch, M.S., Mendelssohn, I.A., 1989. Sulphide as a soil phytotoxin: differential responses in two marsh species. Journal of Ecology 77, 565–578.

Koch, M.S., Mendelssohn, I.A., Mckee, K.L., 1990. Mechanism for the hydrogen sulfide-induced growth limitation in wetland macrophytes. Limnology and Oceanography 35, 399–408.

Komiyama, A., Havanond, S., Srisawatt, W., Mochida, Y., Fujimoto, K., Ohnishi, T., Ishihara, S., Miyagi, T., 2000. Top/root biomass ratio of a secondary mangrove (*Ceriops tagal* (Perr.) CB Rob.) forest. Forest Ecology and Management 139, 127–134.

Komiyama, A., Ong, J.E., Poungparn, S., 2008. Allometry, biomass, and productivity of mangrove forests: a review. Aquatic Botany 89, 128–137.

Koné, Y.-M., Borges, A., 2008. Dissolved inorganic carbon dynamics in the waters surrounding forested mangroves of the Ca Mau Province (Vietnam). Estuarine, Coastal and Shelf Science 77, 409–421.

Koop-Jakobsen, K., Giblin, A.E., 2009. Anammox in tidal marsh sediments: the role of salinity, nitrogen loading, and marsh vegetation. Estuaries and Coasts 32, 238–245.

Krauss, K.W., Doyle, T.W., Twilley, R.R., Rivera-Monroy, V.H., Sullivan, J.K., 2006. Evaluating the relative contributions of hydroperiod and soil fertility on growth of south Florida mangroves. Hydrobiologia 569, 311–324.

Kreuzwieser, J., Buchholz, J., Rennenberg, H., 2003. Emission of methane and nitrous oxide by Australian mangrove ecosystems. Plant Biology 5, 423–431.

Krishnan, K., Bharathi, P.L., 2009. Organic carbon and iron modulate nitrification rates in mangrove swamps of Goa, south west coast of India. Estuarine, Coastal and Shelf Science 84, 419–426.

Kristensen, E., 2008. Mangrove crabs as ecosystem engineers; with emphasis on sediment processes. Journal of Sea Research 59, 30–43.

Kristensen, E., Alongi, D.M., 2006. Control by fiddler crabs (*Uca vocans*) and plant roots (*Avicennia marina*) on carbon, iron, and sulfur biogeochemistry in mangrove sediment. Limnology and Oceanography 51, 1557–1571.

Kristensen, E., Andersen, F.Ø., Kofoed, L.H., 1988. Preliminary assessment of benthic community metabolism in a south-east Asian mangrove swamp. Marine Ecology Progress Series 48, 137–145.

Kristensen, E., Devol, A.H., Ahmed, S.I., Saleem, M., 1992. Preliminary study of benthic metabolism and sulfate reduction in a mangrove swamp of the Indus Delta, Pakistan. Marine Ecology Progress Series 90, 287–297.

Kristensen, E., Jensen, M.H., Banta, G.T., Hansen, K., Holmer, M., King, G.M., 1998. Transformation and transport of inorganic nitrogen in sediments of a southeast Asian mangrove forest. Aquatic Microbial Ecology 15, 165–175.

Kristensen, E., Andersen, F.Ø., Holmboe, N., Holmer, M., Thongtham, N., 2000. Carbon and nitrogen mineralization in sediments of the Bangrong mangrove area, Phuket, Thailand. Aquatic Microbial Ecology 22, 199–213.

Kristensen, E., Bouillon, S., Dittmar, T., Marchand, C., 2008a. Organic carbon dynamics in mangrove ecosystems: a review. Aquatic Botany 89, 201–219.

Kristensen, E., Flindt, M.R., Ulomi, S., Borges, A.V., Abril, G., Bouillon, S., 2008b. Emission of CO_2 and CH_4 to the atmosphere by sediments and open waters in two Tanzanian mangrove forests. Marine Ecology Progress Series 370, 53–67.

Kristensen, E., Mangion, P., Tang, M., Flindt, M.R., Holmer, M., Ulomi, S., 2011. Microbial carbon oxidation rates and pathways in sediments of two Tanzanian mangrove forests. Biogeochemistry 103, 143–158.

Krom, M.D., Berner, R.A., 1980. Adsorption of phosphate in anoxic marine sediments. Limnology and Oceanography 25, 797–806.

Kuenzer, C., Bluemel, A., Gebhardt, S., Quoc, T.V., Dech, S., 2011. Remote sensing of mangrove ecosystems: a review. Remote Sensing 3, 878–928.

Kumar, G., Ramanathan, A.L., 2014. Biogeochemistry of methane emission in mangrove ecosystem - Review. Indian Journal of Marines Sciences 43, 989–997.

Kyaruzi, J., Kyewalyanga, M., Muruke, M., 2003. Cyanobacteria composition and impact of seasonality on their in situ nitrogen fixation rate in a mangrove ecosystem adjacent to Zanzibar town. Western Indian Ocean Journal of Marine Science 2, 35–44.

Lanly, J.P., 1982. Tropical Forest Resources Assesment. Global Forest Resources Assessment. FAO, Rome, Italy.

Lee, S.Y., 1989. The importance of sesarminae crabs *Chiromanthes* spp. and inundation frequency on mangrove (*Kandelia candel* (L.) Druce) leaf litter turnover in a Hong Kong tidal shrimp pond. Journal of Experimental Marine Biology and Ecology 131, 23–43.

Lee, S.Y., 2004. Relationship between mangrove abundance and tropical prawn production: a re-evaluation. Marine Biology 145, 943–949.

Lee, S.Y., 2005. Exchange of organic matter and nutrients between mangroves and estuaries: myths, methodological issues and missing links. International Journal of Ecology and Environmental Sciences 31, 163–176.

Lee, R.Y., Joye, S.B., 2006. Seasonal patterns of nitrogen fixation and denitrification in oceanic mangrove habitats. Marine Ecology Progress Series 307, 127–141.

Lin, B.B., Dushoff, J., 2004. Mangrove filtration of anthropogenic nutrients in the Rio Coco Solo, Panama. Management of Environmental Quality: An International Journal 15 (2), 131–142. http://doi.org/10.1108/14777830410523071.

Linto, N., Barnes, J., Ramachandran, R., Divia, J., Ramachandran, P., Upstill-Goddard, R., 2014. Carbon dioxide and methane emissions from mangrove-associated waters of the Andaman Islands, Bay of Bengal. Estuaries and Coasts 37, 381–398.

Livesley, S.J., Andrusiak, S.M., 2012. Temperate mangrove and salt marsh sediments are a small methane and nitrous oxide source but important carbon store. Estuarine, Coastal and Shelf Science 97, 19–27.

Lovelock, C.E., 2008. Soil respiration and belowground carbon allocation in mangrove forests. Ecosystems 11, 342–354.

Lovelock, C., Feller, I.C., Mckee, K.L., Engelbrecht, B., Ball, M., 2004. The effect of nutrient enrichment on growth, photosynthesis and hydraulic conductance of dwarf mangrovees in Panama. Functional Ecology 18, 25–33.

Lovelock, C., Feller, I.C., Ball, M., Engelbrecht, B., Ewe, M.L., 2006. Difference in plant function in phosphorous- and nitrgoen- limited mangrove ecosystems. New Phytologist 172, 514–522.

Ludovici, K.H., Zarnoch, S.J., Richter, D.D., 2002. Modeling in-situ pine root decomposition using data from a 60-year chronosequence. Canadian Journal of Forest Research 32, 1675–1684.

Lugo, A.E., 1990. Fringe wetlands. In: Lugo, A.E., Brinson, M., Brown, S. (Eds.), Forested Wetlands. Elsevier, Amsterdam, pp. 143–169.

Lugo, A.E., Snedaker, S.C., 1974. The ecology of mangroves. Annual Review of Ecology and Systematics 5, 39–64.

Lugomela, C., Bergman, B., 2002. Biological N_2-fixation on mangrove pneumatophores: preliminary observations and perspectives. AMBIO: A Journal of the Human Environment 31, 612–613.

Lynch, J.C., Meriwether, J.R., Mckee, B.A., Vera-Herrera, F., Twilley, R.R., 1989. Recent accretion in mangrove ecosystems based on ^{137}Cs and ^{210}Pb. Estuaries 12, 284–299.

Macdonald, B., White, I., Åström, M., Keene, A., Melville, M.D., Reynolds, J.K., 2007. Discharge of weathering products from acid sulfate soils after a rainfall event, Tweed River, eastern Australia. Applied Geochemistry 22, 2695–2705.

Macnae, W., 1966. Mangroves in eastern and southern Australia. Australian Journal of Botany 14, 67–104.

Maher, D.T., Santos, I.R., Golsby-Smith, L., Gleeson, J., Eyre, B.D., 2013. Groundwater-derived dissolved inorganic and organic carbon exports from a mangrove tidal creek: the missing mangrove carbon sink? Limnology and Oceanography 58, 475–488.

Mahmood, H., Saberi, O., Japar Sidik, B., Misri, K., 2005. Litter flux in Kuala Selangor nature park mangrove forest, Malaysia. Indian Journal of Forestry 28, 233–238.

Mancera-Pineda, J.E., 2003. The Contribution of Mangrove Outwelling to Coastal Food Webs as a Function of Environmental Settings (Dissertation). University of Louisiana at Lafayette.

Mann, F.D., Steinke, T.D., 1992. Biological nitrogen fixation (acetylene reduction) associated with decomposing *Avicennia marina* leaves in the Beachwood Mangrove Nature Reserve. South African Journal of Botany 58, 533–536.

Marchand, C., Lallier-Verges, E., Baltzer, F., 2003. The composition of sedimentary organic matter in relation to the dynamic features of a mangrove-fringed coast in French Guiana. Estuarine, Coastal and Shelf Science 56, 119–130.

Marchand, C., Alberic, P., Lallier-Verges, E., Baltzer, F., 2006. Distribution and characteristics of dissolved organic matter in mangrove sediment pore waters along the coastline of French Guiana. Biogeochemistry 81, 59–75.

Matthews, E., Fung, I., 1987. Methane emission from natural wetlands: global distribution, area, and environmental characteristics of sources. Global Biogeochemical Cycles 1, 61–86.

Matthijs, S., Tack, J., Van Speybroeck, D., Koedam, N., 1999. Mangrove species zonation and soil redox state, sulphide concentration and salinity in Gazi Bay (Kenya), a preliminary study. Mangroves and Salt Marshes 3, 243–249.

McKee, K.L., 1993. Soil physiochemical patterns and mangrove species distribution - reciprocal effects? Journal of Ecology 81, 477–487.

McKee, K.L., 2011. Biophysical controls on accretion and elevation change in Caribbean mangrove ecosystems. Estuarine, Coastal and Shelf Science 91, 475–483.

McKee, K.L., Faulkner, P.L., 2000. Restoration of biogeochemical function in mangrove forests. Restoration Ecology 8, 247–259.

McKee, K.L., Mendelssohn, J.A., Hester, M.W., 1988. Reexamination of pore water sulfide concentrations and redox potentials near the aerial roots of *Rhizpohora Mangle* and *Avicennia Germinans*. American Journal of Botany 75, 1352–1359.

McKee, K.L., Cahoon, D.R., Feller, I.C., 2007. Caribbean mangroves adjust to rising sea level through biotic controls on change in soil elevation. Global Ecology and Biogeography 16, 545–556.

Meyer, R.L., Risgaard-Petersen, N., Allen, D.E., 2005. Correlation between anammox activity and microscale distribution of nitrite in a subtropical mangrove sediment. Applied and Environmental Microbiology 71, 6142–6149.

Middelburg, J.J., Klver, G., Nieuwenhuize, J., Wielemaker, A., De Haas, W., Vlug, T., Van Der Nat, J.F.W.A., 1996. Organic matter mineralization in intertidal sediments along an estuarine gradient. Marine Ecology Progress Series 132, 157–168.

Middleton, B.A., McKee, K.L., 2001. Degradation of mangrove tissues and implications for peat formation in Belizean island forests. Journal of Ecology 89, 818–828.

Millero, F.J., Hiscock, W.T., Huang, F., Roche, M., Zhang, J.Z., 2001. Seasonal variation of the carbonate system in Florida Bay. Bulletin of Marine Science 68, 101–123.

Miyajima, T., Tsuboi, Y., Tanaka, Y., Koike, I., 2009. Export of inorganic carbon from two Southeast Asian mangrove forests to adjacent estuaries as estimated by the stable isotope composition of dissolved inorganic carbon. Journal of Geophysical Research: Biogeosciences 114.

Mohammed, S.M., Johnstone, R.W., 2002. Porewater nutrient profiles and nutrient sediment-water exchange in a tropical mangrove waterway, Mapopwe creek, Chwaka Bay, Zanzibar. East African Wild Life Society. African Journal of Ecology 40, 172–178.

Morell, J.M., Corredor, J.E., 1993. Sediment nitrogen trapping in a mangrove lagoon. Estuarine, Coastal and Shelf Science 37, 203–212.

Morris, J.T., Bowden, W.B., 1986. A mechanistic, numerical model of sedimentation, mineralization, and decomposition for marsh sediments. Soil Science Society of America Journal 50, 96–105.

Morris, A.W., Bale, A.J., Howland, R.J.M., 1982. The dynamics of estuarine manganese cycling. Estuarine, Coastal and Shelf Science 14 (2), 175–192. http://doi.org/10.1016/S0302-3524(82)80044-3.

Morse, J.W., Zullig, J.J., Bernstein, L.D., Millero, F.J., Milne, P., Mucci, A., Choppin, G.R., 1985. Chemistry of calcium carbonate-rich shallow water sediments in the Bahamas. American Journal of Science (United States) 285.

Muñoz-Hincapié, M., Morell, J.M., Corredor, J.E., 2002. Increase of nitrous oxide flux to the atmosphere upon nitrogen addition to red mangroves sediments. Marine Pollution Bulletin 44, 992–996.

Murray, B., Pendleton, L., Jenkins, W., Sifleet, S., 2011. Green Payments for Blue Carbon: Economic Incentives for Protecting Threatened Coastal Habitats. Nicholas Institute for Environmental Policy Solutions, p. 52.

Myint, A., 1986. Preliminary study of nitrogen fixation in Malasyan mangrove soils. Workshop on mangrove ecosystem dynamics. In: UNDP/UNESCO Regional Project (RAS/79/002), New Delhi, pp. 181–195.

Naidoo, G., 2009. Differential effects of nitrogen and phosphorus enrichment on growth of dwarf *Avicennia marina* mangroves. Aquatic Botany 90, 184–190.

Nath, B., Birch, G., Chaudhuri, P., 2013. Trace metal biogeochemistry in mangrove ecosystems: a comparative assessment of acidified (by acid sulfate soils) and non-acidified sites. The Science of the Total Environment 463, 667–674.

Nedwell, D.B., 1975. Inorganic nitrogen metabolism in a eutrophicated tropical mangrove estuary. Journal of Water Research 9, 221–231 (268).

Nedwell, D., Blackburn, T., Wiebe, W., 1994. Dynamic nature of the turnover of organic carbon, nitrogen and sulphur in the sediments of a Jamaican mangrove forest. Marine Ecology Progress Series 223–231.

Nielsen, L.P., 1992. Denitrification in sediment determined from nitrogen isotope pairing. FEMS Microbiology Letters 86, 357–362.

Nielsen, L.P., Christensen, P.B., Revsbech, N.P., Sørensen, J., 1990. Denitrification and photosynthesis in stream sediment studied with microsensor and whole-core techniques. Limnology and Oceanography 35, 1135–1144.

Nielsen, O.I., Kristensen, E., Macintosh, D.J., 2003. Impact of fiddler crabs (Uca spp.) on rates and pathways of benthic mineralization in deposited mangrove shrimp pond waste. Journal of Experimental Marine Biology and Ecology 289, 59–81.

Nixon, S.W., 1980. Between coastal marshes and coastal waters-a review of twenty years of speculation and research on the role of salt marshes in estuarine productivity and water chemistry. In: Hamilton, P., Macdonald, K.B. (Eds.), Estuarine and Wetland Processes with Emphasis on Modeling. Plenum Press, New York, pp. 437–525.

Odum, W.E., 1984. Dual-gradient concept of detritus transport and processing in estuaries. Bulletin of Marine Science 35, 510–521.

Odum, W.E., Heald, E.J., 1975. The detritus-based food web of an estuarine mangrove community. In: Cronin, L.E. (Ed.), Estuarine Research. Academic Press, New York, pp. 265–286.

Ólafsson, E., Buchmayer, S., Skov, M.W., 2002. The East African decapod crab *Neosarmatium meinerti* (de Man) sweeps mangrove floors clean of leaf litter. AMBIO: A Journal of the Human Environment 31, 569–573.

Osland, M.J., Spivak, A.C., Nestlerode, J.A., Lessmann, J.M., Almario, A.E., Heitmuller, P.T., Russell, M.J., Krauss, K.W., Alvarez, F., Dantin, D.D., Harvey, J.E., From, A.S., Cormier, N., Stagg, C.L., 2012. Ecosystem development after mangrove wetland creation: plant-soil change across a 20-year chronosequence. Ecosystems 15, 848–866.

Ovalle, A.R.C., Rezende, C.E., Lacerda, L.D., Silva, C.A.R., 1990. Factors affecting the hydrochemistry of a mangrove tidal creek, Sepetiba Bay, Brazil. Estuarine, Coastal and Shelf Science 31, 639–650.

Palmer, M.A., Filoso, S., Fanelli, R.M., 2014. From ecosystems to ecosystem services: stream restoration as ecological engineering. Ecological Engineering 65, 62–70.

Parkinson, R.W., Delaune, R.D., White, J.R., 1994. Holocene sea-level rise and the fate of mangrove forests within the wider caribbean region. Journal of Coastal Research 10, 1077–1086.

Pelegri, S.P., Twilley, R.R., 1998. Heterotrophic nitrogen fixation (acetylene reduction) during leaf-litter decomposition of two mangrove species from South Florida, USA. Marine Biology 131, 53–61.

Pelegri, S.P., Rivera-Monroy, V.H., Twilley, R.R., 1997. A comparison of nitrogen fixation (acetylene reduction) among three species of mangrove litter, sediments, and pneumatophores in south Florida, USA. Hydrobiologia 356, 73–79.

Penha-Lopes, G., Kristensen, E., Flindt, M., Mangion, P., Bouillon, S., Paula, J., 2010. The role of biogenic structures on the biogeochemical functioning of mangrove constructed wetlands sediments—A mesocosm approach. Marine Pollution Bulletin 60, 560–572.

Pons, L., Van Breemen, N., Driessen, P., 1982. Physiography of coastal sediments and development of potential soil acidity. Acid Sulfate Weathering 1–18.

Poret, N., Twilley, R.R., Rivera-Monroy, V.H., Coronado-Molina, C., 2007. Belowground decomposition of mangrove roots in Florida Coastal Everglades. Estuaries and Coasts 30, 491–496.

Purvaja, R., Ramesh, R., 2001. Natural and anthropogenic methane emission from coastal wetlands of South India. Environmental Management 27, 547–557.

Purvaja, R., Ramesh, R., Ray, A., Rixen, T., 2008. Nitrogen cycling: a review of the processes, transformations and fluxes in coastal ecosystems. Current Science-Bangalore 94, 1419.

Rajkaran, A., Adams, J.B., 2007. Mangrove litter production and organic carbon pools in the Mngazana Estuary, South Africa. African Journal of Aquatic Science 32, 17–25.

Ralison, O.H., Borges, A.V., Dehairs, F., Middelburg, J., Bouillon, S., 2008. Carbon biogeochemistry of the Betsiboka estuary (north-western Madagascar). Organic Geochemistry 39, 1649–1658.

Ramesh, R., Purvaja, R., Neetha, V., Divia, J., Barnes, J., Upstill-Goddard, R., 2007. CO_2 and CH_4 emissions from Indian mangroves and its surrounding waters. In: Greenhouse Gas and Carbon Balances in Mangrove Coastal Ecosystems. Gendai Tosho, Kanagawa, pp. 139–151.

Reis, S., Bekunda, M., Howard, C.M., Karanja, N., Winiwarter, W., Yan, X., Bleeker, A., Sutton, M.A., 2016. Synthesis and review: tackling the nitrogen management challenge: from global to local scales. Environmental Research Letters 11, 120205.

Reis, C.R.G., Nardoto, G.B., Oliveira, R.S., 2017a. Global overview on nitrogen dynamics in mangroves and consequences of increasing nitrogen availability for these systems. Plant and Soil 410, 1–19.

Reis, C.R.G., Nardoto, G.B., Rochelle, A.L.C., Vieira, S.A., Oliveira, R.S., 2017b. Nitrogen dynamics in subtropical fringe and basin mangrove forests inferred from stable isotopes. Oecologia 183, 841–848.

Ren, H., Jian, S., Lu, H., Zhang, Q., Shen, W., Han, W., Yin, Z., Guo, Q., 2008. Restoration of mangrove plantations and colonisation by native species in Leizhou bay, South China. Ecological Research 23, 401–407.

Rezende, C., Lacerda, L., Ovalle, A., Silva, L., 2007. Dial organic carbon fluctuations in a mangrove tidal creek in Sepetiba bay, Southeast Brazil. Brazilian Journal of Biology 67, 673–680.

Rimstidt, J.D., Vaughan, D.J., 2003. Pyrite oxidation: a state-of-the-art assessment of the reaction mechanism. Geochimica et Cosmochimica Acta 67, 873–880.

Rivera-Monroy, V.H., Twilley, R.R., 1996. The relative role of denitrification and immobilization in the fate of inorganic nitrogen in mangrove sediments (Terminos Lagoon, Mexico). Limnology and Oceanography 41, 284–296.

Rivera-Monroy, V.H., Day, J.W., Twilley, R.R., Veraherrera, F., Coronado-Molina, C., 1995a. Flux of nitrogen and sediment in a fringe mangrove forest in Terminos Lagoon, Mexico. Estuarine, Coastal and Shelf Science 40, 139–160.

Rivera-Monroy, V.H., Twilley, R.R., Boustany, R.G., Day, J.W., Vera-Herrera, F., Ramirez, M.D.C., 1995b. Direct denitrification in mangrove sediments in Terminos Lagoon, Mexico. Marine Ecology Progress Series 126, 97–109.

Rivera-Monroy, V.H., Torres, L.A., Bahamon, N., Newmark, F., Twilley, R.R., 1999. The potential use of mangrove forests as nitrogen sinks of shrimp aquaculture pond effluents: the role of denitrification. Journal of the World Aquaculture Society 30, 12–25.

Rivera-Monroy, V.H., Castañeda-Moya, E., Barr, J.G., Engel, V., Fuentes, J.D., Troxler, T.G., Twilley, R.R., Bouillon, S., Smith, T.J., O'Halloran, T.L., 2013. Current methods to evaluate net primary production and carbon budgets in mangrove forests. In: DeLaune, R.D., Reddy, K.R., Megonigal, P., Richardson, C. (Eds.), Methods in biogeochemistry of wetlands, III. Soil Science Society of America Book Series, pp. 243–288.

Robertson, A.I., Daniel, P.A., 1989. The influence of crabs on litter processing in high intertidal mangrove forests in tropical Australia. Oecologia 78, 191–198.

Rodelli, M.R., Gearing, J.N., Gearing, P.J., Marshall, N., Sasekumar, A., 1984. Stable isotope ratio as a tracer of mangrove carbon in Malaysian ecosystems. Oecologia 326–333.

Romero, I.C., Jacobson, M., Fuhrman, J.A., Fogel, M., Capone, D.G., 2012. Long-term nitrogen and phosphorus fertilization effects on N_2 fixation rates and nifH gene community patterns in mangrove sediments. Marine Ecology 33, 117–127.

Romigh, M.M., Davis, S.E., Rivera-Monroy, V.H., Twilley, R.R., 2006. Flux of organic carbon in a riverine mangrove wetland in the Florida Coastal Everglades. Hydrobiologia 569, 505–516.

Rosenfeld, J.K., 1979. Interstitial water and sediment chemistry of two cores from Florida Bay. Journal of Sedimentary Research 49.

Rovai, A.S., Twilley, R.R., Castañeda-Moya, E., Riul, P., Cifuentes-Jara, M., Manrow-Villalobos, M., Horta, P.A., Simonassi, J.C., Fonseca, A.L., Pagliosa, P.R., 2018. Global controls over organic carbon storage in mangrove soils. Nature Climate Change 8 (6), 534–538. https://doi.org/10.1038/s41558-018-0162-5.

Rützler, K., Feller, C., 1988. Mangrove swamp communities. Oceanus 30, 16–24.

Rysgaard, S., Risgaard-Petersen, N., Peter Sloth, N., Jensen, K., Nielsen, L.P., 1994. Oxygen regulation of nitrification and denitrification in sediments. Limnology and Oceanography 39, 1643–1652.

Saenger, P., Hegerl, E., Davie, J.D., 1983. Global status of mangrove ecosystems. In: International Union for Conservation of Nature and Natural Resources.

Saintilan, N., 2004. Relationships between estuarine geomorphology, wetland extent and fish landings in New South Wales estuaries. Estuarine, Coastal and Shelf Science 61, 591–601.

Saintilan, N., Rogers, K., Mckee, K., 2009. Salt marsh-mangrove interactions in Australasia and the Americas. In: Perillo, G.M.E., Wolanski, E., Cahoon, D., Brinson, M. (Eds.), Coastal Wetlands: An Integrated Ecosystem Approach. Elsevier, Netherlands, pp. 855–884.

Saintilan, N., Wilson, N.C., Rogers, K., Rajkaran, A., Krauss, K.W., 2014. Mangrove expansion and salt marsh decline at mangrove poleward limits. Global Change Biology 20, 147–157.

Saintilan, N., Rogers, K., Mazumber, D., Woodroffe, C., 2013. Allochthonous and autochthonous contributions to carbon accumulation and carbon store in southeastern Australian coastal wetlands. Estuarine, Coastal and Shelf Science 128, 84–92. http://doi.org/10.1016/j.ecss.2013.05.010.

Sammut, J., White, I., Melville, M., 1996. Acidification of an estuarine tributary in eastern Australia due to drainage of acid sulfate soils. Marine and Freshwater Research 47, 669–684.

Sánchez-Carrillo, S., Sánchez-Andrés, R., Alatorre, L.C., Angeler, D., Álvarez-Cobelas, M., Arreola-Lizárraga, J., 2009. Nutrient fluxes in a semi-arid microtidal mangrove wetland in the Gulf of California. Estuarine, Coastal and Shelf Science 82, 654–662.

Santantonio, D., Hermann, R., Overton, W., 1977. Root biomass studies in forest ecosystems. Pedobiologia.

Schmid, M.C., Risgaard-Petersen, N., Van De Vossenberg, J., Kuypers, M.M., Lavik, G., Petersen, J., Hulth, S., Thamdrup, B., Canfield, D., Dalsgaard, T., 2007. Anaerobic ammonium-oxidizing bacteria in marine environments: widespread occurrence but low diversity. Environmental Microbiology 9, 1476–1484.

Sengupta, A., Chaudhuri, S., 1991. Ecology of heterotrophic dinitrogen fixation in the rhizosphere of mangrove plant community at the Ganges river estuary in India. Oecologia 87, 560–564.

Shaiful, A.A.A., 1987. Nitrate reduction in mangrove swamps. Malaysian Applied Biology 16, 361–367.

Sheridan, R.P., 1991. Epicaulous, nitrogen-fixing microepiphytes in a tropical mangal community, Guadeloupe, French West Indies. Biotropica 530–541.

Sheridan, P.F., 1992. Comparative habitat utilization by estuarine macrofauna within the mangrove ecosystem of Rookery Bay, Florida. Bulletin of Marine Science 50, 21–39.

Sherman, R.E., Fahey, T.J., Howarth, R.W., 1998. Soil-plant interactions in a neotropical mangrove forest: iron, phosphorus and sulfur dynamics. Oecologia 115, 553–563.

Shiau, Y.-J., Lin, M.-F., Tan, C.-C., Tian, G., Chiu, C.-Y., 2017. Assessing N_2 fixation in estuarine mangrove soils. Estuarine, Coastal and Shelf Science 189, 84–89.

Silva, C.A.R., Mozeto, A.A., Ovalle, Á.R.C., 1998. Distribution and fluxes as macrodetritus of phosphorus in red mangroves, Sepetiba Bay, Brazil. Mangroves and Salt Marshes 2, 37–42.

Simard, M., Zhang, K.Q., Rivera-Monroy, V.H., Ross, M.S., Ruiz, P.L., Castaneda-Moya, E., Twilley, R.R., Rodriguez, E., 2006. Mapping height and biomass of mangrove forests in Everglades National Park with SRTM elevation data. Photogrammetric Engineering and Remote Sensing 72, 299–311.

Simard, M., Rivera-Monroy, V.H., Mancera-Pineda, J.E., Castaneda-Moya, E., Twilley, R.R., 2008. A systematic method for 3D mapping of mangrove forests based on Shuttle Radar Topography Mission elevation data, ICEsat/GLAS waveforms and field data: application to Cienaga Grande de Santa Marta, Colombia. Remote Sensing of Environment 112, 2131–2144.

Simpson, J., Gong, W., Ong, J., 1997. The determination of the net fluxes from a mangrove estuary system. Estuaries and Coasts 20, 103–109.

Singer, P.C., Stumm, W., 1970. Acidic mine drainage: the rate-determining step. Science 167, 1121–1123.

Sousa, W., Dangremond, E., 2011. Trophic Interactions in Coastal and Estuarine Mangrove Forest Ecosystems. In: Treatise on Estuarine and Coastal Science, vol. 6, pp. 43–93.

Spalding, M., 2010. World Atlas of Mangroves. Routledge.

Spalding, M., Blasco, F., Field, C., 1997. World Mangrove Atlas. International Society for Mangrove Ecosystems, Okinawa, Japan.

Stieglitz, T.C., Clark, J.F., Hancock, G.J., 2013. The mangrove pump: the tidal flushing of animal burrows in a tropical mangrove forest determined from radionuclide budgets. Geochimica et Cosmochimica Acta 102, 12–22.

Suárez-Abelenda, M., Ferreira, T.O., Camps-Arbestain, M., Rivera-Monroy, V.H., Macías, F., Nóbrega, G.N., Otero, X.L., 2014. The effect of nutrient-rich effluents from shrimp farming on mangrove soil carbon storage and geochemistry under semi-arid climate conditions in northern Brazil. Geoderma 213, 551–559. http://doi.org/10.1016/j.geoderma.2013.08.007.

Sutherland, J.P., 1980. Dynamics of the epibenthic community on roots of the mangrove Rhizophora mangle, at Bahia de Buche, Venezuela. Marine Biology 58, 75–84.

Sutula, M.A., Perez, B.C., Reyes, E., Childers, D.L., Davis, S., Day, J.W., Rudnick, D., Sklar, F., 2003. Factors affecting spatial and temporal variability in material exchange between the Southern Everglades wetlands and Florida Bay (USA). Estuarine, Coastal and Shelf Science 57, 757–781.

Thamdrup, B., Dalsgaard, T., Jensen, M.M., Petersen, J., 2004. Anammox and the marine N cycle. Geochimica et Cosmochimica Acta 68, A325.

Thamdrup, B., Dalsgaard, T., Jensen, M.M., Ulloa, O., Farías, L., Escribano, R., 2006. Anaerobic ammonium oxidation in the oxygen-deficient waters off northern Chile. Limnology and Oceanography 51, 2145–2156.

Thom, B.G., 1967. Mangrove ecology and deltaic geomorphology: Tabasco, Mexico. Journal of Ecology 55, 301–343.

Thom, B.G., 1982. Mangrove ecology-a geomorphological perspective. In: Clough, B.F. (Ed.), Mangrove Ecosystems in Australia. Australian National University Press, Canberra, pp. 3–17.

Thom, B.G., 1984. Coastal landforms and geomorphic processes. In: Snedaker, S.C., Snedaker, J.G. (Eds.), The Mangrove Ecosystem: Research Methods. UNESCO, United Kingdom, pp. 3–17.

Thomas, N., Lucas, R., Bunting, P., Hardy, A., Rosenqvist, A., Simard, M., 2017. Distribution and drivers of global mangrove forest change, 1996–2010. PLoS One 12, e0179302.

Thongtham, N., Kristensen, E., 2005. Carbon and nitrogen balance of leaf-eating sesarmid crabs (*Neoepisesarma versicolor*) offered different food sources. Estuarine, Coastal and Shelf Science 65, 213–222.

Tilman, D., 1985. The resource-ratio hypothesis of plant succession. The American Naturalist 125, 827–852.

Troxler, T.G., Gaiser, E., Barr, J., Fuentes, J.D., Jaffe, R., Childers, D.L., Collado-Vides, L., Rivera-Monroy, V.H., Castaneda-Moya, E., Anderson, W., Chambers, R., Chen, M., Coronado-Molina, C., Davis, S.E., Engel, V., Fitz, C., Fourqurean, J., Frankovich, T., Kominoski, J., Madden, C., Malone, S.L., Oberbauer, S.F., Olivas, P., Richards, J., Saunders, C., Schedlbauer, J., Scinto, L.J., Sklar, F., Smith, T., Smoak, J.M., Starr, G., Twilley, R.R., Whelan, K., 2013. Integrated carbon budget models for the Everglades terrestrial-coastal oceanic gradient - current status and needs for inter-site comparisons. Oceanography 26, 98–107.

Troxler, T.G., Barr, J.G., Fuentes, J.D., Engel, V., Anderson, G., Sanchez, C., Lagomasino, D., Price, R., Davis, S.E., 2015. Component-specific dynamics of riverine mangrove CO_2 efflux in the Florida coastal Everglades. Agricultural and Forest Meteorology 213, 273–282.

Twilley, R.R., 1985. The exchange of organic-carbon in basin mangrove forests in a southwest Florida estuary. Estuarine, Coastal and Shelf Science 20, 543–557.

Twilley, R.R., 1988. Coupling of mangroves to the productivity of estuarine and coastal waters. In: Janson, B.O. (Ed.), Coastal-Offshore Ecosystem Interactions. Springer-Verlag, Germany, pp. 155–180.

Twilley, R.R., 1995. Properties of mangrove ecosystems related to the energy signature of coastal environments. In: Hall, C.a.S. (Ed.), Maximum Power: The Ideas and Aplications of H.T. Odum. Niwot, Colorado. University Press of Colorado, pp. 43–62.

Twilley, R.R., 1997. Mangrove wetlands. In: Messina, Connor (Eds.), Southern Forested Wetlands: Ecology and Management. CRC Press, Boca Raton, FL, pp. 445–473.

Twilley, R.R., Chen, R.H., 1998. A water budget and hydrology model of a basin mangrove forest in Rookery Bay, Florida. Marine and Freshwater Research 49, 309–323.

Twilley, R.R., Rivera-Monroy, V.H., 2005. Developing performance measures of mangrove wetlands using simulation models of hydrology, nutrient biogeochemistry, and community dynamics. Journal of Coastal Research 79–93.

Twilley, R.R., Chen, R.H., Hargis, T., 1992. Carbon sinks in mangroves and theirs implications to carbon budget of tropical coastal ecosystems. Water, Air, and Soil Pollution 64, 265–288.

Twilley, R.R., Snedaker, S.C., Yanez-Arancibia, A., Medina, E., 1996. Biodiversity and ecosystem processes in tropical estuaries: perspectives from mangrove ecosystems. In: Mooney, H., Cushman, H., Medina, E., Sala, O.E., Schulze, E.D. (Eds.), Functional Roles of Biodiversity: Global Perspectives. John Wiley and Sons, New York, pp. 327–370.

Twilley, R.R., Pozo, M., Garcia, V.H., Rivera-Monroy, V.H., Zambrano, R., Bodero, A., 1997. Litter dynamics in riverine mangrove forests in the Guayas River estuary, Ecuador. Oecologia 111, 109–122.

Twilley, R.R., Rivera-Monroy, V.H., Chen, R.H., Botero, L., 1998. Adapting an ecological mangrove model to simulate trajectories in restoration ecology. Marine Pollution Bulletin 37, 404–419.

Twilley, R.R., Castañeda-Moya, E., Rivera-Monroy, V.H., Rovai, A., 2017. Productivity and carbon dynamics in mangrove wetlands. In: Rivera-Monroy, V.H., Lee, S.Y., Kristensen, E., Twilley, R.R. (Eds.), Mangrove Ecosystems: A Global Biogeographic Perspective Structure, Function and Services. Springer.

Twilley, R.R., Rovai, A.S., Riul, P., 2018. Coastal morphology explains global blue carbon distributions. Frontiers in Ecology and the Environment 16 (9), 1–6. http://doi.org/10.1002/fee.1937.

Uchino, F., Hambali, G.G., Yatazawa, M., 1984. Nitrogen-fixing bacteria from warty lenticellate bark of a mangrove tree, *Bruguiera gymnorrhiza* (L.) Lamk. Applied and Environmental Microbiology 47, 44–48.

Van Der Valk, A.G., Attiwill, P.M., 1984a. Decomposition of leaf and root litter of *Avicennia marina* at Westernport Bay, Victoria, Australia. Aquatic Botany 18, 205–221.

Van Der Valk, A.G., Attiwill, P.M., 1984b. Acetylene reduction in an *Avicennia marina* community in southern Australia. Australian Journal of Botany 32, 157–164.

Viner, A.B., 1982. Nitrogen fixation and denitrification in sediments of two Kenyan lakes. Biotropica 14, 91–98.

Vovides, A.G., Bashan, Y., López-Portillo, J.A., Guevara, R., 2011. Nitrogen fixation in preserved, reforested, naturally regenerated and impaired mangroves as an indicator of functional restoration in mangroves in an arid region of Mexico. Restoration Ecology 19, 236–244.

Wafar, S., Untawale, A., Wafar, M., 1997. Litter fall and energy flux in a mangrove ecosystem. Estuarine, Coastal and Shelf Science 44, 111–124.

Watson, J.G., 1928. Mangrove Forests of the Malay Peninsula, first ed. Fraser & Neave, Singapore.

Wattayakorn, G., Wolanski, E., Kjerfve, B., 1990. Mixing, trapping and outwelling in the Klong Ngao mangrove swamp, Thailand. Estuarine, Coastal and Shelf Science 31, 667–688.

Webb, E.C., Mendelssohn, I.A., Wilsey, B.J., 1995. Causes for vegetation dieback in a Louisiana salt marsh: a bioassay approach. Aquatic Botany 51, 281–289.

Whigham, D.F., Verhoeven, J.T., Samarkin, V., Megonigal, P.J., 2009. Responses of Avicennia germinans (Black Mangrove) and the soil microbial community to nitrogen addition in a hypersaline wetland. Estuaries and Coasts 32, 926–936.

Woitchik, A.F., Ohowa, B.O., Kazungu, J.M., Rao, R.G., Goeyens, L., Dehairs, F., 1997. Nitrogen enrichment during decomposition of mangrove leaf litter in an east African coastal lagoon (Kenya) Relative importance of biological nitrogen fixation. Biogeochemistry 39, 15–35.

Woodroffe, C.D., 1985. Studies of a mangrove basin, Tuff Crater, New Zealand: II. Comparison of volumetric and velocity-area methods of estimating tidal flux. Estuarine, Coastal and Shelf Science 20, 431–445.

Woodroffe, C., 1992. Mangrove sediments and geomorphology. In: Robertson, A.I., Alongi, D.M. (Eds.), Coastal and Estuarine Studies. American Geophysical Union, Washington, DC, pp. 7–41.

Woodroffe, C.D., 2002. Coasts: Form, Process and Evolution. Cambridge University Press.

Woodroffe, C.D., Bardsley, K.N., Ward, P.J., Hanley, J.R., 1988. Production of mangrove litter in a macrotidal embayment, Darwin Harbour, N.T., Australia. Estuarine, Coastal and Shelf Science 26, 581–598.

Wösten, J.H.M., De Willigen, P., Tri, N.H., Lien, T.V., Smith, S.V., 2003. Nutrient dynamics in mangrove areas of the Red River Estuary in Vietnam. Estuarine, Coastal and Shelf Science 57, 65–72.

Young, M., Gonneea, M.E., Herrera-Silveira, J., Paytan, A., 2005. Export of dissolved and particulate carbon and nitrogen from a mangrove-dominated lagoon, Yucatan Peninsula, Mexico. International Journal of Ecology and Environmental Sciences 31, 189–202.

Zablocki, J.A., Andersson, A.J., Bates, N.R., 2011. Diel aquatic CO_2 system dynamics of a Bermudian mangrove environment. Aquatic Geochemistry 17, 841–859.

Zuberer, D.A., Silver, W.S., 1978. Biological dinitrogen fixation (acetylene reduction) associated with Florida mangroves. Applied and Environmental Microbiology 35, 567–575.

Further Reading

Aizpuru, M., Blasco, F., 2000. Global Assessment of Cover Change of the Mangrove Forests Using Satellite Imagery at Medium to High Resolution. Joint Research Center, Ispra, Italy.

Alongi, D.M., Christoffersen, P., 1992. Benthic infauna and organism-sediment relations in a shallow, tropical coastal area: influence of outwelled mangrove detritus and physical disturbance. Marine Ecology Progress Series 56, 229–245.

Alongi, D.M., Wattayakorn, G., Boyle, S., Tirendi, F., Payn, C., Dixon, P., 2004. Influence of roots and climate on mineral and trace element storage and flux in tropical mangrove soils. Biogeochemistry 69, 105–123.

Ayukai, T., Wolanski, E., 1997. Importance of biologically mediated removal of fine sediments from the Fly River plume, Papua New Guinea. Estuarine, Coastal and Shelf Science 44, 629–639.

Dutta, M.K., Mukherjee, R., Jana, T.K., Mukhopadhyay, S.K., 2015. Biogeochemical dynamics of exogenous methane in an estuary associated to a mangrove biosphere; the Sundarbans, NE coast of India. Marine Chemistry 170, 1–10.

Fatoyinbo, T.L., Feliciano, E.A., Lagomasino, D., Lee, S.K., Trettin, C., 2017. Estimating mangrove aboveground biomass from airborne LiDAR data: a case study from the Zambezi River delta. Environmental Research Letters 13 (2018), 025012. https://doi.org/10.1088/1748-9326/aa9f03.

Kaiser, D., Kowalski, N., Böttcher, M.E., Yan, B., Unger, D., 2015. Benthic nutrient fluxes from mangrove sediments of an anthropogenically impacted estuary in southern China. Journal of Marine Science and Engineering 3, 466–491.

Robertson, A.I., 1986. Leaf-burying crabs: their influence on energy flow and export from mixed mangrove forests (Rhizophora spp) in northeastern Australia. Journal of Experimental Marine Biology and Ecology 102, 237–248.

Robertson, A.I., Alongi, D.M., Daniel, P.A., Boto, K.G., 1988. How much mangrove detritus enters the Great Barrier Reef lagoon? In: Proceedings of the 6th International Coral Reef Symposium, 1988 Australia, pp. 601–606.

PART VI

COASTAL WETLAND RESTORATION AND MANAGEMENT

CHAPTER

22

Tidal Marsh Creation

Stephen W. Broome[1], Christopher B. Craft[2], Michael R. Burchell[3]

[1]Department of Crop and Soil Sciences, North Carolina State University, Raleigh, NC, United States; [2]School of Public and Environmental Affairs, Indiana University, Bloomington, IN, United States; [3]Department of Biological and Agricultural Engineering, North Carolina State University, Raleigh, NC, United States

1. INTRODUCTION

Tidal salt and brackish water marshes are ecotones between land and sea, which occur in the upper intertidal zone of sheltered coastal areas such as estuaries and bays and behind barrier islands. Their hydrology is characterized by regular or irregular inundation by tidewater and subsequent drainage. Tidal effects produce distinct zonation of the herbaceous vegetation, which is related to frequency, duration, and depth of inundation as well as salinity. In addition, the ebb and flow of tides connects the marsh to the adjacent waterbody, and tides are regarded as an energy subsidy that increases primary productivity (Mendelssohn and Kuhn, 2003) and facilitates the exchange of organic carbon, mineral nutrients, sediments, aquatic organisms, and other material. The emergent vegetation consists of a limited number of salt-tolerant species, most commonly grasses, sedges, or rushes. Plant species diversity decreases as salinity increases.

Tidal marshes are productive ecosystems that serve as nursery grounds and habitat for aquatic organisms (Minello et al., 2003, 2008) and food and habitat for wildlife. In many coastal environments, much of the primary production that occurs in tidal marshes is exported to adjacent waters in the form of detritus, which becomes a part of the estuarine food web providing energy for bacteria, fungi, worms, oysters, crabs, shrimp, fish, etc. Other important functions of tidal marshes include buffering of storm surges, storing floodwater, protecting shorelines from erosion, stabilizing dredged material, dampening the effects of waves, trapping waterborne sediments, nutrient cycling and transformations, serving as reservoirs of nutrients, and storage of organic carbon (Mitsch and Gosselink, 2000). Socioeconomic services to humans include esthetics, ecotourism, and education (Peterson et al., 2008). Ecosystem functions and services that benefit people can be

Coastal Wetlands
https://doi.org/10.1016/B978-0-444-63893-9.00022-8

TABLE 22.1 Types of Ecosystem Services Provided by Inland and Coastal Wetlands

PROVISIONING	
	Food
	Freshwater
	Fiber and fuel
	Biochemical products
	Genetic materials
REGULATING	
	Climate regulation
	Hydrologic regimes
	Pollution control and detoxification
	Erosion protection
	Natural hazards
CULTURAL	
	Spiritual and inspirational
	Recreational
	Esthetics
	Educational
SUPPORTING	
	Soil formation
	Nutrient cycling
	Pollination

characterized as provisioning, regulating, cultural, and supporting (Millennium Ecosystem Assessment, 2005) (Table 22.1).

It is estimated that tidal marshes and mangroves contribute more than \$190,000/ ha^{-1} $year^{-1}$ of such services (Costanza et al., 2014), especially from their contributions to disturbance regulation, pollution control, and waste detoxification (Costanza et al., 1997). Services such as pollution control and waste detoxification, recreation, and provision of natural habitat and biodiversity (i.e., esthetics) are highly valued by the public and studies show that the value of ecosystem services increases with anthropogenic pressure (i.e., they are more highly valued in urban vs. rural landscapes) (Ghermandi et al., 2010). Created and restored wetlands, in particular, are highly valued for biodiversity enhancement, water quality improvement, and flood control (Ghermandi et al., 2010). For salt marshes, the median cost of such restorations in the developed world is more than \$67,000 USD/ha in 2010 dollars (Bayraktarov et al., 2016). This value is comparable to restoring equivalent acreage of oyster reef (\$66,000 ha^{-1}), less than restoring seagrass \$106,000 ha^{-1} and coral reef (nearly \$2 million USD ha) but more relative to mangrove restoration (\$39,000 ha^{-1}) (Bayraktarov et al., 2016). However, such costs associated with tidal marsh restoration are less than the value of the benefits they provide.

The area occupied by coastal marshes has been reduced over time, and existing marshes have been degraded because of human activities and natural processes. Anthropogenic sources of marsh loss include dredging, filling, dike and levee construction, drainage, urban and agricultural development, oil and gas exploration, and construction of port facilities, highways, bridges, and airports. Natural forces such as sea level rise, land subsidence, and erosion also result in losses of tidal marshes. Marsh functions and values may be lost because of pollutants such as oil or chemical spills.

The vulnerability of tidal marshes to anthropogenic and natural destruction and degradation has led to an interest in creation of new marshes to replace their lost ecosystem services. In many cases, tidal marsh creation is required by regulatory agencies to mitigate losses resulting from development activities.

2. PRINCIPLES AND TECHNIQUES OF TIDAL MARSH CREATION

Wetland creation is defined as the establishment of wetlands through human-induced changes in the landscape on sites where no wetland existed in the recent past (Lewis, 1990; Streever, 1999; Mitsch and Gosselink, 2000). Wetland creation differs from restoration in that restoration implies reestablishing hydrology and functions of a former wetland. We define tidal marsh creation in this paper as human-induced conversion of upland or subtidal habitat to tidal salt or brackish water marsh ecosystems characterized by emergent vegetation. Tidal freshwater marshes (<0.5 PSU) will not be addressed. Tidal marsh creation is accomplished by manipulating topography, hydrology, and soils, often followed by planting vegetation to create conditions that will, through succession, lead to self-sustaining ecosystems similar in structure and providing ecosystem services similar to natural tidal marshes in the area. Typical objectives of tidal marsh creation include dredged material stabilization (Landin et al., 1989; Streever, 2000), mitigation required by government regulations (Darnell and Smith, 2001; Hough and Robertson, 2009), shoreline erosion control (Maryland Department of the Environment, Wetlands and Waterways Program, 2006; Currin et al., 2010; Gittman et al., 2016; Sharma et al., 2016), accumulation of sediment for sea defense (Moller et al., 2001; Hofstede, 2003), and to create agricultural land (Chung, 2006). Tidal marsh creation may also occur expressly for the purpose of increasing the extent of marsh habitat and ecosystem services that are provided. Ecosystem services provided by tidal marshes are habitat and food web support, buffers against storm waves, shoreline stabilization, storage of floodwater, improving water quality, preservation of biodiversity, carbon storage, and socioeconomic services to humans (Peterson et al., 2008).

In Europe, tidal marshes are created using dredged material (Bernhardt and Handke, 1992), accretion enhancement techniques (Hofstede, 2003), and by realignment of coastal defenses such as clay banks that are used to protect the coast from storm tides and rising sea level (Miren et al., 2001; Crooks et al., 2002; Wolters et al., 2005; Garbutt et al., 2006).

These tidal marsh creation projects are passive in that, rather than planting, vegetation is allowed to naturally colonize the site. In the United Kingdom and elsewhere in Europe (Netherlands, Belgium, Germany), managed and unmanaged realignment of coastal defenses has led to reclamation of large areas of tidal marshes, some of which were reclaimed from the sea for agriculture hundreds of years ago (Wolters et al., 2005). Many agricultural sites were accidentally breached during storm surges. Recently, embankments were deliberately breached to recreate salt marshes for conservation and as a first line of defense against the sea (Wolters et al., 2005).

There are many examples of successful tidal marsh creation, but methods and techniques vary depending on local conditions. Tidal marsh creation requires careful planning that takes into account site selection, hydrology, vegetation, soil, and construction costs. It should be recognized that there are likely to be unknowns at any particular site, requiring an adaptive approach to define goals and identify potential techniques to achieve those goals (Zedler, 2001).

2.1 Site Selection

An important consideration in selecting a site is the relative value or importance of the ecosystem that will be replaced. For example, is converting an upland forest or grassland to a tidal marsh a positive gain for the environment? The same question arises if aquatic habitat is destroyed by fill material or sediment deposited at a river mouth or inlet that raises the elevation to create a tidal marsh. Answers to such questions may be determined to some extent by scientific data or by value judgments usually determined by environmental managers, regulators, or political processes. On the other hand, converting a dredged material disposal site, marginal farmland, flood-prone urban land, or an eroding shoreline to tidal marsh habitat would be seen by most as a positive action. Locating tidal marsh creation sites in areas that are protected from erosion is an important consideration in site selection. Waves and currents as well as severe storms are threats to establishment and long-term stability of coastal marshes.

The probability of successfully utilizing marsh creation techniques for shoreline erosion control is site-specific and depends on factors that include wave energy, currents, tidal amplitude, slope of the shoreline, width of the intertidal zone, and physical and chemical properties of the sediment. Fringe marshes along estuarine shorelines help to mitigate erosion that occurs because of rising sea level and severe storm events, and they provide productive intertidal habitat. Hard structures such as bulkheads, groins, rip rap, revetments, or rock sills are common methods of combating erosion to protect valuable coastal property, but these methods generally result in loss of fringe marshes and associated ecosystem services. The concept of "living shorelines" has emerged as a more ecologically sound method of controlling shoreline erosion while maintaining intertidal habitat (Sharma et al., 2016). Living shorelines range from planting vegetation alone to a combination of hard structures and vegetation (Gittman et al., 2016; O'Donnell, 2016), as well as oyster reef restoration (Currin et al., 2010). This hybrid approach utilizes structures to protect plantings along shorelines where vegetation might not otherwise persist because of high wave energy and currents (Currin et al., 2017) (Fig. 22.1). Studies have shown that the relatively narrow fringe marshes

FIGURE 22.1 (A) A portion of a bulkhead was replaced by creating an intertidal zone, constructing a rock sill, and planting *Spartina alterniflora*. (B) Development of the marsh near the end of the second growing season.

generally associated with living shoreline projects do provide ecosystem services that include habitat for fish and other marine organisms, wave attenuation, sediment trapping, and removal of nitrate in groundwater (Currin et al., 2010).

2.2 Conceptual Design

Natural reference marshes, preferably located in the same general area and landscape setting as the proposed tidal marsh creation, should be identified to serve as models or target ecosystems for each creation project. Observations and studies of reference marshes can be used to determine the dominant plant species, the upper and lower elevation limits of each plant community, the slope and surface topography, hydrology (e.g., tidal amplitude, frequency of inundation), and the depth, width, and sinuosity of creeks if they are present. Determining the precise elevation requirements of plant communities and creating those elevations along with proper slope and drainage in the newly created marsh are critical to success. Marsh creation is more difficult in areas that have narrow tide ranges because the elevation difference between upper and lower elevation limits of plant communities is small and leaves little room for error in preparing the surface to be planted (Broome, 1990). Proper planning should include a period where tidal stage is measured at a nearby reference and in the estuarine waters that will feed the created tidal marsh, which will help determine how much earthwork will be needed to achieve proper marsh surface elevations. Creeks or drainage channels should be installed in large creation projects to maximize tidal exchange and utilization by fish and other marine organisms (Zeff, 1999).

2.3 Hydrology

Hydrology is the dominant factor that determines development of the biological and physical characteristics of a wetland (Mitsch and Gosselink, 2000). Tidal flooding is the most obvious hydrologic factor that affects zonation of plant species, plant growth, soil chemical and physical properties, and biological processes in tidal marshes. Hydrology is largely determined by elevation, slope, and tidal regime, which interact to determine the area of the intertidal zone and the depth and duration of flooding that occurs. Subsurface hydrology may also affect these same physical and biological processes. For example, where marshes occur adjacent to uplands, groundwater seepage and runoff may be important hydrologic factors, adding freshwater and nutrients to the landward edge of the marsh (Harvey and Odum, 1990; Nuttle and Harvey, 1995). Rainfall, river flow, and wind effects also influence the hydrology of tidal marshes. A clear understanding of the hydrology of the site is critical for successful tidal marsh creation. As mentioned earlier, the best way to gain this understanding is through a monitoring phase that precedes the design phase. The fundamental requirement of tidal marsh creation is establishing a surface for plant growth within the intertidal zone at elevations that support the flora and fauna of the target communities. If the rate of sea level rise relative to land elevation at a proposed marsh creation site can be estimated, then grading and planting of vegetation could be planned to create relatively more area with elevations equivalent to the upper end of the current intertidal elevation zone.

Natural development of tidal marsh occurs when sediments accumulate to an elevation that can be colonized by pioneer marsh plant species by seed, rhizomes, marsh plants, or whole marsh sods that may wash up on the site (Boorman et al., 2002). Marsh creation by placement of dredged material is similar to this natural process, whereas marsh creation from upland requires excavation and grading to create a surface within the intertidal zone (Fig. 22.2A and B). The surface drainage or creek systems that occur in reference tidal marshes should be duplicated in created marshes with similar spatial distribution, width, and depth to facilitate tidal exchange (Zeff, 1999; Hood, 2007).

FIGURE 22.2 (A) A tidal marsh was created on land previously used for crop production by grading to intertidal elevations, replacing the stockpiled topsoil, and excavating a drainage system that was connected to the estuary after planting. Greenhouse-grown seedlings of *Juncus roemerianus*, *Spartina patens*, and *Spartina alterniflora* were planted (60 cm spacing) in June 2006. (Photo June 2006). (B) Vegetative cover was nearly complete by July 2007.

2.4 Soil

Soil physical and chemical properties affect construction, plant growth, and functional development of habitat in created marshes (Callaway, 2001). Construction operations such as grading, installing drainage channels, and planting vegetation are generally less difficult on sandy soils than on fine-textured clayey and silty soils or organic material. Sand has a greater bearing capacity than finer sediments, which makes operation of equipment, transplanting vegetation, and even walking on created sites much easier.

A disadvantage of sandy material is its low content of plant-available nutrients, which may result in poor initial growth of vegetation. Over time, tidal flooding will deposit silt, clay, and organic sediments that add nutrients to the developing created marsh. Nitrogen (N) and/or phosphorus (P) are the nutrients that often limit plant growth in created marshes. The other essential plant nutrients are usually supplied in adequate amounts either from the soil or saltwater flooding. Adding N and P fertilizers to enhance plant growth and initiate nutrient cycling is beneficial and may be critical to success on sites that are nutrient deficient (Broome, 1990). Maintenance fertilization may also be necessary until nutrient pools are enhanced and nutrient cycling is self-sustaining. At some locations, natural sources of nutrients from tidal inputs, seepage, runoff from uplands, N fixation, and precipitation may be more than adequate to maximize plant growth. Pollutants such as sewage effluent and nonpoint sources from agricultural land and urban areas may also result in N and P enrichment.

When upland soils are graded to intertidal elevations to create marshes, the new surface that is exposed may have physical and chemical properties that are unfavorable for plant growth and marsh development (Fig. 22.3). Appropriate soil amendments such as organic matter, lime, and fertilizer may be needed to improve plant growth. This problem can be avoided by stockpiling and then replacing the topsoil (A horizon) after grading. At most

FIGURE 22.3 Soil pH values below 3.0 developed in small areas in a tidal wetland created by grading an upland soil in eastern North Carolina to intertidal elevations. The low pH was the result of oxidation of acid sulfate soil material when it was exposed to the surface. *Spartina cynosuroides* seedlings did not survive in the low pH soils (foreground) until the affected areas were limed and replanted.

locations, topsoil is likely to have more favorable physical and chemical properties than subsurface layers and would be expected to produce better plant growth and development of marsh functions. Care should be taken to insure that the correct final surface elevation to support plant growth is maintained after the topsoil is applied. Standard soil testing, as used for agricultural crops and turf grass, can be utilized to determine if available plant nutrients are adequate and if the pH is within an acceptable range in the surface soil that will be planted.

Moderate amounts of sediments transported by tides, waves, longshore drift, and upland erosion are beneficial to created marshes. Nutrients associated with sediments increase plant productivity, and accretion helps marsh surfaces keep pace with rising sea level. However, excessive sediment accumulation can damage marsh vegetation by burying plants and increasing surface elevation above the limits of intertidal vegetation (Zedler and West, 2008). Sand blowing onto created marshes may be a problem in some locations, particularly dredged material disposal sites. Installing sand fences and establishing dune vegetation on the upland portion of the site will intercept and stabilize blowing sand to prevent abrupt elevation changes in the intertidal zone.

2.5 Establishing Vegetation

A self-sustaining plant community is the primary goal of tidal marsh creation because vegetation performs many of the biological functions of the ecosystem (Sullivan, 2001). When the appropriate elevation, hydrology, and soil are in place for the created marsh, establishing the dominant emergent plant species that mimic the target reference marsh accelerates development. Vegetation provides structure and creates an environment favorable for recruitment of invertebrates, microbes, and other flora and fauna that are adapted to the tidal marsh environment. Marsh plants are also important primary producers that are a part of the complex food web of marsh/estuary ecosystems (Wainright et al., 2000).

Strategies for establishing vegetation in created tidal marshes are site-specific, varying with geographic region and with environmental factors that determine the plant species composition. Generally, the greater the salinity at the site, the fewer plant species present. For example, along the Atlantic Coast of the United States in marshes with salinities near seawater strength, the vegetation of the intertidal zone is a monoculture of *Spartina alterniflora* with distinct upper and lower elevation limits. Tidal marshes in estuaries farther inland, which are flooded by less saline water, have greater plant species diversity with less distinct zonation of species. The key to planning vegetation establishment is to utilize reference marshes to determine the target plant species, plant community assemblages, salinity and nutrient status of the soil, and their upper and lower elevation limits. Based on this information, establishment methods can be developed to achieve optimum results.

Methods of establishing vegetation range from waiting for natural recruitment of seeds or other propagules to planting all the species identified in the reference marsh. Methods selected should take into consideration environmental factors at the site, characteristics of the plant species, cost, and practicality. Creation sites in low-energy environments that are protected from erosion by waves and currents may become vegetated naturally in a few years as seed or other propagules are spread by tides, wind, or animals. Sites exposed to waves and currents may remain unvegetated indefinitely. Planting the target vegetation has the advantage of producing a cover in a shorter period of time to prevent erosion and accelerating the process of developing a functioning ecosystem. Rapid establishment of the target vegetation gives it a competitive advantage that helps prevent colonization by undesirable or invasive plant species. In some cases, planting a single species that is easy to propagate and grows rapidly could be the best option for initial cover with the expectation that the natural plant assemblage will develop with time. Introducing plants to quickly establish vegetation is logical because the monetary cost and effort in planting at a tidal marsh creation site is usually small in comparison with the cost of earth moving and other construction (Broome, 1990). If a natural marsh in the area is being destroyed because of construction, plants, chunks of sod, and soils, which may contain seed and rhizomes, can be preserved and transferred to the creation site to speed development (Sullivan, 2001). This method also introduces benthic infauna to the new marsh.

Introduction of nonnative plant species should be avoided because it may have unintended consequences that negatively impact ecosystems. An example of this is the introduction of *S. alterniflora* to the west coast of the United States accidentally in Willapa Bay, Washington from packing material used for shipping eastern US oysters (Major et al., 2003), and planting for shoreline stabilization and salt marsh restoration (Global Invasive Species Database, 2007). *S. alterniflora* grows at both higher and lower elevations than native marsh vegetation. Impacts to the ecosystem include conversion of mudflat habitat to tidal marsh, trapping sediment, raising elevations, and loss of native plant communities through competition in both low and high marshes. In San Francisco Bay, California, *S. alterniflora* competes with and hybridizes with the native *Spartina foliosa* (San Francisco Estuary Invasive Spartina Project, 2001).

In some circumstances, introducing nonnative plant species may be beneficial. In his review of utilizing *Spartina* for ecological engineering in China, Chung (2006) documented many benefits of planting several nonnative plant species to create tidal marshes. *Spartina anglica* was brought to China in 1963 followed by the introduction of other species of *Spartina* including *S. alterniflora* in 1979. Significant economic, social, and ecological benefits derived from *Spartina* plantations include coastal stabilization, land reclamation, control of siltation, primary production in marine food webs, esthetic value for tourism, amelioration of saline soil, green manure, animal fodder, fish feed, and biofuel. Qin et al. (1998) reported the health benefits of two products, biomineral liquid and total flavonoids, extracted from *Spartina* plants.

If the goal is to create tidal marsh ecosystems similar to local reference marshes, it is important to obtain planting material from near the creation site to avoid introducing nonnative plants or ecotypes that may not be adapted to the local environment. For plant species that produce significant amounts of seed, collect seed from or near the reference marsh. Direct seeding of created sites may be an option if seed supplies are sufficient and the planting surface is stable. More often, seed are used to grow plants in containers and then planted at created sites. Commercial production of native marsh plants in nurseries and greenhouses is a common practice in some localities. Alternatively, plants can be dug from natural marshes, but there is risk of negatively impacting the donor marsh. If this method of obtaining planting stock is used, newly established stands on dredged material or recently accreted sediments are often good sources of plants. It is difficult to separate viable individual plants from the thick mat of roots and rhizomes in older established marshes.

Planting techniques must be adapted to conditions at the site and the labor available. Mechanical transplanters such as those used for vegetable plants may be used on large sites with soil material that is free of debris and will support equipment. Spades or tree-planting dibbles work well for opening planting holes where planting is done manually. Spacing of plants is an important consideration, which is a trade-off between the numbers of plants required and how long it takes to achieve complete cover. For example, a plant spacing of 1 m × 1 m requires 10,000 plants per hectare, while a spacing of 60 cm × 60 cm requires 27,778 plants per hectare, and 30 × 30 cm spacing requires 111,111 plants per hectare. The rate of growth of the plant species is a factor that must be considered. *S. alterniflora* spreads rapidly, and our experience has shown that 60 cm × 60 cm plant spacing results in complete cover in one growing season (7 months). Spacing plants 1 m apart is acceptable on sites that are protected from erosion, with complete cover in about 1 year through spread of rhizomes and germination of seed produced at the end of the first growing season.

Fertilization at planting with slow release N and P fertilizers enhances growth of vegetation in created marshes (Broome et al., 1983). Ammonium sources of N should be used rather than nitrate-N. When nitrate is applied to reduced soils, it is subject to atmospheric loss as N_2 gas through denitrification and removal by tides. Plant growth response to fertilization varies from location to location depending on whether the available soil nutrients and inputs from sediment accretion, runoff, birds, etc., are adequate for optimum plant growth. Growth response to fertilizer also varies with plant species. For example, the growth rates of grasses such as *Spartina* spp. are more likely to be increased by application of fertilizers than *Juncus* spp. Maintenance applications of fertilizers may be needed in subsequent growing seasons on sites where nutrients are limiting growth. Surface application of ammonium sulfate and concentrated superphosphate at low tide is a good method to supply N and P until nutrient pools and nutrient cycling are sufficient to maintain a healthy marsh ecosystem. Fertilizer should be applied only when necessary because excessive nutrients from pollution are a problem in many coastal environments.

In Europe, several projects have used seeding and planting to accelerate vegetation colonization and succession. A managed realignment at Tollesbury, Essex, UK, employed seeding at low and high densities as well as plugs and turfs of salt marsh vegetation (Garbutt et al., 2006). However, seeding and transplanting were ineffective at the site because the seeds did not germinate (or they washed away) and the transplants experienced near complete (97%) mortality. Although not a marsh creation project, tidal marsh vegetation was planted along the Brittany coast, France, to revegetate salt marshes damaged by the Amoco Cadiz oil spill and subsequent cleanup effort (Broome et al., 1988). Sprigs and plugs of *Puccinellia maritima* and *Halimione portulacoides* were successfully established.

3. EVALUATING FUNCTIONAL EQUIVALENCE OF CREATED TIDAL MARSHES

If the goal of tidal marsh creation is to create self-sustaining ecosystems that develop similar appearances (Fig. 22.4), structures, values, and functions as natural marshes, criteria for success must be defined and measured. The definition of success is controversial and may mean different things to different people, and determining success is challenging (Kentula, 2000).

FIGURE 22.4 A 3-ha brackish water (10–15 PSU) marsh created in 1983 by excavating an upland soil in a pine woodland. Greenhouse-grown seedlings of *Spartina cynosuroides, Spartina alterniflora,* and *Spartina patens* were planted 60 cm apart. Photographed June 12, 2008, 25 years after the project was completed. *Photo courtesy of Mr. Jeffrey C. Furness, Nutrien - Aurora Phosphate.*

Functional equivalence can be determined by evaluating whether key ecological functions are similar to reference marshes. Success may be defined as replicating all the functions of reference marshes, while a more reasonable approach might consider replacement of some of the functions as acceptable (Kentula et al., 1992). Important functions of tidal marshes that may be assessed include biological productivity, food webs, and biogeochemical cycles.

3.1 Vegetation

Because they are detritus-based ecosystems, development of ecological functions of tidal marshes requires high levels of net primary production (NPP) that, over time, accumulates as soil organic matter. The development of an organic-rich surface soil horizon promotes colonization by detritus feeding benthic invertebrates that supports higher-level consumers, finfish, and wading birds. Soil organic matter is also necessary to support heterotrophic microbial processes, organic matter mineralization, methanogenesis, and denitrification.

Development of biological productivity and food webs first requires establishing vegetation with high NPP. In some cases, saltwater marshes achieve aboveground NPP comparable with natural marshes within 3–5 years following creation (Broome et al., 1986, Fig. 22.5A, Table 22.2). Aboveground biomass develops faster than belowground biomass (Fig. 22.5B). In other cases, the plant community and NPP develop slowly or not at all because of problems associated with recreating tidal hydrology and wetland soil characteristics, especially adequate organic matter and N (Zedler, 2001). NPP of marshes created on graded terrestrial soils may not develop to levels found in natural marshes because of acidity and low fertility of the planting substrate (Broome et al., 1988). On these sites, liming and fertilizing with N and P may be necessary for satisfactory plant growth (Broome et al., 1988).

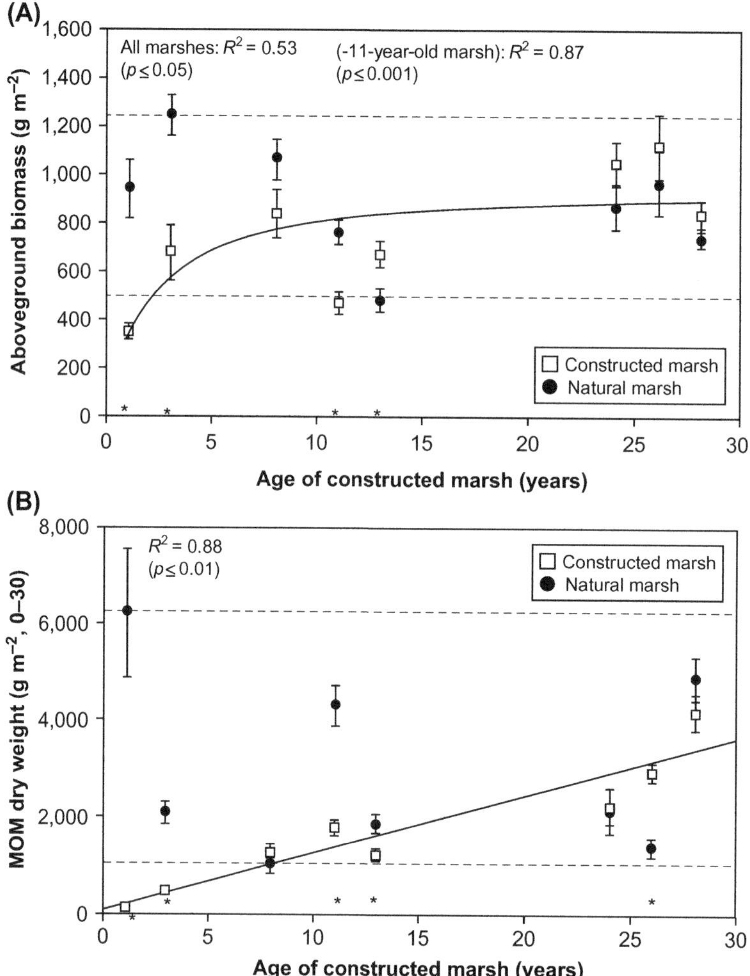

FIGURE 22.5 (A) Aboveground biomass of *Spartina alterniflora* along a chronosequence of created salt marshes and natural reference marshes. *Asterisks* (*) indicate that the created marsh and paired reference marsh are significantly different ($P < .05$) according to Student's t-test. *Dashed lines* represent the range of values measured in the natural marshes. (B) Macroorganic matter (MOM, living and dead root and root/rhizome mat) biomass of *S. alterniflora* along a chronosequence of created salt marshes and natural reference marshes. *Asterisks* (*) indicate that the created marsh and paired reference marsh are significantly different ($P < .05$) according to Student's t-test. *Dashed lines* represent the range of values measured in the natural marshes. *From Craft, C.B., Megonigal, J.P., Broome, S.W., Cornell, J., Freese, R., Stevenson, R.J., Zheng, L., Sacco, J., 2003. The pace of ecosystem development of constructed Spartina alterniflora marshes. Ecological Applications 13, 1417–1432.*

Development of NPP is related to the frequency of tidal inundation, and NPP develops more quickly in salt marshes where tidal inundation is frequent and regular (occurs twice a day) as compared with marshes where tidal inundation is infrequent and irregular. On the marsh levee where inundation occurs frequently, NPP developed to levels similar to

TABLE 22.2 Estimated Rate of Development of Wetland-Dependent Functions Following Saltwater Marsh Creation

	Time Needed to Achieve Equivalence to Natural Marshes[a] (Years)
ECOLOGICAL FUNCTIONS	
Primary production[b]	3–5
Secondary production	
Benthic invertebrates[c]	10–20
Finfish and shellfish[d]	2–>15
Avifauna use	
Waterfowl and wading birds[e]	1–3
Songbirds[e]	10–15
Outwelling[f]	3–5
BIOGEOCHEMICAL FUNCTIONS	
Sedimentation[g]	1–3
N retention[h]	3–5
P retention[i]	1–3
Carbon sequestration[j]	3–5
C mineralization[k]	5–15
Methanogenesis[l]	5–15 years
Denitrification[m]	5–15 years

[a]*See text for an explanation of the rationale for choosing the timescales.*
[b]*Broome et al. (1986, 1988), Craft et al. (2002, 2003).*
[c]*Moy and Levin (1991), Levin et al. (1996), Scatolini and Zedler (1996), Posey et al. (1997), Craft et al. (1999), Craft and Sacco (2003).*
[d]*Moy and Levin (1991), Rulifson (1991), Minello and Zimmerman (1992), Havens et al. (1995), Williams and Zedler (1999), Talley (2000). Rozas et al. (2005, 2007), Zeug et al. (2007), Raposa (2008).*
[e]*Havens et al. (1995), Desrochers et al. (2008).*
[f]*Craft et al. (1989).*
[g]*Craft (1997), Craft et al. (2003).*
[h]*Lindau and Hossner (1981), Craft et al. (1988, 1999, 2002, 2003), Craft (1997).*
[i]*Craft et al. (1988, 1999, 2003), Craft (1997), Poach and Faulkner (1998).*
[j]*Lindau and Hossner (1981), Craft et al. (1988, 1999, 2002, 2003), Craft (1997).*
[k]*Craft et al. (2003), Cornell et al. (2007).*
[l]*C. Craft (unpublished data), Cornell et al. (2007).*
[m]*Thompson et al. (1995), Currin et al. (1996), Fig. 22.8 of this study.*

natural marshes within 3 years (Broome et al., 1988; Craft et al., 2002). At the highest elevations where inundation occurs only during spring tides and storm tides, NPP never consistently achieved equivalence to natural marshes even after 15 years (Craft et al., 2002). Plant species composition also affects the development of NPP. Biomass of C_3 vegetation such as black needle rush (*Juncus roemerianus*), which grows slowly, takes longer to achieve equivalence to natural *Juncus* marshes (10 years) than faster growing C_4 *Spartina* spp. (Broome et al., 1988; Craft et al., 2002).

Macrophyte biomass and NPP do not develop quickly on all sites. In southern California, vegetation development is slowed by low N in soil (Langis et al., 1991) and annual additions of N were needed to maintain high levels of biomass production (Boyer and Zedler, 1998). Species richness, which is greater in southern California salt marshes than in Atlantic coast marshes, also was slow to develop in west coast marshes. While *Salicornia virginica* naturally recruited into created marshes, other native species required seeding or transplanting (Lindig-Cisneros and Zedler, 2002). In southern California, transplanting native species to increase species richness accelerated development of biomass stocks and N accumulation in created marsh soils (Callaway et al., 2003).

In Europe, natural colonization of dredge spoil and areas reclaimed by managed realignment occurs by germination of propagules in the seed bank and by dispersal from nearby natural marshes. Pioneer species consist of *Suaeda maritima*, *Salicornia* sp., and *Atriplex* sp. that colonize the site within 5 years (Bernhardt and Handke, 1992). During succession (5–10 years), pioneer species are replaced by salt marsh species, *Puccinellia maritima*, *Triglochin maritima*, and *Atriplex* sp. Natural colonization of a managed realignment (UK) occurred quickly, and within 6 years after breaching, the site contained 15 different species of plants, and 6 ha of the 21 ha site was colonized by salt marsh vegetation (Garbutt et al., 2006). Nineteen species of benthic invertebrates had also recruited to the site after 6 years, with *Hydrobia ulvae* identified as the dominant species. Breaching of the embankment also led to sedimentation in the newly reclaimed marsh that facilitated soil accretion during the 6-year monitoring period 1995–2001.

The rate of colonization of reclaimed marshes varied depending on size, tide range, and the time since breaching occurred. Species richness was greater in larger sites (>100 ha) than for smaller sites (<30 ha) (Wolters et al., 2005). In addition, sites with the largest elevation range within the tidal prism (mean high water neap to mean high water spring tide) contained more species than sites with less elevation range. Comparison of salt marsh vegetation (*Puccinellia maritima*, *Halimione* [*Atriplex*] *portulacoides*) planted following cleanup of the Amoco Cadiz oil spill (France) revealed that, after 4 years, natural colonization of the site was very slow and that planting was necessary to accelerate succession and restore the site (Broome et al., 1988).

Across Europe, studies indicate that considerable time must pass before reclaimed marshes develop plant communities that are comparable to natural marshes. Miren et al. (2001) observed that while 20- and 35-year-old reclaimed marshes in Spain contained comparable numbers of species (16–17), species richness was much lower than the 36 species recorded in a natural salt marsh.

3.2 Fauna and Food Webs

Establishment of vegetation and NPP are prerequisite to colonization by marsh consumers. Many created marshes contain significantly fewer benthic invertebrates than comparable natural marshes (Moy and Levin, 1991; Sacco et al., 1994; Levin et al., 1996; Scatolini and Zedler, 1996), although there is a trend toward increasing density with marsh age (Fig. 22.6), (LaSalle et al., 1991; Simenstad and Thom, 1996; Craft et al., 1999). Surface deposit feeding

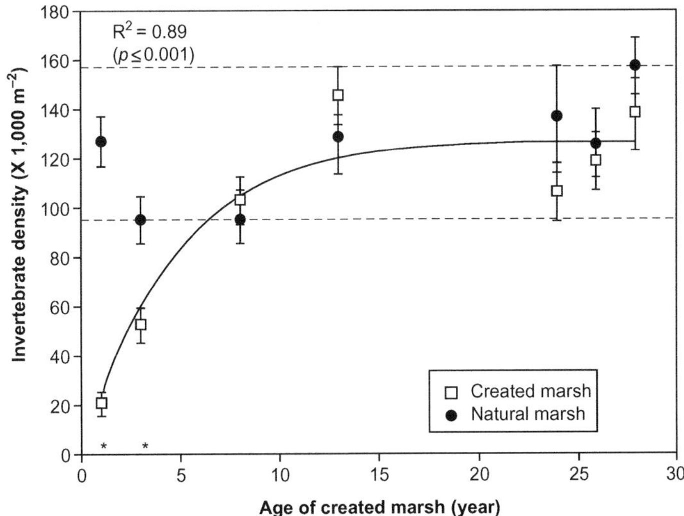

FIGURE 22.6 Density (number m^{-2} [1 SE]) of benthic infauna along a chronosequence of constructed salt marshes and natural reference marshes. *Asterisks* (*) indicate that the created marsh and paired reference marsh are significantly different ($P < .05$) according to Student's t-test. *Dashed lines* represent the range of values measured in the natural marshes. *From Craft. C.B., Sacco, J.N., 2003. Long-term succession of benthic infauna communities on constructed* Spartina alterniflora *marshes. Marine Ecology Progress Series 257, 45–58.*

invertebrates, crabs (*Uca* spp., *Hemigrapsus* spp.) and polychaete worms, colonize created marshes within a few years (Moy and Levin, 1991; Levin et al., 1996; Scatolini and Zedler, 1996; Posey et al., 1997). These organisms feed at the marsh surface and use emergent stems for cover and refuge from predators. However, subsurface deposit feeders such as oligochaete worms and Dipteran larvae were slower to colonize (Moy and Levin, 1991; Levin et al., 1996; Scatolini and Zedler, 1996; Posey et al., 1997). Low densities of subsurface deposit feeders may be the result of insufficient soil organic matter (Moy and Levin, 1991; Levin et al., 1996), and up to 20 years may elapse before created marshes develop an organic-rich surface layer that supports densities of subsurface deposit feeders that are comparable with natural marshes (Craft et al., 1999; Craft and Sacco, 2003).

Studies of shellfish and finfish use of created marshes yield mixed findings. Young (2–5 years old) created salt marshes in Texas contained significantly fewer numbers of grass shrimp (*Palaemonetes pugio*) and brown shrimp (*Penaeus aztecus*) than natural marshes (Minello and Zimmerman, 1992). Likewise, a 5-year-old constructed marsh in Virginia contained fewer blue crabs (*Callinectes sapidus*) than two comparable natural marshes (Havens et al., 1995). Both studies, though, reported no difference in densities of small fish between the constructed and natural marshes. Comparable studies of created and natural marsh creek channels in San Diego Bay and Pamlico River (NC) also revealed no differences in finfish density or species composition (Rulifson, 1991; Williams and Zedler, 1999).

Other studies, though, report significantly lower density of finfish in created versus natural marshes. In southern California, Talley (2000) measured fewer finfish but similar species richness and dominance in 1–3-year-old created marsh creeks relative to natural marsh creeks. A 3-year-old created salt marsh in North Carolina also contained significantly fewer killifish (*Fundulus heteroclitus*) than natural marshes (Moy and Levin, 1991). Differences in trophic dynamics were also noted. Gut analysis revealed that *Fundulus* fed mostly on polychaetes and algae in the created marsh, whereas, in the natural marsh, gut contents were mostly detritus and insects. Minello and Webb (1997) compared finfish and shellfish populations in 5 natural marshes and 10 created marshes (3–15 years old) in Texas and found that created marshes contained significantly fewer finfish (gobies and pinfish, *Lagodon rhomboides*) and crustaceans, including commercially important blue crabs, white shrimp (*Penaeus setiferus*), and brown shrimp (*Penaeus aztecus*). In constructed marshes, lower densities were attributed to higher surface elevation that reduced tidal flooding and limited entry of estuarine nekton. Proper site preparation and grading to create low elevation marsh habitat was recommended for future projects to increase nekton access to the marsh (Minello and Webb, 1997). Other recommendations to increase finfish use of created marshes include increasing channel heterogeneity to create a variety of habitats (Williams and Zedler, 1999) and maximize vegetated edge to facilitate access to the marsh (Minello et al., 1994).

Vegetation provides important habitat for marsh nekton. Rozas et al. (2007) followed populations of brown shrimp (*Farfantepenaeus aztecus*), white shrimp (*Litopenaeus setiferus*), blue crab (*Callinectes sapidus*) over an 11-year period in vegetated and unvegetated (shallow bottom) in Galveston Bay, Texas. They found consistently greater numbers in vegetated habitats. During the 11-year period, marsh nekton declined substantially as the area of marsh and marsh edge declined.

Rozas et al. (2005) compared fisheries use of three different restoration techniques, marsh terracing, mound construction, and marsh island construction, in Galveston Bay, Texas, USA. Terracing contained the greatest amount of marsh edge and also had the greatest use by finfish. Population estimates (per hectare) for the different techniques ranged from 22,200 to 30,800 for brown shrimp (*Farfantepenaeus aztecus*), 21,800 to 33,100 for white shrimp (*Litopenaeus setiferus*), and 17,200 to 24,900 for blue crab (*Callinectes sapidus*). In spite of these large numbers, the created marsh contained lower densities than natural marshes in the region.

Some studies suggest that, even after 30 years, created salt marshes do not support equivalent nekton densities and diversity. Zeug et al. (2007) reported that a 30-year-old created marsh of the Guadalupe Estuary, Texas, USA, contains lower species richness and shrimp and fish biomass compared with a natural marsh. The created marsh, however, contained densities of blue crabs and brown shrimp that were comparable with the natural marsh. It also contained eight nekton species not found in the natural marsh. Zeug et al. (2007) concluded that lower density and diversity of nekton in the created marsh was attributed to reduced habitat structure such as oysters and soil organic matter and that these features should be incorporated into future tidal marsh creation projects. With respect to nekton utilization, it appears that created marshes require anywhere from two to more than 15 years to achieve equivalence to natural marshes.

Use of created marshes by birds has received less attention. A marsh constructed in Virginia attracted similar numbers of wading birds but fewer numbers of songbirds than nearby natural marshes (Havens et al., 1995). In natural marshes, songbird utilization was attributed to the presence of shrubs (*Iva frutescens, Baccharis halimifolia*) that provided more perching habitat. Wading bird and waterfowl utilization of created marshes occur relatively quickly in 1–3 years, whereas songbird use of created marshes takes longer because it depends on establishment of woody shrubs in the marsh (2).

As nesting habitat, created marshes often lack the structural complexity needed to support breeding requirements of certain species. In Virginia, marsh wrens (*Cistothorus palustris*) exhibited a strong preference for nesting in natural marshes versus a constructed marsh (Havens et al., 1995). In southern California, the federally endangered light-footed clapper rail (*Rallus longirostris levipes*) nests only in *S. foliosa* marshes of southern California (Zedler, 1993). Zedler found that *S. foliosa* marsh created to mitigate for wetland loss from highway construction activities did not provide nesting habitat for the rail. In the created marsh, the *Spartina* stems were shorter, 60–90 cm tall, than in the natural marsh where most stems were greater than 90 cm in height. Clapper rails are unable to establish nests in the short *Spartina* canopy because nests are washed out of the lower marsh by high tides (Zedler, 1993). One-time N additions produced 2100 stems m^{-2} and 230 of those stems were taller than the 90 cm height required for nesting (Zedler, 1993; Boyer and Zedler, 1998). However, the following year, fertilized *Spartina* plots produced <30 "tall" stems per m^2, indicating that existing marsh soil and plant N pools were inadequate to sustain improved growth and taller canopies over the long term without annual N additions (Boyer and Zedler, 1998).

Desrochers et al. (2008) compared foraging and breeding use of birds on 11 created salt marshes ranging in age from 9 to 20 years and 11 reference marshes in Virginia. They found that use by obligate marsh species such as the clapper rail (*Rallus longirostris*) did not differ between created and reference marshes. However, during breeding season, avian abundance and species richness was less on created salt marshes, and wetland-dependent species, in particular, were poorly represented during this period (Fig. 22.7).

Creating a diversity of habitats including tidal flat and high marsh (shrub) habitats has been shown to increase avian abundance and species richness. Raposa (2008) reported that

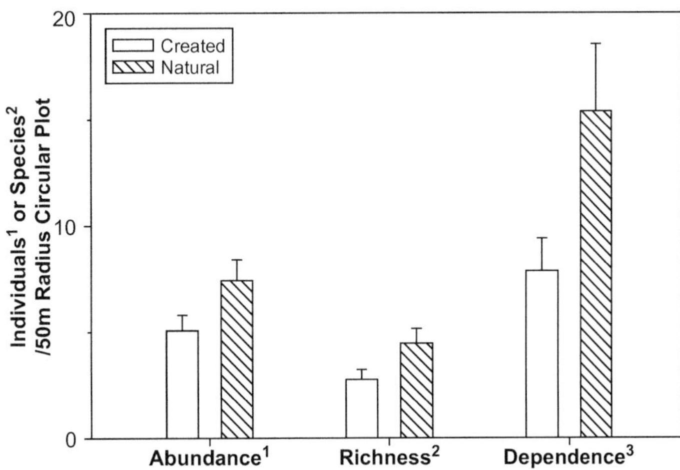

FIGURE 22.7 Avian abundance, richness, and wetland dependence on 11 created and 11 natural salt marshes. *Adapted from Desrochers, D.W., Keagy, J.C., Cristol, D.A., 2008. Created versus natural wetlands: avian communities in Virginia salt marshes. Écoscience 15, 36–74.*

restoration of tidal flows to a tidally restricted marsh led to rapid change in plant community composition and avian use. Three years following reintroduction of tides, open water, the dominant land cover prior to restoration was replaced by tidal mudflat. Percent cover of *Phragmites communis*, the dominant species prior to restoration, decreased by 69% as *Spartina alterniflora* replaced it (a 67% increase). Seven additional bird species were observed 1 year following tidal reintroduction and the number of birds observed increased from 6 to 85 per viewing effort, mostly because of shorebirds using the newly created mudflats.

Like natural marshes, created tidal marshes are a source of detritus to estuarine waters and food webs, which support the outwelling and trophic relay hypotheses. Two created brackish water marshes in North Carolina exported dissolved organic C (DOC) and N (DON) to the tidal creek and imported inorganic N (NH_4) and P (PO_4) (Craft et al., 1989). Overall, the marshes were a source of organic C to the estuary, whereas, with N, they served as transformers by importing NH_4, converting it to organic N and exporting it as DON. One marsh, amended with soil organic matter prior to planting, exported much more DOC and DON than the other marsh, which contained little soil organic matter (Craft et al., 1989).

Natural abundance of stable isotopes of C, N, and S has been used to infer sources of food to marsh invertebrates and nekton following creation and restoration. Several studies using ^{13}C and ^{15}N suggest that, in the early years following marsh creation, benthic microalgae are an important source for marsh invertebrates, whereas, in older natural marshes, *Spartina* and macroalgae make up a larger contribution to their diet (Currin et al., 2003; Moseman et al., 2004).

Llewellyn and La Peyre (2011) employed ^{13}C and ^{15}N to evaluate trophic support of blue crabs (*Callinectes sapidus*) in created and natural marshes of Louisiana. They found that the young (5-year-old) created marsh lacked the trophic niche breadth and dietary diversity found in the natural marsh. However, blue crabs collected from the older (8–24 years old) created marshes exhibited ^{13}C and ^{15}N compositions similar to their natural marsh counterparts, suggesting comparable trophic niche breadth and dietary diversity.

Several studies used multiple stable isotopes (^{13}C, ^{15}N, ^{34}S) to evaluate feeding by resident fish (i.e., *Fundulus* spp.) in created and natural marshes (Talley, 2000; Wosniak et al., 2006). Talley reported differences in ^{13}C, ^{15}N, and ^{34}S in *Fundulus parvipinnis* in 2.5-year-old created tidal marsh and a natural marsh in southern California but was unable to identify the food sources sustaining their diets. Wosniak et al. (2006) used ^{13}C, ^{15}N, and ^{34}S to evaluate changes in *Fundulus* diet following removal of tidal restrictions (e.g., culverts) in Massachusetts. Large differences in ^{13}C were observed between tidal restricted and unrestricted marshes, owing mostly to differences in vegetation: C_3 *Phragmites* in tidally restricted marshes and C_4 *Spartina* in unrestricted marshes. In situations where restoration involves removing tidal restrictions, ^{13}C alone may be useful for inferring dietary shifts as C_3 vegetation (*Phragmites*) is replaced by C_4 *Spartina* (Wosniak et al., 2006).

3.3 Biogeochemical Cycles

Created tidal marshes, as with natural marshes, trap sediment and nutrients (N, P), sequester organic matter and C, and are hotspots for microbial processes such as denitrification, Fe/Mn reduction, and sulfate reduction. Biogeochemical functions develop slowly on wetlands created from upland soils relative to those created on dredged material because of the large amount of oxidized Fe that must be reduced (Craft et al., 1991) before other reducing reactions (e.g., sulfate reduction) will occur.

Created marshes are active sites of sediment deposition. Once vegetation becomes established, aboveground stems dramatically reduce wave energy, which facilitates deposition of inorganic sediment. Created marshes often trap sediment at rates exceeding those of natural marshes. During the 6 months following establishment of vegetation, a created marsh trapped 31 kg sediment per m^2, increasing marsh elevation by 11 mm (C.B. Craft, unpublished data). In a nearby natural marsh, sediment deposition was much lower, 12 kg m^{-2}, during the same period. Sedimentation also deposits substantial amounts of inorganic nutrients. In the same created marsh, sedimentation contributed 7.7 g P m^{-2} to the marsh surface during the first 6 months as compared with 3.5 g P m^{-2} in the nearby natural marsh.

Created tidal marsh soils are also sinks for N (Fig. 22.6). Young created marshes often accumulate N at rates comparable with or exceeding those in natural marshes (Craft et al., 1999). N accumulates at higher rates at more frequently inundated low elevations of the marsh than at higher elevations where inundation is less frequent (Lindau and Hossner, 1981; Craft et al., 2002). Enhanced N accumulation in young created marshes is a result of greater epiphytic (Currin and Paerl, 1998) and benthic (Piehler et al., 1998) N fixation in response to low soil N and N limitation (Craft et al., 1999). In spite of accelerated N accumulation, many created marsh soils still contain less N than natural marshes even after 30 years (Craft et al., 1999).

P accumulation in created and natural tidal marshes is less than N, usually less than 1 $gm^{-2}year^{-1}$ (Table 22.3). Also, like N, recently created marshes often accumulate P at higher rates than older created marshes and natural marshes (Craft, 1997).

Sorption of P to Al or Ca-bearing minerals enhances P accumulation in soil. A brackish water marsh created on a graded terrestrial soil high in Al and Fe sorbed floodwater PO_4–P during tidal inundation (Craft et al., 1989). In the Mississippi River Delta, 15–20-year-old marshes created on dredged material contain more Ca-bound P than younger

TABLE 22.3 Comparison of Selected Biogeochemical Processes in Created and Natural Tidal Salt Marshes

Process (g m^{-2} $year^{-1}$)	Created[a]	Natural[a,b]
Sediment deposition	1000–11,000	160–2700
Soil P accumulation	0–0.9	0–2.3
Soil N accumulation	7.1–17	1.3–23
Soil C sequestration	80–125	21–393
Organic matter quality[d]	15%–20% Lignin	>30% Lignin
Denitrification[c]	<0.1	0–3

[a]*Cammen (1976), Craft (1997), Craft et al. (1999, Table 7), C.B. Craft (unpublished data).*
[b]*Craft et al. (1999, Table 7).*
[c]*Currin et al. (1996), Craft (2001).*
[d]*C.B. Craft (unpublished data).*

(1–3-year-old) marshes because of increased sorption to Ca in floodwaters (Poach and Faulkner, 1998). Salt marshes created on dredged material or shorelines often contain calcareous shell fragments that facilitate P sorption. For example, 10-year-old marsh created on a soil containing numerous shell fragments in North Carolina sorbed 6–8 g P m^{-2} $year^{-1}$ (Craft, 1997). In created marshes, organic forms of P also accrue in accumulating soil organic matter (Poach and Faulkner, 1998), although rates are relatively low, 0.2–1.0 g P $m^{-2}$$year^{-1}$(Craft, 1997).

Carbon sequestration is an important ecosystem service and the C sequestered in coastal wetlands, marshes, mangroves, and seagrasses is known as "blue carbon" (McLeod et al., 2011). Saline tidal wetlands emit little methane compared with freshwater wetlands (Poffenbarger et al., 2011), making saline tidal marsh creation and restoration an attractive instrument for mitigating anthropogenic CO_2 emissions.

The quantity of soil organic matter and sequestered C increases with created marsh age (Fig. 22.8B) as hydroperiod is established and vegetation covers the marsh. Created tidal marshes have been shown to sequester C at rates comparable with natural marshes (Craft et al., 2003) with greater sequestration related to longer hydroperiod (Craft et al., 2002). Like N, soil organic C pools increase with created marsh age but after 30 years still have not achieved equivalence to natural marshes (Fig. 22.6). Thus, while C sequestration in soil begins once tidal inundation and vegetation is established, development of large pools of soil organic C takes time, perhaps 30 years to several hundred years (Craft et al., 2002, Table 22.2).

Carbon sequestration is greater in surface than subsurface soils (Craft, 2000; Havens et al., 2002) and at low elevations in the marsh that are frequently inundated as compared with higher elevations (Lindau and Hossner, 1981; Craft et al., 2002). An interesting difference between created and natural marshes is that in young created marshes the accumulating organic matter is of higher quality. Soil organic matter in young (<10 years old) created marshes contain mostly labile organic compounds such as carbohydrates and water-soluble compounds and relatively little lignin (Craft et al., 2003, Table 22.2), whereas in natural marshes, organic matter contains proportionally more lignin (>30%) that is recalcitrant and not readily decomposable.

Development of heterotrophic microbial activity in created tidal marshes is strongly linked to accumulation of soil organic matter. Comparison of microbial processes and soil organic C stocks in created and natural marshes revealed that C mineralization increased with increasing percent soil organic C ($R^2 = 0.69$, $P < .001$) (Craft et al., 2003). Rates of denitrification ($R^2 = 0.63$, $P < .01$) and methanogenesis ($R^2 = 0.64$, $P < .01$) also increased with increasing soil organic C (Cornell et al., 2007; C. Craft, unpublished data).

In general, denitrification is lower in recently created marshes as compared with natural marshes (Table 22.2). Thompson et al. (1995) measured denitrification rates that were 44 times lower in a 2-year-old created *S. alterniflora* marsh than a nearby natural marsh and Currin et al. (1996) measured low denitrification rates on the same created marsh. In some created marshes, denitrification is inhibited by coarse (sandy) texture of the soil that provides less surface area for microbial populations, greater tidal flushing of pore water nutrients, and more exposure to oxygen relative to fine-textured natural marsh soils (Thompson et al., 1995). However, as created marshes age and soil organic matter accumulates, denitrification increases to levels comparable with levels measured in natural marshes.

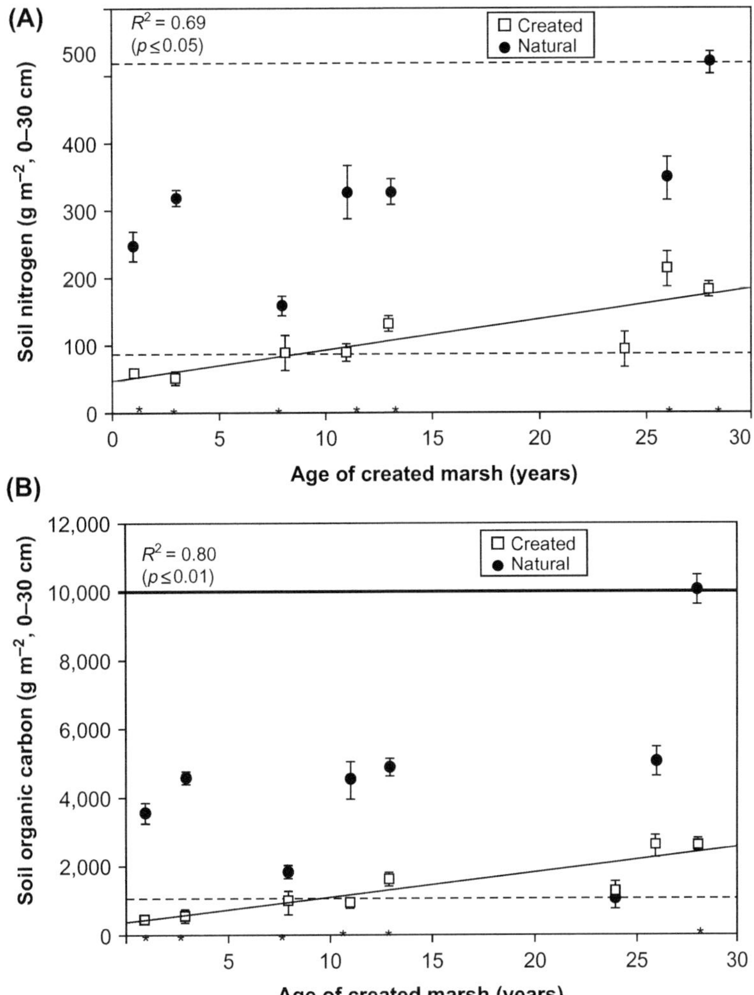

FIGURE 22.8 (A) Soil N pools along a chronosequence of created salt marshes and natural reference marshes. *Asterisks* (*) indicate that the created marsh and paired reference marsh are significantly different ($P < .05$) according to Student's t-test. *Dashed lines* represent the range of values measured in the natural marshes. (B) Soil organic carbon pools along a chronosequence of created salt marshes and natural reference marshes. *Asterisks* (*) indicate that the created marsh and paired reference marsh are significantly different ($P < .05$) according to Student's t-test. *Dashed lines* represent the range of values measured in the natural marshes. *From Craft, C.B., Megonigal, J.P., Broome, S.W., Cornell, J., Freese, R., Stevenson, R.J., Zheng, L., Sacco, J., 2003. The pace of ecosystem development of constructed Spartina alterniflora marshes. Ecological Applications 13, 1417–1432.*

4. MAXIMIZING ECOSYSTEM SERVICES OF CREATED MARSHES

Strategic location and design of tidal marsh creation projects is important in maximizing ecosystem services. For instance, provisioning services are enhanced when tidal marsh creation is designed to include locations for oysters to colonize within associated tidal creeks. Regulating services are enhanced by selecting a site close to agricultural fields to intercept runoff during rain events to retain sediments and nutrients, thereby improving downstream water quality. The hydrologic regime can be designed to minimize greenhouse gas emissions and maximize carbon sequestration. Vegetated coastal habitats, including seagrasses, mangroves, and tidal marshes, are important "blue carbon" sinks. These ecosystems are among the most efficient natural systems for removal of carbon dioxide from the atmosphere and sequestration of organic carbon in marine sediments for long periods of time (Macreadie et al., 2017). Well-designed created marshes with productive vegetation, anaerobic soils, good tidal exchange, and the ability to trap sediment and detritus contribute to sequestration of atmospheric carbon. Cultural services can be planned and coordinated with local stakeholders to ensure that the site is accessible for recreation and education (including both primary schools and university-level research).

An example of this approach was successfully implemented on a 14 ha tidal marsh creation project near Beaufort, NC (USA) in 2007. The brackish marsh was constructed on a historical wetland that had been converted to agriculture through extensive clearing and drainage in the mid-1970s. The site was situated between a large farm, with extensive subsurface and surface drainage infrastructure, and a sensitive estuary designated as a primary shellfishing and fisheries nursery area. Project partners recognized an opportunity to not only reconnect this area with the estuary and surrounding brackish marshes to increase habitat but also to improve downstream water quality by redirecting agricultural drainage water across the created wetland. The site was potentially a significant carbon sink because of the high primary productivity common in tidal marshes. However, carbon losses through emissions of greenhouse gases such as CO_2 and CH_4, which would offset carbon storage, had not been well documented from created sites with salinity ranges >10 PSU.

Retention of nitrate-nitrogen (NO_3−N) from surrounding agricultural drainage was a primary focus. Drainage water diverted across the marsh took twice as long to reach the estuary compared with the original flow through large drainage canals. Because of the complex flow driven by rainfall events and tides (both lunar and wind driven), high-frequency flow and concentration data were essential to determine if the created marsh reduced the mass of NO_3−N exported to the estuary. Measurements of water that moved through the tidal stream and marsh complex were collected at upstream and downstream flumes every 15 min. Flow was recorded using Doppler velocity meters, whereas NO_3−N concentrations within the tidal stream were measured at each station using ultraviolet-visual (UV-Vis) spectrometers calibrated following lab analysis of discrete water samples that were collected concurrent to spectrometer measurements (Etheridge et al., 2015, 2017). The marsh appeared to be a significant sink for NO_3−N during four major events (18%−42% retention) and the majority of the year. However, it was shown that major storm events can result in export of NO_3−N. For example, a large flux of NO_3−N was exported from the marsh following a hurricane. The amount was nearly equivalent to the amount previously retained during the year and therefore reduced the net annual retention of NO_3−N by the marsh to around 10%. This demonstrated that created tidal marshes can be a sink for NO_3−N that enters from the surrounding watershed, although the magnitude of retention will likely fluctuate throughout the year.

Carbon dioxide and methane fluxes, which have not been well studied in natural or created brackish salt marshes with salinity ranges >10 PSU, were measured and coupled with biomass measurements in an attempt to understand the potential for carbon storage at the site. Gas flux samples were collected across seasons over a nearly 3-year period using a replicated static chamber method within three distinct plant zones (*Spartina alterniflora*, *Juncus roemerianus*, and *Spartina patens*) in the created brackish marsh. Data were analyzed to determine flux rates of carbon dioxide (CO_2) and methane (CH_4) in these zones and then combined with biomass harvest and decomposition measurements to estimate whether the marsh had the potential to be a net carbon sink. Additional measurements of marsh hydrology, soil redox, temperature, and pore water chemistry were collected to evaluate their impact on observed results (Shiau et al., 2016). Mean fluxes of carbon losses appeared to be dominated by CO_2 (21.4 ± 66.8 mg C m^{-2} ha^{-1}) rather than CH_4 (0.06 ± 0 0.17 mg C m^{-2} ha^{-1}) in this hydrologic and salinity regime. Soil redox potential (poised above −150 mV) coupled with elevated pore water sulfate (SO_4^{2-}) appeared to create conditions in the created brackish marsh that suppressed microbial production of CH_4, an observation supported by other researchers such as Poffenbarger et al. (2011) in natural brackish marshes. As CH_4 is considered up to 30 times more efficient in trapping heat, low emissions coupled with high biomass production from this created marsh make C sequestration potential for this and other systems highly likely.

This created tidal marsh was designed and constructed to also provide educational, cultural, and recreational opportunities. The site continues to be used for environmental education for university, high school and elementary school students, and outreach to communities and citizens to promote environmental stewardship of coastal ecosystems. The site is also often used for groups of bird-watching enthusiasts.

5. SUMMARY

Successful creation of functionally equivalent tidal marshes in a reasonable length of time requires careful attention to establishing the target biotic communities. Important steps are site selection, assessment of a reference marsh, design and construction, obtaining plant propagules, proper planting, plant spacing, providing adequate nutrients from fertilizers if natural nutrient supplies are low, and maintenance.

Tidal marsh creation methods and basic ecological principles must be adapted to the unique physical conditions and biological communities that occur at each creation site. There is a need for continued research with careful monitoring, documentation of results, and synthesis of information from tidal marsh creation projects in a variety of coastal environments. Some issues that require further investigation include the following: (1) defining design criteria including geomorphic, hydrologic, and biotic factors; (2) developing models for designing tidal creeks and predicting hydrology; (3) the effects of adjacent uplands on hydrology; (4) the value of creating transitional ecotones; (5) the effects of soils, sediments, and plant-available nutrients on success; (6) defining measures of success and determining the time required to achieve structural and functional equivalency to reference wetlands for different marsh types; (7) comparing the benefits of planting native vegetation with natural recruitment; and (8) developing methods for accelerating development of created marsh ecosystems.

Selection of strategic locations and careful attention to design are needed to maximize the ecological benefits of created tidal marshes. Optimizing ecosystem services is the goal that should guide important design, management, policy, and economic considerations for future created marsh projects. When sound principles of ecological engineering are applied, tidal marshes can be created that have the same appearance and, with time, provide many of the ecosystem services of natural marshes.

References

Bayraktarov, E., Saunders, M.I., Abdullah, S., Mills, M., Beher, J., Possingham, H.P., Mumby, P.J., Lovelock, C.E., 2016. The cost and feasibility of coastal marine restoration. Ecological Applications 26, 1055–1074.

Bernhardt, K.G., Handke, P., 1992. Successional dynamics of newly created saline marsh soils. Ekologia 11, 139–152.

Boorman, L., Hazelden, J., Boorman, M., 2002. New salt marshes for old – salt marsh creation and management. In: Littoral 2002, The Changing Coast. Porto, Portugal, pp. 35–45.

Boyer, K.E., Zedler, J.B., 1998. Effects of nitrogen additions on the vertical structure of a constructed cordgrass marsh. Ecological Applications 8, 692–705.

Broome, S.W., 1990. Creation and restoration of wetlands of the southeastern United States. In: Kusler, J.A., Kentula, M.E. (Eds.), Wetland Creation and Restoration. Island Press, Washington, DC, pp. 37–72.

Broome, S.W., Seneca, E.D., Woodhouse Jr., W.W., 1983. The effects of source rate and placement of nitrogen and phosphorus fertilizers on growth of *Spartina alterniflora* transplants in North Carolina. Estuaries 6, 212–226.

Broome, S.W., Seneca, E.D., Woodhouse Jr., W.W., 1986. Long-term growth and development of transplants of the salt marsh grass *Spartina alterniflora*. Estuaries 9, 63–74.

Broome, S.W., Seneca, E.D., Woodhouse Jr., W.W., 1988. Tidal salt marsh restoration. Aquatic Botany 32, 1–22.

Callaway, J.C., 2001. Hydrology and substrate. In: Zedler, J.B. (Ed.), Handbook for Restoring Tidal Wetlands. CRC Press, Boca Raton, FL.

Callaway, J.C., Sullivan, G., Zedler, J.B., 2003. Species-rich plantings increase biomass and nitrogen accumulation in a wetland restoration experiment. Ecological Applications 13, 1626–1639.

Cammen, L.M., 1976. Accumulation rate and turnover time of organic carbon in a salt marsh sediment. Limnology and Oceanography 20, 1012–1015.

Chung, C.H., 2006. Forty years of ecological engineering with Spartina plantations in China. Ecological Engineering 27, 49–57.

Cornell, J.A., Craft, C., Megonigal, J.P., 2007. Ecosystem gas exchange across a created salt marsh chronosequence. Wetlands 27, 240–250.

Costanza, R., d'Arge, R., De Groot, R., Farber, S., Grasso, M., Hannon, B., Limburg, K., Naeem, S., O'Neill, R.V., Paruelo, J., Raskin, R.G., 1997. The value of the world's ecosystem services and natural capital. Nature 387, 253–260.

Costanza, R., de Groot, R., Sutton, P., van der Ploeg, S., Anderson, S.J., Kubiszewski, I., Farber, S., Turner, R.K., 2014. Changes in the global value of ecosystem services. Global Environmental Change 26, 152–158.

Craft, C.B., Broome, S.W., Seneca, E.D., 1988. Nitrogen, phosphorus and organic carbon pools in natural and transplanted marsh soils. Estuaries 11, 272–280.

Craft, C.B., 1997. Dynamics of nitrogen and phosphorus retention during wetland ecosystem succession. Wetlands Ecology and Management 4, 177–187.

Craft, C.B., 2000. Co-development of wetland soils and benthic invertebrate communities following salt marsh creation. Wetlands Ecology and Management 8, 197–207.

Craft, C.B., 2001. Biology of wetland soils. In: Richardson, J.L., Vepraskas, M.J. (Eds.), Wetland Soils: Their Genesis Hydrology, Landscape and Separation into Hydric and Nonhydric Soils. CRC Press, Boca Raton, FL, pp. 107–135.

Craft, C.B., Broome, S.W., Campbell, C.L., 2002. Fifteen years of vegetation and soil development following brackish-water marsh creation. Restoration Ecology 10, 248–258.

Craft, C.B., Broome, S.W., Seneca, E.D., 1989. Exchange of nitrogen, phosphorus and organic carbon between transplanted marshes and estuarine waters. Journal of Environmental Quality 18, 206–211.

Craft, C.B., Sacco, J.N., 2003. Long-term succession of benthic infauna communities on constructed *Spartina alterniflora* marshes. Marine Ecology Progress Series 257, 45–58.

Craft, C.B., Megonigal, J.P., Broome, S.W., Cornell, J., Freese, R., Stevenson, R.J., Zheng, L., Sacco, J., 2003. The pace of ecosystem development of constructed *Spartina alterniflora* marshes. Ecological Applications 13, 1417–1432.

Craft, C.B., Seneca, E.D., Broome, S.W., 1991. Porewater chemistry of natural and created marsh soils. Journal of Experimental Marine Biology and Ecology 152, 187–200.

Craft, C.B., Reader, J.M., Sacco, J.N., Broome, S.W., 1999. Twenty-five years of ecosystem development of constructed *Spartina alterniflora* (Loisel) marshes. Ecological Applications 9, 1405–1419.

Crooks, S., Schutten, J., Sceern, G.D., Pye, K., Davy, A.J., 2002. Drainage and elevation as factors in the restoration of salt marsh in Britain. Restoration Ecology 10, 591–602.

Currin, C.A., Joye, S.B., Paerl, H.W., 1996. Diel rates of N_2-fixation and denitrification in a transplanted *Spartina alterniflora* marsh: implications for N-flux dynamics. Estuarine, Coastal and Shelf Science 42, 597–616.

Currin, C.A., Paerl, H.W., 1998. Epiphytic nitrogen fixation associated with standing dead shoots of smooth cordgrass, *Spartina alterniflora*. Estuaries 21, 108–117.

Currin, C.A., Wainright, S.C., Able, K.W., Weinstein, M.P., Fuller, C.M., 2003. Determination of food web support and trophic position of the mummichog, *Fundulus heteroclitus*, in New Jersey smooth cordgrass (*Spartina alterniflora*), common reed (*Phragmites australis*) and restored salt marshes. Estuaries 26, 495–510.

Currin, C.A., Chappell, W.S., Deaton, A., 2010. Developing alternative shoreline armoring strategies: the living shoreline approach in North Carolina. In: Shipman, H., Dethier, M., Gelfenbaum, G., Fresh, K., Dinicola, R. (Eds.), Puget Sound Shorelines and the Impacts of Armoring—Proceedings of a State of the Science Workshop, May 2009. Reston, VA, U.S. Geological Survey, pp. 91–102. USGS Scientific Investigations Report (2010, 5254).

Currin, C.A., Davis, J., Malhotra, A., 2017. Response of salt marshes to wave energy provides guidance for successful living shoreline implementation. In: Bilkovic, D.M., Toft, J., Mitchell, M., La Peyre, M. (Eds.), Living Shorelines: The Science and Management of Nature-based Coastal Protection. CRC Press, Taylor & Francis Group.

Darnell, T.M., Smith, E.H., 2001. Recommended design for more accurate duplication of natural conditions in salt marsh creation. Environmental Management 29, 813–823.

Desrochers, D.W., Keagy, J.C., Cristol, D.A., 2008. Created versus natural wetlands: avian communities in Virginia salt marshes. Écoscience 15, 36–74.

Etheridge, J.R., Birgand, F., Burchell, M.R., 2015. Quantifying nutrient and suspended solids fluxes in a constructed tidal marsh: insights gained from capturing the rapid changes in flow and concentrations following rainfall. Ecological Engineering 78, 41–52.

Etheridge, J.R., Burchell, M.R., Birgand, F., 2017. Can created tidal marshes reduce nitrate export to downstream estuaries? Ecological Engineering 105, 314–324.

Garbutt, R.A., Reading, C.J., Wolters, M., Gray, A.J., Rothery, P., 2006. Monitoring the development of intertidal habitats on former agricultural land after the managed realignment of coastal defences at Tollesbury, Essex, UK. Marine Pollution Bulletin 53, 155–164.

Ghermandi, A., van den Bergh, J.C.J.M., Brander, L.M., de Groot, H.L.F., Nunes, P.A.L.D., 2010. Values of natural and human-made wetlands: a meta-analysis. Water Resources Research 46, W12516,. https://doi.org/10.1029/2010WR009071.

Gittman, R.K., Peterson, C.H., Currin, C.A., Joel Fodrie, F., Piehler, M.F., Bruno, J.F., 2016. Living shorelines can enhance the nursery role of threatened estuarine habitats. Ecological Applications 26, 249–263.

Global Invasive Species Database, 2007. *Spartina alterniflora* (Grass). http://www.issg.org.

Harvey, J.W., Odum, W.E., 1990. Groundwater inputs to coastal waters. Biogeochemistry 10, 217–236.

Havens, K.J., Varnell, L.M., Bradshaw, J.G., 1995. An assessment of ecological conditions in a constructed tidal marsh and two natural reference tidal marshes in coastal Virginia. Ecological Engineering 4, 117–141.

Havens, K.J., Varnell, L.M., Watts, B.D., 2002. Maturation of a constructed tidal marsh relative to two natural reference marshes over 12 years. Ecological Engineering 18, 305–315.

Hofstede, J.L.A., 2003. Integrated management of artificially created salt marshes in the Wadden Sea of Schleswig-Holstein, Germany. Wetlands Ecology and Management 11, 183–194.

Hood, G.W., 2007. Scaling tidal channel geometry with marsh island area: a tool for habitat restoration, linked to channel formation process. Water Resources Series 43, W03409. https://doi.org/10.1029/2006WR005083.

Hough, P., Robertson, M., 2009. Mitigation under Section 404 of the Clean Water Act: where it comes from, what it means. Wetlands Ecology and Management 17, 15–33.

Kentula, M.E., 2000. Perspectives on setting success criteria for wetland restoration. Ecological Engineering 15, 199–209.

Kentula, M.E., Brooks, R.P., Gwin, S.E., Holland, C.C., Sherman, A.D., Sifneas, J.C., 1992. An Approach to Improving Decision Making in Wetland Creation and Restoration. Island Press, Washington, DC.

Landin, M.C., Webb, J.W., Knutson, P.L., 1989. Long-term Monitoring of Eleven Corps of Engineeers Habitat Development Field Sites Built of Dredged Material, 1974–1987. Technical Report K-89-1. Department of the Army, Waterways Experiment Station, Vicksburg, MS.

Langis, R., Zalejko, M., Zedler, J.B., 1991. Nitrogen assessment in a constructed and natural salt marsh of San Diego Bay. Ecological Applications 1, 40–51.

LaSalle, M.W., Landin, M.C., Sims, J.G., 1991. Evaluation of the flora and fauna of a *Spartina alterniflora* marsh established on dredged material in Winyah Bay, South Carolina. Wetlands 11, 191–208.

Lewis III, R.R., 1990. Wetlands restoration/creation/enhancement terminology: suggestions for standardization. In: Kusler, J.A., Kentula, M.E. (Eds.), Wetland Creation and Restoration: The Status of the Science. Island Press, Washington, DC, pp. 417–422.

Levin, L.A., Talley, D., Thayer, G., 1996. Succession of macrobenthos in a created salt marsh. Marine Ecology Progress Series 141, 67–82.

Lindau, C.W., Hossner, L.R., 1981. Substrate characterization of an experimental marsh and three natural marshes. Soil Science Society of America Journal 45, 1171–1176.

Lindig-Cisneros, R., Zedler, J.B., 2002. Halophyte recruitment in a salt marsh restoration site. Estuaries 25, 1174–1183.

Llewellyn, C., La Peyre, M., 2011. Evaluating ecological equivalence of created marshes: comparing structural indicators with stable isotope indicators of blue crab support. Estuaries and Coasts 34, 172–184.

Macreadie, P.I., et al., 2017. Can we manage coastal ecosystems to sequester more blue carbon? Frontiers in Ecology and the Environment 15, 206–213.

Major III, W.W., Grue, C.E., Grassley, J.M., Conquest, L.L., 2003. Mechanical and chemical control of smooth cordgrass in Willapa Bay, Washington. Journal of Aquatic Plant Management 41, 6–12.

Maryland Department of the Environment Wetlands and Waterways Program, 2006. Shore Erosion Control Guidelines Marsh Creation. http://www.mde.state.md.us/Programs/WaterPrograms/Wetlands_Waterways.

Mcleod, E., Chmura, G.L., Bouillon, S., Salm, R., Björk, M., Duarte, C.M., Lovelock, C.E., Schlesinger, W.H., Silliman, B.R., 2011. A blueprint for blue carbon: toward an improved understanding of the role of vegetated coastal habitats in sequestering CO_2. Frontiers in Ecology and the Environment 9 (10), 552–560.

Mendelssohn, I.A., Kuhn, N.L., 2003. Sediment subsidy: effects on soil-plant responses in a rapidly submerging coastal salt marsh. Ecological Engineering 21, 115–128.

Millennium Ecosystem Assessment, 2005. Ecosystems and Human Well-being: Wetlands and Water Synthesis. World Resources Institute, Washington, DC.

Minello, T.J., Able, K.W., Weinstein, M.P., Hays, C.G., 2003. Salt marshes as nurseries for nekton: testing hypotheses on density, growth, and survival through meta-analysis. Marine Ecology Progress Series 246, 39–59.

Minello, T.J., Matthews, G.A., Caldwell, P.A., Rozas, L.P., 2008. Population and production estimates for decapod crustaceans in wetlands of Galveston Bay, Texas. Transactions of the American Fisheries Society 137, 129–146.

Minello, T.J., Webb Jr., J.W., 1997. Use of natural and created Spartina alterniflora salt marshes by fishery species and other aquatic fauna in Galveston Bay, Texas, USA. Marine Ecology Progress Series 151, 165–179.

Minello, T.J., Zimmerman, R.J., 1992. Utilization of natural and transplanted Texas marshes by fish and decapod crustaceans. Marine Ecology Progress Series 90, 273–285.

Minello, T.J., Zimmerman, R.J., Medina, R., 1994. The importance of edge for natant macrofauna in a created salt marsh. Wetlands 14, 184–198.

Miren, O., Albizu, I., Amezaga, I., 2001. Effect of time on the natural regeneration of salt marsh. Applied Vegetation Science 4, 247–256.

Mitsch, W.J., Gosselink, J.G., 2000. Wetlands, third ed. John Wiley, New York.

Moller, M.A., Phil, M., Spencer, T., French, J.R., Dixon, M., 2001. The sea-defence value of salt marshes: field evidence from North Norfolk. Water and Environment Journal 15, 109–116.

Moseman, S.M., Levin, L.A., Currin, C., Forder, C., 2004. Colonization, succession, and nutrition of macrobenthic assemblages in a restored wetland at Tijuana Estuary, California. Estuarine, Coastal and Shelf Science 60, 755–770.

Moy, L.D., Levin, L.A., 1991. Are Spartina marshes a replaceable resource? A functional approach to evaluation of marsh creation efforts. Estuaries 14, 1–16.

Nuttle, W.K., Harvey, J.W., 1995. Fluxes of water and solute in a coastal wetland sediment. 1. The contribution of regional groundwater discharge. Journal of Hydrology 164, 89–107.

O'Donnell, J.E., 2016. Living Shorelines: a review of literature relevant to New England coasts. Journal of Coastal Research 33, 435–451.

Peterson, C.H., Able, K.W., DeJong, C.F., Piehler, M.F., Simenstad, C.A., Zedler, J.B., 2008. Practical proxies for tidal marsh ecosystem services: application to injury and restoration. Advances in Marine Biology 54, 221–266.

Piehler, M.F., Currin, C.A., Cassanova, R., Paerl, H.W., 1998. Development and N_2 fixing activity of the benthic microbial community in transplanted *Spartina alterniflora* marshes in North Carolina. Restoration Ecology 6, 290–296.

Poach, M.E., Faulkner, S.P., 1998. Soil phosphorus characteristics of created and natural wetlands in the Atchafalaya Delta, Louisiana. Estuarine, Coastal and Shelf Science 46, 195–203.

Poffenbarger, H.J., Needelman, B.A., Megonigal, J.P., 2011. Salinity influence on methane emissions from tidal marshes. Wetlands 31, 831–842.

Posey, M.H., Alpin, T.D., Powell, C.M., 1997. Plant and infaunal communities associated with a created marsh. Estuaries 20, 42–47.

Qin, P., Xie, M., Jiang, Y., 1998. Spartina green food ecological engineering. Ecological Engineering 11, 147–156.

Raposa, K.B., 2008. Early ecological responses to hydrologic restoration of a tidal pond and salt marsh complex in Narragansett Bay, Rhode Island. Journal of Coastal Research 55, 180–192.

Rozas, L.P., Caldwell, P., Minello, T.J., 2005. The fishery value of salt marsh restoration projects. Journal of Coastal Research 40, 37–50.

Rozas, L.P., Minello, T.J., Zimmerman, R.J., Caldwell, P., 2007. Nekton populations, long-term wetland loss, and the recent effect of habitat restoration in Galveston Bay, Texas. Marine Ecology Progress Series 344, 119–130.

Rulifson, R.A., 1991. Finfish utilization of man-initiated and adjacent natural creeks of South Creek estuary, North Carolina using multiple gear types. Estuaries 14, 447–464.

Sacco, J.N., Seneca, E.D., Wentworth, T.R., 1994. Infaunal community development of artificially established salt marshes in North Carolina. Estuaries 17, 489–500.

San Francisco Estuary Invasive Spartina Project, 2001. Introduced *Spartina alterniflora*/Hybrids. Coastal Conservancy. http://www.spartina.org.

Scatolini, S.R., Zedler, J.B., 1996. Epibenthic invertebrates of natural and constructed marshes of San Diego Bay. Wetlands 16, 24–37.

Sharma, S., Goff, J., Cebrian, J., Ferraro, C., 2016. A hybrid shoreline stabilization technique: impact of modified intertidal reefs on marsh expansion and nekton habitat in the northern Gulf of Mexico. Ecological Engineering 90, 352–360.

Shiau, Y., Burchell, M.R., Krauss, K.W., Birgand, F., Broome, S.W., 2016. Greenhouse gas emissions from a created brackish marsh in eastern North Carolina. Wetlands 36, 1009–1024.

Simenstad, C.A., Thom, R.M., 1996. Functional equivalency trajectories of the restored Gog-Le-Hi-Te estuarine wetland. Ecological Applications 6, 38–56.

Streever, W.J., 1999. An International Perspective on Wetland Rehabilitation. Kluwer Academic Publishers, Dorecht, The Netherlands.

Streever, B., 2000. Dredged material marshes: summary of three research projects. Wetland Research Bulletin CRWRP-2 (1), 1–4.

Sullivan, G., 2001. Establishing vegetation in restored and created coastal wetlands. In: Zedler, J.B. (Ed.), Handbook for Restoring Tidal Wetlands. CRC Press, Boca Raton, FL.

Talley, D.M., 2000. Ichthyofaunal utilization of newly-created versus natural salt marsh creeks in Mission Bay, California. Wetlands Ecology and Management 8, 117–132.

Thompson, S.P., Paerl, H.W., Go, M.C., 1995. Seasonal patterns of nitrification and denitrification in a natural and a restored salt marsh. Estuaries 18, 399–408.

Wainright, S.C., Weinstein, M.P., Able, K.W., Currin, C.A., 2000. Relative importance of benthic mcroalgae, phytoplankton and the detritus of smooth cordgrass *Spartina alterniflora* and the common reed Phragmites australis to brackish-marsh food webs. Marine Ecology Progress Series 200, 77–91.

Williams, G.D., Zedler, J.B., 1999. Fish assemblage composition in constructed and natural tidal marshes of San Diego Bay: relative influence of channel morphology and restoration history. Estuaries 22, 702–716.

Wolters, M., Garbutt, A., Bakker, J.P., 2005. Salt-marsh restoration: evaluating the success of de-embankments in north-west Europe. Biological Conservation 123, 249–268.

Wosniak, A.S., Roman, C.T., Wainright, S.G., McKinney, R.A., Jane-Pirr, M., 2006. Monitoring food web changes in tide-restored salt marshes: a stable carbon isotope approach. Estuaries and Coasts 29, 568–578.

Zedler, J.B., 1993. Canopy architecture of natural and planted cordgrass marshes: selecting habitat evaluation criteria. Ecological Applications 3, 123–138.

Zedler, J.B. (Ed.), 2001. Handbook for Restoring Tidal Wetlands. CRC Press, Boca Raton, Florida.

Zedler, J.B., West, J.M., 2008. Declining diversity in natural and restored salt marshes: a 30-year study of Tijuana estuary. Restoration Ecology 16, 249–262.

Zeff, M.L., 1999. Salt marsh tidal channel morphometry: applications for wetland creation and restoration. Restoration Ecology 7, 205–211.

Zeug, S.C., Shervette, V.R., Hoeinghaus, D.J., Davis III, S.E., 2007. Nekton assemblage structure in natural and created marsh edge habitats of the Guadalupe Estuary, Texas, USA. Estuarine, Coastal and Shelf Science 71, 457–466.

CHAPTER

23

Salt Marsh Restoration

Paul Adam

School of Biological Earth and Environmental Science, UNSW, Sydney, NSW, Australia

1. INTRODUCTION

There is a very substantial literature on salt marsh restoration, rehabilitation, and creation, which, as well as numerous site-specific accounts, includes reviews by among others Lewis (1982), Zedler (1995, 2001), Bakker et al. (1997), Niering (1997), Atkinson et al. (2001), Bakker et al. (2002), Zedler and Adam (2002), Boorman (2003), Callaway (2005), Nottage and Robertson (2005), Bakker and Piersma (2006), Wolters (2006), Roman and Burdick (2012), and Thom and Borde (2016). The considerable investment in salt marsh restoration reflects the high degree of public awareness of the ecological values of healthy salt marsh ecosystems and increasing recognition of the extent of absolute loss or degradation of many sites.

A large number of terms have been used in the literature to describe the various approaches applied to improving the state of damaged salt marshes. Elliott (Elliott and Cutts, 2004: Elliott et al., 2007; Elliott et al., 2016) has attempted to establish consistency in the use of terms applied to the various forms of ecological repair conducted in estuaries (in which salt marshes may be a component), and in Elliott et al. (2016) provided a schematic (Fig. 23.1) to display the range of responses to ecosystem degradation. This model is applicable to all ecosystems although developed in the context of estuaries. The degradation axis in this model encompasses both direct and indirect anthropogenic impacts and those caused by natural hazards including major storms, other climatic extremes, and tsunamis, which may be exacerbated by human actions (Elliott et al., 2016). The spatial scale over which both anthropogenic and natural disturbances occur has considerable variation, from the local to the global. The scale of the impact will determine the nature of the response required to manage the damage caused. Where the impact is confined to a single site addressing the problem may be more feasible, even if severe, than it would be dealing with whole estuaries or regions. In the case of damage caused by natural hazards, the response has often been to permit recovery by natural processes. But there is no intrinsic reason why management intervention could not occur, if repairing ecosystems from natural damage was considered appropriate for policy reasons. Whether or not direct intervention is thought desirable will involve consideration of a range of factors, including extent of the

Coastal Wetlands
https://doi.org/10.1016/B978-0-444-63893-9.00023-X

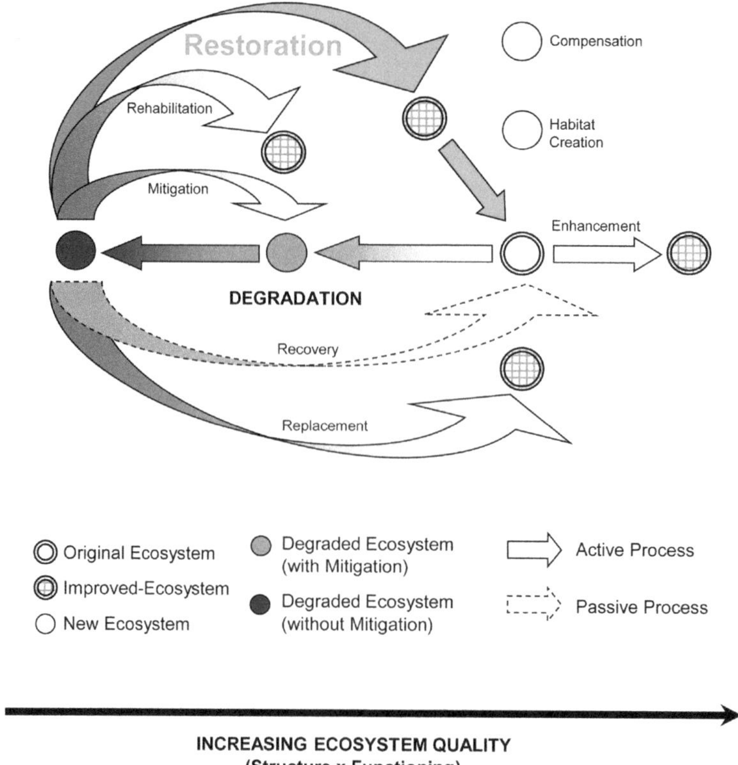

FIGURE 23.1 A typology and terminology of active and passive responses to ecosystem degradation. *From Elliott, M., Mander, L., Masik, K. et al., 2016. Engineering with ecohydrology; success and failures in estuarine restoration. Estuarine, Coastal and Shelf Science 176, 12—35. (Modified from Elliott, M., Burdon, D., Hemingway, K.L., Apitz, S., 2007. Estuarine, coastal and marine ecosystem restoration: confusing management and science - a revision of concepts. Estuarine, Coastal and Shelf Science 74, 349—366.*

damage, the feasibility of the response action, the cost of works required, and whether or not the major disturbance was considered a "normal" event.

The US National Research Council (1992) defined restoration in an ecological context as "the act of bringing an ecosystem back into as nearly as possible its original condition." Craft (2016) added to this the phrase "faster than nature does on its own." However, while achieving an endpoint quickly may be desirable for the reporting purposes of funding agencies and for persuading the public and politicians to continue support of environmental programs to manage coastal wetlands, an important objective is the establishment of self-sustaining (or as near as possible) ecosystems. Demonstrating that the system is on a trajectory to this end, during the reporting cycle of projects (often only a few years), is more likely to be achievable than demonstration of complete success. For many coastal wetlands we do not know, with any degree of certainty, how fast natural succession might proceed.

McDonald et al. (2016) suggest that in heavily modified areas restoration may be neither desirable nor attainable, and in these circumstances the highest and best level of repair is rehabilitation, which they define as "the process of reinstating degrees of ecosystem functionality on degraded sites where restoration is not the aspiration to permit ongoing provision of ecosystem goods and services including support of biodiversity." Many examples from around the world that have been described as restoration should be referred to as rehabilitation, although the distinction is one of degree rather than being absolute.

The aspirational goal of restoration may be unattainable for a number of reasons: change may have been so pervasive over large areas that there are no unaffected reference sites from which to identify the original state or even if there is evidence as to the original state, from fossil or other evidence, the environmental conditions may have changed such that the combination of species and landforms present when a particular marsh started to develop could not be replicated today. Coastal communities are naturally dynamic, so requirements that specify restoration to a particular state, effectively requiring a preservationist rather than conservationist approach, may be both prohibitively expensive and, ultimately, doomed to failure.

Restoration may be undesirable in the case of sites that have been modified as a result of long application of practices regarded as culturally significant or have produced cultural landscapes that are valued for their heritage attributes.

A general aim of both restoration and rehabilitation is an endpoint of self-sustaining ecosystems so that long-term management costs are minimized and environmental gains maximized. This is desirable, but management authorities need to be aware that continuing management to address new and changed environmental circumstances, or to respond to invasions of new species, will be required and is likely to generate continuing need for funding. It is ultimately for society to decide, through the political process, what sorts of environments it wants and what it is prepared to pay for them.

Replacement in Fig. 23.1 may involve habitat creation, but compensation and habitat creation can take place outside the degradation, recovery, and replacement cycles. If sites are converted into urban or industrial development, they are permanently lost as natural or seminatural ecosystems. Government policies may require that in return for development approval, there be compensation in the form of ecological restoration, rehabilitation, or habitat creation at other sites, frequently, though not necessarily always, on a like-for-like basis. One of the challenges for those responsible for approving compensation proposals is determining whether or not the ecological values of the site after the compensatory action will exceed those of the site lost.

Habitat creation involves the creation of ecosystems at localities where systems of that type either did not exist previously or, if they did, the modification to the area in the time since the previous occurrence is such that all continuity has been broken. For example, where original wetlands have been infilled and sometime later the site has been partially excavated to create a new artificial wetland, but not necessarily comparable in its properties to that originally present. The Penrhyn Estuary Project in Botany Bay (Australia) (Sydney Ports n.d.), requiring construction of salt marsh and other intertidal communities as part of the approval process for expansion of Port Botany, can be categorized as creation. One issue with creation is that it must occur somewhere, and that "somewhere" may have unappreciated values. Creating wetlands in old industrial areas may not give rise to concerns, but creation of salt marsh on what would naturally be mudflat requires, unconsciously or otherwise, the assumption that the ecological values of salt marsh are greater than those of mudflat.

Fig. 23.1 also includes the category of enhancement, where a site representative of the original condition or close to it, is subject to interventionist management with the intention of improvement. One example of where this might arise is in the preemptive translocation of species likely to be adversely affected by future climate change. This action would rescue populations that otherwise might be doomed to extinction, and speed up the adjustment to new environmental conditions at the recipient sites. Proposals for translocations for this purpose are controversial (see Ricciardi and Simberloff, 2009, and the opposing view of Thomas, 2011), and the response by regulatory authorities to proposals of this type is difficult to predict.

The number of documented restoration–rehabilitation sites globally is large, and each is in some way unique. Rather than reviewing the extensive case studies, I will focus on issues not always fully considered in project planning. Many of the issues to be discussed are also relevant in freshwater tidal wetlands, and the discussion by Baldwin et al. (2018) provides a complement to this account. The majority of published studies on salt marsh restoration are from North America and Europe, but degraded salt marshes which could be rehabilitated occur widely. Lessons learnt from projects at specific sites are often broadly applicable. Even when the object of restoration is to benefit a component of the fauna (such as birds), projects normally involve manipulation of the physical environment and vegetation, rather than involving direct management of animal populations.

2. SETTING OBJECTIVES

Estuaries have been a focus of human settlement for centuries (Adam, 1990; Marshall, 2004), and globally the pressures on estuaries are increasing in both the developing and developed world. The human population continues to grow, and an increasing proportion of it lives in urban areas, many of them on, or close to estuaries. Historically many estuaries have been reshaped by human activity, and major new port, airport, industrial, and residential developments continue this process, as exemplified in Sydney Harbour (Fig. 23.2). The outer harbor is a drowned river valley (ria), extending in the west into the Parramatta River. Since European colonization in 1788, the shoreline has been very heavily modified as port and other facilities were developed (Lee and Birch, 2013). Intertidal wetlands at the head of embayments have been infilled and lost. The legacy of industrial and urban developments during times with little control and regulation of factory and other discharges is a system with heavily contaminated sediments. The management of activities throughout the estuary system to minimize risks from these sediments will remain a continuing concern for government authorities (Lee and Birch, 2013). Nevertheless, there are still a large number of stands of salt marsh, mostly very small (Kelleway et al., 2007; Williams et al., 2011). Along the Parramatta River, there are some salt marshes which are large in the regional context (Adam, 1996). The Olympic Games in Sydney in 2000 provided an impetus for salt marsh rehabilitation, construction, and ongoing conservation measures in the Homebush Bay section of the Parramatta River estuary (Paul et al., 2012).

Globally large areas of salt marsh have been converted into either infilled developed sites or into agricultural land (both forms of activity previously referred to as reclamation, a misleading and inappropriate term), and others are degraded by past and continuing industrial and waste disposal practices. Few, if any, salt marshes could be regarded as pristine.

Degradation of salt marshes was until recently regarded both as an inevitable consequence of progress and as being economically and socially desirable. Even with the greater

FIGURE 23.2 Sydney Harbour, looking west showing development by European Australians since settlement in 1788.

recognition of the intrinsic, social, and economic values of salt marsh, protection of salt marsh is not guaranteed and proposals for major developments continue to be approved as being in what is perceived by governments as the national interest.

2.1 Context

Recognition of the historic loss and degradation of salt marsh resources and increased understanding of the ecological goods and services provided by salt marshes have provided the impetus for most salt marsh restoration—rehabilitation programs.

The modern conservation movement commenced in the late 19th century and salt marshes were among the early sites dedicated for conservation (for example, see Allison and Morley, 1989), and these early reserves have often been the focus of the research which underpins our current understanding of salt marshes as ecosystems.

The environmental movement developed in the 1960s and promoted a broader approach to holistic environmental management, in which nature conservation reserves are an essential, but not the only, element. Salt marshes were an early beneficiary of the new environmental interest (Nixon, 1980), and the value of the proposed connection between salt marshes and estuarine and inshore coastal water fisheries for both professional and recreational fishers was rapidly seized on as an argument to support marsh preservation in the United States and internationally.

Today a plethora of international treaties and national, state, and local government legislation and policy protect and promote recovery of salt marshes. Some national and state laws came into being as responses to and requirements of international treaties, but much more legislation has evolved within individual countries over extended periods. At the national and subnational level, the amount of legislation can be daunting (Rogers et al., 2016), be administered by a number of separate departments and agencies, and sometimes sets up conflicts as to choice of outcomes. The complexity of international, European, and national law and policy applicable to the marine environment around the United Kingdom has been documented by Boyes and Elliott (2014). Many laws listed in Boyes and Elliott (2014) have application to activities in salt marshes, as do a number of terrestrially based provisions, such as local government planning instruments, reflecting the transitional position of salt marshes between land and sea. Those implementing salt marsh restoration projects need to be aware of the legal framework that applies to each particular project.

Some legislation and policy has a narrow focus, for example, the conservation of threatened species or ecological communities. In the case of salt marsh, there are relatively few species given legal recognition as threatened. However, salt marsh communities have been recognized as being threatened and requiring protection, as, for example, 1330 Atlantic Salt Meadows (Glauco-Puccinellietalia) included under the provisions of the European Commission Habitat Directive (European Commission, 1992; Doody, 2008b) and temperate coastal salt marsh in southern Australia, listed as an endangered ecological community under the *Biodiversity Conservation Act* in New South Wales, and as a vulnerable community nationally under the *Environment Protection and Biodiversity Conservation Act*, even though most stands of these communities do not support any species recognized in their own right as threatened. Management of individual species can be conducted on a local site basis through tailored protective measures or by reintroduction.

Salt marsh may be utilized by migratory shorebirds (a group of species covered by several international treaties) as high tide roosts and, for some species, breeding sites. Although conservation of migratory shorebirds requires protection of individual sites, conservation needs to be considered in the context of entire migratory pathways, and conservation measures applied throughout. The need for geographically wide consideration is particularly apparent along the East Asian–Australasian Flyway where loss of intertidal habitat in the Yellow and Bohai Seas is currently proceeding at unprecedented rates (MacKinnon et al., 2012). Degradation of coastal ecosystems in China driven by economic growth is occurring not just in the northeast but also along the entire coastline and has accelerated since 1978 (He et al., 2014). As China's GDP per capita is still low, continuing economic growth will lead to further threats to coastal biodiversity (He et al., 2014). Although China has introduced coastal conservation measures since the 1980s, these are not always well implemented (He et al., 2014). Given the economic pressures for development, it will be difficult to adequately conserve what natural/seminatural salt marsh remains, let alone make major commitments to restoration and rehabilitation.

2.2 Biodiversity Values

Although legislation may require attention to particular features of salt marshes such as threatened species, there is now greater consideration given to what is sometimes referred to as biodiversity values—the ecosystem goods, services, and functions provided by biodiversity. At the site level, provision of habitat for species (including threatened species) is one of the services, but there are many more, as discussed in Thrush et al. (2013) and Esteves (2014), and in the wider marine context by Beaumont et al. (2007). The identification, assessment, and valuation (in an economic sense) of ecosystem goods and services are still a young science. Broad categories of goods and services have been recognized, and economic analysis has ascribed very large (but realistic) estimates of dollar value to them (for example, Costanza et al., 1997, and for estuarine systems in particular Barbier et al., 2011, Barbier, 2018). Importantly, there is increasing recognition of the less tangible social and cultural values of salt marshes. Study of individual sites and services will lead to further refinement of valuations.

One service that has been given increasing attention in recent years is the sequestration of carbon by intertidal wetlands which is likely to be a major proportion of global "blue carbon." It is suggested that coastal wetlands (salt marsh and mangrove) are responsible for about half of the annual burial of carbon in the ocean (Chmura et al., 2003; Duarte et al., 2005), although the variation between sites in productivity, turnover rates, and allocation of carbon between above- and belowground components of biomass means that current global estimates based on relatively few studies can only be ballpark approximations. Macreadie et al. (2017) have compiled all available estimates of blue carbon in Australian tidal marshes and suggest that the 1.4 million hectares of marshes contain 21.2 million tonnes of organic carbon in the upper meter of sediment. The carbon already present in coastal wetlands may represent hundreds of years of sequestration, and both direct and indirect human impacts can lead to rapid loss of carbon to the atmosphere (McLeod et al., 2011; Pendleton et al., 2012; Macreadie et al., 2013; Coverdale et al., 2014). Promoting carbon sequestration is seen as an important means of combating climate change. The creation and protection of salt marshes with high sequestration capacity as an objective may conflict with other conservation and management objectives, so that decision-makers will need to make choices in what circumstances particular objectives are given priority. In locations where both salt marshes and mangroves occur, mangroves are likely to sequester more carbon (Kelleway et al., 2016a; Owers et al., 2016) so that decision-makers will have to make choices and decide in what circumstances promotion of mangroves will be given priority over salt marsh. In situ accumulation of carbon from root biomass contributes more to long-term sequestration than leaf litter and other allochthonous carbon (Saintilan et al., 2013). However, long-term projections are more uncertain, given the suggestion by Kirwan and Blum (2011) that increased carbon dioxide and higher temperatures might result in increased decomposition resulting in increased release of soil carbon. Kelleway et al. (2016b) have shown that sediment carbon storage is higher within salt marshes in sites with a higher proportion of fine-grained sediment and subject to fluvial inputs, suggesting that sediment characterization will be important "for predicting blue C hotspots for targeted conservation and management activities."

A single site may represent a small contribution to blue carbon stocks, but the aggregate contribution of many rehabilitated or restored sites could be considerable. Crooks et al. (2014) have shown the potential for increased sequestration and reduced emissions from marshes in a whole estuary in a study in the Pacific Northwest. As with the Australian studies of Saintilan et al. (2013), the greatest sequestration benefits came from vegetation with high belowground productivity.

2.3 Mitigation

Restoration of salt marshes frequently occurs with the intention of mitigating or compensating for the damage caused by new development or as part of a more general program to repair past damage and may be mandated by the approval processes for major developments. In these circumstances, objectives are often set in very general or idealistic terms. The objective may be stated simply as the production of a functional (or healthy) salt marsh without "functional" or "healthy" being defined or the type of salt marsh specified. There is a tendency to assume that it is necessary to maximize species and habitat-type diversity, which is justified as increasing biodiversity. However, this is based on a misunderstanding of the concept of biodiversity and its quantification (Adam, 1998) because species-poor communities are still a component of biodiversity. Zedler (1995) points out that more is not necessarily better, and that in the particular case of Californian salt marshes, monotypic stands of *Salicornia bigelovii* are important habitat for Belding's Savannah Sparrow (*Passerculus sandwichensis*), a species protected under the US *Endangered Species Act*. Encouraging the establishment of other plant species in *Salicornia* habitat reduces its value to the sparrow and could constitute a breach of the legislation. Zedler (1995, 1996b) has argued that objectives for individual sites need to be set within a regional context, and that overall the goal should be "to maintain the natural diversity of species and community types." Salt marsh vascular plant communities are often, but not invariably, species poor (Adam, 1990), although the total species richness of the communities, including microalgae and bacteria, is not known. These species-poor plant communities may appear monotonous, even "boring," leading to demand for "more interesting," more species-rich, communities. Many of the ecological functions for which salt marshes are valued, such as provision of fish nursery habitats or absorption of wave energy, are primarily features of the low marsh zone, which supports a small number of vascular plants. In some circumstances, creation of more diverse communities may be possible, even if the resulting assemblage of species is, within its local context, "unnatural," although whether these created communities will be self-sustaining in the longer term is uncertain.

Those setting objectives must be cognizant of the fact that salt marshes are dynamic ecosystems. At individual sites, succession may be occurring, but the present state of a marsh reflects seral changes over the developmental history of the site. Adam (1990) warned against simplistic space–time substitution interpretations of zonation to predict future succession. Creation of "new" salt marshes by establishment of early seral stages and leaving succession to take its course will not necessarily result in mid- and upper marsh communities identical to those previously existing on local sites. Succession can be rapid (Adam, 2000), but at many sites the rate of change appears to be slow; it may be decades before upper marsh develops, probably well after the end of any monitoring programs.

Salt marshes may also be dynamic at larger spatial scales. Adam (2000) documented changes in the distribution and extent of salt marshes in Morecambe Bay in NW England over a quarter of a century. Although some marshes were little changed, others had expanded considerably and yet others had undergone extensive erosion as a result of changes in the position of channels (see also Pringle, 1995). Similar dynamic changes to estuarine wetlands have been reported from the Ebro Delta in Spain by Valdemoro et al. (2007). Restoration projects in regions where dynamic changes are likely can encounter particular problems. Concentration of effort at the local, individual marsh scale may be doomed to failure if the site concerned is destined to erode. If a wider view is adopted, for example, in the Ebro, the whole delta (Valdemoro et al., 2007), a strategy might be developed whereby the loss of some marshes could be compensated by the formation or creation of new ones.

However, while this is conceptually straightforward, implementation would be difficult given the uncertainty of predicting where and when erosion and colonization will occur.

2.4 Monitoring

Many restoration projects have been conducted in an ad hoc fashion with lack of clear goals and with no commitment to evaluation and reporting, indeed some projects were never assessed (Ambrose, 2000). Unless the objectives of the project are clearly articulated, success or failure cannot be determined. Aspirational objectives, such as maintaining or improving ecosystem health, do not lend themselves to assessment unless associated with clearly defined and measurable criteria. Without evaluation, salt marsh restoration will not contribute to the accumulation of basic knowledge required for restoration ecology to become a predictive science, and despite the occasional apparent success there may also be numerous, perhaps expensive, failures. Although there is reluctance from both government agencies and private developers to fund long-term monitoring, without adequate documentation of past experience there will also be reluctance to invest heavily in new projects if the prospects of success are uncertain. The monitoring requirements need to be considered at the same time as objectives are set and a project is planned. Only if monitoring is considered from the beginning, it is likely to be properly designed and funded.

Chapman and Underwood (2000), Chapman (2012), Underwood (2012), and Underwood and Chapman (2016) provide critiques of past practices, inappropriately referred to by their proponents as monitoring, and provide guidance for improvement in setting objectives and protocols for the design and conduct of rigorous monitoring programs. What is monitored will need to be relevant to the objectives. If the requirement is to enhance populations of particular species (e.g., rare or endangered taxa), then assessment of success must involve monitoring of these species, both at restored sites and elsewhere. Success will need to be determined by reference to the starting position at the restoration sites and also against population trends at regional and larger scales. If a species is in general decline across its distribution range, then establishing a sustainable population may be unlikely. The need to sample external reference sites will add to the cost of monitoring but is essential for evaluation of restoration. Reference sites need to be ecologically equivalent to the experimental sites (Chapman and Underwood, 1997); comparing upper estuarine sites with those in more seaward positions will not help assessment. In some circumstances suitable reference sites may not exist, meaning that monitoring design will be suboptimal. If the objectives include restoration of ecosystem functions and processes, then not only do the relevant functions and processes need to be defined but monitoring needs to incorporate direct or indirect measurement of them. Ecosystem process studies are expensive and again require reference sites. In many cases, the appropriate methodologies for assessing functions and processes in particular salt marsh types need to be developed. It is frequently the case that there are a number of objectives, and their assessment will require different sampling designs, both spatially and temporally. There is little background knowledge on the scales of variation in many of the components and processes in salt marshes (Zedler, 1996a) so that pilot studies will frequently be necessary prior to embarking on monitoring. Attempting to include all components within a single sampling design is unlikely to measure any of them adequately (Chapman and Underwood, 2000).

Although the object of restoration is to create fully functional ecosystems, it is unrealistic to require monitoring of ecosystem processes in every restored site. Rather, what is needed is research to establish reliable surrogates of ecosystem processes in different salt marsh types, but surrogates that are appropriate for assessment of *Spartina alterniflora* marshes will not necessarily be valid for marshes dominated by other genera in different geographic regions and vice versa.

The results of monitoring need to feed back into adaptive management (Buchsbaum and Wigand, 2012). The adaptive management approach for responding to climate change proposed by Wigand et al. (2017) is also applicable to restoration sites that will be exposed to climate change. Adaptive management is sometimes perceived by both the public and approval authorities as being simply trial and error, but it needs to be viewed as an experimental approach with defined objectives, the achievement of which can be tested. A technical guide to adaptive management practice is provided by Williams et al. (2009). Adaptive management may lead to modification of the management techniques through an iterative process, but it may also require reassessment of the objectives. If the objective cannot be achieved because of continuing disturbance, and the disturbance factors cannot be controlled, then this needs to be recognized and new goals set. This is not to say that the objectives should be simplified and generalized to triviality, but rather that restoration should be a realistic prospect.

Monitoring should permit change in restored sites to be evaluated against reference sites, so that if the conditions at the reference sites change over time this can be taken into account in determining whether the outcome of restoration is "successful." Given the universality of environmental change and the dynamics of ecosystem responses, reference sites over the longer term will not themselves remain unchanged.

Salt marshes are subjected and respond to a variety of disturbances at a range of spatial and temporal scales (see examples in Li and Pennings, 2016, 2017). Restored salt marshes will experience complex disturbance regimes, as do natural systems. Some disturbances, such as those caused by climatic extremes, may be sufficiently widespread that they will be experienced both at the restoration site and controls, but other disturbances may be of very limited spatial extent and may affect only parts of experimental and control sites. Divergence between control and restored sites does not necessarily indicate failure of restoration—but could indicate the normal response to local disturbance. It is essential that monitoring programs observe the sampling design and protocols established at their commencement. However, those conducting the fieldwork for monitoring programs should nevertheless remain observant and make note of disturbance and other changes across the site so as to place the changes at the sampling locations within their context.

Unfortunately, the legal and regulatory frameworks under which many restoration and rehabilitation projects take place do not recognize environmental change but assume "ecological stationarity." Craig (2010) used this term in his analysis of American environmental law, but it is equally applicable in other jurisdictions. Craig (2010) argued "for a 'principled flexibility' model of climate change adaptation to pursue goals of increasing the resilience and adaptive capacity of socio-ecological systems." Despite the eminent sense of this proposal, flexibility is still an anathema to many regulators and managers as it is perceived as creating uncertainty as to eventual outcomes, so that the requirements for restoration projects may still set targets applicable to the past and not to the future.

2.4.1 *What to Monitor*

Although monitoring programs have included a range of biophysical and biological attributes, the choice of attributes does not always relate to the reasons the activity was undertaken and is often driven by precedent, availability of expertise, and cost. Frequently any regulatory requirement for monitoring is for a limited time, and extended monitoring is a result of longer-term research programs which have been able to capitalize on the existence of the restoration site.

Microbial and invertebrate assemblages are important components of salt marsh biota, with major ecosystem process roles (Christian et al., 1981; Montague et al., 1981; Wiebe et al., 1981). Microbial assemblages in disturbed, polluted salt marshes differ from those in unpolluted marshes (Cao et al., 2006). The development of microbial communities in transplanted *Spartina alterniflora* marshes was studied by Piehler et al. (1998), and after 6 years the community was still not identical to the control site. For assessing ecological process, restoration microbial monitoring may be appropriate but has rarely been attempted.

Development of benthic infauna assemblages has been studied in a number of restored wetlands (e.g., see Levin et al., 1996; Craft, 2000; Craft and Sacco, 2003; Moseman et al., 2004). Although initial colonization may be rapid, convergence to reference assemblages may take many years, 25 years or more in some cases, well beyond the period normally specified for postconstruction monitoring (Craft and Sacco, 2003).

The invertebrate fauna of salt marshes includes a considerable number of species of terrestrial origin, although in general this component has been given far less attention than the marine fauna. David et al. (2016) sampled the abundance and diversity of terrestrial arthropod assemblages in salt marsh on the west coast of the United States. The composition of the assemblages was influenced by the nature of the vegetation and climatic factors. Similar variation in arthropod assemblages was shown by McCall and Pennings (2012a) on the Atlantic and Gulf coasts of North America.

Although a great deal is known about species composition and ecosystem processes in salt marshes in North America, in other parts of the world there is often far less information available from potential reference and degraded sites.

3. PLANNING FOR THE FUTURE

Restoration and rehabilitation projects are undertaken with the intention of providing for long-term survival of functional ecosystems. Projects should therefore be designed with the future in mind. At the most elementary level, this requires consideration of possible future developments or activities which may compromise the viability of the original objectives of restoration. It might be anticipated that given the expense of commitment to restoration that restored sites would be protected, but this is not necessarily the case. The Eve Street Wetland restoration in Sydney (Australia) (Stricker, 1995) resulted in a salt marsh of heritage significance, as part of what had been prior to suburbanization an extensive wetland, and that continued to be a significant habitat for migratory wading birds. Despite its conservation value, a few years after the restoration a major motorway was constructed across part of the site. Although some of the vegetation has survived, the value of the site as bird habitat has been substantially diminished. As population pressures and demand for developable space increase, it is difficult for planning authorities to make commitments which remain constant, so there will continue to be examples of good intentions today being thwarted in the future. The Eve Street example was one where only a single small site was affected. However, it is probable in coming decades that in or close to major urban areas proposals will emerge for new ports, airports, or industrial areas that could have impacts on estuarine habitats, even those habitats that have been valued for many years. In some jurisdictions, the complexity of the planning and appeals process is such that major proposals can remain fiercely contested concepts for decades, but in others, where public participation and environmental values are given less weight, major environmental change can happen very quickly.

The future scenario which has the highest public profile is one of rising sea level as a response to climate change. Over a time scale of a century and beyond this is undoubtedly likely to be a major factor affecting the survival of salt marsh (Reed, 1995). Although much of the public understanding of the issue is based on graphs showing predicted sea level at the start of the 22nd century, it is important to recognize that unless heroic measures are taken in the very near future, the trends expected over the next 100 years will continue and probably accelerate.

Changing sea level is only one of the consequences of an increased greenhouse effect which could impact on salt marsh restoration projects. Osland et al. (2016) have identified the range of possible impacts of consequences of climate change and the factors responsible for it on salt marshes globally. Salt marsh plant species in general have wide geographical distributions. Nevertheless, variation in floristic composition with latitude is observable (Adam, 1990), and temperature is likely to be the major determinant of distribution. Global warming is likely to result in changes to species distributions, although different species are likely to respond at different rates so that assemblages of species (communities) will change. If restoration projects involve planting, should the object be to plant the current regional assemblages or should future change be anticipated so that the species planted are sourced from lower latitudes? Even if such bold experiments are not conducted, the provenance of planting material would still need to be considered. Widespread salt marsh species are often made up of numerous genotypes (Adam, 1990; Zedler and Adam, 2002), and genetic variation is expressed at a range of scales both within and among sites. In restoration projects, use of local provenance material in plantings is often encouraged (Zedler and Adam, 2002), but does climate change require a relaxation of the rules? If environmental conditions are likely to change, should genotypes more adapted to predicted future environments be planted now, rather than geographically local forms which were selected for under previous environmental conditions? There are no simple answers to these questions, but it will be important to treat every project as an experiment, with full documentation of all stages. In terms of planting material it will be necessary to collect and propagate from known provenances, rather than using "off the shelf" stock of unknown origin from commercial nurseries. The complexity of possible phenotypic responses to changes in sea level and temperature has been discussed by Crosby et al. (2015) for *Spartina alterniflora*, where phenology of flowering and biomass allocation (reproduction, aboveground growth, and belowground growth) are likely to change. Predicting the consequences of these changes during the development of a restored marsh will not be easy.

Although the majority of salt marsh plant species have wide distributions, there are a small number of geographically and ecologically restricted taxa. These "rare" species may be particularly vulnerable to climate change. Should these species be given particular attention in restoration projects and be planted outside their current geographical range to anticipate change? Many botanists and conservation authorities would urge caution, and while supporting *ex situ* conservation, would not support planting in the wild outside existing distributions. Such a position would be strengthened by examples of species which are rare in their natural distribution but which have become major weeds as introductions (e.g., *Juncus acutus*). Nevertheless, Ranwell (1981) argued that there was a need to give more consideration to introduction of species, particularly rare species, outside their current natural range. With appropriate monitoring, restoration sites (but not "pristine" natural sites) could provide an opportunity for experimentation with translocation of rare species.

Although temperature is the component of future climate change given the most attention, and for which it is believed the more reliable predictions are available, a number of other aspects of climate are also likely to change. These include the frequency and intensity of extreme climatic events and rainfall seasonal distribution, intensity, and quantity. These changes will have effects on salt marshes. Changed seasonality and quantity of rainfall coupled with temperature increases could result in changes in soil salinity patterns, particularly in upper salt marsh zones. There is too much uncertainty associated with predictions of these parameters for them to be assumed as definitive in planning restoration, but the likelihood of change further strengthens the need for long-term monitoring and adaptive management strategies. Changed climatic conditions may also affect fire regimes. Communities dominated by *Phragmites* or *Juncus* spp. can burn fiercely. Although the likelihood of fire affecting natural or restored salt marshes will remain small, it would be precautionary to build capacity to respond to fires into the management regimes of restored marshes with flammable vegetation.

The underlying driver of anthropogenic climate change is increased atmospheric concentrations of greenhouse gases, particularly carbon dioxide. Changes in carbon dioxide concentration will have effects on vegetation independent of climatic effects. Increased carbon dioxide will differentially favor species utilizing the C_3 photosynthetic pathway over C_4 plants. Salt marsh vegetation is often made up of mixtures of C_3 and C_4 species, so as the carbon dioxide concentration increases there could be a shift to increased abundance of C_3 species. Thus, an aim of retaining existing relative abundances of species in perpetuity sets an unattainable goal. Field experiments involving increasing carbon dioxide in chambers in brackish marshes have shown increased productivity in C_3 Cyperaceae but not in C_4 *Spartina* (Arp and Drake, 1991; Drake, 1992; Drake et al., 1996), and an increase in abundance of the C_3 species over C_4 (Arp et al., 1993). In both C_3 and C_4 species, increased carbon dioxide resulted in decreased transpiration (Drake, 1992), which could limit increases in summer soil salinity. Under increased carbon dioxide, C_3 plants are likely to have lower nitrogen contents. This could not only result in less damage by pathogens (Thompson and Drake, 1994) but also lower attractiveness to herbivores with complex flow-on effects through the food chain. Gray and Mogg (2001) investigated the likely effects of increased carbon dioxide and temperature on the two major low marsh pioneer species, both grasses, in northern Europe—the C_3 *Puccinellia maritima* and the C_4 *Spartina anglica*, a hybrid of recent origin that has an elevation range overlapping with *Puccinellia* but extending lower into the intertidal zone. They suggest that *S. anglica* is likely to extend its geographic range northward in response to rising temperature. But *Puccinellia* growth responded strongly to increased temperature and carbon dioxide as did *Spartina* unexpectedly, possibly due to the greater water use efficiency at higher carbon dioxide levels. *Spartina* showed increased belowground growth (which may increase carbon sequestration). In competition studies, *Puccinellia* was favored over *Spartina*, leading Gray and Mogg (2001) to propose that replacement of *Spartina* by *Puccinellia* in succession might, in the future, occur at lower elevations than at present.

Increased atmospheric carbon dioxide concentrations are a driver of global warming and a major factor leading to sea level rise. Langley et al. (2009) suggest that paradoxically increased carbon dioxide may also contribute to survival of marshes by promoting increased elevation of the marsh surface, as a result of accumulation of belowground biomass. White et al. (2012) found that fine root production decreased in C_3 species with added nitrogen but increased under high carbon dioxide, whereas for C_4 species there was an increase in above- and belowground biomass with high nitrogen but the increase was lessened with increased carbon dioxide. Langley et al. (2013) suggest allocation of plant biomass into above- and belowground components occurs at the individual plant level, rather than through optimizing ecosystem viability. For marshes with highly organic soils increased nitrogen may reduce resource allocation to roots, as found by Turner (2011), but might also lead to changes in community composition.

Similar relative changes in species' ecology are likely to occur globally, suggesting that salt marsh communities in the future will have different composition than those at present. Thus, setting goals for present communities to be identical several decades hence is unrealistic.

One consequence of increased atmospheric carbon dioxide will be ocean acidification. As pH is a log scale, the magnitude of the biological consequences of what is numerically a small decrease in pH has not been sufficiently appreciated by the public. The predicted decreases in pH are unlikely to have any direct effect on macrophytes but will have effects on fauna, particularly those with calcareous exoskeletons (Ross and Adam, 2013).

Climate and associated environmental changes have affected salt marshes throughout geological history. Global warming after the last Ice Age was accompanied by rapid change in the distribution of both individual species and communities in response to both rising sea level and higher temperatures. Sea level reached its present position roughly 6000 years ago, although at high latitudes isostatic rebound following deglaciation continues, raising the level of the land relative to the sea and creating new surface for colonization by salt marsh.

Salt marshes have obviously accommodated past fluctuations by vertical and landward migration, but the ability to respond to future environmental change is severely compromised by the concentration of the world's human population in the coastal zone. Urban, industrial, and agriculture development in the coastal zone limits the opportunities for formation of new habitat for salt marsh colonization and the ability of species to migrate. In planning for salt marsh restoration, the viability of sites under the changed conditions predicted to prevail in decades ahead will need to be assessed. Although the threat posed by climate change–induced sea level rise has been recognized for many years (National Research Council, 1987; Titus, 2000), there is a reluctance in many parts of the world from all levels of government to respond in a coordinated and timely manner.

Climate change–induced increase in the volume of the sea will not translate into a uniform rise in sea level relative to the land. Relative sea level is influenced by other factors at the regional and local scale. For example, the rise in sea level may be balanced or exceeded by continuing isostasy; regional subsidence (as is occurring in southeast England) might increase the effects of global sea level rise, as well as human activities such as groundwater or hydrocarbon extraction that result in local land subsidence. Very rapid changes in land level may result from tectonic activity (Marsden et al., 2016); earthquakes may have regional scale impacts, and associated tsunamis affect even larger areas. Examples include salt marshes elevated above tidal influence in Alaska (Crow, 1971) and forests in Chile lowered into the intertidal and replaced by salt marsh (Valiela, 2006).

3.1 Coastal Squeeze

A rise in relative sea level might be expected to cause loss of the seaward margin of salt marshes. Historically any losses might have been mitigated by retreat of the landward margin and establishment of new salt marsh by colonization of what were previously upland communities. Coastal squeeze is the name given to the loss of habitat when the landward migration of intertidal communities in response to rising relative sea level is prevented. The term coastal squeeze was coined by Doody (2004, 2013), although the concept had been discussed, unnamed, earlier. The phrase has achieved rapid acceptance by the general public as being the most likely consequence on the coast of climate change. The prevention of movement inland is often because of the presence of hard engineering structures. Pontee (2013) and Esteves (2014, 2016a) suggest that coastal squeeze should not be applied to losses due to natural processes, but there are sites where the current landward limit of salt marsh is marked by an abrupt natural change in the topographic gradient (Adam, 2016) that can have the same effect as an artificial seawall.

Diagrams depicting the process (for example, those in Esteves (2014, 2016a)) show an initial intertidal zone of salt marsh that becomes narrower as retreat is impeded. However, an analogous process in the future may result in salt marsh in tropical and subtropical areas being simultaneously squeezed both by running out of retreat space inland and the expansion of mangroves from below. The spread of mangroves into salt marsh is a complex process reflecting responses to a number of factors (Saintilan et al., 2018), and the contribution of sea level rise to date is uncertain, but is likely to be more important in the future.

Coastal squeeze is proposed as one of, if not the, main cause for loss of intertidal habitats, and the model would certainly suggest that this is likely to be the case over the next century and beyond, in some parts of the world. However, it is not clear that the losses of salt marsh over recent decades in some localities are due to coastal squeeze (Hughes and Paramor, 2004). The losses along the south coast of England (Baily and Pearson, 2007) over this period may have a range of causes, but coastal squeeze is most likely to have been a minor component. Hughes and Paramor (2004) implicate bioturbation (in the area they studied principally by the polychaete worm *Nereis*) as a cause of loss, and failure of restoration. But the thesis that bioturbation (including by *Nereis*) is detrimental to salt marsh establishment, including in managed realignment sites, is rejected by Morris et al. (2004). Bioturbation is a normal feature of the pioneer zone in salt marsh succession, and yet marsh establishment has occurred in the past, and there is no reason why there should have been a change at that site. However, increased bioturbation may be a factor inhibiting recovery of marshes experiencing dieback because of increased herbivory following loss of predators (dieback following trophic disruption is a widespread phenomenon in some parts of the world; see Bertness et al., 2014).

In the absence of a barrier to inland migration, replacement of terrestrial communities by salt marsh could occur, as it must have done during previous phases of sea level rise in geological history. Although there are extensive stretches of coastline where movement is now constrained by artificial barriers, as in northern Europe, there are still many parts of the world where this is not the case, as, for example, much of the tropical coasts of Australia, and extensive parts of the coast of the United States. Kirwan et al. (2016) suggest that the likely extent of future loss of salt marsh to rising sea level has possibly been exaggerated as acceleration of soil building may permit marsh survival (see also Kirwan and Megonigal, 2013; Kirwan et al., 2010). Although this is likely to be the case for many marshes, there will still be large areas of salt marsh that will be lost, whereas other possible consequences of climate change, such as changes in severity and frequency of large storms, might in themselves trigger loss of salt marsh. Even where migration inland is possible, there may not be a simple movement of the whole of the intertidal zone. Smith (2013b) investigated the rate of migration of salt marsh inland in the Delaware estuary over more than 100 km of undeveloped coastline. There was loss of the seaward edge to erosion at the same time as there was expansion of salt marsh upward into the retreating deciduous forest, but loss at the eroding edge was greater than upward expansion (see also Kirwan et al., 2016). The lag in change at the salt marsh–forest interface was explained by the expansion of common reed *Phragmites australis*. Smith (2013b) postulated that for forest to be replaced by salt marsh there must first be invasion of dieback forest by *Phragmites* which will then be replaced by salt marsh as salinity increases. This assumes that *Phragmites* is not, under the new environmental conditions, considered a component of salt marsh. Smith (2013b) points out that *Phragmites* performs some of the ecosystem services of salt marsh, but that its expansion relative to salt marsh represents loss of habitat for salt marsh dependent species. Whether *Phragmites* is considered part of the salt marsh is complicated by the fact that the widespread expansion of common reed in North America is by an introduced genotype, which may be more salt tolerant and more responsive to nutrient enrichment than the original native form. Smith (2013b) suggests that establishment of salt marsh would proceed more quickly in locations of high salinity. Whether managers wish to engineer local hydrology to promote development of higher salinity to encourage "true" salt marsh development would be a matter for debate.

Coastal squeeze is not inevitable, even if opportunities for retreat are limited. A rise in sea level could also be countered by accretion of the marsh surface, both from allochthonous sediment input and plant production incorporated into the soil, at a rate greater than the sea level rise, and from isostatic rise of the land, which may be particularly important for the survival of Arctic marshes. Sediment supply derived from erosion in catchments will be affected by the nature of the vegetation cover and its management, and by the presence of artificial structures in rivers impeding the movement of sediment downstream. Kirwan et al. (2011) have shown that in the early European phase of land clearing in the Eastern United States, there was rapid growth of salt marsh, at a rate which could not be achieved today when soil conservation and other measures have reduced sediment supply. Ge et al. (2016) discuss the possible future of salt marsh in the Yangtze estuary, where extensive seawalls limit the potential retreat of salt marsh landward, and the reduction of sediment discharge into the estuary as a result of upstream engineering works will limit the potential for accretion of the marsh surface in the face of rising relative sea level. However, extensive dredging to maintain the deepwater channel for shipping is required, and Ge et al. (2016) suggest that one option to assist the long-term survival of salt marsh would be to relocate the dredged sediment to promote salt marsh development.

Likely changes in relative sea level should be assessed before embarking on restoration.

Cahoon (2015) has provided a valuable discussion of the estimation of relative sea level rise (RSLR) in coastal wetlands. Although it is possible to use tide gauge data to estimate RSLR for upland areas, this is unlikely to provide a general estimate for wetlands in the same region as there will be variation both within and between wetlands in upward growth of marshes from either allochthonous or autochthonous sedimentation. The task is made more difficult by the paucity of long-term data, either in the form of continuous records from tide gauges or measurements from surface elevation tables (SETs) (Cahoon, 2015). Those planning restoration of particular sites may be required to do so without any nearby tide gauge data being available and where no SET has been installed. This adds to the importance of linking subsequent monitoring to adaptive management and emphasizes the importance of applying an ignorance-based worldview (that assumes our current understanding is based on a large field of ignorance with only small areas of certainty) as stressed by Turner (2009).

If the site is likely to be overtaken by the sea within decades, then the appropriateness of the project needs to be considered. If deficiency in sediment supply is identified as a problem, augmentation of effective sedimentation can be considered (Bakker et al., 2002) through sediment recharge, promotion of sediment accretion through construction of a sedimentation field (as in the Schleswig–Holstein method, where a network of brushwood fences is constructed), or by construction of offshore breakwaters to limit erosion (Nottage and Robertson, 2005). These techniques can also be employed to protect and enhance existing marshes so that loss and degradation does not occur. The methods are relatively expensive and labor-intensive, and in the case of sedimentation fields do not offer instant success but may require years of operation and maintenance to achieve the desired result (Nottage and Robertson, 2005). Sediment recharge requires a source of supply, which is often derived from dredging operations. There is a long history of use of dredged material in salt marsh management in the United States, but rather less elsewhere. Dredging to maintain ports and channels is a major and often continuous activity. In the United Kingdom, an estimated 40–50 million cubic meters of sediment is dredged annually, the vast majority of which is dumped at sea. Ausden et al. (2016) argue the case for much greater use of dredged material in salt marsh restoration and rehabilitation, including managed realignment projects. They suggest that in the UK context a major barrier to increase use of dredged material to achieve beneficial conservation outcomes is the labyrinthine complexity of the regulatory regime for gaining approval. However, the sediment available may not have appropriate physical characteristics for salt marsh development and, if derived from an industrialized estuary, may be polluted. Extraction may have environmental impacts which would need to be fully assessed before a project could be approved (Nottage and Robertson, 2005), and, in the context of whole estuaries, the conservation value of the sediment donor site needs to be weighed against the benefits of the particular salt marsh site. Once deposited at the "new" site, the sediment may need to be dewatered, stabilized, and protected from erosion before vegetated marsh could be developed.

3.2 Managed Realignment

In many parts of the world, extensive areas of former marsh have been converted into low-lying agricultural land, protected from the sea by embankments and sluices. With rising relative sea level, the expense of maintaining sea defenses may not be justifiable. Breaching existing sea defenses allows for the creation of new landward defenses and the reestablishment of new salt marsh to replace that lost to erosion and rising sea level outside the original seawalls (Dixon et al., 1998; Nottage and Robertson, 2005; Baldwin et al., 2018). This process has been given various names, including managed retreat, setback, and depolderization, all terms which carry connotations which cause difficulties for political and social acceptance. The currently accepted term is managed realignment (Esteves, 2014, 2016b). Managed realignment has been carried out at many sites in northern Europe, with the majority of examples to date being in England (Esteves, 2014). An online register projects is available as the ABPmer Online Managed Realignment Guide—http://www.apbmer.net/omreg/. Pendle (2013) has provided a review of outcomes of managed realignment projects in England. Managed realignment projects have been initiated outside Europe, but information has not been consolidated. Esteves (2014) lists a number of managed realignments in North America; the approach of restoring salt marshes after breaching embankments has been applied on a large scale in San Francisco Bay (Williams and Faber, 2001; Williams and Associates and Faber, 2004; Williams and Orr, 2002), although in this case the primary objective has been to regenerate salt marsh on former salt production ponds and agriculture lands rather than to mitigate against the effects of rising sea level. Rogers et al. (2014) have identified a potential application of the approach in the Hunter Estuary (NSW, Australia), although their proposals have not yet been adopted. One of the issues which Rogers et al. (2014) identify as a constraint is that the land that would be affected is largely privately owned; the agreement of landholders would be required. However, they emphasize the importance for carbon sequestration of the wetlands that would be created. With appropriate environmental accounting methodologies and credit transfer, the value of extra carbon sequestration might offset the costs of managed realignment.

Managed realignment permits in most cases the establishment of shorter and more economically viable sea defenses, with the reestablished salt marsh in front of the new seawall offering protection to the wall through the attenuation of wave energy. However, land that was embanked long ago, centuries ago in the case of parts of North Europe, will have undergone many changes, including compaction and oxidation of organic matter leading to the land surface now being considerably below that of the current marsh level outside the seawall. In addition, depending on the agricultural practices applied, the soils may be nutrient enriched. The oxidation of soils results in lowering of the land surface and releases previously sequestered carbon (Bu et al., 2015). Former wetland soils when disturbed and oxidized may generate acid, and discharge of acid drainage, accompanied by high levels of mobilized iron and aluminum into estuarine waters, may result in mortality of biota (Stone et al., 1998). If the soil shrinkage has been considerable, then sediment recharge may be necessary before marsh can colonize. If restoration is reliant on natural sedimentation, the process may be slow.

The vegetation that has developed at most managed realignment sites is readily identifiable as salt marsh. However, the question is whether it is comparable with that of reference sites in the general vicinity. Mossman et al. (2012) studied a range of managed realignment sites, natural (accidental) realignment sites (where seawalls or sluices had failed for a variety of reasons, sometimes many years ago), and reference sites. In most cases, the realignment sites contained many of the regional plants species but in combinations and relative abundances that did not match the reference sites. Whether over the long term there will be convergence is unknown, but the current state of the vegetation may not represent "failure"—there may have been an unreasonable expectation of more rapid development than is possible. Quantification of most ecosystem services in managed realignment sites compared with those in reference sites is lacking and there has been primary attention paid to vascular flora. Pétillon et al. (2014), however, have shown relatively rapid colonization by arthropods, a finding they attribute to the greater dispersal potential of arthropods compared with plants.

Morris (2012) argues "that attention must shift away from nature conservation outcomes as the primary policy driver for managed realignment in the UK and in Europe. Instead there are sound engineering and social reasons for greater emphasis on realignment but to make this happen there has to be a reduction in the emphasis on wildlife and a much greater focus on engineering and water management. Relatively small policy adjustments are needed but a much more profound change in presentation is needed so that decision-makers and stakeholders pay more attention to the engineering needs for realignment." Esteves (2013, 2014) has also identified possible conflicts of objectives. If managed realignment cannot, in all instances, meet specific conservation objectives, the ecological values and services provided by salt marshes created through reestablishment of tidal exchange will nevertheless be considerable, and the design and selection of managed realignment sites should seek to maximize ecological outcomes, even if all benefits that might be desirable cannot be delivered.

4. ADDRESSING CAUSES AND NOT SYMPTOMS

Many of the more immediate causes of salt marsh degradation are themselves symptoms of wider problems. Restoration which focuses on ameliorating symptoms without addressing the underlying cause is unlikely to be effective in the long term.

Some forms of degradation can be addressed at the site-specific scale, such as access by vehicles, changes due to too high or too low grazing pressure, changed fire regimes, restoration of tidal flushing and drainage patterns, some forms of pollution, and some species problems; but in many cases action at a much larger spatial scale will be required. If degradation is due to agricultural runoff, the problem needs to be addressed at the catchment scale; for example, the reduction of phosphorus inputs into the Peel–Harvey estuary in Western Australia (Brearley, 2005; Elliott et al., 2016). In this example, the benefits to the fringing salt marsh were secondary to the major objective of increased flushing and improvement in water quality. Some problems will require action at continental scales. In both Eurasia and North America, there have been substantial increases in the populations of migratory geese in recent decades. A major factor in the increase has been greater winter food availability in temperate latitudes because of changes in agricultural practices, combined with significant reduction in hunting pressure. The increased grazing pressure and grubbing of rhizomes has caused disturbance to temperate marshes in winter (Smith and Odum, 1981), but the most extensive damage has been to breeding grounds on Arctic salt marshes, where loss of vegetation and increased soil salinity will be difficult to counter (Jefferies, 1997). Management of the goose populations, which will be an essential prerequisite for any rehabilitation of the marshes, will require international cooperation.

If successful rehabilitation in many cases requires actions that go beyond the site specific, are single site projects a waste of resources? Zedler (1995, 1996b) has argued that restoration ideally should be undertaken in the context of at least regional objectives and strategies, but it is likely that most projects will continue to be site based and often ad hoc. Many projects are initiated, and strongly supported, by the local community. Failure will not only cause disappointment but may also jeopardize future support for conservation management. It is extremely important that there is community endorsement of objectives and strategy, and recognition that some outcomes are only temporary and further remediation to address the same problems will be required. If the need for continuing maintenance is stressed at the outset, it is possible that some projects would not be attempted, but it is likely that often the community will still support rehabilitation of sites in circumstances where the ultimate cause of degradation cannot be addressed but where treatment of symptoms on a recurrent basis will still lead to maintenance of desired features and ecological processes.

5. MANAGING DISTURBANCE

There are many factors that could be managed or modified in the restoration projects. Every site is unique, but there are a number of issues that arise frequently at restoration sites, and these are discussed below.

5.1 Restoring Hydrology

Salt marshes are often topographically complex with numerous creeks and pans. The topography creates habitat diversity which is reflected in the distribution of individual species and communities (Adam, 1990). The creeks provide the major pathway for exchange of nutrients, detritus, and biota between marshes and estuaries and coastal seas. There is considerable variation in the nature of creek and pan systems between marshes, which can be partly related to substrate characteristics, tidal amplitude, and vegetation (Chapman, 1974). In temperate Australia, creeks and pans may be virtually absent from some large salt marshes (Adam, 1997). The basic distribution patterns of creeks and pans are laid down from the earliest stages of marsh colonization, and altered by both erosion and infilling during the course of marsh development. In some mature marshes, the creek system may be thousands of years old. In existing, but degraded, marshes, parts of the creek system may have been obstructed through construction of causeways and culverts, canalized to improve discharge of runoff from the catchments, or augmented by additional ditches to improve drainage.

Given the importance of hydrology to the nature, development, and function of salt marshes, high priority is given to hydrological issues in restoration (Callaway, 2001). Where new wetlands are created it will be necessary to design and construct drainage channels. The process of creation compresses temporally the establishment of processes which in a naturally developed marsh may have taken centuries (Callaway, 2001), and questions arise as to whether the desired complexity can be created instantly or whether a framework should be established with adjustment by natural processes to take place over a period of years.

Callaway (2001) reviewed a large number of restoration projects and identified a number of common problems including low density of creeks, lack of small first-order creeks, generally straight creeks lacking sinuosity, creek bank instability, lack of steep bank slopes, and failure to replicate the natural topography of the creek bank–marsh interface. Many of these problems arise because of the difficulty of constructing replicates of natural geomorphology with bulldozers and excavators.

Further issues arise if the marsh being restored has preexisting fill or where fill is required to bring it to the level needed to support the required community type. If the fill has different physical properties from that found at local reference sites, the ability to create creek banks of the required steepness and stability may be impaired, whereas the chemical properties of fill will influence the development of both vegetation and faunal communities (reviewed by Callaway, 2001). The option of specifying the nature of fill may not be open; often marsh restoration proceeds on the basis of using what is already in place, or moving fill from a predetermined site elsewhere.

Modification of hydrology is one of the major components of restoration which can take place within the confines of a single site, but issues of freshwater input, either from groundwater or as surface flow, will require consideration of the broader catchment. Although there has been consideration given to groundwater providing the route for nitrogen compounds reaching estuaries from agricultural landscapes (Addy et al., 2005), in general groundwater influences on salt marshes have been given little attention.

5.2 Managing Non-Native Plants

The spread of introduced plants in all habitats is considered one of the major threats to biodiversity (Mooney and Hobbs, 2000), although there is an emerging counter opinion that the threat has been exaggerated (Thompson, 2014; Chew, 2015). Salt marshes occupy an environment that is inimical to many plants, and the number of species that can tolerate various combinations of salinity and water logging is relatively small. Additionally, many salt marsh species have very wide geographical distributions (Adam, 1990), and potential vectors for long-distance transport of propagules (currents, tides, and migratory birds) have operated over a far longer time period than humans have been transporting species around the globe. It might be expected that most salt marsh species will already have spread to the tolerance limits of their range, and that the incidence of more recent exotic invaders would be low. However, there are a number of examples of invasive introduced species, and in most cases the time of introduction and subsequent pattern of spread are reasonably well known. The naturally wide distribution of many salt marsh species can give rise to uncertainty as to the status of some species (e.g., is *Spartina maritima* native in southern Africa?—Pierce, 1982, on which continent is *Lampranthus tegens* native?—Adam, 1996)

Removal of introduced species may be part of the restoration of salt marshes but can be an expensive and difficult operation. Although there is often a knee-jerk response by managers to the presence of introduced species that assumes control is essential, this needs to be questioned before expensive programs are initiated.

Firstly, what is the evidence that the introduced species are responsible for "harm"? Introduced invasive weeds could have a variety of impacts including competitive exclusion of native species, changes to the structure and composition of both floral and faunal communities, alteration of physical and chemical characteristics of the environment (such as modification to drainage characteristics, sedimentation rates, or soil properties), or changes in genetic structure of native populations through hybridization. For example, *Spartina townsendii*, and subsequently *Spartina anglica*, arose in southern England following spontaneous hybridization between the European S. *maritima* and the introduced North American S. *alterniflora*. Hybridization between introduced S. *alterniflora* and *Spartina foliosa* in California is regarded as a threat to the survival of the native species (Daehler and Strong, 1994) and to ecosystem processes. These impacts will have effects on biodiversity, although evidence that these extend to hastening extinction of any species is inconclusive (Ranwell, 1981; Valiela, 2006). But there is at least circumstantial evidence for some local extinction of species, and loss of components of biodiversity may render other species more vulnerable to total extinction. In addition to impacts on existing marshes, *Spartina* species may colonize mudflats below the previous lower limit of salt marsh reducing the extent of mudflat habitat (Goss-Custard and Moser, 1988; Adam, 1990; Ayres et al., 2004; Valiela, 2006).

The second issue to be addressed is whether control is likely to succeed. Control will be unsuccessful unless the factors promoting weed invasion are also addressed. These may include alterations or disturbance to the environment (e.g., stormwater discharge promoting growth of *Typha* or *Phragmites*—Zedler et al., 1990) or abundant regional supply of new propagules. There may not be any practical measures for the control of many species because there are considerable constraints on the use of chemical controls, physical removal is labor-intensive and can be onerous in soft mud, and the research and development necessary to obtain regulatory approval for biological control is time consuming, expensive, and not guaranteed of success. Invasive species are often closely related to species indigenous to the marshes being invaded, so that biological control, which requires that the control agent attack only the invasive species, may not be appropriate. Nevertheless, the potential for use of biological methods for the control of S. *alterniflora* on the west coast of the United States (Grevstad et al., 2003) and for *Phragmites australis* (Blossey, 2003) has been suggested.

Some of the issues created by invasive species may, in the longer term, be self-correcting. The original dense monocultures of S. *anglica* will become more species rich during the course of succession (Ranwell, 1981), sometimes relatively quickly (Adam, 2000). The outcome may not be similar in composition to preinvasion communities, but returning to the past may be an impossible dream. In other cases of invasion, such as by *J. acutus* or *P. australis*, there is little indication so far that there will be a decline in dominance and increase in species diversity.

Invasion by introduced species undoubtedly has impacts on some components of biodiversity, but in many cases there are insufficient data to determine the effects on ecosystem processes and services. Does weed invasion change the productivity of salt marshes? What are the impacts on adjacent estuarine and coastal water ecosystems? How important is plant species composition and vegetation structure for the maintenance of the faunal community?

Although there is a very strong argument for being vigilant to detect and control newly arrived plant species and new infestations before they become established, reduction of major long-standing populations may not be possible, necessitating changes to the objectives of restoration projects. A large number of introduced plants are reported from salt marshes, but many are currently relatively local in their impact. The species of most concern, and the subject of active management, are tall-growing competitive dominants, capable of spread by clonal growth. These species are, or are assumed to be, ecosystem engineers.

Species in this category include *Spartina* spp., *P. australis*, and *J. acutus*. *Spartina* species differ from most other salt marsh invaders in that they were deliberately planted at many locations around the world (Ranwell, 1967; Phillips, 1975; Boston, 1981) and have subsequently spread from the point of introduction. *Spartina* planting was still promoted in China (Chung et al., 2004) after the genus had been placed on the official list of harmful invasive plants (Wang et al., 2006). *P. australis*, the common reed, is an almost cosmopolitan species in fresh and brackish wetlands. Change in environmental conditions (such as grazing, drainage, or nutrient regimes) can promote the spread of *Phragmites* (Esselink et al., 2002; Silliman and Bertness, 2004), but the extensive spread in the coastal marshes in the United States in recent decades has been of a European genotype acting as a cryptic alien (Saltonstall, 2002). *J. acutus*, a species of conservation concern in parts of its native northern hemisphere habitats, has become a very aggressive invader in southeast Australia. A range of techniques has been employed to control or eradicate major clonal weeds that are often expensive and labor-intensive. They include physical removal (DPIWE, 2002; Lacambra et al., 2004; Paul and Young, 2006) and herbicide use (Truscott, 1984; DPIWE, 2002; Lacambra et al., 2004), although the effects of herbicide on other estuarine flora and fauna are poorly documented.

Hurst and Boon (2016) reviewed the management options for weed control in salt marshes in southeast Australia. An emerging problem in the area is spread of *Lophopyrum ponticum* (tall wheatgrass) into upper salt marsh. The species had been promoted by the Victorian state government as a salt tolerant pasture grass (salinization of nontidal pastures is an environmental problem in the region), but it has proved to be seriously invasive. Despite numerous examples globally of the use of herbicides in salt marsh, Hurst and Boon (2016) identified potential adverse ecosystem consequences of herbicide use, but these had been poorly studied. Hurst and Boon (2016) suggest that advocacy of herbicide use in salt marshes remains problematic.

In addition to large, obviously aggressive invaders, there are many other nonindigenous species recorded from salt marshes, some currently known from only a few sites. Among these may be sleepers, species that after a latency period can become aggressive invaders. Others may be widespread and abundant, but because they are of relatively small stature, there has been a perception that they do not constitute a problem.

Introduced species are particularly abundant in marshes in Mediterranean climatic zones (Bridgewater and Kaeshagen, 1979; Sullivan and Noe, 2001), often in disturbed areas (such as footpaths) and in the high marsh transition to supralittoral vegetation. This latter zone is where the majority of rare or geographically restricted salt marsh plants occur (Ranwell, 1981). One such species is the endangered hemiparasite *Cordylanthus maritimus* ssp. *maritimus* in upper salt marshes in California. Among the invasive alien species in *Cordylanthus* habitat is the annual European grass *Parapholis incurva*. Fellows and Zedler (2005) have shown that *Parapholis* acts as a "pseudo host" in that haustoria formation occurs between the two species, but flowering of *Cordylanthus* is substantially reduced (average one flower per plant) compared with the native host, the perennial grass *Distichlis spicata* (average 13 flowers per plant). This reduction, if it results in similarly reduced seed production, could over time lead to population decline. This finding warns us of the possibility of subtle adverse consequences of any introduced species on salt marsh.

Another annual grass of European Mediterranean origin which has become established in both California and Australian salt marshes is *Polypogon monspeliensis*. Callaway and Zedler (1998) showed that increased cover of *Polypogon* compared with that of the native succulent perennial chenopod *Sarcocornia perennis* was favored by decreased soil salinity and increased freshwater inputs. In southern California, many salt marshes experience increased freshwater inputs from either agricultural or urban runoff (Callaway and Zedler, 1998), and the impacts are often exacerbated by anthropogenic reductions in tidal exchange (Boland and Zedler, 1996). Spread of annual grasses at the expense of evergreen perennials could alter the vegetation structure and food availability of the salt marsh, with consequent affects up the food chain (Callaway and Zedler, 1998). Kuhn and Zedler (1997) showed that application of salt was an effective means of eradicating small patches of annual grasses. But while this might be an appropriate control technique in the early stages of invasion it is unlikely to be practical at large scales with a well-established population (application of salt, in conjunction with other control techniques, was also shown by Paul and Young (2006) to be effective against *J. acutus*). Restoration of tidal access and reduction in freshwater runoff entering marshes would be likely to control not just *Polypogon* but a suite of other species with similar ecological characteristics.

A number of salt marsh species have become popular in horticulture. This has led to several species of sea lavender (*Limonium*) becoming invasive outside their natural range (Parsons, 2013). Other salt marsh plant genera are also subject to horticultural interest, for example, *Armeria* spp. and *Frankenia* spp., and may also have the potential to become weeds of salt marshes. Managers of salt marshes, restored and natural, should be vigilant for emerging weed problems.

All possible measures should be taken to prevent the introduction and spread of further species. Quarantine can never be perfect, so there is a need for vigilance to detect new arrivals. The control and eradication of existing infestations of many species as part of rehabilitation is likely to be difficult if not impossible, and in some cases may be undesirable.

S. alterniflora, native to the east coast of the United States, was deliberately introduced into San Francisco Bay in the 1970s by the US Army Corps of Engineers. *S. alterniflora* hybridized with the native *S. foliosa*, producing a vigorous form, capable of occupying a wider intertidal range than either of the parents. The spread of the hybrid presented a range of threats to the San Francisco Bay environment (Strong and Ayres, 2016), and intensive control measures (Kerr et al., 2016) were implemented. However, hybrid *Spartina* has been utilized for nesting by the rare Ridgway's Rail (*Rallus obsoletus*). In addition, Strong and Ayres (2016) suggest that tall *Spartina* hybrids could, through attenuation of wave energy, protect developed shorelines. Strong and Ayres (2016) suggest that a barrier of hybrid *Spartina* could protect the shoreline at least for some decades under predicted sea level rise scenario. However, retention of some hybrid *Spartina* would need to be accompanied by active management to prevent reinvasion of habitats where its presence would be undesirable.

The expansion of *Phragmites australis* in upper salt marsh in the United States also gives rise to debate. The invasive form is an introduced genotype. In Australia, the spread of *Phragmites* has given rise to concern among land managers, but there is no evidence of introduced genotypes (Hurry et al., 2013). Spread is a response to changed environmental conditions such as the discharge of stormwater into upper marshes (Zedler et al., 1990). Spread in the United States may be promoted by greater salt tolerance and responses to eutrophication of the introduced form. The development of extensive *Phragmites* monocultures has been viewed as undesirable (Burt, 2007), and large sums have been expended on control, with variable and limited success. Hazelton et al. (2014) advocated that *Phragmites* management should be at a regional scale rather than on an individual site basis, and that there should be greater focus on wetland restoration rather than just *Phragmites* eradication. Hazelton et al. (2014) and Kiviat (2013) also document that *Phragmites*, both native and introduced genotypes, provides a range of ecosystem services. At some localities retention may be preferable to local eradication.

However, where *Phragmites* reduction and control are considered desirable objectives, Silliman et al. (2014) have suggested that use of rotational grazing, in their study by goats in small enclosures at a high stocking rate, could be an economically sustainable method of control when compared with other approaches. They also demonstrated that cattle and horses will also readily consume *Phragmites* and suggest that the long history of grazing by livestock on European marshes may have served to suppress the spread of reeds.

5.3 Introduced Fauna

Ports and estuaries are hot spots for introduced fauna (Valiela, 2006), but relatively little is known about introduced fauna on salt marshes. The Australian burrowing isopod *Sphaeroma quoyanum* is an invasive species in Californian salt marshes (Talley et al., 2001), where it has a serious impact by increasing the erosion of creek banks weakened by its burrows. There is no control method, but the potential for invasion of new sites may be a constraint for future restoration projects.

As with some plants, introduced animals may in some circumstances be advantageous. The European green crab (*Carcinus maenas*) has become one of the most widely distributed estuarine crabs in the world (Grosholz, 2011), and the invasive populations are often perceived to be detrimental to the natural ecosystem. However, where the invaded ecosystem has been degraded, the introduced species may have positive benefits. In the New England (USA) region, depletion of native predators as a result of over exploitation of fisheries has resulted in increases in the herbivorous crab *Sesarma reticulata*. Herbivory on *Spartina alterniflora* by *Sesarma* is a major causal factor in extensive low marsh die-off, and *Sesarma* burrowing increases susceptibility of the sediment to erosion (Holdredge et al., 2009; Coverdale et al., 2012, 2013, 2014; Bertness and Coverdale, 2013: Bertness et al., 2014). Introduced *C. maenas* partially reverses *Spartina* die-off by preying on *Sesarma* and forcing them from their burrows where they are more exposed to desiccation and predation (Bertness and Coverdale, 2013). In the region C. *maenas* is only common in marshes with long histories of *Sesarma* occur induced die-off and is rare on healthy marshes. Deliberate introduction of *Carcinas* should not occur, but its presence in sites proposed for restoration should not be regarded automatically as a negative feature.

In addition to the long history of agricultural use of salt marshes (Adam, 1990; Gedan et al., 2009), a number of introduced vertebrates have become established. Two of the most widespread are the rodents, muskrat (*Ondatra zibethicus*) and coypu (nutria) (*Myocastor coypu*). Both were deliberately introduced for their fur, but escaped from fur farms and established feral populations (Anderson and Valenzuela, 2011). Coypu were also deliberately released in Florida with the aim of controlling invasive aquatic plants (Anderson and Valenzuela, 2011). Both species are ecological engineers and through their burrowing can modify marsh hydrology. In the case of coypu, a combination of grazing and burrowing contributed to the conversion of large areas of salt marsh in Chesapeake Bay and elsewhere into open water (Anderson and Valenzuela, 2011). The presence, outside the natural range, of the species in salt marshes would be an impediment to restoration and rehabilitation projects. However, concerted eradication programs can be successful. Both species escaped from fur farms in the United Kingdom in the 1920s. Fortunately, neither became a problem in salt marsh, but did establish in a range of other wetlands. Initial attention focused on muskrat, and trapping resulted in eradication from Scotland by 1937. Coypu were not originally considered likely to become a problem, but by the 1960s had established large feral populations in southeast England. An extensive eradication program resulted in complete elimination by the 1980s. Anderson and Valenzuela (2011) suggest that local control measures without there being a more widespread program are unlikely to achieve eradication in the long term.

5.4 Mosquitoes

Mosquitoes are a component of the fauna of many salt marshes, although little is known about their role in salt marsh food webs and ecological processes. Salt marsh mosquitoes may, when abundant, be a major irritation to local human populations and, of more significance, the vectors of important human pathogens. Control of mosquitoes on salt marshes has been practiced for many years through habitat manipulation and use of insecticides (Wolfe, 1996). Ditching or runnelling of salt marshes to reduce breeding sites has been a major factor in salt marsh degradation (Daiber, 1986), although increased understanding of the values of salt marshes has led to modified techniques with less adverse consequences (Wolfe, 1996). Filling of historic ditches may appear to be an option in restoration, but Corman et al. (2012) suggest that if sea level is rising, ditch removal may not necessarily achieve the desired end and needs to be carefully considered before a program is commenced. Opposition to salt marsh restoration and rehabilitation projects may arise from concerns of local residents that mosquito populations and incidence of diseases debilitating to humans will increase. In an age of rapid transport of goods and increased volumes of cargo, there is increasing possibility of both mosquito species and pathogens being introduced into new countries. Authorities will need to institute surveillance and development contingency plans for the introduction of emerging disease threats (Webb and Doggett, 2016).

In 1998, the Australian salt marsh mosquito *Aedes camptorhynchus* was recorded for the first time in the North Island of New Zealand. In Australia, the species is known as the vector of the arboviruses responsible for Ross River fever and Barmah Forest disease, neither of which is endemic to New Zealand. Both diseases are debilitating to humans and are notifiable in Australia and subject to long-running surveillance programs, monitoring both mosquitoes and virus incidence (see, for example, the regular reports of the NSW arbovirus and mosquito monitoring program, such as Doggett (2016)).

It is thought that the mosquito was transported to New Zealand in tyres imported for recycling. Its discovery in New Zealand led to a major eradication campaign, involving mobilization of expertise from both New Zealand and Australia. In July 2010, the New Zealand government announced the eradication of the species in the country, the first time this has been achieved. The saga is analyzed in detail in Key and Russell (2013). Continuing surveillance will be required to ensure that further invasions do not occur. The measures applied are not believed to have caused damage to New Zealand salt marshes, but the necessity of acting swiftly meant that there was little opportunity for pretreatment baseline studies.

5.5 Grazing

Waterfowl and mammals are significant herbivores in salt marshes. Although provision of habitat may be one of the long-term objectives of restoration, it may be necessary to limit their access in early stages of development to allow vegetation to become established. Timing of restoration activities will also need to take into account the requirements of fauna. For activities such as sediment recharge or construction of banks and drainage systems that may cause disturbance to birds, there may be only limited windows of opportunity (taking migratory patterns and breeding seasons into consideration) when disturbance will be minimal and conditions are appropriate for establishment of vegetation. Considerable lead time may be required to ensure availability of machinery and other resources during these critical intervals.

In their aboriginal state, many salt marshes would be grazed by a diversity of vertebrate fauna including waterfowl (in the northern hemisphere) and mammals both large (e.g., deer and kangaroos) and small (e.g., voles and hares). This pattern of grazing still continues in many regions, but at many sites natural grazing has been supplemented or replaced by grazing of domestic livestock. Agricultural exploitation of salt marshes has also included haymaking in both Europe and North America (Valiela, 2006), although currently it is a minor use. Both livestock grazing and haymaking result in export of productivity to the hinterland, against the natural flow into estuaries. Agricultural grazing is still an important widespread use of salt marshes in northern Europe and South America but is more local on other continents.

Grazing can affect the composition of vegetation through adverse impacts on sensitive species, and the consequential promotion of grazing-tolerant species. Additionally, trampling by livestock may result in loss of vegetation cover and modification of local drainage patterns.

The effects of grazing have been extensively studied in Europe where grazed salt marshes differ considerably in species composition, structure, and vegetation type from ungrazed marshes (Adam, 1978, 1981; Bakker, 1989). Changes in grazing pressure can result in rapid changes to the composition of salt marsh vegetation (Adam, 2000). An assessment of the condition of salt marshes in Britain and Northern Ireland designated as being of high conservation value (Williams, 2006) concluded that a number were threatened by either overgrazing or undergrazing. The question therefore arises as to whether restoration of salt marsh might, in some circumstances, involve reestablishment of an appropriate grazing regime.

Whether grazing is a desirable management strategy will depend initially on the local context. For very small sites, agricultural grazing would be neither practical nor economically viable. In other cases, changes in local agricultural practice such as from mixed farming to cropping may mean that there are no adjacent farmers able to take up salt marsh grazing. If grazing is thought desirable, then conservation agencies would have to be responsible for management of livestock (Fig. 23.3). If grazing could be practically and ecologically

FIGURE 23.3 An English Longhorn cow, part of a small herd used in the rehabilitation of the Widnes Warth salt marsh, upper Mersey estuary, Cheshire, UK. Use of historic breeds of livestock has become a feature of conservation programs on European salt marshes and achieves the additional benefit of promoting the survival of rare breeds. *Image courtesy of Damian Smith. Copyright Mersey Gateway.*

viable, under what circumstances might it be appropriate? The answer will depend both on the desired end condition of a site, and its prior history.

If the site has never been grazed, then introduction of grazing will result in a change in species composition through loss of sensitive species, although total species diversity might be maximized under moderate grazing pressure (Bakker, 1989). If the site has been grazed, relaxation or cessation of grazing pressure is unlikely to result in a reversion to anything resembling a never grazed state. Rather than reestablishment of sensitive species, the most likely outcome would be a loss of species diversity and the development of rank species-poor plant communities (Adam, 1978, 1981, 1990, 2000; Bakker, 1989) dominated in Northern Europe by *Festuca rubra* or *Elytrigia atherica*. If grazing pressure has only recently been reduced, then reintroduction of the former grazing may prevent the reduction in diversity but once species-poor stands have developed, grazing alone is unlikely to restore the former vegetation.

Manipulation of grazing pressure is one of the tools available to conservation managers to alter vegetation composition and structure. Whether and how this tool should be used at particular sites will require a wider regional perspective to maximize conservation outcomes. If the objective, for example, is to maintain the maximum diversity of salt and brackish marsh birds, then sufficiently large (so as to maintain viable populations) stands of vegetation with different structural characteristics will need to be maintained, restored, or created over a range of sites.

Changes comparable to those shown in response to grazing in northern Europe have not been reported from elsewhere. Although marshes on the Atlantic coast of North America were grazed and used for haymaking from early in European settlement (Teal and Teal, 1969; Valiela, 2006), the overwhelming dominance of *Spartina* spp. may be the reason for the apparent lack of impact from grazing. Outside Europe, grazing is unlikely to come into consideration as a tool in restoration, although following Silliman et al. (2014) use of grazing to control invasive plants might become more widespread.

5.6 Pollution

Many salt marshes have been degraded by pollution. Pollution may be chronic or episodic; local, regional, or global; persistent or ephemeral. In setting objectives for restoration, consideration needs to be given to the impacts of pollution on the ecosystem and whether they can be addressed at a scale appropriate to the restoration project.

Site-specific pollution includes point source discharges and spills. In many urban salt marshes, stormwater discharge has a major impact. The lowering of soil salinity by freshwater inputs can facilitate the invasion of tall growing fresh and brackish water marsh plants which can form dense, often monospecific, stands that out compete the original halophytic vegetation (Zedler et al., 1990). Diversion of flow will permit the normal marsh salinity regime to reestablish, but restoration of the vegetation may require manual removal of the invaders and replanting. Urban stormwater introduces other pollutants into marshes including petrochemicals and heavy metals from road surfaces, nutrients, sediments, weed propagules, and rubbish such as plastic and paper. In new urban development, the drainage system should be designed either not to discharge into wetlands or, if this is not possible, to be treated to an appropriate standard before discharge. Retrofitting existing drainage systems is, however, expensive and may not be practically possible as, for example, space for installation of features such as gross pollutant traps may not be available.

5.6.1 Heavy Metals

In industrialized estuaries, salt marsh sediments may have been long-term sinks for heavy metals; high levels of metals have been reported from both sediments and biota (Beeftink and Nieuwenhuize, 1986; Chenhall et al., 1992; Ohmsen et al., 1995). Although concerns have been expressed about the possible transfer of metals up the food chain (including to humans, Beeftink et al., 1982), and adverse physiological impacts on test species are known from toxicological studies, there have been few studies of the effects of pollution on ecosystem processes in salt marshes. Valiela (2006) reviewed available literature and suggested that despite the clear experimental evidence of physiological effects there was little discernible evidence for population and community level impacts. Chemical loads within organisms, and species composition of assemblages (for example, bacterial assemblages differ in composition in polluted salt marsh sediment—Cao et al., 2006) can be used to monitor pollution, but do not necessarily provide surrogate measures of impacts on functions or processes. Nevertheless, a precautionary approach would suggest that pollution discharges should be reduced and if possible eliminated when planning restoration, but there is little that can be done to address existing pollution loads. Excavation of polluted sediment could mobilize contaminants back into the estuary and create a problem of disposal of contaminated sediment. For these reasons, environmental protection authorities would be unlikely to grant approval for restoration projects which involve disturbance to known contaminated sites. Even if approval were granted to replace contaminated sediment, there would be a need to source the same quantity of clean fill with appropriate physicochemical properties.

The fact that a site is contaminated with heavy metals does not rule it out from being restored, but there would be a need to prevent disturbance of the sediment and possibly long-term restrictions on access and usage, to prevent, for example, human consumption of *Salicornia* spp. Retaining and immobilizing heavy metals is regarded by Smith (2013a) as one of the ecosystem services provided by salt marshes in the upper Mersey estuary, in the heart of one of the oldest industrial areas in Britain. Andrews et al. (2006) and Andrews et al. (2008) have estimated that salt marshes in managed realignment sites in the Humber estuary in eastern England would accumulate and immobilize large amounts of heavy metals every year, and this could be considered a substantial benefit of managed realignment.

5.6.2 Nutrients

A major, insidious cause of salt marsh degradation and loss over the past century has been eutrophication.

Eutrophication arises from increased inputs of nitrogen and phosphorus. Most attention has been given to nitrogen, but in some situations, such as the Peel–Harvey estuary in South West Western Australia, phosphorus loadings have been the most significant drivers of change (Brearley, 2005). Inputs reach salt marshes in a variety of ways. At the scale of individual marshes local stormwater discharge, and small-scale sewage treatment plant outfalls are important and may have initiated significant changes. While addressing these legacies may form part of restoration, the impacts are more restricted than those caused by much larger spatial scale processes, of which catchment runoff from agricultural practices is most important, although increased urbanization and the associated runoff and larger sewage treatment facilities also contribute to greater discharge of nutrients into waterways and inshore waters. Development of cheaper manufacturing processes for fertilizers led to greater availability to farmers, applications greater than could be taken up by crops and consequently greater nutrient runoff.

There have been numerous studies where nutrients have been applied to small plots, often at levels far higher than those in even what are currently the most heavily impacted waterways. Results from such studies are generally similar and have been summarized by (McFarlin et al., 2008). Aboveground biomass increases, belowground biomass decreases, and there may be changes in the plant species composition of communities, although in the low marsh zone, where one species is often mono dominant, opportunity for change is limited. In addition, the more nitrogen-rich, aboveground biomass may be more vulnerable to damage by herbivores. Decomposition of organic material in the soil may increase because added nutrients promote bacterial growth.

Reduction in root and rhizome biomass reduces soil strength, reducing the ability of salt marshes to resist erosive forces (Turner, 2011). Deegan et al. (2012) conducted experimental studies at the marsh creek catchment scale over a period of 9 years and were able to reveal effects that would not have been detectable from studying small quadrats on the marsh surface. In addition to confirming the changes shown in previous studies, these authors showed that the reduction in belowground biomass and greater decomposition of soil organic matter led to collapse of creek banks. This resulted in wider creeks and less vegetated marsh surface (Deegan et al., 2012; Pennings, 2012).

Increased eutrophication can also lead to increased growth of algae (particularly the green alga *Ulva* [formerly *Enteromorpha*] *intestinalis*—Raffaeli et al., 1998). Overgrowth by green algae can smother salt marsh vegetation, reducing its capacity to resist erosion, whereas decaying banks of alga are a nuisance to local human populations.

Addressing eutrophication will require coordinated action at the catchment scale, with cooperation between agencies and in some cases between countries across national boundaries. The process will be slow, and there will be no quick fix.

In the absence of action at the catchment scale approach, the potential for restoration and rehabilitation at the individual salt marsh scale will be limited. There is scope for a policy debate as to whether resources are best concentrated on the catchment approach rather than being expended inefficiently on a site-by-site basis.

If inputs can be lowered, salt marshes may provide part of the long-term solution to reducing the current high levels of nitrogen in estuarine systems because bacteria in salt marsh soils are responsible for denitrification of nitrate, leading to release of nitrogen into the atmosphere.

5.6.3 Oil

Oil pollution is an ever present threat in the marine environment with large numbers of spills, both large and small, being recorded every year (European Environment Agency, 2006). When spills impact on the coast, there is major public and political concern, and pressure for cleanup to occur. If salt marshes are oiled, should there be active rehabilitation or should natural recovery be allowed to take its course? A number of major oil spills have affected salt marshes in different parts of the world, and a review of impacts and treatment methods is provided by Baker et al. (1994). The recent review by Duke (2016), although focused on mangroves, has much that is equally relevant to salt marsh.

A number of techniques can be used to treat surface deposits of oil, but remediating subsurface oil is more difficult. Methods include low pressure water flushing, although care has to be taken not to cause erosion. Rapid deployment of sorbents could reduce penetration into the substrate, but collection and disposal of oiled sorbent material may be a problem. Burning of oiled vegetation may be an option in some circumstances, but burning may increase the hydrocarbon content in the underlying sediment. Cutting of oiled vegetation, particularly if leaving oil might be a threat to wildlife, is an option but the logistics of operating in soft sediment and of removing and disposing of oiled material would need to be considered.

The most extreme treatment approach is combined vegetation and sediment removal. After stripping of the sediment, transplanting and/or seeding is necessary to reestablish habitat and prevent erosion. The approach has been successful in small-scale trials, but serious problems occurred when it was used on a large scale following the "Amoco Cadiz" spill in 1978. This spill deposited large amounts of oil on salt marshes in Brittany (France). Heavy machinery was used to remove as much as 50 cm of sediment, and marsh creeks were widened and straightened. Although vegetation of sites not subject to treatment recovered within a decade, large parts of the treated marshes remained unvegetated in 1990, primarily because the sediment removal had created a surface too low in the intertidal for marsh establishment (Baker et al., 1994).

The immediate impact of an oil spill on salt marshes is dramatic with smothered vegetation and dead or dying fauna very visible. Once the surface oil has weathered or been removed, the marsh surface may appear "normal" to casual observation, but very little is known about the long-term effects of exposure to residual hydrocarbons. Oil residues can be detected in sediments long after pollution incidents, but impacts on the ecosystems from these residues have been harder to establish.

In 1969, the barge "Florida" ran aground in Buzzards Bay, Massachusetts (USA), and spilled oil spread over adjacent salt marshes. Nearly 40 years later, the salt marsh vegetation at the spill site appeared to be the same as that at nearby marshes which were not affected by the spill, but high concentrations of hydrocarbons were detected in the anoxic underlying sediment (Culbertson et al., 2007). One of the major faunal species on the marsh, the fiddler crab *Uca pugnax*, burrows into the sediment. Comparison between the behavior of *U. pugnax* at the contaminated and control sites have shown behavioral differences where crabs at the oiled site avoided burrowing into the contaminated layers, had delayed escape responses, lowered feeding rate, and occurred at lower densities (Culbertson et al., 2007). The oil is thus still biologically active and despite superficial appearance, marsh recovery has not occurred. These results highlight the importance of initial prevention of oiling, but in the context of restoration emphasize the importance of choice of matters to include in monitoring programs. Study of only the vegetation could well have supported an erroneous conclusion that the marsh had fully recovered.

One of the largest recent oil spills was the Deepwater Horizon incident in 2010, following a blowout at a production platform in the Gulf of Mexico. The spill caused extensive contamination of salt marshes in Louisiana and Mississippi (Nixon et al., 2016). Heavily oiled *Spartina alterniflora* died, but lightly oiled plants survived and appeared to be healthy. However, is the total ecological community in lightly oiled sites resilient? McCall and Pennings (2012b) and Pennings et al. (2014) studied the response of terrestrial arthropods to oiling and showed that even when plants were apparently unaffected, there was a decline in arthropods. However, within a year some species had recovered, but others did not, and Pennings et al. (2014) suggested that for sensitive species, oil residues in soil might continue to affect populations for years, again indicating that first visual impressions of recovery based on plants might be misleading at ecosystem level.

The ecological importance of salt marsh has been recognized in national oil spill contingency plans (see, e.g., Carter, 1994), and priority is given to preventing oil reaching marsh surfaces. In designing restoration projects for sites where there is a high probability of oil spills (such as near ports or refineries), it may be appropriate to install permanent anchoring points for the rapid deployment of exclusion.

Although shipping accidents are a source of major oil spills, oil and other chemicals may be spilt into creeks and drains following road or industrial accidents (Fiedler et al., 2007) so that contingency plans to protect salt marshes (both restored and natural) from spills need to consider drainage from the terrestrial catchment.

5.6.4 Agricultural Chemicals

One component of salt marsh biota, which is vital for the restoration of ecosystem functions, is the microalgal film on the sediment surface. These films play a major role in stabilizing the surface, permitting higher plant colonization, and preventing erosion (Coles, 1979; Mason et al., 2003). In addition, they are important food resources for a diversity of fauna and play a considerable role in nutrient fluxes including through nitrogen fixation. Formation of microalgal films in restored marshes has been demonstrated by Underwood (1997), Janousek et al. (2007), and Green et al. (2010). However, Mason et al. (2003) have shown that sublethal concentrations of agricultural triazine herbicides (such as simazine and atrazine) adversely affect growth and photosynthetic efficiency in diatoms. Given the very widespread use of these chemicals, and their presence in runoff into estuaries, they may be affecting colonization success in marsh restoration more widely than has been appreciated.

5.6.5 The Drift Line

An important zone in salt marshes is the drift line, where allochthonous debris—algae, seagrasses, logs, etc.—accumulates. This is a habitat for a large range of invertebrates, and the decay of the wrack is important for nutrient recycling. In small restoration sites close to habitation, the smell of decaying seagrass and algae may be deemed offensive and there may be pressure for removal of the drift material. Even if the drift line is mainly slowly decaying wood, managerial tidiness may lead to its removal.

The drift line is also the resting place for a large amount of artificial material—plastic bottles, fishing tackle, wax paper cartons, coffee cups, and the like (see Fig. 1.22 in Adam (1990)) which may be inert or slow to break down and which might be hazardous to wildlife. Increasingly, a component of the allochthonous material, which may be spread through the marsh as well as being concentrated at the drift line, comprises microplastics derived from artificial fibers by breakdown of larger plastic material or deliberately manufactured for a range of industrial uses. This material is pervasive throughout marine environments and management at site level is not feasible. Managing the problem will require concerted action and regulation at the international level (Rochman et al., 2013).

5.7 Damage Caused by Vehicles

Kelleway (2005) showed that many salt marshes in southern Sydney (Australia) had been extensively damaged by unauthorized vehicle access, a phenomenon repeated in many other urban areas. Although a range of off-road vehicles may be associated with damage, close to urban areas a major issue is use of trail bikes, and BMX and mountain bicycles, which may involve construction of elaborate jumps and other obstacles. Restoration of this damage could require ripping of heavily compacted areas, filling of eroded wheel tracks, reinstatement of original drainage patterns, and replanting. However, such efforts would be to no avail if vehicle access still occurs. Blocking of access will be essential but not sufficient, as barriers and fences can soon be destroyed or circumvented. Enforcement, education, and possibly provision of alternate recreational facilities will be required if recurrence of damage to salt marshes is to be prevented.

6. CONFLICTING PRIORITIES

Given the high conservation values ascribed to salt marshes, the restoration and recreation of degraded salt marshes would be expected to have high priority. Nevertheless, there are circumstances where former salt marsh has developed into habitats of high conservation value in their own right. In northern Europe, embankments and the complex of grazing marshes and brackish ditches behind them may in some instances be hundreds of years old. Embankments may support rich floras, including a number of rare species (Gray, 1977; Ranwell, 1981). Natural unimpounded brackish marshes are now very limited in occurrence, and the grazing marshes and their ditches are the major stronghold for a number of species and communities (Gray, 1977; Doody, 2008a). The habitat values of the grazing marshes are themselves now threatened by changes in agricultural practices and policies and are afforded protected status. Conflicts may arise if proposals for restoration or managed realignment involve loss of grazing marsh (Pethick, 2002). The regulatory and environmental assessment regime required for salt marsh restoration project approval may be complex (Nottage and Robertson, 2005), but ultimately decisions as to whether salt marsh restoration prevails over conservation of some other habitat will be made on the basis of policies, which while they may be informed by science, are made within political fora. There is no absolute criterion for deciding that salt marsh is "better" and that its protection should be favored over conservation of grazing marsh.

However, policy and legislation may provide a basis for addressing issues consistently rather than in an erratic ad hoc manner. Although supporters of salt marsh may press the preeminence of salt marsh restoration (Pethick, 2002), circumstances may arise when protection of other habitat will be given priority. Legislation may set constraints on the determination of objectives for particular restoration projects in that actions which might potentially have adverse impacts on particular species or communities may be forbidden, or there may be additional approval, monitoring, and reporting requirements.

The invasion of salt marshes in southern Australia by mangroves raises similar dilemmas (Saintilan and Williams, 1999; see Saintilan et al., 2018). In the state of New South Wales, salt marsh is legally recognized as an endangered ecological community, but mangroves are afforded special protection because of their contribution to the maintenance of fisheries. If mangroves invade a restored wetland, should this be regarded as an inevitable ecological change or should active removal of mangrove seedlings be undertaken, even though this would require long-term commitments?

7. DISCUSSION

The global trend to increasing urbanization where more than half of the global population now lives in cities (United Nations, 2004) is leading to further expansion of the world's major coastal cities. The related need to further develop transport and industrial facilities on estuaries will inevitably place pressure on intertidal wetlands, continuing the long-term history of coastal wetland being a major focus for development activities (Adam, 2002; Bromberg and Bertness, 2005). Despite increasing recognition of the importance of salt marsh, there will continue to be losses of sites where the social and economic benefits of development are deemed to outweigh those of conservation. However, in return for development approval it is increasingly likely that obligations for mitigation, either in the form of restoration of degraded salt marsh or the creation of new habitat, will be imposed. The overall intent will be expressed as no net loss, but whether this will be achievable either in terms of area or, more importantly, function remains uncertain.

To give greater certainty to future restoration proposals, there will need to be more rigor in setting objectives and in monitoring and evaluating the performance of projects against the objectives. But we must recognize that there is more ignorance than certainty (Turner, 2009), and that over the projected life of restored sites, hopefully decades if not centuries, there will need to be opportunities to amend both objectives and management practices.

As well as providing for mitigation of past and planned damage to natural salt marshes, restoration and recreation projects present other opportunities. They are appropriate venues for public education campaigns, through events, boardwalks, observation sites, citizen science projects, and interpretative signs. These can be designed into restoration projects, so their provision should not entail the disturbance which might result from attempting to provide facilities at natural sites. Additionally, many restoration sites will be in or close to urban areas, thereby increasing accessibility to the public.

An increasingly important component of restoration projects is public consultation and establishment of dialogue between interested parties. A top–down approach of telling the public what is good for them (even if proposals have a strong scientific basis) will be challenged. All the demands of different groups will not be able to be met, indeed some may be incompatible, but scientists must be able to listen to the public and be willing to respond positively. Public consultation may well lead to improvement in project design and implementation and give a sense of ownership to the local community. Smith (2013a) provided an excellent case study of seeking the views of the public including those outside the membership of special interest groups who normally would not be consulted.

Restoration projects themselves need to be treated as experiments, but they provide additional opportunities for manipulative experimentation, which as natural marsh is given increased protection under species, and habitat conservation legislation will become difficult to conduct elsewhere. Not every restoration project will become an experimental research site, but it is an opportune time to identify a range of sites in different geographic regions and with different marsh types that can form a network for long-term ecological research. This has already occurred with sites in southern California (Zedler, 2001), but in many parts of the world there is little interaction between ecological researchers and habitat restoration projects (Chapman and Underwood, 2000), and the opportunities for research are ignored and lost. Monitoring of selected restoration sites could be embedded into broader networks of reference sites, such as that advocated by Osland et al. (2017).

8. CONCLUSIONS

Large numbers of salt marsh restoration projects have been conducted, although the majority have been in the United States and Europe, and have been applied to a relatively restricted range of the world's salt marsh types. Although there is a substantial literature, many projects have not been assessed, and for those that have been, inappropriate setting of objectives and conduct of monitoring limits our ability to draw general conclusions. If the challenge of meeting a goal of no net loss of salt marsh is to be met, then restoration and recreation will have to play a major part, but for success there needs to be a much firmer scientific framework. The task is made more complicated by the uncertainties associated with global climate change and sea level rise and the continuing disturbance associated with development and urbanization. Nevertheless, many of the factors that have caused salt marsh degradation can be identified, and addressing these will be a major component of restoration. However, in some cases, such as the invasion of those exotic species for which there is little prospect of effective control, or rising relative sea level at sites without opportunity for landward retreat, a recasting of objectives for restoration will be required. Recreating the past will rarely be an appropriate option, but restoring ecological functions and habitat for particular species in self-sustaining salt marshes may still be possible. Salt marsh restoration affords unrivaled opportunities for experimental manipulation, adding to our fundamental knowledge of salt marsh ecosystems, permitting improvement in restoration techniques and also better management at the large spatial scale of whole estuaries and their catchments, reducing threats to both natural and restored salt marshes.

References

Adam, P., 1978. Geographical variation in British saltmarsh vegetation. Journal of Ecology 66, 339–366.

Adam, P., 1981. The vegetation of British saltmarshes. New Phytologist 88, 143–196.

Adam, P., 1990. Saltmarsh Ecology. Cambridge University Press, Cambridge, 461 p.

Adam, P., 1996. Saltmarsh vegetation study. In: Homebush Bay Ecological Studies 1993-1995, vol. 2. CSIRO, Melbourne, pp. 113–127.

Adam, P., 1997. Absence of creeks and pans in temperate Australian salt marshes. Mangroves and Salt Marshes 1, 239–241.

Adam, P., 1998. Biodiversity - the biggest of big pictures. In: Lunney, D., Dawson, T., Dickman, C. (Eds.), Is the Biodiversity Tail Wagging the Zoological Dog? Royal Zoological Society of New South Wales, Mosman, pp. 6–14.

Adam, P., 2000. Morecambe Bay saltmarshes: 25 years of change. In: Sherwood, B.R., Gardiner, B.G., Harris, T. (Eds.), British Saltmarshes. Linnean Society of London, London, pp. 81–107.

Adam, P., 2002. Saltmarshes in a time of change. Environmental Conservation 29, 39–61.

Adam, P., 2016. Saltmarshes. In: Kennish, M.J. (Ed.), Encyclopedia of Estuaries. Springer, Dordrecht, pp. 515–535.

Addy, K., Gold, A., Nowicki, B., et al., 2005. Denitrification capacity in a subterranean estuary below a Rhode Island fringing salt marsh. Estuaries 28, 896–908.

Allison, H., Morley, J. (Eds.), 1989. Blakeney Point and Scolt Head Island, fifth ed. The National Trust, Norwich. 116 p.

Ambrose, R.F., 2000. Wetland mitigation in the United States: assessing the success of mitigation policies. Wetlands (Australia) 19, 1–27.

Anderson, C.B., Valenzuela, A.E.J., 2011. Mammals, aquatic. In: Simberloff, D., Reymánek, M. (Eds.), Encyclopedia of Biological Invasions. University of California Press, Berkeley, pp. 445–448.

Andrews, J.E., Burgess, D., Cave, R.R., et al., 2006. Biogeochemical value of managed realignment, Humber estuary, UK. Science of the Total Environment 371, 19–30.

Andrews, J.E., Samways, G., Shimmield, G.B., 2008. Historical storage budgets of organic carbon, nutrient and contaminant elements in saltmarsh sediments: biogeochemical context for managed realignment, Humber Estuary, UK. Science of the Total Environment 405, 1–13.

Arp, W.J., Drake, B.G., 1991. Increased photosynthetic capacity of *Scirpus olneyi* after 4 years of exposure to elevated CO_2. Plant, Cell and Environment 14, 1003–1006.

Arp, W.J., Drake, B.G., Pockman, W.T., Curtis, P.S., Whigham, D.F., 1993. Interactions between C_3 and C_4 salt marsh plant species during four years of exposure to elevated atmospheric CO_2. Vegetatio 104/105, 133–143.

Atkinson, P.W., Crooks, S., Grant, A., Rehfisch, M.M., 2001. The Success of Creation and Restoration Schemes in Producing Intertidal Habitat Suitable for Waterbirds. English Nature Research Reports 425. English Nature, Peterborough, 166 p.

Ausden, M., Dixon, M., Lock, L., et al., 2016. Dredged sediment - still an underused conservation resource. British Wildlife 28, 88–96.

Ayres, D.R., Smith, D.L., Zaremba, K., Klohr, S., Strong, D.R., 2004. Spread of exotic cordgrasses and hybrids (*Spartina* sp.) in the tidal marshes of San Francisco Bay, California, USA. Biological Invasions 6, 221–231.

Baily, B., Pearson, A.W., 2007. Change detection mapping of saltmarsh and analysis of saltmarsh areas of central southern England from Hurst Castle Spit to Pagham Harbour. Journal of Coastal Research 23, 1549–1564.

Baker, J.M., Adam, P., Gilfillan, E., 1994. Biological Impacts of Oil Pollution: Salt Marshes. International Petroleum Industry Environmental Conservation Association, London, 20 p.

Bakker, J.P., Piersma, T., 2006. Restoration of intertidal flats and tidal salt marshes. In: van Andel, J., Aronson, J. (Eds.), Restoration Ecology. Blackwell, Oxford, pp. 174–192.

Bakker, J.P., Esselink, P., van der Wal, R., Dijkema, K.S., 1997. Options for restoration and management of coastal salt marshes in Europe. In: Urbanska, K.M., Webb, N.R., Edwards, P.J. (Eds.), Restoration Ecology and Sustainable Development. Cambridge University Press, Cambridge, pp. 286–322.

Bakker, J.P., Esselink, P., Dijkema, K.S., van Duin, W.E., de Jong, D.J., 2002. Restoration of salt marshes in The Netherlands. In: Nienhuis, P.H., Gulati, R.D. (Eds.), Ecological Restoration of Aquatic and Semi-Aquatic Ecosystems in the Netherlands (NW Europe). Kluwer, Dordrecht, pp. 29–51.

Bakker, J.B., 1989. Nature Management by Grazing and Cutting Regimes Applied to Restore Former Species-Rich Grassland Communities in the Netherlands. Kluwer, Dordrecht, 400 p.

Baldwin, A.H., Hammerschlag, R.S., Cahoon, D.R., 2018. Evaluation of restored tidal freshwater wetlands. In: Perillo, G.M.E., Wolanski, E., Cahoon, D.R., Brinson, M.M. (Eds.), Coastal Wetlands: An Integrated Ecosystem Approach. Elsevier Science, Amsterdam.

Barbier, E.B., Hacker, S.D., Kennedy, C., et al., 2011. The value of estuarine and coastal ecosystem services. Ecological Monographs 81, 169–193.

Barbier, E.B., 2018. The value of coastal wetland ecosystem services. In: Perillo, G.M.E., Wolanski, E., Cahoon, D.R., Brinson, M.M. (Eds.), Coastal Wetlands. An Integrated Ecosystem Approach. Elsevier Science, Amsterdam.

Beaumont, N.J., Austen, M.C., Atkins, J.P., et al., 2007. Identification, definition and quantification of goods and services provided by marine biodiversity: implications for the ecosystem approach. Marine Pollution Bulletin 54, 253–265.

Beeftink, W.G., Nieuwenhuize, J., 1986. Monitoring trace metal contamination in salt marshes of the Westerschelde estuary. Environmental Monitoring and Assessment 7, 233–248.

Beeftink, W.G., Nieuwenhuize, J., Stoeppler, M., Mohl, C., 1982. Heavy-metal accumulation in salt marshes from the western and eastern Scheldt. Science of the Total Environment 25, 119–223.

Bertness, M.D.T., Coverdale, T.C., 2013. An invasive species facilitates the recovery of salt marsh ecosystems on Cape Cod. Ecology 94, 1937–1943.

Bertness, M.D., Brisson, C., Bevil, C., Crotty.S, 2014. Herbivory drives the spread of salt marsh die-off. PLoS One 9 (3), e92916.

Blossey, B., 2003. A framework for evaluating potential ecological effects of implementing biological control of *Phragmites australis*. Estuaries 26, 607–617.

Boland, J., Zedler, J.B., 1996. Maintaining tidal flows in southern California lagoons. In: Zedler, J.B. (Ed.), Tidal Wetland Restoration: A Scientific Perspective and Southern California Focus. California Sea Grant College System. University of California, La Jolla, pp. 11–20.

Boorman, L.A., 2003. Salt Marsh Review. An Overview of Coastal Salt Marshes, Their Dynamic and Sensitivity Characteristics for Conservation and Management. JNCC Report 334. JNCC, Peterborough, 113 p.

Boston, K.G., 1981. Occasional Paper. The Introduction of *Spartina townsendii* (s.l.) to Australia, vol. 6. Melbourne State College, 57 p.

Boyes, S.J., Elliott, M., 2014. Marine legislation - the ultimate 'horrendogram': international law, European directives & national implementation. Marine Pollution Bulletin 86, 39–47.

Brearley, A., 2005. Ernest Hodgkin's Swanland. Estuaries and Coastal Lagoons of South-Western Australia. University of Western Australia Press, Perth, 550pp.

Bridgewater, P.B., Kaeshagen, D., 1979. Changes induced by adventive species in Australian plant communities. In: Williams, O., Tűxen, R. (Eds.), Werden and Vergehen von Pflanzengesellschaften. Cramer, Lehre, pp. 561–579.

Bromberg, K.D., Bertness, M.D., 2005. Reconstructing New England salt marsh losses using historical maps. Estuaries 26, 823–832.

Bu, N.S., Qu, J.F., Li, G., et al., 2015. Reclamation of coastal salt marshes promoted carbon loss from previously-sequestered soil carbon pool. Ecological Engineering 81, 335–339.

Buchsbaum, R.N., Wigand, C., 2012. Adaptive management and monitoring as fundamental tools to effective salt marsh restoration. In: Roman, C.T., Burdick, D.M. (Eds.), Restoring Tidal Wetlands. A Synthesis of Science and Management. Island Press, Washington DC, pp. 213–231.

Burt, W., 2007. Marshes: The Disappearing Eden. Yale University Press, New Haven, 179 p.

Cahoon, D.R., 2015. Estimating relative sea-level rise and submergence potential of a coastal wetland. Estuaries and Coasts 28, 1077–1084.

Callaway, J.C., Zedler, J.B., 1998. Interactions between a salt marsh native perennial *(Salicornia virginica)* and an exotic annual *(Polypogon monspeliensis)* under varied salinity and hydroperiod. Wetlands Ecology and Management 5, 179–194.

Callaway, J.C., 2001. Hydrology and substrate. In: Zedler, J.B. (Ed.), Handbook for Restoring Tidal Wetlands. CRC Press, Boca Raton, pp. 89–117.

Callaway, J.C., 2005. The challenge of restoring functioning salt marsh ecosystems. Journal of Coastal Research 40, 24–36.

Cao, Y., Cherr, G.N., Córdova-Kreylos, A.L., et al., 2006. Relationship between sediment microbial communities and pollutants in two California salt marshes. Microbial Ecology 52, 619–633.

Carter, S., 1994. Coastal Resource Atlas for Oil Spills from Barrenjoey Head to Bellambi Point. Environment Protection Authority of New South Wales, Chatswood, 78 p.

Chapman, M.G., Underwood, A.J., 1997. Concepts and issues in restoration of mangrove forests in urban environments. In: Klomp, N., Lunt, I. (Eds.), Frontiers in Ecology: Building the Links. Elsevier Science, Oxford, pp. 103–114.

Chapman, M.G., Underwood, A.J., 2000. The need for a practical scientific protocol to measure successful restoration. Wetlands (Australia) 19, 28–49.

Chapman, V.J., 1974. Salt Marshes and Salt Deserts of the World, second ed. Cramer, Lehre. 392 p.

Chapman, G., 2012. Rehabilitation of estuarine habitats. In: Sainty, G., Hosking, J., Carr, G., Adam, P. (Eds.), Estuary Plants and What's Happening to Them in South-East Australia. Sainty Books, Sydney, pp. 435–440.

Chenhall, B.E., Yassini, I., Jones, B.G., 1992. Heavy metal concentrations in lagoonal salt marsh species, Illawarra region, southeastern Australia. Science of the Total Environment 125, 203–225.

Chew, M.K., 2015. Ecologists, environmentalists, experts, and the invasion of the 'the second greatest threat'. International Review of Environmental History 1, 7–40.

Chmura, G.L., Anisfeld, S.C., Cahoon, D.R., Lynch, J.C., 2003. Global carbon sequestration in tidal, saline wetland soils. Global Biogeochemical Cycles 17, 1111. https://doi.org/10.1029/2002GB00197.

Christian, R.R., Hanson, R.B., Hall, J.R., Wiebe, W.J., 1981. Aerobic microbes and meiofauna. In: Pomeroy, L.R., Wiegert, R.G. (Eds.), The Ecology of a Salt Marsh. Springer-Verlag, New York, pp. 113–135.

Chung, C.H., Zhuo, R.Z., Xu, G.W., 2004. Creation of *Spartina* plantations for reclaiming Dongtai, China, tidal flats and offshore sands. Ecological Engineering 23, 135–150.

Coles, S.M., 1979. Benthic microalgal populations on intertidal sediments and their role as precursors to salt marsh development. In: Jefferies, R.L., Davy, A.J. (Eds.), Ecological Processes in Coastal Environments. Blackwell Scientific Publishers, Oxford, pp. 25–42.

Corman, S.S., Roman, C.T., King, J.W., Appleby, P.G., 2012. Salt marsh mosquito-control ditches: sedimentation, landscape change, and restoration implications. Journal of Coastal Research 28, 874–880.

Costanza, R., d'Arge, R., de Groot, R., et al., 1997. The value of the world's ecosystem services and natural capital. Nature 387, 253–259.

Coverdale, T.C., Altieri, A.H., Bertness, M.D., 2012. Belowground herbivory increases vulnerability of New England salt marshes to die-off. Ecology 93, 2085–2094.

Coverdale, T.C., Herrmann, N.C., Altieri, A.H., Bertness, M.D., 2013. Latent impacts: the role of historical human activity in coastal habitat loss. Frontiers in Ecology and the Environment 11, 69–74.

Coverdale, T.C., Brisson, C.P., Young, E.W., et al., 2014. Indirect human impacts reverse centuries of carbon sequestration and salt marsh accretion. PLoS One 9 (3), e93296. https://doi.org/10.1371/journal.pone.0093296.

Craft, C., Sacco, J., 2003. Long-term succession of benthic infauna communities on constructed *Spartina alterniflora* marshes. Marine Ecology Progress Series 257, 45–58.

Craft, C., 2000. Co-development of wetland soils and benthic invertebrate communities following salt marsh creation. Wetlands Ecology and Management 8, 197–207.

Craft, C., 2016. Creating and Restoring Wetlands: From Theory to Practice. Elsevier, Amsterdam, 348 p.

Craig, R.K., 2010. "Stationarity is Dead" long live transformation: five principles for climate change adaption law. Harvard Environmental Law Review 34, 9–73.

Crooks, S., Rybczyk, J., O'Connell, K., et al., 2014. Coastal Blue Carbon opportunity assessment for the Snohomish Estuary: the climate benefits of estuary restoration. In: Report by Environmental Science Associates. Western Washington University, EarthCorps, and Restore America's Estuaries, 102 p.

Crosby, S.C., Ivens-Duran, M., Bertness, M.D., et al., 2015. Flowering and biomass allocation in US Atlantic coast *Spartina alterniflora*. American Journal of Botany 102, 669–676.

Crow, J.H., 1971. Earthquake-initiated change in the nesting habitat of the Dusky Canada Goose. In: Foy, J.V., Bishop, E.E., Duggan, M.Y., Olney, H.A. (Eds.), The Great Alaska Earthquake of 1964: Biology. National Academy of Sciences, Washington, DC, pp. 130–136.

Culbertson, J.B., Valiela, I., Peacock, E.E., et al., 2007. Long-term biological effects of petroleum residues on fiddler crabs in salt marshes. Marine Pollution Bulletin 54, 955–962.

Daehler, C.C., Strong, D.R., 1994. Hybridization between introduced smooth cordgrass *(Spartina alterniflora*, Poaceae) and native California cordgrass *(S. foliosa)* in San Francisco Bay, California, U.S.A. American Journal of Botany 84, 604–611.

Daiber, F.C., 1986. Conservation of Tidal Marshes. Van Nostrand Reinhold, New York, 341 p.

David, A.T., Goertler, P.A.L., Munsch, S.H., et al., 2016. Influences of natural and anthropogenic factors and tidal restoration on terrestrial arthropod assemblages in west coast North American estuarine wetlands. Estuaries and Coasts 39, 1491–1504.

Deegan, L.A., Johnson, D.S., Warren, S.W., et al., 2012. Coastal eutrophication as a driver of salt marsh loss. Nature 490, 388–392.

Dixon, A.M., Leggett, D.J., Weight, R.C., 1998. Habitat creation opportunities for landward coastal re-alignment: Essex Case Study. Journal of the Chartered Institution of Water and Environmental Management 12, 107–112.

Doggett, S., 2016. NSW Arbovirus Surveillance & Mosquito Monitoring Program 2016 -2017 Weekly Update 19/Dec/2016.

Doody, J.P., 2004. 'Coastal squeeze' - an historical perspective. Journal of Coastal Conservation 10, 129–138.

Doody, J.P., 2008a. Saltmarsh Conservation, Management and Restoration. Springer, Dordrecht, 217 p.

Doody, J.P., 2008b. Management of Natura 2000 Habitats. 1330 Atlantic Salt Meadows (Glauco-puccinellietalia Maritimae). European Commission, Brussels, 28 p.

Doody, J.P., 2013. Coastal squeeze and managed realignment in southeast England, does it tell us anything about the future? Ocean and Coastal Management 79, 34–41.

DPIWE, 2002. Strategy for the Management of Rice Grass (*Spartina anglica*) in Tasmania, Australia, second ed. Department of Primary Industries, Water and Environment, Hobart. 36 p.

Drake, B.G., 1992. A field study of the effects of elevated CO_2 on ecosystem processes in a Chesapeake Bay wetland. Australian Journal of Botany 40, 579–595.

Drake, B.G., Peresta, G., Beugeling, F., Matamala, R., 1996. Long-term elevated CO_2 exposure in a Chesapeake Bay wetland: ecosystem gas exchange, primary production, and tissue nitrogen. In: Koch, C.W., Mooney, M.A. (Eds.), Carbon Dioxide and Terrestrial Ecosystems. Academic Press, New York, pp. 197–214.

Duarte, C.M., Middelburg, J.J., Caraco, N., 2005. Major role of vegetation on the ocean carbon cycle. Biogeosciences 2, 1–8.

Duke, N.C., 2016. Oil spill impacts on mangroves: recommendations for operational planning and action based on a global review. Marine Pollution Bulletin 109, 700–715.

Elliott, M., Cutts, N.D., 2004. Marine habitats: loss and gain, mitigation and compensation. Marine Pollution Bulletin 49, 671–674.

Elliott, M., Burdon, D., Hemingway, K.L., Apitz, S., 2007. Estuarine, coastal and marine ecosystem restoration: confusing management and science - a revision of concepts. Estuarine, Coastal and Shelf Science 74, 349–366.

Elliott, M., Mander, L., Masik, K., et al., 2016. Engineering with ecohydrology; success and failures in estuarine restoration. Estuarine, Coastal and Shelf Science 176, 12–35.

Esselink, P., Fresco, F.L.M., Dijkema, K.S., 2002. Vegetation change in a man-made salt marsh affected by a reduction in both grazing and drainage. Applied Vegetation Science 5, 17–32.

Esteves, L.S., 2013. Is managed realignment a sustainable long-term coastal management approach? Journal of Coastal Research (Special Issue, 65), 933–938.

Esteves, L.S., 2014. Managed Realignment: A Viable Long-Term Coastal Management Strategy? Springer, New York, 139 p.

Esteves, L.S., 2016a. Coastal squeeze. In: Kennish, M.J. (Ed.), Encyclopedia of Estuaries. Springer, Dordrecht, pp. 158–160.

Esteves, L.S., 2016b. Managed realignment. In: Kennish, M.J. (Ed.), Encyclopedia of Estuaries. Springer, Dordrecht, pp. 390–393.

European Commission, 1992. Council Directive 92/43/EEC of 21 May 1992 on the Conservation of Natural Habitats and Wild Flora and Fauna.

European Environment Agency, 2006. The Changing Faces of Europe's Coastal Areas. EEA Report 6/2006 EEA, Copenhagen, 107 p.

Fellows, M.Q.N., Zedler, J.B., 2005. Effects on the non-native grass, *Parapholis incurva* (Poaceae) on the rare and endangered hemisparasite, *Cordylanthus maritimus* subsp. *maritimus* (Scrophulariaceae). Madroño 52, 91–98.

Fiedler, P.L., Keever, M.F., Grevell, B.I., Partridge, D.I., 2007. Rare plants in the Golden Gate estuary (California): the relationship between scale and understanding. Australian Journal of Botany 55, 206–220.

Ge, Z.-M., Wang, H., Cao, H.-B., et al., 2016. Responses of eastern Chinese salt marshes to sea-level rise combined with vegetative and sedimentary processes. Scientific Reports. https://doi.org/10.1038/srep28466.

Gedan, K.B., Silliman, B.R., Bertness, M.D., 2009. Centuries of human-driven change in salt marsh ecosystems. Annual Reviews of Marine Science 1, 117–141.

Goss-Custard, J.D., Moser, M.E., 1988. Rates of change in the numbers of dunlin *Calidris alpina* wintering in British estuaries in relation to the spread of *Spartina anglica*. Journal of Applied Ecology 25, 95–109.

Gray, A.J., Mogg, R.J., 2001. Climate impacts on pioneer saltmarsh plants. Climate Research 18, 105–112.

Gray, A.J., 1977. Reclaimed land. In: Barnes, R.S.K. (Ed.), The Coastline. Wiley, Chichester, pp. 253–270.

Green, J., Reichelt-Brushett, A., Brushett, D., et al., 2010. Soil algal abundance in a subtropical saltmarsh after surface restoration. Wetlands 30, 87–98.

Grevstad, F.S., Strong, D.R., Garcia-Rossi, D., Switzer, R.W., Wecker, M.S., 2003. Biological control of *Spartina alterniflora* in Willapa Bay, Washington using the planthopper *Prokelisia marginata*: agent specificity and early results. Biological Control 27, 32–42.

Grosholz, E., 2011. Crabs. In: Simberloff, D., Reymánek, M. (Eds.), Encyclopedia of Biological Invasions. University of California Press, Berkeley, pp. 125–128.

Hazelton, E.L.G., Mozder, T.J., Burdick, D.M., et al., 2014. *Phragmites australis* management in the United States: 40 years of methods and outcomes. AoB PLANTS 5, plu001. https://doi.org/10.1093/aobpla/plu001.

He, Q., Bertness, M.D., Bruno, J.F., et al., 2014. Economic development and coastal ecosystem change in China. Scientific Reports 4, 5995. https://doi.org/10.1038/srep05995.

Holdredge, C., Bertness, M.D., Altieri, A.H., 2009. Role of crab herbivory in die-off of New England salt marshes. Conservation Biology 23, 673–679.

Hughes, R.G., Paramor, O.L.A., 2004. The effects of bioturbation and herbivory by the polychaete *Nereis diversicolor* on loss of saltmarsh in south-east England. Journal of Applied Ecology 41, 449–463.

Hurry, C.R., James, E.A., Thompson, 2013. Connectivity, genetic structure and stress response in *Phragmites australis*: Issues for restoration in a salinising wetland system. Aquatic Botany 138–146.

Hurst, T., Boon, P.I., 2016. Agricultural weeds and coastal saltmarsh in south-eastern Australia; an insurmountable problem? Australian Journal of Botany 64, 308–324.

Janousek, C., Currin, C.A., Levin, L.A., 2007. Succession of microphytobenthos in a restored coastal wetland. Estuaries and Coasts 30, 265–276.

Jefferies, R.L., 1997. Long-term damage to sub-arctic coastal ecosystems by geese: ecological indicators and measures of ecosystem dysfunction. In: Crawford, R.M.M. (Ed.), Disturbance and Recovery in Arctic Lands: An Ecological Perspective. Kluwer, Boston, pp. 151–166.

Kelleway, J., Williams, R.J., Allen, C.B., 2007. An Assessment of the Condition of Saltmarsh of the Parramatta River and Sydney Harbour, NSW. NSW Department of Primary Industries. Fisheries Final Report Series No. 90, 100 p.

Kelleway, J.J., Saintilan, N., Macreadie, P.I., et al., 2016a. Seventy years of continuous encroachment substantially increases 'blue carbon' capacity as mangroves replace intertidal salt marshes. Global Change Biology 22, 1097–1109.

Kelleway, J.J., Saintilan, N., Macreadie, P.I., Ralph, R.J., 2016b. Sedimentary factors are key predictors of carbon storage in SE Australian saltmarshes. Ecosystems. https://doi.org/10.1007/s10021-016-9972-3.

Kelleway, J., 2005. Ecological impacts of recreational vehicle use on salt marshes of the Georges River, Sydney. Wetlands (Australia) 22, 52–66.

Kerr, D.W., Hogle, I.B., Orr, B.S., Thornton, W.J., 2016. A review of 15 years of *Spartina* management in the San Francisco Estuary. Biological Invasions 18, 2247–2266.

Key, B.H., Russell, R.C., 2013. Mosquito eradication. The Story of Killing Campto. CSIRO Publishing, Melbourne, 69 p.

Kirwan, M.L., Blum, L.K., 2011. Enhanced decomposition offsets enhanced productivity and soil carbon accumulation in coastal wetlands responding to climate change. Biogeosciences 8, 987–993.

Kirwan, M.L., Megonigal, J.P., 2013. Tidal wetland stability in the face of human impacts and sea-level rise. Nature 504, 53–60.

Kirwan, M.L., Guntenspergen, G.R., D'Alpaos, et al., 2010. Limits on the adaptability of coastal marshes to rising sea level. Geophysical Research Letters 37 (23), L23401. https://doi.org/10.1029/2010GLO45489.

Kirwan, M.L., Murray, A.B., Dunphy, J.B., Corbett, D.R., 2011. Rapid wetland expansion during European settlement and its implications for marsh survival under modern sediment delivery rates. Geology 39, 507–510.

Kirwan, M.L., Temmerman, S., Skeehan, E.E., et al., 2016. Overestimation of marsh vulnerability to sea level rise. Nature Climate Change 6. https://doi.org/10.1038/NCLIMATE2909.

Kiviat, K., 2013. Ecosystem services of *Phragmites* in North America with emphasis on habitat functions. AoB PLANTS 5, plt008. https://doi.org/10.1093/aobpla/plt008.

Kuhn, N.L., Zedler, J.B., 1997. Differential effects of salinity and soil saturation on native and exotic plants of a coastal salt marsh. Estuaries 20, 391–403.

Lacambra, C., Cutts, N., Allen, J., Burd, F., Elliot, M., 2004. Spartina anglica: A Review of its Status, Dynamics and Management. English Nature Research Reports 527. English Nature, Peterborough, 69 p.

Langley, J.A., McKee, K.L., Cahoon, D.R., et al., 2009. Elevated CO_2 stimulates marsh elevation gain, counterbalancing sea-level rise. Proceedings of the National Academy of Sciences 106, 6182–6186.

Langley, J.A., Mozdzer, T.J., Shepard, K.A., Hagerty, S.B., Megonigal, J.P., 2013. Tidal marsh plant responses to elevated CO_2, nitrogen fertilization, and sea level rise. Global Change Biology 19, 1495–1503.

Lee, S.B., Birch, G.F., 2013. Sydney Estuary, Australia. Geology, anthropogenic development and hydrodynamic processes/attributes. In: Wolanski, E. (Ed.), Estuaries of Australia in 2050 and beyond. Springer, Dordrecht, pp. 17–30.

Levin, L.A., Talley, D., Thayer, G., 1996. Succession of macrobenthos in a created salt marsh. Marine Ecology Progress Series 141, 67–82.

Lewis, R.R. (Ed.), 1982. Creation and Restoration of Coastal Plant Communities. CRC Press, Boca Raton, 219 p.

Li, S., Pennings, S.C., 2016. Disturbance in Georgia salt marshes: variation across space and time. Ecosphere 7 (0), e01487. https://doi.org/10.1002/ecs2.1675.

Li, S., Pennings, S.C., 2017. Timing of disturbance affects biomass of a salt marsh plant and attack by stem-boring herbivores. Ecosphere 8 (2), e01675. https://doi.org/10.1002/ecs2.1675.

MacKinnon, J., Verkuil, Y.I., Murray, N., 2012. IUCN situation analysis on East and Southeast Asia intertidal habitats, with particular reference to the Yellow Sea (including the Bohai Sea). In: Occasional Paper of the IUCN Species Survival Commission 47. IUCN, Gland, p. 70.

Macreadie, P.I., Hughes, A.R., Kimbro, D.L., 2013. Loss of 'Blue Carbon' from coastal salt marshes following habitat disturbance. PLoS One 8 (7), e69244. https://doi.org/10.1371/journal.pone.0069244.

Macreadie, P.I., Ollivier, Q.I., Kelleway, J.J., et al., 2017. Carbon sequestration by Australian tidal marshes. Scientific Reports 7, 44071.

Marsden, I.D., Hart, D.E., Reid, G.M., Gomez, C., 2016. Earthquake disturbances. In: Kennish, M.J. (Ed.), Encyclopedia of Estuaries. Springer, Dordrecht, pp. 207–214.

Marshall, S., 2004. The meadowlands before the Commission: three centuries of human use and alteration of the Newark and Hackensack Meadows. Urban Habitats 2, 4–27.

Mason, C., Underwood, G.J.C., Baker, N., Davey, P., Davidson, I., Hanlon, A., Long, S., Oxborough, K., Paterson, D., Watson, A., 2003. The role of herbicides in the erosion of salt marshes in eastern England. Environmental Pollution 122, 41–49.

McCall, B.D., Pennings, S.C., 2012a. Geographic variation in salt marsh structure and function. Oecologia 170, 777–787.

McCall, B.D., Pennings, S.C., 2012b. Disturbance and recovery of salt marsh arthropod communities following BP Deepwater Horizon oil spill. PLoS One 7 (3), e32735. https://doi.org/10.1371/journal.pone.0032735.

McDonald, T., Jonson, J., Dixon, K.W. (Eds.), 2016. National Standards for the Practice of Ecological Restoration in Australasia S1, pp. 1–34.
McFarlin, C.R., Brewer, S.J., Buck, T.L., Pennings, S.C., 2008. Impact of fertilization on a salt marsh food web in Georgia. Estuaries and Coasts 31, 313–325.
McLeod, E., Chmura, G.L., Bouillon, S., et al., 2011. A blueprint for blue carbon: toward an improved understanding of the role of vegetated coastal habitats in sequestering CO_2. Frontiers in Ecology and the Environment. https://doi.org/10.1890/110004.
Montague, C.L., Bunker, S.M., Haines, E.B., Pace, M.L., Wetzel, R.L., 1981. Aquatic macroconsumers. In: Pomeroy, L.R., Wiegert, R.G. (Eds.), The Ecology of a Salt Marsh. Springer-Verlag, New York, pp. 69–85.
Mooney, H.A., Hobbs, R.J., 2000. Introduction. In: Mooney, H.A., Hobbs, R.J. (Eds.), Invasive Species in a Changing World. Island Press, Washington DC, pp. xiii–xv.
Morris, R.K.A., Reach, I.S., Duffy, T.S., et al., 2004. On the loss of saltmarshes in south- east England and the relationship with *Nereis diversicolor*. Journal of Applied Ecology 41, 787–784.
Morris, R.K.A., 2012. Managed realignment: a sediment management perspective. Ocean and Coastal Management 65, 59–66.
Moseman, S., Levin, L., Currin, C., Forder, C., 2004. Colonization, succession and nutrition in a newly restored wetland at Tijuana Estuary, California. Estuarine, Coastal and Shelf Science 60, 755–770.
Mossman, H.L., Davy, A.J., Grant, A., 2012. Does managed coastal realignment create saltmarshes with 'equivalent biological characteristics' to natural reference sites? Journal of Applied Ecology 49, 1446–1456.
National Research Council, 1987. Responding to Changes in Sea Level: Engineering Implications. National Academy Press, Washington, DC, 148 p.
National Research Council, 1992. Restoration of Aquatic Ecosystems: Science, Technology and Public Policy. National Academy Press, Washington, DC, p. 576.
Niering, W.A., 1997. Tidal wetlands restoration and creation along the east coast of North America. In: Urbanska, K.M., Webb, N.R., Edwards, P.J. (Eds.), Restoration Ecology and Sustainable Development. Cambridge University Press, Cambridge, pp. 259–285.
Nixon, Z., Zengel, S., Baker, M., et al., 2016. Shoreline oiling from the Deepwater Horizon oil spill. Marine Pollution Bulletin 107, 170–178.
Nixon, S.W., 1980. Between coastal marshes and coastal waters - a review of twenty years of speculation and research on the role of salt marshes in estuarine productivity and water chemistry. In: Hamilton, P., Macdonald, K.B. (Eds.), Estuarine and Wetland Processes with Emphasis on Modeling. Plenum Press, New York, pp. 437–525.
Nottage, A., Robertson, P., 2005. The Salt Marsh Creation Handbook: A Project Manager's Guide to the Creation of Salt Marsh and Intertidal Mudflat. The RSPB, Sandy and CIWEM, London, 128 p.
Ohmsen, G.S., Chenhall, B.E., Jones, B.G., 1995. Trace metal distributions in two salt marsh substrates, Illawarra region, New South Wales, Australia. Wetlands (Australia) 14, 19–31.
Osland, M.J., Enright, N.M., Day, R.H., et al., 2016. Beyond just sea level change; considering macroclimatic drivers within coastal wetland vulnerability assessments to climate change. Global Change Biology along the northern Gulf of Mexico coast: gaps and opportunities for developing a coordinated regional sampling network. PLoS One. https://doi.org/10.1371/journal.pone.0183431.
Osland, M.J., Griffin, K.T., Larriviere, J.C., et al., 2017. Assessing Coastal Wetland Vulnerability to Sea-Level Rise along the Northern Gulf of Mexico: Gaps and Opportunities for Developing a Coordinated Regional Sampling Network. https://doi.org/10.1371/journal.pone.0183431.
Owers, C.J., Rogers, K., Mazumder, D., Woodroffe, C.D., 2016. Spatial variation in carbon storage; a case study for Currumbene Creek, NSW, Australia. Journal of Coastal Research 75, 1297–1301.
Parsons, R.F., 2013. *Limonium hyblaeum* (Plumbaginaceae), a cushion plant invading coastal southern Australia. Cunninghamia 13, 267–274.
Paul, S., Young, R., 2006. Experimental control of exotic spiny rush, *Juncus acutus*, from Sydney Olympic Park: *I. Juncus* mortality and re-growth. Wetlands (Australia) 23, 1–13.
Paul, S., Darcovich, K., Jack, A., 2012. Saltmarsh conservation at Sydney Olympic Park. In: Sainty, G., Hosking, J., Carr, G., Adam, P. (Eds.), Estuary Plants and What's Happening to Them in South-east Australia. Sainty Books, Sydney, pp. 510–517.
Pendle, M., 2013. Estuarine and coastal managed realignment sites in England. In: A Comparison with Predictions with Monitoring Results for Selected Case Studies. HRRP627. HR Wallingford, Wallingford, 36 p.

Pendleton, L., Donato, D.C., Murray, B.C., et al., 2012. Estimating global "Blue Carbon" emissions from conversion and degradation of vegetated coastal ecosystems. PLoS One. https://doi.org/10.1371/journal.pone.0043542.

Pennings, S.C., 2012. The big picture of marsh loss. Nature 490, 352–353.

Pennings, S.C., McCall, B.D., Hooper-Bui, L., 2014. Effects of oil spills on terrestrial arthropods in coastal wetlands. BioScience 64, 789–795.

Pethick, J., 2002. Estuarine and tidal wetland restoration in the United Kingdom: policy versus practice. Restoration Ecology 10, 431–437.

Pétillon, J., Potier, S., Carpentier, S., Garbutt, A., 2014. Evaluating the success of managed realignment for the restoration of salt marshes: lessons from invertebrate communities. Ecological Engineering 69, 70–75.

Phillips, A.W., 1975. The establishment of *Spartina* in the Tamar estuary, Tasmania. Papers and Proceedings of the Royal Society of Tasmania 109, 66–76.

Piehler, M.F., Currin, C.A., Casanova, R., Paerl, H.W., 1998. Development and N_2-fixing activity of the benthic microalgal community in transplanted *Spartina alterniflora* marshes in North Carolina. Restoration Ecology 6 (3), 290–296.

Pierce, S.M., 1982. What is *Spartina* doing in our estuaries? South African Journal of Marine Science 78, 229–230.

Pontee, N., 2013. Defining coastal squeeze: a discussion. Ocean and Coastal Management 84, 204–207.

Pringle, A.W., 1995. Erosion of a cyclic salt marsh in Morecambe Bay, north-west England. Earth Surface Processes and Landforms 20, 387–405.

Raffaeli, D.G., Raven, J.A., Poole, L.J., 1998. Ecological impact of green macroalgal blooms. Oceanography and Marine Biology an Annual Review 36, 97–126.

Ranwell, D.S., 1967. World resources *of Spartina townsendii (sensu lato)* and economic use *of Spartina* marshland. Journal of Applied Ecology 4, 239–256.

Ranwell, D.S., 1981. Introduced coastal plants and rare species in Britain. In: Synge, H. (Ed.), The Biological Aspects of Rare Plant Conservation. Wiley, Chichester, pp. 413–420.

Reed, D.J., 1995. The response of coastal marshes to sea-level rise: survival or submergence? Earth Surface Processes and Landforms 20, 39–48.

Ricciardi, A., Simberloff, D., 2009. Assisted colonization is not a viable conservation strategy. TREE 24, 248–253.

Rochman, C.M., Browne, M.A., Halpern, B.S., et al., 2013. Policy: classify plastic waste as hazardous. Nature 494, 109–111.

Rogers, K., Saintilan, N., Copeland, C., 2014. Managed retreat of saline coastal wetlands; challenges and opportunities identified from the Hunter River Estuary, Australia. Estuaries and Coasts 37, 67–78.

Rogers, K., Boon, P.I., Branigan, S., et al., 2016. The state of legislation and policy protecting Australia's mangroves and salt marsh and their ecosystem services. Marine Policy 72, 139–155.

Roman, C.T., Burdick, D.M. (Eds.), 2012. Tidal Marsh Restoration. A Synthesis of Science and Management. Island Press, Washington DC, 432 p.

Ross, P.M., Adam, P., 2013. Climate change and intertidal wetlands. Biology 2, 445–480.

Saintilan, N., Williams, R.J., 1999. Mangrove transgression into salt marsh environments in South East Australia. Global Ecology and Biogeography Letters 8, 117–124.

Saintilan, N., Rogers, K., Mazumder, D., 2013. Allochthonous and autochthonous contributions to carbon accumulation and carbon store in Australian coastal wetlands. Estuarine, Coastal and Shelf Science 128, 84–92.

Saintilan, N., Rogers, K., McKee, K., 2018. Salt marsh-mangrove interactions in Australasia and the Americas. In: Perillo, G.M.E., Wolanski, E., Cahoon, D.R., Brinson, M.M. (Eds.), Coastal Wetlands: An Integrated Ecosystem Approach. Elsevier Science, Amsterdam.

Saltonstall, K., 2002. Cryptic invasion by a non-native genotype of the common reed, *Phragmites australis*, in North America. Proceedings of the National Academy of Sciences 99, 2445–2449.

Silliman, B.R., Bertness, M.D., 2004. Shoreline development drives invasion of *Phragmites australis* and the loss of New England salt marsh plant diversity. Conservation Biology 18, 1424–1434.

Silliman, B.R., Mozeder, T., Angelini, C., et al., 2014. Livestock as a potential biological control agent for an invasive wetland plant. PeerJ 2, e567. https://doi.org/10.7717/peerj.567.

Smith, T.J., Odum, W.E., 1981. The effects of grazing by snow geese on coastal salt marshes. Ecology 62, 98–106.

Smith, D.J., 2013a. Changes in Perspectives of the Values and Benefits of Nature. Ph.D. thesis. University of Salford, p. 218.

Smith, J.A.M., 2013b. The role of *Phragmites australis* in mediating inland salt marsh migration in a mid-Atlantic estuary. PLoS One 8 (5), e65091. https://doi.org/10.1371/journal.pone.0065091.

Stone, Y., Ahern, C.R., Blunden, B., 1998. Acid sulfate soils manual. In: Acid Sulfate Soil Management Advisory Committee, Wollongbar. NSW, 277 p.

Stricker, J., 1995. Reviving wetlands. Wetlands (Australia) 14, 20–25.

Strong, D.A., Ayres, D.A., 2016. Control and consequences of *Spartina* spp. invasions with a focus upon San Francisco Bay. Biological Invasions 2237–2246.

Sullivan, G., Noe, G.B., 2001. Coastal wetland plant species in Southern California. Appendix 2. In: Zedler, J.B. (Ed.), Handbook for Restoring Tidal Wetlands. CRC Press, Boca Raton, pp. 369–400.

Sydney Ports and Port Botany expansion, n.d. Penrhyn Estuary Habitat Enhancement Plan. Sydney Ports, Sydney, p. 46.

Talley, T.S., Crooks, J.A., Levin, L.A., 2001. Habitat utilization and alteration by the invasive burrowing isopod, *Sphaeroma quoyanum* in California salt marshes. Marine Biology 138, 561–573.

Teal, J., Teal, M., 1969. Life and Death of the Salt Marsh. Little Brown and Co., Boston, 278 p.

Thom, R.M., Borde, A.M., 2016. Estuarine habitat restoration. In: Kennish, M.J. (Ed.), Encyclopedia of Estuaries. Springer, Dordrecht, pp. 273–283.

Thomas, C.D., 2011. Translocation of species, climate change, and the end of trying to recreate past ecological communities. TREE 26, 218–221.

Thompson, G.B., Drake, B.G., 1994. Insect and fungi on a C_3 sedge and a C_4 grass exposed to elevated atmospheric CO_2 concentrations in open-top chambers in the field. Plant, Cell and Environment 17, 1161–1167.

Thompson, K., 2014. Where Do Camels Belong? the Story and Science of Invasive Species. Profile Books, London, 272 p.

Thrush, S.F., Townsend, M., Hewitt, J.E., et al., 2013. The many uses and values of estuarine ecosystems. In: Dymond, J.R. (Ed.), Ecosystem Services in New Zealand - Conditions and Trends. Maraaki Whenua Press, pp. 226–237.

Titus, J.G., 2000. Does the U.S. government realize that the sea is rising? How to restructure federal programs so that wetlands can survive. Golden Gate University Law Review 30 (4), 717–778.

Truscott, A., 1984. Control *of Spartina anglica* on the amenity beaches of Southport. In: Doody, P. (Ed.), *Spartina Anglica* in Great Britain. Nature Conservancy Council, Attingham Park, pp. 64–69.

Turner, R.E., 2009. Doubt and the values of an ignorance-based world view for restoration: coastal Louisiana wetlands. Estuaries and Coasts 32, 1054–1068.

Turner, R.E., 2011. Beneath the salt marsh canopy: loss of soil strength with increasing nutrient loads. Estuaries and Coasts 34, 1084–1093.

Underwood, A.J., Chapman, M.G., 2016. Ecological monitoring. In: Kennish, M.J. (Ed.), Encyclopedia of Estuaries. Springer, Dordrecht, pp. 223–227.

Underwood, G.J.C., 1997. Microalgal colonization in a salt marsh restoration scheme. Estuarine, Coastal and Shelf Science 44, 471–481.

Underwood, T., 2012. Detecting and understanding environmental impacts. In: Sainty, G., Hosking, J., Carr, G., Adam, P. (Eds.), Estuary Plants and What's Happening to Them in South-east Australia. Sainty Books, Sydney, pp. 464–477.

United Nations, 2004. World Urbanization Prospects; the 2003 Revision. United Nations, New York pp. x+323 (including Annexes).

Valdemoro, H.I., Sánchez-Arcilla, A., Jiménez, J.A., 2007. Coastal dynamics and wetlands stability. The Ebro delta case. Hydrobiologia 577, 17–29.

Valiela, I., 2006. Global Coastal Change. Blackwell Publishing, Oxford, 368 p.

Wang, Q., An, S.Q., Ma, Z.-J., et al., 2006. Invasive *Spartina alterniflora*: biology, ecology and management. Acta Phytotaxonomica Sinica 44, 559–588.

Webb, C.E., Doggett, S.L., 2016. Exotic mosquito threats require strategic surveillance and response planning. Public Health Research and Practice 26 (5), e651656.

White, K.P., Langley, J., Cahoon, D.R., Megonical, J.P., 2012. C_3 and C_4 biomass allocation responses to elevated CO_2 and nitrogen: contrasting resource capture strategies. Estuaries and Coasts 35, 1028–1035.

Wiebe, W.J., Christian, R.R., Hanson, R.B., et al., 1981. Anaerobic respiration and fermentation. In: Pomeroy, L.R., Wiegert, R.G. (Eds.), The Ecology of a Salt Marsh. Springer-Verlag, New York, pp. 136–159.

Wigand, C., Ardito, T., Chaffee, C., et al., 2017. A climate change adaptation strategy for management of coastal marsh systems,. Estuaries and Coasts 40, 682–693.
Williams P. and Associates Faber, P.M., 2004. Design Guidelines for Tidal Wetland Restoration in San Francisco Bay. The Bay Institute and California State Coastal Conservancy, Oakland, 83 p.
Williams, P.B., Faber, P.B., 2001. Salt marsh restoration experience in San Francisco Bay. Journal of Coastal Research 27, 203–211.
Williams, P.B., Orr, M.K., 2002. Physical evolution of restored breached levee salt marshes in the San Francisco Bay estuary. Restoration Ecology 10, 527–542.
Williams, B.K., Szaro, R.C., Shapiro, C.D., 2009. Adaptive Management: The U.S. Department of the Interior Technical Guide. Adaptive Management Working Group, U.S. Department of the Interior, Washington DC.
Williams, R.J., Allen, C.B., Kelleway, J., 2011. Saltmarsh of the Parramatta River - Sydney Harbour: determination of cover and species composition including comparison of ADI and pedestrian survey. Cunninghamia 12, 29–43.
Williams, J.M. (Ed.), 2006. Common Standards Monitoring for Designated Sites: First Six Year Report. Habitats. JNCC, Peterborough, 72 p.
Wolfe, R.G., 1996. Effects of open marsh water management on selected tidal marsh resources: a review. Journal of the American Mosquito Control Association 12, 701–712.
Wolters, H.S., 2006. Restoration of Salt Marshes. Ph.D. thesis. Groningen University.
Zedler, J.B., Adam, P., 2002. Salt marshes. In: Perrow, M.R., Davy, A.J. (Eds.), Handbook of Ecological Restoration, Restoration in Practice, vol. 2. Cambridge University Press, Cambridge, pp. 209–235.
Zedler, J.B., Paling, E., McComb, A., 1990. Differential salinity responses to help explain the replacement of native *Juncus kraussii* by *Typha orientalis* in Western Australian salt marshes. Australian Journal of Ecology 15, 57–72.
Zedler, J.B., 1995. Salt marsh restoration: lessons from California. In: Cairns, J. (Ed.), Rehabilitating Damaged Ecosystems, second ed. CRC Press, Boca Raton, FL, pp. 75–95.
Zedler, J.B., 1996a. Ecological issues in wetland mitigation: an introduction to the forum. Ecological Applications 6, 33–37.
Zedler, J.B., 1996b. Coastal mitigation in Southern California: the need for a regional restoration strategy. Ecological Applications 6, 84–93.
Zedler, J.B. (Ed.), 2001. Handbook for Restoring Tidal Wetlands. CRC Press, Boca Raton, 439 p.

Further Reading

Garbutt, A., Boorman, L.A., 2018. Managed realignment: recreating intertidal habitats on formerly reclaimed land. In: Perillo, G.M.E., Wolanski, E., Cahoon, D.R., Brinson, M.M. (Eds.), Coastal Wetlands. An Integrated Ecosystem Approach. Elsevier Science, Amsterdam.

CHAPTER

24

Methods and Criteria for Successful Mangrove Forest Rehabilitation

Roy R. Lewis III[1], *Benjamin M. Brown*[2], *Laura L. Flynn*[3]

[1]Coastal Resources Group, Inc., Salt Springs, FL, United States; [2]Charles Darwin University, Research Institute for Environment and Livelihoods (RIEL), Darwin, NT, Australia; [3]Coastal Resources Group, Inc., Venice, FL, United States

1. INTRODUCTION

Mangrove forest ecosystems as of 2005 covered 15.2 million ha of the tropical and subtropical shorelines of the world (FAO, 2007 quoted in Spalding et al., 2010). Giri et al., 2011 reported somewhat lower numbers for 2000 coverage (13.8 million ha). These figures represent a decline from 18.8 million ha in 1980 to 16.9 million ha in 1990 (FAO, 2007). These estimates represent about a 2% loss per year from 1980 to 1990 and 1% loss per year from 1990 to 2000 and are currently estimated at an annual loss of 0.66% (Spalding et al., 2010, p. 36). Therefore achieving a goal of no-net-loss of mangroves worldwide would require the successful rehabilitation of approximately 100,000 ha year^{-1}, unless all major losses of mangroves ceased. Increasing the total area of mangroves worldwide would require an even larger-scale effort.

An example of documented losses includes combined losses in the Philippines, Thailand, Vietnam, and Malaysia of 7.4 million ha of mangroves (Spalding, 1997). These figures emphasize the magnitude of the loss and the magnitude of the opportunities that exist to rehabilitate mangrove forests at the landscape scale, represented by the vast area of abandoned, disused, and unproductive aquaculture ponds in Southeast Asia (Stevenson et al., 1999). Prior successful large-scale ecohydrological mangrove (and salt marsh) rehabilitation has been demonstrated in 12,000 ha of mosquito control impoundments in Florida (Brockmeyer et al., 1997; Rey et al., 2012), which serves as an important demonstration of the potential of mangrove forest landscape rehabilitation.

Coastal Wetlands
https://doi.org/10.1016/B978-0-444-63893-9.00024-1

Although great potential exists to reverse the loss of mangrove forests worldwide, most attempts to rehabilitate mangroves often fail completely or fail to achieve the stated goals (Erftemeijer and Lewis, 1999; Lewis, 2000, 2005, 2009; Samson and Rollon, 2008; Brown et al., 2014a,b; Elliott et al., 2016). Previously documented attempts to restore mangroves (Field, 1996, 1999), where successful, have largely concentrated on creation of plantations of mangroves consisting of just a few species, and targeted for harvesting as wood products (Kairo, 2002), or temporarily used to collect eroded soil and raise intertidal areas to usable terrestrial agricultural elevations (Saenger and Siddiqi, 1993).

One of the world's largest attempts at hydrological rehabilitation of mangroves at the landscape scale took place in Cienaga Grande de Santa Marta, Colombia, where hydrologic improvements were employed to attempt to rehabilitate approximately 20,000 ha of dead or stressed mangroves (Perdomo et al., 1998; Twilley et al., 1998). Significant lessons are to be learned from this project, in that failure, during the design phase, to consider hydrological corrections that would result in a self-maintaining system, subsequently led to the need for frequent midcourse hydrological corrections, leaving the system vulnerable to both changing ecohydrological stressors and inadequate governance (Rivera-Monroy et al., 2006).

There are also unverified claims of very large-scale attempts to plant mangroves, claimed to be successful, in Senegal (www.oceaniumdakar.org) where 150 million *Rhizophora* propagules are claimed to have been planted since 2006, supposedly producing almost 12,000 ha of rehabilitated mangroves. However, Alexandris et al. (2013) states that "No, or only a slight change could be detected for the 40 sites analyzed..." using 2010 satellite photography of the project sites. No actual project description or monitoring reports are available from the website, and no known scientific reports have been published. Claims of large plantings, such as this, are common and need to be carefully vetted before being declared "successful" mangrove rehabilitation efforts.

Successful ecological mangrove forest rehabilitation requires careful analyses of a number of variables in advance of undertaking actual rehabilitation. First, for a given area of mangroves or former mangroves, the existing watershed needs to be defined and any changes to the coastal plain hydrology that may have impacted the mangroves documented (Wolanski et al., 2009). Second, careful site selection must take place taking into account the history of the site. This will likely require an investigation of historical maps, aerial and satellite photography, and mapping of changes over time. Third, clearly stated goals and achievable and measurable success criteria need to be defined and incorporated into a proposed monitoring program. Fourth, the restoration methodology must reflect an acknowledgment of the history of routine failure in attempts at mangrove rehabilitation and propose use of proven successful techniques. Fifth, after the initial restoration activities are complete, the proposed monitoring program must be initiated and used to determine if the project is achieving interim measurable success to indicate whether any midcourse corrections are needed. Finally, results should be made available for others to learn from documented successes and failures.

This chapter is an update of Lewis (2009) and reflects progress toward the goal of no-net-loss of mangroves and also the use of "rehabilitation" instead of "restoration" in keeping with the recommendations of Dale et al. (2014).

2. GENERAL SITE SELECTION FOR REHABILITATION

Previous research has documented the general principle that mangrove forests worldwide exist at the down slope end of coastal drainage basins (Kjerfve, 1990). At this down slope location adjacent to the sea, mangroves typically are established on a raised and sloped platform above mean sea level and inundated at approximately 30% or less of the time by tidal waters (Lewis, 2005). Kjerfve (1990) reported inundation times as short as 9% of the time for Klong Ngao on the west coast of Thailand. They are also typically dissected by sinuous tidal streams or rivers that become wider and deeper as they approach their seaward terminus (Figs. 24.1 and 24.2) and channel freshwater drainage from the uplands through the forest within the same tidal channels (Fig. 24.2). Increases in the frequency and duration of flooding more than what is normal for a particular type and location of a mangrove forest induce stress and death of mangroves Lewis et al., 2016. Kjerfve (1990) and Perillo (2009) suggest six key data needs when evaluating the hydrology of a basin and associated mangroves for possible management and rehabilitation:

1. Size and extent of drainage basin
2. Extent and area of mangroves at the down slope (i.e., toward the sea) end of the basin
3. Topography and bathymetry of the mangrove areas including tidal streams
4. Hypsometric characteristics to calculate the current tidal prism of the mangrove areas
5. Rates of terrestrial input of water, sediment, and nutrients
6. Climatic water balance

These analyses will yield one or more distinct areas of mangroves with varying characteristic hydrographic patterns, including more or less natural systems; some with natural

FIGURE 24.1 Oblique aerial photograph of mangroves at Tampa Bay, Florida, USA, illustrating tidal creeks facilitating overall hydrology (*red arrows*) with chief components being tidal exchange (*green double arrow*), upland drainage (*dark blue single arrow*), and rainfall (*light blue single arrow*).

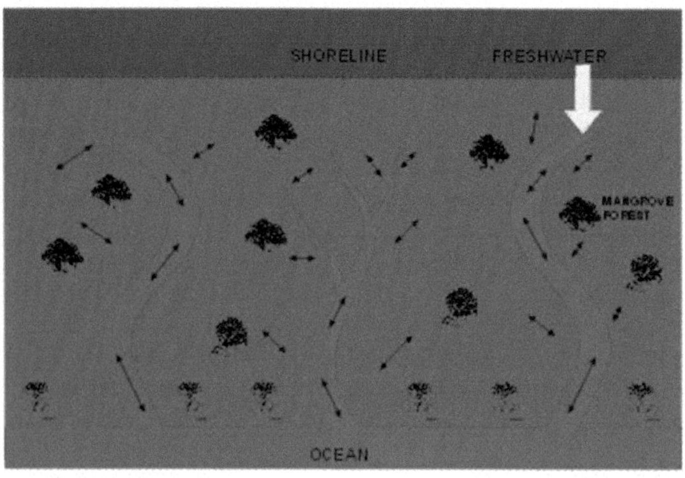

FIGURE 24.2 Graphic illustrating typical tidal creek configurations prior to aquaculture pond construction impacts.

damage such as recent hurricanes, typhoons, or tsunamis; and those impacted by development activities, such as dredging and filling, channelized basin flows, road construction, and restrictions to tidal exchange particularly in lagoonal mangroves. These may be stressed, but functioning at a reduced level, mostly dead, or completely eliminated and replaced by another land use. When combined, this information describes the "ecohydrology system" (Mazda and Wolanski, 2009, p. 256). Such coarse-level analyses are critical to assist countries in assessing their mangrove forest landscape rehabilitation potential to help meet national and international commitments such as the Bonn Challenge, reduction of greenhouse gas emissions, blue carbon, and biodiversity targets (IUCN and WRI, 2014).

3. SPECIFIC SITE SELECTION FOR REHABILITATION

The general site selection process may be applied to more than one coastal drainage basin to yield a list of individual mangrove forest areas and general characteristics of these areas. From this list, those areas showing either some current damage or significant declines in total area of mangroves from historical conditions need to be identified from a reconnaissance-level examination of available maps and aerial or satellite photography. From this effort, a short list of sites warranting further investigation is produced.

Each of these potential restoration sites requires a field-level investigation with maps and aerial or satellite photography in hand to verify vegetation signatures, including areas of stressed, dead, or lost mangroves. There may also be areas of damaged mangroves showing secondary succession or recovery from a previous damage event. The time frame since the damage event needs to be known to answer the key question, which is "does this site need management to support further recovery, or accelerate recovery, or is it likely to recover over time by itself without intervention?" Or as Saenger (2002) emphasizes, "what is the history of the site or area, or more specifically, what prior activities have led to the present conditions?" (p. 229). Characterization of existing and historical tidal courses (tidal streams, tidal rivers), which are critical parts of nearly all mangrove forests, is another critical step.

Lewis (2005) introduced the term "propagule limitation" to define a condition in which natural recovery is slowed or stalled because of a lack of natural mangrove propagules being available to volunteer at a damage site. Propagule limitation may be caused by a large loss of adult trees capable of producing propagules or by hydrologic restrictions or blockages (i.e., dikes), which prevent natural waterborne transport of mangrove propagules to a restoration site. Because propagules are produced at different times of the year by different species in different locations (Tomlinson, 1986), more than one site visit may be necessary to correctly identify a propagule-limited site. Lack of propagules at a single time of year does not necessarily define a propagule-limited site, and therefore careful evaluation of this parameter is important. If a damaged forest is going to recover on its own within an acceptable time frame, any attempt to introduce propagules or plant propagules or plant nursery grown mangroves is likely to be waste of time and money. Recovery is here defined as the recolonization or planting of a site and that site's growth of plant materials to some predefined numerical level (e.g., percent cover, total basal area). Priority should be given to sites that would indeed benefit from intervention by man given always limited time and money to devote to any restoration project.

These suggestions may seem obvious, but the available documentation of successful mangrove forest restoration is very limited, and more commonly former nonmangrove areas, such as mudflats or seagrass meadows seaward of natural mangroves, or damaged areas without a properly documented history, are the primary targets of well-intentioned, but often faulty, mangrove rehabilitation efforts (Field, 1996; Erftemeijer and Lewis, 1999; Lewis, 2000, 2005, 2009; Samson and Rollon, 2008; Brown et al., 2014a; b). The result of unsound evaluations of rehabilitation opportunities has, unfortunately, emphasized first establishing a mangrove nursery and then planting mangroves at a casually selected site as the primary tool in rehabilitation, rather than first assessing the reasons for the loss of mangroves in an area and working with the natural recovery processes (Fig. 24.3).

Rey et al (2012), Stevenson et al. (1999), Lewis et al. (2005), and present examples of successful mangrove rehabilitation following reestablishment of historical tidal connections to adjacent estuaries. This is termed "hydrologic restoration". In the examples discussed, volunteer mangrove and mangrove nurse plant propagules were sufficient to allow for rapid establishment of plant cover. No planting of mangroves was required, although minor amounts of "mangrove nurse plants" such as smooth cordgrass (*Spartina alterniflora)* were used at some sites. Nam et al. (2016) have recently reported similar results with herbicide damaged mangrove forests in Vietnam, where natural recovery (no planting) established equivalent forests 35 years postdamage.

Lewis and Marshall (1998) suggested that five critical steps were necessary to achieve "Ecological Mangrove Rehabilitation (EMR), and these are discussed in more detail in Stevenson et al. (1999), Lewis (2005), and Brown and Lewis (2006). The general approach

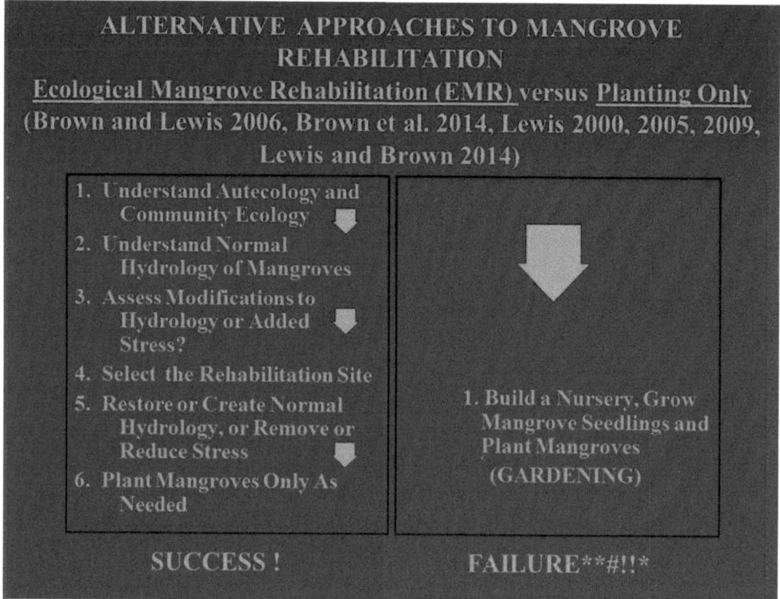

FIGURE 24.3 Alternative approaches to mangrove forest rehabilitation. Plantation establishment often producing failure (right side) versus ecological mangrove forest rehabilitation (left side) more often producing success.

is to emphasize hydrologic rehabilitation opportunities and avoid early emphasis on planting of mangroves (Turner and Lewis, 1997). These steps have been tested in training courses on mangrove restoration in the USA, Nigeria, Indonesia, Thailand, Vietnam, Sri Lanka, and India and have been further modified with review and input by both teachers and students to include one added step for a total of six as shown in Fig. 24.3. These now include the following:

1. Understand the autecology (individual species ecology) of the mangrove species at the site, in particular the patterns of reproduction, propagule distribution, zonation, secondary succession, and successful seedling establishment;
2. Understand the normal hydrologic patterns that control the distribution and successful establishment and growth of targeted mangrove species;
3. Assess the modifications of the previous mangrove environment that occurred that currently prevent natural secondary succession including blocked hydrologic connection, dike construction, overgrazing, and aggressive mangrove tree harvesting;
4. Select appropriate mangrove restoration sites through application of Steps 1–3 above that are both likely to succeed in restoring a sustainable mangrove forest ecosystem, and are cost effective given the available funds and manpower to carry out the projects, including adequate monitoring of their progress towards meeting quantitative goals established prior to restoration. This step includes resolving land ownership/use issues necessary for ensuring long-term access to, and conservation of the site.

5. Design the restoration program at appropriate sites selected in Step 4 above to initially restore the appropriate hydrology and utilize natural volunteer mangrove propagule recruitment for plant establishment.
6. Only utilize actual planting of propagules, collected seedlings, or cultivated seedlings after determining through Steps 1–5 above that natural recruitment will not provide the quantity of successfully established seedlings, rate of stabilization, or rate of growth of saplings established as quantitative goals for the restoration project. This issue may be addressed with small test rehabilitation areas to determine whether volunteer mangrove propagules will colonize a site, or by inspection of previous rehabilitation areas and reading monitoring reports about them.

Step number 2 is often the most overlooked, with essential water courses that previously existed being left out of most design plans and the topography of reference systems largely ignored. This is illustrated in Fig. 24.4 where an oblique aerial photograph of a seemingly successful planting of mangroves in Guyana has in fact some serious problems. First, it is a monospecific planting of only nursery grown *Avicennia germinans* and was "accidently successful" regarding survival of the plantings through saturation plantings regardless of topography. Because of this 41.3% of the plantings (out of a total of 441,599) failed, largely because of improper topography at the planting sites. Those that survived, as seen in Fig. 24.4, resulted in "block" establishment without tidal streams built into the plans and without designed connections to upland drainage ways or even connections to adjacent existing mangroves behind a long seawall separating developed lands from the Atlantic Ocean.

FIGURE 24.4 Oblique aerial photograph of a mangrove plantation established at Chateau Margot/Felicity, East Coast Demerara, Guyana, by the Guyana Mangrove Restoration Project. Only a single species was planted and has established (*Avicennia germinans*). Note the lack of tidal creeks and blockage of upland or riverine drainage due to a prior established seawall running down the right center of the photograph. *Photo credit: National Agricultural Research and Extension Institute, Guyana. Date of Photograph: October 2012.*

FIGURE 24.5 Mangrove plantation established using only a single species of mangrove (*Rhizophora apiculata*) in Myanmar. *Photo courtesy of Toe Aang.*

This is a very typical "mangrove plantation project," not an ecologically sound mangrove rehabilitation project (see Asaeda et al., 2016 for further explanation). Fig. 24.5 shows a similar mangrove plantation project in Myanmar.

An improved design shown in time sequence in Fig. 24.6 is the establishment of a mangrove test rehabilitation area in Naples, Florida, USA, where a tidal stream was restored and persisted (light blue arrows), and no plantings of mangroves took place. Consequently, natural recruitment of the four local species of mangroves (*Rhizophora mangle, Avicennia germinans, Laguncularia racemosa*, and *Conocarpus erectus*) occurred achieving cover, densities, and diversity equivalent to local reference forests and supporting fish populations migrating into and out of the rehabilitated forest utilizing the restored tidal channels. As a result, a larger mangrove rehabilitation project following the same design is designed, permitted, and planned for construction. Test rehabilitation projects of this kind are recommended where rehabilitation methods are not well established for a particular area.

Step number 6 (Fig. 24.3) remains the most controversial application of the EMR process. If natural recruitment fails, that may mean the site has not been adequately rehabilitated to facilitate volunteer mangrove recruitment where propagule limitation does not exist. For example, if the hydrology has not been adequately restored, or at an excavated site, the final topographic grade may be too high or too low. Under these circumstances, planting will not overcome these physical limitations on plant establishment; however, planting does often occur anyway, and failures are common (Brown et al., 2014a,b).

Local communities plant seeds and seedlings even after having undertaken successful EMR for a combination of three reasons: (1) lack of patience, (2) protection of the rehabilitation site because planted areas appear to outsiders (not aware of the project) as an intentional action and provide a measure of protection for that area as it is obvious that there is human activity in the area, and (3) promotion of growth of preferred species such as *Rhizophora* over colonizers such as *Avicennia* or *Sonneratia*.

FIGURE 24.6 Time series photographs of a 1.5 ha mangrove restoration test site at Fruit. Farm Creek, Naples, Florida, USA. The *yellow arrow* in the 2012 photograph shows the disposal site for excavated soil that blocked hydrologic exchange, and the *light blue arrow* in all four photographs shows the restored tidal creek that has persisted without further Intervention. All four of the native mangrove species in Florida have naturally established without planting at this site along with native mangrove associates. *All photographs by Roy R. Lewis III.*

Even with adequate local mangrove recruitment after EMR, planting may serve as an educational process for local communities and encourage local support for the project. It is, however, still important to document natural recruitment during monitoring and report the differential contribution of volunteer and human-planted mangroves to the final measures of total plant cover. It is also important to note that very dense plantings (1 m centers) can result in crowding, lack of functional tidal exchange (Mazda and Wolanski, 2009, p. 248), and prevention of other species of mangroves from volunteering to a site (Salmo et al., 2013; Asaeda et al., 2016). Plantings greater than a density of approximately 2500 ha^{-1} (2 m centers) are rarely needed if conditions are appropriate as survival is typically high (>75%) (see Fig. 24.5).

Through the application of these six simple steps and basic principles of ecological rehabilitation using ecological engineering approaches (Lewis, 2005), including careful cost evaluations prior to final selection of a restoration site and design of a restoration program, the opportunity for a cost-effective and successful restoration effort is maximized. We would caution, however, that there are conflicting recommendations for successful mangrove forest rehabilitation in various publications, and careful reading and comparison of these recommendations should be undertaken before a final decision is made about what rehabilitation method to use. For example, emphasize the EMR approach, whereas Biswas et al. (2009), Global Nature Fund (2015), Primavera and Esteban (2008), Primavera et al. (2011), and Primavera et al. (2014) emphasize that planting is always essential. Some of these differences are based on unique local conditions, but the general recommendation to always approach mangrove rehabilitation as a mangrove planting project, instead of a careful prior examination of the history of hydrologic changes and working with natural recruitment of mangroves (Fig. 24.3), generally defines the existing differences in approach. We maintain that the former approach (e.g., always plant mangroves) is not the correct initial approach to routinely achieving successful and ecologically sound mangrove rehabilitation projects. Salmo et al. (2013) and Asaeda et al. (2016) agree.

Most of the largest attempts to restore mangroves are currently taking place in South and Southeast Asia. While the six-step process of EMR described above has been taught as several dozen short courses and is widely publicized in English and Indonesian in that part of the world, and in Spanish in Central and South America, more education needs to be done and published in native languages (Lewis and Brown, 2014 is available in three languages). The lack of large-scale translations of scientific papers about mangrove rehabilitation into local languages is hampering adoption of the six-step process in these areas. Additionally, the lack of general application of the rule of law in several of these countries limits attempts to protect existing mangroves. Large-scale conversion of existing mangrove forests to aquaculture ponds is still taking place, and more recently, conversion of mangroves to what are perceived as more valuable oil palm plantations and salt-tolerant rice cultivation has accelerated (Giri et al., 2008; Richards and Friess, 2016).

Alternative approaches to mangrove rehabilitation have been proposed and described by Gensac et al. (2015) for the mud coasts of French Guiana and for the mud coasts of Vietnam as described by Albers and Schmitt (2015). It is important to note that both of these approaches emphasize the same general approach as shown in Fig. 24.3, that is, conducting detailed studies of the existing conditions at proposed rehabilitation sites, examining existing topography of mud banks and following accumulation of mud deposits over time until the critical tidal level and flooding and drying periods are reached to allow for natural mangrove propagule recruitment. Actual planting of mangroves seems unnecessary, although installation of T-fences made of bamboo to encourage sediment accumulation is a key component of mangrove rehabilitation in Vietnam. It still remains to be seen if the short life span of bamboo fences (5–7 years, p. 1000, Albers and Schmitt, 2015) results in any loss of mangroves after the T-fences are gone. For this reason, Albers and Schmitt (2015) also emphasize "(M)angrove management is an important element of an area coastal protection strategy" (p. 991), to prevent overharvesting of the new stands of volunteer mangroves.

Ecologists and engineers have not historically understood mangrove hydrology, as Kjerfve (1990) points out. Although a number of papers discuss the science of mangrove hydrology (Kjerfve, 1990; Wolanski et al., 1992; Furukawa et al., 1997, see also reviews in Mazda et al., 2007; Mazda and Wolanski, 2009), their focus has been on tidal and freshwater flows within the forests and not the critical periods of inundation and dryness controlled by the tides and local rainfall, which govern the health of the forest. Kjerfve (1990) does discuss the importance of topography and argues that "...micro-topography controls the distribution of mangroves, and physical processes play a dominant role in formation and functional maintenance of mangrove ecosystems..." Hypersalinty due to year-to-year variations in rainfall can also produce natural mangrove diebacks (Cintron et al., 1978), and disruption of normal freshwater flows that dilute seawater in more arid areas can kill mangroves (Perdomo et al., 1998; Medina et al., 2001) or stress mangroves to the point that they may not be able to keep up with sea level rise through root production and the laying down of a peat layer (Snedaker, 1993).

The point of all of this is that flooding depth, duration, and frequency are critical factors in the survival of both mangrove seedlings and mature trees. Once established, mangroves can be further stressed if the tidal hydrology is changed, for example, by diking (Brockmeyer et al., 1997). Both increased salinity because of reductions in freshwater availability, and flooding stress producing increased anaerobic conditions and free sulfide availability can kill or stress existing stands of mangroves and mangroves at restoration sites (McKee, 1993).

For these reasons, any engineering works constructed near mangrove forests, or in the watershed that drains to mangrove forests, must be designed to allow for sufficient free exchange of seawater with the adjacent ocean or estuary and not interrupt essential upland or riverine drainage into the mangrove forest. Failure to properly account for these essential inputs and exchange of water could result in stress and possible death of the forest, but as seen in Figs. 24.4 and 24.5, as a minimum produce typically very dense, crowded plantings, without ecologically essential tidal courses and thus reduced ecological value as fish and invertebrate habitat. Contrast this with a hydrologic restoration project with NO mangrove plantings (Fig. 24.6), only volunteer seedlings, which over a relatively short period of time have colonized a restored site and establish a diverse mixture of all of the native mangrove species and associated herbaceous salt-tolerant vegetation.

Engineering works such as dikes created to make shrimp aquaculture ponds (Figs. 24.7 and 24.8) disrupt the natural hydrology and produce conditions that prevent natural recovery once these ponds are abandoned because of disease (Stevenson et al., 1999). Rehabilitation of abandoned aquaculture ponds constructed in mangroves potentially has the opportunity to put hundreds of thousands of hectares of damaged mangroves back into natural mangrove forests ecosystems, but the ecological engineering necessary to do this in a cost-effective manner has not been well developed, and ownership issues still prevent some attempts.

Use of a reference mangrove site for examining normal hydrology is one solution for the ecological engineering issues (Lewis, 2005). The installation of tide gages and measurement of the tidal hydrology of a reference mangrove forest in conjunction with the surveyed elevations of a reference mangrove forest floor as a possible surrogate for hydrology is one approach. Establishing those same ranges of elevations at a proposed restoration site or restoring the same hydrology to an impounded mangrove by breaching the dikes in the "right places" can accomplish hydrologic restoration. The "right places" are usually the mouths of historic tidal creeks. These are often visible in vertical (preferred) or oblique aerial photographs.

FIGURE 24.7 Oblique aerial photograph of shrimp aquaculture ponds constructed within former mangrove forests in the Gulf of Thailand. *Photograph by Roy R. Lewis III.*

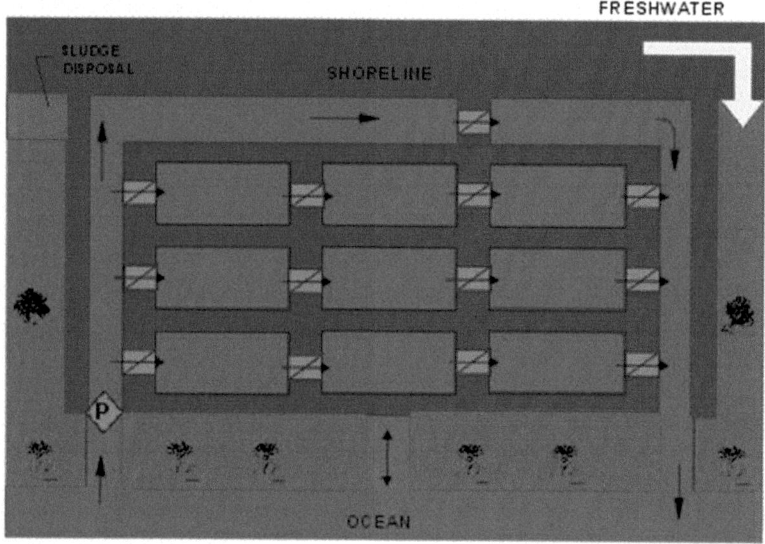

FIGURE 24.8 Graphic illustrating the operation of an aquaculture pond system similar to that shown in Fig. 24.7.

Figs. 24.9 and 24.10 illustrate two possible approaches to such rehabilitation efforts. The first (Fig. 24.9) illustrates the complete regrading of a former pond site through excavation dikes and filling old channels. Such efforts have been estimated to cost USD \$9318 ha^{-1} by Barbier et al. (2011). Our joint professional opinion is that this is a low estimate based on

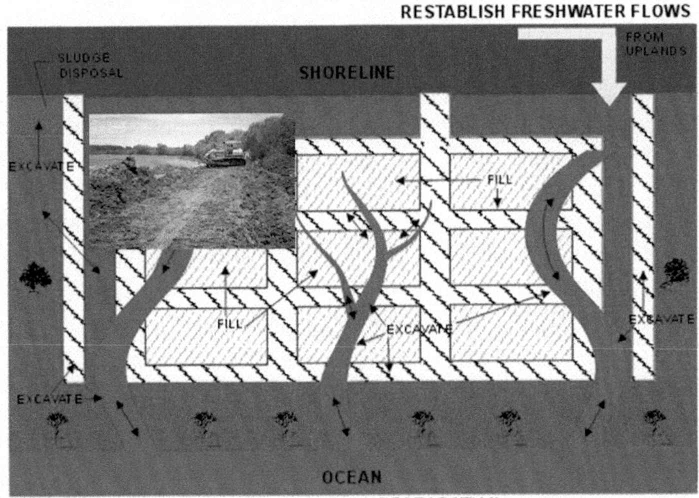

FIGURE 24.9 Graphic illustrating the potential full rehabilitation of a series of aquaculture ponds with heavy construction equipment and targeting full topographic rehabilitation.

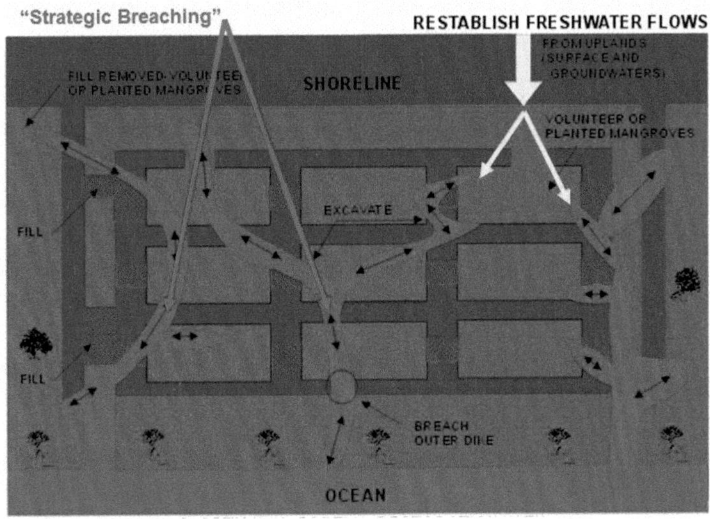

FIGURE 24.10 Strategic breaching (selective breaching of dikes and berms) as an alternative to full rehabilitation of abandoned aquaculture ponds.

similar work in Indonesia, Thailand, and the USA. In any case, Barbier et al. (2011) note that the high costs of restoration of this type can offset any potential ecological and financial values of the restored ecosystem.

Fig. 24.10 illustrates an alternative approach presently being implemented in Indonesia (Brown et al., 2014b) where most of the pond area is left alone and only "strategic breaches" are created, and some channels filled with hand labor or potentially excavating equipment, and full hydrologic functioning restored to the majority of the ponds. This concept was first introduced by Brown and Lewis (2006). Costs for this approach were reported at USD \$1475 ha^{-1} at a 400 ha scale (Brown et al., 2014a,b), and will vary depending on whether or not economies of scale exist and the extent of investment required to effectively engage the support of critical local and regional stakeholders.

4. ESTABLISHING SUCCESS CRITERIA

Once a site is finally chosen for rehabilitation, and a design developed, quantifiable success criteria should be established. Establishing such criteria is important to actually measure progress toward successful rehabilitation. The first step in establishing numeric criteria for success is to prepare a brief narrative goal or set an objective (Saenger, 2002) for the project. This will define the next steps. A goal may be to establish a monotypic plantation of *Rhizophora apiculata* to be harvested after 12 years as poles. This is a typical goal of many mangrove rehabilitation projects. It may be an acceptable goal to local stakeholders in the project such as local villages and fishermen, and harvest of wood products from locally managed forests is a typical goal (see discussion of timber production in the Matang Forest, Malaysia, by Saenger, 2002, pp. 231–234).

If on the other hand the goal is to provide fish and invertebrate habitat to restore local fisheries, a different approach to establishing success criteria is dictated. Maximizing such habitat use usually means maximizing biodiversity of the plant species, and therefore a monotypic stand of mangroves in an area that normally supports 20 or more mangrove tree species is not a logical goal. Establishment and persistence of tidal creeks to assist with entry and exit of juvenile and adult fish and invertebrates may also be needed criteria.

Once narrative success criteria are agreed on, quantitative criteria then need to be established. For the first example above, a certain number of pole size trees per hectare could be established. Such a success criteria would also likely dictate an immediate planting program of collected propagule or nursery grown plants, rather than waiting for volunteer propagules. For the second example, to maximize biodiversity, the restoration site might be left alone and not planted immediately to allow for volunteer colonization of the largest number of different species of mangroves from propagules produced by trees adjacent to a restoration site (see Fig. 24.6).

The next step is to look at available information on both planted and natural recruitment indices of success. Saenger (2002, pp. 256–270) discusses in great detail what is to be expected in terms of biomass and stem density, for example, from typical plantation projects. There has been much work on plantation projects where just a few species of mangroves are managed for, and thus there is a wealth of data to examine. In contrast with this, the availability of data on natural recruitment within a mixed forest is generally not available. McKee and

Faulkner (2000) report on the results of sampling for density and basal area within two restored mangrove forests in Florida, USA, and compared these to two adjacent control areas. Their data show that density and basal area for volunteer mangroves in the restoration areas exceeded that for planted mangroves. Proffit and Devlin (2005) report similar results from one of the same sites sampled by McKee and Faulkner (2000) but in later years as the system matured. Lewis et al. (2005) report on the results of cover sampling over a period of 5 years within a restored mangrove forest in another location in Florida, USA. These studies help define parameters that need to be sampled and sampling methodologies, but provide limited data to apply to local situations in other parts of the world.

5. MONITORING AND REPORTING SUCCESS

Once success criteria have been established and the site restored through hydrologic restoration with or without planting, monitoring and reporting should begin. A typical monitoring schedule would consist of the following 10 reports:

1. Time Zero
2. 0 + 3 months
3. 0 + 6 months
4. 0 + 9 months
5. 0 + 12 months
6. 0 + 18 months
7. 0 + 24 months
8. 0 + 36 months
9. 0 + 48 months
10. 0 + 60 months

A Time Zero report is prepared after all the site restoration changes have been accomplished and any proposed planting completed. It should include photographs taken from fixed stations where future photography will also be taken (see Fig. 24.6). The shorter intervals in the early years of monitoring are designed to insure that any corrective actions necessary due to problems encountered during monitoring are quickly corrected. These are termed "midcourse corrections." A typical real Time Zero report based on the rehabilitation reported in Lewis et al. (2005) can be found as download #301 at www.mangroverestoration.com. A typical Time Zero plus 60 months report, including the graph shown in Fig. 24.11, can be found as download #302 at the same website.

This is a typical schedule of reporting as required by wetland mitigation projects in the United States. As noted by the data of Proffit and Devlin (2005), however, changes in the height, density, and species composition of mangroves on a rehabilitation site will continue over time. Eighteen years had passed since the restoration of the site described in Proffit and Devlin (2005), and changes were still being observed. For example, the dominant species present 7 years after restoration, *Laguncularia racemosa*, experienced "reduced recruitment and apparent density-dependent mortality" through the last sampling 18 years after restoration. Longer-term monitoring in forested restoration areas is recommended where resources are available. The recommended time interval after the last regular monitoring

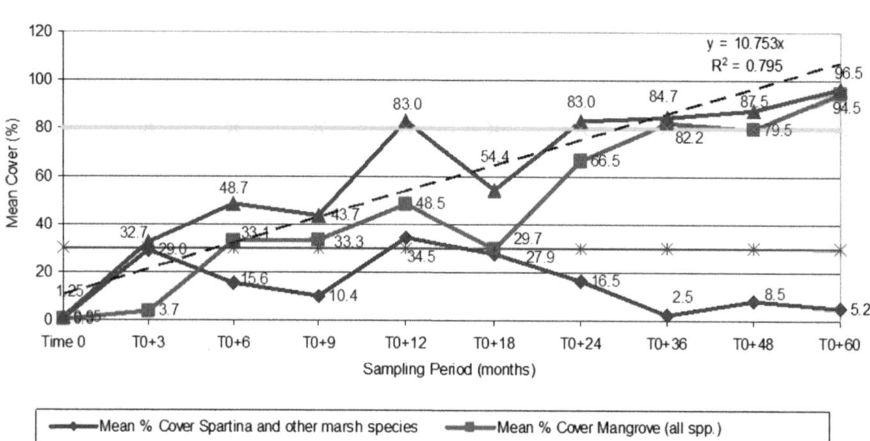

FIGURE 24.11 Graph of mean percent cover over time for all species (*blue triangles*), mangroves only (*pink squares*), and marsh species (*black diamonds*) over a period of 60 months of monitoring after hydrologic restoration. No mangrove plantings took place.

would be 5 year intervals until the time of maturity of the restored forest. Based on the work of Lewis et al. (2005), total cover by mangroves can be expected to occur rapidly (within 3–5 years) (see Fig. 24.6), but basal area equivalency will take much longer.

Ten years after hydrologic restoration, Stevenson et al. (1999) report that the total basal area for all species of mangroves at a hydrologically restored site in Costa Rica (abandoned shrimp aquaculture pond) was 64.2% of that of the control site. Eighteen years after restoration at a site in Florida, USA, Proffit and Devlin (2005) report the total basal area of all mangrove species (42.7 m^2 ha^{-1}) exceeded that of the mean of the two control areas (17.9 m^2 ha^{-1}) by a factor of 2.4. They make an important note, however, in that they encountered a large number of saplings exceeding the 1.3 m height requirement for counting in basal area measurements, but they were less than the minimum of 2.0 cm in diameter at breast height (DBH). Normally these trees are not counted in basal area calculations (Cintron and Novelli, 1984). We believe ignoring them produces a total basal area calculation that does not represent the true value for all trees on the restoration site and recommend they do be counted and DBHs measured, but that category of trees less than 2 cm in DBH be reported separately from the other normally reported classes (i.e., $\geq$2 cm < 10 cm, $\geq$10 cm) to allow for direct comparisons to other data sets like that of Stevenson et al. (1999) and McKee and Faulkner (2000).

We also recommend the establishment of permanent, haphazard, or random plots of 5 × 5m, at a density within the restoration area that allows for stratified sampling over the elevation gradient present within the restoration site. The point-centered quarter method (Cintron and Novelli, 1984) can also be used.

6. FUNCTIONALITY OF REHABILITATED MANGROVE FORESTS

Lewis (2005) notes that ecological rehabilitation of mangroves, where rehabilitated ecosystem functions are the goal, is rarely a prime goal of rehabilitation projects and thus is rarely monitored. As noted above and shown in Figs. 24.4 and 24.5, simple establishment of mangroves, even a dense monoculture, is still considered the definition of success in most cases. Lewis (1991) notes that research on the use by fish of both restored tidal marshes and mangroves in the United States shows that fish populations in these restored plant communities are equivalent in both numbers and species composition within 3–5 years of rehabilitation, if appropriate tidal courses are designed and constructed as part of the project.

McKee and Faulkner (2000) examined the biogeochemical functions of two rehabilitated mangrove forests in Florida (6 and 14 years old). Soil Eh was lower at the rehabilitated sites and porewater sulfide concentrations were significantly higher. Soil carbon and nitrogen levels were greater overall in natural soils and were correlated with soil organic matter content. They concluded that site-specific parameters such as rates of tidal flushing, topography, and salinity played a larger role in primary production and turnover rates of organic material than site age.

Bosire et al. (2008) provide a thorough review of faunal use of rehabilitated mangrove forests, including data for restoration projects in six countries (USA, Thailand, Kenya, Philippines, Qatar, and Malaysia). The data collected for these sites were very variable, however, and no uniform sampling methodology or target faunal groups were the focus of the scattered research efforts. As would be expected, the conclusion of the authors was that functionality depended on what parameters were measured at a given location and generalities were very difficult to make.

For example in Thailand, crab diversity at some of the rehabilitated sites was higher than at an upper shore natural mangrove site, and both biomass and crab numbers were consistently higher in the rehabilitated sites (Macintosh et al., 2002). However, the natural site was characterized by large numbers of sesarmid crabs. Differences in the crab diversity in Thailand were thought to relate to, among other things, inundation zone and differences in the mangrove species present in the rehabilitated sites (Macintosh et al., 2002). However, in Qatar,Al-Khayat and Jones (1999) found lower species richness of crabs at rehabilitated sites compared with natural habitats of *Avicennia marina*. In Kenya, reforested stands of *Rhizophora mucronata* and *Avicennia marina* had higher crab densities than their natural references (Bosire et al., 2004) but with similar species diversity and crab species composition compared with respective bare control with similar site history. In the Philippines, the relative abundance of mud crab (*Scylla olivacea*) and two other noncommercial species was used to separate the effects of habitat from fishing pressure and recruitment limitation. A comparison of mud crab populations in rehabilitated, natural, and degraded sites in the Philippines suggested that 16-year-old rehabilitated *Rhizophora* spp. can support densities of mud crabs equivalent to that of natural mixed species mangroves (Walton et al., 2007). Mollusk diversity showed similar patterns to that of crabs in both previously mentioned studies in Qatar and Thailand, whereas in Kenya, no mollusks were observed in a bare site while the rehabilitated site and natural reference site within a *Sonneratia alba* forest had similar species composition, density, and diversity.

Infauna communities (i.e., polychaete worms, amphipods) showed patterns similar to those described above. Lower diversity of taxa was observed in rehabilitated versus natural sites in Qatar with the data being less clear in Thailand. Studies in Malaysia suggested that 2-year-old rehabilitated mangroves had the greatest biomass and species number followed by the control and 15-year-old stand, although species diversity was highest in the control site and lowest in the 2-year-old site (Sasekumar and Chong, 1998). In Kenya, bare sites had the lowest infauna densities and taxa richness compared with rehabilitated sites with natural reference sites having the highest densities. Taxa richness and composition were similar among respective restored and natural sites (Bosire et al., 2004), suggesting successful fauna recolonization following mangrove rehabilitated.

Studies of mobile fauna in rehabilitated mangroves of varying ages and species composition showed variable patterns. In Qatar, lower diversity of both juvenile and adult fish was observed in rehabilitated sites compared with natural stands of *A. marina* (Al Khayat and Jones, 1999). Studies comparing fish and shrimp density between natural stands of *R. apiculata*, *Avicennia officinalis*, *A. marina*, and a single rehabilitated *R. apiculata* stand (5–6 years old) in the Philippines indicated that density and biomass were primarily influenced by tidal height and mangrove species (Rönnbäck et al., 1999). In *S. alba* plantations in Kenya, there were strong seasonal fluctuations of juvenile fish, showing temporal patterns to be a potentially stronger influence on fish assemblages than type of rehabilitation site or presence of fringing mangroves (Crona and Rönnbäck, 2007). However, the spatial scale of observation is likely a much stronger factor affecting biodiversity studies of more mobile fauna compared with less mobile animal communities described above.

Because most studied rehabilitation sites are small in size, site-specific effects on fish distribution patterns remain largely unknown. The same is true for juvenile shrimp. In Kenya, one species, *Penaeus japonicus*, dominated the community, and lower species richness was observed in a rehabilitated area of *S. alba* than in adjacent clear-felled areas (Crona and Rönnbäck, 2005). Natural forests had higher root complexity and also higher abundances and more even distribution of shrimp species in terms of species composition. Similarly, in the Philippines, higher abundances of juvenile shrimp in a rehabilitated *R. apiculata* site were seemingly related to higher structural root complexity, although more inland stands of mature *Avicennia* spp. and *Rhizophora* spp. showed no such differences (Rönnbäck et al., 1999).

Lewis and Gilmore (2007) discuss fish use of both natural and rehabilitated mangrove forests and report specifically about monitoring of a successful 500 ha mangrove rehabilitation project in Hollywood, Florida, USA, where sampled fish populations in both reference and restored sites were statistically indistinguishable within 3–5 years of restoration. They emphasize three restoration and design goals to ensure functional ecological rehabilitation of mangrove forests:

1. Achieve a biodiverse plant cover similar to that in an adjacent relatively undisturbed control area of mangrove forest.
2. Establish a network of channels that mimic the shape and form of a natural tidal creek system (see Figs. 24.1 and 24.2).
3. Establish a heterogeneous landscape similar to that exhibited by local mangrove ecosystems.

Asaeda et al. (2016) generally recommend the same key aspects for full ecological functional restoration. Few studies exist on trends in biodiversity in restored mangroves, and the range in age, species, and inundation class of restored sites makes generalizations difficult. However, the co-occurrence of many animal species in both restored and comparable natural forests suggests that colonization of restoration sites by both mobile and nonmobile fauna is a rapid process, and equivalent populations of mangrove fauna in both natural controls and restored mangrove sites can typically be found within 5–10 years of restoration.

7. FUTURE STUDIES

Selkoe et al. (2015) have recently outlined some basic principles for managing marine ecosystems prone to tipping points where seemingly healthy marine ecosystems can quickly transition to stressed and potentially severely damaged ecosystems over very short periods of time. This includes mangrove forests. Lewis et al. (2016) have elaborated on this theme with cautionary notes about specifically stressed mangrove forests with the need to identify these sites as part of any mangrove forest management program and have recommended the concept of "preemptive rehabilitation" where actions such as hydrologic restoration are undertaken before the forest is allowed to completely die.

Many advantages accrue to such efforts, including the preservation of carbon stocks within the forest, in belowground biomass, and prevention of soil collapse with widespread death of the forest.

If we know the position of thresholds, the general condition of the system state (conservation phase, renewal phase, stressed, high degraded, converted, etc.), and some of the slow-moving variables that dictate the position of the system with regards to the threshold, we can better understand our management options.

1. Maintain resilience and adaptive capacity
2. Build resilience, enhance adaptive capacity, and identify and remove stressors (preemptive rehabilitation)
3. Restore or rehabilitate
4. Guide intentional transformation

8. SUMMARY

Mangrove forest rehabilitation has not been generally successful except where timber production was the goal and monotypic stands were established and intensely managed. Ecological rehabilitation of mangrove forests, where the goal is the rehabilitation of a mixed species forest cover and functions equivalent to that of an adjacent reference forest, has not typically been a design criteria, and most rehabilitation projects not targeting timber production, but with some general ecological goals, have not been successful (Erftemeijer and Lewis, 1999; Lewis, 2000, 2005, 2009; Samson and Rollon, 2008; Brown et al., 2014a,b). The chosen restoration sites for many of these projects have been mudflats or seagrass beds lying seaward of the outer edge of existing mangrove forests. These sites are typically planted with nursery-grown mangrove seedlings which do not survive because of frequent inundation that exceeds their tolerance for duration of flooding, typically 30% or less of the total annual flooding periods in mangroves (Lewis, 2005).

Although there are relatively few studies on trends in biodiversity in rehabilitated mangroves, the co-occurrence of many animal species in both rehabilitated and comparable natural forests suggests that colonization of rehabilitated sites by both mobile and nonmobile fauna is a rapid process, with equivalent populations of mangrove fauna in both natural control areas and restored mangrove sites typically found within 5–10 years of restoration. The scientific basis for optimum design of restoration projects to meet certain established criteria, such as increased fish production or more use by wading seabirds, is, however, very minimal.

In the future, mangrove rehabilitation projects should be more carefully designed to ensure successful establishment of plant cover at minimal cost over large areas. This can be achieved, for example, by restoring hydrologic connections to impounded mangrove areas as has been done in Florida (Rey et al., 2012), Costa Rica (Stevenson et al., 1999), and the Philippines (Stevenson et al., 1999; Primavera et al., 2014). Agencies typically fund mangrove rehabilitation projects with minimal funds dedicated toward quantitative monitoring and reporting over a reasonable and ecologically based time period (5 years minimum). Both failures and successes thus go undocumented, and mistakes are repeated and lessons learned are lost. Funding agencies and governments need to realize that large amounts of limited rehabilitation funds are now being wasted because of these short-sighted efforts, and at a minimum they should regularly review, publish, and teach the lessons learned from both past successes and failures.

These same funding agencies and governments are very loath to fund careful examination of the ecological functions of rehabilitated mangrove areas. This is somewhat understandable, given the large costs of quantitative monitoring, but at a minimum, attempts should be made to coordinate restoration projects with graduate training programs to provide, at likely minimal costs, opportunities for graduate students and researchers a like access to restoration sites where good research can be done for minimal costs. These efforts to date have been hampered, however, by the lack of application of uniform methodologies of sampling and reporting. We have made some recommendations above but would encourage every researcher to revisit and read carefully the excellent work of Snedaker and Snedaker (1984), where uniform scientific methodologies developed by worldwide consensus for sampling flora, fauna, biochemistry, litter production, and decomposition, among other parameters, were recommended.

In addition, the additional detailed work on mangrove forest hydrology based on recommendations in Kjerfve (1990), Mazda et al. (2007), and Mazda and Wolanski (2009) is essential to round out the beginning point for any mangrove rehabilitation research program to develop a very well thought out and reviewed plan of study. We find most current researchers do not start with these references, producing results and data generation of little value in promoting the advancement of the science of ecological rehabilitation of mangrove forests.

At the present time we believe that greater success at EMR could be achieved with a four step approach that includes the following:

1. General site selection for rehabilitation sites that include examination of multiple coastal basins that contain mangroves.

2. Specific site selection that looks at the history of changes in areal cover of mangroves and changes in hydrology at specific potential rehabilitation sites and initially targets hydrologic restoration of these sites.
3. Establishing quantitative and measurable success criteria and use uniform criteria between study sites (Lewis et al., 2005; Brown et al., 2014b).
4. Monitoring and reporting of progress toward achieving these quantitative success criteria, including reporting on lessons learned from both successes and failures.

References

Albers, T., Schmitt, K., 2015. Dyke design, floodplain restoration and mangrove co-management as parts of an área coastal protection strategy for the mud coasts of the Mekong Delta, Vietnam. Wetlands Ecology and Management 23 (6), 991–1004.

Alexandris, N., Chatenoux, B., Lopez-Torres, L., Peduzzi, P., 2013. Monitoring Mangrove Restoration from Space. UNEP/GRID, Geneva, 42 pp.

Al-Khayat, J.A., Jones, D.A., 1999. A comparison of the macrofauna of natural and replanted mangroves in Qatar. Estuarine, Coastal and Shelf Science 49, 55–63.

Asaeda, T., Barnuevo, A., Sanjaya, K., Fortes, M.D., Kanesaka, Y., Wolanski, E., 2016. Mangrove plantation over a limestone reef – good for the ecology? Estuarine, Coastal and Shelf Science 173, 57–64.

Barbier, E.B., Hacker, S.C., Kennedy, C., Koch, E.W., Stier, A.C., Silliman, B.R., 2011. The value of estuarine and coastal ecosystem services. Ecological Monographs 81 (2), 169–193.

Biswas, S.R., Malik, A.U., Choudhury, J.K., Nishat, A., 2009. A unified framework for the restoration of Southeast Asian mangroves – bridging ecology, society and economics. Wetlands Ecology and Management 17, 365–383.

Bosire, J.O., Dahdough-Guebas, F., Walton, M., Crona, B.I., Lewis III, R.R., Field, C., Kairo, J.G., Koedam, N., 2008. Functionality of restored mangroves: a review. Aquatic Botany 89, 251–259.

Bosire, J.O., Dahdouh-Guebas, F., Kairo, J.G., Cannicci, S., Koedam, N., 2004. Spatial macrobenthic variations in a tropical mangrove Bay. Biodiversity and Conservation 13, 1059–1074.

Brockmeyer Jr., R.E., Rey, J.R., Virnstein, R.W., Gilmore, R.G., Ernest, L., 1997. Rehabilitation of impounded estuarine wetlands by hydrologic reconnection to the Indian River Lagoon, Florida (USA). Wetlands Ecology and Management 4 (2), 93–109.

Brown, B., Lewis, R.R., 2006. In: Lewis, R., Quarto, A., Enright, J., Corets, E., Primavera, J., Ravishankar, T., Stanley, O.D., Djamaluddin, R. (Eds.), Five Steps to Successful Ecological Restoration of Mangroves. Yayasan Akar Rumput Laut (YARL) and the Mangrove Action Project. Yogyakarta, Indonesia, 64 p.

Brown, B., Yuniati, W., Ahmad, R., Soulsby, I., 2014a. Observations of natural recruitment and human attempts at mangrove rehabilitation after seismic (tsunami and earthquake) events in Simulue Island and Singkil Lagoon, Acheh, Indonesia. In: Santiago-Fandino, V., Kontar, Y.A., Kaneda, Y. (Eds.), Post-Tsunami Hazard Reconstruction and Restoration. Springer, pp. 311–327.

Brown, B., Fadilla, R., Nurdin, Y., Soulsby, I., Ahmad, R., 2014b. Community based ecological mangrove rehabilitation (CBEMR) in Indonesia. Surveys and Perspectives Integrating Environment and Society 7 (2), 53–64.

Cintron, G., Lugo, A.E., Pool, D.J., Morris, G., 1978. Mangroves and arid environments in Puerto Rico and adjacent islands. Biotropica 10, 110–121.

Cintron, G., Novelli, Y., 1984. Methods for studying mangrove structure. In: Snedaker, S.C., Snedaker, J.G. (Eds.), The Mangrove Ecosystem: Research Methods. Monographs in Oceanographic Methodology, vol. 8. UNESCO, Paris, pp. 91–113.

Crona, B.I., Rönnbäck, P., 2005. Use of replanted mangroves as nursery grounds by shrimp communities in Gazi Bay, Kenya. Estuarine, Coastal and Shelf Science 65, 535–544.

Crona, B.I., Rönnbäck, P., 2007. Community structure and temporal variability of juvenile fish assemblages in natural and replanted mangroves, *Sonneratia alba* Sm., of Gazi Bay, Kenya. Estuarine, Coastal and Shelf Science 74, 44–52.

Dale, P.E.R., Knight, J.M., Dwyer, P.G., 2014. Mangrove rehabilitation: a review focusing on ecological and institutional issues. Wetlands Ecology and Management 22, 587–604.

Elliott, M., Mander, L., Mazik, K., Simenstad, C., Valesini, F., Whitfield, A., Wolanski, E., 2016. Ecoengineering with ecohydrology: success and failures in estuarine restoration. Estuarine, Coastal and Shelf Science 176, 12–35.

Erftemeijer, P.L.A., Lewis, R.R., 2000. Planting mangroves on intertidal mudflats: habitat restoration or habitat conversion?. In: Proceedings of the ECOTONE VIII Seminar Enhancing Coastal Ecosystems Restoration for the 21st Century, Ranong, Thailand, 23-28 May 1999. Royal Forest Department of Thailand, Bangkok, Thailand, pp. 156–165.

FAO, 2007. The World's Mangroves 1980-2005: A Thematic Study Prepared in the Framework of the Global Forest Resources Assessment 2005. FAO Forestry Paper 153, Rome.

Field, C.D. (Ed.), 1996. Restoration of Mangrove Ecosystems. International Society for Mangrove Ecosystems. Japan, Okinawa.

Field, C.D., 1999. Rehabilitation of mangrove ecosystems: an overview. Marine Pollution Bulletin 37 (8–12), 383–392.

Furukawa, K.E., Wolanski, E., Mueller, H., 1997. Currents and sediment transport in mangrove forests. Estuarine, Coastal and Shelf Science 44, 301–310. Global Nature Fund. 2015.

Gensac, E., Gardel, A., Lesourd, S., Brutier, L., 2015. Morphodynamic evolution of an intertidal mudflat under the influence of Amazon sediment supply – Kourou mud bank, French Guiana, South America. Estuarine, Coastal and Shelf Science 158, 53–62.

Global Nature Fund, 2015. Mangrove Restoration Guide. Radolfzell, Germany, 58 p.

Giri, C., Zhu, Z., Tieszen, L.L., Sigh, A., Gillette, S., Kelmelis, J.A., 2008. Mangrove forest distribution and dynamics (1975-2005) of the tsunami-affected region of Asia. Journal of Biogeography 3, 519–528.

Giri, C., Ochieng, E., Tieszen, L.L., Zhu, Z., Singh, A., Loveland, T., Masek, J., Duke, N., 2011. Status and distribution of mangrove forests of the world using earth observation satellite data. Global Ecology and Biogeography 20, 154–159.

IUCN, WRI, 2014. A Guide to the Restoration Opportunities Assessment Methodology (ROAM): Assessing Forest Landscape Restoration Opportunities at the National or Sub-national Level. Working paper. Gland, Switzerland, 125 p.

Kairo, J.G., 2002. Regeneration status of mangrove forests in Mida Creek, Kenya: a compromised or secured future? Ambio 31 (7–8), 562–568.

Kjerfve, B., 1990. Manual for investigation of hydrological processes in mangrove ecosystems. In: UNESCO/UNDP Regional Project, Research and its Application to the Management of the Mangroves of Asia and the Pacific (RAS/86/120), 79 p.

Lewis, R.R., 1992. Coastal habitat restoration as a fishery management tool. In: Stroud, R.H. (Ed.), Stemming the Tide of Coastal Fish Habitat Loss. Proceedings of a Symposium on Conservation of Coastal Fish Habitat, Baltimore, Md., March 7-9, 1991. National Coalition for Marine Conservation, Inc., Savannah, GA, pp. 169–173.

Lewis, R.R., 2000. Ecologically based goal setting in mangrove forest and tidal marsh restoration in Florida. Ecological Engineering 15 (3–4), 191–198.

Lewis, R.R., 2005. Ecological engineering for successful management and restoration of mangrove forests. Ecological Engineering 24 (4 SI), 403–418.

Lewis, R.R., 2009. Methods and criteria for successful mangrove forest restoration. Chapter 28. In: Perillo, G.M.E., Wolanski, E., Cahoon, D.R., Brinson, M.M. (Eds.), Coastal Wetlands: An Integrated Ecosystem Approach. Elsevier Press, pp. 787–800.

Lewis, R.R., 2014. Mangrove forest restoration and the preservation of mangrove biodiversity. In: Bozzano, M., Jalonen, R., Evert, T., Boshier, D., Gallo, L., Cavers, S., Bordacs, S., Smith, P., Loo, J. (Eds.), Genetic Considerations in Ecosystem Restoration Using Native Tree Species. A Thematic Study for the State of the World's Forest Genetic Resources. United Nations Food and Agriculture Organization, Rome, Italy, pp. 195–200. I-xi + 271 p.

Lewis, R.R., Marshall, M.J., 1998. Principles of Successful Restoration of Shrimp Aquaculture Ponds Back to Mangrove Forests. P. 327 in World Aquaculture Society Book of Abstracts, Aquaculture '98, Las Vegas, Nevada (Abstract).

Lewis, R.R., Hodgson, A.B., Mauseth, G.S., 2005. Project facilitates the natural reseeding of mangrove forests (Florida). Ecological Restoration 23 (4), 276–277.

Lewis, R.R., Gilmore, R.G., 2007. Important considérations to achieve successful mangrove forest restoration with optimum fish habitat. Bulletin of Marine Science 80 (3), 823–837.

Lewis, R.R., Brown, B., 2014. Ecological Mangrove Rehabilitation – a Field Manual for Practitioners. Version 3. Mangrove Action Project Indonesia, Blue Forests. Canadian International Development Agency, and OXFAM, 275 p. www.mangroverestoration.com.

Lewis, R.R., Milbrandt, E.C., Brown, B., Krauss, K.W., Rovai, A.S., Beever III, J.W., Flynn, L.L., 2016. Stress in mangrove forests: early detection and preemptive rehabilitation are essential for future successful worldwide mangrove forest management. Marine Pollution Bulletin 109, 764–771.

Macintosh, D.J., Ashton, E.C., Havanon, S., 2002. Mangrove rehabilitation and intertidal biodiversity: a study in the Ranong mangrove ecosystem, Thailand. Estuarine, Coastal and Shelf Science 55, 331–345.

Mazda, Y., Wolanski, E., Ridd, P.V., 2007. The Role of Physical Processes in Mangrove Environments. Terrapub, Tokyo.

Mazda, Y., Wolanski, E., 2009. Hydrodynamics and modeling of water flow in mangrove areas. Chapter 8. In: Perillo, G.M.E., Wolanski, E., Cahoon, D.R., Brinson, M.M. (Eds.), Coastal Wetlands: An Integrated Ecosystem Approach. Elsevier Press, pp. 231–262.

McKee, K.L., 1993. Soil physiochemical patterns and mangrove species distribution – reciprocal effects? Journal of Ecology 81 (3), 477–487.

McKee, K.L., Faulkner, P.L., 2000. Restoration of biogeochemical function in mangrove forests. Restoration Ecology 8 (3), 247–259.

Medina, E., Fonseca, H., Barboza, F., Francisco, M., 2001. Natural and man-induced changes in a tidal channel mangroves system under tropical semiarid climate at the entrance to the Maracaibo lake (Western Venezuela). Wetlands Ecology and Management 9 (3), 233–243.

Nam, V.N., Sasmito, S.D., Murdiyarso, D., Purbopuspito, J., MacKenzie, R.A., 2016. Carbon stocks in artificially and naturally regenerated mangrove ecosystems in the Mekong Delta. Wetlands Ecology and Management. https://doi.org/10.1007/s11273-015-9479-2.

Perillo, G.M.E., 2009. Tidal courses: classification, origin and functionality. Chapter 6. In: Perillo, G.M.E., Wolanski, E., Cahoon, D.R., Brinson, M.M. (Eds.), Coastal Wetlands: An Integrated Ecosystem Approach. Elsevier Press, pp. 185–210.

Perdomo, L., Ensminger, I., Espinosa, L.F., Elster, C., Wallner-Kersanach, M., Schnetter, M.-L., 1998. The mangrove ecosystem of Cienaga Grande de Santa Marta (Colombia): observations on regeneration and trace metals in sediment. Marine Pollution Bulletin 37 (8–12), 393–403.

Primavera, J.H., Esteban, J.M.A., 2008. A review of mangrove rehabilitation in the Philippines: successes and failures and future prospects. Wetlands Ecology and Management 16 (5), 345–358.

Primavera, J.H., Rollon, R.N., Samson, M.S., 2011. The pressing challenges of mangrove rehabilitation: pond reversion and coastal protection. In: Chicharo, L., Zalewski, M. (Eds.), Chapter 10 in Volume 10: Ecohydrology and Restoration, in the Treatise on Estuarine and Coastal Science. Elsevier, Amsterdam, pp. 217–244. Series eds., E. Wolanski, and D. McLusky.

Primavera, J.H., Yap, W.G., Savaris, J.P., Loma, R.A., Moscoso, A.D.E., Coching, J.D., Montilijao, C.L., Poignan, R.P., Tayo, I.D., 2014. Manual on Mangrove Reversion of Abandoned and Illegal Brackishwater Fishponds – Mangrove Manual Series No. 2. Zoological Society of London, London, UK xii +108 p.

Proffit, C.E., Devlin, D.J., 2005. Long-term growth and succession in restored and natural mangrove forests in southwestern Florida. Wetlands Ecology and Management 13, 531–551.

Rey, J.R., Carlson, D.B., Brockmeyer Jr., R.E., 2012. Coastal wetland management in Florida: environmental concerns and human health. Wetlands Ecology and Management 20 (3), 197–211.

Richards, D.R., Friess, D.A., 2016. Rates and drivers of mangrove deforestation in Southeast Asia, 2000-2012. Proceedings of the National Academy of Sciences 113, 344–349.

Rivera-Monroy, V.H., Twilley, R.R., Mancera, E., Alcantara, A., Castaneda-Moya, E., Casas Monroy, O., Reyes, P., Restrepo, J., Perdomo, L., Campos, E., Cotes, G., Viloria, E., 2006. Aventuras y desventuras en Macondo: rehabilitacion de la Cienega Grande de Santa Marta, Colombia. Ecotropicos 19, 72–93.

Rönnbäck, P., Troell, M., Kautsky, N., Primavera, J.H., 1999. Distribution patterns of shrimps and fish among *Avicennia* and *Rhizophora* microhabitats in the Pagbilao mangroves, Philippines. Estuarine, Coastal and Shelf Science 48, 223–234.

Saenger, P., 2002. Mangrove Ecology, Silviculture and Conservation. Kluwer Academic Publishers, Dordrecht, The Netherlands, 360 p.

Saenger, P., Siddiqi., N.A., 1993. Land from the seas: the mangrove afforestation program of Bangladesh. Ocean and Coastal Management 20, 23–39.

Sasekumar, A., Chong, V.C., 1998. Faunal diversity in Malaysian mangroves. Global Ecology and Biogeography Letters 7, 57–60.

Salmo III, S.G., Lovelock, C., Duke, N.C., 2013. Vegetation and soil characteristics as indicators of restoration trajectories in restored mangroves. Hydrobiologica 720, 1–18.

Samson, M.S., Rollon, R.N., 2008. Growth performance of planted red mangroves in the Philippines: revisiting forest management strategies. Ambio 37 (4), 234–240.

Selkoe, K.A., Blenchner, T., Caldwell, M.R., Crowder, L.B., Erickson, A.L., Essgton, T.E., Estes, J.A., Fujita, R.M., Halpern, B.S., Hunsicker, M.E., Kappel, C.V., Kelly, R.P., Kittinger, J.N., Levin, P.S., Lynham, J.M., Mach, M.E., Martone, R.G., Mease, L.A., Salomon, A.K., Samhouri, J.F., Scarborough, C., Stier, A.C., White, C., Zedler, J., 2015. Principles for managing marine ecosystems prone to tipping points. Ecosystem Health and Systain 1 (5), 17.

Snedaker, S.C., Snedaker, J.G., 1984. The mangrove ecosystem: research methods. In: Monographs on Oceanographic Methodology, vol. 8. UNESCO, Paris.

Snedaker, S.C., 1993. Impact on mangroves. In: Maul, G.A. (Ed.), Climatic Changes in the Intra-American Seas: Implications of Future Climate Change on the Ecosystems and Socio-Economic Structure of the Marine and Coastal Regimes of the Caribbean Sea, Gulf of Mexico, Bahamas and N.E. Coast of S. America. Edward Arnold, London, pp. 282–305.

Spalding, M.D., 1997. The global distribution and status of mangrove ecosystems. Intercoast Network Newsletter Special Edition #1 20–21.

Spalding, M.D., Kainuma, M., Collins, L., 2010. World Atlas of Mangroves. ISME, FAO. Earrthscan, 336 p.

Stevenson, N.J., Lewis, R.R., Burbridge, P.R., 1999. Disused shrimp ponds and mangrove rehabilitation. In: Streever, W.J. (Ed.), An International Perspective on Wetland Rehabilitation. Kluwer Academic Publishers, the Netherlands, pp. 277–297.

Tomlinson, P.B., 1986. The Botany of Mangroves. Cambridge Tropical Biology Series. Cambridge University Press, New York.

Turner, R.E., Lewis, R.R., 1997. Hydrologic restoration of coastal wetlands. Wetlands Ecology and Management 4 (2), 65–72.

Twilley, R.R., Rivera-Monroy, V.H., Chen, R., Botero, L., 1998. Adapting an ecological mangrove model to simulate trajectories in restoration ecology. Marine Pollution Bulletin 37 (8–12), 404–419.

Walton, M.E.M., Le Vay, L., Lebata, J.H., Binas, J., Primavera, J.H., 2007. Assessment of the effectiveness of mangrove rehabilitation using exploited and non-exploited indicator species. Biological Conservation.

Wolanski, E., Brinson, M.M., Cahoon, D.R., Perillo, G.M.E., 2009. Coastal wetlands: a synthesis. Chapter 1. In: Perillo, G.M.E., Wolanski, E., Cahoon, D.R., Brinson, M.M. (Eds.), Coastal Wetlands: An Integrated Ecosystem Approach. Elsevier Press, pp. 1–62.

Wolanski, E., Mazda, Y., Ridd, P., 1992. Mangrove hydrodynamics. In: Robertson, A.I., Alongi, D.M. (Eds.), Tropical Mangrove Ecosystems. American Geophysical Union, Washington, DC, pp. 436–462.

Further Reading

Alongi, D.M., 2009. Paradigm shifts in mangrove biology. Chapter 22. In: Perillo, G.M.E., Wolanski, E., Cahoon, D.R., Brinson, M.M. (Eds.), Coastal Wetlands: An Integrated Ecosystem Approach. Elsevier Press, pp. 615–640.

Friess, D.A., Krauss, K.W., Horstman, E.M., Balke, T., Bouma, T.J., Galli, D., Webb, E.L., 2012. Are all intertidal wetlands created equal? Bottlenecks, thresholds and knowledge gaps to mangrove and saltmarsh ecosystems. Biological Reviews 87, 346–366.

Krauss, K.W., McKee, K.L., Lovelock, C.E., Cahoon, D.R., Saintilan, N., Reef, R., Chen, L., 2014. How mangrove forests adjust to rising sea level. New Phytologist 202, 19–34.

Lewis, R.R., 2010. Mangrove field of dreams: if we build it, will they come? SWS research Brief No. 2009-005. 4 p. Wetland Science and Practice 27 (1), 15–18.

Matsui, N., Suekuni, J., Nogami, M., Havanond, S., Salikui, P., 2010. Mangrove rehabilitation dynamics and soil organic carbon changes as a result of full hydraulic restoration and regrading of a previously intensively managed shrimp pond. Wetlands Ecology and Management 18, 233–242.

López-Portillo, J., Lewis, R., Saenger, P., Rovai, A., Koedam, N., Dahdouh-Guebas, F., Agraz, M., Rivera-Monroy, V., 2017. Mangrove Forest Restoration and Rehabilitation. In: Rivera-Monroy, V.H., Lee, S.Y., Kristensen, E., Twilley, R.W. (Eds.), Mangrove Ecosystems: A Global Biogeographic Perspective: Structure, Function, and Services, pp. 301–345. https://doi.org/10.1007/978-3-319-62206-4_10.

Shafer, D.J., Roberts, T.H., 2008. Long-term development of tidal mitigation wetlands in Florida. Wetlands Ecology and Management 16, 23–31.

Stanley, O.D., Lewis III, R.R., 2009. Strategies for mangrove rehabilitation in an eroded coastline of Selangor, peninsular Malaysia. Journal of Coastal Development 12 (3), 144–156.

Wilkie, M.L., Fortuna, S., 2003. Status and Trends in Mangrove Area Extent Worldwide. Forest Resources Assessment Working Paper 63. Forestry Department, Food and Agriculture Organization of the United Nations, Rome.

Woodroffe, C.D., Davies, G., 2009. The morphology and development of tropical coastal wetlands. Chapter 2. In: Perillo, G.M.E., Wolanski, E., Cahoon, D.R., Brinson, M.M. (Eds.), Coastal Wetlands: An Integrated Ecosystem Approach. Elsevier Press, pp. 65–88.

CHAPTER

25

Evaluating Restored Tidal Freshwater Wetlands

Andrew H. Baldwin[1], *Richard S. Hammerschlag*[2], *Donald R. Cahoon*[2]

[1]Department of Environmental Science and Technology, University of Maryland, College Park, MD, United States; [2]United States Geological Survey, Patuxent Wildlife Research Center, Laurel, MD, United States

1. INTRODUCTION

Tidal freshwater wetlands are recognized as highly productive coastal wetlands that support diverse assemblages of plants and animals and complex biogeochemical cycles (in this book, see Chapter 18 by Whigham et al. and Chapter 19 by Megonigal and Neubauer). Many tidal freshwater wetlands and their associated ecosystem services have been damaged or destroyed by urbanization, agriculture, and other human activities (Baldwin, 2004; Barendregt et al., 2006). Increasing recognition of the value of remaining wetlands and environmental regulations requiring wetland mitigation (i.e., enhancement, creation, or restoration of wetlands to compensate for wetland losses; Kentula, 2000) has driven the restoration of all types of wetlands, including tidal freshwater wetlands. These restoration projects have been increasingly studied by restoration ecologists, with the overarching goal of improving restoration approaches.

In this chapter, we first review characteristics of restored tidal freshwater wetlands in North America and Eurasia, where most studies have been done, including their distribution, general construction methods, and motivating factors for restoration (Section 2). Then we present criteria for evaluating tidal freshwater wetland restoration projects (Section 3). Next we describe a case study of restored tidal freshwater wetlands in the Anacostia River watershed in Washington, DC, USA (Section 4). Finally, we provide conclusions and recommendations to increase the successful restoration of tidal freshwater wetlands (Section 5).

Coastal Wetlands
https://doi.org/10.1016/B978-0-444-63893-9.00025-3

2. MOTIVATION AND CONSTRUCTION METHODS FOR RESTORATION

Tidal freshwater wetlands have been restored to mitigate for wetland losses due to development (roads, buildings), stream channelization, and other hydrologic alterations and to protect shorelines, reduce flooding, control invasive species, and restore the ecosystem functions of converted wetlands (Table 25.1). In the US Pacific Northwest and the Fraser River in British Columbia, Canada, restoring habitat for salmon-dominated fisheries and wildlife has been an important motivation for the restoration of tidal freshwater wetlands (e.g., Simenstad and Thom, 1996). Extensive restoration of tidal freshwater wetlands (and more saline wetland types) in the Mississippi River delta plain in Louisiana has aimed to (1) stabilize or reverse loss of wetlands due to erosion and increasing relative sea level; (2) increase the ratio of land to open water; (3) reduce saltwater intrusion; and (4) promote the development of natural delta features (USACE, 2017). In the more densely populated mid- and southeast Atlantic US regions, restoration has been frequently implemented to mitigate for wetlands lost to urban development, including bridges and roads, or to create mitigation "banks" from which developers can draw credits to offset future losses. On the European Atlantic coast, formerly drained, diked, or "poldered" tidal freshwater wetlands have been reconverted to wetlands to protect shorelines against storms and sea level rise and to create habitat for biodiversity support (Fig. 25.1) (Beauchard et al., 2011).

Restoration of many tidal freshwater wetlands, particularly along the East Coast of the United States, involves excavation of upland soils or placement of dredged sediment in open water areas to create a substrate suitable for plant growth at an intertidal elevation similar to those of naturally or previously occurring tidal freshwater wetlands (Table 25.1; Baldwin, 2009). In locations where former wetlands were surrounded by dikes, levees, or embankments to dewater them for agriculture (forming areas termed "polders" in Europe; e.g., US Pacific and European Atlantic coasts), breaching of embankments is a common method for restoration and that sometimes occurs inadvertently when structures fail (Fig. 25.2) (Hester et al., 2016). In Europe, this approach has been termed "managed retreat," "managed realignment," or "de-embankment." An emerging approach at some locations has been to restore an adequate tidal regime using "controlled reduced tide" structures (Fig. 25.3), which include a high inlet and low outlet at the connection to the estuary (Beauchard et al., 2011; Vandenbruwaene et al., 2011). In the Mississippi River delta, restoration techniques that can be used over vast areas are necessary, including diverting sediment-laden river water into deteriorating areas to increase elevation of subsided former wetlands, restoring historical hydrologic connections to tidal channels, and pumping in dredged material to increase elevation (USACE, 2017).

Most tidal freshwater wetlands projects have focused on creating wetlands dominated by herbaceous plants ("marshes") (Fig. 25.4). However, the McIntyre Tract associated with the Cape Fear River in North Carolina has included planted trees and shrubs, as well as herbaceous plants with the goal of restoring tidal freshwater cypress-gum swamp and marsh/scrub-shrub habitat (Land Management Group, 2004). More recently, interest has increased in restoring tidal freshwater forested wetlands on the US Pacific Coast, an ecosystem type that has been little studied where large woody debris has considerable influence on hydrology and wetland development (Diefenderfer and Montgomery, 2009). In practice, if

TABLE 25.1 Examples of Motivating Factors for Restoration and Construction Methods for Tidal Freshwater Wetlands Restoration Projects

Region	Motivating Factors for Restoration	Construction Methods	References
USA Northeast Atlantic	Remove nonnative plant species	Herbicide and cutting	Findlay et al. (2003), Meyerson et al. (1999)
USA Mid-Atlantic	Mitigate road, airport, bridge construction losses; channel maintenance and dredge material disposal; ecosystem and habitat restoration; restoration of submersed aquatic vegetation	Excavation of uplands; raising elevation with dredged sediment; cutting tidal channels; control invasive plants; reestablish tidal exchange with channels or automated tide gates; planting, seeding, transplanting of submersed plants	Kaminsky and Scelsi (1986), Bartoldus and Heliotis (1989), Bartoldus (1990), Bowers (1995), Syphax and Hammerschlag (1995), Marble and Company (1998), Baldwin and DeRico (1999), Gannett Fleming (2001), Quigley (2001), Verhoeven et al. (2001), Neff (2002), Leck (2003), Neff and Baldwin (2005), Hammerschlag et al. (2006), Neff et al. (2009), Moore et al. (2010), Prasse et al. (2015), DNREC (2016)
USA Southeast Atlantic	Mitigate road construction impacts; restoration for mitigation banking to offset losses	Grading, recontouring, cutting tidal channels, breaching berm along river, removing pump structures	Land Management Group (2004), Hopfensperger et al. (2014)
USA Pacific: Sacramento –San Joaquin Delta, California	Ecosystem restoration	Breach levees surrounding delta islands (similar to "Managed Realignment" used in Europe); excavation to reduce elevation; unrepairable levee failure	Simenstad et al. (2000), Orr et al. (2003), Stillwater Environmental Services (2003), Phillip (2005), Lehman et al. (2010), Whitley and Bollens (2014), Sloey et al. (2015), Hester et al. (2016)
USA Pacific Northwest	Mitigation for development; ecosystem restoration; fisheries habitat restoration	Levee breaching; dike removal; culvert reconnection to estuary; excavation of fill material	Simenstad and Thom (1996), Gray et al. (2002), Tanner et al. (2002), Diefenderfer and Montgomery (2009), David et al. (2014)
USA Gulf of Mexico: Louisiana	Reduce marsh erosion and inundation; reduce salinity; increase land–water ratio; promote natural delta development; fish and wildlife habitat	Hydrologic restoration; shoreline protection; freshwater diversion; dredge material placement; marsh "terraces" to promote sedimentation and reduce erosion of restored wetlands	Lane et al. (1999), Sullivan (2015), USACE (2017)
Canada: Fraser River, British Columbia	Ecosystem restoration for fisheries	Dredge spoil placement; transplanting of vegetation plugs to barren sites	Kistritz (1996), Levings and Nishimura (1996, 1997), Grout et al. (1997)

Continued

TABLE 25.1 Examples of Motivating Factors for Restoration and Construction Methods for Tidal Freshwater Wetlands Restoration Projects—cont'd

Region	Motivating Factors for Restoration	Construction Methods	References
Europe: Various estuaries	Protect shorelines against storms and sea level rise; increase area of coastal wetlands; moderate the "coastal squeeze"	Managed realignment: restoring tidal hydrology to formerly reclaimed land by breaching or removing dikes; "managed retreat; "depoldering"	French (2006)
Europe: Scheldt estuary, Belgium	Flood protection and ecosystem restoration	Controlled reduced tide system to create suitable tidal regimes in managed realignment projects; high inlet and low outlet connection to tidal estuary	Cox et al. (2006), Jacobs et al. (2009), Beauchard et al. (2011, 2013b), Vandenbruwaene et al. (2011, 2012)
China; Yellow River Delta	Restore native plants (including *Phragmites australis*), wildlife, and other wetland functions	Diversion of fresh river water into salinized wetlands to reverse saltwater intrusion cause by dam construction	Wang et al. (2012)

site conditions allow, it may be beneficial from a habitat complexity perspective to create a mosaic of multiple habitats within tidal freshwater wetland restoration projects, including submersed, floating-leaved, and emergent herbaceous vegetation, shrub- or tree-dominated stands, open water, and unvegetated mudflat areas.

Information is lacking on tidal freshwater wetland restoration from many parts of the world, although it appears to be increasing. For example, China began restoring wetlands in 1990 and is dedicating over $100 billion (USD) for coastal and inland wetland restoration by 2030 (Zhao et al., 2016). Some of these are tidal freshwater wetlands (Wang et al., 2012).

3. EVALUATION CRITERIA FOR RESTORED TIDAL FRESHWATER WETLANDS

Despite the difficulties of defining success, several reviews have indicated that wetland restoration projects often do not attain conditions that can be deemed legally or ecologically successful (Mitsch and Wilson, 1996; Zedler and Callaway, 1999; Craft et al., 2003; Baldwin et al., 2009; Suding, 2011). Limited restoration success has been documented for some restored tidal freshwater wetlands. For example, Simenstad and Thom (1996) found that only a few of 16 ecosystem attributes monitored in a restored tidal freshwater wetland were on a functional trajectory approaching that of reference wetlands. Similarly, ecosystem monitoring at restored tidal freshwater wetlands in the US Pacific Northwest, Fraser River in British Columbia, Louisiana coastal zone, US mid-Atlantic region, and the Yellow River Delta in China indicates that persistent differences exist between restored sites and reference wetlands or design goals (Bartoldus, 1990; Levings and Nishimura, 1997; Quigley, 2001;

FIGURE 25.1 Hanöfer Sand, a tidal freshwater restoration site on the Elbe River, Germany, constructed in 2004 by excavating and grading soils and de-embanking to reconnect the area to the Elbe (Kai Jensen, pers. comm.). (A) Overview taken from the adjacent sea wall in spring 2008. Large tidal flats dominate the area of the restoration site. Tidal freshwater marshes have been establishing close to the shore. A band-like stand of willows is found directly in front of the dike in the restored area. (B) In accordance with small differences in elevation, a typical zonation of "pioneer," low, and high tidal freshwater marshes has established at the restoration site. At the highest elevation, a small "island" of willows developed, visible near the center. *Photos by Claudia Mählmann.*

Gray et al., 2002; Neff, 2002; Tanner et al., 2002; Adams and Williams, 2004; Wang et al., 2012). Similarly, tidal freshwater wetlands restored for mitigation purposes met some criteria for legal success (vegetation coverage or survival), but not others (hydrology) (Quigley, 2001; Land Management Group, 2004). In areas where natural sites are degraded, restoration may greatly enhance ecosystem structure and function, becoming more similar to natural wetlands (Beauchard et al., 2013a,b). Taken together, however, these findings suggest that

FIGURE 25.2 Two views of tidal freshwater wetlands at Liberty Island in the Sacramento–San Joaquin Bay Delta, California, USA. The site is a historic tidal freshwater wetland that was diked and drained for agriculture, resulting in oxidation of soil organic matter and land subsidence. The dike failed in 1997 during a high-water event and could not be repaired, reconnecting the area to tidal hydrology. This fostered colonization by wetland plants, primarily "tules" (mostly *Schoenoplectus californicus*, with lesser amounts of *Schoenoplectus acutus*). Lateral expansion of plants is mostly vegetative and on the order of 1 m per year. Areas of highly compacted (high bulk density) legacy agricultural soil tend to have lower rates, and high energy (high wind/wave exposure) shorelines have the lowest rates. Percentage of time the substrate is flooded is also an important driver, with deeper marsh edges tending to have lower rates than less flooded areas. Fish species colonizing the site included the endangered species delta smelt (*Hypomesus transpacificus*). Additional details can be found in Lehman et al. (2010), Whitley and Bollens (2014), Sloey et al. (2015), and Hester et al. (2016). *Photos by M.W. Hester.*

there is considerable room for improvement of techniques and approaches for restoration of tidal freshwater wetlands.

Criteria for evaluating restoration or mitigation success have been developed or applied to many types of wetlands, including coastal wetlands (Weinstein et al., 1997; Zhao et al., 2016), nontidal wetlands (Wilson and Mitsch, 1996; Cole and Shafer, 2002), and mitigation banks (Spieles, 2005). Generally, evaluation efforts have focused on soils, hydrology, geomorphology, vegetation development, and wildlife usage. Some studies in salt marshes (Craft et al., 2003) and tidal freshwater wetlands (Simenstad and Thom, 1996; Baldwin and DeRico, 1999) have gone farther, adding quantitative studies of algae, microbial communities, biogeochemical

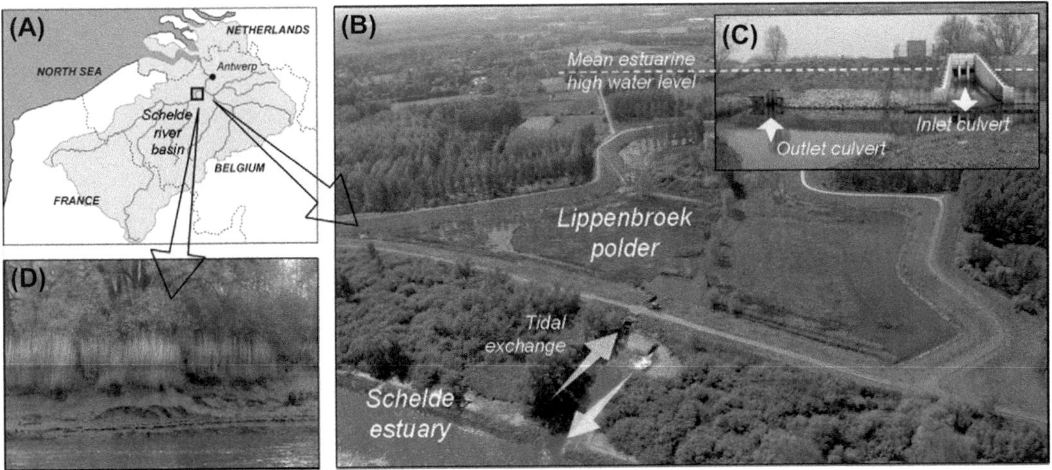

FIGURE 25.3 The controlled reduced tide system on the Scheldt River in Belgium. (A) Location map. (B) The polders, inlet/outlet structures, estuary, and vegetated wetland inside the polders. (C) Close-up of the water exchange system from inside the polder. (D) Typical intertidal gradient in the Scheldt estuary, where erosion prevents vegetation establishment at lower elevations. *Reproduced from Beauchard, O., Jacobs, S., Cox, T.J.S., et al., 2011. A new technique for tidal habitat restoration: evaluation of its hydrological potentials. Ecological Engineering 37, 1849–1858. Used with permission.*

processes, seed banks, invertebrates, fish, or birds as indicators of ecosystem function relative to reference sites. Based on these studies and literature reviewed in Section 2 of this chapter, it is clear that assessments of soil, hydrology, vegetation, and fauna are accepted measures of restoration success used in many types of wetlands, although they are not consistently measured across sites. Here we present success criteria specific to restored tidal freshwater wetlands that address these and related ecosystem attributes (Table 25.2).

As is done for evaluations of other types of restored wetlands, the success criteria we present for restored tidal freshwater wetlands are primarily based on comparisons with reference sites (Craft, 2016). Because success criteria are thus dependent on the condition of the reference site, it is critical to choose reference sites that experience environmental conditions similar to the restored site (Ehrenfeld, 2000a,b). Watershed urbanization or agricultural development may impose landscape constraints on ecosystem components that cannot be overcome through restoration (e.g., Boudell et al., 2015). For this reason, it is unrealistic to expect that restored tidal freshwater wetlands in urbanized landscapes with, for example, high sediment loads, flashy hydrology, fragmented wetlands, and abundant nonnative species will closely resemble those of nonurban landscapes (Baldwin, 2004). The criteria presented in Table 25.2 reflect the need to apply success criteria that can realistically be achieved given environmental constraints imposed by the landscape or watershed surrounding the restored site.

In addition to watershed condition, hydrologic attributes such as tidal range and connectivity to rivers should be similar at the reference and restored sites. If suitable reference sites are not available or evaluated, restored wetlands can be compared with accepted standards

FIGURE 25.4 The Duck Island restored tidal freshwater wetland, Delaware River, Hamilton Township, New Jersey, USA. The 39-ha project site was constructed in 1994 to mitigate highway construction impacts and includes a mix of created tidal and nontidal wetlands and channels and preserved upland areas. (A) Dense and species-rich tidal freshwater herbaceous vegetation in 2001. Purple loosestrife (*Lythrum salicaria*) is visible in clumps in the foreground. (B) Typical view of the wetland in July 2017. Over 2 decades, *Phragmites australis* colonized and formed monocultures across the majority of the site (background), greatly reducing species richness. Tidal channels were planned to be deep enough to be free of emergent vegetation, but sediment eroded from higher elevations increased channel bed elevation and allowed colonization by *Nuphar lutea* (foreground). Detailed descriptions are included in Leck (2003), Leck and Leck (2005), Leck (2012), and Elsey-Quirk and Leck (2015). *(A) Photo by A.H. Baldwin. (B) Photo by M.A. Leck.*

of wetland function (Wilson and Mitsch, 1996) or with literature values for naturally occurring wetlands in similar watersheds.

Some criteria for restoration success may not require comparison with reference sites. Often these are compliance success criteria specified in permit requirements for mitigation projects implemented to replace wetlands lost to development. These may include goals specifying a certain percentage of vegetation cover, a particular hydrologic regime, a preponderance of hydrophytic plant species, use of restored areas by fish and wildlife, or a maximum threshold of nonnative species.

TABLE 25.2 Ecological Criteria for Evaluation of Restored Tidal Freshwater Wetlands

Ecosystem Attribute	Measurements	Evaluation Criteria	Comments
Hydrology	Depth, duration, or percentage of time flooded	Elevation of vegetated high marsh should lie at approximately Mean Sea Level or up to Mean High Water (MHW); vegetated low marsh should lie approximately between Mean Sea Level and Mean Low Water (MLW) (Odum et al., 1984). High marsh should be flooded up to 30 cm depth for 0–4 h per tidal cycle; low marsh should be flooded up to 100 cm depth for 9–12 h per cycle (Simpson et al., 1983; Mitsch and Gosselink, 2000); similar to reference sites.	Differences in elevation of only a few cm (e.g., 3–10) can strongly influence the ability of plants to colonize via seedling recruitment and growth of planted vegetation. Surface elevation relative to Mean Sea Level determines wetland type, for example high marsh vs. low marsh.
Geomorphology	Accretion	Spatially variable; vertical accretion of 5–10 mm year^{-1} (Neubauer et al., 2002), average 6–7 mm year^{-1} (Craft, 2007); similar to reference sites	Restored sites constructed from coarse material (sand and gravel) may not accrete organic matter if elevation is sufficiently high to allow oxidation.
	Elevation change	Little or no net elevation change relative to sea level; similar to reference sites	Naturally occurring tidal freshwater wetlands vary in their ability to accrete vertically at a sufficient elevation to keep pace with rising relative sea level (Craft, 2012; Beckett et al., 2016), and belowground processes of root zone expansion and contraction are particularly important in forested tidal freshwater wetlands, which have hummock-hollow microtopography (Stagg et al., 2016). Excessive erosion, subsidence due to dewatering, or compaction of sediments in restored tidal freshwater wetlands may lead to vegetation species change or dieback; increases in relative sea level may promote vertical increases in elevation by increasing mineral sediment deposition and accumulation of organic matter.
	Channel and pool development	Evidence of small channel formation without excessive erosion or scour of large channels	Large channels cut into restored sites may naturally fill with sediment as small channels form naturally in the wetland (Simenstad and Thom, 1996; Diefenderfer and Montgomery, 2009).
	Topography	Variable within the small range that supports desired vegetation (MLW–MHW)	Naturally occurring tidal freshwater wetlands may exhibit hummocks, particularly in forested systems (15–20 cm height; Baldwin, pers. obs.; Duberstein and Conner, 2009) due to vegetation clumps or fallen trunks, as well as lower areas near channels or in small ponds, but are otherwise very flat.

Continued

TABLE 25.2 Ecological Criteria for Evaluation of Restored Tidal Freshwater Wetlands—cont'd

Ecosystem Attribute	Measurements	Evaluation Criteria	Comments
Soil	Organic matter	Evidence of organic matter accumulation in the surface horizon or streaking into subsurface horizons; average organic carbon concentration for US tidal freshwater marshes is 13%–22% (Craft, 2007)	Sites created by placement of mineral soil or excavation of upland soil to a hydrologically correct elevation may accumulate little organic matter due to oxidation of any material that accumulates vertically; development of organic matter content to the level of nonurban, naturally occurring tidal freshwater wetlands (e.g., 20%–70%, Odum, 1988) may develop extremely slowly (Zedler and Callaway, 1999; Verhoeven et al., 2001; Craft et al., 2003).
		Similar to reference sites	Urban reference sites may have low organic matter content compared with nonurban sites.
	Bulk density	Similar to reference sites	Normally inversely related to organic matter and so tends to be higher in restored than natural tidal freshwater wetlands (e.g., Prasse et al., 2015). Vegetation establishment may reduce bulk density in restored tidal freshwater wetlands (Hester et al., 2016).
	Redox status: redox potential, IRIS tubes, reduced iron test	Similar to reference sites	The degree of soil oxidation and reduction is related to soil hydrology, organic carbon availability, microbial communities, and vegetation, among other variables, and can vary even within a single restored tidal freshwater wetland site (Hester et al., 2016).
	Nutrients, metals, organic contaminants	Similar to reference sites; average nutrient concentrations for US tidal freshwater marshes are 0.9%–1.6% N and 0.9–1.6 mg g −1 P (Craft, 2007)	Concentrations of heavy metals and organic contaminants in dredge material sources should be determined prior to restoration. However, it is unrealistic to try to reduce nutrients or toxins to levels below those of reference sites that experience the same watershed conditions. Metals may accumulate more in woody than in herbaceous vegetation, but overall plants contribute little to phytoremediation (Teuchies et al., 2013).
	Texture	Evidence of surface accumulation of materials of similar texture to reference sites; average bulk density for US tidal freshwater marshes is 0.1–0.3 g cm^{-3} (Craft, 2007)	Sites restored by placing river sediment or excavation of uplands will in general have coarser soil texture (e.g., sand and gravel) than reference sites (silts and clays) (Zedler, 2001).
Salinity	Salinity (parts per 1000, ppt)	Average salinity $\leq$ 0.5 ppt; pulses up to 5 ppt or higher may occur during drought conditions	Saltwater intrusion events that occur only during exceptionally dry years may be important in maintaining plant diversity and may alter fish and invertebrate communities (Odum et al., 1984; Odum, 1988; Baldwin, 2007).

TABLE 25.2 Ecological Criteria for Evaluation of Restored Tidal Freshwater Wetlands—cont'd

Ecosystem Attribute	Measurements	Evaluation Criteria	Comments
Microbial communities and functions	DNA extraction, sequencing	Similar to reference sites	Microbial communities can differ between restored and natural tidal freshwater wetlands, regardless of plant species composition (Prasse et al., 2015). Saltwater intrusion brings in seawater, which contains sulfate. This may shift metabolism toward sulfate reduction and alter decomposition rates (Hopfensperger et al., 2014).
Vegetation	Species composition	The list of perennial species should be similar to those of reference sites, but differences in relative abundance should be accepted	Species such as cattail (*Typha* spp.) are adapted to rapid colonization of exposed, moist substrate such as that created in restored tidal freshwater wetlands; a high abundance of these native highly productive species should be accepted as a natural result of their biology and the environmental conditions created by restoring tidal freshwater wetlands. Species composition of annual species is likely to differ from reference sites because of a lack of seeds of some annual species to colonize; seeding may be required to introduce these species.
		Annual species should comprise 20%–50% of species (Leck et al., 2009), and a peak annual:perennial biomass ratio of 1–5 (Whigham and Simpson, 1992)	Annual species are a key characteristic of many naturally occurring tidal freshwater wetlands (Simpson et al., 1983; Odum et al., 1984). Desired annual species may be slow to colonize restored wetlands sites (Neff and Baldwin, 2005).
	Species richness	Similar to reference sites	Additional plantings or seeding may be necessary if few propagules are present in waterways; nonnative invasive or highly productive native species may contribute substantially to richness in urban areas (Neff, 2002; Rusello, 2006).
	Biomass or total plant cover	Similar to reference sites	Standing biomass and total plant cover are indices of primary production. Belowground biomass may be indicative of adverse physical structure or biochemistry of soils for plant growth. Restoration of submersed aquatic vegetation communities in tidal freshwater areas may require exclosures to prevent transplants from herbivory (Moore et al., 2010). In emergent wetlands, transplanted adults may outperform rhizome plantings and expansion may be altered by hydrology and soil compaction (Sloey et al., 2015).
	Nonnative species abundance	Similar to or less than reference sites	Urban wetlands that might serve as reference sites often contain nonnative plants. Expectations that restored tidal freshwater wetlands in urban areas remain free of nonnative species may be unrealistic (Baldwin, 2004).

Continued

TABLE 25.2 Ecological Criteria for Evaluation of Restored Tidal Freshwater Wetlands—cont'd

Ecosystem Attribute	Measurements	Evaluation Criteria	Comments
Seed banks	Species composition	Dominant species similar to reference sites	Because seed bank composition is related to the composition of vegetation and seeds dispersing to restored sites, watershed characteristics are likely to have a strong influence on seed banks of restored tidal freshwater wetlands and reference sites within the same watershed (Neff et al., 2009).
	Seed density	Similar to reference sites	Higher seed density may occur in restored than in reference tidal freshwater wetland sites (Baldwin and DeRico, 1999; Leck, 2003; Neff et al., 2009).
	Species richness	Similar to reference sites	Richness may be higher in restored than reference sites (Baldwin and DeRico, 1999; Leck, 2003; Neff et al., 2009).
Benthic invertebrates	Density, species composition, species richness	Similar to reference sites	Invertebrate communities, being residential, integrate the water, soil, and vegetation habitat quality functions of the wetland, and are a measure of the capacity of the wetland to support fish, herpetofauna, mammals, and birds. Restoration of tidal freshwater wetlands may improve habitat for macroinvertebrates (Brittingham and Hammerschlag, 2006; Beauchard et al., 2013a).
Fish, birds, mammals, herpetofauna	Density and species composition	Present at restored sites	Restored tidal freshwater wetlands can provide valuable habitat for fish (Whitley and Bollens, 2014), birds (Beauchard et al., 2013b), and other wildlife. It may be useful to document value of wetlands as habitat for particular groups. Comparing restored and reference sites may not be practical because of seasonal and spatial variability in populations. Sampling wetland-dependent guilds may improve resolution.
Ecosystem functions	Nutrient cycling (e.g., mineralization, denitrification)	Similar to reference sites	Removal of invasive plants such as nonnative *Phragmites australis* (in North America) may increase soil ammonium and phosphate concentrations (Meyerson et al., 1999; Findlay et al., 2003).
	Material export and import	Similar to reference sites	Restored tidal freshwater wetlands are net sinks for nitrate and ammonium and net exporters of silica, but different sites vary in import or export of organic N and C, phosphorus, and particulates (Van Damme et al., 2009; Lehman et al., 2010). Restored tidal freshwater wetlands may be important in buffering dissolved silicon loading to estuaries, increasing resilience of diatom communities (Jacobs et al., 2013).

If the restored site is in an urbanized (or agricultural) landscape, reference sites should also be located in an urban (or agricultural) landscape with similar watershed characteristics. If reference sites are not available, restored sites can be compared to accepted standards of wetland function (Wilson and Mitsch, 1996) or literature values for naturally occurring wetlands in similar watersheds or landscapes.

4. RESTORED TIDAL FRESHWATER WETLANDS OF THE ANACOSTIA RIVER, WASHINGTON, DC

4.1 Characteristics of Restored and Reference Tidal Freshwater Wetland Sites

As is the case for many urban rivers, the ecosystems of the Anacostia River in Washington, DC, have been substantially altered by human activities. Development and related projects such as flood control, mandated dredging, sea wall construction, and landfills destroyed more than 3900 ha of forested and herb-dominated wetlands, including about 1000 ha of tidal freshwater wetlands along the Anacostia River (Schmid, 1994; U.S. EPA, 1997), leaving only a few hectares of fragmented tidal freshwater wetlands. Several tidal freshwater wetland restoration projects have been implemented, including Kenilworth Marsh, Kingman Marsh, and Anacostia River "fringe" wetlands (Fig. 25.5). Historically a tidal freshwater wetland existed at the location of Kenilworth Marsh, but the site was dredged to create a recreational lake in the 1940s (Syphax and Hammerschlag, 1995). The US Army Corps of Engineers (USACE) restored wetlands at the site in 1992–93 by pumping about 115,000 m^3 of sediment dredged from the adjacent Anacostia River into containment cells and planting more than 340,000 plants of 16 species (Bowers, 1995; Syphax and Hammerschlag, 1995). At Kingman Marsh, the USACE restored tidal freshwater wetlands in early 2000 in a similar fashion by placing Anacostia River sediment and planting 750,000 plants of seven species. This work created about 13 ha of vegetated wetland (AWRC, 2002). Sediment elevations were designed to be lower than those at Kenilworth Marsh to reduce colonization by nonnative invasive or highly productive native species that form large monoclonal patches (including *Lythrum salicaria*, *Phragmites australis*, and *Typha* spp.).

As part of postconstruction monitoring for the Kingman Marsh project, two natural tidal freshwater wetlands with similar tidal ranges were selected as reference sites to provide a basis for evaluating ecosystem development in the restored wetlands. One of these sites, Dueling Creek Marsh, is a remnant 0.41 ha urban wetland located on a small tributary to the Anacostia River just 0.8 km upstream of Kenilworth Marsh. The other site, Patuxent Wetland Park, is a nonurban tidal freshwater wetland located along the Patuxent River in an adjacent watershed. Furthermore, a multiyear exclosure experiment has been conducted at Kingman Marsh to study the effects of resident Canada geese on marsh vegetation (Krafft et al., 2013). Surface elevation table–marker horizon sampling stations (Boumans and Day, 1993; Cahoon et al., 1995, 1996; 2002a,b) were also installed as a means of tracking sedimentation, elevation change, and geomorphic processes at all of the study locations.

4.2 Evaluation of Success of Restored Tidal Freshwater Wetlands

The goal of ecosystem monitoring at Kingman was to "document both the status and the degree to which the reconstructed marsh achieved a wetland condition similar to reference emergent freshwater tidal wetland habitat" (Hammerschlag et al., 2006). Information from monitoring studies is used here to evaluate the success of restored Anacostia wetlands (see Table 25.3 for details and additional literature references).

Restoration of tidal freshwater wetlands in the Anacostia watershed was successful in creating hydrology similar to that of rural or urban reference sites (Baldwin et al., 2009).

FIGURE 25.5 Views of Anacostia River restored tidal freshwater wetlands. (A) Kingman Marsh prerestoration in 1999; the site was mostly mudflat at low tide and open water at high tide. Restoration was implemented by pumping in river sediment to raise elevation and planting with native species. (B) In 2016, many parts of Kingman remain vegetated; see white building dome landmark also in panels C and A. (C) Exclosure fences demonstrate the complete loss of emergent vegetation in other parts of Kingman due to herbivory by Canada geese (photo: 2016). (D) Anacostia Fringe wetlands relied on sheet piling (visible at right) to prevent erosion and goose fencing (including overhead strings to deter geese) to create desired vegetation (photo: 2008). (E) Kenilworth Marsh achieved dense vegetation but there are large monoculture stands of *Typha* spp. and nonnative *Phragmites australis* (photo: 2007). (F) Some of the restored Heritage Island marshes have been persistently vegetated (2016). Additional details are in Baldwin and DeRico (1999), Neff et al. (2009), and Prasse et al. (2015). *Photos by A.H. Baldwin.*

TABLE 25.3 Evaluation of Restored Tidal Freshwater Wetlands on the Anacostia River, Washington, DC

Parameter	Summary of Monitoring Results	Evaluation
Hydrology	Flooding duration at one of the restored sites (Kingman) was similar to that at a rural reference site (Patuxent), but more prolonged than at an urban reference site (Dueling Creek), which was similar to the older restored site (Kenilworth) (Baldwin et al., 2009).	Because Dueling Creek is in the same watershed as the restored sites, it is probably a more appropriate reference site than the rural reference sites. However, the Kingman restoration was successful in restoring hydrology similar to that in a nonurbanized watershed.
Geomorphology	Surface elevation table and marker horizon measurements were made at five locations at two restored wetlands (Kingman and Kenilworth) over a 3-year period beginning in October 2002, about $2^1/_2$ years after sediment placement. At both wetlands, accretion rates were >20 mm $year^{-1}$ and were at least double the rate of elevation gain, indicating subsurface subsidence but still net elevation gain (Baldwin et al., 2009).	The two restored tidal freshwater wetlands accumulated sediments and grew vertically at rates greater than relative sea level rise (3.22 mm $year^{-1}$; Washington, DC, tide gage 8594900, 1924–2015; tidesandcurrents.gov, accessed 13 January 2017), despite ongoing consolidation of the dredged material substrate (Baldwin et al., 2009). The slower rates of accretion and elevation gain at Kenilworth are consistent with its higher position within the tidal range (less mineral sedimentation and more organic matter decomposition). Rates of elevation increase were 7–8 mm $year^{-1}$ at a natural tidal freshwater wetland on the Patuxent River (Beckett, 2012), but natural tidal freshwater wetlands in other areas are not keeping pace with sea level rise (Craft, 2012; Beckett et al., 2016). Salinity intrusion is likely to have the largest long-lasting impact and may interact with other variables such as atmospheric warming, elevated CO_2, and eutrophication to change the distribution and ecosystem structure and function of tidal freshwater wetlands (Neubauer and Craft, 2009).
Soil	Soil organic matter (SOM) content was only 2.5% at an urban restored tidal freshwater wetland (Kingman), about 12 years following restoration, and 5% at a nearby restored tidal freshwater wetland (Kenilworth), about 20 years after restoration (Prasse et al., 2015).	Although SOM at a rural reference site (Patuxent) was >15%, an urban reference site (Dueling) near the restored sites had only 6% (Prasse et al., 2015), suggesting that accumulation of SOM to higher levels may not be possible in highly urbanized watersheds.
Microbial communities	The two restored sites, Kingman and Kenilworth, had similar community composition and functional gene abundance even though they differed in age (Prasse et al., 2015).	The urban reference site (Dueling) was more similar in composition to the natural reference site (Patuxent) than to the two urban restored sites (Prasse et al., 2015). Thus, microbial communities may require many years to develop.

Continued

TABLE 25.3 Evaluation of Restored Tidal Freshwater Wetlands on the Anacostia River, Washington, DC—cont'd

Parameter	Summary of Monitoring Results	Evaluation
Vegetation	Plant community monitoring revealed significant loss of vegetation cover, species richness, and diversity at a restored site that experienced substantial herbivory from Canada geese (Kingman; Baldwin et al., 2009) but not at any of the other restored or natural wetlands studied (Rusello, 2006; Paul et al., 2006). Most of the dominant species at the urban restored sites (Kingman and Kenilworth) were also dominant at the urban reference site (Dueling; Baldwin et al., 2009). These include the nonnative purple loosestrife *Lythrum salicaria* and the highly productive *Phalaris arundinacea*, neither of which occurred at the rural reference site (Patuxent). *Phragmites australis* and *Typha* spp. were both dominant features of the vegetation community at Kenilworth and have expanded at Kingman.	Populations of resident Canada geese were 3–5 times larger in the area of Kingman than Kenilworth (Paul et al., 2006). The annual species *Impatiens capensis* occurred at both restored and reference sites. However, *Polygonum arifolium* and *Polygonum sagittatum*, which were dominant at one or both reference sites, were rare at the restored sites. The urban reference site also contained nonnative invasive and highly productive native plants occurring at the restored sites, indicating that these are an expected persistent feature of any restored tidal freshwater wetlands in highly urbanized watersheds.
Seed bank	Surface soil samples were collected from restored and reference sites in 2000, 2001, and 2003 for seed bank analysis using the emergence method (Baldwin et al., 1996, 2001; Baldwin and DeRico, 1999; Leck, 2003). The seed bank at one restored site (Kingman) developed rapidly during the first growing season, showing large increases in emerging seedling density and taxa richness between 2000 and 2001 (Neff et al., 2009).	In 2003, all restored and reference sites were found to be similar in density and taxa richness. Significantly higher seedling density and species density were also found at a created tidal wetland in Delaware after 1 year of development (Leck, 2003). Seeds of the nonnative plant *Lythrum salicaria* were important at all urban sites in 2003.
Benthic invertebrates	Benthic macroinvertebrate organisms were collected over a 3-year period (2001–04) using an Ekman bottom grab sampler, sediment corer, dip-net, and Hester-Dendy sampler (Brittingham and Hammerschlag, 2006). Macroinvertebrate density was significantly greater at the newer restored site (Kingman) than at the older restored sited (Kenilworth) due to more numerous chironomids and oligochaetes (Brittingham and Hammerschlag, 2006).	Macroinvertebrate taxa composition at the older restored site (Kenilworth) was similar to that of the urban reference site, although richness was higher. The rural reference site (Patuxent) had more diversity, containing representatives from 30 families, whereas all of the urban restored and reference sites combined had only 23 families.
Birds	A total of 137 bird species were observed at Kingman and 164 at Kenilworth (177 species total); 124 of the species occurred at both wetlands (Paul et al., 2006).	Although birds were not studied at reference sites, results indicate that both restored wetlands provide habitat for numerous species.

The Kingman restored tidal freshwater wetland gained in elevation at rates greater than relative sea level rise (likely due to high sediment levels in the tidal Anacostia), indicating the wetlands are likely to persist as long as the rate of sea level rise does not increase (Baldwin et al., 2009). This outcome is in contrast to natural tidal freshwater wetlands on the US Atlantic Coast, which may not be keeping pace with rising seas. Accumulation of soil organic matter at the restored tidal freshwater wetlands has been slow, as is widely found in wetland restoration projects, but approaching levels at the urban reference site (Prasse et al., 2015). It is unlikely that the restored sites will ever attain soil organic matter similar to tidal freshwater wetlands in rural settings because of constraints imposed by the urban environment (Baldwin et al., 2009). Microbial communities developed slowly in restored sites and remained different from urban and rural reference sites even after 20 years (Prasse et al., 2015). Vegetation has also been constrained by the urban environment, as indicated by heavy grazing pressure from nonmigratory ("resident") geese at one site and by the establishment and expansion of nonnative plant species at all urban restored and reference sites (Rusello, 2006; Paul et al., 2006). The seed bank of restored tidal freshwater wetlands developed rapidly and converged with that of the urban reference site within a few years (Neff et al., 2009). Macroinvertebrate composition was dominated by chironomids and oligochaetes and appeared to be converging with that of the urban reference site but remained at lower taxa richness and had different distribution of dominant species than the rural natural site (Brittingham and Hammerschlag, 2006). Both restored wetlands provided habitat for a variety of bird species, including Canada goose, great blue heron, great egret, American green-winged teal, mallard, greater yellowlegs, song sparrow, killdeer, ring-billed herring, and laughing gulls (Paul et al., 2006).

5. CONCLUSIONS AND IMPLICATIONS

5.1 Restoration of Tidal Freshwater Wetlands in Urban Landscapes and Selection of Urban Reference Sites

Tidal freshwater wetlands typically have higher plant diversity than brackish or saline wetlands (Chapter 18 by Whigham et al.), creating unique challenges for restoration efforts. Monitoring at restored tidal freshwater wetland sites has demonstrated that restoration of elevation, hydrology, vegetation, geomorphological characteristics and processes, and faunal communities is possible and can be considered successful to varying degrees (e.g., Simenstad and Thom, 1996; Beauchard et al., 2013a,b). The case study of the Anacostia River tidal freshwater wetlands in Washington, DC highlights the difficulties of reestablishing wetland structure and function in an urbanized landscape (particularly as overshadowed by the large resident Canada goose population) (Baldwin, 2004). Thus, altered hydrology, environmental pollutants, fragmented landscapes, and nonnative species can override efforts to restore tidal freshwater wetlands to a structure similar to naturally occurring tidal freshwater wetlands in nonurban areas. For urban restoration projects, therefore, it makes sense to be realistic in setting goals and to consider urban reference sites (Ehrenfeld, 2000a,b).

5.2 Establishment of Vegetation

Because restoration efforts typically involve extensive earthmoving (e.g., excavation, dredged material placement, grading) and subsequent rapid changes in geomorphology related to tidal scouring, sedimentation, or compaction, a phased approach to wetland restoration is likely to improve success. Increases in inundation due to erosion or subsidence may reduce survival of plantings over time, but plantings may also help to reduce erosion. By completing sediment placement, excavation, and grading before or during the dormant season, sediment compaction and dewatering can occur for several months prior to the growing season. A topographic survey completed at this time will allow determination of suitability of elevation for plant growth, and additional grading can be performed or sediment placed before or during the early spring. While many species are likely to disperse to restored tidal freshwater wetlands (Neff and Baldwin, 2005), planting or seeding of native species not expected in dispersal pathways may be necessary during the spring to rapidly establish desired species, stabilize sediments against erosion, and possibly reduce establishment of nonnative species.

If herbivores such as resident Canada geese are present at or nearby restoration sites, it may be necessary to reduce those populations through management or protect sites with fencing for several years until vegetation has established. Dense vegetation dominated by native species has established at two other restored tidal freshwater wetland projects in the Anacostia (Heritage Island and River Fringe, Fig. 25.5), where geese have been excluded (Hammerschlag, pers. obs.). A multiyear study at Kingman Marsh has conclusively and strongly demonstrated the strong effect of grazing by geese on marsh vegetation (Fig. 25.5, Krafft et al., 2013). When geese were excluded, sites with planted vegetation persisted and unplanted areas were rapidly colonized and reached about 60% cover in 2 years (Krafft et al., 2013). In unfenced areas, cover of planted vegetation was reduced from about 100% to about 20% in a single year (Krafft et al., 2013). The strong impact of Canada geese on plant communities, particularly annual species, has also been documented in natural tidal freshwater wetlands (Baldwin and Pendleton, 2003; Haramis and Kearns, 2007).

5.3 Control of Nonnative Species

The benefits of controlling nonnative species in restored wetlands should be weighed against the negative environmental impacts of chemical use, as well as labor and materials costs, particularly if the nonnative species also occur in reference wetlands. Furthermore, the beneficial ecological functions of nonnative species should be considered in decisions regarding their control. However, governmental agencies and conservation groups may emphasize establishment of a diversity of local native plants, so it may be necessary to support efforts to suppress nonnative invasive or highly productive native species to promote a habitat that reflects these project goals.

The Anacostia experience suggests that elevations at or just below mean high tide will support a number of native high marsh species but will reduce the vigor of aggressive high marsh species. In contrast, at a restored tidal freshwater wetland in New Jersey, USA, the nonnative lineage of *Phragmites australis* gradually colonized 85%–95% of the wetland over an approximately 20-year period, replacing two earlier invasive colonizers, *Lythrum salicaria* and *Phalaris arundinacea*, and was associated with decreasing species richness (Mary Leck, pers. comm.; Leck, 2012; Elsey-Quirk and Leck, 2015). An exception was low-elevation areas colonized by the flood-tolerant native plant *Nuphar lutea* (Fig. 25.3B; Mary Leck, pers. comm.). *Phragmites*-free patches within the site contained diverse plant communities (Leck, 2012), suggesting that *Phragmites* control would increase diversity in some restored sites.

5.4 Implications for Restoration of Tidal Freshwater Wetlands

In a larger context, this review brings to light a number of considerations that are likely to improve the success of tidal freshwater wetlands restoration:

- *Clear objectives or goals for restoration during the early planning stages.* This need has been stated repeatedly for wetland restoration in general (e.g., Mitsch and Gosselink, 2000; Zedler, 2001), and it applies equally to tidal freshwater wetland restoration.
- *Realistic criteria for success in meeting goals or objectives*, preferably with regard to appropriately chosen reference sites. Planners may envision a pristine, diverse, exotic species-free wetland as the goal, but this may not be possible in a highly urbanized or agricultural landscape (Ehrenfeld, 2000b; Baldwin, 2004).
- *Increased use of adaptive management* for several years following restoration, for example, to fine-tune elevations, introduce additional plantings or seeds, or spot-control nonnative plants.
- *Restoration of tidal freshwater wetlands viewed ecologically as catastrophic landscape disturbances* that create high-light, high-nutrient, moist soil conditions optimal for rapid colonization by native and nonnative wetland species adapted to colonizing disturbed substrates. These species can be viewed as a natural initial phase of vegetation and community development, with the expectation that vegetation development will continue for many years, as influenced by hydrology, geomorphology, seed and propagule supply, and watershed condition.
- *Postconstruction monitoring* not only to document level of success but also to highlight situations that may require adjustment for better outcomes.

Restoration of tidal freshwater wetlands is increasingly practiced in North America and Eurasia. Because of the biological and hydrogeological complexity of tidal freshwater wetlands, outcomes of restoration are often uncertain, although the studies reviewed here demonstrate success in restoring structure and function at many tidal freshwater wetland sites. We hope that the success criteria proposed here stimulate discussion and promote dissemination of information improving the restoration potential of these wetlands.

References

Adams, M.A., Williams, G.L., 2004. Tidal marshes of the Fraser River estuary: composition, structure, and history of marsh creation efforts to 1997. In: Groulx, B.J., Mosher, D.C., Luternauer, J.L., Bilderback, D.E. (Eds.), Fraser River Delta, British Columbia: Issues of an Urban Estuary, Geological Survey of Canada Bulletin 567 Geological Survey of Canada, Vancouver, British Columbia, pp. 147–172.

AWRC (Anacostia Watershed Restoration Committee), 2002. Working Together to Restore the Anacostia Watershed, 2001 Annual Report. Washington, DC.

Baldwin, A.H., 2004. Restoring complex vegetation in urban settings: the case of tidal freshwater marshes. Urban Ecosystems 7, 125–137.

Baldwin, A.H., 2007. In: Conner, W., Doyle, T., Krauss, K. (Eds.), Vegetation and Seed Bank Studies of Salt-pulsed Swamps of the Nanticoke River, Chesapeake Bay. Springer-Verlag, Berlin, Germany, pp. 139–160.

Baldwin, A.H., 2009. Restoration of tidal freshwater wetlands in North America. In: Barendregt, A., Whigham, D.F., Baldwin, A.H. (Eds.), Tidal Freshwater Wetlands. Backhuys Publishers, Leiden, The Netherlands, pp. 207–222.

Baldwin, A.H., DeRico, E.F., 1999. The seed bank of a restored tidal freshwater marsh in Washington, DC. Urban Ecosystems 3, 5–20.

Baldwin, A.H., Pendleton, F.N., 2003. Interactive effects of animal disturbance and elevation on vegetation of a tidal freshwater marsh. Estuaries 26, 905–915.

Baldwin, A.H., McKee, K.L., Mendelssohn, I.A., 1996. The influence of vegetation, salinity, and inundation on seed banks of oligohaline coastal marshes. American Journal of Botany 83, 470–479.

Baldwin, A.H., Egnotovich, M.S., Clarke, E., 2001. Hydrologic change and vegetation of tidal freshwater marshes: field, greenhouse, and seed-bank experiments. Wetlands 21, 519–531.

Baldwin, A.H., Hammerschlag, R.S., Cahoon, D.R., 2009. Evaluation of restored tidal freshwater wetlands. In: Perillo, G.M.E., Wolanski, E., Cahoon, D.R., Brinson, M.M. (Eds.), Coastal Wetlands: An Integrated Ecosystem Approach, first ed. Elsevier, Oxford, UK, pp. 801–832.

Barendregt, A., Whigham, D.F., Baldwin, A.H., et al., 2006. Wetlands in the tidal freshwater zone. In: Bobbink, R., Beltman, B., Verhoeven, J.T.A., Whigham, D.F. (Eds.), Wetlands: Functioning, Biodiversity Conservation, and Restoration. Springer-Verlag, Berlin, pp. 117–148.

Bartoldus, C.C., 1990. Revegetation and Production in a Constructed Freshwater Tidal Marsh (PhD dissertation). George Mason University, Fairfax, VA.

Bartoldus, C.C., Heliotis, F.D., 1989. Factors affecting survival of planted materials in Marley Creek constructed freshwater tidal marsh, Maryland. Transportation Research Record 1224, 1–5.

Beauchard, O., Jacobs, S., Cox, T.J.S., et al., 2011. A new technique for tidal habitat restoration: evaluation of its hydrological potentials. Ecological Engineering 37, 1849–1858.

Beauchard, O., Jacobs, S., Ysebaert, T., et al., 2013a. Sediment macroinvertebrate community functioning in impacted and newly-created tidal freshwater habitats. Estuarine, Coastal and Shelf Science 120, 21–32.

Beauchard, O., Jacobs, S., Ysebaert, T., et al., 2013b. Avian response to tidal freshwater habitat creation by controlled reduced tide system. Estuarine, Coastal and Shelf Science 131, 12–23.

Beckett, L.H., 2012. Subsidence, Accretion, and Elevation Trends in Estuarine Wetlands and Relationships to Salinity and Sediment Stratigraphy (PhD dissertation). University of Maryland, College Park.

Beckett, L.H., Baldwin, A.H., Kearney, M.S., 2016. Tidal marshes across a Chesapeake Bay subestuary are not keeping up with sea-level rise. PLoS One 11, e0159753 doi:10.1371.journal.pone.0159753.

Boudell, J.A., Dixon, M.D., Rood, S.B., Stromberg, J.C., 2015. Restoring functional riparian ecosystems: concepts and applications. Ecohydrology 8, 747–752.

Boumans, R., Day Jr., J.W., 1993. High precision measurements of sediment elevation in shallow coastal areas using a sedimentation-erosion table. Estuaries 16, 375–380.

Bowers, J.K., 1995. Innovations in tidal marsh restoration: the Kenilworth Marsh account. Restoration and Management Notes 13, 155–161.

Brittingham, K.D., Hammerschlag, R.S., 2006. Benthic Macroinvertebrate Communities of Reconstructed Freshwater Tidal Wetlands in the Anacostia River, Washington, D.C. Final Report (2002-2004). U.S. Geological Survey Patuxent Wildlife Research Center, Beltsville, Maryland.

Cahoon, D.R., Reed, D.J., Day Jr., J.W., 1995. Estimating shallow subsidence in microtidal salt marshes of the southeastern United States: Kaye and Barghoorn revisited. Marine Geology 128, 1–9.

Cahoon, D.R., Lynch, J.C., Knaus, R.M., 1996. Improved cryogenic coring device for sampling wetland soils. Journal of Sedimentary Research 66, 1025–1027.

Cahoon, D.R., Lynch, J.C., Hensel, P., et al., 2002a. High-precision measurements of wetland sediment elevation: I. Recent improvements to the Sedimentation-Erosion Table. Journal of Sedimentary Research 72, 730–733.

Cahoon, D.R., Lynch, J.C., Perez, B.C., et al., 2002b. High-precision measurements of wetland sediment elevation: II. The rod surface elevation table. Journal of Sedimentary Research 72, 734–739.

Cole, C.A., Shafer, D., 2002. Section 404 wetland mitigation and permit success criteria in Pennsylvania, USA, 1986-1999. Environmental Management 30, 508–515.

Cox, T., Maris, T., De Vleeschauwer, P., et al., 2006. Flood control areas as an opportunity to restore estuarine habitat. Ecological Engineering 28, 55–63.

Craft, C., 2007. Freshwater input structures soil properties, vertical accretion, and nutrient accumulation of Georgia and U.S. tidal marshes. Limnology and Oceanography 52, 1220–1230.

Craft, C.B., 2012. Tidal freshwater forest accretion does not keep pace with sea level rise. Global Change Biology 18, 3615–3623.

Craft, C., 2016. Creating and Restoring Wetlands: From Theory to Practice. Elsevier, Amsterdam.

Craft, C., Megonigal, P., Broome, S., et al., 2003. The pace of ecosystem development of constructed *Spartina alterniflora* marshes. Ecological Applications 13, 1417–1432.

David, A.T., Ellings, C.S., Woo, I., et al., 2014. Foraging and growth potential of juvenile Chinook Salmon after tidal restoration of a large River delta. Transactions of the American Fisheries Society 143, 1515–1529.

Diefenderfer, H.L., Montgomery, D.R., 2009. Pool spacing, channel morphology, and the restoration of tidal forested wetlands of the Columbia River, U.S.A. Restoration Ecology 17, 158–168.

DNREC (Delaware Department of Natural Resources and Environmental Control), 2016. Northern Delaware Wetland Rehabilitation Program. New Castle, DE.

Duberstein, J.A., Conner, W.H., 2009. Use of hummocks and hollows by trees in tidal freshwater forested wetlands along the Savannah River. Forest Ecology Management 258, 1613–1618.

Ehrenfeld, J.G., 2000a. Evaluating wetlands within an urban context. Urban Ecosystems 4, 69–85.

Ehrenfeld, J.G., 2000b. Defining the limits of restoration: the need for realistic goals. Restoration Ecology 8, 2–9.

Elsey-Quirk, T., Leck, M.A., 2015. Patterns of seed bank and vegetation diversity along a tidal freshwater river. American Journal of Botany 102, 1996–2012.

Findlay, S., Groffman, P., Dye, S., 2003. Effects of *Phragmites australis* removal on marsh nutrient cycling. Wetlands Ecology and Management 11, 157–165.

French, P.W., 2006. Managed realignment – the developing story of a comparatively new approach to soft engineering. Estuarine, Coastal and Shelf Science 67, 409–423.

Gannett Fleming, 2001. Plans of Route 29-Open Water Mitigation, Township of Hamilton, County of Mercer. Gannett Fleming, Inc., Hammonton, NJ.

Gray, A., Simenstad, C.A., Bottom, D.L., et al., 2002. Contrasting functional performance of juvenile salmon habitat in recovering wetlands of the Salmon River estuary, Oregon, U.S.A. Restoration Ecology 10, 514–526.

Grout, J.A., Levings, C.D., Richardson, J.S., 1997. Decomposition rates of purple loosestrife (*Lythrum salicaria*) and Lyngbyei's sedge (*Carex lyngbyei*) in the Fraser River estuary. Estuaries 20, 96–102.

Hammerschlag, R.S., Baldwin, A.H., Krafft, C.C., et al., 2006. Five Years of Monitoring Reconstructed Freshwater Tidal Wetlands in the Urban Anacostia River (2000–2004). U.S. Geological Survey Patuxent Wildlife Research Center, Beltsville, MD.

Haramis, G.M., Kearns, G.D., 2007. Herbivory by resident geese: the loss and recovery of wild rice along the tidal Patuxent river. The Journal of Wildlife Management 71, 788–794.

Hester, M.W., Willis, J.M., Sloey, T.M., 2016. Field assessment of environmental factors constraining the development and expansion of *Schoenoplectus californicus* marsh at a California tidal freshwater restoration site. Wetlands Ecology and Management 24, 33–44.

Hopfensperger, K.N., Burgin, A.J., Schoepfer, V.A., et al., 2014. Impacts of saltwater incursion on plant communities, anaerobic microbial metabolism, and resulting relationships in a restored freshwater wetland. Ecosystems 17, 792–807.

Jacobs, S., Beauchard, O., Struyf, E., et al., 2009. Restoration of tidal freshwater vegetation using controlled reduced tide (CRT) along the Schelde Estuary (Belgium). Estuarine, Coastal and Shelf Science 85, 368–376.

Jacobs, S., Müller, F., Teuchies, J., et al., 2013. The vegetation silica pool in a developing tidal freshwater marsh. Silicon 5, 91–100.

Kaminsky, M., Scelsi, P., 1986. Route 130, Section 9F Rancocas Creek Bridge, Site III Wetland Replacement. Bureau of Environmental Analysis. New Jersey Department of Transportation, Trenton, NJ.

Kentula, M.E., 2000. Perspectives on setting success criteria for wetland restoration. Ecological Engineering 15, 199–209.

Kistritz, R., 1996. Habitat Compensation, Restoration and Creation in the Fraser River Estuary: Are We Achieving a No-Net-Loss of Fish Habitat? R.U. Kistritz Consultants Ltd., Richmond, British Columbia.

Krafft, C.C., Hatfield, J.C., Hammerschlag, R.S., 2013. Effects of Canada Goose Herbivory on the Tidal Freshwater Wetlands in Anacostia Park, 2009–2011. Natural Resource Technical Report, NPS/NCR/NCRO/NRTR—2013/001). United States Department of the Interior, National Park Service, Washington, DC.

Land Management Group, 2004. McIntyre Tract: Year Four Monitoring Report, Brunswick County, North Carolina. Land Management Group, Inc., Wilmington, NC.

Lane, R.R., Day Jr., J.W., Thibodeaux, B., 1999. Water quality analysis of a freshwater diversion at Caernarvon, Louisiana. Estuaries 22, 327–336.

Leck, M.A., 2003. Seed-bank and vegetation development in a created tidal freshwater wetland on the Delaware River, Trenton, New Jersey, USA. Wetlands 23, 310–343.

Leck, M.A., 2012. Dispersal potential of a tidal river and colonization of a created tidal freshwater marsh. AoB Plants 5, pls050. https://doi.org/10.1093/aobpla/pls050.

Leck, M.A., Leck, C.F., 2005. Vascular plants of a Delaware River tidal freshwater wetland and adjacent terrestrial areas: seed bank and vegetation comparisons of reference and constructed marshes and annotated species list. Journal of the Torrey Botanical Society 132, 323–354.

Leck, M.A., Baldwin, A.H., Parker, V.T., et al., 2009. Plant communities of North American tidal freshwater wetlands. In: Barendregt, A., Whigham, D.F., Baldwin, A.H. (Eds.), Tidal Freshwater Wetlands. Backhuys, Leiden, The Netherlands.

Lehman, P.W., Mayr, S., Mecum, L., et al., 2010. The freshwater tidal wetland Liberty Island, CA was both a source and sink of inorganic and organic material to the San Francisco Estuary. Aquatic Ecology 44, 359–372.

Levings, C.D., Nishimura, D.J.H., 1996. Created and Restored Sedge Marshes in the Lower Fraser River and Estuary: An Evaluation of Their Functioning as Fish Habitat. Department of Fisheries and Oceans, West Vancouver, British Columbia.

Levings, C.D., Nishimura, D.J.H., 1997. Created and restored marshes in the lower Fraser River, British Columbia: summary of their functioning as fish habitat. Water Quality Research Journal Canada 32, 599–618.

Marble and Company, 1998. Post-Construction Wetland Monitoring Report, Trenton Complex Mitigation Site, Mercer County, New Jersey. A.D. Marble and Company, Inc., Rosemont, PA.

Meyerson, L.A., Chambers, R.M., Vogt, K.A., 1999. The effects of *Phragmites* removal on nutrient pools in a freshwater tidal marsh ecosystem. Biological Invasions 1, 129–136.

Mitsch, W.J., Gosselink, J.G., 2000. Wetlands, third ed. John Wiley and Sons, NY.

Mitsch, W.J., Wilson, R.F., 1996. Improving the success of wetland creation and restoration with know-how, time, and self-design. Ecological Applications 6, 77–83.

Moore, K.A., Shields, E.C., Jarvis, J.C., 2010. The role of habitat and herbivory on the restoration of tidal freshwater submerged aquatic vegetation populations. Restoration Ecology 18, 596–604.

Neff, K.P., 2002. Plant Colonization and Vegetation Change in a Restored Tidal Freshwater Wetland in Washington, DC (M.S. thesis). University of Maryland, College Park, Maryland.

Neff, K.P., Baldwin, A.H., 2005. Seed dispersal into wetlands: techniques and results for a restored tidal freshwater marsh. Wetlands 25, 392–404.

Neff, K.P., Rusello, K., Baldwin, A.H., 2009. Rapid seed bank development in restored tidal freshwater wetlands. Restoration Ecology 17, 539–548.

Neubauer, S.C., Craft, C.B., 2009. Global change and tidal freshwater wetlands: scenarios and impacts. In: Barendregt, A., Whigham, D.F., Baldwin, A.H. (Eds.), Tidal Freshwater Wetlands. Backhuys, Leiden, The Netherlands.

Neubauer, S.C., Anderson, I.C., Constantine, J.A., et al., 2002. Sediment deposition and accretion in a mid-Atlantic (USA) tideal freshwater marsh. Estuarine, Coastal and Shelf Science 54, 713–727.

Odum, W.E., 1988. Comparative ecology of tidal freshwater and salt marshes. Annual Review of Ecology and Systematics 19, 147–176.

Odum, W.E., Smith III, T.J., Hoover, J.K., et al., 1984. The Ecology of Tidal Freshwater Marshes of the United States East Coast: A Community Profile. FWS/OBS-83/17. U.S. Fish and Wildlife Service, Washington, DC.

Orr, M., Crooks, S., Williams, P.B., 2003. Will restored tidal marshes be sustainable? San Franciso Estuary and Watershed Science 1, 1–33.

Paul, M.M., Krafft, C.C., Hammerschlag, R.S., 2006. Avian Comparisons Between Kingman and Kenilworth Marshes, Final Report 2001–2004. U.S. Geological Survey Patuxent Wildlife Research Center, Beltsville, MD.

Phillip, M., 2005. Decker Island Wildlife Area: Enhancing Delta Wetlands One Phase at a Time. Outdoor California March – April, pp. 4–8.

Prasse, C.E., Baldwin, A.H., Yarwood, S.A., 2015. Site history and edaphic features override the influence of plant species on microbial communities in restored tidal freshwater wetlands. Applied and Environmental Microbiology 81, 3482–3491.

Quigley, P.A., 2001. Wetlands Monitoring Report, Henderson Freshwater Tidal Wetland Restoration, John Heinz National Wildlife Refuge, Monitoring Year 2000. Patricia Ann Quigley, Inc., Norristown, PA.

Rusello, K., 2006. Wetland Restoration in Urban Settings: Studies of Vegetation and Seed Banks in Restored and Reference Tidal Freshwater Marshes (M.S. thesis). University of Maryland, College Park.

Schmid, J.A., 1994. Wetlands in the urban landscape of the United States. In: Platt, R.H., Rowntree, R.A., Muick, P.C. (Eds.), The Ecological City: Preserving and Restoring Urban Biodiversity. The University of Massachusetts Press, Amherst, pp. 106–133.

Simenstad, C.A., Thom, R.M., 1996. Functional equivalency trajectories of the restored Gog-Le-Hi-Te estuarine wetland. Ecological Applications 6, 38–56.

Simenstad, C., Toft, J., Higgins, H., et al., 2000. Sacramento/San Joaquin Delta Breached Levee Wetland Study (BREACH). University of Washington School of Fisheries, Seattle.

Simpson, R.L., Good, R.E., Leck, M.A., et al., 1983. The ecology of freshwater tidal wetlands. BioScience 33, 255–259.

Sloey, T.M., Willis, J.M., Hester, M.W., 2015. Hydrologic and edaphic constraints on *Schoenoplectus acutus, Schoenoplectus californicus*, and *Typha latifolia* in tidal marsh restoration. Restoration Ecology 23, 430–438.

Spieles, D.J., 2005. Vegetation development in created, restored, and enhanced mitigation wetland banks of the United States. Wetlands 25, 51–63.

Stagg, C.L., Krauss, K.W., Cahoon, D.R., et al., 2016. Processes contributing to resilience of coastal wetlands to sea-level rise. Ecosystems 19, 1445–1459.

Stillwater Environmental Services, 2003. Decker Island Phase II Habitat Development and Levee Rehabilitation Project, Initial Study/Mitigated Negative Declaration Public Review Draft. Stillwater Environmental Services, Davis, CA.

Suding, K.N., 2011. Toward an era of restoration in ecology: successes, failures, and opportunities ahead. Annual Review of Ecology, Evolution, and Systematics 42, 465–487.

Sullivan, L.R., 2015. If You Build It, What Will Come? Assessing the Avian Response to Wetland Restoration in the Mississippi River Bird's Foot Delta Through Multiple Measures of Density and Biodiversity (M.S. Thesis). Louisiana State University, Baton Rouge.

Syphax, S.W., Hammerschlag, R.S., 1995. The reconstruction of Kenilworth Marsh, the last tidal marsh in Washington, D.C. Park Science 15, 15–19.

Tanner, C.D., Cordell, J.R., Rubey, J., et al., 2002. Restoration of freshwater intertidal habitat functions at Spencer Island, Everett, Washington. Restoration Ecology 10, 564–576.

Teuchies, J., Jacobs, S., Oosterlee, L., et al., 2013. Role of plants in metal cycling in a tidal wetland: implications for phytoremidiation. The Science of the Total Environment 445–446, 146–154.

USACE (U.S. Army Corps of Engineers), 2017. Louisiana Coastal Wetlands Planning, Protection and Restoration Act. www.lacoast.gov.

US EPA (US Environmental Protection Agency), 1997. An environmental characterization of the District of Columbia – a scientific foundation for setting an environmental agenda. US Environmental Protection Agency – Region 3, Philadelphia, PA.

Van Damme, S., Frank, D., Micky, T., et al., 2009. Tidal exchange between a freshwater tidal marsh and an impacted estuary: the Scheldt estuary, Belgium. Estuarine, Coastal and Shelf Science 85, 197–207.

Vandenbruwaene, W., Maris, T., Cox, T.J.S., et al., 2011. Sedimentation and response to sea-level rise of a restored marsh with reduced tidal exchange: comparison with a natural tidal marsh. Geomorphology 130, 115–126.

Vandenbruwaene, W., Meire, P., Temmerman, S., 2012. Formation and evolution of a tidal channel network within a constructed tidal marsh. Geomorphology 151–152, 114–125.

Verhoeven, J.T.A., Whigham, D.F., van Logtestijn, R., et al., 2001. A comparative study of nitrogen and phosphorus cycling in tidal and non-tidal riverine wetlands. Wetlands 21, 210–222.

Wang, X., Yu, J., Zhou, D., et al., 2012. Vegetative ecological characteristics of restored reed (*Phragmites australis*) wetlands in the Yellow River Delta, China. Environmental Management 49, 325–333.

Weinstein, M.P., Balletto, J.H., Teal, J.M., et al., 1997. Success criteria and adaptive management for a large-scale wetland restoration project. Wetlands Ecology and Management 4, 111–127.

Whigham, D.F., Simpson, R.L., 1992. Annual variation in biomass and production of a tidal freshwater wetland and comparison with other wetland systems. Virginia Journal of Science 43, 5–14.

Whitley, S.N., Bollens, S.M., 2014. Fish assemblages across a vegetation gradient in a restoring tidal freshwater wetland: diets and potential for resource competition. Environmental Biology of Fishes 97, 659–674.

Wilson, R.F., Mitsch, W.J., 1996. Functional assessment of five wetlands constructed to mitigate wetland loss in Ohio, USA. Wetlands 16, 436–451.

Zedler, J.B., 2001. Handbook for Restoring Tidal Wetlands. CRC Press, Boca Raton, FL.

Zedler, J.B., Callaway, J.C., 1999. Tracking wetland restoration: do mitigation sites follow desired trajectories? Restoration Ecology 7, 69–73.

Zhao, Q., Bai, J., Huang, L., et al., 2016. A review of methodologies and success indicators for coastal wetland restoration. Ecological Indicators 60, 442–452.

PART VII

COASTAL WETLAND SUSTAINABILITY

CHAPTER

26

The Shifting Saltmarsh-Mangrove Ecotone in Australasia and the Americas

Neil Saintilan[1], *Kerrylee Rogers*[2], *Karen L. McKee*[3]

[1]Department of Environmental Sciences, Macquarie University, Sydney, NSW, Australia; [2]Geoquest, University of Wollongong, Wollongong, NSW, Australia; [3]U.S. Geological Survey, Wetland and Aquatic Research Center, Lafayette, LA, United States

1. INTRODUCTION

Intertidal coastal wetlands, principally mangrove and saltmarsh, are among the most valuable natural ecological systems in the world (Costanza et al., 2014) providing ecological goods and services ranging from biological habitat, coastal productivity, and protection to carbon sequestration and storage (Lee et al., 2014), However, because of their position at the land–ocean interface, they are sensitive to the impacts of climate change, particularly in the many places in the world where mangroves and saltmarsh coexist and competitively interact. These interactions can be studied over a range of timescales and inform our understanding of the impacts of climate change and other anthropogenic stressors on these critically important natural systems. Stratigraphic and palynological evidence has been used to reconstruct distribution of mangrove and saltmarsh communities over geological timescales with the greatest clarity emerging from the Holocene. At this scale, interactions between mangroves and saltmarshes are governed by geomorphic processes, most notably, patterns of sedimentation following the postglacial marine high stand.

More recently, over the historical time period, archival air photographs and other remotely sensed images have been used to study changes in distribution of these communities over wide geographic areas. The decline in the area of tropical mangrove over recent decades has been well-documented (Pendleton et al., 2012; Friess and Webb, 2014), and though the rate of decline has slowed in recent years, clearance for aquaculture and palm oil agriculture continues in many parts of the world (Murdiyarso et al., 2015). At their poleward limits, the

Coastal Wetlands
https://doi.org/10.1016/B978-0-444-63893-9.00026-5

trend is one of mangrove expansions, particularly where mangroves hitherto have been delimited by cold temperatures (Cavanaugh et al., 2014a). Mangrove expansion at the expense of saltmarsh has been documented in the continental USA, Central America, South America, Australia, South Africa, and China (Saintilan et al., 2014).

Mangroves compete with saltmarshes at the limits of their physiological tolerance. At their latitudinal extreme, the vigor of mangrove growth is constrained by cold, and survival may be dictated by frost. The landward limit of mangroves in the intertidal zone is constrained by the often severe environment of the upper intertidal zone, where low soil moisture and high salinity may prohibit seedling establishment and growth. Changes to environmental conditions at these extremes, including elevated temperatures, decreased frost frequency, elevated atmospheric CO_2, and higher sea levels, have the potential to alter conditions such that the growth of mangrove is favored over saltmarsh (McKee et al., 2012). The encroachment of mangrove into saltmarsh has many parallels in the encroachment of woody vegetation into grasslands in many of the world's biomes (Saintilan and Rogers, 2013).

This chapter reviews a range of environmental variables that govern mangrove and saltmarsh interactions: geomorphic, hydrologic, climatic, physicochemical, and biotic. An emerging theme is the potential importance of climate change as a key control on vegetation shifts and changes in species diversity. For example, higher temperatures may confer a competitive advantage on mangroves, which are tropical plants. Saltmarshes, which predominate at temperate and subarctic latitudes, may increase in diversity with increasing latitude (Saenger et al., 1977). Furthermore, higher sea levels could promote the landward encroachment of mangrove into saltmarsh, as could elevated levels of atmospheric CO_2 (McKee et al., 2012; Saintilan and Rogers, 2015). The changing distribution of mangrove and saltmarsh may serve as an important indicator of climate change impacts, a sentinel of change for the broad range of ecosystem services dependent on these habitats.

2. DISTRIBUTION/GEOMORPHIC SETTINGS—WHERE DO MANGROVE AND SALTMARSH COEXIST?

2.1 Mangrove Distribution

Mangroves are the typical intertidal vegetation of sheltered tropical and subtropical coastlines. However, some mangrove species will readily occupy sheltered temperate coastlines (Saenger, 2002). The global distribution of mangroves has been examined on numerous occasions (Chapman, 1977; Spalding et al., 1997a; Duke et al., 1998a; Duke, 2006; Giri et al., 2011; Fig. 26.1) and is primarily limited by the physiological tolerance of mangrove species to low temperature. Other factors, such as the availability of suitable habitat and climate, dispersal and establishment of propagules, continental drift and tectonic events, are also important (Duke et al., 1998a). Mangroves tend to be restricted to coastlines where mean air temperatures of the coldest month are higher than 20°C and the seasonal range is not greater than 10°C (Walsh, 1974; Chapman, 1977, from Duke et al., 1998a), which appears to correlate with the 20°C isotherm for seawater (Duke et al., 1998a).

In the Americas, distribution of mangroves roughly corresponds to the sea surface winter isotherm of 20°C, except for the North Atlantic coast where the northern limit corresponds to

FIGURE 26.1 Global mangrove and saltmarsh distribution and the average 20°C sea surface temperature isotherm. *From Spalding, M.D., Blasco, F., Field, C.D. (Eds), 1997b. World Mangrove Atlas. International Society for Mangrove Ecosystems. Okinawa, Japan; Long, S.P., Mason, C.F., 1983. Saltmarsh Ecology. Blackie and Son Ltd., Glasgow and London; NOAA, 2007. Sea Surface Temperature (SST) Contour Charts. Prepared by National Oceanic and Atmospheric Administration (NOAA) Satellites and Information. Accessed at: http://www.osdpd.noaa.gov/PSB/EPS/SST/contour.html.*

the 27°C isotherm (i.e., the isotherm of the annual average for monthly maximal sea temperature where areas poleward are always cooler) (see Davy and Costa, 1992). The cold Humboldt and Falkland currents influence sea temperatures along the western and eastern coasts of South America so that the winter 20°C isotherm occurs at different latitudes (Fig. 26.1). Similarly, the distribution of mangroves on the west coast of Africa is more restricted than on the east coast because of the different latitudinal positions of the 20°C isotherm (Fig. 26.1). The reduced occurrence of mangroves on the western coasts of Africa and South America also coincides with the limits of arid regions (Saenger, 2002) and implies that the development of mangroves in these regions is limited by aridity in addition to temperature (Saenger and Moverley, 1985; Smith and Duke, 1987).

However, latitude and its relationship with low temperatures and/or the occurrence of frosts are the primary explanation for the latitudinal distribution of mangroves in the Americas (Sherrod and McMillan, 1985; McMillan and Sherrod, 1986). Mangroves thus extend from about 31°0′N (Baja California; Peinado et al., 1994) to 5°30′S (Peru (Clusener and Breckle, 1987)) on the Pacific coast and from 30°N (St. Augustine, Florida, USA) to 28°30′S (Laguna, Brazil (Schaeffer-Novelli et al., 1990)) on the Atlantic coast. The northernmost mangroves in this region occur in Bermuda (32°18′N), off the east coast of the United States, because of the warm waters of the Gulf Stream and absence of freezing temperatures; species found there include *Rhizophora mangle*, *Avicennia germinans*, and *Conocarpus erectus* (Thomas, 1993). In contrast, mangroves historically extended only to Ponce de Leon Inlet (29°04′N) on the Atlantic coast of Florida and the Mississippi River delta in the Gulf of Mexico (29°12′N) where they are periodically killed by freezing temperatures. Recent reports indicate northernmost individuals of *R. mangle* have extended from 29°40′N (Fort Matanzas, Florida, USA)

(Zomlefer et al., 2006) to 30°01′N, near the northernmost *A. germinans* at 30°02′N (St. Augustine, Florida, USA) (Williams et al., 2014).

Exceptions to the distribution pattern of mangrove in waters cooler than 20°C also occur around Australia and the North Island of New Zealand. In these locations, the species of mangrove with the greatest latitudinal extent is *Avicennia marina*, with its distribution extending to Corner Inlet (38°54′S) in southeastern (SE) Australia. Explanations for the more southerly distribution of *Avicennia marina* in Australia include transmission by ocean currents or that they are relict populations that had greater poleward distributions in the past (MacNae, 1966; Duke et al., 1998a). The latter explanation is more widely accepted, particularly because *Avicennia marina* var. *australasica* is genetically distinct from more tropical varieties, such as var. *marina* and var. *eucalyptifolia* (Duke et al., 1998a,b). Survival at this latitudinal extreme might be explained by the lower incidence of the heavy frosts that cause mass mortality in populations of *A. germinans* and *R. mangle* in the United States. *A. marina* and *Aegiceras corniculatum*, the next most poleward species in Australia, have smaller vessel diameters than *A. germinans* and *R. mangle*, with a consequent higher tolerance of the mild freezing more commonly encountered in southern Australia (Stuart et al., 2007).

Avicennia marina var. *australasica* is the only mangrove species found in New Zealand and is restricted to the North Island with a latitudinal limit at Kutarere (38°03′S) (Lange and de Lange, 1994). Like its Australian counterpart, its southern distribution does not appear to relate to climatic factors (Lange and de Lange, 1994) and may be a relict of greater poleward distributions (Mildenhall and Brown, 1987; Saenger, 2002). Because of the successful establishment of *Avicennia marina* in more southerly locations, the latitudinal distribution of mangroves in New Zealand is hypothesized to be in disequilibrium with climatic factors. Furthermore, low current velocities and large distances between suitable habitats inhibit the distribution of mangrove to the same extent observed in Australia (Lange and de Lange, 1994).

2.2 Saltmarsh Distribution

In contrast, saltmarsh may occur on many of the world's shorelines (Adam 1990; Mendelssohn and McKee, 2000). Information about the global distribution of saltmarsh species is poor and extensive detailed mapping of saltmarsh has not been undertaken for some time (Chapman, 1977; Long and Mason, 1983). At a global scale, saltmarsh establishes on shorelines where mangrove establishment is precluded or development is limited (Kangas and Lugo, 1990). For this reason, saltmarshes are most common in temperate, subarctic, and arctic zones (Long and Mason, 1983; Mitsch and Gosselink, 2000; Mendelssohn and McKee, 2000).

In North America, saltmarshes occur from the southern boundary of Central America to Alaska and Canada, with greatest development in the temperate United States (Mendelssohn and McKee, 2000). The northernmost limits of saltmarsh on that continent occur along the Arctic Ocean at 70°N latitude and extend southward along the Pacific shoreline of Alaska, Canada, the continental United States, Mexico, and Central America. Saltmarsh also occurs along the shoreline of Hudson Bay (55°N) and the Atlantic coasts of Canada and the United States, extending along the Gulf of Mexico into Central America where the distribution overlaps with that of mangrove vegetation. Saltmarsh species can be found in Belize

(16–17°N), Guyana (6–8°N), Peru/Ecuador (0–5°S), Brazil (0–32°S), Argentina (34–51°S), and Chile (40–52°S) (Table 12.2 in Costa and Davy, 1992).

Saltmarsh vegetation can be classed into six biogeographical types: arctic, boreal, temperate, west Atlantic, dry coast, and tropical types (Adam, 1990). A description of saltmarsh vegetation in North America using this classification can be found in Mendelssohn and McKee (2000). Saltmarshes in the Arctic and subarctic regions are dominated by species such as *Puccinellia phryganoides* and *Carex* spp. Boreal saltmarsh along the Atlantic coast is characterized by *Spartina patens*, *Spartina alterniflora* in the low marsh, whereas boreal saltmarsh in the Pacific oceanic region, including British Columbia, contain *Triglochin maritima*, *Salicornia virginica*, and *Distichlis spicata*. Temperate saltmarsh on the Atlantic coast is dominated by *S. alterniflora*, which extends southward into northern Florida. Other temperate species along the Atlantic coast include *Juncus balticus* (boreal), *Juncus gerardii* (boreal to temperate), and *Juncus roemerianus* (temperate), which also occurs along the Gulf of Mexico. Temperate saltmarsh on the Pacific coast contains *Spartina foliosa*, *Salicornia virginica*, and *D. spicata*. In tropical zones, species such as *S. patens*, *Spartina spartinae*, and *D. spicata* can occur but may be replaced in hypersaline settings by succulents such as *Batis maritima*, *Borrichia frutescens*, *Suaeda maritima*, and *Sesuvium portulacastrum*.

2.3 Coexisting Mangrove and Saltmarsh

Saltmarsh typically occurs in settings where mangrove development is limited (Kangas and Lugo, 1990; West, 1977), enabling saltmarsh to establish. Extensive coexisting mangrove and saltmarsh communities can be found in the temperate regions of Australia, New Zealand, South Africa, and southern continental USA. Saltmarsh may also establish behind mangrove communities within tropical and subtropical climates where rainfall is low and soil salinities in these areas become hypersaline (Chapman, 1977; Long and Mason, 1983). In Mexico, Central America, and Florida, for example, saltmarsh may occur on the margins of mangrove forests (either colonizing seaward mudflats or the saline soils on landward edges), within mangrove woodlands with more open canopies, or in disturbed areas (West, 1977; Lopez-Portillo and Ezcurra, 1989).

A distinction must be made between saltmarsh vegetation typical of the low intertidal zone and that restricted to the "high marsh," i.e., areas in the upper intertidal range that are characterized by dryer and more saline conditions. More specifically, it is necessary to identify saltmarsh and mangrove physiographic equivalents when considering interactions between coexisting mangrove and saltmarsh vegetation and future changes in climate and sea level. For example, in North America, extensive stands of the temperate low-marsh–dominant *S. alterniflora* can be found as far south as subtropical Florida, Louisiana, and Texas. Although *S. alterniflora* can be found throughout Latin America, its occurrence is infrequent and mainly limited to mangal fringes (Costa and Davy, 1992). Other saltmarsh species such as *S. spartinae*, *D. spicata*, *B. maritima*, *S. portulacastrum*, and *Sporobolus virginicus* may be more abundant at tropical latitudes (Costa and Davy, 1992), possibly because of their greater tolerance of arid conditions and desiccation. Species differences in tolerance of inundation, salinity, desiccation, and frost as well as ability to acquire resources under differing stresses will determine mangrove versus saltmarsh shifts in response to global drivers. The physiographic equivalent of *S. alterniflora* is *R. mangle*, but

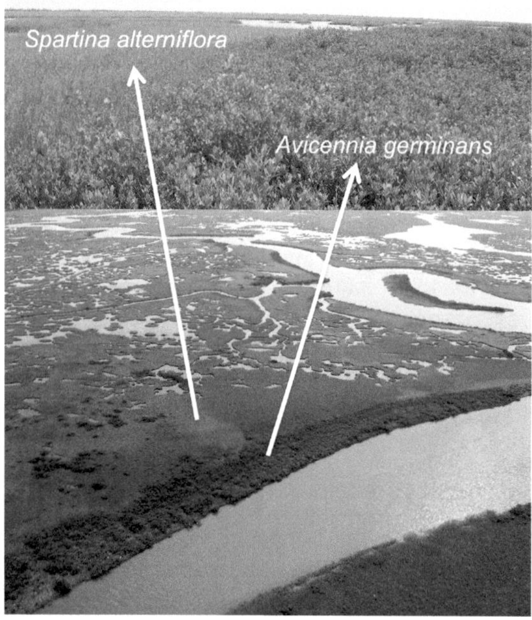

FIGURE 26.2 Ground (top) and aerial (bottom) views of a saltmarsh-mangrove community in coastal Louisiana, USA. Black mangrove (*Avicennia germinans*) occurs along creekbanks and intergrades with smooth cordgrass (*Spartina alterniflora*) in the marsh interior.

the latter cannot tolerate freezing temperatures and does not extend far into subtropical latitudes. Consequently, in the southern USA, *S. alterniflora* often intergrades with *A. germinans*, which is more cold tolerant (Fig. 26.2).

Coexisting mangrove and saltmarsh communities occur in a range of geomorphic settings. Along the coastline of New South Wales (NSW) in Australia, mangrove and saltmarsh are primarily located within drowned river valleys and barrier estuaries (Roy et al., 2001). Although some saltmarsh may be located within saline coastal lagoons, mangroves are generally excluded because mangrove propagule dispersal to these estuaries is restricted by the near permanent closure of the estuary mouth. Periodic extended flooding within coastal lagoons may also lead to the dieback of mangroves, yet allow the survival of saltmarsh (Mbense et al., 2016). Due to high wave energies along the southern Victorian coastline, mangrove and saltmarsh are almost exclusively distributed within three coastal embayments; Port Phillip Bay, Western Port Bay, and Corner Inlet. In New Zealand, mangroves are located within major embayments of the northern part of the North Island (Burns and Ogden, 1985) and may occur with species such as *Apodasmia similis* (jointed rush) (Fig 26.3).

Along the east coast of Florida, mangrove and saltmarsh coexist in low-energy environments such as the Indian River Lagoon (Reimold, 1977; Montague and Wiegert, 1990) and the Ponce de Leon inlet. Along the northern coastline of the Gulf of Mexico, mangrove and

FIGURE 26.3 Mangroves (*Avicennia marina*) invading a mudflat in New Zealand (Coromandel Penninsula). Marsh vegetation in foreground is *Apodasmia similis* (jointed wire rush).

saltmarsh may co-xist within the inactive deltaic environments of the Mississippi River and chenier coastal plains in Louisiana (Patterson and Mendelssohn, 1991), in embayments along the west coast of Florida from Tampa Bay to the Cedar Key (Kangas and Lugo, 1990; Stevens et al., 2006), and in lagoons along the southeast coast of Texas (Sherrod and McMillan, 1985; Everitt et al., 1996). In tropical Mexico, saltmarshes are often associated with mangroves in coastal lagoons or near river deltas with low sediment loads (Olmsted et al., 1993). Elsewhere in Latin America, saltmarsh may develop in settings that promote the development of hypersaline conditions and in disturbed mangrove areas (Costa and Davy, 1992).

The species composition of saltmarsh-mangrove communities in any geographic location may vary substantially because of differences in tide range, local topography, wave energy, and temperature regime. This variation is exemplified along the southern coast of the United States. On the east coast of Florida in the Indian River Lagoon complex, a typical shoreline might be dominated by *R. mangle* or *S. alterniflora* in front of a mid-marsh mixed community of *Salicornia* spp., *D. spicata*, *Borrichia frutescens*, *B. maritima*, and *A. germinans* (Montague and Wiegert, 1990). In south Florida, *S. alterniflora* might form a narrow fringe in front of a well-developed mangrove zone (*R. mangle*, *A. germinans*, *L. racemosa*) and a back-mangal zone dominated by *J. roemerianus* (Davis, 1940). On the west coast of Florida (north of Tampa Bay to the Cedar Key), *S. alterniflora* stands contain small stature, but abundant *A. germinans*. Similarly, the saltmarsh-mangrove communities in coastal Louisiana and south Texas consist of *S. alterniflora* and *A. germinans*. The mangrove *R. mangle* occurs farther south along the Texas coast because of its greater freeze sensitivity. Geographic variation in species co-occurrence and respective inherent environmental tolerances and competitive abilities will likely determine mangrove–saltmarsh interactions in any specific location.

3. LONG-TERM DYNAMICS

3.1 Tropical Northern Australia

Tropical northern Australia is drained by several macrotidal rivers, each receiving seasonal floods in relation to the monsoons (Woodroffe et al., 1989). Mangroves are confined to active depositional environments, including channel banks, the edges of mid-channel islands, and prograding coastlines (Woodroffe et al., 1985). The wide upper intertidal and supratidal flats are bare in lower rainfall areas or covered in grasses, sedges, and, in wetter areas, Melaleuca forest (Woodroffe et al., 1989). These wide estuarine flats are known as black soil plains.

Beneath the black soil plains lay extensive mangrove peat deposits. These have been located in King Sound (Semeniuk, 1980, 1982), the Fitzroy River (Jennings, 1975), the Ord River (Thom et al., 1975), and the Daly and Alligator Rivers (Woodroffe et al., 1985). Radiocarbon dating of these mangrove facies shows a consistency in age across the estuarine plains and also between river systems. For example, in the South Alligator River, 33 radiocarbon dates returned values within a range of 5370–6860 years BP with no spatial trends (Woodroffe et al., 1985). These dates corresponded with dates returned from other systems in northern Australia, including the Fitzroy (5800–7500 years BP) and the Ord (circa 6700 BP Thom et al., 1975). From these dates Woodroffe et al. (1985) hypothesized that extensive mangrove deposits occupied much of the estuarine plains adjacent to the macrotidal rivers near the end of the postglacial marine transgression.

Jennings (1975) had provided a climate interpretation of the extensive cover of mangrove or "big swamp" phase in northern Australian estuaries, postulating higher rainfall as a possible mechanism. In tropical north Queensland, mangroves still occupy the entire estuarine plain in some rivers experiencing high year-round rainfall but can occur together with saltmarsh species (Fig 26.4). By contrast, Woodroffe et al. (1985) presented a geomorphological explanation for this phase. In this model, continued sedimentation following sea level stabilization raised intertidal elevations above those tolerated by mangroves to the point that by 5500 BP most of the mangrove extent was confined to the fringes of channels. In dryer areas, mangroves gave way to hypersaline flats, whereas in wetter areas, upper intertidal and supratidal environments were colonized by salt-tolerant grasses and sedges.

The more detailed investigation by Woodroffe et al. (1985) had shown that the relationship between sedimentation and sea level was the driver of mangrove-saltmarsh/saltflat dynamics at the scale of the Holocene, at least in northern Australia. These studies provided a theoretical context for later work in the estuaries of SE Australia.

3.2 Southeastern Australia

Roy (1984) presented a model of estuarine infill for SE Australia that described phases in the availability of intertidal habitat. Under this model, the deeply incised drowned river valleys characteristic of SE Australia support little in the way of intertidal vegetation immediately following postglacial marine transgression. As these valleys infill with sediment, the fluvial bayhead delta progrades seaward, supporting wide intertidal flats. It is during this "intermediate" stage of infill that estuaries support the greatest extent of mangrove and saltmarsh. With the completion of infill floodplains accrete above intertidal elevations,

FIGURE 26.4 Saltmarsh (*Distichlis* sp.) growing adjacent to a stand of mangroves (*Ceriops tagal*) in Queensland, Australia.

flow is channelized throughout the length of the estuary, and intertidal habitats are restricted to channel fringes and cutoff embayments.

The Roy (1984) model assumes stable sea levels following the cessation of the postglacial marine transgression at approximately 6500 BP. Recent dating of encrusting organisms at several locations in NSW has suggested that mid-Holocene sea levels may have been somewhat higher (50–150 cm) than present levels (Baker and Haworth, 1997, 2000), an observation that might explain the widespread occurrence of potentially acid sulfate soils above current sea levels on estuarine floodplains (Wilson, 2005). Were mangroves more widely distributed in the mid-Holocene than at present?

Early work on the interactions between mangrove and saltmarsh in eastern Australia was dominated by succession theory. Pidgeon (1940) saw mangroves as the initial colonizers of exposed estuarine mudflat under conditions of extreme salinity and saturation. Mangroves were seen as active land builders, which accumulated sediment among aerial roots, inviting invasion by subsequent colonizers.

> Saltmarsh plants, the first of which are *Salicornia australis*, *Suaeda australis*, and *Samolus repens*, invade the landward margin of the mangrove area. With subsequent silt accumulation the mangroves are forced further out into the estuary and in their place a saltmarsh is formed. ***Pidgeon (1940, p. 24)***

Under this model, continued accretion led to colonization by salt-tolerant grasses (*Sporobolus virginicus*), followed by the rush *Juncus* spp., and then *Casuarina glauca*. As freshwater conditions replace saline, Eucalyptus invades, with the "climax" community appearing as a mixed Eucalypt forest (Pidgeon, 1940). Evidence supporting this dynamic succession was seen as the occurrence of "relict species in more advanced zones" and the active invasion of *Juncus* into the *Sarcocornia* meadow.

The succession model can be tested by coring within the saltmarsh in search of relict mangrove peat material. To this end, Mitchell and Adam (1989a) retrieved shallow sediment cores from several saltmarshes in Botany Bay and the Georges River in the Sydney region. Their coring failed to find any evidence of previous occupation of the saltmarsh habitat by mangrove and suggested a model that had saltmarsh species as primary colonizers, followed by invasion by mangrove. This model accorded with observations of initial colonization of newly created intertidal flats by saltmarsh on the northern foreshore of Botany Bay and the spreading of mangroves into the saltmarsh zone at various intertidal locations in the Sydney region (Mitchell and Adam, 1989b).

The Hawkesbury River in central NSW is a large drowned river valley in an intermediate phase of infill. Unlike the macrotidal estuaries of tropical northern Australia, the Hawkesbury supports a diverse array of vegetation on depositional terraces through its tidal length of some 106 km. There is no "big swamp" phase or there does not appear to have been at any point in the Holocene or late Pleistocene. Present-day intertidal flats support widespread mangrove in the central reaches of the estuary. Seaward of these, mudflats exposed at low tide are unvegetated. Headward, intertidal flats are dominated by saltmarsh, principally *Juncus kraussii*. Saintilan (1997) suggested that gradual infill has led to the replacement of mangrove with saltmarsh on successive intertidal flats as the river progrades seaward within its valley.

Evidence for this model of geomorphically driven vegetation succession is found in the stratigraphy of intertidal flats currently occupied by saltmarsh. Using techniques similar to those of Mitchell and Adam (1989b), Saintilan and Hashimoto (1999) found mangrove peats well preserved 20–30 cm beneath the present-day marsh surface, at the approximate elevation of contemporary mangrove root systems. Beneath these peats, estuarine shells dated to approximately 5000 years BP. The age of the mangrove peats varied with distance from the edge of wide intertidal flats, from 1200 to 1700 years BP at the upslope fringes of the flats to 500 years BP close to the current mangrove/saltmarsh boundary, suggesting gradual infill.

Mangrove root material retrieved from beneath saltmarsh in small creeks in southern NSW told a similar story. Mangrove root material dating to 1300 BP in Currambene Creek and 1900 BP in Cararma Inlet was retrieved from beneath the contemporary saltmarsh (Saintilan and Wilton, 2001). The lack of any significant fluvial input into Cararma Inlet suggested tidal reworking of aeolian and washover deposits as the most likely mechanism for infill. Within Currambene, preserved mangrove root material was more sporadic, and the suggestion was made that channel migration had erased evidence of prior mangrove occupation in some locations. At neither site was there evidence of widespread mangrove colonization earlier than 3000 BP, though more headward estuarine floodplains were not sampled.

3.3 Western Atlantic–Caribbean Region

Early work in Florida and the Caribbean region was also influenced by succession theory and the concept of mangroves as "land builders" (Spackman et al., 1966; Cohen and Spackman, 1977). Davis (1940) proposed that mangrove-saltmarsh zonation in Florida reflected ecological succession and seaward progradation by mangroves as sea level rose.

However, Egler (1952), Thom (1967), and others later challenged both the zonation–succession interpretation and the idea that mangroves were geological agents capable of building land. Egler (1952) presented evidence that mangroves had retreated landward during sea level rise, invading freshwater habitats in the Everglades. More recent work in the Caribbean has shown that mangroves growing in sediment-deficient settings can build vertically via peat formation (Woodroffe, 1981, 1983; Cameron and Palmer, 1995; Macintyre et al. 1995, 2004; McKee and Faulkner, 2000; Toscano and Macintyre, 2003; McKee et al., 2007a). In some cases, mangrove peat deposits reach 10 m in thickness, and radiocarbon dating shows that this biogenic accretion has kept pace with sea level rise (e.g., Twin Cays and Tobacco Range, Belize) (Macintyre et al., 2004; McKee et al., 2007a).

Paleostratigraphic evidence of fluctuations in mangrove and saltmarsh dominance is limited for the western Atlantic/Caribbean region and generally does not record the interplay between these specific vegetation types. Early work in Florida showed basal freshwater peat beneath mangrove peat in the modern Everglades-mangrove complex, suggesting mangrove invasion of freshwater marsh during sea level rise (Cohen and Spackman, 1977). Some of the deepest and oldest peat deposits occur in Belize where basal peats have been radiocarbon dated to about 8000 Cal BP (Toscano and Macintyre, 2003; Macintyre et al., 2004; McKee et al., 2007a). Most of these studies identified mangrove-derived peat throughout the stratigraphic profile and no components were identified as "saltmarsh" peat. Work by Woodroffe (1981, 1983) at Grand Cayman also showed predominately mangrove peat in radiocarbon dated cores. These latter observations show the continued dominance of mangroves throughout the Holocene, despite changes in sea level.

Other locations, including Jamaica (Digerfeldt and Hendry, 1987) and Trinidad (Ramcharan, 2004), show fluctuations between mangroves and brackish water sedges (e.g., *Cladium* sp.) or freshwater species, respectively. In the Negril swamp, Jamaica, radiocarbon dated cores show sedge (*Cladium* sp.) peat at the base (c. 7000–8000 Cal BP) and interspersed at intervals with mangrove peat (*Rhizophora* or *Rhizophora-Conocarpus*) or mixtures of *Cladium-Rhizophora* or *Cladium-Conocarpus* peat (Digerfeldt and Hendry, 1987). Pollen records from a late Holocene core on the Mexican Caribbean coast (Quintana Roo) showed mangroves dominating during humid periods and *Conocarpus erectus* and nonmangrove taxa dominating during drier periods, e.g., that coincided with the decline in the Mayan culture (Islebe and Sanchez, 2002).

4. RECENT INTERACTIONS

4.1 Air Photographic Evidence of Mangrove-Saltmarsh Dynamics in Southeastern Australia

On the basis of the stratigraphic evidence described above, the trend in SE Australia over the later Holocene has been one of the saltmarshes replacing mangrove with the infilling of estuaries and mangrove occupying freshly accreted habitat. This pattern has been consistent across a range of settings, from drowned river valleys (Saintilan and Hashimoto, 1999) to smaller barrier estuaries (Saintilan and Wilton, 2001). The pattern presents a contrast to trends identified from more recent times from air photographic records. Numerous studies from SE Australia have demonstrated the encroachment of mangrove into upper intertidal saltmarsh.

Saintilan and Williams (2000) cited 28 surveys that demonstrated saltmarsh loss to mangrove encroachment over the period covered by archival air photographs (usually 1940s–present). The trend is apparent across all east coast bioregions and a range of geomorphic settings. Within southern Queensland, Pleistocene sand barriers protect wide, shallow back-barrier deposits that support extensive mangrove and saltmarsh. Mangrove encroachment into saltmarsh in these environments is well-documented (McTainsh et al., 1986; Morton, 1994; Hyland and Butler, 1988; Manson et al., 2003). In northern NSW, mangroves and saltmarshes occupy the mouths of large rivers. Although losses of mangrove and saltmarsh to agriculture have been extensive (West, 1993), mangroves are encroaching on saltmarsh and some agricultural pastures (Saintilan, 1998). In central coast NSW, widespread losses of saltmarsh to mangrove have been reported from both shallow coastal lakes (Winning, 1990) and drowned river valleys (Mitchell and Adam, 1989a,b; Williams and Watford, 1997; Evans and Williams, 1997; Williams et al., 1999; McLoughlin, 2000; Haworth, 2002; Williams and Meehan, 2004). South coast NSW estuaries, mostly smaller "barrier" estuaries (Roy et al., 2001), have shown similar trends with a median loss of approximately 40% of the saltmarsh to mangrove encroachment (Chafer, 1998; Saintilan and Wilton, 2001; Meehan, 1997).

Within Victoria, saltmarshes and mangroves occupy the shorelines of large coastal embayments. Here, loss of saltmarsh to mangrove encroachment has been consistent, though less dramatic. Declines of 5%–12% of saltmarsh to mangrove encroachment have been reported for the saltmarshes of Western Port Bay (Rogers et al., 2005b) and losses have also been noted for Corner Inlet (Vanderzee, 1988) and the Gulf St. Vincent in South Australia (Burton, 1982). Within New Zealand, the proliferation of mangrove is more commonly described as a seaward colonization (Craggs et al., 2001; Park, 2001; Morrisey et al., 2003), though landward encroachment has been noted (Burns and Ogden, 1985).

4.2 Saltwater Intrusion in Northern Australia

Marked geomorphological changes to estuaries and coastal plains in northern Australia over the past 50 years have been associated with saltwater intrusion in the Alligator River region (Winn et al., 2006) and Mary River (Knighton et al., 1991; Mulrennan and Woodroffe, 1998). The gradual extension of tidal influence along stream channels, the expansion of tidal creeks, and the formation of new tidal creeks (Winn et al., 2006) are linked to the encroachment of mangrove and saline mudflats into freshwater vegetation (Finlayson et al., 1998), localized scour and dieback within Melaleuca forests, accretion of sediment on floodplains (Knighton et al., 1991; Woodroffe and Mulrennan, 1993; Bell et al., 2001), changes in subsurface hydrology (Jolly and Chin, 1992), and land cover changes (Ahmad and Hill, 1995; Bell et al., 2001). Changes since 1950 are significant with bare saline mudflats on the East Alligator River exhibiting a ninefold increase and an associated loss of 64% of Melaleuca forests by 2000 (Winn et al., 2006). More than 17,000 ha of freshwater vegetation have been adversely affected on the Mary River and a further 35%–40% of floodplains are immediately vulnerable to intrusion (Mulrennan and Woodroffe, 1998). A concurrent seaward and landward expansion of mangrove has been observed across the Gulf of Carpentaria, coincident with relatively rapid sea level rise and higher river discharge in the period 1987–2014 (Asbridge et al., 2015). Asbridge et al. (2016) attribute landward mangrove encroachment in this case to freshening of upper intertidal flats as a result of the combined influence of increased tidal inundation and increased wet season inundation from rivers. Mangrove mortality in the region in the period 2016–2017 reversed some to these trends, and corresponded to a period of dryer than normal monsoons and lower sea-levels (Lovelock et al. 2017; Saintilan et al. 2018).

A single cause for saltwater intrusion has not been identified (Mulrennan and Woodroffe, 1998). Instead, it is apparent that factors such as drier-than-average monsoonal conditions, low frequency and low-intensity cyclonic events, and above average ocean water levels (Winn et al., 2006) facilitate extension of tidal influence into freshwater environments, whereas tributary development, large tidal range, small elevation differences over floodplains, and uncontrolled feral buffalo promote the expansion of tidal influence (Knighton et al., 1991). Because of the desiccation of floodplain sediments in the dry seasons, the process of saltwater intrusion now appears to be internally driven and is likely to continue until an equilibrium state is reached between floodplain elevation and tidal influence (Mulrennan and Woodroffe, 1998; Winn et al., 2006).

4.3 Western Atlantic–Gulf of Mexico

Information on historical shifts in saltmarsh and mangrove vegetation in the western Atlantic–Gulf of Mexico region has been highlighted in site-specific observations (Lonard and Judd, 1985; Sherrod and McMillan, 1985; McMillan and Sherrod, 1986; Montague and Wiegert, 1990; Montague and Odum, 1997; McKee et al., 2004; Stevens et al., 2006) and more recently in regional analyses of Landsat and aerial photographic data (Cavanaugh et al., 2014a; Osland et al., 2013; Giri and Long, 2016). Mangroves have expanded by 4.3% across the region since the earliest Landsat records in 1980 (Giri and Long, 2016). In the northern Gulf of Mexico and the Atlantic Florida coast, the persistence and spread of mangroves is controlled mainly by temperature, with dieback and recovery occurring over decadal timescales (Sherrod and McMillan, 1985; Stevens et al., 2006; Osland et al., 2013, 2015; Giri and Long, 2014). Cavanaugh et al. (2014a,b) used a 28-year Landsat chronosequence to demonstrate a doubling of mangrove cover on the Florida east coast between 1984 and 2011 and associated this spread to a reduction in the frequency of days colder than −4°C. This increase has primarily been a landward expansion and "infilling" within poleward limits rather than poleward expansion (Giri and Long, 2016).

Hydroperiod, sedimentation, salinity, and nutrients as well as propagule dispersal and predation also influence local patterns of mangrove establishment and spread (Patterson and Mendelssohn, 1991; Patterson et al. 1993, 1997; Osland et al., 2014). In coastal Louisiana, for example, mangroves tend to dominate shorelines of tidal creeks, and saltmarsh occupies the interior position. The *A. germinans* zone is characterized by higher elevation, salinity, and soil bulk density, whereas the *S. alterniflora* zone is lower in elevation with greater flooding and reducing soils (Patterson and Mendelssohn, 1991).

Periodic freezes have killed or damaged mangroves, allowing expansion of saltmarsh, most recently during the 1980s (McMillan and Sherrod, 1986; Montague and Wiegert, 1990; Montague and Odum, 1997; Stevens et al., 2006). Stevens et al. (2006) review reports of major damage to mangroves in Florida as a result of severe freezes in 1962, 1977, 1981,

1983, 1985, 1989, and 1996. *A. germinans* has some capacity to recover from freeze damage through coppicing (stump sprouting), but periodic damage results in a stunted, scrub-like growth form. In Louisiana, the last severe freeze that caused widespread dieback of mangroves (*A. germinans*) occurred in December 1989 (K.L McKee pers. obs.). Although *A. germinans* can recover from mild freezes, the 1989 event killed mature trees that never recovered. However, propagules that had already dispersed survived this freeze and appeared to be the primary means by which the population reestablished. Thousands of propagules were observed establishing in the months after the freeze, and their viability was further confirmed in greenhouse culture (K.L. McKee unpublished data.). The resistance of *A. germinans* propagules in Louisiana to freezing temperatures is consistent with previous studies of cold tolerance of this species (Markley et al., 1982; Norman et al., 1984).

Most observations indicate that saltmarsh replaces freeze-killed mangroves in the Gulf of Mexico within 4–5 years (Stevens et al., 2006). However, relatively mild winters and the lack of severe freezes in recent years has allowed mangrove expansion in Florida (Stevens et al., 2006; Cavanaugh et al., 2014a,b) and Louisiana (K.L. McKee pers. obs.). Mangroves have expanded (average of 21% increase in cover) at the Cedar Key, Florida (29°08′N), between 1995 and 1999 during a period free from severe freezes (Stevens et al., 2006) and by 74% across the Texas coast between 1990 and 2010 (Armitage et al., 2015). Recent mangrove expansion in coastal Louisiana, USA, has been attributed to a drought-induced, widespread dieback of *S. alterniflora* during 2000 (McKee et al., 2004). Over 40,000 ha of saltmarsh were severely damaged with areas up to 5 km^2 reduced to mudflat. However, *A. germinans* was unaffected and even thrived during and after this marsh dieback event (Fig 26.5).

5. STRESSORS CONTROLLING DELIMITATION OF MANGROVE

Mangroves commonly exhibit a gradual reduction in development with latitude (Lot et al., 1975; Saenger and Snedaker, 1993; Saenger, 2002) and stunted forms at their latitudinal limit may be due to an increasingly stressful environment (Schaeffer-Novelli et al., 1990). Kangas and Lugo (1990) hypothesize that without "stress," in their case frost stress, mangrove vegetation is competitively superior and coastlines would be dominated by mangrove. However, field observations of saltmarsh dieback in coastal Louisiana, USA, and *A. germinans* expansion into areas formerly occupied by *S. alterniflora* (McKee et al., 2004) suggest that at subtropical latitudes some mangrove species are competitively inferior. In addition, greenhouse and field experiments show that *S. alterniflora* competitively inhibits growth of *A. germinans* (Patterson et al., 1993; McKee and Rooth, 2008). Where mangrove and saltmarsh coexist, environmental stressors control the delimitation of mangrove so that these communities typically form three zones: a seaward zone dominated by mangrove, a landward transition zone containing both mangrove (possibly stunted) and saltmarsh, and an interior zone dominated by saltmarsh. Numerous stressors, including geomorphic, hydrological and climatic controls, physicochemical factors, and biotic interactions, may act to delimit mangrove establishment (Patterson et al., 1993, 1997; McKee and Rooth, 2008) and enable saltmarsh development within the interior of marshes. In some extreme environments, however, marsh vegetation may facilitate mangrove recruitment through propagule trapping and/or amelioration of stressful soil conditions (McKee and Rooth, 2008).

FIGURE 26.5 Saltmarsh (*Spartina alterniflora*) dieback in coastal Louisiana, USA (top). Black mangrove (*Avicennia germinans*) was unaffected and later became dominant in some dieback areas (bottom).

5.1 Geomorphic and Hydrological Controls

Inundation by tides is not an ecophysiological requirement of mangroves and saltmarshes, but it is a typical feature and plays an important role by limiting excessive buildup of salts within soils (Saenger, 2002) and distributing mangrove and saltmarsh propagules (Adam, 1990; Clarke, 1993). Indeed, inundation may play a significant role in mangrove and saltmarsh zonation as tidal flushing may act to alter soil salinity (Patterson and Mendelssohn, 1991) and control the transport of propagules along an elevation gradient (Adam, 1990; Patterson et al., 1997; Saenger, 2002). The hydrologic role of inundation is largely controlled by the geomorphology of a marsh so that inundation frequency and water volume are negatively proportional to land elevation. However, the relative positions of saltmarsh and mangrove vegetation along a topographic gradient vary with geographic region and the species involved. In the northern Gulf of Mexico, mangroves coincide and compete with low-marsh species such as *S. alterniflora*, which are highly flood tolerant (Patterson and Mendelssohn, 1991; Patterson et al., 1993, 1997). In these communities, mangroves (*A. germinans*) occupy the higher elevation creekbanks and saltmarsh species dominate the interior, lower elevation areas. In other locations, low elevation sites are commonly inhabited by mangrove, are inundated more frequently, and receive higher water volumes, whereas higher elevations are commonly inhabited by saltmarsh. The latter pattern can be found in coexisting mangrove and saltmarsh communities in Australia and in Florida, USA (Montague and Wiegert, 1990). In the latter case, the low elevation sites along shorelines are dominated by *R. mangle* and higher elevations are occupied by high-marsh species such as *B. maritima*, *D. spicata*, *B. frutescens*, and *Salicornia* spp.; the upland transition may be occupied by *J. roemerianus*, *Spartina bakeri*, or *S. patens* (Montague and Wiegert, 1990).

Adaptations to cope with submergence and high salt environments are evident within both mangrove and saltmarsh species (summarized in Adam, 1990 and Saenger, 2002), and species tolerance to submergence and salinity are hypothesized to cause the zonation of plants observed within marshes. Numerous studies have demonstrated the role of salinity in causing mangrove species to segregate (Ball, 1988a,b; Lin and Sternberg, 1992; Ball and Pidsley, 1995; Ball, 2002; López-Hoffman et al., 2006) and coexist with temporal salinity variation (Ball, 1998). However, for saltmarsh, Adam (1990) argues that because soil salinity varies spatially and temporally, zonation may be more readily explained on the basis of species tolerance to frequency and duration of submergence. Furthermore, although soil salinity and tidal regime are important factors, these factors do not appear to explain observed distributions of saltmarsh species across the intertidal zone (Silvestri et al., 2005). The zonation of coexisting mangrove and saltmarsh communities in Sydney (Clarke and Hannon, 1971) and Brazil (Santos et al., 1997, cited in Saenger, 2002) was found to be salinity based, with species exhibiting definable salt tolerances (Clarke and Hannon, 1971) and their distribution being determined by a salinity model based on topographic level and upland runoff (Santos et al., 1997). Although causes of mangrove and saltmarsh zonation remain incompletely understood (Snedaker, 1982; Adam, 1990), it is evident that zonation, soil salinity, and submergence are all maintained along elevation gradients and is likely to contribute to the distribution of mangrove and saltmarsh vegetation (Snedaker, 1982; Saenger, 2002).

Zonation as a consequence of differential dispersal of mangrove propagules has also been proposed. This is primarily based on the concept of tidal sorting of propagules proposed by Rabinowitz (1975, 1978a,b,c) for mangroves in Panama. However, the relationship between propagule size and zonation has not been supported for mangroves in other locations (Ball, 1980; McKee, 1995a; Sousa et al., 2007) In addition, differential dispersal and survival of propagules with tides has not been established for saltmarsh species, and because many saltmarsh species are able to vegetatively propagate, the contribution of tides to dispersal is likely to be limited (Adam, 1990).

Regardless of the role of tidal sorting of propagules in zonation, tides do distribute mangrove propagules and it is likely that zonation of mangrove to saltmarsh may be related to differential seedling mortality at different tidal levels (Ellison and Farnsworth, 1993). High soil salinity decreases the rate of pericarp shedding by mangrove propagules, thereby limiting survival (Downton, 1982; Clarke and Myerscough, 1991). Thus, higher soil salinities and reduced inundation frequency, common in the high marsh, cause mangrove propagules to desiccate and suffer high mortality (Clarke and Allaway, 1993). However, work in the northern Gulf of Mexico found that mortality of *A. germinans* propagules was associated with excessive flooding and damage by predators in low-marsh areas, resulting in decay and loss of viability (Patterson et al., 1997).

Observed interactions between mangrove and saltmarsh globally have highlighted the role of hydrology and geomorphology in increasing the dispersal and survival of mangrove propagules and seedlings. For example, a strong relationship has been found between relative sea level rise and the upslope migration of mangrove into saltmarsh environments in SE Australia (Rogers, 2004). In this case, relative sea level rise incorporates both geomorphological and hydrological changes and may be altered by eustatic sea level rise, which relates to changes in the volume of water within oceans through the melting of ice caps and thermal expansion of water and/or surface elevation changes of the land through surface processes of vertical accretion or subsurface processes of subsidence and autocompaction. Sea level rise has also been suggested as a primary driver of mangrove expansion within thermal limits across the US Gulf and Atlantic coasts (Krauss et al., 2011; Raabe et al., 2012; Armitage et al., 2015; Giri and Long, 2016).

Hydrological changes due to global warming has caused global sea level to increase by 1.0–2.0 mm/yr; in the 20th century, the rate of sea level rise is predicted to accelerate toward 2100 (Solomon et al., 2007). Recent studies of relative sea level trends for saltmarshes in Tasmania and New Zealand show higher rates over the past century (2.5–3.1 mm/yr) compared with the previous four centuries (Gehrels et al., 2007). Numerous studies predict that global sea level rise will cause a shift in the boundary between mangrove and saltmarsh (Snedaker, 1995; Blasco et al., 1996) and indeed the study of Rogers et al. (2006) related sea level changes to the expansion of mangroves in SE Australia. Changes to the water levels inundating a marsh may also occur because of altered tidal prisms resulting from engineering works, such as dredging of estuary entrances. These hydrological changes may increase the mean water level and/or alter the tidal range within a marsh, thereby altering soil salinity and increasing the distribution potential of mangrove propagules. Both reduced soil salinities and increased inundation promote the survival of mangrove within saltmarsh and increase their competitive potential. Along the US Gulf and Atlantic Coasts, mangrove encroachment is occurring in association with retreat of marsh seaward edges and forest-to-marsh transition, further implicating sea level rise as a major driver of change (Armitage et al., 2015; Raabe and Stumpf, 2016).

Geomorphological changes to marsh elevation alter the inundation frequency and duration within a marsh and may occur as a result of surface processes such as vertical accretion. Increased sediment availability and vertical accretion within estuaries have been hypothesized as mechanisms promoting mangrove encroachment of saltmarsh in SE Australia (summarized by Saintilan and Williams, 1999 and McLoughlin, 2000) by creating more areas suitable for mangrove establishment and providing a soft substrate for mangrove propagule establishment within the saltmarsh. Certainly the spread of mangrove along estuarine foreshores seems related to the availability of new habitat, and commonly the landward encroachment of mangrove is accompanied by seaward encroachment also. Perhaps the overall increase in population of *A. marina* and increased fecundity made possible by higher nutrient loads has increased propagule availability within estuaries, including the upper intertidal environment. However, increased vertical accretion is also more likely to increase marsh elevation, reducing inundation frequency and thereby reducing the likelihood of mangrove survival due to propagule desiccation. Furthermore, the "fertilization" of the saltmarsh from anthropogenic sources does not necessarily lead to improved mangrove colonization. Experimental fertilization of *A. marina* propagules has failed to show an impact on propagule survival (Clarke and Myerscough, 1993; Saintilan, 2003).

Rates of vertical accretion have been used as an estimate for marsh elevation and a surrogate for investigating the cause of mangrove–saltmarsh interaction. This is appropriate at some sites, such as Homebush Bay and Western Port Bay, Australia, where saltmarsh surface elevation is maintained primarily by vertical accretion (Rogers et al., 2005a,b). However, this assumption is not always true. Recent literature has highlighted the role of belowground processes in altering marsh elevations (Cahoon et al., 1999, 2006; McKee et al., 2007a). Processes such as subsidence, soil compaction, plant productivity and dieback, groundwater availability, and tidal flooding interact to alter soil volumes over a range of spatial and temporal scales.

The relationship between marsh elevations and sea level has primarily been explored through the use of surface elevation tables (SETs) (described by Cahoon et al., 2002). By incorporating analyses of surface processes such as vertical accretion, the belowground component of surface elevation change can be differentiated. In SE Australia, SETs were used to show that mangrove surface elevation gain was consistently lower than saltmarsh surface elevation gain, despite vertical accretion being consistently higher in the mangrove zone (Rogers et al., 2005a,b, 2006). Mangrove encroachment of saltmarsh in the same region has been attributed to autocompaction or subsidence, which causes the observed disequilibrium between mangrove surface elevation change and vertical accretion (Rogers et al., 2006). Howard et al. (2016) found a similar association between mangrove encroachment and soil subsidence at the 10000 Island National Wildlife Refuge in Florida. Across the SE Australian sites, those marshes with the highest rates of relative sea level rise were those with the highest rates of mangrove encroachment (Rogers et al., 2005b). Autocompaction of sediments corresponded to a severe El Nino-related drought; and surface elevation change correlated strongly with total monthly rainfall (Rogers et al., 2005a), the Southern Oscillation Index, and groundwater depth (Rogers and Saintilan, 2008). This research is consistent with other studies of surface elevation and groundwater (Schmidt and Burgman, 2003; Watson, 2004; Whelan et al., 2005) and suggests that rainfall and subsequent groundwater recharge play a significant role in determining interannual variability in surface elevation with respect to sea level.

5.2 Climatic Controls

Climatic factors are commonly proposed to promote or limit the survival of mangrove within coexisting mangrove and saltmarsh communities as a result of extreme low temperatures and frost causing the death of mangrove or rainfall influencing soil salinity.

Many tropical and subtropical plants exhibit physiological dysfunction and damage in response to chilling temperatures (van Steenis, 1968) and this appears to be the case for mangroves (McMillan, 1975; Markley et al., 1982; McMillan and Sherrod, 1986; Kangas and Lugo, 1990; Olmsted et al., 1993). Freeze sensitivity is particularly apparent along the Gulf and Atlantic coasts of North America, and the distribution of *Avicennia germinans, Laguncularia racemosa*, and *Rhizophora mangle* corresponds to their tolerance of freezing frequency and intensity under experimental conditions (Cavanaugh et al., 2015). In these locations, mangroves tend to expand during periods of mild winters (Stevens et al., 2006; Giri and Long, 2014; Cavanaugh et al., 2014a) and when weather extremes or other factors stress or eliminate saltmarsh (McKee et al., 2004). When mangroves are periodically damaged or killed by freezes, saltmarsh quickly regains dominance (Stevens et al., 2006). Consequently, future changes in global climate regime and local weather patterns, in particular a reduction in extreme daily minima, will likely influence the relative dominance of mangrove and saltmarsh vegetation in this region. Osland et al. (2013) used downscaled future climate projections for 2070–2100 to suggest that a 2–4°C temperature rise would lead to mangrove replacing 95% of saltmarsh in Louisiana, 100% in Texas, and 60% in Florida. Cavanaugh et al. (2015) projected a northern expansion in Florida over the next 50 years of 160, 110, and 120 km for *A. germinans, R. mangle*, and *L. racemosa*, respectively. Rapid colonization at poleward limits may be facilitated by the "precocious reproduction" of individuals along the invasion front, with observations of younger *R. mangle* reproducing and producing larger propagules at the leading edge of the expanding range in Florida (Dangremond and Feller, 2016).

Mangroves show greater luxuriance and species richness in high rainfall areas (Tomlinson, 1986; Hutchings and Saenger, 1987; Smith and Duke, 1987), and species distribution correlates strongly with freshwater influence from rainfall (Bunt et al., 1982), controlling for latitude. An aridity gradient extending across the Texan coastline from the Louisiana border to Texas–Mexico border strongly influences the degree of vegetative cover, and small changes in freshwater delivery in regions with mean annual rainfall below ~995 mm/yr can profoundly influence the relative extent of mangrove, saltmarsh and hypersaline flat (Osland et al., 2016), and the dominance of graminoid and succulent vegetation in the saltmarsh (Osland et al., 2014). Low rainfall and a shortened rainfall season in Senegal in the 1980s contributed to the decline of mangroves and increased surface areas of bare saltflats and saline grasslands (Diop et al., 1997). In addition, studies in Australia have suggested that freshening of saltmarsh environments from rainfall might promote survivorship of mangroves (McTainsh et al., 1986) and may explain the establishment of mangrove at the saltmarsh-terrestrial boundary in tropical and subtropical areas of Australia (Duke, 2006).

There is other evidence that mangrove encroachment in tropical and subtropical climates may be related to rainfall trends. In these situations, the proportion of mangrove and saltmarsh in estuaries is correlated to annual rainfall (Bucher and Saenger, 1991). Where saltmarsh occurs, it often occupies a middle position between seaward and landward mangrove zones. Increased rainfall in recent decades has been used to explain the encroachment of tall mangrove forest into salt pan in Hinchinbrook Channel (Duke, 1995) and stunted mangrove into saltmarsh in Moreton Bay (Duke, pers. comm.). Conversely, declines in rainfall have resulted in mangrove dieback in the upper intertidal environment in some locations, most notably in the Gulf of Carpentaria during the usually dry 2015–16 monsoon (Duke pers. comm.) The correlation between rainfall and the proportion of mangrove and saltmarsh breaks down in NSW (Saintilan, 2003). Here, the upslope limit of mangrove is more consistently related to frequency of tidal inundation, and there is no landward mangrove fringe.

5.3 Physicochemical Factors

Physicochemical factors have been hypothesized to limit the establishment of mangrove within saltmarsh areas. In particular, mangroves may be excluded from saltmarsh environments because of nutrient deficiency and/or high soil salinity within saltmarshes. Salinity decreases pericarp shed and increases desiccation of mangrove propagules (Downton, 1982; Clarke and Myerscough, 1991) and may contribute to zonation of mangrove and saltmarsh species because of differential species tolerance to soil salinity (summarized in Adam, 1990 and Saenger, 2002) as discussed above.

Experimental fertilization of mangroves indicates that nitrogen and phosphorus limitation is common (Boto, 1983; Boto and Wellington, 1983; McKee et al., 2002) and on fertilization mangroves exhibit increases in growth and biomass (Naidoo, 1987; Clarke and Allaway, 1993; Feller et al. 2003a,b; Lovelock et al. 2004, 2006; Simpson et al., 2013).

However, because the effect of nutrient enrichment does not become apparent until mangrove cotyledons have been exhausted (Naidoo, 1987; Clarke and Allaway, 1993), it has less impact on early mangrove establishment within saltmarsh. Saintilan (2003) confirmed that mangrove mortality within saltmarsh is not reduced through nutrient enrichment of seedlings. In some geographic regions, mangrove mortality within saltmarshes is the result of desiccation of propagules (Clarke and Allaway, 1993; Saintilan, 2003), which is dependent on climatic, geomorphic, and hydrologic factors. However, changes in tissue chemistry by fertilization or light conditions (McKee, 1995a,b) could influence mangrove seedling susceptibility to herbivores.

5.4 Biotic Considerations and Interactions

Several studies suggest that saltmarsh precedes mangrove in successional sequences because of tidal transport of mangrove propagules and shading of saltmarsh by mangrove seedlings (summarized in Kangas and Lugo, 1990). Work by McKee et al. (2007b) found that establishment, survival, and growth of mangrove seedlings were facilitated by herbaceous species, *Sesuvium portulacastrum* and *Distichlis spicata*. Both species promoted mangrove propagule trapping and ameliorated soil conditions (aeration, temperature). Peterson and Bell (2012) demonstrated that *Sporobolus virginicus* had a far greater capacity than *S. portulacastrum* and *B. maritima* to retain propagules of *A. germinans*. Alternatively, dense stands of saltbush (*Tecticornia* spp.) have been suggested to inhibit mangrove establishment in saltmarshes in Western Port Bay, Australia (Rogers et al., 2005b), with similar competition being reported between mangroves and *S. alterniflora* in marshes in Louisiana (Patterson et al., 1993; McKee and Rooth, 2008) and Texas (Guo et al., 2013), though the extent of this suppression may decrease in higher latitudes (Guo et al., 2013). Several mechanisms have been advanced, including shading seedlings, competition for nutrients (Guo et al., 2013; Simpson et al., 2013), or facilitated accretion (Rogers et al., 2005b). Experimental data presented by Peterson and Bell (2015) suggest that succulent saltmarsh vegetation may not be as efficient at trapping and retaining mangrove propagules during the ebb tide as graminoid vegetation.

Plant–animal interactions are also evident within coexisting communities, and differential predation of mangrove propagules by crabs is suggested as a cause of zonation within mangroves (Smith, 1987; Smith et al., 1989), but this hypothesis is not supported in other locations

(McKee, 1995a; Sousa and Mitchell, 1999). A study of *A. germinans* seedlings found that predator damage to propagules (and frequent flooding) leads to decay of propagules within the *Spartina* zone in coastal Louisiana (Patterson et al., 1997).

Once established, tall *A. germinans* are more resistant and resilient to freeze-induced mortality than smaller individuals (Osland et al., 2015). This observation suggests that nonlinear responses of mangrove expansion to declining frost frequency may occur. As poleward individuals grow taller between successive frosts, a greater number reach the required "escape height," providing the basis for subsequent expansion (Osland et al., 2015). Anthropogenic modification of nutrient status of soils and atmospheric CO_2 may enhance this trend as suggested in terrestrial woody encroachment models (Bond and Midgley, 2000; Saintilan and Rogers, 2015).

5.5 Ecosystem Effects

The broad-scale transition of intertidal vegetation from saltmarsh to mangrove represents a change in the physical structure and composition of above- and belowground plant material with potential ecosystem-scale effects. Comeaux et al. (2012) describe two mechanisms whereby mangrove colonization may increase the resistance of coastal marshland to rising sea levels. First, mangroves may facilitate vertical elevation gain more efficiently than saltmarsh, through more efficient trapping of sediment and vertical displacement as the root system expands (Friess et al., 2012). Second, the comparative density of mangrove above- and belowground root mass may facilitate soil cohesive strength while also dampening wave energy (Comeaux et al., 2012).

Comeaux et al. (2012) used radiometric dating of marsh systems in Texas to suggest a higher rate of accretion in the mangrove. This is supported by Surface Elevation Table-Marker Horizon (SET-MH) observations where mangroves and saltmarsh coexist (Rogers et al. 2006, 2013, 2014), though hydroperiod is a confounding factor in SE Australia, where saltmarsh consistently occupies higher elevations. The volume and nature of carbon sequestered by the wetland may also change as saltmarsh transitions to mangrove. In the Gulf of Mexico, studies have focused on saltmarshes recently encroached by mangroves, with some showing an increase in carbon sequestration to root mass (Bianchi et al., 2013; Osland et al., 2012) and others showing no significant change (Perry and Mendelssohn, 2009; Henry and Twilley, 2013; Doughty et al., 2016). In SE Australia, where longer chronosequences of transition may be studied, Kelleway et al. (2016) showed an increase in carbon gain, but only after several decades. This may indicate that the gain is due to the relative recalcitrant nature of mangrove carbon compared with saltmarsh (Bianchi et al., 2013). Relative sequestration potential may be altered by climate change, though the increased mineralization effected by elevated temperatures may be offset by decreased mineralization of carbon resulting from increased hydroperiod (Lewis et al., 2014).

6. CONCLUSIONS

The interaction between mangrove and saltmarsh is of interest to climate change research for several reasons. Having evolved in the tropics (Chapman, 1977; Specht, 1981), mangroves are most prolific in lower latitudes and decrease in diversity and vigor toward the poles

(Lot et al., 1975; Saenger and Snedaker, 1993). By contrast, saltmarshes increase in floristic diversity with increasing latitude within the temperate zone (Adam, 1990) and in the absence of mangroves form the characteristic intertidal vegetation in these latitudes. In the predominantly temperate latitudes where the two community types coexist, saltmarsh distribution is confined in part by the presence of mangrove and in part by the extreme environmental conditions of the upper intertidal zone.

Environmental variability can therefore profoundly influence the competitive interactions between mangrove and saltmarsh. In southern United States, where the proliferation of mangrove is periodically checked by frost, higher temperature and decreased freeze frequency and severity could enhance the survival and growth of species such as *A. germinans*. Wider and more continuous stands of this species may be expected in the northern Gulf of Mexico and the Atlantic coast of Florida. Less cold-tolerant species, such as *R. mangle*, may be able to shift northward along the coasts of Baja, California, Texas, and Florida.

In SE Australia and New Zealand, the growth of the mangrove *A. marina* will be aided not only by increased temperatures toward its southern limit of distribution but also by higher sea levels. Presently, the upslope limit of *A. marina* is defined by the mean high water mark. While there is continued debate over the causes of mangrove encroachment into the saltmarsh of eastern Australia, there is growing evidence for a role of relative sea level rise, evidenced by the relationship between the rate of encroachment and relative sea level trends, and an increased rate of relative sea level rise in temperate Australasian saltmarshes over the past century. The subtle elevation gradients defining the position of mangrove and saltmarsh in these situations provide an early indication of the effects of sea level rise in the coastal wetlands of the region. The concern in these situations is that the landward transition of saltmarsh may be impeded by topographic constraints, both cultural and natural. Where feasible, thought should be given to the designation of landward "accommodation space" in anticipation of projected rates of sea level rise so that the full range of wetland vegetation communities can continue to coexist in these estuaries.

References

Adam, P., 1990. Saltmarsh Ecology. Cambridge University Press, Cambridge.

Ahmad, W., Hill, G.J.E., 1995. Land cover classification of the Mary River floodplain with emphasis on the effect of saltwater intrusion. In: Davenport, C.C., Riley, S.J., Ringrose, S.M. (Eds.), Proceedings of the Second North Australian Remote Sensing and GIS Forum. Australian Government Publishing Service, Canberra, Australia, pp. 76–83.

Armitage, A.R., Highfield, W.E., Brody, S.D., Louchouarn, P., 2015. The contribution of mangrove expansion to salt marsh loss on the Texas Gulf Coast. PLoS One 10 (5), e0125404.

Asbridge, E., Lucas, R., Accad, A., Dowling, R., 2015. Mangrove response to environmental changes predicted under varying climates: case studies from Australia. Current Forestry Reports 1 (3), 178–194.

Asbridge, E., Lucas, R., Ticehurst, C., Bunting, P., 2016. Mangrove response to environmental change in Australia's Gulf of Carpentaria. Ecology and Evolution 6 (11), 3523–3539.

Baker, R.G.V., Haworth, R.J., 1997. Further evidence from relic shellcrust sequences for a late Holocene higher sea level for eastern Australia. Marine Geology 141, 1–9.

Baker, R.G.V., Haworth, R.J., 2000. Smooth or oscillating late Holocene sea-level curve? Evidence from the palaeozoology of fixed biological indicators in east Australia and beyond. Marine Geology 163, 367–386.

Ball, M.C., 1980. Patterns of secondary succession in a mangrove forest of southern Florida. Oecologia 44 (2), 226–235.

Ball, M.C., 1988a. Ecophysiology of mangroves. Trees 2, 129–142.

Ball, M.C., 1988b. Salinity tolerance in the mangroves, *Aegiceras corniculatum* and *Avicennia marina*, I. Water use in relation to growth, carbon partitioning and salt balance. Australian Journal of Plant Physiology 15, 447–464.

Ball, M.C., 1998. Mangrove species richness in relation to salinity and waterlogging: a case study along the Adelaide River Floodplain, northern Australia. Global Ecology and Biogeography Letters 7, 73–82.

Ball, M.C., Pidsley, S.M., 1995. Growth responses to salinity in relation to distribution of two mangrove species, *Sonnertia alba* and *S. lanceolata*, in northern Australia. Functional Ecology 9, 77–85.

Ball, M.C., 2002. Interactive effects of salinity and irradiance on growth: implications for mangrove forest structure along salinity gradients. Trees - Structure and Function 16, 126–139.

Bell, D., Menges, C.H., Bartolo, R.E., 2001. Assessing the extent of saltwater intrusion in a tropical coastal environment using radar and optical remote sensing. Geocarto International 16, 1–8.

Bianchi, T.S., Allison, M.A., Zhao, J., Li, X., Comeaux, R.S., Feagin, R.A., Kulawardhana, R.W., 2013. Historical reconstruction of mangrove expansion in the Gulf of Mexico: linking climate change with carbon sequestration in coastal wetlands. Estuarine, Coastal and Shelf Science 119, 7–16.

Blasco, F., Saenger, P., Janodet, E., 1996. Mangroves as indicators of coastal change. Catena 149, 1–12.

Bond, W.J., Midgley, G.F., 2000. A proposed CO_2-controlled mechanism of woody plant invasion in grasslands and savannas. Global Change Biology 6 (8), 865–869.

Boto, K.G., 1983. Nutrient status and other soil factors affecting mangrove productivity in northeast Australia. Wetlands 3, 45–49.

Boto, K.G., Wellington, J.T., 1983. Phosphorus and nitrogen nutritional status of a northern Australian mangrove forest. Marine Ecology Progress Series 11, 63–69.

Bucher, D., Saenger, P., 1991. An inventory of Australian estuaries and enclosed marine waters: an overview of results. Australian Geographic Studies 29, 370–381.

Bunt, J.S., Williams, W.T., Duke, N.C., 1982. Mangrove distribution in northeast Australia. Journal of Biogeography 9, 111–120.

Burns, B.R., Ogden, J., 1985. The demography of the temperate mangrove (*Avicennia marina* (Forsk.) Vierh.) at its southern limit in New Zealand. Australian Journal of Ecology 10, 125–133.

Burton, T., 1982. Mangrove changes recorded north of Adelaide. Safic 6, 8–12.

Cahoon, D.R., Day Jr., J.W., Reed, D.J., 1999. The influence of surface and shallow subsurface soil processes in wetland elevation: a synthesis. Current Topics in Wetland Biogeography 3, 72–88.

Cahoon, D.R., Lynch, J.C., Hensel, P., Boumans, R., Perez, B.C., Segura, B., Day Jr., J.W., 2002. High-precision measurements of wetland sediment elevation: I. Recent improvements to the Sedimentation-Erosion Table. Journal of Sedimentary Research 72, 730–733.

Cahoon, D.R., Hensel, P.F., Spencer, T., Reed, D.J., McKee, K.L., Saintilan, N., 2006. Wetland vulnerability to relative sea-level rise: wetland elevation trends and process controls. In: Verhoeven, J.T.A., Beltman, B., Bobbink, R., Whigham, D.F. (Eds.), Wetland Conservation and Management. Springer-Verlag, Berlin Heidelberg, pp. 271–292.

Cameron, C.C., Palmer, C.A., 1995. The mangrove peat of the Tobacco range islands, Belize barrier reef, Central America. Atoll Research Bulletin 431, 1–32.

Cavanaugh, K.C., Kellner, J.R., Forde, A.J., Gruner, D.S., Parker, J.D., Rodriguez, W., Feller, I.C., 2014a. Poleward expansion of mangroves is a threshold response to decreased frequency of extreme cold events. Proceedings of the National Academy of Sciences of the United States of America 111 (2), 723–727.

Cavanaugh, K.C., Kellner, J.R., Forde, A.J., Gruner, D.S., Parker, J.D., Rodriguez, W., Feller, I.C., 2014b. Reply to Giri and Long: freeze-mediated expansion of mangroves does not depend on whether expansion is emergence or reemergence. Proceedings of the National Academy of Sciences of the United States of America 111 (15), E1449.

Cavanaugh, K.C., Parker, J.D., Cook-Patton, S.C., Feller, I.C., Williams, A.P., Kellner, J.R., 2015. Integrating physiological threshold experiments with climate modeling to project mangrove species' range expansion. Global Change Biology 21 (5), 1928–1938.

Chafer, C.J., 1998. A Spatio-temporal Analysis of Estuarine Vegetation Change in the Minnamurra River 1938-1997, 47 p., plus appendices. Unpublished report to the Minnamurra Estuary Management Committee.

Chapman, V.J., 1977. Introduction. In: Chapman, V.J. (Ed.), Ecosystems of the World. 1. Wet Coastal Ecosystems. Elsevier Scientific Publications Company, Amsterdam, pp. 1–29.

Clarke, L.D., Hannon, N.J., 1971. The mangrove swamp and saltmarsh communities of the Sydney district. IV. The significance of species interaction. Journal of Ecology 59, 535–553.

Clarke, P.J., 1993. Dispersal of grey mangrove (*Avicennia marina*) propagules in south-eastern Australia. Aquatic Botany 45, 195–204.

Clarke, P.J., Allaway, W., 1993. The regeneration niche of the grey mangrove *Avicennia marina* – effects of salinity, light and sediment factors on establishment, growth and survival in the field. Oecologia 93, 548–556.

Clarke, P.J., Myerscough, P.J., 1991. Buoyancy of *Avicennia marina* propagules in south-eastern Australia. Australian Journal of Botany 39, 77–83.

Clarke, P.J., Myerscough, P.J., 1993. The intertidal distribution of the grey mangrove *(Avicennia marina)* in south-eastern Australia: the effects of physical conditions, interspecific competition, and predation on propagule establishment and survival. Australian Journal of Ecology 18, 307–315.

Clusener, M., Breckle, S.W., 1987. Reasons for the limitation of mangrove along the west-coast of northern Peru. Vegetatio 68, 173–177.

Costanza, R., de Groot, R., Sutton, P., van der Ploeg, S., Anderson, S.J., Kubiszewski, I., Farber, S., Turner, R.K., 2014. Changes in the global value of ecosystem services. Global Environmental Change 26, 152–158.

Cohen, A.D., Spackman, W., 1977. Phytogenic organic sediments and sedimentary environments in the evergladesmangrove complex. Part II. The origin, description and classification of the peats of southern Florida. Palaeontographica, Abteilung B 162, 71–114.

Comeaux, R.S., Allison, M.A., Bianchi, T.S., 2012. Mangrove expansion in the Gulf of Mexico with climate change: implications for wetland health and resistance to rising sea levels. Estuarine. Coastal and Shelf Science 96, 81–95.

Costa, C.S.B., Davy, A.J., 1992. Coastal salt marsh communities of Latin America. In: Seeliger, U. (Ed.), Coastal Plant Communities of Latin America. Academic Press, San Diego, USA, pp. 179–199.

Craggs, R., Hofstra, D., Ellis, J., Schwarz, A., Swales, A., Nicholls, P., Hewitt, J., Ovenden, R., Pickmere, S., 2001. Physiological Responses of Mangroves and Saltmarsh to Sedimentation. NIWA Client Report to Auckland Regional Council. ARC01268/1, 78p.

Dangremond, E.M., Feller, I.C., 2016. Precocious reproduction increases at the leading edge of a mangrove range expansion. Ecology and Evolution 6 (14), 5087–5092.

Davis, J.H.J., 1940. The ecology and geologic role of mangroves in Florida. Papers from Tortugas Laboratory 32, 311–384.

Davy, A.J., Costa, C.S.B., 1992. Development and organization of saltmarsh communities. In: Seeliger, U. (Ed.), Coastal Plant Communities of Latin America. Academic Press, San Diego, pp. 179–199.

Digerfeldt, G., Hendry, M.D., 1987. An 8000 year Holocene sea-Level record from Jamaica - implications for interpretation of Caribbean reef and coastal history. Coral Reefs 5, 165–169.

Diop, E.S., Soumare, A., Diallo, N., Guisse, A., 1997. Recent changes of the mangroves of the Saloum river estuary, Senegal. Mangroves and Salt Marshes 1, 163–172.

Doughty, C.L., Langley, J.A., Walker, W.S., Feller, I.C., Schaub, R., Chapman, S.K., 2016. Mangrove range expansion rapidly increases coastal wetland carbon storage. Estuaries and Coasts 39 (2), 385–396.

Downton, W.J.S., 1982. Growth and osmotic relations of the mangrove *Avicennia marina*, as influenced by salinity. Australian Journal of Plant Physiology 9, 519–528.

Duke, N.C., 1995. Genetic diversity, distributional barriers and rafting continents – more thoughts on the evolution of mangroves. Hydrobiologia 295, 167–181.

Duke, N.C., 2006. Australia's Mangroves. The Authoritative Guide to Australia's Mangrove Plants. University of Queensland, Brisbane, 200 p.

Duke, N.C., Ball, M.C., Ellison, J.C., 1998a. Factors influencing biodiversity and distributional gradients in mangroves. Global Ecology and Biogeography Letters 7, 27–47.

Duke, N.C., Benzie, J.A.H., Goodall, J.A., Ballment, E.R., 1998b. Genetic structure and evolution of species in the mangrove genus *Avicennia* (Avicenniaceae) in the Indo-West Pacific. Evolution 52, 1612–1626.

Egler, F.E., 1952. Southeast saline everglades vegetation, Florida, and its management. Vegetatio Acta Geobotanica 3, 213–265.

Ellison, A.M., Farnsworth, E.J., 1993. Seedling survivorship, growth, and response to disturbance in Belizean mangal. American Journal of Botany 80, 1137–1145.

Evans, M.J., Williams, R.J., 1997. Historical distribution of estuarine wetlands at Kurnell Peninsula, Botany Bay: the need and potential for rehabilitation. Wetlands (Australia) 18, 49–54.

Everitt, J.H., Judd, F.W., Escobar, D.E., Davis, M.R., 1996. Integration of remote sensing and spatial information technologies for mapping black mangrove on the Texas gulf coast. Journal of Coastal Research 12, 64–69.

Feller, I.C., Whigham, D.F., McKee, K.L., Lovelock, C.E., 2003b. Nitrogen limitation of growth and nutrient dynamics in a disturbed mangrove forest, Indian River Lagoon, Florida. Oecologia 134, 405–414.

Feller, I.C., McKee, K.L., Whigham, D.F., O'Neill, J.P., 2003a. Nitrogen vs. phosphorus limitation across an ecotonal gradient in a mangrove forest. Biogeochemistry 62, 145–175.

Finlayson, C.M., Bailey, B.J., Freeland, W.J., Fleming, M.R., 1998. Wetlands of the northern territory. In: McComb, A.J., Lake, P.S. (Eds.), The Conservation of Australian Wetlands. Surrey Beatty and Sons, Sydney, pp. 103–126.

Friess, D.A., Webb, E.L., 2014. Variability in mangrove change estimates and implications for the assessment of ecosystem service provision. Global Ecology and Biogeography 23 (7), 715–725.

Friess, D.A., Krauss, K.W., Horstman, E.M., Balke, T., Bouma, T.J., Galli, D., Webb, E.L., 2012. Are all intertidal wetlands naturally created equal? Bottlenecks, thresholds and knowledge gaps to mangrove and saltmarsh ecosystems. Biological Reviews 87 (2), 346–366.

Gehrels, R., Hayward, B., Grenfell, H., Newnham, R., Hunter, J., Southall, K., Sabaa, A., 2007. Rapid Sea-level Rise in Australia and New Zealand during the Past Century. Abstract, XVII INQUA Congress, Cairns Australia.

Giri, C., Ochieng, E., Tieszen, L.L., Zhu, Z., Singh, A., Loveland, T., Masek, J., Duke, N., 2011. Status and distribution of mangrove forests of the world using earth observation satellite data. Global Ecology and Biogeography 20 (1), 154–159.

Giri, C.P., Long, J., 2014. Mangrove reemergence in the northernmost range limit of eastern Florida. Proceedings of the National Academy of Sciences of the United States of America 111, E1447–E1448.

Giri, C., Long, J., 2016. Is the geographic range of mangrove forests in the conterminous United States really expanding? Sensors 16 (12), 2010.

Guo, H., Zhang, Y., Lan, Z., Pennings, S.C., 2013. Biotic interactions mediate the expansion of black mangrove (*Avicennia germinans*) into salt marshes under climate change. Global Change Biology 19 (9), 2765–2774.

Haworth, R.J., 2002. Changes in mangrove/salt-marsh distribution in the Georges River estuary, southern Sydney, 1930-1970. Wetlands (Australia) 20, 80–103.

Henry, K.M., Twilley, R.R., 2013. Soil development in a coastal Louisiana wetland during a climate-induced vegetation shift from salt marsh to mangrove. Journal of Coastal Research 29 (6), 1273–1283.

Howard, R.J., Day, R.H., Krauss, K.W., From, A.S., Allain, L., Cormier, N., 2016. Hydrologic restoration in a dynamic subtropical mangrove-to-marsh ecotone. Restoration Ecology 25 (3), 471–482.

Hutchings, P., Saenger, P., 1987. Ecology of Mangroves. University of Queensland Press, St Lucia, Queensland, Australia.

Hyland, S.J., Butler, C.T., 1988. The Distribution and Modification of Mangroves and Saltmarsh-claypans in Southern Queensland. Queensland Department of Primary Industries, Brisbane, 74 p.

Islebe, G., Sanchez, O., 2002. History of late holocene vegetation at Quintana Roo, Caribbean coast of Mexico. Plant Ecology 160, 187–192.

Jennings, J.N., 1975. Desert dunes and estuarine fill in the Fitzroy estuary, north-western Australia. Catena 2, 215–262.

Jolly, P., Chin, D., 1992. Rehabilitation of the Mary River floodplains: a review of MSS and TM imagery 1972-1992. In: Davenport, C.C., Riley, S.J., Ringrose, S.M. (Eds.), Proceedings of the GIS and Environmental Rehabilitation Workshop. Australian Government Publishing Service, Canberra, Australia, pp. 43–54.

Kangas, P.C., Lugo, A.E., 1990. The distribution of mangrove and saltmarsh in Florida. Tropical Ecology 31, 32–39.

Kelleway, J.J., Saintilan, N., Macreadie, P.I., Skilbeck, C.G., Zawadzki, A., Ralph, P.J., 2016. Seventy years of continuous encroachment substantially increases 'blue carbon'capacity as mangroves replace intertidal salt marshes. Global Change Biology 22 (3), 1097–1109.

Krauss, K.W., From, A.S., Doyle, T.W., Doyle, T.J., Barry, M.J., 2011. Sea-level rise and landscape change influence mangrove encroachment onto marsh in the Ten Thousand Islands region of Florida, USA. Journal of Coastal Conservation 15 (4), 629–638.

Knighton, A.D., Mills, K., Woodroffe, C.D., 1991. Tidal-creek extension and saltwater intrusion in Northern Australia. Geology 19, 831–834.

Lange, W.P., de Lange, P.J., 1994. An appraisal of factors controlling the latitudinal distribution of mangrove (*Avicennia marina* var. resinifera) in New Zealand. Journal of Coastal Research 10, 539–548.

Lee, S.Y., Primavera, J.H., Dahdouh-Guebas, F., McKee, K., Bosire, J.O., Cannicci, S., Diele, K., Fromard, F., Koedam, N., Marchand, C., Mendelssohn, I., Mukherjee, N., Record, S., 2014. Ecological role and services of tropical mangrove ecosystems: a reassessment. Global Ecology and Biogeography 23, 726–743.

Lewis, D.B., Brown, J.A., Jimenez, K.L., 2014. Effects of flooding and warming on soil organic matter mineralization in *Avicennia germinans* mangrove forests and Juncus roemerianus salt marshes. Estuarine. Coastal and Shelf Science 139, 11–19.

Lin, G.H., Sternberg, L.D.S.L., 1992. Effect of growth form, salinity, nutrients and sulfide on photosynthesis, carbon isotope discrimination and growth of red mangrove (*Rhizophora mangle* L.). Australian Journal of Plant Physiology 19, 509–517.

Lonard, R.I., Judd, F.W., 1985. Effects of a severe freeze on native woody-plants in the lower Rio-Grande valley, Texas. Southwestern Naturalist 30, 397–403.

Long, S.P., Mason, C.F., 1983. Saltmarsh Ecology. Blackie and Son Ltd, Glasgow and London.

López-Hoffman, L., DeNoye, J.L., Monroe, I.E., Shaftel, R., Anten, N.P.R., Martínez-Ramos, M., David, D., Ackerly, D.D., 2006. Mangrove seedling net photosynthesis, growth, and survivorship are interactively affected by salinity and light. Biotropica 38, 606–616.

Lopez-Portillo, Ezcurra, E., 1989. Response of three mangroves to salinity in two geoforms. Functional Ecology 3, 355–361.

Lot, H.A., Vázquez-Yánes, C., Menéndez, F., 1975. Physiognomic and floristic changes near the northern limit of mangroves in the gulf coast of Mexico. In: Walsh, G.E., Snedaker, S.C., Teas, H.J. (Eds.), Proceedings of the International Symposium on Biology and Management of Mangroves, vol. 1. University of Florida, Gainesville, pp. 52–61.

Lovelock, C.E., Feller, I.C., McKee, K.L., Engelbrecht, B.M.J., Ball, M.C., 2004. Experimental evidence for nutrient limitation of growth, photosynthesis, and hydraulic conductance of dwarf mangroves in Panama. Functional Ecology 18, 25–33.

Lovelock, C.E., Feller, I.C., Ball, M.C., Engelbrecht, B.M.J., Ewel, M.L., 2006. Differences in plant function in phosphorus- and nitrogen-limited mangrove ecosystems. New Phytologist 172, 514–522.

Lovelock, C.E., Feller, I.C., Reef, R., Hickey, S., Ball, M.C., 2017. Mangrove dieback during fluctuating sea levels. Scientific Reports 7 (1), 1680.

Macintyre, I.G., Toscano, M.A., Lighty, R.G., Bond, G.B., 2004. Holocene history of the mangrove islands of Twin Cays, Belize, Central America. Atoll Research Bulletin 510, 1–16.

Macintyre, I.G., Little, M.M., Littler, D.S., 1995. Holocene history of Tobacco range, Belize, Central America. Atoll Research Bulletin 430, 1–18.

MacNae, W., 1966. Mangroves in eastern and southern Australia. Australian Journal of Botany 14, 67–104.

Manson, F.J., Loneragan, N.R., Phinn, S.R., 2003. Spatial and temporal variation in distribution of mangroves in Moreton Bay, subtropical Australia: a comparison of pattern metrics and change detection analyses based on aerial photographs. Estuarine, Coastal and Shelf Science 57, 653–666.

Markley, J.L., McMillan, C., Thompson Jr., G.A., 1982. Latitudinal differentiation in response to chilling temperatures among populations of three mangroves, *Avicennia germinans, Laguncularia racemosa*, and *Rhizophora mangle*, from the western tropical Atlantic and Pacific Panama. Canadian Journal of Botany 60, 2704–2715.

Mbense, S., Rajkaran, A., Bolosha, U., Adams, J., 2016. Rapid colonization of degraded mangrove habitat by succulent salt marsh. South African Journal of Botany 107, 129–136.

McKee, K.L., 1995a. Mangrove species distribution patterns and propagule predation in Belize: an exception to the dominance-predation hypothesis. Biotropica 27, 334–345.

McKee, K.L., 1995b. Interspecific variation in growth, biomass partitioning, and defensive characteristics of neotropical mangrove seedlings: response to light and nutrient availability. American Journal of Botany 82, 299–307.

McKee, K.L., Cahoon, D.R., Feller, I.C., 2007a. Caribbean mangroves adjust to rising sea level through biotic controls on soil elevation change. Global Ecology and Biogeography 16, 545–556.

McKee, K.L., Faulkner, P.L., 2000. Mangrove peat analysis and reconstruction of vegetation history at the Pelican Cays, Belize. Atoll Research Bulletin 468, 46–58.

McKee, K.L., Feller, I.C., Popp, M., Wanek, W., 2002. Mangrove isotopic (δ^{15}N and δ^{13} C) fractionation across a nitrogen vs. phosphorous limitation gradient. Ecology 83, 1065–1075.

McKee, K.L., Mendelssohn, I.A., Materne, M.D., 2004. Acute salt marsh dieback in the Mississippi River deltaic plain: a drought-induced phenomenon? Global Ecology and Biogeography 13, 65–73.

McKee, K.L., Rogers, K., Saintilan, N., 2012. Response of salt marsh and mangrove wetlands to changes in atmospheric CO_2, climate, and sea level. In: Middleton, B. (Ed.), Global Change and the Function and Distribution of Wetlands. Springer, pp. 63–96.

McKee, K.L., Rooth, J.E., 2008. Where temperate meets tropical: multi-factorial effects of elevated CO_2, nitrogen enrichment, and competition on a mangrove-salt marsh community. Global Change Biology 14, 1–14.

McKee, K.L., Rooth, J.E., Feller, I.C., 2007b. Mangrove recruitment after forest disturbance is facilitated by herbaceaous species in the Caribbean. Ecological Applications 17 (6), 1678–1693.

McLoughlin, L., 2000. Estuarine wetlands distribution along the Parramatta river, Sydney, 1788-1940: implications for planning and conservation. Cunninghamia 6, 579–610.

McMillan, C., 1975. Adaptive differentiation to chilling in mangrove populations. In: Walsh, G.E., Snedaker, S.C., Teas, H.J. (Eds.), Proceedings of the International Symposium on Biology and Management of Mangroves, vol. 1. University of Florida, Gainesville, pp. 62–68.

McMillan, C., Sherrod, C.L., 1986. The chilling tolerance of black mangrove, Avicennia germinans, from the Gulf of Mexico Coast of Texas, Louisiana and Florida. Contributions in Marine Science 29, 9–16.

McTainsh, G., Iles, B., Saffinga, P., 1986. Spatial and temporal patterns of mangroves at Oyster point Bay, south east Queensland, 1944-1983. Proceedings of the Royal Society of Queensland 99, 83–91.

Meehan, A., 1997. Historical Changes in Seagrass, Mangrove and Saltmarsh Communities in Merimbula Lake and Pambula Lake (Honours thesis). University of Wollongong.

Mendelssohn, I.A., McKee, K.L., 2000. Salt marshes and mangroves. In: Barbour, M.G., Billings, W.D. (Eds.), North American Terrestrial Vegetation. Cambridge University Press, Cambridge, pp. 501–536.

Mildenhall, D.C., Brown, L.J., 1987. An early Holocene occurrence of the mangrove *Avicennia marina* in Poverty Bay, North Island, New Zealand: its climatic and geological implications. New Zealand Journal of Botany 25, 281–294.

Mitchell, M.L., Adam, P., 1989a. The relationship between mangrove and saltmarsh communities in the Sydney region. Wetlands (Australia) 8, 37–46.

Mitchell, M.L., Adam, P., 1989b. The decline of saltmarsh in Botany Bay. Wetlands (Australia) 8, 55–60.

Mitsch, W.J., Gosselink, J.G., 2000. Wetlands, third ed. John Wiley and Sons, New York.

Montague, C.L., Wiegert, R.G., 1990. Salt marshes. In: Myers, R.L., Ewel, J.J. (Eds.), Ecosystems of Florida. University of Central Florida, Orlando, pp. 481–516.

Montague, C.L., Odum, H.T., 1997. The intertidal marshes of Florida's Gulf Coast. In: Coultas, C.L., Hsieh, Y. (Eds.), Ecology and Management of Tidal Marshes. A model from the Gulf of Mexico St Lucie Press, pp. 1–9, 334 pp.

Morrisey, D.J., Skilleter, G.A., Ellis, J.I., Burns, B.R., Kemp, C.E., Burt, K., 2003. Differences in benthic fauna and sediment among mangrove (*Avicennia marina* var. *australasica*) stands of different ages in New Zealand. Estuarine, Coastal and Shelf Science 56, 581–592.

Morton, R.M., 1994. Fluctuations in wetland extent in southern Moreton Bay. In: Greenwood, J.G., Hall, N.J. (Eds.), Future Marine Science in Moreton Bay. School of Marine Science, University of Queensland, Brisbane, pp. 145–150.

Mulrennan, M.E., Woodroffe, C.D., 1998. Saltwater intrusion into the coastal plains of the lower Mary River, northern territory, Australia. Journal of Environmental Management 54, 169–188.

Murdiyarso, D., Purbopuspito, J., Kauffman, J.B., Warren, M.W., Sasmito, S.D., Donato, D.C., Kurnianto, S., 2015. The potential of Indonesian mangrove forests for global climate change mitigation. Nature Climate Change 5 (12), 1089–1092.

Naidoo, G., 1987. Effects of salinity and nitrogen on growth and water relations in the mangrove *Avicennia marina* (Forsk.) Vierh. New Phytologist 107, 317–325.

Norman, H., McMillan, C., Thompson, G.A., 1984. Phosphatidylglycerol molecular species in chilling-sensitive and chilling-resistant populations of *Avicennia germinans* (L.) L. Plant and Cell Physiology 25, 1437–1444.

NOAA, 2007. Sea Surface Temperature (SST) Contour Charts. Prepared by National Oceanic and Atmospheric Administration (NOAA) Satellites and Information. Accessed at: http://www.osdpd.noaa.gov/PSB/EPS/SST/contour.html.

Olmsted, I., Dunevitz, H., Platt, W.J., 1993. Effects of freezes on tropical trees in everglades National Park Florida, USA. Tropical Ecology 34, 17–34.

Osland, M.J., Enwright, N., Day, R.H., Doyle, T.W., 2013. Winter climate change and coastal wetland foundation species: salt marshes vs. mangrove forests in the southeastern United States. Global Change Biology 19 (5), 1482–1494.

Osland, M.J., Enwright, N., Stagg, C.L., 2014. Freshwater availability and coastal wetland foundation species: ecological transitions along a rainfall gradient. Ecology 95 (10), 2789–2802.

Osland, M.J., Day, R.H., From, A.S., McCoy, M.L., McLeod, J.L., Kelleway, J.J., 2015. Life stage influences the resistance and resilience of black mangrove forests to winter climate extremes. Ecosphere 6 (9), 1–15.

Osland, M.J., Enwright, N.M., Day, R.H., Gabler, C.A., Stagg, C.L., Grace, J.B., 2016. Beyond just sea-level rise: considering macroclimatic drivers within coastal wetland vulnerability assessments to climate change. Global Change Biology 22 (1), 1–11.

Osland, M.J., Spivak, A.C., Nestlerode, J.A., Lessmann, J.M., Almario, A.E., Heitmuller, P.T., Russell, M.J., Krauss, K.W., Alvarez, F., Dantin, D.D., Harvey, J.E., 2012. Ecosystem development after mangrove wetland creation: plant–soil change across a 20-year chronosequence. Ecosystems 15 (5), 848–866.

Park, S., 2001. Bay of Plenty Maritime Wetlands Database. Environment of the Bay of Plenty. Environmental Report 2000/21.

Patterson, C.S., Mendelssohn, I.A., 1991. A comparison of physicochemical variables across plant zones in a mangal/salt marsh community in Louisiana. Wetlands 11, 139–161.

Patterson, C.S., Mendelssohn, I.A., Swenson, E.M., 1993. Growth and survival of Avicennia germinans seedlings in a mangal/salt marsh community in Louisiana, U.S.A. Journal of Coastal Research 9, 801–810.

Patterson, S., McKee, K.L., Mendelssohn, I.A., 1997. Effects of tidal inundation and predation on Avicennia germinans seedling establishment and survival in a sub-tropical mangal/salt marsh community. Mangroves and Salt Marshes 1, 103–111.

Peinado, M., Alcaraz, F., Delgadillo, J., De La Cruz, M., Alvarez, J., Aguirre, L.J., 1994. The coastal salt marshes of California and Baja California. Plant Ecology 110, 55–66.

Pendleton, L., Donato, D.C., Murray, B.C., Crooks, S., Jenkins, W.A., Sifleet, S., Fourqurean, J.W., Kauggman, J.B., Marba, N., Megonigal, P., 2012. Estimating global "blue carbon" emissions from conversion and degradation of vegetated coastal ecosystems. PLoS One 7 (9), e43542.

Perry, C.L., Mendelssohn, I.A., 2009. Ecosystem effects of expanding populations of *Avicennia germinans* in a Louisiana salt marsh. Wetlands 29 (1), 396–406.

Peterson, J.M., Bell, S.S., 2012. Tidal events and salt-marsh structure influence black mangrove (*Avicennia germinans*) recruitment across an ecotone. Ecology 93 (7), 1648–1658.

Peterson, J.M., Bell, S.S., 2015. Saltmarsh Boundary Modulates Dispersal of Mangrove Propagules: Implications for Mangrove Migration with Sea-Level Rise. PLoS One 10 (3), e0119128.

Pidgeon, I.M., 1940. The ecology of the central coast area of New South Wales III. Types of Primary Succession. Proceedings of the Linnaean Society of New South Wales 65, 221–249.

Raabe, E.A., Roy, L.C., McIvor, C.C., 2012. Tampa Bay coastal wetlands: nineteenth to twentieth century tidal marsh-to-mangrove conversion. Estuaries and Coasts 35 (5), 1145–1162.

Raabe, E.A., Stumpf, R.P., 2016. Expansion of tidal marsh in response to sea-level rise: Gulf Coast of Florida, USA. Estuaries and Coasts 39 (1), 145–157.

Rabinowitz, D., 1975. Planting experiments in mangrove swamps of Panama. In: Walsh, G.E., Snedaker, S.C., Teas, H.J. (Eds.), Proceedings of the International Symposium on Biology and Management of Mangroves, vol. 1. University of Florida, Gainesville, pp. 385–393.

Rabinowitz, D., 1978a. Dispersal properties of mangrove propagules. Biotropica 10, 47–57.

Rabinowitz, D., 1978b. Mortality and initial propagule size in seedlings in Panama. Journal of Ecology 66, 45–51.

Rabinowitz, D., 1978c. Early growth of mangrove seedlings in Panama, and an hypothesis concerning the relationships of dispersal and zonation. Journal of Biogeography 5, 113–133.

Ramcharan, E.K., 2004. Mid-to-late Holocene sea level influence on coastal wetland development in Trinidad. Quaternary International 120, 145–151.

Reimold, R.J., 1977. Mangals and salt marshes of eastern United States. In: Chapman, V.J. (Ed.), Ecosystems of the World. 1. Wet Coastal Ecosystems. Elsevier Scientific Publications Company, Amsterdam, pp. 157–166.

Rogers, K., 2004. Mangrove and Saltmarsh Surface Elevation Dynamics in Relation to Environmental Variables in Southeastern Australia. A thesis submitted in fulfillment of the requirements of the degree of Doctor of Philosophy from University of Wollongong, Earth and Environmental Sciences, Faculty of Science, 270 p.

Rogers, K., Saintilan, N., 2008. Relationships between surface elevation and groundwater in mangrove forests of SE Australia. Journal of Coastal Research 24, 63–69.

Rogers, K., Saintilan, N., Cahoon, D., 2005a. Surface elevation dynamics in a regenerating mangrove forest, Homebush Bay, Australia. Wetlands Ecology and Management 13, 587–598.

Rogers, K., Saintilan, N., Heijnis, H., 2005b. Monitoring of mangrove and saltmarsh resources in Westernport Bay, Australia. Estuaries 28, 551–559.

Rogers, K., Saintilan, N., Wilton, K., 2006. Vegetation change and surface elevation dynamics of the estuarine wetlands of southeast Australia. Estuarine, Coastal and Shelf Science 66, 559–569.

Rogers, K., Saintilan, N., Howe, A.J., Rodríguez, J.F., 2013. Sedimentation, elevation and marsh evolution in a southeastern Australian estuary during changing climatic conditions. Estuarine, Coastal and Shelf Science 133, 172–181.

Rogers, K., Saintilan, N., Woodroffe, C.D., 2014. Surface elevation change and vegetation distribution dynamics in a subtropical coastal wetland: implications for coastal wetland response to climate change. Estuarine. Coastal and Shelf Science 149, 46–56.

Roy, P.S., 1984. New South Wales estuaries- their origin and evolution. In: Thom, B.G. (Ed.), Development in Coastal Geomorphology in Australia. Academic Press, San Diego CA, pp. 99–121.

Roy, P.S., Williams, R.J., Jones, A.R., Yassini, I., Gibbs, P.J., Coates, B., West, R.J., Scanes, P.R., Hudson, J.P., Nichol, S., 2001. Structure and function of south-east Australian estuaries. Estuarine, Coastal and Shelf Science 53, 351–384.

Saenger, P., 2002. Mangrove Ecology, Silviculture and Conservation. Kluwer Academic Press, Dordrecht, Boston, London.

Saenger, P., Moverley, J., 1985. Vegetative phenology of mangroves along the Queensland coastline. Proceedings of the Ecological Society of Australia 13, 257–265.

Saenger, P., Snedaker, S.C., 1993. Pantropical trends in mangrove above-ground biomass and annual litterfall. Oecologia 96, 293–299.

Saenger, P., Specht, M.M., Specht, R.L., Chapman, V.J., 1977. Mangal and coastal saltmarsh communities in Australasia. In: Chapman, V.J. (Ed.), Ecosystems of the World I: Wet Coastal Ecosystems. Elsevier, Amsterdam, pp. 293–345.

Saintilan, N., 1997. Mangroves as successional stages on the Hawkesbury River Estuary, New South Wales. Wetlands (Australia) 16, 99–107.

Saintilan, N., 1998. Photogrammetric survey of the Tweed River wetlands. Wetlands (Australia) 17, 74–82.

Saintilan, N., 2003. The Influence of Nutrient Enrichment upon Mangrove Seedling Establishment and Growth in the Hawkesbury River Estuary, New South Wales, Australia. Wetlands (Australia) 21, 29–35.

Saintilan, N., Williams, R.J., 1999. Mangrove transgression into saltmarsh environments in south-east Australia. Global Ecology and Biogeography Letters 8, 117–124.

Saintilan, N., Hashimoto, R., 1999. Mangrove-saltmarsh dynamics on a prograding bayhead delta in the Hawkesbury River estuary, New South Wales, Australia. Hydrobiologia 413, 95–102.

Saintilan, N., Williams, R.J., 2000. The decline of saltmarshes in Southeast Australia: Results of recent surveys. Wetlands (Australia) 18, 49–54.

Saintilan, N., Wilton, K., 2001. Changes in the distribution of mangroves and saltmarshes in Jervis Bay, Australia. Wetlands Ecology and Management 9, 409–420.

Saintilan, N., Rogers, K., 2013. The significance and vulnerability of Australian saltmarshes: implications for management in a changing climate. Marine and Freshwater Research 64 (1), 66–79.

Saintilan, N., Wilson, N.C., Rogers, K., Rajkaran, A., Krauss, K.W., 2014. Mangrove expansion and salt marsh decline at mangrove poleward limits. Global Change Biology 20 (1), 147–157.

Saintilan, N., Rogers, K., 2015. Woody plant encroachment of grasslands: a comparison of terrestrial and wetland settings. New Phytologist 205 (3), 1062–1070.

Saintilan, N., Rogers, K., Kelleway, J.J., Ens, E., Sloane, D.R., 2018. Climate Change Impacts on the Coastal Wetlands of Australia, pp. 1–10.

Santos, M.C.F.V., Zieman, J.C., Cohen, R.R.H., 1997. Interpreting the upper mi-littoral zonation patterns of mangroves in Maranhão (Brazil) in response to microtopography and hydrology. In: Kjerfve, B., Lacerda, L.D., Diop, E.H.S. (Eds.), Mangrove Ecosystem Studies in Latin America and Africa. UNESCO, Paris, pp. 127–144.

Schaeffer-Novelli, Y., Cintron-Molero, G., Adaime, R.R., de Camargo, T.M., 1990. Variability of mangrove ecosystems along the Brazilian Coast. Estuaries 13, 204–216.

Schmidt, D.A., Burgman, R., 2003. Time-dependent uplift and subsidence in the Santa Clara Valley, California, from a large interferometric synthetic aperture radar data set. Journal of Geophysical Research – Solid Earth 108, 2416.

Semeniuk, V., 1980. Mangrove zonation along an eroding coastline in King Sound, North-Western Australia. Journal of Ecology 68, 789–812.

Semeniuk, V., 1982. Geomorphology and Holocene history of the tidal flats, King Sound, north-western Australia. Journal of the Royal Society of Western Australia 65, 47–68.

Sherrod, C.L., McMillan, C., 1985. The distributional history and ecology of mangrove vegetation along the northern Gulf of Mexico coastal region. Contributions in Marine Science 28, 129–140.

Silvestri, S., Defina, A., Marani, M., 2005. Tidal regime, salinity and salt marsh plant zonation. Estuarine, Coastal and Shelf Science 62, 119–130.

Simpson, L.T., Feller, I.C., Chapman, S.K., 2013. Effects of competition and nutrient enrichment on *Avicennia germinans* in the salt marsh-mangrove ecotone. Aquatic Botany 104, 55–59.

Smith, T.J., 1987. Seed predation in relation to tree dominance and distribution in mangrove forests. Ecology 68, 266–273.

Smith, T.J., Duke, N.C., 1987. Physical determinants of inter-estuary variation in mangrove species richness around the tropical coastline of Australia. Journal of Biogeography 14, 9–20.

Smith, T.J., Chan, H.T., McIvor, C.C., Robblee, M.B., 1989. Comparisons of seed predation in tropical, tidal forests from three continents. Ecology 70, 146–151.

Snedaker, S.C., 1982. Mangrove species zonation: why?. In: Sen, D.N., Rajpurohit, K.S. (Eds.), Contribution to the Ecology of Halophytes, vol. 2. Dr. W. Junk, The Hague, pp. 111–125.

Solomon, S., Quin, D., Manning, M., Marquis, M., Avery, K., Tignor, M., Miller, H., Chen, Z. (Eds.), 2007. Climate Change 2007- the Physical Science Basis. Contribution of Working Group 1 to the Fourth Assessment Report of the IPCC. Cambridge University Press.

Snedaker, S.C., 1995. Mangroves and climate change in the Florida and Caribbean region: scenarios and hypotheses. Hydrobiologia 295, 43–49.

Sousa, W.P., Kennedy, P.G., Mitchell, B.J., Ordóñez L, B.M., 2007. Supply-side ecology in mangroves: do propagule dispersal and seedling establishment explain forest structure? Ecological Monographs 77 (1), 53–76.

Sousa, W.P., Mitchell, B.J., 1999. The effect of seed predators on plant distributions: is there a general pattern in mangroves? Oikos 86, 55–66.

Spackman, W., Dolsen, C.P., Riegel, W., 1966. Phytogenic organic sediments and sedimentary environments in the Everglades-Mangrove complex. Part 1: Evidence of a transgressing sea and its effects on environments of the Shark River areas of Southwestern Florida. Palaeontographica 117, 135–152.

Spalding, M.D., Blasco, F., Field, C.D., 1997a. World mangrove atlas. International Society for Mangrove Conservation, National Council for Scientific Research, Paris.

Spalding, M.D., Blasco, F., Field, C.D., 1997b. World Mangrove Atlas. International Society for Mangrove Ecosystems, Okinawa, Japan.

Specht, R.L., 1981. Biogeography of halophytic angiosperms (salt-marsh, mangrove and sea-grass). Ecological biogeography of Australia. Dr. W. Junk bv Publishers, The Hague, pp. 575–589.

Stevens, P.W., Fox, S.L., Montague, C.L., 2006. The interplay between mangroves and saltmarshes at the transition between temperate and subtropical climate in Florida. Wetlands Ecology and Management 14, 435–444.

Stuart, S.A., Choat, B., Martin, K.C., Holbrook, N.M., Ball, M.C., 2007. The role of freezing in setting the latitudinal limits of mangrove forests. New Phytologist 173 (3), 576–583.

Thom, B.G., 1967. Mangrove ecology and deltaic geomorphology: Tabasco, Mexico. Journal of Ecology 55, 301–343.

Thom, B.G., Wright, L.D., Coleman, J.M., 1975. Mangrove ecology and deltaic-estuarine geomorphology, Cambridge Gulf-Ord River, Western Australia. Journal of Ecology 63, 203–222.

Thomas, M.L.H., 1993. Mangrove swamps in Bermuda. Atoll Research Bulletin 386, 1–18.

Tomlinson, P.B., 1986. The Botany of Mangroves. Cambridge University Press, Cambridge.

Toscano, M.A., Macintyre, I.G., 2003. Corrected western Atlantic sea-level curve for the last 11,000 years based on calibrated C-14 dates from *Acropora palmata* framework and intertidal mangrove peat. Coral Reefs 22, 257–270.

Van Steenis, C.G.G.J., 1968. Frost in the tropics. In: Misra, R., Gopal, B. (Eds.), Recent Advances in Tropical Ecology, Part I. International Society for Tropical Ecology, Varanasi, India, pp. 154–167.

Vanderzee, M.P., 1988. Changes in saltmarsh vegetation as an early indicator of sea-level rise. In: Pearman, G.I. (Ed.), Greenhouse: Planning for Climate Change. CSIRO, Melbourne.

Walsh, G.E., 1974. Mangroves: a review. In: Reimhold, R.J., Queens, W.H. (Eds.), Ecology of Halophytes. Academic Press, New York, pp. 51–174.

Watson, E.B., 2004. Changing elevation, accretion, and tidal marsh plant assemblages in a South San Francisco Bay tidal marsh. Estuaries 27, 684–698.

West, R.C., 1977. Tidal salt-marsh and mangal formations of Middle and South America. In: Chapman, V.J. (Ed.), Ecosystems of the World. 1. Wet Coastal Ecosystems. Elsevier Scientific Publications Company, Amsterdam, pp. 193–213.

West, R.J., 1993. Estuarine Fisheries Resources of Two South Eastern Australian Rivers (Ph.D. thesis). University of New South Wales, 173 p.

Whelan, K.R.T., Smith, T.J., Cahoon, D.R., Lynch, J.C., Anderson, G.H., 2005. Groundwater control of mangrove surface elevation: Shrink and swell varies with soil depth. Estuaries 28, 833–843.

Williams, A.A., Eastman, S.F., Eash-Loucks, W.E., Kimball, M.E., Lehmann, M.L., Parker, J.D., 2014. Record northernmost endemic mangroves on the United States Atlantic Coast with a note on latitudinal migration. Southeastern Naturalist 13 (1), 56–63.

Williams, R.J., Meehan, A.J., 2004. Focusing management needs at the sub-catchment level via assessments of change in the cover of estuarine vegetation, Port Hacking, NSW, Australia. Wetlands Ecology and Management 12, 499–518.

Williams, R.J., Watford, F.A., 1997. Changes in the Distribution of Mangrove and Saltmarsh in Berowra and Marramarra Creeks, 1941-1992. NSW Fisheries, Fisheries Research Institute, Cronulla, 21 p.

Williams, R.J., Watford, F.A., Balashov, V., 1999. Kooragang Wetland Rehabilitation Project: Changes in Wetland Fish Habitats of the Lower Hunter River. NSW Fisheries, Office of Conservation, Fisheries Research Institute, Cronulla.

Wilson, B.P., 2005. Elevation of sulphurous layers in acid sulphate soils: what do they indicate about sea levels during the Holocene in eastern Australia? Catena 62, 45–56.

Winn, K.O., Saynor, M.J., Eliot, M.J., Eliot, I., 2006. Saltwater intrusion and morphologicvla change at the mouth of the East Alligator River, Northern Territory. Journal of Coastal Research 22, 137–149.

Winning, G., 1990. Lake Macquarie Littoral Habitats Study. The Wetlands Centre, Shortland, 29 p., plus appendices.

Woodroffe, C.D., 1981. Mangrove swamp stratigraphy and holocene transgression, Grand Cayman Island, West Indies. Marine Geology 41, 271–294.

Woodroffe, C.D., 1983. Development of Mangrove forests from a geological perspective. In: Teas, H.J. (Ed.), Tasks for Vegetation Science. Dr. W. Junk Publishers, The Hague, pp. 1–17.

Woodroffe, C.D., Chappell, J., Thom, B.G., Wallensky, E., 1989. Depositional model of a macrotidal estuary and floodplain, South Alligator River, Northern Australia. Sedimentology 36, 737–756.

Woodroffe, C.D., Thom, B.G., Chappell, J., 1985. Development of widespread mangrove swamps in mid-Holocene times in northern Australia. Nature 317, 711–713.

Woodroffe, C.D., Mulrennan, M.E., 1993. Geomorphology of the Lower Mary River Plains, Northern Territory. Australian National University and the Conservation Commission of the Northern Territory, Darwin.

Zomlefer, W.B., Judd, W.S., Giannasi, D.E., 2006. Northernmost limit of *Rhizophora mangle* (red mangrove; Rhizophoraceae) in St. Johns County, Florida. Castanea 71, 239–244.

CHAPTER

27

The Value of Coastal Wetland Ecosystem Services

Edward B. Barbier

Department of Economics, Colorado State University, Fort Collins, Colorado, United States

1. INTRODUCTION

Ecosystems can be viewed as assets that produce a flow of benefits, which are commonly referred to as *ecosystem services* (Barbier, 2011; MEA, 2005). Such benefits are diverse and wide-ranging and generally arise through the natural functioning of ecosystems. For example, as Daily et al. (2000, p. 395) state, "the world's ecosystems are capital assets. If properly managed, they yield a flow of vital services, including the production of goods (such as seafood and timber), life support processes (such as pollination and water purification), and life-fulfilling conditions (such as beauty and serenity)."

However, we are doing a poor job in managing and maintaining the world's ecosystems, especially coastal wetlands, which are some of the most heavily used and threatened natural systems globally (Doney et al., 2012; Halpern et al., 2008; Lotze et al., 2006; Worm et al., 2006). Their deterioration because of human activities is intense and increasing. For example, around one quarter of the world's mangroves have been lost because of human action, mainly through conversion to aquaculture, agriculture, and urban land uses (Barbier and Cox, 2003; Duke et al., 2007; Friess and Webb, 2014; Spalding et al., 2010). As much as 50% of salt marshes have been lost or degraded worldwide over recent decades (Barbier et al., 2011; Doney et al., 2012). This global decline in coastal wetlands is affecting their ability to provide critical ecosystem services, such as raw materials, food, and other products collected by local communities, the provision of nursery and breeding habitats for offshore fisheries, filtering and detoxification services, control of biological invasions, declining water quality, recreational opportunities, shoreline stabilization and control of coastal erosion, and protection from flooding and storm events (Alongi, 2008; Barbier, 2014; Cochard et al., 2008; Spalding et al., 2014; Worm et al., 2006). In addition, the changes in precipitation, temperature, and hydrology accompanying climate change are likely to threaten remaining coastal

Coastal Wetlands
https://doi.org/10.1016/B978-0-444-63893-9.00027-7

and nearshore ecosystems (Dasgupta et al., 2011; Doney et al., 2012; Elliott et al., 2014; Erwin, 2009; Spalding et al., 2014; Webb et al., 2013).

At the core of this problem is the failure to consider the various values, or benefits, provided by coastal wetlands when deciding on their current and future development and use. The purpose of this chapter is to discuss why valuing the goods and services of coastal wetlands is essential to the wise management of these vital ecosystems. For these systems, how we quantify and value ecosystem goods and services can impact significantly our approach to conservation versus development decisions (Barbier et al., 2011). Unfortunately, for many important goods and services provided by coastal wetlands, there are large gaps in our knowledge of these benefits, and as a consequence, there are inadequate estimates of the important values that are needed for making adequate management decisions.

Improving the methodology for estimating the benefits of coastal wetlands is therefore urgently needed, and policy-makers responsible for coastal management are increasingly frustrated with the lack of sufficient progress. A survey of US Environmental Protection Agency wetland regulators revealed that they rarely used monetary estimates of wetland values in their environmental decision-making (Arnold, 2013). The survey respondents cited uncertainty about the scientific validity of estimates and subsequent concerns about the scientific and legal defensibility of estimate use as key reasons for ignoring wetland values. Even for specific services, such as coastal protection from storms, there is concern over the lack of progress in valuation. For example, in the case of US Gulf Coast wetlands, the reliability of estimates of the value of wetlands for storm surge protection has been questioned because the methods used have not taken into account that "the level of storm surge attenuation provided by wetlands depends on many factors including the location, type, extent, and condition of the wetlands and the properties of the storm itself" (Engle, 2011, p. 185).

Yet, in recent years there have been some encouraging signs of progress in the valuation of coastal wetlands. Important challenges still remain, and the reliability of some valuation methods used is questionable. By reviewing both the progress and challenges, this chapter hopes to spur further research into this important aspect of improving coastal wetland management.

The chapter proceeds as follows. The next section provides a brief overview of the unique challenges encountered in valuing ecosystem services and especially those associated with coastal wetlands. The subsequent section provides a brief discussion and survey of past coastal wetland value studies and explains why most studies have been limited to a small range of ecosystem goods and services. The following section focuses on how spatial considerations are becoming increasingly important in valuing coastal wetland services. Equally, as the next section discusses, unique insights can be gained from valuing certain coastal wetland services as part of a broader marine seascape. Finally, the chapter concludes by examining the key areas of future research and progress in the valuation of coastal wetlands.

2. THE CHALLENGE OF VALUING COASTAL WETLANDS

In identifying the ecosystem services provided by natural habitats, such as coastal wetlands, a common practice is to adopt the broad definition of the Millennium Ecosystem Assessment (MEA, 2005) that "ecosystem services are the benefits people obtain from ecosystems." Although this definition has been interpreted in different ways, a consensus is emerging on what "ecosystem services" are and how they arise from ecological processes and functions (Barbier et al., 2011; EPA, 2009).

A wide range of valuable goods and services to humans arise in myriad ways via the structure and functions of coastal wetlands. For example, some of the living organisms found in such ecosystems might be harvested or hunted for food, collected for raw materials, or simply valued because they are esthetically pleasing. Some of the ecosystem functions of coastal wetlands, such as nutrient and water cycling, can also benefit humans through purifying water, controlling floods, reducing pollution, or simply by providing more pleasing environments for recreation. However, although they are the source of ecosystem services, the structure and functions of a coastal ecosystem are not synonymous with such services. Ecosystem structure and functions describe the components of an ecosystem and its biophysical relationship regardless of whether or not humans benefit from them. In contrast, as noted by EPA (2009, p. 12), "ecosystem services are the direct or indirect contributions that ecosystems make to the well-being of human populations." Quantifying these contributions, or "benefits," in terms of human welfare is what constitutes *valuing ecosystem services.*

Fig. 27.1 summarizes why valuing ecosystem services is important for management of coastal wetlands. Human drivers of ecosystem change in these wetlands, such as pollution, resource exploitation, land conversion, species introductions, and habitat fragmentation, affect the structure and functioning of coastal ecosystems. Assessing and quantifying this impact is important, as it alters the ecological production of ecosystem goods and services that benefit humans. The role of economic valuation is to measure explicitly any resulting

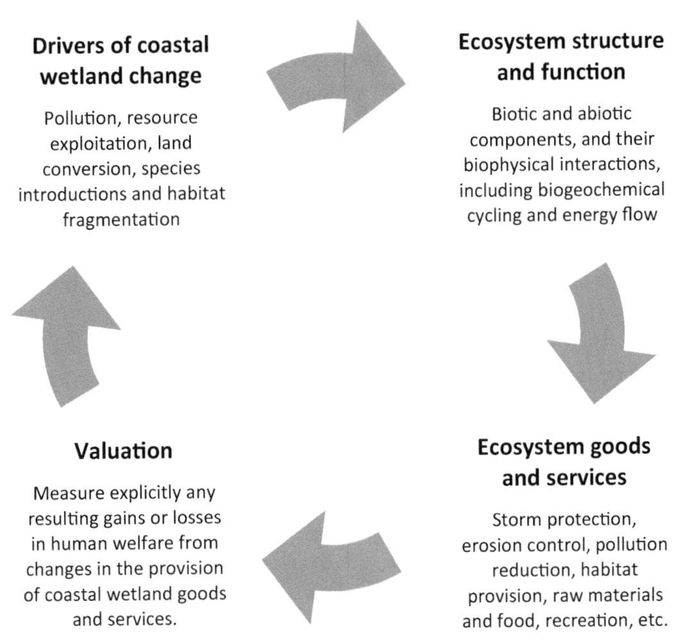
Drivers of coastal wetland change
Pollution, resource exploitation, land conversion, species introductions and habitat fragmentation
Ecosystem structure and function
Biotic and abiotic components, and their biophysical interactions, including biogeochemical cycling and energy flow
Ecosystem goods and services
Storm protection, erosion control, pollution reduction, habitat provision, raw materials and food, recreation, etc.
Valuation
Measure explicitly any resulting gains or losses in human welfare from changes in the provision of coastal wetland goods and services.

FIGURE 27.1 Valuing ecosystem services and coastal wetland management.

gains or losses in human welfare from these changes in benefits. These values can then be used to guide the necessary changes in coastal policies and management to control the human drivers of ecosystem change.

For economists, the term "benefit" has a specific meaning. According to Mendelsohn and Olmstead (2009, p. 326), "The economic benefit provided by an environmental good or service is the sum of what all members of society would be willing to pay for it." In general, there is a consensus that the benefits from ecosystems arise from a combination of ecosystem and human-derived assets. For example, fishing is a combination of ecosystem conditions that support fish stocks along with fishing boats and facilities that allow access to harvesting fish. Thus, the key issue in ecosystem service valuation is to determine the willingness to pay for the specific contribution of ecosystems to the final benefit, such as fishing (Boyd and Banzhaf, 2007; Polasky and Segerson, 2009). Thus, to determine society's willingness to pay for the benefits provided by ecosystem goods and services, one needs to measure and account for their various impacts on human welfare. Or as Bockstael et al. (2000, p. 1385) state, "In economics, valuation concepts relate to human welfare. So the economic value of an ecosystem function or service relates only to the contribution it makes to human welfare, where human welfare is measured in terms of each individual's own assessment of his or her well-being." The key is determining how changes in ecosystem goods and services affect an individual's well-being and then determining how much the individual is either willing to pay for changes that have a positive welfare impact or conversely how much the individual is willing to accept as compensation to avoid a negative effect.

The starting point in identifying coastal wetland goods and services and their values is the consensus economic view outlined above. As long as these habitats make a contribution to human welfare, either entirely on its own or through joint use with other human inputs, we can designate this contribution as an "ecosystem service." Following this approach, for example, in estimating the value of the "intermediate" ecosystem service of coastal wetlands in producing recreational benefits, it is therefore important to assess only the effects of changes in the ecosystem on recreation and not the additional influence of any human inputs. The same approach should be taken for those "final" ecosystem services, such as coastal protection, erosion control, nutrient cycling, water purification, and carbon sequestration of coastal wetlands, which may benefit human well-being with or without any additional human-provided goods and services. Valuation should show how changes in these services affect human welfare, after controlling for the influence of any additional human-provided goods and services.

Although valuing ecosystem goods and services of coastal wetlands seems straightforward, in practice there are a number of challenges to overcome. These difficulties are key to understanding why there are still a large number of coastal wetland goods and services that have yet to be valued or have very unreliable valuation estimates (see next section).

The biggest challenge to valuing ecosystem services is inadequate knowledge to link changes in ecosystem structure and function to the production of valuable goods and services (Barbier, 2011; Barbier et al., 2011; Bateman et al., 2011; NRC, 2005; Polasky and Segerson, 2009). This is certainly true for many coastal wetlands. For these systems, we often do not know how variation in ecosystem structure, functions, and processes gives rise to the change in an ecosystem good or service. For example, in the case of coastal wetlands, the change could be in the spatial area or quality of a particular type of wetland, such as a mangrove, marsh vegetation, or swamp forest. Alternatively, changes in coastal wetlands could be due to variation in the flow of water, energy, or nutrients through the system, such as the variability in tidal surges due to coastal storm events or the influx of organic waste from inland pollution, or the impacts of oil spills and other human-induced hazards.

Another problem encountered in quantifying and valuing ecosystem services is that very few are marketed. Some of the products provided by coastal wetlands, such as raw materials, food, and fish harvests, are bought and sold in markets. Given that the price and quantities of these marketed products are easy to observe, there are numerous ways to estimate the contribution of the environmental input to this production (Barbier, 2007, 2011; McConnell and Bockstael, 2005; Freeman et al., 2014). However, many other key services of coastal wetlands do not lead to observable marketed outputs. These include many services arising from ecosystem processes and functions that benefit human beings largely without any additional input from them, such as coastal protection, nutrient cycling, erosion control, water purification, and carbon sequestration. In recent years, substantial progress has been made by economists working with ecologists and other natural scientists in applying environmental valuation methodologies to assess the welfare contribution of these services (see, for example, Barbier, 2011; EPA, 2009; Freeman et al., 2014; NRC, 2005). The next section reviews some of these studies.

3. VALUATION EXAMPLES

Table 27.1 indicates how specific coastal wetland goods and services are linked to the underlying ecological structure and functions underlying each service. It also cites, where possible, 80 examples from economic studies that have estimated the values arising from the good or service. The list of studies in Table 27.1 is not inclusive of all valuation studies of coastal wetlands; nevertheless, the valuation studies are representative of the literature and thus instructive.

As Table 27.1 indicates, valuation studies of coastal wetlands have largely focused on only a few ecosystem services, such as recreation (16 studies), coastal habitat-fishery linkages (18 studies), and storm protection (19 studies). In recent years, there have been more estimates of the benefits associated with erosion control (4 studies), water pollution and sediment control (7 studies), carbon sequestration (5 studies), raw materials and food production (7 studies), and cultural, existence, and other nonuse values (3 studies). However, there have been no studies of the role of coastal wetlands in maintaining temperature and precipitation and only one comprehensive study of flood control.

The distribution of coastal wetland valuation studies in Table 27.1 shows that, although there is interest in the overall benefits of these important ecosystems, there is a need to have more studies of a wider range of ecosystem goods and services. Moreover, geographical coverage could be improved. On the other hand, recent metaanalyses of coastal wetland valuation suggest that there is growing experience in valuing specific coastal wetland benefits, including those for specific regions such as Europe or developing countries (Brander et al., 2012; Chaikumbung et al., 2016). Although such studies are generally for all wetlands, rather than just coastal ecosystems, such metaanalyses suggest that the number and geographic coverage of valuation estimates are increasing. For example, the study by Chaikumbung et al. (2016) was able to

TABLE 27.1 Examples of Coastal Ecosystem Services and Valuation Studies

Ecosystem Structure and Function	Ecosystem Services	Valuation Examples (80 Studies)
Attenuates and/or dissipates waves, buffers wind	Storm protection	Badola and Hussain (2005), Barbier (2007, 2012), Barbier et al. (2008), Barbier et al. (2013), Barbier and Enchelmeyer (2014), Costanza et al. (2008), Das and Crépin (2013), Das and Vincent (2009), Kim and Petrolia (2013), King and Lester (1995), Landry et al. (2011), Laso Bayas et al. (2011), Mangi et al. (2011), Narayan et al. (2016), Petrolia and Kim (2009, 2011), Petrolia et al. (2014) and Rao et al. (2015). **(19 studies)**
Provides sediment stabilization and soil retention	Erosion control	Huang et al. (2007), Landry et al. (2003), Rulleau et al. (2014) and Sathirathai and Barbier (2001). **(4 studies)**
Water flow regulation and control	Flood control	Turner et al. (2004). **(1 study)**
Provides nutrient and pollution uptake, as well as retention, particle deposition, and clean water	Water pollution and sediment control	Breaux et al. (1995), Byström (2000), Leggett and Bockstael (2000), Massey et al. (2006), Smith (2007), Smith and Crowder (2011) and Turner et al. (2004). **(7 studies)**
Generates biogeochemical activity, sedimentation, biological productivity	Carbon sequestration	Barbier et al. (2011), Luisetti et al. (2011), Mangi et al. (2011), Sikamäki et al. (2012) and Thompson et al. (2014) **(5 studies)**
Climate regulation and stabilization	Maintenance of temperature, precipitation	**(0 studies)**
Generates biological productivity and diversity	Raw materials and food	Janssen and Padilla (1999), King and Lester (1995), Naylor and Drew (1998), Nfotabong Atheull et al. (2009), Pinto et al. (2010), Ruitenbeek (1994) and Sathirathai and Barbier (2001). **(7 studies)**
Provides suitable reproductive habitat and nursery grounds, sheltered living space	Maintains fishing, hunting, and foraging activities	Aburto-Oropeza et al. (2008), Barbier (2003, 2007, 2012), Barbier and Strand (1998), Bell (1997), Freeman (1991), Janssen and Padilla (1999), Johnston et al. (2002), Lange and Jiddawi (2009), Luisetti et al. (2011), Milon and Scrogin (2006), O'Higgins et al. (2010), Plummer et al. (2013), Samonte-Tan et al. (2007), Sanchirico and Mumby (2009), Stål et al. (2008) and Swallow (1994). **(18 studies)**
Provides unique and esthetic landscape, suitable habitat for diverse fauna and flora	Tourism, recreation, education, and research	Bateman and Langford (1997), Birol and Cox (2007), Brouwer and Bateman (2005), Johnston et al. (2002), Kaoru (1993), Kaoru et al. (1995), King and Lester (1995), Kreitler et al. (2013), Landry and Liu (2009), Lange and Jiddawi (2009), Luisetti et al. (2011), Milon and Scrogin (2006), Othman et al. (2004), Smith and Palmquist (1994), Smith et al. (1991) and Turner et al. (2004). **(16 studies)**
Provides unique and esthetic landscape of cultural, historic, or spiritual meaning	Culture, spiritual and religious benefits, existence, and bequest values	Bateman and Langford (1997), Milon and Scrogin (2006) and Naylor and Drew (1998). **(3 studies)**

draw on 1432 estimates of the economic value of various benefits from 379 distinct wetlands from 50 different developing countries.

As our experience and methods of valuing coastal wetland goods and services improve, we are also identifying new areas for further progress. The next sections discuss two such important areas: how spatial considerations are becoming increasingly important in valuing coastal wetland services and how unique insights can be gained from valuing certain coastal wetland services as part of a broader marine seascape.

4. SPATIAL VARIABILITY AND THE VALUE OF COASTAL WETLANDS

Increasingly, it is being recognized that many ecosystem services provided by coastal wetlands are not uniformly distributed across large areas of coastlines, but vary considerably across "coastal landscapes." The reason for this variation may have to do with the structure, functioning, and production processes of the coastal wetlands, such as the marshes and mangroves. But the services provided by these wetlands are also affected by the wider coastal geomorphology, bathymetry, tidal flows, and other physical features and processes of the surrounding coastline in which they are located. Regardless of the cause of the spatial variation, there is increasing recognition that taking into account the spatial variation of benefits across a coastal landscape comprising marshes, mangroves and other wetlands are important to valuation of changes to that landscape.

Recent valuation studies have begun focusing on how the services of coastal wetlands are related to their landscape extent. What is more, such an approach is becoming important to evaluating different conservation versus development scenarios for these ecosystems. Some examples include managing environmental change in the Norfolk and Suffolk Broads of the United Kingdom (Turner et al., 2004), evaluating preferences for alternative restoration options for the Greater Everglades ecosystem in the United States (Milon and Scrogin, 2006), valuing ecosystem services from different land use options for the Peconic Estuary in the United States (Johnston et al., 2002), and assessing different mangrove management options in Malaysia (Othman et al., 2004) and Thailand (Barbier, 2007; Barbier et al., 2008).

Because the functional relationships inherent in many ecological processes are understudied, and there is so little corresponding economic information on the value of important services, estimations of how the value of an ecosystem good or service varies across an ecological landscape are rare. However, studies of coastal wetlands suggest that it is possible to track how the ecological functions underlying some ecosystem benefits vary spatially (Aburto-Oropeza et al., 2008; Aguilar-Perera and Appeldoorn, 2008; Barbier et al., 2008; Gedan et al., 2011; Koch et al., 2009; Meynecke et al., 2008; Peterson and Turner, 1994; Petersen et al., 2003; Rountree and Able, 2007). In particular, storm protection and support for marine fisheries provided by coastal wetland habitats, such as mangroves and salt marshes, tend to decline with the distance inshore from the seaward edge.

For example, the protection against storms provided by mangroves depends on their critical ecological function in terms of "attenuating" or reducing the height of storm waves (Barbier et al., 2008; Gedan et al., 2011; Koch et al., 2009). Ecological and hydrological field studies suggest that mangroves are unlikely to stop storm-induced waves that are greater than 6 m (Alongi, 2008; Cochard et al., 2008; Wolanski, 2007). Mangroves are effective in reducing storm-induced waves less than 6 m in height, and studies suggest that the wave height decreases nonlinearly for each 20–100 m that a mangrove forest extends out to sea (Bao, 2011; Barbier et al., 2008; Mazda et al., 1997). In other words, wave attenuation is greatest for the first 100 m of mangroves but declines as more mangroves are added to the seaward edge.

In addition, mangrove trees also have the capacity to buffer winds (Das and Crépin, 2013; McIvor et al., 2012a). The value of mangroves in providing such protection against high-speed and damaging winds is an often overlooked, but nonetheless very important, benefit. The growing evidence indicating that mangroves have significant wave attenuation and wind buffering functions has led to interest in valuing their storm protection benefit and also provided better understanding of the underlying ecological structure and functions contributing to this benefit, including how it varies across mangrove landscapes and different tide levels (Barbier, 2012; Barbier et al., 2008; Koch et al., 2009; McIvor et al., 2012a,b).

For salt marshes, wave attenuation also diminishes with increasing habitat distance inland from the shoreline (Barbier et al., 2008; Bouma et al., 2010; Gedan et al., 2011; Koch et al., 2009; Shephard et al., 2012; Ysebaert et al., 2011). This wave attenuation function derives from the amount of plant and sedentary animal material obstructing the water column and the bathymetry, or water depth, of the area (Koch et al., 2009). That is, as marsh serves as an important source of friction to moving water, both the presence of marsh wetlands and the amount of vegetation in the wetlands will slow and thus reduce the size of waves generated by hurricanes and storms as they approach coastlines. For example, for 5 mangrove and 10 salt marsh sites, the seaward margin of all the wetlands exhibited greater wave attenuation than equivalent landward distances, and the nonlinear decline in wave attenuation was similar for marsh and mangrove landscapes (Gedan et al., 2011). Salt marshes with two different species occurring in the Yangtze Estuary, China, were both found to reduce heights up to 80% over an initial distance inland of up to 50 m (Ysebaert et al., 2011). In Louisiana, death of salt marsh plants along the shoreline following heavy oiling after the Deepwater Horizon BP oil spill led to a doubling of already high shoreline erosion rates and thus reduced shoreline protection and control of coastal erosion (Silliman et al., 2012).

Mangroves and marshes also strongly influence the abundance, growth, and structure of neighboring marine fisheries by providing nursery, breeding, and other habitat functions for commercially important fish and invertebrate species that spend at least part of their life cycles in coastal and estuarine environments. Evidence of this coastal habitat-fishery linkage indicates that the value of this service is higher at the seaward edge or "fringe" of the vegetated coastal habitat than further inland (Aburto-Oropeza et al., 2008; Aguilar-Perera and Appeldoorn, 2008; Manson et al., 2005; Peterson and Turner, 1994; Rountree and Able, 2007). For example, Peterson and Turner (1994) find that densities of most fish and crustaceans were highest in salt marshes in Louisiana within 3 m of the water's edge compared with the interior marshes. This spatial relationship is confirmed for fish density in mangroves at Montalva Bay, southwestern Puerto Rico, where a study shows that fish density is significantly lower for mangroves located 30–50 m or more inshore compared with the mangroves on the seaward edge and no fish were found more than 50 m inshore from the sea (Aguilar-Perera and Appeldoorn, 2008).

Increasingly, valuation studies of coastal ecosystem services, such as some of the ones listed in Table 27.1, are taking into account the spatial variability in the ecological production functions of coastal landscapes. In the Gulf of California, Mexico, the mangrove fringe with a width of 5–10 m has the most influence on the productivity of nearshore fisheries, with a median value of \$37,500 ha^{-1}. Fishery landings also increased positively with the length of the mangrove fringe in a given location (Aburto-Oropeza et al., 2008). Barbier et al. (2008) show that how much of a mangrove forest coastline should be converted to shrimp aquaculture may depend critically on the spatial variability of coastal storm protection across the mangrove landscape. Barbier (2012) also includes the spatial variability of the support for offshore fisheries in the decision to convert a mangrove wetland to shrimp farms. The results of the latter analysis are depicted in Fig. 27.2.

In the absence of any spatially declining production of storm protection and habitat-linkage benefits, Barbier (2012) finds that the total value of all mangrove ecosystem services (\$18,978 ha^{-1}) easily exceeds either the commercial net returns to shrimp farming (\$9632 ha^{-1}) or the economic net returns (\$1220 ha^{-1}), which are the commercial returns adjusted for the subsidies for shrimp farm operations. If the benefits of mangrove ecosystem services do not vary from the seaward edge to the farthest most inland boundary, then the entire ecosystem should be preserved. However, Fig. 27.2 indicates that the spatial variation in the production of storm protection and habitat-fishery linkages changes the land use outcome significantly. As shown in the figure, when these two mangrove services decline spatially across the mangrove system, it causes the total net benefit of all ecosystem services to diminish significantly from the seaward edge to the landward boundary. This changes, in turn, the land use outcome of how much mangrove area to convert to shrimp farms and where they should be located.

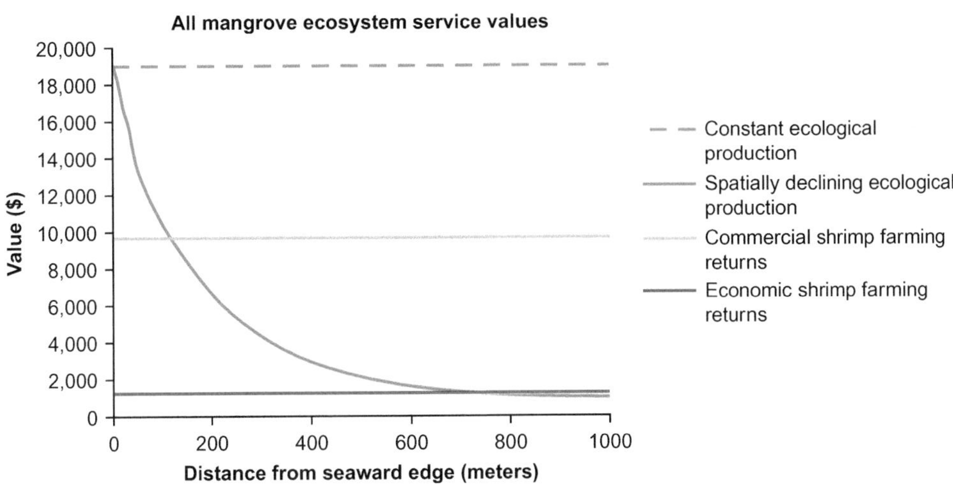

FIGURE 27.2 Spatial variation in mangrove ecosystem services. *Adapted from Barbier, E.B., 2012. A spatial model of coastal ecosystem services. Ecological Economics 78, 70–79.*

For example, Fig. 27.2 indicates that, when the value of preserving the mangroves is compared with the commercial net returns to shrimp farming, it is optimal to conserve only the first 118 m of mangroves from the seaward edge and to convert the rest to aquaculture. When compared with the net economic returns to shrimp farming, mangroves up to 746 m from the seaward edge should be preserved and the remaining mangroves inland converted. Thus, accounting for the spatial production and distribution of ecosystem services across a coastal landscape influences not only the amount of landscape converted but also where the conversion activity takes place, when compared with the scenario when this spatial variation is not accounted for.

5. VALUING COASTAL WETLANDS WITHIN A SEASCAPE

The term *seascape* is now widely used to refer to spatial mosaics of interconnected coastal and nearshore marine habitat types, such as mangroves, salt marsh, seagrasses, and coral reefs (Berkström et al., 2012; Harborne et al., 2006; Moberg and Rönnbäck, 2003; Mumby, 2006; Olds et al., 2016; Pittman et al., 2011). Boström et al. (2011) suggest that typical habitats found in a seascape are mangroves, salt marshes, seagrass meadows, coral reefs, and oyster reefs, as the *connectivity* between and among these coastal and nearshore marine habitats is the most pronounced. As a result of nutrient fluxes and material exchange, including movements of marine fauna, these habitats provide important goods and services both individually and through functional linkages across the seascape (Harborne et al., 2006, Olds et al., 2016).

Increasingly, economists are interested in estimating the multiple benefits arising from interconnected coastal and marine habitats. For example, Johnston et al. (2002) estimate the benefits arising from a wide range of ecosystem services provided by the Peconic Estuary on Long Island, New York, USA. The tidal mudflats, salt marshes, and seagrass (eelgrass) beds of the estuary support the shellfish and demersal (seabed) fisheries. In addition, bird-watching and waterfowl hunting are popular activities. The authors incorporate production function methods to simulate the biological and food web interactions of the ecosystems to assess the marginal value per acre in terms of gains in commercial value for fish and shellfish, bird-watching, and waterfowl hunting. The aggregate annual benefits are estimated to be $67 per acre for intertidal mudflats, $338 for salt marsh, and $1065 for seagrass across the estuary system. Using these estimates, the authors calculate the asset value of protecting existing habitats to be $12,412 per acre for seagrass, $4291 for salt marsh, and $786 for mudflats. In comparison, the asset value of restored habitats is $9996 per acre for seagrass, $3454 for marsh, and $626 for mudflats.

Some economic assessments also consider how ecosystem goods and services are affected by models of the biological connectivity of habitats, food webs, and migration and life cycle patterns across specific seascapes, such as mangrove–seagrass–reef systems (Barbier and Lee, 2014). For example, Plummer et al. (2013) use dynamic simulations in a food web model of central Puget Sound, Washington, USA, to determine how eelgrass beds provide valuable refuge, foraging, and spawning habitat for many commercially and recreationally fished species, including Pacific salmon, Pacific herring, and Dungeness crab. The authors found that a 20% increase in the area of eelgrass beds would lead to increases of $3 km^{-2} (of eelgrass area) in the value of herring, \$15 km^{-2} for crabs, and \$316 km^{-2} for salmon. However, the authors also examine the potential trade-offs of enhanced eelgrass beds with other coastal ecosystem benefits, notably, bird and marine mammal watching. The latter activities were not valued but the quantity of the service was estimated in terms of days of the recreational activity. The trade-off analysis reveals that the positive values to recreational and commercial fishing from increased eelgrass coverage are also associated with a slight rise in marine mammal watching but does lead to a slight decline in bird-watching.

Sanchirico and Mumby (2009) developed an integrated seascape model to illustrate that the presence of mangroves and seagrasses considerably enhances the biomass of coral reef fish communities. A key finding is that mangroves become more important as nursery habitat when excessive fishing effort levels are applied to the reef because the mangroves can directly offset the negative impacts of fishing effort. The results of this study support the development of "ecosystem-based" fishery management and the design of integrated coastal marine reserves that emphasize four key priorities: (1) the relative importance of mangrove nursery sites, (2) the connectivity of individual reefs to mangrove nurseries, (3) areas of nursery habitat that have an unusually large importance to specific reefs, and (4) priority sites for mangrove restoration projects (Mumby, 2006). A further extension of this work maps the life cycle of fish through coral reef–mangrove–seagrass systems to determine the contribution of each habitat to the biological growth and productivity of marine fisheries and to the protection of coastal properties from storm damage (Sanchirico and Springborn, 2012). The authors demonstrate how payments for ecosystem services for a particular habitat with multiple services are interdependent, change over time, and can be greater than the profits associated with fishing activities that degrade the coral reef–mangrove–seagrass system.

Stål et al. (2008) attempt to determine how three coastal habitats—soft sediment bottoms, seagrass beds, and rocky bottoms with macroalgae—influence the value of marine commercial, recreational, and fisheries off the west coast of Sweden. The five species studied are cod, plaice, eel, mackerel, and sea trout. The distributional pattern and habitat dependence varied among fish species. For example, plaice were more highly associated with shallow soft bottoms as a nursery habitat for juveniles but migrate to offshore fisheries as adults. Eel mainly inhabit seagrass beds and rocky bottoms, which is where they are frequently harvested. Cod are primarily caught in offshore fisheries, but their juveniles depend on rocky shores with algae and seagrass beds on soft sediment bottoms. As a consequence, decrease in the seagrass beds had a significant negative impact on the catch, working hours, and profits of eel fisherman. Macroalgae coverage of 30%–50% of rocky bottoms reduces significantly recruitment in the plaice fishery and substantially lowers its net present value. Finally, loss of seagrass habitats on the Swedish west coast may have accounted for around 3% of the total annual cod recruitment in the entire North Sea and Swedish Skagerrak archipelago system.

6. CONCLUSIONS

Although there remain a number of important challenges to valuing coastal wetland goods and services, significant progress has been made in recent years. The number of studies is increasing, more ecosystem goods and services are being valued, and the geographic coverage is improving. Overall, there is growing interest globally in valuing coastal wetland benefits, including for specific regions such as Europe or developing countries (Brander et al., 2012; Chaikumbung et al., 2016). As the world's coastal wetlands continue to disappear because of human population and development pressures, it becomes increasingly essential to assess the benefits provided by these important systems.

In addition, valuation approaches are becoming more advanced. Recently, there has been interest in estimating the multiple benefits arising from interconnected habitats (e.g., Johnston et al., 2002), how ecosystem goods and services are affected by the biological connectivity of habitats, food webs, and migration and life cycle patterns across specific seascapes (e.g., Plummer et al., 2013; Sanchirico and Mumby, 2009; Sanchirico and Springborn, 2012), and how specific ecosystem services are influenced by their spatial variation across a specific seascape habitat or habitats (Aburto-Oropeza et al., 2008; Barbier, 2012; Barbier et al., 2008, 2013; Barbier and Lee, 2014; Smith and Crowder, 2011). Future research should continue to develop such approaches.

The advancement in benefit transfer methods and modeling techniques, including the application of geographical information systems (GIS) and metaregression analysis, means that there are more opportunities to use these methods as a way of extrapolating and transferring estimated ecosystem service values for coastal wetlands from one location, population, and time to other locations, populations, and periods (Brander et al., 2012; Troy and Wilson, 2006). However, this technique is not a substitute for reliability, and there must be strict rules governing its use for coastal wetlands (Barbier, 2007, 2016; Plummer, 2009; Troy and Wilson, 2006). If there is a lack of economic and ecological correspondence between study and policy sites, transferring values between the two sites through GIS and other methods will simply lead to inaccurate valuation estimates (Plummer, 2009; Troy and Wilson, 2006).

Similarly, there are potential drawbacks of applying benefit transfer through metaanalysis regression. This requires knowledge of the values of the independent variables for the policy site of interest and assumes that the statistical relationship between the dependent and independent variables is the same between the study and policy sites (Richardson et al., 2015; Rosenberger and Stanley, 2006). If one can statistically control for these differences in ecological and economic correspondence, this reduces the benefit transfer errors. In addition, there needs to be a sufficient number and variety of reliable policy site valuation studies to make the metaanalysis regression applicable in the first place (Barbier, 2016). For example, as Table 27.1 shows, only a handful of valuation studies of the various benefits of coastal wetlands may serve this purpose. Unfortunately, this suggests that benefit transfers may be less helpful in overcoming the lack of sufficient studies of these benefits for some coastal wetlands and their ecosystem services.

Despite the improvement in valuation methods and the growing number of case study applications globally, there is still skepticism among wetland regulators and coastal managers about how useful such valuation studies are for informing practical policy and management decisions. As discussed previously, for example, US wetland regulators still query the scientific validity of valuation estimates and express concerns about whether they are sufficiently scientifically and legally defensible for decision-making purposes (Arnold, 2013). This would suggest that the most important progress in valuing coastal wetland services would be in providing estimates of value with sufficiently small error bounds that could inform decision-makers on how to rank alternative management options with high confidence.

Improvement in the scientific validity of coastal valuation will be especially important as the field tackles new and more complicated tasks, such as the influence of spatial variation on the benefits across large coastal wetland landscapes and the assessment of benefits, such as support for nearshore fisheries, storm protection, and pollution control, across an interconnected seascape.

References

Aburto-Oropeza, O., Ezcurra, E., Danemann, G., Valdez, V., Murray, J., Sala, E., 2008. Mangroves in the Gulf of California increase fishery yields. Proceedings of the National Academy of Sciences of the United States of America 105, 10456–10459.

Aguilar-Perera, A., Appeldoorn, R.S., 2008. Spatial distribution of marine fishes along a cross-shelf gradient containing a continuum of mangrove-seagrass-coral reefs off southwestern Puerto Rico. Estuarine, Coastal and Shelf Science 76, 378–394.

Alongi, D.M., 2008. Mangrove forests: resilience, protection from tsunamis, and responses to global climate change. Estuarine, Coastal and Shelf Science 76, 1–13.

Arnold, G., 2013. Use of monetary wetland value estimates by EPA Clean Water Act Section 404 regulators. Wetlands Ecology and Management 21, 117–129.

Badola, R., Hussain, S.A., 2005. Valuing ecosystems functions: an empirical study on the storm protection function of Bhitarkanika mangrove ecosystem, India. Environmental Conservation 32, 85–92.

Bao, T.Q., 2011. Effect of mangrove forest structures on wave attenuation in coastal Vietnam. Oceanologia 53, 807–818.

Barbier, E.B., Cox, M., 2003. Does economic development lead to mangrove loss: a cross-country analysis. Contemporary Economic Policy 21, 418–432.

Barbier, E.B., Enchelmeyer, B., 2014. Valuing the storm surge protection service of US Gulf Coast wetlands. Journal of Environmental Economics and Policy 3 (2), 167–185.

Barbier, E.B., Lee, K.D., 2014. Economics of the marine seascape. International Review of Environmental and Resource Economics 7, 35–65.

Barbier, E.B., Strand, I., 1998. Valuing mangrove-fishery linkages: a case study of Campeche, Mexico. Environmental and Resource Economics 12, 151–166.

Barbier, E.B., Koch, E.W., Silliman, B.R., Hacker, S.D., Wolanski, E., Primavera, J.H., Granek, E., Polasky, S., Aswani, S., Cramer, L.A., Stoms, D.M., Kennedy, C.J., Bael, D., Kappel, C.V., Perillo, G.M., Reed, D.J., 2008. Coastal ecosystem-based management with nonlinear ecological functions and values. Science 319, 321–323.

Barbier, E.B., Hacker, S.D., Kennedy, C., Koch, E.W., Stier, A.C., Silliman, B.R., 2011. The value of estuarine and coastal ecosystem services. Ecological Monographs 81 (2), 169–183.

Barbier, E.B., Georgiou, I.Y., Enchelmeyer, B., Reed, D.J., 2013. The value of wetlands in protecting southeast Louisiana from hurricane storm surges. PLoS One 8 (3), e58715. https://doi.org/10.1371/journal.pone.0058715. http://www.plosone.org/article/info:doi/10.1371/journal.pone.0058715.

Barbier, E.B., 2003. Habitat-fishery linkages and mangrove loss in Thailand. Contemporary Economic Policy 21, 59–77.

Barbier, E.B., 2007. Valuing ecosystems as productive inputs. Economic Policy 22, 177–229.

Barbier, E.B., 2011. Capitalizing on Nature: Ecosystems as Natural Assets. Cambridge University Press, Cambridge and New York.

Barbier, E.B., 2012. A spatial model of coastal ecosystem services. Ecological Economics 78, 70–79.

Barbier, E.B., 2014. A global strategy for protecting vulnerable coastal populations. Science 345, 1250–1251.

Barbier, E.B., 2016. The protective service of mangrove ecosystems: a review of valuation methods. Marine Pollution Bulletin 109, 676–681.

Bateman, I., Langford, I.H., 1997. Non-users willingness to pay for a national park: an application of the contingent valuation method. Regional Studies 31, 571–582.

Bateman, I.J., Mace, G.M., Fezzi, C., Atkinson, G., Turner, R.K., 2011. Economic analysis for ecosystem service assessments. Environmental and Resource Economics 48, 177–218.

Bell, F.W., 1997. The economic valuation of saltwater marsh supporting marine recreational fishing in the southeastern United States. Ecological Economics 21 (3), 243–254.

Berkström, C., Gullström, M., Lindborg, R., Mwandya, A.W., Yahya, S.A.S., Kautsky, N., Nyström, M., 2012. Exploring 'knowns' and 'unknowns' in tropical seascape connectivity with insights from East African coral reefs. Estuarine, Coastal and Shelf Science 107, 1–21.

Birol, E., Cox, V., 2007. Using choice experiments to design wetland management programmes: the case of the Severn Estuary Wetland, UK. Journal of Environmental Planning and Management 50 (3), 363–380.

Bockstael, N.E., Freeman III, A.M., Kopp, R.J., Portney, P.R., Smith, V.K., 2000. On measuring economic values for nature. Environmental Science and Technology 34, 1384–1389.

Boström, C., Pittman, S.J., Simenstad, C., Kneib, R.T., 2011. Seascape ecology of coastal biogenic habitats: advances, gaps and challenges. Marine Ecology Progress Series 427, 101–217.

Bouma, T.J., De Vries, M.B., Herman, P.M.J., 2010. Comparing ecosystem engineering efficiency of two plant species with contrasting growth strategies. Ecology 91, 2696–2704.

Boyd, J., Banzhaf, S., 2007. What are ecosystem services? The need for standardized environmental accounting units. Ecological Economics 63, 616–626.

Brander, L., Bräuer, I., Gerdes, H., Ghermandi, A., Kuik, O., Markandya, A., Navrud, S., Nunes, P., Schaafsma, M., Vos, H., Wagendonk, A., 2012. Using meta-analysis and GIS for value transfer and scaling up: valuing climate change induced losses of European wetlands. Environmental and Resource Economics 52, 395–413.

Breaux, A., Farber, S., Day, J., 1995. Using natural coastal wetlands systems for wastewater treatment: an economic benefit analysis. Journal of Environmental Management 44 (3), 285–291.

Brouwer, R., Bateman, I.J., 2005. Temporal stability and transferability of models of willingness to pay for flood control and wetland conservation. Water Resources Research 41, 1–6.

Byström, O., 2000. The replacement value of wetlands in Sweden. Environmental and Resource Economics 16, 347–362.

Chaikumbung, M., Doucouliagos, H., Scarborough, H., 2016. The economic value of wetlands in developing countries: a meta-regression analysis. Ecological Economics 124, 164–174.

Cochard, R., Ranamukhaarachchi, S.L., Shivakotib, G.P., Shipin, O.V., Edwards, P.J., Seeland, K.T., 2008. The 2004 tsunami in Aceh and Southern Thailand: a review on coastal ecosystems, wave hazards and vulnerability. Perspectives in Plant Ecology, Evolution and Systematics 10, 3–40.

Costanza, R., Pérez-Maqueo, O., Martinez, M.L., Sutton, P., Anderson, S.J., Mulder, K., 2008. The value of coastal wetlands for hurricane protection. Ambio 37, 241–248.

Daily, G.C., Söderqvist, T., Aniyar, S., Arrow, K.J., Dasgupta, P., et al., 2000. The value of nature and the nature of value. Science 289, 395–396.

Das, S., Crépin, A.-S., 2013. Mangroves can provide against wind damage during storms. Estuarine, Coastal and Shelf Science 134, 98–107.

Das, S., Vincent, J.R., 2009. Mangroves protected villages and reduced death toll during Indian super cyclone. Proceedings of the National Academy of Sciences of the United States of America 106, 7357–7360.

Dasgupta, S., Laplante, B., Murray, S., Wheeler, D., 2011. Exposure of developing countries to sea-level rise and storm surges. Climatic Change 106, 567–579.

Doney, S.C., Ruckelshaus, M., Duffy, J.E., Barry, J.P., Chan, F., English, C.A., Galindo, H.M., Grebmeier, J.M., Hollowed, A.B., Knowlton, N., Polovina, J.J., Rabalais, N.N., Sydeman, W.J., Talley, L.D., 2012. Climate change impacts on marine ecosystems. Annual Review of Marine Science 4, 11–37.

Duke, N.C., Meynecke, J.-O., Dittmann, S., Ellison, A.M., Anger, K., Berger, U., Cannicci, S., Diele, K., Ewel, K.C., Field, C.D., Koedam, N., Lee, S.Y., Marchand, C., Nordhaus, I., Dahdouh-Guebas, F., 2007. A world without mangroves? Science 317, 41–42.

Elliott, M., Cutts, N.D., Trono, A., 2014. A typology of marine and estuarine hazards and risks as vectors of change: a review of vulnerable coasts and their management. Ocean and Coastal Management 93, 88–99.

Engle, V.D., 2011. Estimating the provision of ecosystem services by Gulf of Mexico Coastal Wetlands. Wetlands 31, 179–193.

Environmental Protection Agency (EPA), 2009. Valuing the Protection of Ecological Systems and Services. A Report of the EPA Science Advisory Board. EPA, Washington, DC.

Erwin, K.L., 2009. Wetlands and global climate change: the role of wetland restoration in a changing world. Wetlands Ecology and Management 17, 71–84.

Freeman III, A.M., Herriges, J.A., Kling, C.L., 2014. The Measurement of Environmental and Resource Values: Theory and Methods. Routledge, London.

Freeman III, A.M., 1991. Valuing environmental resources under alternative management regimes. Ecological Economics 3, 247–256.

Friess, D.A., Webb, E.L., 2014. Variability in mangrove change estimates and implications for the assessment of ecosystem service provision. Global Ecology and Biogeography 23, 715–725.

Gedan, K.B., Kirwan, M.L., Wolanski, E., Barbier, E.B., Silliman, B.R., 2011. The present and future role of coastal wetland vegetation in protecting shorelines: answering recent challenges to the paradigm. Climatic Change 106, 7–29.

Halpern, B.S., Waldbridge, S., Selkoe, K.A., Kappel, C.V., Micheli, F., D'Agrosa, C., et al., 2008. A global map of human impact on marine ecosystems. Science 319, 948–952.

Harborne, A.R., Mumby, P.J., Micheli, F., Perry, C.T., Dahlgren, C.P., Holmes, K.E., Burmbaugh, D.R., 2006. The functional value of Caribbean coral reef, seagrass and mangrove habitats to ecosystem processes. Advances in Marine Biology 50, 59–189.

Huang, J.-C., Poor, P.J., Zhao, M.Q., 2007. Economic valuation of beach erosion control. Marine Resource Economics 32, 221–238.

Janssen, R., Padilla, J.E., 1999. Preservation or conservation? Valuation and evaluation of a mangrove forest in the Philippines. Environmental and Resource Economics 14, 297–331.

Johnston, R.J., Grigalunas, T.A., Opaluch, J.J., Mazzotta, M., Diamantedes, J., 2002. Valuing estuarine resource services using economic and ecological models: the Peconic Estuary system. Coastal Management 30 (1), 47–65.

Kaoru, Y., Smith, V.K., Liu, J.L., 1995. Using random utility models to estimate the recreational value of estuarine resources. American Journal of Agricultural Economics 77, 141–151.

Kaoru, T., 1993. Differentiating use and nonuse values for coastal pond water quality improvements. Environmental and Resource Economics 3, 487–494.

Kim, T.-G., Petrolia, D.R., 2013. Public perceptions of wetland restoration benefits in Louisiana. ICES Journal of Marine Science 70 (5), 1045–1054.

King, S.E., Lester, J.N., 1995. The value of salt marsh as a sea defence. Marine Pollution Bulletin 30 (3), 180–189.

Koch, E.W., Barbier, E.B., Silliman, B.R., Reed, D.J., Perillo, G.M.E., Hacker, S.D., Granek, E.F., Primavera, J.H., Muthiga, N., Polasky, S., Halpern, B.S., Kennedy, C.J., Kappel, C.V., Wolanski, E., 2009. Non-linearity in ecosystem services: temporal and spatial variability in coastal protection. Frontiers in Ecology and the Environment 7, 29–37.

Kreitler, J., Papenfus, M., Byrd, K., Labiosa, W., 2013. Interacting coastal based ecosystem services: recreation and water quality in Puget Sound, WA. PLoS One 8 (2), e56670. https://doi.org/10.1371/journal.pone.0056670.

Landry, C.E., Liu, H., 2009. A semi-parametric estimator for revealed and stated preference application to recreational beach visitation. Journal of Environmental Economics and Management 57, 205–218.

Landry, C.E., Keeler, A.G., Kriesel, W., 2003. An economic evaluation of beach erosion management alternatives. Marine Resource Economics 18, 105–127.

Landry, C.E., Hindsley, P., Bin, O., Kruse, J.B., Whitehead, J.C., Wilson, K., 2011. Weathering the storm: measuring household willingness-to-pay for risk-reduction in post-Katrina New Orleans. Southern Economic Journal 77, 991–1013.

Lange, G.-M., Jiddawi, N., 2009. Economic value of marine ecosystem services in Zanzibar: implications for marine conservation and sustainable development. Ocean and Coastal Management 52, 521–532.

Laso Bayas, J.C., Marohn, C., Dercon, G., Dewi, S., Piepho, H.P., Joshi, L., van Noordwijk, M., Cadisch, G., 2011. Influence of coastal vegetation on the 2004 tsunami wave impact Aceh. Proceedings of the National Academy of Sciences of the United States of America 108, 18612–18617.

Leggett, C.G., Bockstael, N.E., 2000. Evidence of the effects of water quality on residential land prices. Journal of Environmental Economics and Management 39, 121–144.

Lotze, H.K., Lenihan, H.S., Bourque, B.J., Bradbury, R.H., Cooke, R.G., et al., 2006. Depletion, degradation and recovery potential of estuaries and coastal seas. Science 312, 1806–1809.

Luisetti, T., Turner, R.K., Bateman, I.J., Morse-Jones, S., Adams, C., Fonseca, L., 2011. Coastal and marine ecosystem services valuation for policy: managed realignment case studies in England. Ocean and Coastal Management 54, 212–224.

Mangi, S.C., Davis, C.E., Payne, L.A., Austen, M.C., Simmonds, D., Beaumont, N.J., Smyth, T., 2011. Valuing the regulatory services provided by marine ecosystems. Environmetrics 22, 686–698.

Manson, F.J., Loneragan, N.R., Harch, B.D., Skilleter, G.A., Williams, L., 2005. "A broad-scale analysis of links between coastal fisheries production and mangrove extent: a case-study for northeastern Australia." Fisheries Research 74 (1–3), 69–85.

Massey, D.M., Newbold, S.C., Gentner, B., 2006. Valuing water quality changes using a bioeconomic model of a coastal recreational fishery. Journal of Environmental Economics and Management 52, 482–500.

Mazda, Y., Magi, M., Kogo, M., Hong, P.N., 1997. Mangroves as a coastal protection from waves in the Tong King Delta, Vietnam. Mangroves and Salt Marshes 1, 127–135.

McConnell, K.E., Bockstael, N.E., 2005. Valuing the environment as a factor of production (Chapter 14). In: Mäler, K.-G., Vincent, J.R. (Eds.), Handbook of Environmental Economics, vol. 2. Elsevier, Amsterdam, pp. 621–669.

McIvor, A.L., Möller, I., Spencer, T., Spalding, M., 2012a. Reduction of Wind and Swell Waves by Mangroves. Natural Coastal Protection Series: Report 1. Cambridge Coastal Research Unit Working Paper 40. The Nature Conservancy and Wetlands International, Cambridge, UK.

McIvor, A.L., Möller, I., Spencer, T., Spalding, M., 2012b. Storm Surge Reduction by Mangroves. Natural Coastal Protection Series: Report 1 Cambridge Coastal Research Unit Working Paper 41. The Nature Conservancy and Wetlands International, Cambridge, UK.

Mendelsohn, R., Olmstead, S., 2009. The economic valuation of environmental amenities and disamenities: methods and applications. Annual Review of Environment and Resources 34, 325–347.

Meynecke, J.-O., Lee, S.Y., Duke, N.C., 2008. Linking spatial metrics and fish catch reveals the importance of coastal wetland connectivity to inshore fisheries in Queensland, Australia. Biological Conservation 141, 981–996.

Millennium Ecosystem Assessment (MEA), 2005. Ecosystems and Human Well-being: Current State and Trends. Island Press, Washington, DC.

Milon, J.W., Scrogin, D., 2006. Latent preferences and valuation of wetland ecosystem restoration. Ecological Economics 56, 152–175.

Moberg, F., Rönnbäck, P., 2003. Ecosystem services of the tropical seascape: interactions, substitutions and restoration. Ocean and Coastal Management 46, 27–46.

Mumby, P.J., 2006. Connectivity of reef fish between mangroves and coral reefs: algorithms for the design of marine reserves at seascape scales. Biological Conservation 128, 215–222.

Narayan, S., Beck, M.W., Reguero, B.G., Losada, I.J., van Wesenbeeck, B., Pontee, N., Sanchirico, J.N., Ingram, J.C., Lange, G.-M., Burks-Copes, K.A., 2016. The effectiveness, costs and coastal protection benefits of natural and nature-based defences. PLoS One 11 (5), e0154735. https://doi.org/10.1371/journal.pone.0154735.

National Research Council (NRC), 2005. Valuing Ecosystem Services: Toward Better Environmental Decision Making. The National Academies Press, Washington, DC.

Naylor, R., Drew, M., 1998. Valuing mangrove resources in Kosrae, Micronesia. Environment and Development Economics 3, 471–490.

Nfotabong Atheull, A., Din, N., Longonje, S.N., Koedam, N., Dahdouh-Guebas, F., 2009. Commercial activities and subsistence utilization of mangrove forests around the Wouri estuary and the Douala-Edea reserve (Cameroon). Journal of Ethnobiology and Ethnomedicine 5, 35–49.

O'Higgins, T.G., Ferraro, S.P., Dantin, D.D., Jordan, S.J., Chintala, M.M., 2010. Habitat scale mapping of fisheries ecosystem service values in estuaries. Ecology and Society 15 (4), 7. http://www.ecologyandsociety.org/vol15/iss4/art7/.

Olds, A.D., Connolly, R.M., Pitt, K.A., Pittman, S.J., Maxwell, P.S., Huijbers, C.M., Moore, B.R., Albert, S., Rissk, D., Babcock, R.C., Schlacher, T.A., 2016. Quantifying the conservation value of seascape connectivity: a global synthesis. Global Ecology and Biogeography 25, 3–15.

Othman, J., Bennett, J., Blamey, R., 2004. Environmental management and resource management options: a choice modelling experience in Malaysia. Environment and Development Economics 9, 803–824.

Petersen, J.E., Kemp, W.M., Bartleson, R., Boynton, W.R., Chen, C.-C., Cornwell, J.C., Gardner, R.H., Hinkle, D.C., Houde, E.D., Malone, T.C., Mowitt, W.R., Murray, L., Sanford, L.P., Stevenson, J.C., Sundberg, K.L., Suttles, S.E., 2003. Multiscale experiments in coastal ecology: improving realism and advancing theory. BioScience 53, 1181–1197.

Peterson, G.W., Turner, R.E., 1994. The value of salt marsh edge versus interior as habitat for fish and decapods crustaceans in a Louisiana tidal marsh. Estuaries 17, 235–262.

Petrolia, D.R., Kim, T.-G., 2009. What are barrier islands worth? Estimates of willingness to pay for restoration. Marine Resource Economics 24, 131–146.

Petrolia, D.R., Kim, T.-G., 2011. Preventing land loss in coastal Louisiana: estimates of WTP and WTA. Journal of Environmental Management 92, 859–865.

Petrolia, D.R., Interis, M.G., Hwang, J., 2014. America's wetland? A national survey of willingness to pay for restoration of Louisiana's coastal wetlands. Marine Resource Economics 29 (1), 17–37.

Pinto, R., Patricio, J., Neto, J.M., Salas, F., Marques, J.C., 2010. Assessing estuarine quality under the ecosystem services scope: ecological and socioeconomic aspects. Ecological Complexity 7, 389–402.

Pittman, S.J., Kneib, R.T., Simenstad, C.A., 2011. Practicing coastal seascape ecology. Marine Ecology Progress Series 427, 187–190.

Plummer, M.L., Harvery, C.J., Anderson, L.E., Guerry, A.D., Ruckelshaus, M.H., 2013. The role of eelgrass in marine community interactions and ecosystem services: results from an ecosystem-scale food web model. Ecosystems 16, 237–251.

Plummer, M.L., 2009. Assessing benefit transfer for the valuation of ecosystem services. Frontiers in Ecology and the Environment 7 (1), 38–45.

Polasky, S., Segerson, K., 2009. Integrating ecology and economics in the study of ecosystem services: some lessons learned. Annual Review of Resource Economics 1, 409–434.

Rao, N.S., Ghermandi, A., Portela, R., Wang, X., 2015. Global values of coastal ecosystem services: a spatial economic analysis of shoreline protection values. Ecosystem Services 11, 95–105.

Richardson, L., Loomis, J., Kroeger, T., Casey, F., 2015. The role of benefit transfer in ecosystem service valuation. Ecological Economics 115, 51–58.

Rosenberger, R.S., Stanley, T.D., 2006. Measurement, generalization, and publication: sources of error in benefit transfers and their management. Ecological Economics 60 (2), 372–378.

Rountree, R.A., Able, K.W., 2007. Spatial and temporal habitat use patterns for salt marsh nekton: implications for ecological functions. Aquatic Ecology 41, 25–45.

Ruitenbeek, H.J., 1994. Modeling economy-ecology linkages in mangroves: economic evidence for promoting conservation in Bintuni Bay, Indonesia. Ecological Economics 10, 233–247.

Rulleau, B., Rey-Valette, H., Hérivaux, C., 2014. Valuing welfare impacts of climate change in coastal areas: a French case study. Journal of Environmental Planning and Management. http://www.tandfonline.com/action/showCitFormats?doi=10.1080/09640568.2013.862492.

Samonte-Tan, G.P.B., White, A.T., Diviva, M.T.J., Tabara, E., Caballes, C., 2007. Economic valuation of coastal and marine resources: Bohol Marine Triangle, Philippines. Coastal Management 35, 319–338.

Sanchirico, J.N., Mumby, P., 2009. Mapping ecosystem functions to the valuation of ecosystem services: implications of species-habitat associations for coastal land-use decisions. Theoretical Ecology 2, 67–77.

Sanchirico, J.N., Springborn, M., 2012. How to get there from here: ecological and economic dynamics of ecosystem service provision. Environmental and Resource Economics 48, 243–267.

Sathirathai, S., Barbier, E.B., 2001. Valuing mangrove conservation, Southern Thailand. Contemporary Economic Policy 19, 109–122.

Shephard, C.C., Crain, C.M., Beck, M.W., 2012. The protective role of coastal marshes: a systematic review and meta-analysis. PLoS One 6, e27374.

Sikamäki, J., Sanchirico, J.N., Jardine, S.L., 2012. Global economic potential for reducing carbon dioxide emissions from mangrove loss. Proceedings of the National Academy of Sciences of the United States of America 109, 14369–14374.

Silliman, B.R., Diller, J., McCoy, M., Earl, K., Adams, P., von de Koppel, J., Zimmerman, A., 2012. Degradation and resilience in Louisiana salt marshes following the BP-Deepwater Horizon oil spill. Proceedings of the National Academy of Sciences of the United States of America 109, 11234–11239.

Smith, M.D., Crowder, L.B., 2011. Valuing ecosystem services with fishery rents: a lumped-parameter approach to hypoxia in the Neuse Estuary. Sustainability 3, 2229–2267.

Smith, V.K., Palmquist, R.B., 1994. Temporal substitution and the recreational value of coastal amenities. The Review of Economics and Statistics 76, 119–126.

Smith, V.K., Palmquist, R.B., Jakus, P., 1991. Combining farrell frontier and hedonic travel cost models for valuing estuarine quality. The Review of Economics and Statistics 73, 694–699.

Smith, M.D., 2007. Generating value in habitat-dependent fisheries: the importance of fishery management institutions. Land Economics 83, 59–73.

Spalding, M.D., Kanuma, M., Collins, L., 2010. World Atlas of Mangroves. Earthscan, London.

Spalding, M.D., Ruffo, S., Lacambra, C., Meliane, I., Hale, L.Z., Shepard, C.C., Beck, M.W., 2014. The role of ecosystems in coastal protection: adapting to climate change and coastal hazards. Ocean and Coastal Management 90, 50–57.

Stål, J., Paulsen, S., Pihl, L., Rönnbäck, P., Söderqvist, T., Wennhage, H., 2008. Coastal habitat support to fish and fisheries in Sweden: integrating ecosystem functions into fisheries management. Ocean and Coastal Management 51, 594–600.

Swallow, S.K., 1994. Renewable and nonrenewable resource theory applied to coastal agriculture, forest, wetland and fishery linkages. Marine Resource Economics 9, 291–310.

Thompson, B.S., Clubbe, C.P., Primavera, J.H., Curnick, D., Koldewey, H.J., 2014. Locally assessing the economic viability of blue carbon: a case study from Panay Island, the Philippines. Ecosystem Services. http://www.sciencedirect.com/science/article/pii/S2212041614000242.

Troy, A., Wilson, M.A., 2006. Mapping ecosystem services: practical challenges and opportunities in linking GIS and value transfer. Ecological Economics 60, 435–449.

Turner, R.K., Bateman, I.J., Georgiou, S., Jones, A., Langford, I.H., Matias, N.G.N., Subramanian, L., 2004. An ecological economics approach to the management of a multi-purpose coastal wetland. Regional Environmental Change 4, 86–99.

Webb, E.L., Friess, D.A., Krauss, K.W., Cahoon, D.R., Guntenspergen, G.R., Phelps, J., 2013. A global standard for monitoring coastal wetland vulnerability to accelerated sea-level rise. Nature Climate Change 3, 458–465.

Wolanski, E., 2007. Estuarine Ecohydrology. Elsevier, Amsterdam.

Worm, B., Barbier, E.B., Beaumont, N., Duffy, J.E., Folke, C., Halpern, B.S., et al., 2006. Impacts of biodiversity loss on ocean ecosystem services. Science 314, 787–790.

Ysebaert, T., Yang, S.-L., Zhang, L., He, Q., Bouma, T.J., Herman, P.M.J., 2011. Wave attenuation by two contrasting ecosystem engineering salt marsh macrophytes in the intertidal pioneer zone. Wetlands 31, 1043–1105.

CHAPTER

28

Conservation of Blue Carbon Ecosystems for Climate Change Mitigation and Adaptation

Oscar Serrano[1], *Jeffrey J. Kelleway*[2], *Catherine Lovelock*[3,4], *Paul S. Lavery*[1,5]

[1]School of Science & Centre for Marine Ecosystems Research, Edith Cowan University, Joondalup, WA, Australia; [2]Department of Environmental Sciences, Macquarie University, Sydney, NSW, Australia; [3]The School of Biological Sciences, The University of Queensland, St Lucia, QLD, Australia; [4]Global Change Institute, The University of Queensland, St Lucia, QLD, Australia; [5]Centro de Estudios Avanzados de Blanes, Consejo Superior de Investigaciones Científicas, Blanes, Spain

1. WHAT IS BLUE CARBON?

Vegetated coastal ecosystems, specifically tidal marshes, mangroves, and seagrass meadows, have exceptional capacities to sequester carbon dioxide (CO_2). They rank among the most productive habitats on earth, and through photosynthesis they fix CO_2 as organic matter, thus removing CO_2 from the atmosphere (Duarte et al., 2005). The capacity of vegetated coastal ecosystems to store organic carbon (C) in their canopies and soils over centennial to millennial time scales has been termed blue carbon (BC; Nellemann and Corcoran, 2009; Duarte et al., 2013). Recently, marine macroalgae (seaweed), that are the dominant primary producers in the coastal zone, have also been considered to contribute toward global BC sequestration, acting as C donors to other ecosystems where their organic material accumulates (Hill et al., 2015; Krause-Jensen and Duarte, 2016).

The burning of fossil fuels (e.g., coal, gas, and petroleum) is increasing the atmospheric levels of greenhouse gases (GHGs) including CO_2, methane (CH_4), and nitrous oxides (NO_x), which contributes to climate change and global warming. Land use change, and deforestation in particular, is also a major source of GHG emissions, accounting for 8%–20% of all

Coastal Wetlands
https://doi.org/10.1016/B978-0-444-63893-9.00028-9

global emissions (van der Werf et al., 2009). Oceans play a key role in the capture and recycling of atmospheric CO_2 globally, due to the gaseous exchange at the ocean–atmosphere interface (Siegenthaler and Sarmiento, 1993). The oceans have absorbed over one-third of anthropogenic CO_2 emissions, through biological, physical, and chemical processes. In the open ocean, marine living organisms capture C, converting CO_2 into biomass through photosynthesis. However, most of the CO_2 assimilated by photosynthetic organisms in the ocean is recycled near the ocean surface and converted back into CO_2 by marine bacteria, which has been termed the "biological pump" (Longhurst and Harrison, 1989). Yet a significant portion of this biogenic C reaches the seafloor, where it can be buried and effectively locked away from the atmosphere over long time scales (centuries to millennium), constituting a sink of CO_2 and contributing to mitigate climate change (Bowler et al., 2009).

The CO_2 sequestration process in BC ecosystems is similar to that occurring in the open ocean (i.e., conversion of CO_2 into biomass through photosynthesis), but their C storage capacity is exceptional compared with other natural C sinks such as the open ocean and terrestrial forests (Nellemann and Corcoran, 2009, Fig. 28.1). Despite tidal marsh, mangrove, and

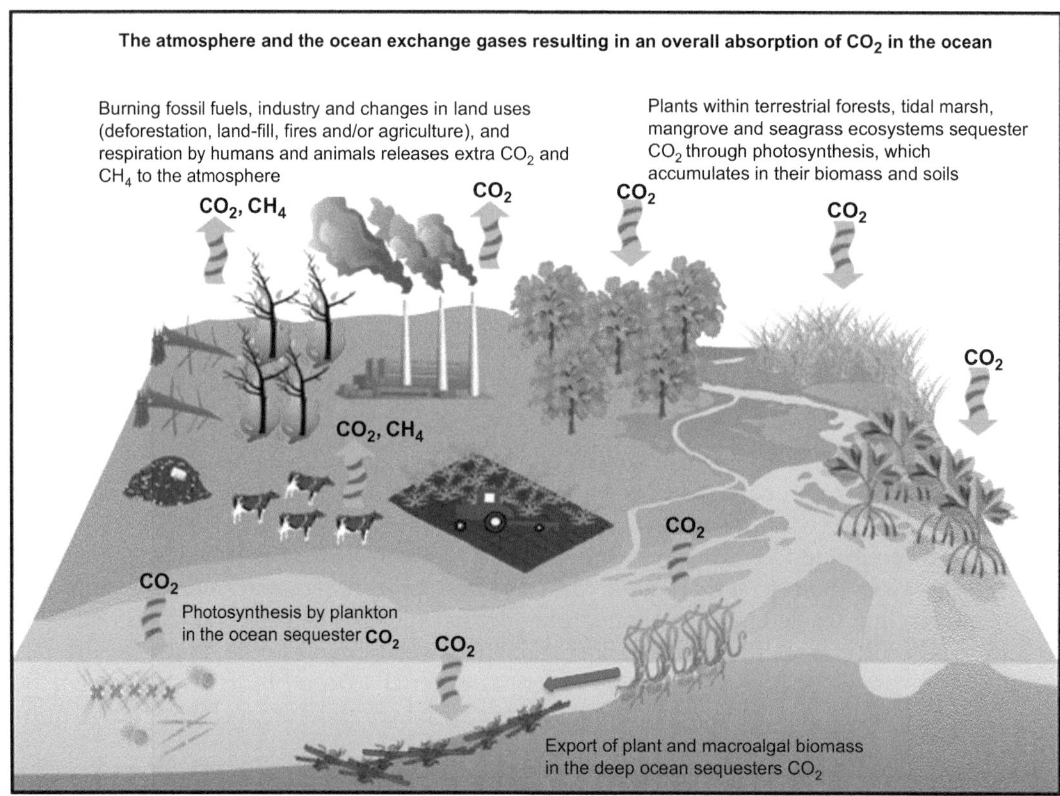

FIGURE 28.1 Conceptual diagram of carbon sequestration by blue carbon ecosystems and some of the activities that influence CO_2 exchange among the atmosphere, soil, and ocean in coastal areas and the open ocean. The major global C pools include the atmosphere, oceans, fossil fuels, vegetation, soils, and detritus. Landfill, smokestacks, cattle farming, and other human activities result in additional methane (CH_4) emissions.

seagrass ecosystems occupying less than 0.2% of the seabed area, they contribute nearly 50% of the CO_2 sequestration in marine sediments, and their C sequestration rates exceed those in the soils of many terrestrial ecosystems by 30- to 50-fold (Chmura et al., 2003; Duarte et al. 2005, 2013; Mcleod et al., 2011). Most macroalgal communities grow on rocky substrate and do not form significant in situ sedimentary C deposits, but the initial estimates of the amount of macroalgae C sequestered in sediments and deep-sea waters suggest that it is comparable to the C sequestered by all other BC ecosystems combined (Krause-Jensen and Duarte, 2016). Furthermore, the C captured by BC ecosystems is stored in marine soils for millennia, rather than the decades or centuries typical of terrestrial forests. This is due in part to the high rates of vertical accretion in tidal marsh, mangrove, and seagrass ecosystems, ranging from 0.4 to 21 mm $year^{-1}$ (Mateo et al., 1997; McKee et al., 2007; Duarte et al., 2013). This process of raising the seafloor is driven partly through the trapping and settling of particles from the water column and partly through organic matter production. This acts to bury the C in anoxic conditions, thereby slowing down its remineralization by microbes (Krauss et al., 2014; Mateo et al., 2006; Pedersen et al., 2011). Globally, tidal marsh, mangrove, and seagrass ecosystems sequester annually a similar amount of C to terrestrial forests, despite their extent being less than 3% of that of forests (Duarte et al., 2013). Unlike terrestrial forests, mangroves and tidal marshes rarely burn in wildfires, although they are exposed to other disturbances (e.g., tropical storms).

BC ecosystems provide important and valuable ecosystem services critical for climate change mitigation and adaptation, including coastal protection from storms and shoreline erosion, regulation of water quality, provision of habitat for commercially important fisheries and enhancing biodiversity, and being globally significant C sinks (Costanza et al., 1997; Duarte et al., 2013; Smale et al., 2013). Despite these recognized values, tidal marshes, mangroves, and seagrasses rank among the most threatened habitats on Earth (Valiela et al., 2001; Waycott et al., 2009; Gedan et al., 2009; Pendleton et al., 2012; Hamilton and Casey, 2016). The degradation, conversion to alternative land uses, and loss of BC ecosystems can result in the release of GHG to the atmosphere from the microbial decomposition of C stocks in their biomass and soil, fueling climate change (Pedersen et al., 2011; Pendleton et al., 2012; Murdiyarso et al., 2015; Kroeger et al., 2017; Lovelock et al., 2017a, Fig. 28.1). Despite the relatively small global decline of macroalgal communities (Byrnes et al., 2016), aquaculture of macroalgae has been identified as a prospective BC resource with great potential (Duarte, 2017).

Knowledge of the role of natural ecosystems in capturing and storing CO_2 is an increasingly important component in developing strategies to mitigate climate change associated with anthropogenic inputs of CO_2 to the earth's atmosphere (Raupach and Canadell, 2008). The loss and degradation of natural ecosystems comprise at least 20%–30% of global CO_2 emissions (UNEP, 2012; Pendleton et al., 2012). Although overall emissions from the burning of fossil fuels need to be severely reduced, mitigating climate change can also be achieved by protecting, restoring, or creating natural ecosystems, which has the simultaneous advantage of returning multiple other benefits (Canadell and Raupach, 2008; Trumper, 2009).

The term BC involves different global initiatives, leaded by United Nations, International Union for Conservation of Nature, and other nongovernmental organizations, with the objective of developing conservation, mitigation, and adaptation strategies for coastal ecosystems (Laffoley and Grimsditch, 2009; Nellemann and Corcoran, 2009). BC strategies comprise a range of management activities for preventing or mitigating CO_2 emissions and/or enhancing CO_2 sequestration through their conservation and restoration (Duarte et al., 2013), similar to those already available for terrestrial forests (i.e., REDD+; Angelsen, 2009).

The potential of managing BC ecosystems as a strategy to mitigate climate change throughout the capture of CO_2 and to adapt to climate change throughout coastal protection is increasingly being recognized by policy-makers (Duarte et al., 2013; Macreadie et al., 2017a). In 2013, the Intergovernmental Panel on Climate Change released a Supplement to its guidelines for national GHG accounting (IPCC, 2014) acknowledging the role of coastal wetlands, including mangroves and tidal marshes in sequestering CO_2. In the interim, many nations, including Australia, India, and Indonesia, have made steps toward including BC strategies in their portfolios of initiatives to abate climate change impacts (Alongi et al., 2016; Kelleway et al., 2017a). At a global level, the International Partnership for BC—comprising nation states, research institutions, and nongovernment organizations—has been established to build awareness, share knowledge, and accelerate action to protect and restore coastal BC ecosystems for climate change mitigation and adaptation.

2. ECOLOGY OF BLUE CARBON ECOSYSTEMS

Tidal marshes, mangroves, and seagrass meadows are widely distributed along low-energy shorelines, occupying intertidal and subtidal coastal areas of every continent except Antarctica (Duarte et al., 2013, Fig. 28.2). The extent of mangroves has been estimated at 152,000 km^2, whereas the extent of seagrasses (ranging from 177,000 to 600,000 km^2) and tidal marsh (22,000 to 400,000 km^2) is poorly constrained (Table 28.1 and references therein). For seagrasses, this is partly due to the difficulties associated with mapping underwater vegetation.

Tidal marsh and mangrove ecosystems are coastal, saline ecosystems typically restricted to the upper–intertidal zones inundated from tidal flows (Hutchings and Saenger, 1987; Adam, 1993, Fig. 28.3). Tidal marshes, vegetated by salt-tolerant plants comprising a diversity of grasses, rushes, and herbs often referred to as salt marshes (Adam, 1993), are currently considered BC ecosystems, though freshwater tidal marshes are not. Mangroves are trees and shrubs adapted to live in coastal saline or brackish water within estuaries and marine shorelines (Duke et al., 1998). Seagrasses are also flowering plants (ranging from large forms with straplike leaves through to small forms with oval- to strap-shaped leaves) that grow in marine and estuarine areas and are common in intertidal and shallow waters to depths of about 30 m, where there is sufficient light for them to grow (Hemminga and Duarte, 2000). Macroalgae or seaweeds are a polyphyletic group of multicellular algae: red, green, and brown algae inhabiting the littoral zone to a depth with sufficient light to drive photosynthesis (Hurd et al., 2014).

BC ecosystems are experiencing a steep global decline at rates four times faster than rain forests (Duarte, 2009). Though quantifications can be problematic, it has been estimated that between 25% and 40% of tidal marsh, mangrove, and seagrass ecosystems extent has been lost since the 1940s, with the rate of loss accelerating in many countries (Valiela et al., 2001; Duarte et al., 2008; Duarte, 2009; Waycott et al., 2009; Nellemann and Corcoran, 2009; Atwood et al., 2017, Fig. 28.4). Major threats to BC ecosystems include pressures

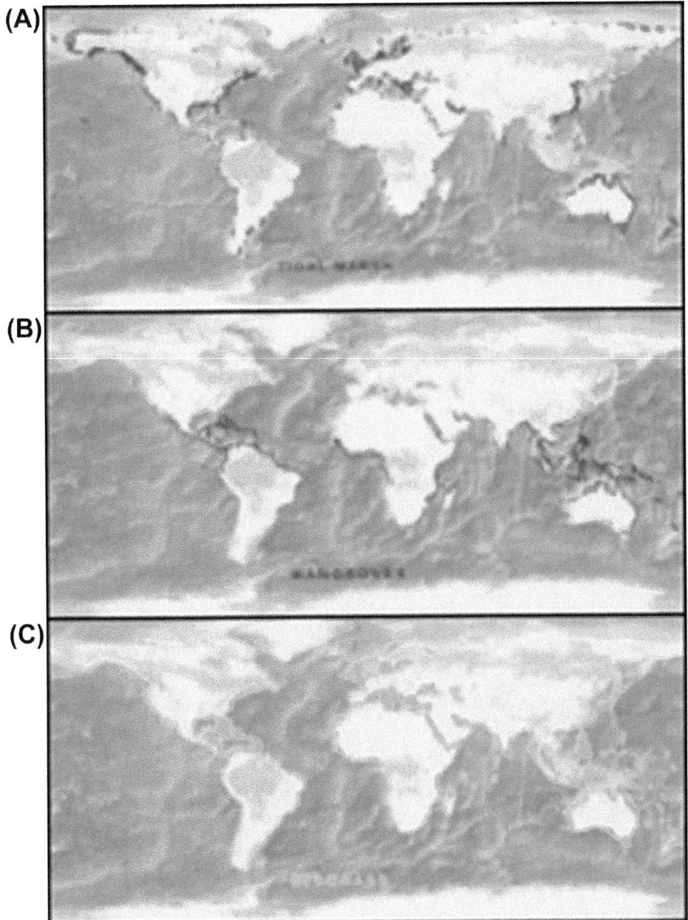

FIGURE 28.2 Global distribution of tidal marsh (A), mangrove (B), and seagrass (C) ecosystems. *Data sources: United Nations Environment Programme World Conservation Monitoring Centre (UNEP-WCMC). References: Hutchison, J., Manica, A., Swetnam, R., Balmford, A., Spalding, M., 2014. Predicting global patterns in mangrove forest biomass. Conservation Letters 7 (3), 233–240; UNEP-WCMC, Short F.T., 2017. Global distribution of seagrasses (version 5.0). In: Fourth Update to the Data Layer Used in Green and Short (2003), Cambridge (UK). UNEP World Conservation Monitoring Centre. http://data.unepwcmc.org/datasets/7; McOwen C., Weatherdon L.V., Bochove J., Sullivan E., Blyth S., Zockler C., Stanwell-Smith D., Kingston N., Martin C.S., Spalding M., Fletcher S., 2017. A global map of saltmarshes. Biodiversity Data Journal 5, e11764. Paper DOI: https://doi.org/10.3897/BDJ.5.e11764. http://data.unepwcmc.org/datasets/43 (v.5).*

associated with coastal development (e.g., conversion of coastal vegetated areas due to land clearance, urbanization, aquaculture, or restriction of tidal flow), deterioration of water quality (e.g., eutrophication), global warming, and sea level rise (SLR) (Nellemann and Corcoran, 2009; Mcleod et al., 2011; Atwood et al., 2017; Arias-Ortiz et al., 2018). Since the 1950s, there has been a global decline in the extent of macroalgal kelp forests of 0.018 $year^{-1}$, attributed to harvesting, pollution, invasive species, and/or temperature (Byrnes et al., 2016). Though the risks facing rain forests and their socioeconomic and ecological benefits are relatively well

TABLE 28.1 Extension and C Stocks and Burial Rates Within the Top 1 m of Soil of Tidal Marsh, Mangrove and Seagrass Ecosystems, and Macroalgae C Buried in the Ocean

Ecosystem	Global Extension (km^2)	Global C Burial Rate ($Tg\ C\ year^{-1}$)	Global C Stock in Soil (Pg C)
Tidal marshes	22,000–400,000[a,b]	4.8–87.3[a]	0.4–6.5[d]
	200,000[c]		
Mangroves	137,760–152,3615[a]	22.5–24.9[e]	5[f]–10.4[d]
	152,308[g]		
Seagrasses	177,000–600,000[a]	48–112[a]	4.2–8.4[j]
Macroalgae	1,400,000[h]–5,700,000[i]	61–268[h]	n/a
	3,540,000[h]	173	

Mean, maximum and minimum (range) estimates. Superscript numbers indicate the sources of data. *n/a*, not applicable. References indicated with a number: [a], Mcleod et al., 2011; [b], McOwen et al., 2017; [c], Cai 2010; [d], Duarte et al., 2013; [e], Breithaupt et al., 2012; [f], Jardine and Siikamäki 2014; [g], Spalding et al., 2010; [h], Krause-Jensen and Duarte 2016; [i], Gattuso et al., 2006; [j], Fourqurean et al., 2012.

FIGURE 28.3 Tidal marshes, mangroves, seagrass meadows, and macroalgal communities. (A) *Sarcocornia* and *Juncus* tidal marsh species along the Peel-Harvey estuary (Western Australia). (B) *Avicennia marina* mangrove forest at Lake Doonella (Australia). (C) Erosional escarpment showing *Posidonia oceanica* C-rich sedimentary deposits at Es Pujols (Spain). (D) *Ecklonia radiata* macroalgal communities at Abrolhos Island. *(D) Australia; Credit: Joan Costa.*

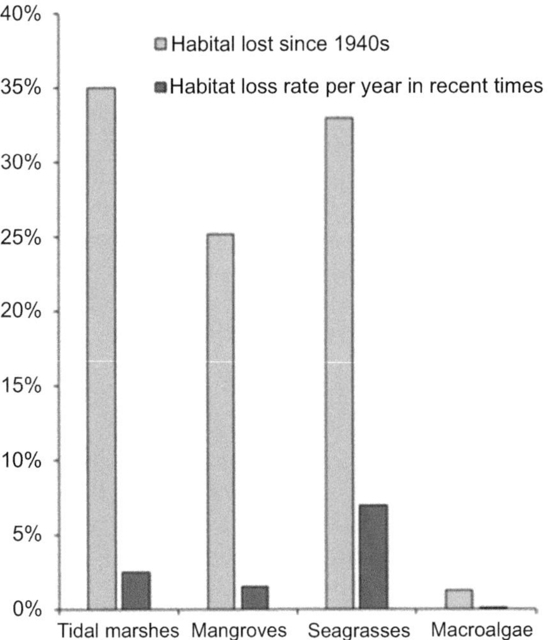

FIGURE 28.4 Loss of the aerial extent of tidal marsh, mangrove, and seagrass ecosystems, and macroalgal communities since the 1940s, and recent loss rates per year. Coastal eutrophication, reclamation, engineering, urbanization, harvesting, and climate change are the main threats to blue carbon ecosystems (Duarte et al., 2008; Duarte, 2009; Nellemann and Corcoran, 2009; UNEP, 2012; Byrnes et al., 2016).

understood, there is a comparatively poor understanding of the status and benefits associated with BC ecosystems (Chmura, 2013; Nellemann and Corcoran, 2009).

3. CARBON CYCLING IN BLUE CARBON ECOSYSTEMS

Tidal marshes, mangroves, seagrass meadows, and macroalgal communities are net autotrophic ecosystems (i.e., fix CO_2 as organic matter photosynthetically in excess of the CO_2 respired back by biota; Duarte and Cebrian, 1996). They constitute hot spots of C cycling, ranking among the most intense C sinks in the biosphere, often with rates of production comparable to the most productive agricultural crops (Nixon, 1980; Gattuso et al., 1998; Duarte et al., 2013). There are several reasons why BC ecosystems are hot spots for C sequestration. First, they are highly productive ecosystems converting CO_2 into plant biomass. Second, they are mainly found in (and/or export C to) depositional environments, which accumulate both autochthonous and allochthonous particulate C (Kennedy et al., 2010; Saintilan et al., 2013). Third, soils within BC ecosystems have high accretion rates resulting in the rapid burial of organic matter in anoxic conditions that slow down microbial decomposition thereby contributing to the formation of organic-rich soils that can exceed 10 m in depth (Donato et al., 2011; Lo Iacono et al., 2008; Serrano et al., 2016a). In addition, their vegetation canopies and/or aerial root networks are complex three-dimensional structures that slow water flow and facilitate the trapping and settling of particles, including C, while protecting the soil C deposits from erosion (Gacia and Duarte, 2001).

Tidal marshes, mangroves, and seagrass meadows store C in two main pools: (1) an aboveground pool in the form of standing biomass (leaves, stems, branches, and trunks), in situ dead biomass such as trees and plant litter, and epiphytes that grow on the surface of these materials and (2) a belowground pool comprising living and dead belowground biomass (roots and rhizomes) and C within the soils. Typically, the majority of the C stocks are found in the soils, with ~90% of total C stocks found in the soils of tidal marshes and seagrasses, and 75% in the soils of mangroves (Nellemann and Corcoran, 2009). A proportion of both the autochthonous and allochthonous C is buried in their soils where it can be preserved thousands of years (Mateo et al., 1997; Filho et al., 2006; Ward et al., 2008), thereby constituting a relevant C sink for climate change mitigation.

In contrast, most macroalgal communities grow on rocky shores and, therefore, do not accumulate sediments. Consequently, within macroalgal habitats, such as kelp forests, the C stock is in the form of standing biomass. However, as with the other BC ecosystems, a large portion of the organic matter and C produced within macroalgal ecosystems is exported as dissolved or particulate organic C to adjacent ecosystems, including sandy shores and the open ocean, where it can also accumulate over time scales relevant for climate change mitigation (Bouillon et al., 2008; Krause-Jensen and Duarte, 2016).

At a global scale, BC ecosystems combined (tidal marsh, mangrove, seagrass, and macroalgae) sequester 130–490 Tg year^{-1} of C (equivalent to 500–1800 Tg of CO_2 per year; Table 28.1 and references therein), whereas current fossil fuel emissions total some 9900 Tg C year^{-1} (36,200 Tg CO_2 per year; CDIAC, 2018). Their annual C sequestration, therefore, is equivalent to 1%–5% of current CO_2 emissions from fossil fuel combustion, making them important and efficient (sequestration rate/area) C sinks. Indeed, over the last millennia and owing to their capacity to raise the seafloor, tidal marsh, mangrove, and seagrass ecosystems accumulated 10–25 Pg C in 1-m thick soils (Table 28.1), equivalent to 2.4%–6.3% of global fossil fuel CO_2 emissions between 1751 and 2014 (estimated at 1474 Pg CO_2; CDIAC, 2018). However, this is likely an underestimate because the long-term preservation and continuous accretion of C in tidal marsh, mangrove, and seagrass soils results in the formation of organic-rich deposits several meters in thickness (Mateo et al., 1997; Donato et al., 2011).

4. FACTORS INFLUENCING CARBON STORAGE IN BLUE CARBON ECOSYSTEMS

BC ecosystems encompass a wide variety of species across a range of depositional environments and water depths, and the variability in the sedimentary C stocks has been found to be up to 18-fold among seagrasses (Lavery et al., 2013) and up to 4-fold in mangroves and tidal marshes (Pendleton et al., 2012). Based on terrestrial analogues and research undertaken on BC ecosystems, it is likely that multiple factors influence C storage, including biotic and abiotic factors acting in the water column, canopy, and the soils, as well as the history of the landscape and past variation in sea level (Fig. 28.5).

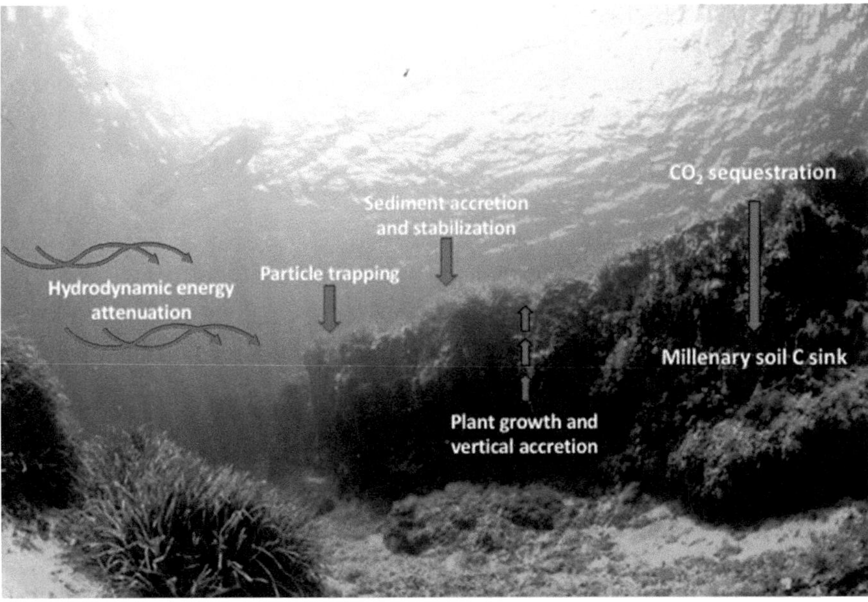

FIGURE 28.5 Processes explaining the capacity of seagrass ecosystems for climate change mitigation (CO_2 sequestration) and adaptation (coastal protection from erosion and sea level rise). The processes occurring in tidal marsh and mangrove ecosystems are similar.

The plants themselves exert a primary control on C storage through production of biomass and nutrient cycling (Lavery et al., 2013; Serrano et al. 2014, 2016a; Miyajima et al., 2015; Kelleway et al., 2016a; Krause-Jensen and Duarte, 2016; Atwood et al., 2017), both of which are highly variable depending on plant species and ecosystem characteristics (Pianka, 1970; Smith, 1981; Lovelock et al., 2014; Saintilan et al., 2013). Plant density, biomass, and productivity are strongly related to the underwater light penetration and soil type in seagrasses (Dennison, 1987; Duarte, 1991), whereas salinity, nutrient supply, and precipitation are important factors for mangroves and tidal marshes (Reef et al., 2010; Wigand et al., 2015; Lovelock et al., 2006). Geomorphological settings (i.e., encompassing variation in landscape and hydrology), soil characteristics (e.g., mineralogy and texture), and biological features (e.g., primary production and remineralization rates) control soil C storage in BC ecosystems (Donato et al., 2011; Adame et al., 2013; Ouyang and Lee, 2014; Kelleway et al., 2016b; Hayes et al., 2017). Both soil C stocks and accumulation rates for BC ecosystems vary with plant community types due to differences in their abilities to trap and retain sediments (Mudd et al., 2009; Lavery et al., 2013; Lovelock et al., 2014; Kelleway et al., 2017b). In addition, tidal transport, fluvial inputs, local hydrodynamics, and seagrass canopy density can all affect how much allochthonous C that is delivered and trapped (Kennedy et al., 2010; Saintilan et al., 2013; Kelleway et al., 2016b).

Once the C is buried in the soil, interactions between biotic and abiotic factors control its accumulation and preservation (Burdige, 2007). The rates of soil accretion, the soil texture, and the biochemical composition of the organic matter buried may strongly influence C accumulation and preservation, parameters that are highly variable among BC ecosystems

(De Falco et al., 2000; Marchand et al., 2003; Kennedy et al., 2010; Saintilan et al., 2013; Ouyang and Lee, 2014; Serrano et al., 2016a). If the accumulated soils are fine-grained, then they are likely to enhance the preservation of C by reducing oxygen exchange thereby reducing remineralization rates (Keil and Hedges, 1993; Serrano et al., 2016a). Finally, while both autochthonous (e.g., plant detritus and epiphytes) and allochthonous (e.g., seston and terrestrial matter) sources contribute to the soil C pool in BC ecosystems (Kennedy et al., 2010; Kelleway et al., 2016b), the proportion of plant-derived C is an important factor controlling their C storage capacity (Serrano et al., 2016a). Plant tissues contain relatively high amounts of degradation-resistant organic compounds (e.g., lignin; Harrison, 1989, Klap et al., 2000, Burdige, 2007, Trevathan-Tackett et al., 2015; Kaal et al., 2016) compared with seston and algal detritus (Laursen et al., 1996), which are more prone to remineralization during early diagenesis (Henrichs, 1992).

The group of macroalgae comprises species with large diversity in forms and size, from small algae with short life spans and high productivity rates (r-selection) to larger algae with slower turnover rates (K-selection; Pianka, 1970). Typically, r-selected species such as the sea lettuce (*Ulva* spp.) hold less structural tissues and are nutrient-enriched compared with K-selected species such as kelp, which are less grazed and more susceptible to contribute to C sequestration (Littler et al., 1983; Steneck and Dethier, 1994; Trevathan-Tackett et al., 2015; Krause-Jensen and Duarte, 2016). A comparison of vascular marine plants (mangrove, tidal marsh, and seagrass species) and macroalgae revealed greater stability in vascular plants related to their lignocellulose matrix (Trevathan-Tackett et al., 2015). In contrast, the presence of specific compounds (e.g., carbonates, long-chain lipids, alginates, xylans, and sulfated polysaccharides) in macroalgae means they may also contribute to long-term sedimentary C storage.

5. ROLE OF BLUE CARBON ECOSYSTEMS FOR CLIMATE CHANGE ADAPTATION

The conservation and creation of BC ecosystems also provide numerous benefits and services that are essential for climate change adaptation along coasts, including protection from storms, prevention of coastal flooding and shoreline erosion, regulation of water quality, provision of habitat for commercially important fisheries and endangered marine species, and food security for coastal communities (Beck et al., 2001; Nellemann and Corcoran, 2009; Barbier et al., 2011; Jones et al., 2012; Duarte et al., 2013; Smale et al., 2013). BC ecosystems function as "ecosystem engineers" by altering light levels, water flow, and near-bed velocity, sedimentation rates, modifying the local environment, and providing shelter from predation for other organisms of significant ecological or socioeconomic importance (Duarte et al., 2013). Through direct provision of food and structural habitat, BC ecosystems support higher levels of biodiversity and biomass than simple, unstructured habitats (Beck et al., 2001; Steneck et al., 2002; Barbier et al., 2011).

The increase in atmospheric CO_2 concentrations not only results in global warming but also entails SLR, and a higher intensity and frequency of severe storms, and extreme sea level events increasing the risks of coastal flooding and erosion, posing the well-being of vulnerable coastal communities under threat (Menéndez and Woodworth, 2010; Hoegh-Guldberg and Bruno, 2010; Young et al., 2011; Church and White, 2011). The melting of ice has doubled SLR globally between 1900s until present (0.17–0.32 cm $year^{-1}$), whereas SLR projections point to peaks of 0.45 cm $year^{-1}$ in 2050 and 1.1 cm $year^{-1}$ by 2100 (Church et al., 2013). The global population under flooding threat increases comparably, from 190 million people in 2000 to 400 million by 2060 (Neumann et al., 2015). The socioeconomic impacts of SLR have been valued in millions of dollars, in particular in tropical and subtropical areas which are exposed to major risks (Hallegatte et al., 2013; Neumann et al., 2015). Tidal marshes, mangroves, seagrass meadows, and macroalgal communities constitute a natural protection against SLR, thereby their conservation and restoration support adaptation to climate change (Duarte et al., 2013). BC ecosystems dissipate hydrodynamic energy and favor particle deposition, and the rate of soil accretion in BC ecosystems is comparable to that of SLR (global estimates ranging from 0.2 to 0.7 cm $year^{-1}$ globally; Duarte et al., 2013 and reference therein), which suggest that tidal marsh, mangrove, and seagrass ecosystems have some capacity to tolerate SLR (Krauss et al., 2014), though there remain uncertainties as to their likely response to future accelerations in SLR.

Over the last 200 years, significant amounts of anthropogenic CO_2 have been released into the atmosphere, reaching a concentration of >400 ppm in the atmosphere, the greater mark since the scientists began to measure it in Mauna Loa Observatory, Hawaii (Showstack, 2013). CO_2 has been continuously absorbed by oceans within a rate of 30% (Millero, 1995), causing a reduction of seawater pH. Carbon dioxide dissolves in seawater and forms carbonic acid (H_2CO_3), which is in turn dissociated in bicarbonate (HCO_3^-) and can also dissociate in carbonate (CO_3^{-2}). The last two reactions also form hydrogen ions (H^+), that lower the seawater pH and decrease carbonate concentration. This process, known as ocean acidification (Caldeira and Wickett, 2003), affects directly the capacity of calcifying organisms to build their own calcium carbonate ($CaCO_3$) shells (Orr et al., 2005; Doney et al., 2009). BC ecosystems, and seagrass meadows and macroalgal communities in particular, modify carbonate chemistry and raise pH in the surrounding water by intense photosynthesis, providing refuge for calcifying organisms from ocean acidification (Middelboe and Hansen, 2007; Fabry et al., 2008; Hendriks et al., 2014). Hence, the conservation and restoration of BC ecosystems contributes to climate change adaptation at local scales, providing direct benefits (i.e., protection against coastal, flooding, erosion, ocean acidification, and food security) for communities inhabiting coastal areas (Duarte et al., 2013).

6. DISTURBANCE OF BLUE CARBON ECOSYSTEMS AND ASSOCIATED EMISSIONS

Terrestrial ecosystems can shift from net sinks of CO_2 to sources of CO_2 as a result of changes in land use and climate effects (e.g., deforestation, afforestation, agricultural practices, changes in precipitation and temperature, fires, and damage by pollution; Solomon et al., 2007). BC ecosystems are no different, and all are currently experiencing global declines in area (Gedan et al., 2009; Waycott et al., 2009; Byrnes et al., 2016; Atwood et al., 2017). The major drivers of these declines include land use conversion (clearing for development, agri(aqua)culture, salt production), altered hydrology and tidal flow, eutrophication, dredging, and other light-reducing activities (Pendleton et al., 2012, Fig. 28.6).

FIGURE 28.6 Examples of healthy (A–D) and damaged (E–H) blue carbon ecosystems that entail CO_2 emissions to the atmosphere. Images are of tidal marshes (A and E), mangroves (B and F), seagrass meadows (C and G), and macroalgal communities (D and H). Tidal marsh ecosystems at Madagascar (A and E) were impacted by land conversion and impounding, restricting landward migration. Mangrove dieback at Shark Bay (Australia; B and F) during fluctuating sea levels (Lovelock et al., 2017b). Eutrophication of coastal water resulted in the dieback of *Posidonia australis* meadows (C and G) at Cockburn Sound (Australia). Kelp communities cleared by sea urchins at Port Phillip Bay (D and H; Australia; Carnell and Keough, 2016). *Credits: (B) Sharyn Hickey; (D and H) Paul Carnell.*

Long-term sea temperature increases, shorter-term heat wave events, and SLR associated with climate change poses additional threats (Warren and Niering, 1993; Lovelock et al., 2017b; Arias-Ortiz et al., 2018).

These pressures have the potential to affect the C storage of BC ecosystems in two ways: through direct impacts on the primary producer and through direct or indirect disturbance of the soils (Fig. 28.7). A particularly unique threat to BC ecosystems is that of coastal squeeze, the phenomenon by which SLR causes the coastal ecosystem to retreat landward, to stay within a suitable water depth, but where this is prevented by the presence of coastal development at the landward margin (Doody, 2004; Torio and Chmura, 2013).

Threats that result in a loss or reduction of primary producer biomass and productivity include light reduction (e.g., through dredging, eutrophication, or increased turbid river flows), altered hydrology and salinity regimes, increased grazing, fire, and heat waves (Lovelock et al., 2017a). These pressures have the initial effect of reducing the production of autochthonous C that can be potentially sequestered and leading to reduced canopy density which may affect the trapping of allochthonous C. Even if the ecosystem suffers no further damage, these changes will likely lead to much-reduced C sequestration. Furthermore, there is likely to be an enhanced GHG emission through the decomposition of the dead biomass in the system.

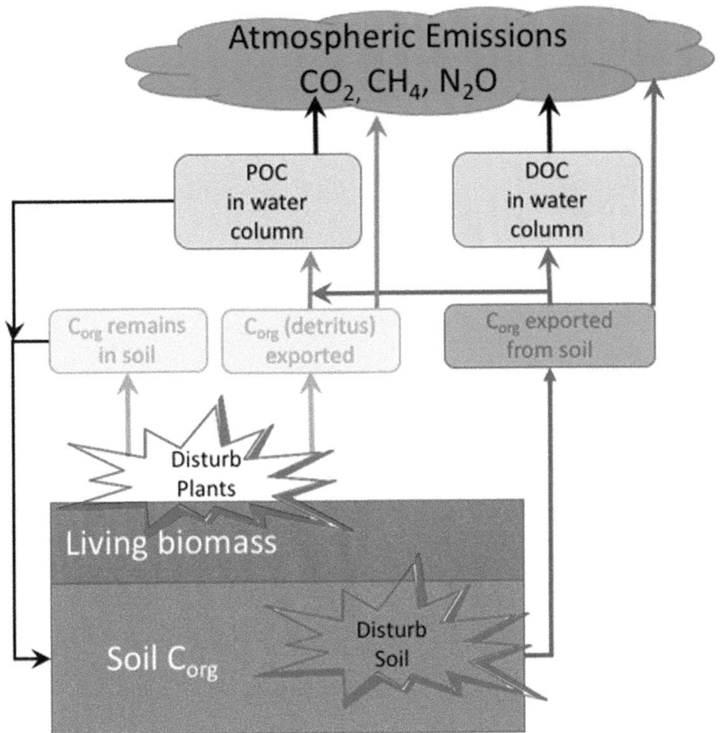

FIGURE 28.7 Conceptual model for the disturbance mechanisms and potential remineralization of organic carbon (C_{org}) following disturbance of blue carbon ecosystems, which could result in carbon dioxide (CO_2), methane (CH_4), and nitrous oxide (N_2O) emissions. *DOC*, dissolved organic carbon; *POC*, particulate organic carbon.

Other threats act on the soil of the BC ecosystems. These physical disturbances include clearing or reclamation (e.g., for housing, timber, or aquaculture development), dredging for mining or infrastructure development, boat moorings, anchor damage, impounding, and erosion driven by altered hydrodynamics. Threats that do not initially lead to physical disturbance of the soils may ultimately do so once the loss of biomass results in an enhanced susceptibility to soil erosion through hydrodynamic forces. These disturbances can lead to previously buried C stocks being resuspended and exposed to oxygenated conditions, enhancing remineralization and CO_2 emissions (Burdige, 2007; Pendleton et al., 2012). Therefore, in these situations, the threats may result not only in reduced sequestration but also enhanced GHG emissions (Hicks et al., 2003; Serrano et al., 2016b; Boxes 28.1 and 28.2).

BOX 28.1

GREENHOUSE GAS EMISSIONS FROM IMPOUNDING COASTAL WETLANDS

In 1976, the mangrove forests and tidal marshes surrounding the East Trinity Inlet (Cairns, Australia) were drained for sugar cane production (Fig. 28.8). The soils within 110 ha of coastal wetland habitat were drained and exposed to erosion (Hicks et al., 2003). This resulted in the loss of 1.3 m of soil elevation and the subsequent loss of 74,800 Mg of soil C per hectare over 23 years after the disturbance occurred. The CO_2 emissions associated with soil organic carbon remineralization were estimated at 0.27 Tg of CO_2, significantly

FIGURE 28.8 Impounded coastal wetlands at Trinity Inlet (Australia). Degradation of tidal marsh and mangrove forest can be observed at both sides of the levee used to drain the wetland (Hicks et al., 2003).

BOX 28.1 *(cont'd)*

contributing to Australia's CO_2 emissions despite the small area of the site. In addition to CO_2 emissions, the exposure of soil C to oxygen resulted in the production of sulfuric acid and highly acidic water (pH of 3.2) that reached the adjacent estuary causing unquantified damage to adjacent ecosystems (Hicks et al., 2003). In 2000, the tidal flow was reintroduced to restore this site, yet the avoided CO_2 emissions and enhanced soil C burial after restoration remains unassessed.

BOX 28.2

CO_2 EMISSIONS FROM MOORING ACTIVITIES IN SEAGRASS MEADOWS

Rottnest Island is a popular holiday destination in Perth (Australia). The first moorings to support boating activities were installed in 1930s, expanding to nowadays reach 893 moorings distributed within Rottnest embayment. As moored boats drift with the currents, they drag a heavy chain across the seafloor removing the seagrass.

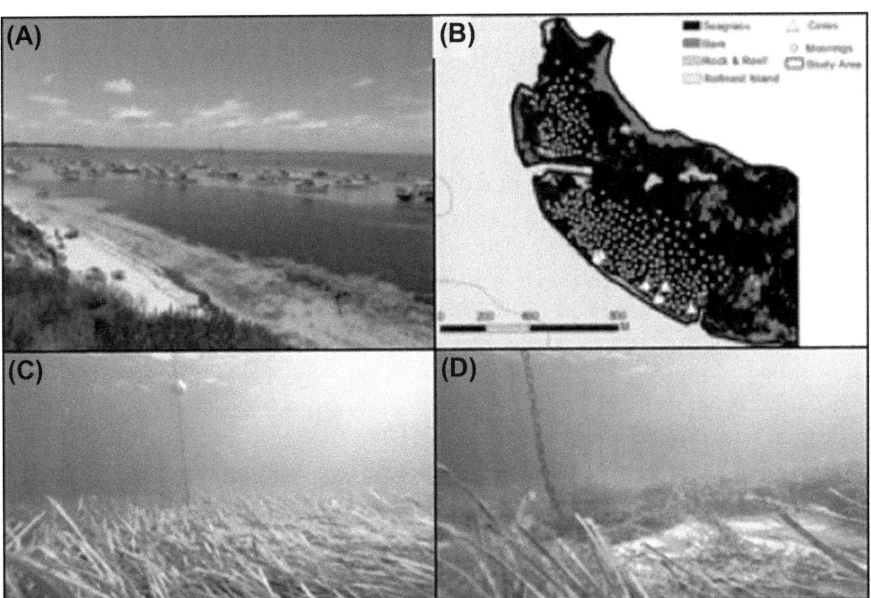

FIGURE 28.9 Rottnest Island is located 18 km off the coast of Perth (Australia). (A) Scouring of seagrass meadows by boating activities can be observed around the moorings; (B) Boating activities triggered the installation of 316 moorings at Thomson Bay; (C and D) Effects of moorings on seagrass meadows exposing soils to erosion (Serrano et al., 2016b).

Continued

BOX 28.2 *(cont'd)*

The soil C stocks underneath seagrass meadows have been compromised by the mooring deployment, which involved both the erosion of existing soil C stocks and the lack of further accumulation of C (Serrano et al., 2016b). The CO_2 emissions associated with the soil C remineralization within the top 50-cm soil accumulated over the last 200 years were estimated at 845 metric tons CO_2. Over the last 25 years, the Rottnest Island Authority has replaced the swing-chain moorings with either pin moorings or secured chain moorings, which reduces the disturbance of seagrasses. However, the presence of barren patches within the meadows at mooring sites can still be seen (Fig. 28.9), indicating that some physical impacts of boating disturbances continues despite conservation measures were undertaken.

Changes in C storage from BC ecosystems has been reported due to land reclamation, chemical and physical disturbances, eutrophication, poleward expansion of mangroves, loss of top–down control by the removal of associated fauna, and marine heat waves (Bu et al., 2015; Deegan et al., 2012; Atwood et al., 2015; Marbà et al., 2015; Kelleway et al., 2016a; Lovelock et al., 2017b; Arias-Ortiz et al., 2018). Given that the soils are typically the greatest sink of C within BC ecosystems and, conversely, present the greatest risk of substantial CO_2 emissions after ecosystem disturbance, the belowground C stocks are the primary interest in BC initiatives (Sutton-Grier et al., 2014) and often dominate potential GHG emissions assessments (Lovelock et al., 2017a).

Estimating the consequences of ecosystem disturbance for GHG emissions can be difficult due to uncertainty around some key factors. These uncertainties can include the size of the C stock, the baseline conditions that postdisturbance conditions need to be compared to, and the fate of the disturbed C sink, the latter of these often the most difficult to resolve (Lovelock et al., 2017a; Duarte, 2017). Although rigorous but fairly routine assessments can determine how much C has been lost from a site following disturbance (e.g., Marbà et al., 2015; Hicks et al., 2003; Coverdale et al., 2014), it is much less clear how much of that stock will be remineralized, and whether the emissions comprise either CO_2, CH_4, and/or NO_x, and whether fluxes will be via the water column or directly across the soil–atmosphere interface (Fig. 28.7). BC ecosystems differ biogeochemically from freshwater wetlands, which may act as substantial natural sources of methane (CH_4) to the atmosphere (Whalen, 2005). This is due to the supply of sulfates from marine waters to BC soils, which acts to suppress CH_4 production by providing a lower energy decomposition pathway (i.e., sulfate reduction; Lovley et al., 1982). In general, it is expected that there is little CH_4 emission from marine ecosystems because the high levels of sulfate generally restrict methanogenesis occurring. However, mangrove and tidal marsh ecosystems which lose the supply of sulfate (e.g., through disconnection of tidal flow) and are converted into freshwater habitat may become major sources of methane emission (Whalen, 2005; Gatland et al., 2014).

Boxes 28.1 and 28.2 show examples of assessments of C stock loss and potential emissions following disturbances, for a mixed mangrove/tidal marsh ecosystem and a seagrass ecosystem, respectively. In both cases, there was a relatively high level of certainty about the stocks that were lost from the site but, in both cases, significant assumptions needed to be made regarding the emission that this represented. Generally, emissions are estimated by assuming a proportion of the lost stock is remineralized under oxic conditions (Lovelock et al., 2017c). The IPCC (IPCC, 2003, 2006 and 2014) provides default values for the size of C stocks in some coastal wetland ecosystems as well as emission factors, using a tiered approach. Progressing from Tier 1 to Tier 3 estimates (e.g., from global values to site-specific estimates) typically reduces the level of uncertainty around the potential GHG emission, though this usually comes at the cost of increased investment in measurements, modeling, and other studies that underpin the improved estimates. Unfortunately, for coastal wetlands, the range of activities (e.g., land use changes) that emission factors are provided for is currently limited and the uncertainties around them quite substantial, reflecting the limited empirical data on GHG emissions from coastal wetlands (Table 28.2).

For activities not covered by the IPCC Wetlands Supplement (IPCC, 2014), other approaches are required to estimate emissions. Lovelock et al. (2017a) proposed a risk assessment approach that combines the size of soil C stocks with assessments of the likelihood of remineralization of soil C. Lovelock et al. (2017c) used a simple model to estimate CO_2 emissions from tidal marsh, mangrove, and seagrass ecosystems based on decomposition rates for organic matter in these ecosystems under either oxic or anoxic conditions combined with assumptions of the proportion of C being exposed to either oxic or anoxic environments following disturbance. Recently, Kelleway et al. (2017a) assessed the potential emission reductions or enhanced C sequestration associated with a range of activities designed to increase C capture and storage by management of BC ecosystems (see Box 28.3). They estimated the size of C stocks from previously reported estimates for BC ecosystems and estimated the potential change in emissions by using the IPCC default values or by using

TABLE 28.2 A Partial Summary of the Possible Avoided Greenhouse Gas (GHG) Emissions and Enhanced GHG Sequestration Associated With Management Activities in Different Blue Carbon Ecosystems Reported by Kelleway et al. (2017a)

Ecosystem	**Activities**	**Avoided Emissions ($Mg\ CO_2\ ha^{-1}\ year^{-1}$)**	**Enhanced Sequestration ($Mg\ CO_2\ ha^{-1}\ year^{-1}$)**
Tidal marsh	Tidal introduction to land without mangroves; enhanced sediment supply; land use planning for future sea level rise (SLR).	0.11–8	0.3–9.2
Mangrove	Tidal introduction to land without mangroves; enhanced sediment supply; land use planning for future SLR.	3.7–161	1.8–73
Seagrass	Avoiding loss; reestablishment of habitat; creation of new habitat; treatment of wrack.	1–1678[a]	0.07–5.2

[a]*One-off emissions estimated assuming that 25%–75% soil C stocks in 1 m soil deposits are Remineralized after disturbance and emitted as* CO_2.

BOX 28.3

BLUE CARBON SCIENCE AND POLICY DEVELOPMENT IN AUSTRALIA

Australia is home to vast areas of mangrove, tidal marsh, and seagrass and has a long history of scientific investigation into the function and management of these ecosystems. Consequently, Australian research has been at the forefront as the concept of BC has developed globally over the past decade. In 2013, Australia's Commonwealth Scientific and Industrial Research Organisation (CSIRO) commenced a 3-year research partnership with multiple local research institutes through the Coastal Carbon Biogeochemistry Cluster. The purpose of this partnership was to build the knowledge base of C cycling in Australian coastal waters and BC ecosystems (mangrove, tidal marsh, and seagrass). Key outcomes included the production of national estimates of BC stocks, accumulation rates, and patterns of BC losses (e.g., Atwood et al., 2017; Hayes et al., 2017; Macreadie et al., 2017b; Arias-Ortiz et al., 2018). At the same time, there has been improved understanding and quantification of key coastal biogeochemical processes relevant to burial and export of BC (e.g., Lavery et al., 2013; Sippo et al., 2016; Krause-Jensen and Duarte, 2016; Kelleway et al., 2017b; Maher et al., 2017).

These new data and insights have underpinned recent considerations of policy mechanisms which might incorporate BC into coastal management and environmental accounts in Australia. In 2017, Australia was also one of the first countries to voluntarily report human-induced GHG emissions and removals associated with coastal wetlands as part of its National Greenhouse Accounts. The Department of the Environment and Energy compiles these accounts annually and reports to the United Nations' Framework Convention on Climate Change (UNFCCC). The inclusion of wetlands in the national inventory enables quantification of how mitigation initiatives (e.g., avoiding loss or degradation of wetlands and/or the restoration or creation of wetland habitat) may contribute to Australia meeting its international GHG commitments.

Although a growing number of countries aspire to report GHG emission and C sequestration changes from these ecosystems under voluntary international reporting requirements, few countries have domestic policy frameworks that specifically support the quantification and financing of C abatement through coastal wetland management. In 2016, the Australian Government also commissioned a technical review of opportunities for including BC activities within the nation's main domestic C accounting mechanism, the Emissions Reduction Fund (ERF) (Kelleway et al., 2017a). This review used a participatory workshop of scientific experts and C industry stakeholders to identify BC management actions that would meet the offsets integrity standards of the ERF (Table 28.3).

The review identified a number of activities which might be suitable for inclusion under the ERF. These include actions which (1) avoid the loss or disturbance of existing mangroves, tidal marshes, and seagrasses (e.g., through clearing and dredging); (2) restore biogeochemistry of BC ecosystems

BOX 28.3 (cont'd)

(e.g., restore tidal flow to coastal wetlands; improve water quality for seagrasses); and/or (3) plan for the creation of new habitat, including in the context of SLR (Kelleway et al., 2017a). Although significant information gaps remain in regard to both BC science and policy-making, the Australian experience to date provides a broad template which other nations may follow as they identify options for C abatement through management of coastal landscapes.

TABLE 28.3 Factors to Consider in the Identification of Potential C Abatement Activities

Integrity Requirement	Description of Requirement
Additional abatement	Undertaking the BC enhancement activity must result in C abatement that is unlikely to occur in the ordinary course of events.
Measurement and verification	Estimating the activity's C removals, reductions or emissions must be achieved using an approach that is measurable and capable of being verified.
Eligible abatement	Carbon abatement using in ascertaining the CO_2 net abatement amount for the activity must be eligible C abatement in accordance with Australia's nationally reported GHG emissions.
Evidence base	The approaches used for the activity must be supported by clear and convincing evidence.
Material amounts considered	Material amounts of greenhouse gases that are emitted as a direct consequence of the activity must be considered.
Conservative estimates	Estimates, projections, or assumptions regarding activity abatement are conservative

These abatement integrity requirements formed part of the assessment of potential Blue Carbon (BC) activities under Australia's Emissions Reductions Fund (Kelleway et al., 2017a).

assumed rates of remineralization based on expert knowledge. This applied example of estimating potential changes in GHG emissions reveals the complexity of the exercise and the large amount of assumptions and mixed methods that may be required to produce estimates.

7. CONSERVATION AND RESTORATION OF BLUE CARBON ECOSYSTEMS AS A STRATEGY FOR CLIMATE CHANGE MITIGATION AND ADAPTATION

Removal of atmospheric CO_2 through sequestration in natural ecosystems is necessary to keep global warming under 2°C by 2050, consistent with the Paris agreement, as reduction of emissions alone may not suffice to achieve this target. Whereas technological approaches to remove atmospheric CO_2 are being sought, biosequestration of CO_2 is already a proven process that can be conserved, strengthened, and incorporated into emission accounting frameworks (IPCC, 2014) and has additional benefits for society.

Actions to prevent these emissions and mitigate climate change have concentrated on the preservation and restoration of terrestrial ecosystems, mainly tropical forests (Agrawal et al., 2011). Determination to reduce CO_2 emissions caused by forest clearance and land degradation led to the elaboration of global climate change mitigation solutions such as the Reducing Emissions from Deforestation and Forest Degradation program (REDD+; IPCC, 2003). The basis of this strategy is to economically compensate nations to manage terrestrial vegetation to enhance C sequestration and avoid CO_2 emissions. With an increased focus on C-trading schemes worldwide, it is important to better understand the C cycle and clearly identify both C sinks and sources of emissions (Ullman et al., 2013).

Coastal regions have constituted strategic points of human settlement through history, causing persistent and severe impacts on BC ecosystems by a wide variety of human activities (Lotze et al., 2006). The loss of BC ecosystems entails the loss of all the ecological services they provide, including C sequestration (Barbier et al., 2011; Kirwan and Megonigal, 2013) and the emission of GHG (Pendleton et al., 2012). Conservation and restoration of BC ecosystems will facilitate the maintenance of the benefits they provide, including fisheries, coastal protection, and related ecosystem services that support coastal communities and their livelihoods (Barbier et al., 2011; Duarte et al., 2013), while contributing to achieve the goal of keeping global warming under 2°C by 2050 (UNFCCC, 2016). As such, the need to identify, accurately value and quantify, and incorporate systems which naturally sequester atmospheric CO_2 into future C inventories and emission frameworks is becoming increasingly paramount. With growth in C markets around nature-based climate mitigation (via biosequestration), there are opportunities to finance the restoration of BC ecosystems using C offset schemes (Thomas, 2014).

As a result, anthropogenic activities that enhance C sequestration and/or avoid emissions of GHG (CO_2, CH_4, and NO_x) form the basis for BC strategies to reduce emissions and mitigate climate change. An important point here, however, is that often it is only activities or interventions that fall outside business-as-usual scenarios (i.e., entailing "additionality") that may be eligible for crediting. These activities may include creating, restoring, and/or managing hydrological conditions, sediments supply, salinity characteristics, water quality, and/or native plant communities (VCS, 2015; Kelleway et al., 2017a). Quantifying the C potential of habitat restoration has been a focal point of BC research. Areas of demonstration include the generation of BC credits through the sustainable management of mangrove ecosystems in Madagascar (Benson et al., 2017; Box 28.4), restoration of aquaculture ponds to mangrove habitat (Matsui et al., 2012), reforestation of logged and degraded mangrove forests in Kenya (Huxham et al., 2015; Box 28.5), the restoration of hydrology to degraded coastal floodplains (Howe et al., 2009), and the revegetation of lost seagrass ecosystems (Marbà et al., 2015; Box 28.6). Managed realignment is an option for recreating or restoring tidal marshes aimed to a more sustainable coastal flood defense together with the provision of other services, including C benefits (Luisetti et al., 2011). The creation of macroalgal belts in southern Korea promotes CO_2 removal, and the management of macroalgal ecosystems could also be considered within BC projects (Chung et al., 2013; Box 28.7).

To quantify net GHG avoided emissions and enhanced sequestration from management activities implemented, there is a need to use approved methodologies for this purpose.

BOX 28.4

BLUE CARBON IN ACTION IN MADAGASCAR

The African island nation of Madagascar contains 2% of the global mangrove distribution, but more than one-fifth of its extent has been deforested since 1990 (Jones et al., 2014). The main drivers of this deforestation have been increased extraction for the production of charcoal, timber, and conversion to agriculture and aquaculture. Under their Blue Forests initiative, the nongovernment organization Blue Ventures is supporting the generation of BC credits through the sustainable management of mangrove ecosystems, to help alleviate poverty and support biodiversity conservation in Madagascar's coastal areas.

BC projects currently under development across Madagascar incorporate multiple C accounting standards, including the Verified Carbon Standard, Climate Community and Biodiversity Alliance, and Plan Vivo. The accurate accounting of BC outcomes and alignment with standards requires significant research and monitoring activity. To undertake this, Blue Ventures works in partnership with local community guides, Malagasy scientists, and international researchers.

Nationwide mapping of historical changes of mangrove forest cover, incorporating remote sensing and field data (Jones et al., 2016), has been central to identifying targeted project areas and defining baseline conditions. Other research initiatives have included the measurement of C stocks (biomass and soil C pools) in healthy, degraded, and restored mangroves, analysis of the drivers of mangrove loss (Scales et al., 2017) and the modeling of future mangrove forest and wetland changes (Benson et al., 2017) (Fig. 28.10).

FIGURE 28.10 (A) Deforestation of mangrove forests in Madagascar for the production of timber and charcoal. (B) Revegetation of mangroves in previously vegetated habitats to enhance carbon sequestration and avoid emissions from soil carbon deposits.

BOX 28.5

NEW POLICY CONCEPT OF CLIMATE COMPATIBLE DEVELOPMENT: A CASE STUDY OF KENYA'S MANGROVE FORESTS

Mangrove forests in Kenya are being degraded for timber, fuelwood, and coastal development. However, the benefits associated with mangrove conservation and restoration (e.g., fisheries, tourism, C sequestration, and protection against coastal erosion and extreme weather events) remain undervalued and are being dismissed (Tamooh et al., 2008). Huxham et al. (2015) developed a new policy concept of Climate Compatible Development (CCD) to predict the economic consequences of plausible alternative future scenarios for Kenya's mangrove forests. The projection of recent mangrove loss trends over the next 20 years suggests a 43% loss of total forest extent over that time, whereas the restoration of the mangrove extent loss between 1992 and 2012 is still plausible. The combination of these scenarios allowed modeling economic outcomes, suggesting that the conservation of Kenya's mangroves entails a net present value of US$20 million. Huxham et al. (2015) provides economic evidence for mangrove conservation and demonstrates a policy tool (CCD) that can be used to convince stakeholders and policy-makers to implement actions for mangrove conservation and restoration (Fig. 28.11).

FIGURE 28.11 Studying the carbon storage potential of mangrove forests in Kenya. *Credit: Gloria Salgado.*

BOX 28.6

IMPACT OF SEAGRASS LOSS AND SUCCEEDING REVEGETATION ON CARBON SEQUESTRATION

The loss of seagrass meadows is a common problem around the world, but thanks to a citizen project, seagrass meadows in Oyster Harbour (Australia) are recovering. Around 90% of the seagrass disappeared in the early 1980's in Oyster Harbour due to extensive land use changes in the catchment area and subsequent eutrophication of the Harbour; however, Geoff Bastyan began transplanting seagrass in 1994. Transplanting seagrass was proved successful, and a recent study by Marbà et al. (2015) demonstrated that soil C stocks erode following seagrass loss but revegetation effectively restored seagrass C sequestration capacity, resulting in avoided CO_2 emissions and enhanced C sequestration. The C burial increased with the age of the restored sites, reaching similar C sequestration capacity than continuously vegetated meadows after 18 years of planting, demonstrating the conservation and restoration of seagrass meadows are effective strategies for climate change mitigation (Fig. 28.12).

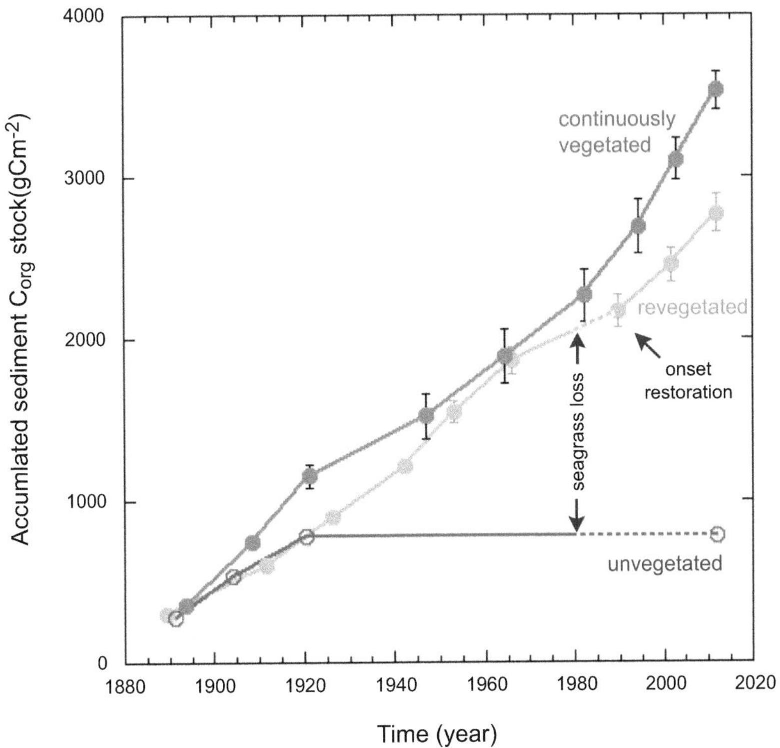

FIGURE 28.12 Reconstructed trajectories of carbon accumulation in resilient seagrass meadows (*black line*), restores meadows (*dark gray line*), and bare sediments (*light gray line*) at Oyster Harbour (Australia) since year 1890 (Marbà et al., 2015).

BOX 28.7

INSTALLATION OF MACROALGAE ECOSYSTEMS FOR MITIGATION AND ADAPTATION AGAINST CLIMATE CHANGE: KOREAN PROJECT OVERVIEW

Macroalgae beds can sequester CO_2 in their biomass while providing food, fuel, pharmaceutical products, and other goods and services for climate change mitigation and adaptation. The Korean Project has established the Coastal CO_2 Removal Belt (CCRB), which encompasses both natural and farmed seaweed communities in southern Korea (Chung et al., 2013). This scheme promotes the CO_2 removal via macroalgae production and estimated that, e.g., *Ecklonia* kelp forests can draw down ~10 tones of CO_2 per ha per year, resulting in a decrease in dissolved organic carbon concentrations in the water column. The conservation and creation of macroalgae ecosystems has also the potential to, e.g., enhance coastal protection from storms and shoreline erosion and food security for coastal communities (Smale et al., 2013). Despite recent evidence that highlights the major direct role of macroalgal C in C sequestration through carbon export and burial beyond the macroalgal habitat (e.g., Krause-Jensen and Duarte, 2016), there is substantial discussion as to whether macroalgae meet the principles to be contemplated within the BC model (Howard et al., 2017; Smale et al., 2018). A research priority is to clearly establish the carbon benefits of macroalgal culture and set a foundation for developing the necessary crediting and regulatory frameworks by which those benefits can be recognized (Chung et al., 2013).

Currently, methodologies available for BC accounting include the (1) Clear Air Development (CDM) methodology for mangroves (AR-AM0014: Afforestation and reforestation of degraded mangrove habitats), which can be used in developing nations; (2) a mangrove methodology for the Mississippi Delta that is available in the American Carbon Registry (ACR) (https://americancarbonregistry.org/carbon-accounting/standards-methodologies/restoration-of-degraded-deltaic-wetlands-of-the-mississippi-delta) (Mack et al., 2012), which is linked to a methodology for accounting for C sequestration with the restoration of tidal marshes in California (https://americancarbonregistry.org/carbon-accounting/standards-methodologies/restoration-of-california-deltaic-and-coastal-wetlands); and (3) the Verified Carbon Standard (VCS), which has an approved methodology for restoration and conservation of tidal marshes, mangroves, and seagrasses (VM0033; http://database.verra.org/methodologies/methodology-tidal-wetland-and-seagrass-restoration-v10), as well as a methodology in development for avoided deforestation for mangroves and other flooded forested wetlands (VM0007, REDD + Methodology Framework (REDD + MF), v1.6, http://database.verra.org/methodologies/redd-methodology-framework-reddmf-v16).

All existing methodologies share common features. First, there is a need to establish the project boundaries, defined by the extent of area potentially affected by the management activity. When the project area is not homogeneous in terms of C storage or habitat characteristics, stratification of the project area is required to enhance the accuracy of avoided GHG emissions or enhanced C sequestration estimates. Prior to the activity, the most plausible baseline scenario in the absence of the project activity must be identified (enhanced C sequestration and/or avoided emissions). An important point here, however, is that often it is only activities or interventions that fall outside business-as-usual scenarios (i.e., entailing "additionality") that may be eligible for crediting. Therefore, the natural C sink capacity of an ecosystem will not mitigate anthropogenic CO_2 emissions except when human-induced changes increase their C sequestration rate over time, such as restoration resulting in increased areal extent or primary productivity. After the activity, the project proponent must reassess the change in C storage every 5–10 years and to quantify enhanced C sequestration and/or avoided emissions.

8. FUTURE OF BLUE CARBON

Despite advances in BC science and governance over the past decade, there remains technical, financial, and policy barriers which limit the upscaling of local initiatives to have impacts that are globally relevant to climate change mitigation. Significant barriers include biases in the geographic coverage of research and data, poor prediction capacity for some C pools (e.g., atmospheric emissions remain poorly quantified outside of near-pristine ecosystems), high transaction costs, and ensuring that equity and justice are achieved. In addition, most demonstrated BC projects are relatively recent actions with little quantification of C mitigation benefits (or societal outcomes) beyond the scale of a few years, thereby the feasibility of BC project (e.g., cost/benefit and success of restoration) remains poorly understood (Bayraktarov et al., 2016).

Importantly, we now have the fundamental knowledge to justify the inclusion of BC ecosystem protection, restoration, and creation in coastal management directions and climate mitigation and adaptation mechanisms. The effective demonstration, monitoring, and reporting of existing and new BC projects will also address many of the outstanding issues. Adoption of standardized methods will also facilitate linking BC to National Determined Contributions (NDCs) and the transfer of information from early adopters of BC projects to those who follow. Currently, successful BC projects involve local stakeholders and consider alternative livelihoods for coastal communities (Wylie et al., 2016). BC projects have been successfully implemented within small scale in Kenya, Madagascar, India, and Vietnam, largely due to simpler requirements for project development, and funding has come from either the voluntary carbon market or alternative financing mechanisms. As more countries seek to respond and adapt to climate change impacts by reducing their carbon footprint, and with BC becoming an emerging issue in the United Nations Framework Convention on Climate Change (UNFCCC), it is likely that more BC projects will be implemented around the world in the near future.

References

Adam, P., 1993. Saltmarsh Ecology. Cambridge University Press.

Adame, M.F., Kauffman, J.B., Medina, I., Gamboa, J.N., Torres, O., Caamal, J.P., et al., 2013. Carbon stocks of tropical coastal wetlands within the karstic landscape of the Mexican Caribbean. PLoS One 8 (2), e56569.

Agrawal, A., Nepstad, D., Chhatre, A., 2011. Reducing emissions from deforestation and forest degradation. Annual Review of Environment and Resources 36, 373–396.

Arias-Ortiz, A., Serrano, O., Masqué, P., Lavery, P.S., Mueller, U., Kendrick, G.A., 2018. A marine heatwave drives massive losses from the world's largest seagrass carbon stocks. Nature Climate Change 1.

Alongi, D.M., Murdiyarso, D., Fourqurean, J.W., Kauffman, J.B., Hutahaean, A., Crooks, S., et al., 2016. Indonesia's blue carbon: a globally significant and vulnerable sink for seagrass and mangrove carbon. Wetlands Ecology and Management 24 (1), 3–13.

Angelsen, A., 2009. Realising REDD+: National Strategy and Policy Options. Cifor.

Atwood, T.B., Connolly, R.M., Almahasheer, H., Carnell, P.E., Duarte, C.M., Lewis, C.J.E., et al., 2017. Global patterns in mangrove soil carbon stocks and losses. Nature Climate Change 7 (7), 523.

Atwood, T.B., Connolly, R.M., Ritchie, E.G., Lovelock, C.E., Heithaus, M.R., Hays, G.C., et al., 2015. Predators help protect carbon stocks in blue carbon ecosystems. Nature Climate Change 5 (12), 1038.

Barbier, E.B., Hacker, S.D., Kennedy, C., Koch, E.W., Stier, A.C., Silliman, B.R., 2011. The value of estuarine and coastal ecosystem services. Ecological Monographs 81 (2), 169–193.

Bayraktarov, E., Saunders, M.I., Abdullah, S., Mills, M., Beher, J., Possingham, H.P., et al., 2016. The cost and feasibility of marine coastal restoration. Ecological Applications 26 (4), 1055–1074.

Beck, M.W., Heck Jr., K.L., Able, K.W., Childers, D.L., Eggleston, D.B., Gillanders, B.M., et al., 2001. The identification, conservation, and management of estuarine and marine nurseries for fish and invertebrates: a better understanding of the habitats that serve as nurseries for marine species and the factors that create site-specific variability in nursery quality will improve conservation and management of these areas. BioScience 51 (8), 633–641.

Benson, L., Glass, L., Jones, T.G., Ravaoarinorotsihoarana, L., Rakotomahazo, C., 2017. Mangrove carbon stocks and ecosystem cover dynamics in Southwest Madagascar and the implications for local management. Forests 8 (6), 190.

Bouillon, S., Connolly, R.M., Lee, S.Y., 2008. Organic matter exchange and cycling in mangrove ecosystems: recent insights from stable isotope studies. Journal of Sea Research 59 (1–2), 44–58.

Bowler, C., Karl, D.M., Colwell, R.R., 2009. Microbial oceanography in a sea of opportunity. Nature 459 (7244), 180.

Breithaupt, J.L., Smoak, J.M., Smith, T.J., Sanders, C.J., Hoare, A., 2012. Organic carbon burial rates in mangrove sediments: strengthening the global budget. Global Biogeochemical Cycles 26 (3).

Bu, N.S., Qu, J.F., Li, G., Zhao, B., Zhang, R.J., Fang, C.M., 2015. Reclamation of coastal salt marshes promoted carbon loss from previously-sequestered soil carbon pool. Ecological Engineering 81, 335–339.

Burdige, D.J., 2007. Preservation of organic matter in marine sediments: controls, mechanisms, and an imbalance in sediment organic carbon budgets? Chemical Reviews 107 (2), 467–485.

Byrnes, J.E.K., Krumhansl, K.A., Okamoto, D., Rassweiler, A., Novak, M., Bolton, J.J., et al., 2016. Linking global patterns of kelp forest change and variation in climate over the past half-century. In: 11th International Temperate Reefs Symposium.

Cai, W.J., 2010. Estuarine and coastal ocean carbon paradox: CO2 sinks or sites of terrestrial carbon incineration? Annual Review of Marine Science 3, 123–145.

Caldeira, K., Wickett, M.E., 2003. Oceanography: anthropogenic carbon and ocean pH. Nature 425 (6956), 365.

Canadell, J.G., Raupach, M.R., 2008. Managing forests for climate change mitigation. Science 320 (5882), 1456–1457.

Carnell, P.E., Keough, M.J., 2016. The influence of herbivores on primary producers can vary spatially and interact with disturbance. Oikos 125 (9), 1273–1283.

CDIAC, 2018. Carbon Dioxide Information Analysis Center. http://cdiac.ess-dive.lbl.gov.

Chmura, G.L., 2013. What do we need to assess the sustainability of the tidal salt marsh carbon sink? Ocean and Coastal Management 83, 25–31.

Chmura, G.L., Anisfeld, S.C., Cahoon, D.R., Lynch, J.C., 2003. Global carbon sequestration in tidal, saline wetland soils. Global Biogeochemical Cycles 1 (4).

Chung, I.K., Oak, J.H., Lee, J.A., Shin, J.A., Kim, J.G., Park, K.S., 2013. Installing kelp forests/seaweed beds for mitigation and adaptation against global warming: Korean project overview. ICES Journal of Marine Science 70 (5), 1038–1044.

Church, J.A., White, N.J., 2011. Sea-level rise from the late 19th to the early 21st century. Surveys in Geophysics 32 (4–5), 585–602.

Church, J.A., Clark, P.U., Cazenave, A., Gregory, J.M., Jevrejeva, S., Levermann, A., et al., 2013. Sea Level Change. PM Cambridge University Press.

Costanza, R., d'Arge, R., De Groot, R., Farber, S., Grasso, M., Hannon, B., et al., 1997. The value of the world's ecosystem services and natural capital. Nature 387 (6630), 253.

Coverdale, T.C., Brisson, C.P., Young, E.W., Yin, S.F., Donnelly, J.P., Bertness, M.D., 2014. Indirect human impacts reverse centuries of carbon sequestration and salt marsh accretion. PLoS One 9 (3), e93296.

da Silva Copertino, M., 2011. Add coastal vegetation to the climate critical list. Nature 473 (7347), 255–256.

De Falco, G., Ferrari, S., Cancemi, G., Baroli, M., 2000. Relationship between sediment distribution and Posidonia oceanica seagrass. Geo-Marine Letters 20 (1), 50–57.

Deegan, L.A., Johnson, D.S., Warren, R.S., Peterson, B.J., Fleeger, J.W., Fagherazzi, S., Wollheim, W.M., 2012. Coastal eutrophication as a driver of salt marsh loss. Nature 490 (7420), 388.

Dennison, W.C., 1987. Effects of light on seagrass photosynthesis, growth and depth distribution. Aquatic Botany 27 (1), 15–26.

Donato, D.C., Kauffman, J.B., Murdiyarso, D., Kurnianto, S., Stidham, M., Kanninen, M., 2011. Mangroves among the most carbon-rich forests in the tropics. Nature Geoscience 4 (5), 293.

Doney, S.C., Fabry, V.J., Feely, R.A., Kleypas, J.A., 2009. Ocean Acidification: The Other CO_2 Problem.

Doody, J.P., 2004. 'Coastal squeeze'—an historical perspective. Journal of Coastal Conservation 10 (1), 129–138.

Duarte, C.M., 1991. Seagrass depth limits. Aquatic Botany 40 (4), 363–377.

Duarte, C.M., Cebrian, J., 1996. The fate of marine autotrophic production. Limnology and Oceanography 41 (8), 1758–1766.

Duarte, C.M., Middelburg, J.J., Caraco, N., 2005. Major role of marine vegetation on the oceanic carbon cycle. Biogeosciences Discussions 1 (1), 659–679.

Duarte, C.M., Dennison, W.C., Orth, R.J., Carruthers, T.J., 2008. The charisma of coastal ecosystems: addressing the imbalance. Estuaries and Coasts 31 (2), 233–238.

Duarte, C.M., 2009. Global Loss of Coastal Habitats: Rates, Causes and Consequences. FBBVA, Madrid, 181 p. ISBN: 978-84-96515-84-0 avaiable as pdf's at: http://www.fbbva.es/TLFU/tlfu/esp/publicaciones/libros/fichalibro/index.jsp?codigo=411.

Duarte, C.M., Losada, I.J., Hendriks, I.E., Mazarrasa, I., Marbà, N., 2013. The role of coastal plant communities for climate change mitigation and adaptation. Nature Climate Change 3 (11), 961–968.

Duarte, C.M., 2017. Reviews and syntheses: hidden forests, the role of vegetated coastal habitats in the ocean carbon budget. Biogeosciences 14 (2), 301.

Duke, N., Ball, M., Ellison, J., 1998. Factors influencing biodiversity and distributional gradients in mangroves. Global Ecology and Biogeography Letters 7 (1), 27–47.

Fabry, V.J., Seibel, B.A., Feely, R.A., Orr, J.C., 2008. Impacts of ocean acidification on marine fauna and ecosystem processes. ICES Journal of Marine Science 65 (3), 414–432.

Filho, P.S., Cohen, M.C.L., Lara, R.J., Lessa, G.C., Koch, B., Behling, H., 2006. Holocene coastal evolution and facies model of the Bragança macrotidal flat on the Amazon Mangrove Coast, Northern Brazil. Journal of Coastal Research 306–310.

Fourqurean, J.W., Duarte, C.M., Kennedy, H., Marbà, N., Holmer, M., Mateo, M.A., et al., 2012. Seagrass ecosystems as a globally significant carbon stock. Nature Geoscience 5 (7), 505–509.

Gacia, E., Duarte, C.M., 2001. Sediment retention by a Mediterranean Posidonia oceanica meadow: the balance between deposition and resuspension. Estuarine, Coastal and Shelf Science 52 (4), 505–514.

Gatland, J.R., Santos, I.R., Maher, D.T., Duncan, T.M., Erler, D.V., 2014. Carbon dioxide and methane emissions from an artificially drained coastal wetland during a flood: implications for wetland global warming potential. Journal of Geophysical Research: Biogeosciences 119 (8), 1698–1716.

Gattuso, J.P., Frankignoulle, M., Wollast, R., 1998. Carbon and carbonate metabolism in coastal aquatic ecosystems. Annual Review of Ecology and Systematics 29 (1), 405–434.

Gattuso, J.P., Gentili, B., Duarte, C.M., Kleypas, J.A., Middelburg, J.J., Antoine, D., 2006. Light availability in the coastal ocean: impact on the distribution of benthic photosynthetic organisms and contribution to primary production. Biogeosciences 3 (4), 489–513.

Gedan, K.B., Silliman, B.R., Bertness, M.D., 2009. Centuries of human-driven change in salt marsh ecosystems. Annual Reviews in Marine Science 1, 117–141.

Hallegatte, S., Green, C., Nicholls, R.J., Corfee-Morlot, J., 2013. Future flood losses in major coastal cities. Nature Climate Change 3 (9), 802.

Hamilton, S.E., Casey, D., 2016. Creation of a high spatio-temporal resolution global database of continuous mangrove forest cover for the 21st century (CGMFC-21). Global Ecology and Biogeography 25 (6), 729–738.

Harrison, P.G., 1989. Detrital processing in seagrass systems: a review of factors affecting decay rates, remineralization and detritivory. Aquatic Botany 35 (3–4), 263–288.

Hayes, M.A., Jesse, A., Hawke, B., Baldock, J., Tabet, B., Lockington, D., Lovelock, C.E., 2017. Dynamics of sediment carbon stocks across intertidal wetland habitats of Moreton Bay, Australia. Global Change Biology 23 (10), 4222–4234.

Hemminga, M.A., Duarte, C.M., 2000. Seagrass Ecology. Cambridge University Press.

Henrichs, S.M., 1992. Early diagenesis of organic matter in marine sediments: progress and perplexity. Marine Chemistry 39 (1–3), 119–149.

Hendriks, I.E., Olsen, Y.S., Ramajo, L., Basso, L., Steckbauer, A., Moore, T.S., et al., 2014. Photosynthetic activity buffers ocean acidification in seagrass meadows. Biogeosciences 11 (2), 333.

Hicks, W., Fitzpatrick, R., Bowman, G., 2003. Managing coastal acid sulfate soils: the East Trinity example. In: Advances in Regolith: Proceedings of the CRC LEME Regional Regolith Symposia. CRC LEME, Bentley, pp. 174–177.

Hill, R., Bellgrove, A., Macreadie, P.I., Petrou, K., Beardall, J., Steven, A., Ralph, P.J., 2015. Can macroalgae contribute to blue carbon? An Australian perspective. Limnology and Oceanography 60 (5), 1689–1706.

Hoegh-Guldberg, O., Bruno, J.F., 2010. The impact of climate change on the world's marine ecosystems. Science 328 (5985), 1523–1528.

Howard, J., Sutton-Grier, A., Herr, D., Kleypas, J., Landis, E., Mcleod, E., et al., 2017. Clarifying the role of coastal and marine systems in climate mitigation. Frontiers in Ecology and the Environment 15 (1), 42–50.

Howe, A.J., Rodríguez, J.F., Saco, P.M., 2009. Surface evolution and carbon sequestration in disturbed and undisturbed wetland soils of the Hunter estuary, southeast Australia. Estuarine, Coastal and Shelf Science 84 (1), 75–83.

Hurd, C.L., Harrison, P.J., Bischof, K., Lobban, C.S., 2014. Seaweed Ecology and Physiology. Cambridge University Press.

Hutchings, P.A., Saenger, P., 1987. Ecology of Mangroves. University of Queensland Press, 388 pages. Maher 44 (10), 4889–4896.

Hutchison, J., Manica, A., Swetnam, R., Balmford, A., Spalding, M., 2014. Predicting global patterns in mangrove forest biomass. Conservation Letters 7 (3), 233–240.

Huxham, M., Emerton, L., Kairo, J., Munyi, F., Abdirizak, H., Muriuki, T., et al., 2015. Applying climate compatible development and economic valuation to coastal management: a case study of Kenya's mangrove forests. Journal of Environmental Management 157, 168–181.

IPCC, 2003. 2003 IPCC Good practice guidance for land use, land-use change and forestry. In: Penman, J., Gytarsky, M., Hiraishi, T., Krug, T., Kruger, D., Pipatti, R., Buendia, L., Miwa, K., Ngara, T., Tanabe, K., Wagner, F. (Eds.). IPCC, Switzerland.

IPCC, 2006. 2006 IPCC guidelines for national greenhouse gas inventories. In: Eggleston, H.S., Buendia, L., Miwa, K., Ngara, T., Tanabe, K. (Eds.), Prepared by the National Greenhouse Gas Inventories Programme. IGES, Japan.

IPCC, 2014. In: Hiraishi, T., Krug, T., Tanabe, K., Srivastava, N., Baasansuren, J., Fukuda, M., Troxler, T.G. (Eds.), 2013 Supplement to the 2006 IPCC Guidelines for National Greenhouse Gas Inventories: Wetlands. IPCC, Switzerland.

Jardine, S.L., Siikamäki, J.V., 2014. A global predictive model of carbon in mangrove soils. Environmental Research Letters 9 (10), 104013.

Jones, H.P., Hole, D.G., Zavaleta, E.S., 2012. Harnessing nature to help people adapt to climate change. Nature Climate Change 2 (7), 504–509.

Jones, T.G., Ratsimba, H.R., Ravaoarinorotsihoarana, L., Cripps, G., Bey, A., 2014. Ecological variability and carbon stock estimates of mangrove ecosystems in northwestern Madagascar. Forests 5 (1), 177–205.

Jones, T.G., Glass, L., Gandhi, S., Ravaoarinorotsihoarana, L., Carro, A., Benson, L., et al., 2016. Madagascar's mangroves: quantifying nation-wide and ecosystem specific dynamics, and detailed contemporary mapping of distinct ecosystems. Remote Sensing 8 (2), 106.

Kaal, J., Serrano, O., Nierop, K.G., Schellekens, J., Cortizas, A.M., Mateo, M.Á., 2016. Molecular composition of plant parts and sediment organic matter in a Mediterranean seagrass (Posidonia oceanica) mat. Aquatic Botany 133, 50–61.

Keil, R.G., Hedges, J.I., 1993. Sorption of organic matter to mineral surfaces and the preservation of organic matter in coastal marine sediments. Chemical Geology 107 (3–4), 385–388.

Kelleway, J.J., Saintilan, N., Macreadie, P.I., Skilbeck, C.G., Zawadzki, A., Ralph, P.J., 2016a. Seventy years of continuous encroachment substantially increases 'blue carbon'capacity as mangroves replace intertidal salt marshes. Global Change Biology 22 (3), 1097–1109.

Kelleway, J.J., Saintilan, N., Macreadie, P.I., Ralph, P.J., 2016b. Sedimentary factors are key predictors of carbon storage in SE Australian saltmarshes. Ecosystems 19 (5), 865–880.

Kelleway, J., Serrano, O., Baldock, J., Cannard, T., Lavery, P., Lovelock, C.E., et al., 2017a. Technical Review of Opportunities for Including Blue Carbon in the Australian Government's Emissions Reduction Fund. Department of Environment and Energy, Canberra, p. 287.

Kelleway, J.J., Saintilan, N., Macreadie, P.I., Baldock, J.A., Ralph, P.J., 2017b. Sediment and carbon deposition vary among vegetation assemblages in a coastal salt marsh. Biogeosciences 14 (16), 3763.

Kennedy, H., Beggins, J., Duarte, C.M., Fourqurean, J.W., Holmer, M., Marbà, N., Middelburg, J.J., 2010. Seagrass sediments as a global carbon sink: isotopic constraints. Global Biogeochemical Cycles 24 (4), 1–8.

Kirwan, M.L., Megonigal, J.P., 2013. Tidal wetland stability in the face of human impacts and sea-level rise. Nature 504 (7478), 53.

Klap, V.A., Hemminga, M.A., Boon, J.J., 2000. Retention of lignin in seagrasses: angiosperms that returned to the sea. Marine Ecology Progress Series 1–11.

Krause-Jensen, D., Duarte, C.M., 2016. Substantial role of macroalgae in marine carbon sequestration. Nature Geoscience 9 (10), 737.

Krauss, K.W., McKee, K.L., Lovelock, C.E., Cahoon, D.R., Saintilan, N., Reef, R., Chen, L., 2014. How mangrove forests adjust to rising sea level. New Phytologist 202 (1), 19–34.

Kroeger, K.D., Crooks, S., Moseman-Valtierra, S., Tang, J., 2017. Restoring tides to reduce methane emissions in impounded wetlands: a new and potent Blue Carbon climate change intervention. Scientific Reports 7 (1), 11914.

Laffoley, D., Grimsditch, G.D. (Eds.), 2009. The Management of Natural Coastal Carbon Sinks. Iucn.

Laursen, A.K., Mayer, L.M., Townsend, D.W., 1996. Lability of proteinaceous material in estuarine seston and subcellular fractions of phytoplankton. Marine Ecology Progress Series 227–234.

Lavery, P.S., Mateo, M.Á., Serrano, O., Rozaimi, M., 2013. Variability in the carbon storage of seagrass habitats and its implications for global estimates of blue carbon ecosystem service. PLoS One 8 (9), e73748.

Littler, M.M., Littler, D.S., Taylor, P.R., 1983. Evolutionary strategies in a tropical barrier reef system: functional-form groups of marine macroalgae. Journal of Phycology 19 (2), 229–237.

Lo Iacono, C., Mateo, M.A., Gracia, E., Guasch, L., Carbonell, R., Serrano, L., et al., 2008. Very high-resolution seismo-acoustic imaging of seagrass meadows (Mediterranean Sea): implications for carbon sink estimates. Geophysical Research Letters 35 (18).

Longhurst, A.R., Harrison, W.G., 1989. The biological pump: profiles of plankton production and consumption in the upper ocean. Progress in Oceanography 22 (1), 47–123.

Lotze, H.K., Lenihan, H.S., Bourque, B.J., Bradbury, R.H., Cooke, R.G., Kay, M.C., et al., 2006. Depletion, degradation, and recovery potential of estuaries and coastal seas. Science 312 (5781), 1806–1809.

Lovelock, C.E., Ball, M.C., Feller, I.C., Engelbrecht, B.M., Ling Ewe, M., 2006. Variation in hydraulic conductivity of mangroves: influence of species, salinity, and nitrogen and phosphorus availability. Physiologia Plantarum 127 (3), 457–464.

Lovelock, C.E., Adame, M.F., Bennion, V., Hayes, M., O'Mara, J., Reef, R., Santini, N.S., 2014. Contemporary rates of carbon sequestration through vertical accretion of sediments in mangrove forests and saltmarshes of South East Queensland, Australia. Estuaries and Coasts 37 (3), 763–771.

Lovelock, C.E., Atwood, T., Baldock, J., Duarte, C.M., Hickey, S., Lavery, P.S., et al., 2017a. Assessing the risk of carbon dioxide emissions from blue carbon ecosystems. Frontiers in Ecology and the Environment 15 (5), 257–265.

Lovelock, C.E., Feller, I.C., Reef, R., Hickey, S., Ball, M.C., 2017b. Mangrove dieback during fluctuating sea levels. Scientific Reports 7 (1), 1680.

Lovelock, C.E., Fourqurean, J.W., Morris, J.T., 2017c. Modeled CO2 emissions from coastal wetland transitions to other land uses: tidal marshes, mangrove forests, and seagrass beds. Frontiers in Marine Science 4, 143.

Lovley, D.R., Dwyer, D.F., Klug, M.J., 1982. Kinetic analysis of competition between sulfate reducers and methanogens for hydrogen in sediments. Applied and Environmental Microbiology 43 (6), 1373–1379.

Luisetti, T., Turner, R.K., Bateman, I.J., Morse-Jones, S., Adams, C., Fonseca, L., 2011. Coastal and marine ecosystem services valuation for policy and management: Managed realignment case studies in England. Ocean and Coastal Management 54 (3), 212–224.

Mack, S.K., Lane, R.R., Day, J.W., 2012. Restoration of Degraded Deltaic Wetlands of the Mississippi Delta v2.0. American carbon Registry (ACR). Winrock international.

Macreadie, P.I., Nielsen, D.A., Kelleway, J.J., Atwood, T.B., Seymour, J.R., Petrou, K., et al., 2017a. Can we manage coastal ecosystems to sequester more blue carbon? Frontiers in Ecology and the Environment 15 (4), 206–213.

Macreadie, P.I., Ollivier, Q.R., Kelleway, J.J., Serrano, O., Carnell, P.E., Lewis, C.E., et al., 2017b. Carbon sequestration by Australian tidal marshes. Scientific Reports 7, 44071.

Maher, D.T., Santos, I.R., Schulz, K.G., Call, M., Jacobsen, G.E., Sanders, C.J., 2017. Blue carbon oxidation revealed by radiogenic and stable isotopes in a mangrove system. Geophysical Research Letters.

Marbà, N., Arias-Ortiz, A., Masqué, P., Kendrick, G.A., Mazarrasa, I., Bastyan, G.R., et al., 2015. Impact of seagrass loss and subsequent revegetation on carbon sequestration and stocks. Journal of Ecology 103 (2), 296–302.

Marchand, C., Lallier-Vergès, E., Baltzer, F., 2003. The composition of sedimentary organic matter in relation to the dynamic features of a mangrove-fringed coast in French Guiana. Estuarine, Coastal and Shelf Science 56 (1), 119–130.

Mateo, M.A., Romero, J., Pérez, M., Littler, M.M., Littler, D.S., 1997. Dynamics of millenary organic deposits resulting from the growth of the Mediterranean seagrass Posidonia oceanica. Estuarine, Coastal and Shelf Science 44 (1), 103–110.

Mateo, M., Cebrián, J., Dunton, K., Mutchler, T., 2006. Carbon flux in seagrass ecosystems. Seagrasses: Biology, Ecology and Conservation 159–192.

Matsui, N., Morimune, K., Meepol, W., Chukwamdee, J., 2012. Ten year evaluation of carbon stock in mangrove plantation reforested from an abandoned shrimp pond. Forests 3 (2), 431–444.

McKee, K.L., Cahoon, D.R., Feller, I.C., 2007. Caribbean mangroves adjust to rising sea level through biotic controls on change in soil elevation. Global Ecology and Biogeography 16 (5), 545–556.

Mcleod, E., Chmura, G.L., Bouillon, S., Salm, R., Björk, M., Duarte, C.M., et al., 2011. A blueprint for blue carbon: toward an improved understanding of the role of vegetated coastal habitats in sequestering CO_2. Frontiers in Ecology and the Environment 9 (10), 552–560.

McOwen, C., Weatherdon, L.V., Bochove, J., Sullivan, E., Blyth, S., Zockler, C., Stanwell-Smith, D., Kingston, N., Martin, C.S., Spalding, M., Fletcher, S., 2017. A global map of saltmarshes. Biodiversity Data Journal 5, e11764. https://doi.org/10.3897/BDJ.5.e11764. http://data.unepwcmc.org/datasets/43 (v.5).

Menéndez, M., Woodworth, P.L., 2010. Changes in extreme high water levels based on a quasi-global tide-gauge data set. Journal of Geophysical Research: Oceans 115 (C10).

Middelboe, A.L., Hansen, P.J., 2007. High pH in shallow-water macroalgal habitats. Marine Ecology Progress Series 338, 107–117.

Millero, F.J., 1995. Thermodynamics of the carbon dioxide system in the oceans. Geochimica et Cosmochimica Acta 59 (4), 661–677.

Miyajima, T., Hori, M., Hamaguchi, M., Shimabukuro, H., Adachi, H., Yamano, H., Nakaoka, M., 2015. Geographic variability in organic carbon stock and accumulation rate in sediments of East and Southeast Asian seagrass meadows. Global Biogeochemical Cycles 29 (4), 397–415.

Mudd, S.M., Howell, S.M., Morris, J.T., 2009. Impact of dynamic feedbacks between sedimentation, sea-level rise, and biomass production on near-surface marsh stratigraphy and carbon accumulation. Estuarine, Coastal and Shelf Science 82 (3), 377–389.

Murdiyarso, D., Purbopuspito, J., Kauffman, J.B., Warren, M.W., Sasmito, S.D., Donato, D.C., et al., 2015. The potential of Indonesian mangrove forests for global climate change mitigation. Nature Climate Change 5 (12), 1089.

Nellemann, C., Corcoran, E. (Eds.), 2009. Blue Carbon: The Role of Healthy Oceans in Binding Carbon: a Rapid Response Assessment. UNEP/Earthprint.

Neumann, B., Vafeidis, A.T., Zimmermann, J., Nicholls, R.J., 2015. Future coastal population growth and exposure to sea-level rise and coastal flooding-a global assessment. PLoS One 10 (3), e0118571.

Nixon, S.W., 1980. Between coastal marshes and coastal waters—a review of twenty years of speculation and research on the role of salt marshes in estuarine productivity and water chemistry. In: Estuarine and Wetland Processes. Springer US, pp. 437–525.

Orr, J.C., Fabry, V.J., Aumont, O., Bopp, L., Doney, S.C., Feely, R.A., et al., 2005. Anthropogenic ocean acidification over the twenty-first century and its impact on calcifying organisms. Nature 437 (7059), 681.

Ouyang, X., Lee, S.Y., 2014. Updated estimates of carbon accumulation rates in coastal marsh sediments. Biogeosciences 5057.

Pedersen, M.Ø., Serrano, O., Mateo, M.Á., Holmer, M., 2011. Temperature effects on decomposition of a Posidonia oceanica mat. Aquatic Microbial Ecology 65 (2), 169–182.

Pendleton, L., Donato, D.C., Murray, B.C., Crooks, S., Jenkins, W.A., Sifleet, S., et al., 2012. Estimating global "blue carbon" emissions from conversion and degradation of vegetated coastal ecosystems. PLoS One 7 (9), e43542.

Pianka, E.R., 1970. On r-and K-selection. The American Naturalist 104 (940), 592–597.

Raupach, M.R., Canadell, J.G., 2008. Observing a vulnerable carbon cycle. In: The Continental-Scale Greenhouse Gas Balance of Europe. Springer, New York, NY, pp. 5–32.

Reef, R., Feller, I.C., Lovelock, C.E., 2010. Nutrition of mangroves. Tree Physiology 30 (9), 1148–1160.

Saintilan, N., Rogers, K., Mazumder, D., Woodroffe, C., 2013. Allochthonous and autochthonous contributions to carbon accumulation and carbon store in southeastern Australian coastal wetlands. Estuarine, Coastal and Shelf Science 128, 84–92.

Scales, I.R., Friess, D.A., Glass, L., Ravaoarinorotsihoarana, L., 2017. Rural livelihoods and mangrove degradation in south-west Madagascar: lime production as an emerging threat. Oryx 1–5.

Serrano, O., Lavery, P.S., Rozaimi, M., Mateo, M.Á., 2014. Influence of water depth on the carbon sequestration capacity of seagrasses. Global Biogeochemical Cycles 28 (9), 950–961.

Serrano, O., Ricart, A.M., Lavery, P.S., Mateo, M.A., Arias-Ortiz, A., Masque, P., et al., 2016a. Key biogeochemical factors affecting soil carbon storage in Posidonia meadows. Biogeosciences 13 (15), 4581.

Serrano, O., Ruhon, R., Lavery, P.S., Kendrick, G.A., Hickey, S., Masqué, P., et al., 2016b. Impact of mooring activities on carbon stocks in seagrass meadows. Scientific Reports 6, 23193.

Showstack, R., 2013. Carbon dioxide tops 400 ppm at Mauna Loa, Hawaii. Eos, Transactions American Geophysical Union 94 (21), 192.

Siegenthaler, U., Sarmiento, J.L., 1993. Atmospheric carbon dioxide and the ocean. Nature 365 (6442), 119.

Sippo, J.Z., Maher, D.T., Tait, D.R., Holloway, C., Santos, I.R., 2016. Are mangroves drivers or buffers of coastal acidification? Insights from alkalinity and dissolved inorganic carbon export estimates across a latitudinal transect. Global Biogeochemical Cycles 30 (5), 753–766.

Smale, D.A., Burrows, M.T., Moore, P., O'Connor, N., Hawkins, S.J., 2013. Threats and knowledge gaps for ecosystem services provided by kelp forests: a northeast Atlantic perspective. Ecology and Evolution 3 (11), 4016–4038.

Smale, D.A., Moore, P.J., Queirós, A.M., Higgs, N.D., Burrows, M.T., 2018. Appreciating interconnectivity between habitats is key to blue carbon management. Frontiers in Ecology and the Environment 16 (2), 71–73.

Smith, S.V., 1981. Marine macrophytes as a global carbon sink. Science 211 (4484), 838–840.

Solomon, S., Qin, D., Manning, M., Averyt, K., Marquis, M. (Eds.), 2007. Climate change 2007-the physical science basis: Working group I contribution to the fourth assessment report of the IPCC (Vol. 4), Cambridge university press.

Spalding, M., Kainuma, M., Collins, L., 2010. World Atlas of Mangroves. A Collaborative Project of ITTO, ISME, FAO, UNEP-WCMC. Earthscan, London, UK.

Steneck, R.S., Dethier, M.N., 1994. A functional group approach to the structure of algal-dominated communities. Oikos 476–498.

Steneck, R.S., Graham, M.H., Bourque, B.J., Corbett, D., Erlandson, J.M., Estes, J.A., Tegner, M.J., 2002. Kelp forest ecosystems: biodiversity, stability, resilience and future. Environmental Conservation 29 (4), 436–459.

Sutton-Grier, A.E., Moore, A.K., Wiley, P.C., Edwards, P.E., 2014. Incorporating ecosystem services into the implementation of existing US natural resource management regulations: operationalizing carbon sequestration and storage. Marine Policy 43, 246–253.

Tamooh, F., Huxham, M., Karachi, M., Mencuccini, M., Kairo, J.G., Kirui, B., 2008. Below-ground root yield and distribution in natural and replanted mangrove forests at Gazi bay, Kenya. Forest Ecology and Management 256 (6), 1290–1297.

Thomas, S., 2014. Blue carbon: knowledge gaps, critical issues, and novel approaches. Ecological Economics 107, 22–38.

Torio, D.D., Chmura, G.L., 2013. Assessing coastal squeeze of tidal wetlands. Journal of Coastal Research 29 (5), 1049–1061.

Trevathan-Tackett, S.M., Kelleway, J., Macreadie, P.I., Beardall, J., Ralph, P., Bellgrove, A., 2015. Comparison of marine macrophytes for their contributions to blue carbon sequestration. Ecology 96 (11), 3043–3057.

Trumper, K., 2009. The Natural Fix?: The Role of Ecosystems in Climate Mitigation: a UNEP Rapid Response Assessment. UNEP/Earthprint.

Ullman, R., Bilbao-Bastida, V., Grimsditch, G., 2013. Including blue carbon in climate market mechanisms. Ocean and Coastal Management 83, 15–18.

UNEP, 2012. The Emissions Gap Report 2012. United Nations Environment Programme (UNEP), Nairobi.

UNEP-WCMC, Short F. T, 2017. Global distribution of seagrasses (version 5.0). In: Fourth Update to the Data Layer Used in Green and Short (2003). UNEP World Conservation Monitoring Centre, Cambridge (UK). http://data.unepwcmc.org/datasets/7.

UNFCCC (United Nations Framework Convention on Climate Change), 2016. Report of the Conference of the Parties on its Twenty-first Session, Held in Paris from 30 November to 13 December 2015. FCCC/CP/2015/10/Add.1. Bonn. UNFCCC. http://unfccc.int/resource/docs/2015/cop21/eng/10a01.pdf.

Valiela, I., Bowen, J.L., York, J.K., 2001. Mangrove forests: one of the world's threatened major tropical environments. BioScience 51 (10), 807–815.

Van der Werf, G.R., Morton, D.C., DeFries, R.S., Olivier, J.G., Kasibhatla, P.S., Jackson, R.B., et al., 2009. CO_2 emissions from forest loss. Nature Geoscience 2 (11), 737–738.

VCS (Verified Carbon Standard), 2015. Methodology for Tidal Wetland and Seagrass Restoration (VM003). http://database.v-c-s.org/sites/vcs.benfredaconsulting.com/files/VM0033%20Tidal%20Wetland%20and%20-Seagrass%20Restoration%20v1.0%2020%20NOV%202015_0.pdf.

Ward, L.G., Zaprowski, B.J., Trainer, K.D., Davis, P.T., 2008. Stratigraphy, pollen history and geochronology of tidal marshes in a Gulf of Maine estuarine system: climatic and relative sea level impacts. Marine Geology 256 (1–4), 1–17.

Warren, R.S., Niering, W.A., 1993. Vegetation change on a northeast tidal marsh: interaction of sea-level rise and marsh accretion. Ecology 74 (1), 96–103.

Waycott, M., Duarte, C.M., Carruthers, T.J., Orth, R.J., Dennison, W.C., Olyarnik, S., et al., 2009. Accelerating loss of seagrasses across the globe threatens coastal ecosystems. Proceedings of the National Academy of Sciences 106 (30), 12377–12381.

Whalen, S.C., 2005. Biogeochemistry of methane exchange between natural wetlands and the atmosphere. Environmental Engineering Science 22 (1), 73–94.

Wigand, C., Davey, E., Johnson, R., Sundberg, K., Morris, J., Kenny, P., et al., 2015. Nutrient effects on belowground organic matter in a minerogenic salt marsh, North Inlet, SC. Estuaries and Coasts 38 (6), 1838–1853.

Wylie, L., Sutton-Grier, A.E., Moore, A., 2016. Keys to successful blue carbon projects: lessons learned from global case studies. Marine Policy 65, 76–84.

Young, I.R., Zieger, S., Babanin, A.V., 2011. Global trends in wind speed and wave height. Science 332 (6028), 451–455.

CHAPTER

29

Toward a Salt Marsh Management Plan for New York City: Recommendations for Strategic Restoration and Protection

Christopher Haight, Marit Larson, Rebecca K. Swadek, Ellen Kracauer Hartig

New York City Department of Parks & Recreation, New York, NY, United States

1. INTRODUCTION

Salt marshes in New York City (NYC) are essential tidal wetland habitats that sustain vibrant ecosystems in our highly urbanized environment while also enhancing the open space opportunities for visitors and nearby residential communities. There are approximately 1600 ha (O'Neil-Dunne et al., 2014) of salt marsh remaining in NYC today, representing less than 20% of the extent of historic tidal wetlands around New York Harbor (USFWS, 1997). Nearly 600 ha of the remaining salt marsh are owned and managed by the New York City Department of Parks & Recreation (NYC Parks). These tidal ecosystems fringe the City's outer boroughs in the Bronx, Brooklyn, Queens, and Staten Island. The salt marshes included in this study alone occupy about 27 km of the City's shoreline. Of the five boroughs of NYC (including the city's center, Manhattan), only the Bronx is a contiguous part of the mainland of the United States; the others are islands, with Brooklyn and Queens forming the westernmost part of Long Island. Given NYC's fundamentally water-bound geography and many low-lying land areas, the remaining salt marshes provide buffers to wave action and sea level rise; improve water quality by filtering pollutants and excess nutrients; provide natural habitat for birds, fish, and other wildlife; and create rare expansive vistas and places for passive recreation, education, and inspiration for New Yorkers. More than 325 species of birds use NYC's salt marshes during some part of their life cycle

Coastal Wetlands
https://doi.org/10.1016/B978-0-444-63893-9.00029-0

(Trust for Public Land and NYC Audubon, 1987, 1993; Sanderson et al., 2016). However, even the largest NYC marshes continue to be threatened by sea level rise, coastal erosion, encroachment, and other human impacts (Hartig et al., 2002; Horton et al., 2015; Kemp et al., 2017).

The wetlands of NYC are part of the nationally significant New York–New Jersey Harbor Estuary. For decades, various government agencies, including NYC Parks, the NYC Department of Environmental Protection (NYCDEP), the New York State Department of Environmental Conservation (NYSDEC), the United States Army Corps of Engineers (USACE), and the United States Environmental Protection Agency (USEPA), have worked to protect, manage, and restore these coastal resources so that they can provide the environmental benefits that are critical for public health, sustainability, and resilience (USACE, 2016). Over the last 20 years, NYC Parks has partnered with these and other agencies, as well as nongovernmental organizations, to restore over 40 ha of salt marsh across over 24 sites. Similar efforts are underway nationally and internationally (Boesch, 2006; Coops and van Geest, 2007; Roman and Burdick, 2012).

Given that locations in the city for salt marsh restoration have become more limited and projects more costly over time, a more strategic and comprehensive approach is needed to ensure that salt marsh ecosystems are retained at the edge of our urban environment. The recommendations outlined in this chapter derive from the synthesis of our field data and desktop analyses of the conditions of and threats to these habitats. While the start of this assessment project predated Hurricane Sandy, which made landfall just south of NYC on October 29, 2012 (Sobel, 2014), the hurricane gave new urgency and relevancy to our project. Our assessment report and recommendations can inform planners, resource managers, and community advocates from public and private entities when they are considering projects along the city's coastline. This report also supports efforts to create a more sustainable waterfront city and build on other NYC planning documents, including NYC Wetlands Strategy (NYC OLTPS, 2012), and One NYC, The Plan for a Strong and Just City (NYC Office of Recovery & Resilience, NYC ORR, 2015).

2. STUDY OBJECTIVES AND APPROACH

The overall goal of this study was to develop recommendations that would guide planning for the protection, restoration, and management of the remaining salt marsh habitat in NYC.

To this end, we focused on the 25 largest naturally occurring salt marshes, representing over two-thirds of the total tidal marsh area on NYC Parks property. First, we assessed the existing conditions at these salt marsh complexes (Fig. 29.1) and evaluated their vulnerability to various stressors through field data collection and desktop analysis. Next, we selected metrics that best represented marsh condition and the threats faced to create a framework for comparison across salt marsh sites. Finally, we identified a select set of actions that we considered best for addressing these threats and preserving existing marsh function. These actions are not all-inclusive but offer innovative forms of interventions that can realistically be implemented by NYC Parks.

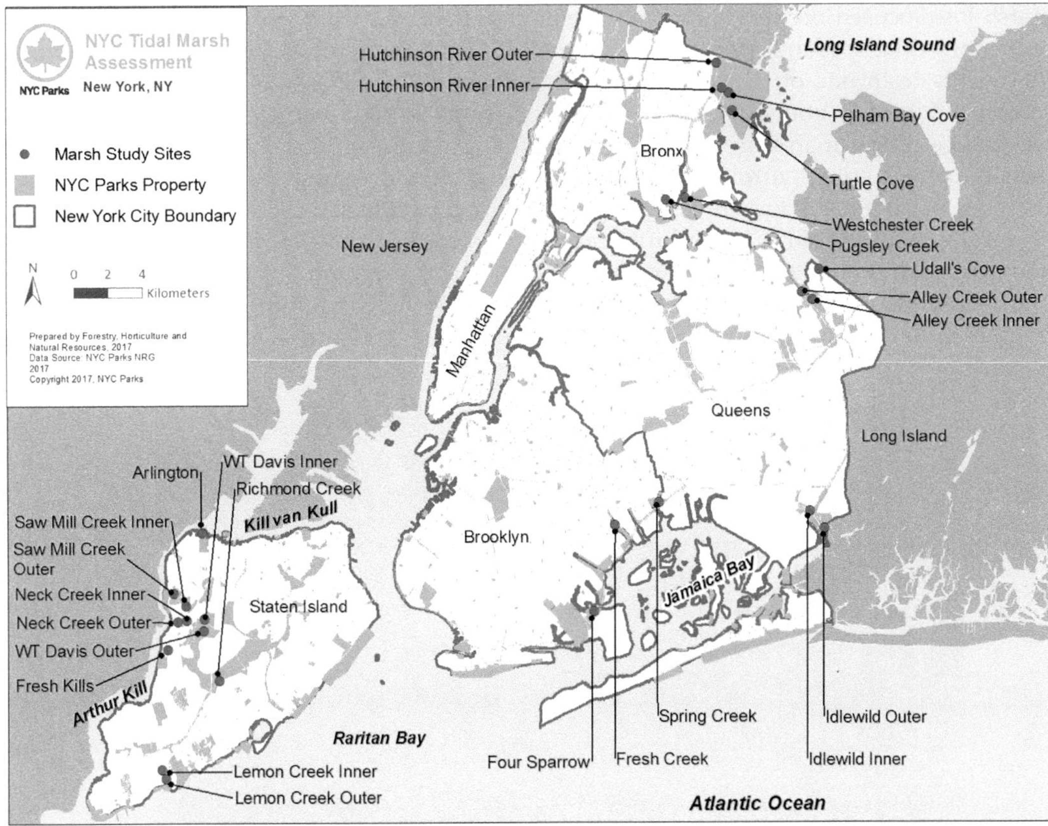

FIGURE 29.1 Map of 25 New York City salt marsh complexes. Sites are located in the Bronx, Brooklyn, Queens, and Staten Island, with receiving waterbodies of the Long Island Sound, Jamaica Bay, Arthur Kill, and Raritan Bay, which connect to the Atlantic Ocean.

2.1 Field Assessments

NYC Parks staff conducted salt marsh field assessments focusing on vegetation and soil data using three different protocols between 2013 and 2014. First, we used the Mid-Atlantic tidal wetland rapid assessment method so that we could compare NYC marshes with marshes in other regions in the US Mid-Atlantic states (Rogerson et al., 2010). Second, we developed a new protocol that involved more extensive measurements to allow comparisons across sites within NYC. Finally, we implemented a long-term intensive monitoring protocol at six of the sites (Partnership for the Delaware Estuary, 2012, 2014).

2.2 Historic and Landscape Analysis

Desktop analyses were conducted to assess historic change at the marshes and to develop landscape-level indicators of salt marsh condition and vulnerability. The analysis of salt marsh loss focused on the change in vegetated marsh area from 1974 (the year of the first regulatory tidal wetlands maps in NYC) to 2012 using aerial photographs. An example of the results from one of the 25 study sites is given in Fig. 29.2. Overlaying the past and recent salt marsh boundaries allowed calculations of total area, average width, and percentage of salt marsh area loss over time. Other landscape-level indicators include the density of mosquito control ditches over total marsh area, marsh perimeter to area ratio, presence or absence of breeding bird species at each marsh since 2000, and the percentage of development within a 200 m buffer adjacent to the marsh. The mosquito control ditch network is an on-the-ground legacy from Great Depression era (1930s) attempts to drain marshes while giving people jobs (Teal and Teal, 1969). These indicators help assess the fragmentation, potential neighboring land use threat, and ecosystem services affecting marshes.

FIGURE 29.2 Aerial image of marsh loss from 1974 to 2012 in Pelham Bay Cove, Pelham Bay Park, Bronx, NY.

2.3 Future Inundation Modeling

Information from a spatial model called Sea Level Affecting Marshes Model (SLAMM) was used to project the impacts of future sea levels on salt marshes and the adjacent upland area (Clough et al., 2014). Its sea level rise projections were based on the NYC Panel on Climate Change report updated after Hurricane Sandy (Horton et al., 2015). SLAMM was applied to estimate how land cover and vegetation cover types would change based on average sea level rise projections by 2085. Undeveloped upland areas next to salt marsh likely to be flooded according to SLAMM were identified as areas for salt marsh migration. Upland areas that are currently developed (roads, parking lots, paved paths) and likely to be flooded by sea level rise were considered separately as sites that could be reclaimed for salt marsh. The model output was used to develop indicators of vulnerability and to identify locations where salt marsh buffer should be protected and where salt marsh could expand in the future.

2.4 Conditions and Vulnerability Indices

The field and spatial data collected in the approach described above were used to develop a conditions index and a vulnerability index for NYC salt marshes. Nine metrics were chosen to represent the condition, or health, of a salt marsh. Six different metrics were selected that served as indicators of the vulnerability of the salt marsh to various threats (Fig. 29.3). Values for each of the metrics were used to develop scores that were then normalized so that each metric for condition and vulnerability could be compared across sites. Summary scores were generated so that each of the 25 sites could be plotted in a general matrix of condition versus vulnerability (Fig. 29.4). Conceptually, this matrix provides a framework for prioritizing sites for protection and restoration; sites in better health and with lower vulnerability are the highest priority for protection because they are the most likely salt marshes to be self-sustaining in the long term. Sites with lower condition and moderate to higher vulnerability are the highest priority for restoration because some intervention is likely needed to increase their viability.

3. SALT MARSH ASSESSMENT RESULTS

3.1 The Condition of Our Wetlands

The overall condition of salt marshes in NYC is lower compared with larger and more rural salt marshes in the Mid-Atlantic region, according to rapid assessment findings. The healthiest NYC marshes, however, potentially provide ecological functions comparable with those of similar size in other developed watersheds. The smallest salt marshes in the study are the most stressed and limited in the ecological and environmental services they provide, but these marshes still provide forage, nursery habitat, and refuge for birds, fish, and other wildlife. They also offer an opportunity for New Yorkers to observe and experience a remnant ecosystem that has largely been replaced by an armored shoreline.

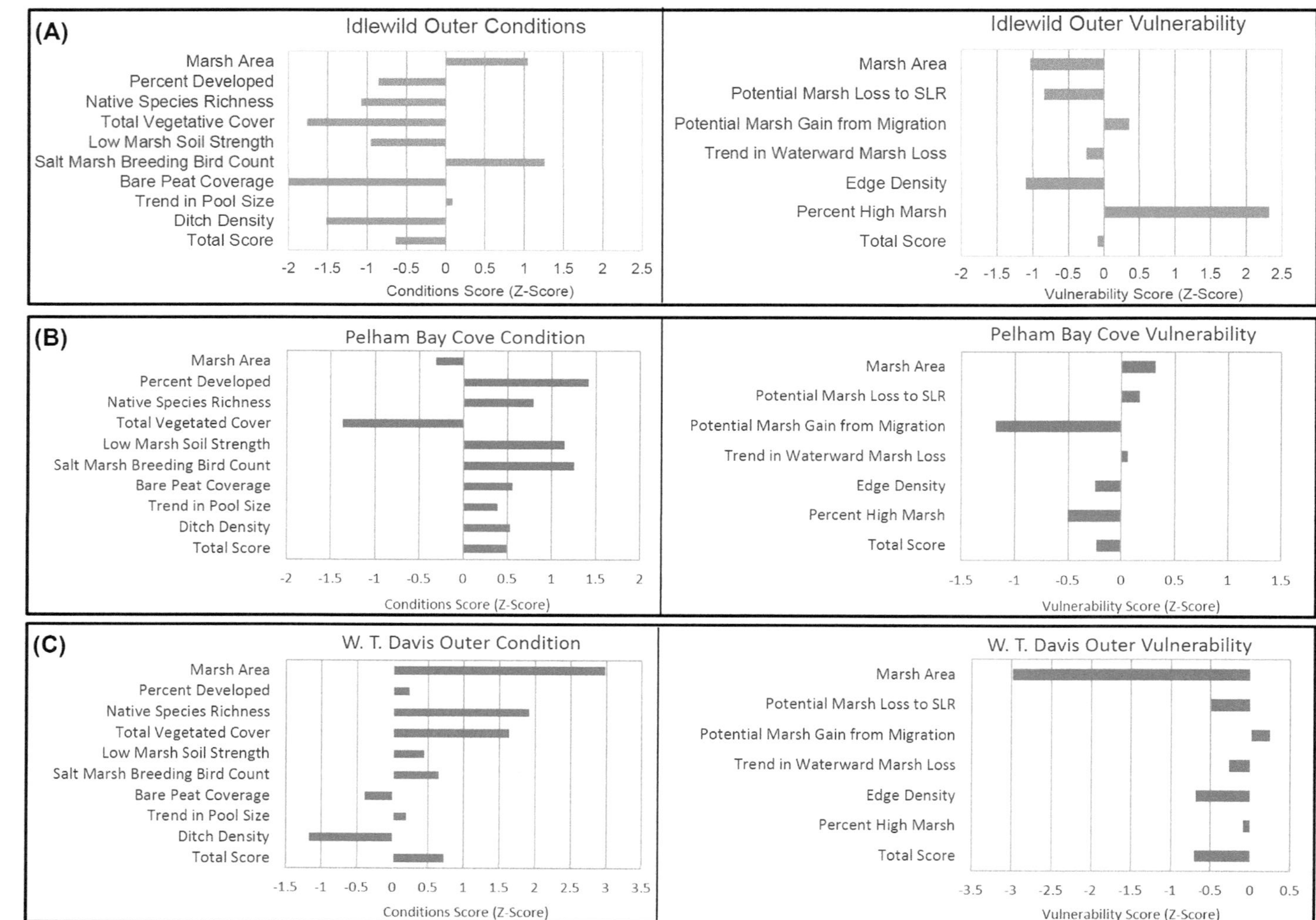

FIGURE 29.3 Conditions and vulnerability metrics and scores for three of the 25 study sites: (A) Idlewild Outer, Queens, NY, with lower condition and moderate vulnerability, (B) Pelham Bay Cove, Bronx, NY, with higher condition and lower vulnerability, and (C) William T. Davis Outer, Staten Island, NY, with higher condition and lower vulnerability. The total condition and vulnerability scores are driven by the individual metric scores, Sea Level Rise (SLR).

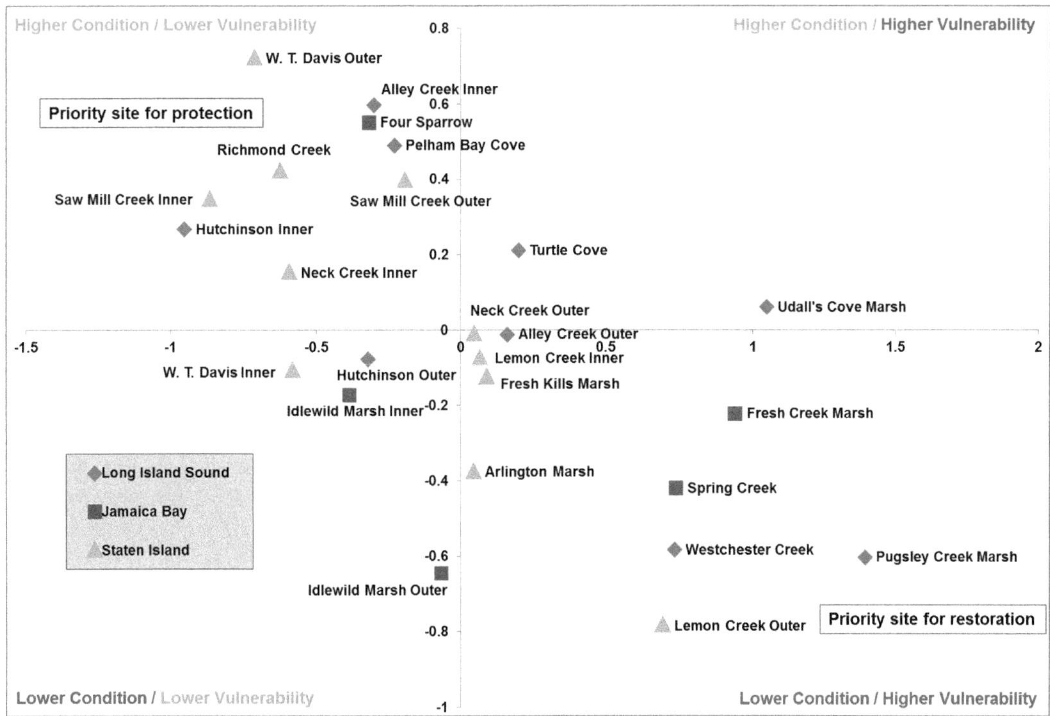

FIGURE 29.4 The distribution of study sites across the condition and vulnerability gradient. Sites in the top left quadrant have higher condition and lower vulnerability and are priority sites for protection to keep the status quo, if not increase health via sound conservation. Sites in the bottom right quadrant have lower condition and higher vulnerability and are priority sites for restoration to improve their condition and lower their vulnerability.

Within NYC, the highest condition salt marshes are found on Staten Island. According to Tiner (2000) approximately one-third of the 1870s salt marsh extent, or 728 ha, remains intact. These marshes are also the most resilient, in part because they are large, have more high marsh area relative to low marsh area (and thus are less susceptible to inundation), and have the greatest potential area for landward migration as sea level rises (allowing for both high marsh to convert to low marsh and upland shrub/scrub to convert to high marsh). This high-quality habitat provides breeding and foraging grounds for birds, fish, and other wildlife including some rare species. Along Jamaica Bay in southern Brooklyn and Queens, the salt marshes are in poorer condition and are less resilient. NYC Parks properties are mainly fringe salt marshes along Jamaica Bay tributaries and include habitat essential to breeding bird species, such as the saltmarsh sparrow, seaside sparrow, clapper rail, and willet (Jonathan Cohen and Alison Kocek SUNY-ESF, personal communication), but these marshes were found to have low total cover of vegetation as indicated by the relatively large areas of bare ground and lower numbers of native plant species. Long Island Sound salt marshes, in the Bronx and northern Queens, vary widely in their condition. On average, sites in Long Island Sound are smaller in size, surrounded by a higher level of development, and have higher cover of invasive *Phragmites australis*. However, they have higher total cover of vegetation and a higher diversity of native plant species compared with Jamaica Bay sites.

3.2 The Threats to Our Wetlands

Threats to NYC salt marshes are multiple and pervasive, but some marshes are more vulnerable than others. In this study, we focus on those threats directly related to or exacerbated by sea level rise. Development pressure is the greatest threat to salt marshes in Staten Island, as this is the least developed borough and has a large amount of privately owned undeveloped land adjacent to wetlands. Soil and water contamination are also threats to these marshes, particularly along the Arthur Kill and Kill van Kull, where oil refineries and shipping operations along the New York/New Jersey shoreline have left a legacy of contaminated soil from industry and frequent oil spills.

The Jamaica Bay fringe marshes have the lowest elevation of all marshes in the city and therefore have the greatest risk of losing interior and shoreline vegetated marsh area due to drowning or excessive inundation. These sites are bordered by both dense urban development and deepened channels, which were historically dredged for navigation and to provide fill material for shoreline development and the construction of John F. Kennedy International Airport. Swanson and Wilson (2008) identify increased tidal range because of large-scale deepening from dredge activity as a contributing factor to marsh loss. This recontouring altered tidal hydrodynamics and, as a result, the marshes are subject to more hours of inundation, promoting low marsh habitat. In addition, given the extensively armored shoreline, there is little influx of mineral sediment from freshwater connections to facilitate accretion. Also, many of these marshes are adjacent to combined sewage outflow (CSO) pipes, which negatively impacts marsh health due to chronic exposition to high nutrient loads and other contaminants (Deegan et al., 2012).

The greatest threat to the marshes along Long Island Sound appears to be shoreline erosion. This type of salt marsh loss at the water's edge has been significant across NYC. The 25 wetland marsh complexes in the study have lost a total of 65 ha (or 15% of their total area) between 1974 and 2012. Staten Island marshes had the greatest total area of loss. However, Long Island Sound marshes had the greatest proportion of loss at 21%. The continued conversion of vegetated marsh to mudflat is likely a result of a combination of factors that include wave action, boat wake, increased inundation due to sea level rise, reduced root density and peat accumulation, predation by herbivorous crabs, and changes in soil chemistry and plant biology due to high nutrient loads. Similar to Jamaica Bay sites, Long Island Sound marshes are surrounded by extensive development, have large areas of fill and marine debris, and are in close proximity to CSO pipes. Another potential threat to Long Island Sound marshes is the crab species *Sesarma reticulatum*, observed at Pelham Bay Park, Bronx; however, their abundance and impact on marshes in NYC is unknown. Elsewhere (e.g., in the US Atlantic coastal marshes of Massachusetts and Rhode Island), extensive marsh vegetation loss has been attributed to *Sesarma* feeding on underground roots of *Spartina* grasses (Coverdale et al., 2012; Bertness et al., 2014).

4. STRATEGIES FOR ADDRESSING WETLAND THREATS

Our ability to influence the long-term viability of salt marshes in NYC depends on a number of factors. These include the following:

- The degree and extent of the threats to the salt marsh systems
- How well we understand and can influence the causes and processes of marsh degradation
- Our ability to create or adapt strategies based on their effectiveness
- Our ability to take timely action on a significant scale

Timely action is critical because shoreline salt marsh loss and threats to salt marsh sustainability, including that from sea level rise, affect all NYC marshes to some degree and are increasing in intensity.

To approach this problem, we identified and focused on three main threats to the long-term viability of salt marsh in NYC (Fig. 29.5):

- Lack of locations to migrate inland
- Inability of the marsh elevation to keep up with sea level rise
- Erosion of the marsh edge

Based on our best understanding of condition and the factors influencing it, we focused on a narrow set of actions and strategies to address threats. These actions and strategies do not necessarily address the causes or processes of degradation; however, they are ones NYC Parks has the capacity to undertake, will provide some level of protection, and will allow us to learn important lessons about what measures will be most effective and feasible at different sites over time.

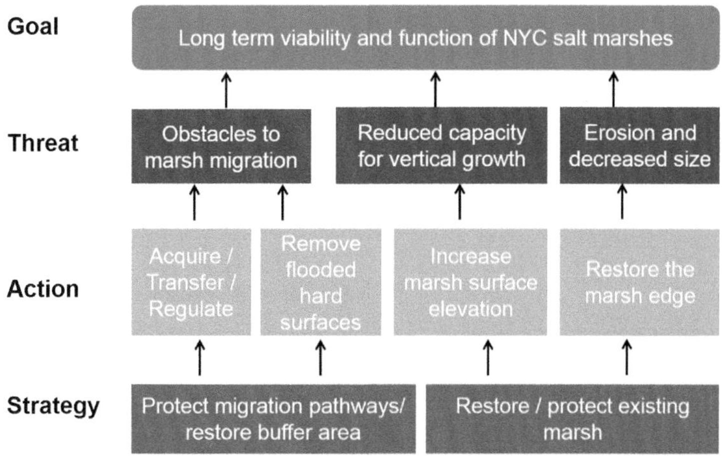

FIGURE 29.5 *Goal, threats, actions,* and *strategies* for salt marsh protection and restoration.

Our first overarching strategy is to protect pathways for landward migration of salt marsh in the future and restore salt marsh buffer or adjacent areas. This strategy is critical for the survival of salt marshes as sea levels rise and development pressure increases on areas adjacent to the marsh. Migration pathways can be protected through land transfer to NYC Parks, acquisition by NYC Parks, or, potentially, establishment of an easement that NYC Parks would manage and use in combination with existing regulations. Where possible, pathways should be ecologically restored and existing and future marsh protected by reclaiming hard surfaces such as paved trails and parking lots that will be regularly flooded in the future.

Our second broad strategy is to restore and protect existing marshes to reduce further degradation and shoreline erosion. There is a risk of marsh loss due to drowning and erosion where there is a predominance of low marsh with relatively low sediment supply, historic high rates of shoreline erosion, or interior marsh degradation.

Our restoration actions are twofold as they relate to this strategy:

1. Selective addition of sand to increase marsh surface elevation in subsided areas and, selectively, where salt pannes have expanded
2. Building out of the marsh edge with a sill or protective toe anchor, where needed, to regain and stem further loss of vegetation along the marsh edge

5. PRIORITIZATION OF RESTORATION AND PROTECTION STRATEGIES

To implement our strategies and actions at all salt marshes across the city would be extremely costly. It would also be risky because some of the restoration actions are still relatively untested and some actions are likely more effective and time sensitive at some sites than at others. Within our two broad strategies, to create and protect buffers around tidal wetlands and to restore marshes in situ, we developed an approach to help prioritize restoration or protection actions across sites. This approach required identifying metrics associated with the site that served as an indicator of the most appropriate action (Table 29.1).

TABLE 29.1 Metric Associated With a Site and Recommended Actions

Metric	Action
Area of land adjacent to marsh that SLAMM projected to be flooded by 2085	Acquire or transfer property to prevent development in potential marsh migration areas through property acquisition or regulations
Area of hard surface land adjacent to marsh that SLAMM projected to be flooded by 2085	Remove hard surfaces that will be flooded (to allow for migration)
Proportion of the marsh dominated by low marsh species and bare ground	Increase marsh surface elevation through sediment application
Proportion of the marsh and area of vegetated marsh loss since 1970s	Restore and protect the marsh edge

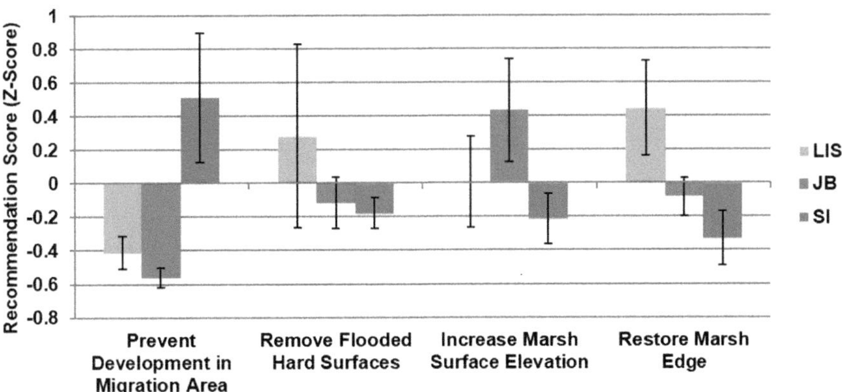

FIGURE 29.6 Z-scores for actions to protect or restore salt marshes, summarized by New York City marsh region. Regions include Long Island Sound (LIS), Jamaica Bay (JB), and Staten Island (SI).

Scores were applied to each metric for each action for each site and z-scores, centered around zero, were calculated to allow comparisons between sites for each action. Each site was then ranked according to its score. Scores were summarized by waterbody and region to determine if there was a greater need for specific actions in one area compared to another (Fig. 29.6). The results suggest that there is a greater need for specific actions in each area. Sites in Staten Island have the greatest need for protection of salt marsh migration pathways (with a z-score of 0.5). Sites in Jamaica Bay have the greatest need for marsh surface elevation increase (with a z-score just over 0.4). Sites in Long Island Sound have the greatest need for restoration of the lost salt marsh edge (with a z-score just over 0.4).

6. RECOMMENDED ACTIONS

Below we describe our recommendations, where and why they should be applied, and the constraints and limitations to implementation.

6.1 Strategy 1: Protect and Create Pathways for Migration

6.1.1 Action: Protect Land in Tidal Wetland Adjacent Area Through Transfer, Acquisition, Easements, and Regulation

By preventing low-lying vegetated uplands adjacent to salt marshes from being developed, salt marshes can potentially migrate landward. Protecting land adjacent to salt marsh from development does not guarantee salt marsh migration; however, it is a necessary first step. Management changes such as ceasing to mow lawns can allow an area to transition to salt marsh if it is subject to even irregular tidal inundation (Anisfeld et al., 2017). Effective protections for wetland migration pathways on land that is now or will become marsh would apply to (1) land owned by other city agencies to be transferred to NYC Parks, (2) privately owned land acquired by NYC Parks, or (3) future potential conservation easements on

TABLE 29.2 Hectares of Current and Additional Future Marsh and Number of Parcels by Ownership

Owner	Current Marsh (Hectares)	Future Marsh (Hectares)	Number of Parcels
NYC Parks	350	+83	228
Private	9	+12	80
Other government agencies	15	+10	18
Total	374	105	326

privately owned land. Simultaneously, NYC Parks can advocate for strict limitations on issuance of NYSDEC tidal wetland permits for fill activity within the 46 m (maximum) jurisdictional adjacent area to wetlands.

To understand where this type of action is most needed and potentially most effective, we identified individual properties that will likely contain marsh by 2085, based on SLAMM. The properties identified at each marsh site were classified by ownership type and prioritized based on the proportion of the parcel that is currently wetland or likely to be future wetland. Within a site, the adjacent parcels recommended for transfer, acquisition, or easements were those with the greatest existing wetland area and/or projected additional future flooded area. A parcel without buildings was designated for acquisition or transfer whenever the combined existing and projected wetland components of the parcel were determined to meet a 45% threshold. Parcels were identified for easement when buildings were present or when less than 45% of a parcel without buildings had existing and/or projected additional future flooded area.

NYC Parks owns most of the property in our study area that is likely to support future salt marsh, but 24 ha of existing wetland are under private and non-NYC Parks public ownership, as are 22 ha of projected future wetland (Table 29.2). Most of these parcels with future wetland also have a large proportion of nonwetland area, especially those adjacent to sites in Staten Island (Fig. 29.7).

6.1.2 *Action: Transfer Public Parcels With Wetlands to New York City Parks*

Properties owned by other city government agencies adjacent to salt marsh that either currently contain marsh or will likely support salt marsh in the future should ideally be transferred to NYC Parks ownership. Across our study area, 18 parcels owned by other agencies currently support 15 ha of existing marsh and are likely to support about 10 additional hectares of marsh in the future (Table 29.2). Some of the identified public parcels were already considered for transfer by the Wetlands Transfer Task Force, a temporary interagency task force formed as a result of 2005 legislation by the NYC Council (Local Law 83) to inventory city-owned properties containing wetlands and to determine the feasibility of their transfer to NYC Parks (NYC Parks, 2007). Because transferring all parcels that will support future salt marsh is not necessarily feasible, we must focus on ensuring maximum regulatory protection in adjacent wetland areas (see Regulation section below).

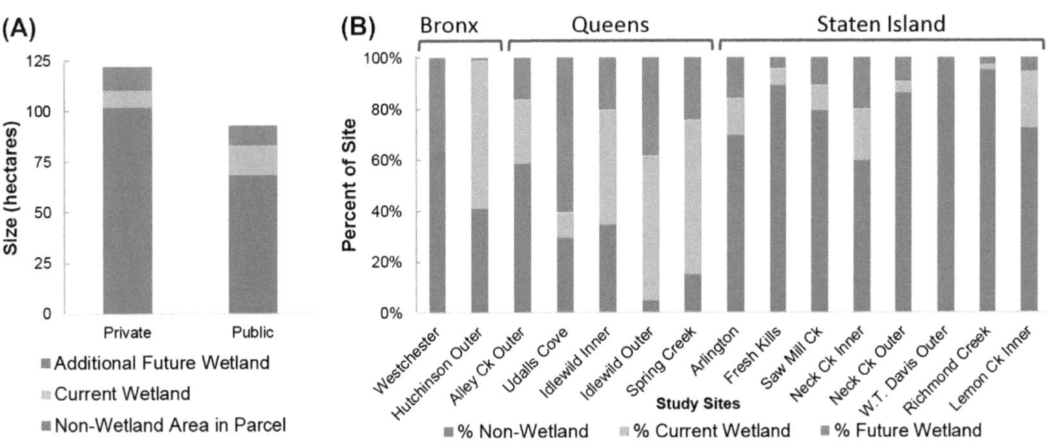

FIGURE 29.7 Current and future wetland and nonwetland area (upland including developed areas) in non–NYC Parks parcels. (A) Areas summarized by private and public land, (B) proportions of parcels summarized by study sites adjacent to parcels in the Bronx, Queens, and Staten Island; no parcels were identified in Brooklyn. Property ownership was determined using information from the New York City Department of Information Technology and Telecommunications (NYC DOITT, 2017) and New York City Open Access Space Information System (OASIS NYC, 2017).

While the transfer cost does not include the cost of the land (the property is already under City ownership), it does include the required cost of a Phase I environmental assessment (e.g., including a site inspection and historic review to assess likelihood of contamination, and if necessary, Phase II site sampling to determine the degree of contamination), remediation or restoration of the site, and installing fencing. We estimated these costs as ranging between $160,000 and $4.9 million per ha depending on the site condition.

6.1.3 Action: Acquire and Establish Conservation Easements

There are a total of 44 parcels of land citywide identified as candidates for acquisition or easement. The parcels should be either purchased outright by the city or secured as conservation easements through purchase or donation. In each case, NYC Parks would manage the land to protect current and future marsh. Conservation easements potentially allow protection from development on a subsection of property while allowing for continued private ownership and management, e.g., on parcels with existing buildings that are not appropriate for NYC Parks to own or manage.

The cost of acquisition was estimated based on the market price of property in NYC and assumed to be $17.3 million per ha on the high end and $3.2 million per ha on the low end. We assumed that easements on private property could cost as much as acquisition, depending on the arrangement with the property owner.

6.1.4 Action: Comply with Wetland Adjacent Area Regulations

In New York State, Article 25 of the Environmental Conservation Law 6 NYCRR Part 661 (Tidal Wetland Land Use Regulation) authorizes State jurisdiction of an adjacent area landward of the tidal wetland boundary (NYSDEC, 2017). In NYC, this area extends 46 m, or to the 3-m elevation contour, or to roads, or to other aboveground structures, whichever distance is shortest. These regulations are implemented by NYSDEC and pertain to activities that could substantially impair or alter the natural condition of the jurisdictional wetland adjacent area, including soil removal, fill activity, and construction of new roads or other structures. Most of the projected future wetland areas identified in our study fall within the 46-m maximum extent of the regulated wetland adjacent area and therefore any development activity proposed for these areas will require a tidal wetland permit from the state agency. However, it is often difficult to track and regulate actions in the wetland adjacent area across the highly developed (and still developing) landscape of NYC; development in these buffer areas continues today through both permitted and illegal activities, though less frequently than in the past.

All of the property we have identified as priorities for acquisition or easements (privately owned property), or transfer (publicly owned property), include wetland adjacent area under State jurisdiction, subjected to wetland protection regulations. NYC Parks will work with NYSDEC at the state level to strengthen protection efforts in these areas, particularly where there may be existing violations and evidence of encroachment or when infrastructure construction projects encroach on wetland adjacent areas.

6.1.5 Action: Reclaim Future Flooded Hard Surfaces That Prevent Migration

Hard surfaces that are likely to be regularly flooded in the future should be removed and restored to native plant communities wherever possible. This action will facilitate salt marsh migration and the long-term viability of the salt marsh. However, in NYC there is very little hard surface adjacent to existing marsh that is not associated with an actively used road or parking lot.

Hard surfaces defined as pavement include roads, parking lots, asphalt pathways, and turf sports fields. Parcels with hard surfaces that cannot be removed include those with buildings or other structures or railroad tracks. Removing hard surfaces entails first breaking up and overexcavating pavement, concrete, and fill, then adding clean sand (or other planting medium) to the desired grade, and planting appropriate native species.

To identify those paved surfaces that can be removed, the snow plow priority status, as listed by NYC DOITT (2017), was used to estimate the level of activity associated with those surfaces. Roads listed as primary and secondary priority for snow plowing were assumed to have high activity or use and thus would be more difficult to remove. Roads listed as tertiary priority were assumed to have low levels of use. Roads without snow plow priority designation were assumed to have the same status as the closest adjacent existing status. Within a site, the hard surfaces we recommend for removal, in the short or long term, are those on NYC Parks property that have low activity or use (tertiary snow plow priority status or those without priority that are adjacent to areas with tertiary priority status). Larger areas of roads and parking lots with higher levels of activity were not included.

Across all of the study sites, the greatest amount of future flooding is projected to occur on parking lots, followed by roads and other hard surfaces (Table 29.3). By far the largest parking lot area to be flooded in the future is the 10-ha lot at Orchard Beach in Pelham Bay Park, the Bronx (Fig. 29.8). This lot is used heavily during the summer months when the beach is open and used as an emergency staging area at other times. To address the flooding problems that will be associated with this parking lot and to protect the surrounding water resources,

TABLE 29.3 Summary of Hard Surface Removal Opportunities Adjacent to All 25 Study Sites

Hard Surface Type	Number of Marsh Sites	Number of Locations of Flooded Hard Surfaces	Area of Flooded Hard Surface (Hectares)
Parking lots	4	7	11.9
Roads	17	47	3.0
Other hard surfaces	7	21	2.4
Total	18	75	17.3

we recommend the consideration of sustainable ways to redesign this paved area, including long-term planning for some expansion of salt marsh habitat.

In total, 3 ha of road adjacent to existing marsh (Table 29.3) are likely to be flooded in the future. These roads include residential streets and heavily used traffic arteries, such as Brookville Boulevard in Queens, NY, and Travis Avenue in Staten Island, NY. Such critical roads cannot feasibly be removed for salt marsh migration in the near term, but long-term planning and design at these sites needs to consider approaches that will protect and sustain the salt marsh. NYC is currently designing projects to raise roads in low-lying areas that are increasingly flooded by tides, such as that portion of Travis Avenue in Staten Island that divides the NYC Parks William T. Davis Wildlife Refuge. Elevating these roads on piers, for example, could maintain the long-term function of the marsh and viability of the road. Although these measures are extremely costly, they provide the opportunity to design and construct for long-term sea level rise. Where this is not feasible, any road raising or flood protection should be sensitive to the potential impacts on adjacent wetlands.

The issue of regular flooding of developed land in the coming decades will need to be addressed and all of the locations identified here should be considered as part of the strategy to adapt to sea level rise.

We assumed the cost of removing hard surfaces, including concrete or pavement removal, excavation of fill, placement of clean sand, planting, and all associated construction costs, ranges from $1.2 to $2.5 million per ha.

6.2 Strategy 2: Protect and Restore Existing Marsh

6.2.1 Action: Apply a Thin Layer of Sediment to Elevate Marshes

To help protect and maintain the function of existing wetlands, we recommend pursuing the addition of sediment to low elevation areas in marshes in target areas. Ideally, there would be sufficient sediment supply in the streams, bays, and estuaries deposited on the marshes to accrete, or vertically build up, at the same rate as sea level rises. However, NYC is a sediment-starved system, especially in Jamaica Bay (Gateway National Recreation Area, 2007; NYCDEP, 2007). Applying clean silt and sand to a bare or partially bare marsh surface is a way to increase the surface elevation of the marsh so that plants can colonize at a higher elevation to keep from drowning (Frame et al., 2006; Ray, 2007).

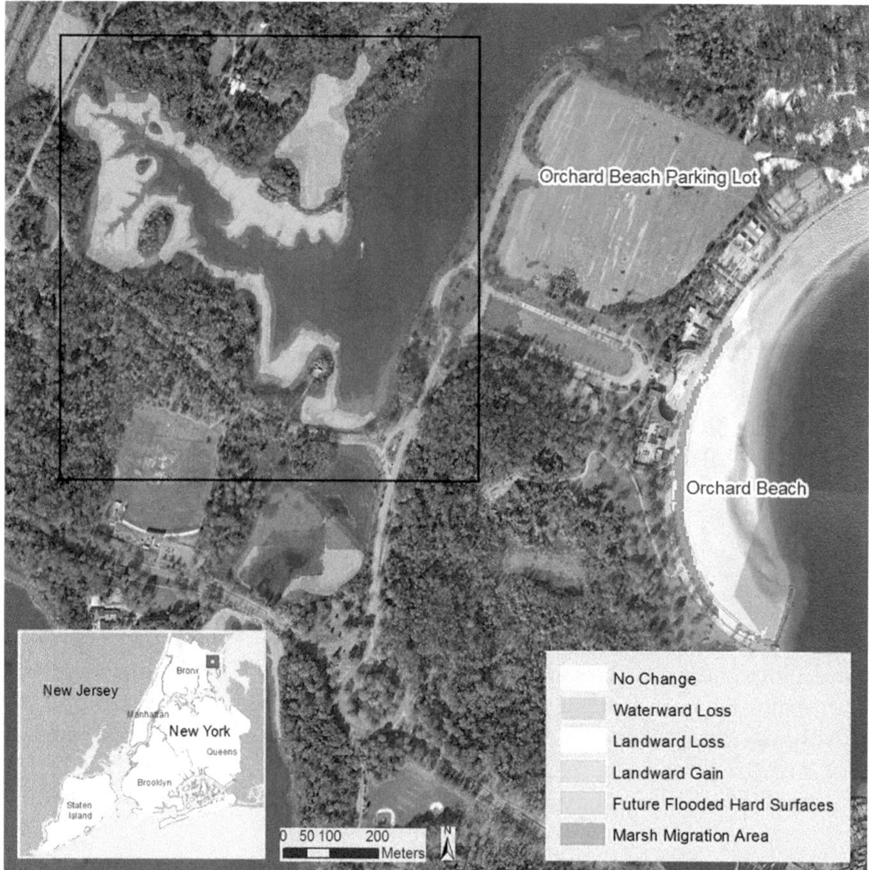

FIGURE 29.8 Aerial image showing the area of current marshes (light greens), future marsh migration areas (dark green) and future flooded hard surfaces including at the Orchard Beach Parking Lot (blue) by 2085 (Clough et al. 2014) in Pelham Bay Park, Bronx, NY. The black frame is the area shown in Fig. 29.2.

Over a quarter of our large marsh systems consist of low marsh. This indicates a great need and opportunity to increase elevation to sustain these ecosystems (Table 29.4 and Fig. 29.9). Marshes where this action should be implemented were prioritized based on the percentage of *Spartina alterniflora* cover and bare ground cover data (used as an indicator of deteriorating marsh), the breeding presence of any of several species of saltmarsh sparrow (these sparrows are especially vulnerable to loss of high marsh habitat and their nesting sites are a priority), future loss due to sea level rise (based on SLAMM results), and feasibility for sediment application (proximity to land or water access points).

Thin sediment layer application has not been part of NYC Parks past practices. However, there are several local precedents; this technique has been used since the early 2000s in NYC on federal lands—at the Jamaica Bay unit of Gateway National Recreation Area, National Park Service (NPS). In 2003, dredged silt and sand from the adjacent channel was used to reconstruct an eroded section of the unit's Big Egg Marsh in Jamaica Bay. From 2007 to

TABLE 29.4 Summary of Low Marsh Area Across the 25 Study Sites. Includes the Total Potential Area of Where Elevation Could Potentially be Increased

Location	Number of Study Marshes	Total Study Marsh Area (Hectares)	Low Marsh in Study Marshes	
			Total Area (%)	Total Area (Hectares)
Long Island Sound	9	94.2	21	19.6
Jamaica Bay	5	103.1	61	62.5
Arthur Kill, Kill van Kull, Raritan Bay	11	214.4	14	29.6
Citywide	25	411.7	27	111.7

FIGURE 29.9 Area of low elevation and interior marsh loss with pooling and hummock formation at Idlewild Park Preserve, Outer, Queens, NY, June 3, 2014. Funding for future pilot project at sites to be provided by New York Department of State's Local Waterfront Revitalization Program.

2012, the USACE, NYSDEC, NYCDEP, and NPS placed clean sand across 63 ha of mudflat to expand or rebuild five of the Jamaica Bay unit's marsh islands (Messaros et al., 2012). The USACE used various planting and seeding techniques to reestablish salt marsh vegetation after the sand was placed.

In 2017, NYC Parks was awarded a state grant to pilot techniques for placing thin layers of sediment in a degrading marsh to build up elevation. The project will take place at Idlewild Park Preserve in Queens, NY (Fig. 29.9), which lies to the north of John F. Kennedy

International Airport. This pilot project will include long-term monitoring and help us evaluate the effectiveness and longevity of this kind of intervention. The pilot project will cover a little over 1.34 ha at an approximate cost of $2.96 million per ha. Construction costs are extremely high in NYC and for this project include mobilization, sand, sand placement, planting, installation of herbivory fencing, erosion control measures, and basic monitoring.

6.2.2 Action: Restore Salt Marsh Where Prolonged Ponding Has Expanded or Created Pools

NYC Parks conducted a pilot project to restore recently expanded areas of vegetation loss, here described as pools, in the interior of the Alley Creek salt marsh with sediment placement and replanting (Fig. 29.10). These pools had increased by 40% in size since 1974, resulting in fragmentation and loss of contiguous marsh. Without this intervention, the Alley Creek salt marsh would become increasingly limited in function as existing pools enlarged and new pools formed. This effort piloted an approach that could become an important management alternative for addressing the ongoing loss of habitat from marsh pool expansion that has been documented in NYC over the last 40 years.

In April 2017, NYC Parks staff placed coir logs along the 1974 boundary of the largest pools in Alley Creek marsh. The 1974 boundary was chosen to acknowledge the value of pools for shorebird foraging habitat, while mitigating further conversion of salt marsh habitat to open water or mudflat. The coir logs were staked and bound in place to stabilize newly applied sediment. Once sediment had settled and elevations were verified, volunteers helped plant *Spartina alterniflora* at the low marsh elevations and *Spartina patens* and *Distichlis spicata* at the higher elevations. Materials were transported on-site using marsh mats consisting of plywood or engineered plastic boards to protect vegetation and minimize soil compaction (Fig. 29.10).

6.2.3 Action: Restore Eroded Marsh Edge

In addition to thin layer interior sediment application, we recommend restoring former marsh edge wherever possible. Building out former marsh may include clean sand placement

FIGURE 29.10 Salt marsh restoration at Alley Creek in Queens, NY, through placement of clean sediment where open water persists (A) at the start by setting coir log along the flooded former marsh, (B) using marsh mats to access the marsh with machinery, and (C) with staff and volunteers planting *Spartina alterniflora*.

FIGURE 29.11 Shoreline edge exhibiting terracing and slumping, or marsh bank erosion, indicative of vegetation loss and the creation of bare peat in formerly vegetated marsh in Udalls Cove Park Preserve, Queens, NY, May 8, 2014.

by itself, like at the Jamaica Bay marsh islands, up to the elevation of the former marsh. More often, however, it will include installing some type of sill, or breakwater, and the placement of clean sand behind that structure. NYC marshes have lost a total of 65 ha (15% of total area) from 1974 to 2012, indicating a great need and opportunity for marsh edge restoration (Figs. 29.2 and 29.11 and Table 29.5). Sites in the Bronx and northern Queens and Brooklyn on Long Island Sound, and southern Queens and Brooklyn in Jamaica Bay, experienced the highest proportion of loss and thus have the greatest need for restoration. The waterward boundary of intertidal (low) marsh as mapped in 1974 provides a baseline for these types of projects.

TABLE 29.5 Summary of Marsh Edge Restoration Opportunities by Borough Across the 25 Study Sites

Study Site Location	Number of Study Sites	Study Site (Hectares)	Salt Marsh Loss Since 1974 (Hectares)	Salt Marsh Loss Since 1974 (%)	Average Width of Salt Marsh Loss Since 1974 (Meters)
Bronx	6	64	14	19	15
Queens	6	116	21	17	20
Brooklyn	2	17	3	17	11
Staten Island	11	214	27	12	8
Citywide	25	411	160	15	54

In addition to the local precedent for sediment placement to build tidal wetlands vertically and horizontally established in Jamaica Bay, this restoration approach is documented elsewhere including, for example, in the United States in Louisiana (CWPPRA, 2017; Reed and Wilson, 2004), in Chesapeake Bay, Maryland, at the Paul S. Sarbanes Ecosystem Restoration Project (also known as Poplar Island) (Stinchcomb et al., 2014; USACE, 2017), in San Francisco Bay (Williams and Faber, 2001), and internationally (Coops and van Geest, 2007; Veenklass et al., 2015).

Data on the area, proportion, and width of marsh loss along the water's edge between 1974 and 2012 were used to identify sites and locations for projects. Sites that experienced the greatest area, proportion, and width of salt marsh loss are assumed to be in the greatest need of restoration. Within sites that have experienced loss, those with lower condition and moderate to higher vulnerability were prioritized for marsh edge restoration. Additional priority criteria include sites with significant marsh loss that are at least 0.4 ha in area and at least 30 m wide, which have lower levels of exposure to wind/wave action and which can be accessed by land or water for sediment application.

Restoring and expanding marsh edge acts both to protect existing marsh and to improve ecosystem condition by creating habitat for marine life such as oysters and ribbed mussels.

Shoreline protection techniques piloted previously include the securing and planting of fibrous logs into the shoreline (2013), the reintroduction and expansion of live oysters (2013), and the installation of gabion baskets to accumulate sediment and to create a growth medium for plants and ribbed mussels (1997), all at Freshkills Park, Staten Island, and the installation of riprap breakwaters at the Bronx Kill Salt Marsh in Randall's Island Park and at the Pier One Salt Marsh in Brooklyn Bridge Park.

NYC Parks estimates that the restoration of eroded marsh edge costs is similar to current average restoration costs in NYC (approximately $2.9 M per ha), including the cost of clean sediment placement, planting, installation of herbivory fencing, erosion control measures, and applicable shoreline protection measures (e.g., riprap breakwaters or sills, armored toe, or coir logs).

6.3 Strategy 3: Other Ongoing Restoration Opportunities, Actions, and Recommendations

Although this study focused on future restoration needs and piloting new interventions at our largest marsh complexes, NYC currently conducts restoration work throughout our wetland properties using similar decision-making approaches. We map opportunities for restoration in the field and use spatial and tabular data to maintain an updated inventory of restoration needs across NYC; this is referred to as the Restoration Opportunities Inventory (Natural Areas Conservancy, 2016).

6.3.1 Action: Remove Debris and Trash

Floatable trash and heavy marine debris collects in our coastal wetlands with the ebb and flow of the tide. This material, along with illegally dumped domestic or industrial garbage, can accumulate in large areas, smother marsh vegetation, and compact soil. NYC Parks employs contractors to remove large objects such as boats or cars (Fig. 29.12). NYC Parks is

FIGURE 29.12 Contractors remove large marine debris from a salt marsh on Long Island Sound in Pelham Bay Park, Bronx, NY, August 4, 2015.

FIGURE 29.13 Example of tire dumping on salt marsh in Idlewild Park Preserve, Queens, NY, September 15, 2014.

currently implementing a project funded by National Oceanic and Atmospheric Administration (NOAA) and the Federal Emergency Management Agency (FEMA) to remove large marine debris from salt marshes across NYC deposited during Hurricane Sandy. We also hold cleanup events with volunteer groups across the city year-round to remove smaller floatable debris, residential garbage, tires, and more (Fig. 29.13). The Restoration Opportunities

Inventory contains information regarding debris removal opportunities, which is used to identify projects, estimate resources, and pursue funding for project implementation.

6.3.2 Action: Excavate Historic Fill

We also restore marshes by removing historical landfill from marsh habitat. Many marshes across NYC were filled during the construction of bridges, roadways, and housing developments in past decades. In some locations, this fill can be excavated to elevations where the tidal hydrologic regime can be reestablished to support salt marsh vegetation. Sometimes the substrate exposed by excavation is historical marsh peat, but sometimes it is still more contaminated fill or material unsuitable for planting, in which case the material is overexcavated and backfilled with clean sand to the desired elevation for sustaining a salt marsh (see Box 29.1).

6.3.3 Action: Remove Tidal Barriers

Removal of tidal barriers, such as dikes, berms, and underdesigned culverts, can allow impounded areas to revert back to salt marsh by the reintroduction of tidal inundation behind the barrier (NOAA, 2010; Roman and Burdick, 2012). In Europe this is referred to as coastal or managed realignment (Coops and van Geest, 2007; Veenklass et al., 2015). In Staten Island's Saw Mill Creek Preserve, a 0.4 ha berm removal allowed tidal inundation to return to formerly connected salt marsh, resulting in a 3 ha restoration. In Pelham Bay Park, Bronx, a pedestrian road built on a dike was replaced by a bridge walkway, restoring an 8 ha salt marsh; however, few feasible opportunities for barrier removal remain in NYC.

7. SUMMARY AND NEXT STEPS

Restoring and protecting the remaining salt marsh in NYC is essential for making a more resilient and livable city in the face of climate change. Salt marsh habitat specifically provides refuge for wildlife, captures and stores carbon, reduces the impact of continuous wave action, and improves water quality. It also provides essential open space for the highly urbanized NYC landscape and an important opportunity for engagement with the natural world through recreation, education, and research. An NYC tidal wetlands conservation plan is critical for the long-term vitality of our salt marshes. The strategies and recommendations in this report are building blocks for this plan.

For any given marsh site, it is essential to understand the existing and historical conditions, the near- and long-term threats, and, to the fullest extent possible, the degree to which the factors or causes of degradation can be mitigated. For any given intervention, all relevant environmental feasibility, cost, and community support parameters must be considered to determine the appropriate protection or restoration strategy. We believe that in the face of threats associated with sea level rise, particularly in the NYC region, the protection of pathways for migration and the restoration of existing marshes in place, through pilot projects and new techniques, are critical. We intend for our recommendations to contribute to a future NYC tidal wetlands management plan that helps to ensure that future generations will experience the ecological and cultural heritage of the once vast salt marsh systems of NYC.

BOX 29.1

EXAMPLE RESTORATION PROJECT INVOLVING HISTORIC FILL EXCAVATION

The Pugsley Creek restoration project at Castle Hill Park, Bronx, completed in 2012, was funded through a New York State Clean Water/Clean Air Bond Act Grant and the City of New York. Habitat improvements included 0.9 ha of salt marsh, scrub shrub, and meadow habitat in a highly urbanized neighborhood along a historic waterfront. The project was designed to expand breeding and foraging habitat for birds and other wildlife, to help improve Long Island Sound water quality by expanding intertidal vegetation that can remove nutrients and trap suspended sediment from the water column, and to offer park visitors a walking trail along the salt marsh.

To construct the salt marsh, contractors excavated historic fill material up to 1 m deep, stockpiled the excavated fill material for off-site removal, and then backfilled with clean sand to design grades (Fig. 29.14). Stockpiled overburden was tested for contaminants and sent to landfills that could handle the material. The on-site excavation activity was largely conducted above the midtide line, leaving the existing gravel–cobble shore and remnant low marsh with tall *Spartina alterniflora* largely undisturbed.

To compensate for projected sea level rise in the absence of existing high marsh at the project site, the highest design elevation grades were determined from the analysis of low and high marsh zones in neighboring marshes within Castle Hill Park. The lower design elevation grades were informed by the upper limit of the remnant low marsh. These remnant shoreline and neighboring marshes continued to be used as reference sites during the postconstruction monitoring period.

Monitoring for vegetation, benthic invertebrates, and other parameters was initiated preconstruction and conducted annually for 5 years postconstruction along permanent transects in both the reference site and the project site, as recommended under the New York State Salt Marsh Restoration and Monitoring Guidelines (Niedowski, 2000). Adaptive management techniques have been applied postconstruction including the reinstallation of temporary exclusion fencing to protect the establishing vegetation from herbivory by Canada geese.

FIGURE 29.14 Salt marsh restoration at Pugsley Creek in the Bronx, NY. Images show the site (A) before, (B) during, and (C) after restoration.

Acknowledgments

This report could not have been completed without the work of numerous staff and partners, most importantly, Leah Beckett, formerly of NYC Parks, Nichole Maher, Stephen Lloyd and Lauren Alleman of The Nature Conservancy, and Helen Forgione of the Natural Areas Conservancy (NAC). The NAC, a nonprofit that works closely with NYC Parks on the conservation of our natural areas, provided NYC Parks staff with technical advice and expertise. Valuable assistance was provided by Maria Amin, Katie Conrad, Kevin Eng, Minona Heaviland, and Ryan Morrison. Additional valuable guidance was provided by Jennifer Greenfeld and Bram Gunther. This report was made possible through funding from the USEPA, Region 2, Wetland Program Development Grant managed by Kathleen Drake.

References

Anisfeld, S.C., Cooper, K.R., Kemp, A.C., 2017. Upslope development of a tidal marsh as a function of upland land use. Global Change Biology 23 (2), 755–766.

Bertness, M.D., Brisson, C.P., Bevil, M.C., Crotty, S.M., 2014. Herbivory drives the spread of salt marsh die-off. PLoS One 9. http://journals.plos.org/plosone/article?id=10.1371/journal.pone.0092916.

Boesch, D.F., 2006. Scientific requirements for ecosystem-based management in the restoration of Chesapeake bay and coastal Louisiana. Ecological Engineering 26, 6–26.

Clough, J., Propato, M., Polaczyk, A., 2014. Application of Sea-Level Affecting Marshes Model (SLAMM) to Long Island, NY, and New York City. Final Report. New York State Energy Research and Development Authority (NYSERDA), Albany, New York. http://www.warrenpinnacle.com/prof/SLAMM/NYSERDA/.

Coops, H., van Geest, G., 2007. Ecological Restoration of Wetlands in Europe: Significance for Implementing the Water Framework Directive in the Netherlands. Report Prepared for: Rijkswaterstaat RIZA. https://www.researchgate.net/profile/Hugo_Coops/publication/235005665_Ecological_restoration_of_wetlands_in_Europe_significance_for_implementing_the_Water_Framework_Directive_in_the_Netherlands/links/0fcfd510442487b115000000/Ecological-restoration-of-wetlands-in-Europe-significance-for-implementing-the-Water-Framework-Directive-in-the-Netherlands.pdf.

Coverdale, T.C., Altieri, A.H., Bertness, M.D., 2012. Belowground herbivory increases vulnerability of New England salt marshes to die-off. Ecology 93, 2085–2094.

Coastal Wetlands Planning, Protection and Restoration Act (CWPPRA), 2017. Restoring Coastal Louisiana Since 1990. https://www.lacoast.gov/new/About/.

Deegan, L.A., Johnson, D.S., Warren, R.S., Peterson, B.J., Fleeger, J.W., Fagherazzi, S., Wollheim, W.M., 2012. Coastal eutrophication as a driver of salt marsh loss. Nature 490, 388–392.

Frame, G.W., Mellander, M.K., Adamo, D.A., 2006. Big Egg marsh experimental restoration in Jamaica bay, New York. In: Harmon, D. (Ed.), People, Places and Parks: Proceedings of the 2005 George Wright Society Conference on Parks, Protected Areas, and Cultural Sites. The George Wright Society, Hancock, Michigan, pp. 123–130. http://www.georgewright.org/0520frame.pdf.

Gateway National Recreation Area, National Park Service, US Department of the Interior, 2007. An Update on the Disappearing Salt Marshes of Jamaica Bay. Jamaica Bay Watershed Protection Plan Advisory Committee, New York. http://www.nytimes.com/packages/pdf/nyregion/city_room/20070802_FinalJamaicaBayReport.pdf.

Hartig, E.K., Gornitz, V., Kolker, A., Mushacke, F., Fallon, D., 2002. Anthropogenic and climate-change impacts on salt marshes of Jamaica Bay, New York City. Wetlands 22 (1), 71–89.

Horton, R., Little, C., Gornitz, V., Bader, D., Oppenheimer, M., 2015. New York city Panel on climate change 2015 report chapter 2: sea Level rise and coastal storms. Annals of the New York Academy of Sciences 1336, 36–44. http://onlinelibrary.wiley.com/doi/10.1111/nyas.12593/full.

Kemp, A.C., Hill, T.D., Vane, C.H., Cahill, N., Orton, P.M., Talke, S.A., Parnell, A.C., Sanborn, K., Hartig, E.K., 2017. Relative sea-level trends in New York City during the past 1500 years. The Holocene 1–18. https://www.researchgate.net/publication/312181480.

Messaros, R.C., Woolley, G.S., Morgan, M.J., Rafferty, P.S., 2012. Chapter 8. Tidal wetlands restoration. In: Ali, M. (Ed.), The Functioning of Ecosystems. InTech, Europe, pp. 149–170. https://www.intechopen.com/books/the-functioning-of-ecosystems/tidal-wetlands-restoration.

Natural Areas Conservancy (NAC), 2016. Inventory of Coastal Wetland Restoration Opportunities in NYC. http://naturalareasnyc.org/content/3-in-print/2-research/roi-project-summary-august-2016_final.pdf.

New York City Department of Environmental Protection (NYCDEP), 2007. Jamaica Bay Watershed Protection Plan. http://home2.nyc.gov/html/dep/html/dep_projects/jamaica_bay.shtml.

New York City Department of Information Technology and Telecommunications (NYC DOITT), 2017. Citywide Street Centerline. https://data.cityofnewyork.us/City-Government/NYC-Street-Centerline-CSCL-/exjm-f27b.

New York City Department of Parks & Recreation (NYC Parks), 2007. Recommendations for the Transfer of City-Owned Properties Containing Wetlands. Interagency Wetlands Transfer Task Force. https://www.nycgovparks.org/greening/natural-resources-group/wetlands-transfer-task-force.

New York City, Office of Long-term Planning & Sustainability (NYC OLTPS), 2012. New York City Wetlands Strategy. http://www.nyc.gov/html/planyc2030/downloads/pdf/nyc_wetlands_strategy.pdf.

New York City, Office of Recovery & Resiliency (NYC ORR), 2015. One New York (OneNYC), The Plan for a Strong and Just City. http://www.nyc.gov/html/onenyc/downloads/pdf/publications/OneNYC.pdf.

New York State Department of Environmental Conservation (NYSDEC), 2017. Article 25, Title 6, Codes, Rules, and Regulations of New York State (6 NYCRR) Part 661 of the Environmental Conservation Law (ECL) Entitled. Tidal Wetlands–Land Use Regulations. https://govt.westlaw.com/nycrr/Browse/Home/NewYork/NewYorkCodesRulesandRegulations?guid=I031f2dc0b5a111dda0a4e17826ebc834&originationContext=documenttoc&transitionType=Default&contextData=(sc.Default).

Niedowski, N.L., 2000. New York State Salt Marsh Restoration and Monitoring Guidelines, Prepared for New York State Department of State and New York State Department of Environmental Conservation (NYSDEC), Albany, New York. http://www.dec.ny.gov/docs/wildlife_pdf/saltmarsh.pdf.

National Oceanic and Atmospheric Administration (NOAA), 2010. In: Craig, L., McCraken, K., Schnabolk, H., Ward, B. (Eds.), Returning the Tide: A Tidal Hydrology Restoration Guidance Manual for the Southeastern United States. NOAA Coastal Services Center and NOAA Restoration Center. http://www.habitat.noaa.gov/toolkits/tidal_hydro/download_all_manual_chapters.pdf.

Open Access Space Information System New York City (OASIS NYC), 2017. Interactive Community Maps for NYC. Center for Urban Research, The Graduate Center/City University of New York (CUNY). http://www.oasisnyc.net/.

O'Neil-Dunne, J.P.M., MacFaden, S.W., Forgione, H.M., Lu, J.W.T., 2014. Urban Ecological Land-cover Mapping for New York City. Final Report to the Natural Areas Conservancy. Spatial Informatics Group, University of Vermont, Natural Areas Conservancy, and New York City Department of Parks & Recreation. http://naturalareasnyc.org/content/3-in-print/2-research/urbanecologicalmap_newyorkcity_report_2014.pdf.

Partnership for the Delaware Estuary (PDE), 2012. Development and Implementation of an Integrated Monitoring and Assessment Program for Tidal Wetlands. PDE Report No. 12-03. https://s3.amazonaws.com/delawareestuary/sites/default/files/Quirk%20PDE%20Jan%202013.pdf.

Partnership for the Delaware Estuary (PDE), 2014. Site Specific Intensive Monitoring of Coastal Wetlands in Dividing Creek New Jersey Watershed, 2012-2013. PDE Report No. 14-04. https://s3.amazonaws.com/delawareestuary/pdf/PDE-Report-14-04_SiteSpecificIntensiveMonitoringofCoastalWetlandsinDividingCreekNewJerseyWatershed20122013.pdf.

Ray, G.L., 2007. Thin Layer Disposal of Dredged Material on Marshes: A Review of the Technical and Scientific Literature. ERDC/EL Technical Notes Collection (ERDC/EL TN-07–1). U.S. Army Engineer Research and Development Center, Vicksburg, MS.

Reed, D.J., Wilson, L., 2004. Coast 2050: a new approach to restoration of Louisiana coastal wetlands. Physical Geography 25 (1), 4–21. http://labs.uno.edu/restoration/Reed%20and%20Wilson%202050.pdf.

Rogerson, A., McLaughlin, E., Havens, K., 2010. Mid-Atlantic Tidal Wetland Rapid Assessment Method (MidTRAM) Version 3.0. Delaware Department of Natural Resources and Environmental Control. http://www.dnrec.delaware.gov/Admin/DelawareWetlands/Documents/Tidal%20Rapid_Protocol%203.0%20Jun10.pdf.

Roman, C.T., Burdick, D.M. (Eds.), 2012. Tidal Marsh Restoration: A Synthesis of Science and Management. Island Press, Washington.

Sanderson, E.W., Solecki, W.D., Waldman, J.R., Parris, A.S. (Eds.), 2016. Prospects for Resilience, Insights from New York City's Jamaica Bay. Island Press, Washington.

Sobel, A., 2014. Storm Surge: Hurricane Sandy, Our Changing Climate, and Extreme Weather of the Past and Future. Harper Publishers, New York.

Stinchcomb, C., Sharrett, R., Army Corps of Engineers, North Atlantic Division Baltimore (CENAB), The Maryland Environmental Service (MES), 2014. Poplar Island Reaching Capacity; Expansion Authorized by WRRD. IDR International Dredging Review; News and Information for the Worldwide Dredging Industry. http://www.dredgemag.com/January-February-2015/P-oplar-Island-Reaching-Capacity-Expansion-Authorized-by-WRRDA-2014/.

Swanson, R.L., Wilson, R.E., 2008. Increased tidal ranges coinciding with Jamaica Bay development contribute to marsh flooding. Journal of Coastal Research 24 (6), 1565–1569.

Teal, J., Teal, M., 1969. Life and Death of the Salt Marsh. Little, Brown and Company, Boston.

Tiner, R.W., 2000. Wetlands of Staten Island, New York, Valuable and Vanishing Urban Wetlands. US Fish and Wildlife Service, Cooperative National Wetlands Inventory Publication, Hadley, Massachusetts. https://www.fws.gov/wetlands/Documents/Wetlands-of-Staten-Island-New-York-Valuable-Vanishing-Urban-Wildlands.pdf.

Trust for Public Land and the New York City Audubon Society, 1987. Buffer the Bay, a Survey of Jamaica Bay's Unprotected Open Shoreline and Uplands. Trust for Public Land, New York.

Trust for Public Land and the New York City Audubon Society, 1993. Buffer the Bay Revisited, an Updated Report on Jamaica Bay's Open Shoreline and Uplands. Capital Cities/ABC, New York.

U.S. Fish and Wildlife Service (USFWS), 1997. Significant Habitats and Habitat Complexes of the New York Bight Watershed. Coastal Ecosystem Program. https://nctc.fws.gov/pubs5/begin.htm.

U.S. Army Corps of Engineers (USACE), 2016. Hudson-Raritan Estuary, Comprehensive Restoration Plan, vol. 1. Version 1.0. http://www.harborestuary.org/watersweshare/pdfs/CRP/FinalReport-0616.pdf.

U.S. Army Corps of Engineers (USACE), 2017. Fact Sheet: Paul S. Sarbanes Ecosystem Restoration Project at Poplar Island, Talbot County, MD. http://cdm16021.contentdm.oclc.org/cdm/ref/collection/p16021coll11/id/521.

Veenklass, R.M., Koppenaal, E.C., Bakker, J.P., Esselink, P., 2015. Salinization during salt-marsh restoration after managed realignment. Journal of Coastal Conservation 19 (4), 405–415. https://link.springer.com/article/10.1007/s11852-015-0390-z.

Williams, P.B., Faber, P.M., 2001. Salt marsh restoration experience in san Francisco bay. Journal of Coastal Research 27, 203–311. Special Issue. http://www.tidalmarshmonitoring.org/pdf-project/Williams2001_SaltMarshRestorationSanFranBay.pdf.

CHAPTER

30

Living Shorelines for Coastal Resilience

Carolyn A. Currin

NOAA, National Centers for Coastal Ocean Science, Beaufort Lab, Beaufort, NC, United States

1. INTRODUCTION TO LIVING SHORELINES

Shorelines are ever-changing features of coastal ecosystems, responding to wave energy, storm events, and changes in sea level and sediment supply. Despite the impermanent nature of shorelines, humans are drawn to live near them for access to transportation corridors and the abundant natural resources available in coastal ecosystems. An estimated one-third of the human population lives within 100 km of a coastline, and this proportion is expected to increase (Gittman et al., 2016b; MEA, 2005). Coastal communities have historically combated erosion by hardening the shoreline, including the construction of seawalls and breakwaters (Dugan et al., 2011). As the human population along the coastline continues to increase and as sea level rise accelerates, there will be an increasing demand for erosion protection by waterfront property owners and communities (Titus et al., 2009; Arkema et al., 2013; Temmerman et al., 2013).

Sheltered coasts occupy smaller bodies of water than the open ocean, and the lower energy of these systems allows coastal wetland habitats to occupy the shoreline. Coastal wetlands currently occupy 48% of the tidal shoreline in the United States (Gittman et al., 2015). A recent analysis of shoreline hardening along sheltered coasts in the conterminous United States concluded that, as of 2005, 14% of the tidal shoreline had been armored (Gittman et al., 2015). Shoreline hardening is correlated with land development and the percentage of hardened shoreline can exceed 50% within a given watershed or county (Gittman et al., 2015; Bilkovic et al., 2016; Kornis et al., 2017).

Coastal wetlands, including salt marshes, are able to maintain themselves over centuries or millennia as a result of a dynamic balance between erosion, sediment supply, and accretion (Fagherazzi et al., 2011; Marani et al., 2011). The focus of this chapter is on salt marshes, as their ability to reduce erosion of estuarine shorelines has long been recognized and

Coastal Wetlands
https://doi.org/10.1016/B978-0-444-63893-9.00030-7

incorporated into shoreline stabilization efforts (Garbisch et al., 1975; Broome et al., 1986). Early assessments of the potential for marsh vegetation to stabilize shorelines concluded that marsh vegetation alone would be sufficient to stabilize many shorelines on which hardened structures have been built (Knutson et al., 1981; Shafer and Streever, 2000). More recently, the term "Living Shoreline" (hereafter Living Shoreline) has been adopted by coastal resource managers and environmental groups to describe shoreline stabilization efforts that incorporate natural vegetation or habitats (NRC, 2007; Currin et al., 2010; NOAA, 2015). Living Shorelines have been advocated as an alternative to traditional hardened structures such as bulkheads, revetments, and breakwaters because of concerns about the adverse impact of shoreline hardening on estuarine habitats and ecosystem function and the desire to conserve wetland habitat threatened by sea level rise and coastal development (Arkema et al., 2013; Sutton-Grier et al., 2015). By 2015, this effort had led to the permitting and construction of over 200 projects identified as Living Shorelines in the United States (Gittman et al., 2016a). In response to information needs expressed by coastal managers and regulatory agencies, over the last decade there has been a dramatic increase in research on both the ecosystem impacts of traditional shoreline hardening and on the ecosystem function and effectiveness of Living Shoreline designs.

Incorporating natural coastal habitats into a shoreline protection plan is gaining interest globally and may include salt marsh, mangrove, seagrass, coral reef, oyster reef, tidal flat, and rocky shoreline habitats (French, 2008; Koch et al., 2009; Shipman, 2010; Arkema et al., 2013; Temmerman et al., 2013; Firth et al., 2014). The Living Shoreline approaches that are commonly used in temperate ecosystems, and which primarily utilize salt marsh in association with oyster reefs, tidal flats, and/or seagrasses to stabilize the shoreline, are the focus of this chapter, along with a summary of studies examining their efficacy and impact on ecosystem services. Factors that affect the planning and implementation of a Living Shoreline approach will be described, including regulatory, policy, and public perceptions. Current recommendations are discussed on the utilization of Living Shorelines and potential incorporation of a Living Shoreline approach into a long-term strategy for coastal communities to respond to accelerated sea level rise.

2. LIVING SHORELINES IN ESTUARINE ECOSYSTEMS

2.1 Definition and Types of Living Shorelines

The Living Shoreline term was first developed, and will be used here, specifically for stabilization of estuarine, embayed, or sheltered coastlines with native vegetation (NRC, 2007). Living Shoreline definitions have been developed by numerous government agencies and environmental groups, and the variety of usages reflects the range of approaches utilized in differing environments. A definition that captures the key elements addressed in this chapter is "A living shoreline has a footprint that is made up mostly of native material. It incorporates vegetation or other living, natural 'soft' elements alone or in combination with some type of harder shoreline structure (e.g., oyster reefs or rock sills) for added stability. Living shorelines maintain continuity of the natural land-water interface and reduce erosion while providing habitat value and enhancing coastal resilience" (NOAA, 2015). Living Shoreline options (Fig. 30.1A–C) incorporate an amount of natural habitat equal to or greater than

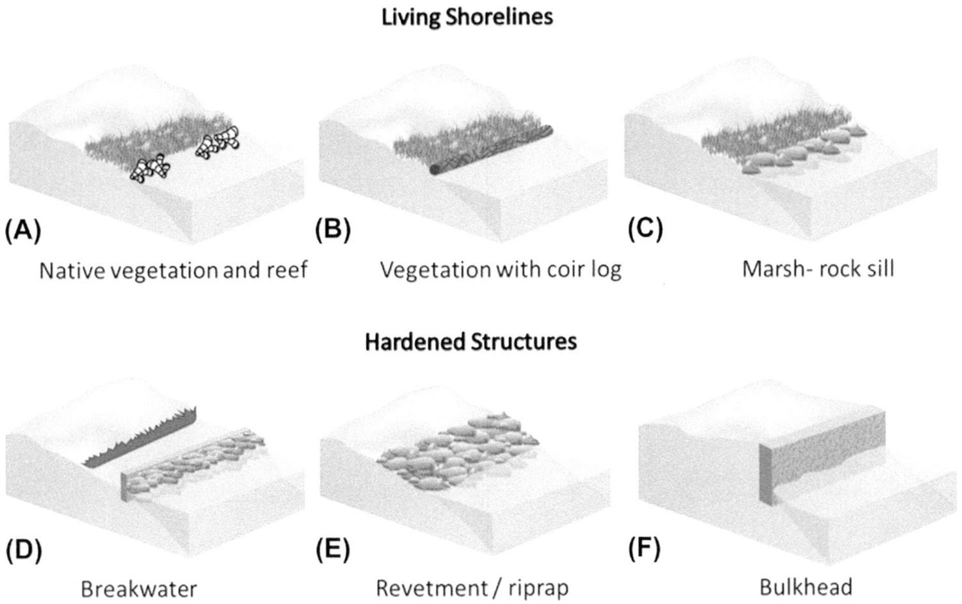

Living Shorelines

(A) Native vegetation and reef

(B) Vegetation with coir log

(C) Marsh- rock sill

Hardened Structures

(D) Breakwater

(E) Revetment / riprap

(F) Bulkhead

FIGURE 30.1 Illustrations of shoreline stabilization approaches, ranging from those that represent Living Shorelines that emphasize native habitats (A–C) to those that utilize primarily hardened structures (D–F); (A) native marsh vegetation and reef; (B) native vegetation with coir log at marsh toe; (C) native marsh with an offshore rock sill; (D) offshore stone breakwater; (E) stone revetment or riprap; (F) bulkhead. *Figure adapted from SAGEsagecoast.org/docs/Living%20Shoreline%20Brochure_Inside.jpg.*

hard structure and maintain the land–water interface. Hardened structures, including breakwaters, revetments, and bulkheads, do not, although they are sometimes found adjacent to vegetation (Fig. 30.1D–F). In addition to describing the scope of this chapter, another reason to define the Living Shoreline term is to distinguish these efforts from shoreline stabilization structures that are primarily composed of hardened structures but may be promoted under the Living Shoreline description (Popkin, 2015). Large breakwaters, wave attenuation devices, and vertical walls can be combined with natural habitats to provide shoreline stabilization (Firth et al., 2014; Bridges et al., 2015), but their resilience to accelerated sea level rise and preservation of ecosystem function will be less than the Living Shorelines as defined and discussed in this chapter.

There are additional terms used to describe estuarine shoreline stabilization approaches consistent with Living Shorelines that vary regionally in their use. In the Pacific Northwest and other areas, the analogous term Green Shores is often used in place of Living Shorelines. This is in part because on high-energy coastlines with steep topography, coastal wetlands do not occupy the estuarine shoreline. An approach to shoreline stabilization that shares many of the goals of the Living Shorelines effort may not incorporate any living plants but instead utilize native features such as logs, bluffs, and beaches (Shipman, 2010; Dethier et al., 2016) (Fig. 30.2A). The term "hybrid structure" is often used to distinguish Living Shoreline approaches that incorporate a structural feature from those that only use natural vegetation and habitats (Fig. 30.2B and C). Natural and nature-based infrastructure (NNBF) is another

FIGURE 30.2 Regional Living Shoreline approaches for shoreline stabilization reflect local habitats and geomorphology; (A) stabilization of bluff with large woody debris in Puget Sound, Washington, (B) hybrid Living Shoreline in microtidal estuary in the Florida panhandle with submerged rock sill, (C) aerial view of hybrid Living Shoreline utilizing rock sill with dropdowns and adjacent to bulkhead and riprap structure in Beaufort, NC, (D) eroding beach on Pivers Island in Beaufort, NC, in March 2001. Site was planted with four rows of *Spartina alterniflora* in 2000, with a deployment of oyster cultch placed offshore of the planted area, (E) Pivers Island marsh in July 2006, showing growth of marsh and oyster reef in the foreground, second deployment of oyster cultch visible in middle of planting site, (F) Pivers Island marsh in May 2014, showing continuing marsh and oyster reef development.

term frequently used to describe the incorporation of natural ecosystems in efforts to increase coastal resilience to erosion, storms and sea level rise. This term was popularized post-Hurricane Sandy (Bridges et al., 2015; Sutton-Grier et al., 2015). Although NNBF efforts encompass Living Shorelines, it is a much broader term that can include designs to protect beach communities from severe storms and tsunamis, which is beyond the capability of Living Shorelines as discussed in this chapter.

It is important to note that managing land use is the preferred alternative to dealing with shoreline erosion; that is, where possible, any shoreline stabilization project should be considered only if it is impossible to relocate infrastructure or otherwise accommodate natural erosion processes (French, 2008; Titus et al., 2009). This priority is often incorporated into natural resource agency policy documents (NRC, 2007). When shoreline stabilization is required, there are settings where a Living Shoreline approach as described above will not be adequate because of high wave energy, proximity to navigation channels, surrounding built infrastructure, or the desired use of a waterfront property. High bluffs and steep shoreline morphology also limit the use of wetland vegetation for erosion control (NRC, 2007). In these cases, although it may be impossible to design a Living Shoreline as defined above, an NNBF approach may be undertaken that incorporates natural vegetation, shellfish habitat, and beach-building processes into large stabilization structures.

The type of Living Shoreline used should be considerate of the wave energy of the site, and the use of structural features in addition to natural habitats should be limited to providing adequate protection for site conditions (Bilkovic et al., 2016; Currin et al., 2017). At low-energy sites, a combination of planted marsh, intertidal flat, and/or oyster reef can provide effective protection (Figs. 30.1A and 30.2D–F). In cases where there is an existing marsh scarp or slightly higher wave energy conditions, fiber coir logs or a marsh toe stone revetment may be added directly adjacent to the marsh edge (Fig. 30.1B). A rock sill or low offshore breakwater may be used in areas with greater wave energy (Figs. 30.1 and 30.2B and C). As noted above, in regions where salt marsh does not occur, mangroves, seagrasses, logs, or other natural material may be used in place of marsh.

2.2 Physical Processes and Natural Shorelines

The presence of even narrow bands of vegetation on natural estuarine shorelines can decrease shoreline erosion and increase resilience to sea level rise (Fig. 30.3A) (Koch et al., 2009; Morgan et al., 2009; Gedan et al., 2011; Shepard et al., 2011; Temmerman et al., 2013). These fringing wetlands, which are defined here as less than 50 m wide from shoreline to upland, provide many of the ecosystem services for which more expansive wetland ecosystems are valued, as many of these functions are associated with the marsh edge (Morgan et al., 2009; Currin et al., 2010; Bilkovic et al., 2016). In particular, the sediment trapping and wave attenuation capability of salt marshes that make them ideal for shoreline protection are maximized within several meters of the lower marsh edge (Knutson et al., 1982; Christiansen et al., 2000; Leonard et al., 2002; Möller and Spencer, 2002; Morgan et al., 2009; Tonelli et al., 2010). Although all sloping shorelines will dissipate wave energy, vegetation increases friction, dampens waves, and reduces turbulent mixing, resulting in increased wave energy attenuation as compared with unvegetated shorelines or mudflats (Möller, 2006; Yang et al., 2012; Möller et al., 2014). Even fairly narrow bands of marsh (10–15 m wide), as might

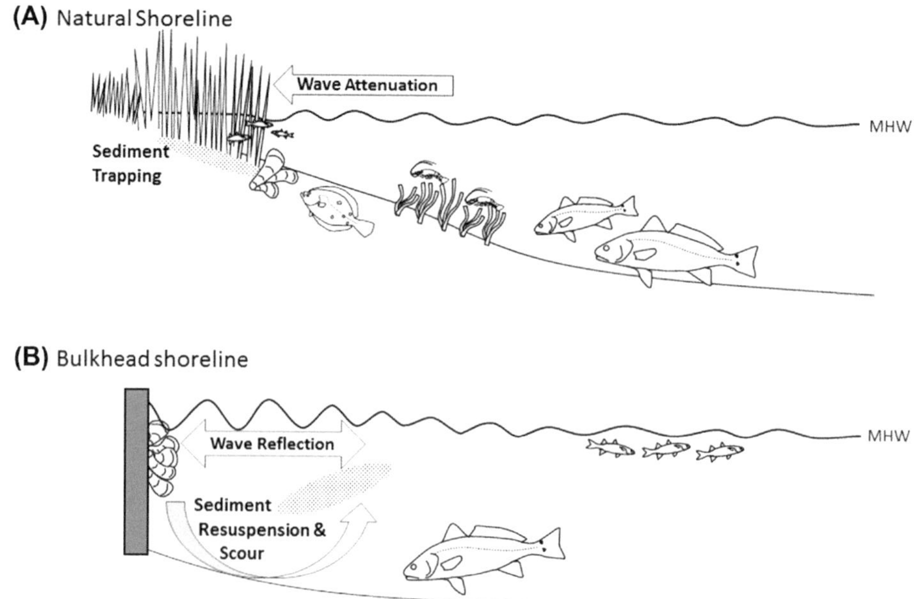

FIGURE 30.3 Natural shoreline and impact of bulkhead on ecosystem properties. (A) Fringing salt marsh attenuates wave energy and traps sediment from incoming tidal water, depositing it on the marsh surface. Shallow water provides refuge for small larval and juvenile fish, and shallow offshore bottom supports seagrasses, benthic fish, and oysters. (B) Bulkhead displaces marsh vegetation, and reflected wave energy promotes scour and sediment resuspension. Deepening of water column eliminates shallow water refuge for larval and juvenile fish.

be incorporated into a Living Shoreline, have been shown to be effective at attenuating up to 80% of incoming wave energy, and wider fringing marshes can attenuate over 90% of incoming wave energy (Leonard and Croft, 2006; Shepard et al., 2011; Yang et al., 2012).

Wave energy reduction increases with increased marsh stem density and/or canopy height (Leonard and Luther, 1995; Möller, 2006; Yang et al., 2012). Therefore, seasonal changes in canopy height and biomass will result in decreased wave attenuation in winter months (Möller, 2006). In tidal systems, it is important to note that wetlands can only attenuate wave energy when submerged and that the ability of wetland vegetation to attenuate wave energy changes with the relationship between canopy height and water depth (Koch et al., 2009; Möller et al., 2014). Marsh scarps are most susceptible to erosion just prior to inundation of the marsh surface, and scarped marsh shorelines are more susceptible to erosion than sloped ones (Tonelli et al., 2010; Theuerkauf et al., 2015).

The response of salt marsh vegetation to storm conditions is of great interest to property owners considering a Living Shoreline. A recent study utilized an outdoor flume, planted with marsh species native to the North Sea (*Puccinella* and *Elymus)*, to measure the wave attenuation and erosion protection of a 40 m marsh over a variety of water levels and wave heights (Möller et al., 2014). The combination of marsh vegetation and soil surface was capable of reducing wave energy by 20% during tests run to simulate storm conditions, with wave heights up to 0.9 m (Möller et al., 2014). Overall, however, a wetland's ability to attenuate wave energy decreases as water levels exceed canopy height, and storm surge reduction by a kilometer of marsh is on the order of tens of centimeters (Gedan et al., 2011). During these events, however, the marsh surface is resistant to erosion (Möller et al., 2014), and numerous observations have been made of hurricanes delivering sediment to the marsh surface, rather than eroding it (Cahoon, 2006; see Section 3.2). These observations are consistent with research concluding that it is frequent lower-energy waves that drive marsh shoreline erosion, rather than isolated large storm events (Leonardi et al., 2016).

As marsh plants attenuate incoming wave energy and slow the movement of water, they cause an increase in the deposition of suspended sediments onto the marsh surface (Leonard and Croft, 2006). Salt marsh vegetation influences sediment deposition both by the reduction of wave energy and by direct interception of particles by plant stems (Mudd et al., 2010; Mudd, 2011). The trapped sediments contribute to both vertical and horizontal maintenance/growth of marshes and enable the salt marsh to maintain surface elevation relative to sea level rise (Morris et al., 2002; Fagherazzi et al., 2011; Kirwan et al., 2016). With adequate sediment supply, natural fringing marshes can also grow horizontally in the seaward direction as they aggrade (Mattheus et al., 2010). More typically, however, the cycle of erosion, slumping, and return of reworked sediment to the marsh surface represents a long-term landward movement of the shoreline during periods of sea level rise (Chauhan, 2009; Fagherazzi et al., 2013). Nonetheless, there is a significant positive effect of vegetation on shoreline sediment accretion, surface elevation increase, and erosion reduction (Shepard et al., 2011; Currin et al., 2015). With adequate sediment supply and low wave energy, salt marsh shorelines can maintain their position even with current and accelerated rates of sea level rise (Mudd, 2011).

Although not a focus of this chapter, seagrasses, oyster reefs, and tidal flats may occur seaward of fringing marshes and contribute many of the same shoreline stabilization functions, including wave attenuation or dissipation, and sediment trapping and binding (Stal, 2003; Koch et al., 2009; Scyphers et al., 2011), and these habitats may be incorporated into Living Shoreline designs. Seagrasses, or submerged aquatic vegetation, do attenuate wave energy, but their flexible stems are not as effective as salt marsh species (Bouma et al., 2005; Koch et al., 2009). However, submerged aquatic vegetation occupies the water column for virtually all of the tidal cycle and can attenuate wave energy during low tide periods when marshes are not inundated (Fonseca and Callahan, 1992; Koch et al., 2009).

2.3 Ecological Function of Natural Shoreline Habitats

Fringing salt marshes also provide valuable fishery habitat. Numerous studies have shown high abundances of nekton at the marsh edge (Rozas and Reed, 1993; Minello et al., 1994; Peterson and Turner, 1994) and that fish abundance in fringing shoreline marshes may be similar to that found in extensive marshes bordering tidal creeks (Hettler, 1989; Currin et al., 2008). Marsh edge habitats are preferentially utilized by blue crabs and other species, which take advantage of the juxtaposition of marsh and seagrass vegetation to find abundant prey and avoid predation themselves (Micheli and Peterson, 1999) (Fig. 30.3A). Tidal flats and the shallow unvegetated bottom adjacent to the marsh edge provide important refuge and settlement for juvenile and larval fish (Burke et al., 1991; Ruiz et al., 1993).

Fringing salt marshes serve as critical buffers against estuarine eutrophication by absorbing or transforming nutrients, particularly nitrogen, before they enter the estuary's open waters (Tobias et al., 2001; Davis et al., 2004; Smyth et al., 2013). Moreover, denitrification rates in fringing salt marshes were found to scale with marsh area so that even narrow fringing marshes afford significant nitrogen removal capability (O'Meara et al., 2015).

Oyster reefs may occur in the intertidal or subtidal areas of estuaries and provide abundant ecosystem services, on par with vegetated coastal habitats (Grabowski et al., 2012). Oysters are effective ecosystem engineers, forming cohesive, structurally complex reefs that grow both horizontally and vertically and are often incorporated into Living Shorelines designs, alone or in conjunction with marsh restoration or creation (Meyer et al., 1997; Rodriguez et al., 2014; La Peyre et al., 2014).

Mangroves replace salt marsh along sheltered coasts in tropical and subtropical regions and are similarly capable of reducing shoreline erosion (Mazda et al., 1997; Gedan et al., 2011). Mangrove restoration and conservation is also driven by their demonstrated ability to reduce tsunami impacts (Kathiresan and Rajendran, 2005). Recently, mangroves have been incorporated into Living Shoreline projects in North America, and their use is likely to become more widespread.

The natural distribution of coastal wetland vegetation and shoreline habitats across the physical, energetic, and geomorphologic gradients found in estuaries serve as a template for the design of Living Shorelines. Similarly, the numerous ecosystem functions and erosion protection provided by naturally vegetated shorelines provide a benchmark for assessing the ecosystem function associated with anthropogenically modified shorelines, including Living Shorelines.

2.4 Impact of Traditional Shoreline Hardening

A wide variety of hardened structures utilizing stone, concrete, wood, vinyl, and sheet metal have been devised to armor shorelines. The most widely used structures on sheltered coasts include offshore breakwaters, revetments or riprap, and bulkheads or vertical seawalls (hereafter referred to as bulkheads) (Fig. 30.1D–F). There are a number of potential ecosystem impacts of traditional shoreline hardening on coastal wetlands. The most immediate is loss of habitat when the footprint of a structure covers existing wetlands. This type of direct loss occurs with marina construction, port expansion, and other types of businesses and infrastructure requiring direct placement on the shoreline.

In other cases, hardened structures are placed at or just above the mean high water line and therefore pose indirect impacts on adjacent wetlands and the aquatic ecosystem (Fig. 30.4). Those impacts include the elimination of connectivity at the land–water interface, wave reflection or refraction and associated scouring of the sediment bottom, increased water depth and loss of intertidal habitat, and resuspension of sediments in the water column. The result can be a reduction in water quality, wetland habitat, fisheries, and wildlife (Fig. 30.3B). Hardened structures also prevent marshes from migrating landward and therefore limit the ability of marshes to contribute long-term natural erosion protection (Titus et al., 2009; Temmerman et al., 2013; Kirwan et al., 2016). The materials used in seawalls or bulkheads may include chemical contaminants, which can reduce the biomass and diversity of the adjacent benthic community (Weis et al., 1998). The introduction of hardened substrate into an estuarine ecosystem, particularly one dominated by soft sediments, can provide an entry point for invasive species (Davis et al., 2002; Bulleri and Chapman, 2010), as can the disturbance from construction (Silliman and Bertness, 2004). Finally, shoreline erosion can

FIGURE 30.4 Impact of bulkhead construction on fringing *Spartina alterniflora* marsh.

be an important source of sediments needed for sustaining coastal wetland elevation, and eliminating all shoreline erosion via hardening may contribute to sediment starvation and drowning of downstream coastal wetlands (Mariotti and Carr, 2014; Currin et al., 2015).

2.4.1 *Hydrodynamics and Sediment*

The impact of hardened structures on adjacent sediments and water depth has been well-documented. As summarized by Dugan et al. (2011), "… any engineered structure placed in a coastal setting will alter hydrodynamics and modify the flow of water, wave regime, sediment dynamics, grain size, and depositional processes." (Fig. 30.3B). Bulkheads accelerate erosion and increase scour and turbidity on estuarine shorelines (Jackson and Nordstrom, 1992; NRC, 2007; Miles et al., 2001; Nordstrom and Jackson, 2012), and revetments are associated with steeper offshore slopes (Bilkovic and Mitchell, 2013). Breakwaters, larger but analogous to the rock sill used in hybrid Living Shoreline designs, can also alter littoral transport of sediments. This impact can significantly alter shoreline profiles and patterns of sediment accretion, including the creation of stagnant lagoons (Martin et al., 2005), and increases in accretion of fine sediments landward of the structure (Airoldi et al., 2005). With adequate longshore sediment transport, offshore breakwaters create tambolos, elevated sandy features which may not support marsh vegetation (NRC, 2007).

2.4.2 *Impacts on Vegetation*

The scour and deepening associated with bulkheads can lead to a reduction in adjacent vegetation. In New England, the high marsh community was significantly reduced in marshes seaward of bulkheads (Bozek and Burdick, 2005), and marshes adjacent to bulkheads occupied a lower elevation than natural fringing marshes in Georgia (Gehman et al., 2018). Marsh loss may be greater adjacent to bulkheads that experience high wave energy and/or where there is minimal sediment supply, as these factors increase erosion and decrease a marsh's ability to accrete sediment.

Seagrasses located adjacent to hardened shorelines can be similarly affected by erosion and are also negatively impacted by any increase in turbidity associated with a hardened structure (Miles et al., 2001). Light is the primary limiting factor for seagrass growth, and both deepening of the water column and an increase in suspended sediments reduce light availability to benthic vegetation. Bulkheads negatively impacted the distribution of *Ruppia maritima* and *Zostera marina* in polyhaline watersheds in the Mid-Atlantic (Patrick et al., 2016). The negative impact of bulkheads was greatest in those watersheds that experienced the least human modification through land use and that contained the most abundant seagrass habitats. Hardened structures were also associated with greater decadal declines in submerged aquatic vegetation cover in the Hudson River (Findlay et al., 2014). The impact of shore-parallel breakwaters and revetments on seagrasses is less than that of bulkheads (Patrick et al., 2016), although modeling efforts demonstrate that when wave energy is sufficient to resuspend sediment, breakwaters have a negative effect on the growth potential of seagrass shoreward of the structure (Smith et al., 2009). An analysis of the relationship between the percentage of hardened shoreline and distribution of wetlands and seagrass was made utilizing 587 sites in 39 subestuaries of the Chesapeake Bay and Delaware Coastal Bays (Kornis et al., 2017). This analysis revealed that bulkhead and riprap shorelines were associated with both increased water depth of the nearshore and with reduced extent of coastal wetland habitat (Kornis et al., 2017).

2.4.3 *Impacts on Fauna*

The increase in water depth, alterations to sediment type, and reduction of vegetation associated with breakwaters, revetments, and bulkheads can lead to changes in biodiversity and a decrease in faunal utilization of nearshore habitats adjacent to hardened structures (Figs. 30.3B and Fig. 30.5) (Bulleri and Chapman, 2010; Gittman et al., 2016b; Kornis et al., 2017). The negative impact of bulkheads on fauna associated with shoreline habitats is usually greater than that of sloped revetments or breakwaters (Gittman et al., 2016a), although an analysis of the relationship between shoreline cover type and faunal communities in Chesapeake Bay found similar negative impacts of the two structure types (Kornis et al., 2017). Benthic infauna and macrofauna are abundant organisms in estuarine sediments and represent an important link in food webs supporting commercially and recreationally important fisheries. Numerous studies comparing sites adjacent to bulkheads with naturally vegetated shorelines demonstrate declines in benthic invertebrate abundance and diversity at sites with bulkheads (Tourtellotte and Dauer, 1983; Bilkovic et al., 2006; Seitz et al., 2006; Seitz and Lawless, 2006). This reduction in benthic organisms may be a result of greater predation, reduced food associated with sandier sediments, and impacts from increased runoff and pollutants in sediment adjacent to bulkheads, which is reflected in a significant decline in the pollution-sensitive species *Limecola balthica* (Seitz et al., 2018). Infaunal abundance and diversity is not altered as consistently or as much by riprap revetments as by bulkheads, consistent with fewer changes to offshore sediments and water depth (Seitz et al., 2006; Lawless and Seitz, 2014), although negative impacts on infauna have been reported (Bilkovic and Mitchell, 2013). Breakwaters that intercept and accrete sediments can have a significant impact on infaunal communities, although site-specific

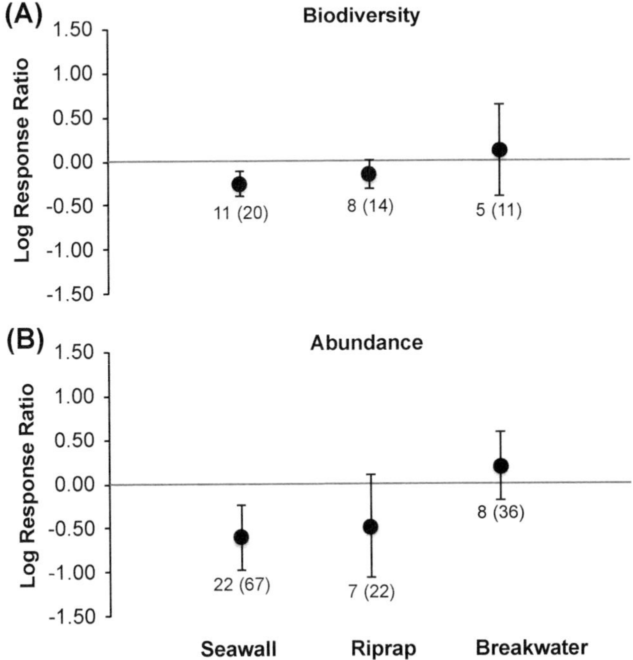

FIGURE 30.5 Overall log response ratios between hardened shorelines (seawall, riprap, breakwater) and natural marsh shorelines for (A) biodiversity and (B) abundance. *Error bars* represent 95% confidence intervals and data labels show the number of studies and the total number of responses from the studies. *From Gittman, R.K., Scyphers, S.B., Smith, C.S., Neylan, I.P., Grabowski, J.H., 2016b. Ecological consequences of shoreline hardening: a meta-analysis. BioScience 66 (9), 763–773, with permission of the author.*

hydrodynamics and sediment transport mechanisms determine whether fine- or coarse-grained sediments accumulate (Airoldi et al., 2005; NRC, 2007).

Bulkheads also impact fish abundance and biodiversity. A recent metaanalysis found that fish abundance was reduced by 45% as compared to natural shorelines and biodiversity reduced by 23% (Gittman et al., 2016b, Fig. 30.5). In some cases, fish species associated with wetlands were entirely absent from bulkheaded shorelines (Bilkovic and Roggero, 2008). In the large study of Chesapeake and Delaware Bay watersheds, nearshore habitat adjacent to hardened shorelines had lower abundances of small-bodied fish and higher abundances of large-bodied fish (Kornis et al., 2017). On a subestuary scale, the percentage of hardened shoreline was negatively correlated with the abundance of 9 of 15 fish species (Kornis et al., 2017). In Mobile Bay, Alabama, the composition of the fish community associated with bulkheaded shorelines was less stable over an 11-year period than that found adjacent to natural shorelines (Scyphers et al., 2015a). Along with alterations to water depth, wave climate, and sediment characteristics, bulkheads have reduced structural complexity compared with natural habitats and provide less refuge for nekton and their prey (Fig. 30.3B) (Strayer et al., 2012; Scyphers et al., 2015a; Gittman et al., 2016b).

Nekton abundance and biodiversity associated with revetments/riprap and breakwaters (see Fig. 30.1D and E) is similar to that of natural shorelines in the majority of studies reported to date (Gittman et al., 2016b), although Kornis et al. (2017) report negative impacts of revetments on fish and mobile crustacean abundance, similar to those observed with bulkheaded shorelines. The stability of fish community composition was reported as greater adjacent to revetments than to bulkheads (Scyphers et al., 2011). Although revetments and breakwaters do not appear to have as negative an impact on nearshore fish communities as do bulkheads, the nekton response does not capture the complete ecosystem impact of these structures. Revetments and breakwaters are placed directly over unvegetated shallow water bottom. These structures displace intertidal and shallow subtidal sandflats and mudflats (Bilkovic and Mitchell, 2013) that have been described as "the Secret Garden" because they provide a unique habitat for algal primary producers and benthic infauna, which combine to stabilize sediments and support fishery food webs (Stal, 2003; Miller et al., 1996).

Bulkheads and revetments greatly reduce the connectivity between land and aquatic ecosystems, with adverse impacts on species that utilize both habitats. These include diamondback terrapins and birds (DeLuca et al., 2008; Dugan et al., 2011; Isdell et al., 2015). In addition to reducing the connectivity between habitats, bulkheads and revetments fragment and narrow fringing marsh habitat, which has detrimental impacts on a number of species (Meyer and Posey, 2009; Isdell et al., 2015).

2.4.4 *Cumulative Impacts*

In addition to the direct impacts of hardened structures on the structure and function of adjacent habitats, there is emerging evidence of cumulative and multistressor impacts of hardened shorelines and upland development on estuarine ecosystems (Bilkovic and Roggero, 2008; Dethier et al., 2016; Kornis et al., 2017; Seitz et al., 2018). Watershed or subestuary reductions in native marsh and seagrass habitat, benthic infauna, nekton, birds, and terrapins have been linked to system-wide shoreline hardening, with significant negative impacts observed when 10%–30% of the shoreline is hardened (Silliman and Bertness, 2004; Bilkovic and Roggero, 2008; DeLuca et al., 2008; Isdell et al., 2015; Kornis et al., 2017). Upland development is often correlated with shoreline hardening and is associated with increased sediment, nutrient, and pollutant inputs (Kennish, 2002; Kornis et al., 2017; Seitz et al., 2018). In highly developed areas with significant nutrient or pollutant inputs, the cumulative impact of shoreline hardening on shoreline habitats may be more difficult to document as overall ecosystem function is reduced (Patrick et al., 2016; Kornis et al., 2017; Seitz et al., 2018).

3. LIVING SHORELINE IMPLEMENTATION AND ASSESSMENT

Most of this section is based on the limited number of published, quantitative assessments of Living Shoreline structure, function, and effectiveness, all from sites in the Gulf of Mexico and the southeast and mid-Atlantic of the United States. These sites are typically microtidal and vegetated by *Spartina alterniflora* and at higher elevations, *S. patens*. As the practice of Living Shorelines increases, unique aspects of other areas will be incorporated into regional Living Shoreline designs. For example, sites in the Northeast will be challenged by ice scour and topographical limitations on marsh landward transgression. The slope of narrow fringing marshes in macrotidal regions will be greater than illustrated here, and Living

Shorelines may require terracing, gabions, or other methods to stabilize steeper shorelines. Farther south, mangroves and seagrasses may be incorporated into Living Shorelines. On the West Coast, where *S. alterniflora* is a nonnative species, steep topography, high-energy shorelines, drought, and an abundance of invasive species will require a modified Living Shoreline approach (Dethier et al., 2016). Continued monitoring, experimentation, and quantitative assessment of Living Shorelines are needed to provide much-needed guidance for these areas (NRC, 2007). Involvement of citizen scientists in this process can yield cost-effective data collection and provide a valuable outreach and educational opportunity (Currin et al., 2008). Given the widespread interest in the subject, it is expected that there will be an increase in the published assessments of Living Shoreline effectiveness and ecological function in the coming years.

The ability of fringing salt marshes to attenuate wave energy, trap sediment, and reduce shoreline erosion provides the basis for Living Shoreline implementation and evaluation. Marsh vegetation, sediment accretion, nutrient cycling, and faunal utilization of Living Shorelines and natural fringing marshes have been compared to assess the development of ecosystem function in Living Shoreline projects. The incorporation of oyster reefs into Living Shoreline designs has been evaluated in regard to erosion protection and impacts on marsh flora and fauna. These assessments of ecosystem function are crucial to efforts to improve and increase the use of Living Shorelines for erosion control. This section summarizes those findings and also presents a summary of the regulatory hurdles and social perceptions of shoreline stabilization that influence Living Shoreline policy and implementation.

3.1 Wave Energy and Geomorphology

The goal of a Living Shoreline is to stabilize the shoreline while preserving the sustainability of natural habitats and the ecosystem services they provide. There is a balancing act between an engineering goal of fail-safe protection and an ecological goal of minimizing adverse impacts of hardened structures. Most nonstabilized salt marsh shorelines exposed to open water erode over time because of the consequences of wave energy and sea level rise, unless there is substantial sediment supply in the system (Marani et al., 2011; Fagherazzi et al., 2013). Studies on the physical setting in which natural habitats occur, and on the performance of hybrid Living Shorelines in different hydrodynamic and geomorphological settings, can provide guidance for implementation of Living Shorelines.

There is a strong linear relationship between wave energy or wave power and marsh erosion rates (Leonardi et al., 2016), and the wave energy a site experiences is the first factor to consider when designing a shoreline stabilization project. There are a number of alternatives for estimating site-specific wave energy. These include a simple measure of fetch or the distance over the water that wind blows (Knutson et al., 1981; Roland and Douglass, 2005); measures of "effective fetch" that incorporate dominant wind direction and speed and can be modified into a Relative Exposure Index (Shafer and Streever, 2000; La Peyre et al., 2015); desktop Geographic Information System (GIS)-based computer models that incorporate bathymetry with fetch, wind speed, and direction to calculate representative wave energy (RWE) (Currin et al., 2015, 2017; Theuerkauf et al., 2016); and more complex computer models that incorporate shoreline morphology and wave refraction (Priestas et al., 2015). The measures of wind-driven wave energy generated from the modeling approaches include measures of significant wave height (m), RWE (Joules (J) m^{-1}), and wave power (J m^{-1} s^{-1} or W m^{-1}). As fetch is the easiest measure to make and one that property owners are familiar with, most guidance documents available from coastal resource managers use fetch to guide Living Shoreline implementation (Bilkovic et al., 2016; Currin et al., 2017). Currently available guidance documents recommend that Living Shorelines composed of natural vegetation or habitats (marsh and/or oyster reefs) be limited to sites with a fetch of <1.5 km (1 mile) and that hybrid structures can be used for fetches between 1.5 and 8 km (1–5 miles) (Currin et al., 2017). However, site orientation to prevailing winds, offshore bathymetry, and shoreline sinuosity can significantly alter the wind-driven wave energy experienced by sites with similar fetch (Knutson et al., 1981; Priestas et al., 2015; Currin et al., 2017). Several of the Living Shoreline studies discussed later in this chapter experience maximum fetches considerably greater than the guidelines described above and have demonstrated the ability to maintain vegetation, elevation, and faunal communities (Meyer et al., 1997; Craft et al., 2003; Bilkovic and Mitchell, 2013; Gittman et al., 2016a; Currin et al., 2017). If available, utilization of wave energy models that include effective fetch and offshore bathymetry to generate measures of significant wave height, wave energy, or wave power will provide clearer guidance on a site's ability to support vegetated habitats associated with Living Shorelines. If those measures are not available, resource managers and project planners should consider the factors other than fetch affecting wave energy, including exposure to prevailing winds, offshore bathymetry, proximity to navigation channels, bank slope, and shoreline geomorphology (Knutson et al., 1981; Broome et al., 1986; Currin et al., 2017). Proximity of existing marsh and oyster reef to a proposed site is one of the clearest indications that a Living Shoreline approach is warranted and is particularly important for incorporation of oyster reefs, which rely on successful recruitment for long-term sustainability.

The distribution and erosion rates of fringing *Spartina alterniflora* marshes are correlated with measures of significant wave height and RWE. A significant wave height of less than 0.3 m was identified as a threshold for fringing marshes in a number of places (Shafer et al., 2003; Roland and Douglass, 2005; Currin et al., 2015). In North Carolina, average natural fringing marsh width was >100 m in sites that experienced an average RWE calculated with the top 20% of wind events (RWE_{20}) of <300 J m^{-1}, whereas fringing marshes narrowed substantially at higher RWE_{20} values and were rarely found at RWE_{20} values >700 J m^{-1} (Currin et al., 2017). The distribution of natural intertidal oyster reefs in North Carolina was limited to RWE_5 (top 5% wind events) values of <500 J m^{-1}, which is similar to the RWE_{20} threshold determined for wide fringing marshes (Theuerkauf et al., 2016; Currin et al., 2017). Oysters attached to hardened structures (riprap and bulkheads) were found at RWE_5 values of up to 2000 J m^{-1}, and the authors recommend hardened substrate be used to create artificial oyster breakwaters at wave energies above the threshold for natural intertidal reefs (Theuerkauf et al., 2016).

Boat wakes from recreational and commercial vessels can cause shoreline erosion and may equal or exceed wind wave energy, particularly in narrow waterways and sheltered settings (Manis et al., 2014). In North Carolina, proximity to navigation channels significantly reduced the width of natural fringing marshes (Currin et al., 2017). Boat wakes have also been associated with loss of live oyster reefs and reduction in seagrass shoot density (Grizzle et al., 2002; Fonseca and Bell, 1998). Boat wake wave heights can vary from 0.1 m for recreational boats to nearly 1 m for large displacement vessels (Manis et al., 2014). The frequency, proximity, and type of boat traffic should be incorporated into estimates of the wave energy experienced by a site.

Shoreline geomorphology also plays a significant role in determining erosion rates, including bank height or slope and shoreline sinuosity or shape. Headland marshes receive wave energy from more directions than do embayed marshes, which may receive greater refracted wave energy than those on headlands (Priestas et al., 2015). Structures on adjacent property, including docks and piers, may also reflect wave energy onto a site and increase erosion (Mattheus et al., 2010). Eroding marshes often have scarped edges, which erode faster than a ramped or sloped border (Tonelli et al., 2010; Theuerkauf et al., 2015). The slope and width of the tidal flat in front of the marsh makes a significant difference in the wave energy reaching the marsh during high tides and is more effective at attenuating wave energy at low tides than a scarped edge. Steeply sloped fringing marshes, such as those that might occur in macrotidal regions, are more subject to erosion and slumping than gently sloped marshes.

The development of sediment organic matter in created and transplanted marshes occurs over a much longer time period than vegetation and other ecosystem attributes, and the same observation has been made in assessments of Living Shorelines (Craft et al., 1999; Davis et al., 2015). Natural fringing marshes often have sediments with sandier, lower-organic matter content than interior or meadow marshes (Craft et al., 1999; Currin et al., 2008; Morgan et al., 2009; Bilkovic and Mitchell, 2013). These shoreline marshes are likely to receive coarser-grained mineral input, as offshore sandflats are eroded and transported landward (Mariotti and Carr, 2014). Laboratory experiments have demonstrated that coarse-grained marsh sediment erodes faster than fine-grained sediment because of the lack of interparticle reactions and reduced biofilms that bind fine-grained sediments (Feagin et al., 2009; Paterson et al., 1990). Over time, the addition of organic matter and fine-grained sediment to Living Shoreline marsh sediment will increase their resistance to erosion (Craft et al., 1999).

Sediment carbon (C) accumulation in natural and transplanted fringing marsh is a result of sediment accretion and the burial of belowground marsh C. Twenty-five years after establishment, the annual organic C accumulation rate was similar between transplanted and natural fringing marshes, although soil nutrient concentrations remained lower in transplanted fringing marshes than in natural marshes (Craft et al., 1999). Carbon stocks and the C sequestration rate of natural and transplanted marshes, including three constructed as Living Shorelines, were compared in North Carolina (Davis et al., 2015). The calculated C sequestration rate declined with transplanted marsh age, ranging from 283 g C m^{-2} $year^{-1}$ in a 14-year-old marsh to 73 g C m^{-2} $year^{-1}$ in a 38-year-old marsh. Carbon stocks in the top 30 cm of sediment were similar between sites, and the decrease in C sequestration rate with marsh age is due to ongoing decomposition of newly deposited material, until a steady state of 70–80 g C m^{-2} $year^{-1}$ is reached (Davis et al., 2015). Long-term C sequestration rates reported in Davis et al. (2015) are on the lower end of average C accumulation rates reported elsewhere for *S. alterniflora*, and this may be because of the well-flushed nature of the sandy sediments. Wider implementation of Living Shorelines for erosion protection will result in the conservation of existing stocks of buried marsh C and increase the area of salt marsh available to contribute to the reduction of atmospheric greenhouse gases (Pendleton et al., 2012).

3.2 Development of Marsh Vegetation and Sediment

Living Shorelines have shown the ability to persist over time and provide both shoreline erosion protection and ecosystem services similar to natural vegetated shorelines. Early publications demonstrated that *Spartina alterniflora* could be transplanted successfully on estuarine shorelines for erosion protection and provided guidelines for site selection and planting protocols (Garbisch et al., 1975; Knutson et al., 1981; Broome et al., 1986). Depending on site conditions and original planting density, aboveground marsh biomass in these transplanted marshes became similar to that in natural marshes within 5–10 years, and fertilizer additions were frequently successful in increasing plant density, particularly on sandy soils (Garbisch et al., 1975; Broome et al., 1986; Chapter 22, this book). Planting density was found to increase transplant success in field experiments especially in more exposed settings (Broome et al., 1986), and higher transplant density increased sediment accretion rates in laboratory flume experiments (Gleason et al., 1979).

Two marshes created for shoreline stabilization, one in front of an existing bulkhead and the other on an island created from dredged material, were monitored periodically for 25 years to assess development of the plant community and associated measures of ecosystem function (Broome et al., 1986; Craft et al., 1999). Once established, the transplanted marshes persisted throughout the monitoring period without additional planting, and measures of the marsh plant community achieved equivalence with natural reference marshes within 5 years (Craft et al., 1999). Loose oyster shell, or cultch, was added to the edge of treatment plots at another three marshes created for dredged material stabilization in North Carolina and evaluated for its effect on marsh erosion and sediment elevation (Meyer et al., 1997). Both marsh and marsh-oyster cultch treatments exhibited a slight ($x = 0.26$ m) net seaward horizontal expansion, and oyster cultch treatment plots had a significantly greater vertical increase ($x = 4.8$ cm) in surface elevation over the 21-month study period. The stability of both treatments was especially noteworthy because of the wave energy at the sites; one site had a maximum fetch of >8 km, two sites experienced significant boat wake wave energy because of proximity to the IntraCoastal Waterway (ICW), and the study period for all the sites bracketed the March 1983 storm, which brought sustained wind speeds of 97 kph to the study area (Meyer et al., 1997).

In addition to these studies, a variety of alternative shoreline stabilization approaches utilizing marsh vegetation were initiated in the 1980 and 1990s, particularly in the states of Virginia, Maryland, and North Carolina (NRC, 2007, and references within). In the United States, the overall success of these efforts, together with a growing awareness of the combined threat to coastal wetlands from coastal development and sea level rise, prompted a call for more scientific study of the ecosystem impacts of stabilized shorelines and for assessment of alternative methods utilizing natural features (NRC, 2007).

The recovery of marsh vegetation at Living Shorelines that utilize rock sills in combination with salt marsh vegetation (sill-marshes) has been examined in several studies in North Carolina and Virginia. The stone sills in these studies conform with recently proposed regulatory requirements for Living Shorelines, in that sill height is <0.3 m above Mean High Water (MHW), sill length is less than 500 m, sills are constructed of granite rock or marl, and longer sills have dropdowns or openings to facilitate tidal exchange and faunal access (Federal Register, 2017; Currin et al., 2008, 2017; Bilkovic and Mitchell, 2013; Gittman et al., 2014, 2016a). In a study of newly established sill-marshes (1–3 years), *Spartina alterniflora* percentage cover was lower in sill sites than natural reference sites throughout the study period, while average stem density became equivalent to the reference marsh at one site after 3 years (Currin et al., 2008). Natural and transplanted sill-marshes had slopes from the marsh edge to the upland border of between 3% and 10%, and measures of *S. alterniflora* vegetation extended landward 15 m from the lower marsh edge. Average stem density in the natural fringing marsh sites was 180 stems m^{-2}. A similar result was reported for newly constructed sill-marshes in Gittman et al. (2016a), where marsh macrophyte stem density became equivalent to natural marshes in 3–8 years. Natural fringing marsh stem density in this study averaged 400 stems m^{-2}, and transect lengths ranged from 5 to 20 m, terminating at the upland boundary.

Stem density of *S. alterniflora* near the lower marsh edge was examined in three paired natural and sill-marsh sites over a period of 5 years (Currin et al., 2017). The sill-marsh sites were 4–5 years old at the beginning of the study and included several sites examined previously (Currin et al., 2008). Marsh stem density in plots established both 1 m seaward and 1 m landward of a defined lower marsh edge showed no significant change over time in natural marshes with and without an adjacent oyster reef (Fig. 30.6B and C). In contrast, stem density in sill-marsh edge plots located seaward of the original marsh edge showed a gain in stem density over the 5-year period, from an average of less than 50 stems m^{-2} to an average of over 200 stems m^{-2} (Fig. 30.6B and C). Stem density 5 m behind the original defined marsh edge was similar in natural and sill-marsh plots and averaged 210 stems m^{-2} (Fig 30.6D). Both low marshes dominated by *S. alterniflora* and high marshes dominated by *S. patens* were examined at paired natural and sill-marsh sites in Virginia (Bilkovic and Mitchell, 2013). The study was conducted 2–6 years postsill construction. Slope ranged from 1.3% to 5.6%, and marsh width ranged from 10 to 21 m. *Spartina alterniflora* stem density behind sill-marshes was similar or less than that of the natural marshes, whereas *S. patens* stem density was significantly higher in sill-marshes than natural marshes (Bilkovic and Mitchell, 2013).

Only one study has compared the macroalgal community in natural and sill-marshes, and it documented reduced macroalgal biomass and diversity behind sills (O'Connor et al., 2011). Sills had no impact on seagrass distribution located within 20 m seaward of the lower sill edge at three sites in North Carolina (Gittman et al., 2016a).

Marsh surface elevation change in Living Shoreline and natural reference marshes provides an indication of a site's long-term sustainability, as marshes must maintain their elevation relative to local sea level rise to survive (Morris et al., 2002; Kirwan et al., 2010). In North Carolina, surface elevation increase was greater behind sill-marshes than natural marshes in three studies (Currin et al., 2008, 2017; Gittman et al., 2016a). Surface elevation tables (SETs; Cahoon et al., 2002) established 1 m landward of the lower marsh edge in natural fringing marshes demonstrated little elevation gain or loss over 4 years. SETs located similarly behind sills exhibited an elevation increase during the study period, with an average increase of 3 mm/year, similar to the local relative sea level rise (Currin et al., 2017). These results are similar to results obtained with surveying techniques, where sill-marshes exhibited gains in elevation while natural marshes exhibited less elevation gain or loss (Currin et al., 2008; Gittman et al., 2016a). The combination of increased sediment accretion, higher surface elevation, and seaward movement of the marsh edge in sill-marshes results in the elimination of an unvegetated border between the sill and marsh edge, and

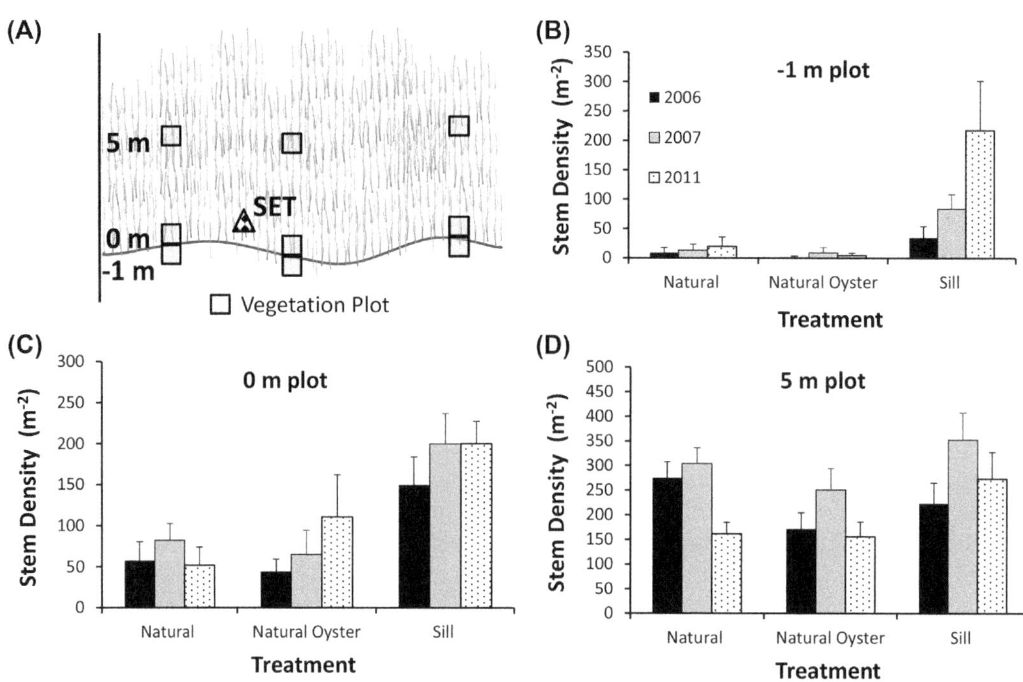

FIGURE 30.6 Density of *Spartina alterniflora* stems collected from permanent plots in natural marshes, natural marshes with fringing oyster reef, and sill-marshes established in 2001 or 2002. (A) Schematic showing relative locations of the −1, 0, and 5 m plots relative to shoreline and surface elevation table (SET). (B–D) Average stem density as a function of treatment and year for the −1, 0, and 5 m plots, respectively. *Figure from Currin, C.A., Davis, J., and Malhotra, A., 2017. Response of salt marshes to wave energy provides guidance for successful living shoreline implementation. In: Bilkovic, D.M., Mitchell, M., La Peyre, M., Toft, J. (Eds.), Living Shorelines: The Science and Management of Nature-based Coastal Protection, CRC Press, Taylor & Francis Group, pp. 211–234.*

marsh growth into the rock sill is often seen (Fig. 30.7). Sediment accretion behind sill-marshes may increase in high-energy settings, as resuspension of sediment from offshore flats increases with higher wave energy (Mariotti and Carr, 2014), potentially increasing the sediment delivered to the marsh during high tide. This was observed in a study of marsh-sill sites in North Carolina where a positive relationship between wave energy and surface elevation change was observed at sill-marsh sites over a 4-year period (Currin et al., 2017). Together, these studies of sediment accretion and marsh vegetation from hybrid Living Shoreline sites demonstrate that the density and biomass of marsh vegetation (particularly *S. alterniflora*) can achieve levels similar to that of natural fringing marshes over a period of 3–5 years. However, because of the alterations sills make on sediment accretion processes, over longer periods differences in the distribution of low and high marsh between natural and Living Shoreline sites may develop (Currin et al., 2017; Bilkovic and Mitchell, 2013).

Several studies have captured the response of natural and stabilized fringing marsh surface elevation and vegetation to storm events (Meyer et al., 1997; Currin et al., 2008, 2017; Gittman et al., 2014). In each of these studies, posthurricane marsh surface elevation measures made behind sill-marshes showed a greater net gain in surface elevation than natural

FIGURE 30.7 Marsh plants encroaching on a rock sill at a Living Shoreline site in North Carolina. The site was established with a 1–2 m unvegetated border between the marsh edge and rock sill. Over time, sediment accretion has raised the surface elevation behind the sill to promote seaward expansion of the marsh.

reference marshes. The passage of Hurricane Isabel over paired marsh sites in North Carolina resulted in a net average elevation gain in both natural and sill-marshes (Currin et al., 2008). The pattern of accretion varied, however, with the natural marsh lower edge showing a net 6 cm erosion and the upper natural marsh gaining over 15 cm in elevation. There was no erosion behind the rock sill, and elevation gain was greatest in the first 5 m of marsh immediately behind the sill, ranging from 7 to 13 cm. The response of three paired natural and sill-marsh sites to Hurricane Irene was reported in Gittman et al. (2014) who also compared storm damage among bulkheads, revetments, and Living Shorelines. Prior to the storm, sill-marsh elevations were greater than natural referenced marshes, and marsh surface elevation was unchanged in both marsh types after the storm. Vegetation density decreased in both marsh types after the storm, but it rebounded to prestorm levels within 2 years at both natural and sill-marsh sites (Gittman et al., 2014). Overall, bulkheads suffered more storm damage than any other type of estuarine shoreline in the study area (Gittman et al., 2014).

3.3 Oyster Reefs and Living Shorelines

Living Shorelines that incorporate salt marsh and oyster reefs have been evaluated at several sites in the Gulf of Mexico. In Louisiana, oyster reefs constructed with either loose shell or precast concrete structures were effective at reducing erosion of the marsh edge behind them, by as much as 1 m $year^{-1}$, whereas unprotected marsh erosion rates ranged from 1.3 to 2.5 m $year^{-1}$ (La Peyre et al., 2015). Although shoreline erosion was greatest at sites with the highest effective fetch, oyster reefs had the greatest impact on reducing shoreline erosion at these sites. In this study, it was noted that subsidence and sea level rise were likely responsible for continued marsh erosion after reef construction and that additional measures may be needed to stabilize the marsh shoreline (La Peyre et al., 2015). Oyster breakwaters protecting three intertidal marshes located in a high-energy area (>10 km fetch) of the Mississippi Sound also significantly reduced marsh erosion over a 3-year study period, although there was a net shoreline loss of >1 m year^{-1} even in reef-protected marshes (Moody et al., 2013). Despite the significant shoreline erosion, landward transgression of the marsh community preserved the marsh species distribution patterns even at the highest energy sites (Moody et al., 2013). In Mobile Bay, subtidal oyster breakwaters were established offshore of eroding marsh, at two sites with maximum fetches of 30 km and erosion rates compared to nearby control sites (Scyphers et al., 2011). Marsh erosion at control and breakwater sites exceeded 1 m year^{-1} during the 2-year study period, although one breakwater did reduce overall edge erosion by 40% (Scyphers et al., 2011). Oyster breakwaters established in these studies reported successful, although variable, recruitment and survival of oysters (Scyphers et al., 2011; Moody et al., 2013; La Peyre et al., 2015). Oyster habitat suitability must be determined prior to incorporating oysters into shoreline stabilization projects to insure long-term sustainability of the reef structure (Theuerkauf et al., 2016; La Peyre et al., 2017). Although the ability of constructed or restored oyster reefs to reduce erosion has been demonstrated in the majority of studies, in most cases oyster reefs have not demonstrated the ability to eliminate or reverse marsh erosion (La Peyre et al., 2017).

3.4 Faunal Utilization of Living Shorelines

Living shorelines have been found to support benthic infauna, epifauna, and nekton communities with similar or greater abundance and biodiversity than found in natural fringing marshes. Benthic infauna communities in transplanted marshes are slower to develop than fish and epifauna, as they respond to gradual changes in marsh sediment organic matter content and grain size (Craft et al., 1999). However, long-term monitoring of marshes constructed for shoreline stabilization in North Carolina demonstrated higher infauna abundance and species richness after 15–25 years of soil development (Craft et al., 1999). A study of benthic infauna associated with marsh-sills <8 years old revealed significant reductions in infaunal abundance and species composition (Bilkovic and Mitchell, 2013). These authors also noted that subtidal habitat offshore of salt marshes supported the greatest abundance, biomass, and diversity of infauna and that the use of rock sills should be minimized to reduce the loss of subtidal habitat. However, the added structure of sills and oyster reefs contributes to observations of greater epifaunal abundance and species richness in Living Shorelines that incorporate built reefs or sills than in natural fringing marshes (Scyphers et al., 2011; Bilkovic and Mitchell, 2013; Gittman et al., 2016a).

Nekton communities quickly utilize habitat associated with sill-marshes (Currin et al., 2008; Gittman et al., 2016a). In North Carolina, fish, crab, and shrimp abundance obtained with fyke nets in fall and spring matched that found in natural reference sites within 3 years (Currin et al., 2008). Similarly, Gittman et al. (2016a) used fyke nets to sample sill-marsh sites 1–8 years postconstruction and found that marshes with sills supported higher abundance and diversity of fish and higher crab abundance after 3 years. The abundance and biomass of fishes caught in traps set along unvegetated edges of natural marsh and sill-marsh habitat also did not differ, and bivalve cover (oysters, ribbed mussels) was greater on sill substrate than the unvegetated marsh edge (Gittman et al., 2016a).

3.5 Regulatory and Social Drivers

Despite clear evidence of the adverse impact of hardened structures on estuarine ecosystems, traditional hardened structures remain the most common approach to shoreline stabilization in coastal communities (NRC, 2007; Sutton-Grier et al., 2015; Gittman et al., 2014). Both regulatory and social hurdles have been identified that limit a broad implementation of the Living Shoreline approach to shoreline stabilization (NRC, 2007; Sutton-Grier et al., 2015; Gittman et al., 2014, 2016b; Scyphers et al., 2015b). A Living Shorelines project consisting of planting vegetation on existing sediment substrate would have few permit requirements in most states. However, in addition to vegetation, many Living Shoreline designs include the placement of sand fill, oyster shell, or a rock sill into intertidal or subtidal waters. In this case, several existing state and federal laws and regulations come into play (Pace, 2017). The line between public and private property along the shoreline varies from state to state and is usually defined by vertical datums based on tides. State to state, the public–private property boundary in the United States occurs somewhere between mean higher high water to mean low water, with the majority of states declaring the boundary at mean high water (NOAA, 2001). Therefore, Living Shoreline projects are usually placed within the public trust resource, as defined by state laws (Maloney and Ausness, 1974). Many states have established comprehensive coastal zone management legislation as a result of the Coast Zone Management Act of 1972 that may impact the Living Shoreline permitting process. At the US federal level, the National Marine Fisheries Service (NMFS) is required by the Magnuson-Stevens Fisheries Conservation Act and additional authorities to review projects that directly or indirectly affect managed fishery resources. This review considers impacts to essential fish habitat (EFH) provided by coastal wetlands. When the use of fill or other alteration of EFH is requested for a Living Shoreline project, the NMFS requires project applicants to justify conversion of one naturally functioning aquatic system at the expense of another. In addition, Section 404 of the Clean Water Act of 1977 regulates fill in wetlands and requires that the US Army Corps of Engineers (USACE) take the lead on developing policy and guidance on the topic, as well as enforcing permit provisions in coordination with the US Environmental Protection Agency, US Fish and Wildlife Service, and/or the NMFS. There are additional local, state, and federal laws that may apply, depending on the design and location of a Living Shorelines project (Pace, 2017).

The net result of the regulations placed on activities within the intertidal zone, which are designed to protect the environment, has often been to slow down the permitting process for Living Shorelines relative to shoreline hardening measures that take place above the mean high water tidal datum (NRC, 2007; Currin et al., 2010; Pace, 2017). In recognition of that, over the past decade both state and federal regulatory agencies have developed policies and general permits to facilitate the utilization of Living Shorelines. For example, several states now require that a Living Shoreline approach must be considered prior to bulkhead or riprap (Bilkovic et al., 2016). In early 2017, the USACE released Nationwide Permit 54 specifically for Living Shorelines in an effort to standardize and minimize the review and permitting process (Federal Register, 2017). The application of that permit may vary regionally to reflect differences in coastal geomorphology and hydrodynamics.

Human attitudes, experience, and perceptions about shoreline erosion and structural protection have a tremendous influence on how or whether a community or property owner chooses to stabilize a shoreline (Titus et al., 2009; Sutton-Grier et al., 2015; Scyphers et al., 2015b). Despite evidence that hardened structures are more expensive, and in some cases less resilient to storms and sea level rise, Living Shoreline approaches represent a small fraction of permitted bank stabilization activities in the United States (Bridges et al., 2015; Sutton-Grier et al., 2015; Gittman et al., 2016a). In Alabama, waterfront property owners reported that effectiveness, cost, and durability were the top criteria driving their shoreline stabilization choice and also recognized the harm that hardened shorelines represented to natural coastal habitats and ecosystem services (Scyphers et al., 2015b). Nonetheless, this study found that the strongest predictor of shoreline condition of a given property was the type of shoreline adjacent to the property, and homeowners where neighboring properties have a bulkhead were twice as likely to prefer that option as those that were not adjacent to a bulkhead (Scyphers et al., 2015b). In addition to cost, durability, and familiarity, other factors that can contribute to shoreline stabilization choice include aesthetics, perceived maintenance costs, availability of knowledgeable contractors, and ease of permitting. These factors are likely to vary regionally, and assembling this information should be a priority for scientists and resource agencies, as the demand for shoreline protection in coastal communities can be expected to increase in coming years (Arkema et al., 2013; Sutton-Grier et al., 2015).

4. CLIMATE ADAPTATION AND LIVING SHORELINES

Wider implementation of Living Shorelines for stabilization of estuarine shorelines can help to address two pressing societal needs: protecting coastal communities from accelerated sea level rise and conserving coastal habitats and the ecosystem services they provide. In an era of accelerating sea level rise, the area of coastal wetland habitat can be maintained via two mechanisms: vertical accretion to maintain marsh surface elevation relative to mean sea level and horizontal transgression onto higher ground (Kirwan et al., 2016, Fig. 30.8A). The former is controlled by tidal inundation, sediment supply, and marsh biomass, whereas the latter is limited by topography and anthropogenic modification of the shoreline (Brinson et al., 1995; Titus et al., 2009; Kirwan et al., 2016). A recent metaanalysis of global marsh elevations concludes that although marshes are currently building elevation at rates similar to historic rates of sea level rise, under accelerated sea level rise, the primary mechanism for marsh survival will be landward transgression (Kirwan et al., 2016). In the United States, tidal wetlands occupy almost half of the remaining, unhardened coastline (Gittman et al., 2015). After a review of state and local land use plans in the United States, Titus et al. (2009) estimated that over 60% of the land below 1 m elevation is expected to be developed in the future. That development would represent a barrier to the ability of salt marsh and other coastal wetlands to migrate landward. The term "coastal squeeze" is used to describe the loss of intertidal habitat seaward of hardened shoreline structures as sea level rises (Pontee, 2013, Fig. 30.8B).

Living Shorelines are utilized to protect homes and infrastructure from marsh edge erosion and to provide some protection from sea level rise through biological and geomorphological

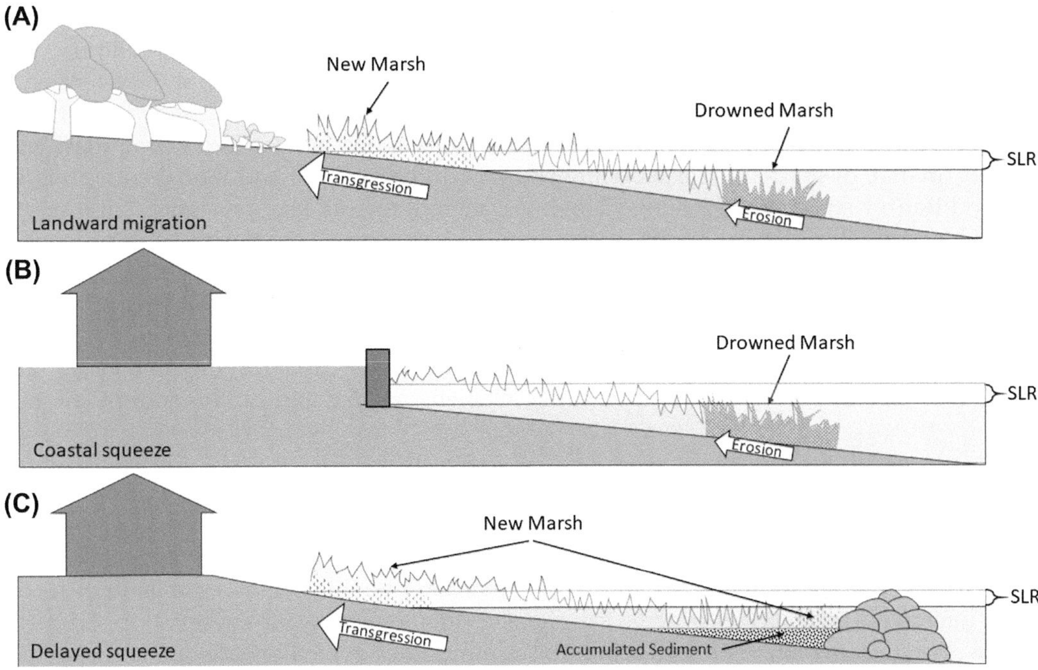

FIGURE 30.8 Fringing marsh response to sea level rise (SLR), shoreline hardening, and coastal development associated with Living Shorelines: (A) marsh landward transgression with SLR allows maintenance of marsh habitat despite seaward erosion; (B) hardened structure protecting property promotes seaward erosion and loss of wetlands, or coastal squeeze (Pontee, 2013), and eliminates landward migration of the marsh; (C) delayed squeeze as a result of a Living Shoreline established to protect coastal property. Erosion protection and enhanced sediment accretion may result in short-term marsh gain and resiliency, but long-term property ownership will limit marsh landward migration response to SLR.

feedback processes that increase marsh and oyster reef elevation (Morris et al., 2002; Rodriguez et al., 2014). Hybrid Living Shorelines that incorporate a low sill at the marsh edge can further increase sediment accretion in the marsh, building elevation capital and providing increased resilience to sea level rise (Morris et al., 2002; Currin et al., 2008, 2017; Gittman et al., 2016a), although this will vary regionally and with project design specifications (Bilkovic et al., 2016). However, Living Shorelines may result in a "delayed coastal squeeze," where natural shoreline habitats are able to persist longer than if traditional hardening took place (Fig. 30.8B), but eventually marsh landward progression will be prevented on low-lying properties where valuable real estate or infrastructure has been built (Fig. 30.8C). Therefore, although Living Shorelines represent a short-term (decadal) preservation of coastal wetlands, they may contribute to wetland loss further in the future, if policy is not developed to accommodate marsh migration or managed retreat (French, 2008; Titus et al., 2009). Regional assessments of areas where future landward migration of marshes can occur, and of areas where topography and coastal development combine to limit the ability of shoreline marshes to migrate landward in the future, can support the development of environmental policy and permitting guidelines that preserve coastal wetlands (Titus et al., 2009; Torio and Chmura, 2013).

Finally, all shoreline stabilization strategies that alter, reduce, or create natural habitat result in a trade-off of ecosystem services (Bilkovic et al., 2016). Hardened shorelines, while protecting uplands and built infrastructure, clearly result in both local and system-wide reductions in wetlands and fauna (Gittman et al., 2016b; Kornis et al., 2017). Even native habitats created as part of Living Shorelines may replace one habitat (beach, mudflat) with another (marsh, oyster reef) and therefore support different floral and faunal communities. The addition of structure to Living Shorelines, including coir logs or stone at the marsh toe, or low rock sills offshore of the marsh, may add structural complexity, which increases faunal use, and sills may increase sediment accretion, which results in greater resilience to sea level rise compared with natural fringing marshes (Gittman et al., 2016a; Currin et al., 2017). However, these structures also reduce benthic fauna via elimination of habitat, and higher marsh elevations can result in less fish utilization (Minello et al., 2012) and reductions in other marsh functions including sediment trapping, nutrient cycling, and carbon accumulation.

5. SUMMARY

This chapter defines Living Shorelines as shoreline stabilization efforts that primarily utilize natural estuarine shoreline habitats and vegetation and that preserve the natural land–water interface. The ability of fringing salt marshes to attenuate wave energy, trap sediment, reduce shoreline erosion, and support faunal communities is well-established and provides the basis for Living Shoreline implementation and evaluation. Hardened structures used to stabilize the shoreline, including bulkheads, riprap revetments, and breakwaters, deepen the nearshore water column and alter sediment properties and have an adverse impact on wetland habitats and nearshore faunal communities. In addition to site-level reductions in faunal abundance and biodiversity, recent studies have demonstrated the cumulative impacts of shoreline hardening on subestuary-scale measures of wetland vegetation and faunal abundance.

Salt marsh vegetation has been used successfully to stabilize shorelines since the 1970s, and general guidelines for the wave energy and geomorphological settings in which vegetation could be utilized have been established. The addition of oyster reefs offshore of fringing salt marshes can effectively reduce shoreline erosion, although in high-energy settings erosion of the marsh edge may still take place. More recently, hybrid Living Shoreline designs, which add structural features to the marsh edge or just offshore to further reduce wave energy, have been developed. Assessment of these designs demonstrates that marsh vegetation is established within 3–8 years, and sediment accretion and organic matter accumulation occur at rates similar to or greater than natural fringing marshes. These designs have also proven resilient to hurricanes. The abundance and biodiversity of fish and crustaceans found in Living Shoreline sites equal or exceed those of natural fringing marshes. Epifauna communities are also similar, especially where sills or oyster reefs offer additional structural complexity. Benthic infauna may be reduced in Living Shorelines compared to natural fringing marshes, as sediment characteristics take longer to develop, and placement of rock sills reduces habitat availability.

Living Shorelines can help to minimize the loss of wetland habitat from "coastal squeeze," where sea level rise and erosion on the seaward edge and hardened structures on the landward side combine to reduce fringing marsh habitat. However, the ability of Living Shorelines to protect coastal property has a limited lifetime, and longer-term protection may not be possible, resulting in a "delayed coastal squeeze." Habitat trade-offs, particularly those associated with hybrid designs, may limit the long-term provision of ecosystem services. Improved design guidelines and regulatory policy are needed for the successful implementation of Living Shorelines to conserve wetland habitats and promote coastal resiliency.

Acknowledgments

I thank Q. Walker for preparation of Figs. 30.1, 30.2 and 30.8 and R. Giannelli for help with the Reference section. R. Gittman kindly permitted our use of an original figure. Reviews by J. Davis, B. Ward, K. Onerevole, and two anonymous reviewers improved the manuscript. This work was supported by funding from the NOAA National Centers for Coastal Ocean Science and the Defense Coastal Estuarine Research Program. The scientific results and conclusions, as well as any views and opinions, expressed herein, are those of the author and do not necessarily reflect the views of NOAA or the Department of Commerce and should not be construed as an official US Department of Defense position or decision unless so designated by other official documentation.

References

Airoldi, L., Abbiati, M., Beck, M.W., Hawkins, S.J., Jonsson, P.R., Martin, D., Moschella, P.S., Sundelöf, A., Thompson, R.C., Aberg, P., 2005. An ecological perspective on the deployment and design of low-crested and other hard coastal defense structures. Coastal Engineering 52 (10–11), 1073–1087.

Arkema, K.K., Guannel, G., Verutes, G., Wood, S.A., Guerry, A., Ruckelshaus, M., Kareiva, P., Lacayo, M., Silver, J.M., 2013. Coastal habitats shield people and property from sea-level rise and storms. National Climate Change 3, 913–918.

Bilkovic, D.M., Roggero, M.M., Hershner, C.H., Havens, K.H., 2006. Influence of land use on microbenthic communities in nearshore estuarine habitats. Estuaries and Coasts 29 (6), 1185–1195.

Bilkovic, D.M., Roggero, M.M., 2008. Effects of coastal development on nearshore estuarine nekton communities. Marine Ecology Progress Series 358, 27–39.

Bilkovic, D.M., Mitchell, M.M., 2013. Ecological tradeoffs of stabilized salt marshes as a shoreline protection strategy: effects of artificial structures on microbenthic assemblages. Ecological Engineering 61, 469–481.

Bilkovic, D.M., Mitchell, M., Mason, P., Duhring, K., 2016. The role of living shorelines as estuarine habitat conservation strategies. Coastal Management 44, 161–174.

Bouma, T.J., De Vries, M.B., Low, E., Peralta, G., Tanczos, I.C., van de Koppel, J., Herman, P.M.J., 2005. Trade-offs related to ecosystem engineering: a case study on stiffness of emerging macrophytes. Ecology 86 (8), 2187–2199.

Bozek, C.M., Burdick, D.M., 2005. Impacts of seawalls on saltmarsh plant communities in the Great Bay Estuary, New Hampshire USA. Wetlands Ecology and Management 13 (5), 553–568.

Bridges, T.S., Wagner, P.W., Burks-Copes, K.A., et al., 2015. Use of Natural and Nature-based Features (NNBF) for Coastal Resilience. US Army Corps of Engineers. ERDC SR-15-1.

Brinson, M.M., Christian, R.R., Blum, L.K., 1995. Multiple states in the sea-level induced transition from terrestrial forest to estuary. Estuaries 18 (4), 648–659.

Broome, S.W., Seneca, E.D., Woodhouse, W.W., 1986. Long-term growth and development of transplants of the salt-marsh grass *Spartina alterniflora*. Estuaries 9, 63–74.

Bulleri, F., Chapman, M.G., 2010. The introduction of coastal infrastructure as a driver of change in marine environments. Journal of Applied Ecology 47, 26–35.

Burke, J.S., Miller, J.M., Hoss, D.E., 1991. Immigration and settlement pattern of *Paralichthys dentatus* and *P. lethostigma* in an estuarine nursery ground, North Carolina, USA. Netherlands Journal of Sea Research 27 (3–4), 393–405.

Cahoon, D.R., Lynch, J.C., Hensel, P., Boumans, R., Perez, B.C., Segura, B., Day Jr., J.W., 2002. High-precision measurements of wetlands sediment elevation: I. Recent improvements to the sedimentation-erosion table. Journal of Sedimentary Research 72 (5), 730–733.

Cahoon, D.R., 2006. A review of major storm impacts on coastal wetlands elevations. Estuaries and Coasts 29 (6), 889–898.

Chauhan, P.P.S., 2009. Autocyclic erosion in tidal marshes. Geomorphology 110 (3–4), 45–57.

Christiansen, T., Wiberg, P.L., Milligan, T.G., 2000. Flow and sediment transport on a tidal salt marsh surface. Estuarine, Coastal and Shelf Science 50 (3), 315–331.

Craft, C., Reader, J., Sacco, J.N., Broome, S.W., 1999. Twenty-five years of ecosystem development of constructed *Spartina alterniflora* (loisel) marshes. Ecological Applications 9 (4), 1405–1419.

Craft, C., Megonigal, P., Broome, S., Stevenson, J., Freese, R., Cornell, J., Zheng, L., Sacco, J., 2003. The pace of ecosystem development of constructed *Spartina alterniflora* marshes. Ecological Applications 13 (5), 1417–1432.

Currin, C.A., Delano, P.C., Valdes-Weaver, L.M., 2008. Utilization of a citizen monitoring protocol to assess the structure and function of natural and stabilized fringing salt marshes in North Carolina. Wetlands Ecology and Management 16, 97–118.

Currin, C.A., Chappell, W.S., Deaton, A., 2010. Developing alternative shoreline armoring strategies: the living shoreline approach in North Carolina. In: Shipman, H., Dethier, M., Gelfanbaum, G., Fresh, K.L., Dinicola, R.S. (Eds.), Puget Sound Shorelines and the Impacts of Armoring-proceedings: USGS Scientific Investigations Report 2010-5254, pp. 91–102.

Currin, C.A., Davis, J., Cowart Baron, L., Malhotra, A., Fonseca, M., 2015. Shoreline change in the new river estuary, North Carolina: rates and consequences. Journal of Coastal Research 31, 1069–1077.

Currin, C.A., Davis, J., Malhotra, A., 2017. Response of salt marshes to wave energy provides guidance for successful living shoreline implementation. In: Bilkovic, D.M., Mitchell, M., La Peyre, M., Toft, J. (Eds.), Living Shorelines: The Science and Management of Nature-based Coastal Protection. CRC Press, Taylor & Francis Group, pp. 211–234.

Davis, J., Levin, L., Walther, S., 2002. Artificial armored shorelines: sites for open-coast species in a southern California bay. Marine Biology 140 (6), 1249–1262.

Davis, J., Nowicki, B., Wigand, C., 2004. Denitrification in fringing salt marshes of Narragansett Bay, Rhode Island, USA. Wetlands 24 (4), 870–878.

Davis, J.L., Currin, C.A., O'Brien, C., Raffenburg, C., Davis, A., 2015. Living shorelines: coastal resilience with a blue carbon benefit. PLoS One. https://doi.org/10.1371/journal.pone.0142595.

DeLuca, W.V., Studds, C.E., King, R.S., Marra, P.P., 2008. Coastal urbanization and the integrity of estuarine waterbird communities: threshold responses and the importance of scale. Biological Conservation 141 (11), 2669–2678.

Dethier, M.N., Raymond, W.W., McBride, A.N., Toft, J.D., Cordell, J.R., Ogston, A.S., Heerhartz, S.M., Berry, H.D., 2016. Multiscale impacts of armoring on Salish Sea shorelines: evidence for cumulative and threshold effects. Estuarine, Coastal and Shelf Science 175, 106–117.

Dugan, J.E., Airoldi, L., Chapman, M.G., Walker, S.J., Schlacher, T., 2011. Estuarine and coastal structures: environmental effects, a focus on shore and nearshore structures. In: Wolanski, E., McLusky, D.S. (Eds.), Treatise on Estuarine and Coastal Science, vol. 8. Academic Press, Waltham, pp. 17–41.

Fagherazzi, S., Kirwan, M.L., Mudd, S.M., et al., 2011. Numerical models of salt marsh evolution: ecological, geomorphic, and climatic factors. Reviews of Geophysics 50. https://doi.org/10.1029/2011RG000359. RG1002.

Fagherazzi, S., Mariotti, G., Wiber, P.L., McGlathery, K.J., 2013. Marsh collapse does not require sea level rise. Oceanography 26, 70–77.

Feagin, R.A., Lozada-Bernard, S.M., Ravens, T.M., Möller, I., Yeager, K.M., Baird, A.H., 2009. Does vegetation prevent wave erosion of salt marsh edges? Proceedings of the National Academy of Sciences 106 (25), 10109–10113.

Federal Register, January 2017. www.federalregister.gov/documents/2017/01/06/2016-31355/issuance-and-reissuance-of-nationwide-permits.

Findlay, S.E.G., Strayer, D.L., Smith, S.D., Curri, N., 2014. Magnitude and patterns of change in submerged aquatic vegetation of the tidal freshwater Hudson River. Estuaries and Coasts 37 (5), 1233–1242.

Firth, L.B., Thompson, R.C., Bohn, K., et al., 2014. Between a rock and a hard place: environmental and engineering considerations when designing coastal defence structures. Coastal Engineering 87, 122–135.

Fonseca, M.S., Callahan, J.A., 1992. A preliminary evaluation of wave attenuation for four species of seagrasses. Estuarine, Coastal and Shelf Science 35, 565–576.

Fonseca, M.S., Bell, S.S., 1998. Influence of physical setting on seagrass landscapes near Beaufort, North Carolina, USA. Marine Ecology Progress Series 171, 109–121.

French, J.R., 2008. Hydrodynamic modelling of estuarine flood defence realignment as an adaptive management response to sea-level rise. Journal of Coastal Research 24, 1–12.

Garbisch Jr., E.W., Woller, P.B., Bostian, W.J., McCallum, R.J., 1975. Biotic techniques for shore stabilization. In: Cronin, L.E. (Ed.), Estuarine Research. Geology and Engineering, vol. 2. Academic Press, Inc., New York, pp. 405–426.

Gedan, K.B., Kirwan, M.L., Wolanski, E., Barbier, E.B., Silliman, B.R., 2011. The present and future role of coastal wetland vegetation in protecting shorelines: answering recent challenges to the paradigm. Climatic Change 106, 7–29.

Gehman, A.M., McLenaghan, N.A., Byers, J.E., Alexander, C.R., Pennings, S.C., Alber, M.A., 2018. Effects of small-scale armoring and residential development on the salt marsh-upland ecotone. Estuaries and Coasts 41 (Suppl 1), S54–S67.

Gittman, R.K., Popowich, A.M., Bruno, J.F., Peterson, C.H., 2014. Marshes with and without sills protect estuarine shorelines from erosion better than bulkheads during a Category 1 hurricane. Ocean and Coastal Management 102 (A), 94–102.

Gittman, R.K., Fodrie, F.J., Popowich, A.M., Keller, D.A., Bruno, J.F., Currin, C.A., Peterson, C.H., Piehler, M.F., 2015. Engineering away our natural defenses: an analysis of shoreline hardening in the US. Frontiers in Ecology and the Environment 13, 301–307.

Gittman, R.K., Peterson, C.H., Currin, C.A., Fodrie, F.J., Piehler, M.F., Bruno, J.F., 2016a. Living shorelines can enhance the nursery role of threatened coastal habitats. Ecological Applications 26, 249–263.

Gittman, R.K., Scyphers, S.B., Smith, C.S., Neylan, I.P., Grabowski, J.H., 2016b. Ecological consequences of shoreline hardening: a meta-analysis. BioScience 66 (9), 763–773.

Gleason, M.L., Elmer, D.A., Pien, N.C., Fisher, J.S., 1979. Effects of stem density upon sediment retention by salt marsh cord grass, *Spartina alterniflora* loisel. Estuaries 2 (4), 271–273.

Grabowski, J.H., Brumbaugh, R.D., Conrad, R.F., Keeler, A.G., Opaluch, J.J., Peterson, C.H., Piehler, M.F., Powers, S.P., Smyth, A.R., 2012. Economic valuation of ecosystem services provided by oyster reefs. BioScience 62 (10), 900–909.

Grizzle, R.E., Adams, J.R., Walters, L.J., 2002. Historical changes in intertidal oyster (*Crassostrea virginica*) reefs in a Florida lagoon potentially related to boating activities. Journal of Shellfish Research 21 (2), 749–756.

Hettler Jr., W.F., 1989. Nekton use of regularly-flooded saltmarsh cordgrass habitat in North Carolina, USA. Marine Ecology Progress Series 56, 111–118.

Isdell, R.E., Chambers, R.M., Bilkovic, D.M., Leu, M., 2015. Effects of terrestrial-aquatic connectivity on an estuarine turtle. Diversity and Distributions 21 (6), 643–653.

Jackson, N.L., Nordstrom, K.F., 1992. Site specific controls on wind and wave processes and beach mobility on estuarine beaches in New Jersey, U.S.A. Journal of Coastal Research 8, 88–98.

Kathiresan, K., Rajendran, N., 2005. Coastal mangrove forests mitigated tsunami. Estuarine, Coastal and Shelf Science 65, 601–606.

Kennish, M.J., 2002. Environmental threats and environmental future of estuaries. Environmental Conservation 29 (1), 78–107.

Kirwan, M.L., Guntenspergen, G.R., D'Alpaos, A., Morris, J.T., Mudd, S.M., Temmerman, S., 2010. Limits on the adaptability of coastal marshes to rising sea level. Geophysical Research Letters. https://doi.org/10.1029/2010GL045489.

Kirwan, M.L., Temmerman, S., Skeehan, E.E., Guntenspergen, G.R., Fagherazzi, S., 2016. Overestimation of marsh vulnerability to sea level rise. Nature Climate Change 6, 253–260.

Knutson, P.L., Ford, J.C., Inskeep, M.R., Oyler, J., 1981. National survey of planted salt marshes (Vegetative stabilization and wave stress). Wetlands 1 (1), 129–157.

Knutson, P.L., Brochu, R.A., Seelig, W.N., Inskeep, M., 1982. Wave damping in *Spartina alterniflora* marshes. Wetlands 2, 87–104.

Koch, E.W., Barbier, E.B., Silliman, B.R., et al., 2009. Non-linearity in ecosystem services: temporal and spatial variability in coastal protection. Frontiers in Ecology and the Environment 7, 29–37.

Kornis, M.S., Breitburg, D., Balouskus, R., et al., 2017. Linking the abundance of estuarine fish and crustaceans in nearshore waters to shoreline hardening and land cover. Estuaries and Coasts 1–23.

La Peyre, M.K., Humphries, A.T., Casas, S.M., La Peyre, J.F., 2014. Temporal variation in development of ecosystem services from oyster reef restoration. Ecological Engineering 63, 34–44.

La Peyre, M.K., Serra, K., Joyner, T.A., Humphries, A., 2015. Assessing shoreline exposure and oyster habitat suitability maximizes potential success for sustainable shoreline protection using restored oyster reefs. PeerJ 3, e1317. https://doi.org/10.771/peerj.1317.

La Peyre, M.K., Miller, L.S., Miller, S., Melancon, E., 2017. Comparison of oyster populations, shoreline protection services, and site characteristics at seven created fringing reefs in Louisiana, Key parameters and responses to consider. In: Bilkovic, D.M., Mitchell, M., La Peyre, M., Toft, J. (Eds.), Living Shorelines: The Science and Management of Nature-based Coastal Protection. CRC Press, Taylor & Francis Group, pp. 363–382.

Lawless, A.S., Seitz, R.D., 2014. Effects of shoreline stabilization and environmental variables on benthic infaunal communities in the Lynnhaven River system of Chesapeake Bay. Journal of Experimental Marine Biology and Ecology 457, 41–50.

Leonard, L.A., Luther, M.E., 1995. Flow hydrodynamics in tidal marsh canopies. Limnology and Oceanography 40, 1474–1484.

Leonard, L.A., Wren, P.A., Beavers, R.L., 2002. Flow dynamics and sedimentation in *Spartina alterniflora* and *Phragmites australis* marshes of the Chesapeake Bay. Wetlands 22 (2), 415–424.

Leonard, L.A., Croft, A.C., 2006. The effect of standing biomass on flow velocity and turbulence in *Spartina alterniflora* canopies. Estuarine, Coastal and Shelf Science 69, 325–336.

Leonardi, N., Ganju, N.K., Fagherazzi, S., 2016. A linear relationship between wave power and erosion determines salt-marsh resilience to violent storms and hurricanes. Proceedings of the National Academy of Sciences 113, 64–68.

Maloney, F.E., Ausness, R.C., 1974. The use of legal significant of the mean high water line in coastal boundary mapping. North Carolina Law Review 53, 185.

Manis, J.E., Garvis, S.K., Jachec, S.M., Walters, L.J., 2014. Wave attenuation experiments over living shorelines over time: a wave tank study to assess recreational boating pressures. Journal of Coastal Conservation 19 (1), 1–11.

Marani, M., D'Alpaos, A., Lanzoni, S., Santalucia, M., 2011. Understanding and predicting wave erosion of marsh edges. Geophysical Research Letters 38, L21401.

Mattheus, C.R., Rodriguez, A., Mckee, B., Currin, C.A., 2010. Impact of land-use change and hard structures on the evolution of fringing marsh shorelines. Estuarine, Coastal and Shelf Science 88, 365–376.

Mariotti, G., Carr, J., 2014. Dual role of salt marsh retreat: long-term loss and short-term resilience. Water Resources Research 50, 2963–2974.

Martin, J.B., Cable, J.E., Jaeger, J., 2005. Quantification of Advective Benthic Processes Contributing Nitrogen and Phosphorus to Surface Waters of the Indian River Lagoon, p. 244.

Mazda, Y., Magi, M., Kogo, M., Hong, P.N., 1997. Mangroves as a coastal protection from waves in the Tong King delta, Vietnam. Mangroves and Salt Marshes 2, 127–135.

Meyer, D.L., Townsend, E.C., Thayer, G.W., 1997. Stabilization and erosion control value of oyster cultch for intertidal marsh. Restoration Ecology 5 (1), 93–99.

Meyer, D.L., Posey, M.H., 2009. Effects of life history strategy on fish distribution and use of estuarine salt marsh and shallow-water flat habitats. Estuaries and Coasts 32 (4), 797–812.

Micheli, F., Peterson, C.H., 1999. Estuarine vegetated habitats as corridors for predator movements. Conservation Biology 13 (4), 869–881.

Miles, J.R., Russell, P.E., Huntly, D.A., 2001. Field measurements of sediment dynamics in front of a seawall. Journal of Coastal Research 17, 195–206.

Millennium Ecosystem Assessment (MEA), 2005. Ecosystems and Human Well-being. Island Press, Washington, DC. http://www.maweb.org/en/index.aspx.

Miller, D.C., Geider, R.J., MacIntyre, H.L., 1996. Microphytobenthos: The ecological role of the "secret garden" of unvegetated, shallow-water marine habitats. II. Role in sediment stability and shallow-water food webs. Estuaries 19 (2), 202–212.

Minello, T.J., Zimmerman, R.J., Medina, R., 1994. The importance of edge for natant macrofauna in a created salt marsh. Wetlands 14 (3), 184–198.

Minello, T.J., Rozas, L.P., Baker, R., 2012. Geographic variability in salt marsh flooding patterns may affect nursery value for fishery species. Estuaries and Coasts 35, 501–514.

Möller, I., Spencer, T., 2002. Wave dissipation over macro-tidal saltmarshes: Effects of marsh edge typology and vegetation change. Journal of Coastal Research 36, 506–521.

Möller, I., 2006. Quantifying saltmarsh vegetation and its effect on wave height dissipation: Results from a UK east coast saltmarsh. Estuarine, Coastal and Shelf Science 69, 337–351.

Möller, I., Kudella, M., Rupprecht, F., et al., 2014. Wave attenuation over coastal salt marshes under storm surge conditions. Nature Geoscience 7, 727–731.

Moody, R.M., Cebrian, J., Kerner, S.M., Heck Jr., K.L., Powers, S.P., Ferraro, C., 2013. Effects of shoreline erosion on salt-marsh floral zonation. Marine Ecology Progress Series 488, 145–155.

Morgan, P.A., Burdick, D.M., Short, F.T., 2009. The functions and values of fringing salt marshes in northern New England, USA. Estuaries and Coasts 32, 483–495.

Morris, J.T., Sundareshwar, P.V., Nietch, C.T., Kjerfve, B., Cahoon, D.R., 2002. Responses of coastal wetlands to rising sea level. Ecology 83, 2869–2877.

Mudd, S.M., D'Alpaos, A., Morris, J.T., 2010. How does vegetation affect sedimentation on tidal marshes? Investigating particle capture and hydrodynamic controls on biologically mediated sedimentation. Journal of Geophysical Research 115 (F3).

Mudd, S.M., 2011. The life and death of salt marshes in response to anthropogenic disturbance of sediment supply. Geology 39 (5), 511–512.

NOAA (National Oceanic and Atmospheric Administration), 2001. Sea Level Variations of the United States 1854-1999. Technical Report NOS CO-OPS 36.

NOAA (National Oceanic and Atmospheric Administration), 2015. Conceptual guidance for considering the use of Living Shorelines. In: NOAA Habitat Conservation Team Living Shorelines Workgroup.

Nordstrom, K.F., Jackson, N.L., 2012. Physical processes and landforms on beaches in short fetch environments in estuaries, small lakes and reservoirs: A review. Earth-Science Reviews 111 (1–2), 232–247.

NRC (National Research Council), 2007. Mitigating Shore Erosion along Sheltered Coasts. National Research Council, National Academies Press, Washington, DC.

O'Connor, M.I., Violin, C., Anton, A., Ladwig, L.M., Piehler, M.F., 2011. Salt marsh stabilization affects algal primary producers at the marsh edge. Wetlands Ecology and Management 19, 131–140. https://doi.org/10.1007/s11273-010-9206-y.

O'Meara, T., Thompson, S.P., Piehler, M.F., 2015. Effects of shoreline hardening on nitrogen processing in estuarine marshes of the U.S. mid-Atlantic coast. Wetlands Ecology and Management 23 (3), 385–394.

Pace, N.L., 2017. Permitting a Living Shoreline: A look at the legal framework governing Living Shoreline projects at the federal, state and local level. In: Bilkovic, D.M., Mitchell, M., La Peyre, M., Toft, J. (Eds.), Living Shorelines: The Science and Management of Nature-based Coastal Protection. CRC Press, Taylor & Francis Group, pp. 33–50.

Paterson, D.M., Crawford, R.M., Little, C., 1990. Sub-aerial exposure and changes in the stability of intertidal estuarine sediment. Estuarine, Coastal and Shelf Science 34, 223–234.

Patrick, C.J., Weller, D.E., Ryder, M., 2016. The relationship between shoreline armoring and adjacent submerged aquatic vegetation in Chesapeake Bay and nearby Atlantic coastal bays. Estuaries and Coasts 39, 158–170.

Pendleton, L., Donato, D.C., Murray, B.C., et al., 2012. Estimating global "blue carbon" emissions from conversion and degradation of vegetated coastal ecosystems. PLoS One. https://doi.org/10.1371/journal.pone.0043542.

Peterson, G.W., Turner, R.E., 1994. The value of salt marsh edge vs interior as a habitat for fish and decapod crustaceans in a Lousisiana tidal marsh. Estuaries 17, 235–262.

Pontee, N., 2013. Defining coastal squeeze: A discussion. Ocean and Coastal Management 84, 204–207.

Popkin, G., 2015. Breaking the waves. Science 350 (6262), 756–759.

Priestas, A.M., Mariotti, G., Leonardi, N., Fagherazzi, S., 2015. Coupled wave energy and erosion dynamics along a salt marsh boundary, Hog Island Bay, Virginia, USA. Journal of Marine Science and Engineering 3, 1041–1065.

Rodriguez, A.B., Fodrie, F.J., Ridge, J.T., et al., 2014. Oyster reefs can outpace sea-level rise. Nature Climate Change 4, 493–497.

Roland, R.M., Douglass, S.L., 2005. Estimating wave tolerance of *Spartina alterniflora* in coastal Alabama. Journal of Coastal Research 21 (3), 453–463.

Rozas, L.P., Reed, D.J., 1993. Nekton use of marsh-surface habitats in Louisiana (USA) deltaic salt marshes undergoing submergence. Marine Ecology Progress Series 96, 147–157.

Ruiz, G.M., Hines, A.H., Posey, M.H., 1993. Shallow water as a refuge habitat for fish and crustaceans in non-vegetated estuaries: an example from Chesapeake Bay. Marine Ecology Progress Series 99, 1–16.

Scyphers, S.B., Powers, S.P., Heck Jr., H.L., Byron, D., 2011. Oyster reefs as natural breakwaters mitigate shoreline loss and facilitate fisheries. PLoS One. https://doi.org/10.1371/journal.pone.0022396.

Scyphers, S.B., Gouhier, T.C., Grabowski, J.H., Beck, M.W., Mareska, J., Powers, S.P., 2015a. Natural shorelines promote the stability of fish communities in an urbanized coastal system. PLoS One. https://doi.org/10.1371/journal.pone.0118580.

Scyphers, S.B., Picou, J.S., Powers, S.P., 2015b. Participatory conservation of coastal habitats: The importance of understanding homeowner decision making to mitigate cascading shoreline degradation. Conservation Letters 8 (1), 41–49.

Seitz, R.D., Lawless, A.S., 2006. Landscape-level impacts of shoreline development on Chesapeake Bay benthos and their predators. In: Erdle, S.Y., Davis, J.L., Sellner, K.G. (Eds.), Management, Policy, Science, and Engineering of Nonstructural Erosion Control in the Chesapeake Bay, CRC Publ. No. 08-164, Chesapeake Bay, pp. 63–70.

Seitz, R.D., Lipcius, R.N., Olmstead, N.H., Seebo, M.S., Lambert, D.M., 2006. Influence of shallow-water habitats and shoreline development on abundance, biomass, and diversity of benthic prey and predators in Chesapeake Bay. Marine Ecology Progress Series 326, 11–27.

Seitz, R.D., Knick, K.E., Davenport, T.M., Saluta, G.G., 2018. Human influence at the coast: Upland and shoreline stressors affect coastal macrofauna and are mediated by salinity. Estuaries and Coasts 41 (Suppl 1), 114. https://doi.org/10.1007/s12237-017-0347-6.

Shafer, D.J., Streever, W.J., 2000. A comparison of 28 natural and dredged material salt marshes in Texas with an emphasis on geomorphological variables. Wetlands Ecology and Management 8, 353–366.

Shafer, D.J., Roland, R., Douglass, S.L., 2003. Preliminary Evaluation of Critical Wave Energy Thresholds at Natural and Created Coastal Wetlands. WRP Technical Notes Collection (No. ERDC TNWRP-HS-CP-2.2). US Army Engineer Research and Development Center, Vicksburg, Mississippi.

Shepard, C.C., Crain, C.M., Beck, M.W., 2011. The protective role of coastal marshes: A systematic review and meta-analysis. PLoS One. https://doi.org/10.1371/journal.pone.0027374.

Shipman, H., 2010. The Geomorphic Setting of Puget Sound: Implications for Shoreline Erosion and the Impacts of Erosion Control Structures. Puget Sound Shorelines and the Impacts of Armoring – Proceedings: USGS Scientific Investigations Report 2010-5254, pp. 91–102.

Silliman, B.R., Bertness, M.D., 2004. Shoreline development drives invasion of *Phragmites australis* and the loss of plant diversity on New England salt marshes. Conservation Biology 18 (5), 1424–1434.

Smith, K.A., North, E.W., Shi, F., Chen, S., Hood, R.R., Koch, E.W., Newell, R.I.E., 2009. Modeling the effects of oyster reefs and breakwaters on seagrass growth. Estuaries and Coasts 32 (4), 748–757.

Smyth, A.R., Thompson, S.P., Siporin, K.N., Gardner, W.S., McCarthy, M.J., Piehler, M.F., 2013. Assessing nitrogen dynamics throughout the estuarine landscape. Estuaries and Coasts 36, 44–55.

Stal, L.J., 2003. Microphytobenthos, their extracellular polymeric substances, and the morphogenesis of intertidal sediments. Geomicrobiology Journal 20 (5), 463–478.

Strayer, D.L., Findlay, S.E.G., Miller, D., Malcom, H.M., Fischer, D.T., Coote, T., 2012. Biodiversity in Hudson River shores zones: influence of shoreline type and physical structure. Aquatic Sciences 74 (3), 597–610.

Sutton-Grier, A.E., Wowk, K., Bamford, H., 2015. Future of our coasts: The potential for natural and hybrid infrastructure to enhance the resilience of our coastal communities, economies and ecosystems. Environmental Science and Policy 51, 137–148.

Temmerman, S., Meire, P., Bouma, T.J., Herman, P.M.J., Ysebaert, T., De Vriend, H.J., 2013. Ecosystem-based coastal defence in the face of global change. Nature 504, 79–83.

Theuerkauf, E.J., Stephens, J.D., Ridge, J.T., Fodrie, F.J., Rodriguez, A.B., 2015. Carbon export from fringing saltmarsh shoreline erosion overwhelms carbon storage across a critical width threshold. Estuarine, Coastal and Shelf Science 164 (5), 367–378.

Theuerkauf, S.J., Eggleston, D.B., Puckett, B.J., Theuerkauf, K.W., 2016. Wave exposure structures oyster distribution on natural intertidal reefs, but not on hardened shorelines. Estuaries and Coasts 40 (2), 376–386.

Titus, J.G., Hudgens, D.E., Trescott, D.L., et al., 2009. State and local governments plan for development of most land vulnerable to rising sea level along the US Atlantic coast. Environmental Research Letters 4 (4). https://doi.org/10.1088/1748-9326/4/044008.

Tobias, C.R., Macko, S.A., Anderson, I.C., Canuel, E.A., Harvey, J.W., 2001. Tracking the fate of a high concentration groundwater nitrate plume through a fringing marsh: A combined groundwater tracer and in situ isotope enrichment study. Limnology and Oceanography 46 (8), 1977–1989.

Tonelli, M., Fagherazzi, S., Petti, M., 2010. Modeling wave impact on salt marsh boundaries. Journal of Geophysical Research 115. https://doi.org/10.1029/2009JC006026.

Torio, D.D., Chmura, G.L., 2013. Assessing coastal squeeze of tidal wetlands. Journal of Coastal Research 29 (5), 1049–1061.

Tourtellotte, G.H., Dauer, D.M., 1983. Macrobenthic communities of the lower Chesapeake Bay. II. Lyn-haven roads, Lynnhaven Bay, Broad Bay, and Linkhorn Bay. Hydrobiology 68, 59–72.

Weis, J.S., Weis, P., Proctor, T., 1998. The extent of benthic impacts of cca-treated wood structures in Atlantic coast estuaries. Archives of Environmental Contamination and Toxicology 34 (4), 313–322.

Yang, S.L., Shi, B.W., Bouma, T.J., Ysebaert, T., Luo, X.X., 2012. Wave attenuation at a salt marsh margin: A case study of an exposed coast on the Yangtze Estuary. Estuaries and Coasts 35, 169–182.

Further Reading

Dahl, T.E., Stedman, S.M., 2013. Status and Trends of Wetlands in the Coastal Watersheds of the Conterminous United States 2004 to 2009. U.S. Department of the Interior, Fish and Wildlife Service and National Oceanic and Atmospheric Administration, National Marine Fisheries Service.

CHAPTER

31

Mangrove Management: Challenges and Guidelines

Daniel O. Suman

Rosenstiel School of Marine and Atmospheric Science, University of Miami, Miami, FL, United States

1. MANGROVE ECOSYSTEM SERVICES

Mangroves colonize intertidal areas and can cope with varying salinities and dynamic sediment movements. Mangrove species are found in 123 countries and include 73 species and hybrids (Spalding et al., 2010). Although they account for less than 1% of the world's tropical forests, they contribute important ecosystem functions and are being significantly altered and lost throughout the world. This chapter discusses some of the challenges and obstacles to effective management of mangrove ecosystems, guidelines for management of mangrove ecosystems that have been accepted by the international community, and various categories of management alternatives.

Mangrove ecosystems provide numerous benefits to local communities and adjacent environments (UNEP-WCMC, 2006). They represent great direct economic importance to the livelihoods of millions of coastal residents throughout the world who harvest marine resources; extract timber for construction, firewood, and charcoal production; and cultivate mangrove honey. Mangrove ecosystems have great linkages to coastal waters; with their export of nutrients, they support biological productivity in coastal waters. Mangroves also serve as nursery and breeding grounds and shelter for many marine species of fish and crustaceans. Shrimp larvae and postlarvae juveniles use the mangrove ecosystem before migrating out to the open ocean as adults. Faced with rising sea levels and tropical storms, tropical and subtropical nations also benefit from the shoreline stabilization and buffering that mangrove trees and root systems provide (Zhang et al., 2012). Mangrove forests reduce coastal erosion rates and help build up land and consolidate sediments. Benefits to the coastal region are also significant. Mangrove wetlands help maintain the coastal hydrological regime potentially reducing inland flooding.

Coastal Wetlands
https://doi.org/10.1016/B978-0-444-63893-9.00031-9

Their fine sediments and abundant populations of filter feeders also help maintain water quality and act as biological filters for some pollutants. Many mangrove areas shelter a rich biological diversity and serve as particularly important roosting, resting, and feeding sites for many species of migratory birds. With the rapid global warming of the anthropocene, mangroves are important areas for carbon sequestration. Mangroves' high standing biomass and soils store a significant amount of organic carbon—perhaps 5000 ± 400 TgC in this relatively stable sink. Thus, they may play an important role in the global carbon cycle and help mitigate climate change (Spalding et al., 2010). Additional nonconsumptive uses—such as recreation, ecotourism, environmental education, scientific research, and cultural and religious uses (i.e. sacred forest sites)—are attracted to the unique characteristics of mangrove ecosystems.

2. MANGROVE LOSSES

The Food and Agriculture Organization of the United Nations (FAO, 2017) estimates that the global area of mangroves in 2000 was 14,653,000 ha with annual losses from 1990 to 2000 at 1.1% (FAO, 2017). According to Spalding et al. (2010), mangrove deforestation rates are 3–5 times greater than global deforestation rates. As the estimate of global mangrove area in 1980 was 19.8 million ha, some 5 million ha of mangrove forests were lost during this 20 year period amounting to about 25% of the mangrove area in 1980. Countries with the largest mangrove losses were Brazil, Indonesia, Australia, Mexico, Papua New Guinea, and Pakistan, whereas countries with the greatest relative losses were the Ivory Coast, Honduras, China, Congo, and Barbados.

The causes of mangrove ecosystem losses are numerous. Habitat degradation and loss through conversion to other activities that may produce a more immediate economic gain are perhaps the most significant. Among these activities are urban and industrial developments, unsustainable aquaculture practices, and cattle ranching and rice and sugar cane cultivation. Water pollution and alteration of hydrological changes also take their toll. While extraction of mangrove resources contributes to the livelihoods of millions of coastal residents, overexploitation of these resources may also degrade the ecosystem. A newly recognized threat to mangroves is climate change/sea level rise. Although mangroves may be able to migrate inland with sea level rise, the existence of coastal developments and infrastructure may make this natural adaptation impossible.

3. CHALLENGES TO MANGROVE MANAGEMENT

Mangrove ecosystems are clearly important centers for biodiversity, providers of numerous ecosystem functions, and sources of natural resources for thousands of small coastal communities throughout the tropics. Despite the importance of mangroves recognized by the scientific community, environmental organizations, and environmental authorities, mangrove losses continue to be high. Numerous factors may partially explain this apparent contradiction (Table 31.1).

TABLE 31.1 Challenges to Effective Management of Mangrove Ecosystems

Threats	Potential Impacts
Population increase	Increased demand for mangrove resources and increased pressure to convert mangrove space to other uses
Conversion of mangrove forests to other uses (urban expansion, ports, agriculture, tourism projects, shrimp aquaculture ponds)	Habitat loss, loss of biodiversity, loss of shoreline erosion protection, loss of carbon sinks, loss of nurseries for marine resources, loss of educational and cultural resources, loss of livelihoods for traditional users of mangrove resources and increasing poverty levels
Water pollution	Increased stress on mangrove fauna and flora
Decreased water flow from upstream diversions	Decreased sediment input may lead to accelerated coastal erosion. Increased salinity may be a stress on mangrove growth.
Sea level rise and climate change	Accelerated erosion and ecosystem migration inland, as well as to higher latitudes
Uncertain land tenure and property regimes in mangrove forests	Traditional mangrove users may be displaced and lose access to mangrove resources; a common pool resource may be usurped by developers
Low level of community input and involvement in mangrove resource management	Lack of support for mangrove conservation efforts; adverse impacts on local community livelihoods
Lack of use of best available science in mangrove rehabilitation efforts	Allowed uses of mangroves that are unsustainable; mangrove restoration efforts that fail
Overexploitation of mangrove resources (wood, fish, and shellfish)	Decreased resource base
Lack of sustainable funding for mangrove management efforts	Mangrove conservation or rehabilitation activities that cannot be maintained
Poor legal framework for mangrove resources	Mangrove resources unprotected; institutional overlaps and contradictions; resource exploitation that cannot be controlled
Lack of enforcement of existing legislation related to mangrove resources	Loss of mangrove habitats, impunity for developers and individuals responsible for mangrove forest conversion
Institutional overlaps and conflicts	Duplication of efforts, wasted resources; mangrove conservation efforts that are undermined
Lack of appreciation of values of mangroves	Absence of rational bases for mangrove conservation and sustainable management
Difficulty of quantifying values of the direct and indirect uses of mangroves	Absence of incentives to protect and sustainably manage mangrove ecosystems
Difficulty of quantifying the ecosystem services that mangrove provide	Devaluation of mangrove ecosystems by decision-makers

3.1 Lack of Awareness of the Value of Mangrove Ecosystems

Long-held perceptions that mangroves are wastelands and breeding grounds for mosquitos and other insects continue to influence actions in many countries. Few people live in mangroves, and those who do tend to be poor, marginalized, and politically powerless. Although some mangrove forests may be declared protected areas, visitor access, and therefore public appreciation, can often be challenging. In short, mangroves have few champions. Decision-makers' lack of awareness of the values of mangrove ecosystems (often nonmarket) compounds this problem.

This lack of awareness of the ecological benefits of mangrove ecosystems has led many governments and international lending organizations (World Bank, FAO, Asian Development Bank, Inter-American Development Bank, among others) to view this coastal space as an optimal site for economic development, particularly through their conversion to shrimp aquaculture ponds. For example, in the Philippines, about a third of international fisheries assistance (approximately $1 billion) has been targeted to aquaculture projects. As a result, aquaculture conversion has been responsible for perhaps half of Philippine mangrove losses (Primavera, 2000). Moreover, the rush to promote shrimp aquaculture and its export earnings has led government institutions to undervalue mangrove ecosystems through often ridiculously low fees for concessions to "develop" mangrove forests through conversion to shrimp ponds (Primavera, 2005). Armitage (2002) also documents the support of governments, multilateral economic development agencies, and well-connected entrepreneurs to expand shrimp aquaculture ponds at the expense of community groups connected to mangrove common pool resources. The privatization of a common pool resource undermines local property rights and resource access and leads to marginalization of traditional mangrove users.

In Guinea, undervalued mangroves are clear-cut, diked, and converted to rice fields that are soon abandoned because of hypersalinity (Wolanski and Cassagne, 2000). Rapid urban expansion has also led to conversion of important mangrove forests in Panama and Singapore (Castellanos-Galindo et al., 2017; Corlett, 1992). Shrimp aquaculture ponds constructed in mangrove forests create numerous adverse impacts that affect hydrology, biota, and soils. When these shrimp ponds or rice fields are abandoned, the resulting soil degradation may make reestablishment of mangrove trees impossible except perhaps where tidal flushing is restored (Wolanski and Cassagne, 2000). Even in sites where mangroves are not overtly converted to different uses, activities in mangrove forests may adversely affect the ecosystem. Excessive harvest of mangrove woods for construction, poles, fuelwood, and charcoal or unsustainable fishing practices, such as use of very fine nets placed across mangrove channels, may degrade mangrove ecosystem functions. Other activities, such as livestock grazing may impact forest structure and recruitment. Grazing of mangrove foliage by cattle, buffalo, and camels is a serious problem particularly in arid environments, such as Orissa State (Eastern India), the Indus Delta (Pakistan), and the Nxaxo Estuary (South Africa) (Hoppe-Spear and Adams, 2015; Pattanaik et al., 2008; Saifullah, 1997).

Coastal space is a limited commodity that is often in high demand. Mangrove ecosystems must compete with powerful economic interests supported by national and local elites, such as urban residential developments, airports, power plants, construction and expansion of ports, aquaculture activities, and coastal tourism projects (Dale et al., 2014). In the eyes of the decision-maker, the immediate economic gain from these conversion activities often outweighs the long-term—often misunderstood and unappreciated—ecosystem benefits that mangrove ecosystems provide. Ecosystem services are extremely difficult to quantify and compare with the immediate economic benefits of development projects. Recent estimates of the economic value of mangrove forests range from $2000 to $9000 per ha (Spalding et al., 2010; Vo et al., 2012). Often these attempts to quantify values of mangrove ecosystems only consider commodities that can be traded, such as wood products and fish. However, it is extremely difficult to quantify nonmarket values for natural resources (shoreline protection and erosion control, fish nurseries, carbon sequestration, biodiversity, water purification, nutrient release, research and education, esthetic and cultural uses), and these efforts always risk cultural bias (Dale et al., 2014).

3.2 Uncertain Property Regimes

Although private and community-based property regimes exist in mangrove forests, in most countries mangrove forests are government-owned lands. Nevertheless, effective government control and presence are often lacking or absent. The majority of mangrove forests are found in developing countries where the regulatory capacity of government institutions may often be weak, social needs are high, and corruption may be problematic. Although these coastal spaces are theoretically government-owned, they are de facto open access areas with minimal control exercised over exploitation and conversion to other uses (Berkes et al., 1989; Primavera and Esteban, 2008; Walters et al., 2008).

In many countries confusion exists over land tenure and exercise of authority in mangrove forests. Old and dubious land titles sometimes arise in areas that are legally public lands. Conflicts with individuals who claim private ownership in mangrove forests even occur in mangroves that have been declared national parks, wildlife refuges, and protected wetlands. Confusion over land tenure and registry, failure to recognize traditional land tenure regimes, and weak government authorities that are often incapable of confronting or even co-opted by strong private interests often create scenarios for conversion of mangrove forests to other uses and alienation of traditional users (Walters et al., 2008).

3.3 Interagency Conflicts and Jurisdictional Overlaps

Located at the land–sea interface, mangrove forests face the dubious honor of existing in a zone of multiple agency jurisdictions. In many countries, the national or central government owns and manages the coastal oceans, whereas the provincial or local government has authority over coastal lands. Mangroves straddle this boundary where governmental management strategies and legislation may be contradictory or duplicative. Similarly, several different agencies at different governmental levels may have regulatory authority over mangrove forests (Iftekhar, 2008; Islam and Wahab, 2005). These may include forestry, fisheries and aquaculture, environmental conservation, wildlife management agencies, and even tourism management authorities. Again, their management objectives often are incompatible and contradictory. Instead of ecosystem-based management of mangrove forests, often the management is fragmented at both sectoral and intergovernmental levels (Dale et al., 2014; Kairo et al., 2001). For example, Primavera (2000) mentions that in the Philippines local governments may eagerly accept property tax payments for land claims in mangrove forests without inspecting the area or evaluating the claim.

3.4 Off-Site Impacts and Linkages

Mangrove forests are interconnected with adjacent terrestrial ecosystems, such as mudflats, salt flats, salt marshes, upstream swamp forests and palm forests, and the upstream watersheds. Upstream water pollution, dams, or water diversion for irrigation may produce serious impacts on downstream coastal wetlands and mangroves (Iftekhar, 2008). These linkages can be physical, ecological, and involve energy flows. Many different faunal species migrate between mangroves and adjacent ecosystems. Fluxes of water and nutrients also occur between mangroves and adjacent ecosystems. Similarly, mangroves have linkages with offshore coastal ocean ecosystems, such as seagrass beds and coral reefs. Wise management recognizes these linkages with adjacent ecosystems. However, often these interconnected ecosystems are not protected comprehensively together with the mangrove ecosystems or at least not managed in an integrated manner with the mangrove forests.

3.5 Climate Change and Sea Level Rise

Sea level rise may be a great stressor for mangrove ecosystems in the future, particularly in subsiding coastal areas or regions with low sediment accretion rates (Gilman et al., 2008; Loucks et al., 2010). Mangrove systems may be able to adapt to sea level rise if they are able to retreat landward with rising seas. However, this may be impossible in areas with developed infrastructure or coastal protective dikes. Proactive management may involve acquisition of inland buffer zones to which mangroves may migrate (Erwin, 2009). On the other hand, rising tides and increased concentrations of atmospheric CO_2 may increase productivity of mangroves. Additionally, this ecosystem may also migrate to higher latitudes.

3.6 Insufficient or Inappropriate Community Involvement

Local communities that live in or near mangroves often experience high levels of poverty and depend on mangroves for their livelihood. Their social and cultural norms and concepts of land tenure have evolved with the adjacent mangrove ecosystem. As a result, local communities with their traditional ecological knowledge should be central to any attempt to sustainably manage or rehabilitate mangrove ecosystems (Iftekhar and Takama, 2008). Moreover, conversation efforts must provide for diversification of livelihood options for adjacent communities. Without their support and involvement, conservation efforts are doomed to fail. Research by Badola et al. (2012) in eastern India corroborates that local communities value mangrove functions that are directly related to their livelihood. Thus, we can expect communities to support mangrove conservation measures but not at the expense of their livelihood.

3.7 Lack of Sustainable Funding for Conservation Efforts

Even in the best scenario where authorities, communities, and/or environmental organizations have developed integrated comprehensive management plans for the mangrove forests and where local community buy-in exists, sustainable funding to implement management plans is often lacking. Sources of sustainable funding for environmental conservation in developing countries are often difficult to tap. However, this is especially the case in mangrove forests that, as described earlier, often lack strong champions.

3.8 Lack of Understanding and Misuse of the Best Available Science

Attempts and investments in mangrove replanting or rehabilitation often fail for lack of understanding of the biophysical environment and the selection of inappropriate mangrove species (Lewis, 2005). Primavera and Esteban (2008) and Barnuevo et al. (2017) identify numerous restoration examples from the Philippines that selected inappropriate sites in the lower intertidal to subtidal zones instead of higher middle to upper intertidal zones. The rationale for the inappropriate site selection was the potential for aquaculture pond development in the middle to upper intertidal zone. Unsuccessful efforts in the Philippines planted *Rhizophora* species instead of the natural colonizers *Avicennia marina* and *Sonneratia alba* because of lack of ecological knowledge or the influence of neotropical restoration efforts that primarily use *Rhizophora mangle*. Planted mangrove stands in abandoned shrimp ponds show low recruitment of nonplanted mangrove species and differ substantially from natural mangrove forests. Implementation of forestry methods, such as thinning to open the canopy and induce enhancement of structural and species diversity, could promote to improve ecological restoration (Barnuevo et al., 2017).

A comprehensive review of success of mangrove planting efforts (largely *Rhizophora* spp.) in Sri Lanka (where "success" was defined as 5-year survival rates) reported that out of 1000–1200 ha of replanted mangroves, only 200–220 ha were successful (Kodikara et al., 2017). Only 3 sites out of 67 showed survival rates greater than 50%. Failures can be explained largely because the projects selected inappropriate sites above the intertidal zone, susceptible to trampling by cattle, with excessive submergence and with extreme exposure to sunlight. The projects' failure to follow scientific approaches and best management practices led to their failures.

4. INTERNATIONAL GUIDELINES FOR MANGROVE MANAGEMENT

Numerous international organizations and non-governmental organizations recognize the importance of mangrove ecosystems for the direct and indirect benefits they proportion to coastal peoples and environments. Through their resolutions, statements, publications, and actions, they have created a body of management guidelines that support mangrove conservation and sustainable use. Although the approaches vary somewhat, good consensus exists among the various organizational guidelines. First, I will describe the approaches of the international organizations that create the body of guidelines for mangrove management. Subsequently, I will summarize the guiding management principles that are shared by the organizations.

4.1 Food and Agriculture Organization of the United Nations

The **FAO of the United Nations** published the first guidelines for mangrove management over 20 years ago (FAO, 1994). The FAO guidelines clearly focus on sustainable forestry management (silviculture), rather than mangrove preservation. The document discusses the advantages and disadvantages of different methods of mangrove harvesting: clear-felling with and without seed bearers, selection systems, and shelterwood systems. It also describes mangrove nurseries and planting, thinning, harvesting and cutting policies, and rotation. Despite focusing on production, the FAO guidelines do touch on the sustainable use of the broad mangrove resources.

4.2 International Tropical Timber Organization

The **International Tropical Timber Organization (ITTO)** is an intergovernmental organization established in 1986 under the International Tropical Timber Agreement. The ITTO promotes the conservation and sustainable management of tropical forest resources through dissemination of data and funding of projects related to sustainable forest management. Sources of ITTO's mangrove management guidelines include its five-year strategic action plans and numerous publications that focus on themes such as the sustainable forestry management, the conservation and sustainable use of biodiversity in tropical production forests, restoration of degraded forests, and indicators for sustainable forestry management. ITTO's *Guidelines for the Restoration, Management and Rehabilitation of Degraded and Secondary Forests* (2002) describes the necessary elements for successful restoration (ITTO, 2002). Effective forest governance and adequate policies, appropriate incentives, and sanctions are essential, as are secure land tenure, clarification of access, and recognition of customary use rights. Local communities and stakeholders must be able to participate in decision-making regarding restoration, as well as share the responsibilities. Natural processes should be favored in restoration where possible. In its *Guidelines for the Conservation and Sustainable Use of Biodiversity in Tropical Timber Production Forests*, ITTO stresses the importance of integrated approaches involving coordination among various agencies that have different agendas in the management of forest ecosystem resources (ITTO/IUCN, 2009). Biodiversity should be an important component of all forest management plans from preparation to implementation. The publication *Voluntary Guidelines for the Sustainable Management of Natural Tropical Forests* describes the broad principles of forestry management—sustainability, integrated management, multiple use management, adaptive management, participation of local communities and stakeholders, inclusion of biodiversity protection, and restoration (ITTO, 2015). Most recently, ITTO's *Criteria and Indicators for the Sustainable Management of Tropical Forests* presents lists that can be used to evaluate forestry management efforts (ITTO, 2016).

4.3 International Society for Mangrove Ecosystems

The **International Society for Mangrove Ecosystems (ISME),** established in 1990 with Secretariat located in Okinawa, Japan, is an international non-governmental organization whose efforts focus on mangrove conservation and research. ISME with funding from the World Bank has published comprehensive guidelines for the conservation and sustainable utilization of mangrove ecosystems (World Bank et al., 2004). ISME also maintains a database of mangrove information and resources—Global Mangrove Database and Information System (GLOMIS). ISME's orientation is more pro-mangrove conservation than those of the other organizations. ISME guidelines stress conservation of mangrove ecosystems and the necessity to strictly regulate activities that may cause degradation. Integrated sustainable management of mangrove ecosystems is an important principle that ISME emphasizes, as is meaningful participation of local communities in all aspects of mangrove management. ISME has also published detailed guidelines and case studies on mangrove ecosystem restoration—*Restoration of Mangrove Ecosystems* (Field, 1995).

4.4 Ramsar Convention

The **Ramsar Convention**, signed in 1971 in Ramsar, Iran, is the only global treaty that focuses specifically on wetlands. Today 170 nations are signatories to the Ramsar Convention. A contracting party agrees to nominate at least one wetland in its territory to the List of Wetlands of International Importance based on enumerated criteria. By August 6, 2018, over 2323 wetland areas were inscribed on the Ramsar List, comprising over 248 million ha (Ramsar Convention Secretariat, 2018). In addition, contracting parties agree to manage all their wetlands based on the concept of "wise use." Wise use means the maintenance of the ecological character of the wetland and allowance of sustainable use for the benefit of people and the environment. The Convention also mandates contracting parties to adopt National Wetland Policies, produce wetland inventories, conduct wetland monitoring and research, raise public awareness of wetlands, and develop integrated management plans for wetlands sites. The Ramsar Secretariat has prepared numerous manuals to assist wetland decision-makers. Ramsar Handbook 18 contains Guidelines for Wetland Management (Ramsar Convention Secretariat, 2010b). Ramsar Handbook 12 concerning Coastal Management also stresses many of the same principles that should guide management of coastal wetlands (Ramsar Convention Secretariat, 2010a). In 2002, the eighth Conference of the contracting parties to the Ramsar Convention (COP-8) adopted Resolution VIII.14—"New Guidelines for Management Planning for Ramsar Sites and Other Wetlands." At the COP-8, contracting parties also adopted specific Ramsar guidelines for mangrove management—Resolution VIII.32—"Conservation, Integrated Management, and Sustainable Use of Mangrove Ecosystems and their Resources."

4.5 Choluteca Declaration

The **Choluteca Declaration** was signed by 21 delegates of environmental and community non-governmental organizations (NGOs) from Latin America, North America, Europe, and Asia at the 1996 Choluteca (Honduras) Forum on Aquaculture and its Impacts. Delegates were extremely concerned about the conversion of mangroves to shrimp aquaculture ponds and the adverse impacts on local communities that depend on mangrove resources for their livelihood. The delegates called for a global moratorium on further establishment of shrimp farming until technologies could be implemented that ensure the long-term survival of wetland ecosystems (particularly mangroves) and the fishers and communities that depend on these resources. The Declaration also called on multilateral developments banks, FAO, and bilateral development agencies not to finance or promote unsustainable shrimp aquaculture projects. All types of coastal development, such as shrimp aquaculture, should occur via integrated plans that include meaningful participation of all interested and affected groups, particularly local communities.

FAO's **Code of Conduct for Responsible Fisheries** (1995) includes many of the principles of the Choluteca Declaration with a softer touch. The Code of Conduct calls on States to ensure the ecological integrity of aquaculture activities and guarantee that these will not adversely affect the livelihoods of local communities or access to their fishing grounds. This FAO document also recommends that States integrate fisheries (and fishery aquaculture) into coastal area management. The sustainable use of coastal resources is the goal that must consider the fragility of coastal ecosystems.

Scientific and policy documents of the organizations and consortia mentioned above present common themes for mangrove ecosystem management that I summarize below.

Conservation and sustainability—Promotion of sustainable use and conservation of mangrove forests is the most basic principle espoused by the four international organizations. Mangrove resources can be used to benefit local populations within limits based on scientific foundations. However, mangrove ecological functions must continue and not be allowed to be degraded so that they will continue to provide future benefits to people and the broader environment.

Multiple use management—Generally, mangrove forests should be managed for multiple uses (forestry, wildlife and biodiversity, fisheries, mariculture, soil and water protection) to provide the maximum benefits to the greatest number of people. The numerous cultural benefits, products, and services should all be respected, but no use should adversely affect another or take priority over others. As a general rule, nondestructive uses of mangrove ecosystems should have preference over uses that convert and degrade mangroves and alter their hydrological regimes. Some mangrove areas of particular environmental quality or importance should be zoned for strict and permanent protection. Balancing uses is difficult, of course, and involves compromises and trade-offs—decisions that should involve all stakeholders.

Integration (sectoral and intergovernmental)—Many different activities occur in mangrove forests (timber extraction, fishing, aquaculture, shellfish collection, honey gathering, tourism, scientific research, education), and these are often managed by different governmental authorities possibly at different administrative levels. Holistic ecosystem-based management of mangroves requires coordination and cooperation among the different sectoral agencies at different levels of government during all of their functions (development of laws, regulations, and policies; planning; research and monitoring; enforcement; public education and outreach). This requires improved agency coordination and avoidance of duplication of efforts, resolution of interagency conflicts, and identification of policy and management gaps.

Ecosystem integration both landward and seaward—It is an error to manage the mangrove ecosystem in isolation. Flows of water, energy, pollutants, and flora and fauna occur between mangrove forests and adjacent ecosystems both landward (salt flats, freshwater wetlands, estuaries, and watersheds) and seaward (seagrass beds and coral reefs), as well as along the coastline. Management plans must include or at least coordinate with management efforts in adjacent ecosystems. This calls for broad land use planning at the local, regional, national, and transnational levels in which mangrove management plays an integral role. Thus, management of mangroves must be included in broader Integrated Coastal Management and Watershed Management programs.

Mangrove management plans—Government authorities together with all mangrove stakeholders should develop realistic management plans for the sustainable use and conservation of the mangrove ecosystem, which are accompanied by sustainable funding mechanisms. The management plan should clarify and quantify activities that are allowed and those that are prohibited. It should also delineate interventions/actions that will help resolve conflicts and minimize threats to the ecosystem. Mangrove ecosystem management must include measures to address climate change impacts (such as using mangrove forests to mitigate against sea level rise and increasing atmospheric CO_2 concentrations) and protection of biodiversity on genetic, species, and landscape scales. The plan should indicate responsible parties (government agencies, local community user groups, environmental NGOs, private sector interests), timelines, contingency plans, personnel and funding requirements for implementation, and indicators to evaluate the success of efforts. Zoning of the mangrove ecosystem into areas that should be strictly and permanently preserved because of their high environmental value or the important functions they provide, sustainable use zones, and recuperation/restoration zones should be a key component of the management plan. Environmental planning is a long-term dynamic process. After a determined number of years of implementation of the interventions/actions of the management plan, stakeholders must evaluate the successes and failures of implementation, as well as changing conditions and circumstances, as well as new knowledge, and revise the management plan to take into account new realities (Adaptive Management).

Effective participation of stakeholders and local communities—All phases of the mangrove management process (inventories and research, planning, implementation, evaluation) must involve the local communities and all stakeholders in an effective manner. Participation and consultation of all stakeholders is essential for transparent and equitable decision-making. The local community must be a partner in management of mangrove resources as they benefit from its ecosystem functions. Mangrove users have a right to participate and also should share in the responsibilities for the sustainable use and conservation of the mangrove ecosystem—perhaps creating a system of co-management between the local community of users and government authorities. This will ensure buy-in and compliance with management actions.

Respect for customary rights—Sustainable management of mangrove forests should support and where possible integrate traditional ecological knowledge, cultural values, and traditional management measures and, also, respect the traditional uses of the local communities. Local users who have traditionally used mangrove resources should have guaranteed access to continue their practices. However, levels of exploitation should be within sustainable levels.

Minimization of harm to the mangrove ecosystem from economic activities—The precautionary approach should be followed when considering approval of development projects that may impact mangrove ecosystems. Activities that have the potential to directly or indirectly harm the ecosystem services that mangroves provide should be carefully and independently evaluated before they are allowed to proceed. This may be accomplished through the use of environmental impact assessments. Post-project assessments and monitoring of the ecological conditions and parameters of the mangrove ecosystem must be followed to determine compliance with permit conditions and any adverse impacts that may result. Mitigation measures must be required to address the adverse impacts of development projects. Mitigation implies avoidance, minimization of harm, or compensation for the damage leading to environmental restoration.

Economic incentives—Often formal concessions or permits for conversion of mangroves to other uses or for extraction of water from mangrove waterways have an extremely low cost and, therefore, serve as incentives for replacement of mangrove ecosystems. These economic incentives that promote mangrove ecosystem conversion and degradation and that are inconsistent with conservation and wise use should be eliminated.

Best management practices for different categories of development projects and activities should be disseminated and followed. Fish and shellfish extraction have scientifically determined limits based on sustainability principles. Aquaculture practices should be consistent with mangrove management and avoid conversion of forests to aquaculture ponds, and integrated mangrove/aquaculture systems should be promoted. Tourism projects should avoid harm to mangroves and instead develop linkages with mangrove conservation efforts and local communities. Land use planners should require buffer zones of at least 200 m between mangroves and adjacent activities, such as tourism projects, housing developments, or aquaculture ponds (Alongi, 2002). Sustainably harvested mangrove products should be marketed using "fair trade" or "green labeling" mechanisms.

Restoration or rehabilitation of mangrove ecosystems—Degraded or converted mangrove ecosystems should be restored or rehabilitated. First, it is essential to clearly define the objectives of the restoration, such as coastline protection, sustainable use and production, or conservation to protect the natural ecosystem and its biodiversity. The objectives will determine the methodology used, as well as the indicators that will measure success of the restoration project.

Before beginning the restoration work, the characteristics of the adjacent mangrove ecosystem must be determined—soil characteristics, salinity, appropriate tidal flushing, freshwater input, winds, shade, availability of seeds, natural regeneration, and local community cooperation. Local community cooperation is essential for project success, and the community must see that restoration benefits their livelihoods and the environment.

Restoration methods may be natural or artificial, but natural regeneration is the preferred method. In natural regeneration, tides transport seeds or propagules to the site. The advantage of natural regeneration is that the vegetation will approximate the natural vegetation and the process is less expensive. Artificial regeneration using native species should be used where natural regeneration is insufficient, impossible, or when time constraints exist. Restoration should only occur in areas where mangroves once existed. Artificial methods control the plant distribution and species composition. Active planting may also provide an educational function and must involve the local community. Regardless of the restoration method selected, the restored area should have connectivity with forested areas to facilitate migration of flora and fauna and natural regeneration.

Seeds should be selected based on the species that naturally occur at the site and be collected from healthy well-developed trees. If the mangrove species reproduces via propagules, these should be mature and ready for collection. When direct planting is difficult or when it is advantageous to plant mangroves of a certain size, mangrove nurseries may be a solution. Spacing between seedlings or propagules is critical. After planting, it is essential to monitor growth, mortality, ecosystem parameters, and success of the project.

Institutional strengthening and capacity building—The capacity of government agencies responsible for management of mangrove resources must be strengthened. They must implement a system of effective governance based on a legal framework that clearly establishes rights and responsibilities. Adequate policies must be implemented that guide activities and encourage users to act sustainably with appropriate incentives and sanctions that, respectively, encourage sustainable use and deter violations of the legal regime.

Public education and awareness—Efforts must be dedicated to raising the awareness of the general public and local communities about the important services that mangrove ecosystems provide. Only then will they comprehend the need for effective conservation of these coastal wetlands.

5. SPECIFIC STRATEGIES

Management strategies can be classified into several broad areas ranging from preservation/conservation at the State and/or community levels, community-based management, regulatory measures, silviculture management, and ecosystem restoration and rehabilitation. These different strategies are not necessarily mutually exclusive.

5.1 Conservation via Protected Areas

Spalding et al. (2010) report that globally over 1200 protected areas that include mangroves have been designated and that a quarter of the world's mangroves are targeted for conservation with these designations. Categories of protected areas may include nature reserves, wilderness areas, national parks, habitat/species management areas (wildlife refuges), national forests, and protected areas with sustainable use of natural resources—each with different conservation and sustainable use objectives and levels of protection. The management authority may be a government agency (protected areas, fisheries, forestry, wildlife) at the national, provincial/state, or local government level; a university or environmental NGO; or a community. Several international networks of protected areas exist that may include mangrove ecosystems—each with its own objectives and management strategies.

The Ramsar Convention on Wetlands, signed in 1971 in Ramsar, Iran, is the only intergovernmental treaty that focuses specifically on wetlands. Signatory states must nominate at least one wetland to the List of Wetlands of International Importance and agree to maintain the ecological characteristics of the site. Broad criteria for inclusion of a site on the List include ecological, botanical, zoological, or hydrological characteristics. The contracting parties have approved specific criteria based on waterbird populations, fish populations, species and ecological communities, and rare and representative wetland types. Inclusion of a site raises the stature of the wetland, provides opportunities for international cooperation and support, and creates a platform for debate about activities that may harm the ecological characteristics of the wetland. By August 2018, 170 countries were contracting parties to the Ramsar Convention. These nations had dedicated some 2323 wetland areas as Ramsar Sites (Ramsar Convention Secretariat, 2018). Of these, about 251 Ramsar Sites in some 71 countries include mangrove ecosystems. Contracting parties also agree to promote the wise use of wetlands in their territories and conduct a national inventory of wetlands. The Ramsar Convention recognizes that wetlands benefit people directly but stresses that the use should be sustainable so that future generations can also benefit from the wetland resources and services.

The Convention concerning the Protection of the World Cultural and Natural Heritage ("World Heritage Convention") was adopted by the United Nations Educational, Scientific and Cultural Organization (UNESCO, 2018c) in 1971. The objective of the Convention is to identify, protect, and preserve the world's cultural and natural heritage. Today 167 states are contracting parties to the World Heritage Convention. UNESCO maintains a list of World Heritage Sites that State Parties have nominated based on satisfaction of specific selection criteria that demonstrate that the site is of outstanding universal value. Natural World Heritage Sites contain natural features of physical or biological formations or habitats for threatened species that are of outstanding universal value from esthetic, scientific, or conservation points of view. The World Heritage List contains 1092 properties that may be cultural, natural, or mixed natural and cultural (UNESCO, 2018c). Of the 209 natural sites and 38 mixed natural and cultural sites, 36 sites in 24 countries include mangrove ecosystems.

The UNESCO launched the Man and the Biosphere Program (MAB) in 1971. This effort promotes interdisciplinary research in the ecological and social aspects of biodiversity conservation and loss. The heart of the MAB Program is the international network of Biosphere Reserves characterized by three zones: a core area of strict protection, a surrounding buffer zone allowing for limited use, and a transition zone with greatest sustainable use. Today 122 countries have designated 686 biosphere reserves (UNESCO, 2018a). Over 91 biosphere reserves of the 686 MAB sites include mangrove ecosystems. UNESCO Global Geoparks is another international network of protected areas (UNESCO, 2018b). The 140 Global Geoparks represent sites of international geological significance and adopt management philosophies of protection, education, community involvement, and sustainable development. Two of these sites containing mangroves are the Langkawi Global Geopark in Malaysia and the Ciletuh-Palabuhanratu Global Geopark in Indonesia.

5.2 Community-Based Management

A growing trend is formal devolution of management of and decision-making regarding mangrove resources to local communities and families that live in or adjacent to mangrove forests and extract natural resources from that ecosystem, such as mollusks, fish, wood, and honey. Some analysts distinguish between "community-based management" where local communities manage mangroves and "co-management" where authorities and local communities share management responsibilities (Nguyen, 2014). Both concepts imply active participation of community members. Community-based management should provide increased benefits to local community members and reduce their economic vulnerability. Communities have no incentive to support mangrove conservation measures if these do not support their livelihoods (Alongi, 2002; Armitage, 2002; Badola et al., 2012; Barbier, 2006; Iftekhar and Takama, 2008; Nguyen, 2014; Rönnbäck et al., 2007). Communities' responsibility or right is often contingent on an approved management plan that emphasizes sustainable use and extraction. Such a system of co-management between local communities of resource extractors and the government—if well-conceived and implemented—could protect mangrove forests from conversion to other uses, such as aquaculture ponds, agriculture, and urban expansion areas. Success of community-based management regimes depends on the existence of legal regimes, such as the Mangrove Stewardship Agreements that exist in the Philippines, which establish the enforceable legal rights of the local community to the mangrove resources, community leadership and social capital, realistic community expectations, a well-conceived management plan with sustainable exploitation limits, and clear delimitation of the area with secure land tenure (Barbier, 2006; Pagdee et al., 2006; Primavera, 2005). Community-based mangrove management must also embrace ecosystem services that mangroves provide in addition to the sustainable use of marketable products. Enforcement should be transparent and egalitarian with respect to all stakeholders—economically powerful or not (Datta et al., 2012). Many examples of co-management regimes in mangroves illustrate this strategy. South Asia provides more examples of community-based mangrove management than Africa or Latin America.

In the late 1990s, Ecuador's Coastal Zone Management Program developed a series of Community User Agreements that granted exclusive harvest rights to users of specific mangrove areas. This initial program has grown into the "Socio Manglar" / Mangrove Associate Program, managed by the Ministry of the Environment, which grants concessions of mangrove areas to local communities. Once communities have obtained a concession and agree to protect their mangroves via an Agreement for the Sustainable Use and Custody of Mangroves (Acuerdo de Uso Sustentable y Custodia del Manglar), they can obtain financial incentives from the government. About 42% of 66,000 ha of Ecuador's remaining mangroves (157,000 ha) in various provinces (Esmeraldas, Manabí, Guayas, and El Oro) have been concessioned to communities in this program (Barona, 2014; Ministerio del Ambiente – Ecuador, 2016).

Kisakasaka Village in Zanzibar, Tanzania, formed a conservation committee that created a set of rules regarding harvesting of adjacent mangrove resources. The rules established cutting periods, closed areas, a rotation system, harvesting limits, penalties for violations, and limited access by outsiders (Lindsay, 1998). Although mangrove forests in Thailand are state-owned, villages in Trang Provincc in southern Thailand gained exclusive rights to manage their mangrove forests and developed governance strategies for mangrove management with the assistance of an NGO (Sudtongkong and Webb, 2008). In 2007, the Government of Kerala State, India, declared a community mangrove reserve to protect community property rights and provide sustainable livelihoods (Hema and Indira Devi, 2013). A six member Community Reserve Management Committee manages the reserve and is composed of five elected members from the community and a representative of the Forestry Department. The Committee's responsibilities include preparation and implementation of a management plan and monitoring.

Primavera and Esteban (2008) examined numerous mangrove replanting projects in the Philippines and noted that survival rates are generally low. Reasons for failures are numerous and include inappropriate siting and species selection. Some of the most successful projects are those that have high levels of community involvement and initiative. One of the most successful projects, located on Pagan Island, Bohol, was a community self-help initiative with no government assistance.

5.3 Regulatory Measures

Permits for development projects that may impact mangroves and are conditioned on mitigation obligations provide "generic protection" for mangrove ecosystems. In many jurisdictions, mangroves—whether publicly or privately owned—are managed with such regulatory methods. The landowner or developer must submit a permit application before undertaking activities involving cutting of mangroves or alteration of the hydrology. The authorities evaluating the permit application determine whether the proposal significantly harms the mangrove ecosystem or its resources and then may request revision of the proposal to avoid or minimize impacts. If the impacts are unavoidable but the project is deemed worthy, the project proponent must compensate the harm through mitigation measures.

Miami-Dade County, Florida, provides a relevant example of these regulatory strategies. Section 24–48 of the Miami-Dade County Code requires that anyone who intends to work in, on, or over tidal waters or coastal wetlands must obtain a Class I Permit for Coastal Construction and Mangrove Trimming from the Department of Regulatory and Economic Resources—Environmental Resources Management (Miami-Dade County). The objective of this regulatory process is to avoid or minimize environmental and esthetic impacts to mangroves and other coastal wetlands. The developer must compensate unavoidable impacts with mitigation, which may involve wetland creation, restoration, or enhancement. Not only may the project proponent have to obtain a permit from Miami-Dade County, but he or she also may have to obtain regulatory permits from the Florida Department of Environmental Protection and the US Army Corps of Engineers/Environmental Protection Agency, depending on the type and size of the Project.

Panama offers a different regulatory model. That country's mangroves are protected (even those outside of designated protected areas) and alteration of ecosystem components and its hydrology is illegal—illegal unless the developer obtains the appropriate permit and pays for the area she is converting to another use (Panama, 2008a,b, 2012). What appears at first blush to be a highly protective regulatory scheme has eventually evolved into an increasingly lenient payment structure for allowed conversion of mangrove forests (Spalding et al., 2015).

5.4 Mosquito Control

Mosquitos that breed in mangroves and coastal wetlands may be vectors for waterborne diseases. Moreover, their existence may weaken the public's support for wetland conservation measures or lead to increased use of pesticides for mosquito control. Dale and Knight (2006, 2012) suggest the potential use of shallow (<0.3 m), narrow (<0.9 m) ditches, called runnels, in coastal wetlands to modify tidal flow and increase flushing, thereby reducing the numbers of mosquito larvae. Use of runnels does not appear to cause significant habitat modification and preserves the ecosystem services that wetlands provide while reducing public health risks from mosquitos.

5.5 Silviculture Management

Mangrove trees are harvested for timber construction products, firewood, and charcoal in many countries on a small scale. Several countries have developed long-standing commercial forestry practices in some of their mangrove forests. These examples claim to be examples of sustainable silviculture—not sustainability in the broader ecological sense. FAO (1994) explored mangrove silviculture management practices over 20 years ago. However, in recent decades, the trend has been to recognize the ecosystem services of mangrove forests rather than viewing them solely as forestry resources.

The Cuban Forestry Institute classified a small percentage of Cuban mangrove forests (6%) as timber production zones to be managed according to principles of sustainable forestry (Padrón Milian, 1999). These areas are not immediately adjacent to coastlines and involve felling of trees in alternating strips that are 25–30 m wide. Species exploited are *Avicennia germinans*, *Laguncularia racemosa*, and *Conocarpus erectus* primarily for fuelwood, charcoal, and construction materials. The harvest is based on 30-year rotation periods. Regeneration by direct planting of seedlings is used if natural regeneration is impractical at the site.

In Bangladesh, large areas of the Sundarbans have been managed since the 18th century as mangrove plantations on a sustainable yield basis. The major species exploited are *Heritiera fomes* and *Excoecaria agallocha*, respectively, used for hardboard and construction and pulp for newspaper. The cutting cycle for commercial purposes decreased from 40 to 20 years (Iftekhar, 2008). Despite goals of sustainable yield management, stocks of both these species have decreased in recent decades because of overexploitation, sea level rise and salinity intrusion, decreasing input of freshwater and riverine sediments, cyclones, conversion to shrimp aquaculture ponds, and oil pollution (Akhtaruzzaman, 2004; Iftekhar and Islam, 2004). Changes in forest policy in the past two decades now prohibit harvesting of *H. fomes* and only allow limited cutting of *E. Agallocha* (Aziz and Paul, 2015). In recent decades, Bangladesh forestry policy has evolved from "sustainable yield management" to "sustainable ecosystem management" with less mangrove silviculture and increased efforts on afforestation for benefits of ecosystem services (Iftekhar and Islam, 2004).

In Malaysia the Matang Mangrove Forest Reserve on the northwestern corner of the Malaysian Peninsula is perhaps the oldest example of mangrove silviculture management with more than a century of scientific management (Alongi, 2002; Arnaud Goessens et al., 2014). The 40,151 ha Matang Reserve has adopted a 30-year rotation cycle for clear-cutting of 1050 ha blocks of mangrove trees that are subsequently used for charcoal production. Thinnings occur at 15 and 20 years. After felling at 30 years, the blocks are replanted with seedlings of *Rhizophora apiculata* and *Rhizophora mucronata*. Similarly, in Indonesia, several mangrove concessions have been granted for commercial forestry in Kalimantan and West Papua provinces. The mangrove timber is used for charcoal and paper pulp. Of the 140,000 ha in concession, 3% are harvested annually with a 20-year rotation (Evans, 2013). The Thai Royal Forest Department used to grant long-term concessions for silviculture systems based on clear-cutting of mangroves with a 30-year rotation. Half of each block was cut every 15 years. Both *Rhizophora apiculata* and *Rhizophora mucronata* were used to make charcoal. However, since 1996 the government has prohibited all mangrove cutting throughout the country (Ajiki, 2004).

During selective exploitation of a mangrove forest, attention must be given to species regeneration. The case of Kenya's Mida Creek mangrove forests, dominated by two species, *Rhizophora mucronata* and *Ceriops tagal*, is illustrative. Woodcutters prefer the *Rhizophora* species because of its high value for the production of building poles. Regeneration tends to favor the economically inferior *Ceriops* species (Kairo et al., 2002). Wise forestry management requires careful thought about reforestation of the exploited species if the objective is sustainable production of the economically valuable species—in this case *Rhizophora mucronata*.

5.6 Restoration and Rehabilitation

Many sites throughout the world with newly accreted sediments or mangrove areas that have been degraded or converted to different uses have become sites of mangrove planting, rehabilitation, restoration, or enhancement. These efforts may be promoted by government authorities, environmental NGOs, and/or local communities that wish to capitalize on the products or ecosystem benefits that mangrove forests provide. In some cases, mangrove restoration may be a requirement of mitigation for a development project. Definition of the objectives of the rehabilitation or restoration project, funding to implement the entire process, and long-term monitoring are all essential (Barnuevo et al., 2017). Extremes might range from restoration of the complex diverse mangrove ecosystem to monoculture mangrove forest for silviculture or coastal defense purposes (Ellison, 2000). Asaeda et al. (2016) provide recommendations for mangrove restoration projects that have an objective of replication of a natural forest, while Lewis and Gilmore (2007) discuss optimization of fish habitat during mangrove restoration projects. They suggest that different species be planted with appropriate spacing to avoid high tree density. The design should also create tidal creeks to recreate natural mangrove hydrology.

Mangrove restoration ecologists recommend active restoration of native mangrove species only when the site has been altered to such an extent that natural restoration cannot occur. If mangroves never grew at a site, it is likely that the hydrologic or soil conditions were not conducive to mangrove establishment. Lewis et al. (2000); Lewis (2009); and Dale et al. (2014) suggest five essential steps for successful mangrove restoration projects: (1) understanding the ecology of the mangrove species present, (2) understanding the hydrologic regime at the site, (3) assessment of variations from the original environment (changes in hydrology, elevation, soil), (4) development of a restoration plan based on restoration of hydrology so that natural recruitment occurs, and (5) planting of propagules or seedlings only if natural recruitment is impossible. The inner portion of accreting mud bars that are sheltered from wave action and are relatively stable because of the migration of biofilm may be ideal sites for mangrove restoration (Gensac et al., 2015). The costs for restoration vary considerably depending on labor requirements, need to excavate filled soil and build canals, and the potential purchase of mangrove propagules or seedlings. Many mangrove restoration projects opt for planting propagules or seedlings instead of allowing for natural recruitment. This may be feasible where labor costs are low (volunteers or students), for educational purposes, or to promote community "buy-in" and a sense of responsibility for the regenerated forest. In addition to understanding the biophysical environment, restoration success usually requires an additional essential element—community involvement and some guarantee that traditional rights and land tenure will be respected (Dale et al., 2014; Walters et al., 2008).

A coalition of groups (Global Nature Fund, Mangrove Action Project—Asia, Foundation Ursula Merz, and the German Ministry for Economic Cooperation and Development) is engaged in mangrove restoration of abandoned fish/shrimp ponds and rice paddies in Trang Province, Thailand (Mangrove Action Project). This community-based restoration project reproduces the natural hydrology by opening dikes and then allowing natural recruitment of mangroves to occur. The restored mangroves are seen as protection against tsunamis and storms.

The Philippines has extensive experience with mangrove restoration. The Zoological Society of London has sponsored a number of projects to rehabilitate mangroves in abandoned fish ponds (Zoological Society of London) and produced field guides for mangrove restoration (Philippine Tropical Forest Conservation Foundation, 2012). Similarly, the US Agency for International Development has promoted mangrove restoration in the Philippines and developed important guidelines for mangrove nurseries and planting, community forestry projects, and organizational strengthening (Melana et al., 2000).

After cutting trees in Cuban silviculture plantations (discussed above), *Rhizophora mangle* propagules are directly planted in rows with spacing between plants, which varies between

1.0 m × 1.0 m and 2.0 m × 1.6 m. In areas where *Avicennia germinans*, *Laguncularia racemosa*, and *Conocarpus erectus* are clear-cut, natural regeneration gives the best results (Padrón Milian, 1999).

The principal objective of mangrove restoration in Vietnam is to mitigate the impacts of sea level rise and storm surges from tropical monsoons. Mangroves (*Rhizophora stylosa*, *Kandelia candel*, and *Sonneratia caseolaris*) have been planted in areas between the protective sea dike and the coastline and appear to be successful in protecting the dikes from storm surges (Marchand, 2008; Powell et al., 2011). Research in northern Vietnam indicates that dissipation of wave energy depends on the vertical structure of mangroves (pneumatophores, leaves, branches), as well as forest density, wave height and period, water depth, and tidal stage (Alongi, 2008). Unique vertical configurations of different mangrove species cause different reduction of sea waves (Wolanski et al., 2001; Mazda et al., 1997, 2006). Therefore, careful thought should be given to the most appropriate mangrove species for the desired protective result. In the Mekong Delta in southern Vietnam bamboo T-groins and fences placed in front of earthen dikes help restore the eroded foreshore as a precondition of natural regeneration of mangroves. Through their function of wave attenuation, these mangroves help protect the sea dikes from storm surges and sea level rise (Albers and Schmitt, 2015). The researchers stress the importance of participatory involvement of local communities in these restoration efforts.

Bangladesh began a major mangrove afforestation program in 1966, and since then between 765 and 1200 km^2 of mangroves have been planted on newly accreted mudflats—largely *Sonneratia apetala* and *Avicennia officinalis* (Iftekhar and Islam, 2004; Saenger and Siddiqi, 1993). The principal objectives of the afforestation program are stabilization of newly accreted lands to minimize risks from cyclones and storm surges, as well as production of fuelwood for local communities and poverty alleviation (Iftekhar and Islam, 2004; Saenger and Siddiqi, 1993). Although Erftemeijer and Lewis (1999) recognize the success of Bangladesh's extensive afforestation projects in stabilizing newly accreted mudflats, they also raise concerns about this practice in other sites in Southeast Asia (Malaysia, Philippines, Thailand, Vietnam) where high mortality rates have been reported. Mudflats possess ecological importance themselves as avian feeding sites and sources of shellfish and other products for local communities. These authors recommend mangrove afforestation in accreting mudflat sites only where shoreline protection from cyclones is a priority.

After Hurricane Andrew destroyed the exotic forest of Australian pines (*Casuarina equisetifolia*) in Cape Florida State Park (Key Biscayne, Miami, Florida) in 1992, the State of Florida implemented a native vegetation restoration project for the entire park. Restoration begun in 1998 included a 25 ha mangrove restoration site. This project involved removal of fill ("scrape down"), creation of a system of interconnected channels and lagoons connected to Biscayne Bay, and subsequent planting of *R. mangle* grown in nurseries. Some 20 years after the planting of the red mangrove seedlings, the trees are 5–6 m tall. Birds, fish, invertebrates, and an occasional crocodile have returned to the site.

6. SUMMARY—MOVING FORWARD

Although mangroves comprise only a small percentage of the global tropical forest area, their contributions to millions of people and the health and security of the coastal zone and coastal ocean are great. Yet despite these contributions, mangrove forest degradation and losses are significant and even surpass those of tropical forests in general. Numerous reasons exist for these losses, but perhaps the most significant is conversion to other uses that provide a more immediate economic benefit to some individuals—usually not the traditional users of mangrove ecosystem resources. This is not surprising because coastal space is limited, houses many different commercial and economic interests, and has higher human population densities than inland areas.

A growing body of principles for mangrove conservation and sustainable use has formed at the international level through the contributions of at least four international organizations, environmental NGOs advocating for sustainable use of natural resources and empowerment of local communities and subsistence users, and the scientific community. Although differences in their orientations exist, their principles illustrate remarkable consensus. Increased protection of mangrove ecosystems via protected areas is clearly a priority. Conservation of remaining mangrove forests should be an important objective of efforts of international organizations, national governments, and local communities. Co-management regimes between government authorities and local communities of mangrove users must be fostered and favored as mangrove management strategies because often the best protectors of mangrove forests are those individuals and groups that have an interest in mangrove ecosystems' continued sustainable use because they obtain benefits that support their basic needs. Successful co-management regimes require much planning and depend on the social capital of the community. On the other hand, increased institutional capacity and planning skills for government institutions that have management responsibilities in mangrove ecosystems are essential components to ensure effective management of mangrove forests. Public awareness and understanding of the important values of mangrove ecosystems is crucial for garnering support for mangrove conservation.

Management of mangrove ecosystems should adopt the principles of integrated coastal management. Mangrove conservation and sustainable use must be coordinated with all institutions that possess authority over mangrove resources or on- and offsite activities that may affect them. Appropriate management also involves coordination among national, provincial, and local governments that may all possess some responsibility for various governmental functions. In regions with transboundary watersheds, international cooperation may be necessary to avoid adverse impacts to the downstream mangrove ecosystems. As described above, integrated management also calls for co-management and sharing of responsibilities between governmental authorities and local communities that use and benefit from mangrove resources.

Restoration and rehabilitation of degraded mangrove forests is also essential and should be a high priority for international and national efforts. Mangrove restoration or rehabilitation projects must first identify the objectives and attempt to remove or lessen the responsible stressors. Success of mangrove rehabilitation hinges on application of proper techniques and knowledge of the biophysical environments (ideal hydrology and soil elevation, correct mangrove species for the site, source of seeds or propagules, consideration of wave action and predators), as well as a monitoring program to evaluate success and identify problems. The ultimate success of these efforts, however, depends on the involvement and empowerment of local communities that understand the direct and indirect benefits that mangrove ecosystems provide them (Dale et al., 2014). Success also requires long-term guarantee of the community's property rights and control over mangrove forests (land tenure), as well as its responsibilities to utilize the resources sustainably.

While aquaculture ponds sited in mangroves have been responsible for a significant percentage of global mangrove losses and should be located outside mangrove communities, "mangrove-friendly aquaculture" or "aquasilviculture" *might* provide an alternative and a "win-win" situation. Acceptable activities would be small-scale, perhaps family-operated, and respect forest structure and water flows. Examples of small-scale mollusk and crab aquaculture activities in mangroves exist. Additional research efforts should be dedicated to developing appropriate techniques that are sustainable, low-impact, and that could provide alternative incomes for local communities.

Increased linkages between ecological research and sustainable management of mangrove ecosystems are essential. Research needs should focus on appropriate techniques for restoration and rehabilitation, particularly in abandoned aquaculture ponds; ecological structural and functional results and endpoints from mangrove restoration projects; fostering the capability of mangrove ecosystems to adapt in light of sea level rise; the best mangrove restoration techniques to protect against storm surges and tsunamis; and the sociocultural factors that optimize community-based management strategies for mangrove ecosystems.

Mangrove ecosystems provide extensive benefits that are nonconsumptive and indirect, but quantification of these nonmarket benefits is extremely difficult. As a result, mangroves are often converted to other uses that provide an immediate short-term economic benefit. Further research that attempts to holistically quantify mangroves' ecosystem services and cultural values is a necessity. The recent interest in Blue Carbon may facilitate new mechanisms to quantify the contribution of mangroves as carbon sinks to mitigate global climate change by reduction of atmospheric concentrations of CO_2. Once better consensus exists in the scientific community regarding mangroves as carbon sinks, we will see application of payments for ecosystem services (PES) and REDD+ (reduction of emissions from deforestation and forest degradation to foster conservation and sustainable forestry management) mechanisms to transfer funds from the Global North to the Global South for mangrove conservation and rehabilitation.

References

Ajiki, K., 2004. Socio-economic study of the utilization of mangrove forests in South-East Asia. In: Vannucci, M. (Ed.), Mangrove Management and Conservation. United Nations University Press, Tokyo and New York, pp. 257–269.

Akhtaruzzaman, A.F.M., 2004. Mangrove forestry research in Bangladesh. In: Vannucci, M. (Ed.), Mangrove Management and Conservation. United Nations University Press, Tokyo and New York, pp. 249–256.

Albers, T., Schmitt, K., 2015. Dyke design, floodplain restoration and mangrove co-management as parts of an area coastal protection strategy for the mud coasts of the Mekong Delta, Vietnam. Wetlands Ecology and Management 23, 991–1004.

Alongi, D.M., 2002. Present state and future of the world's mangrove forests. Environmental Conservation 29, 331–349.

Alongi, D.M., 2008. Mangrove forests: resilience, protection from tsunamis, and response to global climate change. Estuarine, Coastal and Shelf Science 76, 1–13.

Armitage, D.L., 2002. Socio-institutional dynamic and the political ecology of mangrove forest conversion in central Sulawesi, Indonesia. Global Environmental Change 12, 203–217.

Arnaud Goessens, A., Satyanarayana, B., Van der Stocken, T., Quispe Zúñiga, M., Mohd-Lokman, H., Sulong, I., Dahdouh-Guebas, F., 2014. Is Matang Mangrove Forest in Malaysia sustainably rejuvenating after more than a century of conservation and harvesting management? PLoS One 9 (8), e105069. https://doi.org/10.1371/journal.pone.0105069.

Asaeda, T., Barnuevo, A., Sanjaya, K., Fortes, M.D., Kanesaka, Y., Wolanski, E., 2016. Mangrove plantation over a limestone reef – good for the ecology? Estuarine, Coastal and Shelf Science 173, 57–64.

Aziz, A., Paul, A.R., 2015. Bangladesh Sundarbans: present status of the environment and biota. Diversity 7, 242–269. https://doi.org/10.3390/d7030242.

Badola, R., Barthwal, S., Hussain, S.A., 2012. Attitudes of local communities toward conservation of mangrove forests: a case study from the east coast of India. Estuarine, Coastal and Shelf Science 96, 188–196.

Barbier, E.B., 2006. Natural barriers to natural disasters: replanting mangroves after the tsumani. Frontiers in Ecology and the Environment 4, 124–131.

Barnuevo, A., Asaeda, T., Sanjaya, K., Kanesaka, Y., 2017. Drawbacks of mangrove rehabilitation schemes: lessons learned from the large-scale mangrove plantations. Estuarine, Coastal and Shelf Science 198 (B), 432–437.

Barona, J., 2014. Ecuador, un referente en la conservación del manglar con participación comunitaria. Available at: http://www.andes.info.ec/es/noticias/ecuador-referente-conservacion-manglar-participacion-comunitaria.html.

Berkes, F., Feeny, D., McCay, B.J., Acheson, J.M., 1989. The benefits of the commons. Nature 340, 91–93.

Castellanos-Galindo, G.A., Kluger, L.C., Tompkins, P., 2017. Panama's impotent mangrove laws. Science 355, 918–919.

Choluteca Declaration, October 16, 1996. Choluteca Forum on Aquaculture and its Impacts, Choluteca, Honduras. Available at: Darwin.bio.uci.edu/-sustain/shrimpecos/declare1.html.

Corlett, R.T., 1992. The ecological transformation of Singapore, 1819-1990. Journal of Biogeography 19, 411–420.

Dale, P.E.R., Knight, J.M., 2006. Managing salt marshes for mosquito control: impacts of runnelling, open marsh water management and grid-ditching in sub-tropical Australia. Wetlands Ecology and Management 14, 211–220.

Dale, P.E.R., Knight, J.M., 2012. Managing mosquitos without destroying wetlands: an Eastern Australian approach. Wetlands Ecology and Management 20, 233–242.

Dale, P.E.R., Knight, J.M., Dwyer, P.G., 2014. Mangrove rehabilitation: a review focusing on ecological and institutional issues. Wetlands Ecology and Management 22, 587–604.

Datta, D., Chattopadhyay, R.N., Guha, P., 2012. Community based mangrove management: a review on status and sustainability. The Journal of Environmental Management 107, 84–95.

Ellison, A.M., 2000. Mangrove restoration: do we know enough? Restoration Ecology 8, 219–229.

Erwin, K.L., 2009. Wetlands and global climate change: the role of wetland restoration in a changing world. Wetlands Ecology and Management 17, 71–84.

Erftemeijer, P.L.A., Lewis, R.R., 1999. Planting mangroves on intertidal mudflats: habitat restoration or habitat conversion? In: Paper Presented at the ECOTONE-VIII Seminar "Enhancing Coastal Ecosystem Restoration for the 21st Century", Ranong & Phuket, Thailand, 23–28 May 1999.

Evans, K., 2013. Could Sustainable Logging Save Indonesia's Mangroves? Available at: http://blog.cifor.org/14229/could-sustainable-logging-save-indonesias-mangroves?fnl=en.

FAO, 1995. Code of Conduct for Responsible Fisheries. FAO, Rome.

FAO, 1994. Mangrove Forest Management Guidelines. FAO Forestry Paper 117 (Rome).

FAO, 2017. Status and Trends in Mangrove Area Extent Worldwide. At http://www.fao.org/docrep/007/j1533e/J1533E02.htm#P199_6397.

Field, C. (Ed.), 1995. Restoration of Mangrove Ecosystems. International Society for Mangrove Ecosystems, Okinawa, Japan.

Gensac, E., Gardel, A., Lesourd, S., Brutier, L., 2015. Morphodynamic evolution of an intertidal mudflat under the influence of Amazon sediment supply – Kourou mud bank, French Guiana, South America. Estuarine, Coastal and Shelf Science 158, 53–62.

Gilman, E.L., Ellison, J., Duke, N., Field, C., 2008. Threats to mangroves from climate change and adaptation options: a review. Aquatic Botany 89, 237–250.

Hema, M., Indira Devi, P., 2013. Socioeconomic impacts of the community-based management of the mangrove reserve in Kerala, India. Journal of Environmental Professionals Sri Lanka 1, 30–45. Available at: www.sljol.info/index.php/JEPSL/article/view/5146.

Hoppe-Speer, S.C.L., Adams, J.B., 2015. Cattle browsing impacts on stunted Avicennia marina mangrove trees. Aquatic Biology 121, 9–15.
Iftekhar, M.S., 2008. An overview of mangrove management strategies in three South Asian countries: Bangladesh, India and Sri Lanka. International Forestry Review 10, 38–51.
Iftekhar, M.S., Islam, M.R., 2004. Managing mangroves in Bangladesh: a strategy analysis. Journal of Coastal Conservation 10, 139–146.
Iftekhar, M.S., Takama, T., 2008. Perceptions of biodiversity, environmental services, and conservation of planted mangroves: a case study on Nijhum Dwip Island, Bangladesh. Wetlands Ecology and Management 16, 119–137.
International Tropical Timber Organization (ITTO), 2002. Guidelines for the Restoration, Management and Rehabilitation of Degraded and Secondary Forests. Policy Development Series No. 13. ITTO, Yokohama, Japan.
International Tropical Timber Organization (ITTO), 2015. Voluntary Guidelines for the Sustainable Management of Natural Tropical Forests. ITTO Policy Development Series No. 20. ITTO, Yokohama, Japan.
International Tropical Timber Organization (ITTO), 2016. Criteria and Indicators for the Sustainable Management of Tropical Forests. ITTO Policy Development Series No. 21. ITTO, Yokohama, Japan.
International Tropical Timber Organization (ITTO)/IUCN), 2009. Guidelines for the Conservation and Sustainable Use of Biodiversity in Tropical Timber Production Forests. ITTO Policy Development Series No. 17. ITTO, Yokohama, Japan.
Islam, M.S., Wahab, M.A., 2005. A review on the present status and management of mangrove wetland habitat resources in Bangladesh with emphasis on mangrove fisheries and aquaculture. Hydrobiologia 542, 165–190.
Kairo, J.G., Dahdouh-Guebas, F., Bosire, J., Koedam, N., 2001. Restoration and management of mangrove systems – a lesson for and from the East African region. South African Journal of Botany 67, 383–389.
Kairo, J.G., Dahdouh-Guebas, F., Gwada, P.O., Ochieng, C., Koedam, N., 2002. Regeneration status of mangrove forests in Mida Creek, Kenya: a compromised or secured future? Ambio 31, 562–568.
Kodikara, K.A.S., Mukherjee, N., Jayatissa, L.P., Dahdouh-Guebas, F., Koedam, N., 2017. Have mangrove restoration projects worked? An in-depth study in Sri Lanka. Restoration Ecology 25, 705–716.
Lewis, R.R., 2005. Ecological engineering for successful management and restoration of mangrove forests. Ecological Engineering 24, 403–418.
Lewis, R.R., Gilmore, R.G., 2007. Important considerations to achieve successful mangrove forest restoration with optimum fish habitat. Bulletin of Marine Science 80, 823–837.
Lewis, R.R., Streever, B., Theriot, R.F., 2000. Restoration of Mangrove Habitat. WRP Technical Notes Collection (ERDC TN-WRP-VN-RS-3.2). U.S. Army Engineer Research and Development Center, Vicksburg, Mississippi. Available at: http://oai.dtic.mil/oai/oai?verb=getRecord&metadataPrefix=html&identifier=ADA384964.
Lewis, R.R., 2009. Methods and criteria for successful mangrove forest restoration. In: Perillo, G.M.E., Wolanski, E., Cahoon, D.R., Brinson, M.M. (Eds.), Coastal Wetlands: An Integrated Ecosystem Approach. Elsevier, Amsterdam and Oxford, pp. 787–800.
Lindsay, J.M., 1998. Creating Legal Space for Community-based Management: Principles and Dilemmas. FAO Development Law Service, Rome. Available at: http://www.mekonginfo.org/assets/midocs/0001835-environment-creating-legal-space-for-community-based-management-principles-and-dilemmas.pdf.
Loucks, C., Barber-Meyer, S., Houssain, M.A.A., Barlow, A., Chowdhury, R.M., 2010. Sea level rise and tigers: predicted impacts to Bangladesh's Sundarbans mangroves. Climatic Change 98, 291–298.
Mangrove Action Project. Mangrove Restoration and Reforestation in Asia. Available at: http://mangrove actionproject.org/mangrove-restoration-and-reforestation-in-asia/.
Marchand, M., 2008. Mangrove Restoration in Vietnam: Key Considerations and a Practical Guide. Report WRU/TUD. Available at: Mangrove restoration in Vietnam. repository.tudelft.nl.
Mazda, Y., Magi, M., Kogo, M., Hong, P.N., 1997. Mangroves as a coastal protection from waves in the Tong King delta, Vietnam. Mangroves and Salt Marshes 1, 1127–1135.
Mazda, Y., Magi, M., Ikeda, Y., Kurokawa, T., Asano, T., 2006. Wave reduction in a mangrove forest dominated by *Sonneratia* sp. Wetlands Ecology and Management 14, 365–378.
Melana, D.M., Atchue III, J., Yao, C.E., Edwards, R., Melana, E.E., Gonzales, H.I., 2000. Mangrove Management Handbook. Department of Environment and Natural Resources, Manila, Philippines through the Coastal Resource Management Project, Cebu City, Philippines.
Miami-Dade County (Florida). Class I Permit Applications. Available at: https://www.miamidade.gov/permits/library/class-1.pdf.

Ministerio del Ambiente (Ecuador), August 26 , 2016. Primer encuentro de organizaciones que custodian el manglar de Guayas. Available at: http://www.ambiente.gob.ec/primer-encuentro-de-organizaciones-que-custodian-el-manglar-del-guayas/.

Nguyen, H.H., 2014. The relation of coastal mangrove changes and adjacent land-use: a review of Southeast Asia and Kien Giang, Vietnam. Ocean and Coastal Management 90, 1–10.

Padrón Milian, C., 1999. Estudio de caso: manejo integrado de ecosistemas de manglar en Cuba. In: Ammour, T., Imbach, A., Suman, D., Windevoxhel, N. (Eds.), Manejo Productivo de Manglares en América Central. CATIE, Turrialba, Costa Rica, pp. 293–302.

Pagdee, A., Kim, Y., Dougherty, P.J., 2006. What makes community forest management successful: a meta-study from community forests throughout the world. Society and Natural Resources 19, 33–53.

Panama, February 26, 2008a. Autoridad de los Recursos Acuáticos de Panamá (ARAP)/Panamanian Aquatic Resources Authority, ARAP Board of Directors Resolution J.D. No. 1.

Panama, May 23, 2012. Autoridad de los Recursos Acuáticos de Panamá (ARAP)/Panamanian Aquatic Resources Authority, ARAP Board of Directors Resolution J.D. No. 20.

Panama, 29 January 2008b. Autoridad de los Recursos Acuáticos de Panamá (ARAP)/Panamanian Aquatic Resources Authority. Resolution ARAP No. 1.

Pattanaik, C., Reddy, C.S., Prasad, S.N., 2008. Mapping, monitoring and conservation of Mahanadi wetland ecosystem, Orissa, India using remote sensing and GIS. The Proceedings of the National Academy of Sciences, India, Section B: Biological Sciences 78, 81–89.

Philippine Tropical Forest Conservation Foundation and the Zoological Society of London, 2012. Community-based Mangrove Rehabilitation Training Manual. Available at: https://www.zsl.org/sites/default/files/media/2015-06/Mangrove%20Rehab_Training%20Manual.pdf.

Powell, N., Osbeck, M., Tan, S.B., Toan, V.C., 2011. Mangrove Restoration and Rehabilitation for Climate Change Adpatation in Vietnam: World Resources Report Case Study. World Resources Institute, Washington, D.C. Available at: http://www.wri.org/sites/default/files/wrr_case_study_mangrove_restoration_vietnam.pdf.

Primavera, J.H., 2000. Development and conservation of Philippine mangroves: institutional issues. Ecological Economics 35, 91–106.

Primavera, J.H., 2005. Mangroves, fishponds, and the quest for sustainability. Science 310, 57–59.

Primavera, J.H., Esteban, J.M.A., 2008. A review of mangrove rehabilitation in the Philippines: successes, failures and future prospects. Wetlands Ecology and Management 16, 345–358.

Ramsar Convention Secretariat, 2010a. Coastal Management: Wetland Issues in Integrated Coastal Zone Management In: Ramsar Handbooks for the Wise Use of Wetlands, fourth ed., vol. 12. Ramsar Convention Secretariat, Gland, Switzerland.

Ramsar Convention Secretariat, 2010b. Managing Wetlands: Frameworks for Managing Wetlands of International Importance and Other Wetlands Sites In: Ramsar Handbooks for the Wise Use of Wetlands, fourth ed., vol. 18. Ramsar Convention Secretariat, Gland, Switzerland.

Ramsar Convention Secretariat, 2018. The List of Wetlands of International Importance. Available at: http://www.ramsar.org/sites/default/files/documents/library/sitelist.pdf.

Rönnbäck, P., Crona, B., Ingwall, L., 2007. The return of ecosystem goods and services in replanted mangrove forests: perspectives from local communities in Kenya. Environmental Conservation 34, 313–324.

Saenger, P., Siddiqi, N.A., 1993. Land from the sea: the mangrove afforestation program of Bangladesh. Ocean and Coastal Management 20, 23–39.

Saifullah, S.M., 1997. Management of the Indus Delta mangroves. In: Haq, B.U., Haq, S.M., Kullenberg, G., Stel, J.H. (Eds.), Coastal Zone Management Imperative for Maritime Developing Nations. Kluwer Academic Press Publishers, Dordrecht, The Netherlands, pp. 333–346.

Spalding, A.K., Suman, D.O., Mellado, M.E., 2015. Navigating the evolution of marine policy in Panama: current policies and community responses in the Pearl Islands and Bocas del Toro. Marine Policy 62, 161–168.

Spalding, M., Kainuma, M., Collins, L., 2010. World Atlas of Mangroves. Earthscan, London and Washington, D.C.

Sudtongkong, C., Webb, E.L., 2008. Outcomes of state vs. community-based mangrove management in southern Thailand. Ecology and Society 13 (27). Available at: http://mangroves.elaw.org/node/51.

UNEP-WCMC, 2006. In the Front Line: Shoreline Protection and Other Ecosystem Services from Mangroves and Coral Reefs. UNEP-WCMC, Cambridge, UK.

UNESCO, 2018a. Directory of the World Network of Biosphere Reserves (WNBR). Available at: http://www.unesco.org/new/en/natural-sciences/environment/ecological-sciences/biosphere-reserves/world-network-wnbr/wnbr/.

UNESCO, 2018b. List of UNESCO Global Geoparks. Available at: http://www.unesco.org/new/en/natural-sciences/environment/earth-sciences/unesco-global-geoparks/list-of-unesco-global-geoparks/.

UNESCO, 2018c. World Heritage List. Available at: http://whc.unesco.org/en/list/.

Vo, Q.T., Kuenzer, C., Vo, Q.M., Moder, F., Oppelt, N., 2012. Review of valuation methods for mangrove ecosystem services. Ecological Indicators 23, 431–446.

Walters, B.B., Rönnbäck, P., Kovacs, J.M., Crona, B., Hussain, S.A., Badola, R., Primavera, J.H., Barbier, E., Dahdouh-Guebas, F., 2008. Ethnobiology, socio-economic and management of mangrove forests: a review. Aquatic Botany 89, 220–236.

Wolanski, E., Cassagne, B., 2000. Salinity intrusion and rice farming in the mangrove-fringed Konkoure River delta, Guinea. Wetlands Ecology and Management 8, 29–36.

Wolanski, E., Mazda, Y., Furukawa, K., Ridd, P., Kitheka, J., Spagnol, S., Stieglitz, T., 2001. Water circulation in mangroves and its implications for biodiversity. In: Wolanski, E. (Ed.), Oceanographic Processes of Coral Reefs. CRC Press, London, pp. 53–76.

World Bank, ISME, cenTER Aarhus, 2004. Principles for a Code of Conduct for the Management and Sustainable Use of Mangrove Ecosystems.

Zhang, K., Liu, H., Li, Y., Xu, H., Shen, J., Rhome, J., 2012. The role of mangrove in attenuating storm surges. Estuarine, Coastal and Shelf Science 102–103, 11–23.

Zoological Society of London. Rehabiliting Mangroves in the Philippines. Available at: https://www.zsl.org/conservation/regions/asia/rehabilitating-mangroves-in-the-philippines.

Subject Index

Note: 'Page numbers followed by "f" indicate figures, "t" indicate tables and "b" indicate boxes.'

G

R

S

T

U

Y

Z

Geographic Index

Note: 'Page numbers followed by "f" indicate figures, "t" indicate tables.'

P

Q

R

S

T